to the Next Level with MasteringA&P®

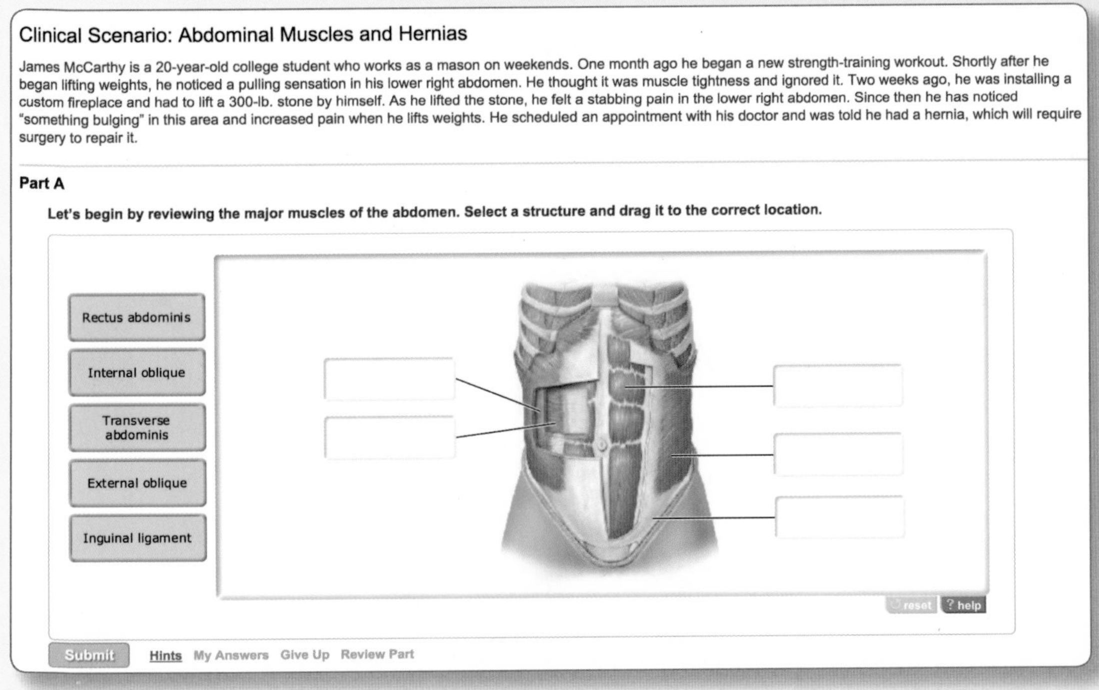

NEW! **Clinical Scenario Coaching Activities** bring the real world into your course.

NEW! The student package for Marieb, Wilhelm, and Mallatt's *Human Anatomy,* Eighth Edition includes:

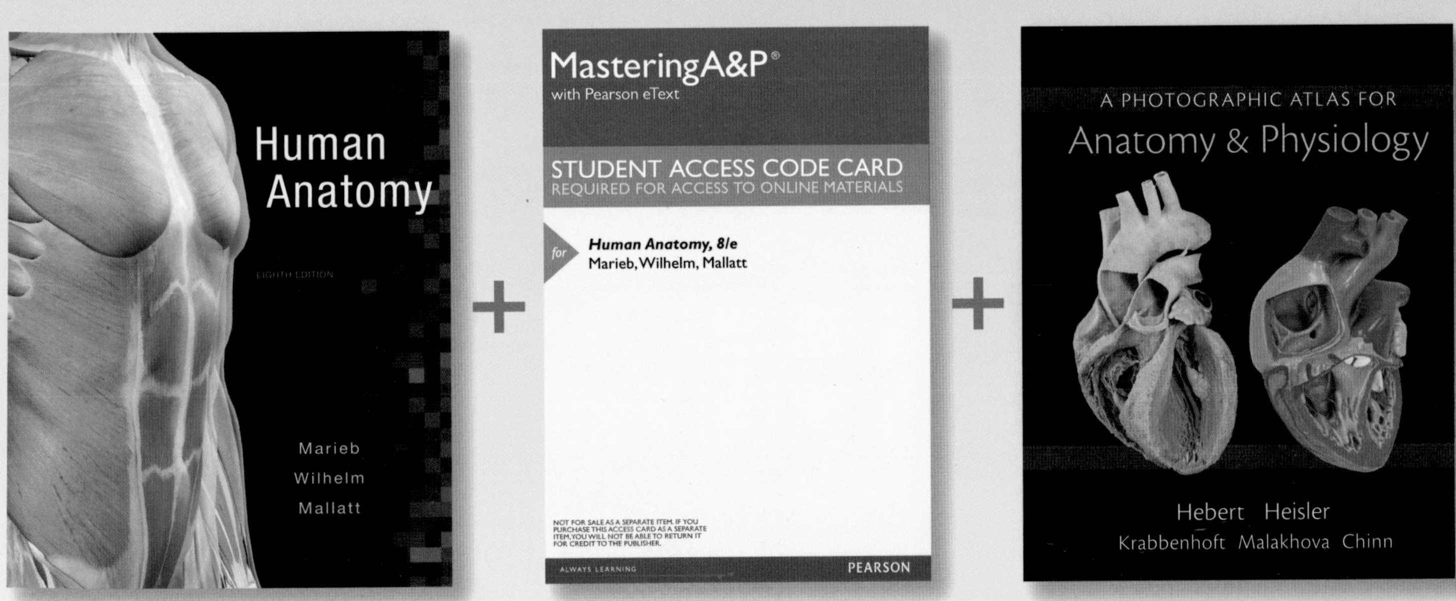

Human Anatomy

EIGHTH EDITION

Elaine N. Marieb, R.N., Ph.D.
Holyoke Community College

Patricia Brady Wilhelm, Ph.D.
Johnson & Wales University

Jon Mallatt, Ph.D.
Washington State University

Editor-in-Chief: Serina Beauparlant
Project and Design Manager: Michele Mangelli
Program Manager: Tiffany Mok
Art Development Manager: Laura Southworth
Editorial Assistant: Nicky Montalvo Romero
Senior Project Manager, Rights and Permissions: Maya Gomez
Program Management Team Lead: Michael Early
Project Management Team Lead: Nancy Tabor
Copyeditor: Anita Wagner Hueftle
Compositor: Cenveo® Publisher Services
Production Coordinator: Karen Gulliver
Art Coordinator: Jean Lake
Proofreader: Betsy Dietrich
Indexers: Kathy Pitcoff and Sallie Steele
Interior Designer: tani hasegawa
Cover Designer: Tandem Creative, Inc.
Illustrators: Imagineering STA Media Services Inc.
Photo Researcher: Kristin Piljay
Senior Procurement Specialist: Stacey Weinberger
Executive Marketing Manager: Allison Rona
Senior Anatomy & Physiology Specialist: Derek Perrigo

Cover image: Full color 3D computer-generated illustration on black background. Hank Grebe/Superstock/Corbis.

Library of Congress Cataloging-in-Publication Data

Names: Marieb, Elaine Nicpon, author. | Wilhelm, Patricia Brady, author. | Mallatt, Jon, author.
Title: Human anatomy / Elaine N. Marieb, R.N., Ph.D., Holyoke Community College, Patricia Brady Wilhelm, Ph.D., Johnson & Wales University Jon Mallatt, Ph.D., Washington State University.
Description: Eighth edition. | San Francisco : Pearson Education, Inc., [2017]
Identifiers: LCCN 2015038748| ISBN 9780134243818 (student edition) | ISBN 0134243811 (student edition) | ISBN 9780134398303 (instructor review copy) | ISBN 0134398300 (instructor review copy)
Subjects: LCSH: Human anatomy.
Classification: LCC QM23.2 .M348 2017 | DDC 577.01—dc23
LC record available at http://lccn.loc.gov/2015038748

ISBN 10: 0-13-424381-1 (student edition)
ISBN 13: 978-0-13-424381-8 (student edition)
ISBN 10: 0-13-439830-0 (instructor review copy)
ISBN 13: 978-0-13-439830-3 (instructor review copy)
3 17

www.pearsonhighered.com

ABOUT THE AUTHORS

Elaine N. Marieb

After receiving her Ph.D. in zoology from the University of Massachusetts at Amherst, Elaine N. Marieb joined the faculty of the Biological Science Division of Holyoke Community College.

While teaching at Holyoke Community College, where many of her students were pursuing nursing degrees, she developed a desire to better understand the relationship between the scientific study of the human body and the clinical aspects of the nursing practice. To that end, while continuing to teach full time, Dr. Marieb pursued her nursing education, which culminated in a Master of Science degree with a clinical specialization in gerontology from the University of Massachusetts. It is this experience that has informed the development of the unique perspective and accessibility for which her publications are known.

Dr. Marieb has given generously to provide opportunities for students to further their education. She funds the E. N. Marieb Science Research Awards at Mount Holyoke College, which promotes research by undergraduate science majors, and has underwritten renovation of the biology labs in Clapp Laboratory at that college. Dr. Marieb also contributes to the University of Massachusetts at Amherst where she generously provided funding for reconstruction and instrumentation of a cutting-edge cytology research laboratory. Recognizing the severe national shortage of nursing faculty, she underwrites the Nursing Scholars of the Future Grant Program at the university. In January 2012, Florida Gulf Coast University named a new health professions facility "Dr. Elaine Nicpon Marieb Hall." With the help of Dr. Marieb's generous donation, this facility contains simulated laboratories in the School of Nursing.

Dr. Marieb is an active member of the Human Anatomy and Physiology Society (HAPS) and the American Association for the Advancement of Science (AAAS).

Patricia Brady Wilhelm

Patricia Brady Wilhelm received her Ph.D. in biological and medical sciences from Brown University and is currently Professor and Chair of Science at Johnson & Wales University, Providence RI. She has taught human anatomy at Brown University, Rhode Island College, Community College of Rhode Island, and currently at the Center for Physician Assistant Studies at Johnson & Wales University.

Dr. Wilhelm's commitment to teaching has been recognized throughout her career. As a doctoral student, she received the Presidential Award for Excellence in Graduate Teaching and in 2011 the Teaching Excellence Award from the Community College of Rhode Island. Dr. Wilhelm embraces innovation in the classroom and laboratory, incorporating project-based learning, Process Oriented Guided Inquiry Learning (POGIL) activities, cooperative team-based dissection, and other active learning strategies. Dr. Wilhelm has shared her techniques, experience, and enthusiasm for student success through professional presentations, including those of the Human Anatomy and Physiology Society (HAPS) and the New England Biology Association of Two-Year Colleges (NEBATYC) conferences.

In addition to teaching, Dr. Wilhelm contributes to the development of media tools for human anatomy instruction and is a reviewer for *Anatomical Sciences Education*. She is a member of Sigma Xi, the Human Anatomy and Physiology Society (HAPS), the American Association of Anatomists (AAA), and the PULSE (Partnership for Undergraduate Life Science Education) Community.

Jon Mallatt

With a Ph.D. in anatomy from the University of Chicago, Dr. Mallatt is currently an Associate Professor of Biological Sciences at Washington State University. He is also a member of the department of Basic Medical Sciences, where he teaches courses in histology and in anatomy of the trunk in the WWAMI Medical Program. WWAMI has honored him numerous times with its Excellence in Teaching Award. Dr. Mallatt is an accomplished researcher with 45 publications in the fields of comparative anatomy and molecular phylogeny to his credit.

Help Your Students Prepare for Lab

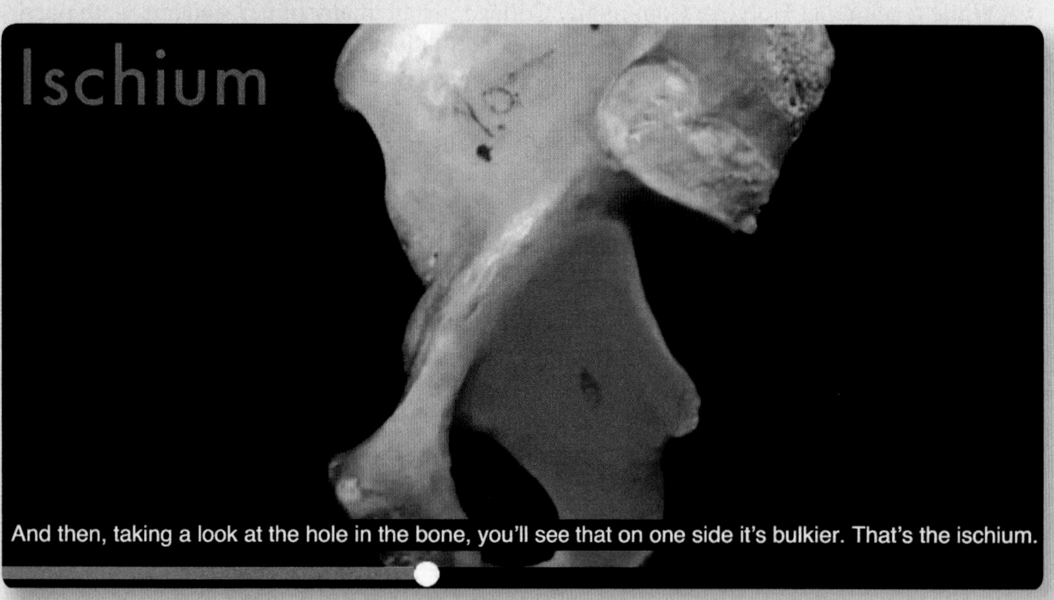

Bone and Organ Dissection Videos cover major bone and organ dissections to help students prepare for lecture and lab.

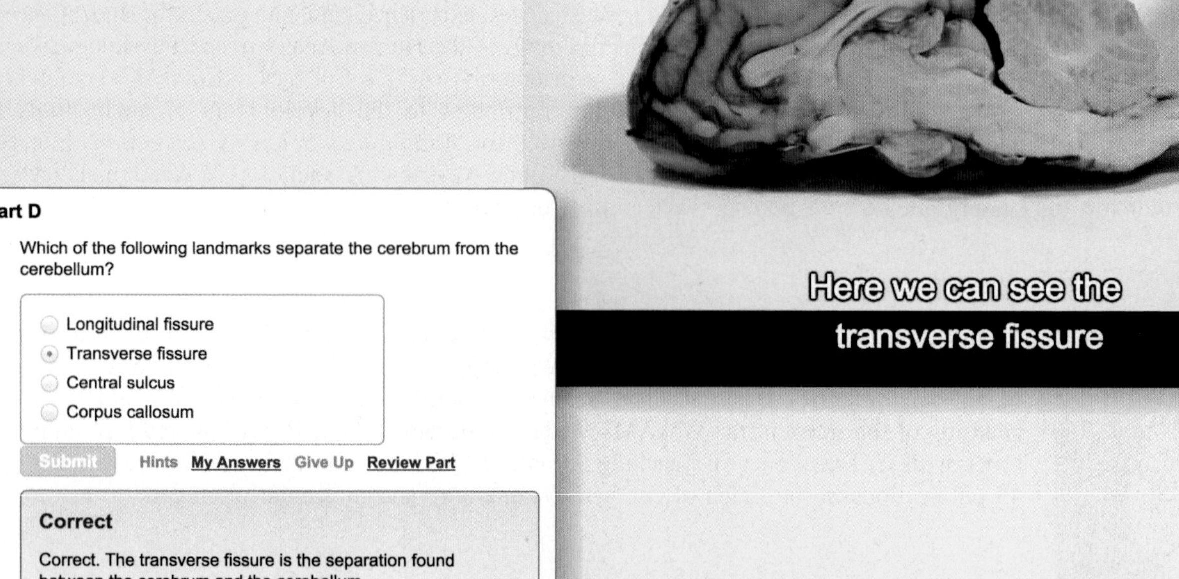

with MasteringA&P® Assignments

NEW! Cat Dissection Videos help students prepare for cat dissection lab and identify key anatomical structures.

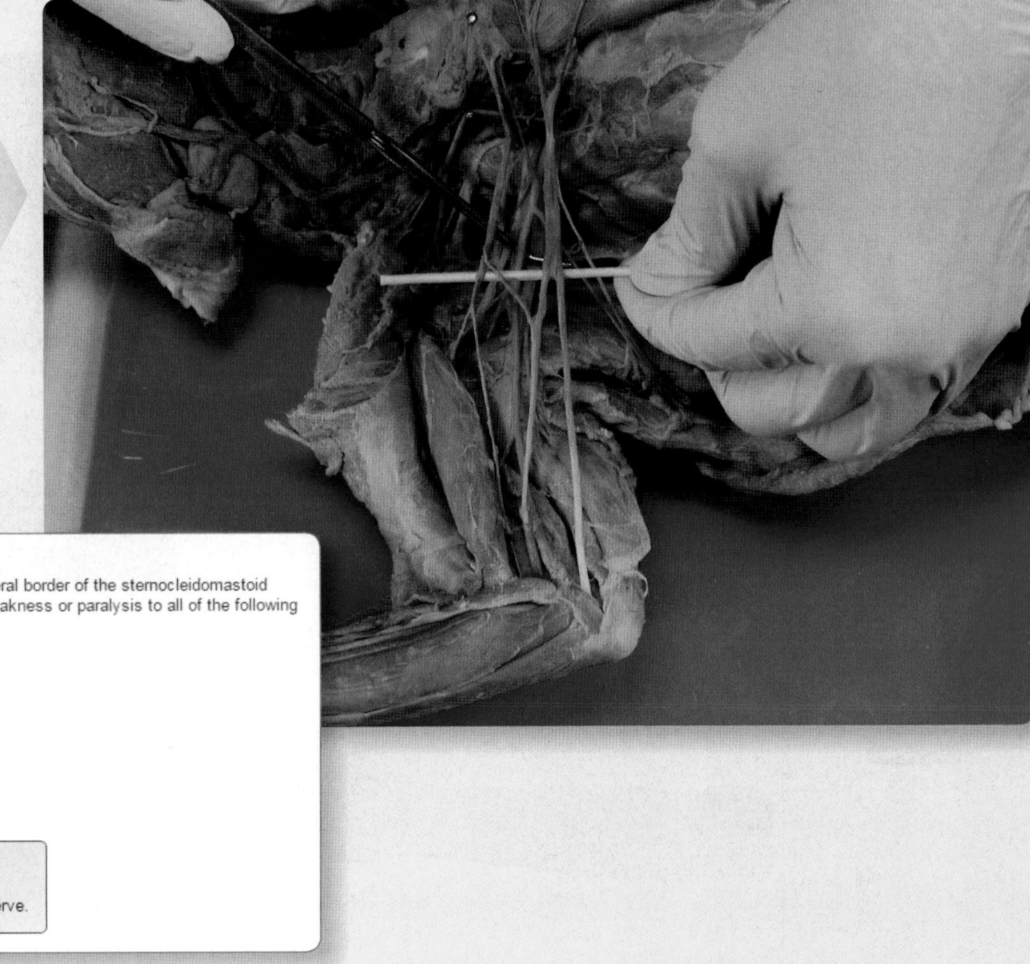

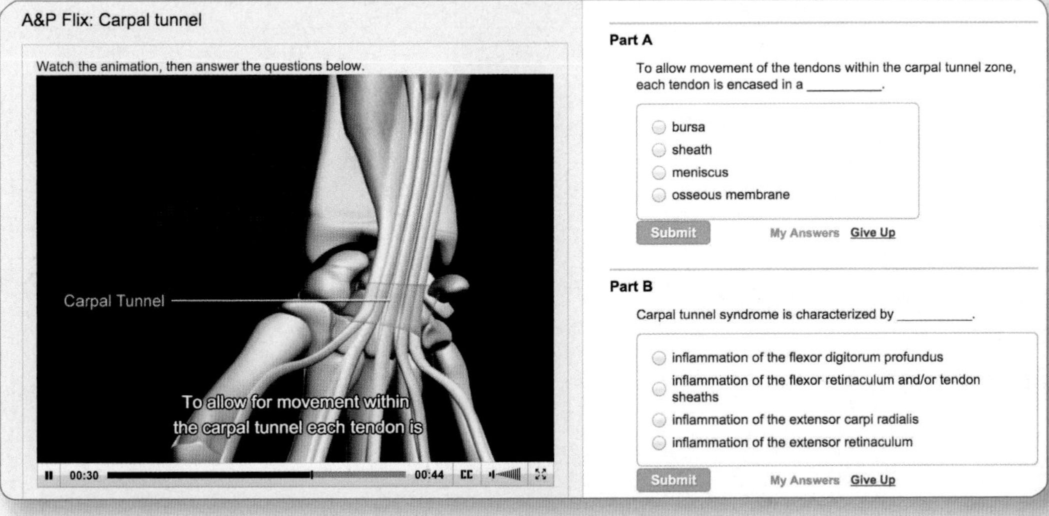

A&P Flix™ Animations are 3D movie-quality animations that help students visualize joint movements and origins, insertions, actions and innervations of over 65 muscles.

Give Your Students Access to Lab

Practice Anatomy Lab™ 3.0 is a virtual anatomy study tool that gives students 24/7 access to the most widely used lab specimens including human cadaver, anatomical models, histology, cat, and fetal pig.

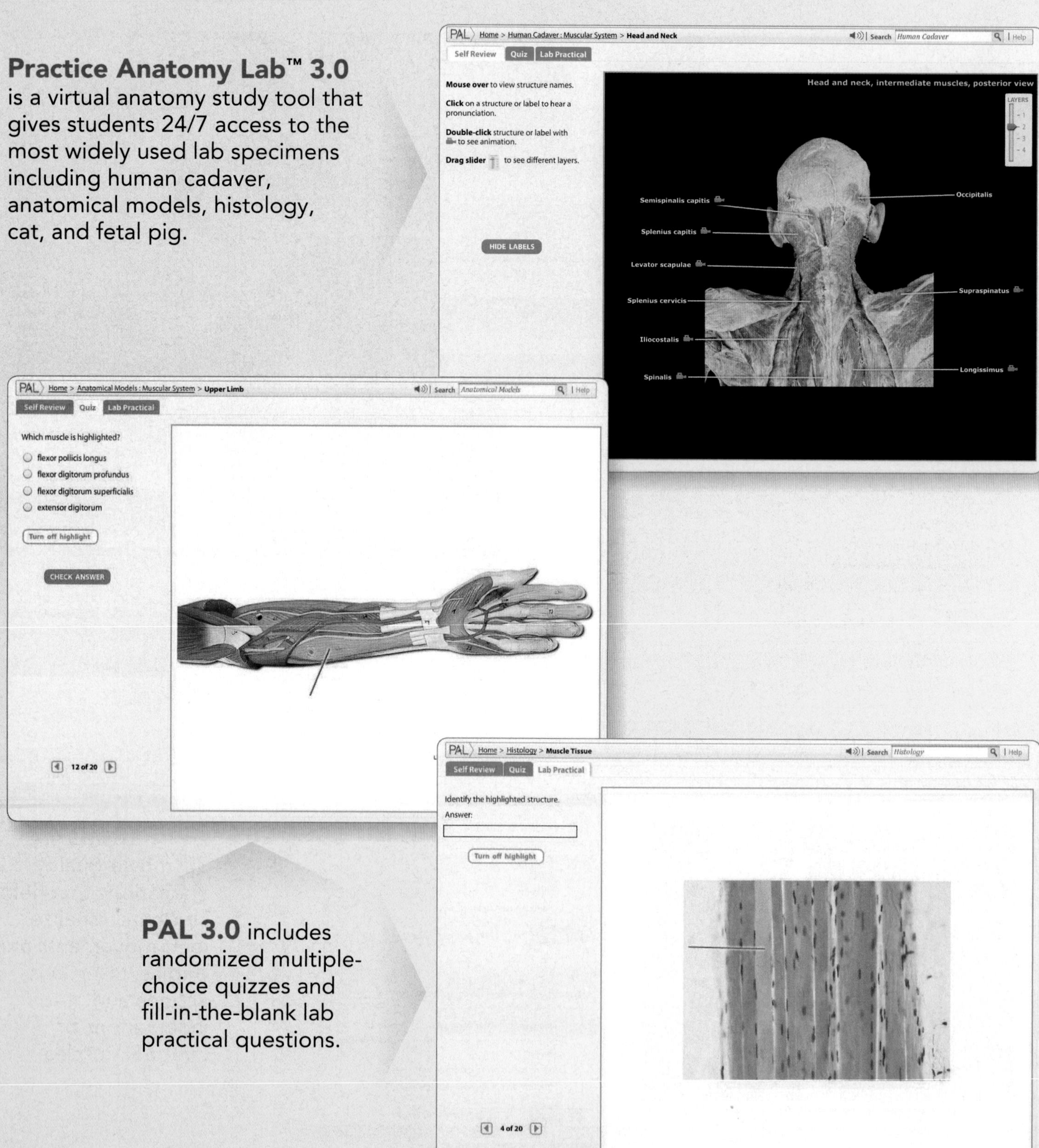

PAL 3.0 includes randomized multiple-choice quizzes and fill-in-the-blank lab practical questions.

Practice 24/7 with MasteringA&P®

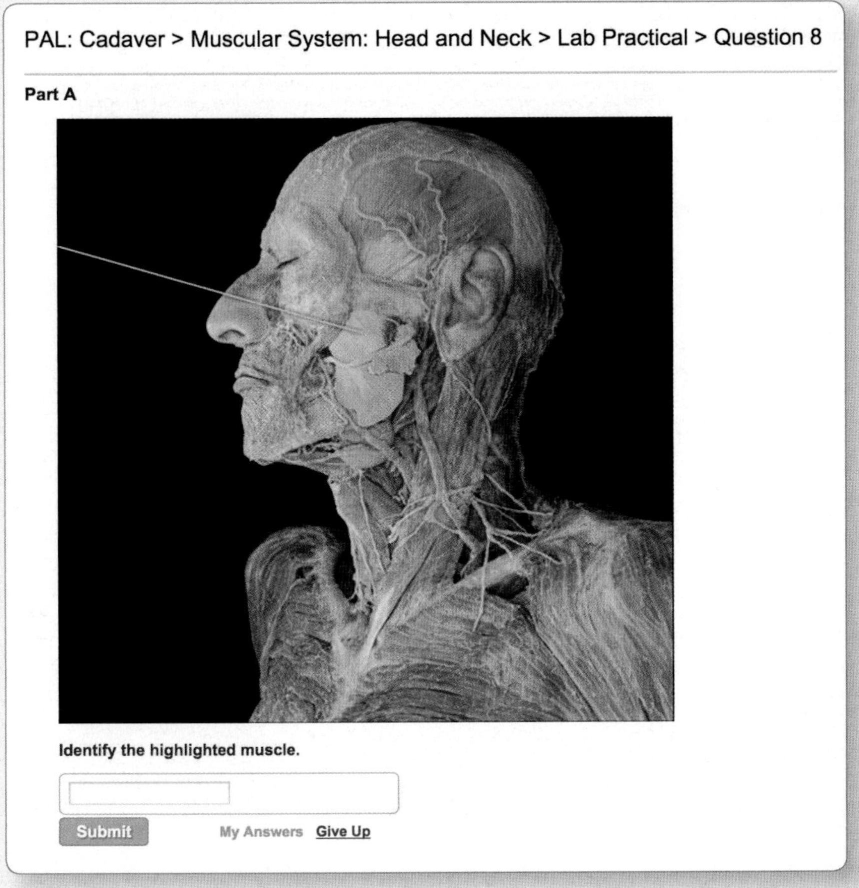

Assign only the structures you want your students to know by using the **PAL™ 3.0 Test Bank**. The PAL 3.0 Test Bank includes over 4,000 customizable questions.

The **PAL 3.0 App** lets students access PAL 3.0 on their iPad or Android tablet. Students can enlarge images, watch animations, and study for lab practicals with multiple-choice and fill-in-the-blank quizzes—all while on the go!

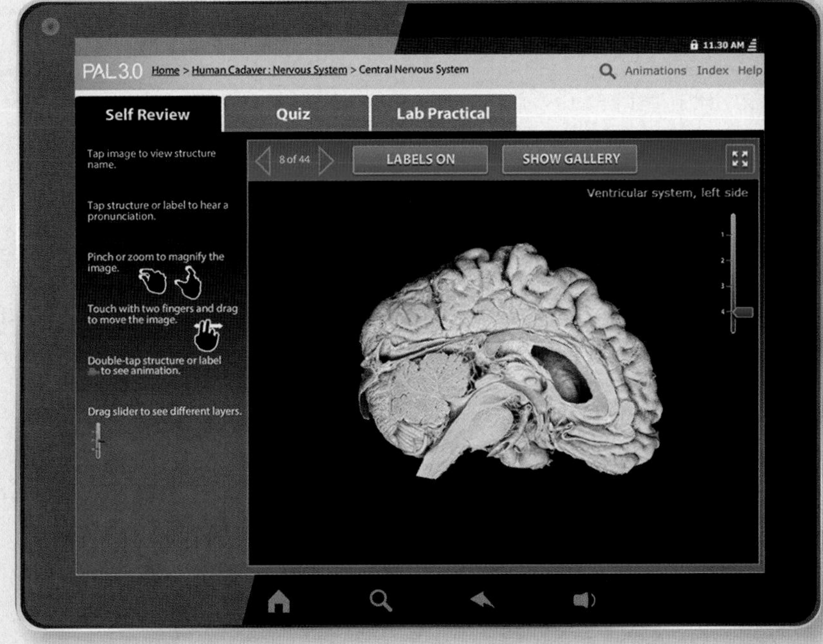

Engage Your Students in Higher-

NEW! Clinical Scenario Coaching Activities use real world examples and art from the book to engage students.

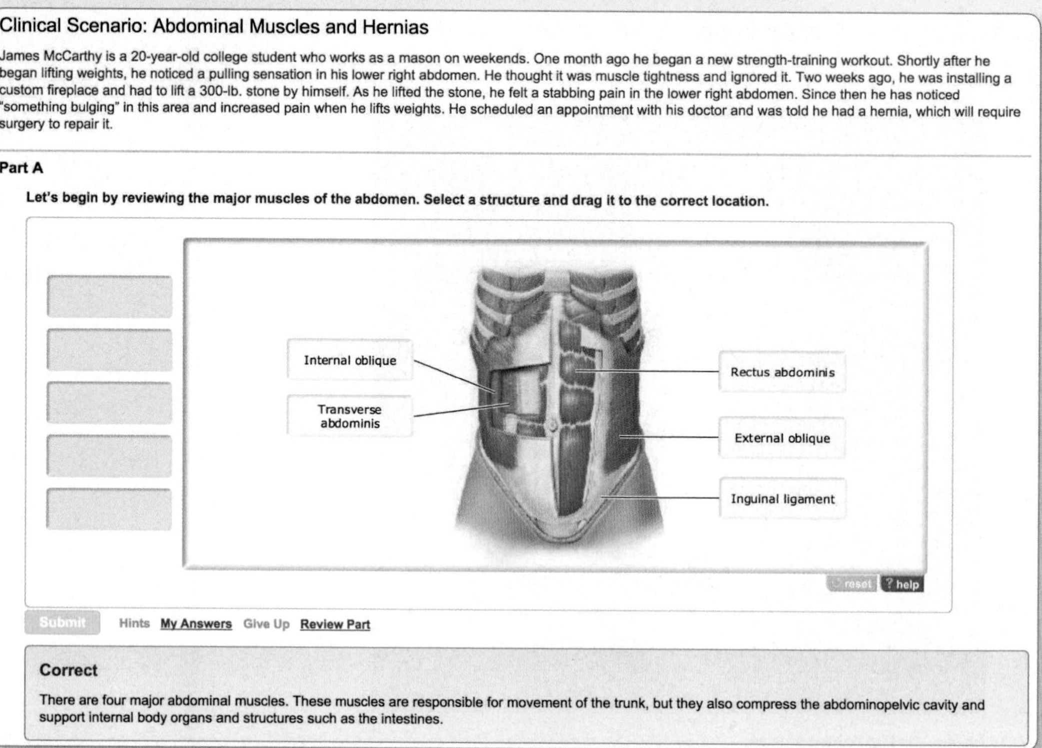

Part F

James needed to have his hernia surgically repaired. This can be done through laparoscopic surgery. The surgeon must cut through to the herniated area and put a Teflon mesh underneath the hernia to close it off and provide more support to the area. Use your knowledge of anatomy to select the correct sequence of structures that would be cut for this procedure. Make sure you select the structures in the order they would be cut by the surgeon.

- ○ Skin, hypodermis, subcutaneous fat, internal oblique aponeurosis, external oblique, transverse abdominis, peritoneum
- ◉ Skin, hypodermis, peritoneum, subcutaneous fat, external oblique aponeurosis, internal oblique, and transverse abdominis
- ○ Skin, hypodermis, subcutaneous fat, external oblique aponeurosis, internal oblique, transverse abdominis, peritoneum
- ○ Skin, hypodermis, peritoneum, subcutaneous fat, transverse abdominis, internal oblique, and the external oblique aponeurosis.

Submit   Hints   My Answers   Give Up   Review Part

Incorrect; Try Again

The peritoneum is the layer of tissue that surrounds the abdominopelvic cavity, and is the deepest structure.

Level Thinking with MasteringA&P®

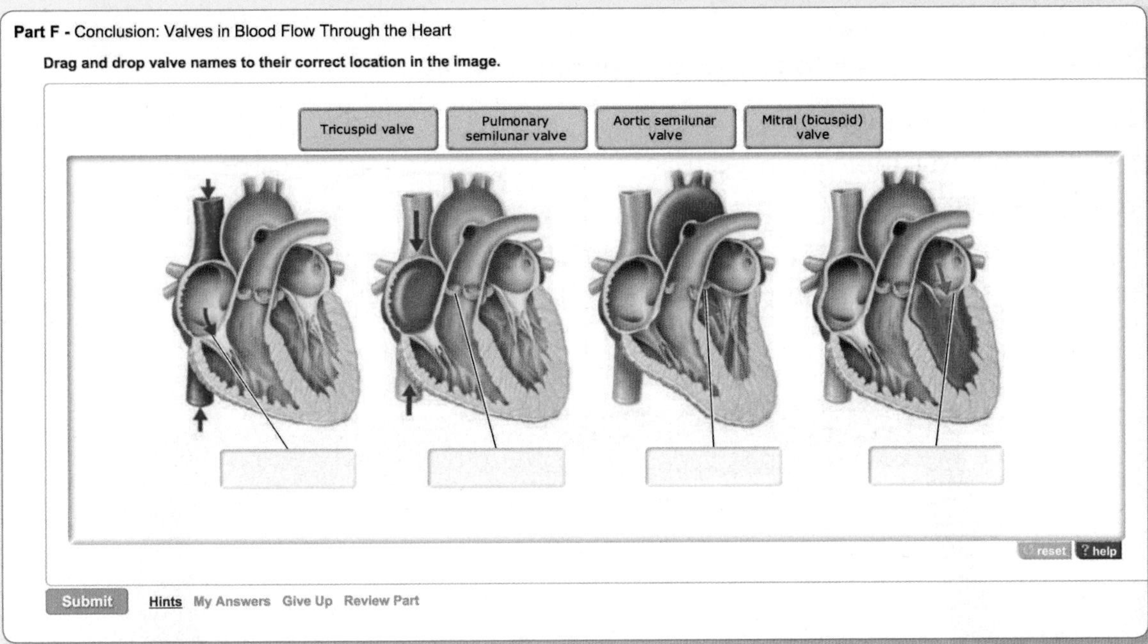

Every Focus Figure has an assignable, multi-step Coaching Activity in MasteringA&P.

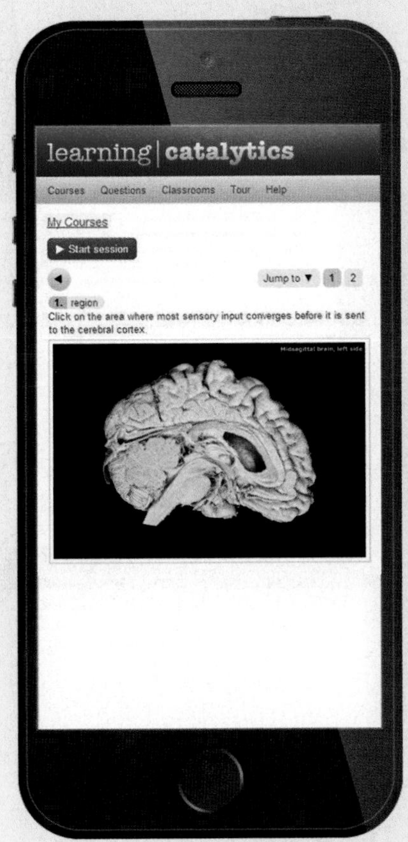

NEW! Learning Catalytics is a "bring-your-own-device" engagement, assessment, and classroom intelligence system. With Learning Catalytics, instructors can flip the classroom and assess students in real time using open-ended tasks to probe their understanding. Students use their smartphone, tablet, or laptop to respond to questions in class.

Help Your Students Study Anywhere,

NEW! eText 2.0 brings the textbook to any web-enabled device.

- Now available on smartphones and tablets.
- Seamlessly integrated videos and other rich media.
- Accessible (screen-reader ready).
- Configurable reading settings, including resizable type and night-reading mode.
- Instructor and student note-taking, highlighting, bookmarking, and search.

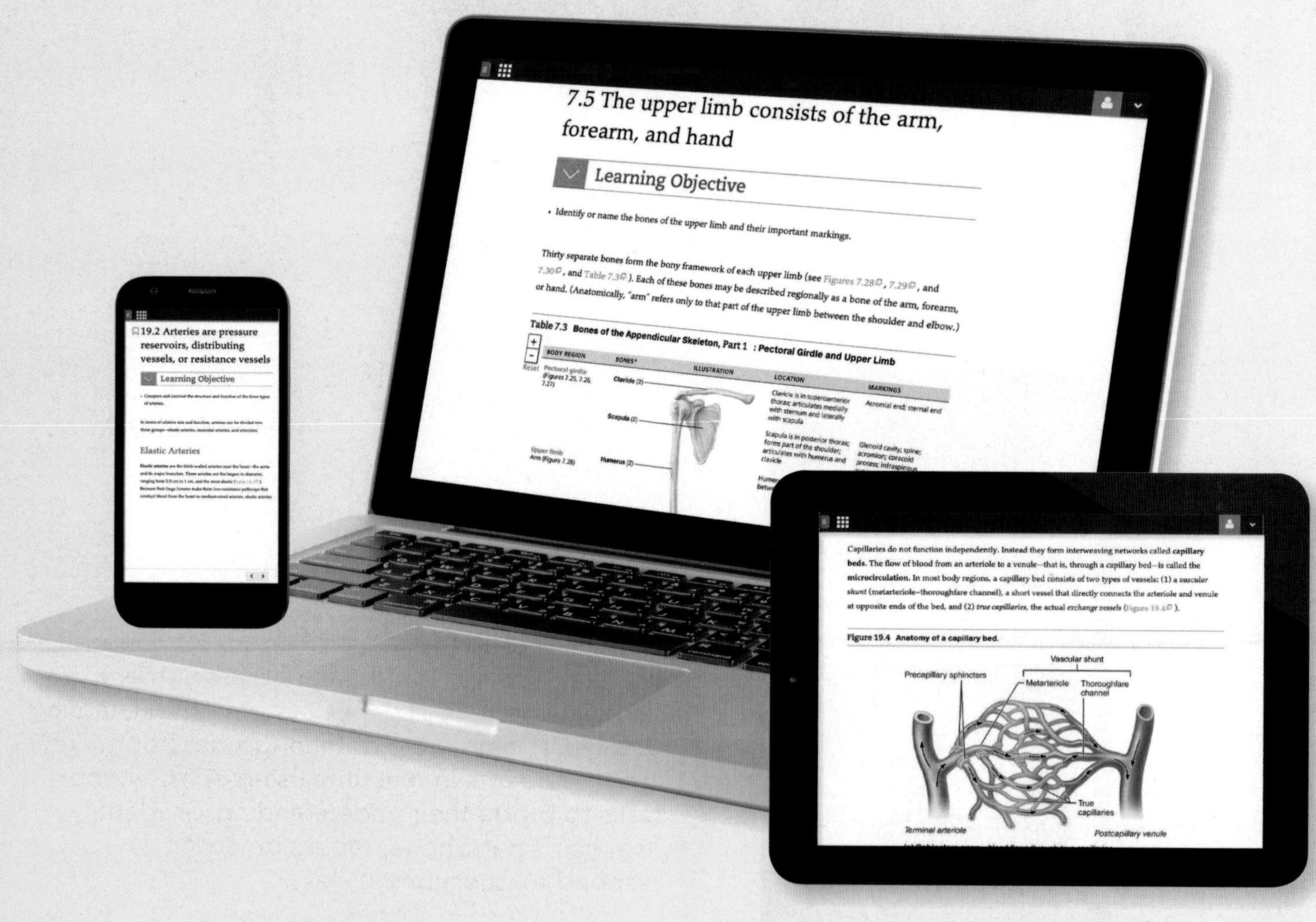

Anytime with Mobile Tools

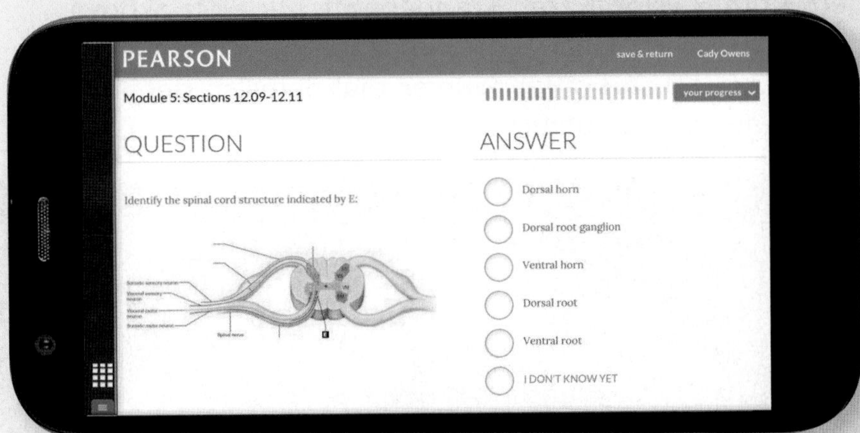

NEW! Dynamic Study Modules help students acquire, retain, and recall information faster and more efficiently than ever before. These mobile-friendly, flashcard-style questions adapt to a student's performance, and include feedback with text and art from the book itself.

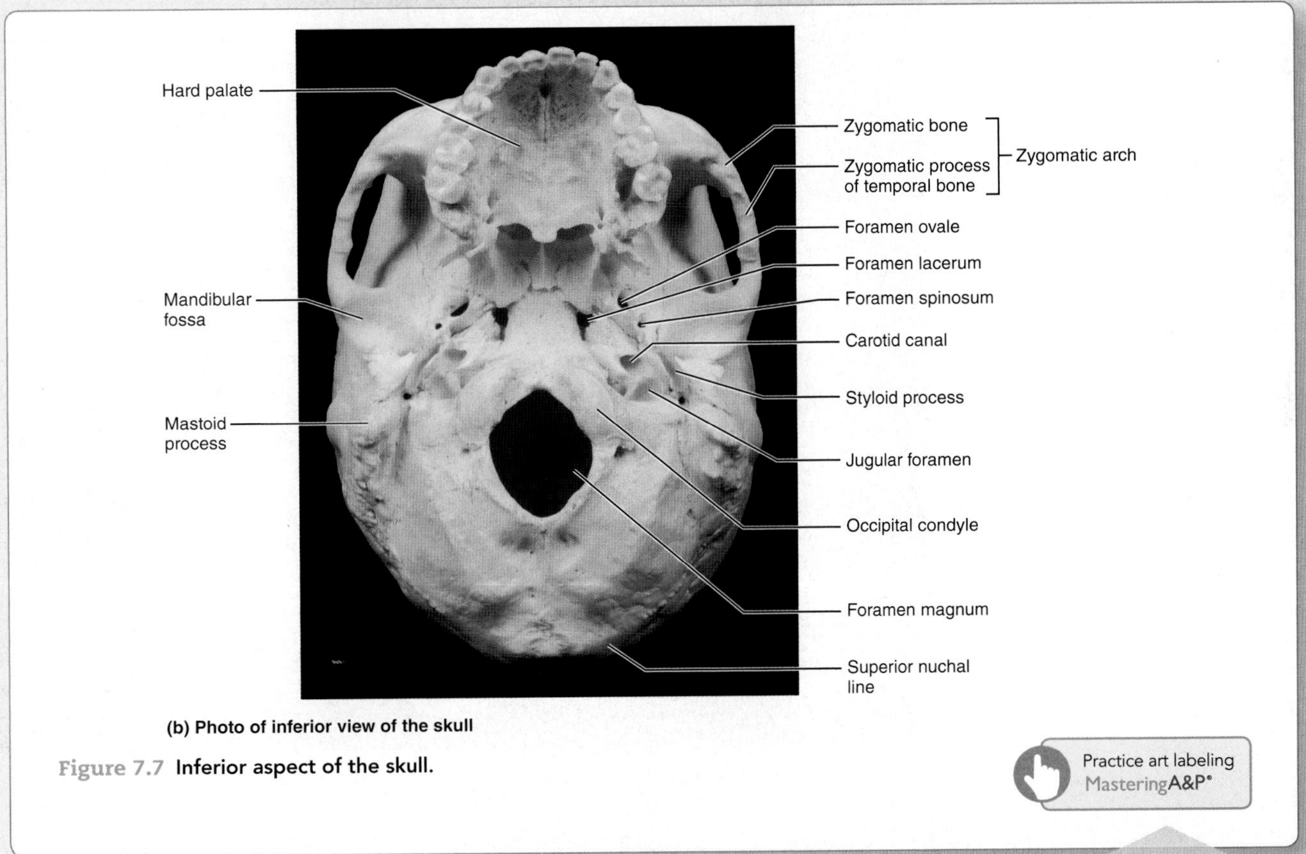

(b) Photo of inferior view of the skull

Figure 7.7 Inferior aspect of the skull.

NEW! Media References in the textbook to PAL 3.0, A&P Flix animations, bone videos, animal organ dissection and cat dissection videos, and art-labeling activities in MasteringA&P help students easily find relevant media resources as they are reading the book.

Maximize Your Students' Learning

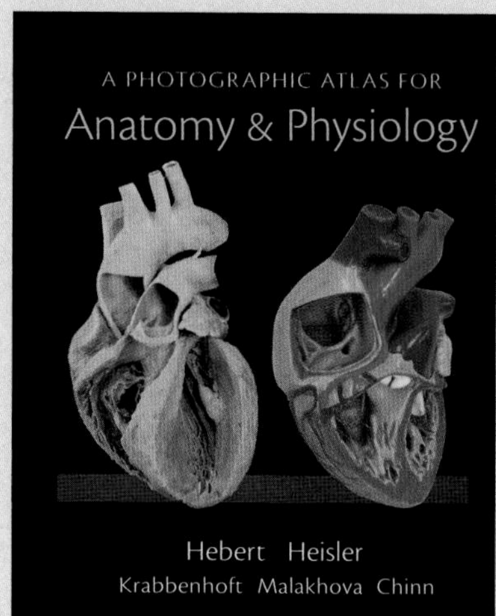

NEW! ***A Photographic Atlas for Anatomy & Physiology,*** by Nora Hebert, Ruth Heisler, Karen Krabbenhoft, Olga Malakhova, and Jett Chinn is a new visual lab study tool featuring photos from Practice Anatomy Lab 3.0 and other sources that helps students learn and identify key anatomical structures.

Available as part of the student package for Marieb, Wilhelm, and Mallatt's *Human Anatomy,* Eighth Edition.

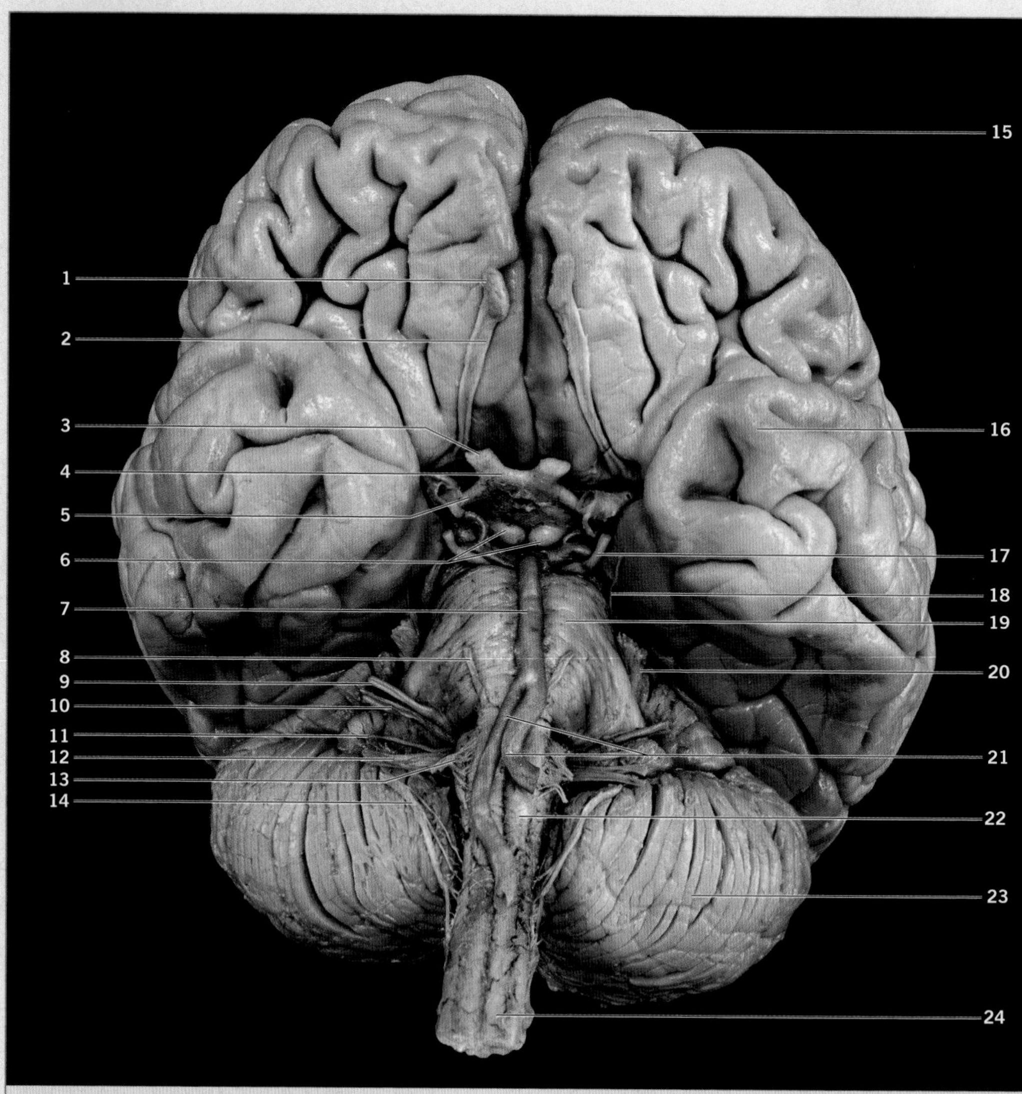

in the Lab

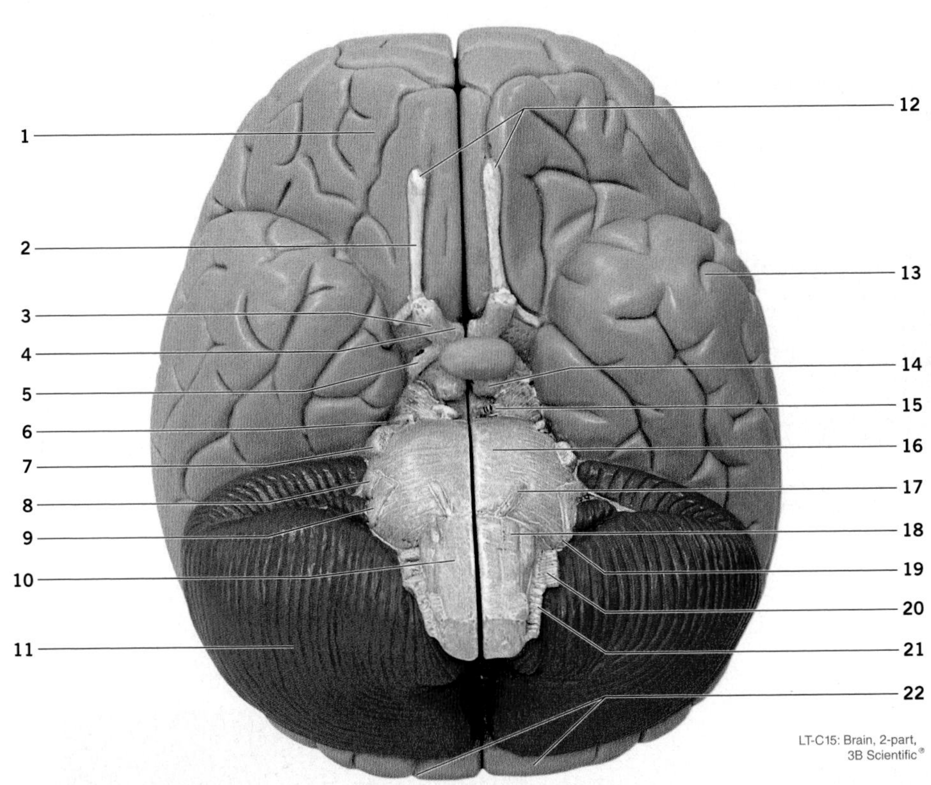

FIGURE 5.7 Brain with Cranial Nerves, Inferior View

1. Frontal lobe
2. Olfactory tract
3. Optic nerve (II)
4. Optic chiasma
5. Optic tract
6. Trochlear nerve (IV)
7. Trigeminal nerve (V)
8. Vestibulocochlear nerve (VIII)
9. Facial nerve (VII)
10. Medulla oblongata
11. Cerebellum
12. Olfactory bulbs
13. Temporal lobe
14. Mammillary body
15. Oculomotor nerve (III)
16. Pons
17. Abducens nerve (VI)
18. Hypoglossal nerve (XII)
19. Glossopharyngeal nerve (IX)
20. Vagus nerve (X)
21. Accessory nerve (XI)
22. Occipital lobes

Everything Your Students Need to Succeed in Lecture and Lab

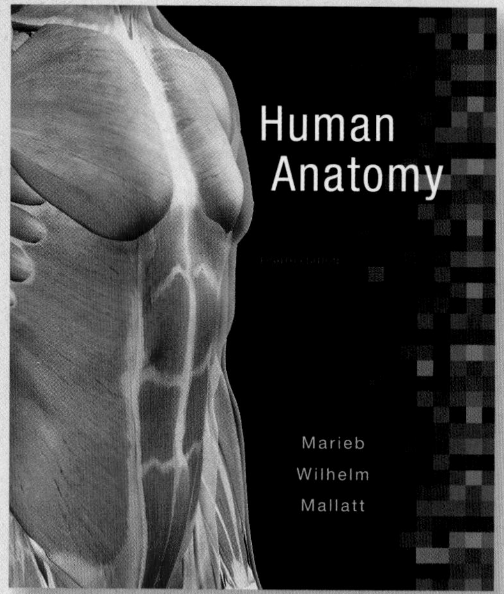

NEW! Student package for Marieb, Wilhelm, and Mallatt's *Human Anatomy*, Eighth Edition now includes MasteringA&P + *A Photographic Atlas for Anatomy & Physiology*.

Marieb/Wilhelm/Mallatt,
Human Anatomy,
Eighth Edition

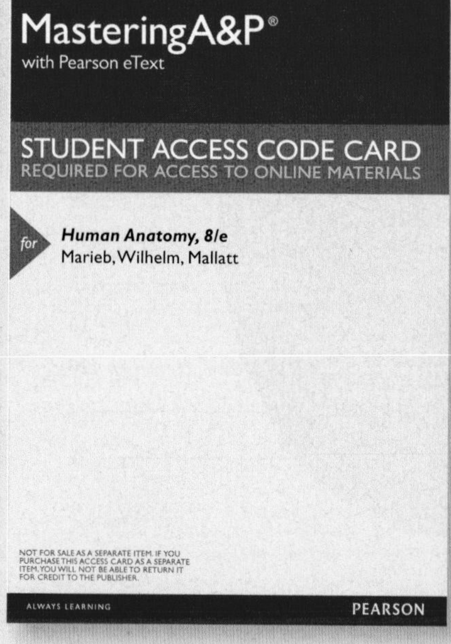

MasteringA&P® Student Access Code for Marieb/Wilhelm/Mallatt, *Human Anatomy,* Eighth Edition

A PHOTOGRAPHIC ATLAS FOR
Anatomy & Physiology
Hebert Heisler
Krabbenhoft Malakhova Chinn

Hebert et al.,
A Photographic Atlas for Anatomy & Physiology

Could Your Students Use a Refresher?

NEW! MyReadinessTest™

for A&P prepares students *before* their Human Anatomy course begins.

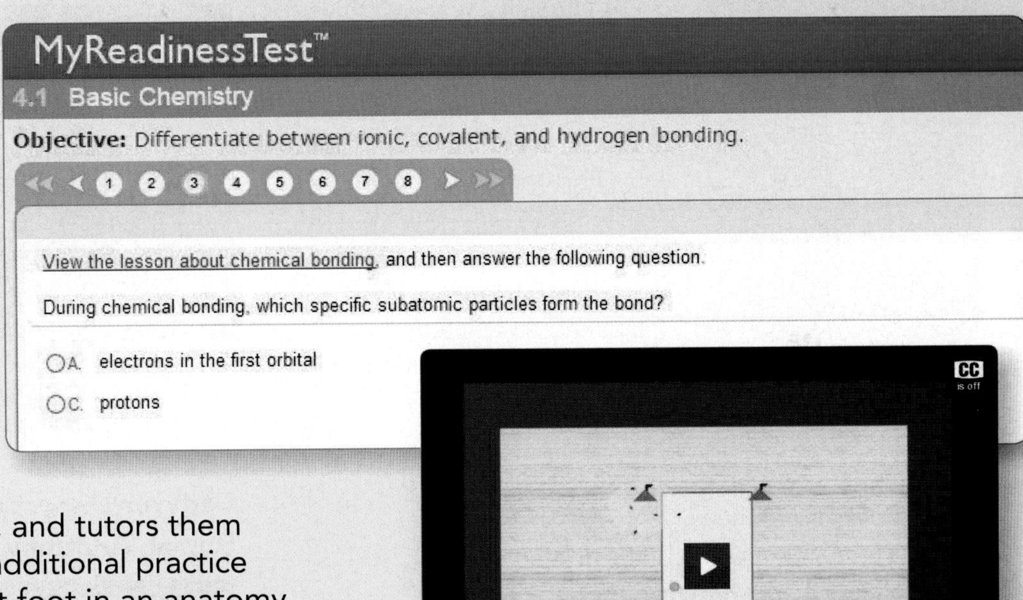

Students can get *free* online access the moment they register for your anatomy course. MyReadinessTest assesses students' proficiency in study skills and foundation concepts in science, and tutors them in core areas where they need additional practice and review, before they even set foot in an anatomy classroom. It offers:

- **Student Online Access** upon registration for their Human Anatomy course.
- **Diagnostic Test and Cumulative Test** based on learning outcomes from a widely used primer, *Get Ready for A&P* by Lori Garrett.
- **Personalized Study Plan** based on students' test results that includes practice questions with tutorials.
- **Flexible Testing** that allows instructors to edit the Diagnostic Test or implement their own placement test or exit exam.
- **Gradebook** that automatically records students' results.

NEW! Five new video tutors on topics such as Learning Styles, Relative Positions, Meiosis, Chemical Reactions, and Concept Mapping.

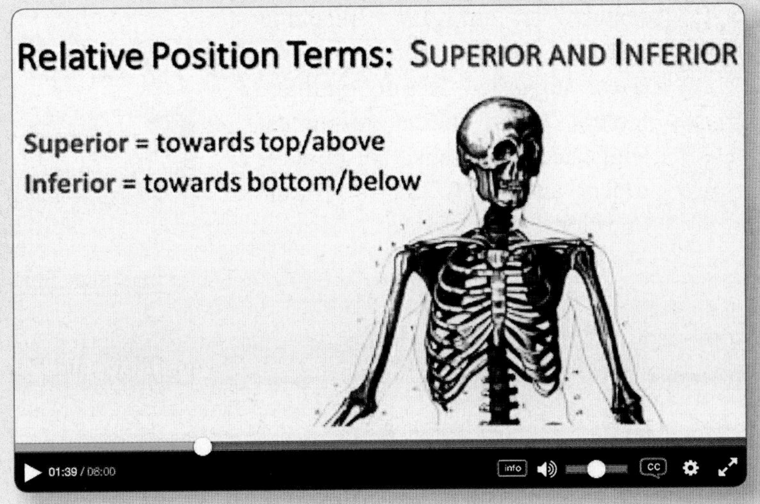

Supplements for Instructors

Instructor Resource DVD

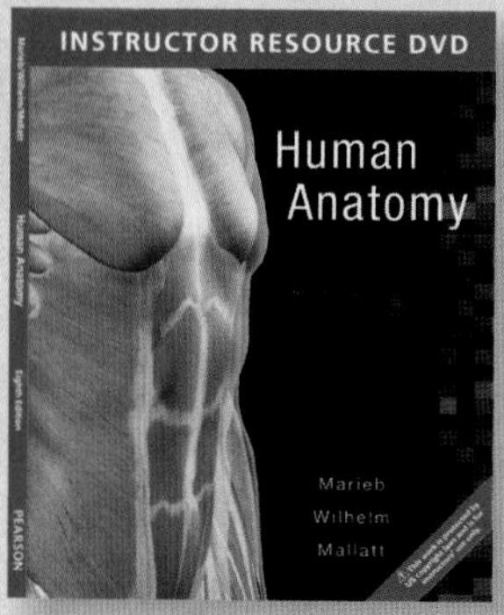

0-134-28663-4 / 978-0-134-28663-1

The Instructor Resource DVD organizes all instructor media resources by chapter into one convenient package that allows you to easily and quickly pull together a lecture.

- Customizable PowerPoint® Lecture Presentations, with labeled and unlabeled images.
- All figures from the book in JPEG format and PowerPoint slides
- Instructor's Resource Guide in Microsoft Word®
- Cat Dissection Videos
- A&P Flix™ Animations and Clicker Questions
- Bone and Dissection Videos
- Images from *A Brief Atlas of the Human Body,* Second Edition
- Complete Test Bank with TestGen® software
- Clicker Questions and Quiz Show Presentations
- PAL 3.0™ Instructor Resource DVD with Test Bank
- Index of anatomical structures covered in PAL 3.0 (PDF format)

Instructor's Resource Guide

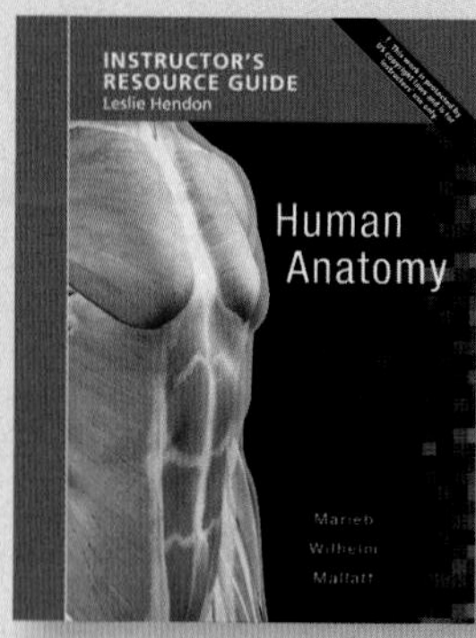

By Leslie Hendon
0-134-28661-8 / 978-0-134-28661-7

The *Instructor's Resource Guide* features an innovative Teaching with Art feature, learning objectives, suggested lecture outlines, lecture hints, media resources, suggested readings, discussion topics, answers to end-of-chapter questions, and more.

Printed Test Bank

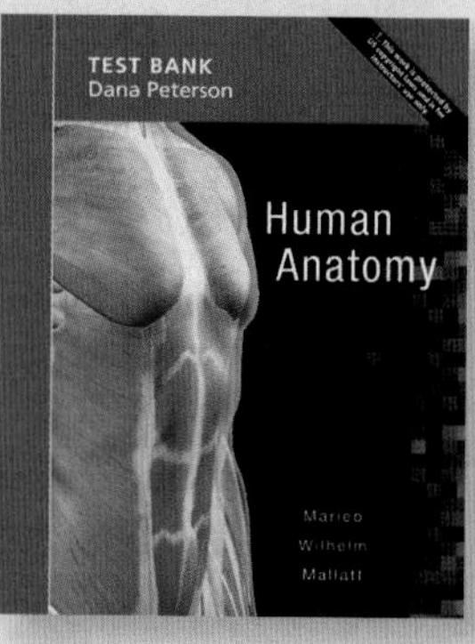

By Dana Peterson
0-134-28659-6 / 978-0-134-28659-4

The Eighth Edition Test Bank covers all major topics at a range of difficulty levels. All questions in the printed Test Bank are available in Microsoft Word and TestGen formats on the IRDVD and in the Instructor Resources section of MasteringA&P®. Both electronic options are cross-platform and allow instructors to easily generate and customize tests.

Cadaver Dissection Video Series for Human Anatomy & Physiology

By Rose Leigh Vines
0-321-85920-0 / 978-0-321-85920-4

These videos reinforce many of the concepts covered in Human Anatomy and are available on a single DVD. Please contact your Pearson sales representative for more information about an Instructor's Review Copy.

PAL™ 3.0 Instructor Resource DVD with Test Bank

By Nora Hebert, Ruth Heisler, Jett Chinn, Karen M. Krabbenhoft, Olga Malakova
0-321-74963-4 / 978-0-321-74963-5

Includes everything an instructor needs to present and assess PAL 3.0 in lecture and lab. The DVD includes images in PowerPoint with editable labels and leader lines, labeled and unlabeled images in JPEG and PowerPoint format, and more.

Supplements for Students

Human Anatomy Laboratory Manual with Cat Dissections, Eighth Edition

By Elaine N. Marieb and Lori A. Smith
0-134-25558-5 / 978-0-134-25558-3

This lab manual contains 30 gross anatomy and histology exercises for all major body systems, featuring 24 cat dissection photos. Illustrated in full color, with convenient spiral binding and an Instructor's Guide, the lab manual is an excellent accompaniment to *Human Anatomy* for lab.

A Photographic Atlas for Anatomy & Physiology

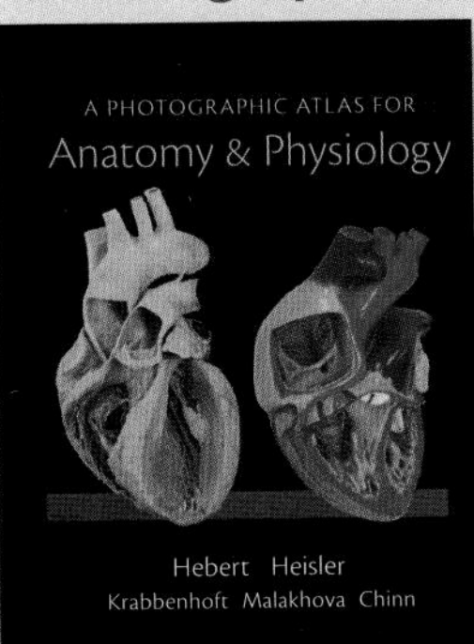

By Nora Hebert, Ruth Heisler, Karen Krabbenhoft, Olga Malakhova, and Jett Chinn
0-321-86925-7 / 978-0-321-86925-8

NEW! Visual lab study tool that helps students learn and identify key anatomical structures. Featuring photos from Practice Anatomy Lab™ 3.0 and other sources, the Atlas includes over 250 cadaver dissection photos, histology photomicrographs, and cat dissection photos plus over 50 photos of anatomical models.

Practice Anatomy Lab™ 3.0 DVD

By Ruth Heisler, Nora Hebert, Jett Chinn, Karen Krabbenhoft, and Olga Malakhova
0-321-68211-4 / 978-0-321-68211-6

PAL 3.0 is an indispensable virtual anatomy study and practice tool that gives students 24/7 access to the most widely used lab specimens, including the human cadaver, anatomical models, histology, cat, and fetal pig.

Practice Anatomy™ (PAL™) 3.0 Lab Guide

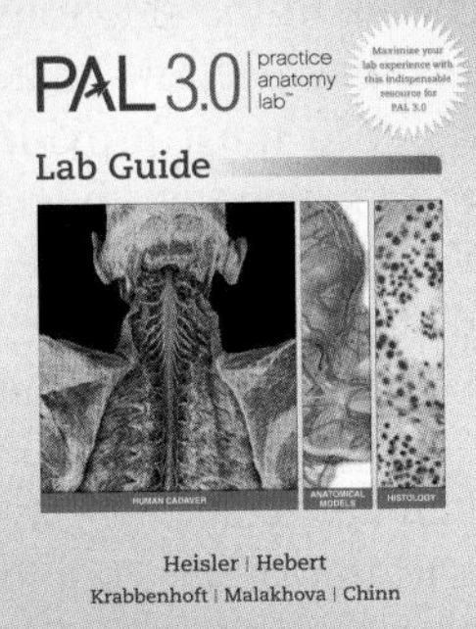

By Ruth Heisler, Nora Hebert, Karen Krabbenhoft, Olga Malakhova, and Jett Chinn
Without PAL 3.0 DVD (0-321-84025-9); with PAL 3.0 DVD (0-321-85767-4)

The PAL 3.0 Lab Guide enhances students' virtual anatomy lab experience by helping them explore anatomical structures through a series of labeling activities and quizzes using the images from PAL.

Get Ready for A&P, Third Edition

By Lori K. Garrett
0-321-81336-7 / 978-0-321-81336-7

This book and online component were created to help students be better prepared for their course. Features include pre-tests, guided explanations followed by interactive quizzes and exercises, and end-of-chapter cumulative tests. Also available in the Study Area of MasteringA&P®.

The Anatomy Coloring Book, Fourth Edition

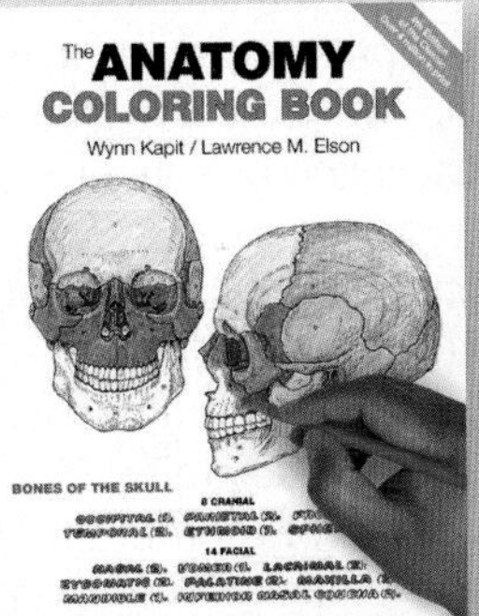

By Wynn Kapit and Lawrence M. Elson
0-321-83201-9 / 978-0-321-83201-6

For over 35 years, The Anatomy Coloring Book has been the #1 best-selling human anatomy coloring book. A useful tool for anyone with an interest in learning anatomical structures, this concisely written text features precise, extraordinary hand-drawn figures that were crafted especially for easy coloring and interactive study. Organized according to body systems, each of the 162 spreads featured in this book includes an ingenious color-key system where anatomical terminology is linked to detailed illustrations of the structures of the body.

PREFACE

The general philosophy behind this Eighth Edition of *Human Anatomy* remains the same as in the previous editions. As an instructor, you know that teaching anatomy is not just the presentation of facts. You must provide information in a framework that encourages genuine understanding, devise new presentations to help students remember large amounts of material, and help students apply what they have learned to new situations. All the while you hope that you inspire in the students a love of the subject.

After many years of teaching human anatomy, we became convinced that new approaches to the subject could excite and challenge the students' natural curiosity. That is why we decided to write this book. We are fortunate to have collaborated with Pearson Education, a publisher that shares our goal: to set a new standard for pedagogical and visual effectiveness in an anatomy text.

This book is designed for one-semester or one-quarter introductory anatomy courses that serve students in pre-nursing, pre-medical, pre-physical therapy, radiological technology, physician assistant training, pre-dentistry, pharmacy, and other allied-health fields, as well as physical education, athletic training, and nutrition.

Unique Approach to Anatomy

Since its inception, we have worked diligently to distinguish *Human Anatomy* from the many other anatomy books currently available. This book explains anatomy thoroughly, and its discussions are not merely brief summaries of the art. We have striven to present the basic concepts of anatomy—gross, microscopic, developmental, and clinical—in a manner that is clearly written, effectively organized, up to date, and well illustrated. We realize that learning anatomy involves assimilating gargantuan amounts of material, and we have tried to make our presentation as logical and accessible as possible. To this end, we present anatomy as a "story" that can be explained and understood—convincing the students that the structure of the body makes sense.

Although descriptive gross anatomy is a relatively static science, knowledge is growing quickly in the subfields of functional anatomy, neuroanatomy, developmental anatomy, and the functional aspects of tissue and cellular anatomy. This text strives to keep up with the knowledge explosion in these subfields and to present anatomy in a way that allows modern biology students, whose training is becoming ever more molecular and cellular, to anchor their biochemical and medical training in the physical context of the human body.

Functional Approach

We strongly emphasize the functional anatomy theme, giving careful consideration to the adaptive characteristics of the anatomical structures of the body. Wherever possible, we explain how the shape and composition of the anatomical structures allow them to perform their functions. Such functional anatomy is not physiology (which focuses on biological mechanisms), but is more akin to "design analysis." This approach is unique for a text at this level.

Microscopic Anatomy

We have worked to provide an especially effective treatment of microscopic anatomy. Many undergraduate texts treat histology as a specialized and minor subfield that takes a back seat to gross anatomy. This is unfortunate, because most physiological and disease processes take place at the cellular and tissue level, and most allied-health students require a solid background in histology and subcellular structure to prepare them for their physiology courses.

Embryology

Our text is designed to present embryology in the most effective and logical way. We are convinced that the fundamentals should be presented early in the text, before the more advanced discussions of the developing organ systems in the relevant chapters. Therefore, we wrote Chapter 3 as a basic introduction to embryology. Because a comprehensive presentation of embryology early in the book could be intimidating to some students, we have used a "velvet glove approach," providing only the most important concepts in a concise, understandable way, visually reinforced with exceptionally clear art.

Life Span Approach

Most chapters in this book close with a "Throughout Life" section that first summarizes the embryonic development of organs of the system and then examines how these organs change across one's life span. Diseases particularly common during certain periods of life are pointed out, and effects of aging are considered. The implications of aging are particularly important to students in the health-related curricula because many of their patients will be older adults.

Helpful Presentation of Terminology

The complex terminology of anatomy is one of the most difficult aspects of the subject to make interesting and accessible. To this end, we highlight important terms in boldfaced type, and we provide the pronunciations of more terms than do many competing texts. Also, we include the Latin or Greek translations of almost every term at the point where the term is introduced in the text. This promotes learning by showing students that difficult terms have simple, logical derivations. The anatomical terms used in this text are consistent with the terms accepted by the International Federation of Associations of Anatomists (IFAA). Clinical terminology is also presented in the Related Clinical Terms section found at the conclusion of most chapters. A helpful glossary, pronunciation guide, and list of word roots and suffixes are located at the end of the text.

NEW TO THE EIGHTH EDITION

The Eighth Edition builds on the book's hallmark strengths—art that teaches better, a student-friendly narrative, and easy-to-use media and assessment tools—and improves on them.

- **Twelve updated body movement photos and seven updated facial movement photos** clearly demonstrate movements allowed by synovial joints, as well as actions of muscles of the face, scalp, and neck.
- **Two updated Focus figures**, Focus Figure 4.11 (Identifying Epithelial and Connective Tissues) and Focus Figure 15.2 (Comparing Somatic Motor and Autonomic Innervation), have been revised to better highlight and teach important, tough-to-understand concepts.
- **New and improved in-text media references** to PAL 3.0, A&P Flix animations, bone videos, animal organ dissection and cat dissection videos, and art-labeling activities in the Study Area of MasteringA&P® help students easily find helpful study tools as they are reading the book.

More Robust MasteringA&P

MasteringA&P now includes:

- **NEW! Clinical Scenario Coaching Activities** that complement lecture and lab, and can be assigned as part of in-class activities or as post-class assignments. Multiple coaching activities for each chapter include an assortment of multiple choice, sorting, labeling, and matching questions.
- **NEW! Cat Dissection Videos**, created by coauthor Patricia Wilhelm, that are assignable in MasteringA&P with hints and wrong-answer feedback. The videos without questions are also available in the Study Area of MasteringA&P.
  Video topics cover:
  - Superficial Muscles of the Trunk, Dorsal View
  - Deep Muscles of the Trunk, Dorsal View
  - Posterior Muscles of the Hip and Thigh
  - Brachial Plexus and Innervation of the Muscles of the Arm and Forearm
  - Digestive Structures of the Head
  - Peritoneum and Mesenteries of the Abdomen
  - Structures That Pass Through Mesenteries
  - Blood Vessels of the Thorax
  - Male Reproductive Structures
  - Female Reproductive Structures
- **NEW! Dynamic Study Modules** that help students study effectively on their own by continuously assessing their activity and performance in real time. Here's how it works: Students complete a set of questions with a unique answer format that also asks them to indicate their confidence level. Questions repeat until the student can answer them all correctly and confidently. Once completed, Dynamic Study Modules explain the concept using materials from the text. These are available as graded assignments prior to class, and accessible on smartphones, tablets, and computers.
- **NEW! eText 2.0**, which seamlessly integrates videos and other rich media right into the reading experience. eText 2.0 is available in MasteringA&P and on smartphones and tablets. It is screen-reader ready, includes note-taking, highlighting, bookmarking, and search capabilities, and features customizable settings such as night reading mode.
- **Bone and Dissection Video Coaching Activities** review all major bones and organ dissections. Each video is supported by activities with hints and specific wrong-answer feedback.
- **UPDATED! Focus Figure Coaching Activities** expand upon the popular Focus figures in the text by guiding students through complex processes step by step with hints and specific wrong-answer feedback. The Coaching Activities for Focus Figures 4.11 and 15.2 have been updated.
- ***Get Ready for A&P* Diagnostic, Learning Styles, and Cumulative Tests** along with ***Get Ready for A&P* Video Tutors** feature award-winning teacher Lori Garrett walking students through key basic concepts needed for students to be successful in A&P. Students can take the assignable Diagnostic Test and/or Learning Styles Test in MasteringA&P to assess their base knowledge at the start of the course. Chapter assessments include Reading Questions and Video Tutor Coaching Activities. The key concepts covered include: Learning Styles, Study Skills, Basic Math Review, Terminology, Body Basics, Chemistry, and Cell Biology.
- **A&PFlix™ Coaching Activities** provide dramatic 3-D animations of key anatomy topics, including individual muscle origins, insertions, actions, and innervations, and key muscle actions and joint movement. Each animation provides practice quizzes and wrong-answer feedback.
- **Drag-and-Drop Art Labeling Activities** and Art-Based Questions
- **Practice Anatomy Lab™ 3.0** is an indispensable virtual anatomy study and practice tool that gives students 24/7 access to the most widely used lab specimens including human cadaver, anatomical models, histology, cat, and fetal pig. PAL™ 3.0 includes built-in pronunciation guides, rotatable bones, multiple choice quizzes, and fill-in-the-blank lab practical exams.
- **Practice Anatomy Lab™ 3.0 Test Bank** includes over 4,000 customizable multiple choice and fill-in-the-blank questions. With this test bank, you can assign only the structures you want your students to know.
- **Learning Catalytics™** is an interactive, classroom tool that uses students' smartphones, tablets, or laptops to engage them in more sophisticated tasks and thinking. Now included with Mastering with eText, Learning Catalytics enables you to generate classroom discussion, guide your lecture, and promote peer-to-peer learning with real-time analytics. Instructors can:
  - Pose a variety of open-ended questions that help your students develop critical thinking skills

- Monitor responses to find out where students are struggling
- Use real-time data to adjust your instructional strategy and try other ways of engaging your students during class
- Manage student interactions by automatically grouping students for discussion, teamwork, and peer-to-peer learning

ACKNOWLEDGMENTS

As we work on each new edition, we are reminded of the great pleasure of working collaboratively with dedicated, competent, and skilled professionals. This experience reinforces the importance of developing collaborative skills in our students. This edition is no different. So many individuals have been involved in the various stages of manuscript preparation, review, and production. Each person mentioned here has directly influenced and improved the final product. More important, each has been a pleasure to work with, and we thank them all.

Serina Beauparlant, Editor-in-Chief, guided the planning and implementation of this new edition and associated electronic media. Serina's understanding of the needs of faculty and students, focused approach, and dedication to producing the best product available have proved invaluable. We thank her for her leadership and her friendship.

Michele Mangelli, Mangelli Productions, was the project manager for this new edition, coordinating the editorial and production aspects of the project. As always, Michele's diligence and attention to detail and flexibility with due dates has facilitated the work on this edition. Michele also organized and managed the production of video content that supplements the textbook, along with Amanda Kaufman, Coproducer; Richard Boghosian, Videographer; and Anthony Saccocio, Sound Engineer. The Content Producers, Nicole Tache and Patrice Fabel, oversaw the development of the book's media program. Stacey Weinberger once again contributed to this text with her manufacturing expertise. Allison Rona and Derek Perrigo, our Marketing Managers, have efficiently kept a finger on the pulse of the marketplace—keeping in touch with professors and students, and providing feedback on what they do or do not like about the text and media products. Thank you all.

Laura Southworth worked on the revision of Focus Figures. Her skill with art and layout, combined with her extensive content knowledge helped to improve select focus figures tremendously.

Kristin Piljay has done an outstanding job as photo researcher. Special thanks to the team at Imagineering for work on the art for this edition. Their skillful work has added significantly to this new edition.

We thank other members of the production team—Karen Gulliver, production coordinator; Jean Lake, art and photo coordinator; Anita Hueftle, copyeditor; and Betsy Dietrich, proofreader—for their excellent work and attention to detail. Many thanks go out to tani hasegawa for a beautiful interior design and to Tandem Creative for the stunning new cover. We also appreciate the fine work of Cenveo in assembling the pages.

Special thanks to Ruth Heisler for writing Clinical Scenario Coaching Activities for MasteringA&P, Leslie Hendon for her work on the Dynamic Study Modules, Justin Shaffer for writing alt-text and completing correlations for questions in MasteringA&P, and Janet Brodsky for her work on art-labeling activities in MasteringA&P. Leif Saul, Anthony Friscia, and Anthony Weinhaus reviewed revised Focus Figures. Their insight and expertise contributed significantly to the effectiveness of these figures. Thank you. The administration at Johnson & Wales University provided teaching release time for work on this revision. We are also grateful to Eric Leaver, Development Editor, for revising the Test Bank.

We also want to thank the following reviewers for their feedback and advice on MasteringA&P and our new media for the eighth edition:

Joslyn Ahlgren, *University of Florida*
Gary Allen, *Dalhousie University*
Kathleen Azevedo, *University of California, Berkeley Extension, Las Positas College*
David Babb, *West Hills College–Lemoore Campus*
Elizabeth Co, *Boston University*
David Conley, *Washington State University*
Lisa Flick, *Monroe Community College*
Jill Harper-Judd, *St. Petersburg College–Clearwater Campus*
Ruth Heisler, *University of Colorado, Boulder*
Leslie Hendon, *University of Alabama at Birmingham*
Kerrie Hoar, *University of Wisconsin–La Crosse*
Steve Hobbs, *University of Colorado, Boulder*
Dawn Hunter, *West Virginia University*
Jeremy Ingraham, *University of North Carolina, Greensboro*
Patricia Phelps, *University of California, Los Angeles*
Carmen Rodriguez, *Virginia Commonwealth University*
Justin Shaffer, *University of California, Irvine*
Judith Tamburlin, *University at Buffalo*
Shelley Thai, *Glendale Community College*

Finally, a note of gratitude for the support and encouragement provided by our families. They have been patient and understanding about the time taken from family while we focused on revisions, and we thank them. Our last acknowledgment is a shout-out to our students, who continue to inspire us.

Elaine N. Marieb

Patricia Brady Wilhelm

Jon Mallatt

BRIEF CONTENTS

DETAILED CONTENTS

The Human Body: An Orientation

1

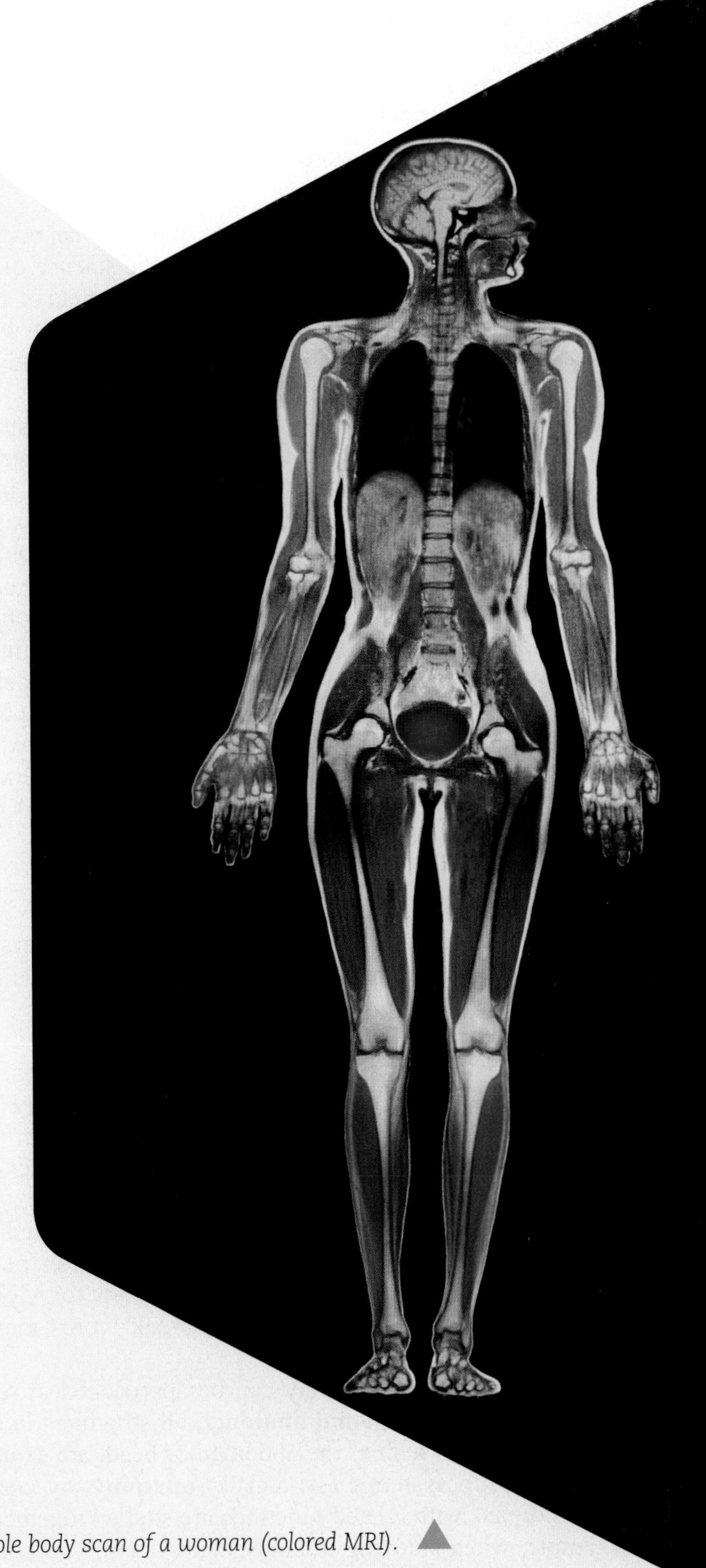

Whole body scan of a woman (colored MRI).

As you read this book, you will learn about a subject that has forever fascinated people—their own bodies. The study of human anatomy is not only an interesting and highly personal experience, but also a timely one. Almost every week, the news media report advances in medical science. Understanding how your body is built and how it works allows you to appreciate newly developed techniques for detecting and treating disease and to apply guidelines for staying healthy. If you are preparing for a career in the health sciences, your knowledge of human anatomy is the foundation of your clinical practice.

AN OVERVIEW OF ANATOMY

learning outcomes

- Define anatomy and physiology, and describe the subdisciplines of anatomy.
- Identify the levels of structural organization in the human body, and explain the interrelationships between each level.
- List the organ systems of the body, and briefly state their functions.
- Use metric units to quantify the dimensions of cells, tissues, and organs.
- Use the meaning of word roots to aid in understanding anatomical terminology.

Anatomy is the study of the structure of the human body. It is also called **morphology** (mor″fol′o-je), the science of form. An old and proud science, anatomy has been a field of serious intellectual investigation for at least 2300 years. It was the most prestigious biological discipline of the 1800s and is still dynamic.

Anatomy is closely related to **physiology,** the study of body function. Although you may be studying anatomy and physiology in separate courses, the two are truly inseparable, because structure supports function. For example, the lens of the eye is transparent and curved; it could not perform its function of focusing light if it were opaque and uncurved. Similarly, the thick, long bones in our legs could not support our weight if they were soft and thin. This textbook stresses the closeness of the relationship between structure and function. In almost all cases, a description of the anatomy of a body part is accompanied by an explanation of its function, emphasizing the structural characteristics that contribute to that function. This approach is called *functional anatomy.*

Subdisciplines of Anatomy

Anatomy is a broad field of science consisting of several subdisciplines, or branches. Each branch of anatomy studies the body's structures in a specialized way.

Gross Anatomy

Gross anatomy (*gross* = large) is the study of body structures that can be examined by the naked eye—the bones, lungs, and muscles, for example. An important technique for studying gross anatomy is **dissection** (dĭ-sek′shun; "cut apart"), in which connective tissue is removed from between the body organs so that the organs can be seen more clearly. Then the organs are cut open for viewing. The term *anatomy* is derived from Greek words meaning "to cut apart."

Studies of gross anatomy can be approached in several different ways. In **regional anatomy,** all structures in a single body region, such as the abdomen or head, are examined as a group. In **systemic** (sis-tem′ik) **anatomy,** by contrast, all the organs with related functions are studied together. For example, when studying the muscular system, you consider the muscles of the entire body. The systemic approach to anatomy is best for relating structure to function. Therefore, it is the approach taken in most college anatomy courses and in this book. Medical schools, however, favor regional anatomy because many injuries and diseases involve specific body regions (sprained ankle, sore throat, heart disease); furthermore, surgeons need extensive and detailed knowledge of each body region.

Another subdivision of gross anatomy is **surface anatomy,** the study of shapes and markings (called *landmarks*) on the surface of the body that reveal the underlying organs. This knowledge is used to identify the muscles that bulge beneath the skin in weight lifters, and clinicians use it to locate blood vessels for placing catheters, feeling pulses, and drawing blood. Clinically useful surface landmarks are described throughout the text in reference to the organ system that they relate to. (Chapter 11 concludes with a section on surface anatomy, which integrates the anatomical relationships between skeletal and muscular structures.)

Microscopic Anatomy

Microscopic anatomy, or **histology** (his-tol′o-je; "tissue study"), is the study of structures that are so small they can be seen only with a microscope. These structures include cells and cell parts; groups of cells, called *tissues;* and the microscopic details of the organs of the body (stomach, spleen, and so on). A knowledge of microscopic anatomy is important because physiological and disease processes occur at the cellular level.

Other Branches of Anatomy

Two branches of anatomy explore how body structures form, grow, and mature. **Developmental anatomy** traces the structural changes that occur in the body throughout the life span and the effects of aging. **Embryology** is the study of how body structures form and develop before birth. A knowledge of embryology helps you understand the complex design of the adult human body and helps to explain birth defects, which are anatomical abnormalities that occur during embryonic development and are evident after birth.

Some specialized branches of anatomy are used primarily for medical diagnosis and scientific research. **Pathological** (pah-tho-loj′ĭ-kal) **anatomy** deals with the structural changes in cells, tissues, and organs caused by disease. (**Pathology** is the study of disease.) **Radiographic** (ra″de-o′graf′ic) **anatomy** is the study of internal body structures by means of X-ray studies and other imaging techniques (see pp. 15–19). **Functional morphology** explores the functional properties of body structures and assesses the efficiency of their design.

The Hierarchy of Structural Organization

The human body has many levels of structural complexity as illustrated in Focus on Levels of Structural Organization (Figure 1.1). At the **chemical level,** *atoms* are tiny building blocks of matter such as carbon, hydrogen, oxygen, and

Levels of Structural Organization

Figure 1.1

Recognizing connections between structural levels leads to better understanding of organismal function.

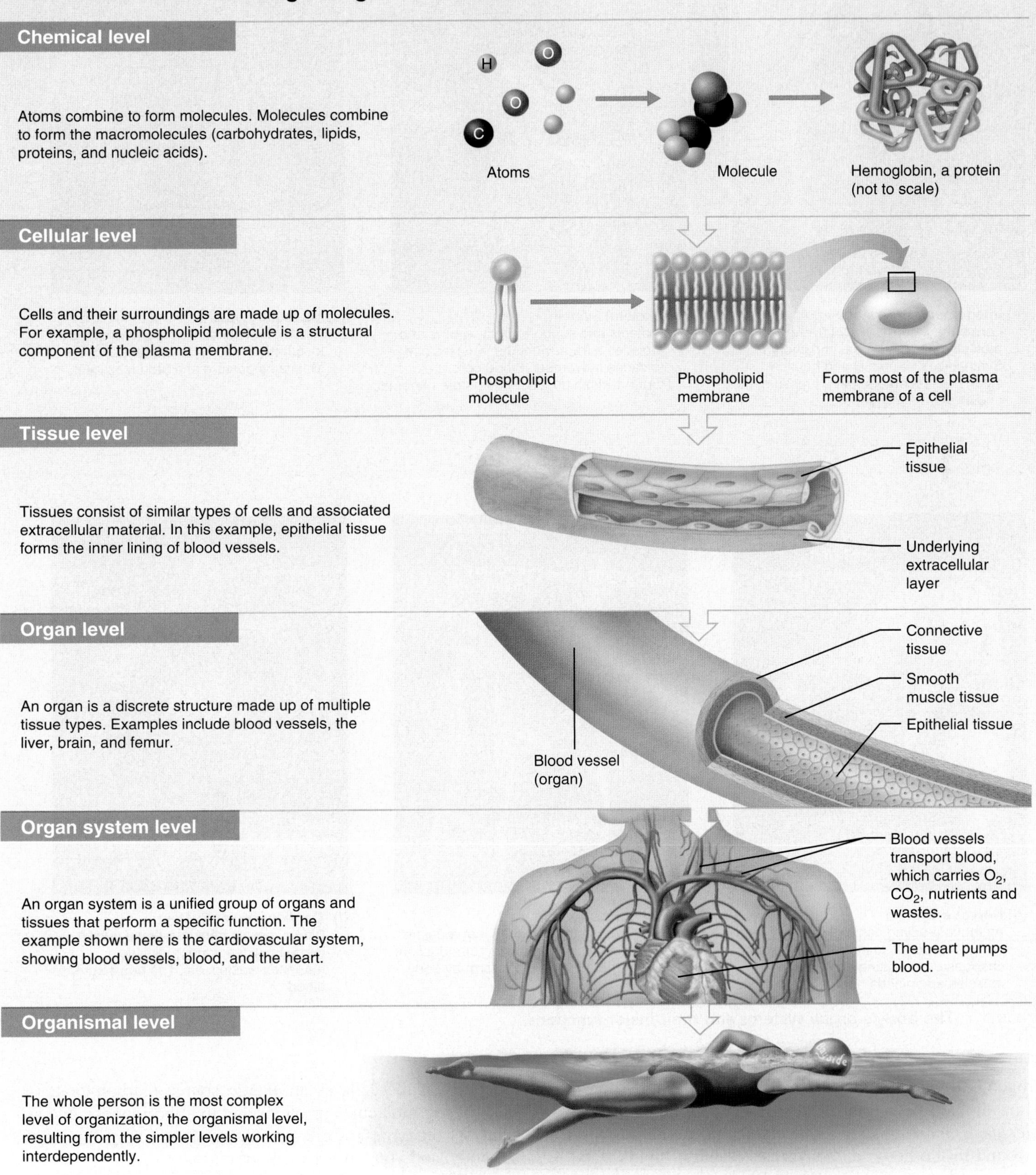

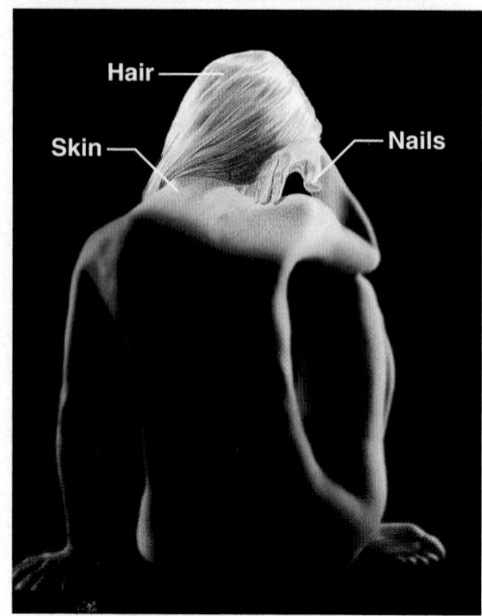

(a) Integumentary System
Forms the external body covering and protects deeper tissues from injury. Synthesizes vitamin D and houses cutaneous receptors (pain, pressure, etc.) and sweat and oil glands.

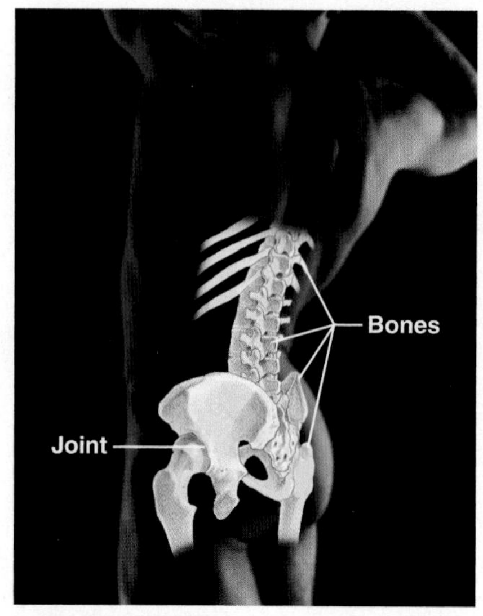

(b) Skeletal System
Protects and supports body organs and provides a framework the muscles use to cause movement. Blood cells are formed within bones. Bones store minerals.

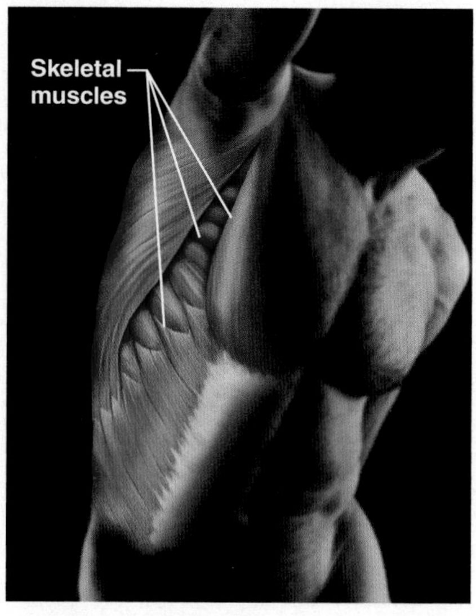

(c) Muscular System
Allows manipulation of the environment, locomotion, and facial expression. Maintains posture and produces heat.

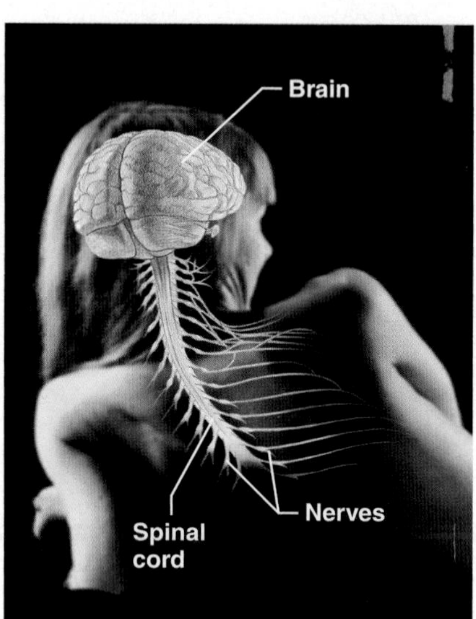

(d) Nervous System
As the fast-acting control system of the body, it responds to internal and external changes by activating appropriate muscles and glands.

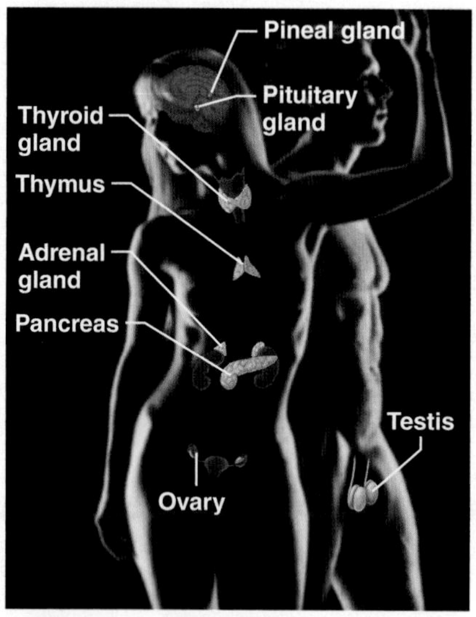

(e) Endocrine System
Glands secrete hormones that regulate processes such as growth, reproduction, and nutrient use (metabolism) by body cells.

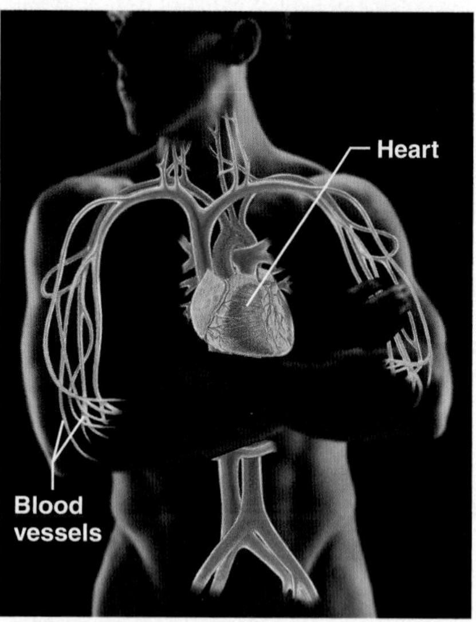

(f) Cardiovascular System
Blood vessels transport blood, which carries oxygen, carbon dioxide, nutrients, wastes, etc. The heart pumps blood.

Figure 1.2 **The body's organ systems and their major functions.**

nitrogen. Atoms combine to form small *molecules,* such as carbon dioxide (CO_2) and water (H_2O), and larger *macromolecules* (*macro* = big). Four classes of macromolecules are found in the body: carbohydrates (sugars), lipids (fats), proteins, and nucleic acids (DNA, RNA). These macromolecules are the building blocks of the structures at the **cellular level:** the *cells* and their functional subunits, called *cellular organelles.* Macromolecules also contribute to the metabolic functions of the cells as an energy source (carbohydrates), as signaling molecules (proteins and lipid hormones), and as catalysts (enzymes). Cells are the smallest living things in the body, and you have trillions of them.

The next level is the **tissue level.** A tissue is a group of cells that work together to perform a common function. Only four tissue types make up all organs of the human body: epithelial tissue (epithelium), connective tissue, muscle tissue, and nervous

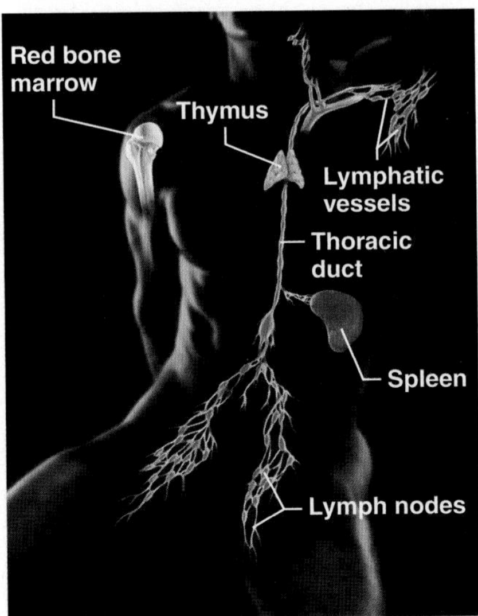

(g) Lymphatic System/Immunity
Picks up fluid leaked from blood vessels and returns it to blood. Disposes of debris in the lymphatic stream. Houses white blood cells (lymphocytes) involved in immunity. The immune response mounts the attack against foreign substances within the body.

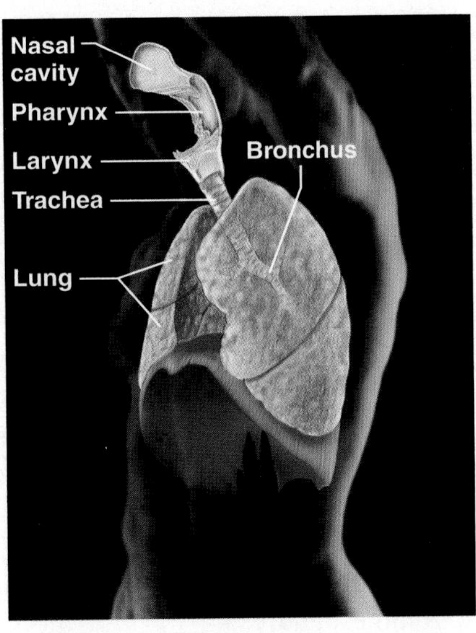

(h) Respiratory System
Keeps blood constantly supplied with oxygen and removes carbon dioxide. The gaseous exchanges occur through the walls of the air sacs of the lungs.

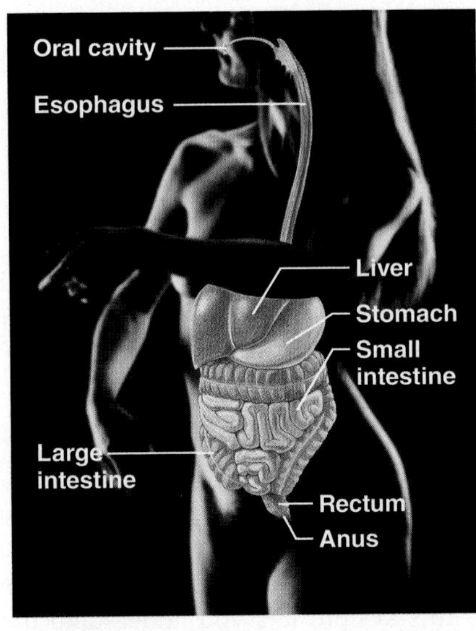

(i) Digestive System
Breaks down food into absorbable units that enter the blood for distribution to body cells. Indigestible foodstuffs are eliminated as feces.

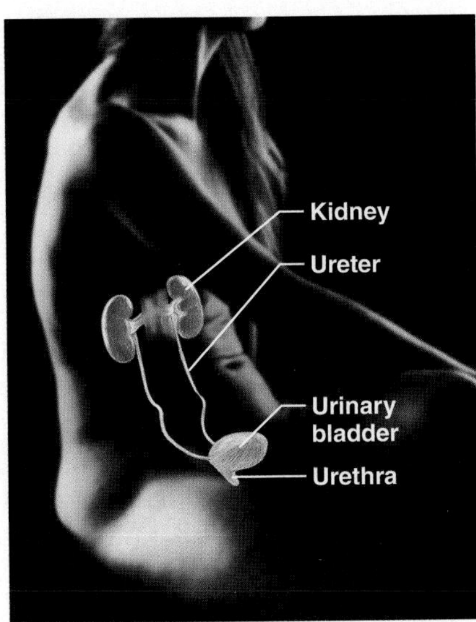

(j) Urinary System
Eliminates nitrogenous wastes from the body. Regulates water, electrolyte, and acid-base balance of the blood.

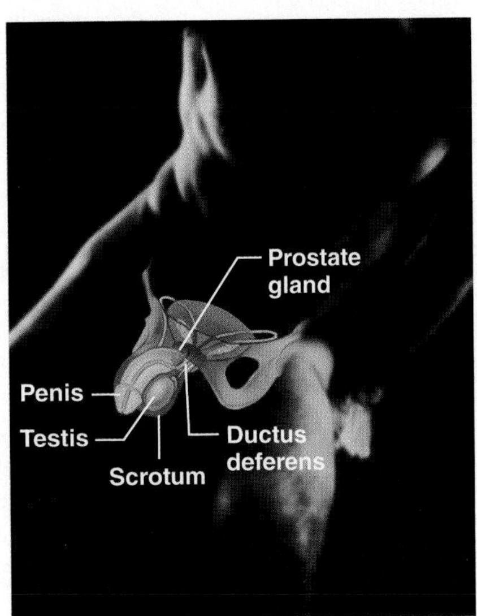

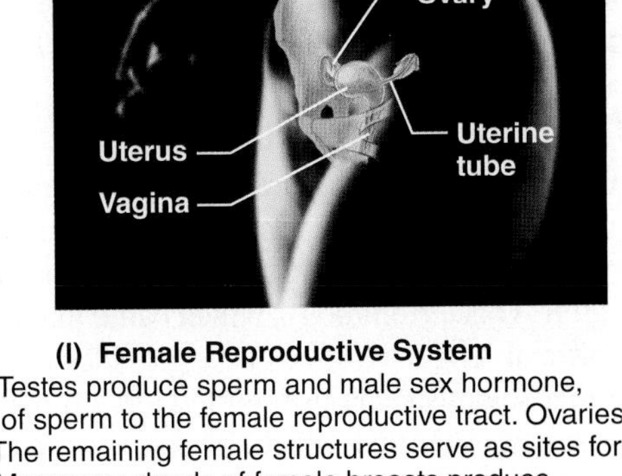

(k) Male Reproductive System **(l) Female Reproductive System**
Overall function is production of offspring. Testes produce sperm and male sex hormone, and male ducts and glands aid in delivery of sperm to the female reproductive tract. Ovaries produce eggs and female sex hormones. The remaining female structures serve as sites for fertilization and development of the fetus. Mammary glands of female breasts produce milk to nourish the newborn.

Figure 1.2 The body's organ systems and their major functions, *continued.*

tissue. Each tissue plays a characteristic role in the body. Briefly, epithelium (ep″ĭ-the′le-um) covers the body surface and lines its cavities; connective tissue supports the body and protects its organs; muscle tissue provides movement; and nervous tissue provides fast internal communication by transmitting electrical impulses.

Extremely complex physiological processes occur at the **organ level.** An organ is a discrete structure made up of more than one tissue. Most organs contain all four tissues. The liver, brain, femur, and heart are good examples. You can think of each organ in the body as a functional center responsible for an activity that no other organ can perform.

Organs that work closely together to accomplish a common purpose make up an **organ system,** the next level (Figure 1.2). For example, organs of the cardiovascular system—the heart and blood vessels—transport blood to all body tissues. Organs of the digestive system—the mouth, esophagus, stomach, intestine, and so forth—break down the food we eat so that we can absorb the nutrients into the blood. The body's organ systems are the *integumentary* (skin), *skeletal, muscular, nervous, endocrine, cardiovascular, lymphatic, immune, respiratory, digestive, urinary,* and *reproductive* systems.*

The highest level of organization is the **organismal level;** for example, the human organism is a whole living person. The organismal level is the result of all of the simpler levels working in unison to sustain life.

Scale: Length, Volume, and Weight

To describe the dimensions of cells, tissues, and organs, anatomists need a precise system of measurement. The **metric system** provides such precision (Appendix A). Familiarity with this system lets you understand the sizes, volumes, and weights of body structures.

An important unit of *length* is the **meter (m),** which is a little longer than a yardstick. If you are 6 feet tall, your height is 1.83 meters. Most adults are between 1.5 and 2 meters tall. A **centimeter (cm)** is a hundredth of a meter (*cent* = hundred). You can visualize this length by remembering that a nickel is about 2 cm in diameter. Many of our organs are several centimeters in height, length, and width. A **micrometer (μm)** is a millionth of a meter (*micro* = millionth). Cells, organelles (structures found inside cells), and tissues are measured in micrometers. Human cells average about 10 μm in diameter, although they range from 5 μm to 100 μm. The human cell with the largest diameter, the egg cell (ovum), is about the size of the tiniest dot you could make on this page with a pencil.

The metric system also measures *volume* and *weight* (mass). A **liter (l)** is a volume slightly larger than a quart; soft drinks are packaged in 1-liter and 2-liter bottles. A **milliliter (ml)** is one-thousandth of a liter (*milli* = thousandth). A **kilogram (kg)** is a mass equal to about 2.2 pounds, and a **gram (g)** is a thousandth of a kilogram (*kilo* = thousand).

Anatomical Terminology

Most anatomical terms are based on ancient Greek or Latin words. For example, the arm is the brachium (bra′ke-um; Greek for "arm"), and the thigh bone is the femur (fe′mer; Latin for "thigh"). This terminology, which came into use when Latin was the official language of science, provides a standard nomenclature that scientists can use worldwide, no matter what language they speak. This text will help you learn anatomical terminology by explaining the origins of selected terms as you encounter them. Dividing an unfamiliar term into its word roots will help you understand its meaning. For example, the word *hepatitis* is made up of *hepata,* "liver," and *itis,* "inflammation"; thus, hepatitis is inflammation of the liver. For further help, see the Glossary in the back of the book, and the list of word roots inside the back cover of the book.**

*The cardiovascular and lymphatic systems are collectively known as the *circulatory system* because of their interrelated roles in circulating fluids (blood and lymph) through the body.

☑ check your understanding

- ☐ 1. What is the difference between histology and radiography?
- ☐ 2. Use the word root definitions located in the end pages of this text to define each of the terms listed: pathology, hepatitis, brachial, leukocyte, pneumonia.
- ☐ 3. Define a tissue. List the four types of tissues in the body, and briefly state the function of each.
- ☐ 4. Name the organ system described in each of the following: (a) eliminates wastes and regulates water and ion balance; (b) fast-acting control system that integrates body activities; (c) supplies blood with oxygen and removes carbon dioxide.

(For answers, see Appendix B.)

GROSS ANATOMY: AN INTRODUCTION

learning outcomes

- ▶ Define the anatomical position.
- ▶ Use anatomical terminology to describe body directions, regions, and planes.
- ▶ Describe the basic structures that humans share with other vertebrates.
- ▶ Locate the major body cavities and their subdivisions.
- ▶ Name the four quadrants of the abdomen, and identify the visceral organs located within each quadrant.

Regional and Directional Terms

To accurately describe the various body parts and their locations, you need to use a common visual reference point. This reference point is the **anatomical position** (Figure 1.3a). In this position, a person stands erect with feet flat on the ground, toes pointing forward, and eyes facing forward. The palms face anteriorly with the thumbs pointed away from the body. It is essential to learn the anatomical position because most of the directional terminology used in anatomy refers to the body in this position. Additionally, the terms *right* and *left* always refer to those sides belonging to the person or cadaver being viewed—not to the right and left sides of the viewer.

Regional terms are the names of specific body areas. The fundamental divisions of the body are the *axial* and *appendicular* (ap″en-dik′u-lar) *regions.* The **axial region,** so named because it makes up the main axis of the body, consists of the *head, neck,* and *trunk.* The trunk, in turn, is divided into the *thorax* (chest), *abdomen,* and *pelvis;* the trunk also includes the region around the anus and external genitals, called the *perineum* (per″ĭ-ne′um; "around the anus"). The **appendicular region** of the body consists of

**For a guide to pronunciation, see the Glossary.

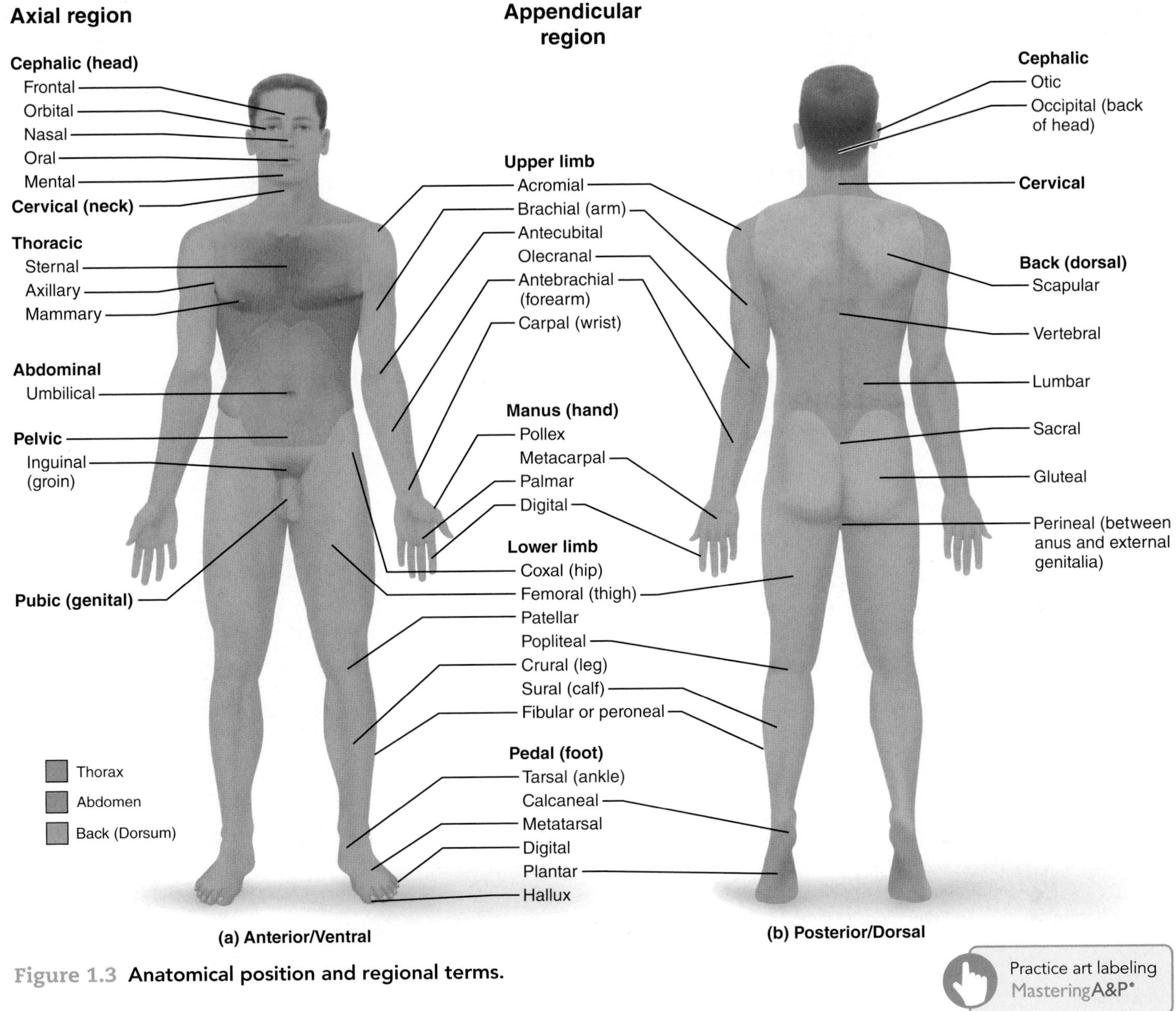

Figure 1.3 **Anatomical position and regional terms.**

the limbs, which are also called *appendages* or *extremities*. The fundamental divisions of the body are subdivided into smaller regions (as shown in Figure 1.3).

Standard directional terms are used by medical personnel and anatomists to explain precisely where one body structure lies in relation to another. For example, you could describe the relationship between the eyebrows and the nose informally by stating, "The eyebrows are at each side of the face to the right and left of the nose and higher than the nose." In anatomical terminology, this is condensed to, "The eyebrows are lateral and superior to the nose." Clearly, the anatomical terminology is less wordy and confusing. Most often used are the paired terms **superior/inferior, anterior (ventral)/posterior (dorsal), medial/lateral,** and **superficial/deep** (Table 1.1).

Body Planes and Sections

In the study of anatomy, the body is often *sectioned* (cut) along a flat surface called a *plane*. The most frequently used body planes are sagittal, frontal, and transverse planes, which lie at right angles to one another (Figure 1.4). A section bears the name of the plane along which it is cut. Thus, a cut along a sagittal plane produces a sagittal section.

A **sagittal plane** (sag′ĭ-tal; "arrow") extends vertically and divides the body into left and right parts (Figure 1.4a). The specific sagittal plane that lies exactly in the midline is the **median plane,** or **midsagittal plane.** All other sagittal planes, offset from the midline, are **parasagittal** (*para* = near). A **frontal (coronal) plane** also extends vertically and divides the body into anterior and posterior parts (Figure 1.4b). A **transverse (horizontal) plane** runs horizontally

Table 1.1 Orientation and Directional Terms

Term	Definition/Example
Superior (cranial)	Toward the head end or upper part of a structure or the body; above *The head is superior to the abdomen.*
Inferior (caudal)	Away from the head end or toward the lower part of a structure or the body; below *The intestines are inferior to the liver.*
Medial	Toward or at the midline of the body; on the inner side of *The heart is medial to the lungs.*
Lateral	Away from the midline of the body; on the outer side of *The thumb is lateral to the pinky.*
Proximal	Closer to the origin of the body part or the point of attachment of a limb to the body trunk *The elbow is proximal to the wrist.*
Distal	Farther from the origin of a body part or the point of attachment of a limb to the body trunk *The knee is distal to the thigh.*
Ipsilateral	On the same side *The right hand and right foot are ipsilateral.*
Contralateral	On opposite sides *The right hand and left foot are contralateral.*
Anterior (ventral)*	Toward or at the front of the body; in front of *The sternum is anterior to the heart.*
Posterior (dorsal)*	Toward or at the back of the body; behind *The vertebra is posterior to the heart.*
Superficial (external)	Toward or at the body surface *The skin is superficial to the skeletal muscles.*
Deep (internal)	Away from the body surface; more internal *The lungs are deep to the skin.*

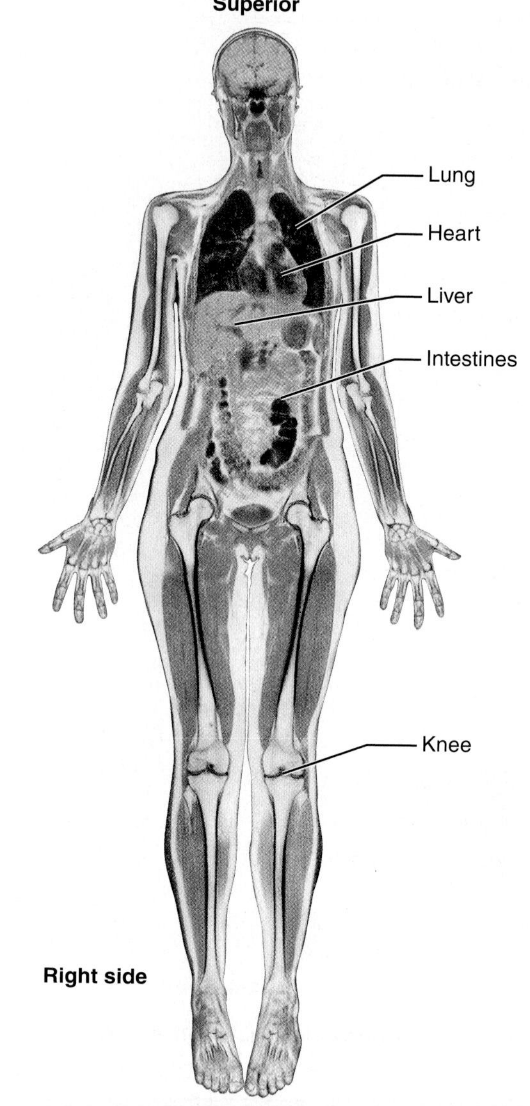

Whole body MRI, frontal section, anterior view

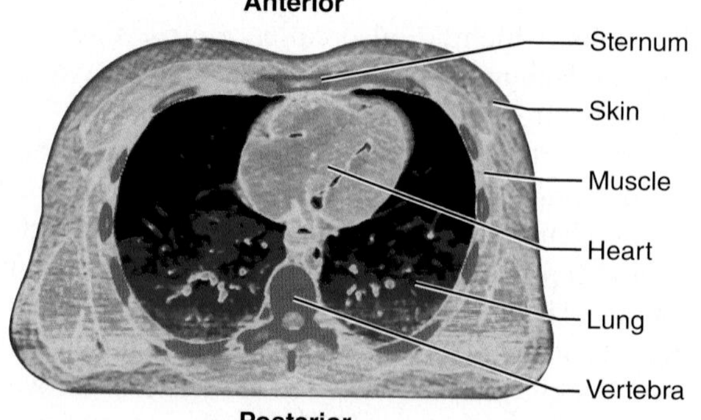

CT scan, transverse section through thorax

*Whereas the terms *ventral* and *anterior* are synonymous in humans, this is not the case in four-legged animals. *Ventral* specifically refers to the "belly" of a vertebrate animal and thus is the inferior surface of four-legged animals. Likewise, although the dorsal and posterior surfaces are the same in humans, the term *dorsal* specifically refers to an animal's back. Thus, the dorsal surface of four-legged animals is their superior surface.

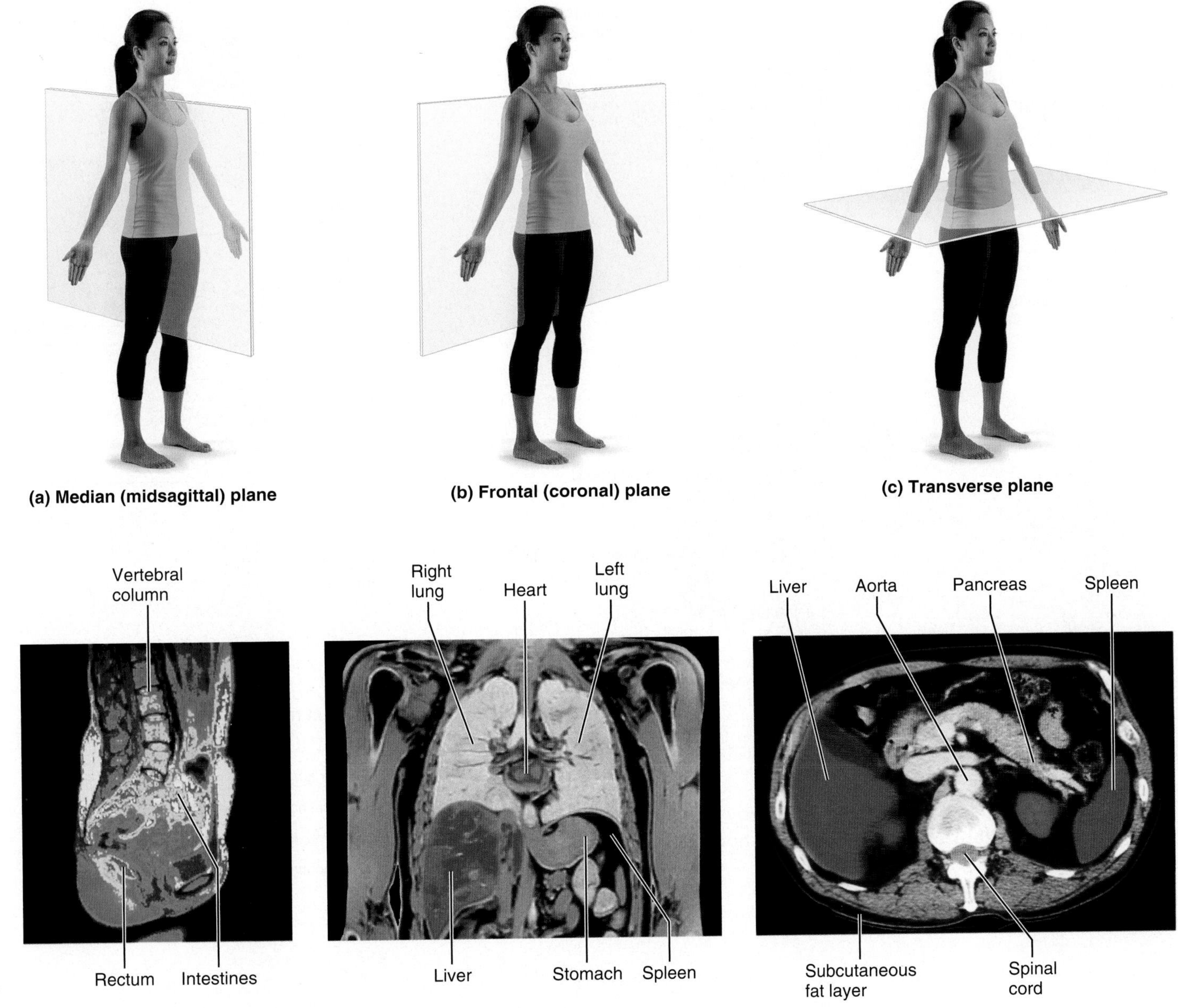

Figure 1.4 Planes of the body with corresponding magnetic resonance imaging (MRI) scans.

from right to left, dividing the body into superior and inferior parts (Figure 1.4c). A transverse section is also called a **cross section.**

Cuts made along any plane that lies diagonally between the horizontal and the vertical are called **oblique sections.** Not frontal, transverse, or sagittal, such oblique sections are difficult to interpret because the orientation of the view is not obvious. For this reason, oblique sections are seldom used.

The ability to interpret sections through the body, especially transverse sections, is increasingly important in the clinical sciences. Many medical imaging devices (described on pp. 16–19) produce sectional images rather than three-dimensional images. It can be difficult, however, to decipher an object's overall shape from a sectional view alone. A cross section of a banana, for example, looks like a circle and gives no indication of the whole banana's crescent shape.

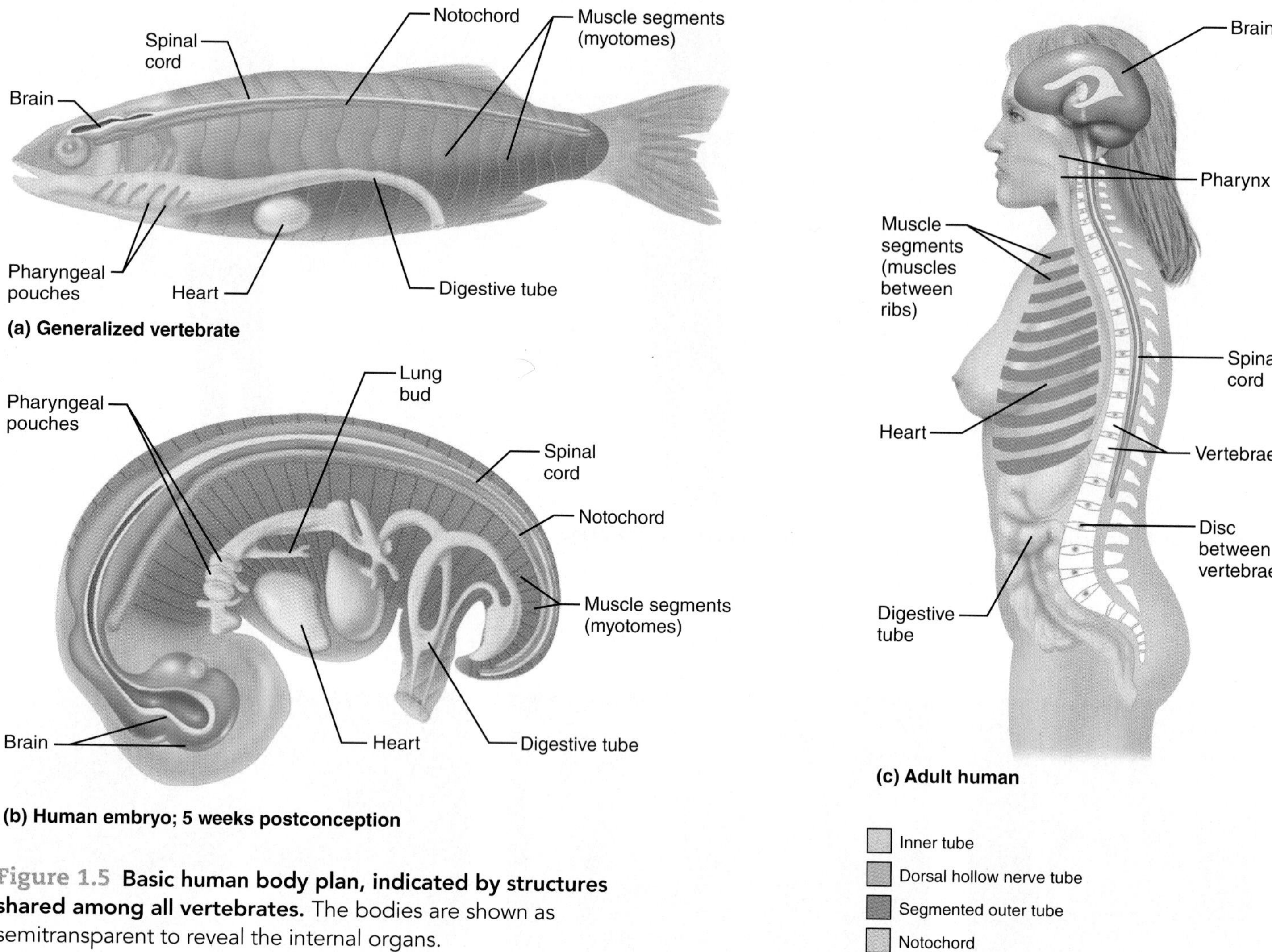

Figure 1.5 Basic human body plan, indicated by structures shared among all vertebrates. The bodies are shown as semitransparent to reveal the internal organs.

Sometimes, you must mentally assemble a whole series of sections to understand the true shape of an object. With practice, you will gradually learn to relate two-dimensional sections to three-dimensional shapes.

The Human Body Plan

Humans belong to the group of animals called *vertebrates*. This group also includes cats, rats, birds, lizards, frogs, and fish. An understanding of the basic vertebrate body plan will aid your understanding of the complexities of human anatomical structure. All vertebrates share the following basic features (Figure 1.5):

1. **Tube-within-a-tube body plan.** The inner tube extends from the mouth to the anus and includes the respiratory and digestive organs (yellow structures in Figure 1.5). The outer tube consists of the axial skeleton and associated axial muscles that make up the outer body wall, and nervous structures.
2. **Bilateral symmetry.** The left half of the body is essentially a mirror image of the right half. Most body structures, such as the right and left hands, eyes, and ovaries, occur in pairs. Structures in the median plane are unpaired, but they tend to have identical right and left sides (the nose is an example).
3. **Dorsal hollow nerve cord.** All vertebrate embryos have a hollow nerve cord running along their back in the median plane. This cord develops into the brain and spinal cord.
4. **Notochord and vertebrae.** The **notochord** (no′to-kord; "back string") is a stiffening rod in the back just deep to the spinal cord. In humans, a complete notochord forms in the embryo, although most of it is quickly replaced by the vertebrae (ver′tĕ-bre), the bony pieces of the vertebral column, or backbone. Still, some of the notochord persists throughout life as the cores of

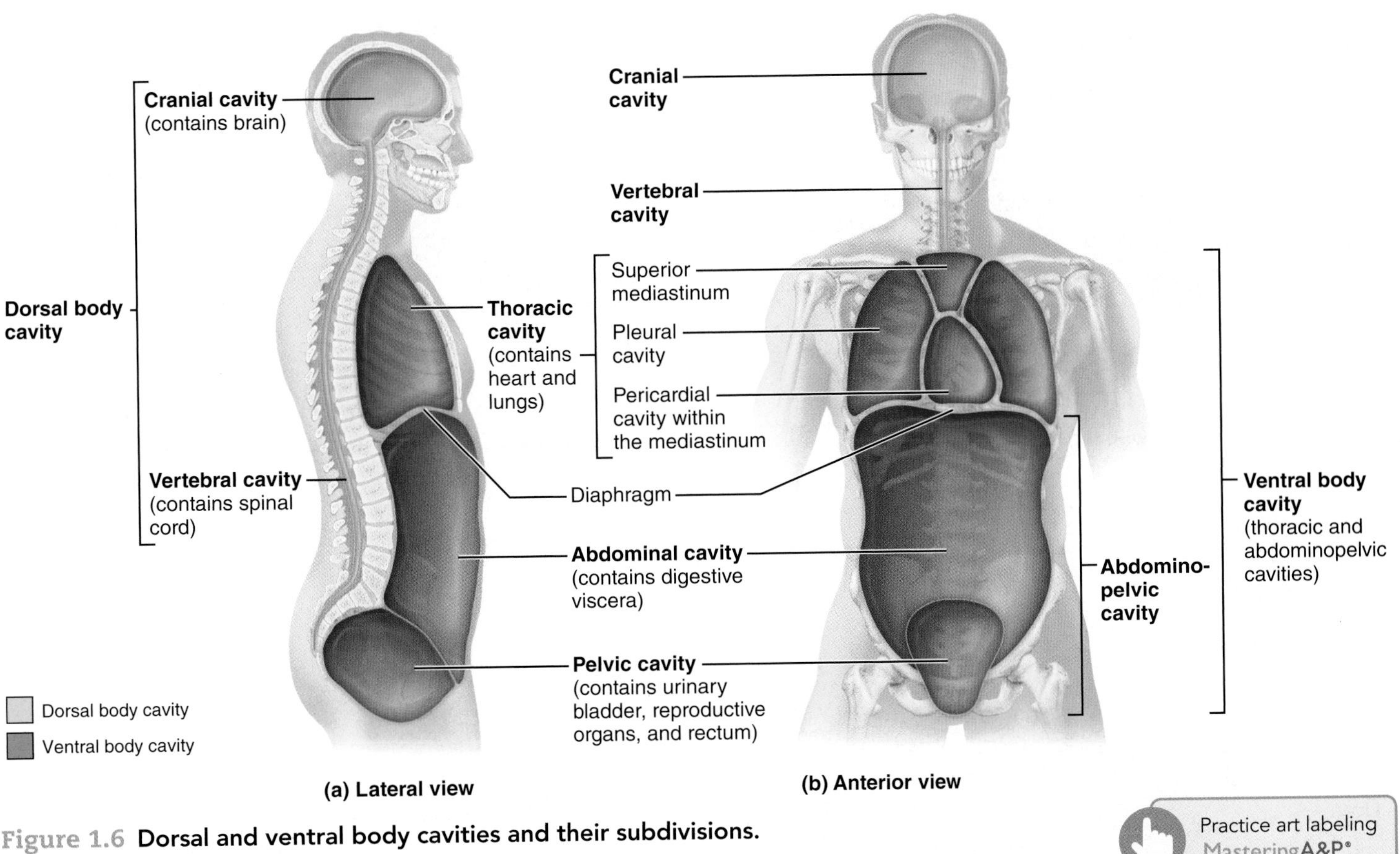

Figure 1.6 **Dorsal and ventral body cavities and their subdivisions.**

the discs between the vertebrae (see the description of nucleus pulposus, p. 171).

5. **Segmentation.** The "outer tube" of the body shows evidence of segmentation. Segments are repeating units of similar structure that run from the head along the full length of the trunk. In humans, the ribs and the muscles between the ribs are evidence of segmentation, as are the many nerves branching off the spinal cord. The bony vertebral column, with its repeating vertebrae, is also segmented.
6. **Pharyngeal pouches.** Humans have a **pharynx** (far′ingks), which is the throat region of the digestive and respiratory tube. In the embryonic stage, the human pharynx has a set of outpocketings called *pharyngeal* (far-rin′je-al) *pouches* that correspond to the clefts between the gills of fish. Such pouches give rise to some structures in the head and neck. An example is the middle ear cavity, which runs from the eardrum to the pharynx.

Body Cavities and Membranes

Within the body are two large cavities called the dorsal and ventral cavities (Figure 1.6). These are closed to the outside, and each contains internal organs. Think of them as filled cavities, like toy boxes containing toys.

Dorsal Body Cavity

The **dorsal body cavity** is subdivided into a **cranial cavity,** which lies in the skull and encases the brain, and a **vertebral cavity,** which runs through the vertebral column to enclose the spinal cord. The hard, bony walls of this cavity protect the contained organs.

Ventral Body Cavity

The more anterior and larger of the closed body cavities is the **ventral body cavity** (Figure 1.6). The organs it contains, such as the lungs, heart, intestines, and kidneys, are called **visceral organs** or **viscera** (vis′er-ah). The ventral body cavity has two main divisions: (1) a superior **thoracic cavity,** surrounded by the ribs and the muscles of the chest wall; and (2) an inferior **abdominopelvic** (ab-dom″ĭ-no-pel′vic) **cavity** surrounded by the abdominal walls and pelvic girdle. The thoracic and abdominal cavities are separated from each other by the diaphragm, a dome-shaped muscle used in breathing.

The *thoracic cavity* has three parts: (a) two lateral parts, each containing a lung surrounded by a **pleural cavity** (ploo′-ral; "the side, a rib"), and (b) a central band of organs called the **mediastinum** (me″de-ah-sti′num; "in the middle"). The mediastinum contains the heart surrounded by a **pericardial cavity** (per″ĭ-kar′de-al; "around the heart"). It also houses other major thoracic organs, such as the esophagus and trachea (windpipe).

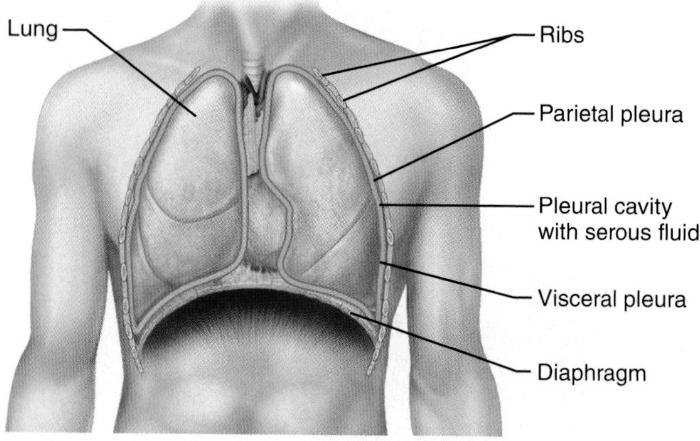

(a) Serosae associated with the lungs: pleura

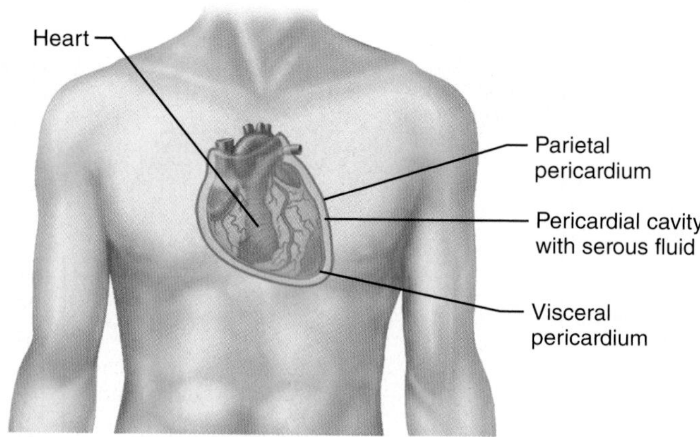

(b) Serosae associated with the heart: pericardium

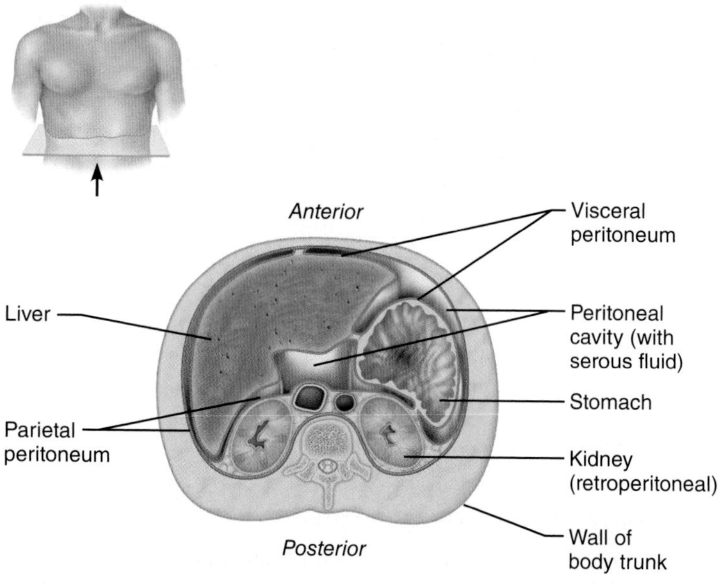

(c) Serosae associated with the abdominal viscera: peritoneum

Outer balloon wall (comparable to parietal serosa)
Air (comparable to serous cavity)
Inner balloon wall (comparable to visceral serosa)

(d) Model of the serous membranes and serous cavity

Figure 1.7 **The serous cavities and their associated membranes.**

The *abdominopelvic cavity* is divided into two parts. The superior part, called the **abdominal cavity,** contains the liver, stomach, kidneys, and other organs. The inferior part, or **pelvic cavity,** contains the bladder, some reproductive organs, and the rectum. These two parts are continuous with each other, not separated by any muscular or membranous partition. Many organs in the abdominopelvic cavity are surrounded by a **peritoneal** (per″ĭ-to-ne′al) **cavity.**

Serous Cavities

The previous section mentioned the *pleural cavity* around the lung, the *pericardial cavity* around the heart, and the *peritoneal cavity* around the viscera in the abdominopelvic cavity. Each of these serous cavities is a slitlike space lined by a **serous** (se′rus) **membrane,** or **serosa** (se-ro′-sah; plural, **serosae**) (Figure 1.7). These serous membranes (indicated by the red lines in Figure 1.7) are named **pleura,** serous **pericardium,** and **peritoneum,** respectively. The part of a serosa that forms the outer wall of the cavity is called the **parietal** (pah-ri′ĕ-tal; "wall") **serosa.** The parietal serosa is continuous with the inner, **visceral serosa,** which covers the visceral organs. You can visualize the relationship of the serous membranes by pushing your fist into a limp balloon (Figure 1.7d):

- The part of the balloon that clings to your fist represents the visceral serosa on the organ's (your fist's) outer surface.
- The outer wall of the balloon represents the parietal serosa.
- The balloon's thin airspace represents the serous cavity itself.

Serous cavities contain a thin layer of **serous fluid** (*serous* = watery). This fluid is produced by both serous membranes. The slippery serous fluid allows the visceral

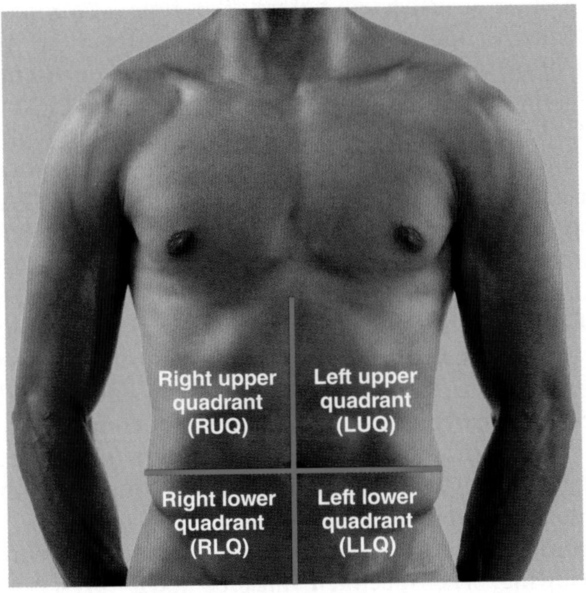

(a) The four abdominopelvic quadrants

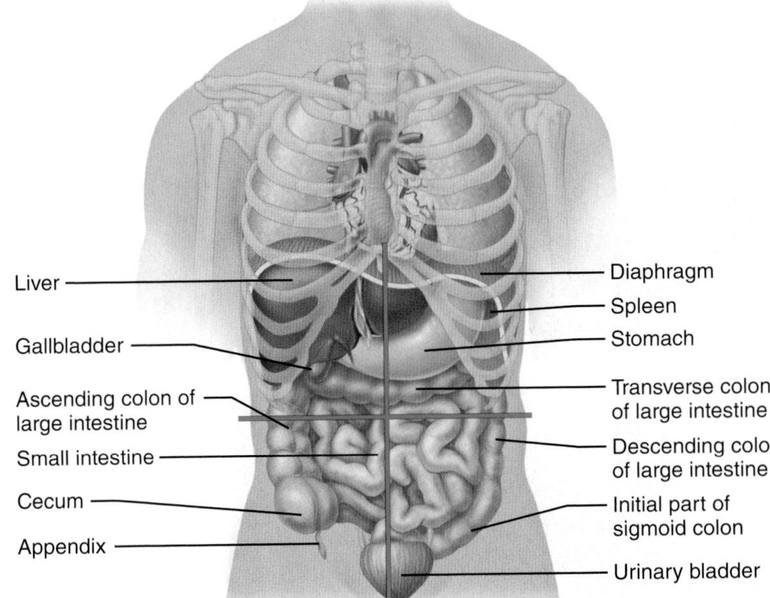

(b) Anterior view of the four quadrants showing the superficial organs

Figure 1.8 **Abdominal quadrants.** In (a), the two planes through the abdominopelvic cavity, one horizontal and one vertical, intersect at the navel.

organs to slide with little friction across the cavity walls as they carry out their routine functions. This freedom of movement is extremely important for organs that move or change shape, such as the pumping heart and the churning stomach.

Abdominal Quadrants

Because the abdominopelvic cavity is large and contains many organs, it is helpful to divide it into smaller areas for study. To localize organs in a general way, the abdomen is divided into four **quadrants** ("quarters") by drawing one vertical and one horizontal plane through the navel (Figure 1.8a). Knowledge of which abdominal organs are located within each quadrant (Figure 1.8b) aids clinicians in diagnosing disorders or injuries.

The rib cage is commonly thought of as protection for the thoracic organs, but it also protects the organs in the most superior part of the abdomen. The liver and the spleen, two blood-rich organs particularly vulnerable to injury, are protected by the surrounding rib cage on the right and left sides, respectively. The kidneys, located along the posterior abdominal wall, are also protected by the inferior ribs.

Anatomical Variability

You know from looking at the faces and body shapes of the people around you that humans differ in their external anatomy. The same kind of variability holds for internal organs as well. Thus, not every structural detail described in an anatomy book is true of all people or of all the cadavers (dead bodies) you observe in the anatomy lab. In some bodies, for example, a certain blood vessel may branch off higher than usual, a nerve or vessel may be somewhat "out of place," or a small muscle may be missing. Such minor variations are unlikely to confuse you, however, because well over 90% of all structures present in any human body match the textbook descriptions. Extreme anatomical variations are seldom seen, because they are incompatible with life. For example, no living person could be missing the blood vessels to the brain.

✓ check your understanding

- ☐ 5. Using directional terms, describe the location of the liver in reference to the heart (see Figure 1.8 and Table 1.1).
- ☐ 6. Which tube of the body shows evidence of segmentation, the outer tube or the inner tube?
- ☐ 7. What is the outer layer of serous membrane that lines the pleural cavity called?

(For answers, see Appendix B.)

MICROSCOPIC ANATOMY: AN INTRODUCTION

learning outcomes

- ▶ Explain how human tissue is prepared and examined for its microscopic structure.
- ▶ Distinguish tissue viewed by light microscopy from that viewed by electron microscopy.

Light and Electron Microscopy

Microscopy is the examination of small structures with a microscope. When microscopes were introduced in the early 1600s, they opened up a tiny new universe whose

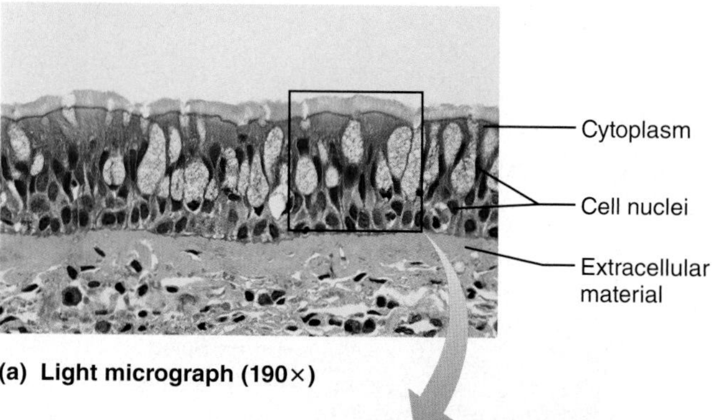

(a) Light micrograph (190×)

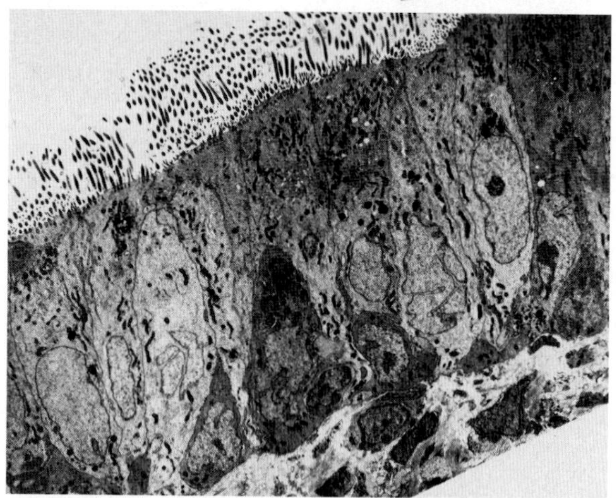

(b) Transmission electron micrograph (2250×)

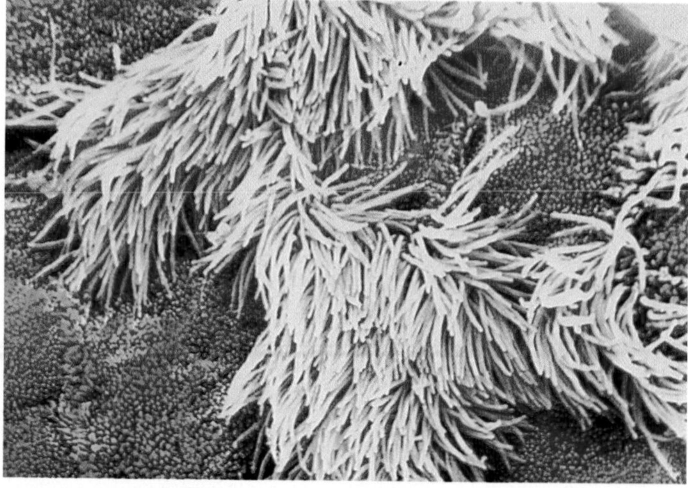

(c) Scanning electron micrograph, artificially colored (2500×)

Figure 1.9 Cells viewed by three types of microscopy. (a) Light micrograph of ciliated epithelium. (b) Transmission electron micrograph showing enlarged area of the cell region that is indicated in the box in part (a). (c) Scanning electron micrograph: surface view of cells lining the trachea, or windpipe. The long, grasslike structures on the surfaces of these cells are cilia, and the tiny knoblike structures are microvilli (Chapter 4, p. 76).

existence was unsuspected before that time. Two main types of microscopes are now used to investigate the fine structure of organs, tissues, and cells: the **light microscope (LM)** and the **transmission electron microscope (TEM** or just **EM).** Light microscopy illuminates body tissue with a beam of light, whereas electron microscopy uses a beam of electrons. LM is used for lower-magnification viewing; EM, for higher magnification (Figure 1.9a and b, respectively). Light microscopy can produce sharp, detailed images of tissues and cells, but not of the small structures within cells (organelles). A light microscope's *low resolution*—its inability to reveal small structures clearly—remains its basic limitation, despite technical advances that have greatly improved the quality of LM images. EM, by contrast, uses electron beams of much smaller wavelength to produce sharp images at much greater magnification, thus revealing the fine details of cells and tissues.

Elaborate steps are taken to prepare tissue for microscopic viewing. The specimen must be **fixed** (preserved) and then cut into **sections** (slices) thin enough to transmit light or electrons. Finally, the specimen must be **stained** to enhance contrast. The stains used in light microscopy are beautifully colored organic dyes, most of which were originally developed by clothing manufacturers in the mid-1800s (Figure 1.9a). These dyes helped to usher in the golden age of microscopic anatomy from 1860 to 1900. The stains come in almost all colors. Many consist of charged molecules (negative or positive molecules) of dye that bind within the tissue to macromolecules of the opposite charge. This electrical attraction is the basis of staining. Dyes with negatively charged molecules stain the positively charged structures of the cell or tissue, and thus they are called **acidic stains.** Positively charged dyes, by contrast, are called **basic stains** because they bind to, and stain, negatively charged structures. Because different parts of cells and tissues take up different dyes, the stains distinguish the different anatomical structures.

One of the most commonly used histological stains is a combination of two dyes, hematoxylin and eosin (H&E stain). Hematoxylin is a basic stain that binds to the acidic structures of the cell (the nucleus, ribosomes, rough ER) and colors them a dark blue to purple hue. Eosin, an acidic stain, binds to basic cytoplasmic structures and extracellular components, coloring them red to pink. Many of the micrographs throughout this text show tissues stained with H&E. (In Figure 1.9a, for example, the dark, almost black, spots are the cell nuclei, the cellular cytoplasm is magenta, and the extracellular material in the bottom half of the image is stained a lighter pink.) A variety of other stains can be used to visualize specific structures. Some of these stains create strikingly beautiful images illuminating detailed histological structure.

For transmission electron microscopy, tissue sections are stained with heavy-metal salts. These metals deflect electrons in the beam to different extents, thus providing contrast in the image. Electron-microscope images contain only shades of gray because color is a property of light, not of electron waves. The image may be artificially colored to enhance contrast (Figure 1.9c).

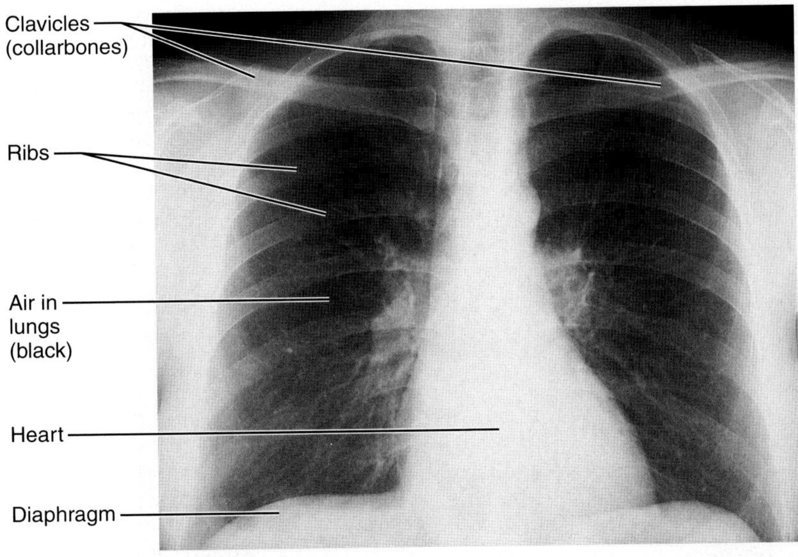

(a) Radiograph of the chest

(b) Lower GI with barium contrast medium, normal

Figure 1.10 **X-ray images.**

Scanning Electron Microscopy

The types of microscopy introduced so far are used to view cells and tissue that have been sectioned. Another kind of electron microscopy, **scanning electron microscopy (SEM),** provides three-dimensional pictures of whole, unsectioned surfaces with striking clarity (Figure 1.9c). First, the specimen is preserved and coated with fine layers of carbon and gold dust. Then, an electron beam scans the specimen, causing other, secondary electrons to be emitted from its surface. A detector captures these emitted electrons and assembles them into a three-dimensional image on a video screen, based on the principle that more electrons are produced by the higher points on the specimen surface than by the lower points. Although artificially constructed, the SEM image is accurate and looks very real. Like all electron-microscopy images, the original is in black and white, although it can be colored artificially to highlight structural details (Figure 1.9c).

Artifacts

The preserved tissue seen under the microscope has been exposed to many procedures that alter its original condition. Because each preparatory step introduces minor distortions, called **artifacts,** most microscopic structures viewed by anatomists are not exactly like those in living tissue. Furthermore, the human and animal corpses studied in the anatomy laboratory have also been preserved, so their organs have a drabber color and a different texture from those of living organs. Keep these principles in mind as you look at the micrographs (pictures taken with a microscope) and the photos of human cadavers in this book.

☑ check your understanding

☐ 8. In tissue stained with H&E stain, what color are the cellular nuclei?

☐ 9. Which type of microscopy produces detailed three-dimensional images of the surface features of a structure?

(For answers, see Appendix B.)

CLINICAL ANATOMY: AN INTRODUCTION TO MEDICAL IMAGING TECHNIQUES

learning outcome

▶ Describe the medical imaging techniques that are used to visualize structures inside the body.

Physicians have long sought ways to examine the body's internal organs for evidence of disease without subjecting the patient to the risks of exploratory surgery. Physicians can identify some diseases and injuries by feeling the patient's deep organs through the skin or by using traditional X rays. Powerful new techniques for viewing the internal anatomy of living people continue to be developed. These imaging techniques not only reveal the structure of functioning internal organs but also can yield information about cellular activity. The new techniques all rely on powerful computers to construct images from raw data transmitted by electrical signals.

X-Ray Imaging

Before considering the newer techniques, you need to understand traditional X-ray images, because these still play the major role in medical diagnosis (Figure 1.10a). Discovered quite by accident in 1895 and used in medicine ever since, X rays are electromagnetic waves of very short wavelength. When X rays are directed at the body, some are absorbed. The amount of absorption depends on the density of the

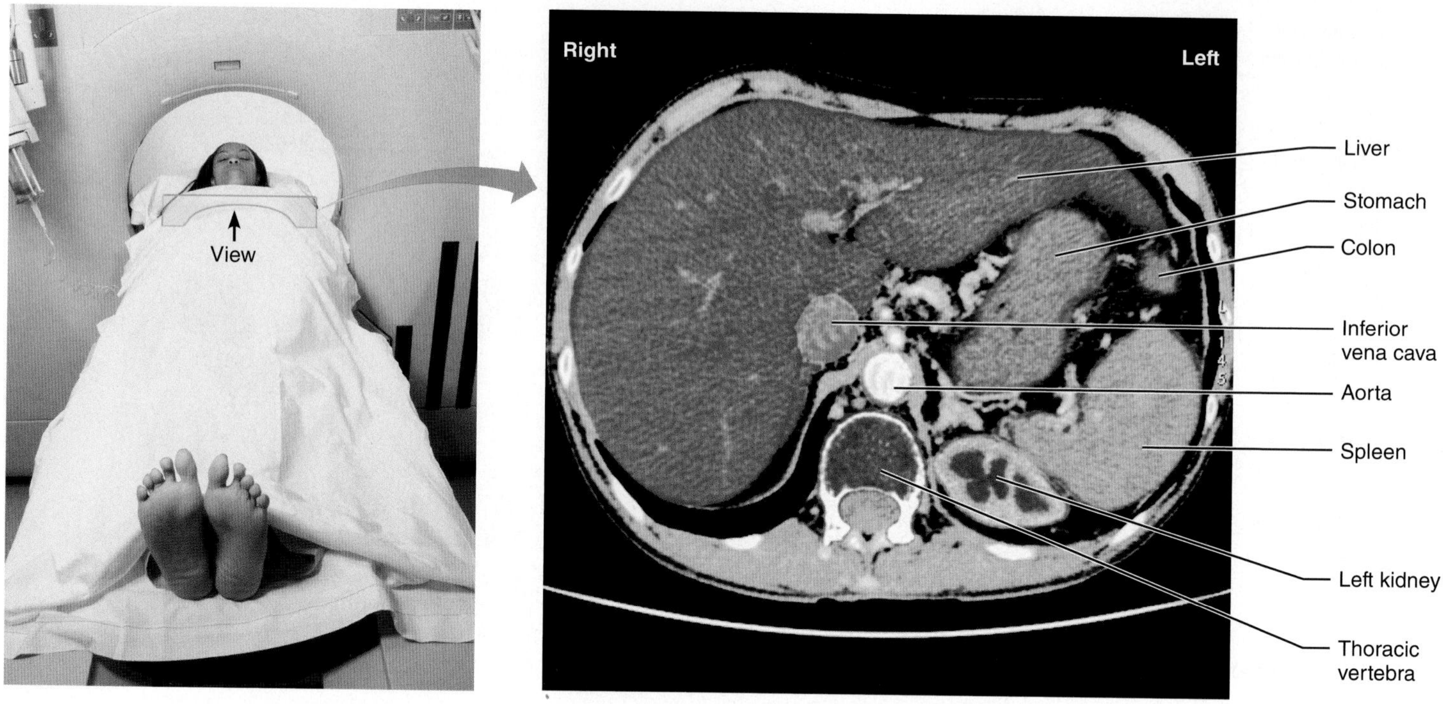

View
Right
Left
Liver
Stomach
Colon
Inferior vena cava
Aorta
Spleen
Left kidney
Thoracic vertebra

Figure 1.11 Computed tomography (CT). CT scan through the upper abdomen. CT sections are conventionally oriented as if viewed from an inferior direction, with the posterior surface of the body directed toward the inferior part of the picture; therefore, the patient's right side is at the left side of the picture.

matter encountered. X rays that pass through the body expose a piece of film behind the patient. The resulting image (radiograph) is a negative: The darker, exposed areas on the film represent soft organs, which are easily penetrated by X rays, whereas light, unexposed areas correspond to denser structures, such as bones, which absorb most X rays.

X-ray images are best for visualizing bones and for locating abnormal dense structures, such as some tumors and tuberculosis nodules in the lungs. Mammography ("breast image") uses low-dose X rays to screen for tumors in the breast, and bone density scans use X rays of the lower back and hip to screen for osteoporosis ("porous bone"). X-ray examination of hollow soft tissue organs is enhanced by the use of a **contrast medium,** a liquid that contains atoms of a heavy element such as barium that absorb more passing X rays. The contrast medium is injected or ingested, depending on the structure to be examined, to fill organs of interest and allow better visualization of these soft tissue structures. The gastrointestinal ("stomach intestine") tract is commonly examined using this procedure (upper and lower GI imaging) to screen for ulcers or tumors (Figure 1.10b).

In many instances, conventional X-ray images are very informative; however, conventional X-ray studies have several limitations that have prompted clinicians to seek more advanced imaging techniques. First, X-ray images, especially those of soft tissues, can be blurry. Second, conventional X-ray images flatten three-dimensional body structures into two dimensions. Consequently, organs appear stacked one on top of another. Even more problematic, denser organs block the less dense organs that lie in the same path. For improved images, particularly of soft tissues, clinicians use computer-assisted imaging techniques that produce sectional images of the body's interior.

Advanced X-Ray Techniques

Computed Tomography

One of the more useful modern imaging techniques is a refined X-ray technology called **computed tomography (CT),** or **computed axial tomography (CAT)** (Figure 1.11). A CT scanner is shaped like a square metal nut (as in "nuts and bolts") standing on its side. The patient lies in the central hole, situated between an X-ray tube and a recorder, both of which are in the scanner. The tube and recorder rotate to take about 12 successive X-ray images around the person's full circumference. Because the fan-shaped X-ray beam is thin, all pictures are confined to a single transverse body plane about 0.3 cm thick. This explains the term *axial tomography,* which literally means "pictures of transverse sections taken along the body axis." Information is obtained from all around the circumference so that every organ is recorded from its best angle, with the fewest structures blocking it. The computer translates all the recorded information into a detailed picture of the body section, which it displays on a viewing screen. CT produces superb images of soft tissue as well as of bone and blood vessels. CT is a fast and relatively inexpensive

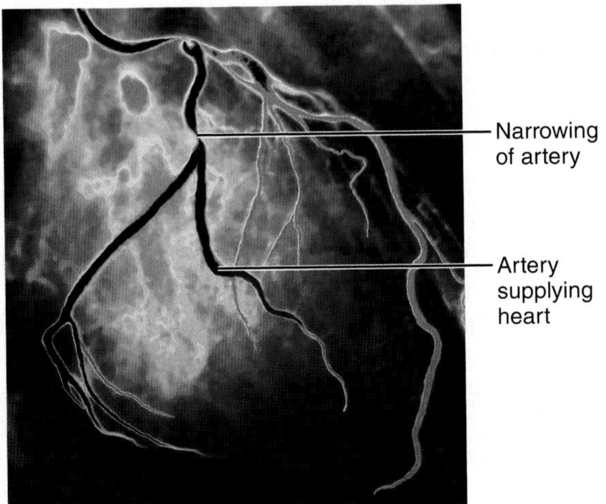

Figure 1.12 **Digital subtraction angiography (DSA).** A DSA image of the arteries that supply the heart.

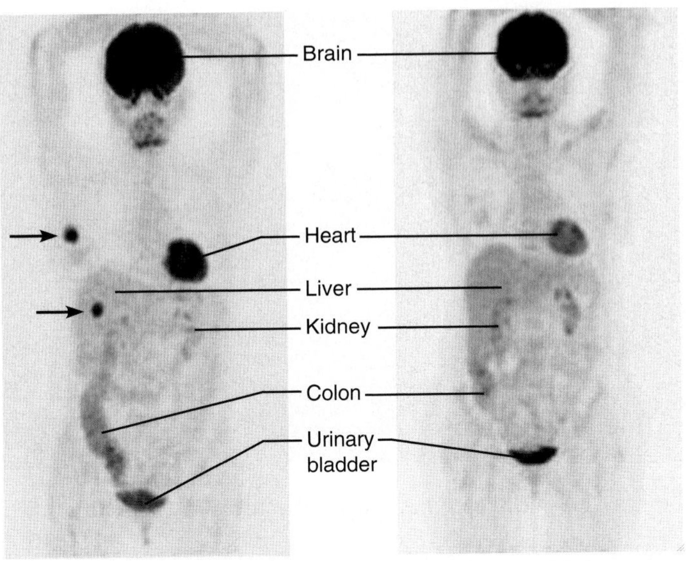

Figure 1.13 **Positron emission tomography (PET).** PET scans are used in oncology to assess tumor size, location, and response to treatment.

test. It can be used quickly and readily in trauma situations to assess internal injury. CT does use X rays to produce images, so it does pose some, although minimal, concern about radiation exposure. CT is less useful for nervous tissue structures and for joint images, particularly the knee and shoulder, because bone can obscure the joint details. However, because it is less costly than magnetic resonance imaging (MRI, described on p. 18), and its use less restrictive, CT is an essential diagnostic tool for clinicians.

Angiography

Angiography ("vessel image") is a technique that produces images of blood vessels. A contrast medium is injected into a vessel and distributed via the vascular system. Images of the vessels of interest are recorded using either conventional radiography or a digitally assisted imaging technique such as a CT scan or an MRI. The contrast medium highlights the vessel structure and allows for clear visualization of blood vessels. This procedure is used for diagnosing aneurisms (ballooning out of a vessel due to weakening of the vessel wall) and atherosclerosis (narrowing of blood vessels due to the buildup of fatty plaques) and for identifying a source of internal bleeding.

An extension of angiography, **digital subtraction angiography (DSA)** provides an unobstructed view of small arteries (Figure 1.12). In this procedure, images of the vessel are taken before and after the injection of contrast medium. The computer subtracts the "before" image from the "after" image, eliminating all traces of body structures that obscure the vessel. DSA is often used to identify blockage of the arteries that supply the heart wall and the brain.

Positron Emission Tomography

Positron emission tomography (PET) (Figure 1.13) is an advanced procedure that produces images by detecting radioactive isotopes injected into the body. The special advantage of PET is that its images indicate regions of cellular activity. For example, radioactively tagged sugar or water molecules are injected into the bloodstream and traced to the body areas that take them up in the greatest quantity. This procedure identifies the body's most active cells and pinpoints the body regions that receive the greatest supply of blood. As the radioactive material decays, it gives off energy in the form of gamma rays. Sensors within the doughnut-shaped scanner pick up the emitted gamma rays, which are translated into electrical impulses and sent to the computer. A picture of the isotope's location is then constructed on the viewing screen.

PET is used to assess the functional flow of blood and areas of high metabolic activity. In the brain, it can depict areas of the normal brain most active during specific tasks (speech, seeing, comprehension), thereby providing direct evidence for the functions of specific brain regions. The resolution of a PET image is low, however, and the image takes a relatively long time to form, so PET cannot record fast changes in brain activity. PET is gradually being eclipsed by functional MRI.

PET scans are used in oncology (cancer treatment) for detecting and staging tumors and for assessing cancer therapy. Because PET measures metabolic activity, it can indicate areas of enhanced cellular activity due to tumor growth. PET may reveal the presence of cancerous growths before they become visible in CT or MRI imaging. In cancer treatment, PET imaging is used to monitor the size and distribution of tumors and the response of cancerous tumors to therapeutic treatment (Figure 1.13). PET imaging is increasingly being used in combination with CT or MRI to correlate metabolic activity of cancerous tissues with alteration of anatomic structure.

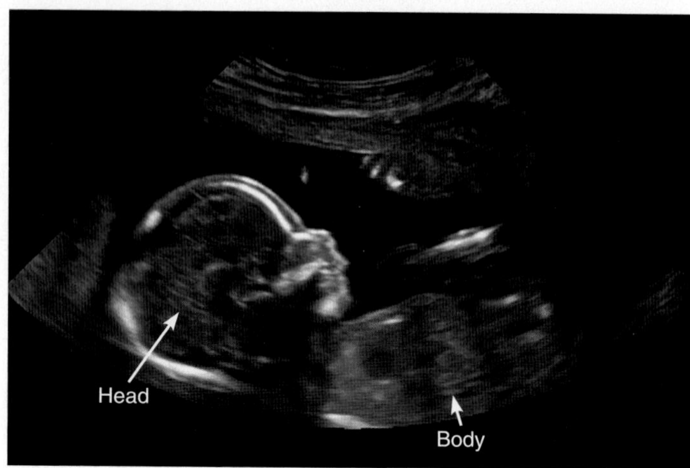

Figure 1.14 Ultrasound image of a fetus in the uterus.

Sonography

In **sonography,** or **ultrasound imaging** (Figure 1.14), the body is probed with pulses of high-frequency (ultrasonic) sound waves that reflect (echo) off the body's tissues. A computer analyzes the echoes to construct sectional images of the outlines of organs. A handheld device that looks something like a microphone emits the sound and picks up the echoes. The device is moved across the surface of the body, allowing organs to be examined from many different body planes.

Sonography has two distinct advantages over other imaging techniques. First, the equipment is relatively inexpensive. Second, ultrasound seems to be safer than ionizing forms of radiation, with fewer harmful effects on living tissues.

Because of its apparent safety, ultrasound is the imaging technique of choice for determining the age and health of a developing fetus. It is also used to visualize the gallbladder and other viscera and, increasingly, the arteries to detect atherosclerosis (thickening and hardening of the arterial walls). Sonography is of little value for viewing air-filled structures (lungs) or structures surrounded by bone (brain and spinal cord) because sound waves do not penetrate hard objects well and rapidly dissipate in air.

Ultrasound images are somewhat blurry, although their sharpness is being improved by using higher-frequency sound waves. Liquid contrast media containing sound-reflecting bubbles can be injected into the bloodstream to better reveal the vessels and heart.

Magnetic Resonance Imaging

Magnetic resonance imaging (MRI) is a technique with tremendous appeal because it produces high-contrast images of soft tissues (Figure 1.15), an area in which X-ray imaging is weak. MRI also does not use radiation for generating an image. MRI primarily detects the levels of the element hydrogen in the body, most of which is in water. Thus, MRI tends to distinguish body tissues from one another on the basis of differences in water content. Because bones contain less water than other tissues do, MRI peers easily through the skull to reveal the brain. MRI can distinguish the fatty white matter from the more watery gray matter of the brain. Many tumors show up distinctly, and MRI has even revealed brain tumors missed by direct observation during exploratory surgery. The soft tissues of the joints, ligaments, and cartilage are also visualized well with MRI.

The technique subjects the patient to magnetic fields up to 60,000 times stronger than that of the earth. The patient lies in a chamber, with his or her body surrounded by a huge magnet. When the magnet is turned on, the nuclei of the body's hydrogen atoms—single protons that spin like tops—line up parallel to the strong magnetic field. The patient is then exposed to a brief pulse of radio waves just below the frequency of FM radio, which knock the spinning protons out of alignment. When the radio waves are turned off, the protons return to their alignment in the magnetic field, emitting their own faint radio waves in the process. Sensors detect these waves, and the computer translates them into images. With the use of advanced volume-rendering techniques, multiple MRI scans can be assembled into three-dimensional reconstructions (Figure 1.15b). The images produced are dramatic views into the body's organs.

In the early 1990s, MRI technology leaped forward with the development of **functional MRI (fMRI).** This technique measures blood oxygen, so it reveals the amount of oxygenated blood flowing to specific body regions. It can therefore show which parts of the brain are active during various mental tasks. Functional MRI can pinpoint much smaller brain areas than PET can, works faster, and requires no messy radioactive tracers. For these reasons, it is replacing PET in the study of brain function.

Despite the advantages of MRI, there are limitations to its use. MRI does not use X rays, so it poses no concern about radiation exposure; however, the large magnets can cause implanted metallic devices to malfunction. MRIs require a longer time to produce an image than a CT scan, and medical devices, such as traction or life support equipment, cannot be used during MRI imaging. For these reasons, MRI is not useful in trauma situations. MRI is also more sensitive to patient movement during the scan.

The images formed by computerized imaging techniques can be quite stunning. Keep in mind, however, that the images are abstractions assembled within a computer. They are artificially enhanced for sharpness and artificially colored to increase contrast or to highlight areas of interest. Although computer-based images are not inaccurate, they are several steps removed from direct visual observation.

✓ check your understanding

☐ 10. What imaging technique is best suited for each of the clinical situations described? (a) Examining gallbladder for possible gallstones in response to a patient's complaints of sharp pain in the upper right quadrant of the abdomen; (b) ruling

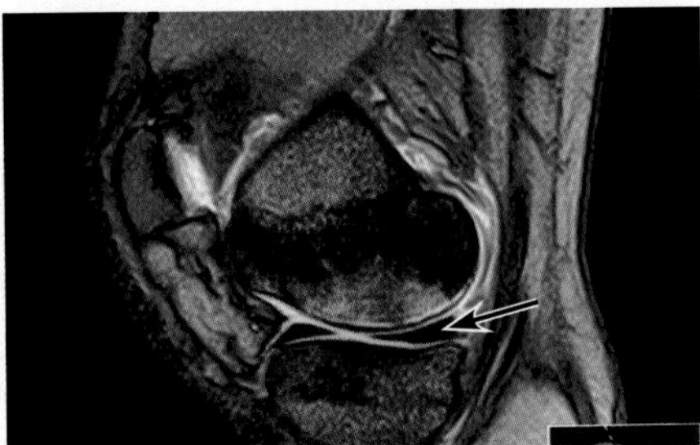

Normal knee, meniscus intact

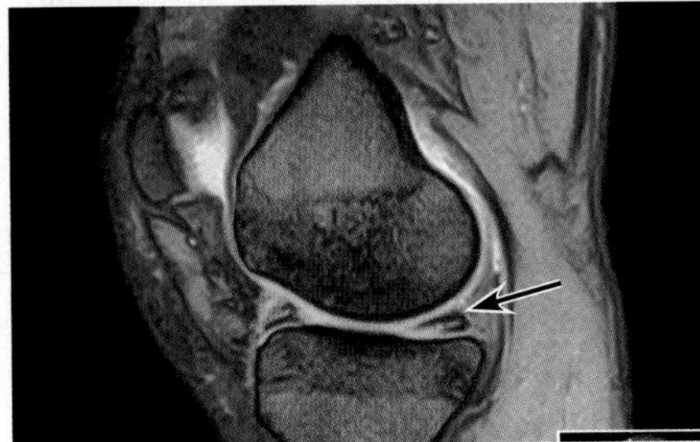

Injured knee, torn meniscus

(a) MRI of knee, sagittal section. Arrow indicates meniscus. Note tear in meniscus in bottom image.

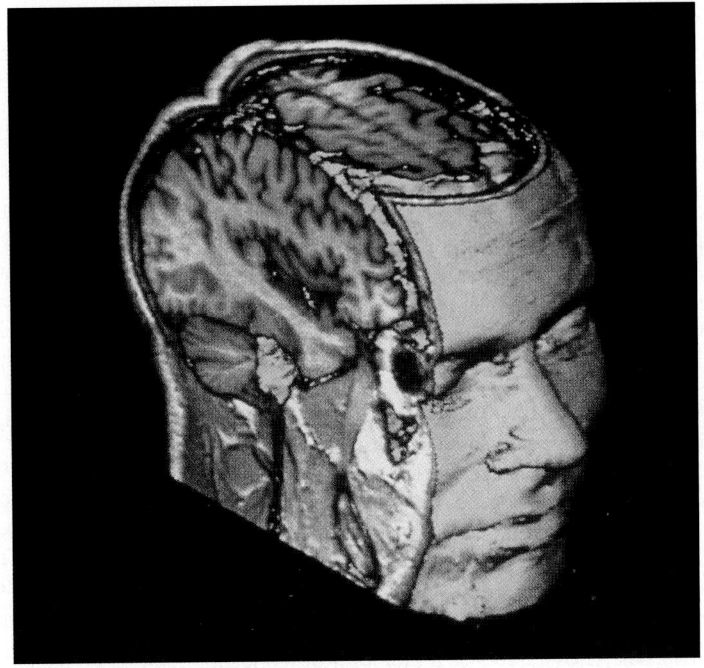

(b) Volume rendering of an MRI of the head

Figure 1.15 **Magnetic resonance image (MRI).** The flat surfaces in (b) show the original MRI data.

out a broken bone in a patient complaining of wrist and forearm pain; (c) examining of the knee of a patient complaining of persistent pain following an injury on the athletic field; (d) assessing possible damage to abdominal viscera resulting from a car accident.

(For answers, see Appendix B.)

CHAPTER SUMMARY

Take full advantage of the invaluable online practice and study tools for this chapter by going to the Study Area of MasteringA&P®. There you will find:

- Art Labeling Activities, Quizzes, Practice Tests, Memory Games
- Animal Organ Dissection, Cat Dissection and Bone Videos
- PAL Cadaver
- PAL Histology
- A&P Flix Animations
- Dynamic Study Modules

AN OVERVIEW OF ANATOMY (pp. 2–6)

1. Anatomy is the study of body structure. In this book, structures are considered in terms of their function.

Subdisciplines of Anatomy (p. 2)

2. Subdisciplines of anatomy include gross anatomy, microscopic anatomy (histology), and developmental anatomy.

The Hierarchy of Structural Organization (pp. 2–6)

3. The levels of structural organization of the body, from simplest to most complex, are chemical, cellular, tissue, organ, organ system, and the human organism itself.
4. The organ systems in the body are the integumentary (skin), skeletal, muscular, nervous, endocrine, cardiovascular, lymphatic, immune, respiratory, digestive, urinary, and reproductive systems.

Scale: Length, Volume, and Weight (p. 6)

5. Important units of length measurement are meters (m) for the organism, centimeters (cm) for the organs, and micrometers (μm) for cells. (For other units of measurement, see Appendix A.)

Anatomical Terminology (p. 6)

6. Because most structures in the body have formal Greek and Latin names, learning the meaning of word roots will help you understand anatomy.

GROSS ANATOMY: AN INTRODUCTION (pp. 6–13)

Regional and Directional Terms (pp. 6–7)

7. In the adult anatomical position, the body stands erect, facing forward with legs together. The arms are at the sides, with the palms forward.
8. Regional terms are used to designate specific areas of the body (Figure 1.3).
9. Directional terms allow anatomists to describe the location of body structures with precision. Important terms include superior/inferior; anterior/posterior (or ventral/dorsal); medial/lateral; proximal/distal; and superficial/deep (Table 1.1).

Body Planes and Sections (pp. 7–10)

10. The body or its organs may be cut along planes to produce different types of sections. The frontal plane divides a structure into anterior and posterior portions. A sagittal plane divides into right and left portions. A transverse plane (cross section) is horizontal to the longitudinal axis and divides a structure into superior and inferior sections.

The Human Body Plan (pp. 10–11)

11. The basic structures we share with all other vertebrate animals are the tube-within-a-tube body plan, bilateral symmetry, a dorsal hollow nerve cord, notochord and vertebrae, segmentation, and pharyngeal pouches.

Body Cavities and Membranes (pp. 11–13)

12. The body contains two major closed cavities: the dorsal body cavity, subdivided into the cranial and vertebral cavities; and the ventral body cavity, subdivided into the thoracic and abdominopelvic cavities.

13. Within the ventral body cavity are the visceral organs (such as the heart, lungs, intestines, and kidneys) and three serous cavities: pleural, pericardial, and peritoneal cavities. These slitlike cavities are lined by thin membranes, the parietal and visceral serosae (Figure 1.7). The serosae produce a thin layer of lubricating fluid that decreases friction between moving organs.

Abdominal Quadrants (p. 13)

14. To map the visceral organs in the abdominopelvic cavity, clinicians divide the abdomen into four quadrants.

MICROSCOPIC ANATOMY: AN INTRODUCTION (pp. 13–15)

Light and Electron Microscopy (pp. 13–15)

15. To illuminate cells and tissues, the light microscope (LM) uses light beams and the transmission electron microscope (TEM or EM) uses electron beams. EM produces sharper images than LM at higher magnification.

16. The preparation of tissues for microscopy involves preservation (fixation), sectioning, and staining. Stains for LM are colored dyes, whereas stains for TEM are heavy-metal salts.

Scanning Electron Microscopy (p. 15)

17. Scanning electron microscopy (SEM) provides sharp, three-dimensional images at high magnification.

CLINICAL ANATOMY: AN INTRODUCTION TO MEDICAL IMAGING TECHNIQUES (pp. 15–19)

X-Ray Imaging (pp. 15–16)

18. In conventional radiographs, X rays are used to produce negative images of internal body structures. Denser structures in the body appear lighter (whiter) on the X-ray film. X-ray images are useful for visualizing bones and abnormal dense structures, such as tumors. A contrast medium injected or ingested into hollow organs enables visualization of soft tissue organs, such as those of the gastrointestinal tract.

Advanced X-Ray Techniques (pp. 16–17)

19. Computed tomography (CT) produces improved X-ray images that are taken in cross section and are computer enhanced for clarity. CT produces excellent images of bone, blood vessels, and soft tissue and is especially useful in trauma situations to assess internal injury. Angiography produces sharp X-ray images of blood vessels injected with a contrast medium.

Positron Emission Tomography (p. 17)

20. PET tracks radioisotopes in the body, locating areas of high energy consumption and high blood flow. PET imaging is used in functional brain studies and in cancer diagnosis and assessment of treatment.

Sonography (p. 18)

21. Sonography provides sonar images of developing fetuses and internal body structures. Ultrasound images allow for immediate, inexpensive visualization of internal organs.

Magnetic Resonance Imaging (pp. 18–19)

22. MRI subjects the body to strong magnetic fields and radio waves, producing high-contrast images of soft body structures. MRI is very useful for visualizing structures surrounded by bone, such as nervous tissue and joints.

REVIEW QUESTIONS

Multiple Choice/Matching Questions

(For answers, see Appendix B.)

1. The correct sequence of levels forming the body's structural hierarchy is (a) organ, organ system, cellular, chemical, tissue; (b) chemical, tissue, cellular, organismal, organ, organ system; (c) chemical, cellular, tissue, organ, organ system, organismal; (d) organismal, organ system, organ, chemical, cellular, tissue.

2. Using the terms listed below, fill in the blank with the proper term.

anterior	superior	medial	proximal	superficial
posterior	inferior	lateral	distal	deep

(a) The heart is located ____ to the diaphragm.
(b) The muscles are ____ to the skin.
(c) The shoulder is ____ to the elbow.
(d) In anatomical position, the thumb is ____ to the index finger.
(e) The vertebral region is ____ to the scapular region.
(f) The nose is ____ to the chin.
(g) The toes are ____ to the heel.

3. Match each anatomical term for body regions listed in column B with the common name listed in column A.

Column A	Column B
____ **(1)** armpit	**(a)** inguinal
____ **(2)** buttocks	**(b)** dorsal
____ **(3)** back	**(c)** axilla
____ **(4)** groin	**(d)** lumbar
____ **(5)** lower back	**(e)** gluteal
____ **(6)** sole of foot	**(f)** plantar

4. Which of these organs would *not* be cut by a section through the midsagittal plane of the body? (a) urinary bladder, (b) gallbladder, (c) small intestine, (d) heart. (Hint: See Figure 1.8.)

5. State whether each structure listed below is part of the inner tube (I) or outer tube (O).

____ (1) intestines
____ (2) lungs
____ (3) backbone (vertebra)
____ (4) abdominal muscles
____ (5) esophagus
____ (6) spinal cord

6. Indicate whether each of the following conditions or statements applies to the dorsal body cavity (D) or the ventral body cavity (V).

____ (1) surrounded by the skull and the vertebral column
____ (2) includes the thoracic and abdominopelvic cavities
____ (3) contains the brain and spinal cord
____ (4) located more anteriorly
____ (5) contains the heart, lungs, and many digestive organs

7. List the following structures, from darkest (black) to lightest (white), as they would appear on an X-ray film. Number the darkest one 1, the next darkest 2, and so on.

____ (a) soft tissue
____ (b) femur (bone of the thigh)
____ (c) air in lungs
____ (d) gold (metal) filling in a tooth

8. The ventral surface of the body is the same as its ____ surface. (a) medial, (b) superior, (c) superficial, (d) anterior, (e) distal.

9. Match each serous membrane in column B with its description in column A.

Column A	Column B
____ (1) covers the surface of the heart	(a) parietal pericardium
____ (2) forms the outer lining of the pericardium	(b) parietal pleura
____ (3) lines the wall of the thoracic cavity	(c) visceral pericardium
____ (4) covers the outer surface of the small intestine	(d) visceral peritoneum

10. Which microscopic technique provides sharp pictures of three-dimensional structures at high magnification? (a) light microscopy, (b) X-ray microscopy, (c) scanning electron microscopy, (d) transmission electron microscopy.

11. Histology is the same as (a) pathological anatomy, (b) ultrastructure, (c) functional morphology, (d) surface anatomy, (e) microscopic anatomy.

Short Answer Essay Questions

12. Describe the anatomical position, and then assume this position.

13. Identify the organ system that each group of organs listed here is part of:
(a) thymus, thyroid, ovary, pancreas
(b) nasal cavity, larynx, bronchi, lungs
(c) kidney, ureter, urethra

14. (a) Define *bilateral symmetry*. (b) Although many body structures are bilaterally symmetrical, much of the *abdomen* is not. Find a picture that demonstrates this lack of symmetry, and name some abdominal organs that are not symmetrical.

15. The following advanced imaging techniques are discussed in the text: CT, angiography, PET, sonography, and MRI. (a) Which of these techniques uses X rays? (b) Which uses radio waves and magnetic fields? (c) Which uses radioactive isotopes? (d) Which displays body regions in sections? You may have more than one answer to each question.

16. Give the formal regional term for each of these body areas: (a) arm, (b) chest, (c) thigh, (d) navel, (e) limbs.

17. How would the appearance of a tissue differ if viewed with light microscopy versus electron microscopy? Consider the types of structures that would be viewable, the level of detail, the color of the image, and whether the view would show surface features or a sectional view.

18. Construct sentences that use the following directional terms: superior/inferior; dorsal/ventral; medial/lateral; and superficial/deep. (For ideas, look at whole-body diagrams that show many structures, such as Figures 1.3 and 1.8.

19. The main cavities of the body include the abdominal and pelvic cavities. List three organs in each of these cavities.

20. Where would you be injured if you pulled a muscle in your axillary region, cracked a bone in your occipital region, or cut a digit?

21. (a) The human body is designed as a tube within a tube. List three organs that are part of the inner tube and three structures that are components of the outer tube. (b) Give an example of segmentation in the human body.

Critical Reasoning & Clinical Application Questions

1. Dominic's doctors strongly suspect he has a tumor in his brain. Which of the following medical imaging techniques—conventional X-ray imaging, angiography, PET, sonography, or MRI—would probably be best for precisely locating the tumor?

2. The Nguyen family was traveling in their van and had a minor accident. The children in the back seat were wearing lap belts, but they still sustained bruises around the abdomen and had some internal injuries. Why is the abdominal area more vulnerable to damage than others?

3. A patient had a hernia in his inguinal region, pain from an infected kidney in his lumbar region, and hemorrhoids in his perineal region. Explain where each of these regions is located.

4. A woman fell off a motorcycle. She tore a nerve in her axillary region, and she broke bones in her coxal, sacral, acromial, and fibular regions. Explain where each of these regions is located.

5. New anatomy students often mix up the terms *spinal cord, spinal column,* and *spine.* Look up these words in the index of this book or in a medical dictionary, define each one, and explain whether they are the same or different.

6. Using the list of word roots located in the inside covers of the text: (1) Look up the meaning of each root listed below. (2) Identify one anatomical term used in this chapter that is derived from each root. (3) Define the anatomical term from your knowledge of the meaning of the word root.

super- , contra- , para- , ante- , mamm- , epi- , peri- , -graph , trans-

2 Cells: The Living Units

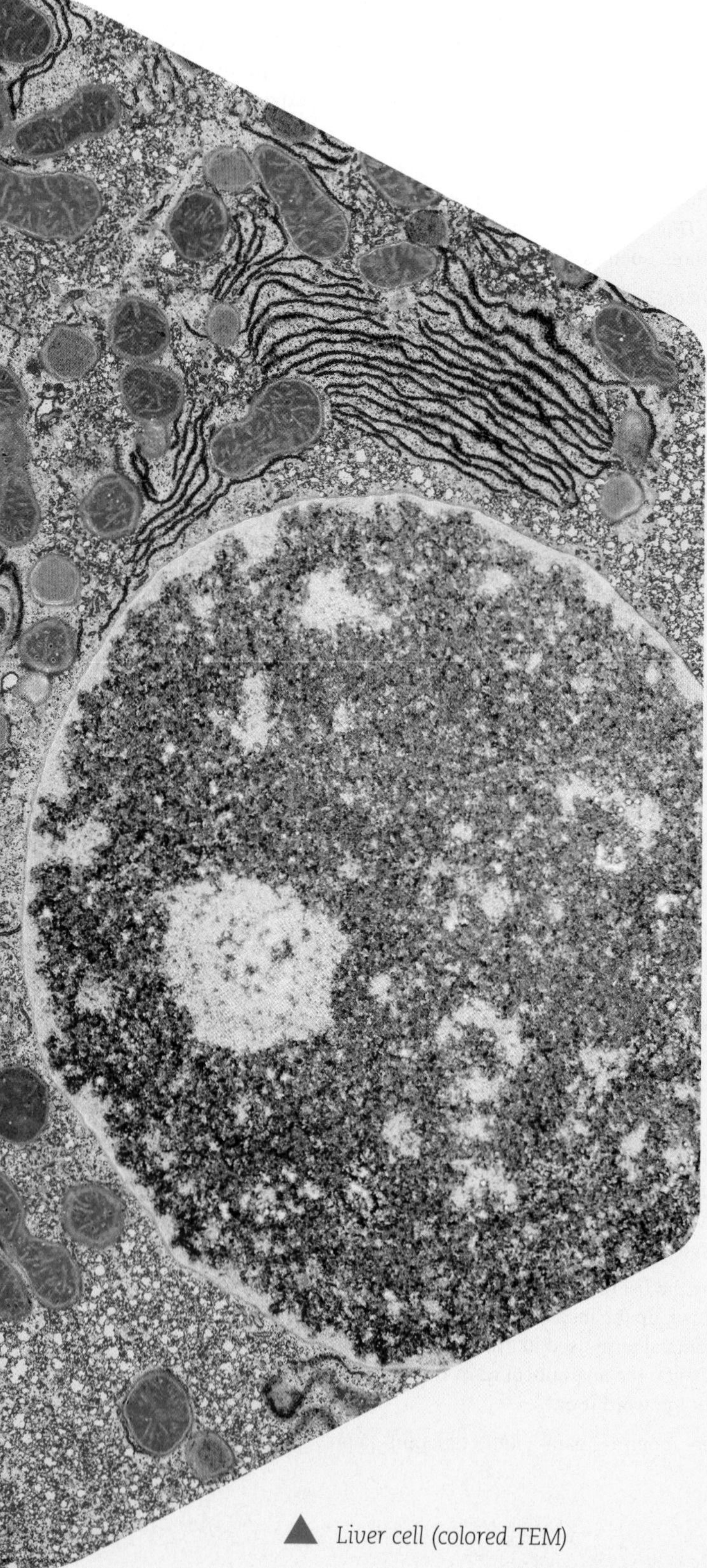

▲ *Liver cell (colored TEM)*

All living organisms are cellular in nature, from one-celled "generalists" such as amoebas to complex multicellular organisms such as trees, dogs, and humans. Just as bricks and timbers are the structural units of a house, cells are structural units of all living things. The human body has about *50 to 100 trillion cells.* This chapter examines the structures and functions that are common to the different cells of the body. Specialized cells and their unique functions are addressed in detail in later chapters.

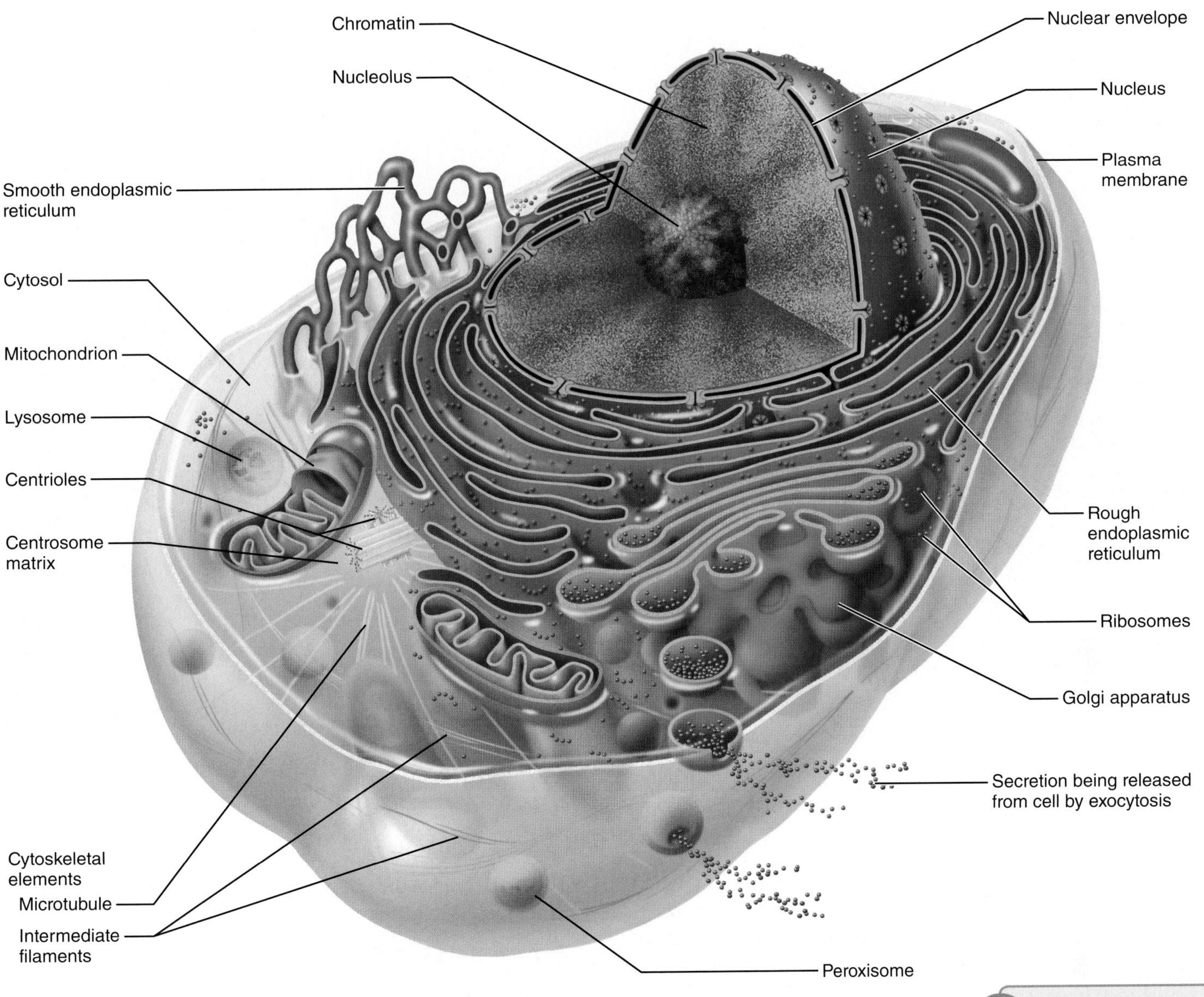

Figure 2.1 **Structure of a generalized cell.** No cell type is exactly like this one, but this composite illustrates features common to animal cells. Not all of the organelles are drawn to the same scale in this illustration.

Practice art labeling MasteringA&P®

OVERVIEW OF CELLS

learning outcome

▶ Define a cell, its basic activities, and its three major regions.

The scientist Robert Hooke first observed plant cells with a crude microscope in the late 1600s. However, it was not until the 1830s that two scientists, Matthias Schleiden and Theodor Schwann, boldly asserted that all living things are composed of cells. Shortly thereafter, the pathologist Rudolf Virchow extended this idea by contending that cells arise only from other cells. Virchow's thesis was revolutionary because it challenged the prevailing theory of spontaneous generation, which held that organisms can arise from nonliving matter. Since the late 1800s, cell research has been exceptionally fruitful. Currently, scientific knowledge about the cell is increasing exponentially.

Cells are the smallest living units in the body. Each cell performs all the functions necessary to sustain life. Each cell can:

- Obtain nutrients and other essential substances from the surrounding body fluids.
- Use these nutrients to make the molecules it needs to survive.
- Dispose of its wastes.
- Maintain its shape and integrity.
- Replicate itself.

These functions are carried out by the cell's many subunits, most of which are called **organelles** ("little organs"). Although different cell types perform different functions, virtually all human cells contain the same basic parts and can be described in terms of a generalized cell (Figure 2.1). Human cells have three main parts: the plasma membrane, the cytoplasm, and

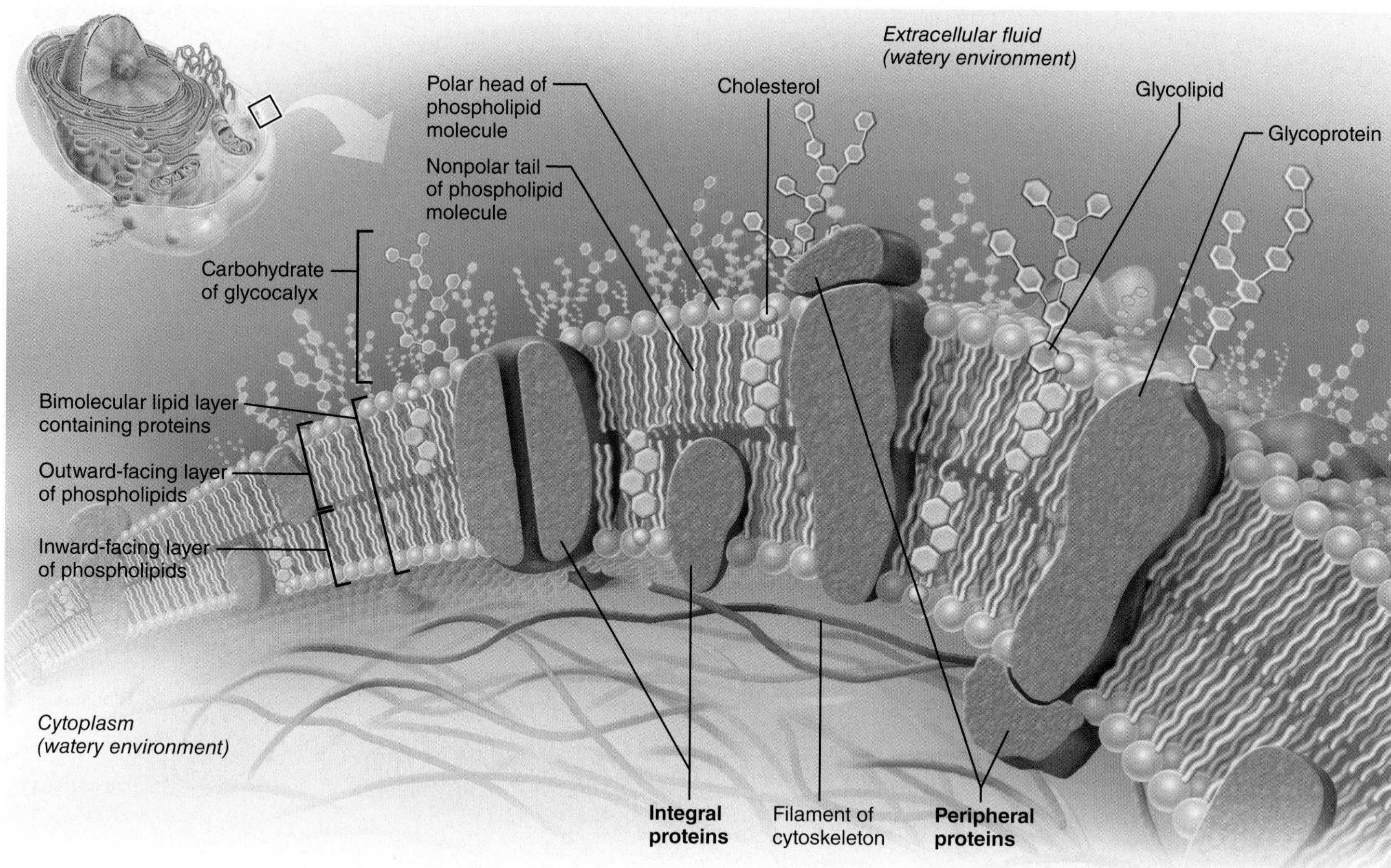

Figure 2.2 **The plasma membrane according to the fluid mosaic model.**

the nucleus (Table 2.1). The *plasma membrane* is the outer boundary. Internal to this membrane is the *cytoplasm* (si′to-plazm), which makes up the bulk of the cell, contains most of the cellular organelles, and surrounds the nucleus. The *nucleus* (nu′kle-us) controls cellular activities and lies near the cell's center. Think of the cell as a manufacturing plant. Each organelle functions like a specific division of the factory:

- The **plasma membrane** is the boundary fence and security gate; it forms the boundary of the cell and selectively allows materials to pass into and out of the cell.
- **Ribosomes** are the assembly line; they produce proteins.
- **Endoplasmic reticulum (ER)** is part of the production division; rough ER produces proteins, and smooth ER metabolizes lipids and stores calcium.
- **Golgi apparatus** is the packaging and shipping division; it packages proteins for use either within or outside of the cell.
- **Lysosomes** are the janitorial crew and recycling center; they break down used proteins and other cellular debris.
- **Mitochondria** are the power plant; they make energy.
- **Peroxisomes** are the toxic waste treatment facility; they neutralize and remove toxic substances within the cell.
- **Cytoskeletal elements** form the framework and infrastructure of the building; they maintain cell shape and structure and transport materials within the cell.
- The **nucleus** is the chief executive officer (CEO); it directs the operation of the cell.

Each of these cell structures is discussed next.

☑ check your understanding

☐ 1. What are the three general regions of a cell?

(For the answer, see Appendix B.)

THE PLASMA MEMBRANE

learning outcomes

▶ Describe the composition and basic functions of the plasma membrane.

▶ Explain the different processes used to move molecules across the plasma membrane.

The outer cell membrane (Figure 2.2) is called the **plasma membrane** or **plasmalemma** (plaz′mah-lem′ah; *lemma* = sheath, husk). This thin, flexible layer defines the extent of the cell, thereby separating two of the body's major fluid compartments: the *intracellular fluid* within the cells and the *extracellular fluid* that lies outside and between cells. To return to our analogy, you can think of the plasma membrane as a security fence surrounding the manufacturing plant (cell). This boundary contains specific checkpoints (receptors) that influence cellular activity in various ways.

Table 2.1 Parts of the Cell: Structure and Function

Cell Part	Structure	Functions
PLASMA MEMBRANE (Figure 2.2)	Membrane made of a double layer of lipids (phospholipids, cholesterol, etc.) within which proteins are embedded; externally facing proteins and some lipids have attached sugar groups	Serves as an external cell barrier; acts in transport of substances into or out of the cell; externally facing proteins act as receptors (for hormones, neurotransmitters, etc.) and in cell-to-cell recognition
CYTOPLASM	Cellular region between the nuclear and plasma membranes; consists of fluid **cytosol** containing dissolved solutes, **inclusions** (stored nutrients, pigment granules), and **organelles**, the metabolic machinery of the cytoplasm	
Cytoplasmic organelles		
• Ribosomes (Figure 2.6)	Dense particles consisting of two subunits, each composed of ribosomal RNA and protein; free or attached to rough ER	The sites of protein synthesis
• Rough endoplasmic reticulum (Figure 2.6)	Membrane system of sacs and tubules externally studded with ribosomes	Makes proteins that are secreted from the cell; makes the cell's membranes
• Smooth endoplasmic reticulum (Figure 2.6)	Membranous system of sacs and tubules; free of ribosomes	Site of lipid and steroid hormone synthesis, lipid metabolism, and drug detoxification
• Golgi apparatus (Figures 2.7, 2.8)	A stack of smooth membrane sacs close to the nucleus	Packages, modifies, and segregates proteins for secretion from the cell, inclusion in lysosomes, and incorporation into the plasma membrane
• Lysosomes (Figure 2.9)	Membranous sacs containing acid hydrolases	Sites of intracellular digestion
• Mitochondria (Figure 2.10)	Rodlike, double-membrane structures; inner membrane folded into projections called cristae	Site of ATP synthesis; powerhouse of the cell
• Peroxisomes (Figure 2.1)	Membranous sacs of oxidase enzymes	The enzymes detoxify a number of toxic substances; the most important enzyme, catalase, breaks down hydrogen peroxide
• Microfilaments (Figure 2.11a)	Fine filaments of the contractile protein actin	Involved in muscle contraction and other types of intracellular movement; help form the cell's cytoskeleton
• Intermediate filaments (Figure 2.11b)	Protein fibers; composition varies	The stable cytoskeletal elements; resist tension forces acting on the cell
• Microtubules (Figure 2.11c)	Cylindrical structures made of tubulin proteins	Support the cell and give it shape; involved in intracellular and cellular movements; form centrioles
• Centrioles (Figure 2.12)	Paired cylindrical bodies, each composed of nine triplets of microtubules	Organize a microtubule network during mitosis to form the spindle and asters; form the bases of cilia and flagella
NUCLEUS (Figure 2.13)	Surrounded by the nuclear envelope; contains fluid nucleoplasm, nucleoli, and chromatin	Control center of the cell; responsible for transmitting genetic information and providing the instructions for protein synthesis
• Nuclear envelope (Figure 2.13)	Double-membrane structure; pierced by the pores; continuous with the cytoplasmic ER	Separates the nucleoplasm from the cytoplasm and regulates passage of substances to and from the nucleus
• Nucleoli (Figure 2.13)	Dense spherical (non-membrane-bounded) bodies	Site of ribosome subunit manufacture
• Chromatin (Figures 2.13, 2.15)	Granular, threadlike material composed of DNA and histone proteins	DNA constitutes the genes

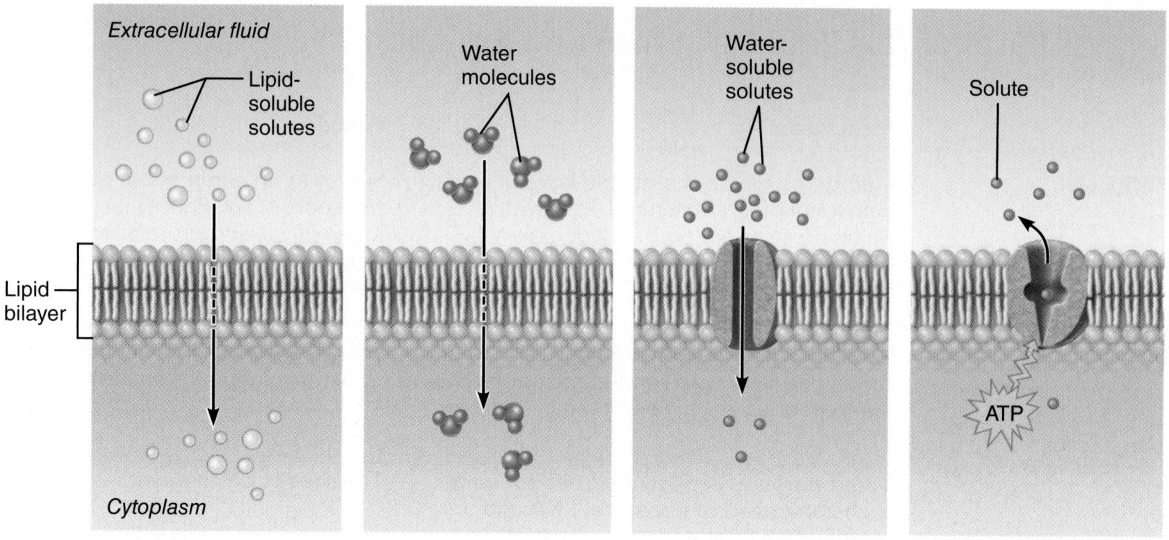

(a) Simple diffusion of fat-soluble molecules directly through the phospholipid bilayer down their concentration gradient

(b) Osmosis, diffusion of water through the lipid bilayer

(c) Facilitated diffusion An integral protein that spans the plasma membrane enables the passage of a particular solute across the membrane.

(d) Active transport Some transport proteins use ATP as an energy source to actively pump substances across the plasma membrane against their concentration gradient.

Figure 2.3 **Membrane transport mechanisms.**

Structure

The *fluid mosaic model* of membrane structure depicts the plasma membrane as a double layer, or bilayer, of lipid molecules with protein molecules embedded within it (Figure 2.2). The most abundant lipids in the plasma membrane are phospholipids. Like a lollipop on two sticks, each phospholipid molecule has a polar "head" that is charged, and an uncharged, nonpolar "tail" made of two chains of fatty acids. The polar heads are attracted to water—the main constituent of both the cytoplasm and the fluid external to the cell—so they lie along the inner as well as the outer face of the membrane. The nonpolar tails avoid water and line up in the center of the membrane. The result is two parallel sheets of phospholipid molecules lying tail to tail, forming the membrane's basic bilayered structure.

The inner and outer layers of the membrane differ somewhat in the kinds of lipids they contain. Sugar groups are attached to about 10% of the outer lipid molecules, making them "sugar-fats," or glycolipids (gli″ko-lip′ids). The plasma membrane also contains substantial amounts of cholesterol, another lipid. Cholesterol makes the membrane more rigid and increases its impermeability to water and water-soluble molecules.

Proteins make up about half of the plasma membrane by weight. The membrane proteins are of two distinct types: integral and peripheral (Figure 2.2). **Integral proteins** are firmly embedded in or strongly attached to the lipid bilayer. Some integral proteins protrude from one side of the membrane only, but most are *transmembrane proteins* that span the whole width of the membrane and protrude from both sides (*trans* = across). **Peripheral proteins,** by contrast, are not embedded in the lipid bilayer at all. Instead, they attach rather loosely to the membrane surface. The peripheral proteins include a network of filaments that helps support the membrane from its cytoplasmic side. Without this strong, supportive base, the plasma membrane would tear apart easily.

Short chains of carbohydrate molecules attach to the integral proteins to form glycoproteins. These sugars project from the external cell surface, forming the *glycocalyx* (gli″-ko-kal′iks; "sugar covering"), or *cell coat.* Also contributing to the glycocalyx are the sugars of the membrane's glycolipids. You can therefore think of your cells as "sugar-coated." The glycocalyx is sticky and may help cells to bind when they come together. Because every cell type has a different pattern of sugars that make up its glycocalyx, the glycocalyx is also a distinctive biological marker by which approaching cells recognize each other. For example, a sperm recognizes the ovum (egg cell) by the distinctive composition of the ovum's glycocalyx.

Functions

The functions of the plasma membrane relate to its location at the interface between the cell's exterior and interior:

1. The plasma membrane provides a protective barrier against substances and forces outside the cell.
2. Some of the membrane proteins act as **receptors;** that is, they have the ability to bind to specific molecules arriving

from outside the cell. After binding to the receptor, the molecule can induce a change in the cellular activity. Membrane receptors act as part of the body's cellular communication system.

3. The plasma membrane controls which substances can enter and leave the cell. The membrane is a selectively permeable barrier that allows some substances to pass between the intracellular and extracellular fluids while preventing others from doing so. The processes involved in moving substances across the plasma membrane are described next.

Membrane Transport

Small, uncharged molecules, such as oxygen, carbon dioxide, and fat-soluble molecules, can pass freely through the lipid bilayer of the plasma membrane through a process called **simple diffusion.** Diffusion is the tendency of molecules in a solution to move down their concentration gradient; that is, the molecules move from a region where they are more concentrated to a region where they are less concentrated (Figure 2.3a). Water, like other molecules, diffuses down its concentration gradient. The diffusion of water molecules across a membrane is called **osmosis** (oz-mo′sis, Figure 2.3b).

However, most water-soluble or charged molecules, such as glucose, amino acids, and ions, cannot pass through the lipid bilayer by simple diffusion. Such substances can cross the plasma membrane only by means of specific transport mechanisms that use the integral proteins to carry or pump molecules across the membrane or to form channels through which specific molecules pass. Some of these molecules move down their concentration gradient, diffusing through the plasma membrane by moving through a specific integral protein. This transport process is called **facilitated diffusion** (Figure 2.3c). Other integral proteins move molecules across the plasma membrane against their concentration gradient, a process called **active transport,** which requires the use of energy (Figure 2.3d).

The largest molecules (macromolecules) and large solid particles are actively transported through the plasma membrane by another set of processes, called *vesicular* or *bulk transport.* Knowledge of the two general types of bulk transport, *exocytosis* and *endocytosis,* is essential to understanding basic functional anatomy.

Endocytosis (en″do-si-to′sis; "into the cell") is the mechanism by which large particles and macromolecules *enter* cells (Figure 2.4). The substance to be taken into the cell is enclosed by an infolding part of the plasma membrane. In the region of invagination, specific proteins may cover the inner surface of the plasma membrane (the purple, tack-shaped structures shown in Figure 2.4c). This protein coat aids in the selection of the substance to be transported and deforms the membrane to form a membrane-walled sac called a **vesicle.** The membranous vesicle is pinched off from the plasma membrane and moves into the cytoplasm, where its contents are digested. Three types of endocytosis are recognized: phagocytosis, pinocytosis, and receptor-mediated endocytosis.

Phagocytosis (fag″o-si-to′sis) is literally "cell eating" (Figure 2.4a). Some cells—most white blood cells, for example—are experts at phagocytosis. Such cells help to police and protect the body by ingesting bacteria, viruses, and other foreign substances. They also "eat" the body's dead and diseased cells.

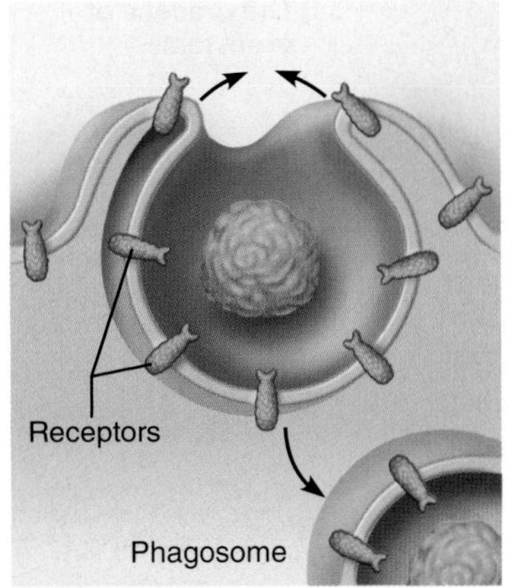

(a) Phagocytosis
The cell engulfs a large particle by forming projecting pseudopods ("false feet") around it and enclosing it within a membrane sac called a **phagosome.** The phagosome then combines with a lysosome, and its contents are digested. Vesicle may or may not be protein-coated but has receptors capable of binding to microorganisms or solid particles.

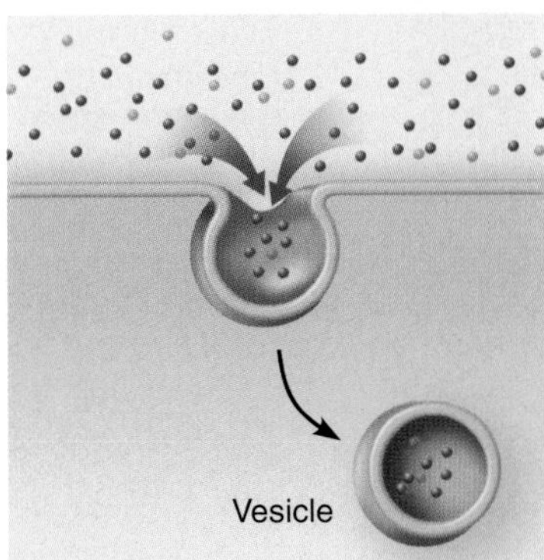

(b) Pinocytosis
Infolding of the plasma membrane carries a drop of extracellular fluid containing solutes into the cell in a tiny membrane-bound vesicle. No receptors are used, so the process is nonspecific.

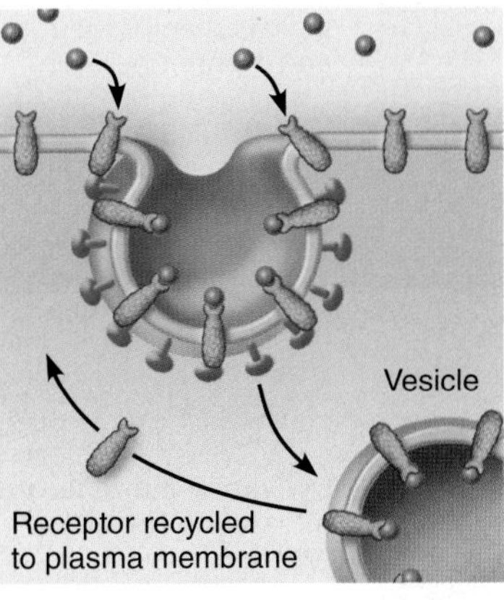

(c) Receptor-mediated endocytosis
Extracellular substances bind to specific receptor proteins in regions of protein-coated pits, enabling the cell to ingest and concentrate specific substances in protein-coated vesicles. The ingested substance may simply be released inside the cell or combined with a lysosome to digest contents. Receptors are recycled to the plasma membrane in vesicles.

Figure 2.4 **The three types of endocytosis.**

Just as cells eat in a manner of speaking, they also drink. **Pinocytosis** (pin″o-si-to′sis) is "cell drinking" (Figure 2.4b). Pinocytosis, a routine activity of most cells, is an unselective way of sampling the extracellular fluid. This process is particularly important in cells that function in nutrient absorption, such as cells that line the intestines.

Some molecules, such as insulin and other hormones, enzymes, and *low-density lipoproteins* (*LDLs,* the molecules that carry cholesterol through the bloodstream to the body's cells) are brought into cells through **receptor-mediated endocytosis,** an exquisitely selective transport process (Figure 2.4c). These substances bind to specific receptors on the cell

(a) The process of exocytosis

① The molecule to be secreted migrates to the plasma membrane in a membrane-bound vesicle.

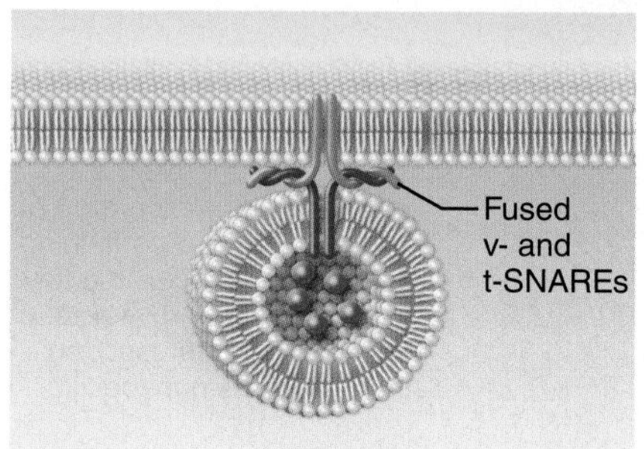

② At the plasma membrane, proteins at the vesicle surface (v-SNAREs, *v* for "vesicle") bind with t-SNAREs (plasma membrane proteins, *t* for "target").

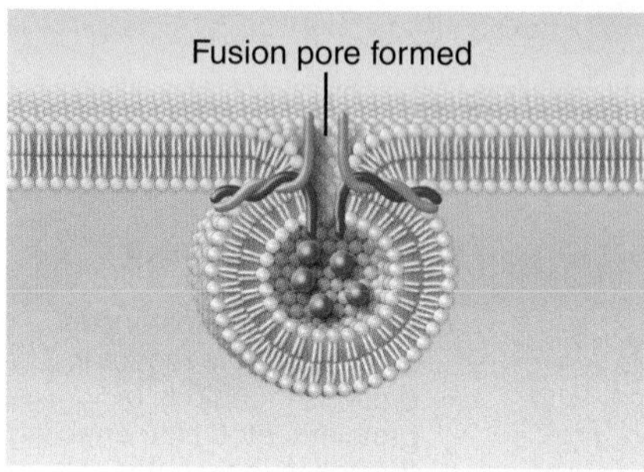

③ The vesicle and plasma membrane fuse and a pore opens up.

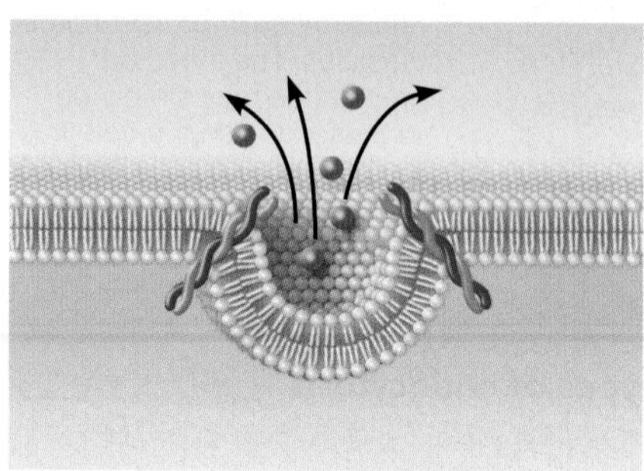

④ Vesicle contents are released to the cell exterior.

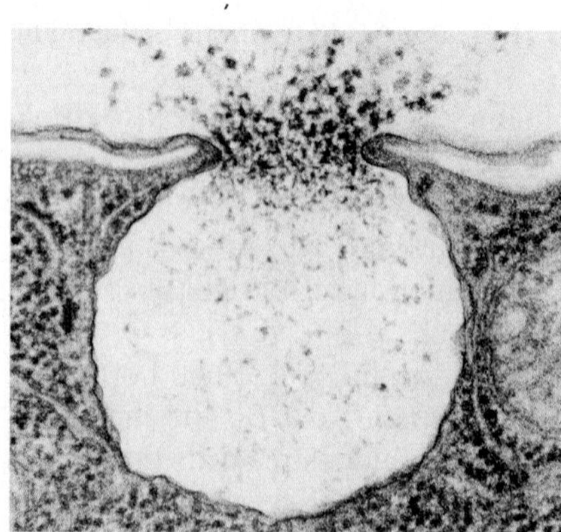

(b) Photomicrograph of a secretory vesicle releasing its contents by exocytosis (110,000×)

Figure 2.5 **Exocytosis.**

membrane for transport into the cell. Unfortunately, harmful substances such as some toxins and viruses also use receptor-mediated endocytosis to enter and attack cells.

CLINICAL APPLICATION

Hypercholesterolemia In an inherited disease called *familial hypercholesterolemia*, the body's cells lack the protein receptors that bind to cholesterol-delivering low-density lipoproteins (LDLs.) As a result, cholesterol cannot enter the cells and accumulates in the blood. If untreated, hypercholesterolemia causes atherosclerosis, also known as "hardening of the arteries," a condition that places the individual at high risk of stroke (blockage of a blood vessel in the brain) or of coronary artery disease and heart attack.

Exocytosis (ek″so-si-to′sis; "out of the cell") is an active mechanism by which substances move from the cytoplasm to the outside of the cell (Figure 2.5). Exocytosis accounts for most secretion processes, such as the release of mucus or protein hormones from the gland cells of the body. (The details of this process are shown in Figure 2.5.)

check your understanding

- ☐ 2. What types of macromolecules compose the plasma membrane?
- ☐ 3. By what process does water enter and leave the cell?
- ☐ 4. Which transport process carries large macromolecules out of the cell?

(For answers, see Appendix B.)

THE CYTOPLASM

learning outcomes

- ▶ Describe the structure and cellular activity of each organelle: ribosomes, endoplasmic reticulum, Golgi apparatus, lysosomes, peroxisomes, mitochondria, cytoskeleton, centrosome, and centrioles.
- ▶ Explain the structure of glycosomes and lipid droplets.

Cytoplasm, literally "cell-forming material," is the part of the cell that lies internal to the plasma membrane and external to the nucleus. Most cellular activities are carried out in the cytoplasm, which consists of three major elements: *cytosol, organelles,* and *inclusions.*

Cytosol

The **cytosol** (si′to-sol), is the jellylike, fluid-containing substance within which the other cytoplasmic elements are suspended (Figure 2.1). It consists of water, ions, and many enzymes. Some of these enzymes start the breakdown of nutrients (sugars, amino acids, and lipids) that are the raw materials and energy source for cell activities. In many cell types, the cytosol makes up about half the volume of the cytoplasm.

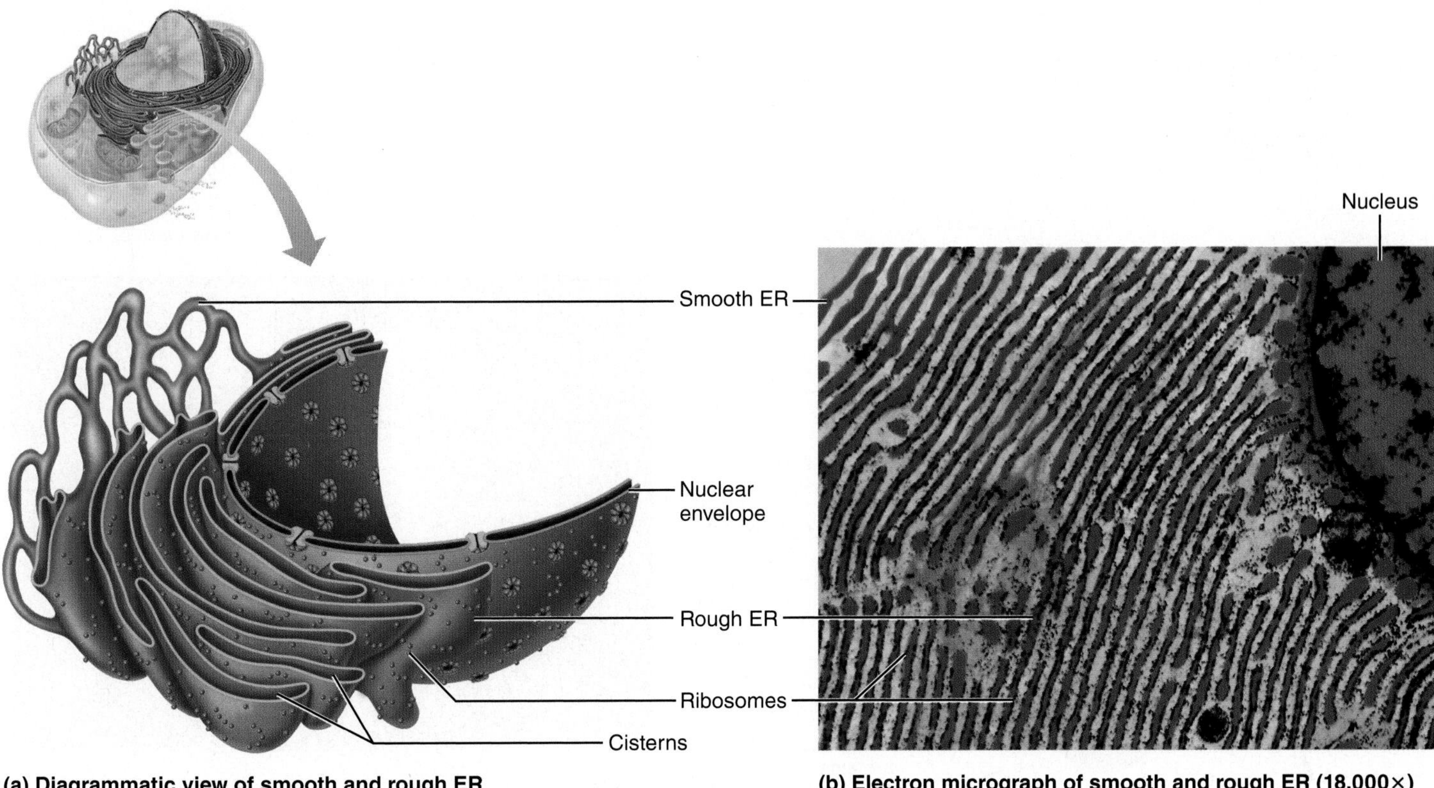

(a) Diagrammatic view of smooth and rough ER

(b) Electron micrograph of smooth and rough ER (18,000×)

Figure 2.6 **The endoplasmic reticulum (ER) and ribosomes.**

Cytoplasmic Organelles

Typically, the cytoplasm contains about nine types of organelles, each with a different function that is essential to the survival of the cell. As separate units, the organelles compartmentalize the cell's biochemical reactions, thus preventing reactions from interfering with one another and promoting functional efficiency. The organelles include mitochondria, ribosomes, rough and smooth endoplasmic reticulum, Golgi apparatus, lysosomes, peroxisomes, the cytoskeleton, and centrioles (Figure 2.1). As you will learn, most organelles are bounded by a membrane that is similar in composition to the plasma membrane but lacks a glycocalyx.

With very few exceptions, all cells of the human body share the same kinds of organelles. However, when a cell type performs a special body function, the organelles that contribute to that function are especially abundant in that cell. Thus, certain organelles are better developed in some cells than in others. You will see examples of this principle as you explore the organelles and their roles.

Ribosomes

Ribosomes (ri′bo-sōmz) are the assembly line of the manufacturing plant, producing proteins for cellular or extracellular function. They are small, dark-staining granules (Figure 2.6). Unlike most organelles, they are not surrounded by a membrane, but are constructed of proteins plus **ribosomal RNA** (*RNA* = ribonucleic acid). Each ribosome consists of two subunits that fit together like the body and cap of an acorn.

Almost all cells make large amounts of protein, and ribosomes are the site of protein synthesis. On the ribosomes, building blocks called amino acids are linked together to form protein molecules. This assembly process is called *translation*. It is dictated by the genetic material in the cell nucleus (DNA), whose instructions are carried to the ribosomes by messenger molecules called **messenger RNA (mRNA).**

Many ribosomes float freely within the cytosol. Such **free ribosomes** make the soluble proteins that function within the cytosol itself. Ribosomes attached to the membranes of the rough endoplasmic reticulum (Figure 2.6), make proteins that become part of the cell membrane or that are exported out of the cell.

Endoplasmic Reticulum

The **endoplasmic reticulum** (en″do-plaz′mik ret-tik′u-lum), or **ER,** is literally the "network within the cytoplasm." The ER is an extensive system of membrane-walled envelopes and tubes that twists through the cytoplasm (Figure 2.6). It accounts for more than half of the membranous surfaces inside an average human cell. There are two distinct types of ER: *rough ER* and *smooth ER*. Either type may predominate in a given cell type, depending on the specific functions of the cell.

Rough Endoplasmic Reticulum The **rough endoplasmic reticulum (rough ER)** consists mainly of stacked membrane-enclosed cavities called **cisterns** ("fluid-filled cavities"). Ribosomes stud the external faces of the membranes of the rough ER, assembling proteins. The ribosomes attach to the membrane when the protein is being made, then detach when the protein is completed.

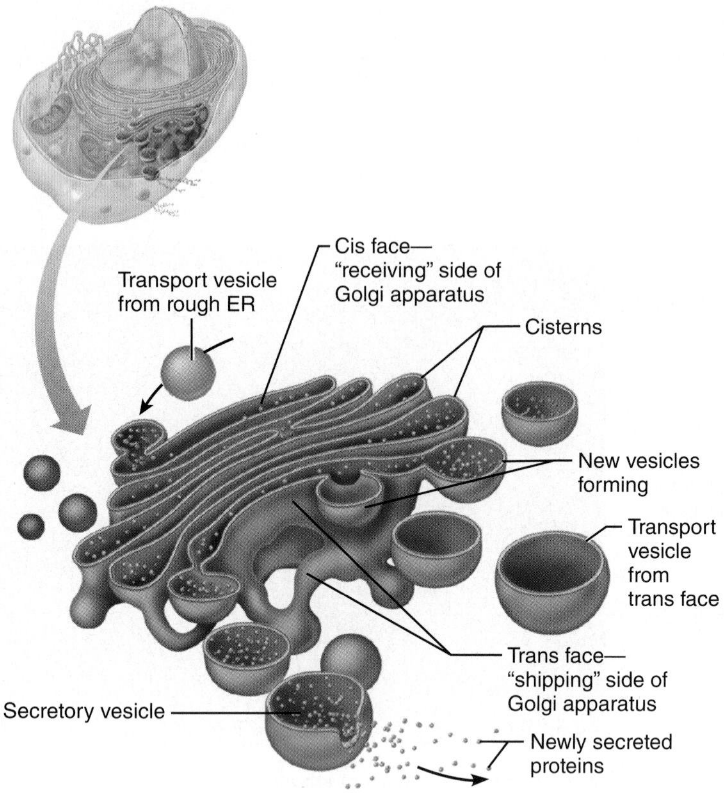

(a) Many vesicles in the process of pinching off from the Golgi apparatus

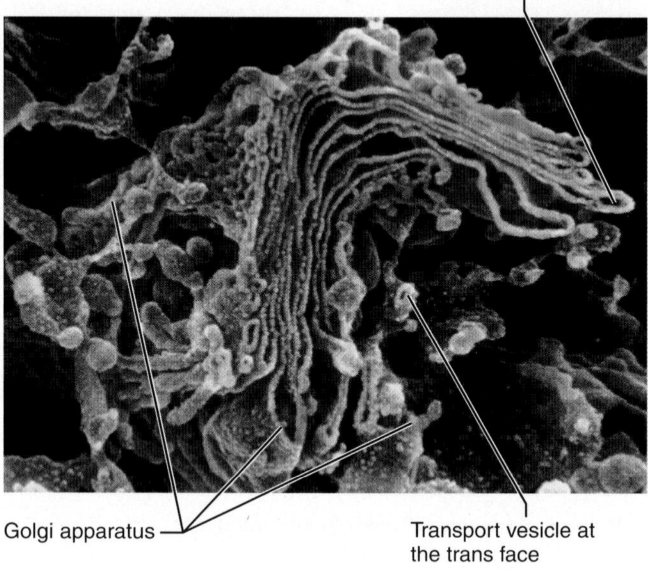

(b) Electron micrograph of the Golgi apparatus (90,000×)

Figure 2.7 **Golgi apparatus.**

The rough ER has several functions. Its ribosomes make all proteins that are secreted from cells; thus, rough ER is especially well developed in gland cells that secrete a large amount of protein (mucous cells, for example). It makes the digestive enzymes that will be contained in lysosomes. The rough ER also makes both the integral proteins and the phospholipid molecules of the cell's membranes. In other words, all cell membranes start out as rough ER membrane. The rough ER can therefore be considered the cell's "membrane factory."

Smooth Endoplasmic Reticulum The **smooth endoplasmic reticulum (smooth ER)** is continuous with the rough ER (Figure 2.6). It consists of tubules arranged in a branching network. Because no ribosomes are attached to its membranes, the smooth ER is not a site of protein synthesis. It performs different functions in different cell types, but most of these relate to lipid metabolism, the making or breaking down of fats. Smooth ER is abundant in cells that make lipid steroid hormones from cholesterol and in liver cells that detoxify lipid-soluble drugs. Most cell types, however, have little smooth ER.

Another important function of smooth ER is storing calcium ions. Ionic calcium is a signal for the beginning of many cellular events, including muscle contraction and glandular secretion. The calcium concentration in the cytosol is kept low when such events are not occurring, because most calcium ions are pumped into the ER and held there until the cell needs them. The ER in muscle cells is very extensive, reflecting this essential function.

Golgi Apparatus

The **Golgi** (gōl′je) **apparatus** is a stack of three to ten disc-shaped cisterns, each bound by a membrane (Figure 2.7). It resembles a stack of hollow saucers, one cupped inside the next. The products of the rough ER move through the Golgi stack from the convex *(cis)* to the concave *(trans)* side. More specifically, the **cis face** receives spherical, membranous **transport vesicles** from the rough ER; new vesicles bud off a **trans face** to leave the apparatus.

The Golgi apparatus sorts, processes, and packages the proteins and membranes made by the rough ER. Activities and products in the Golgi apparatus follow three pathways—A, B, and C (Figure 2.8). In pathway A, which occurs in gland cells, the protein product is contained in **secretory vesicles;** these vesicles ultimately release their contents to the cell's exterior by exocytosis. In pathway B, common in all cells, the membrane of the vesicle fuses to and contributes to the plasma membrane, whose components are constantly being renewed and recycled. In pathway C, also common in all cells, the vesicle leaving the Golgi apparatus is a lysosome, a sac filled with digestive enzymes, that remains inside the cell. Thus, the Golgi apparatus is the packaging and shipping division of the manufacturing plant. It receives product produced by the rough ER, packages it, and ships it to its appropriate destination.

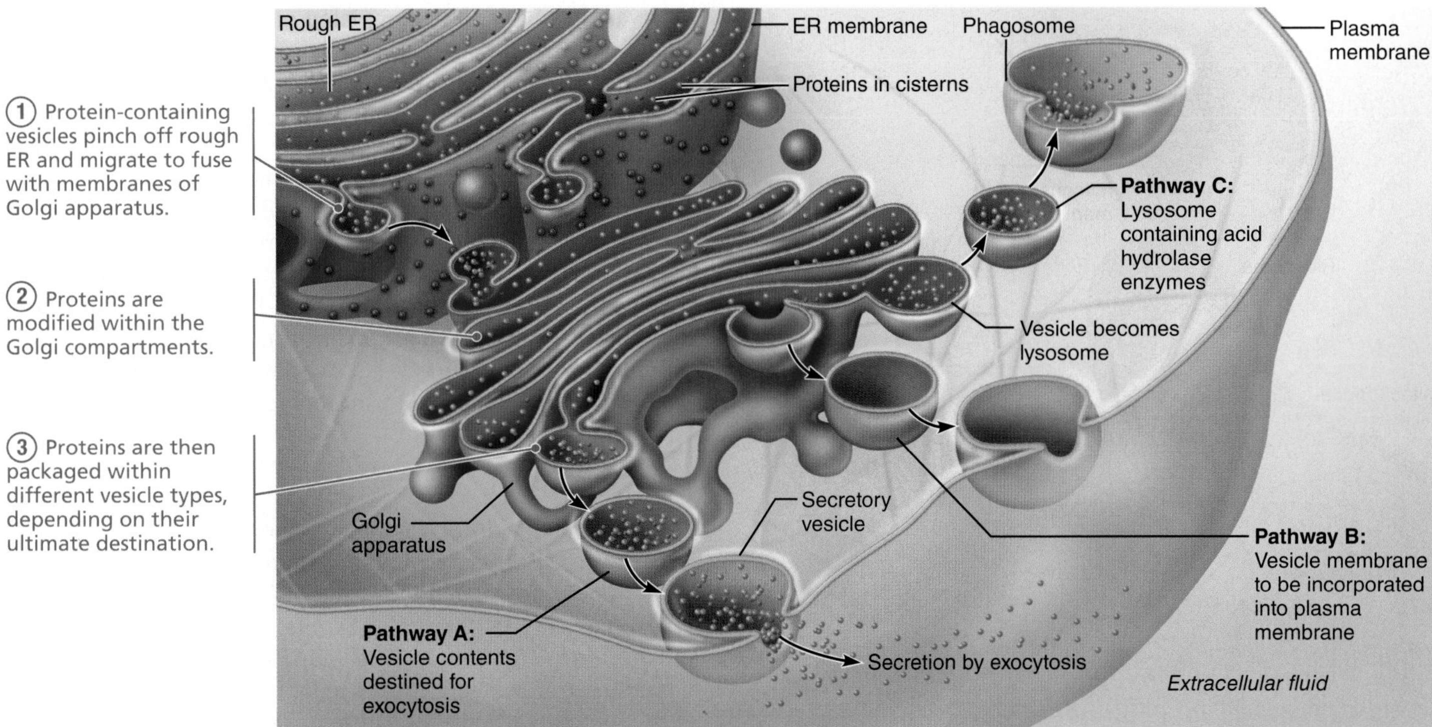

Figure 2.8 The sequence of events from protein synthesis on the rough ER to the final distribution of these proteins.

Lysosomes

Lysosomes are spherical, membrane-walled sacs containing many kinds of digestive enzymes (Figure 2.9). These enzymes, called acid hydrolases, can digest almost all types of large biological molecules. Lysosomes can be considered the cell's "demolition crew," because they break apart and digest unwanted substances. For example, they fuse with phagosomes, emptying their enzymes into these vesicles and breaking down their contents (Figure 2.8, pathway C).

When a cell's own internal membranes, proteins, or organelles are damaged or wear out, they are encircled by a new membrane from the rough ER, forming a vesicle. Then, nearby lysosomes fuse with this vesicle to digest its contents. Within such vesicles, digestion can proceed safely, because the enclosing membrane keeps the destructive enzymes away from other cell components. Phagocytic cells, such as some white blood cells, have an exceptional number of lysosomes to degrade ingested bacteria and viruses.

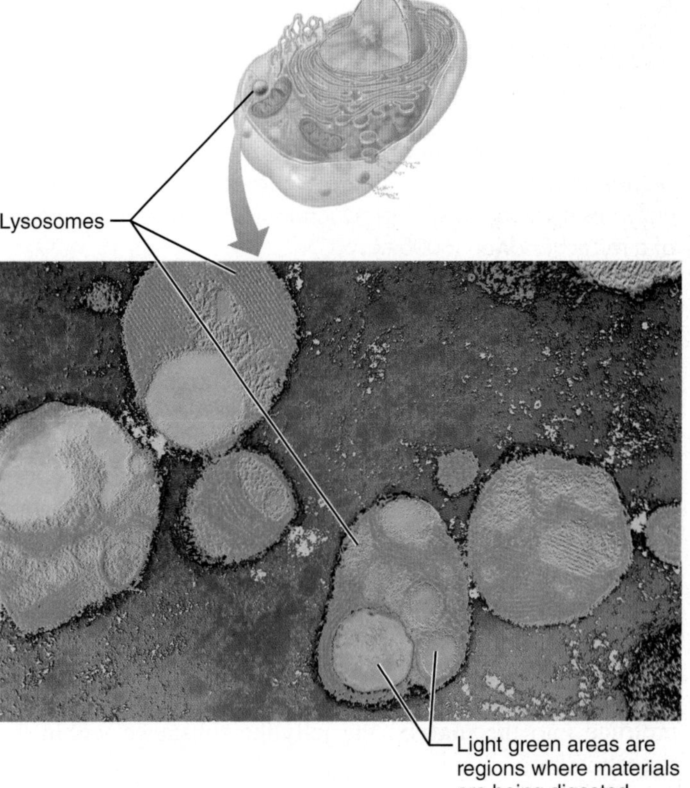

Figure 2.9 Electron micrograph of lysosomes (27,000×), artificially colored.

CLINICAL APPLICATION

Tay-Sachs Disease In an inherited condition called Tay-Sachs disease, an infant's lysosomes lack a specific enzyme that breaks down certain glycolipids in the normal recycling of worn-out cellular membranes. Such glycolipids are especially abundant in the membranes of nerve cells. Accumulation of undigested glycolipids in the lysosomes interferes with nerve cell function, resulting in mental retardation, blindness, spastic movements, and ultimately death before age 5.

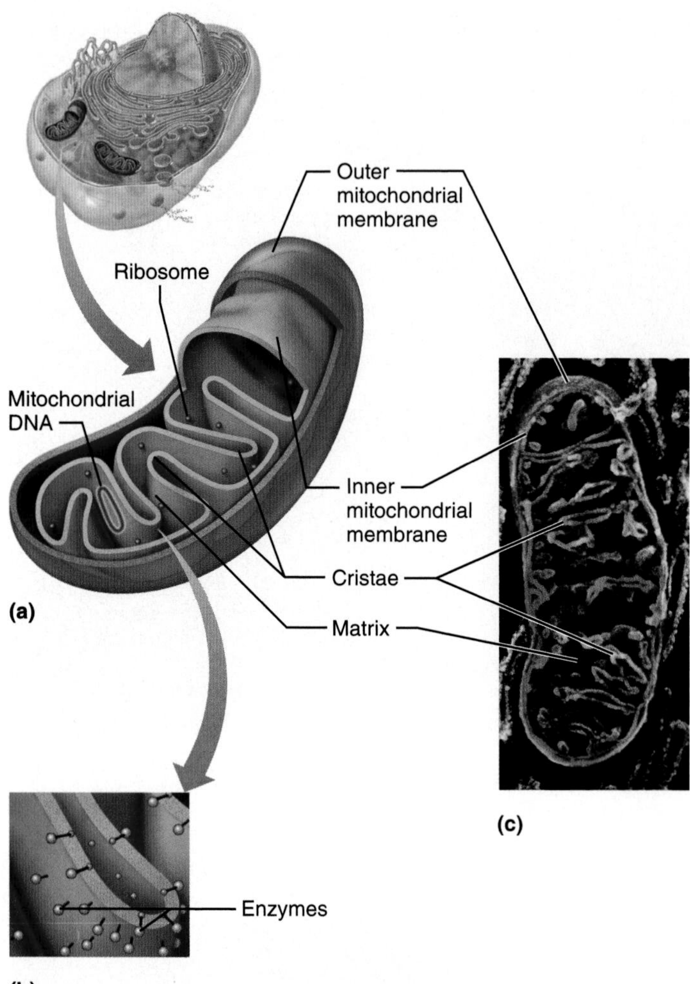

Figure 2.10 **Mitochondria.** (**a**) Diagram of a longitudinally sectioned mitochondrion. (**b**) Enlargement of crista showing enzymes involved in ATP production. (**c**) Electron micrograph of a mitochondrion (6000×).

Mitochondria

Mitochondria (mi′to-kon″dre-ah) are analogous to the power plant of the manufacturing company. These organelles produce the energy for cellular function. They usually are depicted as bean-shaped structures because of their appearance in sections under the microscope (Figure 2.10). In reality, mitochondria are long and threadlike (*mitos* = thread). In living cells, they squirm about and change shape as they move through the cytoplasm. Most organelles are surrounded by a membrane, but mitochondria are enclosed by two membranes: The outer membrane is smooth and featureless, and the inner membrane folds inward to produce shelflike **cristae** (krĭ′ste; "crests"). These protrude into the **matrix,** the jellylike substance within the mitochondrion.

Mitochondria generate most of the energy the cell uses to carry out work. They do this by systematically releasing the energy stored in the chemical bonds of nutrient molecules and then transferring this energy to produce **adenosine triphosphate (ATP),** the high-energy molecules that cells use to power chemical reactions. Within the mitochondrion, the ATP-generating process starts in the matrix (by a process called the citric acid cycle) and is completed on the inner membrane of the cristae (by the processes called oxidative phosphorylation and electron transport). Cell types with high energy requirements, muscle cells for example, have large numbers of mitochondria in their cytoplasm. These types of cells also have large numbers of cristae within their mitochondria.

Mitochondria are far more complex than any other organelle. They even contain some maternally inherited genetic material (DNA) and divide to form new mitochondria, as if they were miniature cells themselves. Intriguingly, mitochondria are very similar to a group of bacteria, the purple bacteria phylum. It is now widely believed that mitochondria arose from bacteria that invaded the ancient ancestors of animal and plant cells.

Peroxisomes

Peroxisomes (pĕ-roks′ĭ-sōmz; "peroxide bodies") are like the toxic waste removal system of the manufacturing plant. They are membrane-walled sacs that resemble small lysosomes (Figure 2.1). They contain a variety of enzymes, most importantly oxidases and catalases. Oxidases use oxygen to neutralize aggressively reactive molecules called **free radicals,** converting these to hydrogen peroxide. Free radicals are normal by-products of cellular metabolism, but if allowed to accumulate they can destroy the cell's proteins, membranes, and DNA. Hydrogen peroxide is also reactive and dangerous, but it is converted by catalase into water and oxygen. This catalase-driven reaction breaks down poisons that have entered the cell, such as alcohol, formaldehyde, and phenol. Peroxisomes are numerous in liver and kidney cells, which play a major role in removing toxic substances from the body. Peroxisomes also perform other metabolic reactions, such as breaking down long chains of fatty acids in lipid metabolism.

Cytoskeleton

The **cytoskeleton,** literally "cell skeleton," is an elaborate network of rods running throughout the cytosol (the framework of the manufacturing building, Figure 2.11). This network acts as a cell's "bones," "muscles," and "ligaments" by supporting cellular structures and generating various cell movements. The three types of rods in the cytoskeleton are *microfilaments, intermediate filaments,* and *microtubules,* none of which is covered by a membrane.

Microfilaments, the thinnest elements of the cytoskeleton, are strands of the protein **actin** (ak′tin) (Figure 2.11a). Also called **actin filaments,** they concentrate most heavily in a layer just deep to the plasma membrane. Actin filaments interact with another protein called **myosin** (mi′o-sin) to generate contractile forces within the cell. The interaction of actin and myosin squeezes one cell into two during cell division (cytokinesis, p. 42) and causes the membrane changes that accompany endocytosis and exocytosis. This interaction also enables some cells to send out and then retract extensions called *pseudopods* (soo′do-pods; "false feet"), in a crawling action called amoeboid motion (ah-me′boid; "changing shape"). Additionally, myosin acts as a motor protein to move some organelles within the cell. Except in muscle cells, where they are stable and permanent, the actin

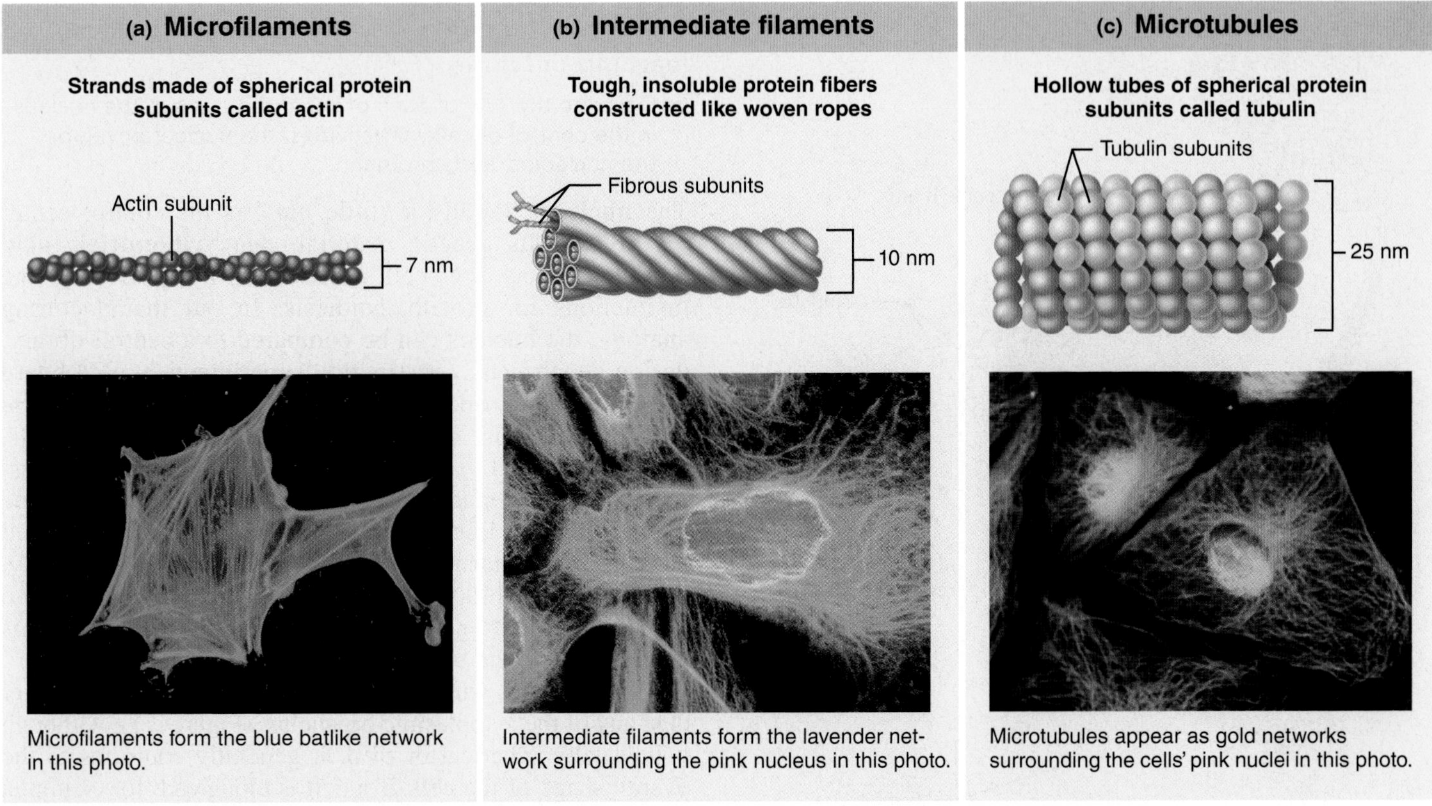

Microfilaments form the blue batlike network in this photo.

Intermediate filaments form the lavender network surrounding the pink nucleus in this photo.

Microtubules appear as gold networks surrounding the cells' pink nuclei in this photo.

Figure 2.11 Cytoskeletal elements. Diagrams (above) and photos (below) of cells treated to fluorescently tag the structure of interest.

microfilaments are unstable, constantly breaking down and re-forming from smaller subunits.

Intermediate filaments (Figure 2.11b) are tough, insoluble protein fibers, with a diameter between those of microfilaments and microtubules. Intermediate filaments are the most stable and permanent of the cytoskeletal elements. Their most important property is high tensile strength; that is, they act like strong guy-wires to resist *pulling* forces that are placed on the cell. They also function to link adjacent cells together by attaching to specific cell junctions called desmosomes (p. 75).

Microtubules, the elements with the largest diameter, are hollow tubes made of spherical protein subunits called tubulins (Figure 2.11c). They are stiff but bendable. All microtubules radiate from a small region of cytoplasm near the nucleus called the *centrosome* ("center body"; Figure 2.1). This radiating pattern of stiff microtubules determines the overall shape of the cell, as well as the distribution of cellular organelles. Mitochondria, lysosomes, and secretory granules attach to the microtubules like ornaments hanging from the limbs of a Christmas tree. Organelles move within the cytoplasm, pulled along the microtubules by small *motor proteins, kinesins* (ki-ne′sinz), and *dyneins* (di′ne-inz), which act like train engines on the microtubular railroad tracks. Microtubules are remarkably dynamic organelles, constantly growing out from the cell center, disassembling, then reassembling.

Centrosome and Centrioles

The **centrosome** (sen′tro-sōm) is a spherical structure in the cytoplasm near the nucleus (Figure 2.12). It contains no membranes. Instead, it consists of an outer cloud of protein called the **centrosome matrix** and an inner pair of **centrioles** (sen′tre-ōlz). The matrix protein seeds the growth and elongation of microtubules, which explains why the long microtubules of the cytoskeleton radiate from the centrosome in nondividing cells (Figure 2.11c) and why a mitotic spindle of microtubules radiates from it in dividing cells (Figure 2.17, pp. 38–39).

In the core of the centrosome, the two barrel-shaped centrioles lie perpendicular to one another. The wall of each centriole consists of 27 short microtubules, arranged in nine groups of three. Unlike most other microtubules, those in centrioles are stable and do not disassemble. Functionally, centrioles act in forming cilia and flagella (p. 76 in Chapter 4) and the mitotic spindle (Figure 2.17).

Cytoplasmic Inclusions

Inclusions are temporary structures in the cytoplasm that may or may not be present in a given cell type. Inclusions include pigments, crystals of protein, and food stores. The food stores, by far the most important kind, are lipid droplets and glycosomes. **Lipid droplets** are spherical drops of stored fat. They can have the same size and appearance as lysosomes but can be distinguished by their lack of a surrounding membrane. Only a few cell types contain lipid droplets: Small lipid droplets are found in liver cells, large ones in fat cells. **Glycosomes** ("sugar-containing bodies") store sugar in the form of glycogen (gli′ko-jen), which is a long branching chain of glucose molecules, the cell's main energy source. Glycosomes also contain enzymes that make and degrade the glycogen into its glucose subunits.

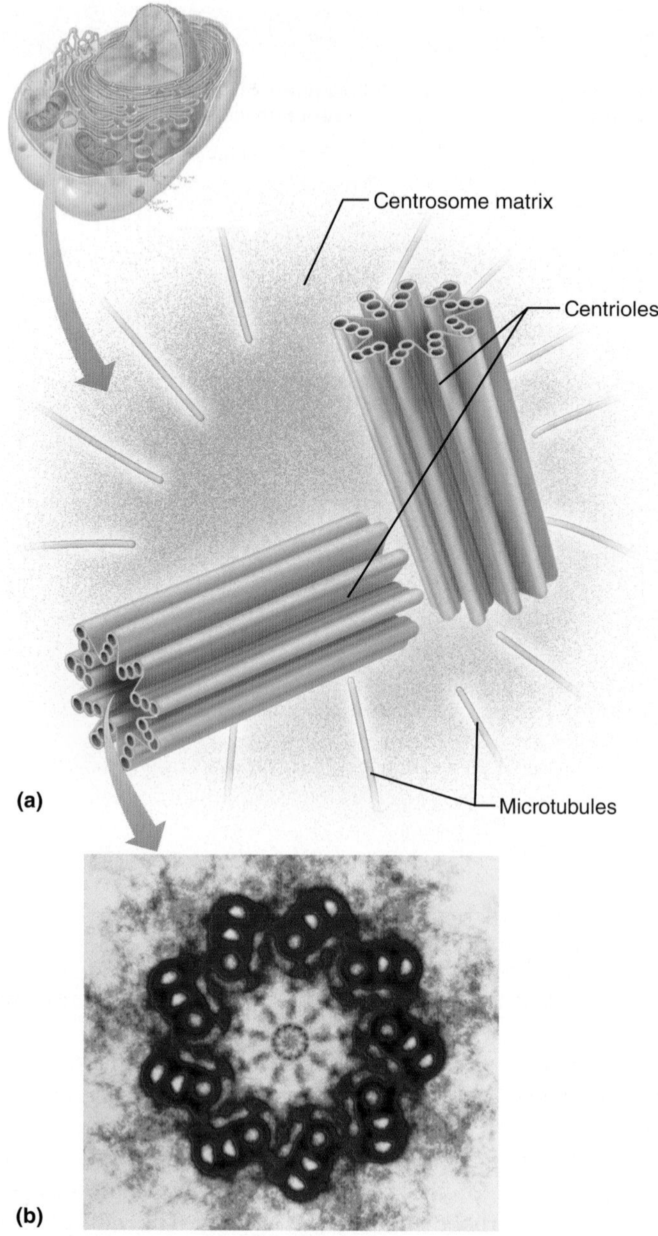

Figure 2.12 **Centrosome and centrioles.** (a) A centrosome. (b) Electron micrograph of a centriole in cross section. Its wall consists of nine groups of three microtubules (124,000×).

Structurally, glycosomes are dense, spherical granules. They resemble ribosomes, but their diameter is twice as large.

check your understanding

- ☐ 5. Which cellular organelles are involved with protein synthesis and packaging?
- ☐ 6. Which organelle produces the energy needed for cellular activity?
- ☐ 7. Which organelle would be prevalent in a cell that specialized in phagocytosis?
- ☐ 8. Which cytoskeletal element functions to resist tension and thus helps to keep the cell intact?

(For answers, see Appendix B.)

THE NUCLEUS

learning outcome

- ▶ Describe the role of each of the three parts of the nucleus in the control of cellular activities: the nuclear envelope, the nucleolus, and chromatin.

The **nucleus,** literally a "little nut," is the control center of the cell. Its genetic material, **deoxyribonucleic acid (DNA),** directs the cell's activities by providing the instructions for protein synthesis. In our manufacturing analogy, the nucleus can be compared to a central library, design department, construction superintendent, and board of directors all rolled into one. Whereas most cells have only one nucleus, some, including skeletal muscle cells, have many; that is, they are *multinucleate* (mul″tĭ-nu′-kle-āt; *multi* = many). The presence of more than one nucleus usually signifies that a cell has a larger-than-usual amount of cytoplasm to regulate. One cell type in the body, the mature red blood cell, is *anucleate;* that is, it has no nucleus at all. Its nucleus normally is ejected before this cell first enters the bloodstream.

The nucleus, which averages 5 μm in diameter, is larger than any of the cytoplasmic organelles (Figure 2.1). Although it is usually spherical or oval, it generally conforms to the overall shape of the cell. If a cell is elongated, for example, the nucleus may also be elongated. The main parts of the nucleus are the *nuclear envelope, nucleolus,* and *chromatin and chromosomes* (Figure 2.13).

Nuclear Envelope

The nucleus is surrounded by a **nuclear envelope** that consists of two parallel membranes separated by a fluid-filled space (Figure 2.13a). The outer membrane is continuous with the rough ER and has ribosomes on its external face. It forms anew from rough ER after every cell division, so it is evidently a specialized part of the rough ER. The inner membrane is lined by protein filaments, the **nuclear lamina,** which maintain the shape of the nucleus (Figure 2.13b).

At various points, the two layers of the nuclear envelope fuse, and **nuclear pores** penetrate the fused regions (Figure 2.13a and b). Each pore is formed by a bracelet-shaped complex of more than 22 proteins, and there are several thousand pores per nucleus. Like other cellular membranes, the membranes of the nuclear envelope are selectively permeable, but the pores allow large molecules to pass in and out of the nucleus as necessary. For example, protein molecules imported from the cytoplasm and RNA molecules exported from the nucleus routinely travel through the pores.

The nuclear envelope encloses a jellylike fluid called *nucleoplasm* (nu′kle-o-plazm″), in which the chromatin and nucleolus are suspended. Like the cytosol, the nucleoplasm contains salts, nutrients, and other essential chemicals.

Nucleolus

The **nucleolus** (nu-cle′o-lus, "little nucleus") is a dark-staining body in the cell nucleus (Figure 2.13). There may be one or several within a cell nucleus. A nucleolus contains parts

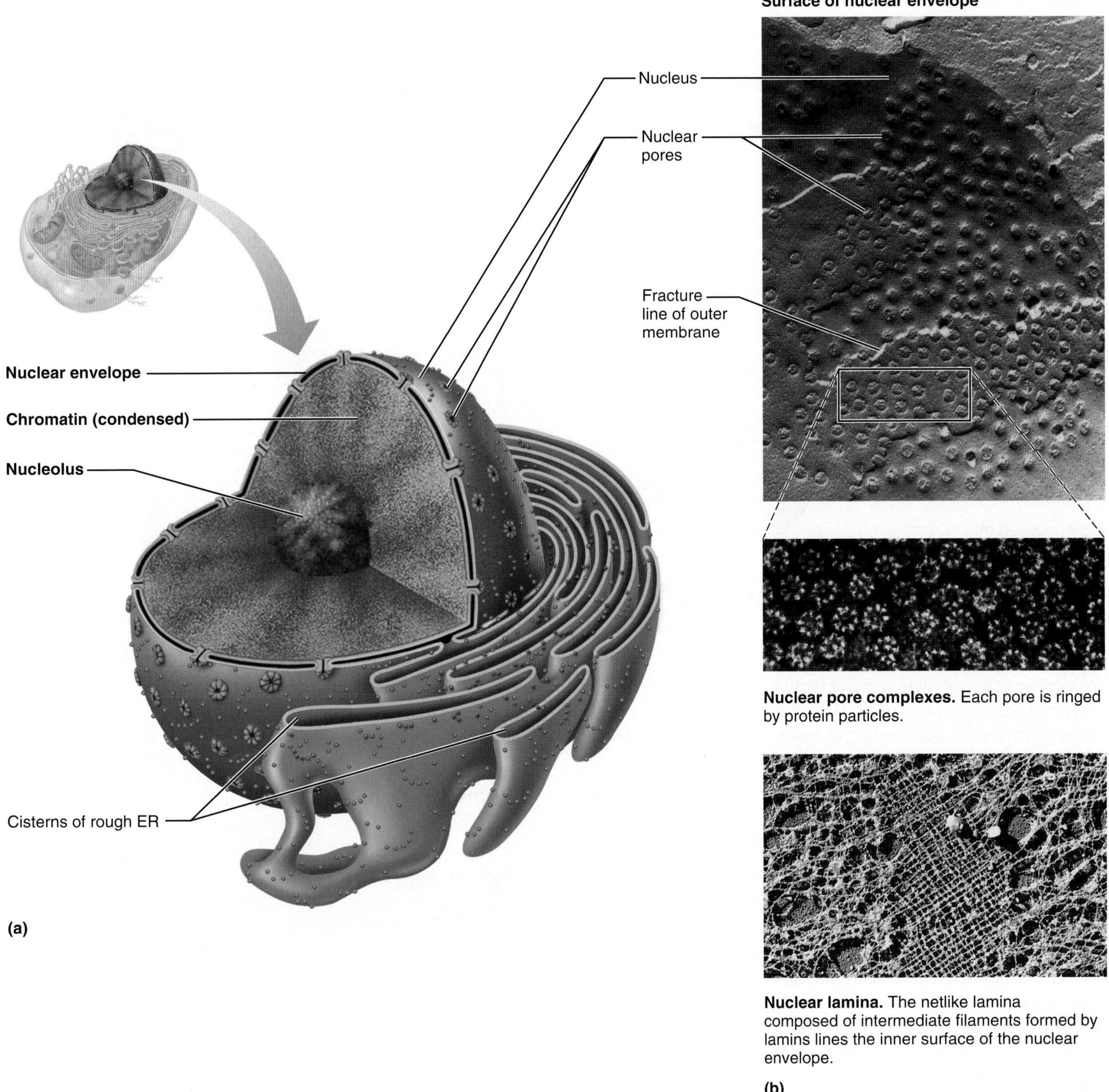

Figure 2.13 The nucleus. (**a**) Three-dimensional illustration of the nucleus, showing the continuity of the nuclear envelope with the rough ER. (**b**) Freeze-fracture micrograph transmission electron micrographs (TEMs).

of several different chromosomes and serves as the cell's "ribosome-producing machine." Specifically, it has hundreds of copies of the genes that code for ribosomal RNA and serves as the site where the large and small subunits of ribosomes are assembled. These subunits leave the nucleus through the nuclear pores and join within the cytoplasm to form complete ribosomes.

Chromatin and Chromosomes

DNA is a long double helix that resembles a spiral staircase (Figure 2.14). This double helix is in turn composed of four kinds of subunits called *nucleotides,* each of which contains a distinct base. These bases—thymine (T), adenine (A), cytosine (C), and guanine (G)—bind to form the "stairs" of the "staircase" and to hold the DNA helix together.

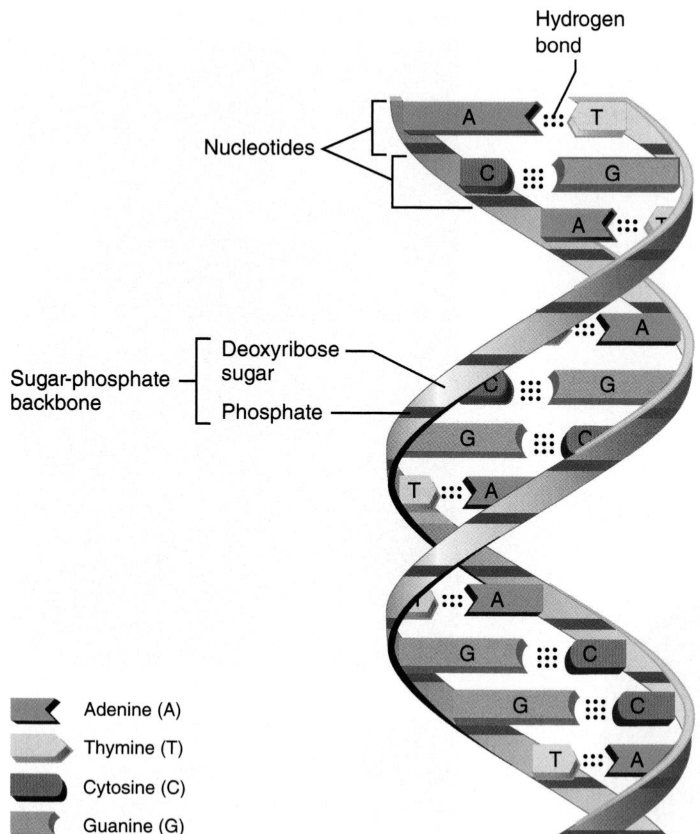

Figure 2.14 Molecular structure of DNA. DNA is a double helix constructed of chains of nucleotide molecules. Each nucleotide consists of a sugar, phosphate, and one of four bases: thymine (T), adenine (A), cytosine (C), or guanine (G).

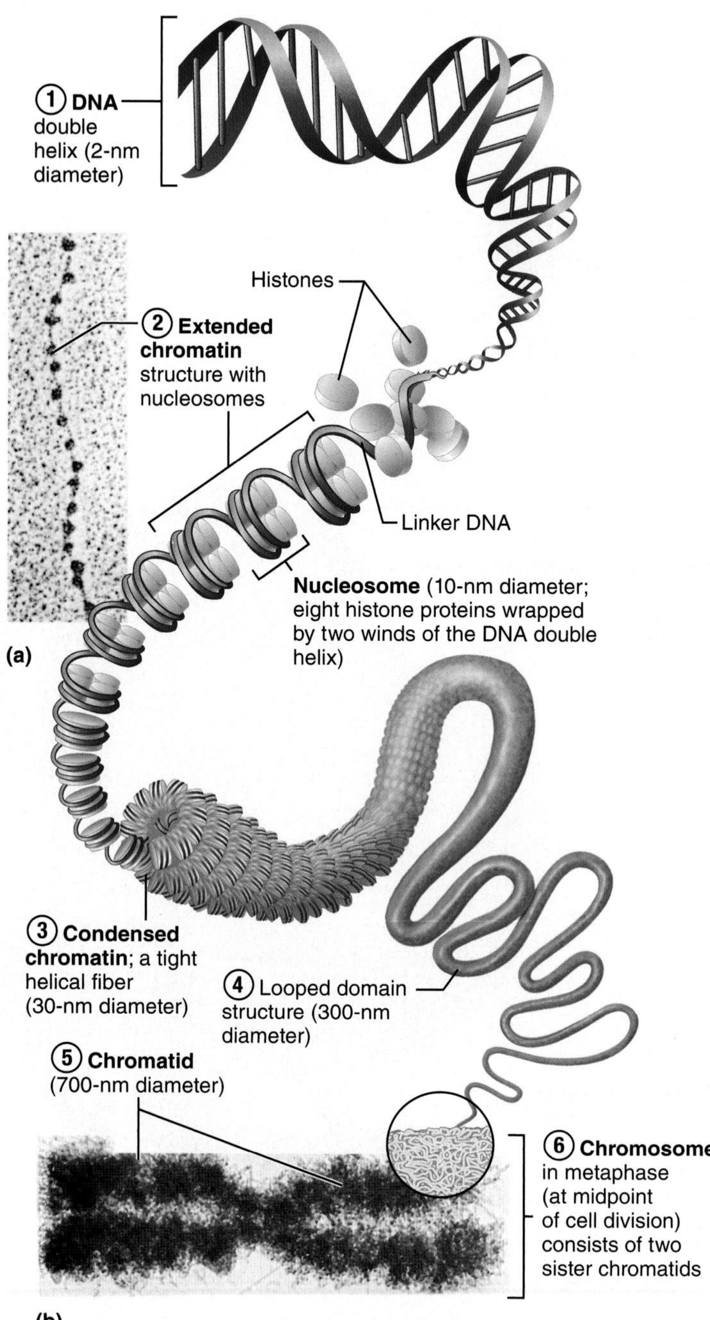

Figure 2.15 Chromatin and chromosome structure. (**a**) Electron micrograph of extended chromatin (36,500×). (**b**) The arrangement of DNA and histones in chromatin, from its most extended state ① to its most condensed state ⑥, in a chromosome.

The double helix of DNA (Figure 2.15 ①) is packed with protein molecules and coiled in strands of increasing structural complexity and thickness. The DNA molecule plus the proteins form **chromatin.** Each two turns of the DNA helix is packed with eight disc-shaped protein molecules called **histones** (his′tōnz, Figure 2.15 ②). Each cluster of DNA and histones is called a **nucleosome.** In an electron micrograph of chromatin, the nucleosomes have the appearance of beads on a string. Chromatin in this form is called **extended chromatin.** Further coiling of the nucleosomes forms a tight helical fiber. These thick fibers of chromatin are called **condensed chromatin.** (Figure 2.15 ③.)

During a cell's nondividing phase, when it is performing its normal activities, the chromatin is in either its extended or condensed form. The tightly coiled DNA of condensed chromatin is inactive. The extended chromatin is the active region of the DNA, directing the synthetic activities of the cell. Specifically, extended chromatin is the site where DNA's genetic code is copied onto messenger RNA molecules in a process called **transcription.** The most active cells in the body have a large amount of extended chromatin and little condensed chromatin.

During cell division, the chromatin is further packed: The helical fibers of nucleosomes are looped (Figure 2.15 ④) and then packed further into the most complex structure, the chromatid (Figure 2.15 ⑤) of a **chromosome** (kro′mo-sōm; "colored body"). Each chromosome contains a single, very long molecule of DNA, and there are 46 chromosomes in a typical human cell. When a cell is dividing, its chromosomes are maximally coiled, so they appear as thick rods (Figure 2.15 ⑥). Chromosomes move extensively during cell division (pp. 40–41), and their compact nature helps to keep the delicate chromatin strands from tangling and breaking as the chromosomes move. When cell division stops, many parts of the chromosome uncoil to form the extended chromatin, thereby allowing transcription to occur.

check your understanding

- ☐ 9. What does the nucleolus produce?
- ☐ 10. Which cytoplasmic organelle is continuous with the nuclear envelope?
- ☐ 11. How does the appearance of extended chromatin differ from that of condensed chromatin? What is the difference in function between these forms of chromatin?

(For answers, see Appendix B.)

THE CELL LIFE CYCLE

learning outcome

- ▶ List the phases of the cell life cycle, and describe a key event of each phase.

The **cell life cycle** is the series of changes a cell undergoes from the time it forms until it reproduces itself. This cycle can be divided into two major periods (Figure 2.16): *interphase,* in which the cell grows and carries on its usual activities; and *cell division,* or the *mitotic phase,* during which it divides into two cells.

Interphase

In addition to carrying on its life-sustaining activities, a cell in **interphase** prepares for the next cell division. Interphase is divided into G_1, S, and G_2 subphases. During **G_1 (gap 1),** the first part of interphase, cells are metabolically active, make proteins rapidly, and grow vigorously. This is the most variable phase in terms of duration. In cells with fast division rates, G_1 lasts several hours; in cells that divide slowly, it can last days or even years. Near the end of G_1, the centrioles start to replicate in preparation for cell division. During the next stage, the **S (synthetic) phase,** DNA replicates itself, ensuring that the two daughter cells will receive identical copies of the genetic material. The final part of interphase, called **G_2 (gap 2),** is brief. In this period, the enzymes needed for cell division are synthesized. Centrioles finish copying themselves at the end of G_2. The cell is now ready to divide. Throughout all three subphases, the cell continues to grow, producing proteins and cytoplasmic organelles, and to carry out its normal metabolic activities.

Checkpoints that evaluate cellular activities such as cell growth, DNA replication, and mitotic spindle formation occur throughout the cell cycle. The G_1 checkpoint assesses cell size before DNA synthesis, and the G_2/M checkpoint checks for DNA damage and accuracy of replication (Figure 2.16). Mitosis can be halted at these checkpoints, thus preventing damaged cells from dividing. Mutation in genes at these checkpoints can cause uncontrolled cell division and lead to tumor growth.

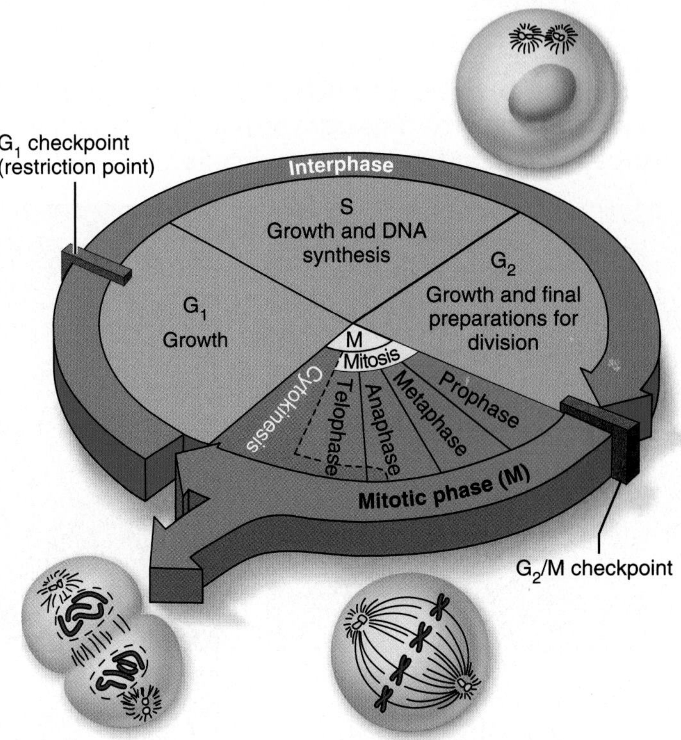

Figure 2.16 **The cell cycle.** The two basic phases in the life and reproduction of each cell are interphase and the mitotic (M) phase. The length of the cell cycle varies in different cell types, but the G_1 stage of interphase tends to be the longest and the most variable in duration.

Cell Division

Cell division is essential for body growth and tissue repair. Short-lived cells that continuously wear away, such as cells of the skin and the intestinal lining, reproduce themselves almost continuously. Others, such as liver cells, reproduce slowly (replacing those cells that gradually wear out) but can divide quickly if the organ is damaged. Cells of nervous tissue and, for the most part, skeletal muscle are unable to divide after they are fully mature; repair is carried out by scar tissue (a fibrous connective tissue).

Cells divide in the **M (mitotic) phase** of their life cycle, which follows interphase (Figure 2.16). In most cell types, division involves two distinct events: *mitosis* (mi-to′sis), or division of the nucleus, and *cytokinesis* (si″to-ki-ne′sis), or division of the entire cell into two cells.

Mitosis

Mitosis is the series of events during which the replicated DNA of the original cell is parceled out into two new cells, culminating in the division of the nucleus. Throughout these events, the chromosomes are evident as thick rods or threads. Indeed, *mitosis* literally means "the stage of threads." Mitosis is described in terms of four consecutive phases: prophase, metaphase, anaphase, and telophase. However, it is actually a continuous process, with each phase merging smoothly into the next. Its duration varies according to cell type, but it typically lasts about 2 hours. Mitosis is described in detail in *Focus on Mitosis* (Figure 2.17).

(Text continues on page 40.)

Figure 2.17

Mitosis is the process of nuclear division in which the chromosomes are distributed to two daughter nuclei. Together with cytokinesis, it produces two identical daughter cells.

Interphase

Early Prophase

Late Prophase

Centrosomes (each has 2 centrioles)
Plasma membrane
Nucleolus
Nuclear envelope
Chromatin

Early mitotic spindle
Aster
Chromosome consisting of two sister chromatids
Centromere

Spindle pole
Nonkinetochore microtubule
Fragments of nuclear envelope
Kinetochore
Kinetochore microtubule

Interphase

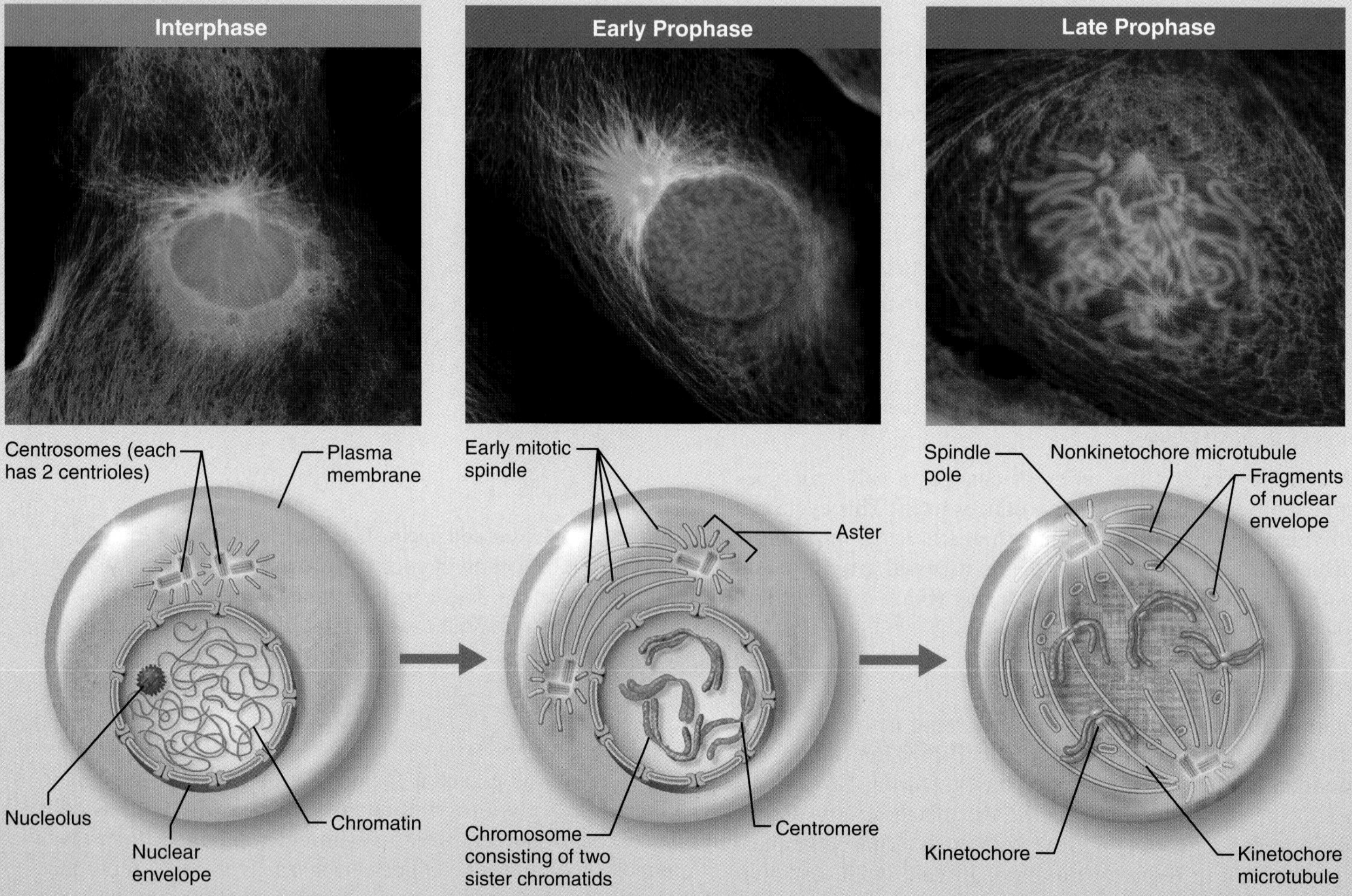

Interphase

Interphase is the period of a cell's life when it carries out its normal metabolic activities and grows. Interphase is not part of mitosis.

- During interphase, the DNA-containing material is in the form of chromatin. The nuclear envelope and one or more nucleoli are intact and visible.
- There are three distinct periods of interphase: G_1, S, and G_2.

The light micrographs show dividing lung cells from a newt. Fluorescent markers color cell structures. The chromosomes appear blue and the microtubules green. (The red fibers are intermediate filaments.) The schematic drawings show details not visible in the micrographs. For simplicity, only four chromosomes are drawn.

Prophase—first phase of mitosis

Early Prophase

- The chromatin coils and condenses, forming barlike *chromosomes*.
- Each duplicated chromosome consists of two identical threads, called *sister chromatids*, held together at the *centromere*. (Later when the chromatids separate, each will be a new chromosome.)
- As the chromosomes appear, the nucleoli disappear, and the two centrosomes separate from one another.
- The centrosomes act as focal points for growth of a microtubule assembly called the *mitotic spindle*. As the microtubules lengthen, they propel the centrosomes toward opposite ends (poles) of the cell.
- Microtubule arrays called *asters* ("stars") extend from the centrosome matrix.

Late Prophase

- The nuclear envelope breaks up, allowing the spindle to interact with the chromosomes.
- Some of the growing spindle microtubules attach to *kinetochores* (ki-ne´ to-korz), special protein structures at each chromosome's centromere. Such microtubules are called *kinetochore microtubules*.
- The remaining (unattached) spindle microtubules are called *nonkinetochore microtubules*. The microtubules slide past each other, forcing the poles apart.
- The kinetochore microtubules pull on each chromosome from both poles in a tug-of-war that ultimately draws the chromosomes to the center, or equator, of the cell.

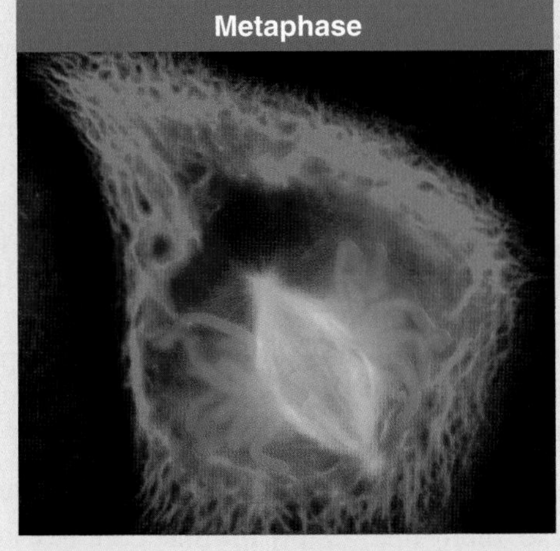

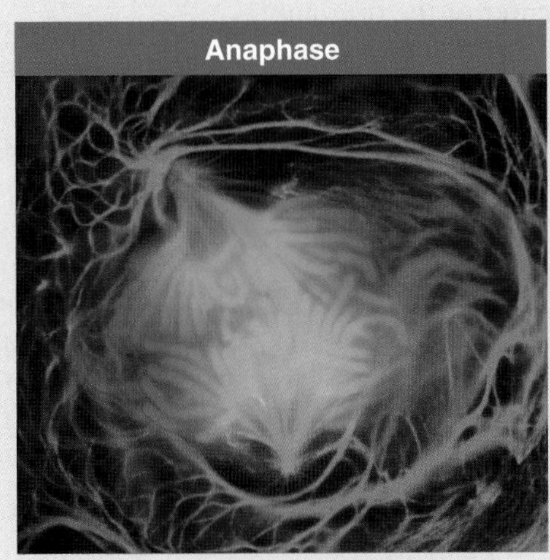

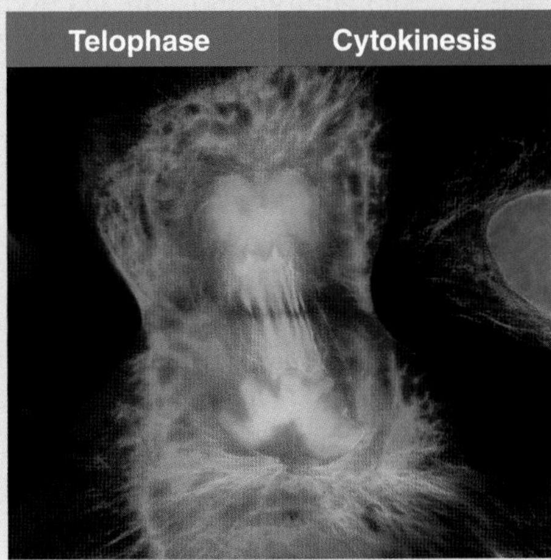

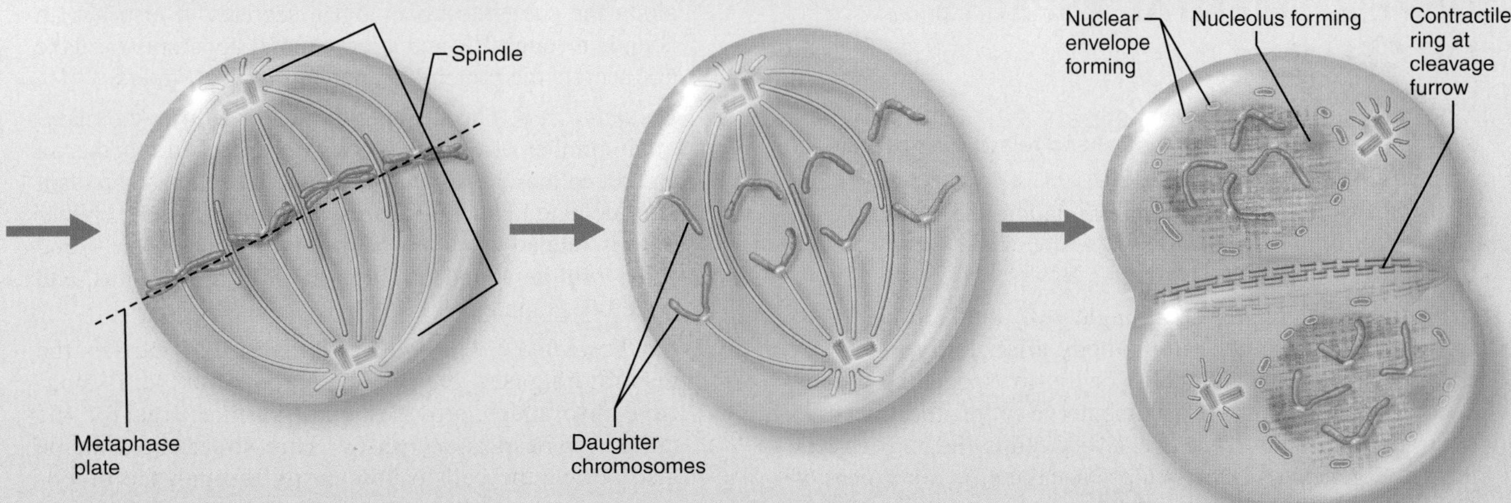

Metaphase—second phase of mitosis

- The two centrosomes are at opposite poles of the cell.
- The chromosomes cluster at the midline of the cell, with their centromeres precisely aligned at the spindle *equator*. This imaginary plane midway between the poles is called the *metaphase plate*.
- At the end of metaphase, enzymes that will act to separate the chromatids from each other are triggered.

Anaphase—third phase of mitosis

The shortest phase of mitosis, anaphase begins abruptly as the centromeres of the chromosomes split simultaneously. Each chromatid now becomes a chromosome in its own right.

- The kinetochore microtubules, moved along by motor proteins in the kinetochores, gradually pull each chromosome toward the pole it faces.
- At the same time, the nonkinetochore microtubules slide past each other, lengthen, and push the two poles of the cell apart.
- The moving chromosomes look V shaped. The centromeres lead the way, and the chromosomal "arms" dangle behind them.
- Moving and separating the chromosomes is helped by the fact that the chromosomes are short, compact bodies. Diffuse threads of chromatin would trail, tangle, and break, resulting in imprecise "parceling out" to the daughter cells.

Telophase—final phase of mitosis

Telophase

Telophase begins as soon as chromosomal movement stops. This final phase is like prophase in reverse.

- The identical sets of chromosomes at the opposite poles of the cell begin to uncoil and resume their threadlike chromatin form.
- A new nuclear envelope forms around each chromatin mass, nucleoli reappear within the nuclei, and the spindle breaks down and disappears.
- Mitosis is now ended. The cell, for just a brief period, is binucleate (has two nuclei) and each new nucleus is identical to the original mother nucleus.

Cytokinesis—division of cytoplasm

Cytokinesis begins during late anaphase and continues through and beyond telophase. A contractile ring of actin microfilaments forms the *cleavage furrow* and pinches the cell apart.

Cytokinesis

The separation of one cell into two at the end of the cell cycle is called **cytokinesis,** literally "cells moving (apart)." It begins during anaphase and is completed after mitosis ends (Figure 2.17). Essentially, a ring of contractile actin and myosin filaments in the center of the original cell constricts to pinch that cell in two. The two new cells, called daughter cells, then enter the interphase part of their life cycle.

check your understanding

- ☐ 12. In which phase of the cell life cycle does the cell spend most of its life?
- ☐ 13. What is the meaning of the root words that form the terms anaphase, metaphase, and telophase? What is happening during each of these phases of mitosis?

(For answers, see Appendix B.)

DEVELOPMENTAL ASPECTS OF CELLS

learning outcomes

- ▶ Name some specific cell types, and relate their overall shape to their special functions.
- ▶ Compare theories of cell differentiation and aging.

Cell Differentiation

All humans begin life as a single cell, the fertilized egg, from which all the cells in the body arise. Early in embryonic development, the cells begin to specialize: Some become liver cells; some become nerve cells; others become the transparent lens of the eye. Every cell in the body carries the same genes. (A *gene,* simply speaking, is a segment of DNA that dictates a specific cell function, usually by coding for a specific protein.) If all our cells have identical genes, how do cells differentiate and take on specialized structures and functions?

Cells in each region of the developing embryo are exposed to different chemical signals that channel the cells into specific pathways of development. The cytoplasm of a fertilized egg contains gradients of maternally produced messenger RNA (mRNA) molecules and proteins. In the early days of development as the fertilized egg divides, the cytoplasm of each daughter cell receives a different composition of these molecules. These maternally derived molecules in the cytoplasm influence the activity of the embryonic genome. In this way, different genes are activated in each cell, leading to cellular differentiation. Once the cell-specific gene expression begins, a cell may produce signaling molecules that influence the development of neighboring cells by switching some of their genes "on" or "off." Some genes are active in all cells; for example, all cells must carry out protein synthesis and make ATP. However, the genes for the synthesis of specialized proteins, such as hormones or mucus, are activated only in certain cell populations. The key to cell specialization lies in the kinds of proteins made and reflects differential gene activation in the different cell types.

Cell specialization, also called *cell differentiation,* leads to *structural* variation among the cell types in the body. Different organelles come to predominate in different cells. For example, muscle cells make tremendous quantities of actin and myosin proteins, and lipid accumulates in fat cells. Phagocytic cells produce more lysosomal enzymes and contain many lysosomes. There are about 200 different cell types in the body, which vary greatly in size, shape, and function. They include sphere-shaped fat cells, disc-shaped red blood cells, branching nerve cells, and cube-shaped cells of kidney tubules. The shapes of cells and their arrangement of organelles relate to the specialized function of these cells (Figure 2.18). Cells fall into these functional groups:

(a) Cells that connect body parts or cover and line organs

Fibroblast. The elongated shape of this cell extends along the cablelike fibers that it secretes. It also has an abundant rough ER and a large Golgi apparatus to make and secrete the protein components of these fibers.

Epithelial cell. The shape of these cells allows the maximum number of epithelial cells to be packed together in a sheet called *epithelium.* An epithelial cell has abundant intermediate filaments that resist tearing when the epithelium is rubbed or pulled. Some epithelial cells are gland cells, with an abundant rough ER, Golgi apparatus, and secretory granules.

Erythrocyte (red blood cell). This cell carries the respiratory gases, oxygen and carbon dioxide. Its concave disc shape provides extra surface area for the uptake of respiratory gases. This streamlined shape also allows the cell to flow easily through the bloodstream. So much oxygen-carrying pigment is packed in erythrocytes that all other organelles have been shed to make room.

(b) Cells that produce movement and move body parts

Skeletal muscle and smooth muscle cells. These cells are elongated and filled with abundant actin and myosin filaments, so they can shorten forcefully.

(c) Cell that stores nutrients

Fat cell. The huge spherical shape of a fat cell is produced by a large lipid droplet in its cytoplasm.

(d) Cell that fights disease

Macrophage (a phagocytic cell). This cell extends long pseudopods to crawl through tissue to reach infection sites. The many lysosomes within the cell digest the infectious microorganisms it takes up.

(e) Cell that gathers information and controls body functions

Nerve cell (neuron). This cell has long processes for receiving messages and transmitting them to other structures in the body. The processes are covered with

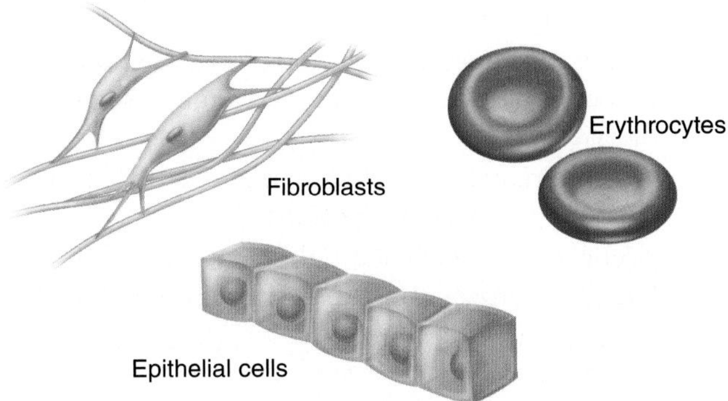

(a) Cells that connect body parts, form linings, or transport gases

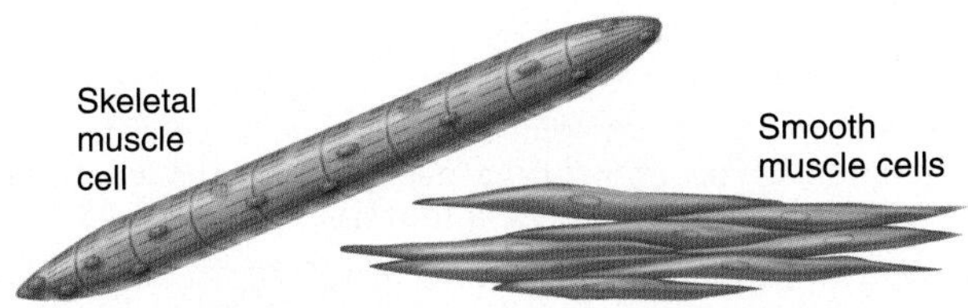

(b) Cells that move organs and body parts

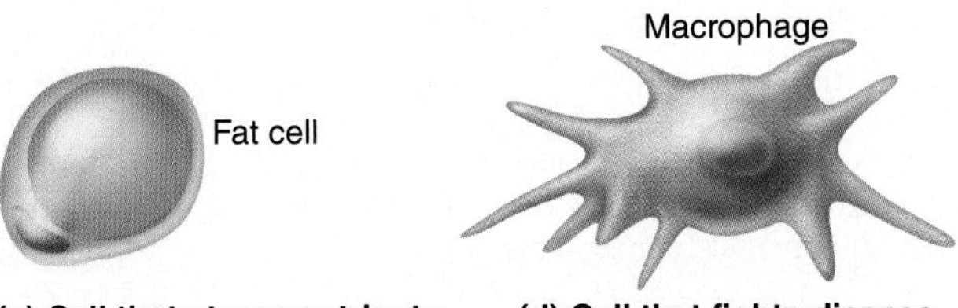

(c) Cell that stores nutrients **(d) Cell that fights disease**

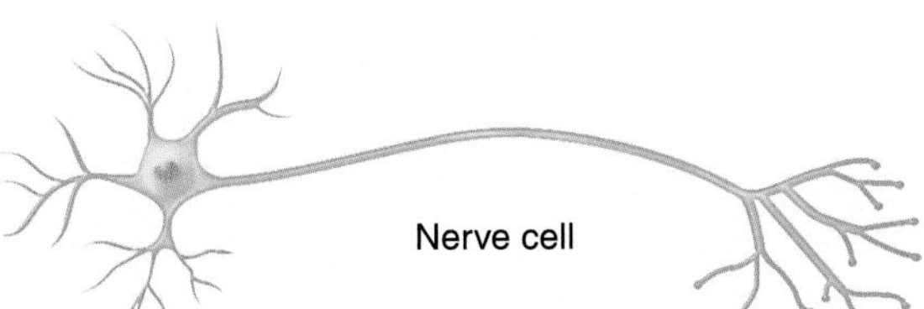

(e) Cell that gathers information and controls body functions

(f) Cell of reproduction

Figure 2.18 **Cellular diversity.** (Note that cells are not drawn to the same scale.)

an extensive plasma membrane, whose components are continually recycled; a large rough ER is present to synthesize membrane components.

(f) Cell of reproduction

Sperm (male). This cell is long and streamlined for swimming to the egg for fertilization. The swimming tail is a motile whip called a *flagellum* (p. 76 in Chapter 4).

Most organs are well formed and functional long before birth, but the body continues to grow by forming more cells throughout childhood and adolescence. Once adult size is reached, cell division slows considerably and occurs primarily to replace short-lived cell types and to repair wounds.

Aging

There is no doubt that cellular aging occurs and that it accounts for most problems associated with old age. Although aging is complex and certainly the result of many mechanisms, the best-documented theory proposes that free radicals play the major role. These highly reactive and thus destructive molecules are primarily by-products of normal cellular metabolism, although they also form in response to external insults, such as radiation and chemical pollutants. The theory proposes that free radicals build up and progressively damage the essential cell molecules. Most evidence for this comes from experiments on less complex animals such as worms and fruit flies, in which neutralizing the free radicals and repairing their damage have been shown to increase life span. Vitamins E and C appear to act as antioxidants in the human body and may help to prevent excessive free-radical formation.

Most free radicals are produced in the mitochondria, the organelle with the highest rate of metabolism. Scientists propose that a decrease in energy production by free-radical-damaged mitochondria weakens and ages the cells. This is called the *mitochondrial theory of aging.* It has long been known that when laboratory rats and mice are slightly undernourished, their life span increases by up to 30%. The same finding has also been demonstrated in primates. Because caloric restriction lowers the metabolic rate and makes metabolism more efficient, fewer of the destructive free radicals are produced, and aging slows.

Genetic theories of aging propose that aging is programmed into our genes. These theories originated from the observation that the body ages in a predictable pattern, as if aging were a normal part of human development (and development is known to be controlled by genes). Rats and fruit flies can be bred to live longer than usual, and genes that increase and decrease longevity have been identified in animals. Although some of these genes work by influencing free radicals, others act in less understood ways.

The best evidence for planned senescence (programmed aging) involves **telomeres,** structures that limit the maximum number of times cells can divide. Telomeres are repeating, seemingly nonsensical stretches of DNA that cap the ends of chromosomes (*telo* = end; *mere* = piece). Although they carry no genes, they appear to be vital for chromosomal survival: Each time DNA is replicated, 50 to 100 of the end nucleotides are lost, and the telomeres get a bit shorter. When the telomeres reach a certain minimum length, the "stop-division" signal is given. *Telomerase* is an enzyme that prevents telomeres from degrading by adding more repeating DNA to the ends. Telomerase occurs in our endlessly replicating germ-line cells and in cancer cells, but not in other cell types.

Many studies show that shortened telomeres are associated with an increased risk of developing age-related illness such as diabetes, cardiovascular disease, Alzheimer's disease, and certain cancers. In the near future, tests to measure telomere length may be used clinically as one measure to assess an individual's health status.

✓ check your understanding

- ☐ 14. Which cellular structures would be abundant in cells that specialize in producing movement, such as muscle cells?
- ☐ 15. Which organelles would be abundant in cells that produce and secrete hormones?
- ☐ 16. According to which aging theory presented here can the aging process be altered by individual behavior?

(For answers, see Appendix B.)

RELATED CLINICAL TERMS

Apoptosis (ap″op-to′sis; "falling away") Programmed cell death. This process of controlled cellular suicide eliminates cells that are stressed, unneeded, excessive, or aged. In response to damaged macromolecules within the cell or to some extracellular signal, a series of intracellular enzymes is activated that destroy the cell's DNA, cytoskeleton, and other structures, producing a quick, neat death. The apoptotic cell shrinks without leaking its contents into the surrounding tissue. It detaches from other cells and is immediately consumed by nearby cells. This tidy sequence avoids inflammation (p. 95), and therefore minimizes tissue injury. Cancer cells fail to undergo apoptosis, but oxygen-starved cells do so excessively (heart-muscle and brain cells during heart attacks and strokes, for example).

Dysplasia (dis-pla′ze-ah) (*dys* = abnormal) A change in cell size, shape, or arrangement due to long-term irritation or inflammation (from infections, for example).

Hyperplasia (hi″per-pla′ze-ah; "excess shape") Excessive cell proliferation. Unlike cancer cells, hyperplastic cells retain their normal form and arrangement within tissues.

Hypertrophy (hi-per′tro-fe; "excess growth") Growth of an organ or tissue due to an increase in the size of its cells. Hypertrophy is a normal response of skeletal muscle cells to exercise. Hypertrophy differs from hyperplasia, the condition in which cells increase in number but not in size.

Necrosis (ne-kro′sis) (*necros* = death; *osis* = process; condition) Death of a cell or group of cells due to injury or disease. Acute injury causes the cells to swell and burst, and they induce an inflammatory response. This is accidental, *uncontrolled* cell death, in contrast to apoptosis.

CHAPTER SUMMARY

Take full advantage of the invaluable online practice and study tools for this chapter by going to the Study Area of MasteringA&P®. There you will find:

- Art Labeling Activities, Quizzes, Practice Tests, Memory Games
- Animal Organ Dissection, Cat Dissection and Bone Videos
- PAL Cadaver
- PAL Histology
- A&P Flix Animations
- Dynamic Study Modules

OVERVIEW OF CELLS (pp. 23–24)

1. Cells are the basic structural and functional units of life.
2. There are 50 to 100 trillion cells in the human body. This chapter emphasizes the features common to all cells.
3. Cells obtain nutrients, make molecules, dispose of wastes, maintain their shape, and replicate.
4. Each cell has three main regions: plasma membrane, cytoplasm, and nucleus.

THE PLASMA MEMBRANE (pp. 24–28)

Structure (p. 26)

5. The plasma membrane defines the cell's outer boundary. The fluid mosaic model interprets this membrane as a flexible bilayer of lipid molecules (phospholipids, cholesterol, and glycolipids) with embedded proteins.
6. Most proteins in the membrane are integral proteins and extend entirely through the membrane. Peripheral proteins, by contrast, are attached to the membrane surface, helping to support the membrane along with other functions.
7. Sugar groups of membrane glycoproteins and glycolipids project from the cell surface and contribute to the cell coat (glycocalyx), which functions in cell-to-cell binding and recognition.

Functions (pp. 26–27)

8. The plasma membrane functions as a fragile barrier to protect the cell contents. It determines what enters and leaves the cell, and some proteins in the plasma membrane function as receptors for extracellular signal molecules.

Membrane Transport (pp. 27–28)

9. Small uncharged molecules pass through the membrane by simple diffusion; water enters and leaves by osmosis; larger or charged molecules pass through by transport mechanisms that involve the integral proteins. The movement of molecules down their concentration gradient by way of an integral protein is called facilitated diffusion. Movement against the concentration gradient is active transport, a process that requires the use of energy.

10. Large particles and macromolecules are actively transported through the membrane by endocytosis and exocytosis. Endocytosis brings large substances into the cell as packets of plasma membrane fold in to form cytoplasmic vesicles. If the substance is a particle, the process is called phagocytosis; if the substance is dissolved molecules in the extracellular fluid, the process is known as pinocytosis. Receptor-mediated endocytosis is selective: Specific molecules attach to receptors on the membrane before being taken into the cell in protein-coated vesicles.

11. In exocytosis, membrane-lined cytoplasmic vesicles fuse with the plasma membrane and release their contents to the outside of the cell.

THE CYTOPLASM (pp. 28–34)

Cytosol (p. 28)

12. The cytosol is a viscous fluid containing water, dissolved ions, and enzymes; cytoplasmic organelles and inclusions are suspended in the cytosol.

Cytoplasmic Organelles (pp. 29–33)

13. Each organelle performs specific functions. The various cell types in the body have different numbers of each organelle type.

14. Ribosomes are dark-staining granules that consist of two subunits, each made of protein and ribosomal RNA. Ribosomes are the sites of protein synthesis (translation). Free ribosomes make proteins used in the cytosol.

15. The rough endoplasmic reticulum is a ribosome-studded system of membrane-walled cavities (cisterns). Its ribosomes make proteins, which enter the cisterns and which may ultimately be secreted by the cell. The rough ER also makes all the cell's membranes.

16. The smooth endoplasmic reticulum, a network of membrane-walled tubules containing no ribosomes, is involved in the metabolism of lipids. Smooth ER also stores calcium ions.

17. The Golgi apparatus is a stack of disc-shaped cisterns that has a cis (convex) and a trans (concave) face. It sorts the products of the rough endoplasmic reticulum and then sends these products, in membrane-bound vesicles, to their proper destination. Lysosomes and secretory granules arise from the Golgi apparatus.

18. Lysosomes are spherical, membrane-walled sacs of digestive enzymes. They digest deteriorated organelles and substances brought into the cell in membrane-bound vesicles.

19. Mitochondria are threadlike organelles covered by two membranes, the inner of which forms shelflike cristae. Mitochondria are the main sites of ATP synthesis, the cell's main energy generators.

20. Peroxisomes are membrane-walled, enzyme-containing sacs that perform several metabolic processes. They convert free radicals to hydrogen peroxide. They also use hydrogen peroxide to break down some organic poisons.

21. The cytoskeleton includes protein rods of three distinct types—actin microfilaments, intermediate filaments, and microtubules—all in the cytosol. Actin microfilaments interact with myosin to produce contractile forces. Intermediate filaments, which act to resist tension placed on the cell, are stable. Microtubules, which radiate out from the centrosome region, give the cell its shape; they also organize the distribution and the transport of various organelles within the cytoplasm. Both microtubules and microfilaments tend to be unstable, breaking down and re-forming.

22. The centrosome is a spherical region of cytoplasm near the nucleus. It consists of a cloudlike matrix surrounding a pair of centrioles. Proteins in the matrix anchor the long microtubules of the cytoskeleton and microtubules of the mitotic spindle. The centrioles are barrel-shaped structures with walls of short microtubules.

Cytoplasmic Inclusions (pp. 33–34)

23. Inclusions are temporary structures in the cytoplasm. Examples include food stores, such as lipid droplets and glycogen-containing glycosomes.

THE NUCLEUS (pp. 34–37)

24. The nucleus contains genetic material (DNA) and is the control center of the cell. Most cells have one centrally located nucleus shaped like a sphere or an egg.

Nuclear Envelope (p. 34)

25. The nucleus is surrounded by a selectively permeable nuclear envelope, which is penetrated by nuclear pores. These pores allow the passage of large molecules such as RNA and proteins into and out of the nucleus. The nuclear envelope is continuous with the rough endoplasmic reticulum.

Nucleolus (pp. 34–35)

26. A nucleolus is a dark-staining body within the nucleus containing copies of genes for ribosomal RNA. Nucleoli make the subunits of ribosomes.

Chromatin and Chromosomes (pp. 35–37)

27. The DNA molecule is a double helix consisting of four types of nucleotides, with bases of thymine, adenine, cytosine, and guanine.

28. Chromatin is strandlike material (DNA and histones) in the nucleus that forms chromosomes. During cell division, all chromatin is highly coiled, making the chromosomes appear as thick rods. In nondividing cells, the chromatin is a mixture of inactive, coiled regions (condensed chromatin) and active, uncoiled regions (extended chromatin).

THE CELL LIFE CYCLE (pp. 37–40)

29. The cell life cycle is the series of changes a cell experiences from the time it forms until it divides.

Interphase (p. 37)

30. Interphase is the nondividing phase of the cell life cycle. It consists of the subphases G_1, S, and G_2. During G_1, the cell grows; during S, DNA replicates; and during G_2, the final preparations for division are made.

Cell Division (pp. 37–40)

31. Cell division, essential for growth and repair of the body, occurs during the M (mitotic) phase. Cell division has two distinct aspects: mitosis and cytokinesis.

32. Mitosis, the division of the nucleus, has four stages: (1) *prophase,* when chromatids appear, the nuclear membrane is lost, and the mitotic spindle forms; (2) *metaphase,* when the chromatids line up at the cell's equator; (3) *anaphase,* when the V-shaped chromatids are pulled apart; and (4) *telophase,* when the chromatin extends and the nucleus reassembles. Mitosis parcels out the replicated chromosomes to two daughter nuclei. Cytokinesis, completed after mitosis, is the division of the cell into two cells.

DEVELOPMENTAL ASPECTS OF CELLS (pp. 40–42)

Cell Differentiation (pp. 40–41)

33. The first cell of a human is the fertilized egg. Cell differentiation begins early in development and is thought to reflect differential gene activation.

34. There are about 200 different cell types in the human body. These cells have a variety of shapes, which reflect their functions; different organelles dominate in different cell types (Figure 2.18).

35. During adulthood, cell numbers remain fairly constant, and cell division occurs primarily to replace lost cells.

Aging (pp. 41–42)

36. Aging of cells (and of the whole body) may reflect accumulated damage from free radicals, or it may be a genetically influenced process, or both. It may also reflect a loss of the capacity for cell division over time.

REVIEW QUESTIONS

Multiple Choice/Matching Questions

(For answers, see Appendix B.)

1. The endocytotic process in which particulate matter is brought into the cell is called (a) phagocytosis, (b) pinocytosis, (c) exocytosis.
2. The nuclear substance composed of histone proteins and DNA is (a) chromatin, (b) the nuclear envelope, (c) nucleoplasm, (d) nuclear pores.
3. Final preparations for cell division are made during this stage of the cell life cycle: (a) G_1, (b) G_2, (c) M, (d) S.
4. The fundamental bilayered structure of the plasma membrane is determined almost exclusively by (a) phospholipid molecules, (b) peripheral proteins, (c) cholesterol molecules, (d) integral proteins.
5. Identify the cell structure or organelle described by each of the following statements.
   (a) A stack of 3–10 membranous discs, with vesicles.
   (b) A continuation of the nuclear envelope forms this ribosome-covered cytoplasmic organelle.
   (c) In nondividing cells, the highly coiled parts of chromosomes form this type of chromatin.
   (d) This type of endoplasmic reticulum is a network of hollow tubes involved in lipid metabolism.
   (e) This is a cluster of eight histone molecules and the DNA strand that wraps them.
   (f) These are the cytoskeletal rods with the thickest diameter (choose from microtubules, microfilaments, intermediate filaments).
   (g) The only organelle with DNA and cristae.
   (h) This energy-producing organelle is probably descended from bacteria.
   (i) Protein synthesis occurs at this organelle.
6. Identify the *false* statement about centrioles. (a) They start to duplicate in G_1. (b) They lie in the centrosome. (c) They are made of microtubules. (d) They are membrane-walled barrels lying parallel to each other.
7. The trans face of the Golgi apparatus (a) is its convex face; (b) is where products leave the Golgi apparatus in vesicles; (c) receives transport vesicles from the rough ER; (d) is in the very center of the Golgi stack; (e) is the same as the cis face.
8. Identify the *false* statement about lysosomes. (a) They have the same structure and function as peroxisomes. (b) They form by budding off the Golgi apparatus. (c) Lysosomal enzymes do not occur freely in the cytosol in healthy cells. (d) They are abundant in phagocytic cells.
9. Name the appropriate stage of mitosis (prophase, metaphase, anaphase, telophase) for each of the following.
   (a) The chromosomes are lined up in the middle of the cell.
   (b) The nuclear membrane fragments.
   (c) The nuclear membrane re-forms.
   (d) The mitotic spindles form.
   (e) The chromosomes (chromatids) are V-shaped.
10. Name the cytoskeletal element (actin microfilaments, intermediate filaments, or microtubules) for each of the following.
   (a) give the cell its shape
   (b) resist tension placed on a cell
   (c) radiate from the centrosome
   (d) interact with myosin to produce contraction force
   (e) are the most stable
   (f) associated with kinesins and dyneins
   (g) associated with the motor protein myosin
11. Different organelles are abundant in different cell types. Match the cell types with their abundant organelles by placing the correct letter from column B into each blank in column A. Follow the hints provided in parentheses.

Column A	Column B
____ **(1)** cell in the adrenal gland that makes steroid hormones	**(a)** mitochondria
____ **(2)** white blood cell (phagocytic)	**(b)** smooth ER
____ **(3)** liver cell (detoxifies poisons)	**(c)** peroxisomes
____ **(4)** muscle cell (highly contractile)	**(d)** microfilaments
____ **(5)** mucous cell (secretes protein product)	**(e)** rough ER
____ **(6)** epithelial cell in the outer layer of skin (withstands tension)	**(f)** intermediate filaments
____ **(7)** kidney tubule cell (makes and uses large amounts of ATP)	**(g)** lysosomes

12. Which of these processes involves specific membrane receptors? (a) phagocytosis, (b) receptor-mediated endocytosis, (c) exocytosis, (d) pinocytosis.

Short Answer Essay Questions

13. List all the cytoplasmic organelles that are composed (at least in part) of lipid-bilayer membranes. Then list all the cytoplasmic organelles that are *not* membranous.
14. Martin missed a point on his anatomy test because he thought *nucleus* and *nucleolus* were the same word and the same structure. Distinguish the nucleus from the nucleolus.
15. In this chapter, we claimed that mitochondria are the most complex organelles in the cytoplasm. What evidence supports this claim?
16. Define *chromosome*. Then compare the arrangement of the chromatin in a dividing versus a nondividing cell.
17. Indicate which cellular organelle or organelles participate in each of the cell functions listed earlier in the chapter (p. 23).

18. Use the analogy of the cell as a manufacturing factory to describe the function of the following cellular structures: plasma membrane, mitochondria, nucleus, Golgi apparatus, ribosomes, lysosomes, peroxisomes.

Critical Reasoning & Clinical Application Questions

1. Chrissy, who is 30 years old, stays 5 kg (11 pounds) underweight because she thinks being thin will slow her aging process. After warning her not to lose any more weight or to become anorexic, her doctor admitted that there is at least some scientific evidence to support her view. What is that evidence?
2. Kareem had a nervous habit of chewing on the inner lining of his lip with his front teeth. The lip grew thicker and thicker from years of continual irritation. Kareem's dentist noticed his greatly thickened lip, then told him to have it checked to see if the thickening was a tumor. A biopsy revealed hyperplasia and scattered areas of dysplasia, but no evidence of tumor. What do these terms mean? Did Kareem have cancer of the mouth?
3. The normal function of one tumor-suppressor gene acting at the G_1 checkpoint is to prevent cells with damaged chromosomes and DNA from "progressing from G_1 to S." Another tumor-suppressor gene, acting at the G_2/M checkpoint, prevents "passage from G_2 to M." When these tumor-suppressor genes fail to work, cancer can result. Explain what the phrases in quotations mean.
4. In their anatomy labs, many students are exposed to the chemical preservatives phenol, formaldehyde, and alcohol. Our cells break down these toxins very effectively. What cellular organelle is responsible for this?
5. The sedative phenobarbital is a lipid-soluble drug. What may happen to the smooth ER in liver cells of people who use this drug for a period of time?
6. The drug vinblastine is used in cancer therapy to stop the runaway division of cancer cells. Vinblastine inhibits the assembly and growth of microtubules. Explain how the action of this drug prevents mitosis (refer to Figure 2.17).

3 Basic Embryology

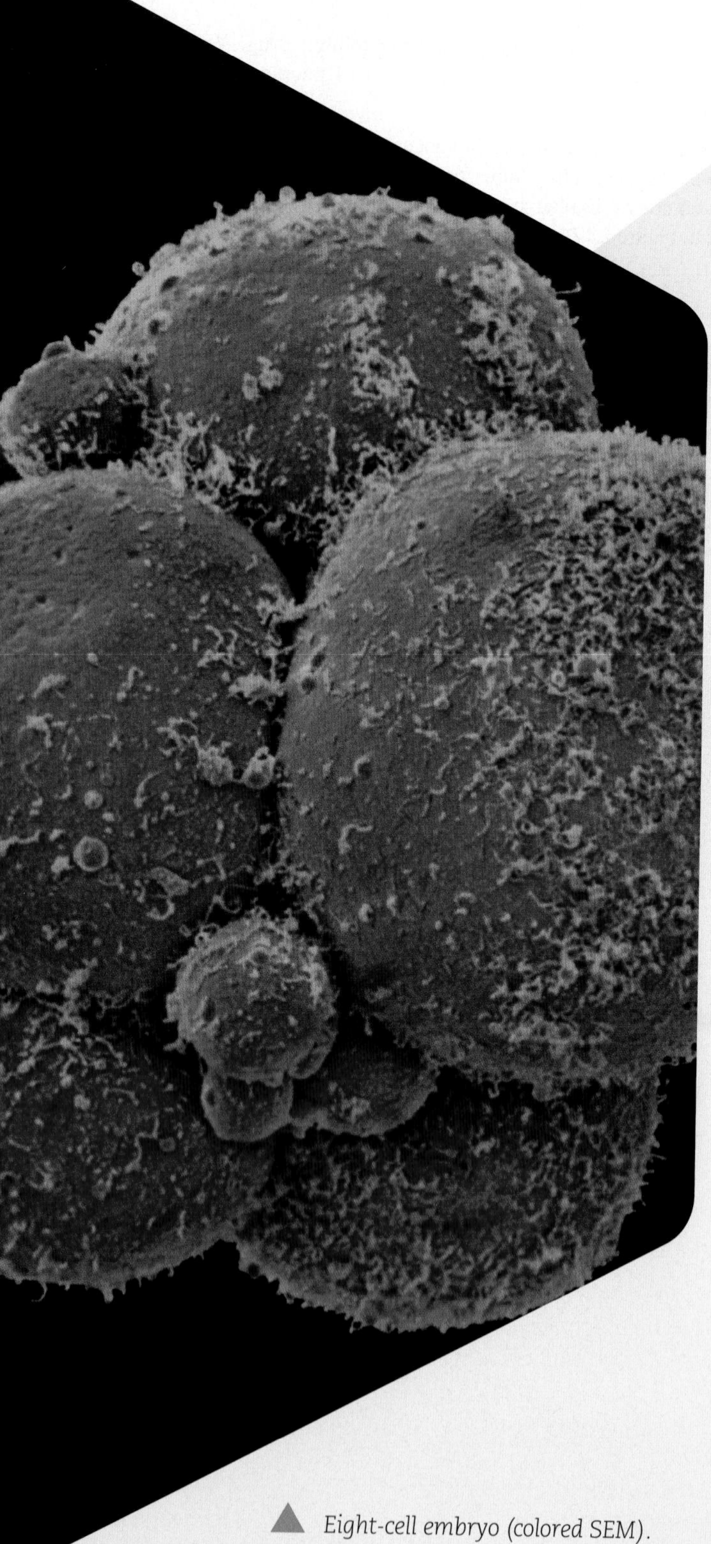

▲ *Eight-cell embryo (colored SEM).*

In just 38 weeks, from conception to birth, a single fertilized egg cell develops into a fully formed human being. The body will not change this much again during its remaining life span of 70 to 90 years. This chapter introduces you to human **embryology,** the study of the origin and development of an individual person. A knowledge of basic embryological events and structures is especially valuable as you begin your study of human anatomy. By knowing how the body methodically assembles itself, you will better understand adult anatomy. Moreover, embryology helps to explain the origin of many birth defects, anatomical abnormalities that are evident in about 3% of live births.

STAGES OF PRENATAL DEVELOPMENT

learning outcomes

- ▶ List the practical and clinical reasons for studying embryology.
- ▶ Distinguish the embryonic period from the fetal period of development.

The **prenatal period** (pre-na′tal; "before birth") is the time between conception (the union of an egg and a sperm cell) and birth. It is divided into two stages (Figure 3.1): The **embryonic period**

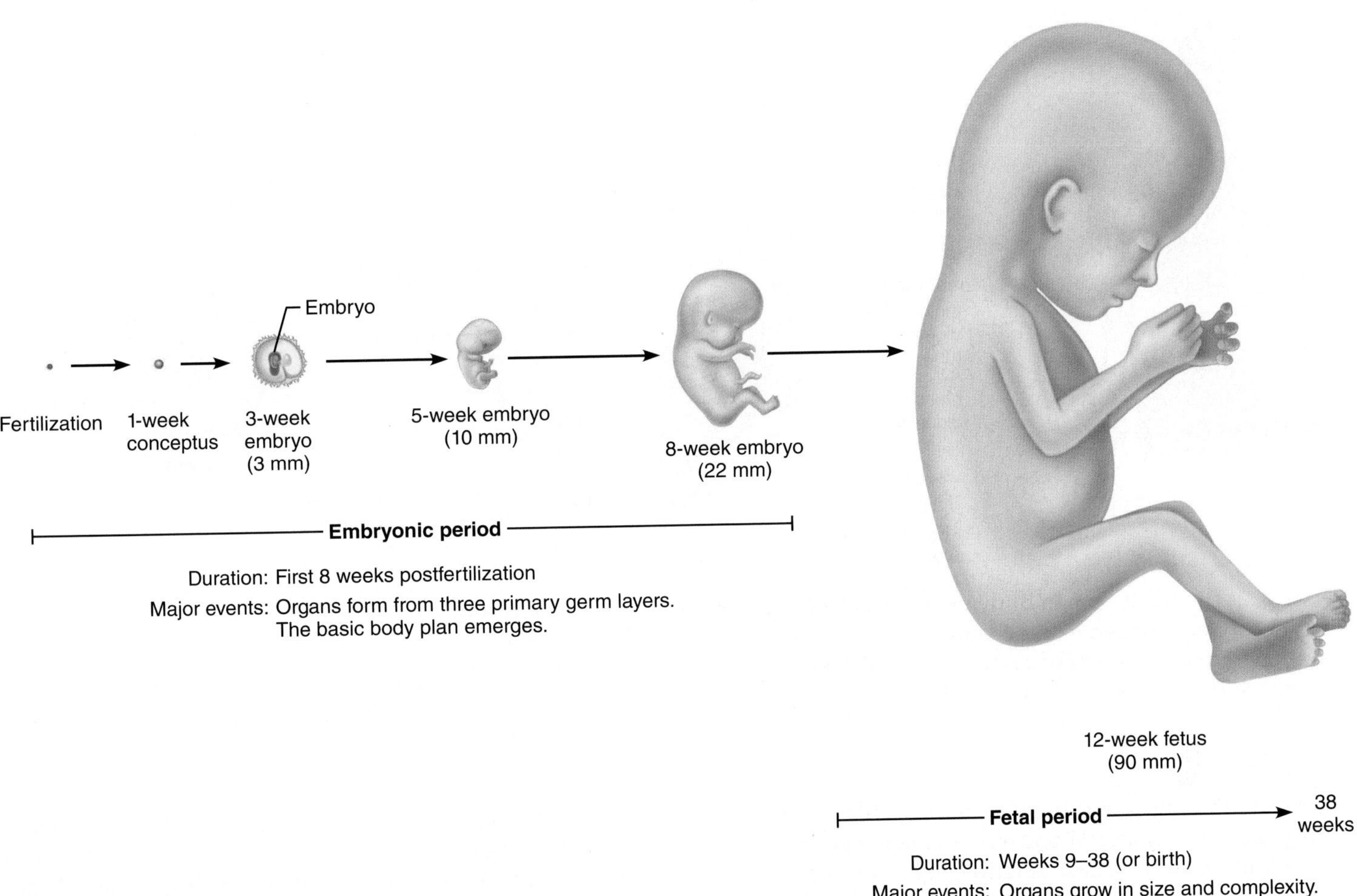

Figure 3.1 **The prenatal period:** All embryos and the fetus are shown in actual size.

spans the first 8 weeks, and the **fetal** (fe′tal) **period** encompasses the remaining 30 weeks. The embryonic period is an exceptionally busy one. By its end, all major organs are in place, and the **embryo,*** the early form of the body, looks distinctly human. In the longer fetal period that follows, the organs of the **fetus** (fe′tus; "the young in the womb") grow larger and more complex.

THE BASIC BODY PLAN

learning outcome

▶ Sketch the basic structural plan of the adult body, which is established during the embryonic period.

To simplify the treatment of embryology, this chapter limits the discussion to the derivation of the basic adult body plan. The basic body plan can be described as a tube within a tube (Chapter 1, p. 10). This basic plan is evident by month 2 of development (Figure 3.2). The outer body wall makes up the outer tube and the inner tube in this section through the abdomen is the digestive tube. These two tubes are separated by a serous cavity. In the abdomen, this cavity is the peritoneal cavity. In the figure, note the following adult structures:

1. **Skin.** The skin has two layers: an outer layer called the *epidermis* and an inner, leathery layer called the *dermis*.
2. **Outer body wall.** The outer body wall consists mostly of trunk muscles. Dorsally, it also contains the vertebral column, through which the spinal cord runs. Ribs attach to each bony vertebra in the thoracic region of the trunk wall.
3. **Body cavity and inner tube.** The inner tube is composed of the respiratory and digestive structures (see Figure 1.5, p. 10). In Figure 3.2, the section through the abdomen shows the peritoneal cavity, lined by visceral and parietal serosae, surrounding the digestive tube (stomach, intestines, and so on). The digestive tube has a muscular wall and is lined internally by a sheet of cells. (This lining is shown in yellow in Figure 3.2.)

In the thoracic region, the body plan is similar. The respiratory structures form from the inner tube. The body cavities in the thorax are the pleural cavity

*Technically, the embryo is the stage in prenatal development between the third or fourth and eighth weeks, inclusive. However, the term *embryo* can be used informally to encompass all stages in the embryonic period.

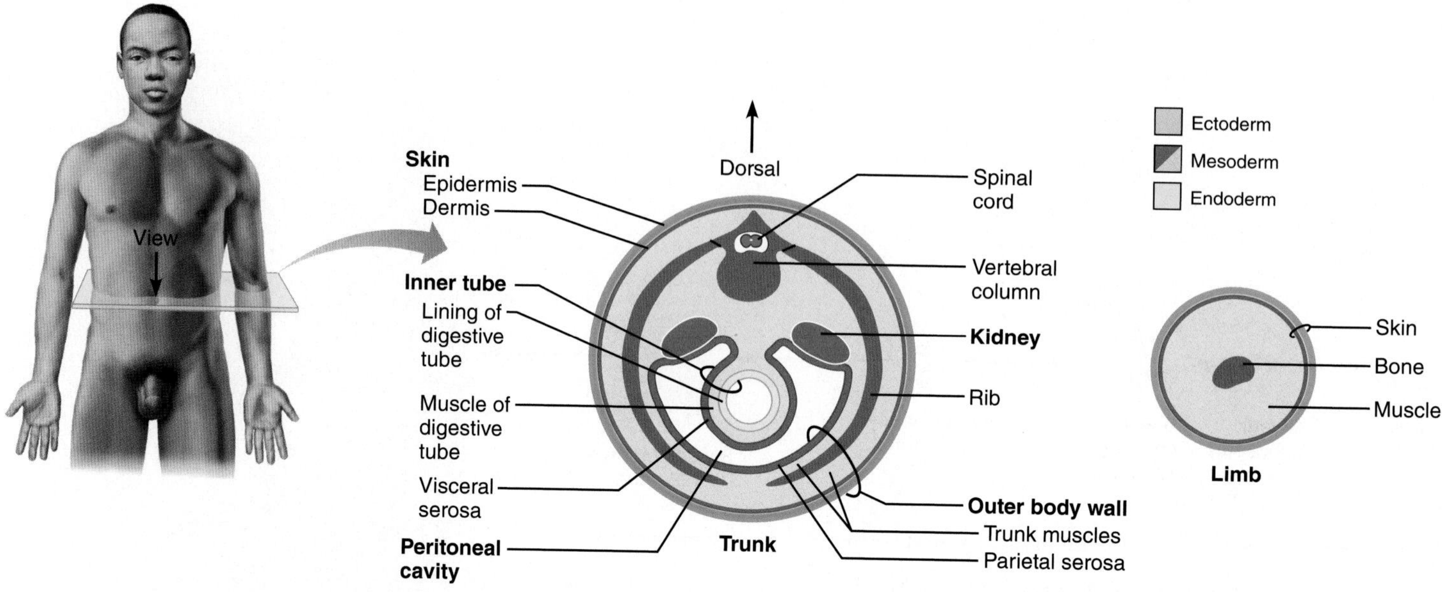

Figure 3.2 **The adult human body plan (simplified cross section).** The development of this body plan is traced in this chapter. The blue, red, and yellow colors denote derivation from the three basic embryonic germ layers. (See the discussion on pp. 50-51 and Figure 3.9 for more information.)

around the lungs and the pericardial cavity surrounding the heart. Parietal and visceral serosae line these cavities as well. (These relationships were introduced in Chapter 1, p. 12.)

4. **Kidneys and gonads.** The kidneys lie directly deep to the dorsal body wall, in the lumbar region of the back posterior to the parietal serosa. The gonads (testes or ovaries) originate in a similar position but migrate to other body regions during the fetal period.
5. **Limbs.** The limbs consist mostly of bone, muscle, and skin.

You can see how this adult body plan takes shape by following the events of month 1 of human development.

✓ check your understanding

- ☐ 1. During which prenatal period is the basic body plan established?
- ☐ 2. Which abdominal structures form from the inner tube?
- ☐ 3. Using directional terms (Table 1.1, p. 8), describe the position of the kidneys in reference to the peritoneal cavity.

(For answers, see Appendix B.)

THE EMBRYONIC PERIOD

Week 1: From Zygote to Blastocyst

learning outcome

▶ Describe the earliest stage of development, from zygote to blastocyst (week 1).

Each month, one of a fertile woman's two ovaries releases an immature egg, called an *oocyte* (Figure 3.3). This cell is normally drawn into a *uterine tube* (fallopian tube), which provides a direct route to the uterus (womb). **Fertilization** of an oocyte by a sperm generally occurs in the lateral third of the uterine tube. The fertilized oocyte, now called a **zygote** (zi′gōt; "a union"), is moved toward the uterus. Along the way, it divides repeatedly to produce two cells, then four cells, then eight cells, and so on. Because there is not much time for cell growth between divisions, the resulting cells become smaller and smaller. This early division sequence, called **cleavage,** provides the large number of cells needed as building blocks for the embryo.

About 72 hours* after fertilization, cleavage has generated a solid cluster of 12–16 cells called a **morula** (mor′u-lah; "mulberry"). During day 4, the late morula—now consisting of about 60 cells—enters the uterus. It takes up fluid, which gathers into a central cavity. This new fluid-filled structure is called a **blastocyst** (blas′to-sist; *blasto* = bud or sprout; *cyst* = *bag*).

Two distinct types of cells are obvious in the blastocyst stage (Figure 3.3e). A cluster of cells on one side of the blastocyst cavity is called the **inner cell mass,** and the layer of cells surrounding the cavity is called the **trophoblast** (trōf′o-blast; *tropho* = nourishment). The inner cell mass will form the embryo, and the trophoblast will help form the placenta, the structure that transfers nutrients from the mother to the fetus.

*All dates given for the developmental events in this chapter are *average* times. Actual dates vary by 1–2 days or more among different pregnancies.

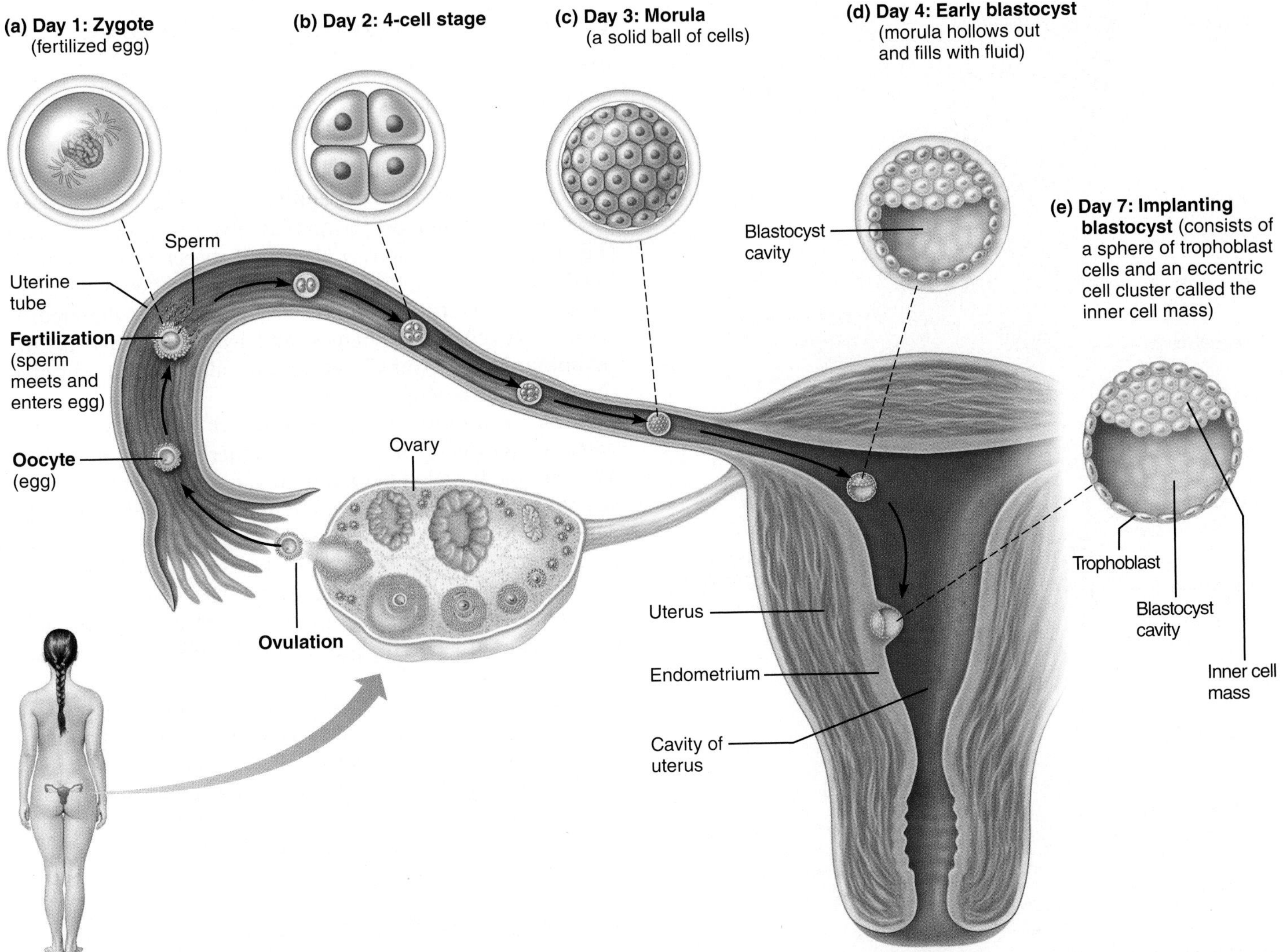

Figure 3.3 Fertilization and the events of the first 6 days of development. The ovary, uterus, and uterine tubes, which lie in the mother's pelvis, are shown in posterior view. An egg (oocyte) is released into the peritoneal cavity (ovulation) and taken up by the uterine tube, where it undergoes fertilization to become a zygote (**a**). Then, as it moves through the uterine tube and into the uterus, it passes through the four-cell stage (**b**) and the morula (**c**) and blastocyst (**d**) stages. Finally, the blastocyst implants into the wall of the uterus (**e**), as shown in detail in the next figure. Parts (d) and (e) show the blastocyst cut in half to reveal the inside.

This chapter focuses on the inner cell mass and the embryo; the trophoblast and placenta are discussed along with the female reproductive system (Chapter 25).

The blastocyst stage lasts about 3 days, from day 4 to day 7. For most of this time, the blastocyst floats freely in the cavity of the uterus, but on day 6, it starts to burrow into the wall of the uterus (Figure 3.4). This process, called **implantation,** takes about a week to complete. In implantation, the trophoblast layer erodes inward until the entire blastocyst is embedded in the uterine wall.

In some pregnancies, the inner cell mass of a single blastocyst splits into two during the early stages of cleavage of week 1 or week 2. This produces identical twins, also called monozygotic twins (*monozygotic* = from one zygote). Worldwide, the birth rate for identical twins is approximately 4 per 1000 births.

CLINICAL APPLICATION

Conjoined (Siamese) Twins Identical twins that are born joined together are called conjoined twins. This phenomenon is caused by incomplete division of the inner cell mass or embryonic disc during the twinning process. The twins may be joined at any body region and often share organs. Some can be separated successfully by surgery; others live fulfilling lives with their conjoined sibling.

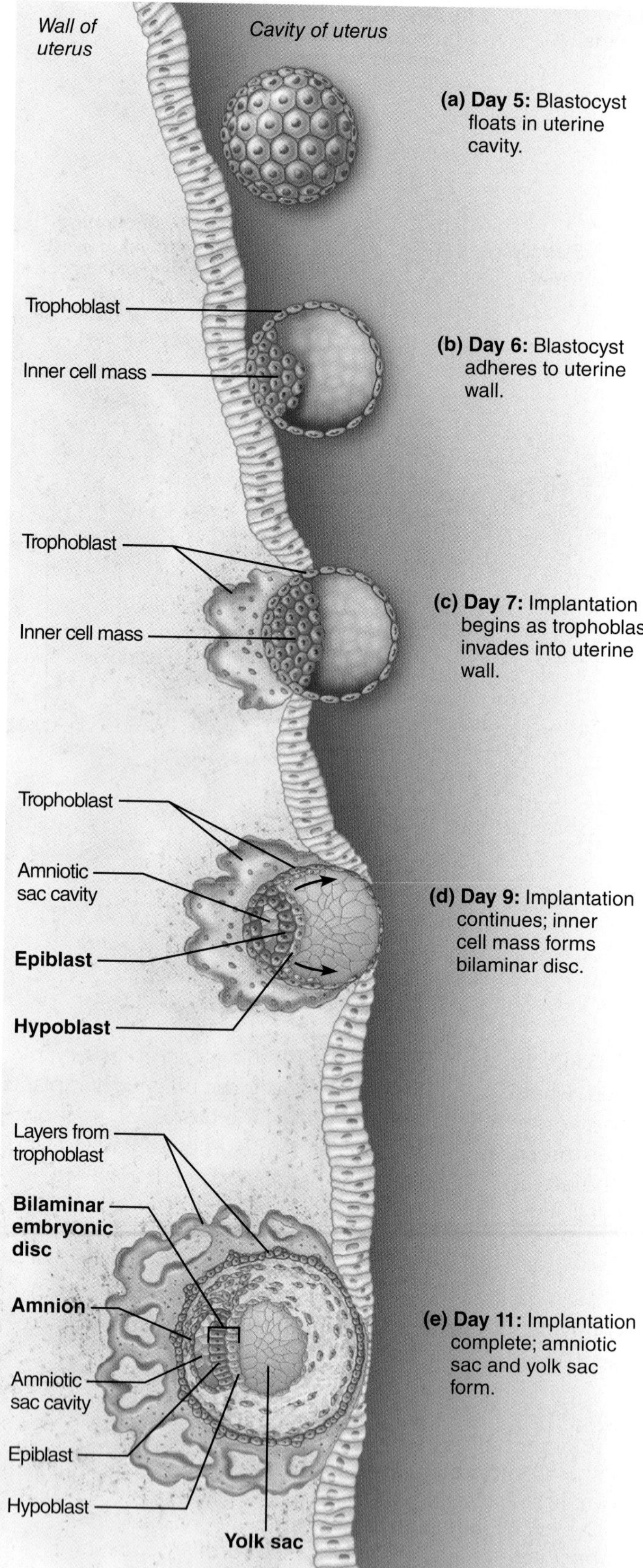

Figure 3.4 Implantation of the blastocyst during week 2 of development. The embryos are shown in section (cut in half).

Week 2: The Two-Layered Embryo

learning outcome

▶ Identify the structures that develop as the inner cell mass differentiates into a two-layered disc.

About 9 days after fertilization, the inner cell mass has divided into two sheets of cells, the **epiblast** (ep′ĭ-blast; *epi* = over, upon) and the **hypoblast** (hi′po-blast; *hypo* = under, beneath) (Figure 3.4d). Extensions of these cell sheets form two fluid-filled sacs (Figure 3.4d–e) resembling two balloons touching one another, with the epiblast and hypoblast at the area of contact. Together, the epiblast and hypoblast make up the **bilaminar** ("two-layered") **embryonic disc,** which will give rise to the whole body.

The sac formed by an extension of the epiblast is the *amniotic* (am″ne-ot′ik) *sac.* The outer membrane of the amniotic sac is called the **amnion,** and the internal **amniotic sac cavity** is filled with *amniotic fluid.* This fluid buffers the developing embryo and fetus from physical shock until the time of birth. You may have heard the expression, "The mother's water broke just before she gave birth." This refers to the rupture of the amniotic sac and release of its amniotic fluid near the start of labor, the process that expels the mature fetus from the uterus.

The **yolk sac,** formed by an extension of the hypoblast, holds a very small amount of yolk, which is insignificant as a food source. The human yolk sac, however, is important because the digestive tube forms from part of it (p. 56). Furthermore, tissue around the yolk sac gives rise to the earliest blood cells and blood vessels (see Chapter 18, "The Blood Throughout Life" section).

Week 3: The Three-Layered Embryo

learning outcome

▶ Explain gastrulation and the formation of the three germ layers (week 3).

The Primitive Streak and the Three Germ Layers

During week 3, the embryo grows from a two-layered disc to a three-layered (trilaminar) disc. This process, called **gastrulation** (gas″troo-la′shun), forms the three *primary germ* layers—ectoderm, mesoderm, and endoderm—the layers from which all body tissues develop. The germ layer begins to form on days 14–15, when a raised groove called the **primitive streak** appears on the dorsal surface of the epiblast (Figure 3.5). Epiblast cells migrate inward at this streak. On days 14–15 the first cells that migrate through the primitive streak displace the cells of the underlying hypoblast to become the **endoderm.** (Figure 3.5f). Then, starting on day 16, the ingressing epiblast cells form a new layer between epiblast and endoderm, the **mesoderm** (Figure 3.5g). The epiblast cells that remain on the embryo's dorsal surface make up the **ectoderm.** In this way, the three primary germ layers of the body are established, all derived from epiblast cells.

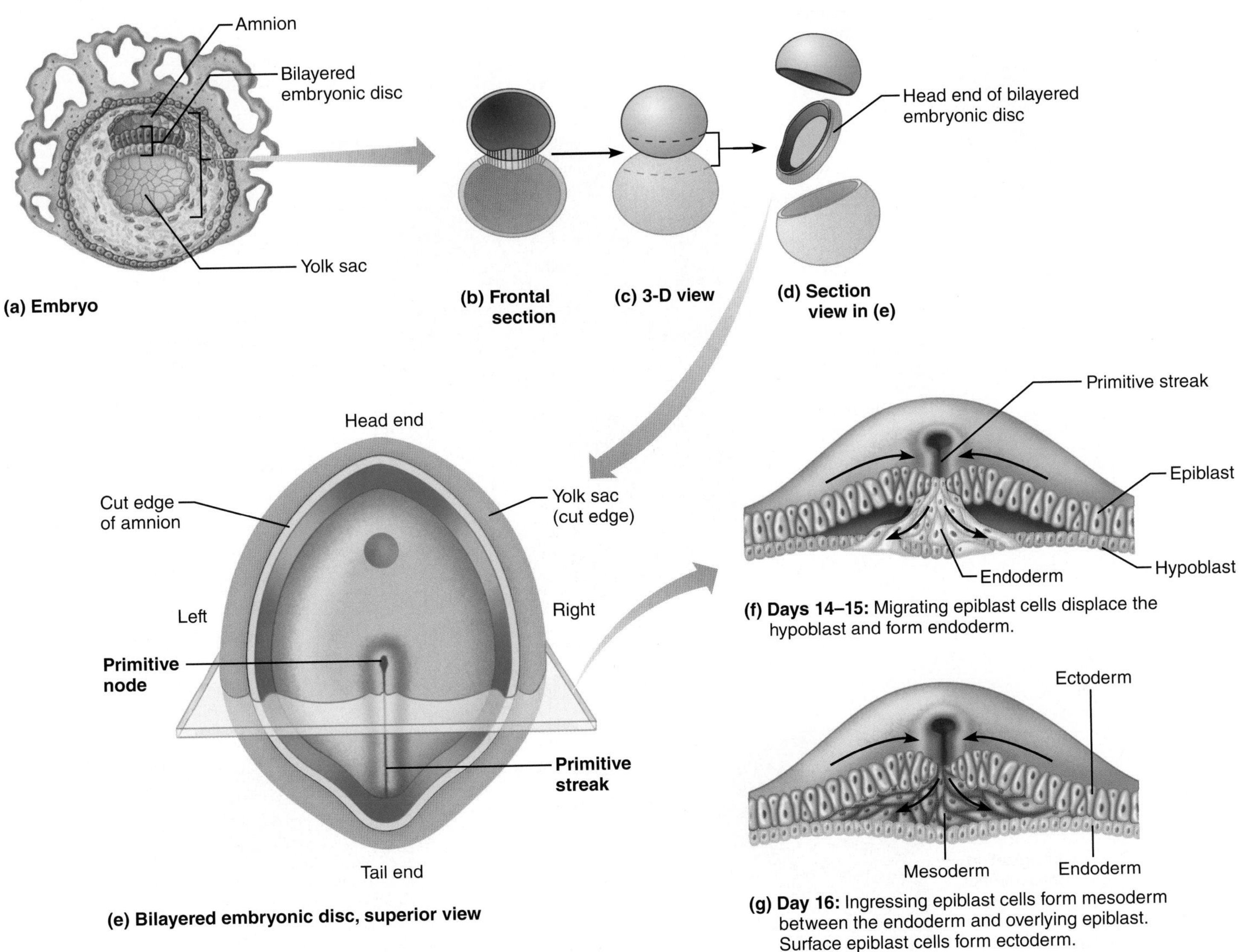

Figure 3.5 The primitive streak stage. Diagrams (**a**) through (**d**) show orientation of the bilaminar disc shown in (e). (**e**) The primitive streak appears on the epiblast on about day 14. (**f, g**) Sections through the embryonic disc at the location shown in (e).

The structures formed from each germ layer are color-coded in the figures throughout this chapter:

- **Ectoderm** ("outside skin"; colored blue in Figure 3.5) forms the outer layer of the skin (epidermis), the brain, and the spinal cord.
- **Mesoderm** ("middle skin"; colored red in Figure 3.5) forms muscle, bone, and connective tissues.
- **Endoderm** ("inner skin"; colored yellow in Figure 3.5) forms the innermost lining of the inner tube (epithelial lining).

The three germ layers differ in their tissue structure. A tissue is a collection of similar cells that perform a common function (Chapter 1). Ectoderm and endoderm are *epithelial tissues,* or *epithelia*—sheets of tightly joined cells (p. 5). Mesoderm, by contrast, is a *mesenchyme tissue* (mes′eng-kīm; *mesen* = middle; *chyme* = fluid). A mesenchyme is any embryonic tissue with star-shaped cells that do not attach to one another. Thus, mesenchyme cells are free to migrate widely within the embryo.

The Notochord

At one end of the primitive streak is a swelling called the **primitive node** (Figure 3.5e). The epiblast cells that move through the primitive node migrate straight anteriorly. These mesodermal cells, along with a few cells from the underlying endoderm, form a rod called the **notochord** (Figure 3.6a). The notochord defines the body *axis* (the midline that divides the left and right sides of the body). It extends the length of the body and is the site of the future vertebral column. The notochord appears on day 16, and by day 18 it reaches the future head region.

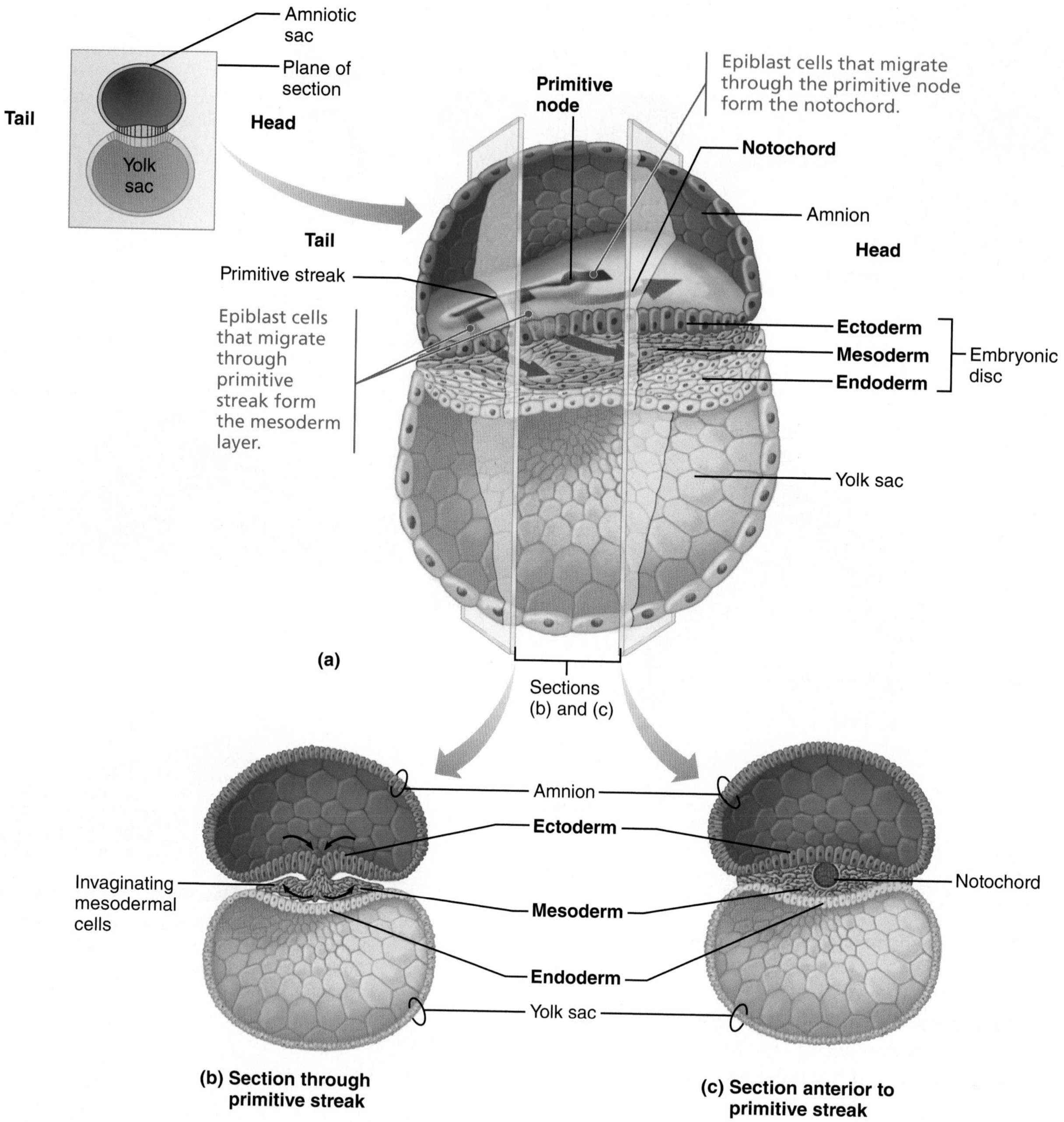

Figure 3.6 Formation of the mesoderm and the notochord in a 16-day embryo. (**a**) The three-layered embryonic disc in sagital section (see the inset above for orientation). (**b**) Cross-sectional view of the embryonic disc, through the primitive streak. (**c**) Cross section taken anterior to the primitive streak, in the future thorax.

Neurulation

As the notochord develops, it signals the overlying ectoderm to start forming the spinal cord and brain, an event called **neurulation** (nu″roo-la′shun) (central column of Figure 3.7). Specifically, the ectoderm in the dorsal midline thickens into a **neural plate,** and then starts to fold inward as a **neural groove.** This groove deepens until a hollow **neural tube** is pinched off into the body. Closure of the neural tube begins at the end of week 3 in the region that will become the neck and then proceeds both cranially (toward the head or crown) and caudally (toward the tail or rump). Complete closure occurs by the end of week 4. The cranial part of this neural tube becomes the brain, and the rest becomes the spinal cord.

Neural crest cells (green in Figure 3.7) are pulled into the body along with the invaginating neural tube. The neural crest cells originate from ectodermal cells on the lateral ridges (neural folds) of the neural plate, and they come to lie just external to the closed neural tube. The neural crest forms

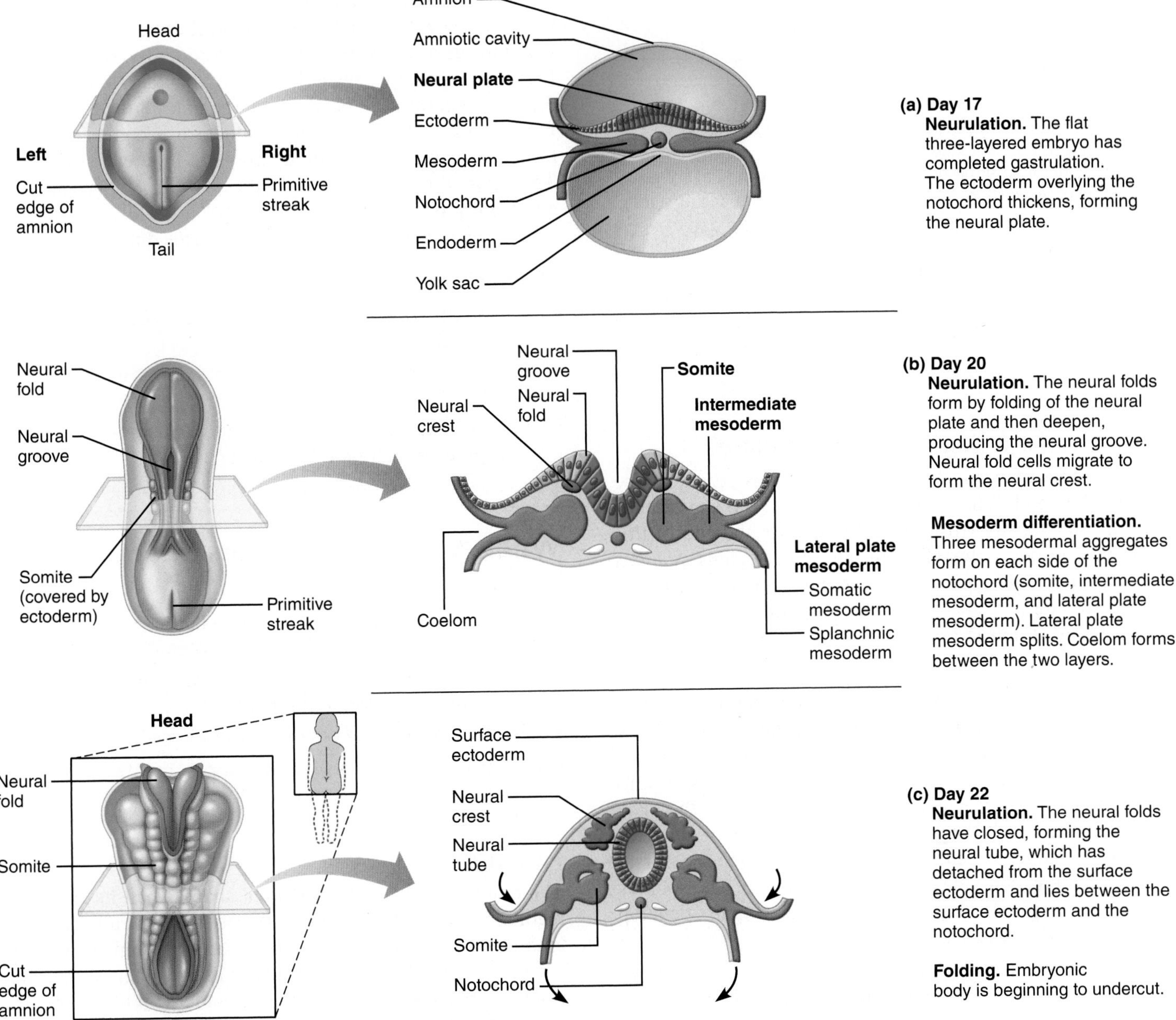

Figure 3.7 **Neurulation and mesoderm differentiation, week 3.** (a–c) At left, dorsal surface views of the embryo. At right, transverse sections through the entire embryo showing formation of the neural tube and mesoderm differentiation.

the sensory nerve cells and some other important structures, as described later in the chapter (p. 55).

The ability of one group of cells to influence the developmental direction of other cells is called **induction.** The influence of the notochord on the formation of the neural tube is an example of induction. In fact, the notochord initiates a chain reaction of inductions that continue throughout development. Disruption in any of these inductive processes can result in developmental abnormalities. The genes and molecules that signal these inductive events are currently being identified.

CLINICAL APPLICATION

Neural Tube Defects Neural tube defects result from failure of the neural tube to close completely. The most common neural tube defects are spina bifida and anencephaly. **Spina bifida** results when the caudal portion of the neural tube does not close completely, causing malformation of the spinal cord, spine, and spinal nerves. Depending on the location and extent of the defect, resulting disabilities range from mild (lower body weakness and pain, bowel and bladder dysfunction) to severe (paralysis below the area of ▶

Neural Tube Defects (continued)

defect and complications from hydrocephalus, excessive fluid buildup on the brain). **Anencephaly** results when the neural tube fails to close cranially. The brain does not develop, and the baby is either stillborn or dies within a few hours after birth. Up to 70% of neural tube defects have been linked to low maternal levels of folic acid. It is recommended that women of childbearing age consume 400 micrograms (0.4 mg) of folic acid daily. In addition to vitamin supplementation, good sources of folic acid include green leafy vegetables, whole-grain cereals, and nuts. Taking supplements prior to pregnancy is important because the neural tube forms in the third week of development, often before a woman knows she is pregnant.

The Mesoderm Begins to Differentiate

In the middle of week 3, the mesoderm lies lateral to the notochord on both sides of the body (Figure 3.7a) and extends cranially to caudally (from head to tail). By the end of this week, the mesoderm has divided into three regions. *Somites* and *intermediate mesoderm* are segmented and form the segmented structures of the outer tube. *Lateral plate mesoderm* is unsegmented and is associated with the developing inner tube organs.

1. **Somites** (so′mītz; "bodies"). The mesoderm closest to the notochord begins as paraxial mesoderm (par″ak′-se-al; "near the body axis"). Starting cranially and proceeding caudally, the paraxial mesoderm divides into a series of blocks called *somites* (Figure 3.7b). The somites are visible in surface view as a row of subectodermal bulges on each side of the back (see the dorsal views on the left side of Figure 3.7b and c). The somites are the first body segments, and about 40 pairs develop by the end of week 4.
2. **Intermediate mesoderm.** This begins as a continuous strip of tissue just lateral to the paraxial mesoderm. Influenced by the segmentation of the somites, the intermediate mesoderm divides into spherical segments in a cranial-to-caudal sequence (Figure 3.7b). Each segment of intermediate mesoderm attaches to a somite.
3. **Lateral plate.** This, the most lateral part of the mesoderm, remains unsegmented (Figure 3.7b and c). The lateral plate begins as one layer, but soon splits into two. A wedge of space is formed between these two sheets. This space is called the **coelom** (se′lum; "cavity"). The two resulting divisions of the lateral plate are the **somatic mesoderm** (so-mat′ik; "body"), next to the ectoderm, and the **splanchnic mesoderm** (splangk′-nik; "viscera"), next to the endoderm. The coelom that intervenes between the splanchnic and somatic mesoderm will become the serous cavities of the ventral body cavity, namely the peritoneal, pericardial, and pleural cavities (see Figure 1.7).

When you compare cross sections of a 3-week embryo (Figure 3.7) with the adult body plan (Figure 3.2), you might begin to see a few similarities. The ectoderm (colored blue in Figure 3.7) becomes the epidermis of the skin (blue in Figure 3.2); the neural tube becomes the spinal cord (blue in both Figures 3.2 and 3.7); and the endoderm becomes the lining of the digestive tube (colored yellow in Figures 3.2 and 3.7). The main difference between the 3-week embryo and the adult body is that the embryo is still a flat disc. The three-dimensional, cylindrical body shape forms during week 4.

☑ check your understanding

- ☐ 4. Describe gastrulation. During which week of embryonic development does it occur?
- ☐ 5. What structure induces the formation of the neural tube?
- ☐ 6. Which type or types of mesoderm cluster into segments along the body axis?

(For answers, see Appendix B.)

Week 4: The Body Takes Shape

learning outcomes

- ▶ Discuss how the body folds from a flat disc into its three-dimensional, tubular shape (week 4).
- ▶ In a cross section of a week 4 embryo, identify the outer tube, the inner tube, and the coelom.
- ▶ List the main derivatives of ectoderm and endoderm.
- ▶ In a cross section of a week 4 embryo, identify the five regions of mesoderm and list the main derivatives of each region.

Folding

The embryo takes on a cylindrical shape when its sides fold medially and it lifts off the yolk sac and protrudes into the amniotic cavity (Figure 3.8). This process resembles the folding of three stacked sheets of paper into a tube. At the same time, the head and tail regions fold under (Figure 3.8b). The embryonic disc bulges upward because it is growing so much faster than the yolk sac below it. Its lateral folding is caused by the fast growth of the somites. The folding at the head and tail is caused by expansion of the brain and lengthening of the spinal cord.

As a result of this folding, the embryo acquires a tadpole shape by day 24. As the embryo becomes cylindrical, it encloses a tubular part of the yolk sac, called the **primitive gut** (future digestive tube and respiratory structures: Figure 3.8c and Figure 1.5 on p. 10). This tube is lined by endoderm. The embryo remains attached to the yolk sac below by a duct located at the future navel. This duct becomes incorporated into the umbilical cord.

Derivatives of the Germ Layers

By day 28, the basic human body plan has been attained (Figure 3.9). Comparing a cross section through the trunk of a 1-month-old embryo (Figure 3.9b) to the adult body section

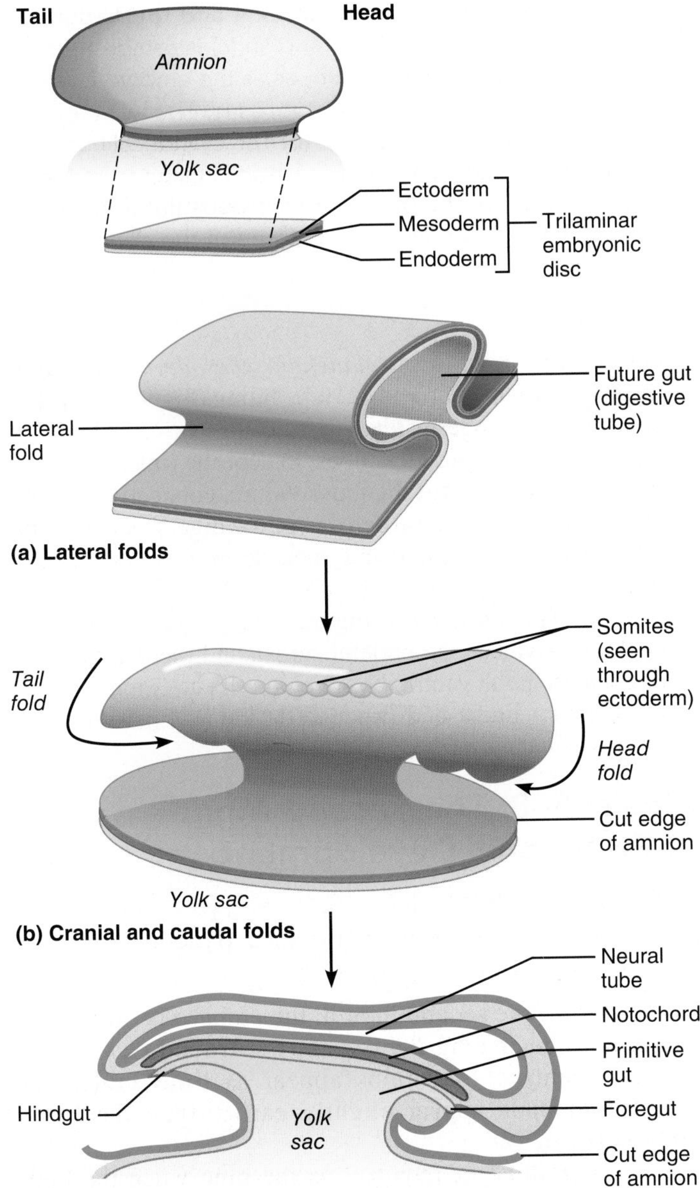

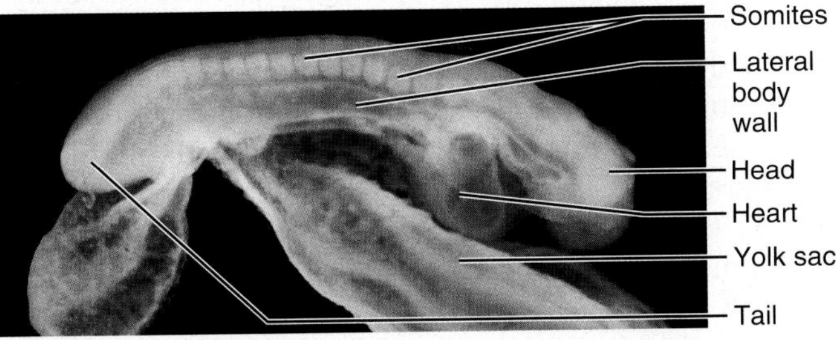

(d) 23-day embryo, lateral view (20×). Lateral folding nearing completion.

Figure 3.8 Folding of the embryo, lateral views.

(Figure 3.9c) will help you understand the adult derivatives of the germ layers (Figure 3.10).

Derivatives of Ectoderm The ectoderm becomes the brain, spinal cord, and epidermis of the skin. The early epidermis, in turn, produces the hair, fingernails, toenails, sweat glands, and the oil glands of the skin. Neural crest cells, from ectoderm, give rise to the sensory nerve cells. Furthermore, much of the neural crest breaks up into a mesenchyme tissue, which wanders widely through the embryonic body. These wandering neural crest derivatives produce such varied structures as the pigment-producing cells in the skin (melanocytes) and the bones of the face.

Derivatives of Endoderm The endoderm becomes the inner epithelial lining of the gut tube and its derivatives: the respiratory tubes, digestive organs, and the urinary bladder. It also gives rise to the secretory cells of the glands that develop from gut-lining epithelium: the thyroid, thymus, and parathyroid glands from the pharynx; and the liver and pancreas from the digestive tract.

Derivatives of the Mesoderm and Notochord The mesoderm has a complex fate, so you may wish to start by reviewing its basic parts: the notochord; the segmented portions, the somites and intermediate mesoderm; and the unsegmented somatic and splanchnic lateral plate mesoderm (Figure 3.9).

- **The notochord.** The *notochord* gives rise to an important part of the spinal column, the springy cores of the discs between the vertebrae. These spherical centers, each called a *nucleus pulposus* (pul-po′sus), give the vertebral column some bounce as we walk.
- **The segmented mesoderm.** Each of the *somites* divides into three parts (Figure 3.9b). One part is the **sclerotome** (skle′ro-tōm; "hard piece"). Its cells migrate medially, gather around the notochord and the neural tube, and produce the vertebra and rib at the associated level. The most lateral part of each somite is a **dermatome** ("skin piece"). Its cells migrate externally until they lie directly deep to the ectoderm, where they form the dermis of the skin in the dorsal part of the body. The third part of each somite is the **myotome** (mi′-o-tōm; "muscle piece"), which stays behind after the sclerotome and dermatome migrate away. Each myotome grows ventrally until it extends the entire dorsal-to-ventral height of the trunk. Myotomes become the segmented trunk musculature of the body wall (Figures 1.5b and 3.11). Additionally, the ventral parts of myotomes grow into the limb buds and form the muscles of the limbs.

  The *intermediate mesoderm,* lateral to each somite, forms the kidneys and the gonads. The intermediate mesoderm lies in the same relative location as the adult kidneys, outside the peritoneal cavity, or **retroperitoneal** (compare Figure 3.9b and c).
- **The unsegmented mesoderm.** The splanchnic and somatic lateral plate mesoderm are separated by the coelom body

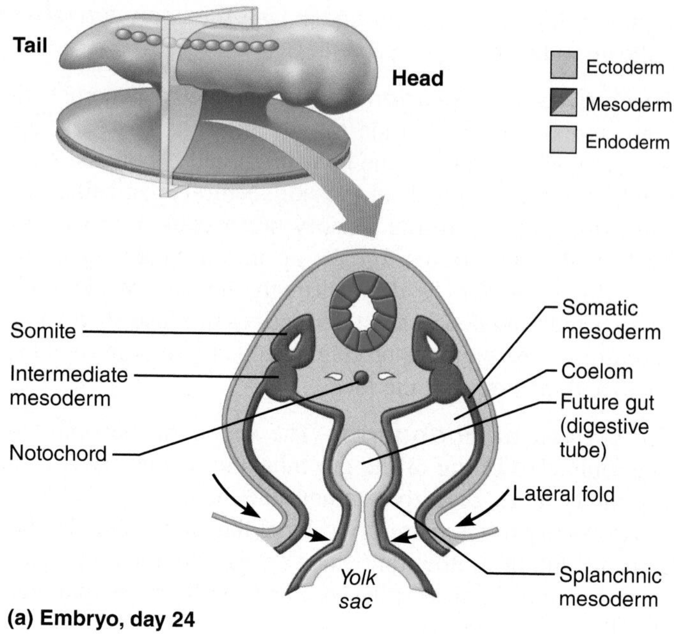

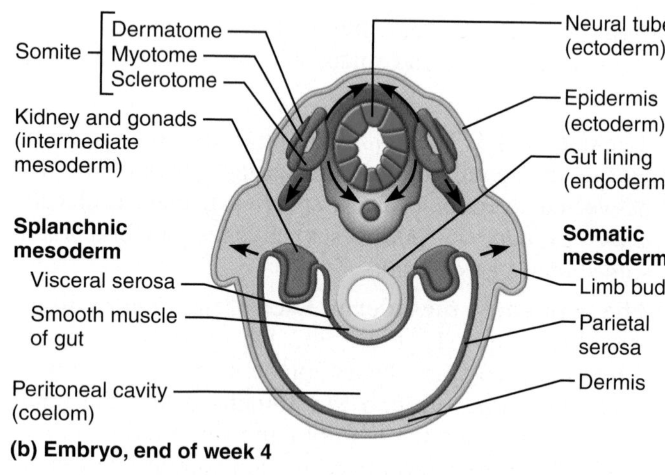

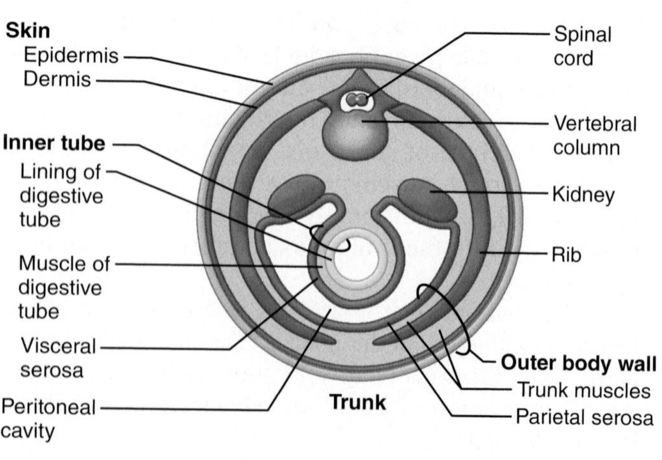

Figure 3.9 **The germ layers in week 4 and their adult derivatives.** The basic adult body plan (c) is established in week 4 (a–b). (**a**) Folding continues as the embryo forms into a cylinder and lifts up off the yolk sac. The right and left lateral folds will join ventrally. (**b**) The cylindrical human body plan on day 28. This cross section is taken from the trunk region, where the posterior limb buds (future legs) attach. (**c**) Simplified cross section through the abdomen of an adult.

cavity. By now, the *splanchnic mesoderm* surrounds the endodermally derived gut tube lining (Figure 3.9b). The splanchnic mesoderm gives rise to the entire wall of the digestive and respiratory tubes, except the inner epithelial lining; that is, it forms the musculature, connective tissues, and the slippery visceral serosae of the digestive and respiratory structures. Splanchnic mesoderm also gives rise to the heart and most blood vessels.

Somatic mesoderm (Figure 3.9b), just external to the coelom, produces the parietal serosa and the dermal layer of the skin in the ventral body region. Its cells migrate into the forming limbs and produce the bone, ligaments, and dermis of each limb.

Weeks 5–8: The Second Month of Embryonic Development

learning outcome

▶ Describe the main events of the second month of development.

At the start of the second month, the embryo is only about a half centimeter long (Figure 3.11). Around day 28, the first rudiments of the limbs appear as **limb buds.** The upper limb buds appear slightly earlier than the lower limb buds.

You can think of month 2 as the time when the body becomes less tadpole-like and starts to become recognizably human (Figure 3.1). The limbs grow from rudimentary buds to fully formed extremities with fingers and toes. The head enlarges quickly and occupies almost half the volume of the entire body. The eyes, ears, and nose appear, and the face gains a human appearance as the embryonic period draws to a close. The protruding tail of the 1-month-old embryo disappears at the end of week 8. All major organs are in place by the end of month 2, at least in rudimentary form. Developmental errors during this time can cause severe malformation (see A CLOSER LOOK, p. 60). As other chapters discuss the development of each organ system, you will often return to the events of month 2.

☑ check your understanding

☐ 7. After folding takes place, which embryonic germ layer forms the outer covering of the embryo?

(Text continues on page 58.)

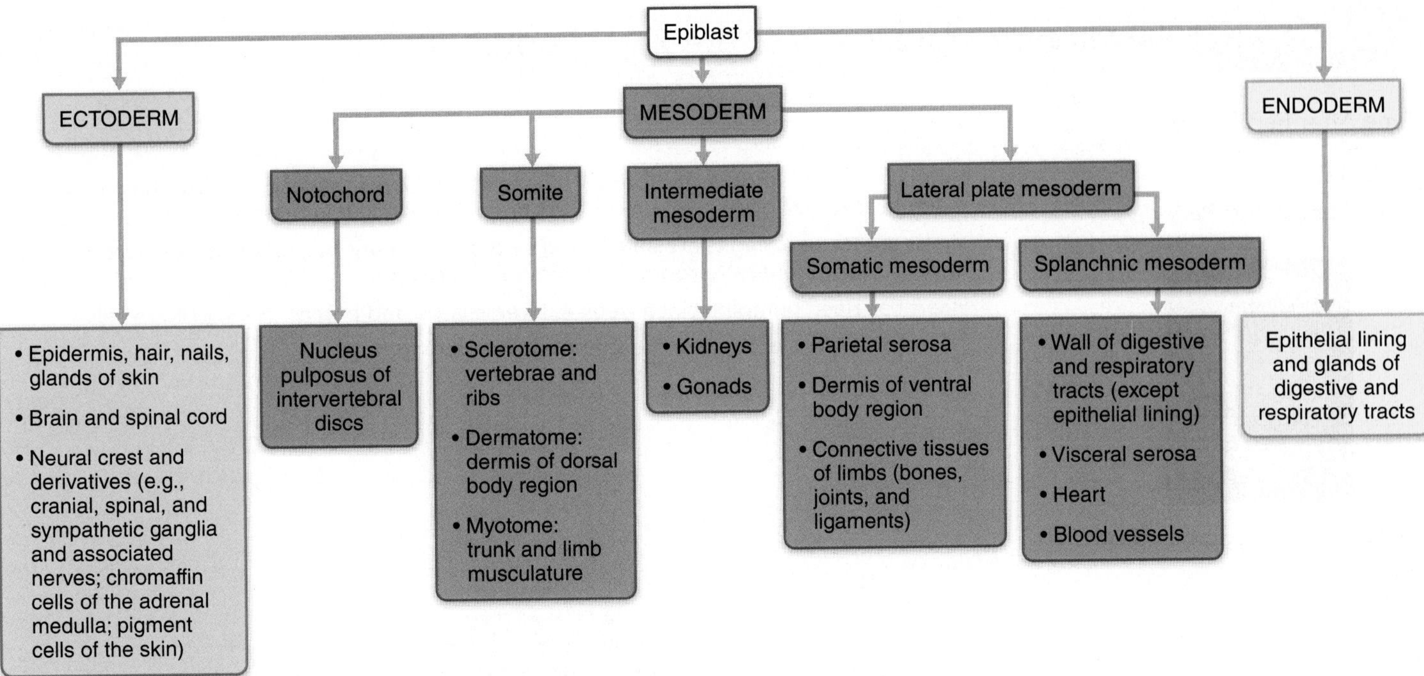

Figure 3.10 **Flowchart showing major derivatives of embryonic germ layers.**

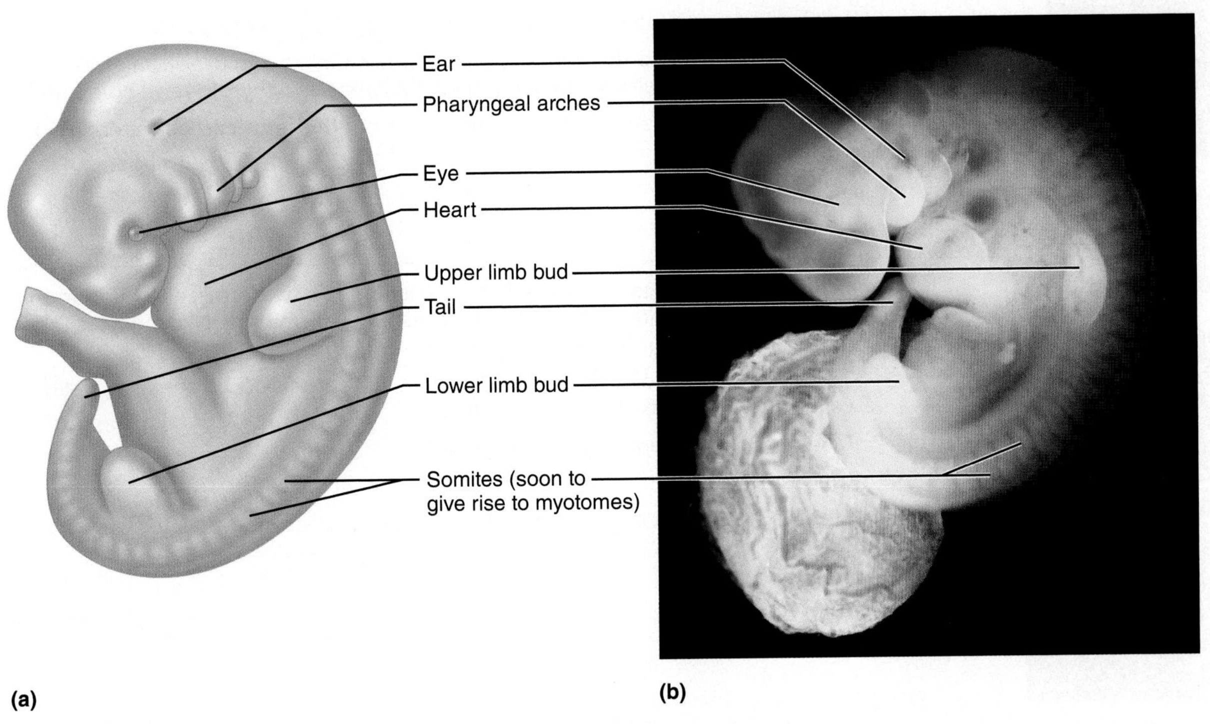

Figure 3.11 **A 29-day embryo.** (**a**) Diagram. (**b**) Fiber-optic photograph. Note the limb buds. Also note the segmented somites, which soon give rise to segmented myotomes.

Table 3.1 Developmental Events of the Fetal Period

Time After Fertilization	Events
8 Weeks (End of Embryonic Period) 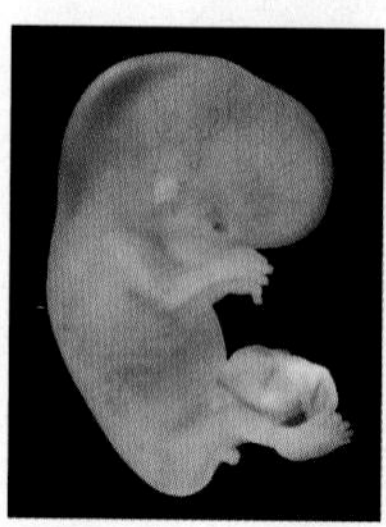 Crown-to-rump length: ~ 3 cm Weight: 2 g at end of period	Head is nearly as large as the body. Nose, ears, and eyes are recognizably human. All major divisions of brain are formed. First brain waves occur in brain stem. Limbs are formed. Digits are initially webbed but separate by end of week 8. Ossification begins in long bones. Vertebrae are formed in cartilage. Heart has been pumping since week 4. Liver is large and begins to form blood cells. All major organ systems are present in rudimentary form.
9–12 Weeks (Month 3) 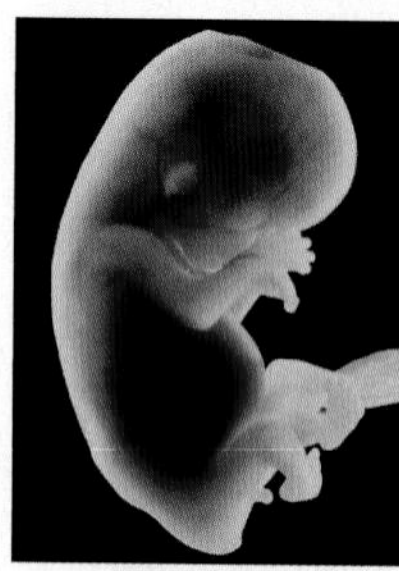 Crown-to-rump length: ~ 6 cm at end of period	Brain continues to enlarge. Cervical and lumbar enlargements are apparent in spinal cord. Retina of eye is present. Trunk and limbs elongate. Palate (roof of mouth) begins to fuse at the midline. Fetus begins to move, but mother does not feel movement. Heartbeat can be detected externally. Blood cell formation begins in bone marrow. Lungs begin to develop. Fetus inhales and exhales amniotic fluid. Intestines move into the abdomen. Liver is prominent and producing bile. Smooth muscle is forming in the walls of hollow organs. Pancreas and thyroid have completely formed. Male and female genitalia are distinctive; sex of the fetus can be determined.
13–16 Weeks (Month 4) 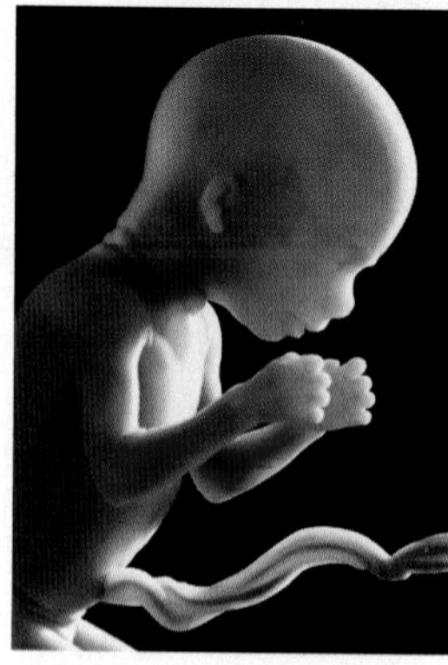 Crown-to-rump length: ~ 11 cm at end of period	Skin development continues with differentiation of the dermis and hypodermis. Epidermis at tips of fingers and toes thickens to initiate nail formation. Melanocytes (pigment cells) migrate into the epidermis. Torso elongates. Bone formation begins in vertebrae. Most bones are distinct, and joint cavities are present. Hard palate is fused. Myelin begins to form around nerve cells. Glands develop in the GI tract. Meconium is collecting. Kidneys attain typical structure. Primary follicles containing oocytes begin to form in the ovary (female).

☐ 8. Does endoderm form the inner lining of the inner tube or the outer tube?

☐ 9. Which part of the somite forms the vertebrae and ribs?

☐ 10. What does the splanchnic lateral plate mesoderm form?

☐ 11. A birth defect in the heart is caused by a developmental error during which weeks of embryonic development?

(For answers, see Appendix B.)

Table 3.1 *continued*

Time After Fertilization	Events
17–20 Weeks (Month 5) Crown-to-rump length: ~ 16 cm at end of period	Hair follicles and sebaceous and sweat glands form. The body is covered with vernix caseosa (fatty secretions of sebaceous glands), and lanugo (silklike hair) covers the skin. Brown fat, a site of heat production, forms in the neck, chest, and groin. Mother can feel fetal movements (quickening). The brain grows rapidly.
21–30 Weeks (Months 6 and 7) Crown-to-rump length: ~ 38 cm at end of period	Period of substantial increase in weight. Fetus has periods of sleep and wakefulness. Fingernails and toenails are complete. Hair is apparent on the head. Distal limb bones begin to ossify. Cerebrum grows, and convolutions develop on brain surface to accommodate the increasing size of the cerebral cortex. Lungs complete development; terminal air sacs and surfactant-secreting cells form at end of month 6. Bone marrow becomes only site of blood cell formation. Testes descend to scrotum in month 7 (males).
30–38 Weeks (Months 8 and 9) 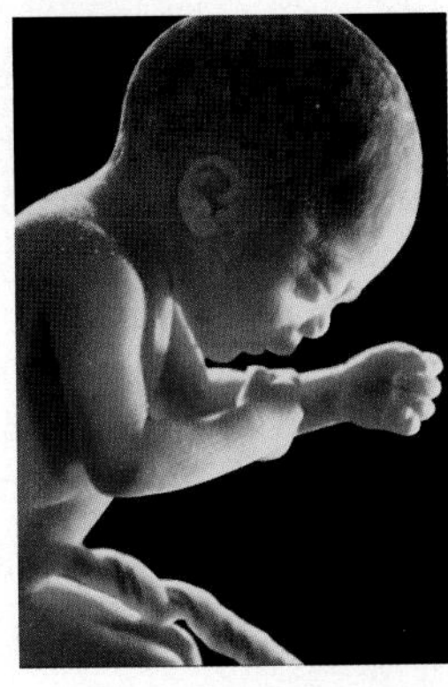 Crown-to-rump length: ~ 47 cm Weight: 2.7 – 4.5 kg (6 – 10 lb) at end of period	Fat accumulates in hypodermis; skin thickens. Surfactant production in the lungs increases. Immune system develops.

THE FETAL PERIOD

learning outcome

▶ List the major events of fetal development.

The fetal period spans weeks 9 through 38 (**Table 3.1**). It is a time of maturation and rapid growth of the body structures that were established in the embryo. The fetal period is more than just a time of growth, however. During the first half of this period, cells are still differentiating into specific cell types to form the body's distinctive tissues and complete the fine details of body structure.

Normal births typically occur 38 weeks after conception. A **premature birth** is one that occurs before that time. Infants born as early as week 30 (7 months) usually survive without requiring life-saving measures. Medical technology can save many fetuses born prematurely, sometimes as early as 22 weeks (in month 6). However, the lungs of premature infants are not fully functional, and death may result from respiratory distress (see Chapter 22, the section "Respiratory Distress Syndrome," for more details). Among those premature infants that survive, rates of visual problems and mental retardation are elevated.

☑ check your understanding

☐ 12. Why is respiratory distress common in babies born in their sixth or seventh month of gestation?

(For the answer, see Appendix B).

Focus on Birth Defects

About 3% of all newborns exhibit a significant birth defect, or **congenital abnormality** (*congenital* = present at birth). This figure increases to 6% by 1 year of age, when initially unrecognized conditions become evident.

Birth defects have many causes, including defective chromosomes or genes, exposure to chemicals and harmful radiation, and viruses passed from the mother to the embryo. Chemical, physical, and biological factors that disrupt development and cause birth defects are called **teratogens** (ter′ah-to-jenz; "monster formers"). Teratogens include many drugs, some organic solvents and pollutants, toxic metals, various forms of radiation such as X rays, and infectious agents, particularly rubella (German measles). Although many teratogens have been identified, most birth defects cannot be traced to any known teratogen or genetic disorder. Most are caused by several factors.

When a pregnant woman drinks alcohol, she increases the likelihood of birth defects. Excessive alcohol consumption can produce **fetal alcohol syndrome (FAS),** typified by microcephaly (small head and jaw), growth deficiencies, and distinctive facial features (see photo). FAS is the most severe of a range of brain and behavioral abnormalities called **fetal alcohol spectrum disorders (FASD).** FASD are estimated to occur in 1% of births and are the most common cause of mental retardation in the United States. Maternal binge drinking (high quantity, regular frequency) greatly increases the risk of FASD. The risks of moderate alcohol consumption are less clear. Maternal factors, such as age, body size, nutrition, socioeconomic status, and genetics, contribute to the incidence of FASD. Given the one constant factor occurring in all instances of FASD—alcohol consumption—pregnant women are advised to refrain from drinking alcohol entirely.

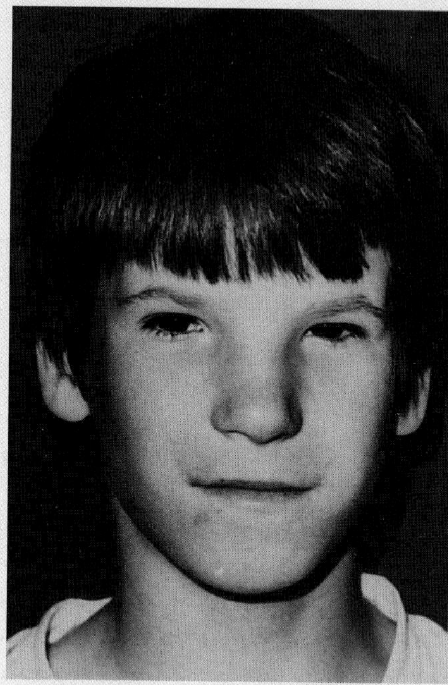

A child with fetal alcohol syndrome, the most common preventable cause of mental retardation in the United States. Typical features are unusually shaped eyes and lips (including a thin upper lip), short nose, receding chin, problems with vision and hearing, and possible heart defects. Behavioral traits include difficulties with learning, attention, problem solving, and memory, as well as problems with mental health and social interaction.

Some prescription drugs can cause catastrophic birth defects. Thalidomide, for example, was used in Europe from 1958 to 1962 as a sedative and to relieve morning sickness. When taken during the period of limb differentiation (days 24–36), it sometimes resulted in tragically deformed infants with short, flipperlike limbs. Similarly, a drug called diethylstilbestrol (DES) was commonly given to pregnant women from the 1940s to 1970, because it was wrongly thought to prevent miscarriages. Instead, it caused cancer of the vagina and uterus in daughters exposed to DES in the womb.

An expectant mother must be particularly careful of her health and chemical intake during the third to eighth weeks after conception. Weeks 3–8 are the "danger period": Insults delivered *before* week 3 tend to destroy critical early cell lines so that development cannot proceed, and a natural abortion occurs. After week 8, almost all organs are already in place, so teratogens may produce only minor malformations. However, the subsequent fetal period is not risk-free; exposure to teratogens can still impair body growth and organ functions, produce abnormalities of the eyes, and lead to mental retardation.

Surprisingly, most human embryos die in the womb because of lethal chromosome defects, severe malformation, or inability to implant in the uterus. Many embryos die in the first 2 weeks and are shed without notice in the next menstrual flow. This natural loss of embryos is a biological selection process that promotes the birth of healthy offspring.

RELATED CLINICAL TERMS

Abortion (*abort* = born prematurely) Premature removal of the embryo or fetus from the uterus. Abortions can be spontaneous (resulting from fetal death) or induced.

Amniocentesis (am″ne-o-sen-te′sis; "puncture of the amnion") A procedure for obtaining a sample of amniotic fluid. A needle is inserted through the abdominal and uterine walls, and a small amount of amniotic fluid is obtained. Shed cells from the fetus can be studied for chromosomal and genetic abnormalities, and chemicals in the fluid can reveal other disorders. Conditions such as Down syndrome and spina bifida can be diagnosed. The procedure can be performed as early as week 11 of pregnancy.

Teratology (ter″ah-tol′o-je) (*terato* = a monster; *logos* = study of) The study of birth defects and of fetuses with congenital deformities.

CHAPTER SUMMARY

Take full advantage of the invaluable online practice and study tools for this chapter by going to the Study Area of MasteringA&P®. There you will find:

- Art Labeling Activities, Quizzes, Practice Tests, Memory Games
- Animal Organ Dissection, Cat Dissection and Bone Videos
- PAL Cadaver
- PAL Histology
- A&P Flix Animations
- Dynamic Study Modules

1. The study of the 38-week prenatal period is called embryology. Knowledge of embryology helps one understand the basic organization of the human body, as well as the origin of birth defects. (Congenital abnormalities are considered in the box on p. 60).

STAGES OF PRENATAL DEVELOPMENT (pp. 46–47)

2. The embryonic period, during which the basic body plan is established, is the first 8 weeks of prenatal development. The fetal period, primarily a time of growth and maturation, is the remaining 30 weeks.

THE BASIC BODY PLAN (pp. 47–48)

3. Basic features of the adult body that are established in the embryonic period include the skin, trunk muscles, vertebral column, spinal cord and brain, digestive and respiratory tubes, serous cavities (peritoneal, pleural, and pericardial cavities), heart, kidneys, gonads, and limbs (Figure 3.2).

THE EMBRYONIC PERIOD (pp. 48–59)

Week 1: From Zygote to Blastocyst (pp. 48–50)

4. After fertilization occurs in the uterine tube, the zygote undergoes cleavage (early cell division without growth) to produce a solid ball of cells called the morula. Around day 4, the morula enters the uterus.

5. The morula gains an internal fluid-filled cavity and becomes a blastocyst, which consists of an inner cell mass and a trophoblast. The inner cell mass becomes the embryo, and the trophoblast forms part of the placenta. The blastocyst is implanted in the uterine wall from about day 6 to day 13.

Week 2: The Two-Layered Embryo (p. 50)

6. By day 7, the inner cell mass has formed two touching sheets of cells, the epiblast and hypoblast. These layers form the bilaminar embryonic disc.

7. The amniotic sac, whose wall is an extension of the epiblast, contains amniotic fluid that buffers the growing embryo from physical trauma. The yolk sac, whose wall is an extension of the hypoblast, later contributes to the digestive tube. The earliest blood cells and blood vessels form from tissue around the yolk sac.

Week 3: The Three-Layered Embryo (pp. 50–54)

8. At the start of week 3, gastrulation occurs. The primitive streak appears on the posterior part of the epiblast layer. Epiblast cells migrate through the primitive streak to form the endoderm and mesoderm. The embryonic disc now has three layers—ectoderm, mesoderm, and endoderm—all derived from epiblast.

9. Ectoderm and endoderm are epithelial tissues (sheets of closely attached cells). Mesoderm, by contrast, is a mesenchyme tissue (embryonic tissue with star-shaped, migrating cells that remain unattached to one another).

10. The primitive node, at the anterior end of the primitive streak, is the site where the notochord originates. The notochord defines the longitudinal body axis and later forms each nucleus pulposus of the vertebral column.

11. The notochord signals the overlying ectoderm to fold into a neural tube, the future spinal cord and brain. This infolding carries other ectoderm cells, neural crest cells, into the body.

12. By the end of week 3, the mesoderm on both sides of the notochord has condensed into three parts. These are, from medial to lateral, somites (segmented blocks derived from the paraxial mesoderm), intermediate mesoderm (segmented clusters attached to the somite), and the lateral plate (an unsegmented layer). The coelom divides the lateral plate into somatic and splanchnic mesoderm layers.

Week 4: The Body Takes Shape (pp. 54–56)

13. The flat embryo folds inward at its sides and at its head and tail, taking on a tadpole shape. As this occurs, a tubular part of the yolk sac becomes enclosed in the body as the primitive gut.

14. The basic body plan is attained as the body achieves this tadpole shape (compare Figure 3.9b and c). The major adult structures derived from ectoderm, endoderm, and mesoderm are summarized (Figure 3.10).

Weeks 5–8: The Second Month of Embryonic Development (pp. 56–58)

15. The limb buds appear around day 28.

16. All major organs are present by the end of month 2. During this month, the embryo becomes less tadpole-like and recognizably human as the eyes, ears, nose, and limbs form.

THE FETAL PERIOD (pp. 59–60)

17. Most cell and tissue differentiation is completed in the first half of the fetal period. The fetal period is a time of growth and maturation of the body (Table 3.1).

REVIEW QUESTIONS

Multiple Choice/Matching Questions

(For answers, see Appendix B).

1. Indicate whether each of the following relates to (a) cleavage or (b) events at the primitive streak.

____ **(1)** period when a morula forms

____ **(2)** period when the notochord forms

____ **(3)** period when the three embryonic germ layers appear

____ **(4)** period when individual cells become markedly smaller

2. The outer layer of the blastocyst, which attaches to the uterine wall, is the (a) yolk sac, (b) trophoblast, (c) amnion, (d) inner cell mass.

3. Most birth defects can be traced to disruption of the developmental events during this part of the prenatal period: (a) first 2 weeks, (b) second half of month 1 and all of month 2, (c) month 3, (d) end of month 4, (e) months 8 and 9.

4. The primary germ layer that forms the trunk muscles, heart, and skeleton is (a) ectoderm, (b) endoderm, (c) mesoderm.

5. Match each embryonic structure in column B with its adult derivative in column A.

Column A	Column B
____ **(1)** kidney	**(a)** notochord
____ **(2)** peritoneal cavity	**(b)** ectoderm (not neural tube)
____ **(3)** pancreas secretory cells	**(c)** intermediate mesoderm
____ **(4)** parietal serosa	**(d)** splanchnic mesoderm
____ **(5)** nucleus pulposus	**(e)** sclerotome
____ **(6)** visceral serosa	**(f)** coelom
____ **(7)** hair and epidermis	**(g)** neural tube
____ **(8)** brain	**(h)** somatic mesoderm
____ **(9)** ribs	**(i)** endoderm

6. Match each date in column B (approximate time after conception) with a developmental event or stage in column A.

Column A	Column B
____ **(1)** blastocyst	**(a)** 38 weeks
____ **(2)** implantation occurs	**(b)** 15 days
____ **(3)** first few somites appear	**(c)** 4–7 days
____ **(4)** flat embryo becomes cylindrical	**(d)** about 21 days
____ **(5)** embryo/fetus boundary	**(e)** about 21–24 days
____ **(6)** birth	**(f)** 2nd of week 8
____ **(7)** primitive streak appears	**(g)** days 6–13

7. It is currently possible to save some premature babies born as early as week (a) 8, (b) 16, (c) 22, (d) 38, (e) 2.

8. Somites are evidence of (a) a structure from ectoderm, (b) a structure from endoderm, (c) a structure from lateral plate mesoderm, (d) segmentation in the human body.

9. Which of the following embryonic structures are segmented? (More than one is correct.) (a) intermediate mesoderm, (b) lateral plate, (c) endoderm, (d) myotomes.

10. Endoderm gives rise to (a) the entire digestive tube, (b) the skin, (c) the neural tube, (d) the kidney, (e) none of these.

11. When it is 1.5 months old, an average embryo is the size of (a) a black bean, (b) an adult fist, (c) its mother's nose, (d) the head of a common pin.

12. Gastrulation is the (a) formation of three germ layers by the epiblast at the primitive streak, (b) formation of the placenta, (c) same as cleavage, (d) folding of the gut into a tube during week 4.

13. The epiblast forms (a) only the ectoderm, (b) the yolk sac, (c) only the mesoderm and ectoderm, (d) all three primary germ layers.

14. The limbs develop during the (a) start of the second month, (b) start of the second trimester, (c) start of the second week, (d) start of the fetal period.

15. The notochord develops primarily from (a) ectoderm, (b) neural crest, (c) mesoderm, (d) endoderm.

16. The fetal period is (a) from weeks 9–38, (b) the time of rapid growth of body organs, (c) a time of cell differentiation, (d) all of the above, (e) none of the above.

Short Answer Essay Questions

17. What important event occurs at the primitive streak? At the primitive node?

18. What is the function of the amniotic sac and the fluid it contains?

19. (a) What is mesenchyme? (b) How does it differ from epithelium?

20. Explain how the flat embryonic disc takes on the cylindrical shape of a tadpole.

21. In anatomy lab, Thaya pointed to the vertebrae of a cadaver and said "sclerotome." He pointed to a kidney and said, "intermediate mesoderm," to the biceps muscle in the arm and said, "splanchnic mesoderm," to the inner lining of the stomach and said, "endoderm," and to the brain and said, "ectoderm." Point out Thaya's *one* mistake.

22. Neural crest is the one embryonic tissue that is unique to vertebrate animals. List some adult structures that develop from the neural crest.

23. Many embryonic events first occur cranially, then proceed caudally. Give three examples of structures that develop craniocaudally.

24. Define *induction,* and give one example of an inductive interaction.
25. Differentiate the outer tube from the inner tube. Identify three structures that are located in the outer tube and three that are located in the inner tube.
26. Before Delta studied embryology in her anatomy course, she imagined a developing human as a shapeless mass of indistinct tissues until about halfway through pregnancy. Was Delta correct? At what stage does the embryo or fetus really start to look like a developing human?

Critical Reasoning & Clinical Application Questions

1. A friend in your dormitory, a freshman, tells you that she just discovered she is 3 months pregnant. You know that since she came to college she has been experimenting with recreational drugs. Circle the best advice you could give her, and explain your choice. (a) She must stop taking the drugs, but they could not have affected her fetus during these first few months of her pregnancy. (b) Harmful substances usually cannot pass from mother to embryo, so she can keep using drugs. (c) There could be defects in the fetus, so she should stop using drugs and visit a doctor as soon as possible. (d) If she has not taken any drugs in the last week, she is OK.
2. Insufficient folic acid in the early weeks of pregnancy can result in neural tube deficits in the developing embryo. Why is it important that a woman take folic acid supplements prior to becoming pregnant?
3. Your cat finds a large ball of string and starts eating it unbeknownst to you. When you discover him, he has string extending from both his mouth and his anus. What primary germ layer(s) is the string in contact with as it passes through the cat?
4. When is the earliest time the sex of a fetus can be determined using ultrasound examination?

4 Tissues

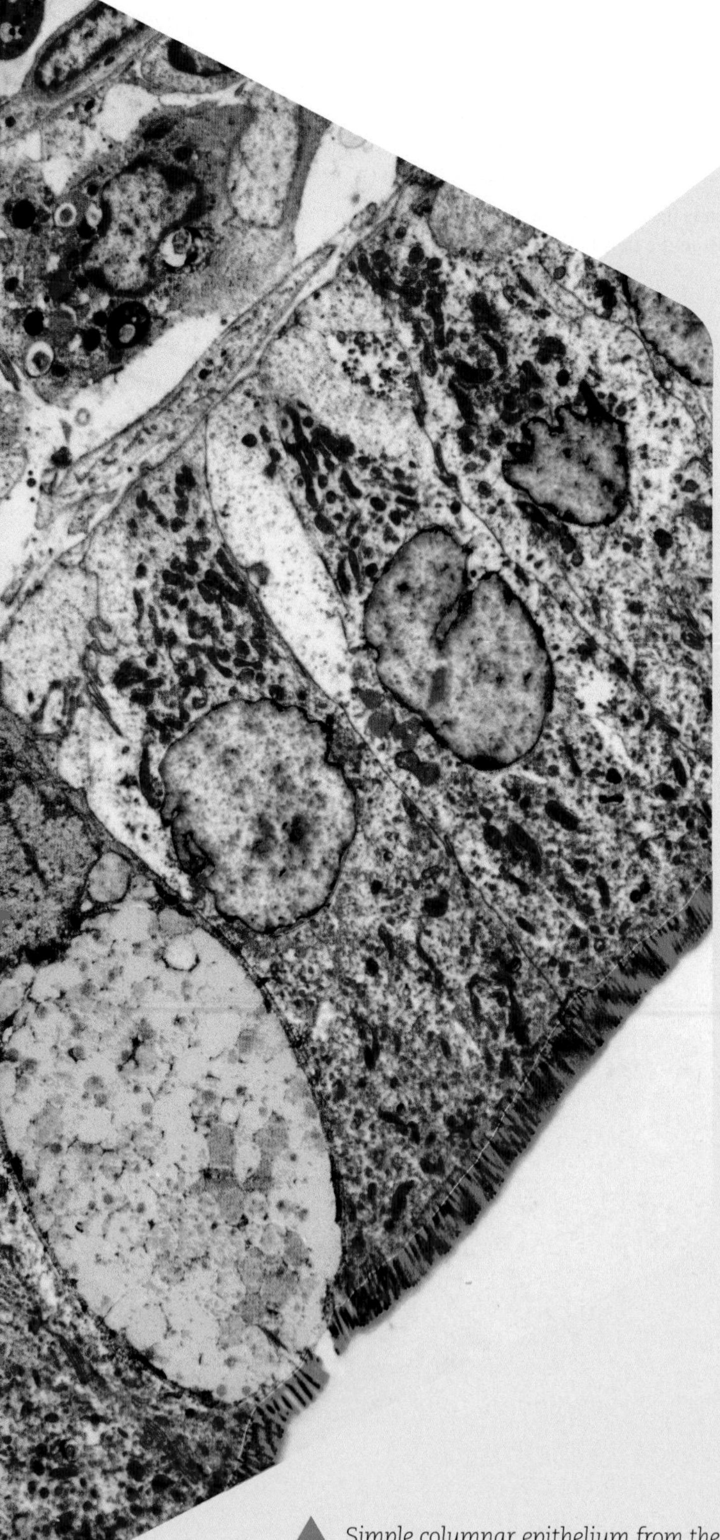

Simple columnar epithelium from the small intestine (colored TEM).

The cells of the human body do not operate independently of one another. Instead, related cells live and work together in cell communities called **tissues.** A tissue is defined as *a group of cells of similar structure that perform a common function*. However, tissues do not consist entirely of cells: Between the cells is a nonliving material, the *extracellular matrix*.

The word *tissue* derives from the Old French word meaning "to weave," reflecting the fact that the different tissues are woven together to form the "fabric" of the human body. The four basic types of tissue are **epithelial tissue, connective tissue, muscle tissue,** and **nervous tissue.** If a single, broad functional term were assigned to each basic tissue, the terms would be *covering* (epithelial tissue), *support* (connective), *movement* (muscle), and *control* (nervous). However, these terms reflect only a fraction of the functions that each tissue performs.

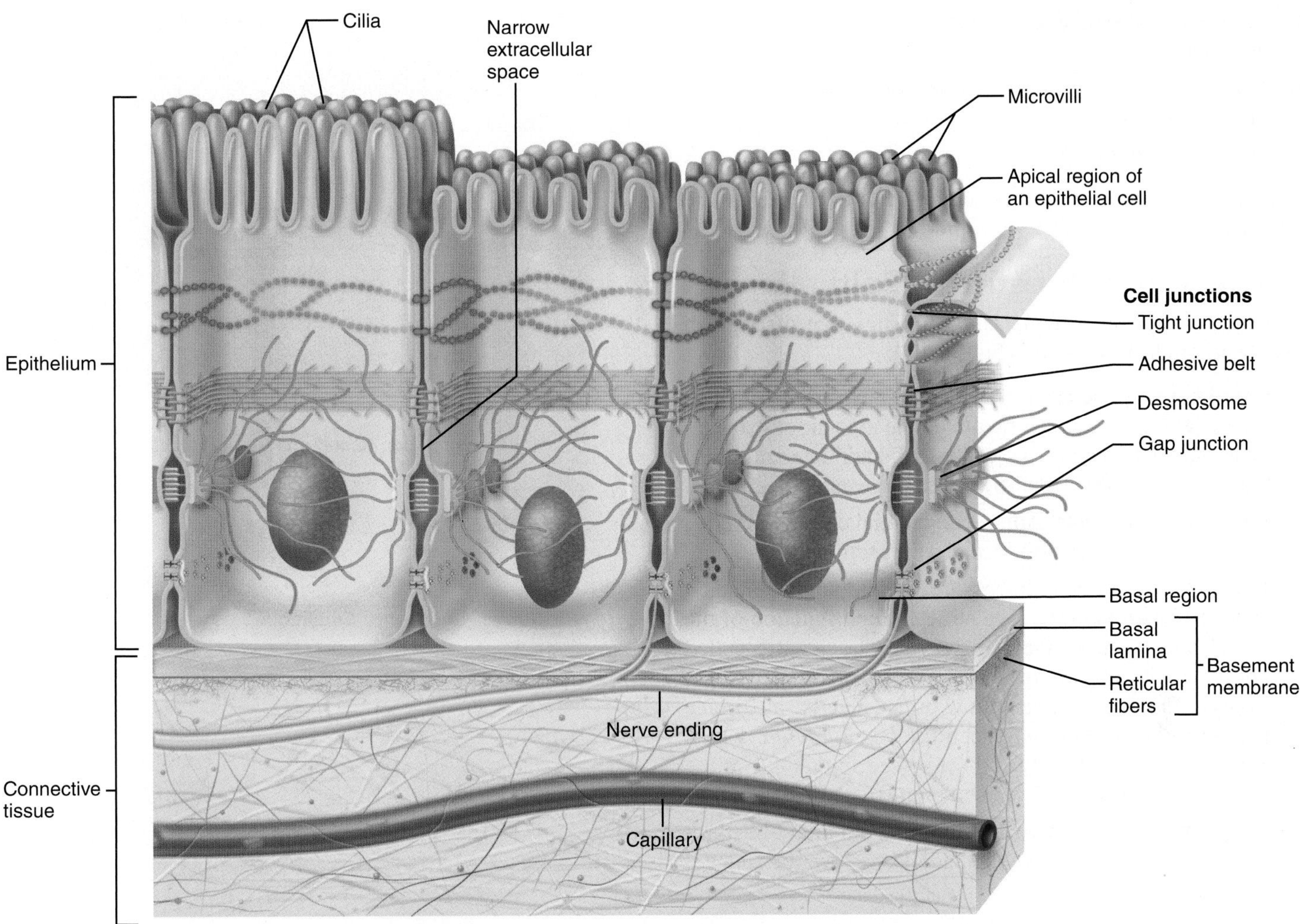

Figure 4.1 Special characteristics of epithelium. A sheet of closely joined epithelial cells rests on connective tissue proper. Epithelia contain nerve endings but no blood vessels. Note the special features on the epithelial cell surfaces: cilia, microvilli, cell junctions, and basal lamina.

Tissues are the building blocks of the body's organs. Because most organs contain all four tissue types, learning about the structure and functions of tissues will provide a strong foundation for your understanding of the structure and functions of the organs discussed in the remaining chapters of this book.

I. EPITHELIAL TISSUE

learning outcomes

- ▶ List the functional and structural characteristics of epithelial tissue.
- ▶ Identify the different epithelia of the body, and describe the chief function(s) and location(s) of each.

An **epithelium** (ep″ĭ-the′le-um; "covering") is a sheet of cells that covers a body surface or lines a body cavity (Figure 4.1).

Epithelial tissue occurs in two different forms:

- *Covering and lining epithelium* covers the outer and inner surfaces of most body organs. Examples include the outer layer of the skin; the inner lining of all hollow viscera, such as the stomach and respiratory tubes; the lining of the peritoneal cavity; and the lining of all blood vessels.
- *Glandular epithelium* forms most of the body glands.

Epithelia occur at the boundary between two different environments. The epidermis of the skin, for example, lies between the inside and the outside of the body. Most substances that enter into the body or are released from the body must pass through an epithelium. Therefore all functions of epithelia reflect their roles as interface tissues. These functions include:

1. *Protection* of the underlying tissues
2. *Secretion* (release of molecules from cells)
3. *Absorption* (bringing small molecules into cells)

4. *Diffusion* (movement of molecules down their concentration gradient)
5. *Filtration* (passage of small molecules through a sieve-like membrane)
6. *Sensory reception*

These functions will be discussed further as we describe the specific types of epithelia.

Special Characteristics of Epithelia

Epithelial tissues have many characteristics that distinguish them from other tissue types (Figure 4.1):

1. **Cellularity.** Epithelia are composed almost entirely of cells. These cells are separated by a minimal amount of extracellular material, mainly projections of their integral membrane proteins into the narrow spaces between the cells.
2. **Specialized contacts.** Adjacent epithelial cells are directly joined at many points by special cell junctions.
3. **Polarity.** All epithelia have a free **apical surface** and an attached **basal surface**. The structure and function of the apical and basal surfaces differ, a characteristic called *polarity*. The apical surface abuts the open space of a cavity, tubule, gland, or hollow organ. The basal surface lies on a thin supporting sheet, the *basal lamina,* which is part of the *basement membrane* (see Figure 4.1; these structures are further explained on p. 76).
4. **Support by connective tissue.** All epithelial sheets in the body are supported by an underlying layer of connective tissue.
5. **Avascular but innervated.** Whereas most tissues in the body are *vascular* (contain blood vessels), epithelium is *avascular* (a-vas′ku-lar), meaning it lacks blood vessels. Epithelial cells receive their nutrients from capillaries in the underlying connective tissue. Although blood vessels do not penetrate epithelial sheets, nerve endings do; that is, epithelium is *innervated.*
6. **Regeneration.** Epithelial tissue has a high regenerative capacity. Some epithelia are exposed to friction, and their surface cells rub off. Others are destroyed by hostile substances in the external environment such as bacteria, acids, and smoke. As long as epithelial cells receive adequate nutrition, they can replace lost cells quickly by mitosis, cell division.

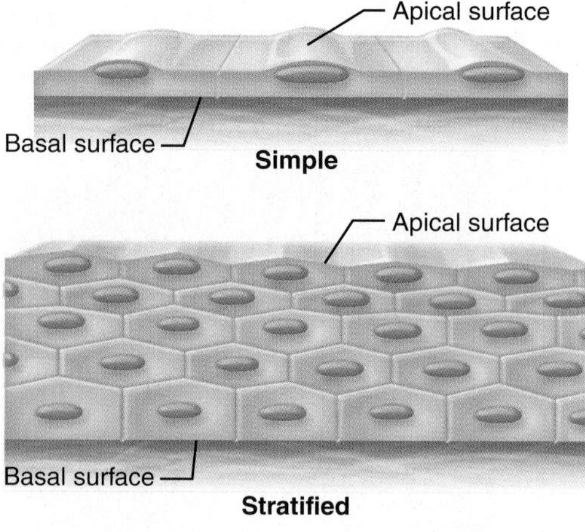

(a) Classification based on number of cell layers

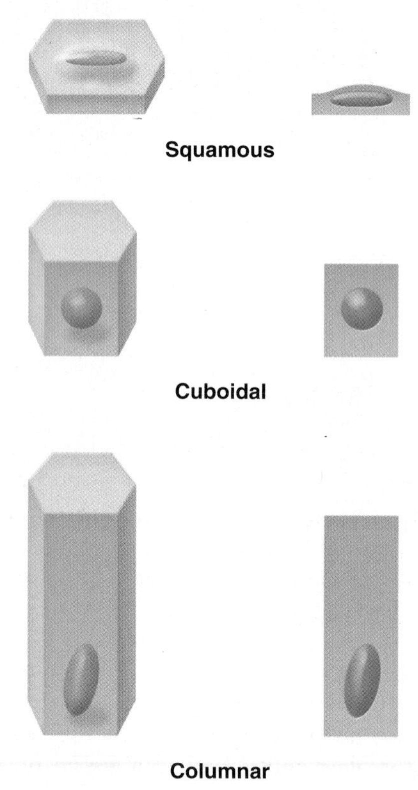

(b) Classification based on cell shape

Figure 4.2 Classification of epithelia.

Classification of Epithelia

Many kinds of epithelia exist in the body. Two features are used to classify and name epithelia: the number of cell layers and the shape of the cells (Figure 4.2). The terms *simple* and *stratified* describe the number of cell layers in an epithelium (Figure 4.2a).

- **Simple epithelia** contain a single layer of cells, with each cell attached to the basement membrane.
- **Stratified epithelia** contain more than one layer of cells. The cells on the basal surface are attached to the basement membrane; those on the apical surface border an open space.

Cell shape is described as *squamous, cuboidal,* or *columnar,* referring to the appearance of the cells in section (Figure 4.2b). In each case, the shape of the nucleus conforms to the shape of the cell. This is an important feature to observe when distinguishing epithelial types.

- **Squamous cells** (sqwa′mus; "scale") are flat cells with flat, disc-shaped nuclei.
- **Cuboidal cells** are cube-shaped cells with spherical, centrally located nuclei.

Table 4.1 Function of Epithelial Tissue Related to Tissue Type

	Number of Layers	
Cell Shape	**One Layer: Simple Epithelial Tissues**	**More Than One Layer: Stratified Epithelial Tissues**
Squamous	Diffusion and filtration	Protection
Cuboidal **Columnar**	Secretion and absorption; ciliated types propel mucus or reproductive cells	Protection; these tissue types are rare in humans (see pp. 69–70)
Transitional		Protection; stretching to accommodate distension of urinary structures

- **Columnar cells** are taller than they are wide, like columns. The nuclei of columnar cells are located near the basal surface and are commonly oval in shape, elongated from top to bottom.

Simple epithelia are easy to classify by cell shape because all cells in the layer usually have the same shape. In stratified epithelia, however, the cell shapes usually differ among the different cell layers. To avoid ambiguity, stratified epithelia are named according to the shape of the cells in the *apical* layer.

It is useful to keep in mind how tissue structure reflects tissue function (Table 4.1). Stratified epithelial tissues function to protect. Multiple layers of cells protect underlying connective tissues in areas where abrasion is common. In simple epithelia, the shape of the cells is indicative of tissue function. Squamous cells are found where diffusion or filtration are important, because these are distance-dependent processes; the thinner the layer, the more quickly the process occurs. Columnar and cuboidal cells are found in tissues involved in secretion and absorption. Larger cells are necessary for the additional cellular machinery needed to produce and package secretions and to produce the necessary energy for these processes. Ciliated epithelia function to propel material, for example, mucus. Keep these generalizations in mind as you study each type of epithelial tissue in detail.

As you view micrographs of different types of epithelium (Figure 4.3), try to pick out the individual cells within each. This is not always easy, because the boundaries between epithelial cells often are indistinct. Furthermore, the nucleus of a particular cell may or may not be visible, depending on the plane of the cut made to prepare the tissue slides. When a pear is sliced in transverse sections, slices from the top of the pear will not contain any seeds, but slices from the middle region will. The same is true for sections through tissues: Some cells may be sliced through the nucleus, whereas others may not.

Simple Epithelia

Simple Squamous Epithelium (Figure 4.3a) A simple squamous epithelium is a single layer of flat cells. When viewed from above, the closely fitting cells resemble a tiled floor. When viewed in lateral section, they resemble fried eggs seen from the side. Thin and often permeable, this type of epithelium occurs wherever small molecules pass through a membrane quickly, by processes of diffusion or filtration. The walls of capillaries consist exclusively of this epithelium, whose exceptional thinness encourages efficient exchange of nutrients and wastes between the bloodstream and surrounding tissue cells. In the lungs, this epithelium forms the thin walls of the air sacs, where gas exchange occurs.

Simple Cuboidal Epithelium (Figure 4.3b) Simple cuboidal epithelium consists of a single layer of cube-shaped cells. This epithelium forms the secretory cells of many glands, the walls of the smallest ducts of glands, and the walls of many tubules in the kidney. Its functions are the same as those of simple columnar epithelium.

Simple Columnar Epithelium (Figure 4.3c) Simple columnar epithelium is a single layer of tall cells aligned like soldiers in a row. It lines the digestive tube from the stomach to the rectum. It functions in the active movement of molecules, namely in absorption and secretion. The structure of simple columnar epithelium is ideal for these functions: It is thin enough to allow large numbers of molecules to pass through it quickly, yet thick enough to house the cellular machinery needed to perform the complex processes of molecular transport.

Some simple columnar epithelia bear *cilia* (sil′e-ah; "eyelashes"), whiplike bristles on the apex of epithelial cells that beat rhythmically to move substances across certain body surfaces (see Figure 4.1). A simple ciliated columnar epithelium lines the inside of the uterine tube. Its cilia help move the ovum to the uterus. (This journey is traced in Chapter 3.) Cilia are considered in detail later in this chapter.

Pseudostratified Columnar Epithelium (Figure 4.3d) The cells of pseudostratified (soo-do-strat′ĭ-fīd) columnar epithelium are varied in height. All of its cells rest on the basement membrane, but only the tall cells reach the apical surface of the epithelium. The short cells are undifferentiated and continuously give rise to the tall cells. The cell nuclei lie at several different levels, giving the false impression that this epithelium is stratified (*pseudo* = false).

(Text continues on page 72.)

(a) Simple squamous epithelium

Description: Single layer of flattened cells with disc-shaped central nuclei and sparse cytoplasm; the simplest of the epithelia.

Function: Allows passage of materials by diffusion and filtration in sites where protection is not important; produces lubricating fluid in serosae.

Location: Kidney glomeruli; air sacs of lungs; lining of heart, blood vessels, and lymphatic vessels; lining of ventral body cavity (serosae).

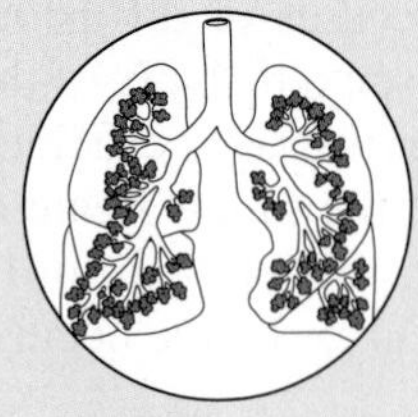

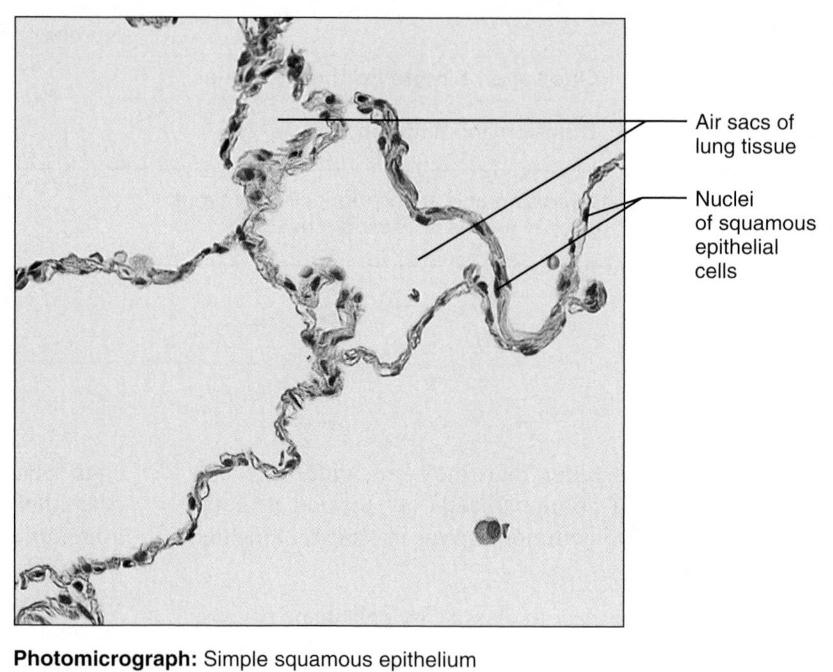

Photomicrograph: Simple squamous epithelium forming part of the alveolar (air sac) walls (140×).

(b) Simple cuboidal epithelium

Description: Single layer of cubelike cells with large, spherical central nuclei.

Function: Secretion and absorption.

Location: Kidney tubules; ducts and secretory portions of small glands; ovary surface.

Simple cuboidal epithelial cells

Basement membrane

Connective tissue

Photomicrograph: Simple cuboidal epithelium in kidney tubules (430×).

Figure 4.3 Epithelial tissues.

View PAL Histology MasteringA&P®

(c) Simple columnar epithelium

Description: Single layer of tall cells with *round* to *oval* nuclei; some cells bear cilia; layer may contain mucus-secreting unicellular glands (goblet cells).

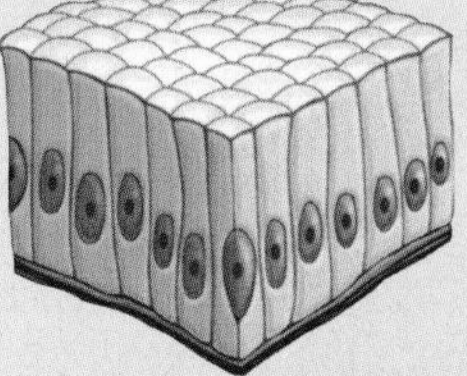

Function: Absorption; secretion of mucus, enzymes, and other substances; ciliated type propels mucus (or reproductive cells) by ciliary action.

Location: Nonciliated type lines most of the digestive tract (stomach to rectum), gallbladder, and excretory ducts of some glands; ciliated variety lines small bronchi, uterine tubes, and some regions of the uterus.

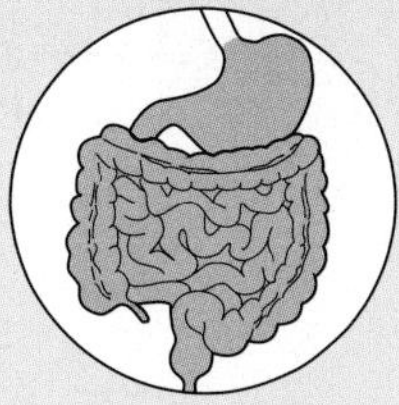

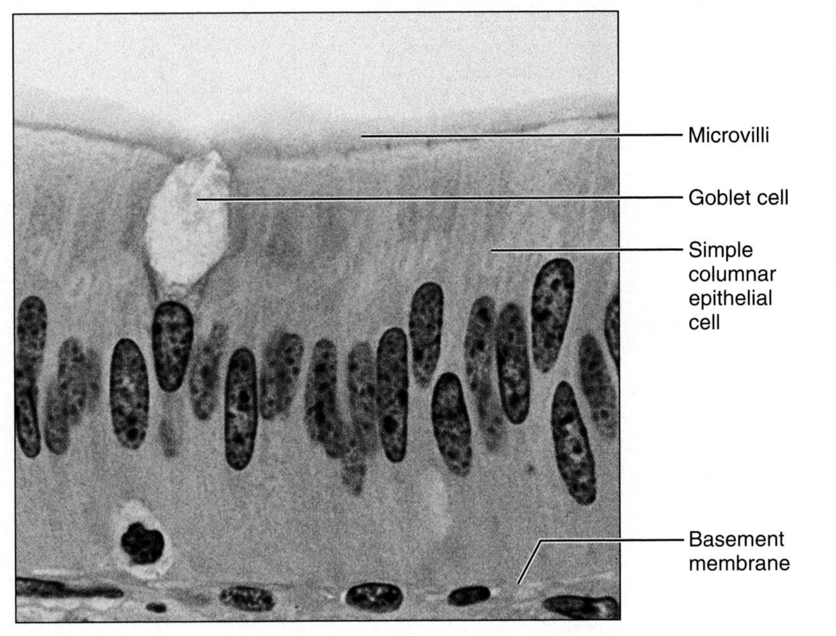

Photomicrograph: Simple columnar epithelium of the small intestine (650×).

(d) Pseudostratified columnar epithelium

Description: Single layer of cells of differing heights, some not reaching the free surface; nuclei seen at different levels; may contain mucus-secreting goblet cells and bear cilia.

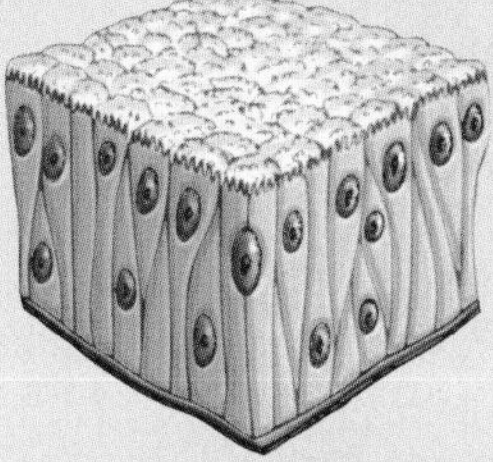

Function: Secretion, particularly of mucus; propulsion of mucus by ciliary action.

Location: Nonciliated type in male's sperm-carrying ducts and ducts of large glands; ciliated variety lines the trachea, most of the upper respiratory tract.

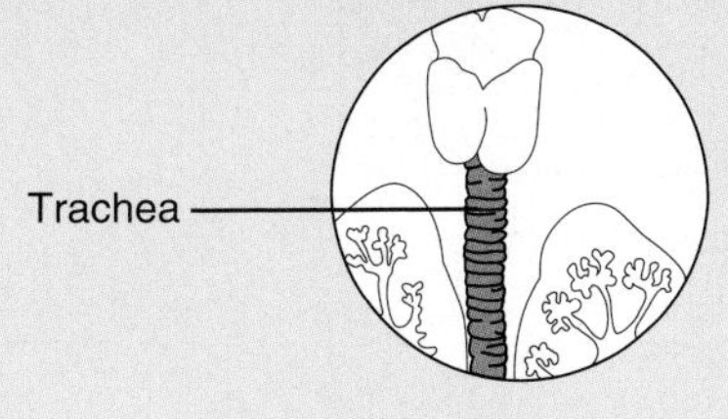

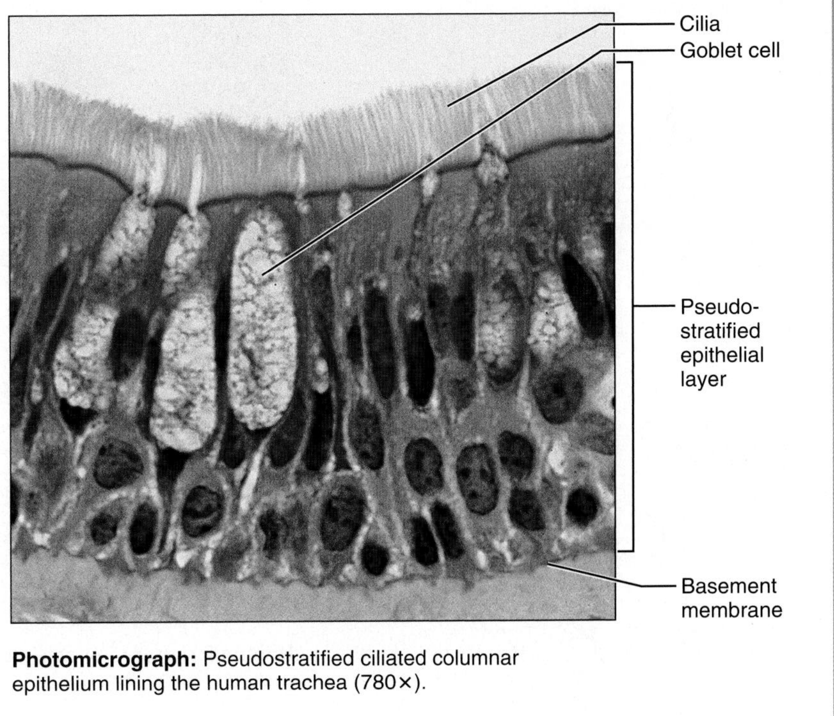

Photomicrograph: Pseudostratified ciliated columnar epithelium lining the human trachea (780×).

Figure 4.3 Epithelial tissues, *continued*

(e) Stratified squamous epithelium

Description: Thick membrane composed of several cell layers; basal cells are cuboidal or columnar and metabolically active; surface cells are flattened (squamous); in the keratinized type, the surface cells are full of keratin and dead; basal cells are active in mitosis and produce the cells of the more superficial layers.

Function: Protects underlying tissues in areas subjected to abrasion.

Location: Nonkeratinized type forms the moist linings of the esophagus, mouth, and vagina; keratinized variety forms the epidermis of the skin, a dry membrane.

Stratified squamous epithelium

Nuclei

Basement membrane

Connective tissue

Photomicrograph: Stratified squamous epithelium lining the esophagus (280×).

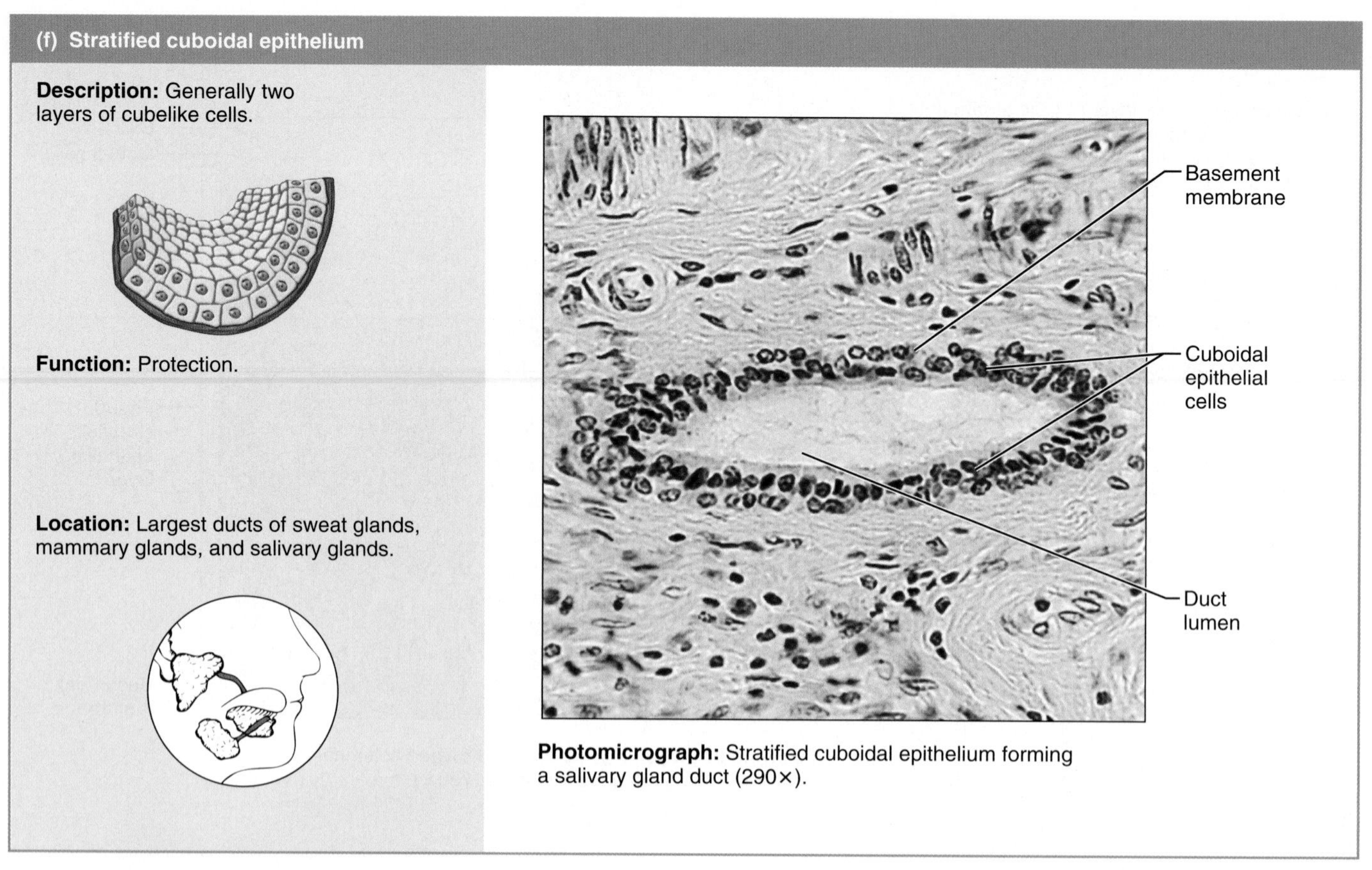

(f) Stratified cuboidal epithelium

Description: Generally two layers of cubelike cells.

Function: Protection.

Location: Largest ducts of sweat glands, mammary glands, and salivary glands.

Photomicrograph: Stratified cuboidal epithelium forming a salivary gland duct (290×).

Figure 4.3 Epithelial tissues, *continued*

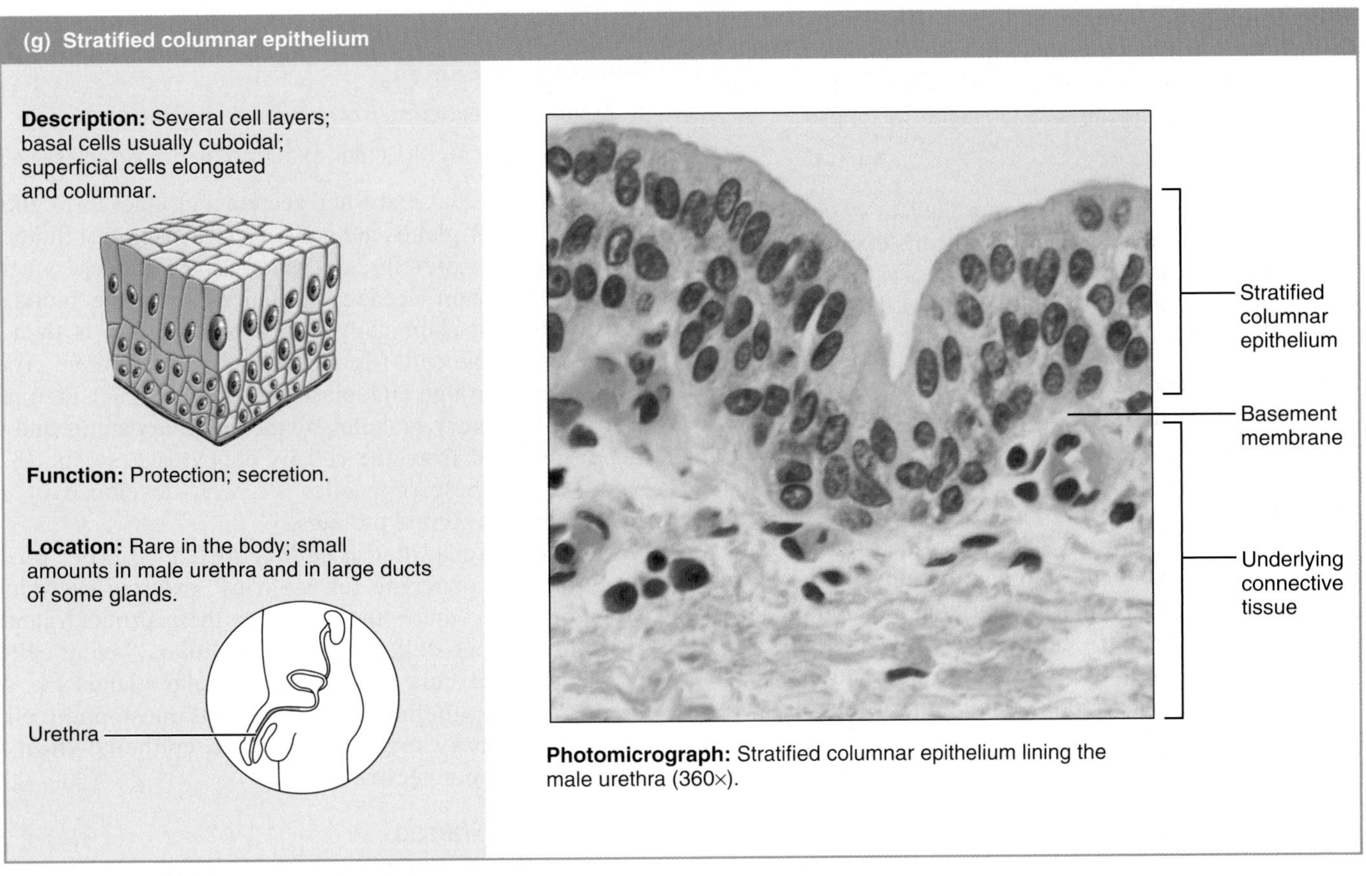

Photomicrograph: Stratified columnar epithelium lining the male urethra (360×).

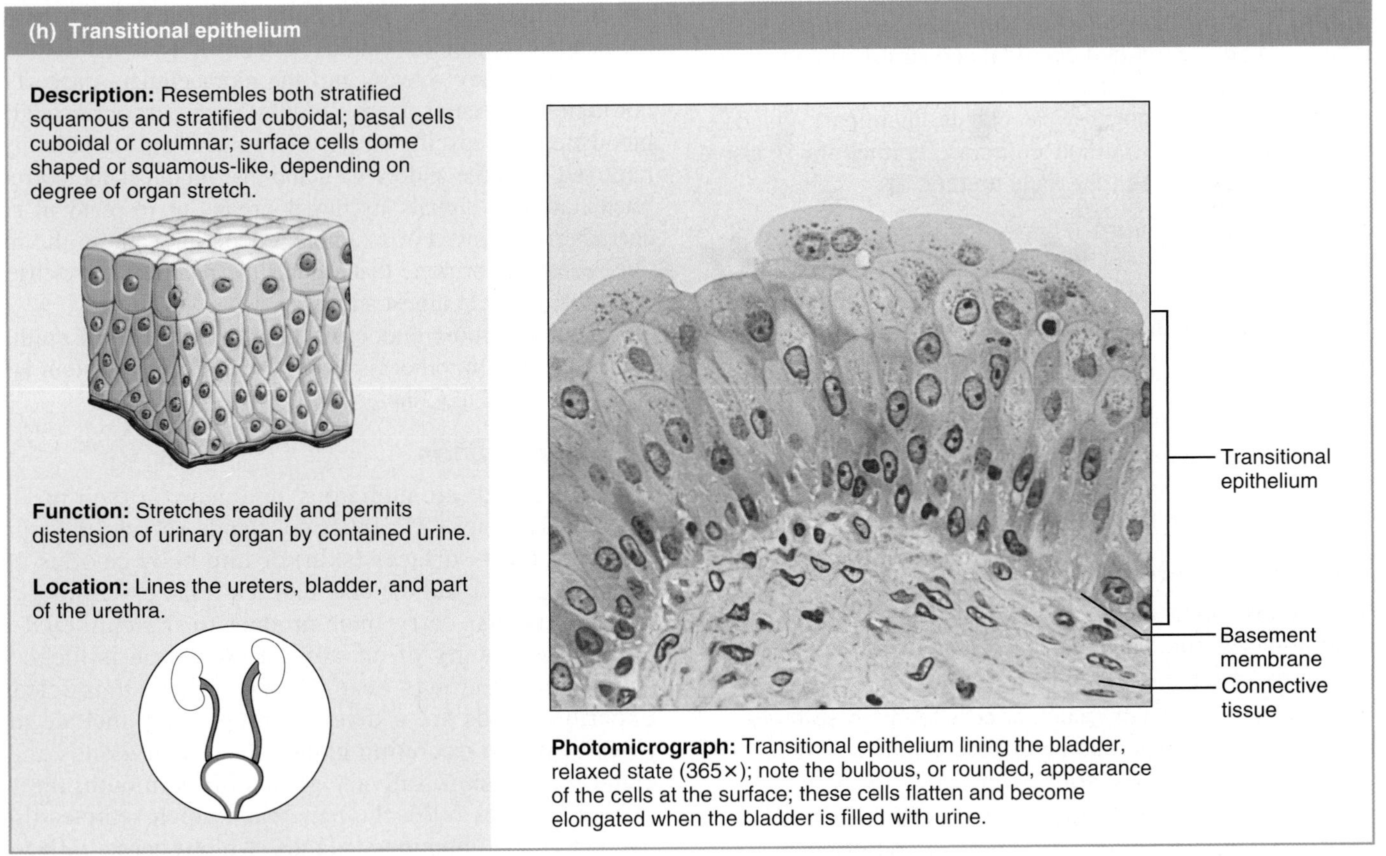

Photomicrograph: Transitional epithelium lining the bladder, relaxed state (365×); note the bulbous, or rounded, appearance of the cells at the surface; these cells flatten and become elongated when the bladder is filled with urine.

Figure 4.3 Epithelial tissues, *continued*

Pseudostratified columnar epithelium, like simple columnar epithelium, functions in secretion or absorption. A ciliated type lines the interior of the respiratory tubes. Here, the cilia propel sheets of dust-trapping mucus out of the lungs.

Stratified Epithelia

Stratified epithelia contain two or more layers of cells. They regenerate from below; that is, the basal cells divide and push apically to replace the older surface cells. Stratified epithelia are more durable than simple epithelia, and their major (but not only) role is protection.

Stratified Squamous Epithelium (Figure 4.3e) As you might expect, stratified squamous epithelium consists of many cell layers whose surface cells are squamous. In the deeper layers, the cells are cuboidal or columnar. Of all the epithelial types, this is the thickest and best adapted for protection. It covers the often-abraded surfaces of our body, forming the epidermis of the skin and the inner lining of the mouth, esophagus, and vagina. To learn the location of stratified squamous epithelium, simply remember that this epithelium forms the outermost layer of the skin and extends a certain distance into every body opening that is directly continuous with the skin.

The epidermis of the skin is *keratinized,* meaning that its surface cells contain an especially tough protective protein called *keratin.* The other stratified squamous epithelia of the body lack keratin and are *nonkeratinized.* (Keratin is explained in detail in Chapter 5, p. 105).

Stratified Cuboidal and Columnar Epithelia (Figure 4.3f and Figure 4.3g) Stratified cuboidal and stratified columnar epithelia are rare types of tissue, located in the large ducts of some glands, for example, sweat glands, mammary glands, and salivary glands. Stratified columnar epithelium is also found in small amounts in the male urethra.

Transitional Epithelium (Figure 4.3h) Transitional epithelium lines the inside of the hollow urinary organs. Such organs (the urinary bladder, for example) stretch as they fill with urine. As the transitional epithelium stretches, it thins from about six cell layers to three, and its apical cells unfold and flatten. When relaxed, portions of the apical surface invaginate into the cell, giving this surface a scalloped appearance. Thus, this epithelium undergoes "transitions" in shape. It also forms an impermeable barrier that keeps urine from passing through the wall of the bladder.

☑ check your understanding

☐ 1. Describe the location of the apical region of an epithelial tissue. How can this region be easily identified in a micrograph?

☐ 2. Why are cuboidal or columnar cells found in epithelia that function in secretion and absorption? List three locations of cuboidal epithelium.

☐ 3. Why is columnar epithelium *not* found in locations where diffusion occurs? What type of epithelium is found in these locations?

(For answers, see Appendix B.)

Glands

learning outcomes

▶ Distinguish exocrine from endocrine glands.

▶ Explain how multicellular exocrine glands are classified.

Epithelial cells that make and secrete a product form **glands.** The products of glands are aqueous (water-based) fluids that usually contain proteins. *Secretion* is the process whereby gland cells obtain needed substances from the blood and transform them chemically into a product that is then discharged from the cell. More specifically, the protein product is made in the rough endoplasmic reticulum (ER), then packaged into secretory granules by the Golgi apparatus and ultimately released from the cell by exocytosis (see p. 28 and Figure 2.8). These organelles are well developed in most gland cells that secrete proteins.

Glands are classified as **endocrine** (en′do-krin; "internal secretion") or **exocrine** (ek′so-krin; "external secretion"), depending on where they release their product, and as **unicellular** ("one-celled") or **multicellular** ("many-celled") on the basis of cell number. Unicellular glands are scattered within epithelial sheets, whereas most multicellular glands develop by invagination of an epithelial sheet into the underlying connective tissue.

Endocrine Glands

Endocrine glands lack ducts, so they are often referred to as *ductless glands.* They secrete directly into the tissue fluid that surrounds them. More specifically, endocrine glands produce messenger molecules called **hormones** (hor′mōnz; "exciters"), which they release into the extracellular space. These hormones then enter nearby capillaries and travel through the bloodstream to specific *target organs,* which are commonly far removed from the endocrine gland that produces the hormone. Each hormone signals its target organs to respond in some characteristic way. For example, endocrine cells in the intestine secrete a hormone that signals the pancreas to release the enzymes that help digest a meal.

Although most endocrine glands derive from epithelia, some derive from other tissues. (The endocrine system is discussed in detail in Chapter 17.)

Exocrine Glands

Exocrine glands are numerous, and many of their products are familiar ones. All exocrine glands secrete their products onto body surfaces (skin) or into body cavities (such as the digestive tube), and multicellular exocrine glands have **ducts** that carry their product to the epithelial surfaces. The activity of an exocrine secretion is local; that is, the secretion acts near the area where it is released. Exocrine glands are a diverse group: They include many types of mucus-secreting glands, the sweat glands and oil glands of the skin, salivary glands of the mouth, the liver (which secretes bile), the pancreas (which secretes digestive enzymes), mammary glands (which secrete milk), and many others.

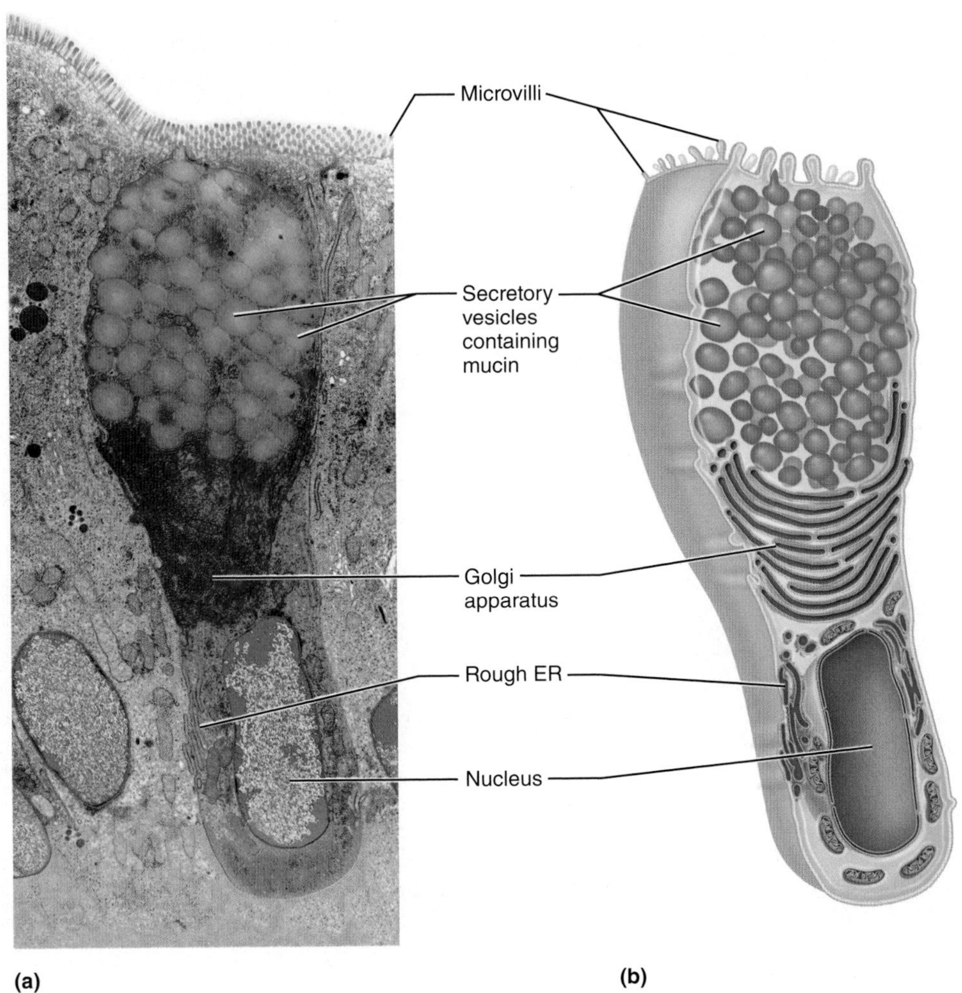

Figure 4.4 Goblet cell (unicellular exocrine gland). (**a**) Photomicrograph of a mucus-secreting goblet cell in the simple columnar epithelium that lines the small intestine (1650×). (**b**) Diagram of a goblet cell. Note the secretory vesicles and the well-developed rough ER and Golgi apparatus.

Unicellular Exocrine Glands The only important example of a one-celled exocrine gland is the **goblet cell** (Figure 4.4). True to its name, a goblet cell is indeed shaped like a goblet, a drinking glass with a stem. Goblet cells are scattered within the epithelial lining of the intestines and respiratory tubes, between columnar cells with other functions. They produce **mucin** (mu′sin), a glycoprotein (sugar protein) that dissolves in water when secreted. The resulting complex of mucin and water is viscous, slimy **mucus.** Mucus covers, protects, and lubricates many internal body surfaces.

Multicellular Exocrine Glands Each multicellular exocrine gland has two basic parts: an epithelium-walled *duct* and a *secretory unit* consisting of the secretory epithelium (Figure 4.5). Also, in all but the simplest glands, a supportive connective tissue surrounds the secretory unit, carrying with it blood vessels and nerve fibers. Often, the connective tissue forms a *fibrous capsule* that extends into the gland proper and partitions the gland into subdivisions called *lobes* (not illustrated).

Multicellular glands are classified by the structure of their ducts (Figure 4.5). **Simple** glands have an unbranched duct, whereas **compound** glands have a branched duct. The glands are further categorized by their secretory units: They are **tubular** if their secretory cells form tubes and **alveolar** (al-ve′o-lar) if the secretory cells form spherical sacs (*alveolus* = a small, hollow cavity). Furthermore, some glands are **tubuloalveolar;** that is, they contain both tubular secretory units and alveolar units. The term *acinar* (as′ĭ-nar; *acinus* = grape or berry) is commonly used as a synonym for *alveolar*.

☑ check your understanding

- ☐ 4. What type of gland are goblet cells? What do they secrete?
- ☐ 5. An ________ gland secretes into the tissue fluid, and its secretion is carried to a target organ by the bloodstream.
- ☐ 6. What feature distinguishes a simple exocrine gland from a compound exocrine gland?

(For answers, see Appendix B.)

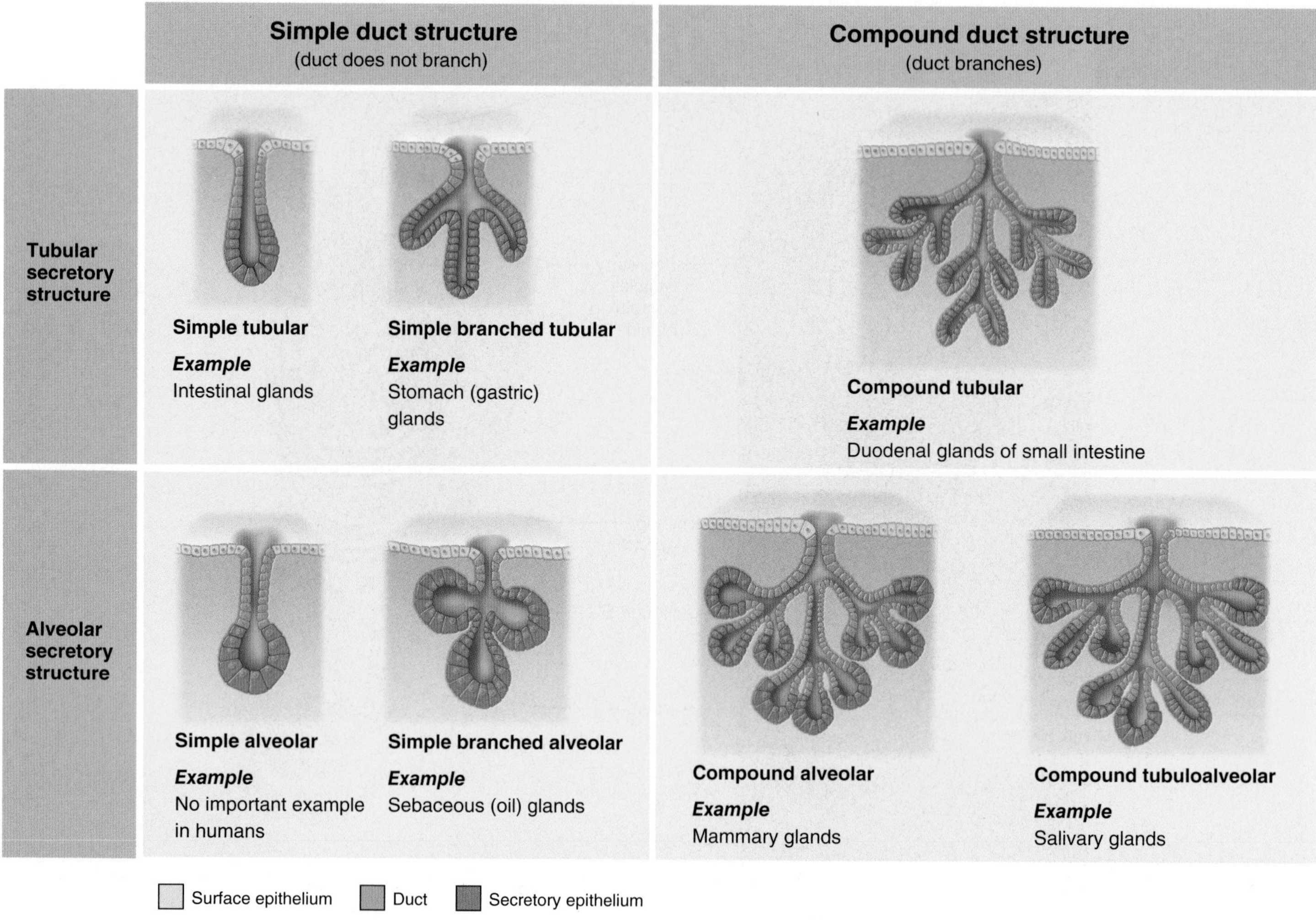

Figure 4.5 **Types of multicellular exocrine glands.** Multicellular glands are classified according to the structure of their ducts (simple or compound) and the structure of their secretory units (tubular, alveolar, tubuloalveolar).

Epithelial Surface Features

learning outcome

▶ Describe apical, lateral, and basal surface features of epithelia and epithelial cells.

As previously described, epithelial tissues are composed of many cells closely joined together by special cell junctions along their lateral walls. Epithelial tissues also have distinct apical and basal regions. The basal surface sits on a specialized boundary with the underlying connective tissue; the apical region of certain epithelia has modifications associated with specific functions. These special features are described next.

Lateral Surface Features: Cell Junctions

Three factors act to bind epithelial cells to one another: (1) *adhesion proteins* in the plasma membranes of the adjacent cells link together in the narrow extracellular space; (2) the wavy contours of the membranes of adjacent cells join in a tongue-and-groove fashion; and (3) there are special cell junctions (Figure 4.6). **Cell junctions,** the most important of the factors, are characteristic of epithelial tissue but are found in other tissue types as well.

Tight Junctions In the apical region of most epithelial tissues, a beltlike junction extends around the periphery of each cell (Figure 4.6a). This is a **tight junction,** or a *zonula occludens* (zōn′u-lah o-klood′enz; "small zone that shuts off"). At tight junctions, the adjacent cells are so close that some proteins in their plasma membranes are fused. This fusion forms a seal that closes off the extracellular space; thus tight junctions prevent molecules from passing between the cells of epithelial tissue. For example, the tight junctions in the epithelium lining the digestive tract keep digestive enzymes, ions, and microorganisms in the intestine from seeping into the bloodstream. Tight junctions need not be entirely impermeable; some are more leaky than others and may let certain types of ions through.

Adhesive Belt Junctions Just below the tight junctions in epithelial tissues are **adhesive belt junctions,** or *zonula*

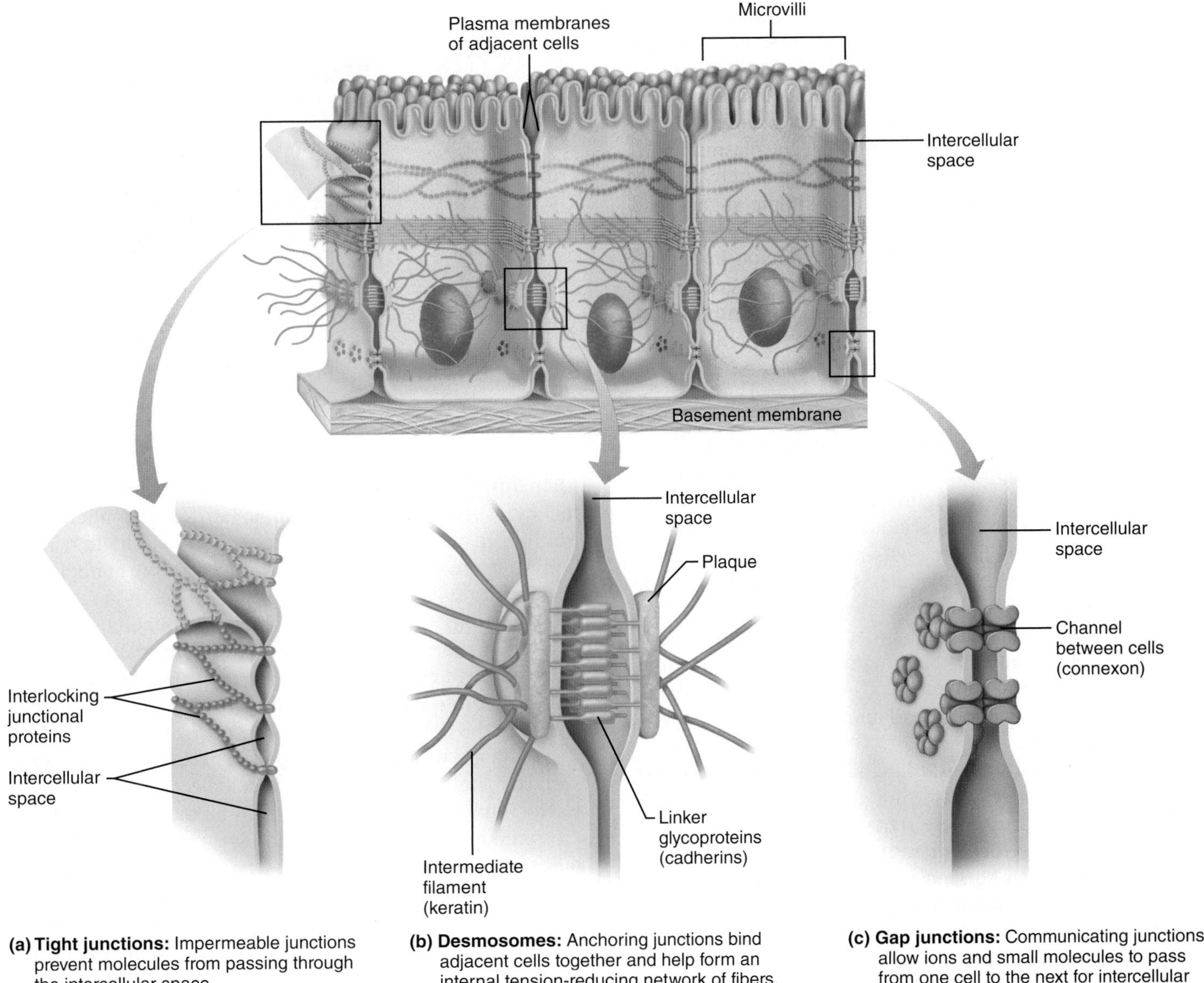

(a) Tight junctions: Impermeable junctions prevent molecules from passing through the intercellular space.

(b) Desmosomes: Anchoring junctions bind adjacent cells together and help form an internal tension-reducing network of fibers.

(c) Gap junctions: Communicating junctions allow ions and small molecules to pass from one cell to the next for intercellular communication.

Figure 4.6 **Cell junctions.** An epithelial cell shown joined to adjacent cells by three common types of cell junction.

adherens (zōn′u-lah ad-hir-ens), a type of anchoring junction (Figure 4.1). Transmembrane linker proteins attach to the actin microfilaments of the cytoskeleton and bind adjacent cells. This junction reinforces the tight junctions, particularly when the tissues are stretched. Together with tight junctions, these form the tight junctional complex around the apical lateral borders of epithelial tissues.

Desmosomes The main junctions for binding cells together are called **desmosomes** (dez′mo-sōmz; "binding bodies"), an *anchoring junction.* These adhesive spots are scattered along the abutting sides of adjacent cells (Figure 4.6b). Desmosomes have a complex structure: On the cytoplasmic face of each plasma membrane is a circular plaque. The plaques of neighboring cells are joined by linker proteins. These project from both cell membranes and interdigitate, like the teeth of a zipper, in the extracellular space. In addition, intermediate filaments (the cytoskeletal elements that resist tension) insert into each plaque from its inner, cytoplasmic side. Bundles of these filaments extend across the cytoplasm and anchor at other desmosomes on the opposite side of the same cell. Overall, this arrangement not only holds adjacent cells together but also interconnects intermediate filaments of the entire epithelium into one continuous network of strong guy-wires. The epithelium is thus less likely to tear when pulled on, because the pulling forces are distributed evenly throughout the sheet.

Desmosomes are found in cardiac muscle tissue as well as in epithelial tissues. In general, these junctions are common in tissues that experience great mechanical stress.

Gap Junctions A **gap junction,** or *nexus* (nek′sus; "bond"), is a tunnel-like junction that can occur anywhere along the lateral membranes of adjacent cells (Figure 4.6c). Gap junctions function in intercellular communication by allowing small molecules to move directly between neighboring cells. At such junctions, the adjacent plasma membranes are very close, and the cells are connected by hollow cylinders of protein (connexons). Ions, simple sugars, and other small molecules pass through these cylinders from one cell to the next. Gap junctions are common in embryonic tissues and in many adult tissues, including connective tissues. They are also prevalent in smooth and cardiac muscle, where the passage of ions through gap junctions synchronizes contraction.

Basal Feature: The Basal Lamina

At the border between the epithelium and the connective tissue deep to it is a supporting sheet called the **basal lamina** (lam′ĭ-nah; "sheet") (see Figure 4.1). This thin, noncellular sheet consists of proteins secreted by the epithelial cells. Functionally, the basal lamina acts as a selective filter; that is, it determines which molecules from capillaries in the underlying connective tissue are allowed to enter the epithelium. The basal lamina also acts as scaffolding along which regenerating epithelial cells can migrate. Luckily, infections and toxins that destroy epithelial cells usually leave the basal lamina in place, for without this lamina, epithelial regeneration is more difficult.

Directly deep to the basal lamina is a layer of reticular fibers (defined shortly) belonging to the underlying connective tissue. Together, these reticular fibers plus the basal lamina form the **basement membrane** (see Figure 4.1). The thin basal lamina can be seen only by electron microscopy, but the thicker basement membrane is visible by light microscopy (see Figure 4.3). Although this text distinguishes *basal lamina* from *basement membrane,* many scientists use these two terms interchangeably.

CLINICAL APPLICATION

Basement Membranes and Diabetes In untreated cases of diabetes mellitus (type 1 or type 2 diabetes), the basement membranes of the epithelial lining of capillaries thicken over time. This thickening is caused by increased amounts of glucose, present in high concentrations in diabetics, binding to the proteins of the basement membrane. This process is referred to as *increased glycosylation of the basement membrane.* Thickening is especially evident in the capillaries in the kidneys and retina of the eye, which can become nonfunctional. For this reason, kidney failure and blindness are major symptoms of advanced diabetes.

Apical Surface Features: Microvilli and Cilia

Microvilli (mi″cro-vĭ′li; "little shaggy hairs") are fingerlike extensions of the plasma membrane of apical epithelial cells (Figure 4.7). Each microvillus contains a core of actin filaments that extend into the actin microfilaments of the cytoskeleton and function to stiffen the microvillus. Microvilli occur on almost every moist epithelium in the body but are longest and most abundant on epithelia that absorb nutrients (in the small intestine) (see Figure 4.3c) or transport ions (in the kidney). In such epithelia, microvilli maximize the surface area across which small molecules enter or leave cells. Microvilli are also abundant on epithelia that secrete mucus, where they help anchor the mucous sheets to the epithelial surface.

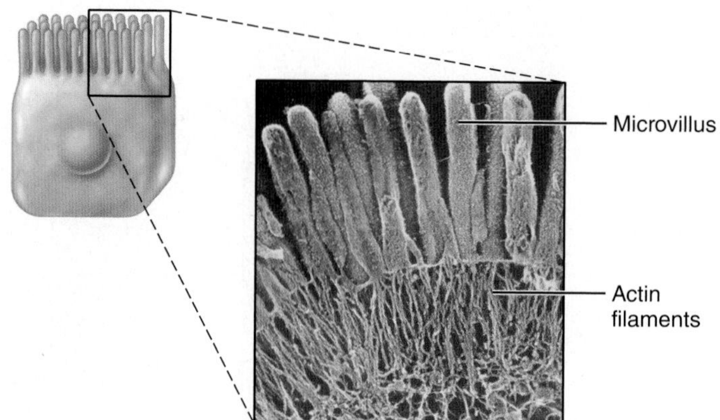

Figure 4.7 **Microvilli.**

Cilia are whiplike, highly motile extensions of the apical surface membranes of certain epithelial cells (see Figure 4.1). Each cilium contains a core of microtubules held together by cross-linking and radial proteins (Figure 4.8). The microtubules are arranged in pairs, called *doublets,* with nine outer doublets encircling one central pair. Ciliary movement is generated when adjacent doublets grip one another with side arms made of the motor protein dynein (p. 33) and these arms start to oscillate. This causes the doublets to slide along the length of each other, like centipedes trying to run over each other's backs. As a result, the cilum bends.

The microtubules in cilia are arranged in much the same way as in the cytoplasmic organelles called centrioles (p. 33). Indeed, cilia originate as their microtubules assemble around centrioles that have migrated from the centrosome to the apical plasma membrane. The centriole at the base of each cilium is called a **basal body** (Figure 4.8a).

The cilia on an epithelium bend and move in coordinated waves, like waves across a field of grass on a windy day. These waves push mucus and other substances over the epithelial surface (Figure 4.8c). Each cilium executes a propulsive *power stroke,* followed by a nonpropulsive *recovery stroke* similar to feathering an oar or a canoe paddle (Figure 4.8b). This sequence ensures that fluid is moved in one direction only. An extremely long, isolated cilium is called a *flagellum* (flah-jel′-um; "whip"). The only flagellated cells in the human body are sperm, which use their flagella to swim through the female reproductive tract.

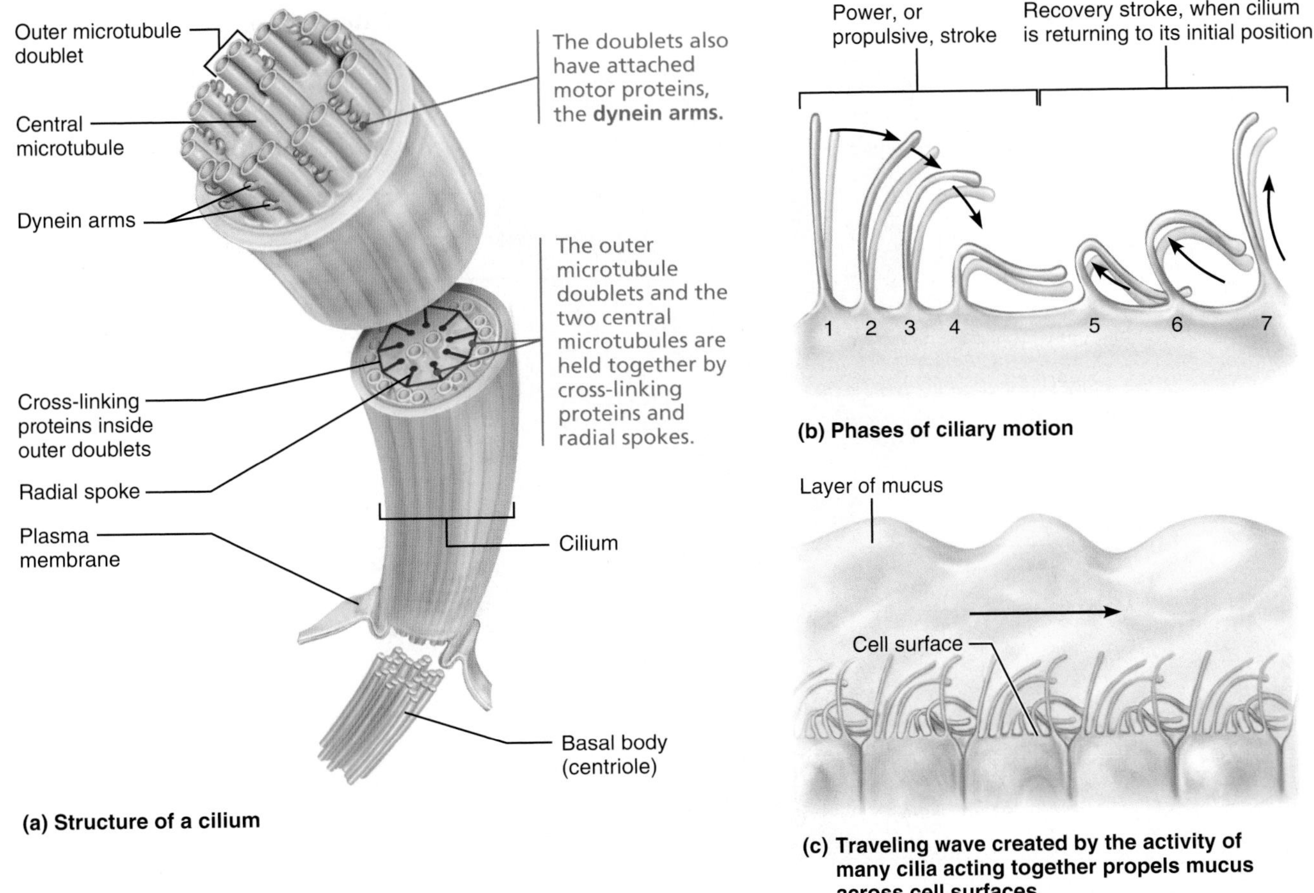

(a) Structure of a cilium

(b) Phases of ciliary motion

(c) Traveling wave created by the activity of many cilia acting together propels mucus across cell surfaces.

Figure 4.8 **Cilia structure and function.**

CLINICAL APPLICATION

Kartagener's Syndrome A type of *immotile cilia syndrome,* Kartagener's syndrome is an inherited disease in which the dynein arms within the cilia fail to form. This condition leads to frequent respiratory infections because the nonfunctional cilia cannot sweep inhaled bacteria out of the respiratory tubes.

check your understanding

☐ 7. How do cilia differ from microvilli?

☐ 8. How do the intermediate filaments within epithelial cells function to bind together the entire sheet of epithelia?

☐ 9. In addition to epithelial tissue, what other types of tissues contain gap junctions?

(For answers, see Appendix B.)

II. CONNECTIVE TISSUE

learning outcomes

- Describe the features that are common to all connective tissues.
- Identify the four main classes of connective tissue.

The second of the four basic types of tissue is **connective tissue,** the most diverse and abundant type of tissue. There are four main classes of connective tissue and many subclasses. (**Table 4.2**, p. 79). The main classes are:

1. *Connective tissue proper,* familiar examples of which are fat tissue and the fibrous tissue of ligaments
2. *Cartilage*
3. *Bone tissue*
4. *Blood*

Connective tissues do far more than just connect the tissues and organs of the body together. They also form the basis of the skeleton (bone and cartilage), store and carry nutrients (fat tissue and blood), surround all the blood vessels and nerves of the body (connective tissue proper), and lead the body's fight against infection.

In this section, we will first discuss the special characteristics of connective tissues and then describe the structural elements found in connective tissues. Finally, we will review the structure, function, and location of the specific types of connective tissues.

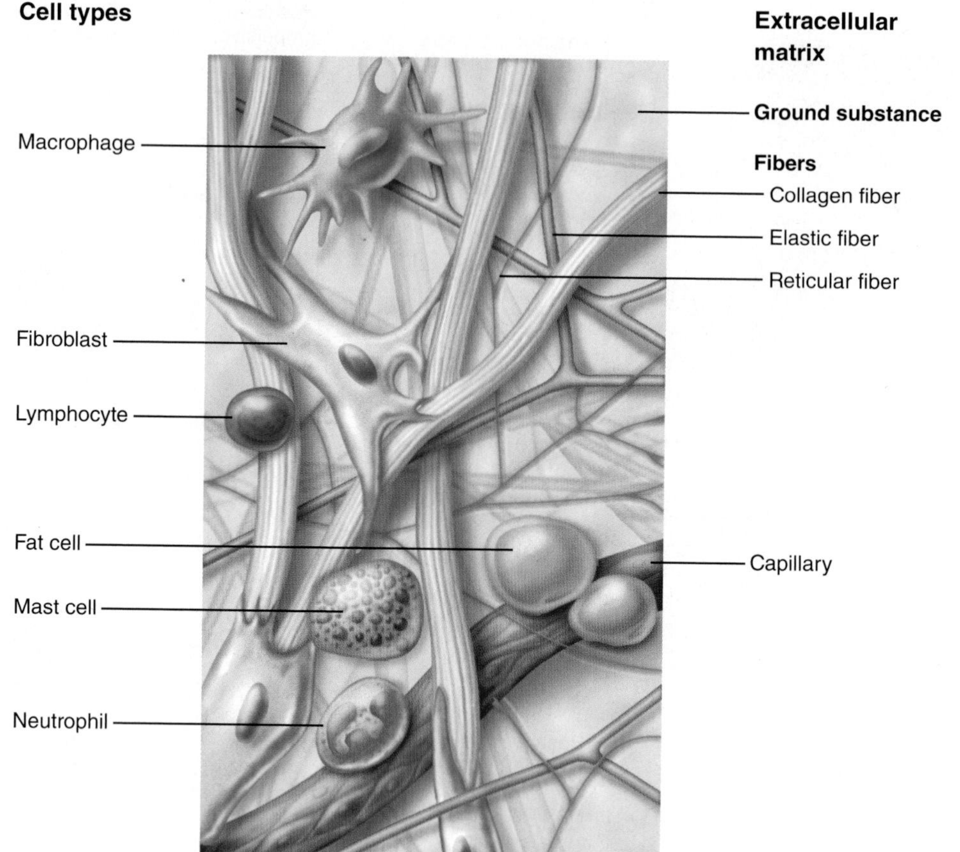

Figure 4.9 **Areolar connective tissue: A model connective tissue.**

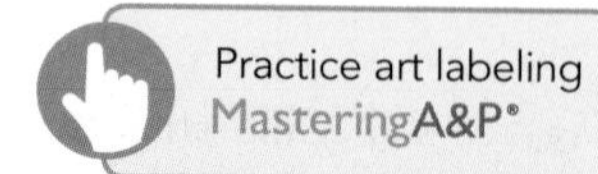

Special Characteristics of Connective Tissues

As different as they are, fat, bone, and blood are all connective tissues. All connective tissues share the same simple structural plan (Figure 4.9).

1. **Relatively few cells, lots of extracellular matrix.** The cells of connective tissues are separated from one another by a large amount of extracellular material called the **extracellular matrix** (ma′triks; "womb"). This differs markedly from epithelial tissue, whose cells crowd closely together.
2. **Extracellular matrix composed of ground substance and fibers.** The extracellular matrix is produced by the cells of the connective tissue. It is composed of some type of **ground substance** embedded with protein **fibers.** The ground substance varies for each class of connective tissue. In many, it is a soft, gel-like substance that holds tissue fluid. In bone, it is hard—calcified by inorganic calcium salts. The fibrous portion of the matrix provides support for the connective tissue. Three types of protein fibers are found in connective tissues: *collagen fibers, reticular fibers,* and *elastic fibers.* The types, density, and distribution of the fibers are distinctive for each type of connective tissue. The differences among the physical properties and functions of each type of connective tissue are due to differences in the composition of the extracellular matrix. The details of the matrix structure will be discussed with each type of connective tissue.
3. **Embryonic origin.** Another feature common to connective tissues is that they all originate from the embryonic tissue called **mesenchyme** (Figure 4.10; see also p. 51).

Structural Elements of Connective Tissues

Connective tissues are composed of cells and a significant amount of extracellular matrix. Connective tissues differ in their physical properties because of differences in the types of cells and the composition of the extracellular matrix. However, they all share similar structural elements: *cells, fibers,* and *ground substance*.

Cells

In most connective tissues, the primary cell type produces the extracellular matrix (see Table 4.2, third column). In connective tissue proper, these cells are called **fibroblasts** (literally, "fiber buds" or "fiber formers"). Fibroblasts make the protein subunits of fibers, which are secreted into the extracellular matrix and assemble into fibers. Fibroblasts

Table 4.2 Comparison of Classes of Connective Tissues

Tissue Class and Example	Subclasses	Components: Cells	Components: Matrix	General Features
Connective Tissue Proper *Dense regular connective tissue*	1. Loose connective tissue • Areolar • Adipose • Reticular 2. Dense connective tissue • Regular • Irregular • Elastic	Fibroblasts Fibrocytes Defense cells Fat cells	Gel-like ground substance All three fiber types: collagen, reticular, elastic	Six different types; vary in density and types of fibers Functions as a binding tissue Resists mechanical stress, particularly tension
Cartilage *Hyaline cartilage*	1. Hyaline cartilage 2. Elastic cartilage 3. Fibrocartilage	Chondroblasts found in growing cartilage Chondrocytes	Gel-like ground substance Fibers: collagen, elastic fibers in some	Resists compression because of the large amounts of water held in the matrix Functions to cushion and support body structures
Bone Tissue *Compact bone*	1. Compact bone 2. Spongy bone	Osteoblasts Osteocytes	Gel-like ground substance calcified with inorganic salts Fibers: collagen	Hard tissue that resists both compression and tension Functions in support
Blood	Blood cell formation and differentiation are quite complex (Details are provided in Chapter 18)	Erythrocytes (RBCs) Leukocytes (WBCs) Platelets	Plasma No fibers	A fluid tissue Functions to carry O_2, CO_2, nutrients, wastes, and other substances (hormones, for example)

also secrete the molecules that form the ground substance of the matrix. In cartilage tissue, the cells that secrete the matrix are **chondroblasts** (kon′dro-blasts; "cartilage formers"), and in bone they are **osteoblasts** (os″te-o-blasts′; "bone formers"). Once these tissue-forming cells are not actively secreting new matrix, they are termed **fibrocytes, chondrocytes,** and **osteocytes** (*cyte* = cell). They function to maintain and repair the tissue matrix and keep the tissue healthy.

The cells found in blood are an exception. These cells do not produce the plasma matrix of blood. The cellular components of blood function to carry respiratory gases (red blood cells), to fight infections (white blood cells), and to aid in blood clotting (platelets).

Additional cell types are found in many connective tissues (shown in Figure 4.9).

- **Fat cells**, or *adipose* (ad′ĭ-pōs) *cells,* store nutrients. Fat cells are egg-shaped, and their cytoplasm is dominated by a single, giant lipid droplet that flattens the nucleus and cytoplasm at one end of the cell. Mature fat cells are among the largest cells in the body and cannot divide.
- **White blood cells** (neutrophils, lymphocytes, eosinophils) respond to and protect against infectious agents.
- **Macrophages** specialize in phagocytosis. They engulf and devour a wide variety of foreign materials, ranging from whole bacteria to foreign molecules, dirt particles, and dead tissue cells.
- **Mast cells** mediate inflammation and promote healing. Mast cells contain many large secretory granules filled with numerous chemicals that are secreted in response to infectious and allergy-inducing agents. The chemicals secreted by mast cells include histamine (increases blood flow and capillary permeability in the local area), heparin (interacts with other mast cell chemicals), and proteases (protein-degrading enzymes).

Fibers

The extracellular matrix of a connective tissue is composed of fibers and ground substance (Figure 4.9). Three types of fibers are found in connective tissues: *collagen fibers, reticular fibers,* and *elastic fibers*. In general, the fibers function in support, yet each type of fiber contributes unique properties to the connective tissue.

Collagen fibers (shown as thick, whitish gray fibers in Figures 4.9 and 4.10), are the strongest and most abundant type of fiber in connective tissues. Collagen fibers resist tension (pulling forces) and contribute strength to a connective tissue. Pulling tests show that collagen fibers are stronger than steel fibers of the same size! The thick collagen fibers that one sees with the light microscope are bundles of thinner collagen fibrils, which consist of still thinner strands that are strongly cross-linked to one another. This cross-linking is the source of collagen's great tensile strength.

Reticular (re-tik′u-lar) **fibers** (shown as thin blue fibers in Figures 4.9 and 4.10), are bundles of a special type of collagen fibril. These short fibers cluster into a meshlike network (*reticulum* = network) that covers and supports the structures bordering the connective tissue. For example, capillaries are coated with fuzzy nets of reticular fibers, and these fibers form part of the basement membrane of epithelia. Individual reticular fibers glide freely across one another when the network is pulled, so they allow more give than collagen fibers do.

The last fiber type found in certain connective tissues is **elastic fibers** (illustrated as gold fibers in Figures 4.9 and 4.10). Elastic fibers contain the rubberlike protein **elastin** (e-last′in), which allows them to function like rubber bands. Although we typically think of rubber bands as being stretchy, the defining feature of an elastic material is its ability to recoil back to its original form after being stretched. Connective tissue can stretch only so much before its thick, ropelike collagen fibers become taut. When the tension is released, the elastic fibers recoil, and the stretched tissue resumes its original shape.

CLINICAL APPLICATION

Scurvy Vitamin C, which is abundant in citrus fruits such as oranges and lemons, is necessary for the proper cross-linking of the molecules that make up collagen fibers. A deficiency of this vitamin in the diet can lead to **scurvy,** a weakening of collagen and connective tissue throughout the body. Strong collagen is necessary for holding teeth in their sockets, reinforcing the walls of blood vessels, healing wounds, and forming scar tissue. Common signs of scurvy include loss of teeth, blood vessel rupture, and poor healing. Scurvy was a common ailment of sea explorers in the sixteenth to eighteenth centuries. In the mid-1700s, a British naval physician observed that consumption of citrus juice prevented scurvy. For that reason, the British Royal Navy began serving lime juice to sailors to prevent the disease—and British sailors came to be called "limeys."

Ground Substance

The other component of the matrix of connective tissue is the ground substance (Figure 4.9). The molecules that compose the ground substance are produced and secreted by the primary cell type of the connective tissue (fibroblasts, chondroblasts, or osteoblasts). The ground substance in most connective tissues is a gel-like material that consists of large sugar and sugar-protein molecules (proteoglycans and glycosaminoglycans). These molecules soak up fluid like a sponge. The fluid-filled ground substance functions to cushion and protect body structures (connective tissue proper), to withstand compressive stresses (cartilage), or to hold the tissue fluid that bathes all the cells in our body (areolar connective tissue). In bone tissue, the secreted ground substance is embedded with calcified mineral salts, which make the matrix hard and contribute to its function in supporting the body.

Here again, blood is an exception. The ground substance of blood, plasma, is not produced by the blood cells. Blood plasma is composed of water (about 90%) and various dissolved proteins, nutrients, ions, gases, and other molecules. These molecules are either produced by cells in other organs and then secreted into blood (for example, plasma proteins and hormones) or transported into blood from an external source (water, air, and nutrients).

A model of connective tissue can be built with a few simple ingredients: gelatin, jellybeans, and licorice. Dissolve the gelatin, and allow it to set. The gelatin represents the ground substance of the connective tissue. Gelatin is composed of proteins (actually hydrolyzed collagen) that hold fluid, very similar to the ground substance of most connective tissues. Once the gelatin thickens (approximately 1.5 hours or so), add a few jellybeans and some licorice. The jellybeans represent the cellular components of connective tissue, and licorice represents the fibers. Different types of licorice can represent different types of fibers: thick red twists represent collagen fibers; thin black laces represent elastic fibers. This model will not mimic the physical properties of connective tissues—strength, elasticity, and resistance to compression—but it does give you a visual analogy of the composition of a connective tissue.

☑ check your understanding

- ☐ 10. How do epithelial tissues differ from connective tissues in the arrangement of the cells in each tissue?
- ☐ 11. Distinguish the matrix of a connective tissue from the ground substance.
- ☐ 12. Which structural element of connective tissue resists tension? Which resists compression? Which allows for recoil? Which element produces the matrix?

(For answers, see Appendix B.)

Classification of Connective Tissues

learning outcome

▶ Differentiate the different types of connective tissues in reference to the types of cells located within the tissue, the structure and composition of the extracellular matrix, and the main function of each.

Connective Tissue Proper—Loose Connective Tissues

Connective tissue proper has two subclasses: **loose connective tissue** (areolar, adipose, and reticular) and **dense connective tissue** (dense irregular, dense regular, and elastic). These subclasses are distinguished by the density of fibers: In loose connective tissues, the fibers are distributed throughout the tissue but are separated from each other by ground substance. There are three types of loose connective tissue: *areolar, adipose,* and *reticular.*

Areolar Connective Tissue (Figure 4.9 and Figure 4.10b) Loose areolar connective tissue is the most widespread type of connective tissue proper. This tissue underlies almost all the epithelia of the body and surrounds almost all the small nerves and blood vessels, including the capillaries. The structure of this tissue reflects its basic functions:

1. Supporting and binding other tissues
2. Holding body fluids
3. Defending the body against infection
4. Storing nutrients as fat

The fibers of areolar connective tissue provide support. Areolar connective tissue has all three types of fibers in its extracelluar matrix: collagen fibers, reticular fibers, and elastic fibers (although the reticular fibers do not show up when the common histological stains hematoxylin and eosin are used).

The ground substance of areolar connective tissue holds fluid. All the cells of the body are bathed in tissue fluid, or **interstitial fluid.** This fluid is derived from leakage of fluid and small molecules from the blood as it travels through the capillaries. Nutrients and oxygen are delivered to cells and waste molecules are carried away from cells via diffusion through this fluid. Areolar connective tissue lies between the capillaries and all other cells and tissues in the body. It soaks up this tissue fluid, much like a sponge. Thus it keeps the body's cells surrounded by fluid and facilitates the passage of nutrients, gases, waste products, and other molecules to and from the cells.

Areolar connective tissue is the body's first line of defense against invading microorganisms, such as bacteria, viruses, fungi, and parasites. Lying just deep to the epithelial tissues that line the body surfaces and surrounding capillaries, it is in an ideal location to destroy microorganisms at their entry site—before they enter the capillaries and use the vascular system to spread to other locations. Areolar connective tissue contains a variety of defense cells (shown in Figure 4.9), which respond to infectious agents.

Cellular defenses are not the only means by which areolar connective tissue fights infection. The viscous ground substance and the dense networks of collagen fibers in the extracellular matrix slow the progress of invading microorganisms. Some bacteria, however, secrete enzymes that rapidly break down ground substance or collagen. Such matrix-degrading bacteria are highly invasive; that is, they spread rapidly through the connective tissues and are especially difficult for the body's defenses to control. An example of such a bacterium is the *Streptococcus* strain responsible for strep throat.

A minor function of areolar connective tissue is to store energy reserves as fat. Fat cells occur singly or in small groups in this tissue (see Figure 4.9).

Adipose Tissue (Figure 4.10c) **Adipose tissue** is similar to areolar connective tissue in structure and function, but its nutrient-storing function is much greater. Correspondingly, adipose tissue is crowded with fat cells, which account for 90% of its mass. These fat cells are grouped into large clusters called lobules. Adipose tissue is richly vascularized, reflecting its high metabolic activity. It removes lipids from the bloodstream after meals and later releases them into the blood, as needed. Without the fat stores in our adipose tissue, we could not live for more than a few days without eating.

Much of the body's adipose tissue occurs in the layer beneath the skin called the *hypodermis.* Adipose tissue also is abundant in the mesenteries, which are sheets of serous membranes that hold the stomach and intestines in place. Fat in this location is called *visceral fat.* Additionally, fat forms cushioning pads around the kidneys and behind the eyeballs in the orbits.

Whereas the abundant fat beneath the skin serves the general nutrient needs of the entire body, smaller depots of fat serve the local nutrient needs of highly active organs. Such depots occur around the hard-working heart and around lymph nodes (where cells of the immune system are furiously fighting infection), within some muscles, and as individual fat cells in the bone marrow (where new blood cells are produced at a frantic rate). Many of these local depots offer special lipids that are highly enriched.

(Text continues on page 84.)

(a) Embryonic connective tissue: mesenchyme

Description: Embryonic connective tissue; gel-like ground substance containing fibers; star-shaped mesenchymal cells.

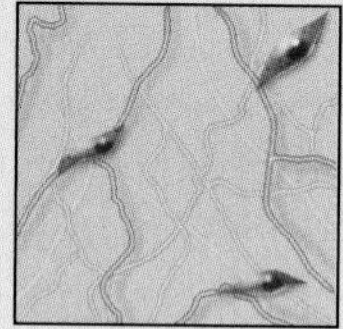

Function: Gives rise to all other connective tissue types.

Location: Primarily in embryo.

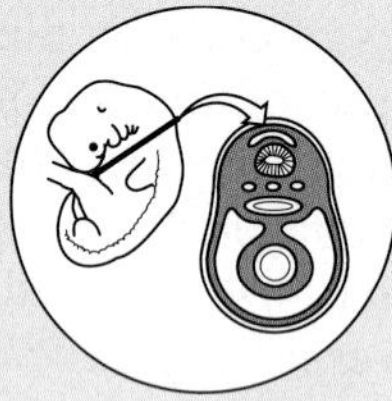

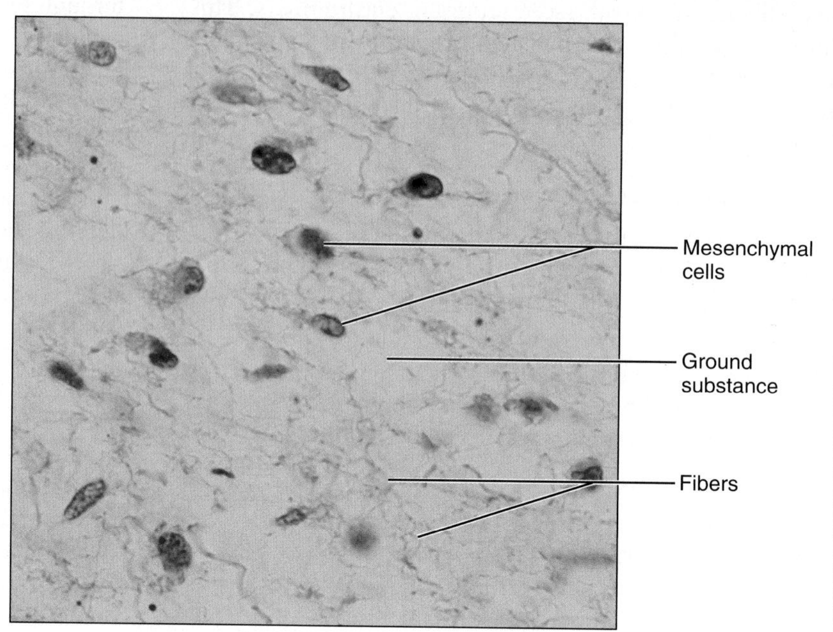

Photomicrograph: Mesenchyme, an embryonic connective tissue (385×). The matrix is composed of the fluid ground substance (clear-appearing background) and fine, sparse fibers.

(b) Connective tissue proper: loose connective tissue, areolar

Description: Gel-like matrix with all three fiber types; cells: fibroblasts, macrophages, mast cells, and some white blood cells.

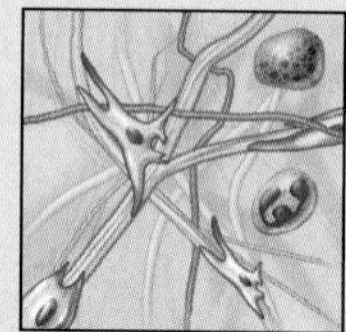

Function: Wraps and cushions organs; its macrophages phagocytize bacteria; plays important role in inflammation; holds and conveys tissue fluid.

Location: Widely distributed under epithelia of body, e.g., forms lamina propria of mucous membranes; packages organs; surrounds capillaries.

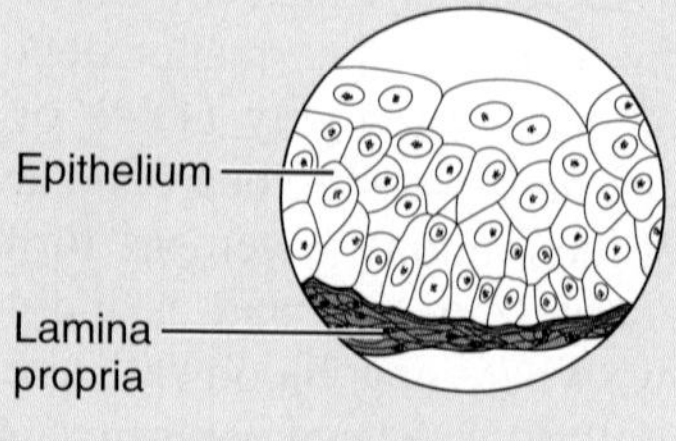

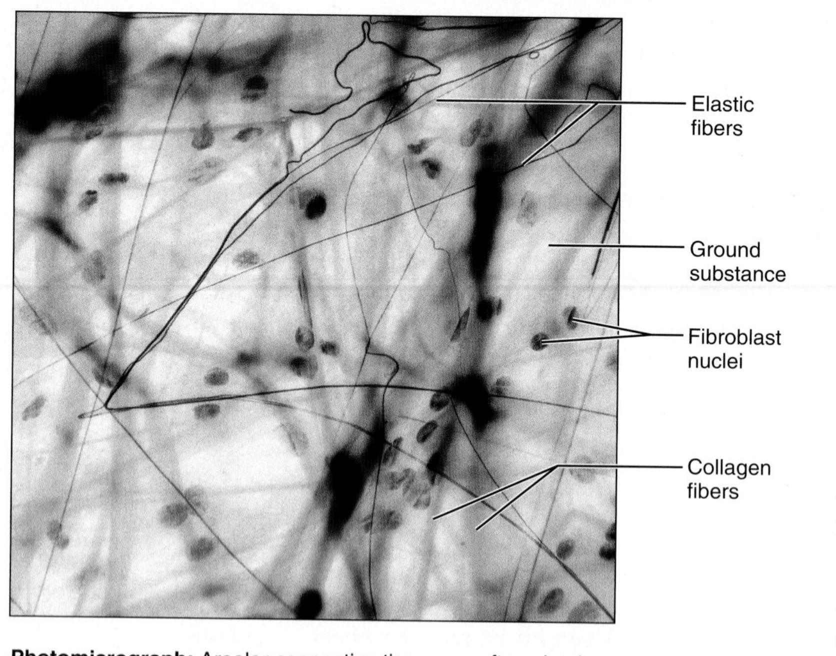

Photomicrograph: Areolar connective tissue, a soft packaging tissue of the body (340×).

Figure 4.10 **Connective tissues.**

View PAL Histology MasteringA&P®

(c) Connective tissue proper: loose connective tissue, adipose

Description: Matrix as in areolar connective tissue, but very sparse; closely packed adipocytes, or fat cells, have nucleus pushed to the side by large fat droplet.

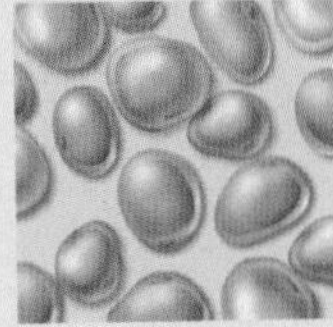

Function: Provides reserve food fuel; insulates against heat loss; supports and protects organs.

Location: Under skin in the hypodermis; around kidneys and eyeballs; within abdomen; in breasts.

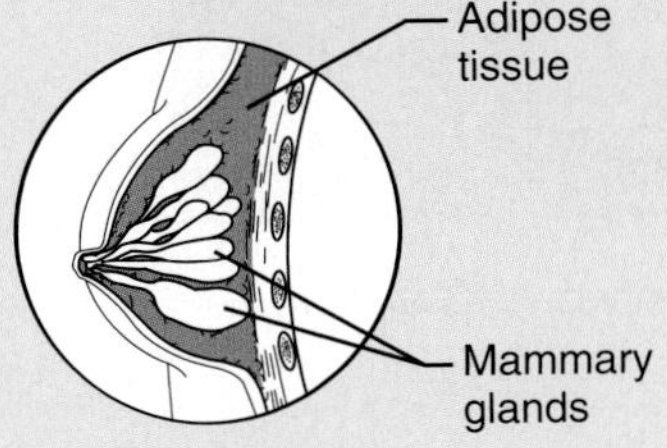

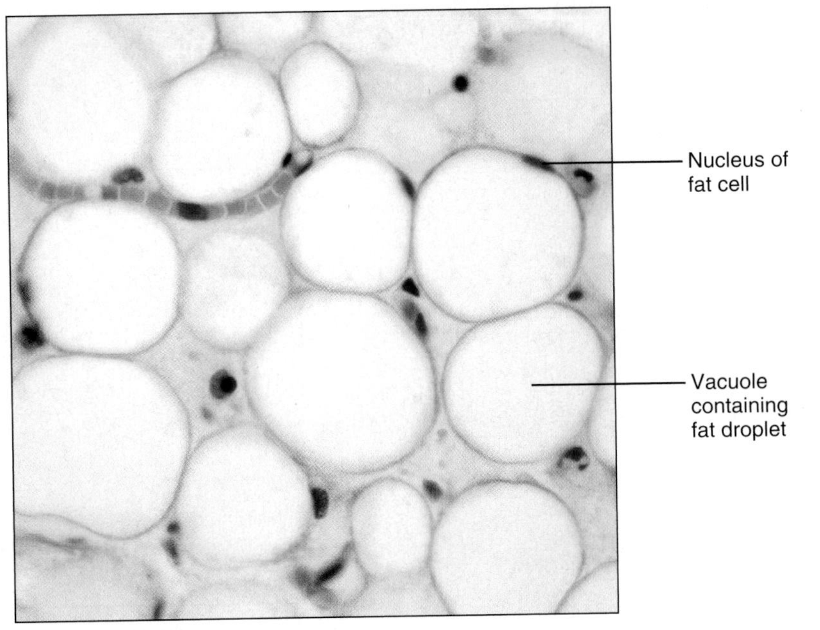

Photomicrograph: Adipose tissue from the subcutaneous layer under the skin (350×).

(d) Connective tissue proper: loose connective tissue, reticular

Description: Network of reticular fibers in a typical loose ground substance; reticular cells lie on the network.

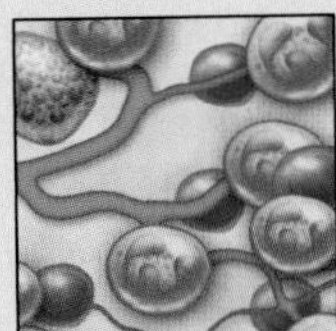

Function: Fibers form a soft internal skeleton (stroma) that supports other cell types including white blood cells, mast cells, and macrophages.

Location: Lymphoid organs (lymph nodes, bone marrow, and spleen).

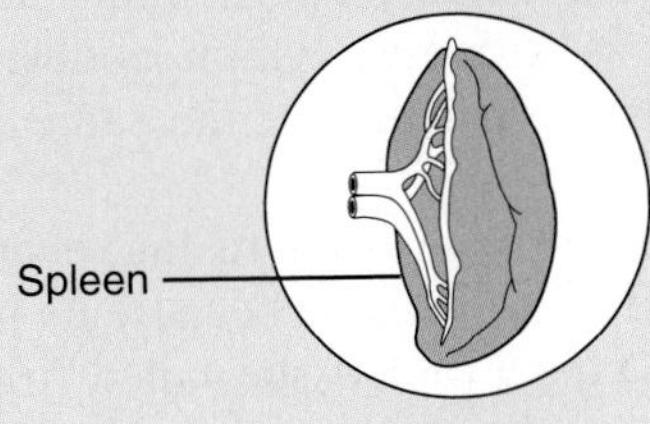

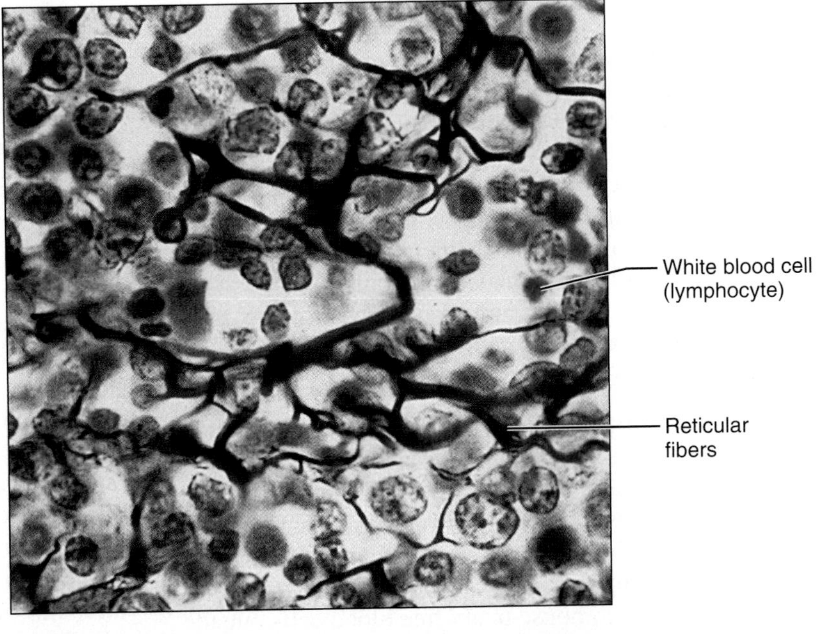

Photomicrograph: Dark-staining network of reticular connective tissue fibers forming the internal skeleton of the spleen (350×).

Figure 4.10 **Connective tissues,** *continued.*

(e) Connective tissue proper: dense connective tissue, dense irregular

Description: Primarily irregularly arranged collagen fibers; some elastic fibers; major cell type is the fibroblast; defense cells and fat cells are also present.

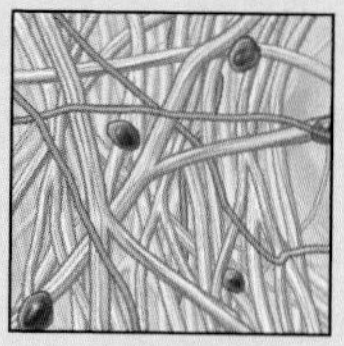

Function: Able to withstand tension exerted in many directions; provides structural strength.

Location: Fibrous capsules of organs and of joints; dermis of the skin; submucosa of digestive tract.

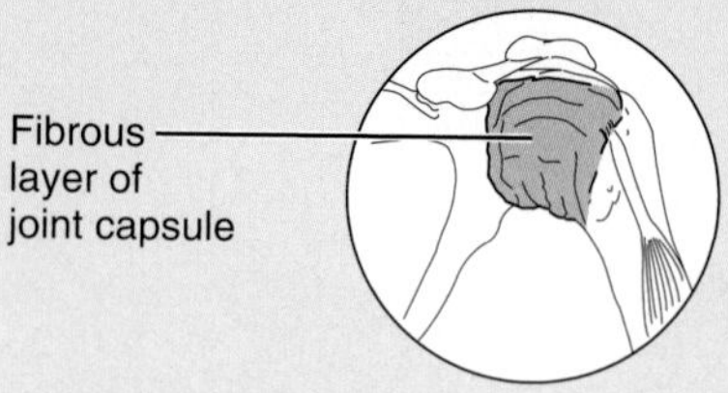

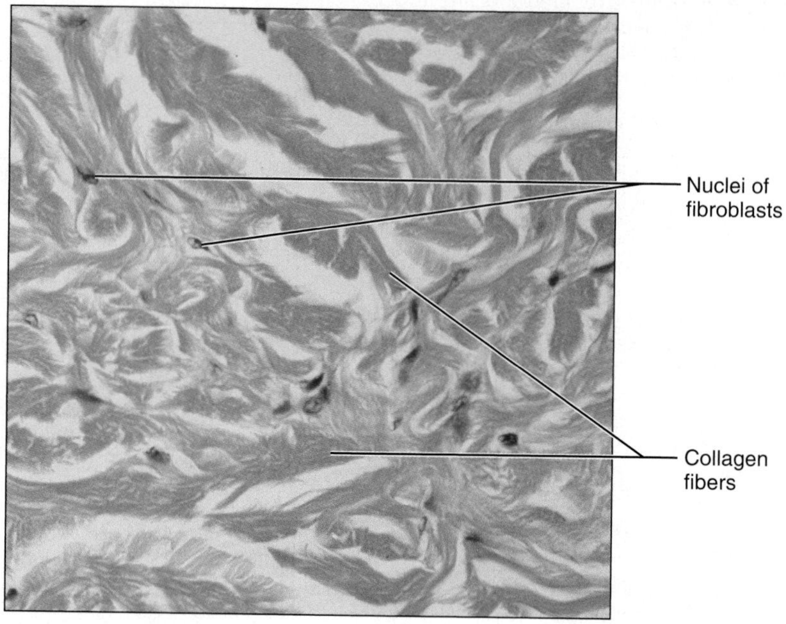

Photomicrograph: Dense irregular connective tissue from the dermis of the skin (300×).

Figure 4.10 **Connective tissues,** *continued.*

The typical, nutrient-storing fat is *white adipose tissue,* or *white fat.* Another type, called **brown adipose tissue,** produces heat and is a nutrient consumer. Brown fat, thought to occur only in babies to aid in thermoregulation, has also been identified in adults. It is located in the hypodermis between the two scapulae (shoulder blades) in the center of the back, on the side of the anterior neck, and on the anterior abdominal wall. It is even more richly vascularized than white fat. Each brown-fat cell contains many lipid droplets and numerous mitochondria, which use the lipid fuel to heat the bloodstream rather than to produce ATP molecules.

Reticular Connective Tissue (Figure 4.10d) **Reticular connective tissue** resembles areolar tissue, but the only fibers in its matrix are reticular fibers. These fine fibers form a broad, three-dimensional network like the frame of a house. The spaces in the framework create a labyrinth of caverns that hold many free cells. The bone marrow, spleen, and lymph nodes, which house many free blood cells outside their capillaries, consist largely of reticular connective tissue. Fibroblasts called **reticular cells** lie along the reticular network of this tissue. (Reticular tissue is discussed further in Chapters 18 and 21.)

Connective Tissue Proper—Dense Connective Tissue

Dense connective tissue, or *fibrous connective tissue,* contains more collagen than areolar connective tissue does. With its thick collagen fibers, it can resist extremely strong pulling (tensile) forces. There are three types of dense connective tissue: *irregular, regular,* and *elastic.*

Dense Irregular Connective Tissue (Figure 4.10e) **Dense irregular connective tissue** is similar to areolar tissue, but its collagen fibers are much thicker. These fibers run in different planes, allowing this tissue to resist strong tensions from different directions. This tissue dominates the leathery dermis of the skin, which is commonly stretched, pulled, and hit from various angles. This tissue also makes up the *fibrous capsules* that surround certain organs in the body, such as kidneys, lymph nodes, and bones. Its cellular and matrix elements are the same as in areolar connective tissue.

Dense irregular connective tissue is confusing, even for experts. The dermis obviously fits the name because its fibers run randomly, as one would expect for a tissue called "irregular." However, the fibrous capsules of organs consist of two

(Text continues on page 86.)

(f) Connective tissue proper: dense connective tissue, dense regular

Description: Primarily parallel collagen fibers; a few elastic fibers; major cell type is the fibroblast.

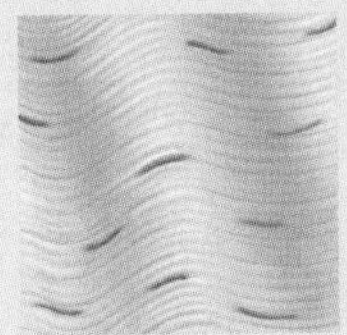

Function: Attaches muscles to bones or to muscles; attaches bones to bones; withstands great tensile stress when pulling force is applied in one direction.

Location: Tendons, most ligaments, aponeuroses.

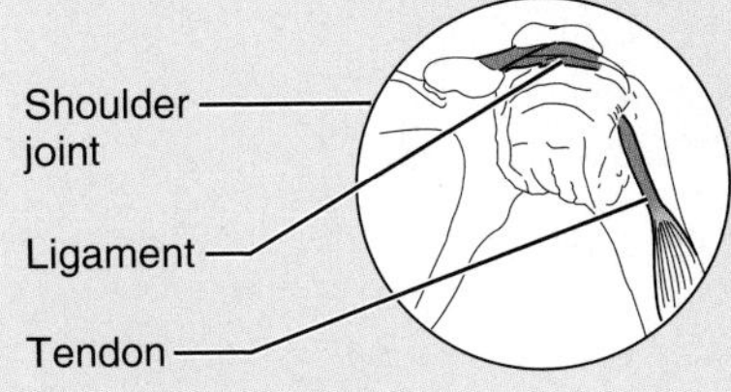

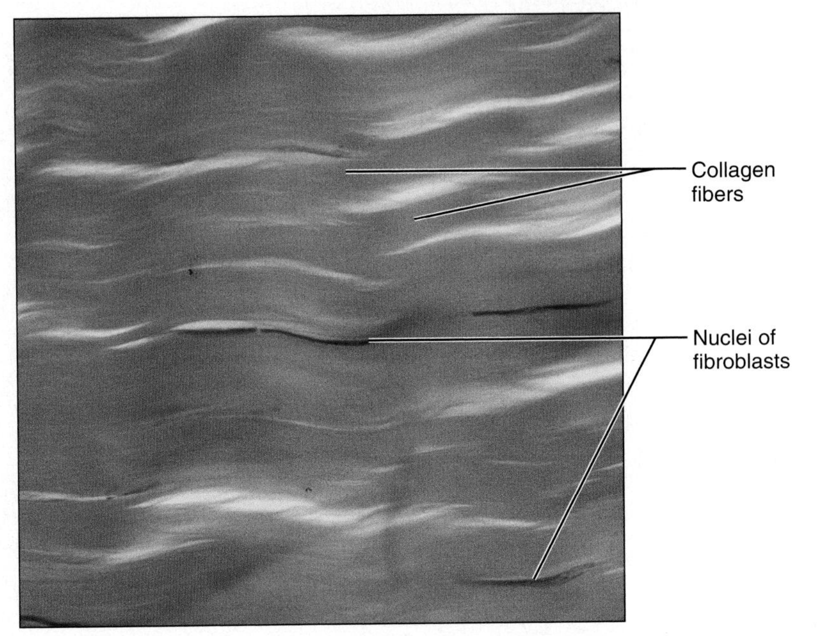

Photomicrograph: Dense regular connective tissue from a tendon (425×).

(g) Connective tissue proper: dense connective tissue, elastic

Description: Dense regular connective tissue containing a high proportion of elastic fibers.

Function: Allows recoil of tissue following stretching; maintains pulsatile flow of blood through arteries; aids passive recoil of lungs following inspiration.

Location: Walls of large arteries; within certain ligaments associated with the vertebral column; within the walls of the bronchial tubes.

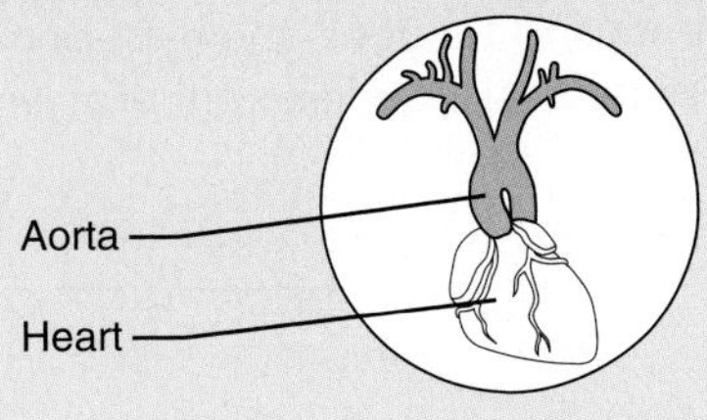

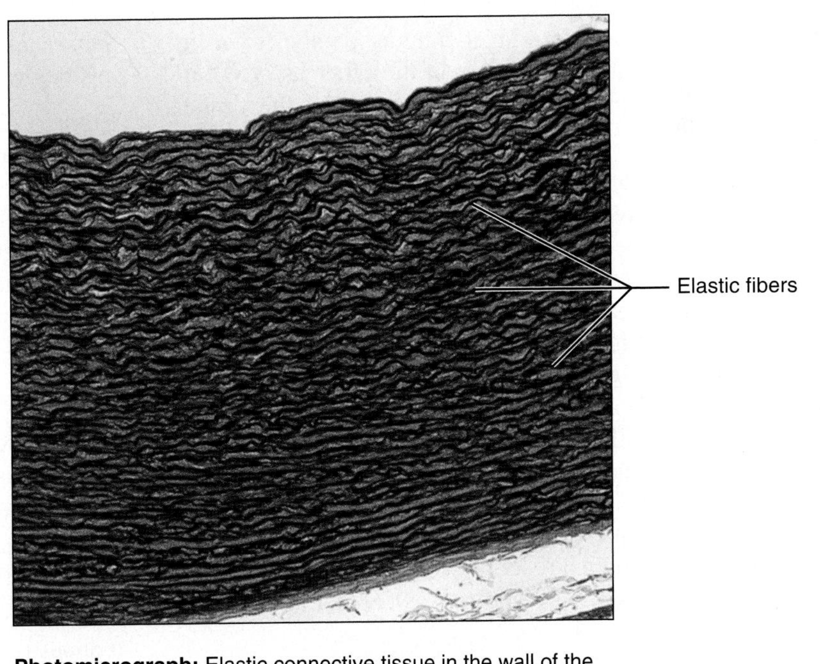

Photomicrograph: Elastic connective tissue in the wall of the aorta (250×).

Figure 4.10 Connective tissues, *continued.*

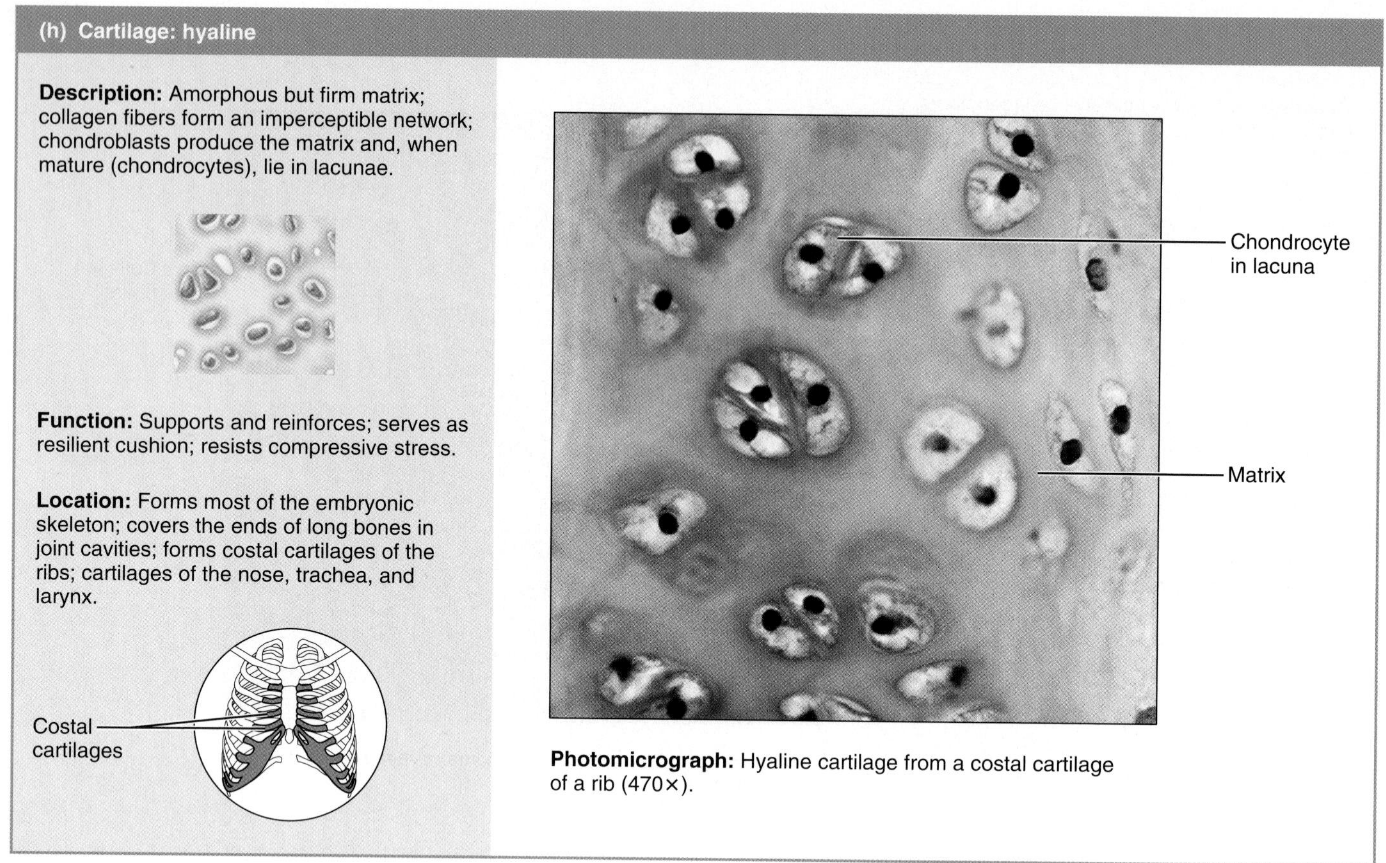

Figure 4.10 Connective tissues, *continued.*

layers, with all the fibers in a layer running parallel to each other but perpendicular to the fibers in the other layer. That is, two perpendicular "regular" layers can make the tissue "irregular."

Dense Regular Connective Tissue (Figure 4.10f) All collagen fibers in **dense regular connective tissue** usually run in the same direction, parallel to the direction of pull. Crowded between the collagen fibers are rows of fibroblasts, which continuously manufacture the fibers and a scant ground substance. When this tissue is not under tension, its collagen fibers are slightly wavy. Unlike areolar connective tissue, dense regular connective tissue is poorly vascularized and contains no fat cells or defense cells.

With its enormous tensile strength, dense regular connective tissue is the main component of *ligaments,* bands or sheets that bind bones to one another. It also is the main tissue in *tendons,* which are cords that attach muscles to bones, and *aponeuroses* (ap″o-nu-ro′sēs), which are sheetlike tendons.

Dense regular connective tissue also forms **fascia** (fash′e-ah; "a band"), a fibrous membrane that wraps around muscles, muscle groups, large vessels, and nerves. Many sheets of fascia occur throughout the body, binding structures together like plastic sandwich wrap. When the word *fascia* is used alone, it is understood to mean *deep fascia. Superficial fascia,* something else entirely, is the fatty hypodermis below the skin.

Elastic Connective Tissue (Figure 4.10g) In **elastic connective tissue,** elastic fibers are the predominant type of fiber, and bundles of elastic fibers outnumber the bundles of collagen fibers. This tissue is located in structures where recoil from stretching is important: within the walls of arteries, in certain ligaments (*ligamentum nuchae* and *ligamentum flavum,* which connect successive vertebrae), and surrounding the bronchial tubes in the lungs.

✓ check your understanding

☐ 13. How does loose connective tissue differ from dense connective tissue?

☐ 14. Which type of connective tissue forms the following structures: ligaments and tendons; the hypodermis of the skin; the tissue that underlies epithelia; lymph nodes?

(For answers, see Appendix B.)

Cartilage

As you have seen, connective tissue proper has the ability to resist tension (pulling). Cartilage and bone are the

(Text continues on page 89.)

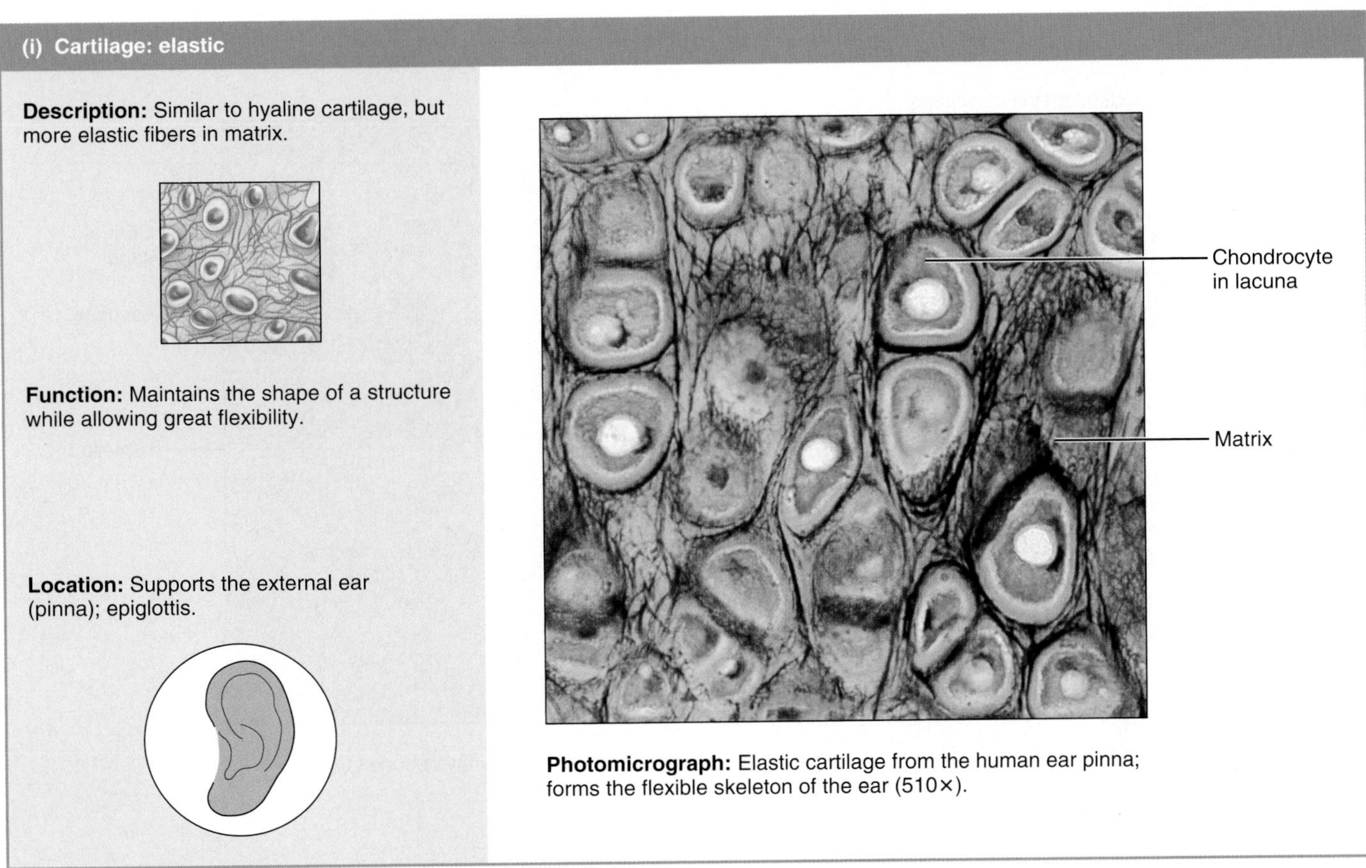

Photomicrograph: Elastic cartilage from the human ear pinna; forms the flexible skeleton of the ear (510×).

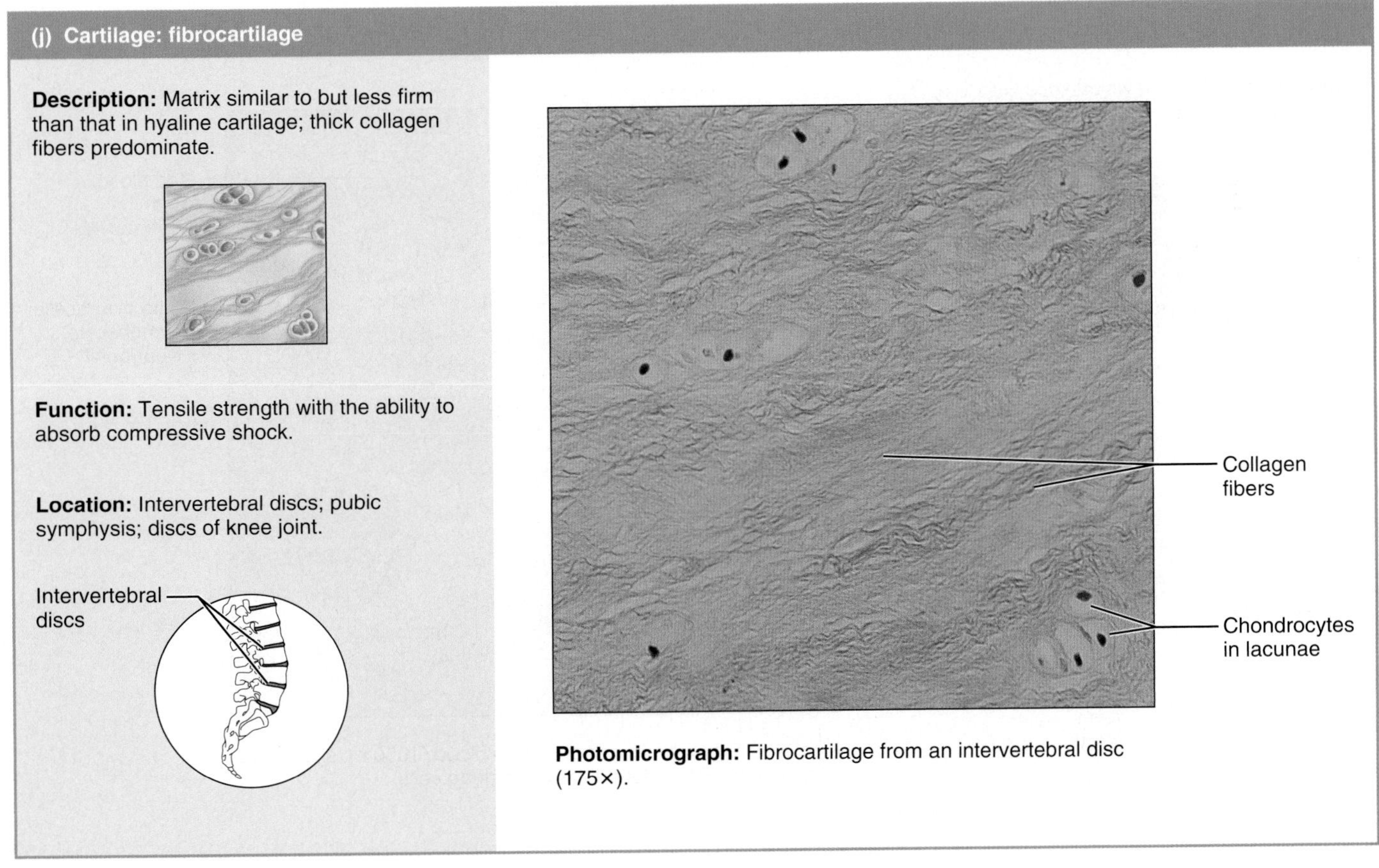

Photomicrograph: Fibrocartilage from an intervertebral disc (175×).

Figure 4.10 Connective tissues, *continued.*

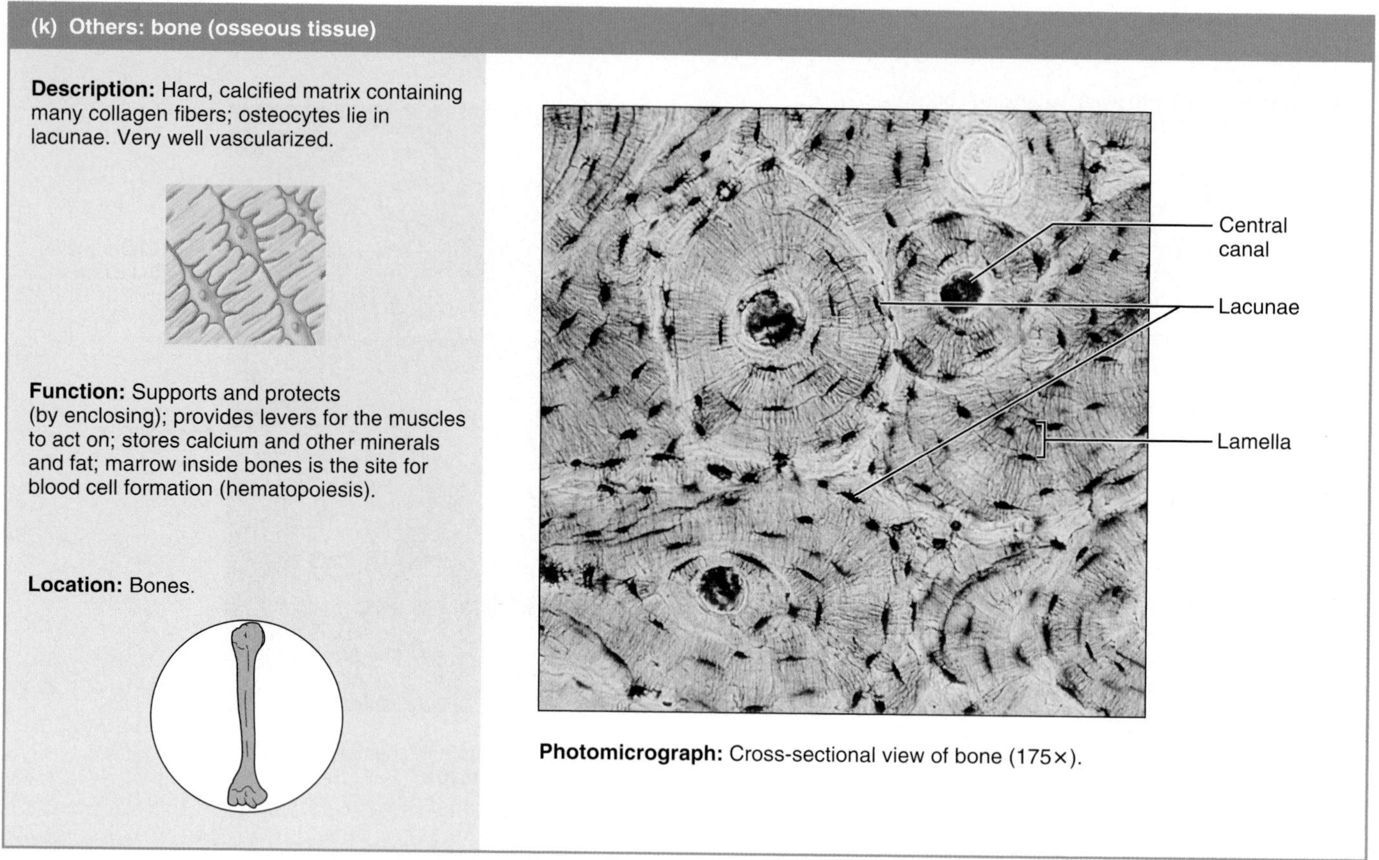

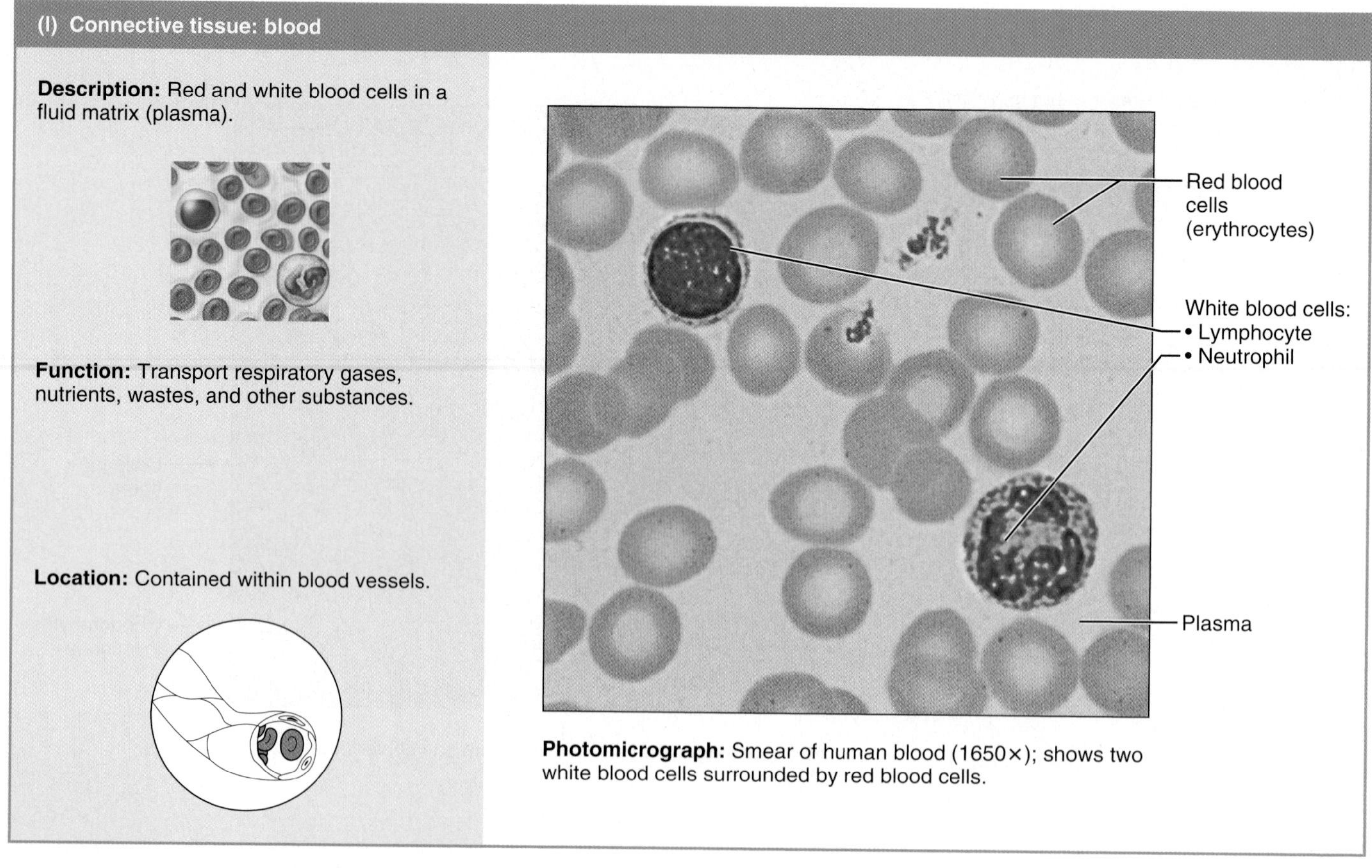

Figure 4.10 **Connective tissues,** *continued.*

firm connective tissues that resist *compression* (pressing) as well as tension. Like all connective tissues, they consist of cells separated by a matrix containing fibers, ground substance, and tissue fluid. However, these skeletal tissues exaggerate the supportive functions of connective tissue and play no role in fat storage or defense against disease.

Cartilage (Figure 4.10h–j), a firm but flexible tissue, occurs in several parts of the skeleton. For example, it forms the supporting rings of the trachea (windpipe) and gives shape to the nose and ears. Like nearly all connective tissues, cartilage consists of cells separated by an abundant extracellular matrix (Figure 4.10h). This matrix contains thin collagen fibrils, a ground substance, and an exceptional quantity of tissue fluid; in fact, cartilage consists of up to 80% water! The arrangement of water in its matrix enables cartilage to spring back from compression (as explained in Chapter 6, p. 126.)

Cartilage is simpler than other connective tissues: It contains no blood vessels or nerves and just one kind of cell, the **chondrocyte.** Each chondrocyte resides within a cavity in the matrix called a **lacuna.** Immature chondrocytes are *chondroblasts,* cells that actively secrete the matrix during cartilage growth. Cartilage is found in three varieties, each dominated by a particular fiber type: *hyaline cartilage, elastic cartilage,* and *fibrocartilage*. (These are introduced in Figure 4.10 h–j and are also discussed in Chapter 6, pp. 125–127.)

Bone

Because of its rocklike hardness, **bone tissue** has a tremendous ability to support and protect body structures. Bone matrix contains inorganic calcium salts, which enable bone to resist compression, and an abundance of collagen fibers, which allow bone to withstand strong tension (Figure 4.10k, p. 88).

Immature bone cells, called *osteoblasts,* secrete the collagen fibers and ground substance of the matrix. Then calcium salts precipitate on and between the collagen fibers, hardening the matrix. The mature bone cells, called **osteocytes,** inhabit cavities (**lacunae;** singular: **lacuna**) in this hardened matrix. Bone is a living and dynamic tissue, well supplied with blood vessels. (It is discussed further in Chapter 6.)

Blood

Blood (Figure 4.10l), the fluid in the blood vessels, is the most atypical connective tissue. It does not bind things together or give mechanical support. It is classified as a connective tissue because it develops from mesenchyme and consists of blood *cells* surrounded by a nonliving matrix, the liquid blood *plasma.* Its cells and matrix are very different from those in other connective tissues. Blood functions as the transport vehicle for the cardiovascular system, carrying defense cells, nutrients, wastes, respiratory gases, and many other substances throughout the body. (Blood is discussed in detail in Chapter 18.)

Epithelial and connective tissues are the most diverse tissue types in the body. For this reason, they are commonly the most difficult for students to understand. *Focus on Distinguishing Between Epithelial and Connective Tissues* (Figure 4.11) presents the distinctions between these types of tissues in reference to structure and function.

☑ check your understanding

- ☐ 15. Which connective tissues contain collagen fibers?
- ☐ 16. In which connective tissues are the cells located in lacunae?
- ☐ 17. Which component of cartilage tissue functions to resist compressive forces?

(For answers, see Appendix B.)

Covering and Lining Membranes

learning outcome

▶ Discuss the structure and function of mucous, serous, and cutaneous membranes.

Now that you have learned about connective and epithelial tissues, you can consider the **covering** and **lining membranes** that combine these two tissue types (Figure 4.12). These membranes, which cover broad areas within the body, consist of an epithelial sheet plus the underlying layer of connective tissue proper. These membranes are of three types: *cutaneous, mucous,* and *serous.*

The **cutaneous membrane** (ku-ta′ne-us; "skin") is the skin, covering the outer surface of the body (Figure 4.12a). Its outer epithelium is the thick epidermis, and its inner connective tissue is the dense dermis. It is a dry membrane. (The skin is discussed further in Chapter 5.)

A **mucous membrane,** or **mucosa** (mu-ko′sah), lines the inside of every hollow internal organ that opens to the outside of the body. More specifically, mucous membranes line the tubes of the respiratory, digestive, reproductive, and urinary systems (Figure 4.12b). Although different mucous membranes vary widely in the types of epithelia they contain, all are wet or moist. As their name implies, many mucous membranes secrete mucus. Not all of them do so, however.

All mucous membranes consist of an epithelial sheet directly underlain by a layer of loose areolar connective tissue called the *lamina propria* (lam′ĭ-nah pro′pre-ah; "one's own layer"). In the mucous membranes of the digestive system, the lamina propria rests on a layer of smooth muscle cells. Later chapters discuss the specific mucous membranes of the body.

Serous membranes, or **serosae** (introduced in Chapter 1), are the slippery membranes that line the closed pleural, pericardial, and peritoneal cavities (Figures 4.12c and 1.7, p. 12). A serous membrane consists of a simple squamous epithelium, called **mesothelium,** lying on a thin layer of areolar connective tissue. This membrane produces a slippery *serous fluid,* beginning as a filtrate of the blood in capillaries in the connective tissue, with the addition of lubricating molecules by the mesothelium.

Identifying Epithelial and Connective Tissues

Figure 4.11 **The structure of a tissue provides clues to its name and location.**

How are the cells distributed in epithelial tissue?

Lots of cells, little extracellular material
Tightly packed cells

Apical and basal regions
Apical region borders open space
Basal region attached to underlying connective tissue

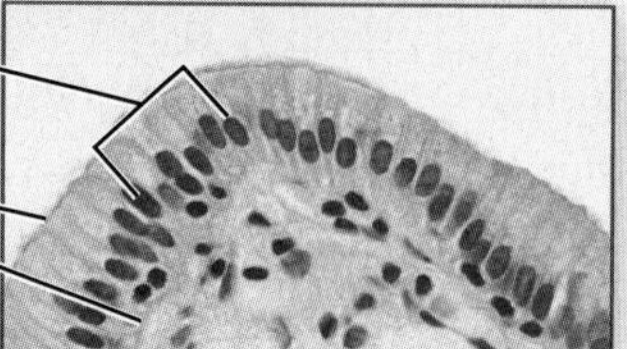

EPITHELIAL TISSUES

Epithelia form selective barriers for the passage of molecules across body surfaces. Epithelia line the external and internal surfaces of many organs and form the secretory portion of most glands.

How are the cell nuclei arranged?

Single row of nuclei:
Simple epithelium

Stacked rows of nuclei:
Stratified epithelium

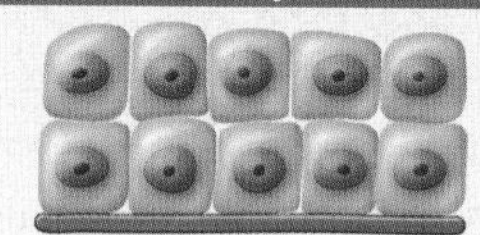

Cell heights differ, nuclei appear stacked; each cell attached to basal region:
Pseudostratified epithelium

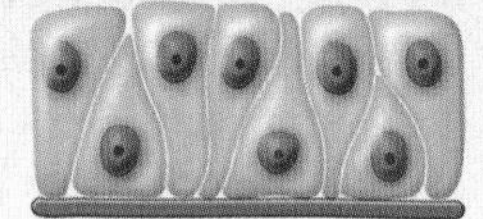

What shape are the cells at the apical region?

Squamous

Simple squamous epithelium

Thin, flat cells
Elongated nucleus

Thin layer of cells allows for rapid transport of molecules. Found where **diffusion** and **filtration** occur (see Figure 4.3a).

Stratified squamous epithelium

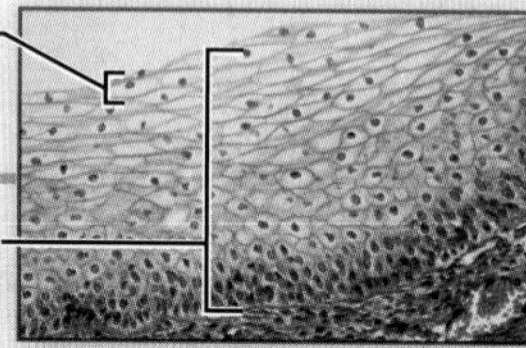

Multiple layers of cells provide **protection** and rapid regeneration. Found in areas that experience abrasion or friction (see Figure 4.3e).

Cuboidal

Simple cuboidal epithelium

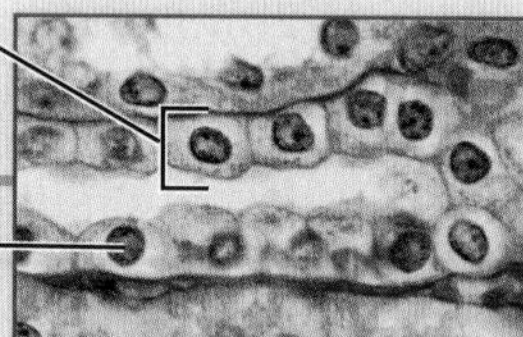

Larger cells house cellular organelles active in **secretion** and **absorption** yet allow rapid passage through epithelial layer (see Figure 4.3b).

Stratified cuboidal epithelium

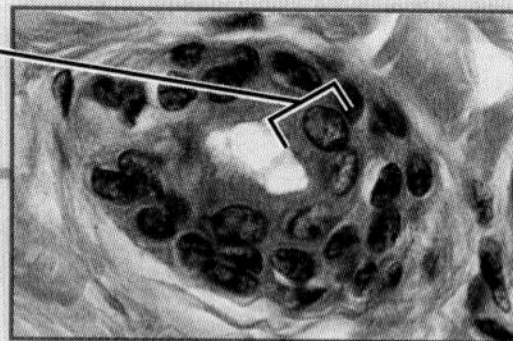

Two layers of cells **protect** underlying tissue from glandular secretions. Forms the ducts of many large glands (see Figure 4.3f).

Columnar

Simple columnar epithelium

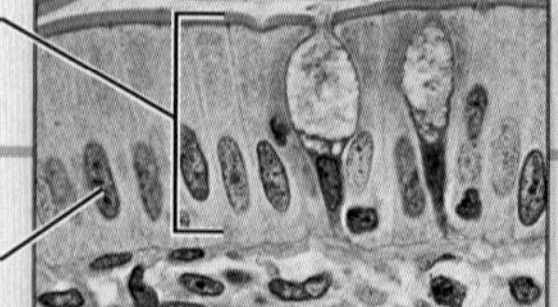

Elongated columnar cells house cellular organelles necessary for a tissue active in **secretion** and **absorption** (see Figure 4.3c).

Stratified columnar epithelium

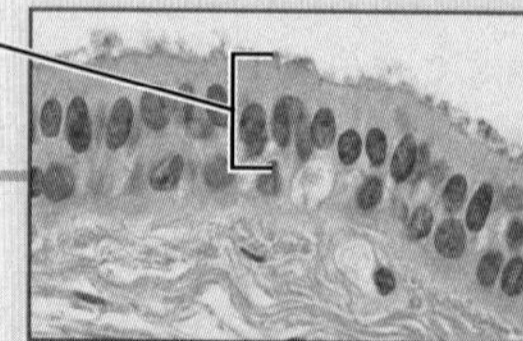

Two layers of cells **protect** underlying tissues; columnar cells have space for organelles needed to produce **secretions** (see Figure 4.3g).

Pseudostratified columnar epithelium

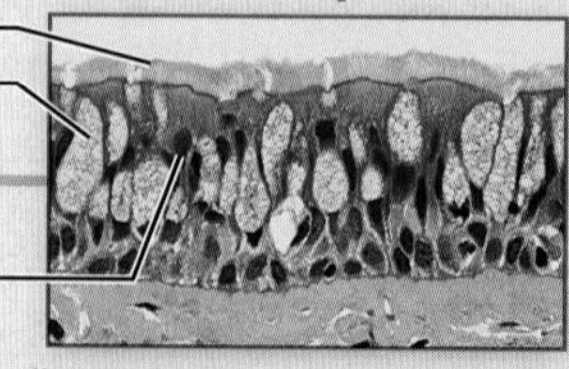

Columnar cells and goblet cells produce **secretions**; cilia **move fluid** along the epithelial surface (see Figure 4.3d).

How are the cells distributed in connective tissue?

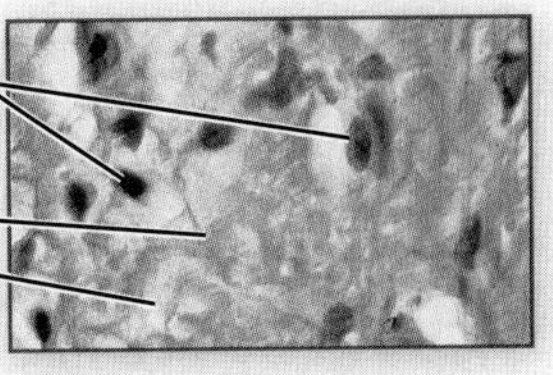

Few cells, widely spaced apart
Fibroblast nuclei

Lots of extracellular matrix
Fibers
Ground substance

CONNECTIVE TISSUES (CT)

Connective tissues are binding and support tissues. They provide structural support, connect body structures, store nutrients, and hold tissue fluids. The physical properties of each connective tissue result from the composition of its matrix.

Where are the cells located within the matrix?

Cells in direct contact with matrix components:

Connective tissue proper

Loose CT
Loosely dispersed individual fibers

Dense CT
Densely packed bundles of fibers

Areolar CT
Fibroblast nucleus
Fibers

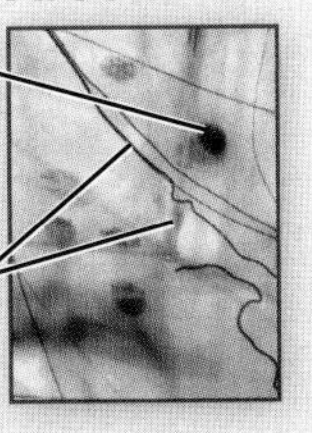

Adipose CT
Fat cell nuclei
Extra-cellular matrix

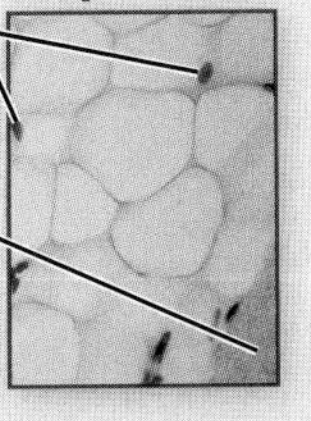

Reticular CT
Fibers

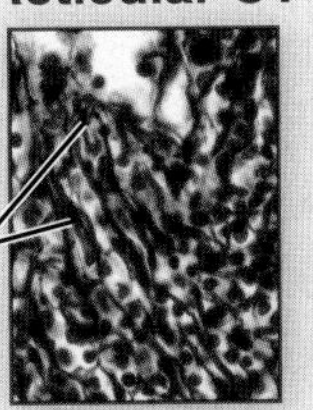

Matrix holds fluid, binds and supports. Cells fight infections and store nutrients. Loose CTs **fill body spaces** and allow for **exchange of materials** with the blood (see Figure 4.10b–d).

Dense irregular CT
Fibroblast nuclei
Thick collagen fibers

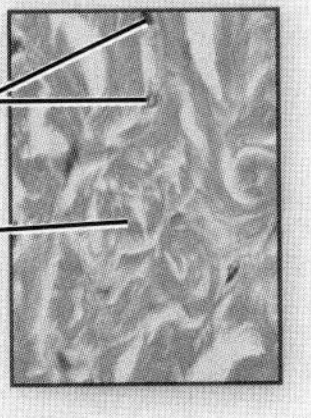

Dense regular CT
Fibroblast nuclei
Thick collagen fibers

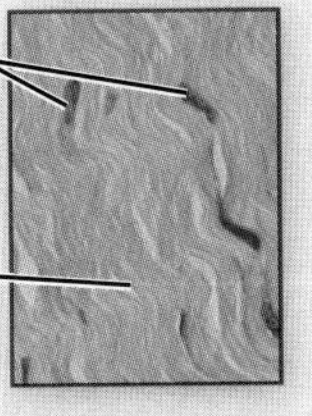

Elastic CT
Thick elastic fibers

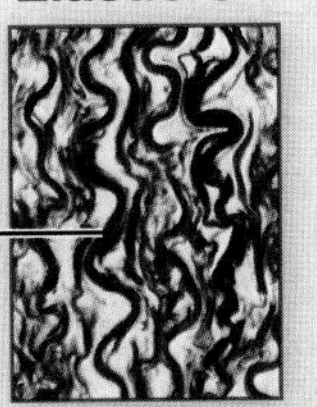

Thick collagen fibers resist tension; elastic fibers enable recoil. Alignment of fibers indicates the direction of applied forces. Dense CTs are strong, resilient **binding** tissues (see Figure 4.10 e–g).

Cells in spaces (lacunae) within the extracellular matrix:

Bone and Cartilage

Bone
Calcified ground substance

Osteocytes in lacunae
Extracellular matrix

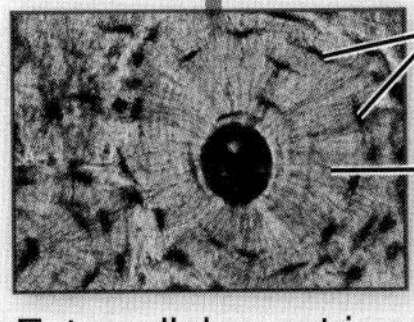

Extracellular matrix contains collagen fibers and calcified ground substance. Strong, stiff, composite tissue functions in **support** and **protection** (see Figure 4.10k).

Cartilage
Gelatinous ground substance

Hyaline cartilage
Chondrocyte
Lacuna
Extracellular matrix

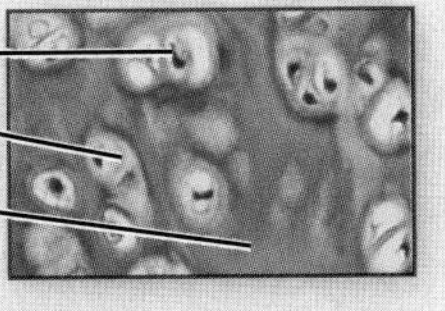

Elastic cartilage
Chondrocyte
Lacuna
Elastic fibers

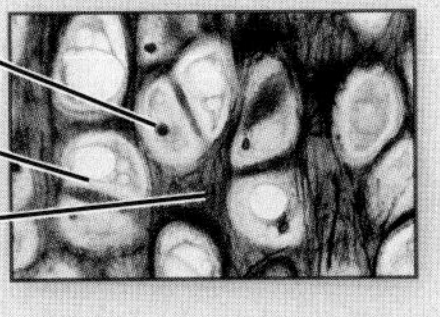

Fibrocartilage
Chondrocyte in lacuna
Collagen fibers

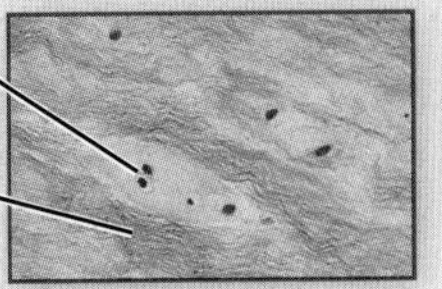

Firm, resilient matrix. Ground substance holds fluid and resists compression; fibers add tensile strength (collagen) and elasticity (elastic). Cartilage contributes to **flexible support** (see Figure 4.10h–j).

Cells suspended in a fluid matrix:

Blood

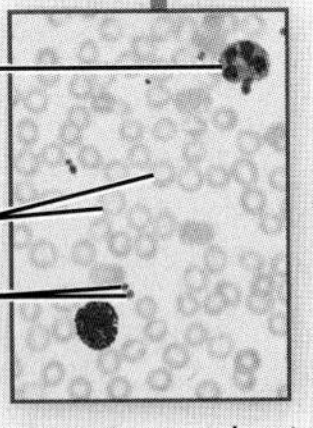

Leukocyte
Erythrocytes
Platelets

Fluid matrix **transports** nutrients, signaling molecules, and waste products. Cellular components transport respiratory gases and **fight infections** (see Figure 4.10l).

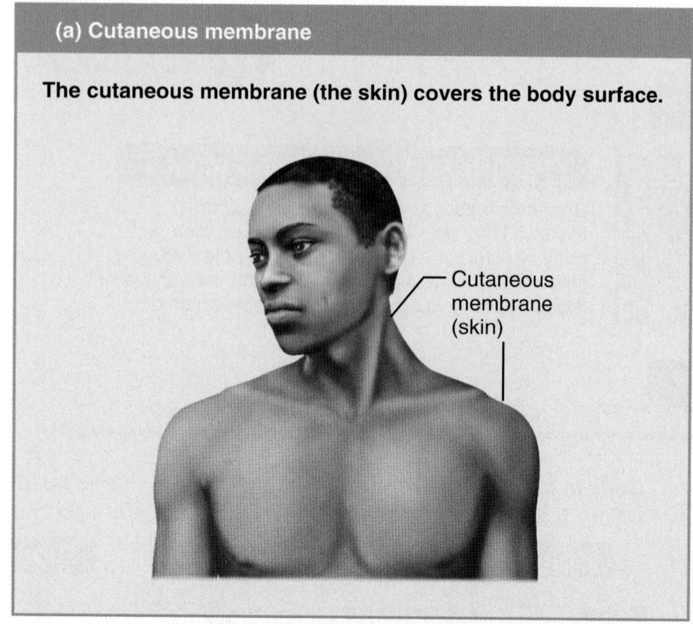

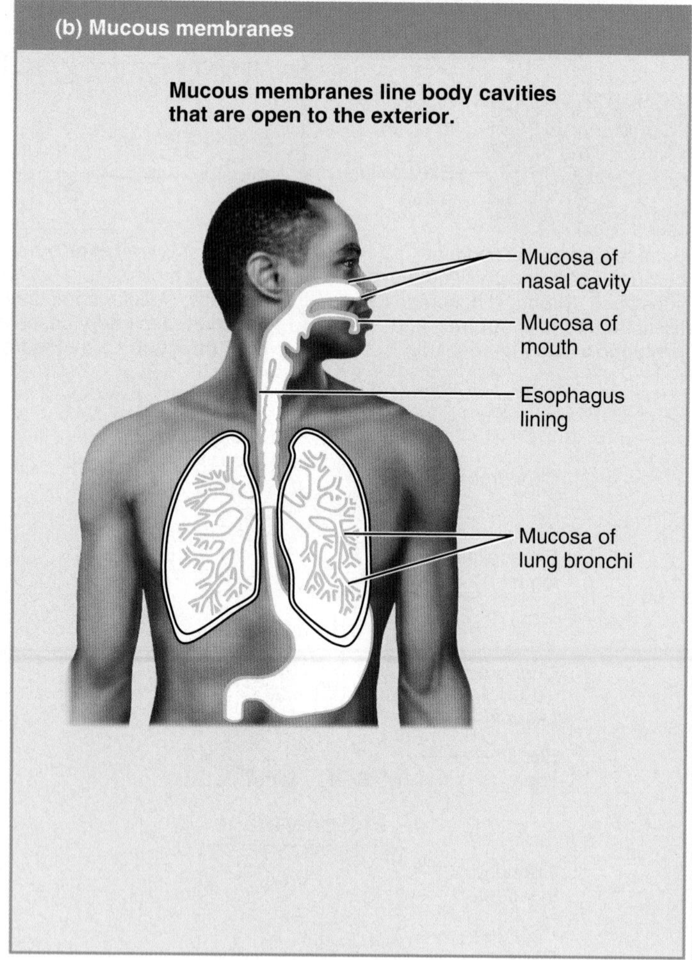

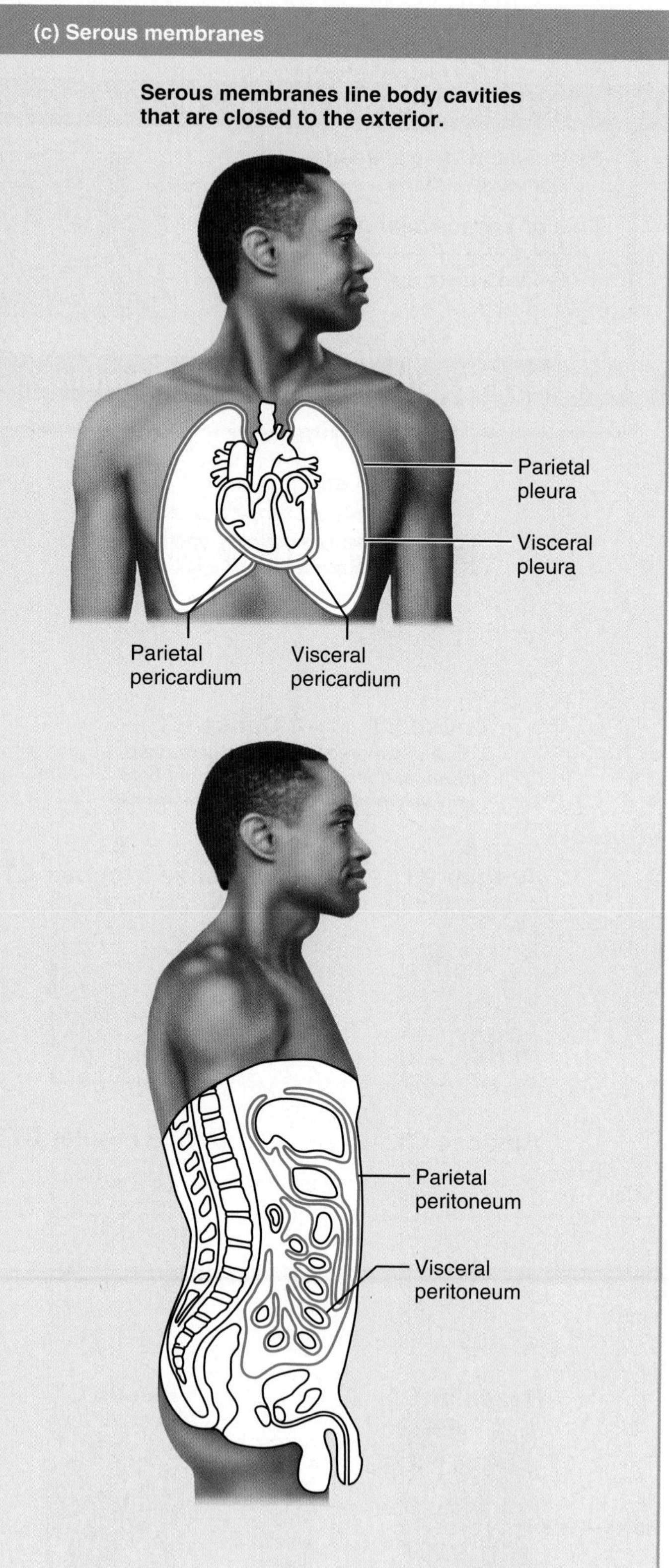

Figure 4.12 Types of membranes.

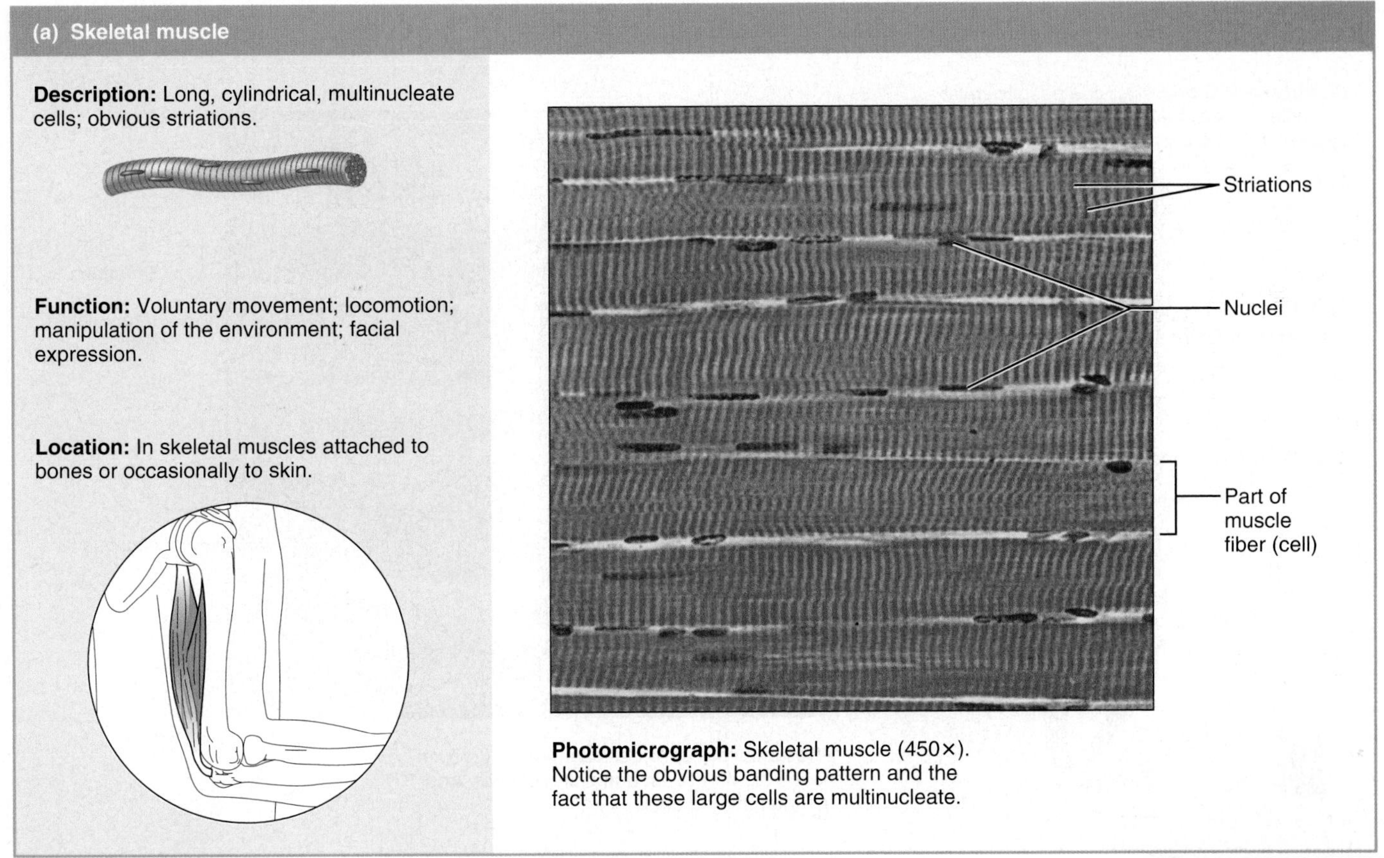

Figure 4.13 **Muscle tissues.**

III. MUSCLE TISSUE

learning outcome

▶ Briefly describe the three types of muscle tissue.

The two remaining tissue types are muscle and nervous tissues. These are sometimes called *composite tissues* because, along with their own muscle or nerve cells, they contain small amounts of areolar connective tissue. (Areolar connective tissue surrounds all blood vessels, and both muscle and nervous tissue are richly vascularized.)

Muscle tissues (Figure 4.13) bring about most kinds of body movements. Most muscle cells are called **muscle fibers.** They have an elongated shape and contract forcefully as they shorten. These cells contain many myofilaments (mi″o-fil′ah-ments; "muscle filaments"), cellular organelles filled with the actin and myosin filaments that bring about contraction in all cell types (p. 32). There are three kinds of muscle tissue: *skeletal, cardiac,* and *smooth*.

Skeletal muscle tissue (Figure 4.13a) is the major component of organs called skeletal muscles, which pull on bones to cause body movements. Skeletal muscle cells are long, large cylinders that contain many nuclei. Their obvious striated, or banded, appearance reflects a highly organized arrangement of their myofilaments. (Skeletal muscle tissue is described in detail in Chapter 10.)

Cardiac muscle tissue (Figure 4.13b) occurs in the wall of the heart. It contracts to propel blood through the blood vessels. Like skeletal muscle cells, cardiac muscle cells are striated. However, they differ in two ways: (1) Each cardiac cell has just one nucleus, and (2) cardiac cells branch and join at special cellular junctions called *intercalated* (in-ter′kah-la″-ted) *discs.* (The details of cardiac muscle tissue are discussed in Chapter 19.)

Smooth muscle tissue (Figure 4.13c) is so named because there are no visible striations in its cells. These cells are elongated with tapered ends and contain one centrally located nucleus. Smooth muscle primarily occurs in the walls of hollow viscera such as the digestive and urinary organs, uterus, and blood vessels. It generally acts to squeeze substances through these organs by alternately contracting and relaxing. (Smooth muscle tissue is described in detail in Chapter 23.)

IV. NERVOUS TISSUE

learning outcome

▶ Distinguish the cell types found in nervous tissue.

Nervous tissue is the main component of the nervous organs—the brain, spinal cord, and nerves—which regulate and control body functions. It contains two types of cells,

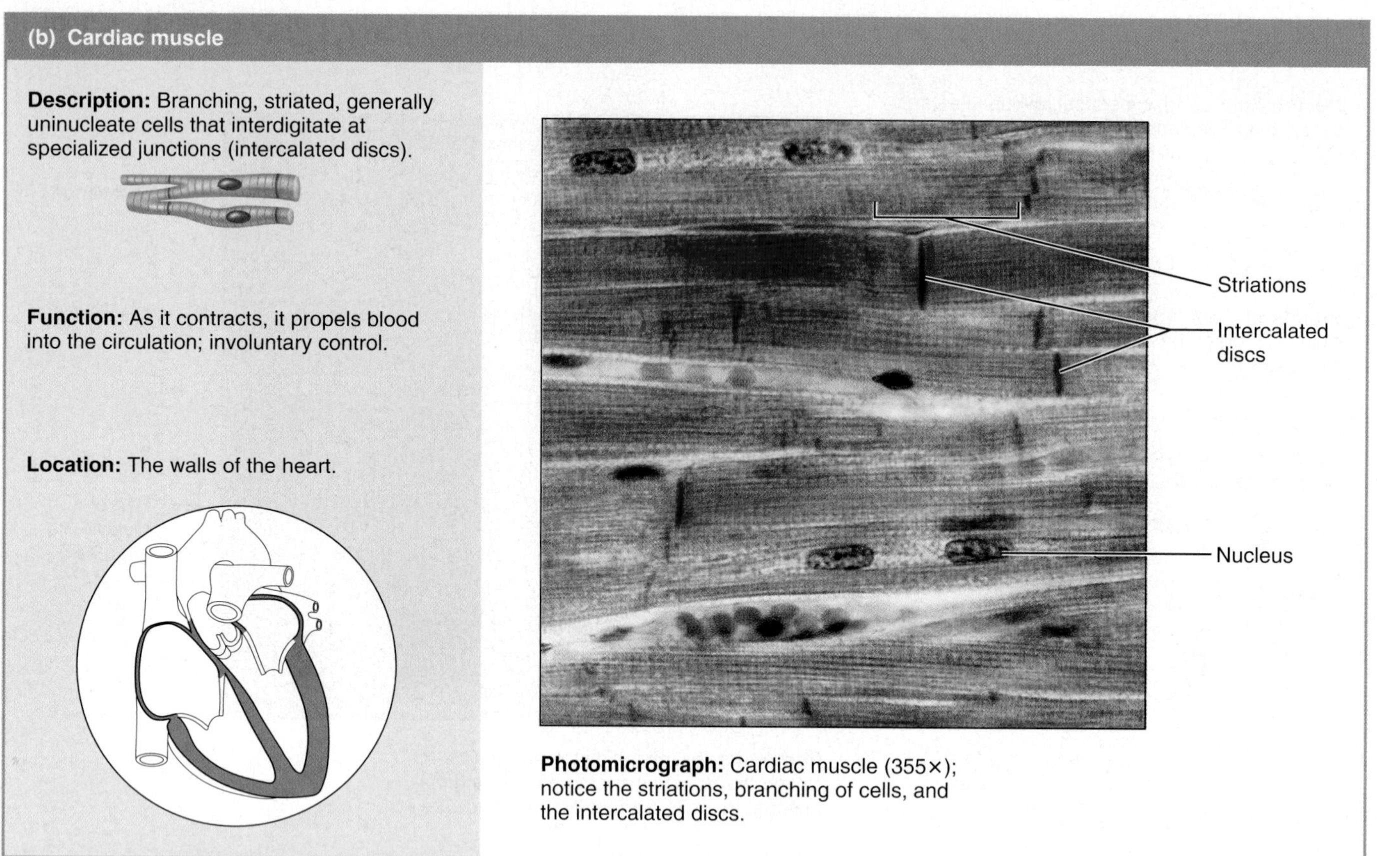

(b) Cardiac muscle

Description: Branching, striated, generally uninucleate cells that interdigitate at specialized junctions (intercalated discs).

Function: As it contracts, it propels blood into the circulation; involuntary control.

Location: The walls of the heart.

Photomicrograph: Cardiac muscle (355×); notice the striations, branching of cells, and the intercalated discs.

(c) Smooth muscle

Description: Spindle-shaped cells with central nuclei; no striations; cells arranged closely to form sheets.

Function: Propels substances or objects (foodstuffs, urine, a baby) along internal passageways; involuntary control.

Location: Mostly in the walls of hollow organs.

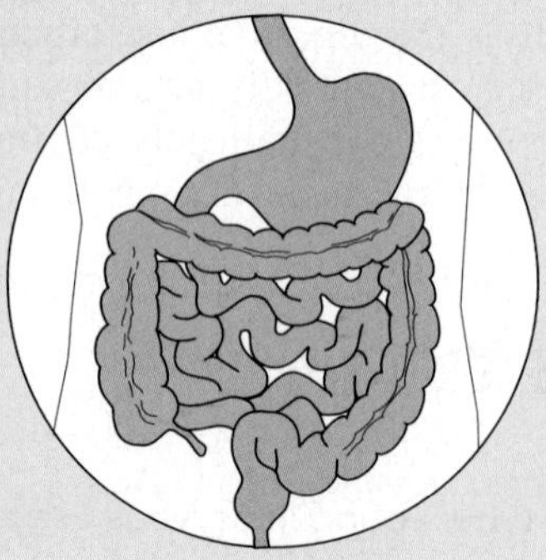

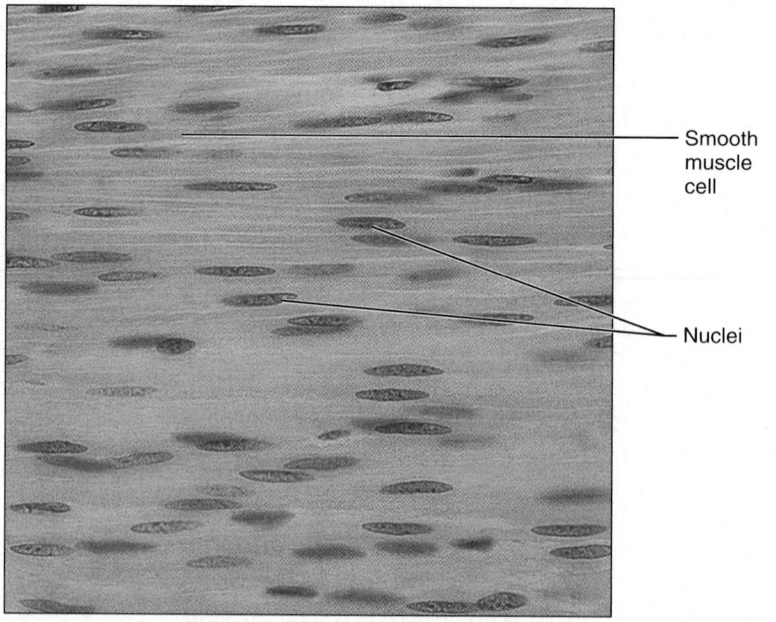

Smooth muscle cell

Nuclei

Photomicrograph: Sheet of smooth muscle from the digestive tract (465×).

Figure 4.13 **Muscle tissues,** *continued.*

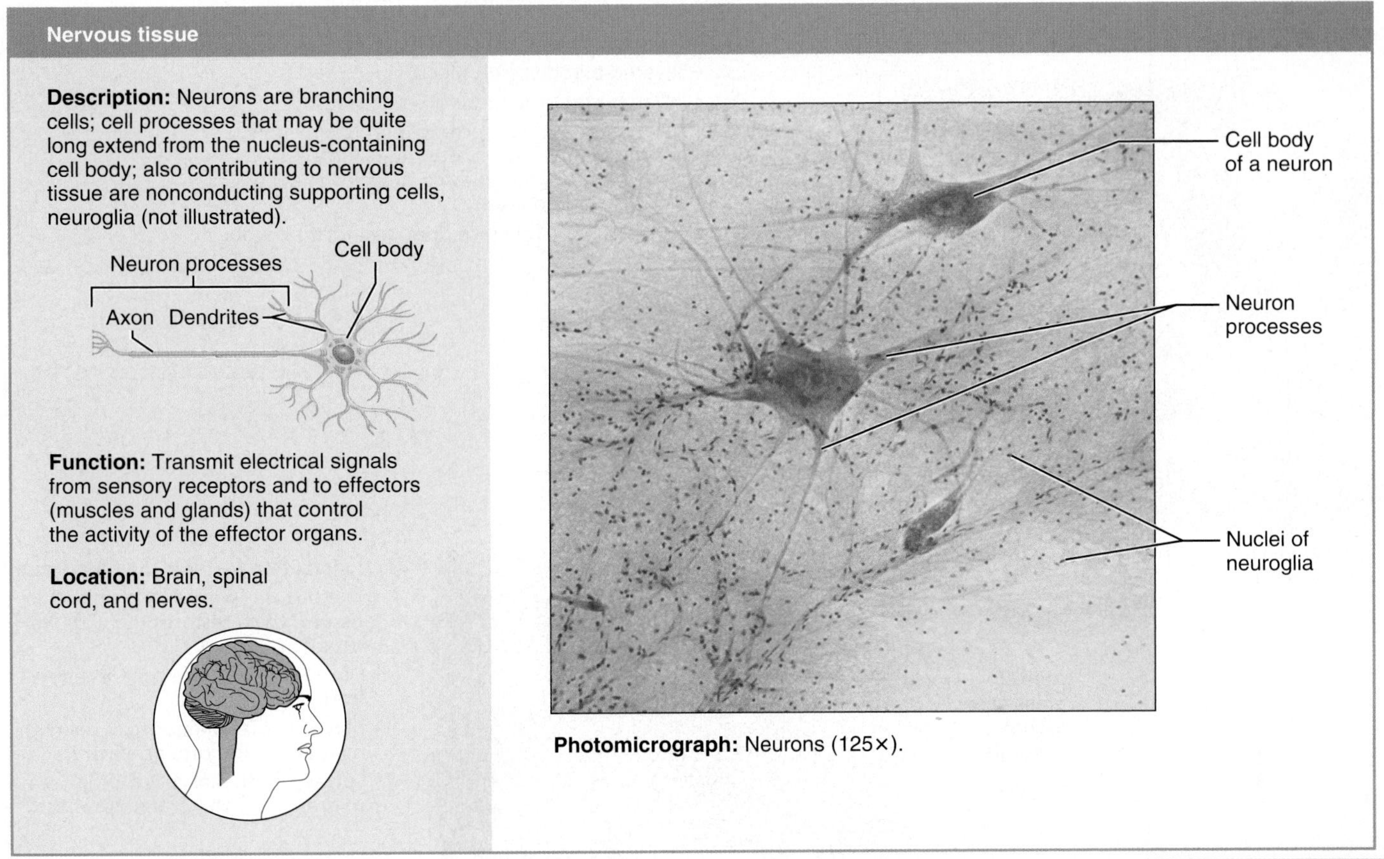

Figure 4.14 Nervous tissue.

neurons and supporting cells. **Neurons** are the highly specialized nerve cells (Figure 4.14) that generate and conduct electrical impulses. They have extensions, or processes, that allow them to transmit impulses over substantial distances within the body. **Dendrites** are cell processes that extend from the cell body of a neuron like branches of a tree. The dendrites, the receptive region of the neuron, transmit signals toward the cell body. A neuron also has a singular, long cell process extending from its cell body, the **axon,** that generates nerve impulses and transmits them away from the cell body. The supporting cells, called **neuroglia,** are nonconducting cells that nourish, insulate, and protect the delicate neurons. (A more complete discussion of nervous tissue appears in Chapter 12.)

☑ check your understanding

- ☐ 18. Look at a photomicrograph of smooth muscle tissue (Figure 4.13c). How can you tell this tissue is *not* stratified squamous epithelium?
- ☐ 19. Which nerve cells function to transmit electrical impulses?
- ☐ 20. Distinguish between the terms *striated* and *stratified.* Which tissues are striated? Which tissues are stratified?

(For answers, see Appendix B.)

TISSUE RESPONSE TO INJURY

learning outcome

- ▶ Describe the inflammatory and repair processes by which tissues recover from injury.

The body has many mechanisms for protecting itself from injury and invading microorganisms. Intact epithelia act as a physical barrier, but once that barrier has been penetrated, protective responses are activated in the underlying connective tissue proper. These are the *inflammatory* and *immune responses.* Inflammation is a nonspecific, local response that develops quickly and limits the damage to the injury site. The immune response, by contrast, takes longer to develop and is highly specific. It destroys particular infectious microorganisms and foreign molecules at the site of infection and throughout the body. This section concentrates on inflammation. (The immune response is discussed in Chapter 21.)

Inflammation

Almost every injury or infection leads to an inflammatory response. For example, assume your skin is cut by a dirty piece of glass, crushed by a blow in football, or has an infected pimple. As short-term, or acute, inflammation develops in the connective tissue, it produces four symptoms: *heat, redness, swelling,* and *pain.* You can trace the source of each symptom.

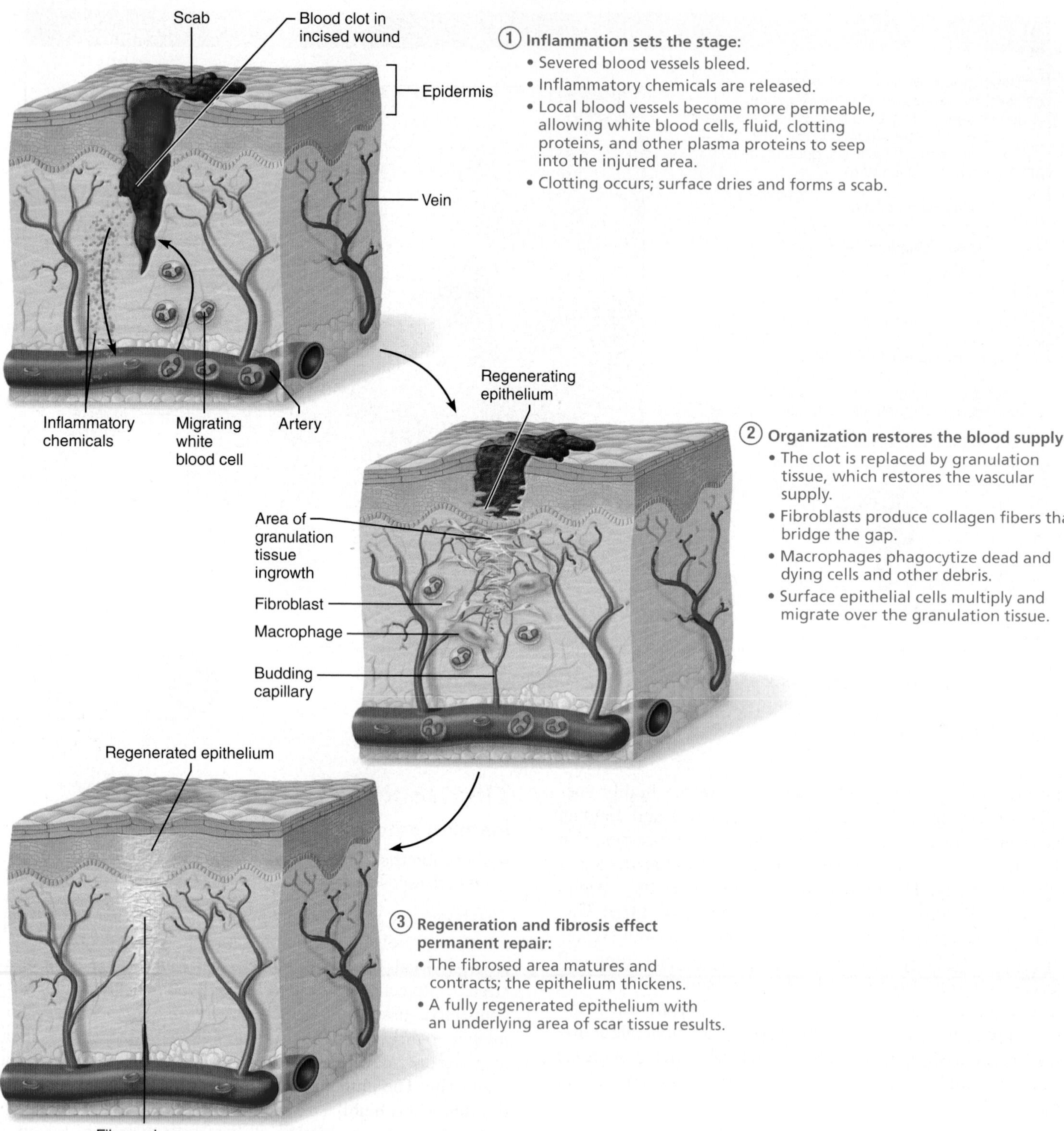

Figure 4.15 Tissue repair of a skin wound.

The initial insult induces the release of *inflammatory chemicals* into the nearby tissue fluid. Injured tissue cells, macrophages, mast cells, and proteins from blood all serve as sources of these inflammatory mediators. These chemicals signal nearby blood vessels to dilate (widen), thus increasing the flow of blood to the injury site. The increase in blood flow is the source of the *heat* and *redness* of inflammation. Certain inflammatory chemicals, such as histamine, increase the permeability of the capillaries, causing large amounts of tissue fluid to leave the bloodstream. The resulting accumulation of fluid in the connective tissue, called **edema** (ĕ-de′mah; "swelling"), causes the *swelling* of inflammation. The excess fluid presses on nerve endings, contributing to the sensation of *pain*. Some of the inflammatory chemicals also cause pain by affecting the nerve endings directly.

At first glance, inflammatory edema seems detrimental, but it is actually beneficial. The entry of blood-derived fluid into the injured connective tissue (1) helps to dilute toxins secreted by bacteria; (2) brings in oxygen and nutrients from the blood, necessary for tissue repair, and (3) brings in antibodies from the blood to fight infection. In very severe infections and in all wounds that sever blood vessels, the fluid leaking from the capillaries contains *clotting* proteins. In these cases, clotting occurs in the connective tissue matrix. The fibrous clot isolates the injured area and "walls in" the infectious microorganisms, preventing their spread.

The next stage in inflammation is *stasis* ("standing"). This is the slowdown in local blood flow that necessarily follows a massive exit of fluid from the capillaries. At this stage, white blood cells begin to leave the small vessels. First to appear at the infection site are neutrophils, then macrophages. These cells devour the infectious microorganisms and the damaged tissue cells as well.

Repair

Even as inflammation proceeds, repair begins. Tissue repair can occur in two major ways: by regeneration and by fibrosis. **Regeneration** is the replacement of a destroyed tissue by new tissue of the same kind, whereas **fibrosis** involves the proliferation of a fibrous connective tissue called **scar tissue.** Tissue repair in a skin wound involves both regeneration and fibrosis. After the blood within the cut has clotted, the surface part of the clot dries to form a scab (Figure 4.15, ①). At this point, the repair begins with a step called organization.

Organization is the process by which the clot is replaced by granulation tissue (Figure 4.15, ②). **Granulation tissue** is a delicate pink tissue made of several elements. It contains capillaries that grow in from nearby areas, as well as proliferating fibroblasts that produce new collagen fibers to bridge the gash. Some of its fibroblasts have contractile properties that pull the margins of the wound together. As organization proceeds, macrophages digest the original clot, and the deposit of collagen continues. As more collagen is made, the granulation tissue gradually transforms into fibrous scar tissue (Figure 4.15, ③).

During organization, the surface epithelium begins to *regenerate,* growing under the scab until the scab falls away. The end result is a fully regenerated epithelium and an underlying area of scar.

The process we have just outlined describes the healing of a wound such as a cut, scrape, or puncture. In pure *infections* (a pimple or sore throat), by contrast, there is usually no clot formation or scarring. Only severe infections lead to scarring.

The capacity for regeneration varies widely among the different tissues. Epithelia regenerate extremely well, as do bone, areolar connective tissue, dense irregular connective tissue, and blood-forming tissue. Smooth muscle and dense regular connective tissue have a moderate capacity for regeneration, but skeletal muscle and cartilage have only a weak capacity. Cardiac muscle and the nervous tissue in the brain and spinal cord have no functional regenerative capacity. However, recent studies have shown that some unexpected cellular division occurs in both these tissues after damage, and efforts are under way to coax them to regenerate better.

In nonregenerating tissues and in exceptionally severe wounds, fibrosis totally replaces the lost tissue. The resulting scar appears as a pale, often shiny area and shrinks during the months after it first forms. A scar consists mostly of collagen fibers and contains few cells or capillaries. Although it is very strong, it lacks the flexibility and elasticity of most normal tissues, and it is unable to perform the normal functions of the tissue it has replaced.

Irritation of visceral organs can cause them to adhere to one another or to the body wall as they scar. Such **adhesions** can prevent the normal churning actions of loops of the intestine, dangerously halting the movement of food through the digestive tube. They can also restrict the movement of the heart and lungs and immobilize the joints. After almost all abdominal surgeries, adhesions form between the body wall and the abdominal viscera, making subsequent surgery in that region more difficult.

✓ check your understanding

☐ 21. Which tissues regenerate easily? Which tissues do not regenerate?

☐ 22. Is the scar tissue that creates a "scar" located in the epithelium or in the underlying connective tissue?

☐ 23. What causes the heat and swelling in an infected tissue?

(For answers, see Appendix B.)

THE TISSUES THROUGHOUT LIFE

learning outcome

▶ Indicate the embryonic derivation of each tissue class.

▶ Briefly describe changes that occur in tissues with age.

The embryonic derivations of the four basic tissues are as follows (Figure 4.16):

- Connective and muscle tissue derive from mesenchyme, mostly of the mesoderm germ layer.
- Most epithelial tissues develop from embryonic epithelium of the ectoderm and endoderm layers. A few epithelia—namely, the epithelium lining the vessels and the mesothelium lining the ventral body cavity—derive from mesodermal mesenchyme.
- Nervous tissue in the brain and spinal cord derives from ectodermal epithelium (recall Chapter 3, p. 52).

By the end of the second month of development, the four primary tissues have formed, and all major organs are in place. In virtually all tissues, the cells continue to divide throughout the prenatal period, providing the rapid body growth that occurs before birth. The division of *nerve* cells, however, stops or nearly stops during the fetal period. After birth, the cells of most other tissues continue to divide until adult body size is reached. In adulthood, only the epithelial tissues and blood cell–forming tissues are highly mitotic. Cellular division in other tissues slows greatly, although many tissues retain a regenerative capacity.

CHAPTER SUMMARY

Take full advantage of the invaluable online practice and study tools for this chapter by going to the Study Area of MasteringA&P®. There you will find:

- Art Labeling Activities, Quizzes, Practice Tests, Memory Games
- Animal Organ Dissection, Cat Dissection and Bone Videos
- PAL Cadaver
- PAL Histology
- A&P Flix Animations
- Dynamic Study Modules

1. Tissues are collections of structurally similar cells with related functions. They also contain nonliving material between their cells. The four primary tissues are epithelial tissue, connective tissue, muscle tissue, and nervous tissue.

I. EPITHELIAL TISSUE (pp. 65–77)

2. Epithelia are sheets of cells that cover body surfaces and line body cavities. Epithelia also form most of the body glands. The functions of epithelia include protection, secretion, absorption, diffusion, filtration, and sensory reception.

Special Characteristics of Epithelia (p. 66)

3. Epithelia exhibit a high degree of cellularity, little extracellular material, specialized cell junctions, polarity, avascularity, and the ability to regenerate. They are underlain by loose connective tissue.

Classification of Epithelia (pp. 66–72)

4. Epithelia are classified by the number of cell layers as simple (one layer) or stratified (more than one layer) and by cell shape as squamous, cuboidal, or columnar. Stratified epithelia are named according to the shape of their apical cells.

5. Simple squamous epithelium is a single layer of flat cells. Its thinness allows molecules to pass through it rapidly by passive diffusion. It lines the air sacs of the lungs, the interior of blood vessels (endothelium), and the ventral body cavity (mesothelium).

6. Simple cuboidal epithelium functions in absorption and secretion. It occurs in kidney tubules, the secretory cells of many glands, and the small ducts of glands.

7. Simple columnar epithelium lines the stomach and intestines, and a ciliated version lines the uterine tubes. Like simple cuboidal epithelium, it is active in secretion and absorption.

8. Pseudostratified columnar epithelium is a simple epithelium that contains both short and tall cells. A ciliated version lines most of the respiratory passages.

9. Stratified squamous epithelium is multilayered and thick. Its apical cells are flat, and it resists abrasion. Examples are the epidermis and the lining of the mouth, esophagus, and vagina.

10. Stratified cuboidal and stratified columnar epithelium, while rare, line the large ducts of some glands.

11. Transitional epithelium is a stratified epithelium that thins when it stretches. It lines the hollow urinary organs.

Glands (pp. 72–73)

12. A gland is one or more cells specialized to secrete a product. Most glandular products are proteins released by exocytosis.

13. Endocrine (ductless) glands secrete hormones, which enter the circulatory vessels and travel to target organs, from which they signal a response.

14. Exocrine glands secrete their products onto body surfaces or into body cavities. Mucus-secreting goblet cells are unicellular exocrine glands. Multicellular exocrine glands are classified by the structure of their ducts as simple or compound and by the structure of their secretory units as tubular, alveolar (acinar), or tubuloalveolar.

Epithelial Surface Features (pp. 74–77)

15. Lateral surface features of epithelia: The main types of cell junctions are tight junctions that close off the extracellular spaces, adhesive belt junctions and desmosomes that bind cells together, and gap junctions through which small molecules pass from cell to cell.

16. Basal surface feature: Epithelial cells lie on a protein sheet called the basal lamina. This acts as a filter and a scaffolding on which regenerating epithelial cells can grow. The basal lamina, plus some underlying reticular fibers, form the thicker basement membrane.

17. Apical surface features: Microvilli occur on most moist epithelia. They increase the epithelial surface area and may anchor sheets of mucus. Cilia are whiplike projections of the plasma membrane that beat to move fluid (usually mucus). Microtubules in the cores of cilia generate ciliary movement.

PAL Histology/Epithelial Tissue

II. CONNECTIVE TISSUE (pp. 77–92)

18. Connective tissue is the most diverse and abundant class of tissues. Its four basic classes are connective tissue proper (loose and dense), cartilage, bone, and blood.

Special Characteristics of Connective Tissues (p. 78)

19. All connective tissues consist of cells separated by abundant extracellular matrix. In all connective tissues except blood, the matrix consists of ground substance, fibers, and tissue fluid. All connective tissues are derived from embryonic mesenchyme.

Structural Elements of Connective Tissues (pp. 78–81)

20. The primary cell type in connective tissue produces the fibers and the ground substance of the extracellular matrix: fibroblasts in connective tissue proper; chondroblasts in cartilage; osteoblasts in bone. Once the tissue is formed, these cells are called fibrocytes, chondrocytes, and osteocytes, and they function to maintain the tissue matrix. Additional cell types found in many connective tissues include fat cells and various defense cells: white blood cells, macrophages, and mast cells.

21. Three types of fibers are found in connective tissues: Collagen fibers function to resist tension, reticular fibers provide structural support, and elastic fibers enable recoil of stretched tissues.

22. The ground substance is a gel-like material that functions to hold fluid. The matrix of bone is hardened with calcified mineral salts.

Classification of Connective Tissues (pp. 81–89)

23. *Loose areolar connective tissue* surrounds capillaries and underlies most epithelia. Its main functions are to (1) support and bind other tissues with its fibers (collagen, reticular, elastic), (2) hold tissue fluid in its jellylike ground substance, (3) fight infection with its many blood-derived defense cells (macrophages, plasma cells, neutrophils, etc.), and (4) store nutrients in fat cells.

24. *Adipose connective tissue* is similar to areolar connective tissue but is dominated by nutrient-storing fat cells. This white fat is plentiful

in the hypodermis below the skin. Brown fat, common in infants but also found in adults, produces heat.

25. *Reticular connective tissue* resembles areolar tissue, except that its only fibers are reticular fibers. These form networks of caverns that hold free blood cells. Reticular tissue occurs in bone marrow, lymph nodes, and the spleen.

26. Dense connective tissue contains exceptionally thick collagen fibers and resists tremendous pulling forces. In *dense irregular connective tissue,* the collagen fibers run in various directions. This tissue occurs in the dermis of the skin and in organ capsules.

27. *Dense regular connective tissue* contains bundles of collagen fibers that all run in the same direction and are separated by rows of fibroblasts. This tissue, which is subject to high tension from a single direction, is the main tissue in tendons, ligaments, and fascia.

28. *Elastic connective tissue* has a high proportion of elastic fibers. This tissue is located in the walls of arteries, around the bronchial tubes, and within certain ligaments.

29. *Cartilage* and *bone* have the basic structure of connective tissue (cells and matrix), but their stiff matrix allows them to resist compression. Their cells are located in lacunae, spaces within the matrix. Cartilage is springy and avascular. Its matrix contains mostly water.

30. Bone tissue has a hard, collagen-rich matrix embedded with calcium salts. This mineral gives bone compressive strength.

31. *Blood* consists of red and white blood cells in a fluid matrix called plasma. It is the most atypical connective tissue.

Covering and Lining Membranes (pp. 89–92)

32. Membranes, each consisting of an epithelium plus an underlying layer of connective tissue, cover broad surfaces in the body. The cutaneous membrane (skin), which is dry, covers the body surface. Mucous membranes, which are moist, line the hollow internal organs that open to the body exterior. Serous membranes, which produce serous fluid, line the pleural, pericardial, and peritoneal cavities.

PAL **Histology/Connective Tissue**

III. MUSCLE TISSUE (p. 93)

33. Muscle tissue consists of long muscle cells that are specialized to contract and generate movement. A scant loose connective tissue separates the muscle cells. There are three types of muscle tissue: *Skeletal muscle* is in the muscles that move the skeleton. Its cells are cylindrical and striated. *Cardiac muscle* is in the wall of the heart, and it pumps blood. Its cells branch and have striations. *Smooth muscle* is in the walls of hollow organs, and it usually propels substances through these organs. Its cells are elongated with tapered ends and lack striations.

IV. NERVOUS TISSUE (pp. 93–95)

34. Nervous tissue, the main tissue of the nervous system, is composed of neurons and supporting cells called neuroglia. Nervous tissue regulates body functions. *Neurons* are branching cells that receive and transmit electrical impulses. *Neuroglia* nourish, insulate, and protect neurons.

PAL **Histology/Nervous Tissue**

TISSUE RESPONSE TO INJURY (pp. 95–97)

Inflammation (pp. 95–97)

35. Inflammation is a response to tissue injury and infection. It is localized to the connective tissue and the vessels of the injury site. The inflammatory response begins with dilation of blood vessels (causing redness and heat), followed by edema (causing swelling and pain). Stasis results, and white blood cells migrate into the injured tissue.

Repair (pp. 97)

36. Tissue repair begins during inflammation and may involve tissue regeneration, fibrosis (scarring), or both. Repair of a cut begins with organization, during which the clot is replaced with granulation tissue. Collagen deposition replaces granulation tissue with scar tissue.

37. Epithelia, areolar and dense irregular connective tissue, bone, and blood-forming tissue are tissues that regenerate well. Tissues with moderate regenerative capabilities are smooth muscle and dense regular connective tissue. Skeletal muscle and cartilage have limited regenerative capability. Cardiac muscle and most nervous tissues do not regenerate naturally and are replaced by scar tissue.

THE TISSUES THROUGHOUT LIFE (pp. 97–99)

38. Muscle and connective tissues are derived from mesenchyme, primarily from the mesoderm germ layer. Epithelial tissue forms from all three embryonic layers: ectoderm, mesoderm, and endoderm. Nervous tissue is derived from ectoderm.

39. Tissue function declines with age. The decrease in mass and viability seen in most tissues during old age partially reflects circulatory deficits or poor nutrition.

REVIEW QUESTIONS

Multiple Choice/Matching Questions

(For answers, see Appendix B.)

1. An epithelium that has several cell layers, with flat cells in the apical layer, is called (choose all that apply): (a) ciliated, (b) columnar, (c) stratified, (d) simple, (e) squamous.

2. The type of gland that secretes products such as milk, saliva, bile, or sweat through a duct is (a) an endocrine gland, (b) an exocrine gland, (c) a goblet cell.

3. Match the epithelial type named in column B with the appropriate location in column A.

Column A	Column B
____ **(1)** lines inside of stomach and most of intestines	**(a)** pseudostratified ciliated columnar
____ **(2)** lines inside of mouth	**(b)** simple columnar
____ **(3)** lines much of respiratory tract (including the trachea)	**(c)** simple cuboidal
____ **(4)** mesothelium	**(d)** simple squamous
____ **(5)** lines inside of urinary bladder	**(e)** stratified columnar
	(f) stratified squamous
	(g) transitional

4. In connective tissue proper, the cell type that secretes the fibers and ground substance is the (a) fibroblast, (b) neutrophil, (c) mast cell, (d) macrophage, (e) chondrocyte.

5. Identify each of the cell surface features described below.
   (a) whiplike extensions that move fluids across epithelial surfaces
   (b) little "fingers" on apical epithelial surfaces that increase cell surface area and anchor mucus
   (c) the cell junction that holds cells together by linking the intermediate filaments of adjacent cells
   (d) a cell junction that closes off the extracellular space
   (e) of the basement membrane and the basal lamina, the one that can be seen by light microscopy

6. Match each epithelial tissue in column B with its function listed in column A.

Column A	Column B
____ (1) functions in diffusion and filtration	(a) stratified epithelium
____ (2) functions in protection	(b) simple columnar epithelium
____ (3) functions in secretion and absorption	(c) transitional epithelium
____ (4) changes shape as the tissue stretches	(d) ciliated epithelium
____ (5) produces movement of material	(e) simple squamous epithelium

7. For each connective tissue (CT) listed, indicate the predominant type or types of fibers found in the extracellular matrix by: (a) collagen fibers, (b) elastic fibers, (c) reticular fibers.

____ (1) bone
____ (2) loose areolar (CT)
____ (3) dense irregular (CT)
____ (4) reticular (CT)
____ (5) fibrocartilage

8. The muscle tissue that is striated is (a) skeletal muscle, (b) cardiac muscle, (c) smooth muscle, (d) cardiac and skeletal muscle, (e) all three types of muscle tissue.

9. Neuroglia (a) conduct electrical impulses, (b) are nerve stem cells, (c) are support cells that aid neurons, (d) are an abnormal growth of neural tissue.

10. Which of the following cells are not found in a connective tissue? (a) fibroblasts, (b) goblet cells, (c) macrophages, (d) mast cells, (e) chondrocytes.

11. The ground substance in connective tissue proper functions to (a) support and strengthen the tissue, (b) hold tissue fluid, (c) fight infection, (d) store nutrients, (e) resist tension.

12. What embryonic layer (ectoderm, mesoderm, or endoderm) primarily forms each of the following major tissues? (a) connective tissue, (b) muscle tissue, (c) nervous tissue.

Short Answer Essay Questions

13. Define tissue.

14. Explain the classification of multicellular exocrine glands, and supply an example of each class.

15. Name four functions of areolar connective tissue, and relate each function to a specific structural part of this tissue.

16. What is fascia?

17. (a) Where does tissue fluid come from? (b) What is the function of tissue fluid?

18. Name the specific type of connective tissue being described:
   (a) around the capillaries
   (b) in the ligamentum flavum between vertebrae
   (c) the original, embryonic connective tissue
   (d) hard tissue of the skull
   (e) main tissue in ligaments
   (f) dominates the dermis
   (g) dominates the hypodermis

19. Name the four classic symptoms of inflammation, and explain what each symptom represents in terms of changes in the injured tissue.

20. (a) Define endocrine gland. (b) What is a hormone?

21. Name two specific tissues that regenerate well and two that regenerate poorly.

22. Of the four basic tissue types, which two develop from mesenchyme?

23. The body does *not* entirely consist of cells, and, in fact, a noncellular material probably makes up more of the body than the cells do. What is this noncellular material?

24. What are the differences between a mucous membrane and a serous membrane?

25. What is the main type of tissue in the following structures? (a) a ligament or tendon; (b) a bone of the leg; (c) a muscle such as the biceps of the arm; (d) the brain; (e) the flexible skeleton in the outer ear; (f) the contractile wall of the heart; (g) kidney tubules.

26. Indicate the name and location of an epithelial tissue formed from each of the embryonic layers: ectoderm, mesoderm, and endoderm.

Critical Reasoning & Clinical Application Questions

1. Systemic lupus erythematosus, or lupus, is a condition that sometimes affects young women. It is a chronic (persistent) inflammation that affects all or most of the connective tissue proper in the body. Suzy is told by her doctor that she has lupus, and she asks whether it will have widespread or merely localized effects within the body. What would the physician answer?

2. Sailors who made long ocean journeys in the time of Christopher Columbus ate only bread, water, and salted meat on their journeys. They often suffered from scurvy. What changes were made to prevent scurvy?

3. Three patients in an intensive care unit have sustained damage and widespread tissue death in three different organs. One patient has brain damage from a stroke, another had a heart attack that destroyed cardiac muscle, and the third injured much of her liver (a gland) in a crushing car accident. All three patients have stabilized and will survive, but only one will gain full functional recovery through tissue regeneration. Which one, and why?

4. In adults, over 90% of all cancers are either adenomas (adenocarcinomas) or carcinomas. In fact, cancers of the skin, lung, colon, breast, and prostate are all in these categories. Which one of the four basic tissue types gives rise to most cancers? Why might this type of tissue be so susceptible to cancer?

5. A patch of scar tissue that forms in the wall of the urinary bladder, the heart, or another hollow organ may severely hamper the function of that organ. Considering the properties of scar tissue, why do you think this is so?

6. Ciliated epithelium is located in the bronchial tubes and in the uterine tubes. What functional similarity do these structures share that is accomplished by ciliated epithelium?

The Integumentary System

5

Would you be enticed by an advertisement for a coat that is waterproof, stretchable, washable, and permanent press, that automatically repairs small rips and burns, and that is guaranteed to last a lifetime with reasonable care? It sounds too good to be true, but you already have such a coat—your skin. The skin and its appendages (sweat glands, oil glands, hair, and nails) serve a number of functions. Together, these organs make up the **integumentary system** (in-teg″u-men′tar-e; "covering").

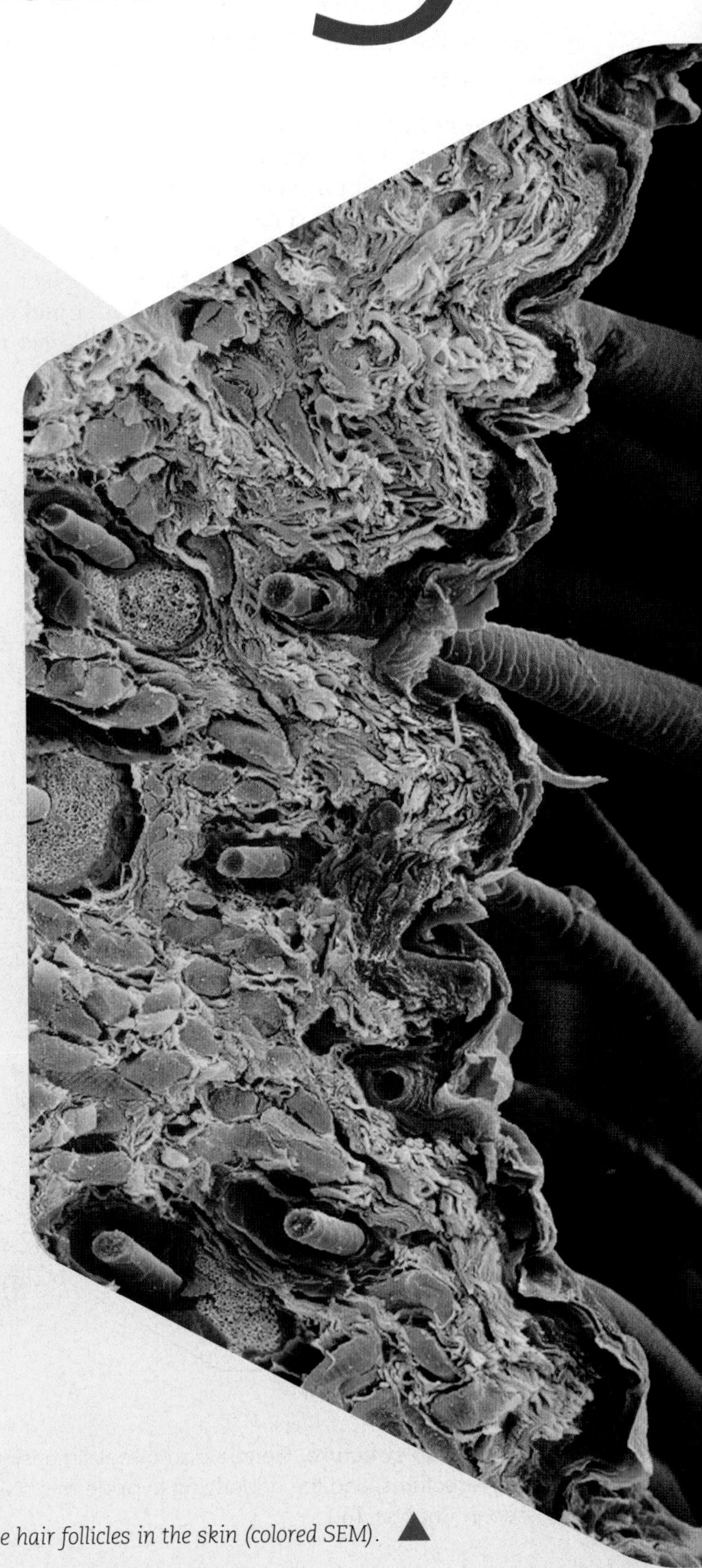

Freeze fracture section through the hair follicles in the skin (colored SEM). ▲

THE SKIN AND THE HYPODERMIS

learning outcomes

- Name the tissue types that compose the epidermis and the dermis, and describe the primary functions of skin.
- Name the layers of the epidermis and the dermis, and describe the structure and functions of each layer.
- Describe the structure and function of the hypodermis.
- Describe the factors that contribute to skin color.

The **skin (integument)** and its appendages are the first *organs* discussed in this book. An organ consists of tissues working together to perform certain functions (see Chapter 1). Although the skin is less complex than most other organs, it is still an architectural marvel. It is also the largest of all the organs, accounting for about 7% of total body weight.

The skin varies in thickness from 1.5 to 4 mm or more in different regions of the body. It has two distinct regions (Figure 5.1 and Figure 5.2): The superficial region is a thick epithelial tissue, the *epidermis*. Deep to the epidermis is the *dermis,* a fibrous connective tissue. Just deep to the skin lies a fatty layer called the *hypodermis,* composed of loose areolar connective tissue and adipose tissue. Although the hypodermis is not part of the integumentary system, it shares some of the skin's functions and is therefore described in this chapter.

The skin performs a variety of functions:

1. **Protection.** Skin cushions and insulates the deeper body organs and protects the body from bumps, scrapes, and cuts. Skin also protects the body from chemicals and invading microorganisms. The epidermis is waterproof, preventing unnecessary loss of water across the body surface, and cells in the epidermal layer produce pigment to protect the skin from the harmful effects of ultraviolet (UV) radiation.
2. **Body temperature regulation.** The skin's rich capillary networks and sweat glands regulate the loss of heat from the body, helping to control body temperature.

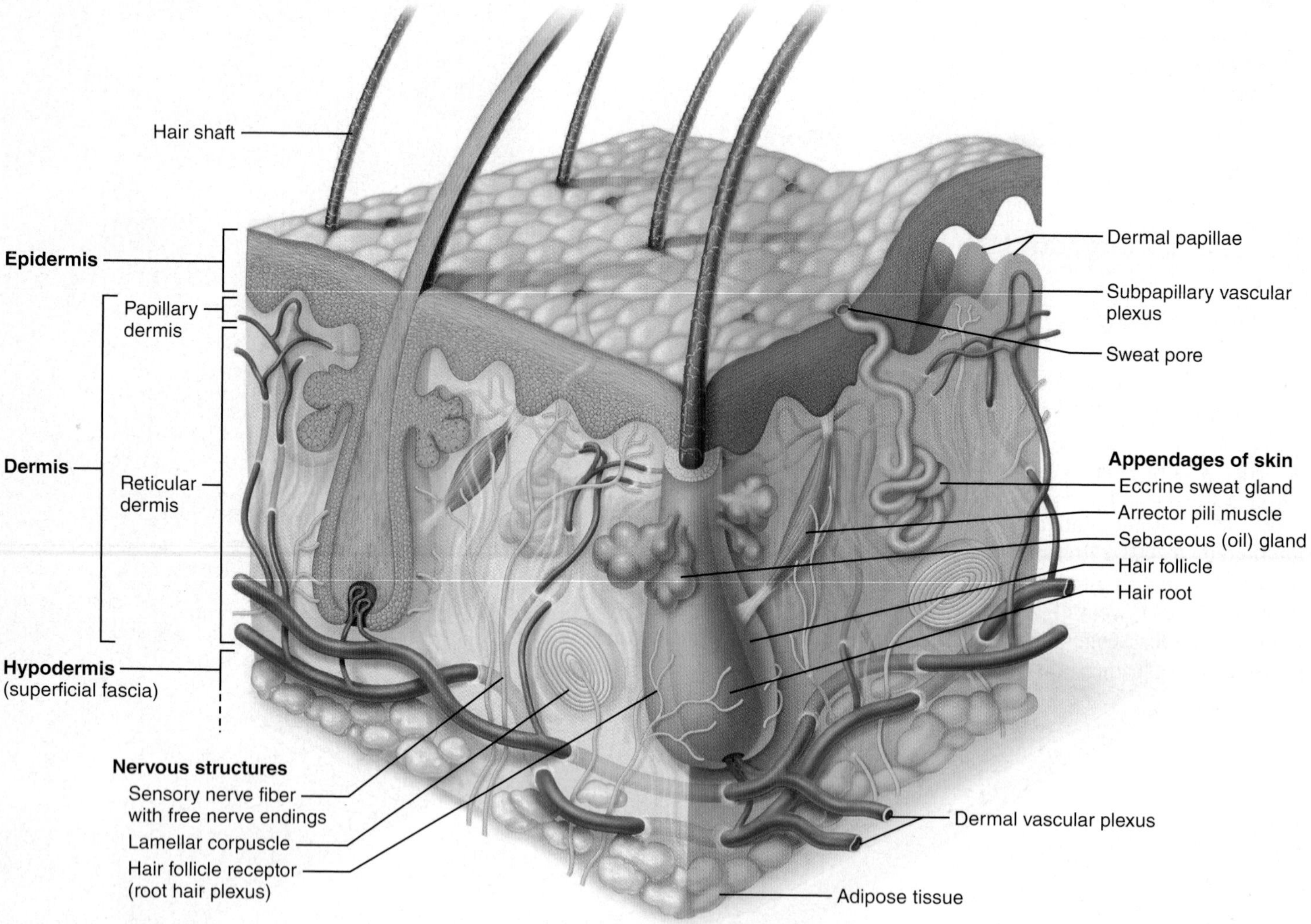

Figure 5.1 Skin structure. Generalized three-dimensional diagram of the skin, accessory structures, and the underlying hypodermis. (Nervous system structures are discussed in Chapter 14.)

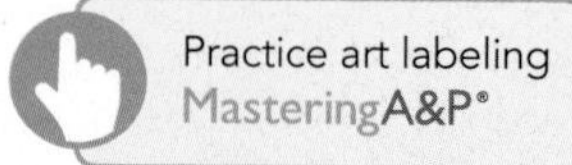

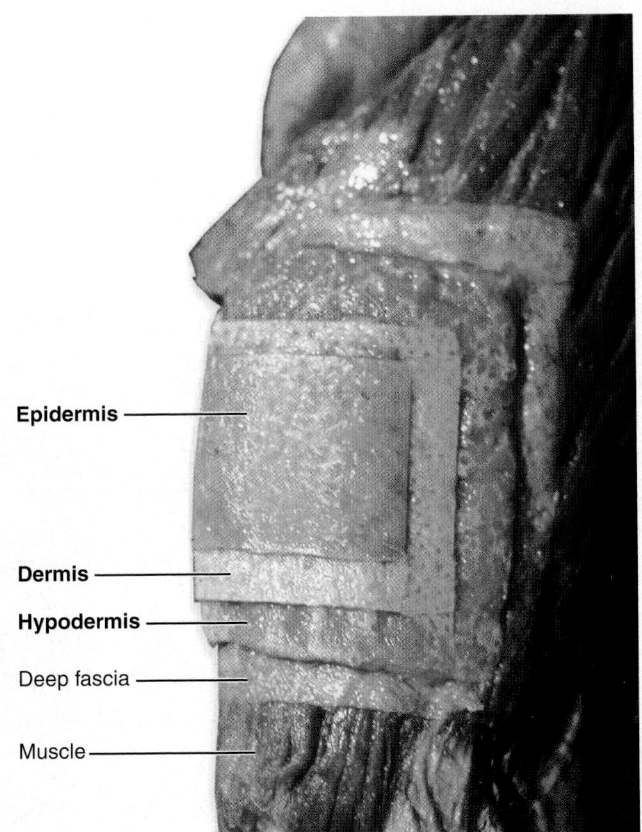

Figure 5.2 **Gross structure of skin and underlying tissues.** Photo of the upper arm of a cadaver.

3. **Excretion.** The skin acts as a miniature excretory system when urea, salts, and water are lost through sweat.
4. **Production of vitamin D.** The epidermal cells use UV radiation to synthesize vitamin D, a molecule necessary for absorbing calcium from the digestive tract.
5. **Sensory reception.** The skin contains sense organs called *sensory receptors* that are associated with nerve endings. By sensing touch, pressure, temperature, and pain, these receptors keep us aware of conditions at the body surface.

You will explore the skin's functions in greater detail as you explore its anatomy.

check your understanding

☐ 1. What is the meaning of the word roots *epi-*, *hypo-*, and *derm*?

☐ 2. Name the five major functions of skin.

(For answers, see Appendix B.)

Epidermis

The **epidermis** (ep″ĭ-der′mis; "on the skin") is a keratinized stratified squamous epithelium that contains four distinct types of cells: *keratinocytes, melanocytes, tactile epithelial cells,* and *dendritic cells* (Figure 5.3).

Keratinocytes (ke-rat′ĭ-no-sīts), the most abundant epidermal cell, produce **keratin,** a tough fibrous protein that gives the epidermis its protective properties. Keratinocytes provide physical and mechanical protection. They also produce antibiotics and enzymes that detoxify the harmful chemicals to which our skin is exposed.

Keratinocytes are tightly connected to one another by a large number of desmosomes. They arise in the deepest part of the epidermis from cells that undergo almost continuous mitosis. As these cells are pushed up by the production of new cells beneath them, they make the keratin that eventually fills their cytoplasm. By the time they approach the skin surface, they are dead, flat sacs completely filled with keratin. Millions of these dead cells rub off every day, giving us an entirely new epidermis every 35 to 45 days—the average time from the birth of a keratinocyte to its final wearing away. Where the skin experiences friction, both cell production and keratin formation are accelerated.

The other epidermal cell types are sparsely distributed among the keratinocytes. We will discuss these cells and their functions as we examine the layers of the epidermis.

Layers of the Epidermis

Variation in the thickness of the epidermis determines whether skin is thick or thin. In **thick skin,** which covers the palms and soles, the epidermis consists of five layers, or *strata* (stra′tah; "bed sheets"). In **thin skin,** which covers the rest of the body, only four strata are present.

Stratum Basale (Basal Layer) The **stratum basale** (stra′tum ba-sal′e), the deepest epidermal layer, is firmly attached to the underlying dermis along a wavy borderline. Also called the *stratum germinativum* (jer-mĭ-na″te′-vum; "germinating layer"), it consists of a single row of cells, mostly stem cells representing the youngest keratinocytes (Figure 5.3). These cells divide rapidly, and many mitotic nuclei are visible. **Tactile epithelial cells,** or Merkel cells, are distributed sparsely among the keratinocytes. Each hemisphere-shaped tactile epithelial cell is intimately associated with a disclike sensory nerve ending and functions as a receptor for touch.

About 10% to 25% of the cells in the stratum basale are spider-shaped **melanocytes** (mel′ah-no-sīts; "melanin cells"), which make the dark skin pigment **melanin** (mel′ah-nin; "black"). Melanin is made in membrane-walled granules and then transferred through the cell processes (the "spider-legs") to nearby keratinocytes. Consequently, the basal keratinocytes contain more melanin than do the melanocytes themselves. This melanin clusters on the superficial side of keratinocytes between the incoming radiation and the cell nuclei, thus shielding the cell nuclei from UV rays, which can damage DNA and cause cancer (see p. 118). In light-skinned people, the melanin is digested by lysosomes in cells a short distance above the basal layer. In dark-skinned people, no such digestion occurs, so melanin occupies keratinocytes throughout the epidermis. Although dark-skinned people

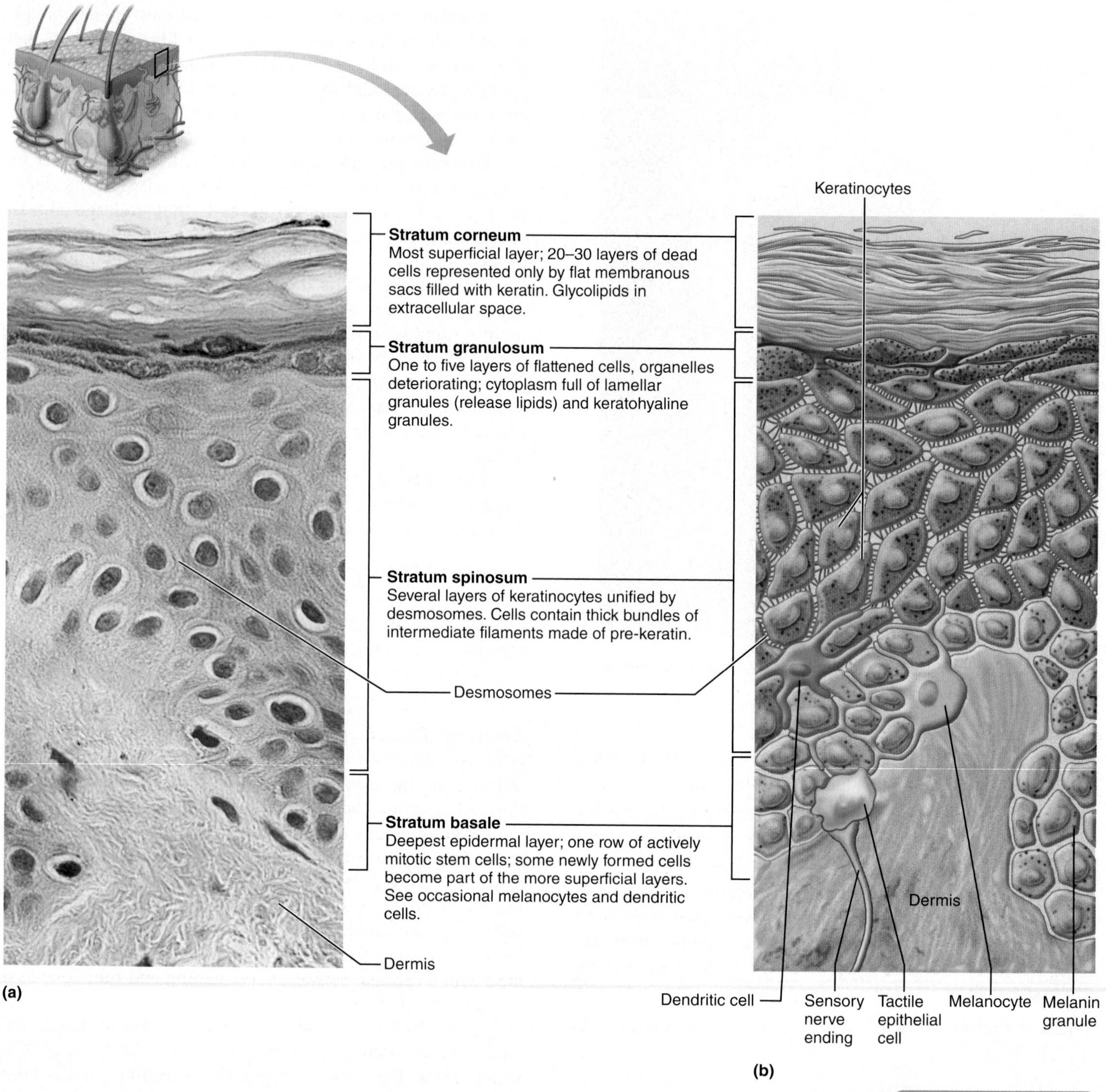

Figure 5.3 Epidermal cells and layers of the epidermis. (**a**) Photomicrograph of the four major epidermal layers in thin skin (500×). (**b**) Diagram showing these four layers and the distribution of different cell types. The stratum lucidum, present in thick skin, is not illustrated here.

View PAL Histology MasteringA&P®

have a darker melanin, more granules, and more pigment in each melanocyte, they do *not* have more melanocytes in their skin. In all but the darkest people, melanocytes respond to ultraviolet radiation (UVR) by increasing the production of melanin and increasing its transfer to keratinocytes, the protective response we know as suntanning.

The role of melanocytes in skin pigmentation has long been recognized. In addition, melanocytes also secrete a variety of signaling molecules in response to ultraviolet radiation that act to modulate the immune response in the skin. These signaling molecules influence the inflammatory response and may have other regulatory functions.

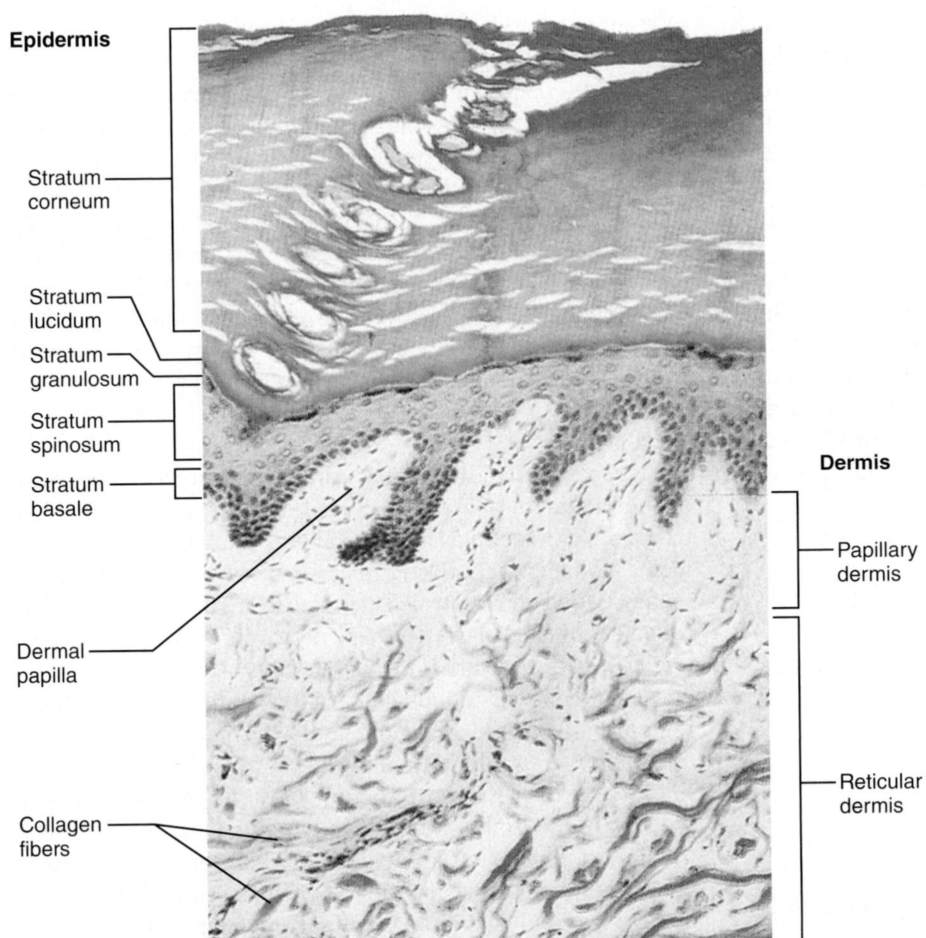

Figure 5.4 Thick skin.
Photomicrograph of the epidermis and dermis of thick skin (185×).

Stratum Spinosum The **stratum spinosum** (spi-no′sum; *spin* = spine) is several cell layers thick (Figure 5.3). Mitosis occurs here, but less often than in the basal layer. This layer gets its name from the many spinelike extensions of its keratinocytes, as seen in typical histological slides. However, these spines do not exist in living cells: They are artifacts created during tissue preparation, when the cells shrink while holding tight at their many desmosomes. Cells of the stratum spinosum contain thick bundles of intermediate filaments, which consist of a tension-resisting protein pre-keratin.

Scattered among the keratinocytes of the stratum spinosum are **dendritic cells.** These star-shaped cells are part of the immune system (see Chapter 21). Dendritic cells police the outer body surface, using receptor-mediated endocytosis (p. 27) to take up foreign proteins (antigens) that have invaded the epidermis. Then they leave the skin and travel to a nearby lymph node and initiate an immune response to all foreign cells that carry the antigen (see the discussion of lymphocyte activation in Chapter 21).

Stratum Granulosum The thin **stratum granulosum** (gran′u-lo″sum; *gran* = grain) consists of one to five layers of flattened keratinocytes. Along with abundant pre-keratin intermediate filaments, these cells also contain *keratohyalin granules* and *lamellar granules,* thus its name—granular layer. The keratohyalin (ker′ah-to-hi″ah-lin) granules help form keratin in the more superficial layers, as described shortly. The lamellar granules (lam″ĭ-la-ted; "plated") contain a waterproofing glycolipid that is secreted into the extracellular space. The glycolipid, along with tight junctions between adjacent cells, plays a major role in slowing water loss across the epidermis. Furthermore, the plasma membranes of the cells thicken so that they become more resistant to destruction. It is as though the keratinocytes "toughen up" to make the outer layer the strongest skin region.

Like all epithelia, the epidermis relies on capillaries in the underlying connective tissue (the dermis) for its nutrients. Above the stratum granulosum, the epidermal cells are too far from the dermal capillaries to receive nourishment, so they die, a completely normal occurrence.

Stratum Lucidum The **stratum lucidum** (lu′sĭ-dum; *luci* = clear) occurs in thick skin (Figure 5.4) but not in thin skin. Appearing through the light microscope as a thin translucent band, the stratum lucidum consists of a few rows of flat, dead keratinocytes. Electron microscopy reveals that its cells are identical to those at the bottom of the next layer, the stratum corneum.

Stratum Corneum (Horny Layer) The most external part of the epidermis, the **stratum corneum** (kor′ne-um; *cornu* = horn), is many cells thick. It is much thicker in thick skin than in thin skin (Figure 5.4). Its dead keratinocytes are flat sacs completely filled with keratin because their nuclei and organelles disintegrated upon cell death. Keratin consists of the pre-keratin intermediate filaments embedded in a "glue" from the keratohyalin granules. Both the keratin and the thickened plasma membranes of cells in the stratum corneum protect the skin against abrasion and penetration. Additionally, the glycolipid between its cells keeps this layer waterproof. It is amazing that a layer of dead cells can still perform so many functions!

The cells of the stratum corneum are shed regularly. These cells are the dandruff shed from the scalp and the flakes that come off dry skin. The average person sheds 18 kg (40 pounds) of these skin flakes in a lifetime. These shed cells are replaced by cells from the deeper layers. The next time you hear the saying "Beauty is only skin deep," consider that nearly everything you see when you look at someone is dead!

CLINICAL APPLICATION

Skin's Response to Friction Persistent friction (from a poorly fitting shoe, for example) causes a thickening of the epidermis called a **callus.** Short-term but severe friction (from using a hoe, for example) can cause a **blister,** the separation of the epidermis from the dermis by a fluid-filled pocket.

check your understanding

- ☐ 3. Which layers of the epidermis contain living cells?
- ☐ 4. Which functions of skin are performed by cells located in the epidermis?
- ☐ 5. How does thick skin differ from thin skin in structure?

(For answers, see Appendix B.)

Dermis

The **dermis,** the second major region of the skin, is a strong, flexible connective tissue. The cells of the dermis are typical of any connective tissue proper: fibroblasts, macrophages, mast cells, and scattered white blood cells (see p. 80). The fiber types—collagen, elastic, and reticular—also are typical. The dermis binds the entire body together like a body stocking. It is your "hide" and corresponds to animal hides used to make leather products.

The dermis has two regions: the *papillary dermis* and the *reticular dermis* (Figure 5.4).

Papillary Dermis

The **papillary** (pap′ĭ-lar-e) **dermis,** the superficial 20% of the dermis, is areolar connective tissue containing very thin collagen and elastic fibers. It includes the **dermal papillae** (pah-pil′le; "nipples"), fingerlike projections that extend into the overlying epidermis. These projections of the dermal papillae into the epidermis increase the surface area for exchange of gases, nutrients, and waste products between these layers. Recall that the epidermis is avascular and depends on the diffusion of these materials from the underlying dermis. The interdigitation of these layers also strengthens the dermal-epidermal junction and thus reduces blister formation.

On the palms of the hands and the soles of the feet, the dermal papillae lie atop larger mounds called *dermal ridges.* These elevate the overlying epidermis into *epidermal ridges* or *friction ridges,* which create fingerprints, palmprints, and footprints (Figure 5.5a). Epidermal ridges increase friction and enhance the gripping ability of the hands and feet. Patterns of these ridges are genetically determined and unique to each person. Because *sweat pores* open along the crests of the friction ridges, they leave distinct fingerprints on almost anything they touch. Thus, fingerprints are "sweat films."

Reticular Dermis

The deeper **reticular dermis,** which accounts for about 80% of the thickness of the dermis, is dense irregular connective tissue. Its extracellular matrix contains thick bundles of interlacing collagen and elastic fibers that run in many different planes. However, most run parallel to the skin surface. The reticular layer is named for its networks of collagen fibers (*reticulum* = network); the name does not imply any special abundance of reticular fibers. Separations or less dense regions between the collagen bundles form the *cleavage lines* or *tension lines* of the skin (Figure 5.5b). These invisible lines occur over the entire body: They run longitudinally in the skin of the limbs and head and in circular patterns around the neck and trunk. A knowledge of cleavage lines is important to surgeons. Incisions made *parallel* to these lines tend to gape less and heal more readily than incisions made *across* cleavage lines.

The collagen fibers of the dermis give skin its strength and resilience. Thus, many jabs and scrapes do not penetrate this tough layer. Furthermore, elastic fibers in the dermis provide the skin with stretch-recoil properties. Extreme stretching of the skin, as occurs in obesity and pregnancy, can tear the collagen in the dermis. Such dermal tearing results in silvery white scars called *striae* (stri′e; "streaks"), which we know as "stretch marks." The dermis is also the receptive site for the pigments used in tattoos (see A CLOSER LOOK, p. 110).

CLINICAL APPLICATION

Decubitus Ulcer Bedsores, or **decubitus ulcers,** are a constant concern for patients who have impaired mobility. Bedridden elderly people are especially susceptible, as are wheelchair-bound individuals, such as those with spinal cord injuries. A decubitus ulcer usually occurs over a bony prominence, such as the hip, sacrum, or heel. Constant pressure from body weight causes a localized breakdown of the skin due to reduction in blood supply. Tissue death can occur in 2–3 hours. Over time, the epidermis and papillary layer of the dermis are lost, and the damaged area shows an increase in collagen fibers. Without the protective covering of the epidermis, infectious agents can easily enter the body and cause serious and fatal complications.

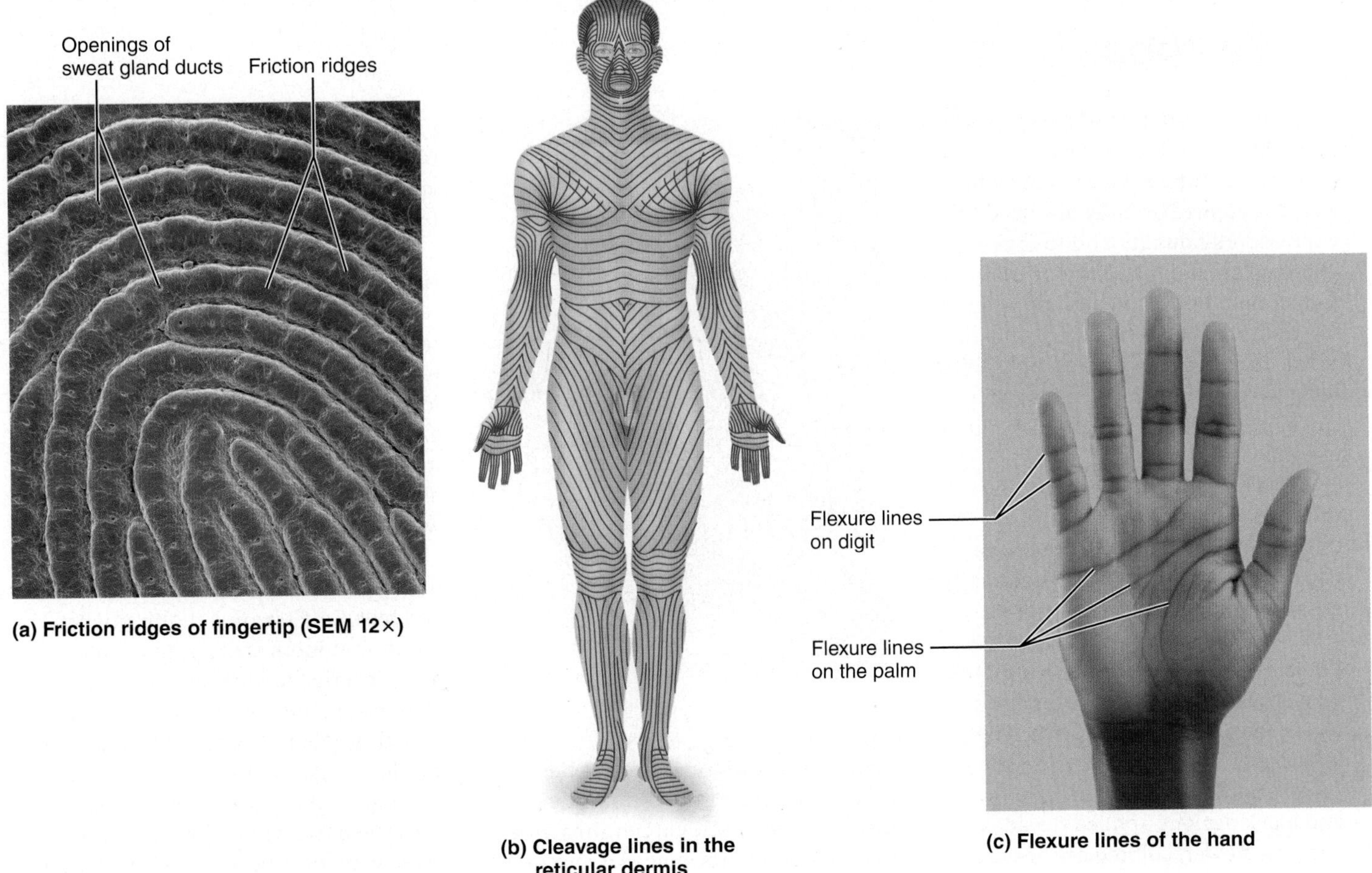

(a) Friction ridges of fingertip (SEM 12×)

(b) Cleavage lines in the reticular dermis

(c) Flexure lines of the hand

Figure 5.5 **Dermal modifications.** (**a**) Friction ridges. (**b**) Cleavage lines represent the separations between bundles of collagen fibers in the reticular layer of the dermis. (**c**) Flexure lines form where dermis is closely attached to the underlying fascia.

From the deep part of the dermis arise the skin surface markings called *flexure lines*. Observe, for example, the deep skin creases on your palm (Figure 5.5c). These result from a continual folding of the skin, often over joints, where the dermis attaches tightly to underlying structures. Flexure lines are also visible on the wrists, soles, fingers, and toes.

The dermis is richly supplied with nerve fibers (described in Chapter 14) and blood vessels. The dermal blood vessels consist of two vascular plexuses (a plexus is a network of converging and diverging vessels) (see Figure 5.1). The deep **dermal plexus** is located between the hypodermis and the dermis. It nourishes the hypodermis and the structures located within the deeper portions of the dermis. The more superficial **subpapillary plexus,** located just below the dermal papillae, supplies the more superficial dermal structures, the dermal papillae, and the epidermis.

Dermal blood vessels do more than just nourish the dermis and overlying epidermis; they also perform a critical role in temperature regulation. These vessels are so extensive that they can hold 5% of all blood in the body. When internal organs need more blood or more heat, nerves stimulate the dermal vessels to constrict, shunting more blood into the general circulation and making it available to the internal organs. By contrast, on hot days the dermal vessels engorge with warm blood, cooling the body by radiating heat away from it.

CLINICAL APPLICATION

The Patch Drug Delivery System The transdermal patch is designed so that drug molecules diffuse through the epidermis to the blood vessels in the dermal layer. A typical patch works well for small lipid-soluble molecules (for example, estrogen, nitroglycerin, and nicotine) that can make their way between epidermal cells. Patches are being developed to deliver larger molecules or molecules that are water soluble, such as insulin and vaccines. In one new patch design, thin, hollow needles extend from the reservoir of the drug, through the epidermis, and down to the dermis, deep enough to deliver drugs to the subpapillary plexus but not so deep as to reach the nerve endings.

Tattoos

Tattooing—using a needle to deposit pigment in the skin dermis—was first practiced around 8000 B.C. Today tattoos are applied as body art, used to express individuality and to make personal statements. Tattoos may signal membership in a group, such as the military, a street gang, or a sports team. They are also increasingly being used for cosmetic purposes, such as the application of permanent eyeliner, eyebrows, and liplines.

But what if the tattoo becomes unfashionable or the pigment migrates? Until very recently, once you had a tattoo, you were essentially stuck with it because attempts at removal—dermabrasion, cryosurgery (freezing the tissue), or applying caustic chemicals—left nasty scars, some worse than the original tattoo. The lasers of today can usually remove "older" tattoos, that is, the blue and black designs applied decades ago. The newer, colored tattoos are another matter entirely. They require the use of several different lasers that emit light at different frequencies to remove each of the different pigments. Tattoo removal requires seven to nine treatments spaced about one month apart, and it can hurt as much as obtaining the tattoo in the first place. Even then, tattoos can't always be eradicated completely: Green and yellow pigments are the most difficult to vaporize.

Tattoos present some other risks. The U.S. Food and Drug Administration does exert some control over the composition of the pigments used, but they are approved for application to the skin surface only—the safety of injecting them under the skin is not well established. Furthermore, state laws vary widely. Some states forbid the practice entirely, and others have no regulations at all. Tattooing involves needles and bleeding, training requirements for practitioners are minimal, and there is a risk of allergic reactions and bloodborne infectious diseases such as hepatitis. (Even if the practitioner uses sterile needles, they may be dipped into pigments that are contaminated with the blood of previous customers.) Even with the availability of laser removal techniques, it makes sense to consider the risks of getting a tattoo.

Embedded within the dermis are glands and hair follicles. These appendages of the skin are derived from the epidermal layer and extend down into the deep dermis and hypodermis.

Hypodermis

Just deep to the skin is the fatty **hypodermis** (hi″po-der′mis), from Greek words meaning "below the skin" (see Figures 5.1 and 5.2). This layer is also called the **superficial fascia** and the *subcutaneous layer* (sub″ku-ta′ne-us; from Latin words meaning "below the skin"). The hypodermis consists of both areolar and adipose connective tissue, but adipose tissue normally predominates. Besides storing fat, the hypodermis anchors the skin to the underlying structures (mostly to muscles), but loosely enough that skin can slide relatively freely over those structures. Such sliding ensures that many blows just glance off our bodies. The hypodermis is also an insulator: Because fat is a poor conductor of heat, it helps prevent heat loss from the body. The hypodermis thickens markedly with weight gain, but this thickening occurs in different body areas in the two sexes. In women, subcutaneous fat accumulates first in the thighs and breasts, whereas in men it first accumulates in the anterior abdomen as a "beer belly."

☑ check your understanding

- ☐ 6. How do the dermal blood vessels regulate body temperature?
- ☐ 7. What type of tissue makes up (a) the papillary layer of the dermis, (b) the reticular layer of the dermis, and (c) the hypodermis?
- ☐ 8. What types of cells are found in the dermis, and how does each contribute to the functions of skin?

(For answers, see Appendix B.)

Skin Color

Three pigments contribute to skin color: melanin, carotene, and hemoglobin. **Melanin,** the most important, is made from an amino acid called tyrosine. Present in several varieties, melanin ranges from yellow to reddish to brown to black. As noted earlier, melanin passes from melanocytes to keratinocytes in the stratum basale of the epidermis. The variations in skin color in humans result from differences in both the amount and type of melanin produced.

> **CLINICAL APPLICATION**
>
> **Freckles and Moles** Both freckles and moles (nevi) are localized accumulations of melanin in the skin. In freckles, the increase in melanin is restricted to the basal layer of the epidermis. Freckles form as a result of repeated exposure to sunlight. Individuals with light complexions are more prone to freckle formation. Moles form when clusters of melanocytes are transformed into melanin-containing cells (nevus cells). These clusters are located in the basal layer of the epidermis and in the top layers of the dermis. Moles form shortly after birth and can appear anytime through young adulthood. They do not form as a result of ultraviolet exposure.

Carotene (kar′o-tēn) is a yellow-orange pigment that the body obtains from vegetable sources such as carrots and tomatoes. It tends to accumulate in the stratum corneum of the epidermis and in the fat of the hypodermis.

The pink hue of Caucasian skin reflects the crimson color of oxygenated **hemoglobin** (he′mo-glo″bin) in the capillaries of the dermis. Because Caucasian skin contains little melanin, the epidermis is nearly transparent and allows the color of blood to show through. Black-and-blue marks represent discolored blood that is visible through the skin. Such bruises, usually caused by blows, reveal sites where blood has escaped from the circulation and clotted below the skin. The general term for a clotted mass of escaped blood anywhere in the body is a **hematoma** (he″mah-to′mah; "blood swelling").

> **CLINICAL APPLICATION**
>
> **Cyanosis** When hemoglobin is poorly oxygenated, both the blood and the skin of Caucasians appear blue, a condition called **cyanosis** (si″ah-no′sis; "dark blue condition"). Skin often becomes cyanotic in people with heart failure or severe respiratory disorders. In dark-skinned people, the skin may be too dark to reveal the color of the underlying vessels, but cyanosis can be detected in the mucous membranes and nail beds.

The reasons for variations in the skin colors of different human populations are not fully understood. Theories explaining this variation reflect the fact that the sun's UV rays are both dangerous and essential. One hypothesis proposes that dark skin coloration evolved to eliminate the danger of skin cancer caused by UV radiation. An alternative hypothesis argues that sunlight's effects on levels of folic acid in the blood was the selective pressure for the evolution of dark skin in tropical areas. In light-skinned people, high levels of sunlight decrease levels of folic acid in the blood. Decreased folic acid levels in pregnant women increase the risk of neural tube defects in the developing fetus. The evolution of dark skin in the tropics protects blood folic acid levels and promotes healthy offspring.

The other factor influencing skin color is vitamin D production. UV rays stimulate the deep epidermis to produce vitamin D, a vital hormone required for the uptake of calcium from the diet and essential for healthy bones. Calcium, and thus adequate vitamin D, is also critical to the developing fetus.

Producing vitamin D poses no problem in the sunny tropics, but Caucasians in northern Europe receive little sunlight during the long, dark winter. Therefore, their colorless epidermis ensures that enough UV rays will penetrate for vitamin D production. Natural exposure to UV rays in these temperate regions is not sufficient to diminish folic acid levels. Most human skin color types evolved at intermediate latitudes (China, the Middle East, and so on) and are characterized by moderately brown skin tones that are neither "white" nor "black." Such skin is dark enough to provide some protection from sunlight's negative effects in the summer, especially when it tans, yet light enough to allow vitamin D production in the moderate winters of intermediate latitudes. The trade-off between protection from the harmful effects of UV radiation and production of vitamin D may have been the selective pressure behind the evolution of skin coloration as early humans moved away from the tropics.

☑ check your understanding

- ☐ 9. Which pigment causes the large variation in human skin color?
- ☐ 10. Should a dark-skinned person living in the northern latitudes have concerns about vitamin D production? Why or why not?

(For answers, see Appendix B.)

APPENDAGES OF THE SKIN

learning outcomes

- ▶ Describe the structure of nails.
- ▶ List the parts of a hair and a hair follicle, and explain the function of each part.
- ▶ Compare the structure and location of oil and sweat glands.
- ▶ Compare eccrine and apocrine sweat glands.

Along with the skin itself, the integumentary system includes several derivatives of the epidermis. These **skin appendages** include nails, hair and hair follicles, sebaceous (oil) glands, and sweat glands. Although they derive from the epithelial cells of the epidermis, they all extend into the dermis.

Nails

A **nail** (Figure 5.6) is a scalelike modification of the epidermis that corresponds to the hoof or claw of other mammals. Fingernails are built-in tools that enable us to pick up small objects and scratch the skin when it itches. Nails are made of dead, keratinized cells. The **hard keratin** that predominates in hairs and nails differs from the **soft keratin** that is found

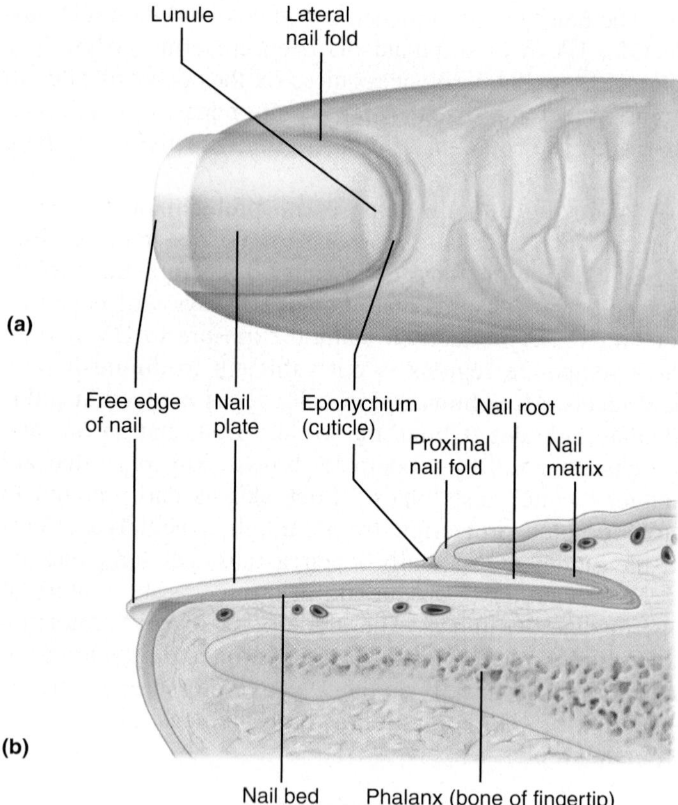

Figure 5.6 Structure of a nail. **(a)** Surface view of the distal part of a finger. **(b)** Sagittal section through a fingertip.

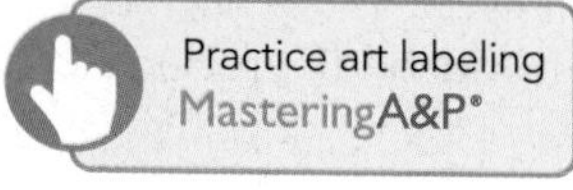

in typical epidermal cells in two ways: (1) It is tougher and more durable, and (2) the cells of hard keratin do not flake off. Each nail has a distal **free edge,** a **nail plate** (the visible attached part), and a **root** (the proximal part embedded in the skin). The nail rests on a bed of epidermis called the **nail bed.** This bed contains only the deeper layers of the epidermis, because the nail itself corresponds to the superficial keratinized layers.

Nails look pink because of the rich network of capillaries in the underlying dermis. At the root and the proximal end of the nail body, the bed thickens to form the **nail matrix,** the actively growing part of the nail. The matrix is so thick that the pink dermis cannot show through it. Instead, we see a white crescent, the **lunule** (lo͞o′no͞ol′; "little moon"), under the nail's proximal region. The lateral and proximal borders of the nail are overlapped by skin folds called **nail folds.** The proximal nail fold projects onto the nail body as the cuticle or **eponychium** (ep″o-nik′e-um; "on the nail").

An *ingrown toenail* is just what its name implies, a nail whose growth pushes it painfully into the lateral nail fold. This happens when the nail grows crookedly, most commonly from the pressure of an ill-fitting shoe.

☑ check your understanding

☐ 11. How does the keratin in hair and nails differ from the keratin in the epidermis?

(For the answer, see Appendix B.)

Hair and Hair Follicles

Together, hairs and their follicles form complex structural units (Figure 5.7). In these units, the *hairs* are the long filaments, and the *hair follicles* are tubular invaginations of the epidermis from which the hairs grow.

Hair

Although hair serves to keep other mammals warm, human body hair is far less luxuriant and less useful for conserving body temperature. Even so, hair is distributed over all our skin surface, except on the palms, soles, nipples, and parts of the external genitalia (the head of the penis, for example). The main function of this sparse body hair is to sense things that lightly touch the skin. The hair on the scalp protects the head against direct sunlight in summer and against heat loss on cold days. Eyelashes shield the eyes, and nose hairs filter large particles such as insects and lint from inhaled air.

A **hair** is a flexible strand made of dead cells filled with hard keratin. The chief parts of a hair are the **root,** the part embedded in the skin, and the **shaft,** the part that projects above the skin surface (Figure 5.7). If its shaft is flat and ribbonlike in cross section, the hair is kinky; if oval in section, the hair is wavy; if perfectly round, the hair is straight.

A hair consists of three concentric layers of keratinized cells (Figure 5.7a and b). Its central core, the **medulla** (mĕ-dul′ah; "middle"), consists of large cells and air spaces. The medulla is absent in fine hairs. The **cortex,** surrounding the medulla, consists of several layers of flattened cells. The outermost **cuticle** is a single layer of cells that overlap one another from below like shingles on a roof. This shingle pattern helps to keep neighboring hairs apart so that the hair does not mat. Hair conditioners smooth out the rough surface of the cuticle to make hair look shinier. The cuticle is the most heavily keratinized part of the hair, providing strength and keeping the inner layers tightly compacted. Because it is subjected to the most abrasion, the oldest part of the hair cuticle tends to wear away at the tip of the hair shaft. The abrasion causes the keratin fibrils in the cortex and medulla to frizz out, creating "split ends."

Hair pigment is made by melanocytes at the base of the hair follicle (shown in Figure 5.7c) and is transferred to the cells of the hair root. Different proportions of two types of melanin (black-brown and yellow-rust in color) combine to produce all the common hair colors—black, brown, red, and blond. Graying or whitening of hair results from a decrease in the production of melanin and from the replacement of melanin by colorless air bubbles in the hair shaft.

Hair Follicles

Hair follicles extend from the epidermal surface into the dermis. The deep end of the follicle is expanded, forming a **hair bulb** (Figure 5.7c). A knot of sensory nerve endings wraps around each hair bulb to form a **hair follicle receptor** or

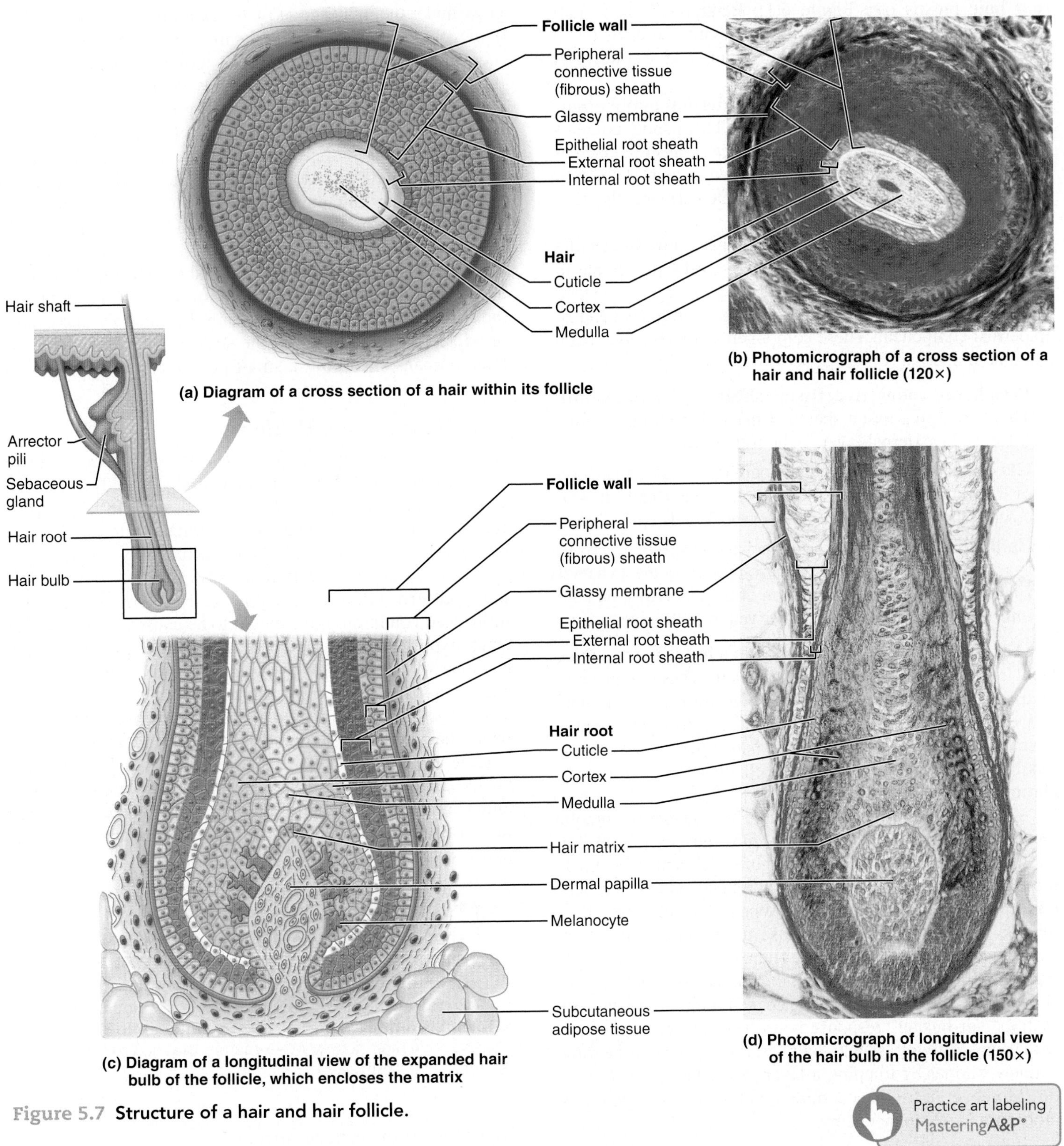

(a) Diagram of a cross section of a hair within its follicle

(b) Photomicrograph of a cross section of a hair and hair follicle (120×)

(c) Diagram of a longitudinal view of the expanded hair bulb of the follicle, which encloses the matrix

(d) Photomicrograph of longitudinal view of the hair bulb in the follicle (150×)

Figure 5.7 **Structure of a hair and hair follicle.**

root hair plexus (see Figure 5.1). Bending the hair shaft stimulates these nerve endings. Hairs are thus exquisite touch receptors. Test this by lightly running your hand over the hairs on your forearm.

A nipplelike bit of the dermis, the **dermal papilla** *(hair papilla)*, protrudes into each hair bulb. This papilla contains a knot of capillaries that deliver substances that stimulate hair growth and supply nutrients to the growing hair. If the hair papilla is destroyed by trauma, the follicle permanently stops producing hair.

The epithelial cells in the hair bulb just above the papilla make up the **hair matrix.** These matrix cells proliferate to form the hair shaft.

The wall of a hair follicle has a dermal component and an epidermal component. These components are described from external to internal (Figure 5.7a–d).

- **Peripheral connective tissue sheath** *(fibrous sheath).* This connective tissue sheath is derived from the dermis. It forms the external layer of the follicle wall.
- **Glassy membrane.** The glassy membrane is at the junction of the fibrous sheath and the epithelial root sheath. It is in essence the basement membrane of the follicle epithelium.
- **Epithelial root sheath.** The epithelial root sheath is derived from the epidermis. It has two components: the **external root sheath,** a direct continuation of the epidermis; and the **internal root sheath,** which is derived from the matrix cells.

Epidermal stem cells are located in a **bulge** in the superficial region of the external root sheath. These stem cells give rise to the hair matrix cells, which form the hair shaft as well as new epidermal cells. Stem cells from this region have been cultured and grown to produce patches of epidermis derived from a patient's own cells that are used to treat chronic wounds.

Associated with each hair follicle is a bundle of smooth muscle cells called an **arrector pili** muscle (ah-rek′tor pi′li; "raiser of the hair") (see Figures 5.1 and 5.7). Each arrector pili runs from the most superficial part of the dermis to a deep-lying hair follicle. When their arrector pili is relaxed, most hairs lie flat because most follicles lie at an oblique angle to the skin surface. Then, when this muscle contracts in response to cold or fear, the hair stands erect and the skin surface dimples, producing goose bumps. Even though such a "hair-raising" response is not very useful to humans, with our relatively sparse hair, it does keep fur-bearing animals warmer by trapping a layer of insulating air in their fur. Moreover, a frightened animal with its hair on end looks more formidable to an enemy.

Types and Growth of Hair

Hairs come in various sizes and shapes, but as a rule they can be classified as either **vellus** (vel′us; *vell* = wool, fleece) or **terminal.** The body hair of children and women is of the fine, short, vellus variety. The longer, coarser hair of everyone's scalp is terminal hair. Terminal hairs also appear at puberty in the axillary (armpit) and pubic regions in both sexes and on the face, chest, arms, and legs of men. These terminal hairs grow under the influence of male sex hormones called *androgens,* of which *testosterone* is the most important type.

Hairs grow an average of 2 mm per week, although this rate varies widely among body regions and with sex and age. Each follicle goes through growth cycles. In each cycle, an active growth phase is followed by a resting phase, in which the hair matrix is inactive and the follicle atrophies somewhat. At the start of each active phase, the newly growing hair pushes out the old hair, causing it to be shed. The life span of hairs varies: On the scalp, the follicles stay active for an average of 4 years, so individual hairs grow quite long before being shed. On the eyebrows, by contrast, the follicles are active for only a few months, so the eyebrows never grow very long. Fortunately, the cycles of adjacent hair follicles on the scalp are not in synchrony; thus, humans shed only a small percentage of the hairs on the head at any one time.

Hair Thinning and Baldness

Given ideal conditions, hair grows fastest from the teen years to the 40s. When hairs are no longer replaced as quickly as they are shed, the hair thins. By age 60 to 65, both sexes usually experience some degree of balding. Coarse terminal hairs are replaced by vellus hairs, and the hair becomes increasingly wispy.

True baldness is different. The most common type, **male pattern baldness,** is a genetically determined, gender-influenced condition. It is thought to be caused by a gene that is not expressed until adulthood, at which time it changes the response of hair follicles to androgens. The hairs respond to androgens by increasingly shortening their growth cycles. The cycles become so short that many hairs never emerge from their follicles before being shed, and those that do emerge are fine vellus hairs that look like peach fuzz in the "bald" area. The drugs used to treat male pattern baldness either inhibit the production of androgens or increase blood flow to the skin and hair follicles. These treatments are only partly successful.

CLINICAL APPLICATION

Chemotherapy and Hair Loss Chemotherapy drugs used to treat cancer target the most rapidly dividing cells in the body, thereby destroying many hair stem cells and causing hair loss. After chemotherapy stops, the hair can recover and regrow. Hair loss due to severe burns, excessive radiation, or other factors that destroy the follicles is permanent.

check your understanding

☐ 12. From what region of skin are the hair and hair follicle derived?

☐ 13. Place the three layers of hair cells in order from deepest to most superficial.

☐ 14. Why is hair loss due to chemotherapy temporary but hair loss resulting from a severe burn permanent?

(For answers, see Appendix B.)

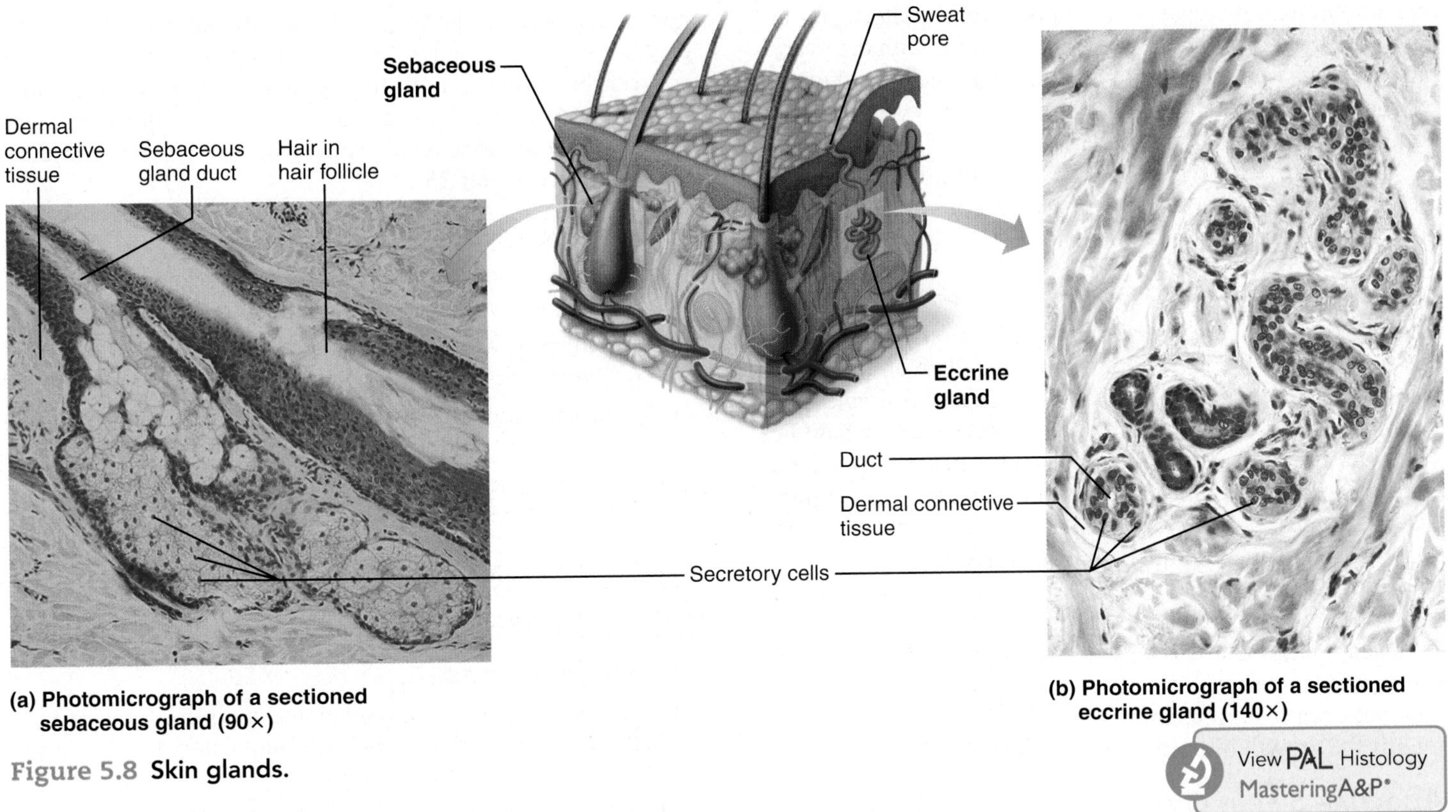

(a) Photomicrograph of a sectioned sebaceous gland (90×)

(b) Photomicrograph of a sectioned eccrine gland (140×)

Figure 5.8 **Skin glands.**

Sebaceous Glands

The **sebaceous glands** (se-ba′shus; "greasy") are the skin's oil glands (Figure 5.1 and Figure 5.8a). They occur over the entire body, except on the palms and soles. They are simple alveolar glands with several alveoli opening into a single duct (see Figure 4.5 p. 74 for their basic structure), but the alveoli are actually filled with cells, so there is no lumen (central cavity). Their oily product, called **sebum** (se′bum; "animal fat"), is secreted in a most unusual way: The central cells in the alveoli accumulate oily lipids until they become engorged and burst apart. This process is called **holocrine secretion** (hol′o-krin; *holos* = whole), because *whole* cells break up to form the product. Most sebaceous glands are associated with hair follicles, emptying their sebum into the upper third of the follicle. From there, the sebum flows superficially to cover the skin. In addition to making our skin and hair oily, sebum collects dirt, softens and lubricates the hair and skin, prevents hair from becoming brittle, and keeps the epidermis from cracking. It also helps to slow water loss across the skin and to kill bacteria.

The secretion of sebum is stimulated by hormones, especially androgens. The sebaceous glands are relatively inactive during childhood but are activated in both sexes during puberty, when the production of androgens begins to rise.

CLINICAL APPLICATION

Acne In some teenagers, so much sebum is produced that it cannot be ducted from the glands quickly enough. When a sebaceous gland becomes blocked by sebum, a whitehead appears on the skin surface. If the material oxidizes and dries, it darkens to form a blackhead (the dark color of a blackhead is *not* due to dirt). Blocked sebaceous glands are likely to be infected by bacteria, producing a pimple. The bacteria break down the sebum into irritating fatty acids. These acids, along with the bacterial products themselves, induce inflammation, especially when the entire infected mass bursts out of the follicle into the surrounding dermis. The *acne* that results can be mild or extremely severe, leading to permanent scarring.

The treatment of acne addresses the different causal factors: increased sebum production and formation of whiteheads, inflammation, and bacterial infection. Topical medications such as benzoyl peroxide treat inflammatory acne; topical retinoids, derivatives from vitamin A, act to prevent the formation of whiteheads; and antibiotics destroy acne-causing bacteria. Often combinations of treatments are used for maximal results.

Sweat Glands

Sweating prevents overheating of the body, because sweat cools the skin as it evaporates. Only mammals have **sweat glands (sudoriferous glands).** Humans have more than 2.5 million sweat glands distributed over the entire skin surface, except on the nipples and parts of the external genitalia. Humans normally produce about 500 ml of sweat per day, but this amount can increase to 12 liters (over 3 gallons) on hot days during vigorous exercise. Hair interferes with the evaporation of sweat and the ability to cool the body, so the need for increased temperature regulation through sweating led to a reduction of hairiness in humans.

There are two types of sweat glands, both of which increase their secretion in response to stress as well as to heat: *eccrine* and *apocrine* glands.

Eccrine Sweat Glands

Eccrine glands (ek′rin; "secreting") are by far the more numerous type (Figure 5.8b). They are most abundant on the palms, soles, and forehead. Each is a coiled version of a simple tubular gland (see Figure 4.5 p. 74). The coiled, secretory base lies in the deep dermis and hypodermis, and the duct runs superficially to open at the skin surface through a funnel-shaped **pore.** (Although most pores on the skin surface are sweat pores, the "pores" seen on the face are openings of hair follicles.)

Sweat is an unusual secretory product in that it is primarily a filtrate of blood that passes through the secretory cells of the sweat glands and is released by exocytosis. Sweat is 99% water, with some salts (mostly sodium chloride) and traces of metabolic wastes (urea, ammonia, uric acid). It is acidic, so it retards the growth of bacteria on the skin.

Apocrine Sweat Glands

Apocrine (ap′o-krin) **glands** are mostly confined to the axillary, anal, and genital areas. They are larger than eccrine glands, and their ducts open into hair follicles. Apocrine glands produce a special kind of sweat consisting of fatty substances and proteins, in addition to the components of true sweat. For this reason, apocrine sweat is viscous and sometimes has a milky or yellow color. This product is odorless when first secreted, but as its organic molecules are decomposed by bacteria on the skin, it takes on a musky smell. This is the source of body odor.

Apocrine glands start to function at puberty under the influence of androgens. Their activity is increased by sexual foreplay, and they enlarge and recede with the phases of a woman's menstrual cycle. The secretions from the apocrine glands were identified as true human pheromones (chemical signals that convey information to a member of the same species) in the late 1990s when it was shown that they are responsible for the synchrony of the menstrual cycle that occurs in females who live together.

Apocrine glands are involved with sexual signaling and appear to function in attractiveness and mate selection. The genes that encode for the immune system, the major histocompatibility complex (MHC), also influence secretions from apocrine glands. Each person has a unique set of these genes. In experiments, the body odor scents that women selected as "sexy" or "attractive" came from men who had immune system genes most different from their own. Mates with complementary immune system genes provide their offspring with greater disease protection and decreased likelihood of recessive disorders.

The skin forms several types of modified sweat glands. *Ceruminous glands* (sĕ-roo′mĭ-nus; "waxy") are modified apocrine glands in the lining of the external ear canal. Their product is a component of earwax. The *mammary glands* are specialized sweat glands highly modified to secrete milk. Although mammary glands are part of the integumentary system, they are discussed along with the female reproductive system (Chapter 25).

✓ check your understanding

☐ 15. Apocrine glands become active in adolescence. Are these also the glands that cause acne?

☐ 16. Which type of tissue forms eccrine glands? (Look at the micrograph of the eccrine gland in Figure 5.8b).

☐ 17. Which functions of skin are performed by eccrine glands?

(For answers, see Appendix B.)

DISORDERS OF THE INTEGUMENTARY SYSTEM

learning outcomes

▶ Describe the layers involved in and the symptoms of first-, second-, and third-degree burns, and explain why serious burns are life-threatening.

▶ Identify the cell type involved, the characteristic appearance, and the degree of malignancy in the three types of skin cancer.

Exposed directly to the dangers and germs of the outside world, skin can develop more than a thousand different conditions and ailments. The most common skin disorders are bacterial, viral, or yeast infections (some of these are described in the Related Clinical Terms at the end of this chapter). The most severe threats to the skin are burns and skin cancer.

Burns

Burns are a devastating threat to the body, primarily because of their effects on the skin. A burn is tissue damage inflicted by heat, electricity, radiation, extreme friction, or certain harmful chemicals.

The immediate threat to life from serious burns is a catastrophic loss of body fluids. Inflammatory edema is severe. As fluid seeps from the burned surfaces, the body quickly loses water and essential salts. This dehydration in turn leads to fatal circulatory shock; that is, inadequate blood circulation caused by the reduction in blood volume. To save the patient, the medical team must replace the lost fluids immediately. After the initial crisis has passed, infection becomes the main threat. Pathogens can easily invade areas where the skin barrier is destroyed.

Burns are classified by their severity (depth) as first-, second-, or third-degree burns (Figure 5.9). Third-degree burns are the *most* severe. In **first-degree burns,** only the epidermis is damaged. Symptoms include redness, swelling, and pain: the typical inflammatory reaction to tissue damage (see p. 95). Generally, first-degree burns heal in a few days without special attention. Sunburn is usually a first-degree burn.

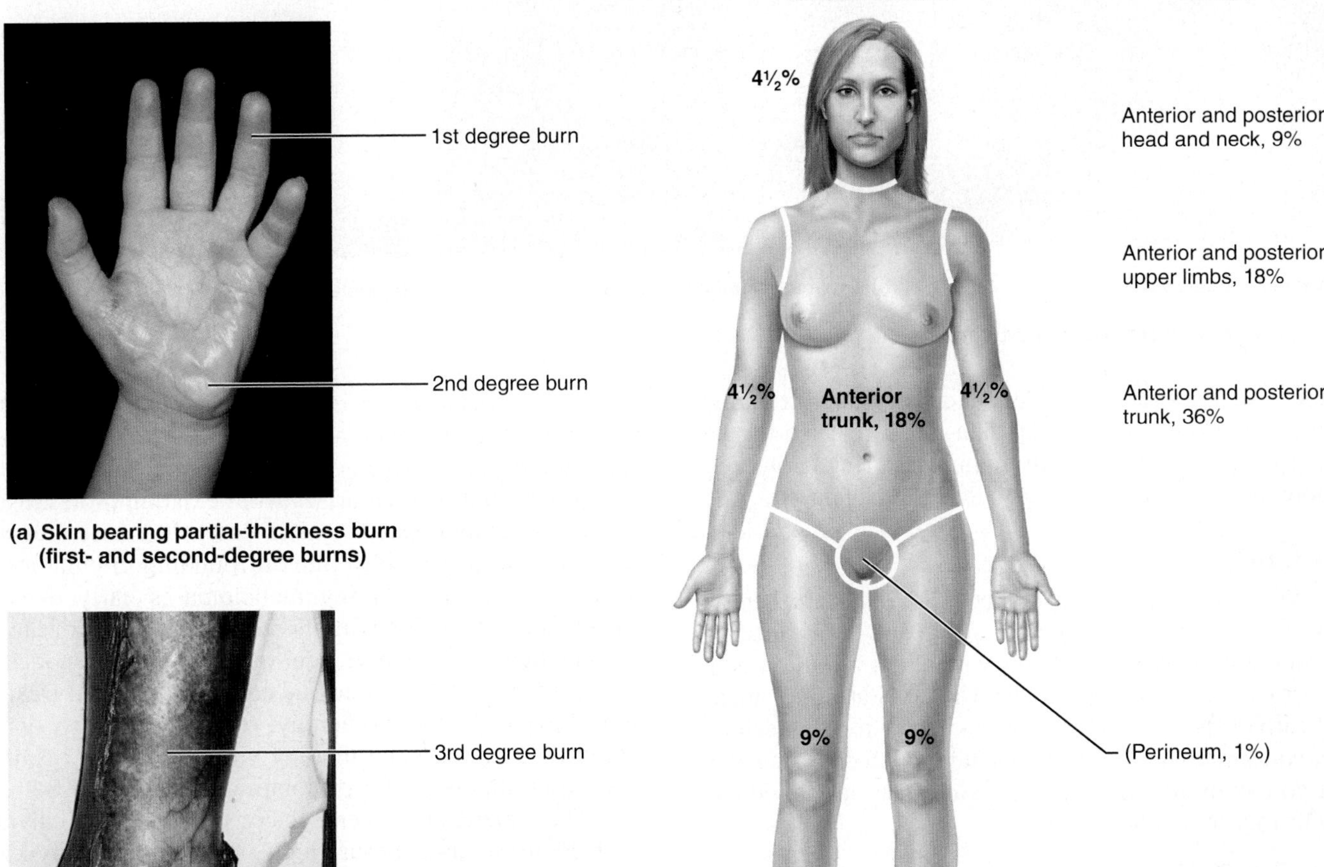

(a) Skin bearing partial-thickness burn (first- and second-degree burns)

(b) Skin bearing full-thickness burn (third-degree burn)

(c) Rule of nines; used to estimate extent of burns

Figure 5.9 **Burns.**

Second-degree burns involve injury to the epidermis and the upper part of the dermis. Symptoms resemble those of first-degree burns, but blisters also appear as fluid accumulates between the epidermal and dermal layers. The skin regenerates with little or no scarring in 3 to 4 weeks if care is taken to prevent infection. First- and second-degree burns are considered **partial-thickness burns.**

Third-degree burns consume the entire thickness of the skin and thus are **full-thickness burns.** The burned area appears white, red, or blackened. Although the skin might eventually regenerate, it is usually impossible to wait for this because of fluid loss and infection. Therefore, skin from other parts of the patient's body must be grafted onto the burned area. Such a graft, in which an individual is both donor and recipient, is called an *autograft.*

In general, burns are considered critical if any of the following conditions exists: (1) Over 10% of the body has third-degree burns; (2) 25% of the body has second-degree burns; (3) there are third-degree burns on the face, hands, or feet. A quick way to estimate how much surface is affected by a burn is to use the **rule of nines** (Figure 5.9c). This method divides the body surface into 11 regions, each accounting for 9% (or a multiple of 9%) of total body area. (The values shown on the human figure in Figure 5.9c indicate surface area values for the anterior body surfaces; total values for each region are indicated to the right of the illustration.) This method is only roughly accurate, so special tables are used when greater accuracy is needed.

For patients whose burns are too large or who otherwise are poor candidates for autografts, good artificial coverings are now available. Artificial skin consists of a "dermis" made from bovine collagen and an "epidermis" made of silicone. The patient's own dermis gradually replaces and reabsorbs the artificial one, and the silicone sheet is later peeled off and

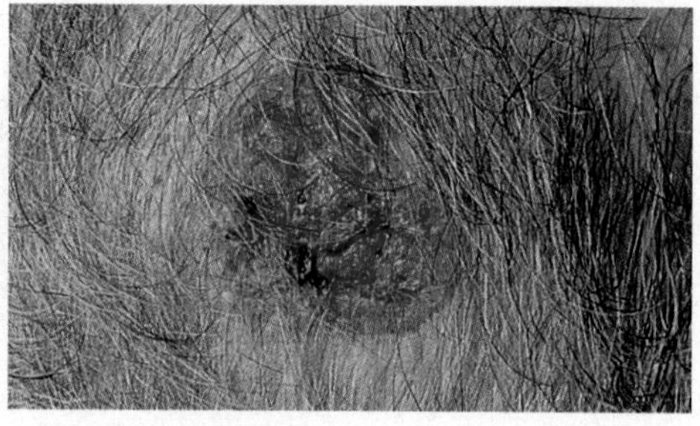

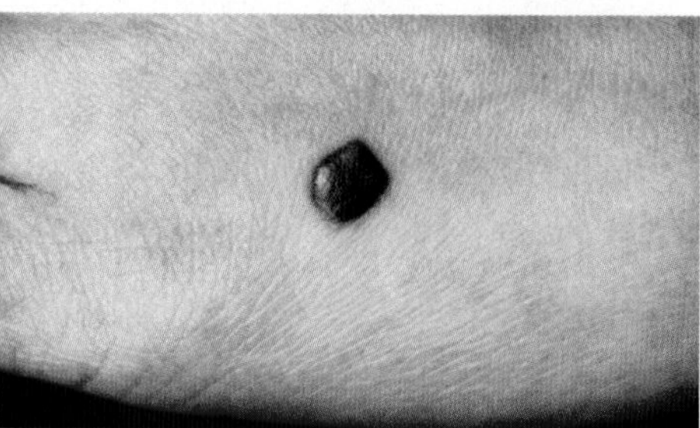

(a) Basal cell carcinoma (b) Squamous cell carcinoma (c) Melanoma

Figure 5.10 Photographs of skin cancers.

replaced with a network of epidermal cells cultured from the patient's own skin. The artificial skin is not rejected by the body, saves lives, and results in minimal scarring. However, it is more likely to become infected than is an autograft.

Skin Cancer

Many types of tumors arise in the skin. Most are benign (warts, for example), but some are malignant. Skin cancer is the most common type of cancer, with about a million new cases appearing each year in the United States. As mentioned earlier, the most important risk factor for skin cancer is overexposure to the UV rays in sunlight. Recent data show increased risk of all three types of skin cancer in individuals with a history of indoor tanning.

Basal Cell Carcinoma

Basal cell carcinoma is the least malignant and most common of the skin cancers (Figure 5.10a). Over 30% of all Caucasians get it in their lifetimes! Cells of the stratum basale proliferate, invading the dermis and hypodermis, and causing tissue erosions there. The most common lesions of this cancer are dome-shaped, shiny nodules on sun-exposed areas of the face. These nodules later develop a central ulcer and a "pearly," beaded edge. They often bleed. Basal cell carcinoma grows relatively slowly, and metastasis seldom occurs. Full cure by surgical excision or other methods of removal is the rule in 99% of cases.

Squamous Cell Carcinoma

Squamous cell carcinoma (Figure 5.10b) arises from the keratinocytes of the stratum spinosum. The lesion appears as a scaly, irregular, reddened papule (small, rounded elevation) that tends to grow rapidly and metastasize if not removed. If treated early, however, the chance of a complete cure is good, and the overall cure rate is 99%. The carcinoma can be removed by radiation therapy, by surgery, or by skin creams containing anticancer drugs.

Melanoma

Melanoma (mel″ah-no′mah), a cancer of melanocytes, is the most dangerous kind of skin cancer (Figure 5.10c). Melanocytes are derived from neural crest cells, which wander widely during embryonic development. This predisposition for migration explains the invasive nature of melanoma. Melanoma accounts for only about 1 of every 20 skin cancers, but it is increasing rapidly in countries with light-skinned populations—by 3% to 8% *per year* in the United States. Melanoma can originate wherever there is pigment, but it often arises from existing moles, usually appearing as an expanding dark patch. Because melanoma cells metastasize rapidly into surrounding circulatory vessels, the key to surviving melanoma is early detection. Survival rates decline with increasing thickness of the melanoma, degree of involvement of nearby lymph nodes, and extent of metastasis. In advanced cases, surgical treatment is followed by immunotherapy, radiation therapy, or more recently, targeted gene therapy with promising results for shrinking tumors and/or prolonging life.

The American Cancer Society suggests that individuals with frequent sun exposure regularly examine their skin for moles and new pigment spots, applying the **ABCD rule** for recognizing melanoma: **A,** *Asymmetry:* The two halves of the spot or mole do not match; **B,** *Border irregularity:* The borders have indentations and notches; **C,** *Color:* The pigment spot contains several colors, including blacks, browns, tans, and sometimes blues and reds; **D,** *Diameter:* larger than 6 mm (larger than a pencil eraser). Some experts have found that adding an **E,** for *Elevation* above the skin surface, improves diagnosis, so they use the **ABCD(E) rule.**

☑ check your understanding

- ☐ 18. Which type of burn causes blisters?
- ☐ 19. What are the two life-threatening concerns resulting from third-degree burns, and how are they treated?
- ☐ 20. Which type of skin cancer is most common? Which one can be fatal?

(For answers, see Appendix B.)

THE SKIN THROUGHOUT LIFE

learning outcomes

- ▶ Identify the primary germ layers that form the skin and its appendages.
- ▶ Describe the changes that occur in the skin from birth to old age.

The epidermis develops from embryonic ectoderm, and the dermis and hypodermis develop from mesoderm (see Chapter 3, pp. 55–57). Melanocytes are derived from neural crest cells that migrate into the ectoderm during the first 3 months of prenatal development. By the end of the fourth month, the skin is fairly well formed: The epidermis has all its layers, dermal papillae are evident, and the epidermal appendages are present in rudimentary form. During the fifth and sixth months, the fetus is covered with a downy coat of delicate hairs called the *lanugo* (lah-nu′go; "wool or down"). This cloak is shed by the seventh month, and vellus hairs make their appearance.

When a baby is born, its skin is covered with *vernix caseosa* (ver′niks ka′se-o-sah; "varnish of cheese"), a cheesy-looking substance produced by sebaceous glands. It protects the skin of the fetus from constant contact with amniotic fluid. A newborn's skin and hypodermis are very thin, but they thicken during infancy and childhood.

With the onset of adolescence, sebaceous glands become more active, and acne may appear (see p. 115). Acne generally subsides in early adulthood, and the skin reaches its optimal appearance in one's 20s and 30s. Thereafter, the skin starts to show the harmful effects of continued environmental assaults, such as abrasion, wind, sun, and chemicals. Scaling and various kinds of skin inflammation, called **dermatitis** (der″mah-ti′tis), become more common.

In middle age, the lubricating and softening substances produced by the glands start to diminish. As a result, the skin becomes dry and itchy. People with naturally oily skin may avoid this dryness, so their skin often ages well.

Most aspects of skin aging are not intrinsic but are caused by sunlight in a process called "photoaging." Aged skin that has been protected from the sun has lost some elasticity and is thinner (as is the hypodermis under it), but it remains unwrinkled and unmarked. Sun-exposed skin, by contrast, is wrinkled, loose, inelastic, leathery, and has pigmented spots called "liver spots." In the dermis of such sun-aged skin, the amount of collagen has declined, and an abnormal, elastin-containing material has accumulated. Much of this change is due to UV-induced activation of enzymes that degrade collagen and other components of the dermis.

By blocking UV rays, large amounts of melanin protect the skin from photoaging. This is why fair-skinned individuals, who have little melanin to begin with, show age-related changes at a younger age than do people with darker skin and hair. For the same reason, dark-skinned people look youthful much longer than Caucasians.

☑ check your understanding

☐ 21. From which embryonic germ layer do the appendages of the skin develop? From which embryonic layer does the hypodermis develop?

☐ 22. How does UV radiation damage the dermal layer of the skin?

(For answers, see Appendix B.)

RELATED CLINICAL TERMS

Actinic keratosis (ak-tin′ik ker-ă-tō′sis) A precancerous lesion of the skin composed of rough, scaly spots that can range in size from a pinhead to larger than a quarter. These lesions may feel like sandpaper and may look like an abrasion that does not heal. They are a precursor to squamous cell carcinoma and are common in sun-damaged skin. Treatment with creams or surgery removes the damaged cells. Once a lesion is diagnosed, screening for further lesions is essential.

Alopecia (al″o-pe′she-ah) Any condition involving absence or loss of hair. Male pattern baldness, the most common type, is technically named *androgenic alopecia.*

Athlete's Foot Itchy, red, peeling condition of the skin between the toes, resulting from fungal infection.

Boils and Carbuncles (kar′bung″kls; "little glowing embers") Infection and inflammation of hair follicles and sebaceous glands that has spread into the underlying hypodermis. A boil can be likened to a giant pimple; carbuncles are composite boils. The common cause is bacterial infection.

Cold Sores (Fever Blisters) Small, painful, fluid-filled blisters that usually occur around the lips and in the mucosa in the mouth. They are caused by the herpes simplex virus, which localizes in nerve cells that supply the skin, where it remains dormant until activated by emotional upset, fever, or UV radiation.

Impetigo (im″pĕ-ti′go; "an attack") Pink, fluid-filled, raised lesions that are common around the mouth and nose. They develop a yellow crust and eventually rupture. This contagious condition, caused by a *Staphylococcus* infection, is common in school-aged children.

Psoriasis (so-ri′ah-sis; "an itching") A chronic inflammatory condition characterized by reddened epidermal papules covered with dry, silvery scales. The scales result from an overproliferation of the epidermis, and the pink color is due to widened capillaries in the dermis. Relatively common, it affects 2% of Americans. It may be an autoimmune condition triggered by bacterial products. One effective treatment for psoriasis employs a drug that slows epidermal growth and that is activated by exposure to UV light.

Rosacea (rō-zā′shē-ah) A chronic skin eruption produced by dilated small blood vessels of the face, particularly those of the nose and cheeks, causing flushing of the face. Papules and acne-like pustules may or may not occur. More common in women, but tends to be more severe when it occurs in men. Cause is unknown, but stress, some endocrine disorders, and anything that produces flushing (hot beverages, alcohol, sunlight, etc.) can aggravate this condition.

Vitiligo (vit″il-i′go; "blemish") An abnormality of skin pigmentation due to a loss of melanocytes, characterized by light spots surrounded by areas of normally pigmented skin. It can be cosmetically disfiguring, especially in dark-skinned people. An autoimmune disorder.

CHAPTER SUMMARY

Take full advantage of the invaluable online practice and study tools for this chapter by going to the Study Area of MasteringA&P®. There you will find:

- Art Labeling Activities, Quizzes, Practice Tests, Memory Games
- Animal Organ Dissection, Cat Dissection and Bone Videos
- PAL Cadaver
- PAL Histology
- A&P Flix Animations
- Dynamic Study Modules

1. The integumentary system consists of the skin and its appendages (hair, sebaceous glands, sweat glands, and nails). The hypodermis, although not part of the integumentary system, is also considered in this chapter.
2. Skin functions to protect our body from bumps and scrapes, chemicals, invading microorganisms, water loss, and ultraviolet (UV) radiation. It also functions in body temperature regulation, excretion, production of vitamin D, and sensory reception.

THE SKIN AND THE HYPODERMIS (pp. 104–111)

Epidermis (pp. 105–108)

3. The epidermis is a keratinized stratified squamous epithelium. Its main cell type is the keratinocyte. From deep to superficial, its strata are the basale, spinosum, granulosum, lucidum, and corneum. The stratum lucidum occurs only in thick skin found on palms and soles.
4. The dividing cells of the stratum basale are the source of new keratinocytes for the growth and renewal of the epidermis.
5. In the stratum spinosum, keratinocytes contain strong pre-keratin intermediate filaments. In the stratum granulosum, keratinocytes contain keratohyalin granules, which combine with the pre-keratin filaments to form tension-resisting keratin. Stratum granulosum cells also secrete an extracellular waterproofing glycolipid. Keratinized cells of the stratum corneum protect the skin and are shed as skin flakes and dandruff.
6. Scattered among the keratinocytes in the deepest layers of the epidermis are pigment-producing melanocytes, tactile epithelial cells, and the dendritic cells of the immune system.

Dermis (pp. 108–110)

7. The leathery dermis is composed of connective tissue proper. The superficial papillary dermis consists of areolar connective tissue, whereas the deep reticular dermis is dense irregular connective tissue. The papillary layer abuts the undulating undersurface of the epidermis and is responsible for the epidermal friction ridges that produce fingerprints. In the reticular layer, less dense regions between the collagen bundles produce cleavage lines. Flexure lines occur over joints, where the dermis attaches tightly to underlying structures.
8. The dermis is well supplied with vessels and nerves. Also located in the dermis are skin glands and hair follicles.

Hypodermis (p. 110)

9. The hypodermis underlies the skin. Composed primarily of adipose tissue, it stores fat, insulates the body against heat loss, and absorbs and deflects blows. During weight gain, this layer thickens most markedly in the thighs and breasts of women and in the anterior abdominal wall of men.

Skin Color (pp. 110–111)

10. Skin color reflects the amounts of pigments (melanin and/or carotene) in the skin and the oxygenation level of hemoglobin in the dermal blood vessels.
11. Melanin, made by melanocytes and transferred to keratinocytes, protects skin and folic acid levels in the blood from the damaging effects of UV radiation. The epidermis produces vitamin D under the stimulus of UV radiation. As ancestral populations moved away from the tropics, the need to balance vitamin D synthesis with UV protection resulted in the evolution of a range of skin colors.

APPENDAGES OF THE SKIN (pp. 111–116)

12. Skin appendages, which develop from the epidermis but project into the dermis, include hairs and hair follicles, sebaceous glands, sweat glands, and nails.

Nails (pp. 111–112)

13. A nail is a scalelike modification of the epidermis that covers the dorsal tip of a finger or toe. The actively growing region is the nail matrix.

Hair and Hair Follicles (pp. 112–114)

14. A hair is produced by a tube-shaped hair follicle and consists of heavily keratinized cells. Each hair has an outer cuticle, a cortex, and usually a central medulla. Hairs have roots and shafts. Hair color reflects the amount and variety of melanin present.
15. A hair follicle consists of an epithelial root sheath that includes the hair matrix and connective tissue root sheath derived from the dermis. Stem cells, which renew the hair matrix and give rise to new epidermal cells, are in the epithelial root sheath. A knot of capillaries in the hair papilla nourishes the hair, and a nerve knot surrounds the hair bulb. Arrector pili muscles pull the hairs erect and produce goose bumps in response to cold or fear.
16. Fine and coarse hairs are called vellus hairs and terminal hairs, respectively. At puberty, under the influence of androgens, terminal hairs appear in the axillae and pubic regions in both sexes.
17. Hairs grow in cycles that consist of growth and resting phases. Hair thinning results from factors that lengthen follicular resting phases, including natural age-related atrophy of hair follicles and expression of the male pattern baldness gene during adulthood.

Sebaceous Glands (p. 115)

18. Sebaceous glands are simple alveolar glands that usually empty into hair follicles. Their oily holocrine secretion is called sebum.
19. Sebum lubricates the skin and hair, slows water loss across the skin, and acts as a bactericidal agent. Sebaceous glands secrete increased amounts of sebum at puberty under the influence of androgens. Blocked and infected sebaceous glands lead to pimples and acne.

Sweat Glands (pp. 115–116)

20. Eccrine sweat glands are simple coiled tubular glands that secrete sweat, a modified filtrate of the blood. Evaporation of sweat cools the skin and the body.
21. Apocrine sweat glands, which produce pheromones, occur primarily in the axillary, anal, and genital regions. Their secretion contains proteins and fatty acids; bacterial action on this secretion causes body odor.

PAL **Histology/Integumentary System**

DISORDERS OF THE INTEGUMENTARY SYSTEM (pp. 116–118)

22. The most common skin disorders result from microbial infections.

Burns (pp. 116–118)

23. In severe burns, the initial threat is loss of body fluids leading to circulatory collapse. The secondary threat is overwhelming infection.
24. The extent of a burn may be estimated by using the rule of nines. The severity of burns is indicated by the terms *first-*, *second-*, and *third-degree.* Third-degree burns harm the full thickness of the skin and require grafting for successful recovery.

Skin Cancer (p. 118)

25. The major risk factor for skin cancer is exposure to UV sunlight.
26. Basal cell carcinoma and squamous cell carcinoma are the most common types of skin cancer and usually are curable if detected early. Melanoma, a cancer of melanocytes, is less common but dangerous. The ABCD or ABCD(E) rule can be used to examine a mole or spot for the possibility of melanoma.

THE SKIN THROUGHOUT LIFE (pp. 118–119)

27. Epidermis develops from embryonic ectoderm; dermis and hypodermis develop from mesoderm; and melanocytes develop from neural crest cells.
28. At 5–6 months, a fetus has a downy lanugo coat. Fetal sebaceous glands produce vernix caseosa that protects the skin within the amniotic sac.
29. In old age, the skin thins and loses elasticity. Damage from sunlight leads to wrinkles, age spots, and loose and leathery skin; this reflects a loss of collagen and accumulation of elastin-containing material in the dermis.

REVIEW QUESTIONS

Multiple Choice/Matching Questions

(For answers, see Appendix B.)

1. Which epidermal cell type functions in the immune system? (a) keratinocyte, (b) melanocyte, (c) dendritic cell, (d) tactile epithelial cell.
2. Match each epidermal layer in column B to its description in column A.

Column A	Column B
____ (1) Desmosomes and shrinkage artifacts give its cells "spiny" projections.	(a) stratum basale
____ (2) Its cells are flat, dead bags of keratin.	(b) stratum corneum
____ (3) Its cells divide, and it is also called the stratum germinativum.	(c) stratum granulosum
____ (4) It contains keratohyalin and lamellated granules.	(d) stratum lucidum
____ (5) It is present only in thick skin.	(e) stratum spinosum

3. The ability of the epidermis to resist rubbing and abrasion is largely due to the presence of (a) melanin, (b) carotene, (c) collagen, (d) keratin.
4. Skin color is determined by (a) melanin, (b) carotene, (c) oxygenation level of the blood, (d) all of these.
5. What is the major factor accounting for the waterproof nature of the skin? (a) desmosomes in stratum corneum, (b) glycolipid between stratum corneum cells, (c) the thick insulating fat of the hypodermis, (d) the leathery nature of the dermis.
6. Circle the *false* statement about vitamin D: (a) Dark-skinned people make no vitamin D. (b) If the production of vitamin D is inadequate, one may develop weak bones. (c) If the skin is not exposed to sunlight, one may develop weak bones. (d) Vitamin D is needed for the uptake of calcium from food in the intestine.
7. Use logic to deduce the answer to this question. Given what you know about skin color and skin cancer, the highest rate of skin cancer on Earth occurs among (a) blacks in tropical Africa, (b) scientists in research stations in Antarctica, (c) whites in northern Australia, (d) Norwegians in the United States, (e) blacks in the United States.
8. An arrector pili muscle (a) is associated with each sweat gland, (b) causes the hair to stand up straight, (c) enables each hair to be stretched when wet, (d) squeezes hair upward so it can grow out.
9. Sebum (a) lubricates the skin and hair; (b) consists of dead cells and fatty substances; (c) collects dirt, so hair has to be washed; (d) all of these.
10. An eccrine sweat gland is a ____ gland: (a) compound tubular, (b) compound tubuloalveolar, (c) simple squamous, (d) simple tubular.
11. Identify which embryonic layer listed in column B forms each of the skin structures in column A.

Column A	Column B
____ (1) hypodermis	(a) ectoderm
____ (2) sebaceous gland	(b) mesoderm
____ (3) epidermis	(c) endoderm
____ (4) dermis	(d) neural crest
____ (5) melanocytes	

12. The reticular layer of the dermis (a) provides strength and elasticity to the skin, (b) is composed of loose connective tissue, (c) insulates to prevent heat loss, (d) forms the dermal papilla.

This chapter focuses on the structure, function, and growth of cartilage and bone. (The individual bones of the skeleton are described in Chapters 7 and 8. The details of joint structure and function are covered in Chapter 9).

CARTILAGES

learning outcomes

- Locate the major cartilaginous structures of the adult human body, and explain the functional properties of cartilage tissue.
- Compare the structure, functions, and locations of the three kinds of cartilage tissue.
- Explain how cartilage grows.

Location and Basic Structure

Cartilaginous structures are found throughout the adult human body. These cartilages (shown in Figure 6.1) include:

1. Cartilage in the external ear
2. Cartilages in the nose

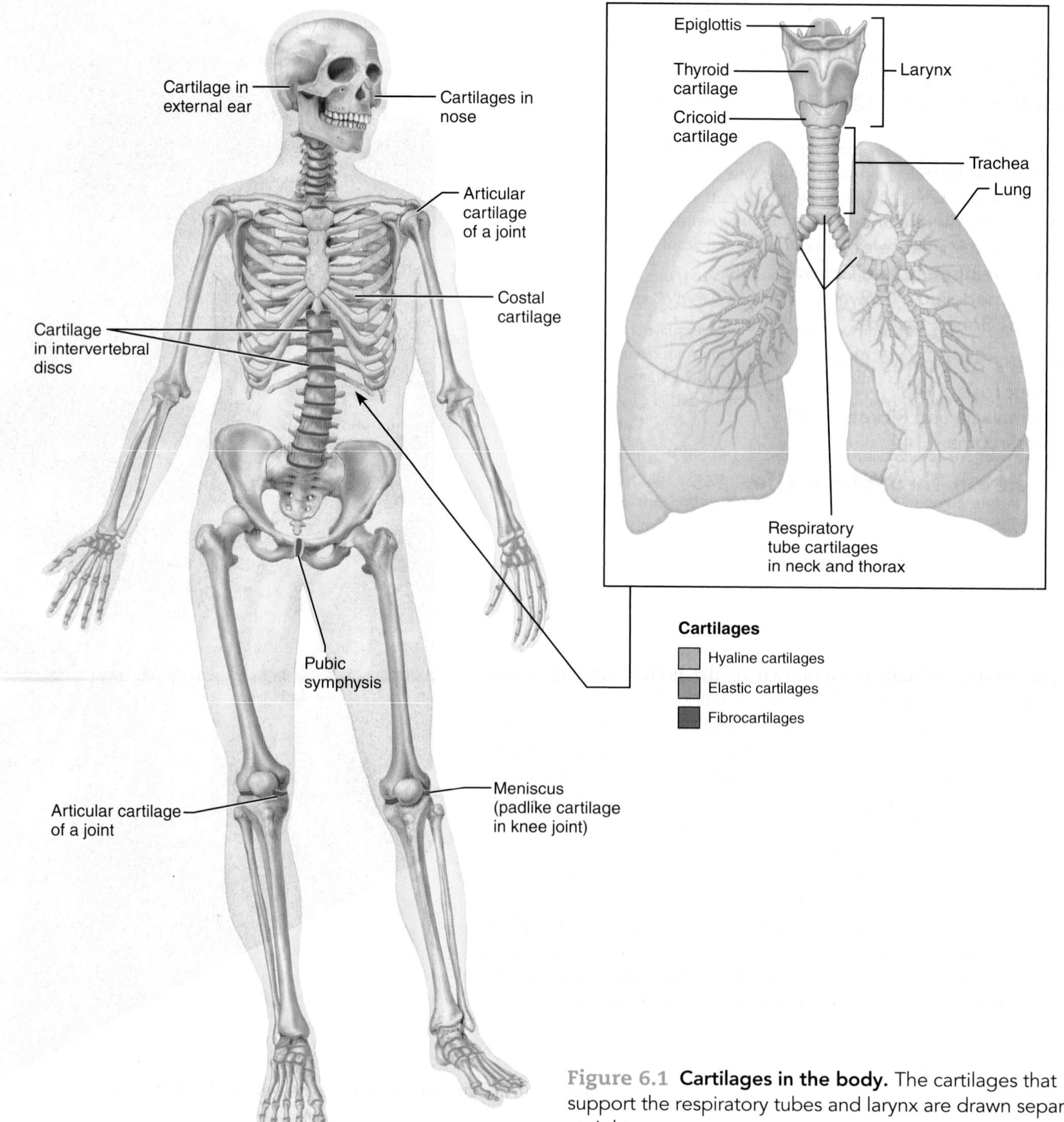

Figure 6.1 **Cartilages in the body.** The cartilages that support the respiratory tubes and larynx are drawn separately at right.

3. **Articular cartilages,** which cover the ends of most bones at movable joints
4. **Costal cartilages,** which connect the ribs to the sternum (breastbone)
5. Cartilages in the larynx (voice box), including the *epiglottis,* a flap that keeps food from entering the larynx and the lungs
6. Cartilages that hold open the air tubes of the respiratory system
7. Cartilage in the discs between the vertebrae
8. Cartilage in the pubic symphysis
9. Cartilages that form the *articular discs* within certain movable joints, the *meniscus* in the knee for example

Cartilage is far more abundant in the embryo than in the adult: Most of the skeleton is initially formed by fast-growing cartilage, which is subsequently replaced by bone tissue in the fetal and childhood periods. (This process is described later in the chapter, pp. 136–138).

A typical cartilaginous structure in the skeleton, such as an articular cartilage, is composed of the connective tissue cartilage, which contains no nerves or blood vessels. The structure is surrounded by a layer of dense irregular connective tissue, the **perichondrium** (per″ĭ-kon′dre-um; "around the cartilage"), which acts like a girdle to resist outward expansion when the cartilage is subjected to pressure (see Figure 6.2a). The perichondrium also functions in the growth and repair of cartilage.

Cartilage tissue consists primarily of water (60% to 80%) and is very resilient; that is, it has the ability to spring back to its original shape after being compressed. (For more information on the unique properties of cartilage tissue, see A CLOSER LOOK on p. 126.)

Types of Cartilage

Three types of cartilage tissue occur in the body: *hyaline cartilage, elastic cartilage,* and *fibrocartilage* (Figure 6.2). While reading about these, keep in mind that cartilage is a connective tissue that consists of cells called **chondrocytes** and an abundant extracellular matrix. In contrast to connective tissue proper, in cartilage tissue each chondrocyte is located in a space in the matrix called a **lacuna** (lah-koo′-nah), literally a "lake" or "cavity." The matrix contains fibers and a jellylike ground substance of complex sugar molecules that attract and hold water (see Chapter 4, p. 81).

Hyaline Cartilage

Hyaline cartilage (hi′ah-līn; "glass"), which looks like frosted glass when viewed by the unaided eye, is the most abundant kind of cartilage (see Figure 6.1). When viewed under the light microscope, its chondrocytes appear spherical (Figure 6.2a). The only type of fiber in the matrix is a collagen unit fibril, which forms networks that are too thin to be seen with a light microscope. The gelatinous ground substance holds large amounts of water; thus, this tissue resists compression well. Hyaline cartilage provides

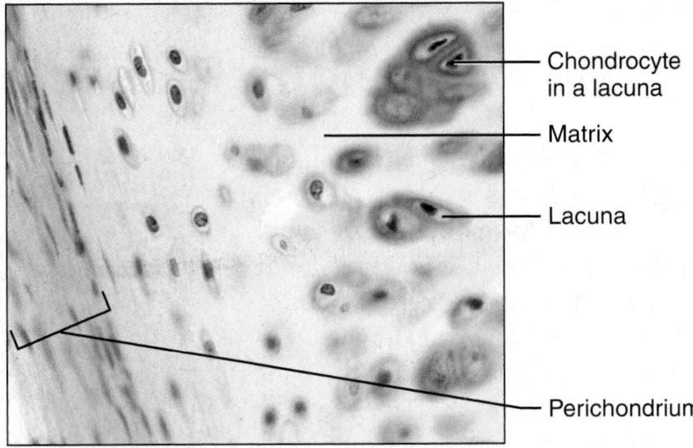

(a) Hyaline cartilage (260×)

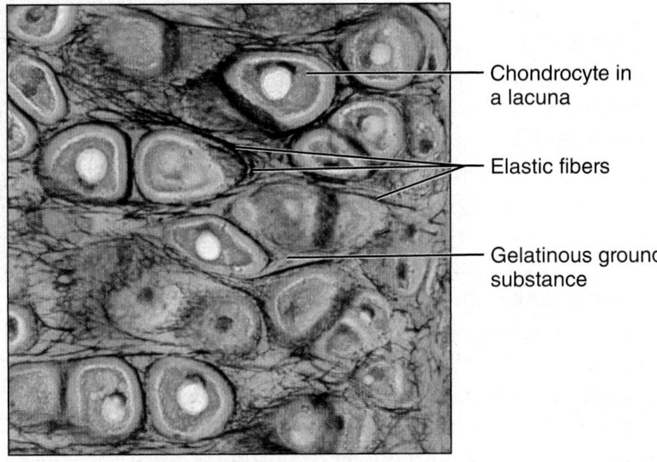

(b) Elastic cartilage (350×)

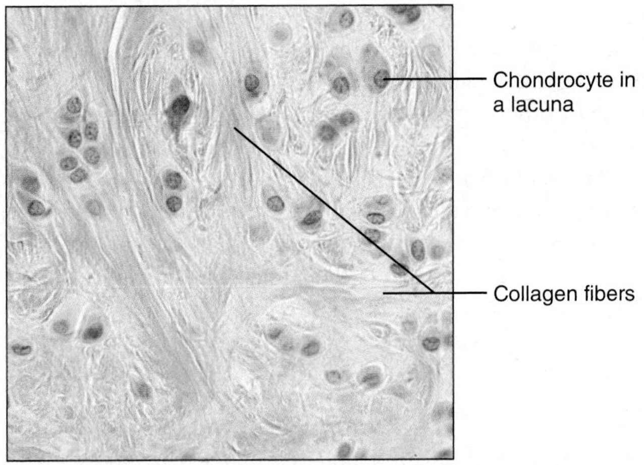

(c) Fibrocartilage (320×)

Figure 6.2 **Types of cartilage tissue.**

A CLOSER LOOK

The Marvelous Properties of Cartilage

Without healthy cartilage to cushion the movable joints, people would find vigorous activities next to impossible. Cartilage is resilient: If you push on it and then ease the pressure, it bounces back. Such resilience allows the articular cartilages at the ends of the bones to cushion the impact of, for example, running, jumping, or pole vaulting.

Another important feature is cartilage tissue's capability for rapid growth, which enables it to keep pace with the rapid growth of the embryo. Most bones originate as cartilage models that are gradually replaced by bone tissue. Throughout childhood and adolescence, growth plates of cartilage remain in most bones. Once adult size is reached, growth plates fuse, and cartilage growth slows. One would expect cartilage to require a rich blood supply for rapid growth, yet cartilage contains no blood vessels. The cells (chondrocytes) in cartilage are specially adapted to survive on the little oxygen that diffuses to them from distant vessels in the surrounding perichondrium. In fact, cartilage and blood vessels are strictly incompatible: Cartilage secretes chemicals that prevent blood vessels from growing into it.

The resilience of cartilage is due to its ability to hold tremendous amounts of water. In fact, cartilage is 60% to 80% water, largely because the complex sugars in its ground substance have so many water-attracting negative charges. When the cartilage is compressed, the water molecules are forced away from the negative charges. As the pressure increases, the negative charges are pushed closer together until they repel each other, resisting further compression.

The cartilage of this pole vaulter and other athletes should be well nourished and well hydrated because of the great physical demands vigorous sports place on the joints.

When the pressure is released, the water molecules rush back into the tissue, causing the cartilage to spring back forcefully to its original shape. These dynamics also play a vital role in the nourishment of cartilage: The flow of fluid during and after compression carries nutrients to the chondrocytes. For this reason, long periods of inactivity can weaken the cartilages of the joints.

Rapid growth is another result of cartilage's capacity to hold fluid. Because cartilage matrix is mostly water, oxygen and nutrients from distant capillaries can diffuse through it quickly enough to supply the metabolic needs of the rapidly dividing cartilage-forming cells (chondroblasts). Furthermore, the ground substance secreted by the chondroblasts attracts so much water that the cartilage expands quickly with little expenditure of materials.

Because its ground substance attracts water, cartilage has a natural tendency to swell. To prevent excessive swelling, its matrix contains a network of thin, unstretchable collagen fibrils. This collagen also gives cartilage its ability to resist tension when pulled by external forces. Even though cartilage is strong in resisting tension and compression, it is weak in resisting shear forces (twisting and bending). Because of this weakness, torn cartilages are a common sports injury.

support through flexibility and resilience. It makes up the articular cartilage that covers the ends of adjoining bones in movable joints. It also forms the cartilaginous attachments of the ribs to the sternum, accounts for most of the cartilage found in the respiratory structures, and forms the embryonic skeleton.

Elastic Cartilage

Elastic cartilage is similar to hyaline cartilage, but its matrix contains many elastic fibers along with the delicate collagen fibrils (Figure 6.2b). This cartilage is more elastic than hyaline cartilage and better able to tolerate repeated bending. The epiglottis, which bends down to cover the glottis (opening) of the larynx each time we swallow, is made of elastic cartilage, as is the highly bendable cartilage in the outer ear (Figure 6.1).

Fibrocartilage

Fibrocartilage is an unusual tissue that resists both strong compression and strong tension (pulling) forces. It occurs in certain ligaments and certain cartilages that experience both of these forces. It is a perfect structural intermediate between hyaline cartilage and dense regular connective tissue. Microscopically, it consists of thick collagen fibers (as in dense regular connective tissue) surrounding the chondrocytes within lacunae (Figure 6.2c). Two specific locations of fibrocartilage are in the *anulus fibrosus* portion of the discs between the vertebrae and in the articular discs of some joints, for example the menisci of the knee.

Growth of Cartilage

A cartilaginous structure grows in two ways. In **appositional** (ap″o-zish′un-al) **growth,** "growth from outside," cartilage-forming cells (chondroblasts) in the surrounding perichondrium produce the new cartilage tissue by actively secreting matrix. In **interstitial** (in″ter-stish′al) **growth,** "growth from within," the chondrocytes within the cartilage divide and secrete new matrix. Cartilage grows rapidly during embryonic development, childhood, and adolescence. Cartilage stops growing in the late teens, when the skeleton itself stops growing, and chondrocytes do not divide again. As a result, cartilage regenerates poorly in adults. The limited healing that occurs within cartilage in adults reflects the ability of the surviving chondrocytes to secrete more extracellular matrix, and, typically, cartilage is repaired with fibrocartilage.

Under certain conditions, crystals of calcium phosphate precipitate in the matrix of cartilage. Such calcification is a sign of aging in an adult, but in a child it is a normal stage in the growth of most bones (see p. 138). Note, however, that **calcified cartilage** is not bone: Bone and cartilage are always distinct tissues.

☑ check your understanding

- ☐ 1. How does the matrix differ in each of the three types of cartilage?
- ☐ 2. Which type of cartilage is most abundant? List three locations where this type of cartilage is found.
- ☐ 3. Where are the chondroblasts located that produce new cartilage by appositional growth?

(For answers, see Appendix B.)

BONES

learning outcomes

- ▶ Describe the functions of the bony skeleton and of bone tissue.
- ▶ Describe the structure of bone tissue and the functions of its organic and inorganic components.

The bones of the skeleton are *organs* because they contain several different tissues. Although bone tissue predominates, bones also contain nervous tissue in nerves, blood tissue in blood vessels, cartilage in articular cartilages, and epithelial tissue lining the blood vessels.

In our discussion of bone, we describe the functions of bone and then examine the composition of bone tissue. We then look at bone as an organ, examining the macroscopic and microscopic structure of bones, bone development and growth, and bone repair.

Functions of Bones

The functions of the skeletal system were briefly mentioned in the introductory paragraph of this chapter. More specifically, bone carries out the following functions:

1. **Support.** Bones provide a hard framework that supports the weight of the body. For example, the bones of the legs are pillars that support the trunk of the body in the standing person.
2. **Movement.** Skeletal muscles attach to the bones by tendons and use the bones as levers to move the body and its parts. As a result, humans can walk, grasp objects, and move the rib cage to breathe. The arrangement of the bones and the structure of the joints determine the types of movement that are possible. Support and movement are mutually dependent functions: The supportive framework is necessary for movement, and the skeletal muscles contribute significantly to the support of body weight.
3. **Protection.** The bones of the skull form a protective case for the brain. The vertebrae surround the spinal cord, and the rib cage helps protect the organs of the thorax.
4. **Mineral storage.** Bone serves as a reservoir for minerals, the most important of which are calcium and phosphate. The stored minerals are released into the bloodstream as ions for distribution to all parts of the body as needed.
5. **Blood cell formation and energy storage.** Bones contain red and yellow *bone marrow.* Red marrow makes the blood cells, and yellow marrow is a site of fat storage, with little or no role in blood cell formation. (Red bone marrow and the production of blood cells are described in detail in Chapter 18.)
6. **Energy metabolism.** Bone-producing cells, *osteoblasts* (described in the next section), secrete a hormone that influences blood sugar regulation and energy metabolism. This hormone, *osteocalcin,* stimulates pancreatic secretions that reduce blood sugar levels (insulin). Osteocalcin also influences fat cells, causing them to store less fat and to secrete a hormone that increases the insulin sensitivity of cells. These results have clinical implications for the treatment of metabolic disorders related to blood sugar regulation, such as type 2 diabetes.

Bone Tissue

Like other connective tissues, bone tissue consists of cells separated by an extracellular matrix. Unlike other connective tissues, bone has both organic and inorganic components. The organic components are the cells, fibers, and ground substance. The inorganic components are the mineral salts that invade the bony matrix, making bone tissue hard. Bone does contains a small amount of tissue fluid, although bone contains less water than other connective tissues.

Extracellular Matrix

As with other connective tissues, it is the unique composition of the matrix that gives bone its exceptional physical properties.

The organic components of bone tissue account for 35% of the tissue mass. These organic substances, particularly collagen, contribute the flexibility and tensile strength that allow bone to resist stretching and twisting. Collagen is remarkably abundant in bone tissue.

The balance of bone tissue, 65% by mass, consists of inorganic hydroxyapatites (hi-drok″se-ap′ah-tītz), or mineral salts, primarily calcium phosphate. These mineral salts are present as tiny crystals that lie in and around the collagen fibrils in the extracellular matrix. The crystals pack tightly, providing bone with its exceptional hardness, which enables it to resist compression. These mineral salts also explain how bones can endure for hundreds of millions of years, providing information on the sizes, shapes, lifestyles, and even some of the diseases (for example, arthritis) of ancient vertebrates.

The composite nature of bone can be compared to reinforced concrete: The collagen fibers, like the steel rods, provide tensile strength; the mineral salts, like the sand and rock in the concrete, provide compressional strength. In fact, bone is stronger than reinforced concrete in resisting compression and almost equal in tensile strength. Neither bone nor reinforced concrete resist torsional forces well. Indeed, most fractures to limb bones are caused by torsional forces.

To demonstrate this composite nature:

1. Soak a bone from a chicken leg or a wishbone in vinegar for several days. The acidic vinegar dissolves away the bone's mineral salts, leaving only the organic component, mainly collagen. The demineralized bone can be tied in a knot, demonstrating the great flexibility provided by the collagen in bone. It also confirms that without its mineral content, bone bends too easily to support weight.
2. Bake a bone at high temperature to destroy the organic portion of bone. What remains is the mineral component, stiff but extremely brittle.

The proper combination of organic and inorganic elements allows bone to be exceedingly durable, strong, and resilient without being brittle.

Cells

Three types of cells in bone tissue produce or maintain the tissue: osteogenic cells, osteoblasts, and osteocytes. **Osteogenic cells** (os′tē-ō-jen′ik; *osteo* = bone, *genic* = producing) are stem cells that differentiate into bone-forming osteoblasts. **Osteoblasts** (os′tē-ō-blasts; *blast* = bud, sprout) are cells that actively produce and secrete the organic components of the bone matrix: the ground substance and the collagen fibers. This bone matrix secreted by osteoblasts is called **osteoid** (os′te-oid, "bonelike"). Within a week, inorganic calcium salts crystallize within the osteoid. Once osteoblasts are completely surrounded by bone matrix and are no longer producing new osteoid, they are called osteocytes. **Osteocytes** (*cyte* = cell) function to keep the bone matrix healthy. In fact, if osteocytes die or are destroyed, the bone matrix is resorbed.

The cells responsible for the resorption of bone are the fourth type of cell found within bone tissue, osteoclasts. **Osteoclasts** (*clast* = break; see Figure 6.14) are derived from a lineage of white blood cells. These multinucleated cells break down bone by secreting hydrochloric acid, which dissolves the mineral component of the matrix, and lysosomal enzymes, which digest the organic components.

Creating and destroying bone tissue is an ongoing process in normal, healthy bones. Destroying old bone tissue and replacing it with new tissue helps to keep our bones strong and enables them to respond to changing stresses. When we are physically active, new bone is formed, strengthening our skeletal support; when we are incapacitated, as with a broken limb or bedridden with illness, bone is resorbed because it is not needed to support the body. We will return to this process, called bone remodeling, later in the chapter.

check your understanding

- ☐ 4. Which component of bone tissue contributes to the strength and flexibility of bone? Which contributes to the hardness of bone?
- ☐ 5. What minerals are stored in bone, and which cells in bone tissue function to remove these minerals from the bone tissue?
- ☐ 6. What is the difference between an osteoblast and an osteocyte?

(For answers, see Appendix B.)

Gross Anatomy of Bones

learning outcomes

- Describe the gross anatomy of a typical long bone and a typical flat bone.
- Explain how bones withstand tension and compression.
- Describe the types of markings found on bones.

Classification of Bones

Bones come in many sizes and shapes. For example, the tiny pisiform bone of the wrist is the size and shape of a pea, whereas the femur (thigh bone) is large and elongated. The shape of each bone reflects its function and formation. The femur, for example, must withstand great weight and pressure, and its hollow cylindrical design provides maximum strength with minimum weight.

Bones are classified by their shape as long, short, flat, or irregular (Figure 6.3).

1. **Long bones.** As their name suggests, long bones are considerably longer than they are wide (Figure 6.3a). A long bone has a shaft plus two distinct ends. Most bones in the limbs are long bones. These bones are named for their elongated shape, not their overall size: The bones of the fingers and toes are long bones, even though they are small.
2. **Short bones.** Short bones are roughly cube-shaped. They occur in the wrist and the ankle (Figure 6.3b).

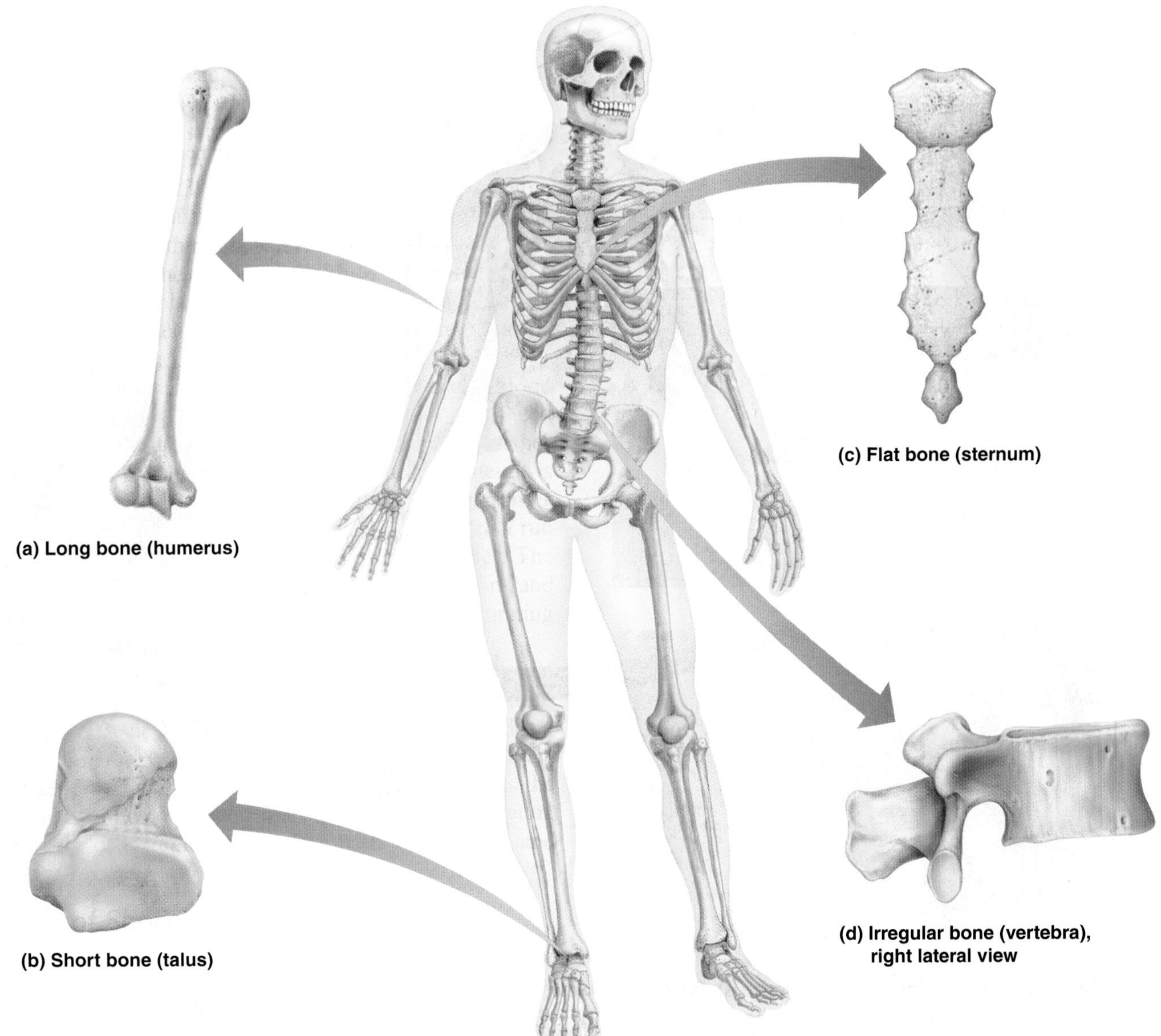

Figure 6.3 **Classification of bones.**

Sesamoid bones (ses′ah-moid; "shaped like a sesame seed") are a special type of short bone that forms within a tendon. A good example is the kneecap, or patella. Sesamoid bones vary in size and number in different people. Some sesamoid bones clearly act to alter the direction of pull of a tendon. Others reduce friction and modify pressure in tendons, thus reducing abrasion or tearing.

3. **Flat bones.** Flat bones are thin, flattened, and usually somewhat curved (Figure 6.3c). Most cranial bones of the skull are flat, as are the ribs, sternum (breastbone), and scapula (shoulder blade).
4. **Irregular bones.** Irregular bones have various shapes that do not fit into the previous categories. Examples are the vertebrae and hip bones (Figure 6.3d).

Compact and Spongy Bone

Almost every bone of the skeleton has a dense outer layer that looks smooth and solid to the naked eye. This external layer is **compact bone** (Figure 6.4 and Figure 6.7). Internal to this is **spongy bone,** also called *trabecular bone,* a honeycomb of small needle-like or flat pieces called **trabeculae** (trah-bek′u-le; "little beams"). In this network, the open spaces between the trabeculae are filled with red or yellow bone marrow.

Structure of a Typical Long Bone

With few exceptions, all the long bones in the body have the same general structure (Figure 6.4).

Diaphysis and Epiphyses The tubular **diaphysis** (di-af′ĭ-sis), or shaft, forms the long axis of a long bone; the **epiphyses** (e-pif′ĭ-sez) are the bone ends (Figure 6.4a). The

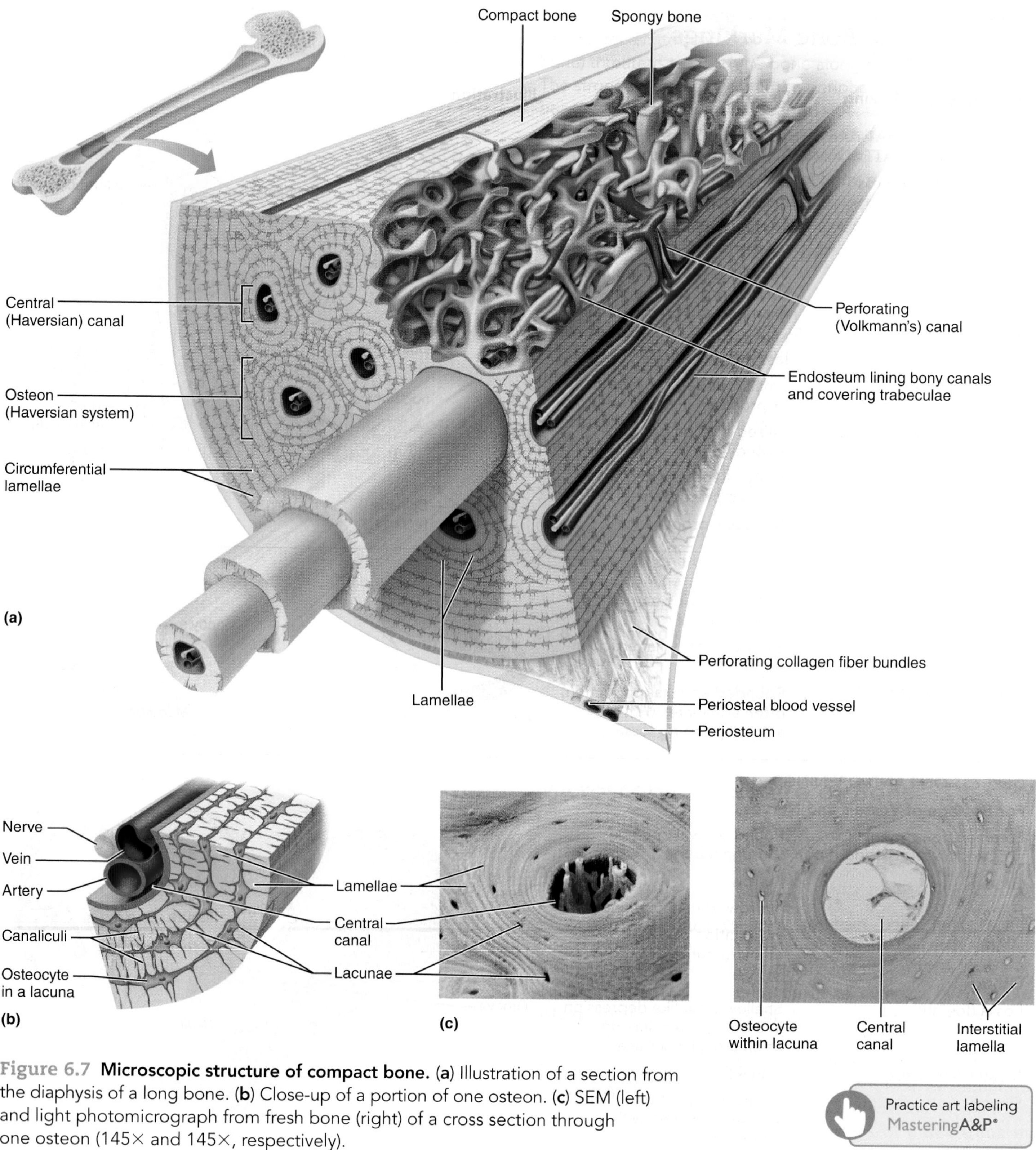

Figure 6.7 Microscopic structure of compact bone. (**a**) Illustration of a section from the diaphysis of a long bone. (**b**) Close-up of a portion of one osteon. (**c**) SEM (left) and light photomicrograph from fresh bone (right) of a cross section through one osteon (145× and 145×, respectively).

Unlike the growth rings in trees, lamellae of bone tissue are added to the inner surface of the osteon, thus decreasing the diameter of the central canal. **Perforating canals,** also called *Volkmann's* (fōlk′mahnz) *canals,* lie at right angles to the central canals and connect the blood and nerve supply of the periosteum to that of the central canals and the marrow cavity.

The mature bone cells, the osteocytes, are spider-shaped (Figure 6.7b). Their bodies occupy small cavities in the solid matrix called lacunae ("little lakes"), and their "spider legs" occupy thin tubes called **canaliculi** (kan″ah-lik′u-li). These "little canals" run through the matrix, connecting neighboring lacunae to one another and to the nearest capillaries, such as

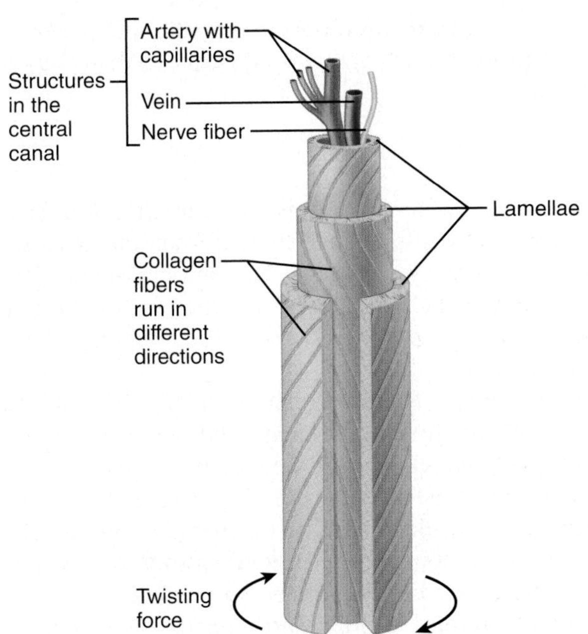

Figure 6.8 A single osteon. To illustrate the individual lamellae, the osteon is drawn as though it were pulled out like a telescope.

those in the central canals. Within the canaliculi, the extensions of neighboring osteocytes touch each other and form gap junctions (see p. 76). Nutrients diffusing from capillaries in the central canal pass across these gap junctions, from one osteocyte to the next, throughout the entire osteon. This direct transfer from cell to cell is the only way to supply the osteocytes with the nutrients they need, because the intervening bone matrix is too solid and impermeable to act as a diffusion medium.

Not all lamellae in compact bone occur within osteons. Lying between the osteons are groups of incomplete lamellae called **interstitial** (in″ter-stish′al) **lamellae** (Figure 6.7c). These are simply the remains of old osteons that have been cut through by bone remodeling. Additionally, **circumferential lamellae** occur in the external and internal surfaces of the layer of compact bone; each of these lamellae extends around the entire circumference of the diaphysis (Figure 6.7a). Functioning like an osteon but on a much larger scale, the circumferential lamellae effectively resist twisting of the entire long bone.

Spongy Bone

The microscopic anatomy of spongy bone (Figure 6.9) is less complex than that of compact bone. Each trabecula contains several layers of lamellae and osteocytes but is too small to contain osteons or vessels of its own. The osteocytes receive their nutrients from capillaries in the endosteum surrounding the trabecula via connections through the canaliculi.

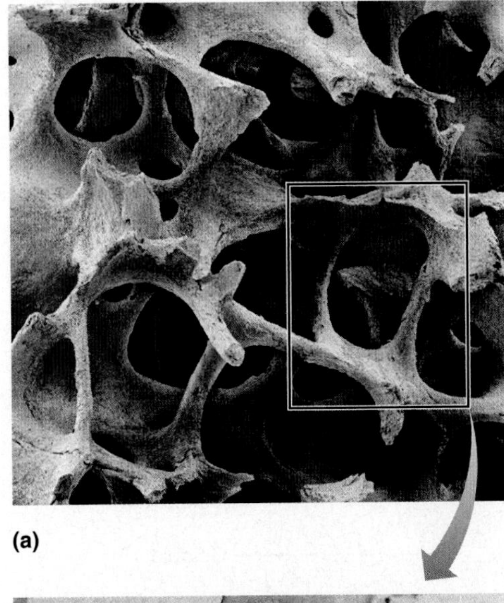

(a)

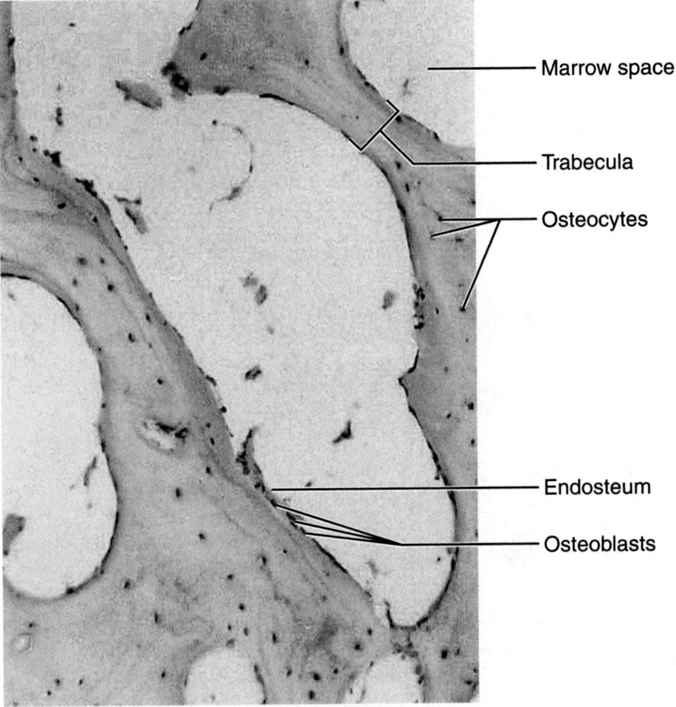

(b)

Figure 6.9 Spongy bone. (a) Micrograph of spongy bone, showing a network of bone trabeculae (32×). (b) Photomicrograph of trabeculae (85×).

✓ check your understanding

- ☐ 10. Differentiate between a central canal, a perforating canal, and the canaliculi.
- ☐ 11. How do the osteocytes in the outer lamella of an osteon receive oxygen and nutrients?
- ☐ 12. What is a trabecula? How is it different from an osteon?

(For answers, see Appendix B.)

Bone Development and Growth

learning outcomes

- ▶ Compare and contrast the two types of bone formation: intramembranous and endochondral ossification.
- ▶ Describe how endochondral bones grow at their epiphyseal plates.

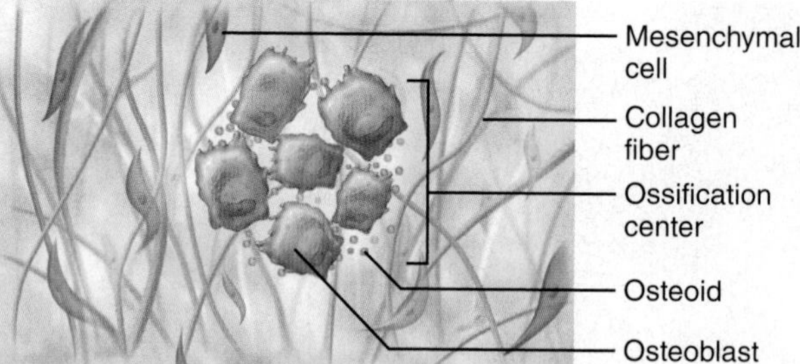

① **Ossification centers appear in the fibrous connective tissue membrane.**
- Selected centrally located mesenchymal cells cluster and differentiate into osteoblasts, forming an ossification center that produces the first trabeculae of spongy bone.

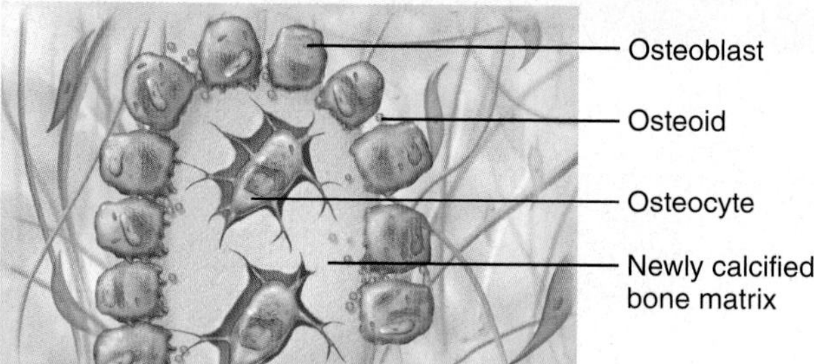

② **Osteoid is secreted within the fibrous membrane and calcifies.**
- Osteoblasts continue to secrete osteoid, which calcifies in a few days.
- Trapped osteoblasts become osteocytes.

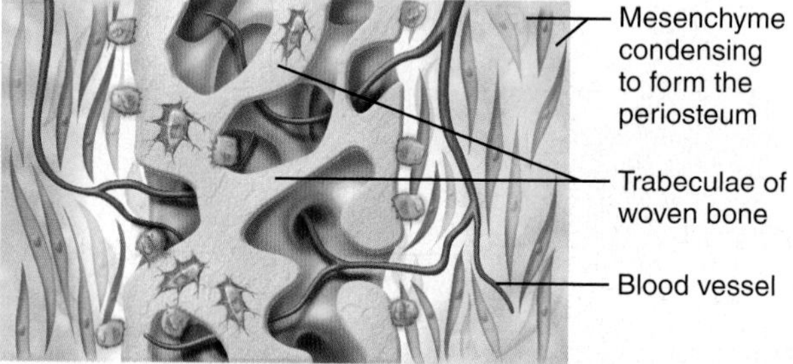

③ **Woven bone and periosteum form.**
- Accumulating osteoid is laid down between embryonic blood vessels in a manner that results in a network (instead of concentric lamellae) of trabeculae called woven bone.
- Vascularized mesenchyme condenses on the external face of the woven bone and becomes the periosteum.

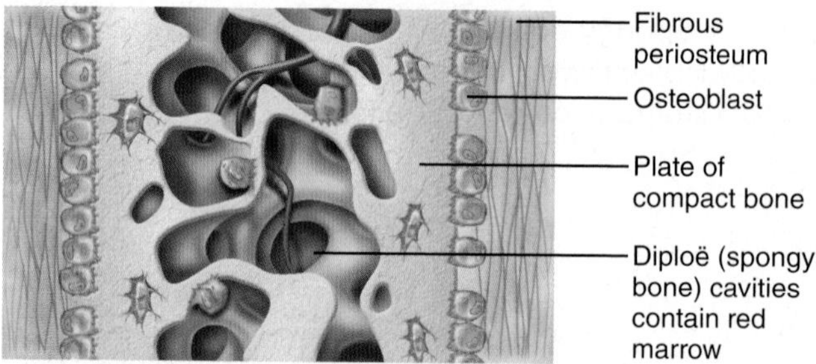

④ **Lamellar bone replaces woven bone, just deep to the periosteum. Red marrow appears.**
- Trabeculae just deep to the periosteum thicken. Mature lamellar bone replaces them, forming compact bone plates.
- Spongy bone (diploë), consisting of distinct trabeculae, persists internally and its vascular tissue becomes red marrow.

Figure 6.10 Intramembranous ossification. Diagrams ① and ② represent much greater magnification than diagrams ③ and ④.

Osteogenesis (os″te-o-jen′ĕ-sis) and **ossification** are both names for the process of bone-tissue formation. Osteogenesis begins in the embryo, proceeds through childhood and adolescence as the skeleton grows, and then occurs at a slower rate in the adult as part of a continual remodeling of the full-grown skeleton.

Before week 8, the skeleton of the human embryo consists only of hyaline cartilage and some membranes of mesenchyme, an embryonic connective tissue (Chapter 4, p. 82). Bone tissue first appears in week 8 and eventually replaces most cartilage and mesenchymal membranes in the skeleton. Some bones, called **membranous bones,** develop from a mesenchymal membrane through a process called **intramembranous ossification** (*intra* = inside). Other bones develop as hyaline cartilage, which is replaced through a process called **endochondral ossification** (*endo* = within; *chondro* = cartilage). These bones are called **endochondral bones** or *cartilage replacement bones.*

Intramembranous Ossification

Membranous bones form directly from mesenchyme without first being modeled in cartilage. All bones of the skull, except a few at the base of the skull, are of this category. The clavicles (collarbones) are the only bones formed by intramembranous ossification that are not in the skull.

Intramembranous ossification proceeds in the following way: During week 8 of embryonic development, mesenchymal cells cluster within the connective tissue membrane and become bone-forming osteoblasts (**Figure 6.10**, ①). These cells begin secreting the organic part of bone matrix, called osteoid, which then becomes mineralized (Figure 6.10, ②). Once surrounded by their own matrix, the osteoblasts are called osteocytes. The new bone tissue forms between embryonic blood vessels, which are woven in a random network (Figure 6.10, ③). The result is **woven bone tissue,** with trabeculae arranged in networks. This embryonic tissue lacks the lamellae that occur in mature spongy bone. During this same stage, more mesenchyme condenses just external to the developing membrane bone and becomes the periosteum.

To complete the development of a membranous bone, the trabeculae at the periphery grow thicker until plates of compact bone are present on both surfaces (Figure 6.10, ④). In the center of the membranous bone, the trabeculae remain distinct, and spongy bone results. The final pattern is that of the flat bone (Figure 6.5).

Endochondral Ossification

All bones from the base of the skull down, except for the clavicles, are endochondral bones. They are first modeled in hyaline cartilage, which then is gradually replaced by bone tissue. Endochondral ossification begins late in the second month of development and is not completed until

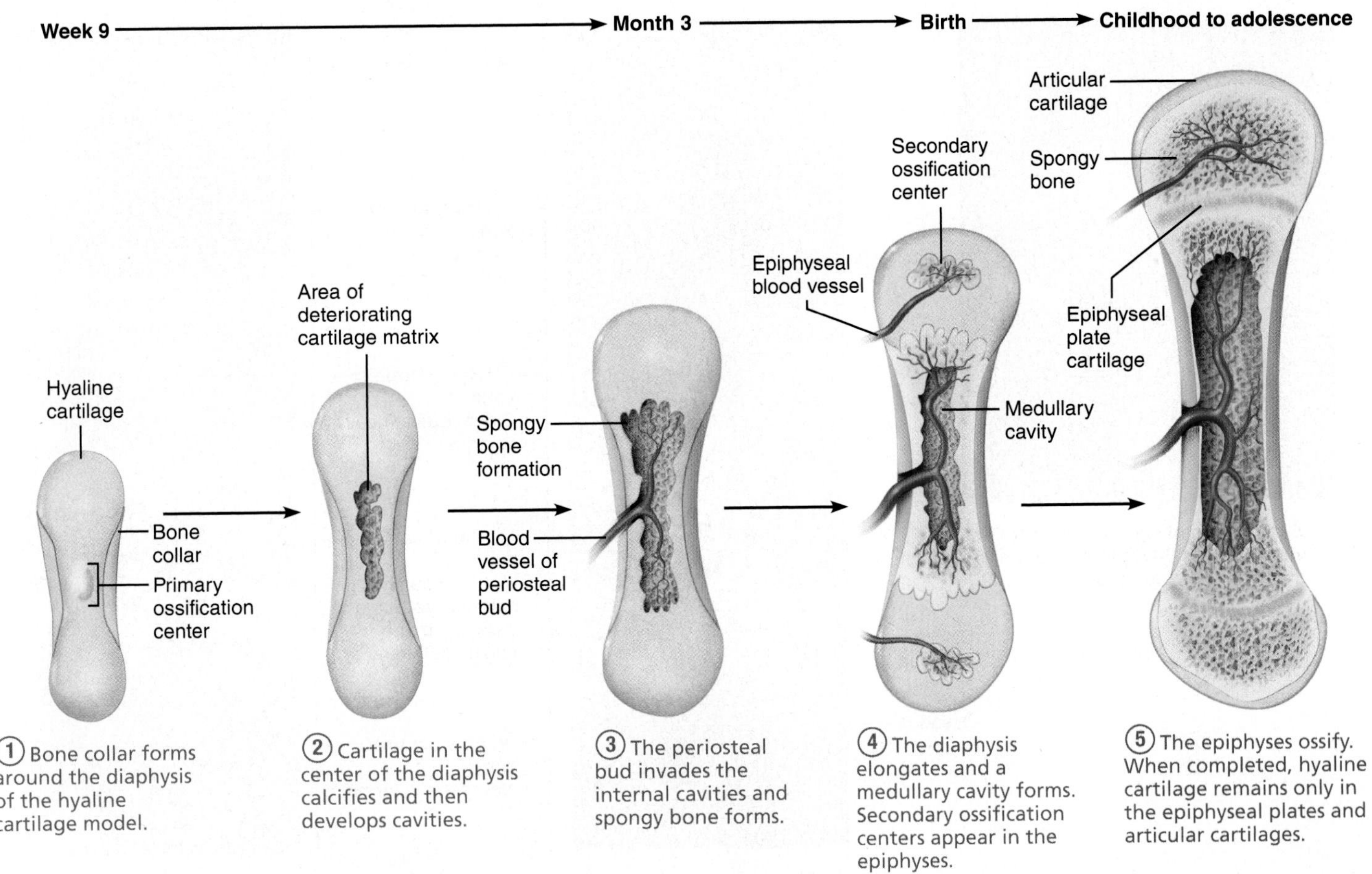

Figure 6.11 **Endochondral ossification of a long bone.**

the skeleton stops growing in early adulthood. Growing endochondral bones increase both in length and in width. The following stages outline only the increase in length, using a large long bone as an example (Figure 6.11).

① **A bone collar forms around the diaphysis.** In the late embryo (week 8), the endochondral bone begins as a piece of cartilage called a *cartilage model.* Like all cartilages, it is surrounded by a perichondrium. Then, at the end of week 8 of development, the perichondrium surrounding the diaphysis is invaded by blood vessels and becomes a bone-forming periosteum. Osteoblasts in this new periosteum lay down a collar of bone tissue around the diaphysis.

② **Cartilage calcifies in the center of the diaphysis.** At the same time the bone collar forms, the chondrocytes in the center of the diaphysis enlarge (hypertrophy) and signal the surrounding cartilage matrix to calcify. The matrix of calcified cartilage is impermeable to diffusing nutrients. Cut off from all nutrients, the chondrocytes die and disintegrate, leaving cavities in the cartilage. No longer maintained by chondrocytes, the cartilage matrix starts to deteriorate. This does not seriously weaken the diaphysis, which is well stabilized by the bone collar around it. These changes affect only the center of the diaphysis. Elsewhere, the cartilage remains healthy and continues to grow, causing the entire endochondral bone to elongate.

③ **The periosteal bud invades the diaphysis, and the first bone trabeculae form.** In the third month of development, the cavities within the diaphysis are invaded by a collection of elements called the **periosteal bud.** This bud consists of a nutrient artery and vein, along with the cells that will form the bone marrow. Most important, the bud contains bone-forming and bone-destroying cells (osteogenic stem cells and osteoclasts). The entering osteoclasts partly erode the matrix of calcified cartilage, and the osteogenic cells differentiate into osteoblasts, which secrete osteoid around the remaining fragments of this matrix, forming bone-covered trabeculae. In this way, the earliest version of spongy bone appears within the diaphysis.

By the third month of development, bone tissue continues to form around the diaphysis from the periosteum and has begun to appear in the center of the diaphysis. This bone tissue of the diaphysis makes up the **primary ossification center.**

④ **Diaphysis elongates, and the medullary cavity forms.** Throughout the rest of the fetal period, the cartilage of the epiphysis continues to grow rapidly, with the part nearest the diaphysis continually calcifying and being replaced by the bone trabeculae, thus elongating the diaphysis. Osteoclasts in turn break down the ends of these bone trabeculae to form a central, boneless medullary cavity.

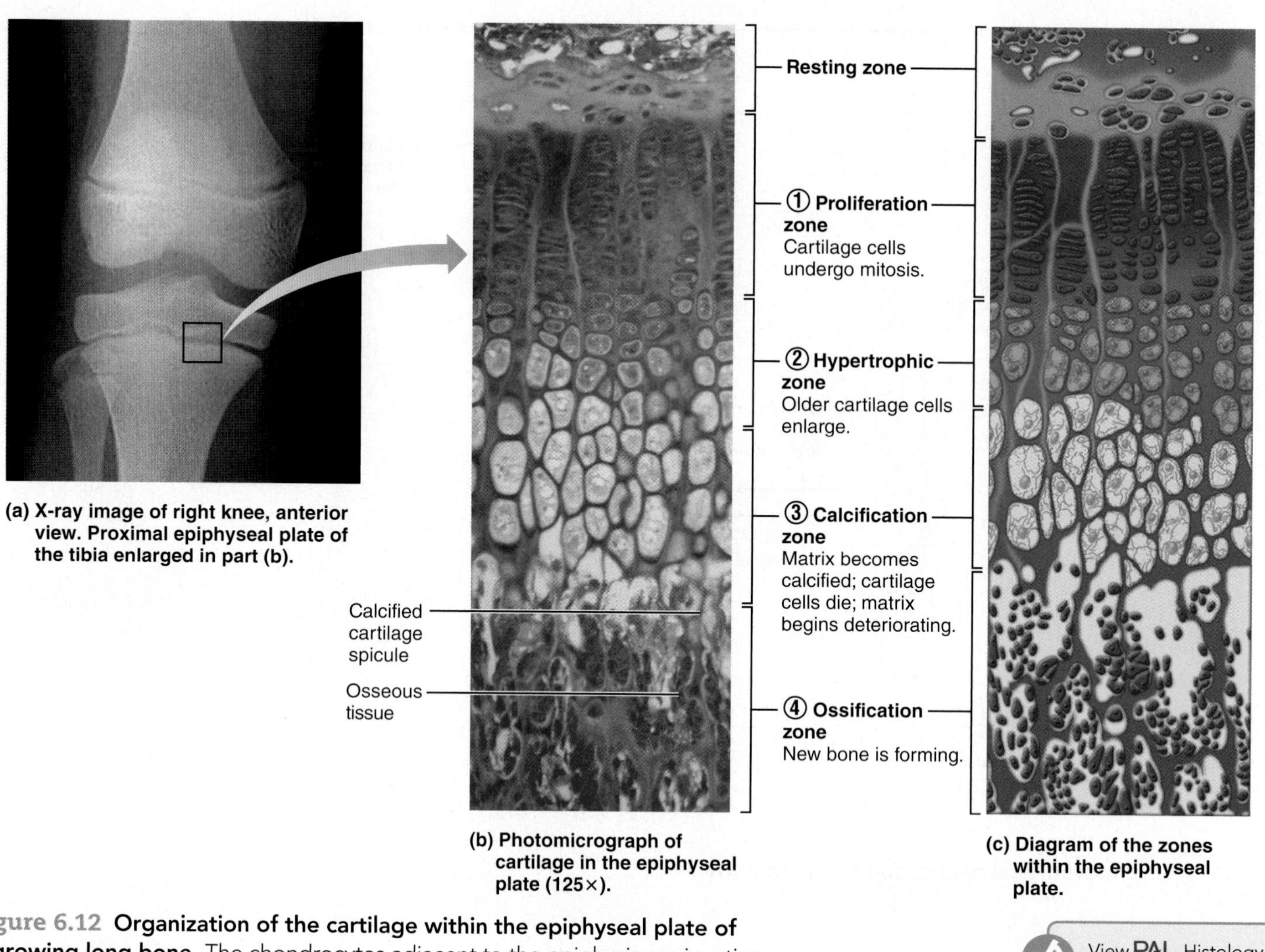

(a) X-ray image of right knee, anterior view. Proximal epiphyseal plate of the tibia enlarged in part (b).

(b) Photomicrograph of cartilage in the epiphyseal plate (125×).

(c) Diagram of the zones within the epiphyseal plate.

Figure 6.12 **Organization of the cartilage within the epiphyseal plate of a growing long bone.** The chondrocytes adjacent to the epiphysis are inactive. This region is the resting zone. The other cartilage cells within the epiphyseal plate are organized into four distinct zones: proliferation, hypertrophic, calcification, and ossification.

View PAL Histology MasteringA&P®

Shortly before or after birth, the epiphyses begin to ossify: First, the cartilage in the center of each epiphysis calcifies and degenerates. Then, a bud containing the epiphyseal vessels invades each epiphysis. Bone trabeculae appear, just as they appeared earlier in the primary ossification center. The areas of bone formation in the epiphyses are called **secondary ossification centers.** The larger long bones of the body can have several ossification centers in each epiphysis.

⑤ **Epiphyses ossify, and cartilaginous epiphyseal plates separate diaphysis and epiphyses.** After the secondary ossification centers have appeared and epiphyses have largely ossified, hyaline cartilage remains at only two places: (1) on the epiphyseal surfaces, where it forms the articular cartilages; and (2) between the diaphysis and epiphysis, where it forms the **epiphyseal plates.** The epiphyseal plates, also called growth plates, are responsible for lengthening the bones during the two decades following birth.

Anatomy of the Epiphyseal Plate

In both the epiphyses of the fetus and the epiphyseal plates of the child, the cartilage is organized in a way that allows it to grow exceptionally quickly and efficiently (Figure 6.12). The cartilage cells nearest the epiphysis are relatively small and inactive. This region is called the **resting** *(quiescent)* **zone.** Below the resting zone, the cartilage cells form tall columns, like coins in a stack. The chondroblasts at the "top" of the stack in the **proliferation zone** divide quickly, pushing the epiphysis away from the diaphysis, thereby causing the entire long bone to lengthen. The older chondrocytes deeper in the stack, in the **hypertrophic zone,** enlarge and signal the surrounding matrix to calcify. In the **calcification zone,** the cartilage matrix becomes calcified, and the chondrocytes die. This process leaves long spicules (trabeculae) of calcified cartilage on the diaphysis side of the epiphysis-diaphysis junction. These spicules are partly eroded by osteoclasts, then covered with bone tissue by osteoblasts, forming spicules of bone. This region is the **ossification zone.** These bony spicules are destroyed from within the diaphysis by the action of osteoclasts at the same rate that they are formed at the epiphysis; thus they stay a constant length and the marrow cavity grows longer as the long bone lengthens.

Postnatal Growth of Endochondral Bones

During childhood and adolescence, the endochondral bones lengthen entirely by growth of the epiphyseal plates. Because its cartilage is replaced with bone tissue on the diaphysis side about as quickly as it grows, the epiphyseal plate maintains a constant thickness while the whole bone lengthens. As adolescence draws to an end, the chondroblasts in the epiphyseal plates divide less often, and the plates become thinner. Eventually, they exhaust their supply of mitotically active cartilage cells, so the cartilage stops growing and is replaced by bone tissue. Long bones stop lengthening when the bone of the epiphyses and diaphysis fuses. This process, called *closure of the epiphyseal plates,* occurs between the ages of 15 and 23. The timing of closure varies among individuals, among bones, and even between different epiphyseal plates of the same bone. In general, the epiphyseal plates of females close earlier than those of males. Once the epiphyseal plates close, a person can grow no taller. The age of a child or adolescent can be estimated by measuring bone length and degree of closure of the epiphyseal plate of a long bone, as shown on an X-ray image. In adults, because no further growth in length occurs after closure of the epiphyseal plates, long bone length can be used to estimate overall height. Both of these techniques are used forensically to help identify unknown individuals.

CLINICAL APPLICATION

Achondroplasia A congenital condition called **achondroplasia** (a-kon″dro-pla′ze-ah) involves defective growth of cartilage and defective endochondral ossification. This condition results from a mutation in a gene on chromosome 4 (fibroblastic growth factor receptor-3 gene or FGFR3). Increased activity of this gene inhibits cartilage proliferation in the epiphyseal plates. As a result, the growth and hypertrophic zones of the epiphyseal plate are narrow and disorganized, and ossification and closure of the epiphyseal plate occurs before normal bone length is reached. Achondroplasia results in typical dwarfism, in which the limbs are short but the trunk and membranous bones are of normal size. Although this condition is genetic, 80% of cases are the result of new mutations. It is also not rare, affecting between 1 in 15,000 and 1 in 40,000 people.

Growing bones must also widen as they lengthen. Osteoblasts in the osteogenic layer of the periosteum add bone tissue in circumferential lamellae to the external face of the diaphysis as osteoclasts in the endosteum remove bone from the internal surface of the diaphysis wall. These two processes occur at about the same rate, so that the circumference of the long bone expands and the bone widens. Growth of a bone by the addition of bone tissue to its surfaces is called **appositional growth.**

Bone growth is regulated by several hormones, primarily growth hormone (produced by the pituitary gland), which stimulates the epiphyseal plates to grow. Thyroid hormones modulate the effects of growth hormone, ensuring that the skeleton retains its proper proportions as it grows. The sex hormones (androgens and estrogens) first promote bone growth in the growth spurt at adolescence and later induce the epiphyseal plates to close, ending growth in length.

This section has focused on the growth and development of large long bones. The other types of endochondral bones grow in slightly different ways. Short bones, such as those in the wrist, arise from only a single ossification center. Most of the irregular bones, such as the hip bone and vertebrae, develop from several distinct ossification centers. Small long bones, such as those in the palm and fingers, form from a primary ossification center (diaphysis) plus a single secondary center; that is, they have just one epiphysis. However, regardless of the number and location of ossification centers, all endochondral bones follow similar steps (shown in Figure 6.11): calcification and deterioration of cartilage internally, invasion of a periosteal bud containing osteoclasts and osteogenic stem cells, and deposition of bone tissue by osteoblasts.

check your understanding

- ☐ **13.** Which bones of the skeleton are membranous bones?
- ☐ **14.** Which portion of the long bones in a 6-month-old fetus is ossified?
- ☐ **15.** As a bone grows in length during childhood, does the thickness of the epiphyseal plate change? In which region of the epiphyseal plate is bone tissue added: the epiphyseal end or the diaphyseal end?

(For answers, see Appendix B.)

Bone Remodeling

learning outcomes

- ▶ Discuss how bone tissue is remodeled within the skeleton.
- ▶ Explain the steps in the healing of bone fractures.

Bones appear to be the most lifeless of body organs when seen in the lab, and once they are formed, they seem set for life. Nothing could be further from the truth. Bone is a dynamic and active tissue. Large amounts of bone matrix and thousands of osteocytes are continuously being removed and replaced within the skeleton, and the small-scale architecture of bones constantly changes. As much as half a gram of calcium may enter or leave the adult skeleton each day.

In the adult skeleton, bone is deposited and removed primarily at the endosteal surface. Together, these two processes constitute **bone remodeling.** The spongy bone in the skeleton, which is covered with endosteum, is entirely replaced

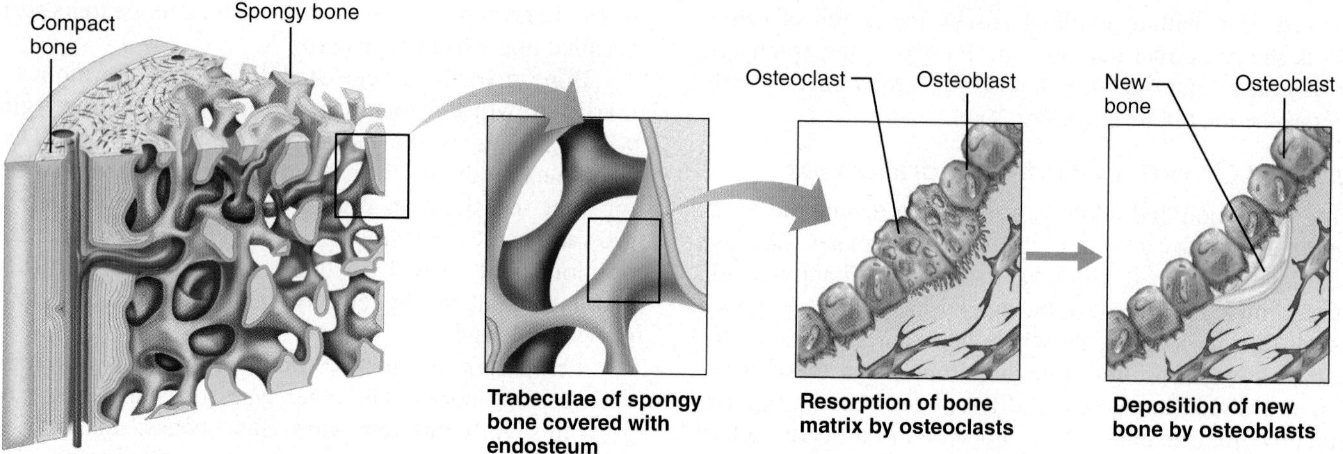

Figure 6.13 Remodeling of spongy bone. Resorption of bony matrix by osteoclasts form a cavity at the bone surface. Osteoblasts from the endosteum fill in the cavity, forming new bone.

every 3 or 4 years. Remodeling in compact bone occurs at the endosteum lining the central canals of the osteons. This process occurs more slowly than in spongy bone; compact bone is completely replaced every 10 years.

Bone remodeling is coordinated by cohorts of adjacent osteoblasts and osteoclasts (Figure 6.13). In healthy young adults, the total mass of bone in the body stays constant, an indication that the rates of deposit and resorption are essentially equal. The remodeling process is not uniform, however. Some bones (or bone parts) are very heavily remodeled; others are not. For example, the distal region of the femur is fully replaced every 5 to 6 months, whereas the diaphysis of the femur changes much more slowly.

Bone resorption is accomplished by osteoclasts (Figure 6.13 and Figure 6.14). Each of these giant cells has many nuclei. Osteoclasts crawl along bone surfaces, essentially digging pits as they break down the bone tissue (Figure 6.13). The part of their plasma membrane that touches the bone surface is highly folded, or ruffled. This expanded membrane forms a tight seal against the bone and secretes concentrated hydrochloric acid, which dissolves the mineral part of the matrix. The liberated calcium ions (Ca^{2+}) and phosphate ions (PO_4^{3-}) enter the tissue fluid and the bloodstream. Lysosomal enzymes are also released across the ruffled membrane and digest the organic part of the bone matrix. Finally, osteoclasts apparently take up collagen and dead osteocytes by phagocytosis.

Bone deposition is accomplished by osteoblasts, these cells lay down organic osteoid on bone surfaces (Figure 6.13), and calcium salts crystallize within this osteoid. This calcification process takes about a week. As stated earlier, the osteoblasts transform into osteocytes when they are surrounded by bone matrix.

Bone-forming osteoblasts derive from mesenchyme cells. In adults, osteoblasts form from mesenchyme-like stem cells located in the periosteum, the endosteum, and the connective tissues of the nearby bone marrow. Osteoclasts,

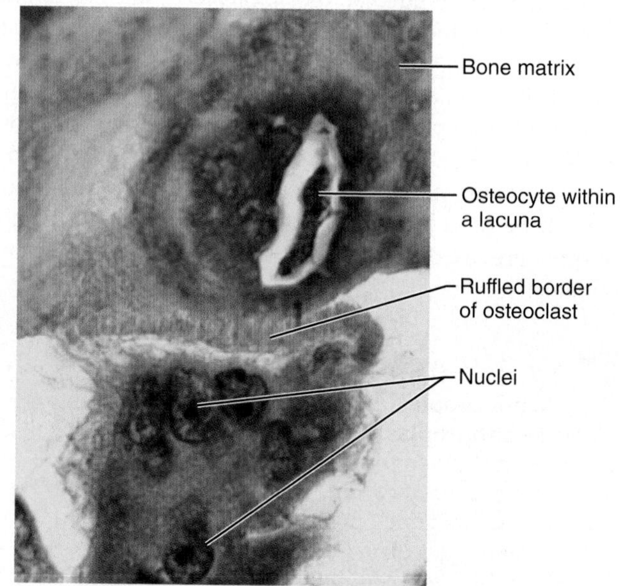

Figure 6.14 An osteoclast. Photomicrograph of an osteoclast destroying bone tissue (520×).

which also form in the bone marrow, arise from immature blood cells called *hematopoietic stem cells,* and they may be related to macrophages. Many of these stem cells fuse together to form each osteoclast, thus their multinucleate structure.

The bones of the skeleton are continually remodeled for two reasons. First, bone remodeling helps maintain constant concentrations of Ca^{2+} and PO_4^{3-} in body fluids. Ca^{2+} levels are strictly controlled because Ca^{2+} is critical for muscle contraction. When the concentration of Ca^{2+} in body

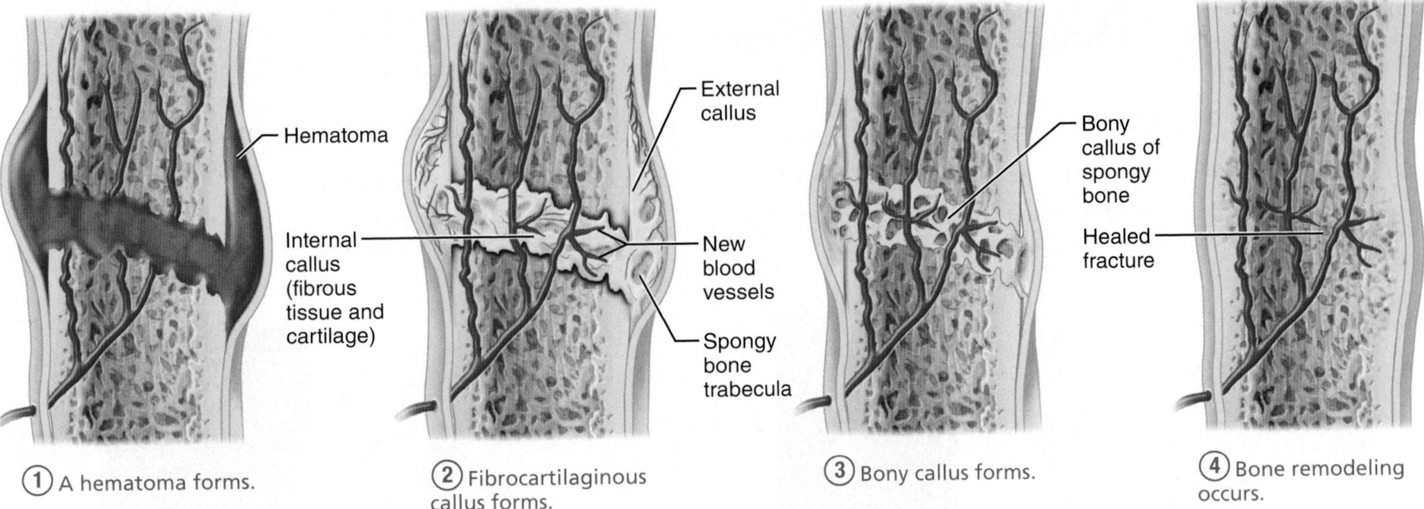

Figure 6.15 Stages in the healing of a bone fracture.

fluids starts to fall, a hormone is released by the parathyroid (par″ah-thi′roid) glands of the neck. This *parathyroid hormone* stimulates osteoclasts to resorb bone, a process that releases more Ca^{2+} into the blood.

Second, bone is remodeled in response to the mechanical stress it experiences. Accordingly, both the osteons of compact bone and the trabeculae of spongy bone are constantly replaced by new osteons and trabeculae that are more precisely aligned with newly experienced compressive and tensile stresses. Furthermore, bone grows thicker in response to the forces experienced during exercise and gains in weight. Conversely, in the absence of mechanical stress, bone tissue is lost, which is why the bones of bedridden people atrophy. A loss of bone under near-zero-gravity conditions is the main obstacle to long missions in outer space. To slow bone loss, astronauts perform isometric exercises during space missions.

CLINICAL APPLICATION

Paget's Disease **Paget's** (paj′ets) **disease** is characterized by excessive rates of bone deposition and bone resorption. The newly formed bone has an abnormally high ratio of immature woven bone to mature compact bone. This, along with reduced mineralization, makes the bones soft and weak. Late in the course of the disease, the activity of osteoblasts outpaces that of osteoclasts. Therefore, the bones can thicken, but in an irregular manner, and the medullary cavities may fill with bone. Paget's disease may affect many parts of the skeleton but is usually a localized and intermittent condition. It rarely occurs before age 40 and affects about 3% of all older people in North America. It progresses slowly, often produces no symptoms, and is seldom life-threatening. Its cause is unknown. Treatment involves inhibiting osteoclasts with bisphosphonate drugs or calcitonin.

Repair of Bone Fractures

Despite their strength, bones are susceptible to **fractures,** or breaks. In young people, most fractures result from trauma (sports injuries, falls, or car accidents, for example) that twists or smashes the bones. In old age, bones thin and weaken, and fractures occur more often. A fracture in which the bone breaks cleanly but does not penetrate the skin is a **simple fracture.** When broken ends of the bone protrude through the skin, the fracture is **compound.** Other common types of fractures include comminuted, compression, spiral, epiphyseal, depressed, and greenstick fractures (**Table 6.2**, p. 142).

A fracture is treated by **reduction,** the realignment of the broken bone ends. In **closed reduction,** the bone ends are coaxed back into position by the physician's hands. In **open reduction,** the bone ends are joined surgically with pins or wires. After the broken bone is reduced, it is immobilized by a cast or traction to allow the healing process to begin. Healing time is about 6 to 8 weeks for a simple fracture, but it is longer for large, weight-bearing bones and for the bones of older people.

The healing of a simple fracture occurs in several phases (**Figure 6.15**).

① **Hematoma formation.** The fracture is usually accompanied by hemorrhaging. Blood vessels break in the periosteum and inside the bone, releasing blood that clots to form a hematoma. The stages of inflammation (described in Chapter 4, p. 96), are evident in and around the clot.

② **Fibrocartilaginous callus formation.** Within a few days, new blood vessels grow into the clot. The periosteum and endosteum near the fracture site show a proliferation of bone-forming cells, which then invade the clot, filling it with repair tissue called *soft callus* (kal′us; "hard skin"). Initially, the soft callus is a fibrous granulation tissue (p. 96). As more fibers are produced, the

Table 6.2 Common Types of Fractures

Fracture Type	Description and Comments
Comminuted	Bone fragments into three or more pieces. Particularly common in the aged, whose bones are more brittle. 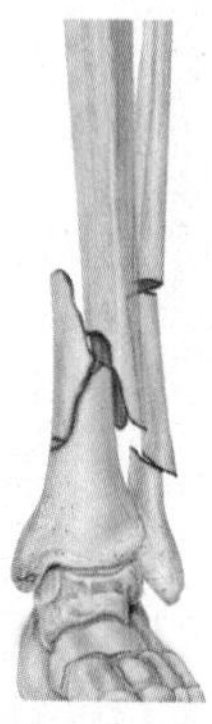 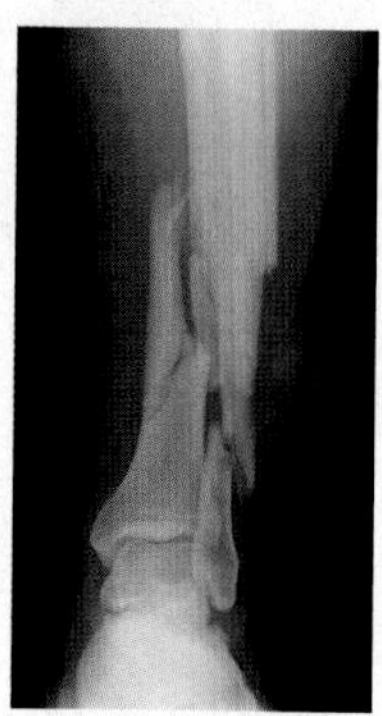
Compression	Bone is crushed. Common in porous bones (i.e., osteoporotic bones) subjected to extreme trauma, as in a fall. 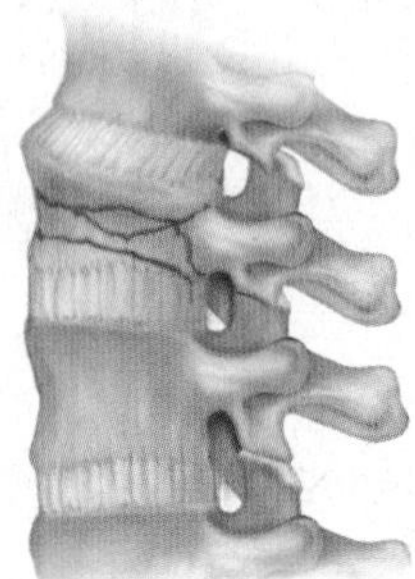 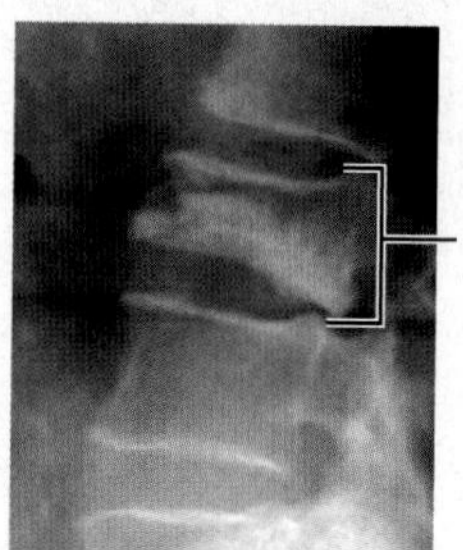
Spiral	Ragged break occurs when excessive twisting forces are applied to a bone. Common sports fracture. 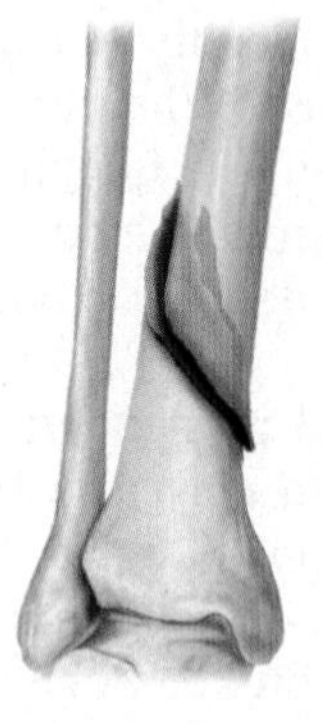 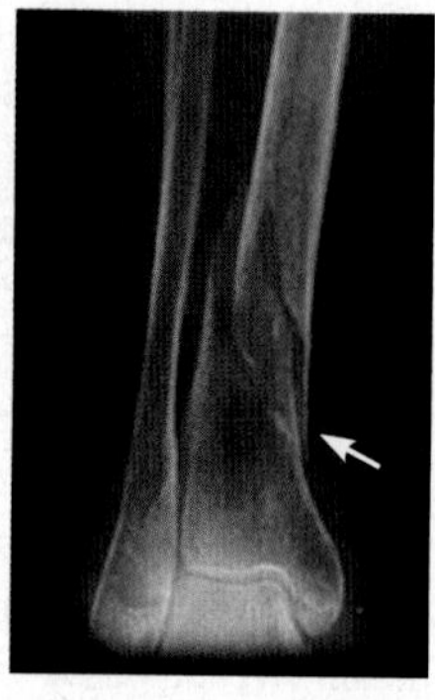
Epiphyseal	Epiphysis separates from the diaphysis along the epiphyseal plate. Tends to occur where cartilage cells are dying and calcification of the matrix is occurring. 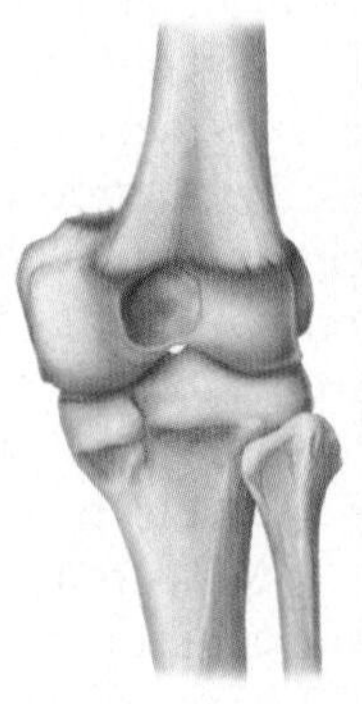 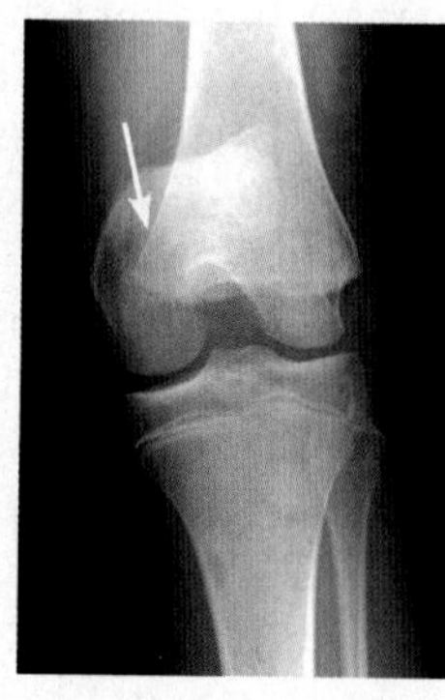
Depressed	Broken bone portion is pressed inward. Typical of skull fracture. 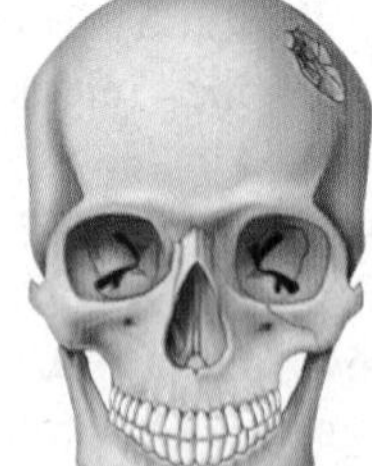 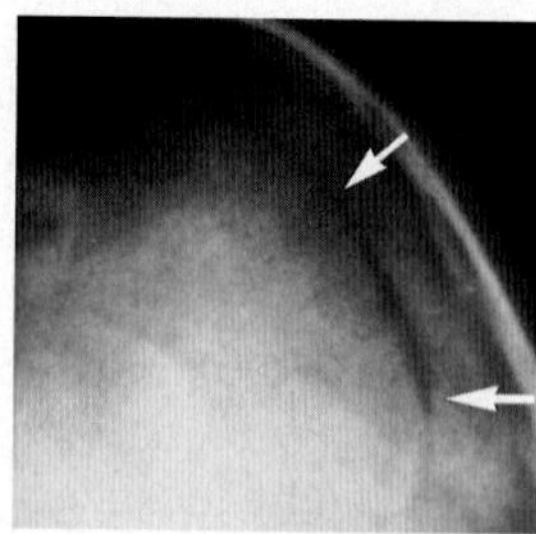
Greenstick	Bone breaks incompletely, much in the way a green twig breaks. Only one side of the shaft breaks; the other side bends. Common in children, whose bones have relatively more organic matrix and are more flexible than those of adults.  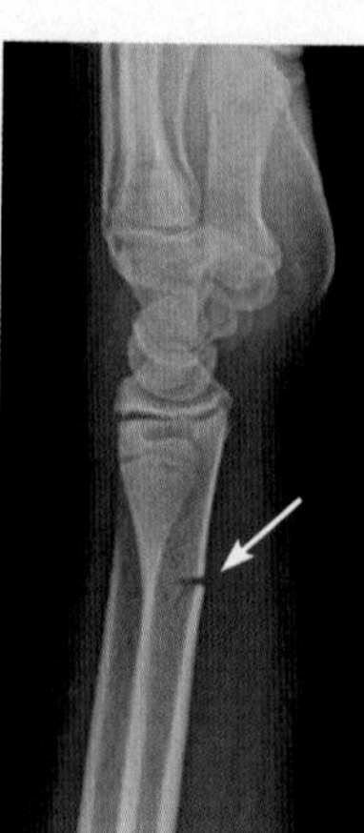

soft callus becomes a dense connective tissue containing fibrocartilage and hyaline cartilage. At this point, the soft callus is also called a **fibrocartilaginous callus.**

③ **Bony callus formation.** Within a week, trabeculae of new bone begin to form in the callus, mostly by endochondral ossification. These trabeculae span the width of the callus and unite the two fragments of the broken bone. The callus is now called a bony callus, or *hard callus,* and its trabeculae grow thicker and stronger and become firm about 2 months after the injury.

④ **Bone remodeling.** Over a period of many months, the bony callus is remodeled. The excess bony material is removed from both the exterior of the bone shaft and the interior of the medullary cavity. Compact bone is laid down to reconstruct the shaft walls. The repaired area resembles the original unbroken bone region, because it responds to the same set of mechanical stresses.

CLINICAL APPLICATION

Traction Traction ("pulling") involves placing a sustained tension on a region of the body to keep the parts of a fractured bone in the proper alignment. Traction also keeps the bone immobilized as it heals. Without traction of the lower limb, for example, strong spasms of the large muscles of the thigh would separate the fracture in a broken femur. Traction is also used to immobilize fractures of the vertebral column, because any movement there could crush the spinal cord.

check your understanding

- ☐ 16. How does exercise affect bone? Why?
- ☐ 17. How does bone remodeling help repair a bone after a fracture?
- ☐ 18. Which types of bone fractures are more common in older individuals (see Table 6.2)?

(For answers, see Appendix B.)

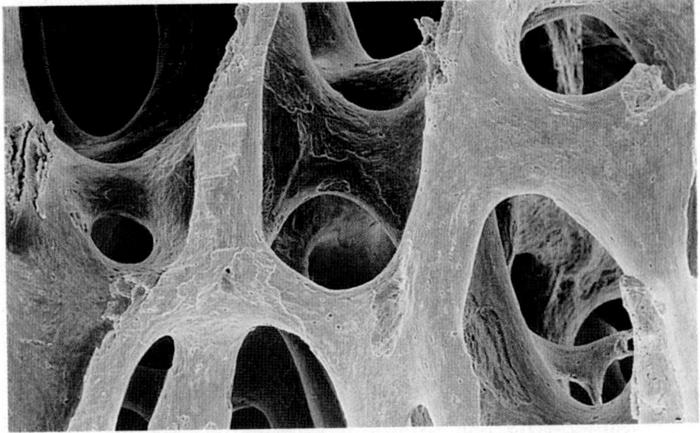

(a) Normal bone

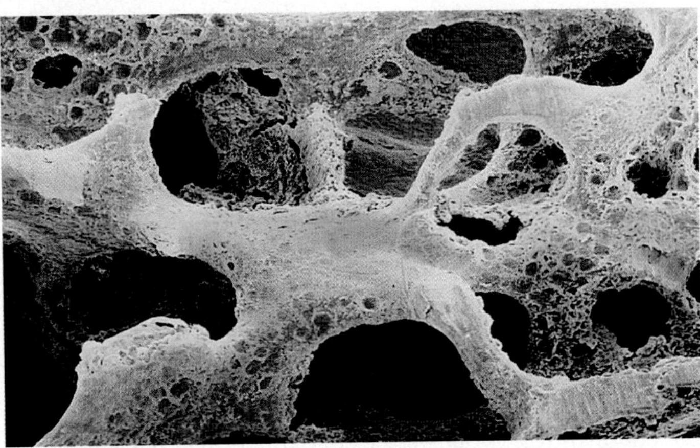

(b) Osteoporotic bone

Figure 6.16 **Osteoporosis.** Scanning electron micrographs, artificially colored (about 75×).

DISORDERS OF BONES

learning outcomes

- Relate the disease processes that cause osteoporosis, osteomalacia, rickets, and osteosarcoma to what you have learned about the structure, composition, and growth of bone tissue.
- Identify the symptoms and treatments for each of these diseases.

As mentioned in the introduction to this chapter, nutritional deficiencies and disease processes are reflected in our skeleton. Studying the skeletal remains of prehistoric human populations can reveal a wealth of information about their nutrition and lifestyle. For modern humans, understanding the growth and development of bone is essential for treating diseases of the skeletal system.

Osteoporosis

Osteoporosis (os″te-o-po-ro′sis; "bone-porous-condition") is characterized by low bone mass and a deterioration of the microscopic architecture of the bony skeleton (Figure 6.16). Although the chemical composition of the bone matrix remains normal, bone resorption outpaces bone deposition, in association with elevated numbers of osteoclasts. Osteoporotic bones become porous and light. The compact bone becomes thinner and less dense than normal, and the spongy bone has fewer trabeculae. The loss of bone mass often leads to fractures. Even though osteoporosis affects the whole skeleton, the vertebral column is most vulnerable, and compression fractures of the vertebrae are frequent. The femur (thigh bone), especially its neck near the hip joint, is also very susceptible to fracture. A break there, called a broken hip, is a common problem in people with osteoporosis.

Osteoporosis occurs most often in the aged, particularly in women who have gone through menopause. Although men develop it to a lesser degree, 30% of American women between the ages of 60 and 70 have osteoporosis; 70% have it by age 80. Moreover, 30% of all Caucasian women (the most susceptible group) will experience a bone fracture due to osteoporosis. Estrogen deficiency is strongly implicated in osteoporosis in older women because the secretion of estrogens, which helps maintain bone density, wanes after menopause. Additional factors that contribute to osteoporosis include insufficient exercise to stress the bones, a diet poor in calcium and protein, and inadequate levels of vitamin D. Because bone deposition rates are high during adolescence and early adulthood, proper diet and exercise during these years is crucial for developing and maintaining healthy bone in later life.

Osteoporosis has traditionally been treated by supplemental calcium and vitamin D (vitamin D is necessary for the absorption of calcium from the digestive tract), increased exercise, and estrogen replacement. Because of the increased risk of heart attack, stroke, and breast cancer associated with estrogen replacement therapy, this treatment is no longer standard for postmenopausal women. A number of newer treatments act to suppress osteoclast activity and slow bone loss. These include (1) the bisphosphonate drugs, such as alendronate (Fosamax), risedronate (Actonel), ibandronate (Boniva), and zoledronate (Reclast); (2) hormones such as calcitonin; and (3) a new treatment, denosumab (Prolia), an injected antibody that blocks a chemical messenger necessary for the formation and function of osteoclasts, thus inhibiting osteoclast formation.

Selective estrogen receptor modulators (SERMs), such as raloxifene (Evista) and tamoxifen, mimic the beneficial effects of estrogen but target bone alone, without any undesired stimulation of tissues of the breast and uterus.

The first treatment to act on the other side of the equation, building bone, is teriparatide (Forteo). Just approved for use in the United States, teriparatide is synthetic parathyroid hormone which influences both osteoblast and osteoclast activity. Intermittent administration of teriparatide increases osteoblast formation, stimulates increased calcium uptake from the intestine, and increases bone density.

These new treatments do not provide a cure, but they do increase bone mass to a moderate degree, slow bone loss, and significantly lessen the risk of fractures.

Osteomalacia and Rickets

The term **osteomalacia** (os″te-o-mah-la′she-ah; "soft bones") applies to a number of disorders in adults in which the bones are inadequately mineralized. Even though osteoid matrix is produced, calcification does not occur, and the bones soften and weaken. The main symptom is pain when weight is put on the affected bone.

Rickets, the analogous disease in children, is accompanied by many of the same signs and symptoms. Because young bones are still growing rapidly, however, rickets is more severe than osteomalacia. Along with weakened and bowed leg bones (Figure 6.17), malformations of the child's head and rib cage are common. Because the epiphyseal plates cannot be replaced with calcified bone, they grow atypically thick, and the epiphyses of the long bones become abnormally long.

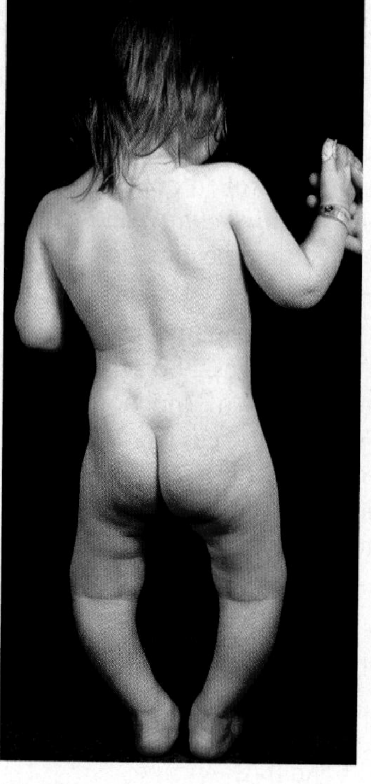

(a)

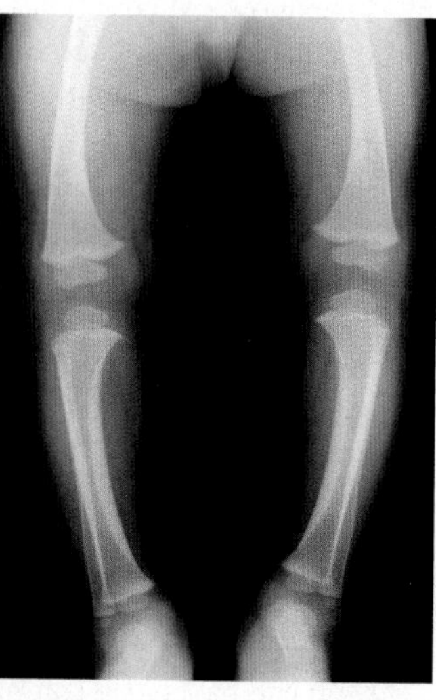

(b)

Figure 6.17 **Rickets.** (**a**) Photograph of a toddler with malformed legs due to rickets. (**b**) X-ray image of legs of a toddler with rickets.

Osteomalacia and rickets are caused by inadequate amounts of vitamin D or calcium phosphate in the diet. They are cured by drinking vitamin D–fortified milk and exposing the skin to sunlight.

It is estimated that in the 1800s, upwards of 90% of children in the industrialized cities of North America and Europe suffered from rickets. The change from an agricultural society to an industrial society resulted in large populations living in smog-filled cities and children working long hours in factories with little exposure to sunlight. Vitamin D was added to milk in the United States in 1930 in response to this epidemic triggered by the Industrial Revolution. Child labor laws followed in 1937.

Rickets is increasing in prevalence in children ages 6–24 months. Several factors have contributed to this increase; breast-feeding without vitamin D supplementation, spending more time indoors, and increased use of sunscreen. Balancing the need for sun protection and vitamin D production is necessary to maintain bone health and prevent damage to skin. A short period (10–15 minutes) of sun exposure three times per week is adequate for vitamin D synthesis.

Osteosarcoma

A *sarcoma* is any cancer arising from a connective tissue cell or muscle cell, and *osteo* means "bone." Clearly, then, **osteosarcoma** is a form of bone cancer.

Osteosarcoma primarily affects young people between 10 and 25 years of age. It usually originates in a long bone of the upper or lower limb, with 50% of cases arising near the knee. The cancer cells derive from osteoblast-like cells of mesenchymal origin in the parts of the diaphyses nearest the epiphyses. Secreting osteoid and growing quickly, the tumor alters the affected bone by eroding the medullary cavity internally and the compact bone externally. The tumor metastasizes, and most deaths result from secondary tumors in the lungs. Most osteosarcomas are recognized by the pain and the visible swelling they produce in the affected bone, and the diagnosis is confirmed by X-ray studies or other medical imaging techniques. Treatment begins by removing the cancerous region of the bone and replacing it with bone grafts or prostheses (although limb amputation is necessary in severe cases). This is followed by chemotherapy and surgical removal of any metastases in the lung. The survival rate is 60% to 70% if the disease is detected early.

☑ check your understanding

- ☐ **19.** Which diseases result from inadequate mineralization of bone?
- ☐ **20.** If you wish to slow bone loss, the activity of which cells in bone tissue should be targeted?
- ☐ **21.** At what age can you best prevent the development of osteoporosis later in life?

(For answers, see Appendix B.)

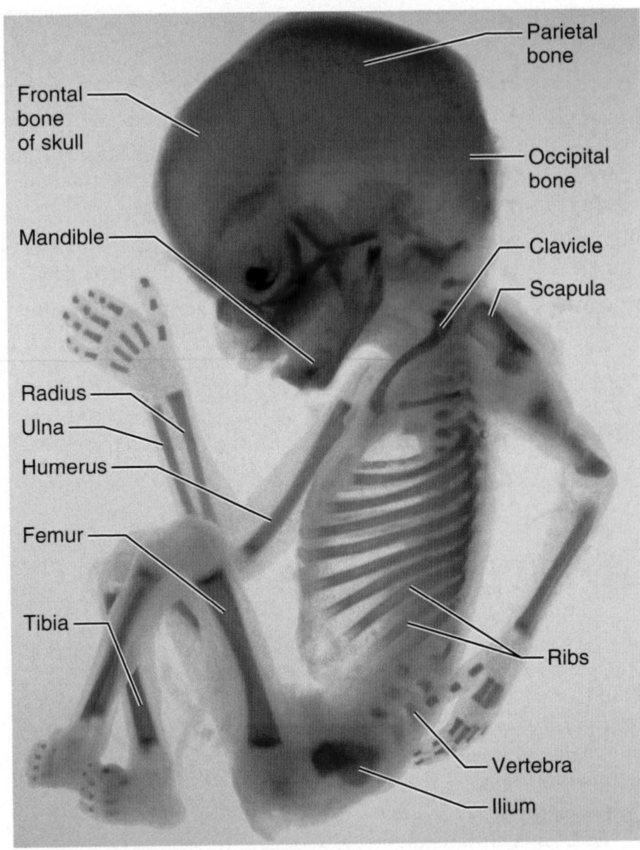

Figure 6.18 Primary ossification centers in the skeleton of a 12-week-old fetus.

THE SKELETON THROUGHOUT LIFE

learning outcome

▶ Describe how bone architecture and bone mass change from the embryonic period to old age.

As previously noted, cartilage grows quickly during youth and then stops growing during early adulthood. In older adults, cartilage contains fewer chondrocytes and shows some degradation and calcification of its matrix, resulting in thinning of articular cartilage.

Bones can be said to be on a timetable from the time they form until death. The mesoderm germ layer and neural crest (in the skull) give rise to embryonic mesenchyme cells, which in turn produce the membranes and cartilages that form most of the embryonic skeleton. These structures then ossify according to a predictable schedule. Although each bone of the skeleton has its own developmental schedule, most long bones start ossifying by week 8 and have obvious primary ossification centers by week 12 (Figure 6.18). So precise is the ossification timetable that the age of a fetus in the womb can be determined from X-ray images or sonograms of the fetal skeleton.

At birth, all bones are relatively smooth and featureless. As the child increasingly uses its muscles, the bones develop projections and other markings (see Table 6.1). Children's bones are not particularly weak, but the cartilage of their epiphyseal plates is not as strong as bone. Thus, childhood injuries often split the epiphyses off from the diaphysis. To treat such injuries, the bone parts are manipulated back into place, then stabilized with metal pins.

As mentioned earlier, the skeleton keeps growing until the age of 18 to 21 years. In children and adolescents, the rate of bone formation exceeds the rate of bone resorption. In young adults, these two processes are in balance. In old age, resorption predominates. Beginning at about age 40, the mass of both compact and spongy bone begins to decline. Among young adults, skeletal mass is generally greater in men than in women. Age-related bone loss is greater in women than in men.

As bone mass declines with age, other changes occur. An increasing number of osteons remain incompletely formed, and mineralization is less complete. The amount of nonliving compact bone increases, reflecting a diminished blood supply to the bones in old age.

☑ check your understanding

- ☐ **22.** At what age do bones begin to ossify? At what age does bone mass begin to decline?
- ☐ **23.** Why is age-related bone loss greater in women than in men?

(For answers, see Appendix B.)

RELATED CLINICAL TERMS

Bone graft Transplantation of a piece of bone from one part of a person's skeleton to another part where bone has been damaged or lost. The graft, often taken from the crest of the iliac bone of the hip, encourages regrowth of lost bone.

Bony spur An abnormal projection on a bone due to bone overgrowth; is common in older individuals with osteoarthritis.

Osteogenesis imperfecta (brittle bone disease) A congenital disease caused by a defect in the gene for collagen (type 1 collagen). As a result, bone tissue is brittle and susceptible to fracture. In mild forms of the disease, high susceptibility to bone fracture and hypermobility of joints is common; more severe forms can result in bone deformities and reduced life span. Treatment with bisphosphonate drugs increases bone strength, reduces fractures, and reduces pain.

Ostealgia (os″te-al′je-ah) (*algia* = pain) Pain in a bone.

Osteomyelitis (os″te-o-mi-ĕ-li′tis; "bone and marrow inflammation"). Bacterial infection of the bone and bone marrow. The pathogen enters bones either from infections in surrounding tissues or through the bloodstream, or follows a compound bone fracture.

Pathologic fracture Fracture occurring in a diseased bone and involving slight or no physical trauma. An example is a broken hip caused by osteoporosis: The hip breaks first, causing the person to fall.

CHAPTER SUMMARY

Take full advantage of the invaluable online practice and study tools for this chapter by going to the Study Area of MasteringA&P®. There you will find:

- Art Labeling Activities, Quizzes, Practice Tests, Memory Games
- Animal Organ Dissection, Cat Dissection and Bone Videos
- PAL Cadaver
- PAL Histology
- A&P Flix Animations
- Dynamic Study Modules

CARTILAGES (pp. 124–127)

Location and Basic Structure (pp. 124–125)

1. Important cartilages in the adult body are cartilage in the external ear and nose, articular cartilages, costal cartilages, the epiglottis, cartilages of the respiratory tubes, cartilage in intervertebral discs, and cartilage discs in certain movable joints. Cartilage makes up most of the embryonic skeleton.

2. Perichondrium is the girdle of dense connective tissue that surrounds a piece of cartilage.

3. Cartilage is resilient: Water squeezed out of its matrix by compression rushes back in as the compression eases, causing the cartilage to spring back.

4. Growing cartilage enlarges quickly because the small amount of matrix it manufactures attracts a large volume of water. Cartilage is avascular and is weak in resisting shearing stresses.

Types of Cartilage (pp. 125–127)

5. Three types of cartilages occur in the body: hyaline, elastic, and fibrocartilage (Figure 6.1).

6. Hyaline cartilage is the most common type. Its matrix contains fine collagen fibrils. Elastic cartilage resembles hyaline cartilage, but its matrix also contains elastic fibers that make it pliable. Fibrocartilage contains thick collagen fibers and can resist both compression and extreme tension.

Growth of Cartilage (p. 127)

7. Cartilages grow from within (interstitial growth) and externally, as chondroblasts in the perichondrium adds cartilage tissue at the periphery (appositional growth). In adults, damaged cartilage regenerates poorly. In the growing and aged skeleton, cartilage calcifies.

BONES (pp. 127–143)

8. Skeletal bones are considered organs because they contain different tissue types.

Functions of Bones (p. 127)

9. Bones support body weight, act as levers for muscles to pull on, protect soft organs, store calcium, and contain bone marrow that makes blood cells. Bone cells also function in energy metabolism by influencing blood sugar regulation.

Bone Tissue (pp. 127–128)

10. Bone consists of cells plus an extracellular matrix. The matrix contains organic substances secreted by osteoblasts, including collagen, which gives bone tensile strength. The organic part of bone matrix is called osteoid. Calcium phosphate salts (hydroxyapatites) crystallize in this matrix, making bone hard and able to resist compression.

11. The cells in bone tissue are osteogenic stem cells; osteoblasts, which secrete osteoid; osteocytes, which maintain the bone matrix; and osteoclasts, which destroy bone tissue.

Gross Anatomy of Bones (pp. 128–132)

12. Bones are classified on the basis of their shape as long, short, flat, or irregular.

13. Bones have an external layer of compact bone and are filled internally with spongy bone, in which trabeculae are arranged in networks.

14. A long bone is composed of a diaphysis or shaft, and epiphyses or ends. Nutrient vessels serve the diaphysis and epiphyseal vessels serve each epiphysis. Bone marrow occurs within the spongy bone and in a central medullary (marrow) cavity. A periosteum covers the outer surface of bones, and an endosteum covers the inner bone surfaces.

15. Flat bones consist of two plates of compact bone separated by a layer of spongy bone.

16. The density of bone material and the magnitude of bending stresses decline from the superficial to the deep regions of the bones. Thus, the strongest forces occur at the periphery, where they are resisted by the strong compact bone. The spaces within bones lighten the skeleton and contain bone marrow.

17. The trabeculae of spongy bone appear to be arranged along the dominant lines of stress experienced by the bone.

18. Bone markings are landmarks that represent sites of muscle attachment, articulation, and passage of blood vessels and nerves (see Table 6.1, p. 133).

Microscopic Structure of Bone (pp. 132–135)

19. An important structural unit in compact bone is the osteon, essentially a pillar consisting of a central canal surrounded by concentric lamellae. The central canals of adjacent osteons are linked by perforating canals. Osteocytes, embedded in lacunae, are connected to each other and to the central canal by canaliculi.

20. Bone lamellae are concentric tubes of bone matrix. The collagen fibers in adjacent lamellae run in roughly opposite directions. This arrangement gives bone tissue great strength in resisting torsion (twisting).

21. Spongy bone consists of trabeculae containing several layers of lamellae and osteocytes, but no osteons.

Bone Development and Growth (pp. 135–139)

22. Flat bones of the skull and the clavicle form by intramembranous ossification of embryonic mesenchyme. A network of bone tissue woven around capillaries appears first and is then remodeled into a flat bone.

23. Most bones develop by endochondral ossification of a hyaline cartilage model, starting in the late embryonic period (week 8). The stages of development of a long bone are (1) formation of a bone collar around the diaphysis; (2) calcification and cavitation in the center of the diaphysis; (3) growth of a periosteal bud into the center of the shaft, and formation of the first bone trabeculae; (4) appearance of the medullary cavity and continued rapid growth throughout the fetal period; near birth, formation of secondary ossification centers in the epiphyses; (5) ossification of the epiphyses and continued growth in length at the epiphyseal plates through adolescence.

24. The growing cartilage of the fetal epiphyses and the postnatal epiphyseal plates is organized into several zones, which allow rapid growth. (These zones are explained in Figure 6.12.)

25. Endochondral bones lengthen during youth through the growth of epiphyseal plate cartilages, which close in early adulthood.

26. Bones increase in width through appositional growth.

Bone Remodeling (pp. 139–141)

27. New bone tissue is continuously deposited and resorbed in response to hormonal and mechanical stimuli. Together, these processes are called bone remodeling. Bone remodeling in adults occurs in the endosteum.

28. Osteoclasts break down bone tissue by secreting digestive enzymes and acid onto bone surfaces. This process releases Ca^{2+} and PO_4^{3-} into the blood. Parathyroid hormone stimulates this resorption of bone.

29. Osteoid is secreted by osteoblasts at areas of bone deposit. Calcium salt is then deposited in the osteoid.

30. Compressive forces and gravity acting on the skeleton help maintain bone strength, because bones thicken at sites of stress.

Repair of Bone Fractures (pp. 141–143)

31. Fractures are treated by open or closed reduction. Healing involves the formation of a hematoma, a fibrocartilaginous callus, and a bony callus, and then a remodeling of the callus into the original bone pattern.

DISORDERS OF BONES (pp. 143–145)

32. Osteoporosis is a condition in which bone breakdown outpaces bone formation, causing bones to weaken. Postmenopausal women are most susceptible.

33. Osteomalacia and rickets occur when bones are inadequately mineralized, making them soft and deformed. The most common cause is inadequate intake or production of vitamin D.

34. Osteosarcoma is the most common form of bone cancer.

THE SKELETON THROUGHOUT LIFE (p. 145)

35. Bone formation in the fetus occurs in a predictable and precisely timed manner.

36. The mass of the skeleton increases dramatically during puberty and adolescence, when bone formation exceeds resorption.

37. Bone mass is constant in young adulthood, but beginning at about age 40, bone resorption exceeds formation.

REVIEW QUESTIONS

Multiple Choice/Matching Questions

(For answers, see Appendix B.)

1. Which is a function of the skeletal system? (a) support, (b) blood cell formation, (c) mineral storage, (d) providing levers for muscle activity, (e) all of these.

2. Articular cartilages are located (a) at the ends of bones, (b) between the ribs and the sternum (breastbone), (c) between the epiphysis and diaphysis, (d) in the nose.

3. The perichondrium of cartilage is similar to the ____ of bone. (a) articular cartilages, (b) spongy bone layer, (c) osteons, (d) marrow, (e) periosteum.

4. Use the key to indicate the type of cartilage that forms each of the structures below.

Key: (a) hyaline cartilage (b) elastic cartilage (c) fibrocartilage

____ (1) articular cartilage covering the surfaces of movable joints

____ (2) the epiglottis

____ (3) the thyroid cartilage

____ (4) the menisci in the knee

____ (5) the embryonic skeleton

5. Indicate whether each of the following statements is true (T) or false (F).

____ (1) Cartilage is more resilient than bone.

____ (2) Cartilage is especially strong in resisting shear (bending and twisting) forces.

____ (3) Cartilage can grow faster than bone in the growing skeleton.

____ (4) In the adult skeleton, cartilage heals and regenerates faster than bone when damaged.

_____ **(5)** Neither bone nor cartilage contains capillaries.

_____ **(6)** Bone tissue contains very little water compared to other connective tissues, while cartilage tissue contains a large amount of water.

_____ **(7)** Nutrients diffuse quickly through cartilage matrix but very poorly through the solid bone matrix.

6. A bone that has essentially the same width, length, and height is most likely (a) a long bone, (b) a short bone, (c) a flat bone, (d) an irregular bone.

7. The shaft of a long bone is properly called the (a) epiphysis, (b) periosteum, (c) diaphysis, (d) compact bone.

8. Match the function of bone markings described in column B with the bone markings listed in column A.

Column A	Column B
_____ **(1)** trochanter	**(a)** attachment site for muscle or ligament
_____ **(2)** condyle	**(b)** forms a joint surface
_____ **(3)** foramen	**(c)** passageway for vessels or nerves
_____ **(4)** process	
_____ **(5)** facet	
_____ **(6)** tuberosity	

9. Which listed feature is found in compact bone but not found in spongy bone? (a) a central canal containing blood vessels, (b) lamellae of matrix, (c) osteocytes in lacunae, (d) canaliculi.

10. The flat bones of the skull develop from (a) areolar tissue, (b) hyaline cartilage, (c) mesenchyme membranes, (d) compact bone.

11. The following events apply to the endochondral ossification process as it occurs in the primary ossification center. Put these events in their proper order by assigning each a number (1–6).

_____ **(a)** Cartilage in the diaphysis calcifies, and chondrocytes die and disintegrate, leaving cavities.

_____ **(b)** A collar of bone is laid down around the hyaline cartilage model just deep to the periosteum.

_____ **(c)** The periosteal bud invades the center of the diaphysis.

_____ **(d)** The perichondrium of shaft becomes more vascularized and becomes a periosteum.

_____ **(e)** Osteoblasts first deposit bone tissue around the cartilage spicules within the diaphysis.

_____ **(f)** Osteoclasts remove the bone from the center of the diaphysis, leaving a medullary cavity that then houses marrow.

12. The remodeling of bone tissue is a function of which cells? (a) chondrocytes and osteocytes, (b) osteoblasts and osteoclasts, (c) chondroblasts and osteoclasts, (d) osteoblasts and osteocytes.

13. Osteogenic cells are located in (a) the lacunae, (b) the fibrous layer of the periosteum, (c) the endosteum, (d) the perichondrium, (e) the growth zone of the epiphysis.

14. A fracture in which the bone penetrates soft tissue and skin is (a) greenstick, (b) compound, (c) simple, (d) comminuted, (e) compression.

15. The disorder in which bones are porous and thin but the chemistry of the bone matrix remains normal is (a) osteomalacia, (b) osteoporosis, (c) osteomyelitis, (d) Paget's disease.

16. Where within an epiphyseal plate is the calcified cartilage located? (a) nearest the diaphysis, (b) in the medullary cavity, (c) farthest from the diaphysis, (d) in the primary ossification center, (e) all of these.

17. Endosteum is in all these places, except: (a) around the exterior of the femur, (b) on the trabeculae of spongy bone, (c) on the lining of the central canal of an osteon, (d) often in direct contact with bone marrow.

18. Match the cells listed in column B with the descriptions in column A. More than one answer may be correct.

Column A	Column B
_____ **(1)** located within lacunae	**(a)** osteoblasts
_____ **(2)** secrete matrix	**(b)** osteocytes
_____ **(3)** maintain bone matrix	**(c)** osteoclasts
_____ **(4)** destroy bone tissue	**(d)** chondrocytes
_____ **(5)** located in the endosteum	

Short Answer Essay Questions

19. Explain (a) why cartilages are resilient and (b) why cartilage can grow so quickly in the developing skeleton.

20. Some anatomy students are joking between classes, imagining what a bone would look like if it had spongy bone on the outside and compact bone on the inside, instead of the other way around. You tell the students that such an imaginary bone would be of poor mechanical design and would break very easily when bent. Explain why this is so.

21. When and why do the epiphyseal plates close?

22. During what period of life does skeletal mass increase dramatically? Begin to decline? Why are fractures most common in older adults?

23. In a piece of cartilage in the young skeleton, interstitial and appositional growth occur together. Compare and contrast interstitial and appositional growth.

24. Photocopy a picture of a skeleton, and then use a red pencil to color its membranous bones and a blue pencil to color its endochondral bones.

25. Differentiate the following: osteoclast, osteocyte, osteoblast, osteon.

26. List three structural features of cartilage and bone tissue that are similar. List three structural features that are different.

Critical Reasoning & Clinical Application Questions

1. Explain why people confined to wheelchairs because of paralyzed lower limbs have thin, weak bones in their thighs and legs.

2. While walking home from class, 52-year-old Ike fell. He broke a bone and damaged cartilage in his knee. Assuming no special tissue grafts are made, which will probably heal faster, the bone or the cartilage? Why?

3. Carlos went to weight-lifting camp in the summer between seventh and eighth grade. He noticed that the trainer put tremendous pressure on participants to improve their strength. After an especially vigorous workout, Carlos's arm felt extremely sore and weak around the elbow. He went to the camp doctor, who took X-ray photos and then told him that the injury was serious because the "end, or epiphysis, of his upper arm bone was starting to twist off." What had happened? Could the same thing happen to Carlos's 23-year-old sister, Selena, who was also starting a weight-lifting program? Why or why not?

4. Ming posed the following question: "If the epiphyseal growth plates are growing so fast, why do they stay thin? Growing things are supposed to get larger or thicker, but these plates remain the same thickness." How would you answer her?

5. Old Norse stories tell of a famous Viking named Egil, an actual person who lived around A.D. 900. His skull was greatly enlarged and misshapen, and the cranial bones were hardened and thickened (6 cm, or several inches, thick). After he died, his skull was dug up, and it withstood the blow of an ax without damage. In life, he had headaches from the pressure exerted by enlarged vertebrae on his spinal cord. So much blood was diverted to his bones to support their extensive remodeling that his fingers and toes always felt cold, and his heart was damaged through overexertion. What bone disorder did Egil probably have?

6. Bernice, a 75-year-old woman, stumbled slightly while walking, then felt a terrible pain in her left hip. At the hospital, X-ray pictures revealed that the hip was broken. Furthermore, the spongy bone throughout her spine was very thin. What condition does Bernice probably have?

7. Why might repeated pregnancies cause a woman to develop osteomalacia?

8. Traditional treatments for osteoporosis address calcium deficiencies in the diet and the importance of weight-bearing exercise. Describe how weight-bearing exercise improves bone mass.

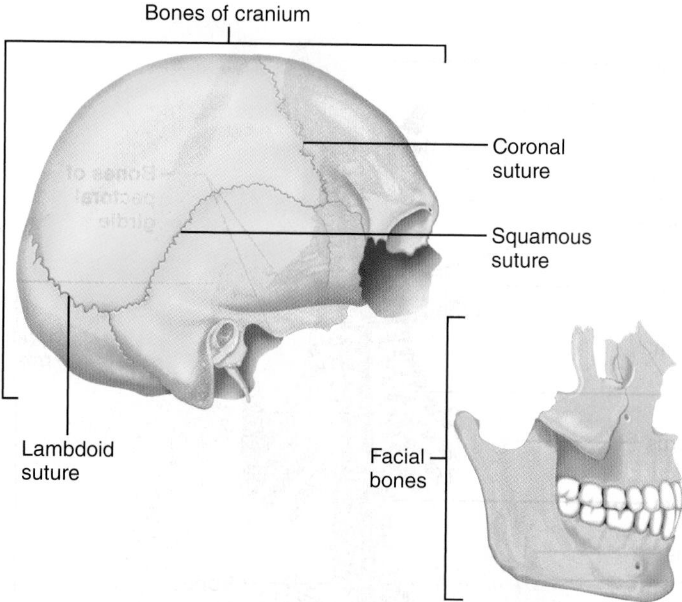

(a) Cranial and facial divisions of the skull

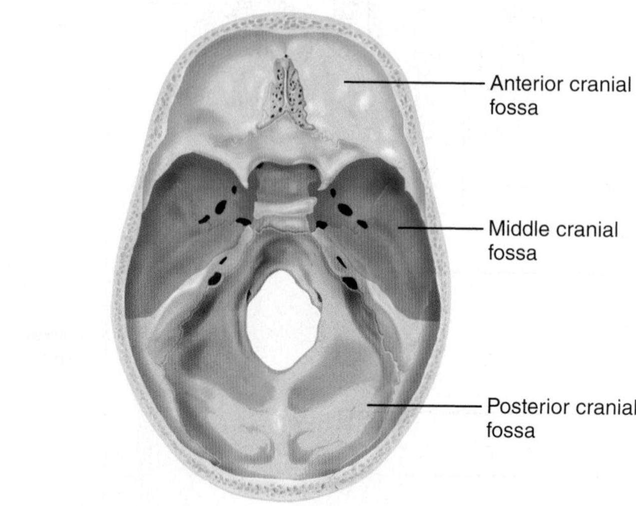

(b) Superior view of the cranial fossae

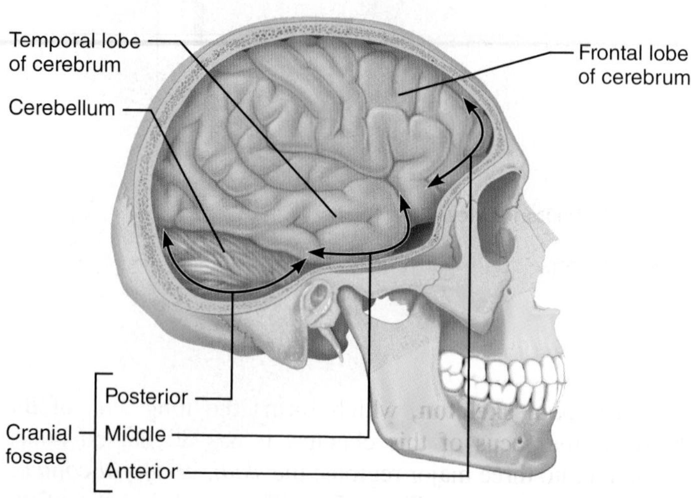

(c) Lateral view of cranial fossae showing the contained brain regions

Figure 7.2 The skull: cranial and facial divisions and fossae.

THE SKULL

learning outcomes

- ▶ Compare the functions of the cranial and facial bones.
- ▶ Name and describe the bones of the skull. Identify their important features.

The **skull** is the body's most complex bony structure. It is formed by cranial and facial bones (Figure 7.2a). The *cranial bones,* or **cranium** (kra′ne-um), enclose and protect the brain and provide attachment sites for some head and neck muscles. The *facial bones* (1) form the framework of the face; (2) form cavities for the sense organs of sight, taste, and smell; (3) provide openings for the passage of air and food; (4) hold the teeth; and (5) anchor the muscles of the face.

Most skull bones are flat bones and are firmly united by interlocking, immovable joints called sutures (soo′-cherz; "seams"). The suture lines have an irregular, saw-toothed appearance. The longest sutures—the *coronal, sagittal, squamous,* and *lambdoid sutures*—connect the cranial bones. Most other skull sutures connect facial bones and are named according to the specific bones they connect.

Overview of Skull Geography

It is worth surveying basic skull "geography" before describing the individual bones. With the lower jaw removed, the skull resembles a lopsided, hollow, bony sphere. The facial bones form its anterior aspect, and the cranium forms the rest. The cranium can be divided into

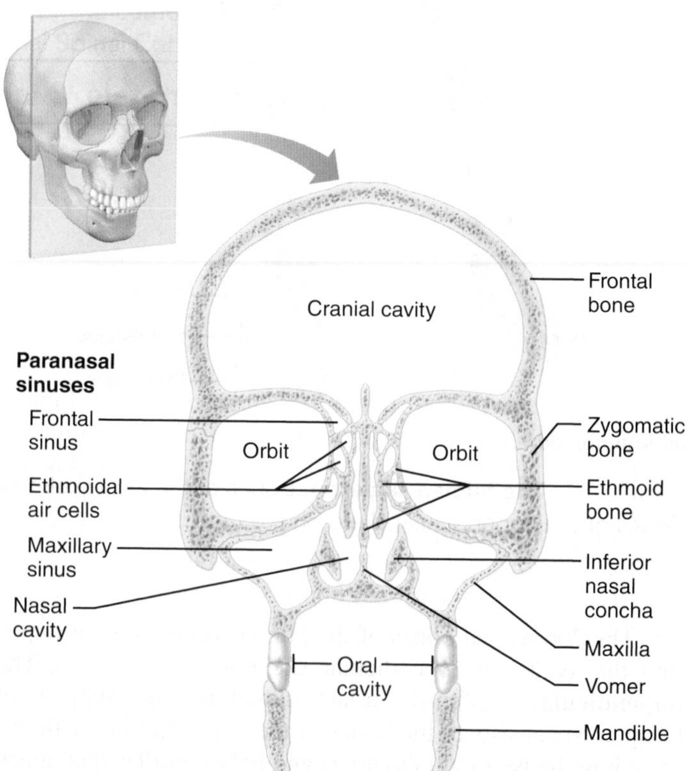

Figure 7.3 Major cavities of the skull, frontal section.

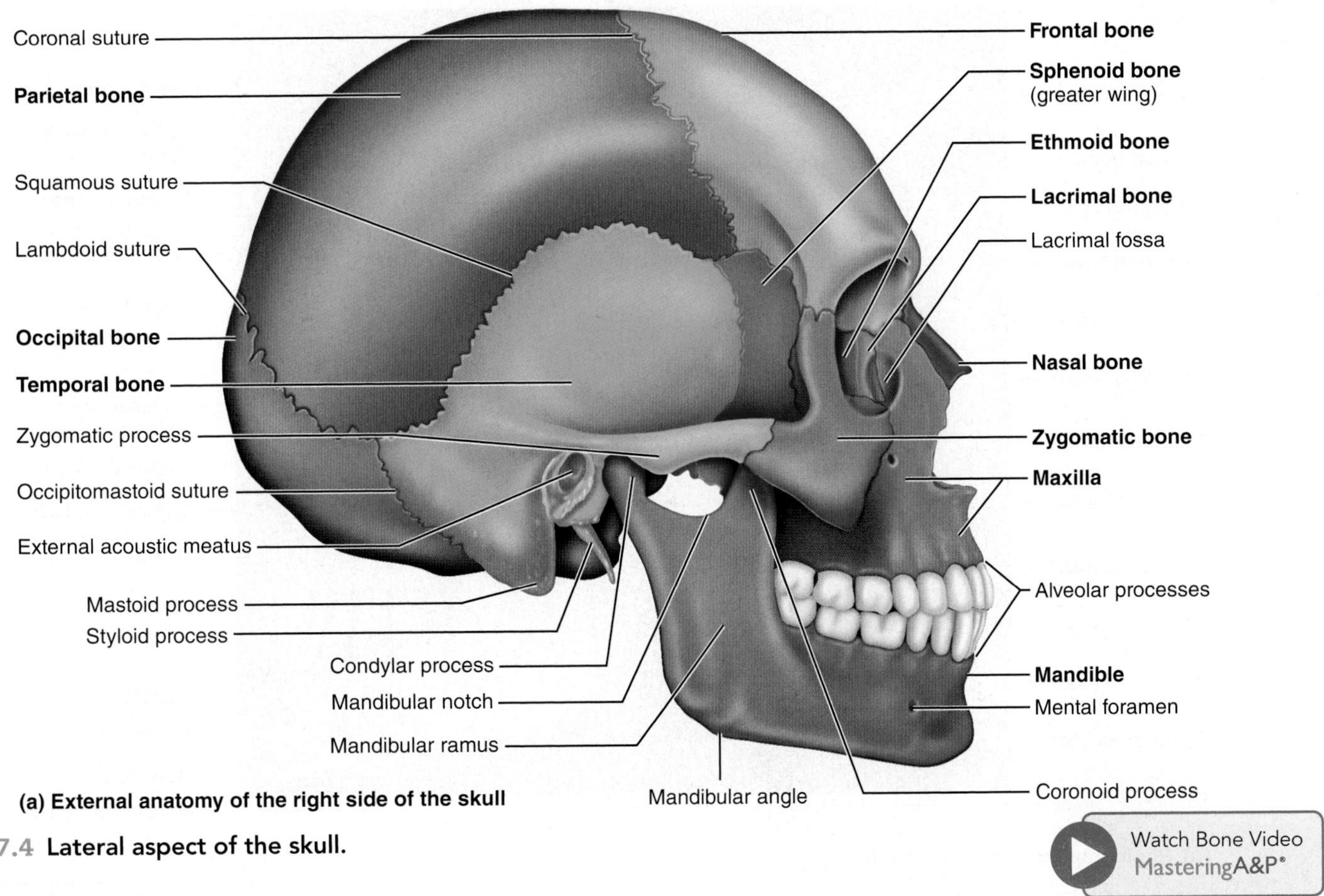

Figure 7.4 Lateral aspect of the skull.

a vault and a base. The cranial vault, called the *calvaria* (kal-va′re-ah; "bald part of skull") or skullcap, forms the superior, lateral, and posterior aspects of the skull, as well as the forehead region. The *cranial base* is the inferior part. Internally, prominent bony ridges divide the cranial base into three distinct "steps," or fossae—the anterior, middle, and posterior *cranial fossae* (Figure 7.2b). The brain sits snugly in these cranial fossae and is completely enclosed by the calvaria (Figure 7.2c). Overall, the brain is said to occupy the *cranial cavity.*

In addition to its large cranial cavity, the skull contains many smaller cavities, including the *middle ear* and *inner ear cavities* (carved into the lateral aspects of the cranial base), the *nasal cavity,* and the *orbits* (Figure 7.3). The nasal cavity lies in and posterior to the nose, and the orbits house the eyeballs. Air-filled sinuses that occur in several bones around the nasal cavity are the *paranasal sinuses.*

Moreover, the skull has about 85 named openings (foramina, canals, fissures). The most important of these provide passageways for the spinal cord, the major blood vessels serving the brain, and the 12 pairs of *cranial nerves,* which conduct impulses to and from the brain. Cranial nerves (which are discussed in Chapter 14, p. 432) are classified by number, using the Roman numerals I through XII. (The skull bones and their features are illustrated in Figures 7.3–7.16 and summarized in Table 7.1, pp. 164–165. The colored box beside each bone's name corresponds to the color of that bone in the figures.)

Cranial Bones

The eight large bones of the cranium are the paired parietal and temporal bones and the unpaired frontal, occipital, sphenoid, and ethmoid bones. Together these bones form the brain's protective "shell." Because its superior aspect is curved, the cranium is self-bracing. This allows the bones to be thin, and, like an eggshell, the cranium is remarkably strong for its weight.

Parietal Bones and the Major Sutures

(Figures 7.4 and 7.5)

The two large **parietal bones,** shaped like curved rectangles, make up the bulk of the calvaria; that is, they form most of the superior part of the skull, as well as its lateral walls (*parietal* = wall) (Figure 7.4). The sites at which the parietal bones articulate (form a joint) with other cranial bones are the four largest sutures:

1. The **coronal suture,** running in the coronal plane, occurs anteriorly where the parietal bones meet the frontal bone.

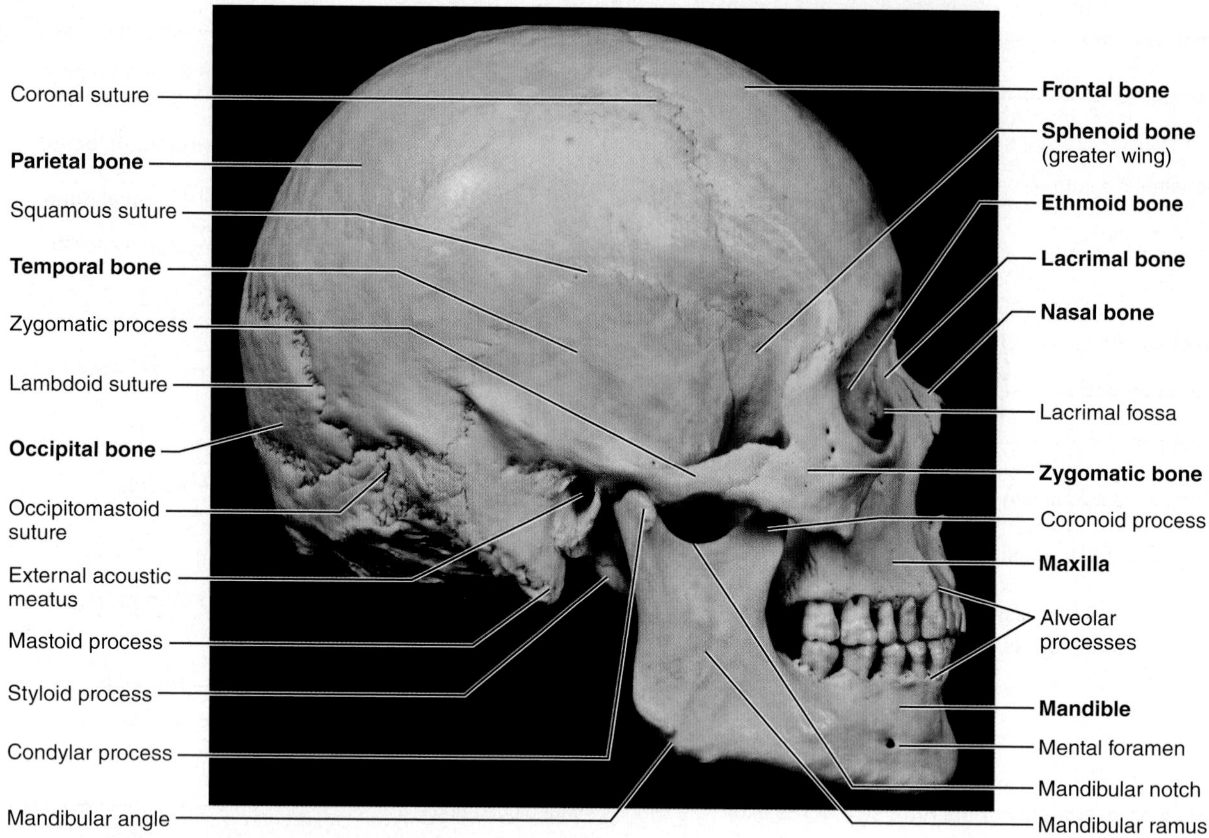

(b) Photograph of right side of skull

Figure 7.4 **Lateral aspect of the skull,** *continued.*

2. A **squamous suture** occurs where each parietal bone meets a temporal bone inferiorly, on each lateral aspect of the skull.
3. The **sagittal suture** occurs where the right and left parietal bones meet superiorly in the midline of the cranium (Figure 7.5).
4. The **lambdoid suture** occurs where the parietal bones meet the occipital bone posteriorly (Figures 7.4 and 7.5). This suture is so named because it resembles the Greek letter lambda (λ).

These sutures vary somewhat in appearance in different skulls. As a person ages, the sutural lines close up, making these sutures less noticeable.

Sutural Bones

Sutural bones are small bones that occur *within* the sutures, especially in the lambdoid suture (Figure 7.5). They are irregular in shape, size, and location, and not all people have them. They develop between the major cranial bones during the fetal period and persist throughout life. The significance of these bones is unknown.

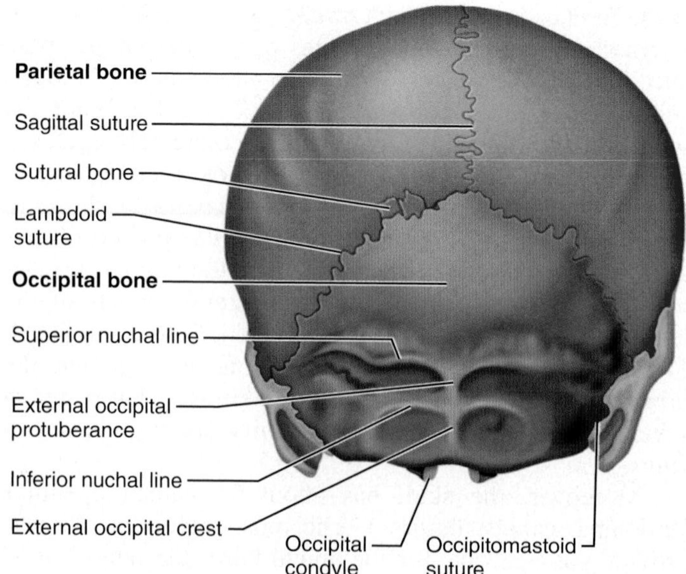

Figure 7.5 **Posterior view of the skull.**

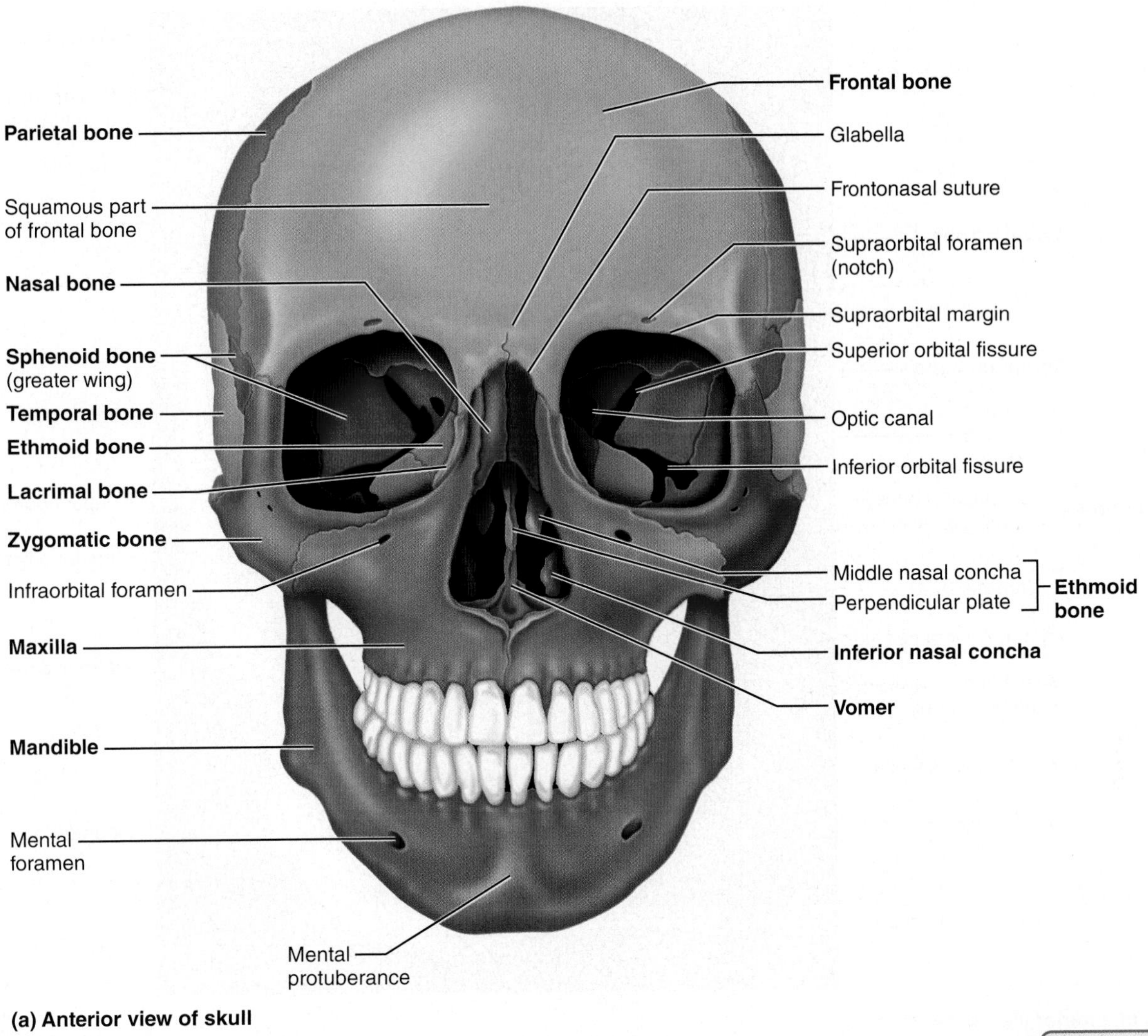

Frontal bone
Glabella
Frontonasal suture
Supraorbital foramen (notch)
Supraorbital margin
Superior orbital fissure
Optic canal
Inferior orbital fissure
Middle nasal concha
Perpendicular plate
Ethmoid bone
Inferior nasal concha
Vomer
Parietal bone
Squamous part of frontal bone
Nasal bone
Sphenoid bone (greater wing)
Temporal bone
Ethmoid bone
Lacrimal bone
Zygomatic bone
Infraorbital foramen
Maxilla
Mandible
Mental foramen
Mental protuberance

(a) Anterior view of skull

Figure 7.6 Anterior view of the skull.

Frontal Bone (Figure 7.6)

The **frontal bone** forms the forehead and the roofs of the orbits (**Figure 7.6**). Just superior to the orbits, it protrudes slightly to form *superciliary* (soo″per-sil′e-a-re; "eyebrow") *arches,* which lie just deep to our eyebrows. The **supraorbital margin,** or superior margin of each orbit, is pierced by a hole or by a notch, respectively called the **supraorbital foramen** or *supraorbital notch.* This opening transmits the supraorbital nerve (a branch of cranial nerve V) and artery, which supply the forehead. The smooth part of the frontal bone between the superciliary arches in the midline is the **glabella** (glah-bel′ah; "smooth, without hair"). Just inferior to it, the frontal bone meets the nasal bones at the *frontonasal suture.* The regions of the frontal bone lateral to the glabella contain the air-filled **frontal sinuses** (see Figure 7.3).

Internally, the frontal bone contributes to the **anterior cranial fossa** (see Figures 7.2b and 7.9), which holds the large frontal lobes of the brain.

Occipital Bone (Figures 7.5 and 7.7)

The **occipital bone** (ok-sip′ĭ-tal; "back of the head") makes up the posterior part of the cranium and cranial base (Figure 7.5). It articulates with the parietal bones at the lambdoid suture and with the temporal bones at the *occipitomastoid sutures* (Figure 7.4). Several features occur on the external surface of the occipital bone (Figures 7.5 and 7.7). The **external occipital protuberance** is a knob in the midline, at the junction of the base and the posterior wall of the skull. The *external occipital crest* extends anteriorly from the protuberance to the foramen magnum. This crest secures the *ligamentum nuchae* (nu′ke; "of the neck"), an elastic, sheet-shaped ligament that lies in the median plane of the posterior neck and connects the neck vertebrae to the skull. Extending laterally from the occipital protuberance are the **superior nuchal** (nu′kal) **lines,** and running laterally from a point halfway along the occipital crest are the **inferior nuchal lines.** The nuchal lines and the bony regions between them anchor many muscles of the neck and back. The superior nuchal line marks the upper limit of the neck.

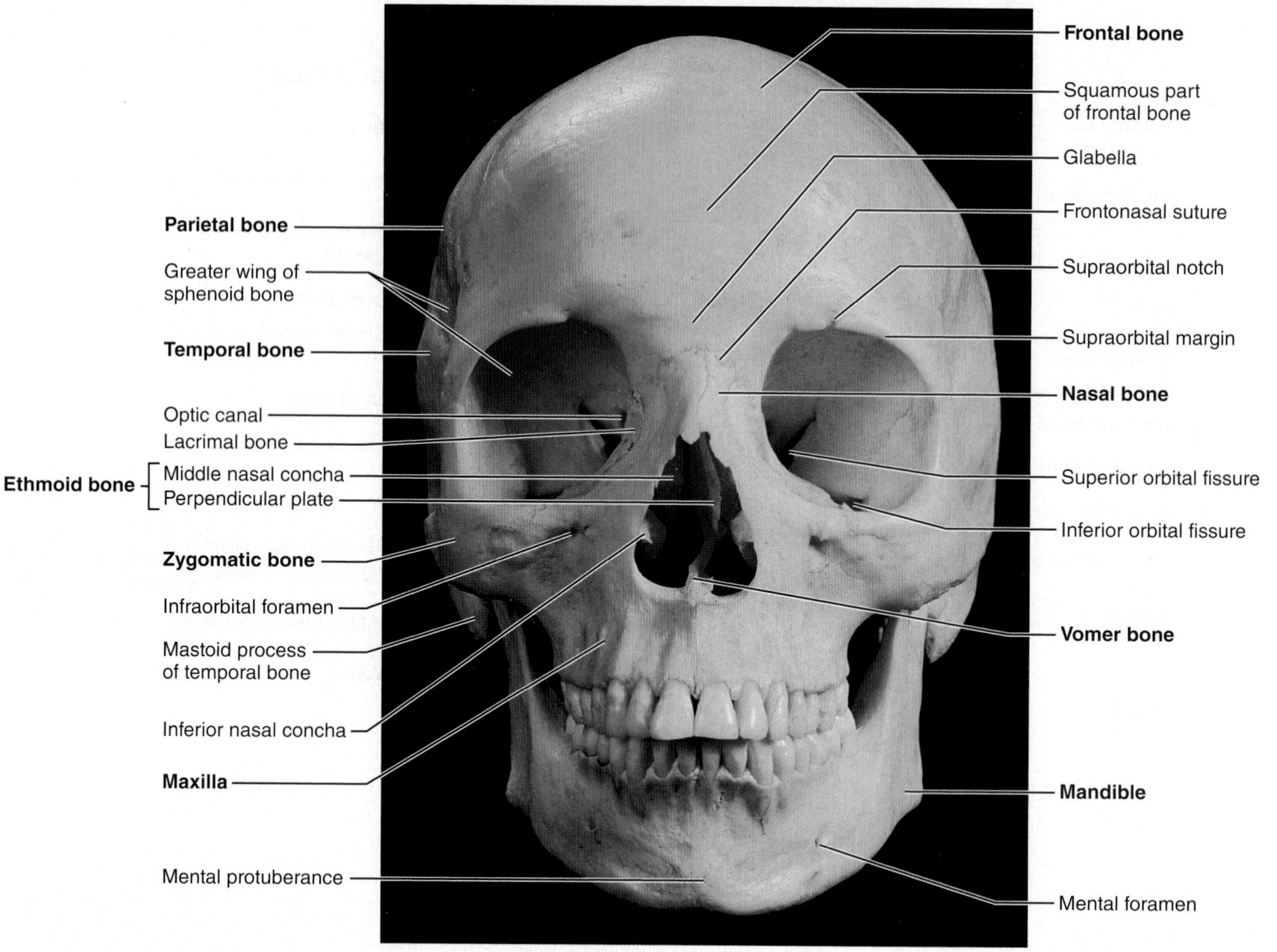

(b) Photo of anterior view of skull

Figure 7.6 **Anterior view of the skull,** *continued.*

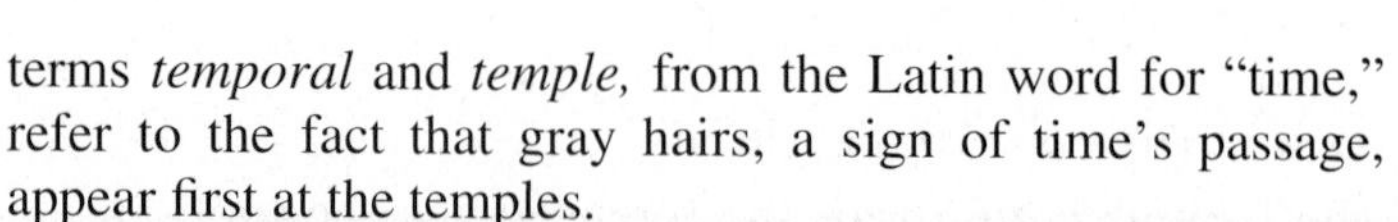

In the base of the occipital bone is the **foramen magnum,** literally, "large hole" (Figure 7.7). Through this opening, the inferior part of the brain connects with the spinal cord. The foramen magnum is flanked laterally by two rockerlike **occipital condyles,** which articulate with the first vertebra of the vertebral column in a way that enables the head to nod "yes." Hidden medial and superior to each occipital condyle is a **hypoglossal** (hi″po-glos′al) **canal,** through which runs cranial nerve XII, the hypoglossal nerve. Anterior to the foramen magnum, the occipital bone joins the sphenoid bone via the **basilar part of the occipital bone.**

Internally, the occipital bone forms the walls of the **posterior cranial fossa,** which holds a part of the brain called the cerebellum (Figure 7.2).

Temporal Bones (Figures 7.4 and 7.8)

The temporal bones are best viewed laterally (Figure 7.4). They lie inferior to the parietal bones and form the inferolateral region of the skull and parts of the cranial floor. The terms *temporal* and *temple,* from the Latin word for "time," refer to the fact that gray hairs, a sign of time's passage, appear first at the temples.

Each temporal bone has an intricate shape and is described in terms of its three major parts: the *squamous, tympanic,* and *petrous parts.* The plate-shaped **squamous part** (Figure 7.8) abuts the squamous suture. It has a barlike **zygomatic process** (zi″go-mat′ik; "cheek") that projects anteriorly to meet the zygomatic bone of the face. Together, these two bony structures form the *zygomatic arch,* commonly called the cheek bone. The oval **mandibular** (man-dib′-u-lar) **fossa** on the inferior surface of the zygomatic process receives the condylar process of the mandible (lower jawbone), forming the freely movable *temporomandibular joint* (jaw joint).

The **tympanic part** (tim-pan′ik; "eardrum") surrounds the **external acoustic meatus,** or external ear canal. It is through this canal that sound enters the ear. The external acoustic meatus and the tympanic membrane (eardrum) at its deep end are parts of the external ear. In a dried skull, the tympanic membrane has been removed. Thus, part of the

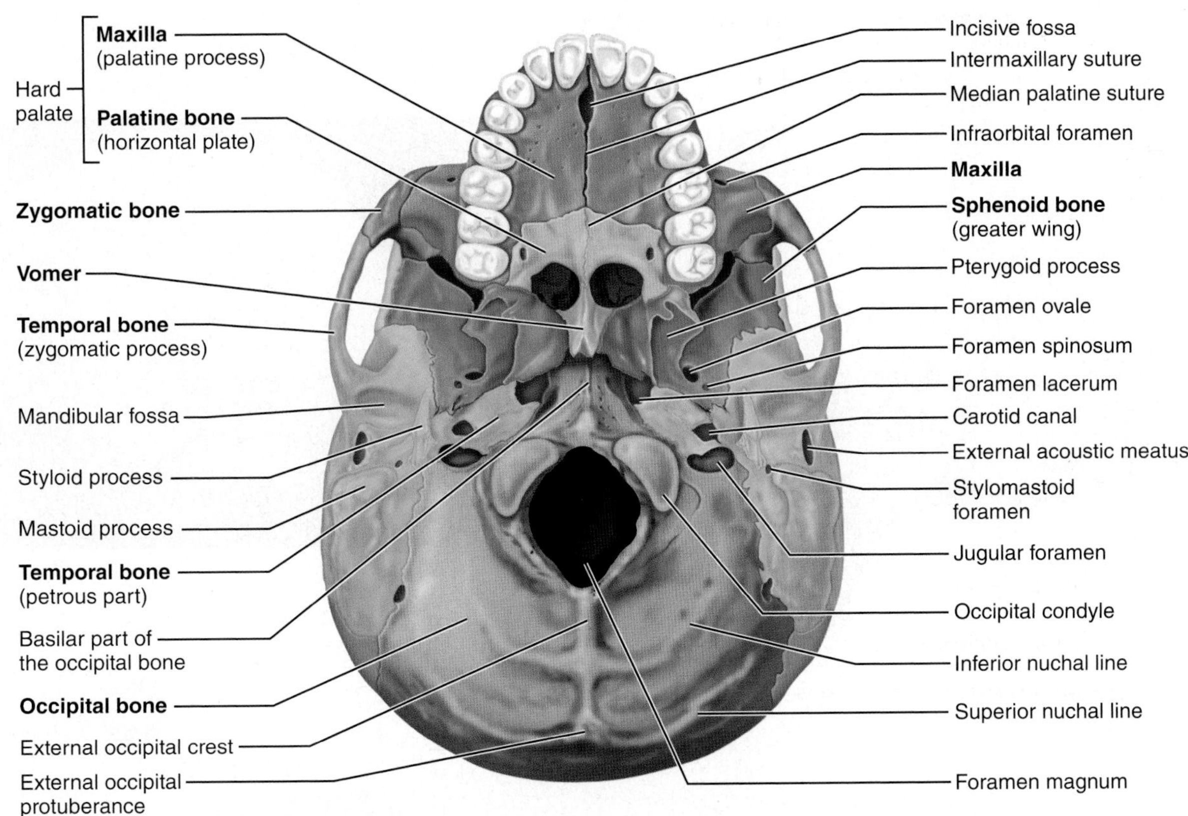

(a) Inferior view of the skull (mandible removed)

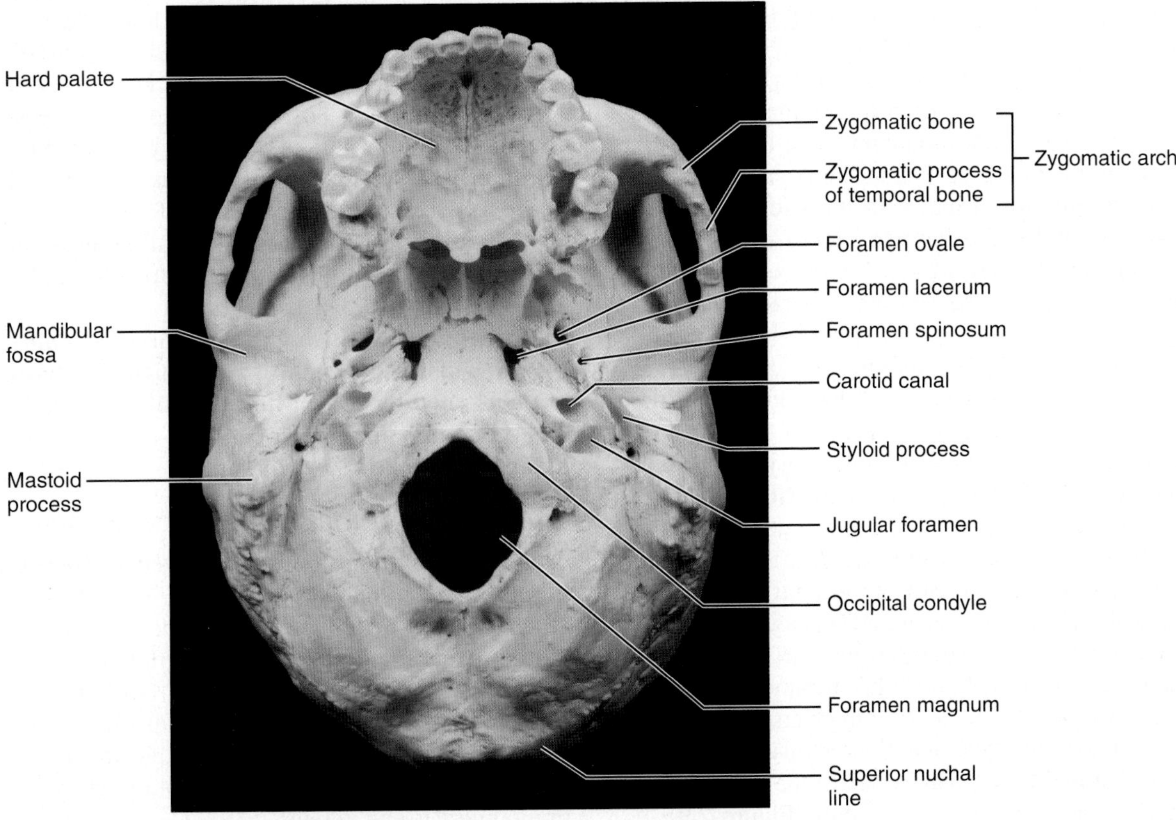

(b) Photo of inferior view of the skull

Figure 7.7 **Inferior aspect of the skull.**

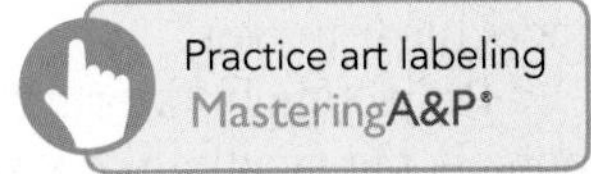

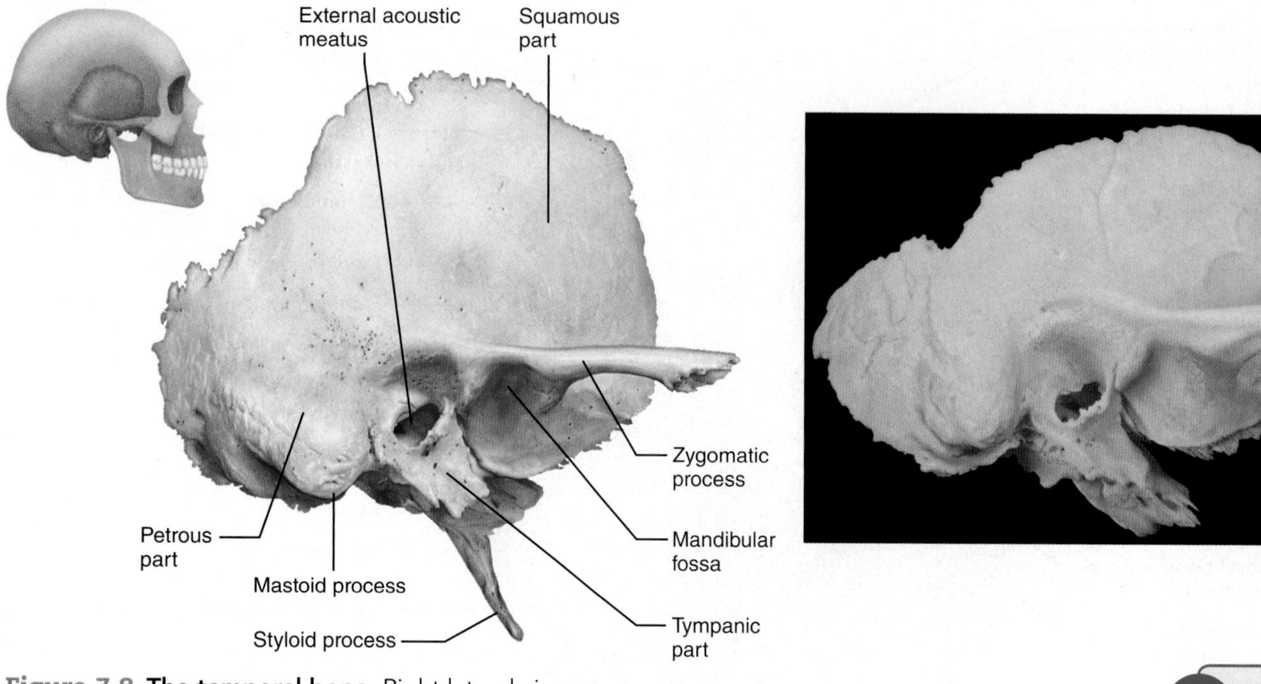

Figure 7.8 The temporal bone. Right lateral view.

middle ear cavity deep to the tympanic region may be visible through the meatus.

The **petrous** (pet′rus; "rocky") **part** of the temporal bone contributes to the cranial base. It forms a bony wedge between the occipital bone posteriorly and the sphenoid bone anteriorly (Figure 7.7a). From within the cranial cavity this very dense region looks like a mountain ridge (**Figure 7.9**). The posterior slope of this ridge lies in the posterior cranial fossa, whereas the anterior slope is in the **middle cranial fossa,** the fossa that holds the temporal lobes of the brain. Housed inside the petrous part are the cavities of the middle and inner ear, which contain the sensory apparatus for hearing and balance.

Several foramina penetrate the petrous part (Figure 7.7a). The large **jugular foramen** is located where the petrous part joins the occipital bone. Through this foramen pass the largest vein of the head, the internal jugular vein, and cranial nerves IX, X, and XI. The **carotid** (ka-rot′id) **canal** opens on the skull's inferior aspect, just anterior to the jugular foramen. The internal carotid artery, the main artery to the brain, passes through it into the cranial cavity. The **foramen lacerum** (la′ser-um; "lacerated") is a jagged opening between the medial tip of the petrous part of the temporal bone and the sphenoid bone. This foramen is almost completely closed by cartilage in a living person, but it is so conspicuous in a dried skull that students usually ask its name. The **internal acoustic meatus** lies in the cranial cavity on the posterior face of the petrous part (Figure 7.9). It transmits cranial nerves VII and VIII, the facial and vestibulocochlear nerves.

Projecting inferiorly from the petrous part of the temporal bone is the needle-like **styloid process** (sti′loid; "stakelike") (Figure 7.8). This process is an attachment point for some muscles of the tongue and pharynx and for a ligament that connects the skull to the hyoid bone of the neck.

Lateral and posterior to the styloid process is the prominent **mastoid process** (mas′toid; "breast-shaped"), an anchoring site for some neck muscles. This process can be felt as a lump just posterior to the ear. The **stylomastoid foramen** (Figure 7.7a) is located between the styloid and mastoid processes. A branch of cranial nerve VII, the facial nerve, leaves the skull through this foramen.

The mastoid process is full of air sinuses called *mastoid air cells,* which lie just posterior to the middle ear cavity. Infections can spread from the throat to the middle ear to the mastoid cells. Such an infection, called *mastoiditis,* can even spread to the brain, from which the mastoid air cells are separated by only a thin roof of bone. This was a serious problem before the late 1940s, when antibiotics became available.

Sphenoid Bone (Figures 7.9 and 7.10)

The **sphenoid bone** (sfe′noid; "wedge-shaped") spans the width of the cranial floor (Figure 7.9) and has been said to resemble a bat with its wings spread. It is considered the keystone of the cranium because it forms a central wedge that articulates with every other cranial bone. It is a challenging bone to study because of its complex shape and orientation: Portions of the sphenoid are viewable from most aspects of the skull. It also has a number of foramina for the passage of cranial nerves and vessels.

The sphenoid consists of a central **body** and three pairs of processes: the *greater wings, lesser wings,* and *pterygoid* (ter′ĭ-goid) *processes* (**Figure 7.10**). The superior surface of the body bears a saddle-shaped prominence, the **sella turcica** (sel′ah ter′sik-ah; "Turkish saddle"). The seat of

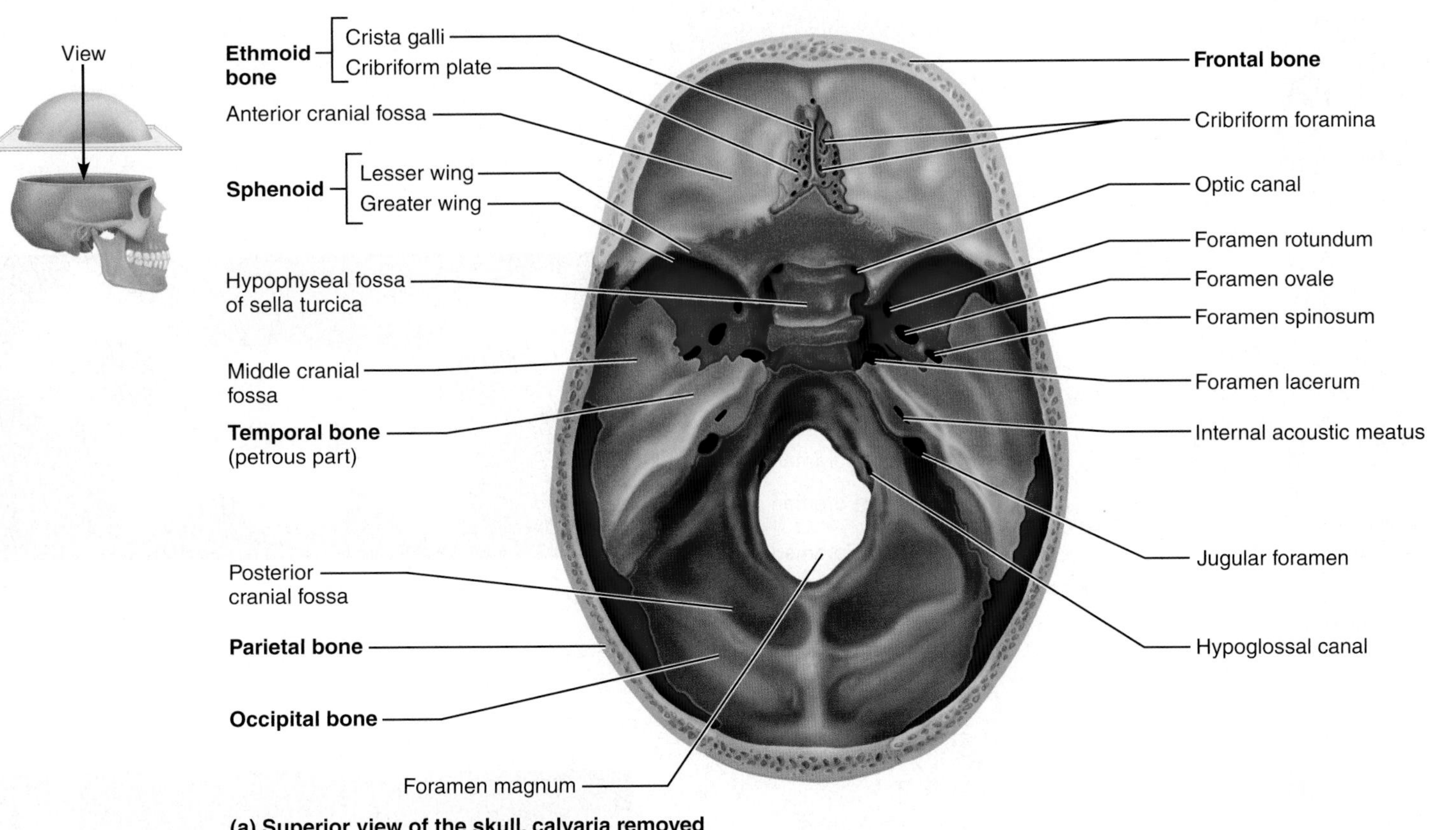

(a) Superior view of the skull, calvaria removed

Ethmoid bone
Crista galli
Cribriform plate
Anterior cranial fossa
Sphenoid
Lesser wing
Greater wing
Hypophyseal fossa of sella turcica
Middle cranial fossa
Temporal bone (petrous part)
Posterior cranial fossa
Parietal bone
Occipital bone
Foramen magnum
Frontal bone
Cribriform foramina
Optic canal
Foramen rotundum
Foramen ovale
Foramen spinosum
Foramen lacerum
Jugular foramen

(b) Superior view of the skull, calvaria removed

Figure 7.9 **Floor of the cranial cavity.**

Watch bone video
MasteringA&P®

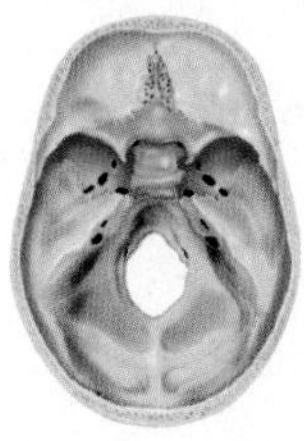

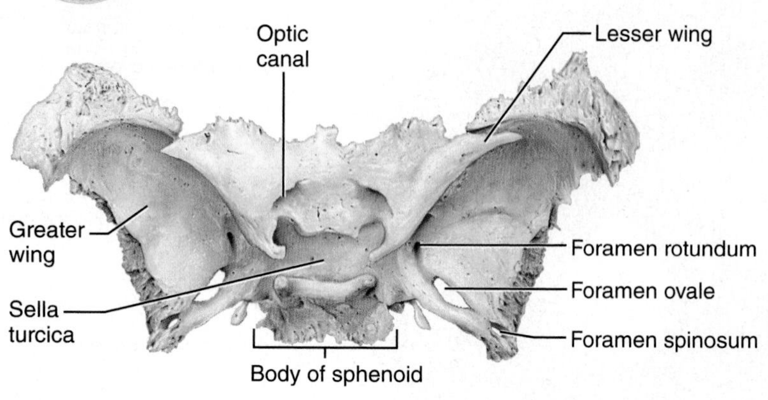

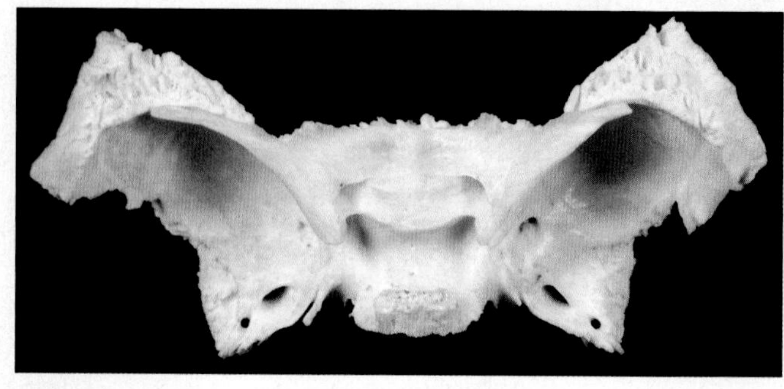

(a) Superior view

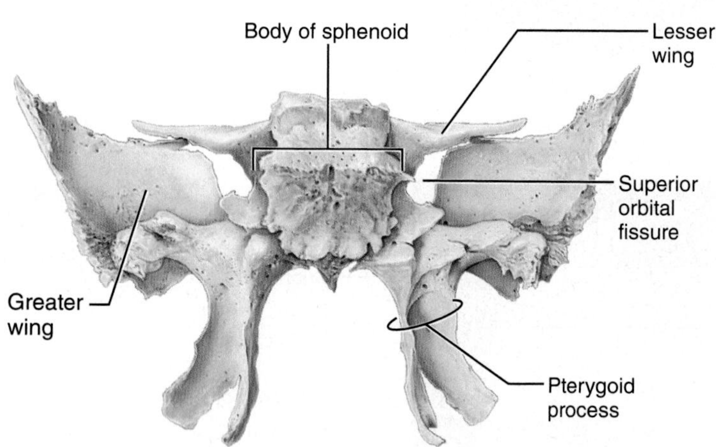

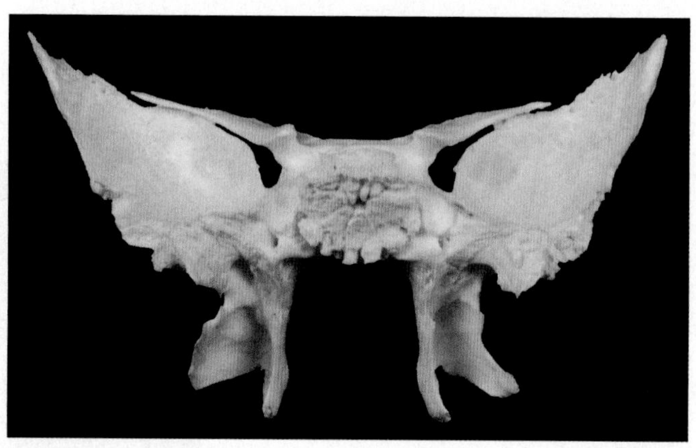

(b) Posterior view

Figure 7.10 **The sphenoid bone.**

this saddle, called the *hypophyseal* (hi″po-fiz′e-al) *fossa,* holds the pituitary gland, or hypophysis. Within the sphenoid body are the paired **sphenoidal sinuses** (Figure 7.11). The **greater wings** project laterally from the sphenoid body, forming parts of the middle cranial fossa (Figure 7.9) and the orbit. Externally, the greater wings form the lateral wall of the skull anterior to the squamous part of the temporal bones (Figure 7.4a). The horn-shaped **lesser wings** form part of the floor of the anterior cranial fossa (Figure 7.9) and a part of the orbit. The trough-shaped **pterygoid** ("winglike") **processes** project inferiorly from the greater wings (see Figures 7.7 and 7.10). These processes, which have both medial and lateral plates, are attachment sites for the pterygoid muscles that help close the jaw in chewing.

The sphenoid bone has five important openings (Figures 7.9 and 7.10) on each side. The **optic canal** lies just anterior to the sella turcica. Cranial nerve II, the optic nerve, passes through this hole from the orbit into the cranial cavity. The other four openings lie in a crescent-shaped row just lateral to the sphenoid body on each side. The most anterior of these openings, the **superior orbital fissure,** is a long slit between the greater and lesser wings (Figure 7.10b). It transmits several structures to and from the orbit, such as the cranial nerves that control eye movements (III, IV, and VI). This fissure is best seen in an anterior view of the orbit (Figure 7.6). The **foramen rotundum** lies in the medial part of the greater wing. It is usually oval, despite its name, which means "round opening." The **foramen ovale** (o-val′e) is an oval hole posterolateral to the foramen

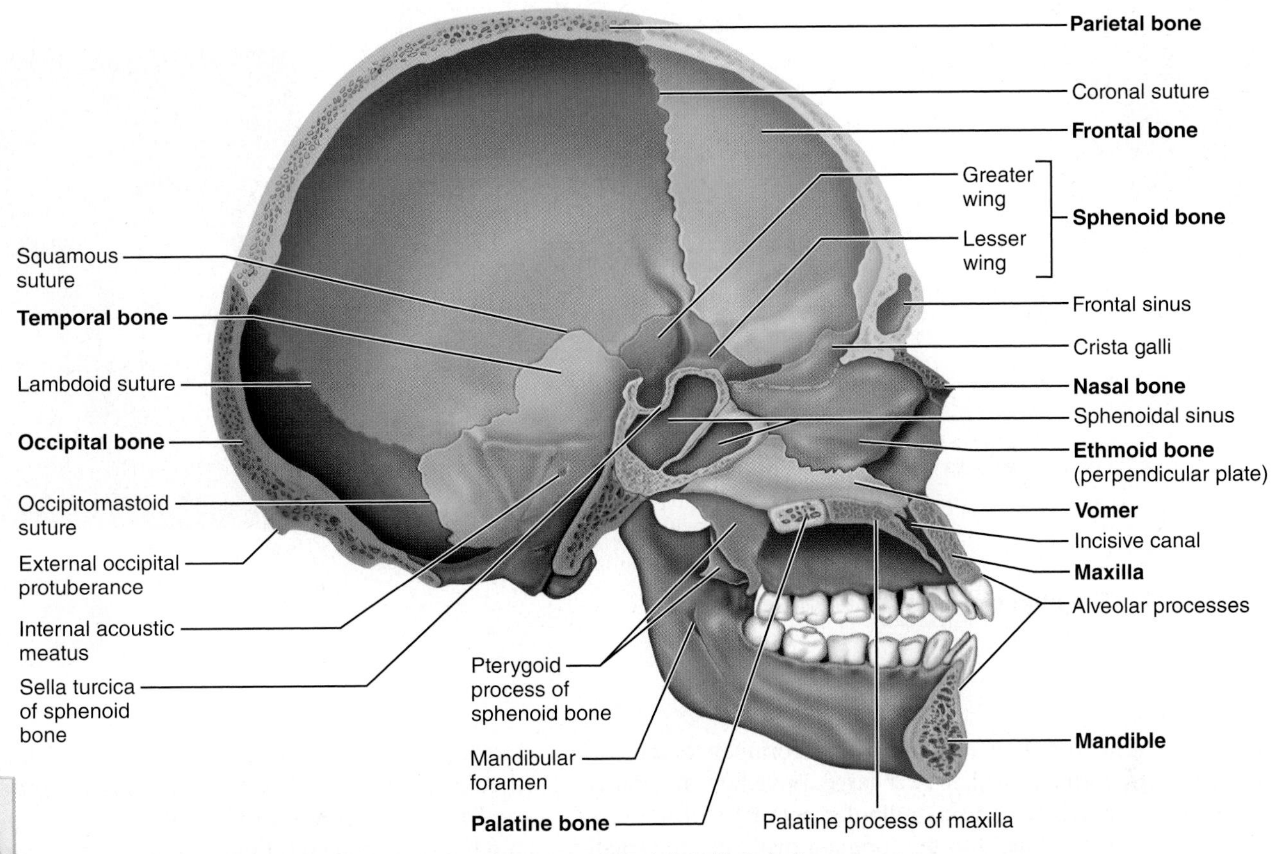

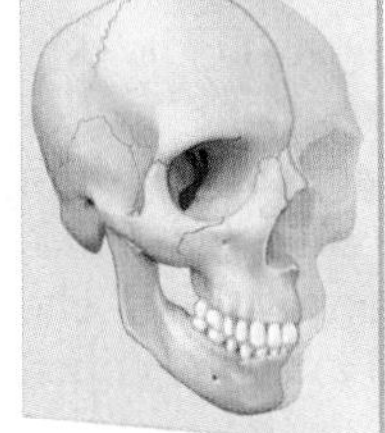

(a) Midsagittal section showing the internal anatomy of the left half of skull

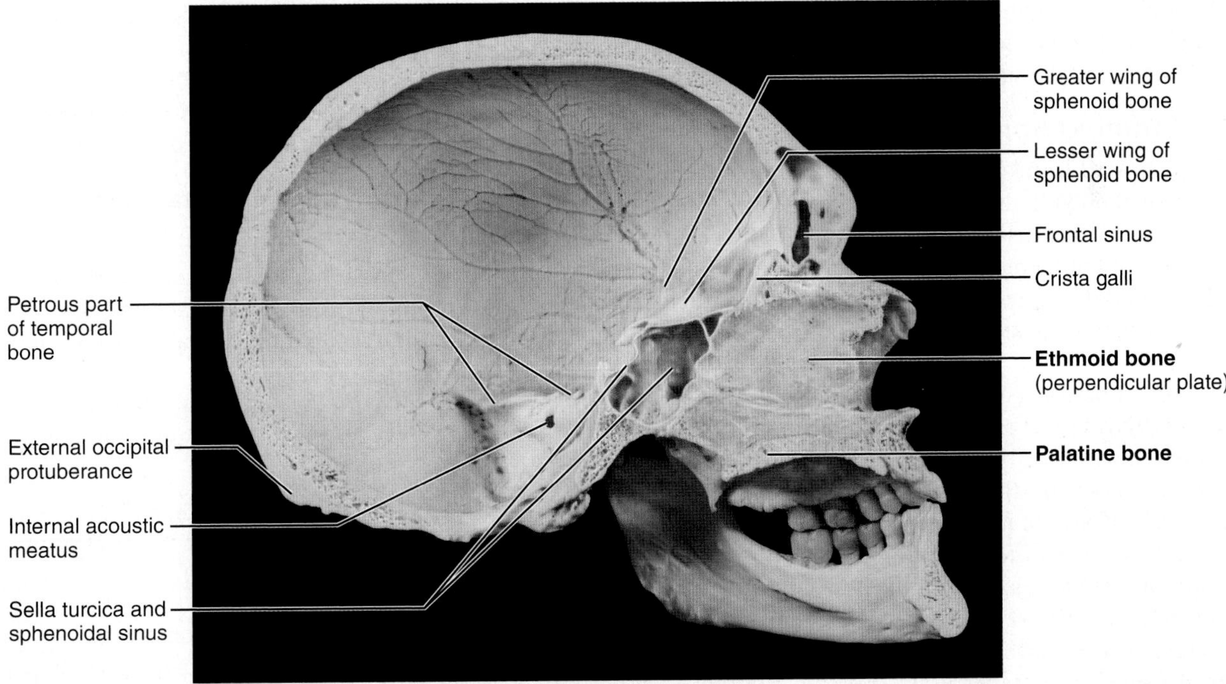

(b) Photo of skull cut through the midline, same view as in (a)

Figure 7.11 **Midsagittal section through the skull.**

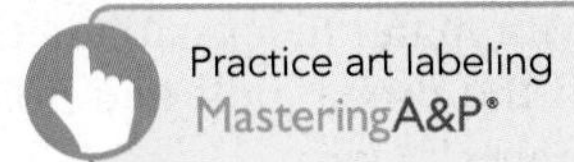

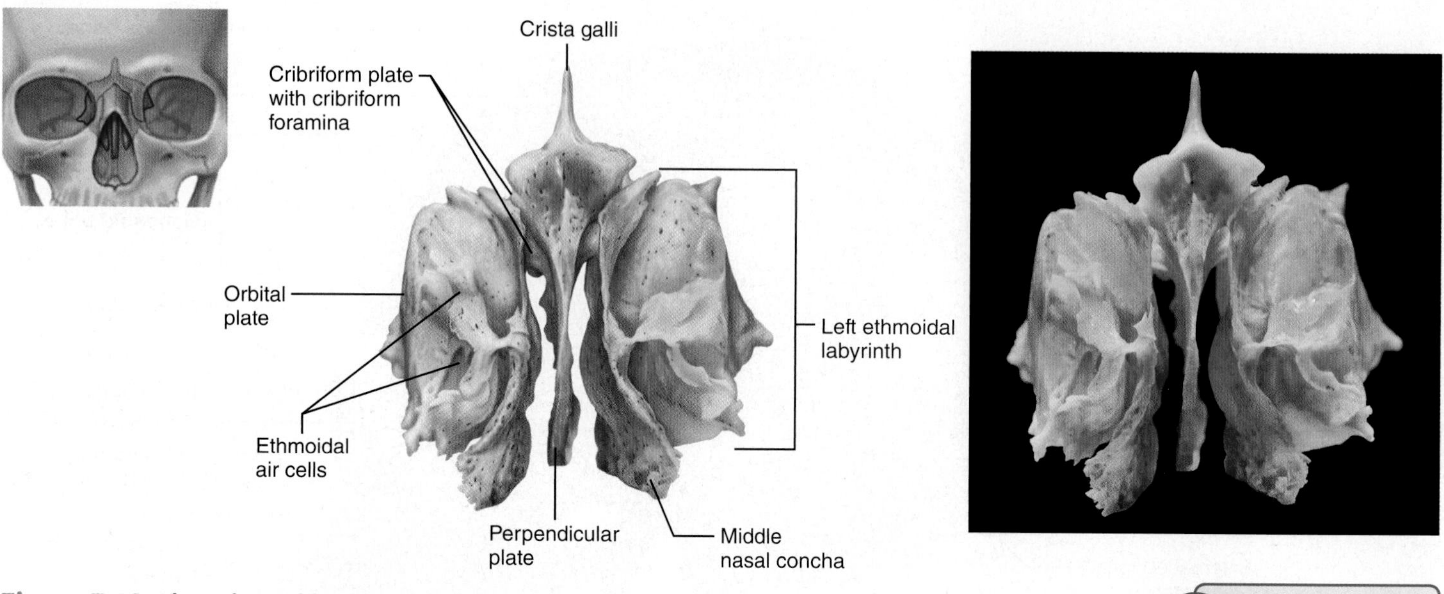

Figure 7.12 The ethmoid bone. Anterior view.

rotundum. The foramen rotundum and foramen ovale are passageways through which two large branches of cranial nerve V (the maxillary and mandibular nerves) exit the cranium. Posterior and lateral to the foramen ovale lies the small **foramen spinosum** (spi-no′sum), named for a short spine that projects from its margin on the inferior aspect of the skull. Through this foramen passes the middle meningeal artery, which supplies blood to the broad inner surfaces of the parietal bones and the squamous part of the temporal bones.

Ethmoid Bone (Figures 7.11 and 7.12)

The **ethmoid bone** (Figure 7.12) is the most deeply situated bone of the skull. It lies anterior to the sphenoid bone and posterior to the nasal bones (Figure 7.11), forming most of the medial bony area between the nasal cavity and the orbits. The ethmoid is a remarkably thin-walled and delicate bone. In the articulated skull, only small portions of the ethmoid are viewable.

Its superior surface is formed by paired, horizontal **cribriform** (krib′rĭ-form; "perforated like a sieve") **plates** that contribute to the roof of the nasal cavities and the floor of the anterior cranial fossa (Figure 7.9). The cribriform plates are perforated by tiny holes called *cribriform foramina.* The filaments of cranial nerve I, the olfactory nerve, pass through these holes as they run from the nasal cavity to the brain. Between the two cribriform plates, in the midline, is a superior projection called the **crista galli** (kris′tah gal′li; "rooster's comb"). A fibrous membrane called the falx cerebri (discussed further in the dura mater discussion in Chapter 13, pp. 403–404) attaches to the crista galli and helps to secure the brain within the cranial cavity.

The **perpendicular plate** of the ethmoid bone projects inferiorly in the median plane. It forms the superior part of the nasal septum, the vertical partition that divides the nasal cavity into right and left halves (Figure 7.6). Flanking the perpendicular plate on each side is a delicate **ethmoidal labyrinth** riddled with **ethmoidal air cells** (ethmoid sinuses). The ethmoid bone is named for these sinuses, as *ethmos* means "sieve" in Greek. Extending medially from the ethmoidal labyrinth are the thin **superior** and **middle nasal conchae** (kong′ke), which protrude into the nasal cavity (see Figure 7.14a). The conchae are curved like scrolls and are named after the conch shells one finds on warm ocean beaches. The lateral surfaces of the labyrinth are called **orbital plates** because they contribute to the medial walls of the orbits.

☑ check your understanding

☐ 1. Name the bones that form the anterior cranial fossa.

☐ 2. For each feature listed, name the bone that contains the feature: crista galli, mastoid process, nuchal line, sella turcica, supraorbital foramen, zygomatic process.

☐ 3. Which four bones articulate with the left parietal bone? Name the sutures that join these bones to the left parietal bone.

(For answers, see Appendix B.)

Facial Bones

The skeleton of the face consists of 14 bones (see Figures 7.4 and 7.6). These are the unpaired mandible and the vomer, plus the paired maxillae, zygomatics, nasals, lacrimals, palatines, and inferior nasal conchae.

Mandible (Figure 7.13a)

The U-shaped **mandible** (man′dĭ-bl), or lower jawbone, is the largest, strongest bone in the face (Figure 7.13a). It has a horizontal *body* that forms the inferior jawline, and

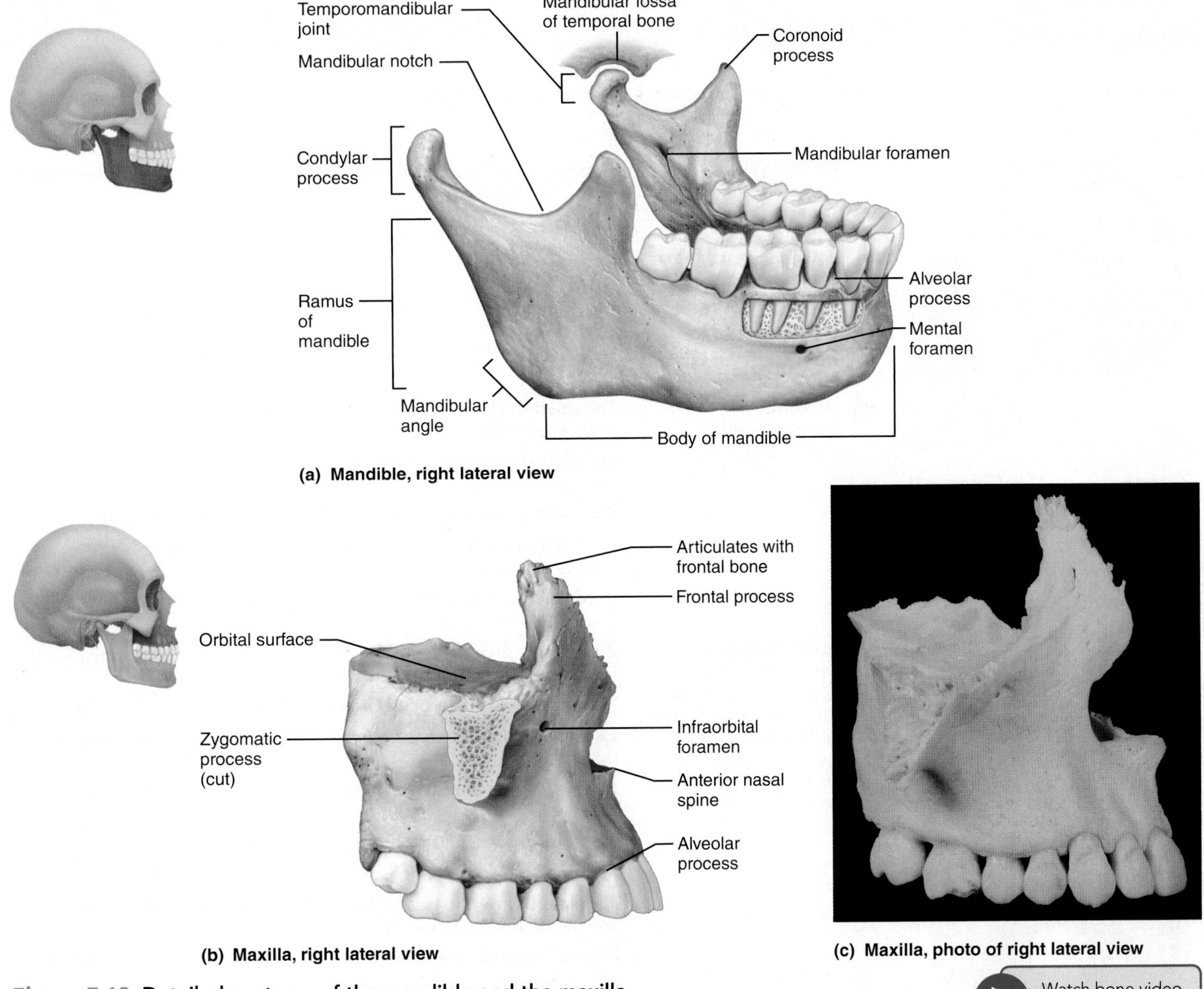

Figure 7.13 **Detailed anatomy of the mandible and the maxilla.**

two upright *rami* (ra′mi; "branches"). Each **ramus** meets the body posteriorly at a **mandibular angle.** At the superior margin of each ramus are two processes. The anterior **coronoid process** (kor′o-noid; "crown-shaped") is a flat, triangular projection. The temporalis muscle, which elevates the lower jaw during chewing, inserts here. The posterior **condylar process** enlarges superiorly to form the **head of the mandible.** It articulates with the temporal bone to form the temporomandibular joint. The coronoid and condylar processes are separated by the **mandibular notch.**

The **body** of the mandible anchors the lower teeth and forms the chin. Its superior border is the **alveolar** (al-ve′-o-lar) **process.** The tooth sockets, called *alveoli,* open onto this surface. Anteriorly, the two halves of the mandible fuse at the mandibular symphysis. The **mental protuberance** at this junction (Figure 7.6) forms the chin.

Several openings pierce the mandible. On the medial surface of each ramus is a **mandibular foramen** (Figure 7.13a), through which a nerve responsible for tooth sensation (inferior alveolar nerve, a branch of cranial nerve V) enters the mandibular body and supplies the roots of the lower teeth. Dentists inject anesthetic into this foramen before working on the lower teeth. The **mental** ("chin") **foramen,** which opens on the anterolateral side of the mandibular body, transmits blood vessels and nerves to the lower lip and the skin of the chin.

Maxillary Bones (Figures 7.6 and 7.13b)

The **maxillary bones,** or **maxillae** (mak-sil′e; "jaws"), form the upper jaw and the central part of the facial skeleton (see Figures 7.6 and 7.13b). They are considered the keystone bones of the face because they articulate with all other facial bones except the mandible.

Table 7.1 Bones of the Skull

View of Skull	Bone (#) with Comments*	Important Markings
	CRANIAL BONES	
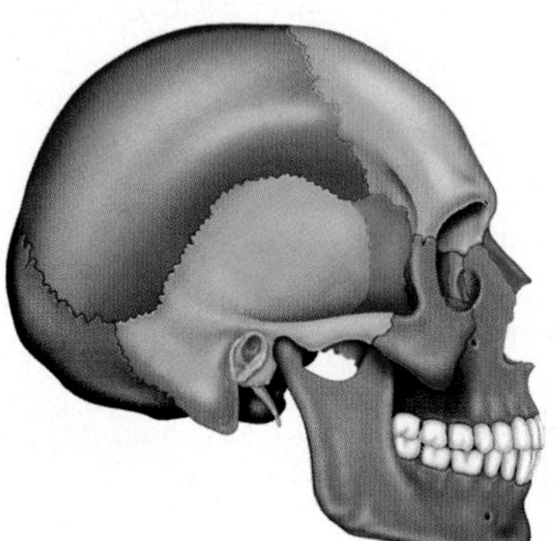	**Frontal (1)** Forms forehead, superior part of orbits, and most of the anterior cranial fossa; contains sinuses	**Supraorbital foramina (notches):** passageway for the supraorbital arteries and nerves
	Parietal (2) Forms most of the superior and lateral aspects of the skull	
Lateral view of skull (Figure 7.4)	**Occipital (1)** Forms posterior aspect and most of the base of the skull	**Foramen magnum:** allows passage of the spinal cord from the brain stem to the vertebral canal **Hypoglossal canals:** passageway for the hypoglossal nerve (cranial nerve XII) **Occipital condyles:** articulate with the atlas (first vertebra) **External occipital protuberance and nuchal lines:** sites of muscle attachment **External occipital crest:** attachment site of ligamentum nuchae
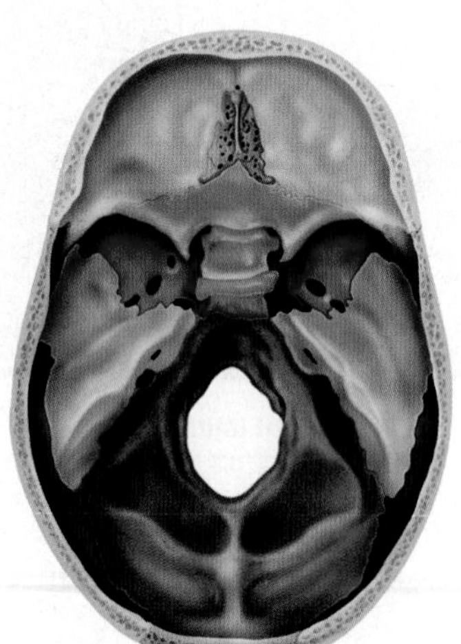	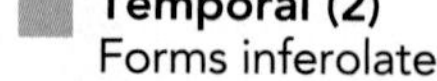 **Temporal (2)** Forms inferolateral aspects of the skull and contributes to the middle cranial fossa; has squamous, tympanic, and petrous parts	**Zygomatic process:** contributes to the zygomatic arch, which forms the prominence of the cheek **Mandibular fossa:** articular point for the head of the mandible **External acoustic meatus:** canal leading from the external ear to the eardrum **Styloid process:** attachment site for several neck and tongue muscles and for a ligament to the hyoid bone **Mastoid process:** attachment site for several neck muscles **Stylomastoid foramen:** passageway for cranial nerve VII (facial nerve) **Jugular foramen:** passageway for the internal jugular vein and cranial nerves IX, X, and XI **Internal acoustic meatus:** passageway for cranial nerves VII and VIII **Carotid canal:** passageway for the internal carotid artery
Superior view of skull, calvaria removed (Figure 7.9)	**Sphenoid (1)** Keystone of the cranium; contributes to the middle cranial fossa and orbits; main parts are the body, greater wings, lesser wings, and pterygoid processes	**Sella turcica:** hypophyseal fossa portion is the seat of the pituitary gland **Optic canals:** passageway for cranial nerve II and the ophthalmic arteries **Superior orbital fissures:** passageway for cranial nerves III, IV, VI, part of V (ophthalmic division), and ophthalmic vein **Foramen rotundum (2):** passageway for the maxillary division of cranial nerve V **Foramen ovale (2):** passageway for the mandibular division of cranial nerve V **Foramen spinosum (2):** passageway for the middle meningeal artery
	Ethmoid (1) Small contribution to the anterior cranial fossa; forms part of the nasal septum and the lateral walls and roof of the nasal cavity; contributes to the medial wall of the orbit	**Crista galli:** attachment point for the falx cerebri, a dural membrane fold **Cribriform plates:** passageway for filaments of the olfactory nerves (cranial nerve I) **Superior** and **middle nasal conchae:** form part of lateral walls of nasal cavity; increase turbulence of air flow

Table 7.1 *(continued)*

View of Skull	Bone (#) with Comments*	Important Markings
	FACIAL BONES	
	 **Nasal (2)** Form the bridge of the nose	
	Lacrimal (2) Form part of the medial orbit wall	**Lacrimal fossa:** houses the lacrimal sac, which helps to drain tears into the nasal cavity
	Zygomatic (2) Form the cheek and part of the orbit	
	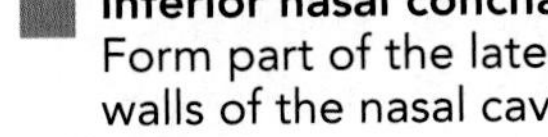 **Inferior nasal concha (2)** Form part of the lateral walls of the nasal cavity	
Anterior view of skull (Figure 7.6)	**Mandible (1)** The lower jaw	**Coronoid processes:** insertion points for the temporalis muscles **Condylar processes:** articulate with the temporal bones to form the jaw (temporomandibular) joints **Mental protuberance:** forms the chin **Alveoli:** sockets for the teeth **Mandibular foramina:** passageway for the inferior alveolar nerves **Mental foramina:** passageway for blood vessels and nerves to the chin and lower lip
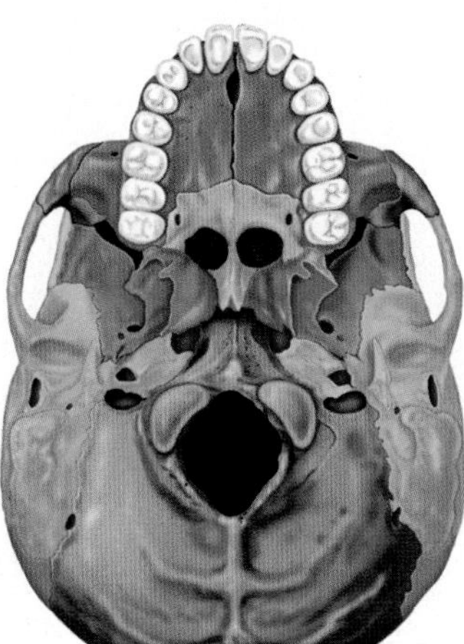 *Inferior view of skull, mandible removed (Figure 7.7)*	**Maxilla (2)** Keystone bones of the face; form the upper jaw and parts of the hard palate, orbits, and nasal cavity walls	**Alveoli:** sockets for teeth **Zygomatic process:** helps form the zygomatic arches **Palatine process:** forms the anterior part of the hard palate; the two processes meet medially in the intermaxillary suture **Frontal process:** forms part of lateral aspect of bridge of nose **Incisive fossa:** passageway for blood vessels and nerves through hard palate (fused palatine processes) **Inferior orbital fissure:** passageway for maxillary branch of cranial nerve V, the zygomatic nerve, and blood vessels **Infraorbital foramen:** passageway for infraorbital nerve to skin of face
	Palatine (2) Form posterior part of the hard palate and a small part of nasal cavity and orbit walls	
	Vomer (1) Inferior part of the nasal septum	
	Auditory ossicles (malleus, incus, and stapes) (2 each) Found in middle ear cavity; involved in sound transmission (see Chapter 16)	

*The color code beside each bone name corresponds to the bone's color in the illustrations (see Figures 7.4 to 7.16). The number in parentheses ( ) following the bone name indicates the total number of such bones in the body.

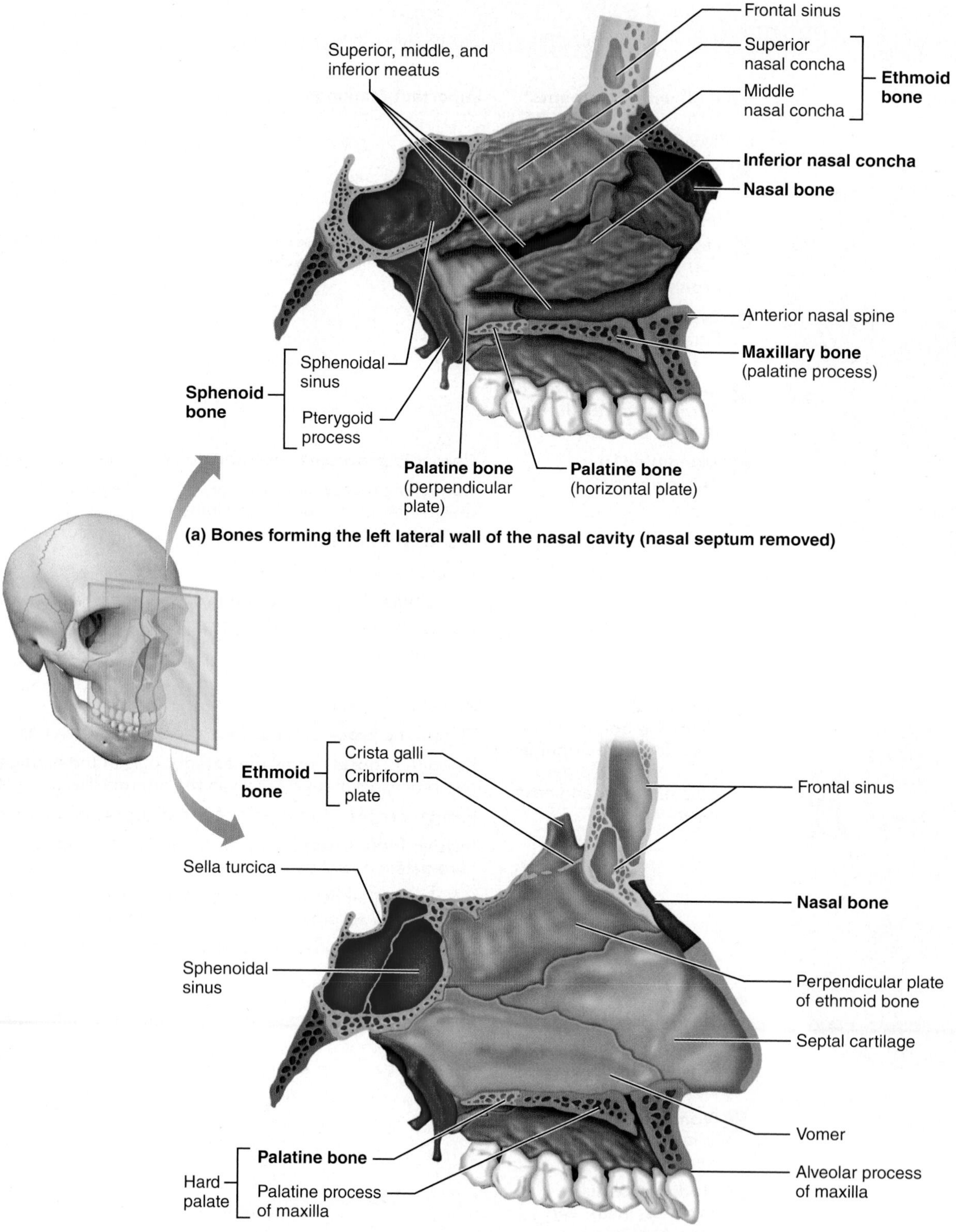

Frontal sinus
Superior, middle, and inferior meatus
Superior nasal concha
Middle nasal concha
Ethmoid bone
Inferior nasal concha
Nasal bone
Anterior nasal spine
Maxillary bone (palatine process)
Sphenoid bone
Sphenoidal sinus
Pterygoid process
Palatine bone (perpendicular plate)
Palatine bone (horizontal plate)

(a) Bones forming the left lateral wall of the nasal cavity (nasal septum removed)

Ethmoid bone
Crista galli
Cribriform plate
Frontal sinus
Sella turcica
Nasal bone
Sphenoidal sinus
Perpendicular plate of ethmoid bone
Septal cartilage
Vomer
Hard palate
Palatine bone
Palatine process of maxilla
Alveolar process of maxilla

(b) Nasal cavity with septum in place showing the contributions of the ethmoid bone, the vomer, and septal cartilage

Figure 7.14 **Bones of the nasal cavity.**

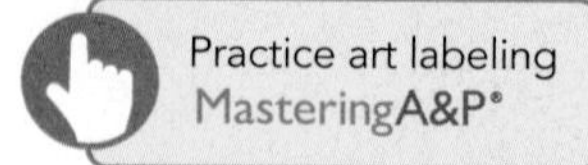

Like the mandible, the maxillae have an **alveolar process** that contains teeth in alveoli. The **palatine** (pal′ah-tēn) **processes** project medially from the alveolar margins to form the anterior region of the **hard palate,** or bony roof of the mouth (Figure 7.7a). The **frontal processes** extend superiorly to reach the frontal bone, forming part of the lateral aspect of the bridge of the nose (Figure 7.6). The maxillae lie just lateral to the nasal cavity and contain the **maxillary sinuses.** These sinuses, the largest of the paranasal air sinuses, extend from the orbit down to the roots of the upper teeth. Laterally, the maxillae articulate with the zygomatic bones at the **zygomatic processes.**

The maxilla, along with several other bones, forms the borders of the **inferior orbital fissure** in the floor of the orbit. This fissure transmits several vessels and nerves (see Table 7.1), including the maxillary nerve (a branch of cranial nerve V) or its continuation, the infraorbital nerve. The infraorbital nerve proceeds anteriorly to enter the face through the **infraorbital foramen.**

Zygomatic Bones (Figures 7.4 and 7.6)

The irregularly shaped **zygomatic bones** are commonly called the cheekbones (*zygoma* = cheekbone). Each joins the zygomatic process of a temporal bone posteriorly, the zygomatic process of the frontal bone superiorly, and the zygomatic process of the maxilla anteriorly (Figure 7.4). The zygomatic bones form the prominences of the cheeks and define part of the margin of each orbit.

Nasal Bones (Figure 7.6)

The paired, rectangular **nasal bones** join medially to form the bridge of the nose. They articulate with the frontal bone superiorly, the maxillae laterally, and the perpendicular plate of the ethmoid bone posteriorly. Inferiorly, they attach to the cartilages that form most of the skeleton of the external nose (see Figure 22.2).

Lacrimal Bones (Figure 7.4)

The delicate, fingernail-shaped **lacrimal** (lak′rĭ-mal) **bones** are located in the medial orbital walls. They articulate with the frontal bone superiorly, the ethmoid bone posteriorly, and the maxilla anteriorly. Each lacrimal bone contains a deep groove that contributes to a *lacrimal fossa.* This fossa contains a *lacrimal sac* that gathers tears, allowing the fluid to drain from the eye surface into the nasal cavity (*lacrima* = tear).

Palatine Bones (Figures 7.7 and 7.14)

The **palatine bones** lie posterior to the maxillae **(Figure 7.14a)**. These paired, L-shaped bones articulate with each other at their inferior **horizontal plates,** which complete the posterior part of the hard palate. The **perpendicular plates** form the posterior part of the lateral walls of the nasal cavity (Figure 7.14a) and a small part of the orbits (see Figure 7.16).

Vomer (Figure 7.14b)

The slender, plow-shaped **vomer** (vo′mer; "plowshare") lies in the nasal cavity, where it forms the inferior part of the nasal septum (Figures 7.6, 7.7, and 7.14b).

Inferior Nasal Conchae (Figure 7.14a)

The paired **inferior nasal conchae** are thin, curved bones in the nasal cavity (see Figures 7.6 and 7.14a). Projecting medially from the lateral walls of the nasal cavity, just inferior to the middle nasal conchae of the ethmoid bone, they are the largest of the three pairs of conchae.

check your understanding

- ☐ 4. Name all the bones that articulate with the maxilla.
- ☐ 5. Which bones or bony processes form the hard palate?
- ☐ 6. What are the alveolar processes, and on which bones are they located?

(For answers, see Appendix B.)

Special Parts of the Skull

learning outcome

- ▶ Define the bony boundaries of the nasal cavity, paranasal sinuses, and orbit.

Next we will examine four special parts of the skull: the nasal cavity and the orbits, which are restricted regions of the skull formed from many bones; the paranasal sinuses, which are extensions of the nasal cavity; and the hyoid bone.

Nasal Cavity (Figure 7.14)

The **nasal cavity** is constructed of bone and cartilage. Its *roof* is the ethmoid's cribriform plates. The *floor* is formed by the palatine processes of the maxillae and the horizontal plates of the palatine bones. These same nasal-floor structures also form the roof of the mouth and are collectively called the hard palate. Contributing to the *lateral walls* of the nasal cavity are the nasal bones, the superior and middle conchae of the ethmoid, the inferior nasal conchae, the frontal process of the maxilla, and the perpendicular plates of the palatine bones (Figure 7.14a). On these lateral walls, each of the three conchae forms a roof over a groove-shaped air passageway called a *meatus* (me-a′tus; "a passage"). Therefore, there are **superior, middle,** and **inferior meatuses.**

The nasal cavity is divided into right and left halves by the nasal septum. The bony part of this septum is formed by the vomer inferiorly and by the perpendicular plate of the ethmoid superiorly (Figure 7.14b). A sheet of cartilage, called the *septal cartilage,* completes the septum anteriorly.

CLINICAL APPLICATION

Deviated Septum The nasal septum in many people is slightly off center. A septum that is markedly off center is referred to as a *deviated septum*. This condition is commonly a result of trauma to the nose and can cause difficulty in breathing through the nose, as well as nasal congestion, frequent nosebleeds, and frequent sinus infections. Surgery to correct the deviation (septoplasty) may be indicated if the symptoms are severe.

Frontal sinus
Ethmoidal air cells (sinus)
Sphenoidal sinus
Maxillary sinus

(a) Anterior aspect

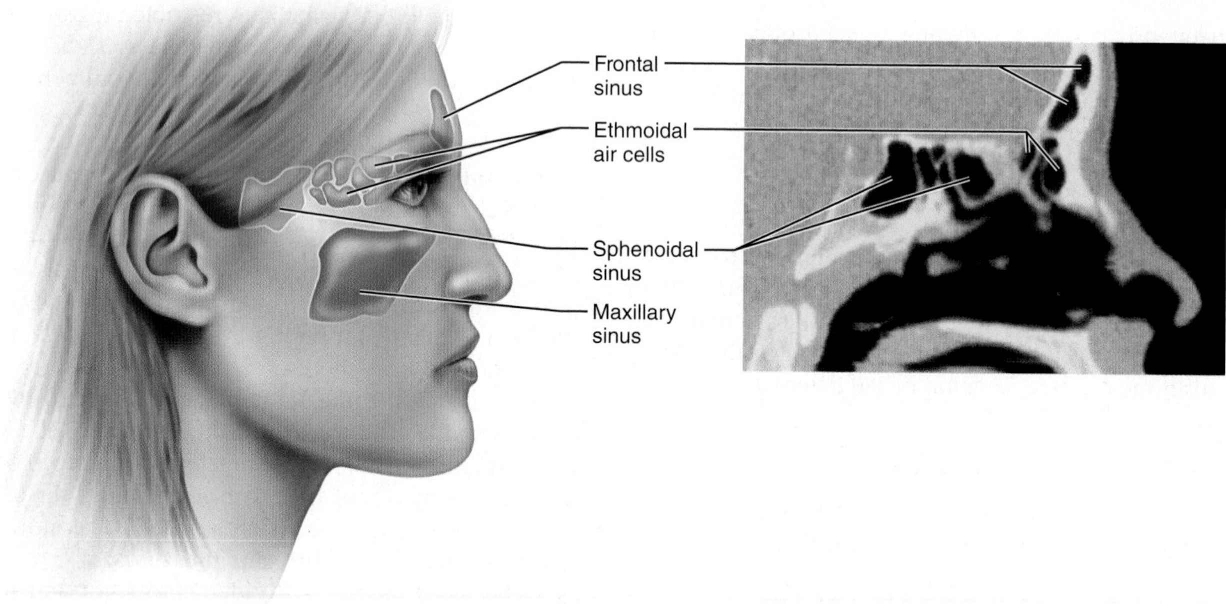

(b) Medial aspect

Figure 7.15 Paranasal sinuses. Illustration on left, CT scan on right.

The walls of the nasal cavity are covered with a mucous membrane that moistens and warms inhaled air. This membrane also secretes mucus that traps dust, thereby cleansing the air of debris. The three pairs of scroll-shaped nasal conchae cause the air flowing through the nasal cavity to swirl. This turbulence increases the contact of inhaled air with the mucous membrane throughout the nasal cavity, such that the air is warmed, moistened, and filtered more efficiently.

Paranasal Sinuses (Figure 7.15)

The bones surrounding the nasal cavity—the frontal, ethmoid, sphenoid, and both maxillary bones—contain hollow spaces internally. These air-filled sinuses are called **paranasal sinuses** (*para* = near) because they cluster around the nasal cavity (Figure 7.15). In fact, they are extensions of the nasal cavity, lined by the same mucous membrane, and help warm, moisten, and filter inhaled air. The paranasal sinuses also lighten the skull, giving the bones they occupy a moth-eaten appearance in an X-ray image. These sinuses connect to the

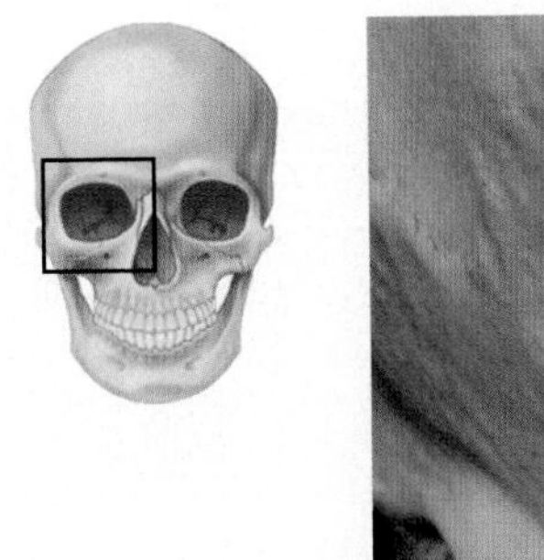

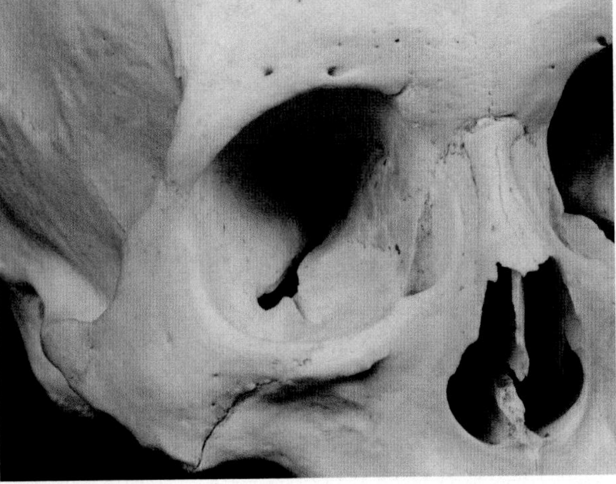

(a) Photograph, right orbit

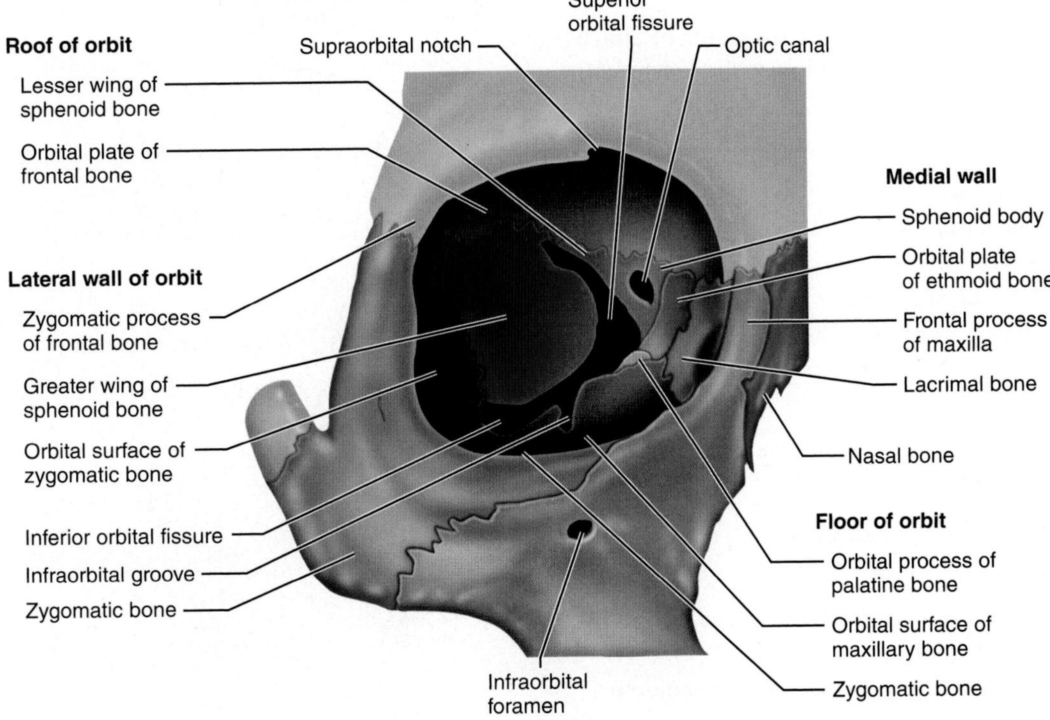

(b) Contribution of each of the seven bones forming the right orbit

Figure 7.16 **Bones that form the orbit.**

nasal cavity through small openings, most of which occur at the meatuses inferior to the conchae. (For more on paranasal sinuses, see Chapter 22).

Orbits (Figure 7.16)

The **orbits** are cone-shaped bony cavities that hold the eyes, the muscles that move the eyes, some fat, and the tear-producing glands. The walls of each orbit are formed by parts of seven bones—the frontal, sphenoid, zygomatic, maxillary, palatine, lacrimal, and ethmoid bones (Figure 7.16). The superior and inferior orbital fissures, optic canal, and lacrimal fossa (described earlier) are also in the orbit.

The Hyoid Bone (Figure 7.17)

Not really part of the skull but associated with it, the **hyoid bone** (hi′oid; "U-shaped") lies just inferior to the mandible in the anterior neck (Figure 7.17). This bone resembles an archer's bow. It has a **body** and two pairs of **horns,** each of which is also called a *cornu,* the Latin word for "horn." The hyoid is the only bone in the skeleton that does not articulate directly with any other bone. Instead, its lesser horns are tethered by thin *stylohyoid ligaments* to the styloid processes of the temporal bones. Other ligaments connect the hyoid to the larynx (voice box) inferior to it. Functionally, the hyoid bone acts as a movable base for the tongue.

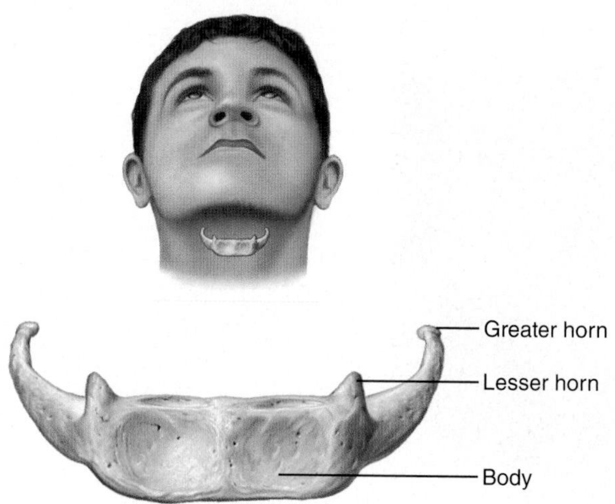

Figure 7.17 The hyoid bone. Anterior view.

Furthermore, its body and greater horns are points of attachment for neck muscles that raise and lower the larynx during swallowing.

check your understanding

- ☐ 7. Which bones form the nasal conchae? What is the function of these structures?
- ☐ 8. Which of the bones that form the orbit are cranial bones? Which are facial bones?
- ☐ 9. Which paranasal sinuses are located along the lateral walls of the nasal cavity? What type of membrane lines all the paranasal sinuses?

(For answers, see Appendix B.)

THE VERTEBRAL COLUMN

learning outcomes

- Name a function performed by both the spinal curvatures and the intervertebral discs.
- Describe the general structure of the vertebral column, and list its components.
- Discuss the structure of a typical vertebra, and describe the special features of cervical, thoracic, and lumbar vertebrae.

The **vertebral column** (Figure 7.18), also called the *spinal column* or *spine,* consists of 26 bones connected into a flexible, curved structure. The main support of the body axis, the vertebral column extends from the skull to the pelvis, where it transmits the weight of the trunk to the lower limbs. It also surrounds and protects the delicate spinal cord and provides attachment points for the ribs and for muscles of the neck and back. In the fetus and infant, the vertebral column consists of 33 separate bones, or **vertebrae** (ver′te-bre). Inferiorly, nine of these eventually fuse to form two composite bones, the sacrum and the tiny coccyx (tailbone). The remaining 24 bones persist as individual vertebrae separated by *intervertebral discs.*

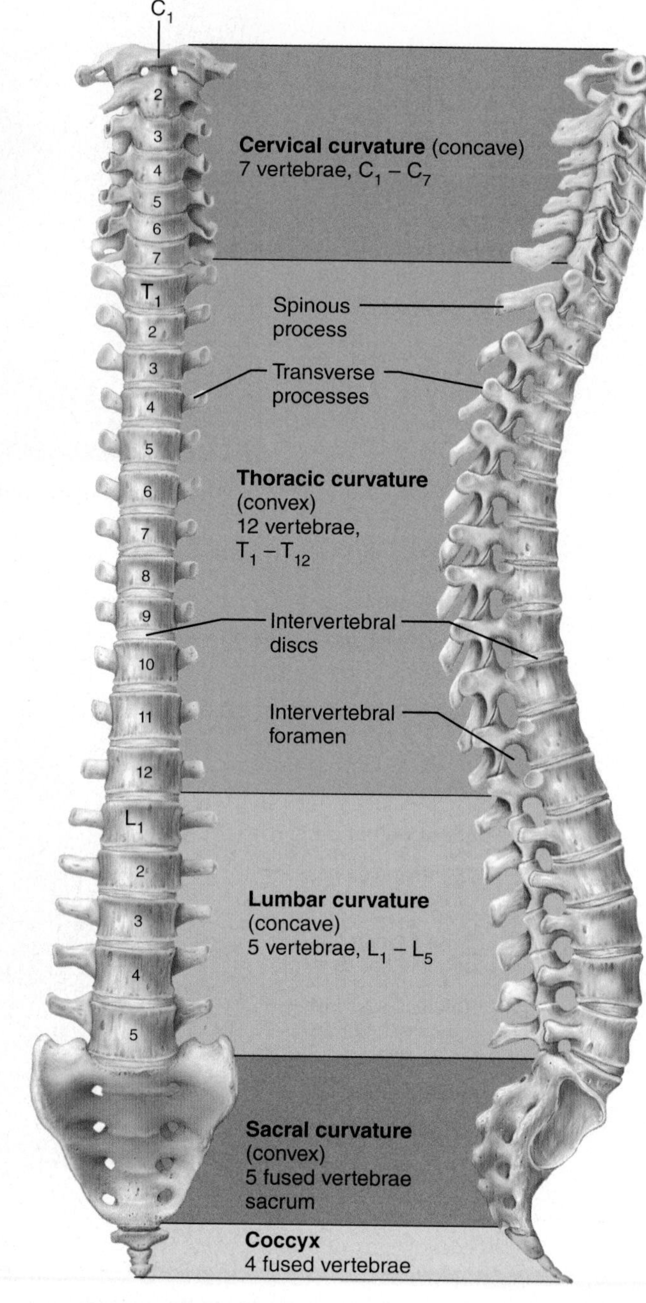

Figure 7.18 The vertebral column. Note the four curvatures in the lateral view at right. The terms *convex* and *concave* are relative to the *posterior* aspect of the column.

Regions and Normal Curvatures

The vertebral column, which is about 70 cm (28 inches) long in an average adult, has five major regions (Figure 7.18). The 7 vertebrae of the neck are the **cervical vertebrae,** the next 12 are the **thoracic vertebrae,** and the 5 that support the lower back are the **lumbar vertebrae.** To remember the number of vertebrae in these three regions, think of the usual meal times of 7:00 A.M., 12:00 noon, and 5:00 P.M.

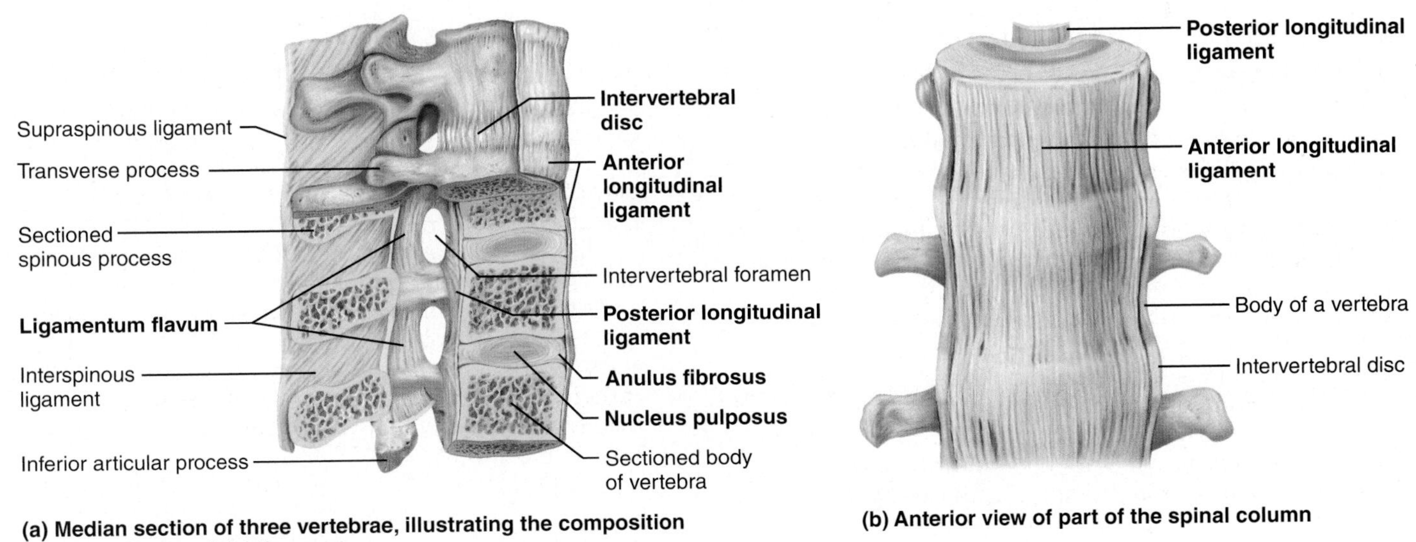

(a) Median section of three vertebrae, illustrating the composition of the discs and the ligaments

(b) Anterior view of part of the spinal column

Figure 7.19 **Ligaments and intervertebral discs of the spine.**

The vertebrae become progressively larger from the cervical to the lumbar region as the weight they must support progressively increases. Inferior to the lumbar vertebrae is the **sacrum** (sa′krum), which articulates with the hip bones of the pelvis. The most inferior part of the vertebral column is the tiny **coccyx** (kok′siks).

All people (and in fact the majority of mammals) have seven cervical vertebrae. Variations in numbers of vertebrae in the other regions occur in about 5% of people.

From a lateral view, four curvatures that give the vertebral column an S shape are visible (Figure 7.18, right). The **cervical** and **lumbar curvatures** are concave posteriorly, whereas the **thoracic** and **sacral curvatures** are convex posteriorly. These curvatures increase the resilience of the spine, allowing it to function like a spring rather than a straight, rigid rod.

Only the thoracic and sacral curvatures are well developed at birth. Both of these *primary curvatures* are convex posteriorly, so that an infant's spine arches (is C-shaped) like that of a four-legged animal. The *secondary curvatures,* the cervical and lumbar curvatures, are concave posteriorly and develop during the first 2 years of childhood as the intervertebral discs are reshaped. The cervical curvature is present before birth but is not pronounced until the baby starts to lift its head at 3 months, and the lumbar curvature develops when the baby begins to walk, at about 1 year. The lumbar curvature positions the weight of the upper body over the lower limbs, providing optimal balance during standing.

Ligaments of the Spine

Like a tremulous telecommunication transmitting tower, the vertebral column cannot stand upright by itself. It must be held in place by an elaborate system of supports. Serving this role are the straplike ligaments of the back and the muscles of the trunk. (The trunk muscles are discussed in Chapter 11, pp. 287–290.) The major supporting ligaments are the **anterior** and **posterior longitudinal ligaments** (Figure 7.19) that run vertically along the anterior and posterior surfaces of the bodies of the vertebrae, from the neck to the sacrum. The anterior longitudinal ligament is wide and attaches strongly to both the bony vertebrae and the intervertebral discs. Along with its supporting role, this thick anterior ligament prevents hyperextension of the back (bending too far backward). The posterior longitudinal ligament, which is narrow and relatively weak, attaches only to the intervertebral discs. This ligament helps to prevent hyperflexion (bending the vertebral column too sharply forward).

Several other posterior ligaments connect each vertebra to those immediately superior and inferior (Figure 7.19a). Among these is the **ligamentum flavum** (fla′vum; "yellow"), which connects the lamina of adjacent vertebrae. It contains elastic connective tissue and is especially strong: It stretches as we bend forward, then recoils as we straighten to an erect position.

Intervertebral Discs

Each **intervertebral disc** (Figure 7.19) is a cushionlike pad composed of an inner sphere, the **nucleus pulposus** (pul-po′sus; "pulp"), and an outer collar of about 12 concentric rings, the **anulus fibrosus** (an′u-lus fi-bro′sus; "fibrous ring"). Each nucleus pulposus is gelatinous and acts like a rubber ball, enabling the spine to absorb compressive stress. In the anulus fibrosus, the outer rings consist of ligament and the inner ones consist of fibrocartilage. The main function of these rings is to contain the nucleus pulposus, limiting its expansion when the spine is compressed. However, the rings also function like a woven strap, binding the successive vertebrae together, resisting tension on the spine, and absorbing compressive forces. Collagen fibers in adjacent rings in the anulus cross like an X, allowing the spine to withstand twisting. This arrangement creates the

CLINICAL APPLICATION

Herniated Disc Severe or sudden physical trauma to the spine—for example, due to lifting a heavy object—may cause one or more **herniated discs** (also called **prolapsed discs** or, in common terms, *slipped discs*). This condition usually involves rupture of the anulus fibrosus followed by protrusion of the nucleus pulposus. Aging is a contributing factor, because the nucleus pulposus loses its cushioning properties over time, and the anulus fibrosus weakens and tears. This mechanical fatigue allows the nucleus to rupture through the anulus. The anulus is thinnest posteriorly, but the posterior longitudinal ligament prevents the herniation from proceeding directly posteriorly, so the rupture proceeds posterolaterally—toward the spinal nerve roots exiting from the spinal cord (see figure a). The resulting pressure on these nerve roots causes pain or numbness. In most cases, the pain eventually resolves, so treatment is usually conservative: moderate exercise, physical therapy, massage, heat therapy, and painkillers. If these treatments fail, the herniated disc may be removed surgically, and bone grafts are used to fuse the adjacent vertebrae.

In many of these cases, back pain results not from herniation of the disc but instead from small nerves that enter tears in the disc on newly invading veins. A new, simple, and low-pain treatment called IDET (intradiscal electrothermal annuloplasty) consists of threading a fine catheter with a heated tip into the disc, burning away the invading nerves, and at the same time sealing the tears in the anulus.

Vertebral spinous process (posterior aspect of vertebra)
Spinal cord
Spinal nerve root
Transverse process
Herniated portion of disc
Anulus fibrosus of disc
Nucleus pulposus of disc

(a) Superior view of a herniated intervertebral disc

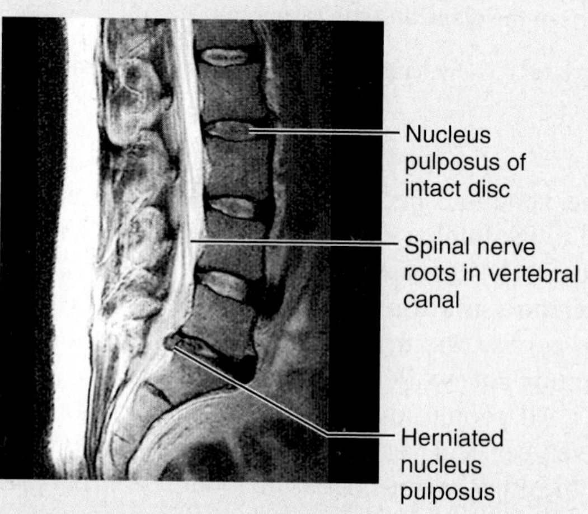

(b) MRI of lumbar region of vertebral column in sagittal section showing normal and herniated discs

same antitwisting design provided by bone lamellae in osteons (see Figure 6.8 on p. 135).

The intervertebral discs act as shock absorbers during walking, jumping, and running. At points of compression, the discs flatten and bulge out a bit between the vertebrae. The discs are thickest in the lumbar (lower back) and cervical (neck) regions of the vertebral column. Collectively, the intervertebral discs make up about 25% of the height of the vertebral column. As a result of compression and loss of fluid from the gelatinous nucleus pulposus, they flatten somewhat by the end of each day. So, you are probably 1 to 2 centimeters shorter at night than when you awake in the morning.

☑ check your understanding

☐ 10. Which portion of the intervertebral disc expands under compression? Which portion resists twisting forces?

☐ 11. When and how do the secondary curvatures of the vertebral column develop?

☐ 12. Why do intervertebral discs usually herniate in the posterolateral direction?

(For answers, see Appendix B.)

General Structure of Vertebrae

Vertebrae from all regions share a common structural pattern. The following features are typical of all vertebrae (Figure 7.20):

- The anterior portion of the vertebra is the disc-shaped **body.** The body is the weight-bearing region of the vertebra.
- The **vertebral arch** forms the posterior portion of the vertebra. It is composed of two pedicles and two laminae. The **pedicles** (ped′ĭ-klz; "little feet") are short, bony walls that project posteriorly from the vertebral body and form the sides of the arch. The two **laminae** (lam′ĭ-ne; "sheets") are flat, bony plates that complete the arch posteriorly, extending from the transverse processes to the spinous process. The vertebral arch protects the spinal cord and spinal nerves located in the vertebral foramen.
- The large hole encircled by the body and vertebral arch is the **vertebral foramen.** Successive vertebral foramina of the articulated vertebrae form the long *vertebral canal,* through which the spinal cord and spinal nerve pass.
- The **spinous process** is the median, posterior projection arising at the junction of the two laminae. It is an attachment

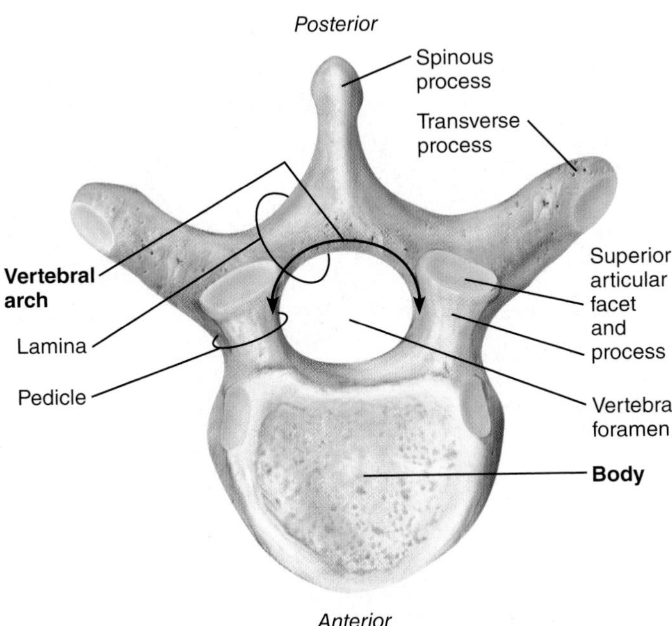

Figure 7.20 **Structure of a typical vertebra.** Superior view. Only features of the bone are illustrated in this figure and subsequent figures in this chapter. Articular cartilage is not depicted.

site for muscles and ligaments that move and stabilize the vertebral column.

- A **transverse process** projects laterally from each pedicle-lamina junction. As with the spinous process, the transverse processes are attachment sites for the muscle and ligaments.
- Articular processes protrude superiorly and inferiorly from the pedicle-lamina junctions and form movable joints between successive vertebrae: The **inferior articular processes** of each vertebra join with the **superior articular processes** of the vertebra immediately inferior. Successive vertebrae are joined by both intervertebral discs and by these articular processes. The smooth joint surfaces of these processes are **facets** ("little faces").
- Notches on the superior and inferior borders of the pedicles form lateral openings between adjacent vertebrae, the **intervertebral foramina** (see Figure 7.18). Spinal nerves from the spinal cord pass through these foramina.

Regional Vertebral Characteristics

The different regions of the spine perform slightly different functions, so vertebral structure shows regional variation (Figure 7.21). In general, the types of movements that can occur between vertebrae are (1) flexion and extension (anterior bending and posterior straightening of the spine), (2) lateral flexion (bending the upper body to the right or left), and (3) rotation, in which the vertebrae rotate on one another in the long axis of the vertebral column. (Refer to Table 7.2 on pp. 175–176 while reading the following discussions of vertebral characteristics.)

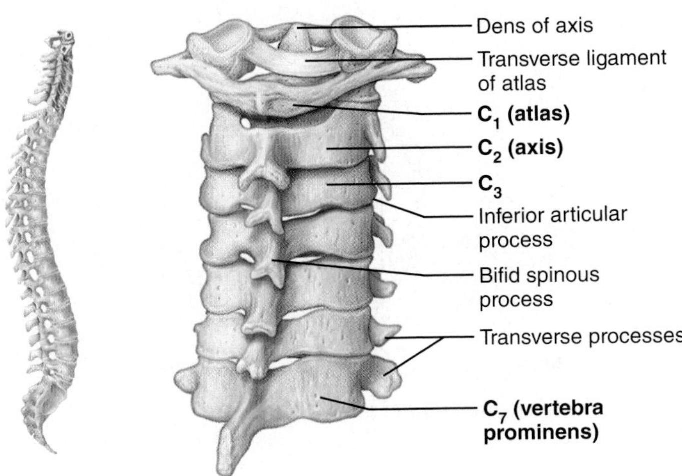

(a) Cervical vertebrae

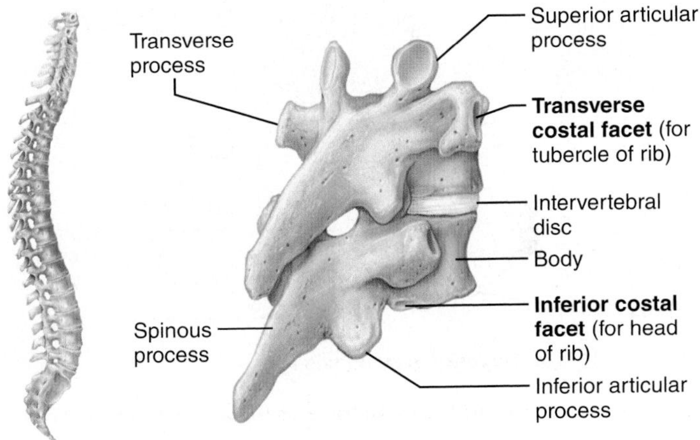

(b) Thoracic vertebrae

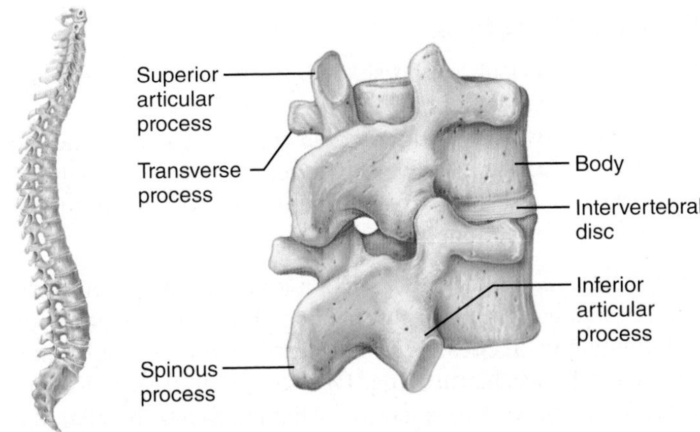

(c) Lumbar vertebrae

Figure 7.21 **Posterolateral views of articulated vertebrae.** In (a), note the prominent spinous process of C_7, the vertebra prominens.

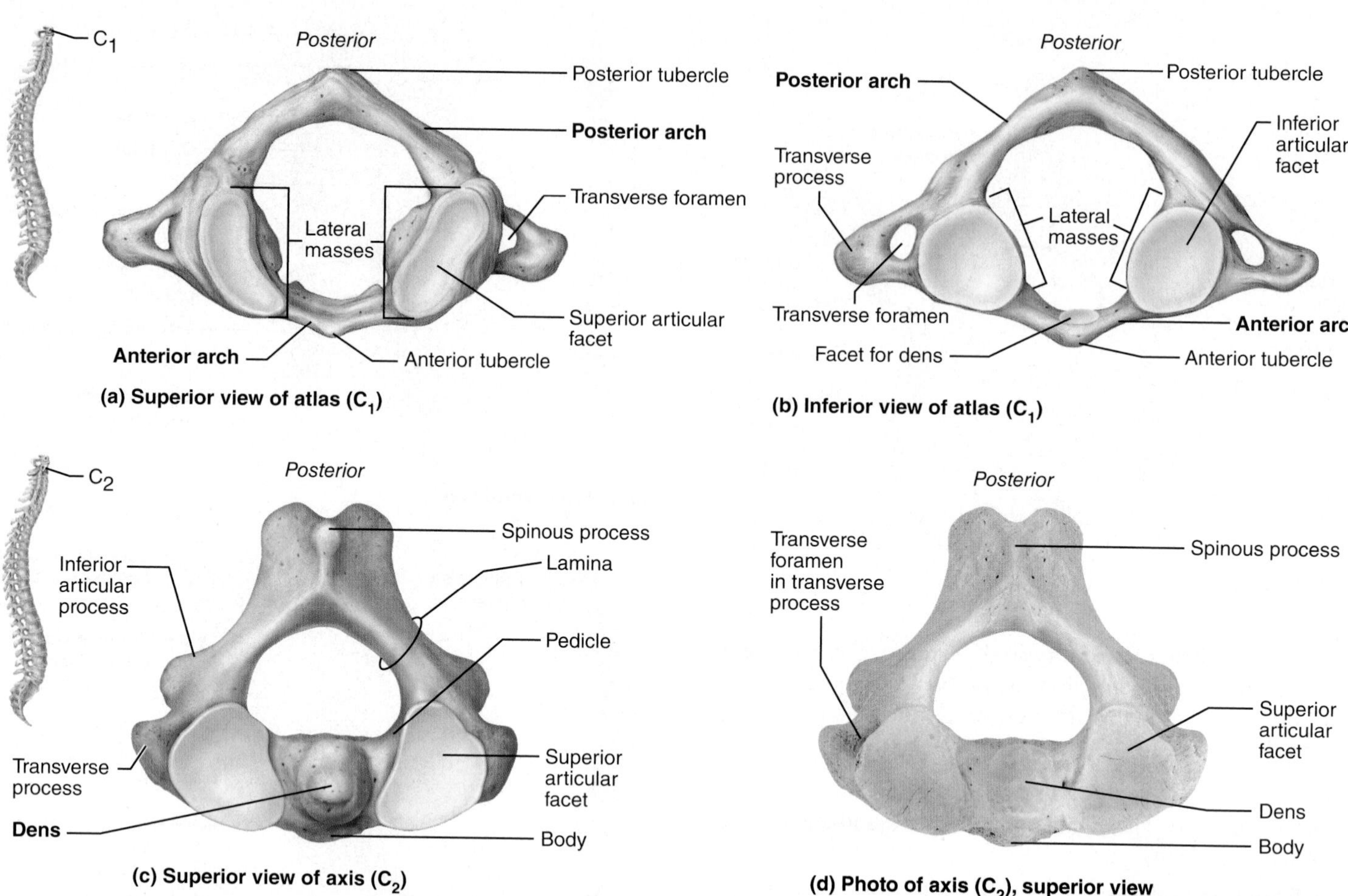

Figure 7.22 The first and second cervical vertebrae.

Watch Bone Video MasteringA&P®

Cervical Vertebrae

The seven cervical vertebrae, identified as C_1–C_7, are the smallest, lightest vertebrae.

The first two cervical vertebrae are the *atlas* (C_1) and the *axis* (C_2) (**Figure 7.22**). These two vertebrae are unusual: No intervertebral disc lies between them, and they have unique structural and functional features.

The Atlas, C_1 The **atlas** lacks a body and a spinous process. Essentially, it is a ring of bone consisting of **anterior** and **posterior arches,** plus a **lateral mass** on each side. Each lateral mass has articular facets on both its superior and inferior surfaces. The superior articular facets receive the occipital condyles of the skull. Thus, they "carry" the skull, just as the giant Atlas supported the heavens in Greek mythology. These joints participate in flexion and extension of the head on the neck, as when you nod "yes." The inferior articular facets form joints with the axis.

The Axis, C_2 The **axis,** which has a body, a spinous process, and the other typical vertebral processes, is not as specialized as the atlas. Its only unusual feature is the knoblike **dens** ("tooth") projecting superiorly from its body. The dens is actually the "missing" body of the atlas that fuses with the axis during embryonic development. Cradled in the anterior arch of the atlas (Figure 7.21a), the dens acts as a pivot for the rotation of the atlas and skull. Hence, this joint participates in rotating the head from side to side to indicate "no." *Axis* is a good name for the second cervical vertebra because its dens allows the head to rotate on the neck's axis.

CLINICAL APPLICATION

The Dens and Fatal Trauma In cases of severe head trauma in which the skull is driven inferiorly toward the spine, the dens may be forced into the brain stem, causing death. Alternatively, if the neck is jerked forward, as in an automobile collision, the dens may be driven posteriorly into the cervical spinal cord. This injury is also fatal.

Typical Cervical Vertebrae, C_3–C_7 The typical cervical vertebrae, C_3–C_7, have the following distinguishing features (**Table 7.2**):

- The body is wider laterally than in the anteroposterior dimension.
- Except in C_7, the spinous process is short, projects directly posteriorly, and is bifid (bi′fid; "cleaved in two"), or forked; that is, it is split at its tip.

Table 7.2 Regional Characteristics of Cervical, Thoracic, and Lumbar Vertebrae

Characteristic	Cervical (3–7)	Thoracic	Lumbar
Body	Small, wide side to side	Larger than cervical; heart-shaped; superior and inferior costal facets near pedicle	Massive; kidney-shaped
Spinous process	Short; forked; projects directly posteriorly	Long; sharp; projects inferiorly	Short; blunt; rectangular; projects directly posteriorly
Vertebral foramen	Triangular	Circular	Triangular
Transverse processes	Contain foramina	Costal facets for tubercle of rib (except T_{11} and T_{12}) on anterior surfaces	Thin and tapered
Superior and inferior articulating processes	Superior facets directed superoposteriorly Inferior facets directed inferoanteriorly	Superior facets directed posteriorly Inferior facets directed anteriorly	Superior facets directed posteromedially (or medially) Inferior facets directed anterolaterally (or laterally)
Movements allowed	Flexion and extension; lateral flexion; rotation; the spine region with the greatest range of movement	Rotation; lateral flexion possible but restricted by ribs; flexion and extension limited	Flexion and extension; some lateral flexion; rotation prevented

SUPERIOR VIEW

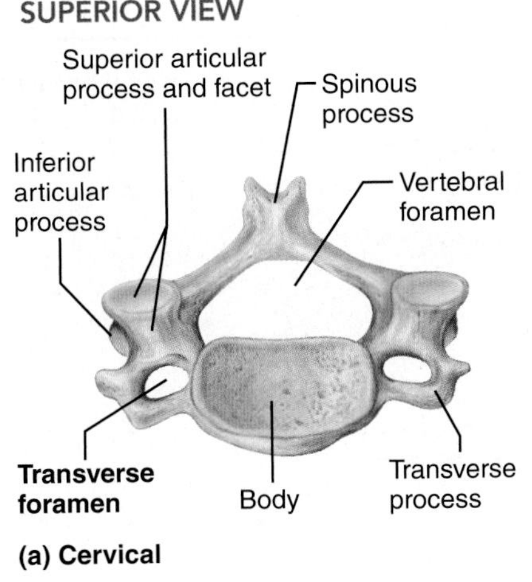

(a) Cervical

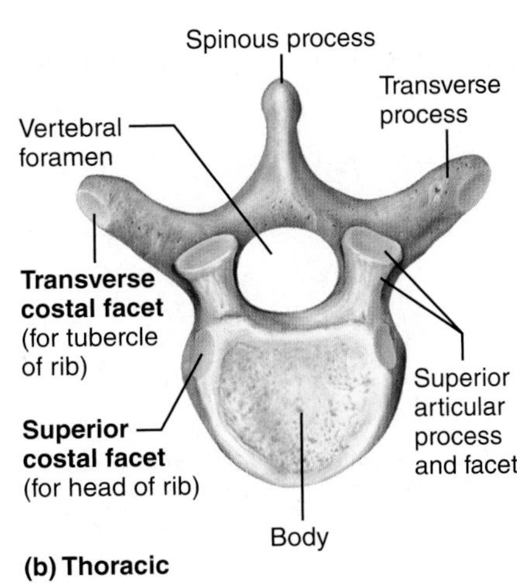

(b) Thoracic

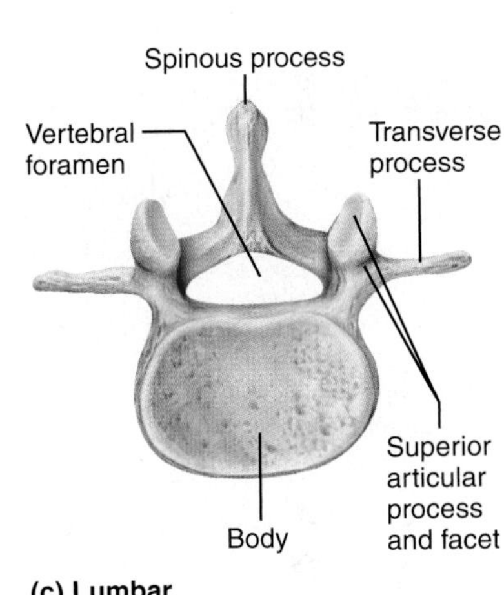

(c) Lumbar

- The vertebral foramen is large and generally triangular.
- Each transverse process contains a hole, a **transverse foramen,** through which the vertebral blood vessels pass. These vessels ascend and descend through the neck to help serve the brain.
- The superior articular facets face superoposteriorly, whereas the inferior articular facets face inferoanteriorly. Thus these articulations lie in an oblique plane. The orientation of these articulations allows the neck to carry out an extremely wide range of movements: flexion and extension, lateral flexion, and rotation.

The spinous process of C_7 is not forked and is much larger than those of the other cervical vertebrae (Figure 7.21a). Because its large spinous process can be seen and felt through the skin, C_7 is called the **vertebra prominens** ("prominent vertebra") and is used as a landmark for counting the vertebrae in living people. To locate this landmark, run your fingers inferiorly along the back of your neck, in the posterior midline, where you can feel the spinous processes of the cervical vertebrae. The spine of C_7 is especially prominent.

Thoracic Vertebrae

The 12 thoracic vertebrae, T_1–T_{12} (Figure 7.21b and Table 7.2), all articulate with ribs. Their other unique characteristics are as follows:

- From a superior view, the vertebral body is roughly heart-shaped. Laterally, each side of the vertebral body has two facets, commonly referred to as demifacets (dem′e-fas″ets). The **superior costal facet** lies at the superior edge, and the **inferior costal facet** lies at the inferior edge (*costa* = rib; *facet* = joint surface). The heads of the ribs articulate

Table 7.2 Regional Characteristics of Cervical, Thoracic, and Lumbar Vertebrae (*continued*)

Characteristic	Cervical (3–7)	Thoracic	Lumbar
RIGHT LATERAL VIEW			

Spinous process
Superior articular process
Body
Inferior articular process and facet
Transverse process
(a) Cervical

Superior articular process and facet
Transverse process
Transverse costal facet (for tubercle of rib)
Superior costal facet (for head of rib)
Body
Inferior notch
Inferior costal facet (for head of rib)
Spinous process
(b) Thoracic

Superior articular process
Transverse process
Body
Inferior notch
Inferior articular process and facet
Spinous process
(c) Lumbar

Superior articular process and facet
Transverse foramen
Inferior articular process and facet

Superior articular process and facet
Superior costal facet
Transverse costal facet
Inferior articular process and facet
Inferior costal facet

Superior articular process
Transverse process
Spinous process
Inferior articular process and facet

with these facets. For most ribs, the head of the rib is attached to the bodies of two vertebrae, the inferior costal facet of the superior vertebra and the superior costal facet of the inferior vertebra. Vertebra T_1 differs from this general pattern in that its body bears a full facet for the first rib and a demifacet for the second rib; furthermore, the bodies of T_{10}–T_{12} have only single facets to receive their respective ribs.

- The spinous process is long and points inferiorly.
- The vertebral foramen is circular.
- With the exception of T_{11} and T_{12}, the transverse processes have facets that articulate with the tubercles of the ribs called **transverse costal facets** (Figure 7.21b).
- The superior and inferior articular facets, which join adjacent vertebrae, lie mainly in the frontal plane; that is, the superior articular facets face posteriorly, whereas the inferior articular facets face anteriorly (Figure 7.21b). Such articulations limit flexion and extension, but they allow rotation between successive vertebrae. Much of the ability to rotate the trunk comes from the thoracic region of the vertebral column. Lateral flexion is also possible but is restricted by the ribs.

Lumbar Vertebrae

The lumbar region of the vertebral column, the area commonly referred to as the small of the back, receives the most stress. The enhanced weight-bearing function of the five lumbar vertebrae (L_1–L_5) is reflected in their sturdy structure: Their bodies are massive and appear kidney-shaped from a superior view (see Table 7.2). Their other characteristics are as follows:

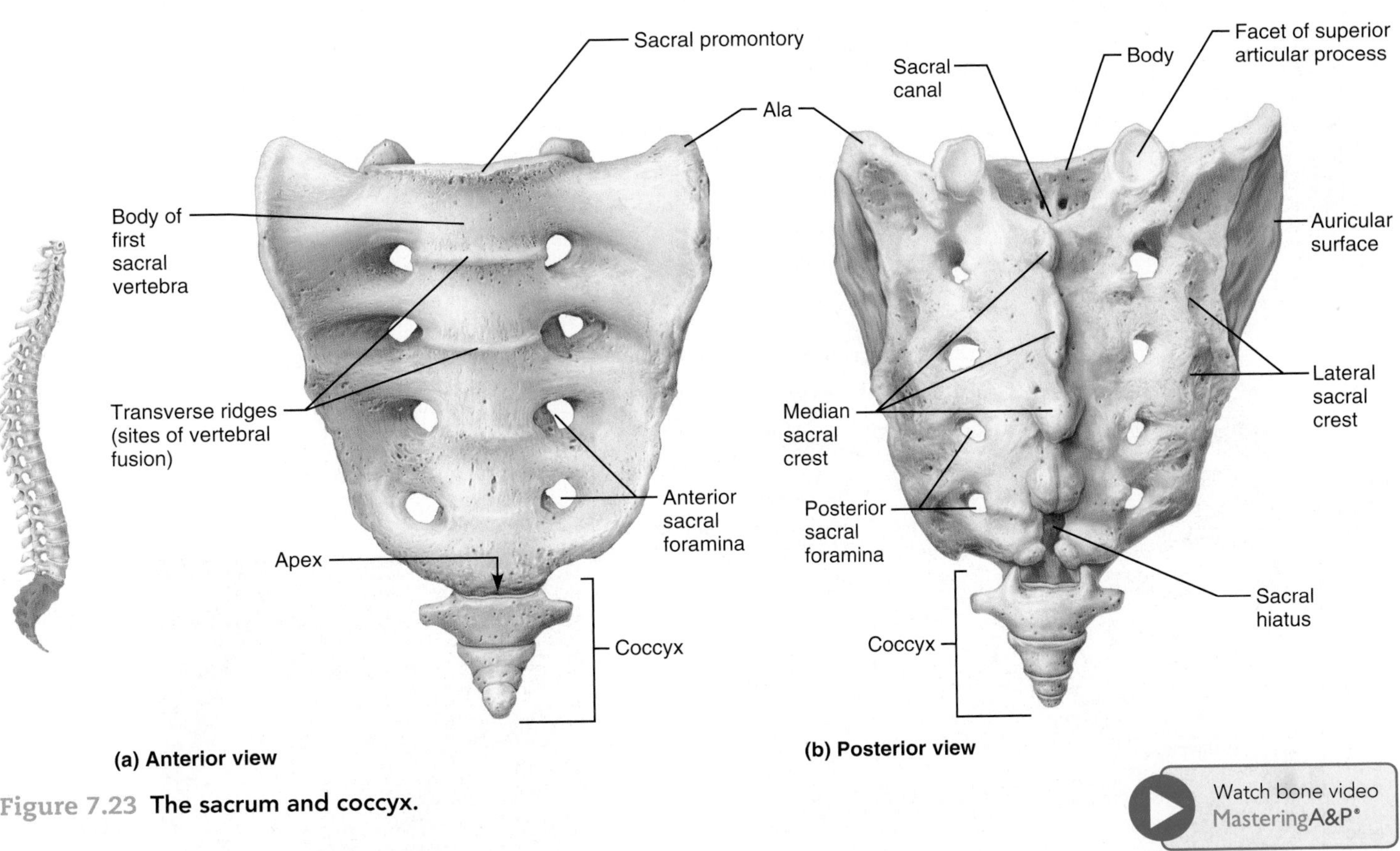

Figure 7.23 **The sacrum and coccyx.**

- The pedicles and laminae are shorter and thicker than those of other vertebrae.
- The spinous processes are short, flat, and hatchet-shaped, and they project straight posteriorly. These processes are robust for the attachment of large back muscles.
- The vertebral foramen is triangular.
- The superior articular facets face posteromedially (or medially), whereas the inferior articular facets face anterolaterally (or laterally) (see Figure 7.21c), oriented approximately in the sagittal plane. Such articulations provide stability by preventing rotation between the lumbar vertebrae while allowing flexion and extension. The lumbar region flexes, for example, when you do sit-ups or bend forward to pick up a coin from the ground. Additionally, lateral flexion is allowed by this spinal region.

Sacrum

The curved, triangular **sacrum** shapes the posterior wall of the pelvis (Figure 7.23). It is formed by five fused vertebrae (S_1–S_5); thus, most sacral features are modifications of the typical vertebral features. Superiorly, the sacrum articulates with L_5 through a pair of **superior articular processes** (Figure 7.23b) and an intervertebral disc. Inferiorly, it joins the coccyx.

The anterosuperior margin of the first sacral vertebra bulges anteriorly into the pelvic cavity as the **sacral promontory** (prom′on-tor-ē, "a high point of land projecting into the sea"). The human body's center of gravity lies about 1 cm posterior to this landmark. Four **transverse ridges** cross the anterior surface of the sacrum, marking the lines of fusion of the sacral vertebrae. The **anterior sacral foramina** transmit the ventral divisions (ventral rami) of the sacral spinal nerves. The large region lateral to these foramina is simply called the *lateral part.* This part expands superiorly into the flaring **ala** (a′lah; "wing"), which develops from fused *rib* elements of S_1–S_5. (Embryonic rib elements form in association with *all* vertebrae, although they only become true ribs in the thorax. Elsewhere, they fuse into the ventral surfaces of the transverse processes.) The alae articulate at the auricular surface with the two hip bones to form the *sacroiliac* (sa″kro-il′e-ak) *joints* of the pelvis.

On the posterior surface, in the midline, is the bumpy **median sacral crest,** which represents the fused spinous processes of the sacral vertebrae. Lateral to it are the **posterior sacral foramina,** which transmit the dorsal rami of the sacral spinal nerves. Just lateral to these is the **lateral sacral crest,** representing the tips of the transverse processes of the sacral vertebrae.

The vertebral canal continues within the sacrum as the **sacral canal.** The laminae of the fifth (and sometimes the fourth) sacral vertebrae fail to fuse medially, leaving an enlarged external opening called the **sacral hiatus** (hi-a′tus; "gap") at the inferior end of the sacral canal.

Coccyx

The **coccyx,** or tailbone, is small and triangular (Figure 7.23). The name *coccyx* is from a Greek word for "cuckoo," and the bone was so named because of a fancied resemblance to a bird's beak. The coccyx consists of three to five vertebrae fused

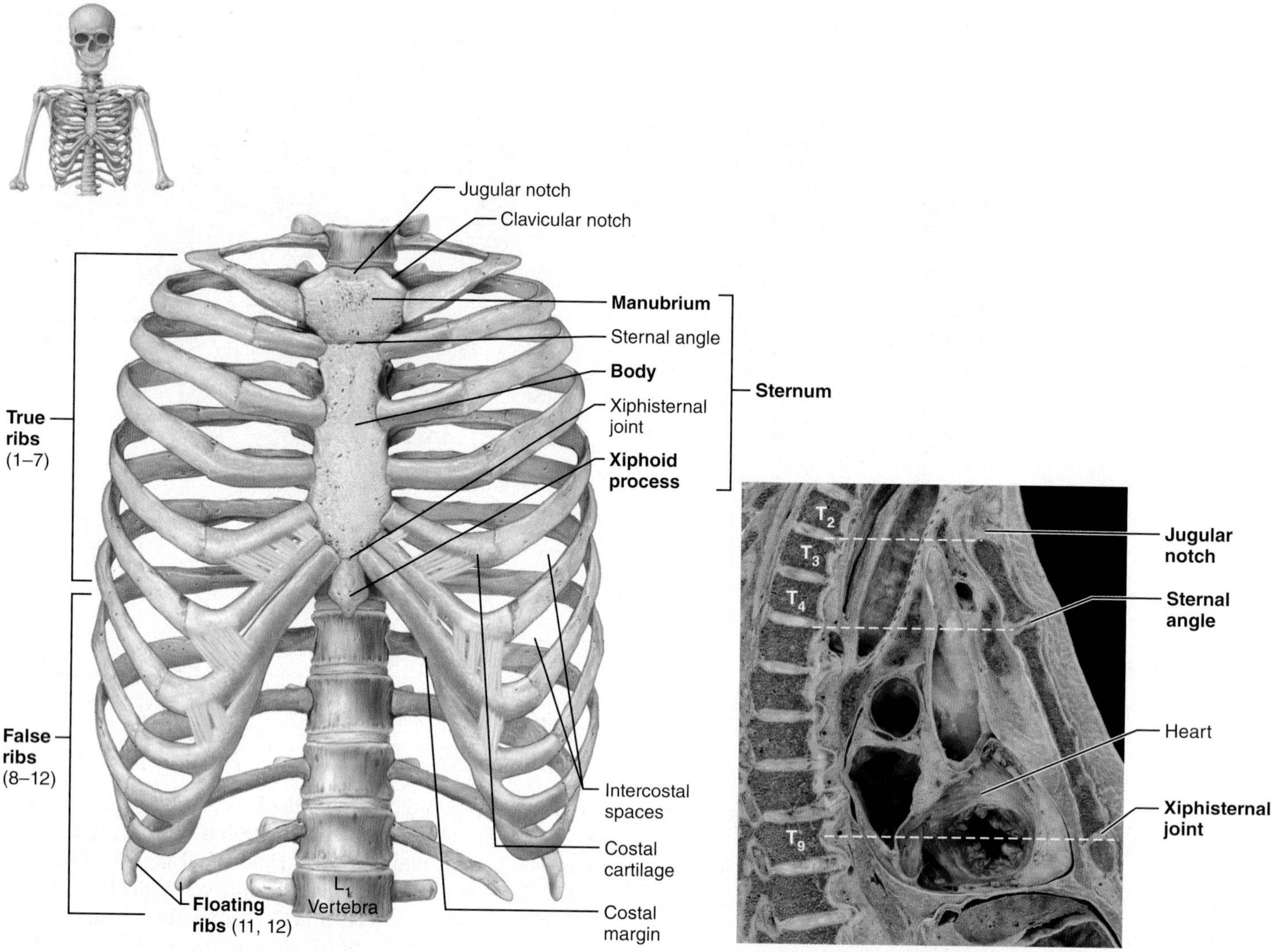

(a) Skeleton of the thoracic cage, anterior view

(b) Midsagittal section through the thorax, showing the relationship of surface anatomical landmarks of the thorax to the vertebral column

Figure 7.24 The thoracic cage.

together. Except for the slight support it affords the pelvic organs, it is an almost useless bone. Injury to the coccyx, such as from an abrupt fall, is extremely painful. Occasionally, a baby is born with an unusually long coccyx. In most such cases, this bony "tail" is discreetly snipped off by a physician.

✓ check your understanding

- ☐ 13. What does the superior articular process of a vertebra articulate with?
- ☐ 14. Name one feature that is unique for all cervical vertebrae.
- ☐ 15. How can you distinguish thoracic vertebra T_{12} from lumbar vertebra L_1?
- ☐ 16. What part of the vertebrae form the median sacral crest?

(For answers, see Appendix B.)

THE THORACIC CAGE

learning outcomes

- ▶ Identify the distinctive features of the ribs and sternum.
- ▶ Differentiate true ribs from false ribs, and explain how they relate to floating ribs.
- ▶ Describe how a typical rib attaches to the vertebral column.

The bony framework of the chest (thorax), called the **thoracic cage,** is roughly barrel-shaped and includes the thoracic vertebrae posteriorly, the ribs laterally, and the sternum and costal cartilages anteriorly (Figure 7.24). The thoracic cage forms a protective cage around the heart, lungs, and other organs. It also supports the shoulder girdles and upper limbs and provides attachment points for many muscles of the back, neck, chest, and shoulders. In addition, the *intercostal spaces* (*inter* = between; *costa* = the ribs) are occupied by the intercostal muscles, which lift and depress the thorax during breathing.

Sternum

The **sternum** (breastbone) lies in the anterior midline of the thorax. Resembling a dagger, it is a flat bone about 15 cm long consisting of three sections: the manubrium, body, and xiphoid process. The **manubrium** (mah-nu′bre-um; "knife handle"), the superior section, is shaped like the knot in a necktie. Its **clavicular notches** articulate with the clavicles (collarbones) superolaterally. Just below this, the manubrium articulates with the first and second ribs. The **body,** or midportion, makes up the bulk of the sternum. It is formed from four separate bones, one inferior to the other, that fuse after puberty. The sides of the sternal body are notched where it articulates with the costal cartilages of the second to seventh ribs. The **xiphoid process** (zif′oid; "sword-like") forms the inferior end of the sternum. This tongue-shaped process is a plate of hyaline cartilage in youth, and it does not fully ossify until about age 40. In some people, the xiphoid process projects posteriorly. Blows to the chest can push such a xiphoid into the underlying heart or liver, causing massive hemorrhage.

The sternum has three important anatomical landmarks that can be palpated: the jugular notch, the sternal angle, and the xiphisternal joint (Figure 7.24a).

- The **jugular notch,** also called the *suprasternal notch,* is the central indentation in the superior border of the manubrium. If you slide your finger down the anterior surface of your neck, it will land in the jugular notch. The jugular notch generally lies in the same horizontal plane as the disc between the second and third thoracic vertebrae (Figure 7.24b).
- Just inferior is the **sternal angle,** a horizontal ridge across the anterior suface of the sternum where the manubrium joins the body. This fibrocartilage joint acts like a hinge, allowing the sternal body to swing anteriorly when we inhale. The sternal angle is in line with the disc between the fourth and fifth thoracic vertebrae. Anteriorly, it lies at the level of the second ribs. It is a handy reference point for finding the second rib. Once the second rib is located, you can count down to identify all the other ribs (except the first and sometimes the twelfth, which are too deep to be palpated). By locating the individual ribs, you attain a series of horizontal lines of "latitude" by which to locate the underlying visceral organs of the thoracic cavity.
- The **xiphisternal** (zif″ĭ-ster′nal) **joint** is where the sternal body and xiphoid process fuse. It lies at the level of the ninth thoracic vertebra. Deep to this joint, the heart lies on the diaphragm.

Ribs

Twelve pairs of **ribs** (Latin: *costa)* form the flaring sides of the thoracic cage (see Figure 7.24a). All ribs attach to the thoracic vertebrae posteriorly and run anteroinferiorly to reach the front of the chest. The superior seven pairs, which attach directly to the sternum by their *costal cartilages,* are the **true ribs,** or *vertebrosternal* (ver″tĕ-bro-ster′nal) *ribs.* The inferior five pairs, ribs 8–12, are called **false ribs** because they attach to the sternum either indirectly or not at all. Ribs 8–10 attach to the sternum indirectly, as each joins the costal cartilage above it; these are called *vertebrochondral* (ver″tĕ-bro-kon′dral) *ribs.* Ribs 11 and 12 are called **floating ribs,** or *vertebral ribs,* because they have no anterior attachments. Instead, their costal cartilages lie embedded in the muscles of the lateral body wall. The ribs increase in length from pair 1 to 7, then decrease in length from pair 8 to 12. The inferior margin of the rib cage, or **costal margin,** is formed by the costal cartilages of ribs 7 to 10. The right and left costal margins diverge from the region of the xiphisternal joint, where they form the **infrasternal angle** (*infra* = below).

A typical rib is a bowed flat bone (Figure 7.25). The bulk of a rib is simply called the **shaft** or **body.** Its superior

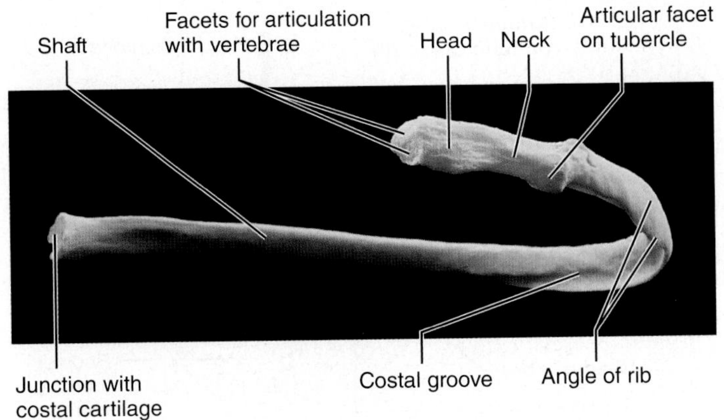

(a) A typical rib (rib 6, right), posterior view

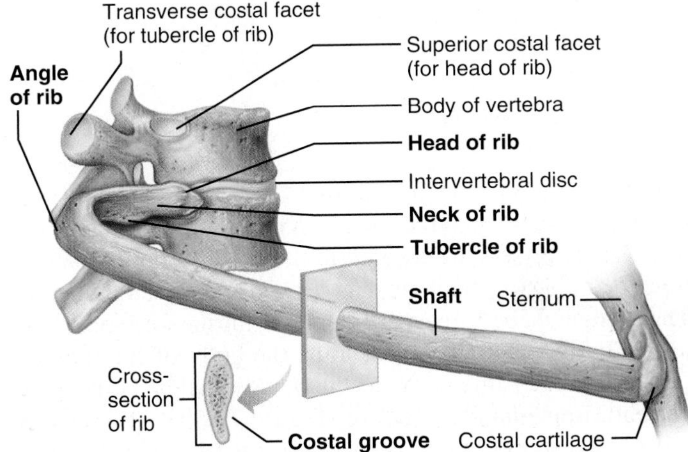

(b) Vertebral and sternal articulations of a typical true rib

Figure 7.25 Ribs. All ribs illustrated in this figure are right ribs.

Articular facet on tubercle of rib
Spinous process
Shaft
Ligaments
Transverse costal facet (for tubercle of rib)
Neck of rib
Head of rib
Body of thoracic vertebra
Superior costal facet (for head of rib)

(c) Superior view of the articulation between a rib and a thoracic vertebra

Figure 7.25 **Ribs,** *continued.*

border is smooth, but its inferior border is sharp and thin and has a **costal groove** on its inner face. The intercostal nerves and vessels are located in the costal groove. In addition to the shaft, each rib has a head, neck, and tubercle. The wedge-shaped **head** articulates with the vertebral bodies by two facets: One facet joins the body of the thoracic vertebra of the same number; the other joins the body of the vertebra immediately superior (Figure 7.25b). The **neck** of a rib is the short, constricted region just lateral to the head. Just lateral to the neck on the posterior surface, the knoblike **tubercle** articulates with the transverse process of the thoracic vertebra of the same number (Figure 7.25c). Lateral to the tubercle at the **angle** of the rib, the shaft curves sharply anteriorly and extends to the costal cartilage anteriorly. The costal cartilages provide secure but flexible attachments of ribs to the sternum and contribute to the elasticity of the thoracic cage.

The first rib is atypical because it is flattened from superior to inferior and is quite broad (Figure 7.24a). The subclavian vessels, the large artery and vein servicing the upper limb, run in a groove along its superior surface. There are other exceptions to the typical rib pattern: Rib 1 and ribs 10–12 articulate with only one vertebral body, and ribs 11 and 12 do not articulate with a vertebral transverse process.

☑ check your understanding

☐ 17. Define the sternal angle. Which rib articulates with the sternum at this landmark?

☐ 18. What specific features of the thoracic vertebrae articulate with the head of a rib? Where does the tubercle of a rib articulate?

(For answers, see Appendix B.)

DISORDERS OF THE AXIAL SKELETON

learning outcomes

▶ Describe the causes and consequences of cleft palate.

▶ List three types of abnormal curvatures of the spinal column, and explain spinal stenosis.

Cleft Palate

Several congenital abnormalities may distort the skull. Most common is **cleft palate,** in which the right and left halves of the palate fail to join medially. This defect leaves an opening between the mouth and the nasal cavity that interferes with sucking and the baby's ability to nurse. Cleft palate can lead to aspiration (inhalation) of food into the nasal cavity and lungs, resulting in pneumonia. Often accompanied by cleft lip (Figure 7.26), it is repaired surgically. Recently it has been found that the likelihood of cleft palate is minimized if the mother takes multivitamins containing folic acid during early pregnancy.

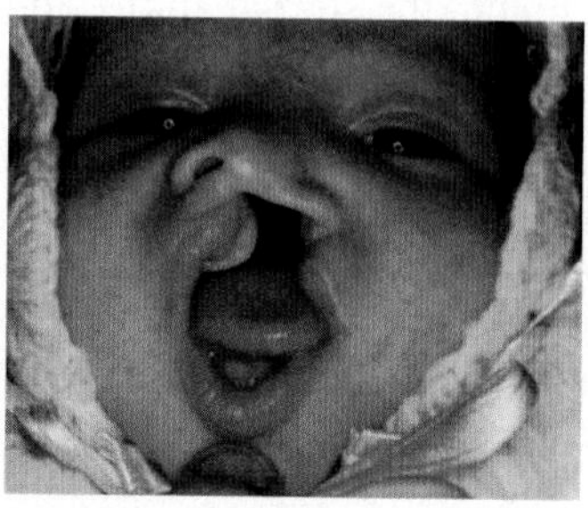

(a) A boy born with a cleft palate and lip

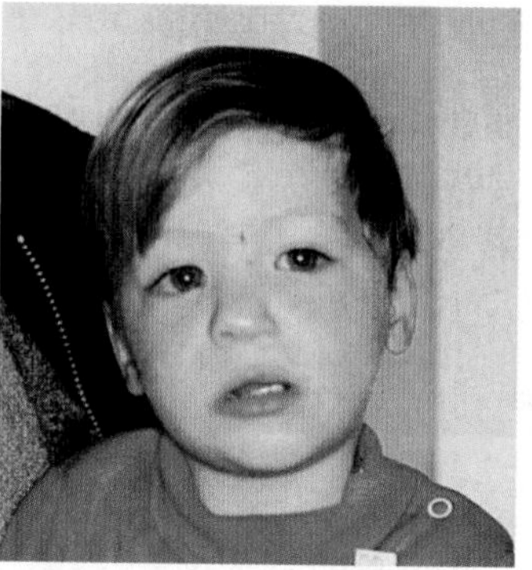

(b) The boy as a toddler, following surgical repair during infancy

Figure 7.26 **Cleft palate.**

Stenosis of the Lumbar Spine

The condition known as **stenosis of the lumbar spine** (ste-no′-sis; "narrowing") is a narrowing of the vertebral canal in the lumbar region. This disorder can compress the roots of the spinal nerves and cause back pain. The narrowing may result from degenerative or arthritic changes in the vertebral joints and ligaments or be caused by bone spurs projecting into the vertebral canal.

Abnormal Spinal Curvatures

There are several types of abnormal spinal curvatures. Some are congenital (present at birth), whereas others result from disease, poor posture, or unequal pull of muscles on the spine.

Scoliosis (sko″le-o′sis; "twisted disease") is an abnormal *lateral* curvature of more than 10 degrees that occurs most often in the thoracic region (Figure 7.27a). One type, of

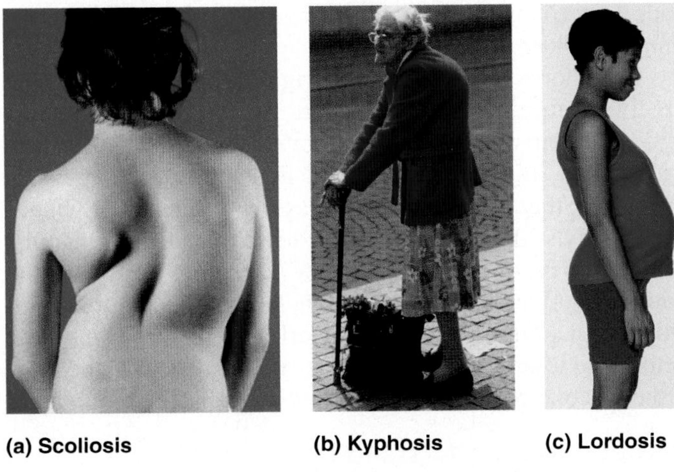

Figure 7.27 **Abnormal spinal curvatures.**

unknown cause, is common in adolescents, particularly girls. Other, more severe cases result from abnormally structured vertebrae, lower limbs of unequal length, or muscle paralysis. If muscles on one side of the back are nonfunctional, those on the opposite side pull unopposed and move the spine out of alignment. To prevent permanent deformity, clinicians treat scoliosis with body braces or surgery before the child stops growing. Severe scoliosis can compress a lung, causing difficulty in breathing.

Kyphosis (ki-fo′sis; "humped disease"), or hunchback, is an exaggerated thoracic curvature that is most common in aged women because it often results from spinal fractures that follow osteoporosis (Figure 7.27b). It may also result from either tuberculosis of the spine or osteomalacia.

Lordosis (lor-do′sis: "bent-backward disease"), or swayback, is an accentuated lumbar curvature (Figure 7.27c). Temporary lordosis is common in people carrying a "large load in front," such as obese men and pregnant women. Like kyphosis, it can result from spinal tuberculosis or osteomalacia.

THE AXIAL SKELETON THROUGHOUT LIFE

learning outcome

▶ Describe how the axial skeleton changes as we grow.

The membranous bones of the skull begin to ossify late in the second month of development. In these flat bones, bone tissue grows outward from **ossification centers** within the mesenchyme membranes. At birth, the skull bones remain incomplete and are separated by still-unossified remnants of the membranes, called **fontanelles** (fon″tah-nelz′) (Figure 7.28). The major fontanelles—the **anterior, posterior, mastoid,** and **sphenoidal fontanelles**—allow the skull to be compressed slightly as the infant passes through the narrow birth canal, and they accommodate brain growth in the baby. A baby's pulse surging in these "soft spots" feels like gushing water (*fontanelle* means "little fountain"). The large, diamond-shaped anterior fontanelle can be felt for 1½ to 2 years after birth. The others usually are replaced by bone by the end of the first year.

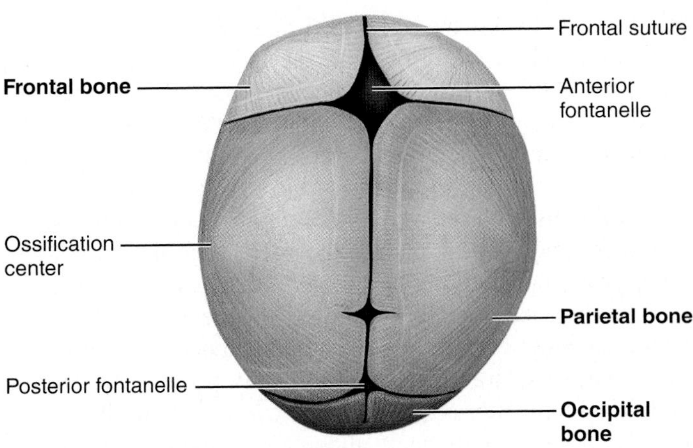

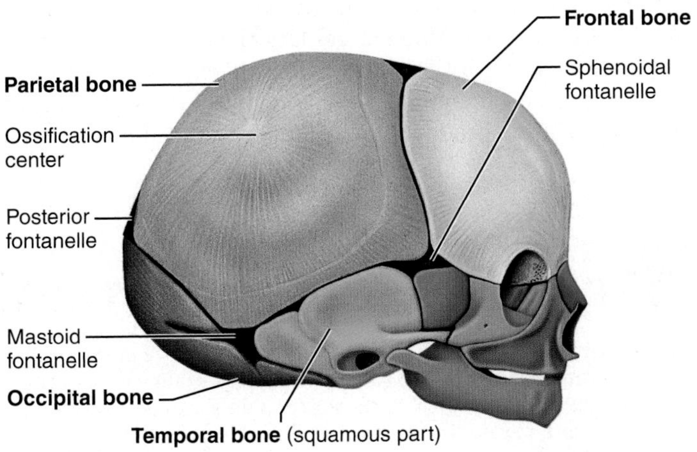

Figure 7.28 **Skull of a newborn.** Note the fontanelles and ossification centers.

Whereas the bones of the calvaria and face form directly from mesenchyme by intramembranous ossification, those at the base of the skull develop from cartilage by endochondral ossification. More specifically, the skull's endochondral bones are the occipital (except for its extreme posterosuperior part), the petrous part of the temporal, most of the sphenoid, and the ethmoid.

At birth, the skull bones are thin. The frontal bone and the mandible begin as paired bones that fuse medially during childhood. The tympanic part of the temporal bone is merely a C-shaped ring in the newborn.

The skull changes throughout life, but the changes are most dramatic during childhood. At birth, the baby's cranium is huge relative to its small face (see Figure 7.28). The maxillae and mandible are comparatively tiny, and the contours of the face are flat. By 9 months of age, the skull is already half its adult size. By 2 years, it is three-quarters of its full size, and by 8 or 9 years, the cranium is almost full-sized.

However, between the ages of 6 and 13, the head appears to enlarge substantially because the face literally grows out from the skull as the jaws, cheekbones, and nose become more prominent. The enlargement of the face correlates with an expansion of the nose, paranasal sinuses, and chewing muscles, plus the development of the large permanent teeth.

Problems with the vertebral column, such as lordosis or scoliosis, may appear during the early school years, when rapid growth of the long limb bones stretches many muscles. Children normally have a slight lordosis because holding the abdomen anteriorly helps to counterbalance the weight of their relatively large heads, held slightly posteriorly. During childhood, the thorax grows wider, but true adult posture (head erect, shoulders back, abdomen in, and chest out) does not develop until adolescence.

Aging affects many parts of the skeleton, especially the spine. The water content of the intervertebral discs declines. As the discs become thinner and less elastic, the risk of herniation increases. By 55 years, a loss of several centimeters from a person's height is common. Further shortening of the trunk can be produced by osteoporosis of the spine that leads to kyphosis, which is sometimes called "dowager's hump" in older adults.

The thorax becomes more rigid with increasing age, largely because the costal cartilages ossify. This loss of elasticity of the rib cage leads to shallow breathing, which in turn leads to less efficient gas exchange in the lungs.

Recall that all bones lose mass with age. Skull bones lose less mass than most, but they lose enough to change the facial contours of the older person. As the bony tissue of the mandible and maxilla decreases, the jaws come to look small and childlike once again. If an elderly person loses his or her teeth, bone loss from the jaws is accelerated because the bone of the alveolar region (tooth sockets) is reabsorbed.

☑ check your understanding

☐ **19.** Which skull bones form as two individual bones that fuse during childhood to form a single bone in the adult?

☐ **20.** Which abnormal spinal curvature is associated with age-related degenerative bone loss?

(For answers, see Appendix B.)

RELATED CLINICAL TERMS

Craniotomy ("cutting the skull") Surgery to remove part of the cranium, usually done to gain access to the brain—for example, to remove a brain tumor, a blood clot, or a sample of brain tissue for a biopsy. The piece of skull is removed by drilling a series of round holes through the bone at regular intervals to outline a square and then cutting between these holes with a stringlike saw. At the end of the operation, the square piece of bone is replaced, and it heals normally.

Laminectomy (lam″ĭ-nek′to-me; "lamina-cutting") Surgical removal of a vertebral lamina. Laminectomies are usually done to reach and relieve a herniated disc.

Orthopedist (or″tho-pe′dist) or **Orthopedic Surgeon** A physician who specializes in restoring lost function or repairing damage to bones and joints.

Spinal fusion Surgical procedure involving insertion of bone chips as a tissue graft to stabilize a portion of the vertebral column, particularly after the fracture of a vertebra or the prolapse of a disc.

Whiplash An injury to the neck region caused by a rapid sequence of hyperextension followed by hyperflexion of the neck. Often results when a car is hit from the rear. Damage and pain usually result from stretching of the ligaments and joints between the neck vertebrae, and torn neck muscles. Treatment typically involves use of a neck brace, painkillers, muscle-relaxing drugs, and physical therapy.

CHAPTER SUMMARY

Take full advantage of the invaluable online practice and study tools for this chapter by going to the Study Area of MasteringA&P®. There you will find:

- Art Labeling Activities, Quizzes, Practice Tests, Memory Games
- Animal Organ Dissection, Cat Dissection and Bone Videos
- PAL Cadaver
- PAL Histology
- A&P Flix Animations
- Dynamic Study Modules

1. The axial skeleton forms the long axis of the body. Its parts are the skull, vertebral column, and thoracic cage. The appendicular skeleton, by contrast, consists of the pectoral (shoulder) and pelvic girdles and the bones of the limbs.

THE SKULL (pp. 152–170)

2. The cranium forms the vault and base of the skull and protects the brain. The facial skeleton provides openings for the eyes, openings for respiratory and digestive passages, and areas of attachment for facial muscles.
3. Almost all bones of the skull are joined by immobile sutures.
4. The eight bones of the cranium are the paired parietal and temporal bones and the unpaired frontal, occipital, sphenoid, and ethmoid bones (see Table 7.1, pp. 164–165).
5. The 14 facial bones are the paired maxillae, zygomatics, nasals, lacrimals, palatines, and inferior nasal conchae, plus the unpaired mandible and vomer (see Table 7.1).

6. The nasal cavity and orbits are complicated regions of the skull, each formed by many bones (see Figures 7.14 and 7.16).

7. The air-filled paranasal sinuses occur in the frontal, ethmoid, sphenoid, and maxillary bones. They are extensions of the nasal cavity, to which they are connected.

8. The bowed hyoid bone in the anterior neck serves as a movable base for the tongue, among other functions.

THE VERTEBRAL COLUMN (pp. 170–178)

9. The vertebral column includes 26 irregular bones: 24 vertebrae plus the sacrum and coccyx.

10. The vertebral column has five major regions: the cervical vertebrae (7), the thoracic vertebrae (12), the lumbar vertebrae (5), the sacrum, and the coccyx. The cervical and lumbar curvatures are concave posteriorly, and the thoracic and sacral curvatures are convex posteriorly.

11. The thoracic and sacral curvatures are primary curvatures, present at birth. The cervical curvature becomes pronounced when a baby starts to lift its head, and the lumbar curvature develops when a toddler starts to walk. These are secondary curvatures.

12. The spinal column is supported by a number of ligaments, including the anterior and posterior longitudinal ligaments, which run along the bodies of the vertebrae from the neck to the sacrum; and the ligamentum flavum, an especially strong elastic ligament running between the lamina of adjacent vertebrae.

13. Intervertebral discs, with their nucleus pulposus cores and anulus fibrosus rings, act as shock absorbers and give flexibility to the spine. Herniated discs usually involve rupture of the anulus followed by protrusion of the nucleus pulposus.

14. With the exception of C_1, all cervical, thoracic, and lumbar vertebrae have a body, two transverse processes, two superior and two inferior articular processes, a spinous process, and a vertebral arch.

15. The atlas and axis (C_1 and C_2) are atypical vertebrae. The ringlike atlas supports the skull and helps make nodding movements possible. The axis has a dens that enables the head to rotate.

16. Special characteristics distinguish the cervical, thoracic, and lumbar vertebrae from one another (see Table 7.2, pp. 175–176). The orientation of the articular facets determines the movements possible in the different regions of the spinal column. The sacrum and coccyx, groups of fused vertebrae, contribute to the posterior wall of the pelvis.

THE THORACIC CAGE (pp. 178–180)

17. The thoracic cage includes the 12 pairs of ribs, the sternum, and the thoracic vertebrae.

18. The sternum consists of the manubrium, body, and xiphoid process. Important landmarks are the jugular notch, sternal angle, and xiphisternal joint.

19. The first seven pairs of ribs are called true ribs; the rest are false ribs. The eleventh and twelfth are called floating ribs. A typical rib consists of a head with facets, a neck, a tubercle, and a shaft. A costal cartilage occurs on the ventral end of each rib.

PAL Human Cadaver/Axial Skeleton

DISORDERS OF THE AXIAL SKELETON (pp. 180–181)

20. Cleft palate, a congenital malformation, results when the bones of the hard palate do not fuse medially. Stenosis of the lumbar spine is a pathological narrowing of the vertebral canal in the lumbar region. Scoliosis, kyphosis, and lordosis are three types of abnormal spinal curvatures.

THE AXIAL SKELETON THROUGHOUT LIFE (pp. 181–182)

21. Fontanelles are membrane-covered regions between the cranial bones in the infant skull. They allow growth of the brain and ease birth passage.

22. Most skull bones are membranous bones. The skull's only endochondral bones are most of the occipital and sphenoid, the petrous part of the temporal, and the ethmoid.

23. Growth of the cranium early in life is closely related to the rapid growth of the brain. Enlargement of the face in late childhood follows tooth development and expansion of the nose, paranasal sinuses, and chewing muscles.

24. In old age, the intervertebral discs thin. This, along with osteoporosis, leads to a gradual decrease in height.

REVIEW QUESTIONS

Multiple Choice/Matching Questions

(For answers, See Appendix B.)

1. Match bones in column B to descriptions in column A. (Some entries in column A require more than one answer from column B.)

Column A	Column B
____ (1) bones connected by the coronal suture	(a) ethmoid
____ (2) keystone bone of the cranium	(b) frontal
____ (3) keystone bone of the face	(c) mandible
____ (4) bones that form the hard palate	(d) maxillary
____ (5) bone that contains the foramen magnum	(e) occipital
____ (6) forms the chin	(f) palatine
____ (7) contain paranasal sinuses	(g) parietal
____ (8) contains mastoid air cells	(h) sphenoid
	(i) temporal

2. Herniated intervertebral discs tend to herniate (a) superiorly, (b) anterolaterally, (c) posterolaterally, (d) laterally.

3. The parts of the sternum that articulate at the sternal angle are (a) xiphoid and body, (b) xiphoid and manubrium, (c) manubrium and body, (d) second ribs.

4. The only rib whose shaft is flattened in the horizontal plane, instead of vertically, is the (a) first rib, (b) seventh rib, (c) eleventh rib, (d) twelfth rib.

5. The name of the first cervical vertebra is (a) atlas, (b) axis, (c) occiput, (d) vertebra prominens.

6. Match the vertebrae listed in column B with the features in column A.

Column A	Column B
____ (1) no spinous process	(a) atlas only
____ (2) transverse foramen	(b) axis only
____ (3) superior articular facets	(c) cervical vertebrae
____ (4) dens process	(d) thoracic vertebrae
____ (5) transverse costal facets	(e) lumbar vertebrae
____ (6) kidney-shaped body	(f) all vertebrae
____ (7) forked spinous process	
____ (8) circular vertebral foramen	
____ (9) transverse process	
____ (10) articular facets directed medially/laterally	

7. Match the foramen in column B with the bone in column A in which it is located.

Column A	Column B
____ (1) ethmoid bone	(a) hypoglossal canal
____ (2) maxillary bone	(b) foramen ovale
____ (3) sphenoid bone	(c) external acoustic meatus
____ (4) occipital bone	(d) infraorbital foramen
____ (5) temporal bone	(e) cribriform foramina

Short Answer Essay Questions

8. In the fetus, how do the relative proportions of the cranium and face compare with those in the skull of a young adult?
9. Name and diagram the four normal vertebral curvatures. Which are primary and which are secondary?
10. List two specific structural characteristics each for cervical, thoracic, and lumbar vertebrae that would enable anyone to identify each type correctly.
11. (a) What is the function of intervertebral discs? (b) Distinguish the anulus fibrosus from the nucleus pulposus of a disc. (c) Which part herniates in the condition called prolapsed disc?
12. Is a floating rib a true rib or a false rib? Explain.
13. Briefly describe the anatomical characteristics and impairment of function seen in cleft palate.
14. Compare the skeleton of a young adult to that of an 85-year-old person, with respect to bone mass in general and the basic bony structure of the skull, thorax, and vertebral column.
15. Identify what types of movement are allowed by the lumbar region of the vertebral column, and compare these with the movements allowed by the thoracic region.
16. List the bones in each of the three cranial fossae.
17. Describe the important features of the sternum.
18. Define the following structures in a way that shows you can tell them apart: vertebral arch, vertebral canal, vertebral foramen, and intervertebral foramen.
19. Describe where the four major fontanelles are located in relation to the major sutures of the skull.
20. Professor Ron Singer pointed to the foramen magnum of the skull and said, "That is where the food goes. The food passes down through this hole when you swallow." Some of the students believed him, but others said this was a big mistake. Can you correct his statement?
21. In your anatomy course, you may be handed an isolated vertebra and asked to determine whether it is cervical, thoracic, or lumbar. Devise a reliable scheme for distinguishing these three types of vertebrae.
22. Describe how a typical true rib (for instance, the fifth rib) articulates with both the vertebral column and the sternum.

Critical Reasoning & Clinical Application Questions

1. Antonio was hit in the face with a bad-hop grounder during baseball practice. An X-ray film revealed multiple fractures of the bones around an orbit. Name the bones that form the margins of an orbit.
2. Lindsey had polio as a child and was partially paralyzed in one lower limb for over a year. Although she is no longer paralyzed, she now has a severe lateral curvature of the lumbar spine. Explain what has happened, and name her condition.
3. Mr. Chester, a heavy beer drinker with a large potbelly, complained of severe lower back pains. After X-ray films were taken, he was found to have displacement of his lumbar vertebrae. What would this condition be called, and what would cause it?
4. Which region of the vertebral column (cervical, thoracic, or lumbar) is most likely to experience a herniated disc?
5. After falling off a horse, Mary complained of pain on the right side of her thorax that intensified when she took a deep breath or coughed. The emergency medical technician suspected a broken rib. Using palpation, how could the EMT identify the precise location of the injury?

Bones, Part 2: The Appendicular Skeleton

8

Limb bones and their girdles are appended, or attached, to the axial skeleton. Thus, they are collectively called the **appendicular skeleton** (see Figure 7.1, p. 151). The pectoral girdles (pek′tor-al; "chest") attach the upper limbs to the trunk, whereas the pelvic girdle secures the lower limbs. Although the bones of the upper and lower limbs differ in their functions, they share the same basic structural plan. That is, each limb is composed of three basic segments: the hand, forearm, and arm in the upper limb; and the foot, leg, and thigh in the lower limb.

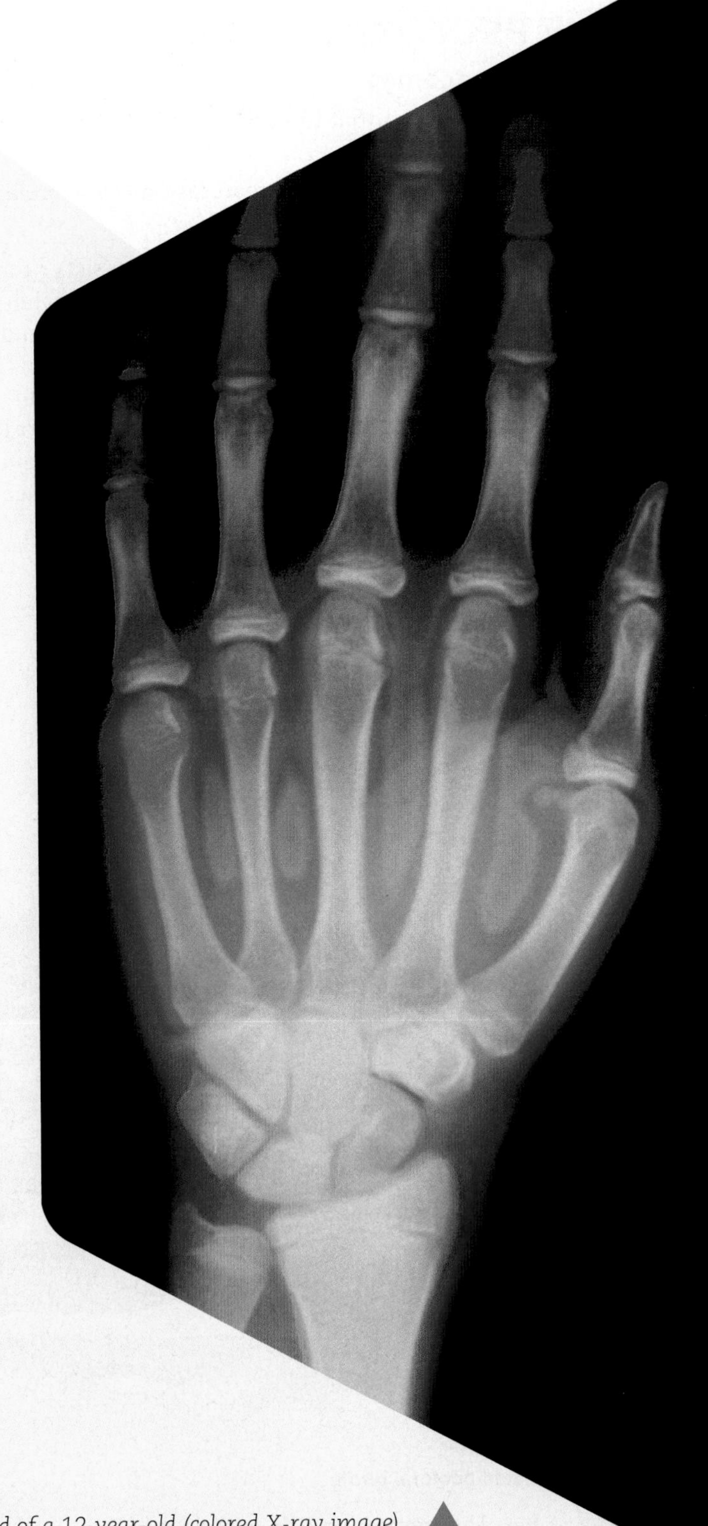

Hand of a 12-year-old (colored X-ray image).

The appendicular skeleton enables people to carry out the wide variety of movements typical of their active, mobile, and object-manipulating lifestyle. Each time you take a step, throw a ball, write with a pencil, or even drive a car, you use your appendicular skeleton.

The bones of the appendicular skeleton are illustrated and described in detail in this chapter. We begin with the pectoral girdle and upper limb.

THE PECTORAL GIRDLE

learning outcomes

- ▶ Identify the bones that form the pectoral girdle, and explain their functions.
- ▶ Identify the important bone features on the scapula and clavicle.

The **pectoral girdle,** or *shoulder girdle,* consists of a *clavicle* (klav′ĭ-kl) anteriorly and a *scapula* (skap′u-lah) posteriorly (Figure 8.1). The paired pectoral girdles and their associated muscles form the shoulders. The term *girdle* implies a belt completely circling the body, but these girdles do not quite satisfy this description: Anteriorly, the medial end of each clavicle joins to the sternum and first rib, and the lateral ends of the clavicles join to the scapulae at the shoulder. However, the two scapulae fail to complete the ring posteriorly, because their medial borders do not join to each other or the axial skeleton.

Besides attaching the upper limb to the trunk, the pectoral girdle provides attachment for many muscles that move the limb. This girdle is light and allows the upper limbs to be quite mobile. This mobility results from two factors:

1. Because only the clavicle attaches to the axial skeleton, the scapula can move quite freely across the thorax, allowing the arm to move with it.
2. The socket of the shoulder joint—the scapula's glenoid cavity—is shallow, so it does not restrict the movement of the humerus (arm bone). Although this arrangement is good for flexibility, it is bad for stability: Shoulder dislocations are fairly common.

Clavicles

The **clavicles** ("little keys"), or collarbones, are slender, S-shaped bones that extend horizontally across the superior thorax on the anterior surface (Figure 8.1a). The cone-shaped **sternal end** attaches to the manubrium medially, and the flattened **acromial** (ah-kro′me-al) **end** articulates with the scapula laterally (Figure 8.1b and c). The medial two-thirds of the clavicle is convex anteriorly; you can

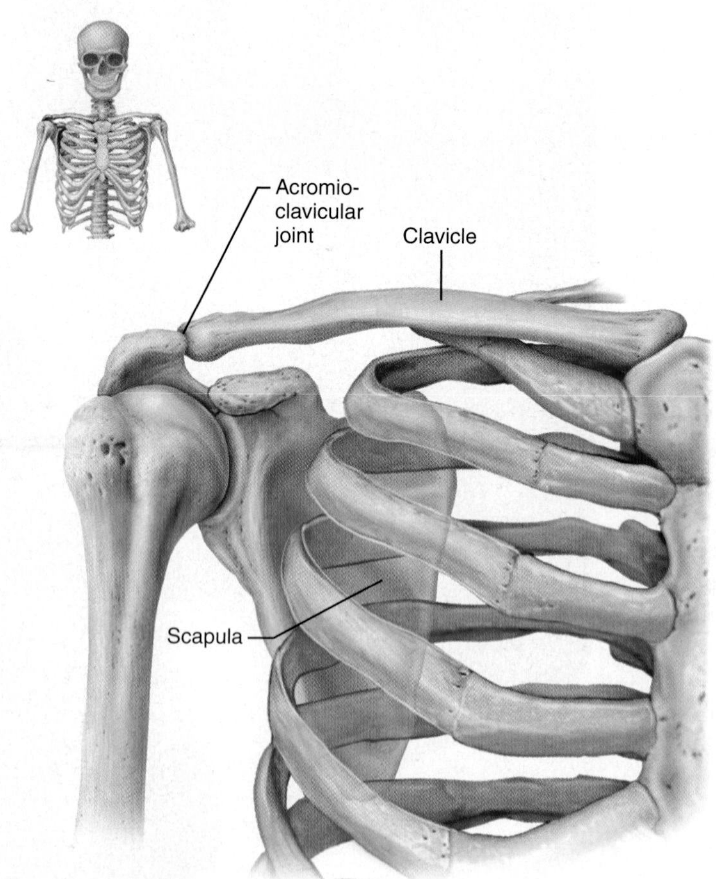

(a) Articulated pectoral girdle

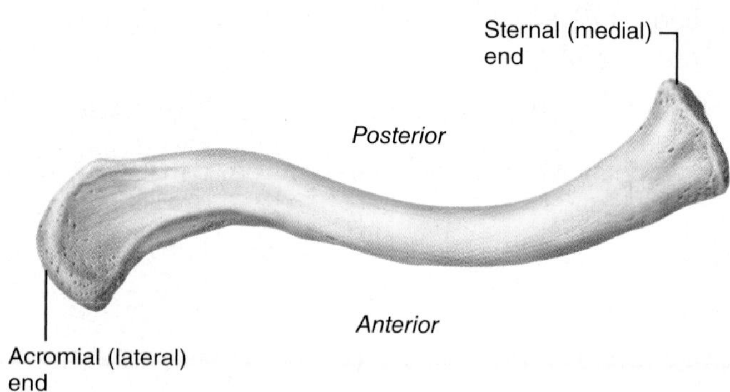

(b) Right clavicle, superior view

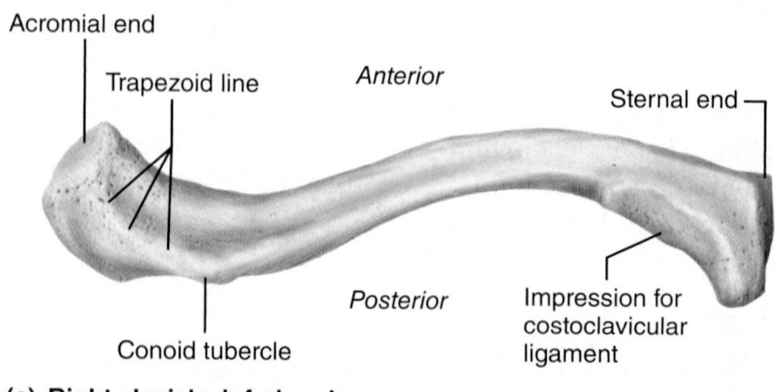

(c) Right clavicle, inferior view

Figure 8.1 The pectoral girdle and clavicle.

feel this anterior projection on yourself when you palpate the clavicle. The lateral third is concave anteriorly. The superior surface is almost smooth, but the inferior surface is ridged and grooved for the ligaments and muscles that attach the clavicle to the rib cage and scapula. The thick **trapezoid line** and the **conoid tubercle** near the acromial end (see Figure 8.1c) provide attachment for a ligament that runs to the scapula's coracoid process (defined below), and a roughened area near the sternal end is the attachment site of the costoclavicular ligament, a ligament that connects the clavicle to the first rib.

The clavicles perform several functions. Besides providing attachment for muscles, they act as braces that hold the scapulae and arms out laterally from the thorax. This function becomes obvious when a clavicle is fractured: The entire shoulder region collapses medially. The clavicles also transmit compression forces from the upper limbs to the axial skeleton, as when someone puts both arms forward and pushes a car to a gas station.

CLINICAL APPLICATION

Fractures of the Clavicle The clavicles are not very strong, and they often fracture. This can occur when a person falls on the lateral border of a shoulder, is hit directly on the clavicle in a contact sport, or uses outstretched arms to break a fall. A fractured clavicle is also a common injury in automobile accidents in which the occupants are wearing seat belts. The curves in the clavicle ensure that it usually fractures anteriorly (outward) at its middle third. If it were to fracture posteriorly (inward), bone splinters would pierce the main blood vessels to the arm, the subclavian vessels, which lie just deep to the clavicle.

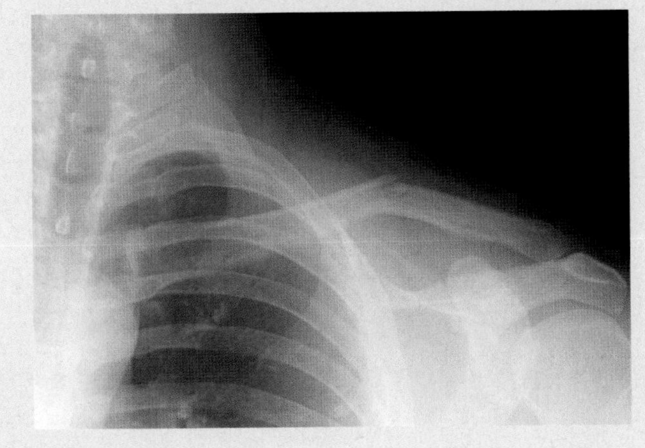

Scapulae

The **scapulae,** or shoulder blades, are thin, triangular flat bones (Figure 8.2) located on the dorsal surface of the rib cage, between rib 2 superiorly and rib 7 inferiorly. Each scapula has three borders. The **superior border** is the shortest and sharpest. The **medial border,** or *vertebral border,* parallels the vertebral column. The thick **lateral border,** or *axillary border,* abuts the axilla (armpit) and ends superiorly in a shallow fossa, the **glenoid cavity** (gle′noid; "pit-shaped") (Figure 8.2c). This cavity articulates with the humerus, forming the shoulder joint.

Like all triangles, the scapula has three corners, or *angles.* The glenoid cavity lies at the scapula's **lateral angle.** The **superior angle** is where the superior and medial borders meet, and the **inferior angle** is at the junction of the medial and lateral borders. The inferior angle moves as the arm is raised and lowered, and it is an important landmark for studying scapular movements.

The anterior, or costal, surface of the scapula is slightly concave and relatively featureless. The **coracoid** (kor′ah-coid) **process** projects anteriorly from the lateral part of the superior scapular border. The root *corac* means "like a crow's beak," but this process looks more like a bent finger. It is an attachment point for the biceps muscle of the arm. Strong ligaments also bind the coracoid process to the clavicle. Just medial to the coracoid process lies the **suprascapular notch** (passageway for the suprascapular nerve), and just lateral to it lies the glenoid cavity.

The posterior surface bears a prominent **spine** that is easily felt through the skin. The spine ends laterally in a flat projection, the **acromion** (ah-kro′me-on; "apex of shoulder"), which articulates with the acromial end of the clavicle.

Several large fossae occur on both surfaces of the scapula and are named according to location. The **infraspinous** ("below the spine") and **supraspinous** ("above the spine") **fossae** lie inferior and superior to the scapular spine, respectively (Figure 8.2b). The **subscapular** ("under the scapula") **fossa** is the shallow concavity formed by the entire anterior surface of the scapula (Figure 8.2a). Lying within these fossae are muscles with similar names, *infraspinatus, supraspinatus,* and *subscapularis.*

check your understanding

- ☐ 1. Which part of the scapula articulates with the clavicle?
- ☐ 2. How is the pectoral girdle attached to the axial skeleton?
- ☐ 3. Name the three fossae of the scapula, and describe their location.

(For answers, see Appendix B.)

THE UPPER LIMB

learning outcome

- ▶ Identify the bones of the upper limb and their important features, and describe how they articulate with each other.

Thirty bones form the skeleton of the upper limb (see Figures 8.3 to 8.6). They are grouped into bones of the arm, forearm, and hand (**Table 8.1**, p. 193).

Arm

Anatomists use the term *arm* or *brachium* (bra′ke-um) to designate the part of the upper limb between the shoulder and elbow only. The **humerus** (hu′mer-us) is the only bone of the

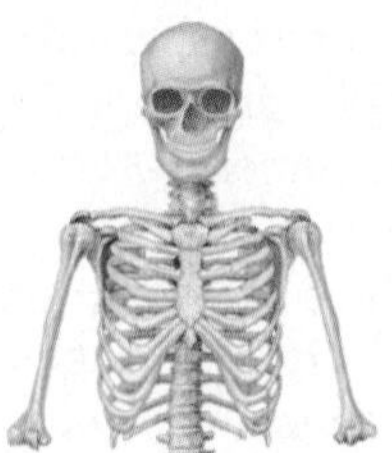

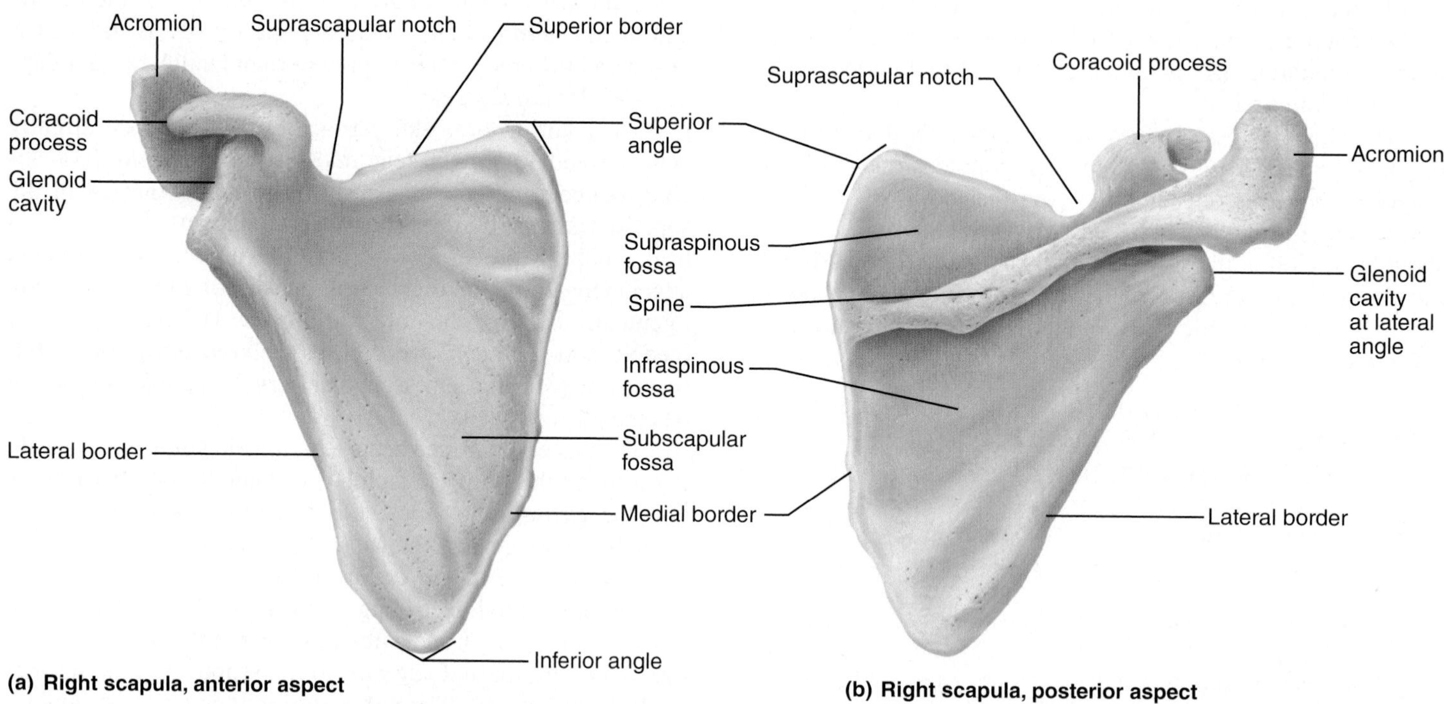

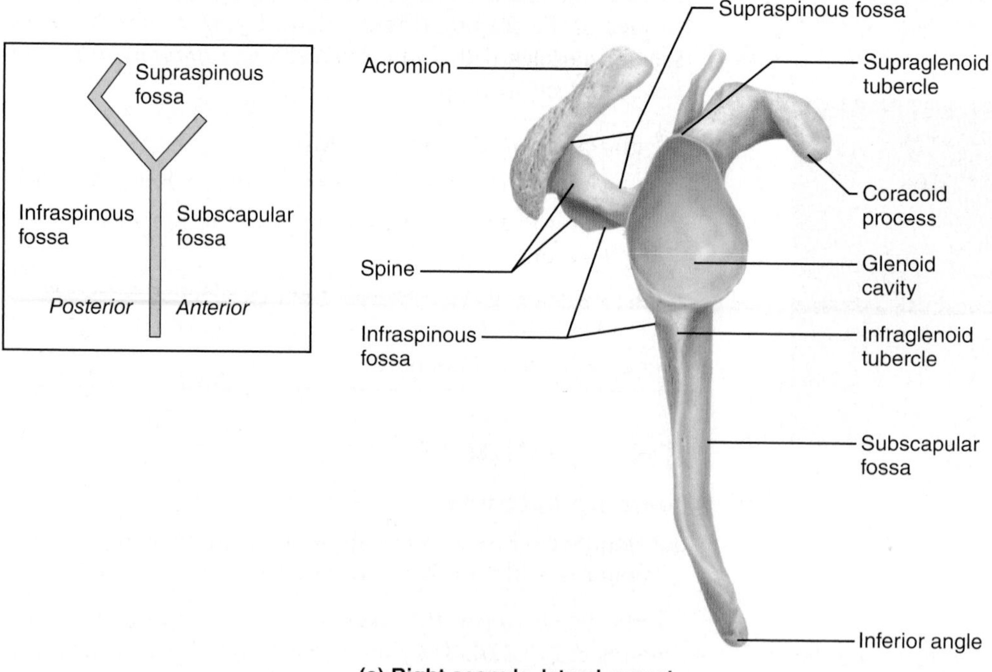

Figure 8.2 **The scapula.**

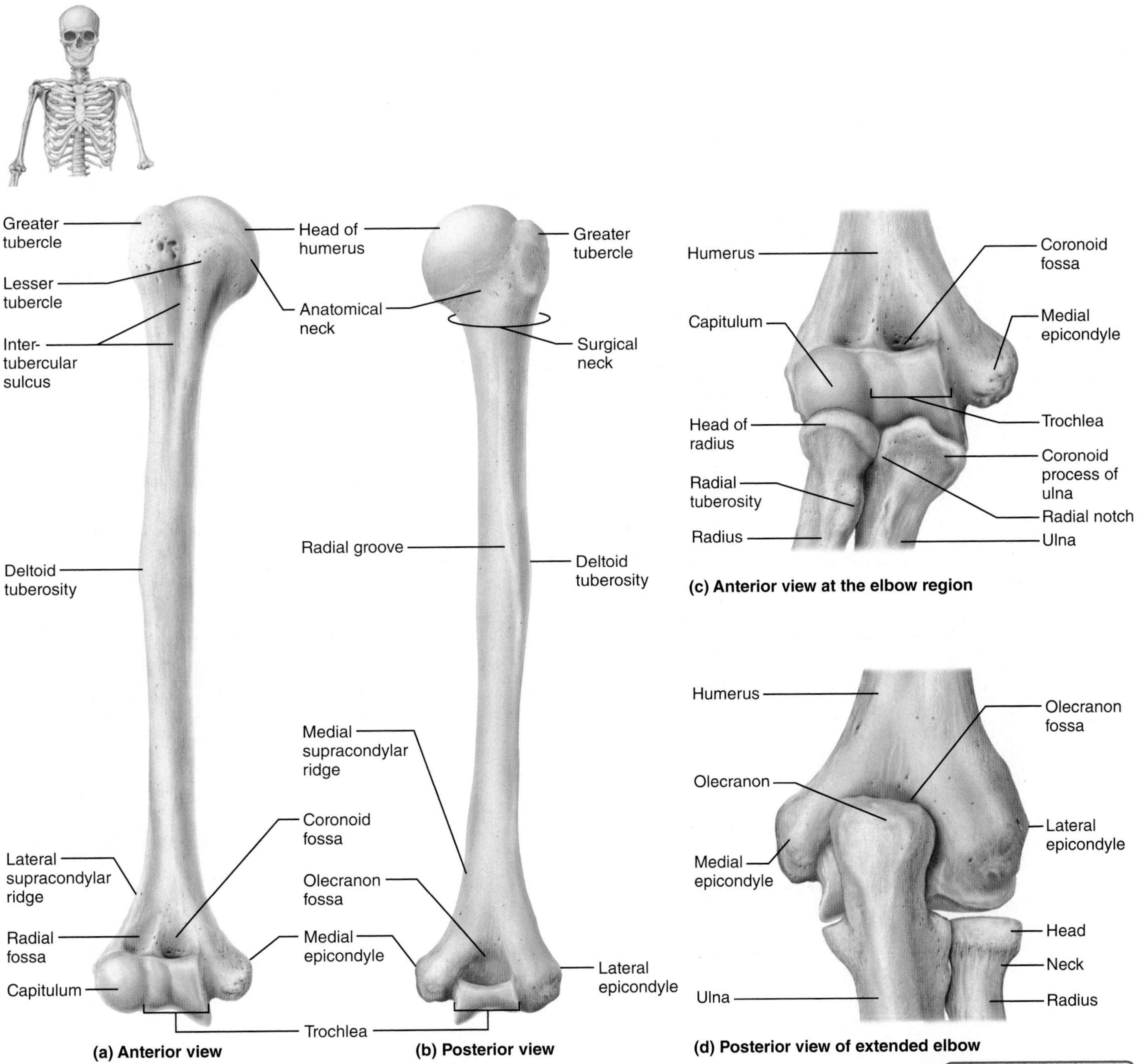

Greater tubercle
Lesser tubercle
Inter-tubercular sulcus
Head of humerus
Anatomical neck
Deltoid tuberosity
Lateral supracondylar ridge
Radial fossa
Capitulum
Coronoid fossa
Medial epicondyle
Trochlea

(a) Anterior view

Greater tubercle
Surgical neck
Radial groove
Deltoid tuberosity
Medial supracondylar ridge
Olecranon fossa
Lateral epicondyle

(b) Posterior view

Humerus
Capitulum
Head of radius
Radial tuberosity
Radius
Coronoid fossa
Medial epicondyle
Trochlea
Coronoid process of ulna
Radial notch
Ulna

(c) Anterior view at the elbow region

Humerus
Olecranon
Medial epicondyle
Ulna
Olecranon fossa
Lateral epicondyle
Head
Neck
Radius

(d) Posterior view of extended elbow

Figure 8.3 The humerus of the right arm and detailed views of articulation at the elbow.

arm (Figure 8.3). The largest and longest bone in the upper limb, it articulates with the scapula at the shoulder and with the radius and ulna (forearm bones) at the elbow.

At the proximal end of the humerus is the hemispherical **head,** which fits into the glenoid cavity of the scapula. Just inferior to the head is a slight constriction, the **anatomical neck.** Inferior to this, the lateral **greater tubercle** and the more medial **lesser tubercle** are separated by the **intertubercular** ("between the tubercles") **sulcus,** or *bicipital* (bi-sip′-ĭ-tal) *groove.* The tubercles are sites of attachment for the rotator cuff muscles (Chapter 11, p. 304). The intertubercular sulcus guides a tendon of the biceps muscle to its attachment point at the rim of the glenoid cavity (the supraglenoid tubercle, Figure 8.2c). The **surgical neck** of the humerus, so named because it is the most frequently fractured part of the humerus, is inferior to the tubercles. About midway down the shaft, on the lateral side, is the **deltoid tuberosity.** This V-shaped, roughened area is an attachment site for the deltoid muscle of the shoulder. Near the deltoid tuberosity along the posterior surface of the shaft, the **radial groove** descends obliquely. It marks the course of the radial nerve, an important nerve of the upper limb.

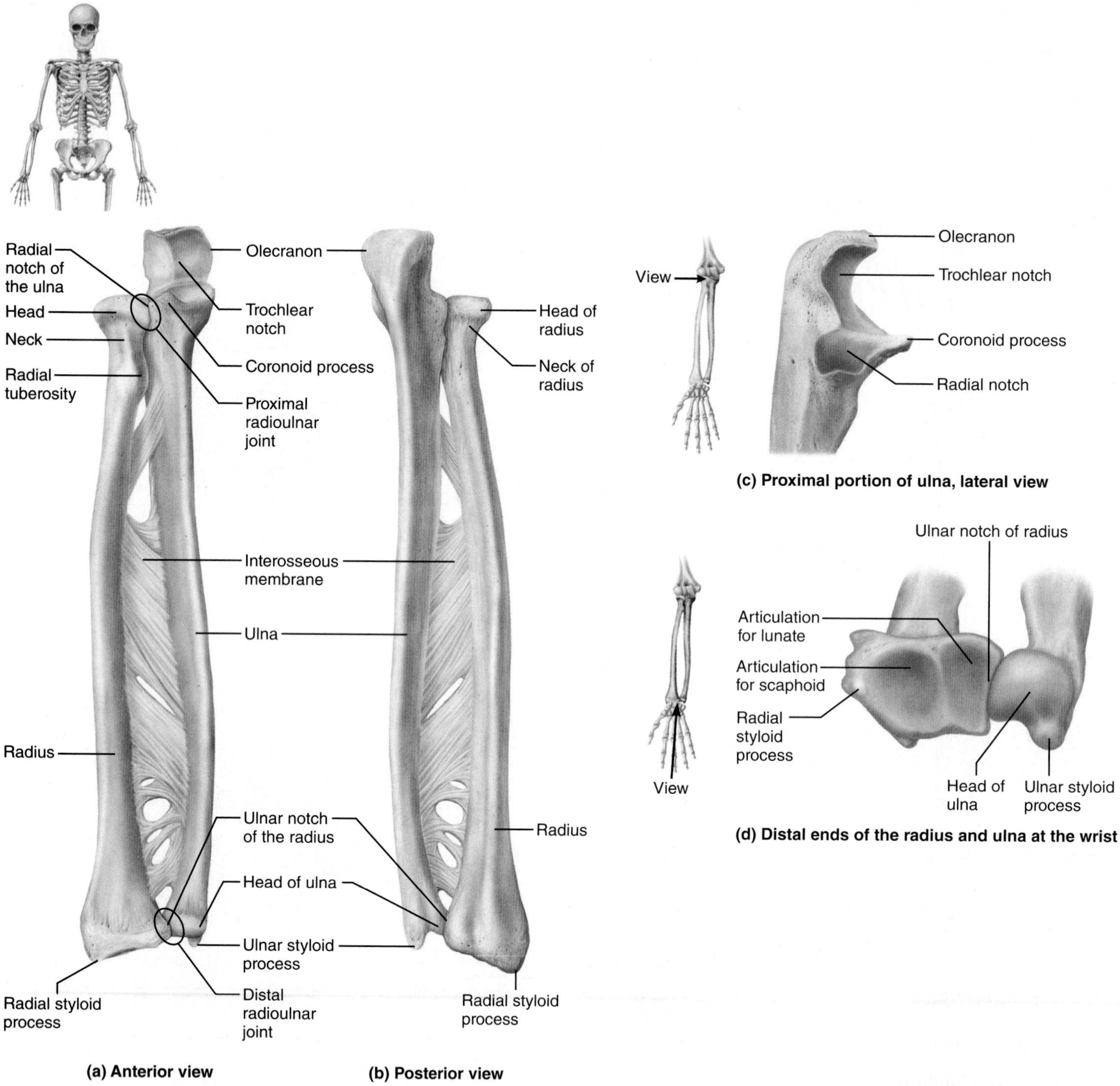

Figure 8.4 Radius and ulna of the right forearm.

Watch Bone Video MasteringA&P®

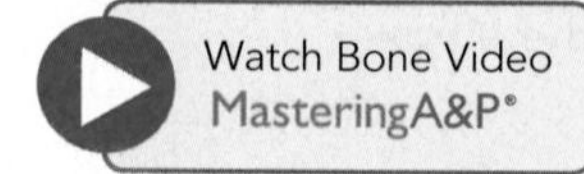

At the distal end of the humerus are two condyles, a medial **trochlea** (trok′le-ah; "pulley") that articulates with the ulna, and a lateral **capitulum** (kah-pit′u-lum; "small head") that articulates with the radius (Figure 8.3c). The trochlea looks like an hourglass turned on its side, and the capitulum is shaped like half a ball. They are flanked by the **medial** and **lateral epicondyles** ("beside the condyles"), which are attachment sites for muscles of the forearm. Directly above these epicondyles are the **medial** and **lateral supracondylar ridges** (Figure 8.3a).

On the posterior surface of the humerus directly proximal to the trochlea is the deep **olecranon** (o-lek′rah-non) **fossa.** In the corresponding position on the anterior surface is a shallower **coronoid** (kor′o-noid) **fossa** medially and **radial fossa** laterally. These fossae receive similarly named projections of the forearm bones during forearm movement (see Figure 8.3c and d).

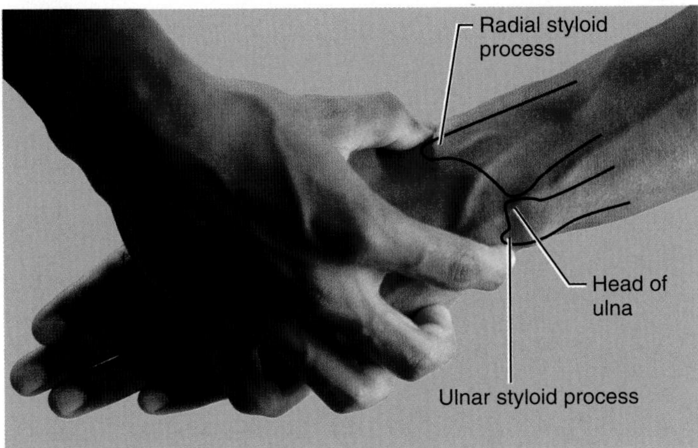

Figure 8.5 Location of styloid processes of radius and ulna. Palpation of normal positioning.

Forearm

Forming the skeleton of the *forearm,* or *antebrachium* (an′-te-bra′ke-um), are two parallel long bones, the radius and ulna (Figure 8.4), that articulate with the humerus proximally and the bones of the wrist distally. The radius and ulna also articulate with each other both proximally and distally at the small *radioulnar* (ra″de-o-ul′nar) *joints.* Furthermore, they are interconnected along their entire length by a flat ligament called the **interosseous membrane** (in″ter-os′e-us; "between the bones"). In the anatomical position, the radius lies laterally (on the thumb side), and the ulna medially. However, when the palm faces posteriorly, the distal end of the radius crosses over the ulna, and the two bones form an X.

Ulna

The **ulna** (ul′nah; "elbow"), which is slightly longer than the radius, is the main bone forming the elbow joint with the humerus. It looks much like a monkey wrench. At its proximal end are two prominent projections, the **olecranon** ("elbow") and **coronoid** ("crown-shaped") **process,** separated by a deep concavity, the **trochlear notch** (Figure 8.4c). Together, these two processes grip the trochlea of the humerus (Figure 8.3c), forming a hinge joint that allows the forearm to bend upon the arm (flex), then straighten again (extend). When the forearm is fully extended, the olecranon process "locks" into the olecranon fossa of the humerus (Figure 8.3d). When the forearm is flexed, the coronoid process of the ulna fits into the coronoid fossa of the humerus. On the lateral side of the coronoid process is a smooth depression, the **radial notch** (Figure 8.4c), where the head of the radius articulates with the ulna.

Distally, the shaft of the ulna narrows and ends in a knoblike **head** that articulates with the radius (Figure 8.4d). Medial to this is the **ulnar styloid** ("stake-shaped") **process,** from which a ligament runs to the wrist. The head of the ulna is separated from the bones of the wrist by a disc of fibrocartilage and plays little or no role in hand movements.

Radius

The **radius** ("spoke" or "ray") is thin at its proximal end and widened at its distal end—the opposite of the ulna (Figure 8.4). The proximal **head** of the radius is shaped like the end of a spool of thread (Figure 8.3c and d). Its superior surface is concave, and it articulates with the capitulum of the humerus. Medially, the head of the radius articulates with the radial notch of the ulna, forming the *proximal radioulnar joint.* Just distal to the head, on the anterior surface in anatomical position, is a rough bump, the **radial tuberosity,** a site of attachment of the biceps muscle. On the distal end of the radius (Figure 8.4d), the medial **ulnar notch** articulates with the head of the ulna, forming the *distal radioulnar joint,* and the lateral **radial styloid process** anchors a ligament that runs to the wrist. The distal articular surface is concave and articulates with carpal bones of the wrist. Whereas the ulna contributes heavily to the elbow joint, the radius is the primary forearm bone contributing to the wrist joint. When the radius rotates, the hand moves with it.

The distal, knoblike radial and ulnar styloid processes can be palpated by grasping the dorsal side of the distal forearm adjacent to the wrist with the thumb and index finger of the opposite hand (Figure 8.5). In this position, the thumb palpates the radial styloid process and the index finger palpates the ulnar styloid process. The radial styloid process lies about 1 cm (0.4 inch) distal to the ulnar styloid process.

CLINICAL APPLICATION

Palpation of Colles' Fracture When presented with a patient who has fallen on an outstretched hand and whose wrist has curves that resemble those on a fork, a clinician palpates for the position of the styloid processes (Figure 8.5). If palpation reveals that the styloid process of the radius has moved proximally from its normal position, the diagnosis is **Colles' fracture,** an impacted fracture in which the distal end of the radius is forced proximally into the shaft of the radius.

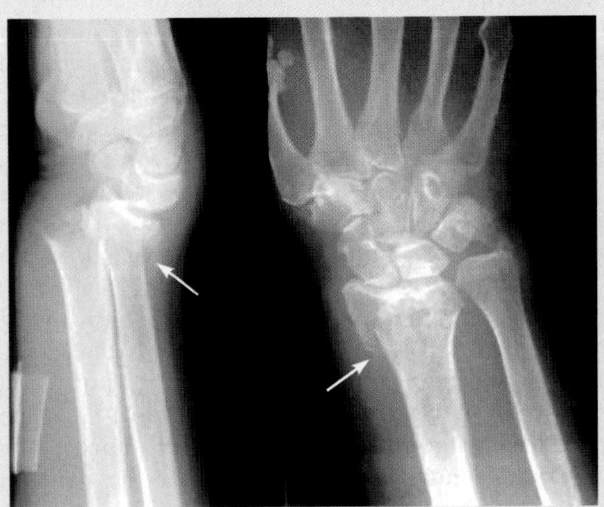

Radiograph of fractured radius, Colles' fracture

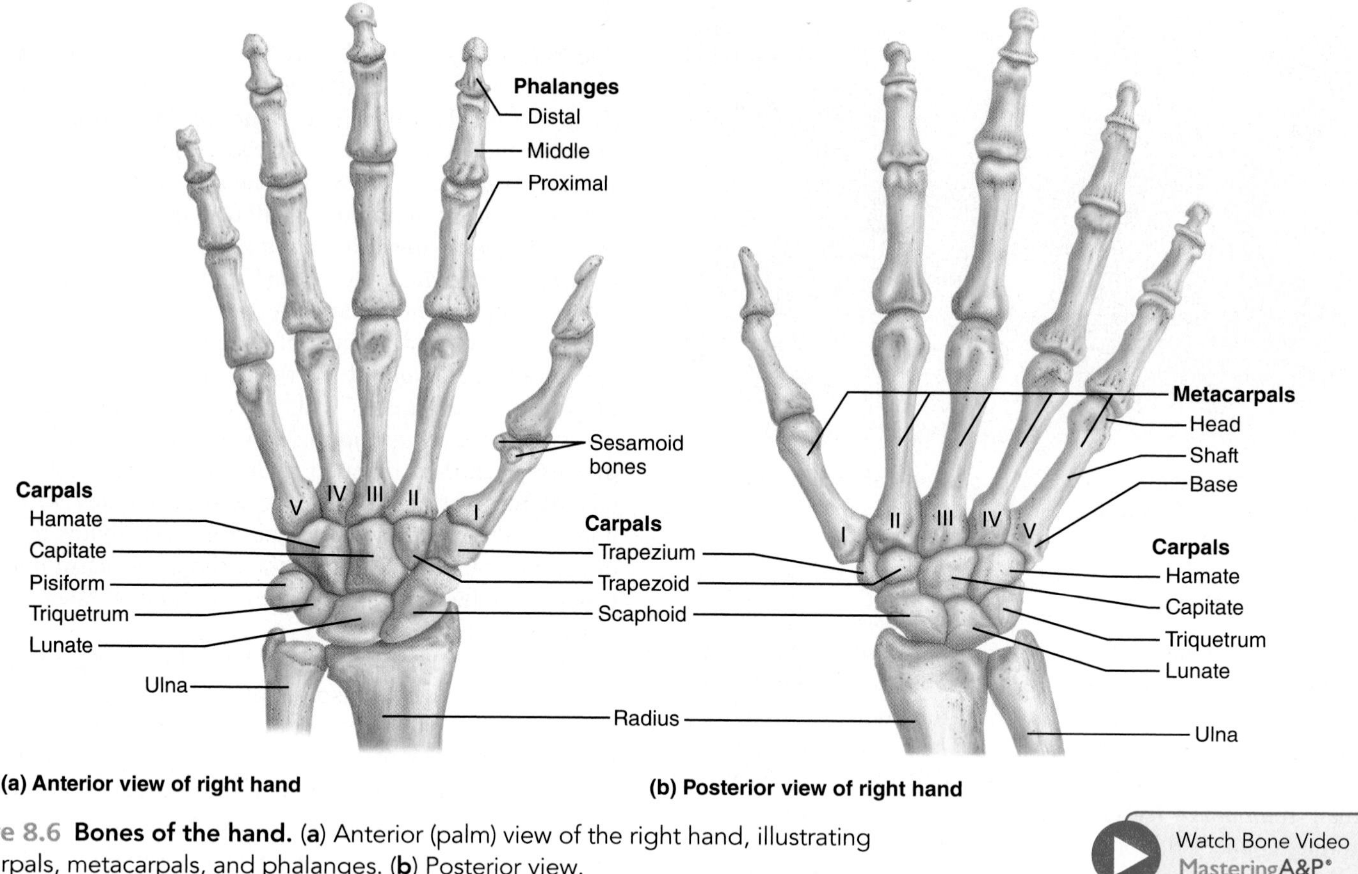

Figure 8.6 Bones of the hand. (**a**) Anterior (palm) view of the right hand, illustrating the carpals, metacarpals, and phalanges. (**b**) Posterior view.

Hand

The skeleton of the hand includes the bones of the *carpus,* or wrist; the bones of the *metacarpus,* or palm; and the *phalanges,* or bones of the fingers (Figure 8.6).

Carpus

A wristwatch is actually worn on the distal forearm, not on the wrist at all. The true wrist, or **carpus** (kar′pus), is the proximal region of the hand, just distal to the wrist joint. The carpus contains eight marble-sized short bones, or **carpals** (kar′palz), closely united by ligaments. Gliding movements occur between the carpals, making the wrist rather flexible. The carpals are arranged in two irregular rows of four bones each (Figure 8.6). In the proximal row, from lateral (thumb side) to medial, are the **scaphoid** (skaf′oid; "boat-shaped"), **lunate** (lu′nāt; "moonlike"), **triquetrum** (tri-kwet′rum; "triangular"), and **pisiform** (pis′i-form; "pea-shaped") bones. Only the scaphoid and lunate bones articulate with the radius to form the wrist joint. The carpals of the distal row, again from lateral to medial, are the **trapezium** (trah-pe′ze-um; "little table"), **trapezoid** ("four-sided"), **capitate** (kap′i-tāt; "head-shaped"), and **hamate** (ham′āt; "hooked") bones. A simple mnemonic may help you remember the names and positions of the carpal bones, starting with the proximal row from lateral to medial, and continuing with the distal row from lateral to medial: **S**ally **L**eft **T**he **P**arty **T**o **T**ake **C**armen **H**ome.

The scaphoid is the most frequently fractured carpal bone, which often results from falling on an outstretched hand. The impact bends the scaphoid, which then breaks at its narrow midregion.

Metacarpus

Five **metacarpals** radiate distally from the wrist to form the **metacarpus,** or palm of the hand (*meta* = beyond). These small long bones are not named individually but instead are numbered I to V, from thumb to little finger (Figure 8.6). The *bases* of the metacarpals articulate with the carpals proximally and with each other on their lateral and medial sides. Distally, the bulbous *heads* of the metacarpals articulate with the proximal phalanges of the fingers to form knuckles. Metacarpal I, associated with the thumb, is the shortest and most mobile.

Phalanges of the Fingers

The digits, or fingers, are numbered I to V beginning with the thumb, or *pollex* (pol′eks). The fingers contain miniature long bones called **phalanges** (fah-lan′jēz). The singular of this term is *phalanx* (fa′langks; "a closely knit row of soldiers"). In most people, the third finger is the longest. With the exception of the thumb, each finger has three phalanges: *proximal, middle,* and *distal* (Figure 8.6). The thumb has no middle phalanx.

Table 8.1 Bones of the Upper Limb

Body Region	Bones*	Location	Markings
PECTORAL GIRDLE (Figures 8.1, 8.2)	**Clavicle** (2)	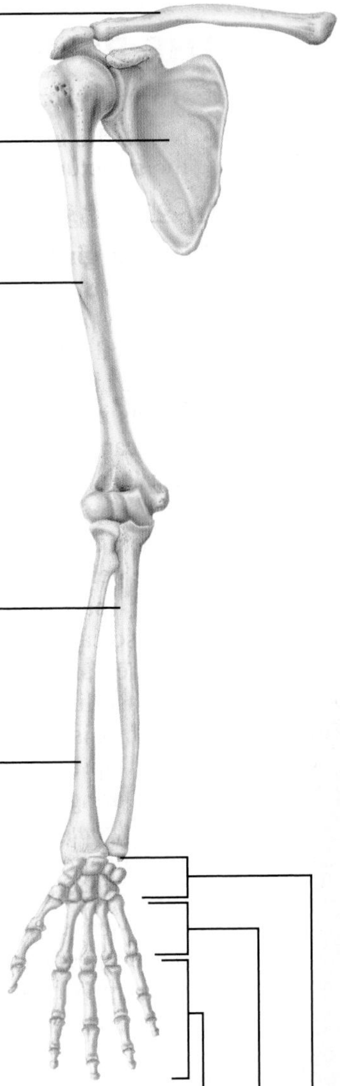 Clavicle is in superoanterior thorax; articulates medially with sternum and laterally with scapula	Acromial end; sternal end; conoid tubercle
	Scapula (2)	Scapula is in posterior thorax; forms part of the shoulder; articulates with humerus and clavicle	Glenoid cavity; spine; acromion; coracoid process; infraspinous, supraspinous, and subscapular fossae
UPPER LIMB Arm (Figure 8.3)	**Humerus** (2)	Humerus is sole bone of arm; between scapula and elbow	Head; greater and lesser tubercles; intertubercular sulcus; radial groove; deltoid tuberosity; trochlea; capitulum; coronoid and olecranon fossae; medial and lateral epicondyles
Forearm (Figure 8.4)	**Ulna** (2)	Ulna is medial bone of forearm between elbow and wrist; forms elbow joint	Coronoid process; olecranon; radial notch; trochlear notch; ulnar styloid process; head
	Radius (2)	Radius is lateral bone of forearm; articulates with proximal carpals to form part of the wrist joint	Head; radial tuberosity; radial styloid process; ulnar notch
Hand (Figure 8.6)	8 **Carpals** (16): Scaphoid, Lunate, Triquetrum, Pisiform, Trapezium, Trapezoid, Capitate, Hamate	Carpals form a bony crescent at the wrist; arranged in two rows of four bones each	
	5 **Metacarpals** (10)	Metacarpals form the palm; one in line with each digit	
	14 **Phalanges** (28): Proximal, Middle, Distal	Phalanges form the fingers; three in digits II–V; two in digit I (the thumb)	

Anterior view of pectoral girdle and upper limb

*The number in parentheses ( ) following the bone name denotes the total number of such bones in the body.

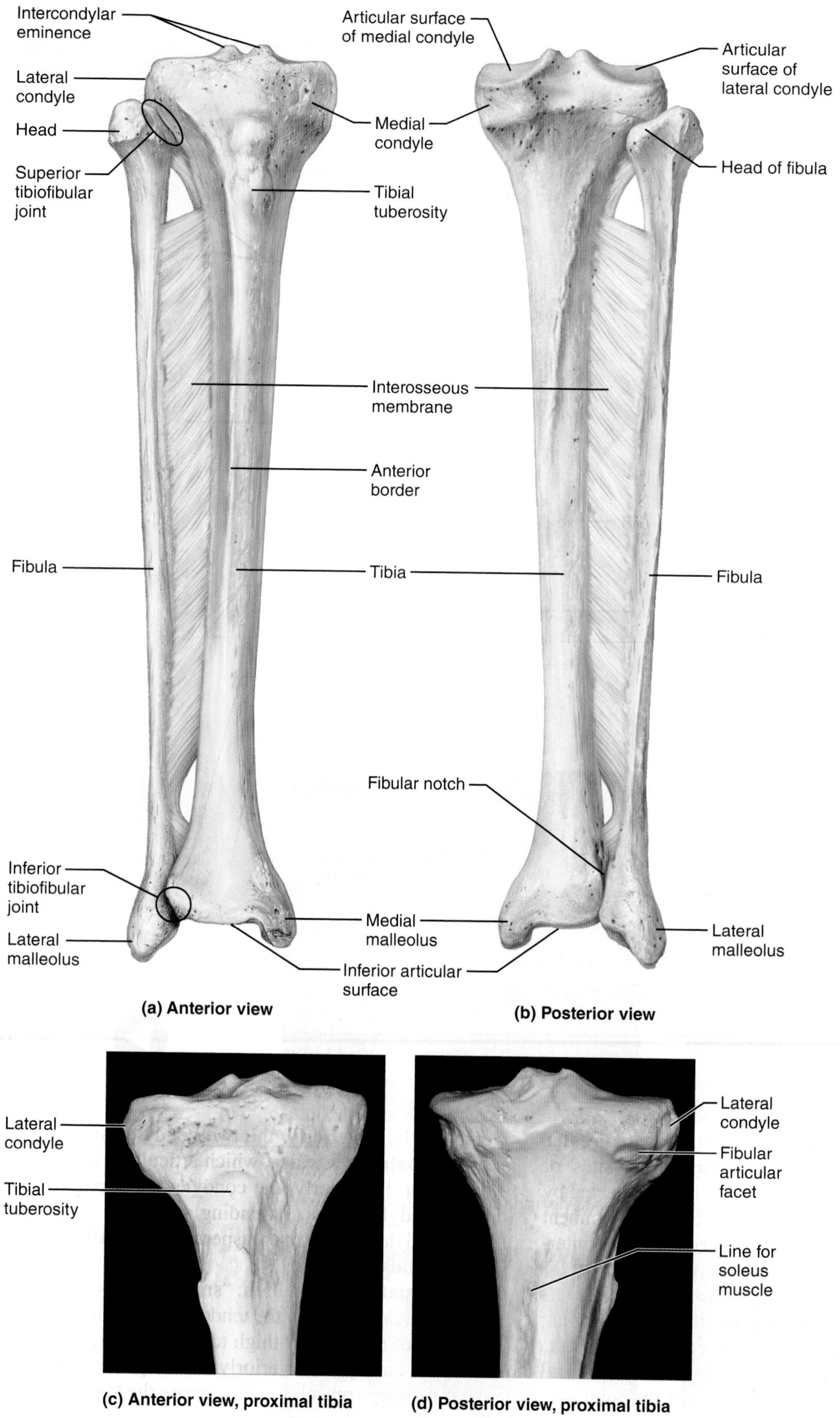

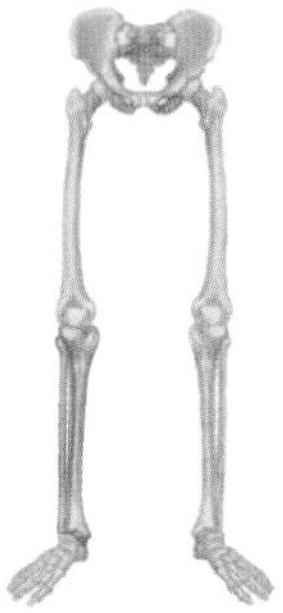

Intercondylar eminence
Lateral condyle
Head
Superior tibiofibular joint
Articular surface of medial condyle
Medial condyle
Tibial tuberosity
Articular surface of lateral condyle
Head of fibula
Interosseous membrane
Anterior border
Fibula
Tibia
Fibula
Fibular notch
Inferior tibiofibular joint
Lateral malleolus
Medial malleolus
Inferior articular surface
Lateral malleolus
(a) Anterior view
(b) Posterior view
Lateral condyle
Tibial tuberosity
Lateral condyle
Fibular articular facet
Line for soleus muscle
(c) Anterior view, proximal tibia
(d) Posterior view, proximal tibia

Figure 8.10 The tibia and fibula of the right leg.

Watch Bone Video
MasteringA&P®

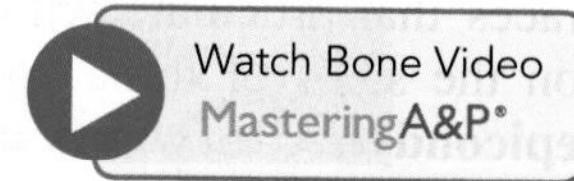

Table 8.3 Bones of the Lower Limbs

Body Region	Bones*	Illustration	Location	Markings
PELVIC GIRDLE (Figure 8.7)	**Hip** (2) (coxal)		Each hip bone is formed by the fusion of an ilium, ischium, and pubis; the hip bones fuse anteriorly at the pubic symphysis and form sacroiliac joints with the sacrum posteriorly; girdle consisting of both hip bones and the sacrum is basinlike	Iliac crest; anterior and posterior iliac spines; auricular surface; greater and lesser sciatic notches; obturator foramen; ischial tuberosity and spine; acetabulum; pubic arch; pubic crest; pubic tubercle
LOWER LIMB Thigh (Figure 8.9)	**Femur** (2)		Femur is the sole bone of thigh; between hip joint and knee; largest bone of the body	Head; greater and lesser trochanters; neck; lateral and medial condyles and epicondyles; gluteal tuberosity; linea aspera
Kneecap (Figure 8.9)	**Patella** (2)		Patella is a sesamoid bone formed within the tendon of the quadriceps (anterior thigh) muscles	
Leg (Figure 8.10)	**Tibia** (2)		Tibia is the larger and more medial bone of leg; between knee and foot	Medial and lateral condyles; tibial tuberosity; anterior border; medial malleolus
	Fibula (2)		Fibula is the lateral bone of leg; sticklike	Head; lateral malleolus
Foot (Figure 8.11)	7 **Tarsals** (14) Talus Calcaneus Navicular Cuboid Lateral cuneiform Intermediate cuneiform Medial cuneiform		Tarsals are seven bones forming the proximal part of the foot; the talus articulates with the leg bones at the ankle joint; the calcaneus, the largest tarsal, forms the heel	
	5 **Metatarsals** (10)		Metatarsals are five bones numbered I–V	
	14 **Phalanges** (28) Proximal Middle Distal		Phalanges form the toes; three in digits II–V, two in digit I (the great toe)	

Anterior view of pelvic girdle and left lower limb

*The number in parentheses ( ) following the bone name denotes the total number of such bones in the body.

and *fibula,* form the skeleton of the leg (Figure 8.10). The tibia, located medially, is more massive than the sticklike fibula. These two bones articulate with each other both superiorly and inferiorly. However, unlike the joints between the radius and ulna of the forearm, the *tibiofibular* (tib″e-o-fib′u-lar) *joints* allow almost no movement. An **interosseous membrane** connects the tibia and fibula along their entire length. The tibia articulates with the femur to form the knee joint, and with the talus bone of the foot at the ankle joint. The fibula does not contribute to the knee joint and merely helps stabilize the ankle joint.

Tibia

The **tibia** (tib′e-ah; "shinbone") receives the weight of the body from the femur and transmits it to the foot. It is second only to the femur in size and strength. At its proximal end the broad **medial** and **lateral condyles,** which resemble two thick checkers lying side by side on the top of the shaft, articulate with the corresponding condyles of the femur. The tibial condyles are separated by an irregular projection, the **intercondylar eminence.** On the inferior part of the lateral tibial condyle is a facet that articulates with the fibula to

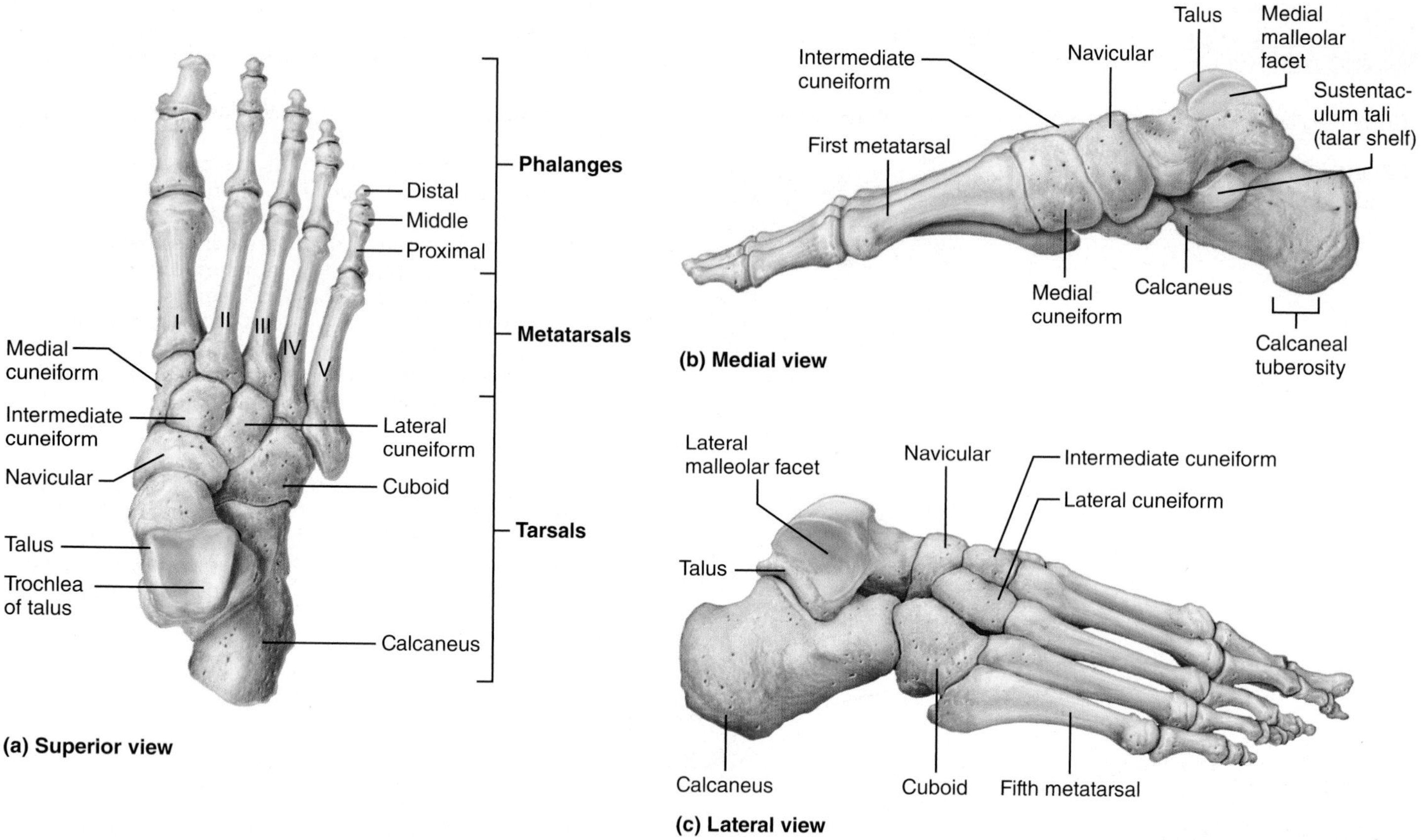

Phalanges
Distal
Middle
Proximal
Metatarsals
I II III IV V
Medial cuneiform
Intermediate cuneiform
Navicular
Talus
Trochlea of talus
Lateral cuneiform
Cuboid
Calcaneus
Tarsals

(a) Superior view

Intermediate cuneiform
First metatarsal
Navicular
Talus
Medial malleolar facet
Sustentaculum tali (talar shelf)
Medial cuneiform
Calcaneus
Calcaneal tuberosity

(b) Medial view

Lateral malleolar facet
Navicular
Intermediate cuneiform
Lateral cuneiform
Talus
Calcaneus
Cuboid
Fifth metatarsal

(c) Lateral view

Figure 8.11 **Bones of the right foot.**

Watch Bone Video
MasteringA&P®

form the *superior tibiofibular joint* (Figure 8.10a and d). Just inferior to the condyles, on the tibia's anterior surface, is the **tibial tuberosity** (Figure 8.10c), attachment site of the patellar ligament.

The shaft of the tibia is triangular in cross section. The sharp **anterior border** lies just below the skin and is easily palpated. Distally, the end of the tibia is flat where it articulates with the talus of the foot. Medial to this joint surface, the tibia has an inferior projection called the **medial malleolus** (mah-le′-o-lus; "little hammer"), which forms the medial bulge of the ankle. The **fibular notch,** on the lateral side of the distal tibia, articulates with the fibula, forming the *inferior tibiofibular joint.*

Fibula

The **fibula** (fib′u-la; "pin"), located lateral to the tibia, is a thin long bone with two expanded ends. Its superior end is its **head,** and its inferior end is the **lateral malleolus.** This malleolus forms the lateral bulge of the ankle and articulates with the talus bone of the foot. The shaft of the fibula is heavily ridged; the fibula does not bear weight, but several muscles originate from it.

Foot

The skeleton of the foot includes the bones of the *tarsus,* the bones of the *metatarsus,* and the *phalanges,* or toe bones (Figure 8.11). The foot has two important functions: It supports the weight of the body, and it acts as a lever to

CLINICAL APPLICATION

Ankle Fractures The medial and lateral malleoli are commonly fractured when the foot is forcefully inverted or everted at the ankle—that is, when one lands either on the lateral side of the foot and twists the sole medially (inversion) or on the medial side and turns the sole laterally (eversion). This type of injury occurs frequently in the general population and in people participating in contact sports.

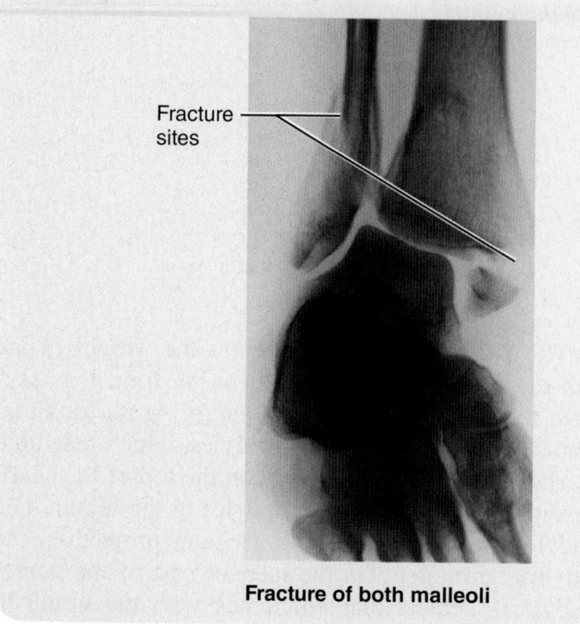

Fracture of both malleoli

propel the body forward during walking or running. A single bone could serve both these purposes but would function poorly on uneven ground. Its multicomponent structure makes the foot pliable, avoiding this problem.

Tarsus

The **tarsus** (tar′sus) makes up the posterior half of the foot and contains seven bones called **tarsals.** It is comparable to the carpus of the hand. The weight of the body is carried primarily by the two largest, most posterior tarsal bones: the **talus** (ta′lus; "ankle"), which articulates with the tibia and fibula superiorly, and the strong **calcaneus** (kal-ka′-ne-us; "heel bone"), which forms the heel of the foot. The tibia articulates with the talus at the **trochlea of the talus** (Figure 8.11a). Inferiorly, the talus articulates with the calcaneus. The thick tendon of the calf muscles attaches to the posterior surface of the calcaneus. The part of the calcaneus that touches the ground is the **calcaneal tuberosity,** and the medial, shelflike projection is the **sustentaculum tali** (sus″-ten-tak′u-lum ta′le; "supporter of the talus") or **talar shelf** (Figure 8.11b). The remaining tarsal bones are the lateral **cuboid** (ku′-boid; "cube-shaped"), the medial **navicular** (nah-vik′u-lar; "boatlike"), and the anterior **medial, intermediate,** and **lateral cuneiforms** (ku-ne′ĭ-form; "wedge-shaped").

Metatarsus

The **metatarsus** of the foot, which corresponds to the metacarpus of the hand, consists of five small long bones called **metatarsals.** These bones are numbered I to V beginning on the medial side of the foot (Figure 8.11a). The first metatarsal at the base of the big toe is the largest, and it plays an important role in supporting the weight of the body. The metatarsals are more nearly parallel to one another than are the metacarpals in the palm. Distally, where the metatarsals articulate with the proximal phalanges of the toes, the enlarged head of the first metatarsal forms the "ball" of the foot.

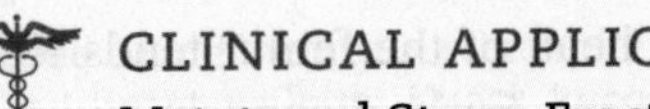

CLINICAL APPLICATION

Metatarsal Stress Fracture One of the most common foot injuries, **metatarsal stress fracture,** results from repetitive stress on the foot, typically as a result of increasing one's running mileage too quickly. The second and third metatarsals are most often affected. Treatment generally involves resting the foot and wearing stiff or well-cushioned shoes.

Phalanges of the Toes

The 14 phalanges of the toes are smaller than those of the fingers and thus are less nimble. Still, their general structure and arrangement are the same: There are three phalanges in each digit except the great toe (the *hallux*), which has only two phalanges. As in the hand, these toe bones are named *proximal, middle,* and *distal phalanges.*

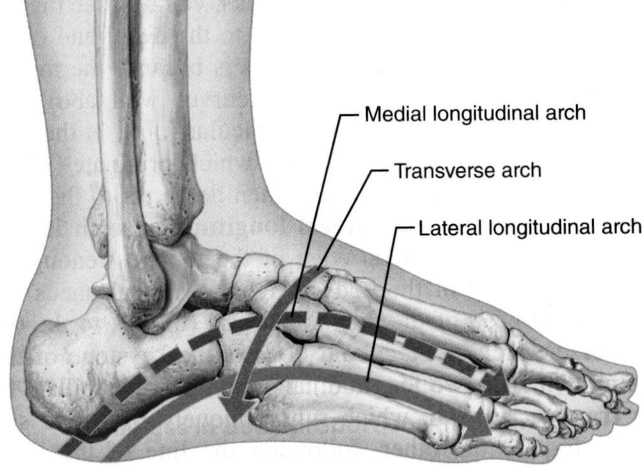

(a) Lateral aspect of right foot

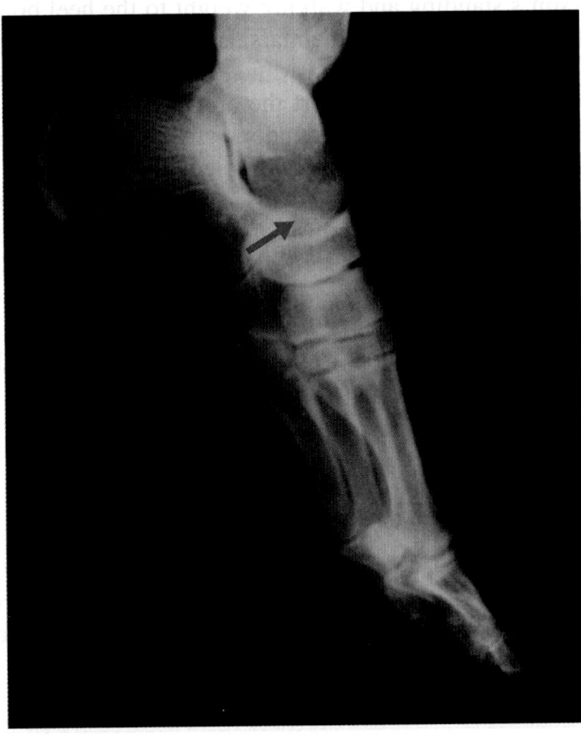

(b) X-ray image, medial aspect of right foot; keystone of medial longitudinal arch at arrow

Figure 8.12 Arches of the foot.

Arches of the Foot

A structure composed of multiple components can support weight only if it is arched. The foot has three arches: the medial and lateral *longitudinal* arches and the *transverse* arch (Figure 8.12). These arches are maintained by the interlocking shapes of the foot bones, by strong ligaments, and by the pull of some tendons during muscle activity; the ligaments and tendons also provide resilience. As a result, the arches "give" when weight is applied to the foot, then spring back when the weight is removed.

Table 9.2 Movements at Synovial Joints

Movement	Definition
GLIDING (Figure 9.6a)	Sliding the flat surfaces of two bones across each other
ANGULAR MOVEMENTS (Figure 9.6b–e)	
Flexion	Decreasing the angle between two bones
Extension	Increasing the angle between bones
Abduction	Moving a limb away from the body midline
Adduction	Moving a limb toward the body midline
Circumduction	Moving a limb or finger so that it describes a cone in space
ROTATION (Figure 9.6f)	Turning a bone around the longitudinal axis
Medial rotation	Rotating toward the medial plane
Lateral rotation	Rotating away from the medial plane

Angular Movements

Angular movements (Figure 9.6b–e) increase or decrease the angle between two bones. These movements include flexion, extension, abduction, adduction, and circumduction.

Flexion and Extension Flexion and extension are movements that occur primarily in the sagittal plane.

Flexion *decreases the angle* between the bones, bringing these bones closer together (Figure 9.6b–d). Examples include flexion of the neck or trunk (Figure 9.6b and c); flexion of the fingers, as in making a fist; and flexion of the forearm toward the arm at the elbow. In less obvious examples, the arm is flexed at the shoulder when moved in an anterior direction (Figure 9.6d), and the hip is flexed when the thigh moves anteriorly.

Extension is the reverse of flexion and occurs at the same joints (see Figure 9.6b–d). It *increases the angle* between the joining bones and is a straightening action. Straightening the fingers after making a fist is an example of extension. At the shoulder and the hip, extension moves the limb posteriorly (see the shoulder extension in Figure 9.6d). Bending a joint back beyond its normal range of motion is called **hyperextension** (literally, "superextension"). Individuals who have loose ligaments that allow a greater range of motion can be capable of hyperextending the joints.

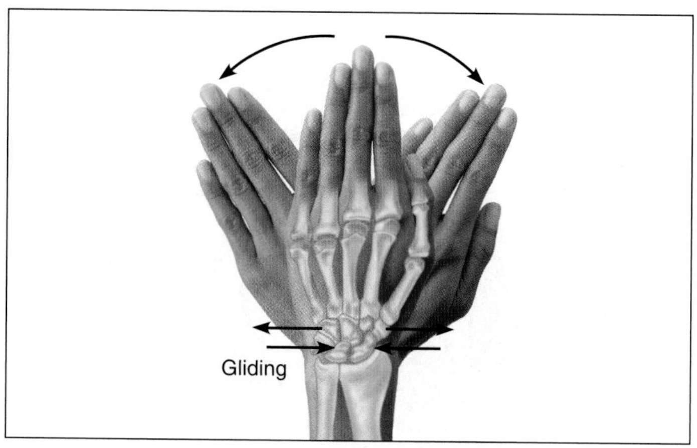

(a) Gliding movements at the wrist

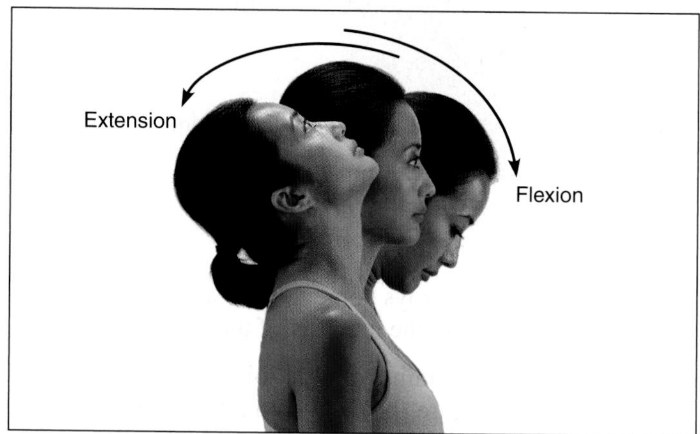

(b) Angular movements: flexion and extension of the neck

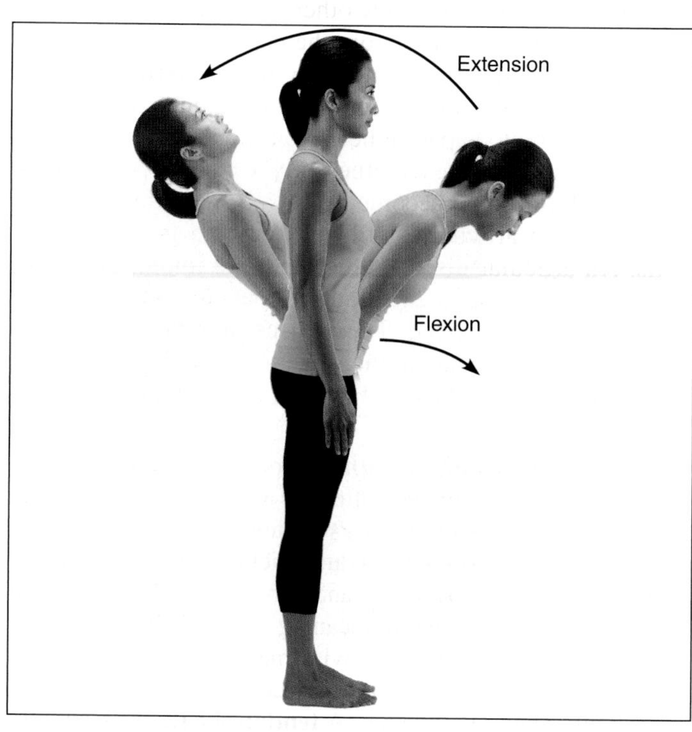

(c) Angular movements: flexion and extension of the trunk

Figure 9.6 **Movements allowed by synovial joints.** The axis around which each movement occurs is indicated by a red dot in (d) and (e), and by a red bar in (f).

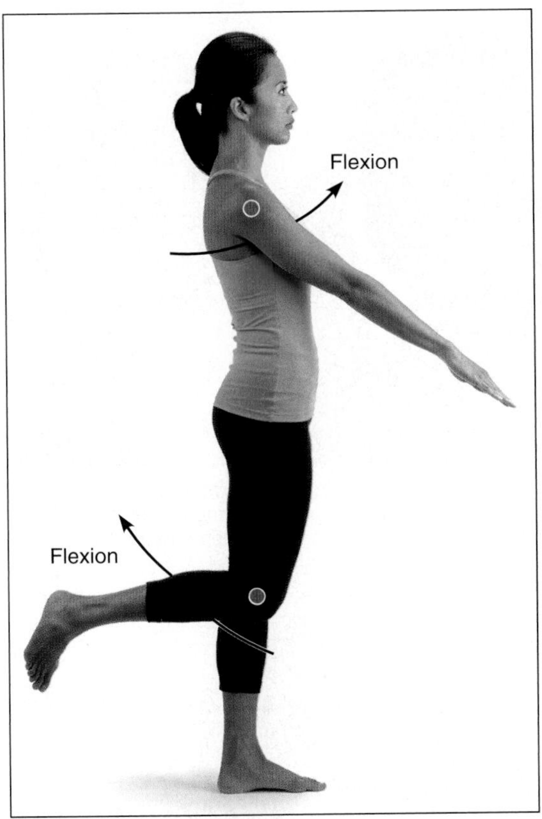

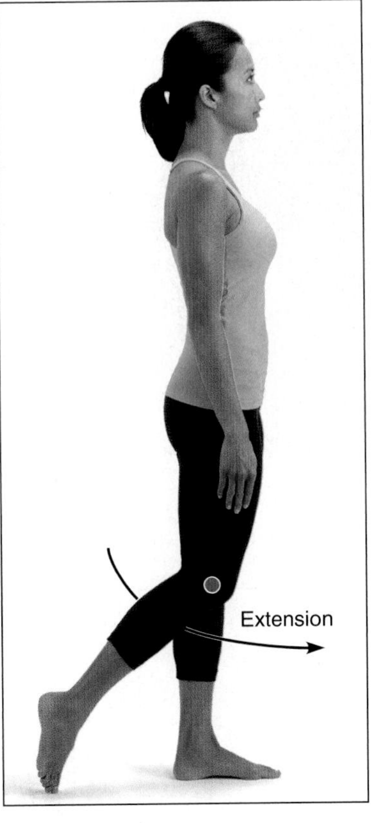

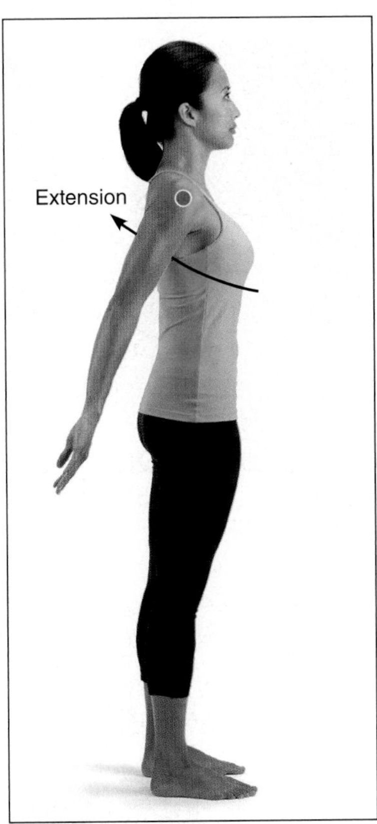

(d) Angular movements: flexion and extension at the shoulder and knee

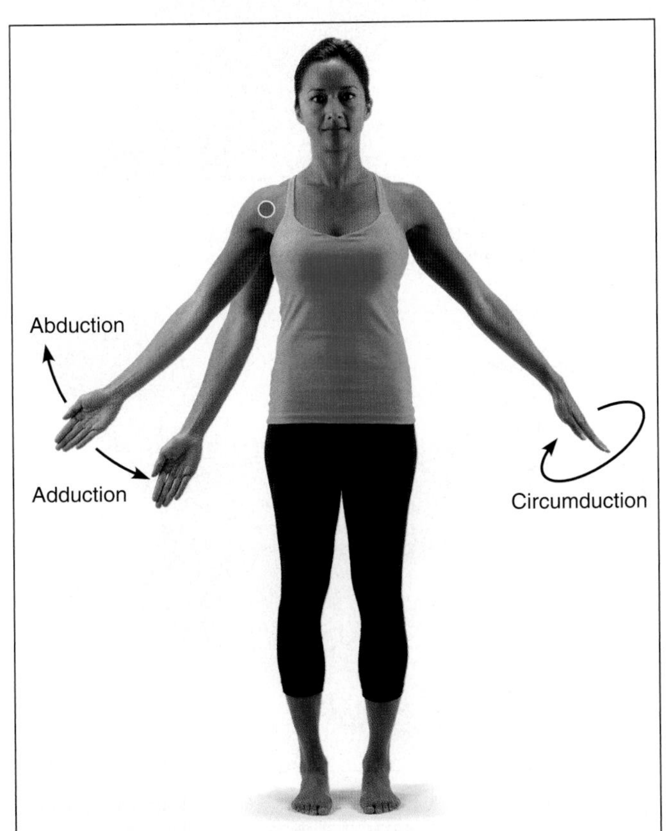

(e) Angular movements: abduction, adduction, and circumduction of the upper limb at the shoulder

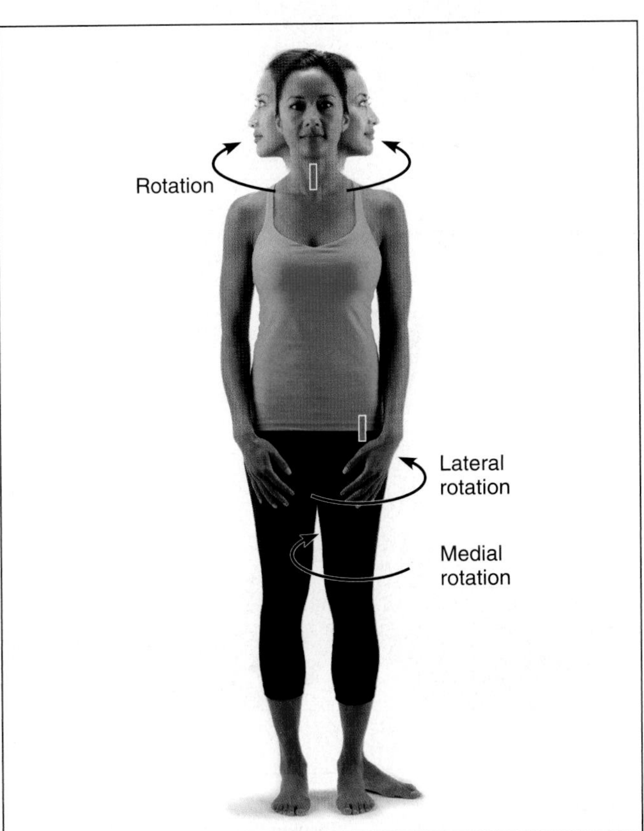

(f) Rotation of the head, neck, and lower limb

Figure 9.6 **Movements allowed by synovial joints,** *continued.*

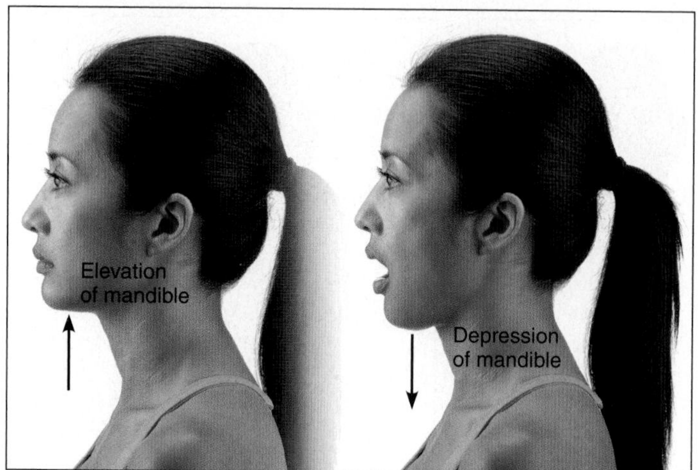

(a) Elevation
Lifting a body part superiorly

Depression
Moving a body part inferiorly

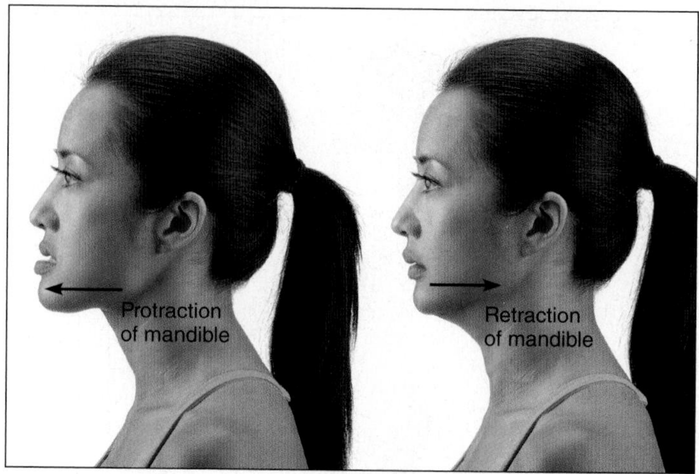

(b) Protraction
Moving a body part in the anterior direction

Retraction
Moving a body part in the posterior direction

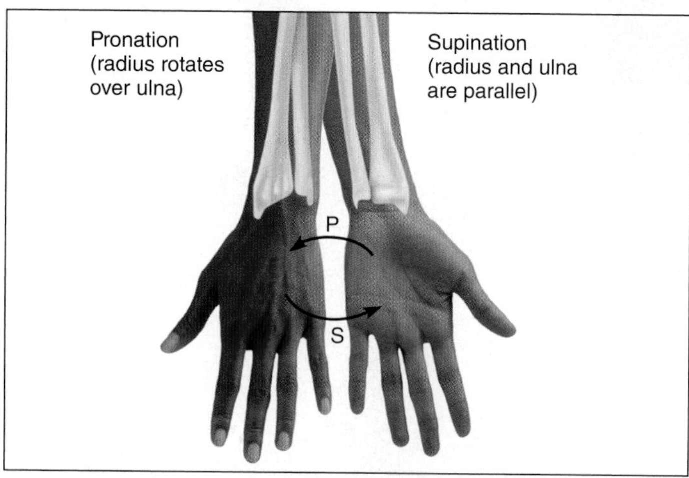

(c) Pronation (P)
Rotating the forearm so the palm faces posteriorly

Supination (S)
Rotating the forearm so the palm faces anteriorly

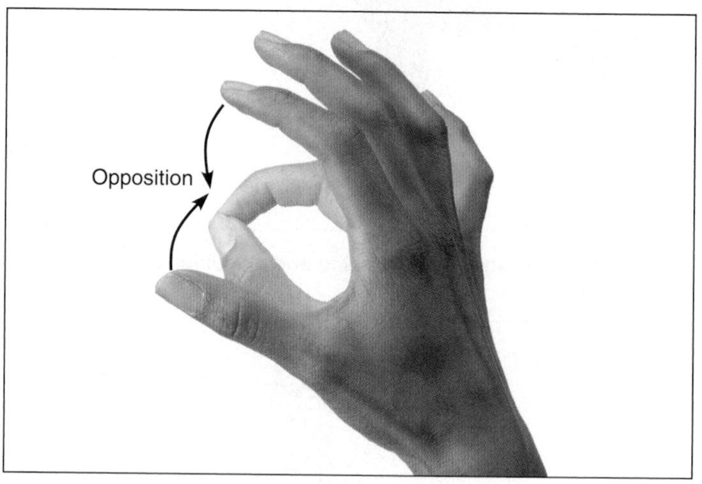

(d) Opposition
Moving the thumb to touch the tips of the other fingers

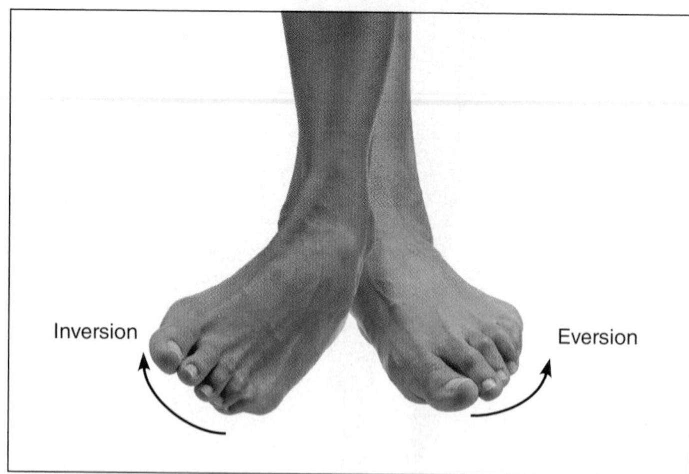

(e) Inversion
Turning the sole of the foot medially

Eversion
Turning the sole of the foot laterally

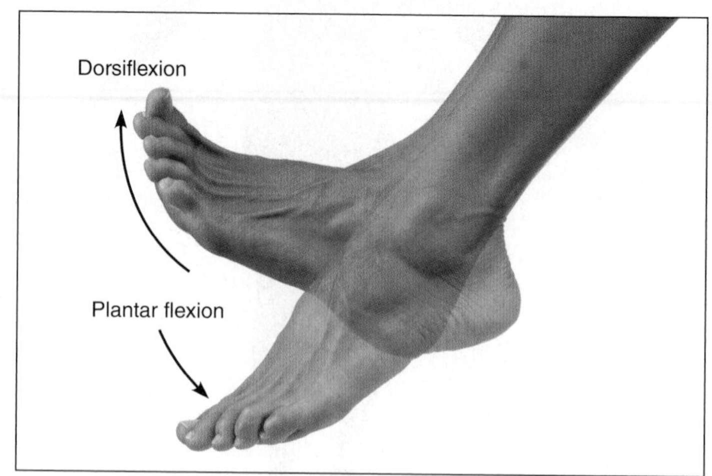

(f) Dorsiflexion
Lifting the foot so its superior surface approaches the shin

Plantar flexion
Depressing the foot elevating the heel

Figure 9.7 **Some special body movements.**

Abduction and Adduction Abduction and adduction move a body part along the frontal plane.

Abduction, from Latin words meaning "moving away," is movement of a limb *away* from the body midline. Raising the arm or thigh laterally is an example of abduction (Figure 9.6e). For the fingers or toes, *abduction* means spreading them apart. In this case, the "midline" is the longest digit: the third finger or the second toe. Note that bending the trunk away from the body midline to the right or left is called *lateral flexion* instead of abduction.

Adduction ("moving toward") is the opposite of abduction: the movement of a limb *toward* the body midline (Figure 9.6e) or, in the case of the digits, toward the midline (longest digit) of the hand or foot.

Circumduction **Circumduction** ("moving in a circle") is moving a limb or finger so that it describes a cone in space (Figure 9.6e). This is a complex movement that combines flexion, abduction, extension, and adduction in succession.

Rotation **Rotation** is the turning movement of a bone around the longitudinal axis. This motion occurs along the transverse plane. Rotation is the only movement allowed between the first two cervical vertebrae. The entire vertebral column also rotates, twisting the whole trunk to the right or left. Rotation also occurs at the hip and shoulder joints (Figure 9.6f). Rotation of the limbs may be directed toward the median plane or away from it. For example, in **medial rotation** of the lower limb, the limb's anterior surface turns toward the median plane of the body; **lateral rotation** is the opposite movement.

Special Movements

Certain movements do not fit into any of the previous categories and occur at only a few joints (Figure 9.7).

Elevation and Depression **Elevation** means lifting a body part superiorly (Figure 9.7a). Moving the elevated part inferiorly is **depression.** During chewing, the mandible is alternately elevated and depressed.

Protraction and Retraction Nonangular movements in the anterior and posterior directions are called **protraction** and **retraction,** respectively (Figure 9.7b). The mandible is protracted when you jut out your jaw and retracted when you bring it back.

Supination and Pronation The terms **supination** (soo″-pĭ-na′shun) and **pronation** (pro-na′shun) refer to movements of the radius around the ulna (Figure 9.7c). Supination occurs when the forearm, specifically the radius, rotates laterally so that the palm faces anteriorly (the hand is lying on its "back," supine). This is standard anatomical position. Pronation occurs when the radius rotates medially so that the palm faces posteriorly (hand lying "belly" side down, as in a prone float). Pronation brings the radius across the ulna so that the two bones form an X.

Opposition In the palm, the saddle joint between metacarpal I and the trapezium allows a movement called *opposition* of the thumb (Figure 9.7d). This is the action by which you move your thumb across the palm enabling it to touch the tips of the other fingers on the same hand. This unique action is what makes the human hand such a fine tool for grasping and manipulating objects.

Inversion and Eversion **Inversion** and **eversion** are special movements of the foot at the intertarsal joints (Figure 9.7e). To invert the foot, turn the sole medially; to evert the foot, turn the sole laterally.

Dorsiflexion and Plantar Flexion Up-and-down movements of the foot at the ankle are also given special names. Lifting the foot so that its superior surface approaches the shin is called **dorsiflexion,** whereas depressing the foot or elevating the heel (pointing the toes) is called **plantar flexion** (Figure 9.7f). Dorsiflexion of the foot corresponds to extension of the hand at the wrist, whereas plantar flexion corresponds to hand flexion.

Types of Synovial Joints

The shapes of the articulating bone surfaces determine the movements allowed at a synovial joint. Synovial joints are described functionally as follows:

- **Nonaxial.** Adjoining bones do not move around a specific axis.
- **Uniaxial.** Movement occurs around a single axis.
- **Biaxial.** Movement can occur around two axes; thus, the joint enables motion along both the frontal and sagittal planes.
- **Multiaxial.** Movement can occur around all three axes and along all three body planes: frontal, sagittal, and transverse.

Based on joint shape, synovial joints are structurally classified as *plane, hinge, pivot, condylar, saddle,* and *ball-and-socket.* This classification is presented in detail in *Focus on Synovial Joints* (Figure 9.8). Most of the synovial joints of the body are summarized by structural and functional type and the movements that occur at each joint (Table 9.3 pp. 220–222) .

Factors Influencing the Stability of Synovial Joints

learning outcome

▶ List three factors that influence the stability of synovial joints.

Because joints are stressed during movement, they must be stabilized to prevent dislocation (misalignment). The stability of a synovial joint depends on three factors: the shapes of the articular surfaces, the number and position of stabilizing ligaments, and muscle tone.

Articular Surfaces

The articular surfaces of the bones in a joint fit together in a complementary manner. Although their shapes determine

Figure 9.8

The shapes of the joint surfaces define the types of movements that can occur at a synovial joint; they also determine the classification of synovial joints into six structural types.

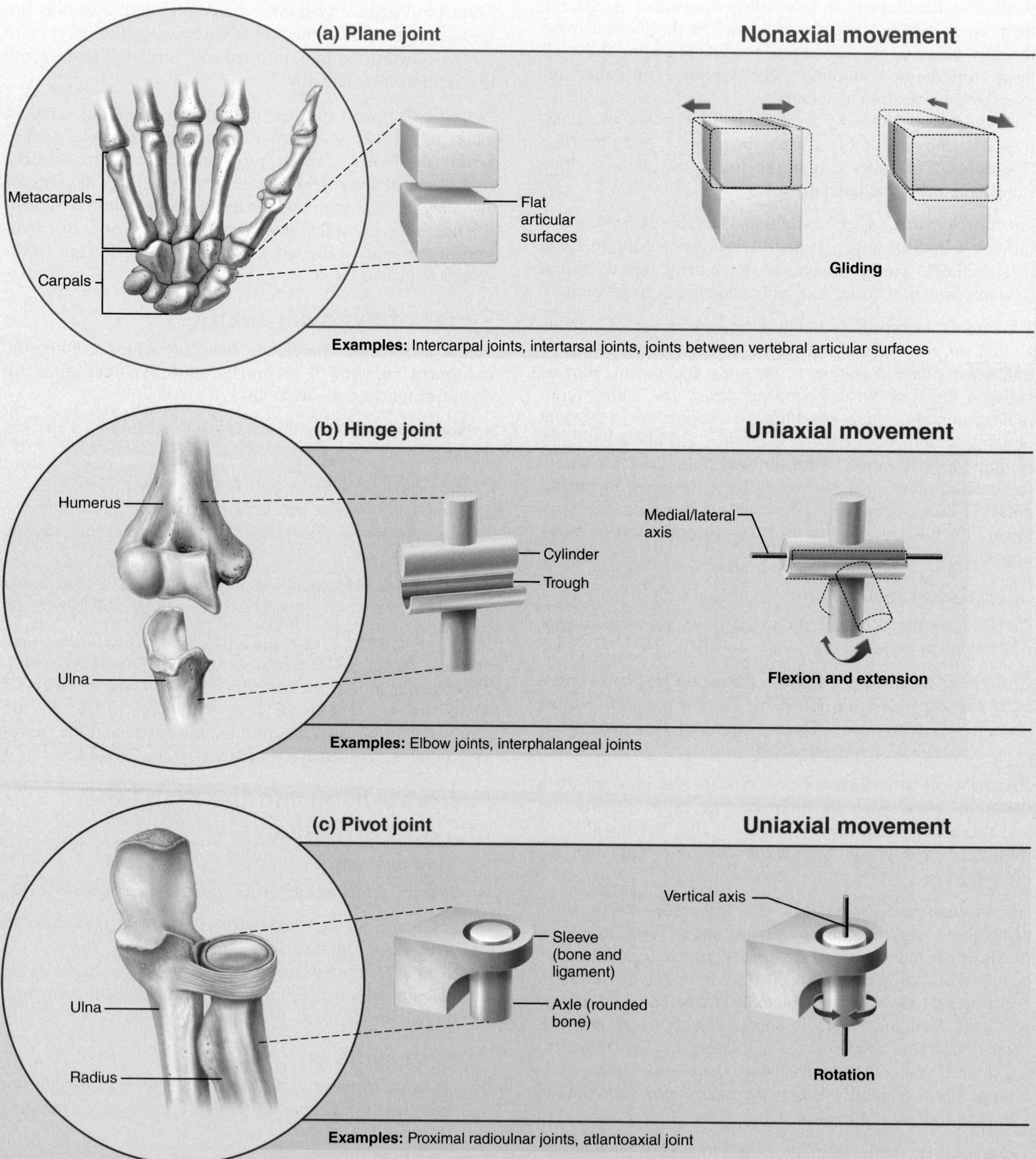

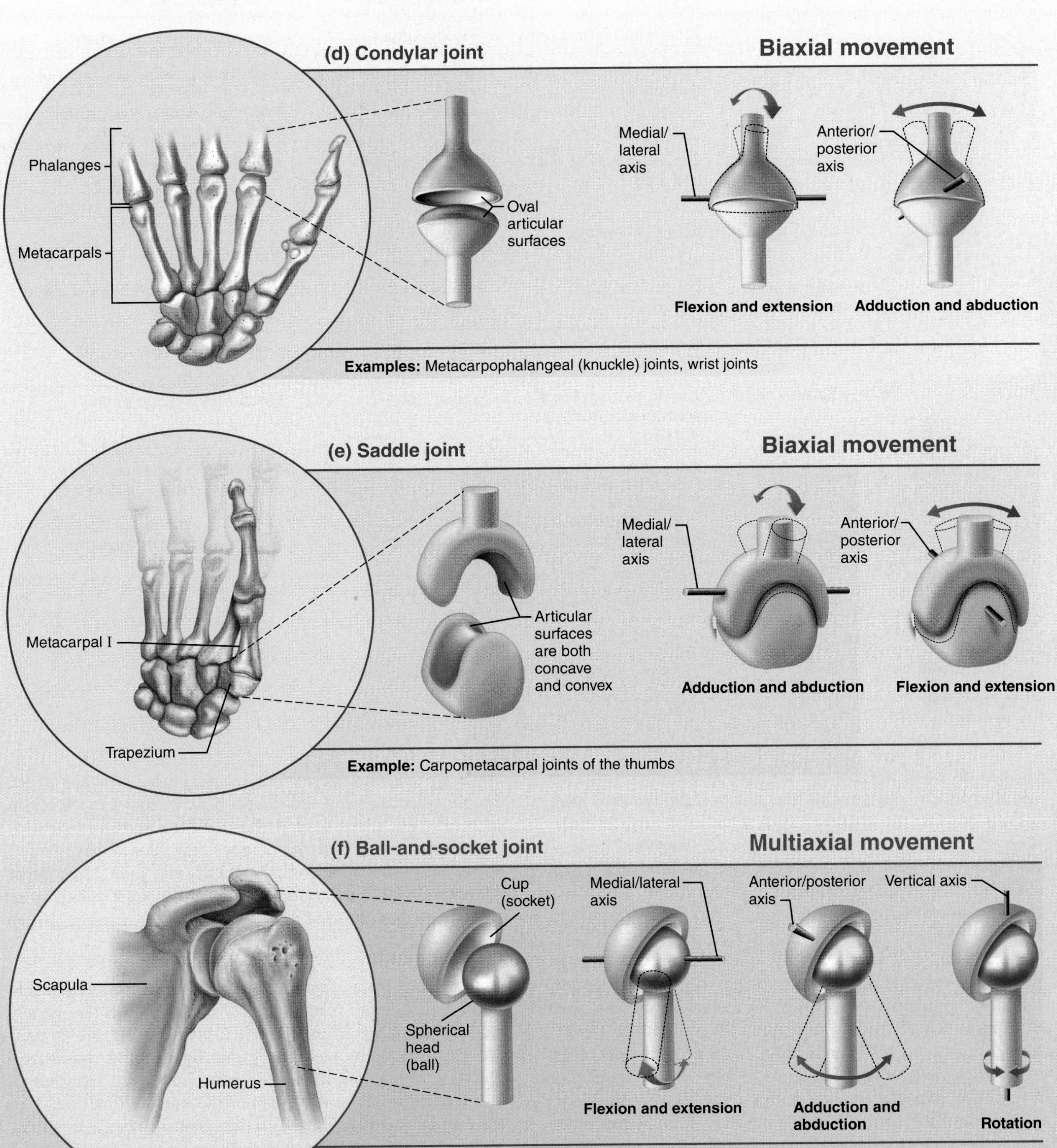

(d) Condylar joint
Biaxial movement
Phalanges
Metacarpals
Oval articular surfaces
Medial/ lateral axis
Anterior/ posterior axis
Flexion and extension
Adduction and abduction
Examples: Metacarpophalangeal (knuckle) joints, wrist joints
(e) Saddle joint
Biaxial movement
Metacarpal I
Trapezium
Articular surfaces are both concave and convex
Medial/ lateral axis
Anterior/ posterior axis
Adduction and abduction
Flexion and extension
Example: Carpometacarpal joints of the thumbs
(f) Ball-and-socket joint
Multiaxial movement
Scapula
Humerus
Cup (socket)
Spherical head (ball)
Medial/lateral axis
Anterior/posterior axis
Vertical axis
Flexion and extension
Adduction and abduction
Rotation
Examples: Shoulder joints and hip joints

Table 9.3 Structural and Functional Characteristics of Body Joints

Illustration	Joint	Articulating Bones	Structural Type	Functional Type; Movements Allowed
	Skull	Cranial and facial bones	Fibrous; suture	Synarthrotic; no movement
	Temporo-mandibular	Temporal bone of skull and mandible	Synovial; modified hinge* (contains articular disc)	Diarthrotic; gliding and uniaxial rotation; slight lateral movement, elevation, depression, protraction, and retraction of mandible
	Atlanto-occipital	Occipital bone of skull and atlas	Synovial; condylar	Diarthrotic; biaxial; flexion, extension, lateral flexion, circumduction of head on neck
	Atlantoaxial	Atlas (C_1) and axis (C_2)	Synovial; pivot	Diarthrotic; uniaxial; rotation of the head
	Intervertebral	Between adjacent vertebral bodies	Cartilaginous; symphysis	Amphiarthrotic; slight movement
	Intervertebral	Between articular processes	Synovial; plane	Diarthrotic; gliding
	Costovertebral	Vertebrae (transverse processes or bodies) and ribs	Synovial; plane	Diarthrotic; gliding of ribs
	Sternoclavicular	Sternum and clavicle	Synovial; shallow saddle (contains articular disc)	Diarthrotic; multiaxial (allows clavicle to move in all axes)
	Sternocostal (first)	Sternum and rib 1	Cartilaginous; synchondrosis	Synarthrotic; no movement
	Sternocostal	Sternum and ribs 2–7	Synovial; double plane	Diarthrotic; gliding

- Fibrous joints
- Cartilaginous joints
- Synovial joints

*These modified hinge joints are structurally bicondylar.

what kinds of movement are possible at the joint, articular surfaces seldom play a major role in joint stability: Most joint sockets are just too shallow. Still, some joint surfaces have deep sockets or grooves that do provide stability. The best example is the ball and deep socket of the hip joint; other examples are the elbow and ankle joints.

Ligaments

The capsules and ligaments of synovial joints help hold the bones together and prevent excessive or undesirable motions. Ligaments located on the medial or inferior side of a joint resist excessive abduction; lateral and superiorly located ligaments resist adduction. Anterior ligaments resist excessive extension and lateral rotation; posterior ligaments resist excessive flexion and medial rotation. As a rule, the more ligaments a joint has, the stronger it is. When other stabilizing factors are inadequate, undue tension is placed on the ligaments, and they fail. Once stretched, ligaments, like taffy, stay stretched. However, a ligament can stretch only about 6% beyond its normal length before it snaps apart. People who can bend the thumb back to touch the forearm or place both heels behind the neck are sometimes called "double-jointed," but of course they don't have more joints than usual. The joint ligaments and joint capsules of "double-jointed" individuals are simply looser and more stretchable than those of most people.

Muscle Tone

Another important factor in joint stabilization is **muscle tone,** a constant, low level of contractile force generated by a muscle even when it is not causing movement. Muscle tone helps stabilize joints by keeping tension on the muscle tendons that cross over joints just external to the joint capsule. In this manner the muscle functions like a ligament holding the adjoining bone surfaces together. This stabilizing factor is especially important in reinforcing the shoulder and knee joints and in supporting the joints in the arches of the foot.

Table 9.3 *continued*

Illustration	Joint	Articulating Bones	Structural Type	Functional Type; Movements Allowed
	Acromio-clavicular	Acromion of scapula and clavicle	Synovial; plane (contains articular disc)	Diarthrotic; gliding and rotation of scapula on clavicle
	Shoulder (glenohumeral)	Scapula and humerus	Synovial; ball-and-socket	Diarthrotic; multiaxial; flexion, extension, abduction, adduction, circumduction, rotation of humerus
	Elbow	Ulna (and radius) with humerus	Synovial; hinge	Diarthrotic; uniaxial; flexion, extension of forearm
	Proximal radioulnar	Radius and ulna	Synovial; pivot	Diarthrotic; uniaxial; rotation of radius around long axis of forearm to allow pronation and supination
	Distal radioulnar	Radius and ulna	Synovial; pivot (contains articular disc)	Diarthrotic; uniaxial; rotation (convex head of ulna rotates in ulnar notch of radius)
	Wrist	Radius and proximal carpals	Synovial; condylar	Diarthrotic; biaxial; flexion, extension, abduction, adduction, circumduction of hand
	Intercarpal	Adjacent carpals	Synovial; plane	Diarthrotic; gliding
	Carpometacarpal of digit I (thumb)	Carpal (trapezium) and metacarpal I	Synovial; saddle	Diarthrotic; biaxial; flexion, extension, abduction, adduction, circumduction, opposition of metacarpal I
	Carpometacarpal of digits II-V	Carpal(s) and metacarpal(s)	Synovial; plane	Diarthrotic; gliding of metacarpals
	Metacarpo-phalangeal	Metacarpal and proximal phalanx	Synovial; condylar	Diarthrotic; biaxial; flexion, extension, abduction, adduction, circumduction of fingers
	Interphalangeal	Adjacent phalanges	Synovial; hinge	Diarthrotic; uniaxial; flexion, extension of fingers

● Fibrous joints
● Cartilaginous joints
● Synovial joints

check your understanding

☐ **6.** Name all the movements that occur at these joints: (a) elbow, (b) hip, (c) ankle, (d) atlantoaxial joint, (e) metacarpophalangeal joint.

☐ **7.** Classify each of the joints named in question 6 according to joint shape. For each joint, indicate whether it is uniaxial, biaxial, or multiaxial.

☐ **8.** Define pronation and supination. At which joints do these movements occur?

(For answers, see Appendix B.)

SELECTED SYNOVIAL JOINTS

learning outcome

▶ Describe the key features of the temporomandibular, sternoclavicular, shoulder, elbow, wrist, hip, knee, and ankle joints.

Important synovial joints include the temporomandibular, sternoclavicular, shoulder, elbow, wrist, hip, knee, and ankle joints. While reading about them, keep in mind that they all contain a joint cavity, articular cartilages, articular capsules, and synovial membranes.

Table 9.3 Structural and Functional Characteristics of Body Joints *continued*

Illustration	Joint	Articulating Bones	Structural Type	Functional Type; Movements Allowed
	Sacroiliac	Sacrum and coxal bone	Synovial in childhood; increasingly fibrous in adult	Diarthrotic in childhood; modified amphiarthrotic in adult (more during pregnancy)
	Pubic symphysis	Pubic bones	Cartilaginous; symphysis	Amphiarthrotic; slight movement (enhanced during pregnancy)
	Hip	Hip bone and femur	Synovial; ball and socket	Diarthrotic; multiaxial; flexion, extension, abduction, adduction, rotation, circumduction of femur
	Knee (tibiofemoral)	Femur and tibia	Synovial; modified hinge* (contains articular discs)	Diarthrotic; biaxial; flexion, extension of leg, some rotation allowed
	Knee (femoropatellar)	Femur and patella	Synovial; plane	Diarthrotic; gliding of patella
	Superior tibiofibular	Tibia and fibula (proximally)	Synovial; plane	Diarthrotic; gliding of fibula
	Inferior tibiofibular	Tibia and fibula (distally)	Fibrous; syndesmosis	Synarthrotic; slight "give" during dorsiflexion of foot
	Ankle	Tibia and fibula with talus	Synovial; hinge	Diarthrotic; uniaxial; dorsiflexion and plantar flexion of foot
	Intertarsal	Adjacent tarsals	Synovial; plane	Diarthrotic; gliding; inversion and eversion of foot
	Tarsometatarsal	Tarsal(s) and metatarsal(s)	Synovial; plane	Diarthrotic; gliding of metatarsals
	Metatarso-phalangeal	Metatarsal and proximal phalanx	Synovial; condylar	Diarthrotic; biaxial; flexion, extension, abduction, adduction, circumduction of great toe
	Interpha-langeal	Adjacent phalanges	Synovial; hinge	Diarthrotic; unaxial; flexion, extension of toes

- Fibrous joints
- Cartilaginous joints
- Synovial joints

*These modified hinge joints are structurally bicondylar.

Temporomandibular Joint

The **temporomandibular joint (TMJ),** or jaw joint, is a modified hinge joint. It lies just anterior to the ear (Figure 9.9). At this joint, the condylar process of the mandible articulates with the inferior surface of the squamous temporal bone. The mandible's condylar process is egg-shaped, whereas the articular surface on the temporal bone has a more complex shape: Posteriorly, it forms the concave **mandibular fossa.** Anteriorly, it forms a dense knob called the **articular tubercle.** Enclosing the joint is a loose articular capsule, the lateral aspect of which is thickened into a **lateral ligament.** Within the capsule is an articular disc, which divides the synovial cavity into superior and inferior compartments (Figure 9.9b). The two surfaces of the disc allow distinct kinds of movement at the TMJ. First, the concave inferior surface receives the condylar process of the mandible and allows the familiar hingelike movement of depressing and elevating the mandible. Second, the superior surface of the disc glides anteriorly with the condylar process when the mouth is opened wide. This anterior movement braces the condylar process against the dense bone of the articular tubercle, so that the mandible is not forced superiorly through the thin roof of the mandibular fossa when one bites hard foods such as nuts or hard candies. To demonstrate the anterior gliding of your mandible, place a

(a) Location of the joint in the skull

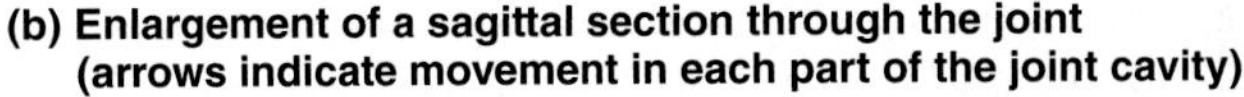

(b) Enlargement of a sagittal section through the joint (arrows indicate movement in each part of the joint cavity)

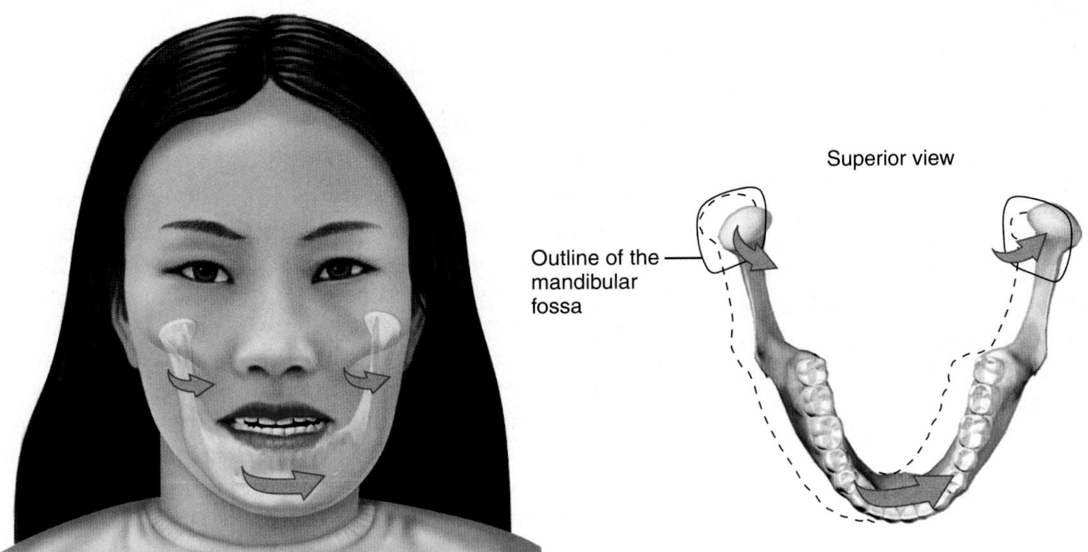

(c) Lateral excursion: lateral (side-to-side) movements of the mandible

Figure 9.9 **The temporomandibular (jaw) joint.**

CLINICAL APPLICATION

Temporomandibular Disorders Because of its shallow socket, the TMJ is the most easily dislocated joint in the body. Even a deep yawn can dislocate it. This joint almost always dislocates anteriorly; the condylar process of the mandible glides anteriorly, ending up in the infratemporal fossa of the skull (see Figure 9.9a) leaving the mouth wide open and unable to close. To realign, the physician places his or her thumbs in the patient's mouth between the lower molars and the cheeks, and then pushes the mandible inferiorly and posteriorly.

Temporomandibular disorders, usually caused by painful spasms of the chewing muscles, can result from stress-induced teeth grinding, an injury to the TMJ, or from poor occlusion of the teeth. The most common symptoms are pain in the ear and face, tenderness of the jaw muscles, popping or clicking sounds when the mouth opens, and stiffness of the TMJ. Treatment usually focuses on getting the jaw muscles to relax using massage, stretching the muscles, applying moist heat or ice, administering muscle-relaxant drugs, bite guards for sleeping, and adopting stress-management techniques.

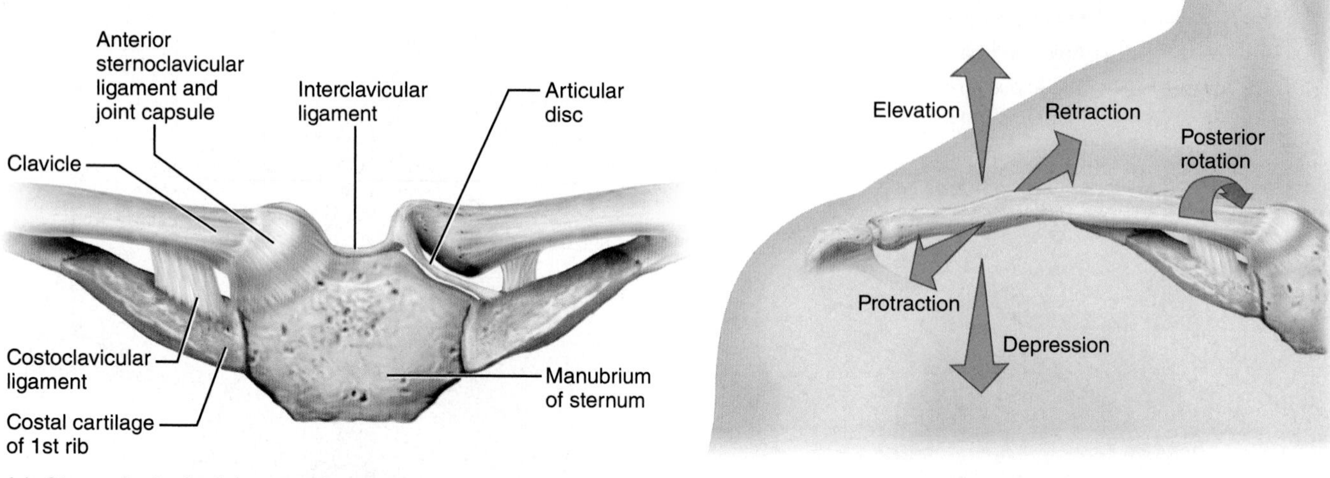

Figure 9.10 **The sternoclavicular joint.**

finger on the condylar process of the mandible just anterior to your ear opening, and yawn. The superior compartment also allows for the side-to-side gliding movements of this joint. As the posterior teeth are drawn into occlusion during grinding, the mandible moves with a side-to-side motion called *lateral excursion* (Figure 9.9c). This lateral joint movement is unique to the masticatory apparatus of mammals and is readily apparent in horses and cows as they chew.

Sternoclavicular Joint

The **sternoclavicular joint (SC)** is a saddle joint (Figure 9.10a). This unusual type of joint is found in only two locations: the sternoclavicular joint and the joint between the trapezium and metacarpal I (the thumb). The saddle-shaped clavicular facet of the sternum and the superior surface of the first costal cartilage articulate with the medial surface of the clavicle. An articular disc within the joint cavity divides the cavity. Four ligaments surround the joint: the **anterior** and **posterior sternoclavicular ligaments;** the **interclavicular ligament,** extending between the medial end of the left and right clavicles; and the **costoclavicular ligament,** extending from the first costal cartilage to the inferior surface of the clavicle. Muscles originating from the sternum, sternocleidomastoid, sternohyoid, and sternothyroid also contribute to the stability of this joint.

This uniquely shaped joint allows for multiple complex movements (Figure 9.10b) and is critical for the mobility of the upper extremity. To demonstrate the three planes of movement of the sternoclavicular joint, place one hand on the junction of the sternum and the clavicle. First, shrug your shoulders to feel elevation and depression of the sternoclavicular joint; second, reach your arm forward and backward maximally to feel protraction and retraction; and finally, abduct or flex your arm to feel posterior rotation (this movement is quite subtle). The SC joint also forms the only bony attachment of the axial skeleton to the pectoral girdle. The sternoclavicular joint is a well reinforced, extremely stable joint. Forceful blows directed medially more commonly result in fracture of the clavicle than dislocation of the SC joint.

check your understanding

- ☐ 9. Both the sternoclavicular and temporomandibular joints contain an articular disc. What is the function of this disc in each of these joints?
- ☐ 10. Which other joint described in this chapter contains an articular disc?

(For answers, see Appendix B.)

Shoulder (Glenohumeral) Joint

In the **shoulder joint** (Figure 9.11), stability has been sacrificed to provide the most freely moving joint of the body. This ball-and-socket joint is formed by the head of the humerus and the shallow glenoid cavity of the scapula. Even though the glenoid cavity is slightly deepened by a rim of fibrocartilage called the **glenoid labrum** (*labrum* = lip) (Figure 9.11a and d), this shallow cavity contributes little to joint stability. The articular capsule (Figure 9.11c) is remarkably thin and loose (qualities that contribute to the joint's freedom of movement) and extends from the margin of the glenoid cavity to the anatomical neck of the humerus. The only strong thickening of the capsule is the superior **coracohumeral ligament,** which helps support the weight of the upper limb. The anterior part of the capsule thickens slightly into three rather weak **glenohumeral ligaments** (Figure 9.11d).

Muscle tendons that cross the shoulder joint contribute most to the joint's stability. One of these is the tendon of the long head of the biceps brachii muscle (see Figure 9.11a and d). This tendon attaches to the superior margin of the glenoid labrum, travels within the joint cavity, and runs in the intertubercular sulcus of the humerus, in the process securing the head of the humerus tightly against the glenoid cavity. Four other tendons and the associated muscles make up the **rotator cuff** (see Figure 9.11e), which encircles the shoulder joint and merges with the joint capsule. The rotator cuff muscles include the subscapularis, supraspinatus, infraspinatus, and teres minor (see Chapter 11, p. 304). Moving the arm vigorously can severely stretch or tear the rotator cuff. Baseball pitchers who throw too hard and too often can easily injure the rotator cuff.

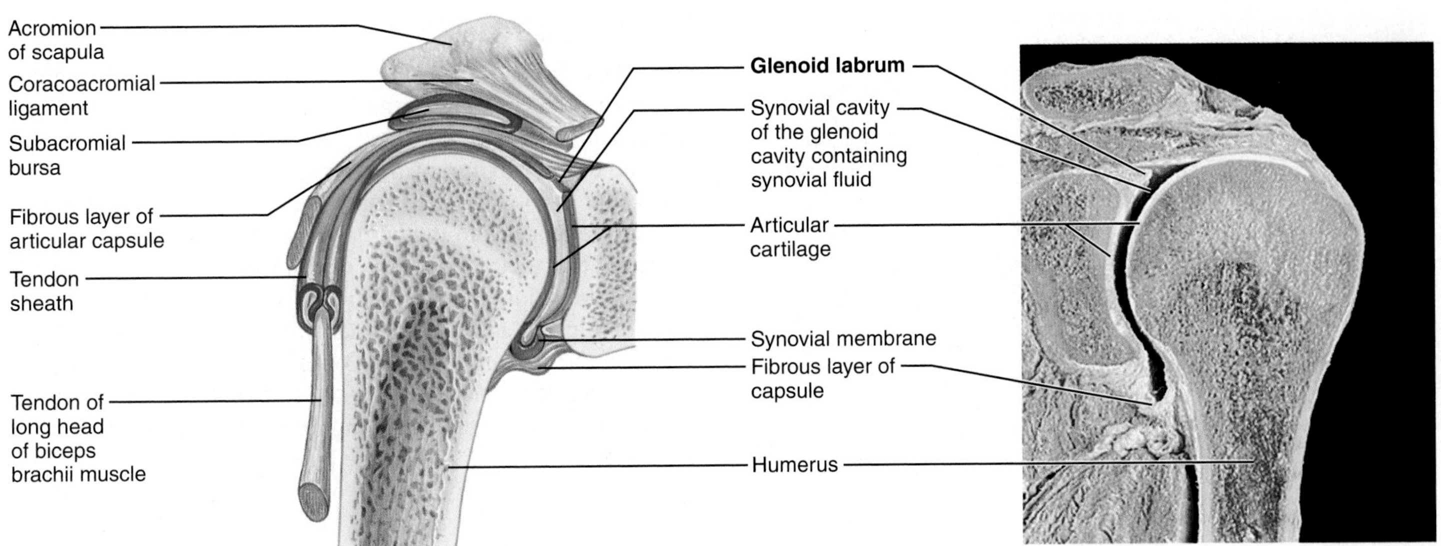

(a) Frontal section through right shoulder joint

(b) Cadaver photo corresponding to (a)

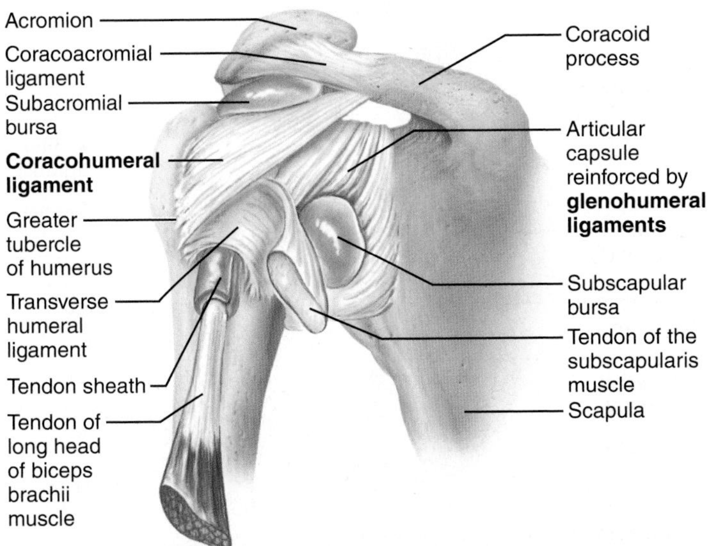

(c) Anterior view of right shoulder joint capsule

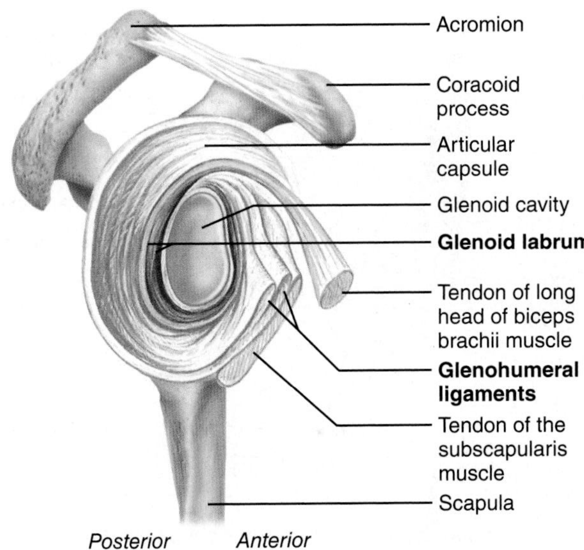

(d) Lateral view of socket of right shoulder joint, humerus removed

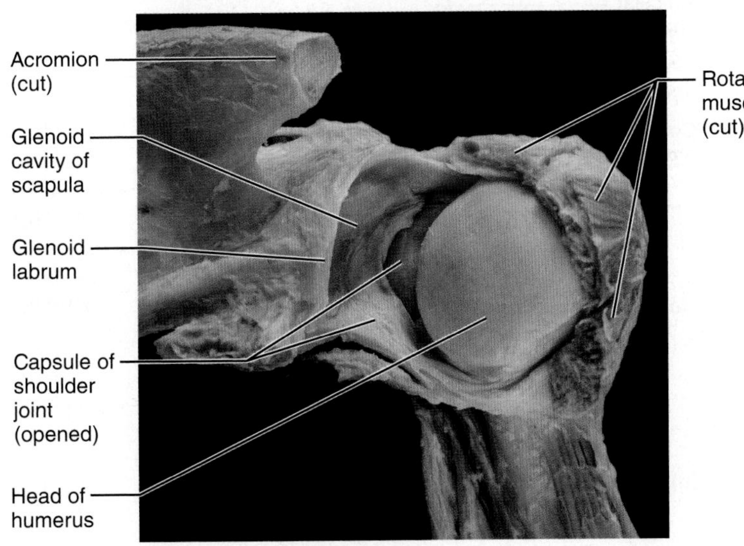

(e) Posterior view of an opened right shoulder joint

Figure 9.11 **The shoulder joint.**

> **CLINICAL APPLICATION**
>
> **Shoulder Dislocations** One price of mobility in the shoulder is that **shoulder dislocations** are common injuries. Because the structures reinforcing this joint are weakest anteriorly and inferiorly, the head of the humerus easily dislocates forward and downward. The glenoid cavity provides poor support when the humerus is rotated laterally and abducted, as when a football player uses the arm to tackle an opponent or a baseball fielder hits the ground diving for a ball. These situations cause many shoulder dislocations, as do blows to the top and back of the shoulder. A **shoulder separation** is a dislocation of the acromioclavicular joint. It results from falling onto an outstretched hand or onto the side of the shoulder.

Elbow Joint

The **elbow joint** *(humeroulnar joint)* (Figure 9.12) is a hinge that allows only extension and flexion. Even though both the radius and ulna articulate with the condyles of the humerus, it is the close gripping of the humerus by the ulna's trochlear notch that forms the hinge and stabilizes the joint. The articular capsule attaches to the humerus and ulna and to the **anular ligament** (an′u-lar; "ringlike") **of the radius,** a ring around the head of the radius (Figure 9.12b–d). Laterally and medially, the capsule thickens into strong ligaments that prevent lateral and medial movements: the **radial collateral ligament,** a triangular band on the lateral side (Figure 9.12b), and the **ulnar collateral ligament** on the medial side (Figure 9.12c and d). Tendons of several arm muscles, such as the biceps brachii and triceps brachii, cross the elbow joint and provide stability (Figure 9.12a).

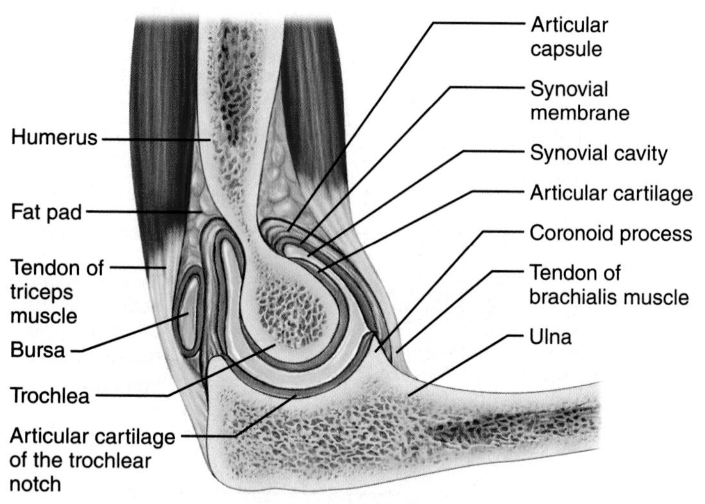

(a) Median sagittal section through right elbow (lateral view)

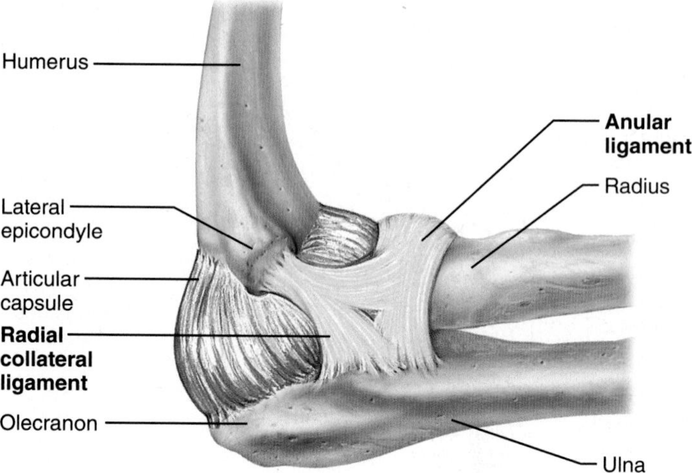

(b) Lateral view of right elbow joint

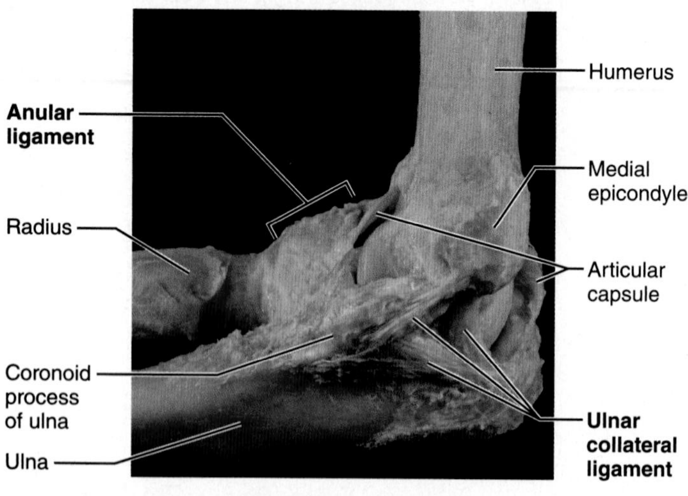

(c) Cadaver photo of medial view of right elbow

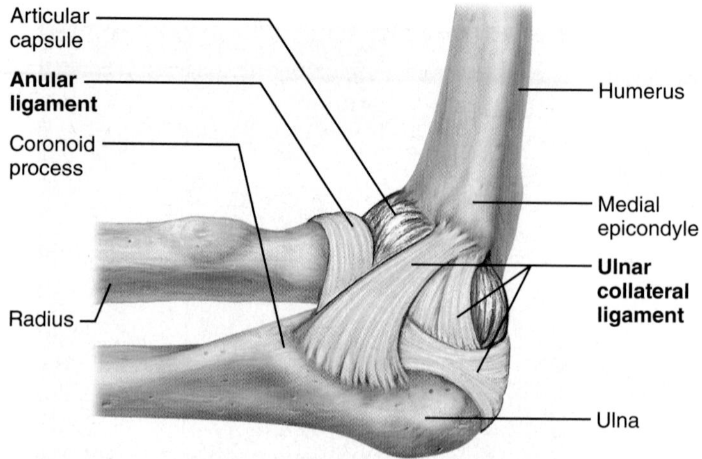

(d) Medial view of right elbow

Figure 9.12 **The elbow joint.**

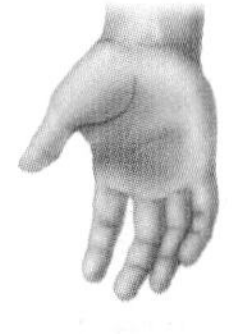

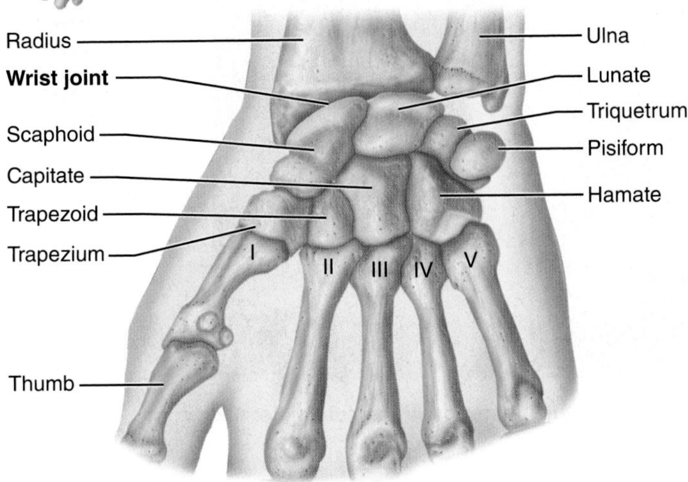

(a) Right wrist, anterior (palmar) view

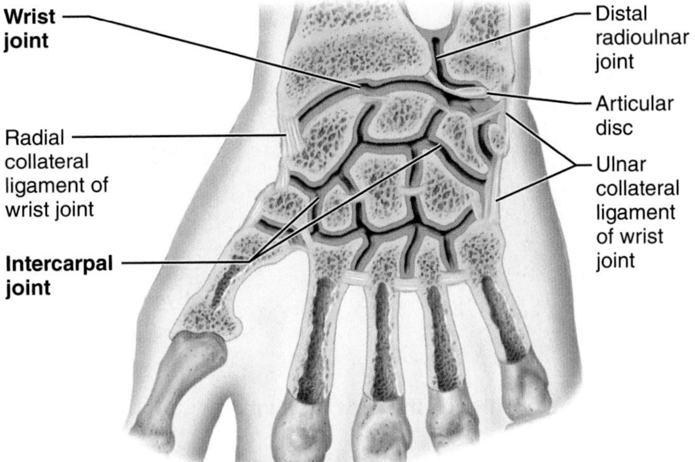

(b) Wrist joints, coronal section

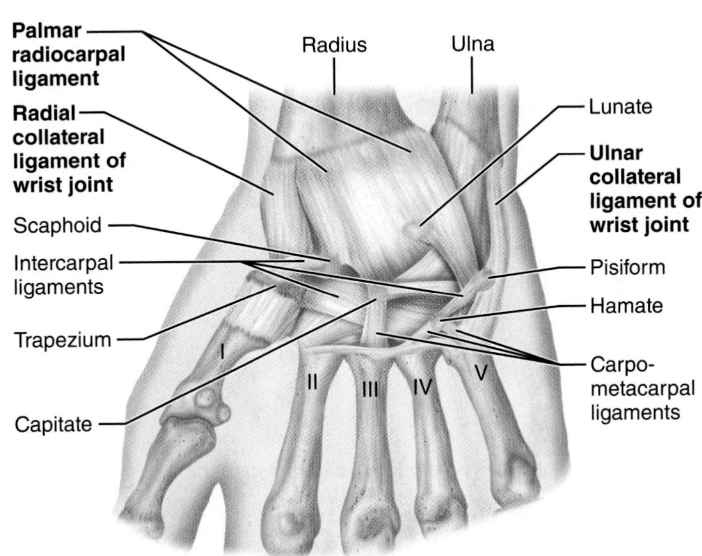

(c) Ligaments of the wrist, anterior (palmar) view

Figure 9.13 **The wrist joints.**

CLINICAL APPLICATION

Elbow Trauma Although the elbow is a very stable joint, it often experiences trauma. Only the jaw and shoulder are dislocated more frequently. Elbow dislocation is usually caused by falling onto an outstretched arm and breaking the fall with the hand. This fall pushes the ulna posteriorly. Such a fall may dislocate the radius posteriorly as well, and it often fractures the articulating elements.

Athletes who repeatedly swing a racket or throw a ball subject their forearms to strong outward-bending forces that can weaken and ultimately rupture the ulnar collateral ligament.

Surgical reconstruction of the ulnar collateral ligament, commonly called "Tommy John surgery" after the Los Angeles Dodgers pitcher who first underwent this procedure, involves an autograft replacement with a tendon from the patient's own forearm, knee, or foot. Full recovery is likely after 6–12 months of rehabilitation.

Wrist Joint

The wrist has two major joint surfaces: the **wrist joint** (radiocarpal) and the **intercarpal,** or **midcarpal, joint.** The wrist joint is the joint between the radius and the proximal carpals, the scaphoid and lunate (Figure 9.13a and b). This joint is a condylar joint, permitting movements of flexion, extension, adduction, abduction, and circumduction. The intercarpal joint is located between the proximal and distal rows of carpals (Figure 9.13b). Gliding occurs at this joint as the adjacent carpals slide by each other.

The wrist is stabilized by several ligaments (Figure 9.13c). Four major ligaments extending from the forearm bones to the carpals reinforce this joint: the **palmar radiocarpal ligament** anteriorly, the **dorsal radiocarpal ligament** posteriorly, the **radial collateral ligament of the wrist joint** laterally, and the **ulnar collateral ligament of the wrist joint** medially. There are multiple smaller ligaments extending between carpal bones that connect the carpals to each other and to the metacarpals (Figure 9.13c).

✓ check your understanding

☐ 11. Of the shoulder, elbow, or wrist, which joint is the most stable? Which is the least stable?

☐ 12. What structures contribute most to stability of the shoulder joint?

☐ 13. Which forearm bone forms part of the elbow joint? Which forms part of the wrist joint?

(For answers, see Appendix B.)

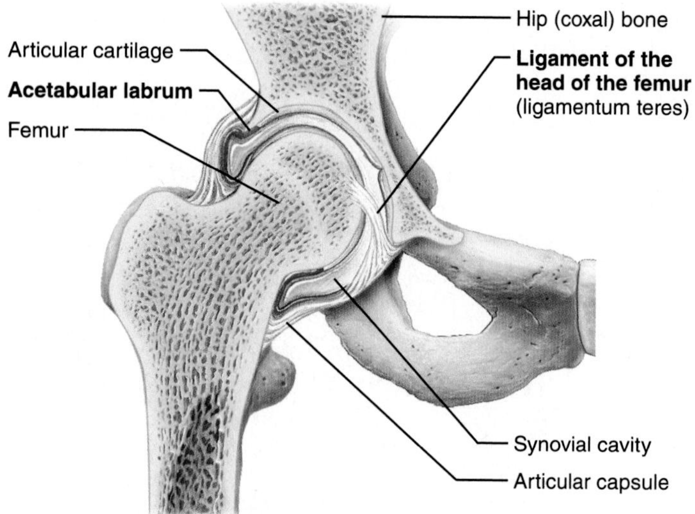

(a) Frontal section through the right hip joint

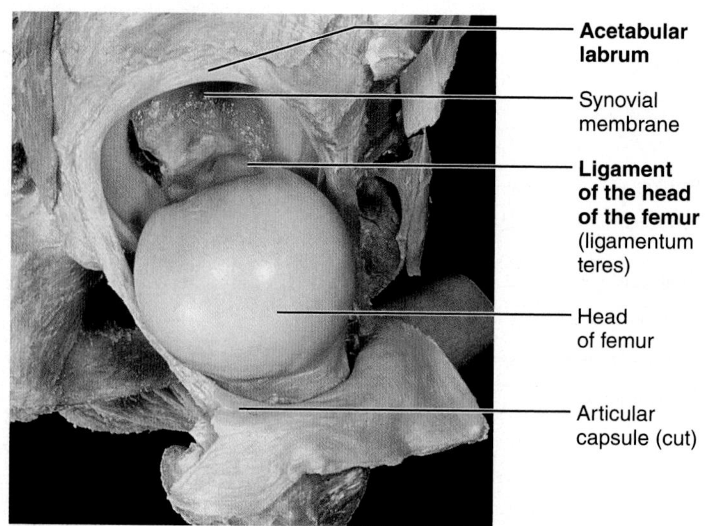

(b) Photo of the interior of the hip joint, lateral view

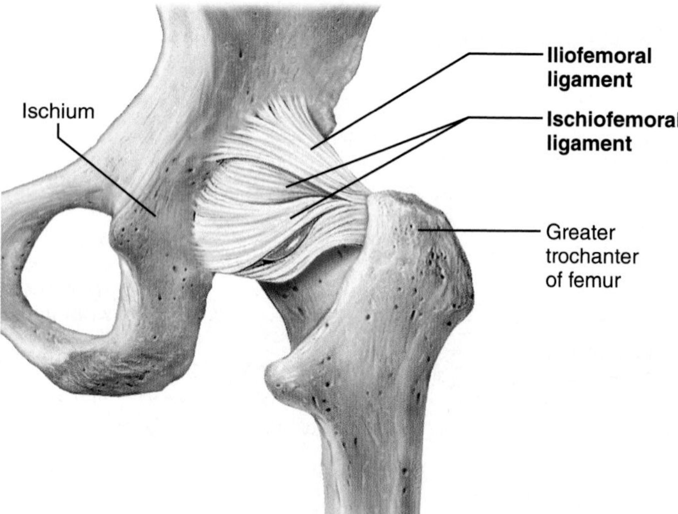

(c) Posterior view of right hip joint, capsule in place

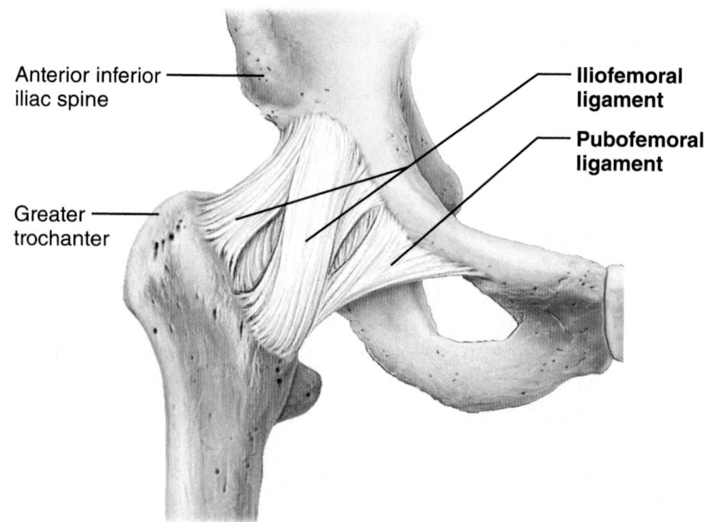

(d) Anterior view of right hip joint, capsule in place

Figure 9.14 The hip joint.

Hip Joint

The **hip joint,** like the shoulder joint, has a ball-and-socket structure (Figure 9.14). It has a wide range of motion, but not nearly as wide as that of the shoulder joint. Movements occur in all possible axes but are limited by the joint's ligaments and deep socket.

The hip joint is formed by the spherical head of the femur and the deeply cupped acetabulum of the hip bone. The depth of the acetabulum is enhanced by a circular rim of fibrocartilage called the **acetabular labrum** (Figure 9.14a and b). Because the diameter of this labrum is smaller than that of the head of the femur, the femur cannot easily slip out of the socket, and hip dislocations are rare. The joint capsule runs from the rim of the acetabulum to the neck of the femur (Figure 9.14c and d).

Three external ligamentous thickenings of this capsule reinforce the joint: the **iliofemoral ligament,** a strong, V-shaped, anteriorly located ligament; the **pubofemoral ligament,** a triangular thickening of the capsule's inferior region (Figure 9.14d); and the **ischiofemoral ligament,** a spiraling, posteriorly located ligament (Figure 9.14c). These three ligaments are arranged in such a way that they "screw" the head of the femur into the acetabulum when a person stands erect, thereby increasing the stability of the joint.

The **ligament of the head of the femur** (Figure 9.14a and b) is a flat, intracapsular band that runs from the head of the femur to the inferior region of the acetabulum. This ligament remains slack during most hip movements, so it is not important in stabilizing the joint. Its mechanical function is unknown, but it does contain an artery that helps supply the head of the femur. Damage to this artery may lead to arthritis of the hip joint.

Muscle tendons that cross the hip joint contribute to its stability, as do the fleshy parts of many hip and thigh muscles that surround the joint. In this joint, however, stability comes chiefly from the cupped socket and the capsular ligaments.

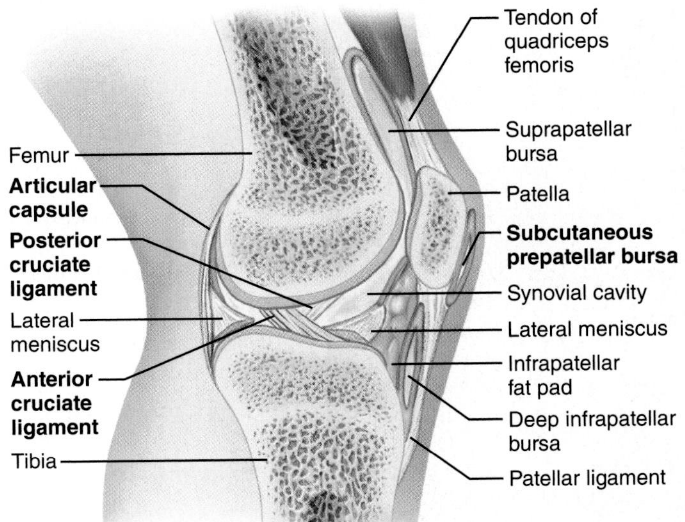

(a) Sagittal section through the right knee joint

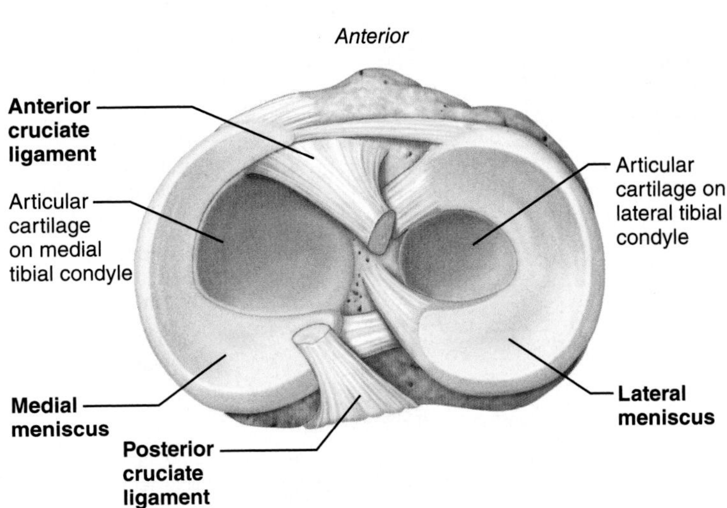

(b) Superior view of the right tibia in the knee joint, showing the menisci and cruciate ligaments

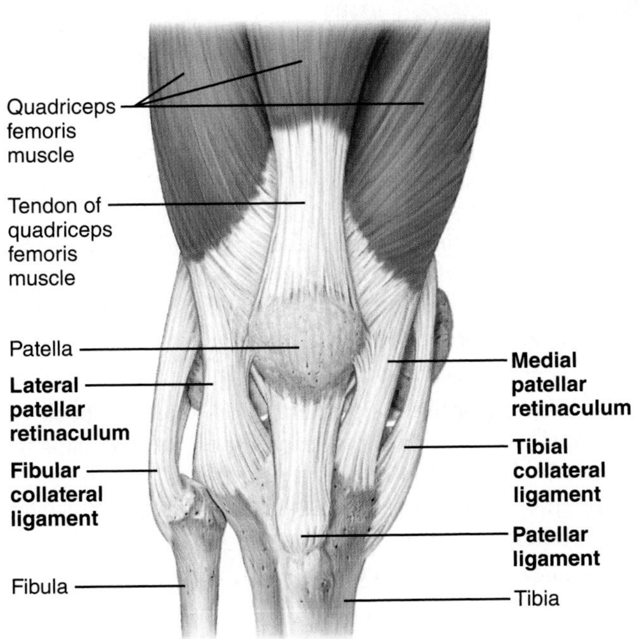

(c) Anterior view of right knee

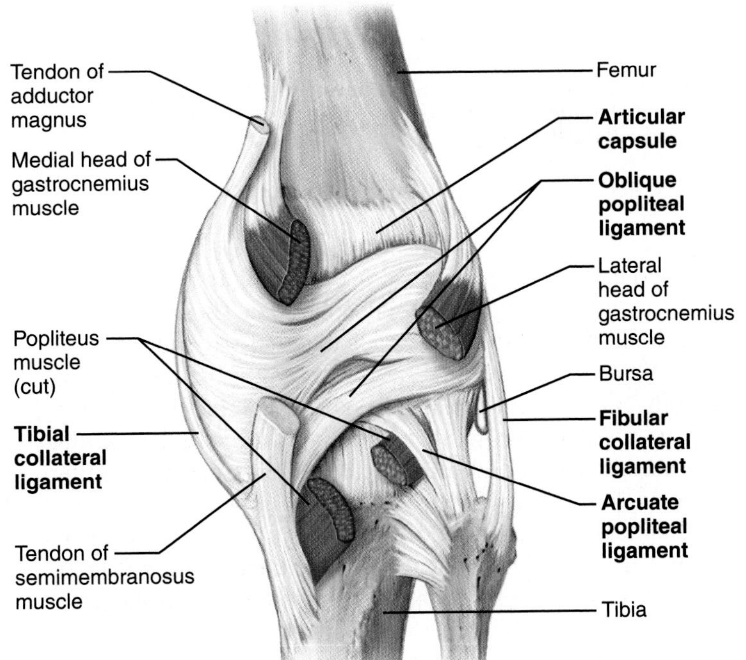

(d) Posterior view of the joint capsule, including ligaments

Figure 9.15 **The knee joint.**

Knee Joint

The **knee joint,** the largest and most complex joint in the body (Figure 9.15), primarily acts as a hinge. However, it also permits some medial and lateral rotation when in the flexed position and during the act of leg extension. Structurally, it is compound and bicondyloid, because both the femur and tibia have two condylar surfaces. In this joint, the wheel-shaped condyles of the femur roll along the almost-flat condyles of the tibia like tires on a road. Sharing the knee cavity is an articulation between the patella and the inferior end of the femur (Figure 9.15a); this *femoropatellar joint* is a plane joint that allows the patella to glide across the distal femur as the knee bends.

The synovial cavity of the knee joint has a complex shape (Figure 9.15a), with several incomplete subdivisions and several extensions that lead to "blind alleys." At least a dozen bursae are associated with this joint, some of which are shown in the figure. The **subcutaneous prepatellar bursa** is often injured when the knee is bumped anteriorly.

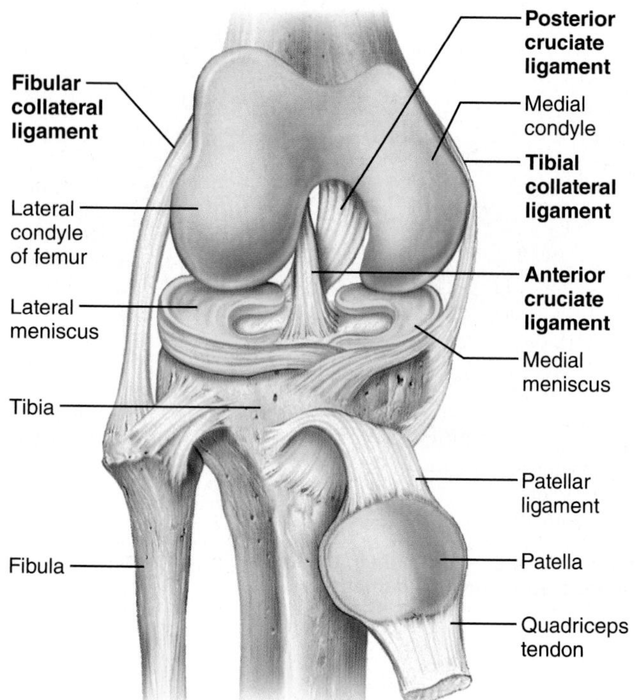

(e) Anterior view of flexed knee, showing the cruciate ligaments (articular capsule removed, and quadriceps tendon cut and reflected distally)

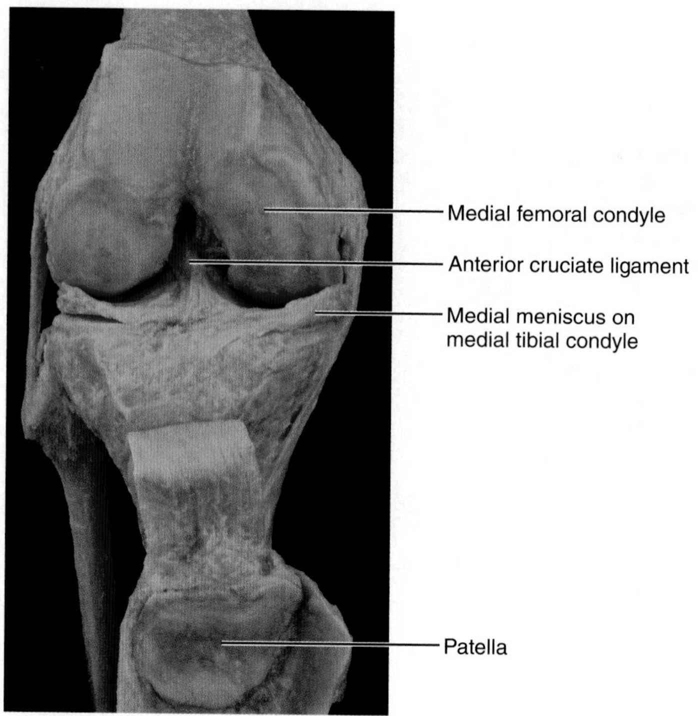

(f) Photograph of an opened knee joint; view similar to (e)

Figure 9.15 **The knee joint,** *continued.*

Two fibrocartilage menisci occur within the joint cavity, between the femoral and tibial condyles. These C-shaped **lateral** and **medial menisci** attach externally to the condyles of the tibia (Figure 9.15b). Besides evening the distribution of both compressive load and synovial fluid, the menisci help to stabilize the joint by guiding the condyles during flexion, extension, and rotation movements and preventing side-to-side rocking of the femur on the tibia.

The articular capsule of the knee joint can be seen on the posterior and lateral aspects of the knee (Figure 9.15d), where it covers most parts of the femoral and tibial condyles. Anteriorly, however, the capsule is absent. Instead, this anterior area is covered by three broad ligaments that run inferiorly from the patella to the tibia (Figure 9.15c): the **patellar ligament,** flanked by the **medial** and **lateral patellar retinacula** (ret″ĭ-nak′u-lah; "retainers"). The patellar ligament is actually a continuation of the tendon of the main muscles on the anterior thigh, the quadriceps femoris. Physicians tap the patellar ligament to test the knee-jerk reflex.

The joint capsule of the knee is reinforced by several capsular and extracapsular ligaments, all of which become taut when the knee is extended to prevent hyperextension of the leg at the knee.

1. The extracapsular **fibular** and **tibial collateral ligaments** are located on the lateral and medial sides of the joint capsule, respectively (Figure 9.15c–e). The fibular collateral ligament descends from the lateral epicondyle of the femur to the head of the fibula. The tibial collateral ligament runs from the medial epicondyle of the femur to the medial condyle of the tibia. Besides halting leg extension and preventing hyperextension, these collateral ligaments prevent the leg from moving laterally and medially at the knee.
2. The **oblique popliteal ligament** (pop′lĭ-te″al; "back of the knee") crosses the posterior aspect of the capsule (Figure 9.15d). Actually it is a part of the tendon of the semimembranosus muscle that fuses with the joint capsule and helps stabilize the joint.
3. The **arcuate popliteal ligament** arcs superiorly from the head of the fibula over the popliteus muscle to the posterior aspect of the joint capsule (Figure 9.15d).

In addition, the knee joint is stabilized by two strong *intracapsular* ligaments called *cruciate ligaments* (kru′-she″āt) because they cross each other like an X (*crus* = cross) (Figure 9.15a, b, and e). Each runs from the tibia to the femur and is named for its site of attachment to the tibia. The **anterior cruciate ligament** attaches to the *anterior* part of the tibia, in the intercondylar area. From there, it passes posteriorly to attach to the femur on the medial side of the lateral condyle. The **posterior cruciate ligament** arises from the *posterior* intercondylar area of the tibia and passes anteriorly to attach to the femur on the lateral side of the medial condyle.

Functionally, the cruciate ligaments act as restraining straps to prevent undesirable movements at the knee joint (Figure 9.16). The anterior cruciate helps prevent anterior sliding of the tibia. The posterior cruciate, which is even stronger than the anterior cruciate, prevents forward sliding

① During movement of the knee, the anterior cruciate prevents anterior sliding of the tibia; the posterior cruciate prevents posterior sliding of the tibia.

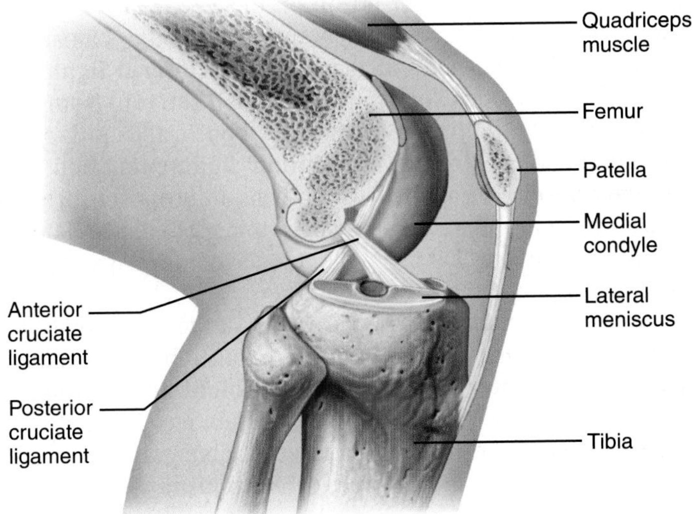

② When the knee is fully extended, both cruciate ligaments are taut and the knee is locked.

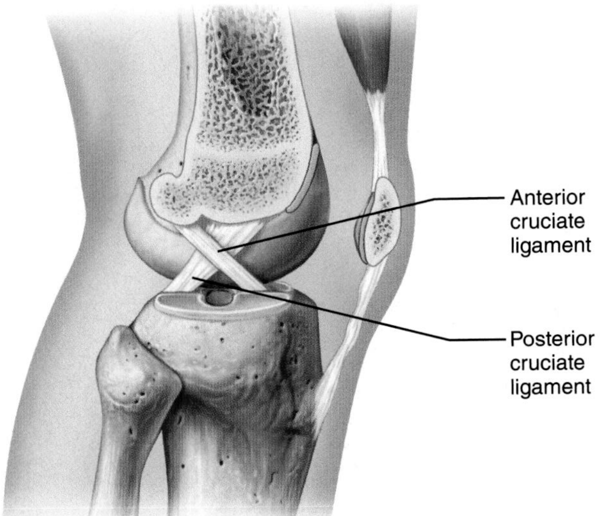

Figure 9.16 Stabilizing function of the cruciate ligaments.

of the femur or backward displacement of the tibia. The two cruciates also function together to lock the knee when one stands (discussed shortly).

The tendons of many muscles reinforce the joint capsule and act as critical stabilizers of the knee joint. Most important are the tendons of the quadriceps femoris and semimembranosus muscles (see Figure 9.15c and d). The greater the strength and tone of these muscles, the less the chance of knee injury.

CLINICAL APPLICATION

Knee Injuries Knee injuries occur frequently in contact sports because even though ligaments and muscles strongly hold the knee together, its articular surfaces offer no stability. That is, the nearly flat tibial surface has no socket to secure the femoral condyles. Thus, the knee joint is especially vulnerable to *horizontal blows*, such as occur in tackling and body blocks in football. Most dangerous are *lateral* blows, which tear the tibial collateral ligament and the medial meniscus attached to it, as well as the anterior cruciate ligament (ACL), an injury coined the "unhappy triad."

Injuries to the ACL alone are increasing rapidly in noncontact sports. Of particular interest is the greater incidence of ACL injuries in women athletes as women's sports become more vigorous and competitive. Numerous factors are implicated in the higher incidence of ACL injuries in women: wider pelvis, narrower intercondylar fossa, effects of female hormones on joint laxity, and slower muscle reaction times. This injury is a common noncontact injury in soccer, basketball, and volleyball. Most ACL injuries result when a runner stops and changes direction quickly, twisting a hyper-extended leg. A torn ACL heals poorly, so it must be replaced surgically, usually with a graft from either the patellar ligament, the calcaneal tendon, or the semitendinosus tendon (see pp. 317, 327, and 320). Training regimens are being developed to decrease the incidence of ACL injuries in women athletes. Injuries to the *posterior* cruciate ligament, caused by posteriorly directed blows to the upper tibia, are less common and more able to heal on their own.

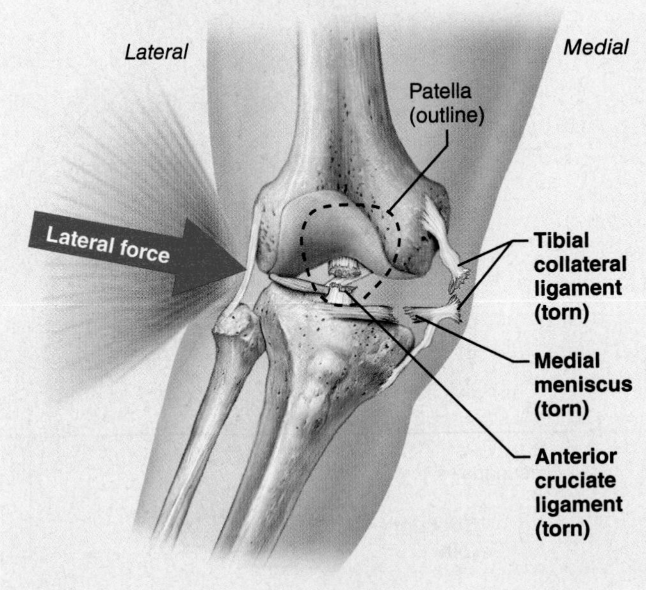

The "unhappy triad": ruptured ACL, ruptured tibial collateral ligament, and torn meniscus. A common injury in American football.

The knees have a built-in locking device that provides steady support for the body in the standing position. As a person stands up, the flexed leg begins to extend at the knee, and the femoral condyles roll like ball bearings on the tibial condyles. Then, as extension nears completion, the lateral femoral condyle stops rolling before the medial condyle stops. This causes the femur to rotate medially on the tibia until both cruciate and both collateral ligaments stretch tight and halt all movement (Figure 9.16b). The tension in these ligaments locks the knee into a rigid structure that cannot be flexed again until it is unlocked by a muscle called the popliteus (see Table 11.16, pp. 326–328), which rotates the femur laterally on the tibia.

Ankle Joint

The **ankle joint** is a hinge joint between (1) the united inferior ends of the tibia and fibula, and (2) the talus of the foot (Figure 9.17). This joint allows only dorsiflexion and plantar flexion. Inversion and eversion occur at the intertarsal joints (Figure 9.17a). The ankle joint has a capsule that is thin anteriorly and posteriorly, but thickened with ligaments medially and laterally. The strong **medial (deltoid) ligament** (Figure 9.17b) runs from the medial malleolus of the tibia down to a long line of insertion on the navicular and talus bones, and on the sustentaculum tali of the calcaneus bone. On the other side of the ankle, the **lateral ligament** (Figure 9.17c) consists of three bands that run from the fibula's lateral malleolus to the foot bones: the horizontal **anterior** and **posterior talofibular ligaments,** and the **calcaneofibular ligament,** which runs inferoposteriorly to reach the calcaneus. Functionally, the medial and lateral ligaments prevent anterior and posterior slippage of the talus and the foot.

Forming the deep, U-shaped socket of the ankle joint, the inferior ends of the tibia and fibula are joined by ligaments of their own: the **anterior** and **posterior tibiofibular ligaments** and the lower part of the interosseus membrane (Figure 9.17c and d). These ligaments, called the tibiofibular syndesmosis, stabilize the socket so that forces can be transmitted to it from the foot.

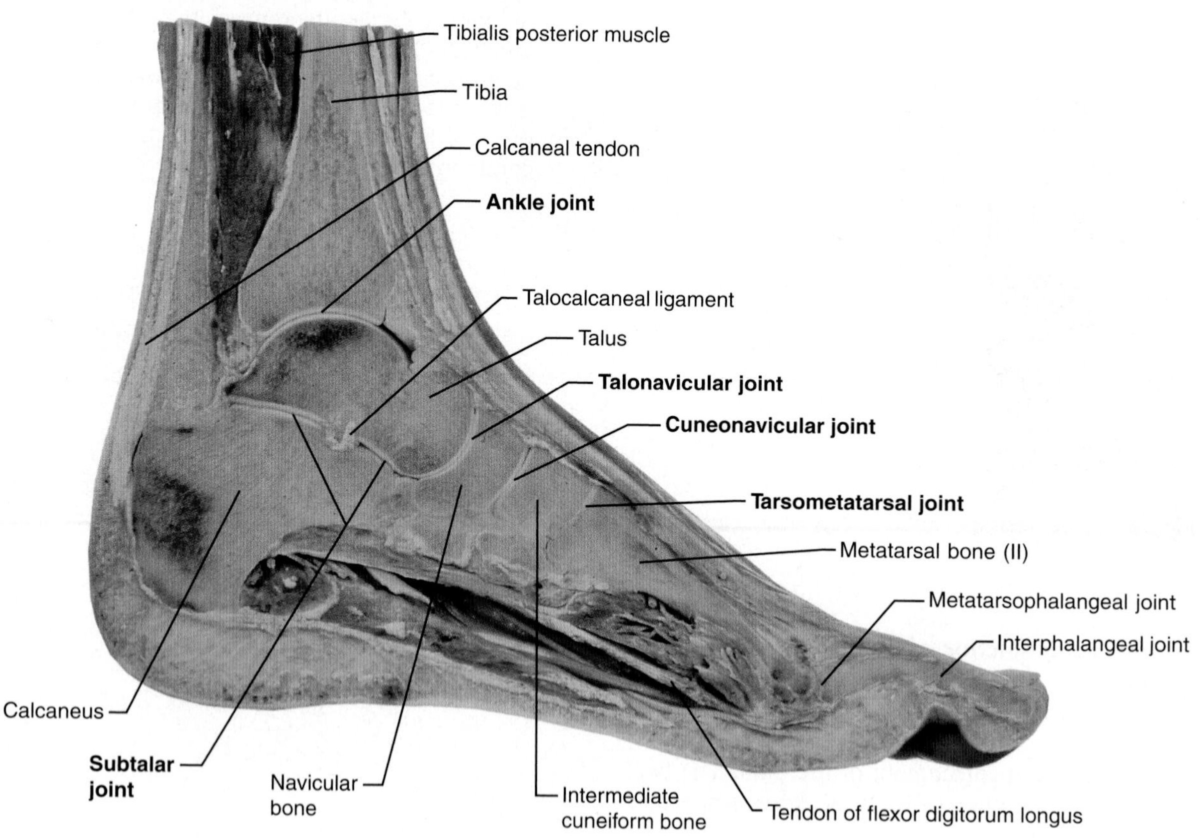

(a) Cadaver photo of ankle and foot, sagittal section

Figure 9.17 **The ankle joint.**

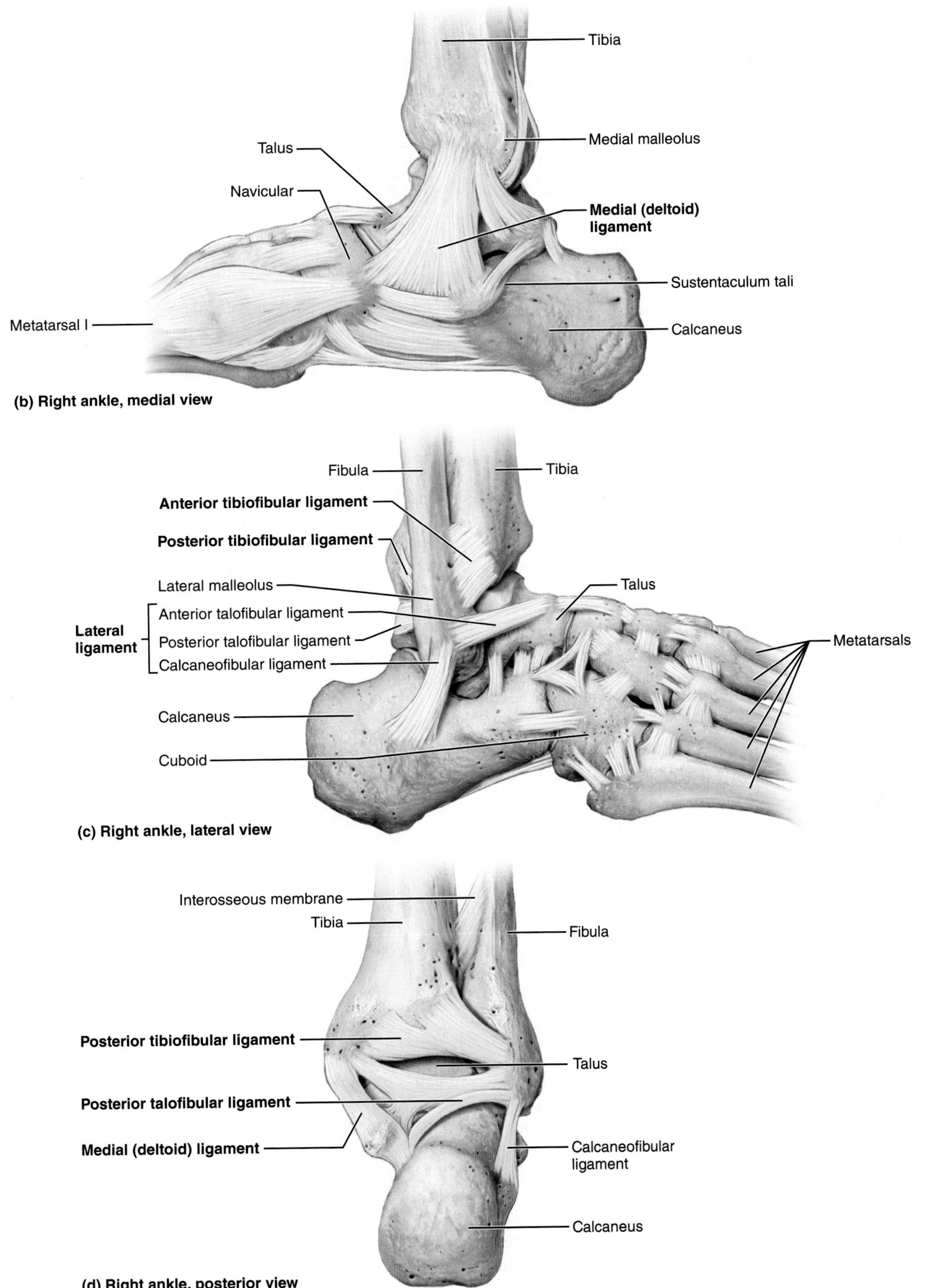

Figure 9.17 **The ankle joint,** *continued.*

CLINICAL APPLICATION

Ankle Sprains The ankle is the most frequently injured joint in the lower limb, and ankle sprains are the most common sports injuries. (A sprain is a stretched or torn ligament in a joint; see the next section, "Disorders of Joints.") Most ankle sprains are caused by excessive inversion of the foot and involve the lateral ligament. *Syndesmosis ankle sprains,* by contrast, are caused by extreme dorsiflexion, internal rotation, or external rotation of the foot; all these actions move the talus in ways that wedge the tibia away from the fibula and stretch the tibiofibular ligaments. Ankle sprains are treated with the **RICE** regimen: rest, ice, compression, and elevation, followed by ankle-strengthening exercises as recovery begins.

check your understanding

☐ 14. Name the intracapsular ligaments found in the hip and the knee.

☐ 15. Which ligament is injured in an ankle sprain resulting from forceful inversion?

☐ 16. The articular surfaces of the knee contribute little to the stability of this joint. What additional structural features aid in stabilizing the knee?

(For answers, see Appendix B.)

DISORDERS OF JOINTS

learning outcomes

- ▶ Name the most common injuries to joints, and discuss the problems associated with each.
- ▶ Name and describe the types of arthritis.

The structure and function of joints make them especially vulnerable to a variety of disorders. Because they experience the same strong forces as the hard skeleton yet consist of soft tissue, joints are prone to *injuries* from traumatic stress. And because their function in movement subjects them to friction and wear, joints can be afflicted by *inflammatory and degenerative processes.*

Joint Injuries

Three common types of joint injuries are torn cartilage, sprains, and dislocations.

Torn Cartilage

Even though most cartilage injuries involve tearing of the meniscus in the knee, tears and overuse injuries in the articular cartilages of other joints are becoming increasingly common in competitive athletes, particularly gymnasts.

Cartilage tears in the knee often happen when a meniscus is simultaneously subjected to both high compression and shear stresses. For example, a tennis player lunging to return a ball can rotate the flexed knee medially with so much force that it tears both the joint capsule and the medial meniscus attached to it (Figure 9.18). Because cartilage is avascular, it can rarely repair itself; thus, torn cartilage usually stays torn.

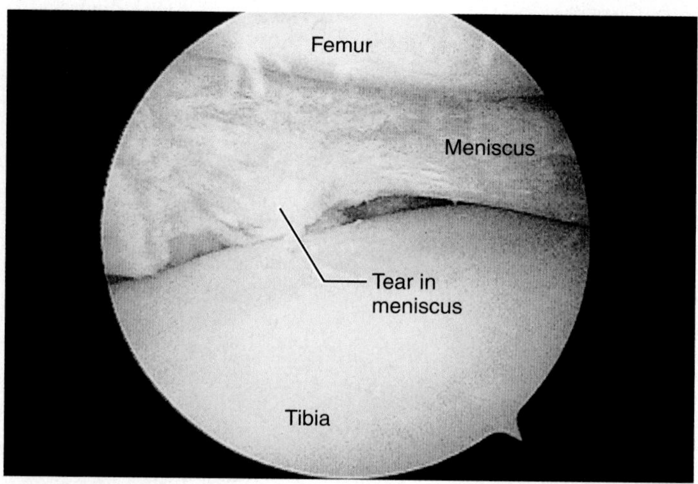

Figure 9.18 Arthroscopic photograph of a torn medial meniscus.

Because cartilage fragments (also called *loose bodies*) can cause joints to lock or bind, the recommended treatment often is surgical removal of the damaged cartilage.

Repair is achieved using a remarkable outpatient procedure called **arthroscopic surgery** (ar-thro-skop′ik; "looking into joints"). Using miniaturized video and surgical equipment inserted into the joint through a single small incision, the surgeon removes the cartilage fragments and repairs the ligaments, in the process minimizing scarring and tissue damage and speeding healing. Arthroscopic surgery has applications in many joints, not just the knee.

CLINICAL APPLICATION

Autologous Cartilage Implantation Damaged joint cartilages heal poorly (p. 127), and coaxing them to heal better is a major goal of medical research. Although much progress remains to be made, it has been found that a patient's own chondrocytes, when cultured in the lab and then grafted onto joint surfaces, can produce enough new hyaline cartilage to fill small holes or breaks in the articular cartilages of the knee. This is called *autologous cartilage implantation,* and its long-term goal is to grow thicker sheets of cartilage to cover entire joint surfaces.

Sprains

In a **sprain,** the ligaments reinforcing a joint are stretched or torn. Common sites of sprains are the lumbar region of the spine, the ankle, and the knee. Partly torn ligaments eventually repair themselves, but they heal slowly because ligaments are poorly vascularized. Sprains tend to be painful and immobilizing. Completely ruptured ligaments, by contrast, require prompt surgical repair because inflammation in the joint breaks down the neighboring tissues and turns the injured ligament to "mush." Surgical repair can be a difficult task: A ligament consists of hundreds of fibrous strands, and sewing a ligament back together has been compared to sewing two hairbrushes together. When important ligaments are too severely damaged to be repaired, they must be removed and replaced with grafts or substitute ligaments.

Dislocations

A **dislocation (luxation)** occurs when the bones of a joint are forced out of alignment. This injury is usually accompanied by sprains, inflammation, pain, and difficulty in moving the joint. Dislocations may result from serious falls or blows and are common sports injuries. The jaw, shoulder, finger, and thumb joints are most commonly dislocated. Like fractures, dislocations must be reduced; that is, the bone ends must be returned to their proper positions by a physician. **Subluxation** is a partial or incomplete dislocation of a joint. In a subluxation, the bone ends return to their proper position on their own. A joint that has been dislocated once is susceptible to repeated injury because the initial dislocation stretches the joint capsule and ligaments. After the capsule is loosened, the joint is more likely to dislocate again. Injured ligaments eventually shorten to their original length, but this aspect of healing can take years.

CLINICAL APPLICATION

Nursemaid's Elbow A common injury to the forearm in young children (less than 5 years of age) is subluxation of the proximal radioulnar joint. When the forearm is pulled with the arm outstretched, as in swinging a child by her hands, the head of the radius can be pulled away from the anular ligament (see Figure 9.12b–d), resulting in entrapment of the anular ligament in the humeroradial joint. The immediate pain that occurs upon injury subsides and is followed by a reluctance to use the injured arm. A history of pulling on the forearm is critical for diagnosing this injury accurately. Treatment is a simple reduction that results in immediate return to use of the arm.

Inflammatory and Degenerative Conditions

Inflammatory conditions that affect joints include inflammations of bursae and tendon sheaths and various forms of arthritis.

Bursitis, Tendonitis, and Tenosynovitis

Bursitis, inflammation of a bursa, usually results from a physical blow or friction, although it may also be caused by arthritis or bacterial infection. In response, the bursa swells with fluid. Falling on one's knee can cause a painful bursitis of the subcutaneous prepatellar bursa (see Figure 9.15a), known as **housemaid's knee.** Resting and rubbing one's elbows on a desk can lead to **student's elbow,** or **olecranon bursitis,** the swelling of a bursa just deep to the skin of the posterior elbow. Severe cases of bursitis may be treated by injecting inflammation-reducing drugs into the bursa. Excessive fluid accumulation may require fluid removal by needle aspiration.

Tendonitis is an inflammation of a tendon. **Tenosynovitis** is inflammation of a tendon sheath. These two conditions commonly occur together. Causes (overuse injury or infection), symptoms (pain, swelling, tenderness) and treatments (rest, ice, anti-inflammatory drugs) mirror those of bursitis.

Arthritis

The term **arthritis** describes over 100 kinds of inflammatory or degenerative diseases that damage the joints. In all its forms, arthritis is the most widespread crippling disease in the United States: One of every five Americans suffers from its effects. All forms of arthritis have, to a greater or lesser degree, the same initial symptoms: pain, stiffness, and swelling of the joint.

Osteoarthritis (Degenerative Joint Disease) The most common type of arthritis is **osteoarthritis (OA),** a chronic (long-term) degenerative condition that is often called "wear-and-tear arthritis." It is most common in the aged and is probably related to the normal aging process. OA affects women more often than men, but most of us will develop this condition by the age of 80. OA affects the articular cartilages, causing them to soften, fray, crack, and erode.

The cause of OA is unknown. According to current theory, normal use causes joints to release metalloproteinase enzymes that break down the cartilage matrix (especially the collagen fibrils); meanwhile, the chondrocytes continually repair the damage by secreting more matrix. Whenever the strain on a joint is repeated or excessive, too much of the cartilage-destroying enzyme is thought to be released, causing OA. Because this process occurs most where an uneven orientation of forces causes extensive microdamage, badly aligned or overworked joints are likely to develop OA.

The bone directly below the articular cartilage is also affected, becoming dense and stiff. As the disease progresses, bone spurs tend to grow around the margins of the damaged cartilages, encroaching on the joint cavity and perhaps restricting joint movement (Figure 9.19). Patients complain of stiffness upon waking in the morning, but this decreases within a half hour. However, there is always joint pain during use. The affected joints may make a crunching noise (called *crepitus*) as their roughened surfaces rub together during movement.

The joints most commonly affected in OA are those of the fingers, knuckles, hips, and knees. The nonsynovial joints between the vertebral bodies are also susceptible, especially in the cervical and lumbar regions of the spine.

The course of OA is slow and irreversible. It is not usually crippling, except for some severe cases involving the hip and knee joints. Inflammation may or may not accompany the degeneration of the joints, but it is usually not severe. In many cases the symptoms of OA can be controlled with a pain reliever such as aspirin or acetaminophen plus a program of low-impact exercise. Rubbing a hot-pepper-like substance called capsaicin on the skin over the joint also helps lessen the pain of OA. As more aging individuals are enjoying active lifestyles, joint replacement has become a common treatment for joints severely damaged by arthritis (Figure 9.19c).

Rheumatoid Arthritis Another kind of arthritis, **rheumatoid** (roo″mah-toid) **arthritis (RA)** is a chronic inflammatory disorder. Its onset usually occurs between the ages of

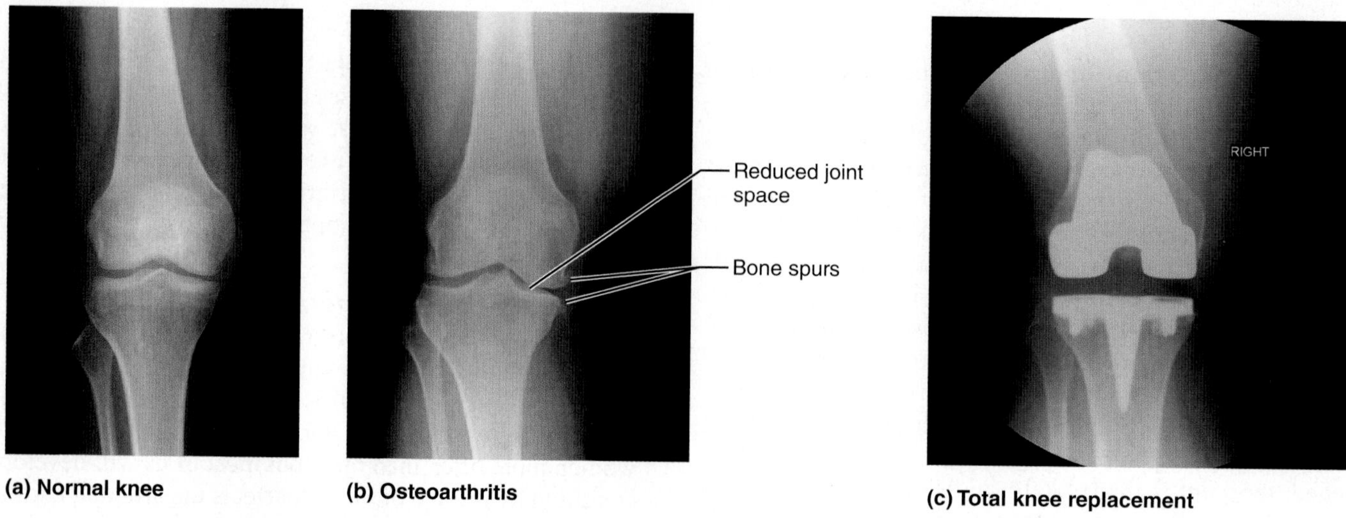

Figure 9.19 Osteoarthritis of the knee.

30 and 50, but it may arise at any age. The course of RA is variable: It may develop gradually or in spurts that are years apart. It is marked by flare-ups and remissions. RA tends to affect many joints simultaneously and bilaterally (on both sides of the body), especially the small joints of the fingers, wrists, ankles, and feet. Along with pain and swelling of the joints, other manifestations of RA include osteoporosis, muscle weakness, and cardiovascular problems.

RA is an **autoimmune disease**—a disorder in which the body's immune system attacks its own tissues. The cause of this reaction is unknown, but RA might follow infection by certain bacteria and viruses that bear surface molecules similar to molecules normally present in the joints. When the body is stimulated to attack the foreign molecules, it inappropriately destroys its own joint tissues as well.

RA begins with an inflammation of the synovial membrane. Capillaries in this membrane leak tissue fluid and white blood cells into the joint cavity. This excess synovial fluid then causes the joint to swell. The chronic inflammation and swelling can deteriorate the connective tissues surrounding the joint. In the hand, the loss of integrity of the joint capsules in the metacarpophalangeal and interphalangeal joints leads to the characteristic deformities due to RA (Figure 9.20). With time, the inflamed synovial membrane thickens into a *pannus* ("rag"), a coat of granulation tissue that clings to the articular cartilages. The pannus erodes the cartilage and, often, the underlying bone. As the cartilage is destroyed, the pannus changes into a fibrous scar tissue that interconnects the bone ends. This scar tissue eventually ossifies, and the bone ends fuse together, immobilizing the joint. This end condition, called **ankylosis** (ang″kĭ-lo′sis; "stiff condition"), often produces bent, deformed fingers.

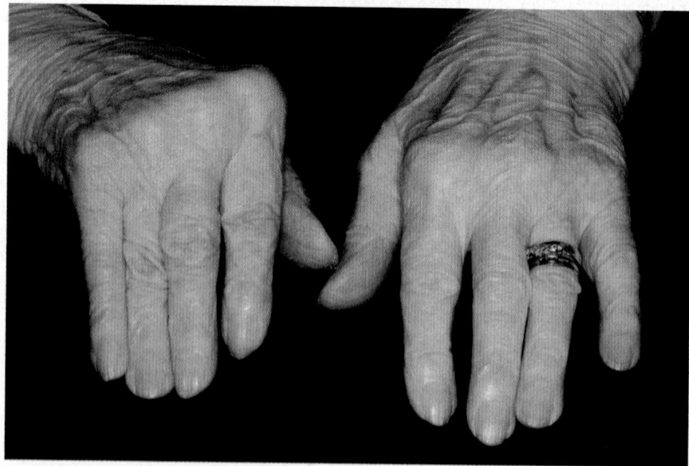

Figure 9.20 Rheumatoid arthritis. Classic signs of RA in the hand on the left: prominent ulnar deviation of digits and swelling in the metacarpophalangeal joints.

Most drugs used to treat RA are either anti-inflammatory drugs or immunosuppressants. These must be given for long periods and are only partly successful; there is no cure. Joint prostheses are the last resort for severely crippled patients. Some RA sufferers have over a dozen artificial joints.

Gouty Arthritis (Gout) Ordinarily, people maintain proper blood levels of uric acid, a normal waste product of nucleic acid metabolism, by excreting it in the urine. If the rate of excretion is low, uric acid levels rise abnormally in the blood and body fluids, and the acid precipitates as solid crystals of urate in the synovial membranes. An inflammatory response follows as the body tries to attack and digest the crystals, producing an agonizingly painful attack of **gouty** (gow′te) **arthritis,** or **gout.** The initial attack involves a single joint, usually in the lower limb, often at the base of the big toe. Other attacks usually follow, months to years later. Gout is far more common in men than in women because men naturally have higher levels of uric acid in their blood (perhaps because estrogens increase the rate of its excretion).

Untreated gout can cause the ends of articulating bones to fuse, immobilizing the joint. Fortunately, effective treatment is available. For acute attacks, nonsteroidal anti-inflammatory drugs such as ibuprofen are used. For the long term, urate-lowering drugs and dietary measures (avoidance of alcohol and red meat) are effective.

☑ check your understanding

☐ 17. Identify the type of arthritis described in each case: (a) crystallization of uric acid in synovial membranes; (b) erosion of articular cartilage; (c) autoimmune response causing inflammation of the synovial membrane.

☐ 18. Why is an injured joint more susceptible to repeat of the injury following a sprain or subluxation?

(For answers, see Appendix B.)

THE JOINTS THROUGHOUT LIFE

learning outcome

▶ Describe how joints develop and how their function may be affected by aging.

Synovial joints develop from mesenchyme that fills the spaces between the cartilaginous "bone models" in the late embryo. The outer region of this intervening mesenchyme condenses to become the fibrous layer of the joint capsule, whereas the inner region hollows to become the joint cavity. By week 8, these joints resemble adult joints in form and arrangement; that is, their synovial membranes are developed, and synovial fluid is being secreted into the joint cavities. These basic structural features are genetically determined, but after birth a joint's size, shape, and flexibility are modified by use. Active joints grow larger and have thicker capsules, ligaments, and bony supports than they would if they were never used.

During youth, many injuries to joints tear or knock the epiphysis off the shaft, a vulnerability that ends after the epiphyseal plates close in early adulthood. From that time on, comparable injuries merely result in sprains.

Osteoarthritis is the most common joint problem associated with advancing age. Just as bones must be stressed to maintain their strength, joints must be exercised to keep their health. Exercise squeezes synovial fluid in and out of articular cartilages, providing the cartilage cells with the nourishment they need. And although exercise cannot prevent osteoarthritis, it strengthens joints and slows degeneration of the articular cartilages. It also strengthens the muscles that stabilize the joints. Overexercising, however, worsens osteoarthritis. Because the buoyancy of water relieves much of the stress on weight-bearing joints, people who exercise in pools often retain good joint function throughout life.

☑ check your understanding

☐ 19. By what age of development are synovial joints formed?

(For the answer, see Appendix B.)

RELATED CLINICAL TERMS

Ankylosing spondylitis (ang″kĭ-lo′sing spon″dĭ-li′tis) (*ankyl* = stiff joint; *spondyl* = vertebra) A distinctive kind of rheumatoid arthritis that mainly affects men. It usually begins in the sacroiliac joints and progresses superiorly along the spine. The vertebrae become interconnected by so much fibrous tissue that the spine becomes rigid ("poker back").

Arthroplasty ("joint reforming") Replacing a diseased joint with an artificial joint (see Figure 9.19).

Chondromalacia patellae ("softening of cartilage of the patella") Damage and softening of the articular cartilages on the posterior surface of the patella and the anterior surface of the distal femur. This condition, seen most often in adolescent athletes, produces a sharp pain in the knee when the leg is extended (in climbing stairs, for example). Chondromalacia may result when the quadriceps femoris, the main group of muscles on the anterior thigh, pulls unevenly on the patella, persistently rubbing it against the femur in the knee joint. Chondromalacia can often be corrected by exercises that strengthen weakened parts of the quadriceps muscles.

Lyme disease An inflammatory disease that often results in joint pain and arthritis, especially in the knee joint. It is caused by spirochetes, bacterial organisms transmitted by the bites of ticks that live on deer and mice. Lyme disease is first characterized by a skin rash and flulike symptoms. It is treatable with antibiotics, especially if detected early.

Patellofemoral pain syndrome Persistent pain in the region behind the patella; results from rubbing pressure between the femoral condyles and the patella, as may occur with overuse of the quadriceps femoris muscles or an abnormally shaped patella. Is distinguished from chondromalacia patellae (see above) by usual absence of damage to the cartilage and generally younger age at onset. Treatment is exercise that strengthens the different quadriceps muscles evenly and gradually.

Synovitis Inflammation of the synovial membrane of a joint. Can result from an injury, an infection, or arthritis. Synovitis leads to the production of excess fluid, causing the joints to swell. Such accumulation of fluid in the joint cavity is called *effusion*.

Valgus and varus injuries *Valgus* means "bent outward, away from the body midline," such as in abduction of the leg at the knee or the forearm at the elbow. *Varus* means "bent inward," such as in adduction of these elements. Because the knee and elbow are not designed for such movements, strong valgus and varus bending injures them.

CHAPTER SUMMARY

Take full advantage of the invaluable online practice and study tools for this chapter by going to the Study Area of MasteringA&P®. There you will find:

- Art Labeling Activities, Quizzes, Practice Tests, Memory Games
- Animal Organ Dissection, Cat Dissection and Bone Videos
- PAL Cadaver
- PAL Histology
- A&P Flix Animations
- Dynamic Study Modules

1. Joints (articulations) are sites where elements of the skeleton meet. They hold bones together and allow various degrees of movement.

CLASSIFICATION OF JOINTS (p. 209)

2. Joints are classified functionally as synarthrotic (no movement), amphiarthrotic (slight movement), or diarthrotic (free movement). They are classified structurally as fibrous, cartilaginous, or synovial.

FIBROUS JOINTS (pp. 209–210)

3. In fibrous joints, the bones are connected by fibrous connective tissue. No joint cavity is present. Nearly all fibrous joints are synarthrotic (immovable).

4. The types of fibrous joints are sutures (between skull bones), syndesmoses ("ligament joints"), and gomphoses (articulation of the teeth with their sockets).

CARTILAGINOUS JOINTS (p. 210)

5. In cartilaginous joints, the bones are united by cartilage. No joint cavity exists.

6. Synchondroses are immovable joints of hyaline cartilage, such as epiphyseal plates. Symphyses are amphiarthrotic (slightly movable) fibrocartilage joints, such as intervertebral discs and the pubic symphysis.

SYNOVIAL JOINTS (pp. 210–221)

7. Most joints in the body are synovial. Synovial joints are diarthrotic (freely movable).

General Structure of Synovial Joints (pp. 211–213)

8. Synovial joints have a fluid-containing joint cavity and are covered by an articular capsule. The capsule has an outer fibrous layer, often reinforced by ligaments, and an inner synovial membrane that produces synovial fluid. The articulating ends of bone are covered with impact-absorbing articular cartilages. Nerves in the capsule provide a sense of "joint stretch."

9. Synovial fluid is mainly a filtrate of the blood, but it also contains molecules that make it a friction-reducing lubricant. The cartilage-covered bone ends glide on a slippery film of synovial fluid squeezed out of the articular cartilages. This mechanism is called weeping lubrication.

10. Some joints contain fibrocartilage discs (menisci), which distribute loads evenly and may allow two movements at one joint.

11. Bursae and tendon sheaths are often associated with synovial joints. A bursa is a fibrous sac lined by a synovial membrane and containing synovial fluid. Tendon sheaths, which are similar to bursae, wrap around certain tendons. They act as lubricating devices that allow adjacent structures to move smoothly over one another.

Movements Allowed by Synovial Joints (pp. 213–217)

12. Contracting muscles produce three common kinds of bone movements at synovial joints: gliding, angular movements (flexion, extension, abduction, adduction, and circumduction), and rotation.

13. Special movements include elevation and depression, protraction and retraction, supination and pronation of the forearm, opposition of the thumb, inversion and eversion of the foot, and dorsiflexion and plantar flexion of the foot.

Types of Synovial Joints (p. 217)

14. The shapes of the articular surfaces reflect the kinds of movements allowed at a joint. Joints are classified by shape as plane (nonaxial), hinge or pivot (uniaxial), condylar or saddle (biaxial), or ball-and-socket (multiaxial) joints.

Factors Influencing the Stability of Synovial Joints (pp. 217–221)

15. Joints are the weakest part of the skeleton. Factors that stabilize joints are the shapes of the articulating surfaces, ligaments, and the tone of muscles whose tendons cross the joint.

SELECTED SYNOVIAL JOINTS (pp. 221–234)

16. The temporomandibular joint is a modified hinge joint formed by (1) the condylar process of the mandible and (2) the mandibular fossa and articular tubercle of the temporal bone. This joint allows both a hingelike opening of the mouth and an anterior gliding of the mandible. It often dislocates anteriorly and is frequently the site of stress-induced temporomandibular disorders.

17. The sternoclavicular joint is a saddle joint between the medial end of the clavicle and the manubrium of the sternum. This joint allows for elevation, depression, protraction, retraction, and slight rotation. It is an extremely stable joint.

18. The shoulder (glenohumeral) joint is a ball-and-socket joint between the glenoid cavity and the head of the humerus. It is the body's most freely movable joint. Its socket is shallow, and its capsule is lax and reinforced by ligaments superiorly and anteriorly. The tendons of the biceps brachii and the rotator cuff help stabilize the joint. The humerus often dislocates anteriorly and inferiorly at the shoulder joint.

19. The elbow is a hinge joint in which the ulna and radius articulate with the humerus. It allows flexion and extension. Its articular surfaces are highly complementary and help stabilize it. Radial and ulnar collateral ligaments prevent side-to-side movement of the forearm.

20. The wrist is composed of a condylar joint between the distal end of the radius and the proximal carpals (scaphoid, lunate) that allows for flexion, extension, abduction, and adduction of the hand. The intercarpal joint between the proximal row of carpals and the distal row of carpals is a gliding joint allowing sliding movements between the carpals.

21. The hip joint is a ball-and-socket joint between the acetabulum of the hip bone and the head of the femur. It is adapted for weight bearing. Its articular surfaces are deep and secure, and its capsule is strongly reinforced by three ligamentous thickenings.

22. The knee is a complex and shallow joint formed by the articulation of the tibial and femoral condyles (and anteriorly by the patella with the femur). Extension, flexion, and some rotation are allowed.

C-shaped menisci occur on the articular surfaces of the tibia. The capsule, which is absent anteriorly, is reinforced elsewhere by several capsular and extracapsular ligaments. For example, the tibial and fibular collateral ligaments help prevent hyperextension, abduction, and adduction of the leg. The intracapsular cruciate ligaments help prevent anterior-posterior displacement of the joint surfaces and help lock the knee when one stands. The tone of the muscles crossing this joint is important for knee stability. In contact sports, lateral blows are responsible for many knee injuries. In noncontact sports, the anterior cruciate ligament is a frequent site of injury caused by quick stopping and change of direction.

23. The ankle joint is a hinge joint between (1) the distal tibia and fibula and (2) the talus of the foot. Its medial and lateral ligaments allow plantar flexion and dorsiflexion but prevent anterior and posterior displacements of the foot. Inversion and eversion of the foot occur through gliding movements at the intertarsal joints.

PAL **Human Cadaver/Joints**

DISORDERS OF JOINTS (pp. 234–237)

Joint Injuries (pp. 234–235)

24. Trauma can tear joint cartilages. Sports injuries to the knee menisci are common and can result from twisting forces as well as from direct blows to the knee. Arthroscopic surgery is used to repair knee and other joint injuries.

25. Sprains are the stretching and tearing of joint ligaments. Because ligaments are poorly vascularized, healing is slow.

26. Joint dislocations move the surfaces of articulating bones out of alignment. Such injuries must be reduced.

Inflammatory and Degenerative Conditions (pp. 235–237)

27. Bursitis and tenosynovitis are inflammation of a bursa and a tendon sheath, respectively. Tendonitis is inflammation of a tendon.

28. Arthritis is the inflammation or degeneration of joints accompanied by pain, swelling, and stiffness. Arthritis includes many different diseases and affects millions of people.

29. Osteoarthritis is a degenerative condition that first involves the articular cartilages. It follows wear or excessive loading and is common in older people. Weight-bearing joints are most often affected.

30. Rheumatoid arthritis, the most crippling kind of arthritis, is an autoimmune disease involving severe inflammation of the joints, starting with the synovial membranes.

31. Gout is joint inflammation caused by a deposition of urate salts in the synovial membranes.

THE JOINTS THROUGHOUT LIFE (p. 237)

32. Joints develop from mesenchyme between the cartilaginous "bone models" in the embryo.

33. Joints usually function well until late middle age, when osteoarthritis almost always appears.

REVIEW QUESTIONS

Multiple Choice/Matching Questions

(For answers, see Appendix B.)

1. Match the joint types in the key to the descriptions that apply to them (More than one joint type might apply.)

Key: (a) fibrous joints
(b) cartilaginous joints
(c) synovial joints

____ (1) have no joint cavity

____ (2) types are sutures and syndesmoses

____ (3) dense connective tissue fills the space between the bones

____ (4) almost all joints of the skull

____ (5) types are synchondroses and symphyses

____ (6) all are diarthroses

____ (7) the most common type of joint in the body

____ (8) nearly all are synarthrotic

____ (9) shoulder, hip, knee, and elbow joints

2. Synovial joints have (a) articular cartilage, (b) a joint cavity, (c) a lubricant, (d) an articular capsule, (e) all of these.

3. In general, the most important factor(s) determining the stability of synovial joints is (a) interlocking shapes of the articular surfaces; (b) reinforcing ligaments; (c) ligaments and muscle tone; (d) synovial fluid, which acts like glue; (e) the body's wrapping of skin, which holds the limbs together.

4. Characteristics of a symphysis include (a) presence of fibrocartilage, (b) ability to resist large compression and tension stresses, (c) presence of a joint cavity, (d) very high mobility, (e) both a and b.

5. Most nerve fibers to synovial joints (a) monitor how much the capsule is stretched, (b) supply the articular cartilages, (c) cause the joint to move, (d) monitor pain when the capsule is injured.

6. Match the parts of a synovial joint listed in the key to their functions below. (More than one part may apply.)

Key: (a) articular cartilage (b) ligaments and fibrous layer of capsule
(c) synovial fluid (d) muscle tendon

____ (1) keeps bone ends from crushing when compressed; resilient

____ (2) resists tension placed on joints

____ (3) lubricant that minimizes friction and abrasion of joint surfaces

____ (4) helps prevent dislocation

7. Indicate the joint (or joints) from the list in column B that contains the structure listed in column A.

Column A	Column B
____ (1) medial meniscus	(a) sternoclavicular joint
____ (2) anular ligament of the radius	(b) temporomandibular joint
____ (3) saddle joint	(c) wrist joint
____ (4) articular disc	(d) shoulder joint
____ (5) rotator cuff	(e) elbow joint
____ (6) anterior cruciate ligament	(f) hip joint
____ (7) deltoid ligament	(g) knee joint
____ (8) ulnar collateral ligament	(h) ankle joint
____ (9) fibular collateral ligament	

8. Classify each of the synovial joints listed as one of the following: (a) plane joint, (b) hinge joint, (c) pivot joint, (d) condylar joint, (e) saddle joint, (f) ball-and-socket joint.

____ **(1)** proximal radioulnar joint

____ **(2)** trapezium and metacarpal I

____ **(3)** knee (tibiofemoral)

____ **(4)** metacarpophalangeal joint

____ **(5)** wrist joint

____ **(6)** atlanto-occipital joint

____ **(7)** atlantoaxial joint

____ **(8)** sternocostal joints, ribs 2–7

____ **(9)** intervertebral joints (between articular processes)

____ **(10)** acromioclavicular

Short Answer Essay Questions

9. Define joint.
10. Where does synovial fluid come from?
11. Explain weeping lubrication of the synovial joint surfaces.
12. Compare and contrast (1) flexion and extension with (2) adduction and abduction.
13. Name two specific examples of each: hinge joint, plane joint, condylar joint, ball-and-socket joint.
14. What are the functions of the menisci of the knee? Of the anterior and posterior cruciate ligaments?
15. Why are sprains and injuries to joint cartilages particularly troublesome?
16. Name the most common direction in which each of the following joints tends to dislocate: (a) shoulder, (b) elbow.
17. Examine the thorax using a skeleton or an illustration. Classify the various joints you see as containing either hyaline cartilage, fibrocartilage, or fibrous tissue (fibrous joints).
18. What are the technical names of and movements allowed at the following joints? (a) joints in the toes, (b) wrist joint, (c) jaw joint, (d) joint between sacrum and hip bones, (e) knuckle joint in the hand.
19. List the functions of the following parts of a synovial joint: (a) fibrous layer of the capsule, (b) synovial fluid, (c) articular disc.
20. Compare a bursa and a tendon sheath in terms of their structure, function, and locations.
21. Provide the anatomically proper names of the joints between (a) the scapula and clavicle, (b) the articular processes of successive vertebrae, (c) the ribs and sternum, (d) the ribs and vertebrae, (e) the various tarsal bones. (Refer to Table 9.3 if needed.)
22. Name the joint that contains the (a) glenoid cavity and labrum; (b) cruciate ligaments and menisci; (c) anular ligament and head of radius; (d) coronoid process, trochlea, and radial collateral ligament; (e) medial and lateral ligaments and talus.

Critical Reasoning & Clinical Application Questions

1. Harry was jogging down the road when he tripped in a pothole. As he fell, his ankle twisted violently to the side. The diagnosis was severe dislocation and spraining of the ankle. The surgeon stated that she would perform a reduction of the dislocation and then attempt to repair the injured ligament using arthroscopy. (a) What kinds of movements can normally occur at the ankle joint? (b) Was the doctor telling Harry that his bones were broken? (c) What is reduction of a joint? (d) Why is it often necessary to repair ligaments surgically? (e) How will the use of arthroscopic surgery minimize Harry's recovery time and his suffering?
2. Dan Park, an exhausted anatomy student, was attending a lecture. After 30 minutes he began to doze. He woke up suddenly (the professor's voice was too loud to permit a good nap) and yawned widely. To his great distress, he couldn't close his mouth—his lower jaw was stuck open. What had happened?
3. Mrs. Estevez, who is 37 years old, visited her physician and complained of unbearable pain in several joints of both hands. The joints were very red, swollen, and warm to the touch. When asked if she had ever had such an episode in the past, she said she had had a similar attack 2 years earlier that had disappeared as suddenly as it had come. Her diagnosis was arthritis. (a) What type of arthritis does she have? (b) What is the cause of joint inflammation in this type of arthritis?
4. At work, a box fell from a shelf onto Ming's acromial region. In the emergency room, the physician could feel that the head of her humerus had moved into the axilla. What had happened to Ming?
5. At his ninety-fourth birthday party, Jim was complimented on how good he looked and was asked about his health. He answered, "I feel good, except that some of my joints ache and are stiff, especially my knees and hips and lower back, and especially when I wake up in the morning." A series of X-ray studies and an MRI scan taken a few weeks earlier showed that the articular cartilages of these joints were rough and flaking off and that bone spurs were present at the ends of some of Jim's bones. What is Jim's probable condition?
6. On the evening news, Samantha heard that the deer population in her state had increased greatly in the past few years, and she knew that deer were often seen walking the streets of her suburban community. Suddenly, she exclaimed, "So that's why several children in my son's class got Lyme disease this year!" Explain her reasoning. (See "Related Clinical Terms")
7. Describe how the structure of the joints in each limb reflects the function of each limb (mobility in the upper limb, stability in the lower limb). What movements are unique to the upper limb? Which joints enable these unique movements?

Skeletal Muscle Tissue 10

Muscle is from a Latin word meaning "little mouse," a name given because flexing muscles look like mice scurrying beneath the skin. Indeed, the rippling skeletal muscles of weight lifters are often the first thing that comes to mind when one hears the word *muscle.* However, muscle is also the main tissue in the heart and in the walls of other hollow organs. Muscle tissue is a composite tissue composed of muscle cells and surrounding connective tissues. In all its forms, muscle tissue makes up nearly half the body's mass.

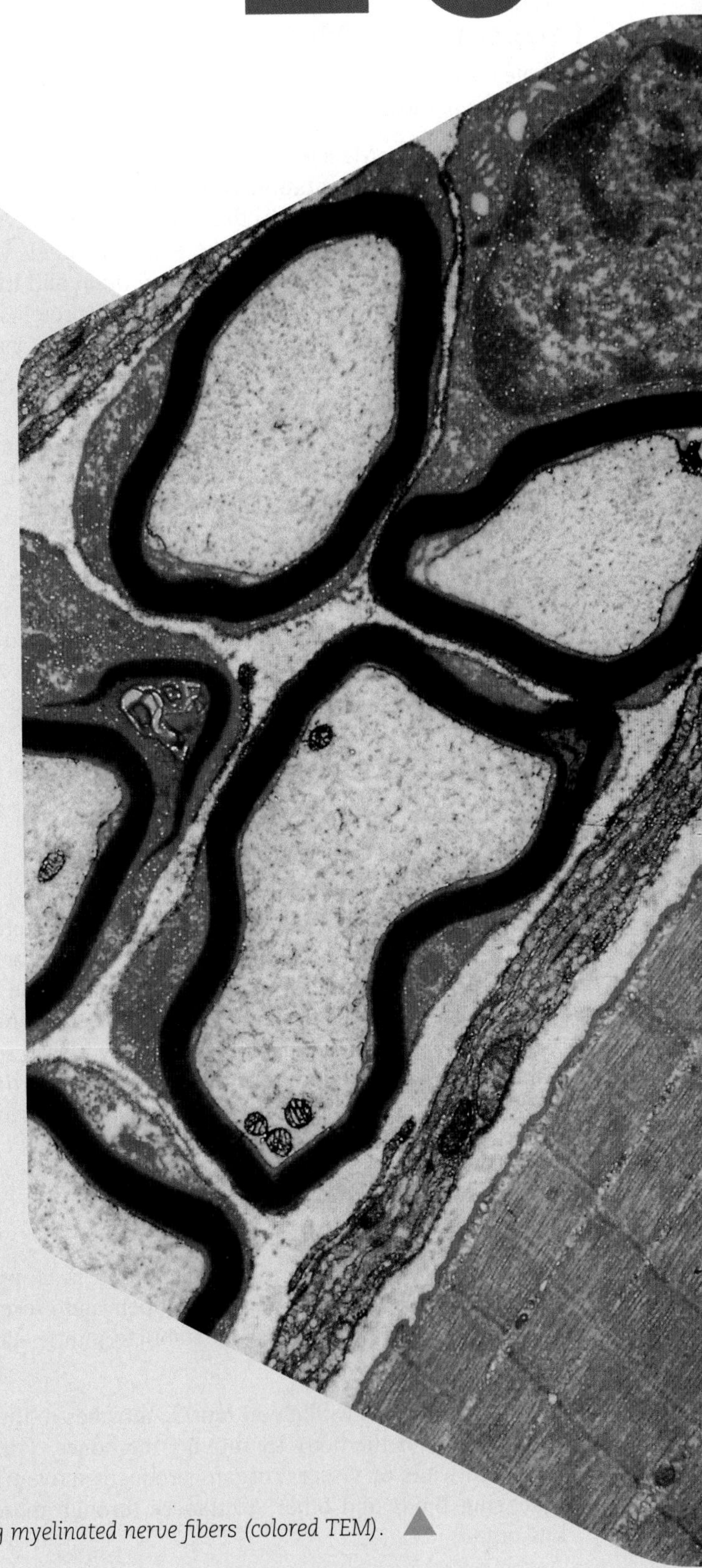

Contracted striated muscle fibers containing myelinated nerve fibers (colored TEM).

OVERVIEW OF MUSCLE TISSUE

learning outcomes

- List four functional properties that distinguish muscle tissue from other tissues.
- Compare and contrast skeletal, cardiac, and smooth muscle tissue.

Properties of Muscle Tissue

Muscle tissue has four functional properties that distinguish it from other tissues:

1. **Contractility.** Muscle tissue contracts forcefully. Muscle cells contain **myofilaments** (*myo* = muscle), specific types of microfilaments that are responsible for the shortening of muscle cells. There are two kinds of myofilaments, one containing the protein *actin* and the other containing the protein *myosin*. These two proteins generate contractile force in every cell in the body (Chapter 2, p. 32). This contractile property is most highly developed in muscle cells.
2. **Excitability.** Nerve signals or other stimuli excite muscle cells, causing electrical impulses to travel along the cells' plasma membrane. These impulses initiate contraction in muscle cells.
3. **Extensibility.** Muscle tissue can be stretched. Contraction of one skeletal muscle will stretch an opposing muscle. The muscular wall of a hollow organ is stretched by the substances contained within that organ—a bolus of food in the digestive tract or urine in the urinary bladder, for example.
4. **Elasticity.** After being stretched, muscle tissue recoils passively and resumes its resting length.

Terminology Specific to Muscle Tissue

The prefixes **myo**- or **mys**- (both mean "muscle") and **sarco**- ("flesh") refer to muscle. There are some specific terms for structures in muscle cells that use these word roots. For example, the plasma membrane of muscle cells is called the **sarcolemma** (sar″ko-lem′ah, "flesh sheath"). The cytoplasm is the **sarcoplasm,** and the endoplasmic reticulum, which is specialized for the storage of calcium, is the *sarcoplasmic reticulum.* The microfilaments containing the contractile proteins are *myofilaments.* You will learn additional terms in this chapter that use these word roots.

Functions of Muscle Tissue

The most obvious function of muscle tissue is to produce movement. Muscle tissue also acts to open and close body passageways, maintain posture and stabilize joints, and generate heat.

- **Produce movement.** Skeletal muscle attaches to the skeleton and moves the body by moving the bones. The muscle in the walls of visceral organs produces movement by squeezing fluids and other substances through these hollow organs.
- **Open and close body passageways.** *Sphincter muscles* encircle many body openings and function as valves: relaxing to allow passage of a substance, contracting to close the passageway. This function is especially apparent in the digestive and urinary tracts. Muscle tissue around the mouth and eyes opens and closes these body orifices. Muscle fibers in the iris of the eye constrict and relax to change pupil diameter.
- **Maintain posture and stabilize joints.** When we are awake, skeletal muscles contract continuously to maintain posture, enabling the body to remain in a standing or sitting position. Muscle tone, the constant low-level contraction of muscles, helps stabilize and strengthen many synovial joints.
- **Generate heat.** Muscle contraction produces heat. This plays a vital role in maintaining normal body temperature at 37°C (98.6°F). This function is clear when we exercise; excess heat generated during exercise stimulates sweating to cool us down.

Types of Muscle Tissue

There are three types of muscle tissue: *skeletal, cardiac,* and *smooth.* Each type can be characterized by two distinctive features: (1) the presence or absence of light and dark stripes, called *striations,* in the muscle cells; and (2) whether control of contraction is *voluntary* or *involuntary.*

Skeletal Muscle Tissue

Skeletal muscle tissue is located in the **skeletal muscles,** discrete organs that attach to and move the skeleton. This tissue makes up 40% of body weight. The muscle cells of skeletal muscle tissue are **striated muscle.** Striated muscle tissue has dark and light stripes extending transversely across its muscle cells. These striations are visible when the muscle tissue is viewed histologically (see Figure 4.13a, p. 93). The elongated, cylindrical skeletal muscle cells are called **muscle fibers.** Skeletal muscle is innervated by the **voluntary** division of the nervous system and is subject to conscious control; you can control this muscle tissue at will. The details of the structure and function of skeletal muscle tissue is the focus of this chapter (**Table 10.1**).

Cardiac Muscle Tissue

Cardiac muscle tissue occurs only in the wall of the heart. The muscle cells of cardiac muscle are **striated muscle** (see Figure 4.13b, p. 94) but its contraction is **involuntary.** In fact, cardiac muscle can contract with no nervous stimulation. The involuntary division of the nervous system regulates contraction of cardiac muscle tissue; we have no direct, conscious control over how fast our heart beats. (The details of the structure and function of cardiac muscle tissue are described in Chapter 19.)

Smooth Muscle Tissue

Most **smooth muscle** tissue in the body is found in the walls of hollow internal organs other than the heart, such as

Table 10.1 Structure and Organizational Levels of Skeletal Muscle

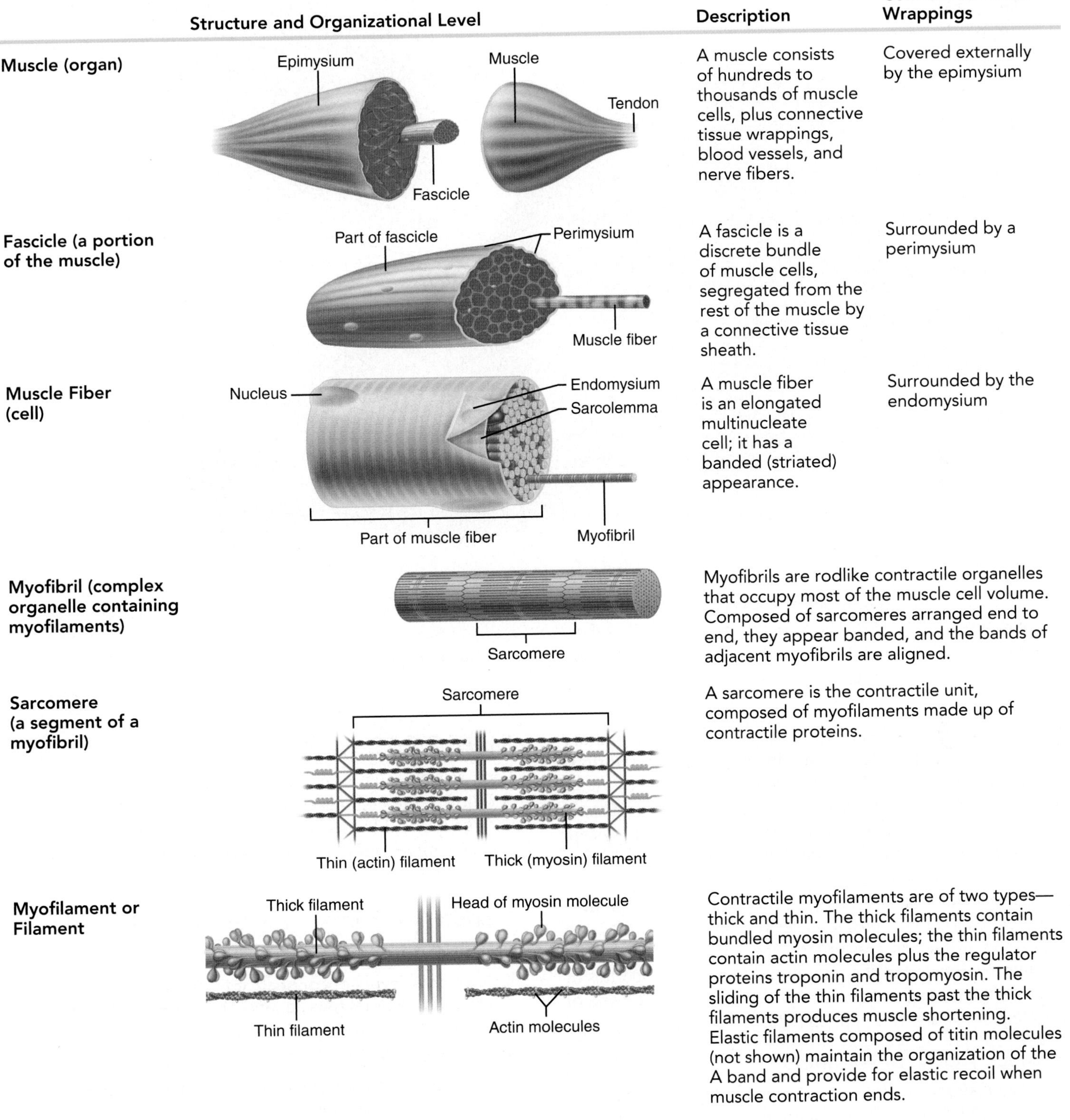

Structure and Organizational Level	Description	Connective Tissue Wrappings
Muscle (organ)	A muscle consists of hundreds to thousands of muscle cells, plus connective tissue wrappings, blood vessels, and nerve fibers.	Covered externally by the epimysium
Fascicle (a portion of the muscle)	A fascicle is a discrete bundle of muscle cells, segregated from the rest of the muscle by a connective tissue sheath.	Surrounded by a perimysium
Muscle Fiber (cell)	A muscle fiber is an elongated multinucleate cell; it has a banded (striated) appearance.	Surrounded by the endomysium
Myofibril (complex organelle containing myofilaments)	Myofibrils are rodlike contractile organelles that occupy most of the muscle cell volume. Composed of sarcomeres arranged end to end, they appear banded, and the bands of adjacent myofibrils are aligned.	
Sarcomere (a segment of a myofibril)	A sarcomere is the contractile unit, composed of myofilaments made up of contractile proteins.	
Myofilament or Filament	Contractile myofilaments are of two types—thick and thin. The thick filaments contain bundled myosin molecules; the thin filaments contain actin molecules plus the regulator proteins troponin and tropomyosin. The sliding of the thin filaments past the thick filaments produces muscle shortening. Elastic filaments composed of titin molecules (not shown) maintain the organization of the A band and provide for elastic recoil when muscle contraction ends.	

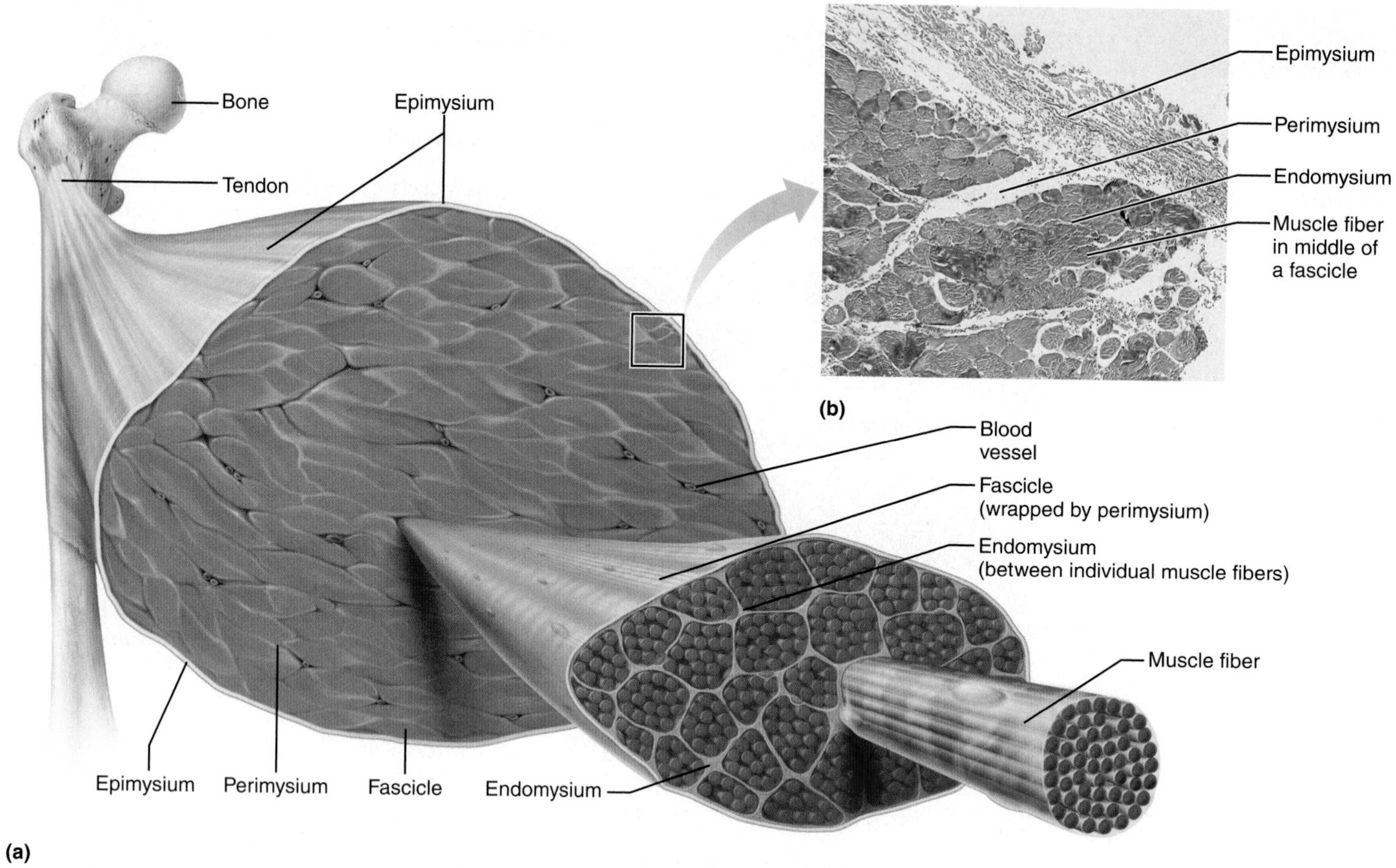

Figure 10.1 Connective tissue sheaths in skeletal muscle: epimysium, perimysium, and endomysium. (**a**) Illustration. (**b**) Photomicrograph of skeletal muscle in cross-section. Connective tissue sheaths are colored blue/purple by trichrome staining. (40×).

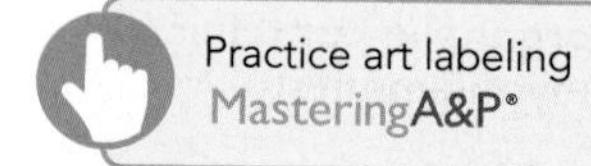

the stomach, urinary bladder, blood vessels, and respiratory tubes. The muscle cells of smooth muscle **lack striations** (see Figure 4.13c, p. 94). Like skeletal muscle, these cells are elongated and referred to as *muscle fibers*. As with cardiac muscle, the **involuntary** division of the nervous system innervates smooth muscle. (The detailed structure and function of smooth muscle tissue are covered in Chapter 23.)

Cardiac muscle and smooth muscle tissues are collectively called *visceral muscle,* a term reflecting the fact that both occur in the visceral organs and are innervated by the involuntary division of the nervous system. The properties of the three types of muscle tissue are summarized at the end of this chapter (**Table 10.2**, pp. 254–255).

check your understanding

- ☐ 1. What structural similarities are shared by all muscle tissue? What are the unique functional properties of this tissue?
- ☐ 2. Which types of muscle tissue are striated? Which types are called visceral muscle? In which types are the muscle cells called fibers?

(For answers, see Appendix B.)

SKELETAL MUSCLE

Each muscle is an organ made of several kinds of tissue: In addition to skeletal muscle tissue, a muscle also contains connective tissue, blood vessels, and nerves. The following sections examine skeletal muscle anatomy from the gross level to the microscopic level.

Gross Anatomy of a Skeletal Muscle

learning outcomes

- Name the layers of connective tissue that occur in and around a skeletal muscle, and briefly describe a muscle's blood and nerve supply.
- Define muscle fascicles.
- Describe the various ways in which muscles attach at their origins and insertions.

Connective Tissue and Fascicles

Several sheaths of connective tissue hold the fibers of a skeletal muscle together. These sheaths, from external to internal, are as follows (**Figure 10.1**):

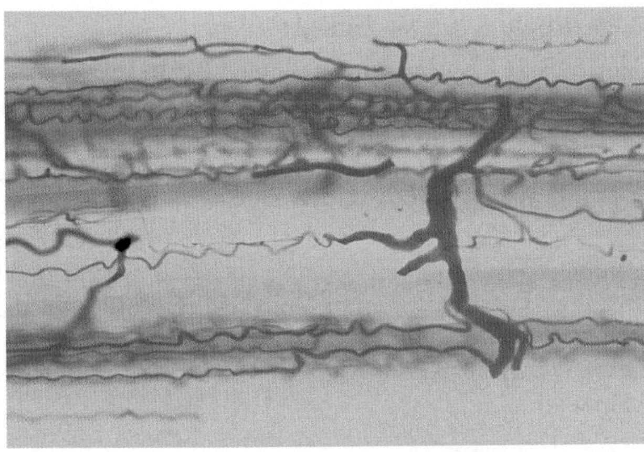

Figure 10.2 **Photomicrograph of the capillary network surrounding skeletal muscle fibers.** The arterial supply was injected with dark red gelatin to demonstrate the capillary bed. The muscle fibers, which run horizontally across the photograph, are stained orange. Note the wavy appearance of the thinnest capillaries (75×).

1. **Epimysium.** An outer layer of dense, irregular connective tissue surrounds the whole skeletal muscle. This layer is the **epimysium** (ep″ĭ-mis′e-um), a name that means "outside the muscle." Sometimes the epimysium blends with the deep fascia that lies between neighboring muscles.
2. **Perimysium.** Within each skeletal muscle, the muscle fibers are separated into groups. Each group, which resembles a bundle of sticks tied together, is called a **fascicle** (fas′ĭ-kl; "bundle"). Surrounding each fascicle is a layer of fibrous connective tissue called **perimysium** (per″ĭ-mis′e-um; "around the muscle [fascicle]").
3. **Endomysium.** Within a fascicle, each muscle fiber is surrounded by a fine sheath of loose connective tissue consisting mostly of reticular fibers. This layer is the **endomysium** (en″do-mis′e-um; "within the muscle").

These fibrous connective tissues bind muscle fibers together and hold them in parallel alignment so they can work together to produce force. These sheaths are continuous with each other: The endomysium merges with the perimysium, which in turn is continuous with the epimysium (Figure 10.1b). All three sheaths converge to form the **tendon,** the connective tissue structure that joins skeletal muscles to bones (Figure 10.1a). When muscle fibers contract, they pull on the surrounding endomysium. Because of the continuity between sheaths, this pull is then exerted on the perimysium, epimysium, and tendon, a sequence that transmits the force of contraction to the bone being moved. The sheaths also provide a muscle with much of its natural elasticity and carry the blood vessels and nerves that serve the muscle fibers.

Nerves and Blood Vessels

In general, each skeletal muscle is supplied by one nerve, one artery, and one or more veins—all of which enter or exit the muscle near the middle of its length. The nerves and vessels branch repeatedly in the intramuscular connective tissue, with the smallest branches serving individual muscle fibers.

The rich blood supply to muscles reflects the high demand that contracting muscle fibers have for nutrients and oxygen. Capillaries in the endomysium form a network (Figure 10.2). These long capillaries are wavy when the muscle fibers contract and stretched straight when the muscle extends.

The innervation of skeletal muscle is described later in this chapter (p. 252).

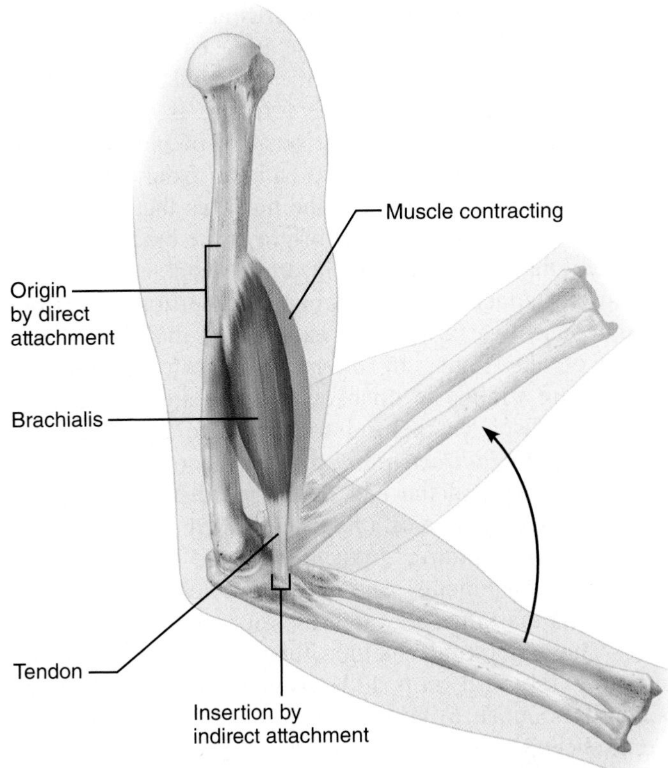

Figure 10.3 **Muscle attachments (origin and insertion).** When a skeletal muscle contracts, its insertion is pulled toward its origin.

Muscle Attachments

A muscle attachment is the location on a bone where a muscle connects to the bone. Each skeletal muscle extends from one bone to another, crossing at least one movable joint. When the muscle contracts, it causes one of the bones to move while the other bone usually remains fixed. The attachment of the muscle on the less movable bone is called the **origin** of the muscle, whereas the attachment on the more movable bone is called the muscle's **insertion** (Figure 10.3). Thus, when the muscle contracts, its insertion is pulled toward its origin. In the muscles of the limbs, the origin is by convention the more proximal attachment of the muscle, and the insertion is the more distal attachment. However, it is important to realize that the functions of the origin and the insertion may switch, depending on body position and the movement produced when the muscle contracts. For example, the conventional attachments for the brachialis muscle in the arm are origin on the shaft of the humerus and insertion on the proximal ulna (Figure 10.3). These are accurate terms when referring to the usual position of the limb and the function of this muscle, which is flexing the forearm, as when you lift an object. Consider, though, the position of the upper limb and its movement when you are hanging from a bar doing chin-ups. In this situation, it is the humerus that moves, not the ulna. Thus the conventional origin of the brachialis is the attachment that moves, and the insertion is stable.

Muscles attach to their origins and insertions via strong fibrous connective tissues that extend into the fibrous periosteum of the bone. In *direct,* or *fleshy, attachments,* the attaching strands of connective tissue are so short that the muscle fascicles themselves appear to attach directly to the bone. In *indirect attachments,* the connective tissue extends well beyond the end of the muscle fibers to form either a cordlike tendon or a flat sheet called an **aponeurosis** (ap″o-nu-ro′sis). Indirect attachments are more common than direct attachments, and most muscles have tendons. Raised bone markings are often present where tendons meet bones. These markings include tubercles, trochanters, and crests (see Table 6.1 on p. 133). Although most tendons and aponeuroses attach to bones, a few attach to skin, to cartilage, to sheets of fascia, or to a seam of fibrous tissue called a **raphe** (ra′fe; "seam").

Many muscles span two or more joints. Such muscles are called *biarticular* ("two-joint") *muscles* or *multijoint muscles.* For example, imagine a muscle that originates on the humerus, runs along the forearm without attaching to the radius or ulna, and inserts on the carpals. Contraction of such a muscle would cause movements at two joints: the elbow and the wrist.

☑ check your understanding

- ☐ 3. Name the connective tissue that surrounds a fascicle.
- ☐ 4. What are the functional definitions of the origin and insertion of a muscle? In the limbs, what are the conventional definitions of a muscle's origin and insertion?

(For answers, see Appendix B.)

Microscopic and Functional Anatomy of Skeletal Muscle Tissue

learning outcomes

- ▶ Describe the microscopic structure of a skeletal muscle fiber and the arrangement of its contractile thick and thin filaments into sarcomeres and myofibrils.
- ▶ Describe the structure and functions of sarcoplasmic reticulum and T tubules in muscle fibers.
- ▶ Explain the sliding filament mechanism of muscle contraction, including the role of titin.
- ▶ Describe the innervation of skeletal muscle fibers at neuromuscular junctions. Define a motor unit.
- ▶ Compare and contrast the three kinds of skeletal muscle fibers.

The Skeletal Muscle Fiber

Skeletal muscle fibers are long, cylindrical cells (**Figure 10.4a** and **b**). They are huge cells, relatively speaking. Their diameter is 10–100 μm (up to ten times that of an average body cell), and their length is phenomenal: from several centimeters in short muscles to dozens of centimeters in long muscles. It would be inaccurate to call skeletal muscle fibers the biggest cells in the body, however, because each one was formed by the fusion of hundreds of embryonic cells. Because the fibers develop this way, they contain many nuclei. These nuclei lie at the periphery of each fiber, just deep to the sarcolemma (Figure 10.4b).

Myofibrils and Sarcomeres

Under the light microscope, the light and dark striations in skeletal muscle fibers are clearly visible. These striations result from the internal structure of long, rod-shaped organelles called **myofibrils.** Myofibrils are unbranched cylinders that are present in large numbers, making up more than 80% of the sarcoplasm. They are specialized contractile organelles unique to muscle tissue. Myofibrils contain myofilaments (Figure 10.4c), the contractile proteins described earlier and discussed in detail shortly. The myofibrils in a fiber are separated from one another by other components of the sarcoplasm. Among those components are mitochondria and glycosomes, both of which supply energy for muscle contraction. It is difficult to distinguish individual myofibrils in a light micrograph (Figure 10.4a) because the striations of adjacent myofibrils line up almost perfectly.

A myofibril is composed of repeating segments called **sarcomeres** (sar′ko-mērz; "muscle segments") (Figure 10.4c and 10.4d). The sarcomere is the basic unit of contraction in skeletal muscle. The boundaries at the two ends of each sarcomere are called **Z discs** (or sometimes **Z lines**). Attached to each Z disc and extending toward the center of the sarcomere are many fine myofilaments called **thin (actin) filaments.** The thin filaments are composed primarily of the contractile protein **actin.** Two regulatory proteins, **troponin** and **tropomyosin**, are also found on the thin filament. Tropomyosin forms a thin strand that spirals around the actin molecule. Troponin is a globular protein with three binding sites: one for actin, one for tropomyosin, and one for calcium. Troponin attaches the tropomyosin strand to the actin molecule. These two regulatory proteins control the interaction of actin with the thick filaments. In the center of the sarcomere and overlapping the inner ends of the thin filaments is a cylindrical bundle of **thick (myosin) filaments.** Thick filaments consist largely of myosin molecules. They also contain ATPase enzymes that split ATP (energy-storing molecules) to release the energy required for muscle contraction. Both ends of a thick filament are studded with knobs called *myosin heads* (Figure 10.4d).

The sarcomere structure explains the pattern of striations in skeletal muscle fibers. The dark bands are created by the full length of the thick filaments in the sarcomeres, along with the inner ends of the thin filaments, which overlap the thick filaments. This region of each sarcomere is called the **A band** (Figure 10.4c). The central part of an A band, where no thin filaments reach, is the **H zone.** The **M line** in the center

(a) Photomicrograph of **portions of two isolated muscle fibers** (725×). Notice the obvious striations (alternating dark and light bands).

Nuclei
Dark A band
Light I band
Fiber

(b) Diagram of **part of a muscle fiber** showing the myofibrils. One **myofibril** is extended from the cut end of the fiber.

Sarcolemma
Mitochondrion
Myofibril
Dark A band
Light I band
Nucleus

(c) Small part of one **myofibril enlarged to show the myofilaments** responsible for the banding pattern. Each **sarcomere** extends from one Z disc to the next.

Thin (actin) filament
Z disc
H zone
Z disc
Thick (myosin) filament
I band
A band
I band
M line
Sarcomere

(d) Enlargement of one sarcomere (sectioned lengthwise). Notice the myosin heads on the thick filaments.

Z disc
M line
Z disc
Thin (actin) filament
Elastic (titin) filaments
Thick (myosin) filament
Myosin heads

(e) Cross-sectional view of a sarcomere cut through in different locations.

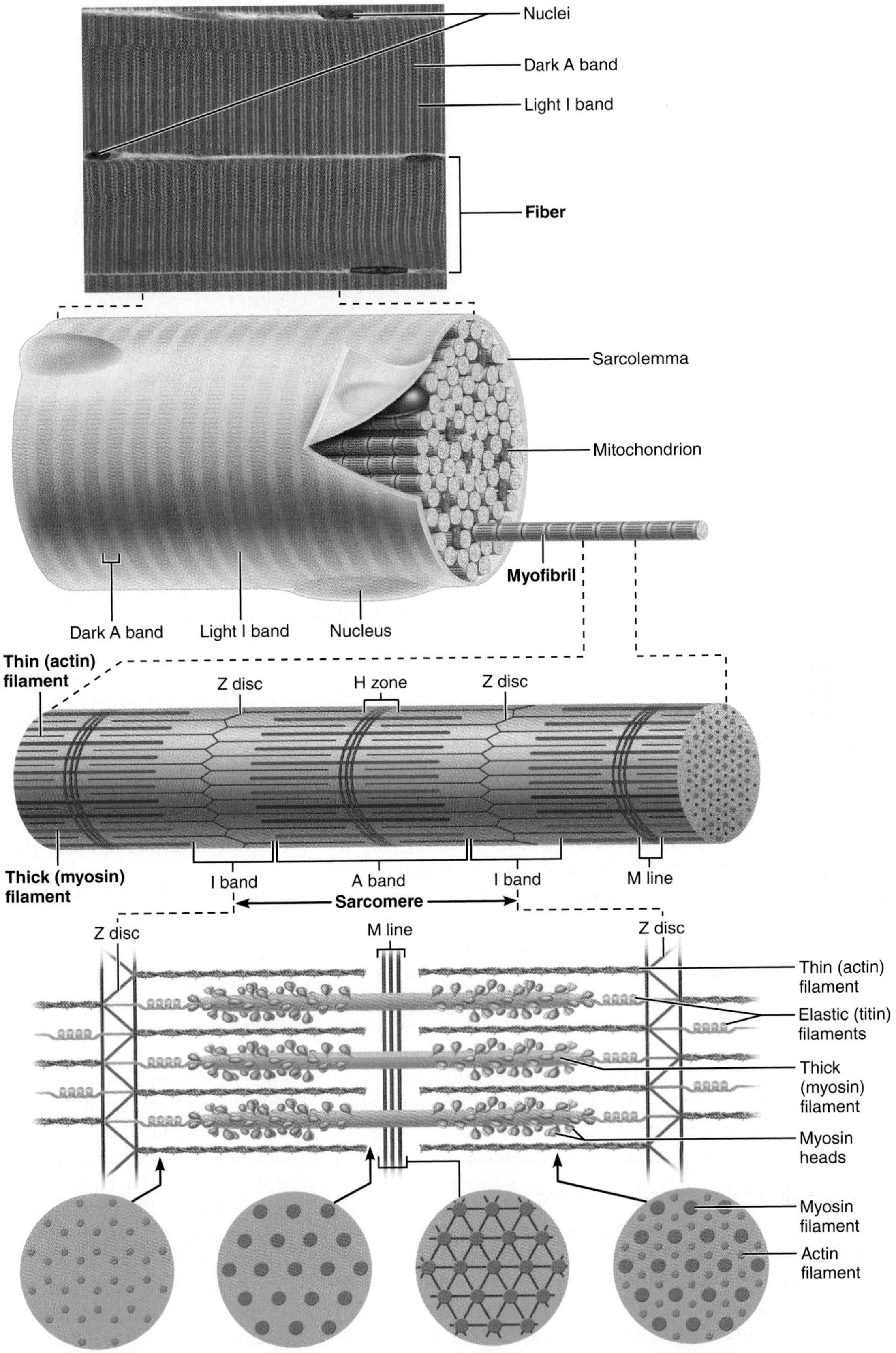

Myosin filament
Actin filament

I band thin filaments only

H zone thick filaments only

M line thick filaments linked by accessory proteins

Outer edge of A band thick and thin filaments overlap

Figure 10.4 Microscopic anatomy of the skeletal muscle fiber (cell).

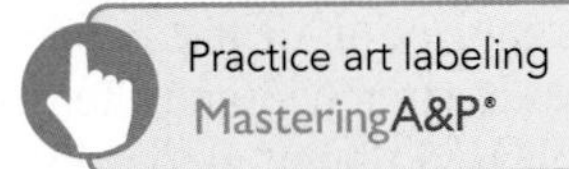

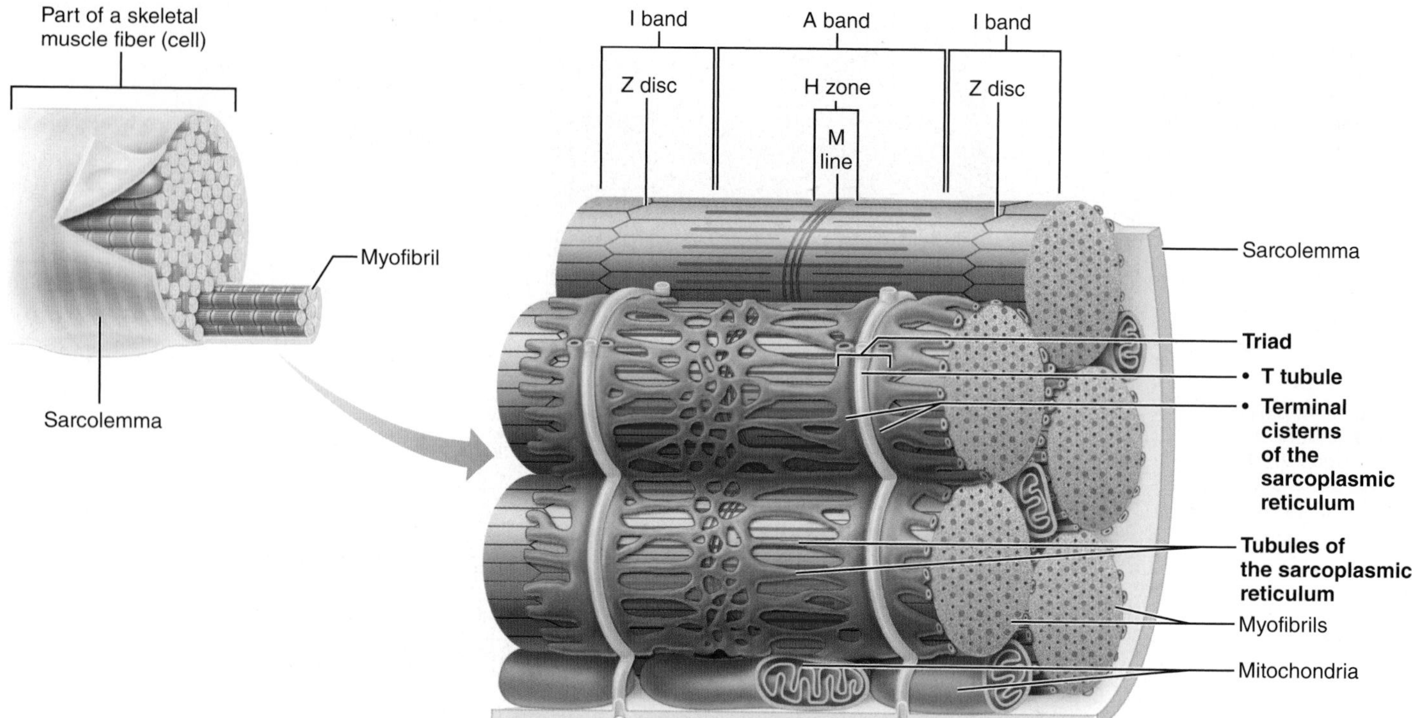

Figure 10.5 **Sarcoplasmic reticulum and T tubules in the skeletal muscle fiber.**

of the H zone contains tiny rods that hold the thick filaments together (Figure 10.4d). The two regions on either side of the A band, regions that contain only thin filaments, are called the **I bands.** It is the I bands of the sarcomeres that create the light portions of the light-dark pattern of striations seen along the length of any skeletal muscle fiber. Each I band is part of two adjacent sarcomeres and has a Z disc running through its center (Figure 10.4c). Recall that the striations of adjacent myofibrils align perfectly, allowing us to see the various bands in a muscle fiber (Figure 10.4a and 10.4b). To help you remember which band is which, remember there is an "*A*" in *dark;* thus, the dark bands are the A bands; there is an "*I*" in *light;* thus, the light bands are I bands. These notations actually refer to how these regions refract polarized light: the A bands are *anisotropic,* and the I bands are *isotropic.*

Cross sections through different regions of the sarcomere (Figure 10.4e) give a three-dimensional view of the types of myofilaments and their arrangement in each region.

Titin and Other Myofibril Proteins

Titin (ti′tin) is a springlike molecule in sarcomeres that resists overstretching. The titin molecules in a sarcomere are found in the elastic filaments. They extend from the Z disc to the thick filament and run within the thick filament to attach to the M line (see Figure 10.4d). When first identified, titin generated much excitement because it is the largest protein ever discovered. It has two basic functions: (1) It holds the thick filaments in place in the sarcomere, thereby maintaining the organization of the A band; and (2) it unfolds when the muscle is stretched and then refolds when the stretching force is released, thereby contributing to muscle elasticity. Titin does not resist stretching in the ordinary range of extension, but it becomes stiffer the more it uncoils; therefore, it strongly resists excessive stretching that tries to pull the sarcomere apart.

In each myofibril, several proteins surround the Z discs. These proteins bind adjacent sarcomeres together.

(Table 10.1, p. 243, provides a quick review of the macroscopic and microscopic structure of skeletal muscle.)

Sarcoplasmic Reticulum and T Tubules

Each skeletal muscle fiber contains two sets of tubules that participate in the regulation of muscle contraction: sarcoplasmic reticulum and T tubules (Figure 10.5). **Sarcoplasmic reticulum (SR)** is an elaborate smooth endoplasmic reticulum whose interconnecting tubules surround each myofibril like the sleeve of a loosely crocheted sweater surrounds your arm. Most SR tubules run longitudinally along the myofibril. Other SR tubules, called **terminal cisterns** ("end sacs"), form larger, perpendicular cross-channels over the junction between each A band in a myofibril and its adjacent I bands (A-I junctions). The sarcoplasmic reticulum and the terminal cisterns store large quantities of calcium ions (Ca^{2+}). These ions are released when the muscle is stimulated to contract.

T tubules (transverse tubules) are deep invaginations of the sarcolemma that run between each pair of terminal cisterns (Figure 10.5). The complex of the T tubule flanked by two terminal cisterns at the A-I junction is called a **triad** (tri′ad; "group of three").

Contraction in skeletal muscle is ultimately controlled by nerve-generated impulses that travel along the sarcolemma of the muscle fiber. Because the T tubules are continuations of the sarcolemma, they conduct each impulse to the deepest regions of the muscle fiber, thus ensuring that the deep-lying myofibrils contract at the same time as the superficial ones. Impulses traveling down the T tubules stimulate the release of calcium from the terminal cistern. Calcium diffuses through the cytosol to the thin filaments and triggers muscle contraction. After contraction, calcium is pumped back into the sarcoplasmic reticulum for storage.

check your understanding

- ☐ 5. Place the following structures in order from smallest to largest, and define each: myofibril, muscle fiber, myofilament, sarcomere.
- ☐ 6. Which myofilaments are found only in the A band?
- ☐ 7. What are the functions of the terminal cistern and the T tubules?

(For answers, see Appendix B.)

Mechanism of Contraction

There are two types of muscle contraction involved in producing movement: concentric contraction and eccentric contraction. *Concentric contraction* is the more familiar type, in which the muscle shortens and does work—picking up a book or kicking a ball. *Eccentric contraction* occurs when a muscle generates force as it *lengthens.*

Eccentric Contraction When contracting eccentrically, a muscle generates force while lengthening. This type of contraction is essential for controlled movement and resistance to gravity. The mechanism for eccentric contraction is less well understood than the mechanism for concentric contraction. Both types of contraction occur during push-ups. During the up portion of the exercise, concentric contractions in the pectoralis muscles of the chest raise the torso off the floor. During the down portion of the movement, these same muscles contract eccentrically and by doing so resist gravity and control the downward motion of the torso. Eccentric contraction occurs in many movements that resist gravity: going down stairs, running downhill, landing from a jump. Whenever muscles are acting as a brake, they are contracting eccentrically.

CLINICAL APPLICATION

Delayed-Onset Muscle Soreness Soreness that begins 8–24 hours after an activity is called **delayed-onset muscle soreness** or **post-exercise muscle soreness.** Starting an exercise regimen, occasional participation in a physically demanding activity, or increasing the duration or intensity of a regular workout can all result in sore muscles. This soreness is caused by microscopic tears in the muscle fibers and is most common after eccentric exercise. The inflammatory response to these small tears results in swelling in the connective tissues surrounding the muscle fibers. The swelling compresses sensory nerve endings in the muscle, causing the characteristic soreness. Delayed-onset muscle soreness does not last long (3–7 days). Low-level aerobic activity increases blood supply to the muscle and speeds recovery. These microscopic tears stimulate increased production of myofibrils and myofilaments, resulting in increased muscle strength.

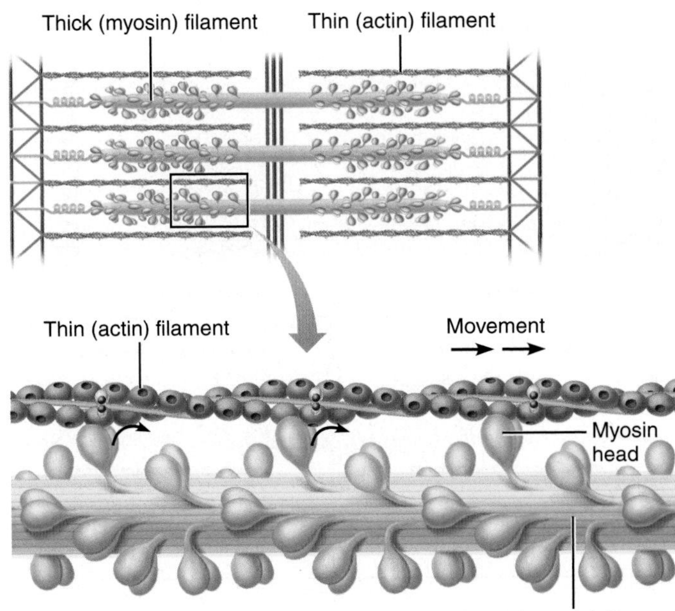

(a) Myosin heads attach to actin in the thin filaments, then pivot to pull the thin filaments towards the M line.

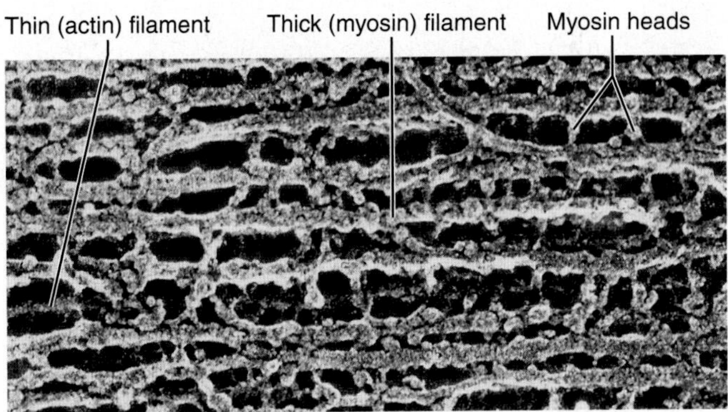

(b) Freeze fracture TEM through the A band of a sarcomere showing myosin heads attached to the thin filaments.

Figure 10.6 Sliding filament mechanism for concentric contraction in a skeletal muscle.

Concentric Contraction When contracting concentrically, a muscle generates force while shortening. Concentric contraction of skeletal muscle is explained by the **sliding filament mechanism** (Figure 10.6). The sliding filament mechanism is initiated by the release of calcium ions from the sarcoplasmic reticulum and the binding of those ions to the troponin molecule on the thin filament. This results in a change of shape of the troponin, which moves the tropomyosin molecule and exposes the binding sites on the actin filament for the myosin heads. Contraction results as the myosin heads of the thick filaments attach to the thin filaments at both ends of the sarcomere and pull the thin filaments toward the center of the sarcomere by pivoting inward. After a myosin head pivots at its "hinge," it lets go, returns to its original position, binds to the thin filament farther along its length,

① Fully relaxed sarcomere of a muscle fiber

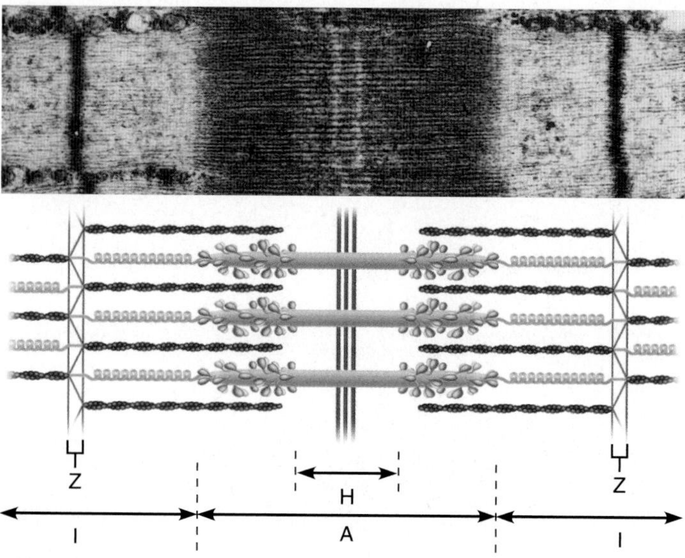

② Fully contracted sarcomere of a muscle fiber

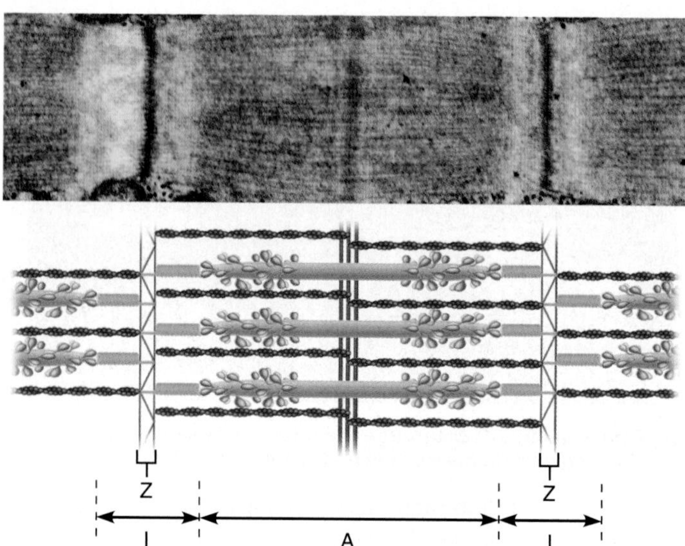

Figure 10.7 Changes in striations as skeletal muscle contracts. The numbers indicate the sequence of events from ① relaxed to ② fully contracted. Electron micrographs (top view in each case) show enlargements of 32,000×.

and pivots again. This ratchet-like cycle is repeated many times during a single contraction. ATP powers this process. It should be emphasized that the thick and thin filaments themselves do not shorten: The thin filament merely slides over the thick filament.

Contraction affects the striation pattern of skeletal muscle (Figure 10.7). In a fully relaxed sarcomere (Figure 10.7 ①), the thin filaments partially overlap the thick filaments. Note the length of the A band, the I band, and the position of the Z discs in the relaxed sarcomere. When the muscle is stimulated to contract, as shown in (Figure 10.7 ②), the action of the thick filaments forcefully pulls the two Z discs closer together, causing each sarcomere to shorten. As the Z discs move closer together, the I bands shorten, and the H zones disappear completely. Notice that the decrease in length of the I band and the loss of the H zone are due to the increased amount of overlap of the thin and thick filaments. The length of the thin filament has not changed. The A bands stay the same length because the length of the thick filaments also does not change.

Muscle Extension

You know that muscle tissue is extensible and that muscle fibers are stretched (extended) back to their original length after they contract. What is responsible for this stretching? Basically, a skeletal muscle—and its contained fibers—are stretched by a movement that is the *opposite* of the movement the muscle normally produces. For example, a muscle that normally *abducts* the arm at the shoulder is stretched by *adducting* the arm at this joint.

When a muscle is stretched rather than contracted, the amount of overlap between the thin and thick filaments decreases: The I bands and H zones lengthen as the Z discs move apart. Again, there is no change in the width of the A bands.

Muscle Fiber Length and the Force of Contraction

The optimal resting length for skeletal muscle fibers is the length that will generate the greatest pulling force when the muscle is contracted. This optimal length occurs when a fiber is slightly stretched, so that its thin and thick filaments overlap to only a moderate extent (Figure 10.7 ①). Under these conditions, the myosin heads can move and pull along the whole length of the thin filaments. Under other conditions, contraction is suboptimal: If a muscle fiber is stretched so much that the thick and thin filaments do not overlap at all, the myosin heads have nothing to attach to, and no pulling force can be generated. Alternatively, if the sarcomeres are so compressed that the thick filaments are touching the Z discs, little further shortening can occur (Figure 10.7 ②).

What is true for a muscle fiber is true for an entire muscle. Whole skeletal muscles have a range of optimal operational length that runs from about 80% of their normal resting length to about 120% of that length. The sites of muscle attachments tend to keep muscles within that optimal range; that is, the joints normally do not let any bone move so widely that its attached muscles could shorten or stretch beyond their optimal range.

Innervation of Skeletal Muscle

The release of calcium ions from the sarcoplasmic reticulum and the subsequent contraction of skeletal muscle is initiated by nervous stimulation. The nerve cells that innervate muscle fibers are called **motor neurons.** A neuron has cell processes that extend from the cell body: *Dendrites* are receptive regions of the neuron; an *axon* is a long, singular cell process that initiates and transmits nerve impulses (Chapter 4, p. 95).

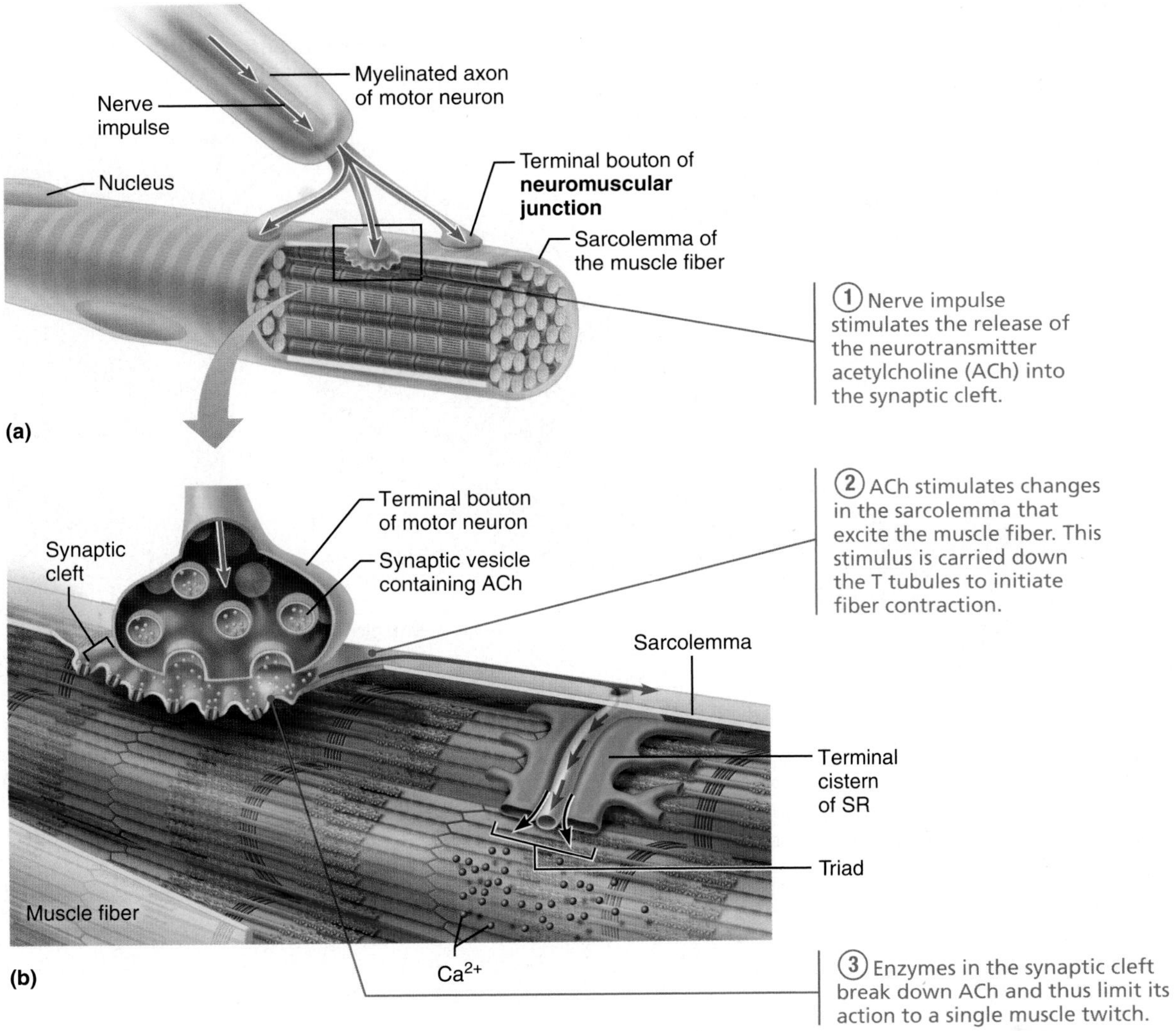

① Nerve impulse stimulates the release of the neurotransmitter acetylcholine (ACh) into the synaptic cleft.

② ACh stimulates changes in the sarcolemma that excite the muscle fiber. This stimulus is carried down the T tubules to initiate fiber contraction.

③ Enzymes in the synaptic cleft break down ACh and thus limit its action to a single muscle twitch.

Figure 10.8 The neuromuscular junction. (**a**) An axon of a motor neuron forming three neuromuscular junctions with a skeletal muscle fiber. (**b**) Enlargement of a single neuromuscular junction contacting a muscle cell.

Neuromuscular Junction Each muscle fiber in a skeletal muscle is served by a nerve ending, which signals the fiber to contract. The point at which the nerve ending and fiber meet is called a **neuromuscular junction** or a *motor end plate* (Figure 10.8). The nerve part of the junction is a cluster of enlargements at the end of the axonal process that stores chemical messenger molecules, neurotransmitters. These enlargements are called **terminal boutons** or *axon terminals.* The terminal boutons are separated from the sarcolemma of the muscle fiber by a space called the **synaptic cleft** (Figure 10.8b). The terminal boutons contain vesicles that release neurotransmitter when a nerve impulse reaches the terminals (Figure 10.8 ①). The neurotransmitter at the neuromuscular junction—acetylcholine—diffuses across the synaptic cleft and binds to receptor molecules on the sarcolemma, where it induces an impulse that initiates fiber contraction (Figure 10.8 ②).

The neuromuscular junction has several unique features. Each terminal bouton lies in a trough-like depression of the sarcolemma, which in turn has its own invaginations (Figure 10.8b). The invaginations of the sarcolemma are covered with a basal lamina (not illustrated). This basal lamina contains the enzyme acetylcholinesterase (as″ĕ-til-ko″lin-es′ter-ās), which breaks down acetylcholine immediately after the neurotransmitter signals a single contraction (Figure 10.8 ③). This ensures that each nerve impulse to the muscle fiber produces just one twitch of the fiber, preventing any undesirable additional twitches that would result if acetylcholine were to linger in the synaptic cleft.

Motor Unit The axon of a motor neuron branches to innervate a number of fibers in a skeletal muscle. A motor neuron and all the muscle fibers it innervates are called a **motor unit.** Two simplified motor units are illustrated (Figure 10.9a). When a motor neuron fires, all the skeletal muscle fibers in the motor unit contract together. Although the average number of muscle fibers in a motor unit is 150, the number may run as high as several hundred or as low as four. Muscles that require very fine control (such as the muscles moving the fingers and eyes) have few muscle fibers per motor unit,

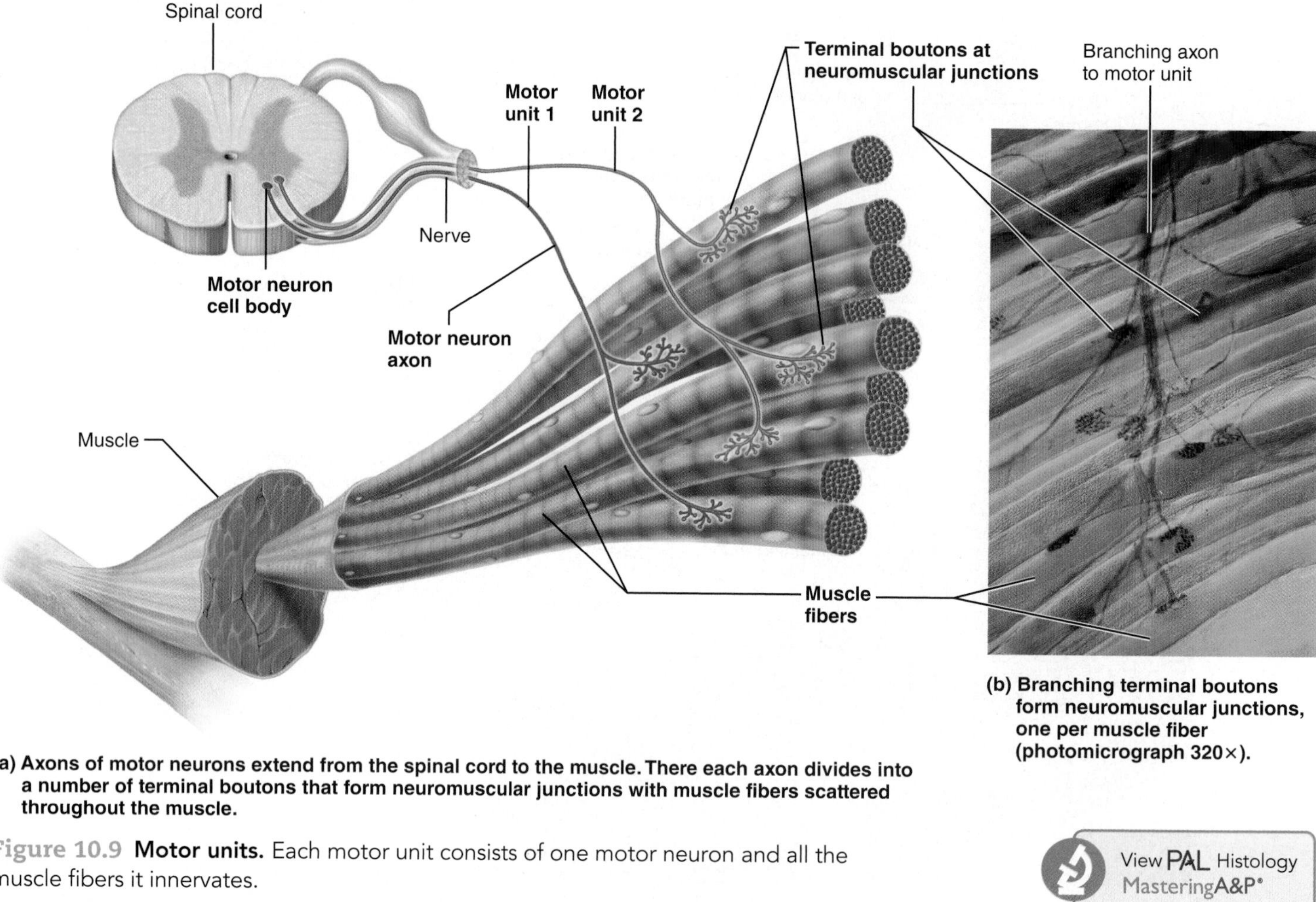

(a) Axons of motor neurons extend from the spinal cord to the muscle. There each axon divides into a number of terminal boutons that form neuromuscular junctions with muscle fibers scattered throughout the muscle.

(b) Branching terminal boutons form neuromuscular junctions, one per muscle fiber (photomicrograph 320×).

Figure 10.9 **Motor units.** Each motor unit consists of one motor neuron and all the muscle fibers it innervates.

View PAL Histology MasteringA&P®

whereas bulky, weight-bearing muscles, whose movements are less precise (such as the hip muscles), have many muscle fibers per motor unit. The muscle fibers of a single motor unit are not clustered together but rather are spread throughout the muscle. As a result, stimulation of a single motor unit causes a weak contraction of the entire muscle.

In addition to variation in the size of motor units in different muscles, each individual muscle contains many motor units. The addition of motor units to accomplish a movement is called **recruitment.** If a small force is required, a small number of motor units are stimulated. As more force is needed, additional motor units are recruited. Thus the same muscle is capable of lifting a pencil off the desk or lifting this textbook.

CLINICAL APPLICATION

Muscle Twitch The uncontrollable contraction of a small portion of a muscle, usually a single motor unit, is a **muscle twitch,** also called a **muscle fasciculation.** Occasional twitches of the eyelid, thumb, or calf are normal. These relatively common twitches may be related to anxiety or stress. Abnormal muscle twitches may be a symptom of certain nervous disorders, dietary deficiencies, or drug side effects.

check your understanding

- ☐ 8. Why is overlap of the thin and thick filaments essential for muscle contraction?
- ☐ 9. Which region of the myofibril changes in length during contraction: the A band, the I band, or the Z disc?
- ☐ 10. Differentiate a neuromuscular junction from a motor unit.

(For answers, see Appendix B.)

Types of Skeletal Muscle Fibers

The various types of skeletal muscle fibers are categorized according to two characteristics: (1) how they manufacture energy (ATP) and (2) how quickly they contract. Some muscle fibers predominantly produce ATP aerobically (using oxygen) and are thus called **oxidative fibers.** Others make ATP anaerobically (without oxygen) via glycolysis and are referred to as **glycolytic fibers.** The speed of contraction, fast versus slow, depends on how quickly a fiber breaks down ATP to gain the energy needed for contraction. Based on these characteristics, muscle fibers are divided into three general classes: *slow oxidative fibers (SO), fast glycolytic fibers (FG),* and *fast oxidative fibers (FO)* (Figure 10.10). Most of the muscles in the body contain all three fiber types, but the proportions differ from one muscle to another.

Slow Oxidative Fibers (SO) As their name implies, slow oxidative fibers obtain their energy from aerobic metabolic reactions; thus, they have a relatively large number of mitochondria (the sites of aerobic metabolism) and a rich supply of capillaries. These relatively thin fibers are red because of their abundant content of *myoglobin* (mi″o-glo′bin), an oxygen-binding pigment in their sarcoplasm. Slow oxidative fibers contract slowly, are extremely resistant to fatigue as long as enough oxygen is present, and deliver prolonged contractions. There are many of these fibers in the postural muscles of the lower back, muscles that must contract continuously to keep the spine straight and maintain posture. Because they are thin, slow oxidative fibers do not generate much power.

CLINICAL APPLICATION

Rhabdomyolysis Although oxygen-binding myoglobin increases endurance, it can cause problems if it leaks from muscle. Damage to skeletal muscles caused by crushing or by extreme, destructive exercise may lead to a condition called **rhabdomyolysis** (rab″do-mi-ol′i-sis; "disintegration of rod-shaped [skeletal] muscle"). The condition was originally recognized in survivors of bombings and earthquakes who had been pinned under fallen buildings. In rhabdomyolysis, myoglobin pours into the bloodstream and clogs the blood-filtering kidneys, causing kidney failure, which eventually leads to heart failure.

Fast Glycolytic Fibers (FG) Fast glycolytic fibers are pale because they contain little myoglobin. They are about twice the diameter of slow oxidative fibers, contain more myofilaments, and thus generate much more power. Because these fibers depend on anaerobic pathways to make ATP, they contain few mitochondria or capillaries but have many glycosomes containing glycogen as a fuel source. Fast glycolytic fibers contract rapidly and tire quickly. They are common in the muscles of the upper limbs, which often lift heavy objects for brief periods.

Fast Oxidative Fibers (FO) Fast oxidative muscle fibers are intermediate in many of their characteristics in comparison with the other two fiber types. Like fast glycolytic fibers, they contract quickly; like slow oxidative fibers, they are oxygen dependent and have a high myoglobin content, a large number of mitochondria, and a rich supply of capillaries. Because fast oxidative fibers depend largely on aerobic metabolism, they are fatigue resistant but less so than slow oxidative fibers. The speed of contraction of fast oxidative fibers is between that of the other two fiber types. The diameter of the fiber is also intermediate; thus, these fibers are more powerful than slow oxidative fibers but not as powerful as fast glycolytic fibers. They are abundant in the muscles of the lower limbs, which must move the body for long periods during locomotion.

Because individual muscles contain a mixture of the three fiber types, each muscle can perform different tasks at different times. A muscle in the calf of the leg, for example, uses its glycolytic fibers to propel the body in a short sprint, its fast oxidative fibers in long-distance running, and its slow oxidative fibers in maintaining a standing posture.

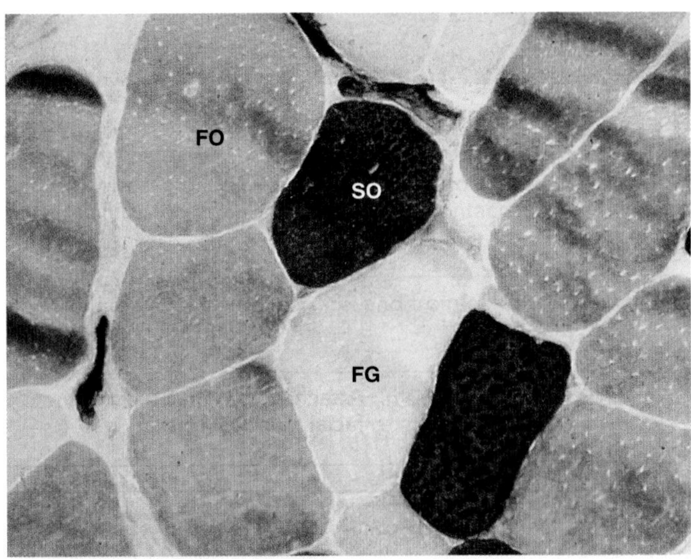

Figure 10.10 Cross section of the three types of fibers in skeletal muscle. The staining technique used in this preparation differentiates the fibers into slow oxidative fibers (SO), fast oxidative fibers (FO), and fast glycolytic fibers (FG) by the abundance of their mitochondria, mitochondrial enzymes, and other features (600×).

Training Effects Although everyone's muscles contain mixtures of the three fiber types, some people have relatively more of one type. These differences are genetically controlled and no doubt influence the athletic capabilities of endurance versus strength. It is possible to transform muscle fiber types through training, however. Specifically, intense resistance training can convert fast glycolytic fibers to fast oxidative fibers. Any fibers converted in this way do revert to their original type when the training stops. In fact, an interesting overshoot phenomenon has been observed. After training ends, the percentage of fast glycolytic fibers increases significantly from the pretraining level. This phenomenon is the physiological basis for tapering off training prior to a major competitive event, a common practice among athletes.

Weight training also increases the diameter and strength of fast muscle fibers. Weight lifting increases the production of the contractile proteins actin and myosin, of the myofilaments containing these proteins, and of the myofibril organelles these myofilaments form. As the number and size of the myofibrils increase, the fibers enlarge. Skeletal muscle is multinucleated, and the enlarged fibers need additional nuclei to direct and support the formation of new proteins. Small immature muscle cells, called satellite cells, are scattered in the muscle tissue outside the muscle fibers. These cells fuse with the fibers, contributing the additional nuclei needed as the fibers enlarge. Thus, muscle fibers do not

REVIEW QUESTIONS

Multiple Choice/Matching Questions

(For answers, see Appendix B.)

1. The connective tissue that lies just outside the sarcolemma of an individual muscle cell is called the (a) epimysium, (b) perimysium, (c) endomysium, (d) endosteum.
2. A fascicle is (a) a muscle, (b) a bundle of muscle cells enclosed by a connective tissue sheath, (c) a bundle of myofibrils, (d) a group of myofilaments.
3. Thick and thin myofilaments have different properties. For each phrase below, indicate whether the filament described is thick or thin (write *thick* or *thin* in the blanks).

____ **(1)** contains actin

____ **(2)** contains myosin heads

____ **(3)** contains myosin

____ **(4)** does *not* lie in the H zone

____ **(5)** does *not* lie in the I band

____ **(6)** attaches to a Z disc

4. Write *yes* or *no* in each blank below to indicate whether each of the following narrows when a skeletal muscle fiber contracts.

____ **(1)** H band

____ **(2)** A band

____ **(3)** I band

____ **(4)** M line

5. Match the level of skeletal muscle organization given in the key with its description:

Key:

(a) muscle (b) fascicle (c) fiber

(d) myofibril (e) myofilament

____ **(1)** rod-shaped organelle; made of sarcomeres

____ **(2)** an organ

____ **(3)** a bundle of cells

____ **(4)** a group of large molecules

____ **(5)** a cell

6. The function of T tubules in muscle contraction is to (a) make and store glycogen, (b) release Ca^{2+} into the cell interior and then pick it up again, (c) transmit an impulse deep into the muscle cell, (d) make proteins.
7. Which fiber type would be the most useful in the leg muscles of a long-distance runner? (a) fast glycolytic, (b) slow glycolytic, (c) fast oxidative.
8. The ions that first enter a muscle cell when an impulse passes over its sarcolemma and then trigger muscle contraction are (a) calcium, (b) chloride, (c) sodium, (d) potassium.
9. Fill in each blank with the correct answer from the key. More than one answer may be correct.

Key:

(a) skeletal muscle (b) cardiac muscle (c) smooth muscle

____ **(1)** striated and involuntary

____ **(2)** striated and voluntary

____ **(3)** not striated and involuntary

____ **(4)** is present in wall of bladder

____ **(5)** is located only in the heart

____ **(6)** its fibers are giant, multinucleate cells

____ **(7)** the individual muscle cells are called muscle fibers

____ **(8)** has no A or I bands

____ **(9)** it is located in the walls of hollow body organs

____ **(10)** its extranuclear materials are called sarcoplasm instead of cytoplasm, and the plasma membrane is called the sarcolemma

Short Answer Essay Questions

10. Name and explain the four special functional properties of muscle tissue.
11. (a) Distinguish a tendon from an aponeurosis and a fleshy attachment. (b) Define origin and insertion, and explain how they differ.
12. Explain the sliding filament theory of contraction by drawing and labeling a relaxed and a contracted sarcomere.
13. Define motor unit.
14. List the structural differences between the three distinct types of skeletal muscle fibers.
15. Cindy Wong was a good anatomy student, but she realized she was mixing up the following "sound-alike" structures in skeletal muscle: myofilaments, myofibrils, fibers, and fascicles. Therefore, she compiled a brief table to define and differentiate these four structures. Construct a table like hers.
16. What is the function of the sarcoplasmic reticulum in a skeletal muscle cell?
17. Define sarcolemma and sarcoplasm.
18. Where is titin located, and what are its functions?
19. What is the general distribution of skeletal muscle fiber types in various body regions (trunk, upper limb, lower limb), and how is fiber type related to function of that body region?

Critical Reasoning & Clinical Application Questions

1. A certain anatomy student decided that his physique left much to be desired. He joined a health club and began to "pump iron" three times a week. After 3 months of training, during which he was able to lift increasingly heavy weights, his arm and chest muscles became much larger. Explain what happened to the fibers in these muscles.
2. Diego, who had not kept in shape, went out to play a game of touch football. As he was running, the calf region of his leg began to hurt. The next day he went to a doctor, who told him he had a strain. Diego kept insisting that no joints hurt. Clearly, he was confusing a *strain* with a *sprain.* Explain the difference.
3. Chickens are capable of only brief bursts of flight, and their flying muscles consist of fast glycolytic fibers. The breast muscles of

ducks, by contrast, consist of slow and fast oxidative fibers. What can you deduce about the flying abilities of ducks?

4. Takashi, an osteopathic physician, saw that Mrs. and Mr. Rogers were suffering because their son was fighting a long battle with Duchenne muscular dystrophy. To comfort them, Takashi said that one day physicians hope to be able to cure this disease by injecting healthy myoblasts into the weakened muscles. What are myoblasts?

5. Why are muscle infections relatively rare (compared to respiratory or skin infections, for example)?

6. As a sprinter, Lateesha knew that the best way to treat pulled muscles was through "RICE" (see Chapter 9, p. 234). What does that mean?

7. After her first day of skiing for the season, Janine woke up with stiff, achy muscles. What caused the sore muscles, and should Janine head out to the slopes for a second day of skiing?

8. Given what you have learned about increasing muscle strength through weight training, why do athletes taper off their training regimen prior to a major competitive event?

9. Skeletal muscle cells cannot divide. How does skeletal muscle repair itself when injured?

Figure 11.2 **Lever systems operating at a mechanical advantage and a mechanical disadvantage.** The equation at the top expresses the relationships among the forces and distances in any lever system.
(**a**) Mechanical advantage with a power lever. When a jack is used, the load lifted is greater than the applied muscular effort: Only 10 kg of force (the effort) lifts a 1000-kg car (the load).
(**b**) Mechanical disadvantage with a speed lever. When a shovel is used to lift soil, the muscular force is greater then the load lifted: A muscular force (effort) of 100 kg lifts 50 kg of soil (the load). Levers operating at a mechanical disadvantage are common in the body.

Effort × length of effort arm = load × length of load arm
(force × distance) = (resistance × distance)

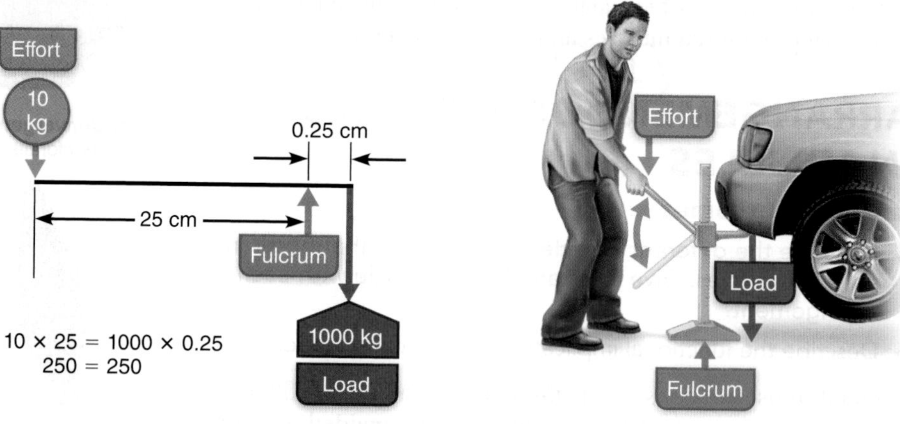

(a) Mechanical advantage with a power lever

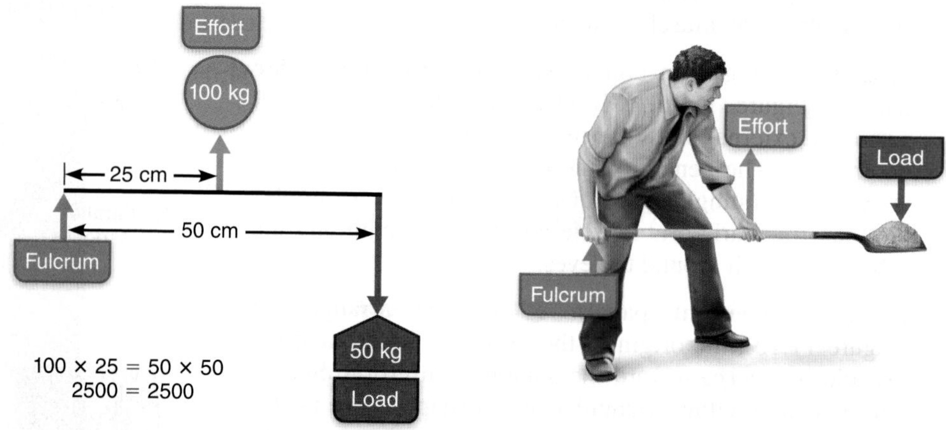

(b) Mechanical disadvantage with a speed lever

LEVER SYSTEMS: BONE-MUSCLE RELATIONSHIPS

learning outcome

▶ Describe the three types of lever systems by which muscles function, and indicate the relative positions of effort, fulcrum, and load in each.

The operation of most skeletal muscles involves *leverage*—the use of a lever to aid the movement of some object. A **lever** is a rigid bar that moves on a fixed point, the **fulcrum,** when a force is applied to the lever. The applied force, called the **effort,** is used to move a resistance, or **load.** A simple lever system that you are certainly familiar with is a see-saw. The middle of the see-saw is the fulcrum; the two levers extend outward. The effort is the body weight of one person on the see-saw; the load is the weight of the other person. The distance from the fulcrum where each is sitting is the effort arm and load arm, respectively.

A lever allows a given effort either to (1) move a heavier load or (2) move a load farther and faster. In the first case, if the load to be moved is close to the fulcrum and the effort is applied far from the fulcrum, a small effort can move a large load (Figure 11.2a). Such a lever is called a **power lever.** Power levers operate at a **mechanical advantage,** enabling a small effort to move a large load. In a power lever, the load is moved only a small distance, but the effort required is also small. For example each large downward push of the jack handle lifts a car only a little, but little muscular effort is required (Figure 11.2a).

In the second case, if the effort is applied closer to the fulcrum than the load to be moved, greater effort is required to move the load. This type of lever operates at a **mechanical disadvantage,** requiring a large effort to move a small load. However, because the load arm is longer than the effort arm, the load is moved over a greater distance and also at a greater speed. Thus, this type of lever is called a **speed lever.** Wielding a shovel is an example of a speed lever (Figure 11.2b).

Although the shovel and the jack are different types of levers, both follow the same universal **law of levers:** When the effort arm is longer than the load arm, the lever operates at a mechanical advantage; when the effort arm is shorter than the load arm, the lever operates at a mechanical disadvantage. Going back to the see-saw example, think about the alterations you would make to balance a see-saw with someone lighter than you. If you can do this, you intuitively understand the law of levers. (The equation at the top of Figure 11.2 expresses this relationship.)

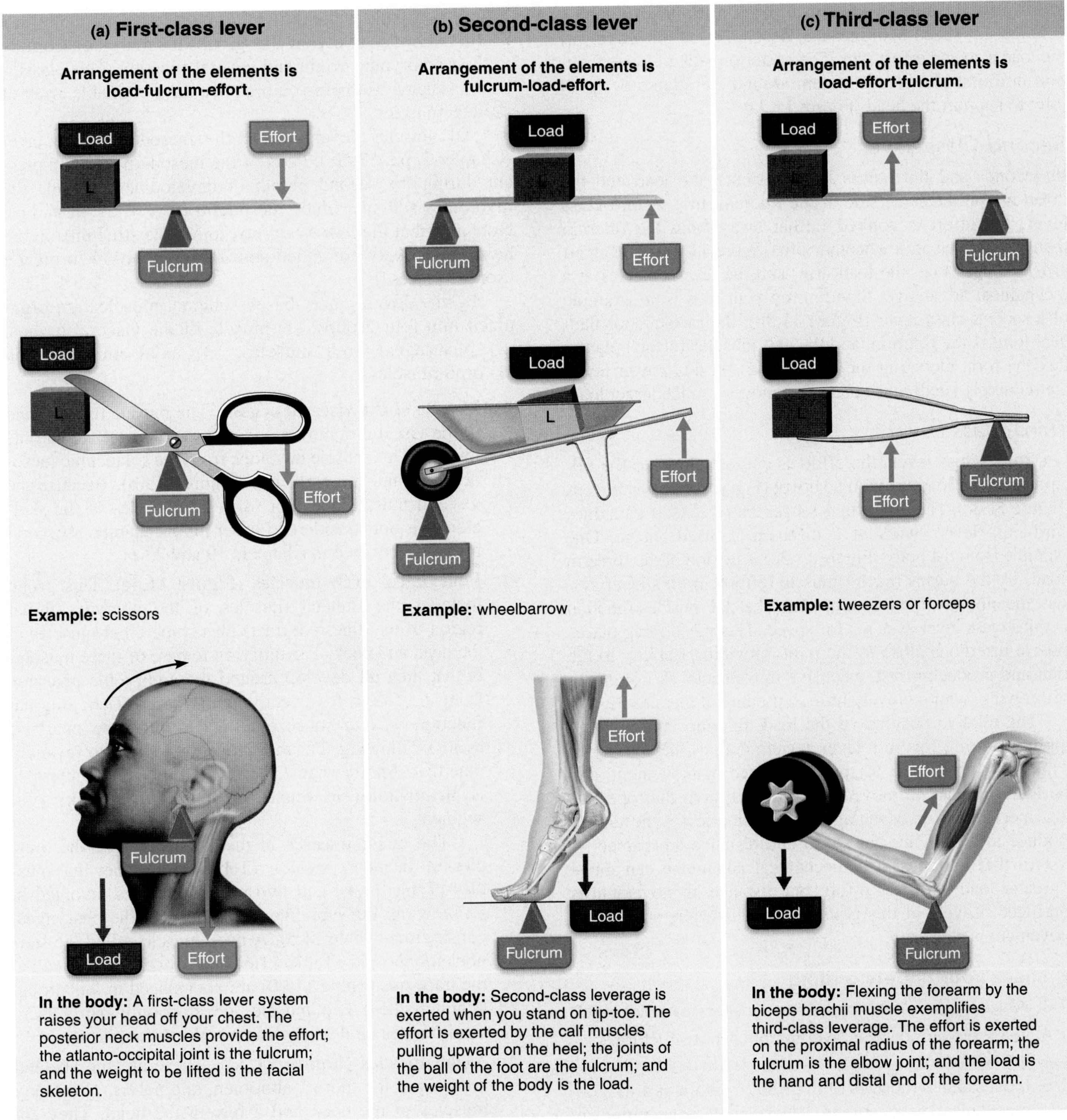

Figure 11.3 **Lever systems.**

So what does all this have to do with skeletal muscles? In the skeletal system, the bones of the skeleton act as *levers*, and the joints act as *fulcrums*. Muscle contraction is the *effort*, which is applied at the point where the muscle attaches to the bone. The *load* is the body part that is moved, along with anything connected to that body part, such as a barbell or suitcase.

First-Class Lever

Levers are divided into three classes, depending on the relative positions of the effort, fulcrum, and load.

In **first-class levers,** the fulcrum is located between the load and the point at which the effort is applied (Figure 11.3a). The see-saw and car jack previously described are examples of first-class levers. First-class levers can operate either at a mechanical advantage, for power, or at a mechanical disadvantage, for speed and distance, depending on the lengths of the load arm and effort arm. The posterior neck muscles function in a first-class lever system around the atlanto-occipital joint to support the head (Figure 11.3a).

Second-Class Lever

In second- and third-class lever systems, the load and the effort are on the same side of the fulcrum. In a **second-class lever,** the effort is applied farther away from the fulcrum than the load, as in a wheelbarrow (Figure 11.3b). The effort arm is longer than the load arm, and the lever operates at a mechanical advantage. Standing on your toes is an example of a second-class lever (Figure 11.3b); the metatarsophalangeal joint is the fulcrum, and the calf muscles (effort) plantar flex the foot, elevating the body (load). In this power lever, the relatively small muscular effort moves a much larger load.

Third-Class Lever

In a **third-class lever,** the effort is applied closer to the fulcrum than the load, as with a shovel (Figure 11.2b) or forceps (Figure 11.3c). The load arm is longer than the effort arm; thus, third-class levers work at a mechanical disadvantage. One example from the body (Figure 11.3c) is flexion of the forearm (load) by the biceps brachii muscle (effort). In this lever system, the elbow is the fulcrum. Most skeletal muscles function as third-class lever systems for speed. This positioning places muscle insertions close to the joint, providing stability to the joint and producing fast, extensive movements, as in running or throwing, with relatively little shortening of the muscle.

The relative position of the load, fulcrum, and point of application of effort in a lever system defines the activity of a muscle with respect to speed, distance of movement, and the load that can be moved. In a lever system that operates at a mechanical disadvantage, a greater effort is needed to produce a movement, but speed and distance are gained. A system that operates at a mechanical advantage can move a greater load with less effort, but distance of movement is sacrificed. Levers of this type are located where strength of movement is a priority.

☑ check your understanding

☐ 3. Which types of levers operate at a mechanical advantage?

☐ 4. In a skeletal/muscular lever system, what structure is the fulcrum?

☐ 5. Most skeletal muscles of the body operate at a mechanical disadvantage. What are the advantages of this arrangement?

(For answers, see Appendix B.)

ORGANIZATIONAL SCHEME BASED ON EMBRYONIC DEVELOPMENT

learning outcome

▶ Organize the body's muscles into four functional groups based on their developmental origin.

Before you look at the individual muscles of the body, it's useful to have an overview that groups the muscles according to their embryonic origin and general function. This classification scheme can help you organize the formidable array of skeletal muscles.

All muscles develop from the mesoderm germ layer (Chapter 3, pp. 57–58). Parts of the mesoderm that are present during the second month of development include the myotomes and splanchnic mesoderm (**Figure 11.4a** and **b**). Note also that the first seven myotome-like structures in the head collectively are called **somitomeres** (so-mit′o-mērz″; "somite pieces").

In this development-based scheme, muscles are organized into four groups: (1) muscle of the visceral organs, (2) pharyngeal arch muscles, (3) axial muscles, and (4) limb muscles.

1. **Muscle of the visceral organs.** The muscle tissue found in the visceral organs is either smooth muscle or cardiac muscle. This muscle develops from the splanchnic mesoderm around the early gut (Figure 11.4b). Because the visceral muscle does not form the muscles of the skeleton, it is not considered further in this chapter. (Visceral muscle is covered in Chapters 19 and 23.)
2. **Pharyngeal arch muscles (Figure 11.4c).** This group includes the skeletal muscles of the pharynx (throat region of the digestive tract) plus some other muscles in the head and neck. The unifying feature of these muscles is that they all develop around the embryonic pharynx, from the fourth to seventh somitomeres. Their original function was to squeeze things through the pharynx, as in swallowing, but they have diversified to do more than that. Another name for the pharyngeal arch muscles is **branchiomeric muscles** (brang″ke-o-mer′ik; "gill segment").

   The major muscles in this group include the muscles of facial expression (Table 11.3), chewing muscles (Table 11.4), and two sets of muscles involved in swallowing: the suprahyoid muscles and the pharyngeal constrictors (Table 11.5). A neck muscle called the sternocleidomastoid (Table 11.6) and a back muscle called the trapezius (Table 11.10) are also placed in this group, although these two muscles develop only partly from somitomeres (and partly from nearby myotomes).
3. **Axial muscles (Figure 11.4d).** These are the skeletal muscles of the thorax, abdomen, and pelvis, plus many muscles of the neck and a few in the head. They are called axial muscles because they lie anterior and posterior to the body axis (vertebral column), and their main functions are moving the trunk and maintaining posture. They develop from the myotomes and some somitomeres.

   The dorsal regions of the myotomes become the deep muscles of the back (Table 11.6), which extend the spine as we stand erect. The ventral regions of the myotomes become the muscles of the anterior and lateral parts of the trunk and neck, which flex the spine

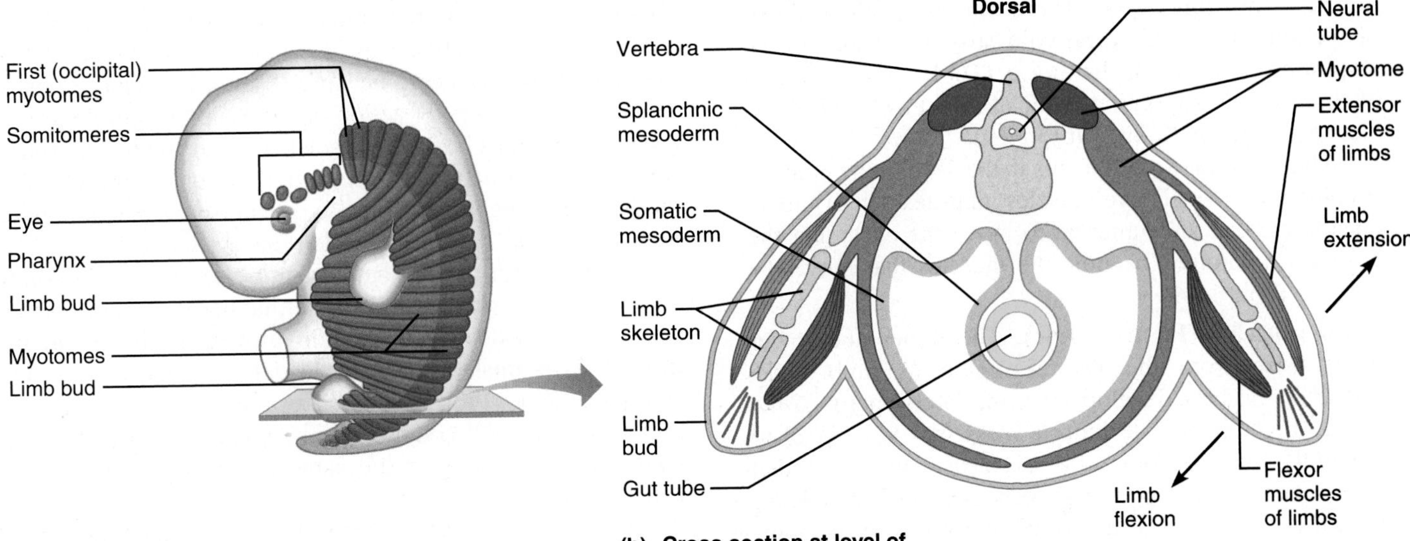

(a) 6-week embryo

(b) Cross section at level of lower limb buds

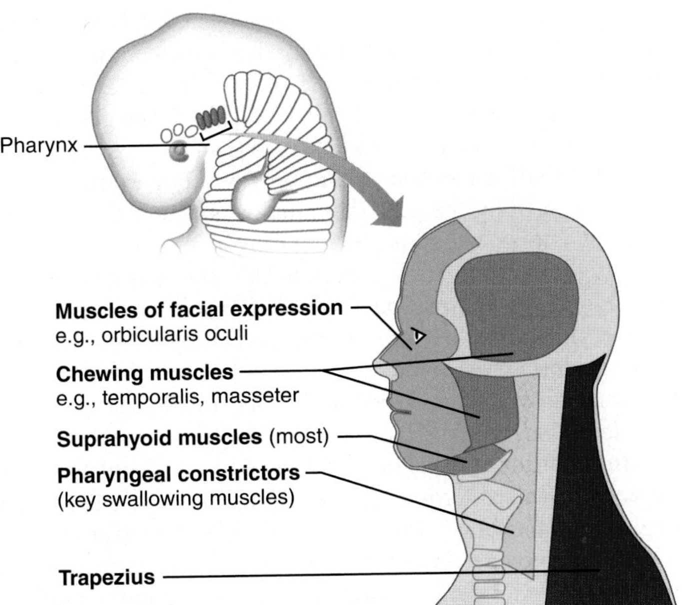

(c) Pharyngeal arch (branchiomeric) muscles: 4th–7th somitomeres

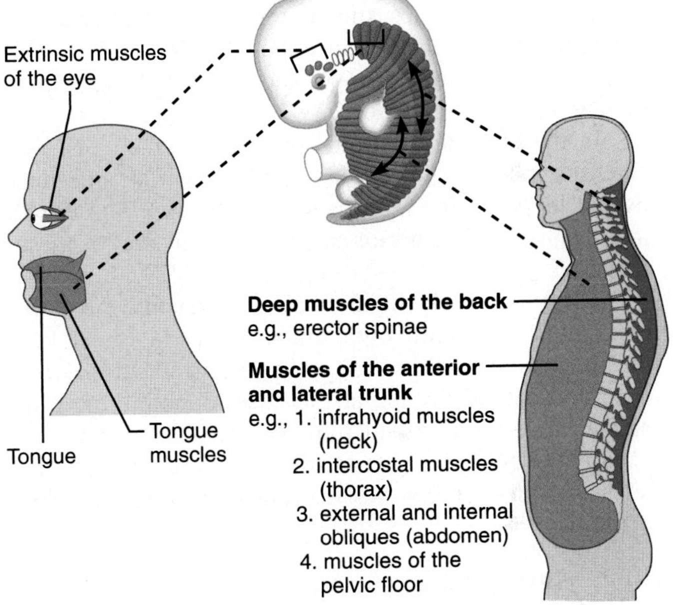

(d) Axial muscles: 1st–3rd somitomeres and from myotomes

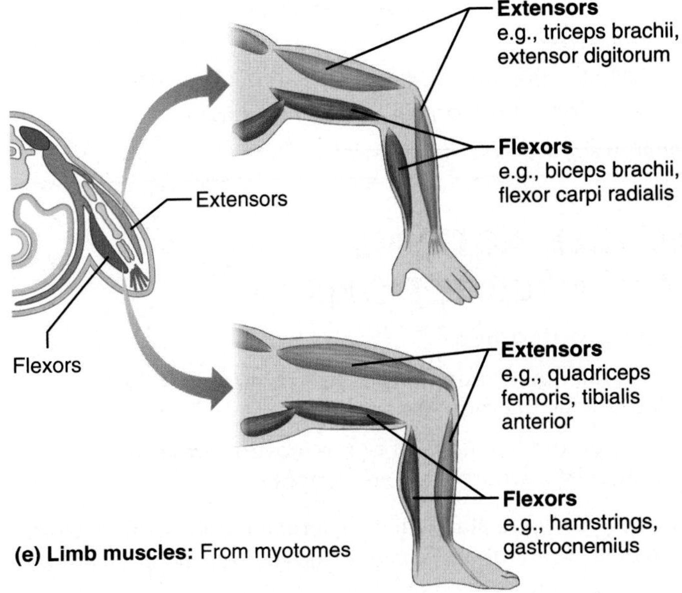

(e) Limb muscles: From myotomes

Figure 11.4 **Development and basic organization of the muscles.** (**a**) A 6-week embryo, showing the myotomes and somitomeres that give rise to muscles. (**b**) Cross section through the trunk and lower limbs of the 6-week embryo, showing divisions of the muscle-forming mesoderm. The plane of section is shown in (a). (**c–e**) Three of the functional classes of muscles, based on developmental origin: (**c**) pharyngeal arch muscles, (**d**) axial muscles, (**e**) limb muscles. (The fourth functional class, visceral muscle, is not represented here.)

(among other functions). These muscles include those of the anterior neck (infrahyoid muscles; Table 11.5), respiratory muscles of the thorax (Table 11.7), muscles of the anterior abdominal wall (Table 11.8), and muscles of the pelvic floor (Table 11.9).

The first three somitomeres in the developing embryo become the extrinsic eye muscles, muscles that originate outside the eye and function to move the eyes, and the occipital myotomes become the muscles that move the tongue (Table 11.4).

4. **Limb muscles (Figure 11.4e).** The upper and lower limbs arise from the ventral region as limb buds, and limb muscles develop from the lateral parts of the nearby myotomes. The muscle mass that lies dorsal to the limb bones becomes the extensor muscles of that limb, whereas the ventral mass becomes the limb's flexor muscles. The muscles that attach the limbs to their girdles and the muscles that attach the girdles to the trunk (Tables 11.10, 11.11, and 11.15) are also part of this group. These muscles function in locomotion and in manipulating objects. Because they develop in the same way, the muscles of the upper and lower limbs are similar and directly comparable.

    In the adult upper limbs (Tables 11.11–11.14), the extensor muscles lie on the limb's posterior side and extend its parts (forearm, hand, and fingers). The flexor muscles lie on the anterior side of the limb and flex these parts.

    The lower limb rotates during embryonic development; thus in the adult lower limbs (Tables 11.15–11.17), the extensor muscles occupy the anterior side of the limb and extend the leg at the knee, dorsiflex the foot at the ankle, and extend the toes. The flexor muscles occupy the posterior side of the lower limb and flex the leg at the knee, plantar flex the foot at the ankle, and flex the toes.

☑ check your understanding

☐ 6. To which developmental group does each of the following belong: (a) the muscles that move the thigh, (b) the abdominal muscles (the "abs"), (c) the chewing muscles?

☐ 7. What is the movement produced by muscles that develop on the dorsal side of the limb?

(For answers, see Appendix B.)

MUSCLE ACTIONS AND INTERACTIONS

learning outcomes

- ▶ Describe how a muscle's position as it crosses a joint determines the muscle's action.
- ▶ Describe the functions of prime movers (agonists), antagonists, synergists, and fixators.

It is easy to understand that different muscles often work together to bring about a single movement. What is not so obvious, however, is that muscles often work *against* one another in a productive way. No single muscle can reverse the motion it produces, because a muscle cannot "push." Therefore, for whatever action one muscle (or muscle group) can do, there must be other muscles that can "undo" the action (or do the opposite action). In general, groups of muscles that produce opposite movements lie on opposite sides of a given joint. See *Focus on Muscle Action* (Figure 11.5).

With respect to their actions, muscles are classified into several *functional* types. A muscle that has the major responsibility for producing a specific movement is the **prime mover,** or **agonist** (ag′o-nist; "leader"), of that motion. For example, the pectoralis major is a prime mover for flexing the arm at the shoulder (Figure 11.5a). Sometimes, two muscles contribute so heavily to the same movement that both are called agonists.

Muscles that oppose or reverse a particular movement act as **antagonists** (an-tag′o-nists; "against the leader"). When a prime mover is active, it is possible for its antagonists to be stretched or remain relaxed. Usually, however, the antagonists contract slightly during the movement to keep the movement from overshooting its mark or to slow it near its completion. Antagonists can also be prime movers in their own right; that is, an antagonist for one movement can serve as an agonist for the opposite movement. For example, *flexion* of the arm by the pectoralis major is antagonized by the latissimus dorsi, which is a prime mover for *extension* of the arm (Figure 11.5b). It is important that the two members of any agonist/antagonist pair be challenged and developed evenly; otherwise the more developed muscle will put constant tension on the joint, reducing its flexibility and causing awkward movements. A person in this situation is said to be *muscle bound.*

In addition to prime movers and antagonists, most movements also involve one or more muscles called **synergists** (sin′er-jist; "together-worker"). Synergists help the prime movers, either by adding a little extra force to the movement being carried out or by reducing undesirable extra movements that the prime mover may produce. This second function of synergists deserves more explanation. Some prime movers cross several joints and can cause movements at all of them, but synergists act to cancel some of these movements. For example, the muscles that flex the fingers cross both the wrist and finger joints, but you can make a fist without flexing your wrist because synergists stabilize the wrist.

Additionally, some prime movers can cause several kinds of movement at the *same joint.* For example, latissimus dorsi (shown in Figure 11.5b), functions in adduction as well as extension of the shoulder because it crosses the shoulder both posteriorly and medially. Synergists prevent the particular movement that is inappropriate at a given time: Pectoralis major (Figure 11.5a), a flexor of the shoulder, will act as a synergist to latissimus dorsi to prevent extension and produce straight adduction.

Some synergists hold a bone firmly in place so that a prime mover has a stable base on which to move a body

Figure 11.5

The action of a muscle can be inferred by the position of the muscle relative to the joint it crosses. (Examples given relate to the shoulder joint.)

(a) A muscle that crosses on the anterior side of a joint produces flexion*

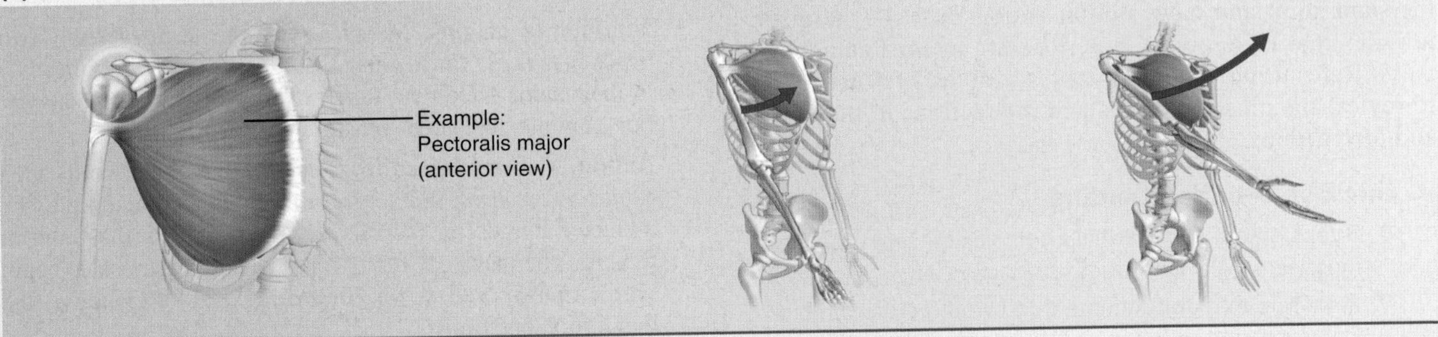

(b) A muscle that crosses on the posterior side of a joint produces extension*

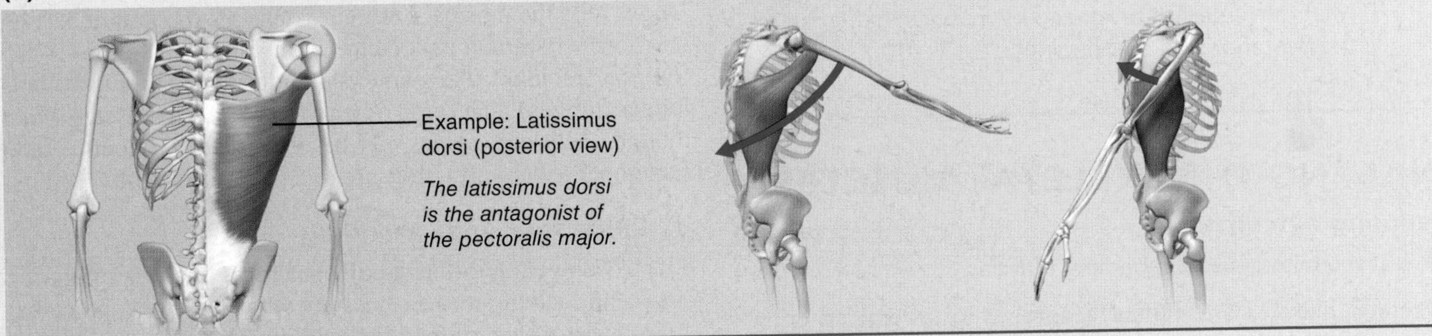

(c) A muscle that crosses on the lateral side of a joint produces abduction

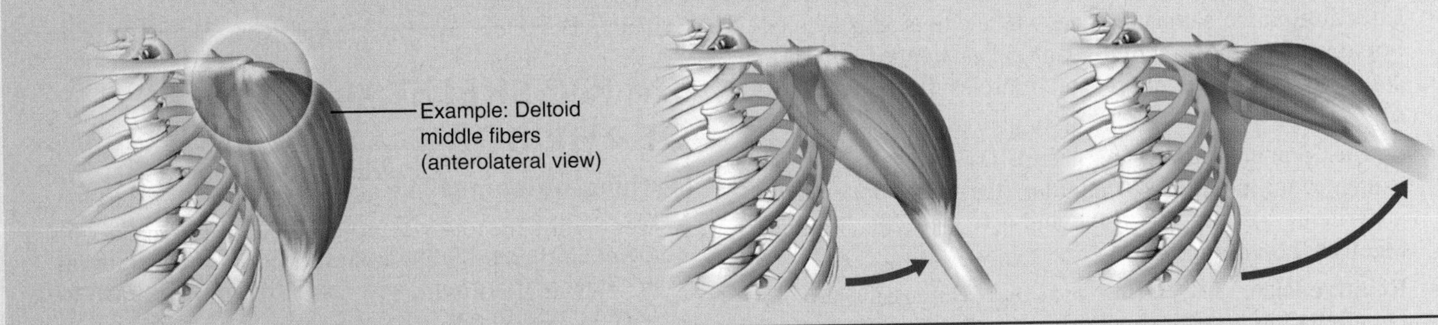

(d) A muscle that crosses on the medial side of a joint produces adduction

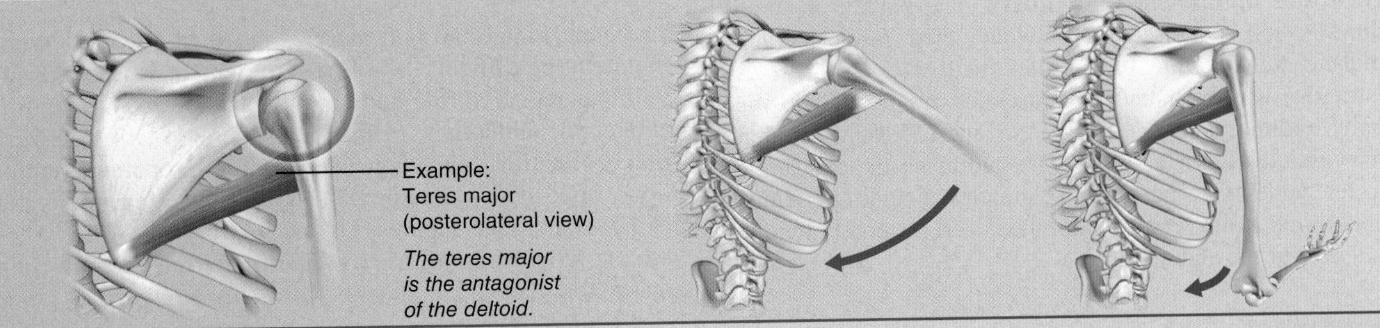

*These generalities do not apply to the knee and ankle because the lower limb is rotated during development. The muscles that cross these joints posteriorly produce flexion, and those that cross anteriorly produce extension.

part. Such synergists are called **fixators.** An example is the muscles that fix the scapula when the arm moves. Muscles that maintain posture and stabilize joints also act as fixators.

In summary, although prime movers get all the credit for causing movements, the actions of antagonistic and synergistic muscles are also important. Muscles that cross the same joint and have similar actions act as synergists; muscles that have opposite actions are agonist/antagonist pairs. Refer to the summary tables (Tables 11.1 and 11.2) to review the muscle actions and interactions in the upper and lower limbs.

☑ check your understanding

- ☐ 8. Biceps brachii and brachialis both flex the forearm. These muscles, which work together, function as ________.
- ☐ 9. A muscle that abducts the thigh would cross the hip on which side of the joint? What action would the antagonist muscles produce?
- ☐ 10. At most joints (the knee and ankle are exceptions), what movement is produced by muscles that cross the posterior side of the joint?

(For answers, see Appendix B.)

NAMING THE SKELETAL MUSCLES

learning outcome

▶ List the criteria used in naming muscles.

Skeletal muscles are named according to several criteria, each of which describes the muscle in some way. Learning these criteria can simplify the task of learning muscle names.

- **Location.** Some names indicate where a muscle is located. For example, the brachialis muscle is in the arm (*brachium* = arm), and intercostal muscles lie between the ribs (*costa* = rib).
- **Shape.** Some muscles are named for their shapes. For example, the deltoid is triangular (the Greek letter *delta* is written Δ), and the right and left trapezius muscles together form a trapezoid.
- **Relative size.** The terms *maximus* (largest), *minimus* (smallest), *longus* (long), and *brevis* (short) are part of the names of some muscles—such as the gluteus maximus and gluteus minimus muscles of the buttocks.
- **Direction of fascicles and fibers.** The names of some muscles tell the direction in which their fascicles (and muscle fibers) run. In muscles with the term *rectus* (straight) in their name, the fascicles are parallel to the body midline, whereas *transversus* and *oblique* mean that the fascicles lie at a right angle and at an oblique angle to the midline, respectively. Examples include the rectus abdominis, transversus abdominis, and external oblique muscles of the abdomen.
- **Location of attachments.** The names of some muscles reveal their points of origin and insertion. Recall that the origin is the less movable attachment of a muscle and that the insertion is the more movable attachment. The origin is always named first. For instance, the brachioradialis muscle in the forearm originates on the bone of the brachium, the humerus, and inserts on the radius.
- **Number of origins.** Muscles with the term *biceps* ("two heads"), *triceps* ("three heads"), or *quadriceps* ("four heads") in their name have two, three, or four origins, respectively. For example, the biceps brachii has two origins.
- **Action.** Muscles named for their action include words such as *flexor, extensor, adductor,* or *abductor*. For example, the adductor longus on the medial thigh adducts the thigh at the hip. The names of many forearm and leg muscles begin with *extensor* and *flexor,* indicating how they move the hand, foot, and digits.

Often, several different criteria are used to name a muscle. For instance, the name *extensor carpi radialis longus* tells the muscle's action (extensor), the joint it acts on (*carp* = wrist), and its location, lying along the radius in the forearm (radialis). The name also indicates that the muscle is longer (longus) than some other wrist extensor muscles. Even though such long names can be difficult to pronounce (and remember accurately), they are extremely informative.

☑ check your understanding

- ☐ 11. Use your knowledge of word roots and the criteria for naming muscles to state what the name indicates about each of the following muscles: (a) latissimus dorsi, (b) sternocleidomastoid, (c) serratus anterior, (d) adductor magnus.

(For the answer, see Appendix B.)

MAJOR SKELETAL MUSCLES OF THE BODY

learning outcomes

▶ Identify the muscular compartments of the upper and lower limb. Identify the general action of the muscles in each compartment, and name the muscles located in each compartment.

▶ Identify the muscles described in the tables in this section (Tables 11.3–11.17). State the origin, insertion, and primary action of each.

There are more than 600 muscles in the body, and learning them can be difficult. The first requirement is to make sure you understand all the body movements: flexion, extension, abduction, adduction, medial and lateral rotation, protraction, retraction, elevation, depression, pronation, supination, opposition, dorsiflexion, plantar flexion, inversion, and eversion (see Figures 9.6 and 9.7, pp. 214–216). These are the terms you will use to describe muscle actions. Some instructors have their students learn only a few major groups of muscles (Figure 11.6), whereas others require a more detailed study.

Summary tables of the limb muscles (Tables 11.1 and 11.2) are presented first for easy reference and review of limb muscle actions. More detailed tables follow (Tables 11.3–11.17) that describe the origin, insertion, action, and innervation of most muscles of the body.

Muscle Compartments of the Limbs

Dense fibrous connective tissue divides the muscles of the limbs into anatomical **compartments.** These fascial compartments group muscles of similar developmental origin and function. Muscles in the same compartment have similar actions and often function synergistically to produce a particular movement. Muscles in opposite compartments function as agonist/antagonist pairs. In most cases, each compartment is innervated by a single named nerve.

Upper Limb

The upper limb has two compartments: anterior and posterior. The muscles in the anterior compartment of the arm flex the arm at the shoulder or flex the forearm at the elbow and are innervated by the musculocutaneous nerve (Figure 11.7a). The muscles of the anterior compartment of the forearm are wrist and digital flexors and pronators innervated by either the median nerve or the ulnar nerve (Figure 11.7b).

The muscles of the posterior compartments of the arm and forearm extend the elbow and wrist, respectively. These muscles, plus the *supinator* and *brachioradialis* (developmentally posterior compartment muscles), are all innervated by branches off the radial nerve. Functionally, the supinator supinates the forearm, and the brachioradialis, which crosses the anterior side of the elbow, acts as a forearm flexor.

(The actions of the muscles that move the arm, forearm, and hand are summarized in **Table 11.1**, pp. 274–275.)

Lower Limb

The thigh portion of the lower limb has three compartments: posterior, anterior, and medial (Figure 11.8a). The muscles of the posterior compartment extend the thigh at the hip and flex the leg at the knee and are innervated by the tibial branch of the sciatic nerve. The muscles of the anterior compartment flex the thigh at the hip and extend the leg at the knee and are innervated by the femoral nerve. The muscles of the medial compartment adduct the thigh and are innervated primarily by the obturator nerve.

The leg also has three compartments: posterior, anterior, and lateral (Figure 11.8b). The posterior compartment contains the digital flexor muscles and plantar flexors innervated by the tibial nerve. The anterior compartment contains the digital extensor muscles and dorsiflexors, all innervated by the deep fibular nerve. The lateral compartment contains the fibularis muscles that plantar flex and evert the foot and are innervated by the superficial fibular nerve.

(The actions of the muscles of the thigh and leg are summarized in **Table 11.2**, pp. 276–277.)

CLINICAL APPLICATION

Compartment Syndrome Any injury to a limb muscle, either traumatic or chronic, can result in swelling of the muscle. Because the inelastic fascia surrounding the muscle compartment prohibits the compartment from expanding, pressure in the compartment increases and can compress the vessels and nerves, resulting in incredible pain. This increased pressure impedes venous drainage from the compartment, further increasing intracompartmental pressure. **Acute compartment syndrome** results from traumatic injury and requires immediate medical attention to minimize the risks of tissue damage resulting from ischemia (decreased blood supply). Surgical treatment, cutting through the skin and fascia of the affected compartment (a procedure called a fasciotomy), may be necessary to reduce pressure in the compartment. **Chronic compartment syndrome** is an overuse injury. When muscle is exercised, intracompartmental pressure increases. When a muscle is overused, as when an athlete trains excessively, microtears in the muscle can cause swelling, increasing the pressure in the muscle compartment. The most common injuries of this type occur in the anterior and lateral compartments of the leg. The characteristic symptom of chronic compartment syndrome is pain during activity that is relieved when exercise stops. Diagnosis is confirmed by measuring intracompartmental pressure before and after exercise. Chronic compartmental syndrome is usually treated conservatively with RICE (rest, ice, compression, elevation) and low-impact activities that do not result in pain.

☑ check your understanding

- ☐ **12.** In what muscle compartment are the biceps brachii and brachialis muscles located? In what compartment is their antagonist muscle located?
- ☐ **13.** What are the actions of the muscles in the posterior compartment of the thigh? What muscle group functions as an antagonist to these muscles?
- ☐ **14.** List two muscles that act in synergy as prime movers for plantar flexion.

(For answers, see Appendix B.)

(Text continues on page 278.)

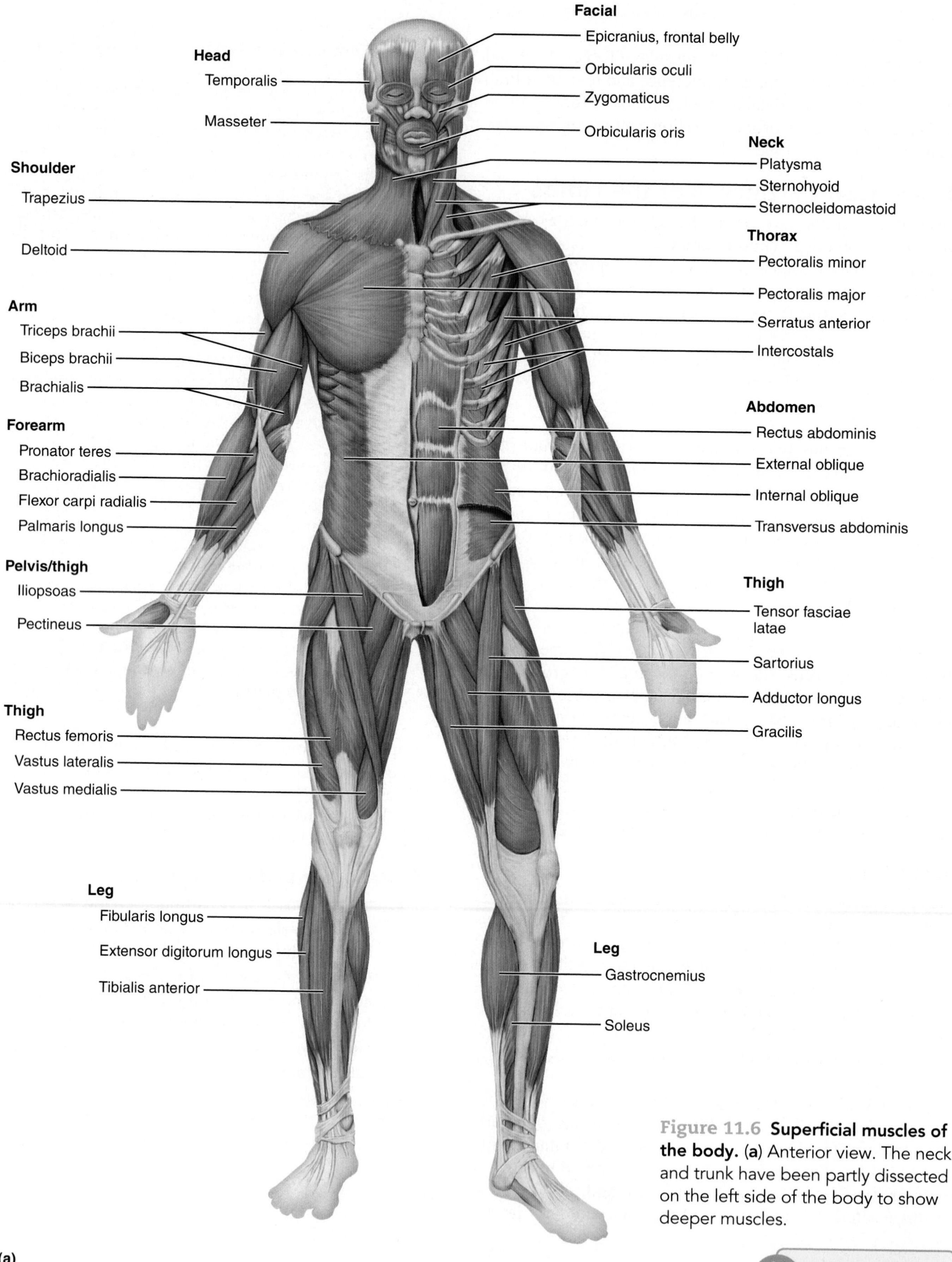

Figure 11.6 **Superficial muscles of the body.** (**a**) Anterior view. The neck and trunk have been partly dissected on the left side of the body to show deeper muscles.

Practice art labeling MasteringA&P®

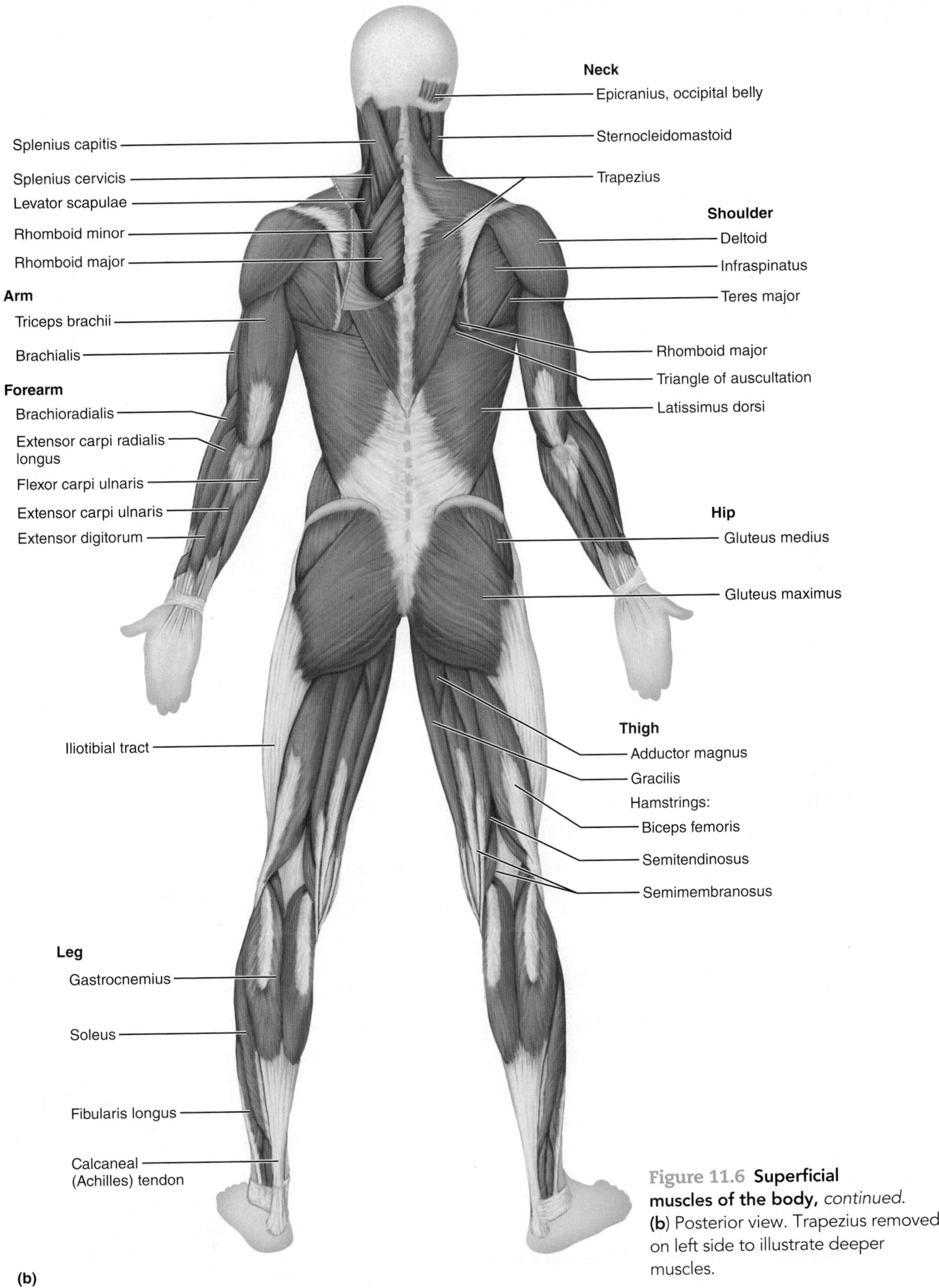

Figure 11.6 Superficial muscles of the body, *continued.* **(b)** Posterior view. Trapezius removed on left side to illustrate deeper muscles.

Table 11.1 Summary of Actions of Muscles Acting on the Arm, Forearm, and Hand (Figure 11.7)

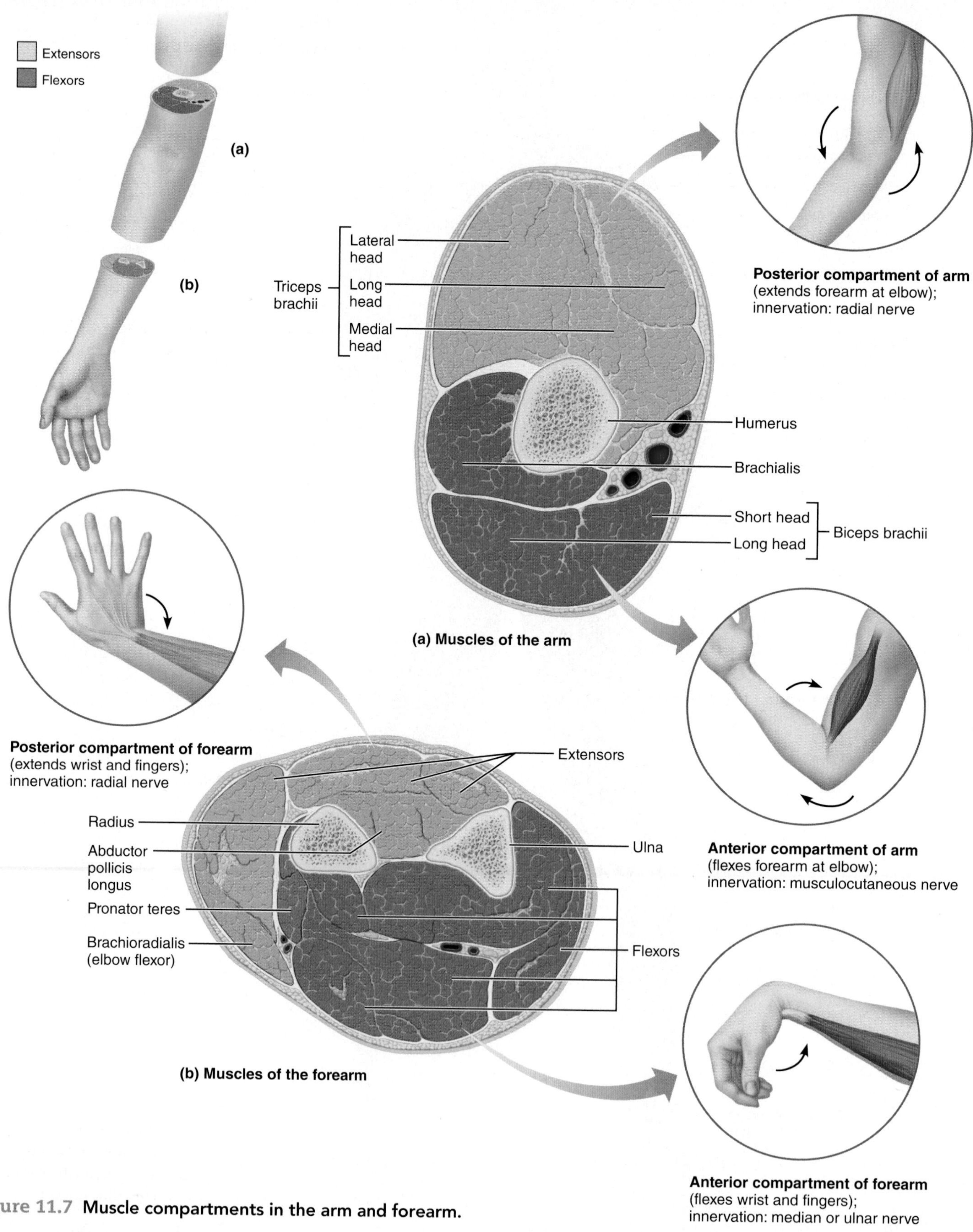

Figure 11.7 Muscle compartments in the arm and forearm.

Table 11.1 *continued*

PART I: Muscles Acting on the Arm (Humerus) (PM = Prime Mover)	Actions at the Shoulder: Flexion	Extension	Abduction	Adduction	Medial Rotation	Lateral Rotation
Pectoralis major	X (PM)			X (PM)	X	
Latissimus dorsi		X (PM)		X (PM)	X	
Teres major		X		X	X	
Deltoid	X (PM) (anterior fibers)	X (PM) (posterior fibers)	X (PM)		X (anterior fibers)	X (posterior fibers)
Subscapularis					X (PM)	
Supraspinatus			X			
Infraspinatus						X (PM)
Teres minor				X (weak)		X (PM)
Coracobrachialis	X			X		
Biceps brachii	X					
Triceps brachii				X		

PART II: Muscles Acting on the Forearm	Actions on the Forearm: Elbow Flexion	Elbow Extension	Pronation	Supination
Biceps brachii	X (PM)			X
Brachialis	X (PM)			
Triceps brachii		X (PM)		
Anconeus		X		
Pronator teres	X (weak)		X	
Pronator quadratus			X (PM)	
Brachioradialis	X			
Supinator				X

PART III: Muscles Acting on the Wrist and Fingers	Actions at the Wrist: Flexion	Extension	Abduction	Adduction	Actions on the Fingers: Flexion	Extension
Anterior Compartment						
Flexor carpi radialis	X (PM)		X			
Palmaris longus	X (weak)					
Flexor carpi ulnaris	X (PM)			X		
Flexor digitorum superficialis	X (PM)				X	
Flexor pollicis longus					X (thumb)	
Flexor digitorum profundus	X				X	
Posterior Compartment						
Extensor carpi radialis longus and brevis		X	X			
Extensor digitorum		X (PM)				X (and abducts)
Extensor carpi ulnaris		X		X		
Abductor pollicis longus			X		(abducts thumb)	
Extensor pollicis longus and brevis						X (thumb)
Extensor indicis						X (index finger)

Table 11.2 Summary of Actions of Muscles Acting on the Thigh, Leg, and Foot (Figure 11.8)

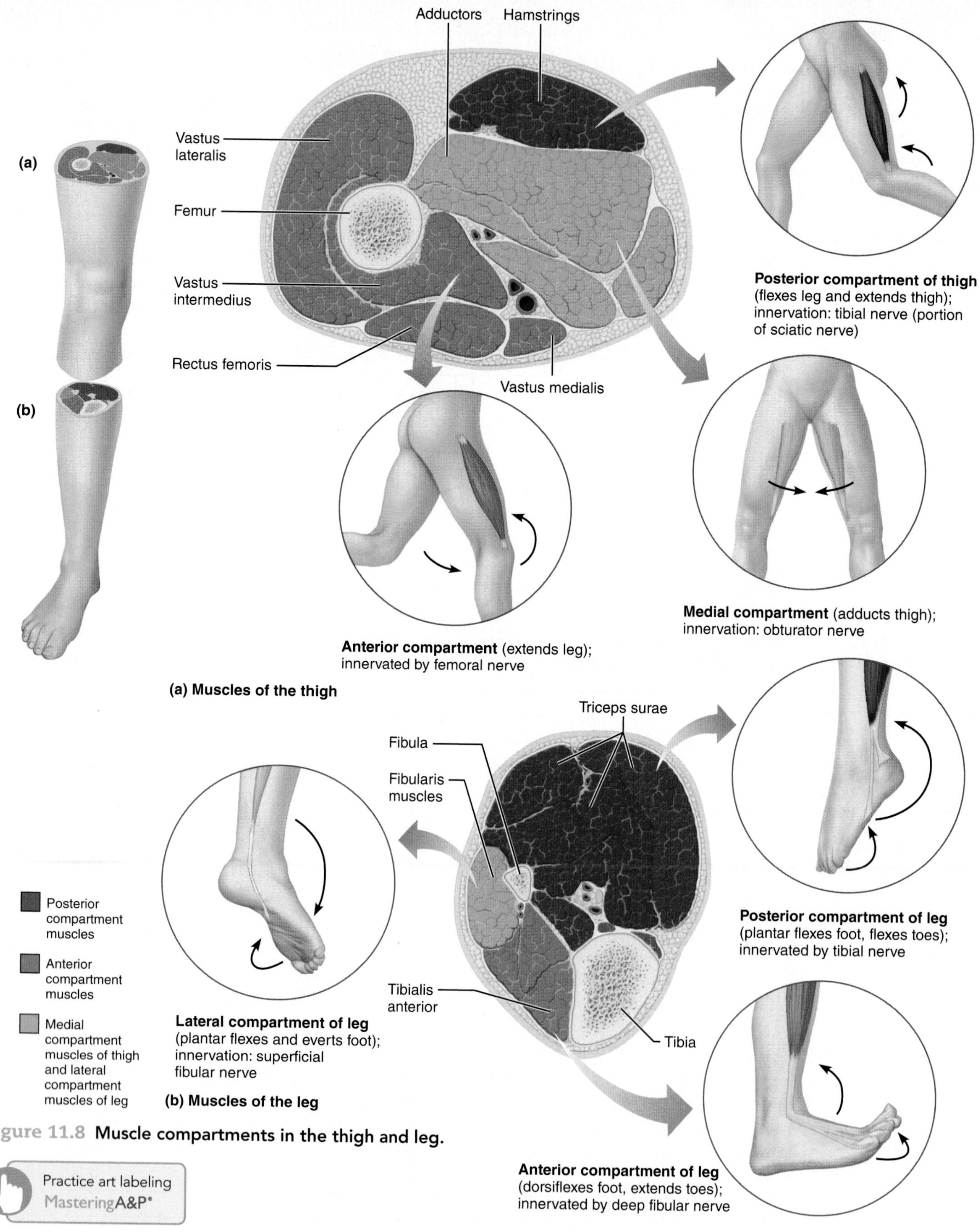

Figure 11.8 Muscle compartments in the thigh and leg.

Practice art labeling MasteringA&P®

Table 11.2 *continued*

PART I: Muscles Acting on the Thigh and Leg (PM = Prime Mover)	Actions at the Hip Joint						Actions at the Knee	
	Flexion	Extension	Abduction	Adduction	Medial Rotation	Lateral Rotation	Flexion	Extension
Posterior Muscles								
Gluteus maximus		X (PM)	X			X		
Gluteus medius			X (PM)		X			
Gluteus minimus			X		X			
Piriformis			X			X		
Obturator internus						X		
Obturator externus						X		
Gemelli						X		
Quadratus femoris						X		
Biceps femoris		X (PM)					X (PM)	
Semitendinosus		X					X (PM)	
Semimembranosus		X					X (PM)	
Gastrocnemius							X	
Plantaris							X	
Popliteus							X (and rotates leg medially)	
Anterior and Medial Muscles								
Iliopsoas	X (PM)							
Tensor fasciae latae	X		X		X			X
Sartorius	X		X			X	X	
Rectus femoris	X							X (PM)
Vastus muscles								X (PM)
Adductor magnus	X	X		X	X			
Adductor longus	X			X	X			
Adductor brevis				X	X			
Gracilis				X	X		X	
Pectineus	X			X	X			

PART II: Muscles Acting on the Ankle and Toes	Actions at the Ankle				Actions on the Toes	
	Plantar Flexion	Dorsiflexion	Inversion	Eversion	Flexion	Extension
Posterior Compartment						
Gastrocnemius	X (PM)					
Soleus	X (PM)					
Plantaris	X					
Flexor digitorum longus	X		X		X (PM)	
Flexor hallucis longus	X		X		X (great toe)	
Tibialis posterior	X		X (PM)			
Anterior Compartment						
Tibialis anterior		X (PM)	X			
Extensor digitorum longus		X				X (PM)
Fibularis tertius		X		X		
Extensor hallucis longus		X	X (weak)			X (great toe)
Lateral Compartment						
Fibularis longus and brevis	X			X		

The Muscle Tables

In the detailed muscle tables that follow (Tables 11.3–11.17), the muscles are grouped by function and by location, roughly from head to foot. Every table starts by introducing a muscle group in the opening paragraphs. The details of shape, location, attachments, actions, and innervation for each muscle in the group follow in table format. All muscles are illustrated in accompanying figures. As you study each muscle, look at its attachments and the direction of its fascicles, and try to understand how these features determine its action. Many muscles have multiple actions; in these tables, the primary action of each muscle is indicated in blue text. A good way to learn muscle actions is to act out the movements yourself and then feel the contracting muscles bulge beneath your skin.

The sequence of tables is as follows:

- Muscles of the Head, Part I: Facial Expression (**Table 11.3**, **Figure 11.9**); pp. 278–281
- Muscles of the Head, Part II: Mastication and Tongue Movement (**Table 11.4**, **Figure 11.10**); pp. 282–283
- Muscles of the Anterior Neck and Throat: Swallowing (**Table 11.5**, **Figure 11.11**); pp. 284–286
- Muscles of the Neck and Vertebral Column: Head Movements and Trunk Extension (**Table 11.6**, **Figure 11.12**); pp. 287–290
- Deep Muscles of the Thorax: Breathing (**Table 11.7**, **Figure 11.13**); pp. 291–292
- Muscles of the Abdominal Wall: Trunk Movements and Compression of Abdominal Viscera (**Table 11.8**, **Figure 11.14**); pp. 293–295
- Muscles of the Pelvic Floor and Perineum: Support of Abdominopelvic Organs (**Table 11.9**, **Figure 11.15**); pp. 296–297
- Superficial Muscles of the Anterior and Posterior Thorax: Movements of the Scapula (**Table 11.10**, **Figure 11.16**); pp. 298–301
- Muscles Crossing the Shoulder Joint: Movements of the Arm (Humerus) (**Table 11.11**, **Figure 11.17**); pp. 302–304
- Muscles Crossing the Elbow Joint: Flexion and Extension of the Forearm (**Table 11.12**, Figure 11.17); p. 305
- Muscles of the Forearm: Movements of the Wrist, Hand, and Fingers (**Table 11.13**, **Figures 11.18**, **11.19**, and **11.20**); pp. 306–311
- Intrinsic Muscles of the Hand: Fine Movements of the Fingers (**Table 11.14**, **Figure 11.21**); pp. 312–314
- Muscles Crossing the Hip and Knee Joints: Movements of the Thigh and Leg (**Table 11.15**, **Figures 11.22**, **11.23**, and **11.24**); pp. 315–322
- Muscles of the Leg: Movements of the Ankle and Toes (**Table 11.16**, **Figures 11.25**, **11.26**, and **11.27**); pp. 323–328
- Intrinsic Muscles of the Foot: Toe Movement and Foot Support (**Table 11.17**, **Figure 11.28**); pp. 329–332

(Text continues on page 332.)

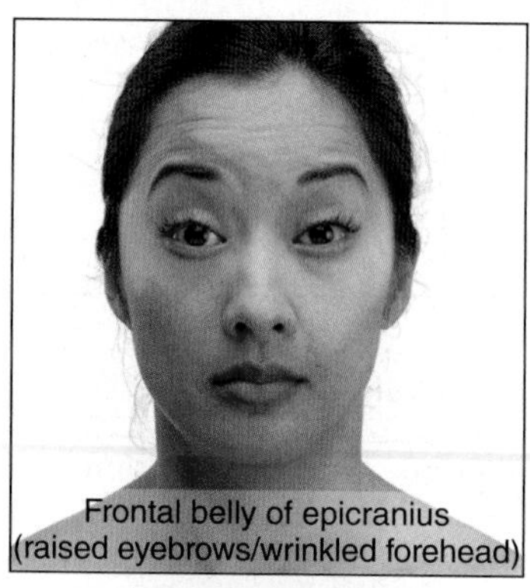
Frontal belly of epicranius (raised eyebrows/wrinkled forehead)

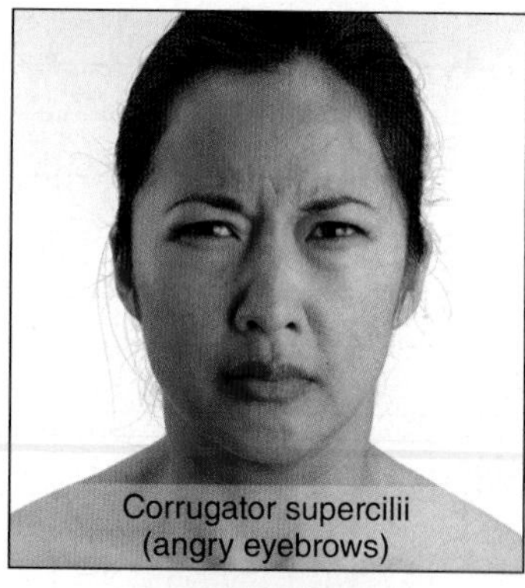
Corrugator supercilii (angry eyebrows)

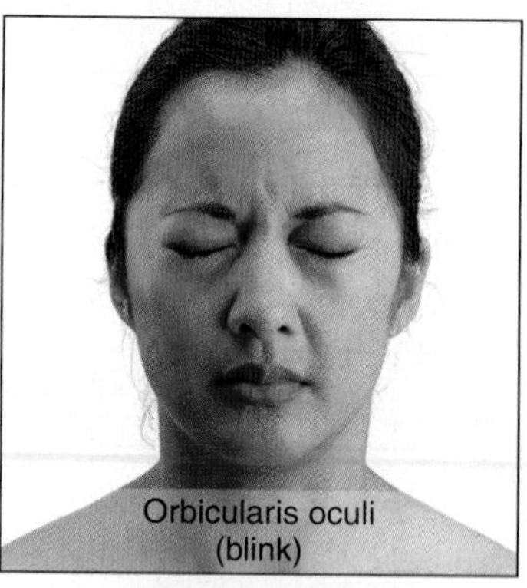
Orbicularis oculi (blink)

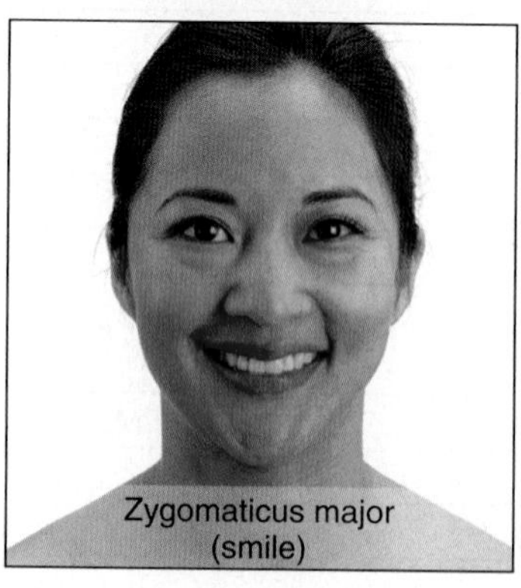
Zygomaticus major (smile)

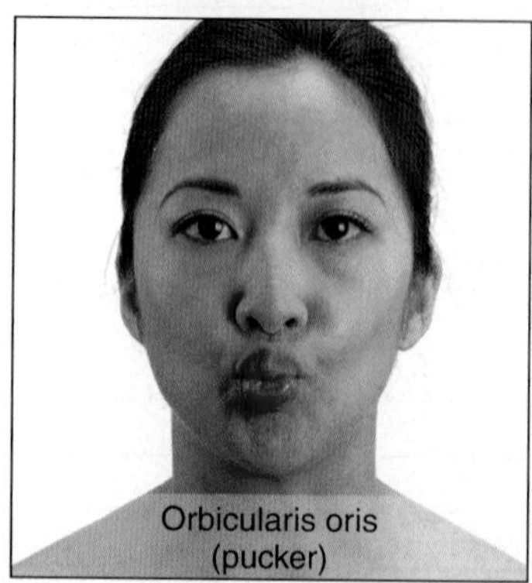
Orbicularis oris (pucker)

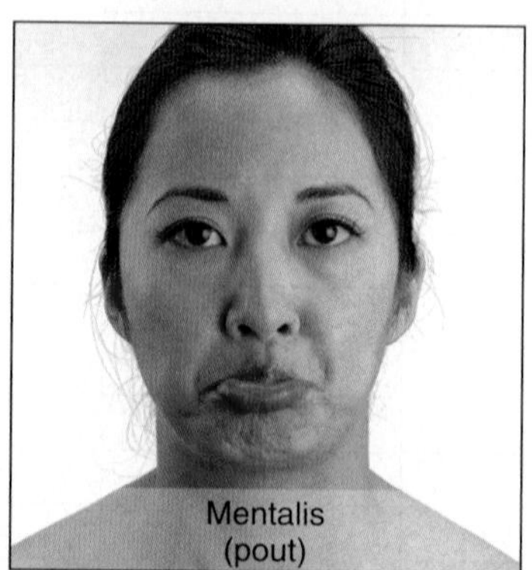
Mentalis (pout)

Platysma (tense neck)

(a)

Figure 11.9 Muscles of facial expression. (a) Actions of select facial muscles. *(Figure continues on page 280.)*

Table 11.3 Muscles of the Head, Part I: Facial Expression (Figure 11.9)

The muscles that promote facial expression lie in the face and scalp, just deep to the skin. They are thin and vary in shape, and adjacent muscles in this group tend to be fused. Unlike other skeletal muscles, facial muscles insert on the skin, not on the bones. In the scalp, the main muscle is the **epicranius,** which has distinct anterior and posterior parts. In the face, the muscles covering the facial bones lift the eyebrows and flare the nostrils, close the eyes and lips, and provide one of the best tools for influencing others—the smile. The great importance of the facial muscles in nonverbal communication becomes clear when they are paralyzed, as in some stroke victims. All muscles listed in this table are innervated by *cranial nerve VII, the facial nerve.* The muscles that move the eyeballs as one looks in various directions are described with the discussion of the eye, special senses.

Muscle	Description	Origin (O) and Insertion (I)	Action	Nerve Supply
MUSCLES OF THE SCALP				
Epicranius (Occipitofrontalis) ep″ĭ-kra′ne-us; ok-sip″ĭ-to-fron-ta′lis) (*epi* = over; *cran* = skull)	Bipartite muscle consisting of frontal and occipital bellies connected by the epicranial aponeurosis. The alternate actions of these two muscles pull scalp anteriorly and posteriorly.			
• **Frontal belly**	Covers forehead and dome of skull; no bony attachments	O—epicranial aponeurosis I—skin of eyebrows and root of nose	With aponeurosis fixed, elevates the eyebrows (as in surprise); wrinkles forehead skin horizontally	Facial nerve (cranial VII)
• **Occipital belly**	Overlies posterior occiput; by pulling on the galea, fixes origin of frontalis	O—occipital and temporal (mastoid) bones I—epicranial aponeurosis	Fixes aponeurosis and pulls scalp posteriorly	Facial nerve
MUSCLES OF THE FACE				
Corrugator supercilii (kor′ah-ga-ter soo″per-sĭ′le-i) (*corrugo* = wrinkle; *supercilium* = eyebrow)	Small muscle to eyebrow; acts with orbicularis oculi	O—arch of frontal bone above nasal bone I—skin of eyebrow	Draws eyebrows medially and inferiorly; wrinkles skin of forehead vertically	Facial nerve
Orbicularis oculi (or-bik′u-lar-is ok′u-li) (*orb* = circular; *ocul* = eye)	Thin, flat sphincter muscle of eyelid; surrounds rim of the orbit	O—frontal and maxillary bones and ligaments around orbit I—tissue of eyelid	Closes eye; produces blinking, squinting, and draws eyebrows inferiorly	Facial nerve
Zygomaticus (zi-go-mat′ĭ-kus), major and minor (*zygomatic* = cheekbone)	Muscle pair extending diagonally from cheekbone to corner of mouth	O—zygomatic bone I—skin and muscle at corner of mouth	Raises lateral corners of mouth (smiling muscle)	Facial nerve
Orbicularis oris (*or* = mouth)	Multilayered muscle of the lips with fibers that run in many different directions; most run circularly	O—arises indirectly from maxilla and mandible; fibers blended with fibers of other facial muscles associated with the lips I—encircles mouth; inserts into muscle and skin at angles of mouth	Closes lips; purses and protrudes lips; kissing and whistling muscle	Facial nerve
Mentalis (men-ta′lis) (*ment* = chin)	One of the muscle pair forming a V-shaped muscle mass on chin	O—mandible below incisors I—skin of chin	Wrinkles chin; protrudes lower lip	Facial nerve
Platysma (plah-tiz′mah) (*platy* = broad, flat)	Unpaired, thin, sheetlike superficial neck muscle; not strictly a head muscle, but plays a role in facial expression	O—fascia of chest (over pectoral muscles and deltoid) I—lower margin of mandible, and skin and muscle at corner of mouth	Tenses skin of neck (e.g., during shaving); helps depress mandible; pulls lower lip back and down	Facial nerve

▶

Table 11.3 Muscles of the Head, Part I: Facial Expression (Figure 11.9) *continued*

Muscle	Description	Origin (O) and Insertion (I)	Action	Nerve Supply
MUSCLES OF THE FACE *(continued)*				
Risorius (ri-zor′e-us) (*risor* = laughter)	Slender muscle inferior and lateral to zygomaticus	O—lateral fascia associated with masseter muscle I—skin at angle of mouth	Draws corner of lip laterally; tenses lips; synergist of zygomaticus	Facial nerve
Levator labii superioris (lĕ-va′tor la′be-i soo-per″e-or′is) (*leva* = raise; *labi* = lip; *superior* = above, over)	Thin muscle between orbicularis oris and inferior eye margin	O—zygomatic bone and infraorbital margin of maxilla I—skin and muscle of upper lip	Opens lips; elevates and furrows the upper lip	Facial nerve
Depressor labii inferioris (de-pres′or la′be-i in-fer″e-or′is) (*depressor* = depresses; *infer* = below)	Small muscle running from mandible to lower lip	O—body of mandible lateral to its midline I—skin and muscle of lower lip	Depresses lower lip (as in a pout)	Facial nerve

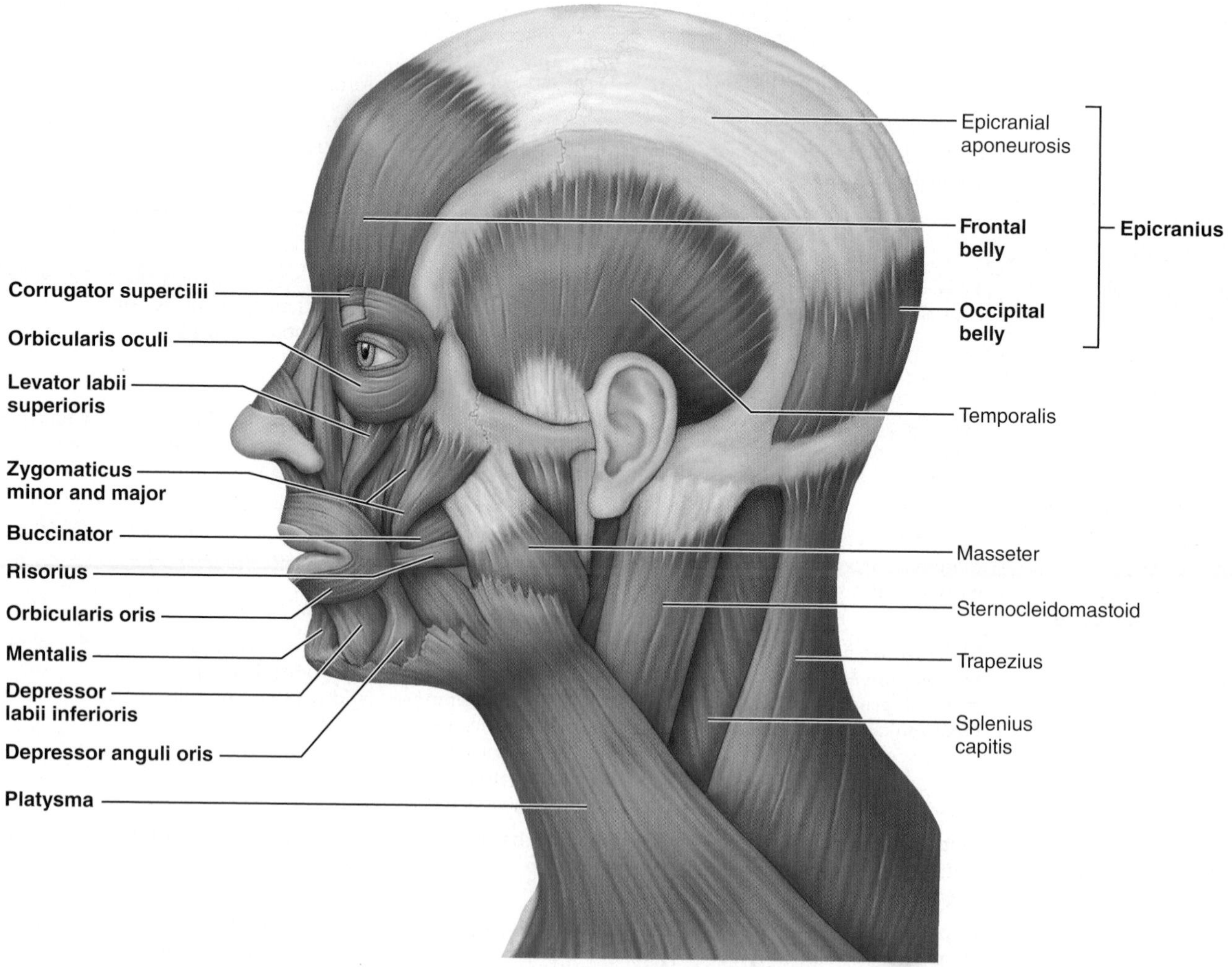

Figure 11.9 Muscles of facial expression, *continued.* (**b**) Illustration, left lateral view.

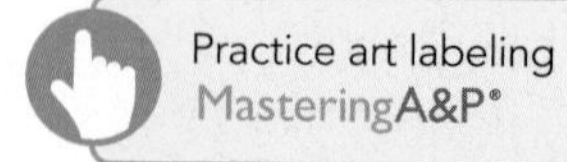

Table 11.3 *continued*

Muscle	Description	Origin (O) and Insertion (I)	Action	Nerve Supply
MUSCLES OF THE FACE *(continued)*				
Depressor anguli oris (ang′gu-li or′is) (*angul* = angle, corner; *or* = mouth)	Small muscle lateral to depressor labii inferioris	O—body of mandible below incisors I—skin and muscle at angle of mouth below insertion of zygomaticus	Draws corners of mouth inferiorly and laterally (as in a frown); zygomaticus antagonist	Facial nerve
Buccinator (bu′si-na″ter) (*bucc* = cheek or "trumpeter")	Thin, horizontal cheek muscle; principal muscle of cheek; deep to masseter (see also Figure 11.10)	O—molar region of maxilla and mandible I—orbicularis oris	Compresses cheek, (as in whistling and sucking); draws corner of mouth laterally: holds food between teeth during chewing; well developed in nursing infants	Facial nerve

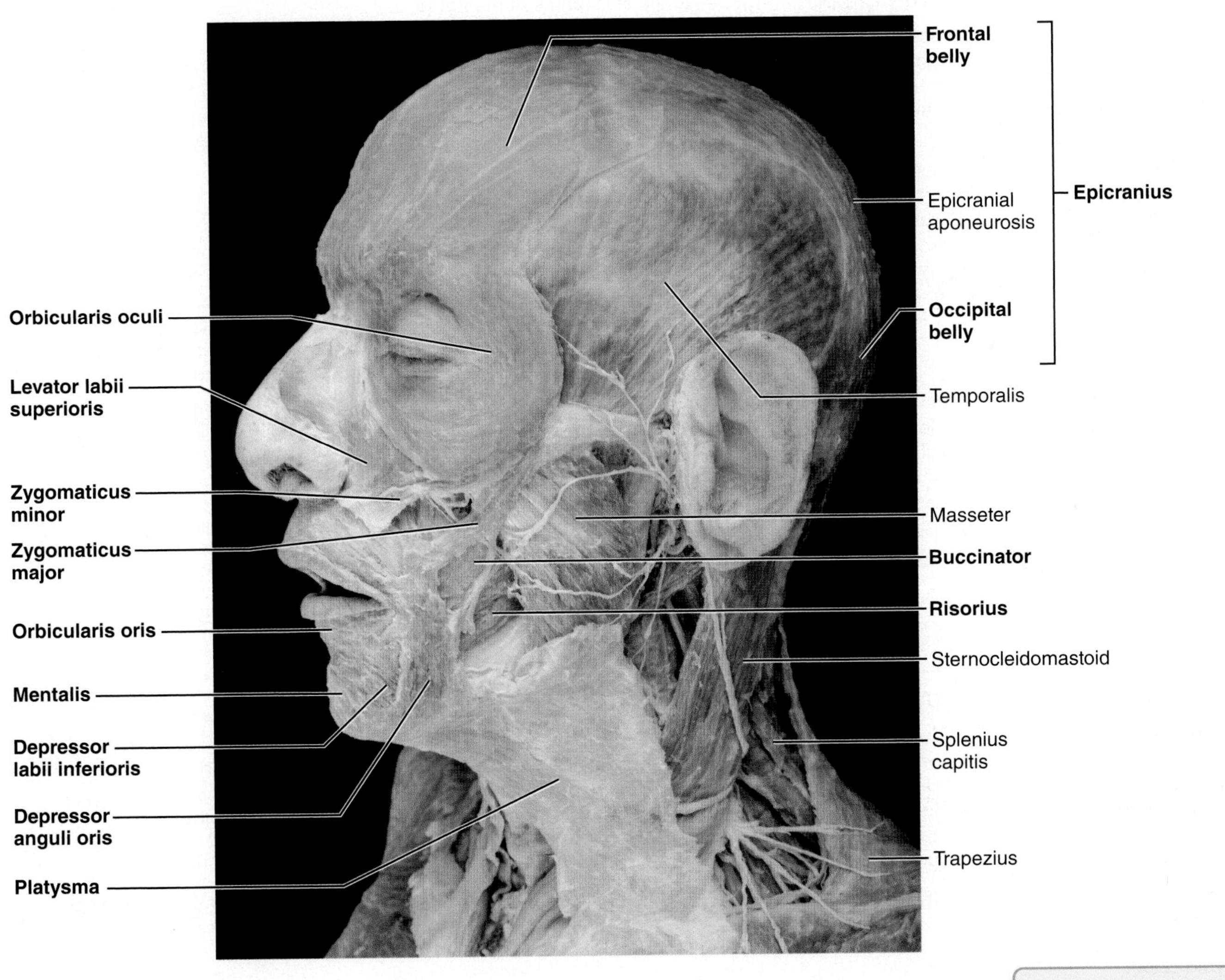

Figure 11.9 *continued.* **(c)** Cadaver dissection, left lateral view.

Table 11.4 Muscles of the Head, Part II: Mastication and Tongue Movement (Figure 11.10)

Four main pairs of muscles are involved in mastication (chewing), and all are innervated by the mandibular division of *cranial nerve V, the trigeminal nerve*. The prime movers of jaw closure and biting are the powerful **masseter** and **temporalis** muscles, which can be felt bulging through the skin when the teeth are clenched. Side-to-side grinding movements are brought about by the **pterygoid** muscles. The **buccinator** muscles in the cheeks (see Table 11.3) also play a role in chewing. For lowering the mandible, gravity is usually sufficient, but if there is resistance to jaw opening, the jaw-opening muscles are activated (**digastric** and **mylohyoid** muscles; see Table 11.5).

The tongue consists of muscle fibers that curl, squeeze, and fold the tongue during speaking and chewing. These **intrinsic tongue muscles,** which change the tongue's shape but do not really move it, are considered with the digestive system (Chapter 23). Only the **extrinsic tongue muscles** are covered in this table. These move the tongue laterally, anteriorly, and posteriorly. All the tongue muscles are innervated by *cranial nerve XII, the hypoglossal nerve*.

Muscle	Description	Origin (O) and Insertion (I)	Action	Nerve Supply
MUSCLES OF MASTICATION				
Masseter (mah-se′ter) (*maseter* = chewer)	Powerful muscle that covers lateral aspect of ramus of mandible	O—zygomatic arch and zygomatic bone I—angle and ramus of mandible	Closes jaw; elevates mandible	Mandibular branch of trigeminal nerve (cranial V)
Temporalis (tem″por-ă′lis) (*tempora* = time; pertaining to the temporal bone)	Fan-shaped muscle that covers parts of the temporal, frontal, and parietal bones	O—temporal fossa I—coronoid process of mandible via a tendon that passes deep to zygomatic arch	Closes jaw; elevates and retracts mandible; maintains position of the mandible at rest; deep anterior part may help protract mandible	Mandibular branch of trigeminal nerve
Medial pterygoid (me′de-ul ter′ĭ-goid) (*medial* = toward median plane; *pterygoid* = winglike)	Deep two-headed muscle that runs along internal surface of mandible and is largely concealed by that bone	O—medial surface of lateral pterygoid plate of sphenoid bone, maxilla, and palatine bone I—medial surface of mandible near its angle	Acts with lateral pterygoid muscle to protract mandible and to produce side-to-side (grinding) movements; synergist of temporalis and masseter muscles in elevation of the mandible	Mandibular branch of trigeminal nerve
Lateral pterygoid (*lateral* = away from median plane)	Deep two-headed muscle; lies superior to medial pterygoid muscle	O—greater wing and lateral pterygoid plate of sphenoid bone I—condylar process of mandible and capsule of temporomandibular joint	Protracts mandible and produces side-to-side grinding movements of the lower teeth	Mandibular branch of trigeminal nerve
Buccinator	(See Table 11.3)	(See Table 11.3)	Compresses the cheek; helps keep food between grinding surfaces of teeth during chewing	Facial nerve (cranial VII)
MUSCLES PROMOTING TONGUE MOVEMENTS (EXTRINSIC MUSCLES)				
Genioglossus (je″ne-o-glah′sus) (*geni* = chin; *glossus* = tongue)	Fan-shaped muscle; forms bulk of inferior part of tongue; its attachment to mandible prevents tongue from falling backward and obstructing respiration	O—internal surface of mandible near symphysis I—inferior aspect of the tongue and body of hyoid bone	Protracts tongue; can depress or, in concert with other extrinsic muscles, retract tongue	Hypoglossal nerve (cranial XII)
Hyoglossus (hi′o-glos″us) (*hyo* = pertaining to hyoid bone)	Flat, quadrilateral muscle	O—body and greater horn of hyoid bone I—inferolateral tongue	Depresses tongue and draws its sides inferiorly	Hypoglossal nerve
Styloglossus (sti-lo-glah′sus) (*stylo* = pertaining to styloid process)	Slender muscle running superiorly to and at right angles to hyoglossus	O—styloid process of temporal bone I—lateral inferior aspect of tongue	Retracts and elevates tongue	Hypoglossal nerve

Table 11.4 continued

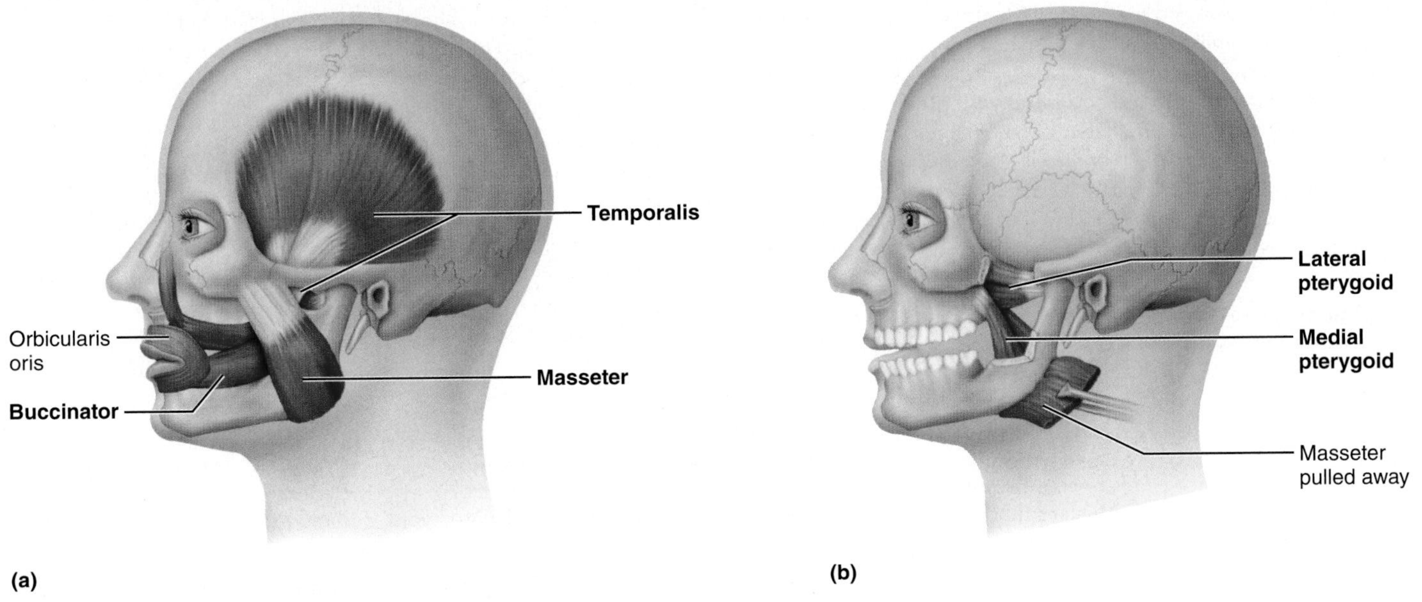

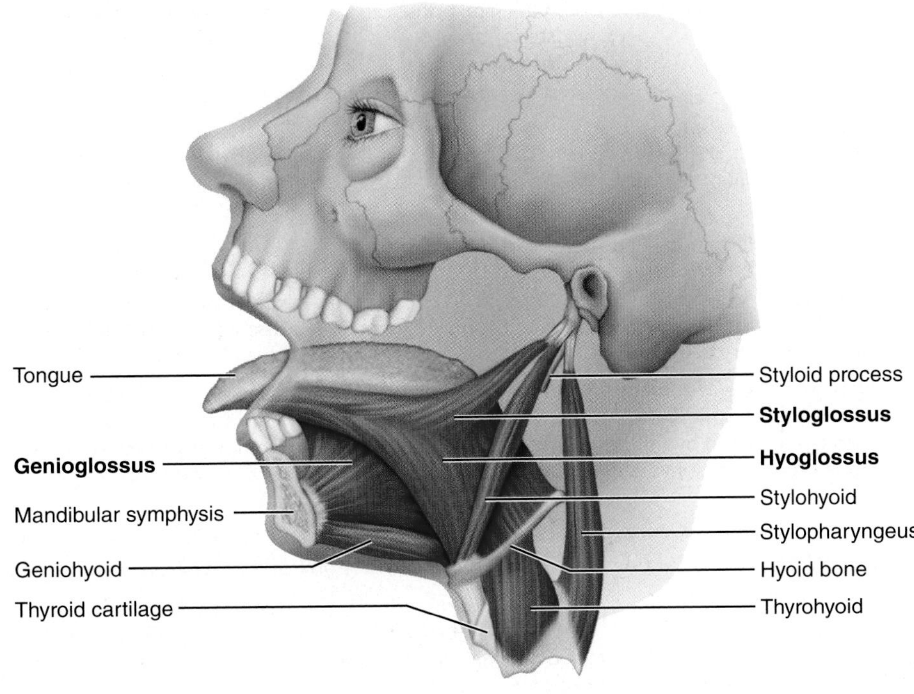

Figure 11.10 Muscles of mastication and tongue movement, left lateral view. (**a**) Superficial muscles of mastication. (**b**) Deep muscles of mastication. (**c**) Extrinsic muscles of the tongue and associated suprahyoid muscles.

Table 11.5 Muscles of the Anterior Neck and Throat: Swallowing (Figure 11.11)

The neck is divided into two triangles (anterior and posterior) by the sternocleidomastoid muscle (Figure 11.11). This table considers the muscles of the anterior triangle of the neck. These muscles are divided into **suprahyoid** and **infrahyoid** groups, which lie superior and inferior to the hyoid bone, respectively. Most of these muscles participate in swallowing.

Swallowing begins when the tongue and buccinator muscles squeeze food posteriorly along the roof of the mouth toward the pharynx. Then, many muscles in the pharynx and neck contract in sequence to complete the swallowing process:

1. The suprahyoid muscles aid in closing the air passageway of the larynx—so that food is not inhaled into the lungs—by lifting the larynx superiorly and anteriorly, under the cover of a protective flap (the epiglottis). By pulling anteriorly on the hyoid bone, suprahyoid muscles also widen the pharynx to receive the food.
2. The muscles of the wall of the pharynx, the **pharyngeal constrictors,** squeeze the food inferiorly into the esophagus.
3. As swallowing ends, the infrahyoid muscles pull the hyoid bone and larynx inferiorly to their original position.

Swallowing also includes mechanisms that prevent food from being squeezed superiorly into the nasal cavity. These mechanisms are considered in the respiratory system chapter.

Muscle	Description	Origin (O) and Insertion (I)	Action	Nerve Supply
SUPRAHYOID MUSCLES (soo″prah-hi′oid)	Muscles that help form floor of oral cavity, anchor tongue, elevate hyoid, and move larynx superiorly during swallowing; lie superior to hyoid bone			
Digastric (di-gas′trik) (*di* = two; *gaster* = belly)	Consists of two bellies united by an intermediate tendon, forming a V shape under the chin	O—lower margin of mandible (anterior belly) and mastoid process of the temporal bone (posterior belly) I—by a connective tissue loop to hyoid bone	Open mouth and depress mandible; acting together, the digastric muscles elevate hyoid bone and steady it during swallowing and speech	Mandibular branch of trigeminal nerve (cranial V) for anterior belly; facial nerve (cranial VII) for posterior belly
Stylohyoid (sti″lo-hi′oid) (also see Figure 11.10)	Slender muscle below angle of jaw; parallels posterior belly of digastric muscle	O—styloid process of temporal bone I—hyoid bone	Elevates and retracts hyoid, thereby elongating floor of mouth during swallowing	Facial nerve
Mylohyoid (mi″lo-hi′oid) (*myle* = molar)	Flat, triangular muscle just deep to digastric muscle; this muscle pair forms a sling that forms the floor of the anterior mouth	O—medial surface of mandible I—hyoid bone and median raphe	Elevates hyoid bone and floor of mouth, enabling tongue to exert backward and upward pressure that forces food bolus into pharynx	Mandibular branch of trigeminal nerve
Geniohyoid (je′ne-o-hy″oid) (also see Figure 11.10) (*geni* = chin)	Narrow muscle in contact with its partner medially; runs from chin to hyoid bone deep to mylohyoid	O—inner surface of mandibular symphysis I—hyoid bone	Elevates and protracts the hyoid, shortening floor of mouth and widening pharynx for receiving food during swallowing	First cervical spinal nerve via hypoglossal nerve (cranial XII)
INFRAHYOID MUSCLES	Straplike muscles that depress the hyoid bone and larynx during swallowing and speaking			
Sternohyoid (ster″no-hi′oid) (*sterno* = sternum)	Most medial muscle of the neck; thin; superficial except inferiorly, where covered by sternocleidomastoid	O—manubrium and medial end of clavicle I—lower margin of hyoid bone	Depresses larynx and hyoid bone if mandible is fixed	C_1–C_3 through ansa cervicalis (slender nerve root in cervical plexus)
Sternothyroid (ster″no-thi′roid) (*thyro* = thyroid cartilage)	Lateral and deep to sternohyoid	O—posterior surface of manubrium of sternum I—thyroid cartilage	Depresses larynx and hyoid bone	As for sternohyoid
Omohyoid (o″mo-hi′oid) (*omo* = shoulder)	Straplike muscle with two bellies united by an intermediate tendon; lateral to sternohyoid	O—superior surface of scapula I—hyoid bone, lower border	Depresses and retracts hyoid bone	As for sternohyoid
Thyrohyoid (thi″ro-hi′oid) (also see Figure 11.10)	Appears as a superior continuation of sternothyroid muscle	O—thyroid cartilage I—hyoid bone	Depresses hyoid bone or elevates larynx if hyoid is fixed	First cervical nerve via hypoglossal

Table 11.5 *continued*

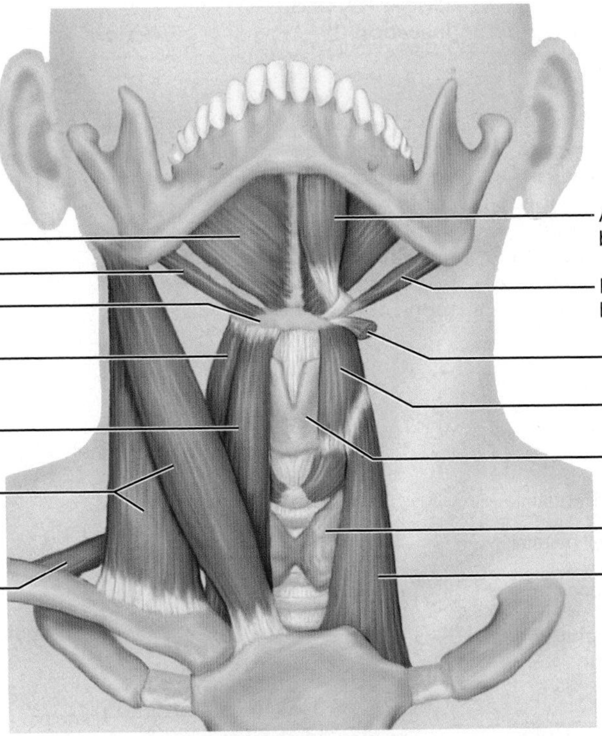

Mylohyoid
Stylohyoid
Hyoid bone
Omohyoid
(superior belly)
Sternohyoid
Sternocleido-
mastoid
Omohyoid
(inferior belly)
Anterior
belly
Posterior
belly
Digastric
Stylohyoid (cut)
Thyrohyoid
Thyroid cartilage
of the larynx
Thyroid gland
Sternothyroid

(a)

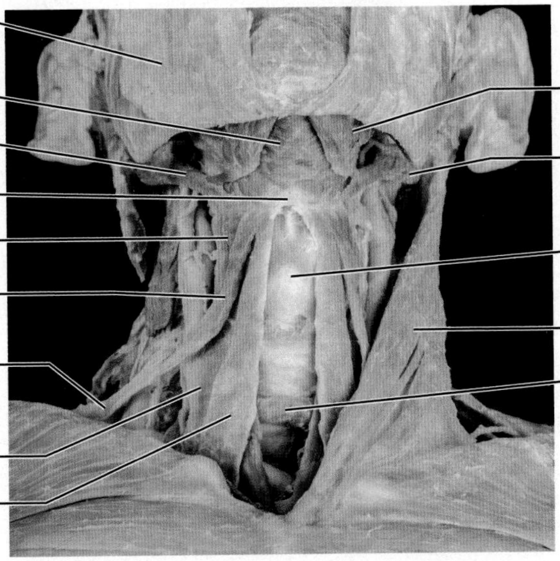

Platysma (cut
and reflected)
Mylohyoid
Stylohyoid
Hyoid bone
Thyrohyoid
Omohyoid
Superior
belly
Inferior
belly
Sternothyroid
Sternohyoid
Anterior
belly
Posterior
belly
Digastric
Thyroid cartilage
of the larynx
Sternocleidomastoid
Thyroid gland

(b)

Figure 11.11 **Muscles of the anterior neck and throat used in swallowing.** (**a**) Anterior view of the suprahyoid and infrahyoid muscles. The sternocleidomastoid muscle (not involved in swallowing) is shown at the left to provide an anatomical landmark. Deeper muscles are shown on the right of the illustration. (**b**) Cadaver dissection of suprahyoid and infrahyoid muscles.

Explore PAL Cadaver
MasteringA&P®

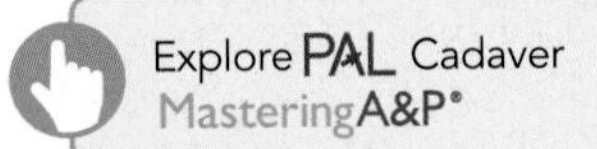

Table 11.5 Muscles of the Anterior Neck and Throat: Swallowing (Figure 11.11) *continued*

Muscle	Description	Origin (O) and Insertion (I)	Action	Nerve Supply
PHARYNGEAL CONSTRICTOR MUSCLES				
Superior, middle, and inferior pharyngeal constrictors (far-in′je-al)	Composite of three paired muscles whose fibers run circularly in pharynx wall; arranged so that the superior muscle is innermost and inferior one is outermost; substantial overlap	O—attached anteriorly to mandible and medial pterygoid plate (superior), hyoid bone (middle), and laryngeal cartilages (inferior) I—posterior median raphe of pharynx	Constrict pharynx during swallowing, which propels a food bolus to esophagus	Pharyngeal plexus [branches of vagus nerve (cranial X)]

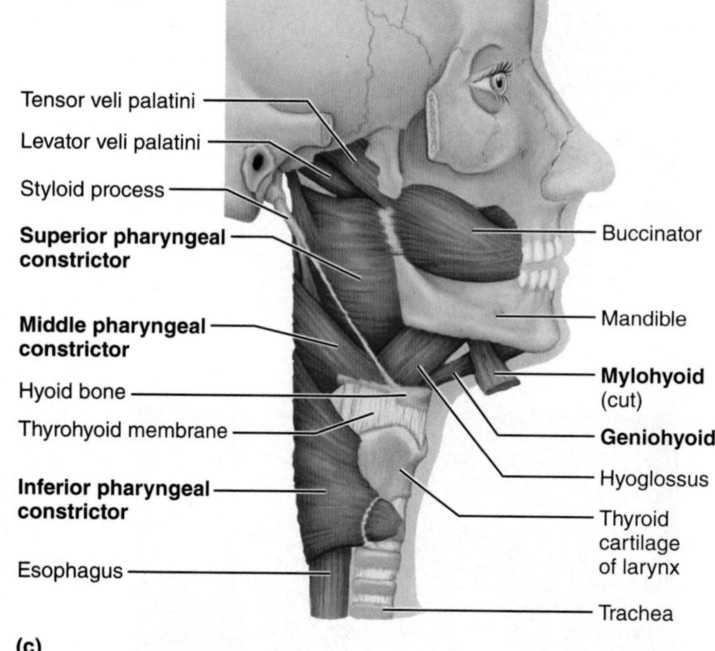

Figure 11.11 **Muscles of the anterior neck and throat used in swallowing,** *continued.* (**c**) The pharyngeal constrictor muscles are shown in their proper anatomical relationship to the buccinator (a chewing muscle) and to the hyoglossus muscle (which promotes tongue movements).

Table 11.6 Muscles of the Neck and Vertebral Column: Head Movements and Trunk Extension (Figure 11.12)

Head Movements. The head is moved by muscles that originate on the axial skeleton inferiorly. *Flexion* of the head is mainly brought about by the **sternocleidomastoid** muscles (Figure 11.12a), with some help from the suprahyoid and infrahyoid muscles (see Table 11.5). *Lateral flexion* and *rotation* of the head and neck result when muscles contract on one side of the neck only. These actions are performed by the sternocleidomastoids and the deeper neck muscles considered in this table. *Extension* of the head is aided by the trapezius muscle of the back (see Figure 11.6b), but the main extensors of the head are the **splenius** muscles deep to the trapezius (Figure 11.12b).

Trunk Extension. Trunk extension is brought about by the *deep, or intrinsic, muscles of the back* (Figure 11.12c and d). These muscles also maintain the normal curvatures of the spine, acting as postural muscles. As you consider these back muscles, keep in mind that they are deep; the superficial back muscles that cover them (see Figure 11.8b) primarily run to the bones of the upper limb (see Tables 11.10 and 11.11).

The deep muscles of the back form a broad, thick column running from the sacrum to the skull (see Figure 11.12c). Many muscles of varying length contribute to this mass. It helps to regard each of these muscles as a string, which when pulled causes one or many vertebrae to extend or rotate on the vertebrae inferior to it. The largest of the deep back muscles is the **erector spinae** group. Many of the deep back muscles are long, so that large regions of the spine can be extended at once. When these muscles contract on just one side of the body, they help bend the spine laterally. Such lateral flexion is automatically accompanied by some rotation of the vertebral column. During vertebral movements, the articular facets of the vertebrae glide on each other.

There are also many short muscles that extend from one vertebra to the next. These small muscles act primarily as synergists in the extension and rotation of the spine and as stabilizers of the spine. They are not described in this table, but you can deduce their actions by examining their origins and insertions (see Figure 11.12d).

The trunk muscles considered in this table are extensors. *Flexion* of the trunk is brought about by muscles that lie anterior to the vertebral column (see Table 11.8).

CLINICAL APPLICATION

Torticollis The condition in which the neck stays rotated to one side, keeping the head tilted in that direction, is called *torticollis* (tor″ti-kol′is; "turned neck"). All of the various forms of this condition result from problems in a sternocleidomastoid muscle on one side of the neck only. The adult-onset form, or spasmotic torticollis, is caused by sustained spasms in the muscle. In congenital forms, either the developing fetus twists into an awkward position because of a lack of amniotic fluid, or excessive stretching of the muscle occurs during birth.The resulting damage prevents the muscle from lengthening as the child grows. Early identification and treatment with physical therapy are effective in resolving this condition.

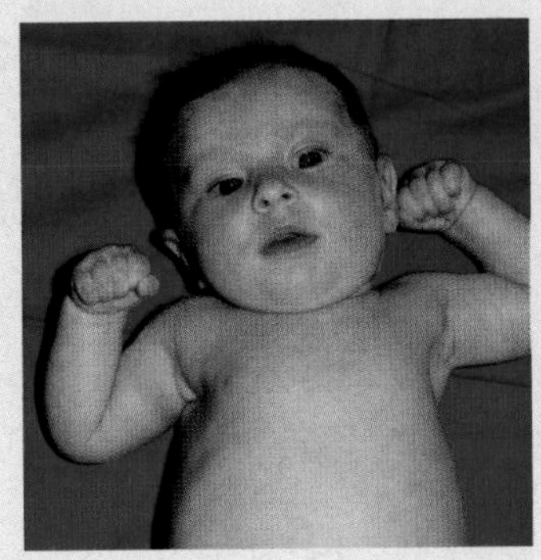

Muscle	Description	Origin (O) and Insertion (I)	Action	Nerve Supply
ANTEROLATERAL NECK MUSCLES (FIGURE 11.12a)				
Sternocleidomastoid (ster″no-kli″do-mas′toid) (*sterno* = breastbone; *cleido* = clavicle; *mastoid* = mastoid process) (also see Figure 11.11a and b)	Two-headed muscle located deep to platysma on anterolateral surface of neck; delineate limits of anterior and posterior triangles of the neck	O—manubrium of sternum and medial portion of clavicle I—mastoid process of temporal bone and superior nuchal line of occipital bone	Flexes and laterally rotates the head; simultaneous contraction of both muscles causes neck flexion; acting alone, each muscle rotates head toward shoulder on opposite side and tilts or laterally flexes head to its own side	Accessory nerve (cranial nerve XI); cervical spinal nerves C_2 and C_3 (ventral rami)
Scalenes (ska′lēnz)—anterior, middle, and posterior (*scalene* = uneven)	Located more laterally than anteriorly on neck; deep to platysma and sternocleidomastoid	O—transverse processes of cervical vertebrae I—anterolaterally on first two ribs	Elevate first two ribs (aid in inspiration); flex and rotate neck	Cervical spinal nerves

Table 11.6 Muscles of the Neck and Vertebral Column: Head Movements and Trunk Extension (Figure 11.12) *continued*

Muscle	Description	Origin (O) and Insertion (I)	Action	Nerve Supply
INTRINSIC MUSCLES OF THE BACK (FIGURE 11.12b–d)				
Splenius (sple′ne-us)—capitis and cervicis portions (kah-pit′us; ser-vis′us) (*splenion* = bandage; *caput* = head; *cervi* = neck) (Figures 11.12b and 11.9)	Broad bipartite superficial muscle (capitis and cervicis parts) extending from upper thoracic vertebrae to skull	O—ligamentum nuchae.* spinous processes of vertebrae C_7–T_6 I—mastoid process of temporal bone and occipital bone (capitis); transverse processes of C_2–C_4 vertebrae (cervicis)	Extend the head; when splenius muscles on one side are activated, head is rotated and bent laterally toward same side	Cervical spinal nerves (dorsal rami)

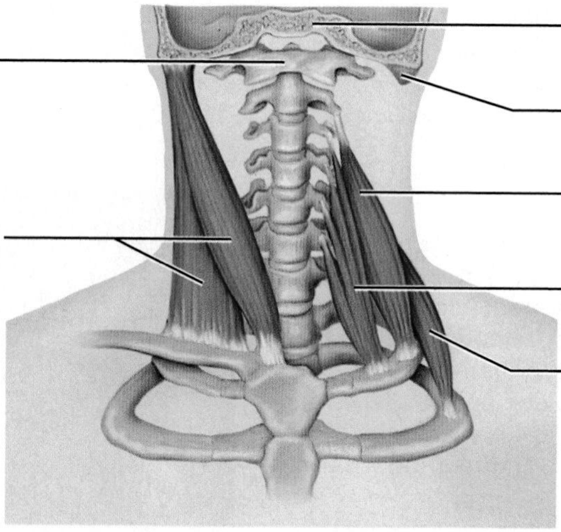

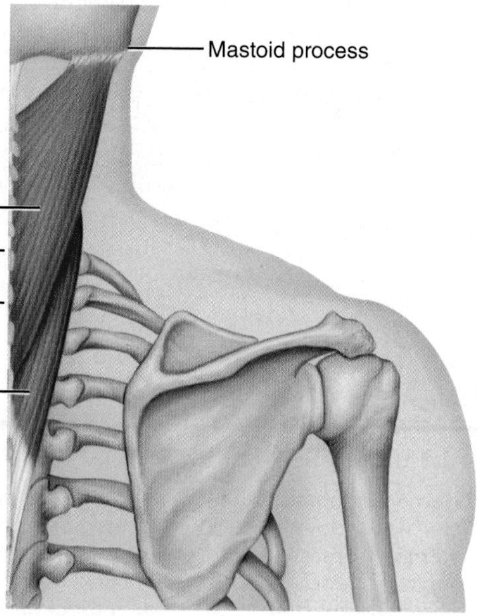

Figure 11.12 Muscles of the neck and vertebral column causing movements of the head and trunk. (**a**) Muscles of the anterolateral neck; superficial platysma muscle and the deeper neck muscles have been removed. (**b**) Deep muscles of the posterior neck; superficial muscles have been removed.

*The ligamentum nuchae (lig″ah-men′tum noo′ke) is a strong, elastic ligament in the midsagittal plane extending from the occipital bone of the skull along the tips of the spinous processes of the cervical vertebrae (Figure 11.12c). It binds the cervical vertebrae together and inhibits excessive head and neck flexion, thus preventing damage to the spinal cord in the vertebral canal.

Table 11.6 *continued*

Muscle	Description	Origin (O) and Insertion (I)	Action	Nerve Supply
INTRINSIC MUSCLES OF THE BACK (FIGURE 11.12b–d) *(continued)*				
Erector spinae (e-rek′tor spi′ne) (Figure 11.12c, left side)	Prime mover of back extension. Erector spinae on each side consist of three columns—the iliocostalis, longissimus, and spinalis muscles—forming the intermediate layer of intrinsic back muscles. Erector spinae provide resistance that helps control action of bending forward at the waist, and act as powerful extensors to promote return to erect position. During full flexion (i.e., when touching fingertips to floor), erector spinae are relaxed and strain is borne entirely by ligaments of back. On reversal of the movement, these muscles are initially inactive, and extension is initiated by hamstring muscles of thighs and gluteus maximus muscles of buttocks. As a result of this peculiarity, lifting a load or moving suddenly from a bent-over position is potentially injurious to muscles and ligaments of back and intervertebral discs. Erector spinae muscles readily go into painful spasms following injury to back structures.			
• **Iliocostalis** (il″e-o-kos-tă′lis)—lumborum, thoracis, and cervicis portions (lum′bor-um; tho-ra′sis) (*ilio* = ilium; *cost* = rib)	Most lateral muscle group of erector spinae muscles; extend from pelvis to neck	O—iliac crests (lumborum); inferior 6 ribs (thoracis); ribs 3 to 6 (cervicis) I—angles of ribs (lumborum and thoracis); transverse processes of cervical vertebrae C_6–C_4 (cervicis)	Extend and laterally flex the vertebral column, maintain erect posture; acting on one side, bend vertebral column to same side	Spinal nerves (dorsal rami)
• **Longissimus** (lon-jis′ĭ-mus)—thoracis, cervicis, and capitis parts (*longissimus* = longest)	Intermediate tripartite muscle group of erector spinae; extend by many muscle slips from lumbar region to skull; mainly pass between transverse processes of the vertebrae	O—transverse processes of lumbar through cervical vertebrae I—transverse processes of thoracic or cervical vertebrae and to ribs superior to origin as indicated by name; capitis inserts into mastoid process of temporal bone	Thoracis and cervicis act to extend and laterally flex vertebral column; acting on one side causes lateral flexion; capitis extends and rotates the head toward same side	Spinal nerves (dorsal rami)
• **Spinalis** (spi-nă′lis)—thoracis and cervicis parts (*spin* = vertebral column, spine)	Most medial muscle column of erector spinae; cervicis usually rudimentary and poorly defined	O—spines of upper lumbar and lower thoracic vertebrae I—spines of upper thoracic and cervical vertebrae	Extends vertebral column	Spinal nerves (dorsal rami)
Semispinalis (sem′e-spĭ-nă′lis)—thoracis, cervicis, and capitis regions (*semi* = half; *thorac* = thorax) (Figure 11.12c, right side)	Composite muscle forming part of deep layer of intrinsic back muscles; extends from thoracic region to head	O—transverse processes of C_7–T_{12} I—occipital bone (capitis) and spinous processes of cervical (cervicis) and thoracic vertebrae T_1 to T_4 (thoracis)	Extends vertebral column and head and rotates them to opposite side; acts synergistically with sternocleidomastoid muscle of opposite side	Spinal nerves (dorsal rami)
Quadratus lumborum (kwod-ra′tus lum-bor′um) (*quad* = four-sided; *lumb* = lumbar region) (also see Figure 11.22a)	Fleshy muscle forming part of posterior abdominal wall	O—iliac crest and lumbar fascia I—transverse processes of upper lumbar vertebrae and lower margin of 12th rib	Flexes vertebral column laterally when acting separately; when pair acts jointly, lumbar spine is extended, and 12th rib is fixed; maintains upright posture; assists in forced inspiration	T_{12} and upper lumbar spinal nerves (ventral rami)

Table 11.6 Muscles of the Neck and Vertebral Column: Head Movements and Trunk Extension (Figure 11.12) *continued*

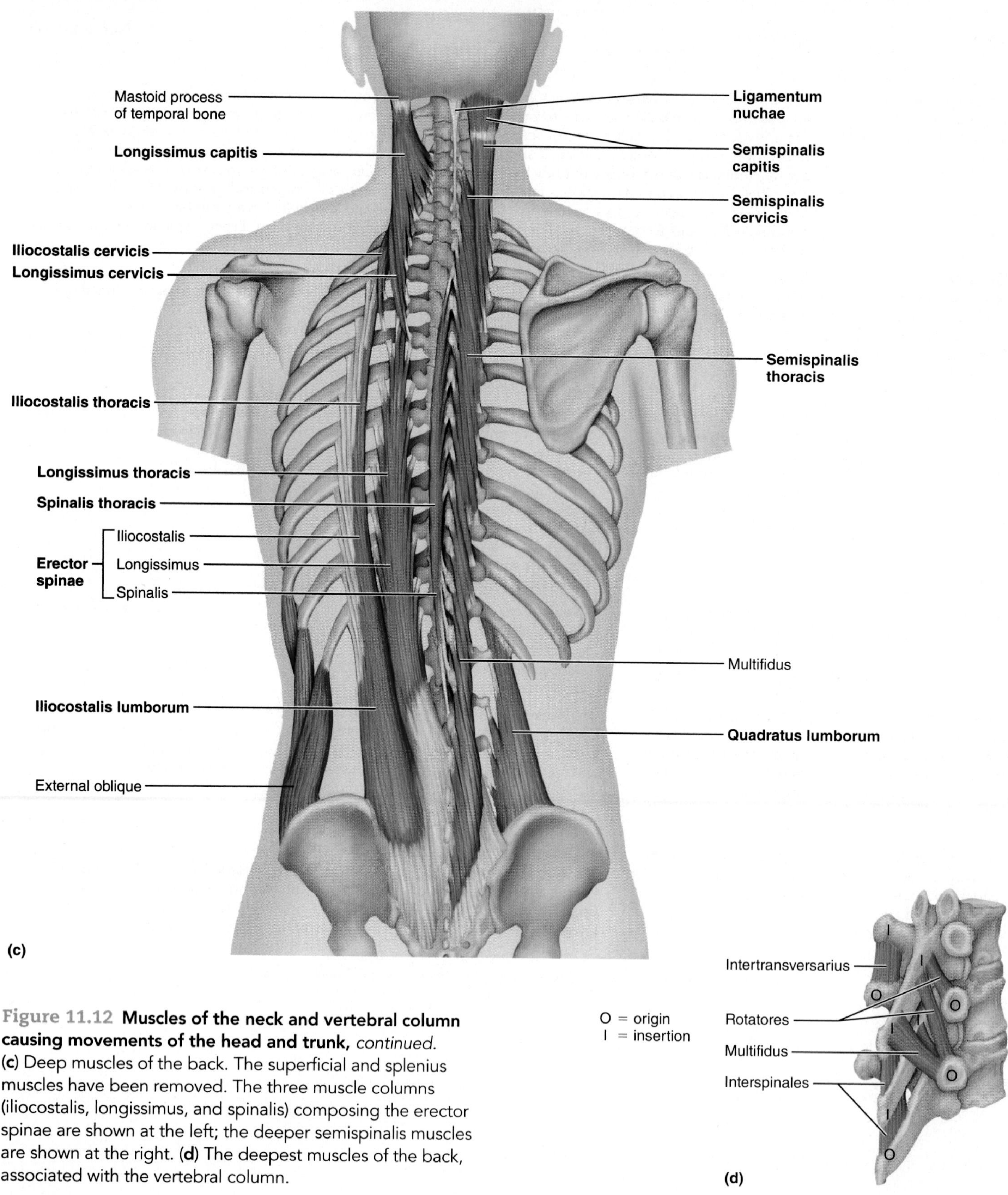

Figure 11.12 Muscles of the neck and vertebral column causing movements of the head and trunk, *continued.* **(c)** Deep muscles of the back. The superficial and splenius muscles have been removed. The three muscle columns (iliocostalis, longissimus, and spinalis) composing the erector spinae are shown at the left; the deeper semispinalis muscles are shown at the right. **(d)** The deepest muscles of the back, associated with the vertebral column.

Table 11.7 Deep Muscles of the Thorax: Breathing (Figure 11.13)

A primary function of the deep muscles of the thorax is to provide the movements necessary for ventilation, or breathing. Breathing has two phases—inspiration, or inhaling, and expiration, or exhaling—caused by cyclical changes in the volume of the thoracic cavity.

The thoracic muscles are very short: Most run only from one rib to the next. They form three layers in the wall of the thorax. The **external intercostal muscles** form the most superficial layer. They lift the rib cage, which increases its anterior-posterior and lateral dimensions. Thus, the external intercostals function during inspiration. The **internal intercostals** form the intermediate muscle layer. They may aid forced expiration by depressing the rib cage. The third and deepest muscle layer of the thoracic wall attaches to the internal surfaces of the ribs. It has three discontinuous parts (from posterior to anterior): the *subcostals, innermost intercostals,* and *transversus thoracis*. These are small, and their function is unclear, so they are not listed in this table.

The **diaphragm,** the most important muscle of respiration (Figures 11.13b and c), forms a complete partition between the thoracic and abdominopelvic cavities. In the relaxed state, the diaphragm is dome-shaped. When it contracts, it moves inferiorly and flattens, increasing the volume of the thoracic cavity. Thus, the diaphragm is a powerful muscle of inspiration. It contracts rhythmically during respiration, but one can also contract it voluntarily to push down on the abdominal viscera and increase the pressure in the abdominopelvic cavity. This pressure helps to evacuate the contents of the pelvic organs (feces, urine, or a baby). It also helps in lifting heavy weights: When one takes a deep breath to fix the diaphragm, the abdomen becomes a firm pillar that will not buckle under the weight being lifted. The muscles of the anterior abdominal wall also help to increase the intraabdominal pressure (see Table 11.8).

With the exception of the diaphragm, which is innervated by *phrenic nerves* from the neck, all muscles described in this table are served by nerves running between the ribs, called *intercostal nerves*.

When breathing is forced and heavy, as during exercise, additional muscles become active in ventilation. For example, in forced inspiration, the scalene and sternocleidomastoid muscles of the neck help lift the ribs. Forced expiration is aided by abdominal wall muscles that pull inferiorly on the ribs and push the diaphragm superiorly by compressing the abdominal organs.

Muscle	Description	Origin (O) and Insertion (I)	Action	Nerve Supply
External intercostals (in″ter-kos′talz) (*external* = toward the outside; *inter* = between; *cost* = rib)	11 pairs lie between ribs; fibers run obliquely (down and forward) from each rib to rib below; in lower intercostal spaces, fibers are continuous with external oblique muscle forming part of abdominal wall	O—inferior border of rib above I—superior border of rib below	With first ribs fixed by scalene muscles, the external intercostals pull ribs toward one another to elevate rib cage; aid in inspiration; synergists of diaphragm	Intercostal nerves
Internal intercostals (*internal* = toward the inside, deep)	11 pairs lie between ribs; fibers run deep to and at right angles to those of external intercostals (i.e., run downward and posteriorly); lower internal intercostal muscles are continuous with fibers of internal oblique muscle of abdominal wall	O—superior border of rib below I—inferior border (costal groove) of rib above	With 12th ribs fixed by quadratus lumborum, muscles of posterior abdominal wall, and oblique muscles of the abdominal wall, the internal intercostals draw ribs together and depress rib cage; aid in forced expiration; antagonistic to external intercostals	Intercostal nerves
Diaphragm (di′ah-fram) (*dia* = across; *phragm* = partition)	Broad muscle pierced by the aorta, inferior vena cava, and esophagus; forms floor of thoracic cavity; in relaxed state is dome-shaped; fibers converge from margins of thoracic cage toward a boomerang-shaped central tendon	O—inferior internal surface of rib cage and sternum, costal cartilages of last six ribs, and lumbar vertebrae I—central tendon	Prime mover of inspiration; flattens on contraction increasing vertical dimensions of thorax; when strongly contracted, dramatically increases intra-abdominal pressure	Phrenic nerves

Table 11.7 Deep Muscles of the Thorax: Breathing (Figure 11.13) *continued*

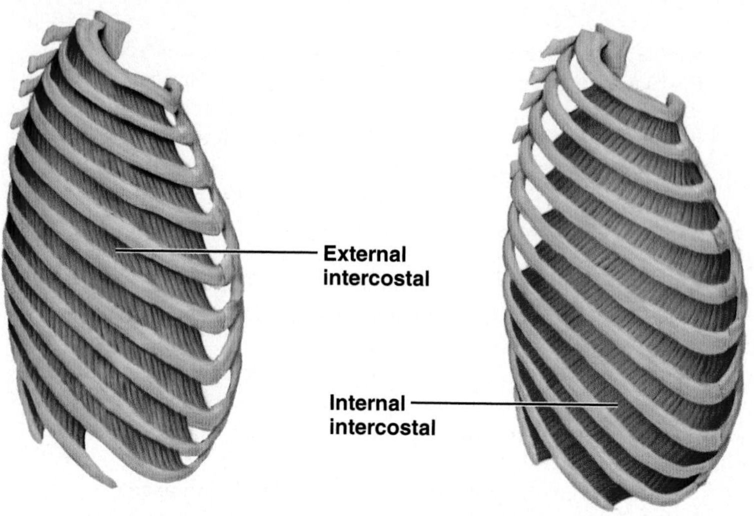

Figure 11.13 Muscles of respiration. (**a**) Deep muscles of the thorax, right lateral view: the external intercostals are superficial to the internal intercostals. These two muscle layers run obliquely and at right angles to each other. (**b**) Inferior view of the diaphragm, the prime mover of inspiration. Its muscle fibers converge toward a central tendon, an arrangement that causes the diaphragm to flatten and move inferiorly as it contracts. The diaphragm and its tendon are pierced by the great vessels (aorta and inferior vena cava) and the esophagus. (**c**) Photograph of the diaphragm, superior view at vertebral level T_9.

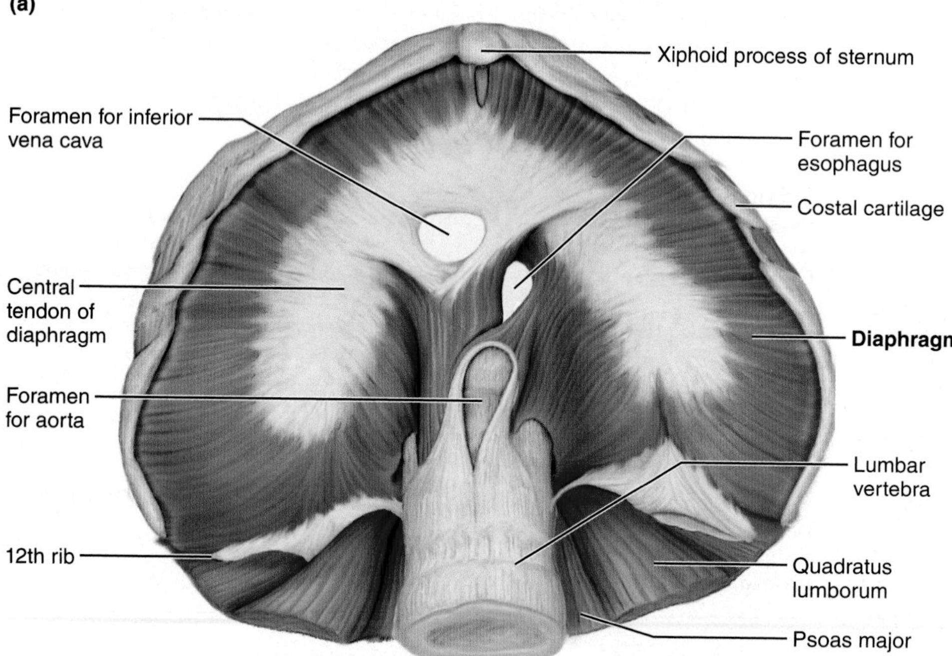

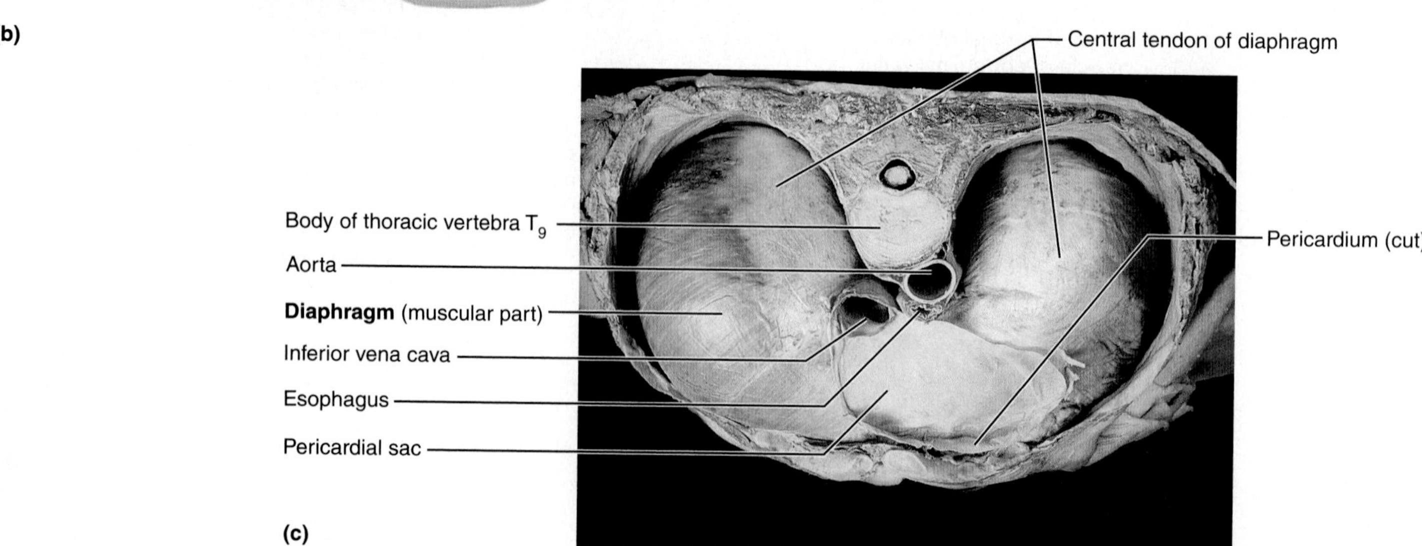

Table 11.8 Muscles of the Abdominal Wall: Trunk Movements and Compression of Abdominal Viscera (Figure 11.14)

The walls of the abdomen have no bony reinforcements (no ribs). Instead, the abdominal wall is a composite of sheetlike muscles, their contained fasciae, and aponeuroses. Forming the anterior and lateral parts of the abdominal wall are three broad, flat muscle sheets, layered one on the next: the **external oblique, internal oblique,** and **transversus abdominis.** These are direct continuations of the three intercostal muscles in the thorax. The fascicles of the external oblique run inferomedially, at right angles to fascicles of the internal oblique muscle (which primarily run superomedially), whereas the fascicles of the deep transversus abdominis are strictly horizontal. This arrangement of different fascicle directions lends great strength to the abdominal wall, just as plywood gets its strength from sheets of wood whose grains run in different directions. These three muscles end anteriorly in broad, white aponeuroses (tendonlike sheets). The aponeuroses enclose a fourth muscle pair, the straplike **rectus abdominis,** and extend medially to insert on the **linea alba** ("white line"), a tendinous raphe (seam) that runs vertically from the sternum to the pubic symphysis. The tight enclosure of the rectus abdominis within aponeuroses, called the **rectus sheath,** ensures that this vertical muscle cannot snap forward like a bowstring when it contracts to flex the trunk.

Functions. The four muscles of the abdominal wall function in anterior flexion (sit-ups), lateral flexion, and rotation of the trunk. When they contract simultaneously, they pull the ribs inferiorly and compress the abdominal contents. This action pushes the diaphragm superiorly and aids in forced expiration. When the abdominal muscles contract with the diaphragm, and the airway is closed (the Valsalva maneuver) the increased intra-abdominal pressure helps promote micturition (voiding urine), defecation, vomiting, childbirth, sneezing, coughing, laughing, burping, screaming, and nose blowing. Next time you perform one of these activities, feel the abdominal wall muscles contract under your skin. These muscles also help support the abdominal viscera and strengthen the trunk. They contract during heavy lifting—sometimes so forcefully that hernias result.

CLINICAL APPLICATION

Hernia An abdominal hernia is an abnormal protrusion of abdominal contents out of the abdominal cavity through a weak point in the muscles of the abdominal wall. The most commonly herniated elements are coils of small intestine and parts of the greater omentum (a large membrane, or mesentery, in the abdominal cavity). A hernia can result from increased intra-abdominal pressure during lifting and straining. An element that herniates from the abdominal cavity usually lies deep to the skin or superficial fascia, where it can form a visible bulge on the body surface. The most common types of hernias occur in the groin in the region of the aponeurosis of the external oblique muscle (inguinal hernia), in the superior thigh (femoral hernia), through the foramen for the esophagus in the diaphragm (hiatal hernia), and in or near the navel (umbilical hernia).

Muscle	Description	Origin (O) and Insertion (I)	Action	Nerve Supply
MUSCLES OF THE ANTERIOR AND LATERAL ABDOMINAL WALL				
	Four paired flat muscles; function in support and protection of abdominal viscera; flexion and lateral flexion of vertebral column			
External oblique (o-blēk′) (*external* = toward outside; *oblique* = running at an angle)	Largest and most superficial of the three lateral muscles; fibers run downward and medially (same direction outstretched fingers take when hands put into pants pockets); aponeurosis turns under inferiorly, forming inguinal ligament	O—by fleshy strips from outer surfaces of lower eight ribs I—most fibers insert into linea alba via a broad aponeurosis; some insert into pubic crest and tubercle and iliac crest	Flex vertebral column and compress abdominal wall when pair contracts simultaneously; acting individually, aid muscles of back in trunk rotation and lateral flexion; used in oblique curls	Intercostal nerves (T_7–T_{12})
Internal oblique (*internal* = toward the inside; deep)	Most fibers run superiorly and medially; the muscle fans such that its inferior fibers run inferiorly and medially	O—lumbar fascia, iliac crest, and inguinal ligament I—linea alba, pubic crest, last three or four ribs and costal margin	As for external oblique	Intercostal nerves (T_7–T_{12}) and L_1
Transversus abdominis (trans-ver′sus) (*transverse* = running straight across or transversely)	Deepest (innermost) muscle of abdominal wall; fibers run horizontally	O—inguinal ligament, lumbar fascia, cartilages of last six ribs; iliac crest I—linea alba, pubic crest	Compresses abdominal contents	Intercostal nerves (T_7–T_{12}) and L_1

Table 11.8 Muscles of the Abdominal Wall: Trunk Movements and Compression of Abdominal Viscera (Figure 11.14) *continued*

Muscle	Description	Origin (O) and Insertion (I)	Action	Nerve Supply
MUSCLES OF THE ANTERIOR AND LATERAL ABDOMINAL WALL *(continued)*				
Rectus abdominis (rek′tus ab-dom′ĭ-nis) (*rectus* = straight; *abdom* = abdomen)	Medial muscle pair; extend from pubis to rib cage; ensheathed by aponeuroses of lateral muscles; segmented by three tendinous intersections	O—pubic crest and symphysis I—xiphoid process and costal cartilages of ribs 5–7	Flex and rotate lumbar region of vertebral column; fix and depress ribs, stabilize pelvis during walking, increase intra-abdominal pressure; used in sit-ups/curls	Intercostal nerves (T_6 or T_7–T_{12})

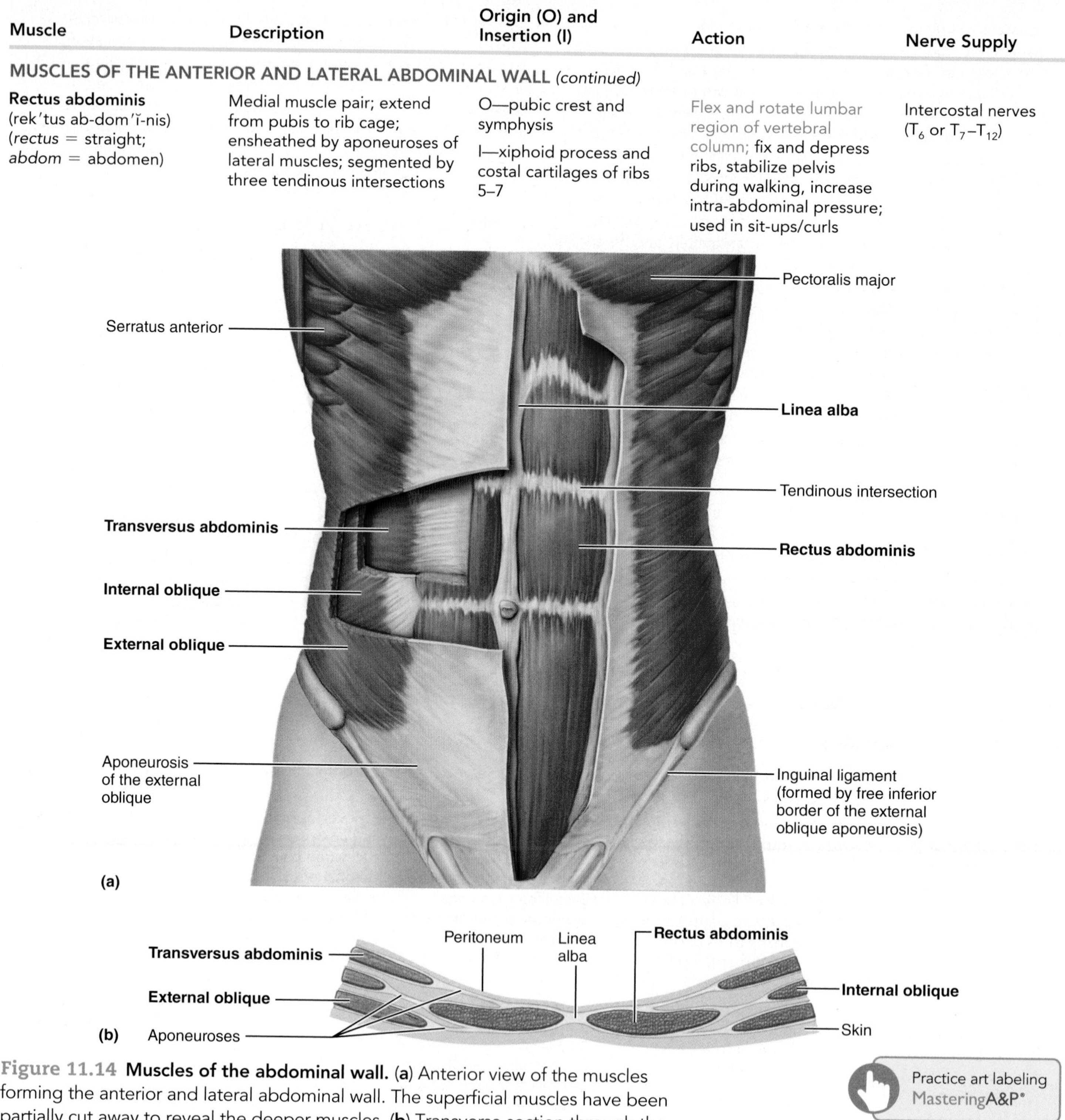

Figure 11.14 Muscles of the abdominal wall. (**a**) Anterior view of the muscles forming the anterior and lateral abdominal wall. The superficial muscles have been partially cut away to reveal the deeper muscles. (**b**) Transverse section through the anterior abdominal wall (midregion), showing how the aponeuroses of the lateral abdominal muscles contribute to the rectus abdominis sheath.

Practice art labeling MasteringA&P®

Table 11.8 continued

Figure 11.14 **Muscles of the abdominal wall,** *continued.* (**c**) Dissection showing the muscles of the abdominal wall in left lateral view. (**d**) Left lateral views of the trunk, illustrating the fiber directions and attachments of the external oblique, internal oblique, and transversus abdominis muscles.

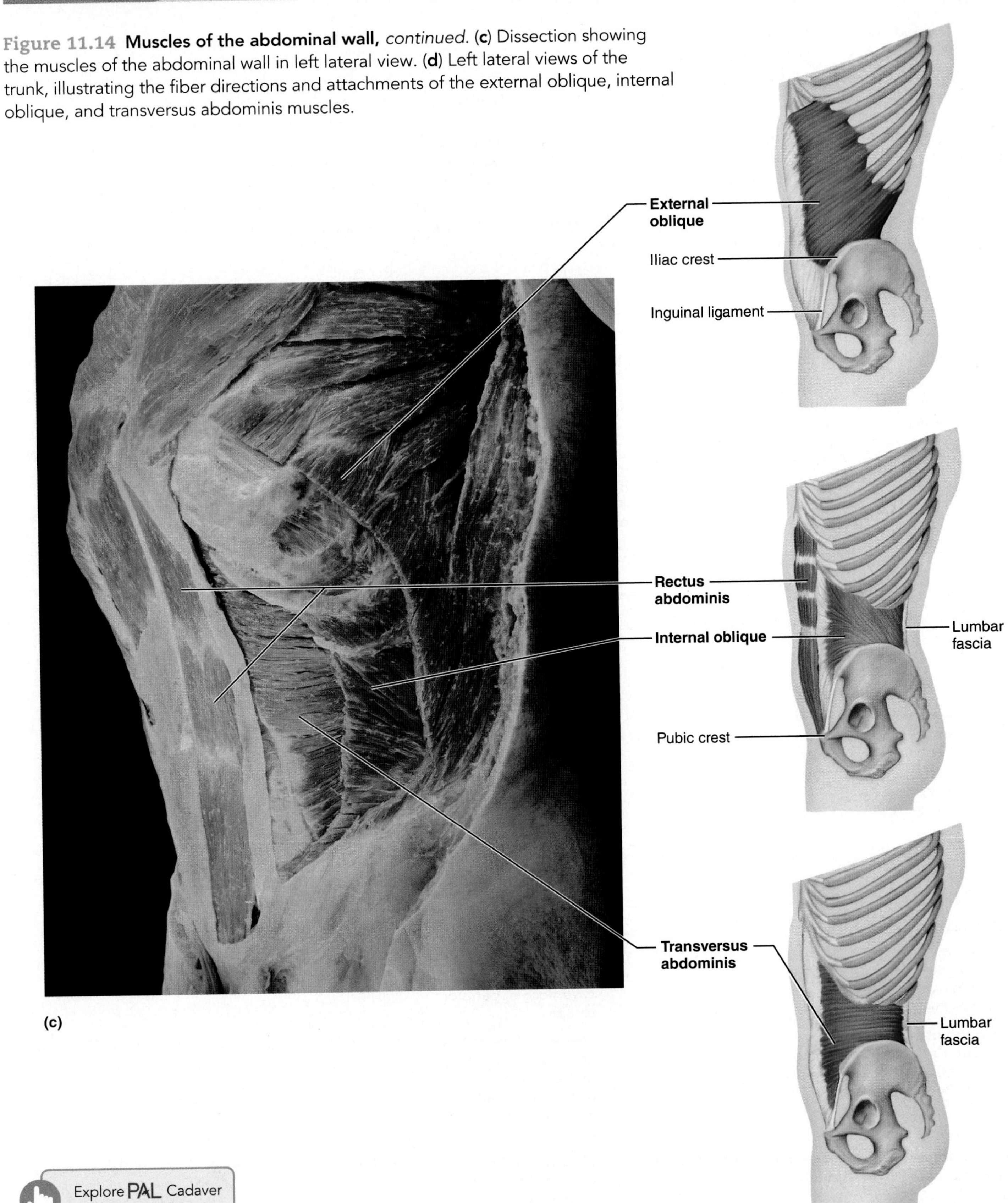

Explore PAL Cadaver
Mastering A&P®

Table 11.9 Muscles of the Pelvic Floor and Perineum: Support of Abdominopelvic Organs (Figure 11.15)

The pelvic floor, or **pelvic diaphragm** (Figure 11.15a), is a sheet consisting of two muscles, the **levator ani** and the small **coccygeus.** This diaphragm supports the pelvic organs, seals the inferior opening of the bony pelvis, and lifts superiorly to help release feces during defecation. The pelvic diaphragm is pierced by the rectum and urethra (the tube for urine) and (in females) by the vagina.

The body region inferior to the pelvic diaphragm is the *perineum* (Figure 11.15b and c). In the anterior half of the perineum is a triangular sheet of muscle called the **urogenital diaphragm.** It contains the **external urethral sphincter** muscle, which surrounds the urethra. One uses this muscle voluntarily to prevent urination. Just inferior to the urogenital diaphragm is the superficial perineal space, which contains muscles **(bulbospongiosus, ischiocavernosus)** that help maintain erection of the penis and clitoris. In the posterior half of the perineum, circling the anus, lies the **external anal sphincter** (Figure 11.15b). This muscle is used voluntarily to prevent defecation. Just anterior to this sphincter, at the exact midpoint of the perineum, is the **central tendon.** Many perineal muscles insert on this strong tendon and, in so doing, are able to support the heavy organs in the pelvis.

Muscle	Description	Origin (O) and Insertion (I)	Action	Nerve Supply
MUSCLES OF THE PELVIC DIAPHRAGM (FIGURE 11.15a)				
Levator ani (lĕ-va′tor a′ne) (*levator* = raises; *ani* = anus)	Broad, thin, tripartite muscle (pubococcygeus, puborectalis, and iliococcygeus parts); its fibers extend inferomedially, forming a muscular "sling" around the prostate (or vagina), urethra, and anorectal junction before meeting in the median plane	O—extensive linear origin inside pelvis from pubis to ischial spine I—inner surface of coccyx, levator ani of opposite side, and (in part) into the structures that penetrate it	Supports and maintains position of pelvic organs; resists downward thrusts that accompany rises in intrapelvic pressure during coughing, vomiting, and expulsive efforts of abdominal muscles; forms sphincters at anorectal junction and vagina; lifts anal canal during defecation	S_3, S_4, and inferior rectal nerve (branch of pudendal nerve)
Coccygeus (kok-sij′e-us) (*coccy* = coccyx)	Small triangular muscle lying posterior to levator ani; forms posterior part of pelvic diaphragm	O—spine of ischium I—sacrum and coccyx	Supports pelvic organs; supports coccyx and pulls it forward after it has been reflected posteriorly by defecation and childbirth	S_4 and S_5
MUSCLES OF THE UROGENITAL DIAPHRAGM (FIGURE 11.15b)				
Deep transverse perineal muscle (per″ĭ-ne′al) (*deep* = far from surface; *transverse* = across; *perine* = near anus)	Together the pair spans distance between ischial rami; in females, lies posterior to vagina	O—ischial rami I—midline central tendon of perineum; some fibers into vaginal wall in females	Supports pelvic organs; steadies central tendon	Pudendal nerve
External urethral sphincter (*sphin* = squeeze)	Muscle encircling urethra and vagina (female)	O—ischiopubic rami I—midline raphe	Constricts urethra; allows voluntary inhibition of urination; helps support pelvic organs	Pudendal nerve
MUSCLES OF THE SUPERFICIAL PERINEAL SPACE (FIGURE 11.15c)				
Ischiocavernosus (is′ke-o-kav′ern-o′sus) (*ischi* = hip; *caverna* = hollow chamber)	Runs from pelvis to base of penis or clitoris	O—ischial tuberosities I—crus of corpus cavernosa of penis or clitoris	Retards venous drainage and maintains erection of penis or clitoris	Pudendal nerve
Bulbospongiosus (bul″bo-spun″je-o′sus) (*bulbon* = bulb; *spongio* = sponge)	Encloses base of penis (bulb) in males and lies deep to labia in females	O—central tendon of perineum and midline raphe of penis I—anteriorly into corpus cavernosa of penis or clitoris	Empties male urethra; assists in erection of penis and clitoris	Pudendal nerve
Superficial transverse perineal muscle (*superficial* = closer to surface)	Paired muscle bands posterior to urethral (and in females, vaginal) opening; variable; sometimes absent	O—ischial tuberosity I—central tendon of perineum	Stabilizes and strengthens central tendon of perineum	Pudendal nerve

Table 11.9 continued

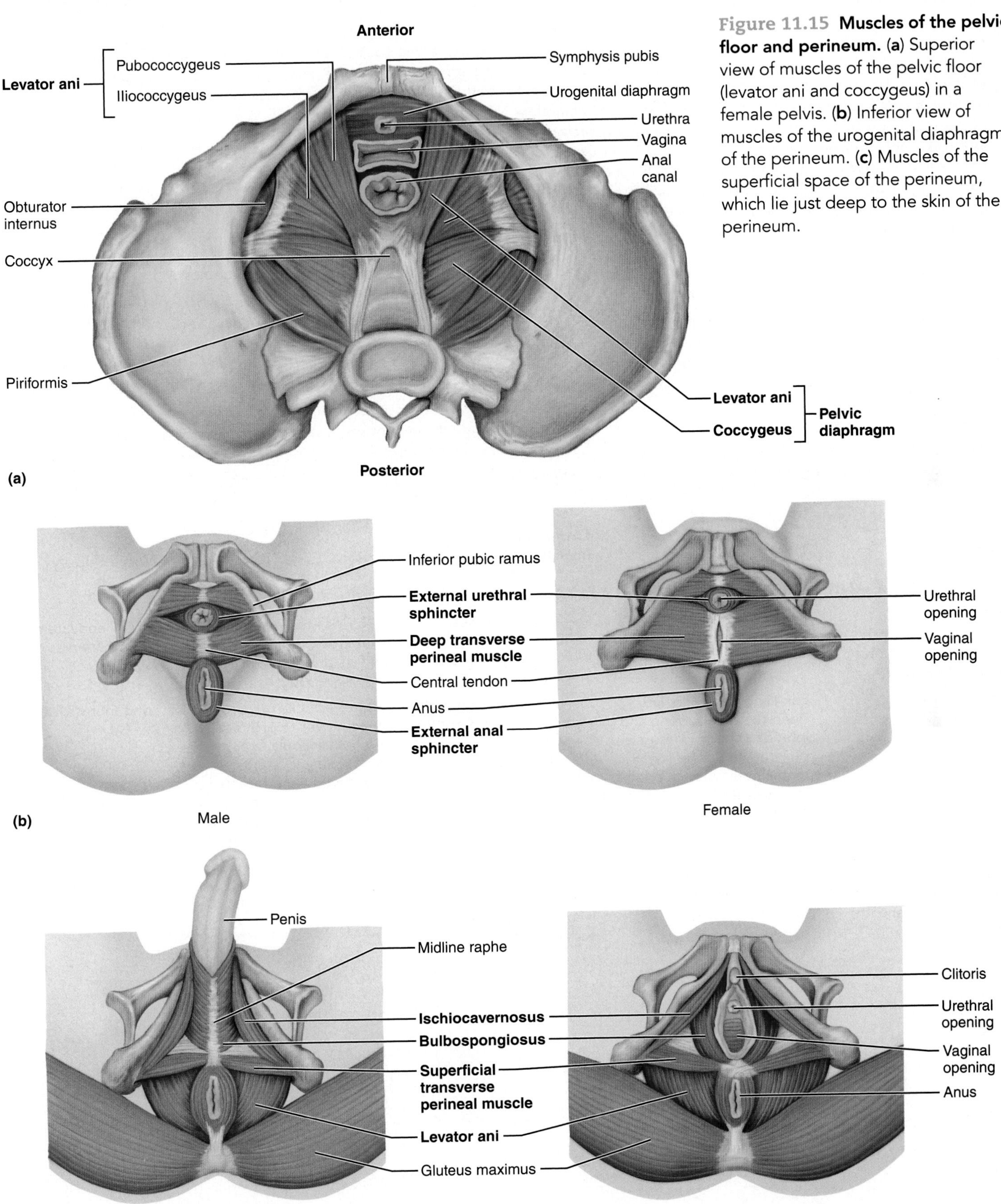

Figure 11.15 Muscles of the pelvic floor and perineum. (**a**) Superior view of muscles of the pelvic floor (levator ani and coccygeus) in a female pelvis. (**b**) Inferior view of muscles of the urogenital diaphragm of the perineum. (**c**) Muscles of the superficial space of the perineum, which lie just deep to the skin of the perineum.

Table 11.10 Superficial Muscles of the Anterior and Posterior Thorax: Movements of the Scapula (Figure 11.16)

Most superficial muscles on the thorax are *extrinsic shoulder muscles*, which run from the ribs and vertebral column to the shoulder girdle. They both fix the scapula in place and move it to increase the arm's range of movements. The *anterior* muscles of this group are **pectoralis major, pectoralis minor, serratus anterior,** and **subclavius** (Figure 11.16a and b). The *posterior muscles* are **trapezius, latissimus dorsi,** superficially and the underlying **levator scapulae,** and **rhomboids** (Figure 11.16c and d). The pectoralis major and latissimus dorsi both insert into the humerus. They function primarily in movement of the arm (Table 11.11).

The scapula undergoes a variety of movements on the posterior rib cage.

Elevation. In *elevation* (superior movement) of the scapula, as in shrugging the shoulders, the prime movers are the levator scapulae and the superior fascicles of the trapezius; the rhomboids are synergists. Note that the fascicles of all these muscles run inferolaterally (see Figure 11.16c and d); thus, they are ideal elevators of the scapula.

Depression. *Depression* (inferior movement) of the scapula is mostly due to gravity (weight of the arm). When the scapula must be depressed against resistance, however, this action is done by the serratus anterior, pectoralis minor, and the inferior part of the trapezius. The fascicles of all these muscles run superiorly to insert on the scapula.

Protraction/Abduction. *Protraction* or *abduction* of the scapula moves it laterally and anteriorly, as in punching. This action is mainly carried out by the serratus anterior, whose horizontal fibers pull the scapula anterolaterally.

Retraction/Adduction. *Retraction* or *adduction* of the scapula moves it medially and posteriorly. This is brought about by the mid part of the trapezius and the rhomboids.

Upward Rotation. *Upward rotation* of the scapula (in which the scapula's inferior angle moves laterally) allows one to lift the arm above the head: The serratus anterior swings the inferior angle laterally while the superior part of the trapezius, gripping the scapular spine, pulls the superior edge of the scapula medially.

Downward Rotation. In *downward rotation* of the scapula (where the inferior angle swings medially, as in paddling a canoe), the rhomboids pull the inferior part of the scapula medially while the levator scapulae steadies the superior part.

Muscle	Description	Origin (O) and Insertion (I)	Action	Nerve Supply
MUSCLES OF THE ANTERIOR THORAX (FIGURE 11.16a AND b)				
Pectoralis minor (pek″to-ra′lis mi′nor) (*pectus* = chest, breast; *minor* = lesser)	Flat, thin muscle deep to pectoralis major	O—anterior surfaces of ribs 3–5 (or 2–4) I—coracoid process of scapula	With ribs fixed, protracts and rotates scapula downward; with scapula fixed, draws rib cage superiorly	Medial and lateral pectoral nerves (C_6–C_8)
Serratus anterior (ser-a′tus) (*serratus* = saw)	Fan-shaped muscle; lies deep to scapula, deep and inferior to pectoral muscles on lateral rib cage; forms medial wall of axilla; origins have serrated, or sawtooth, appearance	O—by a series of muscle slips from ribs 1–8 (or 9) I—entire anterior surface of vertebral border of scapula	Rotates scapula upward; prime mover to protract and hold scapula against chest wall; important role in abduction and raising of arm and in horizontal arm movements (pushing, punching); called "boxer's muscle"	Long thoracic nerve (C_5–C_7)
Subclavius (sub-kla′ve-us) (*sub* = under, beneath; *clav* = clavicle)	Small cylindrical muscle extending from rib 1 to clavicle	O—costal cartilage of rib 1 I—groove on inferior surface of clavicle	Helps stabilize and depress pectoral girdle	Nerve to subclavius (C_5 and C_6)

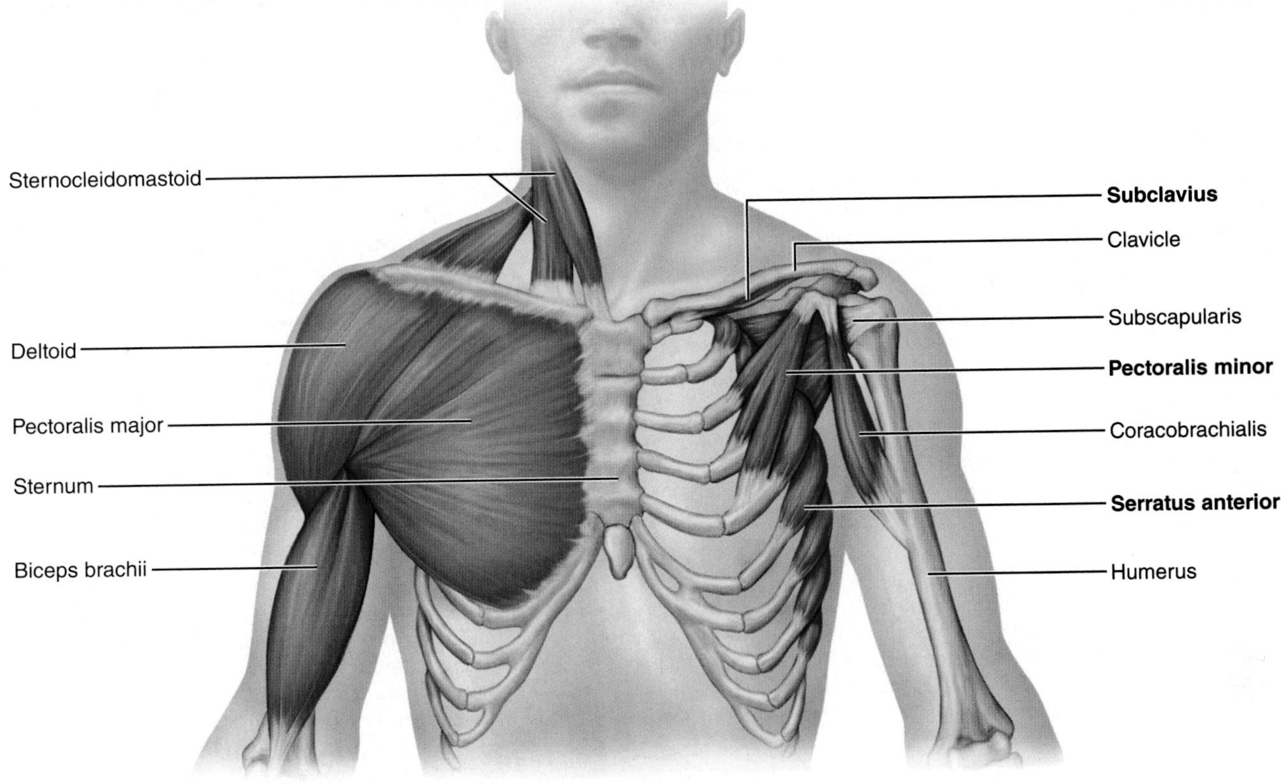

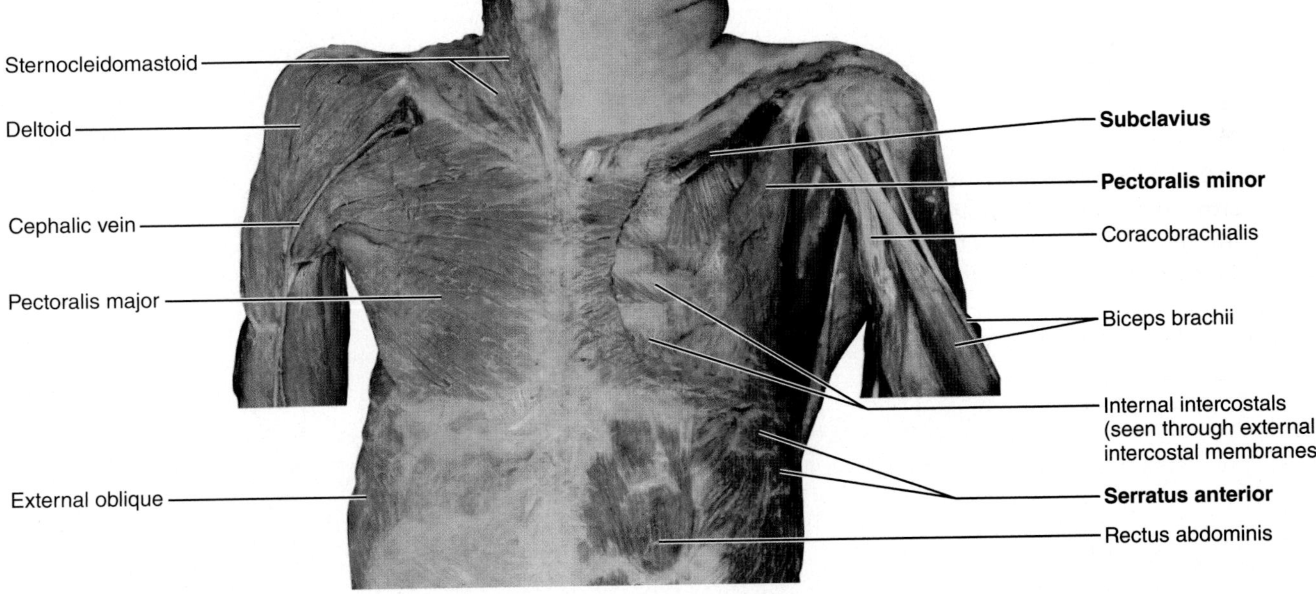

Figure 11.16 **Superficial muscles of the thorax and shoulder acting on the scapula and arm.** (**a**) Anterior view. The superficial muscles, which move the arm, are shown on the left in the illustration. The deep muscles, which act on the pectoral girdle, are shown on the right side of the illustration. (**b**) Cadaver dissection, anterior view.

Play A&P Flix Animation
MasteringA&P®

Table 11.10 Superficial Muscles of the Anterior and Posterior Thorax: Movements of the Scapula (Figure 11.16) *continued*

Muscle	Description	Origin (O) and Insertion (I)	Action	Nerve Supply
MUSCLES OF THE POSTERIOR THORAX (FIGURE 11.16c AND d)				
Trapezius (trah-pe′ze-us) (*trapezion* = irregular four-sided figure)	Most superficial muscle of posterior thorax; flat and triangular; upper fibers run inferiorly to scapula; middle fibers run horizontally to scapula; lower fibers run superiorly to scapula	O—occipital bone, ligamentum nuchae, and spines of C_7 and all thoracic vertebrae I—a continuous insertion along acromion and spine of scapula and lateral third of clavicle	Stabilizes, elevates, retracts, and rotates scapula; middle fibers retract (adduct) scapula; superior fibers elevate scapula (as in shrugging shoulders) or can help extend head when scapula is fixed; inferior fibers depress scapula (and shoulder)	Accessory nerve (cranial nerve XI); C_3 and C_4
Levator scapulae (skap′u-le) (*levator* = raises)	Located at back and side of neck, deep to trapezius; thick, straplike muscle	O—transverse processes of C_1–C_4 I—medial border of the scapula, superior to the spine	Elevates and adducts scapula in synergy with superior fibers of trapezius; rotate scapulae downward; when scapula is fixed, flexes neck to same side	Cervical spinal nerves and dorsal scapular nerve (C_3–C_5)
Rhomboids (rom′boidz)—major and minor (*rhomboid* = diamond-shaped)	Two rectangular muscles lying deep to trapezius and inferior to levator scapulae; rhomboid minor is the more superior muscle	O—spinous processes of C_7 and T_1 (minor) and spinous processes of T_2–T_5 (major) I—medial border of scapula	Stabilize scapula; act together (and with middle trapezius fibers) to retract scapula, thus "squaring shoulders"; rotate scapulae downward (as when arm is lowered against resistance; e.g., paddling a canoe)	Dorsal scapular nerve (C_4 and especially C_5)

Figure 11.16 **Superficial muscles of the thorax and shoulder acting on the scapula and arm,** *continued.* (**c**) Posterior view. The superficial muscles of the back are shown on the left side of the illustration. Superficial muscles are removed on the right side to reveal the deeper muscles acting on the scapula and the rotator cuff muscles that help stabilize the shoulder joint. (**d**) Cadaver dissection, posterior view.

Table 11.10 *continued*

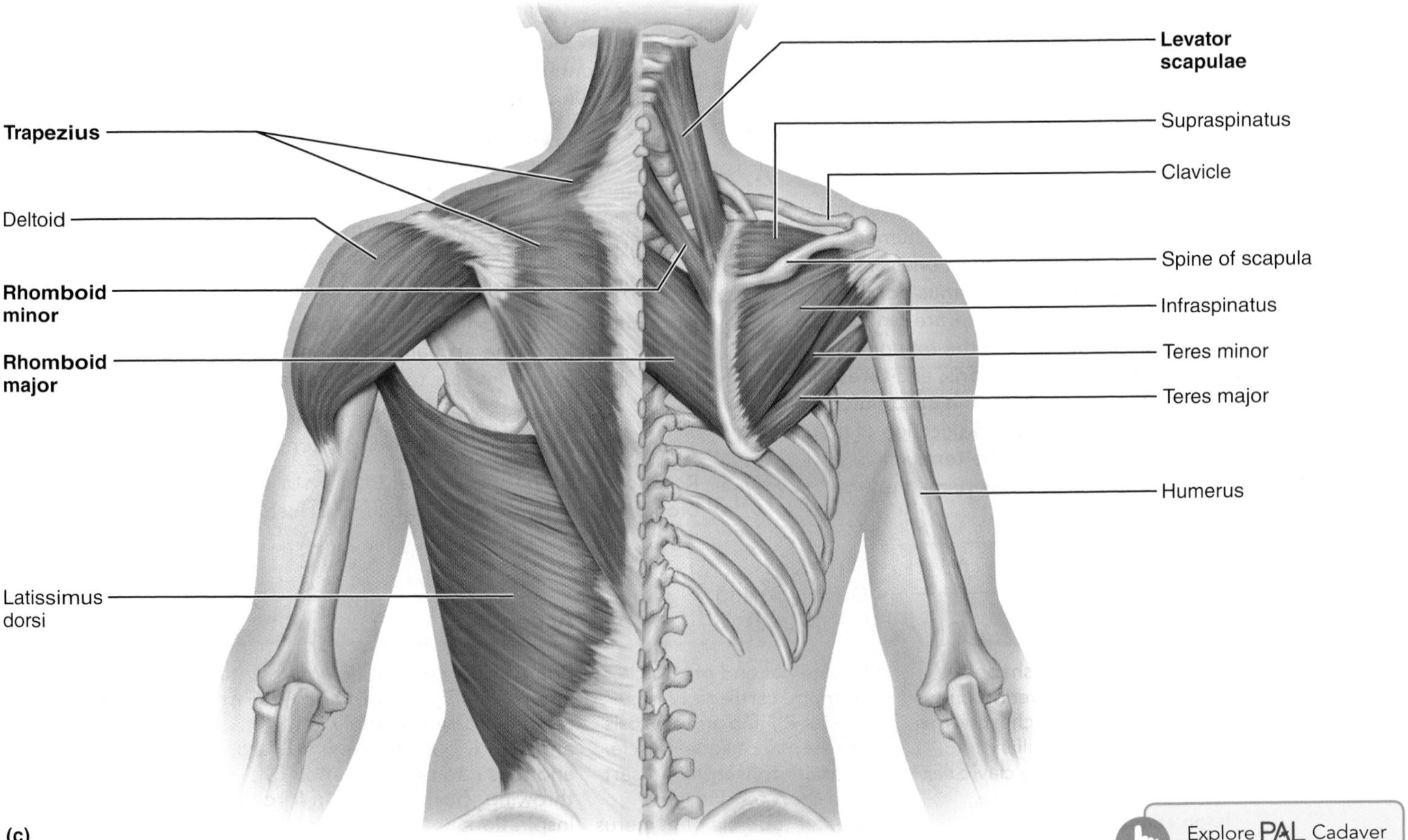

Trapezius
Deltoid
Rhomboid minor
Rhomboid major
Latissimus dorsi
Levator scapulae
Supraspinatus
Clavicle
Spine of scapula
Infraspinatus
Teres minor
Teres major
Humerus

(c)

Explore PAL Cadaver MasteringA&P®

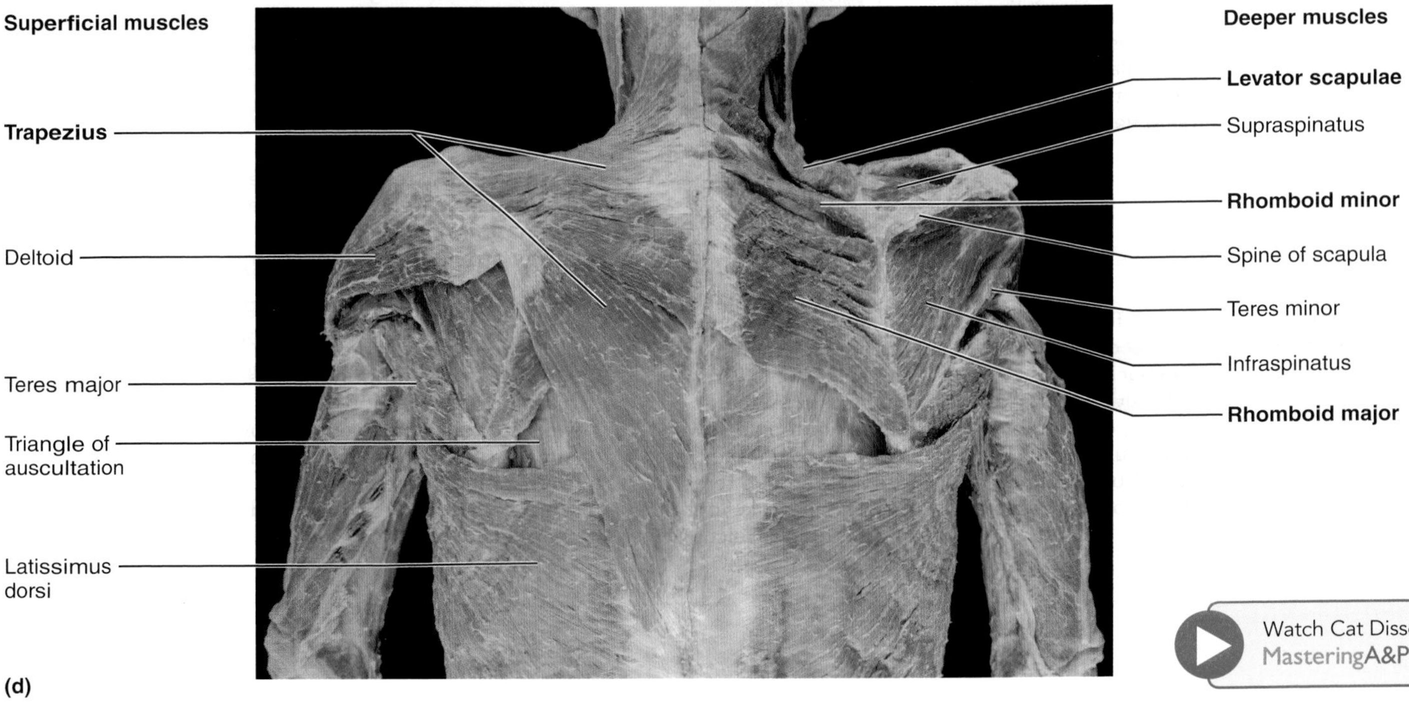

Superficial muscles

Trapezius
Deltoid
Teres major
Triangle of auscultation
Latissimus dorsi

Deeper muscles

Levator scapulae
Supraspinatus
Rhomboid minor
Spine of scapula
Teres minor
Infraspinatus
Rhomboid major

(d)

Watch Cat Dissection MasteringA&P®

Table 11.15 Muscles Crossing the Hip and Knee Joints: Movements of the Thigh and Leg (Figures 11.22, 11.23, and 11.24) *continued*

Muscle	Description	Origin (O) and Insertion (I)	Action	Nerve Supply
LATERAL ROTATORS *(continued)*				
Obturator internus (in-ter′nus) (*internus* = inside)	Surrounds obturator foramen within pelvis; leaves pelvis via lesser sciatic notch and turns acutely forward to insert on femur	O—inner surface of obturator membrane, greater sciatic notch, and margins of obturator foramen I—greater trochanter in front of piriformis	As for piriformis	L_5 and S_1
Gemellus (jĕ′mĕ′lis)—superior and inferior (*gemin* = twin, double; *superior* = above; *inferior* = below)	Two small muscles with common insertions and actions; considered extrapelvic portions of obturator internus	O—ischial spine (superior); ischial tuberosity (inferior) I—greater trochanter of femur	As for piriformis	L_5 and S_1
Quadratus femoris (*quad* = four-sided square)	Short, thick muscle; most inferior of lateral rotator muscles; extends laterally from pelvis	O—ischial tuberosity I—intertrochanteric crest of femur	Rotates thigh laterally and stabilizes hip joint	L_5 and S_1

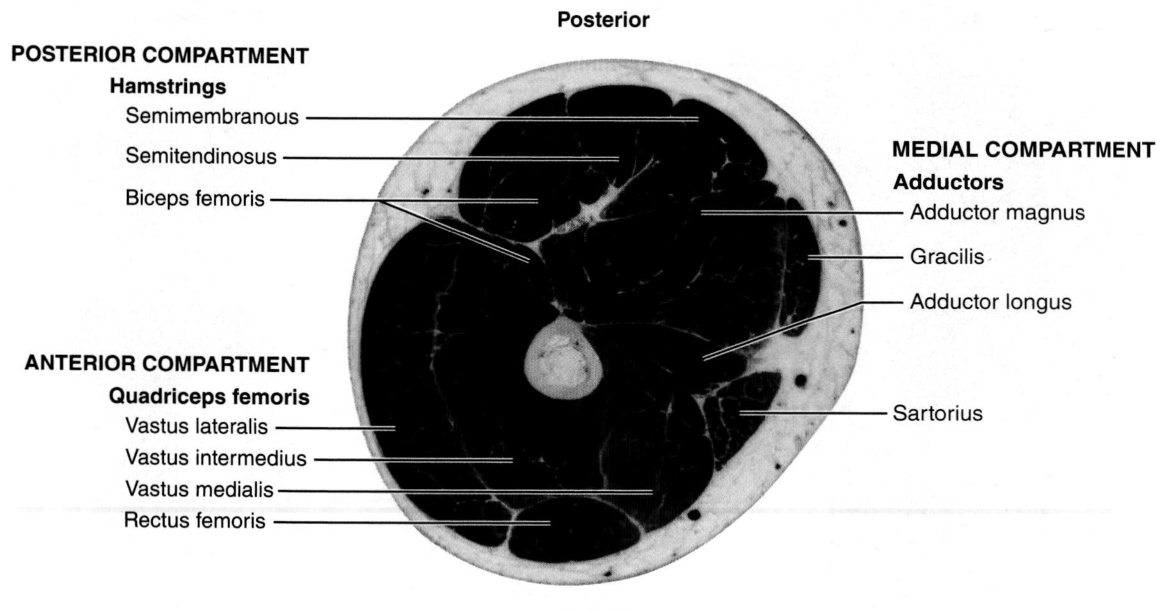

Figure 11.24 **Cross section through the mid-thigh.**

Table 11.16 Muscles of the Leg: Movements of the Ankle and Toes (Figures 11.25, 11.26, and 11.27)

The deep fascia of the leg is continuous with the fascia lata that surrounds the thigh. Like a "knee sock" deep to the skin, this leg fascia surrounds the leg muscles and binds them tightly, preventing excess swelling of these muscles during exercise and also aiding venous return. Inward extensions from the leg fascia divide the leg muscles into *anterior, lateral,* and *posterior compartments* (Figure 11.8), each with its own nerve and blood supply. Distally, the leg fascia thickens to form the **extensor, fibular,** and **flexor retinacula,** "ankle bracelets" that hold the tendons in place where they run to the foot. As in the wrist and hand, the distal tendons are covered with slippery tendon sheaths.

The various muscles of the leg cause movements at the ankle joint (dorsiflexion and plantar flexion), at the intertarsal joints (inversion and eversion of the foot), or at the toes (flexion, extension).

Anterior Extensor Compartment. The muscles in the *anterior extensor compartment* of the leg (Figure 11.25) are directly comparable to the extensor muscle group of the forearm: They extend the toes and dorsiflex the foot. Although dorsiflexion is not a powerful movement, it keeps the toes from dragging during walking.

Lateral Compartment. The *lateral compartment muscles* (Figure 11.26) are the *fibularis* (formerly *peroneal; peron* = the fibula) muscles. They evert and plantar flex the foot.

Posterior Flexor Compartment. Muscles of the *posterior flexor compartment* (Figure 11.27) are comparable to the flexor muscle group of the forearm: They flex the toes and plantar flex the foot. Plantar flexion is the most powerful movement at the ankle: It lifts the weight of the entire body. Plantar flexion is necessary for standing on tiptoe and provides the forward thrust in walking and running.

(The actions of the muscles in this table are summarized in Table 11.2, Part II, p. 277.)

Muscle	Description	Origin (O) and Insertion (I)	Action	Nerve Supply
PART I: MUSCLES OF THE ANTERIOR COMPARTMENT OF LEG (FIGURES 11.25 AND 11.26)				
	All muscles of the anterior compartment of the leg are dorsiflexors of the ankle and have a common innervation, the deep fibular nerve. Paralysis of the anterior muscle group causes *foot drop,* which requires that one lift the leg unusually high during walking to prevent tripping over one's toes. "Shin splints" (p. 344) is a painful inflammatory condition of the muscles of the anterior compartment.			
Tibialis anterior (tib″e-a′lis) (*tibial* = tibia; *anterior* = toward the front)	Superficial muscle of anterior leg; laterally parallels sharp anterior margin of tibia	O—lateral condyle and upper two-thirds of tibial shaft; interosseous membrane I—by tendon into inferior surface of medial cuneiform and first metatarsal bone	Prime mover of dorsiflexion; inverts foot; assists in supporting medial longitudinal arch of foot	Deep fibular nerve (L_4 and L_5)
Extensor digitorum longus (*extensor* = increases angle at a joint; *digit* = finger or toe; *longus* = long)	Unipennate muscle on anterolateral surface of leg; lateral to tibialis anterior muscle	O—lateral condyle of tibia; proximal three-fourths of fibula; interosseous membrane I—middle and distal phalanges of toes II–V via extensor expansion	Prime mover of toe extension (acts mainly at metatarsophalangeal joints); dorsiflexes foot	Deep fibular nerve (L_5 and S_1)
Fibularis (peroneus) tertius (fib-u-lah′ris ter′shus) (*perone* = fibula; *tertius* = third)	Small muscle; usually continuous and fused with distal part of extensor digitorum longus; not always present	O—distal anterior surface of fibula and interosseous membrane I—tendon inserts on dorsum of fifth metatarsal	Dorsiflexes and everts foot	Deep fibular nerve (L_5 and S_1)
Extensor hallucis (hal′u-sis) **longus** (*hallux* = great toe)	Deep to extensor digitorum longus and tibialis anterior; narrow origin	O—anteromedial fibula shaft and interosseous membrane I—tendon inserts on distal phalanx of great toe	Extends great toe; dorsiflexes foot	Deep fibular nerve (L_5 and S_1)
PART II: MUSCLES OF THE LATERAL COMPARTMENT OF LEG (FIGURES 11.26 AND 11.27)				
	These muscles have a common innervation, the superficial fibular nerve. Besides plantar flexion and foot eversion, these muscles stabilize the lateral ankle and lateral longitudinal arch of the foot.			
Fibularis (peroneus) longus (also see Figure 11.25)	Superficial lateral muscle; overlies fibula	O—head and upper portion of lateral side of fibula I—by long tendon that curves under foot to first metatarsal and medial cuneiform	Plantar flexes and everts foot	Superficial fibular nerve (L_5 and S_1)

▶

Table 11.16 Muscles of the Leg: Movements of the Ankle and Toes (Figures 11.25, 11.26, and 11.27) *continued*

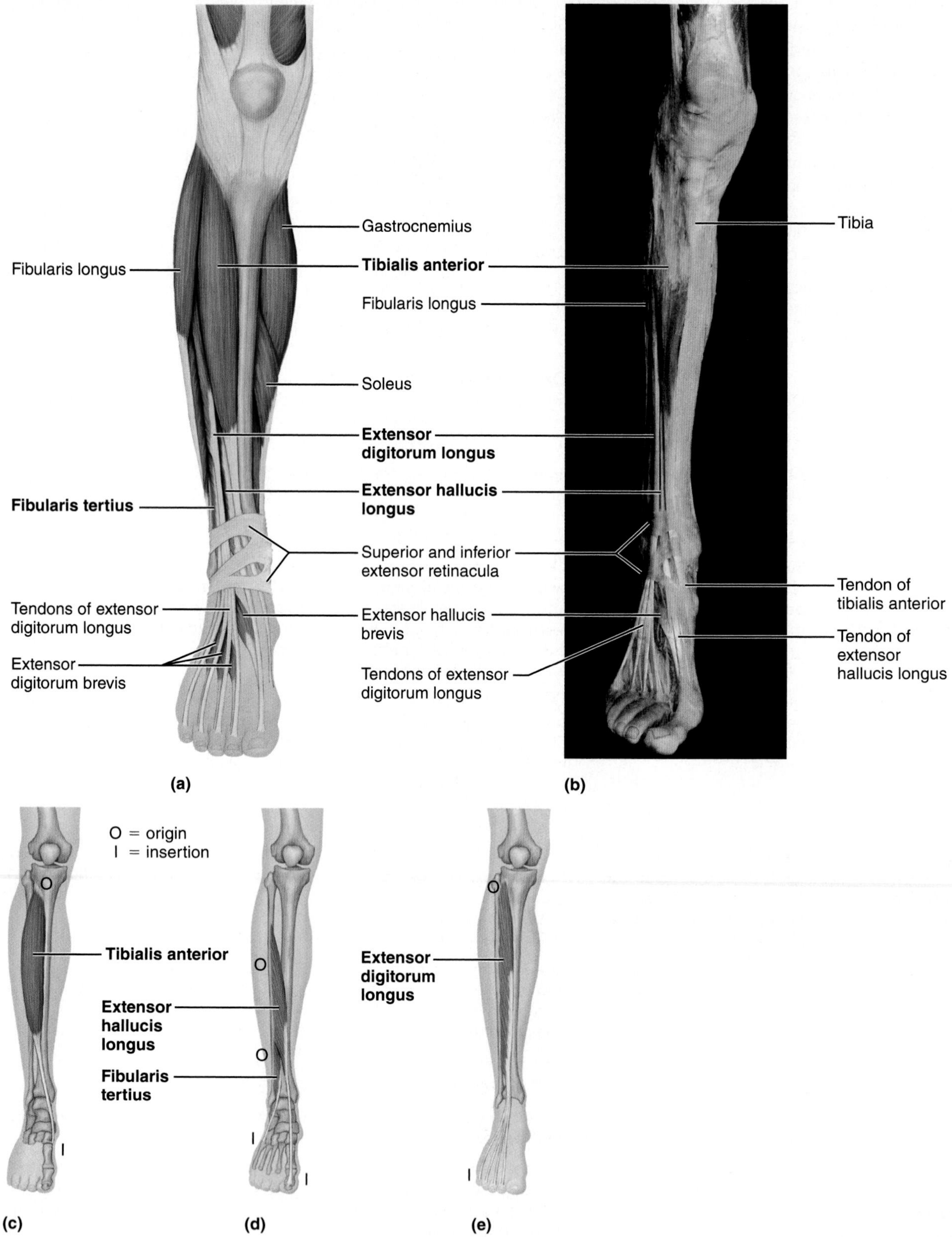

Figure 11.25 Muscles of the anterior compartment of the right leg. (**a**) Superficial view of anterior leg muscles. (**b**) Cadaver dissection of view similar to (a). (**c–e**) Anterior leg muscles shown in isolation to show origins and insertions.

Play A&P Flix Animation MasteringA&P®

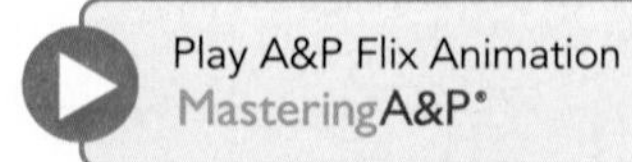

Table 11.16 *continued*

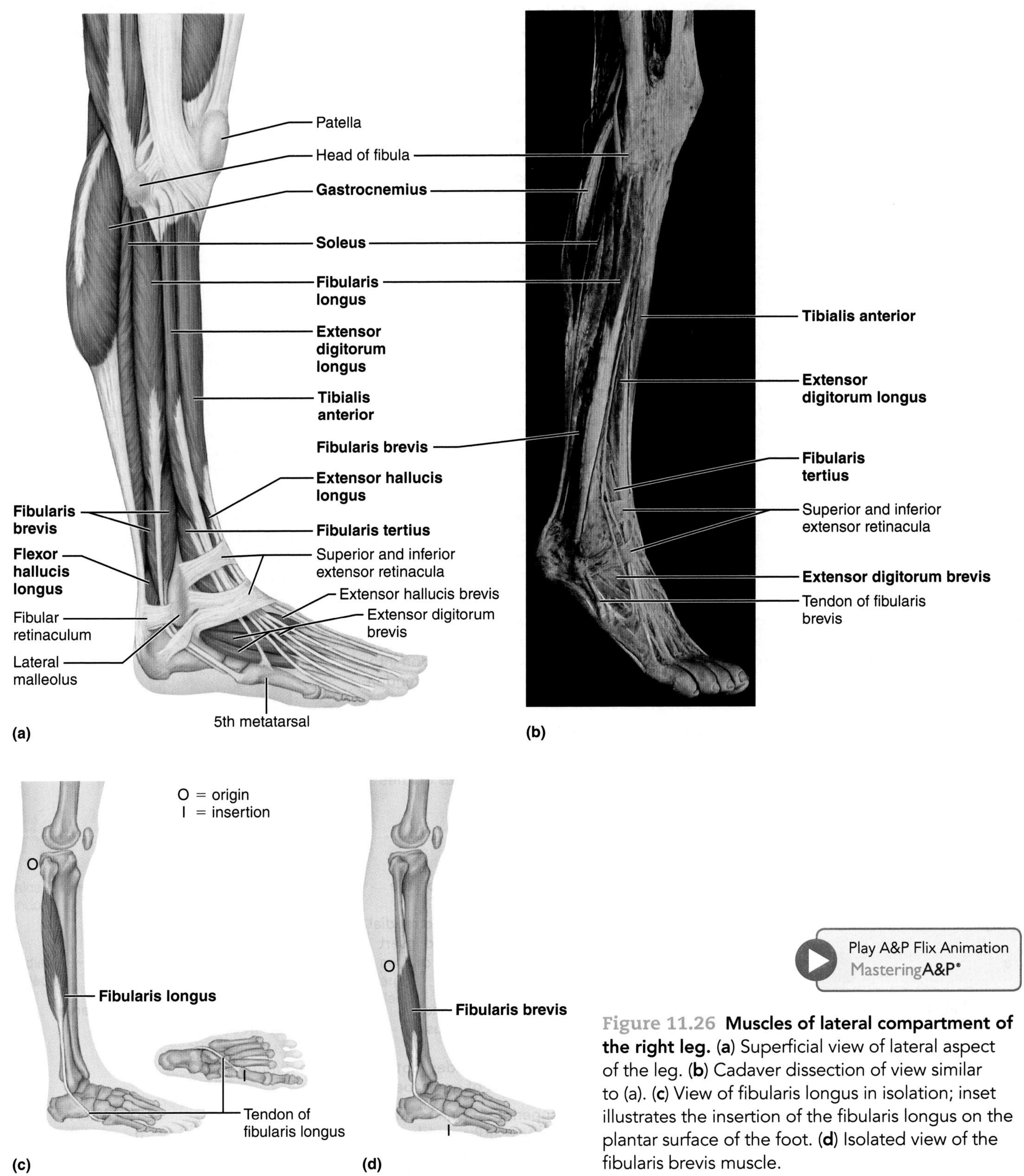

Figure 11.26 **Muscles of lateral compartment of the right leg.** (**a**) Superficial view of lateral aspect of the leg. (**b**) Cadaver dissection of view similar to (a). (**c**) View of fibularis longus in isolation; inset illustrates the insertion of the fibularis longus on the plantar surface of the foot. (**d**) Isolated view of the fibularis brevis muscle.

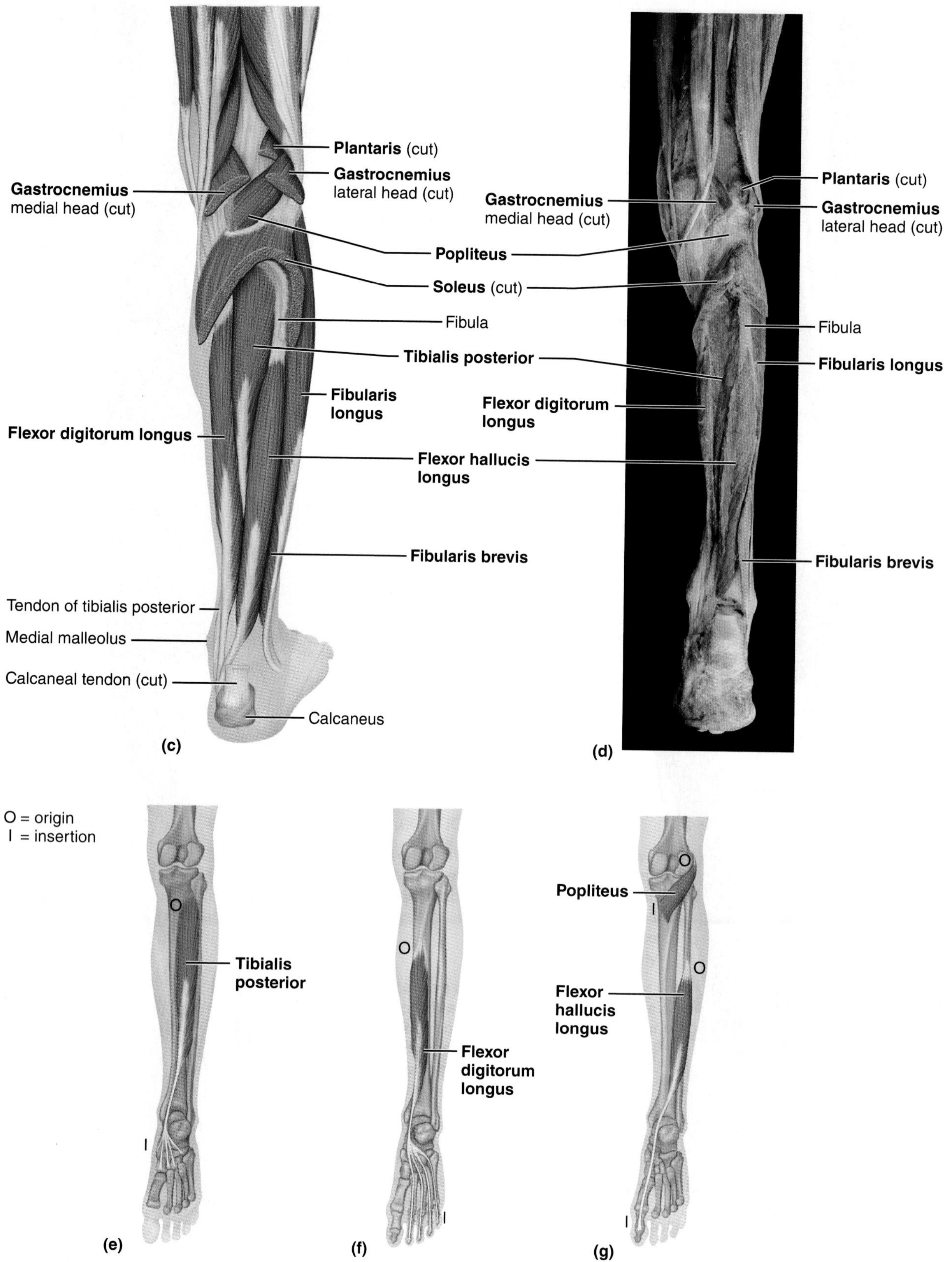

Figure 11.27 Muscles of the posterior compartment of the right leg, *continued.* **(c)** The triceps surae has been removed to show the deep muscles of the posterior compartment. **(d)** Cadaver dissection of view similar to (c). **(e–g)** Individual deep muscles are shown in isolation so that their origins and insertions may be seen.

Table 11.17 Intrinsic Muscles of the Foot: Toe Movement and Foot Support (Figure 11.28)

The intrinsic muscles of the foot help to flex, extend, abduct, and adduct the toes. Furthermore, along with the tendons of some leg muscles that enter the sole, the foot muscles support the arches of the foot. There is a single muscle on the foot's dorsum (superior aspect), and many muscles on the plantar aspect (the sole). The plantar muscles occur in four layers, from superficial to deep. Overall, the foot muscles are remarkably similar to those in the palm of the hand.

Muscle	Description	Origin (O) and Insertion (I)	Action	Nerve Supply
MUSCLE ON DORSUM OF FOOT				
Extensor digitorum brevis (Figure 11.26a)	Small, four-part muscle on dorsum of foot; deep to the tendons of extensor digitorum longus; corresponds to the extensor indicis and extensor pollicis muscles of forearm	O—anterior part of calcaneus bone; extensor retinaculum I—base of proximal phalanx of big toe; extensor expansions on toes II–IV	Helps extend toes at metatarsophalangeal joints	Deep fibular nerve (L_5 and S_1)
MUSCLES ON SOLE OF FOOT (FIGURE 11.28)				
First Layer (Most Superficial)				
• **Flexor digitorum brevis**	Bandlike muscle in middle of sole; corresponds to flexor digitorum superficialis of forearm and inserts into digits in the same way	O—calcaneal tuberosity I—middle phalanx of toes II–IV	Flexes toes	Tibial nerve (medial plantar nerve, a branch of tibial nerve, S_1 and S_2)
• **Abductor hallucis** (hal′u-sis) (*hallux* = the big toe)	Lies medial to flexor digitorum brevis (recall the similar thumb muscle, abductor pollicis brevis)	O—calcaneal tuberosity and flexor retinaculum I—proximal phalanx of great toe, medial side, through a tendon shared with flexor hallucis brevis (see below)	Abducts great toe	Tibial nerve (medial plantar nerve, S_1 and S_2)
• **Abductor digiti minimi**	Most lateral of the three superficial sole muscles (recall similar abductor muscle in palm)	O—calcaneal tuberosity I—lateral side of base of little toe's proximal phalanx	Abducts and flexes little toe	Tibial nerve (lateral plantar nerve, a branch of tibial nerve, S_2, and S_3)
Second Layer				
• **Flexor accessorius (quadratus plantae)**	Rectangular muscle just deep to flexor digitorum brevis in posterior half of sole; two heads (see Figure 11.28b and c)	O—medial and lateral sides of calcaneus I—tendon of flexor digitorum longus in midsole	Straightens out the oblique pull of flexor digitorum longus	Tibial nerve (lateral plantar nerve, S_2 and S_3)
• **Lumbricals**	Four little "worms" (like lumbricals in hand)	O—from each tendon of flexor digitorum longus I—extensor expansion on proximal phalanx of toes II–V, medial side	By pulling on extensor expansion, flex toes at metatarsophalangeal joints and extend toes at interphalangeal joints	Tibial nerve (medial plantar nerve to first lumbrical and lateral plantar nerve to second through fourth lumbricals, L_5–S_2)

Table 11.17 Intrinsic Muscles of the Foot: Toe Movement and Foot Support (Figure 11.28) *continued*

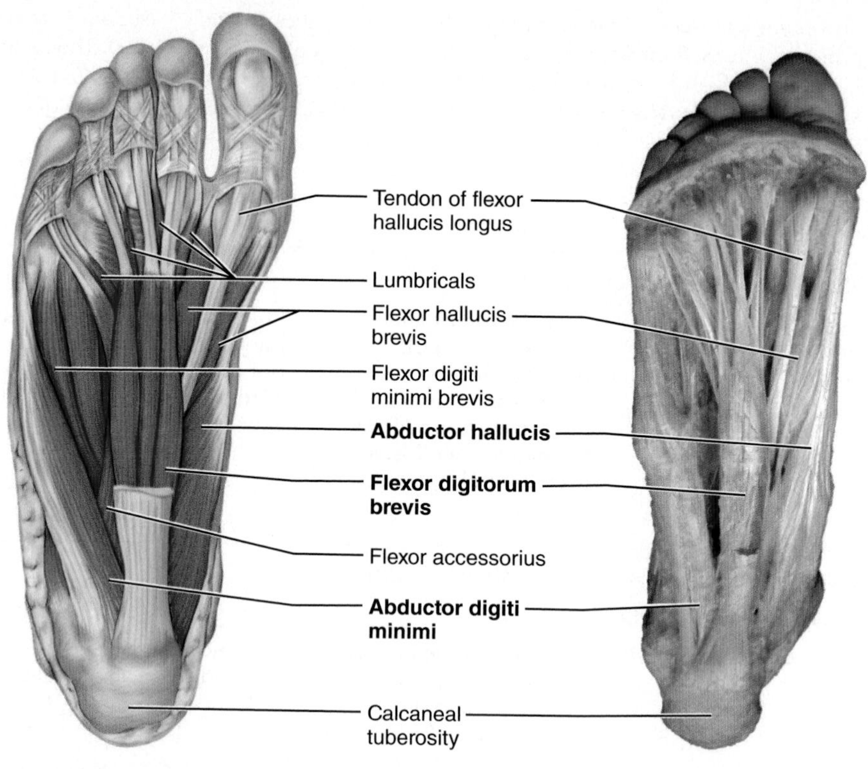

(a) First layer

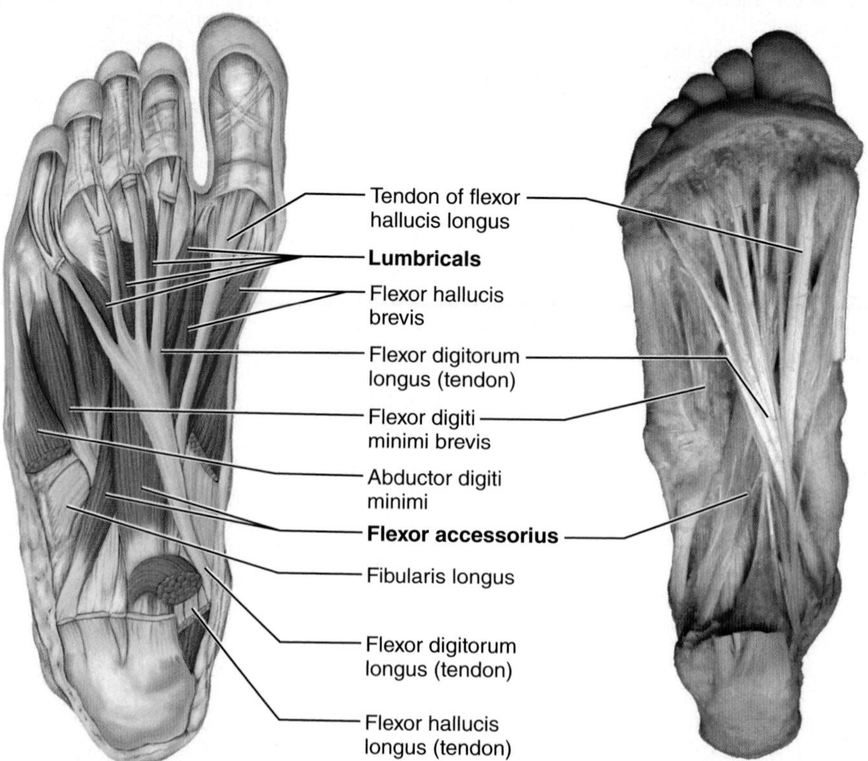

(b) Second layer

Figure 11.28 **Muscles of the right foot.** (**a, b**) Plantar aspect. Dissections of the two superficial layers also shown.

Table 11.17 *continued*

Muscle	Description	Origin (O) and Insertion (I)	Action	Nerve Supply
MUSCLES ON SOLE OF FOOT (FIGURE 11.28) ***(continued)***				
Third Layer				
• **Flexor hallucis brevis**	Covers metatarsal I; splits into two bellies—recall flexor pollicis brevis of thumb (see Figure 11.28c)	O—lateral cuneiform and cuboid bones I—via two tendons onto both sides of the base of the proximal phalanx of great toe; each tendon has a sesamoid bone in it	Flexes great toe at metatarsophalangeal joint	Tibial nerve (medial plantar nerve, S_1 and S_2)
• **Adductor hallucis**	Oblique and transverse heads; deep to lumbricals (recall adductor pollicis in thumb)	O—from bases of metatarsals II–IV and from sheath of fibularis longus tendon (oblique head); from a ligament across metatarsophalangeal joints (transverse head) I—base of proximal phalanx of great toe, lateral side	Helps maintain the transverse arch of foot; weak adductor of great toe	Tibial nerve (lateral plantar nerve, S_2 and S_3)
• **Flexor digiti minimi brevis**	Covers metatarsal V (recall same muscle in hand)	O—base of metatarsal V and sheath of fibularis longus tendon I—base of proximal phalanx of toe V	Flexes little toe at metatarsophalangeal joint	Tibial nerve (lateral plantar nerve, S_2 and S_3)
Fourth Layer (Deepest)				
• **Plantar and dorsal interossei**	Three plantar and four dorsal; similar to palmar and dorsal interossei of hand in locations, attachments, and actions; however, the long axis of foot around which these muscles orient is the second digit, not third	See palmar and dorsal interossei (Table 11.14)	See palmar and dorsal interossei (Table 11.14)	Tibial nerve (lateral plantar nerve, S_2 and S_3)

Table 11.17 Intrinsic Muscles of the Foot: Toe Movement and Foot Support (Figure 11.28) *continued*

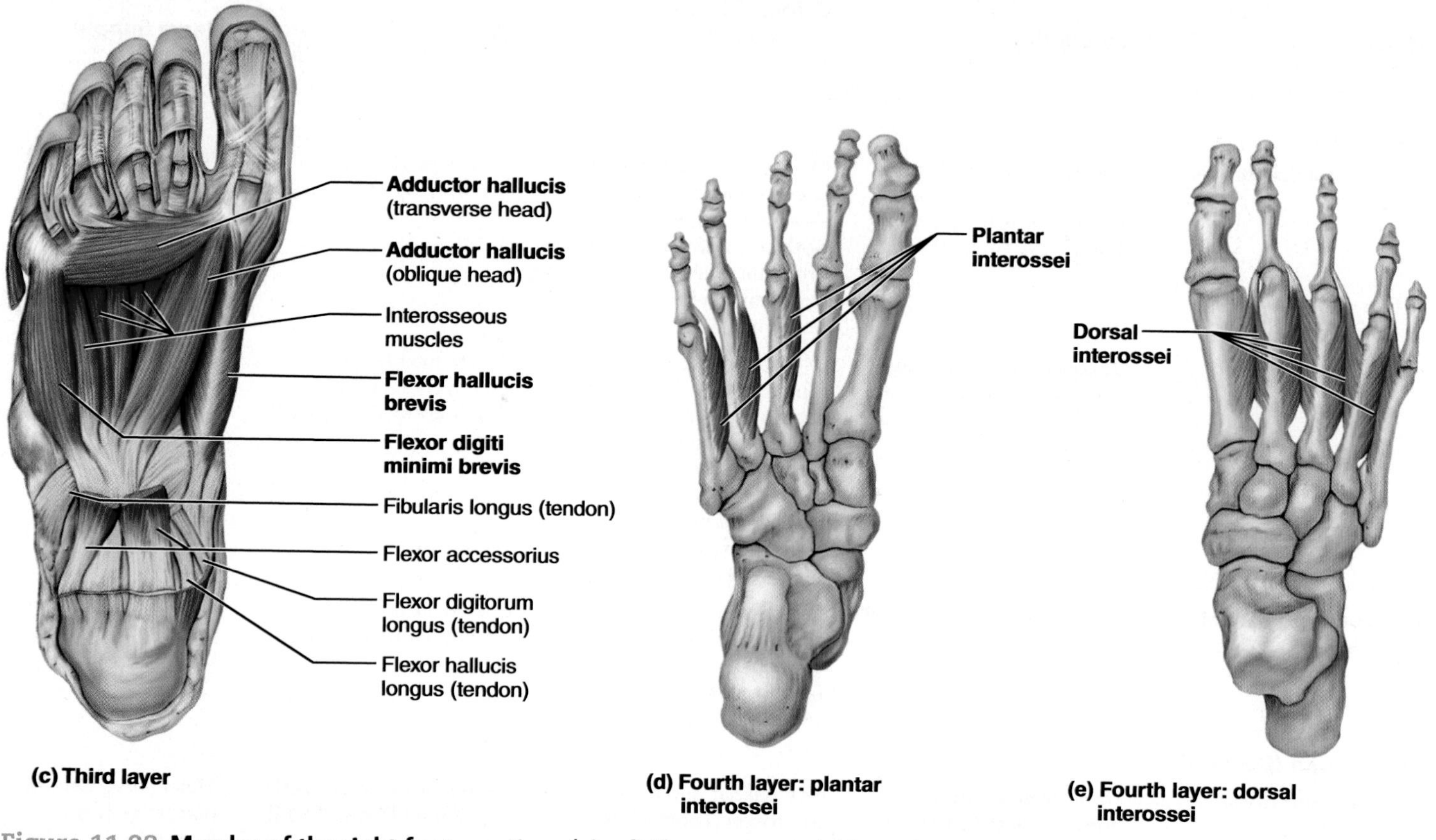

Figure 11.28 **Muscles of the right foot,** *continued.* **(c, d)** Plantar aspect. **(e)** Dorsal view of right foot, deepest muscles.

check your understanding

☐ **15.** What muscle is the most important muscle for respiration? What additional muscles aid in inspiration and forceful expiration?

☐ **16.** Describe the six movements possible at the shoulder joint. List one muscle that performs each movement.

☐ **17.** What muscle is the prime mover of dorsiflexion? Why is this motion important during walking?

☐ **18.** How do the lesser gluteals—gluteus medius and gluteus minimus—function during walking?

(For answers, see Appendix B.)

REGIONAL SURFACE ANATOMY

learning outcome

▶ Use surface landmarks to locate and identify underlying muscles.

Thus far we have studied the skeletal and muscular systems independently, but it is important to think of these systems as an integrated unit that functions to support and move the body. In many clinical situations, an understanding of skeletal and muscular components of a particular region, as well as the nervous and vascular structures of the region, is necessary to assess functionality.

Surface anatomy, the study of the external surface of the body, gives insight into the internal organs. Feeling internal structures through the skin with the fingers is called **palpation** (pal-pa′shun; "touching"). Many features of skeletal and muscular anatomy are either easily viewed from the body surface or can be located by palpation. This section reinforces much of what you learned about skeletal and muscular structure and highlights how these elements relate to one another anatomically. This section takes a *regional* approach to surface anatomy, focusing on the anatomical relationships between body structures in a particular region, exploring the head first and proceeding to the trunk and the limbs. Let your own body structure help you remember the anatomical details. Even better, whenever possible, have a study partner assume the role of the patient as you make your observations and palpations.

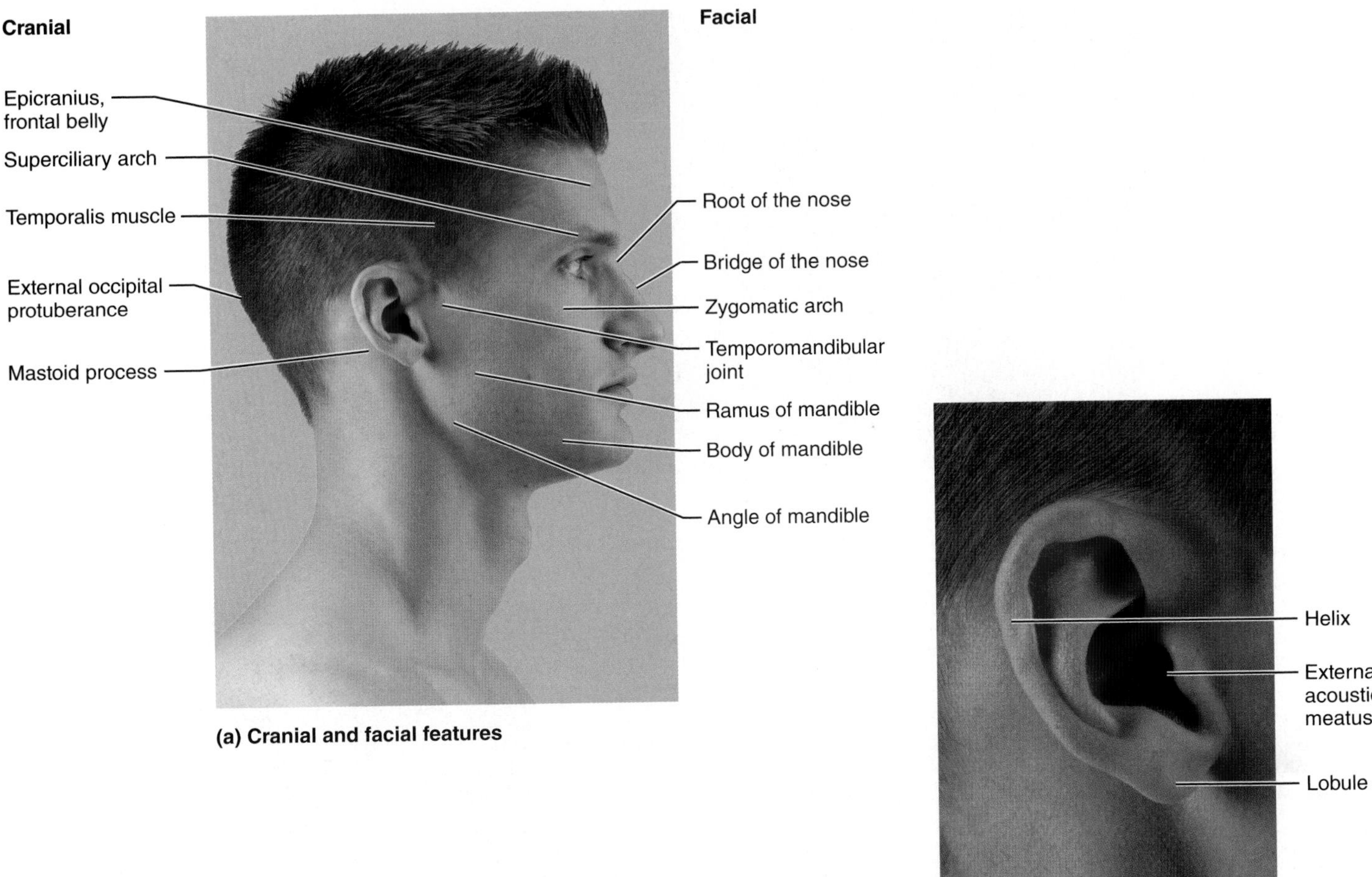

(a) Cranial and facial features

(b) Auricle of the right ear

Figure 11.29 Surface anatomy of the lateral aspect of the head.

The Head (Figure 11.29)

The Cranium

Run your fingers over the superior surface of your head, and notice how the cranial bones under the skin lie very near the surface. Proceed to your forehead, and palpate the *superciliary arches* ("brow ridges") directly superior to your orbits (Figure 11.29a). Then move your hand to the posterior surface of the skull, where you can feel the knoblike *external occipital protuberance.* Running your fingers laterally from this protuberance, feel the ridge-like *superior nuchal line* on the occipital bone. This line marks the superior extent of the muscles of the posterior neck and is a boundary between the head and the neck. With your fingers on the superior nuchal line, extend your head to feel the contraction of the muscles of the posterior neck, the *trapezius* and *splenius.* Now feel the prominent *mastoid process* on each side of the cranium just posterior to your ear.

Place a hand on your temple, and clench your teeth together. You should be able to feel the *temporalis muscle* bulge as it contracts. Next, raise your eyebrows, and feel your forehead wrinkle as the frontal belly of the *epicranius muscle,* a muscle of facial expression, contracts. This portion of the epicranius originates superiorly from the *epicranial aponeurosis,* which covers the superior surface of the cranium and binds tightly to the overlying subcutaneous tissue and skin to form the true scalp. Push on your scalp to confirm that it slides freely over the underlying cranial bones. Because the scalp is only loosely bound to the skull, people can easily be "scalped" (in industrial accidents, for example). Most arteries of the body constrict and close after they are cut or torn, but the vessels in the richly vascularized scalp are unable to do so because they are held open by the dense connective tissue surrounding them. As a result, scalp wounds bleed profusely. However, because the scalp is so well vascularized, these wounds also heal quickly.

The Face

The surface of the face is divided into four regions: orbital (eye), nasal, auricular (ear), and oral (mouth). In the orbital region of your own face, trace the bony margin of an orbit. The groove known as the *lacrimal fossa,* which contains the tear-gathering lacrimal sac, may be felt on the medial side of the eye socket. Next, touch the most superior part of your nose, which is its *root,* between the eyebrows. Just inferior to this, between your eyes, is the *bridge* of the nose, formed by the nasal bones (Figure 11.29a).

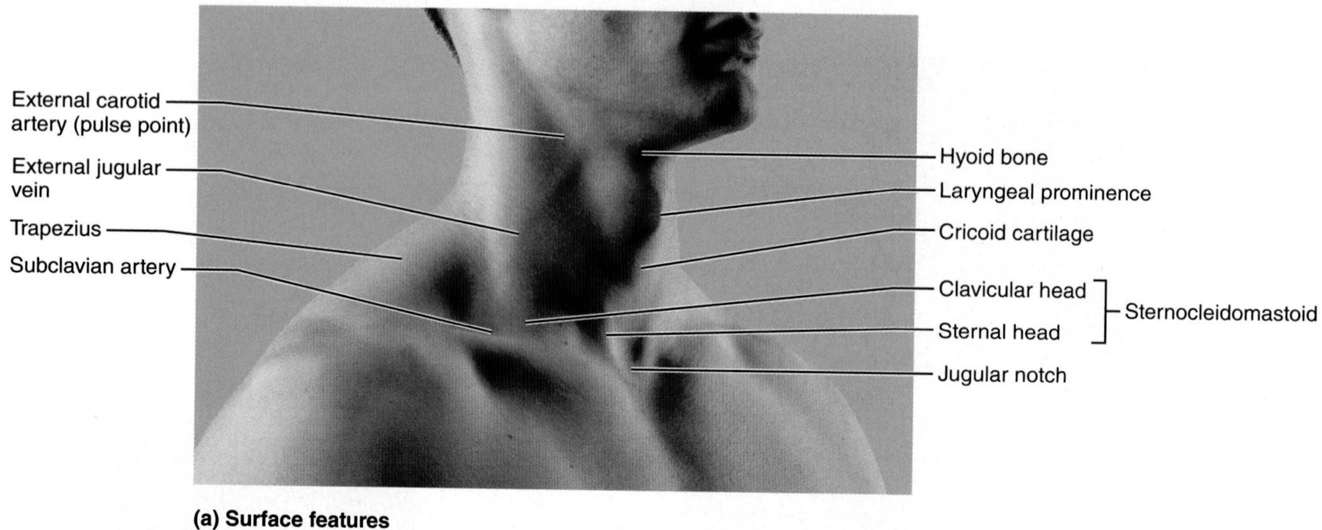

(a) Surface features

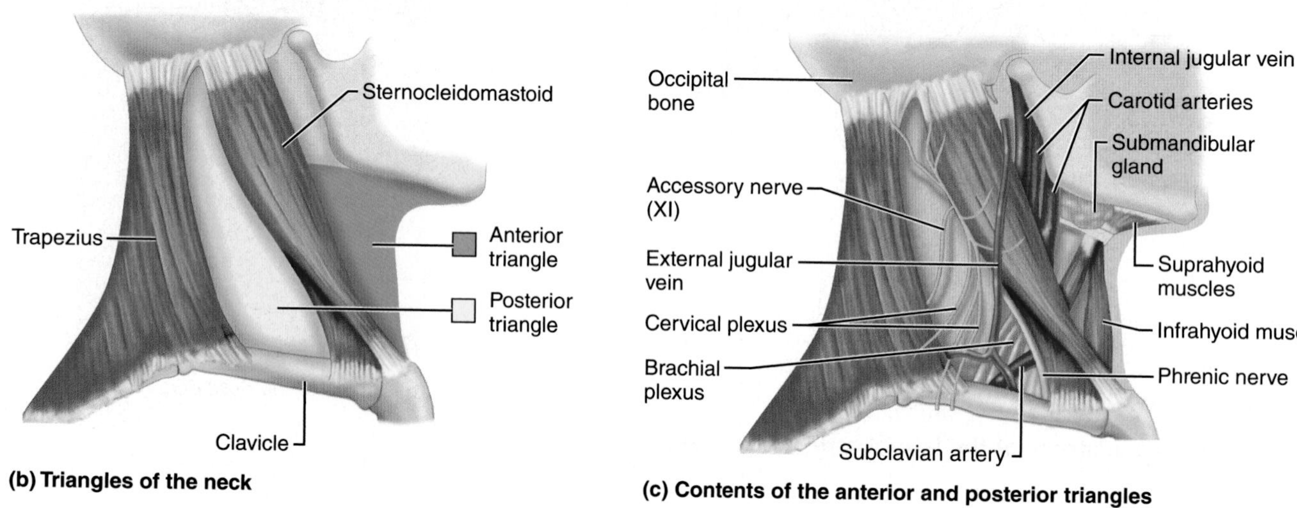

(b) Triangles of the neck

(c) Contents of the anterior and posterior triangles

Figure 11.30 Surface anatomy and regions of the neck.

Grasp your *auricle,* the part of the external ear that surrounds the opening of the *external acoustic meatus* (Figure 11.29b). Trace the ear's outer rim, or *helix,* to the *lobule* (ear lobe) inferiorly. The lobule is easily pierced, and because it is not highly sensitive to pain, it provides a convenient place to obtain a drop of blood for clinical analysis. Additionally, a *pulse oximeter,* a device that measures blood oxygen levels across the surface of the skin, may be clamped to the ear lobe.

Run your hand anteriorly from the ear's external acoustic meatus toward the orbit, and feel the *zygomatic arch* ("cheek bone") just deep to the skin. This bone can easily be broken by blows to the face. Next, place your fingers on your face, and feel it contort as you smile, frown, and grimace; you are now monitoring the action of the subcutaneous *muscles of facial expression* (Table 11.3, pp. 279–281).

In the region of the lower jaw, palpate the parts of the bony *mandible:* its anterior *body* and its posterior ascending *ramus* (Figure 11.29a). Press on the skin over the mandibular ramus, and feel the *masseter muscle* bulge when you bite down. Feel for the anterior border of the masseter, and trace it to its insertion on the mandible's inferior margin and the mandibular angle. Finally, to feel the *temporomandibular joint,* place a finger directly anterior to the external acoustic meatus, and open and close your mouth several times. The bony structure you feel moving is the *head of the mandible.*

The Neck (Figure 11.30)

Skeletal Landmarks

Run your fingers inferiorly along the back of your neck, in the posterior midline, where you can feel the *spinous processes* of the cervical vertebrae. The spine of C_7 is especially prominent. Anteriorly, place a finger on your chin and then run the finger inferiorly along the neck's anterior midline. The first hard structure you encounter will be the U-shaped *hyoid bone* (Figure 11.30a), which lies in the angle between the floor of the mouth and the vertical part of the neck. Directly inferior to this, you will feel the *laryngeal prominence* (Adam's apple) of the thyroid cartilage; this prominence begins with a V-shaped notch superiorly and continues inferiorly as a sharp vertical ridge. Just inferior to the prominence, your finger will sink into a soft depression (formed by the *cricothyroid ligament*) and then will proceed onto the rounded surface of the *cricoid cartilage*.

Muscles of the Neck

The *sternocleidomastoid,* the most prominent muscle in the neck and the neck's most important surface landmark, can best be seen and felt when your head is turned to the side. If you stand in front of a mirror and turn your face sharply from right to left several times, you will be able to see both heads of this muscle, the *sternal head* medially and the *clavicular head* laterally (Figure 11.30).

Another large muscle in the neck, on the posterior aspect, is the *trapezius.* You can feel this muscle contract just deep to the skin as you shrug your shoulders.

Triangles of the Neck

The sternocleidomastoid muscles divide each side of the neck into a posterior and an anterior triangle (Figure 11.30b). The **posterior triangle** is defined by the sternocleidomastoid anteriorly, the trapezius posteriorly, and the clavicle inferiorly. The **anterior triangle** is defined by the inferior margin of the mandible superiorly, the midline of the neck anteriorly, and the sternocleidomastoid posteriorly.

The posterior triangle contains many important nerves and blood vessels (Figure 11.30c), including the *accessory nerve* (cranial nerve XI), most of the *cervical plexus,* and the *phrenic nerve* (discussed in Chapter 14). In the inferior part of the triangle are the *external jugular vein,* the trunks of the *brachial plexus,* and the *subclavian artery*—all of which are relatively superficial.

In the neck's anterior triangle (Figure 11.30c), the important structures include the *submandibular gland,* the *suprahyoid* and *infrahyoid muscles,* and the parts of the carotid arteries and jugular veins that lie superior to the sternocleidomastoid. The *common carotid artery* and *internal jugular vein* lie just deep to the sternocleidomastoid. Just anterior to the sternocleidomastoid, superior to the level of your larynx, you can feel a carotid pulse—the pulsations of the *external carotid artery* (Figure 11.30a).

> **CLINICAL APPLICATION**
>
> **Lacerations to the Neck** The vessels and nerves in the triangles of the neck are all relatively superficial and thus susceptible to injury by a laceration to the neck. A laceration through any of these vessels can result in significant blood loss. A wound to the posterior triangle of the neck can damage nerves located in this region, resulting in long-term loss of sensation in the skin of the neck and shoulder, as well as partial paralysis of the sternocleidomastoid and trapezius muscles.

The Thorax (Figure 11.31)

Skeletal Landmarks

The surface features of the thoracic cage (Figure 11.31) are palpable. The sternum is divided into three sections: manubrium, body, xiphoid process (Chapter 7, p. 179). The *jugular notch* forms a soft depression at the superior edge of the manubrium. The *sternal angle,* a slight bump indicating the junction of the manubrium and the body, is an important landmark for locating the second ribs. The ribs provide a series of horizontal lines that can be used to map the location of the underlying organs of the thoracic cavity. Such mapping also requires longitudinal lines on the wall of the trunk. Two defined vertical lines are the **midaxillary line,** which extends inferiorly from the center of the axilla onto the lateral thoracic wall, and the **midclavicular line,** which runs from the midpoint of the clavicle and extends inferiorly to the groin, passing about 1 cm medial to the nipple.

The *costal margin* marks the inferior boundary of the thoracic wall. The **infrasternal angle** is the location of the *xiphisternal joint,* the junction of the body of the sternum and the xiphoid process.

Muscles of the Thorax

The main superficial muscles of the anterior thoracic wall are the *pectoralis major* and the anterior slips of the *serratus anterior.* (Using Figure 11.31 as a guide, try to palpate these two muscles on your chest.) They both contract during push-ups, and you can confirm this by pushing yourself up from your desk with one upper limb while palpating the muscles with your opposite hand.

The Abdomen (Figure 11.31)

Skeletal Landmarks

The anterior abdominal wall extends from the costal margin down to an inferior boundary that is defined by the following landmarks (all shown in Figure 11.31):

1. **Iliac crest** (superior margins of the iliac bones). You can locate the iliac crests, the superior margins of the iliac bones, by resting your hands on your hips.
2. **Anterior superior iliac spine.** Representing the most anterior point of the iliac crest, this spine is a prominent landmark that can be palpated in everyone, even those who are overweight.
3. **Inguinal ligament.** This ligament, indicated by a groove on the skin of the groin, runs medially from the anterior superior iliac spine to the *pubic tubercle* of the pubic bone.
4. **Pubic crest.** You will have to press deeply to feel this crest on the pubis near the median *pubic symphysis.* The *pubic tubercle,* the most lateral point of the pubic crest, is easier to palpate, but you must still push deeply.

Muscles and Other Abdominal Surface Features

The central landmark of the anterior abdominal wall is the *umbilicus* (navel). Running superiorly and inferiorly

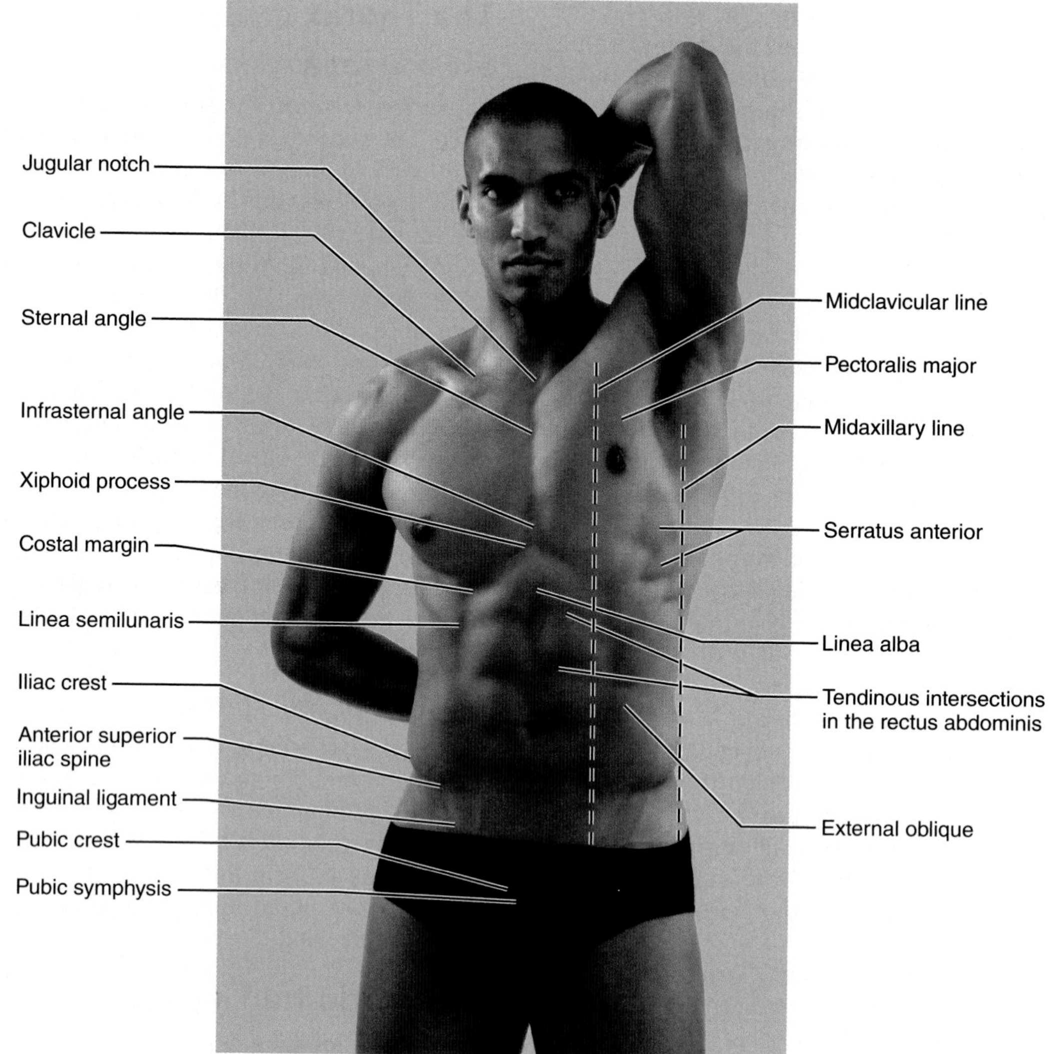

Figure 11.31 The anterior thorax and abdomen.

from the umbilicus is the *linea alba* ("white line"), represented in the skin of lean people by a vertical groove (Figure 11.31). This tendinous line that extends from the xiphoid process to the pubic symphysis, just medial to the rectus abdominis muscles, is the insertion of the abdominal muscles—the obliques and transversus abdominis (see Table 11.8, pp. 293–295). The linea alba is a favored site for surgical entry into the abdominal cavity because a long cut through it produces no muscle damage and a minimum of bleeding.

Flanking the linea alba on the abdominal wall are the vertical, straplike *rectus abdominis* muscles. Feel these muscles contract just deep to your skin as you do a bent-knee sit-up or as you bend forward after leaning back in your chair. In the skin of lean people, the lateral margin of each rectus muscle makes a groove known as the **linea semilunaris** (lin′e-ah sem″ĭ-lu-nar′is; "half-moon line"). In muscular people, three horizontal grooves in the skin covering the rectus abdominis represent the *tendinous intersections,* fibrous bands that subdivide the rectus muscle. Because of these subdivisions, each vertical rectus abdominis muscle presents four distinct horizontal bulges commonly (and inaccurately) referred to as the "six pack" by bodybuilders.

The only other major muscles that can be seen through the anterior abdominal wall are the lateral *external obliques.* Feel these muscles contract as you cough, strain, or in some other way raise your intra-abdominal pressure.

The Back (Figure 11.32)

Bones of the Back

The vertical groove in the center of the back is called the **posterior median furrow** (Figure 11.32). The *spinous processes* of the vertebrae are visible in this furrow when the spinal column is flexed. Palpate a few of these processes on your back (C_7 and T_1 are the most prominent and the easiest to find). Although locating structures on your own back is difficult, you should be able to palpate the posterior parts of some ribs, as well as the prominent *spine of the scapula* and the scapula's long *medial border.* The scapula lies superficial to ribs 2 through 7, its *inferior angle* is at the level of the spinous process of vertebra T_7,

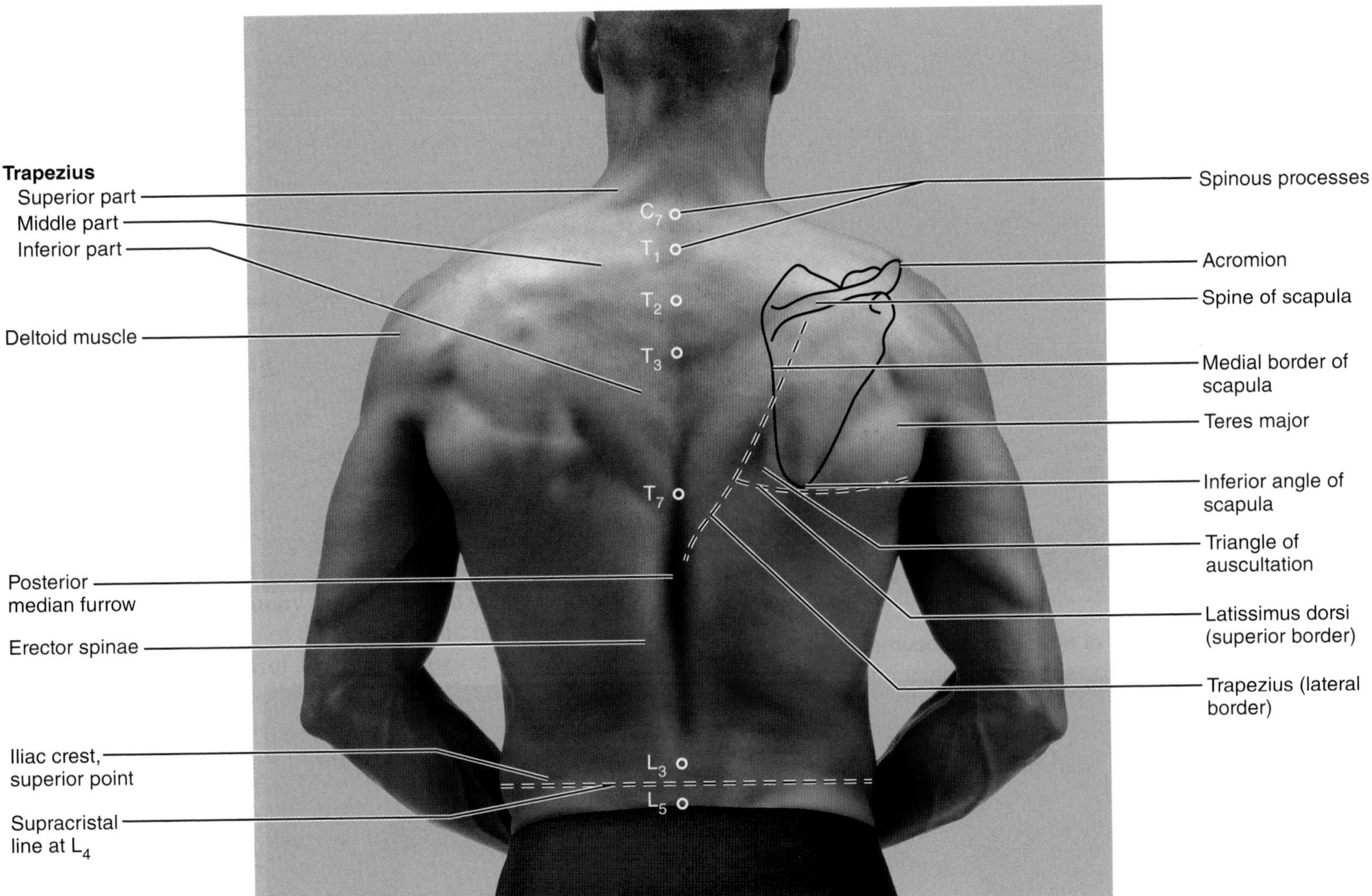

Figure 11.32 **Surface anatomy of the back.**

and the medial end of the scapular spine lies opposite the spinous process of T_3.

Recall that you can locate the iliac crests in your lower back by resting your hands on your hips. Locate the most superior point of each crest, a point roughly halfway between the posterior median furrow and the lateral side of the body (Figure 11.32). A horizontal line through the right and left superior points, the **supracristal line,** intersects L_4, providing a simple way to locate that vertebra. The ability to locate L_4 is essential for performing a *lumbar puncture,* a procedure in which the clinician inserts a needle into the vertebral canal directly superior or inferior to L_4.

The *sacrum* is easy to palpate just superior to the cleft in the buttocks, and one can feel the *coccyx* in the extreme inferior part of that cleft, just posterior to the anus.

Muscles of the Back

The largest superficial muscles of the back are the *trapezius* superiorly and *latissimus dorsi* inferiorly (Figure 11.32). Furthermore, the deeper *erector spinae* muscles are very evident in the lower back, flanking the vertebral column like thick vertical cords. Feel your erector spinae muscles contract and bulge as you extend your spine.

> **CLINICAL APPLICATION**
>
> **The Triangle of Auscultation** The superficial muscles of the back do not cover a small triangular area of the rib cage just medial to the inferior part of the scapula. This area, called the **triangle of auscultation** (aw″skul-ta′shun; "listening"), is bounded by the trapezius medially, the latissimus dorsi inferiorly, and the medial border of the scapula laterally (Figure 11.32). This is the site at which a clinician places a stethoscope to listen to lung sounds.

The Upper Limb and Shoulder

The Axilla (Figure 11.33)

The **base of the axilla** is the groove in which the underarm hair grows (Figure 11.33a). This base forms a "valley" between two thick, rounded ridges of muscle called the **axillary folds.** Just anterior to the base, clutch your **anterior axillary fold,** a fold formed by the pectoralis major muscle. Then grasp your **posterior axillary fold,** which is formed by the latissimus dorsi and teres major muscles of the back as they course toward their insertions on the humerus.

check your understanding

- ☐ 19. Where is the linea alba located, and what muscles insert into it?
- ☐ 20. Define the boundaries of the triangle of auscultation, and describe the importance of this landmark.
- ☐ 21. Describe three locations for giving intramuscular injections and the landmarks that help define these locations.
- ☐ 22. Identify the body region where the following surface landmarks are found: (a) supracristal line; (b) linea semilunaris; (c) cubital fossa; (d) median bicipital groove.
- ☐ 23. What are the boundaries of the posterior triangle of the neck? Where can you locate the carotid pulse?
- ☐ 24. When making a fist, two tendons are prominent in the distal forearm. Name these tendons.

(For answers, see Appendix B.)

RELATED CLINICAL TERMS

Adductor strain Pulled adductor muscle in the thigh, usually the adductor longus. Common sports injury, caused by extreme (or forced) abduction when the thigh is adducting (as when the planted limb slips sideways when a soccer player is kicking a ball). Can also occur if one's hip joint is unstable and the hip-stabilizing adductors strain from continually compensating for the instability. Imaging or other techniques are used to distinguish this from other injuries to the groin region, such as hernias and cracked neck of femur.

Charley horse Tearing of a muscle, followed by bleeding into the tissues (hematoma) and severe pain; a common sports injury. Charley horse of the quadriceps femoris occurs frequently in football players.

Electromyography The recording, study, and interpretation of graphic records of the electrical activity of contracting muscles. Electrodes inserted into the muscles detect the electrical impulses that cross muscle cell membranes and stimulate contraction; related equipment provides a visual display of the impulses. This is the best and most important technique for determining the functions of the muscles and muscle groups of the body.

Hallux valgus (hal'uks val'gus; *valgus* = bent away from body midline) Permanent displacement of the great toe toward the other toes, often caused by tight shoes. In this condition, the sesamoid bones that underlie the head of the first metatarsal have been pushed laterally, away from the great toe. Thus, the abductor hallucis muscle of the sole, whose insertion tendon contains one of the sesamoids, cannot abduct the displaced toe back to its proper position.

Mallet finger Extreme, persistent flexion of the distal phalanx of a finger, caused by forceful bending, as when a baseball hits the tip of a finger. Part of the bony insertion of the extensor digitorum muscles is stripped off, so the finger flexors flex the joint unopposed.

Rupture of calcaneal tendon The calcaneal (Achilles) tendon is the strongest tendon in the body, but its rupture is surprisingly common, particularly in older people as a result of stumbling, and in sprinters who tense the tendon too severely while starting a race. The rupture leads to a gap just superior to the heel, and the calf bulges as the triceps surae muscles are released from their insertion. As a result, plantar flexion is very weak, but dorsiflexion is exaggerated. The tendon must be repaired surgically.

Shin splints Common term for pain in the anterior leg caused by swelling of the tibialis anterior muscle. This muscle swells after heavy exercise in the absence of adequate prior conditioning. Because it is tightly wrapped by fascia, the tibialis anterior cuts off its own circulation as it swells and presses painfully on its own nerves.

CHAPTER SUMMARY

Take full advantage of the invaluable online practice and study tools for this chapter by going to the Study Area of MasteringA&P®. There you will find:

- Art Labeling Activities, Quizzes, Practice Tests, Memory Games
- Animal Organ Dissection, Cat Dissection and Bone Videos
- PAL Cadaver
- PAL Histology
- A&P Flix Animations
- Dynamic Study Modules

ARRANGEMENT OF FASCICLES IN MUSCLES

(p. 263)

1. The arrangement of fascicles in skeletal muscle can be circular, convergent, parallel, or pennate. The long muscle fibers found in convergent- and parallel-arranged fascicles produce a large range of movement; the shorter, yet more densely packed muscle fibers in pennate-arranged fascicles produce more power.

LEVER SYSTEMS: BONE-MUSCLE RELATIONSHIPS (pp. 264–266)

2. A lever is a rigid bar that moves on a fulcrum. When an effort is applied to the lever, a load is moved. In the body, bones are the levers, joints are the fulcrums, and the effort is exerted by skeletal muscles pulling on their insertions.

3. When the effort is farther from the fulcrum than is the load, the lever works at a mechanical advantage, a power lever. When the effort is closer to the fulcrum than is the load, the lever works at a mechanical disadvantage, a speed lever.

4. First-class levers (fulcrum located between load and effort) may operate at a mechanical advantage or disadvantage. Second-class levers (load between fulcrum and effort) all work at a mechanical advantage. Third-class levers (effort between fulcrum and load) all work at a mechanical disadvantage. Most muscles of the body are in third-class lever systems and therefore provide speed of movement.

ORGANIZATIONAL SCHEME BASED ON EMBRYONIC DEVELOPMENT (pp. 266–268)

5. Based on embryonic origin and general functions, the muscles of the body can be divided into four main groups: muscle of the visceral organs, pharyngeal arch muscles, axial muscles, and limb muscles.

6. *Pharyngeal arch muscles* are skeletal muscles that develop from somitomeres and surround the pharynx; some migrate into the head and back during development. The *axial muscles* of the neck and trunk derive from myotomes. They include dorsal muscles that extend the spine and ventral muscles that flex the spine. Of the axial muscles in the head, the eye muscles and the tongue muscles develop from somitomeres and occipital myotomes, respectively. *Limb muscles* derive from the ventral-lateral portions of myotomes and exhibit the same arrangement in the upper and lower limbs: extensors on one side, flexors on the other.

MUSCLE ACTIONS AND INTERACTIONS (pp. 268–270)

7. Skeletal muscles are arranged in opposing groups across movable joints such that one group can reverse or modify the action of the other.

8. Prime movers (agonists) bear the main responsibility for a particular movement. Antagonists produce the opposite movement. Synergists help the prime movers, prevent undesirable movements, or stabilize joints (as fixators).

9. Muscles that cross on the anterior side of a joint generally produce flexion. Those that cross the joint posteriorly generally are extensors. (Exceptions, are at the knee and ankle.) Adductors cross on the medial side of a joint, and abductors cross on the lateral side.

NAMING THE SKELETAL MUSCLES (p. 270)

10. Muscles are named for their location, shape, relative size, fascicle direction, location of attachments, number of origins, and action.

MAJOR SKELETAL MUSCLES OF THE BODY (pp. 270–332)

Muscle Compartments of the Limbs (pp. 271–277)

11. The muscles of the limbs are organized into compartments: anterior (flexor) and posterior (extensor) compartments in the arm and the forearm; posterior, anterior, and medial compartments in the thigh; and posterior, anterior, and lateral compartments in the leg. Muscles located in the same compartment generally have similar function and innervation. Summary tables of the muscles of the upper and lower limbs (Tables 11.1 and 11.2) provide an excellent review of muscle actions, interactions, and compartments.

The Muscle Tables (pp. 278–332)

12. The muscles of facial expression (Table 11.3) are thin and insert primarily on the facial skin. They close the eyes and lips, compress the cheeks, and allow people to smile and show feelings.

13. Chewing muscles include the masseter and temporalis that elevate the mandible and the two pterygoids that assist in grinding food. Extrinsic muscles of the tongue anchor the tongue and control its movements (Table 11.4).

14. Deep muscles in the anterior neck aid swallowing by elevating and then depressing the larynx (suprahyoid and infrahyoid groups). Pharyngeal constrictors squeeze food into the esophagus (Table 11.5).

15. Flexion and rotation of the head and neck are brought about by the sternocleidomastoid and scalenes in the anterior and lateral neck. Extension of the spine and head is brought about by deep muscles of the back and the posterior neck (Table 11.6).

PAL **Human Cadaver/Muscular System/Head and Neck**

16. Movements of breathing are promoted by the diaphragm and by the intercostal muscles of the thorax (Table 11.7). Contraction of the diaphragm increases abdominal pressure during straining, an action that is aided by the muscles of the abdominal wall.

17. The sheetlike muscles of the anterior and lateral abdominal wall (Table 11.8) are layered like plywood. This layering forms a strong wall that protects, supports, and compresses the abdominal contents. These muscles can also flex the trunk.

18. Muscles of the pelvic floor and perineum (Table 11.9) support the pelvic organs, inhibit urination and defecation, and aid erection.

19. The superficial muscles of the thorax attach to the pectoral girdle and fix or move the scapula during arm movements (Table 11.10).

PAL **Human Cadaver/Muscular System/Trunk**

20. The muscles that move the arm at the shoulder joint originate from the shoulder girdle (Table 11.11) or from the axial skeleton. Four muscles form the musculotendinous rotator cuff, which stabilizes the shoulder joint. As a rule, muscles located on the anterior thorax flex the arm (pectoralis major), whereas the posterior muscles extend it (latissimus dorsi). Abduction is mostly brought about by the deltoid. Adduction and rotation are brought about by many muscles on both the anterior and posterior sides.

21. Most muscles in the arm move the forearm (Table 11.12). Anterior arm muscles are flexors of the forearm (biceps, brachialis), and posterior arm muscles are extensors of the forearm (triceps).

22. Forearm muscles primarily move the hand (Table 11.13). Anterior forearm muscles flex the hand and fingers (but include two pronators), whereas the posterior muscles are mostly extensors (but include the strong supinator).

23. The intrinsic muscles of the hand allow for fine movements of the fingers (Table 11.14) and opposition, which aid in gripping objects in the palms. These small muscles are divided into thenar, hypothenar, and midpalmar groups.

PAL **Human Cadaver/Muscular System/Upper Limb**

24. Muscles that bridge the hip and knee joints move the thigh and the leg (Table 11.15). Anterior muscles mostly flex the thigh and extend the leg (quadriceps femoris), medial muscles adduct the thigh (adductors), and posterior muscles (gluteals, hamstrings) primarily abduct and extend the thigh and flex the leg.

25. Muscles in the leg act on the ankle and toes (Table 11.16). Leg muscles in the anterior compartment are mostly dorsiflexors of the foot (extensor group). Muscles in the lateral compartment are everters and plantar flexors of the foot (fibular group), and posterior leg muscles are strong plantar flexors (gastrocnemius and soleus, flexor group).

26. The intrinsic muscles of the foot (Table 11.17) support the foot arches and help move the toes. Most occur in the sole, arranged in four layers. They resemble the small muscles in the palm of the hand.

PAL **Human Cadaver/Muscular System/Lower Limb and Muscular System/All Regions**

REGIONAL SURFACE ANATOMY (pp. 332–344)

27. Surface anatomy, the study of the surface landmarks and their relationship to internal organs, is an excellent tool for reviewing and reinforcing the skeletal and muscular anatomy you are learning. Identifying surface landmarks via palpation is an important clinical skill for assessing structure and function.

The Head (pp. 333–334)

28. In the cranium, the bony and muscular structures that can be felt through the skin include the superciliary arches, external occipital protuberance, superior nuchal line, mastoid process, temporalis muscles, and epicranius muscle.

29. The scalp covers the superior surface of the cranium and consists of skin, a highly vascular subcutaneous layer, and the epicranial aponeurosis.

30. The face is divided into orbital, nasal, auricular, and oral regions. The bony parts of the nose are the root and the bridge. The two principal parts of the auricle are its helix and lobule.

31. Palpable features of the face include the zygomatic arch, the mandibular body, and the mandibular ramus, masseter muscle, and temporomandibular joint in the oral region.

The Neck (pp. 334–335)

32. By running a finger inferiorly along the anterior midline of the neck, you can palpate the hyoid bone, laryngeal prominence, and cricoid cartilage.

33. The sternocleidomastoid muscle is an important surface landmark of the neck. The large trapezius muscle is located on the posterior aspect of the neck.

34. The posterior triangle of the neck is defined by the sternocleidomastoid anteriorly, the trapezius posteriorly, and the clavicle inferiorly. The anterior triangle is defined by the inferior margin of the mandible superiorly, the midline of the neck anteriorly, and the sternocleidomastoid posteriorly.

The Thorax (p. 335)

35. The jugular notch is on the superior surface of the manubrium; the sternal angle lies at the junction of the manubrium and the body of the sternum. By finding the second rib at the sternal angle, you can count inferiorly to identify the other ribs. The midaxillary and midclavicular lines are imaginary vertical lines that extend inferiorly from the center of the axilla and clavicle, respectively. The apex of the inverted V-shaped costal margin (infrasternal angle) is at the xiphisternal joint.

The Abdomen (pp. 335–336)

36. Surface landmarks on the anterior abdominal wall include the iliac crest, anterior superior iliac spine, groove for the inguinal ligament, pubic tubercle, pubic crest, pubic symphysis, umbilicus, groove for linea alba, rectus abdominis muscle, linea semilunaris, and external oblique muscle.

The Back (pp. 336–337)

37. In the posterior median furrow of the back, you can count and identify the vertebral spinous processes. The most easily located vertebral spines are C_7 and T_1 (the most prominent spines). Also palpable are T_3 (at the level of the medial end of the spine of the scapula), T_7 (at the level of the inferior scapular angle), and L_4 (at the level of the supracristal line).

38. Superficial muscles of the back include the trapezius and latissimus dorsi, which, along with the medial border of the scapula, define the triangle of auscultation. The deeper erector spinae muscles flank the vertebral column.

The Upper Limb and Shoulder (pp. 337–340)

39. Important superficial structures associated with the axilla include the base of the axilla and the anterior and posterior axillary folds.

40. Skeletal landmarks that can be palpated in the shoulder region include the acromion of the scapula, clavicle, acromioclavicular joint, and the greater tubercle of the humerus.

41. The principal muscle of the shoulder is the deltoid muscle.

42. On the arm, you can palpate the humerus and the biceps and triceps brachii muscles. The medial bicipital groove is located on the medial side of the arm.

43. Bony landmarks in the elbow region include the two epicondyles of the humerus and the olecranon of the ulna, all lying in the same horizontal plane.

44. The cubital fossa is the triangular depression on the anterior surface of the elbow region bounded laterally by the brachioradialis and pronator teres muscles.

45. In the forearm, the palpable bony structures include the entire posterior margin of the ulna and the distal half and head of the radius. At the wrist, the styloid process of the radius lies distal to the styloid process of the ulna.

46. One can palpate the flexor and extensor muscle groups on the anterior and posterior forearm, respectively.

47. On the anterior forearm near the wrist, the tendons of the flexor carpi radialis, palmaris longus (absent in some people), and flexor carpi ulnaris can be identified.

48. The tendons of the thumb extensor muscles define the anatomical snuff box on the posterior base of the thumb. The styloid process of the radius and the scaphoid bone are located here.

The Lower Limb and Gluteal Region (pp. 340–344)

49. The posterior point of the iliac crest is the posterior superior iliac spine, which is represented by a dimple in the skin and indicates the location of the sacroiliac joint. Other bony landmarks in the gluteal region are the ischial tuberosities and the greater trochanter of the femur.

50. The right and left gluteal prominences, separated by the natal cleft, consist of gluteus maximus muscles and subcutaneous fatty tissue. The gluteal fold is the inferior margin of the gluteus maximus. The ischial tuberosities support body weight during sitting.

51. The main groups of thigh muscles are the quadriceps femoris muscles, the adductor muscles, and the hamstrings. Bony landmarks that can be palpated in the thigh include the patella and the condyles of the femur.

52. The femoral triangle on the anterior superior thigh is formed by the inguinal ligament superiorly and by the sartorius and adductor longus muscles laterally.

53. The popliteal fossa is a diamond-shaped hollow on the posterior aspect of the knee, defined by four muscular boundaries: biceps femoris, semitendinosus, semimembranosus, and gastrocnemius.

54. Bony landmarks that can be palpated in the leg include the tibial tuberosity and tibial condyles, the anterior border and medial surface of the tibia, the head of the fibula, and the lateral and medial malleolus.

55. The main muscles of the leg, which can be palpated as they contract, are the gastrocnemius and soleus on the calf, the muscles in the anterior compartment, and the fibularis muscles on the fibula.

56. On the dorsum of the foot, one can observe the dorsal venous arch and the tendons of the extensor digitorum longus and extensor hallucis longus muscles.

REVIEW QUESTIONS

Multiple Choice/Matching Questions

(*For answers, see Appendix B.*)

1. A muscle whose fascicles are arranged at an angle to a central longitudinal tendon has this arrangement: (a) circular, (b) longitudinal, (c) parallel, (d) pennate, (e) convergent.

2. A muscle that helps an agonist by causing a similar movement or by stabilizing a joint over which the agonist acts is (a) an antagonist, (b) a prime mover, (c) a synergist, (d) a fulcrum.

3. Match the muscles in column B to their embryonic origins in column A.

Column A	Column B
____ (1) from head (occipital) myotomes	(a) biceps brachii
____ (2) from trunk (nonlimb) myotomes	(b) erector spinae
____ (3) limb extensor group	(c) tongue muscles
____ (4) limb flexor group	(d) chewing muscles
____ (5) pharyngeal arch muscles	(e) triceps brachii

4. The muscle that closes the eyes is (a) the occipitalis, (b) the zygomaticus, (c) the corrugator supercilii, (d) the orbicularis oris, (e) none of these.

5. The arm muscle that flexes the forearm at the elbow and supinates it is the (a) brachialis, (b) brachioradialis, (c) biceps brachii, (d) triceps brachii.

6. The chewing muscles that protract the mandible and grind the teeth from side to side are (a) the buccinators, (b) the masseters, (c) the temporalis, (d) the pterygoids, (e) none of these.

7. Muscles that depress the hyoid bone and larynx include all of the following *except* the (a) sternohyoid, (b) omohyoid, (c) geniohyoid, (d) sternothyroid.

8. The quadriceps femoris muscles include all of the following except the (a) vastus lateralis, (b) biceps femoris, (c) vastus intermedius, (d) rectus femoris.

9. A prime mover of thigh flexion at the hip is the (a) rectus femoris, (b) iliopsoas, (c) vastus intermedius, (d) gluteus maximus.

10. Someone who sticks out a thumb to hitch a ride is ____ the thumb. (a) extending, (b) abducting, (c) adducting, (d) opposing.

11. The major muscles used in the up portion of a chin-up are the (a) triceps brachii and pectoralis major, (b) infraspinatus and biceps brachii, (c) serratus anterior and external oblique, (d) latissimus dorsi and brachialis.

12. The anatomical snuff box (a) is in the nose, (b) contains the styloid process of the radius, (c) is defined by tendons of the flexor carpi radialis and palmaris longus, (d) is on the ventral side of the hand.

13. A muscle that contributes to the posterior axillary fold is the (a) pectoralis major, (b) latissimus dorsi, (c) trapezius, (d) infraspinatus, (e) pectoralis minor.

Short Answer Essay Questions

14. Define and distinguish first-, second-, and third-class levers. Which types of levers operate at a mechanical advantage?

15. (a) Name the four pairs of muscles that act together to compress the abdominal contents. (b) How do their fiber (fascicle) directions contribute to the strength of the abdominal wall? (c) Of these four muscles, which is the strongest flexor of the vertebral column? (d) Name the prime mover of inhalation (inspiration) in breathing.

16. List all six possible movements of the humerus that can occur at the shoulder joint, and name the prime mover(s) of each movement.

17. (a) Name two forearm muscles that are both powerful extensors and abductors of the hand at the wrist. (b) Name the only forearm muscle that can flex the distal interphalangeal joints. (c) Name the hand muscles responsible for opposition.

18. (a) Name the lateral rotators of the hip. (b) Explain the main function of each of the three gluteal muscles.

19. What is the functional reason why the muscle group on the dorsal leg (calf) is so much larger than the muscle group in the ventral region of the leg?

20. Define (a) a fascial compartment in a limb and (b) a retinaculum in a limb.

21. Name two muscles in each of the following compartments or regions: (a) thenar eminence (ball of thumb), (b) posterior compartment of forearm, (c) anterior compartment of forearm—deep muscle group, (d) anterior muscle group in the arm, (e) muscles of mastication, (f) third muscle layer of the foot, (g) posterior compartment of leg, (h) medial compartment of thigh, (i) posterior compartment of thigh.

22. (a) What muscles develop from the embryonic structures called somitomeres? (b) What are the major groups of pharyngeal arch (branchiomeric) muscles? (c) What three constrictor muscles squeeze swallowed food through the pharynx to the esophagus?

23. Describe the location and function of the levator ani muscle.

24. Deduce some characteristics of the following muscles based on their names: tibialis anterior; temporalis; erector spinae.

25. For the following muscles, list one action of each, one muscle that acts as an antagonist, and one muscle that functions as a synergist: pectoralis major; biceps femoris; flexor carpi ulnaris; adductor magnus; rectus abdominis.

26. (a) Define palpation. (b) Why is a knowledge of surface anatomy valuable in clinical situations and for evaluating athletic injuries?

27. Explain how one locates the proper site for intramuscular injections into (a) the ventral gluteal site and (b) the deltoid muscle.

28. Esperanza, a pre–physical therapy student, was trying to locate the vertebral spinous processes on the flexed back of her friend Chao, but she kept losing count. Chao told her to check her count against several reliable "guideposts" along the way: the spinous processes of C_7, T_3, T_7, and L_4. Describe how to find each of these four vertebrae without having to count any vertebrae.

29. Name the borders of the triangle of auscultation.

30. Tell where to locate the following thigh structures: (a) greater trochanter of the femur, (b) adductor muscles, (c) vastus lateralis, (d) bony origin of the hamstring muscles.

Critical Reasoning & Clinical Application Questions

1. Rosario, a student nurse, was giving Mr. Graves a back rub. What two broad superficial muscles of his back were receiving most of her attention?

2. When Mrs. O'Brian returned to her doctor for a follow-up visit after childbirth, she complained that she was having problems controlling urine flow (was incontinent) when she sneezed. The physician gave Mrs. O'Brian instructions in performing exercises to strengthen the muscles of the pelvic floor. What are these specific muscles?

3. Assume you are trying to lift a heavy weight off the ground with your right hand. Explain why it will be easier to flex your forearm at the elbow when your forearm is supinated than when it is pronated.

4. Chao-Jung abducted her thigh very strongly while riding a horse and pulled some groin muscles in her medial upper thigh. What muscles did she probably strain, and what is the medical name for her condition? (Hint: See this chapter's Related Clinical Terms.)

5. When bodybuilders develop the muscles on one side of a limb more than on the other side, they are said to be "muscle bound." Describe the symptoms of this condition, and explain how they are produced.

6. Based on its fascicle orientation, determine whether each of the following muscles is built for large range of movement or for power: sartorius; rectus femoris; latissimus dorsi; subscapularis; gluteus maximus.

7. A baseball pitcher injured a rotator cuff muscle. When he abducts and laterally rotates his shoulder, it dislocates anteriorly. What muscle is injured and thus making the joint unstable?

8. What class of lever is described by the following activities: (1) the soleus plantar flexing the foot; (2) the deltoid abducting the arm; (3) the triceps brachii extending the forearm during push-ups?

9. Walking to her car after her 65th birthday party, Mrs. Schultz tripped on ice and fell forward on her outstretched palms. When she arrived at the emergency room, her right wrist and hand were bent like a fork. Dr. Jefferson felt that the styloid process of her radius was outside of the anatomical snuff box and slightly proximal to the styloid process of the ulna. When he checked her elbow, he found that the olecranon lay 2 cm proximal to the two epicondyles of the right humerus. Explain all these observations, and describe what had happened to Mrs. Schultz's limb.

10. An athletic trainer was examining a college basketball player to find the site of a pulled muscle. The trainer asked the athlete to extend her thigh forcefully at the hip, but she found this action too painful to perform. Then the trainer palpated her posterior thigh and felt some swelling of the muscles there. In the simplest terms, which basic muscle group was injured?

Fundamentals of the Nervous System and Nervous Tissue

12

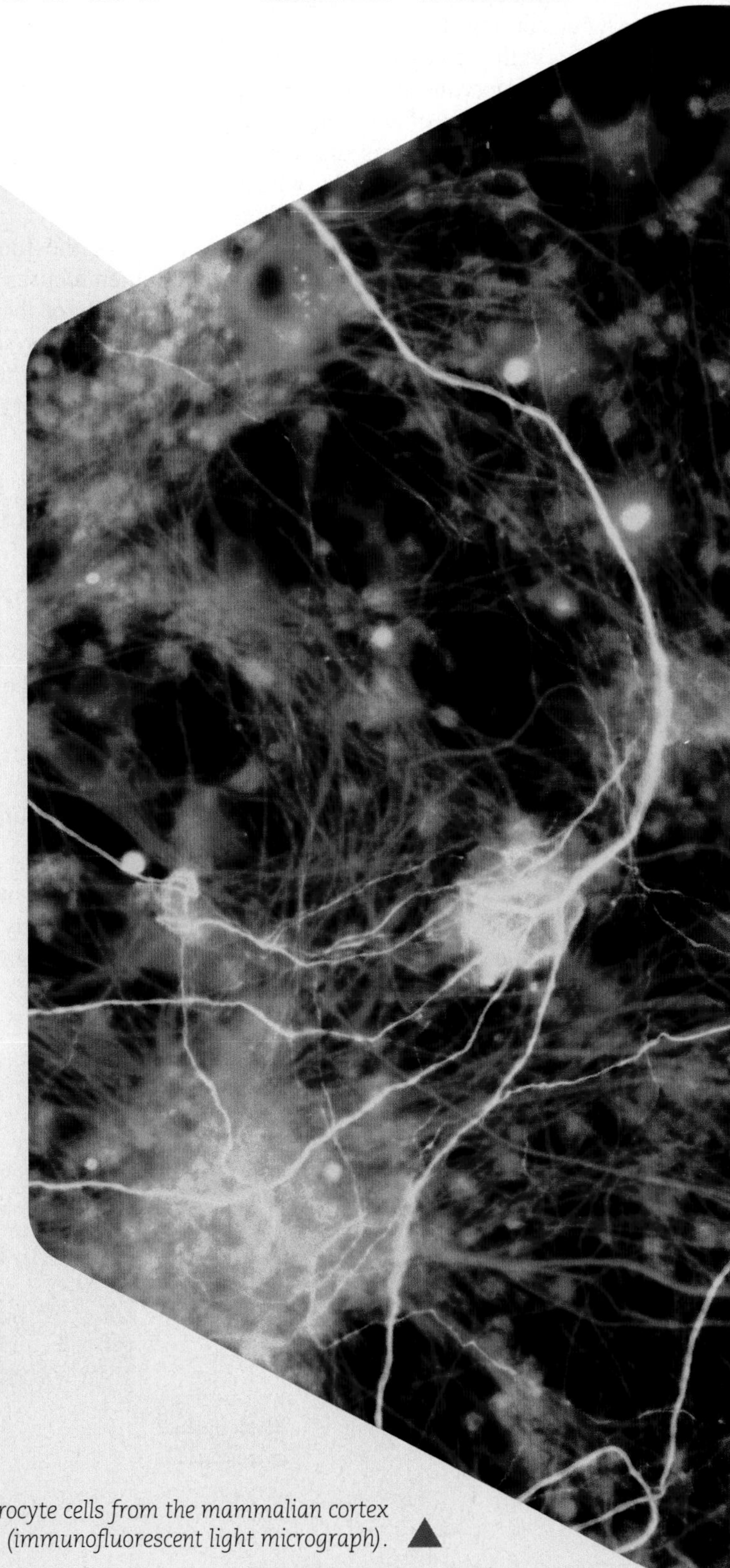

Branching nerve fiber and astrocyte cells from the mammalian cortex (immunofluorescent light micrograph). ▲

As you are driving down the freeway, a horn blares on your right, and immediately you swerve to your left. Charlie's note on the kitchen table reads, "See you later—I will have the stuff ready at six"; you know that the "stuff" is chili with taco chips, and your mouth waters in anticipation. While you are dozing, your infant son cries out softly, and instantly you awaken. What do these events have in common? All are everyday examples of the function of your nervous system, which has the cells in your body humming with activity nearly all the time.

The nervous system is the master control and communications system of the body. Every thought, action, instinct, and emotion reflects its activity. Its cells communicate through electrical signals, which are rapid and specific and usually produce almost immediate responses.

This chapter begins with an overview of the functions and the divisions of the nervous system. It then focuses on the functional anatomy of nervous tissue, especially of the nerve cells, or *neurons,* which are the key to the efficient system of neural communication. Finally, the chapter addresses how the arrangement of neurons determines the structural organization of the nervous system.

THE FUNCTIONAL ORGANIZATION OF THE NERVOUS SYSTEM

learning outcomes

- List the main functions of the nervous system.
- Explain the structural and functional divisions of the nervous system.

Functions of the Nervous System

The nervous system has three overlapping functions (Figure 12.1):

1. It uses its millions of *sensory receptors* to monitor changes occurring both inside and outside the body. Each of these changes is called a *stimulus,* and the gathered information is called **sensory input.**
2. It processes and interprets the sensory input and makes decisions about what should be done at each moment—a process called **integration.**
3. It dictates a response by activating the *effector organs,* our muscles or glands; the response is called **motor output.**

Some examples illustrate how these functions work together. When you are driving and hear a horn to your right, your nervous system integrates this information (a horn signifies danger), and your arm muscles contract to turn the wheel to the left (motor output). As another example, when you taste food your nervous system integrates this sensory information and signals your salivary glands to secrete more saliva into your mouth.

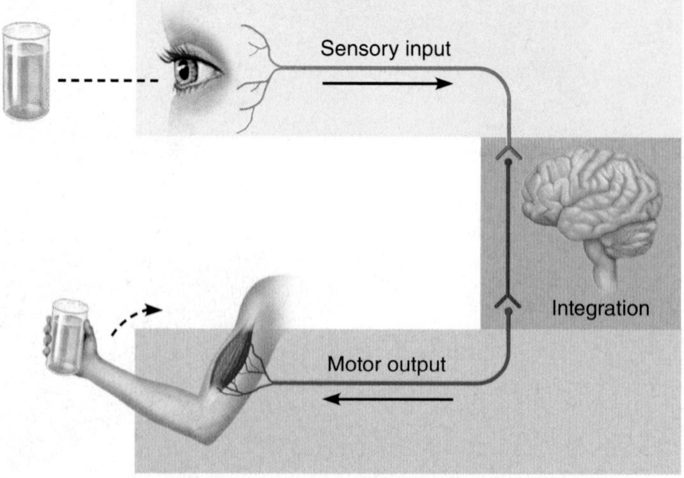

Figure 12.1 Functions of the nervous system.

Basic Divisions of the Nervous System

Humans have only one, highly integrated nervous system. However, for the sake of convenience you can think of it as having two anatomical parts: the central nervous system and the peripheral nervous system (Figure 12.2). The **central nervous system (CNS)** consists of the *brain* and the *spinal cord,* which occupy the cranium and the vertebral canal, respectively. The CNS is the integrating and command center of the nervous system: It receives incoming sensory signals, interprets these signals, and dictates motor responses based on past experiences, reflexes, and current conditions.

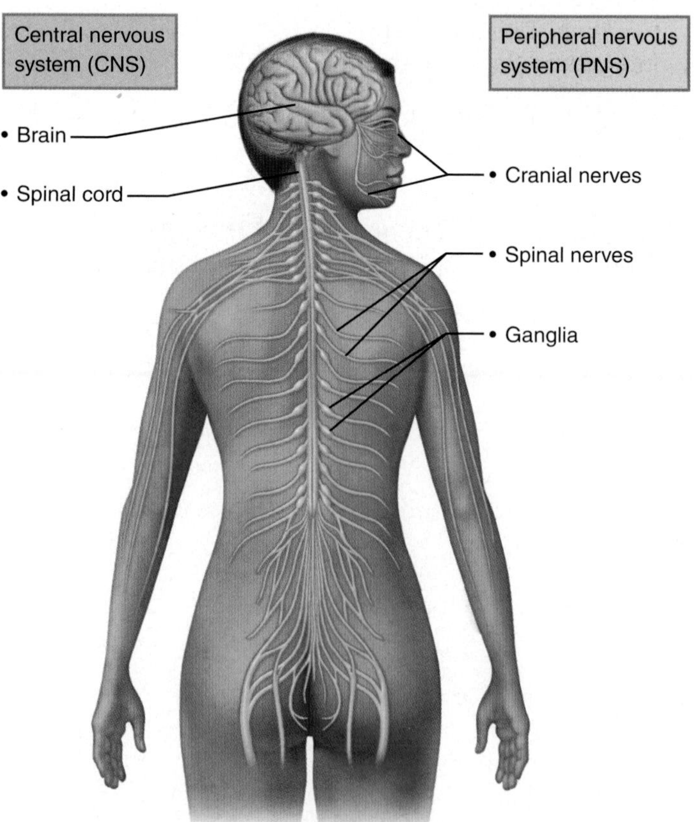

Figure 12.2 Divisions of the nervous system: the central nervous system (CNS) and the peripheral nervous system (PNS).

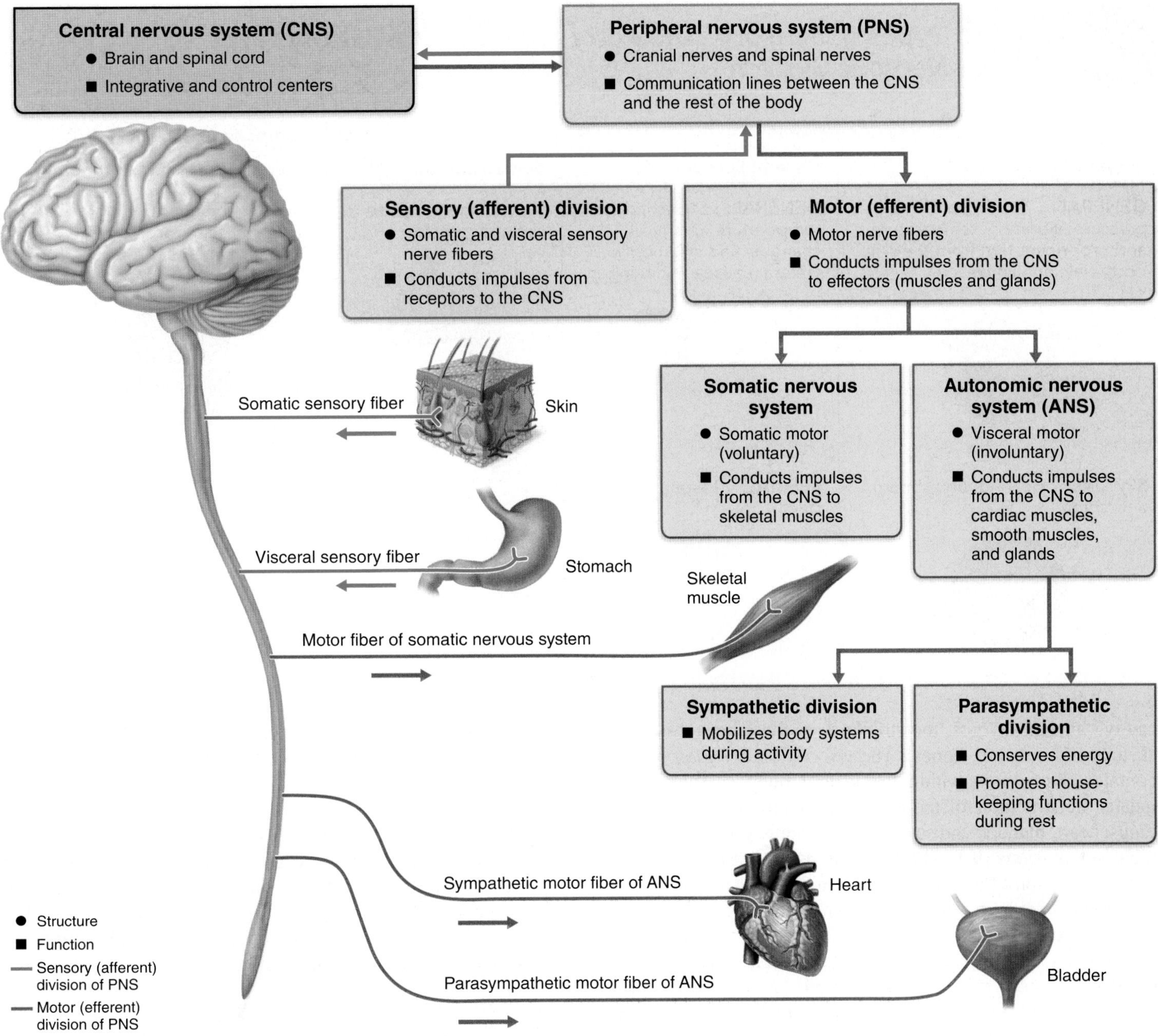

Figure 12.3 Schematic of levels of organization in the nervous system. Visceral organs (located primarily in the ventral body cavity) are innervated by visceral sensory fibers and by motor fibers of the autonomic nervous system. The somatic body regions (limbs and body wall) are innervated by motor fibers of the somatic nervous system and by somatic sensory fibers. Arrows indicate the direction of nerve impulses. (Connections to the spinal cord are not anatomically accurate.)

The **peripheral nervous system (PNS),** the part of the nervous system outside the CNS, consists mainly of the *nerves* that extend from the brain and spinal cord. *Cranial nerves* carry signals to and from the brain, whereas *spinal nerves* carry signals to and from the spinal cord. These peripheral nerves serve as communication lines that link all regions of the body to the central nervous system. Also included in the PNS are **ganglia,** areas where the cell bodies of neurons are clustered.

As mentioned, the nervous system receives sensory inputs and dictates motor outputs (Figure 12.3). In the **sensory,** or **afferent** (af′er-ent), **division,** signals are picked up by sensory receptors located throughout the body and carried by nerve fibers of the PNS into the CNS (*afferent* means "carrying toward"). In the **motor,** or **efferent** (ef′er-ent), **division,** signals are carried away from the CNS by nerve fibers of the PNS to innervate the muscles and glands, causing these organs either to contract or to secrete (*efferent* means "carrying away"). Both the sensory inputs and the motor outputs are further divided according to the body regions they serve: The *somatic body region* consists of the structures external to the ventral body

Table 12.1 Types of Sensory and Motor Fibers of the Peripheral Nervous System

Sensory Components		Motor Components	
Somatic Sensory (SS)	**Visceral Sensory (VS)**	**Somatic Motor (SM)**	**Visceral Motor (VM; Autonomic)**
GENERAL: Touch, pain, pressure, vibration, temperature, and proprioception from the skin, body wall, and limbs	**GENERAL:** Stretch, pain, temperature, chemical changes, and irritation in viscera; nausea and hunger	Motor innervation to skeletal muscles	Motor innervation to smooth muscle, cardiac muscle, and glands
SPECIAL: Hearing, equilibrium, and vision	**SPECIAL:** Taste and smell		

cavity—in other words, the structures of the outer tube (skin, skeletal musculature, bones). The *visceral body region* mostly contains the viscera within the ventral body cavity—which means the structures of the body's inner tube (digestive tube, lungs, heart, bladder, and so on). This scheme results in the four main subdivisions of the PNS: (1) **somatic sensory** (the sensory innervation of the outer tube: skin, body wall, and limbs); (2) **visceral sensory** (the sensory innervation of the viscera); (3) **somatic motor,** or voluntary motor (the motor innervation of the outer tube, specifically skeletal muscles); and (4) **visceral motor,** also called the **autonomic** (aw-to-nom′ik) **nervous system (ANS)** (the involuntary motor innervation of the inner tube, specifically smooth muscle, cardiac muscle, and glands, as well as some outer tube structures: arrector pili muscles, smooth muscle in the vessels, and sweat glands). Because these subdivisions are essential to understanding the nervous system, we will examine each in greater detail (Table 12.1).

Somatic Sensory (SS)

The **general somatic senses** are the senses whose receptors are spread widely throughout the outer tube of the body (in this context, the term *general* means "widespread"). These include the many senses experienced on the skin and in the body wall, such as touch, pain, pressure, vibration, and temperature. The gritty texture of sandpaper, the heat from a mug of hot chocolate, and the pain from a bruise or a pulled muscle are all examples of general somatic sensation.

Another type of general somatic sensation is **proprioception** (pro″pre-o-sep′shun; "sensing one's own body"), a sense that detects the amount of stretch in muscles, tendons, and joint capsules. Proprioception informs you of the position and movement of your body in space, giving you a "body sense." To demonstrate proprioception, flex and extend your fingers without looking at them—you will be able to feel exactly which joints are moving.

The **special somatic senses** are the somatic senses whose receptors are confined to relatively small areas rather than spread widely throughout the body (in this context, the term *special* means "localized"). Most special senses are confined to the head, including hearing and balance, or **equilibrium** (with receptors in the inner ear), and vision (with receptors in the eye).

Visceral Sensory (VS)

The **general visceral senses** include stretch, pain, and temperature, which can be felt widely in the digestive and urinary tracts, reproductive organs, and other viscera. Sensations such as hunger and nausea are also general visceral sensations.

Taste and smell are considered **special visceral senses.** These senses, referred to also as the chemical senses, have their sensory receptors localized to the tongue and nasal cavity, respectively.

Somatic Motor (SM)

The **somatic motor** part of the PNS stimulates contraction of the skeletal muscles in the body. Because we have voluntary control over the contraction of our skeletal muscles, the somatic motor system is often called the *voluntary nervous system.*

Visceral Motor (VM)

The **visceral motor** part of the PNS regulates the contraction of smooth and cardiac muscle and secretion by the body's many glands. General visceral motor neurons make up the *autonomic nervous system (ANS),* which controls the function of the visceral organs (see Figure 12.3). Because we generally have no voluntary control over such activities as the pumping of the heart and movement of food through the digestive tract, the ANS is also called the *involuntary nervous system.* The autonomic nervous system is divided into the sympathetic division, which gets the body ready for activity ("fight or flight"); and the parapsympathetic division, which conserves energy and promotes digestion ("rest and digest").

You will find it useful to return to this basic discussion of the organization of the nervous system as you work through the details of structure and function of this body system.

✓ check your understanding

- ☐ 1. In which direction are afferent signals carried? What type of information do these signals contain?
- ☐ 2. What subdivision of the nervous system regulates contraction of the muscle tissue in the heart?
- ☐ 3. What type of sensation is (a) pain from a pulled muscle, (b) nausea, (c) taste?

(For answers, see Appendix B.)

NERVOUS TISSUE

The nervous system consists mostly of **nervous tissue,** whose cells are densely packed and tightly intertwined. Although exceedingly complex, nervous tissue is made up of just two main types of cells: (1) *neurons,* the excitable nerve cells that transmit electrical signals, and (2) *neuroglia,* nonexcitable supporting cells that surround and wrap the neurons. Both of these cell types develop from the same embryonic tissues: neural tube and neural crest (see p. 52).

The Neuron

learning outcomes

- ▶ Define neuron, describe its structural components, and relate each structure to its functional role.
- ▶ Describe the structure of a synapse.
- ▶ Classify neurons both structurally and functionally.

The human body contains many billions of **neurons,** or **nerve cells** (Figure 12.4), which are the basic structural units of the nervous system. Neurons have a number of special functional characteristics.

- Neurons are highly specialized cells that *conduct electrical signals* from one part of the body to another. These signals are transmitted along the plasma membrane, or **neurilemma,** in the form of **nerve impulses,** or **action potentials.** Basically, an impulse is a reversal of electrical charge that travels rapidly along the neuronal membrane.
- Neurons have *extreme longevity.* They can live and function for a lifetime, over 100 years.
- Neurons *do not divide.* As the fetal neurons assume their roles as communication links in the nervous system, they lose their ability to undergo mitosis. There can be a high price for this characteristic of neurons, for they cannot replace themselves if destroyed. There are some exceptions to this rule; neural stem cells have been identified in certain areas of the CNS.
- Neurons have an exceptionally *high metabolic rate,* requiring continuous and abundant supplies of oxygen and glucose. Neurons cannot survive for more than a few minutes without oxygen.

Neurons typically are large, complex cells. Although they vary in structure, they all have a *cell body* from which one or more *processes* project (Figure 12.4).

The Cell Body

The **cell body** is also called a *soma* (= body). Although the cell bodies of different neurons vary widely in size (from 5 to 140 μm in diameter), all consist of a single nucleus surrounded by cytoplasm. In all but the smallest neurons, the nucleus is spherical and clear and contains a dark nucleolus near its center (Figure 12.4a and b).

The cytoplasm contains all the usual cellular organelles as well as distinctive **chromatophilic substance** *(Nissl bodies).* Chromatophilic (kro-mah′to-fil-ic; "color-loving") substance is large clusters of rough endoplasmic reticulum and free ribosomes that stain darkly with basic dyes. These cellular organelles continually renew the membranes of the cell and the protein components of the cytosol. **Neurofibrils** are bundles of intermediate filaments *(neurofilaments)* that run in a network between the chromatophilic substance. Like all other intermediate filaments (see p. 33), neurofilaments keep the cell from being pulled apart when it is subjected to tensile forces.

The cell body is the focal point for the outgrowth of the neuron processes during embryonic development. In most neurons, the plasma membrane of the cell body acts as a receptive surface that receives signals from other neurons.

Most neuron cell bodies are located within the CNS. However, clusters of cell bodies called **ganglia** (singular: **ganglion,** gang′le-on; "knot in a string"), lie along the nerves in the PNS (see Figure 12.2).

Neuron Processes

Armlike **processes** extend from the cell bodies of all neurons. These processes are of two types, *dendrites* and *axons* (Figure 12.4), which differ from each other both in structure and in functional properties. The cell processes of neurons are described here using a motor neuron as an example of a typical neuron. Motor neurons do indeed resemble most neurons in the arrangement of their processes, but sensory neurons and some tiny neurons differ from the "typical" pattern presented here.

Dendrites Most neurons have numerous **dendrites,** processes that branch from the cell body like the limbs on a tree (*dendro* = tree). Virtually all organelles that occur in the cell body also occur in dendrites, and chromatophilic substance extends into the basal part of each dendrite. Dendrites function as *receptive sites,* providing an enlarged surface area for receiving signals from other neurons. By definition, dendrites conduct electrical signals *toward* the cell body.

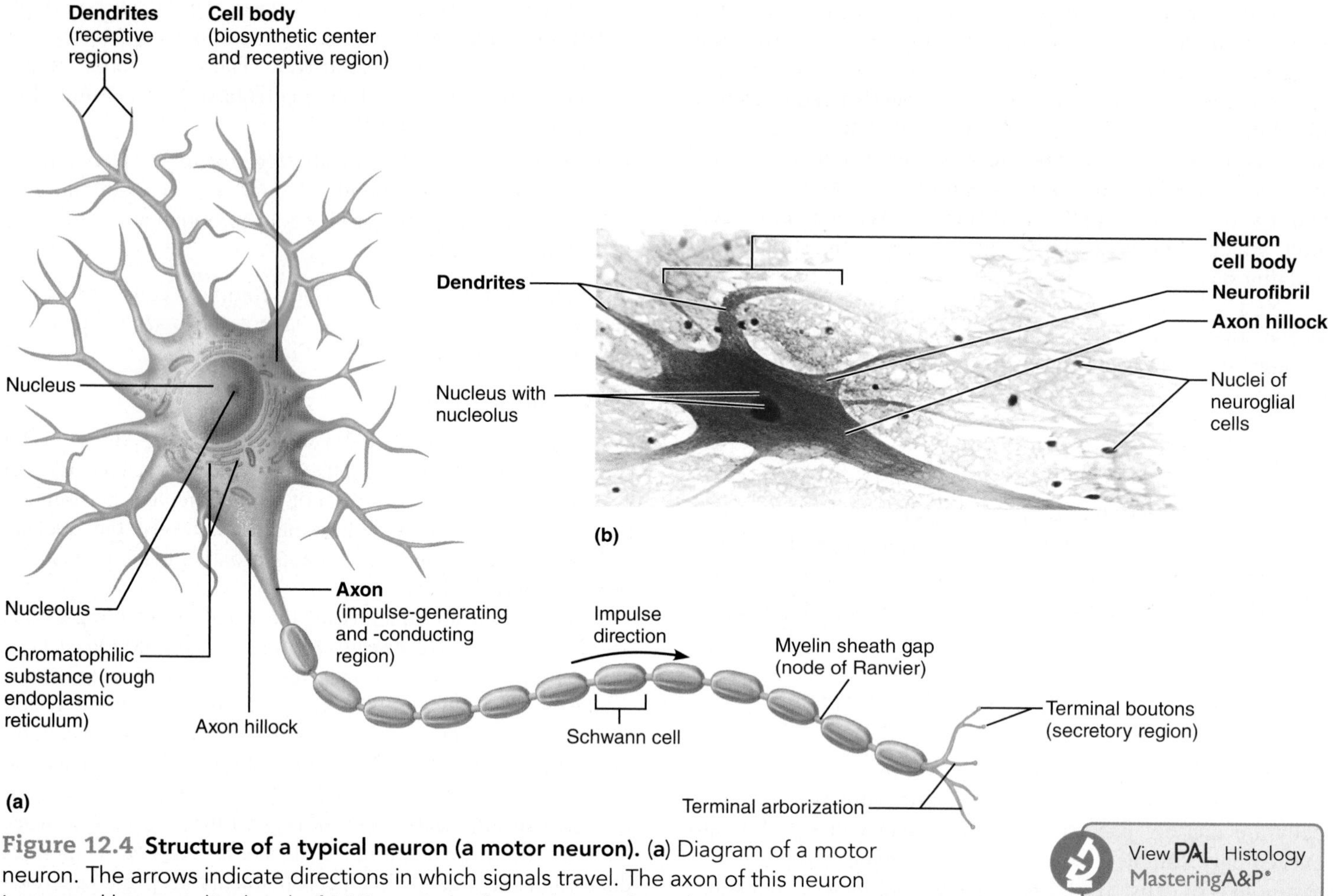

Figure 12.4 Structure of a typical neuron (a motor neuron). (a) Diagram of a motor neuron. The arrows indicate directions in which signals travel. The axon of this neuron is covered by a myelin sheath. (b) Micrograph of neural tissue from the spinal cord showing neuronal cell bodies and surrounding neuroglial cells.

View PAL Histology MasteringA&P®

Axons A neuron has only one **axon** (ak′son; "axis, axle"), which arises from a cone-shaped region of the cell body called the **axon hillock** ("little hill"). Axons are thin processes of uniform diameter throughout their length. By definition, axons are impulse generators and conductors that transmit nerve impulses *away* from their cell body.

Chromatophilic substance and the Golgi apparatus are absent from the axon and the axon hillock alike. In fact, axons lack ribosomes and all organelles involved in protein synthesis, so they must receive their protein components from the cell body. Neurofilaments, actin microfilaments, and microtubules are especially evident in axons, where they provide structural strength. These cytoskeletal elements also aid in the transport of substances to and from the cell body as the axonal cytoplasm is continually recycled and renewed. This movement of substances along axons is called axonal transport.

The axon of some neurons is short, but in others it can be extremely long. For example, the axons of the motor neurons that control muscles in the foot extend from the sacral region of the spinal cord to the sole of the foot, a distance of a meter or more (3–4 feet). Any long axon is called a **nerve fiber.**

Axon diameter varies considerably among the different neurons of the body. Axons with larger diameters conduct impulses faster than those with smaller diameters because of a basic law of physics: The resistance to the passage of an electrical current decreases as the diameter of any "cable" increases.

Although axons branch far less frequently than dendrites, occasional branches do occur along their length. These branches, called **axon collaterals,** extend from the axon at more or less right angles. Whether an axon remains undivided or has collaterals, it usually branches profusely at its end, called the **terminal arborization.** For there to be ten thousand of these *branches* per neuron is not unusual. They end in knobs called **terminal boutons** (boo-tonz′; "buttons") (Figure 12.4a) or *axon terminals.*

A nerve impulse is typically generated where the axon extends from the axon hillock and is conducted along the axon to the terminal boutons, where it causes the release of chemicals called *neurotransmitters* into the extracellular space. The neurotransmitters excite or inhibit the neurons or target organ that are in close contact with the axon boutons.

Synapses

The site at which neurons communicate is called a **synapse** (sin′aps; "union") (Figure 12.5a). Most synapses in the nervous system transmit information through chemical messengers. However, some neurons in certain areas of the CNS transmit signals electrically through gap junctions. Because signals pass across most synapses in one direction only, synapses determine the direction of information flow through the nervous system.

The neuron that conducts signals toward a synapse is called the **presynaptic neuron;** the neuron that transmits signals away from the synapse is called the **postsynaptic neuron** (Figure 12.5a). Most neurons in the CNS function as both presynaptic (information-sending) and postsynaptic (information-receiving) neurons, getting information from some neurons and dispatching it to others.

There are two main types of synapses. Most synapses occur between the terminal boutons of one neuron and the dendrites of another neuron; these are called **axodendritic synapses.** However, many synapses also occur between axons and neuron cell bodies; these are called *axosomatic synapses*.

Structurally, synapses are elaborate cell junctions. This section focuses on the axodendritic synapse (Figure 12.5b) because its structure is representative of both types of synapses. On the presynaptic side, the terminal bouton contains **synaptic vesicles.** These are membrane-bound sacs filled with neurotransmitters, the molecules that transmit signals across the synapse. Mitochondria are abundant in the terminal bouton because the secretion of neurotransmitters requires a great deal of energy. At the synapse, the plasma membranes of the two neurons are separated by a **synaptic cleft.**

How does a synapse function? When an impulse travels along the axon of the presynaptic neuron, the impulse stimulates the synaptic vesicles to fuse with the presynaptic membrane. The fused area then ruptures, causing the vesicles to release their neurotransmitter molecules, which diffuse across the synaptic cleft and bind to the postsynaptic membrane. This binding changes the membrane charge on the postsynaptic neuron, influencing the membrane's ability to generate a nerve impulse.

Classification of Neurons

Neurons may be classified both by structure and by function.

Structural Classification of Neurons Neurons are grouped structurally according to the number of processes that extend from the cell body. By this classification, neurons are *multipolar, bipolar,* or *unipolar* (*polar* = ends, poles) (Table 12.2).

Multipolar neurons have more than two processes (Table 12.2, first column). Usually, multipolar neurons have numerous dendrites and a single axon. However, some small multipolar neurons have no axon and rely strictly on their dendrites for conducting signals. Well over 99% of neurons in the body belong to the multipolar class.

Bipolar neurons have two processes that extend from opposite sides of the cell body (Table 12.2, middle column). These very rare neurons occur in some of the special sensory organs (inner ear, olfactory epithelium of the nose, retina of the eye), where they mostly serve as sensory neurons.

Unipolar neurons have a short, single process that emerges from the cell body and divides like an inverted T into two long branches (Table 12.2, last column). Most unipolar neurons start out as bipolar neurons whose two

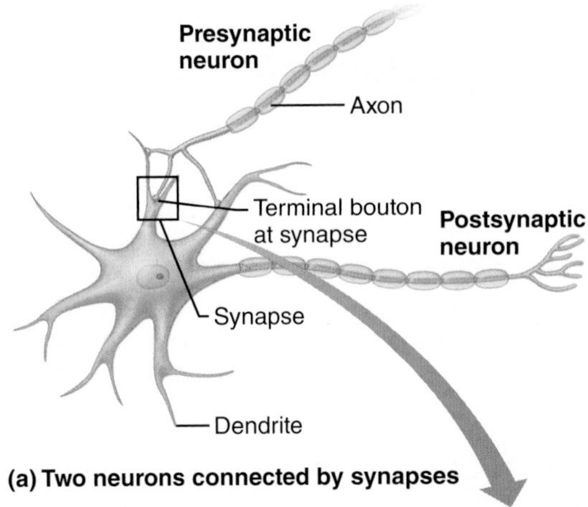

(a) Two neurons connected by synapses

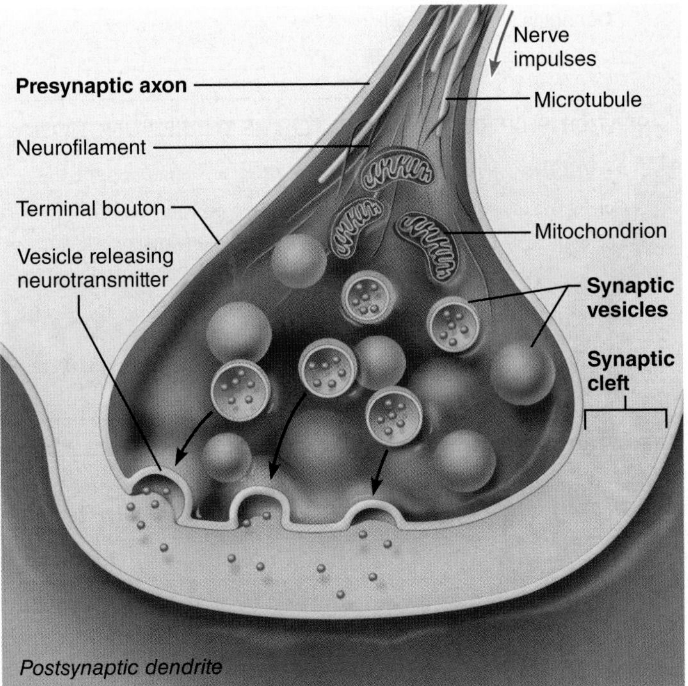

(b) Enlarged view of the synapse

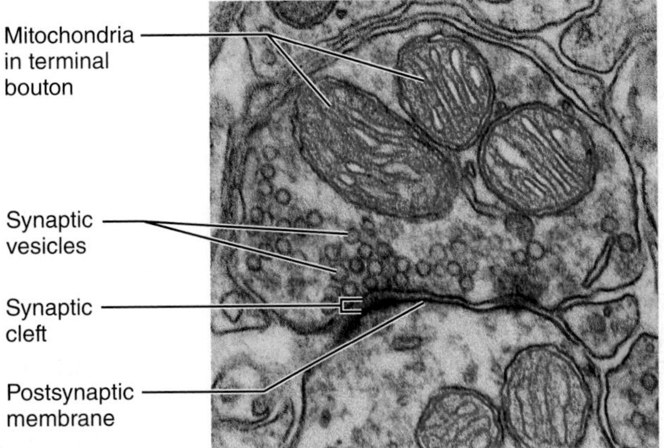

(c) TEM of the synapse (17,000×, artificially colored)

Figure 12.5 Structure of an axodendritic synapse.

Table 12.2 Comparison of Structural Classes of Neurons

Neuron Type

Multipolar	**Bipolar**	**Unipolar (Pseudounipolar)**

STRUCTURAL CLASS: NEURON TYPE ACCORDING TO THE NUMBER OF PROCESSES EXTENDING FROM THE CELL BODY

Multipolar: Many processes extend from the cell body; all are dendrites except for a single axon.

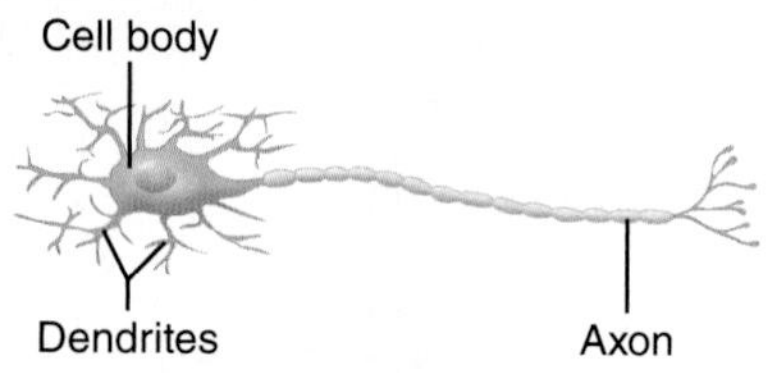

Bipolar: Two processes extend from the cell body: One is a fused dendrite; the other is an axon.

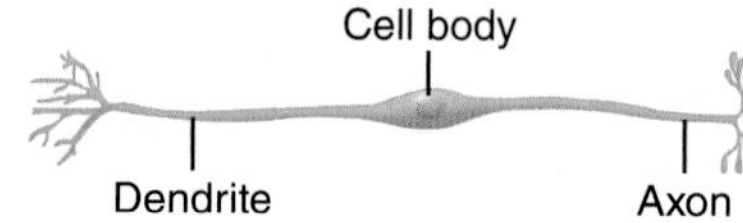

Unipolar (Pseudounipolar): One process extends from the cell body and forms central and peripheral processes, which together comprise an axon.

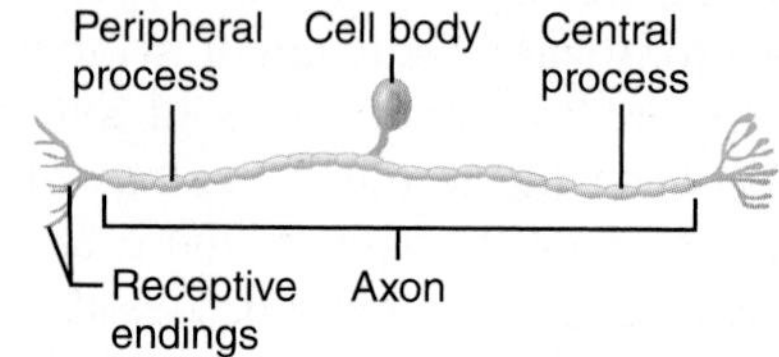

RELATIONSHIP OF ANATOMY TO THE THREE FUNCTIONAL REGIONS

- Receptive region (receives stimulus)
- Conducting region (generates/transmits action potential)
- Secretory region (terminal boutons release neurotransmitters)

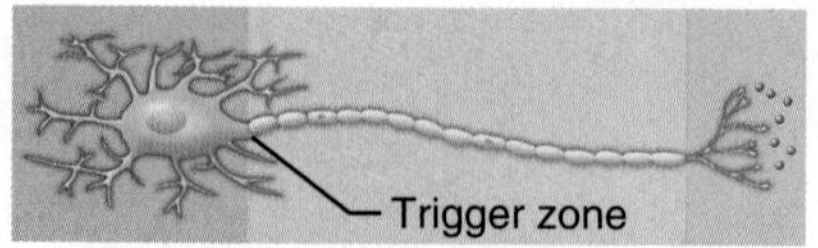

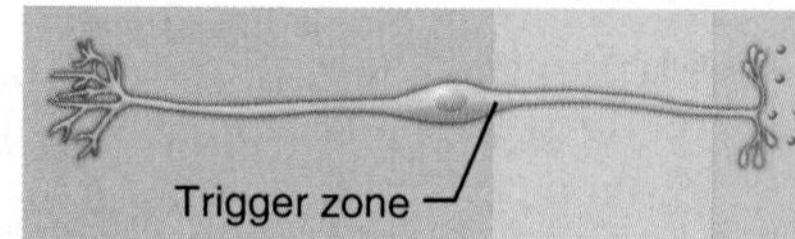

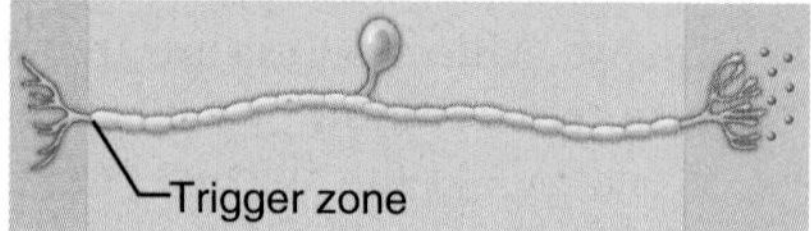

RELATIVE ABUNDANCE AND LOCATION IN HUMAN BODY

Multipolar	Bipolar	Unipolar (Pseudounipolar)
Most abundant in body. Major neuron type in the CNS.	Rare. Found in some special sensory organs (olfactory mucosa, eye, ear).	Found mainly in the PNS. Common only in dorsal root ganglia of the spinal cord and sensory ganglia of cranial nerves.

STRUCTURAL VARIATIONS

Multipolar

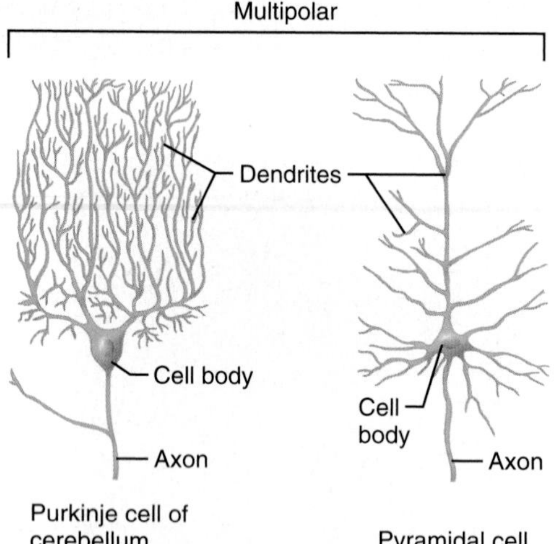

Purkinje cell of cerebellum

Pyramidal cell

Bipolar

Cell body
Dendrite
Axon
Dendrite
Cell body
Axon

Olfactory cell

Retinal cell

Unipolar

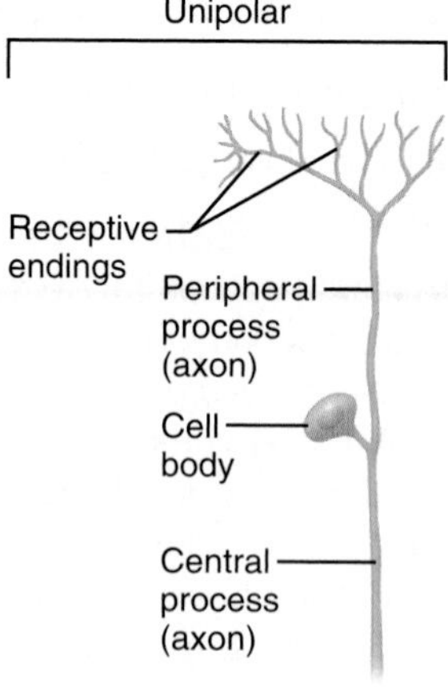

Dorsal root ganglion cell

Table 12.2 continued

Neuron Type		
Multipolar	**Bipolar**	**Unipolar (Pseudounipolar)**
FUNCTIONAL CLASS: NEURON TYPE ACCORDING TO DIRECTION OF IMPULSE CONDUCTION		
1. Most multipolar neurons are **interneurons** that conduct impulses within the CNS, integrating sensory input or motor output; may be one of a chain of CNS neurons, or a single neuron connecting sensory and motor neurons. 2. Some multipolar neurons are **motor neurons** that conduct impulses along the efferent pathways from the CNS to an effector (muscle/gland).	Essentially all bipolar neurons are **sensory neurons** that are located in some special sense organs. For example, bipolar cells of the retina are involved with the transmission of visual inputs from the eye to the brain (via an intermediate chain of neurons).	Most unipolar neurons are **sensory neurons** that conduct impulses along afferent pathways to the CNS for interpretation. (These sensory neurons are called primary or first-order sensory neurons.)
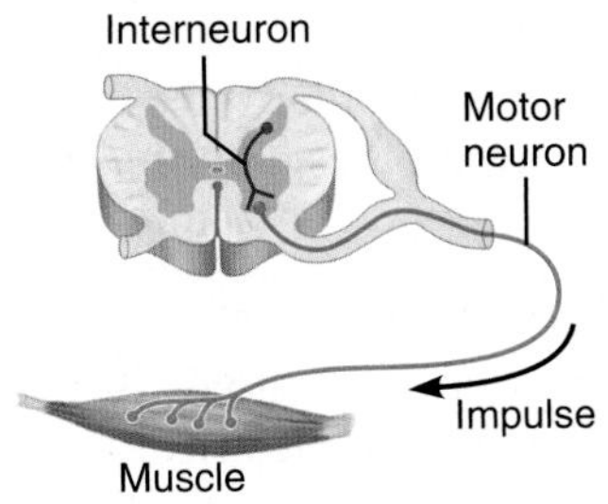	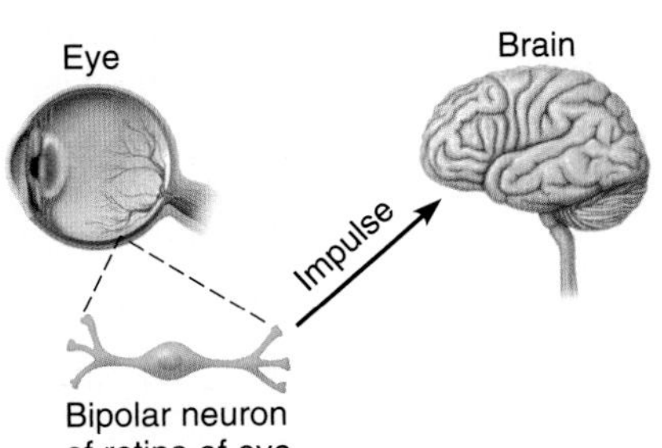	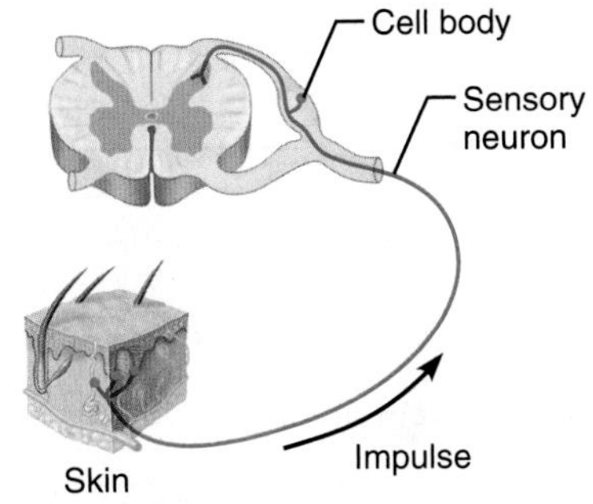

processes fuse together near the cell body during development. Therefore, they are more properly called *pseudounipolar neurons* (*pseudo* = false). Unipolar neurons are found in sensory ganglia in the PNS, where they function as sensory neurons. The short, single process near the neuron cell body divides into two longer branches. One of these branches runs centrally into the CNS and is called the **central process,** whereas the other branch extends peripherally to the receptors and is called the **peripheral process.**

Are the processes of unipolar neurons dendrites, or are they axons? The central process is clearly an axon because it (1) carries a nerve impulse and (2) carries that impulse away from the cell body—the two criteria that define an axon (p. 354). The peripheral process, by contrast, is ambiguous: It generates and carries nerve impulses, as does an axon, but these signals travel *toward* the cell body, a fundamental feature of dendrites. Despite this functional similarity to dendrites, the peripheral process is called an *axon* because its fine structure is identical to that of true axons. Unipolar neurons, therefore, do not have dendrites; receptive endings at the terminal end of the peripheral axon receive sensory stimuli and transmit signals through the peripheral axon toward the cell body.

Functional Classification of Neurons Neurons are grouped functionally according to the direction the nerve impulse travels relative to the CNS. Based on this criterion, there are *sensory neurons, motor neurons,* and *interneurons* (Table 12.2, last row).

Sensory neurons, or *afferent neurons,* make up the sensory division of the PNS (see Figure 12.3). They transmit impulses *toward* the CNS from sensory receptors in the PNS. Most sensory neurons are pseudounipolar, and their cell bodies are in ganglia outside the CNS. The peripheral process extends from a sensory receptor; the central process terminates in the CNS. These two processes function as one, carrying impulses directly from the peripheral receptors to the CNS. Some sensory neurons are bipolar in structure. These neurons are restricted to some of the special sense organs.

Motor neurons or *efferent neurons,* make up the motor division of the PNS (see Figure 12.3). These neurons carry impulses *away* from the CNS to effector organs (muscles and glands). Motor neurons are multipolar, and their cell bodies are located in the CNS (except for some neurons of the autonomic nervous system). Motor neurons form junctions with effector cells, stimulating muscles to contract or glands to secrete.

Interneurons lie between motor and sensory neurons. These multipolar neurons are confined entirely to the CNS (Table 12.2). Interneurons link together into chains that form complex neuronal pathways. The fact that interneurons make up 99.98% of the neurons of the body reflects the vast amount of information processed in the human CNS. These multipolar neurons show great diversity in size and in the branching patterns of their processes (see Table 12.2, "Structural Variations").

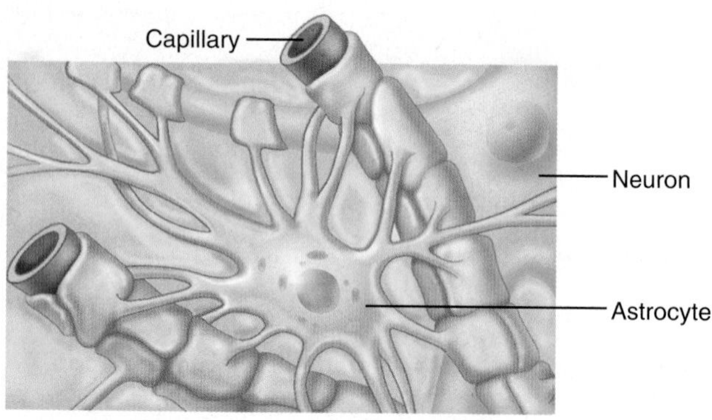

(a) Astrocytes are the most abundant CNS neuroglia.

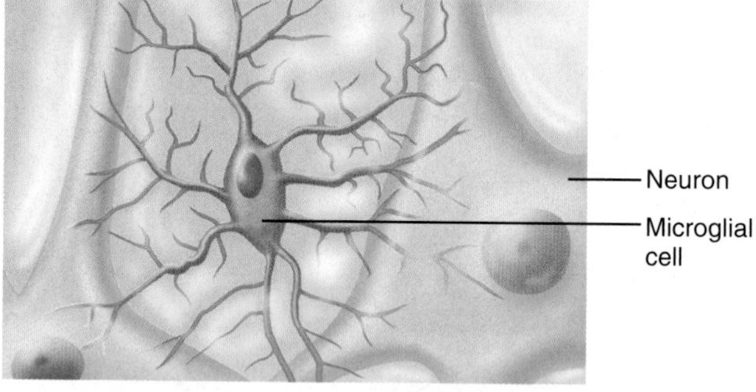

(b) Microglial cells are defensive cells in the CNS.

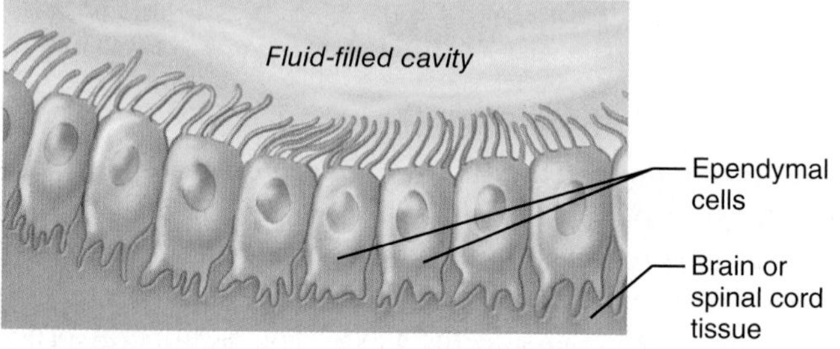

(c) Ependymal cells line cerebrospinal fluid–filled cavities.

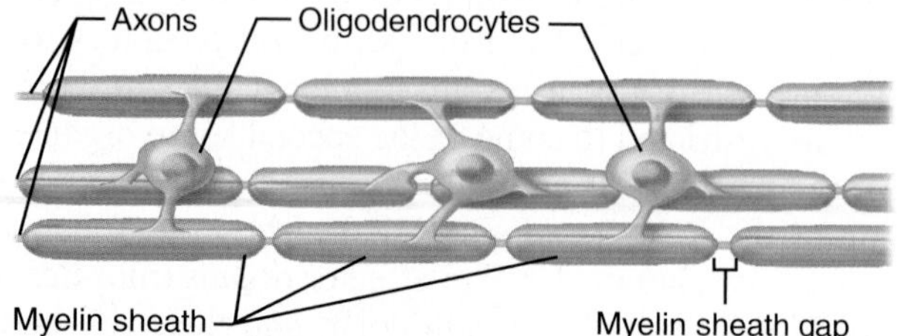

(d) Oligodendrocytes have processes that form myelin sheaths around long axons in the CNS.

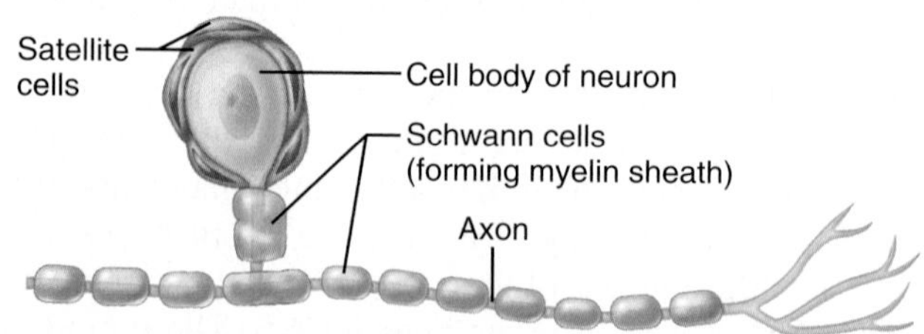

(e) Satellite cells and Schwann cells (which form myelin) surround neurons in the PNS.

Figure 12.6 **Neuroglia.** (**a–d**) The four types of neuroglial cells in the CNS. (**e**) Neuroglia of the PNS.

check your understanding

- ☐ 4. Which type of neuron process receives stimuli?
- ☐ 5. Describe how the electrical impulse from one neuron is passed to another neuron.
- ☐ 6. What is the structural type of most sensory neurons?
- ☐ 7. Which structural type of neuron is most abundant? Which neurons in the PNS are of this type?

(For answers, see Appendix B.)

Neuroglia

learning outcomes

- ▶ List the six types of neuroglia in nervous tissue, and distinguish them by location and function.
- ▶ Describe the structure of myelin sheaths.

All neurons associate closely with non-nervous supporting cells called **neuroglia** (nu-rog′le-ah; "nerve glue"), or **glial** (gle′al) **cells.** There are six types of neuroglia—four in the CNS and two in the PNS (Figure 12.6). Each type has a unique function, yet in general, these cells:

- Provide a supportive scaffolding for neurons.
- Cover all nonsynaptic parts of the neurons, thereby insulating the neurons and keeping the electrical activities of adjacent neurons from interfering with each other.

Neuroglia in the CNS

Like neurons, most neuroglial cells of the CNS have branching processes and a central cell body (Figure 12.6a–d). Neuroglia can be distinguished from neurons, however, by the much smaller size of neuroglia and by their darker-staining nuclei (see Figure 12.4). They outnumber neurons in the CNS by about 10 to 1, and they make up about half the mass of the brain. Unlike neurons, glial cells can divide throughout life.

Astrocytes Star-shaped **astrocytes** (as′tro-sītz; "star cells") are the most abundant glial cells of the CNS (Figure 12.6). They have many radiating processes with bulbous ends. Some of these bulbs cling to neurons (including the axon terminals), whereas others cling to capillaries. The functions of astrocytes are numerous and not fully understood. Known functions include (1) regulating neurotransmitter levels by increasing the rate of neurotransmitter uptake in regions of high neuronal activity; (2) signaling increased blood flow through capillaries in active regions of the brain; and (3) controlling the ionic environment around neurons. Astrocytes also help synapses form in developing neural tissue, produce molecules necessary for neural growth (brain-derived trophic factor, BDTF), and propagate calcium signals that may be involved with memory. No longer are these cells considered passive supportive cells for neurons; rather, they appear to have an active role in neural activity. Understanding the activities of these abundant glial cells is an area of ongoing research.

Microglial Cells **Microglial cells,** the smallest and least abundant neuroglia of the CNS, have elongated cell bodies and cell processes with many pointed projections, like a thorny bush (Figure 12.6b). They are phagocytes, the macrophages of the

CNS. They migrate to, and then engulf, invading microorganisms and injured or dead neurons. Unlike other neuroglial cells, microglial cells do not originate in nervous tissue; like the other macrophages of the body, they are derived from blood cells called monocytes. The monocytes that become microglial cells migrate to the CNS during the embryonic and fetal periods.

Ependymal Cells The CNS originates in the embryo as a hollow neural tube and retains a central cavity throughout life (Chapter 3, p. 52). **Ependymal cells** (ĕ-pen′dĭ-mal; "wrapping garment") form a simple epithelium that lines the central cavity of the spinal cord and brain (Figure 12.6c). Here these cells provide a fairly permeable layer between the *cerebrospinal fluid* that fills this cavity and the tissue fluid that bathes the cells of the CNS. Ependymal cells bear cilia that help circulate the cerebrospinal fluid.

Oligodendrocytes **Oligodendrocytes** (ol″ĭ-go-den′dro-sītz; "few-branch cells") (Figure 12.6d) have fewer branches than astrocytes, as their name implies. They line up in small groups and wrap their cell processes around the thicker axons in the CNS, producing insulating coverings called *myelin sheaths.*

CLINICAL APPLICATION

Gliomas Because glial cells can divide, they accumulate the "mistakes" in DNA replication that may transform them into cells that proliferate abnormally. Such an accumulation does not occur in neurons, most of which do not divide. Therefore, most tumors that originate in the brain (60%) are **gliomas,** tumors formed by uncontrolled proliferation of glial cells. Two percent of all cancers are gliomas, and their incidence is increasing. These are difficult cancers to treat, and the one-year survival rate is under 50%.

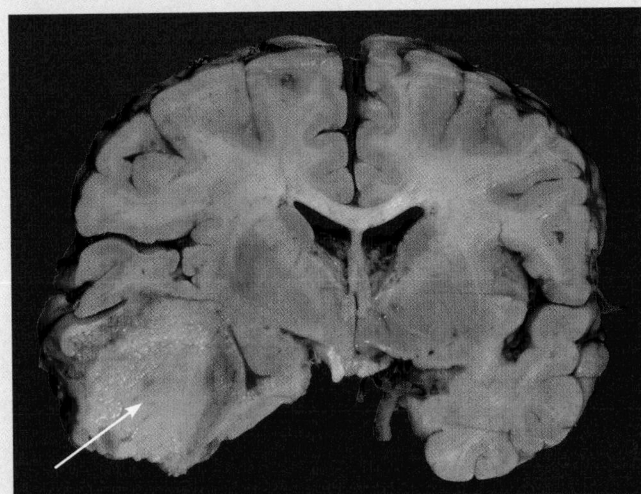

Glioma in temporal lobe of cerebrum

Neuroglia in the PNS

The two kinds of neuroglia in the PNS are *satellite cells* and *Schwann cells* (Figure 12.6e), very similar cell types that differ mainly in location. **Satellite cells** surround neuron cell bodies within ganglia. Their name comes from a fancied resemblance to the moons, or satellites, around a planet. **Schwann cells** surround all axons in the PNS and form myelin sheaths around many of these axons.

Myelin Sheaths

Myelin sheaths are produced by oligodendrocytes in the CNS and Schwann cells in the PNS. These sheaths are segmented structures that are composed of the lipoprotein **myelin** and surround the thicker axons of the body. Each segment of myelin consists of the plasma membrane of a glial cell rolled in concentric layers around the axon (Figure 12.7a). Myelin sheaths form an insulating layer that prevents the leakage of electrical current from the axon, increases the speed of impulse conduction along the axon, and makes impulse propagation more energy-efficient.

Myelin Sheaths in the PNS The myelin sheaths in the PNS are formed by Schwann cells (see Figure 12.6e). Myelin sheaths develop during the fetal period and the first year or so of postnatal life. To form the myelin sheath, the Schwann cells first indent to receive the axon and then wrap themselves around the axon repeatedly, in a jellyroll fashion (Figure 12.7a ① and ②). Initially the wrapping is loose, but the cytoplasm of the Schwann cell is gradually squeezed outward from between the membrane layers. When the wrapping process is finished, many concentric layers of Schwann cell plasma membrane ensheathe the axon in a tightly packed coil of membranes that is the true myelin sheath (Figure 12.7a ③). The nucleus and most of the cytoplasm of the Schwann cell end up just external to the myelin layers.

Because the adjacent Schwann cells along a myelinated axon do not touch one another, there are gaps in the myelin sheath (see Figure 12.4a). These gaps, called **myelin sheath gaps** *(nodes of Ranvier),* occur at regular intervals about 1 mm apart. In myelinated axons, nerve impulses do not travel along the myelin-covered regions of the axonal membrane but instead jump from the membrane of one myelin sheath gap to the next in a way that greatly speeds impulse conduction.

Only thick, rapidly conducting axons are sheathed with myelin. In contrast, thin, slowly conducting axons lack a myelin sheath and are called **nonmyelinated axons** (Figure 12.7b). Schwann cells surround such axons but do not wrap around them in concentric layers of membrane (Figure 12.7b ①). A single Schwann cell can partly enclose 15 or more nonmyelinated axons, each of which occupies a separate tubular recess in the surface of the Schwann cell (Figure 12.7b ②). Nonmyelinated axons are found in portions of the autonomic nervous system (Chapter 15) and in some sensory fibers.

CLINICAL APPLICATION

Tic Douloureux The main sensory nerve of the face, the trigeminal nerve, can be affected by an extremely painful disorder called **tic douloureux** (tik doo″loo-roo; "wincing in pain"), or **trigeminal neuralgia.** Compression of the trigeminal nerve by an adjacent blood vessel causes degeneration and loss of the myelin sheath that surrounds the sensory nerve fibers. Because of the loss of insulation in the nonmyelinated region, impulses in nerve fibers that carry touch sensations stimulate pain fibers in the same nerve (called *cross-talk*), leading to the perception of pain by the brain. Even the softest touch to the face can produce excruciating pain. Although not fatal, this condition is debilitating: The pain is unpredictable and extreme.

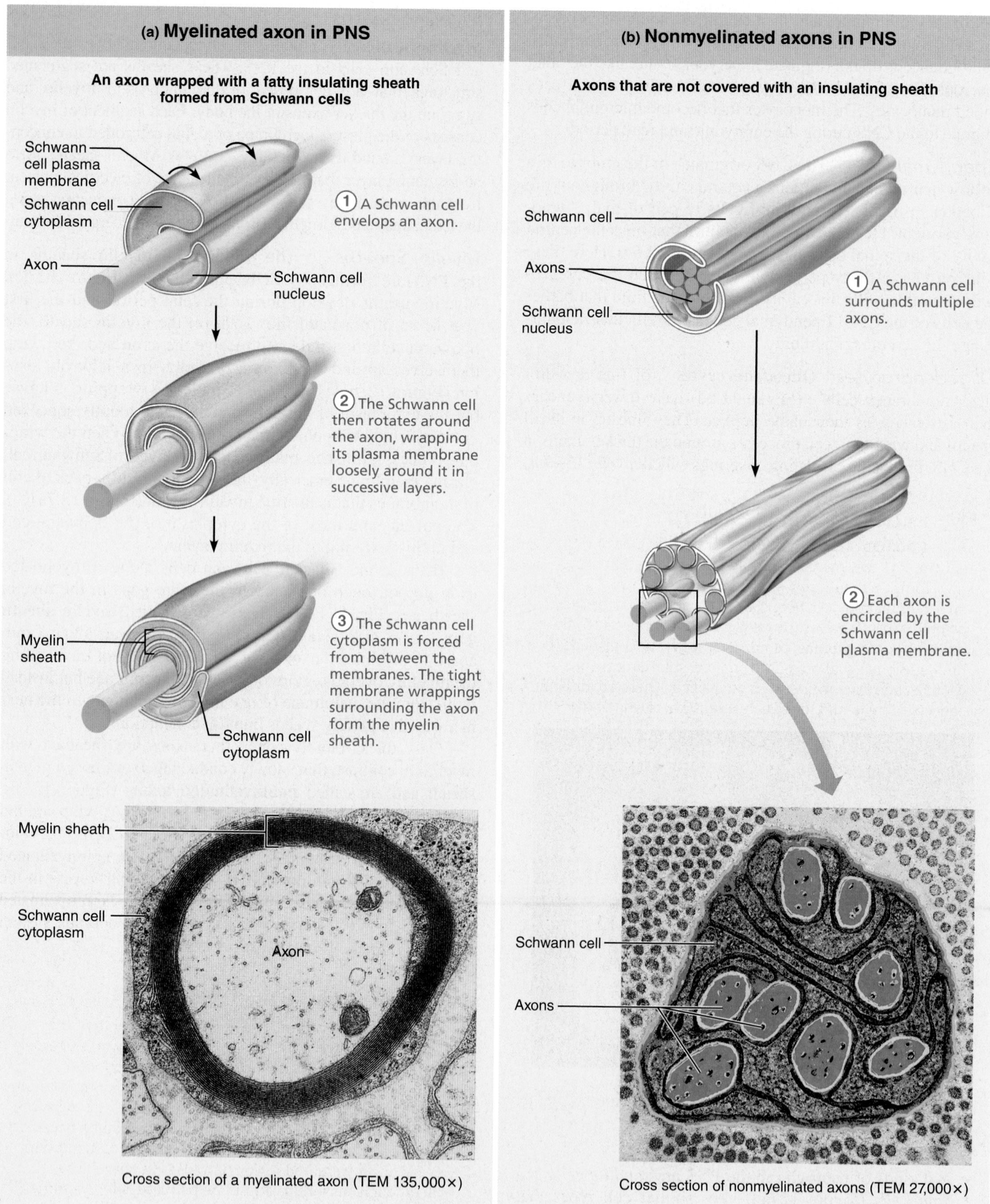

Figure 12.7 Schwann cells on myelinated and nonmyelinated axons in the PNS.

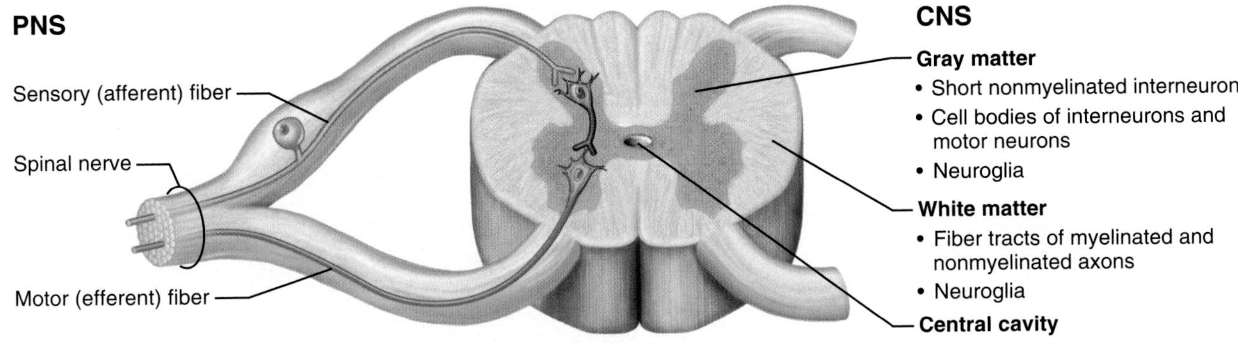

Figure 12.8 **Cross section through the spinal cord.** Linkage between PNS and CNS is illustrated on the left.

Myelin Sheaths in the CNS Oligodendrocytes form the myelin sheaths in the brain and spinal cord (Figure 12.6d). In contrast to Schwann cells, each oligodendrocyte has multiple processes that coil around several different axons. Myelin sheath gaps are present, although they are more widely spaced than those in the PNS.

As in the PNS, the thinnest axons in the CNS are nonmyelinated. These nonmyelinated axons are covered by the many long processes of glial cells, such as astrocytes, that are so abundant in the CNS.

check your understanding

☐ 8. Which neuroglia make myelin in the CNS? In the PNS?

☐ 9. Which neuroglia are common in regions where synapses occur?

☐ 10. Do Schwann cells cover nonmyelinated axons in the PNS?

(For answers, see Appendix B.)

GROSS ANATOMY OF THE NERVOUS SYSTEM: AN OVERVIEW

learning outcomes

- Distinguish gray matter from white matter in the CNS.
- Define nerve and describe the structural components of nerves.

Thus far we have discussed the structure and function of the cells that make up nervous tissue: neurons and neuroglia. Nervous tissue, along with other types of tissues, forms the nervous system: the brain and spinal cord of the CNS and the nerves and ganglia of the PNS. The basic structural organization of the CNS and the structure of nerves are described next. (The detailed structure and function of the organs of the CNS are covered in Chapter 13, and those of the PNS are covered in Chapters 14 and 15.)

Gray and White Matter of the CNS

The brain and spinal cord have distinct regions of gray and white matter that reflect the arrangement of their neurons. The **gray matter** is a gray-colored zone that surrounds the hollow central cavity of the CNS. In the spinal cord it is a butterfly-shaped region in which the dorsal half contains cell bodies of interneurons and the ventral half contains cell bodies of motor neurons (Figure 12.8). Thus, *gray matter is the site where neuron cell bodies are clustered.* More specifically, the gray matter of the CNS is a mixture of neuron cell bodies; dendrites; short, nonmyelinated neurons; and neuroglia. Synapses occur in the gray matter.

External to the gray matter is **white matter,** which contains no neuron cell bodies but millions of axons and neuroglia. Its white color comes from the myelin sheaths around many of the axons. Most of these axons either ascend from the spinal cord to the brain or descend from the brain to the spinal cord, allowing these two regions of the CNS to communicate with each other. Thus, *white matter consists of axons running between different parts of the CNS.* Within the white matter, axons traveling to similar destinations form axon bundles called **tracts.**

Throughout the CNS white matter is external to gray matter, which surrounds the hollow central cavity. In two regions of the brain (the cerebrum and cerebellum), there is an additional layer of gray matter located superficially, the **cortex.** (The anatomy of the CNS and the distribution of gray and white matter are described more completely in Chapter 13.)

Nerves

A **nerve** is a cablelike organ in the peripheral nervous system (see Figure 12.2). Each nerve consists of many axons (nerve fibers) arranged in parallel bundles and enclosed by successive wrappings of connective tissue (Figure 12.9). Almost all nerves contain both myelinated and nonmyelinated sensory and motor axons.

Within a nerve, each axon is surrounded by Schwann cells. Covering the Schwann cells is a delicate layer of loose connective tissue called **endoneurium** (en″do-nu′re-um). Groups of axons are bound into bundles called nerve **fascicles** by a

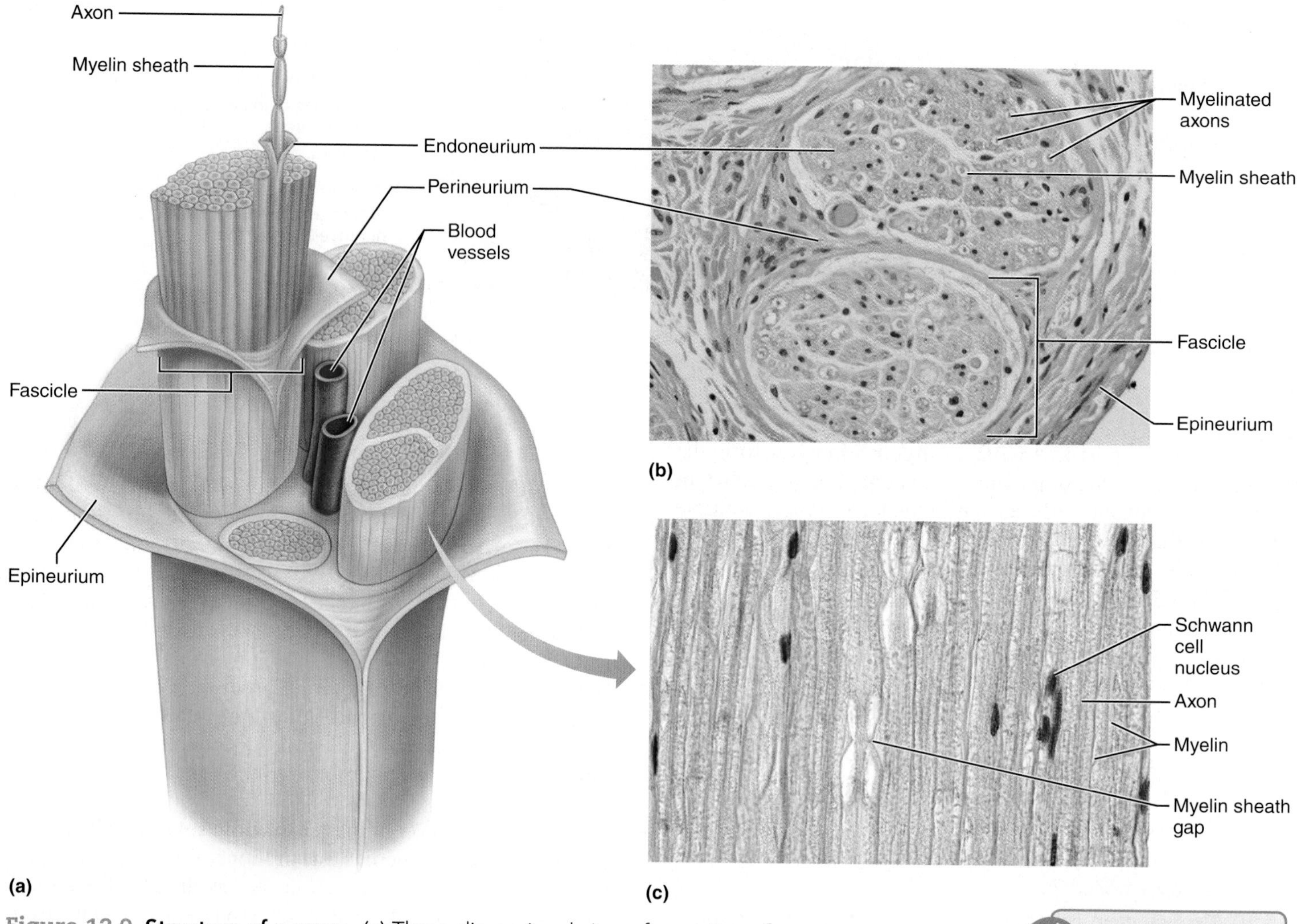

Figure 12.9 Structure of a nerve. (**a**) Three-dimensional view of a portion of a nerve, showing connective tissue wrappings. (**b**) Light micrograph of a cross section of a portion of a nerve (H&E stained, 200×). (**c**) Longitudinal section of a nerve, as viewed by light microscopy (265×).

View PAL Histology Mastering A&P®

wrapping of connective tissue called the **perineurium.** Finally, the whole nerve is surrounded by a tough fibrous sheath, the **epineurium.** The three layers of connective tissue in nerves correspond exactly to those in skeletal muscle: the endomysium, perimysium, and epimysium (see p. 245). As in skeletal muscle, the connective tissue in a nerve contains the blood vessels that nourish the axons and Schwann cells (Figure 12.9).

The terms *neuron, nerve fiber,* and *nerve* are easy to confuse. Keep in mind:

- A **neuron** is a nerve cell.
- A **nerve fiber** is a long axon.
- A **nerve** is a collection of axons in the PNS.

☑ check your understanding

☐ 11. Name the connective tissue wrapping that encloses a bundle of axons into a fascicle.

☐ 12. Where do synapses occur in the CNS, in white matter or in gray matter?

☐ 13. Why is white matter white?

(For answers, see Appendix B.)

NEURONAL INTEGRATION

learning outcomes

- Describe the structural link between the peripheral nervous system and the central nervous system.
- Define reflex, and list the basic components of a reflex arc. Distinguish monosynaptic reflexes from polysynaptic reflexes.

We divide the nervous system into the PNS and CNS to simplify the discussion of each component. Yet it is important to remember that these two divisions of the nervous system are functionally and structurally linked.

The PNS is composed of the axons of sensory (afferent) and motor (efferent) neurons (nerve fibers) bundled together as nerves (Figure 12.8). These nerves function as information pathways to and from the periphery of the body: The afferent fibers respond to sensory stimuli and carry that information toward the CNS; the efferent fibers transmit motor stimuli from the CNS to muscles and glands, causing them to function (contract or secrete).

The CNS is composed of interneurons that (1) receive sensory information, (2) direct and transport this information to specific regions of the CNS, and (3) initiate an appropriate motor response.

The structural link between the PNS and the CNS occurs in the gray matter of the CNS. The simplest example of this neuronal integration is the reflex arc.

Reflex Arcs

Reflex arcs are simple chains of neurons that cause our simplest, reflexive behaviors and reflect the basic structural plan of the nervous system. Reflex arcs account for **reflexes,** which are defined as rapid, automatic *motor* responses to stimuli. Reflexes are unlearned, unpremeditated, and involuntary. Examples are jerking away your hand after it accidentally touches a hot stove and vomiting in response to some food that irritates your stomach. As you can see from these examples, reflexes are either *somatic reflexes* resulting in the contraction of skeletal muscles or *visceral reflexes* activating smooth muscle, cardiac muscle, or glands.

Every reflex arc has five essential components, each of which activates the next (Figure 12.10):

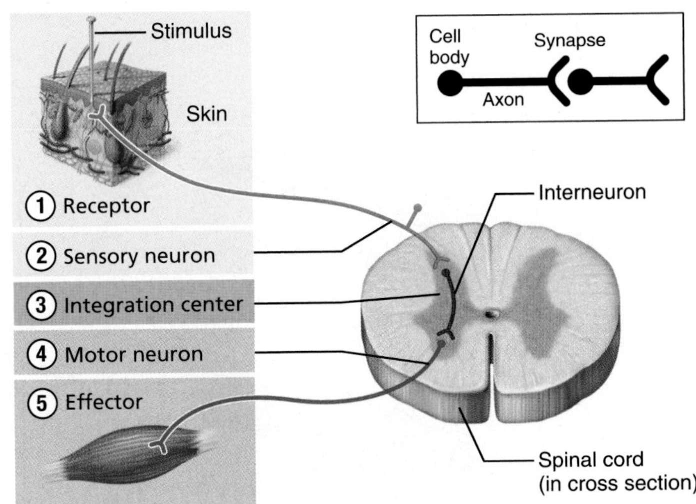

Figure 12.10 **Components of a reflex arc.** Receptors detect changes in the internal or external environment. Effectors are muscles or glands.

① The *receptor* is the site where the stimulus acts. Receptors are located at the terminal end of the peripheral process of a sensory neuron.

② The *sensory neuron* transmits the afferent impulses to the CNS.

③ The *integration center* consists of one or more synapses in the gray matter of the CNS (represented by the spinal cord in Figure 12.10). In the simplest reflex arcs, the integration center is a single synapse between a sensory neuron and a motor neuron. In more complex reflexes, it can involve multiple synapses that send signals through long chains of interneurons to other portions of the CNS, for instance, to portions of the brain.

④ The *motor neuron* conducts efferent impulses from the integration center to an effector.

⑤ The *effector* is the muscle or gland cell that responds to the efferent impulses by contracting or secreting.

Monosynaptic Reflex

The simplest of all reflexes is the **monosynaptic reflex** ("one synapse"; Figure 12.11). In a monosynaptic reflex there is no interneuron between the sensory neuron and the motor neuron; thus, as its name implies, there is only one synapse in this reflex arc. An example is the familiar "knee-jerk" reflex: The impact of a hammer on the patellar ligament stretches the quadriceps muscles of the thigh. This stretching initiates an impulse in a sensory neuron that directly activates a motor neuron in the spinal cord, which then signals the quadriceps muscle to contract. This contraction counteracts the original stretching caused by the hammer.

Many skeletal muscles of the body participate in such monosynaptic *stretch reflexes,* which help maintain equilibrium and upright posture. In these reflexes, the sensory neurons sense the stretching of muscles that occurs when the body starts to sway, and then the motor neurons activate muscles that adjust the body's position to prevent a fall. Because they contain just one synapse, stretch reflexes are the fastest of all reflexes—and speed is essential to keep one from falling to the ground.

Polysynaptic Reflex

Far more common than monosynaptic reflexes are **polysynaptic reflexes,** in which one or more interneurons are part of the reflex pathway between the sensory and motor neurons. The presence of even one interneuron means that there have to be at least two synapses in this type of reflex arc, thus the name polysynaptic. Most of the simple reflex arcs in the body contain a single interneuron and therefore have a total of

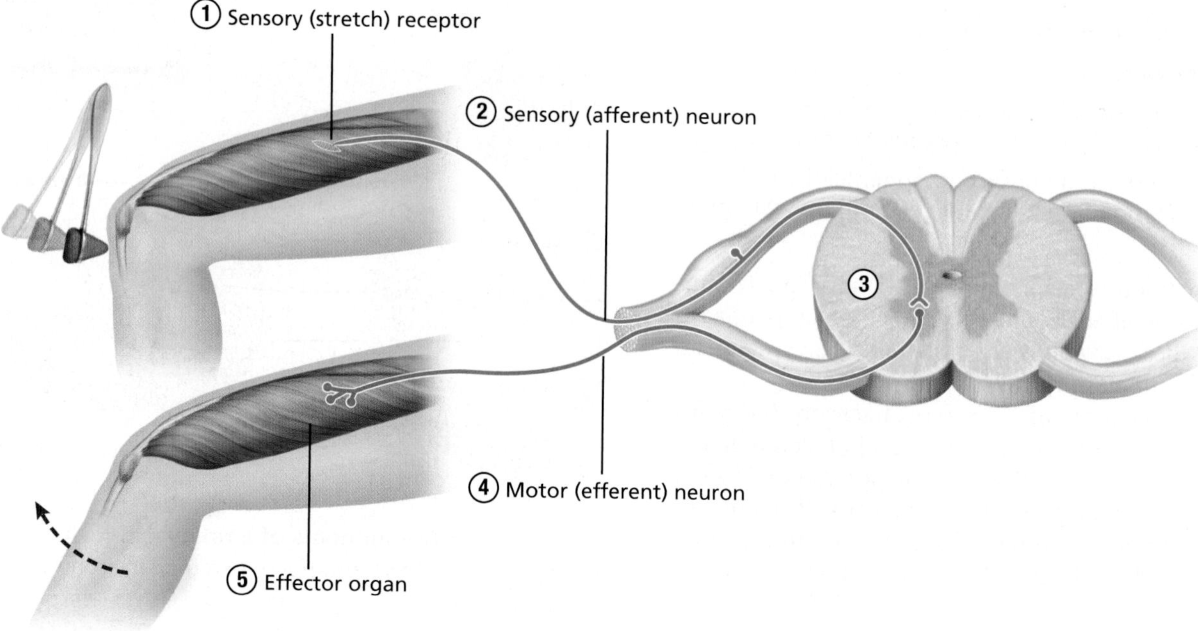

(a) Monosynaptic stretch reflex

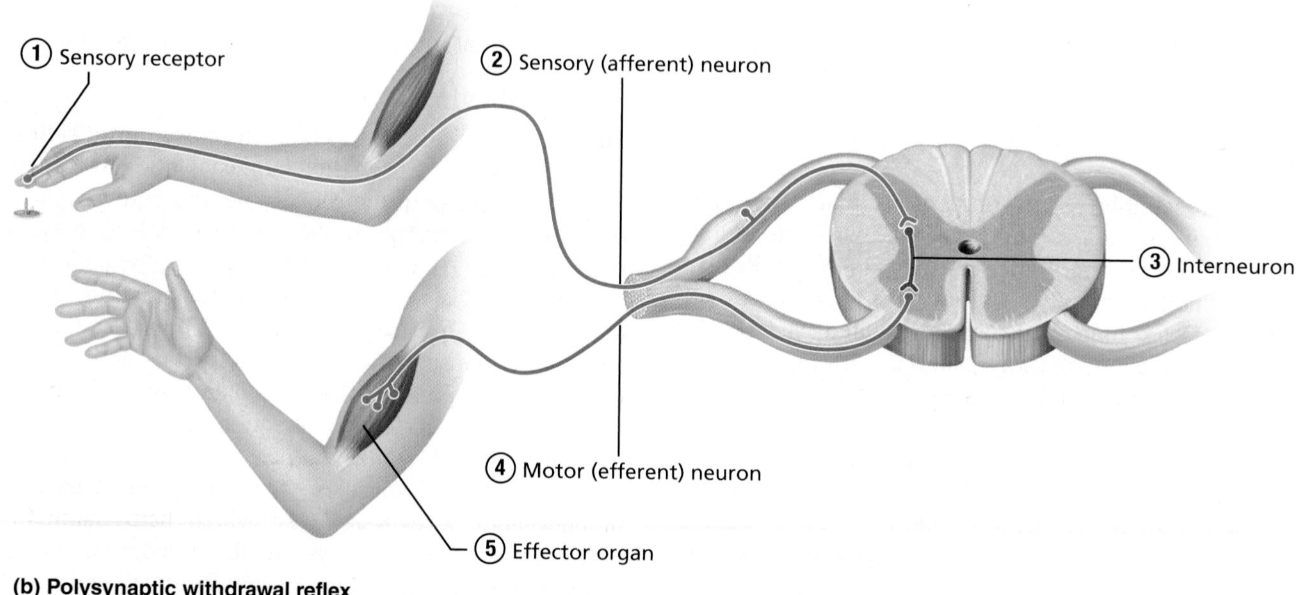

(b) Polysynaptic withdrawal reflex

Figure 12.11 **Types of reflex arcs.** (**a**) A monosynaptic reflex arc has two neurons and a single synapse. (**b**) A polysynaptic reflex arc has more than two neurons (in this case, three) and therefore has at least two synapses. The five components of a reflex arc are indicated by number.

three neurons. *Withdrawal reflexes,* by which we pull away from danger, are three-neuron polysynaptic reflexes (Figure 12.11b). Pricking a finger with a tack initiates an impulse in the sensory neuron, which activates the interneuron in the CNS. The interneuron then signals the motor neuron to contract the muscle that withdraws the hand.

Neuronal Circuits

learning outcomes

- ▶ Describe the roles of interneurons in the CNS.
- ▶ Differentiate serial processing and parallel processing.
- ▶ Define convergent and divergent circuits and give an example of each.

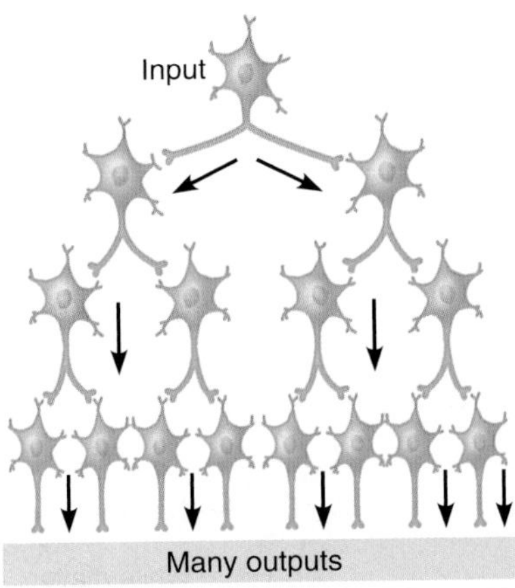

(a) Diverging circuit to multiple pathways

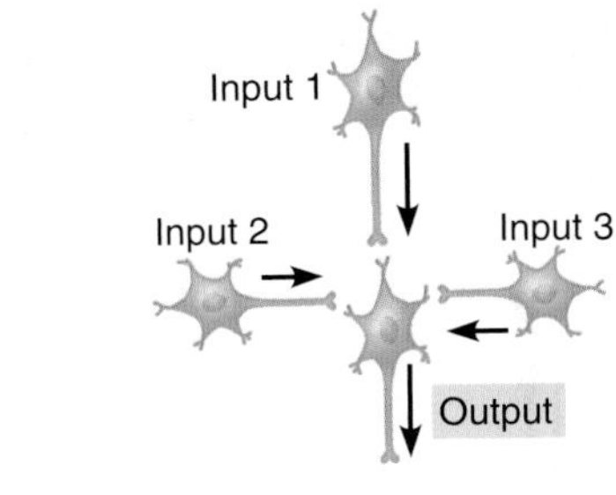

(b) Converging circuit

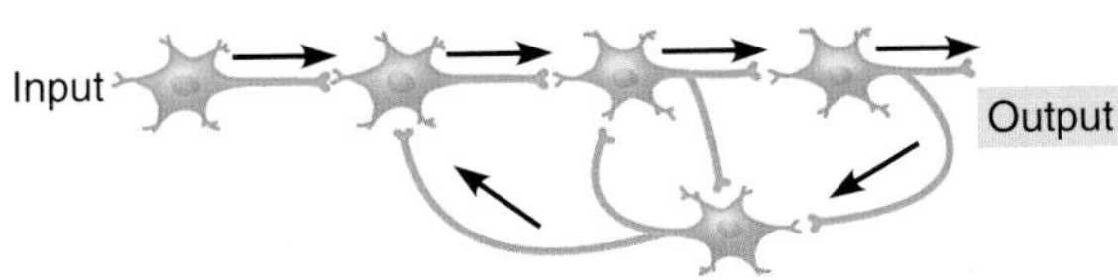

(c) Reverberating circuit

Figure 12.12 **Types of neuronal circuits.**

Even though reflex arcs reflect the basic organization of the human nervous system, it is obviously more than just a series of simple reflex arcs. To appreciate its complexity, you must understand the role that interneurons play. More than 99% of all neurons are interneurons, which include not only the intermediate neurons of reflex arcs but also all the neurons that are confined entirely within the CNS. The complexity of the CNS arises from the organization of this vast number of interneurons. Typically, a single neuron synapses with many other neurons. The interneurons of the CNS may be interconnected in multiple ways. These are referred to as **neuronal circuits** (Figure 12.12).

Diverging Circuit

In a **diverging circuit** (Figure 12.12a), one presynaptic neuron synapses with several other neurons. An example of neuronal **divergence** is seen in the stretch reflex previously described. The stretch of a muscle stimulates numerous sensory neurons. Each sensory neuron synapses directly with 100–150 neurons in the gray matter of the spinal cord. Some of these neurons are motor neurons that directly innervate the stretched muscle and stimulate contraction. Others are interneurons that act to inhibit the activity of the antagonistic muscle group, and still others are interneurons that project sensory information to higher regions of the CNS. As a result of divergence, information is distributed through multiple neuronal pathways.

Converging Circuit

When many neurons synapse on a single postsynaptic neuron, the circuit is a **converging circuit** (Figure 12.12b). For example, **convergence** occurs when a single motor neuron receives both excitatory and inhibitory impulses from many other neurons. These impulses are integrated by the target motor neuron and influence whether it will initiate a nerve impulse. Divergence and convergence are apparent throughout the central nervous system as neurons integrate information.

Reverberating Circuit

In a **reverberating circuit,** one neuron in the circuit receives feedback from another neuron in the same circuit. A branch off the axon of one neuron circles back and synapses with a previous neuron in the circuit (Figure 12.12c). In this pathway, the signal continues to be sent until either synaptic fatigue or inhibition by another signal interrupts the circuit. Reverberating circuits are involved in the control of many rhythmic activities, such as breathing and swinging the arms when walking.

Serial Processing

Information is processed in two ways in the nervous system. Neurons that synapse one-on-one in a sequence are said to be joined **in series,** and the processing is called **serial processing.** Neurons linked this way pass their signal along a single pathway from one neuron to the next, like links in a chain. A reflex arc is one example (see Figure 12.11).

Parallel Processing

The second way of processing information is along neurons joined **in parallel.** Information from a single neuron is sent along two or more parallel pathways. This is called **parallel processing.** Parallel processing occurs when a single sensory stimulus results in multiple perceptions. For example, as you watch a friend walking toward you, multiple pathways process visual stimuli from your retina in parallel, evaluating the color, shape, spatial location, and movement of your friend. The stimuli are also processed by the parts of your brain associated with recognition and memory from past experiences, enabling you to recognize your friend, recall his name, and remember when you last saw him. For all its complexity, parallel processing occurs almost instantaneously. It allows the brain to evaluate stimuli with incredible speed and enables information to be integrated along numerous pathways.

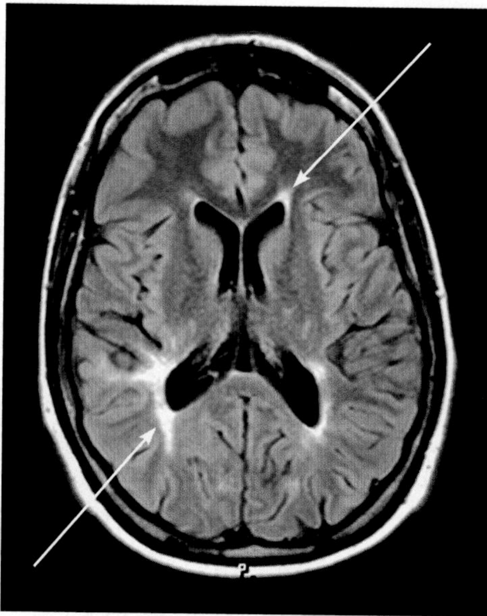

Figure 12.14 Multiple sclerosis. MRI of the brain of a patient with MS, transverse section. Lesions appear as bright white spots within the white matter.

of the immune response, which gives axons time to repair themselves and to recover their functions. However, after 10–15 years of periodic relapses, the accumulated damage to the axons is often too great, and 50% of MS sufferers decline rapidly in their second decade of the disease.

MS is not always easy to recognize. A definitive but invasive test for MS identifies MS-specific antibodies in cerebrospinal fluid taken from the space around the spinal cord. A less invasive approach tests for a delay in the time it takes visual signals to travel from the eye through the brain, as measured by monitoring brain waves detected on the outside of the skull. Finally, magnetic resonance imaging can reveal the disease's characteristic lesions (Figure 12.14).

Treatment is directed toward relieving the symptoms. Anti-inflammatory steroid drugs lessen the severity and duration of relapses, but they do not reduce the frequency of attacks. Other drugs, such as interferons and glatiramer acetate, act to control the immune system to decrease the frequency of relapses as well as the appearance of new lesions. Mitoxantrone, a cancer drug that suppresses lymphocyte activity, has been successful in treating more advanced stages of the disease.

Neuronal Regeneration

Because there is no effective replacement of dead or damaged neurons after the fetal period, neural injuries tend to cause permanent dysfunction in children and adults. If axons alone are destroyed, however, the cell bodies often survive, and the axons may regenerate (Figure 12.15). If a nerve is severed in the PNS, macrophages invade and destroy the axon distal to the injury (Figure 12.15 ②). Axon filaments can grow peripherally from the injured site at an approximate rate of 1.5 mm a day within regeneration tubes formed by the surviving Schwann cells that surrounded the original axons (Figure 12.15 ③). Thus, eventual reinnervation of the target organ, with partial recovery of function, is sometimes possible in the PNS (Figure 12.15 ④). In the CNS, the neuroglia never form bands to guide the regrowing axons and even hinder such axons by secreting growth-inhibiting chemicals. Therefore, there is no effective regeneration after injury to the spinal cord or brain.

CLINICAL APPLICATION

Regeneration and Spinal Cord Injuries

Trauma to the spine has left 200,000 living Americans paralyzed with spinal cord injury. To help these people, extensive medical research is being conducted on spinal cord regeneration. The main goals of this research are to (1) identify and block the chemicals that inhibit axonal growth in the injured CNS; (2) add neurotrophins to induce axonal sprouting and remyelinization; (3) make the damaged region of the spinal cord more like the PNS by implanting Schwann cells or segments of peripheral nerves, along which the CNS axons may grow; and (4) isolate neuronal stem cells and add them to the damaged spinal cord. Currently, the only helpful treatment for spinal injury relies on the fact that much of the axonal damage is secondary—caused by inflammation, production of free radicals, and apoptosis that develop in the cord after the injury. Administering the anti-inflammatory drug methylprednisolone within 8 hours after injury minimizes this secondary damage and improves the prognosis of many patients. (For more on spinal cord injuries, see Chapter 13, p. 419).

☑ check your understanding

☐ 17. Both peripheral nerves and the white matter of the spinal cord are composed of axonal processes. Why is it possible to regain function after an injury to a peripheral nerve, but not after an injury to the spinal cord?

☐ 18. From your understanding of the functions of myelin and functions of the white matter tracts of the CNS, explain how the loss of myelination in the CNS, as occurs in multiple sclerosis, can cause the signs and symptoms typical of this disease.

(For answers, see Appendix B.)

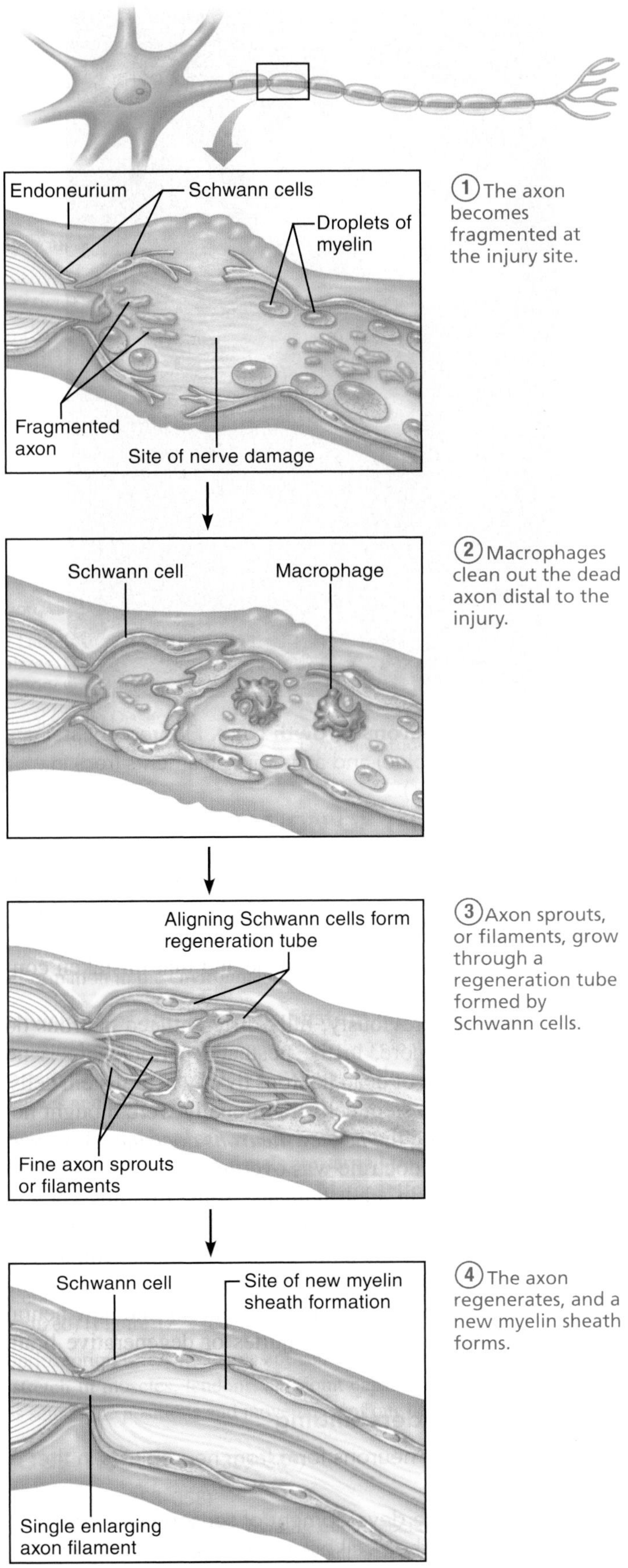
Endoneurium
Schwann cells
Droplets of myelin
Fragmented axon
Site of nerve damage
① The axon becomes fragmented at the injury site.
Schwann cell
Macrophage
② Macrophages clean out the dead axon distal to the injury.
Aligning Schwann cells form regeneration tube
Fine axon sprouts or filaments
③ Axon sprouts, or filaments, grow through a regeneration tube formed by Schwann cells.
Schwann cell
Site of new myelin sheath formation
Single enlarging axon filament
④ The axon regenerates, and a new myelin sheath forms.

Figure 12.15 **Regeneration of an axon in a peripheral nerve.**

NERVOUS TISSUE THROUGHOUT LIFE

learning outcome

▶ Describe the development of the nervous system in the embryo.

Embryonic Development of Nervous Tissue

The nervous system develops from the dorsal ectoderm, which invaginates to form the neural tube and the neural crest (Figure 12.16; see also Figure 3.7 on p. 53). The neural tube, whose walls begin as a layer of pseudostratified **neuroepithelial cells,** becomes the CNS. These cells divide, migrate externally, and become neuroblasts (future neurons), which never again divide, and neuroglial cells (Figure 12.16b). The earliest of these glial cells extend outward from the neuroepithelium and provide pathways along which young neurons migrate to reach their final destinations. Just external to the neuroepithelium, the neuroblasts cluster into an **alar plate** and a **basal plate,** the future gray matter (Figure 12.16c). Dorsally, the neuroblasts of the alar plate become interneurons, which remain in the CNS. Ventrally, the neuroblasts of the basal plate become motor neurons and sprout axons that grow out to the effector organs. (Some interneurons also form from the basal plate and remain among the motor neurons.) Axons that sprout from the young interneurons form the white matter by growing outward along the length of the CNS. These events occur in both the spinal cord and the brain.

Most of the events described so far take place in the second month of development, but neurons continue to form rapidly until about the sixth month. At that time, neuron formation slows markedly, although it may continue at a reduced rate into childhood. Just before neuron formation slows, the early neuroglial cells differentiate into astrocytes and oligodendrocytes. As the division of its cells slows, the neuroepithelium differentiates into the layer of ependymal cells.

Sensory neurons arise not from the neural tube but from the neural crest (Figure 12.16c). This explains why the cell bodies of sensory neurons lie *outside* the CNS. Like motor neurons and interneurons, sensory neurons stop dividing during the fetal period.

Neuroscientists are actively investigating how forming neurons meet and form synapses with one another. As the growing axons elongate at *growth cones* (Figure 12.17), they are attracted by chemical signals such as *neurotrophins* (nu-ro-tro′finz) released from other neurons and astrocytes. At the same time, the receiving dendrites send out thin, wiggling extensions to reach the approaching axons and form synapses. Which synaptic connections are made, and which persist, are determined by two factors: (1) the amount of neurotrophin initially received by an axon and (2) the degree to which a synapse is used after being established. Neurons that make "bad" connections are signaled to die via apoptosis

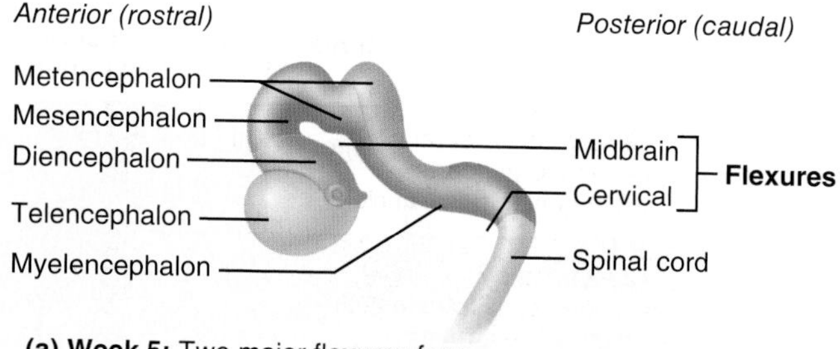

(a) Week 5: Two major flexures form.

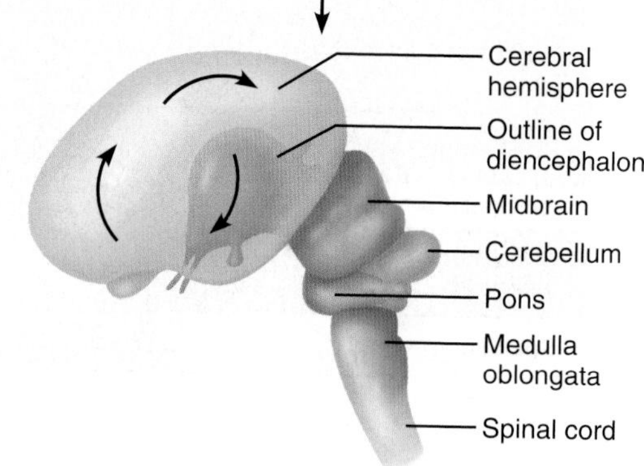

(b) Week 13: Cerebral hemispheres grow posterolaterally, enclosing the diencephalon and the rostral brain stem.

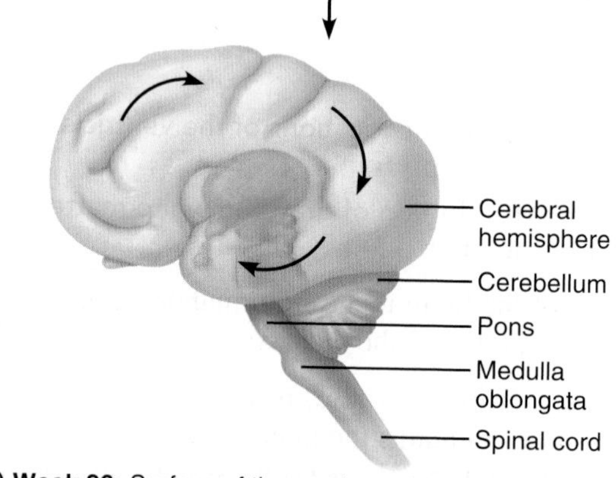

(c) Week 26: Surface of the cerebrum begins to fold.

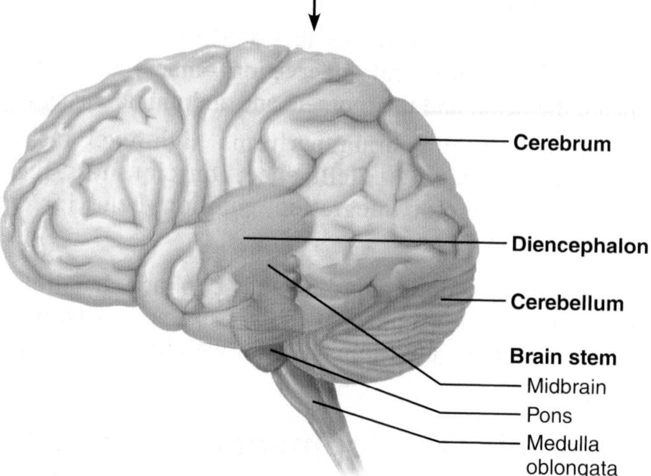

(d) Birth: Brain shows adult pattern of structures and convolutions.

Figure 13.2 Brain development from week 5 to birth. (**a**) Formation of two major flexures by week 5 of development causes the telencephalon and diencephalon to angle toward the brain stem. (**b–d**) Development of the cerebral hemispheres at 13 weeks, 26 weeks, and birth, respectively. The posterolateral growth of the cerebral hemispheres ultimately encloses the diencephalon and the superior aspect of the brain stem, as shown in this see-through view of the cerebral hemisphere. Folding of the surface of the cerebrum begins by week 26, and the convolutions become more distinct as development continues.

appear (Figure 13.1b). These three vesicles are the **prosencephalon** (pros″en-sef′ah-lon), or **forebrain;** the **mesencephalon** (mes″en-sef′ah-lon), or **midbrain;** and the **rhombencephalon** (romb″en-sef′ah-lon), or **hindbrain.** (Note that the word stem *encephalos* means "brain.") The caudal portion of the neural tube becomes the spinal cord.

In week 5, the three primary vesicles give rise to five **secondary brain vesicles** (Figure 13.1c). The prosencephalon divides into the **telencephalon** ("endbrain") and the **diencephalon** ("through-brain"). The mesencephalon remains undivided, but the rhombencephalon divides into the **metencephalon** ("afterbrain") and the **myelencephalon** ("brain most like the spinal cord"). During this period the brain also develops two major bends, or *flexures*—a *midbrain flexure* and a *cervical flexure* (Figure 13.2a).

Each secondary brain vesicle develops rapidly to produce the major structures of the adult brain (Figure 13.1d).

- The telencephalon develops two lateral swellings that look like large mouse ears. These become the large **cerebral hemispheres,** together called the **cerebrum** (ser-e′brum).
- The **diencephalon** develops three main divisions: the thalamus, the hypothalamus, and the epithalamus.
- The mesencephalon forms the **midbrain.**
- The ventral part of the metencephalon becomes the **pons,** and the dorsal roof of the metencephalon develops into the **cerebellum** (ser″e-bel′um).
- The myelencephalon forms the **medulla oblongata** (mĕ-dul′ah ob″long-gah′tah).

The midbrain, pons, and medulla together constitute the **brain stem.** Finally, the central cavity of the neural tube enlarges in certain regions to form the hollow **ventricles** (ven′trĭ-klz; "little bellies") of the brain (Figure 13.1e).

During the late embryonic and the fetal periods, the brain continues to grow rapidly, and changes occur in the relative positions of its parts (Figure 13.2). Space restrictions force the cerebral hemispheres to grow posteriorly over the rest of the brain, and soon these hemispheres completely envelop the diencephalon and midbrain. As each cerebral hemisphere grows, it bends into a horseshoe shape (illustrated by the arrows in Figure 13.2b and c). By week 26, the continued growth of the cerebral hemispheres causes their surfaces to crease and fold (Figure 13.2c), until at birth the hemispheres are wrinkled like a walnut (Figure 13.2d). This infolding allows more neurons to fit in the limited space.

Basic Parts and Organization of the Brain

Anatomists often classify the brain according to four parts (Figure 13.2d): the (1) *brain stem* (medulla oblongata, pons, and midbrain), (2) *cerebellum,* (3) *diencephalon,* and (4) *cerebrum,* composed of the two cerebral hemispheres.

Gray and white matter have a unique distribution in the brain. The gray matter of the CNS contains short, nonmyelinated neurons and neuron cell bodies. The white matter is composed of myelinated and nonmyelinated axons. Like the spinal cord, the brain has an inner region of gray matter adjacent to the hollow ventricles. This inner gray region is surrounded by white matter. However, the brain also has additional regions of gray matter that are not present in the spinal cord (Figure 13.3). This difference occurs because during brain development, certain groups of neurons migrate externally to form collections of gray matter in regions that otherwise consist of white matter. The most extreme examples of this outward migration occur in the cerebellum and cerebrum, where sheets of gray matter lie at the surface of the brain. Each of these external sheets of gray matter is called a **cortex** ("bark on a tree"); the sheet covering the cerebellum is the **cerebellar cortex,** and the one covering the cerebrum is the **cerebral cortex.** All other gray matter in the brain is in the form of clusters of neuron cell bodies called **brain nuclei** (not to be confused with cell nuclei). Functionally, the large amount of gray matter in the brain allows this part of the CNS to perform very complex neural functions—because gray matter contains many small interneurons that process information.

You can visualize the spatial relationships among the basic parts of the brain most easily by first considering the hollow cavities that lie deep in the brain. This knowledge will help you as we explore the four parts of the brain next, our discussion moving from the most caudal region (brain stem) to the most rostral (cerebrum). Therefore, let us begin our survey with a brief look at the brain's deep-lying hollow cavities, the ventricles.

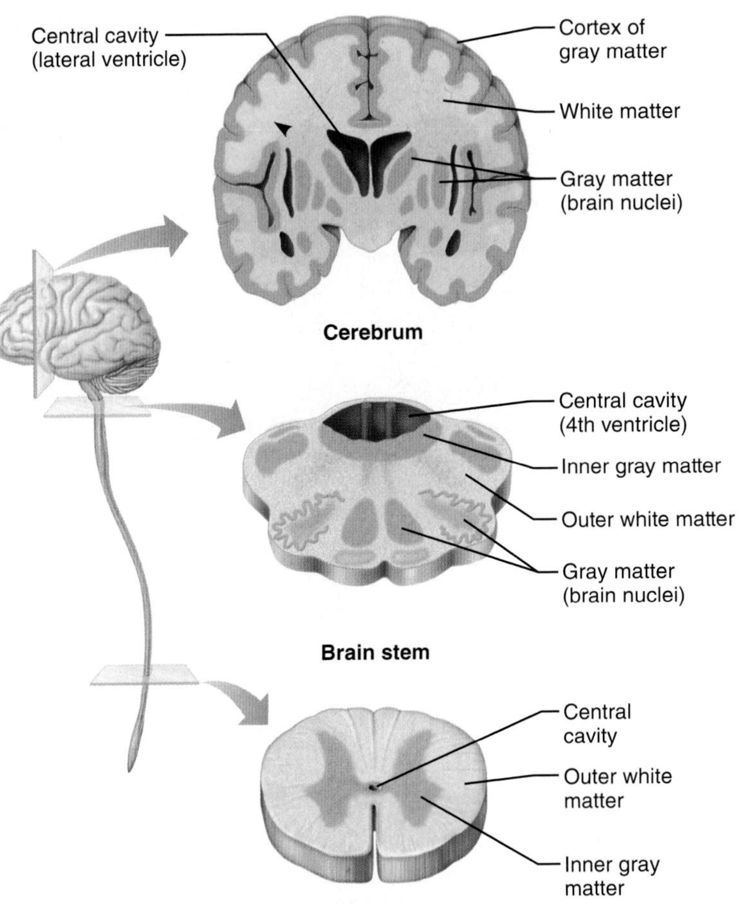

Figure 13.3 **Pattern of gray and white matter in the CNS.** In each section, the dorsal aspect is on top. In general, gray matter surrounds the central cavity of the neural tube, and white matter is located external to the gray matter. In the developing brain, collections of gray matter migrate externally into the white matter. Both the cerebrum (shown) and the cerebellum develop an external cortex of gray matter.

Ventricles of the Brain

The ventricles are expansions of the brain's central cavity, filled with cerebrospinal fluid and lined by ependymal cells (Figure 13.4). They are continuous with one another and with the central canal of the spinal cord.

- The paired **lateral ventricles,** once called the *first* and *second ventricles,* lie in the cerebral hemispheres. Their horseshoe shape reflects the bending of the hemispheres during development. Anteriorly, the two lateral ventricles lie close together, separated by only a thin median membrane called the *septum pellucidum* (sep′tum pĕ-lu′si-dum, "transparent wall"; Figure 13.4a).
- The **third ventricle** lies in the diencephalon. Anteriorly, it connects to each lateral ventricle through an **interventricular foramen.**
- In the midbrain, the thin tubelike central cavity is called the **cerebral aqueduct,** which connects the third and fourth ventricles.
- The **fourth ventricle** lies in the brain stem, dorsal to the pons and the superior half of the medulla oblongata. Three openings occur in the walls of the fourth ventricle: the paired **lateral apertures** and the **median aperture** in its roof. These holes connect the ventricles with the *subarachnoid space,* which surrounds the whole CNS (see p. 406). This connection allows cerebrospinal fluid to fill both the ventricles and the subarachnoid space.
- The fourth ventricle connects caudally to the central canal of the inferior medulla and spinal cord.

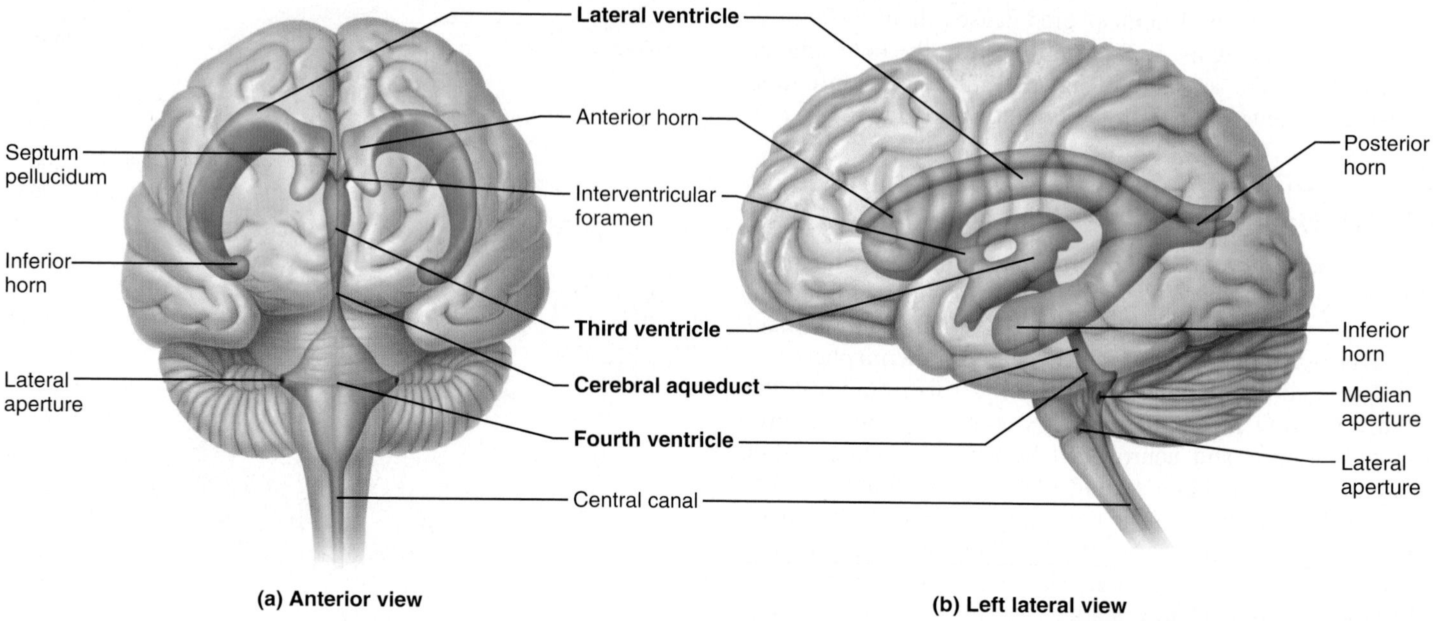

(a) Anterior view **(b) Left lateral view**

Figure 13.4 Ventricles of the brain. The regions of the lateral ventricles, the anterior horn, posterior horn, and inferior horn, are labeled in parts (a) and (b).

check your understanding

- ☐ 1. Which part of the adult brain develops from the telencephalon? From the mesencephalon?
- ☐ 2. What are the brain nuclei?
- ☐ 3. Name the structure that connects the third ventricle to the fourth ventricle. In what part of the brain stem is this structure located?

(For answers, see Appendix B.)

The Brain Stem

learning outcomes

- ▶ Identify the three subdivisions of the brain stem, and list the major structures in each.
- ▶ Relate the structures of the brain stem to the general functions of this part of the brain.

The most caudal of the four major parts of the brain is the brain stem. From caudal to rostral, the three regions of the brain stem are the medulla oblongata, pons, and midbrain (Figures 13.5, 13.6, 13.7). Each region is roughly an inch long, and together they make up only 2.5% of total brain mass. Lying in the posterior cranial fossa of the skull, on the basilar part of the occipital bone, the brain stem has four general functions:

1. It acts as a passageway for all the fiber tracts running between the cerebrum and the spinal cord.
2. It is heavily involved with the innervation of the face and head; 10 of the 12 pairs of cranial nerves attach to it.
3. It produces the rigidly programmed, automatic behaviors necessary for survival.
4. It integrates auditory reflexes and visual reflexes.

The brain stem has the same structural plan as the spinal cord, with outer white matter surrounding an inner region of gray matter (Chapter 12, p. 361). However, brain nuclei of gray matter are also located in the white matter of the brain stem. The three regions of the brain stem are discussed in reference to the general functions of this portion of the brain.

The Medulla Oblongata

The conical medulla oblongata, or simply medulla ("marrow"), is the most caudal part of the brain stem (Figure 13.6). It is continuous with the spinal cord at the level of the foramen magnum of the skull. As mentioned previously, a part of the fourth ventricle lies dorsal to the superior half of the medulla.

White matter fiber tracts connecting the more rostral regions of the brain with the spinal cord must pass through the medulla oblongata.

- Externally, two longitudinal ridges called **pyramids** (Figure 13.6a) flank the ventral midline of the medulla. These ridges are formed by the *pyramidal tracts,* large fiber tracts that originate from pyramid-shaped neurons in the cerebrum and descend through the brain stem and spinal cord carrying voluntary motor output to the spinal cord. In the caudal part of the medulla, 70–90% of these pyramidal fibers cross over to the opposite side of the brain at a point called the **decussation of the pyramids** (de″kus-sa′shun; "a crossing"). The result of this crossover is that each cerebral hemisphere controls the voluntary movements of the opposite side of the body (see Figure 13.17, p. 397).
- Just lateral to each pyramid is the **olive** (Figure 13.6b). This enlargement contains the **inferior olivary nucleus,** a large wavy fold of gray matter viewable in cross section (Figure 13.7c). This brain nucleus is a relay station for

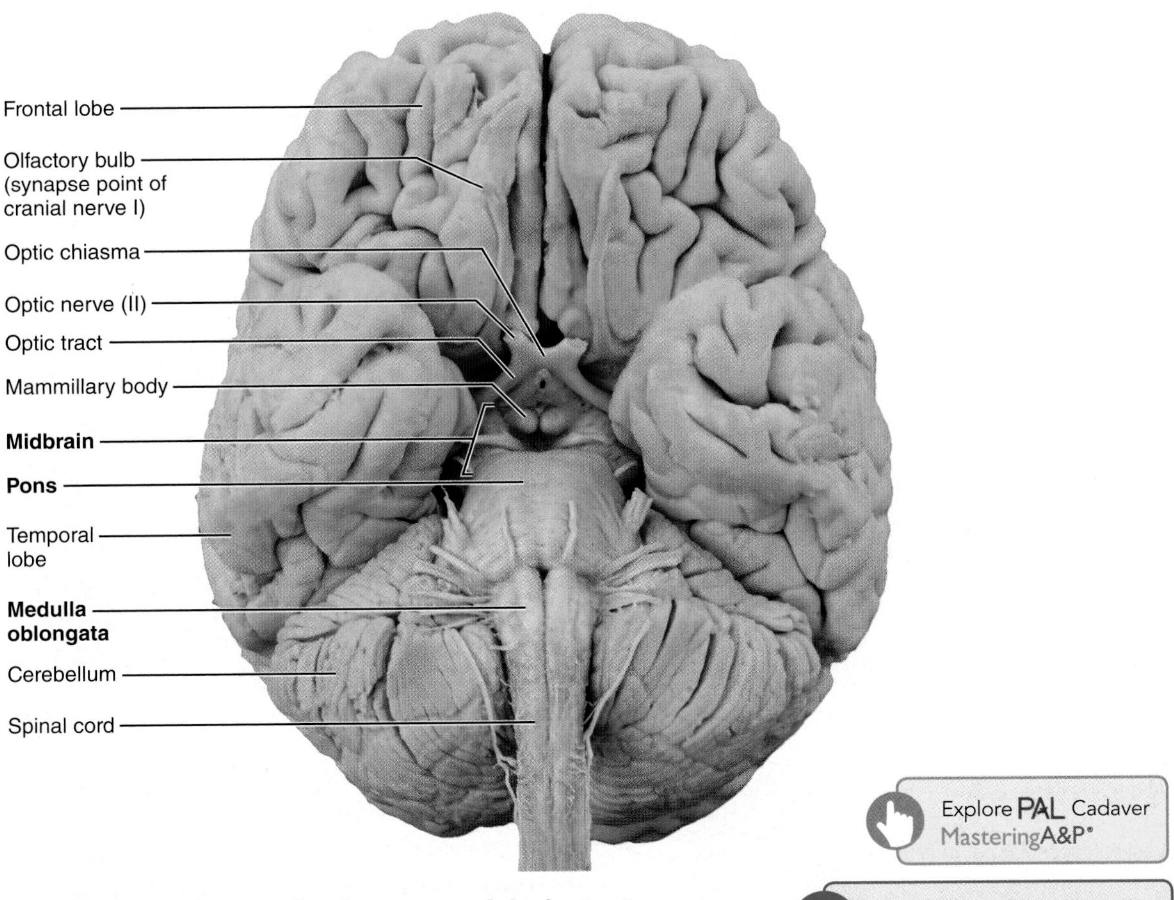

Figure 13.5 Ventral view of the brain showing the three parts of the brain stem: medulla oblongata, pons, and midbrain. Only a small portion of the midbrain is visible in this view.

sensory information traveling to the cerebellum, especially for proprioceptive information ascending from the spinal cord. **Relay nuclei** such as this process and edit information before sending it along. The **inferior cerebellar peduncles** (Figure 13.6c and d) are fiber tracts that connect the medulla to the cerebellum dorsally.

- Internally, two other important relay nuclei are located in the caudal medulla, **nucleus fasiculatus** and **nucleus cuneatus.** Ascending fibers carrying general sensation from the discriminative senses (touch, pressure, limb/joint position) from the skin and proprioceptors synapse in these medullary nuclei along their pathway to the cerebrum.

Four pairs of cranial nerves (discussed in detail in Chapter 14) attach to the medulla (Figure 13.6). The brain nuclei associated with these cranial nerves may be *sensory nuclei* that receive sensory input from the cranial nerve or *motor nuclei* that initiate a motor response which is transmitted on the cranial nerve. The sensory and motor nuclei associated with these cranial nerves lie near the fourth ventricle and can be viewed in a cross section through the medulla (Figure 13.7c).

- The vestibulocochlear nerve (cranial nerve VIII) attaches at the junction of the medulla and the pons and is the sensory nerve of hearing and equilibrium. The *vestibular* and *cochlear nuclei* relay sensory input from each of these nerves to other regions of the brain. They are located on the dorsolateral portion of the medulla.
- The glossopharyngeal nerve (cranial nerve IX) innervates part of the tongue and pharynx. The brain nuclei associated with the glossopharyngeal nerve that are illustrated are the *nucleus ambiguus,* a motor nucleus, and the *solitary nucleus,* a sensory nucleus.
- The vagus nerve (cranial nerve X) innervates many visceral organs in the thorax and abdomen. There are three brain nuclei associated with the vagus nerve: the *dorsal motor nucleus of the vagus,* the *solitary nucleus,* and the *nucleus ambiguus.*
- The hypoglossal nerve (cranial nerve XII) innervates tongue muscles. The *hypoglossal nucleus,* a motor nucleus, is located dorsomedially, just deep to the fourth ventricle.

Running through the core of the brain stem is a loose cluster of brain nuclei called the **reticular formation** (Figure 13.7). The brain nuclei in the reticular formation form three columns on each side that extend the length of the brain stem: (1) the midline **raphe** (ra′fe) **nuclei,** which are flanked laterally by (2) the **medial nuclear group** and then (3) the **lateral nuclear group.** Reticular formation nuclei that are clustered around the motor nuclei of the cranial nerves function to coordinate reflexes and autonomic

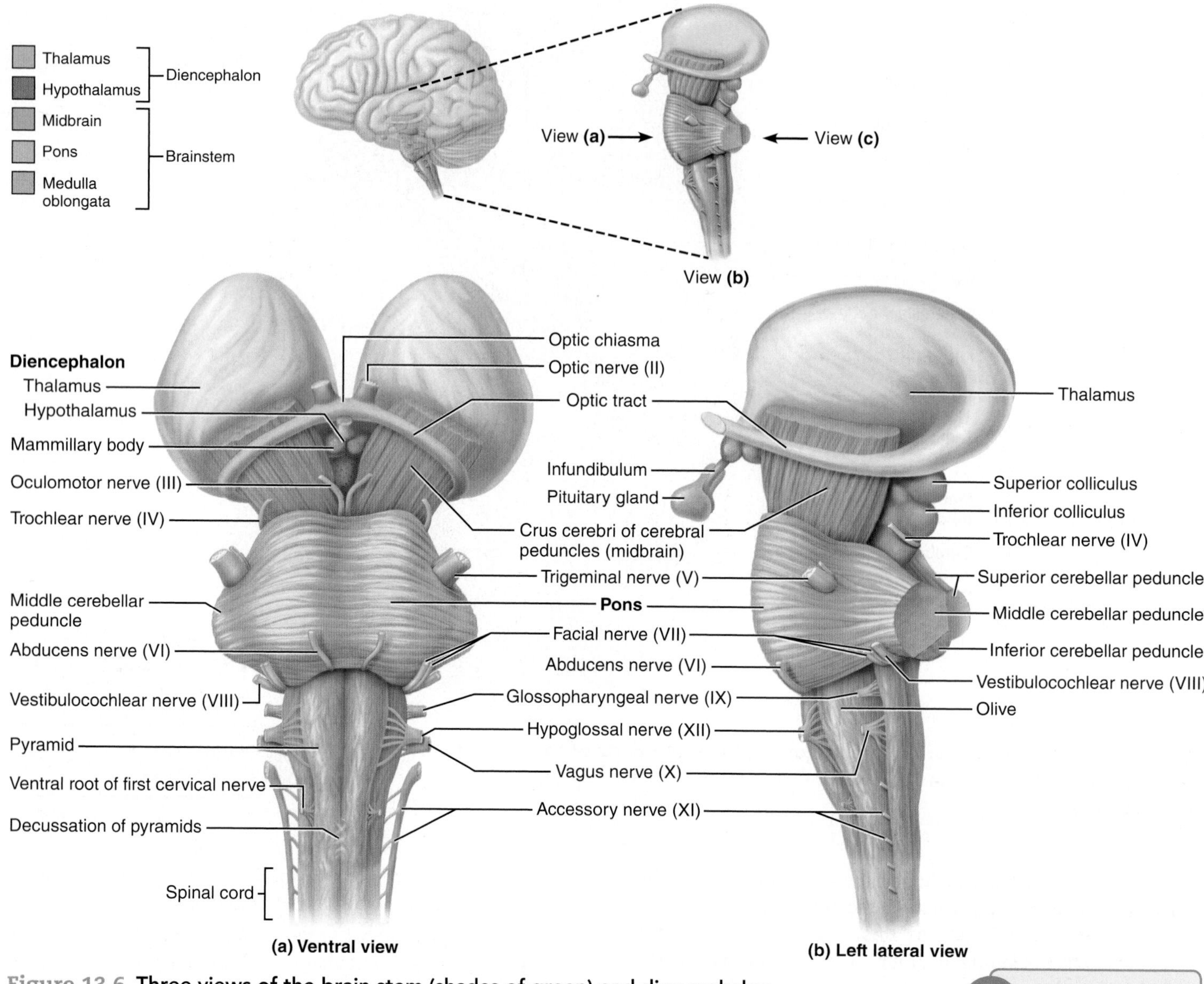

Figure 13.6 Three views of the brain stem (shades of green) and diencephalon (purple).

behaviors involving the cranial nerves. The most important nuclei in the medulla's reticular formation involved with visceral activities are the following:

- The *cardiac center* adjusts the force and rate of the heartbeat.
- The *vasomotor center* regulates blood pressure by stimulating or inhibiting the contraction of smooth muscle in the walls of blood vessels, thereby constricting or dilating the vessels. Constriction of arteries throughout the body causes blood pressure to rise, whereas dilation reduces blood pressure.
- The *medullary respiratory center* controls the basic rhythm and rate of breathing.

The Pons

The second region of the brain stem, the pons, is a bulge wedged between the midbrain and the medulla oblongata (Figure 13.6). Dorsally, it is separated from the cerebellum by the fourth ventricle. As its name suggests, the pons ("bridge") forms a ventral bridge between the brain stem and the cerebellum.

The thick *pyramidal motor tracts* descending from the cerebral cortex pass through the pons ventrally. Interspersed among the fibers of these tracts are numerous **pontine nuclei,** which are relay brain nuclei in a path that connects a portion of the cerebral cortex with the cerebellum (Figure 13.7b). This pathway is involved with the coordination of voluntary movements. The pontine nuclei send axons to the cerebellum in the thick **middle cerebellar peduncles** (Figure 13.6c and d).

Several cranial nerves attach to the pons (Figure 13.6a).

- The trigeminal (cranial nerve V) innervates the skin of the face and the chewing muscles.
- The abducens (cranial nerve VI) innervates an eye-moving muscle.
- The facial (cranial nerve VII) supplies the muscles of facial expression, among other functions.

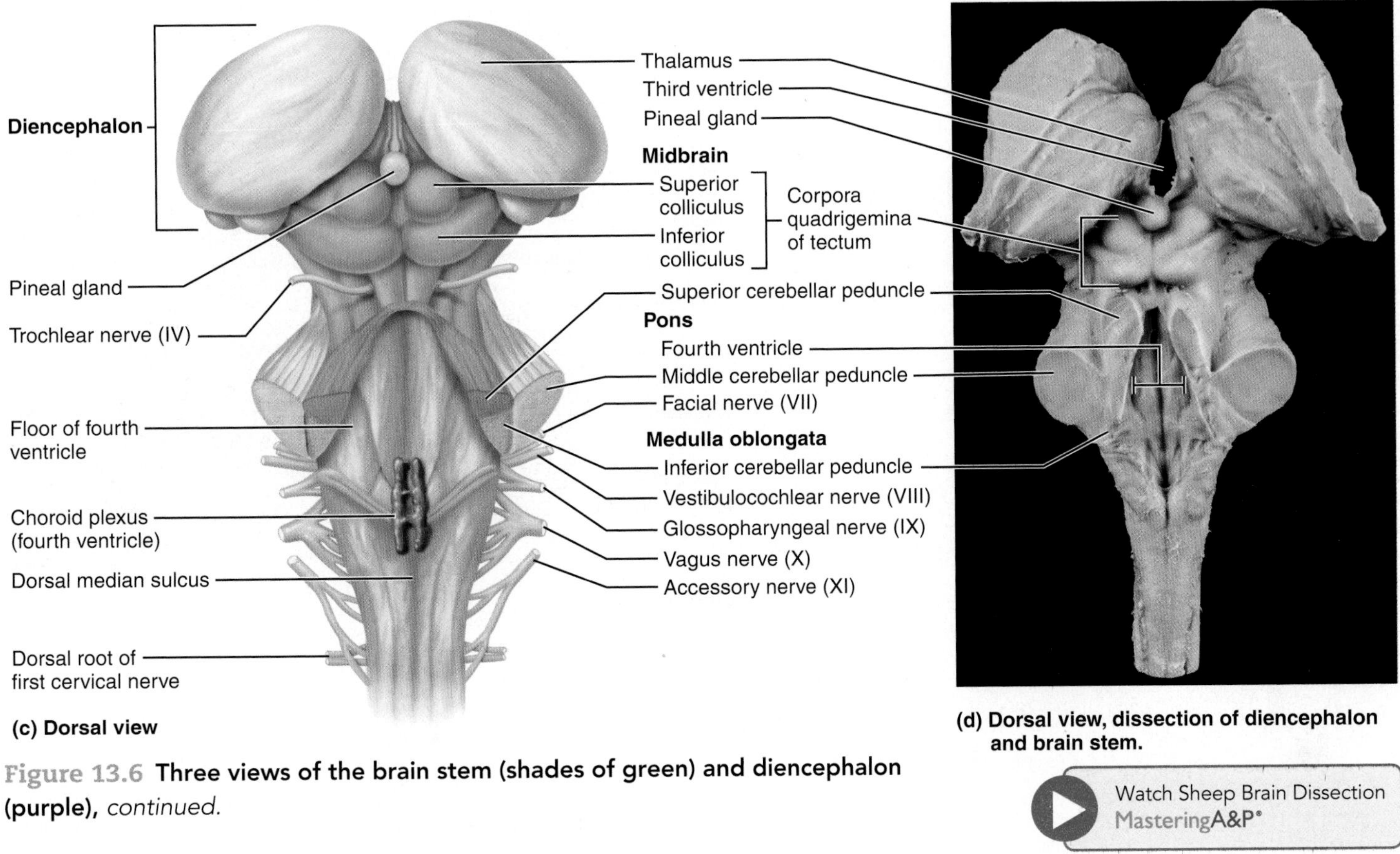

Figure 13.6 Three views of the brain stem (shades of green) and diencephalon (purple), *continued.*

Reticular formation nuclei in the pons, viewable in cross section (Figure 13.7b), lie ventral to the cranial nerve nuclei and function in the control of autonomic behaviors.

The Midbrain

The most rostral of the three regions of the brain stem, the midbrain, lies between the diencephalon and the pons. The central cavity of the midbrain is the cerebral aqueduct, which divides the midbrain into a **tectum** ("roof") dorsally and paired **cerebral peduncles** ventrally (Figure 13.7a).

White fiber tracts and relay nuclei in the midbrain connect regions of the CNS.

- The **cerebral peduncles** on the ventral surface of the brain (Figure 13.6a) form vertical pillars that appear to hold up the cerebrum, hence their name, which means "little feet of the cerebrum." These peduncles contain the *pyramidal motor tracts* descending from the cerebrum toward the spinal cord. The ventral part of each peduncle that contains this tract is the **crus cerebri** ("leg of the cerebrum").
- The **superior cerebellar peduncles,** located dorsally (Figure 13.6c and d), contain fiber tracts that connect the midbrain to the cerebellum.
- In cross section, two pigmented brain nuclei embedded in the white matter of the midbrain are viewable (Figure 13.7a). The bandlike **substantia nigra** (sub-stan′she-ah ni′-grah), whose neuronal cell bodies contain dark melanin pigment, is deep to the pyramidal tracts in the cerebral peduncle. This brain nucleus is functionally linked to the deep gray matter of the cerebrum, the *basal nuclei (ganglia),* and is involved in controlling voluntary movement (pp. 398–399). Degeneration of the neurons in the substantia nigra is the cause of Parkinson's disease.
- The oval **red nucleus** lies deep to the substantia nigra. Its reddish hue is due to a rich blood supply and to the presence of iron pigment in the cell bodies of its neurons. It has a minor motor function: helping to bring about flexion movements of the limbs. The red nucleus is closely associated with the cerebellum.

The cell bodies of motor neurons that contribute to two cranial nerves, the oculomotor (cranial nerve III) and the trochlear (cranial nerve IV) nuclei are located in the ventral part of the gray matter surrounding the cerebral aqueduct, an area called the **periaqueductal gray matter** (Figure 13.7a). These cranial nerves control most of the muscles that move the eyes.

Nuclei that contribute to autonomic behaviors are viewable in a cross section through the midbrain (Figure 13.7a). Around the cerebral aqueduct is the *periaqueductal gray matter,* which has two somewhat related functions. First, it is involved in the "fight-or-flight" (sympathetic) reaction, constituting the midbrain link between the part of the cerebrum that perceives fear and the autonomic pathway that produces the physiological reactions associated with fear: increase in heart rate, skyrocketing blood pressure,

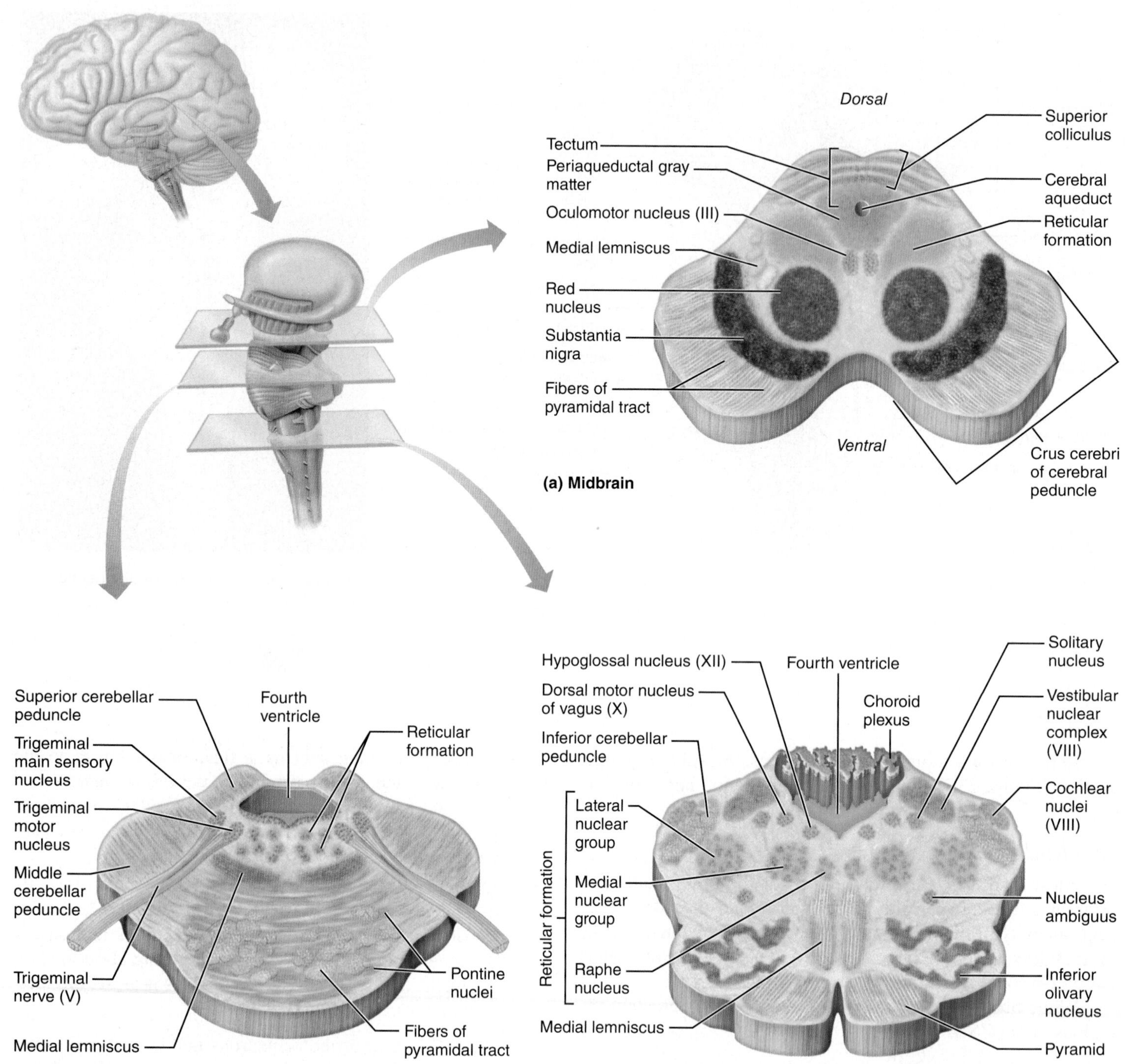

Figure 13.7 Cross sections through the brain stem.

wild fleeing or defensive freezing, and the suppression of pain when the person is injured. Second, the periaqueductal gray matter seems to mediate the response to visceral pain (for instance, nausea), during which it decreases heart rate and blood pressure, produces a cold sweat, and discourages movement.

Auditory and visual reflexes are integrated by brain nuclei in the **corpora quadrigemina** (kwod″rĭ-jem′ĭ-nah; "quadruplets"), which make up the tectum (Figure 13.6c and 13.6d). These brain nuclei form four bumps on the dorsal surface of the midbrain. They are named the *superior* and *inferior colliculi.* The two **superior colliculi** (ko-lik′u-li; "little hills") act in visual reflexes, for example, when the eyes track and follow moving objects even if the person is not consciously looking at the objects. The two **inferior colliculi,** by contrast, belong to the auditory system. Among other functions, they act in reflexive responses to sounds. For instance, when you are startled by a loud noise, your head and eyes reflexively turn toward the sound because of the function of the inferior colliculi.

CLINICAL APPLICATION

Localization of a Brain Stem Injury The location of a brain stem injury is determined by assessing both cranial nerve function and peripheral sensory and motor function. The cranial nerves exit from the brain stem in groups: III and IV from the midbrain; V and VI from the pons; VII and VIII from the junction of the pons and medulla; IX, X, and XII from the medulla. Specific nerve deficits can indicate the region of injury or lesion. Also running through the brain stem in very specific locations are the ascending and descending fiber tracts carrying sensory and motor stimuli between the cerebrum and the periphery. Deficits in sensory and motor function in the periphery can localize an injury more precisely.

check your understanding

☐ 4. Of the general functions of the brain stem described, which does each of the following structures perform? (a) pyramids, (b) vestibular and cochlear nuclei, (c) cerebral peduncles, (d) periaqueductal gray matter, (e) trigeminal nuclei.

☐ 5. In which part of the brain stem are each of the above structures located?

☐ 6. What are the corpora quadrigemina?

(For answers, see Appendix B.)

The Cerebellum

learning outcome

▶ Describe the structure and function of the cerebellum.

The cauliflower-like cerebellum, the second of the brain's major parts as we move caudally to rostrally, makes up 11% of the mass of the brain. Only the cerebrum is larger. The cerebellum is located dorsal to the pons and medulla oblongata, from which it is separated by the fourth ventricle (Figure 13.8). Functionally, the cerebellum smooths and coordinates body movements that are directed by other brain regions, and it helps maintain posture and equilibrium.

The cerebellum consists of two expanded **cerebellar hemispheres** connected medially by the wormlike **vermis** (Figure 13.8c). The surface of the cerebellum is folded into many platelike ridges called **folia** (fo′le-ah; "leaves"), which are separated by deep grooves called **fissures.** Each cerebellar hemisphere is subdivided into three lobes: the large **anterior** and **posterior lobes,** and the small **flocculonodular** (flok″u-lo-nod′u-lar) **lobe.** The anterior and posterior lobes coordinate trunk and limb movements. The flocculonodular lobes are hidden ventral to the posterior lobe (Figure 13.8b). The flocculonodular lobes adjust posture to maintain equilibrium and coordinate head and eye movements.

The cerebellum has three regions: an outer cortex of gray matter, internal white matter called the **arbor vitae** (ar′bor vi′te; "tree of life"), and deeply situated gray matter called **deep cerebellar nuclei** (Figure 13.8d). The cerebellar cortex is a neuron-rich calculator whose function is to smooth our body movements. The arbor vitae consists of axons that carry information to and from the cortex. The deep cerebellar nuclei give rise to axons that relay the instructions from the cerebellar cortex to other parts of the brain.

Information is processed by the cerebellum in the following way:

1. **The cerebellum receives information from the cerebrum on the movements being planned.** The area of the cerebrum that initiates voluntary movements, the *motor cortex* of the cerebrum, passes information from the cerebral cortex through the pontine nuclei in the pons to the lateral portion of the anterior and posterior lobes of the cerebellum.
2. **The cerebellum compares these planned movements with current body position and movements.** Information on equilibrium and head position is relayed from receptors in the inner ear through the vestibular nuclei in the medulla oblongata to the flocculonodular lobe. Information on the current movements of the limbs, neck, and trunk travels from the proprioceptors in muscles, tendons, and joints up the spinal cord to the vermis and medial portions of the anterior and posterior lobes.
3. **The cerebellum sends instructions back to the cerebral cortex on how to resolve any differences between the intended movements and current position.** Using this feedback from the cerebellum, the motor cortex of the cerebrum continuously readjusts the motor commands it sends to the spinal cord, fine-tuning movements so that they are well coordinated.

The cerebellum is also involved in some higher cognitive functions. When learning a new motor skill, the cerebellum refines movements to correct for errors. After practice, the skill is mastered, and it can be performed automatically—for example, riding a bike, playing a musical instrument, or typing on a keyboard. This retention of learned motor skills is called *motor memory.* The cerebellum also plays some role in language, problem solving, and task planning. Overall, its cognitive function may be to recognize, use, and predict sequences of events that we experience or perceive.

The Cerebellar Peduncles

The superior, middle, and inferior cerebellar peduncles are thick fiber tracts that connect the cerebellum to the brain stem (Figure 13.6c and d and Figure 13.8). These fiber tracts carry the information that travels from and to the cerebellum.

- The **superior cerebellar peduncles** connect the cerebellum to the midbrain, carrying primarily efferent instructions from the cerebellum toward the cerebral cortex.

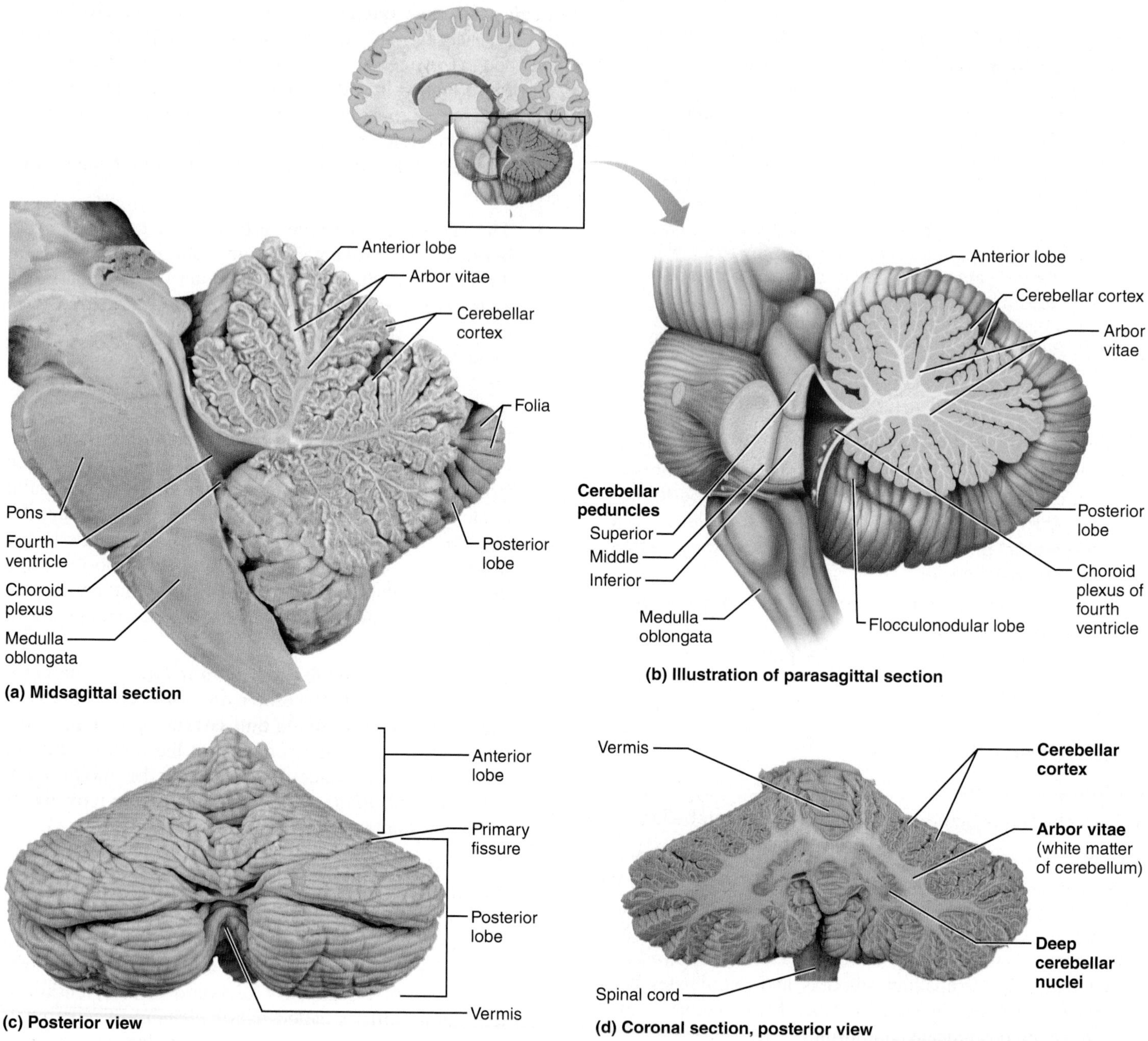

Anterior lobe
Arbor vitae
Cerebellar cortex
Folia
Pons
Fourth ventricle
Choroid plexus
Medulla oblongata
Posterior lobe

(a) Midsagittal section

Anterior lobe
Cerebellar cortex
Arbor vitae
Cerebellar peduncles
Superior
Middle
Inferior
Medulla oblongata
Flocculonodular lobe
Posterior lobe
Choroid plexus of fourth ventricle

(b) Illustration of parasagittal section

Anterior lobe
Primary fissure
Posterior lobe
Vermis

(c) Posterior view

Vermis
Cerebellar cortex
Arbor vitae (white matter of cerebellum)
Deep cerebellar nuclei
Spinal cord

(d) Coronal section, posterior view

Figure 13.8 Cerebellum.

- The **middle cerebellar peduncles** connect the pons to the cerebellum and carry efferent information from the cerebral cortex and the pontine nuclei into the cerebellum.
- The **inferior cerebellar peduncles** arise from the medulla and carry primarily afferent fibers from the vestibular nuclei (equilibrium) and from the spinal cord (proprioception) into the cerebellum.

All fibers that enter and leave the cerebellum are **ipsilateral** (ip″sĭ-lat′er-al; *ipsi* = same), which means that they run to and from the *same* side of the body.

CLINICAL APPLICATION

Injury to the Cerebellum Damage to the anterior and posterior lobes of the cerebellar hemispheres leads to disorders of coordination, characterized by slow or jerky movements that tend to overreach their targets. People with such damage are unable to touch their finger to their nose with their eyes closed. Speech may be slurred, slow, and singsong. Damage to the flocculonodular nodes, by contrast, leads to disorders of equilibrium. Affected individuals have a wide stance and an unsteady drunken gait that predisposes them to falling. In all cases, damage in one hemisphere of the cerebellum (left or right) affects that same side of the body.

☑ **check your understanding**

☐ 7. Name the structure that connects the two cerebellar hemispheres.

☐ 8. What type of sensory information does the flocculonodular lobe receive?

☐ 9. Name the three white fiber tracts that connect the cerebellum to the brain stem. Which tract connects to which portion of the brain stem?

(For answers, see Appendix B.)

The Diencephalon

learning outcome

▶ Name the three divisions of the diencephalon and their functions.

The diencephalon, the third of the four main parts of the brain as we move from caudal to rostral, forms the central core of the forebrain and is surrounded by the cerebral hemispheres (Figure 13.9). The diencephalon consists largely of three paired structures—the thalamus, the hypothalamus, and the epithalamus. These border the third ventricle and consist primarily of gray matter.

The Thalamus

The egg-shaped **thalamus** is a paired structure (see Figure 13.6c, p. 381) that makes up 80% of the diencephalon and forms the superolateral walls of the third ventricle (Figure 13.9b). *Thalamus,* a Greek word meaning "inner room," describes this deep brain region well. Usually the right and left parts of the thalamus are joined by a small midline connection, the **interthalamic adhesion** (*intermediate mass*) (Figure 13.9a).

The thalamus contains about a dozen major nuclei, each of which sends axons to a particular portion of the cerebral cortex. A Y-shaped sheet of white matter, the *internal medullary lamina* divides the thalamic nuclei into three groups: the anterior nuclei, the medial group including the medial dorsal nucleus, and the large lateral nuclear group (Figure 13.10a). Afferent impulses from all the conscious senses except olfaction (sense of smell) converge on the thalamus and synapse in at least one of its nuclei. For example,

- The ventral posterolateral nuclei act as relay stations for the sensory information ascending to the primary sensory areas of the cerebral cortex.
- The medial geniculate body receives auditory input and links to the auditory cortex.
- The lateral geniculate body receives visual input and transmits to the visual cortex.

Sensory inputs are not the only type of information relayed through the thalamus. *Every part of the brain that communicates with the cerebral cortex must relay its signals through a nucleus of the thalamus.* The thalamus can therefore be thought of as the "gateway" to the cerebral cortex.

The thalamus not only relays information to the cerebral cortex but also processes the information as it passes through. The thalamic nuclei organize and then either amplify or "tone down" the signals headed for the cerebral cortex. This is why, for example, you can focus on a conversation with a single person in large, noisy cafeteria.

The Hypothalamus

The **hypothalamus** ("below the thalamus") is the inferior portion of the diencephalon (Figure 13.9). It forms the inferolateral walls of the third ventricle. On the underside of the brain (see Figure 13.5, p. 379), the hypothalamus lies between the **optic chiasma** (point of crossover of cranial nerves II, the optic nerves) and the posterior border of the **mammillary bodies,** rounded bumps that bulge from the hypothalamic floor (*mammillary* = "little breast"). Projecting inferiorly from the hypothalamus is the *pituitary gland* (discussed in Chapter 17).

The hypothalamus, like the thalamus, contains about a dozen brain nuclei of gray matter (Figure 13.10b). Functionally, the hypothalamus is the main visceral control center of the body, regulating many activities of the visceral organs. Its functions include the following:

- **Control of the autonomic nervous system.** Recall that the autonomic nervous system is composed of the peripheral motor neurons that regulate contraction of smooth and cardiac muscle and secretion from glands (p. 352). The hypothalamus is one of the major brain regions involved in directing the autonomic neurons. In doing so, the hypothalamus regulates heart rate and blood pressure, activity of the digestive tract, the secretion from sweat glands and salivary glands, and many other visceral activities. The hypothalamus exerts its control over visceral functions by relaying its instructions through the periaqueductal gray matter of the midbrain (p. 381) and the reticular formation of the brain stem (p. 379), which then carry out those instructions.
- **Regulation of body temperature.** The body's thermostat is in the hypothalamus. Some hypothalamic neurons monitor blood temperature and receive input from peripheral thermoreceptors. The hypothalamus initiates the body's cooling or heating mechanisms as needed (sweating or shivering, respectively). Hypothalamic centers also induce fever.
- **Regulation of hunger and thirst sensations.** By sensing the concentrations of nutrients and salts in the blood, certain hypothalamic neurons mediate feelings of hunger and thirst and thus aid in maintaining the proper concentrations of these substances.
- **Regulation of sleep-wake cycles.** Acting with other brain regions, the hypothalamus helps regulate the complex phenomenon of sleep. The **suprachiasmatic** (soo″-prah-ki-az-mat′ik) **nucleus** (Figure 13.10b) is the body's biological clock. It generates the daily circadian rhythms and synchronizes these cycles in response to dark-light information sensed via the optic nerve. In response to such signals, the **preoptic nucleus** induces sleep. Other hypothalamic nuclei near the mammillary body mediate arousal from sleep.
- **Control of the endocrine system.** The hypothalamus controls the secretion of hormones by the pituitary gland,

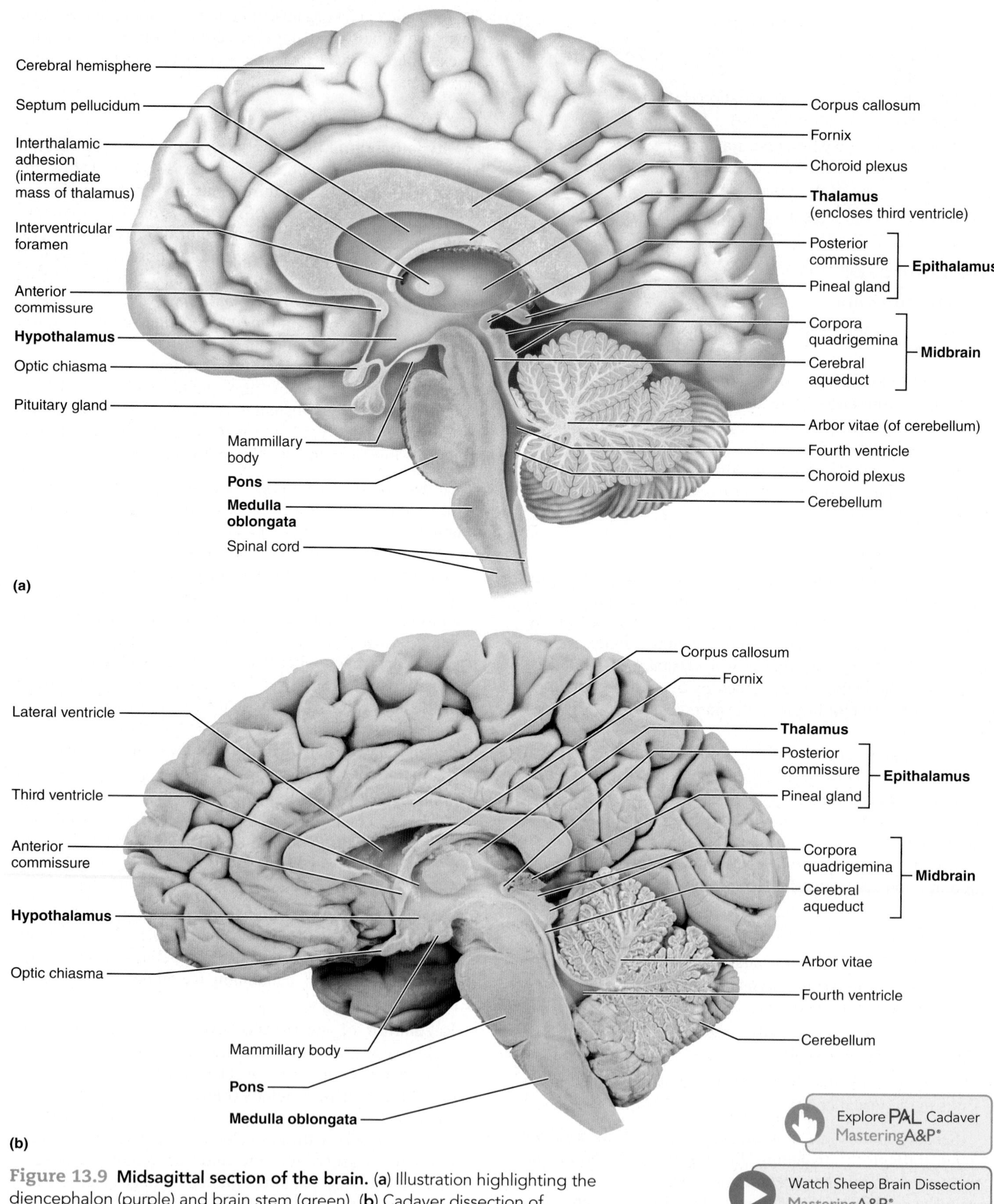

Figure 13.9 Midsagittal section of the brain. (**a**) Illustration highlighting the diencephalon (purple) and brain stem (green). (**b**) Cadaver dissection of same view.

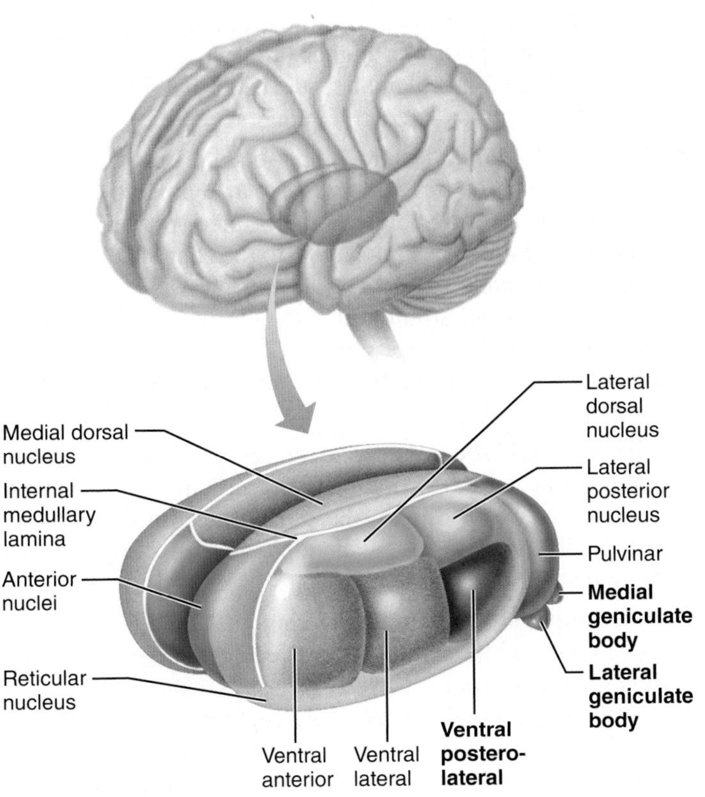

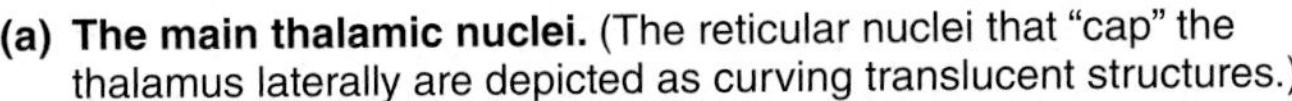
(a) **The main thalamic nuclei.** (The reticular nuclei that "cap" the thalamus laterally are depicted as curving translucent structures.)

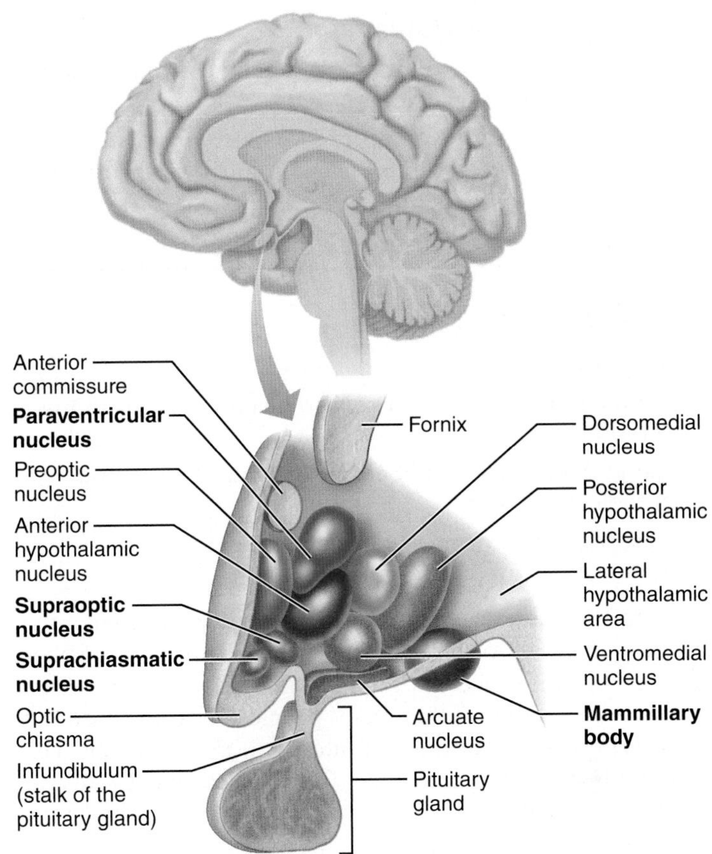

(b) The main hypothalamic nuclei

Figure 13.10 **Nuclei of the thalamus and hypothalamus.**

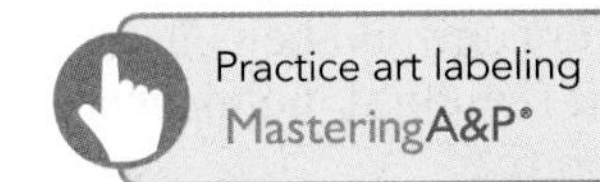

which in turn influences the activity of many other endocrine organs (discussed in Chapter 17).

- **Control of emotional responses.** The hypothalamus lies at the center of the emotional part of the brain, the *limbic system* (discussed later in this chapter). Regions involved in pleasure, rage, and fear are located in the hypothalamus.
- **Control of motivational behavior.** The hypothalamus controls behavior that is rewarding. For example, the hypothalamus influences our motivation for feeding, thereby determining how much we eat, and also influences sex drive and sexual behavior.
- **Formation of memory.** The brain nucleus in the mammillary body receives many inputs from the major memory-processing structure of the cerebrum, the *hippocampal formation*.

Lesions of the hypothalamus cause disorders in visceral functions and in emotions. Thus, injuries to the hypothalamus can result in severe weight loss or obesity, sleep disturbances, dehydration, and a broad range of emotional disorders.

The Epithalamus

The **epithalamus,** the third and most dorsal part of the diencephalon (Figure 13.9), forms part of the roof of the third ventricle. It consists of one tiny group of brain nuclei and a small, unpaired knob called the **pineal gland** (pin′e-al; "pine-cone-shaped"). This gland, which derives from ependymal glial cells, is a hormone-secreting organ. Under the influence of the hypothalamus, the pineal gland secretes the hormone **melatonin** (mel″ah-to′nin), which signals the body to prepare for the nighttime stage of the sleep-wake cycle.

CLINICAL APPLICATION

Why Won't Teenagers Sleep at Night?

Elevated melatonin levels get the body ready for sleep. The timing of the rise in melatonin levels changes during a lifetime. In youth and childhood, melatonin levels rise in the early evening hours, and thus young children tire early in the night. In adolescence, the time at which melatonin levels rise shifts to much later in the night. As a result, teens are not sleepy until quite late in the night. This physiological shift makes it difficult for teenagers to fall asleep at a reasonable hour. The sleep deficit is compounded by the fact that most secondary schools start early in the morning. A few school systems have recognized this biological constraint and have changed the start of high school to later in the morning, allowing their students to get their needed sleep.

check your understanding

☐ 10. To which region of the brain do the thalamic neurons send their axonal fibers?

☐ 11. What part of the diencephalon functions as the main visceral control center? What is meant by the phrase *visceral control*?

(For answers, see Appendix B.)

The Cerebrum

learning outcomes

- List the major lobes, fissures, and functional areas of the cerebral cortex.
- Name three classes of fiber tracts in the white matter of the cerebrum.
- Describe the form and function of the basal nuclei (ganglia).

As noted earlier, the cerebrum, the most rostral portion of the brain, is made up of two cerebral hemispheres and together these hemispheres account for 83% of total brain mass. They so dominate the brain that many people mistakenly use the word *brain* when referring specifically to the *cerebrum.* The cerebral hemispheres cover the diencephalon and the rostral brain stem in much the same way a mushroom cap covers the top of its stalk.

The various fissures evident on and around the cerebral hemispheres separate the major portions of the brain from one another. The **transverse cerebral fissure** separates the cerebral hemispheres from the cerebellum inferiorly (Figure 13.11a), whereas the median **longitudinal fissure** separates the right and left cerebral hemispheres from each other (Figure 13.11b). The cerebrum is composed of a superficial *cerebral cortex* of gray matter, the *cerebral white matter* internal to it, and the *deep gray matter of the cerebrum* within the white matter. Let us now examine each of these regions.

Lobes of the Cerebral Cortex

There are many shallow grooves on the surface of the cerebral hemispheres called **sulci** (sul′ki; singular: **sulcus,** "furrow"). Between the sulci are twisted ridges of brain tissue called **gyri** (ji′ri; singular: **gyrus,** "twister"; Figure 13.11c). The more prominent gyri and sulci are similar in all people and are important anatomical landmarks. Some of the deeper sulci divide each cerebral hemisphere into five major lobes: the *frontal, parietal, occipital,* and *temporal lobes,* and the *insula* (Figure 13.11, Table 13.1). Most of these lobes are named for the skull bones overlying them.

- The **frontal lobe** is located deep to the frontal bone and fills the anterior cranial fossa. It extends posteriorly to the **central sulcus,** which separates the frontal lobe from the parietal lobe. The **precentral gyrus** containing the *primary motor cortex* lies just anterior to the central sulcus. The frontal lobe contains functional areas that plan, initiate, and enact motor movement including eye movement and speech production. The most anterior region of the frontal cortex performs higher-order cognitive functions, such as thinking, planning, decision making, working memory, and other executive functions.
- The **parietal lobe,** deep to the parietal bones, extends posteriorly from the central sulcus to the **parieto-occipital sulcus.** The **lateral sulcus** forms its inferior boundary. The **postcentral gyrus,** just posterior to the central sulcus, contains the *primary somatosensory cortex.* The parietal lobe processes sensory stimuli allowing (1) conscious awarness of general somatic sensation; (2) spacial awarness of objects, sounds and body parts; and (3) understanding of speech.
- The **occipital lobe** lies deep to the occipital bone and forms the most posterior portion of the cerebrum. It is separated from the parietal lobe by the **parieto-occipital sulcus** on the medial surface of the hemiphere. The occipital lobe contains the visual cortex.
- The **temporal lobe,** on the lateral side of the hemisphere, lies in the middle cranial fossa deep to the temporal bone. It is separated from the overlying parietal and frontal lobes by the deep **lateral sulcus.** The temporal lobe contains the auditory cortex and the olfactory cortex. It also functions in the recognition of objects, words, and faces; in language comprehension; and in emotional response and memory.
- The **insula** (in′su-lah; "island") is buried deep within the lateral sulcus and forms part of its floor (Figure 13.11d). The insula is covered by parts of the temporal, parietal, and frontal lobes. The visceral sensory cortex for taste and general visceral sensations are in the insula.

Functional Areas of the Cerebral Cortex

The cerebral cortex is the "executive suite" of the nervous system, the home of the conscious mind. It enables people to be aware of themselves and their sensations, to initiate and control voluntary movements, and to communicate, remember, and understand. Because it is composed of gray matter, the cerebral cortex contains neuron cell bodies, dendrites, and very short unmyelinated axons but no fiber tracts. Even though it is only 2–4 mm thick, its many gyri and sulci triple its surface area to approxmately 2500 cm^2, about the size of a large desk calendar, and it accounts for about 40% of the total mass of the brain.

The cerebral cortex contains billions of neurons arranged in six layers. In 1909, a German neurologist, Korbinian Brodmann, mapped the cerebral cortex into 47 structural areas based on subtle variations in the thickness of the six layers. With Brodmann's structural map emerging, early neurologists were eager to localize *functional* regions of the cerebral cortex. Historically, the methods for determining the functions of brain regions were quite crude: studying the mental defects of people and animals that had sustained localized brain injuries, recording neuronal activity in the brains of uninjured animals, or stimulating and recording electrical activity from the surface of the exposed cerebrum during brain surgery. Since about

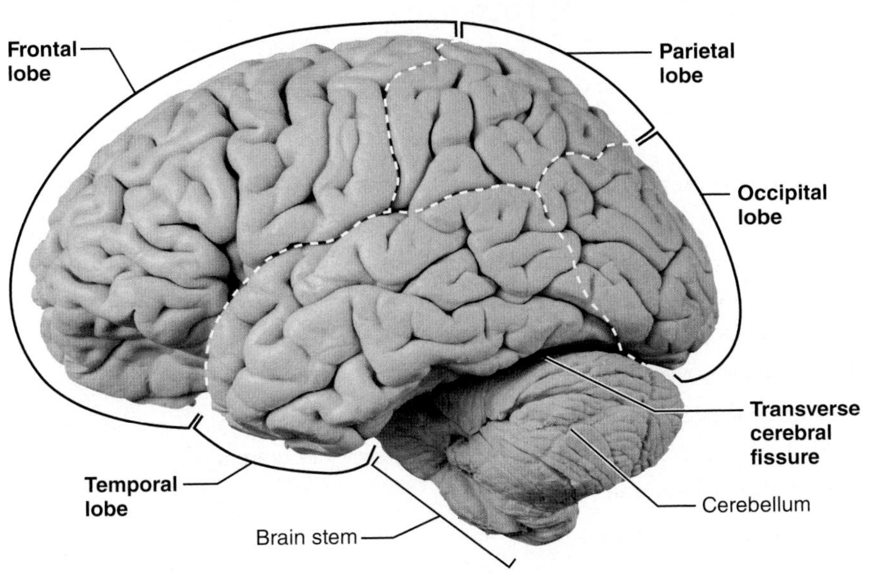

(a) Left lateral view of brain

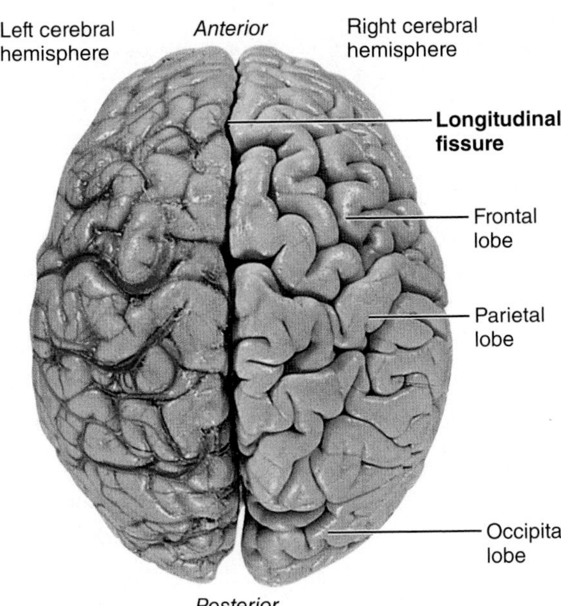

(b) Superior view

Precentral gyrus
Central sulcus
Postcentral gyrus
Frontal lobe
Parietal lobe
Parieto-occipital sulcus
Lateral sulcus
Occipital lobe
Temporal lobe
Fissure (a deep sulcus)
Gyrus
Cortex (gray matter)
Sulcus
White matter

(c) Lobes and sulci of the cerebrum

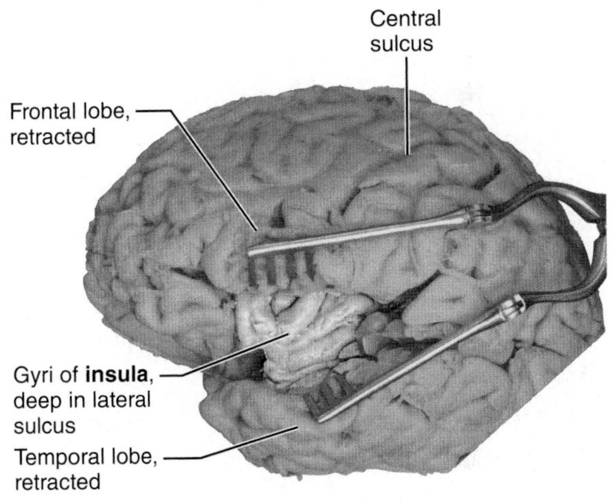

(d) Location of the insula lobe

Figure 13.11 Lobes, sulci, and fissures of the cerebral hemispheres.

Table 13.1 Structures and Functions of the Cerebrum

Structure	Function (functional area)
CEREBRAL CORTEX	
Frontal lobe	Voluntary movement (primary motor cortex) Planning movement (premotor cortex) Eye movement (frontal eye field) Speech production (Broca's area) Executive cognitive functions (anterior association area) Emotional response (limbic association area)
Parietal lobe	General somatic sensation (somatosensory cortex and association area) Spatial awareness of objects, sounds, body parts (posterior association area) Understanding speech (Wernicke's area)
Occipital lobe	Vision (visual cortex and association areas)
Temporal lobe	Hearing (auditory cortex and association area) Smell (olfactory cortex) Object identification (posterior association area) Emotional response, memory (limbic association area)
Insula	Taste (gustatory cortex)
CEREBRAL WHITE MATTER	
Commissural fibers	Connect the corresponding cortices of the two hemispheres
Association fibers	Connect the cortex of the different parts of same hemisphere
Projection fibers	Connect the cortex to more caudal parts of the CNS
DEEP CEREBRAL GRAY MATTER	
Basal nuclei (ganglia)	Control movements in conjunction with the motor cortex
Basal forebrain nuclei	Perform major role in arousal, learning, memory, and motor control; rich in cholinergic fibers
Claustrum	Function unclear; may integrate information between the cerebral cortex and the limbic system

1990, two **functional neuroimaging techniques—PET** and **fMRI** (pp. 17–19)—have revolutionized brain science. These techniques reveal areas of maximal metabolic activity and blood flow to the brain, which are presumed to be areas participating in the mental task being performed. A fMRI obtained during a variety of tasks (Figure 13.12) indicates areas of increased blood flow, as shown by the red-yellow color, in the regions of the cerebral cortex associated with each task.

Three general kinds of functional areas are recognized in the cerebral cortex (see Figure 13.13): *sensory areas,* which allow conscious awareness of sensation; *association areas,* which integrate diverse information to enable purposeful action; and *motor areas,* which control voluntary motor functions. There is a sensory area for each of the major senses. Each region is called a **primary sensory cortex.** Each primary sensory cortex has association areas linked to it that process the sensory information. These areas are **sensory association areas.** Other association areas receive and integrate input from multiple regions of the cerebral cortex. These regions are called **multimodal association areas.** Finally, the regions of the cortex that plan and initiate voluntary motor functions are called the **motor areas.** Information is processed through these regions of the cerebral cortex in the following hierarchical manner.

1. Sensory information is received by the primary sensory cortex, and the arrival of this information results in awareness of the sensation.
2. The information is relayed to the sensory association area that gives meaning to the sensory input.
3. The multimodal association areas receive input in parallel from multiple sensory association areas, integrating all of the sensory input to create a complete understanding of the sensory information. These regions also integrate sensory input with past experience and develop a motor response.
4. The motor plan is enacted by the motor cortex.

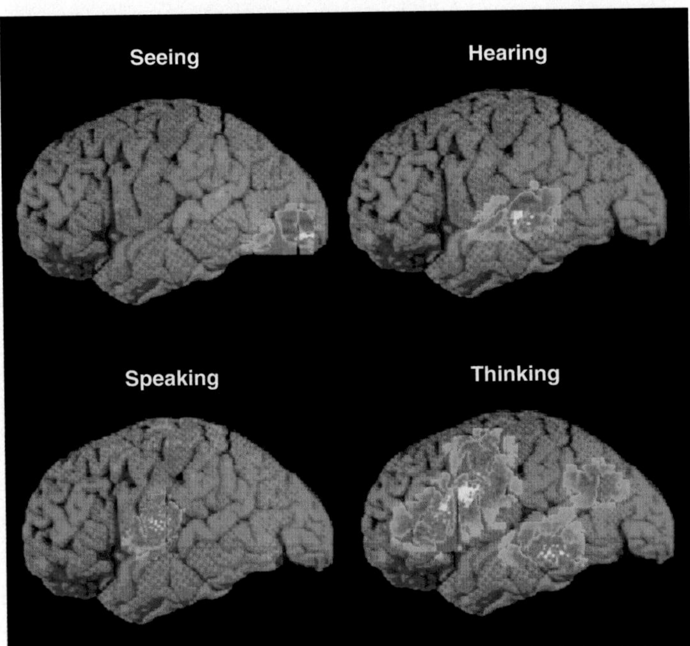

Figure 13.12 **Functional neuroimaging (fMRI) of the cerebral cortex.** Red and orange areas show increased blood flow in the cortical regions responsible for carrying out each activity.

For example, a loud blaring horn interrupts class:

1. The primary sensory cortex enables you to hear the horn.
2. The auditory association cortex indicates that the horn is a fire alarm.
3. The multimodal association areas integrate the sound with past experience (it is a beautiful spring day, the usual time for fire drills) and with other sensory information (there is no smell of smoke, nor sound of any fire engine sirens approaching the school). These areas then develop a motor response (pick up your bag and walk out the nearest exit).
4. The motor cortex initiates the movements necessary to enact the motor plan and you exit the classroom.

As you know, all of this processing occurs in rapid sequence, resulting in a motor response within a fraction of a second.

The functional areas of the cerebral cortex are discussed below in greater detail. As you read, do not confuse motor and sensory *areas* with motor and sensory *neurons.* All the neurons in the cerebral cortex are interneurons, which by definition are confined entirely to the CNS.

Sensory Areas The areas of the cerebral cortex involved with conscious awareness of sensation occur in parts of the parietal, temporal, and occipital lobes. These areas are shown in the dark violet (Figure 13.13). There is a distinct cortical area, a primary sensory cortex, for each of the major senses: the general somatic senses and the special senses of vision, hearing, balance, olfaction, and taste. The primary sensory cortex makes you aware of sensory stimuli: You hear a sound, see an image, or feel pain in your leg. The sensory association areas adjacent to the primary sensory area interpret the sensory stimulus and gives meaning to the sensation: The sound is a bell, the image is my daughter, the pain is from a bruise. These sensory association areas are included in this discussion and are shown in teal (Figure 13.13a).

1. **Somatosensory areas.** The **primary somatosensory cortex** receives information from the general somatic senses (touch, pressure, vibration, pain, and temperature from the skin and proprioception from the muscles and joints) and enables conscious awareness of these sensations. It is located along the postcentral gyrus of the parietal lobe (Figure 13.13). General somatic sensations are picked up by sensory receptors in the periphery of the body and relayed through the spinal cord, brain stem, and thalamus to the primary somatosensory cortex. There, cortical neurons process the information and identify the precise area of the body being stimulated.

   Each region of the somatosensory cortex receives sensory stimuli from a specific area of the body; thus, sensory input from the body can be "mapped" onto the postcentral gyrus. This map of the primary sensory cortex (Figure 13.14, right half) is called a **sensory homunculus** (ho-mung′ku-lus; "little man"). Notice that the body is represented upside down, with the face on the inferolateral part of the postcentral gyrus and the toes at the superomedial end. The amount of somatosensory cortex devoted to a body region is related to the sensitivity of that region—that is, to the number of its sensory receptors. The lips and fingertips are our most sensitive body parts and, hence, the largest parts of the sensory homunculus.

   The primary somatosensory cortex exhibits **contralateral projection** from the sensory receptors to the sensory cortex. This means that the right cerebral hemisphere receives its sensory input from the left side of the body, and the left cerebral hemisphere receives its sensory input from the right side of the body. As you will see, many areas of the brain exhibit contralateral projection.

   Damage to the primary somatosensory cortex destroys the conscious ability to feel and localize touch, pressure, and vibrations on the skin. Most ability to feel pain and temperature is also lost, although these two sensations can still be felt in a vague, poorly localized way.

CLINICAL APPLICATION

Phantom Limb Pain People who have had a limb amputated commonly perceive pain in the "phantom" body part that is no longer present. The pain is often induced by touching the skin of the limb stump or the face. The amputation removes all axonal input to the limb representation on the primary somatosensory cortex. This denervated cortical area is reinnervated by growing axons from the nearby cortical regions that contain face or stump representations; thus, touches on the face or stump are interpreted as coming from the now-missing limb.

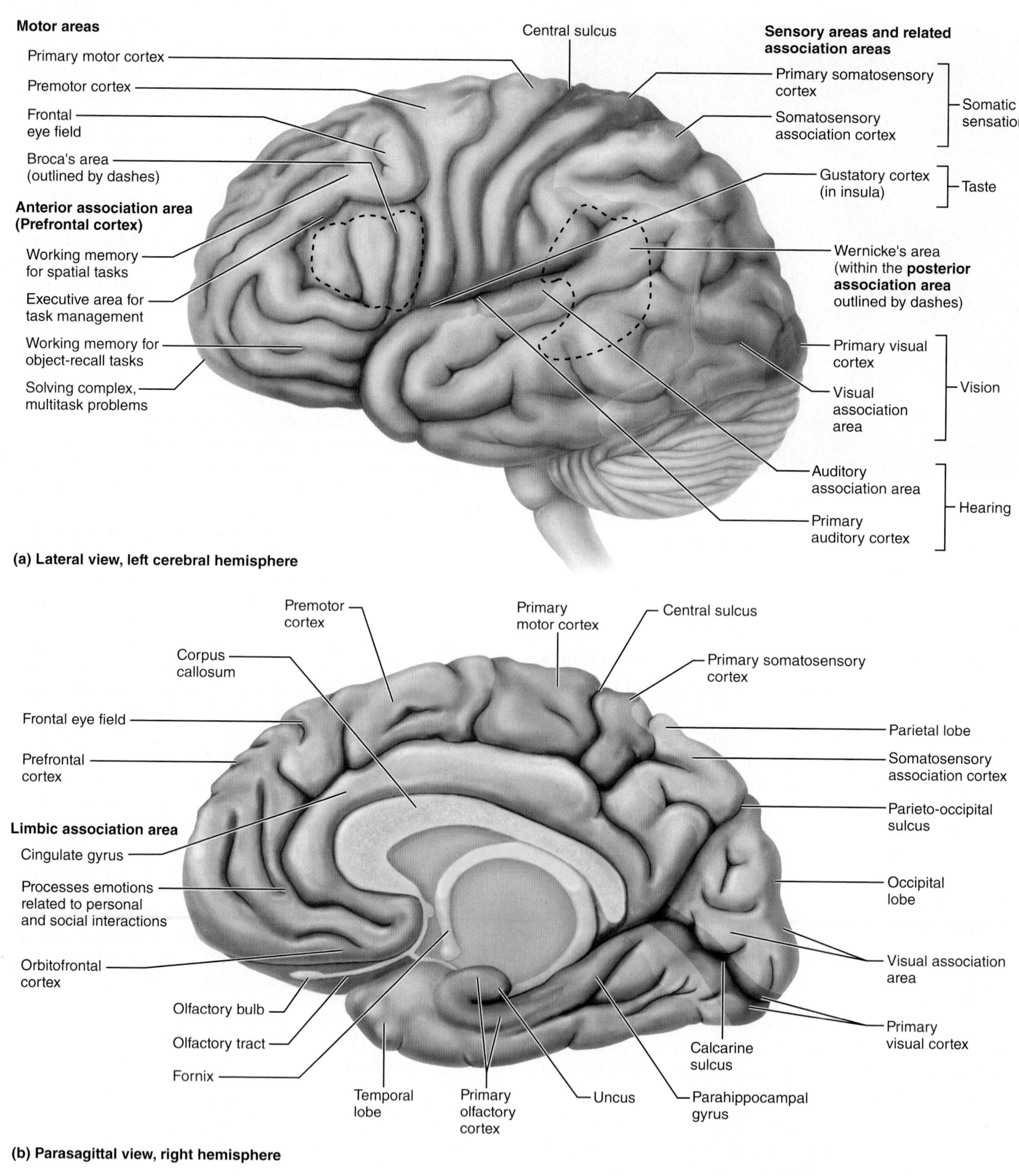

Figure 13.13 **Functional areas of the cerebral cortex.**

Practice art labeling MasteringA&P®

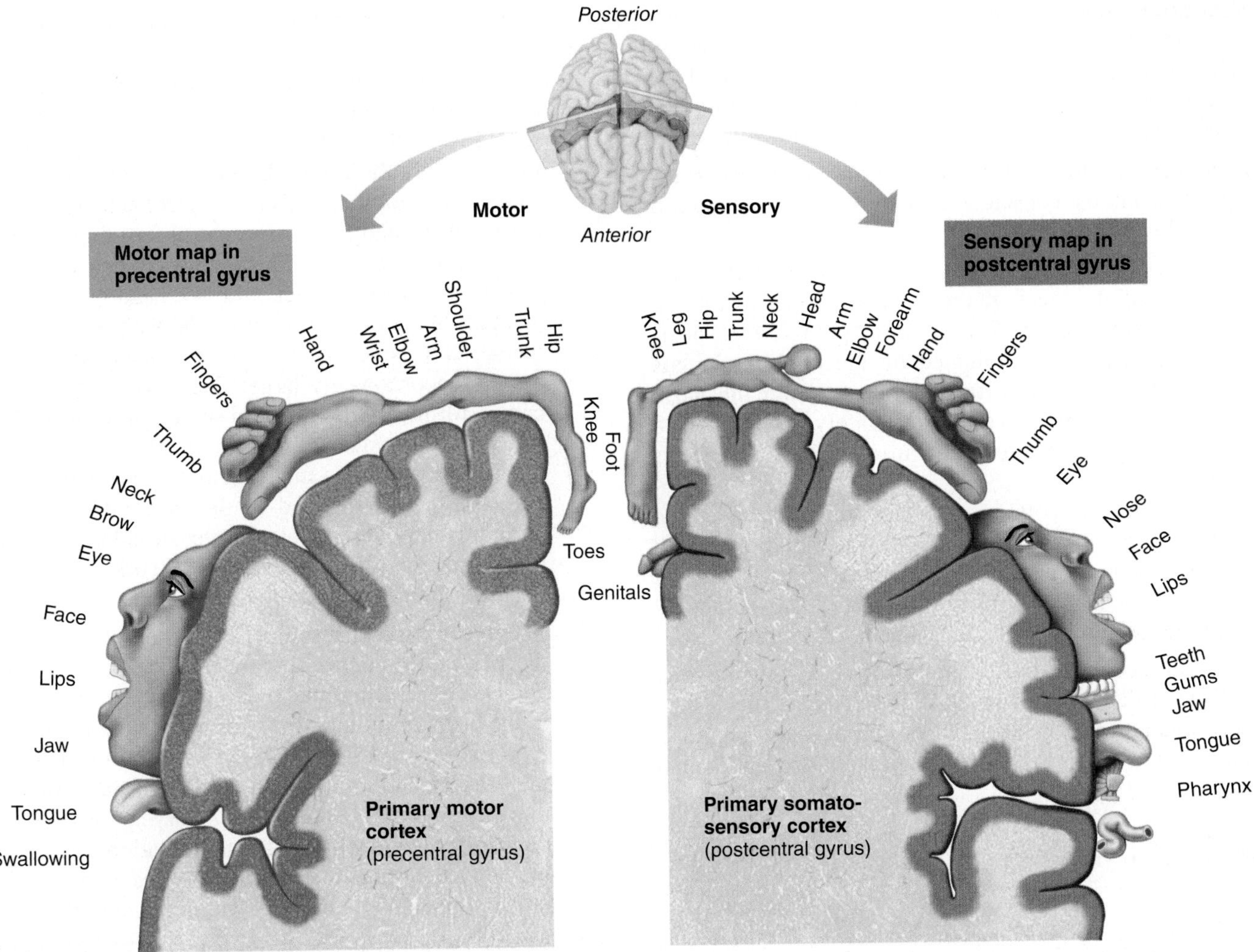

Figure 13.14 Body maps in the primary motor cortex and somatosensory cortex of the cerebrum. The relative amount and location of cortical tissue devoted to each function is proportional to the distorted body diagrams (homunculi).

The **somatosensory association cortex** lies posterior to the primary somatosensory cortex (Figure 13.13). Communicating with the primary somatosensory cortex, this association area integrates sensory inputs (touch, pressure, and others) into a comprehensive understanding of what is being felt. For example, when you reach into your pocket, the somatosensory association area draws on stored memories of past sensory experiences and perceives the objects you feel as, say, coins or keys.

2. **Visual areas.** The **primary visual cortex** is on the posterior and medial part of the occipital lobe (Figure 13.13a and b), much of it buried within the deep **calcarine sulcus** (kal′kar-in; "spur-shaped"). The largest of all cortical sensory areas, the primary visual cortex receives visual information that originates from the retina of the eye. The visual cortex exhibits contralateral projection: The right half of visual space is represented on the left visual cortex, and the left half is on the right cortex. Processing here is at a comparatively low level—noting the orientation of objects being viewed and merging the sensory stimuli from the two eyes. If this cortical area is damaged, the person has no conscious awareness of what is being viewed and is functionally blind.

   The **visual association area** surrounds the primary visual cortex and covers much of the occipital lobe. Communicating with the primary visual cortex, the visual association area continues the processing of visual information by analyzing color, form, and movement.

3. **Auditory areas.** The **primary auditory cortex** functions in conscious awareness of sound. It is located on the superior edge of the temporal lobe, primarily inside the lateral sulcus (Figure 13.13a). When sound waves excite the sound receptors of the inner ear, impulses are transmitted to the primary auditory cortex, where this information is related to loudness, rhythm, and pitch (high and low notes).

   The **auditory association area** lies just posterior and lateral to the primary auditory area. This area permits the evaluation of a sound as, say, a screech, thunder, or music.

4. **Vestibular (equilibrium) cortex.** The vestibular cortex is responsible for conscious awareness of the sense of balance, specifically the awareness of the position of the head in space. This region of the cortex is located in the posterior part of the insula, deep to the lateral sulcus.
5. **Gustatory cortex.** The gustatory (gus′tah-to″re) cortex is involved in the conscious awareness of taste stimuli. It lies in the insula.
6. **Olfactory cortex.** The primary olfactory cortex lies on the medial aspect of the temporal lobe in a small region called the *piriform cortex* (pir′ĭ-form; "pear-shaped"), which is dominated by the hooklike **uncus** (Figure 13.13b). The olfactory nerves from the nasal cavity transmit impulses that ultimately are relayed to the olfactory cortex, resulting in conscious awareness of smells.

   The olfactory cortex is part of a brain area called the *rhinencephalon* (ri″nen-sef′ah-lon; "nose brain"), which includes all parts of the cerebrum that directly receive olfactory signals: the piriform cortex, the **olfactory tract,** the **olfactory bulb,** and some nearby structures.
7. **Visceral sensory area.** The visceral sensory cortex is located deep within the lateral sulcus on the insula. This area receives general sensory input (pain, pressure, hunger, and so forth) from the thoracic and abdominal organs.

CLINICAL APPLICATION

Agnosia Damage to a sensory association area causes **agnosia,** the inability to comprehend sensory stimuli from the sense receptors of the PNS. Thus an individual with damage to the visual association area could have excellent vision but have no understanding of what she is seeing. If the auditory association cortex is damaged, one could hear sounds but not distinguish a siren from a trumpet. Damage to the somatosensory association area would disable you from identifying the item in your pocket as keys by touch alone; you would have to take them out and look at them to determine what they are.

Motor Areas The cortical areas that control motor functions—the *primary motor cortex, premotor cortex, frontal eye field,* and *Broca's area*—all lie in the posterior part of the frontal lobe.

1. **Primary motor cortex.** The **primary motor cortex** is located along the precentral gyrus of the frontal lobe, just anterior to the primary sensory cortex (Figure 13.13). In this area are large neurons called **pyramidal cells.** The long axons of pyramidal cells form the massive **pyramidal tracts** that descend through the brain stem and spinal cord. The axons in the pyramidal tracts synapse on motor neurons to bring about precise voluntary movements of the body, especially of the forearms, fingers, and facial muscles. The projection of the pyramidal axons is contralateral.

   The human body is represented spatially in the primary motor cortex of each hemisphere. This representation of the body mapped onto the motor cortex is illustrated by the **motor homunculus** (Figure 13.14, left side). The pyramidal cells that control hand movement are in one place, and those that control foot movement are in another. As with the primary sensory cortex, the body is represented upside down, with the head on the inferolateral part of the precentral gyrus and the toes at the superomedial end. Note that the face and hand representations are disproportionately large: The face and hand muscles can perform very delicate and skilled movements because so many pyramidal cells control them.

   The concept of a motor map does not mean that each body muscle is mapped in the precentral gryus. Each pyramidal cell in the motor cortex stimulates the contraction of several related muscles that work together to produce an often-executed movement.
2. **Premotor cortex.** Just anterior to the precentral gyrus is the **premotor cortex** (Figure 13.13). This region plans and coordinates complex movements and relays the plan to the primary motor cortex for implementation. The premotor cortex receives highly processed information from the sensory and multimodal regions of the cortex. Using this information, it controls voluntary actions that depend on sensory feedback about spatial relations, such as moving an arm through a maze of obstacles to grasp a hidden object or catching a high fly ball in the outfield.
3. **Frontal eye field.** The **frontal eye field** lies anterior to the premotor cortex (Figure 13.13). It controls voluntary movements of the eyes, especially when one looks quickly at something, as when the eyes follow a moving object.
4. **Broca's area.** The motor area known as **Broca's area** (Bro′kahz) lies anterior to the inferior part of the premotor cortex in the left, or *language-dominant,* cerebral hemisphere (Figure 13.13a). It controls the motor movements necessary for speaking. This area is extensively connected to the language comprehension areas in the posterior association area. The corresponding region in the right, or *intuitive-emotional,* hemisphere controls the emotional overtones given to spoken words. People with damage to Broca's area exhibit deliberate, nonfluent speech with impaired articulation but can understand most aspects of the speech of others.

Multimodal Association Areas **Multimodal association areas** are large regions of the cerebral cortex that receive sensory input from multiple sensory modalities and from the sensory association areas. A multimodal association area ties together, or makes *associations* between, various kinds of sensory information. The multimodal area also associates new sensory inputs with memories of past experiences and plans appropriate motor responses.

There are three multimodal association areas: the **posterior association area,** the **anterior association area** (Figure 13.13a, lavender regions), and the **limbic association area** (Figure 13.13b, lavender region).

1. **Posterior association area.** This multimodal area is located at the interface of the visual, auditory, and somatosensory association areas in the parietal and temporal lobes (Figure 13.13a). It integrates all these types of sensory information to form a unified perception of the sensory input. One of the major functions of the posterior association area is awareness of the spatial location of the body in reference to itself and to the outside world. This "body sense" requires ongoing sensory input from the proprioceptive senses, the visual system, and the vestibular apparatus to determine where the body is at any moment. This information is integrated to determine how to move one's limbs through space and then is relayed to the anterior association area (a motor region), which dictates these movements.

CLINICAL APPLICATION

Neglect Syndrome Damage to the posterior association area can lead to neglect of the opposite side of the body. This is particularly likely when there is injury to this multimodal area on the right side, because in most people the right cerebral hemisphere is more dominant in integrating body awareness. Without the ability to integrate sensory information from and about the left side of the body, the injured individual doesn't perceive that the parts of the body on the left exist and therefore neglects these parts—he will not wash, clothe, or recognize the neglected side. An individual with neglect syndrome will wonder what his left arm or leg is doing in bed with him.

Complex visual processing extends from the occipital lobe into the posterior association area in the temporal and parietal lobes. Visual information proceeds anteriorly through these association areas in two streams (Figure 13.15). The **dorsal stream** extends through the posterior parietal cortex, to the postcentral gyrus and perceives spatial relationships among various objects. This is referred to as the *"where" pathway,* identifying the location of objects. The **ventral stream** extends through the inferior part of the temporal lobe and is responsible for recognizing objects, words during reading, and faces (facial recognition involves the right hemisphere only). It is referred to as the *"what" pathway* because it identifies what things are.

Just as visual stimuli are processed by two streams in parallel, auditory stimuli also are processed both serially and in parallel along two pathways from the auditory association area through the multimodal association areas (Figure 13.16). The posterolateral pathway through the parietal lobe to the lateral part of the frontal lobe evaluates the location of a sound stimulus—the "where" pathway. The anterolateral pathway from the anterior temporal lobe to the inferior region of the frontal lobe processes information related to sound identification—the "what" pathway.

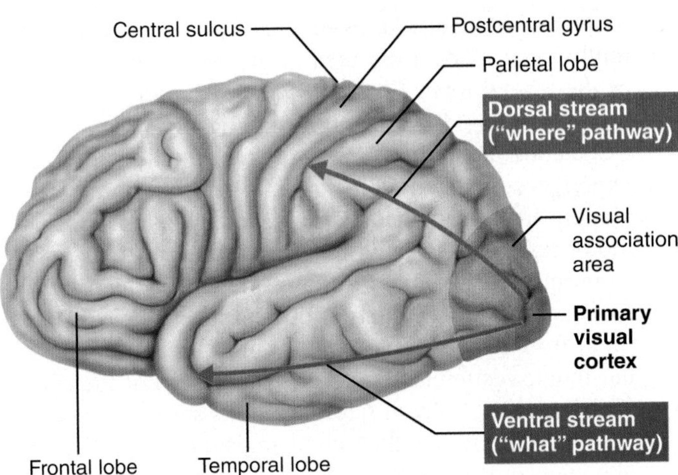

Figure 13.15 Visual pathways through the multimodal association areas.

Another major function of the posterior association area is related to language comprehension and speech. In the posterior association area on the left side (in most people) are regions that function in speech comprehension (*Wernicke's area,* Figure 13.13a) and in the coordination of the auditory and visual aspects of language, such as naming viewed objects and reading words (the "what" pathways through the temporal lobe, Figures 13.15 and 13.16). There is extensive linkage between these regions and the motor regions that produce language (Broca's area). Damage to these regions results in difficulties with language comprehension but does not impair speech production.

Corresponding areas in the right hemisphere function in the creative interpretation of words and in controlling the emotional overtones of speech.

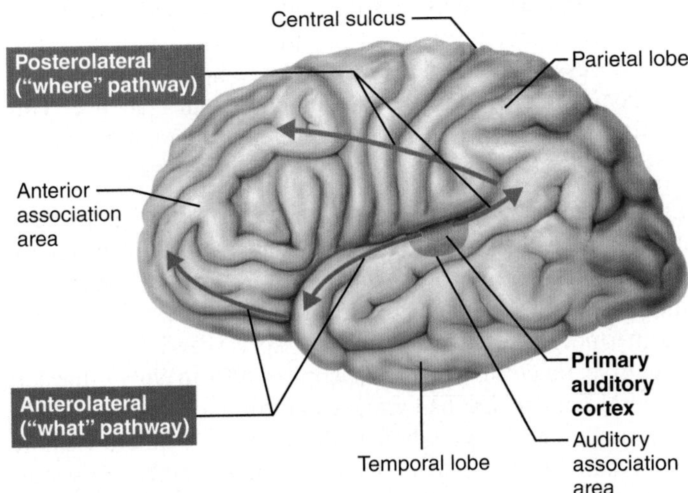

Figure 13.16 Auditory pathways through the multimodal association areas.

2. **Anterior association area (prefrontal cortex).** This multimodal association area is the large anterior region of the frontal lobe (Figure 13.13a). This region receives highly processed sensory information from the posterior association area, integrates this information with past experience through connection with the limbic association area (described below), and plans and initiates motor responses through linkage with the motor regions. Coordinated movement through the environment results from an ongoing evaluation of, and response to, sensory input from proprioceptors in the muscles and joints, visual input, vestibular input, and knowledge of the spatial location of the body in reference to its surroundings.

   In addition to this comparatively straightforward, although complex, motor function, cognitive functions such as thinking, perceiving, and intentionally remembering and recalling information occur in this anterior association area. The proper functioning of this region is necessary for processing abstract ideas, reasoning and judgment, impulse control, long-term planning, complex problem solving, mental flexibility, social skills, appreciating humor, empathy, and conscience. It has close links to the emotional (limbic) part of the brain and thus is also related to mood. The tremendous elaboration of the anterior association area distinguishes humans from other animals.

   Tumors or injury to this region can cause mental and personality disorders. Wide swings in mood may occur, and there may be a loss of attentiveness, of inhibitions, and of judgment. Developmentally, the anterior association area is one of the last parts of the brain to mature. It is not fully formed until early adulthood—a fact that explains why adolescents commonly exhibit poor judgment.

3. **Limbic association area.** This multimodal association area is located on the medial side of the cerebral hemispheres in portions of the temporal, parietal and frontal lobes (Figure 13.13b) and includes the cingulate gyrus, the hippocampus, and parahippocampal gyrus. It is involved in both memory and emotion. It integrates sensory and motor behavior with past experience, helps form memory, and uses this past experience to influence future motor response. The limbic association area also processes emotions involved in complex personal and social interactions and guides emotional response. This cortical region is part of the limbic system, a functional brain system (discussed more fully on p. 399).

Lateralization of Cortical Functioning You have already seen that some division of labor exists between the right and left cerebral hemispheres, as, for example, in the fact that the two hemispheres control opposite sides of the body. Furthermore, the hemispheres are specialized for different cognitive functions: In most people (90% to 95%), the left cerebral hemisphere has greater control over language abilities, math, and logic, whereas the right hemisphere is more involved with visual-spatial skills, reading facial expressions, intuition, emotion, and artistic and musical skills. Whereas the right hemisphere deals with the big picture, the left deals with the details, which it then interprets logically. In the remaining 5% to 10% of the population, either these roles of the hemispheres are reversed, or the two hemispheres share their cognitive functions equally.

Despite their differences, the two cerebral hemispheres are nearly identical in structure and share most functions and memories. Such sharing is possible because the hemispheres communicate through white fiber tracts (*commissural fibers*).

☑ check your understanding

☐ 12. What is the difference in function between a primary sensory cortex and a sensory association area?

☐ 13. Which functional area of the cerebral cortex plans complex movements? Which area signals the execution of these movements?

☐ 14. Define contralateral projection.

☐ 15. What deficits may result from injury to the occipital lobe?

(For answers, see Appendix B.)

White Matter of the Cerebrum

Underlying the gray matter of the cerebral cortex is the cerebral white matter. It is via the many fibers that form the cerebral white matter that the various areas of the cerebral cortex communicate both with one another and with the brain stem and spinal cord. Most of these fibers are myelinated and bundled into large tracts. The fibers are classified as *commissural fibers, association fibers,* or *projection fibers,* according to where they run (Figure 13.17 and Table 13.1, p. 390).

- **Commissural fibers** cross from one side of the CNS to the other. The commisural fibers of the cerebrum interconnect corresponding gray areas of the right and left cerebral hemispheres, allowing the two hemispheres to function together as a coordinated whole. The largest commissure is the **corpus callosum** (kor′pus kah-lo′sum, "thickened body"), a broad band that lies superior to the lateral ventricles, deep within the longitudinal fissure (Figure 13.17a).
- **Association fibers** connect different parts of the same hemisphere. Short association fibers connect neighboring cortical areas, and long association fibers connect widely separated cortical lobes.
- **Projection fibers** either descend from the cerebral cortex to more caudal parts of the CNS or ascend to the cortex from lower regions. It is through projection fibers that sensory information reaches the cerebral cortex and motor instructions leave it. These fibers run vertically, whereas most commissural and association fibers run horizontally.

Deep to the cerebral white matter, the projection fibers form a compact bundle called the **internal capsule,** which passes between the thalamus and some of the deep gray matter, the *basal nuclei (ganglia)* (Figure 13.17). Superior to the internal capsule, the projection fibers running to and from the cerebral cortex fan out to form the **corona radiata** (kŏ-ro′nah ra-de-ah′tah; "radiating crown").

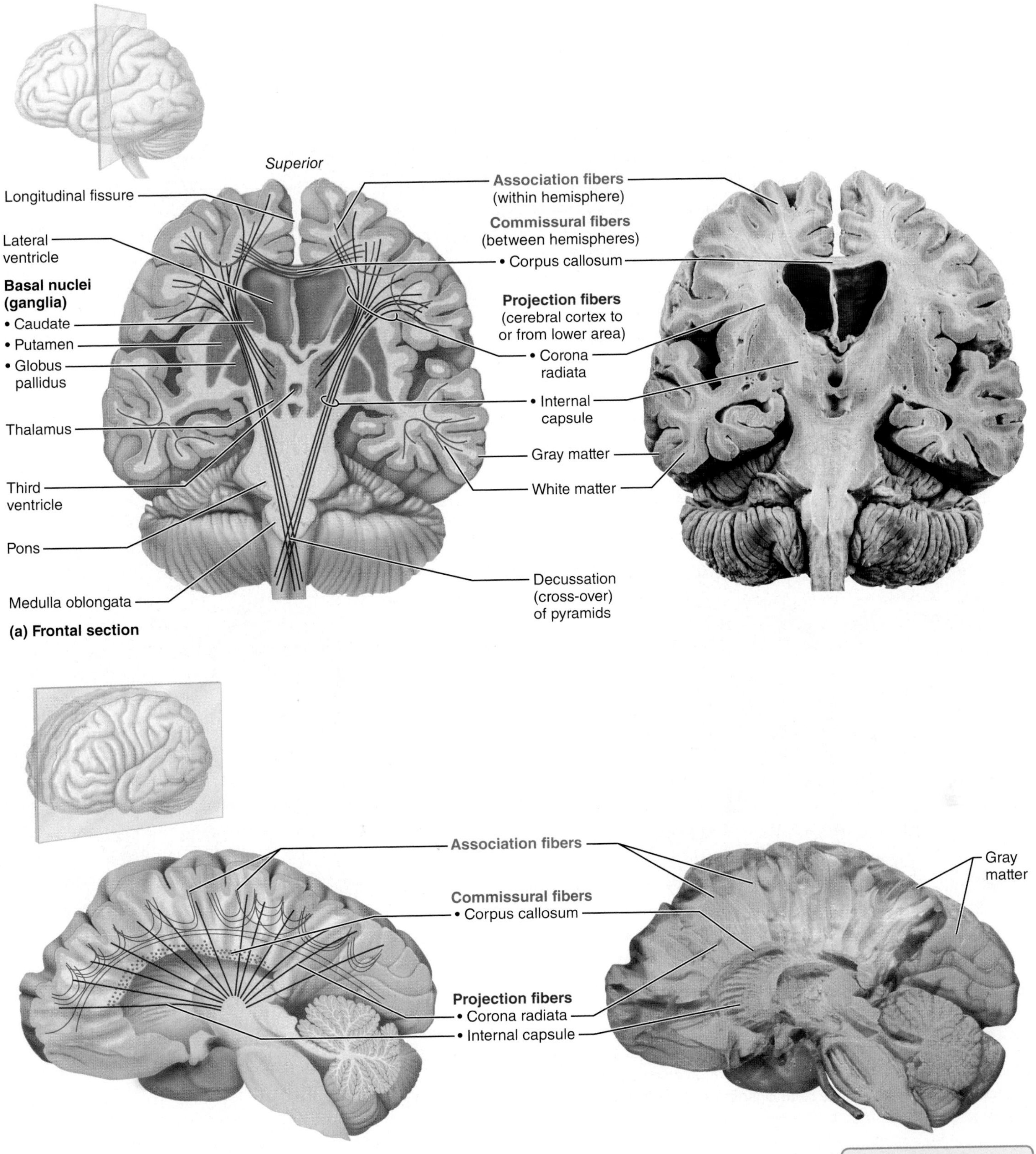

Superior
Longitudinal fissure
Lateral ventricle
Basal nuclei (ganglia)
• Caudate
• Putamen
• Globus pallidus
Thalamus
Third ventricle
Pons
Medulla oblongata
Association fibers (within hemisphere)
Commissural fibers (between hemispheres)
• Corpus callosum
Projection fibers (cerebral cortex to or from lower area)
• Corona radiata
• Internal capsule
Gray matter
White matter
Decussation (cross-over) of pyramids
(a) Frontal section
Association fibers
Commissural fibers
• Corpus callosum
Projection fibers
• Corona radiata
• Internal capsule
Gray matter

(b) Parasagittal section and dissection

Figure 13.17 White fiber tracts of the cerebral hemispheres. Commissural, association, and projection fibers run within the cerebrum and between the cerebrum and lower portions of the CNS.

Explore PAL Cadaver MasteringA&P®

Watch Sheep Brain Dissection MasteringA&P®

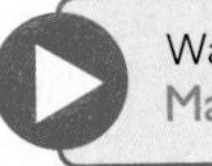

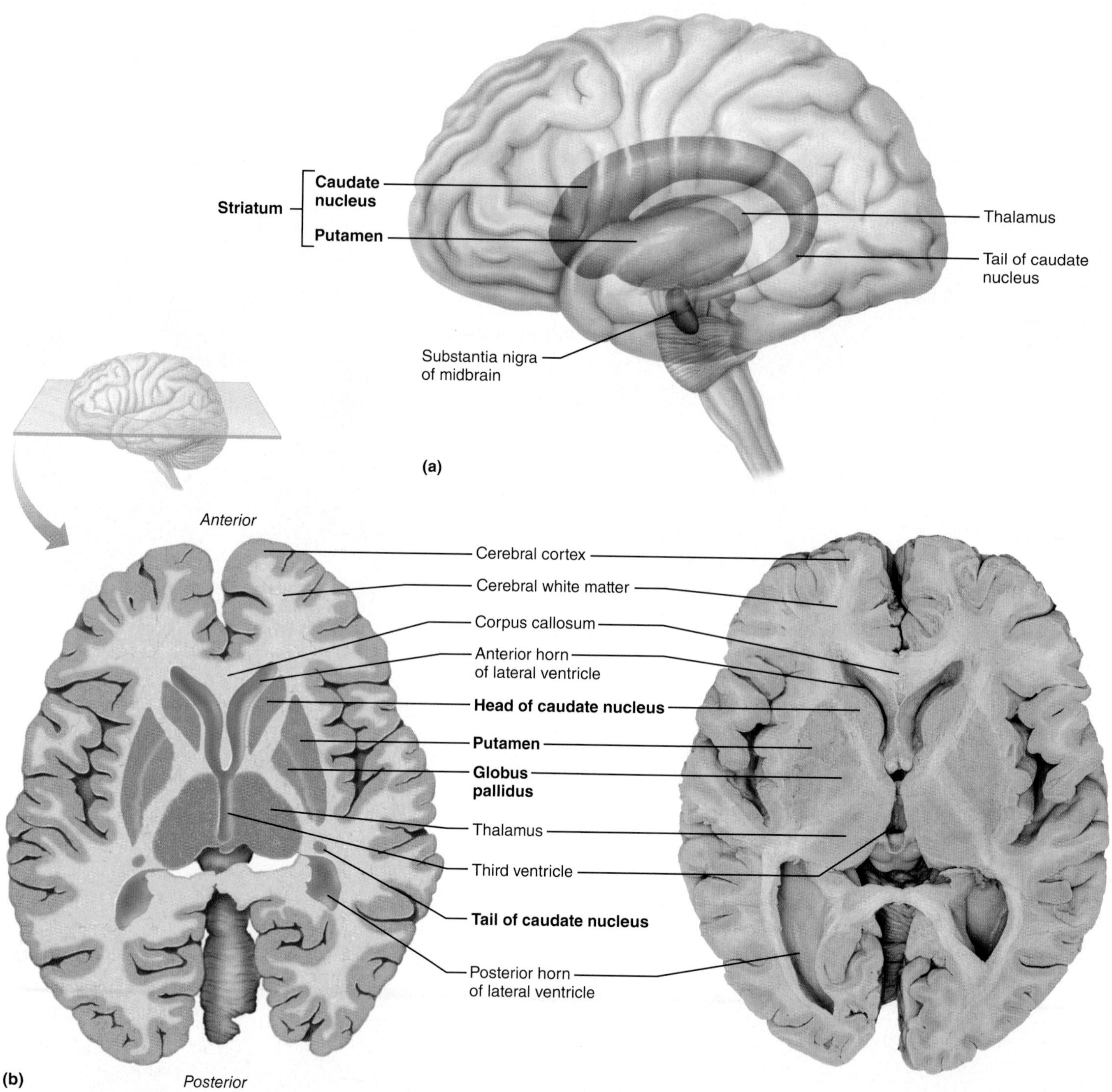

Figure 13.18 Basal nuclei (ganglia). (**a**) Three-dimensional view of the basal nuclei deep in the cerebrum. The substantia nigra of the midbrain is included as it has close functional links to the basal nuclei. (**b**) Transverse section through the cerebrum and diencephalon, showing the relationship of the basal nuclei to the thalamus and the lateral and third ventricles.

Deep Gray Matter of the Cerebrum

The deep gray matter of the cerebrum consists of the *basal nuclei (ganglia),* involved in motor control; the *basal forebrain nuclei,* associated with memory; and the *claustrum,* a brain nucleus of unknown function (Table 13.1, p. 390). The *amygdaloid body* is also deep gray matter in the cerebrum. It is part of the limbic system and is included in the discussion of that functional brain system.

Basal Nuclei (Ganglia) Deep within the cerebral white matter lies a group of subcortical brain nuclei collectively called the **basal nuclei (ganglia*):** the **caudate** ("tail-like") **nucleus,** the **putamen** (pu-ta′men; "pod"), and the **globus pallidus** ("pale globe") (Figure 13.18). The caudate nucleus and the putamen together are called the striatum (stri-a′-tum; "striated") because some fibers of the internal capsule passing through them create a striated appearance (Figure 13.17a). The caudate nucleus arches superiorly over the thalamus and lies medial to the internal capsule. The globus pallidus and the putamen are located lateral to the internal capsule (Figure 13.17a).

Functionally, the basal nuclei can be viewed as complex neural calculators that cooperate with the cerebral cortex in controlling movements. They receive input from many cortical areas and send almost all their output back to the motor cortex through a relay in the thalamus. The substantia nigra in the midbrain (shown in Figure 13.18a) also influences the activity of the basal nuclei.

The basal nuclei function to start, stop, and regulate the intensity of the voluntary movements ordered and executed by the cerebral cortex. They may select the appropriate muscles or movements for a task and inhibit the relevant antagonistic muscles. They also control rhythmic, repetitive tasks and participate in the learning of habits. Finally, in a nonmotor role, the basal nuclei in some way estimate the passage of time.

CLINICAL APPLICATION

Dyskinesia Degenerative conditions of the various basal nuclei produce *dyskinesia* (dis-ki-ne′ze-ah), literally, "bad movements." Two dyskinetic conditions are Parkinson's disease and Huntington's disease.

Parkinson's disease is characterized by slow and jerky movements; tremor of the face, limbs, or hands; muscle rigidity; and great difficulty in starting voluntary movements. This condition results from degeneration of a portion of the substantia nigra in the midbrain that normally sends inhibitory input to the basal nuclei. This loss of inhibitory input from the substantia nigra leads to an overactive globus pallidus, which inhibits the motor cortex and has the effect of a stuck brake on an automobile or a bicycle. The ultimate cause of Parkinson's disease is unknown. It may be caused by accumulations of environmental toxins, the buildup of iron in the brain, or an accumulation of cell and tissue damage from free radicals.

Huntington's disease involves an overstimulation of motor activities, such that the limbs jerk uncontrollably. The signs of the disease are caused by degeneration of the basal nuclei pathway that inhibits motor activity, and the cerebral cortex eventually degenerates as well. The condition is inherited, and the precise nature of the genetic defect in the DNA has been discovered.

*Historically, the deep gray matter of the cerebrum was referred to as the "basal ganglia." These structures are now referred to as "basal nuclei" because the term *ganglia* is currently used to refer to collections of cell bodies in the PNS. Many sources still use the term *ganglia,* so we have included this term for clarity. The *basal nuclei (ganglia),* described here, are distinct from the *basal forebrain nuclei* (part of the cerebral cholinergic system), which are discussed shortly.

Basal Forebrain Nuclei The **basal forebrain nuclei** are four important structures in the cerebrum: the *septal nuclei,* the *diagonal band (of Broca),* the *horizontal band (of Broca),* and the *basal nucleus (of Meynert).* These structures are part of the basal forebrain cholinergic system (neurons that synthesize and release acetylcholine) and are located anterior and dorsal to the hypothalamus. The main functions of this system are related to arousal, learning, memory, and motor control. Degeneration of the basal forebrain nuclei, and the resultant reduction in cholinergic activity, is associated with Alzheimer's disease, a form of dementia. Early symptoms of Alzheimer's disease are forgetfulness and diminished attention span. As the disease progresses, memory loss becomes more severe; language skills deteriorate; and abstract thinking, judgment, personality, and emotion can be affected (for further discussion of Alzheimer's disease, see p. 421).

check your understanding

- ☐ 16. Name the white fiber tracts that connect the cerebral cortex to more caudal regions of the CNS.
- ☐ 17. Where is the caudate nucleus located in reference to the lateral ventricles?

(For answers, see Appendix B.)

Functional Brain Systems

learning outcome

- ▶ Describe the locations and functions of the limbic system and the reticular formation.

Functional brain systems are networks of neurons that function together despite spanning large distances in the brain. This section focuses on two important functional systems: the *limbic system* and the *reticular formation.*

The Limbic System

The limbic system is referred to as the "emotional brain." It is responsible for the emotional impact actions, behavior, and situations have on us (our "feelings"); it directs our response to these emotions; and it functions in creating, storing, and retrieving memories, particularly those that elict strong emotions. This functional brain system consists of a group of structures located in the medial aspect of each cerebral hemisphere and the diencephalon (Figure 13.19). In the cerebrum, the limbic structures form a broad ring (*limbus* = headband) that includes the *cingulate gyrus, septal nuclei,* part of the *amygdaloid body,* and the *hippocampal formation.* In the diencephalon, the main limbic structures are the *anterior thalamic nuclei* and the *hypothalamus.* The **fornix** ("arch") and other fiber tracts link the limbic system together. The limbic system also overlaps the rhinencephalon, the portions of the cortex that process olfactory signals. This explains why smells often trigger emotions.

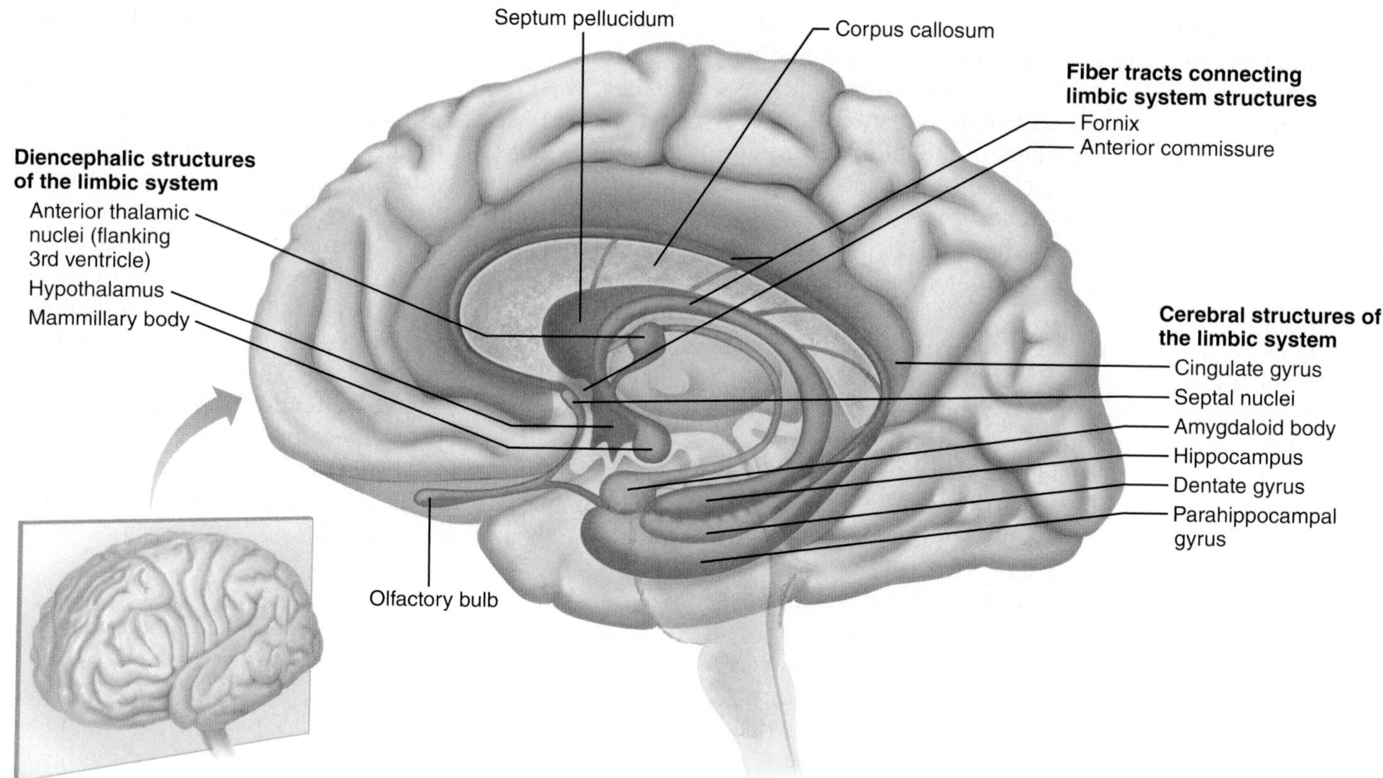

Figure 13.19 The limbic system. Medial view of the cerebrum and diencephalon illustrating the structures of the limbic system, the emotional-visceral brain.

The major cerebral components of the limbic system are described below.

- The **cingulate gyrus** (sing′gu-lāt; "belt-shaped") is part of the cerebral cortex located superior to the corpus callosum (Figure 13.19). It is linked to all other regions of the cerebral cortex. Through its connections to the sensory regions, the cingulate gyrus mediates the emotional response to these stimuli, such as experiencing painful stimuli as unpleasant. Its connections with the hypothalamus and prefrontal cortex function to generate and control visceral and behavorial responses to emotions.
- The **hippocampal formation,** located in the temporal lobe, consists of the **hippocampus** (hip″o-kam′pus; "sea horse") and the **parahippocampal gyrus.** These regions encode, consolidate, and later retrieve memories of facts and events. The hippocampal formation receives information to be remembered from the rest of the cerebral cortex; it processes these data and returns them to the cortex, where they are stored as long-term memories.
- The **amygdaloid body** is subcortical gray matter that contains the key brain nuclei for processing fear and stimulating the appropriate sympathetic response to fear. The amygdaloid body also forms memories of experiences based entirely on their emotional impact, especially those related to fear. If people are reminded of these experiences later, the amygdaloid body retrieves the memories and causes them to reexperience the original emotion. This is beneficial because it lets people make informed decisions about difficult and risky situations, based on memories of past emotional experiences.

Hyperactivity in the amygdaloid body and dysfunction in the medial prefrontal cortex (limbic association area) and the hippocampus are involved in the extreme response to triggered memory experienced by individuals suffering **post-traumatic stress disorder (PTSD).**

The limbic system communicates with many other regions of the brain. Most output from the limbic system is relayed through the hypothalamus and the reticular formation (discussed next), which control visceral responses. This is why people under emotional stress may experience visceral illnesses, such as high blood pressure and heartburn. The limbic system also interacts extensively with the prefrontal cortex of the cerebrum. Thus, feelings (mediated by the emotional brain) interact closely with thoughts (mediated by the thinking brain).

The Reticular Formation

The **reticular formation** runs through the central core of the brain stem (Figure 13.20). The neurons of the reticular formation have long, branching axons that project to widely separated regions of the thalamus, cerebellum, and spinal cord. Such widespread connections make reticular neurons ideal for governing the arousal of the brain as a whole. Certain reticular neurons send a continuous stream of impulses to the cerebrum (through relays in the thalamus), thereby maintaining

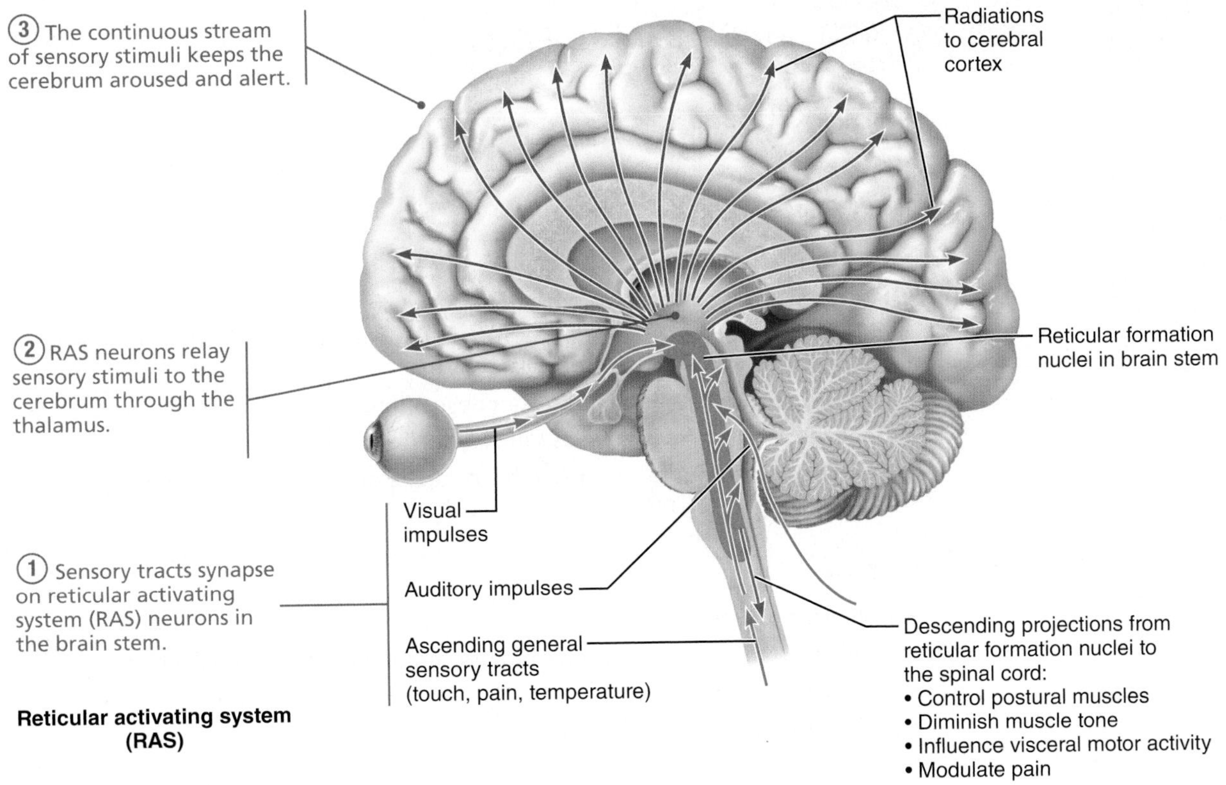

Figure 13.20 The reticular formation. This functional brain system extends the length of the brain stem. Part of this formation, the reticular activating system (RAS), maintains alert wakefulness of the cerebral cortex.

the cerebral cortex in an alert, conscious state (left side of Figure 13.20). This arm of the reticular formation, which maintains consciousness and alertness, is called the **reticular activating system (RAS).** It is located mainly in the medial nuclear group of the pons and medulla in the brain stem (see Figure 13.7).

Axons from all the major ascending sensory tracts synapse on RAS neurons, keeping these reticular neurons active and enhancing their arousing effect on the cerebrum. The fact that visual, auditory, and touch stimuli keep people awake and mentally alert explains why many students like to study in a crowded room: They are stimulated by the bustle of such an environment. The RAS also functions in sleep and in arousal from sleep. General anesthesia, alcohol, tranquilizers, and sleep-inducing drugs depress the RAS and lead to a loss of alertness or consciousness. Severe injury to the RAS is a cause of coma.

Descending pathways of the reticular formation (right side of Figure 13.20) send axons to the spinal cord via the reticulospinal tracts. Certain descending pathways from the reticular formation influence somatic motor neurons to postural muscles. Others diminish muscle tone by inhibiting motor neurons, as when the person is sleeping. Axons from the reticular formation influence the autonomic neurons that regulate visceral functions such as heart rate, blood pressure, and respiration (discussed on pp. 379–380, lateral nuclear group in Figure 13.7c). Another group of descending fibers from the reticular fromation influence our perception of pain by inhibiting the transmission of pain impulses.

The major parts of the brain and their functions are presented in summary for your review (Table 13.2).

✓ check your understanding

☐ 18. From where do the reticular nuclei receive input? To where do they project?

☐ 19. What emotional response does the amygdaloid body stimulate?

(For answers, see Appendix B.)

Protection of the Brain

learning outcomes

- Describe the meningeal coverings surrounding the brain, and explain how the meninges and cerebrospinal fluid protect the structures of the central nervous system.
- Explain how cerebrospinal fluid is formed, and describe its pattern of circulation.

Nervous tissue is delicate, and the irreplaceable neurons can be injured by even a minor impact. The brain is protected from injury due to external forces by the skull, by surrounding connective tissue membranes called the *meninges,* and by a watery cushion of *cerebrospinal fluid* produced in the ventricles of the brain. The *blood brain barrier* protects the brain

Table 13.2 Functions of Major Parts of the Brain

Part	Function

BRAIN STEM (pp. 378–383)

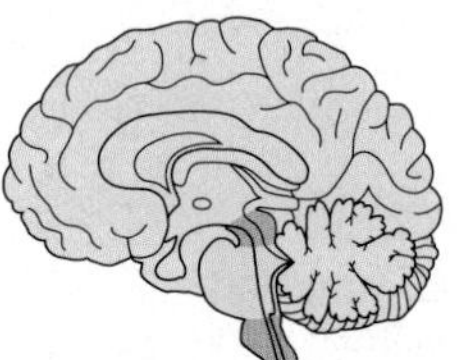

Medulla oblongata:

- Contains projection fibers
- Site of decussation of the pyramids
- Relays ascending sensory pathways transmitting impulses from skin and proprioceptors through nuclei cuneatus and gracilis
- Relays sensory information to the cerebellum through inferior olivary nuclei
- Contains nuclei of cranial nerves VIII–X and XII
- Contains visceral nuclei controlling heart rate, blood vessel diameter, respiratory rate, vomiting, coughing, etc.

Pons:

- Contains projection fibers.
- Pontine nuclei relay information from the cerebrum to the cerebellum
- Contains nuclei of cranial nerves V–VII
- Contains reticular formation nuclei

Midbrain:

- Contains projection fibers (e.g., cerebral peduncles contain the fibers of the pyramidal tracts)
- Contains subcortical motor centers (substantia nigra and red nuclei)
- Contains nuclei for cranial nerves III and IV
- Contains visual (superior colliculi) and auditory (inferior colliculi) reflex centers

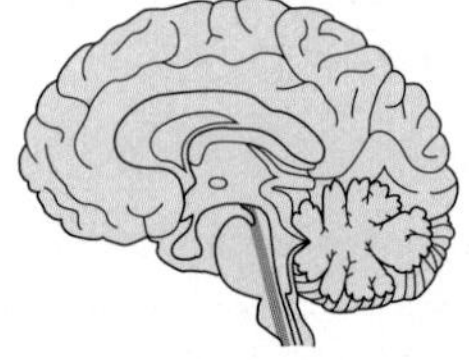

Reticular formation (pp. 400–401)—A functional system:

- Maintains cerebral cortical alertness (reticular activating system)
- Filters out repetitive stimuli
- Helps regulate skeletal and visceral muscle activity and modulate pain

CEREBELLUM (pp. 383–385)

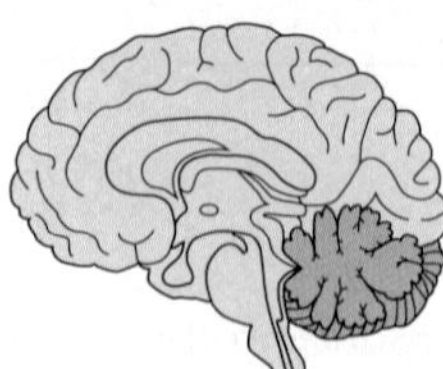

Cerebellum:

- Processes input from cerebral motor cortex, proprioceptors, and visual and equilibrium pathways
- Provides output to cerebral motor cortex and subcortical motor centers that result in smooth, coordinated skeletal muscle movements
- Responsible for balance and posture

Table 13.2 continued

Part	Function
DIENCEPHALON (pp. 385–388) 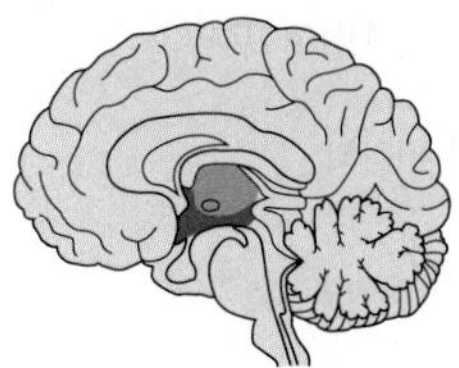	**Thalamus:** • Relays sensory impulses to cerebral cortex for interpretation • Relays impulses between cerebral cortex and subcortical motor centers, including basal nuclei (ganglia) and cerebellum • Involved in memory processing **Hypothalamus:** • Chief integration center of autonomic (involuntary) nervous system • Regulates body temperature, food intake, water balance, thirst, and biological rhythms and drives • Regulates hormonal output of anterior pituitary gland • Acts as an endocrine organ producing posterior pituitary hormones ADH and oxytocin
CEREBRAL HEMISPHERES (pp. 388–399) 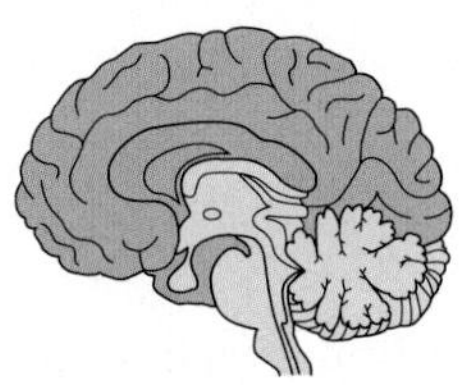	**Cortical gray matter:** • Localizes and interprets sensory inputs • Controls voluntary and skilled skeletal muscle activity • Functions in intellectual and emotional processing **Basal nuclei (ganglia):** • Subcortical motor centers help control skeletal muscle movements
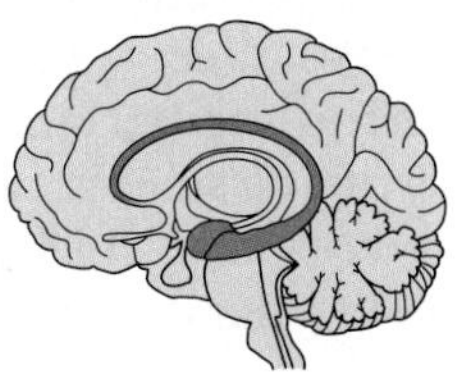	**Limbic system** (pp. 399–400)—A functional system: • Includes cerebral and diencephalon structures (cingulate gyrus, hippocampal formation, amygdaloid body, hypothalamus, and anterior thalamic nuclei) • Mediates emotional response • Forms and retrieves memories

internally from harmful substances carried in the blood, such as toxic chemicals or microorganisms.

Meninges

The **meninges** (mĕ-nin′jēz; "membranes") are three connective tissue membranes that lie just external to the brain and spinal cord. Their functions are to:

- Cover and protect the CNS.
- Enclose and protect the blood vessels that supply the CNS.
- Contain the cerebrospinal fluid.

From external to internal, the meninges are the *dura mater, arachnoid mater,* and *pia mater* (Figure 13.21). This section describes the meninges around the brain. The meninges extend around the spinal cord (described on p. 412) and are continuous throughout the CNS.

The Dura Mater

The leathery *dura mater* (du′rah ma′ter; "tough mother") is the strongest of the meninges. Where it surrounds the brain, the dura mater is a two-layered sheet of dense fibrous connective tissue (Figure 13.21). The more superficial **periosteal layer** attaches to the internal surface of the skull bones (it is the periosteum); the deeper **meningeal layer** forms the true external covering of the brain. These two layers of dura mater are fused together, except where they separate to enclose the blood-filled *dural venous sinuses* (see discussion of *dural venous sinuses* in Chapter 20). The dural sinuses collect blood from the brain and conduct it to the large internal jugular veins of the neck. The largest dural sinus is the *superior sagittal sinus* in the superior midline.

In several places, the meningeal dura mater extends inward to form flat partitions that subdivide the cranial cavity (Figure 13.22) and limit movement of the brain within the cranium. These partitions include the following:

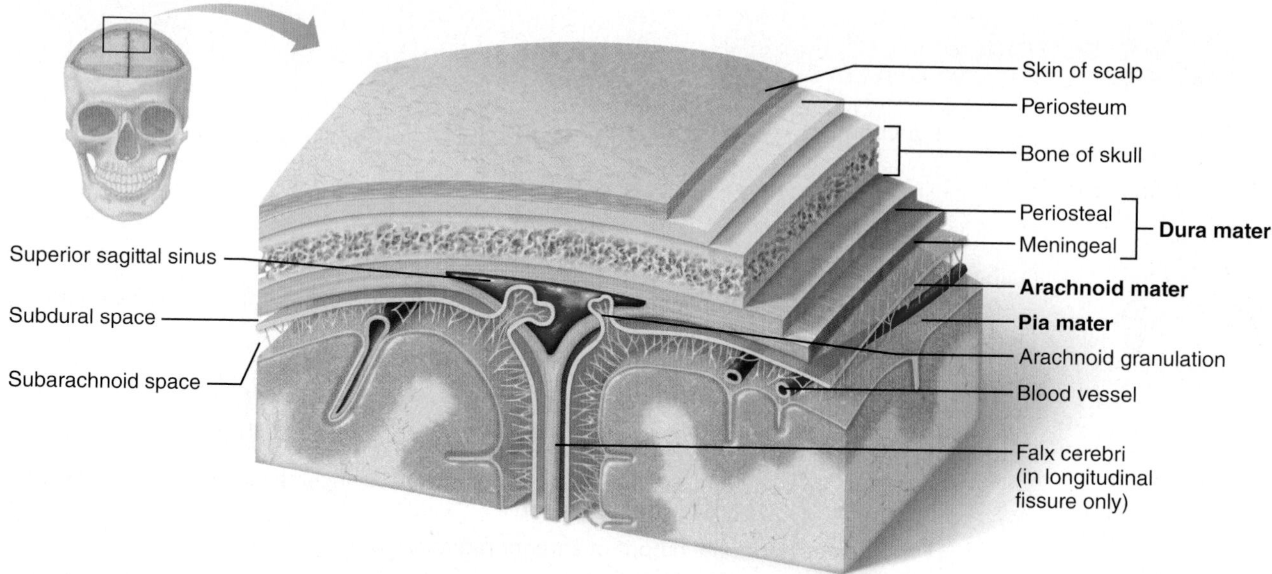

Figure 13.21 Meninges around the brain. The three meninges: dura mater, arachnoid mater, and pia mater, shown in frontal view. The meningeal and periosteal layers of the dura mater separate to form the superior sagittal sinus, illustrated here. The meningeal layers from the right and left sides fuse and extend into the longitudinal fissure, forming the falx cerebri. Arachnoid granulations, which return cerebrospinal fluid to the dural sinus, are also shown.

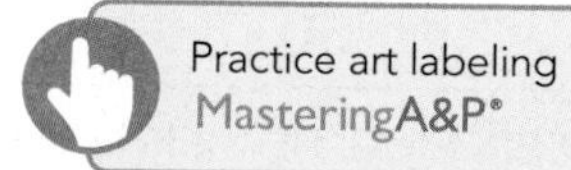

- The **falx cerebri** (falks ser′ĕ-bri). This large, sickle-shaped (*falx* = sickle) vertical sheet lies in the median plane in the longitudinal fissure between the cerebral hemispheres. It attaches anteriorly to the crista galli of the ethmoid bone.
- The **falx cerebelli** (ser″ĕ-bel′li). Continuing inferiorly from the posterior part of the falx cerebri, the falx cerebelli is a vertical partition that runs along the vermis of the cerebellum in the posterior cranial fossa.
- The **tentorium** (ten-to′re-um) **cerebelli.** Resembling a tent over the cerebellum (*tentorium* means "tent"), this almost horizontal sheet lies in the transverse fissure between the cerebrum and cerebellum. The *transverse sinus* is enclosed by the tentorium cerebelli.

The Arachnoid Mater The **arachnoid** (ah-rak′noid) **mater** lies just deep to the dura mater (Figure 13.21). Between these two layers is a thin potential space called the **subdural space,** which contains only a film of fluid. The subdural space is referred to as a *potential space* because although normally it is a thin space, it has the potential to fill with substantial amounts of fluid or blood as a result of disease or trauma. The dura and arachnoid surround the brain loosely, never dipping into the sulci on the cerebral surface.

Deep to the arachnoid membrane is the wide **subarachnoid space,** which is spanned by weblike threads that hold the arachnoid mater to the underlying pia mater. This web is the basis of the name *arachnoid,* which means "spiderlike." The subarachnoid space is filled with cerebrospinal fluid, and it also contains the largest blood vessels that supply the brain. Because the arachnoid is fine and elastic, these vessels are rather poorly protected.

Over the superior part of the brain, the arachnoid forms knoblike projections called **arachnoid granulations,** or *arachnoid villi*. The arachnoid granulations project superiorly through the dura mater into the superior sagittal sinus (see Figure 13.21) and into some other dural sinuses, as well. They act as valves that allow cerebrospinal fluid to pass from the subarachnoid space into the dural venous sinuses.

The Pia Mater The **pia mater** (pi′ah; "gentle mother") is a layer of delicate connective tissue richly vascularized with fine blood vessels. Unlike the other meninges, it clings tightly to the brain surface, following every convolution. As its arteries enter the brain tissue, they carry ragged sheaths of pia mater internally for short distances.

Cerebrospinal Fluid

Cerebrospinal fluid (CSF) is a watery broth that fills the subarachnoid space and the central hollow cavities of the brain and spinal cord. It aids in protecting and nourishing the neural tissue.

- CSF provides a liquid medium that surrounds and gives buoyancy to the central nervous system. The brain and spinal cord actually float in the CSF, which prevents these delicate organs from being crushed under their own weight.
- The layer of CSF surrounding the central nervous system resists compressive forces and cushions the brain and spinal cord from blows and jolts.

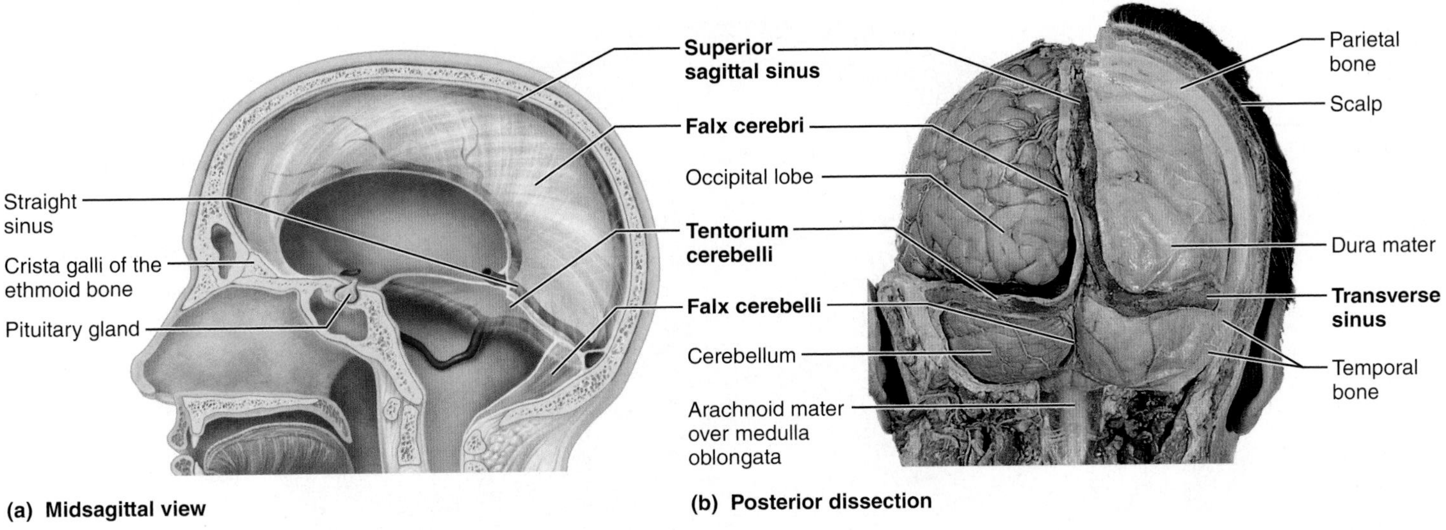

(a) Midsagittal view

(b) Posterior dissection

Explore PAL Cadaver
MasteringA&P®

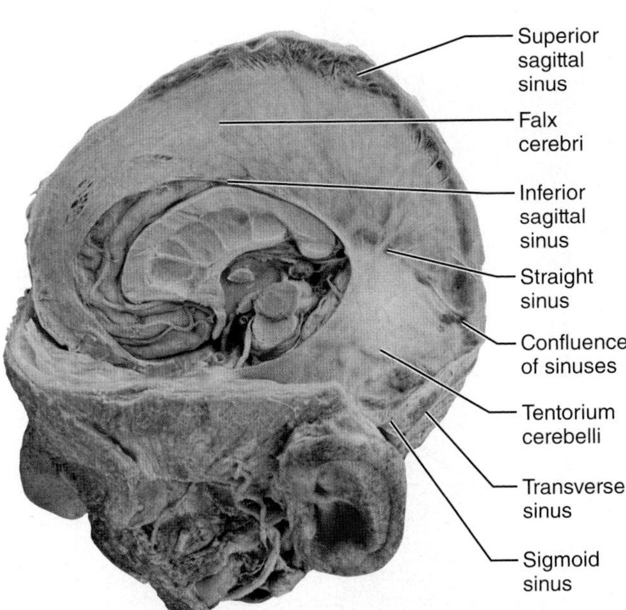

(c) Medial view of the dural partitions and the dural venous sinuses

Figure 13.22 **Partitions of dura mater in the cranial cavity and the dural venous sinuses.**

- CSF helps to nourish the brain, to remove wastes produced by neurons, and to carry chemical signals such as hormones between different parts of the central nervous system. Although similar in composition to the blood plasma from which it arises, CSF contains more sodium and chloride ions and less protein.

Amazingly, very little CSF is needed to perform these functions: Only 100–160 ml, about half a cup, is present at any time.

CSF is produced in the ventricles of the brain, circulates through the hollow central cavities of the CNS and the subarachnoid space, and is reabsorbed back into the blood in the dural venous sinuses (Figure 13.23).

① CSF is made in the **choroid plexuses,** capillary-rich membranes located in the roofs of the four brain ventricles (Figure 13.23a). Each plexus consists of a layer of ependymal cells covered externally by a capillary-rich pia mater (Figure 13.23b). CSF continuously forms from blood plasma by filtration from the capillaries and then passes through the ependymal cells into the ventricles.

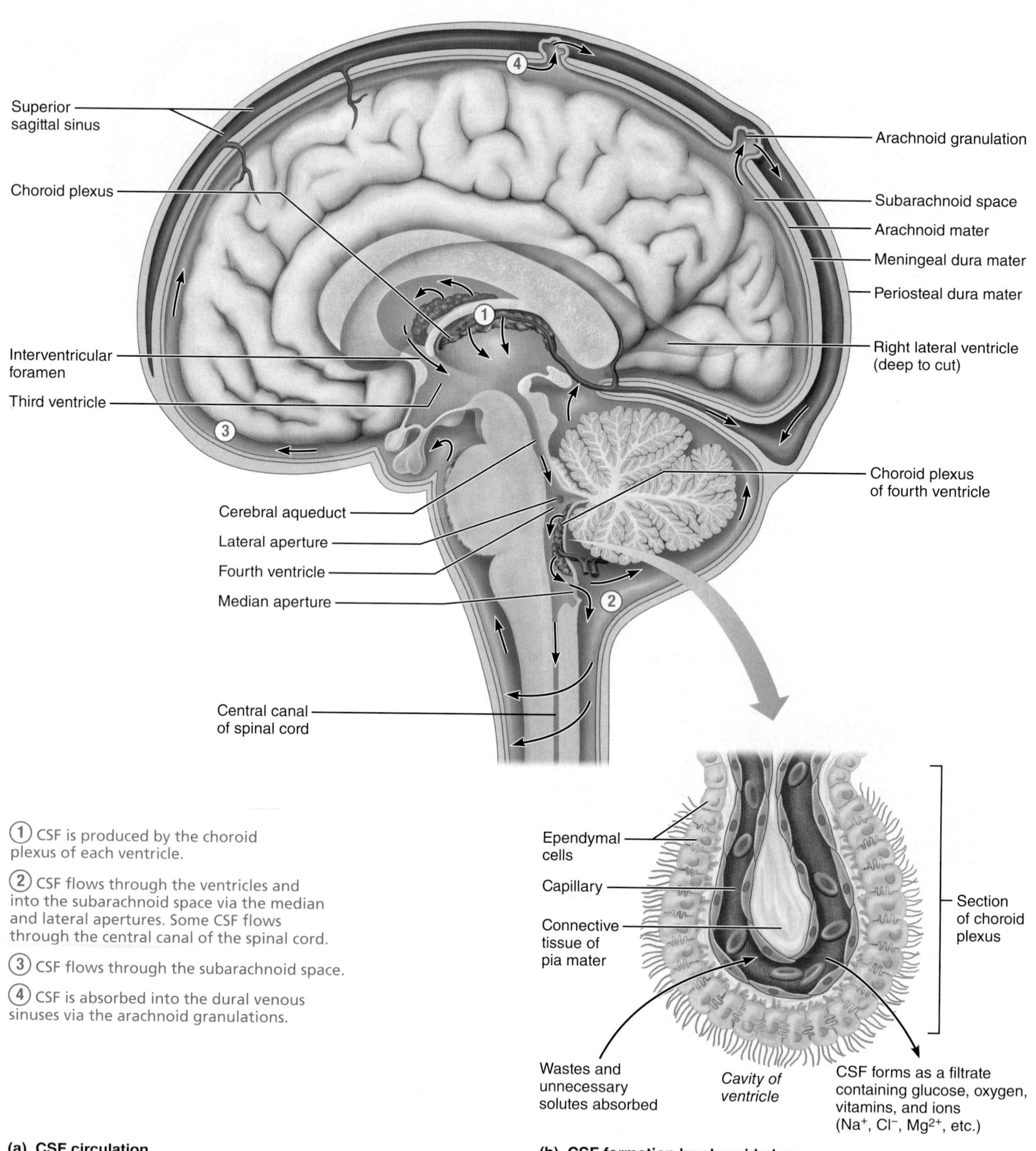

Figure 13.23 Formation, location, and circulation of CSF. (a) Location and circulatory pattern of cerebrospinal fluid (CSF). Arrows indicate the direction of flow. (b) Each choroid plexus consists of a knot of porous capillaries surrounded by a single layer of ependymal cells joined by tight junctions and bearing long cilia. Fluid leaking from porous capillaries is processed by the ependymal cells to form the CSF in the ventricles.

Practice art labeling MasteringA&P®

After entering the ventricles, the CSF moves freely through these chambers.

② Some CSF runs into the central canal of the spinal cord, but most enters the subarachnoid space through the lateral and median apertures in the walls of the fourth ventricle.

③ In the subarachnoid space, CSF bathes the outer surfaces of the brain and spinal cord.

④ CSF then passes through the arachnoid granulations and enters the blood of the dural venous sinuses.

Cerebrospinal fluid arises from the blood and returns to it at a rate of about 500 ml a day.

CLINICAL APPLICATION

Hydrocephalus Excessive accumulation of CSF in the ventricles or subarachnoid space can exert a crushing pressure on the brain. This condition is called **hydrocephalus** (hi″dro-sef′ah-lus; "water on the brain"). Its causes include a tumor or inflammatory swelling that closes off the cerebral aqueduct or the fourth ventricle; meningitis, which causes scarring that closes the subarachnoid space and arachnoid granulations; and, in infants, an overdeveloped choroid plexus that secretes too much CSF. Hydrocephalus in newborns results in an enlarged cranium because the young skull bones have not yet fused (see photograph). In adults, fluid pressure quickly builds up and damages the brain because the cranium is unyielding. Hydrocephalus is treated surgically by inserting a shunt (plastic tube) into the ventricles to drain excess fluid into the abdominal cavity.

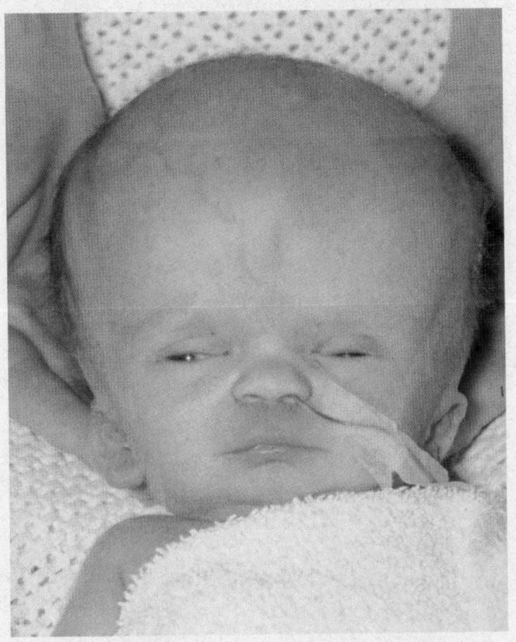

Hydrocephalus in the newborn.

Blood Brain Barrier

The brain has a rich supply of capillaries that provide its nervous tissue with nutrients, oxygen, and all other vital molecules. However, some bloodborne molecules that can cross other capillaries of the body cannot cross the brain capillaries. Bloodborne toxins, such as urea, mild toxins from food, and bacterial toxins, are prevented from entering brain tissue by the **blood brain barrier,** which protects the neurons of the CNS.

The blood brain barrier results primarily from special features of the epithelium that make up the walls of the brain capillaries. The epithelial cells in brain capillaries are joined together around their entire perimeters by tight junctions, making these capillaries the least permeable capillaries in the body. Even so, the blood brain barrier is not an absolute barrier. All nutrients (including oxygen) and ions needed by the neurons pass through, some by special transport mechanisms in the plasma membranes of the capillary epithelial cells. Furthermore, because the barrier is ineffective against fat-soluble molecules, which easily diffuse through all cell membranes, the barrier allows alcohol, nicotine, and anesthetic agents to reach brain neurons. (More information on the blood brain barrier is provided in Chapter 20.)

check your understanding

☐ 20. Name the dura mater extension that lies in the longitudinal fissure.

☐ 21. Where is cerebrospinal fluid produced? How is it returned into the blood?

(For answers, see Appendix B.)

THE SPINAL CORD

learning outcomes

- ▶ Describe the gross structure of the spinal cord.
- ▶ Identify the regions of white matter and gray matter in a cross section through the spinal cord.
- ▶ Describe the organization of the neurons in the gray matter of the spinal cord.

The **spinal cord** functions in three ways:

1. Through the spinal nerves that attach to it, the spinal cord is involved in the *sensory and motor innervation* of the entire body inferior to the head.
2. Through the ascendng and descending tracts traveling within its white matter, the spinal cord provides a *two-way conduction pathway* for signals between the body and the brain.
3. Through sensory and motor integration in its gray matter, the spinal cord is a *major center for reflexes* (see Figure 12.13, pp. 366–367).

The spinal cord runs through the vertebral canal of the vertebral column. The vertebral canal is formed from the

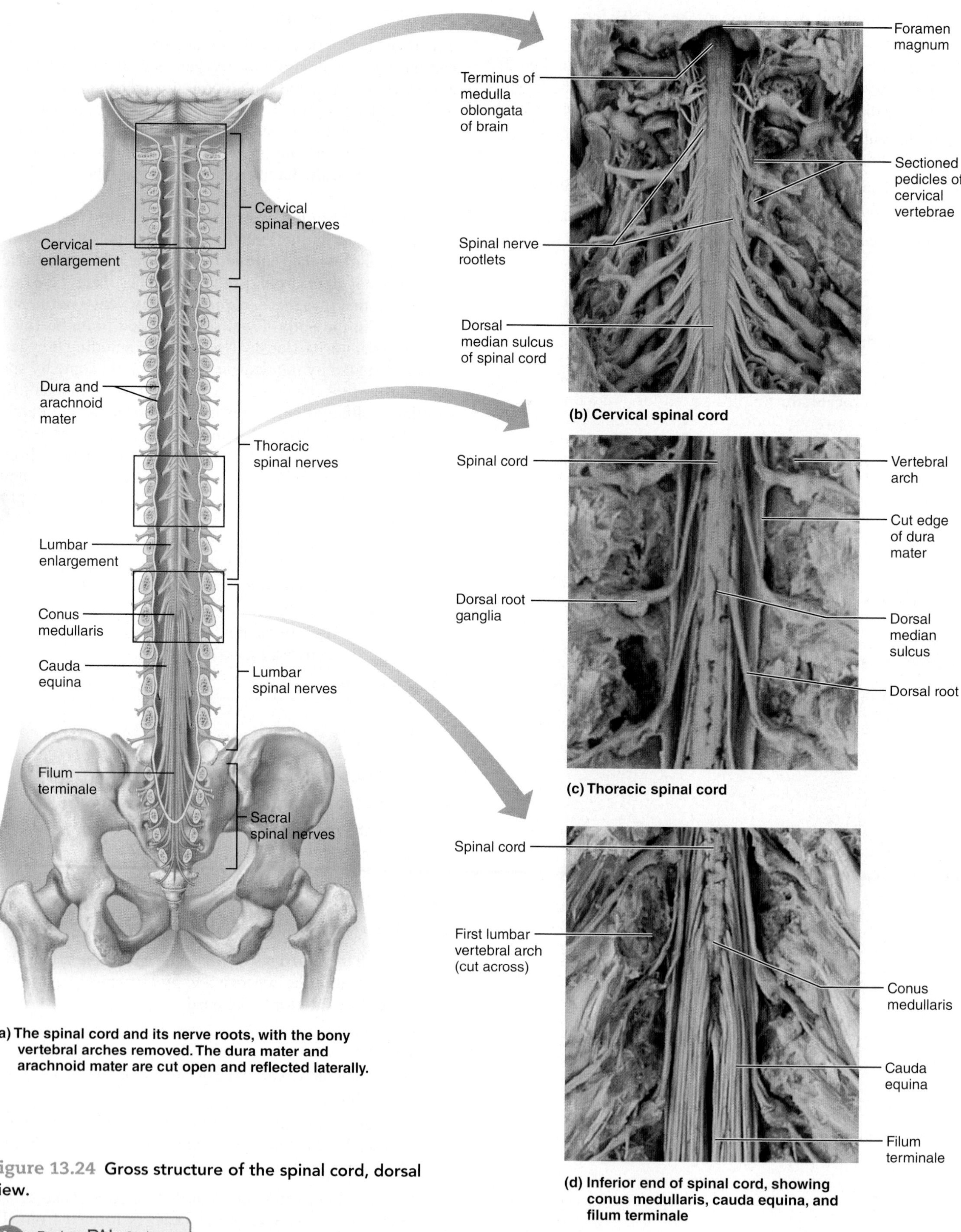

(a) The spinal cord and its nerve roots, with the bony vertebral arches removed. The dura mater and arachnoid mater are cut open and reflected laterally.

(b) Cervical spinal cord

(c) Thoracic spinal cord

(d) Inferior end of spinal cord, showing conus medullaris, cauda equina, and filum terminale

Figure 13.24 **Gross structure of the spinal cord, dorsal view.**

Explore PAL Cadaver MasteringA&P®

successive vertebral foramina of the articulated vertebrae. The spinal cord extends from the foramen magnum at the base of the skull's occipital bone superiorly to the level of the first or second lumbar vertebra (L_1 or L_2) inferiorly (Figure 13.24). At its inferior end, the spinal cord tapers into the **conus medullaris** (ko′nus med′u-lar″is; "cone of the spinal cord") (Figure 13.24a and d). A long filament of connective tissue, the **filum terminale** ("end filament"), extends from the conus medullaris and attaches to the coccyx inferiorly, anchoring the spinal cord in place so that it is not jostled by body movements.

Thirty-one pairs of *spinal nerves* (PNS structures) attach to the spinal cord through dorsal and ventral nerve roots (Figure 13.24a and c). The spinal nerves lie in the intervertebral foramina, from which they send lateral branches throughout the body. The spinal nerves are named based on the vertebral locations in which they lie. The 31 spinal nerves are divided into cervical (8), thoracic (12), lumbar (5), sacral (5), and coccygeal (1) groups. There are eight cervical nerves despite there being only seven cervical vertebrae because the first cervical spinal nerve exits above the first cervical vertebra. Each subsequent cervical spinal nerve exits inferior to a cervical vertebra. The thoracic, lumbar, and sacral spinal nerves exit below the vertebra for which they are named. (The spinal nerves are discussed in detail in Chapter 14.)

In the cervical and lumbar regions of the spinal cord, where the nerves to the upper and lower limbs arise, the spinal cord shows obvious enlargements called the **cervical** and **lumbar enlargements** (Figure 13.24a). Because the cord does not reach the inferior end of the vertebral column, the lumbar and sacral nerve roots must descend within the vertebral canal for some distance before reaching their corresponding intervertebral foramina. This collection of nerve roots at the inferior end of the vertebral canal is the **cauda equina** (kaw′dah e-kwi′-nah; "horse's tail").

The spinal cord does not extend the full length of the vertebral canal because of prenatal events. Until the third month of development, the spinal cord does extend all the way to the coccyx, but thereafter it grows more slowly than the caudal vertebral column. As the vertebral column elongates, the spinal cord assumes a progressively more rostral position. By the time of birth, the spinal cord ends at the level of the lumbar vertebra L_3. During childhood, it attains its adult position, terminating at the level of the intervertebral disc between lumbar vertebrae L_1 and L_2. This is merely the average level; the location of the end of the spinal cord varies from one person to another, from the inferior margin of thoracic vertebra T_{12} to the superior margin of lumbar vertebra L_3.

The spinal cord forms from neuroectoderm, and its segmented appearance reflects the pattern of the adjacent somites. The term **spinal cord segment** is used clinically to indicate the region of the spinal cord from which the axonal processes that form a given spinal nerve emerge (Figure 13.25). Each spinal cord segment is designated by the spinal nerve that arises from it—for example, the first thoracic cord segment (spinal cord segment T_1) is where the first thoracic nerve (spinal nerve T_1) emerges from the spinal cord, the fifth thoracic cord segment (spinal cord segment T_5) is where the fifth spinal nerve (spinal nerve T_5) emerges, and so forth.

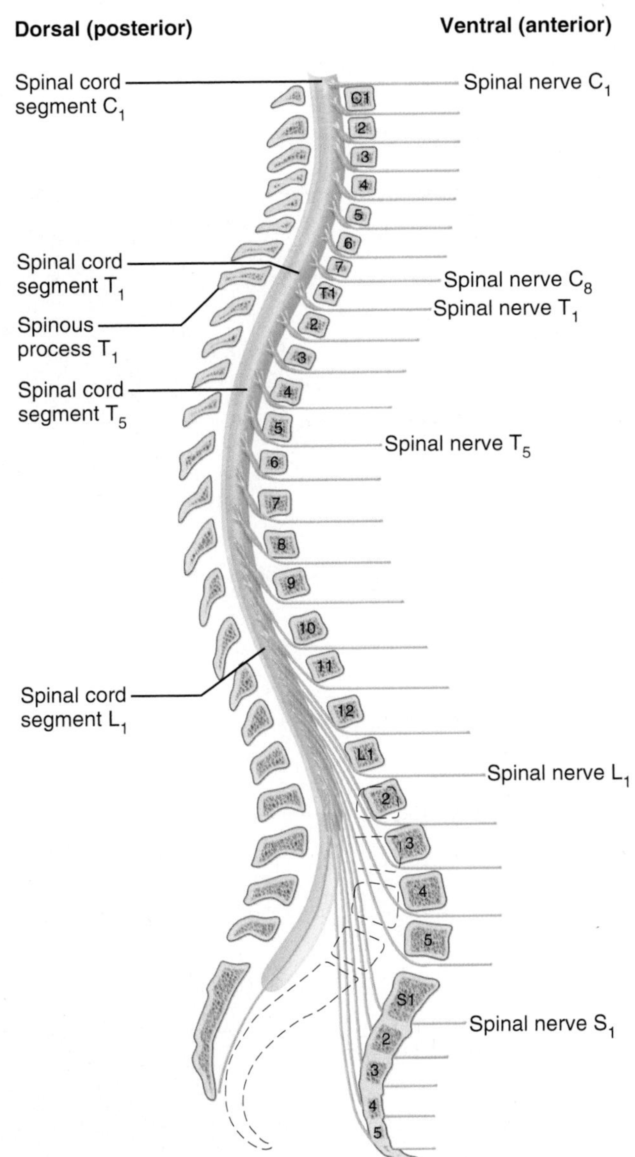

Figure 13.25 Spinal cord segments. Diagram of the spinal cord in sagittal view showing the position of spinal cord segments and the spinal nerves. The vertebral bodies, shown on the right, are numbered. Those in the lumbar and sacral regions are displaced anteriorly to illustrate the emergence of the spinal nerves.

Because the spinal cord does not extend to the end of the spinal column, the spinal cord segments are located superior to where their corresponding spinal nerves emerge through the intervertebral foramina. For example, spinal cord segment T_5 is located at the level of vertebra T_4. This discrepancy is most pronounced in the lumbar and sacral regions of the cord: The lumbar cord segment L_1 is located at vertebra T_{11}, and the sacral cord segment S_1 is at vertebra L_1.

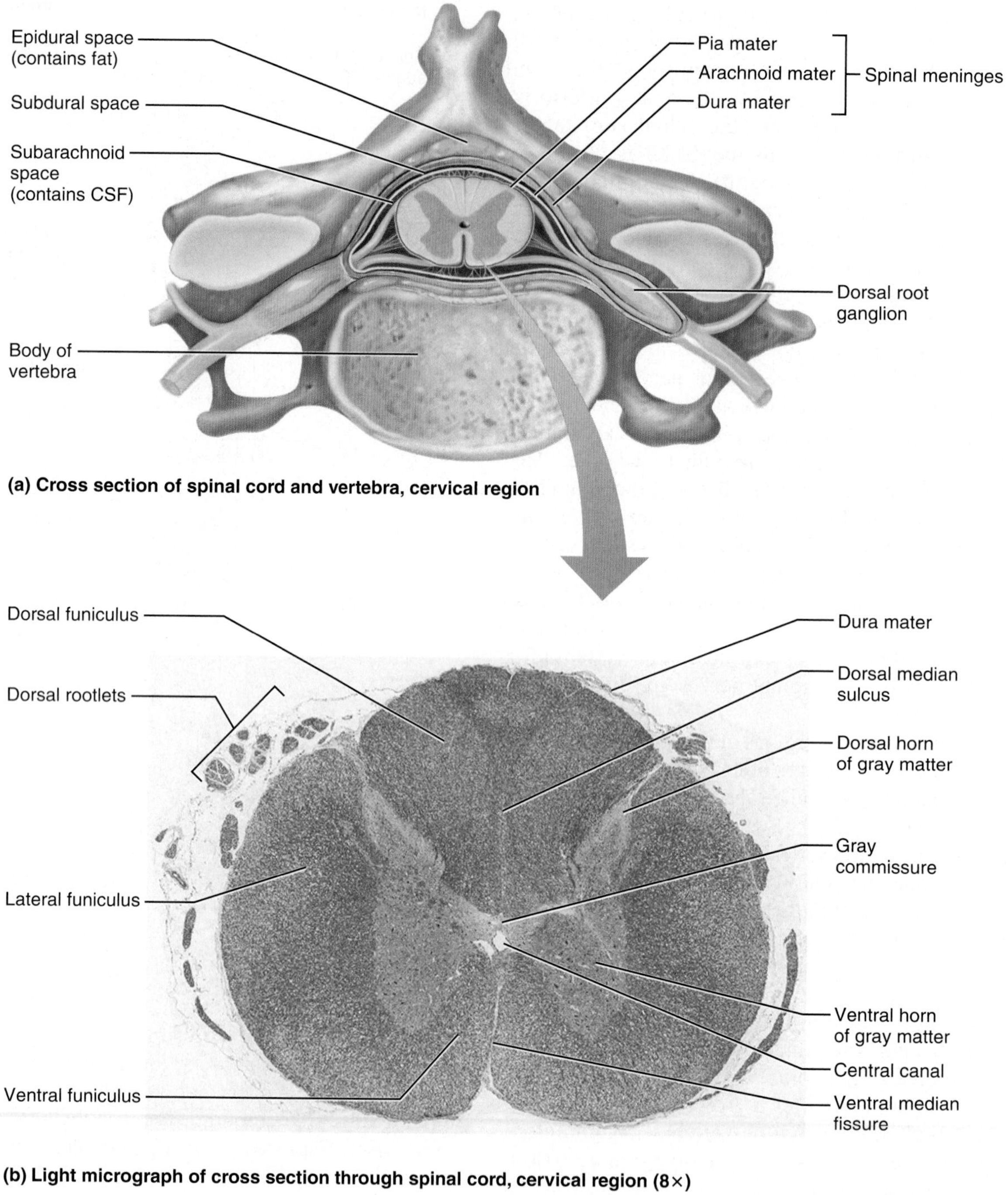

Epidural space (contains fat)
Subdural space
Subarachnoid space (contains CSF)
Body of vertebra
Pia mater
Arachnoid mater
Dura mater
Spinal meninges
Dorsal root ganglion

(a) Cross section of spinal cord and vertebra, cervical region

Dorsal funiculus
Dorsal rootlets
Lateral funiculus
Ventral funiculus
Dura mater
Dorsal median sulcus
Dorsal horn of gray matter
Gray commissure
Ventral horn of gray matter
Central canal
Ventral median fissure

(b) Light micrograph of cross section through spinal cord, cervical region (8×)

Figure 13.26 Anatomy of the spinal cord.

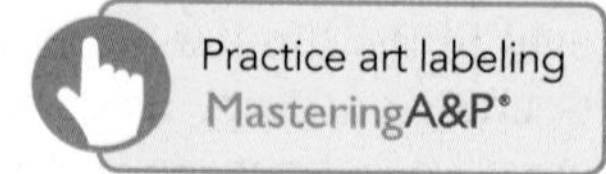

The spinal cord is wider laterally than anteroposteriorly (Figure 13.26). Two grooves—the **dorsal (posterior) median sulcus** and the wider **ventral (anterior) median fissure**—run the length of the cord and partly divide it into right and left halves (Figure 13.26b and c).

White Matter of the Spinal Cord

The spinal cord consists of an outer region of white matter and an inner region of gray matter. The white matter of the spinal cord is composed of myelinated and unmyelinated axons that allow communication between different parts of the spinal cord and between the cord and brain. These fibers are classified as being one of three types, according to the direction in which they carry nerve impulses:

- **Ascending.** Most of the ascending fibers in the spinal cord carry sensory information from the sensory neurons of the body up to the brain.
- **Descending.** Most descending fibers carry motor instructions from the brain to the spinal cord, to stimulate contraction of the body's muscles and secretion from its glands.

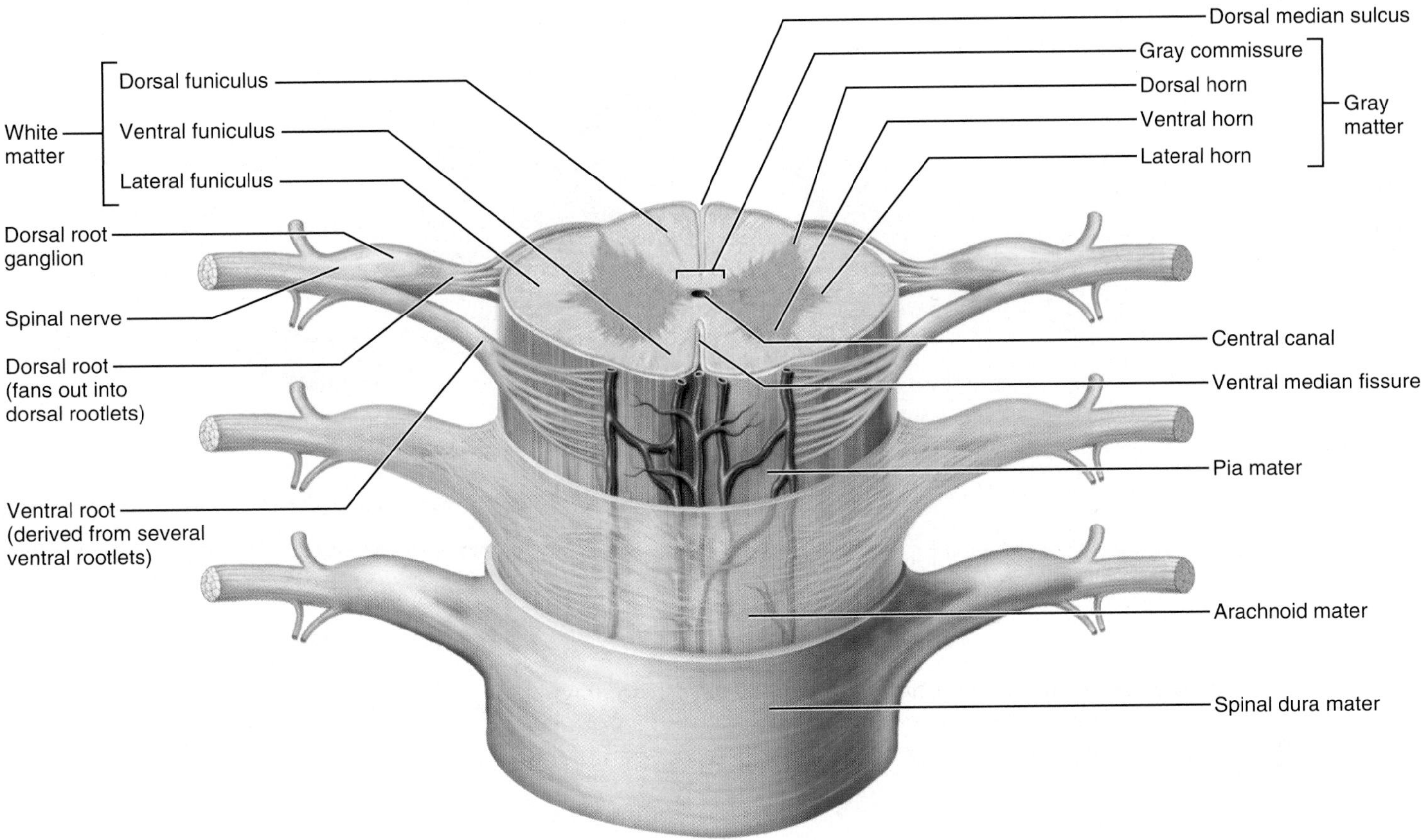

(c) The spinal cord and its meningeal coverings, thoracic region

Figure 13.26 **Anatomy of the spinal cord,** *continued.*

- **Commissural.** Commissural fibers are white-matter fibers that carry information from one side of the spinal cord to the other.

The white matter on each side of the spinal cord is divided into three **funiculi** (fu-nik′u-li; "long ropes"): the **dorsal (posterior) funiculus, ventral (anterior) funiculus,** and **lateral funiculus** (Figure 13.26b and c). The ventral and lateral funiculi are continuous with each other and are divided by an imaginary line that extends out from the ventral horn of gray matter. The three funiculi contain many fiber tracts. Each fiber tract is composed of axons that all have similar destinations and functions. These sensory and motor pathways, described later in the chapter, integrate the PNS with the CNS (p. 413).

Gray Matter of the Spinal Cord and Spinal Roots

As in all other parts of the CNS, the gray matter of the cord consists of a mixture of neuron cell bodies; short, nonmyelinated neurons; and neuroglia. In cross section, the gray matter of the spinal cord is shaped like the letter H (Figure 13.26). The crossbar of the H is called the **gray commissure** and is composed of unmyeliated axons that cross from one side of the CNS to the other. Within the gray commissure is the narrow central cavity of the spinal cord, the **central canal.** The two posterior arms of the H are the **dorsal (posterior) horns,** and the two anterior arms are the **ventral (anterior) horns.** In three dimensions, these arms form columns of gray matter that run the entire length of the spinal cord. In the thoracic and superior lumbar segments of the spinal cord, small lateral gray matter columns—called **lateral horns**—are present (Figure 13.26c).

The neurons in the spinal gray matter are organized functionally (Figure 13.27). The dorsal horns consist entirely of interneurons. These interneurons receive information from sensory neurons whose cell bodies lie outside the spinal cord in *dorsal root ganglia* and whose axons reach the spinal cord via the *dorsal roots.* The ventral (and the lateral) horns contain cell bodies of motor neurons that send their axons out of the spinal cord via the *ventral roots* to supply muscles and glands. Interneurons are also present in the ventral horns (but not emphasized in Figure 13.27). The size of the ventral motor horns varies along the length of the spinal cord, reflecting the amount of skeletal musculature innervated at each level: The ventral horns are largest in the cervical and lumbar

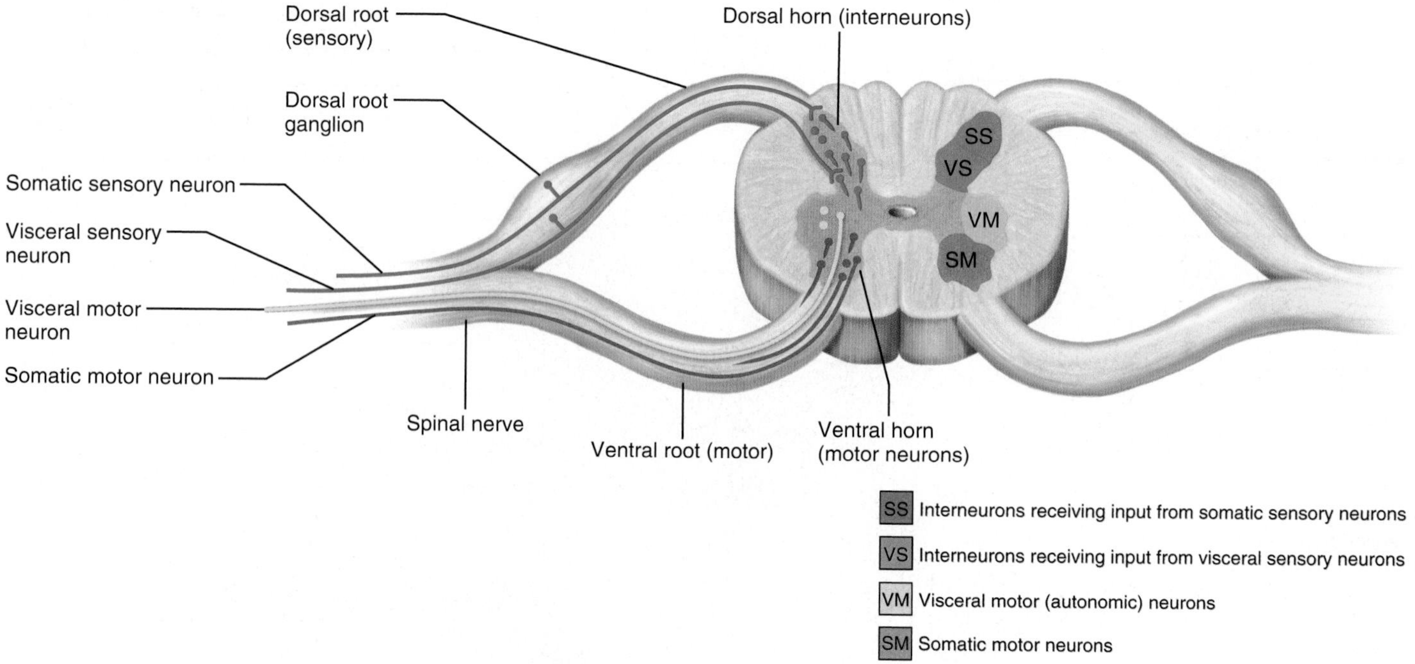

Figure 13.27 **Organization of the gray matter of the spinal cord.** The gray matter of the spinal cord can be divided into a dorsal sensory half and a ventral motor half. The dorsal roots and ventral roots are part of the PNS, not part of the spinal cord.

segments of the cord, which innervate the upper and lower limbs, respectively. These segments form the previously described cervical and lumbar enlargements of the spinal cord.

The spinal gray matter can be further classified according to the innervation of the somatic and visceral regions of the body. This scheme recognizes four zones of spinal cord gray matter: **somatic sensory (SS), visceral sensory (VS), visceral motor (VM),** and **somatic motor (SM).** These zones are equivalent to the functional divisions of the PNS (see Table 12.1, p. 352).

CLINICAL APPLICATION

Flaccid And Spastic Paralysis Any localized damage to the spinal cord or the spinal roots leads to some functional loss, either *paralysis* (loss of motor function) or *paresthesia* (abnormal or lost sensation). Severe damage to the ventral horn or to the ventral motor roots destroys the somatic motor neurons in the region of the injury and results in a complete, or **flaccid, paralysis** of the skeletal muscles served. Because the muscles are no longer stimulated by neurons, they shrink and waste away, and spinal reflexes are absent. By contrast, damage to only the descending fiber tracts in the white matter of the spinal cord leaves the spinal cord motor neurons and spinal reflexes intact. In this case, the muscles remain healthy, but their movements are no longer under voluntary control because their connection to the brain has been lost. This condition is termed **spastic paralysis.**

Protection of the Spinal Cord

Like the brain, the neural tissue of the spinal cord is protected by bone (the vertebrae), meninges, and cerebrospinal fluid (Figure 13.26a). The tough dura mater, called the **spinal dural sheath,** differs somewhat from the dura around the brain: It does not attach to the surrounding bone and corresponds only to the meningeal layer of the brain's dura mater. Just external to this spinal dura is a rather large **epidural space** filled with cushioning fat and a network of veins. Anesthetics are often injected into the epidural space to block nerve impulses in the spinal cord and thereby relieve pain in body regions inferior to the injection site. The subdural space, arachnoid mater, subarachnoid space, and pia mater around the spinal cord are continuous with the corresponding elements around the brain.

Inferiorly, the dura and arachnoid extend to the level of S_2, well beyond the end of the spinal cord. The pia mater extends into the coccyx, covering the filum terminale (Figure 13.24a and d). Lateral extensions of the pia mater called the **denticulate ligaments** (den-tik′u-lāt; "toothed") anchor the spinal cord laterally to the dura mater throughout the length of the cord.

CLINICAL APPLICATION

Meningitis An inflammation of the meninges caused by a bacterial or viral infection is called **meningitis.** The infection can spread to the underlying nervous tissue and cause brain inflammation, or **encephalitis** (en″sef-ah-li′tis). Meningitis is usually diagnosed by taking a sample of cerebrospinal fluid from the subarachnoid space and examining it for the presence of microbes. Because the adult spinal cord ends at the level of vertebra L_1 or L_2, a needle can safely enter the subarachnoid space inferior to that location, a procedure that is called either a **lumbar puncture** or a **spinal tap.** The patient bends forward, and the needle is inserted between the spines and laminae of successive vertebrae, either between L_3 and L_4 or between L_4 and L_5. Lumbar punctures are also used to sample the CSF for chemical analysis, to measure the pressure of the CSF, and to inject antibiotics, anesthetics, or X-ray contrast media into the subarachnoid space.

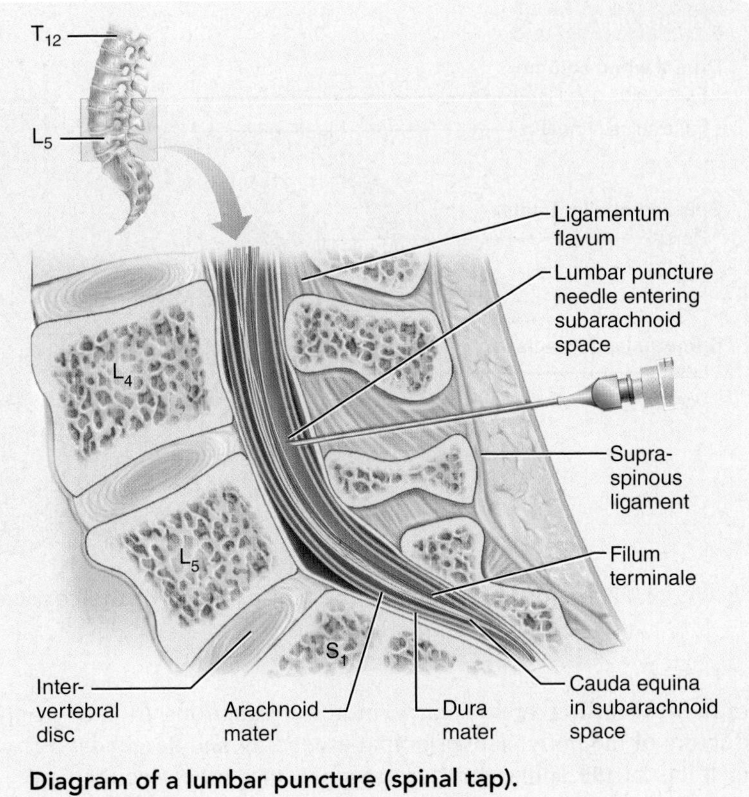

Diagram of a lumbar puncture (spinal tap).

check your understanding

- ☐ 22. What neural structures pass through the vertebral foramen in the region of lumbar vertebra L_3?
- ☐ 23. Which portion of the spinal cord, gray matter or white matter, provides the two-way conduction pathway for signals between the body and the brain?
- ☐ 24. What deficits result if the ventral horn of the spinal cord in the lower cervical segments is damaged?
- ☐ 25. Which two meninges border the space that is filled with cerebrospinal fluid?

(For answers, see Appendix B.)

SENSORY AND MOTOR PATHWAYS IN THE CENTRAL NERVOUS SYSTEM

learning outcomes

- ▶ Identify the major ascending and descending tracts in the spinal cord, and indicate the types of stimuli carried on each.
- ▶ Trace the major neuronal pathways to and from the brain.

Now that we have looked at the anatomy and functioning of both the brain and the spinal cord, we turn to how these two parts of the central nervous system are connected to the rest of the body and how communication between the CNS and the body periphery is accomplished.

The brain is connected to the body periphery by multineuron pathways that travel through the white matter of the CNS. These pathways are composed of interconnected fiber tracts that relay information from one portion of the CNS to another. The pathways that carry information to more rostral regions of the CNS are called **ascending pathways;** those that carry information to more caudal regions of the CNS are called **descending pathways.**

The following statements summarize important features of the ascending and descending pathways:

- Most pathways cross from one side of the CNS to the other, or *decussate,* at some point along their course.
- Most pathways consist of a chain of two or three serially linked neurons that contribute to successive tracts along a given pathway.
- Most pathways are spatially arranged in a specific way, according to the body region they supply. For example, in one ascending tract, the axons transmitting impulses from the superior parts of the body lie lateral to the axons carrying impulses from the inferior body parts.
- All pathways are bilaterally symetrical, occuring on both the right and left side of the brain or spinal cord.

The segments of these pathways that travel through the spinal cord are called ascending and descending **spinal tracts: Ascending tracts** carry sensory information to the

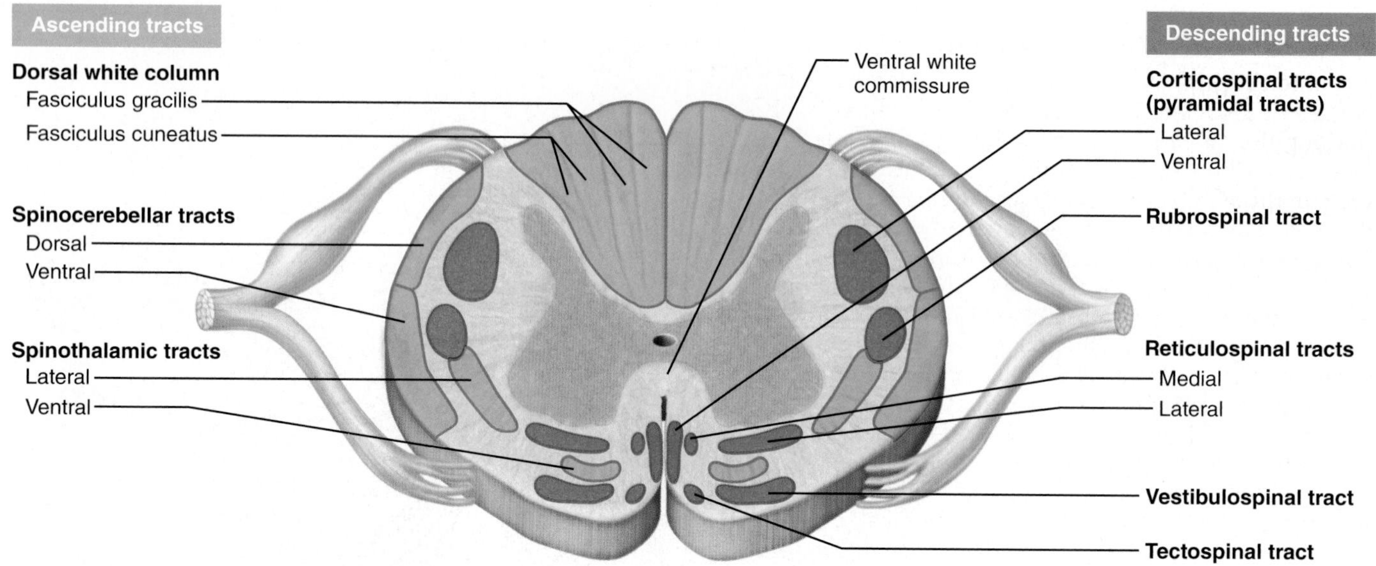

Figure 13.28 Major fiber tracts in the white matter of the cervical spinal cord.

brain; **descending tracts** carry motor instructions to the effectors of the body. The principal ascending and descending tracts of the spinal cord are mapped in a cross section through the cervical spinal cord (Figure 13.28). For the most part, these spinal tracts are named according to their origin and destination.

Ascending Pathways

The ascending pathways conduct general somatic sensory impulses superiorly through chains of two or three neurons to various regions of the brain (Figure 13.29). The first neuron in the pathway, called the *first-order neuron,* is the sensory neuron which extends from the sensory receptor into the spinal cord (bottom of Figure 13.29a and b). The first-order neuron synapses in the CNS with another neuron in the pathway, the *second-order neuron* (middle of 13.29a and b). In some ascending pathway, the second-order neuron synapses with a *third-order neuron* (top of Figure 13.29a and b).

The three main ascending pathways are the *spinocerebellar pathway,* the *dorsal column pathway,* and *spinothalamic pathway* (illustrated in Figure 13.29 and summarized in Table 13.3).

Spinocerebellar Pathway

The **spinocerebellar pathway** arises from second-order neurons in the dorsal horn of the spinal cord and terminates on the cerebellum (left side of Figure 13.29a). This pathway carries information on proprioception from the lower limbs and trunk to the cerebellum, which uses this information to coordinate body movements. These fibers either do not decussate or cross twice, undoing the decussation; thus, they project ipsilaterally.

Dorsal Column–Medial Lemniscal Pathway

The **dorsal column pathway** (right side of Figure 13.29a) carries information on fine touch, pressure, and conscious aspects of proprioception. These are **discriminative senses**—senses that can be localized very precisely on the body surface. In the dorsal column pathway,

- The axons of first-order neurons, the sensory neurons, enter the spinal cord and send an axonal branch up one of the dorsal white column tracts, either the medial **fasciculus gracilis** ("slender bundle") or the lateral **fasciculus cuneatus** ("wedge-shaped bundle"). These axons ascend in the spinal tract to the medulla oblongata.
- In the medulla oblongata, these axons synapse with second-order neurons in the **nucleus gracilis** or **nucleus cuneatus.** Axons from these brain nuclei form a tract called the **medial lemniscus tract** (lem-nis′kus; "ribbon") (Figure 13.29a), which decussates in the medulla and then ascends through the pons and midbrain to the thalamus.
- Third-order neurons originating in the thalamus send axons to the primary somatosensory cortex on the postcentral gyrus, where the sensory information is processed, resulting in awareness of precisely localized sensations.

Spinothalamic Pathway

The **spinothalamic pathway** (Figure 13.29b) carries information on pain, temperature, deep pressure, and nondiscriminative touch—stimuli we are aware of but cannot localize precisely on the body surface. In the spinothalamic pathway:

- The axons of first-order sensory neurons enter the spinal cord, where they synapse on interneurons in the dorsal gray horn.

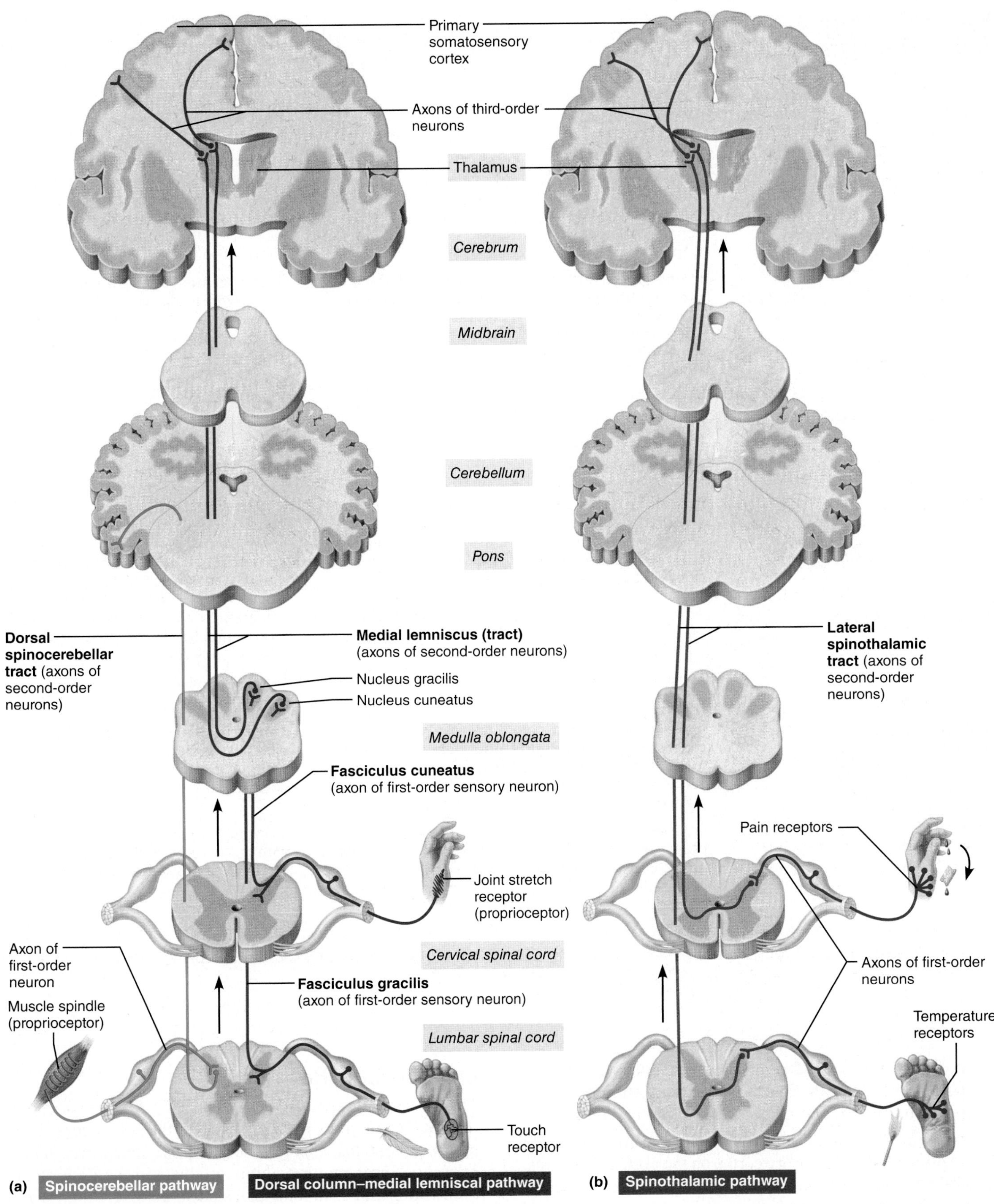

Figure 13.29 Pathways of selected ascending spinal cord tracts. Cross section up to the cerebrum, which is shown in frontal section. (**a**) The spinocerebellar pathway (left) transmits proprioceptive information to the cerebellum, and so it is subconscious. The dorsal column–medial lemniscal pathway transmits discriminative touch and conscious proprioception signals to the cerebral cortex. (**b**) The spinothalamic pathway transmits nondiscriminative sensations (pain, temperature, pressure) to the primary somatosensory cortex.

Table 13.3 Major Ascending (Sensory) Pathways and Spinal Tracts

Spinal Cord Tract	Location (Funiculus)	Origin	Termination	Function
SPINOCEREBELLAR PATHWAYS				
Dorsal spinocerebellar*	Lateral (dorsal part)	Interneurons (second-order neurons) in dorsal horn on same side of cord. Fibers ascend without crossing.	By synapse in cerebellum by way of the inferior cerebellar peduncle	Transmits impulses from trunk and lower limb proprioceptors on one side of body to ipsilateral side of cerebellum for subconscious proprioception.
Ventral spinocerebellar*	Lateral (ventral part)	Interneurons (second-order neurons) of dorsal horn. Contains crossed fibers that cross back to the original ipsilateral side in the pons.	By synapse in cerebellum by way of the superior cerebellar peduncle	Transmits impulses from the trunk and lower limb on one side of body to ipsilateral side of cerebellum for subconscious proprioception.
DORSAL COLUMN–MEDIAL LEMNISCUS PATHWAYS				
Fasciculus cuneatus and fasciculus gracilis	Dorsal	Central processes of sensory (first-order) neurons enter dorsal root of the spinal cord and branch. Branches enter dorsal white column on ipsilateral side without synapsing.	By synapse with second-order neurons in nucleus cuneatus and nucleus gracilis in medulla. Fibers of medullary neurons cross over and ascend in medial lemniscus to thalamus where they synapse with third-order neurons. Thalamic neurons then transmit impulses to somatosensory cortex.	Both tracts transmit general sensory impulses from receptors of skin and proprioceptors, which are interpreted as discriminative touch, pressure, and "body sense" (limb and joint position) in the contralateral somatosensory cortex. Cuneatus transmits afferent impulses from upper limbs, upper trunk, and neck; it is not present in spinal cord below level of T_6. Gracilis carries impulses from lower limbs and inferior body trunk.
SPINOTHALAMIC PATHWAYS				
Lateral spinothalamic	Lateral	Interneurons (second-order neurons) in dorsal horn. Fibers cross to contralateral side before ascending.	By synapse with third-order neurons in thalamus. Thalamic neurons then convey impulses to somatosensory cortex.	Transmits impulses concerned with pain and temperature to contralateral somatosensory cortex.
Ventral spinothalamic	Ventral	Interneurons (second-order neurons) in dorsal horns. Fibers cross to contralateral side before ascending.	By synapse with third-order neurons in thalamus. Thalamic neurons convey impulses to somatosensory cortex.	Transmits impulses concerned with crude touch and pressure to contralateral somatosensory cortex.

*These spinocerebellar tracts carry information from the lower limb and trunk only. The corresponding tracts for the upper limb and neck (called *rostral spinocerebellar* and *cuneocerebellar tracts*) are beyond the scope of this book.

- Axons of the second-order neurons decussate in the spinal cord, enter the lateral and ventral funiculi as the **spinothalamic tract,** and ascend to the thalamus.
- Axons from third-order neurons in the thalamus project to the primary somatosensory cortex on the postcentral gyrus, where the information is processed into conscious sensations. The brain interprets the sensory information carried by the spinothalamic pathway as unpleasant—pain, burns, cold, and so on.

Descending Pathways

The descending spinal tracts (Figure 13.30) deliver motor (output) instructions from the brain to the spinal cord. Each pathway contains *upper motor neurons* and *lower motor neurons.* Upper motor neurons originate in the gray matter of the brain and send long axons down a descending tract of the spinal cord. These axons synapse with lower motor neurons in the ventral horn of the spinal cord. The axons of the

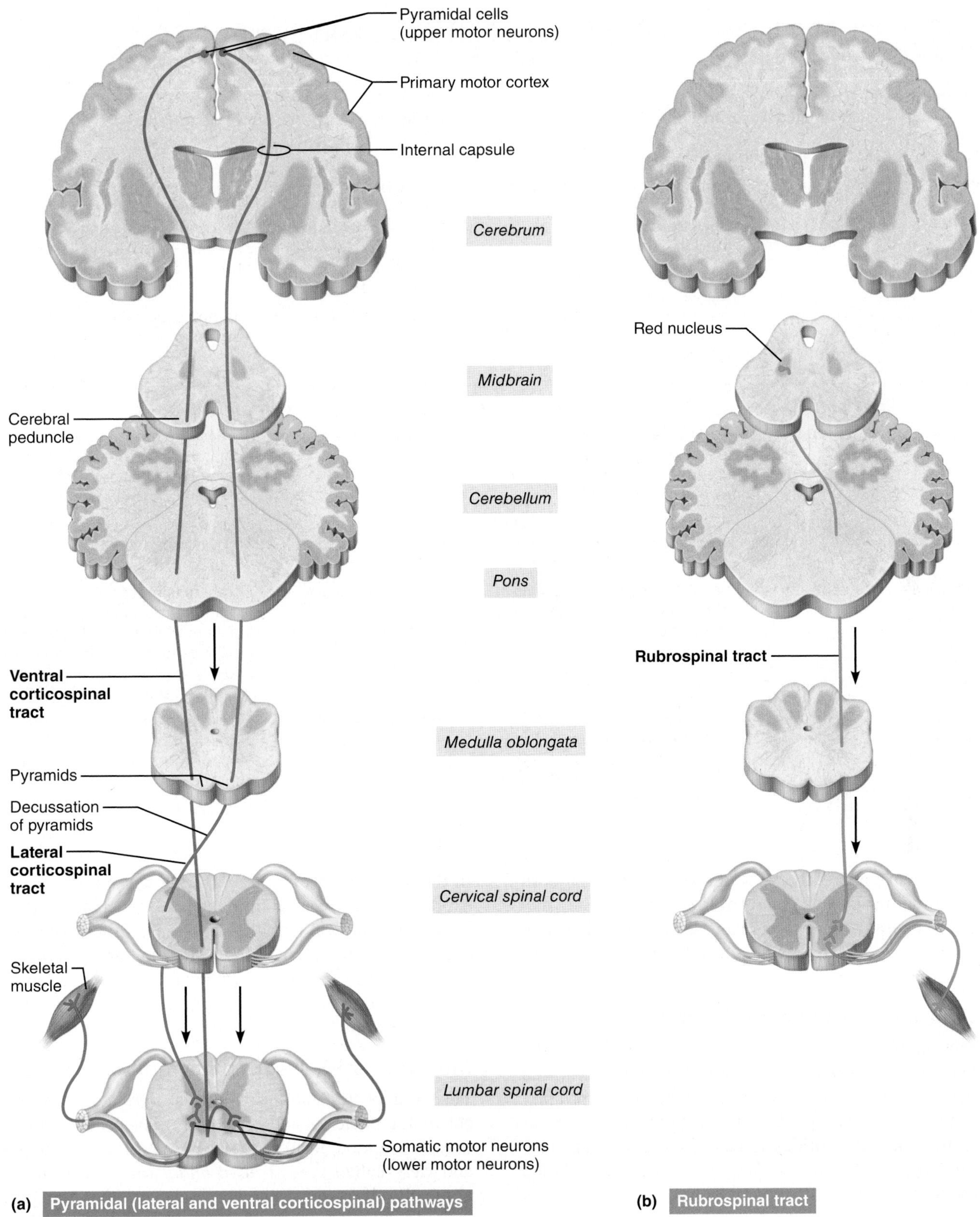

Figure 13.30 **Three descending pathways by which the brain influences movement.** Cross sections up to the cerebrum, which is shown in frontal section. (**a**) Pyramidal (lateral and ventral corticospinal pathways) originate from the cerebral cortex and control skilled, voluntary movements. (**b**) The rubrospinal tract, one of the indirect pathways, helps regulate muscle tone.

Table 13.4 Major Descending (Motor) Pathways and Spinal Tracts

Spinal Cord Tract	Location (Funiculus)	Origin	Termination	Function
DIRECT (PYRAMIDAL) PATHWAYS				
Lateral corticospinal	Lateral	Pyramidal neurons of motor cortex of the cerebrum; decussate in pyramids of medulla	By synapse with ventral horn interneurons that influence motor neurons, and occasionally with ventral horn motor neurons directly	Transmits motor impulses from cerebrum to contralateral spinal cord motor neurons (which activate skeletal muscles); voluntary motor tract to limb muscles
Ventral corticospinal	Ventral	Pyramidal neurons of motor cortex; fibers cross over at the spinal cord level	Ventral horn (as above)	Same as lateral corticospinal tract but to axial muscles
INDIRECT PATHWAYS				
Rubrospinal	Lateral	Red nucleus of midbrain of brain stem (fibers cross to contralateral side just inferior to the red nucleus)	Ventral horn (as above)	Transmits motor impulses concerned with muscle tone of distal limb muscles (mostly flexors) on contralateral side of body
Tectospinal	Ventral	Superior colliculus of midbrain of brain stem (fibers cross to contralateral side of cord)	Ventral horn (as above)	Turns neck so eyes can follow a moving object
Vestibulospinal	Ventral	Vestibular nuclei in medulla of brain stem (fibers descend without crossing)	Ventral horn (as above)	Transmits motor impulses that maintain muscle tone and activate ipsilateral limb and trunk extensor muscles and muscles that move head; helps maintain balance during standing and moving
Reticulospinal (ventral, medial, and lateral)	Ventral and lateral	Reticular formation of brain stem (medial nuclear group of pons and medulla); both crossed and uncrossed fibers	Ventral horn (as above)	Transmits impulses concerned with muscle tone and many visceral motor functions; may control most unskilled movements

lower motor neurons exit the ventral root of the spinal cord to innervate the muscles or glands of the body.

The descending spinal tracts can be classified into two groups: (1) *direct pathways,* which are the pyramidal tracts, and (2) *indirect pathways,* essentially all others (Table 13.4).

Direct Pathways (Pyramidal Tracts)

The direct pathways extend without synapsing from the pyramidal cells in the cerebral cortex to the spinal cord. The pyramidal cells are the large neurons found in the primary motor cortex of the brain (p. 394). The long axons of pyramidal cells form the **pyramidal tracts** (Figure 13.30a), also called **corticospinal tracts,** which control precise and skilled voluntary movements. In the pyramidal tracts:

- The axons of pyramidal cells, the upper motor neurons, descend from the cerebral motor cortex through the brain stem to the spinal gray matter—mostly to the ventral horns.
- In the ventral horn, the axons either synapse with short interneurons that activate somatic motor neurons or synapse directly on somatic motor neurons, the lower motor neurons.

In this way, signals that travel along the pyramidal pathways exert influence over the limb muscles, especially muscles that move the hand and fingers. The axons of the pyramidal tracts decussate along their course: In the *lateral corticospinal tract,* this occurs in the medulla within the decussation of the pyramids; in the *ventral corticospinal tract,* the axons decussate in the spinal cord.

Indirect Pathways (Extrapyramidal Tracts)

The indirect descending pathways originate in the subcortical motor nuclei of the brain stem. Historically, these pathways were thought to be independent of the pyramidal cells and were called the *extrapyramidal tracts,* a term still used clinically. It is now known, however, that the pyramidal cells do project to and influence these pathways; thus, these pathways are referred to as *indirect* or *multineuronal pathways*. The indirect motor pathways include the **tectospinal tract** from the superior colliculus (the tectum of the midbrain), the **vestibulospinal tract** from the vestibular nuclei, the **rubrospinal tract** from the red nucleus (*rubro* = red) (Figure 13.30b), and the **reticulospinal tract** from the reticular formation. These tracts stimulate body movements that are subconscious, coarse, or postural (see Table 13.4 for details).

This discussion of sensory and motor pathways has focused on the innervation of body regions inferior to the head. In general, the pathways to and from the head are similar to those for the trunk and limbs, except that in the tracts servicing the head, the axons are located in cranial nerves and the brain stem, not in the spinal cord.

> **CLINICAL APPLICATION**
>
> **Amyotrophic Lateral Sclerosis (ALS)** Also known as *Lou Gehrig's disease,* **amyotrophic lateral sclerosis** (ah-mi″o-trof″ik, "muscle degeneration" or "muscle atrophy"; skle-ro′sis, "hardening") or ALS, most commonly involves degeneration of the pyramidal tracts with the resultant wasting of the skeletal muscles. As the pyramidal tracts deteriorate, scar tissue hardens the lateral parts of the spinal cord where these tracts lie. Lower motor neurons also degenerate, causing muscle weakness and atrophy. Initial symptoms include weakness in using hands or difficulty in swallowing or speaking. ALS appears after age 40, and its progress is relentless and always fatal; currently, half of people affected die within 3 to 5 years. However, new intensive-care programs may extend survival time to 15 years. The cause of ALS is unclear, but it may result from an excess of the neurotransmitter glutamate in the primary motor cerebral cortex, where the glutamate first overstimulates and then kills the neurons.

check your understanding

- ☐ 26. Which sensory pathway carries discriminative touch sensations?
- ☐ 27. Of the sensory pathways described, which pass through the thalamus? Which pathway does not?
- ☐ 28. Which descending fiber tract originates from the primary motor cortex?
- ☐ 29. Which of the pathways illustrated here (ascending and descending) do not decussate?

(For answers, see Appendix B.)

DISORDERS OF THE CENTRAL NERVOUS SYSTEM

learning outcomes

- Explain the effects of severe injuries to the spinal cord.
- Describe the signs and symptoms of concussions (traumatic brain injuries), strokes, and Alzheimer's disease.

Many disorders can affect the central nervous system. This discussion begins with spinal cord injury and proceeds to disorders of the brain.

Spinal Cord Damage

Spinal cord injuries—resulting from traffic accidents, skiing or diving accidents, falls, and gunshot wounds—can crush, tear, or cut through (transect) the spinal cord. Transection leads to a total loss of voluntary movements and of conscious sensation in body regions inferior to the cut because the connections between the brain and the body, the spinal tracts, have been severed. If the damage occurs between the T_1 and L_2 segments of the spinal cord, the lower limbs are affected but not the upper limbs. This condition is called **paraplegia** (par″ah-ple′je-ah). If the damage is in the cervical region of the spinal cord, all four limbs are affected, a condition known as **quadriplegia** (*quadri* = four; *plegia* = a blow). Spinal cord transection superior to the midneck eliminates the person's ability to breathe because the motor neurons to the diaphragm are located in spinal cord segments C_3–C_5. Such injuries are fatal unless the victim is kept alive by artificial resuscitation at the accident site and by a respirator thereafter.

In evaluating spinal cord injuries, it is important to remember that the spinal cord segments are slightly more rostral than the corresponding vertebral levels (see p. 409).

Brain Dysfunction

Brain dysfunctions are varied and extensive. Traumatic brain injuries are discussed in A CLOSER LOOK on p. 420, and two important degenerative brain diseases—cerebrovascular accidents and Alzheimer's disease—are discussed here in the main text.

Cerebrovascular Accidents

Cerebrovascular accidents (CVAs), commonly called *strokes,* are the most common disorders of the nervous system and the third leading cause of death in the United States. A CVA occurs when either blockage or interruption of the flow of blood to a brain area causes brain tissue to die from lack of oxygen. Such a deprivation of blood to a tissue is called **ischemia** (is-ke′me-ah; "to hold back blood").

CVAs are caused by burst or torn cranial vessels (hemorrhagic stroke) or by a blood clot blocking cranial vessels (ischemic stroke). The neurological deficits that ensue depend on which artery is affected and on which area of the brain the affected artery supplies. People who survive a CVA may be paralyzed on one side of the body or have sensory or language deficits. Some survivors partially recover their lost abilities, especially with rehabilitative treatment, because undamaged neurons sprout new branches that spread into the damaged brain area and take over lost functions. This is an example of **neural plasticity,** the regenerative process by which the damaged CNS rewires itself. However, because neuronal spread is limited to only about 5 mm, recovery is never complete if the damaged area is large.

Strategies for treating CVAs include (1) early administration of a clot dissolver, such as *tissue-type plasminogen activator (tPA),* for strokes caused by clots; (2) earlier detection of CVAs in progress using faster neuroimaging techniques; and (3) administration of neuroprotective drugs (currently in development) to keep damaged neurons from dying.

Traumatic Brain Injury

Each year, 1.4 million **traumatic brain injuries (TBIs)** occur in the United States from falls (28%), traffic accidents (20%), sports injuries (21%), and violence (30%).

Injury to the neural tissue is caused by three mechanisms: (1) a direct blow that causes bruising of neural tissue, (2) shearing forces that tear neuronal fibers, and (3) bleeding and swelling in the cranial cavity that increase intracranial pressure and compress neural tissue.

Being surrounded by cerebrospinal fluid, the brain is essentially floating in the skull. Any outside force impacting the skull causes the brain to move in the skull, slamming the brain into the cranium and injuring the neural tissue. If the force of the blow is great enough, it may also cause the brain to hit the opposite side of the skull. This is referred to as a **contrecoup injury.**

As the brain is being forcefully moved in the cranial cavity, the long, projecting axons are stretched. Shearing forces may tear the axons, resulting in **diffuse axonal injury.** These forces affect widely dispersed axons and disrupt their ability to send information—thus disrupting neuronal circuits. Given time, axonal connections may regrow to reestablish functional neural circuits. However, connections may misroute, resulting in long-term functional deficits.

Both contrecoup injury and diffuse axonal injury are typical in shaken baby syndrome and in the whiplash caused by the rapid deceleration typical of an automobile crash.

Internal bleeding at the site of impact compresses the neural tissue. The swelling that results from edema and water uptake by the brain leads to elevated intracranial pressure, which further compresses neural tissue, causing additional injury. Of great concern are injuries that result in bleeding into the subdural or subarachnoid spaces that is undiagnosed at the time of the injury. Immediately following a blow to the head, a victim may be lucid and act normal. However, as blood and other liquids accumulate around and in the brain tissue, neurological signs may develop, such as trouble with balance, coordination problems, and sensory deficits. The resulting increased intracranial pressure can affect brain stem functions controlling respiration, heart rate, and blood pressure, potentially causing death. Thus, individuals with any type of head injury must be monitored for the development of neurological signs: abnormal pupil reflexes, imbalance, incoherent speech, and abnormal behavior.

Mild brain injuries (concussions) are the most common type of brain injury. These injuries cause a transient change in mental state: confusion, disorientation, and memory loss. A brief loss of consciousness may or may not occur. Symptoms of a mild brain injury include headache, dizziness, nausea, fatigue, and irritability. A **moderate brain injury** is defined by loss of consciousness for an interval ranging from a few minutes to a few hours. A moderate brain injury causes confusion that may last for days or weeks. Neural deficits result in physical, cognitive, and behavioral impairments that may last for an extended period or may be permanent. A **severe brain injury** results in prolonged unconsciousness or coma lasting days, weeks, or months.

Around 75% of all TBIs are classified as mild, including most of the brain injuries that occur during sport activities. However, a mild injury is not an insignificant injury. Mild brain injuries are often ignored or minimized, particularly by competitive athletes.

Neural tissue requires at least 2 weeks to repair itself following a mild injury. If the brain is not allowed to heal, further injury may result from another impact. A second impact that occurs prior to the complete healing of an initial impact can result in brain swelling so acute that it proves fatal. This condition is referred to as **second-impact syndrome.** Although neither impact is severe, the extreme swelling produced by the second injury can alter brain stem function, and the cumulative effect of the injuries can have fatal consequences.

Repetitive brain injuries do combine to cause significant neurological deficits. Research shows that athletes with repeat head injuries perform more poorly on tests for attention, memory, organization, and complex problem solving than individuals who have never suffered a concussion.

Preventing head trauma is critical. For sports that involve contact, protective head gear and intelligent decisions by coaches, athletic trainers, and athletes concerning when to return to play following a head injury are essential.

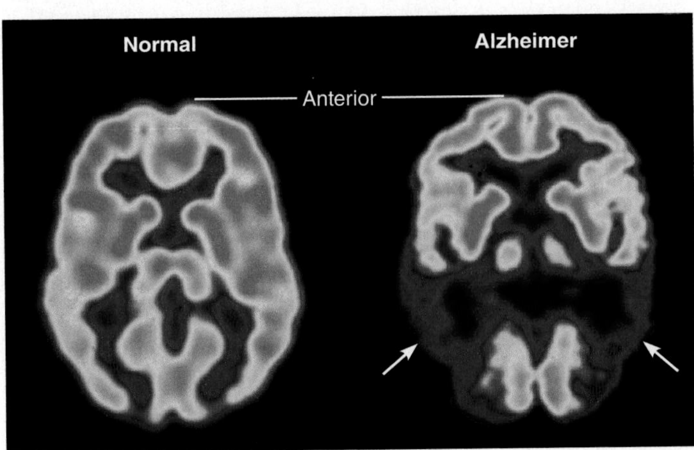

Figure 13.31 **Brain activity differences with Alzheimer's disease.** PET scan of a normal individual on the left and of a patient with Alzheimer's disease on the right. Red and yellow indicate the highest levels of glucose metabolism, thus high neural activity. Metabolic deficits in the parietal cortex are indicated by the arrows in the scan of the Alzheimer's patient.

Alzheimer's Disease

Alzheimer's (Altz′hi-merz) **disease** is a progressive degenerative disease of the brain that ultimately results in dementia (mental deterioration). Twelve percent of Americans over age 65 develop this condition, and up to half of all people over age 85 have Alzheimer's. Although usually seen in the elderly, Alzheimer's disease may begin in middle age. Individuals with Alzheimer's exhibit a wide variety of mental defects, including loss of memory (particularly for recent events), shortened attention span, depression, and disorientation. The disease worsens over a period of years until, eventually, cortical damage is so severe that patients become completely incapacitated, unable to respond to their environment or control movement.

Alzheimer's disease begins with structural changes in the basal forebrain nuclei, the hippocampus, and the association areas of the cerebral cortex, all areas involved in memory and thought. As the disease advances, damage speads to other areas of the cerebrum. The cerebral gyri shrink, the ventricles enlarge, and many neurons ultimately are lost. Functional neuroimaging has documented a large loss of neuronal activity in the cerebral cortex of the parietal lobe (Figure 13.31).

The cause of the disease is unclear. Microscopically, brain tissue of Alzheimer's patients contains *amyloid plaques* between the neurons and *neurofibrillar tangles* within the neurons. The characteristic plaques are formed from amyloid-beta protein (A-beta), a component of a larger amyloid precursor protein found in all cells of healthy brains. A-beta has long been hypothesized as a leading causal factor in the disease. It causes diminished synapse activity and disrupts nerve cell function. However, PET scans have shown A-beta accumulation in some healthy brains as well as in the brains of Alzheimer's patients, suggesting there is more to this disease than A-beta plaques. The roles other factors may play in the development of Alzheimer's diesease are being researched, specifically the effects of inflammation and stress on nerve cells resulting from head trauma or mild strokes. As our population ages and the number of individuals with Alzheimer's increases, the social and economic impact of this disease on victims, caregivers, and society in general becomes increasingly devastating.

THE CENTRAL NERVOUS SYSTEM THROUGHOUT LIFE

learning outcomes

- Describe causes and consequences of the congenital disorders anencephaly, cerebral palsy, and spina bifida.
- Explain the effects of aging on brain structure.

Embryonic Development and Congenital Conditions

Established during the first month of development, the brain and spinal cord continue to grow and mature throughout the prenatal period. A number of congenital malformations originate during this period, the most serious of which are congenital hydrocephalus (see p. 407), neural tube defects, and cerebral palsy.

Neural tube defects result from a delay in the closure of the neural tube (see Figure 3.7). The neural tube defect known as **anencephaly** (an′en-sef′ah-le; "without a brain") is caused by the failure of the rostral part of the tube to close and form a complete brain. As a result, the cerebrum and cerebellum never develop; only the brain stem and traces of the basal nuclei (ganglia) form. The anencephalic newborn is totally vegetative, and mental life, as usually defined, is absent. Death occurs soon after birth. About 1 of 5000 newborns are anencephalic.

Spina bifida (spi′nah bif′ĭ-dah; "forked spine") encompasses a variety of neural tube defects. These defects result from either the failure of the caudal portion of the neural tube to close or the incomplete formation of the bony vertebral arches (absence of vertebral laminae). The delay in formation of the spinal cord and meninges prevents the formation of the bony vertebral laminae. In extremely delayed cases, the neural groove remains unclosed and functionless, but in most cases a complete spinal cord forms but then becomes damaged. About 1 of every 2500 newborns have spina bifida, and defects are most common in the lumbosacral region.

In **spina bifida cystica,** the most common variety, the meninges around the spinal cord are expanded into a baglike cyst called a **meningocele** (mĕ-ning′go-sēl) that protrudes dorsally from the infant's spine. If the spinal cord moves into the dorsal part of the cyst, it is called a **myelomeningocele** (mi″ĕ-lo-mĕ-ning′go-sēl), a "spinal cord in a sac of meninges" (Figure 13.32). Because the wall of the cyst is thin, irritants such as urea in the surrounding amniotic fluid can pass through the wall and harm the spinal cord during the fetal period, as can physical jostling and infections after

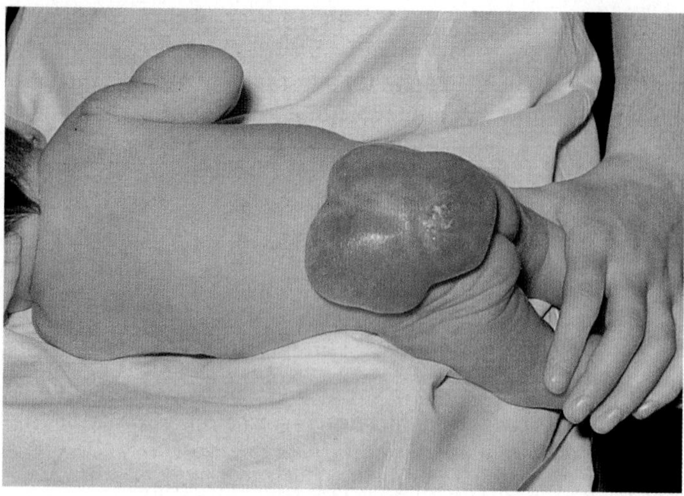

Figure 13.32 **Newborn with a lumbar myelomeningocele.**

birth. Commonly, children with this defect also have a small posterior cranial fossa, which causes the medulla oblongata and parts of the cerebellum to extend through the foramen magnum of the skull, constricting the brain ventricles and resulting in hydrocephalus. Immediate surgical intervention can relieve the hydrocephalus and may sometimes prevent paralysis.

Babies with spina bifida should be delivered by cesarean section to prevent compression of the meningocele and damage to the spinal cord during passage through the narrow birth canal. Of course, planning for a cesarean delivery requires that the spina bifida be recognized prenatally—either by a fetal ultrasound examination or by a test called *maternal serum alpha-fetoprotein (AFP) screening* run before midpregnancy. Inadequate amounts of the B vitamin folate in the maternal diet increase the risk of a neural tube defect in a developing fetus. Folic acid, the supplemental form of this vitamin, is now added to all bread, flour, and pasta products sold in the United States. Since the fortification of these products with folic acid became manditory in 1998, the prevalence of neural tube defects has decreased 27%.

The congenital condition known as **cerebral palsy** is a lifelong CNS disorder in which the voluntary muscles are either poorly controlled or paralyzed. This condition results from damage either to the cerebral cortex or, less often, to the cerebellum or basal nuclei (ganglia). Infections of the placenta and other factors, such as trauma during birth, that interfere with the blood supply to the fetal brain can lead to cerebral palsy. In addition to spastic paralysis, speech difficulties, and other motor impairments, the symptoms can also include seizures, mental retardation, deafness, and visual impairments. Cerebral palsy is the largest single cause of crippling in children. It occurs in 2 of every 1000 full-term births and is 70 times more frequent among premature infants.

Postnatal Changes in the Brain

After birth, different parts of the normal brain complete their development at different times, and the neural connections made in the brain of the growing baby are strongly molded by experiences. PET scans show that the thalamus and the somatosensory cortex are active in a 5-day-old baby but that the visual cortex is not. This finding supports the observation that infants of this age respond to touch but have poor vision. By 11 weeks, more of the cortex is active, and the baby can reach for objects. By 8 months, the cortex is highly active; the child can think about and remember what she or he sees and is starting to understand the sounds of language. At 10 to 18 months, the prefrontal lobe starts to communicate with the limbic regions, so that the child can begin to control emotional urges. Maturation of the nervous system continues through childhood and reflects the progressive myelination and thickening of its axons.

Despite the great plasticity of young brains, there are critical periods after which learning is no longer possible. For example, no one has ever learned a first language after about age 5, and childhood abuse can so greatly retard the maturation of the frontal lobes that normal cognitive functioning is not attained. Recent studies suggest these critical periods are not as rigid as once was believed, providing more hope for deprived children.

Formerly it was thought that the brain's circuitry is fully formed by late childhood, but new findings show that such maturation is not complete until people are in their early 20s. In teenagers, the "rational" frontal lobe is still organizing itself at the same time that "emotional" limbic structures are in developmental overdrive from the effects of sex hormones. This may explain why some teenagers seem moody or abandon good judgment for risky, thrill-seeking behaviors.

The brain reaches its maximum weight in the young adult, but with age it shrinks, and its cognitive abilities slowly decline. Traditionally, this decline has been attributed to neuronal death, but improved techniques now suggest that the number of brain neurons does not decrease with age—at least in the cerebral cortex and hippocampus. The reason for declining brain function with age may instead lie in changes in neural circuitry or in lower amounts of neurotransmitters being released at synapses. Studies show that the elderly compensate for this decline by using more regions of the cerebral cortex and that keeping mentally active can slow the rate of brain shrinkage. Finally, the choroid plexus calcifies in the elderly, drastically reducing the production of cerebrospinal fluid late in life.

☑ check your understanding

☐ 30. Individuals who have suffered a stroke generally have functional deficits on the side of the body opposite from the injured brain tissue. Explain this in reference to your knowledge of ascending and descending pathways in the CNS.

☐ 31. Which regions of the cerebral cortex are active in a newborn baby? Which region of the cortex is not fully developed until early adulthood?

(For answers, see Appendix B.)

RELATED CLINICAL TERMS

Encephalopathy (en-sef"ah-lop'ah-the) Any disease or disorder of the brain.

Microcephaly (mi"kro-sef'ah-le) Congenital condition involving the formation of a small brain and skull. May result from damage to the brain before birth or from premature fusion of the sutures of the skull. Most microcephalic children have severe mental deficits.

Myelitis (mi"ĕ-li-tis; *myel* = spinal cord) Inflammation of the spinal cord.

Myelogram (*gram* = recording) X-ray study of the spinal cord after injection of a contrast medium.

Tourette's syndrome A brain disorder that affects about 1 of every 3000 Americans, mostly males. Characterized by tics (sudden, fast movements or sounds) such as blinks, grimaces, barks, and yelps. May involve obscene vocalizations, repeating the words and actions of others, and obsessive-compulsive behavior. In this condition, the basal nuclei (ganglia) do not properly regulate a neurotransmitter called dopamine and thus cannot interact with the motor cerebral cortex to inhibit stereotypical behaviors. This syndrome is treated with psychotropic drugs. After the preteen years, the severity of the symptoms usually declines with age.

CHAPTER SUMMARY

Take full advantage of the invaluable online practice and study tools for this chapter by going to the Study Area of MasteringA&P®. There you will find:

- Art Labeling Activities, Quizzes, Practice Tests, Memory Games
- Animal Organ Dissection, Cat Dissection and Bone Videos
- PAL Cadaver
- PAL Histology
- A&P Flix Animations
- Dynamic Study Modules

THE BRAIN (pp. 375–407)

1. The brain initiates voluntary and involuntary movements, interprets and integrates sensory stimuli, and is responsible for consciousness and cognitive function. It controls basic life-sustaining activities, and it is also involved with the innervation of the head through the cranial nerves.

Embryonic Development of the Brain (pp. 375–377)

2. The brain develops from the rostral part of the embryonic neural tube. By week 5, the early forebrain has become the telencephalon and diencephalon; the early midbrain has become the mesencephalon; and the early hindbrain has become the metencephalon and myelencephalon. As development continues after birth, the telencephalon becomes the cerebrum; the mesencephalon becomes the midbrain portion of the brain stem; the metencephalon becomes the pons portion of the brain stem and the cerebellum; and the myelencephalon becomes the medulla oblongata portion of the brain stem.

3. The cerebral hemispheres grow rapidly, enveloping the diencephalon and rostral brain stem. The expanding surfaces of the hemispheres fold into a complex pattern of sulci and gyri.

Basic Parts and Organization of the Brain (p. 377)

4. A widely used classification scheme divides the brain into the cerebrum, diencephalon, brain stem (midbrain, pons, medulla), and cerebellum.

5. Throughout the CNS, white matter is external to central gray matter. In the brain stem, clusters of gray matter, called brain nuclei, are embedded within the white matter. The cerebrum and cerebellum have an external cortex of gray matter.

Ventricles of the Brain (pp. 377–378)

6. The two lateral ventricles are in the cerebral hemispheres; the third ventricle is in the diencephalon; the cerebral aqueduct is in the midbrain; and the fourth ventricle is in the pons and medulla regions of the brain stem.

The Brain Stem (pp. 378–383)

7. The medulla oblongata contains the pyramids and their decussation, all formed by the pyramidal tracts. The olives contain relay nuclei to the cerebellum; nucleus cuneatus and nucleus gracilis relay discriminitive sensory stimuli to the thalamus. Nuclei of cranial nerves VIII–XII lie near the fourth ventricle. Centers in the medullary reticular formation regulate respiration, heart rate, blood pressure, and other visceral functions.

8. The ventral region of the pons contains the pyramidal tracts plus the pontine nuclei that project to the cerebellum. The nuclei of cranial nerves V–VII lie near the fourth ventricle.

9. The midbrain is divided into a tectum and paired cerebral peduncles, with the latter containing the pyramidal motor tracts in the crus cerebri. In the tectum, the superior and inferior colliculi mediate visual and auditory reflexes. The red nucleus and substantia nigra participate in motor functions. The periaqueductal gray matter elicits the fear response. The midbrain also contains motor nuclei of cranial nerves III and IV, nerves that control eye muscles.

The Cerebellum (pp. 383–385)

10. The cerebellum smooths and coordinates body movements and helps maintain posture and equilibrium. Its main divisions—the paired cerebellar hemispheres and the vermis—are divided transversely into three lobes: anterior, posterior, and flocculonodular. The cerebellar surface is covered with folia (ridges) and fissures.

11. From superficial to deep, the main regions of the cerebellum are the cortex, the arbor vitae, and the deep cerebellar nuclei.

12. The cerebellar cortex receives sensory information on current body movements and then compares this information to how the body should be moving. In this way, it in effect carries out a dialogue with the cerebral cortex on how to coordinate precise voluntary movements.

13. The cerebellum connects to the brain stem by the superior, middle, and inferior cerebellar peduncles, thick fiber tracts that carry information to and from the cerebellum. All these fibers are ipsilateral.

The Diencephalon (pp. 385–388)

14. The diencephalon, which consists of the thalamus, hypothalamus, and epithalamus, encloses the third ventricle.
15. The thalamus, a paired egg-shaped group of brain nuclei, is the gateway to the cerebral cortex. It is a major relay station for sensory impulses ascending to the sensory cortex and for impulses from all brain regions that communicate with the cerebral cortex.
16. The hypothalamus, a cluster of brain nuclei, is the brain's most important visceral control center. It regulates sleep cycles, hunger, thirst, body temperature, secretion by the pituitary gland, the autonomic nervous system, and some emotions and behaviors.
17. The small epithalamus contains the pineal gland, which secretes a hormone, called melatonin, that is involved in the nighttime stage of the sleep-wake cycle.

The Cerebrum (pp. 388–399)

18. The two large cerebral hemispheres that make up the cerebrum have surface gyri, sulci, and fissures. The five lobes of each hemisphere are the frontal, parietal, occipital, and temporal lobes and the insula.
19. The cerebral cortex is the site of conscious sensory perception, voluntary initiation of movements, and higher thought functions. Structural regions of the six-layered cerebral cortex were identified by Brodmann and others.
20. Functionally, the cerebral cortex is divided into three areas. Sensory areas in the parietal, occipital, and temporal lobes are the primary somatosensory cortex, primary visual cortex, primary auditory cortex, gustatory cortex, vestibular cortex, and olfactory cortex. Individual sensory association areas are adjacent to their primary sensory cortices. Motor areas in the frontal lobe are the primary motor cortex, premotor cortex, frontal eye field, and Broca's area for speech production.
21. The cortex of each cerebral hemisphere receives sensory information from, and sends motor instructions to, the opposite (contralateral) side of the body. The body is mapped in an upside-down "homunculus" on the somatosensory and primary motor areas.
22. The pyramidal tracts descend from the primary motor cortex and extend through the brain stem and the spinal cord, where they signal motor neurons to perform skilled voluntary movements.
23. Multimodal association areas overlap the parietal, temporal, and frontal lobes. These areas integrate sensory input from many modalities with past experience and plan a motor response. Higher-order visual and auditory processing occurs in the temporal lobes (ventral stream for recognizing objects or identifying sounds) and the posterior parietal lobes (dorsal stream for identifying spatial relationships or the location of sounds). The language areas occupy a large part of the left cerebral cortex around the lateral sulcus.
24. In most people, the left cerebral hemisphere is specialized for language and math skills, and the right hemisphere is devoted more to visual-spatial and creative abilities.
25. The three types of cerebral white matter are commissural fibers between hemispheres, association fibers between cortical areas within a hemisphere, and projection fibers to and from the brain stem and spinal cord.
26. The basal nuclei (ganglia) are one component of the deep gray matter of the cerebrum. The structures of the basal nuclei include the caudate nucleus, the putamen, and the globus pallidus. Functionally, the basal nuclei work with the cerebral motor cortex to control complex movements, among other functions.
27. The basal forebrain nuclei, located anterior and dorsal to the hypothalamus, are part of the basal forebrain cholinergic system, whose functions relate to arousal, learning, memory, and motor control.

Functional Brain Systems (pp. 399–401)

28. The limbic system consists of many cerebral and diencephalic structures on the medial aspect of each cerebral hemisphere. It is the "emotional-visceral" part of the brain, and it also forms and retrieves memories. The cingulate gyrus controls the behavioral response to emotions, the hippocampus consolidates memories, and the amygdaloid body mediates our response to fearful stimuli.
29. The reticular formation includes diffuse brain nuclei that span the length of the brain stem. The reticular activating system of the reticular formation maintains the conscious state of the cerebral cortex, and the reticulospinal tracts, which are descending tracts, influence postural muscles, diminish muscle tone, and influence visceral motor function.

Protection of the Brain (pp. 401–407)

30. The three meninges form continuous coverings around the spinal cord and brain. Internal projections of the dura mater surrounding the brain form partitions (falx cerebri, falx cerebelli, tentorium cerebelli) that secure the brain in the skull.
31. Cerebrospinal fluid is derived from blood plasma at the choroid plexuses in the roof of the ventricles. It circulates through the ventricles and into the subarachnoid space, and it returns to the blood through arachnoid granulations at the superior sagittal sinus.
32. The blood brain barrier derives from the relative impermeability of brain capillaries (complete tight junctions between epithelial cells). The barrier lets water, oxygen, nutrients, and fat-soluble molecules enter the neural tissue but prevents entry of most harmful substances.

THE SPINAL CORD (pp. 407–413)

33. The spinal nerves that innervate the neck, limbs, and trunk attach to the spinal cord. The spinal cord also acts as a reflex center and contains axon tracts running to and from the brain. It extends from the foramen magnum to the level of the first or second lumbar vertebra. The terminal end of the spinal cord is the conus medullaris.
34. Thirty-one pairs of spinal nerve roots issue from the spinal cord. The most inferior bundle of roots resembles a horse's tail (cauda equina). The spinal cord is enlarged in its cervical and lumbar regions, reflecting the innervation of the limbs.

White Matter of the Spinal Cord (pp. 410–411)

35. The white matter of the cord is divided into dorsal, lateral, and ventral funiculi containing ascending and descending fibers.

Gray Matter of the Spinal Cord and Spinal Roots (pp. 411–412)

36. The H-shaped gray matter of the spinal cord has two ventral horns containing motor neurons and two dorsal horns containing interneurons. The dorsal horns are subdivided into somatic and visceral sensory regions; the ventral horns, into visceral and somatic motor regions.

37. The roots of the spinal nerves—dorsal sensory roots and ventral motor roots—are PNS structures that attach to the spinal cord.

Protection of the Spinal Cord (pp. 412–413)

38. The three meninges enclose and protect both the brain and the spinal cord: the tough outer dura mater, the arachnoid, and the inner vascularized pia mater.

39. Cerebrospinal fluid both floats and cushions the structures of the CNS. It fills the subarachnoid space and the central cavities of the brain and spinal cord.

PAL Human Cadaver/Nervous System/Central Nervous System

SENSORY AND MOTOR PATHWAYS IN THE CENTRAL NERVOUS SYSTEM (pp. 413–419)

Ascending Pathways (pp. 414–416)

40. The ascending spinal tracts in the funiculi of the spinal cord belong to sensory pathways that run from the body periphery to the brain. These include the spinocerebellar pathway (for proprioception) to the cerebellum and two pathways to the somatosensory cortex: the dorsal column–medial lemniscus pathway (for discriminative touch and proprioception) and the spinothalamic pathway (for pain, temperature, and nondiscriminative touch).

Descending Pathways (pp. 416–419)

41. The descending tracts in the funiculi of the spinal cord belong to motor pathways that connect the brain to the body muscles. These include the direct pathways, the pyramidal (corticospinal) tracts, that control skilled movements, and indirect pathways, the extrapyramidal tracts, that control subconscious and coarse movements.

DISORDERS OF THE CENTRAL NERVOUS SYSTEM (pp. 419–421)

Spinal Cord Damage (p. 419)

42. Transection, tearing, or crushing of the spinal cord permanently eliminates voluntary movements and sensation inferior to the damage site.

Brain Dysfunction (pp. 419–421)

43. Blows to the head may cause concussions. These injuries can be aggravated by intracranial bleeding and swelling of the brain.

44. Cerebrovascular accidents (strokes) result either when the blood vessels to the brain tissue are blocked or when blood flow to the brain is interrupted, causing the tissue to die.

45. Alzheimer's disease is a degenerative brain disease affecting the cerebral cortex and hippocampus. Mental functions diminish progressively.

THE CENTRAL NERVOUS SYSTEM THROUGHOUT LIFE (pp. 421–422)

46. Birth defects that involve the brain include congenital hydrocephalus, neural tube defects (anencephaly, spina bifida), and cerebral palsy. Spina bifida cystica is associated with meningocele and/or myelomeningocele cysts.

47. The brain forms many neuronal connections during childhood, all based on early life experiences. Growth of the brain stops in young adulthood, and cognitive functions decline slowly with age.

REVIEW QUESTIONS

Multiple Choice/Matching Questions

(For answers, see Appendix B.)

1. Choose the correct brain structure from the key for each of the following descriptions. Some letters are used more than once.

Key:

(a) cerebellum
(b) superior and inferior colliculi
(c) striatum
(d) corpus callosum
(e) hypothalamus
(f) medulla
(g) midbrain
(h) pons
(i) thalamus

____ **(1)** basal nuclei (ganglia) involved in motor activities; related to Huntington's disease

____ **(2)** region where there is a crossover of fibers of pyramidal tracts

____ **(3)** control of temperature, autonomic nervous system, hunger, and water balance

____ **(4)** contains the substantia nigra and red nucleus

____ **(5)** involved in visual and auditory reflexes; found in midbrain

____ **(6)** part of diencephalon with vital centers controlling heart rate, some aspects of emotion, and blood pressure

____ **(7)** all inputs to cerebral cortex must first synapse in one of its nuclei

____ **(8)** brain area that has folia and coordinates movements

____ **(9)** brain region that contains the cerebral aqueduct

____ **(10)** associated with fourth ventricle and contains nuclei of cranial nerves V–VII

____ **(11)** thick fiber tract connecting the two cerebral hemispheres

2. A patient suffered a cerebral hemorrhage that damaged the postcentral gyrus in her right cerebral hemisphere. As a result, she (a) cannot voluntarily move her left arm or leg, (b) feels no sensation on the left side of her body, (c) feels no sensation on the right side of her body, (d) can no longer play the violin.

3. Destruction of the ventral horn cells of the spinal cord results in loss of (a) intelligence, (b) all sensation, (c) pain, (d) motor control.

4. For each of the following brain structures, write G (for gray matter) or W (for white matter) as appropriate.

____ **(1)** cortex of cerebellum

____ **(2)** pyramids

____ **(3)** internal capsule and corona radiata

____ **(4)** red nucleus

____ **(5)** medial lemniscus

____ **(6)** cranial nerve nuclei

____ **(7)** cerebellar peduncle

5. Which of the following areas is most likely to store visual memories? (a) visual association area, (b) Wernicke's area, (c) premotor cortex, (d) primary somatosensory cortex.

Short Answer Essay Questions

6. Make a diagram showing the five secondary brain vesicles of an embryo, and then list the basic adult brain regions derived from each.
7. (a) Make a rough sketch of a lateral view of the left cerebral hemisphere. (b) You may be thinking, "But I just can't draw!" So, name the hemisphere involved in most people's ability to draw. (c) On your drawing, locate the following areas, and identify their major functions: primary motor cortex, premotor cortex, somatosensory association area, primary visual area, anterior association area (prefrontal cortex), Broca's area.
8. (a) Name the brain nuclei that make up the basal nuclei (ganglia). (b) What is the basic function of the basal nuclei (ganglia)? (c) Which nucleus arches over the thalamus?
9. Describe the role of the cerebellum in maintaining smooth, coordinated skeletal muscle activity.
10. (a) Where is the limbic system located? (b) What structures make up this system? (c) How is the limbic system important in behavior?
11. (a) Describe the location of the reticular formation in the brain. (b) What does RAS mean, and what is its function?
12. (a) List four ways in which the CNS is protected. (b) How is cerebrospinal fluid formed and drained? Describe its pathway within and around the brain.
13. (a) What are the superior and inferior boundaries of the spinal cord in the vertebral canal? (b) A correctly executed lumbar spinal tap could poke the cauda equina or filum terminale, but not the conus medullaris. Explain.
14. (a) Which of the ascending tracts in the spinal cord carry sensory impulses concerned with discriminative touch and pressure? (b) Which are concerned with proprioception only?
15. (a) In the spinothalamic pathway, where are the cell bodies of the first-order neurons located? Of the third-order neurons? (b) Trace the descending pathway that is concerned with skilled voluntary movements.
16. List the changes in brain structure that occur with aging.
17. Distinguish between the sensory and motor homunculi. Where is each located, what does each illustrate, and how is the body represented in each?
18. A brain surgeon removed a piece of a woman's skull and cut through all the meninges to reach the brain itself. Name all the layers that were cut, from skin to brain.
19. What are some of the functional differences between the right and the left cerebral hemispheres?
20. Sketch the spinal cord in cross section, and label the following structures: (a) lateral funiculus of white matter, (b) ventral root, (c) ventral horn of gray matter, (d) central canal, (e) dorsal root ganglion, (f) dorsal column tracts: fasciculus gracilis and fasciculus cuneatus, (g) ventral corticospinal tract, (h) lateral corticospinal tract.
21. Describe the location and function of the periaqueductal gray matter.
22. Describe the location and function of the ventral stream of visual-processing areas in the cerebral cortex. Then explain why the dorsal and ventral visual streams are called the "where" and "what" streams, respectively.

Critical Reasoning & Clinical Application Questions

1. Kimberly learned that the basic design of the CNS is gray matter on the inside and white matter on the outside. When her friend L'Shawn said that there is gray matter on the outside of the cerebrum of the brain, Kim became confused. Clear up her confusion.
2. When Ralph had brain surgery to remove a small intracranial hematoma (blood mass), the operation was done under local anesthesia because he was allergic to general anesthetic agents. The surgeon removed a small part of the skull, and Ralph remained conscious. The operation went well, and when Ralph asked the surgeon to stimulate his (unharmed) postcentral gyrus mildly with an electrode, the surgeon did so. Indicate which of the following happened, and explain why: (a) Ralph was seized with uncontrollable rage. (b) He saw things that were not there. (c) He asked to see what was touching his hand, but nothing was. (d) He started to kick. (e) He heard his mother's voice from 30 years ago.
3. When their second child was born, Kiko and Taka were told their baby had spina bifida and would be kept in intensive care for a week. After that time, a "shunt" would be put in. Also, immediately after the birth, an operation was performed on the infant's lower back. The parents were told that this operation went well but that their son would always be a "little weak in the ankles." Explain the statements in quotation marks in more informative and precise language.
4. Cesar, a brilliant computer analyst, was hit on the forehead by a falling rock while mountain climbing. It was soon obvious to his coworkers that his behavior had changed dramatically. He had been a smart dresser, but now he was unkempt, and he started revealing the most secret office gossip to everyone. When he was discovered barefoot, washing his socks in the employee lounge, his supervisor ordered Cesar to report to the company doctor. What region of Cesar's brain was affected by the cranial blow?
5. One war veteran was quadriplegic, and another was paraplegic. Explain the relative locations of their injuries.
6. Every time Spike went to a boxing match, he screamed for a knockout. Then his friend Rudy explained what happens when someone gets knocked out. What is the explanation?
7. A spinal cord injury at C_2 results not only in paralysis but also in a lifetime on a respirator. Why does an injury this high affect ability to breathe?
8. What parts of the brain are still developing during adolescence, and how does this lag contribute to the types of behavior typical of this age: mood swings, risky behavior, and poor judgment?
9. Strokes, tumors, or wounds can destroy limited areas of the brain. What are the effects of localized lesions to the primary motor cortex?

The Peripheral Nervous System 14

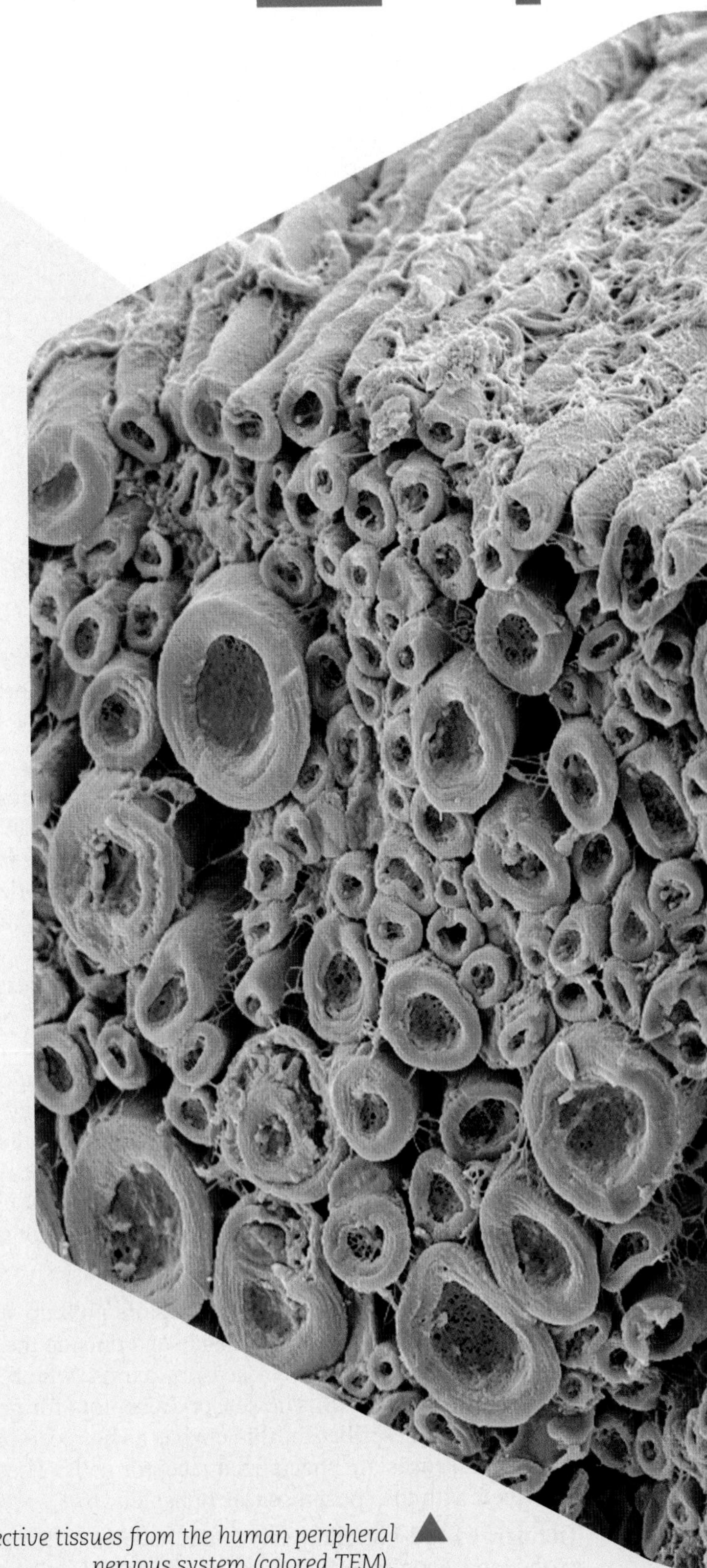

The human brain, for all its sophistication, would be useless without its sensory and motor connections to the outside world. Our very sanity depends on a continual flow of sensory information from the environment. When blindfolded volunteers were suspended in a tank of warm water (a sensory-deprivation tank), they began to hallucinate: One saw pink and purple elephants, another heard a singing chorus, and others had taste hallucinations. Our sense of well-being also depends on our ability to carry out motor instructions sent from the CNS. Many victims of spinal cord injuries experience despair at being unable to move or take care of their own needs. The **peripheral nervous system (PNS)**—the nervous system structures outside the brain and spinal cord— provides these vital links to the body and the outside world.

Myelinated axons and associated connective tissues from the human peripheral nervous system (colored TEM). ▲

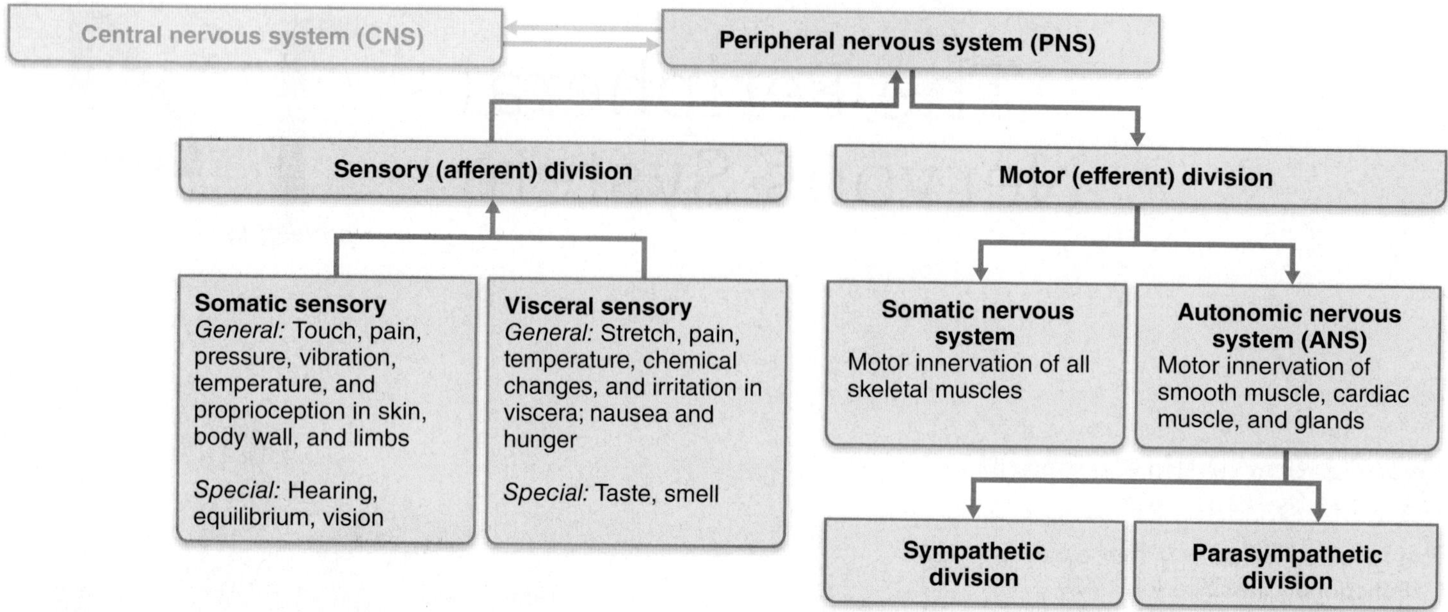

Figure 14.1 **Functional organization of the peripheral nervous system (PNS) and its relation to central nervous system (CNS).**

ORGANIZATION OF THE PERIPHERAL NERVOUS SYSTEM

learning outcomes

- ▶ Outline the organization of the peripheral nervous system.
- ▶ Describe the structural components of the peripheral nervous system.

Nerves thread through almost every part of the body, allowing the CNS to receive information and to initiate action. The PNS is functionally divided into sensory and motor divisions (Figure 14.1; see also Chapter 12, pp. 351–353). The sensory inputs and motor outputs of the PNS are subcategorized as either *somatic* (innervating the outer tube) or *visceral* (innervating the visceral organs or inner tube). Within the sensory division, sensory inputs are differentiated as *general* (widespread) or *special* (localized, i.e., the special senses). The visceral motor component of the PNS is the *autonomic nervous system (ANS)*. The autonomic nervous system has *parasympathetic* and *sympathetic* divisions (discussed in detail in Chapter 15).

It is important to recognize that the structures of the PNS reflect this functional organization—nerve impulses from each modality (somatic sensory, visceral sensory, somatic motor, visceral motor) are carried on distinct neurons. The following are the structures of the PNS:

- The **sensory receptors.** Sensory receptors pick up stimuli (environmental changes) from inside and outside the body and then initiate impulses in sensory axons, which carry the impulse to the CNS. The sensory receptors for general sensations are described in this chapter. The receptors for the special senses are specialized receptor cells. These are described with the special senses (Chapter 16).
- The **nerves** and **ganglia.** Nerves are bundles of peripheral axons, and ganglia are clusters of peripheral cell bodies, such as the cell bodies of the sensory neurons (Chapter 12). Most nerves contain both sensory and motor axons and are called *mixed nerves.* Certain cranial nerves (the nerves attached to the brain) contain only sensory axons and thus are purely sensory in function; certain others contain primarily motor axons and are motor in function. The *cranial nerves* and the *spinal nerves* (the nerves attached to the spinal cord) are described in this chapter, focusing on their somatic functions. Visceral innervation is discussed with the autonomic nervous system (Chapter 15).
- The **motor endings.** The motor endings are the axon terminals of motor neurons that innervate the effector organs, muscles, and glands. Motor endings to the muscles and glands of the body are described with the appropriate organ system: skeletal muscle innervation (Chapter 10); cardiac muscle innervation (Chapter 19); and the innervation of smooth muscle and glands (Chapter 23).

☑ check your understanding

- ☐ 1. List two examples of general somatic sensations. List one example of a special visceral sensation. (Refer to Table 12.1, p. 352, if needed.)
- ☐ 2. Do all nerves contain both sensory and motor fibers?

(For answers, see Appendix B.)

PERIPHERAL SENSORY RECEPTORS

learning outcome

- ▶ Classify sensory receptors according to body location, stimulus detected, and structure.

Peripheral sensory receptors are structures that pick up sensory stimuli and then initiate signals in the sensory axons. Most receptors fit into two main categories: (1) *free nerve endings* *of sensory neurons* and (2) complete *receptor cells,* which are specialized epithelial cells or small neurons that transfer sensory information to sensory neurons. Free nerve endings monitor most types of general sensory information (such as touch, pain, pressure, temperature, and proprioception), whereas specialized receptor cells monitor most types of special sensory information (taste, vision, hearing, and equilibrium).

Sensory receptors may be categorized based on either function or structure.

Functional Classification

Functional classifications group receptors according to their location or the type of stimulus they detect.

Location of Receptors

Sensory receptors monitor both the external and the internal environment. Sensory receptors can be described by their location in the body or the location of the stimuli to which they respond.

- **Exteroceptors** (eks″ter-o-sep′torz) are sensitive to stimuli arising outside the body. Accordingly, most exteroceptors are located at or near the body surface and include receptors for touch, pressure, pain, and temperature in the skin and most receptors of the special sense organs.
- **Interoceptors** (in″ter-o-sep′torz), also called *visceroceptors,* receive stimuli from the internal viscera, such as the digestive tube, bladder, and lungs. Different interoceptors monitor a variety of stimuli, including changes in chemical concentration, taste stimuli, the stretching of tissues, and temperature. Their activation causes us to feel visceral pain, nausea, hunger, or fullness.
- **Proprioceptors** are located in the musculoskeletal organs, such as skeletal muscles, tendons, joints, and ligaments. Proprioceptors monitor the degree of stretch of these locomotory organs and send input on body movements to the CNS (see p. 352).

Stimulus Type

Sensory receptors monitor many types of stimuli. The type of stimulus that most readily activates a sensory receptor can be used to categorize it functionally.

- **Mechanoreceptors** respond to mechanical forces such as touch, pressure, stretch, and vibrations. One type of mechanoreceptor, called a **baroreceptor,** monitors blood pressure.
- **Thermoreceptors** respond to temperature changes.
- **Chemoreceptors** respond to chemicals in solution (such as molecules tasted or smelled) and to changes in blood chemistry.
- **Photoreceptors** in the eye respond to light.
- **Nociceptors** (no″se-sep′torz) respond to harmful stimuli that result in pain (*noci* = harm).

Structural Classification

This section considers only the **general sensory receptors.** (The special senses are discussed in Chapter 16). All these widely distributed receptors are nerve endings of sensory

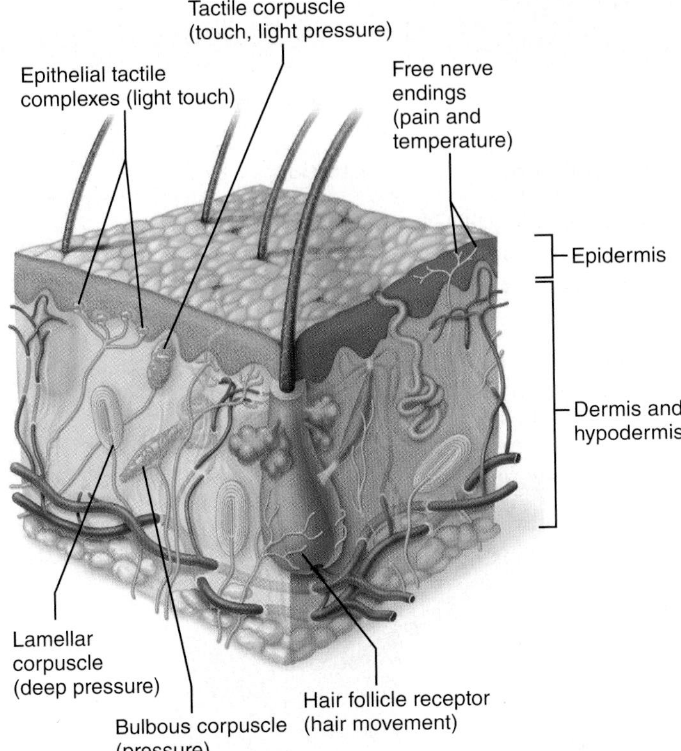

Figure 14.2 **Structure of free and encapsulated general sensory receptors.** Tactile corpuscles are not common in hairy skin but are included here for illustrative purposes.

neurons that monitor touch, pressure, vibration, stretch, pain, temperature, and proprioception. Structurally, sensory receptors are divided into two broad groups (Table 14.1): (1) *free* nerve endings and (2) *encapsulated* nerve endings surrounded by a capsule of connective tissue.

As you learn about the structural types of general sensory receptors, it is important to remember that there is no one-to-one correspondence between structural type and stimulus detected. Each type of receptor can respond to several different kinds of stimuli. In addition, similar stimuli can activate different types of receptors.

Free Nerve Endings

Free nerve endings of sensory neurons invade almost all tissues of the body but are particularly abundant in epithelia and in the connective tissue that underlies epithelia (Figure 14.2). These receptors are primarily nociceptors and thermoreceptors, responding to pain and temperature (though some respond to tissue movements caused by pressure). One way to characterize free nerve endings functionally is to say that they monitor the *affective* senses, those to which people have an emotional response—and people certainly respond emotionally to pain!

Itch receptors consist of free nerve endings in the dermis. These receptors escaped detection until 1997 because of their thin diameter. Their discovery startled scientists, who had not realized that itch is a distinct sense but rather had believed it to be a mild form of pain. Itch receptors are activated by chemicals present at inflamed sites, such as histamine.

Table 14.1 General Sensory Receptors Classified by Structure and Function

Structural Class	Illustration	Functional Class According to Location (L) and Stimulus Type (S)	Body Location
FREE NERVE ENDINGS			
Free nerve endings of sensory neurons	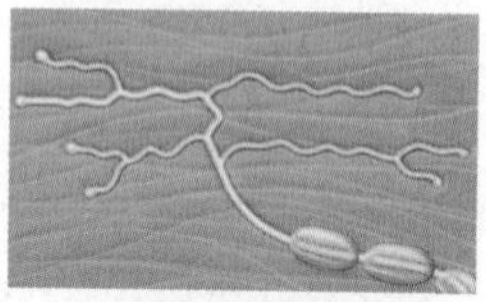	L: Exteroceptors, interoceptors, and proprioceptors S: Nociceptors (pain), thermoreceptors (heat and cold), mechanoreceptors (pressure), chemoreceptors	Most body tissues; most dense in connective tissues (ligaments, tendons, dermis, joint capsules, periosteum) and epithelia (epidermis, cornea, mucosae, and glands)
Modified free nerve endings: Epithelial tactile complexes (Merkel discs)	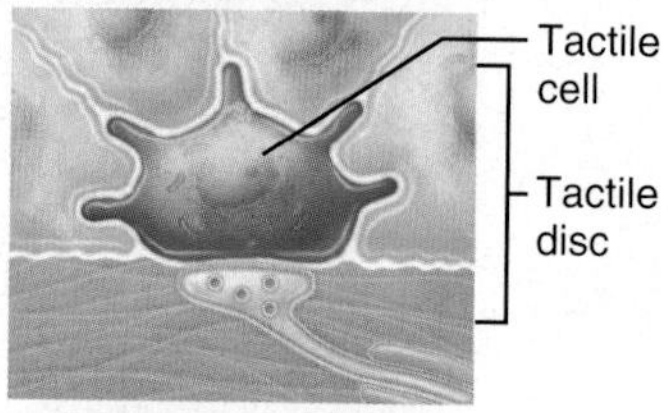 Tactile cell; Tactile disc	L: Exteroceptors S: Mechanoreceptors (light pressure), slowly adapting	Basal layer of epidermis
Hair follicle receptors	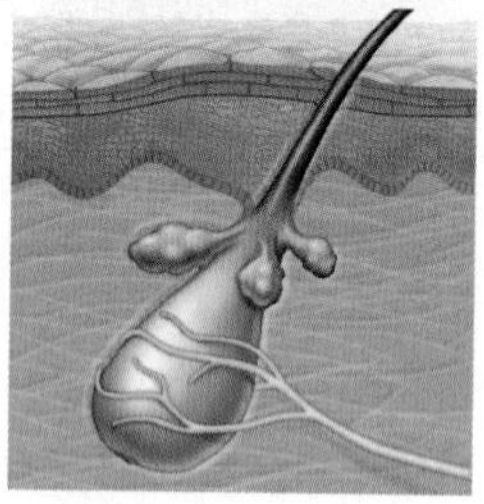	L: Exteroceptors S: Mechanoreceptors (hair deflection), rapidly adapting	In and surrounding hair follicles
ENCAPSULATED			
Tactile (Meissner's) corpuscles		L: Exteroceptors S: Mechanoreceptors (light pressure, discriminative touch, vibration of low frequency), rapidly adapting	Dermal papillae of hairless skin, particularly nipples, external genitalia, fingertips, eyelids
Lamellar (Pacinian) corpuscles	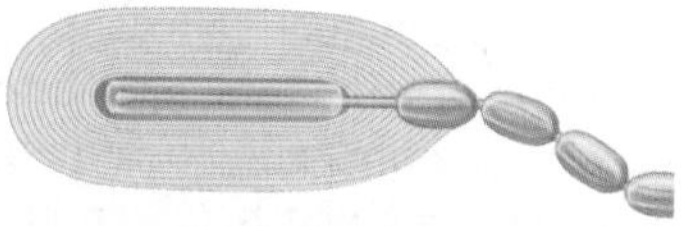	L: Exteroceptors, interoceptors, and some proprioceptors S: Mechanoreceptors (deep pressure, stretch, vibration of high frequency); rapidly adapting	Dermis and hypodermis; periosteum, mesentery, tendons, ligaments, joint capsules, most abundant on fingers, soles of feet, external genitalia, nipples
Bulbous corpuscle (Ruffini endings)	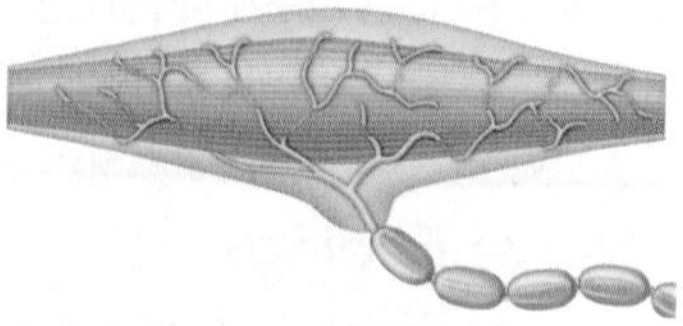	L: Exteroceptors and proprioceptors S: Mechanoreceptors (deep pressure and stretch); slowly adapting or nonadapting	Deep in dermis, hypodermis, and joint capsules
PROPRIOCEPTORS			
Muscle spindles	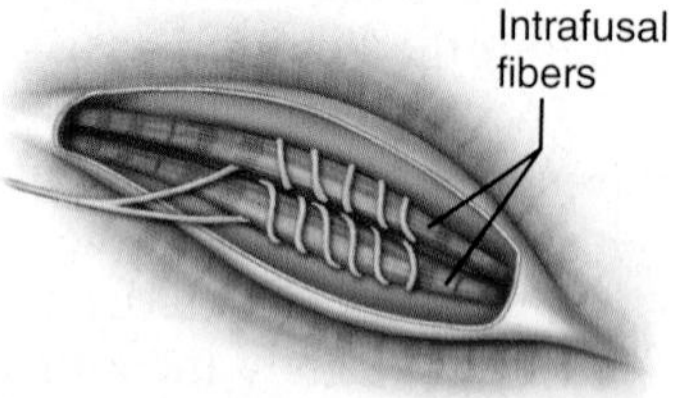 Intrafusal fibers	L: Proprioceptors S: Mechanoreceptors (muscle stretch)	Skeletal muscles, particularly those of the extremities
Tendon organs		L: Proprioceptors S: Mechanoreceptors (tendon stretch)	Tendons
Joint kinesthetic receptors	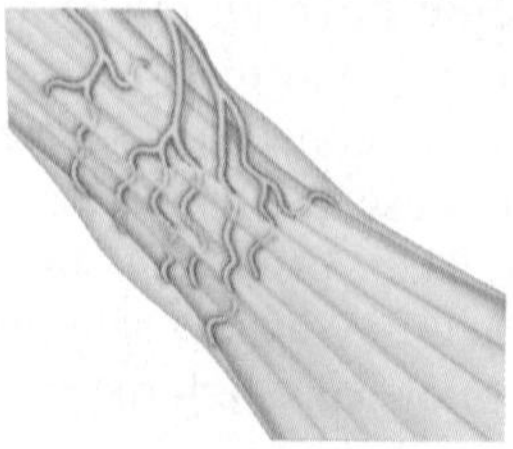	L: Proprioceptors S: Mechanoreceptors and nociceptors	Joint capsules of synovial joints

Certain free nerve endings contribute to **epithelial tactile complexes** (*Merkel discs*), which lie in the epidermis of the skin. Each consists of a disc-shaped tactile epithelial cell (see p. 105) innervated by a sensory nerve ending. These complexes are **slowly adapting mechanoreceptors** for light touch; that is, they continue to respond and send out action potentials even after a long period of continual stimulation. **Hair follicle receptors,** free nerve endings that wrap around hair follicles, are mechanoreceptors for light touch that monitor the bending of hairs. Unlike epithelial tactile complexes, they are **rapidly adapting,** meaning that the sensation disappears quickly even if the stimulus is maintained. The tickle of a mosquito landing on your forearm is mediated by hair follicle receptors.

Encapsulated Nerve Endings

All **encapsulated nerve endings** consist of one or more end fibers of sensory neurons enclosed in a capsule of connective tissue. All seem to be mechanoreceptors, and their capsules serve either to amplify the stimulus or to filter out the wrong types of stimuli. Encapsulated receptors vary widely in shape, size, and distribution in the body. The main types are *tactile (Meissner's) corpuscles, lamellar (Pacinian) corpuscles, bulbous corpuscles (Ruffini endings)* (see Figure 14.2), and *proprioceptors.*

Tactile Corpuscles In a **tactile corpuscle** (*Meissner's corpuscle*), a few spiraling nerve endings are surrounded by Schwann cells, which in turn are surrounded by an egg-shaped capsule of connective tissue. These corpuscles, which occur in the dermal papillae beneath the epidermis, are rapidly adapting receptors for fine, discriminative touch. They mainly occur in sensitive and hairless areas of the skin, such as the soles, palms, fingertips, nipples, and lips. Tactile corpuscles perform the same "light touch" function in hairless skin that hair follicle receptors perform in hairy skin.

Lamellar Corpuscles Scattered throughout the deep connective tissues of the body are **lamellar corpuscles** (*Pacinian corpuscles*). They occur, for example, in the hypodermis deep to the skin. Although they are sensitive to deep pressure, they respond only to the initial application of that pressure before they tire and stop firing. Therefore, lamellar corpuscles are rapidly adapting receptors that are best suited to monitor *vibration,* an on/off pressure stimulus. These corpuscles are large enough to be visible to the unaided eye—about 0.5–1 mm wide and 1–2 mm long. In section, a lamellar corpuscle resembles a cut onion: Its single nerve ending is surrounded by up to 60 layers of flattened Schwann cells, which in turn are covered by a capsule of connective tissue.

Bulbous Corpuscles Located in the dermis and elsewhere are **bulbous corpuscles** (*Ruffini endings*), which contain an array of nerve endings enclosed in a thin, flattened capsule. Like lamellar corpuscles, they respond to pressure and touch. However, they adapt slowly and thus can monitor *continuous pressure* placed on the skin.

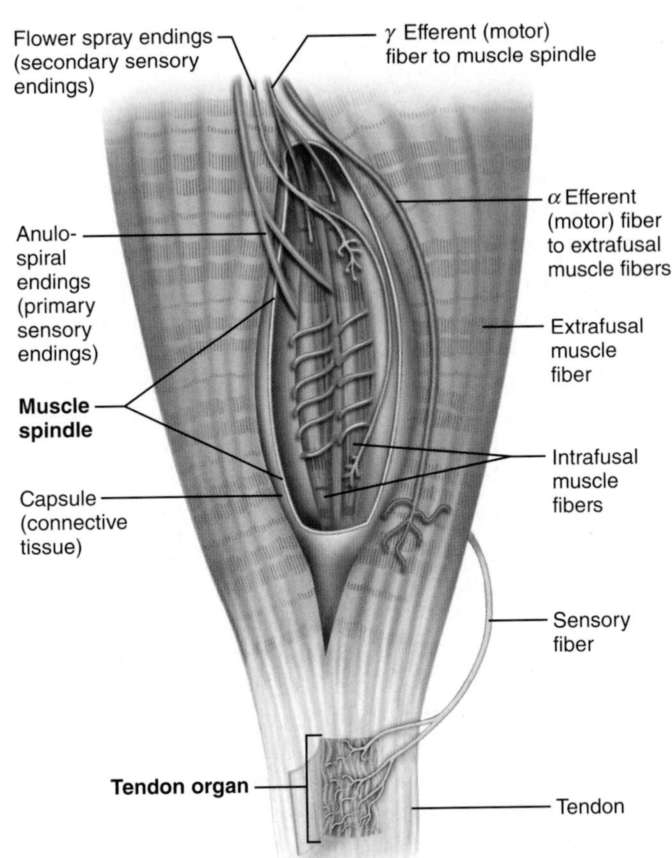

Figure 14.3 **Structure of proprioceptors: muscle spindles and tendon organ.** Myelin has been omitted from all nerve fibers for clarity.

Proprioceptors Virtually all proprioceptors are encapsulated nerve endings that monitor stretch in the locomotory organs. Proprioceptors include *muscle spindles, tendon organs,* and *joint kinesthetic receptors.*

Muscle spindles *(neuromuscular spindles)* measure the changing length of a muscle as that muscle contracts and is stretched back to its original length (Figure 14.3). An average muscle contains some 50 to 100 muscle spindles, which are embedded in the perimysium between the fascicles. Structurally, each spindle contains several modified skeletal muscle fibers called **intrafusal muscle fibers** (*intra*within; *fusal*the spindle) surrounded by a connective tissue capsule. Intrafusal muscle fibers have fewer striations than the **extrafusal** (*extra*outside) **muscle fibers,** that is, the ordinary muscle cells outside the spindles. The intrafusal fibers are innervated by two types of sensory endings: **Anulospiral endings,** or *primary sensory endings,* twirl around the noncontractile middle of the intrafusal fibers innervating the spindle center. These receptors are stimulated by the rate and degree of stretch of the muscle spindle. *Secondary sensory endings* called **flower spray endings,** monitor the spindle ends (the only contractile parts of the spindle) and respond only to degree of stretch.

Muscles are stretched by the contraction of antagonist muscles and also by the movements that occur as a person begins to lose balance. The muscle spindles sense this lengthening in the following way:

- When a whole muscle is stretched, its intrafusal fibers are also stretched. This stretching activates the primary and secondary sensory endings that innervate the spindle, causing them to fire off impulses to the spinal cord and brain.
- The CNS then activates spinal motor neurons called **α (alpha) efferent neurons** (Figure 14.3) that cause the entire muscle (extrafusal fibers) to generate contractile force and resist further stretching. This response can be initiated by a monosynaptic spinal reflex that rapidly prevents a fall; alternatively, the response can be controlled by the cerebellum regulating *muscle tone,* the steady force generated by noncontracting muscles to resist stretching.

Also innervating the intrafusal fibers of the muscle spindle are spinal *motor* neurons called **γ (gamma) efferent neurons** (Figure 14.3). These neurons preset the sensitivity of the spindle to stretch. When the brain stimulates the gamma motor neurons to fire, the intrafusal muscle fibers contract and become tense so that very little stretch is needed to stimulate the sensory endings, making the spindles highly sensitive to applied stretch. Gamma motor neurons are most active when balance reflexes must be razor sharp, as for a gymnast on a balance beam or a rock climber on a vertical face.

Tendon organs (Golgi tendon organs) are proprioceptors located near the muscle-tendon junction, where they monitor tension within tendons (Figure 14.3). Each consists of an encapsulated bundle of tendon fibers (collagen fibers) within which sensory nerve endings are intertwined. When a contracting muscle pulls on its tendon, tendon organs are stimulated, and their sensory neurons send this information to the cerebellum. These receptors also induce a spinal reflex that both relaxes the contracting muscle and activates its antagonist. This relaxation reflex is important in motor activities that involve rapid alternation between flexion and extension, such as running.

Joint kinesthetic receptors (kin″es-thet′ik; "movement feeling") are proprioceptors that monitor stretch in the synovial joints. Specifically, they are sensory nerve endings within the joint capsules. Four types of joint kinesthetic receptors are present within each joint capsule:

1. *Lamellar (Pacinian) corpuscles:* These rapidly adapting stretch receptors are ideal for measuring acceleration and rapid movement of the joints.
2. *Bulbous corpuscles (Ruffini endings):* These slowly adapting stretch receptors are ideal for measuring the positions of nonmoving joints and the stretch of joints that undergo slow, sustained movements.
3. *Free nerve endings:* May be pain receptors.
4. Receptors resembling *tendon organs:* Their function in joints is not known.

Joint receptors, like the other two classes of proprioceptors, send information on body movements to the cerebellum and cerebrum, as well as to spinal reflex arcs.

CLINICAL APPLICATION

Paresthesia An abnormal sensation of numbness, burning, or tingling is referred to as **paresthesia** (par″-es-the′zha; "faulty sensation"). Temporary paresthesia commonly occurs in the foot, arm, or hand as a result of compression of the peripheral nerve serving the region. This is commonly referred to as a body part "going to sleep" or as the sensation of "pins and needles." Once nerve compression is relieved, normal sensation returns. Chronic paresthesia, ongoing numbness or tingling, can occur in any body region. It is usually symptomatic of some other neurological disease such as stroke, multiple sclerosis, a nerve entrapment syndrome such as carpal tunnel syndrome, or a traumatic nerve injury.

check your understanding

- ☐ 3. Name the functional type of the sensory receptors that respond to painful stimuli. What is the structure of these receptors?
- ☐ 4. Are proprioceptors part of the somatic or visceral sensory system?
- ☐ 5. What type of stimulus do encapsulated receptors respond to?

(For answers, see Appendix B.)

CRANIAL NERVES

learning outcome

▶ Name the 12 pairs of cranial nerves, and describe the structures innervated by each.

The remainder of this chapter is devoted to the specific nerves of the body, beginning with the 12 pairs of **cranial nerves** that attach to the brain and pass through various foramina in the skull (Figure 14.4). (For a review of cranial foramina, see Figure 7.9, p. 159.) These nerves are numbered from I through XII in a rostral to caudal direction. The first two pairs attach to the forebrain, the rest to the brain stem. Except for the vagus nerve (X), which extends into the abdomen, the cranial nerves innervate only head and neck structures. The following mnemonic phrase can help you remember the first letters of the names of the 12 cranial nerves in their proper order: "**O**h, **O**h, **O**h, **T**o **T**ouch **A**nd **F**eel **V**ery **G**ood **V**elvet, **AH**!"

The twelve cranial nerves and the origins of their names are briefly described below.

I. **Olfactory.** This is the sensory nerve of smell.

II. **Optic.** Because it develops as an outgrowth of the brain, this sensory nerve of vision is not a true nerve at all. It is more correctly called a brain tract.

III. **Oculomotor.** The name *oculomotor* means "eye mover." This nerve innervates four of the *extrinsic eye muscles*—muscles that move the eyeball in the orbit.

IV. **Trochlear.** The name *trochlear* means "pulley." This nerve innervates an extrinsic eye muscle that hooks through a pulley-shaped ligament in the orbit.

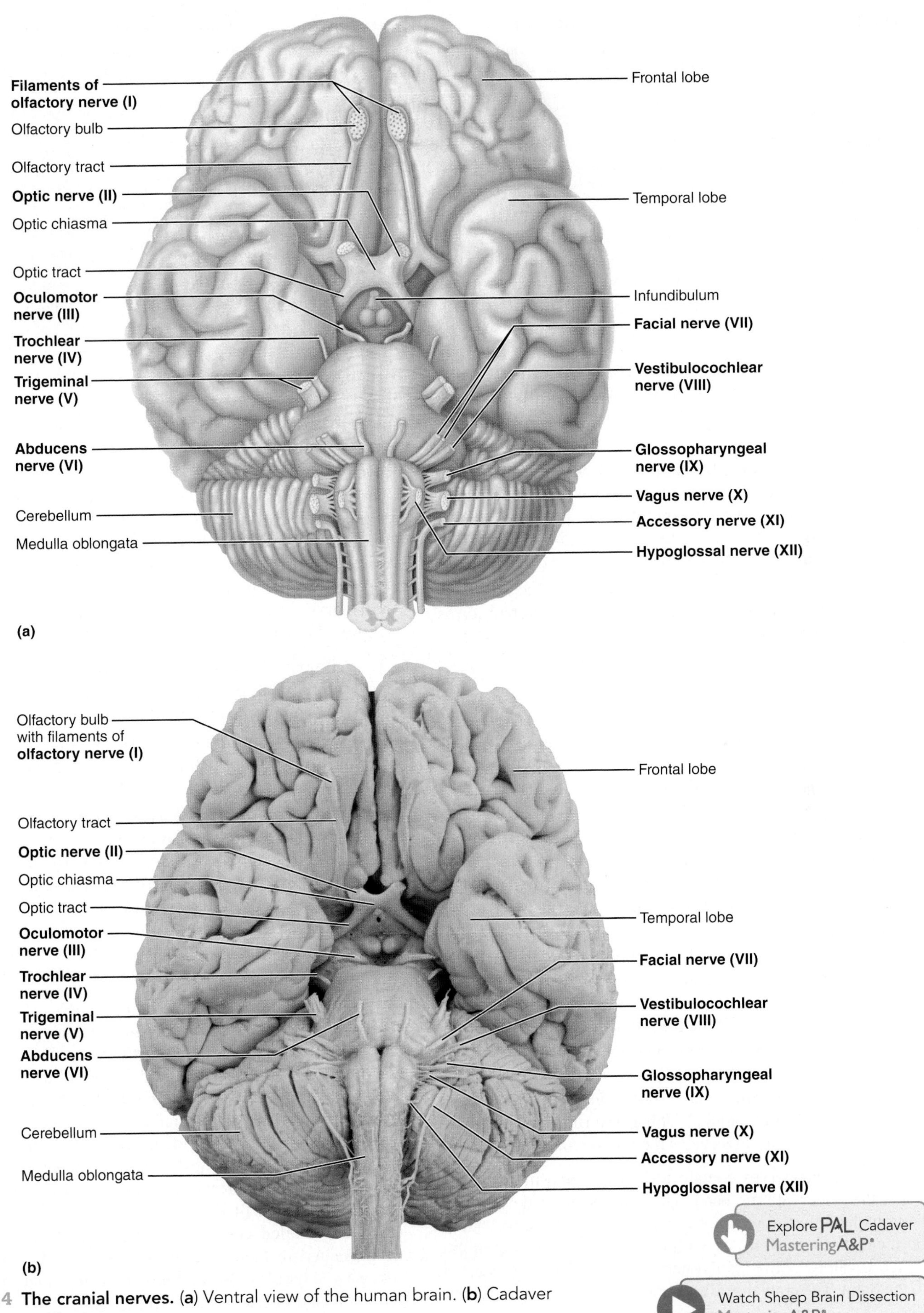

Figure 14.4 **The cranial nerves.** (**a**) Ventral view of the human brain. (**b**) Cadaver dissection, ventral brain.

Cranial nerves	Sensory function		Motor function	
	Somatic sensory (SS)	Visceral sensory (VS)	Somatic motor (SM)	Visceral motor: parasympathetic (VM)
I Olfactory		Smell		
II Optic	Vision			
III Oculomotor			SM	VM
IV Trochlear			SM	
V Trigeminal	General		SM	
VI Abducens			SM	
VII Facial	General	General; taste	SM	VM
VIII Vestibulocochlear	Hearing; equilibrium		Some	
IX Glossopharyngeal	General	General; taste	SM	VM
X Vagus	General	General; taste	SM	VM
XI Accessory			SM	
XII Hypoglossal			SM	

(c)

Figure 14.4 **The cranial nerves,** *continued.* **(c)** Summary of the cranial nerves by function. Cranial nerves that have somatic motor function also contain proprioceptive (sensory) fibers.

V. **Trigeminal.** The name *trigeminal* means "threefold," which refers to this nerve's three major branches. The trigeminal nerve provides general sensory innervation to the face and motor innervation to the chewing muscles.

VI. **Abducens.** This nerve was so named because it innervates the muscle that *abducts* the eyeball (turns the eye laterally).

VII. **Facial.** This nerve innervates the muscles of facial expression as well as other structures.

VIII. **Vestibulocochlear.** This sensory nerve of hearing and equilibrium was once called the *auditory nerve.*

IX. **Glossopharyngeal.** The name *glossopharyngeal* means "tongue and pharynx," structures that this nerve helps to innervate.

X. **Vagus.** The name *vagus* means "vagabond" or "wanderer." This nerve "wanders" beyond the head into the thorax and abdomen.

XI. **Accessory.** This nerve was once called the *spinal accessory nerve.* It originates from the cervical region of the spinal cord, enters the skull through the foramen magnum, and exits the skull with the vagus nerve. The accessory nerve carries motor innervation to the trapezius and sternocleidomastoid muscles.

XII. **Hypoglossal.** The name *hypoglossal* means "below the tongue." This nerve runs inferior to the tongue and innervates the tongue muscles.

The cranial nerves contain the sensory and motor nerve fibers that innervate the head. The cell bodies of the *sensory* neurons lie either in receptor organs (e.g., the nose for smell, or the eye for vision) or within **cranial sensory ganglia,** which lie along some cranial nerves (V, VII–X) just external to the brain. The cranial sensory ganglia are directly comparable to the dorsal root ganglia on the spinal nerves (see p. 411). The cell bodies of most cranial *motor* neurons occur in cranial nerve nuclei in the ventral gray matter of the brain stem, (Figure 13.7, p. 382)—just as cell bodies of spinal motor neurons occur in the ventral gray matter of the spinal cord.

Based on the types of fibers they contain, the 12 cranial nerves can be classified into three functional groups (Figure 14.4c):

1. **Primarily or exclusively sensory nerves** (I, II, VIII) that contain *special sensory* fibers for smell (I), vision (II), and hearing and equilibrium (VIII).
2. **Primarily motor nerves** (III, IV, VI, XI, XII) that contain *somatic motor* fibers to skeletal muscles of the eye, neck, and tongue.
3. **Mixed (motor and sensory) nerves** (V, VII, IX, X). These mixed nerves supply sensory innervation to the face (through *general somatic sensory* fibers) and to the mouth and viscera *(general visceral sensory),* including the taste buds for the sense of taste *(special visceral sensory).* These nerves also innervate pharyngeal arch muscles *(somatic motor),* such as the chewing muscles (V) and the muscles of facial expression (VII).

Additionally, four of the cranial nerves (III, VII, IX, X) contain *visceral motor* fibers that regulate visceral muscle and glands throughout much of the body. These motor fibers belong to the parasympathetic division of the autonomic nervous system (ANS) (see Figure 14.1). The ANS innervates body structures through chains of two motor neurons. The cell bodies of the second neurons occupy *autonomic motor ganglia* in the PNS. The location of these peripheral autonomic ganglia are described in the pathway of these four nerves (see Table 14.2).

The detailed coverage of the individual cranial nerves is presented in table format **(Table 14.2)**. In this table, the foramina of the skull through which each cranial nerve passes are indicated in blue text for emphasis.

(Text continues on page 444.)

Table 14.2 Cranial Nerves

CN I OLFACTORY (ol-fak′to-re) NERVES

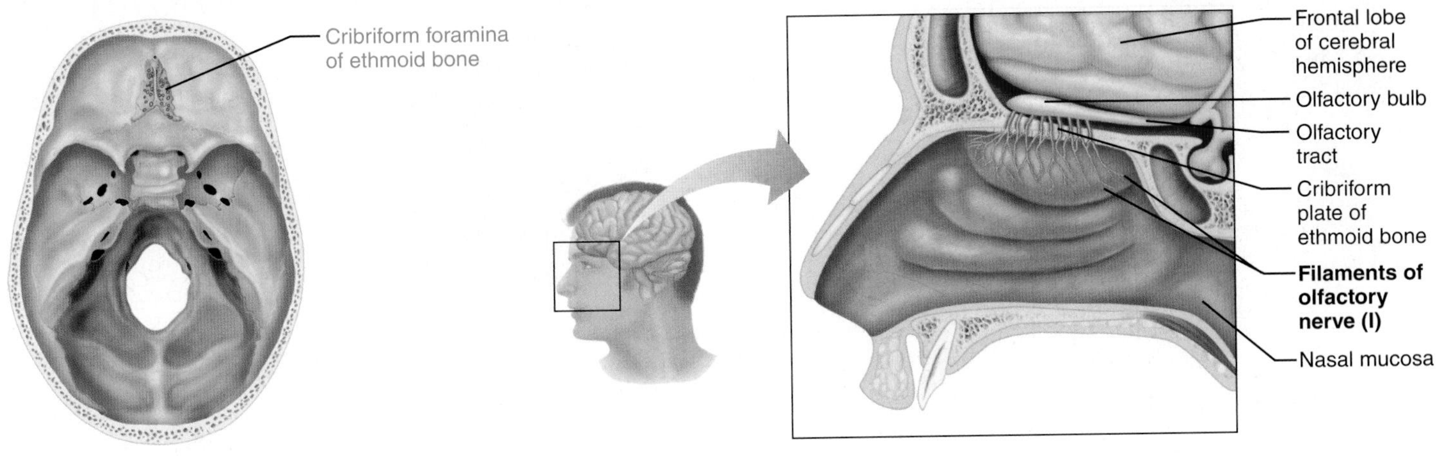

Sensory function (VS)	Special visceral sensory, sense of smell.
Origin	Olfactory receptor cells (bipolar neurons) in the olfactory epithelium of the nasal cavity.
Pathway	Pass through the cribriform foramina of the ethmoid bone to synapse in the olfactory bulb. Fibers of olfactory bulb neurons extend posteriorly beneath the frontal lobe as the olfactory tract. Terminate in the primary olfactory cortex of the cerebrum. (See also Figure 16.3, p. 491.)

CLINICAL APPLICATION **Anosmia** Fracture of the ethmoid bone or lesions of olfactory fibers may result in partial or total loss of smell, a condition known as anosmia (an-oz′me-ah).

CN II OPTIC NERVES

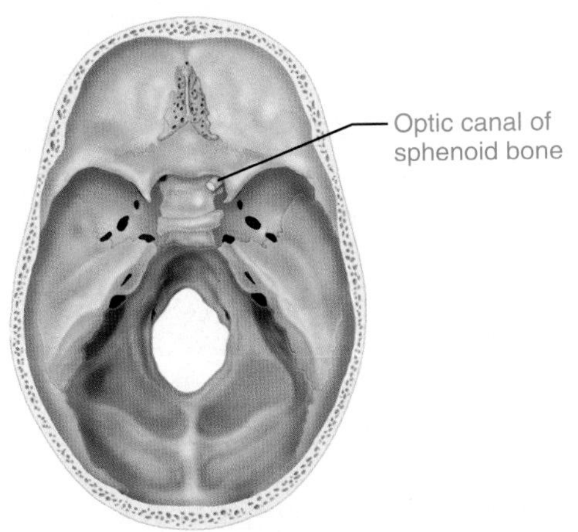

Sensory function (SS)	Special somatic sensory, vision.
Origin	Retina of the eye.
Pathway	Pass through the optic canal of the sphenoid bone. Optic nerves converge to form the optic chiasma (ki-az′-mah), where fibers partially cross over, then continue as the optic tracts to synapse in the thalamus. Thalamic fibers project to and terminate in the primary visual cortex in the occipital lobe. (See also Figure 16.14, p. 504.)

CLINICAL APPLICATION **Optic Nerve Damage** Damage to an optic nerve results in blindness in the eye served by the nerve; damage to the visual pathway distal to the optic chiasma results in partial visual losses; visual defects are called *anopsias* (an-op′se-as).

CN III OCULOMOTOR (OK″U-LO-MO′TER) NERVES

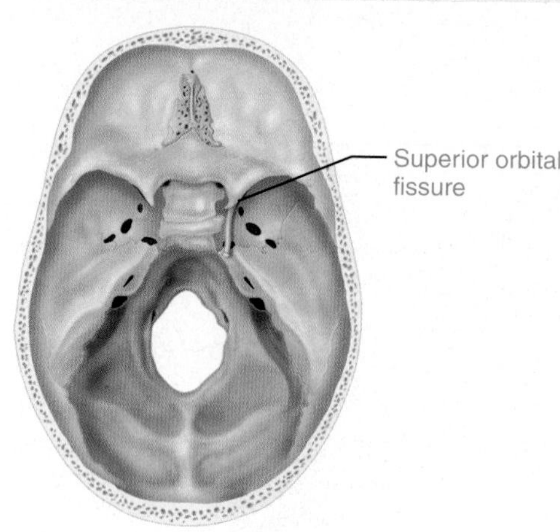

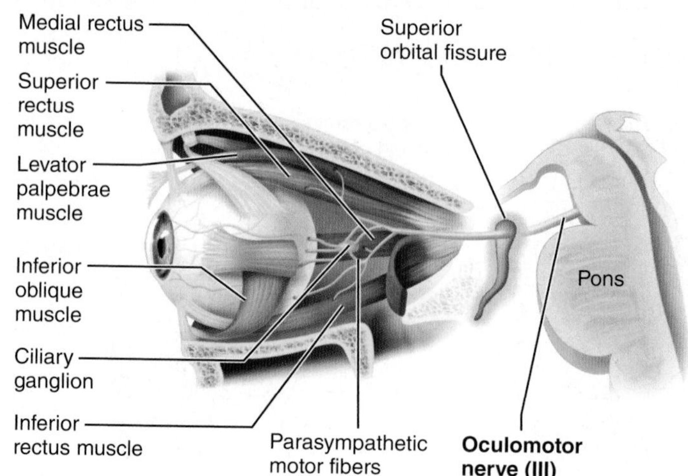

Somatic motor function (SM)	Innervate four extrinsic eye muscles that direct the eyeball: **superior rectus, medial rectus, inferior rectus, inferior oblique muscles.** Innervate levator palpebrae superioris muscle that elevates the upper eyelid (p. 492). Afferent proprioceptor fibers return from the extrinsic eye muscles.
Visceral motor function (VM) (parasympathetic)	Constrictor muscles of the iris constrict the pupil. Ciliary muscle controls lens shape.
Origin	Oculomotor nuclei in the ventral midbrain.
Pathway	Pass through the superior orbital fissure to enter the orbit. Parasympathetic fibers from the brain stem synapse with post ganglionic neurons in the *ciliary ganglion* that innervate the iris and ciliary muscle.

 CLINICAL APPLICATION **Oculomotor Nerve Paralysis** Because the actions of the two extrinsic eye muscles not served by cranial nerve III are unopposed, the eye cannot be moved up or inward, and at rest the eye turns laterally (external strabismus [strah-biz′mus]). The upper eyelid droops (ptosis), and the person has double vision.

CN IV TROCHLEAR (trok′le-ar) NERVES

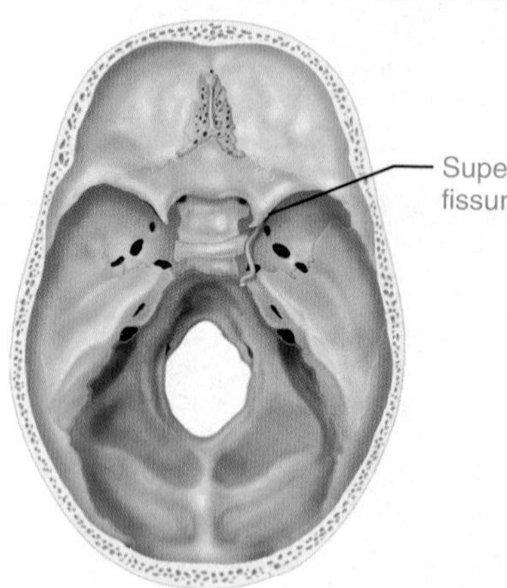

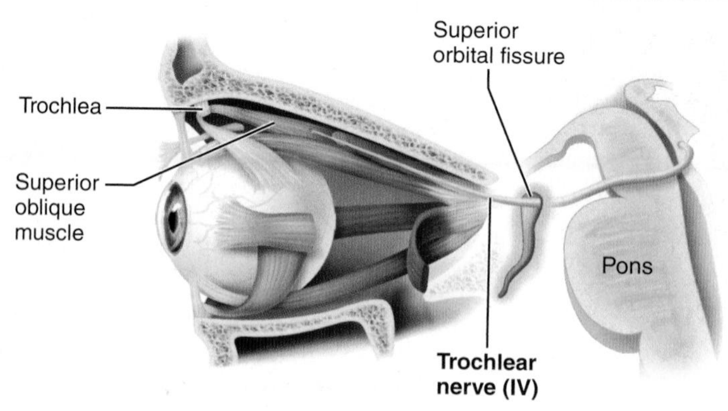

Somatic motor function (SM)	Innervate the **superior oblique muscle.** This muscle passes through a ligamentous pulley at the roof of the orbit, the trochlea, from which its name is derived. Afferent proprioceptor fibers return from the superior oblique.
Origin	Trochlear nuclei in the dorsal midbrain.
Pathway	Pass ventrally around the midbrain; pass through the superior orbital fissure to enter the orbit.

 CLINICAL APPLICATION **Trochlear Nerve Damage** Damage to a trochlear nerve results in double vision and reduced ability to rotate the eye inferolaterally.

Table 14.2 *continued*

CN V TRIGEMINAL NERVES

The large trigeminal nerve forms three divisions (*trigeminal* = threefold): ophthalmic (V_1), maxillary (V_2), and mandibular (V_3) divisions.

This mixed nerve is the general somatic sensory nerve of the face for touch, temperature, and pain. The mandibular division supplies somatic motor innervation to the chewing muscles. Details on the next page.

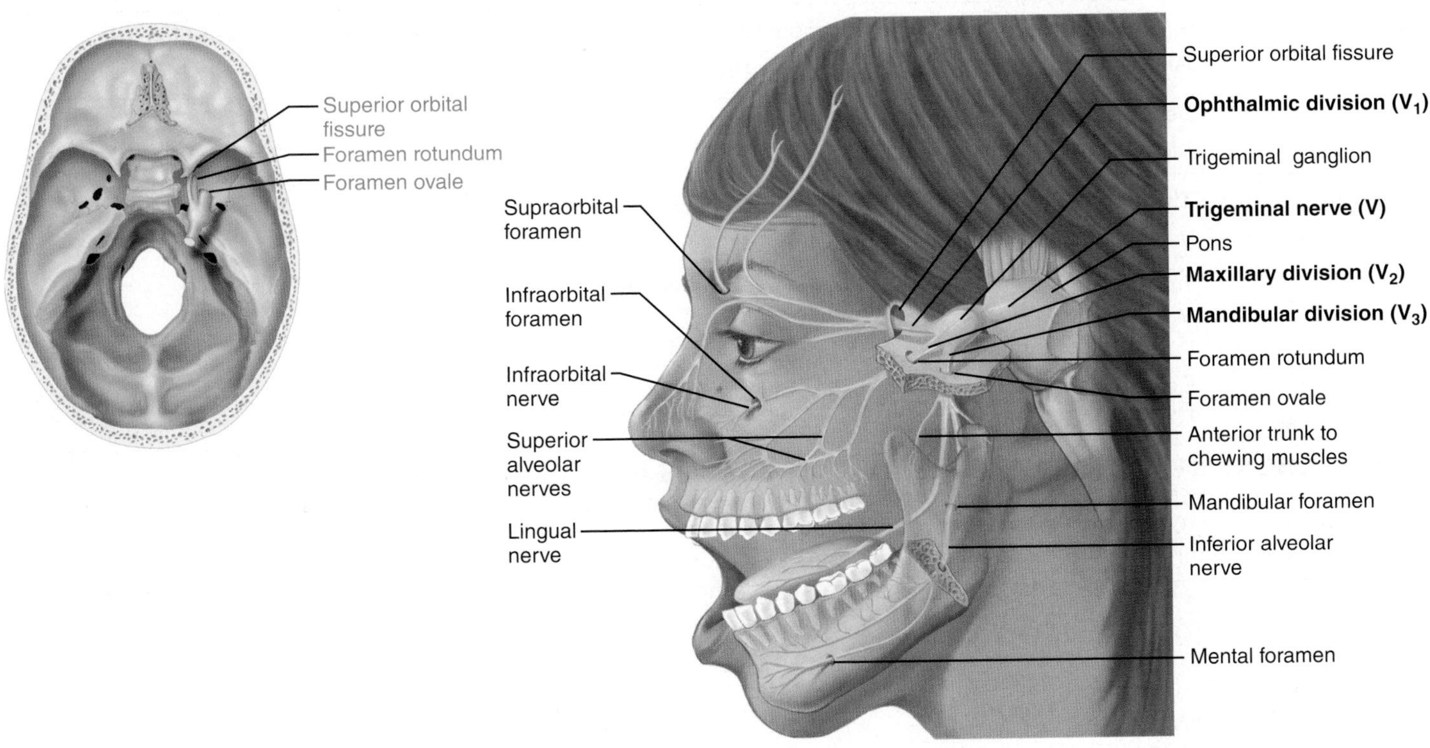

(a) Distribution of the trigeminal nerve

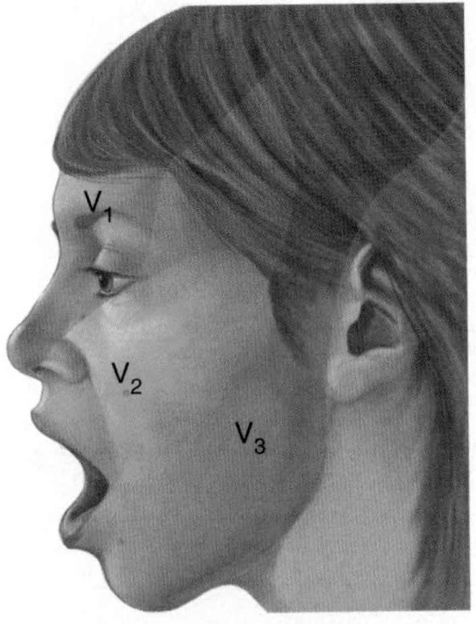

(b) Distribution of sensory fibers of each division

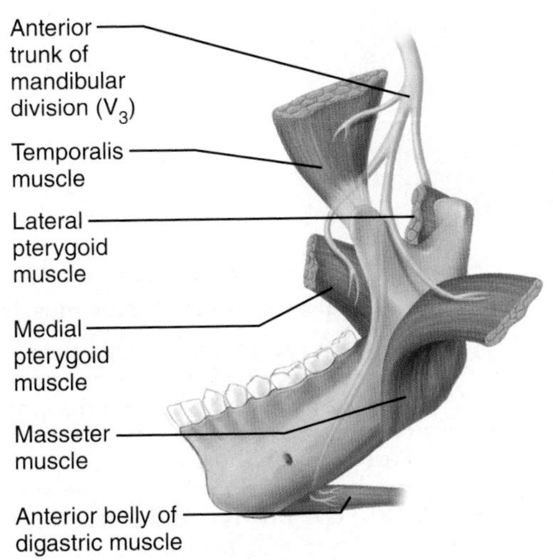

(c) Motor branches of the mandibular division (V_3)

Table 14.2 Cranial Nerves *continued*

CN V TRIGEMINAL NERVES *(continued)*

Sensory function (SS)	V_1 General somatic sensation from skin of anterior scalp and forehead, upper eyelid and nose, nasal cavity mucosa, cornea, and lacrimal gland. V_2 General somatic sensation from skin of cheek, upper lip, and lower eyelid, nasal cavity mucosa, palate, upper teeth. V_3 General somatic sensation from skin of chin and temporal region of scalp, anterior tongue and lower teeth.
Somatic motor function (SM)	V_3 Innervate the muscles of mastication: **temporalis, masseter, pterygoids, anterior belly of digastric.** Afferent proprioceptor fibers return from these muscles.
Origin	Sensory receptors in skin and mucosa of face. Motor fibers from trigeminal motor nucleus in pons.
Pathway	**Through the Skull** / **Cutaneous Branch** V_1 Superior orbital fissure / Supraorbital foramen V_2 Foramen rotundum / Infraorbital foramen V_3 Foramen ovale, Mandibular foramen / Mental foramen Cell bodies of sensory neurons of all three divisions located in the large *trigeminal ganglion.* Fibers extend to trigeminal nuclei in the pons.

CLINICAL APPLICATION **Anesthesia for Upper and Lower Jaws** Dentists desensitize upper and lower jaws by injecting local anesthetic (such as Novocain) into alveolar branches of the maxillary and mandibular divisions of the trigeminal nerve, respectively. This blocks pain-transmitting fibers from the teeth, and the surrounding tissues become numb.

CN VI ABDUCENS (ab-du'senz) NERVES

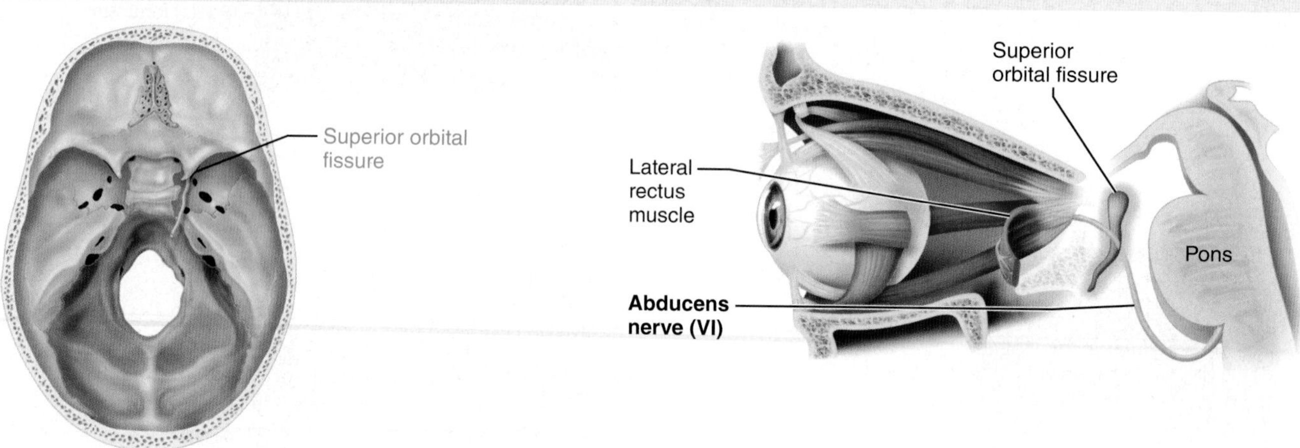

Somatic motor function (SM)	Innervate the **lateral rectus muscle.** This muscle abducts the eye. Afferent proprioceptor fibers return from the lateral rectus.
Origin	Abducens nuclei in the inferior pons.
Pathway	Pass through the superior orbital fissure to enter the orbit.

CLINICAL APPLICATION **Abducens Nerve Paralysis** In abducens nerve paralysis, the eye cannot be moved laterally; at rest, affected eyeball turns medially *(internal strabismus)*.

Table 14.2 *continued*

CN VII FACIAL NERVES

A mixed nerve: Chief somatic motor nerve to the facial muscles; parasympathetic innervation to glands; special sensory taste from the tongue. Details on the next page.

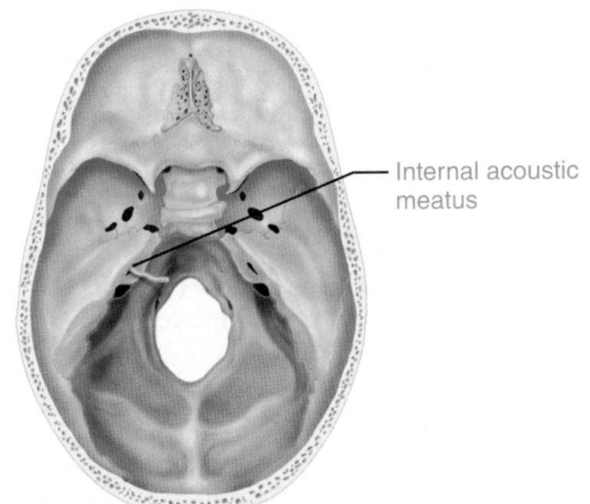

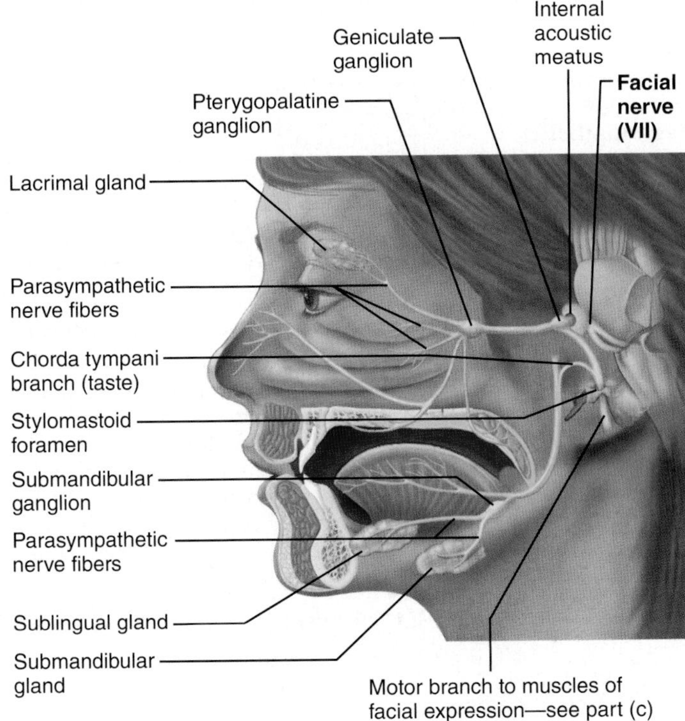

(a) Parasympathetic efferents and sensory afferents

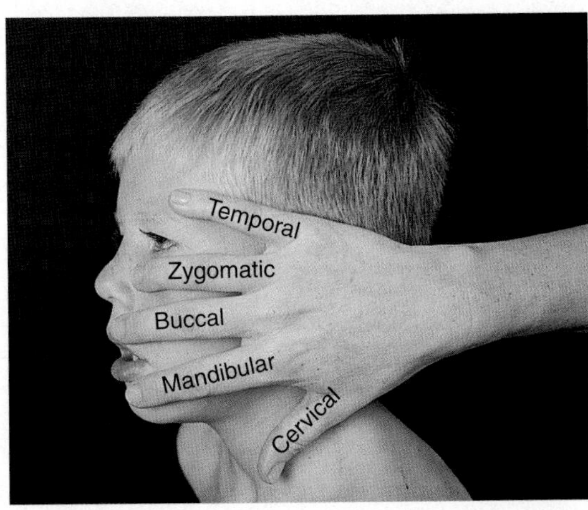

(b) A simple method of remembering the courses of the five major motor branches of the facial nerve

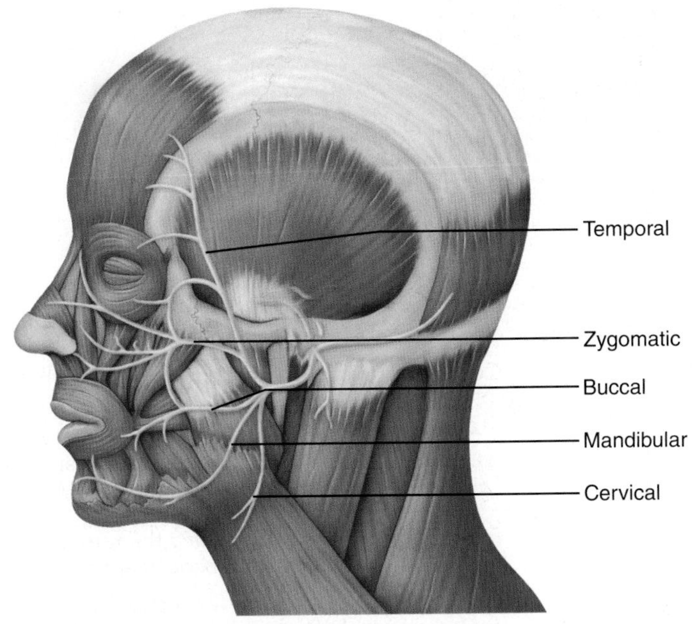

(c) Motor branches to muscles of facial expression and scalp muscles

CN VII FACIAL NERVES *(continued)*

Sensory function (VS, SS)	Special visceral sensory from taste buds on anterior two-thirds of tongue. General somatic sesory from small patch of skin on the ear.
Somatic motor function (SM)	Five major branches on face: temporal, zygomatic, buccal, mandibular, and cervical, to innervate the **facial muscles** (see parts c and d of the figure). Also innervates the **posterior belly of digastric.** Afferent proprioceptor fibers return from these muscles.
Visceral motor function (VM) (parasympathetic)	Innervate the lacrimal (tear) glands, nasal and palatine glands, and the submandibular and sublingual salivary glands.
Origin	Fibers emerge from the pons, just lateral to abducens.
Pathway	Fibers enter the temporal bone via the internal acoustic meatus. Chorda tympani branches off to innervate the two salivary glands and tongue. Branch to facial muscles emerges from the temporal bone through the stylomastoid foramen and courses to lateral aspect of face. Cell bodies of sensory neurons are in *geniculate ganglion.* Cell bodies of postganglionic parasympathetic neurons are in *pterygopalatine* (ter″i-go-pal′ah-tin) and *submandibular ganglia* on the trigeminal nerve (see part a of the figure).

CLINICAL APPLICATION **Bell's Palsy** *Bell's palsy,* characterized by paralysis of facial muscles on affected side and partial loss of taste sensation, may develop rapidly (often overnight). It is caused by herpes simplex (viral) infection, which produces inflammation and swelling of the facial nerve. The lower eyelid droops, the corner of the mouth sags (making it difficult to eat or speak normally), and the eye constantly drips tears and cannot be completely closed. The condition may disappear spontaneously without treatment.

CN VIII VESTIBULOCOCHLEAR (ves-tib″u-lo-kok′le-ar) NERVES

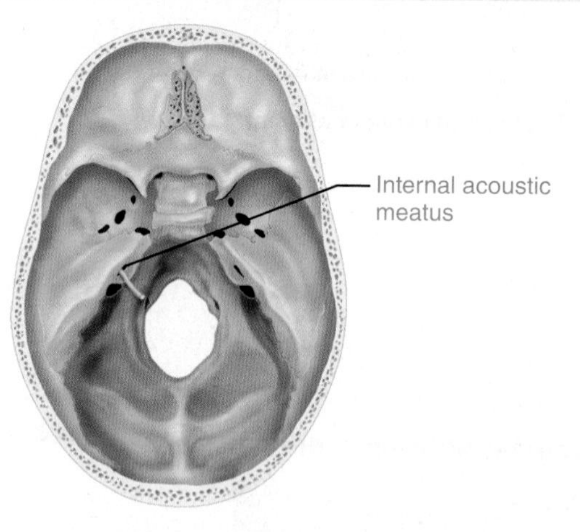

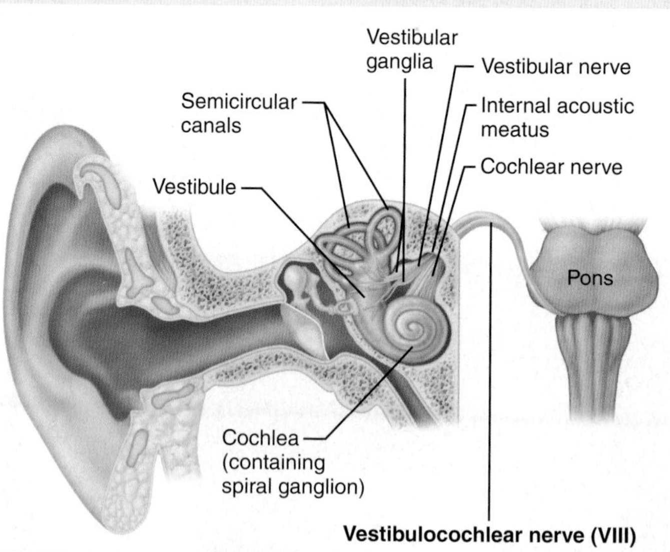

Sensory function (SS)	Vestibular branch: Special somatic sensory, equilibrium. Cochlear branch: Special somatic sensory, hearing. Small motor component adjusts the sensitivity of the sensory receptors
Origin	Sensory receptors in the inner ear for hearing (within the cochlea) and for equilibrium (within the semicircular canals and vestibule).
Pathway	From the inner ear cavity within the temporal bone, fibers pass through the internal acoustic meatus, merge to form the vestibulocochlear nerve and enter the brain stem at the pons. Sensory nerve cell bodies for vestibular branch located in *vestibular ganglia* (see Figure 16.18, p. 510); for the cochlear branch, in the *spiral ganglia* within the cochlea. (See Figure 16.19, p. 511.)

CLINICAL APPLICATION **Vestibulocochlear Nerve Damage** Lesions of cochlear nerve or cochlear receptors result in *central* or *nerve deafness,* whereas damage to vestibular division produces dizziness, rapid involuntary eye movements, loss of balance, nausea, and vomiting.

Table 14.2 *continued*

CN IX GLOSSOPHARYNGEAL (glos″o-fah-rin′je-al) NERVES

Mixed nerve innervating the tongue (general and special sensory), the pharynx, and the parotid salivary gland.

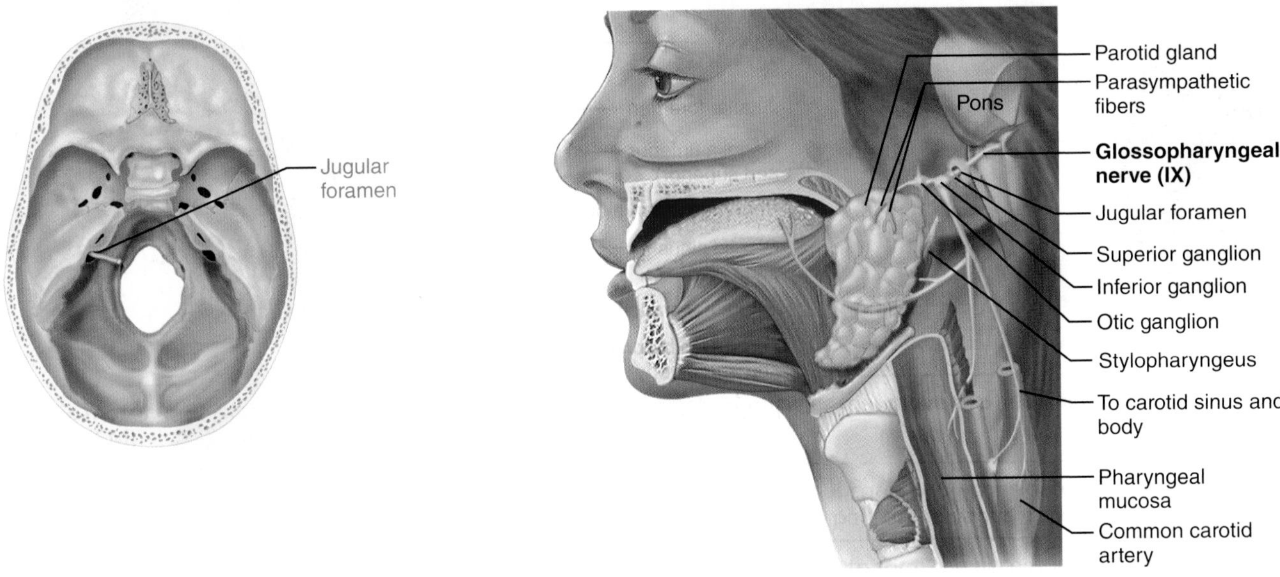

Sensory function (VS, SS)	Special visceral sensory from taste buds on posterior third of tongue. General visceral sensory from posterior third of tongue, pharyngeal mucosa, chemoreceptors in the carotid body (which monitor O_2 and CO_2 in the blood and regulate respiratory rate and depth), and baroreceptors of carotid sinus (regulate blood pressure). General somatic sensory from small area of skin on external ear.
Somatic motor function (SM)	Innervate a pharyngeal muscle, **stylopharyngeus,** which elevates the pharynx during swallowing. Afferent proprioceptor fibers return from this muscle.
Visceral motor function (VM) (parasympathetic)	Innervate the parotid salivary gland.
Origin	Fibers emerge from the medulla oblongata.
Pathway	Fibers pass through the jugular foramen and travel to the pharynx. Cell bodies of sensory neurons are located in the *superior* and *inferior ganglia.* Cell bodies of postganglionic parasympathetic neurons are in *otic ganglion* on the trigeminal nerve.

CLINICAL APPLICATION **Glossopharyngeal Nerve Damage** Injury or inflammation of glossopharyngeal nerves impairs swallowing and taste on the posterior third of the tongue.

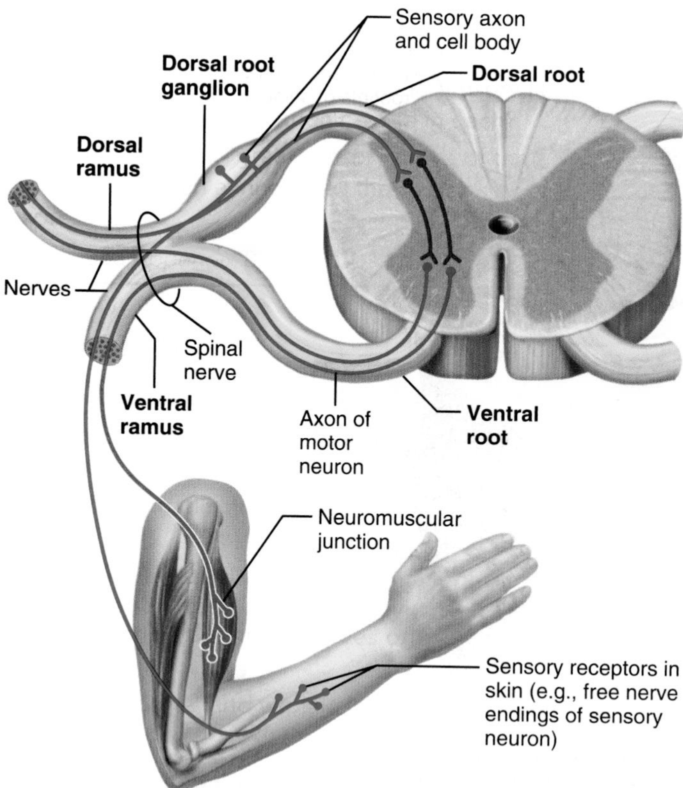

Figure 14.7 The functional components of a spinal nerve, anterior view. The branches and connections to the spinal cord are illustrated.

It is important to review the difference between the *roots* and *rami* (Figure 14.7). Do not confuse these similarly named structures.

- The roots lie medial to the spinal nerves and are either strictly sensory (dorsal root) or strictly motor (ventral root).
- The rami are lateral branches of the spinal nerves, and each contains both sensory fibers and motor fibers.

The following sections explore how the rami and their branches innervate various regions of the body. We will first consider the innervation of the back and the anterior thoracic and abdominal wall. Then we discuss the anatomy and roles of the structures called *nerve plexuses.* Finally, we will examine the innervation of the neck, the upper limbs, the lower limbs, the joints, and the skin. The text throughout refers to the innervation of major muscle groups (specific information on muscle innervation is found in Tables 11.3–11.17 on pp. 279–332).

Innervation of the Back

The innervation of the back (posterior part of the trunk and neck) by the dorsal rami follows a neat, segmented pattern. Each dorsal ramus branches to innervate the intrinsic muscles of the back and a horizontal strip of skin in line with its emergence point from the vertebral column (Figure 14.6b). This pattern of innervation is far simpler than the innervation of the rest of the body by the ventral rami.

Innervation of the Anterior Thoracic and Abdominal Wall

Only in the thorax are the ventral rami arranged in a simple and segmented pattern. The thoracic ventral rami run anteriorly, one deep to each rib, as the **intercostal nerves** (Figures 14.5 and 14.6b). These nerves supply the intercostal muscles, the skin of the anterior and lateral thorax, and most of the abdominal wall inferior to the rib cage. Along its course, each intercostal nerve gives off **lateral** and **anterior cutaneous branches** to the adjacent skin (Figure 14.6b).

Two nerves in this thoracic series are unusual. The last (T_{12}) lies inferior to the twelfth rib and thus is called a *subcostal* ("below the ribs") *nerve* rather than an intercostal nerve. The first (most superior) intercostal nerve is exceptionally small because most fibers of T_1 enter the brachial plexus (discussed shortly).

☑ check your understanding

- ☐ 9. What deficits would result from a lesion that damaged the dorsal root of spinal nerve T_4? What deficits would occur if a lesion damaged the dorsal ramus of T_4?
- ☐ 10. Below which vertebra does each of the following nerves exit: (a) cervical nerve C_5; (b) lumbar nerve L_3?

(For answers, see Appendix B.)

Nerve Plexuses

A **nerve plexus** is a network of nerves. The ventral rami of all spinal nerves except T_2–T_{12} branch and join one another lateral to the vertebral column, forming nerve plexuses (see Figure 14.5). These interlacing networks occur in the cervical, brachial, lumbar, and sacral regions and primarily serve the limbs. Note that these plexuses are formed by *ventral* rami only. Within the plexuses, fibers from the different ventral rami converge and diverge and redistribute so that (1) each end branch of the plexus contains fibers from several different spinal nerves, and (2) fibers from each ventral ramus travel to the body periphery via several different routes or branches. Therefore, each muscle in a limb receives its nerve supply from more than one spinal nerve. As a result of this arrangement, the destruction of a single spinal nerve cannot completely paralyze any limb muscle.

The Cervical Plexus and Innervation of the Neck

The **cervical plexus** is buried deep in the neck, under the sternocleidomastoid muscle, and extends into the posterior triangle of the neck (shown in Figure 11.30, p. 334). It is formed by the ventral rami of the first four cervical nerves (Figure 14.8). The plexus forms an irregular series of

Table 14.3 Branches of the Cervical Plexus (Figure 14.8)

Nerves	Ventral Rami	Structures Innervated
CUTANEOUS BRANCHES (SUPERFICIAL)		
Lesser occipital	C_2, C_3	Skin on posterolateral aspect of neck
Greater auricular	C_2, C_3	Skin of ear, skin over parotid gland
Transverse cervical	C_2, C_3	Skin on anterior and lateral aspect of neck
Supraclavicular (anterior, middle, and posterior)	C_3, C_4	Skin of shoulder and clavicular region
MOTOR BRANCHES (DEEP)		
Ansa cervicalis (superior and inferior roots)	C_1–C_3	Infrahyoid muscles of neck (omohyoid, sternohyoid, and sternothyroid)
Segmental and other muscular branches	C_1–C_5	Deep muscles of neck (geniohyoid and thyrohyoid) and portions of scalenes, levator scapulae, trapezius, and sternocleidomastoid muscles
Phrenic	C_3–C_5	Diaphragm (sole motor nerve supply)

interconnecting loops from which branches arise. Most branches of the cervical plexus are **cutaneous nerves** (*cutane* = skin) that carry sensory impulses from the skin of the neck, the back of the head, and the most superior part of the shoulder. Other branches carry motor innervation to muscles in the anterior neck region (Table 14.3).

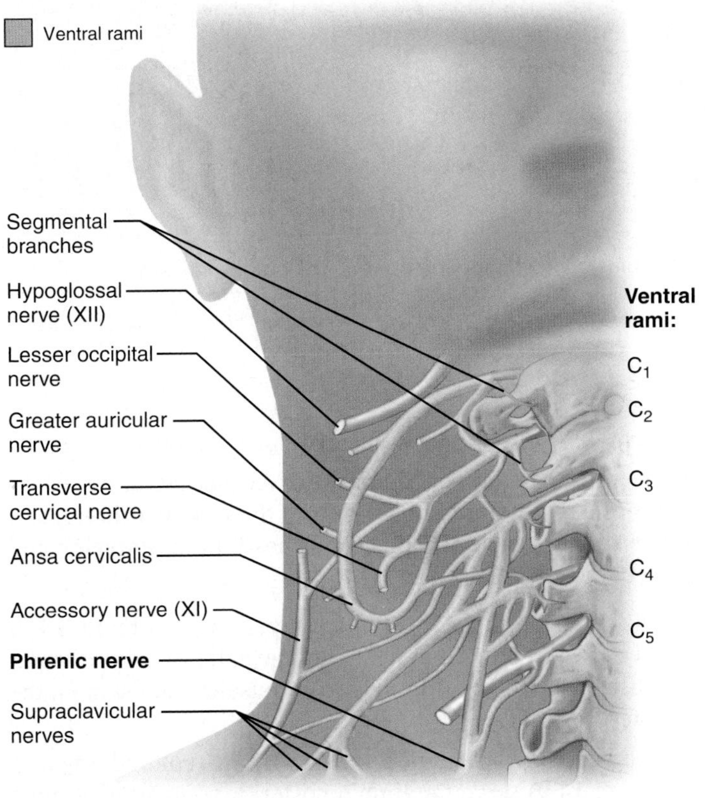

Figure 14.8 The cervical plexus, anterior view. The nerves that are depicted in gray connect to the cervical plexus but do not belong to it. (See Table 14.3 for areas of innervation.)

The most important nerve from this plexus is the **phrenic** (fren′ik) **nerve,** which receives fibers from spinal nerves C_3, C_4, and C_5. The phrenic nerve courses inferiorly through the thorax and innervates the diaphragm (*phren* = diaphragm), providing somatic motor and sensory innervation to this most vital respiratory muscle. If both phrenic nerves are cut (or if the spinal cord is damaged superior to C_3–C_5), respiratory arrest will occur.

CLINICAL APPLICATION

Hiccups Hiccups result when the phrenic nerve induces abrupt, rhythmic contractions of the diaphragm. This reflexive response commonly originates from sensory irritation of the diaphragm or stomach. Swallowing spicy food or acidic soda pop can lead to hiccups via this reflex. Painful stimuli from the diaphragm, carried on the somatic sensory fibers of the phrenic nerve, are perceived as coming from the skin of the shoulder area, which is innervated by the same spinal segments (see Figure 14.15).

The Brachial Plexus and Innervation of the Upper Limb

The **brachial plexus** (Figure 14.9a) lies partly in the neck and partly in the axilla (armpit) and gives rise to almost all of the nerves that supply the upper limb. The brachial plexus can sometimes be felt in the posterior triangle of the neck just superior to the clavicle at the lateral border of the sternocleidomastoid muscle (refer to Figure 11.30, p. 334).

The brachial plexus is formed by the intermixing of the ventral rami of cervical nerves C_5–C_8 and most of the ventral ramus of T_1. Additionally, it may receive small contributions from C_4 or T_2.

Because the brachial plexus is very complex, some consider it the anatomy student's nightmare, yet its arrangement reflects

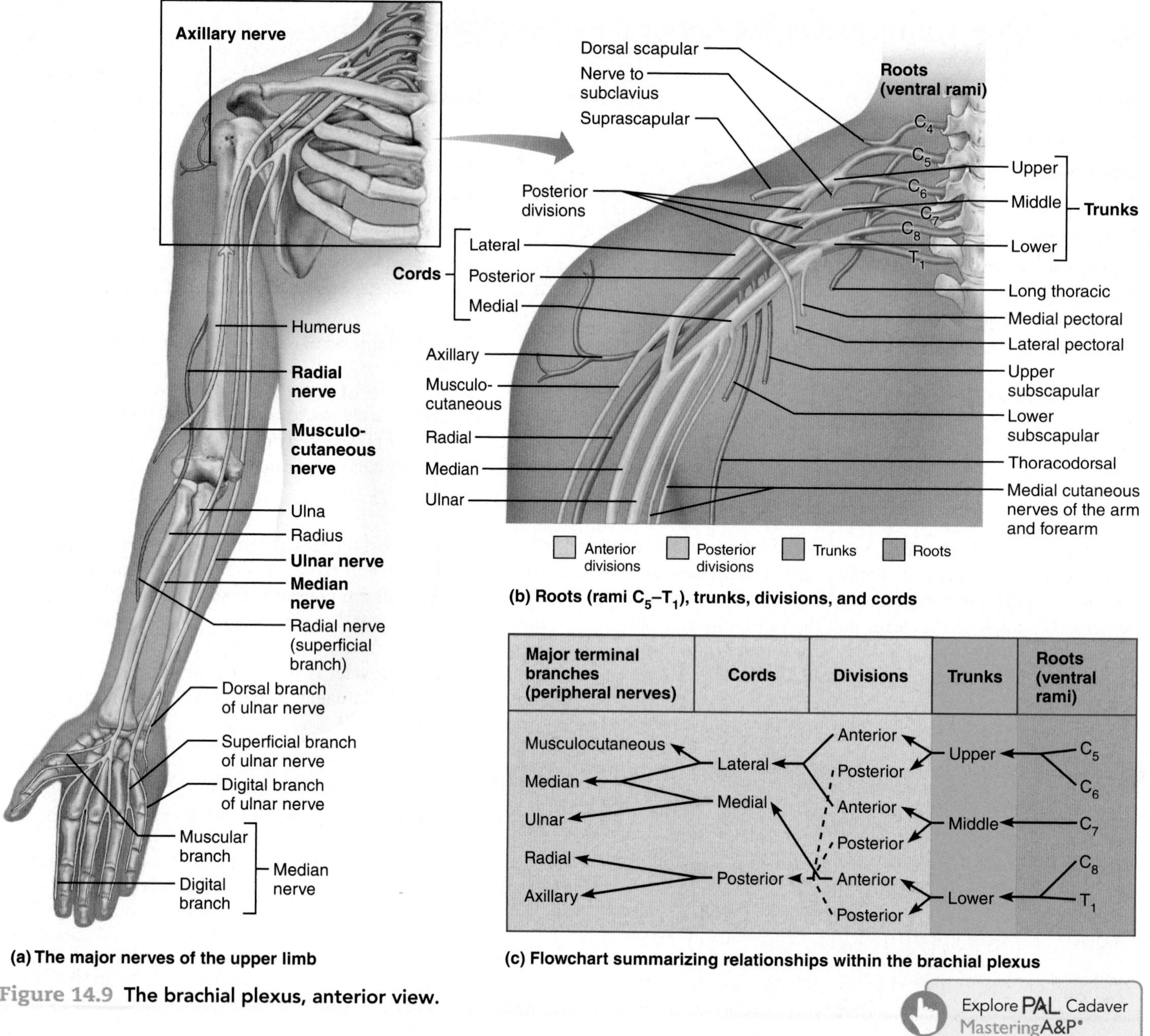

Figure 14.9 The brachial plexus, anterior view.

the functional organization of the structures it innervates. The components of the brachial plexus, from medial to lateral, are:

- **Ventral rami.** The ventral rami from spinal segments C_5–T_1 form the *roots** of the brachial plexus.
- **Trunks.** The ventral rami merge to form three *trunks*.
- **Divisions.** Each trunk splits into two *divisions,* anterior and posterior.
- **Cords.** These six divisions then converge to from three *cords.*

The phrase "**R**eally **T**ired? **D**rink **C**offee" can help you remember the sequence of roots, trunks, divisions, and cords (Figure 14.9b and c).

*Do not confuse these roots of the brachial plexus with the dorsal and ventral roots of spinal nerves.

The five **roots** of the brachial plexus, ventral rami C_5–T_1, lie deep to the sternocleidomastoid. At the lateral border of this muscle, these rami unite to form the **upper, middle,** and **lower trunks,** each of which branches into an **anterior** and **posterior division.**

- The anterior divisions (colored yellow in Figure 14.9) give rise to nerves that innervate the anterior compartment muscles in the upper limb (flexor muscles) and skin on the anterior surface.
- The nerves from the posterior divisions (colored green in Figure 14.9) serve the limb's posterior compartment muscles (extensor muscles) and skin on the posterior surface.

The divisions pass deep to the clavicle and enter the axilla. The anterior divisions from the upper and middle

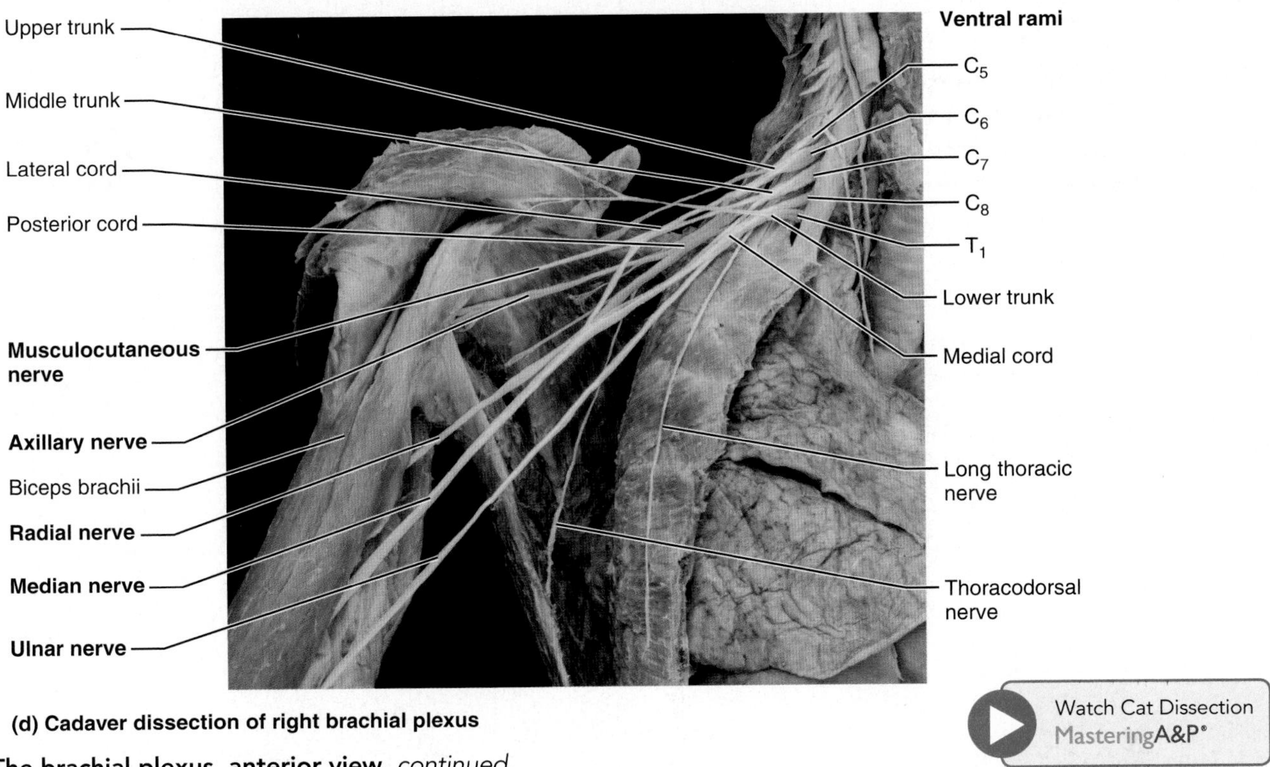

(d) Cadaver dissection of right brachial plexus

Figure 14.9 **The brachial plexus, anterior view,** *continued.*

trunks give rise to the **lateral cord,** and the anterior division from the lower trunk forms the **medial cord.** The **posterior cord** is composed of the posterior divisions of all three trunks. The cords are named for their position in reference to the axillary artery. Some nerves branch off the brachial plexus here to supply muscles of the superior thorax and shoulder (and some skin of the shoulder as well) (Figure 14.9b).

The brachial plexus ends in the axilla, where its three cords give rise to the main nerves of the upper limb (see Figure 14.9). *Focus on Innervation of the Upper Limb* (Figure 14.10) shows the distribution and targets of five of the most important of these peripheral nerves—the *musculocutaneous, median,* and *ulnar nerves* from the anterior divisions of the plexus, and the *axillary* and *radial nerves* from the posterior divisions.

CLINICAL APPLICATION

Brachial Plexus Injuries Injuries to the brachial plexus are common and can be serious. These injuries usually result from stretching the plexus, as can occur in football when a tackler yanks the arm of a ball carrier, or when a blow depresses a shoulder and laterally flexes the head away from that shoulder. Falls, as from a horse or motorcycle, or carrying a heavy backpack can result in overstretching or crushing the brachial plexus and can lead to weakness or even paralysis of the upper limb. Transient injuries to this plexus eventually heal.

Musculocutaneous Nerve The **musculocutaneous nerve** (Figures 14.9d and 14.10), the main terminal branch of the lateral cord, courses within the anterior arm. It passes through the coracobrachialis muscle and descends between the muscles biceps brachii and brachialis, innervating these three arm flexors. Distal to the elbow, it becomes cutaneous and enables skin sensation on the lateral forearm.

Median Nerve The **median nerve** (Figures 14.9a and 14.10) innervates most muscles of the anterior forearm and the lateral palm. After originating from the lateral and medial cords, it descends through the arm without branching, lying medial and posterior to the biceps brachii muscle. Just distal to the elbow region, the median nerve gives off branches to most muscles in the flexor compartment of the forearm (but not the flexor carpi ulnaris and the medial part of the flexor digitorum profundus). Passing through the carpal tunnel deep to the tendon of the palmaris longus and reaching the hand, it innervates five intrinsic muscles in the lateral part of the palm, including the thenar muscles used to oppose the thumb.

Ulnar Nerve The **ulnar nerve** (Figures 14.9a and 14.10) branches off the medial cord of the brachial plexus, then descends along the medial side of the arm. At the elbow it passes posterior to the medial epicondyle of the humerus; then it follows the ulna along the forearm, where it supplies the flexor carpi ulnaris and the medial (ulnar) part of the flexor digitorum profundus. The ulnar nerve then continues into the hand, where it innervates most of the intrinsic hand muscles and the skin on the medial side of the hand. A dorsal branch supplies the skin of the dorsal medial hand.

(Text continues on page 452.)

Figure 14.10

The nerves from the anterior division of the brachial plexus innervate the muscles in the anterior compartments: the flexor muscles of the arm, forearm, and hand. These nerves continue distally to supply cutaneous innervation to the forearm and hand.

Watch Cat Dissection MasteringA&P®

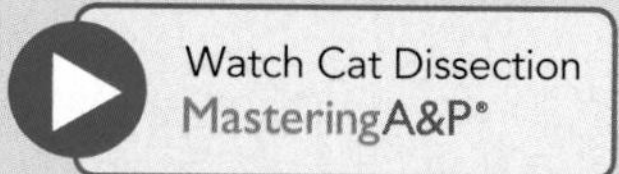

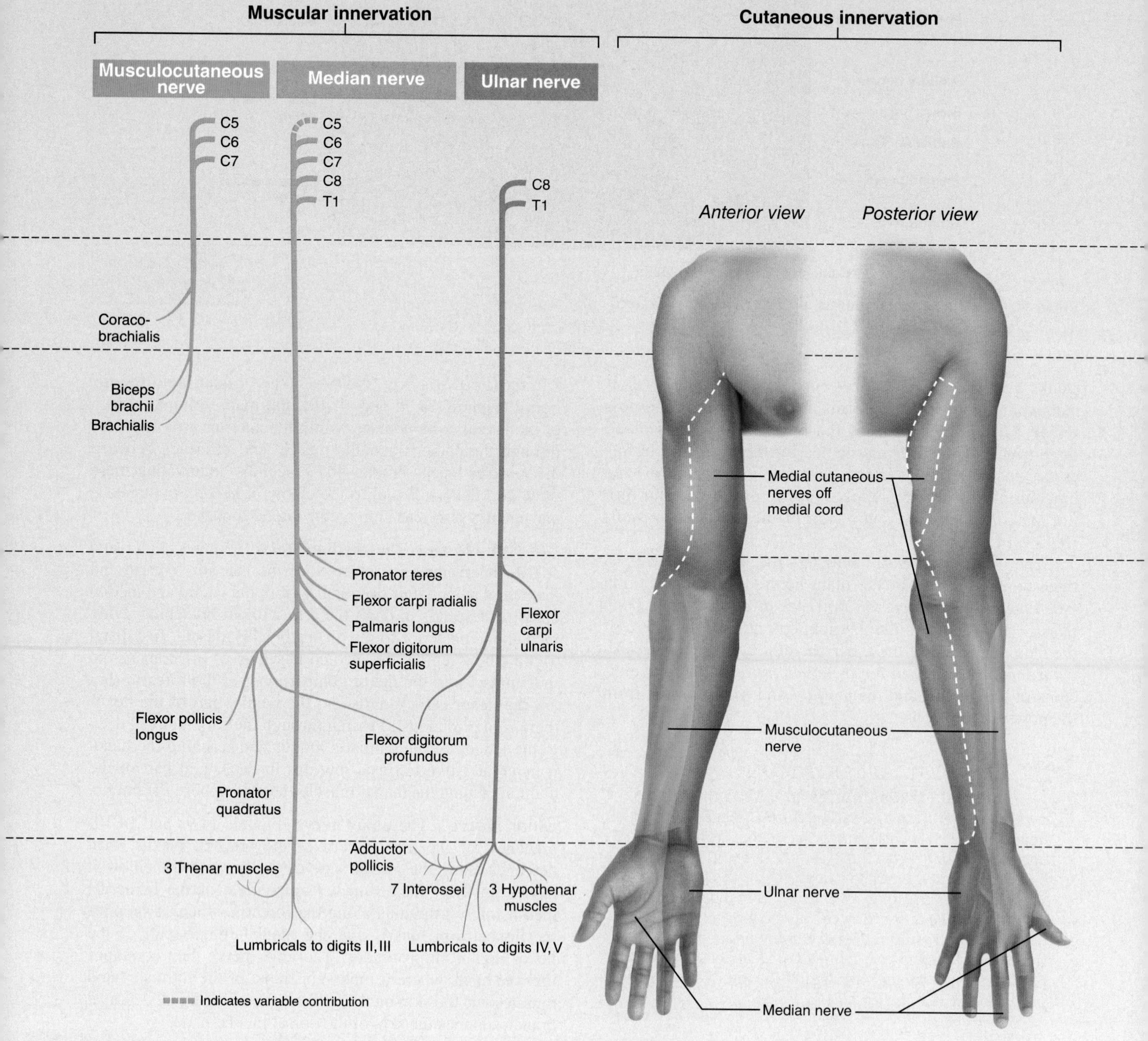

The nerves from the **posterior division of the brachial plexus** innervate the muscles in the posterior compartments: the extensor muscles of the upper limb. These nerves continue distally to supply cutaneous innervation to much of the posterior arm, forearm, and hand.

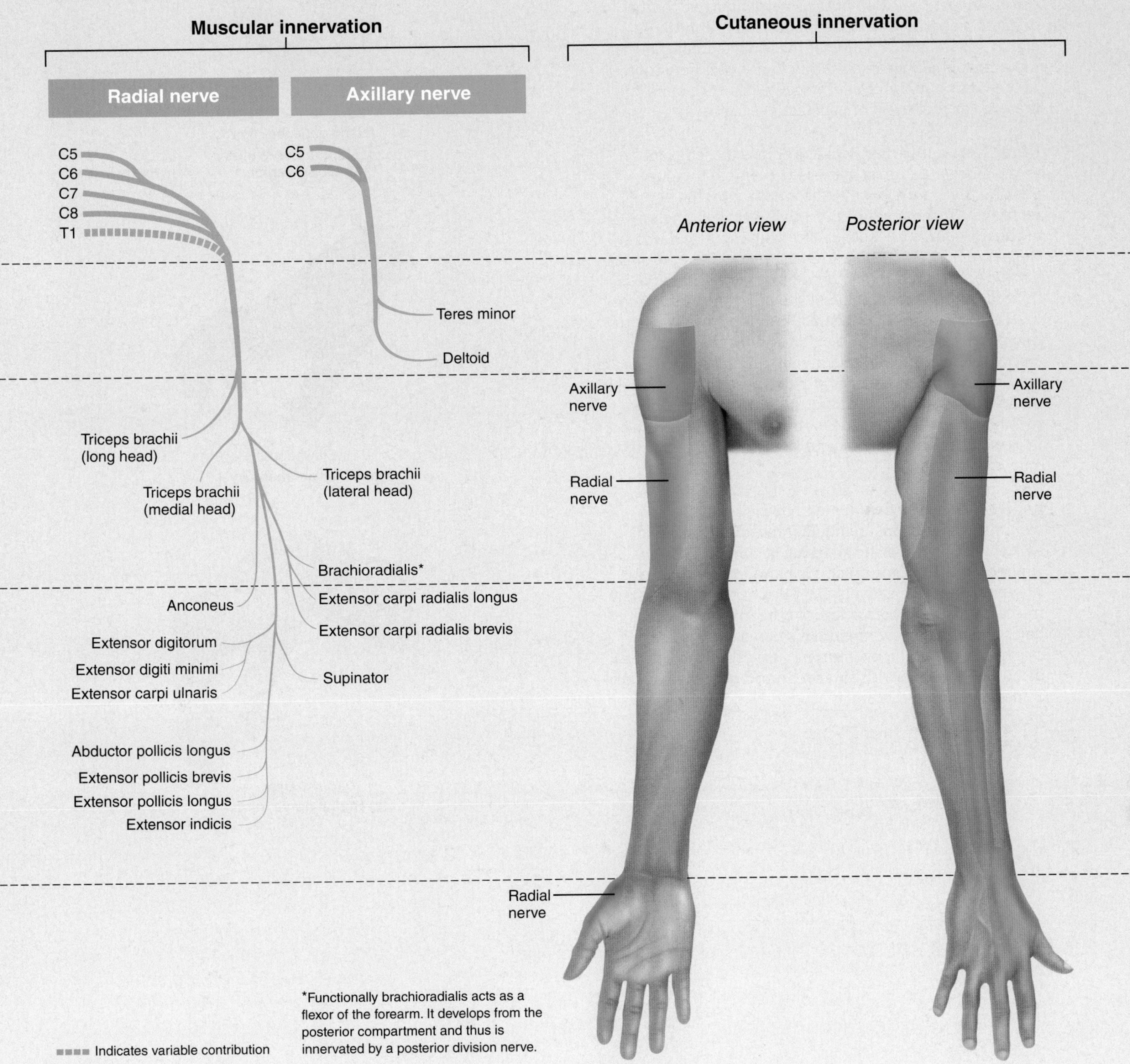

*Functionally brachioradialis acts as a flexor of the forearm. It develops from the posterior compartment and thus is innervated by a posterior division nerve.

CLINICAL APPLICATION

Median Nerve Injury Because the median nerve descends through the exact middle of the forearm and wrist, it is often severed during wrist-slashing suicide attempts. Destruction of the median nerve makes it difficult to oppose the thumb toward the little finger; thus, the ability to pick up small objects is diminished. This deficit is referred to as "ape hand."

An injury to the median nerve in the elbow region results in loss of flexion at the interphalangeal joints of digits II and III (due to deficit of innervation to the digital flexors) and inability of these digits to flex at the metacarpophalangeal joint (due to loss of innervation to the lateral lumbricals). Clinically, this type of injury presents as the "hand of benediction" (figure a).

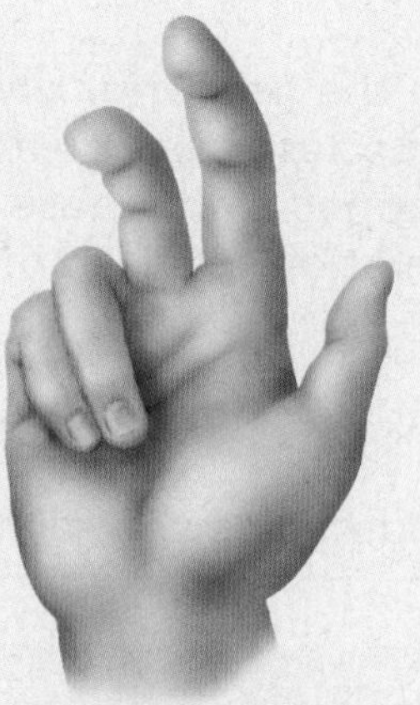

"Hand of benediction"
(a) Median nerve injury in elbow region

Ulnar Nerve Injuries Because of its exposed position at the elbow, the ulnar nerve is vulnerable to injury. Striking the "funny bone"—the spot where this nerve rests against the medial epicondyle—causes tingling of the little finger. Repeated throwing can also damage the nerve here because it stresses the medial part of the elbow region. Even more commonly, the ulnar nerve is injured in the carpal region of the palm, another place where it is superficially located. People with damage to this nerve can neither adduct and abduct their fingers (because the interossei and medial lumbrical muscles are paralyzed) nor form a tight grip (because the hypothenar muscles are useless). Without the interossei muscles to extend the fingers at the interphalangeal joints, the fingers flex at these joints, and the hand contorts into a *clawhand* (figure b).

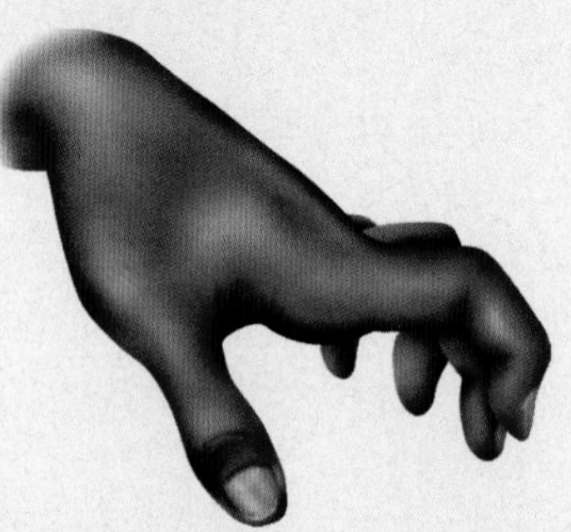

Clawhand
(b) Ulnar nerve injury

Radial Nerve Injuries Trauma to the radial nerve results in **wrist-drop,** an inability to extend the hand at the wrist (figure c). If the lesion occurs far enough superiorly, the triceps muscle is paralyzed, so that the forearm cannot be actively extended at the elbow. Many fractures of the humerus follow the radial groove and harm the radial nerve there. This nerve can also be crushed in the axilla by improper use of a crutch or if a person falls asleep with an arm draped over the back of a chair.

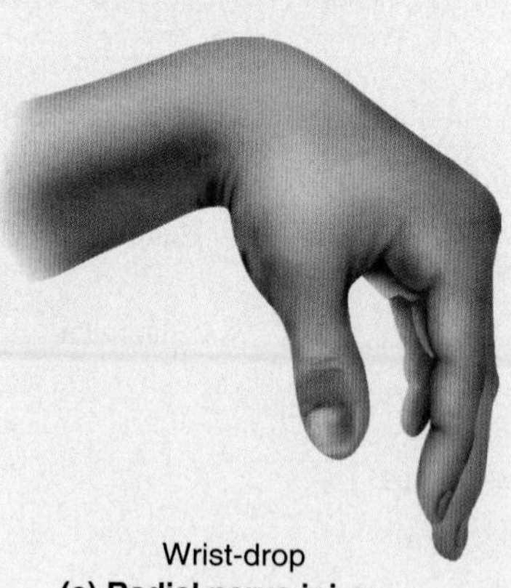

Wrist-drop
(c) Radial nerve injury

Table 14.4 Branches of the Brachial Plexus to Trunk and Shoulder Muscles

Nerves	Cord and Spinal Roots (Ventral Rami)	Structures Innervated
Dorsal scapular	Branches of C_5 rami	Rhomboid muscles and levator scapulae
Long thoracic	Branches of C_5–C_7 rami	Serratus anterior muscle
Subscapular	Posterior cord; branches of C_5 and C_6 rami	Teres major and subscapularis muscles
Thoracodorsal	Posterior cord; branches of C_6–C_8 rami	Latissimus dorsi
Suprascapular	Upper trunk (C_5, C_6)	Shoulder joint; supraspinatus and infraspinatus muscles
Pectoral (lateral and medial)	Branches of lateral and medial cords (C_5–T_1)	Pectoralis major and minor muscles

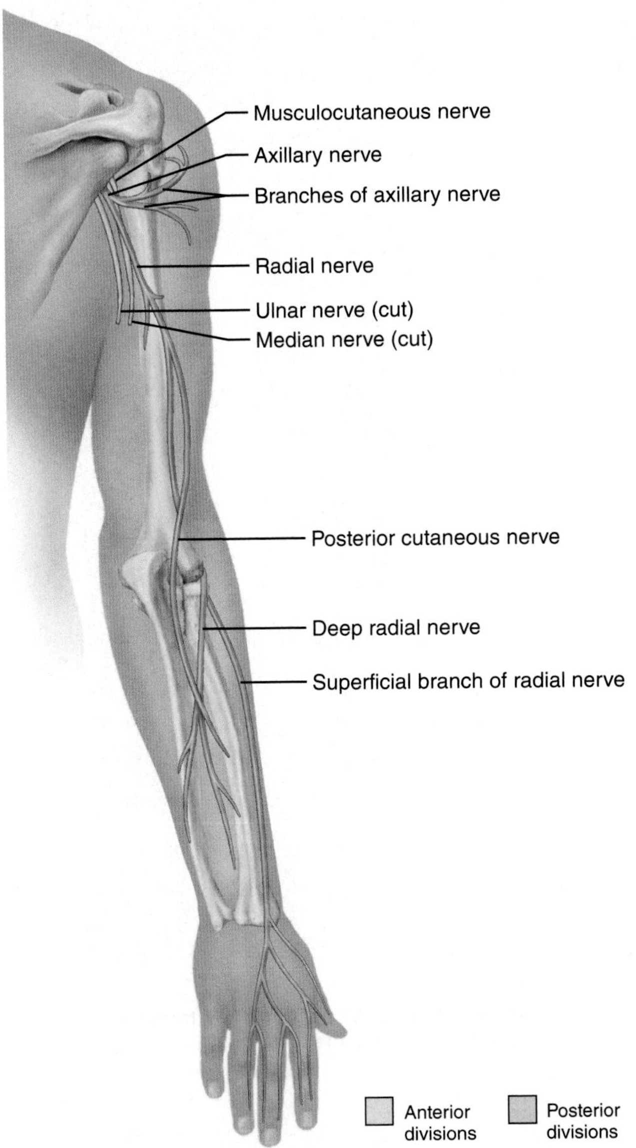

Figure 14.11 Course of the axillary and radial nerves. Posterior view.

Axillary Nerve The **axillary nerve,** a branch of the posterior cord (Figure 14.10 and Figure 14.11), runs posterior to the surgical neck of the humerus and innervates the deltoid and teres minor muscles. Its sensory fibers supply the capsule of the shoulder joint and the skin covering the inferior half of the deltoid muscle.

Radial Nerve The **radial nerve** (Figures 14.10 and 14.11), a continuation of the posterior cord, is the largest branch of the brachial plexus, innervating almost the entire posterior side of the upper limb, including all the limb extensor muscles. As it descends through the arm, it wraps around the posterior humerus in the radial groove, sending branches to the triceps brachii. It then curves anteriorly around the lateral epicondyle at the elbow, where it divides into a superficial and a deep branch. The superficial branch descends along the lateral edge of the radius under cover of the brachioradialis muscle; distally, it supplies the skin on the dorsolateral surface of the hand. The deep branch of the radial nerve runs posteriorly to supply the extensor muscles on the forearm.

Smaller branches off the brachial plexus innervate muscles of the trunk and scapula that move the shoulder **(Table 14.4)**.

✓ check your understanding

- ☐ 11. Would a patient with a spinal injury at C_5 be able to breathe independently?
- ☐ 12. Which nerve innervates all of the extensor muscles of the arm and forearm?
- ☐ 13. A crushing blow to the elbow injured the ulnar nerve where it crosses the medial epicondyle. What sensory and motor deficits will be apparent on examination?
- ☐ 14. Would the patient with the injury at C_5 (described above) have any function in the muscles of the upper limb?

(For answers, see Appendix B.)

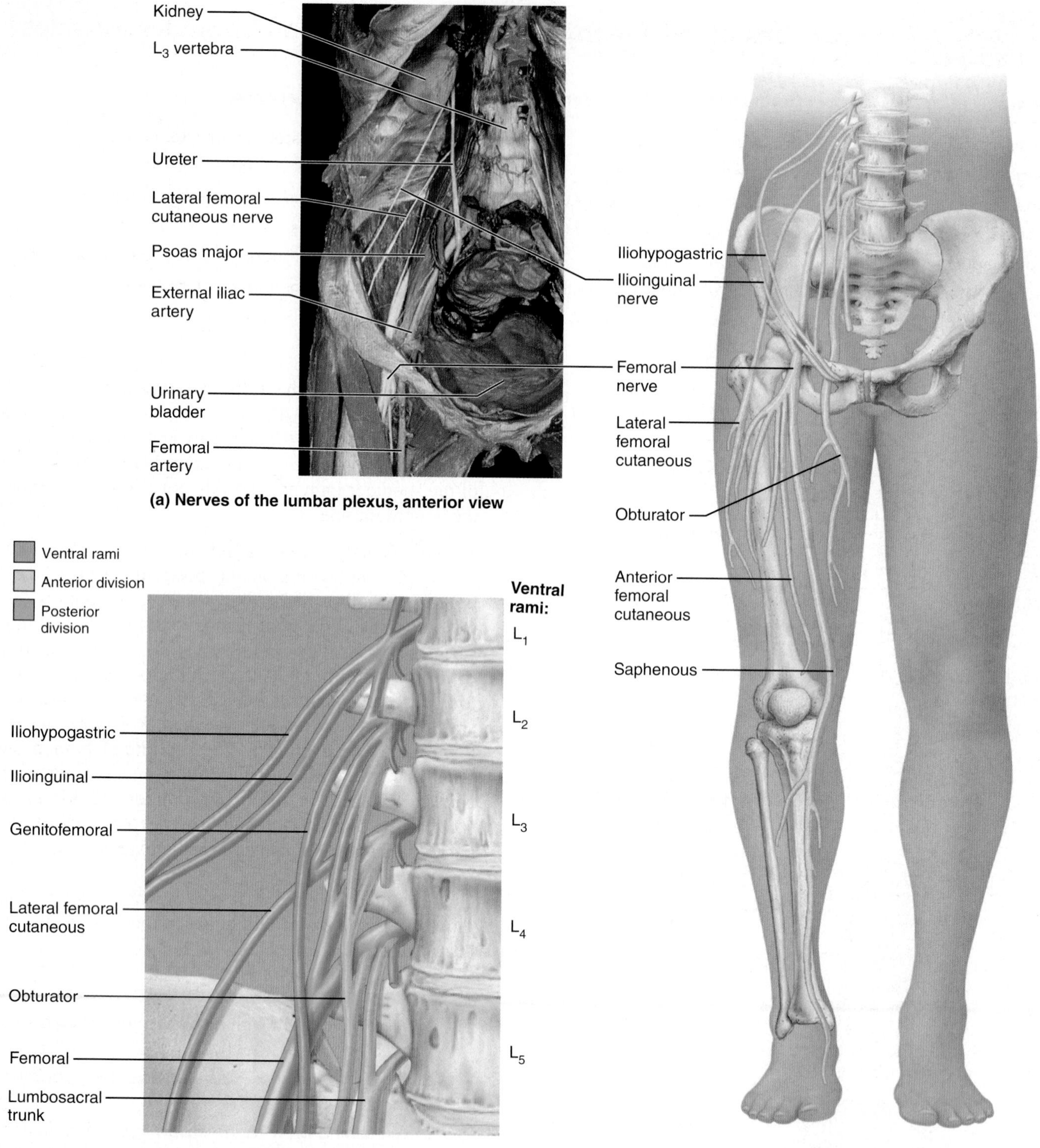

Kidney
L_3 vertebra
Ureter
Lateral femoral cutaneous nerve
Psoas major
External iliac artery
Urinary bladder
Femoral artery

(a) Nerves of the lumbar plexus, anterior view

Ventral rami
Anterior division
Posterior division

Ventral rami:
L_1
L_2
L_3
L_4
L_5

Iliohypogastric
Ilioinguinal
Genitofemoral
Lateral femoral cutaneous
Obturator
Femoral
Lumbosacral trunk

(b) Ventral rami and major branches of the lumbar plexus

Iliohypogastric
Ilioinguinal nerve
Femoral nerve
Lateral femoral cutaneous
Obturator
Anterior femoral cutaneous
Saphenous

(c) Distribution of the major nerves from the lumbar plexus to the lower limb

Figure 14.12 **The lumbar plexus.**

Table 14.5 Branches of the Lumbar Plexus (Figure 14.12)

Nerves	Ventral Rami	Structures Innervated
Lateral femoral cutaneous	L_2–L_3	Skin of lateral thigh; some sensory branches to peritoneum
Iliohypogastric	L_1	Skin on side of buttock and skin on pubis; proprioceptor and motor to the most inferior parts of the internal oblique and transversus abdominis muscles
Ilioinguinal	L_1	Skin of external genitalia and proximal medial aspect of the thigh
Genitofemoral	L_1, L_2	Motor to internal oblique and transversus abdominis muscles. Skin of scrotum in males, of labia majora in females, and of anterior thigh inferior to middle portion of inguinal region; cremaster muscle in males

The Lumbar Plexus and Innervation of the Lower Limb

The **lumbar plexus** (Figure 14.12) arises from the first four lumbar spinal nerves (L_1–L_4) and lies within the psoas major muscle in the posterior abdominal wall. Its smaller branches innervate parts of the abdominal wall and the psoas muscle itself, but the main branches descend to innervate the anterior thigh.

Femoral Nerve The **femoral nerve,** the largest terminal branch, runs deep to the inguinal ligament to enter the thigh. It descends vertically through the center of the femoral triangle (shown in Figure 11.38a, p. 342). From there it divides into several large branches. Motor branches of the femoral nerve innervate the muscles of the anterior compartment of the thigh, including the important quadriceps femoris. Cutaneous branches of the femoral nerve serve the skin of the anterior thigh and the medial surface of the leg from the knee to the foot.

CLINICAL APPLICATION

Compression of Lumbar Spinal Nerves

Compression of the spinal roots of the lumbar plexus by a herniated disc causes a major disturbance in gait. This is because the femoral nerve innervates muscles that flex the thigh at the hip (rectus femoris and iliacus), and muscles that extend the leg at the knee (the entire quadriceps femoris group). Other symptoms are pain or anesthesia of the anterior thigh and of the medial thigh if the obturator nerve is impaired.

Obturator Nerve The **obturator nerve** passes through the large obturator foramen of the pelvis, enters the medial compartment of the thigh, and innervates the adductor muscle group plus some skin on the superomedial thigh.

These major branches of the lumbar plexus are summarized in *Focus on Innervation of the Lower Limb* (Figure 14.13). Smaller branches from the lumbar plexus supply cutaneous innervation to portions of the thigh and buttock, and structures of the perineum (Table 14.5).

The Sacral Plexus and Innervation of the Lower Limb

The **sacral plexus** (Figure 14.14, p. 458) arises from spinal nerves L_4–S_4 and lies immediately caudal to the lumbar plexus. Because some fibers from the lumbar plexus contribute to the sacral plexus via the **lumbosacral trunk,** the two plexuses are often considered together as the *lumbosacral plexus.* Of the dozen named branches of the sacral plexus, about half serve the buttock and lower limb, whereas the rest innervate parts of the pelvis and perineum. The most important branches are described next and summarized in *Focus on Innervation of the Lower Limb* (see Figure 14.13).

Sciatic Nerve The largest branch is the **sciatic** (si-at′ik) **nerve,** the thickest and longest nerve in the body (Figures 14.13 and 14.14c). It supplies all of the lower limb except the anterior and medial regions of the thigh. Actually, the sciatic is two nerves—the *tibial* and *common fibular nerves*—wrapped in a common sheath. It leaves the pelvis by passing through the greater sciatic notch, then courses deep to the broad gluteus maximus muscle and enters the thigh just medial to the hip joint (*sciatic* literally means "of the hip"). From there it descends through the posterior thigh deep to the hamstrings, which it innervates. Superior to the knee region, it branches into the tibial nerve and the common fibular nerve.

Tibial Nerve The **tibial nerve** (Figures 14.13 and 14.14c) courses through the popliteal fossa (the region just posterior to the knee joint, shown in Figure 11.39, p. 343), and then descends through the calf deep to the soleus muscle. At the ankle, it passes posterior to the medial malleolus and divides into the two main nerves of the sole of the foot, the medial and lateral **plantar nerves.** The tibial nerve and its branches supply almost all muscles in the posterior region of the lower limb and supply cutaneous innervation to (1) the skin of the sole of the foot, through the plantar nerves, and (2) a vertical strip of skin along the posterior leg through a **sural nerve,** to which the common fibular nerve also contributes.

Common Fibular Nerve The **common fibular nerve** (Figures 14.13 and 14.14c), or *common peroneal nerve* (*perone* = *fibula*), supplies most structures on the anterolateral

(Text continues on page 459.)

Figure 14.13

The nerves from the lumbar plexus innervate the muscles in the anterior and medial compartments of the thigh, and supply cutaneous innervation to anterior and medial portions of the thigh and leg.

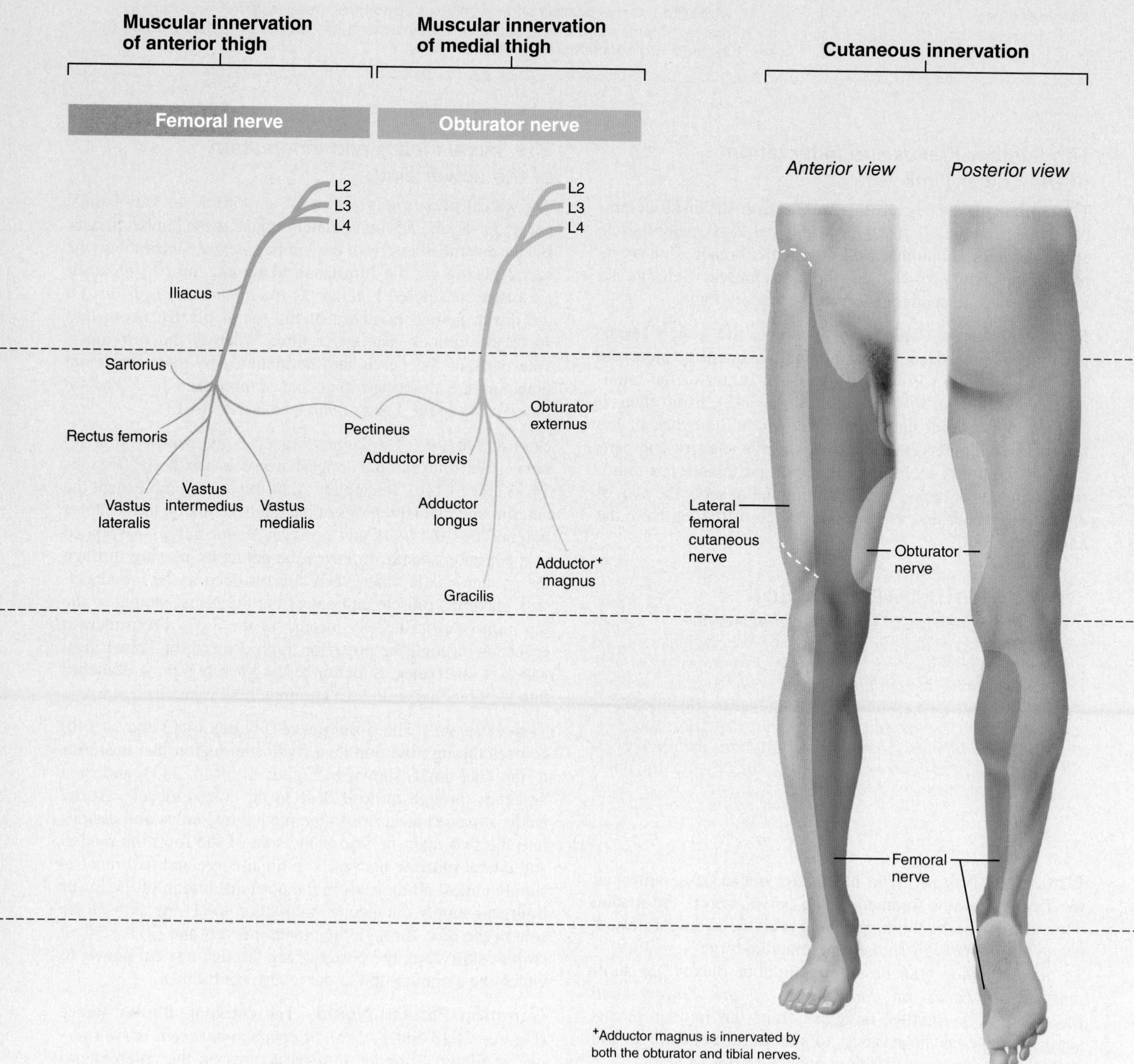

The nerves from the sacral plexus innervate the muscles in the posterior compartment of the thigh and all of the muscles in the leg, and supply cutaneous innervation to the lateral and posterior portions of the leg and to the foot.

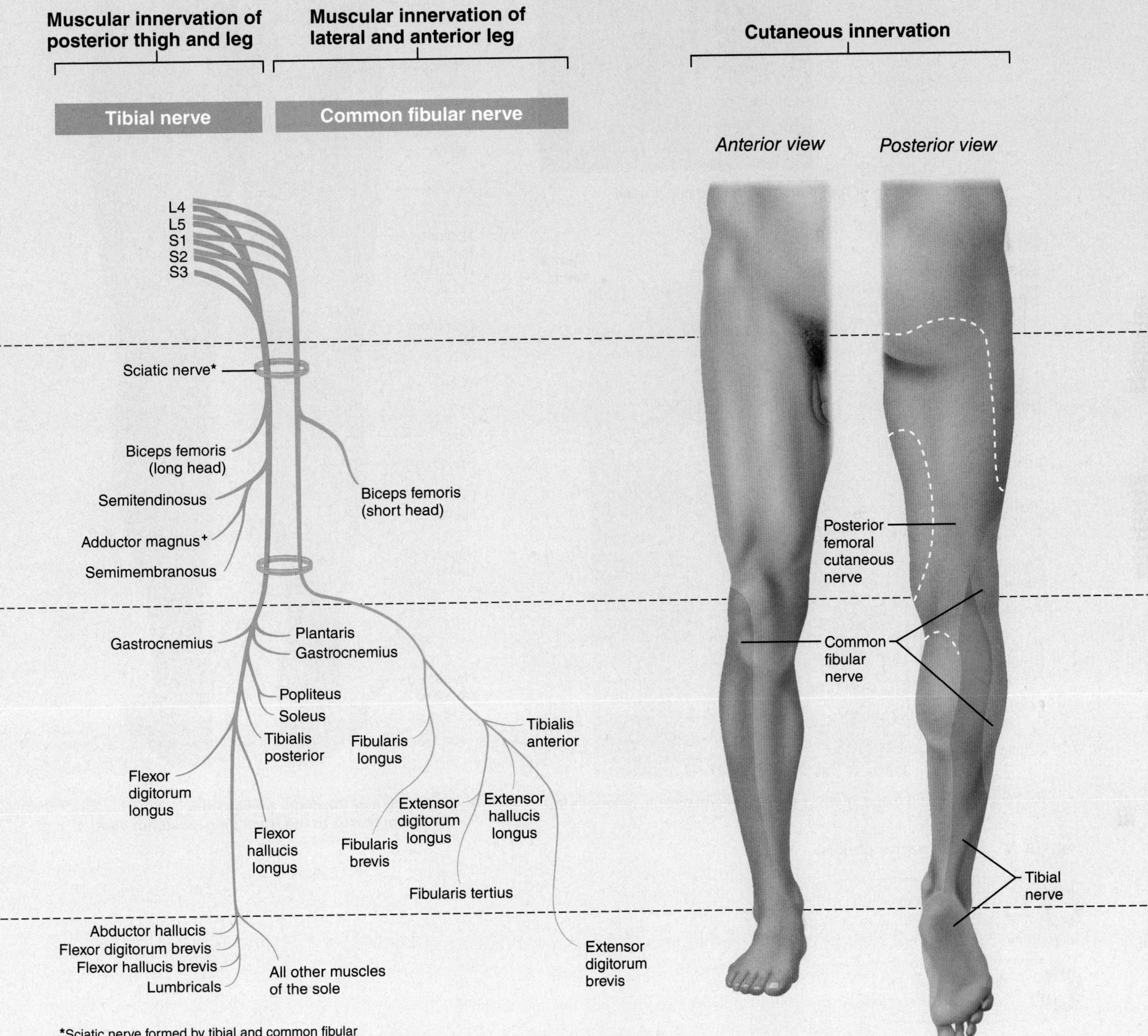

*Sciatic nerve formed by tibial and common fibular nerves wrapped by common sheath.

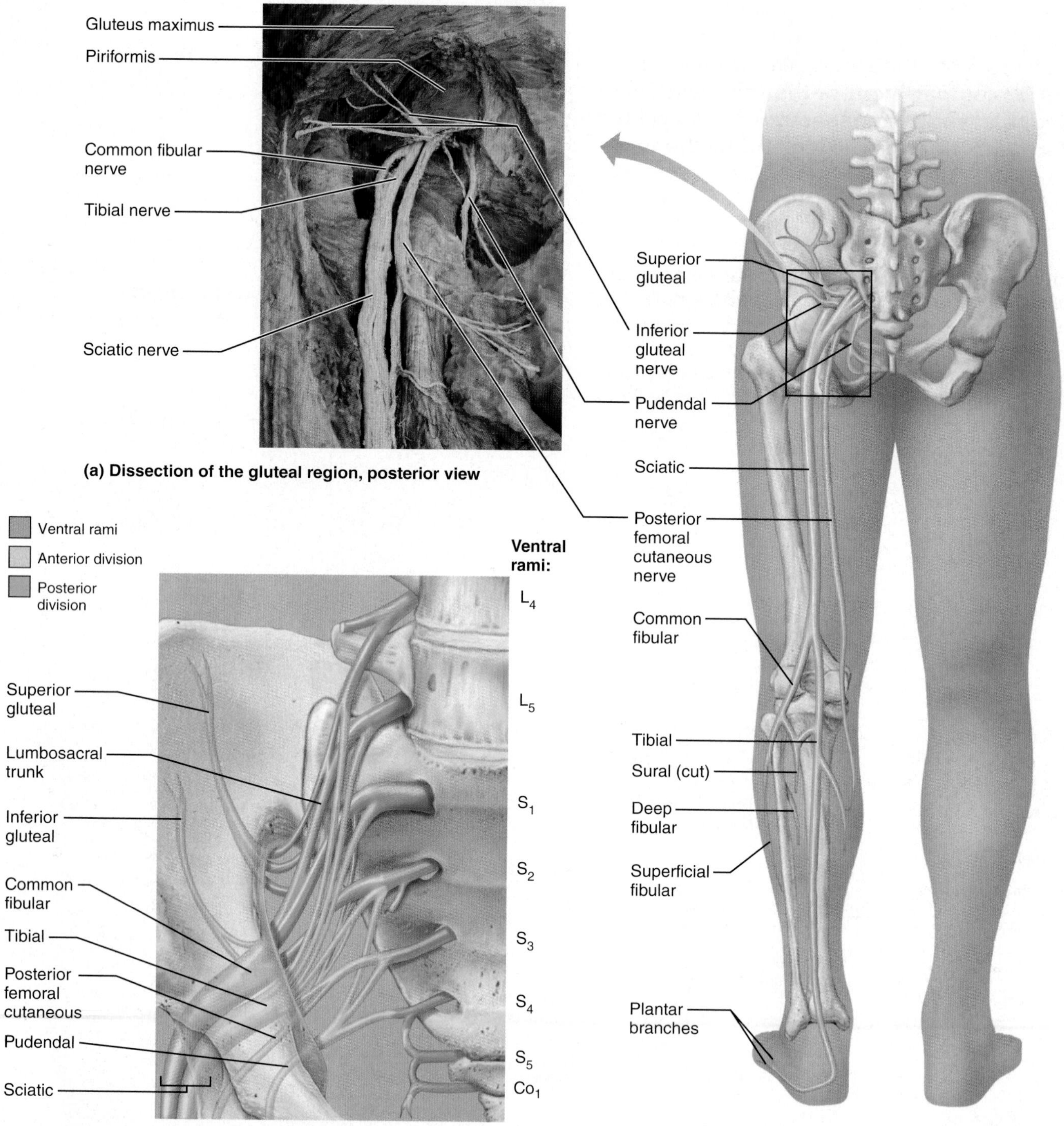

(a) Dissection of the gluteal region, posterior view

(b) Ventral rami and major branches of the sacral plexus, anterior view

(c) Distribution of the major nerves from the sacral plexus to the lower limb, posterior view

Figure 14.14 **The sacral plexus.**

Explore PAL Cadaver
MasteringA&P®

Watch Cat Dissection
MasteringA&P®

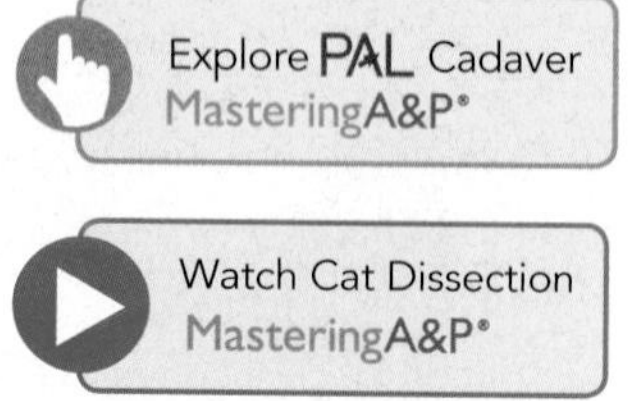

Table 14.6 Additional Branches of the Sacral Plexus (Figure 14.13)

Nerves	Ventral Rami	Structures Innervated
Superior gluteal	L_4, L_5, S_1	Motor branches to gluteus medius and minimus and tensor fasciae latae
Inferior gluteal	L_5–S_2	Motor branches to gluteus maximus
Posterior femoral cutaneous	S_1–S_3	Skin of most inferior part of buttock, posterior thigh, and popliteal region; length variable; may also innervate part of skin of calf and heel
Pudendal	S_2–S_4	Supplies most of skin and muscles of perineum (region encompassing external genitalia and anus and including clitoris, labia, and vaginal mucosa in female, and scrotum and penis in males); external anal sphincter

aspect of the leg. It descends laterally from its point of origin in the popliteal fossa and enters the superior part of the leg, where it can be palpated as it wraps around the neck of the fibula. It then divides into deep and superficial branches. The **superficial fibular (peroneal) nerve** supplies the fibular muscles in the lateral compartment of the leg and most of the skin on the superior surface of the foot. The **deep fibular (peroneal) nerve** serves the muscles of the anterior compartment of the leg—the extensors that dorsiflex the foot.

CLINICAL APPLICATION

Sciatic Nerve Injuries The sciatic nerve can be injured in many ways. **Sciatica** (si-at′ĭ-kah), characterized by a stabbing pain over the course of this nerve, often occurs when a herniated lumbar disc presses on the sacral dorsal roots within the vertebral canal. Furthermore, wounds to the buttocks or posterior thigh can sever the sciatic nerve, in which case the leg cannot be flexed at the knee (because the hamstrings are paralyzed) and the foot cannot be moved at the ankle. When only the tibial nerve is injured, the paralyzed muscles in the calf cannot plantar flex the foot, and a shuffling gait results. When the common fibular nerve is damaged, dorsiflexion is lost, and the foot drops into plantar flexion, a condition called **footdrop.** The common fibular nerve is susceptible to injury in the superolateral leg, where it is superficially located and easily crushed against the neck of the fibula.

Additional Branches of the Sacral Plexus Except for the sciatic nerve, the largest branches of the sacral plexus are the **superior** and **inferior gluteal nerves,** which innervate the gluteal muscles (Figure 14.14). Other branches of the sacral plexus supply the lateral rotators of the thigh and the muscles of the pelvic floor and perineum (Table 14.6). The **pudendal nerve** (pu-den′dal; "shameful") innervates the muscles and skin of the perineum, helps stimulate erection, and is responsible for voluntary inhibition of defecation and urination (Table 11.9, p. 296). As it passes anteriorly to enter the perineum, the pudendal nerve lies just medial to the ischial tuberosity. The pudendal nerve may be blocked with anesthetic, either to help block pain in the perineum during childbirth, or as preparation for surgery on the anal or genital regions. In a **pudendal nerve block,** the injection needle is inserted just medial to the ischial tuberosity.

Innervation of Joints of the Body

Because injuries to joints are common, health professionals should know which nerves serve which synovial joints. Memorizing so much information is almost impossible, however, because every joint capsule receives sensory branches from several different nerves.

The most helpful guideline for deducing which nerves supply a given joint is provided by **Hilton's law:** *Any nerve that innervates a muscle producing movement at a joint also innervates the joint itself (and the skin over it).* Applied to the knee joint, for example, the process of deduction would be as follows: Because the knee is crossed by the quadriceps, hamstring, and gracilis muscles (see Table 11.15, p. 315), and because the nerves to these muscles are the femoral, obturator, and branches of the sciatic nerves (see Figure 14.13), then all these nerves innervate the knee joint as well.

Innervation of the Skin: Dermatomes

The area of skin innervated by the cutaneous branches from a single spinal nerve is called a **dermatome,** literally a "skin segment." All spinal nerves except C_1 participate in dermatomes. The map of dermatomes (Figure 14.15) was constructed by recording areas of numbness in patients who had experienced injuries to specific spinal roots. In the *trunk* region, the dermatomes are almost horizontal, relatively uniform in width, and in direct line with their spinal nerves. In the *limbs,* however, the pattern of dermatomes is less straightforward. In the upper limb, the skin is supplied by the nerves participating in the brachial plexus: the ventral rami of C_5 to T_1 or T_2. In the lower limb, lumbar nerves supply most of the anterior surface, whereas sacral nerves supply much of the posterior surface. This distribution basically reflects the areas supplied by the lumbar and sacral plexuses, respectively.

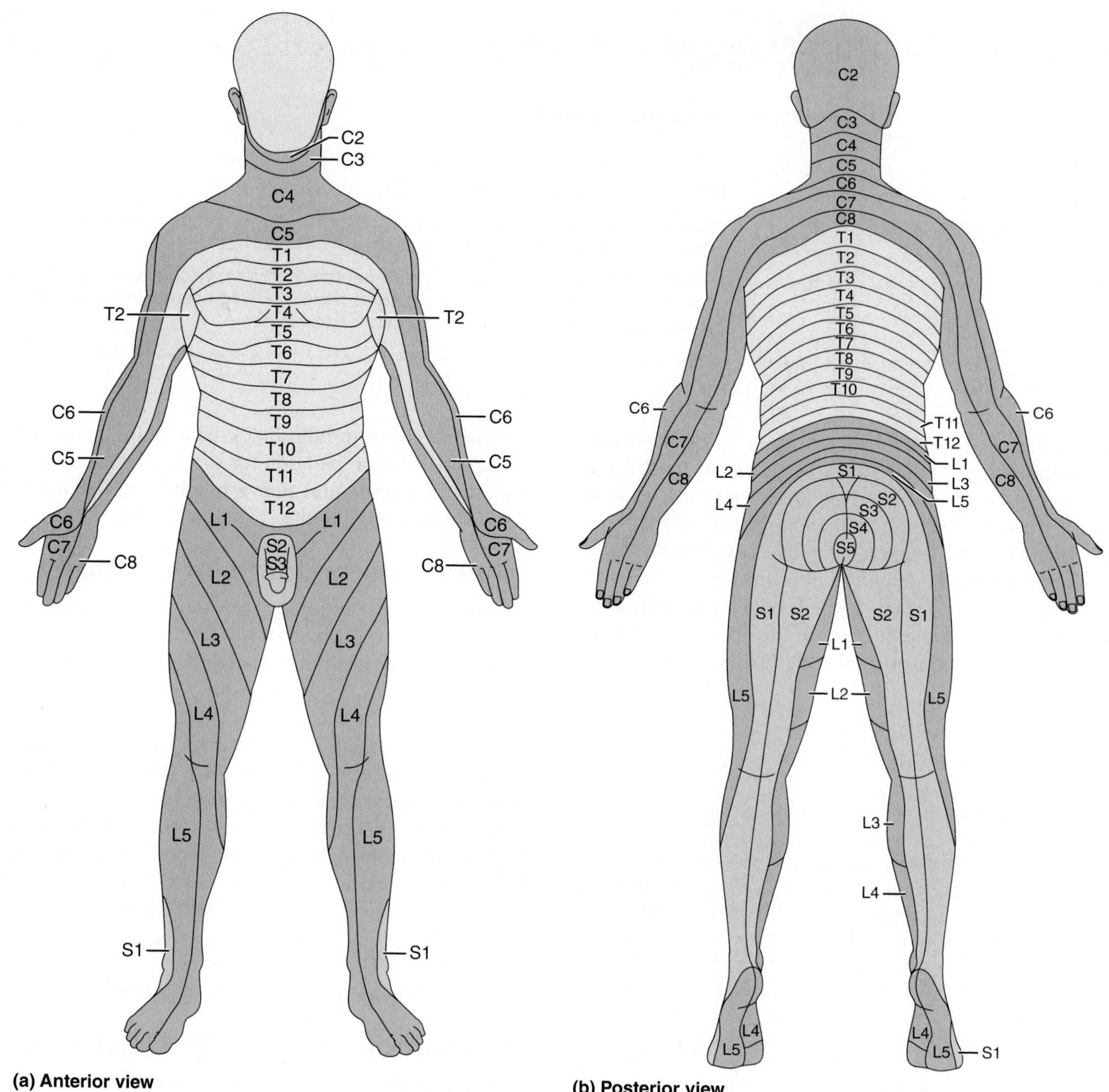

Figure 14.15 Map of dermatomes. A dermatome is an area of skin innervated by the sensory fibers of a single spinal nerve (both ventral and dorsal rami). In the limbs there is considerable overlap between adjacent dermatomes.

Adjacent dermatomes are not as clearly demarcated from one another as a dermatome map indicates. On the trunk, neighboring dermatomes overlap each other by a full 50%; as a result, destruction of a single spinal nerve does not produce complete numbness anywhere. On the limbs, by contrast, the overlap is not complete: Some skin patches are innervated by one spinal nerve only.

☑ check your understanding

☐ 15. Name the three nerves that innervate the muscles of the thigh, and identify the muscle compartment innervated by each.

☐ 16. Using Hilton's law, identify the nerves that innervate the elbow joint.

☐ 17. Explain why a disc herniation at the level of lumbar vertebra L_5 can cause loss of sensation on the anterolateral surface of the leg.

☐ 18. Shingles blisters that encircle the abdomen at the level of the navel indicate viral infection in which spinal nerve?

(For answers, see Appendix B.)

CLINICAL APPLICATION

Clinical Importance of Dermatomes Clinicians use dermatomes in at least two ways. First, dermatomes can help pinpoint the level of spinal injuries. Numbness in all dermatomes below T_1, for example, indicates a spinal cord injury at T_1. Additionally, because the varicella-zoster virus that produces shingles (see next section) resides in a single dorsal root ganglion, a flare-up of this infection may produce skin blisters along the entire course of the associated dermatome (see photo). Finally, dermatomes may also have surgical applications, because anesthetic agents may be injected into specific spinal nerves or roots to desensitize specific skin regions during surgery.

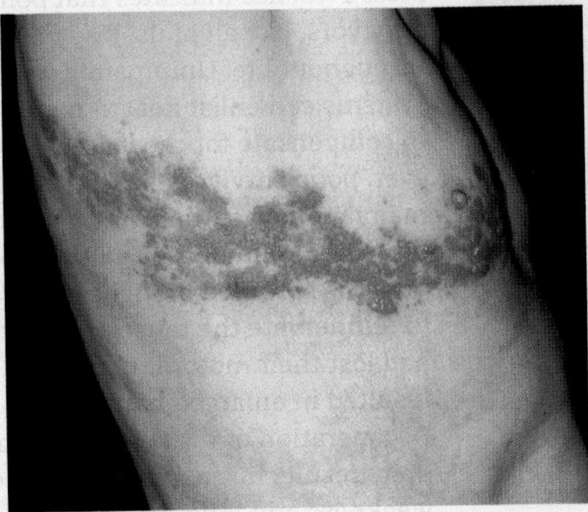

Outbreak of shingles along the T_4 dermatome.

DISORDERS OF THE PERIPHERAL NERVOUS SYSTEM

learning outcome

▶ Explain the causes and symptoms of shingles, migraine headaches, and peripheral neuropathy.

This section considers two disorders that involve sensory neurons (shingles and migraine headaches) and one disorder that can affect both sensory and motor innervation (peripheral neuropathy). For information on other selected motor disorders—polio and postpolio syndrome—see A CLOSER LOOK.

Shingles

Shingles (herpes zoster) is a varicella-zoster viral infection of sensory neurons innervating the skin. It is characterized by a rash of scaly, painful blisters usually confined to a narrow strip of skin on one side of the trunk, usually to one or several adjacent dermatomes (see Clinical Application: Importance of Dermatomes). The disease stems from a childhood infection of chicken pox, during which the viruses are transported from the skin lesions, through the peripheral processes of the sensory neurons, to the cell bodies in a sensory ganglion. There, in the nuclei of the nerve cells, the viruses remain dormant for many years, held in check by the immune system. When the immune system is weakened as a result of age, stress, or another medical condition (HIV infection, cancer treatment, organ transplant), the viruses can multiply and travel back through the sensory axons to the skin, producing the inflammatory rash of shingles. The attacks last for several weeks. In a minority of cases, the attacks recur, separated by periods of healing and remission. Shingles affects 1–2 of every 1000 people, mostly those over 50 years of age. Treatment is to administer pain-relieving drugs and, under certain conditions, antiviral medicines.

Following a shingles outbreak, some patients experience **postherpetic neuralgia,** a condition that occurs if the zoster virus damages the sensory nerve fibers and disrupts the sensory pathways. Sensory signals from the skin are exaggerated, and even light touch (as from clothing) or a slight temperature change (such as stepping outdoors) causes excruciating pain. This condition can last for months and can be debilitating, as the pain is severe and unpredictable.

A zoster virus vaccine, similar to the chicken pox vaccine used for children, has been shown to be effective in reducing the incidence of shingles in people over age 60 by 50% compared to a control group.

Migraines

Migraine headaches are extremely painful, episodic headaches that affect 29.5 million Americans. The cause of migraines was long debated, but they are now known to relate to the sensory innervation of the brain's cerebral arteries by the trigeminal nerve: A signal from the brain stem causes the sensory nerve endings of the ophthalmic division to release chemicals onto the cerebral arteries that they innervate, signaling these arteries to dilate (widen). This dilation then compresses and irritates these sensory nerve endings, causing the headache. Triptans and ergot derivative drugs relieve migraines by constricting blood vessels to the brain. These may be prescribed in addition to over-the-counter pain relief, such as acetaminophen or naproxen.

Peripheral Neuropathy

Peripheral neuropathy refers to any pathologic condition of the peripheral nerves that disrupts nerve function. If an individual nerve is affected, the condition is called a *mononeuropathy;* if multiple nerves are involved, it is termed a *polyneuropathy*. Symptoms of a neuropathy vary according to the nerve fibers affected. Symptoms of sensory nerve involvement include paresthesia, severe pain, burning, or loss of feeling. If motor fibers are affected, muscle weakness and paralysis result. Peripheral neuropathies may be caused by trauma to individual nerves or by repetitive-use injuries that compress nerves. Many systemic disorders (diabetes, autoimmune disorders, HIV, vitamin B deficiency, and alcoholism) can also cause peripheral neuropathy. In fact, 70% of all diabetics have some nerve dysfunction. Treatment of peripheral neuropathy is directed toward treating the cause of the nerve dysfunction and relief of symptoms, most notably pain relief.

and some smooth muscle within the eye; also carry proprioceptive impulses from the skeletal muscles served.

IV. The trochlear nerves: primarily motor; emerge from the dorsal midbrain and carry motor and proprioceptive impulses to and from superior oblique muscles of the eyes.

V. The trigeminal nerves: contain both sensory and motor fibers; emerge from the pons; the main general sensory nerves of the face, nasal cavity, and mouth; three divisions: ophthalmic, maxillary, and mandibular; the mandibular branch contains motor fibers that innervate the chewing muscles.

VI. The abducens nerves: primarily motor; emerge from the pons and supply motor and proprioceptive innervation to the lateral rectus muscles of the eyes.

VII. The facial nerves: contain both sensory and motor fibers; emerge from the pons; motor nerves to muscles of facial expression, sublingual and submandibular salivary glands, and the lacrimal glands; also carry sensory impulses from the taste buds of anterior two-thirds of the tongue.

VIII. The vestibulocochlear nerves: primarily sensory; enter brain at the pons-medulla border; transmit impulses from the hearing and equilibrium receptors of the inner ears.

IX. The glossopharyngeal nerves: contain both sensory and motor fibers; issue from the medulla; transmit sensory impulses from the posterior tongue (including taste) and the pharynx; innervate the stylopharyngeus muscle and parotid gland.

X. The vagus nerves: contain both sensory and motor fibers; arise from the medulla; most of its motor fibers are autonomic; carry motor fibers to, and sensory fibers from, the pharynx, larynx, and visceral organs of the thorax and abdomen.

XI. The accessory nerves: primarily motor; arise from the cervical spinal cord; supply the trapezius and sternocleidomastoid.

XII. The hypoglossal nerves: primarily motor; issue from the medulla; supply motor innervation to the tongue muscles.

10. The cranial nerves can be grouped according to function. Some are *sensory*, containing special somatic sensory fibers (I, II, VIII); some are primarily *motor*, containing motor fibers (III, IV, VI, XI, and XII); and some are *mixed*, containing both sensory and motor fibers (V, VII, IX, X).

SPINAL NERVES (pp. 444–461)

11. The 31 pairs of spinal nerves are numbered according to the region of the vertebral column from which they issue: cervical, thoracic, lumbar, sacral, and coccygeal. There are 8 cervical, 12 thoracic, 5 lumbar, 5 sacral, and 1 coccygeal spinal nerves.
12. Each spinal nerve forms from the union of a dorsal sensory root and a ventral motor root in an intervertebral foramen. On each dorsal root is a dorsal root ganglion containing the cell bodies of sensory neurons. The branches of each spinal nerve are the dorsal and ventral rami (somatic branches) and rami communicantes (visceral branches).
13. Dorsal rami serve the muscles and skin of the posterior trunk, whereas ventral rami serve the muscles and skin of the lateral and anterior trunk. Ventral rami also serve the limbs.
14. The intrinsic back muscles and overlying skin, from neck to sacrum, are innervated by the dorsal rami in a neatly segmented pattern.
15. The anterior and lateral wall of the thorax (and abdomen) are innervated by thoracic ventral rami—the segmented intercostal and subcostal nerves. Thoracic ventral rami do not form nerve plexuses.
16. The major nerve plexuses are networks of successive ventral rami that intermix fibers. They primarily innervate the limbs.
17. The cervical plexus (C_1–C_4) innervates the muscles and skin of the neck and shoulder. Its phrenic nerve (C_3–C_5) serves the diaphragm.
18. The brachial plexus serves the upper limb, including the muscles of the shoulder girdle. It arises primarily from C_5–T_1. From proximal to distal, the brachial plexus consists of "roots" (ventral rami), trunks, divisions, and cords.
19. The main nerves arising from the brachial plexus are the musculocutaneous (to flexors on the arm), median (to anterior forearm muscles and the lateral palm), ulnar (to anteromedial muscles of the forearm and the medial hand), axillary (to the deltoid and teres minor muscles), and radial (to the posterior part of the upper limb).
20. The lumbar plexus (L_1–L_4) innervates the anterior and medial muscles of the thigh, through the femoral and obturator nerves, respectively. The femoral nerve also innervates the skin on the anterior thigh and medial leg.
21. The sacral plexus (L_4–S_4) supplies the muscles and skin of the posterior thigh and almost all of the leg. Its main branch is the large sciatic nerve, which consists of the tibial nerve (to most of the hamstrings and all muscles of the calf and sole) and the common fibular nerve (to muscles of the anterior and lateral leg and the overlying skin). Other branches of the sacral plexus supply the posterior and lateral muscles of the pelvic girdle (such as the gluteal muscles) and the perineum (pudendal nerve).
22. The sensory nerves to a joint are branches of the nerves innervating the muscles that cross that joint (Hilton's law).
23. A dermatome is a segment of skin innervated by the sensory fibers of a single spinal nerve. Adjacent dermatomes overlap to some degree, more so on the trunk than on the limbs. Loss of sensation in specific dermatomes reveals sites of damage to spinal nerves or the spinal cord.

PAL Human Cadaver/Nervous System/Peripheral Nervous System

DISORDERS OF THE PERIPHERAL NERVOUS SYSTEM (pp. 461–462)

Shingles (p. 461)

24. Shingles is a painful disorder in the sensory nerve fibers caused by the varicella-zoster virus (the virus that causes chicken pox). Reactivation of the childhood virus causes painful blisters along the dermatome of the infected spinal nerve.

Migraines (p. 461)

25. Migraines are extremely painful, prolonged, and episodic headaches. They are caused by sensory innervation to the cerebral arteries causing vasodilation, which compresses sensory nerve endings.

Peripheral Neuropathy (p. 461)

26. Peripheral neuropathy is any pathological condition of the peripheral nerves that disrupts nerve function. Trauma or systemic disorders can cause sensory or motor neuropathy.

THE PERIPHERAL NERVOUS SYSTEM THROUGHOUT LIFE (pp. 462–463)

27. During embryonic development, each spinal nerve grows out between newly formed vertebrae to provide the motor innervation of an adjacent myotome (future trunk muscle) and the sensory innervation of the adjacent skin region (dermatome).

REVIEW QUESTIONS

Multiple Choice/Matching Questions

(For answers, see Appendix B.)

1. Proprioceptors include all of the following except (a) muscle spindles, (b) tendon organs, (c) epithelial tactile complexes, (d) lamellar corpuscles in joint capsules.

2. The large, onion-shaped pressure receptors in deep connective tissues are (a) epithelial tactile complexes, (b) lamellar corpuscles, (c) free nerve endings, (d) bulbous corpuscles.

3. Choose the correct receptor type from column B for each description given in column A.

Column A	Column B
____ **(1)** pain, itch, and temperature receptors	(a) bulbous corpuscles
____ **(2)** contain intrafusal fibers and flower spray endings	(b) tendon organs
____ **(3)** discriminative touch receptors in hairless skin (fingertips)	(c) muscle spindles
____ **(4)** contain nerve endings wrapped around thick collagen bundles in a tendon	(d) free nerve ending
____ **(5)** rapidly adapting deep-pressure mechanoreceptors	(e) lamellar corpuscles
____ **(6)** slowly adapting deep-pressure mechanoreceptors	(f) tactile corpuscles

4. Choose the correct cranial nerves from the key for each description below.

Key:

(a) abducens	(g) olfactory
(b) accessory	(h) optic
(c) facial	(i) trigeminal
(d) glossopharyngeal	(j) vestibulocochlear
(e) hypoglossal	(k) vagus
(f) oculomotor	(l) trochlear

____ **(1)** innervates four extrinsic eye muscles
____ **(2)** is the major sensory nerve of the face
____ **(3)** innervates the sternocleidomastoid and trapezius
____ **(4)** are primarily or exclusively sensory (three nerves)
____ **(5)** innervates the tongue muscles
____ **(6)** allows chewing of food
____ **(7)** is anesthetized during dental work
____ **(8)** helps to regulate heart activity
____ **(9)** sensory innervation for hearing and equilibrium
____ **(10)** contain parasympathetic motor fibers (four nerves)
____ **(11)** innervates muscles of facial expression

5. For each of the following muscles or body regions, identify the plexus and the peripheral nerve (or branch of one) involved. Use one choice from Key A followed by one choice from Key B.

Key A: Plexuses

(a) brachial
(b) cervical
(c) lumbar
(d) sacral

Key B: Nerves

(1) common fibular	**(6)** phrenic
(2) femoral	**(7)** radial
(3) median	**(8)** tibial
(4) musculocutaneous	**(9)** ulnar
(5) obturator	

Structure Innervated

____, ____ **(1)** the diaphragm
____, ____ **(2)** muscles of the posterior compartments of thigh and leg
____, ____ **(3)** anterior compartment thigh muscles
____, ____ **(4)** medial compartment thigh muscles
____, ____ **(5)** anterior arm muscles that flex the forearm
____, ____ **(6)** muscles that flex the wrist and fingers (two nerves)
____, ____ **(7)** muscles that extend the wrist and fingers
____, ____ **(8)** skin and extensor muscles of the posterior arm
____, ____ **(9)** fibular muscles, tibialis anterior, and toe extensors

6. Which one of the following contains only motor fibers? (a) dorsal root, (b) dorsal ramus, (c) ventral root, (d) ventral ramus.

7. The trigeminal nerve contains which class(es) of nerve fibers? (a) somatic sensory only, (b) somatic motor and proprioceptor, (c) general sensory and somatic motor.

8. Whereas *bronchial* refers to the airways in the lungs, *brachial* refers to the (a) back, (b) pharynx, (c) arm, (d) thigh.

9. Which cranial nerves contain somatic motor axons to pharyngeal arch muscles (see p. 434)? (a) I, II, VII; (b) V, VII, IX, X; (c) III, IV, VI, XI, XII.

10. Which of the following components occur in all of the mixed cranial nerves V, VII, IX, and X? Choose as many as apply: (a) general somatic sensory; (b) special somatic sensory; (c) general visceral sensory; (d) special visceral sensory (taste), (e) somatic motor to pharyngeal arch muscles.

Short Answer Essay Questions

11. In the sensory receptors called "encapsulated nerve endings," what is the "capsule" made of?

12. (a) Describe the roots to and the composition of a spinal nerve. (b) Are dorsal and ventral roots in the CNS or the PNS? (c) Name the branches of a spinal nerve (other than the rami communicantes), and identify what basic region of the body each branch supplies.

13. (a) Define nerve plexus. (b) List the spinal nerves of origin of the four major nerve plexuses, and name the general body regions served by each plexus.

14. In the brachial plexus, what specific rami make up each of the three trunks?

15. Adrian and Abdul, two anatomy students, were arguing about the facial nerve. Adrian said that it innervates all of the skin of the face, and that it is called the facial nerve for this reason. Abdul claimed the facial nerve does not innervate facial skin at all. Who was correct?

16. Choose the correct answer, and explain why it is correct. Through which pair of intervertebral foramina do spinal nerves L_5 leave the vertebral canal? (a) holes in the sacrum, (b) just superior to vertebra L_5, (c) just superior to S_1, (d) just inferior to S_4.
17. There are 40 roots in the cauda equina. To what spinal nerves do they belong? (See pp. 408–409.)

Critical Reasoning & Clinical Application Questions

1. As Harry was falling off a ladder, he reached out and grabbed a tree branch with his hand. His overstretched upper limb became weak and numb. What major nervous structure had he injured?
2. Frita, in her early 70s, had problems chewing. When she was asked to stick out her tongue, it deviated to the right, and its right half was quite wasted. What nerve was injured?
3. Ted is a war veteran who was hit in the back with small pieces of shrapnel. His skin is numb in the center of his buttocks and along the entire posterior side of a lower limb, but there is no motor problem. Indicate which of the following choices is the most likely site of his nerve injury, and explain your choice: (a) a few dorsal roots of the cauda equina, (b) spinal cord transection at C_6, (c) spinal cord transection at L_5, (d) femoral nerve transected in the lumbar region.
4. A quarterback suffered torn menisci in his right knee joint when he was tackled from the side. The same collision crushed his common fibular nerve against the neck of his right fibula. What locomotor problems can he expect to experience from this nerve injury?
5. In a patient who developed carpal tunnel syndrome, the nerve injury began as a neurapraxia and then became an axonotmesis. (1) Recall what nerve is affected in carpal tunnel syndrome (see Chapter 8, p. 194). (2) Explain the terms *neurapraxia* and *axonotmesis.* (3) Describe the sensory and motor losses experienced. (4) Estimate how long full recovery from this nerve injury will take. (See Related Clinical Terms.)
6. Using Hilton's law, (1) deduce which four nerves send branches that innervate the capsule of the elbow joint, and (2) identify the nerves that supply the hip joint.
7. After suffering from a broken humerus, a patient was unable to extend her wrist and complained of paresthesia on the back of her hand. What nerve must have been injured in the humeral break?
8. Name all of the nerves that innervate the tongue.

The Autonomic Nervous System and Visceral Sensory Neurons 15

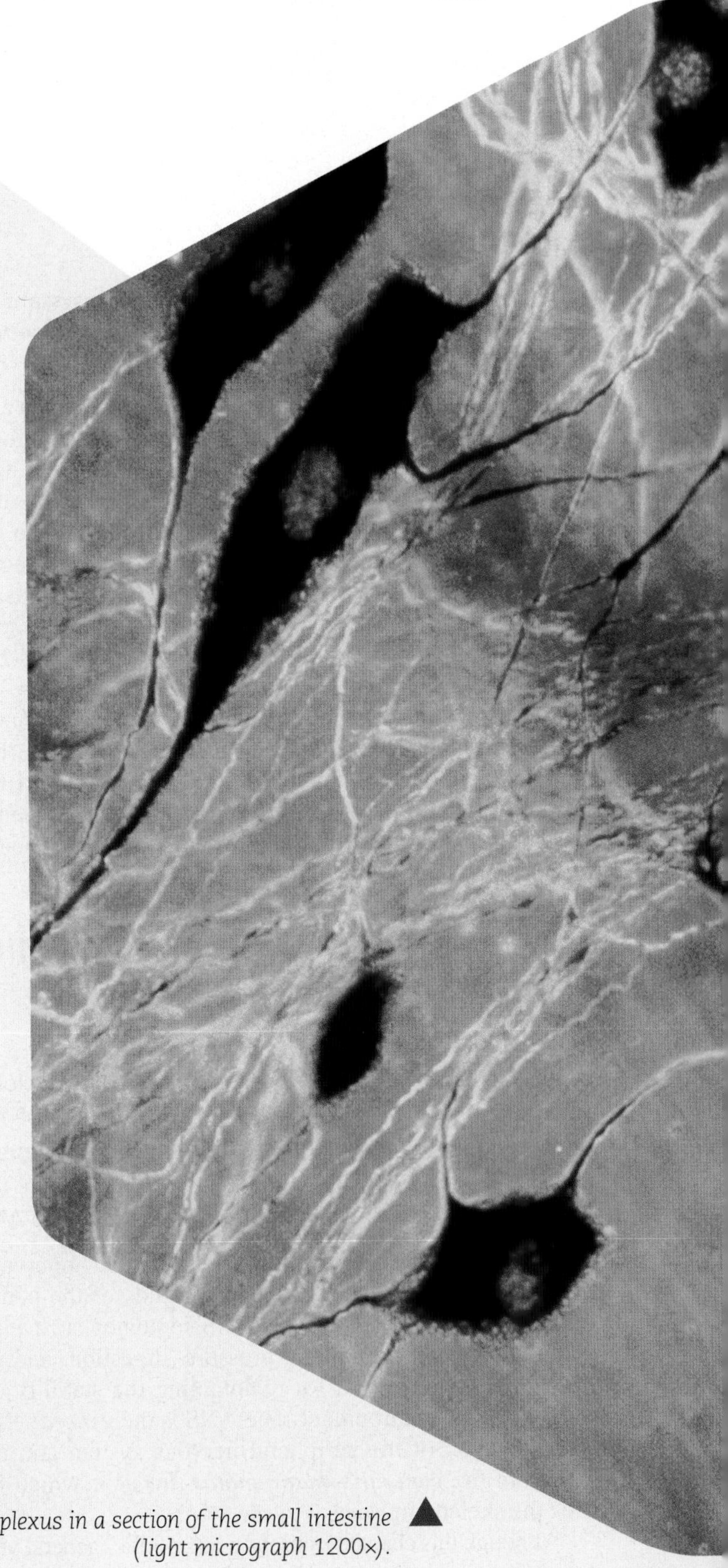

Neurons of the myenteric plexus in a section of the small intestine (light micrograph 1200×).

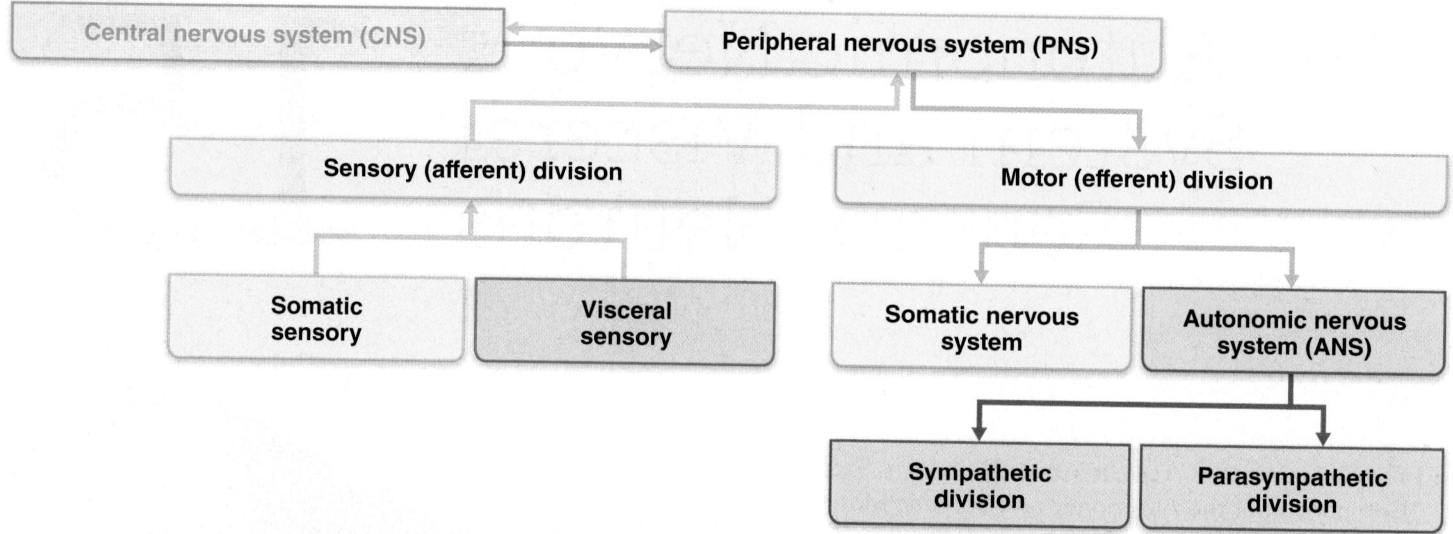

Figure 15.1 Place of the autonomic nervous system (ANS) and visceral sensory components in the structural organization of the nervous system.

Consider the following situations: You wake up at night after having eaten at a restaurant where the food did not taste quite right, and you find yourself waiting helplessly for your stomach to "decide" whether it can hold the food down. A few days later, you are driving to school after drinking too much coffee and wish in vain that your full bladder would stop its uncomfortable contractions. Later that day, your professor asks you a hard question in front of the class, and you try not to let them see you sweat—but the sweat runs down your face anyway. All of these are examples of visceral motor functions that are not easily controlled by the conscious will and that sometimes seem to "have a mind of their own." These functions are performed by the **autonomic nervous system (ANS),** a motor system that does indeed operate with a certain amount of independence (*autonomic* = self-governing).

OVERVIEW OF THE AUTONOMIC NERVOUS SYSTEM

learning outcomes

- Define the autonomic nervous system (ANS), and explain its relationship to the peripheral nervous system as a whole.
- Compare autonomic neurons to somatic motor neurons.
- Describe the basic differences between the parasympathetic and sympathetic divisions of the ANS.

The ANS is the system of motor neurons that innervate the smooth muscle, cardiac muscle, and glands of the body. By controlling these effectors, the ANS regulates such visceral functions as heart rate, blood pressure, digestion, and urination, which are essential for maintaining the stability of the body's internal environment. The ANS is the *general visceral motor* division of the peripheral nervous system and is distinct from the *general somatic motor division,* which innervates the skeletal muscles (Figure 15.1).

Although this chapter focuses on autonomic (general visceral motor) functions, it also considers the **general visceral senses.** The general visceral sensory system continuously monitors the activities of the visceral organs so that the autonomic motor neurons can make adjustments as necessary to ensure optimal performance of visceral functions.

A third component of visceral innervation, the *enteric nervous system,* also innervates smooth muscle and glands, specifically those within the digestive tract. The enteric nervous system regulates the activity of the digestive tract and functions completely independently of the CNS. Autonomic neurons to the digestive tract can influence enteric neurons by either stimulating or inhibiting their activity. This ANS influence acts as a "volume control" rather than as an "on/off" switch. The details of the enteric nervous system and its interaction with the ANS are described with the innervation of the digestive system (Chapter 23).

Before describing the ANS, we need to review two terms. A *synapse* is a junction between two neurons that communicates the message from one neuron, called the *presynaptic neuron,* to another neuron, the *postsynaptic neuron.* A *ganglion* (plural: *ganglia*) is a cluster of neuronal cell bodies in the PNS. (See Chapter 12 for review of these terms.)

Comparison of the Autonomic and Somatic Motor Systems

Other discussions of motor innervation focus largely on the *somatic* motor system, which innervates skeletal muscles. Each somatic motor neuron runs from the central nervous system all the way to the muscle being innervated, and each motor unit consists of a single neuron plus the skeletal muscle cells that it innervates (Figure 15.2). Typical somatic motor axons are thick, heavily myelinated fibers that conduct nerve impulses rapidly.

By contrast, the comparable motor unit in the ANS includes a chain of *two* motor neurons (Figure 15.2). The first of these is called a **preganglionic neuron.** The cell body of this neuron lies within the CNS. Its axon, the **preganglionic axon** (also called a *preganglionic fiber*), synapses with the second motor neuron, the **postganglionic neuron,** in a peripheral **autonomic ganglion.** The **postganglionic axon**

FOCUS Comparing Somatic Motor and Autonomic Innervation

Figure 15.2

Somatic motor and autonomic innervations differ anatomically in two major ways: (1) the target organs innervated and (2) the number of neurons in the pathway.

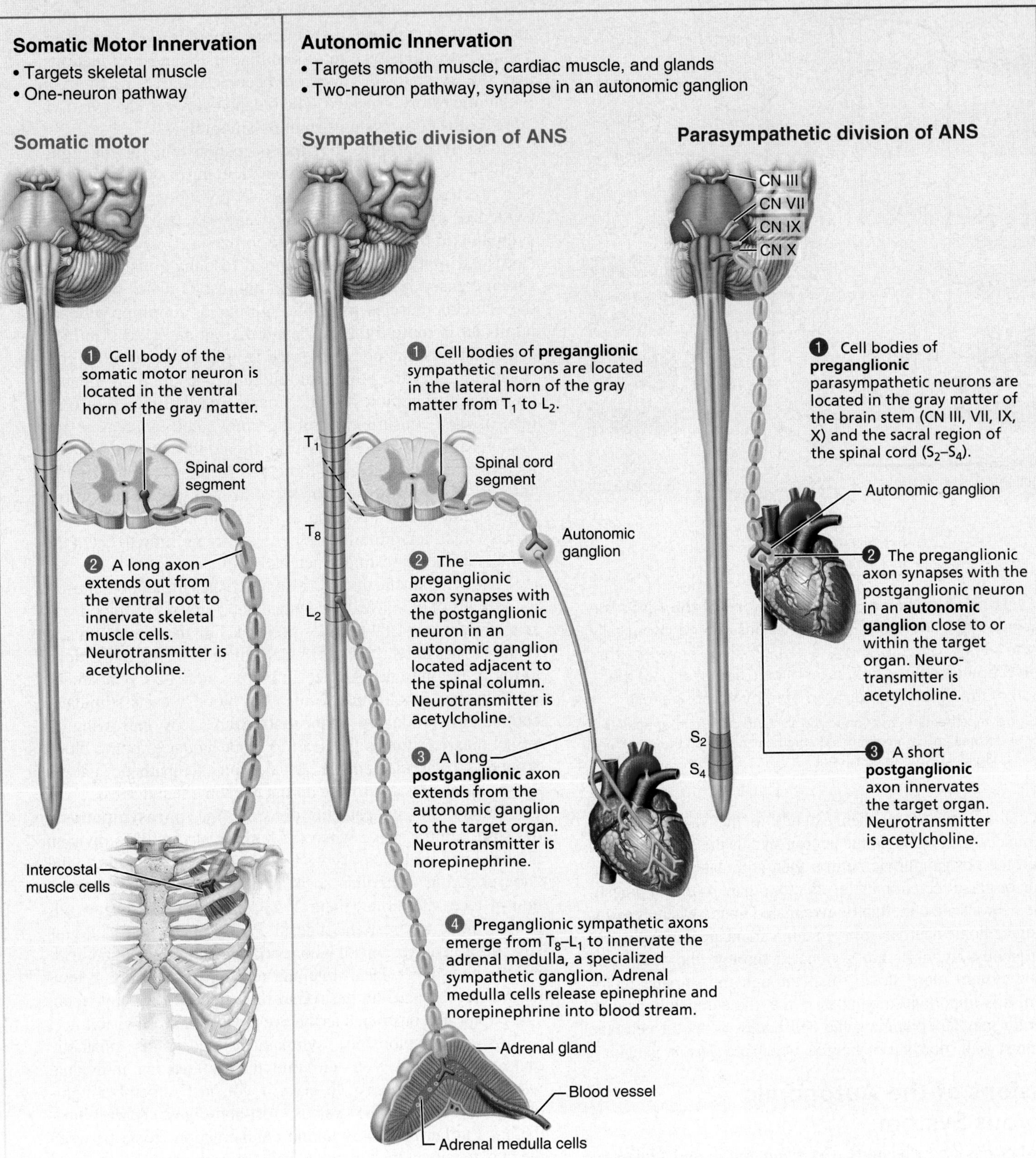

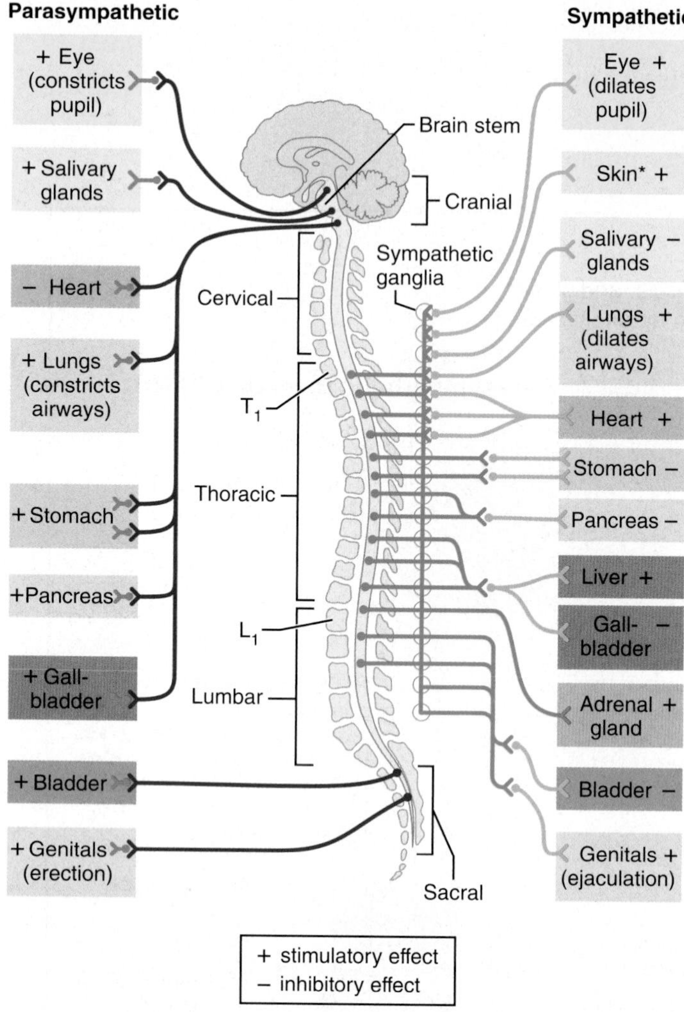

Figure 15.3 Overview of the subdivisions of the ANS. The parasympathetic and sympathetic divisions differ anatomically in (1) the sites of origin of their nerves, (2) the relative lengths of their preganglionic and postganglionic fibers, and (3) the location of their ganglia (indicated here by synapse sites).

*Although sympathetic innervation to the skin and peripheral structures is mapped to the cervical area in this diagram, all nerves to the periphery carry postganglionic sympathetic fibers.

(or *postganglionic fiber*) then extends to the visceral organs. Functionally, the preganglionic neuron signals the postganglionic neuron; the postganglionic neuron then stimulates muscle contraction or gland secretion in the effector organ. Axons of preganglionic neurons are thin, lightly myelinated fibers, whereas axons of postganglionic neurons are even thinner and are unmyelinated. Consequently, impulses are conducted through the autonomic nervous system more slowly than through the somatic motor system. It is important to emphasize that the autonomic ganglia are *motor* ganglia containing the cell bodies of motor neurons, in contrast to the dorsal root ganglia, which are *sensory* ganglia.

Divisions of the Autonomic Nervous System

The ANS has two divisions, the *sympathetic* and *parasympathetic* (par″ah-sim″pah-thet′ik) *divisions,* (Figure 15.2). Both divisions have chains of two motor neurons that mostly innervate the same visceral organs, but they cause opposite effects: One division stimulates some smooth muscle to contract or a gland to secrete; the other division inhibits that action. The sympathetic division mobilizes the body during extreme situations such as fear, exercise, or rage. The parasympathetic division enables the body to unwind and relax and works to conserve body energy. In other words, the parasympathetic division controls routine maintenance functions, and the sympathetic division becomes active when extra metabolic effort is needed. The balance between the two divisions keeps body systems running smoothly.

The **sympathetic division** is responsible for the fight-or-flight response. Its activity is evident during vigorous exercise, excitement, or emergencies. A pounding heart, dilated (widened) eye pupils, and cold, sweaty skin are signs that the sympathetic division has been mobilized (Figure 15.3, right side). All of them help us respond to dangerous situations: The increased heart rate delivers more blood and oxygen to the skeletal muscles used for fighting or running; widened pupils let in more light for clearer vision; and cold skin indicates that blood is being diverted from the skin to more vital organs, such as the brain. Additionally, the small air tubes in the lungs (bronchioles) dilate, increasing the uptake of oxygen; oxygen consumption by the body's cells increases; and the liver releases more sugar into the blood to provide for the increased energy needs of the cells. In this way, the body's "motors are revved up" for vigorous activity. Temporarily nonessential functions, such as digestion and motility of the urinary tract, are inhibited: When you are running to catch the last bus home, digesting lunch can wait.

The sympathetic division also innervates the smooth muscle in the walls of blood vessels. Sympathetic input to the blood vessels servicing skeletal muscles rises, causing the smooth muscle of the vessels to relax. These vessels dilate, bringing more blood to the active muscles. At the same time, increased sympathetic input to the smooth muscle in other blood vessels stimulates contraction, producing **vasoconstriction.** This narrowing of vessel diameter forces the heart to work harder to pump blood around the vascular circuit. As a result, sympathetic activity causes blood pressure to rise during excitement and stress.

Unlike the sympathetic division, the **parasympathetic division** is most active when the body is at rest. This division is concerned with conserving body energy and directing vital "housekeeping" activities such as digestion and the elimination of feces and urine (Figure 15.3, left side). The buzzwords to remember are "rest and digest." Parasympathetic function is best illustrated by a person who is relaxing after dinner and reading the newspaper. Heart rate and respiratory rates are at low-normal levels, and the gastrointestinal tract is digesting food. The pupils are constricted as the eyes focus for close vision.

As you explore the sympathetic and parasympathetic divisions in detail, you will find their effects on individual visceral organs are easy to learn if you just remember fight-or-flight (sympathetic) versus rest-and-digest (parasympathetic). Furthermore, a dynamic counteraction exists between the two divisions, such that they balance each other during times when a person is neither highly excited nor completely at rest.

Table 15.1 Anatomical and Physiological Differences Between the Parasympathetic and Sympathetic Divisions

Characteristic	Sympathetic	Parasympathetic
Origin	Thoracolumbar outflow; lateral horn of gray matter of spinal cord segments T_1–L_2	Craniosacral outflow: brain stem nuclei of cranial nerves III, VII, IX, and X; spinal cord segments S_2–S_4
Location of ganglia	Ganglia close to CNS: alongside vertebral column (sympathetic trunk ganglia) and anterior to vertebral column (collateral ganglia)	Ganglia in or close to visceral organ served
Relative length of pre- and postganglionic axons	Short preganglionic (splanchic nerves are exceptions); long postganglionic	Long preganglionic; short postganglionic
Rami communicantes (see p. 476)	Gray and white rami communicantes; white contain myelinated preganglionic axons; gray contain unmyelinated postganglionic axons	None
Degree of branching of preganglionic axons	Extensive	Minimal
Functional role	Prepares body to cope with emergencies and intense muscular activity; fight-or-flight response	Maintenance functions; conserves and stores energy; rest and digest response
Neurotransmitters	All preganglionic axons release ACh; most postganglionic axons release norepinephrine (adrenergic axons); postganglionic axons to sweat glands and blood vessels of skeletal muscles release ACh; neurotransmitter activity augmented by release of adrenal medullary hormones (epinephrine and norepinephrine)	All axons, preganglionic and postganglionic, release ACH (cholinergic axons)

CLINICAL APPLICATION

Autonomic Neuropathy Damage to the autonomic nerves, which may occur as a complication of diabetes, is called **autonomic neuropathy.** Damage to these nerves results in the inability to control heart rate, blood pressure, and blood sugar levels. Digestion and respiratory functions, urination, sexual response, and vision are also affected. This insidious condition may go untreated because its symptoms are widespread and commonly associated with other conditions. It can be detected via a noninvasive heart rate variability test (HRV).

In addition to these functional differences, there are anatomical and biochemical differences between the sympathetic and parasympathetic divisions (**Table 15.1**).

1. The two divisions originate from different regions of the CNS (Figure 15.2 and Figure 15.3). The sympathetic division can also be called the **thoracolumbar division** because its fibers emerge from the thoracic and superior lumbar parts of the spinal cord. The parasympathetic division can also be termed the **craniosacral division** because its fibers emerge from the brain *(cranial part)* and the sacral spinal cord *(sacral part).*
2. The sympathetic pathways have long postganglionic axons, whereas the postganglionic axons for parasympathetic pathways are comparatively short (Figure 15.2 and Figure 15.3). All sympathetic ganglia lie near the spinal cord and vertebral column; postganglionic axons extend from these ganglia and travel to their target organs. Parasympathetic ganglia lie far from the CNS, in or near the organs innervated; therefore, the postganglionic axons are quite short.
3. Sympathetic fibers branch profusely, whereas parasympathetic fibers do not. Such extensive branching allows each sympathetic neuron to influence a number of different visceral organs, enabling many organs to mobilize simultaneously during the fight-or-flight response. Indeed, the literal translation of *sympathetic,* "experienced together," reflects the bodywide mobilization it produces. Parasympathetic effects, by contrast, are more localized and discrete.
4. The main biochemical difference between the two divisions of the ANS involves the neurotransmitter released by the *post*ganglionic axons (Figure 15.2). In the sympathetic division, most postganglionic axons release *norepinephrine* (also called *noradrenaline*); these fibers are termed *adrenergic.** The postganglionic neurotransmitter in the parasympathetic division is *acetylcholine (ACh);* these fibers are termed *cholinergic*. The *pre*ganglionic axon terminals of both divisions are always cholinergic (release ACh).

*Not all postganglionic fibers in the sympathetic division are adrenergic: Those that innervate the sweat glands and blood vessels in skeletal muscle are cholinergic.

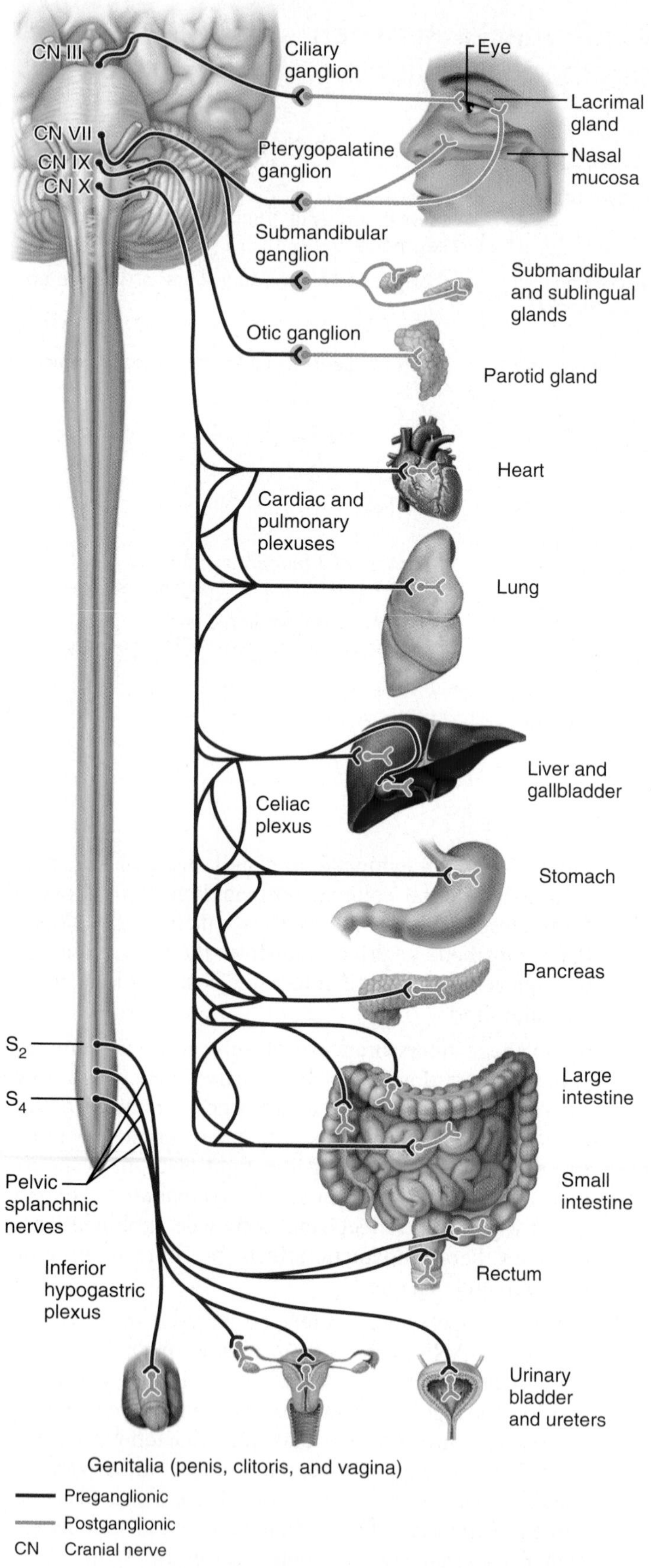

Figure 15.4 Parasympathetic division of the ANS.

The next sections give a more detailed discussion of the anatomical organization of each division.

☑ check your understanding

- ☐ 1. Which fibers in the motor division of the PNS are not myelinated?
- ☐ 2. As you are driving home from school, a car suddenly swerves toward you, forcing you to hit the brakes quickly. You feel your heart pump, and you begin to sweat a bit. Which division of the ANS has been activated?
- ☐ 3. Where are the sympathetic ganglia located? Where are most parasympathetic ganglia located?
- ☐ 4. Are visceral sensory fibers considered part of the ANS?

(For answers, see Appendix B.)

THE PARASYMPATHETIC DIVISION

learning outcomes

- ▶ Describe the pathway of parasympathetic innervation through cranial nerves and sacral spinal nerves to the visceral organs.
- ▶ Describe the effect of parasympathetic innervation on each visceral organ innervated by this division of the ANS.

The *cranial outflow* of the parasympathetic division originates from the brain and innervates organs in the head, neck, thorax, and most of the abdomen, whereas the *sacral outflow* originates from the sacral spinal cord (S_2, S_3, S_4) and supplies the distal portions of the digestive tract and the pelvic organs (**Figure 15.4**).

Cranial Outflow

The preganglionic axons run in the oculomotor (III), facial (VII), glossopharyngeal (IX), and vagus (X) nerves (top of Figure 15.4). The cell bodies of these preganglionic neurons are located in motor cranial nerve nuclei in the gray matter of the brain stem.

Oculomotor Nerve (III)

The parasympathetic fibers, the visceral motor component (VM), of the oculomotor nerve innervate smooth muscles in the eye that cause the pupil to constrict and the lens of the eye to bulge—actions that allow focusing on close objects in the field of vision. In this two-neuron pathway, the preganglionic axons in the oculomotor nerve originate from cell bodies in the *accessory oculomotor nucleus* in the midbrain (Figure 13.7a, p. 382); the postganglionic cell bodies lie in the **ciliary ganglion,** in the posterior part of the orbit just lateral to the optic nerve (see Table 14.2, p. 436).

Facial Nerve (VII)

The parasympathetic fibers (VM) of the facial nerve stimulate the secretion of many glands in the head, including the lacrimal (tear) gland above the eye; mucus-secreting glands in the nasal cavity; and two salivary glands inferior to the mouth (the submandibular and sublingual glands). In the pathway leading to the lacrimal and nasal glands, the preganglionic neurons originate in the *lacrimal nucleus* in the pons and synapse with postganglionic neurons in the **pterygopalatine ganglion,** just posterior to the maxilla (Table 14.2, p. 439). In the pathway leading to the submandibular and sublingual glands, the preganglionic neurons originate in the *superior salivatory nucleus* in the pons and synapse with postganglionic neurons in the **submandibular ganglion,** deep to the mandibular angle.

Glossopharyngeal Nerve (IX)

The parasympathetic fibers (VM) of the glossopharyngeal nerve stimulate secretion of a large salivary gland, the parotid gland, which lies anterior to the ear. The preganglionic neurons originate in the *inferior salivatory nucleus* in the medulla and synapse with postganglionic neurons in the **otic ganglion** inferior to the foramen ovale of the skull (Table 14.2, p. 441).

The three cranial nerves considered so far (III, VII, IX) supply the entire parasympathetic innervation of the head. Note, however, that only the *pre*ganglionic axons run within these three nerves. These axons synapse in the ganglia described above, which are located along the path of the trigeminal nerve (V), and then the postganglionic axons travel via the trigeminal to their final destinations. This routing by way of the trigeminal nerve occurs because the trigeminal has the widest distribution within the face.

Vagus Nerve (X)

Parasympathetic fibers (VM) from the vagus innervate the visceral organs of the thorax and most of the abdomen (Figure 15.4). Note that this does not include innervation of the pelvic organs and that the vagal innervation of the digestive tube ends halfway along the large intestine. The vagus is an extremely important part of the ANS, containing nearly 90% of the preganglionic parasympathetic fibers in the body. Functionally, the parasympathetic fibers in the vagus nerve bring about typical rest-and-digest activities in visceral muscle and glands—stimulation of digestion (secretion of digestive glands and increased motility of the smooth muscle of the digestive tract), reduction in the heart rate, and constriction of the bronchi in the lungs, for example.

The preganglionic cell bodies are mostly in the *dorsal motor nucleus of vagus* in the medulla (see Fig. 13.7c, p. 382), and the preganglionic axons run through the entire length of the vagus nerve. Most postganglionic neurons are confined within the walls of the organs being innervated, and their cell bodies form *intramural ganglia* (in″trah-mu′ral; "within the walls").

The vagus nerve is essential to the functioning of many organs. As the vagus descends through the neck and trunk, it sends branches through many *autonomic nerve plexuses* to the organs being innervated (**Figure 15.5**). (Recall from Chapter 14 that a nerve plexus is a network of nerves.) Specifically, the vagus sends branches through the **cardiac plexus** to the heart, through the **pulmonary plexus** to the lungs, through the **esophageal plexus** to the esophagus and into the stomach wall, and through the **celiac plexus** and the **superior mesenteric plexus** to the other abdominal organs (intestines, liver, pancreas, and so on). Fibers from both divisions of the ANS, parasympathetic and sympathetic, travel to the thoracic and abdominal organs through these plexuses.

Sacral Outflow

The sacral part of the parasympathetic outflow emerges from segments S_2–S_4 of the spinal cord (bottom part of Figure 15.4). Continuing where the vagus ends, it innervates the organs in the pelvis, including the distal half of the large intestine, the bladder, and reproductive organs such as the uterus, and the erectile tissues of the external genitalia. Parasympathetic effects on these organs include stimulation of defecation, voiding of urine, and erection.

The preganglionic cell bodies of the sacral parasympathetics lie in the visceral motor region of the spinal gray matter (Figure 13.27, p. 412). The axons of these preganglionic neurons run in the ventral roots to the ventral rami, from which they branch to form **pelvic splanchnic nerves** (see Figure 15.4). These nerves then run through an autonomic plexus in the pelvic floor, the **inferior hypogastric plexus** (or *pelvic plexus;* Figure 15.5) to reach the pelvic organs. Some preganglionic axons synapse in ganglia in this plexus, but most synapse in intramural ganglia in the organs. This plexus also contains fibers from both divisions of the ANS.

The specific effects of parasympathetic innervation on various organs are presented in comparison with the effects of sympathetic innervation (**Table 15.2**, p. 475).

☑ check your understanding

- ☐ 5. Which spinal nerves carry parasympathetic preganglionic fibers?
- ☐ 6. What is the result of vagal stimulation of (a) the heart, (b) the small intestine, (c) the salivary glands?
- ☐ 7. Cranial nerves III, VII, and IX carry preganglionic parasympathetic fibers. Which cranial nerve carries the postganglionic fibers to the target organs innervated by these three nerves?

(For answers, see Appendix B.)

THE SYMPATHETIC DIVISION

learning outcomes

- ▶ Describe the pathways of sympathetic innervation from the spinal cord to the effector organs in the body periphery, the head, and the visceral organs.
- ▶ Describe the effect of sympathetic innervation on each effector organ.
- ▶ Explain the sympathetic function of the adrenal medulla.

Basic Organization

The sympathetic division exits from the thoracic and superior lumbar part of the spinal cord, from segments T_1 to L_2 (Figure 15.3). Its preganglionic cell bodies lie in the visceral motor region of the spinal gray matter, where they form the lateral gray horn (Figure 13.27, p. 412).

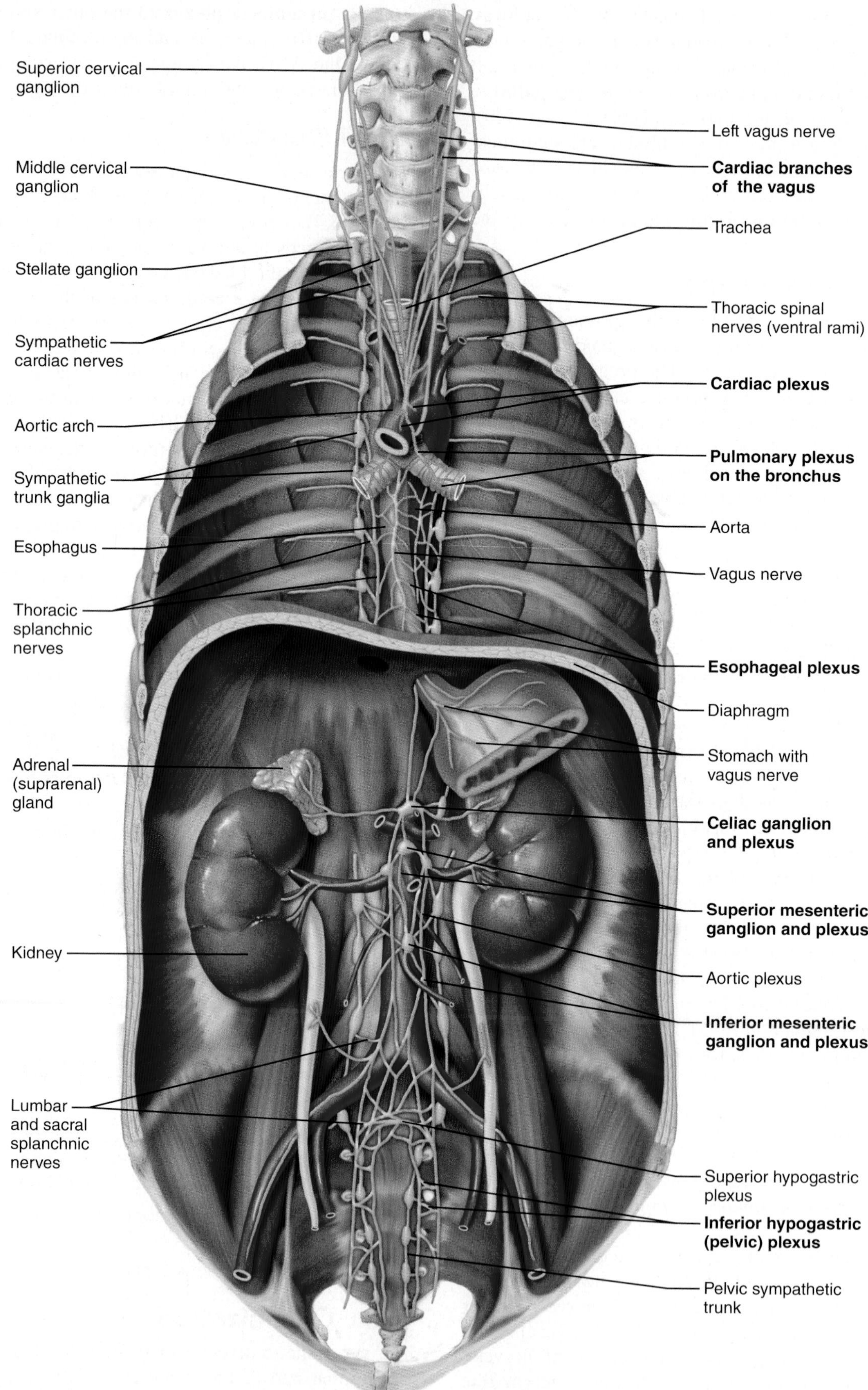

Figure 15.5 **Autonomic nerves, plexuses, and ganglia.** All autonomic plexuses contain both parasympathetic and sympathetic axons. The ganglia are almost exclusively sympathetic.

Table 15.2 Effects of the Parasympathetic and Sympathetic Divisions on Various Organs

Target Organ/System/Activity	Parasympathetic Effects	Sympathetic Effects
Eye (iris)	Stimulates constrictor muscles; constricts eye pupils	Stimulates dilator muscles, dilates eye pupils
Eye (ciliary muscle)	Stimulates ciliary muscles, which results in bulging of the lens for accommodation and close vision	Weakly inhibits ciliary muscles, which flatten the lens for distance vision
Glands (nasal, lacrimal, salivary, gastric, pancreas)	Stimulates secretory activity	Inhibits secretory activity; causes vasoconstriction of blood vessels supplying the glands
Sweat glands	No innervation	Stimulates copious sweating (cholinergic fibers)
Arrector pili muscles attached to hair follicles	No innervation	Stimulates to contract (erects hairs and produces goose bumps)
Heart muscle	Decreases rate; slows and steadies heart	Increases rate and force of heartbeat
Heart: coronary blood vessels	Causes vasoconstriction	Causes vasodilation
Lungs	Constricts bronchioles	Dilates bronchioles and mildly constricts blood vessels
Digestive tract organs	Increases motility (peristalsis) and amount of secretion by digestive organs; relaxes sphincters to allow movement of foodstuffs along tract	Decreases activity of glands and muscles of digestive system and constricts sphincters (e.g., anal sphincter); causes vasoconstriction
Liver	No effect	Epinephrine stimulates liver to release glucose to blood
Gallbladder	Stimulates activity (gallbladder contracts to expel bile)	Inhibits activity (gallbladder is relaxed)
Adrenal medulla	No innervation	Stimulates medulla cells to secrete epinephrine and norepinephrine into bloodstream
Kidney	No effect	Causes vasoconstriction; decreases urine output
Bladder, urethra	Causes contraction of smooth muscle of bladder wall; relaxes urethral sphincter; promotes voiding	Causes relaxation of smooth muscle of bladder wall; constricts urethral sphincter; inhibits voiding
Penis	Causes erection (vasodilation)	Causes ejaculation
Uterus	Inhibits contraction of smooth muscle of uterine wall; causes vasodilation of vessels	Stimulates contraction of smooth muscle of uterine wall; causes vasoconstriction of vessels
Vagina, clitoris	Causes erection (vasodilation) of clitoris	Causes contraction of vagina
Blood vessels	Little or no effect	Constricts most vessels and increases blood pressure; constricts vessels of abdominal viscera and skin to divert blood to muscles, brain, and heart when necessary; dilates vessels of the skeletal muscles (cholinergic fibers) during exercise
Blood coagulation	No innervation	Increases coagulation
Cellular metabolism	No innervation	Increases metabolic rate
Adipose tissue	No innervation	Stimulates lipolysis (fat breakdown)
Mental activity	No innervation	Increases alertness

The sympathetic division is more complex than the parasympathetic, in part because it innervates more organs. It supplies not only all the visceral organs in the internal body cavities but also all visceral structures in the superficial regions of the body: the sweat glands, the hair-raising arrector pili muscles of the skin, and the smooth musculature in the walls of all arteries and veins. It is easy to remember that the sympathetic system alone innervates these structures: You know that you sweat when under stress, your hair stands up when you are terrified, and your blood pressure skyrockets (because of widespread vasoconstriction) when you get excited. The *parasympathetic* division does *not* innervate sweat glands, arrector pili, or (with minor exceptions) blood vessels.

Another reason the sympathetic division is more complex is that it has more ganglia. The sympathetic ganglia fall into two classes: (1) *sympathetic trunk ganglia* and (2) *collateral ganglia*.

Sympathetic Trunk Ganglia

The numerous **sympathetic trunk ganglia,** located along both sides of the vertebral column from the neck to the pelvis, are linked by short nerves into long **sympathetic trunks** *(sympathetic chains)* that resemble strings of beads (Figure 15.6). The sympathetic trunk ganglia are also called *chain ganglia* and *paravertebral* ("near the vertebrae") *ganglia.* These ganglia are joined to the ventral rami of nearby spinal nerves by **white** and **gray rami communicantes** (singular: **ramus communicans,** "communicating arm"). White rami communicantes lie lateral to gray rami communicantes.

There is approximately one sympathetic trunk ganglion for each spinal nerve. However, the number of sympathetic trunk ganglia and spinal nerves is not identical, because some adjacent ganglia fuse during development. Such fusion is most evident in the neck region, where there are eight spinal nerves but only three sympathetic trunk ganglia: the **superior, middle,** and **inferior cervical ganglia** (Figure 15.5). Furthermore, the inferior cervical ganglion usually fuses with the first thoracic ganglion to form the **stellate** ("star-shaped") **ganglion** in the superior thorax (Figure 15.5).

Overall, there are 22–24 sympathetic trunk ganglia per side, and a typical person may have 3 cervical, 11 thoracic, 4 lumbar, 4 sacral, and 1 coccygeal ganglia (Figure 15.5). The cervical ganglia lie just anterior to the transverse processes of the cervical vertebrae; the thoracic ganglia lie on the heads of the ribs; the lumbar ganglia lie on the anterolateral sides of the vertebral bodies; the sacral ganglia lie medial to the sacral foramina; and the coccygeal ganglion is anterior to the coccyx.

Take care not to confuse the sympathetic trunk ganglia with the *dorsal root ganglia.* The dorsal root ganglia are sensory and lie along the dorsal roots of spinal nerves in the intervertebral foramina; the sympathetic trunk ganglia are motor and lie anterior to the ventral rami of spinal nerves.

Collateral Ganglia

The **collateral ganglia,** or *prevertebral ganglia,* differ from the sympathetic trunk ganglia in at least three ways: (1) They are not paired and are not segmentally arranged; (2) they occur only in the abdomen and pelvis; and (3) they all lie anterior to the vertebral column (hence the name *prevertebral*), mostly on the large artery called the abdominal aorta. The main collateral ganglia, the *celiac, superior mesenteric, inferior mesenteric,* and *inferior hypogastric ganglia* (Figure 15.5), lie within the autonomic nerve plexuses of the same names.

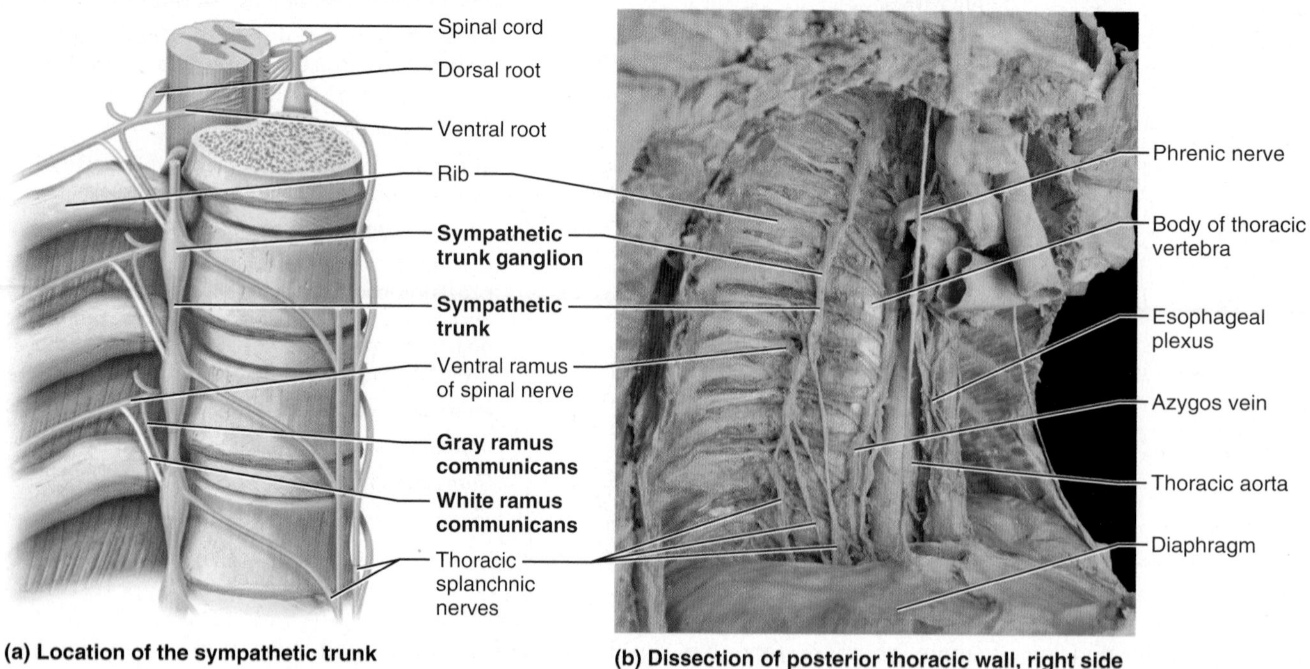

(a) Location of the sympathetic trunk

(b) Dissection of posterior thoracic wall, right side

Figure 15.6 The sympathetic trunk, thoracic region.

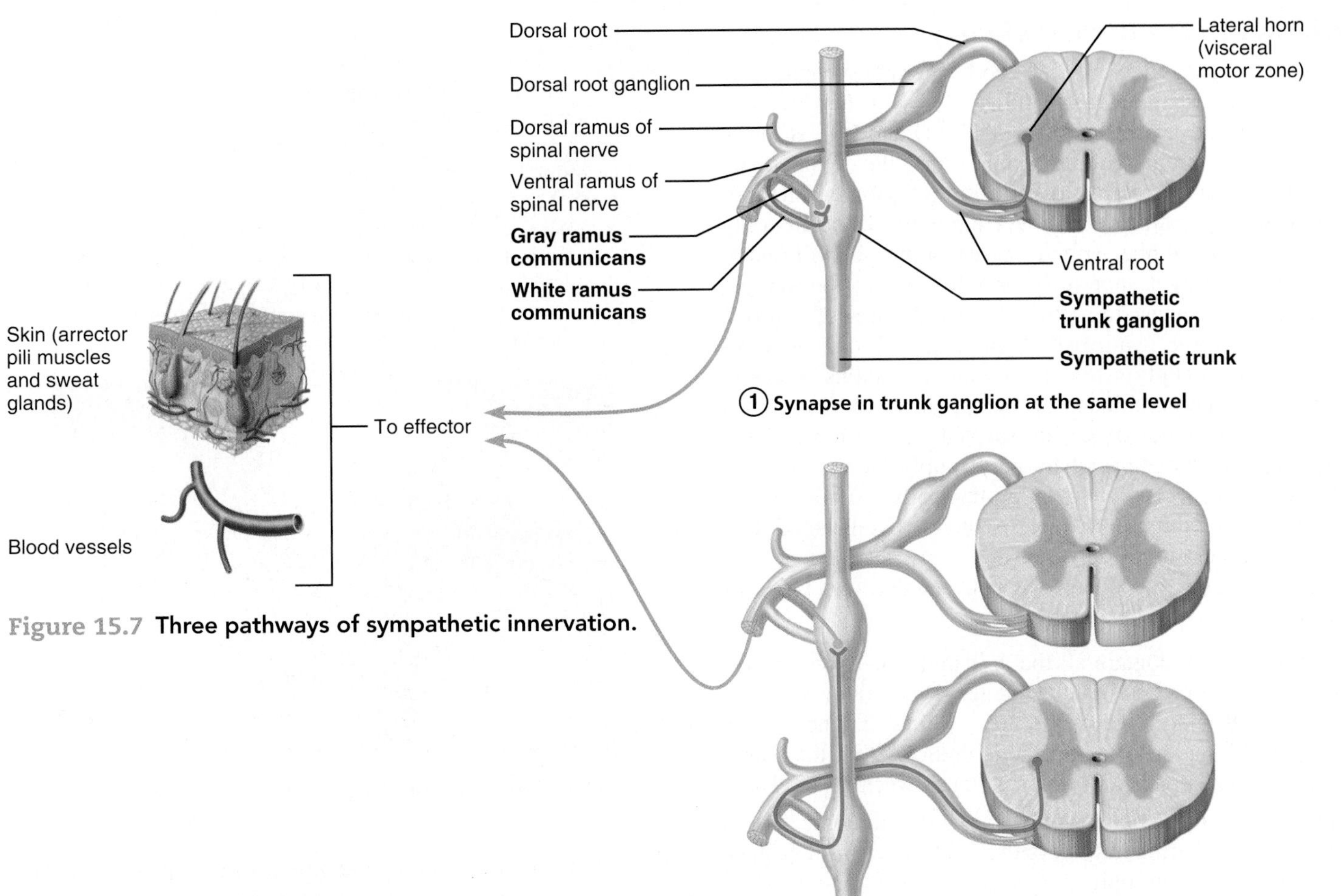

Dorsal root
Dorsal root ganglion
Dorsal ramus of spinal nerve
Ventral ramus of spinal nerve
Gray ramus communicans
White ramus communicans
Lateral horn (visceral motor zone)
Ventral root
Sympathetic trunk ganglion
Sympathetic trunk
Skin (arrector pili muscles and sweat glands)
Blood vessels
To effector

① Synapse in trunk ganglion at the same level

② Synapse in trunk ganglion at a higher or lower level

Figure 15.7 **Three pathways of sympathetic innervation.**

Sympathetic Pathways

In the sympathetic pathways to all body regions (Figure 15.7), preganglionic neurons in the thoracolumbar spinal cord send their motor axons through the adjacent ventral root into the spinal nerve, white ramus communicans, and associated sympathetic trunk ganglion. From there, these preganglionic axons follow one of three pathways:

① The preganglionic axon synapses with a postganglionic neuron in the sympathetic trunk ganglion at the same level and exits via the gray ramus communicans into the spinal nerve at that level.

② The preganglionic axon ascends or descends in the sympathetic trunk to synapse in another trunk ganglion. The postganglionic fiber exits the sympathetic trunk via the gray ramus communicans at the level of the synapse.

③ The preganglionic axon passes through the sympathetic trunk, exits on a splanchnic nerve, and synapses in a collateral ganglion. The postsynaptic fiber extends from the collateral ganglion to the visceral organ via an autonomic nerve plexus.

Keep this overview in mind as you consider the pathways to the specific body regions in more detail.

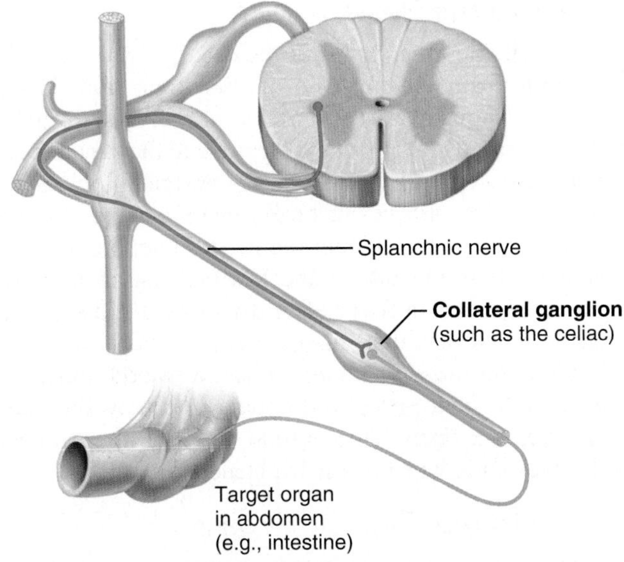

③ Pass through sympathetic trunk to synapse in a collateral ganglion anterior to the vertebral column

Pathways to the Body Periphery

Sympathetic pathways to the body periphery innervate the sweat glands, the arrector pili muscles in the skin, and peripheral blood vessels (Figure 15.7 ① and ②). In these pathways, the preganglionic axons enter the sympathetic trunk ganglia and synapse there with ganglionic cell bodies. Because sympathetic outflow from the CNS occurs only on thoracic and lumbar spinal nerves, some of the preganglionic axons travel superiorly or inferiorly in the sympathetic trunk to synapse in regions outside the area of outflow, thus supplying the head, pelvis, and limbs. From the sympathetic trunk ganglion, the postganglionic axons travel in a *gray* ramus communicans to the spinal nerve. The axons then follow branches of the dorsal and ventral rami of the spinal nerve to the skin, where they supply the arrector pili and sweat glands. Anywhere along this path, the postganglionic axons can "jump onto" nearby blood vessels—and then follow and innervate those vessels.

At this point, the difference between the two types of rami communicantes becomes clear.

- The *white* rami contain all the *pre*ganglionic axons traveling to the sympathetic trunk ganglion. White rami are white because preganglionic axons are myelinated. They occur only in the region of sympathetic outflow from the spinal cord — on sympathetic trunk ganglia between T_1 and L_2.
- The *gray* rami contain only the *post*ganglionic axons headed for peripheral structures. Gray rami are gray because postganglionic axons are nonmyelinated. They occur on all the sympathetic trunk ganglia because sweat glands, arrector pili, and blood vessels must be innervated in *all* body segments.

Pathways to the Head

In the sympathetic innervation of the head (Figure 15.8), the preganglionic axons originate in the first four thoracic segments of the spinal cord (T_1–T_4). From there, these axons ascend in the sympathetic trunk to synapse in the superior cervical ganglion. From this ganglion, the postganglionic axons associate with large arteries that carry them to glands, smooth muscle, and vessels throughout the head. Functionally, these sympathetic fibers (1) inhibit the lacrimal, nasal, and salivary glands (the reason fear causes dry mouth), (2) stimulate the muscle in the iris that dilates the pupil in the eye, and (3) stimulate the *superior tarsal muscle,* a smooth muscle in the upper eyelid that prevents drooping of the eyelid whenever the eyes are open. This muscle also lifts the eyelid to open the eyes wide when one is frightened.

Pathways to the Thoracic Organs

In the sympathetic innervation of thoracic organs (Figure 15.8), the preganglionic axons originate at spinal levels T_1–T_6. Some of these axons synapse in the nearest sympathetic trunk ganglion, and the postganglionic axons run directly to the organ being supplied. Fibers to the lungs and esophagus take this direct route, as do some axons to the heart. Along the way, the postganglionic axons pass through the pulmonary, esophageal, and cardiac plexuses (Figure 15.5).

CLINICAL APPLICATION

Horner's Syndrome *Horner's syndrome* is a condition that follows damage to the sympathetic trunk in the inferior neck region on one side of the body. The symptoms, which occur on the affected side only, result from a loss of sympathetic innervation to the head and the dominance of parasympathetic effects. They include drooping of the upper eyelid (ptosis), pupil constriction, flushing of the face, and inability to sweat. Horner's syndrome can indicate the presence of disease or infection in the neck.

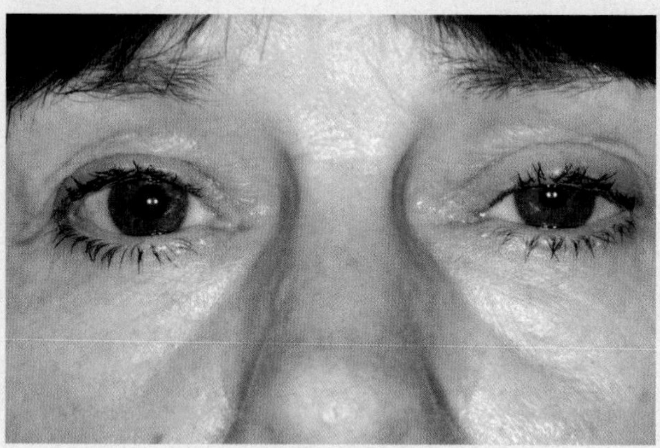

Left eye showing typical signs of Horner's syndrome

Many of the sympathetic fibers to the *heart,* however, take a less direct path. The preganglionic axons ascend in the sympathetic trunk to synapse in the cervical ganglia of the sympathetic trunk. From there, the postganglionic axons descend through the cardiac plexus and into the heart wall. Many nerves to the heart (a thoracic organ) arise in the neck because the heart develops in the neck region of the embryo (discussed in Chapter 19, "Development of the Heart" section).

Functionally, the thoracic sympathetic nerves increase the heart rate and dilate the blood vessels that supply the heart wall (so that the heart muscle itself receives more blood). They also dilate the respiratory air tubes and inhibit the muscle and glands in the esophagus, effects that are integral to the fight-or-flight response.

Pathways to the Abdominal Organs

In the sympathetic innervation of abdominal organs, the preganglionic axons originate in the inferior half of the thoracolumbar spinal cord (T_5–L_2; Figure 15.8). From there, these axons pass through the adjacent sympathetic trunk ganglia and travel in **thoracic splanchnic nerves** (greater, lesser, and least) to synapse in collateral ganglia in the large plexuses on the abdominal aorta. These ganglia include the **celiac** and **superior mesenteric ganglia,** along with some smaller ones. Postganglionic axons from these ganglia then follow the main branches of the aorta to the stomach, liver, kidney, spleen, and intestines (through the proximal half of the large intestine). For the most part, the sympathetic

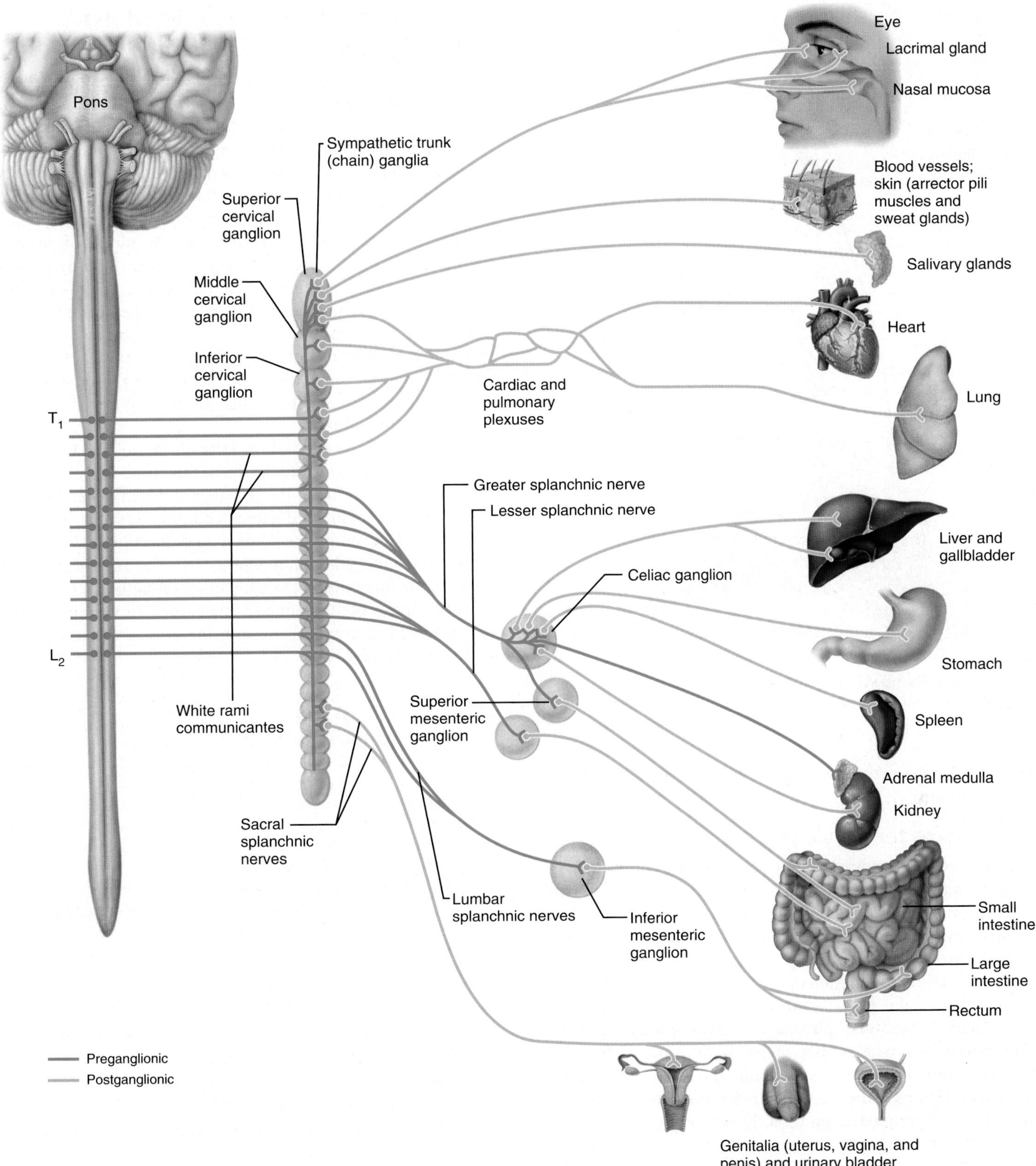

Figure 15.8 Sympathetic division of the ANS. Sympathetic innervation to peripheral structures (blood vessels, glands, and arrector pili muscles) occurs in all areas but is shown only in the cervical area.

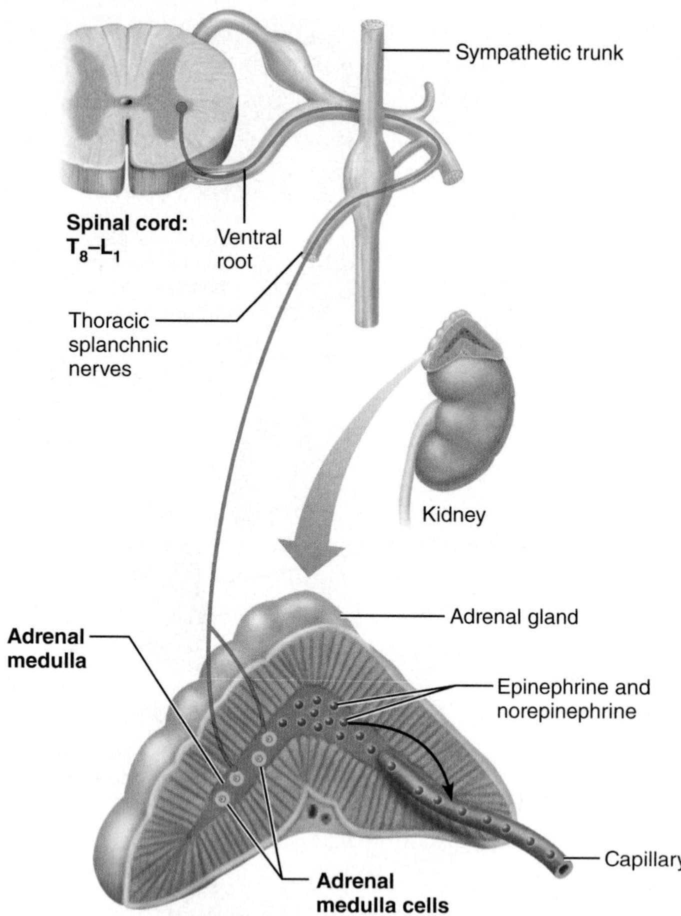

Figure 15.9 Sympathetic innervation of the adrenal medulla.

fibers *inhibit* the activity of the muscles and glands in these visceral organs.

Pathways to the Pelvic Organs

In the sympathetic innervation of pelvic organs (Figure 15.8), the preganglionic axons originate in the most inferior part of the thoracolumbar spinal cord (T_{10}–L_2), then descend in the sympathetic trunk to the lumbar and sacral ganglia of the sympathetic trunk. Some axons synapse there, and the postganglionic axons run in *lumbar* and *sacral splanchnic nerves* to plexuses on the lower aorta and in the pelvis—namely, the **inferior mesenteric plexus,** the **aortic plexus,** and the **hypogastric plexuses.** Other preganglionic axons, by contrast, pass directly to these autonomic plexuses and synapse in collateral ganglia there—the **inferior mesenteric ganglia** and **inferior hypogastric ganglia.** Postganglionic axons proceed from these plexuses to the pelvic organs, including the bladder, the reproductive organs, and the distal half of the large intestine. These sympathetic fibers inhibit urination and defecation and promote ejaculation.

The Role of the Adrenal Medulla in the Sympathetic Division

On the superior aspect of each kidney lies an **adrenal (suprarenal) gland** (Figure 15.9). The internal portion of this gland—the **adrenal medulla**—is a major organ of the sympathetic nervous system. The adrenal medulla is a specialized sympathetic ganglion containing a collection of modified postganglionic neurons that completely lack nerve processes. These neuron-derived cells secrete great quantities of two excitatory hormones into the blood of nearby capillaries during the fight-or-flight response. The hormones secreted are norepinephrine (the chemical secreted by other postganglionic sympathetic neurons as a neurotransmitter) and greater amounts of a related excitatory molecule called **epinephrine (adrenaline).** Once released, these hormones travel throughout the body in the bloodstream, producing the widespread excitatory effects that we have all experienced as a "surge of adrenaline."

The cells of the adrenal medulla are stimulated to secrete by preganglionic sympathetic fibers that arise from cell bodies in the T_8–L_1 region of the spinal cord. From there, they run in the thoracic splanchnic nerves and pass through the celiac plexus before reaching the adrenal medulla (Figure 15.8). Not surprisingly, the adrenal medulla has a more concentrated sympathetic innervation than any other organ in the body.

The effects of sympathetic innervation to various organs are presented in comparison with the effects of parasympathetic innervation (Table 15.2, p. 475).

CLINICAL APPLICATION

Stress-Induced Hypertension Continual stress can promote overactive sympathetic stimulation causing vasoconstriction that results in **hypertension,** or high blood pressure, a circulatory condition that can have a variety of contributing factors. Hypertension is always serious because (1) it increases the workload on the heart, possibly precipitating heart disease, and (2) it increases the wear and tear on artery walls. Stress-induced hypertension is treated with drugs that prevent smooth muscle cells in the walls of blood vessels from binding norepinephrine and epinephrine.

check your understanding

- ☐ 8. Why are white rami communicantes located only on sympathetic trunk ganglia between T_1 and L_2, and gray rami communicantes branch off each sympathetic trunk ganglion?
- ☐ 9. What is the general effect of sympathetic innervation to the abdominal organs?
- ☐ 10. Which types of autonomic fibers (preganglionic, postganglionic, sympathetic, parasympathetic) are located in the thoracic and abdominal autonomic plexuses?

(For answers, see Appendix B.)

VISCERAL SENSORY NEURONS

learning outcomes

- Describe the role and location of visceral sensory neurons relative to autonomic neurons.
- Explain the concept of referred pain.

The visceral division of the PNS contains sensory as well as motor (autonomic) neurons (Figure 15.1). General visceral sensory neurons monitor stretch, temperature, chemical changes, and irritation within the visceral organs. The brain interprets this visceral information as feelings of hunger, fullness, pain, or nausea. Almost all of the receptors for these visceral senses are free (nonencapsulated) nerve endings widely scattered throughout the visceral organs. Visceral sensations tend to be difficult to localize with precision: For example, people usually cannot distinguish whether gas pains originate in the stomach or in the intestine, or whether a pain in the lower abdomen originates from the uterus or the appendix.

Like somatic sensory neurons, the cell bodies of visceral sensory neurons are located in dorsal root ganglia and in the sensory ganglia of cranial nerves. The long peripheral processes of these sensory neurons accompany the autonomic motor fibers to the visceral organs. Many visceral sensory fibers accompany the parasympathetic fibers in the vagus nerve and monitor visceral sensations in the many organs served by this nerve. Other visceral sensory fibers accompany the sympathetic fibers, running from the visceral organs into the autonomic plexuses and then through the splanchnic nerves, sympathetic trunk, rami communicantes, spinal nerves, and dorsal roots. From the dorsal roots, the central processes of these sensory neurons enter the spinal cord. *Most visceral pain fibers follow this sympathetic route to the CNS.*

The pathways by which visceral sensory information is relayed through the spinal cord to the cerebral cortex are not yet fully understood. Most visceral inputs travel along the spinothalamic (and spinoreticular) pathways to the thalamus. Neurons in the thalamus relay visceral sensory information to the visceral sensory cortex in the insula lobe (p. 388) for conscious perception. Visceral sensory information also reaches and influences the visceral control centers in the hypothalamus and medulla oblongata.

Visceral *pain* exhibits an unusual feature that is worth noting: Surprisingly, one often feels no pain when a visceral organ is cut or scraped. When pieces of mucous membrane are snipped from the uterus or the intestines to be examined for cancer, for example, most patients report little discomfort. Visceral pain does result, however, from chemical irritation or inflammation of visceral organs, from spasms of the smooth muscle in these organs (cramping), and from excessive stretching of the organ. In these cases, visceral pain can be severe.

People suffering from visceral pain often perceive this pain to be somatic in origin—that is, as if it originated from the skin or outer body. This phenomenon is called **referred pain.** For example, heart attacks can produce referred pain in cutaneous areas of the superior thoracic wall and the medial aspect of the left arm (Figure 15.10). The cause of referred pain is not fully understood. It is known that both the affected organ and the region of the body wall to which the pain is referred are innervated by the same spinal segments. For example, both the heart and the skin area to which heart pain projects are innervated by sensory neurons from T_1 to T_5. Referred pain may result from the convergence of somatic and visceral sensory neurons on the same second-order neurons in the dorsal horn of the spinal cord.

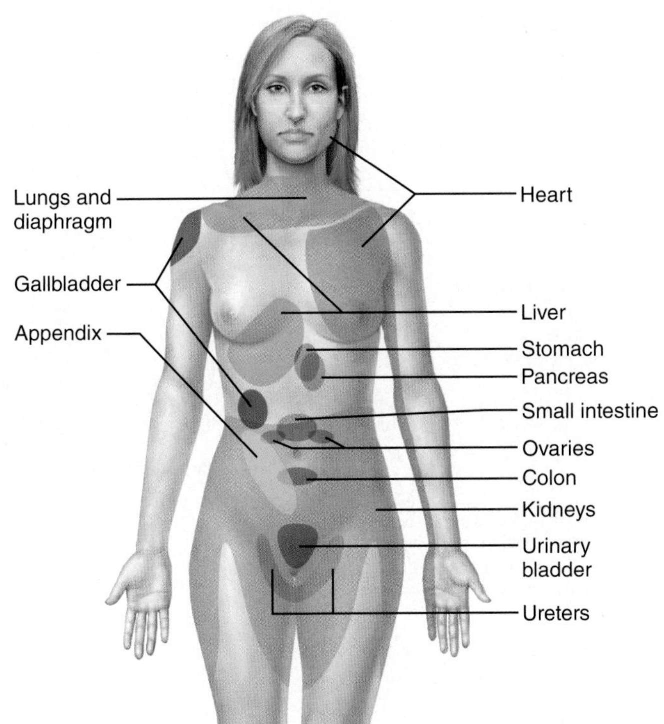

Figure 15.10 A map of referred pain. This map shows the anterior skin areas to which pain is referred from certain visceral organs.

VISCERAL REFLEXES

learning outcome

- Explain how visceral reflexes regulate some functions of visceral organs.

Visceral sensory and autonomic neurons participate in **visceral reflex arcs** (Figure 15.11) including the *defecation reflex* (illustrated in Figure 23.22, p. 703), in which the rectum is stretched by feces and the smooth muscle of the large intestine responds by contracting, and the *micturition reflex* (shown in Figure 24.15, p. 735), in which the smooth muscle in the urine-filled bladder contracts. Many visceral reflexes are simple spinal reflexes in which sensory neurons activate spinal interneurons, which in turn activate preganglionic autonomic neurons.

Other visceral reflexes involve cranial nerves and are integrated in the brain stem. One example is the baroreceptor reflex,

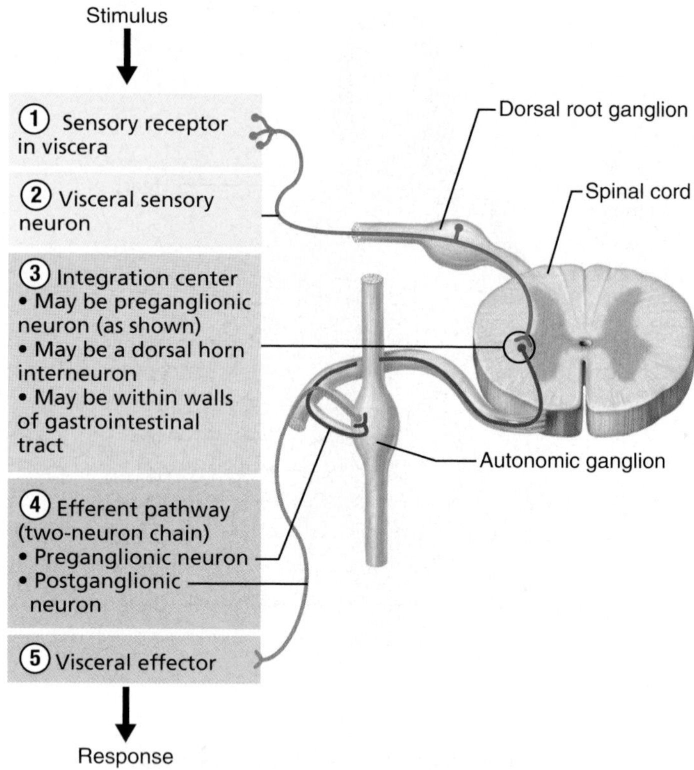

Figure 15.11 Visceral reflexes. Visceral reflex arcs have the same five elements as somatic reflex arcs.

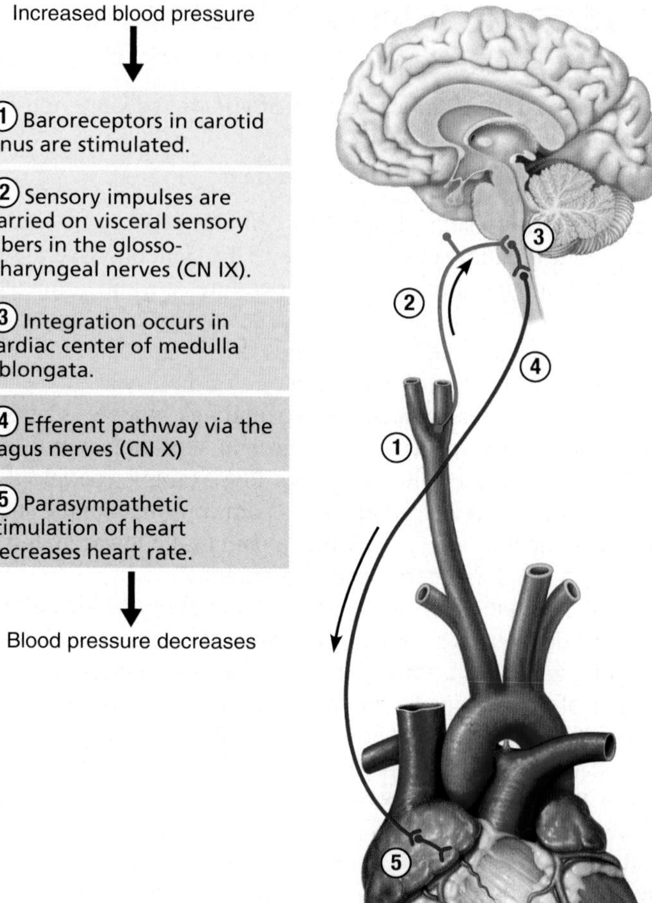

Figure 15.12 Baroreceptor reflex.

a visceral reflex that regulates blood pressure (Figure 15.12). When blood pressure is elevated, baroreceptors in the carotid sinus, located at the junction of the internal and external carotid arteries, stimulate visceral sensory neurons in the glossopharyngeal nerve (cranial nerve IX). Integration in the cardiac center in the medulla stimulates the vagus nerve (cranial nerve X). Vagal stimulation of the heart decreases heart rate, and subsequently blood pressure falls.

Some visceral reflex arcs do not involve the central nervous system at all—they are strictly *peripheral reflexes.* In some of these peripheral reflexes, branches from visceral sensory fibers synapse with postganglionic motor neurons *within sympathetic ganglia.* Furthermore, complete three-neuron reflex arcs (with small sensory, motor, and intrinsic neurons) exist *entirely within the wall of the digestive tube;* these neurons are part of the *enteric nervous system* (see Chapter 23, p. 686). The peripheral reflexes carry out highly localized autonomic responses involving either small segments of an organ or a few different visceral organs. These reflexes enable the peripheral part of the visceral nervous system to control some of its own activity, making it partly independent of the brain and spinal cord. This fact further illustrates the general concept that the autonomic nervous system operates partly on its own.

CLINICAL APPLICATION

Mass Reflex Reaction The **mass reflex reaction** affects quadriplegics and paraplegics with spinal cord injuries above the level of T_6. The cord injury is followed by a temporary loss of all reflexes inferior to the level of the injury. When reflex activity returns, it is exaggerated because of the lack of inhibitory input from higher (brain) centers. The ensuing episodes of mass reflex reaction involve surges of both visceral and somatic motor output from large regions of the spinal cord. The usual trigger for such an episode is a strong stimulus to the skin or the overfilling of a visceral organ, such as the bladder. During the mass reflex episode, the body goes into flexor spasms, the limbs move wildly, the colon and bladder empty, and profuse sweating occurs. Most seriously, sympathetic activity raises blood pressure to life-threatening levels.

CENTRAL CONTROL OF THE AUTONOMIC NERVOUS SYSTEM

learning outcome

▶ Explain how various regions of the CNS help to regulate the autonomic nervous system.

Though the ANS is not considered to be under direct voluntary control, many of its activities are regulated by the central nervous system. Several sources of central control exist—the brain stem and spinal cord, the hypothalamus and amygdala, and the cerebral cortex (Figure 15.13).

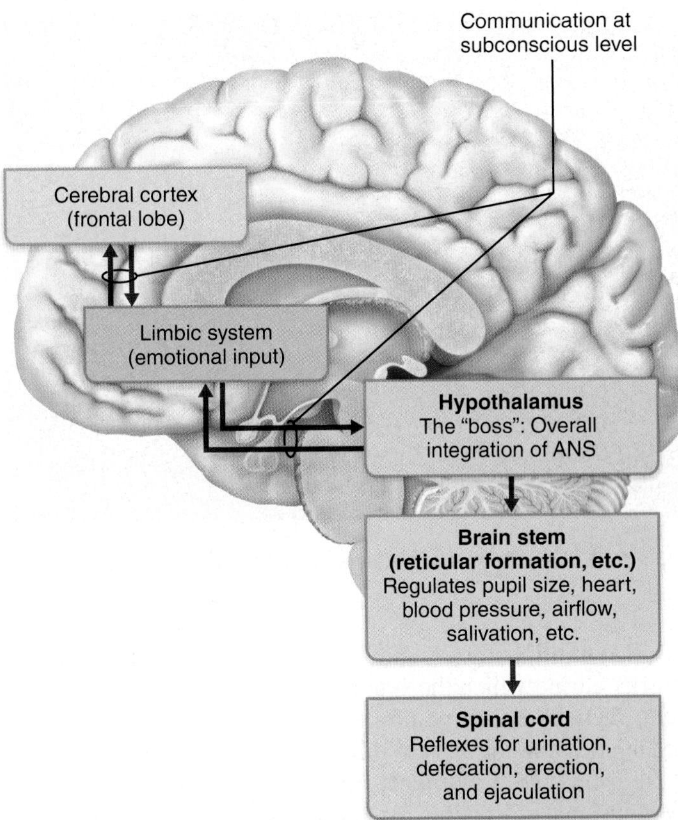

Figure 15.13 **Levels of ANS control.** The hypothalamus is the main integration center of the ANS. Inputs from the cerebral cortex via the limbic lobe can influence hypothalamic functioning.

Control by the Brain Stem and Spinal Cord

The reticular formation of the brain stem appears to exert the most direct influence over autonomic functions. Centers in the medulla oblongata (pp. 379–380) regulate heart rate (*cardiac centers;* Figure 15.12), the diameter of blood vessels *(vasomotor center),* and many digestive activities. Also, the periaqueductal gray matter of the midbrain controls many autonomic functions, especially the sympathetic fear response during a threatening encounter (pp. 381–382).

Control of autonomic functions at the level of the spinal cord involves the spinal visceral reflexes (Figure 15.11). Note, however, that even though the defecation and urination reflexes are integrated by the spinal cord, they are subject to conscious inhibition from the brain. This enables conscious control over when and where to eliminate wastes.

Control by the Hypothalamus and Amygdaloid Body

The main *integration center* of the autonomic nervous system is the hypothalamus (Figure 15.13). In general, the medial and anterior parts of the hypothalamus direct parasympathetic functions, whereas the lateral and posterior parts direct sympathetic functions. These hypothalamic centers influence the preganglionic autonomic neurons in the spinal cord and brain, both through direct connections and through relays in the reticular formation and periaqueductal gray matter. It is through the ANS that the hypothalamus controls heart activity, blood pressure, body temperature, and digestive functions.

Recall that the amygdaloid body is the main limbic region for emotions, including fear (see p. 400). Through communications with the hypothalamus and periaqueductal gray matter, the amygdaloid body stimulates sympathetic activity, especially previously learned fear-related behavior.

Control by the Cerebral Cortex

Even though it was once thought that the autonomic nervous system was not subject to voluntary control by the cerebral cortex, people can exert some conscious control over some autonomic functions by developing control over their thoughts and emotions. For example, feelings of extreme calm achieved during meditation are associated with cerebral cortex influence on the parasympathetic centers in the hypothalamus via various limbic structures. Voluntary sympathetic activation can occur when people decide to recall a frightful experience; in this case the cerebral cortex acts through the amygdaloid body.

☑ check your understanding

☐ 11. Via which pathway do most visceral pain fibers travel to reach the CNS?

☐ 12. What is a peripheral reflex, and how does it differ from a spinal reflex?

☐ 13. Which region of the CNS is the main control center for the ANS?

(For answers, see Appendix B.)

DISORDERS OF THE AUTONOMIC NERVOUS SYSTEM

learning outcome

▶ Briefly describe some diseases of the autonomic nervous system.

Because the ANS is involved in nearly every important process that occurs within the body, it is not surprising that abnormalities of autonomic functioning can have far-reaching effects. Such abnormalities can impair elimination processes and blood delivery, and can even threaten life.

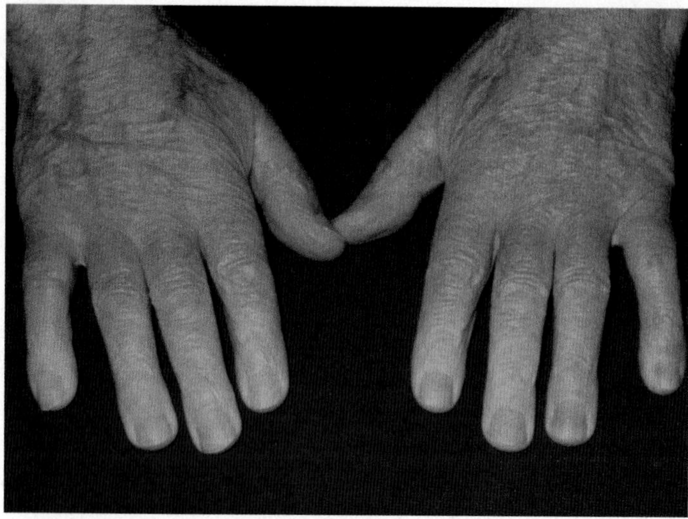

Figure 15.14 **Raynaud's disease in an elderly man.**

Raynaud's disease is characterized by intermittent attacks during which the skin of the fingers and toes becomes pale, then blue and painful (Figure 15.14). When such an attack ends and the vessels dilate, the fingers refill with blood and become red. Commonly provoked by exposure to cold or by emotional stress, this disease is thought to be an exaggerated, sympathetic vasoconstriction response in the affected body regions. The severity of Raynaud's disease ranges from mere discomfort to such extreme constriction of vessels that gangrene (tissue death) results. Treatment typically involves administering drugs that inhibit vasoconstriction, but in severe cases it may be necessary to remove ganglia or to cut the preganglionic sympathetic fibers serving the affected region. Raynaud's disease is rather common among the elderly, affecting 9% of all elderly women and 3% to 5% of elderly men.

Achalasia (ak″ah-la′ze-ah) **of the cardia** is a condition in which some defect in the autonomic innervation of the esophagus results in a loss of that organ's ability to propel swallowed food inferiorly. Additionally, the smooth muscle surrounding the inferior end of the esophagus (cardiac sphincter) remains contracted, preventing the passage of food into the stomach. (*Achalasia* means "failure to relax.") The accumulation of food stretches the esophagus to enormous width, and meals cannot be kept down. The cause of this condition, which usually appears in young adults, is not precisely understood. The preferred treatment is a longitudinal surgical incision through the muscle at the inferior end of the esophagus.

Congenital megacolon, or **Hirschsprung's disease,** is a birth defect in which the parasympathetic and enteric innervation of the distal region of the large intestine fails to develop normally because migrating neural crest cells fail to reach this region (discussed in the next section). Feces and gas accumulate proximal to the immobile bowel segment, greatly distending this area (*megacolon* = enlarged large intestine). The condition is corrected surgically by removing the inactive part of the infant's intestine.

THE AUTONOMIC NERVOUS SYSTEM THROUGHOUT LIFE

learning outcomes

- ▶ Describe the embryonic development of the ANS.
- ▶ List some effects of aging on autonomic functions.

The developmental origins of the structures of the PNS can be summarized in two statements:

- All neurons with cell bodies in the CNS are derived from the neural tube.
- All neurons with cell bodies in the PNS are derived from the neural crest.

Thus, somatic motor neurons and *preganglionic autonomic neurons* form from neuroblasts of the basal plate of the neural tube (Figure 12.16, p. 370); sensory neurons and *postganglionic autonomic neurons* form from the neural crest.

In the development of the sympathetic division (Figure 15.15), some cells migrate ventrally from the neural crest to form the sympathetic trunk ganglia. From there, other cells migrate ventrally to form both the collateral ganglia on the aorta and the adrenal medulla. The sympathetic trunk ganglia and collateral ganglia receive axons from spinal preganglionic neurons, and they in turn send postganglionic axons to the visceral organs.

In the development of the parasympathetic division, the postganglionic neurons also derive from the neural crest and reach the visceral organs by migrating along the growing preganglionic axons.

During youth, impairments of visceral nervous function are usually due to injuries to either the spinal cord or autonomic nerves. With advancing age, the efficiency of the ANS

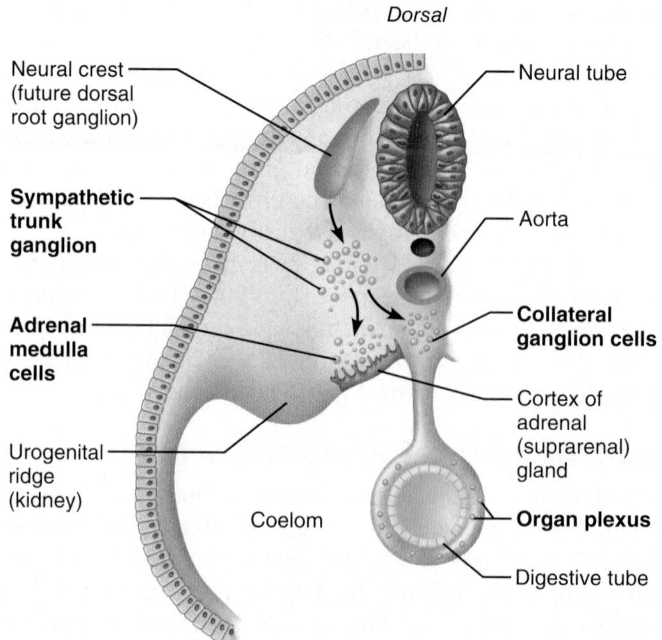

Figure 15.15 **Embryonic development of some sympathetic structures, including the adrenal medulla.** Transverse section through a 5-week embryo.

begins to decline. Elderly people are commonly constipated because the autonomically controlled motility of their gastrointestinal tract is reduced. Frequent eye infections can result from diminished formation of tears, which contain bactericidal enzymes, and the pupils cannot dilate as widely or as quickly. Whenever a young and healthy person rises to a standing position, the sympathetic division induces bodywide vasoconstriction, raising blood pressure so that blood can be pumped to the head and brain. The response becomes sluggish with age, so elderly people may faint if they stand up too quickly. Although these age-related problems are distressing, they usually are not life-threatening and can be easily alleviated. For example, standing up slowly gives the sympathetic nervous system time to adjust blood pressure, eye drops are available for dry eyes, and drinking ample fluid helps alleviate constipation.

☑ check your understanding

☐ 14. Which embryonic tissue forms the postganglionic neurons of the ANS?

☐ 15. Which division of the ANS is deficient in achalasia of the cardia? Which division malfunctions in Raynaud's disease?

(For answers, see Appendix B.)

RELATED CLINICAL TERMS

Atonic bladder (a-tahn'ik; "without tone") A condition in which the bladder becomes flaccid and overfills, allowing urine to dribble out. Atonic bladder results from the temporary loss of the micturition reflex following injury to the spinal cord.

Vagotomy (va-got'o-me) Cutting or severing of a vagus nerve, often to decrease the secretion of stomach acid and other caustic digestive juices that aggravate ulcers.

CHAPTER SUMMARY

Take full advantage of the invaluable online practice and study tools for this chapter by going to the Study Area of MasteringA&P®. There you will find:

- Art Labeling Activities, Quizzes, Practice Tests, Memory Games
- Animal Organ Dissection, Cat Dissection and Bone Videos
- PAL Cadaver
- PAL Histology
- A&P Flix Animations
- Dynamic Study Modules

1. The ANS is the system of motor neurons that innervates smooth muscle, cardiac muscle, and glands. It is the general visceral motor division of the PNS. The ANS largely operates below the level of consciousness.

OVERVIEW OF THE AUTONOMIC NERVOUS SYSTEM (pp. 468–472)

Comparison of the Autonomic and Somatic Motor Systems (pp. 468–470)

2. In the somatic motor division of the nervous system, a single motor neuron forms the pathway from the CNS to the skeletal muscle cells. Autonomic motor pathways, by contrast, consist of chains of two neurons: the preganglionic neuron, whose cell body is in the CNS, and the postganglionic neuron, whose cell body is in an autonomic ganglion. The postganglionic axon innervates the visceral organs.

Divisions of the Autonomic Nervous System (pp. 470–472)

3. The ANS has parasympathetic and sympathetic divisions. Both innervate many of the same organs but produce opposite effects (Table 15.2, p. 475). The sympathetic division prepares the body for "fight or flight," whereas the parasympathetic division is active during "rest and digest."

4. The parasympathetic division exits the CNS on cranial and sacral nerves and has comparatively long preganglionic axons. The sympathetic division exits on thoracic and lumbar nerves and has comparatively long postganglionic axons.

5. The two divisions differ in the neurotransmitters they release at the effector organ. Acetylcholine is released by parasympathetic postganglionic fibers, whereas norepinephrine is released by most sympathetic postganglionic fibers.

THE PARASYMPATHETIC DIVISION (pp. 472–473)

Cranial Outflow (pp. 472–473)

6. Cranial parasympathetic fibers arise in the brain stem nuclei of cranial nerves III, VII, IX, and X and synapse in ganglia in the head, thorax, and abdomen. Axons in cranial nerve (CN) III serve smooth musculature in the eye via a synapse in the ciliary ganglion. Axons in CN VII serve the submandibular, sublingual, lacrimal, and nasal glands and synapse in the submandibular and pterygopalatine ganglia. Axons in CN IX serve the parotid gland via a synapse in the otic ganglion.

7. Parasympathetic fibers in the vagus nerve (CN X) innervate organs in the thorax and most of the abdomen, including the heart, lungs, esophagus, stomach, liver, and most of the intestines. The axons in the vagus are preganglionic. Almost all postganglionic neurons are located in intramural ganglia within the organ walls.

Sacral Outflow (p. 473)

8. Sacral parasympathetic pathways innervate the pelvic viscera. The preganglionic axons exit from the visceral motor region of the gray matter of the spinal cord (S_2–S_4) and form the pelvic splanchnic nerves. Most of these axons synapse in intramural ganglia in the pelvic organs.

THE SYMPATHETIC DIVISION (pp. 473–480)

Basic Organization (pp. 473–476)

9. The preganglionic sympathetic cell bodies are in the lateral horn of the spinal gray matter from the level of T_1 to L_2.

10. The sympathetic division supplies some peripheral structures that the parasympathetic division does not: arrector pili, sweat glands, and the smooth muscle of blood vessels.

11. Sympathetic ganglia include 22–24 pairs of sympathetic trunk ganglia (also called chain ganglia and paravertebral ganglia), which are linked together to form sympathetic trunks on both sides of the vertebral column, and unpaired collateral ganglia (prevertebral ganglia), most of which lie on the aorta in the abdomen.

PAL Human Cadaver/Nervous System/Autonomic Nervous System

Sympathetic Pathways (pp. 477–480)

12. Every preganglionic sympathetic axon leaves the lateral gray horn of the thoracolumbar spinal cord through a ventral root and spinal nerve. From there, it runs in a white ramus communicans to a sympathetic trunk ganglion or collateral ganglion, where it synapses with the postganglionic neuron that extends to the visceral effector. Many preganglionic axons ascend or descend in the sympathetic trunk to synapse in a ganglion at another body level.

13. In the sympathetic pathway to the *body periphery* (to innervate arrector pili, sweat glands, and peripheral blood vessels), the preganglionic axons synapse in the sympathetic trunk ganglia, and the postganglionic axons run in gray rami communicantes to the dorsal and ventral rami of the spinal nerves for peripheral distribution.

14. In the sympathetic pathway to the *head,* the preganglionic axons ascend in the sympathetic trunk and synapse in the superior cervical ganglion. From there, most postganglionic axons associate with a large artery that distributes them to the glands and smooth musculature of the head.

15. In the sympathetic pathway to *thoracic organs,* most preganglionic axons synapse in the nearest sympathetic trunk ganglion, and the postganglionic axons run directly to the organs (lungs, esophagus). Many postganglionic axons to the heart, however, descend from the cervical ganglia in the neck.

16. In the sympathetic pathway to *abdominal organs,* the preganglionic axons run in splanchnic nerves to synapse in collateral ganglia on the aorta. From these ganglia, the postganglionic axons follow large arteries to the abdominal viscera (stomach, liver, kidney, and most of the large intestine).

17. In the sympathetic pathway to *pelvic organs,* the preganglionic axons synapse in sympathetic trunk ganglia or in collateral ganglia on the aorta, sacrum, and pelvic floor. Postganglionic axons travel through the most inferior autonomic plexuses to the pelvic organs.

The Role of the Adrenal Medulla in the Sympathetic Division (p. 480)

18. The adrenal glands, one superior to each kidney, contain a medulla of modified postganglionic sympathetic neurons that secrete the hormones epinephrine (adrenaline) and norepinephrine into the blood. The result is the "surge of adrenaline" felt during excitement.

19. The cells of the adrenal medulla are innervated by preganglionic sympathetic neurons, which signal them to secrete the hormones.

VISCERAL SENSORY NEURONS (p. 481)

20. General visceral sensory neurons monitor temperature, pain, irritation, chemical changes, and stretch in the visceral organs.

21. Visceral sensory fibers run within the autonomic nerves, especially within the vagus and the sympathetic nerves. Most pain fibers from the visceral organs of the body follow the sympathetic pathway to the CNS and are carried in the spinothalamic tract to the thalamus and then to the visceral sensory cortex.

22. Visceral pain is induced by stretching, infection, and cramping of internal organs but seldom by cutting or scraping these organs. Visceral pain is referred to somatic regions of the body that receive innervation from the same spinal cord segments.

VISCERAL REFLEXES (pp. 481–482)

23. Many visceral reflexes are spinal reflexes, such as the defecation reflex. Others, such as the baroreceptor reflex, involve integration centers in the brain stem. Some visceral reflexes, however, involve only peripheral neurons.

CENTRAL CONTROL OF THE AUTONOMIC NERVOUS SYSTEM (p. 483)

24. Visceral motor functions are influenced by the medulla oblongata, the periaqueductal gray matter, spinal visceral reflexes, the hypothalamus, the amygdaloid body, and the cerebral cortex. Some people can voluntarily regulate some autonomic activities by gaining extraordinary control over their emotions.

DISORDERS OF THE AUTONOMIC NERVOUS SYSTEM (pp. 483–484)

25. Most autonomic disorders reflect problems with the control of smooth muscle. Raynaud's disease and hypertension result from abnormalities in vascular control.

THE AUTONOMIC NERVOUS SYSTEM THROUGHOUT LIFE (pp. 484–485)

26. Preganglionic neurons develop from the neural tube. Postganglionic neurons and visceral sensory neurons develop from the neural crest.

27. The efficiency of the ANS declines in old age: Gastrointestinal motility and production of tears decrease, and the vasomotor response to standing slows.

REVIEW QUESTIONS

Multiple Choice/Matching Questions

(For answers, see Appendix B.)

1. Which of the following does *not* characterize the ANS? (a) two-neuron motor chains, (b) preganglionic cell bodies in the CNS, (c) presence of postganglionic cell bodies in ganglia, (d) innervation of skeletal muscle.

2. For each of the following terms or phrases, write either S (for *sympathetic*) or P (for *parasympathetic*) division of the autonomic nervous system.

____ (1) short preganglionic axons, long postganglionic axons

____ (2) intramural ganglia

____ (3) craniosacral outflow

____ (4) adrenergic fibers

____ (5) cervical ganglia of the sympathetic trunk

____ (6) otic and ciliary ganglia

____ (7) more widespread response

____ (8) increases heart rate, respiratory rate, and blood pressure

____ (9) increases motility of stomach and secretion of lacrimal and salivary glands

____ (10) innervates blood vessels

____ (11) most active when you are lolling in a hammock

____ (12) most active when you are running in the Boston Marathon

____ (13) gray rami communicantes

____ (14) synapse in celiac ganglion

____ (15) relates to fear response induced by the amygdaloid body

3. The thoracic splanchnic nerves contain which kind of axons? (a) preganglionic parasympathetic, (b) postganglionic parasympathetic, (c) preganglionic sympathetic, (d) postganglionic sympathetic.
4. The collateral ganglia contain which kind of cell bodies? (a) preganglionic parasympathetic, (b) postganglionic parasympathetic, (c) preganglionic sympathetic, (d) postganglionic sympathetic.
5. Preganglionic sympathetic neurons develop from (a) neural crest, (b) neural tube, (c) alar plate, (d) endoderm.
6. Which of the following is the best way to describe how the ANS is controlled? (a) completely under control of voluntary cerebral cortex, (b) entirely controls itself, (c) completely under control of brain stem, (d) little control by cerebrum, major control by hypothalamus and amygdaloid body, and major control by spinal and peripheral reflexes.
7. The white rami communicantes contain what kind of axons? (a) preganglionic parasympathetic, (b) postganglionic parasympathetic, (c) preganglionic sympathetic, (d) postganglionic sympathetic.
8. Collateral sympathetic ganglia are involved with the innervation of the (a) abdominal organs, (b) thoracic organs, (c) head, (d) arrector pili, (e) all of these.

Short Answer Essay Questions

9. Is the visceral sensory nervous system part of the autonomic nervous system? Explain your answer.
10. (a) Describe the anatomical relationship of the white and gray rami communicantes to a spinal nerve, and to the dorsal and ventral rami. (b) Why are gray rami communicantes gray?
11. What effect does sympathetic activation have on each of the following structures? Sweat glands, pupils of the eye, adrenal medulla, heart, lungs, blood vessels of skeletal muscles, blood vessels of digestive viscera, salivary glands.
12. Which of the effects listed in question 11 would be reversed by parasympathetic activity?
13. A friend asks you how the parasympathetic division, which comes only from cranial and sacral regions, is able to innervate organs in the thorax and abdomen. How would you answer that question?
14. What manifestations of decreased ANS efficiency are seen in elderly people?
15. The students in anatomy class were having difficulty understanding what kind of cell bodies are located in the autonomic ganglia. Frustrated by being asked this same question repeatedly, the exasperated teaching assistant stood on a desk and yelled, "Autonomic ganglia always contain the *postganglionic* cell bodies!" Describe how that information enabled the students to work out which type of neuron has cell bodies in the (a) sympathetic trunk ganglia, (b) intramural ganglia, and (c) head ganglia (ciliary, pterygopalatine, submandibular, and otic).
16. Describe the sympathetic pathways to the (a) submandibular salivary gland, (b) bladder, (c) sweat glands in the skin, (d) adrenal medulla, (e) heart, (f) stomach.
17. Describe the parasympathetic pathways to the (a) smooth muscles in the eye, (b) bladder, (c) small intestine.
18. Describe the path of the vagus through the neck, thorax, and abdomen.
19. What structures are innervated by the *sacral* part of the parasympathetic division?
20. How do the somatic motor and visceral motor (ANS) pathways differ anatomically and functionally?

Critical Reasoning & Clinical Application Questions

1. A 1-year-old infant has a swollen abdomen and is continually constipated. On examination, a mass is felt over the distal colon, and X-ray findings show the colon to be greatly distended in that region. From the information in this chapter, identify one possible cause for these symptoms.
2. Roweena, a high-powered marketing executive, develops a stomach ulcer. She complains of a deep abdominal pain that she cannot quite locate, plus superficial pain on her abdominal wall. Exactly where on the abdominal wall is the superficial pain most likely to be located? (Refer to Figure 15.10.)
3. Constipated people often have hardened feces that can tear the inner lining of the rectum and anus during defecation. Those with tears in the skin of the anus complain of sharp pain, whereas those with tears higher up in the intestinal mucosa do not complain of such pain. How do you explain this difference?
4. A woman with a history of kidney infections was feeling pain in the skin of the lower part of her abdomen and even in her thighs. Does this make sense? Why or why not?
5. Which clinical condition has the classical symptoms of blue fingertips that later turn bright red?
6. Why does a person with a spinal cord injury have no deficits in digestive function?
7. Suppose that a mad scientist seeking to invent a death ray produces a beam that destroys a person's ciliary, pterygopalatine, and submandibular ganglia (and nothing else). List all the signs that would be apparent in the victim. Would the victim die, or would the scientist have to go back to the laboratory to try again?
8. Job-related stress is a concern for people in many occupations. One response to stress is an increase in sympathetic nerve activity. What are some health-related problems that can result from chronic sympathetic stimulation?

16 The Special Senses

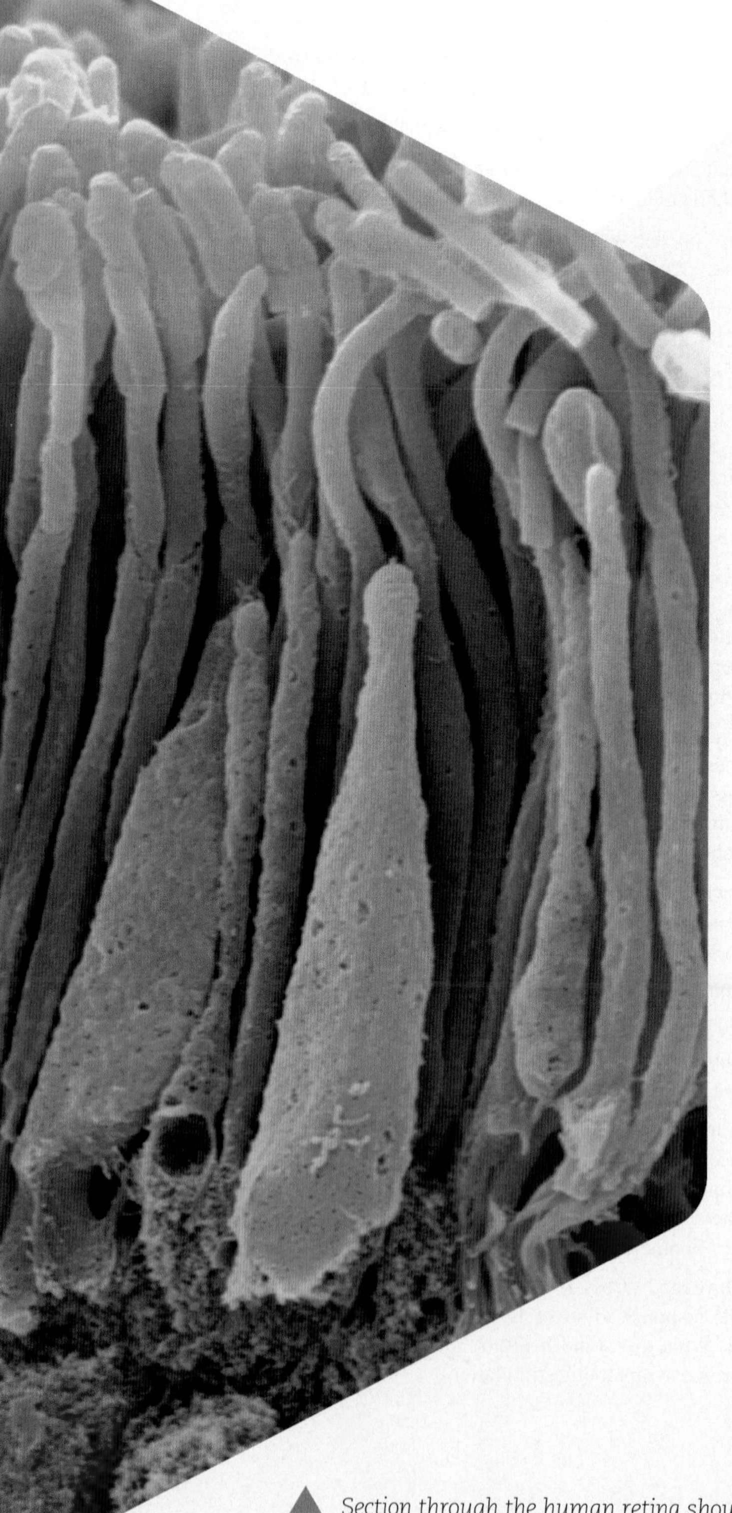

People are responsive creatures. The aroma of baking bread makes our mouths water; lightning makes us blink; thunder makes us recoil. Many such sensory stimuli reach us each day and are processed by the nervous system.

We are usually told that we have five senses: touch, taste, smell, sight, and hearing. Actually, touch is a large group of general senses (considered in Chapter 14); the other four traditional senses—*smell, taste, sight,* and *hearing*—are called *special senses*. Receptors for a fifth special sense—*equilibrium,* or the sense of balance—are located in the ear.

▲ *Section through the human retina showing bipolar neurons (red), rod cells(white), and yellow cone cells (colored SEM).*

The receptors for the general senses are widely distributed; in contrast, the **special sensory receptors** are localized and confined to the head region. The receptors for the special senses are not free endings of sensory neurons but distinct **receptor cells.** These are neuronlike epithelial cells or small peripheral neurons that transfer sensory information to other neurons in afferent pathways to the brain. The special sensory receptors are housed either in complex sensory organs (eye or ear) or in distinctive epithelial structures (taste buds or olfactory epithelium). Their sensory information travels to the brain via cranial nerves.

This chapter explores the functional anatomy of the five special senses: the chemical senses of taste and smell, which are special visceral senses (VS) (p. 352), and the special somatic senses (SS) of vision in the eye, and hearing and equilibrium in the ear.

THE CHEMICAL SENSES: TASTE AND SMELL

learning outcomes

- Describe the receptors for taste and smell.
- Describe the paths by which sensory information from these receptors travels to the brain.

The receptors for taste (gustation) and smell (olfaction) are classified as **chemoreceptors** because they respond to chemical substances—to food chemicals dissolved in saliva and to airborne chemicals that dissolve in fluids on the nasal membranes, respectively.

Taste (Gustation)

Taste Buds

The taste receptors occur in **taste buds** in the mucosa of the mouth and pharynx. The majority of the 10,000 or so taste buds are on the surface of the tongue; a few others occur on the posterior region of the palate (roof of the mouth), on the inner surface of the cheeks, on the posterior wall of the pharynx, and on the epiglottis (a leaf-shaped flap behind the tongue).

Most taste buds occur in peglike projections of the tongue mucosa called **papillae** (pah-pil′e) (Figure 16.1a). Taste buds occur within the epithelium that covers the papillae. In the small **fungiform papillae** scattered over the entire surface of the tongue, the taste buds are on the apical surface. In the large **vallate papillae** arranged in an inverted V near the back of the tongue, and in the **foliate papilla** on the posterolateral surface of the tongue, the taste buds are in the side walls (Figure 16.1b).

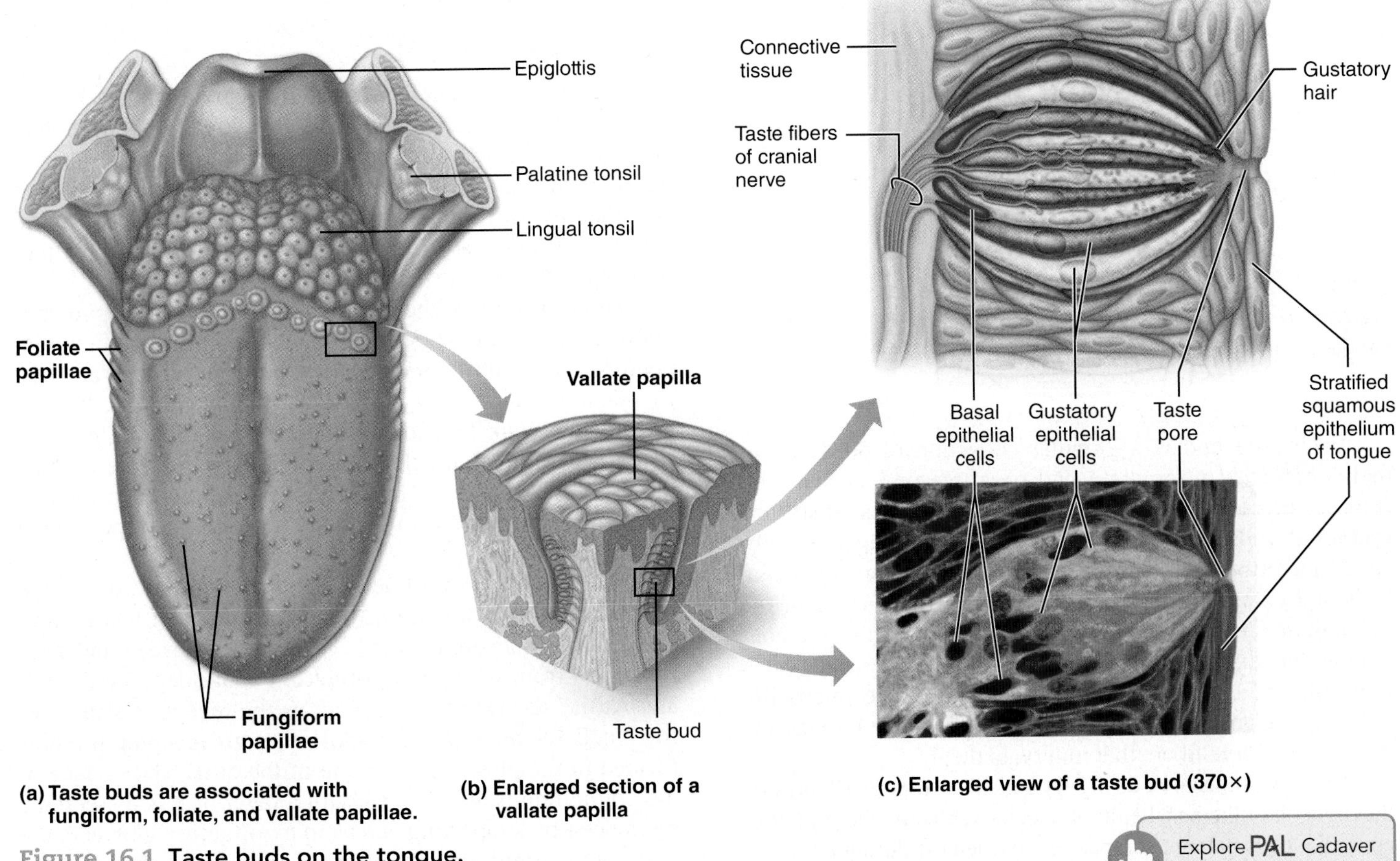

(a) Taste buds are associated with fungiform, foliate, and vallate papillae.

(b) Enlarged section of a vallate papilla

(c) Enlarged view of a taste bud (370×)

Figure 16.1 **Taste buds on the tongue.**

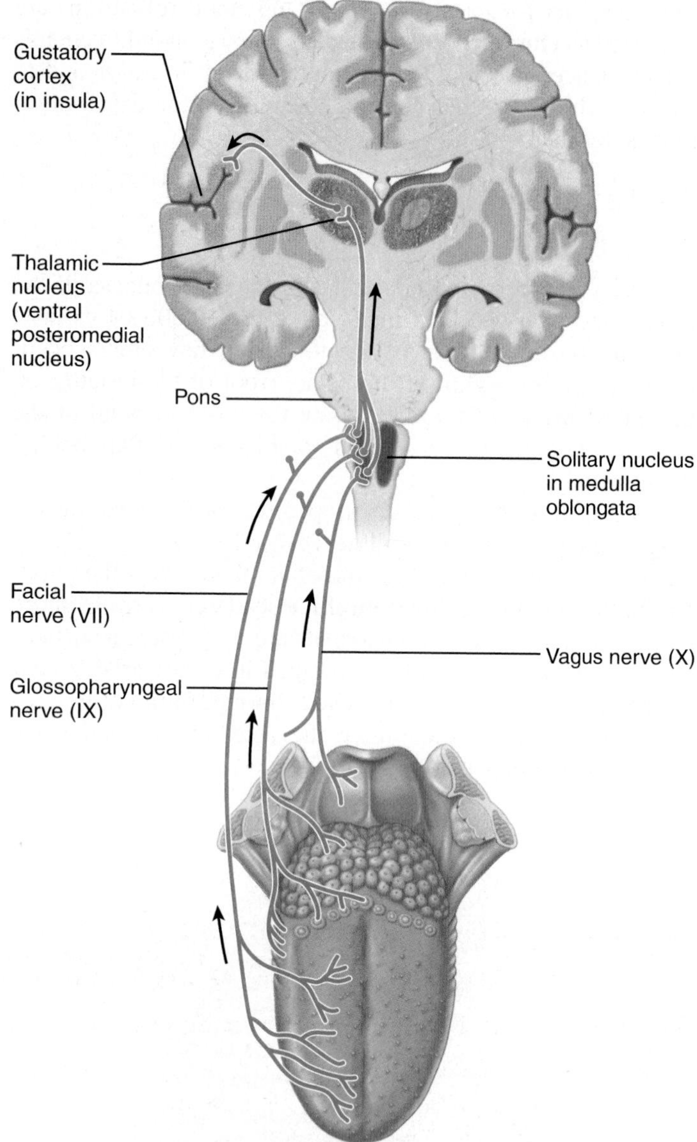

Figure 16.2 The gustatory pathway. Taste sensations are carried from the taste buds to the gustatory area of the cerebral cortex.

Each taste bud is a globular collection of 50–100 epithelial cells that resemble a bud on a tree or a closed tulip (Figure 16.1c). Each contains two major cell types: **gustatory epithelial cells** and **basal epithelial cells.** Long microvilli called **gustatory hairs** project from the gustatory epithelial cells and extend through a **taste pore** to the surface of the epithelium. There, these microvilli are bathed in saliva containing the dissolved molecules that stimulate taste. Such molecules bind to the plasma membrane of the microvilli, inducing the gustatory epithelial cells to generate impulses in the sensory nerve fibers that innervate them.

The cells in taste buds are replenished every 7–10 days by the division of the basal epithelial cells, replacing the gustatory epithelial cells that are scraped and burned off during eating. If an entire taste bud is destroyed, a new one will form after its nerve ending grows back into the regenerating epithelium.

Taste Sensations and the Gustatory Pathway

Taste sensations can be described in terms of five basic qualities: sweet, sour, salty, bitter, and umami. Umami (Japanese for "deliciousness") was recognized as a distinct fifth basic taste in the 1980s. It is elicited by a substance called glutamate that is found naturally in meat, aged cheeses, and tomatoes. Although maps that assign specific taste sensations to specific areas of the tongue are common, researchers have long known that these mapped areas are dubious and that all modalities of taste can be elicited from all areas that contain taste buds.

Taste information reaches the brain stem and cerebral cortex through the **gustatory pathway** (Figure 16.2). Sensory fibers carrying taste information occur in three cranial nerves:

- The *facial nerve* (VII) transmits impulses from taste receptors in the anterior two-thirds of the tongue.
- The *glossopharyngeal nerve* (IX) carries sensations from the tongue's posterior third, as well as from the few buds in the pharynx.
- The *vagus nerve* (X) carries taste impulses from the few taste buds on the epiglottis and lower pharynx.

All the sensory neurons that carry taste information synapse in a nucleus in the medulla called the *solitary nucleus.* From there, impulses are transmitted to the thalamus and ultimately to the gustatory area of the cerebral cortex in the insula lobe.

Smell (Olfaction)

The receptors for smell lie in a patch of epithelium on the roof of the nasal cavity (Figure 16.3a). Specifically, the receptors are a part of the **olfactory epithelium** that covers the superior nasal concha and the superior part of the nasal septum and is bathed by swirling air that has been inhaled into the nasal cavity. Sniffing draws more air across the olfactory epithelium and thus intensifies the sense of smell.

The olfactory epithelium is a pseudostratified columnar epithelium (Figure 16.3b) that contains millions of bipolar neurons called **olfactory sensory neurons.** These are surrounded by columnar **supporting epithelial cells.** At the base of the epithelium lie short **olfactory stem cells,** undifferentiated neuroepithelial cells that continually form new olfactory sensory neurons. Thus, olfactory sensory neurons are among the few neurons in the body that undergo replacement throughout adult life.

The cell bodies of olfactory sensory neurons are located in the olfactory epithelium. Each receptor cell has an apical dendrite that projects to the epithelial surface and ends in a knob from which long **olfactory cilia**, or "hairs," radiate. At the surface these cilia act as the receptive structures for smell by binding odor molecules to receptor proteins located in the plasma membrane of the cilia. The surface of the epithelium is also coated with a layer of mucus secreted by the nearby supporting cells and by olfactory glands in the underlying connective tissue (lamina propria). This mucus, which captures and dissolves odor molecules from the air, is renewed continuously, flushing away old odor molecules

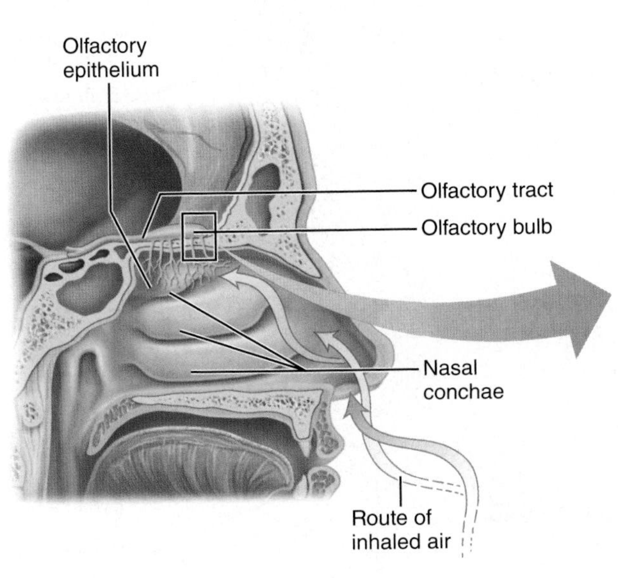

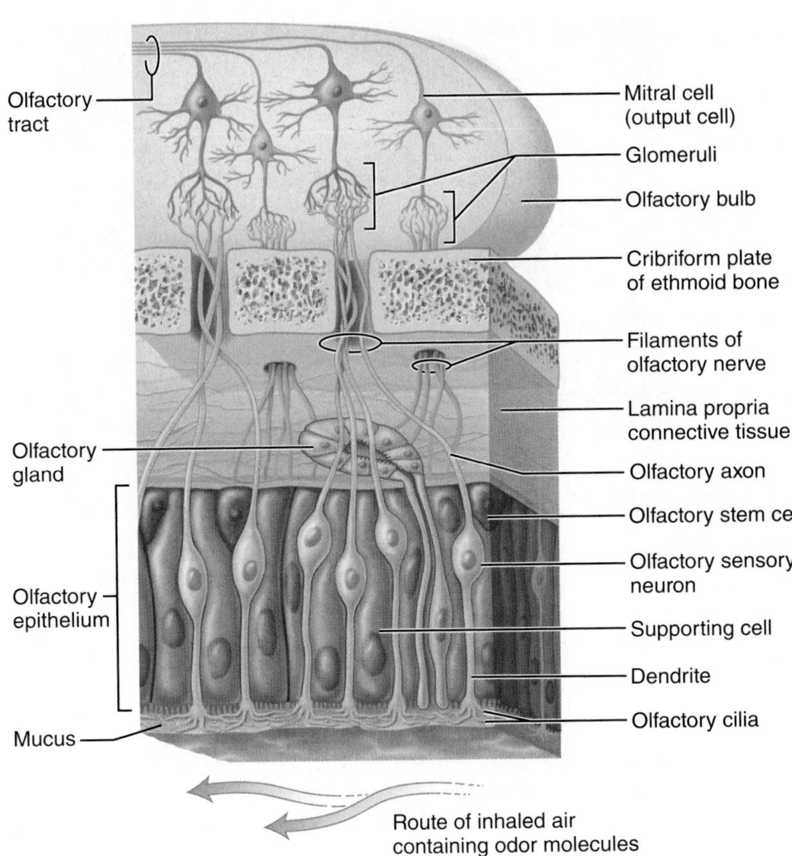

Figure 16.3 Olfactory receptors. (**a**) Site of the olfactory epithelium in the superior nasal cavity. (**b**) Enlarged view of the olfactory epithelium, showing the course of the filaments of the olfactory nerve.

so that new odors always have access to the olfactory cilia. Unlike other cilia in the body, olfactory cilia are largely immotile.

Each olfactory sensory neuron has an axon that enters the connective tissue of the lamina propria (Figure 16.3b). There, axons gather into nerve bundles called **filaments of the olfactory nerve** (cranial nerve I), which penetrate the cribriform plate of the ethmoid bone and enter the overlying **olfactory bulb** of the forebrain. In this bulb, the olfactory nerve axons branch profusely and synapse with neurons called **mitral cells** (mi′tral; "cap-shaped") in complex synaptic clusters called **glomeruli** (glo-mer′u-li; "balls of yarn"). The mitral cells then relay the olfactory information to other parts of the brain.

Upon receiving stimuli at synapses with olfactory sensory neurons, mitral cells transmit the impulses along the olfactory tract to

1. The limbic region, where smells elicit emotions, and
2. The primary olfactory cortex (Figure 13.13b, p. 392).

The olfactory cortex processes olfactory information into a conscious perception of odor; it also sends this information through a thalamic relay to the orbitofrontal cortex (also seen in Figure 13.13b), where the smells are analyzed and compared to other smells.

Embryonic Development of the Chemical Senses

The development of the olfactory epithelium and taste buds is straightforward. Olfactory epithelium derives from paired olfactory placodes, which are platelike thickenings of the surface ectoderm on the embryonic snout region (see Figure 22.19, p. 670). Taste buds develop, upon stimulation by gustatory nerves, from the epithelium lining the embryonic mouth and pharynx, an epithelium that is derived from ectoderm and endoderm.

Disorders of the Chemical Senses

Disorders of both taste and olfaction exist, but olfactory disorders are more common. Absence of the sense of smell, called **anosmia** (an-oz′me-ah; "without smell"), typically results from blows to the head that tear the olfactory nerves, from colds or allergies that produce excessive mucus in the nasal cavity, or from physical blockage of the nasal cavity (by polyps, for example). Surprisingly, the cause in one-third of all cases of loss of chemical senses is zinc deficiency, and a prescribed dietary zinc supplement effects a rapid cure.

Brain disorders can distort the sense of smell. Some people have **uncinate** (un′sĭ-nāt) **fits,** olfactory hallucinations in which they perceive some imaginary odor, such as that of gasoline or rotting meat. Uncinate fits are so called because the primary

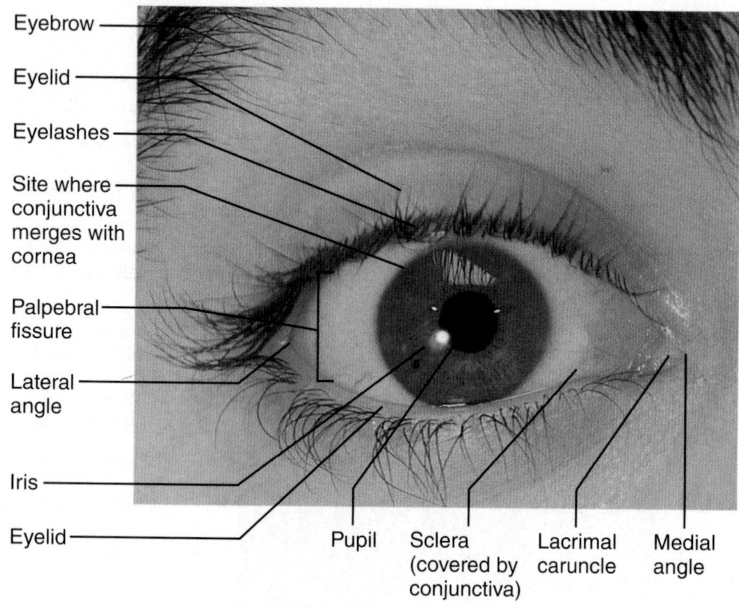

(a) Surface anatomy of the right eye

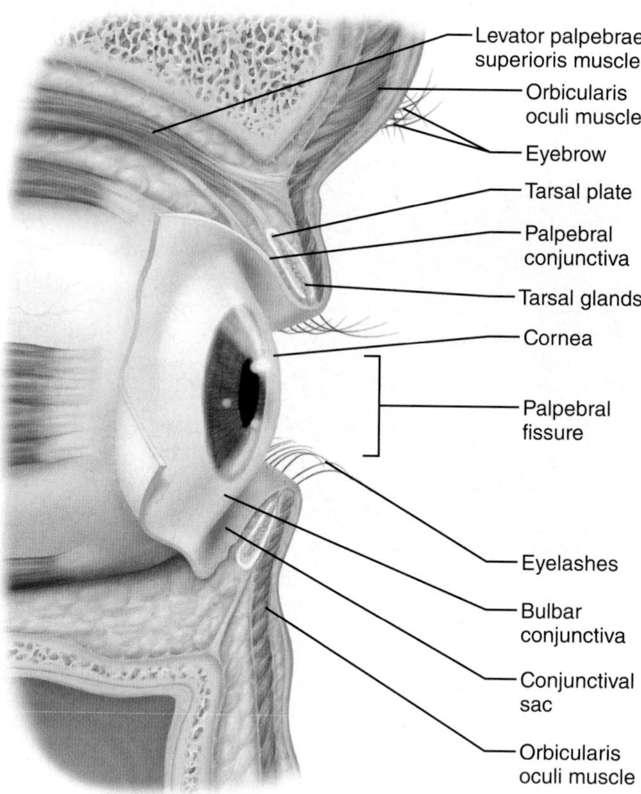

(b) Lateral view; some structures shown in sagittal section

Figure 16.4 **Accessory structures of the eye.**

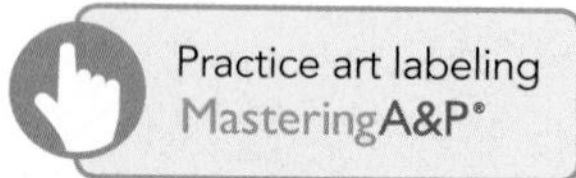

olfactory cortex is in the uncinate region, or uncus, of the cerebrum (see p. 394). These fits may result from irritation of the olfactory pathway by brain surgery or head trauma. Olfactory auras—smells imagined by some epileptics just before they go into a seizure—are brief uncinate fits.

check your understanding

- ☐ 1. What type of tissue forms the receptor cells for taste?
- ☐ 2. What type of cell are the olfactory receptors? When these cells are damaged, are they replaced?

(For answers, see Appendix B.)

THE EYE AND VISION

Vision is the dominant sense in humans: Approximately 70% of the sensory receptors in the body are in the eyes, and 40% of the cerebral cortex is involved in processing visual information (see p. 393). The visual receptor cells (photoreceptors) sense and encode the patterns of light that enter the eye; subsequently, the brain invests these signals with meaning, fashioning visual images of the world.

The visual organ is the **eye,** a spherical structure with a diameter of about 2.5 cm (1 inch). Only the anterior one-sixth of the eye's surface is visible; the rest of the eye lies in the cone-shaped bony *orbit,* where it is surrounded by a protective cushion of fat. Behind the eye, the posterior half of the orbit contains the optic nerve, the arteries and veins to the eye, and the extrinsic eye muscles.

Accessory Structures of the Eye

learning outcomes

- Describe the anatomy and function of the accessory structures of the eye.
- Describe the movements produced by each of the six extrinsic eye muscles and the cranial nerve that innervates each.

Eyebrows

The **eyebrows** consist of coarse hairs in the skin on the superciliary arches (brow ridges of the skull). They shade the eyes from sunlight and prevent perspiration running down the forehead from reaching the eyes.

Eyelids

Anteriorly, the eyes are protected by the mobile **eyelids,** or **palpebrae** (pal′pĕ-bre). The upper and lower lids are separated by the **palpebral fissure** (eye slit) and meet each other at the medial and lateral angles, or eye corners (Figure 16.4a). The medial angle contains a reddish elevation called the **lacrimal caruncle** (kar′ung-k′l; "a bit of flesh"). Glands here produce the gritty "eye sand" reputedly left by the legendary Sandman at night. In most Asian people, a vertical fold of skin called the *epicanthic fold* occurs on both sides of the nose and sometimes covers the medial angle.

The eyelids are thin, skin-covered folds supported internally by connective tissue structures called **tarsal plates** (Figure 16.4b). These stiff plates give the eyelids their curved shape and serve as attachment sites for the eye-closing muscle, the *orbicularis oculi* (see p. 279).

The **levator palpebrae superioris** ("lifter of the upper eyelid") is the skeletal muscle that voluntarily opens the eye. It runs anteriorly from the posterior roof of the orbit, enters the upper eyelid, and inserts on the tarsal plate. The inferior part of the aponeurosis of this muscle contains fibers of smooth muscle, called the *superior tarsal muscle,* an involuntary muscle that prevents the upper eyelid from drooping (discussed previously on p. 478).

Projecting from the free margin of each eyelid are the **eyelashes.** Because the follicles of these hairs are richly innervated by nerve endings, even slight pressure on the eyelashes will trigger reflexive blinking.

Several types of glands occur in the eyelids. **Tarsal glands** are modified sebaceous glands embedded in the tarsal plates (Figure 16.4b). About 25 of these vertical glands line up side by side in the upper eyelid; fewer than this line up in the lower lid. The tarsal gland ducts open along the edge of the eyelids. They release an oil that lubricates the surface of the eye. Other glands are associated with the hair follicles of the eyelashes: These *ciliary glands* (*cilium* = eyelash) include typical sebaceous glands whose ducts open into the hair follicles, and modified sweat glands that lie between the follicles. Infection of a tarsal gland results in an unsightly cyst called a **chalazion** (kah-la′ze-on; "swelling"), whereas infection of the ciliary glands is called a **sty.** Home treatment for both includes warm compresses and medicated ointment or eye wash.

Conjunctiva

The **conjunctiva** (con″junk-ti′vah; "joined together") is a transparent mucous membrane that covers the inner surfaces of the eyelids as the **palpebral conjunctiva** and folds back over the anterior surface of the eye as the **bulbar conjunctiva** (Figure 16.4b). The bulbar conjunctiva, which covers the white of the eye but not the cornea (the transparent tissue over the iris and pupil; Figure 16.4a), is a very thin membrane, and blood vessels are clearly visible beneath it. (These vessels are responsible for "bloodshot" eyes.) When an eye is closed, the slitlike space that forms between the eye surface and the eyelids is the **conjunctival sac,** which is where a contact lens would lie.

Microscopically, the conjunctiva consists of a stratified columnar epithelium underlain by a thin lamina propria of loose connective tissue. Its epithelium contains scattered goblet cells that secrete a lubricating mucus that prevents the eyes from drying. A deficiency of vitamin A, a vitamin required for maintaining epithelia throughout the body, prevents the conjunctiva from secreting mucus. As a result, the conjunctiva dries up and becomes scaly, impairing vision.

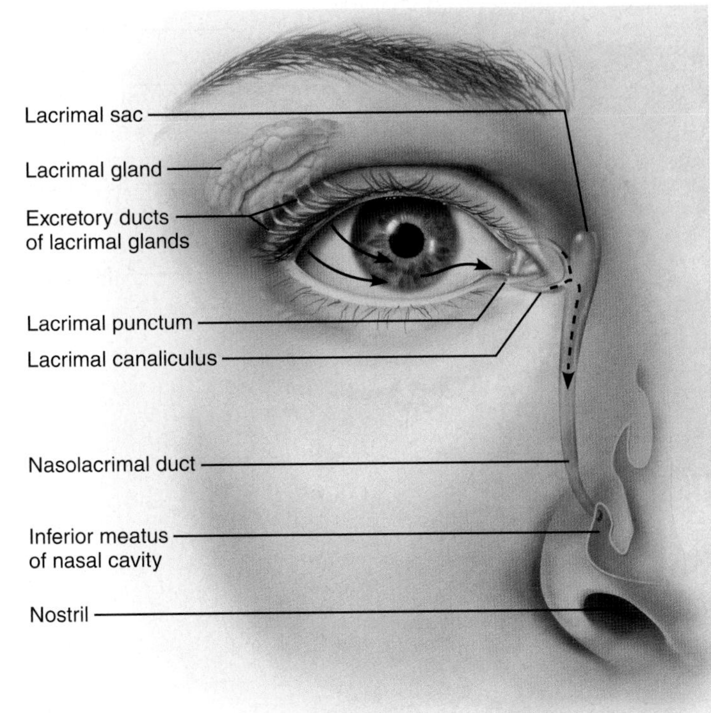

Figure 16.5 **The lacrimal apparatus.** Arrows indicate the direction of flow of lacrimal fluid (tears) from the lacrimal gland to the nasal cavity.

Inflammation of the conjunctiva, called **conjunctivitis,** is a relatively common condition. Conjunctivitis irritates the eyes and makes them red. A highly contagious form of conjunctivitis caused by bacteria or viruses is called **pinkeye.**

Lacrimal Apparatus

The **lacrimal apparatus** (lak′rĭ-mal; "tear"), which keeps the surface of the eye moist with lacrimal fluid (tears), consists of a gland and ducts that drain the lacrimal fluid into the nasal cavity (Figure 16.5). The **lacrimal gland,** lying in the orbit superolateral to the eye, produces lacrimal fluid, which enters the superior part of the conjunctival sac through several small excretory ducts. Blinking the eye spreads this fluid inferiorly across the eyeball to the medial angle. At the medial angle, each lid contains a tiny opening called the **lacrimal punctum** ("puncture") which empties into a small tube, the **lacrimal canaliculus** ("small canal"). From these canals, the fluid drains into the **lacrimal sac** in the medial orbital wall. Finally, the fluid enters the **nasolacrimal duct,** which empties into the nasal cavity at the inferior nasal meatus. Because lacrimal fluid ultimately empties into the nasal cavity, people sniffle when they cry. In people with colds, the viral infection in the nasal passages can spread into the nasolacrimal duct and lacrimal sac, producing inflammation that causes these passages to swell shut. The resulting blockage prevents the drainage of lacrimal fluid from the eye surface; the fluid accumulates, and the eyes water.

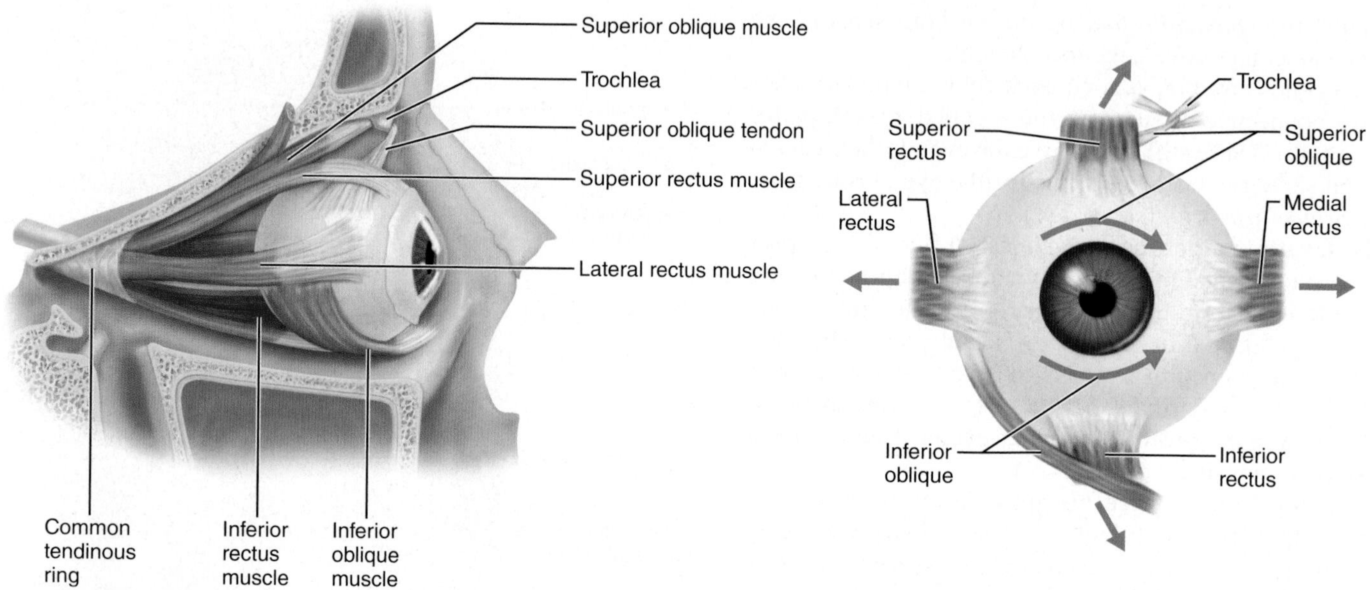

(a) Lateral view of the right eye

(b) Anterior view of the right eye

Muscle	Action	Controlling cranial nerve
Lateral rectus	Moves eye laterally	VI (abducens)
Medial rectus	Moves eye medially	III (oculomotor)
Superior rectus	Elevates eye and turns it medially	III (oculomotor)
Inferior rectus	Depresses eye and turns it medially	III (oculomotor)
Inferior oblique	Elevates eye and turns it laterally	III (oculomotor)
Superior oblique	Depresses eye and turns it laterally	IV (trochlear)

(c) Summary of muscle actions and innervating cranial nerves

Figure 16.6 **Extrinsic eye muscles.** In (b), arrows indicate eye movement resulting from contraction of each muscle.

Lacrimal fluid contains mucus, antibodies, and **lysozyme,** an enzyme that destroys bacteria. When the eye surface is irritated by dust or fumes (from an onion, for example), lacrimal secretion increases to wash away the irritant.

Extrinsic Eye Muscles

Six straplike **extrinsic** (outer) **eye muscles** (Figure 16.6), which originate from the walls of the orbit and insert onto the outer surface of the eyeball, control the movement of each eye and hold the eyes in the orbits.

Four of the extrinsic eye muscles are *rectus* muscles (*rectus* = straight). These originate from the **common tendinous ring,** or **anular ring** (Figure 16.6a), at the posterior point of the orbit. From there, they run straight to their insertions on the anterior half of the eyeball. It is easy to deduce the actions of the rectus muscles from their names and locations (Figure 16.6b and c). The **lateral rectus muscle** turns the eye laterally (outward), whereas the **medial rectus muscle** turns it medially (inward). The **superior rectus muscle** turns the eye superiorly and medially. The **inferior rectus muscle** turns the eye inferiorly and medially.

The actions of the two *oblique* muscles are not so easily deduced because they take indirect paths through the orbit. The **superior oblique muscle** (Figure 16.6a) originates posteriorly near the common tendinous ring, runs anteriorly along the medial orbit wall, and then loops through a ligamentous sling, the **trochlea** ("pulley"), which is suspended from the frontal bone in the anteromedial part of the orbit roof. From there, its tendon runs posteriorly and inserts on the eye's posterolateral surface. Because its tendon approaches from an anterior and medial direction, the superior oblique depresses the eye and turns it laterally (down and out) (Figure 16.6b and c). The **inferior oblique muscle** originates on the anteromedial part of the orbit floor and angles back to insert on the posterolateral part of the eye. The inferior oblique elevates the eye and turns it somewhat laterally (up and out).

The lateral pull of the oblique muscles counteracts the medial pull of the superior and inferior recti to produce strict elevation and depression of the eye. Also, the superior and inferior recti can neither elevate nor depress an eye that has been turned far medially to look inward, so the two oblique muscles must do that.

> **CLINICAL APPLICATION**
>
> **Strabismus** The extrinsic muscles of both eyes are closely controlled by centers in the midbrain so that the eyes move together in unison when we look at objects. If this coordination is disrupted, double vision results, because the two eyes do not look at the same point in the visual field. This misalignment of the eyes is called **strabismus** (strah-biz′mus), meaning "cross-eyed" or "squint-eyed." In strabismus, the affected eye is turned either medially or laterally with respect to the normal eye. Strabismus results from weakness or paralysis of extrinsic eye muscles caused by damage to the nerves that innervate them or other problems. Immediate surgical correction is recommended because over time the brain comes to disregard the image from the affected eye; as a result, the entire visual pathway from that eye degenerates, making the eye functionally blind.

check your understanding

- ☐ 3. What is the conjunctiva, and where is it located?
- ☐ 4. What muscle is not functioning in a person whose eye turns medially? What nerve innervates this muscle?

(For answers, see Appendix B.)

Anatomy of the Eyeball

learning outcomes

- ▶ Describe the structure and function of the three layers of the eye, of the lens, and of the humors of the eye.
- ▶ Explain the structure of the retina and the photoreceptors.

The eye is a complex organ whose many components not only protect and support the delicate photoreceptor cells but also gather, focus, and process light into precise images (Figure 16.7). Because the eyeball is shaped roughly like a globe, it is said to have poles. Its most anterior point is the **anterior pole,** and its most posterior point is the **posterior pole.** Its external wall consists of three *layers,* and its internal cavity contains fluids called *humors.* The *lens,* a structure that helps to focus light, is supported vertically within the internal cavity, dividing it into anterior and posterior segments. The anterior segment is filled with the liquid *aqueous humor,* whereas the posterior segment is filled with the jellylike *vitreous humor.*

Three layers form the external wall of the eye: the fibrous layer, the vascular layer, and the sensory layer (retina).

The Fibrous Layer

The **fibrous layer** is the most external layer. It consists of dense connective tissue arranged into two different regions: sclera and cornea (Figure 16.7). The opaque white, tough **sclera** (skle′rah; "hard") forms the posterior five-sixths of the fibrous layer. Seen anteriorly as the "white of the eye," the sclera protects the eyeball and provides shape and a sturdy anchoring site for the extrinsic eye muscles. The sclera corresponds to the dura mater that covers the brain.

The anterior sixth of the fibrous layer is the transparent **cornea,** through which light enters the eye. This round window bulges anteriorly from its junction with the sclera. The cornea consists of a thick layer of dense connective tissue sandwiched between a superficial corneal epithelium and a deep corneal endothelium. At the junction of the cornea and sclera, between the corneal epithelium and the conjunctiva, are epithelial stem cells that continually renew the corneal epithelium. The cornea's connective tissue layer contains hundreds of sheets of collagen fibers stacked like the pages in a book; the transparency of the cornea is due to this regular alignment of collagen fibers. The cornea not only lets light into the eye but also forms part of the light-bending apparatus of the eye (see pp. 501–502).

The cornea is avascular—it receives oxygen from the air in front of it, and oxygen and nutrients from the aqueous humor that lies posterior to it.

The cornea is richly supplied with nerve endings, most of which are pain receptors. (This is why some people can never adjust to wearing contact lenses.) Touching the cornea causes reflexive blinking and an increased secretion of tears. Even with these protective responses, the cornea is vulnerable to damage by dust, slivers, and other objects. Fortunately, its capacity for regeneration and healing is extraordinary.

> **CLINICAL APPLICATION**
>
> **Corneal Transplants** "Eye banks" are institutions that receive and store corneas for use in the surgical replacement of severely damaged corneas, a procedure called a **corneal transplant.** The cornea is one of the few structures in the body that can be transplanted from one person to another with only minimal risk of rejection. Because it has no blood vessels, it is beyond the reach of the body's immune system. In the 10% of transplanted corneas that are rejected, the damage to the recipient's original cornea was so extensive that it destroyed the corneoscleral junction. Without that region, the corneal epithelium cannot regenerate, so *conjunctival* epithelium overgrows the new cornea, in the process attracting new blood vessels carrying the immune cells that cause rejection. This problem can be alleviated by grafting corneal epithelium from the patient's healthy eye onto the newly transplanted cornea.

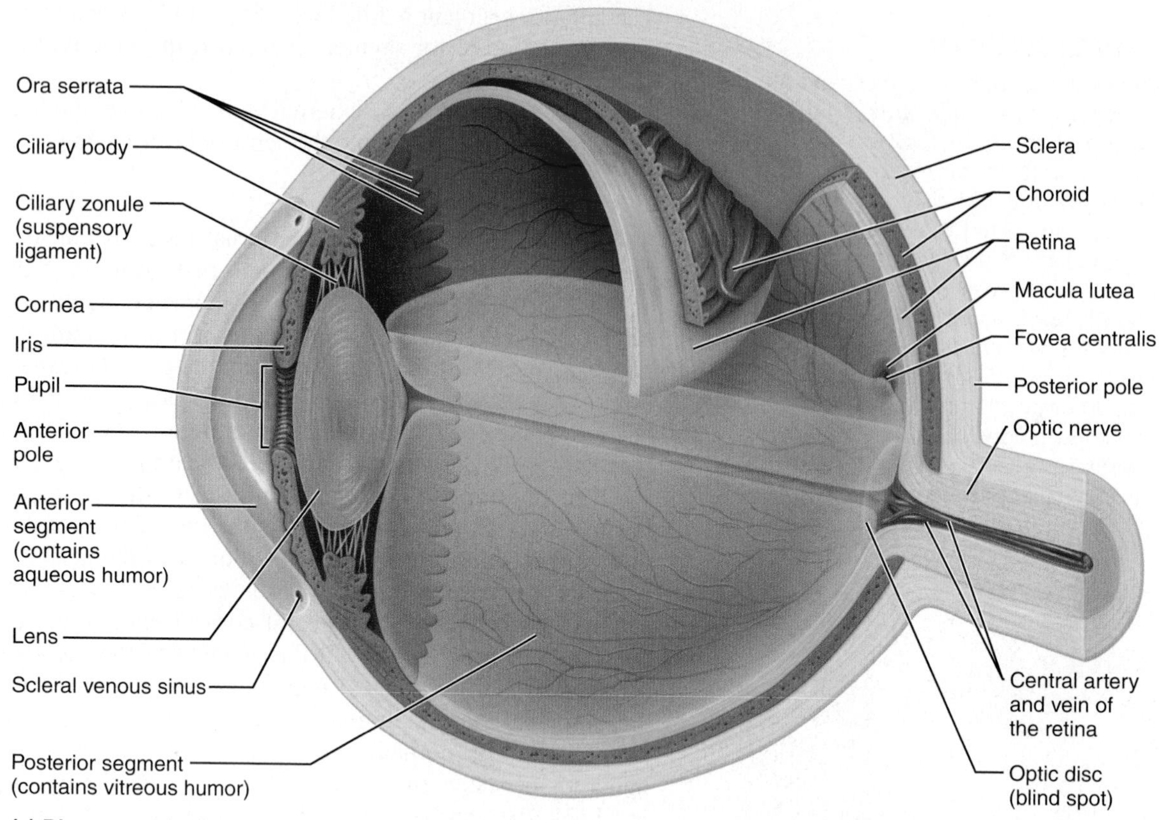

(a) Diagrammatic view. The vitreous humor is illustrated only in the bottom part of the eyeball.

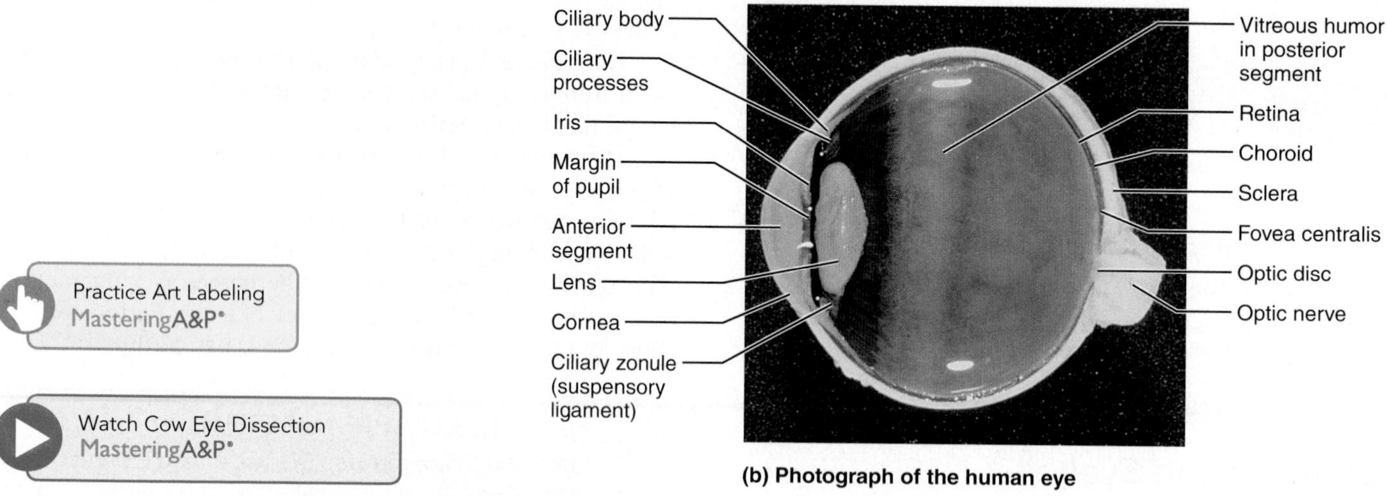

(b) Photograph of the human eye

Figure 16.7 **Internal structure of the eye (sagittal section).**

The Vascular Layer

The **vascular layer,** the middle coat of the eyeball, has three parts: the *choroid,* the *ciliary body,* and the *iris* (Figure 16.7).

The **choroid** (ko′roid; "membrane") is a highly vascular, darkly pigmented membrane that forms the posterior five-sixths of the vascular layer. Its many blood vessels nourish the other layers of the eye. The brown color of the choroid is produced by melanocytes, whose pigment, melanin, helps absorb light, thereby preventing light from scattering within the eye and creating visual confusion. The choroid layer of the eye corresponds to the arachnoid and pia mater around the brain.

Anteriorly, the choroid is continuous with the **ciliary body,** a thickened ring of tissue that encircles the lens. The ciliary body consists chiefly of smooth muscle called the **ciliary muscle,** which acts to focus the lens. Nearest the lens, the posterior surface of the ciliary body is thrown into radiating folds called **ciliary processes.** The halo of fine fibrils that extends from around the entire circumference of the lens and attaches to the ciliary processes is called the **ciliary zonule.**

The **iris** ("rainbow") is the visible, colored part of the eye. It lies between the cornea and lens, and its base attaches to the ciliary body. Its round central opening, the **pupil,** allows light to enter the eye. The iris contains both circularly arranged and radiating smooth muscle fibers, the *sphincter* and *dilator pupillae* muscles, that act to vary the size of the pupil (Figure 16.8). In bright light and for close vision, the sphincter pupillae contracts to constrict the pupil. In dim light and for distant vision, the dilator pupillae contracts to widen the pupil, allowing more light to enter the eye. Constriction and dilation of the pupil are controlled by parasympathetic and sympathetic fibers, respectively. The constriction of the pupils that occurs when a bright light is flashed in the eye is a protective response called the **pupillary light reflex.**

Although irises come in many colors, they contain only brown pigment. Variation in eye color reflects the amount of pigmentation in the iris. All people except albinos have a layer of pigmented cells on the posterior surface of the iris. Brown-eyed people have many pigment cells on the anterior surface of the iris as well. Blue-eyed people, by contrast, do not have pigment on the anterior surface. Blue eyes result from the reflection of light off the pigmented posterior surface. Hazel-eyed people have some pigment in the anterior portion of the iris.

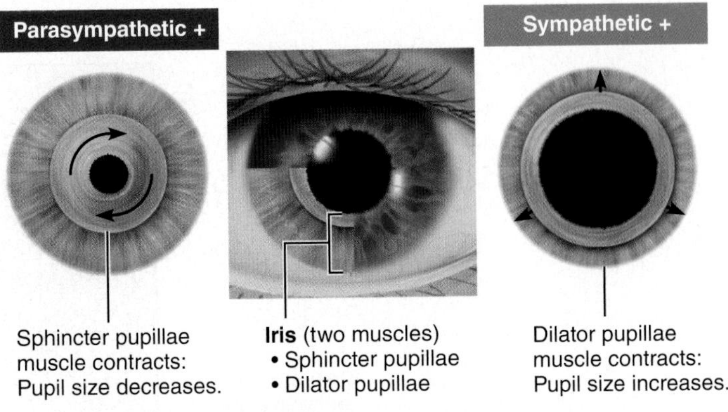

Figure 16.8 **Pupil dilation and constriction, anterior view.** (+ means activation.)

The Inner Layer

The **inner layer** contains the *retina* and the *optic nerve.* The **retina** consists of two layers: a thin pigmented layer and a far thicker neural layer (Figure 16.9). The outer **pigmented layer,** which lies against the choroid, is a single layer of flat-to-columnar melanocytes. Like the choroid, it functions to absorb light and prevent it from scattering within the eye. The much thicker inner **neural layer** is a sheet of nervous tissue that contains the light-sensitive photoreceptor cells. The neural and pigmented layers of the retina are held together by a thin film of extracellular matrix, but they are not tightly fused. Only the neural layer plays a direct role in vision. The pigmented layer supports the photoreceptive cells by removing damaged portions of those cells, maintaining the proper ionic concentration in the fluid surrounding them, recycling the vitamin A derivative used for light detection, and transporting nutrients from the choroid vessels to the photoreceptor cells.

The neural layer contains three main types of neurons. From external to internal, these are the **photoreceptor cells, bipolar cells,** and **ganglion cells** (Figure 16.9b and c). When stimulated by light, the photoreceptor cells signal the bipolar cells, which then signal the ganglion cells to generate nerve impulses potentials. Axons from the ganglion cells run along the internal surface of the retina and converge posteriorly to form the **optic nerve,** which runs from the eye to the brain (Figure 16.9a).

The retina also contains interneurons—including amacrine cells and horizontal cells (Figure 16.9b)—that process and modify visual information before it is sent to higher brain centers for further processing.

Photoreceptors The photoreceptor cells are of two types: *rod cells* and *cone cells.* The more numerous **rod cells** are more sensitive to light and permit vision in dim light. Because rod cells provide neither sharp images nor color vision, things look gray and fuzzy when viewed in dim light. **Cone cells,** by contrast, operate best in bright light and enable high-acuity color vision. Three subtypes of cone cells are sensitive to blue, red, and green light, respectively.

Photoreceptor cells are considered neurons, but they also resemble tall epithelial cells turned upside down, with their "tips" immersed in the pigmented layer (Figure 16.10). Both rod cells and cone cells have an *outer segment* joined to an *inner segment* by a connecting cilium. In each rod cell, the inner and outer segments together form a rod-shaped structure, which connects to the nucleus-containing *cell body* by an *outer fiber.* In each cone cell, the inner and outer segments form a cone-shaped structure, which joins to the cell body directly. In both cell types, the cell body is continuous with an *inner fiber* that synapses with the bipolar cell.

The outer segments are the receptor regions of the rod and cone cells. Each outer segment is a modified cilium whose plasma membrane has folded inward to form hundreds of membrane-covered discs. Light-absorbing *visual pigments* are present within the vast membrane of these discs. The extensive folding of the plasma membrane into discs greatly magnifies the surface area available for trapping light. When light particles hit the visual pigment, the pigment is modified, and a nerve impulse is generated. The stimulated photoreceptor cells signal the bipolar neurons with which they synapse. This reaction initiates the flow of visual information to the brain.

The photoreceptors are highly vulnerable to damage by intense light or heat. These cells cannot regenerate if destroyed, but they continually renew and replace their outer segments through the addition of new discs. In this normal recycling process, as new discs are added to one end of the stack, old discs are removed at the other end by retinal pigment cells phagocytizing the tips of the rods and cones (Figure 16.10).

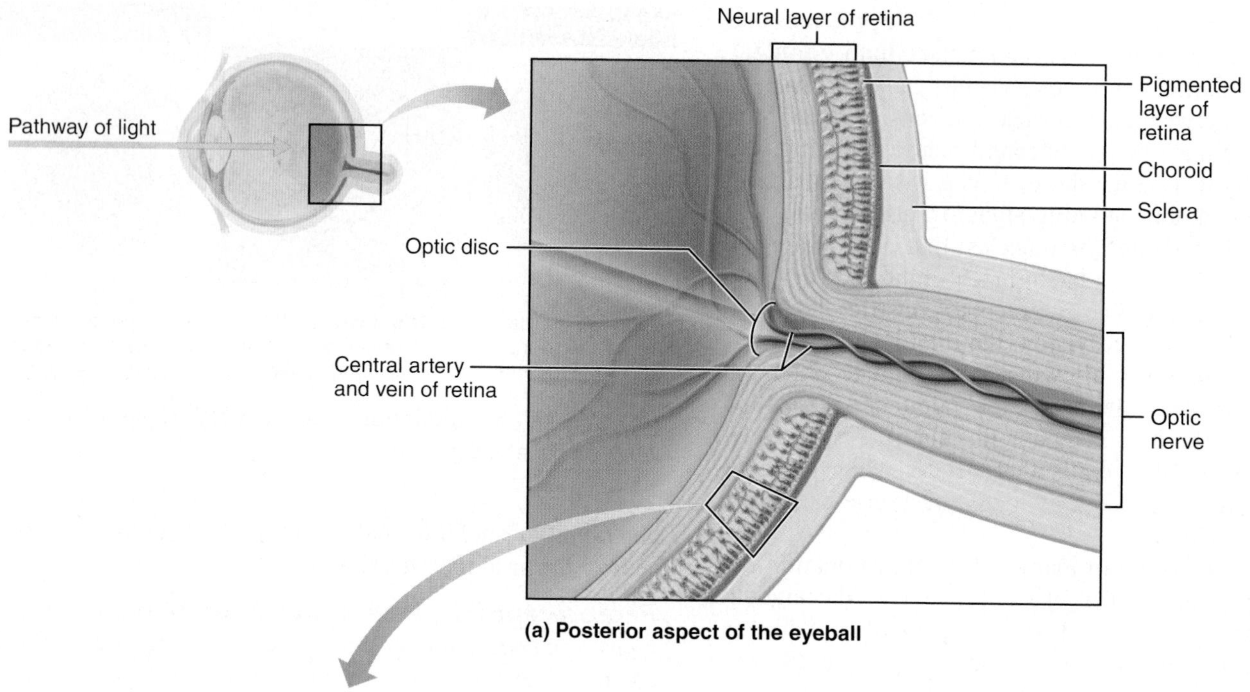

(a) Posterior aspect of the eyeball

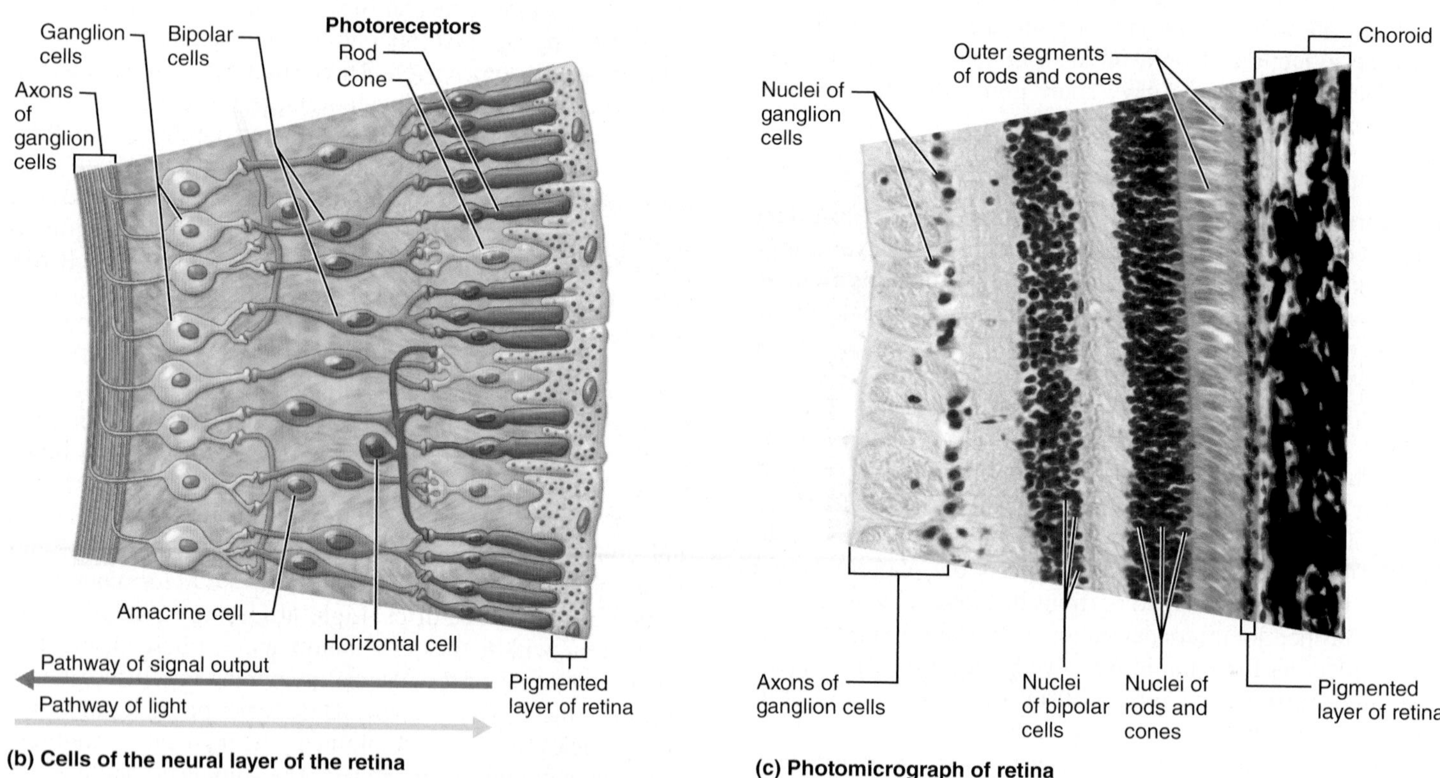

(b) Cells of the neural layer of the retina

(c) Photomicrograph of retina

Figure 16.9 **Microscopic anatomy of the retina.** (**a**) The axons of the ganglion cells form the optic nerve, which leaves the back of the eye at the optic disc. (**b**) Light (indicated by the yellow arrow) passes through the retina to the photoreceptor cells (rods and cones). Information (output signals) flows in the opposite direction via bipolar and ganglion cells. (**c**) Photomicrograph (140×).

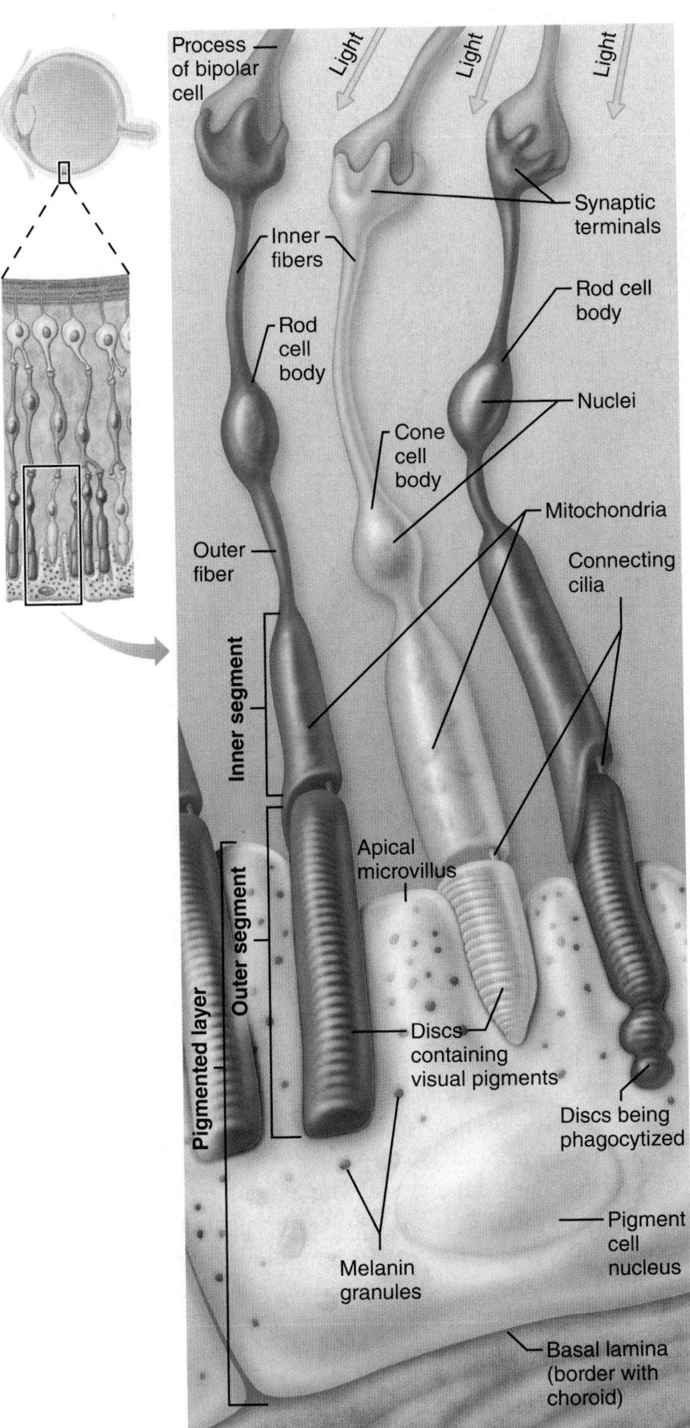

Figure 16.10 **Photoreceptors of the retina.** The outer segments of the rods and cones are embedded in the pigmented layer of the retina.

Regional Specializations of the Retina In certain regions of the eye, the retina differs from its "typical" structure just described.

In the anterior part of the eye, the neural layer ends at the posterior margin of the ciliary body. This junction is called the **ora serrata** (o′rah se-rah′tah; "sawtoothed mouth") (see Figure 16.7a). The pigmented layer extends anteriorly beyond the ora serrata to cover the ciliary body and to form the pigmented layer of the posterior iris.

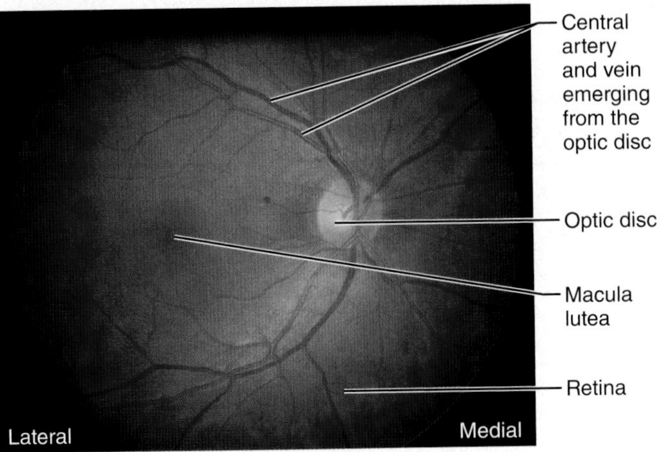

Figure 16.11 **The posterior wall of the right eye as seen with an ophthalmoscope.**

The posterior part of the eye contains several special areas of the retina (Figure 16.7). Lying precisely at the eye's posterior pole is the **macula lutea** (mak′u-lah lu′te-ah; "yellow spot"). At the center of the macula lutea is a tiny pit called the **fovea centralis** (fo′ve-ah sen-trah′lis; "central pit"). The fovea contains only cones and provides maximal visual acuity. Because the fovea lies directly in the anterior-posterior axis of the eye, we see things most clearly when we look straight at them. The macula contains mostly cones and the density of cones declines with increasing distance from the macula. For this reason, peripheral vision is not as sharp as central vision. A few millimeters medial to the fovea is the **optic disc** (Figure 16.7), a circular elevation where the axons of ganglion cells converge to exit the eye as the optic nerve. The optic disc is called the *blind spot* because it lacks photoreceptors, and light focused on it cannot be seen.

CLINICAL APPLICATION

Age-Related Macular Degeneration (AMD) is a progressive deterioration of the retina that affects the macula lutea and leads to loss of central vision. It is the main cause of vision loss in people over age 65, and 200,000 Americans develop it each year. Early stages or mild forms of AMD involve the buildup of visual pigments in the macula caused by loss of cells in the pigmented layer of the retina that normally remove the damaged visual pigments. Continued accumulation of the pigment is associated with the "dry" form of AMD, in which many of the macular photoreceptors die. Less common is the "wet" form, in which new blood vessels grow into the retina from the choroid layer, then bleed and cause scarring and detachment of the retina. The ultimate cause is unknown. AMD is largely untreatable. Wet AMD can be treated with injections or laser treatments that distroy the growing vessels.

Blood Supply of the Retina The retina receives its blood from two different sources. The outer third of the retina, containing the photoreceptors, is supplied by capillaries in the choroid, whereas its inner two-thirds is supplied by the **central artery** and **vein of the retina** (Figure 16.11), which

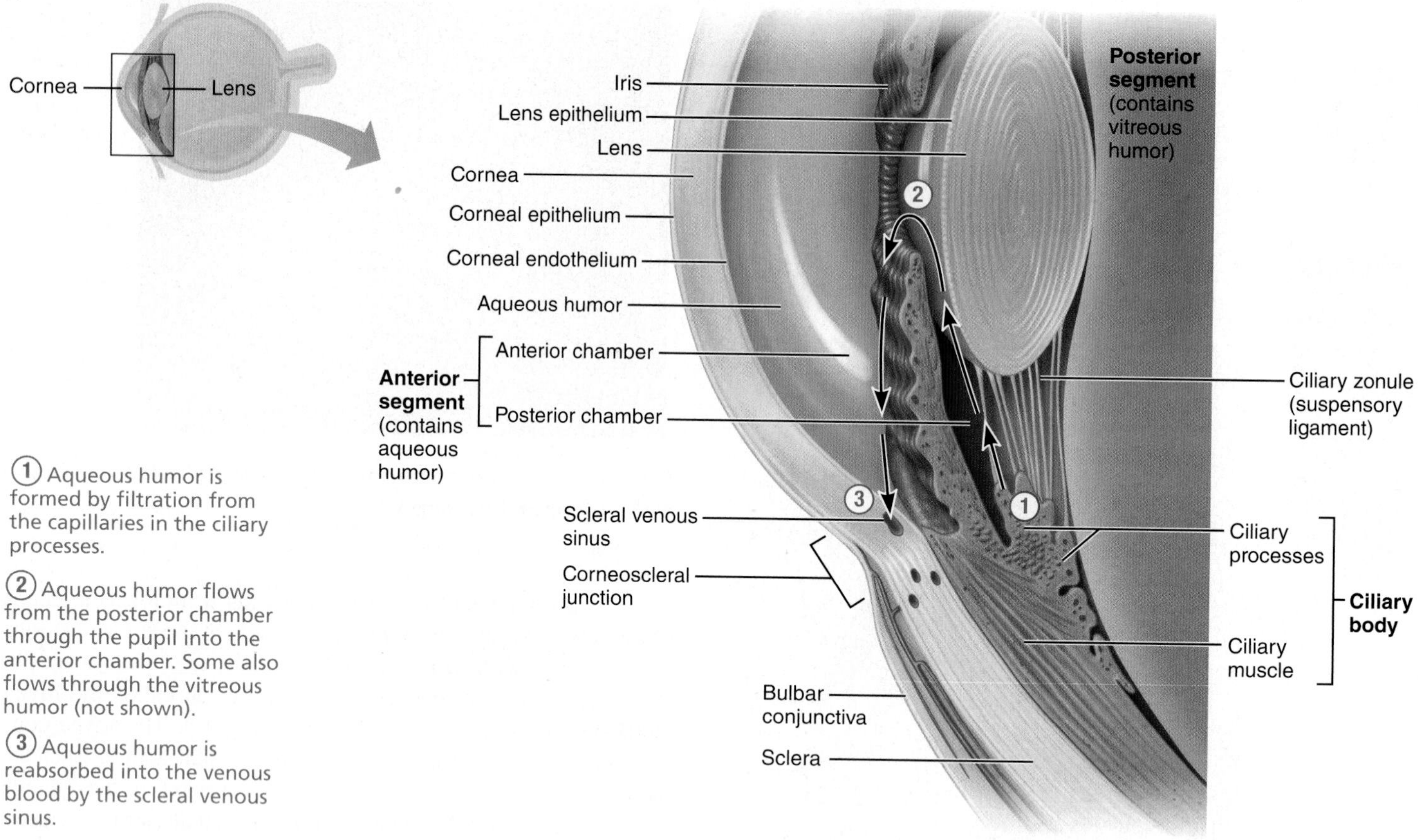

Cornea
Lens
Iris
Lens epithelium
Lens
Cornea
Corneal epithelium
Corneal endothelium
Aqueous humor
Anterior segment (contains aqueous humor)
Anterior chamber
Posterior chamber
Posterior segment (contains vitreous humor)
Ciliary zonule (suspensory ligament)
Scleral venous sinus
Corneoscleral junction
Ciliary processes
Ciliary muscle
Ciliary body
Bulbar conjunctiva
Sclera

(1) Aqueous humor is formed by filtration from the capillaries in the ciliary processes.

(2) Aqueous humor flows from the posterior chamber through the pupil into the anterior chamber. Some also flows through the vitreous humor (not shown).

(3) Aqueous humor is reabsorbed into the venous blood by the scleral venous sinus.

Figure 16.12 **The anterior segment of the eye and circulation of aqueous humor.** The arrows indicate the circulation pathway.

CLINICAL APPLICATION

Detached Retina The retina's pattern of vascularization contributes to a potentially blinding condition called **retinal detachment,** in which the loosely joined neural and pigmented layers of the retina separate ("detach") from one another. The detachment begins with a tear in the retina, which may result from a small hemorrhage, a blow to the eye, or age-related degeneration. The tear allows the jellylike vitreous humor from the eye's interior (see the next section) to seep between the two retinal layers. With the neural layer detached, the photoreceptors, now separated from their blood supply in the choroid, are kept alive temporarily by nutrients obtained from the inner retinal capillaries. However, because these capillaries are too distant to supply them permanently, the photoreceptors soon die. Early symptoms of retinal detachment include seeing flashing lights or spots that float across the field of vision, and having objects appear as if seen through a veil. If the detachment is diagnosed early, blindness may be prevented by reattaching the retina with laser surgery before permanent damage to the photoreceptors occurs.

enter and leave the eye by running through the center of the optic nerve (see Figure 16.7 and Figure 16.9a). These vessels radiate from the optic disc, giving rise to a rich network of tiny vessels that weave among the axons on the retina's inner face. This vascular network is clearly visible through an ophthalmoscope, a handheld instrument that shines light through the pupil and illuminates the retina. Physicians observe the tiny retinal vessels for signs of hypertension, diabetes, and other diseases that damage the smallest blood vessels.

Internal Chambers and Fluids

The lens and its halolike ciliary zonule divide the eye into posterior and anterior segments (Figure 16.7). The **posterior segment** is filled with the clear **vitreous humor** (*vitreus* = glassy), a jellylike substance that contains fine fibrils of collagen and a ground substance that binds tremendous amounts of water. Indeed, water constitutes over 98% of its volume. The vitreous humor functions to:

1. Transmit light
2. Support the posterior surface of the lens and hold the neural retina firmly against the pigmented layer
3. Help maintain *intraocular pressure* (the normal pressure within the eye), thereby counteracting the pulling forces of the extrinsic eye muscles.

The **anterior segment** of the eye (Figure 16.12) is divided into an **anterior chamber** between the cornea and iris, and a **posterior chamber** between the iris and lens. The entire anterior segment is filled with **aqueous humor,** a clear fluid similar to blood plasma (*aqueous* = watery). Unlike the vitreous humor, which forms in the embryo and lasts a lifetime, aqueous humor is renewed continuously and is in constant motion (Figure 16.12). After being formed as a filtrate of the blood from capillaries in the ciliary processes (Figure 16.12 ①), the aqueous humor enters the posterior chamber, flows through the pupil into the anterior chamber (Figure 16.12 ②), and drains into a large vessel at the corneoscleral junction, the **scleral venous sinus,** which returns it to the blood (Figure 16.12 ③). An equilibrium in the rates at which the aqueous humor forms and drains results in a constant intraocular pressure, which supports the eyeball internally. Furthermore, the aqueous humor supplies nutrients and oxygen to the avascular lens and cornea.

CLINICAL APPLICATION

Glaucoma When the aqueous humor drains more slowly than it forms, the result is **glaucoma** (glaw-ko′mah), a disease in which intraocular pressure increases to dangerous levels, causing compression of the retina and the optic nerve. Glaucoma results from an obstruction of outflow, usually from a clogging of the permeable net through which aqueous humor drains into the scleral venous sinus. The resulting destruction of the optic nerve eventually causes blindness. Even though vision can be saved if the condition is detected early, glaucoma often steals sight so slowly and painlessly that most people do not realize they have a problem until the damage is done. Late signs include blurred vision, seeing halos around lights, and headaches. The examination for glaucoma involves testing for high intraocular pressure. A puff of air is directed at the cornea and the amount of deformation of the sclera is measured. A glaucoma exam should be done yearly after age 40 because glaucoma affects fully 2% of people over that age. Early glaucoma is treated with eye drops that increase drainage or decrease production of aqueous humor.

The Lens

The **lens** is a thick, transparent, biconvex disc that changes shape to allow precise focusing of light on the retina (Figure 16.12). It is enclosed in a thin elastic capsule and is held in place posterior to the iris by its ciliary zonule. Like the cornea, it lacks blood vessels, which would interfere with transparency.

The lens has two components: the lens epithelium and the lens fibers. The **lens epithelium,** confined to the anterior surface, consists of cuboidal cells. The subset of epithelial cells around the edge of the lens disc transforms continuously into the elongated **lens fibers** that form the bulk of the lens. These fibers, which are packed together like the layers in an onion, contain no nuclei and few organelles. They do, however, contain precisely folded proteins that make them transparent. New lens fibers are added continuously, so the lens enlarges throughout life. It becomes denser, more convex, and less elastic with age. As a result, its ability to focus light is gradually impaired.

CLINICAL APPLICATION

Cataract A **cataract** (kat′ah-rakt; "waterfall") is a clouding of the lens that causes the world to appear distorted, as if seen through frosted glass. Some cataracts are congenital, but most result from age-related changes in the lens. Evidence indicates that excessive exposure to sunlight, heavy smoking, and certain medications including oral steroids, long-term aspirin use, and tamoxifen (used to treat breast cancer), are linked to cataract formation. No matter the cause, cataracts seem to result from an inadequate delivery of nutrients to the deeper lens fibers. Fortunately, surgical removal of the damaged lens and its replacement by an artificial lens can save a cataract patient's sight.

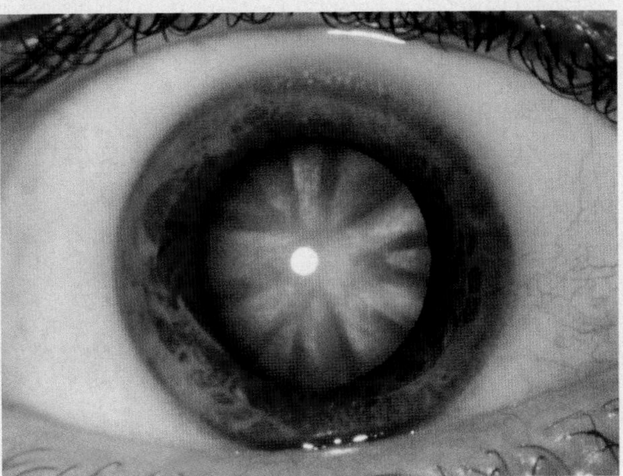

Photograph of a cataract. The lens is milky and opaque, not the cornea.

☑ check your understanding

- ☐ **5.** Differentiate the cornea from the choroid with reference to the location, structure, and function of each.
- ☐ **6.** What portion of the visual field is lost in a person with degeneration of the macula lutea (macular degeneration)?
- ☐ **7.** What is the aqueous humor, where is it located, and what are its functions?

(For answers, see Appendix B.)

The Eye as an Optical Device

learning outcome

▶ Explain how light is focused for close vision.

From each point on an object, light rays radiate in every direction. Some of these light rays enter the eye of the viewer (Figure 16.13). Rays from a distant point are parallel to one another as they reach the eye, whereas rays from a nearby point diverge markedly as they enter the eye. If one is to see clearly, the eye must be able to bend all these light rays so that they converge on the retina at a single *focal point.* The light-bending parts of the eye, called **refractory media,** are

(a) The lens flattens for distant vision.
Sympathetic input relaxes the ciliary muscle. This tightens the ciliary zonule and flattens the lens.

Relaxed ciliary muscle
Tightened ciliary zonule
Flattened lens
Nearly parallel rays from distant object
Image

(b) The lens bulges for close vision.
Parasympathetic input contracts the ciliary muscle. This loosens the ciliary zonule and allows the lens to bulge.

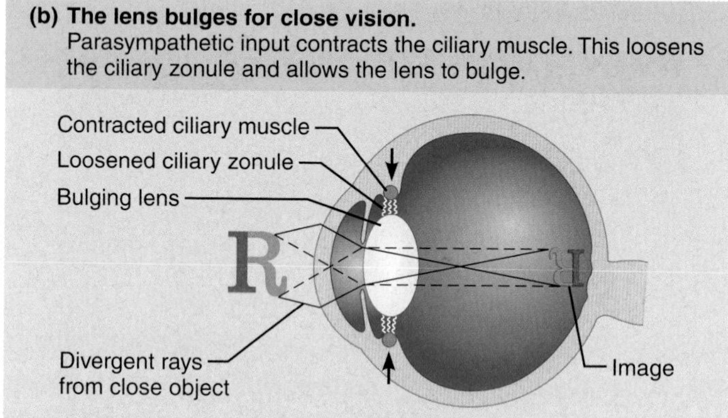

Figure 16.13 Focusing for distant and close vision.

> **CLINICAL APPLICATION**
>
> **Focusing Disorders** Focusing disorders of the eye are common.
>
> **Myopia,** commonly referred to as nearsightedness, occurs when the shape of the eye or the bending of the lens results in a focal point for distant objects that is in front of the retina, creating a blurry image on the retina. Concave lenses, which diverge the incoming rays of light and thus move the focal point posteriorly, correct for this disorder (see illustration on next page).
>
> **Hyperopia,** or farsightedness, results when the eye is short, causing the focal point to occur behind the retina. For viewing distant objects, the lens of the eye can adequately correct for this, enabling clear vision; for viewing close objects, convex corrective lenses, which converge the light rays as they approach the eye and move the focal point anteriorly, are needed (see illustration).
>
> **Presbyopia** affects people as they reach middle age. The lens becomes thicker and less elastic and thus less able to accommodate for near vision. As with hyperopia, convex corrective lenses, commonly called reading glasses, correct for this disorder.
>
> **Astigmatism** results when the cornea or lens has unequal curvatures in different regions. Light entering the eye is bent such that points of light do not converge in one spot on the retina, causing blurred vision. Corrective lenses can easily resolve this common eye disorder. Surgical techniques using lasers (LASIK, LASEK, or photorefractive keratectomy) are readily available to correct for this refractive disorder. In these procedures, an outer portion of the cornea is lifted and the deeper layers are sculpted to change the shape of the cornea and to correct vision abnormalities.

the cornea, the lens, and the humors. The cornea does most of the light bending, the lens does some of it, and the humors do a minimal amount.

Although the lens is not as powerful as the cornea in bending light, its curvature is adjustable. This adjustability allows the eye to focus on nearby objects—a process called **accommodation.** A resting eye, with its lens stretched along its long axis by tension in the ciliary zonule, is "set" to focus the almost-parallel rays from distant points. Therefore, distance vision is the natural state (Figure 16.13a). The diverging rays from *nearby* points must be bent more sharply if they are to focus on the retina. To accomplish this, the lens is made rounder: The ciliary muscle contracts in a complex way that releases most of the tension on the ciliary zonule. No longer stretched, the lens becomes rounder as a result of its own elastic recoil (Figure 16.13b). Accommodation is controlled by the parasympathetic fibers that signal the ciliary muscle to contract.

Focusing on nearby objects is accompanied by pupillary constriction (Figure 16.8), which prevents the most divergent light rays from entering the eye and passing through the extreme edges of the lens. Such rays would not be focused properly and would cause blurred vision.

For simplicity, this discussion of eye focusing is confined to single-point images; how the eye focuses the "multiple-point" images of large objects is beyond the scope of this text. However, it must be noted that the convex lens of the eye, just like the convex lens of a camera, produces images that are upside down and reversed from right to left. Therefore, an inverted and reversed image of the visual field is projected onto each retina (Figure 16.13). The cerebral cortex then "flips" the image back, so that we see things as they are actually oriented.

Visual Pathways

learning outcome

▶ Trace the pathway of nerve impulses from the retina to the cerebral cortex.

Visual information leaves the eye and travels to the brain for complex processing. Most of this visual information goes to the cerebral cortex, which is responsible for conscious "seeing" (Chapter 13, p. 393), but some goes to nuclei in the midbrain and diencephalon, which control reflexes and subconscious behaviors that require visual input.

CLINICAL APPLICATION **Focusing Disorders** *(continued)*

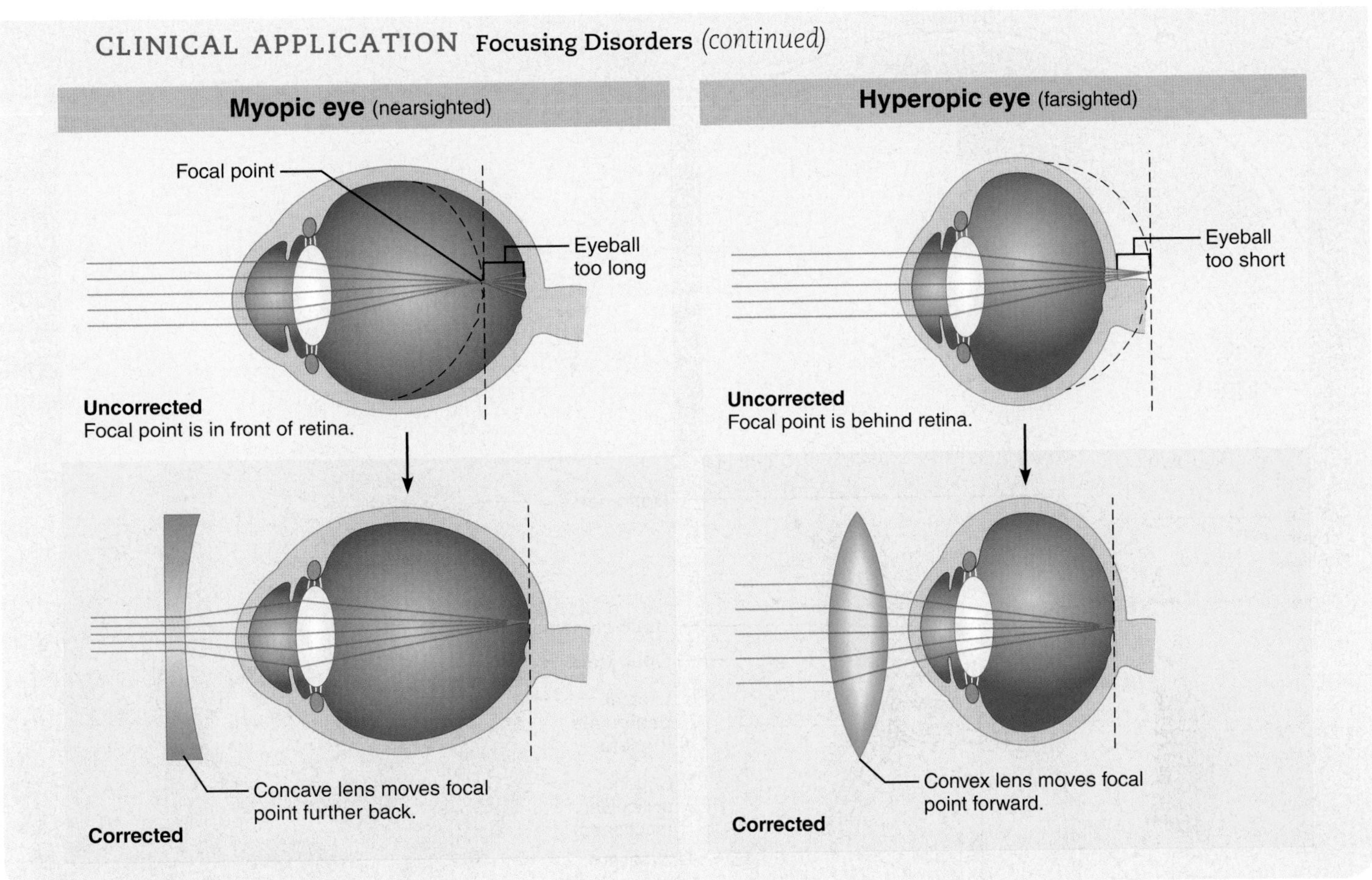

Visual Pathway to the Cerebral Cortex

Visual information travels to the cerebral cortex through the main **visual pathway** (Figure 16.14). Axons of the ganglion cells exit the eye in the **optic nerve** (cranial nerve II). At the X-shaped **optic chiasma** (ki-as′mah; "cross"), which lies anterior to the hypothalamus, the axons from the medial half of each eye decussate and then continue in an **optic tract.** However, axons from the area of the retina lateral to the fovea do not cross at the optic chiasma; they continue to the ipsilateral optic tract. The paired optic tracts sweep posteriorly around the hypothalamus and send most of their axons to the **lateral geniculate nucleus of the thalamus,** where they synapse with thalamic neurons. Axons of those neurons then project through the internal capsule to form the **optic radiation** of fibers in the cerebral white matter. These fibers reach the **primary visual cortex** in the occipital lobe, where conscious perception of visual images occurs.

The partial decussation of axons in the optic chiasma relates to depth perception, which is also called stereoscopic, or three-dimensional, vision. To understand this, you can visualize the retina of each eye as divided into a medial and a lateral half (see the dashed lines extending from the fixation point that bisect the eyes in Figure 16.14a). Recall that the lens system of each eye reverses all images. Because of this reversal, the medial half of each retina receives light rays from the lateral (peripheral) part of the visual field, that is, from objects that lie either to the left or to the right rather than straight ahead. Correspondingly, the lateral half of each retina receives an image of the central part of the visual field. Only those axons from the *medial* halves of the two retinas cross over at the optic chiasma. The result is that all information from the left half of the visual field (shown in yellow in Figure 16.14a) is directed through the right optic tract to be perceived by the right cerebral cortex. Likewise, the right half of visual space (shown in blue) is perceived by the left visual cortex. Each cerebral cortex receives an image of half the visual field, as viewed by the two different eyes from slightly different angles. The cortex then compares these two similar but different images and, in doing so, creates the perception of depth.

These relationships explain patterns of blindness that follow damage to different visual structures. Destruction of one eye (for example, "left eye" in Figure 16.14a) or one optic nerve eliminates true depth perception and causes a loss of peripheral vision on the side of the damaged eye the (visual area stippled yellow in Figure 16.14a, in this example). However, if damage occurs beyond the optic chiasma—in an optic tract, the thalamus, or the visual cortex—then the entire opposite half of the visual field is lost. For example, a stroke affecting the left visual cortex leads to blindness (blackness) throughout the right half of the visual field.

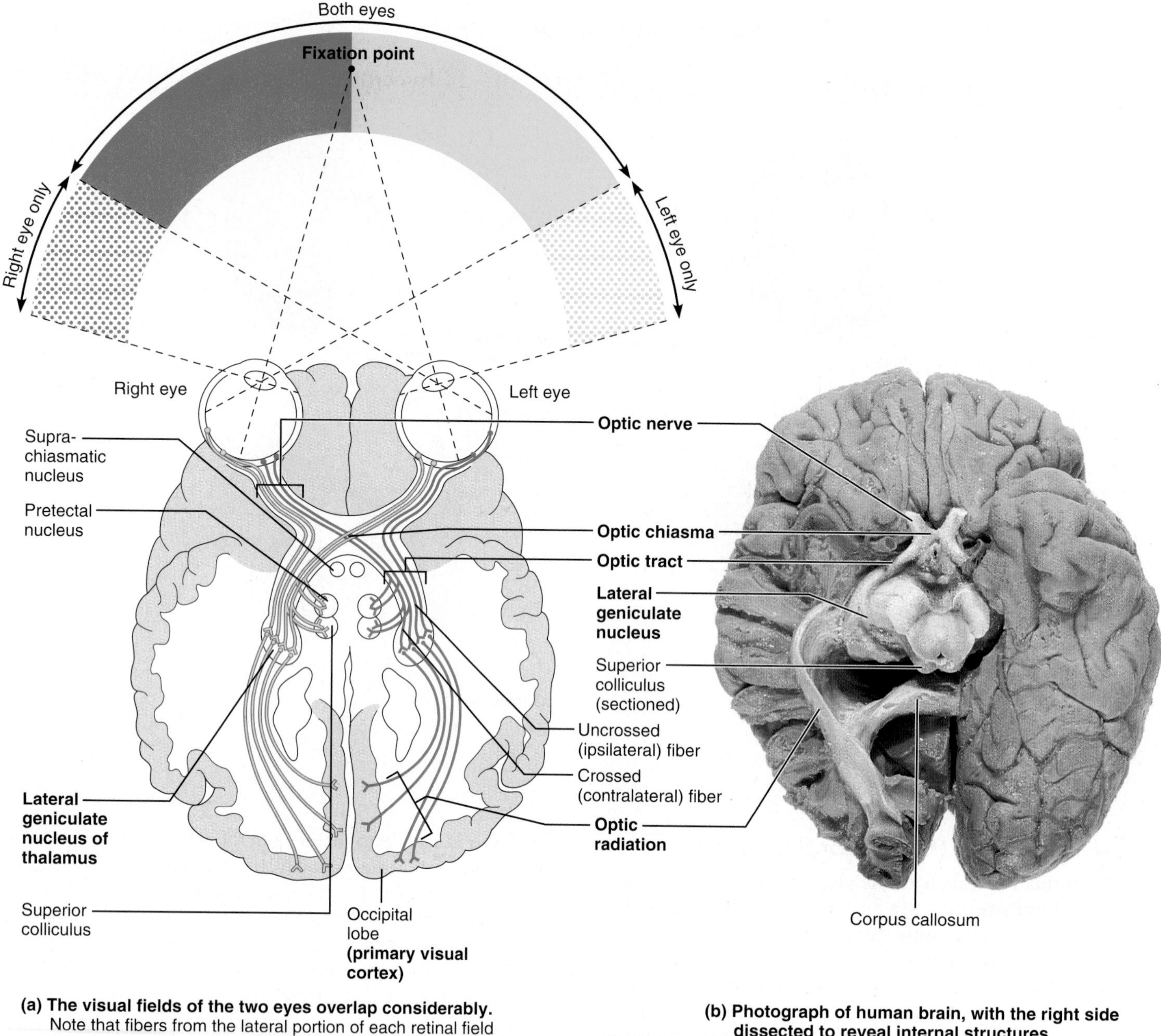

(a) The visual fields of the two eyes overlap considerably. Note that fibers from the lateral portion of each retinal field do not cross at the optic chiasma.

(b) Photograph of human brain, with the right side dissected to reveal internal structures

Figure 16.14 **Visual pathway to the brain and visual fields, inferior view.**

Visual Pathways to Other Parts of the Brain

Some axons from the optic tracts send branches to the midbrain (Figure 16.14a). These branches go to the **superior colliculi,** reflex nuclei controlling the extrinsic eye muscles (discussed on p. 381), and to the **pretectal nuclei,** which mediate the pupillary light reflexes. Other branches from the optic tracts run to the **suprachiasmatic nucleus** of the hypothalamus, which is the "timer" that runs our daily biorhythms and requires visual input to keep it in synchrony with the daylight-darkness cycle.

Embryonic Development of the Eye

learning outcome

▶ Describe the embryonic development of the eye.

The eyes develop as outpocketings of the brain. By week 4, paired lateral outgrowths called **optic vesicles** protrude from the diencephalon (Figure 16.15a). Soon, these hollow vesicles indent to form double-layered **optic cups** (Figure 16.15b). The medial parts of the outgrowths, called the **optic stalks,** form the basis of the optic nerves.

Once a growing optic vesicle reaches the overlying surface ectoderm, it signals the ectoderm to thicken and form

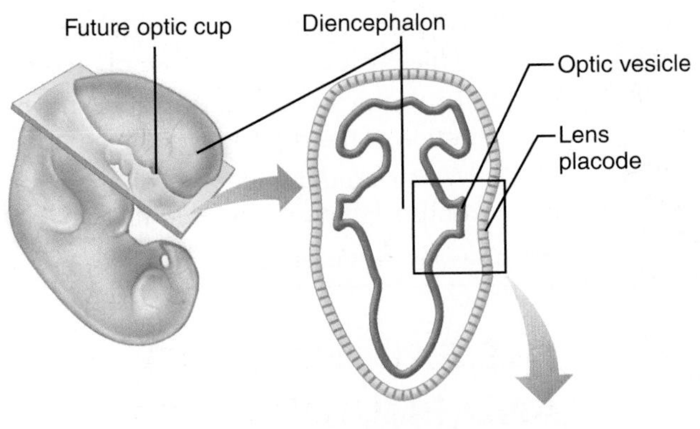

(a) Week 4, early. Outpocketing of the diencephalon forms the optic vesicles.

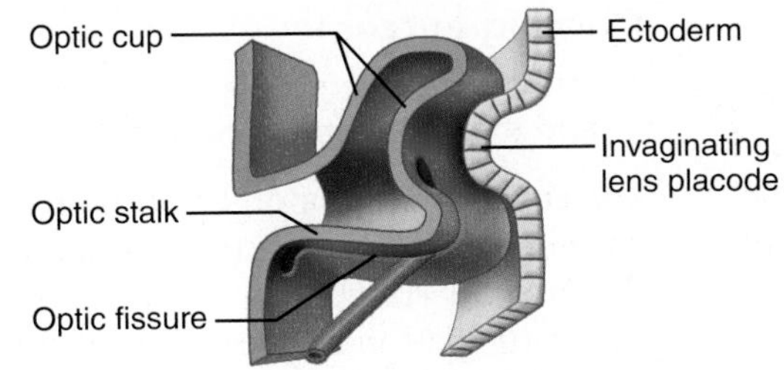

(b) Week 4, late. Optic vesicles invaginate to form the optic cups. The overlying surface ectoderm thickens to form the lens placode.

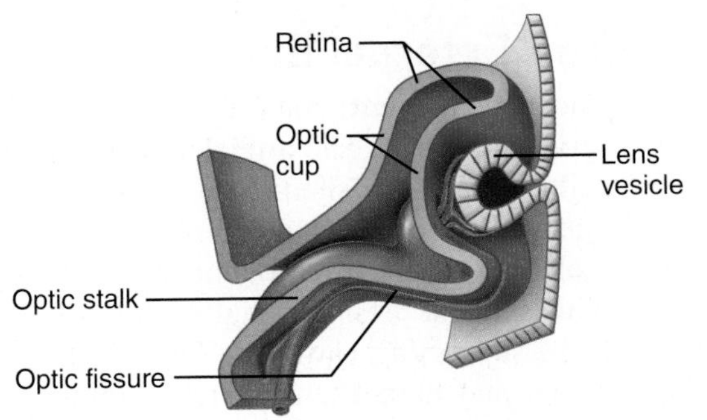

(c) Week 5. Lens placode invaginates and forms the lens vesicle.

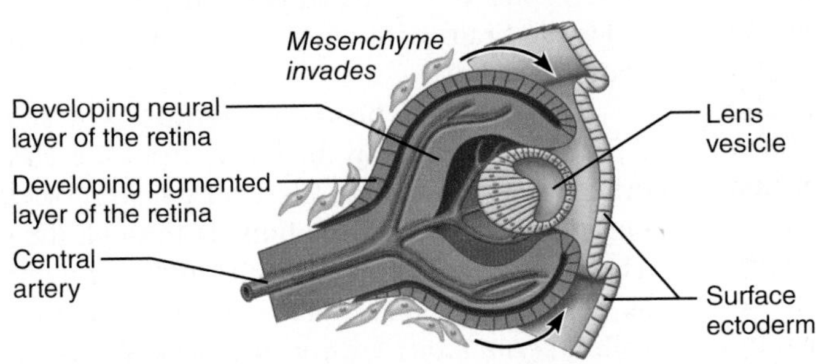

(d) Week 6. The neural and pigmented layers of the retina differentiate from the optic cup. Central artery reaches the interior of the eye. Mesenchyme derived from neural crest invades.

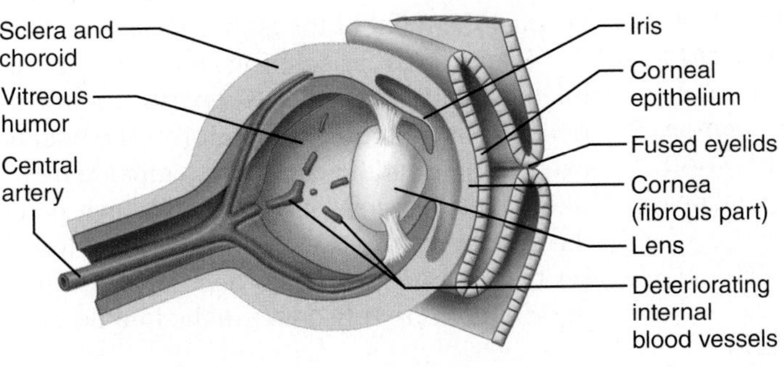

(e) Week 7. Mesenchyme surrounds and invades the optic cup to form the fibrous and vascular layers and the vitreous humor. Lens vesicle forms the lens. Surface ectoderm forms the corneal epithelium and the conjunctiva.

Figure 16.15 **Embryonic development of the eye.**

a **lens placode.** By week 5, this placode has invaginated to form a **lens vesicle** (Figure 16.15c). Shortly thereafter, the lens vesicle pinches off into the optic cup, where it becomes the lens.

The internal layer of the optic cup differentiates into the neural retina, whereas the external layer becomes the pigmented layer of the retina (Figure 16.15d). The **optic fissure,** a groove on the underside of each optic stalk and cup, serves as a direct pathway for blood vessels to reach and supply the interior of the developing eye. When this fissure closes, the optic stalk becomes a tube through which the optic nerve fibers grow from the retina toward the diencephalon. The blood vessels that were originally within the optic fissure now lie in the center of the optic nerve.

The fibrous layer, vascular layer, and vitreous humor form from neural-crest-derived head mesenchyme that surrounds the early optic cup and invades the cup's interior. The central interior of the eyeball has a rich blood supply during development, but these blood vessels degenerate, leaving only those in the vascular layer and retina (Figure 16.15e).

Disorders of the Eye and Vision

The most common of the visual disorders—age-related macular degeneration, glaucoma, cataracts, and disorders of accommodation—have already been discussed. This section covers two other common and important eye disorders, namely *retinopathy of prematurity* and *trachoma.*

Retinopathy of prematurity is a visual impairment that affects many infants born so prematurely that they need to receive oxygen in an oxygen tent. When the infant is weaned from the high concentrations of oxygen, new blood vessels start to grow extensively within the eyes. These abnormal vessels have weak walls, and they hemorrhage, leading to retinal detachment and then blindness. Two treatments to kill many of the growing vessels at their source, laser and cryosurgery (applying cold needles to a circle of points around the eyeball just external to the ora serrata), have been only slightly successful in preventing loss of vision in this disorder.

Trachoma (trah-ko′mah; "rough growth") is a highly contagious infection of the conjunctiva and cornea, caused by the bacterium *Chlamydia trachomatis.* It is transmitted by hand-to-eye contact, by flies that go from eye to eye, or by placing contaminated objects in or near the eye (towels, eye liner, etc.). Symptoms begin with an inflammation of the conjunctiva of the upper eyelid; then the conjunctiva and cornea become highly vascularized, and finally, scarred. The corneal scarring reduces vision and causes blindness. Common worldwide, trachoma blinds millions of people in third-world countries. It is effectively treated with eye ointments containing antibiotic drugs.

☑ check your understanding

☐ 8. In a normal eye, when the lens accommodates for viewing close objects, does it become rounder or more oval in shape?

☐ 9. Why do middle-aged to older adults develop difficulty focusing on close objects?

☐ 10. Trace the visual pathway from the medial half of the retina of the right eye to the cerebral cortex.

☐ 11. Which structure of the eye forms from the outpocketing of the neural tube?

(For answers, see Appendix B.)

THE EAR: HEARING AND EQUILIBRIUM

learning outcome

▶ List the basic structures of the external and middle ear and their corresponding functions.

The **ear,** the receptor organ for both hearing and equilibrium, has three main regions: the *external ear,* the *middle ear,* and the *internal ear* (Figure 16.16a). The external and middle ears participate in hearing only, whereas the internal ear functions in both hearing and equilibrium.

The External Ear

The **external ear** consists of the auricle and the external acoustic meatus. The **auricle,** or **pinna,** is what most people call the ear—the shell-shaped projection that surrounds the opening of the external acoustic meatus. Most of the auricle, including the **helix** (rim), consists of elastic cartilage covered with skin. Its fleshy, dangling **lobule** ("earlobe"), however, lacks supporting cartilage. The function of the auricle is to gather and funnel (and thereby amplify) sound waves coming into the external acoustic meatus. Moreover, the way that sound bounces off the ridges and cavities of the auricle provides the brain with clues about whether sounds come from above or below.

The **external acoustic meatus** is a short tube (about 2.5 cm long) running medially, from the auricle to the eardrum. Near the auricle, its wall consists of elastic cartilage, but its medial two-thirds tunnels through the temporal bone. The entire canal is lined with skin that contains hairs, as well as sebaceous glands and modified apocrine sweat glands called *ceruminous* (sĕ-roo′mĭ-nus) *glands.* The ceruminous and sebaceous glands secrete yellow-brown *cerumen,* or earwax (*cere* = wax). Earwax traps dust and repels insects, keeping them out of the auditory canal.

Sound waves entering the external acoustic meatus hit the thin, translucent **tympanic membrane,** or eardrum (*tympanum* = drum), which forms the boundary between the external and middle ears. The tympanic membrane is shaped like a flattened cone, the apex of which points medially into the middle ear cavity. Sound waves that travel through the air set the eardrum vibrating, and the eardrum in turn transfers the vibrations to tiny bones in the middle ear (discussed next).

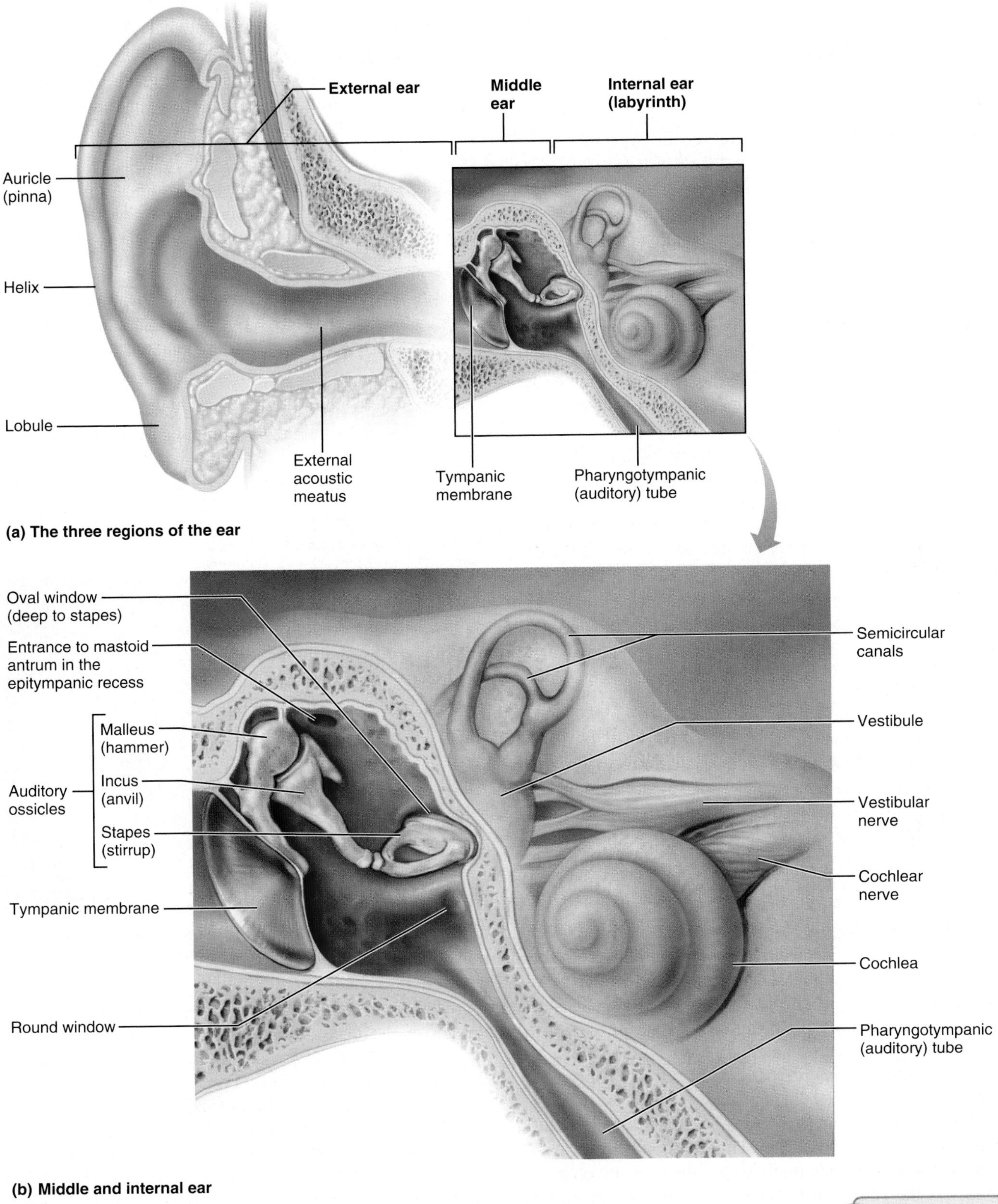

External ear
Middle ear
Internal ear (labyrinth)
Auricle (pinna)
Helix
Lobule
External acoustic meatus
Tympanic membrane
Pharyngotympanic (auditory) tube

(a) The three regions of the ear

Oval window (deep to stapes)
Entrance to mastoid antrum in the epitympanic recess
Auditory ossicles
Malleus (hammer)
Incus (anvil)
Stapes (stirrup)
Tympanic membrane
Round window
Semicircular canals
Vestibule
Vestibular nerve
Cochlear nerve
Cochlea
Pharyngotympanic (auditory) tube

(b) Middle and internal ear

Figure 16.16 **Structure of the ear.** In (b) the bony labyrinth of the internal ear is illustrated.

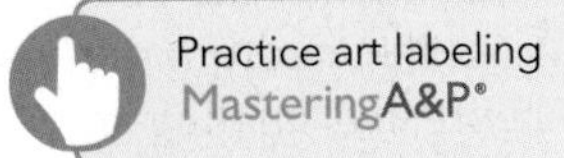

CLINICAL APPLICATION

Perforated Eardrum Undue pressure from a cotton swab or sharp object in the external acoustic meatus can tear the tympanic membrane, a condition called a **perforated eardrum.** A more common cause, however, is a middle ear infection, in which the accumulation of pus medial to the eardrum exerts pressure that bursts the thin membrane. Perforated eardrums heal well, but small amounts of scarring can permanently diminish hearing acuity.

The Middle Ear

The **middle ear,** or *tympanic cavity,* is a small, air-filled space inside the petrous part of the temporal bone. It is lined by a thin mucous membrane and is shaped like a hockey puck standing on its side (Figure 16.16b). Its *lateral boundary* is the tympanic membrane; its *medial boundary* is a wall of bone that separates it from the internal ear. Two small holes penetrate this medial wall: a superior **oval window** and an inferior **round window.** Superiorly, the middle ear arches upward as the **epitympanic recess** (*epi* = over); its *superior boundary* is the roof of the petrous portion of the temporal bone, which is so thin here that middle ear infections can spread to the overlying meninges and brain. The *posterior wall* of the middle ear opens into the **mastoid antrum,** a canal leading to the *mastoid air cells* in the mastoid process. Infections can spread from the middle ear to the mastoid air cells, using the antrum as a passageway. The *anterior wall* of the middle ear lies just behind the internal carotid artery, the main artery to the brain, and it also contains the opening of the pharyngotympanic tube (discussed shortly). The *inferior boundary* of the middle ear is a thin, bony floor, under which lies the important internal jugular vein. Middle ear infections may burst through this floor and clot the blood in this vein.

The **pharyngotympanic tube,** or *auditory tube* (formerly named the *eustachian tube*), links the middle ear to the pharynx (Figure 16.16). About 4 cm (1.5 inches) long, it runs medially, anteriorly, and inferiorly. Its lateral third consists of bone and occupies a groove on the inferior surface of the skull; its medial two-thirds is cartilage and opens into the side wall of the superior pharynx behind the nasal cavity. This normally flattened and closed tube can be opened briefly by swallowing or yawning so that the air pressure in the middle ear equalizes with the outside air pressure. This is important because the eardrum does not vibrate freely unless the pressure on both its surfaces is the same. Differences in air pressure build up across the eardrum during rapid changes in altitude (as during takeoff and landing in an airplane). The next time your ears "pop" in such a situation, remember that yawning is the best way to open your pharyngotympanic tubes and equalize the air pressures.

CLINICAL APPLICATION

Middle Ear Infections Infection and inflammation of the middle ear, called **otitis media** (o-ti′tis; "ear inflammation"), usually starts as a throat infection that spreads to the middle ear through the pharyngotympanic tube. Fluid and pus can build up in the middle ear cavity and exert painful pressure within this enclosed space. More children than adults develop otitis media because the child's pharyngotympanic tube is shorter and enters the pharynx at a less acute angle. Extremely common, otitis media accounts for one-third of all visits to pediatricians in the United States and is frequently treated with antibiotics. However, the overuse of antibiotics has led to bacterial resistance, making persistent and recurrent cases increasingly difficult to treat.

Children with persistent otitis media sometimes have their eardrums lanced and have ear tubes inserted through the eardrum. Such a **myringotomy** (mir″ing-got′o-me; "lancing the eardrum") allows the middle ear to drain and relieves the pressure. The tiny tube that is inserted through the eardrum during myringotomy permits the pus to drain into the external ear. This tube is left in the eardrum and falls out by itself within a year.

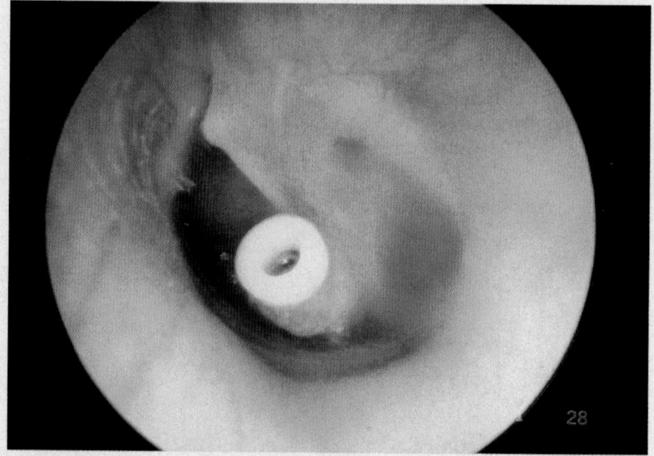

Tube inserted into eardrum to aid drainage and equalize pressure between the external ear and middle ear cavity.

The tympanic cavity is spanned by the three smallest bones in the body, the **auditory ossicles** (Figure 16.16b), which transmit the vibrations of the eardrum across the cavity to a fluid in the internal ear. From lateral to medial, the auditory ossicles are the **malleus** (mal′e-us), or hammer, which looks like a club with a knob on top; the **incus** (ing′kus), or anvil, which resembles a tooth with two roots; and the **stapes** (sta′pēz), which looks like the stirrup of a saddle. The handle of the malleus attaches to the eardrum, and the base of the stapes vibrates against the *oval* window. Most people have trouble remembering whether the stapes fits into the oval window or into the round window inferior to it: To remember, think of the footplate of a saddle stirrup, which is usually oval, not round.

> **CLINICAL APPLICATION**
>
> **Otosclerosis** Excessive growth of bone tissue in the walls of the middle ear cavity can cause the fusion of the footplate of the stapes to the oval window. As a result, the stapes cannot move, and deafness results. This condition, called **otosclerosis** (o″to-sklĕ-ro′sis; "hardening of the ear"), is a common age-related problem that affects 1 in every 200 people. It can be treated by a delicate surgery that removes the stapes and replaces it with a prosthetic (artificial) one.

Tiny ligaments suspend the ossicles in the middle ear, and tiny synovial joints link the ossicles into a chain. By concentrating the vibrations of the eardrum onto the much smaller oval window, the ossicles amplify the pressure of the sound vibrations about 20-fold. Without the ossicles, people could hear only loud sounds.

Two tiny skeletal muscles occur in the middle ear cavity (Figure 16.17). The **tensor tympani** (ten′sor tim′pah-ni) originates on the cartilage part of the pharyngotympanic tube and inserts on the malleus. The **stapedius** (stah-pe′de-us) runs from the posterior wall of the middle ear to the stapes. When the ears are assaulted by very loud sounds, these muscles contract reflexively to limit the vibration of the ossicles and thus prevent damage to the hearing receptors (discussed shortly).

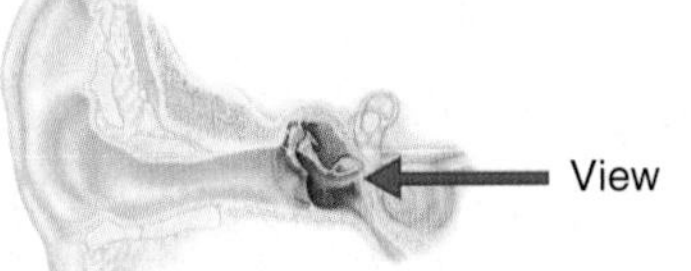

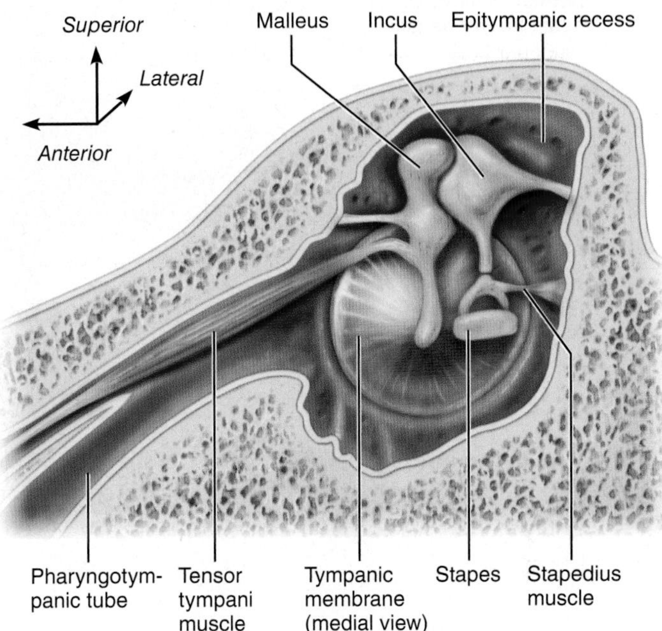

Figure 16.17 **Skeletal muscles associated with the auditory ossicles (right ear, medial view).**

check your understanding

- ☐ 12. What structure separates the external acoustic meatus from the middle ear cavity?
- ☐ 13. Name the four openings or holes into the middle ear cavity.
- ☐ 14. Which auditory ossicle abuts the tympanic membrane?

(For answers, see Appendix B.)

The Internal Ear

learning outcomes

- ▶ Name the parts of the bony and membranous labyrinths in the internal ear.
- ▶ Describe the receptors for hearing and equilibrium.

The **internal ear,** also called the **labyrinth** ("maze") because of its mazelike, complex shape (Figure 16.16), lies within the thick, protective walls of the petrous part of the temporal bone. The internal ear consists of two main divisions: the bony labyrinth and the membranous labyrinth (Figure 16.18 and Table 16.1).

The **bony labyrinth** is a *cavity* in the petrous part of the temporal bone consisting of a system of twisting channels that has three parts. From posterolateral to anteromedial, these parts are the *semicircular canals,* the *vestibule,* and the *cochlea* (Figure 16.16). Textbooks often picture the bony labyrinth as though it were a solid object, but it is actually a cavity.

The **membranous labyrinth** is a continuous series of membrane-walled sacs and ducts that fit loosely within the bony labyrinth and more or less follow its contours (Figure 16.18). The wall of the membranous labyrinth—its "membrane"—is a thin layer of connective tissue lined by a simple squamous epithelium. Parts of this epithelium are thickened and contain the receptors for equilibrium and hearing.

The main parts of the membranous labyrinth are:

- The *semicircular ducts,* one inside each semicircular canal. The semicircular ducts contain the sensory receptors for turning movements of the head.
- The *utricle* and *saccule,* both in the vestibule. The sensory receptors that monitor position and linear acceleration of the head are located in these portions of the membranous labyrinth.
- The *cochlear duct* located within the cochlea. The cochlear duct contains the sensory receptors for hearing.

The membranous labyrinth is filled with a clear fluid called **endolymph** (en′do-limf; "internal water"). External to the membranous labyrinth, the bony labyrinth is filled with another clear fluid called **perilymph** (per′ĭ-limf; "surrounding water"). The perilymph is continuous with the cerebrospinal fluid that fills the subarachnoid space. Nowhere are the perilymph and endolymph continuous with one another.

You will explore the basic parts of the bony and membranous labryinth next.

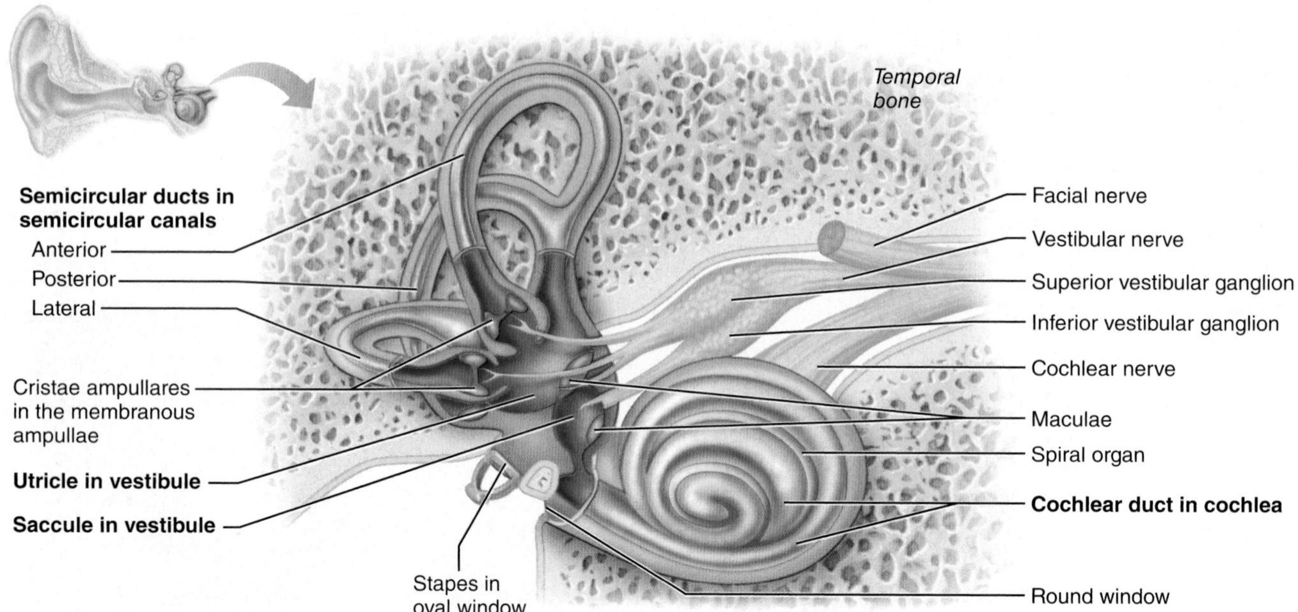

Figure 16.18 The internal ear. The membranous labyrinth (blue) lies within the chambers of the bony labyrinth (tan). The locations of the sensory organs for hearing (spiral organ) and equilibrium (maculae and cristae ampullares) are shown in purple.

The Cochlea

The **cochlea** (kok′le-ah; "snail shell") is a spiraling chamber located inferiorly in the bony labyrinth (Figures 16.16 and 16.18). It is about the size of a split pea. From its attachment to the vestibule at its base, it coils for about two and a half turns around a pillar of bone called the **modiolus** (mo-di′o-lus) (Figure 16.19a). The modiolus is shaped like a screw whose tip lies at the *apex of the cochlea,* pointing anterolaterally. Just as screws have threads, the modiolus has a spiraling projection of bone called the **osseous spiral lamina.** Running through the bony core of the modiolus is the **cochlear nerve,** which is the cochlear division of the vestibulocochlear nerve (VIII).

The coiled part of the membranous labyrinth within the cochlea is the *cochlear duct* (Figure 16.19a and b). This duct winds through the cochlea and ends blindly in the cochlear apex. Within the cochlea, the endolymph-filled cochlear duct (or *scala media*) lies between two perilymph-filled chambers of the bony labyrinth, the **scala vestibuli** and **scala tympani** (*scala* = ladder) (Figure 16.19b). The scala vestibuli is continuous with the vestibule near the base of the cochlea where it abuts the oval window (see Figure 16.20). The scala tympani ends at the round window at the base of the cochlea. The scala vestibuli and scala tympani are continuous with each other at the apex of the cochlea, in a region called the **helicotrema** (hel″ĭ-ko-tre′mah; "the hole in the spiral").

Cochlear Duct The **cochlear duct** or **scala media,** is the part of the membranous labyrinth that contains the sensory receptors for hearing (Figure 16.19b). The "roof" of the cochlear duct, separating it from the scala vestibuli, is the **vestibular membrane.** The external wall of this duct is the *stria vascularis* ("vascularized streak"), an unusual epithelium that contains capillaries and secretes the endolymph of the internal ear. The floor of the cochlear duct consists of the osseous spiral lamina plus an attached sheet of fibers called the **basilar membrane.** The basilar membrane supports the **spiral organ,** the receptor epithelium for hearing. This tall epithelium (Figure 16.19c) consists of columnar *supporting cells* and one row of **inner** and three rows of **outer hair cells,** which are the receptor cells. At the cell's apex, the tips of the hairs, the *stereocilia,* are embedded in a gel-like

Table 16.1 The Internal Ear: Basic Structures of the Bony and Membranous Labyrinths

Bony Labyrinth	Membranous Labyrinth (Within Bony Labyrinth)	Functions of the Membranous Labyrinth
Semicircular canals	Semicircular ducts	Equilibrium: rotational (angular) acceleration of the head
Vestibule	Utricle and saccule	Equilibrium: static equilibrium and linear acceleration of the head.
Cochlea	Cochlear duct	Hearing

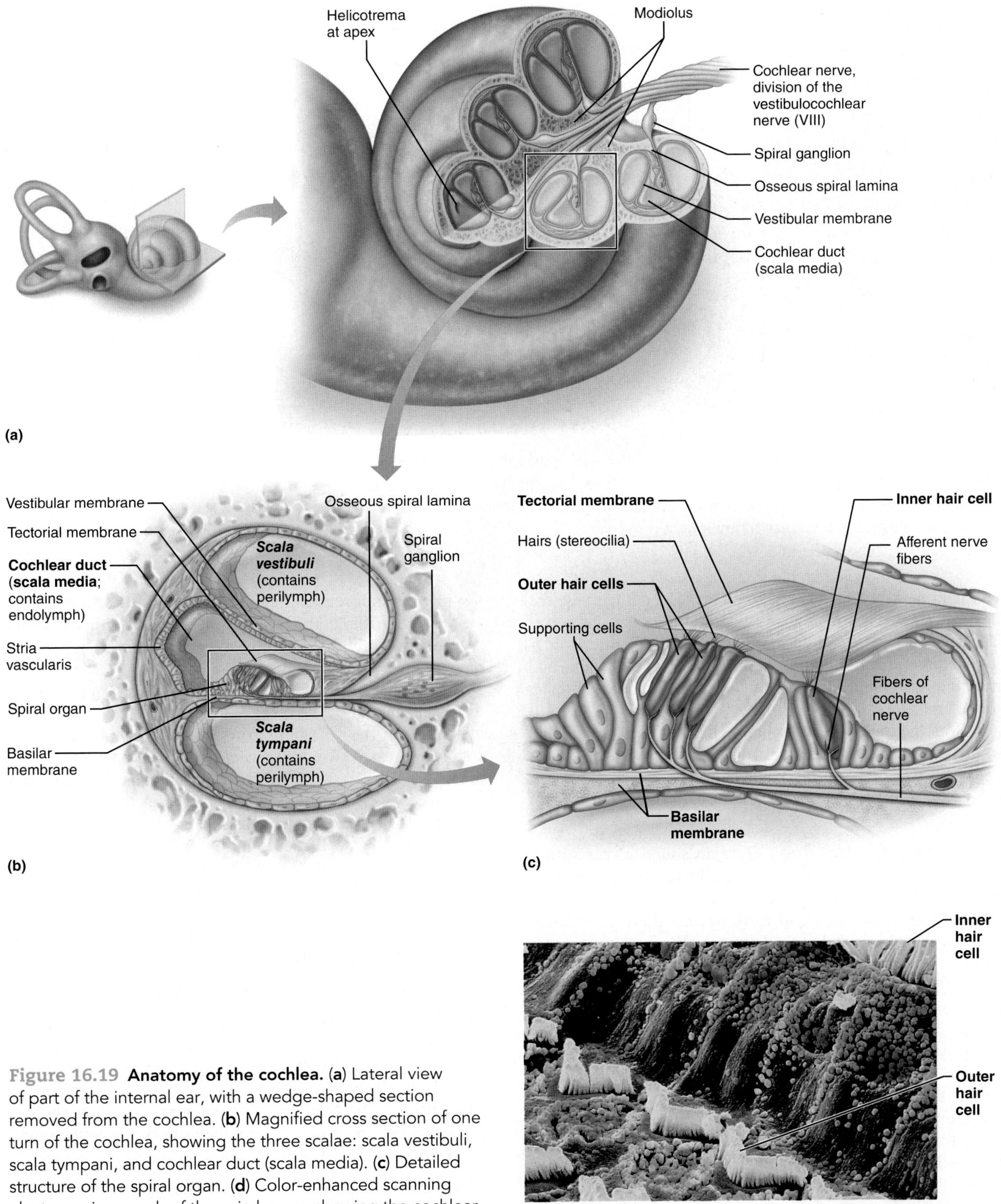

Figure 16.19 Anatomy of the cochlea. (**a**) Lateral view of part of the internal ear, with a wedge-shaped section removed from the cochlea. (**b**) Magnified cross section of one turn of the cochlea, showing the three scalae: scala vestibuli, scala tympani, and cochlear duct (scala media). (**c**) Detailed structure of the spiral organ. (**d**) Color-enhanced scanning electron micrograph of the spiral organ showing the cochlear hair cells (1740×).

View PAL Histology MasteringA&P®

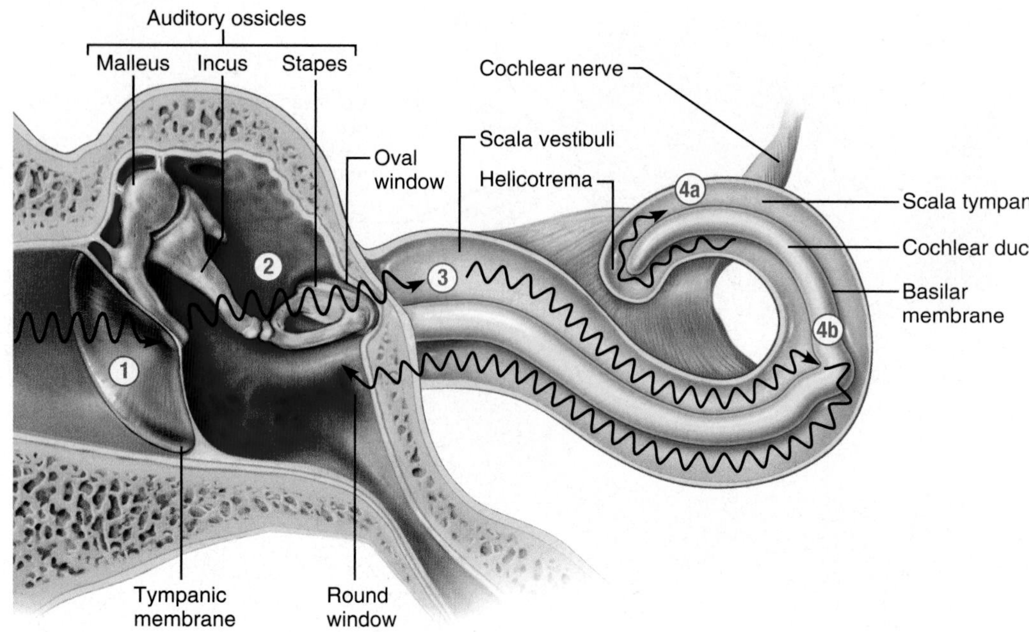

① Sound waves vibrate the tympanic membrane.

② Auditory ossicles vibrate. Pressure is amplified.

③ Pressure waves created by the stapes pushing on the oval window move through fluid in the scala vestibuli.

④a Sounds with frequencies below hearing travel through the helicotrema and do not excite hair cells.

④b Sounds in the hearing range go through the cochlear duct, vibrating the basilar membrane and deflecting hairs on inner hair cells.

Figure 16.20 Role of the cochlea in hearing. The cochlea is drawn as if uncoiled. The structure of the basilar membrane is such that it segregates sound according to frequency—its basal part vibrates in response to high-pitched sounds, whereas its apical part vibrates in response to low-pitched sounds.

tectorial membrane ("roofing membrane"); at their base the hair cells synapse with sensory fibers of the cochlear nerve. These nerve fibers belong to bipolar neurons, whose cell bodies occupy a **spiral ganglion** in the osseous spiral lamina and modiolus (see Figure 16.19b) and whose central fibers project to the brain.

Both the inner and outer hair cells have three stereocilia of increasing length extending from the apical surface of each cell. The stereocilia in the inner hair cells are arranged linearly; those in the outer hair cells form a W pattern (Figure 16.19d).

Mechanism for Hearing How do sound waves stimulate the hair cells in the spiral organ? First, sound vibrations travel from the eardrum through the ossicles, causing the stapes to oscillate back and forth against the oval window (**Figure 16.20** ① and ②). This oscillation sets up pressure waves in the perilymph of the scala vestibuli (Figure 16.20 ③), which are transferred to the endolymph of the cochlear duct (Figure 16.20 ④b). These waves cause the basilar membrane to vibrate up and down. The hair cells in the spiral organ move along with the basilar membrane (Figure 16.19c), but the overlying tectorial membrane (in which the hairs are anchored) does not move. Therefore, the movements of the hair cells cause their hairs to bend. Each time such bending occurs in a specific direction, the hair cells release neurotransmitters that excite the cochlear nerve fibers, which carry the vibratory (sound) information to the brain. The vibrations of the basilar membrane set the perilymph vibrating in the underlying scala tympani. These vibrations then travel to the round window, where they push on the membrane that covers that window, thereby dissipating their remaining energy into the air of the middle ear cavity. Without this release mechanism, echoes would reverberate within the rigid cochlear box, disrupting sound reception.

The inner and outer hair cells in the spiral organ have different functions. The *inner* hair cells are the true receptors that transmit the vibrations of the basilar membrane to the cochlear nerve. The *outer* hair cells are involved with actively tuning the cochlea and amplifying the signal. The outer hair cells receive efferent fibers from the brain that cause these cells to stretch and contract, enhancing the responsiveness of the inner hair cell receptors. Overall, this active mechanism amplifies sounds some 100 times, so that we can hear the faintest sounds.The mobility of the outer hair cells is also responsible for producing ear sounds *(otoacoustic emissions)*. Detection of spontaneous otoacoustic emissions is used to test hearing in newborns.

The Vestibule and the Utricle and Saccule

The **vestibule** is the central cavity of the bony labyrinth (Figure 16.16). It lies just medial to the middle ear, and the oval window is in its lateral bony wall. Suspended within its perilymph are the two egg-shaped parts of the membranous labyrinth, the **utricle** (u′tri-k′l; "leather bag") and the **saccule** ("little sac"). The utricle is continuous with the semicircular ducts; the saccule, with the cochlear duct.

Maculae The utricle and saccule each house a spot of sensory epithelium called a **macula** (mak′u-lah; "spot") (Figure 16.18 and **Figure 16.21a**). Both the macula of the utricle and the macula of the saccule contain receptor cells that monitor the position of the head when the head is held still. This aspect of the sense of balance is called *static equilibrium.* These receptor cells also monitor straight-line changes in the speed and the direction of head movements—that is, *linear acceleration*—but not rotational movements of the head.

Each macula is a patch of epithelium containing columnar **supporting cells** and scattered receptors called **hair cells** (see Figure 16.21a). The hair cells synapse with sensory fibers of the **vestibular nerve,** which is the vestibular division of the vestibulocochlear nerve (VIII). Named for the hairy look of its free surface, each hair cell has many stereocilia (long microvilli) and a single kinocilium (a true cilium) protruding from its apex. The tips of these stiff hairs are embedded in an overlying **otolith** (o″to-lith′ik) **membrane,** which is actually a jellylike disc that contains heavy crystals of calcium carbonate called **otoliths** ("ear stones").

It is easy to understand how the maculae and otoliths contribute to the sense of static equilibrium. The macula of the *utricle* has a *horizontal* orientation within the ear (Figure 16.21a). When one holds the head in a tilted position (Figure 16.21b), the heavy otolithic membrane pulls downward, bending the receptor hairs and signaling the vestibular nerve to tell the brain that the head is tilted. The macula of the *saccule,* in contrast, has a *vertical* orientation within the ear (Figure 16.21a), so its heavy otoliths pull downward on the hairs whenever the head is upright, signaling the brain that the head is in this untilted position. It is also easy to understand how both maculae monitor the movements of linear acceleration: Whenever the body jolts forward, upward, or sideways in a straight line, the heavy otolithic membrane lags behind, again bending the hairs and signaling the brain.

The maculae are innervated by two branches of the vestibular nerve (Figure 16.18). The sensory neurons in this nerve are bipolar neurons, with cell bodies located in the *superior* and *inferior vestibular ganglia.* These ganglia lie in the internal acoustic meatus of the petrous temporal bone. The pathway of the vestibular nerve fibers within the brain is discussed shortly (p. 515).

The Semicircular Canals and Semicircular Ducts

Whereas the vestibule houses the receptors for static equilibrium and linear acceleration, the semicircular canals house receptors for *rotational* acceleration of the head. The three **semicircular canals** of the bony labyrinth lie posterior and lateral to the vestibule (Figure 16.16). Each of these canals actually describes about two-thirds of a circle and has an expansion at one end called an **ampulla** ("flask"). Each canal lies in one of the three planes of space: The **anterior** and **posterior semicircular canals** lie in vertical planes at right angles to each other, whereas the **lateral semicircular canal** *(horizontal canal)* lies almost horizontally. Snaking through each semicircular canal is part of the membranous labyrinth, the **semicircular duct.** Each semicircular duct has a swelling called a **membranous ampulla** within the corresponding bony ampulla.

Cristae Ampullares Each membranous ampulla houses a small crest called a **crista ampullaris,** or "crest of the ampulla"

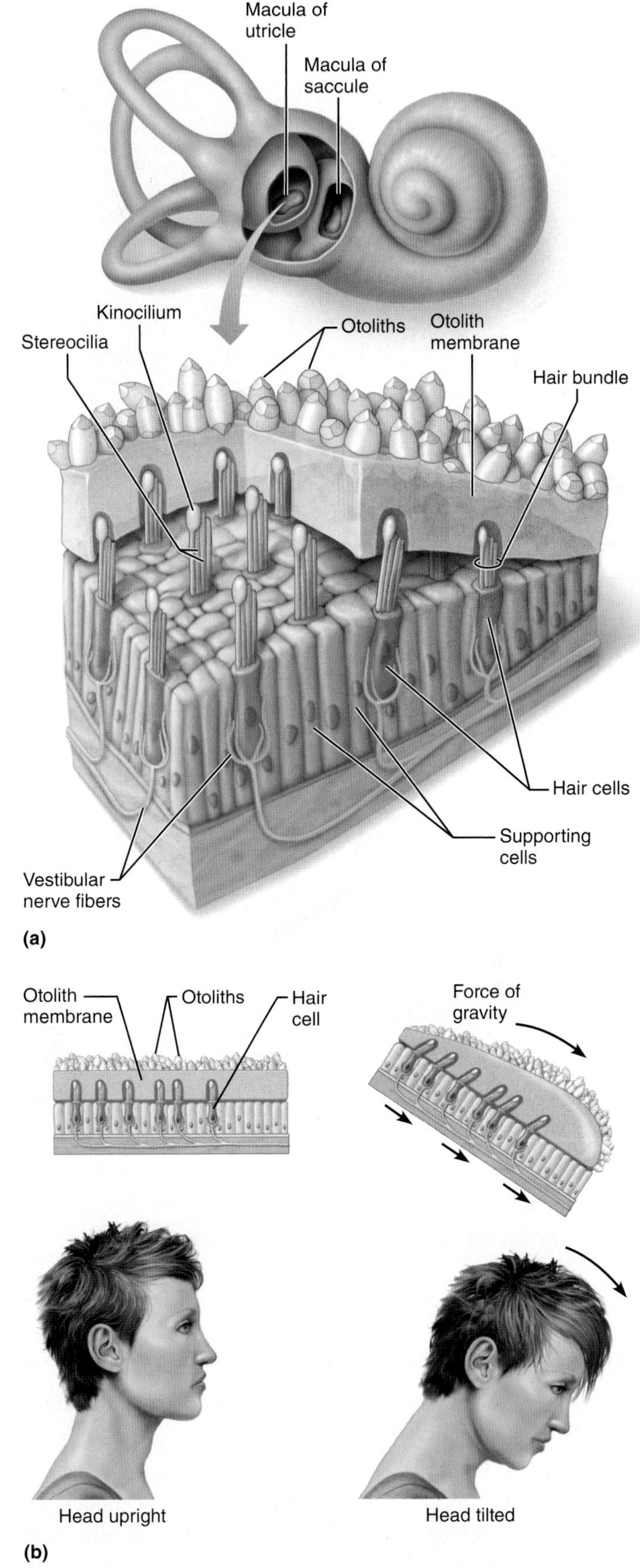

Figure 16.21 **The maculae in the internal ear.** (**a**) Structure of a macula. (**b**) Function of the macula of the utricle in signaling the position of the head.

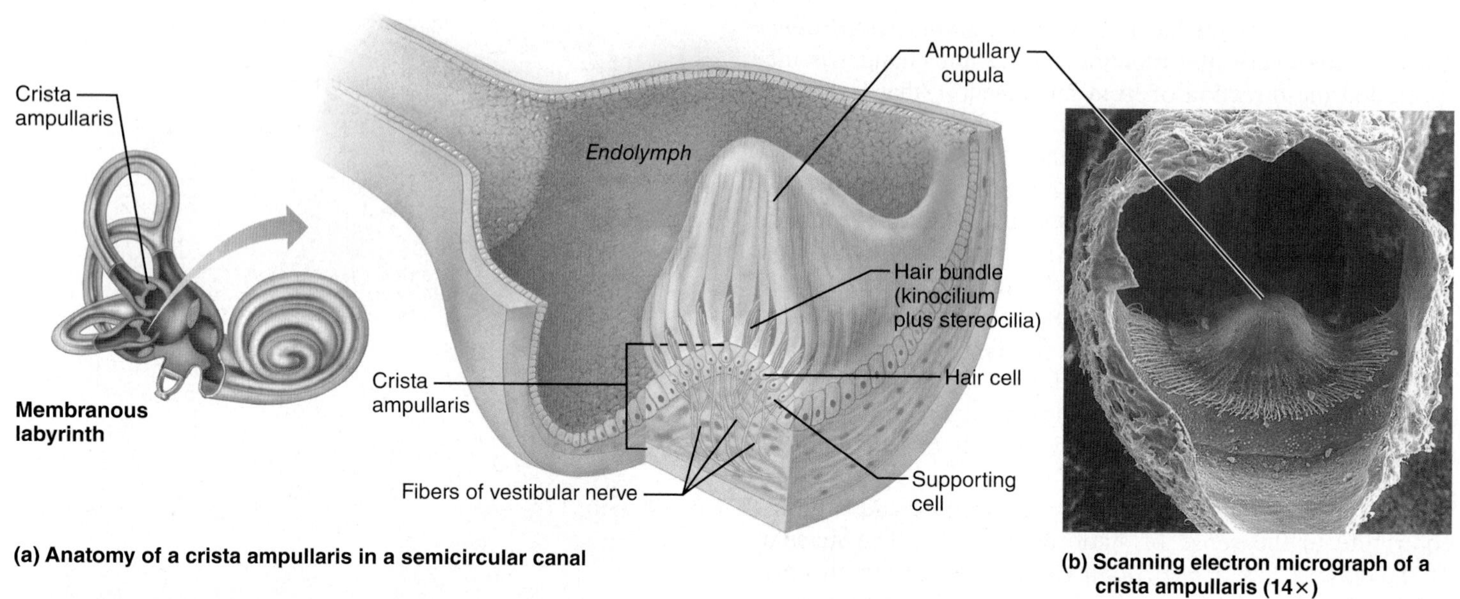

(a) Anatomy of a crista ampullaris in a semicircular canal

(b) Scanning electron micrograph of a crista ampullaris (14×)

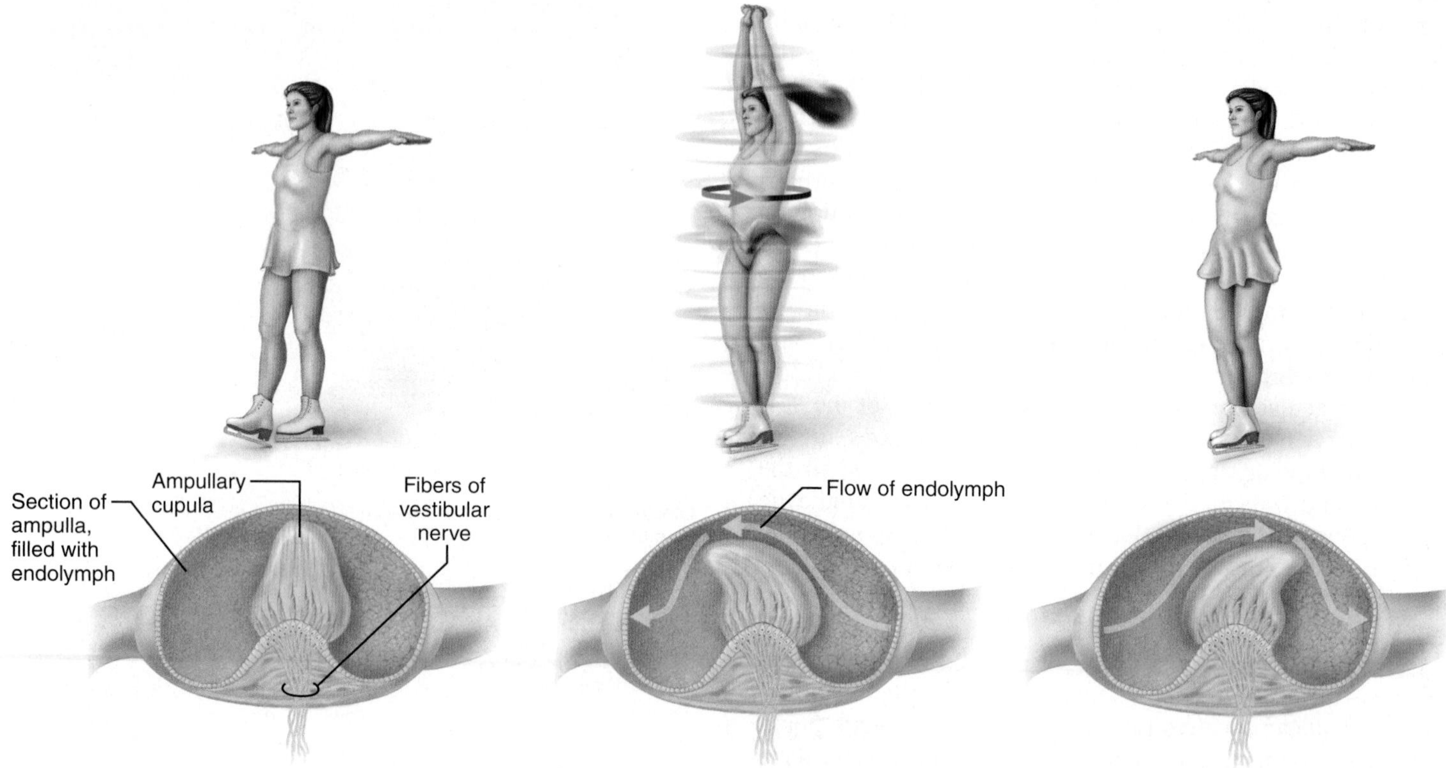

At rest, the ampullary cupula stands upright.

During rotational acceleration, endolymph moves inside the semicircular canals in the direction opposite the rotation (it lags behind because of inertia). Endolymph flow bends the ampullary cupula and excites the hair cells.

As rotational movement slows, endolymph keeps moving in the direction of the rotation, bending the ampullary cupula in the opposite direction from acceleration and inhibiting the hair cells.

(c) Movement of the ampullary cupula during rotational acceleration and deceleration

Figure 16.22 Location, structure, and function of the crista ampullaris in the internal ear.

(Figure 16.18 and Figure 16.22). The three cristae ampullares contain the receptor cells that measure *rotational (angular) acceleration* of the head, as occurs when a figure skater spins or a gymnast does a flip. Each crista has an epithelium on its top that, like the maculae, contains *supporting cells* and receptor *hair cells*. The "hairs" of these hair cells project into a tall, jellylike mass that resembles a pointed cap, the **ampullary cupula** (ku′pu-lah; "little barrel"); the basal parts of the hair cells synapse with fibers of the vestibular nerve.

Because the three semicircular ducts lie in three different planes, each crista responds to head rotation in a different plane of space. When the head starts to rotate, the endolymph in the semicircular duct lags behind at first, pushing on the cupula and bending the hairs (Figure 16.22c). As their hairs bend, the hair cells depolarize and change the pattern of impulses carried by vestibular nerve fibers to the brain.

check your understanding

- ☐ 15. What is the difference between the membranous labyrinth and the bony labyrinth of the internal ear?
- ☐ 16. Where is perilymph located within the cochlea?
- ☐ 17. Where in the cochlear duct is each of these membranes located: vestibular membrane, the basilar membrane, the tectorial membrane? Vibrations in which membrane stimulate the hair cells of the spiral organ?
- ☐ 18. Which sensory receptors monitor stationary head position and linear movements of the head? Where are these receptors located?

(For answers, see Appendix B.)

Auditory and Equilibrium Pathways

learning outcome

▶ Describe the pathways taken by auditory and equilibrium information through the brain.

As is the case for all sensory information, information on equilibrium and hearing travels to the brain for processing and integration.

The ascending **auditory pathway** transmits auditory information primarily from the cochlear receptors of the inner hair cells to the cerebral cortex (Figure 16.23). First, impulses pass through the cochlear nerve to the **cochlear nuclei** in the medulla. From there, some neurons project to the **superior olivary nuclei,** which lie at the junction of the medulla and pons. Beyond this, the axons ascend in the **lateral lemniscus** (a fiber tract) to the **inferior colliculus** (the auditory reflex center in the midbrain), which projects to the **medial geniculate nucleus of the thalamus.** Axons of the thalamic neurons then project to the primary auditory cortex, which provides conscious awareness of sound. The auditory pathway is unusual in that not all of its fibers cross over to the other side of the brain. Therefore, each **primary auditory cortex** receives impulses from both ears. Clinically, this phenomenon makes identifying damage to the primary auditory cortex on one side difficult, because such damage produces only a minimal loss of hearing.

The superior olivary nuclei and inferior colliculus (p. 382) are not merely relay stations along the auditory pathway, but perform important functions of their own. For example, both these structures participate in the localization of sounds.

The **equilibrium pathway** transmits information on the position and movements of the head via the vestibular nerve to the brain stem. Equilibrium is the only special sense for which most information goes to the *lower* brain centers—which are primarily reflex centers—rather than to the "thinking" cerebral cortex. This pathway reflects the fact that the

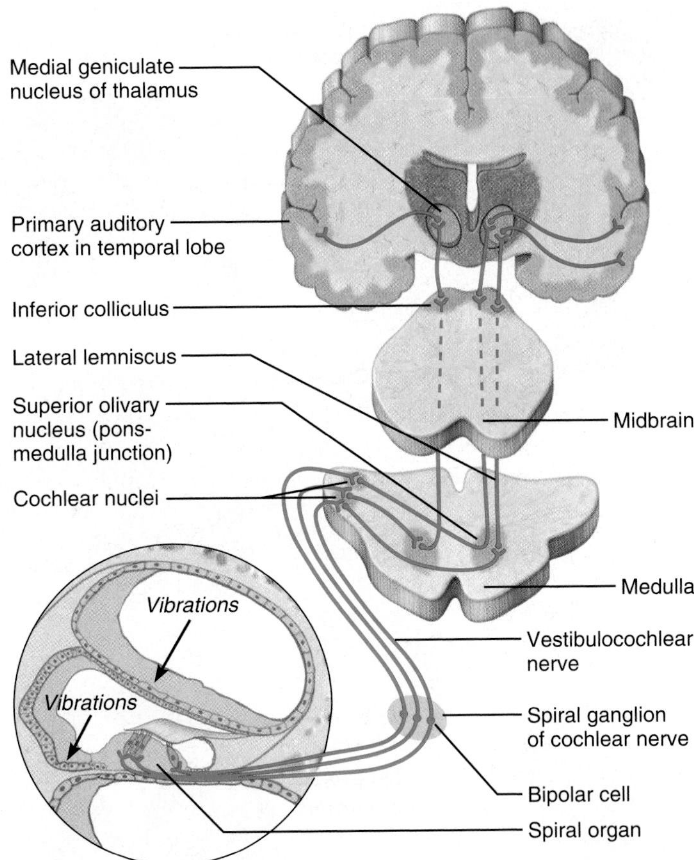

Figure 16.23 The auditory pathway. This simplified diagram shows only the pathway from the right ear.

responses to a loss of balance, such as stumbling, must be rapid and reflexive: In the time it takes you to "think about" correcting a fall, you would hit the ground. The vestibular nuclei in the medulla (shown in Figure 13.7c and discussed on p. 379) and the cerebellum (p. 383) are the major brain centers for processing information on equilibrium. A minor pathway to the cerebral cortex provides conscious awareness of the position and movements of the head. In this minor pathway, vestibular nerve fibers project to the vestibular nuclei, then to the thalamus, and then to the posterior insula of the cerebrum.

Embryonic Development of the Ear

learning outcome

▶ Compare and contrast the embryonic derivations of the external ear, middle ear, and internal ear.

Development of the ear begins in the fourth week after conception (Figure 16.24). First the internal ear begins to form from a thickening of the surface ectoderm called the **otic placode,** which lies lateral to the hindbrain on each side of the head (Figure 16.24a). This placode invaginates to form the **otic pit** (Figure 16.24b). Then its edges fuse to form the **otic vesicle,** which detaches from the surface epithelium, as shown on the right side of Figure 16.24c. The otic vesicle takes on a complex shape and becomes the membranous labyrinth. The mesenchyme tissue around the otic vesicle

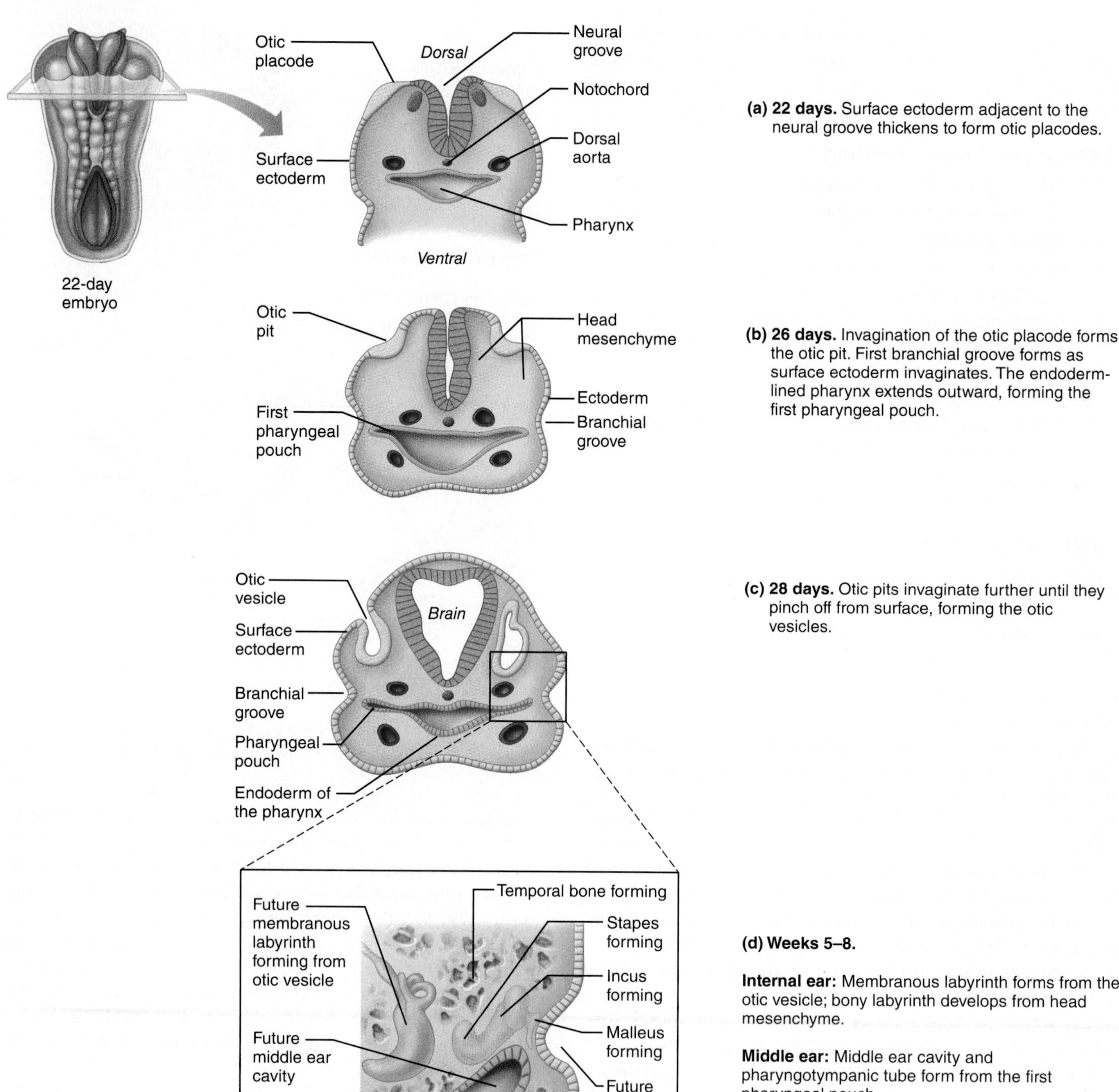

Figure 16.24 Embryonic development of the ear. At left, a surface view of the embryo shows the plane of the sections.

becomes the petrous part of the temporal bone—that is, the walls of the bony labyrinth.

As the internal ear develops, the middle ear starts to form. Lateral outpocketings called *pharyngeal pouches* form from the endoderm-lined pharynx (Figure 16.24b–d). The middle ear cavity and the pharyngotympanic tube develop from the first of the pharyngeal pouches. The ossicles, which will bridge the middle ear cavity, develop from cartilage bars associated with the first and second pharyngeal pouches.

Turning to the external ear, the external acoustic meatus differentiates from the first **branchial groove,** an indentation of the surface ectoderm (Figure 16.24b–d). The auricle of the external ear grows from a series of bulges around this branchial groove.

Disorders of Equilibrium and Hearing

learning outcome

- List the causes and symptoms of motion sickness, Ménière's syndrome, and deafness.

Motion Sickness

Motion sickness is a common disorder of equilibrium in which particular motions (such as riding in a car or aboard a ship) lead to nausea and vomiting. The cause of this condition has been difficult to determine. The most popular theory is that it arises from a mismatch of sensory inputs. For example, if you are in a rocking ship, visual inputs indicate that your body is fixed with reference to a stationary environment (your cabin), but your vestibular apparatus detects movement. The brain's resulting confusion somehow leads to motion sickness. Another theory is that motion sickness occurs because the vestibular nuclei lie near, and may project to, the centers in the medulla that control vomiting. Antimotion drugs can relieve the symptoms of motion sickness by blocking signals from the internal ear to the vomiting center.

Ménière's Syndrome

In a disorder called **Ménière's syndrome,** the membranous labyrinth is apparently distorted by excessive amounts of endolymph. Affected individuals experience a variety of symptoms: equilibrium so disturbed that standing is nearly impossible; transient but repeated attacks of vertigo, nausea, and vomiting; and tinnitus, roaring or buzzing sounds in the ears, such that hearing is impaired and perhaps ultimately lost. Although less severe cases can be managed by antimotion drugs, more debilitating attacks may require diuretics (drugs that increase urine output) and restriction of dietary intake of salt—both of which decrease the volume of extracellular fluid and consequently that of endolymph. Severe cases may require surgery either to drain excess endolymph from the internal ear or to cut the vestibular nerve to relieve the vertigo. A last resort, usually deferred until all hearing is lost, is removal of the entire membranous labyrinth.

Deafness

Any hearing loss, no matter how slight, is considered deafness. The two types of deafness, *conduction deafness* and *sensorineural deafness,* have different causes.

Conduction deafness occurs when sound vibrations cannot be conducted to the internal ear. It can be caused by earwax blocking the external acoustic meatus, a ruptured eardrum, otitis media, or otosclerosis.

Sensorineural deafness results from damage to the hair cells or to any part of the auditory pathway to the brain. Most often it results from the normal, gradual loss of hearing receptor cells that occurs throughout life. In other cases, hair cells can be destroyed at an earlier age by a single, explosively loud noise or by repeated exposure to loud music, factory noise, or airport noise. Strokes and tumors that damage the auditory cortex can also cause sensorineural deafness. When deafness reflects damage to hair cells, hearing aids can help. Traditional hearing aids simply amplify sounds, an effective strategy if the loss of hair cells is not too great. For complete sensorineural deafness, **cochlear implants** are available. Placed in the temporal bone, these devices convert sound energy into electrical signals and deliver these signals directly to the cochlear nerve fibers. Modern models have up to 24 electrodes, each responding to a different frequency, and are so effective that even children who were born deaf can hear well enough to learn to speak.

☑ check your understanding

- ☐ 19. What type of deafness results from damage to the cells of the spiral ganglion?
- ☐ 20. Which embryonic germ layer (ectoderm, mesoderm, or endoderm) forms the membranous labyrinth?
- ☐ 21. What brain regions receive input from the vestibular nerves and process information on equilibrium?

(For answers, see Appendix B.)

THE SPECIAL SENSES THROUGHOUT LIFE

learning outcome

- Describe the changes in the special senses that occur with aging.

Olfaction and Taste

All special senses are functional, to a greater or lesser degree, at birth. Smell and taste are sharp in newborns, and the likely reason that infants relish food that adults consider bland is that children have more taste buds than adults do. Most people experience no difficulties with their chemical senses throughout childhood and young adulthood, but starting in the fourth decade of life, the ability to taste and smell declines. This decline reflects a gradual loss of the chemoreceptors, which are replaced more slowly than in younger years.

Vision

Even though the fetus cannot see in the darkness of the uterus, the photoreceptors are fully formed in the posterior retina by 25 weeks after conception, and the neuronal connections of the visual pathway have developed even earlier. Visual experiences during the first 8 months after birth fine-tune these synaptic connections.

Congenital problems of the eyes are relatively rare, although maternal rubella (German measles) during the first trimester of pregnancy may cause congenital blindness or cataracts. In newborns, the eyeballs are foreshortened, so all infants are hyperopic (farsighted); only gray tones are perceived, eye movements are uncoordinated, and often only one eye is used at a time. By 3 months, however, infants can focus an image on the retina's fovea centralis for sharp vision, and babies can follow moving objects with their eyes. By the age of 6 months, depth perception is present, color vision is well developed, and the earlier hyperopia is almost gone.

As a person ages, the lens loses its clarity and becomes discolored, and it begins to scatter light (the resulting glare is why older people should look slightly away from oncoming headlights while driving at night). The dilator muscles of the iris become less efficient, so the pupils stay partly constricted. All these changes decrease the amount of light reaching the retina, such that visual acuity is dramatically lower in people over 70. In addition, elderly individuals are susceptible to conditions that cause blindness, such as glaucoma, retinal detachment, diabetes mellitus, and age-related macular degeneration.

Hearing

Newborn infants can hear, but their early responses to sounds are mostly reflexive—for example, crying and clenching the eyelids in response to a startling noise. Infants can hear low-pitched and middle-pitched sounds at birth, but the ability to hear high-pitched sounds is a postnatal development. By the third or fourth month, infants can localize sounds and will turn to the voices of family members. By 12 months, babies know all the sounds of their native language, and critical listening begins in toddlers as they learn to speak.

Congenital structural abnormalities of the external ear, including partly or fully missing pinnae and closed or absent auditory canals, are fairly common; less common is sensorineural deafness due to maternal rubella infections during pregnancy. Except for the common ear infections of childhood, few problems usually affect the ears until old age. By about age 60, however, deterioration of the spiral organ becomes noticeable. Humans are born with about 20,000 hair cells in each ear, but they are gradually lost. The ability to hear high-pitched sounds fades first. This gradual loss of hearing with age is called **presbycusis** (pres″bĭ-ku′sis), literally, "old hearing." It is the most common type of sensorineural deafness. Although it is considered a disability of old age, it is becoming more common in younger people as the modern world grows noisier and headphone use pervasive.

In mammals, the hair cells of the internal ear of both hearing and equilibrium do not regenerate naturally. However, gene therapy techniques have induced new hair cell formation. The gene *Math1,* introduced into the internal ear of fluid of adult guinea pigs, resulted in the growth of new hair cells. *Math1* encodes for a protein that induces an immature ear cell to become a hair cell. This research could lead to the treatment of many types of hearing loss.

RELATED CLINICAL TERMS

Congenital deafness One of every 500–1000 children is born deaf because of such physical factors as an immobile stapes, overproduction of perilymph, and various malformations of the middle and internal ear. The ultimate causes of congenital deafness are less well understood, though it can result from maternal mumps, syphilis, or rubella during pregnancy. Many congenital cases are genetically based, and at least four different genes have been identified that, when mutated, cause deafness.

Ophthalmology (of″thal-mol′o-je; "eye study") The study of the eye and eye diseases. An *ophthalmologist* is a medical doctor whose specialty is treating eye disorders. By contrast, an *optometrist* is a licensed nonphysician who measures vision and prescribes corrective lenses.

Otitis externa (swimmer's ear) Inflammation and infection of the external acoustic meatus, caused by bacteria or fungi that enter the canal from outside, especially when the canal is moist.

Otorhinolaryngology (o″to-ri″no-lar″ing-gol′o-je) The study of the ear, nose, and throat and the diseases of these body regions.

Scotoma (sko-to′mah; *scoto* = darkness) A blind spot in the visual field other than the normal blind spot caused by the optic disc; often reflects the presence of a brain tumor pressing on nerve fibers along the visual pathway.

Tinnitus (tĭ-ni′tus) Persistent noise—a ringing, whistling, humming, buzzing, or screeching—that seems to come from the ears in 10% to 20% of all elderly people, causing great distress and annoyance. It often first appears after a loud noise or an injury to the head or cochlea. Recent evidence suggests that tinnitus is analagous to phantom limb pain (see p. 391)—it is a "phantom cochlear noise" caused by destruction of some neurons along the auditory pathway. With the other neurons of this pathway now deprived of their normal input, nearby axons grow in and reinnervate these neurons, and the CNS interprets background signals from the new axons as noise. Treatments include masking the noise with soothing sounds, counseling, and biofeedback; drugs are largely unsuccessful.

CHAPTER SUMMARY

Take full advantage of the invaluable online practice and study tools for this chapter by going to the Study Area of MasteringA&P®. There you will find:

- Art Labeling Activities, Quizzes, Practice Tests, Memory Games
- Animal Organ Dissection, Cat Dissection and Bone Videos
- PAL Cadaver
- PAL Histology
- A&P Flix Animations
- Dynamic Study Modules

THE CHEMICAL SENSES: TASTE AND SMELL (pp. 489–492)

Taste (Gustation) (pp. 489–490)

1. Most taste buds are on the tongue, in the epithelium of fungiform, foliate, and circumvallate papillae.
2. Taste buds contain gustatory epithelial cells and basal epithelial cells that replace damaged gustatory cells. The gustatory epithelial cells are excited when taste-stimulating chemicals bind to their microvilli.

3. The five basic qualities of taste are sweet, sour, salty, bitter, and umami.
4. The sense of taste is served by cranial nerves VII, IX, and X, which send impulses to the medulla. From there, impulses travel to the thalamus and the taste area of the cerebral cortex.

Smell (Olfaction) (pp. 490–491)

5. The olfactory epithelium is located in the roof of the nasal cavity. This epithelium contains olfactory sensory neurons, supporting epithelial cells, and olfactory stem cells.
6. The olfactory sensory neurons are ciliated bipolar neurons. Odor molecules bind to the cilia, exciting the neurons. Axons of these neurons form the filaments of the olfactory nerve (cranial nerve I).
7. Olfactory nerve axons transmit impulses to the olfactory bulb. Here, these axons synapse with mitral cells in structures called glomeruli.
8. After receiving input from the olfactory sensory neurons, the mitral cells send this olfactory information through the olfactory tract to the olfactory cortex and limbic system.

Embryonic Development of the Chemical Senses (p. 491)

9. The olfactory epithelium and taste buds develop from the epithelia on the face and mouth/pharynx, respectively.

Disorders of the Chemical Senses (pp. 491–492)

10. Disorders of smell include anosmia (inability to smell) and uncinate fits (smell hallucinations).

THE EYE AND VISION (pp. 492–506)

11. The eye is located in the bony orbit and is cushioned by fat. The cone-shaped orbit also contains nerves, vessels, and extrinsic muscles of the eye.

Accessory Structures of the Eye (pp. 492–495)

12. Eyebrows shade and protect the eyes.
13. Eyelids protect and lubricate the eyes by reflexive blinking. Each eyelid contains a supporting tarsal plate, the roots of the eyelashes, and tarsal and ciliary glands. Muscles in the eyelids include the levator palpebrae superioris, which opens the eye, and the orbicularis oculi, which closes the eye.
14. The conjunctiva is a mucous membrane that covers the inner surface of the eyelids (palpebral conjunctiva) and the white of the eye (bulbar conjunctiva). Its mucus lubricates the eye surface.
15. The lacrimal gland secretes lacrimal fluid (tears), which spreads medially across the eye surface and drains into the nasal cavity through the lacrimal canaliculi, lacrimal sac, and nasolacrimal duct.
16. The six extrinsic eye muscles are the lateral and medial rectus (which turn the eye laterally and medially, respectively); the superior and inferior rectus (which elevate and depress the eye, respectively, but also turn it medially); and the superior and inferior obliques (which depress and elevate the eye, respectively, but also turn it laterally).

Anatomy of the Eyeball (pp. 495–501)

17. The wall of the eye has three layers. The most external, fibrous layer consists of the posterior sclera and the anterior cornea. The tough sclera protects the eye and gives it shape. The cornea is the clear window through which light enters the eye.
18. The middle, pigmented vascular layer consists of the choroid, the ciliary body, and the iris. The choroid provides nutrients to the retina's photoreceptors and prevents the scattering of light within the eye. The ciliary body contains smooth ciliary muscles that control the shape of the lens and ciliary processes that secrete aqueous humor. The iris contains smooth muscle that changes the size of the pupil.
19. The sensory layer contains the retina and the optic nerve. The retina consists of an outer pigmented layer and an inner neural layer. The neural layer contains photoreceptors (rod and cone cells) and other types of neurons. Light influences the photoreceptors, which signal bipolar cells, which signal ganglion cells. The axons of ganglion neurons run along the inner retinal surface toward the optic disc, forming the optic nerve.
20. The outer segments of the rods and cones contain light-absorbing pigment in membrane-covered discs. Light modifies this pigment to initiate the flow of signals through the visual pathway.
21. Two important spots on the posterior retinal wall are (1) the macula lutea with its fovea centralis (area of highest visual acuity) and (2) the optic disc (blind spot), where axons of ganglion cells form the optic nerve.
22. The outer third of the retina (photoreceptors) is nourished by capillaries in the choroid, whereas the inner two-thirds is supplied by the central vessels of the retina.
23. The posterior segment of the eye, posterior to the lens, contains the gel-like vitreous humor. The anterior segment, anterior to the lens, is divided into anterior and posterior chambers by the iris. The anterior segment is filled with aqueous humor, which continually forms at the ciliary processes in the posterior chamber, flows into the anterior chamber, and drains into the scleral venous sinus.
24. The biconvex lens helps to focus light. It is suspended in the eye by the ciliary zonule attached to the ciliary body. Tension in the zonule resists the lens's natural tendency to round up.

The Eye as an Optical Device (pp. 501–502)

25. As it enters the eye, light is bent by the cornea and the lens and focused on the retina. The cornea accounts for most of this refraction, but the lens allows focusing on objects at different distances.
26. The resting eye is set for distance vision. Focusing on near objects requires accommodation (allowing the lens to round as ciliary muscles release tension on the ciliary zonule). The pupils also constrict. Both these actions are controlled by parasympathetic fibers in the oculomotor nerve.
27. Eye-focusing disorders include myopia (nearsightedness), hyperopia (farsightedness), presbyopia (loss of lens elasticity with age), and astigmatism (irregular curvature of the cornea or lens).

Visual Pathways (pp. 502–504)

28. The visual pathway to the brain begins with some processing of visual information in the retina. From there, ganglion cell axons carry impulses via the optic nerve, optic chiasma, and optic tract to the lateral geniculate nucleus of the thalamus. Thalamic neurons project to the primary visual cortex.
29. At the optic chiasma, axons from the medial halves of the retinas decussate. This phenomenon provides each visual cortex with information on the opposite half of the visual field as seen by both eyes. The visual cortex compares the views from the two eyes and generates depth perception.

Embryonic Development of the Eye (pp. 504–506)

30. Each eye starts as an optic vesicle, a lateral outpocketing of the embryonic diencephalon. This vesicle then invaginates to form the optic cup, which becomes the retina. The overlying ectoderm folds to form the lens. The fibrous and vascular layers derive almost entirely from mesenchyme around the optic cups.

Disorders of the Eye and Vision (p. 506)

31. Two disorders that result in blindness were considered, one that damages the retina (retinopathy of prematurity) and one that damages the cornea (trachoma).

THE EAR: HEARING AND EQUILIBRIUM (pp. 506–517)

The External Ear (pp. 506–508)

32. The auricle and external acoustic meatus constitute the external ear, which acts to gather sound waves. The tympanic membrane (eardrum) transmits sound vibrations to the middle ear.

The Middle Ear (pp. 508–509)

33. The middle ear is a small cavity within the temporal bone. Its boundaries are the eardrum laterally, the bony wall of the internal ear medially, a bony roof, a thin bony floor, a posterior wall that opens into the mastoid antrum, and an anterior wall that opens into the pharyngotympanic tube.

34. The pharyngotympanic tube, which consists of bone and cartilage, runs to the pharynx and equalizes air pressure across the eardrum.

35. The auditory ossicles (malleus, incus, and stapes), which help to amplify sound, span the middle ear cavity and transmit sound vibrations from the eardrum to the oval window. The tiny tensor tympani and stapedius muscles dampen the vibrations of very loud sounds.

The Internal Ear (pp. 509–515)

36. The internal ear consists of the bony labyrinth (semicircular canals, vestibule, and cochlea), which is a chamber that contains the membranous labyrinth (semicircular ducts, utricle and saccule, and cochlear duct). The bony labyrinth contains perilymph, whereas the membranous labyrinth contains endolymph.

37. The coiled cochlea is divided into three parts (scalae). Running through its center is the cochlear duct (scala media), which contains the spiral organ. The spiral organ is an epithelium that lies on the basilar membrane and contains the hair cells (receptors for hearing). The other two parts of the cochlea are the scala vestibuli and the scala tympani parts of the bony labyrinth.

38. In the mechanism of hearing, sound vibrations transmitted to the stapes vibrate the fluids in the cochlea. This vibrates the basilar membrane and spiral organ, and in turn bends the hairs of the receptor cells, whose tips are anchored in a nonmoving tectorial membrane. Bending of the hairs produces impulses in the cochlear nerve.

39. The saccule and utricle each contain a macula, a spot of receptor epithelium that monitors static equilibrium and linear acceleration. A macula contains hair cells whose "hairs" are anchored in an overlying otolith membrane. Forces on the otolithic membrane, caused by gravity and linear acceleration of the head, bend the hairs and initiate impulses in the vestibular nerve.

40. The semicircular ducts lie in three planes of space (anterior vertical, posterior vertical, and lateral horizontal). Their cristae ampullares contain hair cells that monitor rotational acceleration. The "hairs" of these cells are anchored in an overlying ampullary cupula. Forces on the cupula, caused by rotational acceleration of the head, bend the hairs and initiate impulses in the vestibular nerve.

Auditory and Equilibrium Pathways (p. 515)

41. Impulses generated by the hearing receptors travel along the cochlear nerve to the cochlear nuclei in the medulla. From there, auditory information passes through several nuclei in the brain stem (superior olivary, inferior colliculus) to the medial geniculate nucleus of the thalamus and cerebral auditory cortex.

42. Impulses generated by the equilibrium receptors travel along the vestibular nerve to the vestibular nuclei and the cerebellum. These brain centers initiate responses that maintain balance. There is also a minor equilibrium pathway to the posterior insula of the cerebral cortex.

Embryonic Development of the Ear (pp. 515–516)

43. The membranous labyrinth develops from the otic placode, a thickening of ectoderm superficial to the hindbrain.

44. A pouch from the pharynx becomes the middle ear cavity and the pharyngotympanic tube. The outer ear is formed by an external branchial groove (external acoustic meatus) and by swellings around this groove (auricle).

Disorders of Equilibrium and Hearing (p. 517)

45. Motion sickness, brought on by particular movements, causes nausea and vomiting. Ménière's syndrome is an overstimulation of the hearing and equilibrium receptors caused by an excess of endolymph in the membranous labyrinth.

46. Conduction deafness results from interference with the conduction of sound vibrations to the internal ear. Sensorineural deafness reflects damage to auditory receptor cells or neural pathways.

THE SPECIAL SENSES THROUGHOUT LIFE (pp. 517–518)

47. The chemical senses are sharpest at birth and decline with age as the replacement of receptor cells slows.

48. The eye is foreshortened (farsighted) at birth. Depth perception, eye coordination, and color vision develop during the first 6 months. With age, the lens loses elasticity and clarity, and visual acuity declines. Eye problems that may develop with age are presbyopia, cataracts, glaucoma, retinal detachment, and age-related macular degeneration.

49. Initially, infants respond to sound only in a reflexive manner. By 5 months, infants can locate sound. Critical listening develops in toddlers. Obvious age-related loss of hearing (presbycusis) occurs in a person's 60s and 70s.

PAL **Human Cadaver/Nervous System/Special Senses**

REVIEW QUESTIONS

Multiple Choice/Matching Questions

(For answers, see Appendix B.)

1. Sensory impulses transmitted over the facial, glossopharyngeal, and vagus nerves are involved in the special sense of (a) taste, (b) vision, (c) equilibrium, (d) smell.
2. The part of the fibrous layer of the eye that is white, tough, and opaque is the (a) choroid, (b) cornea, (c) retina, (d) sclera.
3. The transmission of sound vibrations through the internal ear occurs chiefly through (a) nerve fibers, (b) air, (c) fluid, (d) bone.
4. Of the neurons in the retina, which form the optic nerve? (a) bipolar neurons, (b) ganglion neurons, (c) cone cells, (d) horizontal neurons.
5. Blocking the scleral venous sinus might result in (a) a sty, (b) glaucoma, (c) conjunctivitis, (d) Ménière's syndrome, (e) a chalazion.
6. Conduction of sound from the middle ear to the internal ear occurs via vibration of the (a) malleus against the tympanic membrane, (b) stapes in the oval window, (c) incus in the round window, (d) tympanic membrane against the stapes.
7. The structure that allows the air pressure in the middle ear to be equalized with that of the outside air is the (a) cochlear duct, (b) mastoid air cells, (c) endolymph, (d) tympanic membrane, (e) pharyngotympanic tube.
8. The receptors for static equilibrium that report the position of the head in space relative to the pull of gravity are in the (a) spiral organ, (b) maculae, (c) crista ampullaris, (d) ampullary cupulae, (e) joint kinesthetic receptors.
9. Paralysis of a medial rectus muscle would affect (a) accommodation, (b) refraction, (c) depth perception, (d) pupil constriction.
10. A light ray passes through the refractory structures of the eye in this order: (a) vitreous humor, lens, aqueous humor, cornea; (b) cornea, aqueous humor, lens, vitreous humor; (c) cornea, vitreous humor, lens, aqueous humor; (d) lens, aqueous humor, cornea, vitreous humor.
11. The optic disc is the site where (a) more rods than cones occur, (b) the macula lutea is located, (c) only cones occur, (d) the optic nerve exits the eye.
12. Which of these is not a basic taste sensation? (a) bitter, (b) tsunami, (c) sour, (d) sweet.

Short Answer Essay Questions

13. An anatomy student was arguing with his grandfather. Granddad, who believed in folk wisdom, insisted that there are only five senses. The student, however, said that there are at least ten senses. Decide who was right, and then list all the senses you know. (See Table 12.1, p. 352.)
14. (a) What is the precise location of the olfactory epithelium? (b) Trace the pathway of olfactory stimuli from the olfactory epithelium to the cerebral cortex.
15. What and where is the fovea centralis, and why is it important?
16. Name two special senses whose receptor cells are replaced throughout life, and two special senses whose receptor cells are replaced so slowly that there can be no functional regeneration.
17. (a) Describe the embryonic derivation of the retina. (b) Explain how the middle ear cavity forms.
18. Describe some effects of aging on the eye and the ear.
19. Trace the auditory pathway to the cerebral cortex.
20. Compare and contrast the functions of the inferior oblique and superior rectus muscles.
21. (a) What is the difference, if any, between a semicircular canal and semicircular duct? Between the cochlea and cochlear duct? (b) Name the three parts of the membranous labyrinth of the internal ear. Which of these parts is for hearing, and which are for balance?
22. Describe the function and the innervation of both the sphincter and dilator muscles of the pupil (in the iris of the eye).

Critical Reasoning & Clinical Application Questions

1. Enrique's uncle tells the physician that 3-year-old Enrique gets many "earaches." Upon questioning, the uncle reveals that Enrique has not had a sore throat for a long time and is learning to swim. Does Enrique have otitis media or otitis externa, and does he need ear tubes? Explain your reasoning.
2. Nine children attending the same day-care center developed inflamed eyes and eyelids. What is the most likely cause and name of this condition?
3. Dr. Nakvarati used an instrument to blow a puff of air on Mr. Jefferson's eye during his annual physical examination on his 60th birthday. The eye deformed very little, indicating that the intraocular pressure was too high. What was Mr. Jefferson's probable condition?
4. Lionel suffered a ruptured artery in his middle cranial fossa, and a pool of blood compressed his left optic tract, destroying its axons. What part of the visual field was blinded?
5. Ming, a student in optometry school, felt very sad when she saw some premature babies at the hospital who were born before 25 weeks after conception. She knew that many of these children would soon be blind for life. Explain why.
6. Jan, a senior citizen, developed a constant buzzing sound in her ear that would not stop. It was very annoying and stressful, and she had to go to counseling to learn how to live with this awful noise. What was her condition called, and what are some other treatments she might receive for it? (See this chapter's Related Clinical Terms.)
7. Describe the effect on vision of a tumor in the hypothalamus or pituitary gland that compresses the optic chiasma, destroying the axons crossing through this structure.

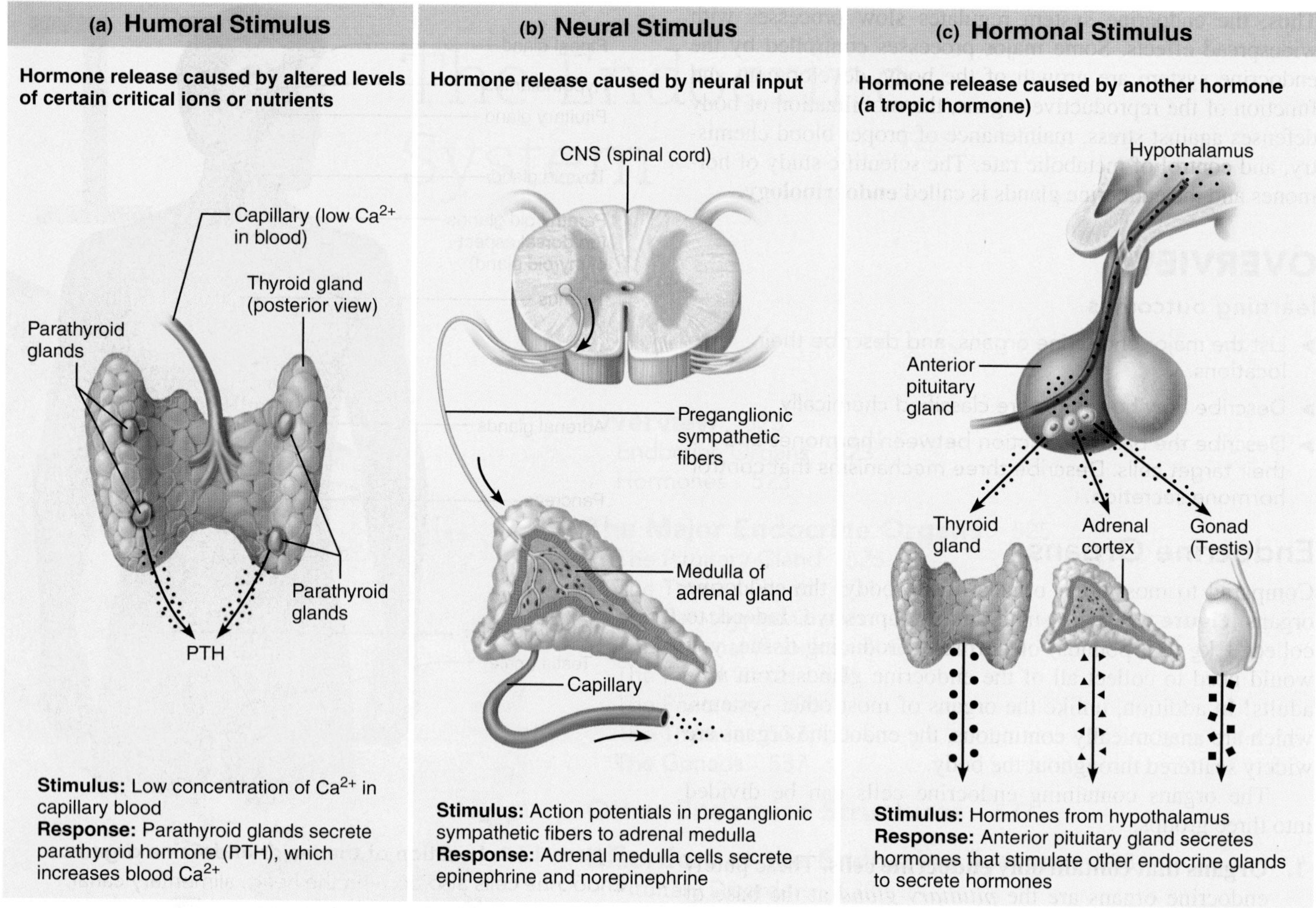

Figure 17.2 **Three types of endocrine gland stimuli.**

smooth ER, which produces the steroid molecules, and abundant lipid droplets, which contain the raw material from which steroids are made. These cells lack secretory granules; steroid hormones are secreted by diffusion across the plasma membrane.

Basic Hormone Action

All major hormones circulate throughout the entire body, leave the bloodstream at the capillaries, and encounter virtually all tissues. Nevertheless, a given hormone influences only specific tissue cells, called its **target cells.** The ability of a target cell to respond to a hormone depends on the presence of specific receptor molecules on the target cell to which that particular hormone can bind. Once binding has occurred, the target cell reacts in a preprogrammed way.

Each kind of hormone produces its own characteristic effects within the body. These effects result from the preprogrammed responses of the target cells, not from any information contained in the hormone molecule itself. In fact, hormones with similar molecular structures often have very dissimilar functions, and the same hormone can have different effects on different target cells. Hormones are just molecular triggers—they do not carry any coded information.

Control of Hormone Secretion

The various endocrine cells of the body are stimulated to make and secrete their hormones by three major types of stimuli (Figure 17.2): *humoral, neural,* and *hormonal* stimuli.

Humoral Stimuli Endocrine glands that secrete their hormones in direct response to changing levels of ions or nutrients in the blood are said to be controlled by **humoral stimuli** (*humoral* = relating to the blood and other body fluids). This is the simplest endocrine control mechanism. For example, the cells of the parathyroid gland (Figure 17.2a) directly monitor the concentration of calcium ions (Ca^{2+}) in the blood and then respond to any decline in this concentration by secreting a hormone that acts to increase blood calcium levels.

Neural Stimuli The secretion of a few endocrine glands is controlled by **neural stimuli** (Figure 17.2b). For example, sympathetic nerve fibers stimulate cells in the adrenal medulla to release epinephrine and norepinephrine during fight-or-flight situations.

Hormonal Stimuli Finally, many endocrine glands secrete their hormones in response to **hormonal stimuli** received from other endocrine glands; that is, certain hormones have the sole purpose of promoting the secretion of other hormones (Figure 17.2c). For example, the hypothalamus of the brain secretes some hormones that stimulate the anterior part of the pituitary gland to secrete its hormones, which in turn stimulate hormonal secretion by other endocrine glands: the thyroid, the adrenal cortex, and the gonads. As you will see, the hypothalamus controls many functions of the endocrine system, through hormonal and other mechanisms.

Feedback Loops No matter how it is stimulated, hormone secretion is always controlled by *feedback loops.* In a negative feedback loop, more hormone is secreted if its blood concentration declines below a minimum set point; then hormone production is halted if the maximum set point is exceeded. This ensures that hormone concentrations stay within a narrow, "desirable" range in the blood. Secretion of certain hormones is regulated by a positive feedback loop: As blood concentrations of the hormone increase, the response of the effector organ stimulates further secretion. The hormone that controls the progression of labor in childbirth, oxytocin, operates via positive feedback—oxytocin stimulates the uterus to contract, and contraction of the uterus stimulates further secretion of oxytocin.

☑ check your understanding

☐ 1. How do hormones reach their target cells?

☐ 2. What are the three types of stimuli that regulate secretion from endocrine cells?

(For answers, see Appendix B.)

THE MAJOR ENDOCRINE ORGANS

The major endocrine organs are the pituitary gland, thyroid gland, parathyroid glands, adrenal (suprarenal) glands, pineal gland, pancreas, thymus, and gonads (ovary and testis).

The Pituitary Gland

learning outcomes

- Name the basic divisions of the pituitary gland.
- List the hormones secreted by the anterior lobe and the cell type that secretes each. State the basic functions of each hormone.
- Explain how the hypothalamus controls secretion of anterior lobe hormones. Define releasing hormones, and trace their path through the pituitary gland.
- Describe the structure of the posterior lobe and the functions of the hormones it releases.

Gross Anatomy

The **pituitary gland,** or **hypophysis** (hi-pof′ĭ-sis; "undergrowth [from the brain]"), is an important endocrine organ that secretes at least nine major hormones. It sits just inferior to the brain in the hypophyseal fossa, a depression in the sella turcica of the sphenoid bone (Figure 17.3). The pituitary gland closely resembles a golf club: The gland itself forms the head of the club, and the stalk of the pituitary, called the **infundibulum** ("funnel"), forms the shaft of the club. The infundibulum connects superiorly to a part of the hypothalamus called the tuber cinereum, which lies between the optic chiasma anteriorly and the mammillary bodies posteriorly.

The pituitary gland has two basic divisions, an **anterior lobe** or **adenohypophysis** (ad″ĕ-no-hi-pof′ĭ-sis; "glandular hypophysis"), composed of glandular tissue, and a **posterior lobe** or **neurohypophysis** (nu″ro-hi-pof′ĭ-sis; "neural hypophysis"), composed of neural tissue and a part of the brain. The anterior lobe has three subdivisions (Figure 17.3c). The largest is the anteriormost **pars distalis.** Just posterior to this lies the **pars intermedia,** and just superior to it lies the **pars tuberalis,** which wraps around the infundibulum like a tube. The posterior lobe has two subdivisions, the **pars nervosa,** inferiorly, and the **infundibulum.**

Arterial blood reaches the pituitary gland through two branches of the internal carotid artery, one of the large arteries that supplies blood to the brain. The **superior hypophyseal artery** supplies the entire anterior lobe and the infundibulum (see Figure 17.4a), whereas the **inferior hypophyseal artery** supplies the pars nervosa (see Figure 17.4b). The veins from the extensive capillary beds in the pituitary gland drain into the cavernous sinus and other nearby dural sinuses (p. 610).

The Anterior Lobe

This description of the anterior lobe concentrates on its largest division, the *pars distalis,* which contains five different types of endocrine cells that make and secrete at least seven different hormones (Table 17.1 on p. 530). Four of the hormones secreted by the anterior lobe—thyroid-stimulating hormone, adrenocorticotropic hormone, follicle-stimulating hormone, and luteinizing hormone—regulate the secretion of hormones by other endocrine glands. These hormones are called **tropic hormones** (tro′pik; "changing"). The remaining three adenohypophysis hormones—growth hormone, prolactin, and melanocyte-stimulating hormone—act directly on nonendocrine target tissues.

Thyroid-Stimulating Hormone (TSH) **Thyroid-stimulating hormone (TSH)** is produced by thyrotropic cells. TSH signals the thyroid gland to secrete its own hormone, thyroid hormone, and thus ultimately controls metabolic rate.

Adrenocorticotropic Hormone (ACTH) **Adrenocorticotropic** (ad-re″no-kor″tĭ-ko-trō′pik; "adrenal cortex changing") **hormone (ACTH)** and melanocyte-stimulating hormone, described shortly, are split from a common parent molecule produced by corticotropic cells in the anterior lobe. ACTH stimulates the adrenal cortex (as its name implies) to secrete hormones that help the body cope with stress.

Gonadotropins **Follicle-stimulating hormone (FSH)** and **luteinizing hormone (LH)** are produced by gonadotropic cells in the anterior lobe and together are referred to as **gonadotropins** (go-nad″o-trō′pinz). These hormones act on the gonads, stimulating maturation of the sex cells and inducing the secretion of sex hormones. In females, FSH and LH stimulate the maturation of the egg-containing ovarian follicles

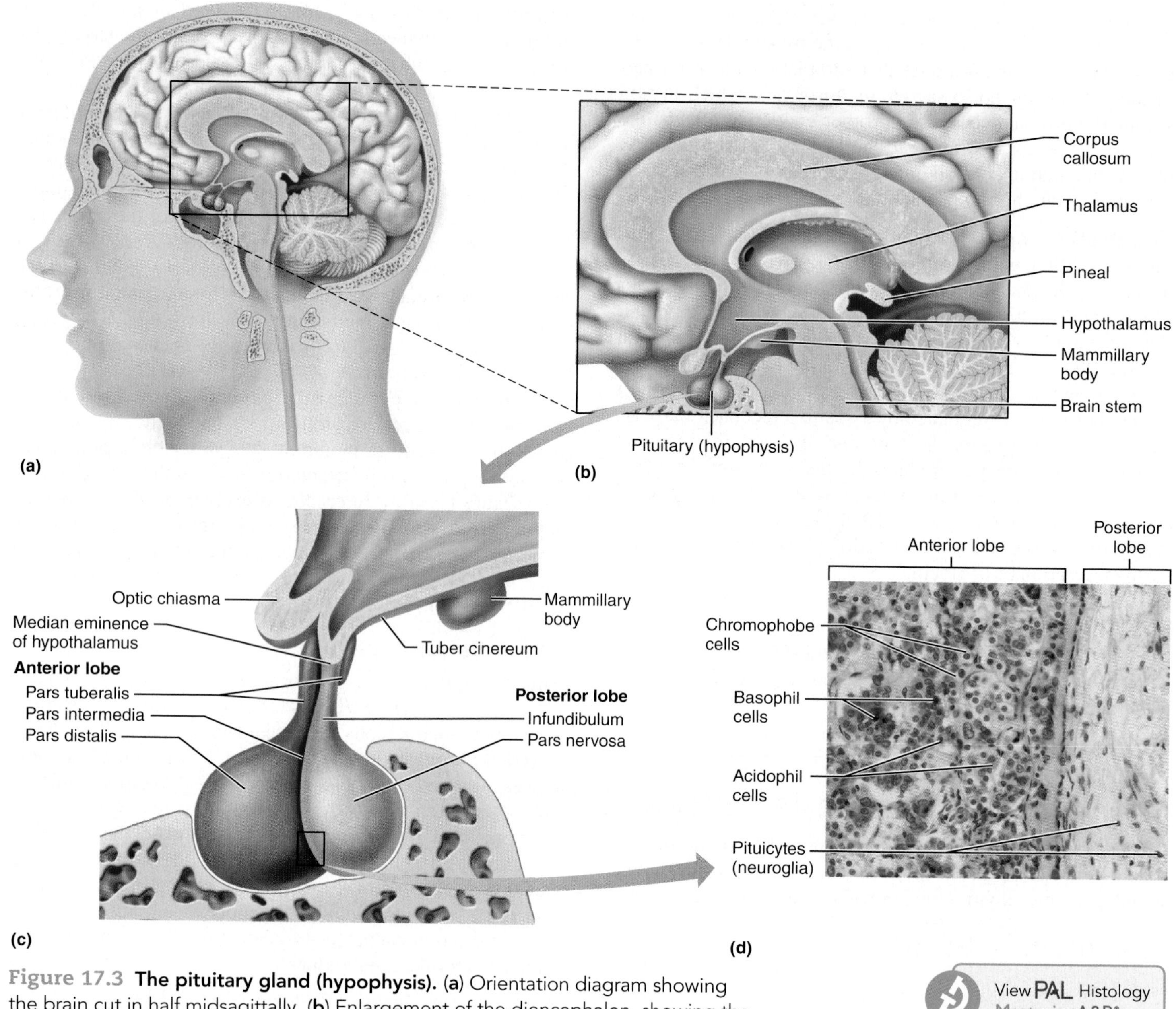

Figure 17.3 The pituitary gland (hypophysis). (**a**) Orientation diagram showing the brain cut in half midsagittally. (**b**) Enlargement of the diencephalon, showing the location of the pituitary gland. (**c**) Basic regions of the pituitary gland. Note that the anterior aspect is to the left. (**d**) Histology of the anterior and posterior lobes of the pituitary gland (160×).

and the secretion of androgens, estrogens, and progesterone from cells in the ovary. Furthermore, a large amount of LH is secreted in the middle of the menstrual cycle to induce ovulation. In males, LH signals the secretion of androgens (primarily testosterone) by interstitial cells in the testes, and FSH stimulates the maturation of sperm cells and the production of androgen-binding protein by cells in the sperm-forming tubules. (See Chapter 25 for a full discussion of these actions.)

Prolactin (PRL) **Prolactin (PRL)** (pro-lak′tin) is produced by prolactin cells in the anterior lobe. PRL targets the milk-producing glands in the breast, the lactiferous glands, and stimulates the manufacture of milk.

Melanocyte-Stimulating Hormone (MSH) **Melanocyte-stimulating hormone (MSH)** is formed from the precursor molecule produced by corticotropic cells. In more primitive vertebrates, amphibians and reptiles, MSH stimulates melanocytes to produce more melanin, the pigment responsible for skin coloration. In humans, MSH functions in the central nervous system (CNS) in appetite suppression; its actions outside the CNS are not well understood.

Growth Hormone (GH) **Growth hormone (GH),** also called **somatotropic hormone** (so″mah-to-tro′pik; "body changing"), is produced in the somatotropic cells, the most abundant cell type in the anterior lobe. This hormone stimulates growth of the entire body by stimulating body cells to increase their production of proteins and by stimulating growth of the epiphyseal plates of the skeleton. It stimulates these actions directly, and also indirectly by signaling the liver to secrete *insulin-like growth factor-1,* which acts together with GH.

CLINICAL APPLICATION

Therapeutic Uses of Growth Hormone Growth hormone is approved for treating a number of disorders that impact growth, enabling individuals with these disorders to reach normal, expected adult height. These disorders include **growth hormone deficiencies** in adults and children, **idiopathic short stature** (ISS, small stature from unknown cause), **chronic kidney disease,** and **Turner's syndrome** (females with a single X chromosome). In addition to stimulating skeletal growth, GH also decreases body fat, increases lean muscle mass and bone density, and decreases blood levels of low-density lipoprotein cholesterol. Because of these actions, GH is used for treating **muscle wasting disease** associated with HIV/AIDS and may be useful in rehabilitation following injury. These properties have led to abuse of GH by athletes hoping to improve their performance; it is a banned substance in competitive sports. The risks associated with GH during the period of use are relatively low: fluid retention and associated complications such as carpal tunnel syndrome. Little data currently exists for the long-term risks and benefits of GH use, particularly for people treated in childhood; further studies are warranted.

The endocrine cells of the pars distalis are clustered into spheres and branching cords separated by capillaries. When tissue of the pars distalis is stained with typical histological dyes and viewed by light microscopy, its five cell types can be classified in three categories (Figure 17.3d):

1. **Acidophils,** which stain with acidic stains and include the somatotropic and prolactin cells
2. **Basophils,** which stain with basic stains and include the thyrotropic, corticotropic, and gonadotropic cells
3. **Chromophobes** ("color avoiders"), which stain poorly. Chromophobes are either immature cells or cells whose supply of hormone has been depleted.

The pars distalis is by far the most important division of the anterior lobe; the other divisions—the *pars intermedia* and *pars tuberalis*—are not as well understood. The pars intermedia contains corticotropic cells that secrete more MSH than ACTH. The pars tuberalis contains gonadotropic cells, thyrotropic cells, and unique, *pars tuberalis–specific endocrine cells* that have receptors for melatonin, a hormone that is secreted by the pineal gland and regulates daily (circadian) rhythms according to light/dark cycles (photoperiod). Therefore, the pars tuberalis may be where sexual functions and metabolic rate are brought under the influence of circadian rhythms and photoperiod. The function of the pineal gland and the role of melatonin are discussed on p. 536.

Hypothalamic Control of Hormone Secretion from the Anterior Lobe

The secretion of hormones by the anterior lobe is controlled by the hypothalamus of the brain. The hypothalamus exerts its control by secreting peptide hormones called **releasing hormones (releasing factors),** which then prompt the cells in the anterior lobe to release their hormones. The hypothalamus also secretes **inhibiting hormones,** which *turn off* the secretion of hormones by the anterior lobe when necessary. There are distinct releasing and inhibiting hormones for almost every anterior lobe hormone.

The role of the hypothalmus in regulating secretion from the pituitary is illustrated in *Focus on Hypothalamus and Pituitary Interactions* (Figure 17.4). Releasing hormones from the hypothalamus signal the secretion of anterior lobe hormones. Releasing hormones made in hypothalamic neurons are secreted like neurotransmitters from the axon terminals of these neurons (Figure 17.4a, ①). In this case, neurons are serving as endocrine cells. The secreted releasing hormones enter a **primary capillary plexus** in the median eminence of the hypothalamus and then travel inferiorly in **hypophyseal portal veins** to a **secondary capillary plexus** in the anterior lobe. The releasing hormones leave the bloodstream and attach to the anterior lobe cells and stimulate these cells to secrete hormones (GH, LH, TSH, PRL, and so on; Figure 17.4a, ②). The hormones secreted from the anterior lobe cells enter the secondary plexus. From there the newly secreted anterior lobe hormones proceed into the general circulation and travel to their target organs throughout the body (Figure 17.4a, ③).

Inhibiting hormones secreted by the hypothalamus follow the same route but function to inhibit hormone secretion by the cells of the anterior lobe.

The primary and secondary capillary plexuses in the pituitary gland, plus the intervening hypophyseal portal veins, constitute the **hypophyseal portal system** (Figure 17.4a). A *portal system* refers to a system of blood vessels in which blood is collected from one capillary bed and travels in veins to a second capillary bed. In the first capillary bed, blood picks up molecules (in this case, releasing or inhibiting hormones) that are then carried to the tissues supplied by the second capillary bed (in this case, the anterior pituitary).

In summary, the hypothalamus controls the secretion of hormones by the anterior lobe, which in turn controls the secretion of hormones by the thyroid gland, the adrenal cortex, and the gonads. In this way, the brain controls these important endocrine glands (Figure 17.2c).

The Posterior Lobe

The posterior lobe, which secretes two hormones, is structurally a part of the brain (Figure 17.4b). It consists of nervous tissue that contains unmyelinated axons and neuroglial cells called **pituicytes** (Figure 17.3d). Its axons form the **hypothalamohypophyseal tract,** which arises from neuron cell bodies in the supraoptic and paraventricular nuclei

Figure 17.4

The hypothalamus controls release of hormones from the pituitary gland in two different ways.

(a) Anterior Pituitary: *Hypothalamic hormones* released into special blood vessels (the hypophyseal portal system) control the release of anterior pituitary hormones.

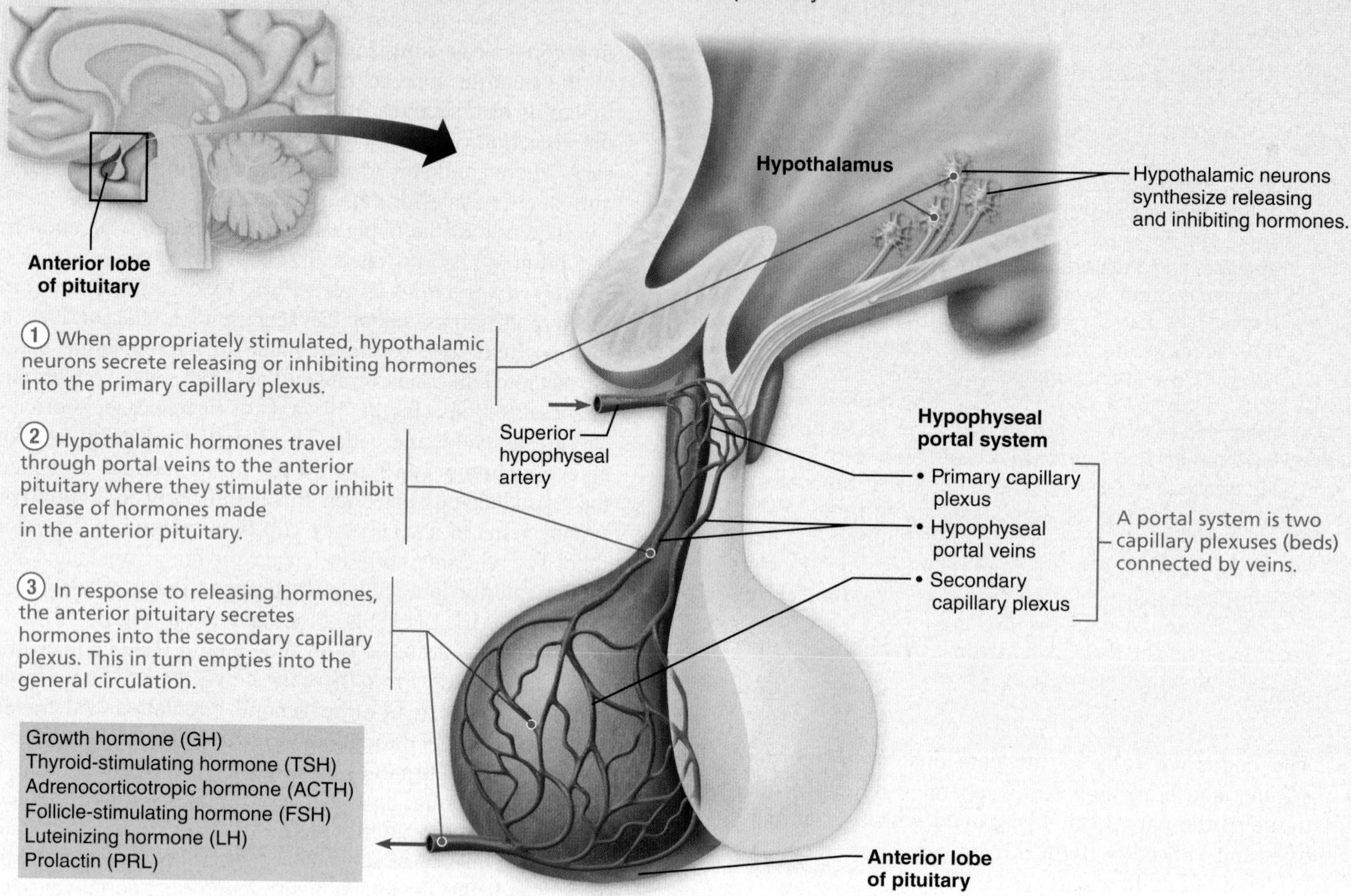

in the hypothalamus and ends in axon terminals in the pars nervosa. The posterior lobe hormones are made in the neuron cell bodies, transported along the axons, and stored in dilated axon terminals called *neurosecretory bodies* (Figure 17.4b ① through ③). When the neurons fire, they release the stored hormones into a capillary bed in the pars nervosa for distribution throughout the body (Figure 17.4b ④). Therefore, the posterior lobe does not make hormones but only stores and releases hormones produced in the hypothalamus. The posterior lobe releases two peptide hormones (Table 17.1, p. 531): *antidiuretic hormone* (an″tĭ-di″u-ret′ik; "inhibiting urination") and *oxytocin* (ok″sĭ-to′sin; "childbirth hormone").

Antidiuretic Hormone Made in the neurons of the supraoptic nucleus, **antidiuretic hormone (ADH),** also called *vasopressin,* targets the kidneys. The kidneys respond by resorbing more water and returning it to the bloodstream. The

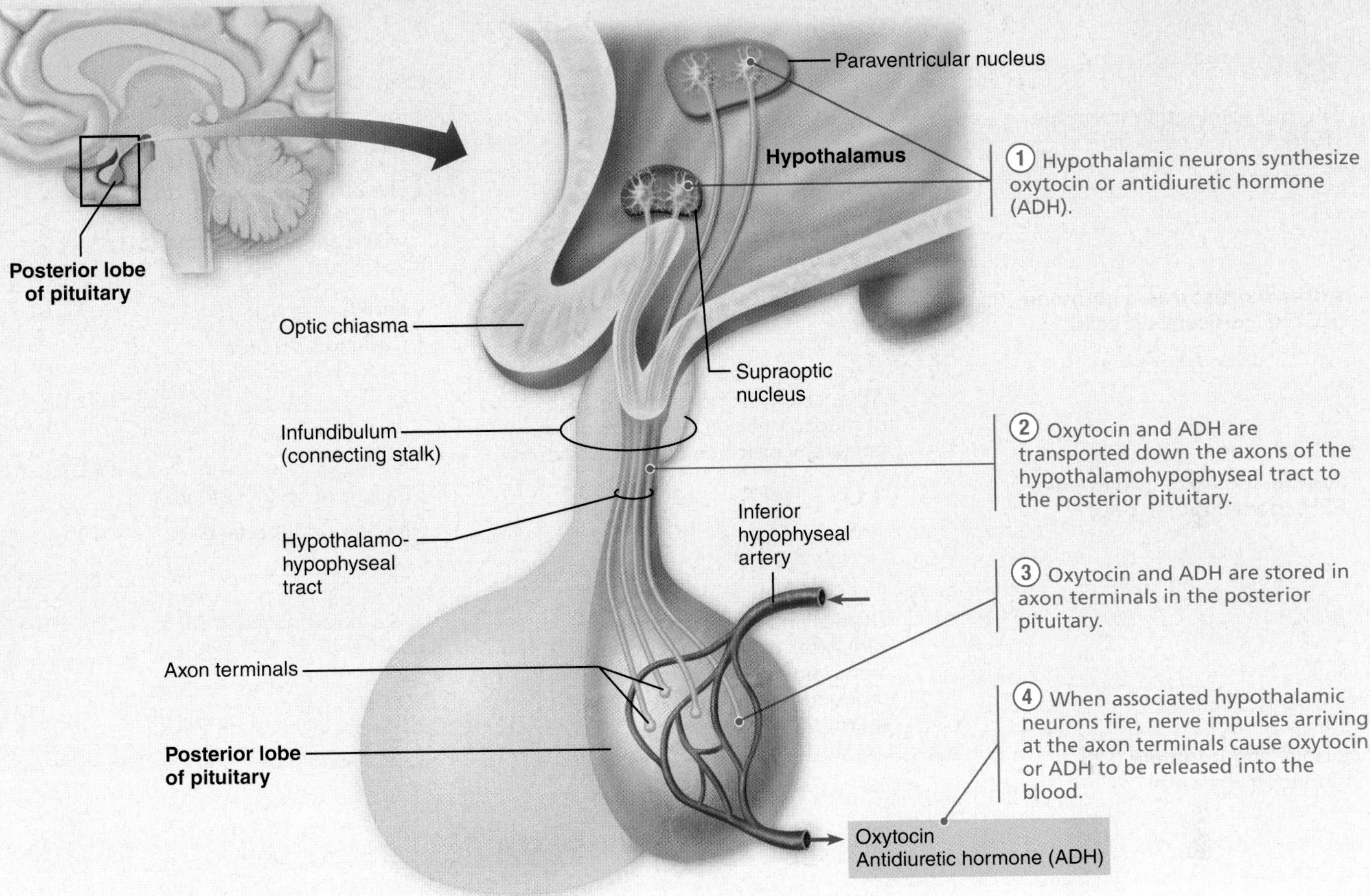

details of this process are described with the urinary system, (Chapter 24). In this way, ADH helps the body retain as much fluid as possible when thirst (dehydration) or fluid loss (severe bleeding) occurs. Also, when fluid loss lowers blood pressure, ADH signals the peripheral arterioles to constrict, thus raising blood pressure to normal (*vasopressin* = vessel constrictor).

Oxytocin **Oxytocin,** produced in the paraventricular nucleus, induces contraction of the smooth musculature of reproductive organs in both males and females. Most important, it signals the smooth muscle in the wall of the uterus to contract, expelling the infant during childbirth; it also induces contraction of muscle-like cells (myoepithelial cells) around the secretory alveoli in the breast to eject milk during breastfeeding. Finally, in monogamous mammals both oxytocin and ADH cause a desire to cuddle, groom, and bond with a mate. Their roles in human socialization are under investigation.

Table 17.1 Pituitary Hormones: Summary of Target Organs and Effects

Hormone (Cell Type)	Target Organ and Effects	Effects of Hyposecretion ↓ and Hypersecretion ↑
ANTERIOR LOBE HORMONES 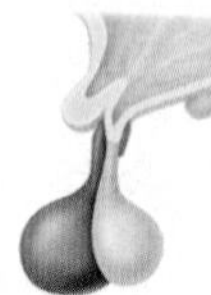		
Thyroid-stimulating hormone (TSH) (thyrotropic cells)	 Thyroid gland: stimulates thyroid gland to release thyroid hormones	↓ Cretinism in children; myxedema in adults ↑ Hyperthyroidism; effects similar to those of Graves' disease, in which antibodies mimic TSH
Adrenocorticotropic hormone (ACTH) (corticotropic cells)	 Adrenal cortex: promotes release of glucocorticoids and androgens (mineralocorticoids to a lesser extent)	↓ Rare ↑ Cushing's disease
Follicle-stimulating hormone FSH) (gonadotropic cells)	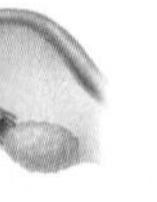 Ovaries and testes: in females, stimulates ovarian follicle maturation and estrogen production; in males, stimulates sperm production	↓ Failure of sexual maturation ↑ No important effects
Luteinizing hormone (LH) (gonadotropic cells)	  Ovaries and testes: in females, triggers ovulation and stimulates ovarian production of estrogen and progesterone; in males, promotes testosterone production	As for FSH
Growth hormone (GH) (somatotropic cells)	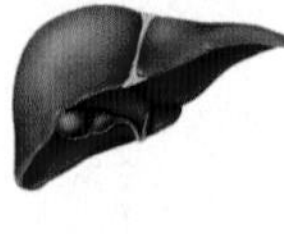  Liver, muscle, bone, cartilage, and other tissues: stimulates protein synthesis and somatic growth; mobilizes fats; increases blood glucose	↓ Pituitary dwarfism in children ↑ Gigantism in children; acromegaly in adults
Prolactin (PRL) (prolactin cells)	 Breast secretory tissue: promotes lactation	↓ Poor milk production in nursing women ↑ Inappropriate milk production (galactorrhea); cessation of menses in females; impotence in males

Table 17.1 *continued*

Hormone (Cell Type)	Target Organ and Effects	Effects of Hyposecretion ↓ and Hypersecretion ↑
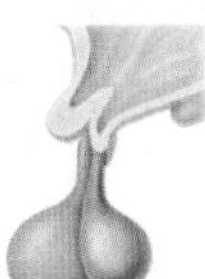 **POSTERIOR LOBE HORMONES (MADE BY HYPOTHALAMIC NEURONS AND STORED IN POSTERIOR LOBE)**		
Antidiuretic hormone (ADH), or **vasopressin** (from neurons in supraoptic nucleus of hypothalamus)	 Kidneys: stimulates kidney tubule cells to resorb water	↓ Diabetes insipidus ↑ Syndrome of inappropriate ADH secretion (SIADH)
Oxytocin (from neurons in paraventricular nucleus hypothalamus)	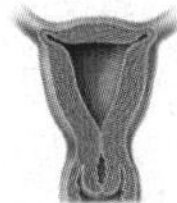  Uterus: stimulates uterine contractions; initiates labor; breast: initiates milk ejecton	Unknown

CLINICAL APPLICATION

Therapeutic Uses of Oxytocin When a pregnancy has lasted well beyond the estimated due date, a physician may deem it necessary to induce labor. The most effective way of inducing labor is to inject the mother with natural or synthetic oxytocin, which initiates uterine contractions. Oxytocin also functions to stop the bleeding following delivery (by causing constriction of the ruptured blood vessels at the site of placental detachment) and to stimulate the ejection of milk by the mammary glands. This dual function of oxytocin on both uterine and mammary tissue is the cause of the intense and painful uterine contractions that can occur during breast-feeding in the early postpartum weeks.

check your understanding

- ☐ 3. How do tropic hormones secreted by the pituitary differ in action from the other pituitary hormones?
- ☐ 4. Where are the hormones produced that are secreted by the posterior lobe of the pituitary? Where are the hormones produced that are secreted by the anterior lobe?
- ☐ 5. Name the target organ(s) for each pituitary hormone listed: (a) oxytocin, (b) ACTH, (c) FSH, (d) ADH.

(For answers, see Appendix B.)

The Thyroid Gland

learning outcomes

- ▶ Describe the anatomy of the thyroid gland.
- ▶ Define and describe the effects of thyroid hormone (TH) and calcitonin, and explain how TH is secreted.

Gross Anatomy

The butterfly-shaped **thyroid gland** is located in the anterior neck, on the trachea just inferior to the larynx. It has two lateral **lobes** (the butterfly's "wings") connected by a median bridge called the **isthmus** (Figure 17.5a). You may be able to feel the isthmus of the thyroid gland as a spongy cushion over the second to fourth tracheal ring. Then, try to palpate the two soft lateral lobes of your thyroid gland along the sides of the trachea. The thyroid gland is the largest purely endocrine gland in the body and has a copious blood supply from the superior and inferior thyroid arteries.

Microscopic Anatomy

Internally, the thyroid gland is composed of hollow, approximately spherical **follicles** separated by an areolar connective tissue rich in capillaries (Figure 17.5b). The walls of each follicle are formed by a layer of cuboidal or squamous epithelial cells called **follicular cells,** and the central lumen is filled with a jellylike substance called **colloid** (kol′oid; "gluelike") consisting of **thyroglobulin** (thi″ro-glob′u-lin), a protein from which thyroid hormone is ultimately derived. Lying within the follicular epithelium are **parafollicular (C) cells,** which appear to project into the surrounding connective tissue. The thyroid produces two hormones: thyroid hormone and calcitonin.

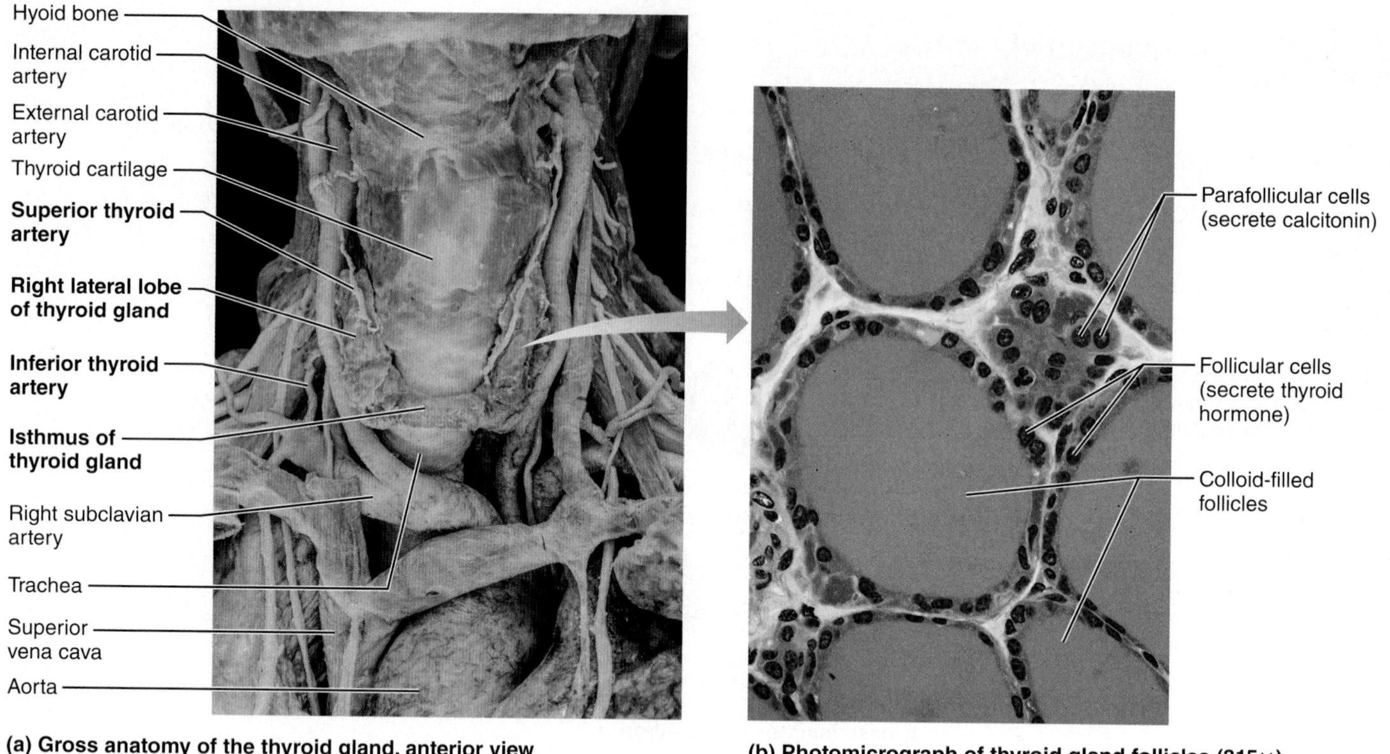

(a) Gross anatomy of the thyroid gland, anterior view

(b) Photomicrograph of thyroid gland follicles (315×)

Figure 17.5 **The thyroid gland.**

Thyroid Hormone The follicular cells of the thyroid gland secrete **thyroid hormone (TH),** a name that actually applies to two similar molecules called thyroxine (thi-rok′-sin) or T_4, and triiodothyronine (tri″i-o″do-thi′ro-nēn) or T_3. Each of these hormone molecules is constructed from a pair of amino acids and contains the element iodine, which is essential to the function of the hormone. Thyroid hormone affects many target cells throughout the body. Its main function is to increase the basal metabolic rate (the rate at which the body uses oxygen to transform nutrients into energy). Individuals who secrete an excess of TH have a high activity level, are fidgety, and continually feel warm, whereas those who do not produce enough TH are sluggish and feel cold. (These and other disorders of the endocrine glands are covered in more detail on pp. 538–539.)

TH is made and secreted by the thyroid follicle cells. The follicular cells continuously synthesize the precursor protein thyroglobulin and secrete it into the center of the follicle for iodination and storage. The thyroid gland is the only endocrine gland that stores its hormone extracellularly and in large quantities—it stores enough TH to last several months. To initiate secretion of the stored TH into the blood, the *pituitary gland* releases thyroid-stimulating hormone (TSH), which signals the follicular cells to reclaim the thyroglobulin by endocytosis. Next, TH is cleaved off the thyroglobulin molecules by lysosomal enzymes in the cytoplasm of the follicular cell. TH then diffuses out of the follicular cells into the capillaries around the follicle.

Calcitonin The parafollicular cells of the thyroid secrete the protein hormone **calcitonin** (kal″sĭ-to′nin) when blood calcium levels are high. Calcitonin lowers blood levels of Ca^{2+} by slowing the calcium-releasing activity of osteoclasts in bone and increasing calcium secretion by the kidney. Calcitonin has no demonstrable function in adults; it seems to act mostly during childhood, when the skeleton grows quickly and osteoclast activity must be slowed to allow bone deposition and bone growth.

CLINICAL APPLICATION

Osteoporosis Treatment Calcitonin is one option in the treatment of osteoporosis to reduce bone loss. One challenge for the therapeutic use of this hormone has been its delivery method. It is a protein-based hormone, so if taken orally, it would be digested before reaching the target tissue, bone. Delivery via a nasal spray has been successful. Treatment with calcitonin results in a modest increase in bone mass and relieves the pain from spinal fractures caused by osteoporosis.

The Parathyroid Glands

learning outcome

▶ Describe the anatomy of the parathyroid gland and the function of parathyroid hormone.

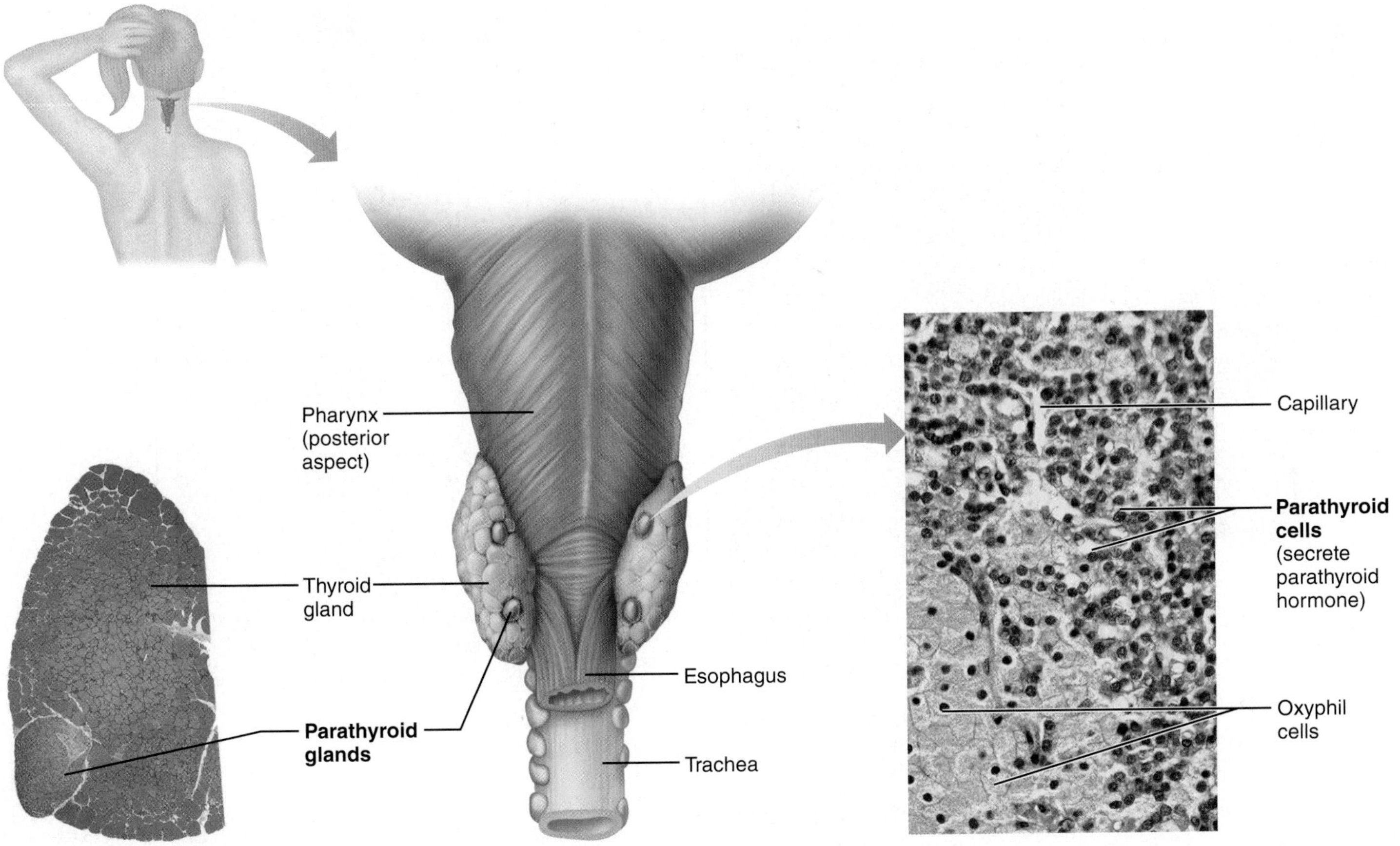

(a) Photomicrograph of parathyroid gland embedded within the thyroid gland (2×)

(b) Illustration of parathyroid glands embedded within the thyroid gland, posterior view

(c) Photomicrograph of parathyroid gland (205×)

Figure 17.6 **The parathyroid glands.** The parathyroid glands may be more inconspicuous than illustrated in (b).

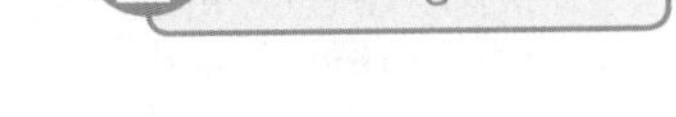

Gross Anatomy

The small, yellow-brown **parathyroid glands** lie on the posterior surface of the thyroid gland (Figure 17.6a). Even though they may be embedded in the substance of the thyroid, the parathyroids always remain distinct organs surrounded by their own connective tissue capsules. Most people have two pairs of parathyroid glands, but the precise number varies among individuals. As many as eight glands have been reported, and some may be located in other regions of the neck or even in the thorax.

Microscopic Anatomy

Histologically, the parathyroid gland contains thick, branching cords composed of two types of endocrine cells (Figure 17.6b): Small, abundant **parathyroid cells** and rare, larger parathyroid **oxyphil cells** (ok′sĭ-fil; "acid = loving" = "stains with acidic dyes"). The function of oxyphil cells is unknown. The parathyroid cells produce a small protein hormone called **parathyroid hormone** (**PTH,** or **parathormone**), which increases the blood concentration of Ca^{2+} whenever it falls below some threshold value. PTH raises blood calcium by:

1. Stimulating osteoclasts to release more Ca^{2+} from bone
2. Decreasing the excretion of Ca^{2+} by the kidney
3. Activating vitamin D, which stimulates the uptake of Ca^{2+} by the intestine

PTH is essential to life because low Ca^{2+} levels lead to lethal neuromuscular disorders. Before PTH was discovered, it was observed that some patients recovered uneventfully after partial (or even total) removal of the thyroid gland, but others exhibited uncontrolled muscle spasms and severe pain and soon died. The lethal neuromuscular disorders resulted when surgeons unknowingly removed the parathyroid glands along with the thyroid.

Note that parathyroid hormone and calcitonin have *opposite,* or antagonistic, effects: PTH raises blood calcium, whereas calcitonin lowers it.

☑ check your understanding

☐ 6. Would a patient with hypothyroidism (low thyroid hormone) have concerns about blood calcium regulation? Explain why or why not.

(For the answer, see Appendix B.)

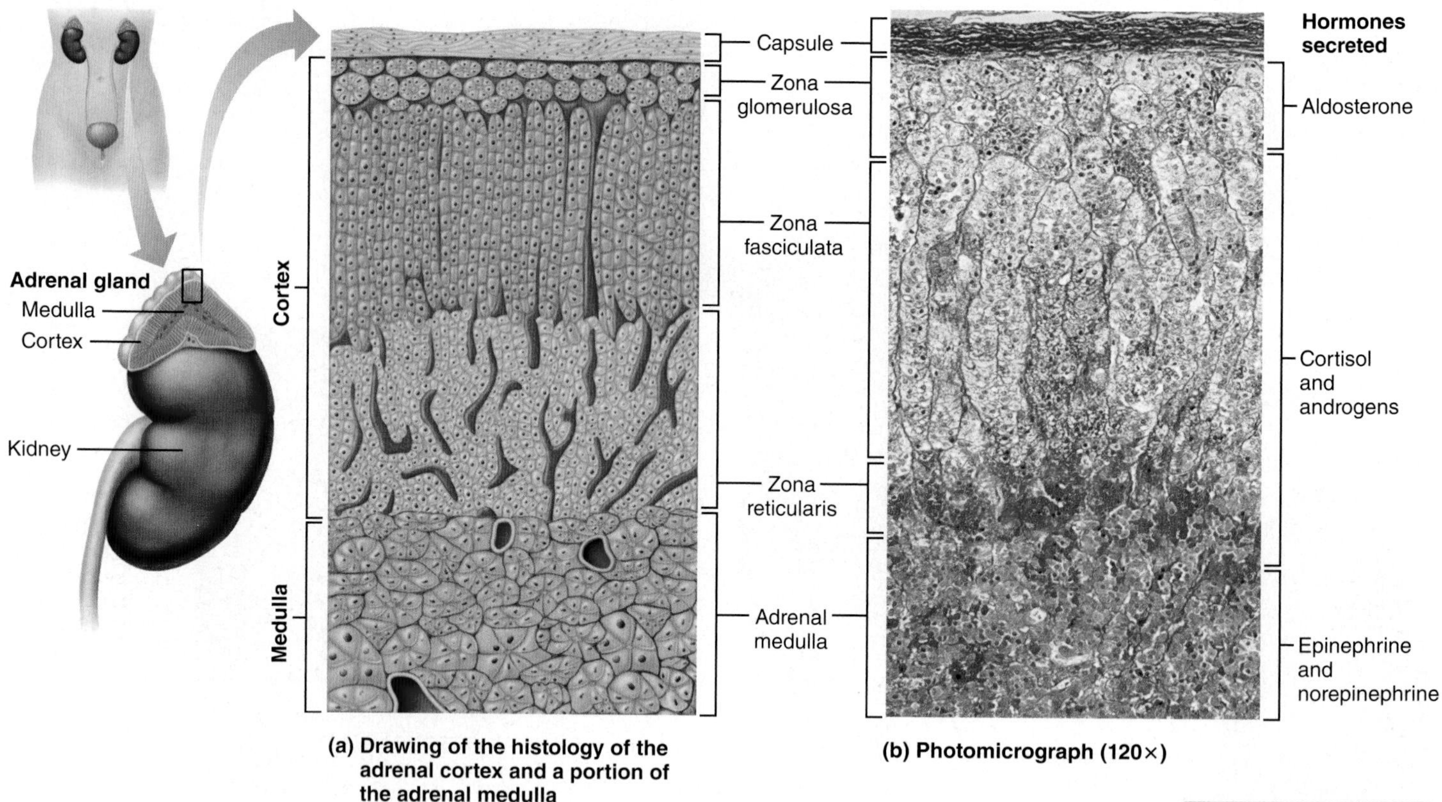

Figure 17.7 The adrenal gland, gross and microscopic structure.

The Adrenal (Suprarenal) Glands

learning outcome

▶ Name the two divisions of the adrenal gland, and compare and contrast them in terms of their structure and the hormones they secrete.

Gross Anatomy

The paired **adrenal (suprarenal) glands** are pyramidal or crescent-shaped organs perched on the superior surface of the kidneys (Figure 17.7). Each adrenal gland is supplied by up to 60 small *suprarenal arteries,* which form three groups: the superior suprarenal arteries from the inferior phrenic artery, the middle suprarenal arteries from the aorta, and the inferior suprarenal arteries from the renal artery. The left *suprarenal vein* drains into the renal vein, whereas the *right suprarenal vein* drains into the inferior vena cava (diagrammed on p. 614). The nerve supply consists almost exclusively of sympathetic fibers to the adrenal medulla (discussed on p. 480).

Each adrenal gland is two endocrine glands in one (Figure 17.7a). The internal *adrenal medulla* is more like a cluster of neurons than a gland. It is derived from the neural crest and it acts as part of the sympathetic nervous system. The external *adrenal cortex,* surrounding the medulla and forming the bulk of the gland, is derived from somatic mesoderm. The cortex and medulla secrete hormones of entirely different chemical types, but all adrenal hormones help people cope with extreme situations associated with danger, terror, or stress.

The Adrenal Medulla

The centrally located **adrenal medulla** is part of the autonomic nervous system (Chapter 15 p. 480), so it is covered only briefly here. Its spherical **medullary chromaffin** (kro-maf′in) **cells** are modified postganglionic sympathetic neurons that secrete the amine hormones epinephrine and norepinephrine into the blood to enhance the fight-or-flight response. These hormones are stored in secretory vesicles within the cell that can be stained with salts containing chromium metal. (Indeed, *chromaffin* literally means "with an affinity for chromium.") Within the adrenal medulla, the chromaffin cells are arranged in spherical clusters (Figure 17.7b) with some branching cords.

The Adrenal Cortex

The thick **adrenal cortex** secretes a variety of hormones, all of which are lipid-based *steroid hormones.* Microscopically, this cortex exhibits three distinct layers, or zones (Figure 17.7). From external to internal, these are as follows:

1. **Zona glomerulosa** (glo-mer-u-lōs′ah), which contains cells arranged in spherical clusters (*glomerulus* = ball of yarn)
2. **Zona fasciculata** (fah-sik″u-lah′tah), whose cells are arranged in parallel cords (*fascicle* = bundle of parallel sticks) and contain an abundance of lipid droplets
3. **Zona reticularis** (rĕ-tik′u-lar″is), whose cells are arranged in a branching network (*reticulum* = network) and stain intensely with the pink dye eosin

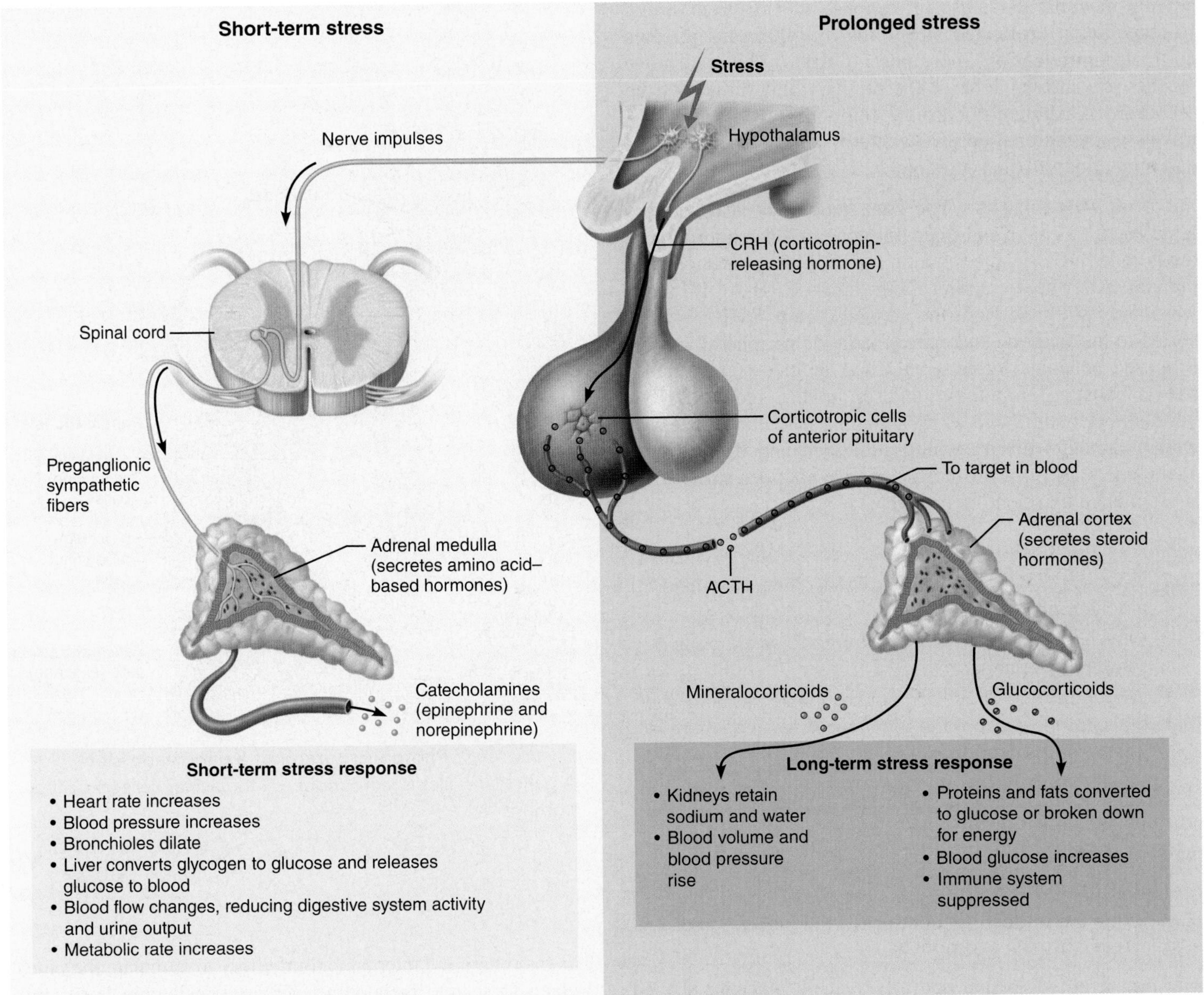

Figure 17.8 Stress and the adrenal gland.

The hormones secreted by the adrenal cortex are **corticosteroids** (kor″tĭ-ko-ste′roids), which, along with the sex hormones, are the body's major steroid hormones. Adrenal corticosteroids are of two main classes: *mineralocorticoids* (min″er-al-o-kor′tĭ-koids) and *glucocorticoids* (gloo″-ko-kor′tĭ-koids).

Mineralocorticoids The main **mineralocorticoid,** called **aldosterone** (al-dos′ter-ōn), is secreted by the zona glomerulosa in response to a decline in either blood volume or blood pressure, as occurs in severe hemorrhage. Aldosterone is the terminal hormone secreted by a complex cascade of hormonal secretions that is initiated in the kidneys in response to low blood volume or low blood pressure (the renin-angiotensin mechanism). To compensate for either decline, aldosterone prompts the ducts in the kidney to resorb more sodium into the blood; water passively follows, thus increasing blood volume.

Glucocorticoids **Glucocorticoids,** of which **cortisol** is the main type, are secreted by the zona fasciculata and zona reticularis to help the body deal with stressful situations such as fasting, anxiety, trauma, crowding, and infection. In essence, glucocorticoids keep blood glucose levels high enough to support the brain's activities while forcing most other body cells to switch to fats and amino acids as energy sources. Glucocorticoids also redirect circulating lymphocytes to lymphoid and peripheral tissues, where most pathogens are. When present in large quantities, however, glucocorticoids depress the inflammatory response and inhibit the immune system. Indeed, glucocorticoids, such as cortisone and prednisone, are administered as anti-inflammatory drugs to treat rheumatoid arthritis, severe allergies, tendonitis, joint injuries, and other inflammatory disorders.

The adrenal gland responds to both short-term and prolonged stress (Figure 17.8). Sympathetic innervation of the adrenal medulla activates the response to short-term stress. In response to prolonged stress, the hypothalamus releases corticotropin-releasing hormone (CRH). This hormone reaches the anterior lobe of the pituitary and stimulates the secretion of adrenocorticotropic hormone (ACTH). ACTH travels to the adrenal cortex, where it signals glucocorticoid and mineralocorticoid secretion.

Adrenal Androgens The zona reticularis also secretes large quantities of an androgen hormone called dehydroepiandrosterone (DHEA), the function of which remains unclear in humans. After its secretion (via the stress-CRH-ACTH pathway) and its release from the adrenal cortex, DHEA is converted to testosterone and estrogens in the peripheral tissues. Proposed beneficial effects of DHEA include counteracting stress, boosting immunity, and improving mood. DHEA has also been suggested as an antiaging molecule. Levels of DHEA are high during young adulthood and decline with age. Clinical uses for this hormone have not been sufficiently substantiated by scientific research.

check your understanding

- ☐ 7. What specific region of the adrenal gland produces aldosterone?
- ☐ 8. Why is the adrenal medulla referred to as a modified sympathetic ganglion?
- ☐ 9. What hormone directly stimulates the secretion of the glucocorticoids from the adrenal cortex, and where is this hormone produced?

(For answers, see Appendix B.)

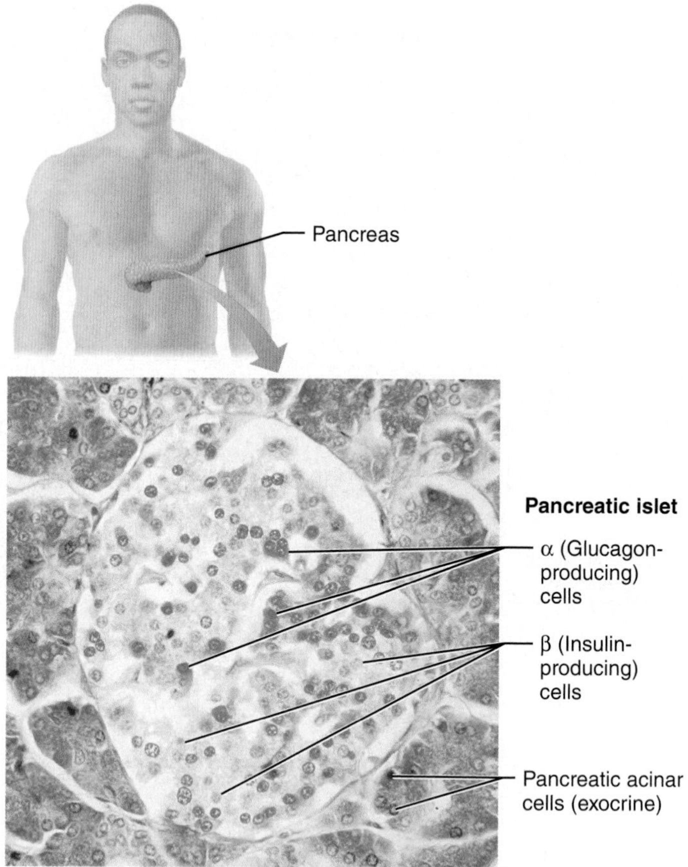

Figure 17.9 Photomicrograph of a pancreatic islet. A pancreatic islet is surrounded by the acinar cells (exocrine portion).

View PAL Histology MasteringA&P®

The Pineal Gland

learning outcome

▶ Describe the endocrine functions of the pineal gland, pancreas, thymus, and gonads.

The **pineal gland** (pineal body) is a small, pinecone-shaped structure at the end of a short stalk on the roof of the diencephalon (see Figure 17.1 and Figure 17.3b). Its endocrine cells, called *pinealocytes,* are arranged in both spherical clusters and branching cords. Pinealocytes are star-shaped cells with long, branching cell processes. Within the adult pineal gland, dense particles of calcium lie between the cell clusters, forming the "brain sand," or "pineal sand." The pineal gland is easy to locate in X-ray images of the brain because the dense calcium minerals in the pineal sand are radiopaque (X rays cannot penetrate them). Thus it stands out in a radiograph and is used as a landmark for identifying other brain structures.

Via a remarkably indirect route, the brain signals the pinealocytes to secrete the hormone **melatonin,** which helps regulate circadian rhythms (see Why Won't Teenagers Sleep at Night, p. 387). During the darkness of night, the "clock" that controls daily rhythms and melatonin secretion—the suprachiasmatic nucleus of the hypothalamus (shown on p. 387)—responds to a lack of visual input from the retina by sending signals to the preganglionic sympathetic neurons in the upper thoracic spinal cord. The signals then go to postganglionic neurons in the superior cervical ganglion, whose axons run on the internal carotid artery to stimulate the pineal gland to secrete melatonin. Increased melatonin levels promote sleepiness.

The Pancreas

Located in the posterior wall of the abdominal cavity, the tadpole-shaped **pancreas** contains both exocrine and endocrine cells. The exocrine *acinar cells,* forming most of the gland, secrete digestive enzymes into the small intestine during digestion of food.

The endocrine cells of the pancreas are contained in spherical bodies called **pancreatic islets** or *islets of Langerhans* (Figure 17.9), about a million of which are scattered among the exocrine acini. In each islet, the endocrine cells are arranged in twisted, branching cords separated by capillaries. The main cell types in the islets are alpha and beta cells.

- **Alpha (α) cells (A cells)** secrete **glucagon** (gloo′kah-gon), a protein hormone that signals *liver* cells to release glucose from their glycogen stores, thus raising blood sugar levels whenever they fall too low.
- **Beta (β) cells (B cells)** secrete **insulin** ("hormone from the islets"), a protein hormone that signals most cells of

the body to take up glucose from the blood and promotes the storage of glucose as glycogen in the liver, thus lowering excessive blood sugar levels (after the digestion of a sugary snack, for example).

The pancreatic islets also contain two rare cell types (not illustrated): **Delta (δ) cells** secrete **somatostatin** (so″mah-to-stat′in), a peptide hormone that inhibits the secretion of glucagon and insulin by the nearby alpha and beta cells; **PP cells** secrete **pancreatic polypeptide,** a hormone that may inhibit the exocrine activity of the pancreas.

The Thymus

Located in the lower neck and anterior thorax is the lobulated **thymus** (see Figure 17.1). The thymus is an important immune organ, the site at which the white blood cells called T lymphocytes arise from precursor cells. This transformation is stimulated by **thymic hormones,** which are secreted by the structural cells of the thymus, the epithelial reticular cells. Thymic hormones are a family of peptide molecules, including thymopoietin (thi″mo-poi′e-tin) and thymosin (thi′-mo-sin). The thymus is described in detail with the lymphatic system (Chapter 21, p. 637).

The Gonads

The gonads—**testes** and **ovaries**—are the main source of the steroid sex hormones. In the testes, **interstitial endocrine cells** between the sperm-forming tubules secrete **androgens** (primarily **testosterone**), which maintain the reproductive organs and the secondary sex characteristics of males and help promote the formation of sperm. In the ovaries, androgens are secreted and directly converted into *estrogens* by ovarian follicle cells. The follicle cells also produce *progesterone*. Following ovulation, estrogens and progesterone continue to be secreted by the remnant of the ovarian follicle, the *corpus luteum*. **Estrogens** maintain the reproductive organs and secondary sex characteristics of females, whereas **progesterone** signals the uterus to prepare for pregnancy. Full details are presented with the reproductive system (Chapter 25, pp. 747, 763, and 764).

OTHER ENDOCRINE STRUCTURES

learning outcomes

- ▶ Name a hormone produced by the heart, and define diffuse neuroendocrine system (DNES).
- ▶ Briefly explain the endocrine functions of the placenta, kidney, and skin.

Other endocrine cells occur within various organs of the body, including the following:

1. **The heart.** The atria of the heart contain some specialized cardiac muscle cells that secrete **atrial natriuretic peptide (ANP)** (na″tre-u-ret′ik; "producing salty urine"), a hormone that decreases excess blood volume, high blood pressure, and high blood sodium concentration, primarily by stimulating the kidney to increase its secretion of salt and its production of salty urine.
2. **The gastrointestinal tract and its derivatives.** *Enteroendocrine cells* are hormone-secreting cells scattered within the epithelial lining of the digestive tract. Related endocrine cells occur within organs that derive from the embryonic gut, such as the respiratory tubes, pancreas, prostate, and thyroid gland. Collectively, all these scattered epithelial cells make up the **diffuse neuroendocrine system (DNES).** To give some concrete examples from this chapter, the parafollicular cells in the thyroid belong to this system, as do the cells of the pancreatic islets. Over 35 kinds of DNES cells secrete amine and peptide hormones that perform such functions as regulating digestion, controlling aspects of blood chemistry, and adjusting local blood flow. Many of these hormones are chemically identical to neurotransmitter molecules, and some of them signal nearby target cells without first entering the bloodstream. These are neuron-like characteristics, explaining why the DNES epithelial cells are called neuroendocrine cells.
3. **The placenta.** Besides sustaining the fetus during pregnancy, the placenta secretes several steroid and protein hormones that influence the course of pregnancy. These hormones include estrogens, progesterone, corticotropin-releasing hormone, and human chorionic gonadotropin (see pp. 772–773). *Human chorionic gonadotropin (hCG)* is the hormone that is tested for in pregnancy tests. It is produced as the fertilized egg implants into the uterus. A positive test indicates the presence of this hormone, thus a positive test for pregnancy.
4. **The kidneys.** Various cells in the kidneys produce hormones. The specialized muscle cells of the juxtaglomerular complex, the granular cells (see p. 731), secrete the protein hormone *renin,* which indirectly signals the adrenal cortex to secrete aldosterone. Other kidney cells—either in the interstitial connective tissue between the kidney tubules or in the endothelial cells of the peritubular capillaries—secrete *erythropoietin* (e-rith″ro-poi′e-tin), which signals the bone marrow to increase the production of red blood cells.
5. **The skin.** When exposed to ultraviolet radiation, keratinocytes in the epidermis convert a modified cholesterol molecule into a precursor of *vitamin D*. Vitamin D is a steroid hormone that is essential for calcium metabolism (see p. 11). The precursor molecules then enter the bloodstream through the dermal capillaries, undergo chemical modification in the liver, and become fully activated vitamin D in the proximal tubules of the kidney. Vitamin D signals the intestine to absorb Ca^{2+} from the diet. Without this vitamin, the bones weaken from insufficient calcium content (see discussion of rickets on p. 144).

☑ check your understanding

- ☐ 10. Where are the endocrine cells of the pancreas located?
- ☐ 11. What structure, other than the ovaries, produces estrogens and progesterone?

(For answers, see Appendix B.)

Figure 17.10 **Disorders of pituitary growth hormone.** An individual exhibiting gigantism (center), is flanked by a pituitary dwarf (left) and a woman of normal height (right).

DISORDERS OF THE ENDOCRINE SYSTEM

learning outcome

- Describe the effects of excessive and inadequate hormonal secretion by the pituitary, thyroid, and adrenal glands; define diabetes mellitus.

Most disorders of endocrine glands involve either a hypersecretion (oversecretion) or a hyposecretion (undersecretion) of a given hormone. Hypersecretion often results from a tumor in an endocrine gland, in which the rapidly proliferating tumor cells secrete hormone at an uncontrolled rate. Hyposecretion, by contrast, typically results from damage to the endocrine gland by infection, autoimmune attack, or physical trauma.

Pituitary Disorders

Some disorders of the anterior pituitary affect the secretion of growth hormone (GH) (Figure 17.10). A tumor that causes hypersecretion of GH in children causes **gigantism,** in which the child grows exceptionally fast and becomes extremely tall, often reaching 2.4 meters (8 feet). If excessive amounts of GH are secreted after the bones' epiphyseal plates have closed, the result is **acromegaly** (ak-ro-meg′ah-le; "enlargement of the extremities"), characterized by enlargement of bony areas that still have active growth areas and are still responsive to GH—the hands, feet, and face.

Hyposecretion of GH in children produces **pituitary dwarfs,** who have bodies of normal proportions but rarely reach 1.2 meters (4 feet) in height. Children with this condition can reach nearly normal stature if given GH injections.

A posterior pituitary disorder, **diabetes insipidus** (di″ah-be′tēz in-sĭ′pĭ-dus; "passing of dilute [urine]"), is caused by insufficient production or secretion of antidiuretic hormone (ADH), or, more rarely, by the kidney's lack of response to this hormone. Because individuals with this condition produce large volumes of dilute urine, they compensate by drinking large quantities of water. Diabetes insipidus can be caused by either a blow to the head that damages the posterior pituitary, a tumor that compresses the pars nervosa, or kidney damage.

A Disorder of the Pancreas: Diabetes Mellitus

Diabetes mellitus affects about 7% of Americans and has a strong hereditary component. It is caused either by insufficient secretion of insulin or by resistance of body cells to the effects of insulin. As a result, glucose cannot enter most cells, so blood sugar remains high and large volumes of urine containing glucose are excreted. Because glucose is unavailable as fuel, the body's cells metabolize fats, whose acidic breakdown products, ketones, accumulate in the blood. Left untreated, the increased urination depletes the body of water and electrolytes, and the ketone acidosis depresses almost all physiological functions and leads to coma.

Type 1 Diabetes

Of the two types of diabetes mellitus, the more serious is **type 1 diabetes** (formerly called **insulin-dependent diabetes**), which develops suddenly, usually before age 15. Because an autoimmune response destroys the insulin-secreting beta cells in the pancreas, insulin must be administered to type 1 diabetics several times daily to control blood glucose levels. After 20–30 years, health-threatening complications can develop. The high level of lipids in their blood predisposes type 1 diabetics to atherosclerosis (hardening of the arteries, discussed on p. 619), and the excessive sugar in their body fluids disrupts capillary functions (see *microangiopathy of diabetes* on p. 618). Research on type 1 diabetes has demonstrated that regular exercise and careful management of diet and blood sugar level can delay the onset of complications.

Type 2 Diabetes

Type 2 (non-insulin-dependent) diabetes develops more slowly (usually appearing after age 40) and accounts for over 90% of all cases of diabetes. Most type 2 diabetics produce some insulin, but their cells have a reduced sensitivity to the effects of insulin. More easily managed than type 1, type 2 diabetes can usually be controlled by dietary modification (such as losing weight and avoiding high-calorie, sugar-rich foods) and regular exercise. If such measures fail, the condition is treated with insulin injections or oral medications that raise blood insulin levels or lower blood glucose levels.

Type 2 diabetes is increasing at an alarming rate in the United States. Of great concern is the rise of type 2 diabetes in children. The increased incidence of this disease is associated

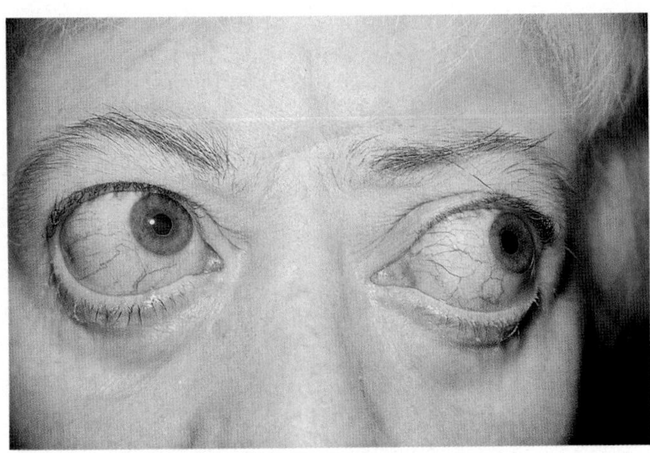

(a) Protrusion of the eyeballs; symptom of hyperthyroidism

(b) An enlarged thyroid (goiter); due to iodine deficiency

Figure 17.11 **Thyroid disorders.**

with decreased activity levels and increased rates of obesity. Regular exercise has been found not only to help control type 2 diabetes but also to decrease the likelihood of its development. One component of this link between exercise and blood sugar regulation is a hormone produced by osteoblasts, the bone-producing cells. Osteoblast activity increases in response to weight-bearing exercise. *Osteocalcin,* a hormone secreted by osteoblasts, stimulates pancreatic secretion of insulin and stimulates fat cells to secrete a hormone that increases the insulin sensitivity of cells. Thus, increased activity of osteoblasts also functions to control blood sugar levels.

Disorders of the Thyroid Gland

The most common form of hyperthyroidism is **Graves' disease,** an autoimmune disease in which the immune system makes abnormal antibodies that mimic TSH and stimulate the oversecretion of TH by follicular cells of the thyroid. Typical signs and symptoms of Graves' disease include elevated metabolic rate, rapid heart rate, sweating, nervousness, and weight loss despite normal food intake. Additionally, the eyeballs may protrude because of edema and inflammation in the orbital tissue behind the eyes and the extrinsic eye muscles (Figure 17.11a). Graves' disease develops most often in middle-aged women and affects 1 of every 20 women overall.

Insufficient secretion of TH produces different effects at different stages of life. In adults, such hyposecretion of TH results in **adult hypothyroidism,** or **myxedema** (mik″se-de′-mah; "mucous swelling"), typically an autoimmune disease in which antibodies attack and destroy thyroid tissue. Signs and symptoms of this condition, which occurs in 7% of women and 3% of men, include a low metabolic rate, weight gain, lethargy, constant chilliness, puffy eyes, edema, and mental sluggishness (but not retardation).

Hypothyroidism can also result from an insufficient amount of iodine in the diet. In such cases, the thyroid gland enlarges, producing in the anterior neck a large, visible lump called an **endemic goiter** (Figure 17.11b). Because the cells of the thyroid follicles produce colloid but cannot iodinate it or make functional hormones, the pituitary gland secretes increasing amounts of TSH in a futile attempt to stimulate the thyroid to produce TH. This action causes the follicles to accumulate ever more colloid, and the thyroid gland swells. The commercial marketing of iodized salts has significantly reduced the incidence of goiters in regions that have iodine-poor soils or lack access to iodine-rich shellfish.

In children, hypothyroidism leads to **cretinism** (cre′tĭ-nizm), a condition characterized by a short, disproportionate body, a thick tongue and neck, and mental retardation.

Disorders of the Adrenal Cortex

Hypersecretion of glucocorticoid hormones leads to **Cushing's disease (Cushing's syndrome),** caused either by an ACTH-secreting pituitary tumor or (rarely) by a tumor of the adrenal cortex. This condition is characterized by high levels of glucose in the blood, loss of protein from muscles, and lethargy. The so-called "cushingoid signs" include a swollen face and the redistribution of fat to the posterior neck, causing a "buffalo hump" (Figure 17.12), as well as

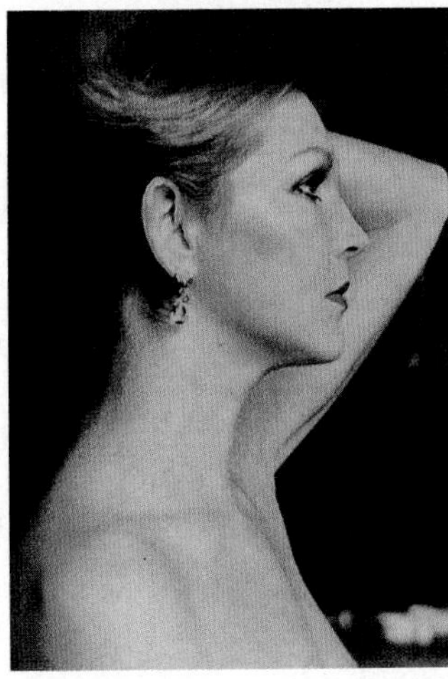

(a) Patient before onset

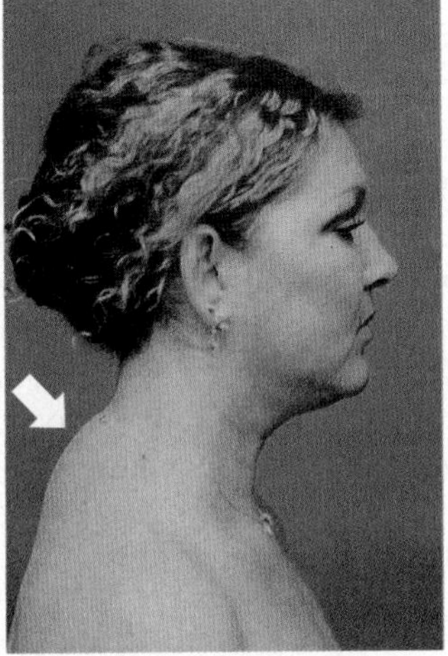

(b) Same patient with Cushing's syndrome. The white arrow shows the characteristic "buffalo hump" of fat on the upper back.

Figure 17.12 **The effects of excess glucocorticoid.**

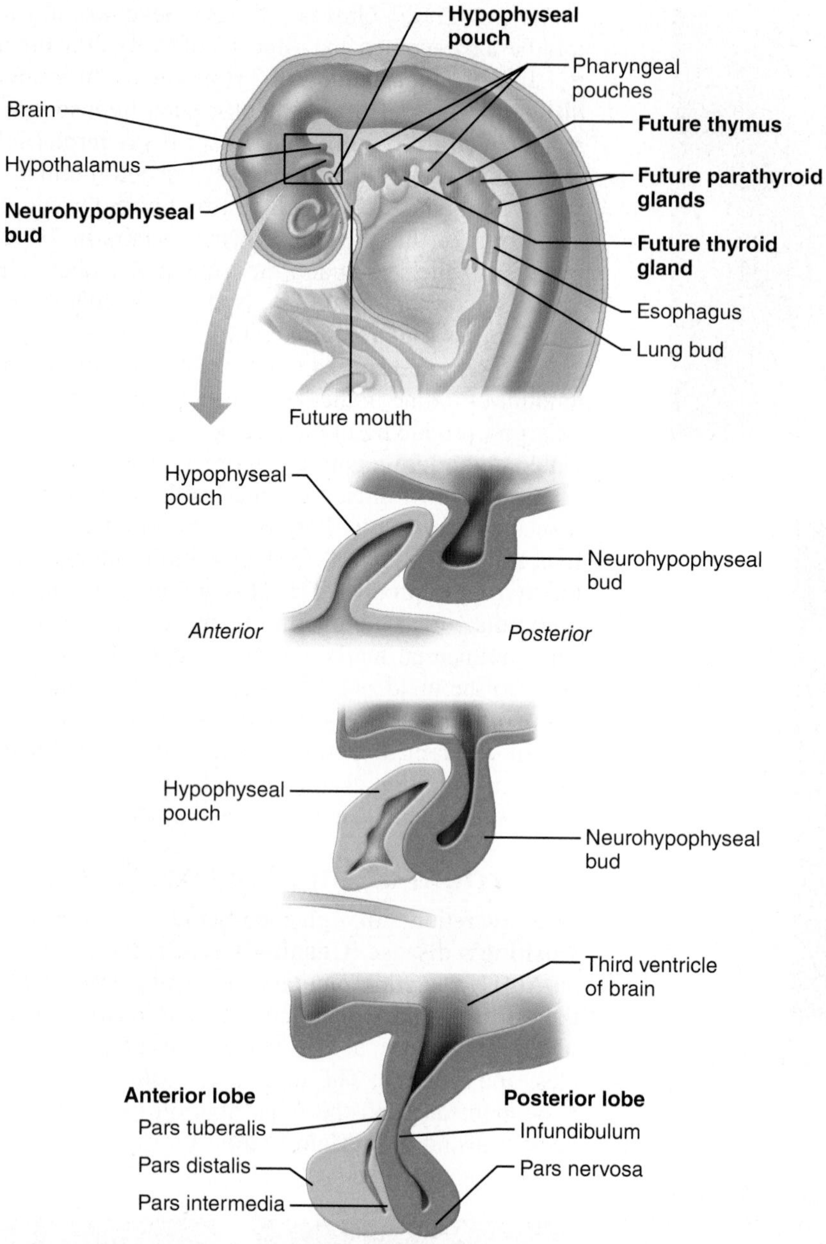

(a) Week 5. Thyroid, thymus, and parathyroid glands form from pharyngeal endoderm. Hypophyseal pouch extends superiorly from ectoderm in the roof of the mouth.

(b) Week 6. Inferior extension of the floor of the diencephalon forms the neurohypophyseal bud.

(c) Week 7. Hypophyseal pouch pinches off the surface ectoderm and is closely associated with the neurohypophyseal bud.

(d) Week 8. Hypophyseal pouch forms the anterior lobe of pituitary; neurohypophyseal bud forms the posterior lobe. Distinct portions of each differentiate.

Figure 17.13 Embryonic development of some major endocrine organs.

depression of the immune and inflammatory responses. Mild cases can result from large doses of glucocorticoids, such as cortisone, prescribed as drugs to suppress inflammation.

Addison's disease, the major hyposecretory disorder of the adrenal cortex, usually involves deficiencies of both glucocorticoids and mineralocorticoids. Blood levels of glucose and sodium drop, and severe dehydration and low blood pressure are common. Other symptoms include fatigue, loss of appetite, and abdominal pain.

✓ check your understanding

☐ 12. Addison's, Graves', and Cushing's (sounds like a law firm): Name the endocrine gland affected by each disorder, whether hypersecretion or hyposecretion is the cause, and the typical symptoms of each.

(For the answer, see Appendix B.)

THE ENDOCRINE SYSTEM THROUGHOUT LIFE

learning outcomes

- Describe the development of the major endocrine glands.
- Describe the effect of aging on some endocrine organs.

The diverse and widely distributed endocrine organs arise from all three germ layers.

The *thyroid gland* forms from a thickening of endoderm on the floor of the pharynx (Figure 17.13a) that first appears on the posterior part of the future tongue and then migrates caudally into the neck. Both the *parathyroid glands* and the *thymus* arise from the endoderm lining the pharyngeal pouches and then migrate to their final positions in the neck and thorax. The parafollicular cells arise in the wall of the last pharyngeal pouch and then migrate caudally, into the thyroid gland. The *pineal gland* arises from ectodermally derived ependymal cells that line the roof of the embryonic diencephalon. (Remember ependymal cells are the glial cells that line the hollow neural tube.)

The *pituitary gland* has a dual origin (Figure 17.13b and c). The anterior lobe arises from the roof of the mouth as a pouch of ectoderm—the **hypophyseal pouch,** or *Rathke's pouch.* This pouch contacts the future posterior lobe, which grows inferiorly from the floor of the brain as the **neurohypophyseal bud.**

The adrenal gland also has a dual embryonic origin (see Figure 15.15, p. 484). Whereas its medulla originates from the neural crest cells of nearby sympathetic trunk ganglia, the cortex develops from somatic mesoderm lining the coelom on the dorsal abdominal wall.

Barring hypersecretory and hyposecretory disorders of the endocrine glands, most endocrine organs operate smoothly throughout life until old age. Then, a number of changes become evident. In the anterior pituitary, the amounts of connective tissue and aging pigment (lipofuscin) increase, vascularization decreases, and the number of hormone-secreting cells declines. These changes may or may not affect hormone production: For example, blood levels of TSH decline slightly with normal aging, ACTH remains constant, and gonadotropins actually *increase* with age.

The adrenal cortex also shows structural changes with age, but normal rates of glucocorticoid secretion appear to persist as long as an individual is healthy. No age-related changes in the release of the amine hormones epinephrine and norepinephrine by the adrenal medulla have been found.

The synthesis and release of thyroid hormones diminish somewhat with normal aging. Typically, the thyroid follicles are loaded with colloid in the elderly, and fibrosis of the gland occurs. Autoimmune diseases of the thyroid become common, affecting about 5% of older women.

The parathyroid glands change little with age, but a tendency for the concentration of PTH in the blood to increase compensates for a lowered intake of calcium and vitamin D if the quality of the diet declines. The decline in estrogen secretion after menopause makes women more susceptible to the bone-demineralizing effects of PTH.

Only growth hormone (GH) from the anterior pituitary, dehydroepiandrosterone (DHEA) from the adrenal cortex, and the sex hormones (estrogens and testosterone) show marked drops in secretion with age. The deterioration of the musculoskeletal, cardiovascular, immune, and other organs that accompany these hormonal declines is discussed elsewhere (see pp. 258, 623, and 641). So closely do the effects of withdrawing these hormones parallel the symptoms of aging that all three have been commercialized as "antiaging drugs" in recent years.

RELATED CLINICAL TERMS

Hypercalcemia Elevated concentrations of calcium ions in the blood caused by either primary hyperparathyroidism or malignancies. In primary hyperparathyroidism, a parathyroid gland spontaneously (usually as a result of a tumor) secretes large amounts of calcium-raising parathyroid hormone. Many different kinds of malignancies also raise blood calcium levels by secreting chemicals that either stimulate osteoclasts to dissolve bone or stimulate calcium resorption in the kidney.

Hypophysectomy Surgical removal of the pituitary gland.

Pancreatic transplantation A procedure performed on people with type 1 diabetes whose lives are threatened by gradually failing kidneys as *microangiopathy of diabetes* (p. 618) destroys the kidney capillaries. The transplanted pancreas secretes insulin and halts the diabetes, and a healthy kidney may be transplanted at the same time to take over for the recipient's failing kidneys. The donated pancreas is placed in the pelvis and sutured to the bladder so that the digestive enzymes it produces can drain into the bladder and be flushed from the body in the urine, instead of irritating the pelvic peritoneum and causing peritonitis. The recipient must take immunosuppressive drugs for life. The success rate for pancreatic transplantation is about 70%. Techniques to transplant only the pancreatic islets have been developed. Islets from a deceased donor are injected into the recipient's liver. Transplanted islets become vascularized and start secreting insulin. This treatment has been moderately successful; only about 50% of patients need no insulin injections after two years, and recipients must take immunosuppressive drugs for life.

Pheochromocytoma (fe″o-kro″mo-si″to′mah; "dusky color tumor") Tumor of the medullary chromaffin cells of the adrenal medulla; results in excessive secretion of the hormones epinephrine and norepinephrine, which produces the symptoms of prolonged sympathetic response (especially hypertension).

Prolactinoma (*oma* = tumor) The most common type of pituitary gland tumor, characterized by a hypersecretion of prolactin, excessive secretion of milk, and menstrual disturbances in women. In men, it causes milk production, a loss of libido, and impotence.

Psychosocial dwarfism Dwarfism (and failure to thrive) resulting from childhood stress and emotional disorders that suppress the hypothalamic release of growth hormone–releasing hormone and thus anterior pituitary secretion of GH.

Thyroid cancer The most common malignancy of the endocrine glands, affecting about 1% of all people, typically between the ages of 25 and 65; characterized by firm, fixed nodules in the thyroid. Incidence is increasing, especially among individuals who had external irradiation of the head and neck when young. Treatment of this highly curable cancer involves removing some or all of the thyroid, then administering a radioactive form of iodine, which is actively taken up by the remaining thyroid follicular cells (normal and cancerous). The sequestered radioactive iodine then kills the cancer cells with a minimum of harm to nonthyroidal tissues.

Thyroid storm (thyroid crisis) A sudden and dangerous increase in the effects of thyroid hormone resulting from excessive amounts of TH in the circulation; metabolic rate is greatly increased, as indicated by fever, rapid heart rate, high blood pressure, nervousness, and tremors. Precipitating factors include stressful situations, excessive intake of TH supplements, and trauma to the thyroid gland.

CHAPTER SUMMARY

Take full advantage of the invaluable online practice and study tools for this chapter by going to the Study Area of MasteringA&P®. There you will find:

- Art Labeling Activities, Quizzes, Practice Tests, Memory Games
- Animal Organ Dissection, Cat Dissection and Bone Videos
- PAL Cadaver
- PAL Histology
- A&P Flix Animations
- Dynamic Study Modules

1. Endocrine organs are ductless glands that release hormones into the blood or lymph.
2. Hormones are messenger molecules that travel in the circulatory vessels and signal physiological changes in target cells.
3. Hormonally regulated processes include reproduction, growth, mobilization of body defenses against stress, maintenance of the proper chemistry of the blood and body fluids, and regulation of cellular metabolism.

OVERVIEW (pp. 523–525)

Endocrine Organs (p. 523)

4. The endocrine organs are small and widely separated from one another within the body. The pure endocrine organs are the pituitary, thyroid, parathyroid, adrenal, and pineal glands. Other organs that contain endocrine cells are the gonads, pancreas, kidney, alimentary canal, heart, thymus, and skin. The hypothalamus of the brain is a neuroendocrine organ.
5. Endocrine organs are richly vascularized.
6. Although most endocrine cells are modified epithelial cells, others are neurons, muscle cells, or fibroblast-like cells.

Hormones (pp. 523–525)

7. Most hormones are either amino acid derivatives (amines, peptides, proteins) or steroids (lipid-based molecules derived from cholesterol).
8. Hormones produce their effects by leaving the capillaries and binding to specific receptor molecules in or on their target cells. Such binding triggers a preprogrammed response in the target cell.
9. Endocrine organs are stimulated to release their hormones by humoral, neural, or hormonal stimuli. Hormonal secretion is controlled by feedback loops.
10. The hypothalamus of the brain regulates many functions of the endocrine system through the hormones it secretes.

THE MAJOR ENDOCRINE ORGANS (pp. 525–537)

The Pituitary Gland (pp. 525–531)

11. The golf club–shaped pituitary gland is suspended from the diencephalon of the brain by its stalk (infundibulum) and lies in the hypophyseal fossa of the sella turcica of the sphenoid bone. It consists of an anterior lobe (adenohypophysis) and a posterior lobe (neurohypophysis).
12. The anterior lobe has three parts: pars distalis, pars intermedia, and pars tuberalis. The posterior lobe has two parts: pars nervosa and infundibulum.
13. The pituitary gland receives its rich blood supply from the superior and inferior hypophyseal arteries.
14. The largest part of the anterior lobe is the pars distalis. Its five cell types secrete seven protein hormones. Four of the hormones secreted are tropic hormones (TSH, ACTH, FSH, LH) that influence the secretion of other endocrine glands; three others directly influence nonendocrine target tissues.
15. The hormones secreted by the anterior lobe and their functions are as follows. Thyroid-stimulating hormone (TSH), produced by thyrotropic cells, stimulates secretion of the thyroid gland. Adrenocorticotropic hormone (ACTH), produced by corticotropic cells, stimulates the secretion of hormones from the adrenal cortex. Follicle-stimulating hormone (FSH) and luteinizing hormone (LH), produced by gonadotropic cells, signal the maturation of sex cells and the secretion of sex hormones. Prolactin, secreted by prolactin cells, stimulates milk production in breast tissue. Melanocyte-stimulating hormone (MSH), whose function is uncertain, is also secreted by corticotropic cells. Growth hormone (GH), secreted by somatotropic cells, stimulates growth of the body and skeleton.
16. The hypothalamus of the brain controls the secretion of hormones from the anterior lobe. First, certain hypothalamic neurons make releasing hormones and inhibiting hormones, which they secrete into a primary capillary plexus in the median eminence. These hormones then travel through hypophyseal portal veins to a secondary capillary plexus in the pars distalis. They leave this plexus to signal the anterior lobe cells to secrete their hormones, which then enter the secondary capillary plexus and travel to their target cells throughout the body.
17. The posterior lobe, which consists of nervous tissue, contains the hypothalamohypophyseal tract. The cell bodies of the neurons that form this tract are located in the paraventricular and supraoptic nuclei of the hypothalamus. These neurons synthesize oxytocin and ADH (vasopressin), respectively, and store them in their axon terminals in the pars nervosa. Here, the stored hormones are released into capillaries when the neurons fire.
18. The posterior lobe hormones have the following functions: ADH increases resorption of water from the urine and raises blood pressure, and oxytocin induces labor and ejection of milk from the breasts. Both these hormones are involved with social bonding.

The Thyroid Gland (pp. 531–532)

19. The thyroid gland, which lies on the superior trachea, consists of spherical follicles covered by epithelial follicle cells and separated by a capillary-rich connective tissue. The follicles are filled with a colloid of thyroglobulin, a storage protein containing thyroid hormone.
20. Thyroid hormone (TH), which contains iodine and increases basal metabolic rate, is made continuously by follicular cells and stored within the follicles until TSH from the pituitary gland signals the follicular cells to reclaim the TH and secrete it into the extrafollicular capillaries.
21. Parafollicular cells protrude from the thyroid follicles and secrete the hormone calcitonin, which can lower blood calcium concentrations.

The Parathyroid Glands (pp. 532–533)

22. Several pairs of parathyroid glands lie on the dorsal aspect of the thyroid gland. Their parathyroid cells are arranged in thick, branching cords and secrete parathyroid hormone, which raises low blood calcium levels.

The Adrenal (Suprarenal) Glands (pp. 534–536)

23. The paired adrenal glands lie on the superior surface of each kidney. Each adrenal gland has two distinct parts, an outer cortex and an inner medulla.

24. The adrenal medulla consists of spherical clusters (and some branching cords) of medullary chromaffin cells. Upon sympathetic stimulation, these cells secrete epinephrine and norepinephrine into the blood—the surge of adrenaline that is experienced during fight-or-flight situations.

25. The adrenal cortex has three layers: outer zona glomerulosa, middle zona fasciculata, and inner zona reticularis; the name of each zone describes its histological structure.

26. The steroid hormones secreted by the adrenal cortex (corticosteroids) include mineralocorticoids (mostly from the zona glomerulosa), glucocorticoids (mostly from the zona fasciculata and reticularis), and the androgen DHEA (from the zona reticularis). Mineralocorticoids (mainly aldosterone) conserve water and sodium by increasing resorption of these substances by the kidney. Glucocorticoids (mainly cortisol) help the body cope with stress by stabilizing blood glucose levels; in large quantities they also inhibit inflammation and the immune system. The functions of DHEA in humans are unclear.

The Pineal Gland (p. 536)

27. The pineal gland, on the roof of the diencephalon, contains pinealocytes, which cluster into spherical clumps and cords separated by dense particles of calcium called "brain sand."

28. Pinealocytes secrete the hormone melatonin, which helps regulate circadian rhythms. This secretion is signaled by the suprachiasmatic nucleus of the hypothalamus through a sympathetic pathway.

The Pancreas (pp. 536–537)

29. The endocrine structures in the pancreas are the spherical pancreatic islets. These islets consist of alpha (α), beta (β), delta (Δ) and PP cells arranged in twisting cords.

30. Alpha cells secrete glucagon, which raises blood sugar levels, whereas beta cells secrete insulin, which lowers blood sugar levels.

The Thymus (p. 537)

31. The thymus, an important organ of the immune system, secretes thymic hormones that are essential for the production of T lymphocytes.

The Gonads (p. 537)

32. Various cells in the ovaries and testes secrete steroid sex hormones, estrogens and androgens.

PAL Human Cadaver/Endocrine System

OTHER ENDOCRINE STRUCTURES (p. 537)

33. Some muscle cells in the atria of the heart secrete atrial natriuretic peptide (ANP), which stimulates loss of body fluids and salts through the production of a sodium-rich urine.

34. Endocrine cells are scattered within the epithelium of the digestive tube and other gut-derived organs (respiratory tubes and so on). These epithelial cells, which have some neuronlike properties, make up the diffuse neuroendocrine system (DNES). There are many classes of DNES cells, some of which secrete hormones that regulate digestion.

35. The placenta secretes hormones of pregnancy, the kidney secretes renin and erythropoietin, and the skin produces vitamin D.

DISORDERS OF THE ENDOCRINE SYSTEM (pp. 538–540)

36. Most disorders of endocrine glands involve either hypersecretion or hyposecretion of a hormone. *Hyper*secretion of GH leads to gigantism, of TH leads to Graves' disease, and of ACTH or glucocorticoids leads to Cushing's disease. *Hypo*secretion of GH leads to pituitary dwarfism, of TH leads to adult hypothyroidism or cretinism, of glucocorticoids and mineralocorticoids leads to Addison's disease, and of insulin leads to type 1 diabetes mellitus.

THE ENDOCRINE SYSTEM THROUGHOUT LIFE (pp. 540–541)

37. The endocrine glands have diverse developmental origins from all three germ layers. The anterior pituitary arises from ectoderm on the roof of the mouth, and the posterior pituitary arises from the floor of the brain. The endocrine organs in the neck (thyroid, parathyroids, thymus) derive from the endoderm of the pharynx. The pineal gland forms from the ectodermally derived ependyma of the roof of the brain, the adrenal medulla develops from sympathetic trunk ganglia, and the adrenal cortex arises from the mesoderm of the dorsal abdominal wall.

38. The efficiency of some endocrine organs gradually declines as the body ages. The hormones whose secretions decline most are GH, DHEA, estrogens, and testosterone.

REVIEW QUESTIONS

Multiple Choice/Matching Questions

(For answers, see Appendix B.)

1. The major stimulus for the release of estrogens is (a) hormonal, (b) humoral, (c) nervous.

2. Choose the correct hormone from the key for each description.

Key:

(a) melatonin
(b) antidiuretic hormone (ADH)
(c) growth hormone (GH)
(d) luteinizing hormone (LH)
(e) thyroid hormone (TH)
(f) thyroid-stimulating hormone (TSH)
(g) prolactin (PRL)
(h) oxytocin
(i) cortisol
(j) parathyroid hormone (PTH)

____ (1) stimulates cell division in the epiphyseal plates of growing bones
____ (2) involved in water balance; causes the kidneys to conserve water
____ (3) stimulates milk production
____ (4) stimulates milk ejection
____ (5) tropic hormone that stimulates the gonads to secrete sex hormones
____ (6) increases basal metabolic rate
____ (7) tropic hormone that stimulates the thyroid gland to secrete thyroid hormone
____ (8) adjusts blood sugar levels and helps the body cope with stress
____ (9) is secreted by the pineal gland
____ (10) increases blood calcium levels
____ (11) is secreted by the posterior lobe of the pituitary (two possible choices)
____ (12) the only steroid hormone in the list

3. The pars distalis of the anterior pituitary does *not* secrete (a) antidiuretic hormone, (b) growth hormone, (c) gonadotropins, (d) thyroid-stimulating hormone.

4. Endocrine cells secrete either protein hormones or steroid hormones. For each endocrine cell described, indicate whether it produces (a) a protein hormone or (b) a steroid hormone.

____ (1) any endocrine cell in the pars distalis
____ (2) interstitial cell in the testis
____ (3) parathyroid cells in the parathyroid gland
____ (4) zona fasciculata cell
____ (5) follicle cell in the ovary that secretes sex hormones
____ (6) parafollicular cells in the thyroid gland

5. From the key, choose the best description of histological structure for each of the following glands. More than one answer may be correct.

Key: **(a)** spherical clusters of cells **(b)** parallel cords of cells **(c)** branching cords of cells **(d)** follicles **(e)** nervous tissue

____ (1) pars nervosa of pituitary gland
____ (2) zona glomerulosa of adrenal gland
____ (3) pars distalis of pituitary gland
____ (4) thyroid gland
____ (5) zona fasciculata of adrenal gland

6. The divisions of the posterior pituitary are the (a) anterior lobe and posterior lobe; (b) pars emphasis, metropolis, and hypothesis; (c) pars distalis, tuberalis, and intermedia; (d) pars nervosa and infundibulum; (e) pars glomerulosa, fasciculata, and reticularis.

7. Which of the following cells secrete releasing or inhibiting hormones? (a) hypothalamic neurons, (b) medullary chromaffin cells, (c) cells in the anterior lobe of the pituitary, (d) parafollicular cells.

8. Medullary chromaffin cells occur in the (a) parathyroid gland, (b) pars distalis, (c) pituitary gland, (d) adrenal gland, (e) pineal gland.

9. The anterior lobe of the pituitary gland is the same as the (a) neurohypophysis, (b) pars nervosa, (c) adenohypophysis, (d) hypothalamus.

10. Many endocrine glands produce multiple hormones. Indicate the two hormones from column B that are secreted by each endocrine gland listed in column A.

Column A

____, ____ (1) thyroid gland
____, ____ (2) pancreas
____, ____ (3) adrenal gland
____, ____ (4) posterior lobe of the pituitary
____, ____ (5) ovaries

Column B

(a) glucagon
(b) calcitonin
(c) cortisol
(d) oxytocin
(e) progesterone
(f) norepinephrine
(g) insulin
(h) antidiuretic hormone
(i) estrogen
(j) thyroid hormone

11. Indicate with a check mark which of the organs listed below have some endocrine function.

____ (a) kidney, ____ (b) mammary glands, ____ (c) pancreas,
____ (d) stomach, ____ (e) heart, ____ (f) urinary bladder,
____ (g) respiratory tubes, ____ (h) skin, ____ (i) salivary glands,
____ (j) thymus.

Short Answer Essay Questions

12. (a) Describe where in the body each of the following endocrine glands is located: anterior and posterior lobe of the pituitary, pineal, thyroid, parathyroids, and adrenals. (b) List the hormones secreted by each of these glands.

13. The anterior lobe of the pituitary secretes so many hormones that it is often called the master endocrine organ, but it too has a "master." What structure controls the release of anterior pituitary hormones?

14. When Joshua explained to his classmate Jennifer that the thyroid gland contains parathyroid cells in its follicles and that the parathyroid cells secrete parathyroid hormone and calcitonin, Jennifer told him he was all mixed up again. Correct Josh's mistakes.

15. (a) Define *hormone*. (b) Name a hormone secreted by a muscle cell and a hormone secreted by a neuron.

16. On a realistic drawing of the endocrine glands in the body (such as a photocopy of Figure 17.1), indicate the gland associated with (a) cretinism, (b) diabetes mellitus, (c) acromegaly, (d) secreting

thyroid-stimulating hormone, (e) secreting a hormone that regulates the nightly activities of our circadian rhythms, (f) secreting estrogens, (g) secreting DHEA.

17. On a realistic drawing of the endocrine glands in the body, trace the following hormones from their glands of origin all the way to their target organs: (a) vitamin D, (b) glucagon, (c) erythropoietin, (d) oxytocin.

18. On a realistic drawing of the endocrine glands in the body, mark and label the endocrine organs that develop from the (a) roof of the embryonic mouth, (b) floor of the diencephalon, (c) endoderm on the floor of the pharynx, (d) endoderm of the pharyngeal pouches (two answers here), (e) neural crest of early sympathetic trunk ganglia.

19. Compare and contrast the functions of both sets of capillaries in the hypophyseal portal system.

20. List the hormones secreted by each of the three zonas of the adrenal cortex.

Critical Reasoning & Clinical Application Questions

1. The brain senses when a person is in a stressful situation, and the hypothalamus responds by secreting a releasing hormone called corticotropin-releasing hormone, which, through a sequence of events, helps the body to deal with the stressful situation. Outline this entire sequence, starting with corticotropin-releasing hormone and ending with the release of cortisol. (As you do this, be sure to trace the hormones through the hypophyseal portal system and out of the pituitary gland.)

2. Jeremy, a 5-year-old boy, has been growing by leaps and bounds such that his height is 70% above normal for his age group. A CT scan reveals a pituitary tumor. (a) What hormone is being secreted in excess? (b) What condition will Jeremy exhibit if corrective measures are not taken?

3. An accident victim who had not been wearing a seat belt received trauma to his forehead when he was thrown against the windshield. The physicians in the emergency room worried that his brain stem may have been driven inferiorly through the foramen magnum. To help assess this, they quickly took a standard X-ray film of his head and searched for the position of the pineal gland. How could anyone expect to find this tiny, boneless gland in a radiograph?

4. Mrs. Giardino had an abnormally high concentration of calcium in her blood, and her physicians were certain she had a tumor of the parathyroid gland. However, when surgery was performed on her neck, the surgeon could not find the parathyroid glands at all. Where should the surgeon look next to find the tumorous parathyroid gland?

5. Explain how endocrine disorders produced the physical characteristics described: (a) obesity in a man with hypothyroidism, (b) small stature and gigantism in people with pituitary disorders, (c) facial hair on a woman with an adrenal tumor, and (d) protrusion of the eyes in a person with Graves' disease.

6. For what therapeutic purposes would pharmaceutical companies seek to design drugs that either mimic atrial natriuretic peptide (ANP) or slow the rate at which ANP is broken down in the body?

7. Which nonsteroid hormone have competitive athletes abused in an effort to improve their performance? What effects of this hormone promote this abuse?

differentiation occurs (Figure 18.8) as the cells experience several final rounds of division. The structural changes that occur in each blood cell line are discussed next, beginning with the line that generates erythrocytes.

Genesis of Erythrocytes

In the line that forms erythrocytes, the committed cells are **proerythroblasts** (pro″ĕ-rith′ro-blasts; "earliest red-formers"), which avidly accumulate iron for future production of hemoglobin. Proerythroblasts give rise to **basophilic erythroblasts,** which act as ribosome-producing factories. Hemoglobin is made on these ribosomes and accumulates during the next two stages: the **polychromatic erythroblast** and the **orthochromatic erythroblast.** The staining properties of the cytoplasm change during these stages, as blue-staining ribosomes become masked by pink-staining hemoglobin. When the orthochromatic erythroblast stage is reached, cell division stops. When the cytoplasm is almost filled with hemoglobin, the nucleus stops directing the cell's activities and shrinks. Then, the nucleus and almost all organelles are ejected, and the cell collapses and assumes its biconcave shape. The cell is now a **reticulocyte,** a young erythrocyte that contains a network of blue-staining material (*reticulum* = network) representing clumps of ribosomes that remain after the other organelles are extruded. Reticulocytes enter the bloodstream and begin their task of transporting oxygen. Erythrocytes remain in the reticulocyte stage for their first day or two in the circulation, after which their ribosomes are degraded by intracellular enzymes and lost.

Formation of Leukocytes and Platelets

The committed cells in each *granulocyte* line are called **myeloblasts** (mi′ĕ-lo-blasts). They accumulate lysosomes and become **promyelocytes** (pro-mi′ĕ-lo-sīts″). The distinctive granules of each granulocyte appear next, in the **myelocyte** stage. When this stage is reached, cell division ceases. In the ensuing **metamyelocyte** stage, the nucleus stops functioning and bends into a thick "horseshoe." Neutrophils with such horseshoe nuclei are called **band cells.** The granulocytes then complete their differentiation and enter the bloodstream.

Not much structural differentiation occurs in the cell lines leading to monocytes and lymphocytes, as these cells look much like the stem cells from which they arise (Figure 18.8). In the line leading to monocytes, committed **monoblasts** enlarge and obtain more lysosomes as they become **promonocytes** and then monocytes. In the line leading to lymphocytes, the chromatin in the nucleus condenses and the amount of cytoplasm declines.

Other cells in the bone marrow become platelet-forming cells (Figure 18.8). In this line, immature **megakaryoblasts** (meg″ah-kar′e-o-blasts) undergo repeated mitoses; however, no cytoplasmic division occurs, and their nuclei never completely separate after mitosis. The result is a giant cell called a **megakaryocyte** (meg″ah-kar′e-o-sīt; "big nucleus cell") that has a large, multilobed nucleus containing many times the normal number of chromosomes. From their sites within the reticular connective tissue of red bone marrow just outside the blood sinusoids (Figure 18.7c), megakaryocytes send cytoplasmic extensions through the walls of the sinusoids and into the bloodstream. These extensions then break apart into platelets, like postage stamps torn from a perforated sheet.

CLINICAL APPLICATION

Abnormal Numbers of Immature Blood Cells

Reticulocytes (immature erythrocytes) make up 1% to 2% of all circulating erythrocytes in most healthy people. Percentages of reticulocytes outside this range indicate that a person is producing erythrocytes at an accelerated or decreased rate. Reticulocyte numbers greater than 2% might indicate that the person is adapting to life at high altitudes (where low oxygen levels stimulate erythrocyte production), whereas numbers less than 1% might indicate a degenerative disease of the bone marrow. To detect disorders of erythrocyte production, clinicians routinely obtain a **reticuloctye count** in blood workups.

Band cells (immature neutrophils) normally make up 1% to 2% of the neutrophils in the blood. This percentage increases dramatically during acute bacterial infections, when the bone marrow releases more immature neutrophils. Thus, detection of **elevated numbers of band cells** in differential WBC counts is considered an indicator of infection.

check your understanding

- ☐ 6. Which bones contain blood stem cells in adults?
- ☐ 7. What type of tissue forms the fibrous network within red bone marrow?
- ☐ 8. Which leukocytes do not form from myeloid stem cells?

(For answers, see Appendix B.)

DISORDERS OF THE BLOOD

learning outcome

- ▶ Consider some common disorders of erythrocytes, leukocytes, and platelets.

Disorders of Erythrocytes

Polycythemia (pol″e-si-the′me-ah; "many blood cells") is an abnormal excess of erythrocytes in the blood. One variety, *polycythemia vera,* results from a cancer of the bone marrow that generates too many erythrocytes. Severe polycythemia causes an increase in the viscosity of the blood, which slows or blocks the flow of blood through the smallest vessels. It is treated by dilution—by removing some blood and replacing it with sterile physiological saline.

Anemia (ah-ne′me-ah; "lacking blood") is any condition in which erythrocyte levels or hemoglobin concentrations are low, such that the blood's capacity for carrying oxygen is diminished. Anemia can be caused by blood loss, iron deficiency, destruction of erythrocytes at a rate that exceeds their replacement, vitamin B_{12} or folic acid deficiency, or a genetic defect of hemoglobin. Anemic individuals are constantly tired and often pale, short of breath, and chilly because their tissues are receiving low amounts of oxygen.

Sickle cell disease, formerly called sickle cell anemia, is a common inherited condition exhibited primarily in people of central African descent; this disease occurs in approximately 1 of every 400 African Americans. Sickle cell disease results from a defect in the hemoglobin molecule that causes the abnormal hemoglobin to crystallize when the concentration of oxygen in the blood is low or the erythrocytes become dehydrated, as during exercise or anxiety. This causes the circulating erythrocytes to distort into the shape of a crescent, thus the name "sickle cell" (Figure 18.9). These deformed erythrocytes are rigid, fragile, and easily destroyed. Because they do not pass through capillaries easily, the sickled erythrocytes block these vessels, causing painful attacks of ischemia. Sickle cell patients experience severe bone and chest pain, infections, and strokes. The disease used to be inevitably fatal during childhood, but current treatments allow many patients to survive into adulthood. A new drug, hydroxyurea, greatly reduces the frequency of attacks and eases the symptoms by increasing the proportion of erythrocytes that contain a normal, fetal form of hemoglobin, which prevents these cells from sickling. Other treatments include drugs that keep the erythrocytes hydrated and repeated blood transfusions. Bone marrow transplants offer a complete cure but their risks—a 10% death rate and a 20% rejection rate—are too great for them to be performed routinely.

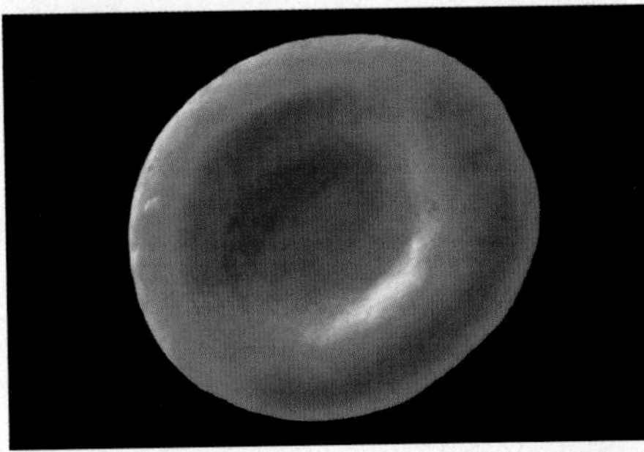

(a) Normal erythrocyte

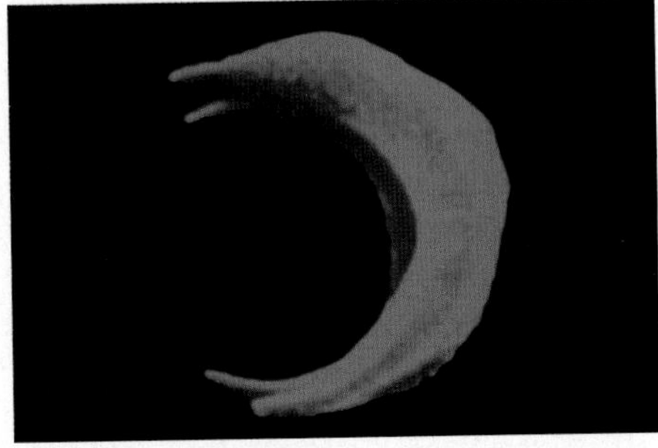

(b) Sickled erythrocyte

Figure 18.9 **Comparison of a normal erythrocyte with a sickled erythrocyte.** (Scanning electron micrographs, artificially colored, 4100×.)

Disorders of Leukocytes

Leukemia is a form of cancer resulting from the uncontrolled proliferation of a leukocyte-forming cell line in the bone marrow. Leukemias are classified according to (1) the cell line involved, as either *lymphoblastic* (from immature lymphocytes) or *myeloblastic* (from immature cells of the myeloid line) (see Figure 18.8); and (2) the rate of progression, as either *acute* (rapidly advancing) or *chronic* (slowly advancing). In all forms of leukemia, immature and cancerous leukocytes flood into the bloodstream. More significantly, however, the cancer cells take over the bone marrow, crowding out the normal blood cell lines and slowing the production of normal blood cells. Therefore, patients in late stages of leukemia suffer from anemia and devastating infections and from internal hemorrhaging due to clotting defects. Infections and hemorrhaging are the usual causes of death in people who succumb to leukemia. For information concerning new treatments for leukemia, see A CLOSER LOOK, (p. 558).

Disorders of Platelets

Thrombocytopenia (throm″bo-si″to-pe′ne-ah; "lack of platelets") is an abnormally low concentration of platelets in the blood. Characterized by diminished clot formation and by internal bleeding from small vessels, thrombocytopenia may result from damage to the bone marrow, chemotherapy, vitamin B_{12} deficiency, leukemia, autoimmune destruction of the platelets, or overactivity of the spleen (an organ that functions to remove and destroy platelets as well as other blood cells).

THE BLOOD THROUGHOUT LIFE

learning outcomes

- ▶ Describe the embryonic origin of blood cells.
- ▶ List four different organs that form blood cells in the fetus.
- ▶ Name some blood disorders that become more common as the body ages.

The first blood cells develop with the earliest blood vessels in the mesoderm around the yolk sac of the 3-week-old embryo. After mesenchymal cells cluster into groups called *blood islands,* the outer cells in these clusters flatten and become the endothelial cells that form the walls of the earliest vessels; the inner cells become the earliest blood cells. Soon vessels form within the embryo itself, providing a route for blood cells to travel throughout the body.

Throughout the first 2 months of development, all blood cells form in the blood islands of the yolk sac, with some contribution from the floor of the aorta. These sources form not only the blood stem cells that will last a lifetime but also the primitive nucleated erythrocytes that carry oxygen in the embryo. Late in the second month, circulating stem cells from the yolk sac become established in the liver and spleen, which take over the blood-forming function and are the major hematopoietic organs until month 7. These stem cells in the liver and spleen produce the first leukocytes and the first platelet-forming cells, plus nucleated and non-nucleated erythrocytes. The bone marrow receives stem cells and begins low-level hematopoiesis during month 3. Bone marrow becomes the major hematopoietic organ in month 7 of development and is the only hematopoietic organ postnatally. Should a severe need for blood cell production arise, however, the liver and spleen may resume their blood-cell-forming roles, even in adults.

Transplants of Bone Marrow and Cord Blood

Typically, treatment for leukemia patients involves transplanting healthy blood stem cells so that these patients can produce normal blood cells. First, however, the patient must undergo chemotherapy or radiation to kill both the cancer cells and blood-forming cells and to destroy the patient's immune system (to reduce the chances of transplant rejection).

In the most common procedure—a **bone marrow transplant**—marrow cells are aspirated from the iliac crest or sternum of a donor and transferred into the bloodstream of the recipient, where they circulate to and repopulate the recipient's marrow. An autologous transplant, in which the patient's own marrow is harvested while the leukemia is in remission, is the least risky approach, especially for older patients. However, in most transplant procedures bone marrow obtained from another person (an allogeneic transplant) must be used.

Allogeneic marrow transplants offer a hope of survival, but they have drawbacks. Perhaps paramount is the difficulty of finding a compatible donor to prevent rejection. Marrow transplants have the smallest margin of error in this respect: Not only can any of the recipient's surviving T cells attack the donor cells, but T cells in the donated marrow can attack the recipient's tissues—a reaction called **graft-versus-host disease.** Potential transplant recipients seeking donors have a 25% chance of finding a suitable match in a parent or sibling, and an even smaller chance among unrelated donors. Patients forced to wait for suitable donors often deteriorate in the meantime, and even those who receive a compatible transplant must take immunosuppressive drugs for life to lessen the risk of rejection. Furthermore, graft-versus-host disease afflicts more than half of all recipients and kills up to a third shortly after the procedure. The overall cure rate for leukemia patients who receive marrow from an unrelated donor is less than 25%.

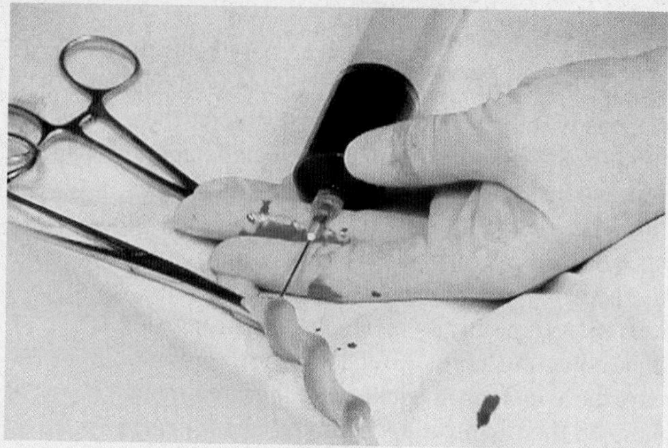

A clinician uses a syringe to extract blood from an umbilical cord (the twisted gray structure at lower center).

A newer source of blood stem cells for transplants is blood from the placenta, a pancake-shaped organ through which the fetus obtains oxygen and nutrients from the mother. The placenta contains stem cells until the very end of pregnancy, and a few tablespoons of placental blood are easily obtainable from the umbilical cord shortly after birth (see photo). Properly stored in liquid nitrogen, stem cells last indefinitely and are alive when thawed. So called **cord-blood** (or **placental-blood) transplants,** first performed in 1988, may soon render marrow transplants obsolete, for the following reasons:

- *Ready availability.* Obtaining cord blood only requires permission from the newborn's parents to harvest blood from a source that is otherwise disposed of as medical waste.
- *Safety.* Cord blood is less likely than adult donor marrow to contain microbes that could infect the recipient.
- *Long shelf life.* Unlike marrow, which must be obtained from the donor immediately before transplantation, cord blood can be stored indefinitely in one of many cord-blood banks worldwide.
- *Reduced risk of rejection.* Because the T cells in cord blood are immature, they are less likely to trigger graft-versus-host disease. Studies document that cord-blood recipients experience lower rates of rejection, regardless of whether they are related to their donors. As a result, the tissue match between donor and recipient need not be so close, which increases the pool of potential donors.

Medical researchers are currently developing new strategies for increasing the success rates of both bone marrow and cord-blood transplants. One approach is to stimulate stem cells in cord blood to multiply before being transplanted, and another is to remove all T cells from donor tissue before the procedure. Ultimately, researchers hope to grow large quantities of nonantigenic blood stem cells in the laboratory for subsequent transplantation.

The most common diseases of the blood that appear with aging are chronic leukemias, anemias, and clotting disorders. However, these and most other age-related blood disorders are usually precipitated by disorders of the heart, blood vessels, or immune system. For example, the increased incidence of leukemias in old age is believed to result from the waning ability of the immune system to destroy cancer cells, and the formation of abnormal thrombi and emboli reflects the progression of atherosclerosis, which roughens the linings of arterial walls.

✓ check your understanding

- ☐ 9. List the symptoms of anemia. What component of blood is affected by this disorder?
- ☐ 10. What are the major hematopoietic organs in the fetus before month 7?
- ☐ 11. Why do cord-blood transplant recipients experience lower rates of tissue rejection?

(For answers, see Appendix B.)

RELATED CLINICAL TERMS

Bone marrow biopsy Procedure for obtaining a sample of bone marrow, usually using a needle to aspirate marrow from the sternum or the iliac bone. The marrow is examined to diagnose disorders of blood cell formation, leukemia, infections, and types of anemia resulting from damage to, or failure of, the marrow.

Embolus Any abnormal mass carried freely in the bloodstream; may be a blood clot, air bubbles, fat masses, clumps of cells, or pieces of tissue. Blood clots are the most common kind of emboli, but *fat emboli,* which enter the blood from the bone marrow following a bone fracture, are also common. *Bacterial emboli* (clusters of bacteria) can occur during blood poisoning. Air bubbles can enter the bloodstream when a central intravenous line is accidentally disconnected, producing an air embolism.

Hemophilia An inherited disease caused by the lack of, or a reduced amount of, a clotting factor, resulting in blood that does not clot normally. People with hemophilia bleed longer and thus can suffer from excessive blood loss when injured. Replacement clotting factor may be administered intravenously following an injury or as a preventive treatment.

Hemorrhage (hem′ŏ-rij; "blood bursting forth") Any abnormal discharge of blood out of a vessel; bleeding.

CHAPTER SUMMARY

Take full advantage of the invaluable online practice and study tools for this chapter by going to the Study Area of MasteringA&P®. There you will find:

- Art Labeling Activities, Quizzes, Practice Tests, Memory Games
- Animal Organ Dissection, Cat Dissection and Bone Videos
- PAL Cadaver
- PAL Histology
- A&P Flix Animations
- Dynamic Study Modules

1. The circulatory system is subdivided into the cardiovascular system (heart, blood vessels, and blood) and the lymphatic system (lymphatic vessels and lymph).
2. Blood carries respiratory gases and nutrients to body tissues and picks up waste products from body tissues. Blood also transports hormones and defense cells to target tissues and regulates body temperature.

COMPOSITION OF BLOOD (pp. 547–553)

3. Blood consists of plasma and formed elements (erythrocytes, leukocytes, and platelets). Erythrocytes make up about 45% of blood volume, plasma about 55%, and leukocytes and platelets under 1%.
4. The volume percentage of erythrocytes is the hematocrit.

Blood Plasma (pp. 547–548)

5. Plasma is a fluid that is 90% water. The remaining 10% consists of nutrients, respiratory gases, salts, hormones, and plasma proteins (mainly albumin, globulins, and fibrinogen). Serum is plasma from which the clotting factors have been removed.

Formed Elements (pp. 548–553)

6. Blood cells are short-lived and are replenished continuously by new cells from the bone marrow.
7. A blood smear is prepared by putting a drop of blood on a glass slide; spreading it to form a film; then drying, preserving, and staining the film. Blood smears are stained with mixtures of two dyes, one acidic, such as eosin, and one basic, such as methylene blue.
8. Erythrocytes, the most abundant blood cells, are anucleate, biconcave discs with a diameter of 7–8 μm. Essentially, they are bags of hemoglobin, the oxygen-transporting protein. The main function of erythrocytes is to transport oxygen between the lungs and the body tissues. Erythrocytes live about 120 days in the circulation.
9. Leukocytes fight infections in the loose connective tissues outside capillaries, using the bloodstream only as a transport system. They leave capillaries by diapedesis and crawl through the connective tissues to the infection sites. There are only 4800–11,000 leukocytes, compared to about 5 million erythrocytes, in a cubic millimeter of blood.
10. The five distinct types of leukocytes are grouped into granulocytes and agranulocytes, according to whether they have distinctive cytoplasmic granules. Granulocytes, which include neutrophils, eosinophils, and basophils, are short-lived cells with distinctive cytoplasmic granules and lobed nuclei. All are phagocytic cells. The agranulocytes are lymphocytes and monocytes.
11. Neutrophils, the most abundant leukocytes, have multilobed nuclei. Two types of small granules give their cytoplasm a light purple color in stained blood smears. The function of neutrophils is phagocytosis of bacteria.

12. Eosinophils have bilobed nuclei and large, red-staining granules that contain digestive enzymes. They destroy and digest parasites and stop allergic reactions.

13. Basophils are rare leukocytes with bilobed nuclei and large, purple-staining granules full of chemical mediators of inflammation. They mediate the late stages of allergic reactions. Although a distinct cell type, they are functionally similar to mast cells.

14. Lymphocytes or their products attack antigens in the specific immune response. These cells have a sparse, blue-staining cytoplasm and a dense, purple-staining, spherical nucleus. T lymphocytes destroy foreign cells directly, whereas B lymphocytes secrete antibodies that mark foreign cells for phagocytosis.

15. Monocytes, the largest leukocytes, resemble large lymphocytes but have a lighter-staining nucleus that may be kidney-shaped. They transform into macrophages in connective tissues.

16. Platelets are disc-shaped, membrane-enclosed fragments of megakaryocyte cytoplasm that contain several kinds of secretory granules. They plug tears in blood vessels, signal vasoconstriction, help initiate clotting, and then retract the clot and close the tear.

BLOOD CELL FORMATION (pp. 553–556)

Bone Marrow as the Site of Hematopoiesis (pp. 553–554)

17. Collectively, bone marrow is the body's second largest organ. Red marrow, located between the trabeculae of spongy bone in the axial skeleton, girdles, and proximal epiphysis of each humerus and femur of adults, actively makes blood cells. Yellow marrow, in the other regions of limb bones of adults, is dormant.

18. Microscopically, bone marrow consists of wide capillaries (sinusoids) snaking through reticular connective tissue. The latter contains reticular fibers, fibroblasts (reticular cells), macrophages, fat cells, and immature blood cells in all stages of maturation. New blood cells enter the blood through the sinusoid walls.

Cell Lines in Blood Cell Formation (pp. 554–556)

19. All blood cells continuously arise from blood stem cells. As these cells divide, there is an early separation into lymphoid stem cells (future lymphocytes) and myeloid stem cells (precursors to all other blood cell classes). These stem cells commit to the specific blood cell lines and structural differentiation begins.

20. Erythrocytes start as proerythroblasts and proceed through various stages, during which hemoglobin accumulates and the organelles and nucleus are extruded. For their first 1 or 2 days in the circulation, erythrocytes are reticulocytes.

21. The three distinct classes of granular leukocytes begin as myeloblasts and then proceed through various stages in which they gain specific granules and their nucleus shuts down, distorting into a horseshoe shape. The cells then mature and enter the bloodstream. Some neutrophils enter the circulation as immature band cells.

22. Monocytes and lymphocytes look somewhat like stem cells. Structural changes are minimal in the developmental pathways of these cells.

23. In the platelet line, immature megakaryoblasts become giant megakaryocytes with multilobed nuclei. The cytoplasm of megakaryocytes breaks up to form platelets in the blood.

DISORDERS OF THE BLOOD (pp. 556–557)

24. Polycythemia is an excess of erythrocytes in the blood, and anemia is any condition in which the blood's capacity for carrying oxygen is diminished. Sickle cell disease is an inherited condition in which the erythrocytes distort into a sickle shape and block the capillaries. Leukemia is cancer of leukocyte-forming cells in bone marrow, and thrombocytopenia is abnormally low concentration of platelets in the blood.

THE BLOOD THROUGHOUT LIFE (pp. 557–559)

25. The blood stem cells develop in blood islands of the yolk sac in the 3-week-old embryo and then travel to the hematopoietic organs.

26. The hematopoietic organs in the fetus are the liver, spleen, and bone marrow.

27. The incidence of leukemia, anemia, and clotting disorders increases with age.

REVIEW QUESTIONS

Multiple Choice/Matching Questions

(For answers, see Appendix B.)

1. Which of the following descriptions of erythrocytes is *false*? Erythrocytes (a) are shaped like biconcave discs, (b) have a life span of about 120 days, (c) contain hemoglobin, (d) have lobed nuclei.

2. The white blood cell that releases histamine and other mediators of inflammation is the (a) basophil, (b) neutrophil, (c) monocyte, (d) eosinophil.

3. Which of the following blood cells are phagocytic? (a) lymphocytes, (b) erythrocytes, (c) neutrophils, (d) lymphocytes and neutrophils.

4. The blood cell that can attack a specific antigen is (a) a lymphocyte, (b) a monocyte, (c) a neutrophil, (d) a basophil, (e) an eryth-rocyte.

5. Rank the following leukocytes in order of their relative abundance in the blood of a normal, healthy person, from 1 (most abundant) to 5 (least abundant).

____ **(a)** lymphocytes

____ **(b)** basophils

____ **(c)** neutrophils

____ **(d)** eosinophils

____ **(e)** monocytes

6. Match the names of the blood cells in column B with their descriptions in column A. Some names are used more than once.

Column A	Column B
____ (1) destroys parasites	(a) erythrocyte
____ (2) has two types of granules	(b) neutrophil
____ (3) does not use diapedesis	(c) eosinophil
____ (4) stops the allergic response	(d) basophil
____ (5) the only cell that is not spherical	(e) lymphocyte
____ (6) granulocyte that phagocytizes bacteria	(f) monocyte
____ (7) the largest blood cell	
____ (8) granulocyte with the smallest granules	
____ (9) a young one is a reticulocyte	
____(10) lives for about 4 months	
____(11) most abundant blood cell	
____(12) includes T cells and B cells	

7. In typically stained monocytes, the blue cytoplasm and the purple nuclei are colored by this dye: (a) eosin, (b) brilliant cresyl blue, (c) basin (as in basophilic), (d) methylene blue, (e) none of the above—you need two dyes for two colors.

Short Answer Essay Questions

8. Which class or classes of formed elements form the buffy coat in a centrifuged blood sample?

9. (a) What is the basic functional difference between red bone marrow and yellow bone marrow? (b) Where is yellow marrow located?

10. (a) Describe the steps of erythrocyte formation in the bone marrow. (b) What name is given to the immature type of red blood cell that is newly released into the circulation?

11. Describe the structure of platelets, and explain their functions.

12. What is the difference between a hematopoietic stem cell and a committed cell in the bone marrow?

13. What is the relationship between megakaryocytes and platelets?

14. Looking at a blood smear under the microscope, Tina became confused by the unusual nuclei of granulocytes. She kept asking why each eosinophil had two nuclei and why neutrophils had four or five nuclei. How would you answer her?

15. Compare and contrast each of the following pairs of terms: (a) *circulatory system* and *cardiovascular system,* (b) *complete blood count* and *complete blood count with differential.*

16. On her test, Floris wrote that the function of platelets was "clotting." The instructor, however, demanded a more complete answer. Indicate what the instructor had in mind.

17. When leukocytes are examined in a blood count, is it possible to tell B lymphocytes from T lymphocytes? Explain your answer.

18. Compare an eosinophil to a basophil, both in structure and in function.

19. What are the advantages of placental-blood transplants over bone marrow transplants?

20. What are the functions of the following plasma proteins: globulins, albumin, fibrinogen?

21. How does serum differ from plasma?

Critical Reasoning & Clinical Application Questions

1. After young Janie was diagnosed as having acute lymphocytic leukemia, her parents could not understand why infection was a major problem for Janie when her WBC count was so high. Provide an explanation for Janie's parents.

2. Freddy could tell that a neuron cell body on his slide was as wide as ten erythrocytes. How wide was the neuron cell body (in micrometers)?

3. The Jones family, who raise sheep on their ranch, let their sheepdog Rooter lick their faces, and they would sometimes kiss him on the mouth. The same day that the veterinarian diagnosed Rooter as having tapeworms, a blood test indicated that both the family's daughters had blood eosinophil levels of over 3000 per cubic millimeter. What is the connection?

4. A reticulocyte count indicated that 5% of Tyler's red blood cells were reticulocytes. His blood test also indicated he had polycythemia and a hematocrit of 65%. Explain the connections among these three facts.

5. Cancer patients being treated with chemotherapy drugs, which are designed to destroy rapidly mitotic cells, are monitored closely for changes in their RBC and WBC counts. Why?

6. Your child has had a moderate fever for 2 days. On the third day, you take her to the pediatrician. After an examination, blood is drawn and a CBC with diff is performed. How does this information aid the pediatrician in determining whether the cause of the infection is viral or bacterial?

19 The Heart

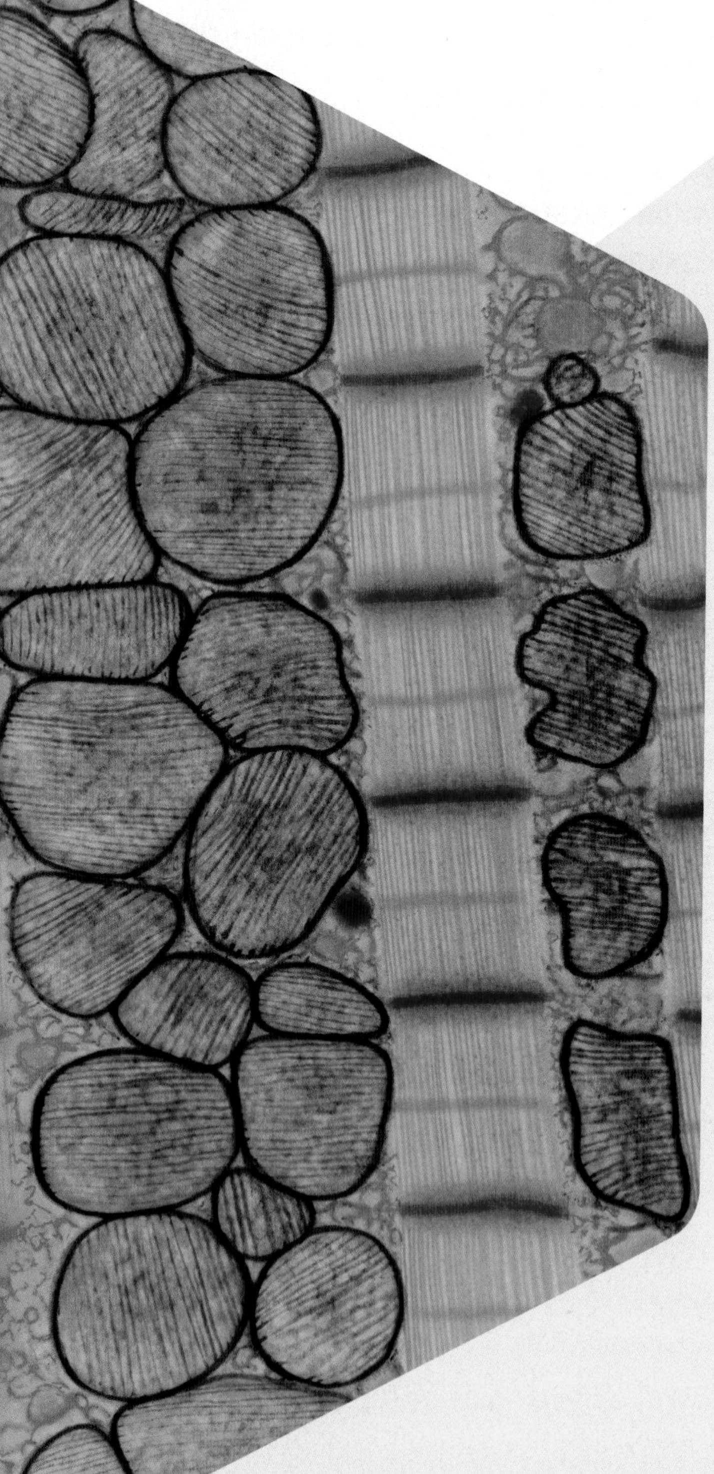

▲ *Healthy cardiac muscle (colored TEM, longitudinal section).*

The ceaselessly beating heart in the thorax has intrigued people for thousands of years. The ancient Greeks believed the heart was the seat of intelligence. Others thought it was the source of emotions. Although these theories have proved false, emotions certainly affect heart rate. Only when your heart pounds or occasionally skips a beat do you become acutely aware of this dynamic organ.

The heart is a muscular double pump with two functions (Figure 19.1):

1. Its right side receives oxygen-poor blood from the body tissues and then pumps this blood to the lungs to pick up oxygen and dispel carbon dioxide. The blood vessels that carry blood to and from the lungs form the **pulmonary circuit** (*pulmonos* = lung).
2. Its left side receives the oxygenated blood returning from the lungs and pumps this blood throughout the body to supply oxygen and nutrients to the body tissues. The vessels that transport blood to and from all body tissues and back to the heart form the **systemic circuit.**

The heart has two receiving chambers, the *right atrium* and *left atrium* (*atrium* = entranceway), that receive blood returning from the systemic and pulmonary circuits. The heart also has two main pumping chambers, the *right ventricle* and *left ventricle* ("hollow belly"), that pump blood around the two circuits.

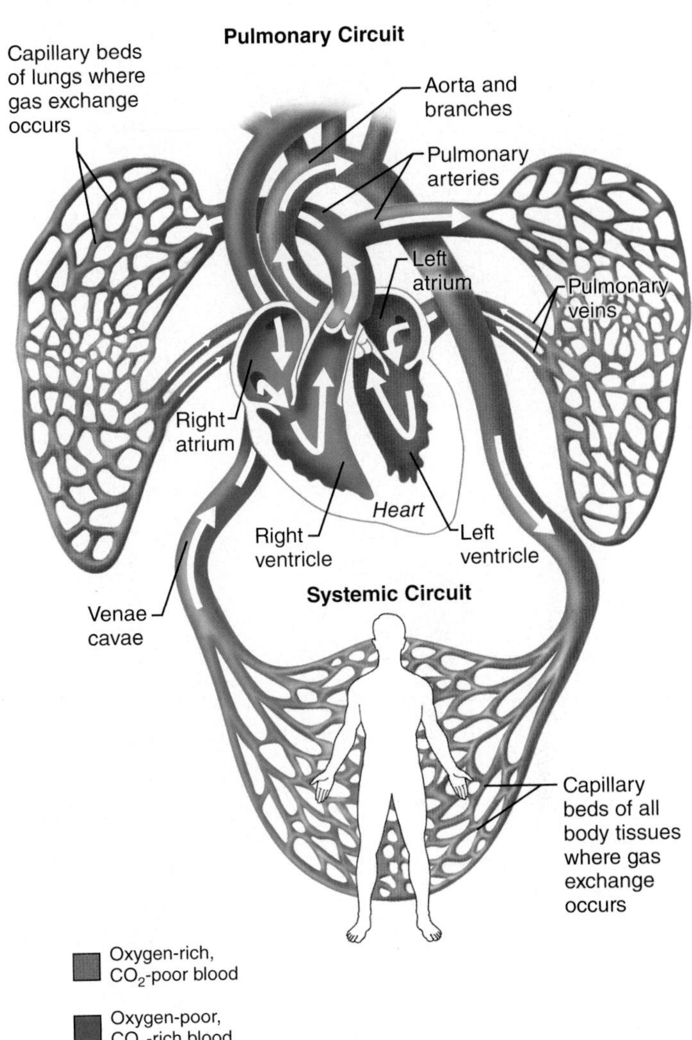

Figure 19.1 **The heart as a double pump.** The right side of the heart pumps blood through the pulmonary circuit. The left side of the heart pumps blood to all the body tissues via the systemic circuit.

LOCATION AND ORIENTATION WITHIN THE THORAX

learning outcome

▶ Describe the orientation, location, and surface anatomy of the heart in the thorax.

The heart's modest size belies its incredible strength and durability. Only about the size of a fist, this hollow, cone-shaped organ looks enough like the popular valentine image to satisfy the sentimentalists among us. Typically it weighs between 250 and 350 grams—less than a pound.

The heart lies in the thorax posterior to the sternum and costal cartilages and rests on the superior surface of the diaphragm (Figure 19.2a). It is the largest organ in the mediastinum, which is the region between the two lungs (and pleural cavities) (Figure 19.2b and c). The heart assumes an oblique position in the thorax, with its pointed **apex** lying to the left of the midline and anterior to the rest of the heart (Figure 19.2c and d). If you press your fingers between the fifth and sixth ribs just inferior to the left nipple, you may feel the beating of your heart where the apex contacts the thoracic wall. Cone-shaped objects have a base as well as an apex, and the heart's **base** is its broad posterior surface.

The heart is said to have four corners defined by four points projected onto the anterior thoracic wall (Figure 19.2a). The second rib is easily palpated just lateral to the sternal angle. Use this landmark to help you locate the four corners of the heart.

- The *superior right* point lies where the costal cartilage of the third rib joins the sternum.
- The *superior left* point lies at the costal cartilage of the second rib, a finger's breadth lateral to the sternum.
- The *inferior right* point lies at the costal cartilage of the sixth rib, a finger's breadth lateral to the sternum.
- Finally, the *inferior left* point (the apex point) lies in the fifth intercostal space at the midclavicular line—that is, at a line extending inferiorly from the midpoint of the left clavicle.

The imaginary lines that connect these four corner points delineate the normal size and location of the heart. Clinicians must know these normal landmarks, because an enlarged or displaced heart as viewed on an X-ray or other medical image can indicate heart disease or other disease conditions.

☑ check your understanding

☐ 1. Which side of the heart receives and pumps deoxygenated blood?

☐ 2. Where is the apex of the heart located in reference to the anterior thoracic wall?

(For answers, see Appendix B.)

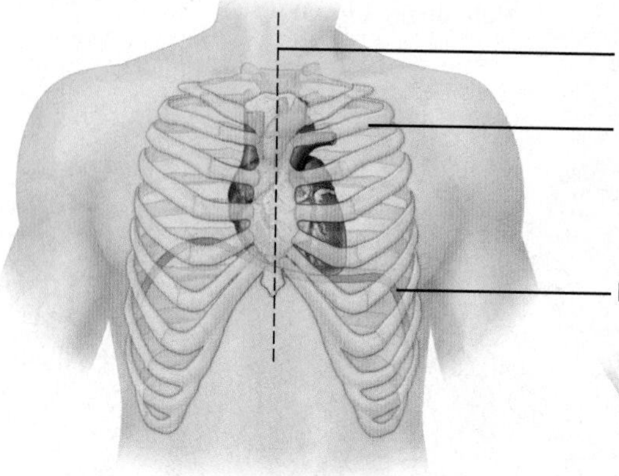

Midsternal line
Rib 2
Diaphragm
(a)

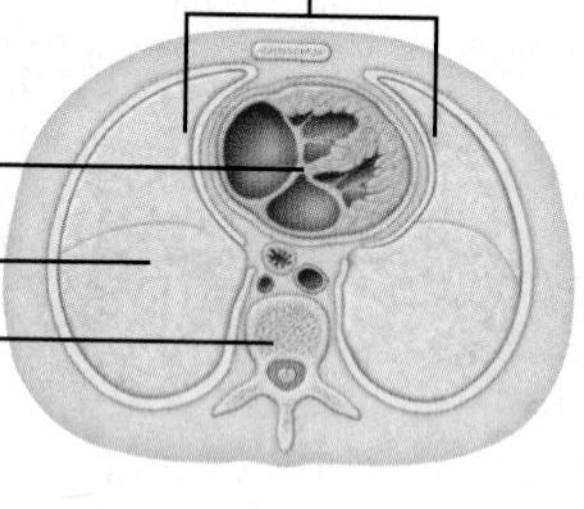

Mediastinum
Heart
Right lung
T_8
Posterior
(b)

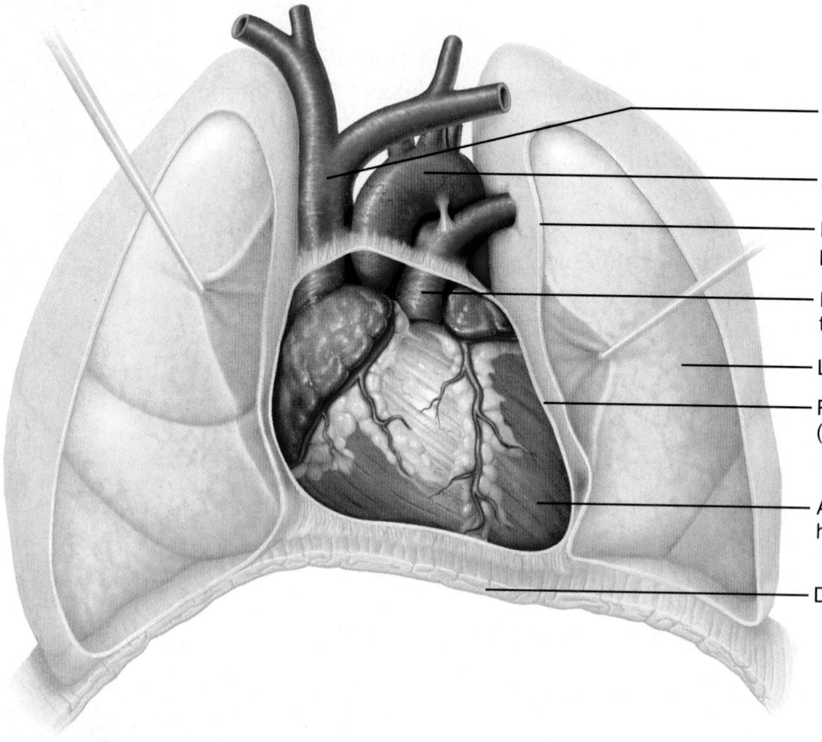

Superior vena cava
Aorta
Parietal pleura (cut)
Pulmonary trunk
Left lung
Pericardium (cut)
Apex of heart
Diaphragm
(c)

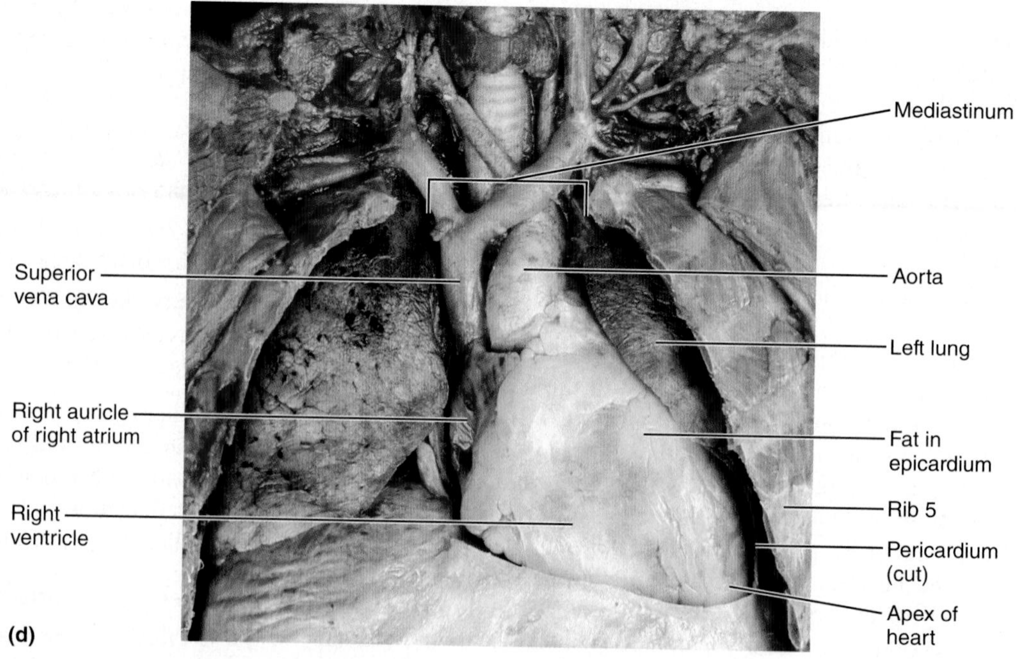

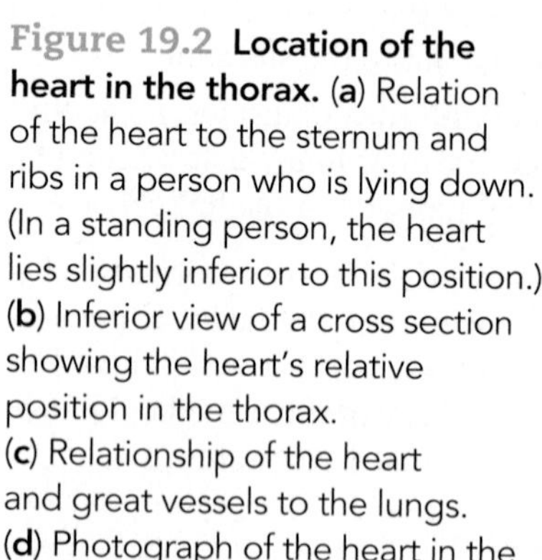

Figure 19.2 Location of the heart in the thorax. (**a**) Relation of the heart to the sternum and ribs in a person who is lying down. (In a standing person, the heart lies slightly inferior to this position.) (**b**) Inferior view of a cross section showing the heart's relative position in the thorax. (**c**) Relationship of the heart and great vessels to the lungs. (**d**) Photograph of the heart in the mediastinum.

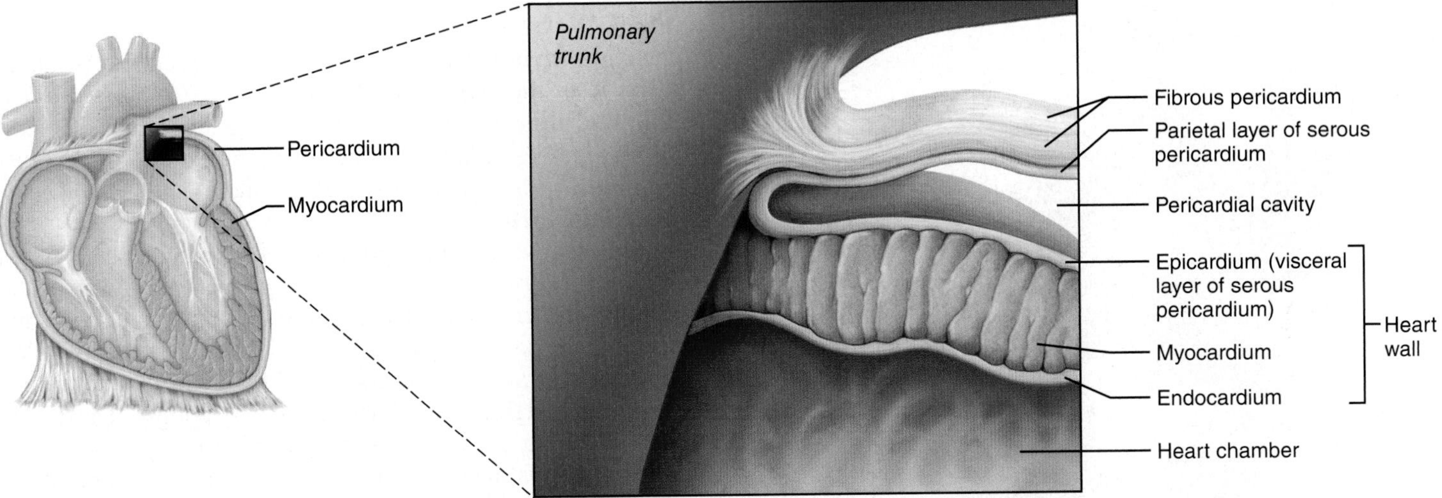

Figure 19.3 **The layers of the pericardium and of the heart wall.**

STRUCTURE OF THE HEART

learning outcomes

- ▶ Describe the layers of the pericardium and the tissue layers of the heart wall.
- ▶ List the important structural features of each heart chamber: right and left atria, and right and left ventricles.

Coverings

The **pericardium** (per″ĭ-kar′de-um; "around the heart") is a triple-layered sac that encloses the heart (Figure 19.3). The outer layer of this sac, called the **fibrous pericardium,** is a strong layer of dense connective tissue. It adheres to the diaphragm inferiorly, and superiorly it is fused to the roots of the great vessels that leave and enter the heart. The fibrous pericardium acts as a tough outer coat that holds the heart in place and keeps it from overfilling with blood.

Deep to the fibrous pericardium is the double-layered **serous pericardium,** a closed sac sandwiched between the fibrous pericardium and the heart (see Chapter 1, "Serous Cavities," and Figure 1.7, p. 12). The outer, **parietal layer of the serous pericardium** adheres to the inner surface of the fibrous pericardium. The parietal layer is continuous with the **visceral layer of the serous pericardium,** or **epicardium,** which lies on the heart and is considered a part of the heart wall (discussed shortly). Between the parietal and visceral layers of serous pericardium is a slitlike space, called the **pericardial cavity,** a division of the embryonic coelom (Chapter 3, p. 54). The epithelial cells of the serous pericardium that line the pericardial cavity produce a lubricating film of serous fluid into the pericardial cavity. This fluid reduces friction between the beating heart and the outer wall of the pericardial sac.

CLINICAL APPLICATION

Pericarditis and Cardiac Tamponade

Infection and inflammation of the pericardium, or **pericarditis,** can lead to a roughening of the serous lining of the pericardial cavity. As a consequence, the beating heart produces a creaking sound called *pericardial friction rub,* which can be heard with a stethoscope. Pericarditis is characterized by pain behind the sternum. Over time, it can lead to adhesions of the heart to the outer pericardial wall, or the pericardium can scar and thicken inhibiting the heart's movements.

In severe acute cases of pericarditis, large amounts of fluid resulting from the inflammatory response exude into the pericardial cavity. Because the fibrous pericardium is a tough, inflexible tissue, the excess fluid compresses the heart, limiting the expansion of the heart between beats and diminishing its ability to pump blood. This condition is called **cardiac tamponade** (tam″po-nād″; "a heart plug"). Physicians treat it by inserting a hypodermic needle into the pericardial cavity to drain the excess fluid. Cardiac tamponade also results if blood accumulates inside the pericardial cavity, as occurs when a penetrating wound to the heart (such as a stab wound) allows blood to leak out of the heart and into the pericardial cavity.

Layers of the Heart Wall

The wall of the heart has three layers: a superficial *epicardium,* a middle *myocardium,* and a deep *endocardium* (Figure 19.3). All three layers are richly supplied with blood vessels.

- The **epicardium** ("upon the heart") is the visceral layer of the serous pericardium, as previously mentioned. This serous membrane is often infiltrated with fat, especially in older people (see Figure 19.2d).
- The **myocardium** ("muscle heart") forms the bulk of the heart. It consists of cardiac muscle tissue and is the layer that

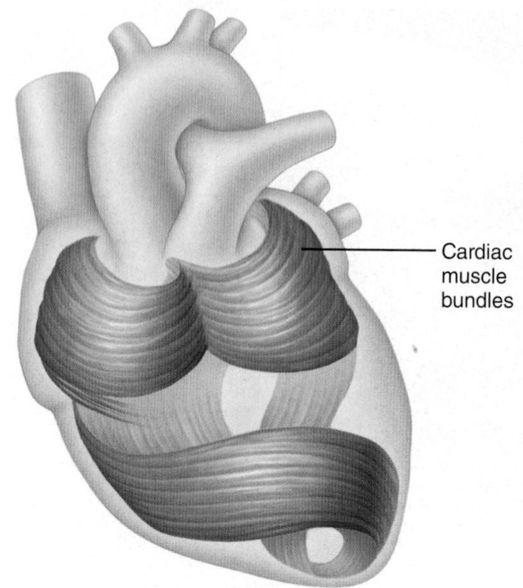

Figure 19.4 The circular and spiral arrangement of cardiac muscle bundles in the myocardium of the heart.

actually contracts. Surrounding the cardiac muscle cells in the myocardium are connective tissues that bind these cells together into elongated, circularly and spirally arranged networks called *bundles* (Figure 19.4). These bundles function to squeeze blood through the heart in the proper directions: inferiorly through the atria and superiorly through the ventricles. The connective tissues of the myocardium form the *cardiac skeleton,* which reinforces the myocardium internally and anchors the cardiac muscle fibers. The histological structure and function of cardiac muscle tissue is covered in detail later in this chapter (pp. 575–578).

- The **endocardium** ("inside the heart"), located deep to the myocardium, is a sheet of simple squamous epithelium resting on a thin layer of connective tissue. Endocardium lines the heart chambers and covers the heart valves.

Heart Chambers

The four heart chambers are the *right* and *left atria* (singular, *atrium*) superiorly, and the *right* and *left ventricles* inferiorly (Figure 19.5). Internally, the heart is divided longitudinally by a partition called the **interatrial septum** between the atria (Figure 19.5c) and the **interventricular septum** between the ventricles (Figure 19.5e and f). Externally, the boundaries of the four chambers are marked by two grooves. The **coronary sulcus** (Figure 19.5b) extends horizontally, circling the boundary between the atria and the ventricles (*corona* = crown). The **anterior interventricular sulcus** (Figure 19.5a and b) extends vertically, marking the anterior position of the interventricular septum, and the **posterior interventricular sulcus** (Figure 19.5d) separates the two ventricles on the heart's inferior surface. Recall that the heart is oriented obliquely within the thorax (Figure 19.2d); the "posterior" of the heart lies against the diaphragm and is thus its inferior surface.

Right Atrium

The **right atrium** forms the entire right border of the human heart. It is the receiving chamber for oxygen-poor blood returning from the systemic circuit (Figure 19.1). The right atrium receives blood via three veins: the *superior vena cava* and *inferior vena cava* (Figure 19.5b and d) and the *coronary sinus* (Figure 19.5c and d).

Externally, the **right auricle,** a small flap shaped like a dog's ear (*auricle* = little ear) projects anteriorly from the superior corner of the atrium (Figure 19.5a). Internally, the right atrium has two parts (Figure 19.5c): a smooth-walled posterior part and an anterior part lined by horizontal ridges called the **pectinate muscles** (*pectin* = comb). These two parts of the atrium are separated by a large, C-shaped ridge called the **crista terminalis** ("terminal crest"). The crista is an important landmark in locating the sites where veins enter the right atrium: The superior vena cava opens into the atrium just posterior to the superior bend of the crista; the inferior vena cava opens into the atrium just posterior to the inferior bend of the crista; and the coronary sinus opens into the atrium just anterior to the inferior end of the crista. Additionally, just posterior to this end of the crista is the **fossa ovalis,** a depression in the interatrial septum that marks the spot where an opening existed in the fetal heart: the *foramen ovale* (see discussion of fetal circulation on (p. 622.)

Inferiorly and anteriorly, the right atrium opens into the right ventricle through the *tricuspid valve (right atrioventricular valve)* (Figure 19.5e and f).

Right Ventricle

The **right ventricle** forms most of the anterior surface of the heart. It receives blood from the right atrium and pumps it into the pulmonary circuit via an artery called the **pulmonary trunk** (Figure 19.5a and e). Internally, the ventricular walls are marked by irregular ridges of muscle called **trabeculae carneae** (trah-bek′u-le kar′ne-e; "little beams of flesh"). Additionally, cone-shaped **papillary muscles** project from the walls into the ventricular cavity (*papilla* = nipple).

Thin, strong bands called **chordae tendineae** (kor′de ten′dĭ-ne-e; "tendinous cords"), the "heart strings," project superiorly from the papillary muscles to the flaps (cusps) of the tricuspid (right atrioventricular) valve. Superiorly, the opening between the right ventricle and the pulmonary trunk contains the *pulmonary semilunar valve* (or simply, *pulmonary valve*).

Left Atrium

The **left atrium** makes up most of the heart's posterior surface, or base. It receives oxygen-rich blood returning from the lungs through two right and two left **pulmonary veins** (Figure 19.5d). The only part of the left atrium visible anteriorly is its triangular left auricle (Figure 19.5a and b). Internally, most of the atrial wall is smooth, with pectinate muscles lining the auricle only. The left atrium opens into the left ventricle through the *mitral valve (left atrioventricular valve)* (Figure 19.5e).

(Text continues on page 571.)

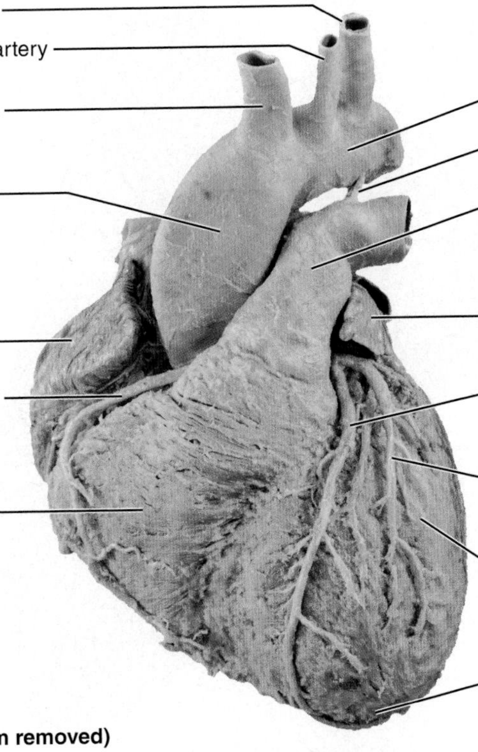

Left subclavian artery
Left common carotid artery
Brachiocephalic trunk
Ascending aorta
Right atrium
Right coronary artery (in coronary sulcus)
Right ventricle
Aortic arch
Ligamentum arteriosum
Pulmonary trunk
Auricle of left atrium
Anterior interventricular artery (in anterior interventricular sulcus)
Additional branch off left coronary artery, normal variation
Left ventricle
Apex of heart (left ventricle)

(a) Anterior aspect (pericardium removed)

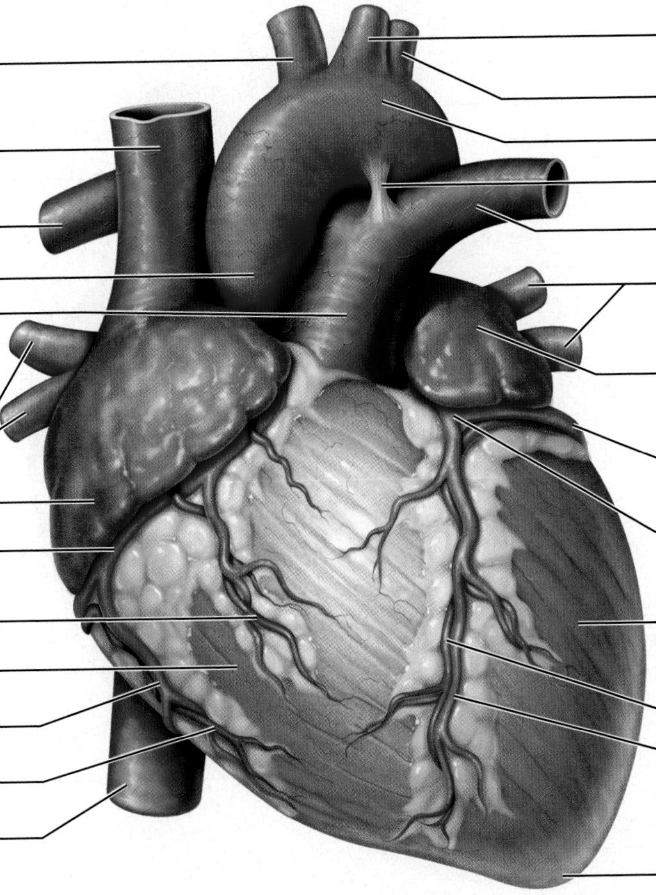

Brachiocephalic trunk
Superior vena cava
Right pulmonary artery
Ascending aorta
Pulmonary trunk
Right pulmonary veins
Right atrium
Right coronary artery (in coronary sulcus)
Anterior cardiac vein
Right ventricle
Right marginal artery
Small cardiac vein
Inferior vena cava
Left common carotid artery
Left subclavian artery
Aortic arch
Ligamentum arteriosum
Left pulmonary artery
Left pulmonary veins
Auricle of left atrium
Circumflex artery
Left coronary artery (in coronary sulcus)
Left ventricle
Great cardiac vein
Anterior interventricular artery (in anterior interventricular sulcus)
Apex

(b) Anterior view

Practice art labeling MasteringA&P®

Figure 19.5 Gross anatomy of the heart. In diagrammatic views, vessels carrying oxygen-rich blood are red; those carrying oxygen-poor blood are blue.

Watch Sheep Heart Dissection MasteringA&P®

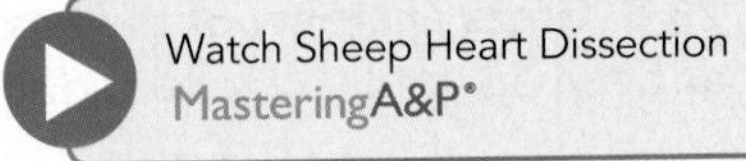

Superior vena cava
Ascending aorta
Pectinate muscles
Crista terminalis
Interatrial septum
Fossa ovalis
Opening of inferior vena cava (cut)
Opening of coronary sinus
Myocardium of left ventricle
Interventricular sulcus

(c) Internal aspect of the right atrium, anterior view

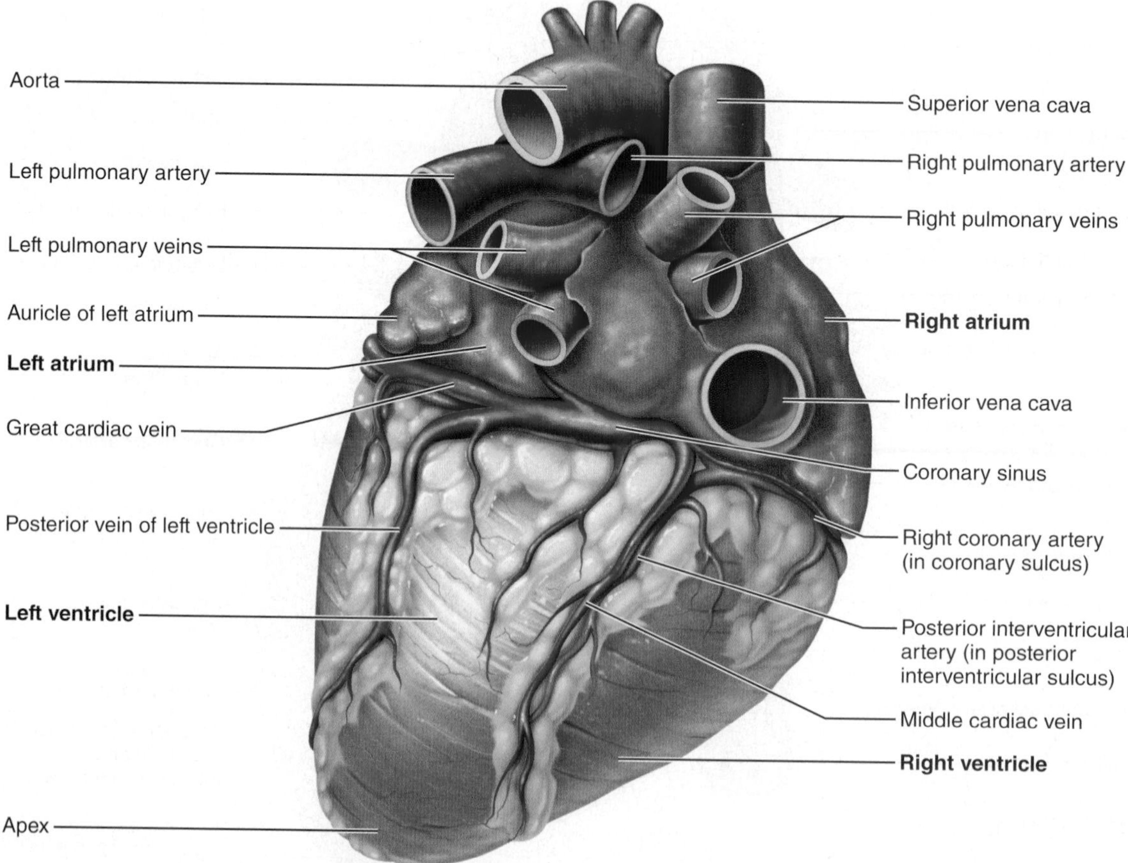

(d) Inferior view; surface shown rests on the diaphragm

Figure 19.5 **Gross anatomy of the heart,** *continued.* In (c), the anterior wall of the atrium has been opened and reflected superiorly.

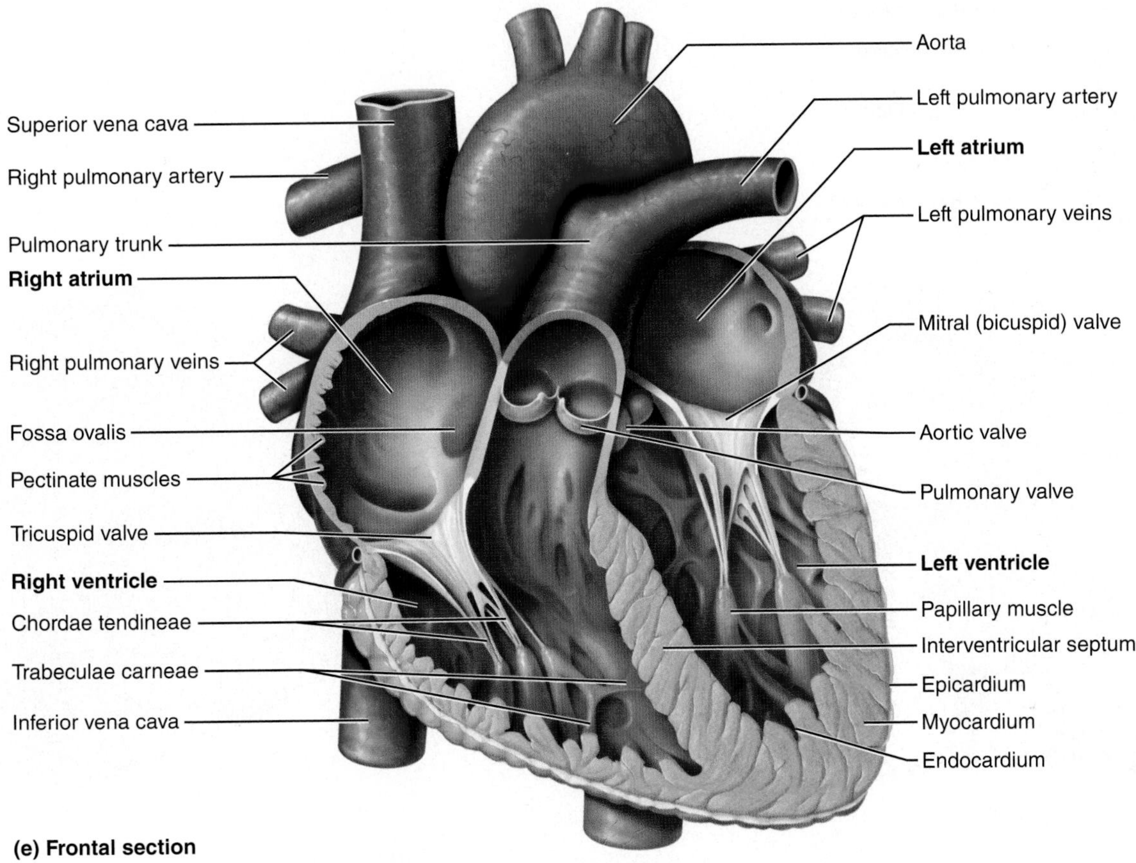

(e) Frontal section

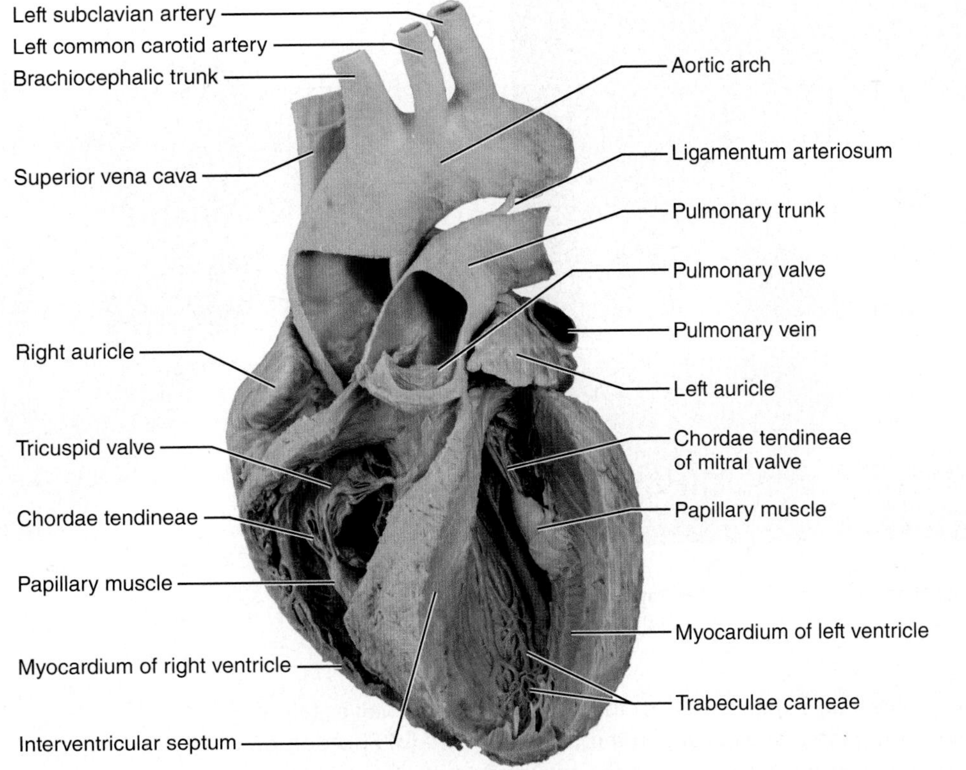

(f) Internal aspect of ventricles; dissection of view similar to (e)

Figure 19.5 **Gross anatomy of the heart,** *continued.*

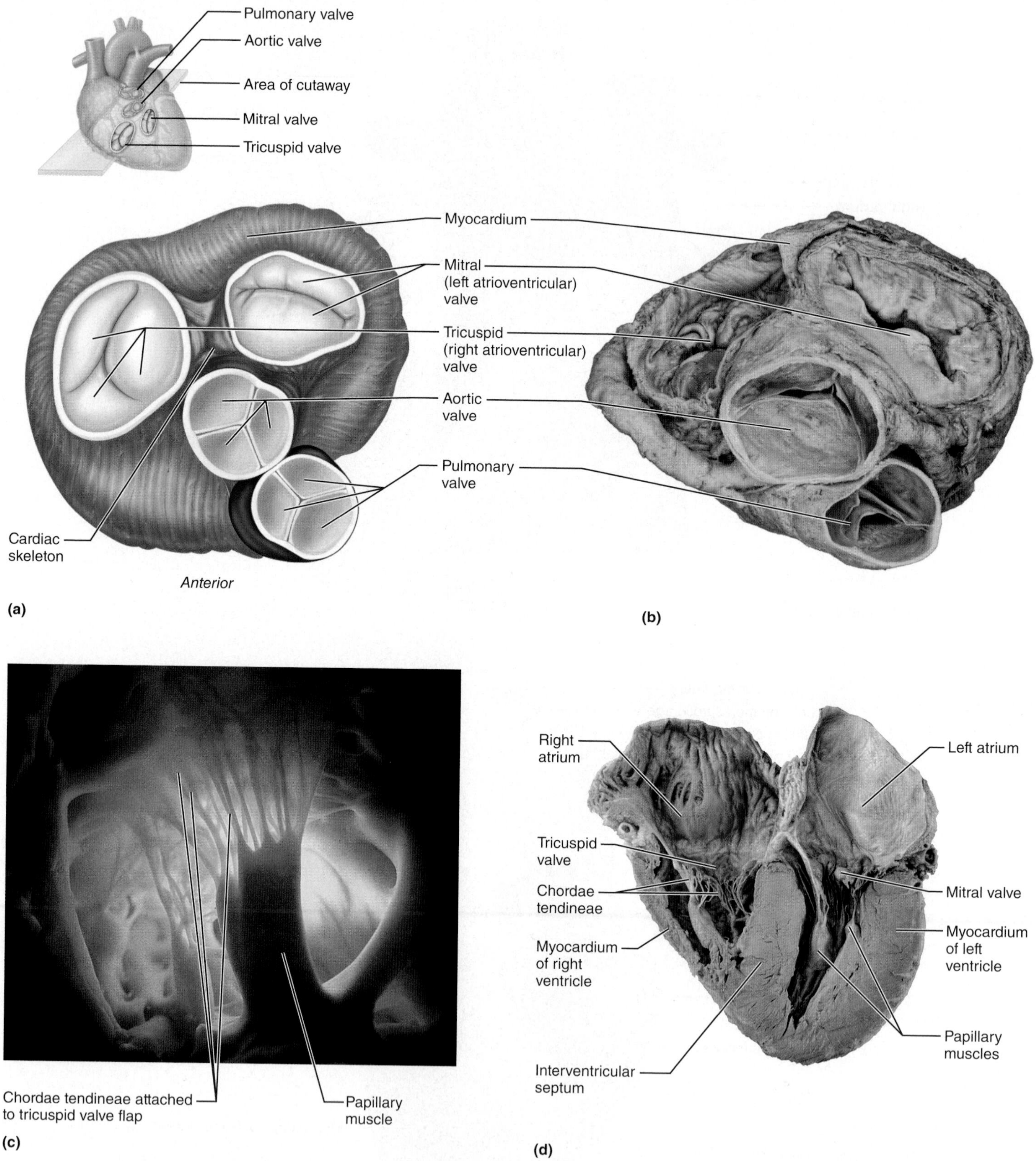

Pulmonary valve
Aortic valve
Area of cutaway
Mitral valve
Tricuspid valve

Myocardium
Mitral (left atrioventricular) valve
Tricuspid (right atrioventricular) valve
Aortic valve
Pulmonary valve
Cardiac skeleton
Anterior

(a)

(b)

Chordae tendineae attached to tricuspid valve flap
Papillary muscle

(c)

Right atrium
Left atrium
Tricuspid valve
Chordae tendineae
Mitral valve
Myocardium of right ventricle
Myocardium of left ventricle
Papillary muscles
Interventricular septum

(d)

Figure 19.6 Heart valves. (**a**) Superior view of the two sets of heart valves (atria removed). The inset shows the plane of section through the heart. (**b**) Photograph of heart valves, superior view. (**c**) Photograph of the tricuspid valve. This inferior-to-superior view shows the valve as seen from the right ventricle. (**d**) Frontal section of the heart.

Left Ventricle

The **left ventricle** forms the apex of the heart and dominates the heart's inferior surface (Figure 19.5d and e). It pumps blood into the systemic circuit. Like the right ventricle, it contains trabeculae carneae, papillary muscles, chordae tendineae, and the cusps of an atrioventricular *(mitral)* valve. Superiorly, the left ventricle opens into the stem artery of the systemic circulation (the aorta) through the *aortic semilunar valve* (or simply, *aortic valve*).

check your understanding

- 3. What is another name for the epicardium?
- 4. Identify the heart chamber or chambers that contain the listed structures: (a) pectinate muscles; (b) papillary muscles; (c) fossa ovalis; (d) trabeculae carneae.
- 5. Name three vessels that empty into the right atrium. Are these vessels arteries or veins?

(For answers, see Appendix B.)

HEART VALVES

learning outcomes

- Name the heart valves, and describe their locations and functions. Indicate where on the chest wall each of the valves is heard.
- Describe the cardiac skeleton, and explain its functions.

Valve Structure

The heart valves—the paired atrioventricular (AV) and semilunar valves—enforce the one-way flow of blood through the heart, from the atria to the ventricles and into the great arteries that leave the superior part of the heart. Each heart valve consists of two or three *cusps,* which are flaps of endocardium reinforced by cores of dense connective tissue (Figure 19.6).

- Located at the junctions of the atria and their respective ventricles are the atrioventricular valves: the **right atrioventricular (tricuspid) valve,** which has three cusps, and the **left atrioventricular (bicuspid) valve,** which has only two cusps. The latter is also called the **mitral** (mi′tral) **valve** because its cusps resemble the two sides of a bishop's hat, or miter.
- Located at the junctions of the ventricles and the great arteries are the **aortic** and **pulmonary (semilunar) valves,** each of which has three pocketlike cusps shaped roughly like crescent moons (*semilunar* = half moon) (Figures 19.6a and b, and 19.5e and f).

The **cardiac skeleton** lies in the plane between the atria and the ventricles and surrounds all four heart valves rather like handcuffs (Figure 19.6a). Composed of dense connective tissue, it has four functions:

1. It anchors the valve cusps.
2. It prevents overdilation of the valve openings as blood pulses through them.
3. It is the point of attachment for the bundles of cardiac muscle in the atria and ventricles (see Figure 19.4).
4. It blocks the direct spread of electrical impulses from the atria to the ventricles (discussed shortly). This function is critical for the proper coordination of atrial and ventricular contractions.

Valve Function

Heart valves open (to allow blood flow) and close (to prevent the backflow of blood) in response to differences in blood pressure on each side of the valves. The two *atrioventricular valves* prevent the backflow of blood into the atria during contraction of the ventricles (Figure 19.7). When the ventricles are relaxed, the cusps of the AV valves hang limply into the ventricular chambers while blood flows into the atria and down through the open AV valves into the ventricles (Figure 19.7a). When the ventricles start to contract, the pressure within them rises and forces the blood superiorly against the valve cusps, pushing the edges of the cusps together and closing the AV valves. The chordae tendineae and papillary muscles that attach to these valves look like the cords of an open parachute, limiting the closed cusps so they cannot fly up and allow reflux of ventricular blood into the atria. The papillary muscles begin to contract slightly before the rest of the ventricle contracts, pulling on the chordae tendineae and preventing the AV valves from everting (Figure 19.7b). If the cusps were not anchored in this manner, they would be forced superiorly into the atria.

The two *semilunar valves* prevent backflow from the great arteries into the ventricles (Figure 19.8, p. 573). When the ventricles contract and raise the intraventricular pressure, the semilunar valves are forced open, and their cusps are flattened against the arterial walls as the blood rushes past them. When the ventricles relax, blood that tends to flow back toward the heart fills the cusps of the semilunar valves and forces them shut.

Heart Sounds

The closing of the valves causes vibrations in the adjacent blood and heart walls that account for the familiar "lub-dup" sounds of each heartbeat: The "lub" sound is produced by the closing of the AV valves at the start of ventricular contraction; the "dub" is produced by the closing semilunar valves at the end of ventricular contraction. The mitral valve closes slightly before the tricuspid closes, and the aortic valve generally closes just before the pulmonary valve closes. Because of these slight differences in timing, all four valve sounds are discernible when the clinician listens through a stethoscope placed on the anterior chest wall.

Even though all four heart valves lie in roughly the same plane (see Figure 19.6a)—the plane of the coronary sulcus—the clinician does not listen directly over the respective valves, because the sounds take oblique paths through the heart chambers to reach the chest wall. Each valve is best heard near a different heart corner: the pulmonary valve near the superior left point, the aortic valve near the superior right point, the mitral valve at the apex point, and the tricuspid near the inferior right point (Figure 19.9, p. 573). Abnormal heart sounds often indicate disorders of the heart valves.

(Text continues on page 573.)

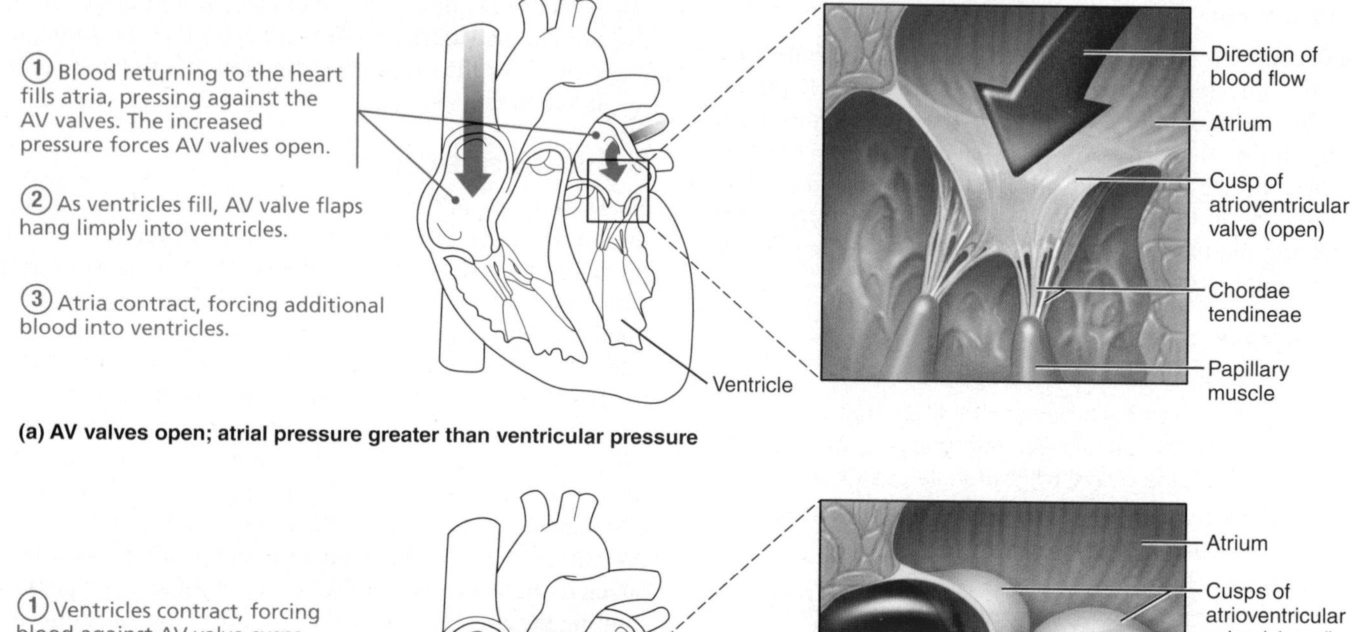

(a) AV valves open; atrial pressure greater than ventricular pressure

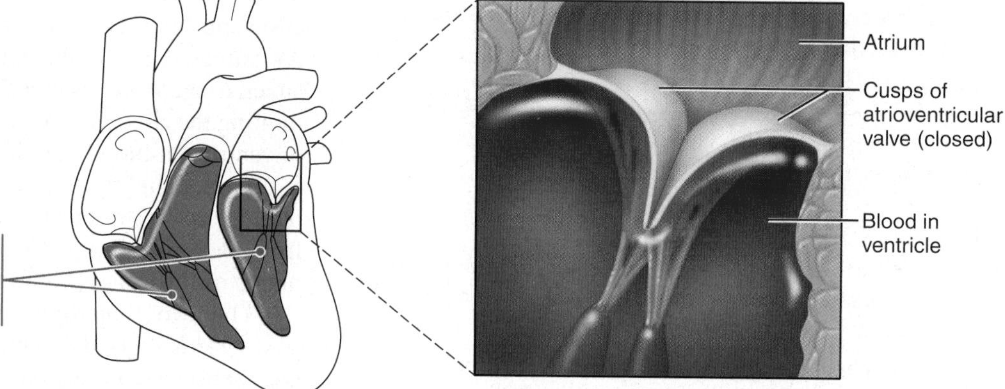

(b) AV valves closed; atrial pressure less than ventricular pressure

Figure 19.7 Function of the atrioventricular valves.

CLINICAL APPLICATION

Valve Disorders Defects of the heart valves are common and have a number of causes, ranging from congenital and genetic abnormalities, to an inadequate blood supply to the valves (due to a heart attack), to bacterial infection of the endocardium. Valvular defects lead to valvular disorders, most of which are classified as one of two types. Valves that leak because they fail to close properly are considered **incompetent** (or are said to exhibit **insufficiency**). An incompetent valve produces a distinct blowing sound after the valve closes. By contrast, valves with narrowed openings, such as occur when cusps have fused or become stiffened by calcium deposits, are termed **stenotic.** Stiffened valves cannot open properly. Stenosis of the aortic valve produces a distinctive "click" sound during ventricular systole as blood passing through the constricted opening becomes turbulent and vibrates. Both incompetent and stenotic valves lessen the heart's efficiency and thus increase its workload; ultimately the heart may weaken severely.

The mitral valve, and the aortic valve to a lesser extent, are most often involved in valve disorders because they are subjected to the great forces resulting from contraction of the powerful left ventricle. In **mitral valve prolapse,** an inherited weakness of the collagen in the valve and chordae tendineae allows one or both cusps of this valve to "flop" into the left atrium during ventricular systole. This is the most common heart valve disorder (affecting 2.5% to 5% of the population); it is characterized by a distinctive heart sound—a click followed by a swish (backflow of blood). Most cases are mild and harmless, but severe cases may lead to heart failure or a disruption of heart rhythm.

Treatment of valve disorders typically involves surgical repair of damaged valves. When this approach is unsuccessful, valves are replaced with either mechanical or biological replacement valves. Tissue for biological valves comes from either cadavers, pigs, or cows. Replacement techniques are neither permanent nor problem-free.

(a) Mechanical valve

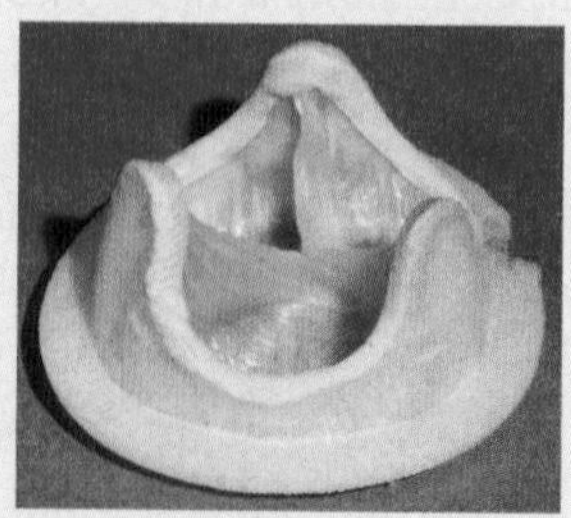

(b) Biological valve

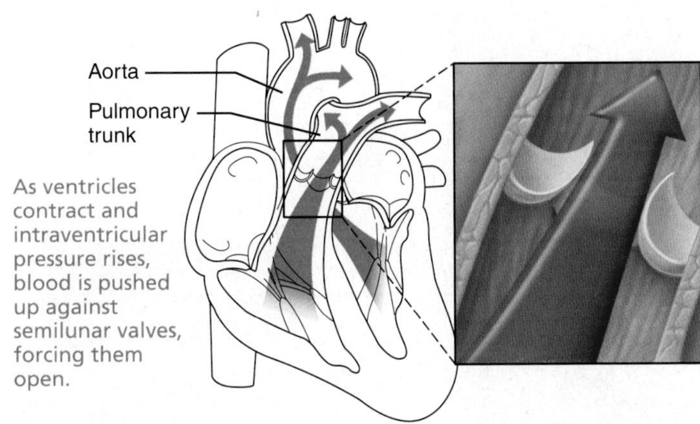

(a) Semilunar valves open

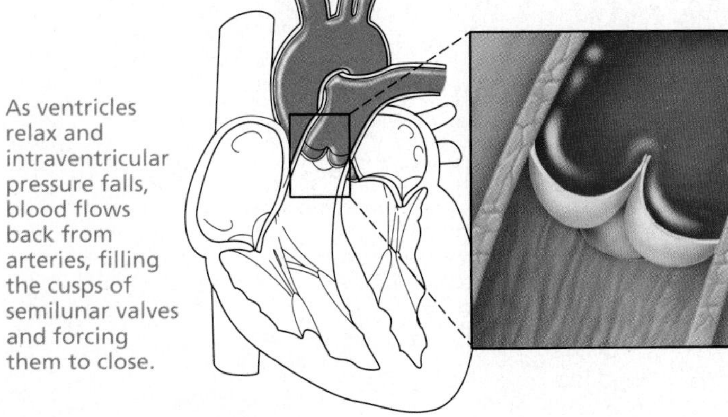

(b) Semilunar valves closed

Figure 19.8 **Function of the semilunar valves.**

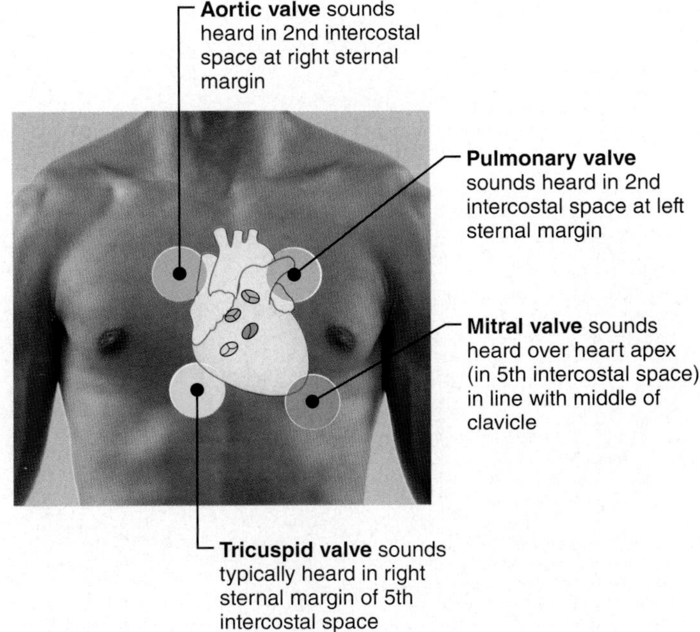

Figure 19.9 **Areas on the thoracic surface where heart sounds are heard most clearly.**

PATHWAY OF BLOOD THROUGH THE HEART

learning outcome

- Describe the path of a drop of blood through the four chambers of the heart and the systemic and pulmonary circuits.

Focus on Blood Flow Through the Heart (Figure 19.10) illustrates the path of blood around the pulmonary and systemic circuits, beginning with oxygen-poor systemic blood as it arrives at the right side of the heart. Blood coming from body regions superior to the diaphragm (excluding the heart wall) enters the right atrium via the superior vena cava (SVC); blood returning from body regions inferior to the diaphragm enters via the inferior vena cava (IVC); and blood draining from the heart wall itself is collected by and enters the right atrium through the coronary sinus. The blood passes from the right atrium through the tricuspid valve to the right ventricle, propelled by gravity and the contraction of the right atrium. Then, the right ventricle contracts, propelling the blood through the pulmonary semilunar valve into the pulmonary trunk and to the lungs through the pulmonary circuit for oxygenation.

The freshly oxygenated blood returns via the four pulmonary veins to the left atrium and passes through the mitral valve to the left ventricle, propelled by gravity and the contraction of the left atrium. The left ventricle then contracts and propels the blood through the aortic semilunar valve into the aorta and its branches. After delivering oxygen and nutrients to the body tissues through the systemic capillaries, the oxygen-poor blood returns through the systemic veins to the right atrium—and the whole cycle repeats continuously.

Although a given drop of blood passes through the heart chambers sequentially (one chamber after another), the four chambers do not contract in that order. Rather, the two atria always contract *together,* followed by the simultaneous contraction of the two ventricles.

A single sequence of atrial contraction followed by ventricular contraction is called a **heartbeat,** and the heart of an average person at rest beats 70–80 times per minute. The term that describes the contraction of a heart chamber is **systole** (sis′to-le; "contraction"); the time during which a heart chamber is relaxing and filling with blood is termed **diastole** (di-as′to-le; "expansion"). Thus, both atria and ventricles experience systole and diastole. Be aware, however, that in common medical usage, diastole and systole most often refer to ventricular filling and contraction, respectively. For example, a blood pressure reading measures the ventricular systolic pressure over the ventricular diastolic pressure (systolic/diastolic) in the systemic circuit, that is, the left ventricle. Because the ventricle is relaxed during diastole, elevation of pressure during this portion of the heartbeat is a symptom of hypertension (high blood pressure).

Now that you know how the heart pumps blood, you can readily understand why the muscular walls of its different

FOCUS Blood Flow Through the Heart

Figure 19.10

The heart is a double pump, each side supplying its own circuit.

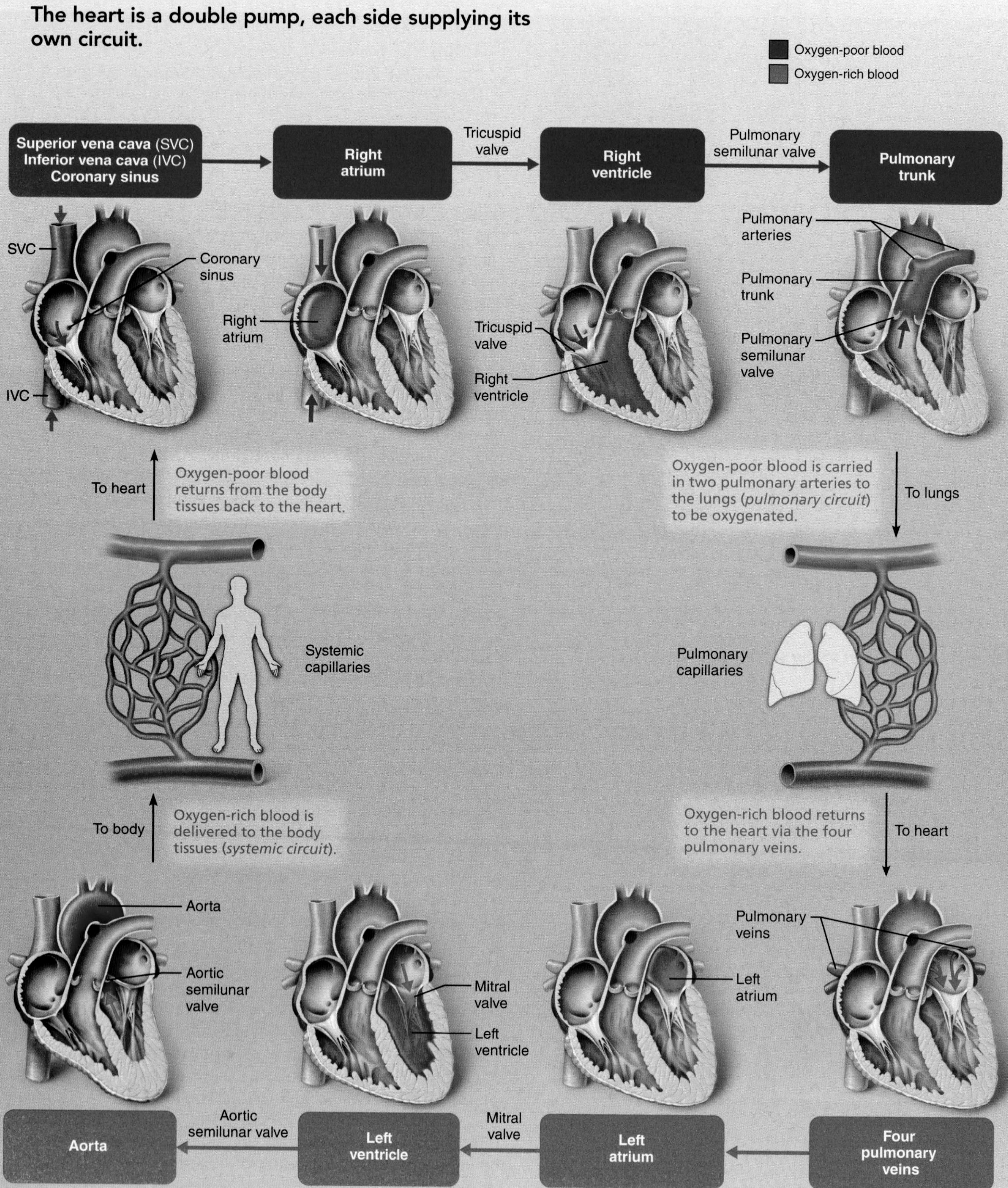

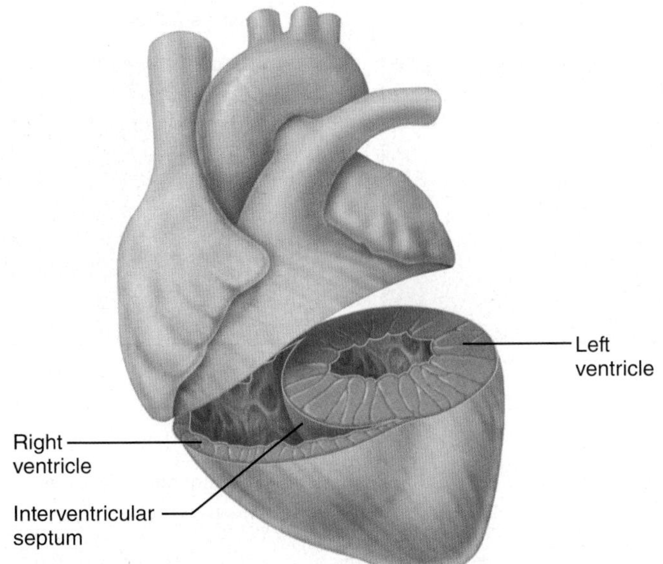

Figure 19.11 Anatomical differences between the right and left ventricles. The thicker, circular wall of the left ventricle encroaches on the cavity of the thinner-walled right ventricle, making the cavity of the right ventricle crescent-shaped.

chambers differ in thickness. The walls of the atria are much thinner than those of the ventricles (Figure 19.5e) because much of ventricular filling is done by gravity, and thus the atria exert little effort to propel blood inferiorly into the ventricles. The wall of the left ventricle (the systemic pump) is at least three times as thick as that of the right ventricle (the pulmonary pump) (Figure 19.11). Consequently, the left ventricle can generate much more force than the right and pumps blood at a much higher pressure. The higher pressure in the systemic circuit reflects the fact that the systemic circuit is much longer than the pulmonary circuit and offers greater resistance to blood flow. The thick wall of the left ventricle gives this chamber a circular shape and flattens the cavity of the adjacent right ventricle into the shape of a crescent.

CLINICAL APPLICATION

Hypertrophic Cardiomyopathy *Hypertrophic cardiomyopathy* is an inherited condition in which the wall of the left ventricle (especially the interventricular septum) is abnormally thick and composed of disorganized muscle cells. Because the lumen of the left ventricle is too small to hold all the atrial blood returning to it, blood pressure in the pulmonary circuit is elevated. Additionally, the grossly thickened myocardium becomes ischemic because it cannot get enough oxygen from the coronary vessels that supply it. Exercising can lead to fainting, chest pain, or even sudden death.

check your understanding

☐ 6. Where does blood travel after passing through the aortic semilunar valve? Is this blood oxygenated or deoxygenated?

☐ 7. During ventricular systole, are the AV valves open or closed? Are the semilunar valves open or closed during this period?

☐ 8. Differentiate a stenotic valve from an incompetent valve.

(For answers, see Appendix B.)

CARDIAC MUSCLE TISSUE

learning outcomes

- Describe the structure of cardiac muscle tissue.
- Describe the structure of intercalated discs, and discuss their importance in the contraction of cardiac muscle.

Thus far, we have described the gross anatomy of the heart and how these structures function to move blood through the heart. To understand how the heart produces forceful contractions, we must examine the structure of cardiac muscle tissue.

Structure of Cardiac Muscle

Cardiac muscle tissue forms the thick myocardium of the heart wall. It contains cardiac muscle cells and the connective tissues that surround these cells. The contractions of cardiac muscle cells pump blood through the heart and into and through the blood vessels of the circulatory system. Cardiac muscle tissue, like skeletal muscle tissue, is striated, and it contracts by the sliding filament mechanism.

Unlike a skeletal muscle cell, which is long, multinucleated, and cylindrically shaped, a cardiac muscle cell is a short, branching cell (Figure 19.12a) with one or two large, centrally located nuclei. Each cell averages about 25 μm in diameter and 120 μm in length. Adjacent cardiac muscle cells are joined together at their ends to form cellular networks. These branching networks of cardiac muscle cells are called *cardiac myofibers.*

The complex junctions that join cardiac muscle cells are called **intercalated discs** (in-ter′kah-la″ted; "inserted between"). At these junctions, the sarcolemmas of adjacent cells interlock through meshing "fingers," like one empty egg carton stacked inside another (Figure 19.12b). Intercalated discs have two distinct regions: Transverse regions contain desmosome-like junctions called **fasciae adherens** (singular: **fascia adherens**) that function to bind adjacent cells together and transmit the contractile force to adjacent cells. Longitudinal regions contain gap junctions that allow ions to pass between cells, transmitting the contractile signal to adjacent cells. The free movement of ions between cells allows the direct transmission of an electrical impulse through the entire network of cardiac muscle cells. This impulse in turn stimulates all the muscle cells in a heart chamber (atria or ventricles) to contract at the same time.

In the intercellular spaces around each cardiac myofiber is a loose fibrous connective tissue, the *endomysium,* which aids to bind adjacent cardiac fibers together and contains the vessels and nerves that serve the muscle cells. Groups of cardiac myofibers form the cardiac muscle bundles in

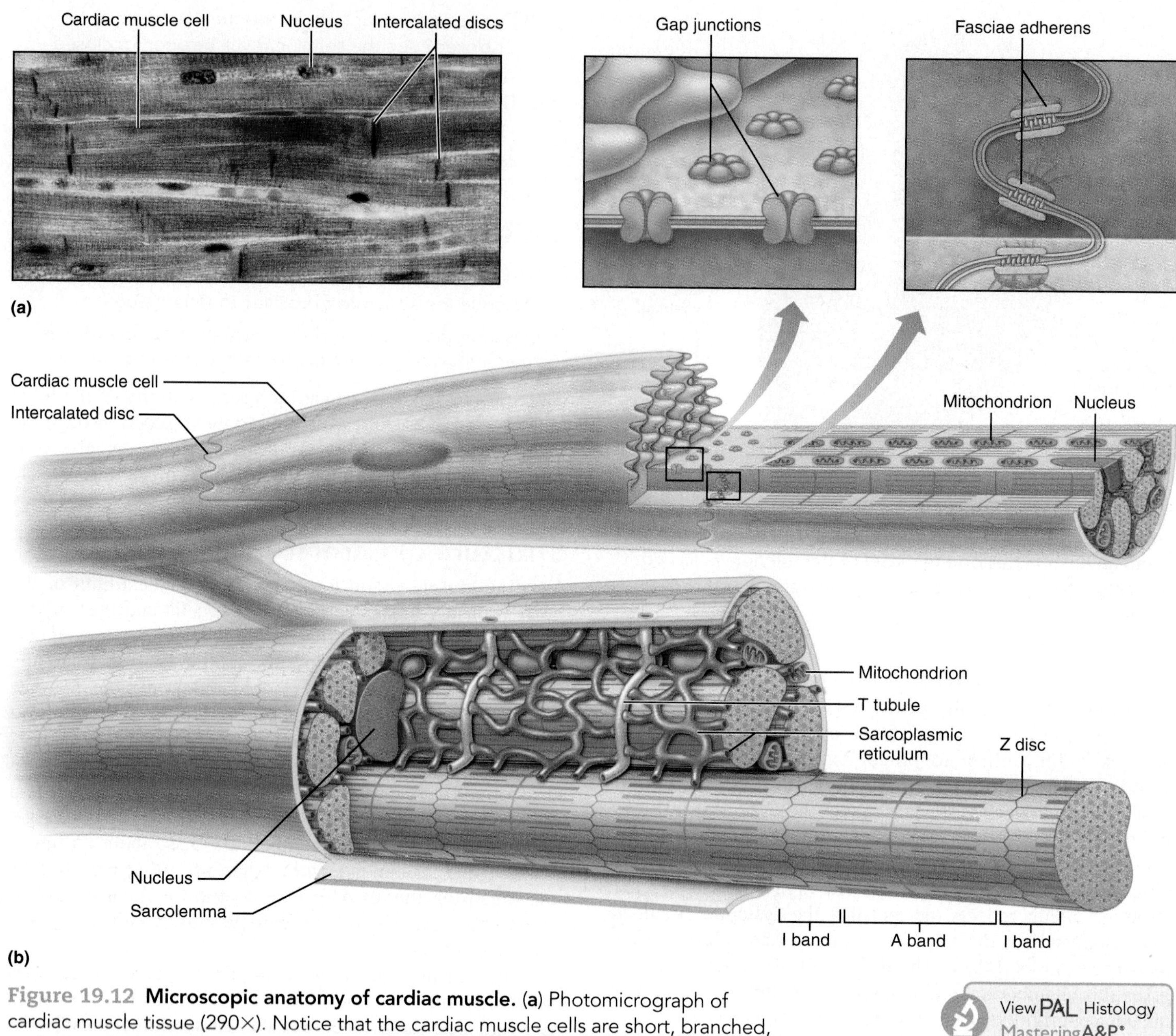

Figure 19.12 Microscopic anatomy of cardiac muscle. (**a**) Photomicrograph of cardiac muscle tissue (290×). Notice that the cardiac muscle cells are short, branched, and striated. The dark-staining areas are intercalated discs, junctions between adjacent cells. (**b**) Diagram of a cardiac muscle cell, partially sectioned, and enlargements of the intercalated disc.

View PAL Histology MasteringA&P®

the myocardium (Figure 19.4). The connective tissues surrounding the cardiac myofibers merge with the cardiac skeleton and thus function to anchor the muscle cells and transmit the contractile forces produced by the muscle cells. In this way, the connective tissues of the myocardium are similar to the tendinous origins and insertions of skeletal muscles.

The striated cardiac muscle cells contain myofibrils with typical sarcomeres composed of A and I bands, H zones, titin, and Z discs and M lines (Figure 19.12b and Figure 19.13). Striations are less apparent in cardiac muscle tissue than in skeletal muscle tissue, especially when viewed by light microscopy, because of the branching of the myofibrils and the great abundance of mitochondria surrounding them (Figure 19.13). The abundant mitochondria make large amounts of ATP by aerobic metabolism; thus, cardiac muscle is highly resistant to fatigue.

Mechanism of Contraction

The molecular mechanism for contraction in cardiac muscle is similar to that in skeletal muscle. Cardiac muscle cells are triggered to contract by ionic calcium (Ca^{2+}) entering the sarcoplasm. In response to an action potential, a small amount

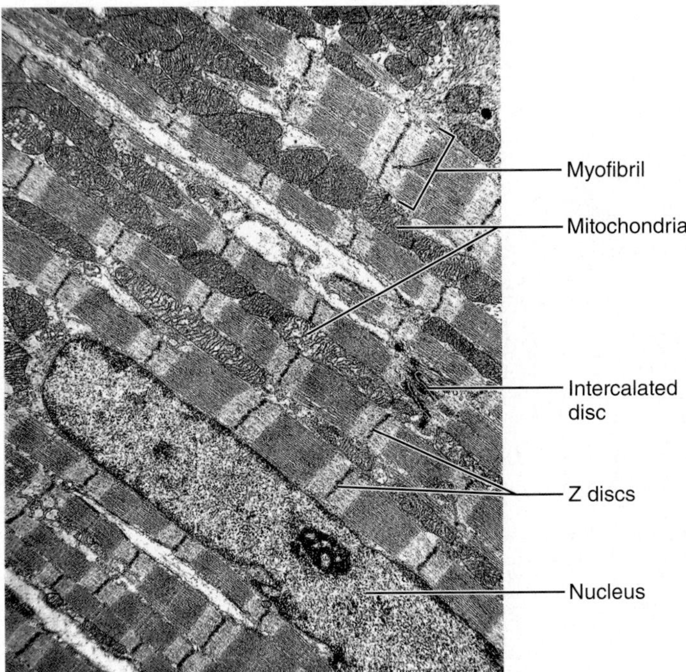

Figure 19.13 **Electron micrograph of a portion of a cardiac muscle cell.** Notice the large number of mitochondria between each myofibril and the irregular size of the myofibrils (7500×).

of Ca^{2+} from the extracellular tissue fluid enters the cardiac muscle cell through the sarcolemma. This rise in intracellular calcium signals the sarcoplasmic reticulum to release its stored Ca^{2+}. These ions diffuse into the sarcomeres and trigger the sliding of the filaments. Reuptake of calcium by the sarcoplasmic reticulum ends contraction. Compared to skeletal muscle, the sarcoplasmic reticulum of cardiac muscle cells is less complex, and the T tubules (recall from skeletal muscle that T tubules are invaginations of the sarcolemma that carry the electrical signal deep into the cell) are less abundant—occurring at Z discs instead of at the A-I junctions (Figure 19.12b and Table 10.2, pp. 254–255).

As with skeletal muscle, the amount of force that cardiac muscle cells can generate depends on their length. Significantly, these cells normally remain slightly *shorter* than their optimal length. Therefore, when they are stretched by a greater volume of blood returning to the heart, their contraction force increases, and they can pump the additional blood.

Unlike cells of skeletal muscle tissue, not all cells of cardiac muscle tissue are innervated. In fact, an isolated cardiac muscle cell will contract rhythmically without any innervation at all. This inherent rhythmicity of cardiac muscle cells is the basis of the rhythmic heartbeat, as explained next.

CONDUCTING SYSTEM AND INNERVATION

learning outcome

- Name the components of the conducting system of the heart, and describe the conduction pathway.

Conducting System

Cardiac muscle cells have an intrinsic ability to generate and conduct electrical impulses that stimulate these same cells to contract rhythmically. These properties are intrinsic to the heart muscle itself and do not depend on extrinsic nerve impulses. Even if all nerve connections to the heart are severed, the heart continues to beat rhythmically. Perhaps you remember removing a beating heart from a dissected frog in your high school biology class.

The **conducting system** of the heart is a series of specialized cardiac muscle cells that carries impulses throughout the heart musculature, signaling the heart chambers to contract in the proper sequence. It also initiates each contraction sequence, thereby setting basic heart rate.

The impulse that signals each heartbeat begins at the **sinoatrial (SA) node** (Figure 19.14, ①), a crescent-shaped mass of muscle cells that lies in the wall of the right atrium, just inferior to the entrance of the superior vena cava. The SA node sets the basic heart rate by generating 70–80 electrical impulses per minute. It is the heart's pacemaker. The signal initiated by the SA node spreads throughout the myocardium through the gap junctions in the intercalated discs.

From the SA node, impulses spread in a wave along the cardiac muscle fibers of the atria, signaling the atria to contract. Some of these impulses travel along an **internodal pathway** to the **atrioventricular (AV) node** in the inferior part of the interatrial septum, where they are delayed for a fraction of a second (Figure 19.14, ②). After this delay, the impulses race through the **atrioventricular (AV) bundle** (formerly, *bundle of His*), which enters the interventricular septum and divides into right and left **bundle branches,** or *crura* ("legs"). About halfway down the septum, the bundle branches terminate in the **subendocardial conducting network (Purkinje fibers),** which approach the apex of the heart and then turn superiorly into the ventricular walls (Figure 19.14, ③–⑤). This arrangement ensures that the contraction of the ventricles begins at the apex of the heart and travels superiorly, so that ventricular blood is ejected superiorly into the great arteries. The brief delay of the contraction-signaling impulses at the AV node enables the ventricles to fill completely before they start to contract. Because the fibrous cardiac skeleton between the atria and ventricles is nonconducting, it prevents impulses in the atrial wall from proceeding directly to the ventricular wall. As a result, only those signals that go through the AV node can continue on.

Examination of the microscopic anatomy of the heart's conducting system reveals that the cells of the nodes and AV bundle are small, but otherwise typical, cardiac muscle cells. The subendocardial conducting network by contrast, is a long row of special, large-diameter, barrel-shaped cells. These muscle cells contain relatively few myofilaments because they are adapted more for conduction than for contraction. Their large diameter maximizes the speed of impulse conduction. This network of conducting cells is located in the deepest part of the ventricular endocardium, between the endocardium and myocardium layers.

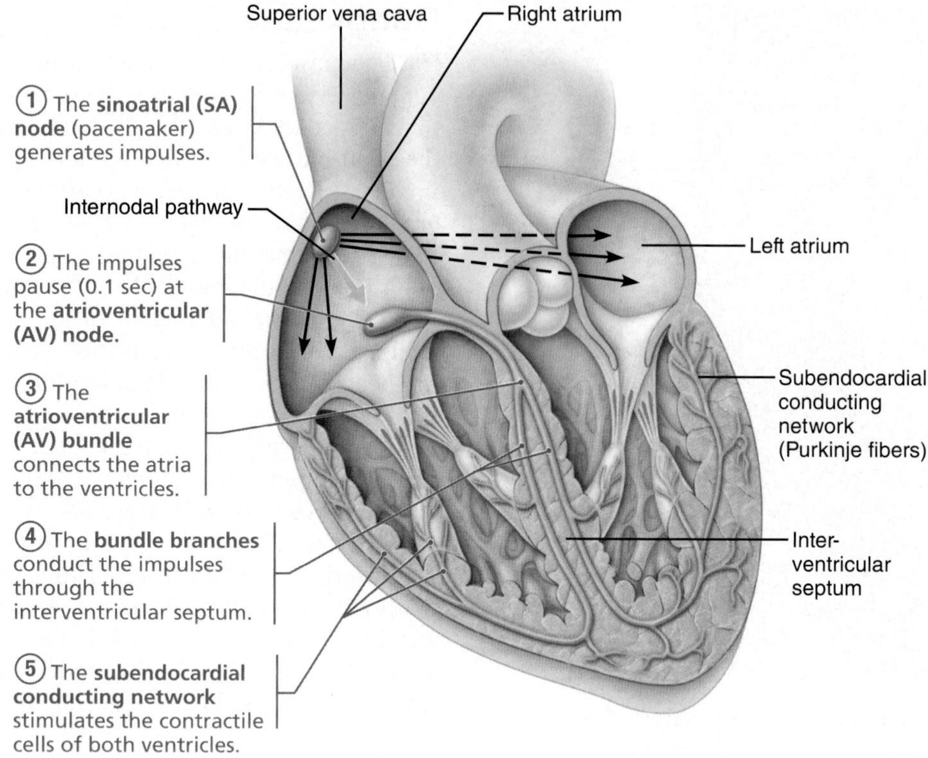

Figure 19.14 The intrinsic conducting system of the heart. This system consists of specialized cardiac muscle cells, not nerves.

CLINICAL APPLICATION

Damage to the Conducting System Because the atria and ventricles are insulated from each other by the electrically inert cardiac skeleton, the only route for impulse transmission from the atria to the ventricles is through the AV node and AV bundle. Therefore, damage to either of these, called a **heart block,** interferes with the ability of the ventricles to receive the pacing impulses.

Without these signals, the ventricles beat at an intrinsic rate that is slower than that of the atria—and too slow to maintain adequate circulation. In such cases, an artificial pacemaker set to discharge at the appropriate rate is usually implanted. For other conditions involving damage to the conducting system, see the section on disorders of the heart (p. 580).

Innervation

Although the heart's inherent rate of contraction is set by the SA node, this rate can be altered by extrinsic neural controls (Figure 19.15). The nerves to the heart consist of *visceral sensory* fibers, *parasympathetic* fibers, and *sympathetic* fibers. All nerves serving the heart pass through the cardiac plexus on the trachea before entering the heart (Figure 15.5, p. 474).

- The **parasympathetic** nerves to the heart arise as branches of the vagus nerve in the neck and thorax. Parasympathetic innervation **decreases heart rate** and is restricted to the SA and AV nodes and the coronary arteries.
- The **sympathetic** nerves travel to the heart from the cervical and upper thoracic chain ganglia (Figure 15.8, p. 479). Sympathetic fibers innervate the same structures as parasympathetic fibers—the SA node, AV node, and coronary arteries. In addition, sympathetic fibers project to the cardiac musculature throughout the heart. Sympathetic innervation of the heart **increases heart rate and strength of contraction.**

The autonomic input to the heart is controlled by *cardiac centers* in the reticular formation of the medulla of the brain (pp. 379–380). In the medulla, the **cardioinhibitory center** influences parasympathetic neurons, whereas the **cardioacceleratory center** influences sympathetic neurons. These medullary cardiac centers, in turn, are influenced by such higher brain regions as the hypothalamus, periaqueductal gray matter, amygdaloid, and insular cortex (discussed in Chapter 13).

check your understanding

- ☐ 9. What is the significance of the gap junctions in the intercalated discs?
- ☐ 10. What is the pacemaker of the heart, and where is it located? What type of tissue forms the structures of the conducting system of the heart?
- ☐ 11. How do the autonomic nerves that innervate the heart influence heart function?

(For answers, see Appendix B.)

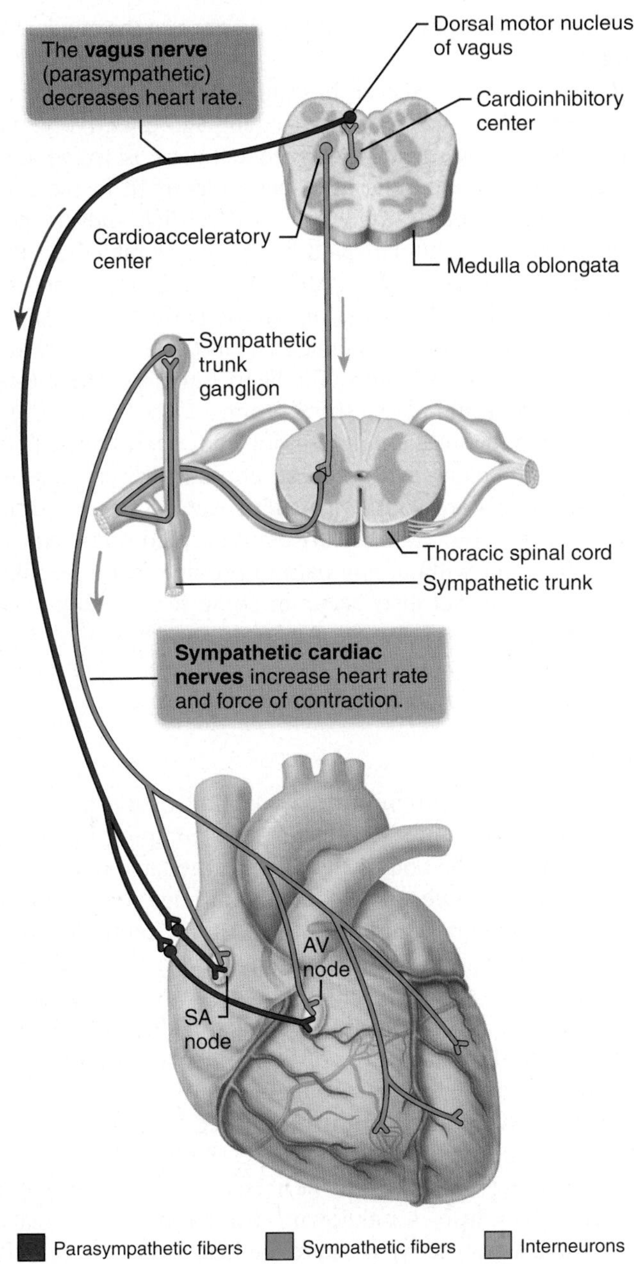

Figure 19.15 Autonomic innervation of the heart.

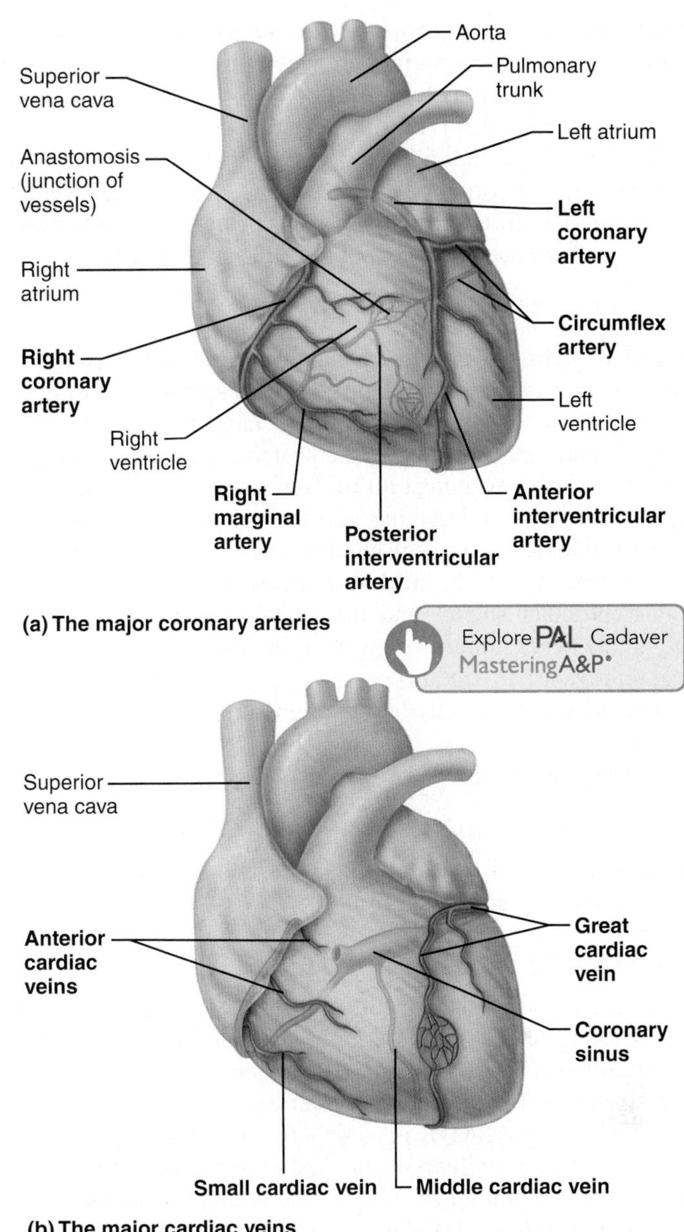

Figure 19.16 Coronary circulation. In both illustrations, the lighter-colored vessels lie more posteriorly on the heart.

BLOOD SUPPLY TO THE HEART

learning outcome

- Describe the locations of the coronary arteries and cardiac veins on the heart surface.

Coronary Arteries

Although the heart is filled with blood, the heart walls are too thick to obtain much nutrition by diffusion from this contained blood. Instead, blood supply to the muscular walls and tissues of the heart is delivered by the right and left *coronary arteries* (Figure 19.16a). These systemic arteries arise from the base of the aorta and run in the coronary sulcus.

The **left coronary artery (LCA)** arises from the left side of the aorta, passes posterior to the pulmonary trunk, then divides into two branches: the *anterior interventricular* and *circumflex arteries.* The **anterior interventricular artery,** commonly referred to clinically as the *left anterior descending artery (LAD),* descends in the anterior interventricular sulcus toward the apex of the heart. This vessel sends branches into and supplies the interventricular septum and anterior walls of both ventricles. The **circumflex artery (Cx)** follows the coronary sulcus posteriorly and supplies the left atrium and the posterior part of the left ventricle.

The **right coronary artery (RCA)** emerges from the right side of the aorta and descends in the coronary sulcus on the anterior surface of the heart, between the right atrium and right ventricle. At the inferior border of the heart, it branches to form the **marginal artery.** Continuing into the posterior part of the coronary sulcus, the right coronary artery gives off a large branch in the posterior interventricular sulcus, the **posterior interventricular artery.** Clinically this vessel is commonly called the *posterior descending artery* (PDA). The right coronary artery and its branches supply the right atrium and much of the right ventricle.

The arrangement of the coronary arteries varies considerably among individuals. For example, in about 15% of people, the left coronary artery gives rise to *both* interventricular arteries. In other people (4%), a single coronary artery emerges from the aorta and supplies the entire heart.

Cardiac Veins

Cardiac veins, which carry deoxygenated blood from the heart wall into the right atrium, also occupy the sulci on the heart surface (Figure 19.16b). The largest of these veins, the **coronary sinus,** occupies the posterior part of the coronary sulcus and returns almost all the venous blood from the heart to the right atrium. Draining into the coronary sinus are three large tributaries: the **great cardiac vein** in the anterior interventricular sulcus, the **middle cardiac vein** in the posterior interventricular sulcus, and the **small cardiac vein** running along the heart's inferior right margin. The anterior surface of the right ventricle contains several horizontal **anterior cardiac veins** that empty directly into the right atrium.

DISORDERS OF THE HEART

learning outcome

- Define coronary artery disease, heart failure, and atrial and ventricular fibrillation.

Coronary Artery Disease

Atherosclerosis is an accumulation of fatty deposits in the inner lining of the body's arteries that can block blood flow through these arteries (see p. 619 for details). When atherosclerosis affects the coronary arteries, it leads to **coronary artery disease (CAD),** in which the arteries supplying the heart wall are narrowed or blocked (See Figure 1.12, p. 17). A common symptom of this disease is **angina pectoris** (an-ji′nah pek′tor-us; "choked chest"), thoracic pain caused by inadequate oxygenation of heart muscle cells, which weaken but do not die. Although the pain of angina usually results directly from tissue hypoxia, it can also result from stress-induced spasms of the atherosclerotic coronary arteries. Angina attacks occur most often during exercise, when the vigorously contracting heart may demand more oxygen than the narrowed coronary arteries can deliver. The onset of angina should be considered a warning sign of other, more serious conditions. Prevention and treatments for CAD are described in A CLOSER LOOK.

If the blockage of a coronary artery is more complete or prolonged, the oxygen-starved cardiac muscle cells die—a condition called **myocardial infarction,** or a heart attack. Few experiences are more frightening than a heart attack: A sharp pain strikes with lightning speed through the chest (and sometimes the left arm and left side of the neck) and does not subside. Death from cardiac arrest occurs almost immediately in about one-third of cases. Heart attacks kill either directly (due to severe weakening of the heart) or indirectly (due to heart-rhythm disruptions caused by damage to the conducting system).

Cardiac muscle tissue has no satellite cells as skeletal muscle does. It does not regenerate effectively, and thus myocardial damage from a heart attack is irreversible. Cardiac muscle stem cells have been identified in the myocardium, and there is evidence that cardiac muscle cells undergo renewal in humans, but at a very low rate—less than 1% annually. Injections of cardiac stem cells and bone marrow stem cells into the hearts of cardiac patients have shown some success in replacing damaged heart tissue with new cardiac muscle tissue and improving heart function. These treatments, in early trial phases, show promise.

For many people, a painless but fatal heart attack is the first and only clear symptom of coronary artery disease. Such individuals are victims of **silent ischemia,** a condition in which blood flow to the heart is interrupted often, exactly as in angina, but without any pain to provide warning. Silent ischemia, which can also occur in some heart-attack survivors, can be detected by measuring heart rhythm through electrocardiography (ECG) during exercise, a procedure called a graded exercise test (or treadmill test).

Heart Failure

Heart failure is a progressive weakening of the heart as it fails to keep pace with the demands of pumping blood and thus cannot meet the body's need for oxygenated blood. This condition may be due to weakened ventricles (damaged, for example, by a heart attack), to failure of the ventricles to fill completely during diastole (for example, as a result of mitral valve stenosis), or to overfilling of the ventricles (for example, because of stenosis or insufficiency of the aortic valve). Most commonly, however, congestive heart failure is the cause of this progressive weakening.

In **congestive heart failure,** the heart enlarges greatly while its pumping efficiency progressively declines. This condition affects 5 million Americans and is increasing in frequency. Its cause is unknown. One hypothesis is that it may involve a destructive positive feedback loop: An initially weakened heart causes the sympathetic nervous system to stimulate the heart to pump harder. This increased demand further weakens the heart, which again causes stimulation of the sympathetic system, and so on.

A condition called **pulmonary arterial hypertension** is the enlargement and sometimes ultimate failure of the right ventricle resulting from elevated blood pressure in the pulmonary circuit. A blockage or constriction of the vessels in the lungs increases resistance to blood flow, which increases blood pressure and forces the right ventricle to work harder. Whereas acute cases of pulmonary hypertension may develop suddenly from an embolism in the pulmonary vessels, chronic cases are usually associated with chronic lung diseases such as emphysema.

Disorders of the Conduction System

This section covers a type of *arrhythmia*—a variation from the normal rhythm of the heartbeat—called *fibrillation*.

Coronary Artery Disease

Out for her morning walk, Barbara, 63, started to feel sharp, squeezing pains in her chest. The pain subsided when she sat down and rested. Her immediate thought was heart attack. A trip to the emergency room showed that fortunately there was no damage to the heart muscle. Barbara had suffered an angina attack, a clear sign of coronary artery disease. Coronary artery disease (CAD) is caused by a buildup of fatty plaque (atherosclerosis) in the coronary arteries, resulting in decreased blood supply to the myocardium of the heart and diminished heart function.

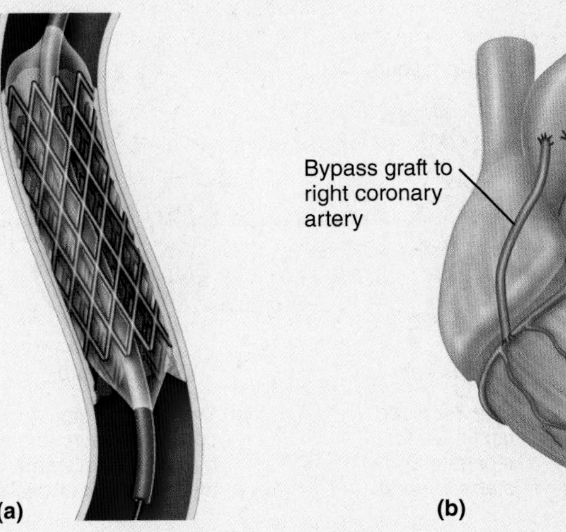

a Angioplasty with insertion of a stent. **b** Coronary artery bypass graft (CABG) performed on two vessels; double bypass surgery.

Once coronary artery disease is detected, the first step is to reduce risk factors through lifestyle modification. Risk factors include high blood pressure, smoking, high cholesterol levels, diabetes, inactivity, and family history of the disease. Changes in diet and exercise routine are primary. Medication may be required to control risk factors, decrease workload on the heart, and increase blood flow through coronary vessels.

When medication and lifestyle changes are inadequate to maintain blood supply to the heart, surgical interventions may be needed. For many years, coronary artery bypass graft (CABG) was the only intervention available. The first bypass occurred in 1967 and was a standard, although risky, treatment requiring major surgery. Other, less invasive treatments have been developed in the last 25 years.

Angioplasty (percutaneous transluminal coronary angioplasty, PTCA), developed in the late 1970s, involves inserting a catheter into the patient's arm or leg and threading it through the arteries until it enters the blocked coronary artery. When the catheter reaches the blockage, a balloon located on the tip of the catheter is inflated to exert pressure on the plaque-filled wall of the vessel. This pressure compresses the plaque and increases the diameter of the vessel lumen, thus improving blood flow to the myocardium.

This procedure revolutionized the treatment of coronary artery disease. However, in some cases angioplasty weakened the wall of the coronary vessel, causing collapse of the coronary artery after the balloon was deflated. In 1993 the use of stents to prevent arterial collapse and reblockage was approved by the FDA. Following the opening of the artery with balloon angioplasty, a metal mesh tube called a bare metal stent (BMS) is inserted into the artery at the site of blockage (see figure a). The stent serves as scaffolding to hold the arterial lumen open. Anticlotting medications are prescribed for at least 6 months after stent insertion to decrease the risk of thrombosis.

Stents improved the success of angioplasty; however, restenosis, the thickening of the vessel wall in response to the stent, continues to be a challenge for some patients. The endothelial cells and smooth muscle cells of the wall grow and divide. This, along with inflammation, results in thickening of the arterial wall and diminished blood flow through the artery.

Drug-eluting stents, first approved in 2003, slowly release drugs that disrupt various portions of the cell cycle. These drugs prevent cells in the vessel wall from proliferating in response to the stent and thus prevent restenosis.

These catheterization procedures are much less invasive than coronary bypass surgery and are appropriate treatments for many individuals. However, some patients may have a better outcome with a coronary artery bypass graft (CABG) procedure. In coronary bypass, vessels from another part of the body are used to reroute blood to the heart. A portion of a vessel is removed from its original location and grafted between the aorta and the heart wall, thus supplying an alternate route for blood to the heart muscle (see figure b). Vessels commonly used for bypass surgery are the internal thoracic artery, which normally serves the anterior body wall; the saphenous vein from the leg; and the radial artery in the forearm.

Most important, the long-term success of any of these treatments depends on lifestyle changes by the patient. Better knowledge of lifestyle factors and improved medical intervention, both pharmacological and surgical, have influenced the prevention of and the prognosis for living with coronary artery disease.

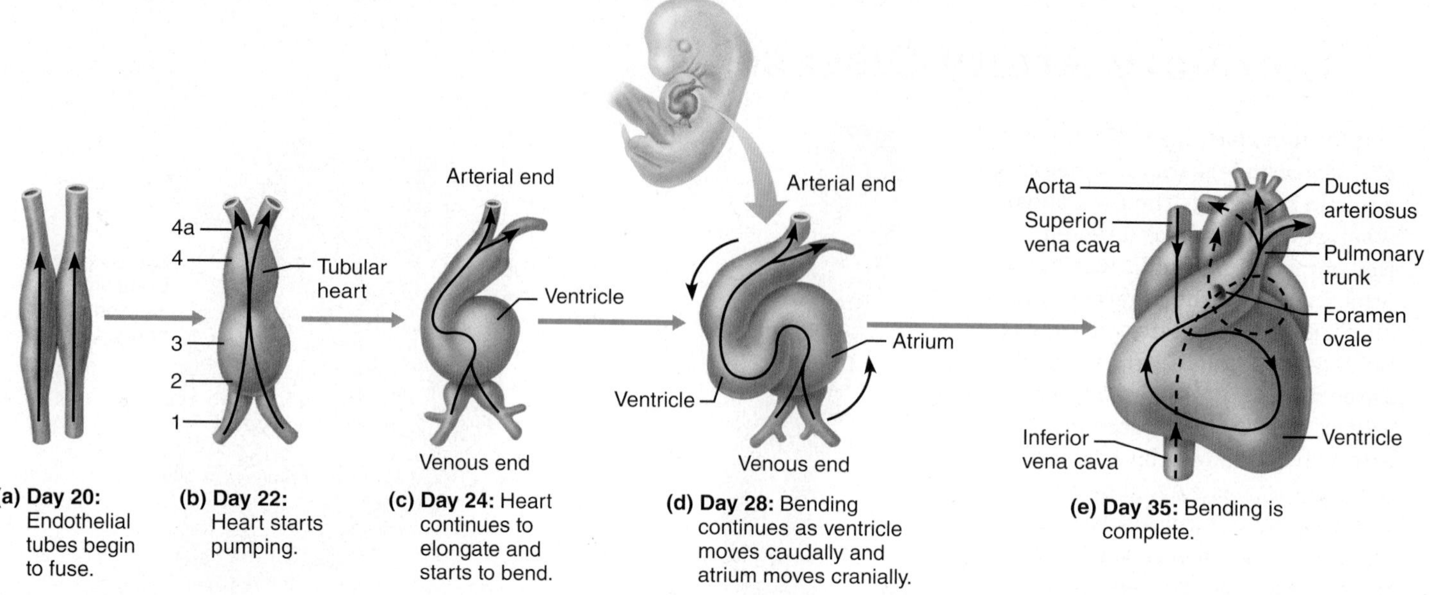

Figure 19.17 Heart development. Ventral views, with the cranial direction toward the top of the figures. Arrows show the direction of blood flow. Days are approximate. In (b), 1 is the sinus venosus; 2, the atrium; 3, the ventricle; 4, the bulbus cordis; and 4a, the truncus arteriosus.

In **ventricular fibrillation,** the ventricles are unable to pump blood into the arteries because rapid, random firing of electrical impulses within ventricular cardiac muscle prevents coordinated contraction of the ventricle. The fibrillating ventricles can be likened to a quivering bag of worms. Ventricular fibrillation results from a crippled conducting system and is the most common cause of cardiac arrest and sudden death in patients with hearts damaged by coronary artery disease.

In **atrial fibrillation,** multiple waves of impulses circle within the atrial myocardium, randomly stimulating the AV node, which then signals the ventricles to contract quickly and irregularly. The resulting lack of smooth movement of blood through the heart can promote the formation of clots, parts of which can break off, reach the brain, and cause strokes. The fibrillations are usually discontinuous, occurring in episodes characterized by palpitations (sensations of an unduly quick and irregular heartbeat), anxiety, fatigue, and shortness of breath. The cause of this condition, which affects 5% of individuals over age 65 and 10% over 75, is unknown, but it is often associated with coronary artery or heart-valve disease. Intermittent rhythm disturbances can be identified by ECG monitoring during normal activity, a procedure called an ambulatory ECG or Holter monitoring. Treatment typically involves drug therapy and administration of anticoagulants to prevent clotting.

☑ check your understanding

- ☐ 12. Which vessel supplies blood to the left ventricle?
- ☐ 13. What are the risk factors for coronary artery disease? What is one common symptom of this disorder?

(For answers, see Appendix B.)

THE HEART THROUGHOUT LIFE

learning outcomes

- ▶ Explain how the heart develops, and describe some congenital heart defects.
- ▶ List some effects of aging on the heart.

Development of the Heart

An understanding of heart development is clinically important because congenital abnormalities of the heart account for nearly half of all deaths from birth defects. One of every 150 newborns has some congenital heart defect.

All blood vessels begin as condensations of mesodermal mesenchyme called blood islands (Chapter 18, p. 557). Subsequently, the blood islands destined to become the heart form in the splanchnic mesoderm around the future head and neck of the embryonic disc. The heart folds neatly into the thorax region when the flat embryonic disc lifts up off the yolk sac to assume its three-dimensional body shape around day 20 or 21 (this folding is described on p. 54).

When the embryonic heart first reaches the thorax, it is a pair of endothelial tubes in the body midline. These tubes fuse into a single tube on about day 20 (Figure 19.17a). The heart starts pumping about day 22, by which time four bulges have developed along the heart tube (Figure 19.17b). These bulges are the earliest heart chambers and are unpaired. From tail to head, following the direction of blood flow, the four chambers are the *sinus venosus, atrium, ventricle,* and *bulbus cordis:*

1. **Sinus venosus** (ven-o′sus; "of the vein"). This chamber, which initially receives all blood from the veins of the embryo, will become the smooth-walled part of the right

atrium and the coronary sinus; it also gives rise to the sinoatrial node. Furthermore, recent evidence indicates that the sinus venosus also contributes to the back wall of the *left* atrium (which mostly derives from the bases of the developing pulmonary veins).

2. **Atrium.** This embryonic chamber eventually becomes the ridged parts of the right and left atria—specifically, the parts lined by pectinate muscles.
3. **Ventricle.** The strongest pumping chamber of the early heart, the embryonic ventricle gives rise to the *left* ventricle.
4. **Bulbus cordis.** This chamber and its most cranial extension, the *truncus arteriosus,* give rise to the pulmonary trunk and first part of the aorta. The bulbus cordis also gives rise to the *right* ventricle.

At the time these four chambers appear, the heart starts bending into an S shape (Figure 19.17c and d). The ventricle moves caudally and the atrium cranially, assuming their adult positions. This bending occurs because the ventricle and bulbus cordis grow quickly, and the heart is unable to accommodate elongation within the confines of the pericardial sac.

During month 2 of development, the heart divides into its four definitive chambers by the formation of its midline septum and valves. These structures develop from cardiac cushions, which are regional thickenings of the endocardium, the inner lining of the heart wall. For example, most of the interatrial septum forms by growing caudally from the heart's roof, and most of the interventricular septum forms by growing cranially from the heart's apex. Additionally, neural crest cells (pp. 52 and 55) migrate into the area where atrium meets ventricle. These cells contribute to the developing heart valves and to the bases of the pulmonary trunk and ascending aorta, the two great arteries that form by the splitting of the bulbus cordis. These month-2 events are so complex that perfect orchestration does not always occur, and developmental defects result.

Most other details of heart development are beyond the scope of this book, but one thing should be mentioned: The two atria remain interconnected by a hole in the interatrial septum—the foramen ovale—until birth, at which time this hole closes to become the fossa ovalis (Figure 19.5c). The foramen ovale plays an important role in blood circulation before birth, and is discussed in the section on fetal circulation in the next chapter (p. 622).

Most of the common congenital heart defects can be traced to month 2 of development. The most common of these is a **ventricular septal defect** (Figure 19.18a), in which the superior (cranial) region of the interventricular septum fails to form, leaving a hole between the two ventricles. As you study this and the other defects presented (Figure 19.18b and c), note that they produce two basic kinds of effects in newborns: Either inadequately oxygenated blood reaches the body tissues because oxygen-poor systemic blood mixes with oxygenated pulmonary blood; or the ventricles labor under an increased workload because of narrowed valves and vessels; or both effects occur. Modern surgical techniques can usually correct these congenital defects.

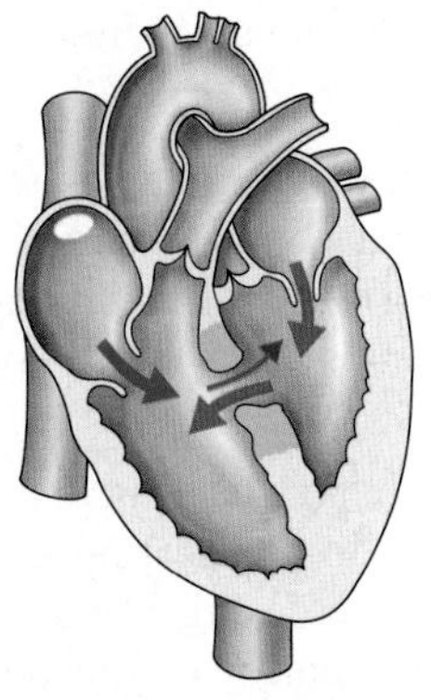
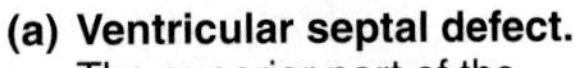

(a) Ventricular septal defect. The superior part of the interventricular septum fails to form; thus, blood mixes between the two ventricles. More blood is shunted from left to right because the left ventricle is stronger.

Occurs in about 1 in every 500 births

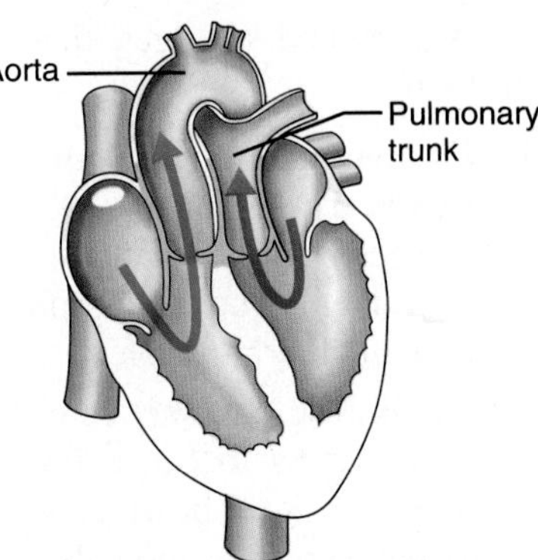

(b) Transposition of the great vessels. Aorta comes from right ventricle; pulmonary trunk from left. Results when the bulbus cordis does not divide properly. Unoxygenated blood passes repeatedly around systemic circuit, while oxygenated blood recycles around the pulmonary circuit.

Occurs in about 1 in every 1000 births

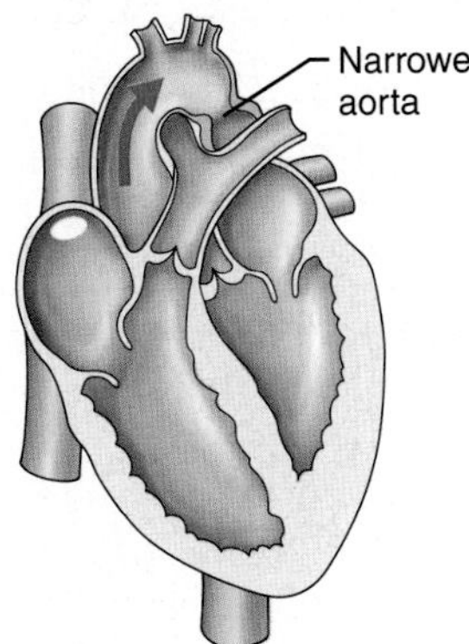

(c) Coarctation of the aorta. A part of the aorta is narrowed, increasing the workload of the left ventricle.

Occurs in about 1 in every 1500 births

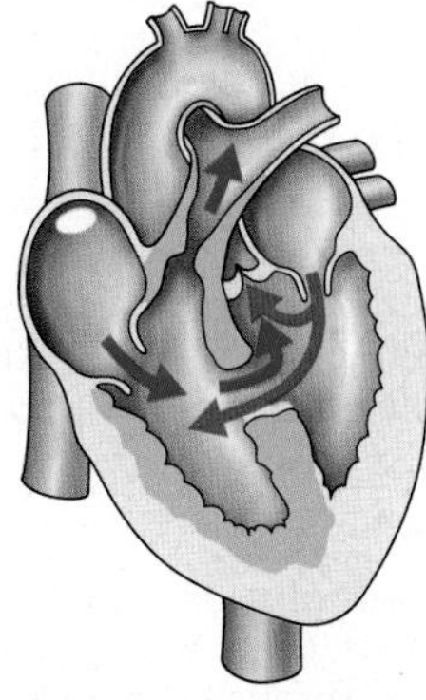

(d) Tetralogy of Fallot. Multiple defects (*tetra* = four): (1) Pulmonary trunk too narrow and pulmonary valve stenosed, resulting in (2) hypertrophied right ventricle; (3) ventricular septal defect; (4) aorta opens from both ventricles.

Occurs in about 1 in every 2000 births

Figure 19.18 **Congenital Heart Defects.** The defects are shown according to relative frequency of occurrence, from the most frequent to the least. Tan areas indicate the locations of the defects.

The Heart in Adulthood and Old Age

In the absence of congenital heart problems, the resilient heart usually functions well throughout life. In individuals who exercise regularly and vigorously, the heart gradually adapts to the increased demand by increasing in strength and size. Aerobic exercise also helps clear fatty deposits from the walls of the coronary vessels, thereby retarding the process of atherosclerosis. Barring some chronic illness, this beneficial response to exercise persists into old age.

Age-related changes that affect the heart include the following:

1. **Hardening and thickening of the cusps of the heart valves.** These changes occur particularly in those valves subjected to the highest blood pressures (mitral and aortic valves). Thus, abnormal heart sounds are more common in elderly people.
2. **Decline in cardiac reserve.** Although the passage of years seems to cause few changes in the resting heart rate, the aged heart is slower at increasing its output to pump more blood. Sympathetic control of the heart becomes less efficient, and heart rate gradually becomes more variable. Maximum heart rate declines, although this problem is much less severe in older adults who are physically active.
3. **Fibrosis of cardiac muscle.** As one ages, more and more cardiac muscle cells die and are replaced by fibrous scar tissue. This fibrosis, which is much more extensive in men than in women, lowers the maximum amount of blood that the heart can pump per unit time. Also, in conjunction with the aging of muscle-cell membranes, fibrosis hinders the initiation and transmission of contraction-signaling impulses, leading to abnormal heart rhythms and other conduction problems.

✓ check your understanding

☐ 14. How would incomplete formation of the interventricular septum alter blood flow through the heart?

☐ 15. Which chamber of the heart is formed from the embryonic ventricle?

☐ 16. What is the single most important factor for maintaining a healthy heart throughout life?

(For answers, see Appendix B.)

RELATED CLINICAL TERMS

Asystole (a-sis′to-le; *a* = without) Failure of the heart to contract.

Cardiac catheterization Diagnostic procedure in which a fine catheter (tube) is passed from a blood vessel at the body surface to the heart. Blood-oxygen content, blood pressure, and blood flow are measured, and heart structures can be visualized. Findings help detect problems with the heart valves, heart deformities, and other cardiac malfunctions.

Cardiomyopathy (kar′de-o-mi-op′ah-the) Any disease of the myocardium that weakens the heart's ability to pump.

Echocardiography (ek″o-kar″de-og′rah-fe) Ultrasound imaging of the heart; used not only for imaging but also for measuring blood flow through the heart.

Endocarditis (en″do-kar-di′tis) Inflammation of the endocardium, usually confined to the endocardium of the heart valves. Endocarditis often results from infection by bacteria that have entered the bloodstream, but may result from fungal infections or an autoimmune response. Drug addicts may develop endocarditis by injecting themselves with contaminated needles. Additionally, the bacteria can enter during routine dentistry and ear-piercing procedures.

Myocarditis (mi′o-kar-di′tis) Inflammation of the heart's myocardium. Sometimes follows an untreated streptococcal infection in children; may be extremely serious because it can weaken the heart and impair its ability to pump blood.

Percussion Tapping the thorax or abdomen wall with the fingertips and using the nature of the resulting sounds to estimate the location, density, and size of the underlying organs. Percussion of the thoracic wall can be used to estimate the size of a patient's heart.

CHAPTER SUMMARY

Take full advantage of the invaluable online practice and study tools for this chapter by going to the Study Area of MasteringA&P®. There you will find:

- Art Labeling Activities, Quizzes, Practice Tests, Memory Games
- Animal Organ Dissection, Cat Dissection and Bone Videos
- PAL Cadaver
- PAL Histology
- A&P Flix Animations
- Dynamic Study Modules

1. The heart is a double pump whose right side pumps blood to the lungs for oxygenation and whose left side pumps blood throughout the body to nourish body tissues; that is, the right side is the pump of the pulmonary circuit, and the left side is the pump of the systemic circuit.
2. The four chambers of the heart are right and left atria (receiving chambers) and right and left ventricles (main pumping chambers).

LOCATION AND ORIENTATION WITHIN THE THORAX (pp. 563–564)

3. The cone-shaped human heart, about the size of a fist, lies obliquely within the mediastinum. Its apex points anteriorly and to the left. Its base is its posterior surface.
4. From an anterior view, the heart is said to have four corners. (The locations of its four corner points are described on p. 564.)

STRUCTURE OF THE HEART (pp. 565–571)

Coverings (p. 565)

5. The pericardium encloses the heart. It consists of a superficial layer (fibrous pericardium and parietal layer of the serous pericardium),

and a deeper layer that covers the heart surface (visceral layer of the serous pericardium, or epicardium). The pericardial cavity, between the two layers, contains lubricating serous fluid.

Layers of the Heart Wall (pp. 565–566)

6. The layers of the heart wall, from external to internal, are the epicardium, the myocardium (which consists of cardiac muscle tissue), and the endocardium (endothelium and connective tissue).

Heart Chambers (pp. 566–571)

7. On the external surface of the heart, several sulci separate the four heart chambers: the coronary sulcus and the anterior and posterior interventricular sulci. Internally, the right and left sides of the heart are separated by the interatrial and interventricular septa.

8. The right atrium has the following features: right auricle and pectinate muscles anteriorly; a smooth-walled posterior part; crista terminalis; openings of the coronary sinus and the superior and inferior venae cavae; and fossa ovalis.

9. The right ventricle contains the following features: trabeculae carneae, papillary muscles, chordae tendineae, and right atrioventricular (tricuspid) valve. The pulmonary valve lies at the base of the pulmonary trunk.

10. The left atrium has a large, smooth-walled, posterior region into which open the four pulmonary veins. Anteriorly, its auricle is lined by pectinate muscles.

11. The left ventricle, like the right ventricle, contains papillary muscles, chordae tendineae, trabeculae carneae, and an atrioventricular valve (mitral). The aortic valve lies at the base of the aorta.

HEART VALVES (pp. 571–573)

Valve Structure (p. 571)

12. The four heart valves are the atrioventricular valves (tricuspid and mitral) and the semilunar valves (aortic and pulmonary). All except the mitral value have three cusps.

13. The cardiac skeleton surrounds the valves between the atria and ventricles. It anchors the valve cusps, is the point of insertion of the heart musculature, and blocks the direct spread of electrical impulses from the atria to the ventricles.

Valve Function (p. 571)

14. The atrioventricular valves prevent backflow of blood into the atria during contraction of the ventricles. The pulmonary and aortic semilunar valves prevent backflow into the ventricles during relaxation of the ventricles.

Heart Sounds (pp. 571–573)

15. Using a stethoscope, physicians can listen to the sounds produced by the closing atrioventricular and semilunar valves. Each valve is best heard near a different heart corner on the anterior chest wall (see Figure. 19.9).

PAL Human Cadaver/Cardiovascular System/Heart

PATHWAY OF BLOOD THROUGH THE HEART (pp. 573–575)

16. A drop of blood circulates along the following path: right atrium to right ventricle to pulmonary circuit to left atrium to left ventricle to systemic circuit to right atrium.

17. In each heartbeat, both atria contract together, followed by simultaneous contraction of both ventricles.

18. The wall of the left ventricle is thicker than that of the right ventricle, resulting in higher pressure in the systemic circuit.

CARDIAC MUSCLE TISSUE (pp. 575–577)

Structure of Cardiac Muscle (pp. 575–576)

19. Cardiac muscle tissue forms the thick myocardium of the heart wall. It contains cardiac muscle cells and connective tissues that surround these cells.

20. Cardiac muscle cells are short, branching, striated cells with one or two central nuclei. They contain myofibrils made of typical sarcomeres. Cardiac muscle contracts by the sliding filament mechanism.

21. Adjacent cardiac cells are connected by intercalated discs to form cardiac myofibers. Intercalated discs contain two types of junctions: desmosome-like fasciae adherens and gap junctions. Gap junctions between cells allow the direct transmission of an electrical impulse through the entire network of cardiac muscle cells.

22. Cardiac muscle cells have many mitochondria and depend on aerobic respiration to form ATP. They are very resistant to fatigue.

PAL Histology/Cardiovascular System

Mechanism of Contraction (pp. 576–577)

23. Contraction of cardiac muscle is triggered by Ca^{2+}. Calcium enters from the extracellular fluid and signals the release of additional calcium from the sarcoplasmic reticulum. The sarcoplasmic reticulum and T tubules are somewhat simpler than those of skeletal muscle.

24. Cardiac muscle cells contract with their own inherent rhythm, the basis of the heartbeat.

PAL Human Cadaver/Cardiovascular System/Heart

CONDUCTING SYSTEM AND INNERVATION (pp. 577–579)

Conducting System (pp. 577–578)

25. The conducting system of the heart is an interconnected series of specialized cardiac muscle cells that initiates each heartbeat, sets the basic rate of the heartbeat, and coordinates the contraction of the heart chambers. The impulse that signals heart contraction travels from the SA node (the pacemaker) through the atrial myocardium and internodal pathway to the AV node, to the atrioventricular bundle, bundle branches, subendocardial conducting network and the ventricular musculature.

Innervation (pp. 578–579)

26. Heart innervation consists of visceral sensory fibers, heart-slowing vagal parasympathetic fibers, and sympathetic fibers that increase heart rate and force of contraction.

BLOOD SUPPLY TO THE HEART (pp. 579–580)

Coronary Arteries (pp. 579–580)

27. The coronary vessels that supply the heart wall are located in the coronary sulcus and interventricular sulci. The right and left coronary arteries branch from the base of the aorta. The right coronary artery and its branches supply the right atrium and much of the right ventricle. The left coronary artery and its branches supply the left atrium, left ventricle, and the anterior wall of the right ventricle.

Cardiac Veins (p. 580)

28. Venous blood, collected by the cardiac veins (great, middle, small, and anterior), empties into the coronary sinus, which opens into the right atrium.

DISORDERS OF THE HEART (pp. 580–582)

29. Coronary artery disease is caused by atherosclerotic blockage of the coronary arteries, and can lead to angina pectoris and myocardial

infarction (heart attack). Heart failure is a weakening of the heart, resulting in its inability to pump blood fast enough to meet the body's needs. Atrial and ventricular fibrillations are rapid, irregular, and uncoordinated contractions of the respective heart chambers, resulting from damage to the conducting system.

THE HEART THROUGHOUT LIFE (pp. 582–584)

Development of the Heart (pp. 582–583)

30. The heart develops from splanchnic mesoderm around the head and neck of the embryonic disc. When the heart folds into the thorax, it is a double tube that fuses into one and starts pumping blood (day 22). It bends into an S shape around day 24. Its four earliest chambers are the sinus venosus, atrium, ventricle, and bulbus cordis.

31. The four final heart chambers are defined during month 2 through the formation of valves and dividing walls. Failures in this complex process account for most congenital defects of the heart (see Figure 19.18).

The Heart in Adulthood and Old Age (pp. 583–584)

32. Age-related changes in the heart include hardening and thickening of the valve cusps, decline in cardiac reserve, fibrosis of cardiac muscle, and atherosclerosis.

REVIEW QUESTIONS

Multiple Choice/Matching Questions

(For answers, see Appendix B.)

1. The most external part of the pericardium is the (a) parietal layer of serous pericardium, (b) fibrous pericardium, (c) visceral layer of serous pericardium, (d) pericardial cavity.

2. Which heart chamber forms most of the heart's inferior surface? (a) right atrium, (b) right ventricle, (c) left atrium, (d) left ventricle.

3. How many cusps does the right atrioventricular valve have? (a) two, (b) three, (c) four.

4. The sequence of contraction of the heart chambers is (a) random, (b) left chambers followed by right chambers, (c) both atria followed by both ventricles, (d) right atrium, right ventricle, left atrium, left ventricle.

5. The middle cardiac vein runs with which artery? (a) marginal artery, (b) aorta, (c) coronary sinus, (d) anterior interventricular artery, (e) posterior interventricular artery.

6. The base of the heart (a) is its posterior surface, (b) lies on the diaphragm, (c) is the same as its apex, (d) is its superior border.

7. Which of the following is an *incorrect* statement about the crista terminalis of the right atrium? (a) It separates the smooth-walled part from the part with pectinate muscles. (b) It is shaped like the letter C. (c) The coronary sinus and inferior vena cava open near its inferior part. (d) It lies mostly in the interatrial septum.

8. The aortic valve closes (a) at the same time the mitral valve closes, (b) just after the atria contract, (c) just before the ventricles contract, (d) just after the ventricles contract.

9. The ventricle of the embryonic heart gives rise to what adult structure(s)? (a) bulbus cordis, (b) both ventricles, (c) left ventricle, (d) the aorta, (e) none of these.

10. Which layer of the heart wall is the thickest? (a) endocardium, (b) myocardium, (c) epicardium, (d) endothelium.

11. The inferior left corner of the heart is located at the (a) second rib slightly lateral to the sternum, (b) third rib at the sternum, (c) sixth rib slightly lateral to the sternum, (d) fifth intercostal space at the midclavicular line.

Short Answer Essay Questions

12. Ben was annoyed when the teaching assistant who ran the discussion section of his anatomy class said that the atria pump blood to the lungs and that the ventricles pump blood throughout the body. Correct this error.

13. Describe the location of the heart within the thorax.

14. Trace a drop of blood through all the heart chambers and heart valves, and through the basic vascular circuits, from the time it enters the left atrium until it returns to the left atrium again.

15. (a) Name the elements of the heart's conducting system in order, beginning with the pacemaker. (b) Is the conducting system made of nerves? Explain. (c) What are the functions of this conducting system?

16. Sketch the heart and draw all the coronary vessels in their correct locations. (Alternatively, you could locate these vessels on a realistic diagram of the heart.)

17. On a diagram of a frontally sectioned heart, indicate the location of the cardiac skeleton.

18. When you view a heart's *anterior* surface, which of its four chambers appears largest?

19. Make a drawing of the adult heart and the associated large vessels. Then color and label the adult regions that derive from each *embryonic* heart chamber: (a) sinus venosus, (b) embryonic atrium, (c) embryonic ventricle, (d) bulbus cordis.

20. How do the right and left ventricles differ structurally, and how do these structural differences reflect functional differences?

21. Which is more resistant to fatigue, cardiac muscle or skeletal muscle? What is the anatomical basis for this difference, and why is it important?

22. Describe the structure and function of an intercalated disc.

23. Compare and contrast the structure of cardiac muscle tissue with skeletal muscle tissue.

Critical Reasoning & Clinical Application Questions

1. Classify the three congenital heart defects—ventricular septal defect, coarctation of the aorta, and tetralogy of Fallot (Figure 19.18)—according to whether they produce (1) mixing of oxygenated and unoxygenated blood, (2) increased workload for the ventricles, or (3) both of these problems.
2. Ms. Hamad, who is 73 years old, is admitted to the coronary care unit of a hospital with a diagnosis of left ventricular failure resulting from a myocardial infarction. Her heart rhythm is abnormal. Explain what a myocardial infarction is, how it is likely to have been caused, and why the heart rhythm is affected.
3. You have been called on to demonstrate where to listen for heart sounds. Explain where on the chest wall you would place a stethoscope to listen for (a) incompetence of the aortic valve and (b) stenosis of the mitral valve.
4. After a man was stabbed in the chest, his face became blue and he lost consciousness from lack of oxygenated blood to the brain. The diagnosis was cardiac tamponade rather than severe blood loss through internal bleeding. What is cardiac tamponade, how did it cause the observed symptoms, and how is it treated?
5. A heroin addict felt tired, weak, and feverish, with vague pains and aches. Finally, he went to a walk-in clinic, terrified that he had AIDS. Instead, the doctor found an abnormal heart sound, and the final diagnosis was not AIDS but endocarditis. What is the most likely way that this patient contracted endocarditis?
6. Another patient had an abnormal heart sound that indicated a stenotic valve. Define this condition, and contrast it with an incompetent heart valve.
7. Assume that the four corner points of a patient's heart, as revealed on an X-ray film, relate to the rib cage as follows: second right rib at sternum, tip of xiphoid process, seventh intercostal space lateral to the left nipple, and second left rib at sternum. Are the size and position of this patient's heart normal?
8. During a lethal heart attack, a blood clot lodges in the first part of the circumflex branch of the left coronary artery, blocking blood flow through this vessel. What regions of the heart will become ischemic and die?
9. Blood from the ventricles is "squeezed" out of the heart from the apex toward the great arteries. What features of the heart contribute to this sequence of contraction?

20 Blood Vessels

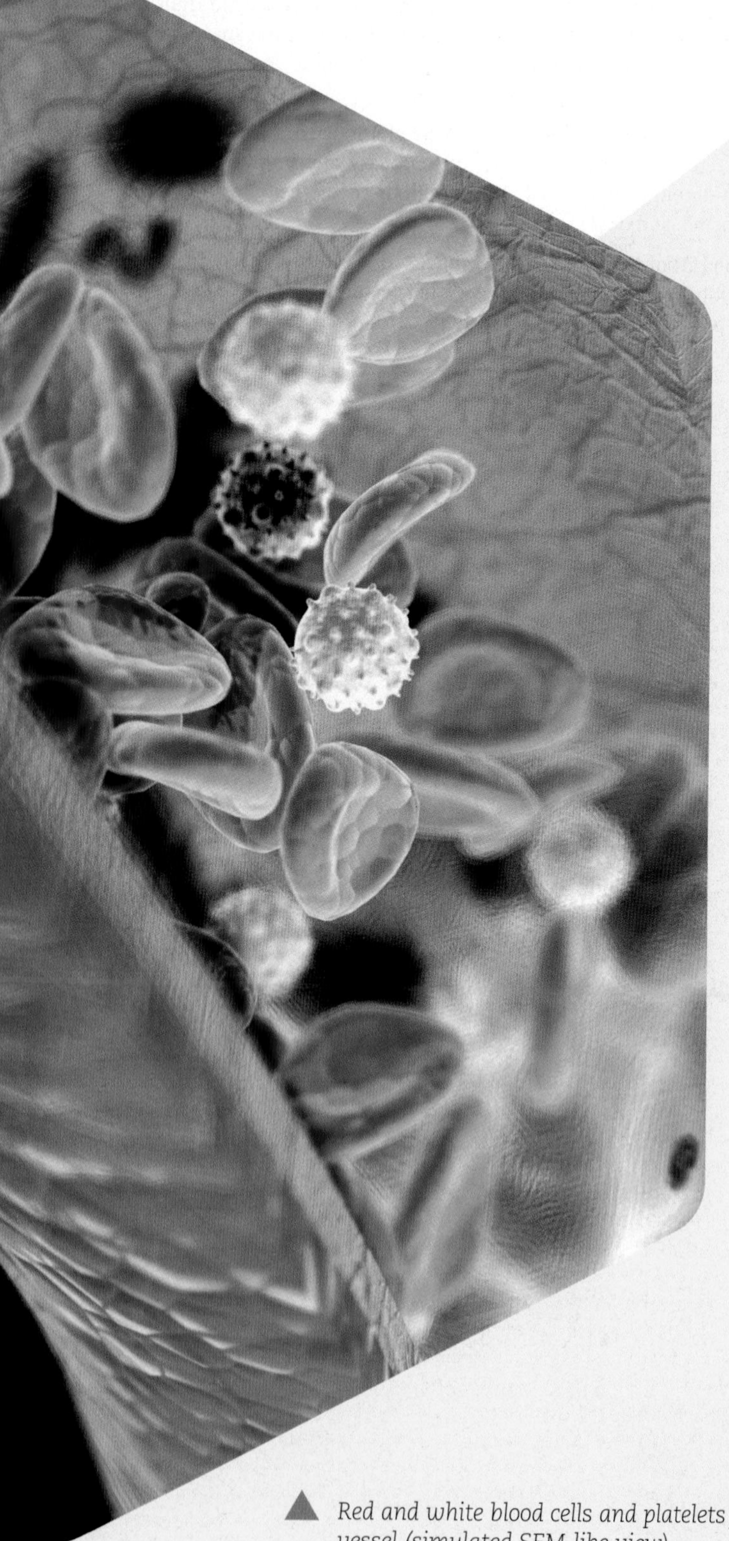

The blood vessels of the body form a closed delivery system powered by the pumping heart. The realization that blood is pumped in a circle dates back to the 1620s and is based on the careful experiments of William Harvey, an English physician. Before that time, it was taught—as proposed by the ancient Greek physician Galen—that blood moved through the body like an ocean tide, first moving out from the heart, then ebbing back into the heart via the same vessels.

Blood vessels are not rigid tubes that simply direct the flow of blood but rather are dynamic structures that pulsate, constrict and relax, and even proliferate, according to the changing needs of the body. This chapter examines the structure and function of these important circulatory pathways through the body.

The three major types of blood vessels are *arteries, capillaries,* and *veins.* When the heart contracts, it forces blood into the large arteries that leave the ventricles. The blood then moves into successively smaller branches of arteries, finally reaching the smallest

▲ *Red and white blood cells and platelets flowing through a blood vessel (simulated SEM-like view).*

branches, the *arterioles* (ar-te′re-ōlz; "little arteries"), which feed into the *capillaries* of the organs. Blood leaving the capillaries is collected by *venules,* small veins that merge to form larger veins that ultimately empty into the heart. This pattern of vessels applies to both the pulmonary and systemic circuits. Altogether, the blood vessels in an adult human body stretch for 100,000 km or 60,000 miles, a distance equivalent to almost 2½ times around the Earth.

Notice that arteries are said to "branch," "diverge," or "fork" as they carry blood *away* from the heart. Veins, by contrast, are said to "join," "merge," "converge," or "serve as tributaries" as they carry blood *toward* the heart.

This chapter is divided into two parts: Part One considers the general characteristics of blood vessels, and Part Two discusses the specific vessels in the pulmonary and systemic circulations.

PART 1 GENERAL CHARACTERISTICS OF BLOOD VESSELS

The first part of this chapter begins by examining the structure of the blood vessel walls and then describes the structure and function of each type of blood vessel.

STRUCTURE OF BLOOD VESSEL WALLS

learning outcomes

- Describe the three tunics that form the wall of an artery or vein.
- Define vasa vasorum.

The walls of all blood vessels, except the very smallest, are composed of three distinct layers, or *tunics* ("cloaks")—the tunica intima, tunica media, and tunica externa—that surround the central blood-filled space, the **lumen** (Figure 20.1). The description of these tunics covers features that are common to arteries as well as veins.

The innermost tunic of a vessel wall is the **tunica intima** (too′nĭ-kah in′tĭ-mah), which is in "intimate" contact with the blood in the lumen. This tunic contains the **endothelium,** the simple squamous epithelium that lines the lumen of all vessels. The flat endothelial cells form a smooth surface that minimizes the friction of blood moving across them. In vessels larger than about 1 mm in diameter, a thin layer of loose connective tissue, the **subendothelial layer,** lies just external to the endothelium.

The middle tunic, or **tunica media,** consists primarily of *circularly* arranged sheets of smooth muscle fibers, between which lie circular sheets of elastin and collagen fibrils. Contraction of the smooth muscle cells decreases the diameter of the vessel, a process called *vasoconstriction.* Relaxation of the smooth muscle cells, a process called *vasodilation,* increases the vessel's diameter. Both activities are regulated by *vasomotor nerve fibers* of the sympathetic division of the autonomic nervous system. The elastin and collagen contribute elasticity and strength for resisting the blood pressure that each heartbeat exerts on the vessel wall. The tunica media is thicker in arteries than in veins. In arteries, which function to maintain blood pressure, the tunica media is the thickest layer.

The outermost layer of the vessel wall is the **tunica externa** or *tunica adventitia* (ad″ven-tish′e-ah). This tunic is a layer of connective tissue that contains many collagen and elastic fibers. Its cells and fibers run *longitudinally.* Functionally, the tunica externa protects the vessel, further strengthens its wall, and anchors the vessel to surrounding structures. The larger arteries and veins have tiny arteries, capillaries, and veins in their tunica externa. These small vessels, the **vasa vasorum** (va′sah va-sor′um; "vessels of the vessels"), arise either as tiny branches from the same vessel or as small branches from other, nearby vessels and nourish the outer half of the wall of the larger vessel (see Figure 20.2a). The inner half, by contrast, gets its nutrients by diffusion from the blood in the vessel's own lumen. Small blood vessels need no vasa vasorum because their walls are entirely supplied by luminal blood.

TYPES OF BLOOD VESSELS

You will learn about the three types of blood vessels—arteries, capillaries, and veins—in the order in which a drop of blood would encounter them as it passes through the systemic or pulmonary circuit. Accordingly, the discussion begins with arteries.

Arteries

learning outcome

- Compare and contrast the structure and functions of elastic arteries, muscular arteries, and arterioles.

Arteries are vessels that carry blood away from the heart. It is a common misconception that *all* arteries carry oxygen-rich blood, whereas *all* veins carry oxygen-poor blood. This statement is correct for the systemic circuit, but it is not correct for the pulmonary circuit, whose arteries carry oxygen-poor blood to the lungs for oxygenation (see Figure 19.1, p. 563).

The passage of blood through the arteries proceeds from elastic arteries, to muscular arteries, to arterioles.

Elastic Arteries

Elastic arteries are the largest arteries near the heart—the aorta and its major branches—with diameters ranging from 2.5 cm (about the width of the thumb) to 1 cm (slightly less than the width of the little finger). Their large lumen allows them to serve as low-resistance conduits for conducting blood between the heart and the medium-sized muscular arteries. For this reason, elastic arteries are sometimes called *conducting arteries.* More elastin occurs in the walls of these arteries than in any other type of vessel, and the sheets of elastin in the tunica media are remarkably thick (Figure 20.2a).

The high elastin content of conducting arteries dampens the surges of blood pressure resulting from the rhythmic contractions of the heart. When the heart forces blood into

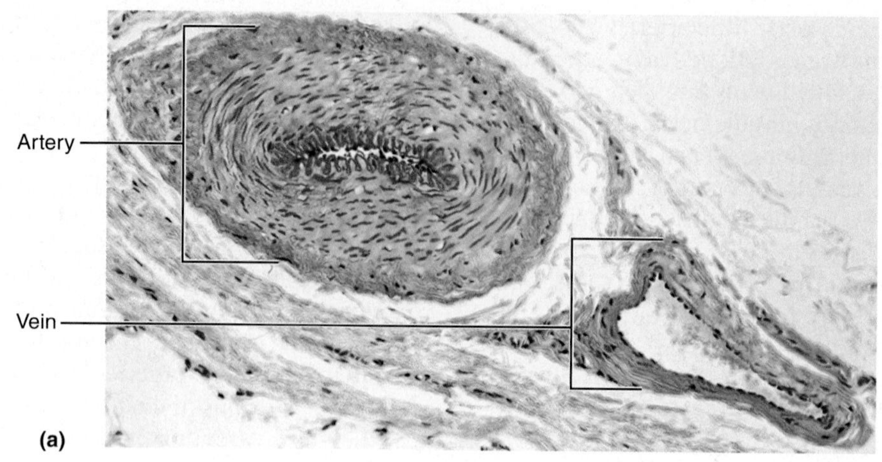

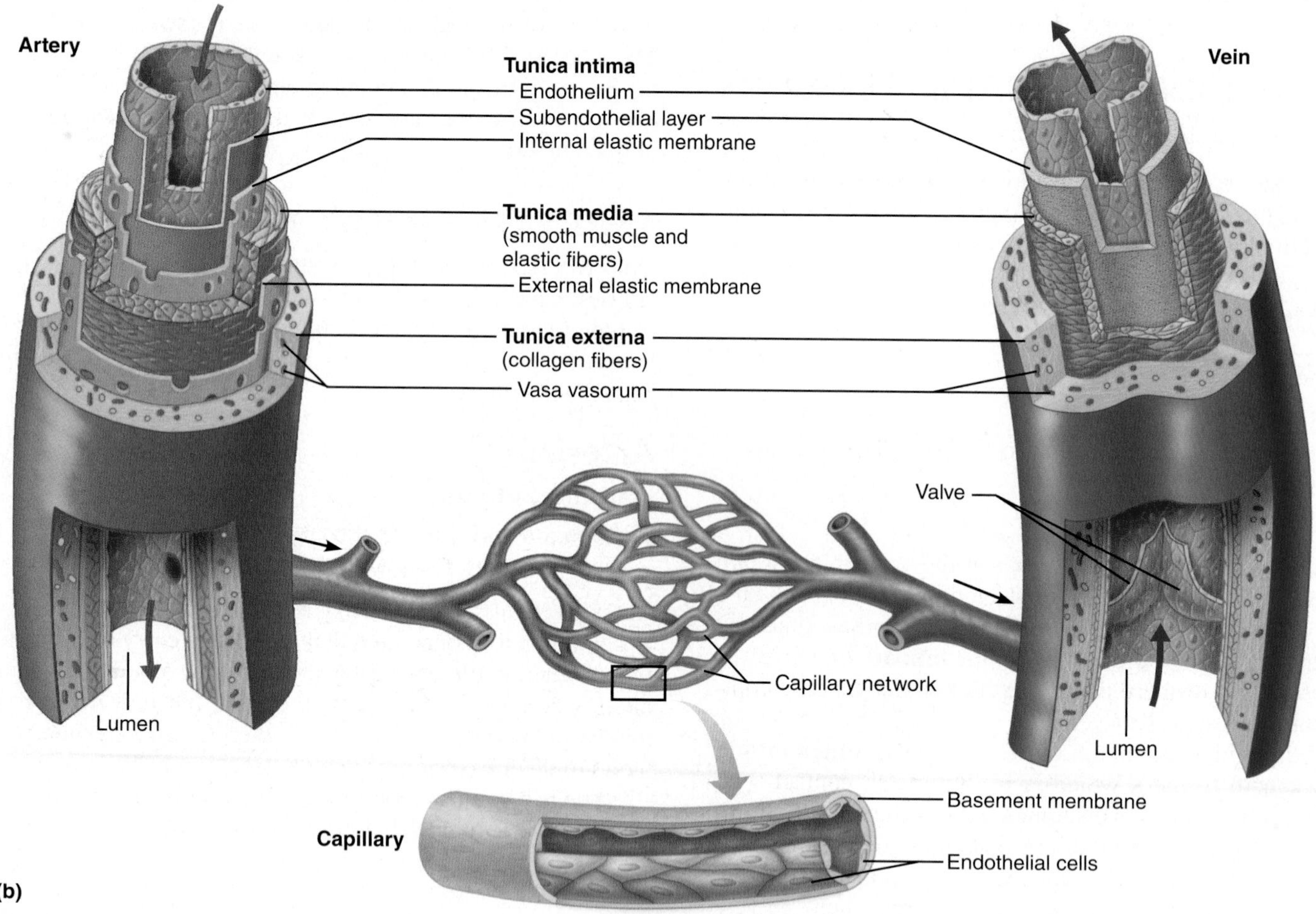

Figure 20.1 Generalized structure of arteries, veins, and capillaries.
(**a**) Photomicrograph of a muscular artery and a vein, in cross section (105×).
(**b**) Comparison of wall structure of arteries, veins, and capillaries. The artery has a thicker wall, a thicker tunica media, a narrower lumen than the similarly sized vein, and thickened elastic membranes not present in the vein. The vein, by contrast, has a thicker tunica externa, a wider lumen, and valves.

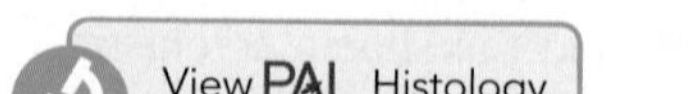

Figure 20.2 **Comparison of arterial wall structure.**
(a) Section through the wall of the aorta, showing its layers, its abundant elastin (stained brown), and the vasa vasorum in the tunica externa. **(b)** The elastic tissue in muscular arteries forms two distinct layers, the internal elastic membrane in the tunica intima, and the external elastic membrane in the tunica media. **(c)** In an arteriole, the tunica media is composed of only a few layers of smooth muscle cells, and the tunica intima is reduced to flattened endothelial cells.

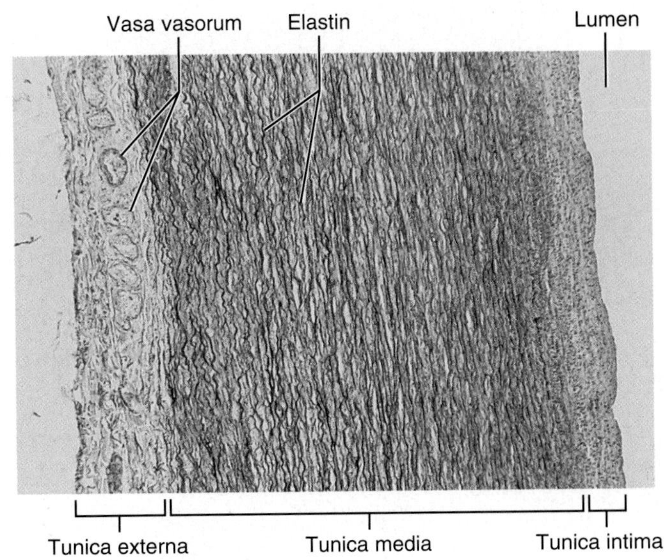

(a) Elastic artery (aorta, 21×)

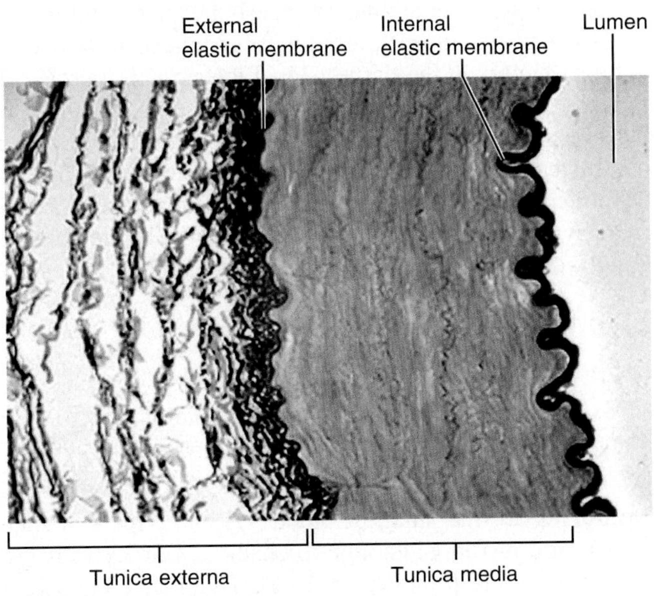

(b) Muscular artery (65×)

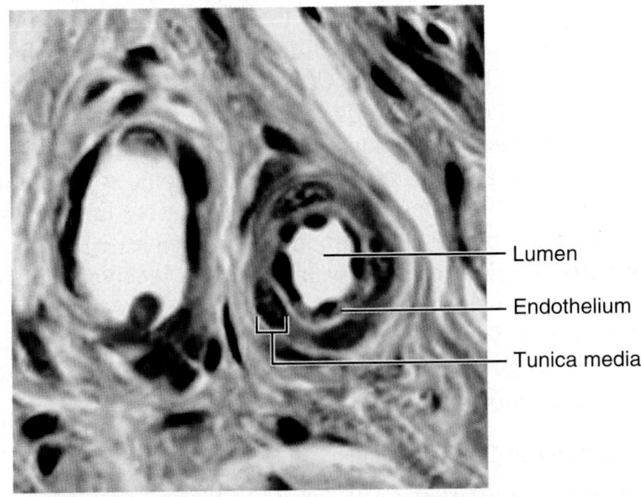

(c) Small arteriole (520×)

the arteries, the elastic elements in these vessels expand in response to increased blood pressure, in effect storing some of the energy of the flowing fluid; then, when the heart relaxes, the elastic elements recoil, propelling the blood onward. As the blood proceeds through smaller arteries, there is a marked decline both in its absolute pressure (due to resistance imparted by the arterial walls) and in the size of the pressure vacillations (due to the elastic recoil of the arteries just described). By the time the blood reaches the thin-walled capillaries, which are too fragile to withstand strong surges in blood pressure, the pressure of the blood is considerably lower and completely steady.

Muscular Arteries

Muscular (distributing) arteries lie distal to the elastic arteries and supply groups of organs, individual organs, and parts of organs. These "middle-sized" arteries constitute most of the named arteries seen in the anatomy laboratory. They range in diameter from about 1 cm to 0.3 mm (as thick as a sharpened pencil lead).

The following features distinguish muscular arteries:

- The tunica media of muscular arteries is thicker relative to the size of the lumen than that of any other type of vessel (see Figure 20.1). By actively changing the diameter of the artery, this muscular layer regulates the amount of blood flowing to an organ according to the specific needs of that organ.
- The smooth muscle of the tunica media of muscular arteries is sandwiched between two thick sheets of elastin: A wavy **internal elastic membrane** forms the outer layer of the tunica intima, and an **external elastic membrane** forms the outer layer of the tunica media (Figure 20.2b). These elastic membranes, in addition to the thin sheets of elastin found within the tunica media, help to dampen the pulsatile pressure produced by the heartbeat.

Arterioles

Arterioles are the smallest arteries, with diameters ranging from about 0.3 mm to 10 μm. Their tunica media contains only one or two layers of smooth muscle cells. Larger arterioles have all three tunics plus an internal elastic network in the tunica intima. Smaller arterioles (Figure 20.2c), which lead into

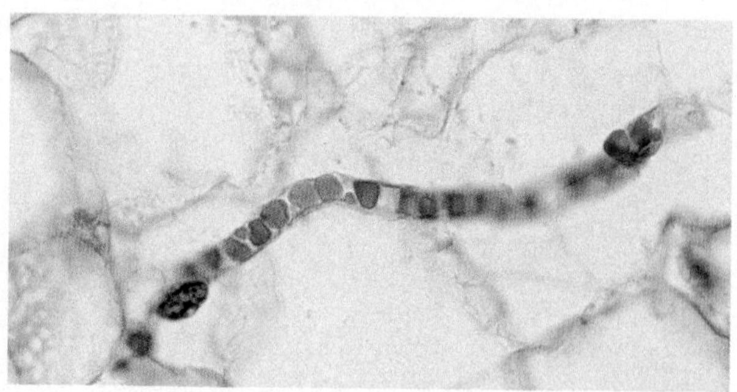

Figure 20.3 **Red blood cells passing through a capillary (512×).**

the capillary beds, are little more than a single layer of smooth muscle cells spiraling around an underlying endothelium.

The diameter of each arteriole is regulated in two ways: (1) Local factors in the tissues signal the smooth muscle cells to contract or relax, thus regulating the amount of blood sent downstream to each capillary bed; and (2) the sympathetic nervous system adjusts the diameter of arterioles throughout the body to regulate systemic blood pressure. For example, a widespread sympathetic vasoconstriction raises blood pressure during fight-or-flight responses (discussed in Chapter 15, p. 470).

Capillaries

learning outcome

- Describe the structure and function of capillaries, sinusoids, and capillary beds, and explain the structural basis of capillary permeability.

Capillaries are the smallest blood vessels, with a diameter of 8–10 μm, just large enough to enable erythrocytes to pass through in single file (Figure 20.3). They are composed of only a single layer of endothelial cells surrounded by a basement membrane (see Figure 20.1b). They are the body's most important blood vessels because they renew and refresh the surrounding tissue fluid (interstitial fluid; p. 81) with which all body cells are in contact. Capillaries deliver to this fluid the oxygen and nutrients cells need, and they remove the carbon dioxide and nitrogenous wastes that cells deposit into the fluid. Along with these universal functions, some capillaries also perform site-specific functions. In the lungs, oxygen enters the blood (and carbon dioxide leaves it) through capillaries. Capillaries in the small intestine receive digested nutrients; those in the endocrine glands pick up hormones; and those in the kidneys remove nitrogenous wastes from the body. There are three types of capillaries: *continuous, fenestrated,* and *sinusoids.*

Continuous Capillaries

Continuous capillaries are the most common type of capillary, occurring in most organs of the body such as skeletal muscles, skin, and the central nervous system (Figure 20.4a).

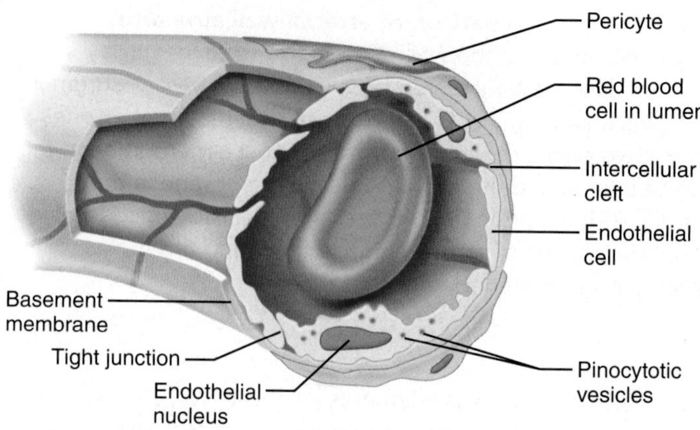

(a) Continuous capillary. Least permeable, and most common (e.g., skin, muscle).

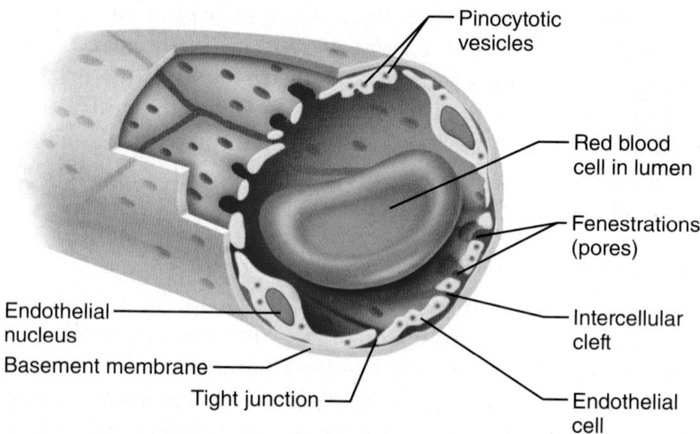

(b) Fenestrated capillary. Large fenestrations (pores) increase permeability. Occurs in areas of active absorption or filtration (e.g., kidney, small intestine).

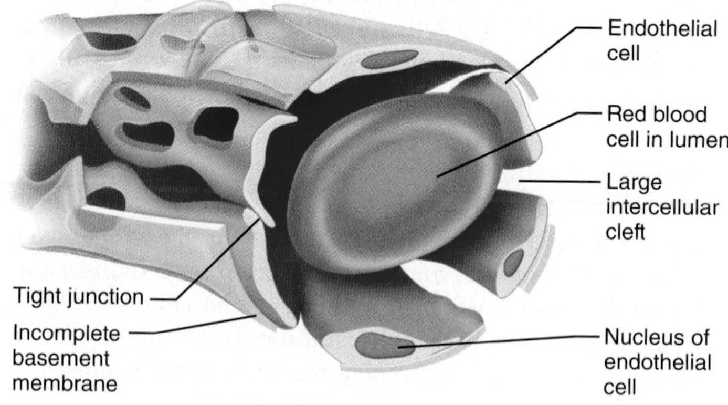

(c) Sinusoid capillary. Most permeable. Occurs in special locations (e.g., liver, bone marrow, spleen).

Figure 20.4 **Capillary structure.**

Tight junctions and occasional desmosomes hold the capillary endothelial cells together. These tight junctions block the passage of small molecules (p. 74), but they do not surround the whole perimeter of the endothelial cells. There are gaps of unjoined membrane called **intercellular clefts** that allow small molecules to pass into and out of the capillary. External to the endothelial cells, the delicate capillary is strengthened and stabilized by scattered **pericytes** (per′ĭ-sīts; "surrounding cells"), spider-shaped contractile stem cells whose thin processes form a widely spaced network around the capillary. Pericytes help control capillary permeability and can give rise to new vessels.

Fenestrated Capillaries

Like continuous capillaries, the endothelial cells of fenestrated capillaries (fen′is-tra-ted) are joined by tight junctions and contain intercellular clefts. In addition, fenestrated capillaries have pores (fenestrations, or "windows") spanning the endothelial cells (Figure 20.4b). Fenestrated capillaries occur only where there are exceptionally high rates of exchange of small molecules between the blood and the surrounding tissue fluid. For example, capillaries in the small intestine, which receive the digested nutrients from food; capillaries in the glomeruli of the kidneys, which filter blood; capillaries in the endocrine glands, which pick up secreted hormones; and capillaries in the synovial membranes of joints, where many water molecules exit the blood to contribute to the synovial fluid.

Sinusoid Capillaries (Sinusoids)

Some organs contain wide, leaky capillaries called **sinusoids** or **sinusoid capillaries** (Figure 20.4c). Each sinusoid follows a twisted path and has both expanded and narrowed regions. Sinusoids are usually fenestrated, and their endothelial cells have fewer cell junctions than do other capillaries. In some sinusoids, in fact, the intercellular clefts are wide open. Sinusoids occur wherever there is an extensive exchange of *large* materials, such as proteins or cells, between the blood and surrounding tissue. For example, they occur in the bone marrow and spleen, where many blood cells move through their walls. The large diameter and twisted course of sinusoids ensure that blood slows when flowing through these vessels, allowing time for the many exchanges that occur across their walls.

Capillary Permeability

Molecules pass into and out of capillaries through four routes.

1. **Direct diffusion through the endothelial cell membranes.** Carbon dioxide and oxygen seem to be the only important molecules that diffuse directly through endothelial cells, because these uncharged molecules easily diffuse through the lipid-containing membranes of cells.
2. **Intercellular clefts.** Most small molecules are exchanged through the intercellular clefts. In sinusoids, larger molecules and cells are exchanged through the wide intercellular clefts.
3. **Fenestrations.** In fenestrated capillaries, the pores in the endothelial cells allow passage of many small molecules.
4. **Pinocytotic vesicles.** Pinocytotic vesicles invaginate from the plasma membrane and migrate across the endothelial cells, transporting dissolved gases, nutrients, and waste products into the capillary.

Low-Permeability Capillaries: The Blood Brain Barrier

The *blood brain barrier* (introduced on p. 407), which prevents all but the most vital molecules (and normally even leukocytes) from leaving the blood and entering brain tissue, derives from the structure of capillaries in the brain. These capillaries *lack* the structural features that account for capillary permeability: Brain capillaries are continuous capillaries with complete tight junctions; intercellular clefts are absent. The vital molecules, such as glucose, that must cross brain capillaries are "ushered through" by highly selective transport mechanisms in the plasma membranes of the endothelial cells. The blood brain barrier is *not* a barrier against uncharged and lipid-soluble molecules such as oxygen, carbon dioxide, and some anesthetics, which diffuse unhindered through the endothelial cells and freely enter brain tissue.

During prolonged emotional stress, the tight junctions between the endothelial cells of brain capillaries are opened, so that the blood brain barrier fails and toxic substances in the blood can enter brain tissue. This mechanism has been implicated in the neurological symptoms associated with Gulf War syndrome—a syndrome marked by numerous chronic symptoms, including chronic fatigue, dizziness, memory loss, and depression, seen in some soldiers who served in the Persian Gulf War in 1991. Further study of the disruption of the blood brain barrier could one day help medical researchers who are seeking ways to deliver beneficial drugs—antibiotics and chemicals to kill brain tumors—into the brain.

Capillary Beds

Capillaries supply body tissues through structures called capillary beds. A *capillary bed* is a network of the body's smallest vessels (Figure 20.5). Capillary beds run throughout almost all tissues, especially the loose connective tissues. A terminal arteriole leads to a **metarteriole**—a vessel that is structurally intermediate between an arteriole and a capillary—from which branch true capillaries. The metarteriole continues into a **thoroughfare channel,** a vessel structurally intermediate between a capillary and a venule. True capillaries merge into the thoroughfare channel, which then joins a venule. Smooth muscle cells called **precapillary sphincters** wrap around the root of each true capillary where it leaves the metarteriole.

The precapillary sphincters regulate the flow of blood to the tissue according to that tissue's needs for oxygen and nutrients. When the tissue is functionally active, the sphincters are relaxed, enabling blood to flow through the wide-open capillaries and supply the surrounding tissue cells. When the tissue has lower demands (such as when nearby tissue cells already have adequate oxygen), the precapillary sphincters contract, closing off the true capillaries and forcing blood to flow straight from the metarteriole into the thoroughfare channel and venule—thereby bypassing the true capillaries. In this way, capillary beds precisely control the amount of blood supplying a tissue at any time.

Most tissues and organs have a rich capillary supply, but not all do. Tendons and ligaments are poorly vascularized. Epithelia and cartilage contain no capillaries; instead, they receive nutrients indirectly via diffusion from nearby vascularized connective tissues. The cornea and the lens have no capillary supply at all and are nourished by the aqueous humor and other sources (described on pp. 501 and 495).

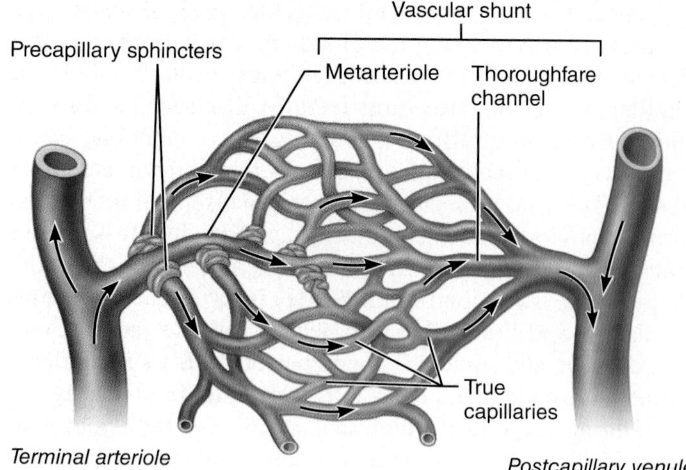

(a) Sphincters open—blood flows through true capillaries.

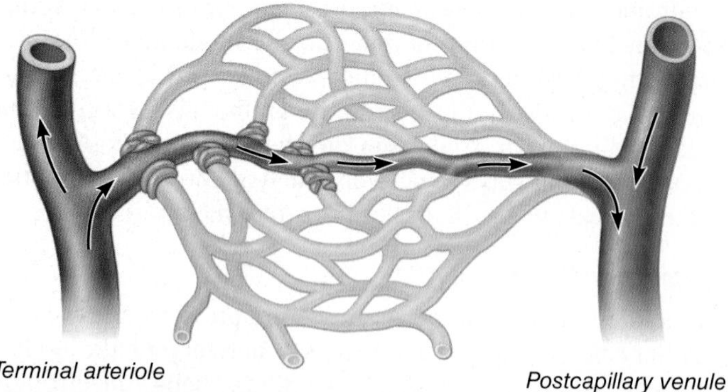

(b) Sphincters closed—blood flows through metarteriole–thoroughfare channel and bypasses true capillaries.

Figure 20.5 Anatomy of a capillary bed.

Venous Vessels

learning outcomes

- Explain how to distinguish a vein from an artery in histological sections.
- Describe the structural features of arteries and veins that help maintain the flow of blood through these vessels.

Veins are the blood vessels that conduct blood from the capillaries toward the heart. Veins in the systemic circuit carry blood that is relatively oxygen-poor, but the pulmonary veins carry oxygen-rich blood returning from the lungs.

Venules

The smallest veins are called **venules** and are 8–100 μm in diameter. The smallest venules, called **postcapillary venules** (Figure 20.5), consist of an endothelium on which lie pericytes. These venules function very much like capillaries. In fact, during inflammatory responses, more fluid and leukocytes leave the circulation through postcapillary venules than through capillaries. Larger venules have a tunica media that consists of one or two layers of smooth muscle cells and a thin tunica externa.

Table 20.1 Summary of Blood Vessel Anatomy

Vessel Type/ Illustration*	Average Lumen Diameter (D) and Wall Thickness (T)	Relative Tissue Makeup			
		Endothelium	Elastic Tissues	Smooth Muscles	Fibrous (Collagenous) Tissues
Elastic artery	D: 1.5 cm T: 1.0 mm				
Muscular artery	D: 6.0 mm T: 1.0 mm				
Arteriole	D: 37.0 μm T: 6.0 μm				
Capillary	D: 9.0 μm T: 0.5 μm				
Venule	D: 20.0 μm T: 1.0 μm				
Vein	D: 5.0 mm T: 0.5 mm				

*Size relationships are not proportional. Smaller vessels are drawn relatively larger so detail can be seen. See column 2 for actual dimensions.

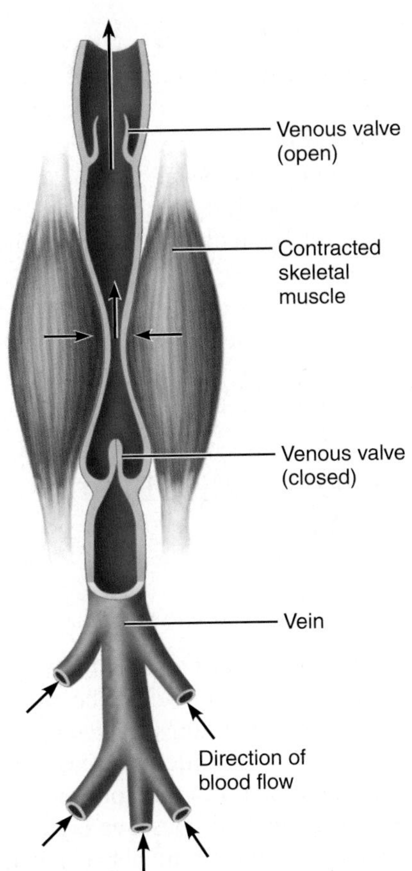

Figure 20.6 **Blood flow through veins.** Valves keep venous blood moving in one direction; the valves are opened by blood flowing toward the heart and are closed by backflow. The skeletal muscular pump aids venous return: Contracting skeletal muscles press against a vein and propel blood toward the heart, forcing valves proximal to the muscles to open and valves distal to close.

Veins

Venules join to form veins. Structurally, veins differs from arteries in the following ways (Table 20.1):

- The lumen of a vein is larger than that of an artery of comparable size (Figure 20.1a). At any given time, veins hold fully 65% of the body's blood.
- In a vein, the tunica externa is thicker than the tunica media. In an artery, the tunica media is the thicker layer. In the body's largest veins—the venae cavae, which return systemic blood to the heart—longitudinal bands of smooth muscle further thicken the tunica externa.
- Veins have less elastin in their walls than do arteries because veins do not need to dampen any pulsations (all of which are smoothed out by arteries before the blood reaches the veins).
- The wall of a vein is thinner than that of a comparable artery. Blood pressure declines substantially while blood passes through the high-resistance arterioles and capillary beds; thus, blood pressure in the veins is much lower than in the arteries.

Several mechanisms counteract the low venous blood pressure and help move the blood back to the heart. One structural feature of some veins is **valves** that prevent the backflow of blood away from the heart (Figure 20.1 and Figure 20.6). Each of these valves has several cusps formed from the tunica intima. The flow of blood toward the heart pushes the cusps apart, opening the valve, and any backflow pushes the cusps together, closing the valve. Valves are most abundant in veins of the limbs, where the superior direction of venous flow is most directly opposed by gravity. A few valves occur in the veins of the head and neck, but none are located in veins of the thoracic and abdominal cavities.

One functional mechanism that aids the return of venous blood to the heart is the normal movement of the body and limbs, for instance, during walking. Swinging a limb moves the blood in the limb, and the venous valves ensure that this blood moves only in the proper direction. Another mechanism aiding venous return is the **skeletal muscular pump,** in which contracting skeletal muscles press against the thin-walled veins, forcing valves proximal to the area of contraction to open and propelling blood toward the heart (Figure 20.6). Valves distal to the contracting muscles are closed by backflowing blood.

The effectiveness of venous valves in preventing the backflow of blood is easily demonstrated. Hang one hand by your side until the veins on its dorsal surface become distended with blood. Next, place two fingertips against one of the distended veins, and, pressing firmly, move the superior finger proximally along the vein, and then release that finger. The vein stays flat and collapsed despite the pull of gravity. Finally, remove the distal fingertip, and watch the vein refill rapidly with blood.

VASCULAR ANASTOMOSES

learning outcome

▶ Define vascular anastomoses and explain their functions.

Where vessels unite or interconnect, they form **vascular anastomoses** (ah-nas″to-mo′sēz; "coming together"). Most organs receive blood from more than one arterial branch, and neighboring arteries often connect with one another to form *arterial anastomoses.* Arterial anastomoses provide alternative pathways, or *collateral channels,* for blood to reach a given body region. If one arterial branch is blocked or cut, the collateral channels can often provide the region with an adequate blood supply. Arterial anastomoses occur around joints, where active body movements may hinder blood flow through one channel (see Figure 20.11 for examples), as well as in the abdominal organs, brain, and heart. Because of the many anastomoses among the smaller branches of the coronary artery in the heart wall, a coronary artery can be 90% occluded by atherosclerosis before a myocardial infarction occurs. By contrast, because arterial anastomoses are poorly developed in the kidneys, spleen, parts of bone diaphyses nearest the epiphyses, and central artery of the retina, blockage of such arteries causes severe tissue damage.

Veins anastomose much more freely than arteries. You may be able to see *venous anastomoses* through the skin on the dorsal surface of your hand. Because of the abundant anastomoses, occlusion of a vein rarely blocks blood flow or leads to tissue death.

CLINICAL APPLICATION

Varicose Veins When the valves in veins weaken and fail, the result is **varicose veins.** The veins twist and swell with pooled blood, and venous drainage slows considerably. Fifteen percent of all adults suffer from varicose veins, usually in the lower limbs. Women are affected more often than men. The left lower limb is more susceptible to varicose veins than the right. Where the aorta divides into the right and left common iliac arteries, the right common iliac artery crosses over the left common iliac vein, the vessel that drains the left lower limb. The higher pressure in the artery can compress the deeper left common iliac vein and impede venous drainage from the left lower limb.

Varicose veins can be hereditary but also occur in people whose jobs require prolonged standing in one position, such as store clerks, hairdressers, dentists, and nurses. In nonmoving legs, the venous blood drains so slowly that it accumulates, stretches the venous walls and valves, and causes the valves to fail. Obesity and pregnancy can cause or worsen the problem, because increased weight constricts the leg-draining veins in the superior thigh.

Straining to deliver a baby or to have a bowel movement raises the intra-abdominal pressure and elevates venous pressure, preventing drainage of blood from the veins of the anal canal at the inferior end of the large intestine. The resulting varicosities in these anal veins are called **hemorrhoids.**

In severe cases, varicose veins slow the circulation through a body region so greatly that the tissues in that region die of oxygen starvation. To prevent this, physicians either remove the affected veins or inject them with an irritating solution that scars them and fuses them shut. Drainage of the body region then proceeds normally, through alternative pathways via vascular anastomoses.

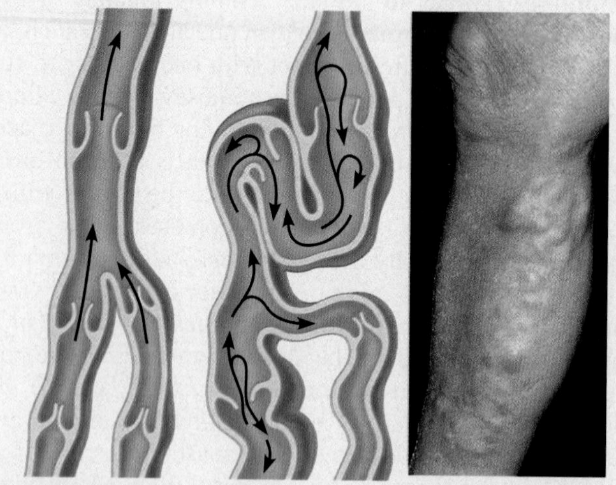

(a) Normal veins **(b) Varicose veins** **(c) Varicosities in great saphenous vein, right leg**

check your understanding

- ☐ 1. What structural features of capillaries make them well suited for their function of nourishing body tissues and removing waste products?
- ☐ 2. What components in the wall of a muscular artery help to move blood through these vessels? What mechanisms aid in maintaining the movement of venous blood?
- ☐ 3. Define each of the following: (a) vasa vasorum, (b) arterial anastomoses, (c) varicose veins, (d) artery.

(For answers, see Appendix B.)

PART 2 BLOOD VESSELS OF THE BODY

learning outcome

▶ Define pulmonary and systemic circuits.

The complex system of blood vessels in the body called the *vascular system* has two basic circuits: The *pulmonary circuit* carries blood to and from the lungs for the uptake of oxygen and the removal of carbon dioxide, whereas the *systemic circuit* carries oxygenated blood throughout the body and picks up carbon dioxide from body tissues (see Figure 19.1, p. 563). Blood vessels in the systemic circuit also (1) pick up nutrients from the digestive tract and deliver them to cells throughout the body, (2) receive nitrogenous wastes from body cells and transport them to the kidneys for elimination in the urine, and (3) pick up hormones or other signaling molecules and transport them to their target organ.

As you read about the vessels, note that arteries and veins tend to run together, side by side (see Figure 20.1a). In many places, these vessels also run with nerves. Note also that by convention, vessels carrying oxygen-rich blood are depicted as red, whereas those transporting oxygen-poor blood are depicted as blue, regardless of vessel type.

THE PULMONARY CIRCULATION

learning outcome

▶ Name the major vessels of the pulmonary circuit.

The pulmonary circulation begins as oxygen-poor blood leaves the right ventricle of the heart via the **pulmonary trunk** (Figure 20.7). This large artery exits the ventricle anterior to the aorta, ascends to the aorta's left, and reaches the concavity of the aortic arch, where it branches at a T-shaped divergence into the **right** and **left pulmonary arteries.** Each pulmonary artery penetrates the medial surface of a lung and then divides into several **lobar arteries** serving the lobes of the lung, three in the right lung and two in the left lung. Within the lung, the arteries branch along with the lung's air passageways (bronchi). As the branching arteries decrease in size, they become arterioles and finally the pulmonary capillaries that surround the delicate air sacs (lung alveoli). Gas exchange occurs across these capillaries, and the newly

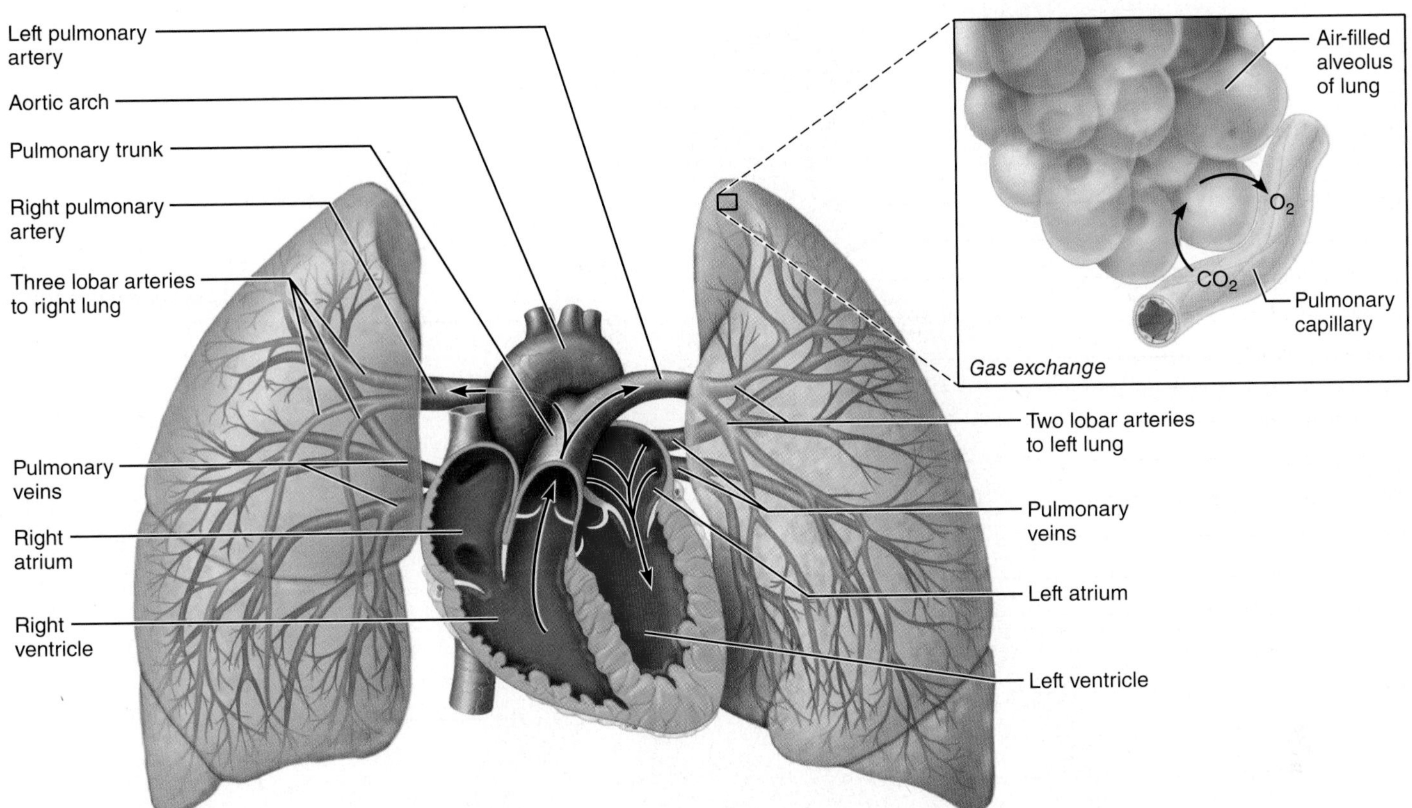

The pulmonary arterial system is shown in blue to indicate that the blood carried is oxygen-poor. The pulmonary venous drainage is shown in red to indicate that the blood transported is oxygen-rich.

Figure 20.7 **Pulmonary circulation.**

oxygenated blood enters venules and then progressively larger veins. The largest venous tributaries form the superior and inferior **pulmonary veins,** which exit the medial aspect of each lung. In the mediastinum posterior to the heart, the four pulmonary veins run horizontally, just inferior to the pulmonary arteries, and empty into the left atrium.

The arteries and veins of the pulmonary circuit have thinner walls than do systemic vessels of comparable diameter, reflecting the fact that the maximum arterial pressure in the pulmonary circuit is much lower, only one-sixth of that in the systemic circuit.

check your understanding

☐ 4. Why is the pulmonary artery illustrated in blue (Figure 20.7)? Is the blood in this vessel traveling toward the heart or away from the heart?

(For the answer, see Appendix B.)

THE SYSTEMIC CIRCULATION

Before you examine the systemic vessels, it is worth noting that the vessels on the right and left sides of the body are not always mirror images of each other. Some of the large, deep vessels of the trunk region are asymmetrical (their initial symmetry is lost during embryonic development). In the head and limbs, by contrast, almost all vessels are bilaterally symmetrical.

Systemic Arteries

learning outcomes

- Name the three vessels that arise from the aortic arch. Describe the routes of arterial blood supply from these vessels to the head and neck, the brain, the thorax, and the upper limb.
- Describe the pathways and the organs supplied by the midline arteries and by the paired arteries branching off the abdominal aorta.
- Describe the pathway of arterial supply to the pelvis and the lower limb.
- Identify the location of pulse points in the limbs, head, and neck.

The systemic arteries (Figure 20.8a) carry oxygenated blood from the heart to the capillaries of organs throughout the body. The elastin in the walls of these arteries maintains the pulsatile flow. Arterial pulses can be palpated in the muscular arteries at numerous body locations (Figure 20.8b) and can be used to determine heart rate and to assess blood flow to a body region after trauma, surgery, or disease. Deep pressure at a pulse point is a first aid technique used to limit blood flow through a vessel that is hemorrhaging and thus limit blood loss.

As you examine the arterial vessels, you will begin with the aorta and then consider the systemic arteries in a generally superior-to-inferior direction.

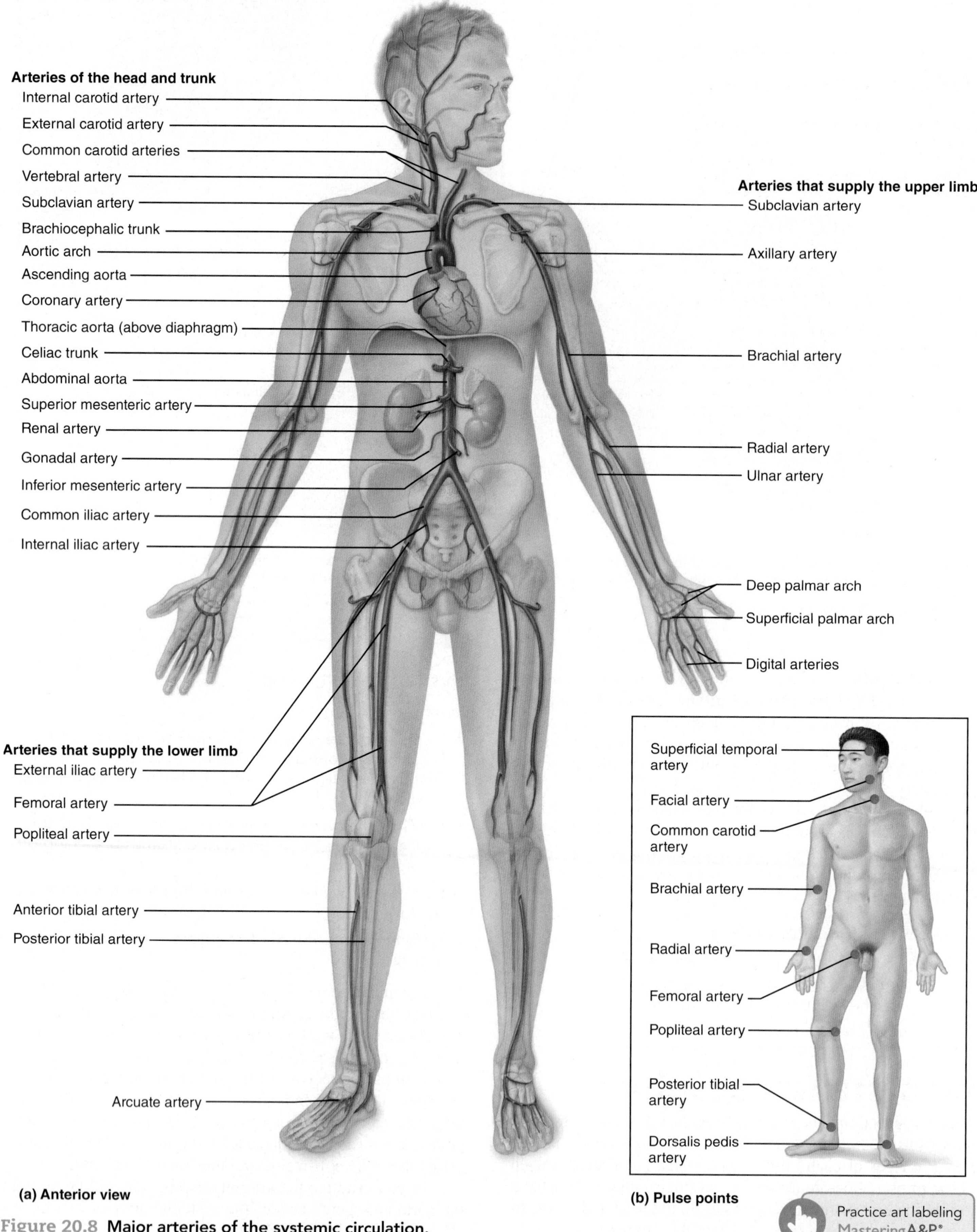

Figure 20.8 Major arteries of the systemic circulation.

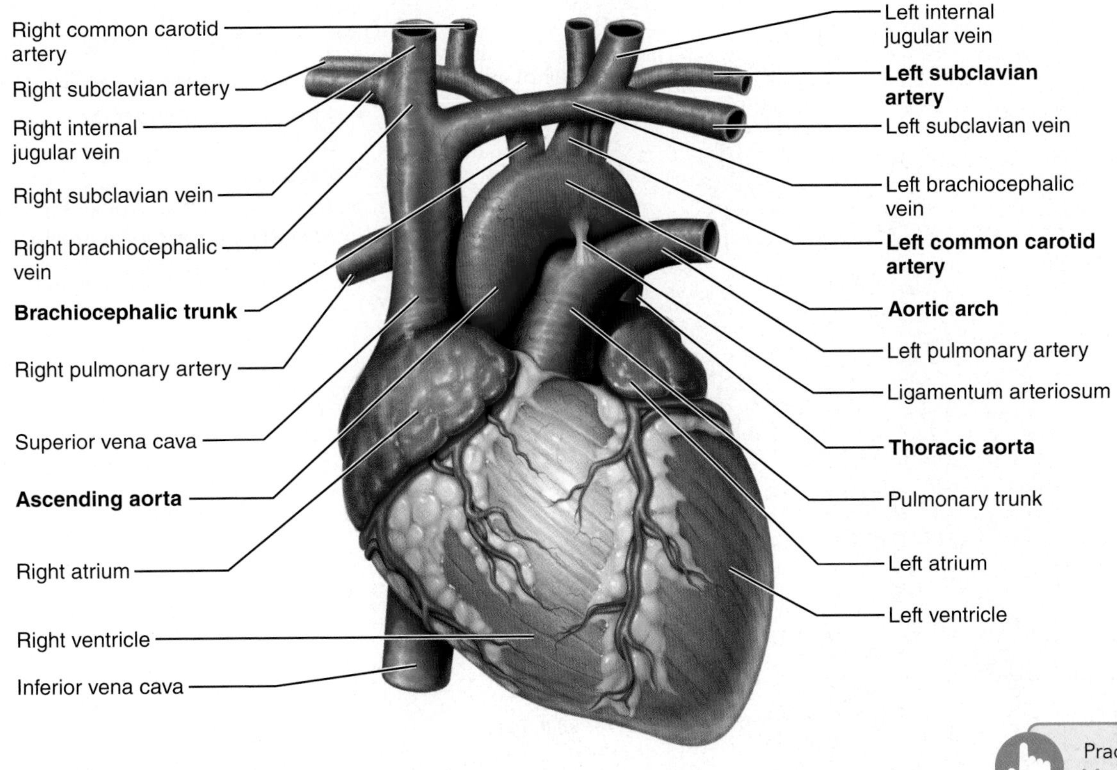

Figure 20.9 The great vessels that exit and enter the heart.

Aorta

The **aorta,** the largest artery in the body, leaves the heart, arcs superiorly and then descends along the bodies of the vertebrae to the inferior part of the abdomen (see Figure 20.8). Along this course, the aorta is divided into the following three parts.

Ascending Aorta The **ascending aorta** (Figure 20.9), one of the great vessels leaving the heart, arises from the left ventricle and ascends for only about 5 cm. It begins posterior to the pulmonary trunk, passes to the right of that vessel, and then curves left to become the aortic arch. The only branches of the ascending aorta are the **right** and **left coronary arteries** that supply the wall of the heart (described on p. 579).

Aortic Arch Arching posteriorly and to the left, the **aortic arch** lies posterior to the manubrium of the sternum. The **ligamentum arteriosum,** a fibrous remnant of a fetal artery called the *ductus arteriosus,* connects the aortic arch and the pulmonary trunk (Figure 20.9).

Three arteries branch from the aortic arch and run superiorly (Figure 20.9).

- The first and largest branch is the **brachiocephalic trunk** (bra″ke-o-sĕ-fal′ik; "arm-head"). This vessel ascends to the right toward the base of the neck where it divides into the **right common carotid artery** and the **right subclavian artery.**
- The second branch of the aortic arch is the **left common carotid artery.**
- The third branch is the **left subclavian artery.**

These three branches of the aorta supply the head and neck, upper limbs, and the superior part of the thoracic wall. Note that the brachiocephalic trunk on the right has no corresponding artery on the left because the left common carotid and subclavian arteries arise directly from the aorta.

This is the typical branching pattern of these vessels off the aortic arch; however, as with all vessels, there is some variability from this pattern. The most frequent variation is the branching of the left common carotid artery from the brachiocephalic trunk. Less commonly, four large arteries (the right and left common carotids, and the right and left subclavian arteries) arise separately from the aortic arch, or the left common carotid artery and the left subclavian artery arise from a left brachiocephalic trunk.

Descending Aorta Continuing from the aortic arch, the **descending aorta** runs posterior to the heart and inferiorly on the bodies of the thoracic and lumbar vertebrae. It has two parts, the *thoracic aorta* and the *abdominal aorta.*

The **thoracic aorta** (Figure 20.9) descends on the bodies of the thoracic vertebrae (T_5–T_{12}) just to the left of the midline. Along the way, it sends many small branches to the thoracic organs and body wall. (These branches are described on p. 602).

The thoracic aorta passes through the diaphragm at the level of vertebra T_{12} and enters the abdominal cavity as the **abdominal aorta** (Figure 20.8a), which lies on the lumbar vertebral bodies in the midline. The abdominal aorta ends at the level of vertebra L_4, where it divides into the right and left *common iliac arteries,* which supply the pelvis and lower limbs.

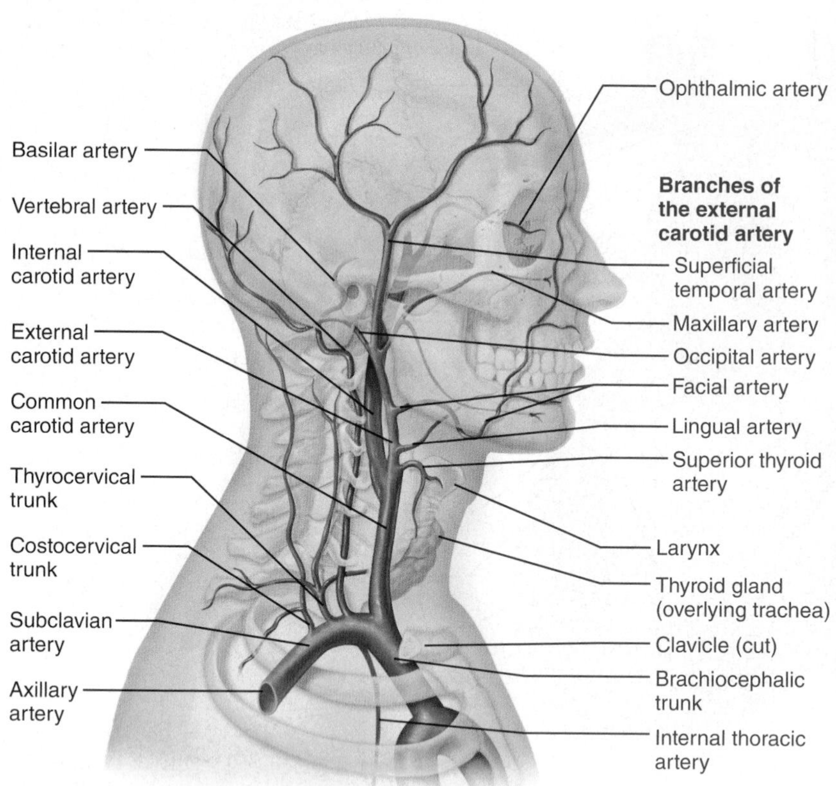

(a) Arteries of the head and neck, right aspect

CLINICAL APPLICATION

Pulse Points in the Head and Neck

- **External carotid artery**
  In the anterior triangle of neck
- **Facial artery**
  At the inferior margin of the mandible
- **Superficial temporal artery**
  Anterior to the auricle

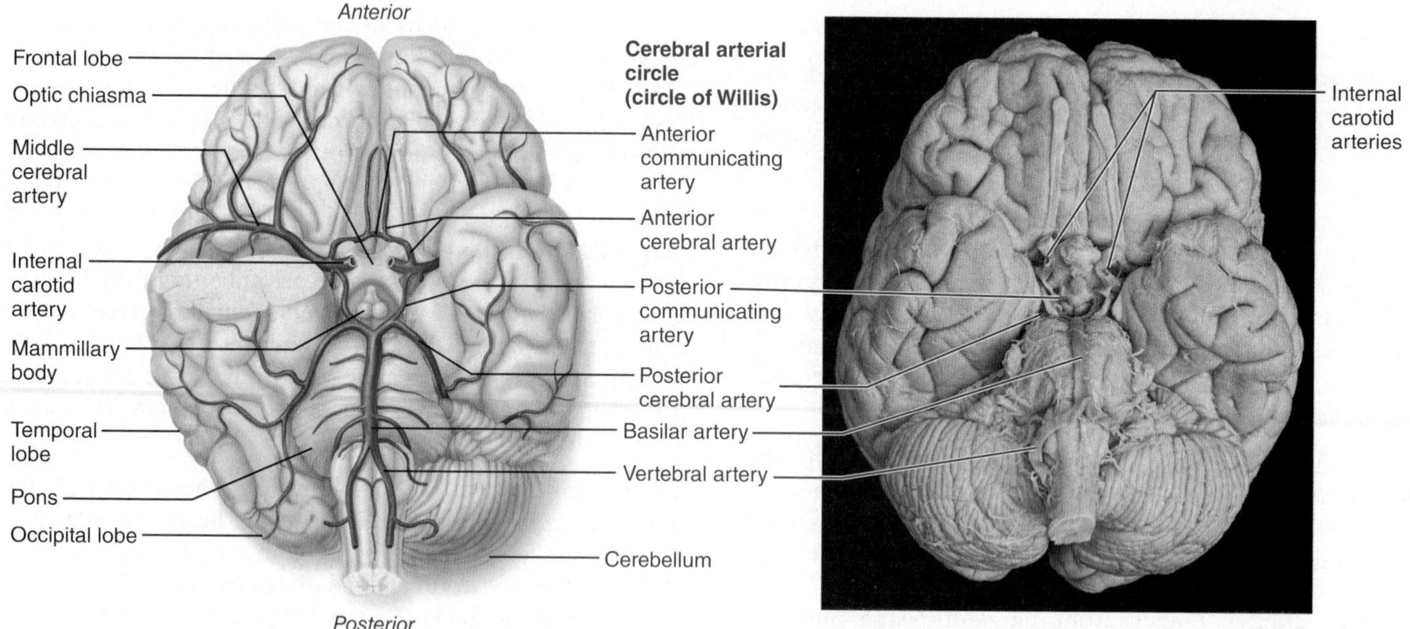

(b) Major arteries serving the brain, inferior view, right side of cerebellum and part of right temporal lobe removed

(c) Cerebral arterial circle, inferior view of brain

Figure 20.10 Arteries of the head, neck, and brain.

Arteries of the Head and Neck

Four pairs of arteries supply the head and neck: the *common carotid arteries* plus three branches from each subclavian artery—the *vertebral artery,* the *thyrocervical trunk,* and the *costocervical trunk* (Figure 20.10a).

Common Carotid Arteries Most parts of the head and neck receive their blood from the right and left **common carotid arteries,** which ascend through the anterior neck just lateral to the trachea. The neck is divided into two triangles by the sternocleidomastoid muscle (shown in

Figure 11.30, p. 334). The common carotid arteries are located in the anterior triangle just deep to the sternocleidomastoid and relatively thin infrahyoid muscles. These vessels are more vulnerable than most other arteries in the body because their relatively superficial location makes them vulnerable to slashing wounds. If a common carotid artery is cut, the victim can bleed to death in minutes. At the superior border of the larynx—the level of the "Adam's apple"—each common carotid ends by dividing into an *external and internal carotid artery.*

The right and left **external carotid arteries** supply most tissues of the head external to the brain and orbit (Figure 20.10a). As each external carotid ascends, it sends a branch to the thyroid gland and larynx **(superior thyroid artery),** to the tongue **(lingual artery),** to the skin and muscles of the anterior face **(facial artery),** to the posterior part of the scalp **(occipital artery),** and to the region around the ear **(posterior auricular artery).** Near the temporomandibular joint, each external carotid ends by splitting into the superficial temporal and maxillary arteries. The **superficial temporal artery** ascends just anterior to the ear and supplies most of the scalp. Branches of this vessel bleed profusely in scalp wounds. The **maxillary artery** runs deep to the ramus of the mandible and anteriorly into the maxillary bone, passing through the chewing muscles. Along the way, it sends branches to the upper and lower teeth, the cheeks, nasal cavity, and muscles of mastication.

A pulse can be palpated at some of these vessels (Figure 20.8b). The external carotid artery pulse can be felt where it branches from the common carotid artery, just anterior to the sternocleidomastoid and superior to the level of the larynx. A pulse from the facial artery can be felt at the inferior margin of the mandible just anterior to the masseter muscle, and pulsations in the superficial temporal artery can be felt on the temple just anterior to the auricle (external ear).

A clinically important branch of the maxillary artery is the *middle meningeal artery* (not illustrated). This vessel enters the skull through the *foramen spinosum* in the sphenoid bone and supplies the broad inner surfaces of the parietal bone and squamous part of the temporal bone, as well as the underlying dura mater. Blows to the side of the head can tear this artery, producing an intracranial hematoma that can compress the cerebrum and disrupt brain function.

The **internal carotid arteries** supply the orbits and most of the cerebrum. Each internal carotid ascends through the superior neck directly lateral to the pharynx and enters the skull through the *carotid canal* in the temporal bone. From there, it runs medially through the petrous part of the temporal bone, runs forward along the body of the sphenoid bone, and bends superiorly to enter the sella turcica just posterior to the optic canal. Here, it gives off the **ophthalmic artery** to the eye and orbit (Figure 20.10a) and divides into the *anterior* and *middle cerebral arteries* (Figure 20.10b).

Each **anterior cerebral artery** anastomoses with its partner on the opposite side through a short **anterior communicating artery** (Figure 20.10b) and supplies the medial and superior surfaces of the frontal and parietal lobes. Each **middle cerebral artery** runs through the lateral fissure of a cerebral hemisphere and supplies the lateral parts of the temporal and parietal lobes. Together, the anterior and middle cerebral arteries supply over 80% of the cerebrum; the rest of the cerebrum is supplied by the posterior cerebral artery (described shortly).

Subclavian Arteries The **right subclavian artery** originates from the brachiocephalic trunk, following the branching of the right common carotid artery. The **left subclavian artery** arises as the third vessel branching from the aortic arch. Branches from the subclavian arteries supply the brain (*vertebral arteries*) and supply the neck and thoracic wall (*thyrocervical trunks, costocervical trunks,* and *internal thoracic arteries*).

Vertebral Arteries The blood supply to the posterior brain comes from the right and left **vertebral arteries,** which arise from the subclavian arteries at the root of the neck (Figure 20.10a and c). The vertebral arteries ascend through the foramina in the transverse processes of cervical vertebrae C_6 to C_1 and enter the skull through the *foramen magnum.* Along the way, they send branches to the vertebrae and cervical spinal cord. Within the cranium, the right and left vertebral arteries join to form the unpaired **basilar** (bas′ĭ-lar) **artery** (Figure 20.10b and c), which ascends along the ventral midline of the brain stem, sending branches to the cerebellum, pons, and inner ear. At the border of the pons and midbrain, it divides into a pair of **posterior cerebral arteries,** which supply the occipital lobes plus the inferior and medial parts of the temporal lobes of the cerebral hemispheres.

Short **posterior communicating arteries** connect the posterior cerebral arteries to the middle cerebral arteries anteriorly. The two posterior communicating arteries and the single anterior communicating artery complete the formation of an arterial anastomosis called the **cerebral arterial circle** (formerly the **circle of Willis**). This circle forms a loop around the pituitary gland and optic chiasma, and it unites the brain's anterior and posterior blood supplies provided by the internal carotid and vertebral arteries. By interconnecting the arteries that supply the anterior, posterior, left, and right parts of the brain, this anastomosis provides alternate routes for blood to reach brain areas that are affected if either a carotid or vertebral artery becomes occluded.

Thyrocervical and Costocervical Trunks The rest of the neck receives its blood from two smaller branches of the subclavian arteries, the *thyrocervical* and *costocervical trunks* (Figure 20.10a).

The **thyrocervical** (thi″ro-ser′vĭ-kal) **trunk,** which arises first, sends two branches posteriorly over the scapula to help supply the scapular muscles, and one branch anteriorly to the inferior part of the thyroid gland *(inferior thyroid artery).* The ascending neck branch (Figure 20.10a) comes off the base of the inferior thyroid artery and helps to supply the cervical vertebrae and spinal cord.

The **costocervical trunk** sends a branch superiorly into the deep muscles of the neck and a branch inferiorly to supply the two most superior intercostal spaces.

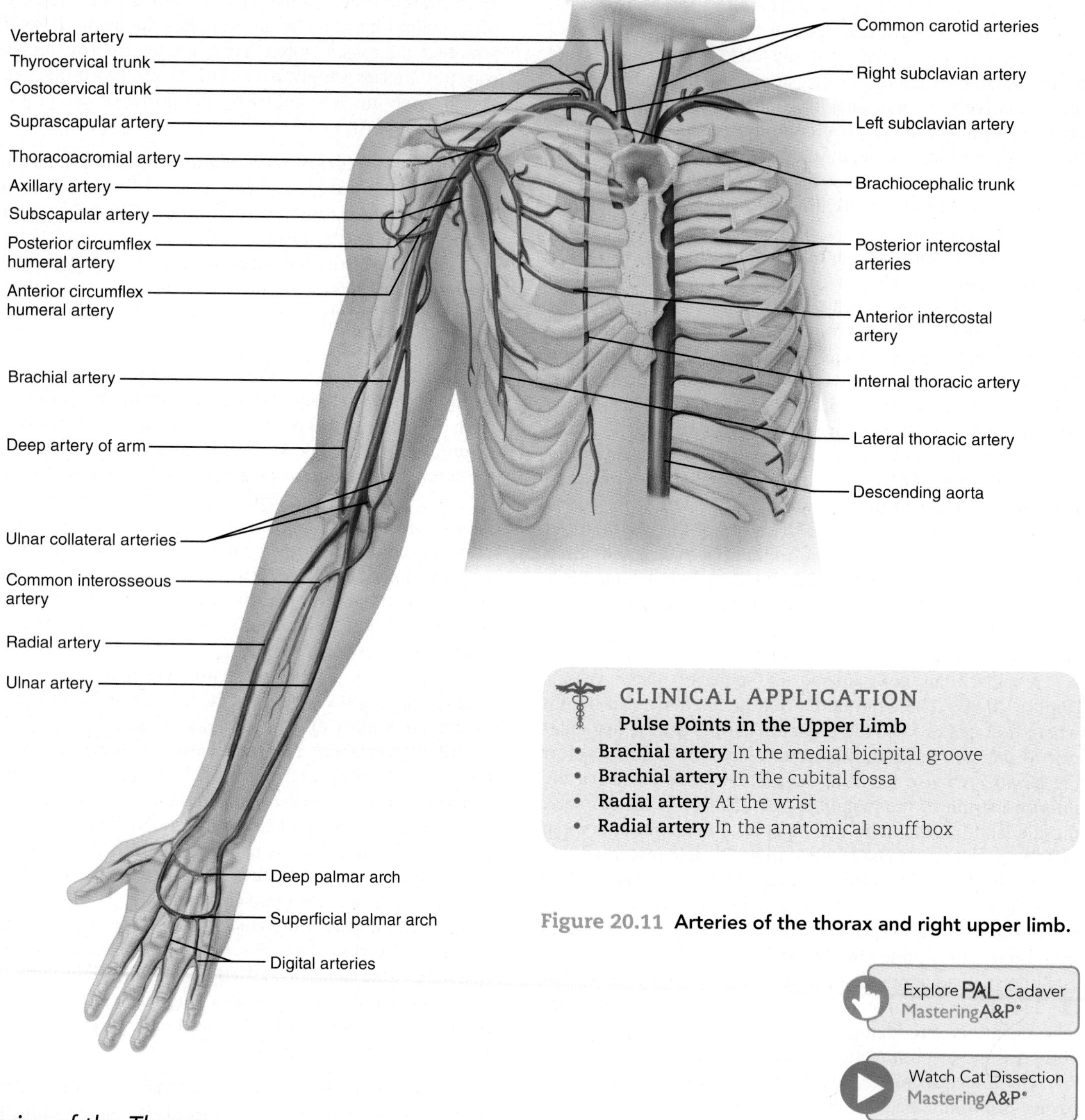

CLINICAL APPLICATION

Pulse Points in the Upper Limb

- **Brachial artery** In the medial bicipital groove
- **Brachial artery** In the cubital fossa
- **Radial artery** At the wrist
- **Radial artery** In the anatomical snuff box

Figure 20.11 Arteries of the thorax and right upper limb.

Explore PAL Cadaver MasteringA&P®

Watch Cat Dissection MasteringA&P®

Arteries of the Thorax

Arteries of the Thoracic Wall The anterior thoracic wall receives its blood from the **internal thoracic artery** (formerly the *internal mammary artery*) (Figure 20.11). This vessel branches from the subclavian artery superiorly, then descends just lateral to the sternum and just deep to the costal cartilages. **Anterior intercostal arteries** branch off at regular intervals and run horizontally to supply the ribs and the structures in the intercostal spaces. The internal thoracic artery also sends branches superficially to supply the skin and mammary gland. It ends inferiorly at the costal margin, where it divides into a branch to the anterior abdominal wall and a branch to the anterior part of the diaphragm.

The posterior thoracic wall receives its blood from the **posterior intercostal arteries.** The superior two pairs arise from the costocervical trunk, whereas the inferior nine pairs issue from the thoracic aorta. All of the posterior intercostal arteries run anteriorly in the costal grooves of their respective ribs. In the lateral thoracic wall, they form anastomoses with the anterior intercostal arteries. Inferior to the twelfth rib, one pair of *subcostal arteries* (not illustrated) branches from the thoracic aorta. Finally, a pair of *superior phrenic arteries* (not illustrated) leaves the most inferior part of the thoracic aorta to supply the posterior, superior part of the diaphragm.

Arteries of the Thoracic Visceral Organs Many thoracic viscera receive their functional blood supply from small branches off the thoracic aorta. Because these vessels are so small, they are not illustrated. The *bronchial arteries* supply systemic (oxygenated) blood to the lung structures. Usually, two bronchial arteries serve the left lung, and one serves the right lung; in some people they arise from posterior intercostal arteries instead of the aorta. Bronchial arteries enter the lung's medial surface along with the large pulmonary vessels.

The thoracic aorta also sends several small branches to the esophagus directly anterior to it, as well as to the posterior part of the mediastinum and pericardium.

Arteries of the Upper Limbs

The upper limb is supplied by arteries that arise from the subclavian artery (Figure 20.11). After giving off its branches to the neck and thorax (the vertebral artery, the thyrocervical and costocervical trunks, and the internal thoracic artery), each subclavian artery runs laterally onto the first rib, where it underlies the clavicle. From here, the subclavian artery enters the axilla as the *axillary artery.*

Axillary Artery The **axillary artery** descends through the axilla, giving off the following branches: the **thoracoacromial** (tho″rah-ko-ah-kro′me-al) **artery,** which arises just inferior to the clavicle and branches to supply much of the pectoralis and deltoid muscles; the **lateral thoracic artery,** which descends along the lateral edge of pectoralis minor supplying the pectoral muscles and serratus anterior, and sends important branches to the breast; the **subscapular artery,** which serves the dorsal and ventral scapular regions and the latissimus dorsi muscle; and **anterior** and **posterior circumflex humeral arteries,** which wrap around the surgical neck of the humerus and help supply the deltoid muscle and shoulder joint. The axillary artery continues into the arm as the *brachial artery.* The boundary for this transition is the inferior border of the teres major muscle.

Brachial Artery The **brachial artery** descends along the medial side of the humerus deep to the biceps muscle in the **medial bicipital groove** (shown in Figure 11.33a, p. 338) and supplies the anterior arm muscles. The brachial pulse can be felt at this location, and firm pressure here can stop bleeding from a hemorrhage in more distal parts of the limb. The brachial artery is used in measuring blood pressure with a sphygmomanometer (sfig″mo-mah-nom′ĕ-ter), a device whose cuff is wrapped around the arm superior to the elbow. One major branch, the **deep artery of the arm** (also called *profunda brachii* = deep brachial), wraps around the posterior surface of the humerus with the radial nerve and serves the triceps muscle. As the brachial artery nears the elbow, it sends several small branches inferiorly, the **ulnar collateral arteries,** that form anastomoses with branches ascending from arteries in the forearm to supply the elbow joint. These vessels also provide collateral circulation to the more distal regions of the limb when the elbow is bent. The brachial artery crosses the anterior aspect of the elbow joint deep to the bicipital aponeurosis in the midline of the arm. At this site the pulse is easily felt. It is the location where one listens when measuring blood pressure. Immediately beyond the elbow joint, the brachial artery splits into the *radial* and *ulnar arteries,* which descend through the anterior aspect of the forearm.

Radial Artery The **radial artery** descends along the medial margin of the brachioradialis muscle, supplying muscles of the lateral anterior forearm, the lateral part of the wrist, and the thumb and index finger. At the root of the thumb just lateral to the tendon of flexor carpi radialis (shown in Figure 11.36, p. 340), the radial artery lies very near the surface and provides a convenient site for taking the pulse. A branch of the radial artery continues into the anatomical snuff box and a radial pulse can be detected there also.

Ulnar Artery The **ulnar artery,** which descends along the medial side of the anterior forearm, lies between the superficial and deep flexor muscles and sends branches to the muscles that cover the ulna. Proximally, it gives off a major branch called the **common interosseous artery,** which splits immediately into *anterior* and *posterior interosseous arteries.* These vessels descend along the respective surfaces of the interosseous membrane between the radius and the ulna. The anterior interosseous artery supplies the deep flexor muscles, whereas the posterior interosseous artery and its branches supply all the extensors on the posterior forearm. The ulnar artery continues into the hand, crossing the wrist just lateral to the tendon of flexor carpi ulnaris.

Palmar Arches In the palm, branches of the radial and ulnar arteries join to form two horizontal arches, the **superficial** and **deep palmar arches.** The superficial arch underlies the skin and fascia of the hand, whereas the deep arch lies against the metacarpal bones. The **digital arteries,** which supply the fingers, branch from these arches. (The radial and ulnar arteries also form a *carpal arch* on the dorsal side of the wrist. Branches from this dorsal arch run distally along the metacarpal bones.)

Arteries of the Abdomen

The arteries to the abdominal organs arise from the abdominal aorta (Figure 20.8a and Figure 20.12). In a person at rest, about half of the entire arterial flow is present in these vessels. Three midline branches (the *celiac trunk, superior mesenteric artery,* and *inferior mesenteric artery*) bring blood to the digestive tract, the inner tube. Paired branches supply structures of the outer tube (adrenal glands, kidneys, gonads, and abdominal body wall). You will consider these arteries from superior to inferior in the order they arise from the aorta.

Inferior Phrenic Arteries The paired **inferior phrenic arteries** branch from the abdominal aorta at the level of vertebra T_{12}, just inferior to the aortic opening (hiatus) of the diaphragm (Figure 20.12). These arteries supply the inferior surface of the diaphragm.

Celiac Trunk The short, wide, unpaired **celiac** (se′le-ak) **trunk** (Figure 20.13a supplies the viscera in the superior part of the abdominal cavity (*coelia* = abdominal cavity). Specifically, it sends branches to the stomach, liver, gallbladder, pancreas, spleen, and a part of the small intestine (duodenum).

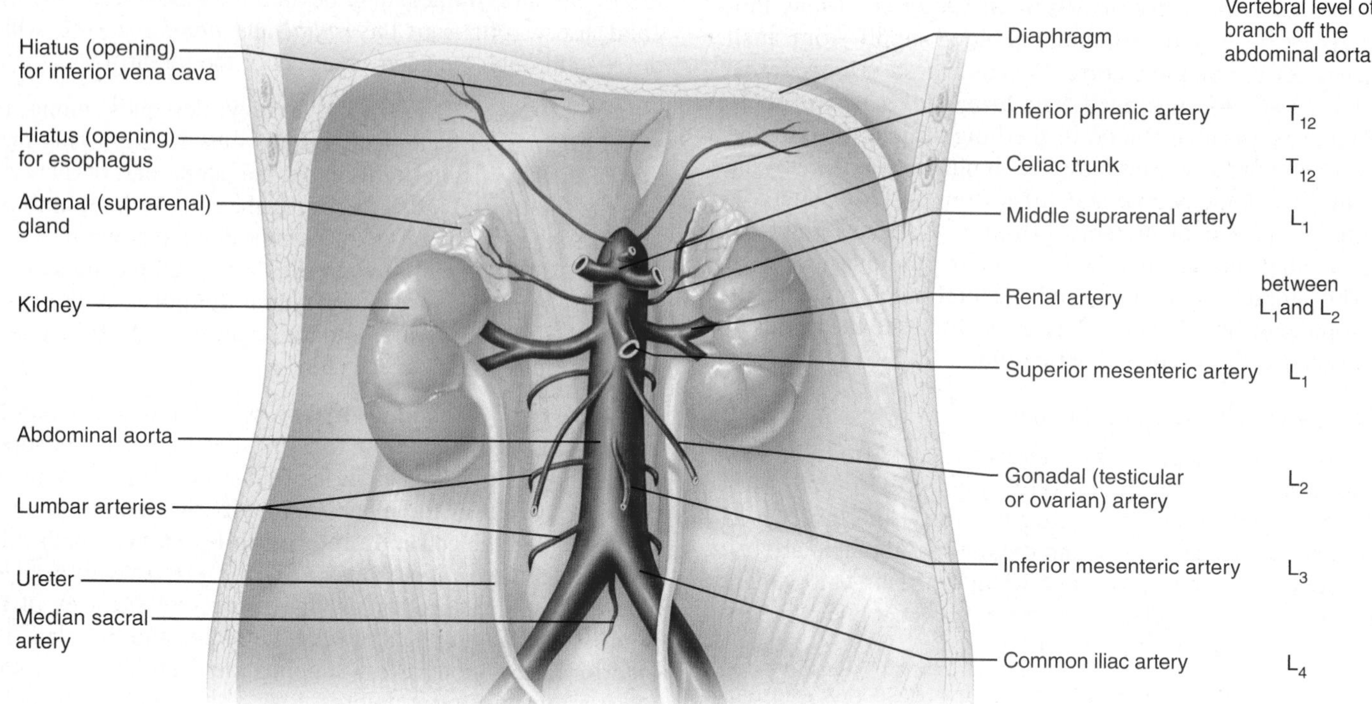

Figure 20.12 Major branches of the abdominal aorta.

It emerges midventrally from the aorta at the level of T_{12} and divides almost immediately into three branches: the *left gastric, splenic,* and *common hepatic arteries.*

The **left gastric artery** (*gaster* = stomach) runs superiorly and to the left, to the junction of the stomach with the esophagus, where it gives off several esophageal branches and descends along the right (lesser) curvature of the J-shaped stomach.

The large **splenic artery** runs horizontally and to the left, posterior to the stomach, to enter the spleen. It passes along the superior border of the pancreas, sending branches to this organ. Near the spleen, it sends several short branches superiorly to the stomach's dome *(short gastric arteries)* and sends a major branch along the stomach's left (greater) curvature—the **left gastroepiploic** (gas″tro-ep″ĭ-plo′ik) **artery.**

The **common hepatic artery** (*hepar, hepat* = liver) is the only branch of the celiac trunk that runs to the right. At the junction of the stomach with the small intestine (duodenum), this artery divides into an ascending branch, the *hepatic artery proper,* and a descending branch, the *gastroduodenal artery.* The **hepatic artery proper** divides into *right* and *left branches* just before entering the liver; the *cystic artery* to the gallbladder usually arises from the right branch of the hepatic artery. The **right gastric artery,** which can arise either from the hepatic artery proper or from the common hepatic artery, runs along the stomach's lesser curvature from the right. The **gastroduodenal artery** (gas″tro-du″o-de′nal), the descending branch of the common hepatic artery, runs inferiorly between the duodenum and the head of the pancreas. One branch, the **superior pancreaticoduodenal artery,** helps supply the pancreas, plus the nearby duodenum. The other branch, the **right gastroepiploic artery,** runs along the stomach's greater curvature from the right.

Superior Mesenteric Artery The large, unpaired **superior mesenteric** (mes″en-ter′ik) **artery** serves most of the intestines (Figure 20.13b). It arises midventrally from the aorta, posterior to the pancreas at the level of L_1. From there, it runs inferiorly and anteriorly to enter the mesentery, a drape-like membrane that supports the long, coiled parts of the small intestine (the jejunum and the ileum). The superior mesenteric artery angles gradually to the right as it descends through the mesentery. From its left side arise many **intestinal arteries** to the jejunum and ileum. From its right side emerge branches that supply the proximal half of the large intestine: the ascending colon, cecum, and appendix via the **ileocolic artery;** part of the ascending colon via the **right colic artery;** and part of the transverse colon via the **middle colic artery.**

Suprarenal Arteries The paired **middle suprarenal arteries** (Figure 20.12), which emerge from the sides of the aorta at L_1, supply blood to the adrenal (suprarenal) glands on the superior poles of the kidneys. The adrenal glands also receive *superior suprarenal* branches from the nearby inferior phrenic arteries, and *inferior suprarenal* branches (not illustrated) from the nearby renal arteries.

Renal Arteries The paired **renal arteries** to the kidneys (*ren* = kidney) stem from the sides of the aorta, between vertebrae L_1 and L_2 (Figure 20.12). The kidneys remove nitrogenous wastes from the blood delivered via the renal arteries. As mentioned previously, the transportation of cellular wastes

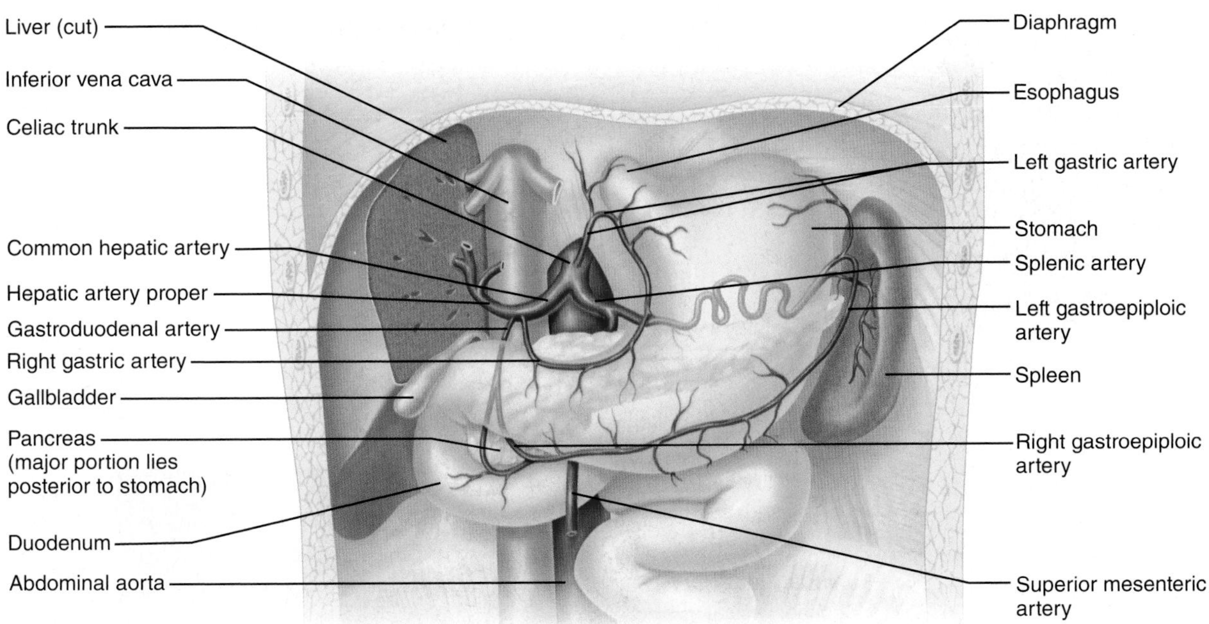

(a) The celiac trunk and its major branches. The left half of the liver has been removed.

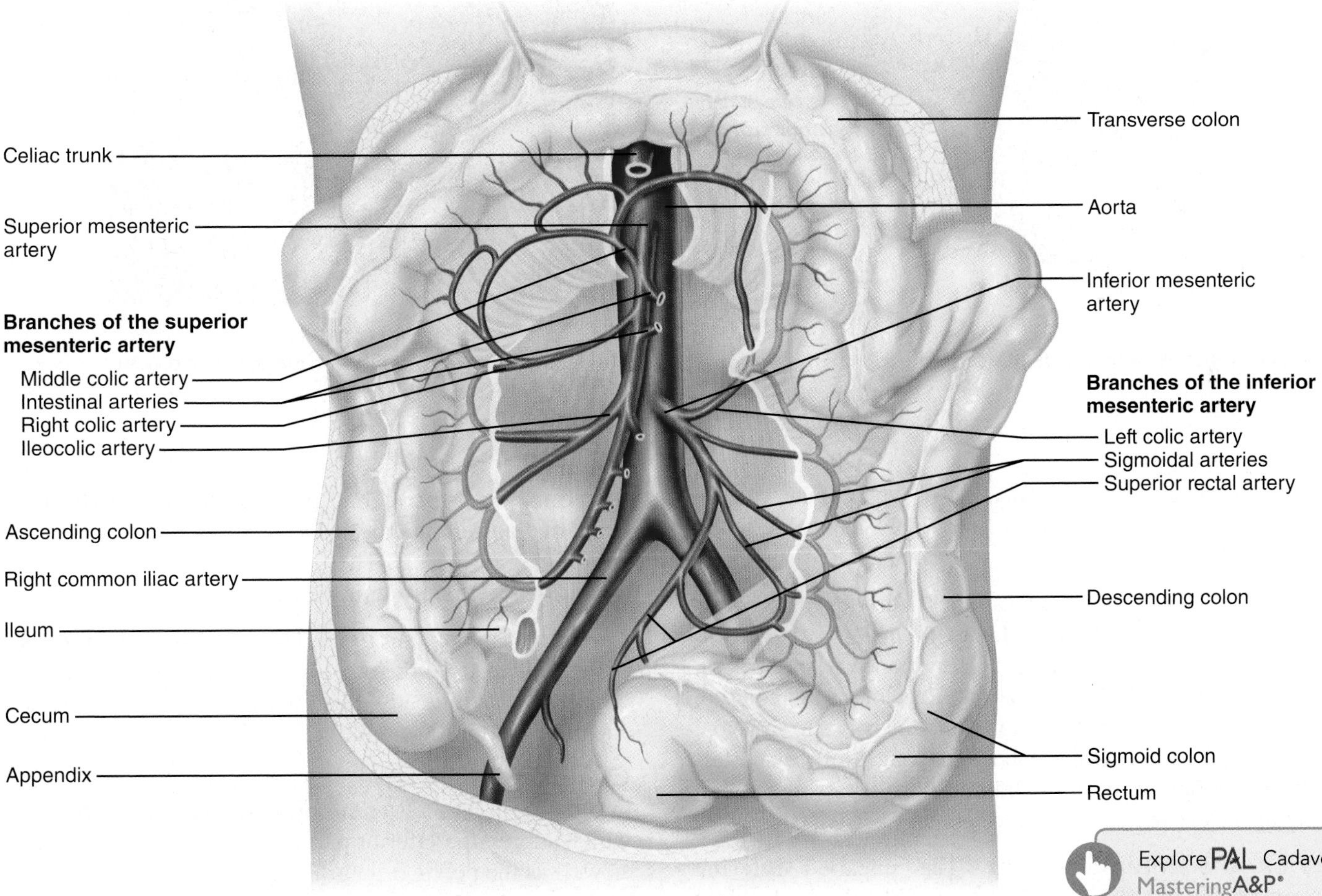

Celiac trunk
Superior mesenteric artery
Branches of the superior mesenteric artery
Middle colic artery
Intestinal arteries
Right colic artery
Ileocolic artery
Ascending colon
Right common iliac artery
Ileum
Cecum
Appendix
Transverse colon
Aorta
Inferior mesenteric artery
Branches of the inferior mesenteric artery
Left colic artery
Sigmoidal arteries
Superior rectal artery
Descending colon
Sigmoid colon
Rectum

(b) Distribution of the superior and inferior mesenteric arteries. The transverse colon has been pulled superiorly.

Figure 20.13 **Midline branches off the abdominal aorta supplying the organs of the digestive tract.**

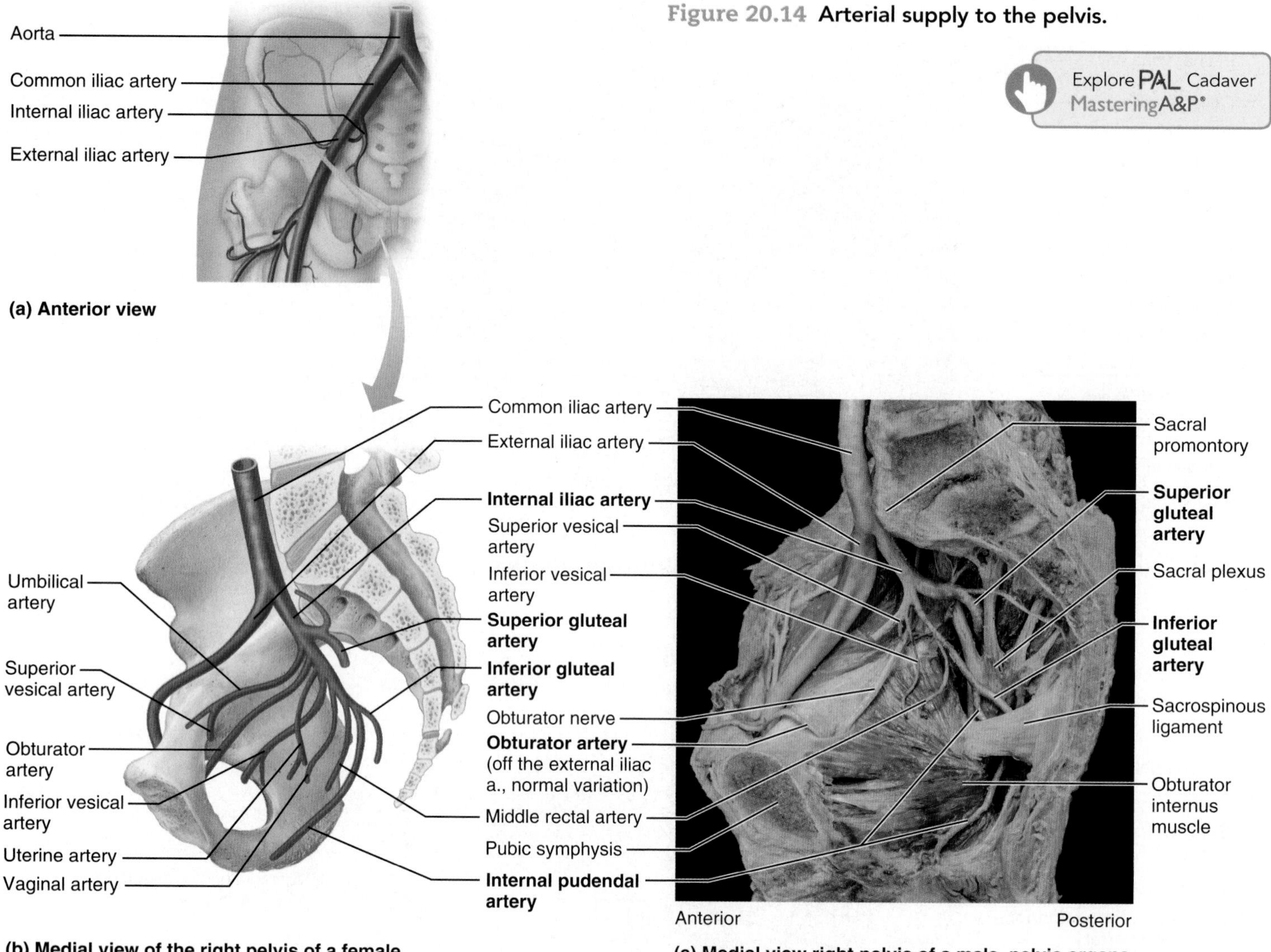

Figure 20.14 Arterial supply to the pelvis.

to the kidney is an important function of the vascular system, so the renal circulation is a major functional subdivision of the systemic circuit.

Gonadal Arteries The paired arteries to the gonads are more specifically called **testicular arteries** in males and **ovarian arteries** in females (Figure 20.12). They branch from the aorta at L_2, the level where the gonads first develop in the embryo. The ovarian arteries extend inferiorly into the pelvis to serve the ovaries and part of the uterine tubes. The longer testicular arteries extend through the anterior abdominal wall, passing through the inguinal canal to the scrotum, where they serve the testes.

Inferior Mesenteric Artery The unpaired **inferior mesenteric artery** (Figures 20.12 and 20.13b) is the final major branch of the abdominal aorta, arising midventrally at the level of L_3. It serves the distal half of the large intestine, from the last part of the transverse colon to the middle part of the rectum. Its branches are the **left colic** (which joins with the middle colic artery on the transverse colon), **sigmoidal,** and **superior rectal arteries.**

Lumbar Arteries Four pairs of **lumbar arteries** arise from the posterolateral surface of the aorta in the lumbar region (Figure 20.12). These segmental arteries run horizontally to supply the posterior abdominal wall.

Median Sacral Artery The unpaired **median sacral artery** issues from the most inferior part of the aorta (Figure 20.12). As it descends, this thin artery supplies the sacrum and coccyx along the midline.

Common Iliac Arteries At the level of L_4, the aorta splits into the right and left **common iliac arteries** (Figure 20.12), which supply the inferior part of the anterior abdominal wall, as well as the pelvic organs and the lower limbs.

Arteries of the Pelvis and Lower Limbs

At the level of the sacroiliac joint on the pelvic brim, each common iliac artery forks into two branches: the *internal iliac artery,* which mainly supplies the pelvic organs, and the *external iliac artery,* which supplies the lower limb (Figure 20.14a).

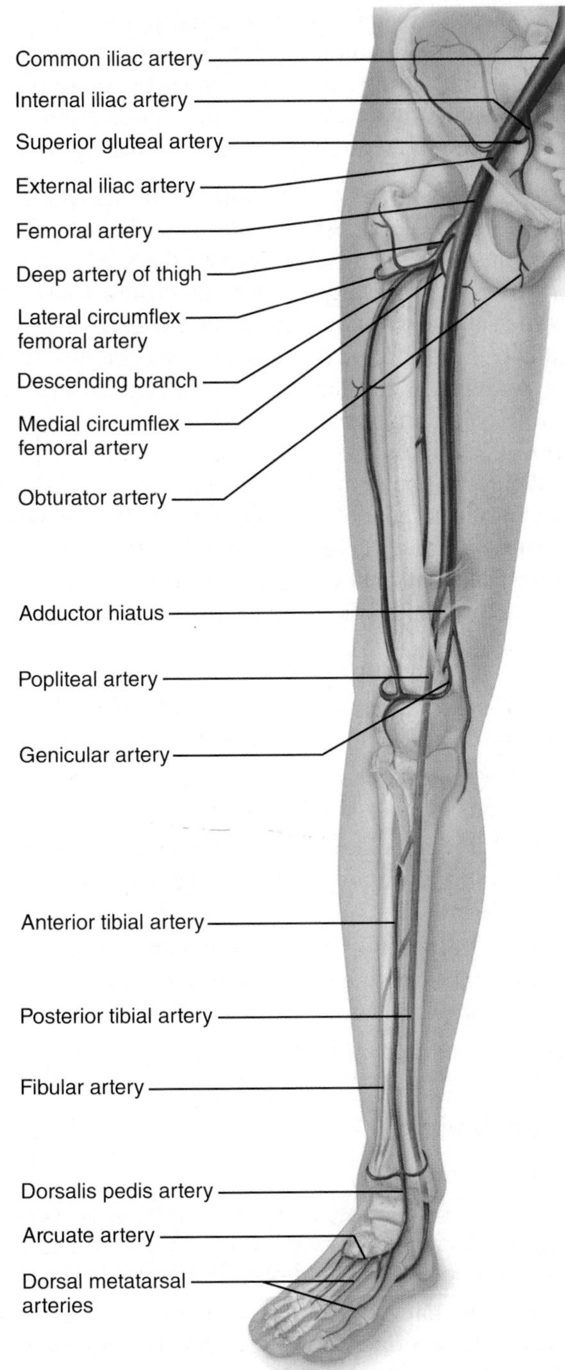

(a) Anterior view

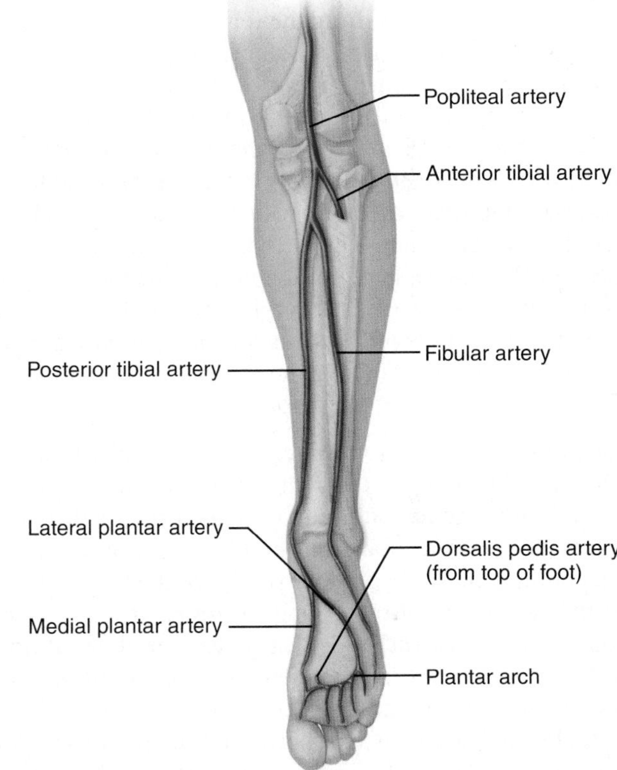

(b) Posterior view of leg

Figure 20.15 **Arteries of the right pelvis and lower limb.**

CLINICAL APPLICATION

Pulse Points in the Lower Limb

- **Femoral artery** In the femoral triangle (proximal thigh)
- **Popliteal artery** In the popliteal fossa (posterior knee)
- **Dorsalis pedis artery** Between the metatarsals I and II
- **Posterior tibial artery** Posterior to the medial malleolus

Internal Iliac Arteries The branches of **internal iliac arteries** (Figure 20.14b and c) supply blood to the pelvic walls, pelvic viscera, buttocks, medial thighs, and perineum. Among the most important branches are the **superior** and **inferior gluteal arteries,** which run posteriorly through the greater sciatic notch to supply the gluteal muscles; the **internal pudendal artery,** which leaves the pelvic cavity to supply the perineum and external genitalia; and the **obturator artery,** which descends through the obturator foramen into the thigh adductor muscles. Other branches run to the bladder (*superior* and *inferior vesical arteries*), the rectum (*middle rectal artery*), the uterus and vagina in females (*uterine* and *vaginal arteries*), and the pelvic reproductive glands in males (branches from the inferior vesical and middle rectal arteries).

External Iliac Artery The right and left **external iliac arteries** carry blood to the lower limbs (Figure 20.15). Originating from the common iliac arteries in the pelvis, each external iliac artery descends along the arcuate line of the ilium bone, sends some small branches to the anterior abdominal wall, and enters the thigh by passing deep to the midpoint of the inguinal ligament. At this point, the external iliac artery is called the femoral artery.

Femoral Artery The **femoral artery** descends vertically through the thigh medial to the femur and along the anterior surface of the adductor muscles. Superiorly, the artery descends through the femoral triangle, a region in the proximal thigh bordered by the sartorius muscle laterally and the adductor longus muscle medially. Inferiorly, the femoral artery passes through a gap in the adductor magnus muscle (the adductor hiatus) and emerges posterior to the distal femur as the *popliteal artery.*

Even though the superior part of the femoral artery is enclosed in a tube of dense fascia, it is relatively superficial and not protected by any overlying musculature. This lack of protection makes the proximal femoral artery a convenient place to take a pulse or apply pressure to stop bleeding from a hemorrhage in the distal limb, but it also makes it susceptible to injury.

Several arteries arise from the femoral artery in the thigh (Figure 20.15a). The largest branch, which arises superiorly and is called the **deep artery of the thigh** (or *profunda femoris* = deep femoral), is the main supplier of the thigh muscles: adductors, hamstrings, and quadriceps. Proximal branches of the deep femoral artery are the **medial** and **lateral circumflex femoral arteries,** which circle the neck and upper shaft of the femur. The medial circumflex artery is the major vessel to the head of the femur. If a fracture of the hip tears this artery, the bone tissue of the head of the femur dies. A long, **descending branch of the lateral circumflex artery** (see Figure 20.15a) runs along the anterior aspect of the vastus lateralis muscle, which it supplies.

Popliteal Artery The **popliteal artery** (pop″lĭ-te′al), the inferior continuation of the femoral artery, lies within the popliteal fossa (the region posterior to the knee), a deep location that offers protection from injury. You may be able to feel a popliteal pulse if you flex your leg at the knee and push your fingers firmly into the popliteal fossa. If a clinician is unable to feel a patient's popliteal pulse, the femoral artery may be narrowed by atherosclerosis. The popliteal artery gives off several **genicular arteries** (je-nik′u-lar; "knee") that circle the knee joint like horizontal hoops. Just inferior to the head of the fibula, the popliteal artery splits into the *anterior* and *posterior tibial arteries.*

Anterior Tibial Artery The **anterior tibial artery** runs through the anterior muscular compartment of the leg, descending along the interosseous membrane lateral to the tibia and sending branches to the extensor muscles along the way (Figure 20.15a). At the ankle, it becomes the **dorsalis pedis artery** ("artery of the dorsum of the foot"). At the base of the metatarsal bones, the **arcuate artery** branches from the dorsalis pedis and sends smaller branches distally along the metatarsals. The end part of the dorsalis pedis penetrates into the sole, where it forms the medial end of the plantar arch (described shortly).

The dorsalis pedis artery is superficial, and the pulse from this artery can be palpated in the proximal space between the first and second metatarsals (the pedal pulse point). The absence of this pulse can indicate that the blood supply to the leg is inadequate. Routine checking of the pedal pulse is indicated for patients known to have impaired circulation to the legs and for those recovering from surgery to the leg or thigh.

Posterior Tibial Artery The **posterior tibial artery** (Figure 20.15b), which descends through the posteromedial part of the leg, lies directly deep to the soleus muscle. Proximally, it gives off a large branch, the **fibular (peroneal) artery,** which descends along the medial aspect of the fibula. Together, the posterior tibial and fibular arteries supply the flexor muscles in the leg, and the fibular arteries send branches to the fibularis muscles.

Inferiorly, the posterior tibial artery passes behind the medial malleolus of the tibia, where its pulse can be palpated. On the medial side of the foot, it divides into **medial** and **lateral plantar arteries.** These serve the sole, and the lateral plantar artery forms the lateral end of the **plantar arch.** Metatarsal and digital arteries to the toes arise from the plantar arch.

check your understanding

☐ 5. Name the three vessels that branch off the aortic arch and the body region each serves.

☐ 6. Name the four major vessels that supply blood to the brain. From what vessel does each arise?

☐ 7. Which vessel do you palpate to feel a pulse in each of the following locations: (a) thigh, (b) arm, (c) wrist, (d) foot, (e) neck?

☐ 8. Name the vessels that branch off the abdominal aorta to supply blood to the digestive organs, and indicate which organs each supplies.

(For answers, see Appendix B.)

Systemic Veins

learning outcomes

- Trace the veins that empty into the superior vena cava as you describe the routes of venous return from the brain, the head and neck, and the thorax and upper limbs.
- Trace the veins that empty into the inferior vena cava as you trace venous return from the abdominal organs and from the pelvis and lower limbs.
- Describe the structure and special function of the hepatic portal system and explain the significance of portal-systemic anastomoses.

Having considered the arteries of the body, you now turn to the veins (Figure 20.16). Although most veins run with corresponding arteries, there are some important differences in the distributions of arteries and veins:

- Whereas just one systemic artery leaves the heart—the aorta exiting the left ventricle, three major veins enter the right atrium of the heart: the superior and inferior venae cavae and the coronary sinus.
- All large and medium-sized arteries have deep locations and are accompanied by deep veins, commonly of similar

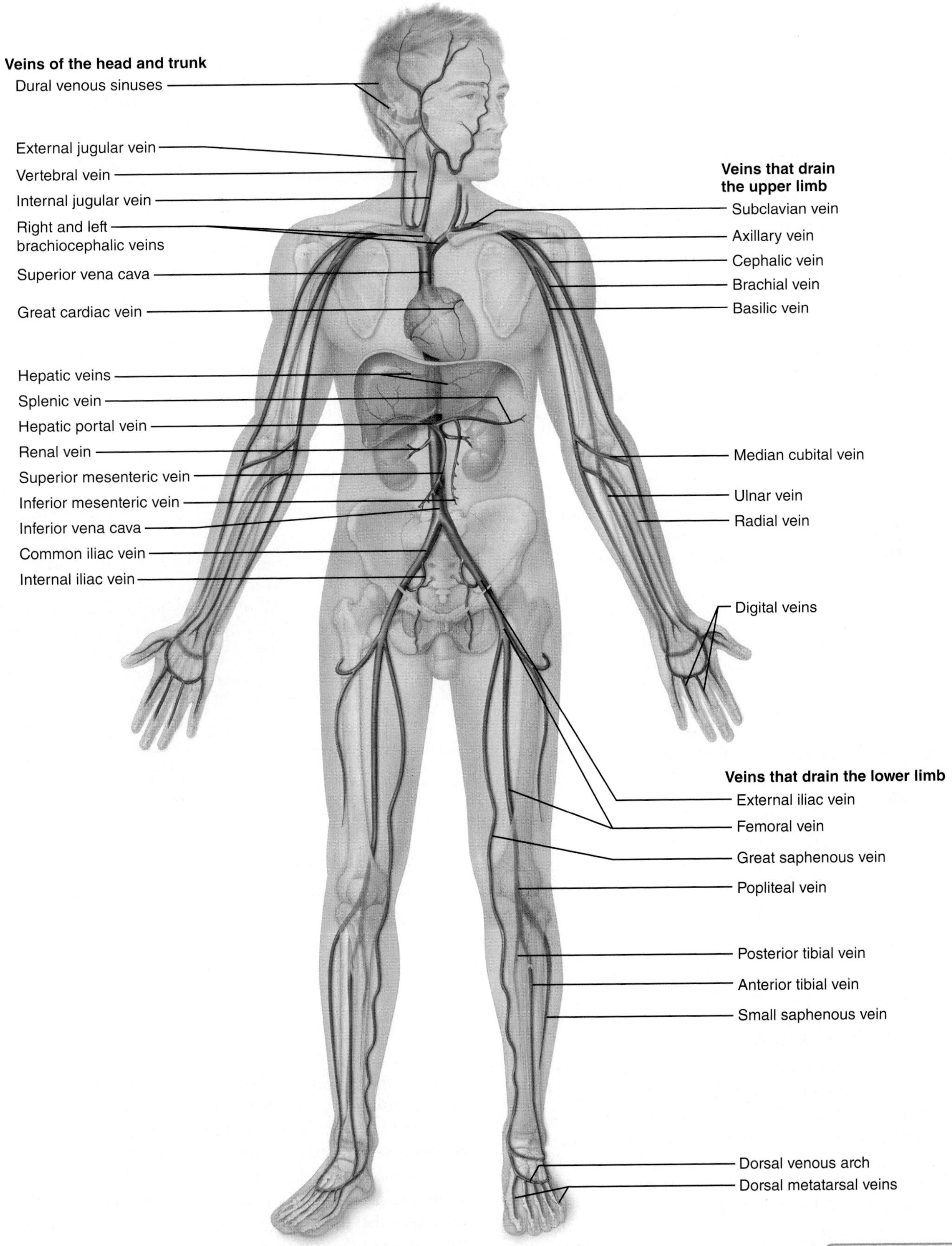

Veins of the head and trunk
Dural venous sinuses
External jugular vein
Vertebral vein
Internal jugular vein
Right and left brachiocephalic veins
Superior vena cava
Great cardiac vein
Hepatic veins
Splenic vein
Hepatic portal vein
Renal vein
Superior mesenteric vein
Inferior mesenteric vein
Inferior vena cava
Common iliac vein
Internal iliac vein
Veins that drain the upper limb
Subclavian vein
Axillary vein
Cephalic vein
Brachial vein
Basilic vein
Median cubital vein
Ulnar vein
Radial vein
Digital veins
Veins that drain the lower limb
External iliac vein
Femoral vein
Great saphenous vein
Popliteal vein
Posterior tibial vein
Anterior tibial vein
Small saphenous vein
Dorsal venous arch
Dorsal metatarsal veins

Figure 20.16 Major veins of the systemic circulation, anterior view.

Practice art labeling
MasteringA&P®

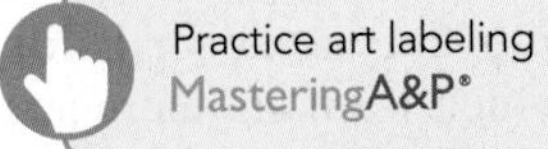

name. In addition, veins are also found just beneath the skin unaccompanied by any arteries. These *superficial veins* are important clinically because they provide sites for drawing blood or placing an intravenous line. Their superficial location also makes them susceptible to cuts or injuries.

- Commonly, two or more parallel veins drain a body region rather than a single larger vein. In some regions, multiple veins anastomose to form a *venous plexus*. Observe the venous plexus draining the dorsum of your hand.
- The brain and digestive tract have unusual patterns of venous drainage. Veins from the brain drain into *dural venous sinuses,* which are not typical veins but endothelium-lined channels supported by walls of dura mater. Venous blood draining from the digestive organs enters a special subcirculation, the *hepatic portal system,* and passes through capillaries in the liver before the blood reenters the general systemic circulation. The dural venous sinuses and hepatic portal system are considered shortly.

Venae Cavae and Their Major Tributaries

The unpaired *superior* and *inferior venae cavae,* the body's two largest veins, empty directly into the right atrium of the heart (see Figure 20.9). The name *vena cava* means "cavelike vein."

Superior Vena Cava The **superior vena cava** (Figure 20.16) receives the systemic blood from all body regions superior to the diaphragm excluding the heart wall. This vein arises from the union of the **left** and **right brachiocephalic veins** posterior to the manubrium and descends to join the right atrium. Of the two brachiocephalic veins, the left is longer and nearly horizontal, whereas the right is vertical (see Figure 20.9). Each brachiocephalic vein is formed by the union of an *internal jugular vein* and *a subclavian vein.*

Inferior Vena Cava The **inferior vena cava,** which ascends along the posterior wall of the abdominal cavity and is the widest blood vessel in the body, returns blood to the heart from all body regions inferior to the diaphragm (Figure 20.16). It begins inferiorly at the union of the two common iliac veins on the body of vertebra L_5, and ascends on the right side of the vertebral bodies to the right of the abdominal aorta. Upon penetrating the central tendon of the diaphragm at T_8, the inferior vena cava joins the right atrium.

Veins of the Head and Neck

Most blood draining from the head and neck enters three pairs of veins: (1) the internal jugular veins that receive blood from the dural venous sinuses, (2) the external jugular veins, and (3) the vertebral veins. Even though most of the extracranial veins have the same names as the extracranial arteries (facial, ophthalmic, occipital, and superficial temporal), their courses and interconnections differ substantially.

Dural Venous Sinuses Most veins of the brain drain into the intracranial **dural venous sinuses** (Figure 20.17), which form an interconnected series of channels in the skull and lie between the two layers of cranial dura mater. The **superior** and **inferior sagittal sinuses** lie in the falx cerebri between the cerebral hemispheres. The inferior sagittal sinus drains posteriorly into the **straight sinus.** The superior sagittal and straight sinuses then drain posteriorly into the **transverse sinuses,** which run in shallow grooves on the internal surface of the occipital bone. Each transverse sinus drains into an S-shaped sinus **(sigmoid sinus),** which becomes the **internal jugular vein** as it leaves the skull through the *jugular foramen.*

The paired **cavernous sinuses** border the body of the sphenoid bone laterally, and each has an internal carotid artery *running within it.* The following cranial nerves also run within the cavernous sinus (or in its wall of dura mater) on

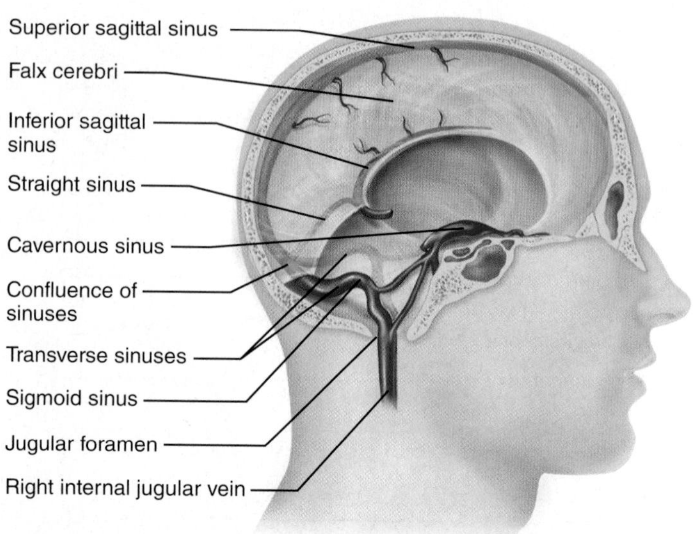

(a) Dural venous sinuses

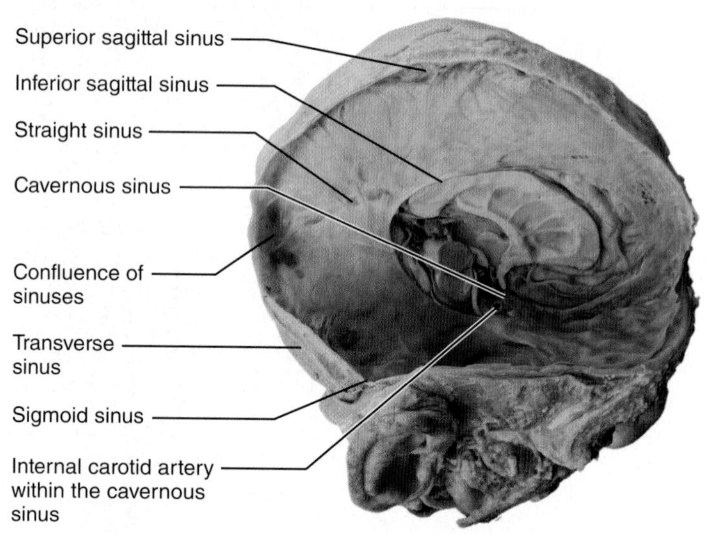

(b) Dural venous sinuses, cadaver, right side

Figure 20.17 Venous drainage of the head, neck, and brain.

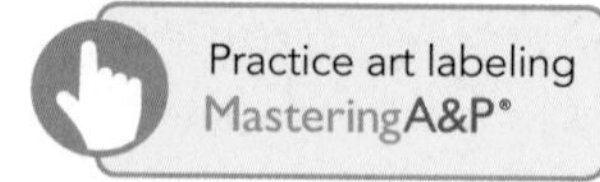

their way to the orbit and the face: the oculomotor, trochlear, abducens, and the maxillary and ophthalmic divisions of the trigeminal nerve. The cavernous sinus drains posterolaterally into the transverse sinus and the internal jugular vein but also communicates with the **ophthalmic vein** of the orbit (see Figure 20.18), which in turn communicates with the facial vein, which drains the nose and upper lip.

Several other dural sinuses exist, but they are small and are not considered here.

CLINICAL APPLICATION

Medical Importance of the Cavernous Sinus

Squeezing pimples on the nose or upper lip can spread infection through the facial vein into the cavernous sinus and, from there, through the other dural sinuses in the skull. For this reason, the nose and upper lip are called the *danger triangle of the face*.

Blows to the head can rupture the internal carotid artery within the confines of the cavernous sinus. Leaked blood then accumulates within this sinus and exerts crushing pressure on the contained cranial nerves, disrupting their functions. For example, damage to the oculomotor, trochlear, and abducens nerves leads to a loss of ability to move the eyes.

Internal Jugular Veins The large **internal jugular veins** drain almost all of the blood from the brain (Figure 20.18). From its origin at the base of the skull, each internal jugular vein descends through the neck deep to the sternocleidomastoid muscle. Superiorly, the internal jugular vein lies lateral to the internal carotid artery and inferiorly it is lateral to the common carotid artery (shown in Figure 11.11b, p. 285). Along the way, the internal jugular vein receives blood from some deep veins of the face and neck—branches of the **facial** and **superficial temporal veins.** At the base of the neck, the internal jugular vein joins the subclavian vein to form the brachiocephalic vein. The jugular veins are named for their end point, as *jugulum* means "the throat just above the clavicle."

Just as wounds to the neck can cut the carotid arteries, they can also sever the internal jugular veins. Cut veins do not bleed as quickly as arteries of comparable size, because the blood pressure in them is lower than in arteries. For this reason, the chances of surviving are greater if a neck wound affects the vein and not the artery.

External Jugular Veins The **external jugular vein** is a superficial vein that descends vertically through the neck on the surface of the sternocleidomastoid muscle. To make this vein visible on your neck, stand before a mirror and gently compress the skin superior to your clavicle with your fingers. Superiorly, its tributaries drain the posterior scalp, lateral scalp, and some of the face (see Figure 20.18). The external jugular vein, which is not accompanied by any corresponding artery, empties inferiorly into the subclavian vein.

Vertebral Veins Unlike the vertebral arteries, the **vertebral veins** do not serve much of the brain, instead draining only the cervical vertebrae, cervical spinal cord, and small muscles in the superior neck. Originating inferior to the occipital condyle, each vertebral vein descends through the transverse foramina of vertebrae C_1–C_6 in the form of a venous plexus. Emerging from C_6 as a single vein, the vertebral vein continues inferiorly to join the brachiocephalic vein in the root of the neck.

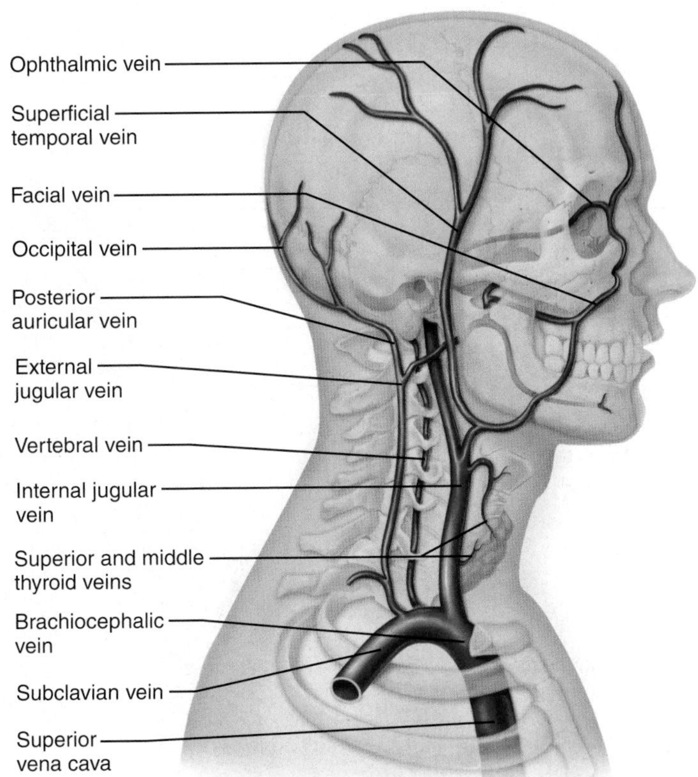

Figure 20.18 Veins of the head and neck, right superficial aspect.

Veins of the Thorax

Blood draining from the first few intercostal spaces enters the brachiocephalic veins. Blood from the other intercostal spaces, as well as from some of the thoracic viscera, drains into a group of veins called the *azygos* (āz′ĭ-gos, or a-zi′gus) *system* (Figure 20.19). This group of veins, which flank the vertebral column and ultimately empty into the superior vena cava, consists of the *azygos vein,* the *hemiazygos vein,* and the *accessory hemiazygos vein.*

Azygos Vein The **azygos vein,** whose name means "unpaired," ascends along the right or the center of the thoracic vertebral bodies. It receives all of the right **posterior intercostal veins** (except the first), plus the subcostal vein. Superiorly, at the level of T_4, the azygos arches over the great vessels that run to the root of the right lung and joins the superior vena cava.

Hemiazygos Vein The **hemiazygos vein** (hem″ĭ-āz′ĭ-gos), which ascends on the left side of the vertebral column, corresponds to the inferior half of the azygos on the right (*hemiazygos* = half the azygos). It receives the ninth through eleventh left posterior intercostal veins and the subcostal vein.

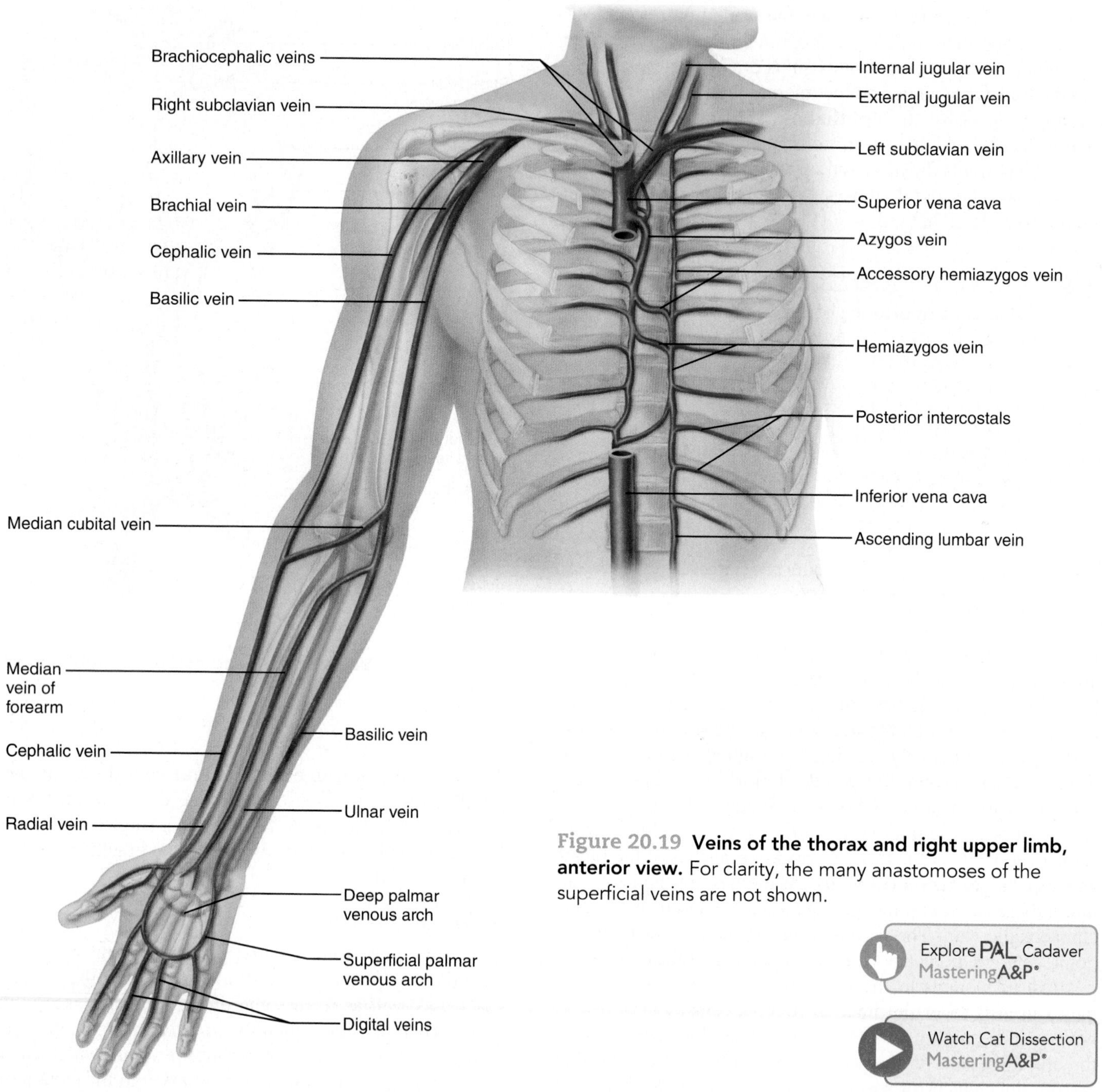

Figure 20.19 Veins of the thorax and right upper limb, anterior view. For clarity, the many anastomoses of the superficial veins are not shown.

Explore PAL Cadaver MasteringA&P®

Watch Cat Dissection MasteringA&P®

At about midthorax, the hemiazygos runs roughly horizontally across the vertebrae and joins the azygos vein.

Accessory Hemiazygos Vein The **accessory hemiazygos vein** can be thought of as a superior continuation of the hemiazygos, receiving the fourth (or fifth) through the eighth left posterior intercostal veins; it also courses to the right to join the azygos.

The superior parts of the azygos and accessory hemiazygos veins also receive the small *bronchial veins,* which drain poorly oxygenated systemic blood from the bronchi in the lungs. Veins from the thoracic part of the esophagus also enter the azygos system.

Veins of the Upper Limbs

The veins of the upper limbs (Figure 20.19) are either deep or superficial.

Deep Veins The deep veins of the upper limbs (shown in Figure 20.19 as light blue vessels) follow the paths of their companion arteries and have the same names. All except the largest, however, are actually two parallel veins that flank their artery on both sides. The **deep** and **superficial palmar venous arches** of the hand empty into the **radial** and **ulnar veins** of the forearm, which unite just inferior to the elbow joint to form the **brachial vein** of the arm. As the brachial vein enters the axilla, it empties into the **axillary vein,** which becomes the **subclavian vein** at the first rib.

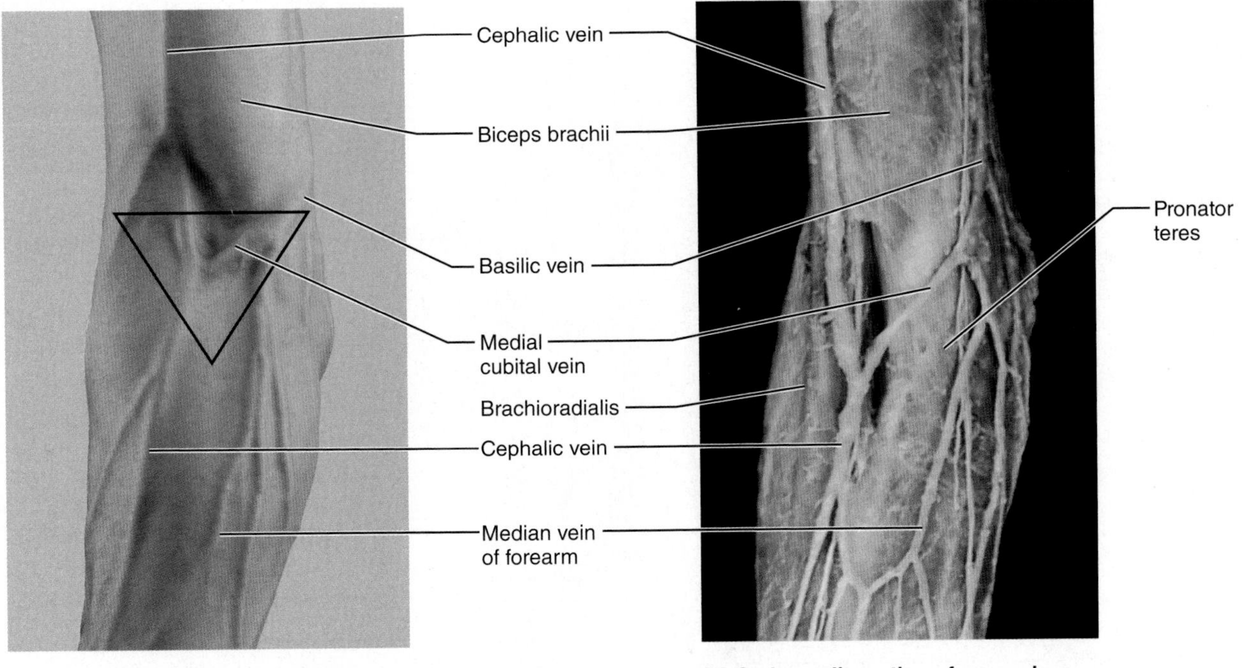

(a) Surface view, right upper limb

(b) Cadaver dissection of same view

Figure 20.20 **The superficial veins of the right upper limb.** The cubital fossa is outlined by the triangle.

Superficial Veins The superficial veins of the upper limb (shown in Figure 20.19 as dark blue vessels) are larger than the deep veins and are visible beneath the skin. They form anastomoses frequently along their course. They begin with the *dorsal venous network* (not illustrated). These veins are readily apparent on the dorsal surface of the hand because of the thinness of the skin. This network provides a preferred site for inserting intravenous catheters. The dorsal venous network drains superiorly into the **cephalic vein,** which starts at the lateral side of this network, then bends around the distal radius to enter the anterior forearm. From there, this vein ascends through the anterolateral side of the entire limb and ends inferior to the clavicle, where it joins the axillary vein. The **basilic vein** arises from the medial aspect of the hand's dorsal venous network, then ascends along the posteromedial forearm and the anteromedial surface of the arm. In the axilla, the basilic vein joins the brachial vein to become the axillary vein. On the anterior aspect of the elbow joint, in the region called the cubital fossa, the **median cubital vein** connects the basilic and cephalic veins (Figure 20.20). The median cubital vein is easy to find in most people and is used to obtain blood or to administer substances intravenously. Because the large brachial artery, along with the median nerve and the tendon of insertion for the biceps brachii, lie just deep to the median cubital vein, a needle must be inserted into the vein at a shallow angle, almost parallel to the skin, to avoid puncturing these structures. The **median vein of the forearm** ascends in the center of the forearm; its termination point at the elbow is highly variable.

Veins of the Abdomen

Blood returning from the abdominopelvic viscera and the abdominal wall reaches the heart via the inferior vena cava (Figure 20.21). Most venous tributaries of this great vein share the names of the corresponding arteries. The veins from the paired abdominal organs, the pelvis, and the abdominal wall drain directly into the inferior vena cava. Blood from the digestive organs returns via the hepatic portal system (discussed below).

Lumbar Veins Several pairs of **lumbar veins** drain the posterior abdominal wall, running horizontally with the corresponding lumbar arteries.

Gonadal (Testicular or Ovarian) Veins The right and left **gonadal veins** ascend along the posterior abdominal wall with the gonadal arteries. The right gonadal vein drains into the anterior surface of the inferior vena cava at L_2. The left gonadal vein drains into the left renal vein (Figure 20.21).

Renal Veins The right and left **renal veins** drain the kidneys; each lies just anterior to the corresponding renal artery.

Suprarenal Veins Although each adrenal gland has several main arteries, it has just one **suprarenal vein.** The right suprarenal vein empties into the nearby inferior vena cava; the left suprarenal vein drains into the left renal vein.

Hepatic Veins The right and left **hepatic veins** exit the liver superiorly and empty into the most superior part of the inferior vena cava. These robust veins carry all the blood that originated in the digestive organs in the abdominal and pelvic cavities and arrived via the hepatic portal system (discussed next).

Hepatic Portal System The **hepatic portal system** (Figure 20.22) is a specialized part of the vascular circuit that serves a function unique to digestion: It picks up digested

(Text continues on page 616.)

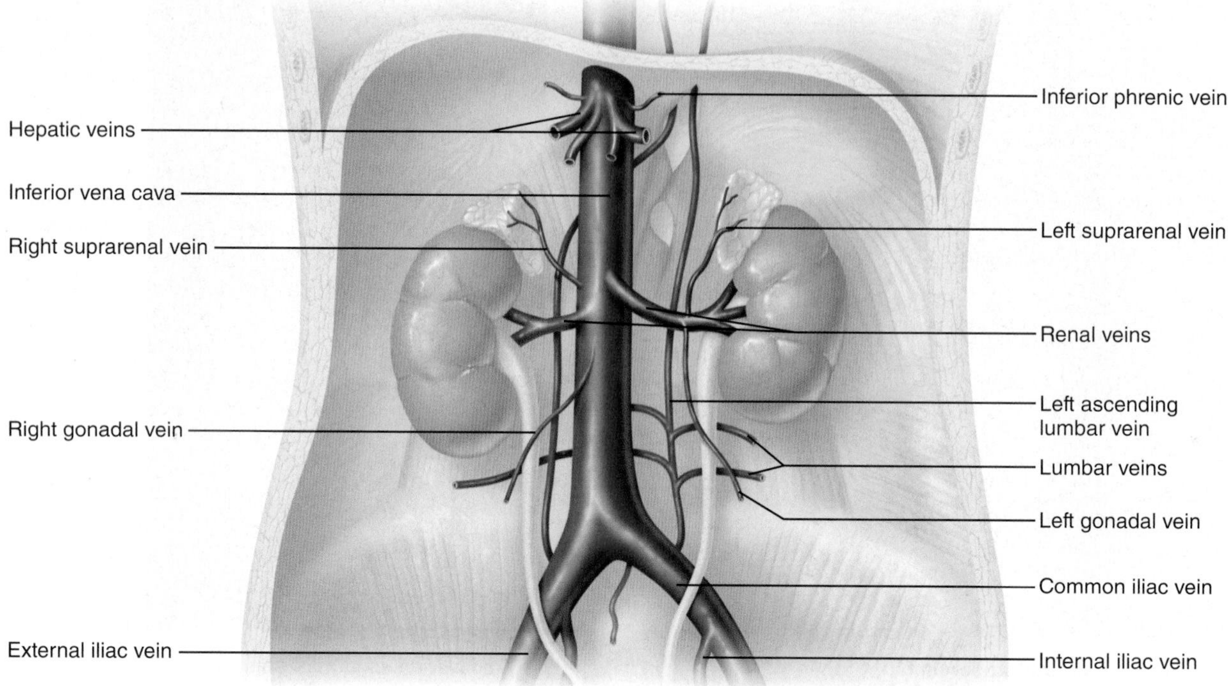

(a) Tributaries of the inferior vena cava; venous drainage of the paired abdominal organs

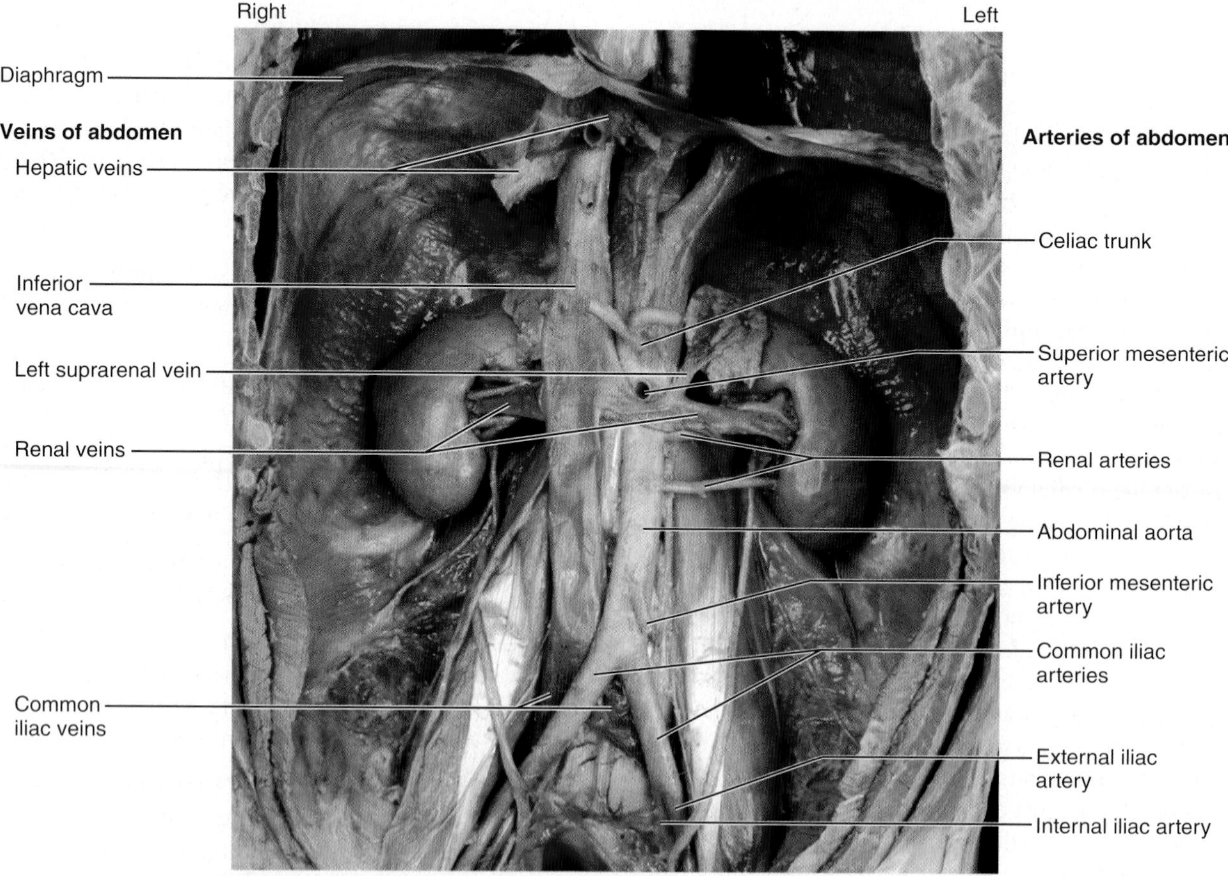

(b) Dissection of the posterior abdominal wall illustrating abdominal vessels

Figure 20.21 **Tributaries of the inferior vena cava.** In (a), notice the asymmetry in venous return: the left gonadal and suprarenal veins drain into the left renal vein. On the right side, these vessels drain directly into the inferior vena cava.

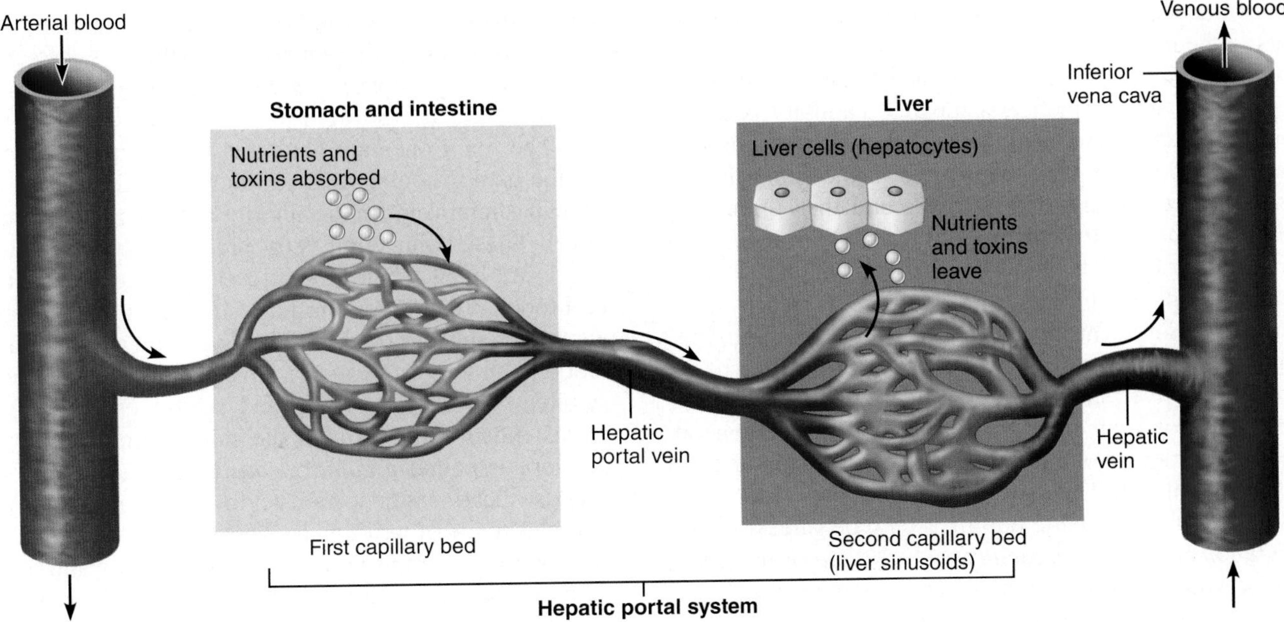

(a) Schematic of the hepatic portal system

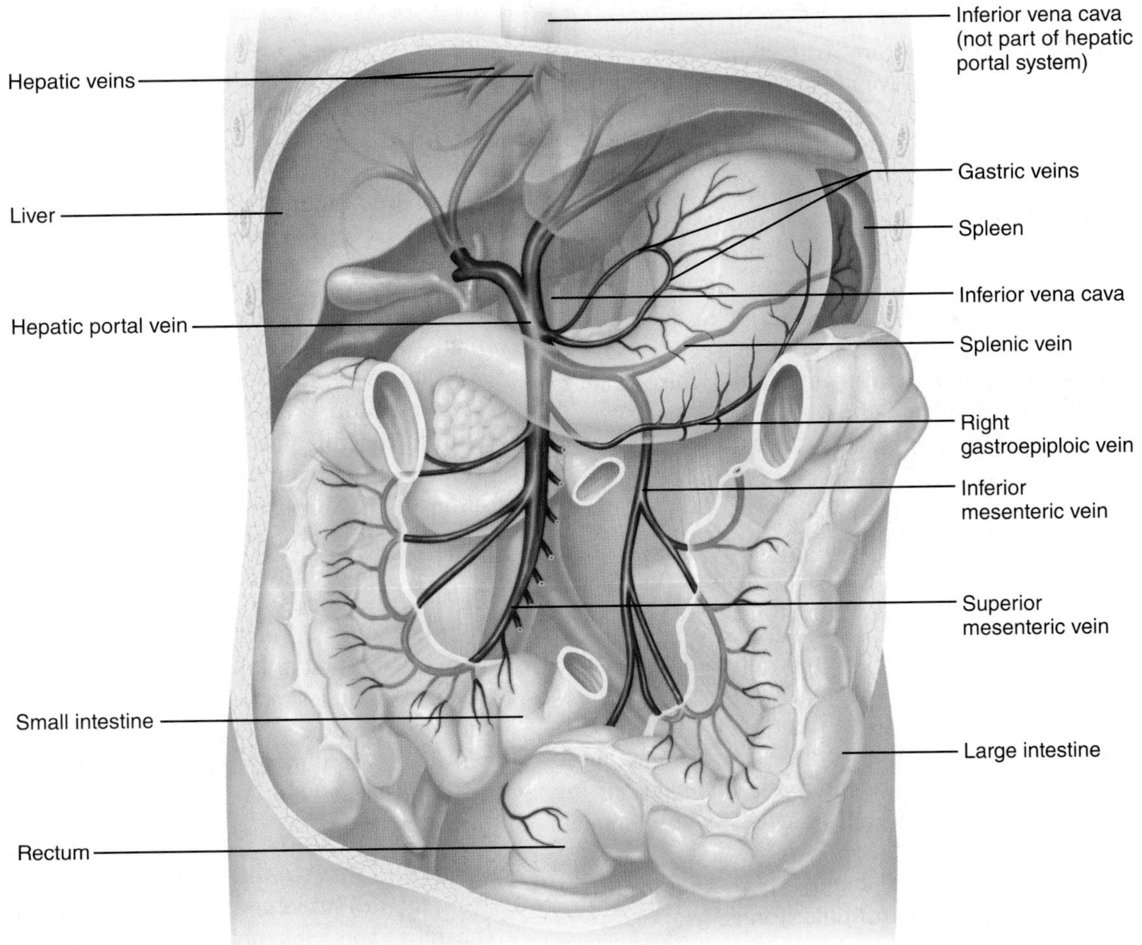

(b) The veins of the hepatic portal system

Figure 20.22 **The hepatic portal system.**

nutrients from the stomach and intestines and delivers these nutrients to the liver for processing and storage.

Like all portal systems, the hepatic portal system is a series of vessels in which two separate capillary beds lie between the arterial supply and the final venous drainage (Figure 20.22a). In this case, capillaries in the stomach and intestines receive the digested nutrients and then drain into the tributaries of the **hepatic portal vein.** This vein then delivers the nutrient-rich blood to a second capillary bed—the *liver sinusoids*—through which nutrients reach liver cells for processing. The liver cells also break down toxins that enter the blood through the digestive tract. After passing through the liver sinusoids, the blood enters the hepatic veins and inferior vena cava, thereby reentering the general systemic circulation. (For a more complete discussion of liver functions, see Chapter 23.)

As you study the hepatic portal system, be careful not to confuse the *hepatic veins,* which carry blood *out* of the liver, with the *hepatic portal vein,* which carries venous blood *into* the liver. The following veins of the hepatic portal system are tributaries of the hepatic portal vein (Figure 20.22b):

1. **Superior mesenteric vein.** This large vein ascends just to the right of the superior mesenteric artery. It drains the entire small intestine, the first half of the large intestine (ascending and transverse colon), and some of the stomach. Its superior part lies posterior to the stomach and pancreas.
2. **Splenic vein.** Even though the spleen is not a digestive organ, venous blood leaving it drains through the hepatic portal system. As a result, any microbes that escape the spleen's infection-fighting activities (discussed in Chapter 21) are carried to the liver for destruction. The splenic vein runs horizontally, posterior to the stomach and pancreas, and joins the superior mesenteric vein to form the hepatic portal vein. Its tributaries correspond to the branches of the splenic artery.
3. **Inferior mesenteric vein.** This vein ascends along the posterior abdominal wall, well to the left of the inferior mesenteric artery. Its tributaries drain the organs that are supplied by that artery—namely, the distal region of the colon and the superior rectum. The inferior mesenteric vein empties into the splenic vein posterior to the stomach and pancreas.

The **hepatic portal vein** is a short, vertical vessel that lies directly inferior to the liver and anterior to the inferior vena cava (Figure 20.22b). Inferiorly, it begins posterior to the pancreas as the union of the superior mesenteric and splenic veins. Superiorly, it enters the underside of the liver and divides into right and left branches, whose smaller branches reach the liver sinusoids.

On its way to the liver, the hepatic portal vein receives the right and left gastric veins from the stomach.

Portal-Systemic Anastomoses

In conditions that lead to scarring and degeneration (cirrhosis) of the liver, especially chronic alcoholism, blood flow through the liver sinusoids is blocked. This blockage raises the blood pressure throughout the hepatic portal system and results in **portal hypertension.** Fortunately, some veins of the portal system form anastomoses with veins that drain into the venae cavae, providing emergency pathways through which the "backed up" portal blood can return to the heart. These pathways are the *portal-systemic (portal-caval) anastomoses.* The main ones are (1) veins in the inferior esophagus, (2) the hemorrhoidal veins in the wall of the anal canal, and (3) superficial veins in the anterior abdominal wall around the navel. These connecting veins are small, however, and they swell and burst when forced to carry large volumes of portal blood. When these anastomosing veins start to fail, as in alcoholics with cirrhosis of the liver, the person may vomit blood from torn esophageal veins, develop hemorrhoids from swollen hemorrhoidal veins, and exhibit a snakelike network of distended veins through the skin around the navel. This network is called a *caput medusae* (kap′ut me-du′se)—the Medusa head—after a monster in Greek mythology whose hair was made of writhing snakes.

Of all the symptoms of cirrhosis, swelling and bursting of veins in the inferior esophagus is the most serious. Bleeding of esophageal veins is associated with a 50% mortality rate, and if the bleeding recurs, another 30% die. Varicose esophageal veins are treated by injecting a hardening agent into them or by tying them off with bands. Also, an implanted metal tube can be run from the inferior vena cava behind the liver into the portal vein, creating a new portalcaval shunt that relieves portal hypertension entirely.

Veins of the Pelvis and Lower Limbs

Veins draining the pelvis and lower limbs (Figure 20.23) are either deep or superficial.

Deep Veins Like those in the upper limb, most deep veins in the lower limbs share the names of the arteries they accompany, and all but the largest are actually two parallel veins. Arising on the sole of the foot from the union of the medial and lateral **plantar veins,** the **posterior tibial vein** ascends deep within the calf muscles and receives the **fibular (peroneal) vein.** The **anterior tibial vein,** which is the superior continuation of the **dorsalis pedis vein** of the foot, ascends to the superior part of the leg, where it unites with the posterior tibial vein to form the **popliteal vein.** The popliteal vein passes through the popliteal fossa and ascends to become the **femoral vein,** which drains the thigh. The femoral vein is located with the femoral artery and nerve in the femoral triangle. This vessel continues superiorly deep to the inguinal ligament and becomes the **external iliac vein.** In the pelvis, the external iliac vein unites with the **internal iliac vein** to form the **common iliac vein.**

Superficial Veins Two large superficial veins, the *great* and *small saphenous veins* (sah-fe′nus; "obvious"), arise from the **dorsal venous arch** located on the dorsal surface of the foot. The saphenous veins frequently form anastomoses with each other and with deep veins along their course. The **great saphenous vein,** the longest vein in the body, ascends along the medial side of the entire limb to empty into the femoral vein just distal to the inguinal ligament. The **small saphenous vein** runs along the lateral side of the foot and then along the posterior calf (Figure 20.23b and c). Posterior to the knee, it empties into the popliteal vein.

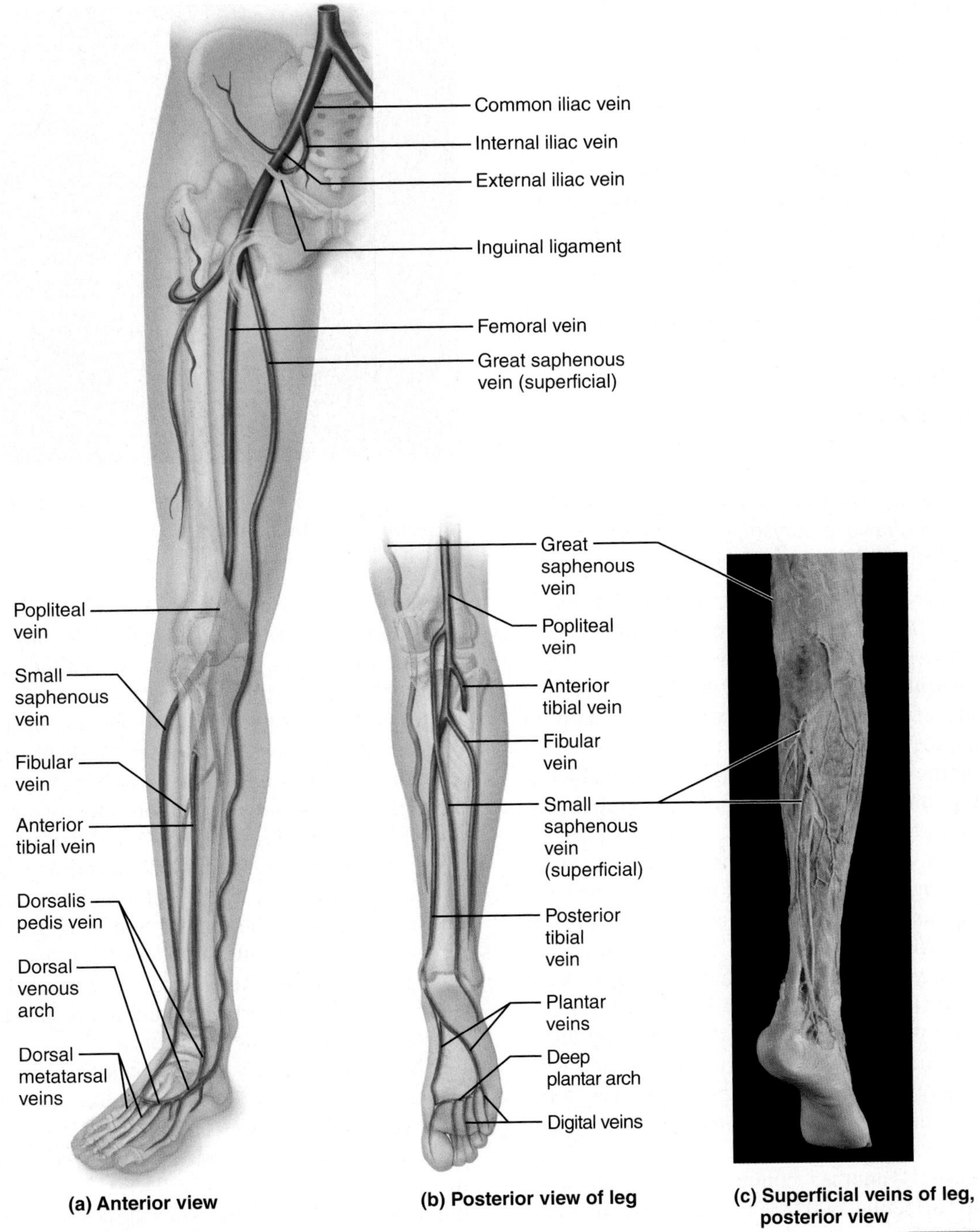

Common iliac vein
Internal iliac vein
External iliac vein
Inguinal ligament
Femoral vein
Great saphenous vein (superficial)
Popliteal vein
Small saphenous vein
Fibular vein
Anterior tibial vein
Dorsalis pedis vein
Dorsal venous arch
Dorsal metatarsal veins

(a) Anterior view

Great saphenous vein
Popliteal vein
Anterior tibial vein
Fibular vein
Small saphenous vein (superficial)
Posterior tibial vein
Plantar veins
Deep plantar arch
Digital veins

(b) Posterior view of leg

Great saphenous vein
Small saphenous vein

(c) Superficial veins of leg, posterior view

Figure 20.23 **Veins of the right lower limb and pelvis.**

CLINICAL APPLICATION

Medical Importance of the Saphenous Veins

Veins The great saphenous vein is the vessel most commonly used in coronary artery bypass operations (see A Closer Look, Chapter 19, on p. 581). Here, it should be noted that when suturing this vein onto a coronary artery, the surgeon must orient it so that blood flow will open, rather than close, its valves. The saphenous veins are more likely to weaken and become varicose than any other veins in the lower limb, because they are poorly supported by surrounding tissue. Furthermore, when valves begin to fail in veins throughout a lower limb, the normal contractions of the leg muscles can squeeze blood out of the deep veins into the superficial veins through the anastomoses between these two groups of veins. This influx of blood engorges and weakens the saphenous veins even further. Even when the saphenous veins do not fail, they are often involved in *venous disease of the lower limb* (see "Disorders of the Blood Vessels" on p. 618).

check your understanding

☐ 9. Consider the hepatic veins and the hepatic portal vein. Which organ or organs does each vein drain? Into which vessel does each vein empty?

☐ 10. In which body region are each of the following veins located: (a) cephalic vein, (b) popliteal vein, (c) transverse sinus, (d) saphenous vein, (e) azygos vein?

☐ 11. Describe the clinical importance of the median cubital vein and the great saphenous vein.

(For answers, see Appendix B.)

DISORDERS OF THE BLOOD VESSELS

learning outcome

▶ Define atherosclerosis, deep vein thrombosis and venous disease of the lower limb, aneurysm, microangiopathy of diabetes, and arteriovenous malformation.

The primary disorder of blood vessels, atherosclerosis, is covered in A CLOSER LOOK. This section considers some other common and serious vessel disorders.

Deep vein thrombosis of the lower limb is the formation of clots in the veins of the lower extremity (usually in the thigh). In half the patients with this condition, the clot detaches, travels to the heart and pulmonary trunk, then blocks a branch of the pulmonary artery (pulmonary embolism). Extremely common, deep vein thrombosis affects 1% to 2% of hospital patients or 2 million Americans per year. It is usually caused by the sluggish flow of blood in veins of inactive and bedridden patients, so that thrombi form on the cusps of the venous valves. Alternatively, it can result from abnormal clotting chemistry or from inflammatory damage to the venous endothelium (in this latter case it is called *thrombophlebitis*). Deep vein thrombosis can be diagnosed by ultrasonagraphy of the veins in the lower limb and is treated with anticlotting drugs. Another treatment involves inserting a filter into the inferior vena cava to prevent emboli from reaching the lungs.

Venous disease is another common venous disorder of the lower limb, affecting 600,000 people, most of them elderly, in the United States. This disease is characterized by inadequate drainage of venous blood from the limb, whose tissues become ischemic and vulnerable to damage and ulceration. It is caused by the failure of valves in interconnecting veins that normally function to prevent blood in the limb's deep veins from flowing outward into the superficial saphenous veins. When these valves fail, the skeletal muscle pump propels the deep venous blood into the saphenous veins, which cannot drain so much blood quickly enough. Blood backs up in the saphenous veins, and so much fluid pours out of the capillaries serving these veins that the leg tissue develops edema and becomes ischemic. The slightest trauma to this oxygen-starved tissue leads to tissue damage, so ulcers form in the lower limb. Venous disease can lead to varicose saphenous veins (p. 596) when the overloaded saphenous veins fail and tear. Treating this disease involves elevating and compressing the lower limb to speed the drainage of blood.

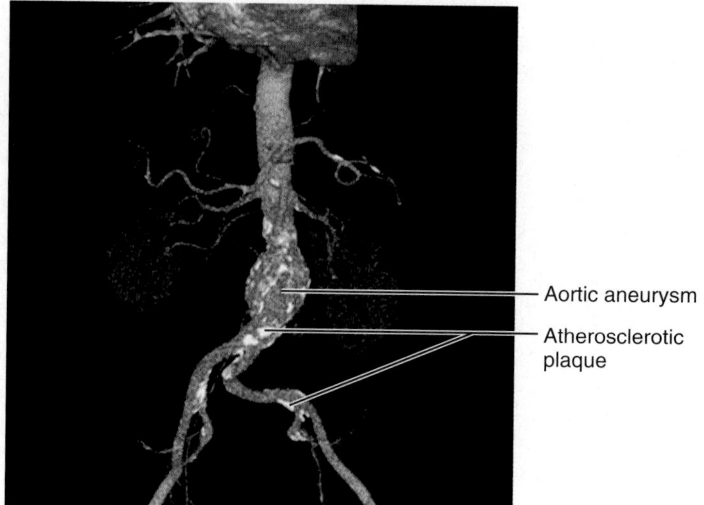

Figure 20.24 Sixty-four-slice CT angiogram of an abdominal aortic aneurysm.

An **aneurysm** (an′u-rizm; "widening") is a saclike widening or outpocketing of an artery (or vein) that places the vessel at risk of rupturing. The aneurysm may result from a congenital weakness of an artery wall or, more often, from a gradual weakening of the vessel by hypertension or arteriosclerosis. The most common sites of aneurysms are the abdominal aorta (Figure 20.24) and the arteries to the brain and kidneys. Aortic aneurysms are present in 1% of women and 8% of men over age 65, and their rupture causes 10,000 deaths per year in the United States. If detected before rupture—by palpation, ultrasound, or CT scans—they are treated by replacing the affected section of the vessel with a synthetic graft or by placing a strong-walled tube inside the part of the vessel where the aneurysm is present.

Microangiopathy of diabetes (mi″kro-an″je-op′ah-the; "small vessel disease") is a common complication of long-term diabetes mellitus. The elevated blood sugar levels of diabetes lead to the deposit of glycoproteins in the basement membrane of the body's capillaries. This results in thickened but leaky capillary walls, and a slowed rate of turnover of the tissue fluid upon which tissue cells rely for oxygen and nutrients. The organs most affected and damaged by this microangiopathy are the kidneys, retina, peripheral nerves, and feet. Foot ulcers commonly develop.

Arteriovenous malformation is a congenital condition in which capillaries fail to develop in a certain location, so that an artery continues directly into a vein. Relatively common among birth defects, it affects 1 in 700 people and usually occurs in the cerebrum of the brain. Without an intervening bed of capillaries to lower the high pressure of arterial blood before it reaches the vein, the vein weakens and forms a bulging aneurysm, which can compress nearby structures or burst to cause a stroke. Arteriovenous malformation is treated by surgically closing or removing the damaged segment of vein.

(Text continues on page 622.)

Atherosclerosis? Get Out the Cardiovascular Drāno

When pipes get clogged, it is usually because we've dumped something down the drain that shouldn't be there—a greasy mass or a hairball. Sometimes, pipes get blocked when something is growing inside them (tree roots, for example), trapping the normal sludge coming through (see top photo). In **arteriosclerosis,** the walls of our arteries become thicker and stiffer, and hypertension results. In **atherosclerosis,** the most common form of arteriosclerosis, small, patchy thickenings called *atheromas* form that can intrude into the vessel lumen, making it easy for arterial spasms or a roaming blood clot to close the vessel completely.

Although all blood vessels are susceptible to atherosclerosis, the aorta and the coronary arteries are most often affected.

Atherosclerosis: Onset and Stages

Atherosclerosis begins with damage to the tunica intima caused by bloodborne chemicals, hypertension, or viral or bacterial infections. Researchers suspect that almost any type of chronic infection—including respiratory, periodontal, and urinary tract problems—could set the stage for atherosclerosis.

Irritation of the endothelium prompts an inflammatory response. The injured endothelial cells release inflammatory agents and growth (mitosis-inducing) factors. Low-density lipoproteins (LDLs, the so called "bad cholesterol") are attracted to the inflamed area. LDLs deliver cholesterol to tissue cells. In the inflamed vessel, they invade the tunica intima and accumulate in the vessel wall. Monocytes, a type of white blood cell, respond to the area of inflammation, enter the tunica intima, and differentiate into macrophages, which phagocytose the LDLs within the vessel wall. The endothelial cells and the macrophages continue to secrete inflammatory chemicals and growth factors, thus stimulating growth of the vessel wall. As the fatty plaque develops, the macrophages become so engorged with LDLs that they are transformed into lipid-laden *foam cells* and lose their scavenging ability.

Smooth muscle cells migrating from the tunica media also take up lipids and become foam cells. The accumulating foam cells initiate the **fatty streak stage.** The smooth muscle cells also deposit collagen and elastin fibers, thickening the intima and producing fibrous lesions with a core of dead and dying foam cells called **fibrous** or **atherosclerotic plaques.** At first the vessel walls accommodate the growing plaque by expanding outward, but eventually these fatty mounds begin to protrude into the vessel lumen, producing full-blown atherosclerosis (see bottom photo).

As enlarging plaques hinder diffusion of nutrients from the blood to deeper tissues of the artery wall, smooth muscle cells in the tunica media die and the elastic fibers deteriorate. These elements are replaced by nonelastic scar tissue, and calcium salts are deposited in the lesions. These events constrict the vessel and cause the arterial walls to fray and ulcerate, conditions that encourage blood sludging and backup, platelet adhesion, and thrombus formation. The vessels' increased rigidity leads to hypertension. Together, these events increase the risk of myocardial infarcts, strokes, and aneurysms and are responsible for the pain (angina) that occurs when heart muscle is ischemic.

The inflammatory process in the vessel wall makes these cholesterol-rich plaques susceptible to rupture, exploding off fragments that trigger massive clots that can cause lethal heart attacks. The victim appears perfectly healthy until he or she drops dead!

Treatment and Prevention

Surgical interventions, angioplasty, and coronary bypass surgery (A Closer Look in Chapter 19, p. 581) have been used for decades to open clogged vessels or reroute blood around a clot to restore myocardial circulation. Anticlotting drugs, such as low-dose aspirin taken regularly, can reduce clot formation. Cholesterol-lowering drugs (statins) not only lower blood cholesterol but also act as antinflammatory agents.

When an atheroma ruptures and induces clot formation, *thrombolytic (clot-dissolving) agents* are used. These revolutionary drugs include *tissue plasminogen activator (tPA)*. Injecting tPA directly into the heart restores blood flow quickly and puts an early end to many heart attacks in progress.

Of course, it's best to prevent atherosclerosis from progressing in the first place—by quitting smoking, losing weight to reduce blood lipid (triglyceride) levels, eating a healthy diet rich in antioxidants, and exercising regularly.

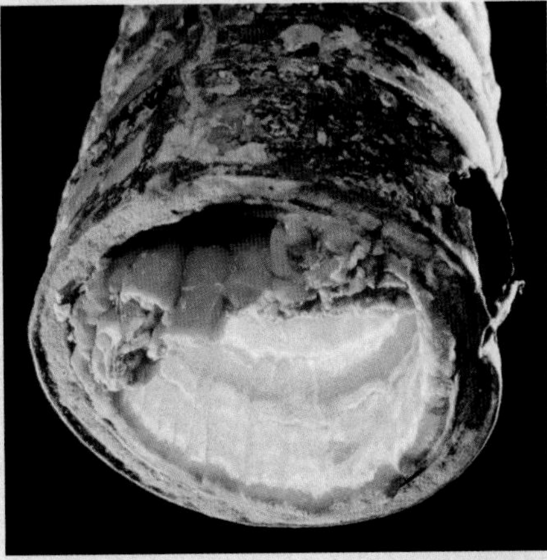

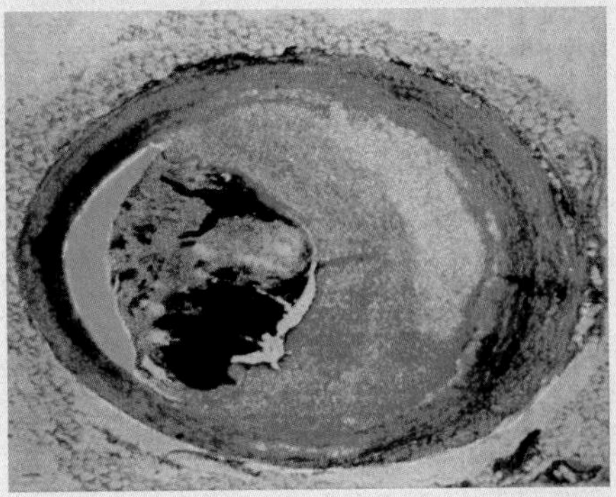

Top A pipe clogged by accumulated deposits. **Bottom** Atherosclerotic plaques nearly close a human artery.

Practice art labeling MasteringA&P®

Figure 20.25

Fetal Circulation. Blood is oxygenated at the placenta; the fetal lungs are not functioning. Fetal circulation has two routes to bypass the pulmonary circuit: the foramen ovale, an opening in the interatrial septum, and the ductus ateriosus, a shunt between the pulmonary trunk and the aorta.

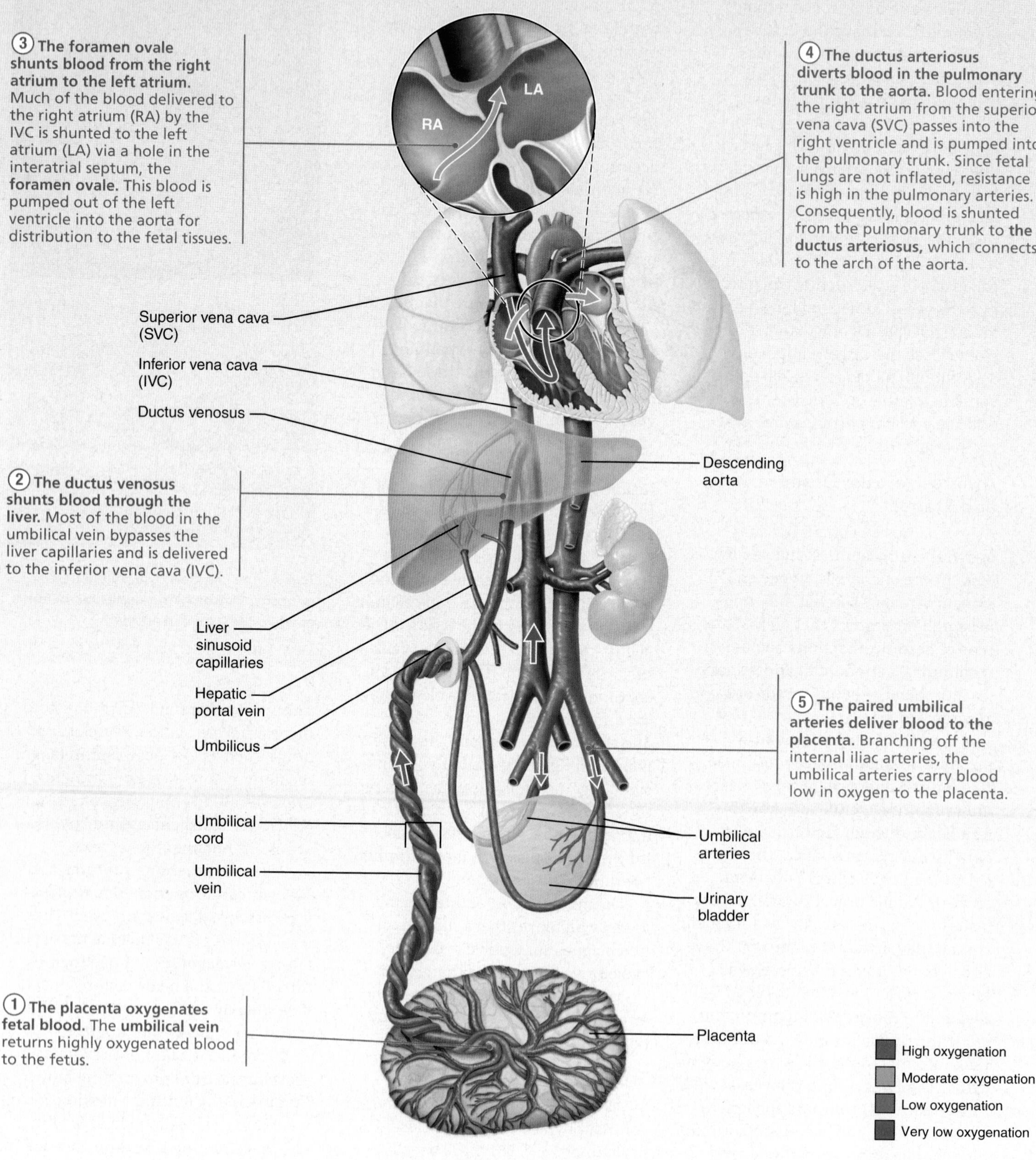

Newborn Circulation. **Blood is oxygenated in the lungs. The heart becomes functionally divided with the first breaths. The right side of the heart receives and pumps poorly oxygenated blood; the left side of the heart receives and pumps highly oxygenated blood.**

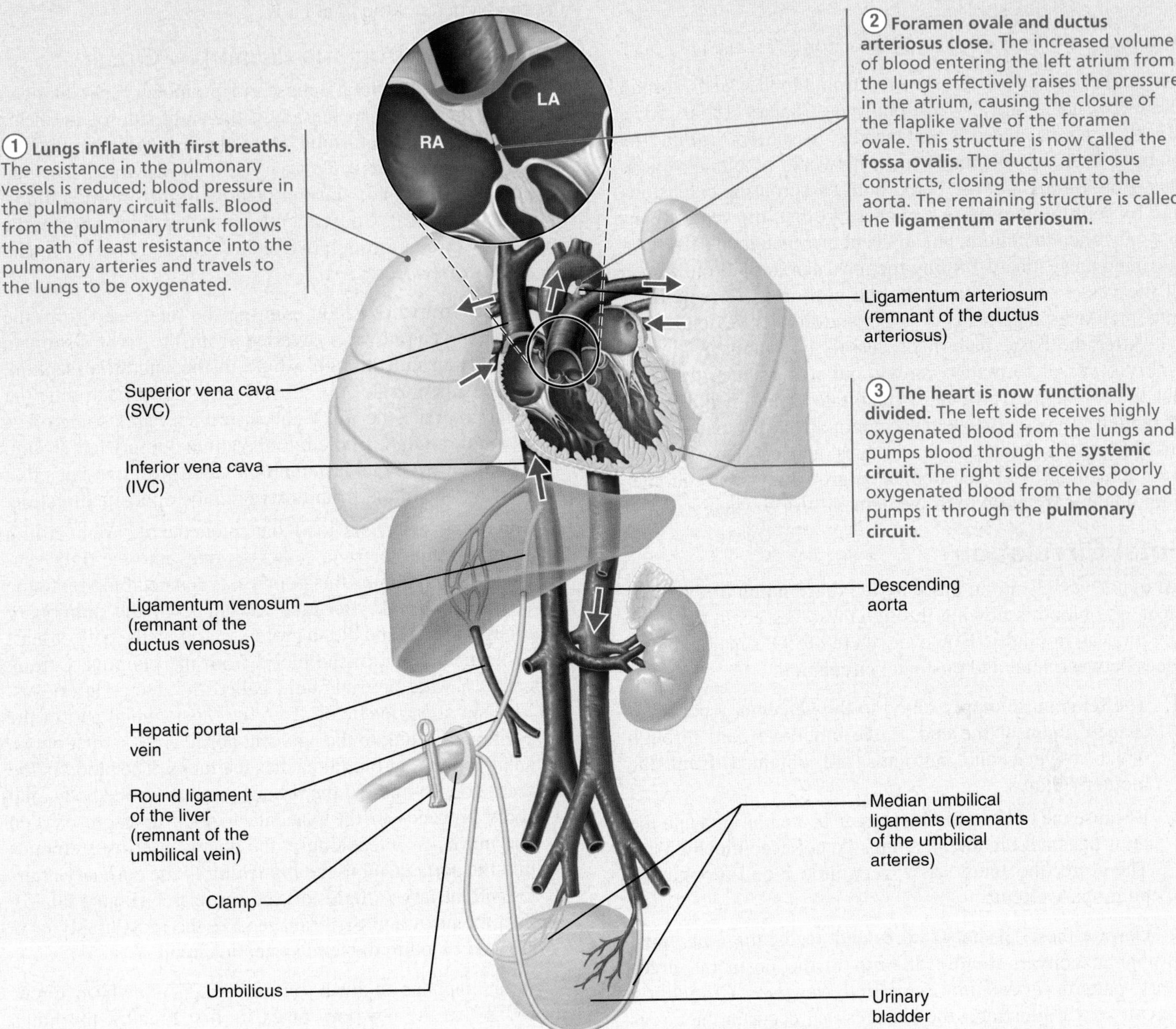

Fetal Structure	Postnatal Structure	Function in Fetus
Foramen ovale	Fossa ovalis	Diverts blood from the right atrium to the left atrium
Ductus arteriosus	Ligamentum arteriosum	Diverts blood from the pulmonary trunk to the aorta
Ductus venosus	Ligamentum venosum	Carries blood from the umbilical vein through the liver into the IVC
Umbilical arteries	Median umbilical ligaments	Paired vessels carry blood from the fetus to the placenta
Umbilical vein	Round ligament of the liver (ligamentum teres)	Single vessel carries oxygenated blood from the placenta to the fetus

BLOOD VESSELS THROUGHOUT LIFE

learning outcomes

- Trace the cardiovascular circuit in the fetus, and explain how it changes at birth.
- List some effects of aging on the blood vessels.

The earliest blood vessels develop from blood islands around the yolk sac in the 3-week embryo (Chapter 18, p. 557). By the end of week 3, splanchnic mesoderm *within* the embryo itself has begun to form networks of blood vessels that grow throughout the body, both by sprouting extensions and by splitting into more branches. At first, the vessels consist only of endothelium, but adjacent mesenchymal cells soon surround these tubes, forming the muscular and fibrous tunics of the vessel walls. Ultimately, the networks are remodeled into treelike arrangements of larger and smaller vessels.

After the basic pattern of vessels is established in the embryo, vessel formation slows, but it continues throughout life: to support growth, to facilitate wound healing, and to rebuild the vessels lost each month after a woman's menstrual period. In adults, as in embryos, new capillaries sprout from existing vessels as tubular outgrowths of the endothelium, in direct relation to tissue demands for oxygen.

Fetal Circulation

All major vessels are in place by the third month of development, and blood is flowing through these vessels in the same direction as in adults. However, there are two major differences between fetal and postnatal circulation:

1. The fetus must supply blood to the *placenta,* a pancake-shaped organ at the end of the umbilical cord through which oxygen and nutrients are obtained from the mother's uterus.
2. Because the fetal respiratory organ is the placenta and the fetus does not breathe, its lungs do not need much blood. Therefore, the fetus sends very little blood around the pulmonary circuit.

Despite these special features and needs, the fetus must be able to convert rapidly at birth to the postnatal circulatory pattern. *Focus on Fetal and Newborn Circulation* (Figure 20.25) illustrates the changes that occur in the circulatory pathways during the fetal and neonatal periods.

Vessels to and from the Placenta

The fetal vessels that carry blood to and from the placenta are called *umbilical vessels* because they run in the umbilical cord (Figure 20.25, Fetal circulation). The paired **umbilical arteries** branch from the internal iliac arteries in the pelvis and carry blood to the placenta to pick up oxygen and nutrients. The unpaired **umbilical vein** returns this blood to the fetus, delivering some of it to the hepatic portal vein so that its nutrients can be processed by the liver cells. However, there is too much returning blood for the liver to process, so most of the blood is diverted through a shunt called the **ductus venosus** ("venous duct"). Whether it goes through the portal system or the ductus venosus, all returning blood proceeds into the hepatic veins, inferior vena cava, and right atrium of the heart. In the inferior vena cava, this oxygen-rich placental blood mixes with the oxygen-poor blood returning to the heart from the caudal parts of the fetal body.

Shunts Away from the Pulmonary Circuit

As mentioned, the fetal lungs and pulmonary circuit need very little blood. As in the adult, the right side of the heart receives all blood returning from the systemic circuit and pumps blood into the pulmonary trunk. However, the fetal lungs are not inflated, and thus the resistance in the pulmonary circuit is great. Consequently, blood is diverted from the pulmonary circuit through two shunts: the *foramen ovale* and the *ductus arteriosus*.

1. **Foramen ovale.** Blood entering the fetal heart from the inferior vena cava is diverted from the right atrium to the left atrium through a hole in the interatrial septum, the foramen ovale (introduced on p. 566). Each time the atria contract, the full right atrium squeezes some of its blood through this hole into the almost-empty left atrium. The foramen ovale is actually a valve with two flaps that prevent any blood from flowing in the opposite direction.
2. **Ductus arteriosus.** Blood that enters the fetal right atrium from the superior vena cava continues into the right ventricle, which pumps this poorly oxygenated blood into the pulmonary trunk. Because resistance in the pulmonary arteries is high and blood prefers to follow the path of least resistance, much of the blood from the pulmonary trunk enters a wide arterial shunt called the ductus arteriosus. This shunt carries blood from the most cranial part of the pulmonary trunk to the adjacent aortic arch, so that only a small amount of blood reaches the lungs. The blood in the aorta goes to nourish the tissues throughout the body, and some proceeds to the placenta to pick up more oxygen and nutrients. Interestingly, the ductus arteriosus empties into the aorta distal to the branching of the coronary arteries and the large arteries off the aortic arch (Figure 20.25), thus the heart and the brain receive the most highly oxygenated blood in the fetal systemic circuit.

What happens at birth (Figure 20.25, Newborn circulation)? When the newborn takes its first breaths, the lungs inflate. There is no longer high resistance in the pulmonary vessels, and the lungs receive more blood. The ductus arteriosus constricts and closes. For the first time, oxygenated pulmonary blood begins pouring into the left atrium, raising the pressure within this chamber. This pressure pushes the two valve flaps of the foramen ovale together, closing it. Both the foramen ovale and ductus arteriosus are now functionally closed, and the postnatal circulatory plan is established.

Although the foramen ovale and ductus arteriosus close shortly after birth, they do not immediately fuse shut. It takes about 3 months for the ductus arteriosus to become the solid **ligamentum arteriosum,** and about a year for the flaps of the foramen ovale to fuse together as the **fossa ovalis.**

Most portions of the umbilical vessels are discarded after birth, when the umbilical cord is cut. The parts of these vessels remaining in the baby's body constrict rapidly and then gradually degenerate into fibrous bands called *ligaments* (Figure 20.25, Fetal circulation). Throughout postnatal life, the remnant of the umbilical vein is the **round ligament of liver** or **ligamentum teres.** The ductus venosus becomes the **ligamentum venosum** on the liver's inferior surface, and the umbilical arteries become **medial umbilical ligaments** in the anterior abdominal wall inferior to the navel.

Blood Vessels in Adulthood

The most important aspect of the aging of the vascular system is the progression of atherosclerosis (discussed on p. 619). Because of the high fat content in the American diet, almost everyone in the United States develops atherosclerotic plaques in the major arteries before the age of 40. Although the degenerative process of atherosclerosis begins in youth, its consequences are rarely apparent until middle to old age, when it may lead to myocardial infarction or stroke.

Gender has an effect on how atherosclerosis develops with age. Until puberty, the blood vessels of boys and girls look alike, but from puberty to about the age of 45 years, women have strikingly less atherosclerosis than men, probably because of protective effects of estrogen. This "gap between the sexes" closes between the ages of 45 and 65, and after age 65 the incidence of heart disease is slightly higher in women. Furthermore, when a woman does experience a heart attack, she is more likely to die from it than is a man. Some of the reasons for this are unpreventable (the average woman having a heart attack is 10 years older than the average male victim, and the vessel-damaging aspects of diabetes seem to affect women more), but others are preventable. Some women may not recognize the symptoms of angina or a heart attack in time to seek help.

✓ check your understanding

☐ 12. What lifestyle factors contribute to the development of atherosclerosis?

☐ 13. How does elevated blood glucose associated with diabetes mellitus affect the delivery of oxygen and nutrients to body tissues?

☐ 14. Which vessel in the fetus carries the most highly oxygenated blood?

☐ 15. Where is the ductus arteriosus located, and how does it function to shunt blood from the pulmonary circuit?

(For answers, see Appendix B.)

RELATED CLINICAL TERMS

Angiography (an-jē-og′ra-fē; *angio* = vessel; *graphy* = writing) Diagnostic technique involving the infusion of a radiopaque substance into the bloodstream for X-ray examination of specific blood vessels. Angiography is the major way of diagnosing occlusion of coronary arteries and risk of heart attack. The images obtained with this procedure are called *angiograms*.

Angiosarcoma Cancer originating from the endothelium of a blood vessel; may develop in liver vessels following exposure to chemical carcinogens.

Blue baby A baby with cyanosis (skin appears blue) due to relatively low levels of oxygen in the blood. This condition is caused by any congenital defect that leads to low oxygenation of the systemic blood, including patent (open) foramen ovale and patent ductus arteriosus, failure of the lungs to inflate at birth, and other conditions (see p. 583).

Carotid endarterectomy Surgical procedure for scraping away atherosclerotic plaques that block the base of the internal carotid artery; done to decrease the risk of a stroke.

Phlebitis (fle-bi′tis; *phleb* = vein; *itis* = inflammation) Inflammation of a vein, accompanied by painful throbbing and redness of the skin over the inflamed vein; most often caused by bacterial infection or local physical trauma.

Phlebotomy (fle-bot′o-me; *tomy* = cut) An incision made in a vein for withdrawing blood.

CHAPTER SUMMARY

Take full advantage of the invaluable online practice and study tools for this chapter by going to the Study Area of MasteringA&P®. There you will find:

- Art Labeling Activities, Quizzes, Practice Tests, Memory Games
- Animal Organ Dissection, Cat Dissection and Bone Videos
- PAL Cadaver
- PAL Histology
- A&P Flix Animations
- Dynamic Study Modules

PART 1: GENERAL CHARACTERISTICS OF BLOOD VESSELS (pp. 589–596)

1. The main types of blood vessels are arteries, capillaries, and veins. Arteries carry blood away from the heart and toward the capillaries; veins carry blood away from the capillaries and toward the heart. Arteries "branch," whereas veins "serve as tributaries."

STRUCTURE OF BLOOD VESSEL WALLS (p. 589)

2. All but the smallest blood vessels have three tunics: tunica intima (with an endothelium), tunica media (smooth muscle and elastin and collagen fibrils), and tunica externa (external connective tissue). The tunica media, which changes the diameter and strengthens the wall of the vessel, is thicker in arteries than in other types of blood vessels. Vasa vasorum are small arteries, capillaries, and veins in the tunica externa that supply the outer part of the wall of larger blood vessels.

TYPES OF BLOOD VESSELS (pp. 589–595)

3. The three classes of arteries, from large to small, are elastic arteries, muscular arteries, and arterioles. Elastic (conducting) arteries, the largest arteries near the heart, contain more elastin than does any other vessel type. The elastin in all arteries helps dampen the pressure pulses produced by the heartbeat.
4. Muscular (distributing) arteries carry blood to specific organs and organ regions. These arteries have a thick tunica media bordered by elastic membranes and are most active in vasoconstriction.
5. Arterioles, the smallest arteries, have one or two layers of smooth muscle cells in their tunica media. They regulate the flow of blood into capillary beds.
6. Capillaries, which have diameters slightly larger than erythrocytes, are endothelium-walled tubes. Tight junctions and intercellular clefts occur between adjacent endothelial cells. Spider-shaped pericytes help support the capillary wall. Continuous capillaries are the most common and least permeable type of capillary; fenestrated capillaries have pores in their endothelial cells and are highly permeable; sinusoidal capillaries are extremely permeable capillaries with wide intercellular clefts.
7. The four routes involved in capillary permeability are (1) by direct diffusion of respiratory gases across the endothelium, (2) through intercellular clefts between endothelial cells, (3) through the pores of fenestrated capillaries, and (4) through endothelial pinocytotic vesicles.
8. Capillaries are arranged in networks called capillary beds. Sometimes, blood is shunted straight through a capillary bed (from arteriole to metarteriole to thoroughfare channel to venule) and does not enter the true capillaries. Precapillary sphincters determine how much blood enters the true capillaries.
9. The smallest veins, postcapillary venules, function like capillaries.
10. Because venous blood is under less pressure than arterial blood, the walls of veins are thinner than those of arteries of comparable size. Veins also have a wider lumen, a thinner tunica media, and a thicker tunica externa.
11. Many veins have valves that prevent the backflow of blood. The skeletal muscle pump aids to move blood through veins. Varicose veins are veins whose valves have failed.

PAL Histology/Cardiovascular System

VASCULAR ANASTOMOSES (pp. 595–596)

12. The joining together of arteries serving a common organ, called an arterial anastomosis, provides collateral channels for blood to reach the same organ. Anastomoses between veins are more common than anastomoses between arteries.

PART 2: BLOOD VESSELS OF THE BODY (pp. 596–623)

THE PULMONARY CIRCULATION (pp. 596–597)

13. The pulmonary trunk splits inferior to the aortic arch into the right and left pulmonary arteries. These arteries divide into lobar arteries, then branch repeatedly in the lung. Arising from the capillary beds, pulmonary venous tributaries empty newly oxygenated blood into the superior and inferior pulmonary veins of each lung. The four pulmonary veins extend from the lungs to the left atrium.

THE SYSTEMIC CIRCULATION (pp. 597–618)

14. The systemic arteries begin with the aorta exiting the heart and branch to supply all of the body regions (Figure 20.8). Systemic veins (Figure 20.16) all empty into one of the three vessels that return blood to the heart: the two large venae cavae and the coronary sinus.

PAL Human Cadaver/Cardiovascular System/Arteries, and Human Cadaver/Cardiovascular System/Veins

15. A portal system is a set of vessels in which *two* capillary beds, interconnected by a vein, lie between the initial artery and final vein. The hepatic portal system is a special subcirculation that drains the digestive organs of the abdomen and pelvis (see Figure 20.22).
16. Cirrhosis of the liver leads to portal hypertension, in which blood backs up through the portal-systemic anastomoses, overloading these delicate veins. This overload leads to esophageal bleeding, hemorrhoids, and caput medusae.

DISORDERS OF THE BLOOD VESSELS (pp. 618–619)

17. *Atherosclerosis* is the most important vascular disorder (see A Closer Look on p. 619). *Deep vein thrombosis of the lower limb* is the formation of clots in veins of the lower extremity. In *venous disease,* the failure of venous valves in some lower-limb veins results in inadequate drainage of venous blood from the limb, whose tissues consequently become ischemic and vulnerable to ulceration. An *aneurysm* is a widening or outpocketing of an artery (or vein). In *microangiopathy of diabetes,* the capillaries are leaky because of sugar deposits in their basement membrane. In *arteriovenous malformation,* the capillaries fail to develop between some important artery and the vein that drains it, often leading to a rupture or an aneurysm of the vein.

BLOOD VESSELS THROUGHOUT LIFE (pp. 620–623)

18. The first blood vessels develop from blood islands on the yolk sac in the 3-week embryo. Soon, vessels begin forming from mesoderm inside the embryo. The formation of new vessels, whether in the embryo or adult, occurs mainly through the sprouting of endothelial buds from existing vessels.
19. The pattern of blood vessels is the same in both the fetus and the newborn, and blood flows through these vessels in the same directions. However, the fetus also has vessels to and from the placenta (umbilical arteries, umbilical vein, and ductus venosus) and two shunts that bypass the nearly nonfunctional pulmonary circuit (the foramen ovale in the interatrial septum, and the ductus arteriosus between the pulmonary trunk and aortic arch). Fetal shunts and vessels close shortly after birth.
20. The most important age-related vascular disorder is the progression of atherosclerosis.

REVIEW QUESTIONS

Multiple Choice/Matching Questions

(For answers, see Appendix B.)

1. Which of the following statements does *not* correctly describe veins? (a) They have less elastic tissue and smooth muscle than arteries. (b) They are subject to lower blood pressures than arteries. (c) They have larger lumens than arteries of similar size. (d) They always carry deoxygenated blood.

2. In atherosclerosis, which of the following layers of the artery wall thickens most? (a) tunica media, (b) tunica intima, (c) tunica externa.

3. Blood flow through the capillaries is steady despite the rhythmic pumping action of the heart because of the (a) elasticity of the large arteries only, (b) elasticity of all the arteries, (c) ligamentum arteriosum, (d) venous valves.

4. Fill in the blanks with the name of the appropriate vascular tunic (intima, media, or externa).

____ (1) contains endothelium

____ (2) the tunic with the largest vasa vasorum

____ (3) the thickest tunic in veins

____ (4) the outer tunic; is mostly connective tissue

____ (5) is mostly smooth muscle

____ (6) forms the valves of veins

5. Which of the following vessels is bilaterally symmetrical (that is, has an identical member of a pair on either side of the body)? (a) internal carotid artery, (b) brachiocephalic trunk, (c) azygos vein, (d) superior mesenteric vein.

6. Identify which artery is *missing* from the following sequence, which traces the flow of arterial blood to the right hand: Blood leaves the heart and passes through the aorta, the right subclavian artery, the axillary and brachial arteries, and through either the radial or ulnar artery to a palmar arch. (a) left coronary, (b) brachiocephalic, (c) cephalic, (d) right common carotid.

7. Which of the following veins do not drain directly into the inferior vena cava? (a) lumbar veins, (b) hepatic veins, (c) inferior mesenteric vein, (d) renal veins.

8. A stroke that blocks a posterior cerebral artery will most likely affect (a) hearing, (b) sight, (c) smell, (d) higher thought processes.

9. The following sequence traces the flow of arterial blood to the left parietal and temporal lobes of the brain. Tell which artery is *missing:* Blood leaves the heart and passes through the aorta, the left common carotid artery, and the middle cerebral artery. (a) vertebral, (b) brachiocephalic, (c) internal carotid, (d) basilar.

10. Tell which two veins are *missing* from the following sequence: Tracing the drainage of *superficial* venous blood from the leg, blood enters the great saphenous vein, femoral vein, inferior vena cava, and right atrium. (a) coronary sinus and superior vena cava, (b) posterior tibial and popliteal, (c) fibular (peroneal) and popliteal, (d) external and common iliacs.

11. The inferior mesenteric artery supplies the (a) rectum and part of the colon, (b) liver, (c) small intestine, (d) spleen.

12. Tell which two vessels are *missing* from the following sequence: Tracing the drainage of venous blood from the small intestine, blood enters the superior mesenteric vein, hepatic vein, inferior vena cava, and right atrium. (a) coronary sinus and left atrium, (b) celiac and common hepatic veins, (c) internal and common iliac veins, (d) hepatic portal vein and liver sinusoids.

13. Tell which vessel is *missing* from the following vascular circuit: Tracing the circulation to and from the placenta, a drop of blood travels in the fetus from the left ventricle to the aorta, to a common iliac artery, to an internal iliac artery, to an umbilical artery, to capillaries in the placenta, to the ductus venosus, to a hepatic vein, to the inferior vena cava, to the right atrium, to the foramen ovale, to the left atrium, and to the left ventricle. (a) umbilical vein, (b) ligamentum arteriosum, (c) internal carotid artery, (d) capillaries in the small intestine.

14. A pressure point is a place where one presses on an artery through the body surface to stop bleeding more distally. Which of the following sites is *not* a pressure point for any major artery? (a) the medial bicipital furrow on arm, (b) the sciatic artery in middle of gluteus maximus, (c) the midinguinal point in the femoral triangle.

15. Which of the following is *not* a pulse point? (a) anatomical snuff box, (b) inferior margin of mandible anterior to masseter muscle, (c) center of distal forearm at palmaris longus tendon, (d) medial bicipital furrow on arm, (e) dorsum of foot between the first two metatarsals.

16. A nurse missed a patient's median cubital vein while trying to draw blood and then inserted the needle far too deeply into the cubital fossa. This error could cause any of the following problems *except* (a) paralysis of the ulnar nerve, (b) paralysis of the median nerve, (c) bruising the insertion tendon of the biceps brachii muscle, (d) blood spurting from the brachial artery.

Short Answer Essay Questions

17. (a) What structural features are responsible for the permeability of capillary walls? (b) Which of these features are absent from brain capillaries in the blood-brain barrier?

18. Distinguish among elastic arteries, muscular arteries, and arterioles relative to their location, histology, and functions.

19. (a) Name an organ containing fenestrated capillaries. (b) What is a sinusoid?

20. Sketch the arterial circle at the base of the brain, and label the important arteries that branch off this circle. Begin with the right and left vertebral arteries merging to form the basilar artery.

21. (a) What are the two large tributaries that form the hepatic portal vein? (b) What is the function of the hepatic portal circulation?

22. State the location and basic body regions drained by the azygos vein.

23. Name all four pulmonary veins.

24. What are the ductus venosus and ductus arteriosus, where are they located, what are their functions, and how are they different?

25. Differentiate between arteriosclerosis and atherosclerosis.

26. A pulse can be felt in the following arteries: superficial temporal, facial, common carotid, brachial, femoral, popliteal, posterior tibial, and dorsalis pedis (Figure 20.8b). Name the artery from which each artery listed branches. Here is an example to get you started: The superficial temporal artery *is a branch of the external carotid artery.*

THE LYMPHATIC SYSTEM

learning outcomes

- ▶ Describe the structure and distribution of lymphatic vessels.
- ▶ Explain how lymph forms and the mechanisms by which it is transported.
- ▶ List and explain the important functions of the lymphatic vessels.
- ▶ Describe how lymph nodes function as lymphatic organs.

All blood capillaries are surrounded by areolar connective tissue that contains tissue fluid. Tissue fluid arises from blood filtered through the capillary walls and consists of the small molecules of blood plasma, including water, various ions, nutrient molecules, and respiratory gases. Even though fluid is continuously leaving and reentering the blood capillaries, for complex reasons slightly more fluid exits from the arteriole end of each capillary bed than reenters the blood at the venule end. The **lymphatic vessels** function to collect this excess tissue fluid from the loose connective tissue around blood capillaries and return it to the bloodstream. Once inside the lymphatic vessels, this fluid is called **lymph** (*lympha* = clear water). Any blockage of the lymphatic vessels causes the affected body region to swell with excess tissue fluid, a condition called *edema* (see p. 96).

The lymphatic vessels also perform another, related function. Blood proteins leak slowly but steadily from blood capillaries into the surrounding tissue fluid. The lymphatic vessels return these leaked proteins to the bloodstream. The proteins in blood generate osmotic forces that are essential for keeping water in the bloodstream (see p. 547). If leaked proteins were not returned to the bloodstream, a massive outflow of water from the blood to the tissues would soon follow, and the entire cardiovascular system would collapse from insufficient volume.

Because lymph flows only *toward* the heart, the lymphatic vessels form a one-way system rather than a full circuit (Figure 21.1a). There are several orders of vessels that collect and carry lymph. The smallest vessels, those that first receive lymph, are the *lymphatic capillaries.* These vessels drain into larger *collecting lymphatic vessels,* along which are scattered *lymph nodes.* The collecting vessels then drain into *lymph trunks,* which unite to form *lymph ducts,* which empty into the veins at the root of the neck.

Lymphatic Capillaries

Lymphatic capillaries, the highly permeable vessels that collect the excess tissue fluid, are located near blood capillaries in the loose areolar connective tissue (Figure 21.1b). Like blood capillaries, their wall consists of a single layer of endothelial cells. Their permeability results from the structure and arrangement of the endothelial cells: They have few intercellular junctions, and the edges of adjacent cells overlap, forming easily opened minivalves. Bundles of fine collagen filaments anchor the endothelial cells to the surrounding connective tissue. As a result, any increase in the volume of the tissue fluid separates the minivalve flaps, opening gaps

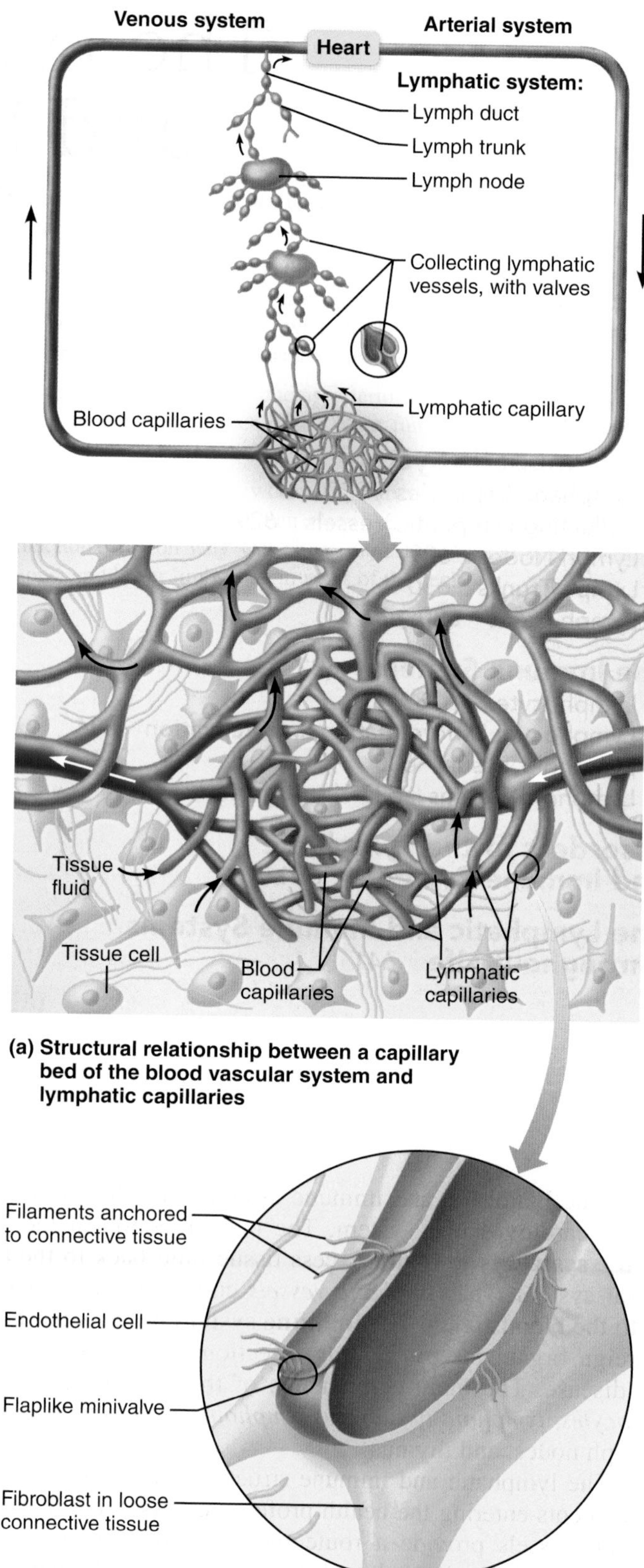

(a) Structural relationship between a capillary bed of the blood vascular system and lymphatic capillaries

(b) Lymphatic capillaries are blind-ended tubes in which adjacent endothelial cells overlap each other, forming flaplike minivalves.

Figure 21.1 **Distribution and special features of lymphatic capillaries.** Arrows in (a) indicate direction of fluid movement.

in the wall and allowing the fluid to enter. Once this fluid enters the lymphatic capillaries, it is called *lymph.* Lymph cannot leak out of the lymphatic capillary because backflow forces the minivalve flaps together.

Although the high permeability of lymphatic capillaries allows the uptake of large quantities of tissue fluid and large protein molecules, it also allows any bacteria, viruses, or cancer cells in the loose connective tissue to enter these capillaries with ease. These pathogenic agents can then travel throughout the body via the lymphatic vessels. However, this threat is averted in part by the lymph nodes, which destroy most pathogens in the lymph (see p. 638).

Lymphatic capillaries are widespread, occurring almost everywhere blood capillaries occur. However, lymphatic capillaries are absent from bone and teeth, from bone marrow, and from the entire central nervous system, where excess tissue fluid drains through the nervous tissue into the cerebrospinal fluid. The cerebrospinal fluid then returns this tissue fluid to the blood at the superior sagittal sinus (see pp. 404–407).

One set of lymphatic capillaries, called *lacteals* (lak′te-alz), has a unique function. Located in the villi of the mucosa of the small intestine, lacteals absorb digested fats from the intestine, which causes the lymph draining from the digestive viscera to become milky white (*lacte* = milk). This fatty lymph is called *chyle* (kīl; "juice"), and, like all lymph, it is carried to the bloodstream.

Collecting Lymphatic Vessels

From the lymphatic capillaries, lymph enters **collecting lymphatic vessels,** which accompany blood vessels: In general, the superficial collecting lymphatic vessels in the skin travel with superficial *veins,* whereas the deep collecting lymphatic vessels of the trunk and digestive viscera travel with the deep *arteries.*

Collecting lymphatic vessels are narrow and delicate, so they usually are not seen in the dissecting laboratory. They have the same tunics as blood vessels (tunica intima, tunica media, and tunica externa), but their walls are always much thinner. This reflects the fact that lymph flows under very low pressure, because lymphatic vessels are not connected to the pumping heart. To direct the flow of lymph, collecting lymphatic vessels contain more valves than do veins (Figure 21.1). At the base of each valve, the vessel bulges, forming a pocket in which lymph collects and forces the valve shut. Because of these bulges, each collecting lymphatic vessel resembles a string of beads. This distinctive appearance, which characterizes the larger lymph trunks and lymph ducts as well, allows physicians to recognize lymphatic vessels in X-ray films taken after these vessels are injected with radiopaque dye. This radiographic procedure is called **lymphangiography** (lim′-fan″je-og′rah-fe; "lymph vessel picturing").

Lymph Transport

Unaided by the force of the heartbeat, lymph is propelled through lymphatic vessels by a series of weaker mechanisms.

- The bulging of contracting skeletal muscles and the pulsations of nearby arteries push on the lymphatic vessels, squeezing lymph through them.
- The muscular tunica media of the lymphatic vessels contracts to help propel the lymph.
- The normal movements of the limbs and trunk help to keep the lymph flowing.

Despite these propulsion mechanisms, the transport of lymph is sporadic and slow, which explains why people who stand for long times at work may develop severe edema around the ankles by the end of the workday. The edema usually disappears if the legs are exercised, as, for example, by walking home. The seemingly useless nervous habit of people who bounce and wiggle their legs while sitting actually performs the important function of moving lymph up the legs.

CLINICAL APPLICATION

The Lymph Vessels and Edema A body region whose lymphatic collecting vessels have been blocked or removed will become swollen and puffy with edema. Edema of the arm often follows a mastectomy (removal of a cancerous breast) in which the lymphatic vessels and nodes that drain the arm are removed from the axilla. Severe edema may continue for several months until the lymphatic vessels grow back. (Lymphatic vessels regenerate quite well and can even grow through scar tissue.) Because this condition, called *lymphedema of the arm,* is very uncomfortable and difficult to treat, surgeons who perform mastectomies are now encouraged to abandon the standard, aggressive procedure of removing all the axillary lymph nodes and, instead, to remove only those nodes to which the cancer is likely to have spread.

Lymph Nodes

Lymph nodes, which cleanse the lymph of pathogens, are bean-shaped organs situated along collecting lymphatic vessels (see Figure 21.1). The popular term "lymph glands" is not correct, because they are not glands at all. There are about 500 lymph nodes in the human body, ranging in diameter from 1 to 25 mm (up to 1 inch). Large clusters of superficial lymph nodes are located in the cervical, axillary, and inguinal regions; deep lymph nodes are found in the thorax, abdomen, and pelvis (**Figure 21.2**).

- The superficial *cervical nodes* along the jugular veins and carotid arteries receive lymph from the head and neck.
- *Axillary nodes* in the armpit and the *inguinal nodes* in the superior thigh filter lymph from the upper and lower limbs, respectively.
- Nodes in the mediastinum, such as the deep *tracheobronchial nodes,* receive lymph from the thoracic viscera.
- Deep nodes along the abdominal aorta, called *aortic nodes,* filter lymph from the posterior abdominal wall.
- Deep nodes along the iliac arteries, called *iliac nodes,* filter lymph from pelvic organs and the lower limbs.

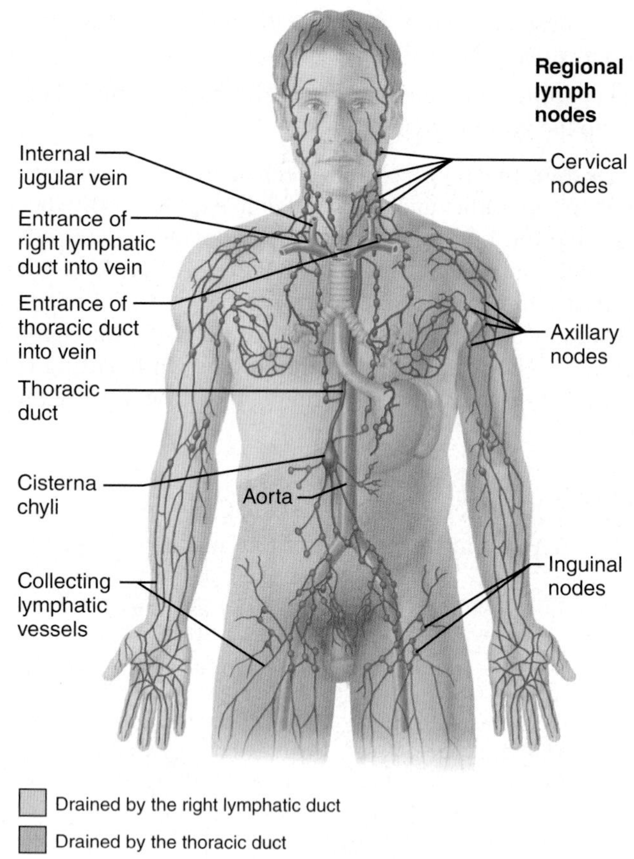

Figure 21.2 General distribution of collecting lymphatic vessels and regional lymph nodes.

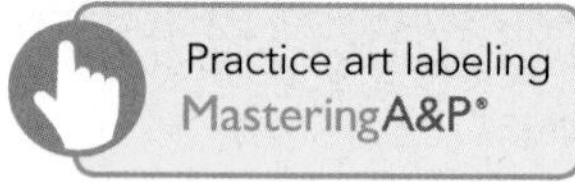

Microscopic Anatomy

A lymph node is surrounded by a fibrous **capsule** of dense connective tissue, from which fibrous strands called **trabeculae** (trah-bek′u-le; "beams") extend inward to divide the node into compartments (Figure 21.3). Lymph enters the convex aspect of the node through several **afferent lymphatic vessels** and exits from the indented region on the other side, the **hilum** (hi′lum), through **efferent lymphatic vessels.** Within the node, between the afferent and efferent vessels, lymph percolates through **lymph sinuses** (*subcapsular, cortical,* and *medullary sinuses*). These large lymph sinuses are spanned internally by a crisscrossing network of reticular fibers covered by endothelial cells. Many macrophages live on this fiber network, consuming pathogens and foreign particles in the lymph that flows through the sinuses (Figure 21.3c). Because most lymph passes through several nodes, it is usually free of pathogens by the time it leaves its last node and enters the lymph trunks on its way to the great veins of the neck.

> **CLINICAL APPLICATION**
>
> **Swollen Lymph Nodes** Sometimes, lymph nodes are overwhelmed by the very agents they are trying to destroy. In one instance, when large numbers of undefeatable bacteria or viruses are trapped by the nodes but are not destroyed, the nodes become enlarged, inflamed, and very tender to the touch. Such infected lymph nodes are called *buboes* (bu′bōz). In bubonic plaque, buboes are the most obvious symptom. In another case, metastasizing cancer cells that enter lymphatic vessels and are trapped in the local lymph nodes continue to multiply there. The fact that cancer-infiltrated lymph nodes are swollen but not painful helps to distinguish cancerous nodes from those infected by microorganisms. (Pain results from inflammation, and cancer cells do not induce the inflammatory response.) Potentially cancerous lymph nodes can be located by palpation, as when a physician examining a patient for breast cancer feels for swollen axillary lymph nodes. Physicians can also locate enlarged, cancerous lymph nodes by using CT and MRI scans.

Lymph Trunks

After leaving the lymph nodes, the largest collecting lymphatic vessels converge to form **lymph trunks** (see Figure 21.2 and Figure 21.4). These trunks drain large areas of the body and are large enough to be found by a skilled dissector. The five major lymph trunks, from inferior to superior, are as follows:

1. **Lumbar trunks.** These paired trunks, which lie along the sides of the aorta in the inferior abdomen, receive all lymph draining from the lower limbs, the pelvic organs, and from some of the anterior abdominal wall.
2. **Intestinal trunk.** This *unpaired* trunk, which lies near the posterior abdominal wall in the midline, receives fatty lymph (chyle) from the stomach, intestines, and other digestive organs.
3. **Bronchomediastinal** (brong″ko-me″de-ah-sti′nal) **trunks.** Ascending near the sides of the trachea, these paired trunks collect lymph from the thoracic viscera and thoracic wall.
4. **Subclavian trunks.** Located near the base of the neck, these paired trunks receive lymph from the upper limbs; they also drain the inferior neck and the superior thoracic wall.
5. **Jugular trunks.** Located at the base of each internal jugular vein, these paired trunks drain lymph from the head and neck.

Lymph Ducts

The lymph trunks drain into the largest lymphatic vessels, the **lymph ducts** (Figures 21.2 and 21.4). Whereas some individuals have two lymph ducts, others have just one.

Thoracic Duct

The **thoracic duct** is present in all individuals. Its most inferior part, located at the union of the lumbar and intestinal trunks, is the **cisterna chyli** (sis-ter′nah ki′li; "sac of chyle"), which lies on the bodies of vertebrae L_1 and L_2 (Figure 21.4). From there, the thoracic duct ascends along the vertebral

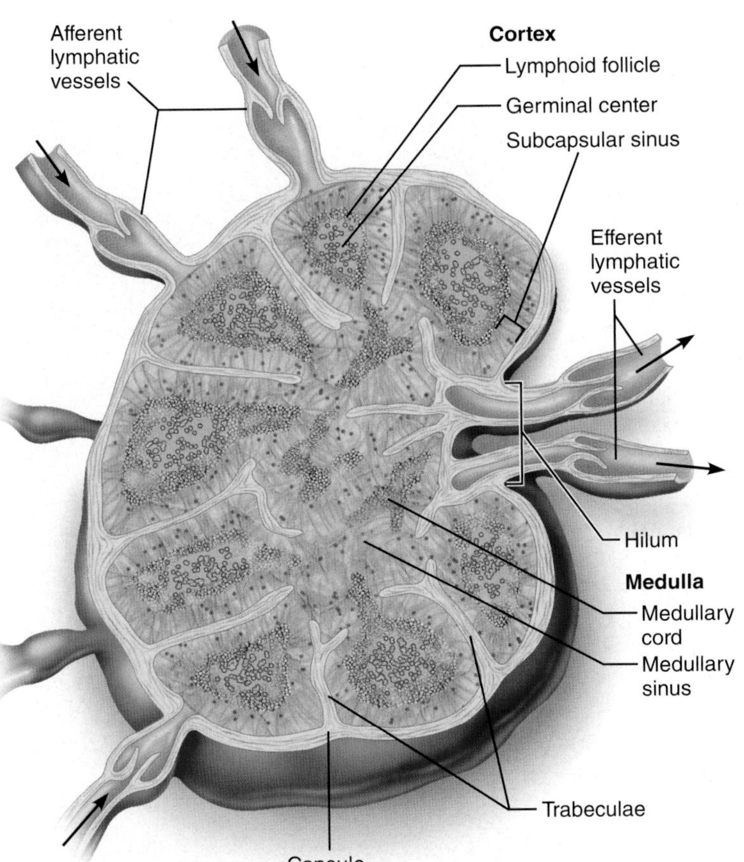

(a) Longitudinal view of the internal structure of a lymph node and associated lymphatics

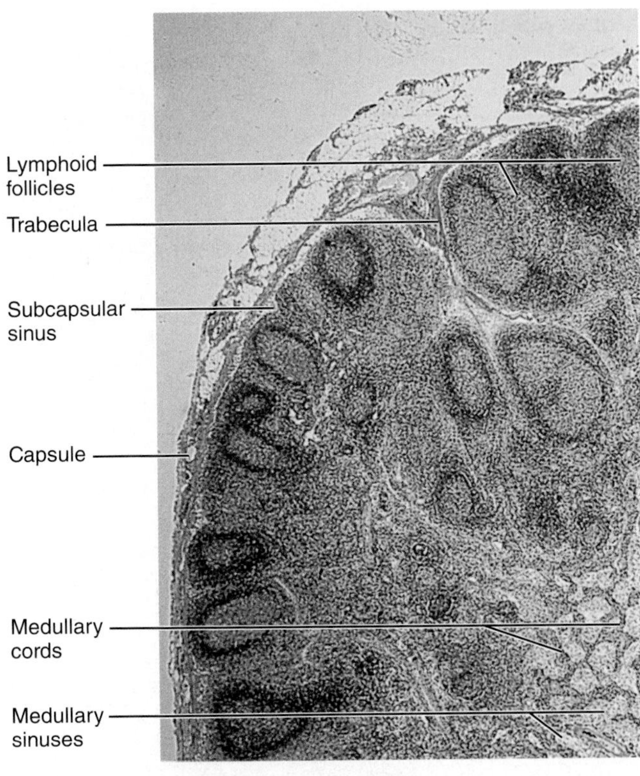

(b) Photomicrograph of part of a lymph node (72×)

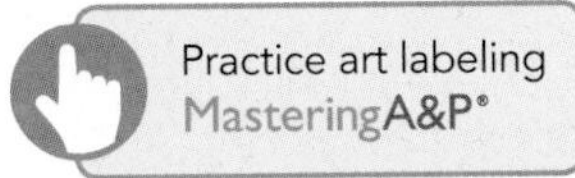

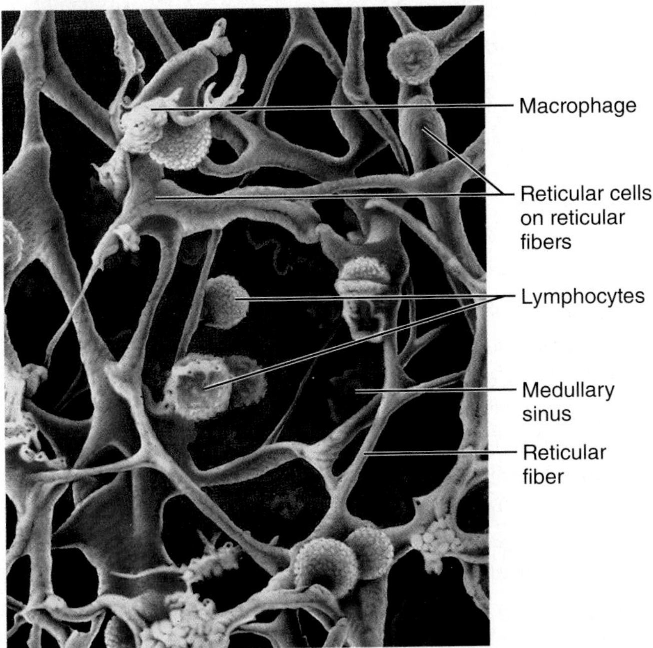

(c) Reticular tissue within the medullary sinus (690×)

Figure 21.3 Structure of a lymph node. In (a), arrows indicate the direction of the lymph flow into, through, and out of the node.

bodies. In the superior thorax, it turns left and empties into the venous circulation at the junction of the left internal jugular and left subclavian veins. The thoracic duct is often joined by the left jugular, subclavian, and/or bronchomediastinal trunks just before it joins with the venous circulation. Alternatively, any or all of these three lymph trunks can empty separately into the nearby veins. When it is joined by the three trunks, the thoracic duct drains three-quarters of the body: the left side of the head, neck, and thorax; the left upper limb; and the body's entire lower half (see Figure 21.2).

Right Lymphatic Duct

The upper right quadrant of the body is drained by the right jugular, subclavian, and bronchomediastinal trunks. In about 20% of people, these ducts join to form a short **right lymphatic duct.** When present, this duct empties into the neck veins at or near the junction of the right internal jugular and subclavian veins (Figure 21.2). More commonly, the three trunks open independently into the neck veins.

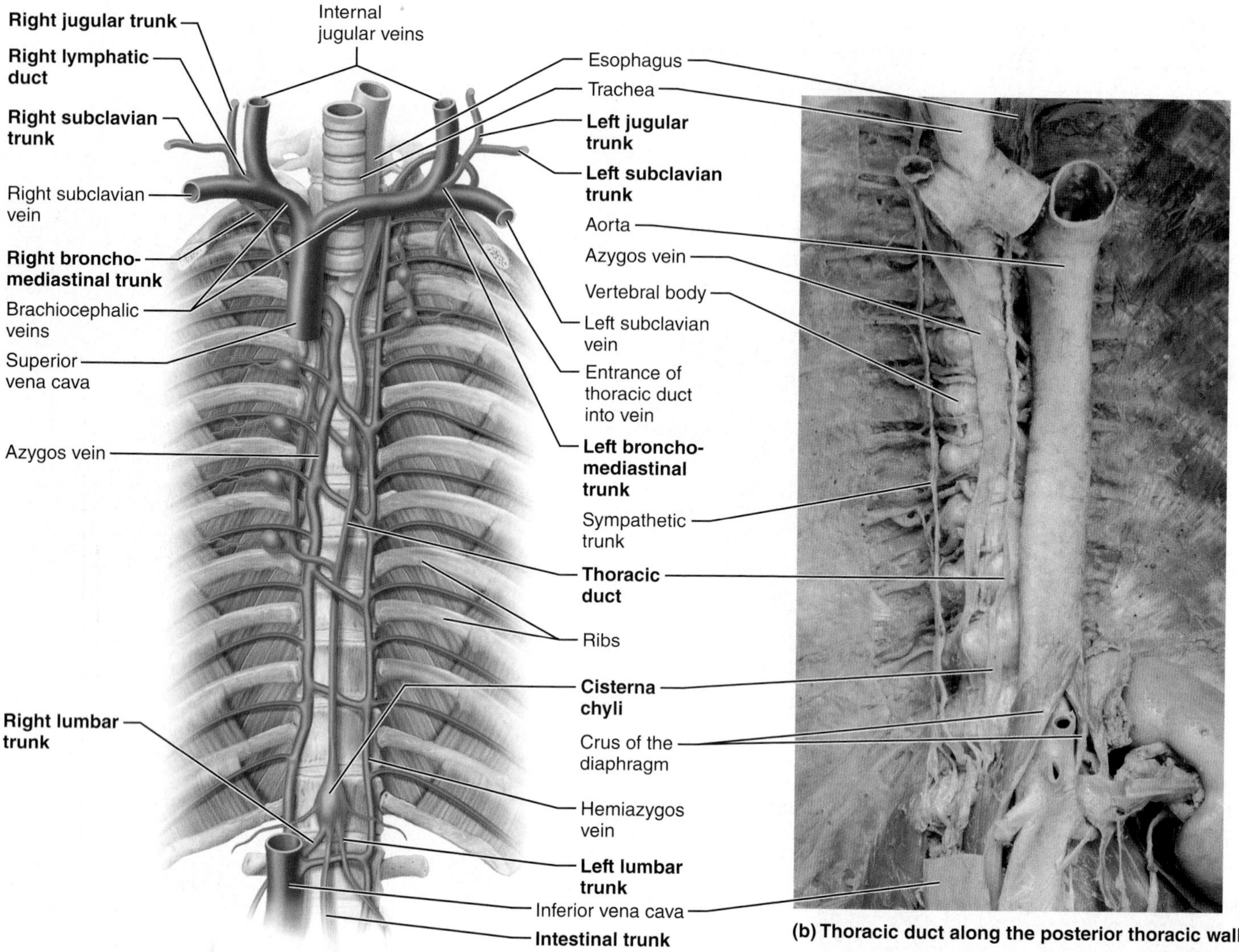

Right jugular trunk
Right lymphatic duct
Right subclavian trunk
Right subclavian vein
Right broncho-mediastinal trunk
Brachiocephalic veins
Superior vena cava
Azygos vein
Right lumbar trunk
Internal jugular veins
Esophagus
Trachea
Left jugular trunk
Left subclavian trunk
Aorta
Azygos vein
Vertebral body
Left subclavian vein
Entrance of thoracic duct into vein
Left broncho-mediastinal trunk
Sympathetic trunk
Thoracic duct
Ribs
Cisterna chyli
Crus of the diaphragm
Hemiazygos vein
Left lumbar trunk
Inferior vena cava
Intestinal trunk

(a) Major lymphatic trunks and ducts in relation to veins and surrounding structures, anterior view

(b) Thoracic duct along the posterior thoracic wall

Figure 21.4 **The lymph trunks and ducts.**

This completes the description of the lymphatic vessels. To summarize their functions, they

1. Return excess tissue fluid to the bloodstream.
2. Return leaked proteins to the blood.
3. Carry absorbed fat from the intestine to the blood (through lacteals).

Moreover, in addition to their roles as lymph filters, the lymph nodes along lymphatic collecting vessels fight disease in their roles as lymphoid organs of the immune system.

check your understanding

☐ 1. What structural feature of lymphatic capillaries makes them permeable to tissue fluid? How does this differ from the features that make blood capillaries permeable?

☐ 2. A sentinel lymph node is the first lymph node to receive drainage from the site of a tumor. This lymph node is tested to determine whether a cancer has spread from its initial location. In which group of lymph nodes would the sentinel node be located in a patient with throat cancer? A patient with breast cancer? A patient with cervical cancer?

☐ 3. Fatty lymph, picked up by the lacteals in the small intestine, travels through the ________ trunk into the ________ duct and empties into venous circulation at the junction of the ________ and ________ veins.

(For answers, see Appendix B.)

THE IMMUNE SYSTEM

The **immune system** is central to the body's fight against disease. Unlike the body's other defense systems, the immune system recognizes and attacks *specific* foreign molecules, and destroys pathogens more and more effectively with each new exposure. The immune system centers around the key

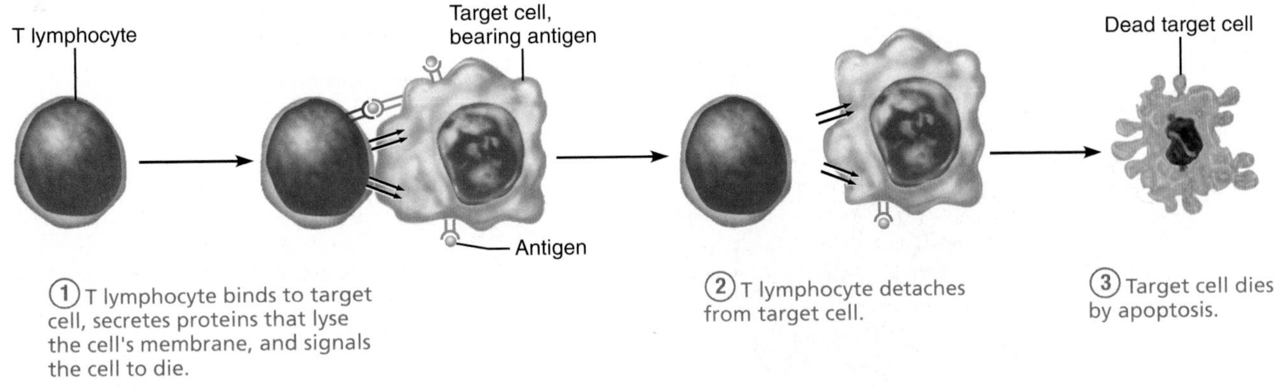

(a) Action of cytotoxic T lymphocyte

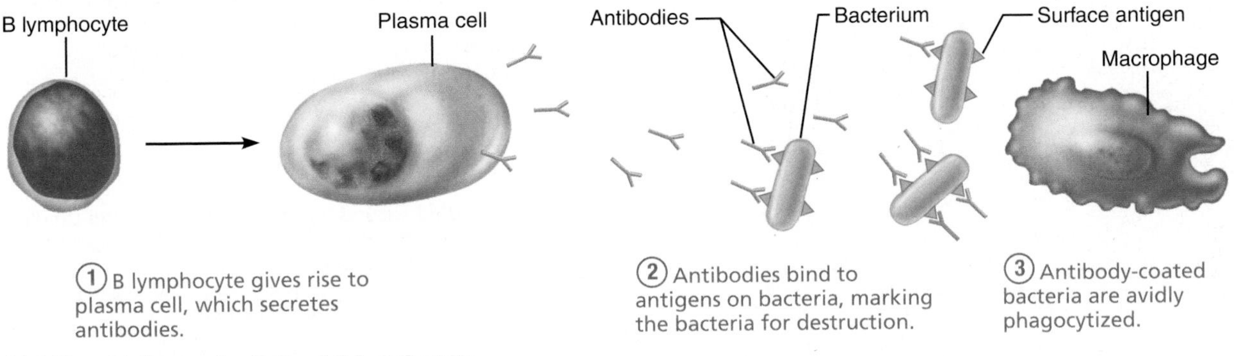

(b) Differentiation and activity of B lymphocyte

Figure 21.5 Lymphocyte function.

defense cells called *lymphocytes.* Lymphocytes are found in *lymphoid tissue* and the *lymphoid organs.* Lymphoid organs include *bone marrow, thymus, lymph nodes, spleen, tonsils,* and *aggregated lymphoid nodules in the small intestine* and *appendix.* The following sections consider these immune components one by one, from the cellular level to the organ level.

Lymphocytes

learning outcomes

- Describe the function, recirculation, and activation of lymphocytes.
- Relate the structure of lymphoid tissue to its infection-fighting function.

Infectious microorganisms that penetrate the epithelial barriers of the body enter the underlying areolar connective tissues. These infectious agents trigger an inflammatory response (described on p. 95) and are attacked by macrophages and by **lymphocytes,** a type of white blood cell of the immune system (p. 552). Lymphocytes are effective in fighting infectious organisms because each lymphocyte recognizes and acts against a *specific* foreign molecule. Any such molecule that induces a response from a lymphocyte is called an *antigen.* Most antigens are either proteins or glycoproteins in the plasma membranes of foreign cells or in the cell walls of bacteria, or proteins secreted by foreign cells (bacterial toxins, for example).

The two main classes of lymphocytes—T cells and B cells—attack antigens in different ways. A major type of T cell, called a **cytotoxic, killer,** or **$CD8^+$ T lymphocyte,** attacks foreign cells directly (Figure 21.5a). It does so by binding to such an antigen-bearing cell, secreting proteins that perforate the foreign cell's membrane, then signaling the cell to undergo programmed cell death (apoptosis; see p. 42). **T cells** bind to antigens that are presented by special proteins that occur only on the membranes of eukaryotic cells (cells that are complex enough to have a nucleus and organelles). These proteins are called the *major histocompatibility complex (MHC).* T cells recognize and respond only to foreign antigens. Thus, T cells target "alien" cells—they reject transplanted organs, destroy our own cells that have been infected with viruses or other pathogens, and kill some cancer cells. All these cells are treated as foreign because they have altered (antigenic) proteins on their surfaces.

B cells, by contrast, differentiate into **plasma cells** that secrete **antibodies** (Figure 21.5b). Antibodies are proteins that bind to specific antigens and mark them for destruction by, for example, making them more recognizable to phagocytic cells. In this way, B cells "flag" cells for destruction by macrophages. B lymphocytes and antibodies respond primarily to bacteria and bacterial toxins in our body fluids. T and B lymphocytes, despite their different actions, cannot be distinguished from one another structurally, even under the electron microscope.

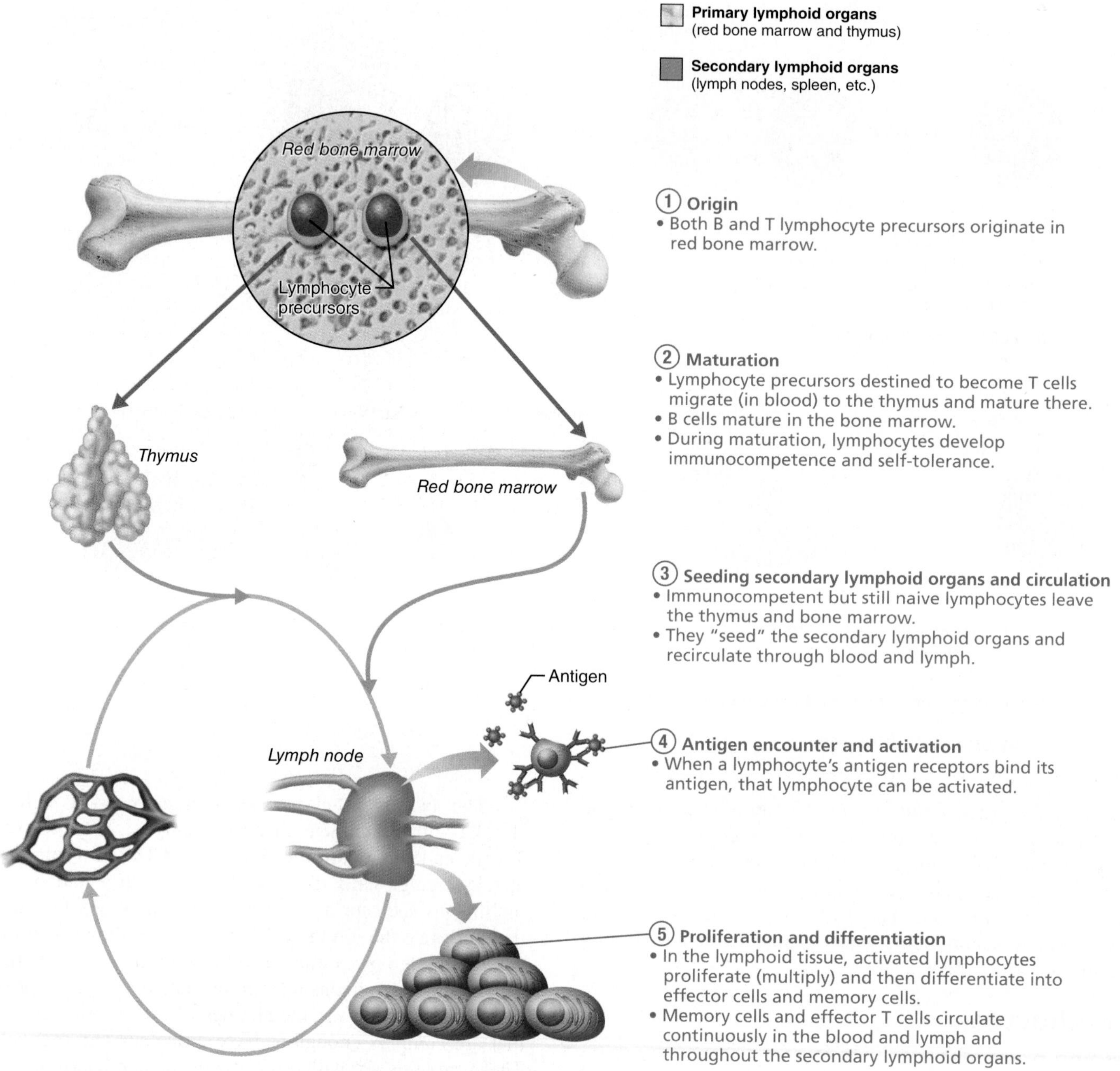

Primary lymphoid organs (red bone marrow and thymus)

Secondary lymphoid organs (lymph nodes, spleen, etc.)

Red bone marrow

Lymphocyte precursors

Thymus

Red bone marrow

Antigen

Lymph node

① **Origin**
- Both B and T lymphocyte precursors originate in red bone marrow.

② **Maturation**
- Lymphocyte precursors destined to become T cells migrate (in blood) to the thymus and mature there.
- B cells mature in the bone marrow.
- During maturation, lymphocytes develop immunocompetence and self-tolerance.

③ **Seeding secondary lymphoid organs and circulation**
- Immunocompetent but still naive lymphocytes leave the thymus and bone marrow.
- They "seed" the secondary lymphoid organs and recirculate through blood and lymph.

④ **Antigen encounter and activation**
- When a lymphocyte's antigen receptors bind its antigen, that lymphocyte can be activated.

⑤ **Proliferation and differentiation**
- In the lymphoid tissue, activated lymphocytes proliferate (multiply) and then differentiate into effector cells and memory cells.
- Memory cells and effector T cells circulate continuously in the blood and lymph and throughout the secondary lymphoid organs.

Figure 21.6 **Differentiation, activation, and recirculation of lymphocytes.** Immature lymphocytes arise in red bone marrow. (Note that red bone marrow is not found in the medullary cavity of the diaphysis of long bones in adults.) Plasma cells (antibody-secreting effector B cells) usually do not circulate.

A third class of lymphocytes, **natural killer cells (NK cells),** do not recognize specific antigens, but instead act when they detect a lack of "self" cell surface molecules or the presence of certain sugars on a target cell. They then rapidly attack tumor cells and virus-infected cells before the immune response is activated. They destroy cells in the same way that cytotoxic T cells do, that is, they lyse them (see Figure 21.5a).

B and T cells continuously travel in the blood and lymph streams to reach infected connective tissues throughout the body, where they fight infection. They repeatedly enter and exit these connective tissues, including the often-infected lymphoid tissue, by squeezing through the walls of capillaries and venules. This repeated movement of activated lymphocytes between the circulatory vessels and the connective tissues, called *recirculation,* ensures that lymphocytes reach all infection sites quickly.

Lymphocyte Differentiation and Activation

Immature lymphocytes go through several stages before they are able to attack antigens (Figure 21.6). Most lymphocytes pass through these stages during a person's infancy and childhood, but many do so in adulthood as well.

1. All lymphocytes originate in the red bone marrow from lymphoid stem cells.
2. Some lymphocytes leave the bone marrow and travel in the bloodstream to mature in the thymus in the thorax and become T lymphocytes (*T* stands for *thymus*). Others stay in the bone marrow to mature and become B lymphocytes. These new T and B lymphocytes divide rapidly and generate many lymphocyte families, each of which has surface receptors able to recognize one unique type of antigen. This is called gaining *immunocompetence.*
3. The mature yet nonactivated (naive) T or B lymphocyte recirculates between the bloodstream and the lymphoid tissue.
4. In the lymphoid tissue, a naive lymphocyte may meet and bind to its specific antigen. Upon this *antigenic encounter,* the lymphocyte becomes fully activated (gains the ability to attack its antigen).
5. In the lymphoid tissue, the activated lymphocyte proliferates rapidly and produces mature lymphocytes. Activated T lymphocytes recirculate throughout the body seeking the same pathogens to attack. Activated B lymphocytes remain in the lymphoid tissue and differentiate into plasma cells, which produce antibodies.

During the activation process, a lymphocyte is presented its antigen by a cell such as a macrophage that has recently phagocytized the antigen, or by a **dendritic cell,** a "professional" antigen gatherer that patrols the body seeking antigens (see Figure 5.3, p. 106). The dendritic cell or macrophage carries the antigen to the lymphoid tissues where lymphocytes are established. The specific response of the activating lymphocyte to the antigen differs for each type of lymphocyte; however, both activating B and T cells produce clones of *effector cells* and *memory cells.*

Effector lymphocytes are short-lived lymphocytes that respond to the pathogen immediately and then die.

- **B cells** divide rapidly to produce plasma cells. These plasma cells remain in the lymphoid tissue and secrete antibodies into the bloodstream. Antibodies bind with soluble antigens at the site of infection, marking them for phagocytosis (Figure 21.5b).
- **Cytotoxic ($CD8^+$) T cells** travel to the infected region via the bloodstream and directly lyse the "foreign" cell (for example, a virus-infected cell or cancer cell, Figure 21.5a).
- **Helper ($CD4^+$) T cells** enter the circulation and stimulate the cells of the immune system by secreting chemicals called cytokines. Cytokines stimulate the proliferation of activated B cells, cytotoxic T cells, and macrophages and amplify and fine-tune the immune response. The importance of helper T cells is illustrated by *acquired immune deficiency syndrome (AIDS),* a viral disease in which a drastic decline in the body's helper T cells greatly weakens the immune system. AIDS is considered in more detail in A CLOSER LOOK on (p. 636).

The other clones produced during activation, **memory lymphocytes,** also called **memory cells,** wait within the lymphoid tissues until the body encounters the specific antigen again—maybe decades later. When a memory lymphocyte finally encounters its antigen, its proliferative response and its attack are most vigorous and rapid. Memory lymphocytes are the basis of acquired immunity; that is, they guard against subsequent infections and prevent people from getting certain diseases more than once.

> **CLINICAL APPLICATION**
>
> **Vaccination** Vaccination mimics acquired immunity by presenting the body with a weakened or inactive dose of a pathogen or toxin. The result of exposure to the infectious agent is the activation of lymphocytes specific to that pathogen and the development of memory lymphocytes. Thus if the body is infected by the same agent again, it can produce a rapid response to eliminate the pathogen before it can cause illness.

Lymphoid Tissue

Lymphoid tissue is a specialized type of connective tissue in which vast quantities of lymphocytes gather to fight invading microorganisms. This tissue has two general locations: (1) in the frequently infected mucous membranes

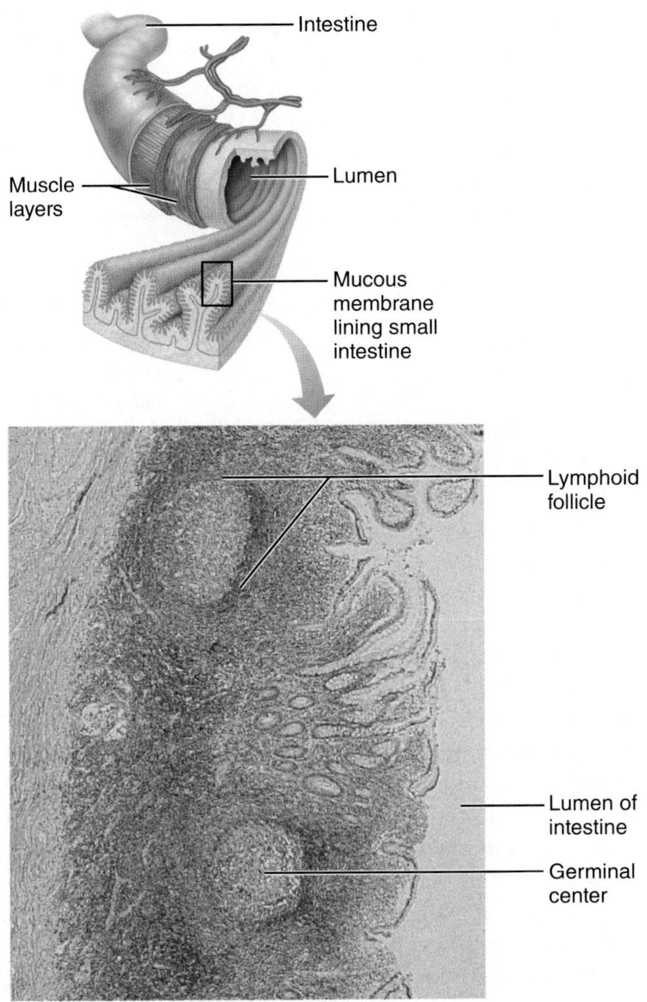

Figure 21.7 Mucosa associated lymphoid tissue (MALT).

AIDS: The Modern-Day Plague?

AIDS is a viral disease that progresses through three stages: (1) An acute stage, which develops within several weeks of infection and lasts about 2 weeks, is characterized by flulike symptoms such as fever, fatigue, rash, headache, sore throat, swollen lymph nodes, muscle and joint pain, night sweats, and diarrhea. (2) Next comes a long period without symptoms that lasts an average of 10 years. (3) Finally, full-blown AIDS develops, characterized by collapse of the immune system that results in increasingly frequent opportunistic infections. These can include tuberculosis, a rare type of fungal pneumonia (*Pneumocystis jirovecii*), and the distinctive purple skin lesions of *Kaposi's sarcoma* (see p. 642). Furthermore, there is wasting (weight loss) accompanied by diarrhea. About 10% to 15% of AIDS patients develop dementia. Untreated AIDS ends in death from wasting or overwhelming infection. This final stage lasts a few months to several years.

Cause

The agent that causes AIDS, the human immunodeficiency virus (HIV), is transmitted solely through body secretions—blood, semen, vaginal secretions, and breast milk. It is not transmitted by casual contact, because the virus dries and dies when exposed to air. Most commonly, HIV enters the body (1) during sexual contact in which the mucosa is torn and bleeds, or where open lesions due to other sexually transmitted diseases give the virus access to the blood; or (2) through blood-contaminated needles.

After entering the body, HIV travels to the lymphoid tissues, where it infects and destroys helper ($CD4^+$) T cells, severely depressing immunity.

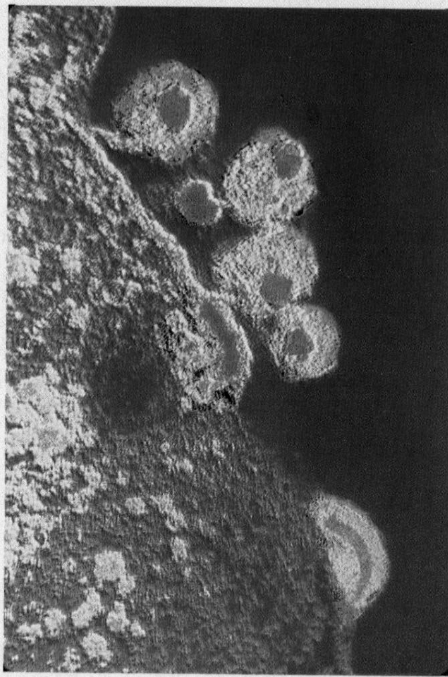

New HIV viruses (red and yellow dots) emerge from an infected human cell.

Helper T cells are the cornerstone of the immune system, because they regulate the populations of B cells and cytotoxic T cells, and without them these lymphocytes cannot be activated or maintained. HIV also enters dendritic cells, macrophages, and microglia cells, all of which share a surface protein called CD4, to which HIV attaches. The virus invades the brain, probably in infected microglia, and induces destruction of neurons, accounting for the dementia of some AIDS patients.

In the initial, acute stage, rapid virus multiplication stimulates the immune system. Antibody levels rise, and cytotoxic ($CD8^+$) T cells fight the infection. During the long, asymptomatic stage, HIV replicates and mutates rapidly, producing deadly strains that evade the immune system by hiding in the $CD4^+$ cells of lymphoid tissues. Ultimately, the immune system is so weakened that the exhausted $CD8^+$ cells can no longer contain the virus. The number of helper T cells declines, and the terminal AIDS stage begins.

Today, AIDS is entering its third decade as a global epidemic. In 2011, the number of people worldwide with HIV infection or AIDS reached 34 million. Of the 2.5 million newly infected individuals, 1.8 million (72%) are in sub-Saharan Africa.

Treatment

Researchers and clinicians had hoped that antiviral drugs would effectively treat HIV. A combination of drugs that prevent viral replication and assembly diminishes viral loads to undetectable levels and reduces transmission to noninfected partners. This treatment does not eliminate HIV from the body, and short breaks in treatment allow viral levels to soar.

Another complication is HIV's ability to mutate. Since 1996, when antiviral drugs were first used against it, HIV has developed resistance to every antiviral medication tested. The quest for a vaccine against HIV, either as prevention or treatment, has been equally challenging as viral mutations form new resistant subtypes.

Despite the grim outlook for a cure, clinicians note a trend in many countries toward high-risk sexual practices, such as casual encounters with multiple partners and falling rates of condom use. The Centers for Disease Control and Prevention estimates that 1.2 million U.S. residents are infected with HIV. The only certain way to avoid AIDS is to avoid contact with HIV-infected body fluids and to practice abstinence. Short of that, the best defense is to practice safe sex by using latex condoms and to know one's sexual partner well.

(p. 89) of the digestive, respiratory, urinary, and reproductive tracts, where it is called **mucosa associated lymphoid tissue,** or **MALT** (Figure 21.7); and (2) in all lymphoid organs except the thymus. Besides serving as the main battleground in the fight against infection, lymphoid tissue is also where most lymphocytes become activated and most effector and memory lymphocytes are generated.

Lymphoid tissue is a reticular connective tissue whose basic framework is a network of reticular fibers secreted by reticular cells (fibroblasts) (Figure 21.3c). Within the spaces of this network reside the many T and B lymphocytes that arrive continuously from venules coursing through this tissue. Macrophages on the fiber network kill invading microorganisms by phagocytosis and, along with dendritic cells, activate nearby lymphocytes by presenting them with antigens.

Evident within lymphoid tissue are scattered, spherical clusters of densely packed lymphocytes, called **lymphoid follicles** or **nodules** (Figure 21.7). These follicles often exhibit lighter-staining centers, called **germinal centers,** of dividing lymphocytes. Each follicle derives from the activation of a single B cell, whose rapid proliferation generates the thousands of lymphocytes in the follicle. Newly produced B cells migrate away from the follicle to become plasma cells.

☑ check your understanding

- ☐ **4.** Where do cytotoxic T lymphocytes gain immunocompetence?
- ☐ **5.** Which immune cells will respond to an infection caused by a bacterial agent?
- ☐ **6.** Which type of lymphocyte is found in lymphoid follicles? What occurs at the germinal center of a lymphoid follicle?

(For answers, see Appendix B.)

Lymphoid Organs

learning outcome

▶ Describe the locations, histological structure, and immune functions of the following lymphoid organs: thymus, lymph nodes, spleen, tonsils, aggregated lymphoid nodules in the intestine, and appendix.

The **lymphoid organs** (Figure 21.8) are grouped into two categories.

- The **primary lymphoid organs** are the bone marrow and the thymus. They produce B and T lymphocytes, respectively (Figure 21.6). The structure of the bone marrow is described in the discussion of blood cell formation (Chapter 18, p. 553). The structure of the thymus is described below.
- The **secondary lymphoid organs** are the lymph nodes, the spleen, and the collections of mucosa associated lymphoid tissue (MALT) that form the tonsils, aggregated lymphoid nodules in the intestine, and the appendix. These organs store immunocompetent lymphocytes and memory lymphocytes, and they gather and destroy infectious microorganisms within their lymphoid tissue. The structure of each of these organs is described next.

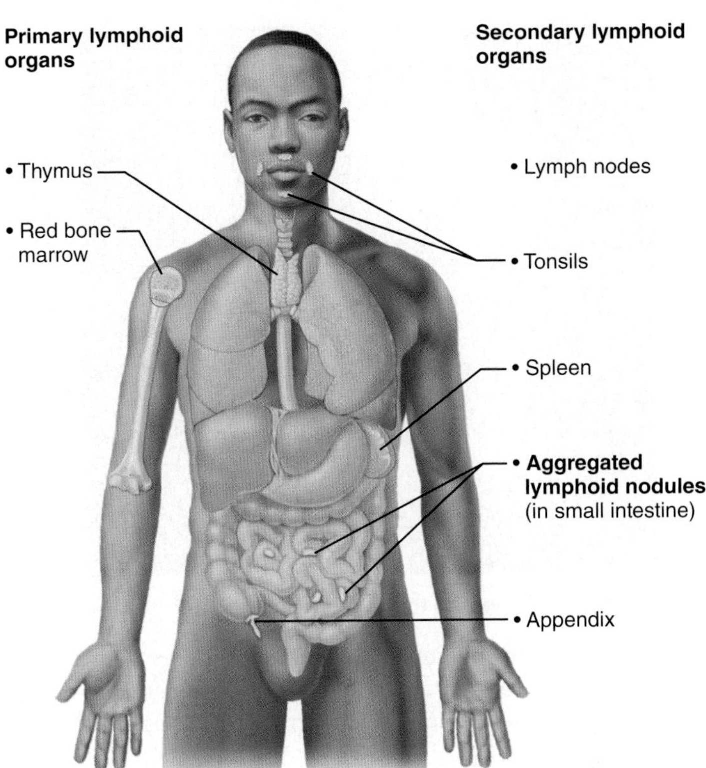

Figure 21.8 **Lymphoid organs.** (Lymph nodes are illustrated in Figure 21.2.)

Thymus

The two-lobed **thymus** lies in the superior thorax and inferior neck, just posterior to the sternum (Figure 21.9a). As noted earlier, the thymus is the site at which immature lymphocytes develop into T lymphocytes. Specifically, it secretes thymic hormones such as thymosin and thymopoietin, which cause T lymphocytes to gain immunocompetence. Prominent in newborns, the thymus continues to increase in size during childhood, when it is most active. During late adolescence, it begins to atrophy gradually as its functional tissue is slowly replaced with fibrous and fatty tissue. At age 20 it still has about 80% of its functional tissue, but at age 40 it typically retains only 5%. By old age, only 2% remains, and the thymus is a fatty mass that is difficult to distinguish from the surrounding connective tissue. Even as it atrophies over time, the thymus continues to produce immunocompetent cells throughout adulthood at a reduced rate.

The thymus contains numerous **lobules** arranged like the florets in a head of cauliflower. Each lobule in turn contains an outer cortex and an inner medulla (Figure 21.9b). The **cortex** stains dark because it is packed with rapidly dividing T lymphocytes gaining immunocompetence; the **medulla** contains fewer lymphocytes and stains lighter. Also in the medulla are **thymic corpuscles,** which are composed of clusters of epithelial cells. The thymic corpuscles function in the development of regulatory T cells, a type of T lymphocyte that prevents autoimmune responses.

The thymus differs from the other lymphoid organs in two basic ways: First, it functions strictly in lymphocyte maturation and thus is the only lymphoid organ that does not directly fight antigens. In fact, the *blood thymus barrier,*

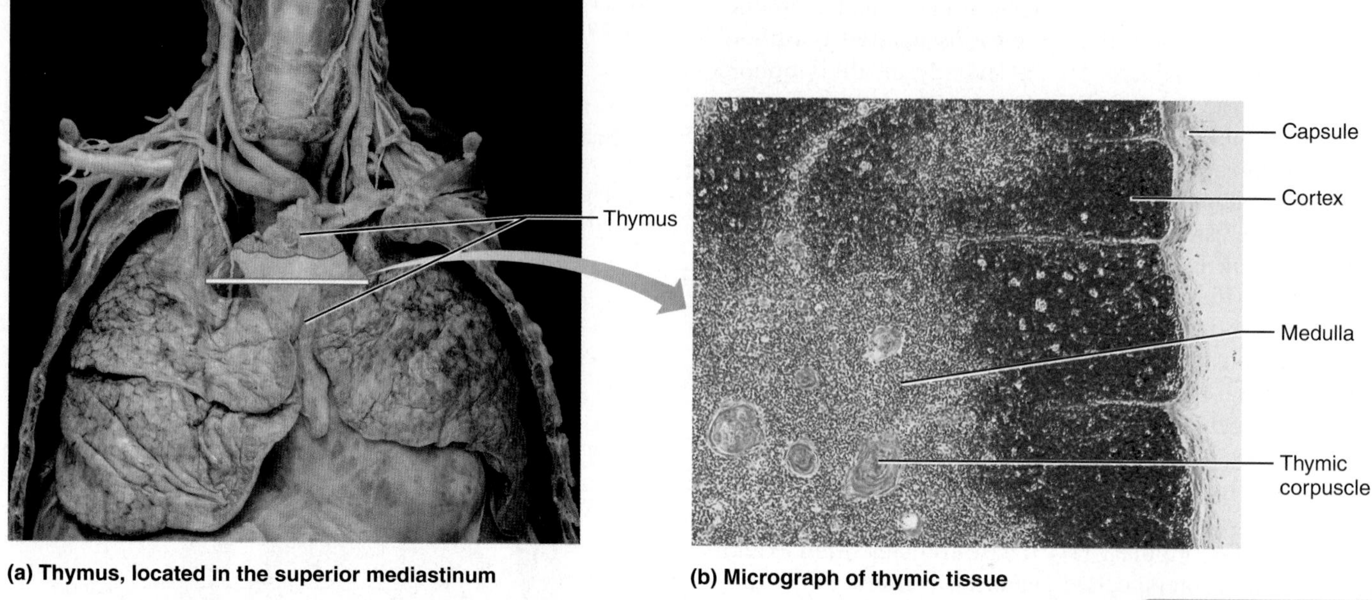

Figure 21.9 The thymus. (**a**) Dissection showing the location of the thymus in the superior thorax. (**b**) Photomicrograph of thymic tissue, showing part of a lobule with cortex and medulla regions, and thymic corpuscles (35×).

analogous to the blood brain barrier, keeps bloodborne antigens from leaking out of thymic capillaries and prematurely activating the immature thymic lymphocytes. Second, the tissue framework of the thymus is not a true lymphoid connective tissue. Because the thymus arises like a gland from the epithelium lining the embryonic pharynx, its basic tissue framework consists of star-shaped epithelial cells rather than reticular fibers. These *epithelial reticular cells* secrete the thymic hormones that stimulate T cells to become immunocompetent. Note as well that the thymus has no lymphoid follicles because it lacks B cells.

Lymph Nodes

The lymph-filtering components of lymph nodes have already been considered (pp. 629–630), but lymph nodes are more than just lymph filters. The lymph nodes are the organs at which the lymphatic and immune systems intersect. Between the lymph sinuses are masses of lymphoid tissue (see Figure 21.3). As the lymph percolates through the lymph sinuses, some of the contained antigens leak out through the sinus walls into this lymphoid tissue. Most antigenic challenges in the human body occur here in the lymph nodes, where the antigens not only meet their destruction but also activate B and T lymphocytes, adding to the body's valuable supply of memory lymphocytes that offer long-term immunity.

Lymph nodes have two histologically distinct regions, an external **cortex** ("outer bark") and a **medulla** ("middle") near the hilum (Figure 21.3). All the lymphoid follicles and most B cells occupy the lymphoid tissue of the most superficial part of the cortex. Deeper in the cortex, the lymphocytes are primarily T cells, especially helper T cells that increase the activity of B cells in the nearby follicles. Clusters of lymphocytes within the medulla, called the medullary cords, contain T and B lymphocytes and plasma cells.

Spleen

The soft, blood-rich **spleen** (Figure 21.10) is the largest lymphoid organ. Its size varies greatly among individuals, but on average it is the size of a fist. This unpaired organ, which lies in the left superior quadrant of the abdominal cavity just posterior to the stomach (see Figure 21.8), is shaped like a jellyfish and has a concave anterior surface. The large *splenic vessels* enter and exit the anterior surface along a line called the **hilum** (Figure 21.10a).

The spleen has two main blood-cleansing functions: (1) the removal of bloodborne antigens (its immune function), and (2) the removal and destruction of aged or defective blood cells. Additionally, the spleen is a site of hematopoiesis in the fetus and stores blood platelets and monocytes throughout life.

The spleen, like lymph nodes, is surrounded by a fibrous capsule from which trabeculae extend inward (Figure 21.10b). The larger branches of the splenic artery run in the trabeculae and send smaller arterial branches into the substance of the spleen. These branches are called **central arteries** because they are enclosed by (lie near the *center* of) thick sleeves of lymphoid tissue that collectively constitute the **white pulp** of the spleen. Bloodborne antigens enter this lymphoid tissue and are destroyed as they activate the immune response. The white pulp provides the immune function of the spleen.

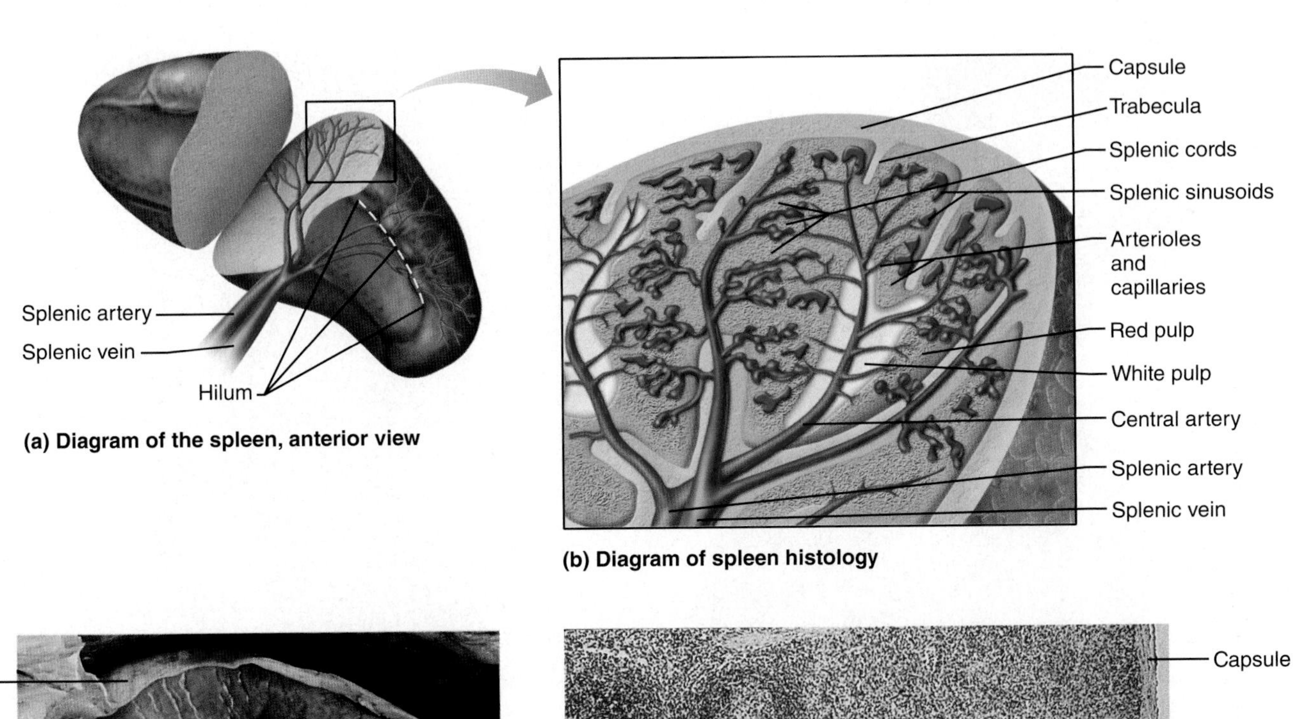

(a) Diagram of the spleen, anterior view

(b) Diagram of spleen histology

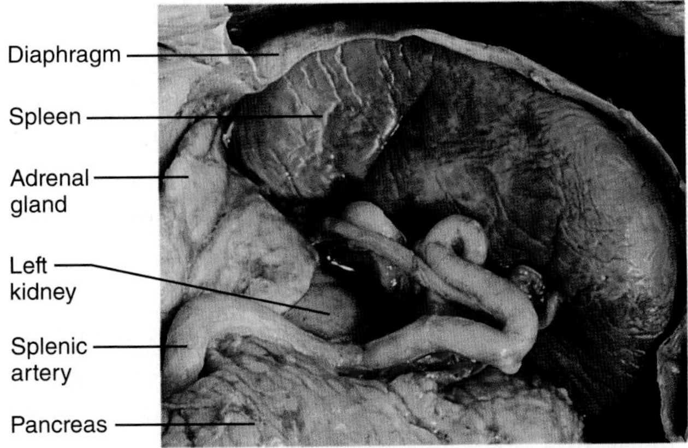

(c) Photograph of the spleen in its normal position in the abdominal cavity, anterior view

Capsule
White pulp
Red pulp

(d) Photomicrograph of spleen tissue (30×). The white pulp, a lymphoid tissue with many lymphocytes, is surrounded by red pulp containing abundant erythrocytes.

Figure 21.10 Structure of the spleen.

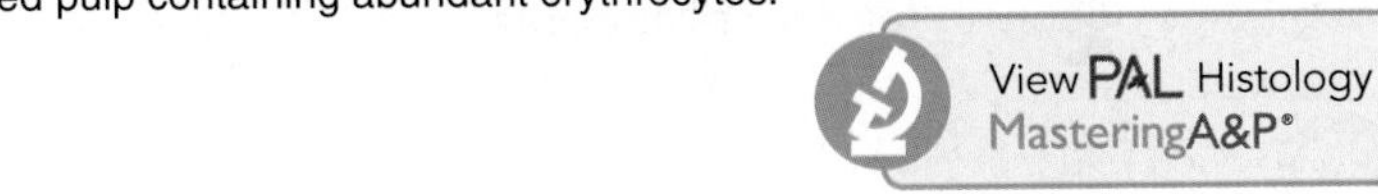

Surrounding the white pulp is **red pulp,** which has two parts: (1) **splenic sinusoids,** blood sinusoids that arise from the distal branches of the central arteries outside of the white pulp; and (2) **splenic cords,** which consist of a reticular connective tissue that is exceptionally rich in macrophages. Whole blood leaks from the sinuses into this connective tissue, where macrophages then phagocytize any defective blood cells. Hence, red pulp is responsible for the spleen's ability to dispose of worn-out blood cells.

In histological sections (Figure 21.10d), the white pulp appears as islands in a sea of red pulp. Note that the naming of the pulp regions reflects their appearances in fresh spleen tissue rather than in stained histological sections; with many stains, the white pulp actually appears darker than the red pulp.

CLINICAL APPLICATION

Splenectomy Because the capsule of the spleen is relatively thin, physical injury or a serious infection may cause the spleen to rupture, leading to severe loss of blood due to hemorrhage into the peritoneal cavity. In such cases, the spleen must be removed quickly and its artery tied off, a surgical procedure called a **splenectomy.** A person can live a relatively healthy life without a spleen, because macrophages in the bone marrow and liver can take over most of the spleen's functions. Such a person will be more susceptible to infections, however, so surgeons performing splenectomies now leave some of the spleen in place whenever possible. Alternatively, healthy fragments of the spleen are surgically reattached immediately after the operation. In some such cases, the spleen can regenerate.

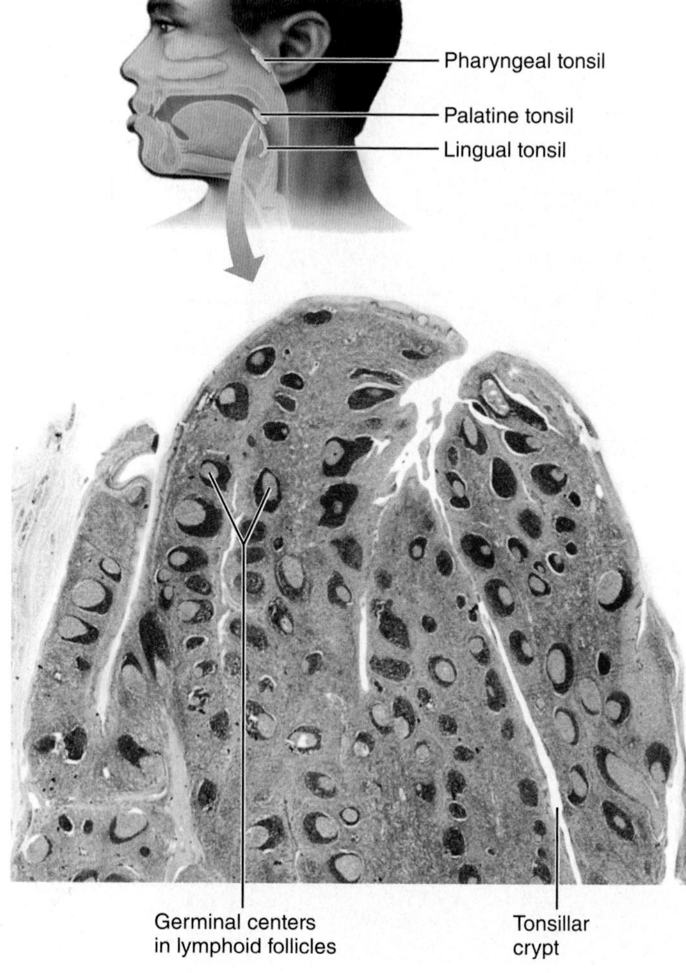

Figure 21.11 Histology of the palatine tonsil. The exterior surface of the tonsil is covered by epithelium, which invaginates deeply to form tonsillar crypts.

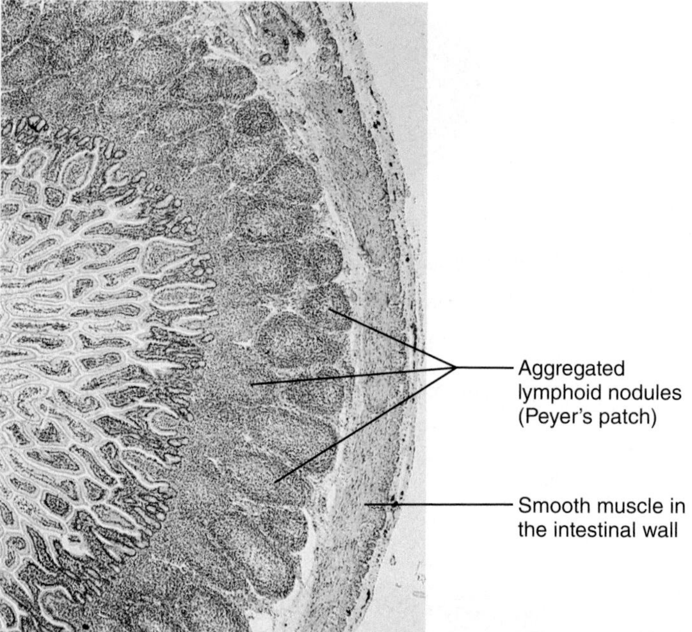

Figure 21.12 Aggregated lymphoid nodules. The photomicrograph shows the histological structure of the aggregated lymphoid nodules (Peyer's patches) in the wall of the ileum of the small intestine. The lumen of the intestine is seen at the left of the photomicrograph (5×).

Tonsils

The **tonsils** are mere swellings of the mucosa lining the pharynx. There are four groups of tonsils (Figure 21.11).

- The paired **palatine tonsils** lie directly posterior to the mouth and palate on the lateral sides of the pharyngeal wall. These are the largest tonsils and the ones most often infected and removed during childhood, in a surgical procedure called *tonsillectomy*.
- The **lingual tonsil** lies on the posterior surface of the tongue.
- The **pharyngeal tonsil** (adenoids) lies on the pharyngeal roof.
- The **tubal tonsils** (illustrated in Figure 22.3, p. 649) are just behind the openings of the pharyngotympanic tubes into the pharynx. The four groups of tonsils are arranged in a ring around the entrance to the pharynx to gather and remove many pathogens that enter the pharynx in inspired air and swallowed food. The tonsils process the antigens and then initiate immune responses.

Like all mucosae, the tonsils consist of an epithelium underlain by a connective tissue lamina propria (Figure 21.11). In the tonsils, the underlying lamina propria consists of abundant mucosa associated lymphoid tissue (MALT) packed with lymphocytes and scattered lymphoid follicles. The overlying epithelium invaginates deep into the interior of the tonsil, forming blind-ended **crypts** that trap bacteria and particulate matter. The trapped bacteria work their way through the epithelium to the underlying lymphoid tissue, causing the activation of lymphocytes. Such trapping of bacteria leads to many tonsil infections during childhood, but it also generates a great variety of memory lymphocytes for long-term immunity.

Aggregated Lymphoid Nodules and the Appendix

Many bacteria permanently inhabit the hollow interior of the intestines and are constantly infecting the intestinal walls. To fight these invaders, MALT is especially abundant in the intestine. In fact, in two parts of the intestine, MALT is large, permanent, and densely packed with lymphocytes. These intestinal structures are the aggregated lymphoid nodules and the appendix.

Aggregated lymphoid nodules *(Peyer's patches)* are clusters of lymphoid follicles in the walls of the distal part (ileum) of the small intestine (Figure 21.12). About 40 of these nodules are present, averaging about a centimeter long and a centimeter wide.

Lymphoid tissue is also heavily concentrated in the wall of the **appendix,** a tubular offshoot of the first part (cecum) of the large intestine (see Figure 21.8). Histological sections reveal that dense lymphoid tissue uniformly occupies over half the thickness of the appendix wall.

Besides destroying the microorganisms that invade them, the aggregated lymphoid nodules and appendix sample many different antigens from within the digestive tube and generate a wide variety of memory lymphocytes to protect the body.

✓ check your understanding

- ☐ 7. What is the function of red pulp in the spleen?
- ☐ 8. Where are T lymphocytes located in lymph nodes?
- ☐ 9. Which part of the intestine contains aggregated lymphoid nodules?

(For answers, see Appendix B.)

DISORDERS OF THE LYMPHATIC AND IMMUNE SYSTEMS

learning outcome

▶ Describe the basic characteristics of two disorders of lymphatic vessels: chylothorax, lymphangitis; and three disorders of lymphocytes and lymphoid organs: mononucleosis, Hodgkin's lymphoma, and non-Hodgkin's lymphoma.

Chylothorax ("chyle in the thorax") is the leakage of the fatty lymph, chyle, from the thoracic duct into a pleural cavity in the thorax. It is caused by tearing or a blockage of the thoracic duct due to chest trauma or to compression by a nearby tumor. Complications can result from (1) large amounts of lymph in the pleural cavity compressing and collapsing the lungs, (2) metabolic problems caused by the loss of fatty nutrients from the circulation, or (3) diminished blood volume following the loss of fluid from the circulation. Treatment involves draining the lymph from the pleural cavity and surgically repairing any damage to the thoracic duct.

Lymphangitis (lim′fan-ji′tis; "lymph vessel inflammation") is inflammation of a lymphatic vessel. Like the large blood vessels, the lymphatic vessels are supplied with blood by the vasa vasorum. When lymphatic vessels are infected and inflamed, the vasa vasorum become congested with blood. The superficial lymphatic vessels then become visible through the skin as red lines that are tender to the touch.

Mononucleosis is a viral disease common in adolescents and young adults. Its symptoms include fatigue, fever, sore throat, swollen lymph nodes, and enlargement of the spleen. It is caused by the Epstein-Barr virus, which specifically attacks B lymphocytes. This attack leads to a massive activation of T lymphocytes, which in turn attack the virus-infected B cells. The large numbers of oversized T lymphocytes that circulate in the bloodstream were originally misidentified as monocytes (*mononucleosis* = condition of monocytes). Mononucleosis is transmitted in saliva ("kissing disease") and usually lasts 4 to 6 weeks. The most serious risk of this disease is rupture of the enlarged spleen, which can cause massive hemorrhaging.

Hodgkin's lymphoma is a malignancy of the lymph nodes characterized by swollen, nonpainful nodes; fatigue; and often, persistent fever and night sweats. It is characterized by distinctive giant cells in the nodes called Reed-Sternberg cells, which are an abnormal type of B lymphocyte. Hodgkin's disease is treated with chemotherapy and radiation therapy, and the cure rate is high relative to that of other cancers.

Non-Hodgkin's lymphoma includes all cancers of lymphoid tissues except Hodgkin's lymphoma. It involves the uncontrolled multiplication and metastasis of undifferentiated lymphocytes, usually B cells (85% of cases), and less often T cells (15% of cases). Lymph nodes become swollen, the spleen may enlarge, gut-lymphoid tissue may be affected, and many organs may eventually be involved. The incidence of non-Hodgkin's lymphoma has remained stable for the last 10 years; it is the seventh most common type of cancer. A low-grade type, which affects the elderly and grows slowly, is often fatal because it is resistant to chemotherapy. An intermediate or high-grade type, which affects young people and grows quickly, can respond to chemotherapy with a 60% remission rate.

THE LYMPHATIC AND IMMUNE SYSTEMS THROUGHOUT LIFE

learning outcome

▶ Outline the development of the lymphatic vessels and lymphoid organs.

The lymphatic system develops from a number of sources. Lymphatic vessels and the main clusters of lymph nodes grow from *lymphatic sacs,* which are projections from the large veins in the embryo. The thymus originates as an outgrowth of the endoderm lining the embryonic pharynx (see Figure 17.13, p. 540). It detaches from the pharynx and migrates caudally into the thorax. Starting early in the fetal period, the thymus receives the first progenitors of T lymphocytes from the blood-forming organs (yolk sac, liver, then bone marrow). The first B lymphocytes are produced in the bone marrow at this time. All the nonthymic lymphoid organs and tissues (spleen, lymph nodes, and MALT) arise from mesodermal mesenchyme. The spleen and tonsils develop before birth. The other secondary lymphoid organs, however, are poorly developed before birth. Shortly after birth, they become heavily populated by circulating lymphocytes and start to gain their functional properties.

The immune system of newborns was long thought to be too immature to attack invading pathogens. However, newborns respond to new antigens just as vigorously as do adults, with both T cells and antibodies. Throughout infancy and childhood, many memory lymphocytes are formed, and a wide range of immunity is attained. After childhood, some immune organs become less active and begin to shrink. The tonsils are regressing by age 14 (except the lingual tonsil, which does not shrink until about age 30). As previously mentioned, the thymus regresses after late adolescence but keeps producing T cells throughout life.

The lymphoid organs and immune system normally serve people well until late in life, when their efficiency begins to wane and their ability to fight infection declines. This reduced effectiveness seems mostly due to a decrease in the production and responsiveness of T cells, which in turn causes similar declines in B cells. Old age is accompanied by an increased susceptibility to disease. The greater incidence of cancer in the elderly is assumed to be another example of the declining ability of the immune system to destroy harmful cells.

☑ check your understanding

- ☐ 10. A few days after a cat scratched his hand, John noticed thin red lines extending from his hand up through his forearm. What is the cause of these marks?
- ☐ 11. From which embryonic layer is the thymus derived? From which embryonic layer are the tonsils, spleen, and lymph nodes derived?

(For answers, see Appendix B.)

RELATED CLINICAL TERMS

Kaposi's sarcoma Tumorlike lesions of the skin and some internal organs, seen in some AIDS patients. The lesions are vascularized and arise from proliferating capillary endothelial cells. The capillaries are so permeable that they leak erythrocytes, giving the lesions a purple color. Kaposi's sarcoma is now known to be caused by a previously unrecognized type of herpes virus that infects the endothelium. Confusion exists as to whether the lesions are merely due to cellular hyperplasia or are true tumors. Treatment is with chemotherapy.

Lymphadenopathy (lim-fad′ĕ-nop-ah-the) (*adeno* = gland; *pathy* = disease) Any disease of the lymph nodes.

Lymphoma Any neoplasm (tumor) of the lymphoid tissue, whether benign or malignant.

Sentinel lymph node in cancer The first node that receives lymph draining from a body area suspected of having a tumor. When examined for the presence of cancer cells, this node gives the best indication of whether metastasis (spread) through the lymph vessels has occurred.

Splenomegaly (sple″no-meg′ah-le; "spleen big") Enlargement of the spleen, usually resulting from blood diseases such as mononucleosis, malaria, leukemia, or polycythemia vera. A way that physicians diagnose spleen enlargement is by feeling for the spleen's notched superior border through the skin of the abdominal wall just anterior to the costal margin. A healthy spleen never reaches this far anteriorly.

Tonsillitis Congestion of the tonsils, typically with infecting bacteria, which causes them to become red, swollen, and sore.

CHAPTER SUMMARY

Take full advantage of the invaluable online practice and study tools for this chapter by going to the Study Area of MasteringA&P®. There you will find:

- Art Labeling Activities, Quizzes, Practice Tests, Memory Games
- Animal Organ Dissection, Cat Dissection and Bone Videos
- PAL Cadaver
- PAL Histology
- A&P Flix Animations
- Dynamic Study Modules

1. The lymphatic system consists of lymphatic circulatory vessels that carry lymph. The immune system contains the lymphocytes, lymphoid tissue, and lymphoid organs, which are involved in the body's fight against disease.

THE LYMPHATIC SYSTEM (pp. 628–632)

2. Lymph is excess tissue fluid, which originates because slightly more fluid leaves blood capillaries than returns there. Lymphatic vessels pick up this excess fluid and return it to the great veins at the root of the neck.
3. Lymphatic vessels also retrieve blood proteins that leak from capillaries and return these proteins to the bloodstream.
4. The vessels of the lymphatic system, from smallest to largest, are lymphatic capillaries, collecting lymphatic vessels (with lymph nodes), lymph trunks, and lymph ducts.

Lymphatic Capillaries (pp. 628–629)

5. Lymphatic capillaries weave through the loose connective tissues of the body. These closed-end tubes are highly permeable to entering tissue fluid and proteins because their endothelial cells are loosely joined. Disease-causing microorganisms and cancer cells also enter the permeable lymphatic capillaries and spread widely through the lymph vessels.
6. Lymphatic capillaries called lacteals absorb digested fat from the small intestine.

Collecting Lymphatic Vessels (p. 629)

7. Collecting lymphatic vessels run alongside arteries and veins but have thinner walls and many more valves than do veins. The collecting vessels resemble a string of beads.
8. Lymph flows very slowly through collecting lymphatic vessels. Flow is maintained by normal body movements, contractions of

skeletal muscles, arterial pulsations, and contraction of smooth muscle in the wall of the lymphatic vessel. Lymphatic valves prevent backflow.

Lymph Nodes (pp. 629–630)

9. Clustered along the collecting lymphatic vessels, bean-shaped lymph nodes remove infectious agents and cancer cells from the lymph stream. Lymph enters the node via afferent lymphatic vessels and exits via efferent vessels at the hilum. In between, the lymph percolates through lymph sinuses, where macrophages remove lymph-borne pathogens. Lymph nodes are located superficially and deep (Figure 21.2).

Lymph Trunks (p. 630)

10. The lymph trunks (lumbar, intestinal, bronchomediastinal, subclavian, and jugular) each drain a large body region. All except the intestinal trunk are paired.

Lymph Ducts (pp. 630–632)

11. The right lymphatic duct (and/or the nearby trunks) drains lymph from the superior right quarter of the body. The thoracic duct (and/or the nearby trunks) drains lymph from the rest of the body. These two ducts empty into the junction of the internal jugular and subclavian veins. The thoracic duct starts at the cisterna chyli at L_1–L_2 and ascends along the thoracic vertebral bodies.

PAL Human Cadaver/Lymphatic System

THE IMMUNE SYSTEM (pp. 632–641)

12. Lymphoid organs and lymphoid tissues house millions of lymphocytes, important cells of the immune system that recognize specific antigens.

Lymphocytes (pp. 633–634)

13. B and T lymphocytes fight infectious microorganisms in the areolar and lymphoid connective tissues of the body—B cells by producing antibody-secreting plasma cells, and cytotoxic ($CD8^+$) T cells by directly killing antigen-bearing cells. B cells and antibodies are best at destroying bacteria and bacterial products, whereas T cells are best at destroying eukaryotic cells that express surface antigens, such as virus-infected cells and grafted and tumor cells. Natural killer lymphocytes do not recognize specific antigens but rapidly attack and kill tumor cells and virus-infected cells.

14. Mature but nonactivated lymphocytes patrol connective tissues throughout the body by passing in and out of the circulatory vessels (recirculation).

Lymphocyte Differentiation and Activation (pp. 634–635)

15. Lymphocytes arise from stem cells in the bone marrow. T cells develop immunocompetence in the thymus, whereas B cells develop immunocompetence in the bone marrow. Immunocompetent lymphocytes then circulate to the loose and lymphoid connective tissues, where antigen binding (the antigenic encounter) leads to lymphocyte activation.

16. The antigenic encounter involves an interaction among the lymphocyte being activated, an antigen-presenting cell (dendritic cell or macrophage), and a helper ($CD4^+$) T lymphocyte. A newly activated T or B cell divides quickly to produce many short-lived effector lymphocytes and some long-lived memory lymphocytes. Recirculating memory lymphocytes provide long-term immunity.

Lymphoid Tissue (pp. 635–637)

17. Lymphoid tissue is reticular connective tissue in which many B and T lymphocytes gather to fight pathogens or become activated. It is located in the mucous membranes (as MALT) and in the lymphoid organs (except the thymus).

18. Lymphoid tissue contains lymphoid follicles (nodules) with germinal centers. Each follicle contains thousands of B lymphocytes, all derived from one activated B cell.

Lymphoid Organs (pp. 637–641)

19. The bone marrow and the thymus are primary lymphoid organs. The lymph nodes, spleen, tonsils, aggregated lymphoid nodules, and the appendix are secondary lymphoid organs.

20. The thymus, located in the superoanterior thorax and neck, is a primary lymphoid organ that is most active during youth. Its hormones, secreted by epithelial reticular cells, signal the contained T lymphocytes to gain immunocompetence.

21. The thymus has lobules, each with an outer cortex packed with maturing T cells and an inner medulla containing fewer T cells and degenerative thymic corpuscles.

22. Within a lymph node, masses of lymphoid tissue lie between the sinuses. This lymphoid tissue receives some of the antigens that pass through the node, leading to lymphocyte activation and memory-lymphocyte production. The masses of lymphoid tissue help divide each node into an outer cortex and an inner medulla.

23. The spleen lies in the superior left part of the abdominal cavity. The splenic vessels enter and exit the hilum on the anterior surface.

24. The spleen has two main functions: (1) removing antigens from the blood and (2) destroying worn-out blood cells. The first function is performed by the white pulp; the second, by the red pulp. White pulp consists of sleeves of lymphoid tissue, each surrounding a central artery. Red pulp consists of splenic sinusoids and strips of blood-filled reticular connective tissue called splenic cords, whose macrophages remove worn-out blood cells.

25. The tonsils in the pharynx, aggregated lymphoid nodules in the small intestine, and the wall of the appendix are parts of MALT in which the lymphoid tissue contains an exceptionally high concentration of lymphocytes and follicles.

PAL Histology/Lymphatic System

PAL Cadaver/Lymphatic System

DISORDERS OF THE LYMPHATIC AND IMMUNE SYSTEMS (p. 641)

26. Chylothorax is the leakage of chyle from the thoracic duct into the pleural cavity. Lymphangitis is inflammation of a lymphatic vessel. Mononucleosis is a viral infection of B lymphocytes, which are attacked by T lymphocytes, leading to flulike symptoms. Hodgkin's lymphoma is a type of lymph node cancer, and non-Hodgkin's lymphoma includes all other cancers of lymphoid tissue.

THE LYMPHATIC AND IMMUNE SYSTEMS THROUGHOUT LIFE (pp. 641–642)

27. Lymphatic vessels develop from lymphatic sacs attached to the embryonic veins. The thymus develops from pharyngeal endoderm, and the other lymphoid organs derive from mesodermal mesenchyme.

28. Lymphoid organs become populated by lymphocytes, which arise from hematopoietic tissue.

29. With aging, the immune system becomes less responsive. Thus, the elderly suffer more often from infections and cancer.

REVIEW QUESTIONS

Multiple Choice/Matching Questions

(For answers, see Appendix B.)

1. Lymphatic capillaries (a) are open-ended like drinking straws, (b) have continuous tight junctions like those of the capillaries of the blood brain barrier, (c) have endothelial cells separated by flap-like minivalves that open wide, (d) have special barriers that stop cancer cells from entering, (e) all of the above.
2. The basic structural framework of most lymphoid organs consists of (a) areolar connective tissue, (b) hematopoietic tissue, (c) reticular connective tissue, (d) adipose tissue.
3. Lymph nodes cluster in all the following body areas except the (a) brain, (b) axillae, (c) groin, (d) neck.
4. The germinal centers in lymph nodes are sites of (a) the lymph sinuses, (b) proliferating B lymphocytes, (c) T lymphocytes, (d) a and c, (e) all of the above.
5. The red pulp of the spleen (a) contains splenic sinusoids and a macrophage-rich connective tissue, (b) is another name for the fibrous capsule, (c) is lymphoid tissue containing lymphocytes, (d) is the part of the spleen that destroys worn-out erythrocytes, (e) a and d.
6. Lymphocytes that develop immunocompetence in the thymus are (a) B lymphocytes, (b) T lymphocytes.
7. Which of the following lymphoid organs have a cortex and a medulla? (More than one choice is correct.) (a) lymph nodes, (b) spleen, (c) thymus, (d) aggregated lymphoid nodules, (e) tonsils.
8. Which one of the following lymphoid organs does *not* contain lymphoid follicles or germinal centers? (a) lymph nodes, (b) spleen, (c) thymus, (d) aggregated lymphoid nodules, (e) tonsils.
9. Developmentally, the embryonic lymphatic vessels are most closely associated with the (a) veins, (b) arteries, (c) nerves, (d) thymus.
10. It sometimes is difficult to distinguish the different lymphoid organs from one another in histological sections. How would you tell the thymus from a lymph node? (a) Only the thymus has a cortex and medulla; (b) lymphocytes are far less densely packed in the thymus than in the lymph node; (c) the thymus contains no blood vessels; (d) only the thymus has distinct lobules and thymic corpuscles.
11. In some people, the thoracic duct does not receive any of the lymph trunks at the base of the neck (jugular, subclavian, or bronchomediastinal). In such people, what part of the body does the thoracic duct drain? (a) upper right quarter, (b) upper left quarter, (c) upper half, (d) lower half.

Short Answer Essay Questions

12. Compare the basic functions of a lymph node to those of the spleen.
13. If you saw a blood vessel and a collecting lymphatic vessel running side by side, how could you tell them apart?
14. Trace the entire course of the thoracic duct. Name the locations of all the lymph trunks.
15. List and briefly explain three important functions of the lymphatic vessels.
16. Which three of the six groups of lymph nodes (described on p. 629) are easily felt (palpated) through the skin during a physical examination?
17. Explain the basic functional differences between B and T lymphocytes.
18. As George was reading this chapter for the second time, he suddenly had an insight, exclaiming, "Lymph comes from the blood, and then it returns to the blood!" Is this insight correct? Explain.
19. Billie Jo is looking at histological slides of the spleen under a microscope. She says that she sees lymphoid follicles in the white pulp. Is she correct? Explain.
20. Aparna, a gross anatomist who has been teaching human anatomy for 35 years, told her class that in adults the thymus is atrophied and has no function. Is she correct? Explain.

Critical Reasoning & Clinical Application Questions

1. A friend tells you that he has tender, swollen "glands" along the left side of the front of his neck. You notice that he has a bandage on his left cheek that is not fully hiding a large infected cut there. The glands are not really glands. Exactly what are his swollen "glands," and how did they become swollen?
2. When young Joe went sledding, the runner of a friend's sled hit him in the left side and ruptured his spleen. Joe almost died before he got to a hospital. What is the immediate danger of a ruptured spleen?
3. The man in the hospital bed next to Joe is an alcoholic with cirrhosis of the liver and portal hypertension. His spleen is seriously enlarged. How could portal hypertension lead to splenomegaly (review Chapter 20, p. 616)?
4. Mrs. Roselli has undergone a left radical mastectomy. Her left arm is severely swollen and painful, and she is unable to raise it higher than her shoulder. (a) Explain the origin of her signs and symptoms. (b) Is she likely to have relief from these symptoms in time? Explain.
5. Traci arrives at the clinic complaining of pain and redness of her right ring finger. The finger and the dorsum of her hand have edema, and red streaks are apparent on her right forearm. Antibiotics are prescribed, and the nurse applies a sling to the affected arm. Why is it important that Traci not move the affected arm excessively?
6. Simi realized she was very ill. During a recent trip to the tropics, she had contracted both malaria and tuberculosis, diseases whose microorganisms travel throughout the bloodstream. Back home for only a week, she then contracted mononucleosis. When she went to the doctor, he was able to feel the notched superior border of her spleen projecting far anterior to the left costal margin in her abdominal wall. Was something wrong with her spleen? Explain.

The Respiratory System 22

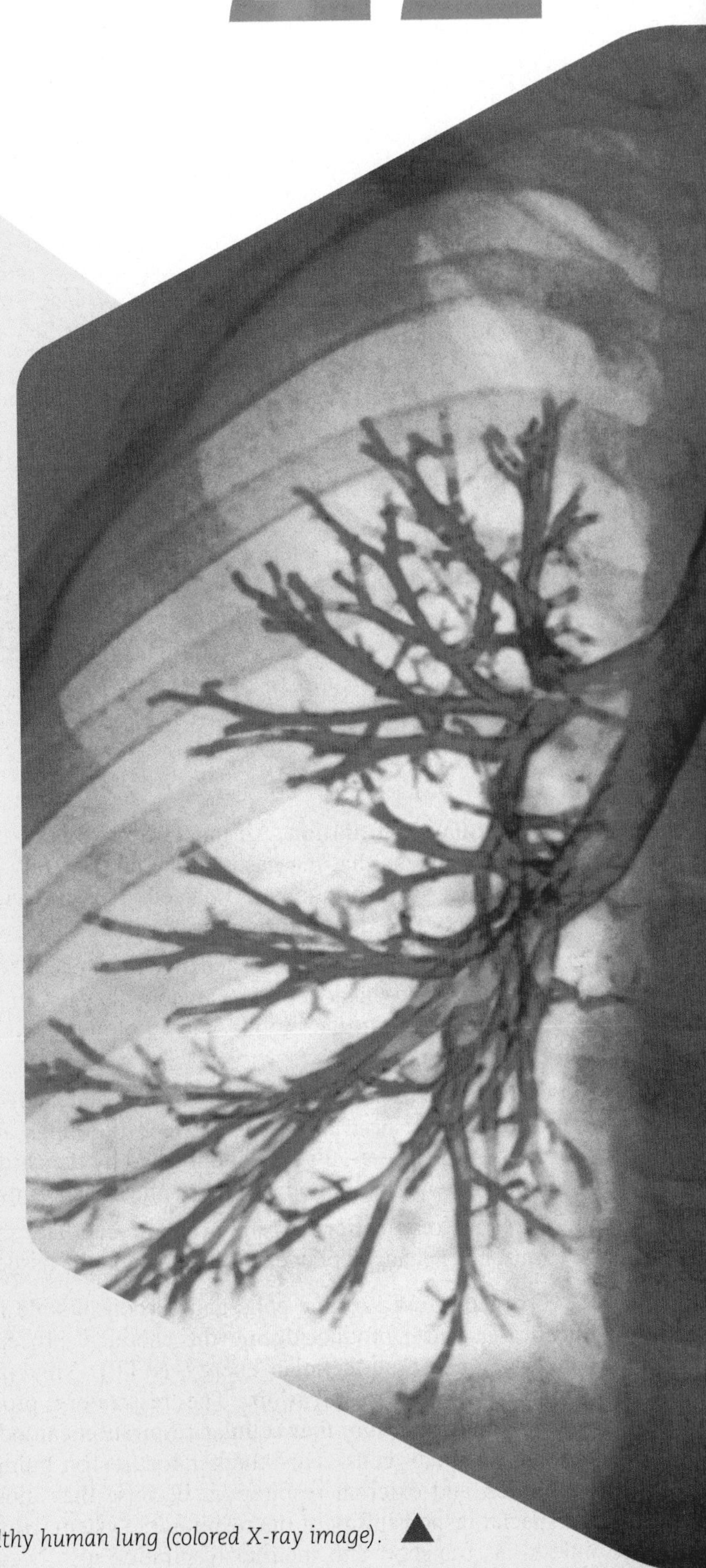

Bronchial tree of a healthy human lung (colored X-ray image). ▲

Humans can live without water for days and without food for weeks, but they cannot live without oxygen for even a few minutes. Breathing is our most urgent need. The trillions of cells in the body need a continuous supply of oxygen to produce the energy needed to carry out their vital functions. Furthermore, as the cells use oxygen, they produce carbon dioxide, a waste product the body must eliminate. The major function of the **respiratory system** is to fulfill these needs—that is, to supply the body with oxygen and dispose of carbon dioxide (CO_2). To accomplish this, the following processes, collectively called **respiration,** must occur:

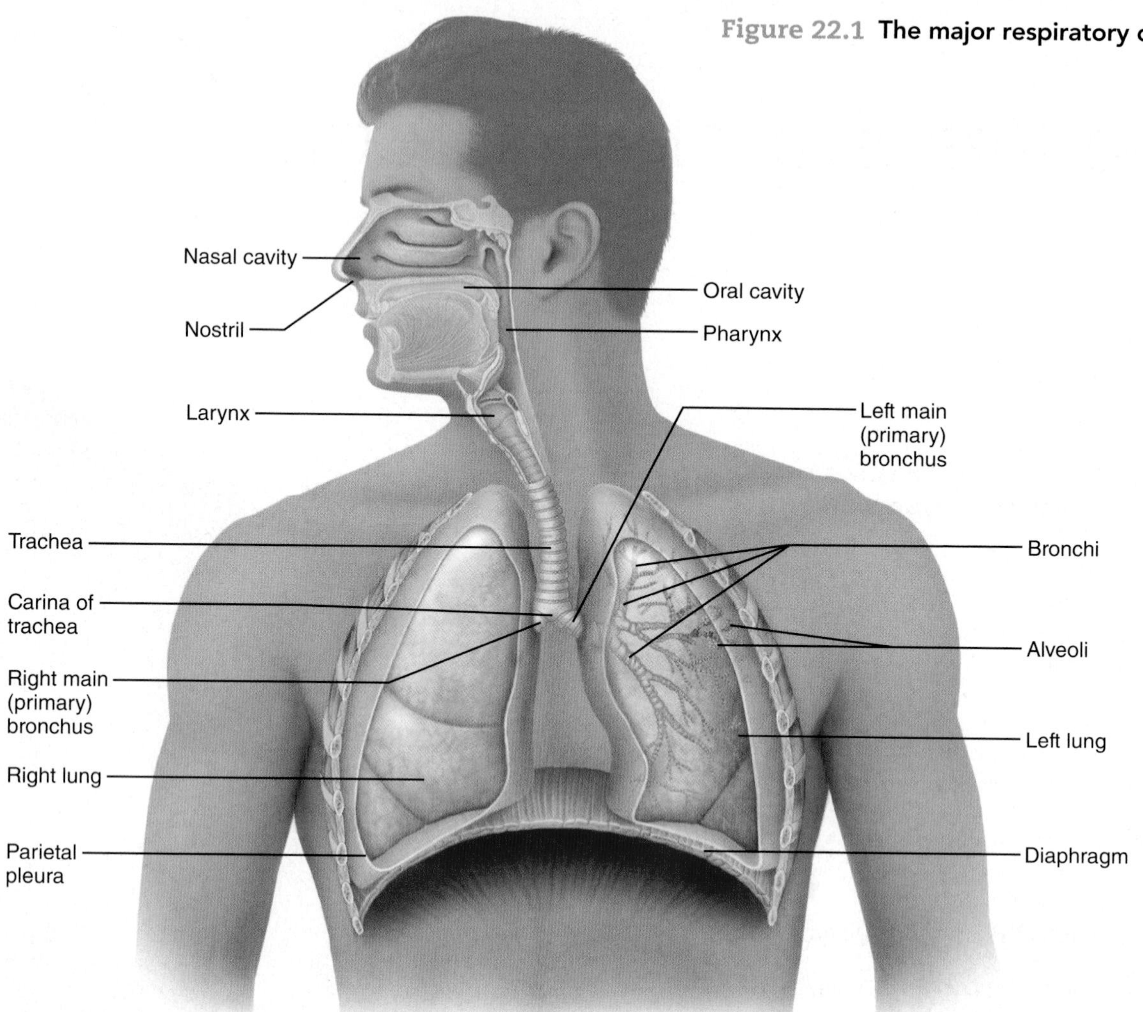

Figure 22.1 The major respiratory organs.

1. **Pulmonary ventilation.** Air must be moved into and out of the lungs so that the gases in the air sacs (alveoli) of the lungs are continuously replaced. This movement is commonly called **ventilation,** or breathing.
2. **External respiration.** Gas exchange must occur between the blood and air at the lung alveoli. Oxygen in the air sacs diffuses into the blood; CO_2 in blood diffuses into the air sacs.
3. **Transport of respiratory gases.** Oxygen and carbon dioxide must be transported between the lungs and the cells of the body. This is accomplished by the cardiovascular system, with blood serving as the transporting fluid.
4. **Internal respiration.** At the systemic capillaries, gases must be exchanged between the blood and the tissue cells.

Oxygen is used by the cells and carbon dioxide is produced as a waste product during the chemical process that converts glucose to cellular energy (ATP). This process is called **cellular respiration.** The respiratory processes described above ensure that cellular respiration can occur in virtually all body cells. This chapter focuses on pulmonary ventilation and external respiration, because they alone are the special responsibility of the respiratory system. However, unless gas transport and internal respiration also occur, the respiratory system cannot accomplish its main goal of supplying oxygen to the cells and removing carbon dioxide. Thus, the respiratory and cardiovascular systems are closely coupled, and if either system fails, the body's cells begin to die from oxygen starvation.

Because the respiratory system moves air into and out of the body, it is also involved in the sense of smell and with the vocalizations of speech, which are briefly discussed along with respiration.

FUNCTIONAL ANATOMY OF THE RESPIRATORY SYSTEM

learning outcomes

- Identify the respiratory passageways in order, from the nose to the alveoli in the lungs. Distinguish the structures of the conducting zone from those of the respiratory zone.
- List and describe the numerous structures that protect the respiratory system from dust, bacteria, food particles, and other foreign matter.

The organs of the respiratory system include the *nose, nasal cavity,* and *paranasal sinuses;* the *pharynx;* the *larynx;* the

Table 22.1 Principal Organs of the Respiratory System

Structure	Description, General and Distinctive Features	Function
Nose	External portion supported by bone and cartilage; internal nasal cavity divided in half by midline nasal septum and lined with respiratory mucosa	Produces mucus; filters, warms, and moistens incoming air; resonance chamber for speech
	Roof of nasal cavity contains olfactory mucosa	Receptors for sense of smell
Paranasal sinuses	Mucosa-lined hollow cavities within the sphenoid, ethmoid, maxillary, and frontal bones	Sinuses function the same as nasal cavity; also lighten skull
Pharynx	Passageway connecting nasal cavity to larynx and oral cavity to esophagus; three subdivisions: nasopharynx, oropharynx, and laryngopharynx	Passageway for air and food
	Houses tonsils	Tonsils are lymphoid tissue that responds to inhaled or ingested antigens
Larynx	Connects pharynx to trachea; framework of cartilage and dense connective tissue; opening (rima glottidis) can be closed by epiglottis or vocal folds	Air passageway; prevents food from entering lower respiratory tract
	Houses true vocal cords	Voice production
Trachea	Flexible tube running from larynx and dividing inferiorly into two main (primary) bronchi; walls contain C-shaped cartilages that are incomplete posteriorly where trachealis muscle occurs	Air passageway; filters, warms, and moistens incoming air
Bronchial tree	Consists of right and left main bronchi, which subdivide within the lungs to form lobar (secondary) and segmental (tertiary) bronchi, smaller bronchi, and bronchioles; bronchiolar walls contain complete layer of smooth muscle; constriction of this muscle impedes expiration	Air passageways connecting trachea with alveoli; warms and moistens incoming air
Alveoli	Microscopic chambers at end of bronchial tree; walls of simple squamous epithelium—type I alveolar cells—underlain by thin basement membrane; external surfaces intimately associated with pulmonary capillaries	Main sites of gas exchange
	Simple cuboidal epithelium—type II alveolar cells—produce surfactant	Surfactant reduces surface tension; helps prevent lung collapse
Lungs	Paired composite organs located within pleural cavities of thorax; composed primarily of alveoli and respiratory passageways; stroma is fibrous elastic connective tissue, allowing lungs to recoil passively during expiration	House passageways smaller than main bronchi
Pleurae	Serous membranes; parietal pleura lines thoracic cavity; visceral pleura covers external lung surfaces	Produce lubricating fluid and compartmentalize lungs

trachea; the *bronchi* and their smaller branches; and the *lungs,* which contain the terminal air sacs, or *alveoli* (**Figure 22.1**, **Table 22.1**). Functionally, these respiratory structures are divided into conducting and respiratory zones. The *conducting zone* includes the respiratory passageways that carry air to the sites of gas exchange. The structures of the conducting zone also filter, humidify, and warm the incoming air. Thus, the air reaching the lungs contains much less dust than it did when it entered the nose and is warm and damp. The *respiratory zone,* the actual site of gas exchange in the lungs, is composed of the terminal respiratory passageways that contain alveoli—namely, the *respiratory bronchioles, alveolar ducts,* and *alveolar sacs.*

The Nose and the Paranasal Sinuses

learning outcome

▶ Identify and describe the functions of the respiratory structures located in the skull.

The Nose

The **nose** is the only externally visible part of the respiratory system. Although the nose is often referred to in negative connotations, as in "brown-nosing" your instructor or being "nosy," its many important functions make it worthy of higher esteem. The nose:

- Provides an airway for respiration.
- Moistens and warms entering air.
- Filters inhaled air to cleanse it of foreign particles.

(a) Surface anatomy

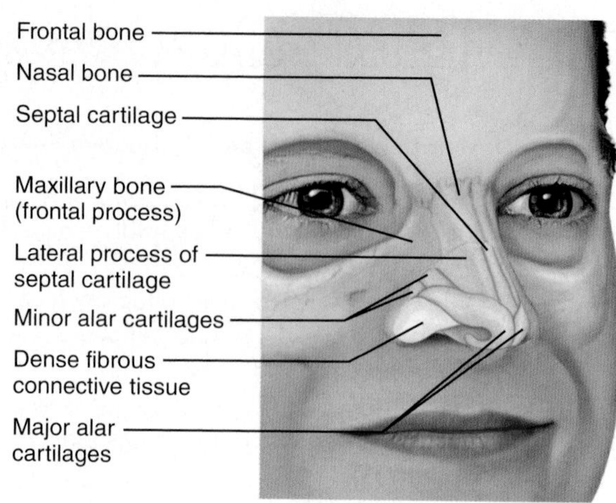

(b) External skeletal framework

Figure 22.2 External nose.

- Serves as a resonating chamber for speech.
- Houses the olfactory (smell) receptors.

The structures of the nose are divided into the *external nose* and the internal *nasal cavity*. The skeletal framework of the **external nose** consists of the frontal and nasal bones superiorly (forming the root and bridge, respectively), the maxillary bones laterally, and flexible plates of hyaline cartilage inferiorly (the lateral, septal, and alar cartilages) (Figure 22.2). The septal cartilage forms the anterior margin of the nose, called the dorsum nasi. The tip of the nose, its apex, is formed by the major alar cartilages, and the lateral border of the nostril, the ala, is formed from dense fibrous connective tissue. The great variation in nose size and shape is largely due to differences in the nasal cartilages. The skin covering the nose's anterior and lateral surfaces is thin and contains many sebaceous glands that open into some of the largest skin pores on the face.

The Nasal Cavity

The **nasal cavity** (Figure 22.1 and Figure 22.3) lies in and posterior to the external nose. During breathing, air enters this cavity by passing through the external **nares** (na′rēz), or *nostrils*. The nasal cavity is divided into right and left halves by the *nasal septum* in the midline; this septum is formed by the perpendicular plate of the ethmoid bone, the vomer, and a septal cartilage (see Figure 7.14 on p. 166), all covered by a mucous membrane. Posteriorly, the nasal cavity is continuous with the nasal part of the pharynx (nasopharynx) through the **posterior nasal apertures,** also called the *choanae* (ko-a′ne; "funnels") or *internal nares.*

The roof of the nasal cavity is formed by the ethmoid and sphenoid bones; its floor is formed by the **palate** (pal′at), which separates the nasal cavity from the mouth inferiorly and keeps food out of the airways. Anteriorly, where the palate contains the horizontal processes of the palatine bones and the palatine process of the maxillary bone, it is called the **hard palate** (see Figure 7.14, p. 166); the posterior part is the muscular **soft palate** (see Figure 22.3).

The part of the nasal cavity that lies just superior to the nostrils, within the flared wings of the external nose, is the **vestibule** ("porch, entranceway"). This is lined with skin containing sebaceous and sweat glands and numerous hair follicles. The nose hairs, or *vibrissae* (vi-bris′e) (*vibro* = to quiver), filter large particles, such as insects and lint, from the inspired air. The rest of the nasal cavity is lined with two types of mucous membrane: (1) the small patch of *olfactory mucosa* near the roof of the nasal cavity, which houses the receptors for smell (see pp. 490–491), and (2) the *respiratory mucosa,* a mucous membrane that lines the vast majority of the respiratory passageway.

Respiratory Mucosa The respiratory mucosa consists of a pseudostratified ciliated columnar epithelium containing scattered goblet cells, and the underlying connective tissue lamina propria. This lamina propria is richly supplied with compound tubuloalveolar glands (see Figure 4.5g, p. 74) that contain mucous cells and serous cells. (*Mucous cells* secrete mucus, whereas *serous cells* in glands secrete a watery fluid containing digestive enzymes.) Each day, the nasal glands and the epithelial goblet cells secrete about a quart of mucus containing lysozyme, an enzyme that digests and destroys bacteria. The sticky mucus forms a sheet that covers the surface of the mucosa and traps inhaled dust, bacteria, pollen, viruses, and other debris from the air. Thus, an important function of the respiratory mucosa is to filter the inhaled air. The ciliated cells in the epithelial lining create a gentle current that moves the sheet of contaminated mucus posteriorly to the pharynx, where it is swallowed. In this way, particles filtered from the air are ultimately destroyed by digestive juices in the stomach. In addition, the sheet of mucus is a wet film that moistens the inhaled air.

CLINICAL APPLICATION

Rhinitis Inflammation of the nasal mucosa, called **rhinitis** (ri-ni′tis; "nose inflammation"), is caused by such things as cold viruses, streptococcal bacteria, and various allergens. This inflammation is accompanied by an excessive production of mucus, which results in nasal congestion, a runny nose, and postnasal drip.

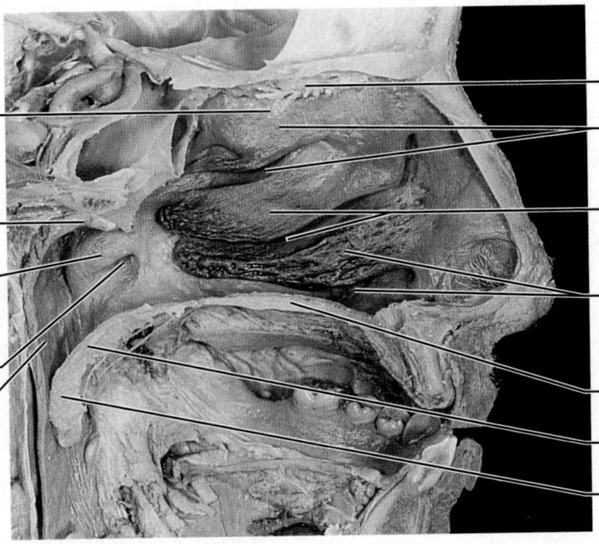

Olfactory epithelium
Mucosa of pharynx
Tubal tonsil
Pharyngotympanic (auditory) tube
Nasopharynx
Olfactory nerves
Superior nasal concha and superior nasal meatus
Middle nasal concha and middle nasal meatus
Inferior nasal concha and inferior nasal meatus
Hard palate
Soft palate
Uvula

(a) Photograph

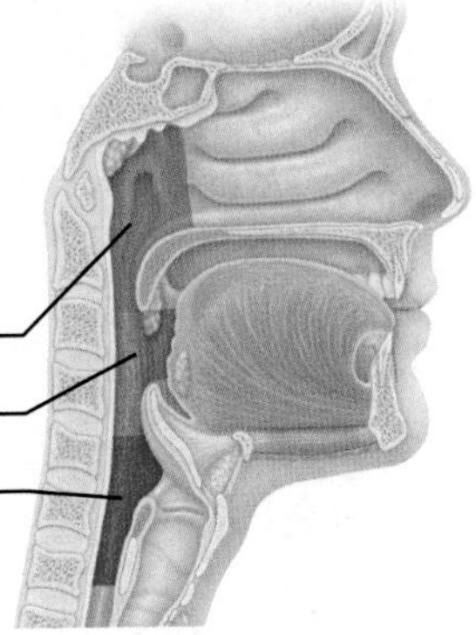

Pharynx
Nasopharynx
Oropharynx
Laryngopharynx

(c) Regions of the pharynx

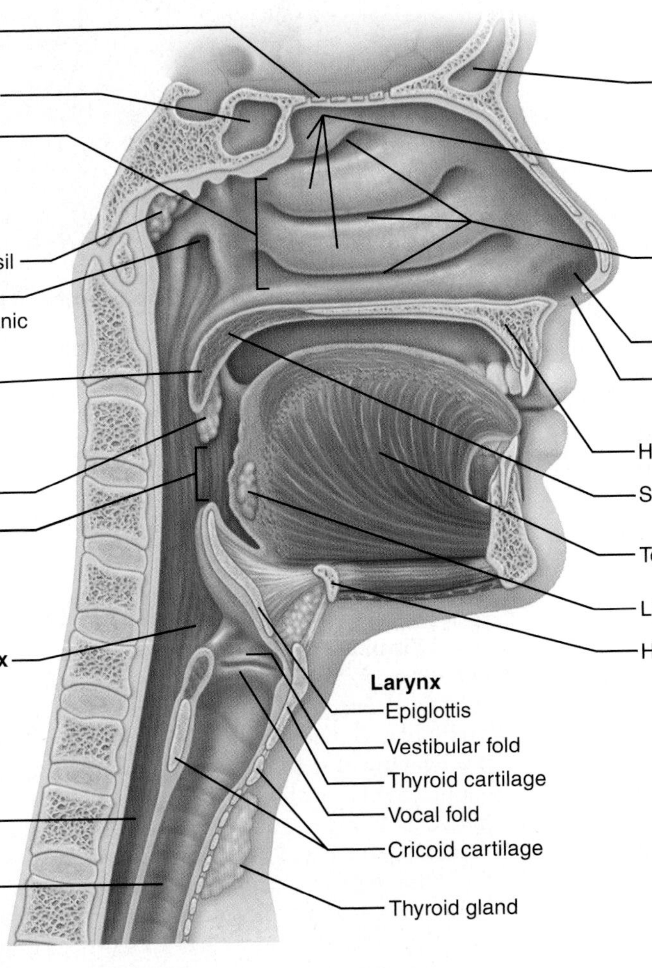

Cribriform plate of ethmoid bone
Sphenoidal sinus
Posterior nasal aperture
Nasopharynx
Pharyngeal tonsil
Opening of pharyngotympanic tube
Uvula
Oropharynx
Palatine tonsil
Isthmus of the fauces
Laryngopharynx
Esophagus
Trachea
Frontal sinus
Nasal cavity
Nasal conchae (superior, middle and inferior)
Nasal meatuses (superior, middle, and inferior)
Nasal vestibule
Nostril
Hard palate
Soft palate
Tongue
Lingual tonsil
Hyoid bone
Larynx
Epiglottis
Vestibular fold
Thyroid cartilage
Vocal fold
Cricoid cartilage
Thyroid gland

(b) Illustration

Figure 22.3 **The upper respiratory tract.** Midsagittal section of the head and neck.

The nasal mucosa is richly supplied with sensory nerve endings from the trigeminal nerve, CNV. A sneeze reflex is stimulated when irritating particles (dust, pollen, and so on) contact this sensitive mucosa. The resulting sneeze propels air outward in a violent burst, expelling the irritant from the nose.

Rich plexuses of capillaries and thin-walled veins occupy the lamina propria of the nasal mucosa and warm the incoming air that flows across the mucosal surface. When the temperature of inhaled air drops, as when you step outside on a cold day, the vascular plexuses respond by engorging with warm blood, thereby intensifying the air-heating process. Because of the abundance and the superficial location of these vessels, nosebleeds are common and often profuse.

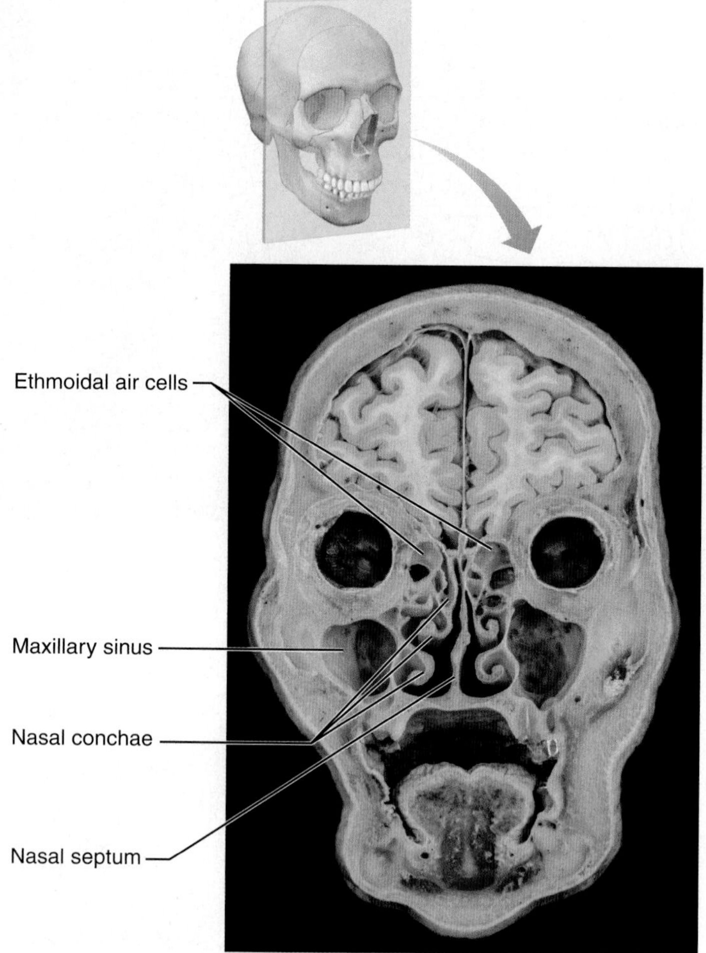

Figure 22.4 Nasal conchae and paranasal sinuses. Frontal section through the face, photograph.

CLINICAL APPLICATION

Epistaxis Epistaxis (ep″ĭ-stak′sis; "to bleed from the nose") is a nose-bleed, or nasal hemorrhage. It commonly follows trauma to the nose, excessive nose blowing, or drying of the mucous membrane lining the nasal cavity, as can occur in dry climates or during the winter heating months in cold climates. Most nasal bleeding is from the highly vascularized *anterior* part of the nasal septum and can be stopped by pinching the nostrils together or packing them with cotton.

The Nasal Conchae Projecting medially from each lateral wall of the nasal cavity are three mucosa-covered, scroll-like structures: the *superior* and *middle conchae* of the ethmoid bone, and the *inferior concha,* which is a separate bone (Figure 22.4 and Figure 22.3). The groove inferior to each concha is a *meatus.* As inhaled air rushes over the curved conchae, the resulting turbulence greatly increases the amount of contact between the nasal mucosa and this inspired air. The gases in the inhaled air swirl through the twists and turns of the conchae, but the air's particulate matter is deflected onto the mucus-coated surfaces, where it becomes trapped. As a result, few particles larger than 4 μm get past the nasal cavities.

The conchae and nasal mucosa function during inhalation to filter, heat, and moisten the air. During exhalation, they reclaim this heat and moisture. The inhaled air cools the conchae, and then during exhalation these cooled conchae precipitate moisture and extract heat from the humid air flowing over them. This reclamation mechanism minimizes the amount of moisture and heat lost from the body through breathing, helping people to survive in dry and cold climates.

The Paranasal Sinuses

The nasal cavity is surrounded by a ring of air-filled cavities called *paranasal sinuses* located in the frontal, sphenoid, ethmoid, and maxillary bones (Figure 22.4; see also p. 168). These sinuses open into the nasal cavity, are lined by the respiratory mucosa, and perform the same air-processing functions as the nasal cavity. Their mucus drains into the nasal cavities, and the suctioning effect caused by nose blowing helps to drain them.

CLINICAL APPLICATION

Sinusitis Inflammation of the paranasal sinuses, a condition called **sinusitis,** is caused by viral, bacterial, or fungal infections. When the passages that connect the paranasal sinuses to the nasal cavity become blocked by swelling of the inflamed nasal mucosa, air in the sinus cavities is absorbed into the blood vessels of the mucosal lining, resulting in a partial vacuum and a sinus headache localized over the inflamed areas. Later, if the infection persists, inflammatory fluid oozes from the mucosa, fills the obstructed sinus, and exerts painful positive pressure. Most cases of sinusitis originate just inferior and lateral to the middle concha, where the openings of the frontal and maxillary sinuses and one of the ethmoidal air cells lie in close proximity and can be blocked simultaneously. Serious cases of sinusitis are treated by promoting drainage and with antibiotics.

check your understanding

- ☐ 1. Which respiratory structures are lined with respiratory mucosa? List the functions of this mucous membrane.
- ☐ 2. Differentiate external respiration, internal respiration, and cellular respiration.

(For answers, see Appendix B.)

The Pharynx

learning outcome

▶ Identify the boundaries of the three regions of the pharynx and the structures located in each region.

The **pharynx** (far′ingks) is the funnel-shaped passageway that connects the nasal cavity and mouth superiorly to the larynx and esophagus inferiorly (Figure 22.3). It descends from the base of the skull to the level of the sixth cervical vertebra and serves as a common passageway for both food and air. In the context of the digestive tract, the pharynx is commonly called the *throat.*

On the basis of location and function, the pharynx is divided into (from superior to inferior) the *nasopharynx, oropharynx,* and *laryngopharynx* (lah-ring″go-far′ingks) (Figure 22.3b). The muscular wall of the pharynx consists of skeletal muscle throughout its length, but the nature of the mucosal lining varies among the three pharyngeal regions.

The Nasopharynx

The **nasopharynx** lies directly posterior to the nasal cavity, inferior to the sphenoid bone and superior to the level of the soft palate (Figure 22.3b and c). Because it is superior to the point where food enters the body, the nasopharynx serves only as an air passageway. During swallowing, the soft palate and its pendulous **uvula** (u′vu-lah; "little grape") reflect superiorly, an action that closes off the nasopharynx and prevents food from entering the nasal cavity. When a person giggles while drinking, this sealing action fails, and fluids can spray from the nose.

The nasopharynx is continuous with the nasal cavity through the posterior nasal apertures, and its ciliated pseudostratified epithelium takes over the job of propelling mucus where the nasal mucosa leaves off, such that dusty mucus is moved downward through the nasopharynx. High on the posterior nasopharyngeal wall is the midline **pharyngeal tonsil,** or *adenoids,* a lymphoid organ that destroys pathogens entering the nasopharynx in the air.

CLINICAL APPLICATION

Infection of the Adenoids Infected and enlarged adenoids, a condition especially common in children, can obstruct the flow of air through the nasopharynx. Because the nasal airway is blocked, breathing through the mouth becomes necessary.

A *pharyngotympanic (auditory) tube,* which drains the middle ear, opens into each lateral wall of the nasopharynx. A ridge of pharyngeal mucosa posterior to each opening constitutes the **tubal tonsil** (Figure 22.3a), whose location provides the middle ear some protection against infections that may spread from the pharynx.

The Oropharynx

The **oropharynx** lies posterior to the oral cavity (mouth); its archlike entranceway, directly behind the mouth, is the **fauces** (faw′sēz; "throat") (Figure 22.3b). The oropharynx extends inferiorly from the level of the soft palate to the level of the epiglottis (a flap posterior to the tongue). Both swallowed food and inhaled air pass through the oropharynx.

As the nasopharynx blends into the oropharynx, the lining epithelium changes from pseudostratified columnar to a thick, protective *stratified squamous epithelium.* This structural adaptation reflects the increased friction and greater chemical trauma accompanying the passage of swallowed food through the oropharynx.

Two kinds of tonsils are embedded in the mucosa of the oropharynx: The paired **palatine tonsils** lie in the lateral walls of the fauces, and the single **lingual tonsil** covers the posterior surface of the tongue.

The Laryngopharynx

Like the oropharynx superior to it, the **laryngopharynx** serves as a common passageway for food and air and is lined with a stratified squamous epithelium. The laryngopharynx lies directly posterior to the larynx and is continuous with both the esophagus, which conducts food and fluids to the stomach, and the larynx, which conducts air to the respiratory tract. The laryngopharynx extends to the inferior edge of the cricoid cartilage.

The Larynx

learning outcome

▶ Describe the structure and functions of the larynx.

The **larynx** (lar′ingks), or voice box, extends from the level of the fourth to the sixth cervical vertebra. Superiorly, it attaches to the *hyoid bone* and opens into the laryngopharynx (see Figure 22.3c); inferiorly, it is continuous with the trachea (windpipe). The larynx has three functions:

- Producing vocalizations.
- Providing an open airway.
- Acting as a switching mechanism to route air and food into the proper channels. During swallowing, the inlet (superior opening) to the larynx is closed; during breathing, it is open.

The framework of the larynx is an intricate arrangement of nine cartilages connected by membranes and ligaments (Figure 22.5). The large, shield-shaped **thyroid cartilage** is formed by two cartilage plates joined in the midline. It resembles an upright open book, with the book's "spine" lying in the anterior midline of the neck. This "book spine" is the ridgelike **laryngeal** (lah-rin′je-al) **prominence,** which is obvious externally as the **Adam's apple.** The thyroid

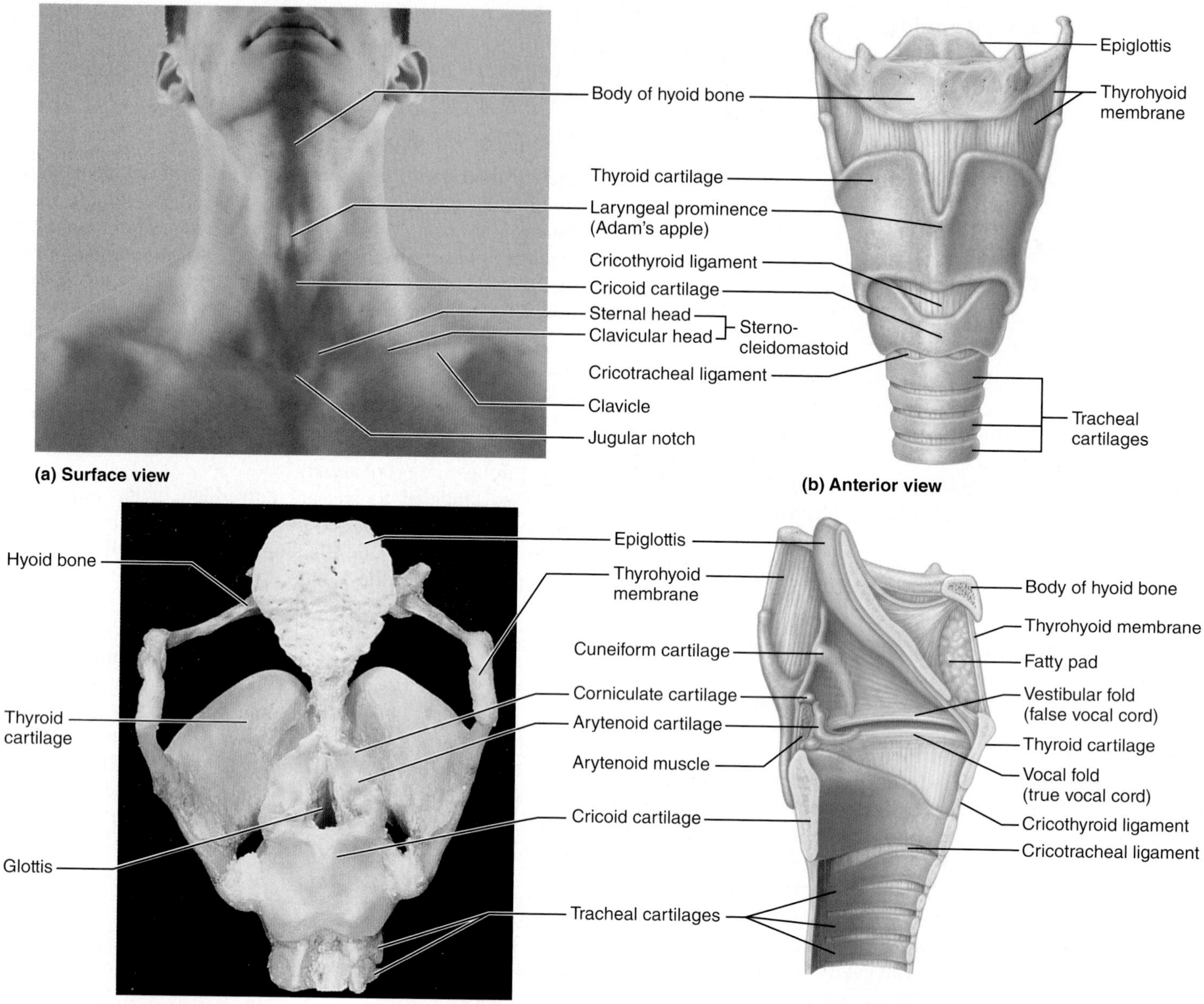

(a) Surface view

(b) Anterior view

(c) Photograph of cartilaginous framework of the larynx, posterior view

(d) Sagittal section (anterior on the right)

Figure 22.5 **Anatomy of the larynx.**

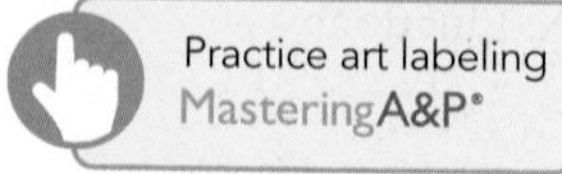

cartilage is larger in men than in women because male sex hormones stimulate its growth during puberty. Inferior to the thyroid cartilage is the **cricoid cartilage** (kri′-koid; "circle"), the only laryngeal cartilage that forms a complete ring. The shape of the cricoid cartilage is likened to a signet ring, a ring historically used to imprint wax seals on letters and official documents. Today, rings of this style are commonly referred to as initial rings. The comparison does accurately describe the cricoid; it is wide posteriorly and narrow anteriorly. The cricoid cartilage is located on top of the trachea.

Three pairs of small cartilages lie just superior to the cricoid cartilage in the posterior part of the larynx (Figure 22.5c and d): the **arytenoid cartilages** (ar″ĭ-te′noid; "ladle-like"), the **corniculate cartilages** (kor-nik′u-lāt; "little horn"), and the **cuneiform cartilages** (ku-ne′ĭ-form; "wedge-shaped"). The most important of these cartilages are the pyramid-shaped arytenoids, which anchor the vocal cords.

The ninth cartilage of the larynx, the leaf-shaped **epiglottis** (ep″ĭ-glot′is), is composed of elastic cartilage and is almost entirely covered by a mucosa. Its stalk attaches anteriorly to the internal aspect of the angle of the thyroid cartilage (Figure 22.5d). From there, the epiglottis projects superoposteriorly and attaches to the posterior aspect of the tongue (*epiglottis* means "upon the tongue"). During swallowing, the entire larynx is pulled superiorly, and the epiglottis tips inferiorly to cover and

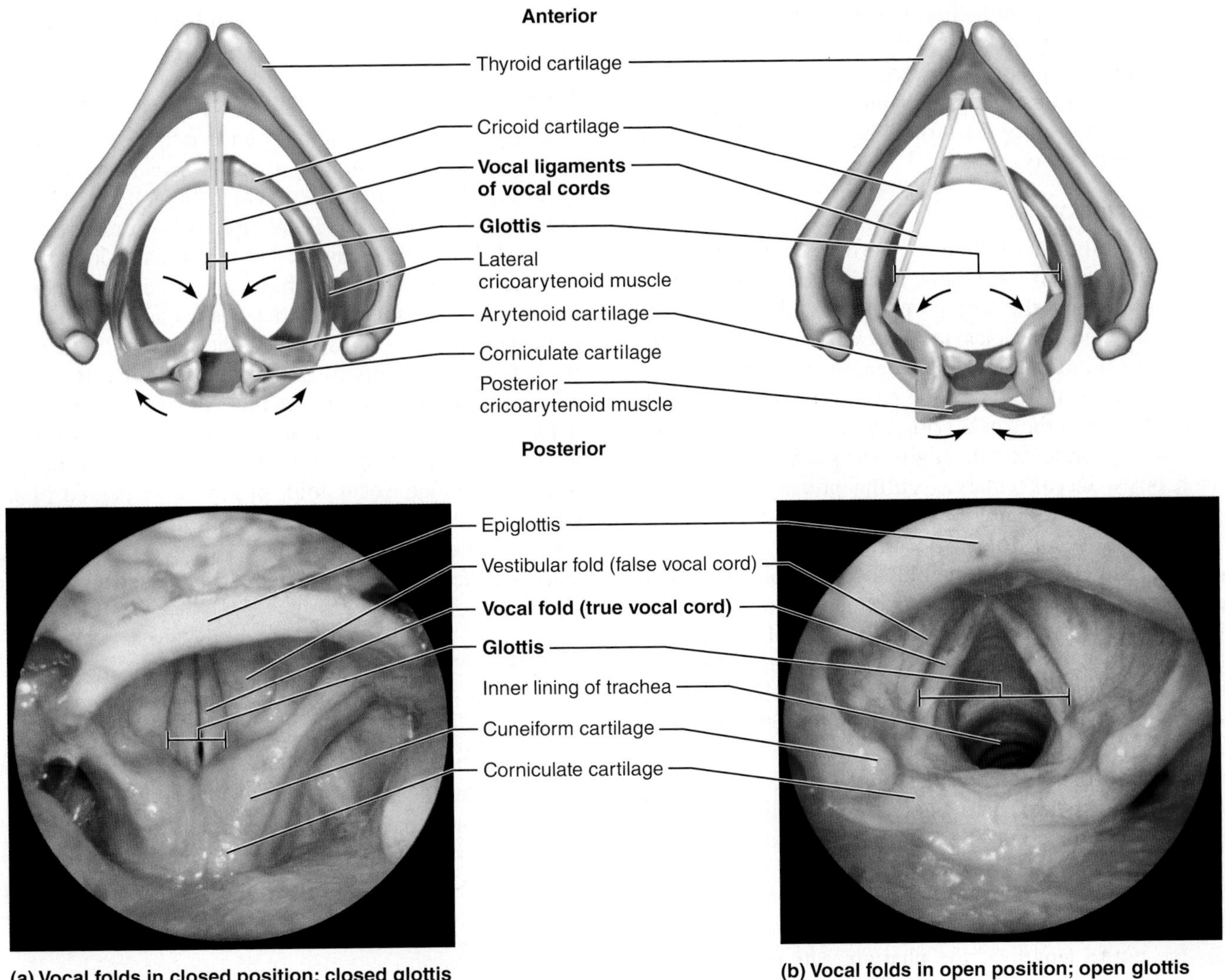

Figure 22.6 Movement of the vocal cords. Superior views of the larynx and vocal cords. The top illustrations show the skeletal elements and two of the many pairs of laryngeal muscles that rotate the arytenoid cartilages. The epiglottis has been left out so that the other laryngeal cartilages can be seen. Lower images show the view of the larynx and vocal folds through a laryngoscope.

seal the laryngeal inlet. This action keeps food out of the lower respiratory tubes. The entry into the larynx of anything other than air initiates the cough reflex, which expels the substance and prevents it from continuing into the lungs.

Because the larynx lies inferiorly in the neck and must ascend so far during swallowing, it sometimes cannot reach the protective cover of the epiglottal lid before food enters the laryngeal inlet. Although the low position of the larynx makes us more susceptible to choking, it is essential to humans' ability to talk. Its inferior location allows for greater movement of the tongue in shaping sounds and for an exceptionally long pharynx, which acts as a resonating chamber for speech. This latter feature greatly improves the quality of the vowel sounds we produce.

Within the larynx, paired *vocal ligaments* run anteriorly from the arytenoid cartilages to the thyroid cartilage. These ligaments, composed largely of elastic fibers, form the core of a pair of mucosal folds called the **vocal folds** or **(true) vocal cords** (see Figure 22.5d and Figure 22.6). Because the mucosa covering them is avascular, the vocal folds appear pearly white. Air exhaled from the lungs causes these folds to vibrate in a wave motion and to clap together, producing the basic sounds of speech. The medial opening between the vocal folds through which air passes is called the **rima glottidis** (*rima* = fissure), and the vocal folds together with this rima compose the **glottis.** Another pair of horizontal mucosal folds that lies directly superior to the vocal folds, the **vestibular folds,** or **false vocal cords,** play no part in sound production. However, they define a slitlike cavity between themselves and the true vocal cords (see Figure 22.5d) that enhances high-frequency sounds, functioning like the tweeter on stereo speakers.

The epithelium lining the superior part of the larynx, an area subject to food contact, is stratified squamous. Inferior to the vocal folds, the epithelium is pseudostratified ciliated columnar that entraps dust. In this epithelium, the power stroke of the cilia is directed upward, toward the pharynx, so that dust-trapping mucus is continuously moved superiorly from the lungs. "Clearing the throat" helps move this mucus up and out of the larynx.

Voice Production

Speech involves the intermittent release of exhaled air and the opening and closing of the glottis. In this process, the length of the vocal folds and the size of the rima glottidis are varied by intrinsic laryngeal muscles, most of which move the arytenoid cartilages (Figure 22.6). As the length and tension of the vocal folds change, the pitch of the produced sound changes. Generally, the tenser the vocal folds, the faster the exhaled air causes them to vibrate, and the higher the pitch.

As a boy's larynx enlarges during puberty, his laryngeal prominence grows anteriorly into a large Adam's apple, lengthening the vocal folds. Because longer vocal folds vibrate more slowly than short folds do, the voice becomes deeper, just as a cello has a lower tone than a violin. For this reason, most men have lower voices than females or young boys. The voices of adolescent boys frequently "crack," alternating between high-pitched and low-pitched sounds, because the boys have not yet learned to control the action of their longer vocal folds.

Loudness of the voice depends on the force with which air rushes across the vocal folds. The greater the force, the stronger the vibrations and the louder the sound. The vocal folds do not move at all when we whisper, but they vibrate vigorously when we yell.

Although the vocal folds produce the basic speech sounds, the entire length of the pharynx acts as a resonating chamber to amplify the quality of sound. The oral cavity, nasal cavity, and paranasal sinuses also contribute to vocal resonance. In addition, normal speech and good enunciation depend on the "shaping" of sounds into recognizable consonants and vowels by the pharynx, tongue, soft palate, and lips.

CLINICAL APPLICATION

Laryngitis Infection from a bad cold stimulates inflammation of the larynx, or **laryngitis,** causing the vocal folds to swell. The swelling interferes with their ability to vibrate, producing hoarseness. Hoarseness is also caused by overuse of the voice, growths on the vocal cords, inhalation of irritating chemicals (as in tobacco smoking), paralysis of some laryngeal muscles, or compression of a recurrent laryngeal nerve in the lung apex.

Sphincter Functions of the Larynx

Under certain conditions, the vocal folds act as a sphincter that prevents the passage of air. During abdominal straining, such as occurs when one strains to defecate, the abdominal muscles contract and the glottis closes to prevent exhalation, raising intrathoracic and intra-abdominal pressure. These events, collectively called **Valsalva's maneuver,** help to evacuate the rectum; they also stabilize the trunk of the body when one lifts a heavy load.

Innervation of the Larynx

The larynx receives its sensory and motor innervation through a superior laryngeal branch of each vagus nerve and from the *recurrent laryngeal nerves,* which branch off the vagus in the superior thorax and loop superiorly to reascend through the neck. The left recurrent laryngeal nerve loops under the aortic arch, whereas the right recurrent laryngeal nerve loops under the right subclavian artery. The backtracking course of these nerves is so unusual that the ancient Greeks mistook them for slings supporting the great arteries. Because these nerves supply innervation to most of the laryngeal muscles, damage to them, as can occur during surgery in this region, disrupts speech. Transection of one recurrent laryngeal nerve immobilizes one vocal fold, producing a degree of hoarseness. In such cases the other vocal fold can compensate, and speech remains almost normal. However, if both recurrent laryngeal nerves are transected, speech (except for whispering) is lost entirely.

✓ check your understanding

- ☐ **3.** Differentiate the pharynx from the larynx.
- ☐ **4.** What structure forms the inferior boundary of each region of the pharynx (nasopharynx, oropharynx, and laryngopharynx)?
- ☐ **5.** To which laryngeal cartilages do the vocal folds attach? What parameters influence the pitch and loudness of our voice? What nerve innervates our vocal apparatus?

(For answers, see Appendix B.)

The Trachea

learning outcome

- ▶ Describe the macroscopic and microscopic structure of the trachea.

The flexible **trachea** (tra′ke-ah), or windpipe, descends from the larynx through the neck and into the mediastinum; it ends by dividing into the two main bronchi (primary bronchi) in the midthorax (see Figure 22.1). Early anatomists mistook the trachea for a rough-walled artery (*trachea* = rough).

The tracheal wall contains 16 to 20 C-shaped rings of hyaline cartilage **(Figure 22.7a)** joined to one another by intervening membranes of fibroelastic connective tissue. Consequently, the trachea is flexible enough to permit bending and elongation, but the cartilage rings prevent it from collapsing and keep the airway open despite the pressure changes that occur during breathing. The open posterior parts of the cartilage rings, which abut the esophagus, contain smooth muscle fibers of the **trachealis** and soft connective tissue. Because the posterior wall of the trachea is not rigid, the esophagus can expand anteriorly as swallowed food passes through it. Contraction of the trachealis muscle decreases the diameter of the trachea: During coughing and sneezing, this action helps to expel irritants from the trachea by accelerating the exhaled air to a speed of 165 km/h (100 mph).

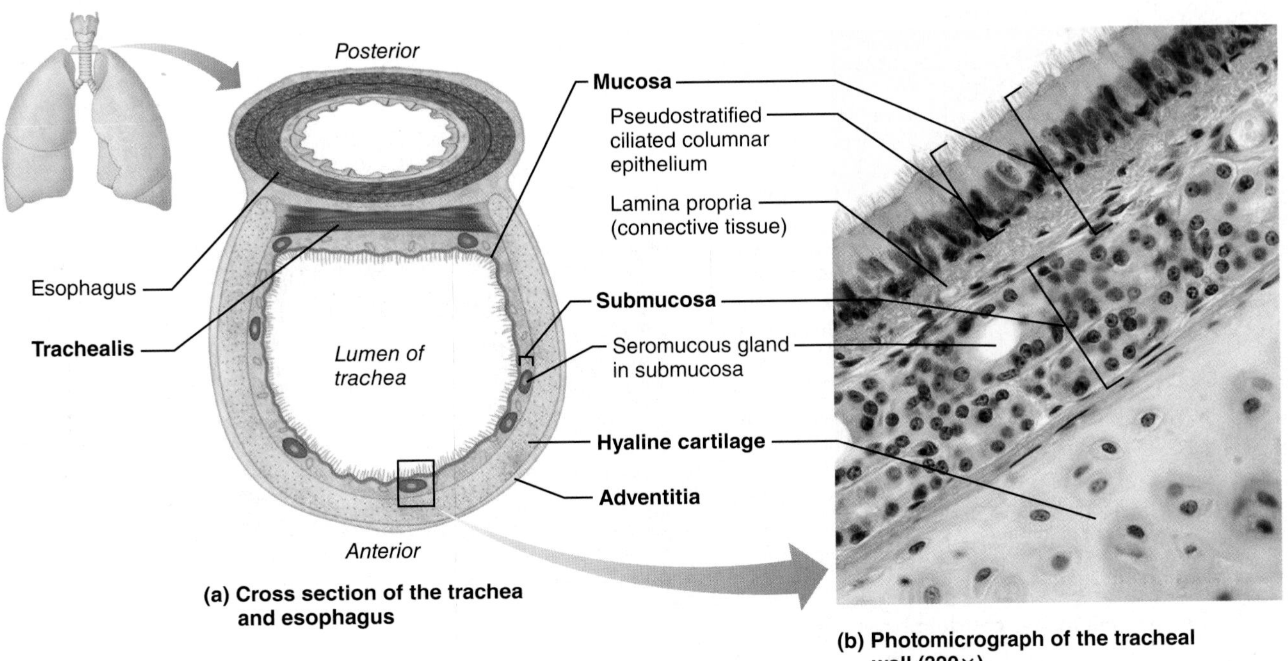

(a) Cross section of the trachea and esophagus

(b) Photomicrograph of the tracheal wall (320×)

(c) Scanning electron micrograph of cilia in the trachea (2500×)

Figure 22.7 **Tissue composition of the tracheal wall.** In the scanning electron micrograph in (c), the cilia appear as yellow, grasslike projections. Mucus-secreting goblet cells (orange) with short microvilli are interspersed between the ciliated cells.

A ridge on the internal aspect of the last tracheal cartilage, called the **carina** (kah-ri′nah; "keel"), marks the point where the trachea branches into the *two main (primary) bronchi* (see Figure 22.1). The mucosa that lines the carina is highly sensitive to irritants, and the cough reflex often originates here.

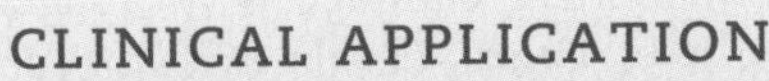

CLINICAL APPLICATION

Tracheotomy (Tracheostomy) When the upper respiratory tract is obstructed, a **tracheotomy,** or **tracheostomy,** is performed to open the airway. In a tracheotomy, the surgeon makes a vertical incision between the second and third tracheal rings in the anterior neck; a tube is then inserted to keep the airway open. In the emergency procedure called cricothyroid tracheotomy, a breathing tube is inserted through an incision made through the cricothyroid ligament of the larynx (see Figure 22.5b).

The microscopic structure of the wall of the trachea (Figure 22.7b) consists of several layers common to many tubular organs of the body: the *mucosa, submucosa,* and *adventitia.* The **mucosa,** a mucous membrane, consists of an inner epithelium and a lamina propria. The epithelium is the same air-filtering pseudostratified epithelium that occurs throughout most of the respiratory tract; its cilia continuously propel dust-laden sheets of mucus superiorly toward the pharynx (Figure 22.7c). The lamina propria contains many elastic fibers and is separated from the submucosa by a distinct sheet of elastin (not illustrated). This elastin, which occurs in all smaller air tubes as well, enables the trachea to stretch during inhalation and recoil during

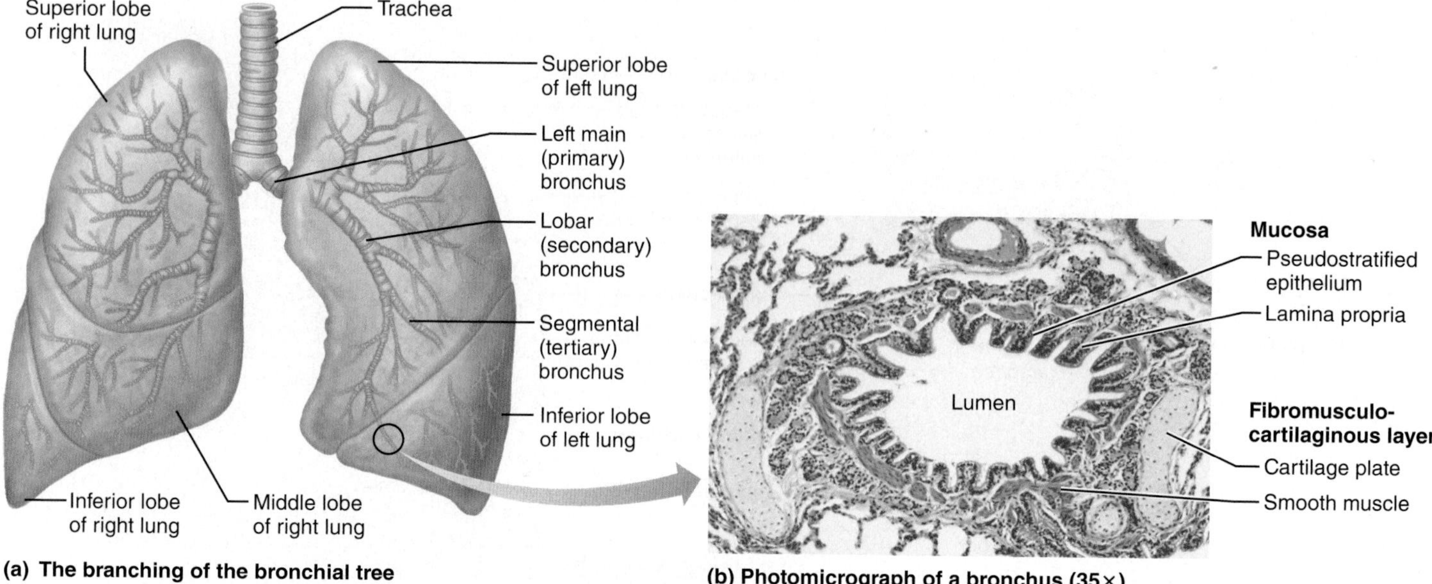

Figure 22.8 **The bronchi in the conducting zone.**

exhalation. The **submucosa** ("below the mucosa"), another layer of connective tissue, contains glands with both serous and mucous cells, called *seromucous glands,* which help produce the sheets of mucus within the trachea. The cartilaginous rings, the fibroelastic connective tissue connecting adjacent rings, and the trachealis lie external to the submucosa and form the **fibromusculocartilaginous layer** of the trachea. The external layer of connective tissue is the **adventitia** (ad″ven-tish′e-ah).

The Bronchial Tree

learning outcomes

- Explain how the walls of the upper respiratory passages differ histologically from those in the lower parts of the respiratory tree.
- Describe the structure of a lung alveolus and of the respiratory membrane.

Bronchi in the Conducting Zone

The right and left **main bronchi** (brong′ki), also called *primary bronchi,* are the largest conduits in the *bronchial tree,* a system of respiratory passages that branches extensively within the lungs (Figure 22.8). The two main bronchi are branches of the trachea in the mediastinum. This bifurcation occurs at the level of the sternal angle (T_4) in the cadavers studied in the anatomy laboratory, but in living, standing individuals, it typically occurs more inferiorly, approximately at T_7. Each main bronchus runs obliquely through the mediastinum before plunging into the medial depression (hilum) of a lung. The main bronchi lie directly posterior to the large pulmonary vessels that supply the lungs (see Figure 22.11d and e). Because the right main bronchus is wider, shorter, and more vertical than the left (see Figure 22.12), an accidentally inhaled object, such as a button or marble, is more likely to lodge in the right main bronchus.

As they approach and enter the lungs, the main bronchi divide into secondary or **lobar bronchi**—three on the right and two on the left—each of which supplies one lung lobe. The lobar bronchi branch into tertiary or **segmental bronchi,** which in turn divide repeatedly into smaller bronchi: fourth-order, fifth-order, and so on. Overall, there are about 23 orders of air tubes in the lungs, the tiniest almost too small to be seen without a microscope. The tubes smaller than 1 mm in diameter are called **bronchioles** ("little bronchi"), and the smallest of these, the **terminal bronchioles,** are less than 0.5 mm in diameter.

The tissue composition of the wall of each main bronchus mimics that of the trachea, but as the conducting tubes become smaller, the following changes occur:

- **The supportive connective tissues change.** The cartilage rings are replaced by irregular *plates* of cartilage as the main bronchi enter the lungs (see Figure 22.8b). By the level of the bronchioles, supportive cartilage is no longer present in the tube walls. By contrast, elastin, which occurs in the walls throughout the bronchial tree, does not diminish.
- **The epithelium changes.** The mucosal epithelium thins as it changes from pseudostratified columnar to simple columnar and then to simple cuboidal epithelium in the terminal and respiratory bronchioles. Neither cilia nor mucus-producing cells are present in these small bronchioles, where the sheets of air-filtering mucus end. Any inhaled dust particles that travel beyond the bronchioles are not trapped in mucus but instead are removed by macrophages in the alveoli (discussed shortly).

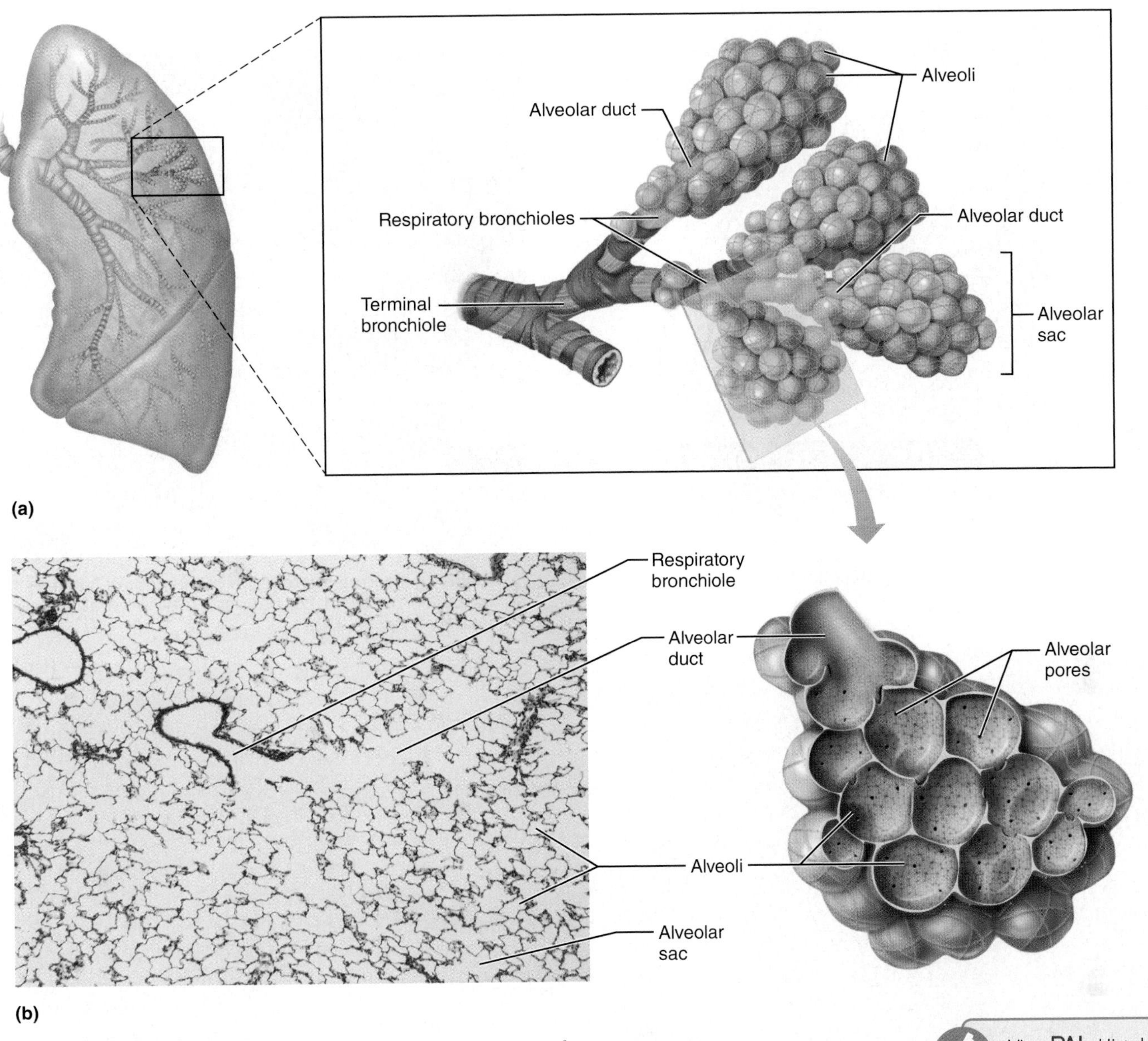

Figure 22.9 **Structures of the respiratory zone.** (a) Diagram of respiratory bronchioles, alveolar ducts, alveolar sacs, and alveoli. (b) Photomicrograph of a part of the lung (55×).

View PAL Histology MasteringA&P®

- **Smooth muscle becomes important.** A layer of smooth muscle first appears in the posterior wall of the trachea, the trachealis muscle, and continues into the large bronchi. This layer forms helical bands that wrap around the smaller bronchi and bronchioles and regulate the amount of air entering the alveoli. The musculature relaxes to widen the air tubes during sympathetic stimulation, thus increasing airflow when respiratory needs are great, and it constricts the air tubes under parasympathetic direction when respiratory needs are low. Strong contractions of the bronchial smooth muscles narrow the air tubes during asthma attacks, as discussed in "Disorders of the Respiratory System" section of this chapter. The smooth muscle thins as it reaches the terminal end of the bronchiole tree and is absent around the alveoli.

The Respiratory Zone

The respiratory zone is the end part of the respiratory tree in the lungs. The respiratory zone consists of structures that contain air-exchange chambers called **alveoli** (Figure 22.9). The first respiratory zone structures, which branch from the terminal bronchioles of the conducting zone, are **respiratory bronchioles.** These can be recognized by the scattered alveoli protruding from their walls. The respiratory bronchioles lead into **alveolar ducts,** straight ducts whose walls consist almost entirely of alveoli. The alveolar ducts then lead into terminal clusters of alveoli called **alveolar sacs.** Note that alveoli and alveolar sacs are not the same things: An alveolar sac is analogous to a bunch of grapes; the individual grapes are the alveoli.

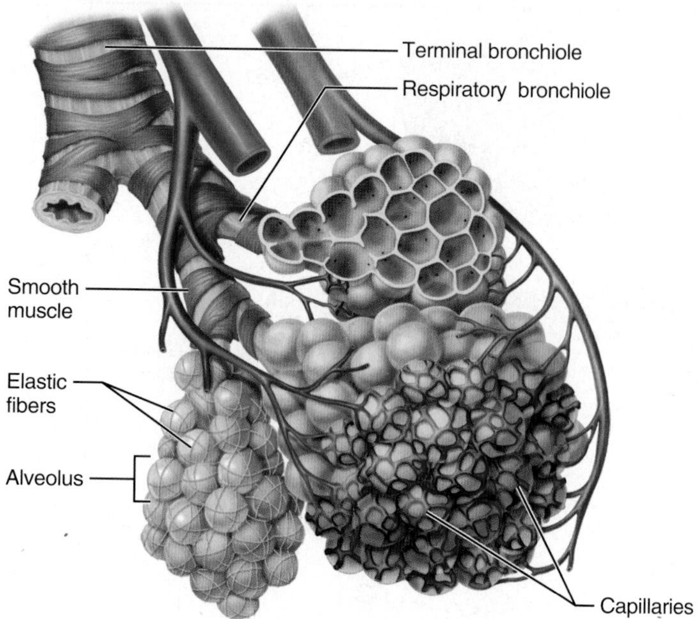

(a) Diagrammatic view of capillary-alveoli relationships

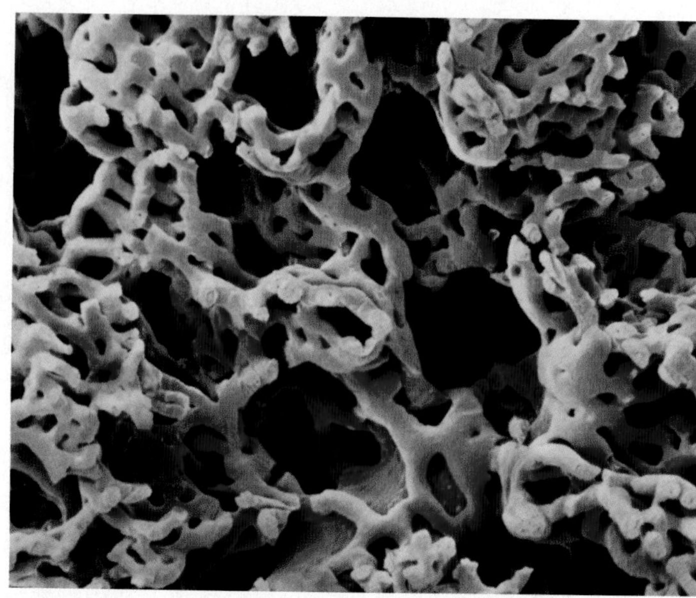

(b) Scanning electron micrograph of pulmonary capillary casts. Tissue forming the alveoli has been removed, leaving only the capillary network (300×).

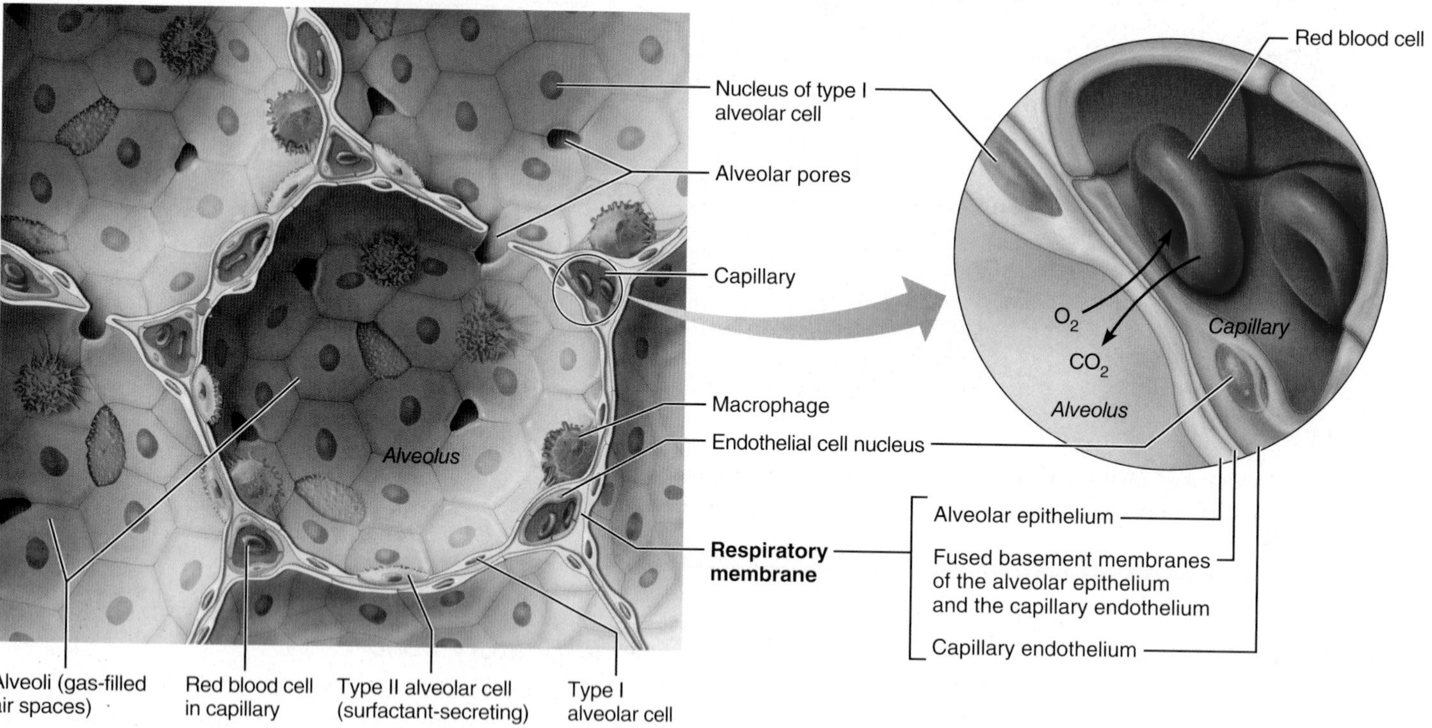

(c) Detailed anatomy of the respiratory membrane

Figure 22.10 Alveoli and the respiratory membrane. Elastic fibers and capillaries surround all alveoli, but for clarity they are shown only on some alveoli in (a).

About 400 million air-filled alveoli crowd together within the lungs, accounting for most of the lung volume and providing a tremendous surface area for gas exchange. The total area of all alveoli in an average pair of lungs is 140 square meters, or 1500 square feet, which is 40 times greater than the surface area of the skin!

The wall of each alveolus consists of a single layer of squamous epithelial cells called **type I alveolar cells** surrounded by a delicate basal lamina (Figure 22.10). The extreme thinness of this wall—0.5 μm—is hard to imagine. It is 15 times thinner than a sheet of tissue paper. The external surfaces of the alveoli are densely covered with a "cobweb" of pulmonary capillaries (Figure 22.10a and b), each of which is surrounded by a thin sleeve of the finest areolar connective tissue. Together, the alveolar and capillary walls and their fused basal laminae form the **respiratory membrane,** where oxygen and carbon dioxide are exchanged between the alveolus and the blood (Figure 22.10c). Air is present on the alveolar side of the membrane, and blood flows past on the capillary side. Gases pass easily through this thin membrane: Oxygen diffuses from the alveolus into the blood, and carbon dioxide diffuses from the blood to enter the air-filled alveolus.

Scattered among the type I squamous cells in the alveolar walls are cuboidal epithelial cells called **type II alveolar cells** (Figure 22.10c), which secrete a fluid that coats the internal alveolar surfaces. This fluid contains a detergent-like substance called **surfactant** (ser-fak′tant) that reduces surface tension within the alveoli. Without this surfactant, the inner walls of an alveolus would stick together during exhalation.

Lung alveoli also have the following significant features:

- Alveoli are surrounded by fine *elastic fibers* of the same type that surround structures along the entire respiratory tree (Figure 22.10a).
- Adjacent alveoli interconnect via **alveolar pores** (Figure 22.10c), which allow air pressure to be equalized throughout the lung and provide alternative routes for air to reach alveoli whose bronchi have collapsed because of disease.
- Internal alveolar surfaces provide a site for the free movement of **alveolar macrophages,** which actually live in the air space and remove the tiniest inhaled particles that were not trapped by mucus. Dust-filled macrophages migrate superiorly from the "dead-end" alveoli into the bronchi, where ciliary action carries them into the pharynx to be swallowed. This mechanism removes over 2 million debris-laden macrophages each hour.

check your understanding

☐ 6. What features of the trachea and larger bronchi trap and remove foreign particles from inhaled air? What performs this function in the alveoli?

☐ 7. At what level of the bronchial tree are cartilage plates no longer found? What is the extent of elastin in the bronchial tree?

☐ 8. Describe the structure of the respiratory membrane.

(For answers, see Appendix B.)

The Lungs and Pleurae

learning outcome

▶ Describe the gross structure of the lungs and the pleurae.

Gross Anatomy of the Lungs

The paired **lungs** and their pleural sacs occupy all the thoracic cavity lateral to the mediastinum (Figure 22.11). Each lung is roughly cone-shaped. The anterior, lateral, and posterior surfaces of a lung contact the ribs and form a continuously curving *costal surface.* Just deep to the clavicle is the **apex,** the rounded, superior tip of the lung. The concave inferior surface that rests on the diaphragm is the **base.** On the *medial (mediastinal) surface* of each lung is an indentation, the **hilum,** through which blood vessels, bronchi, lymphatic vessels, and nerves enter and exit the lung. Collectively, these structures attach the lung to the mediastinum and are called the **root** of the lung. The largest components of this root are the pulmonary artery and veins and the main (primary) bronchus (Figure 22.11b).

Because the heart is tilted slightly to the left of the median plane of the thorax, the left and right lungs differ slightly in shape and size. The left lung is somewhat smaller than the right and has a **cardiac notch,** a deviation in its anterior border that accommodates the heart (see Figure 22.11a). Several deep fissures divide the two lungs into different patterns of **lobes.** The left lung is divided into two lobes, the **superior lobe** and the **inferior lobe,** by the **oblique fissure.** The right lung is partitioned into three lobes, the **superior, middle,** and **inferior lobes,** by the **oblique** and **horizontal fissures.** As previously mentioned, each lung lobe is served by a lobar (secondary) bronchus and its branches.

Each of the lobes, in turn, contains a number of **bronchopulmonary segments** (Figure 22.12) separated from one another by thin partitions of dense connective tissue. Each segment receives air from an individual segmental (tertiary) bronchus. There are approximately ten bronchopulmonary segments arranged in similar, but not identical, patterns in each of the two lungs.

The bronchopulmonary segments have clinical significance in that they limit the spread of some diseases within the lung, because infections do not easily cross the connective-tissue partitions between them. Furthermore, because only small veins span these partitions, surgeons can neatly remove segments without cutting any major blood vessels.

The smallest subdivision of the lung that can be seen with the naked eye is the **lobule.** Appearing on the lung surface as hexagons ranging from the size of a pencil eraser to the size of a penny (see Figure 22.11b), each lobule is served by a bronchiole and its branches. In most city dwellers and in smokers, the connective tissue that separates the individual lobules is blackened with carbon.

As previously mentioned, the lungs consist largely of air tubes and spaces. The balance of the lung tissue, its *stroma* (stro′mah; "mattress"), is a framework of connective tissue containing many elastic fibers. As a result, the lungs are light, soft, spongy, elastic organs that each weigh only about 0.6 kg (1.25 pounds). The elasticity of healthy lungs helps to reduce the effort of breathing, as described shortly.

Blood Supply and Innervation of the Lungs

The **pulmonary arteries** (see p. 596) deliver oxygen-poor blood to the lungs for oxygenation (see Figure 22.11d and e). In the lung, these arteries branch along with the bronchial tree, generally lying *posterior* to the corresponding bronchi. The smallest arteries feed into the **pulmonary capillary networks** around the alveoli (Figure 22.10a). Oxygenated blood is carried from the alveoli of the lungs to the heart by the **pulmonary veins,** whose tributaries generally lie *anterior* to the corresponding bronchi within the lungs. To remind you of the position of the pulmonary vessels around the bronchus, remember **V B A, V**ein **B**ronchus **A**rtery. Also, some venous

(Text continues on page 662.)

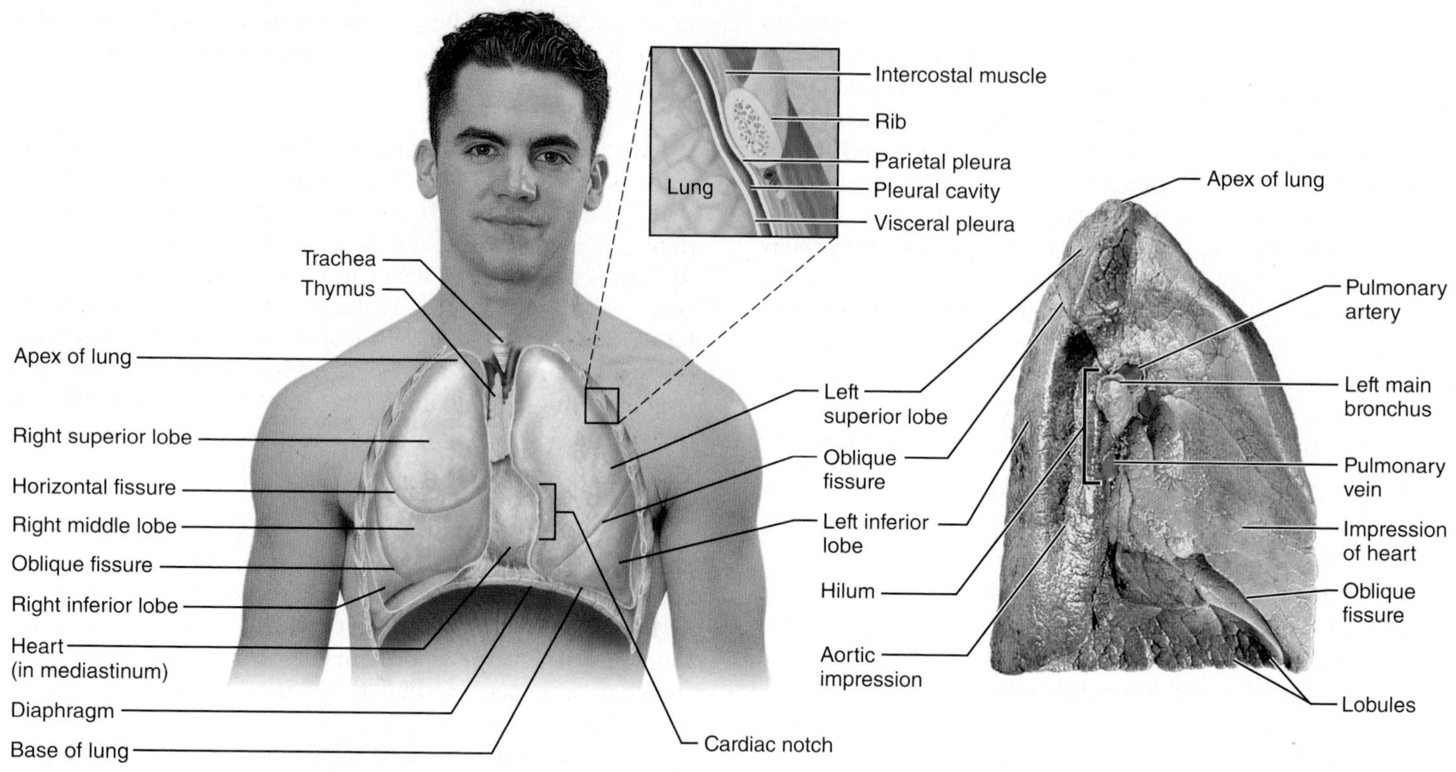

(a) Anterior view. The lungs flank mediastinal structures laterally.

(b) Photograph of medial view of the left lung

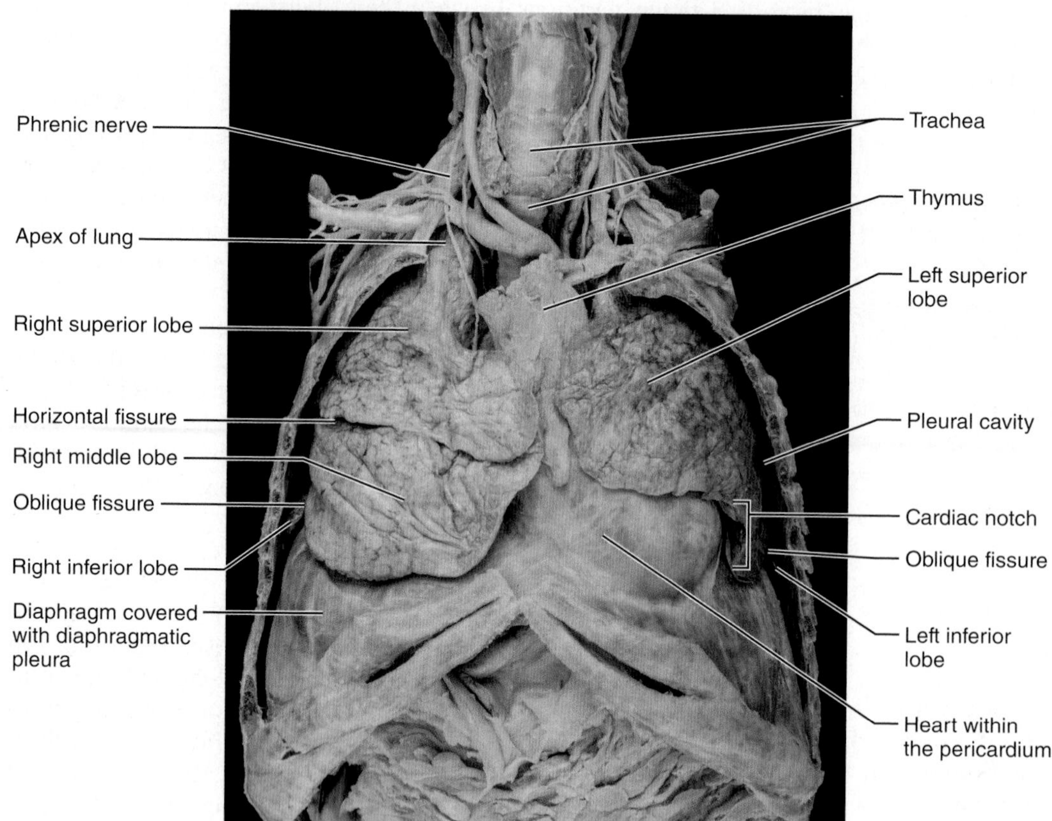

(c) Dissection of the thoracic viscera, anterior view

Figure 22.11 **Anatomical relationships of organs in the thoracic cavity.** The inset in (a) shows details of the pleurae and pleural cavity.

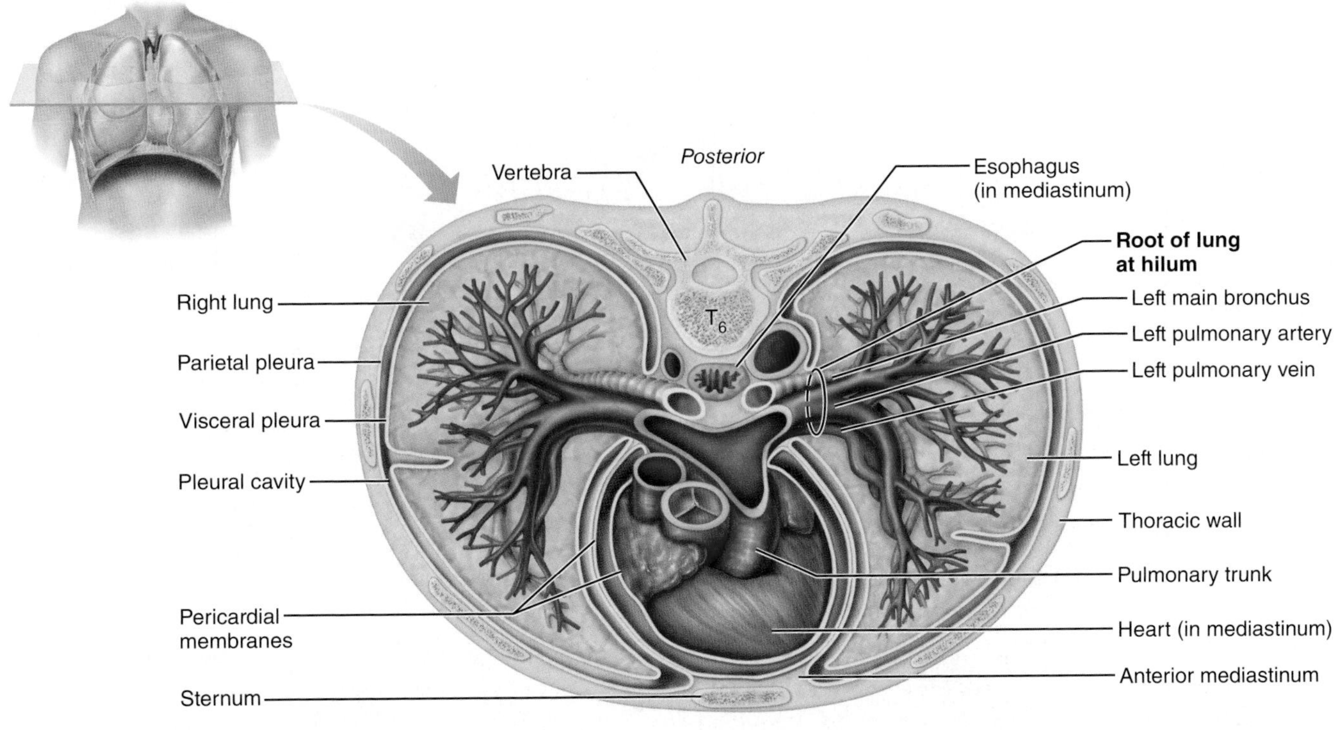

(d) Transverse section through the thorax at level of T_6, viewed from above. Lungs, pleural membranes, and major organs in the mediastinum are shown.

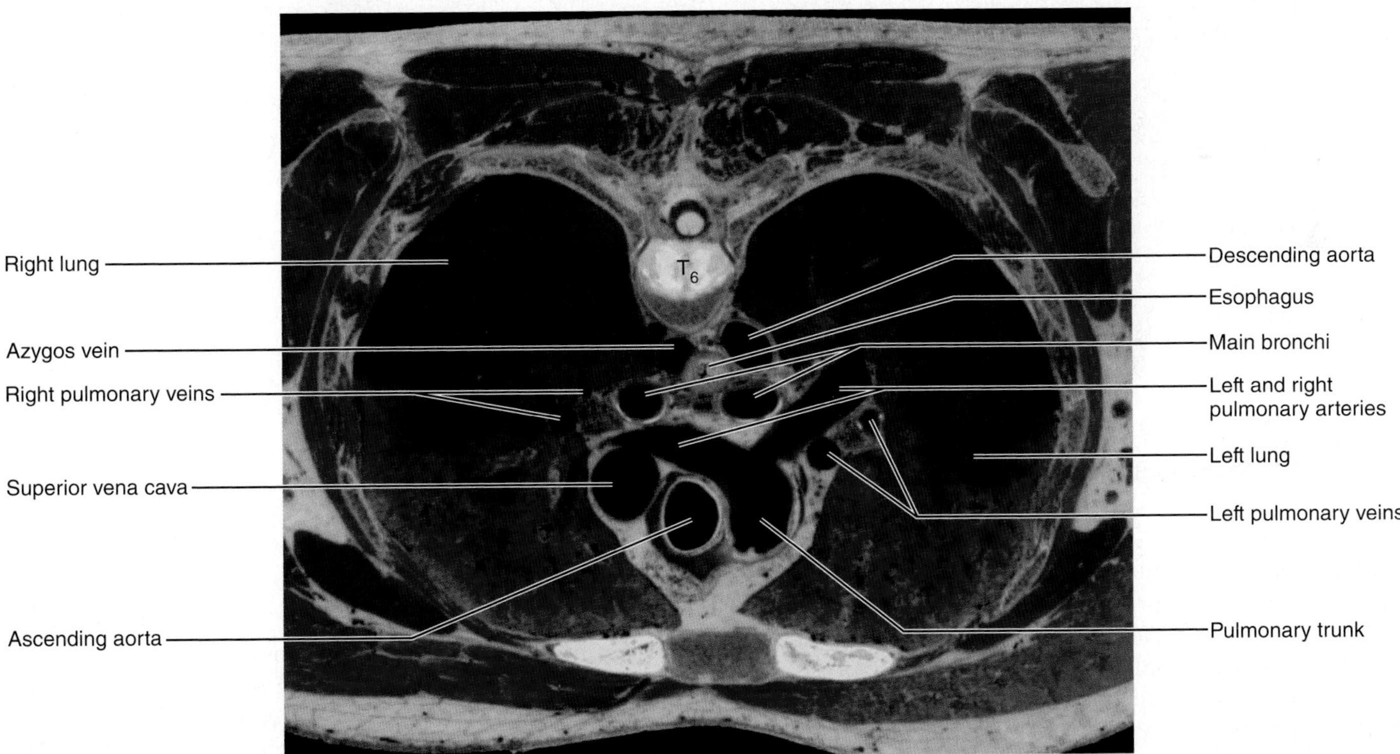

(e) Cross section through the thorax at T_6, view as in (d)

Figure 22.11 *continued.* In (d), the size of the pleural cavity is exaggerated for clarity.

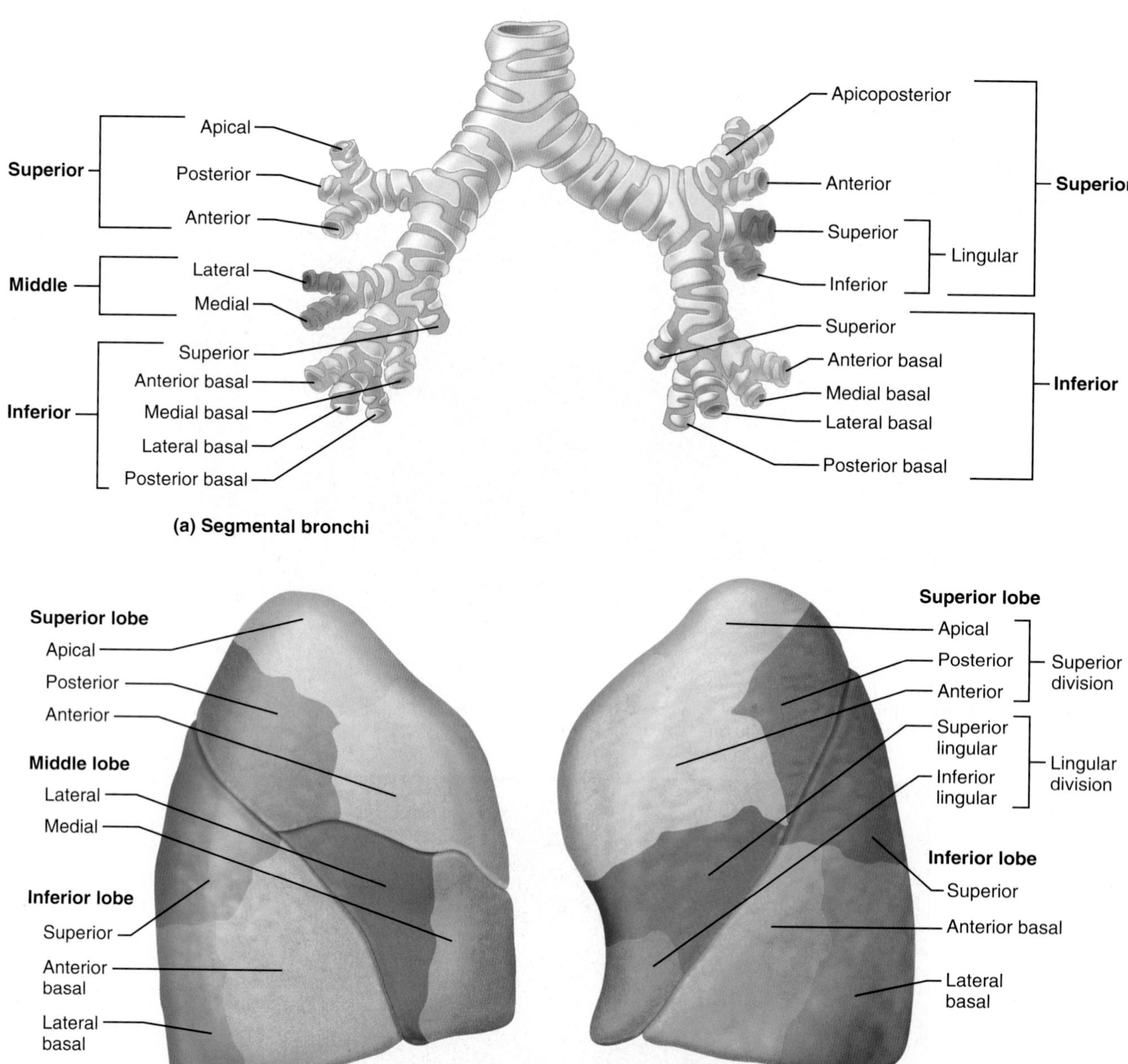

Figure 22.12 **Bronchopulmonary segments.**

tributaries run in the connective tissue partitions between the lung lobules and between the bronchopulmonary segments.

The *bronchial arteries* and *veins* provide and drain systemic blood to and from the lung tissues (Chapter 20, pp. 603, 612). These small vessels enter and exit the lungs at the hilum, and within the lung they lie on the branching bronchi.

The lungs are innervated by sympathetic, parasympathetic, and visceral sensory fibers that enter each lung through the *pulmonary plexus* on the lung root (see Figure 15.5). From there, these nerve fibers lie along the bronchial tubes and blood vessels within the lungs. Parasympathetic innervation causes bronchoconstriction and vasodilation; sympathetic stimulation results in bronchodilation and vasocontriction.

The Pleurae

Around each lung is a flattened sac whose walls consist of a serous membrane called **pleura** (ploo′rah; "the side"). The outer layer of this sac is the *parietal pleura,* whereas the inner layer, directly on the lung, is the *visceral pleura* (Figure 22.11a; see also Chapter 1, p. 12, "Serous Cavities"). The **parietal pleura** covers the internal surface of the thoracic wall, the superior surface of the diaphragm, and the lateral surfaces of the mediastinum. From the mediastinum, it reflects laterally to enclose the great vessels running to the lung (root of lung, Figure 22.11d). In the area where these vessels enter the lung, the parietal pleura is continuous with the **visceral pleura,** which covers the external lung surface.

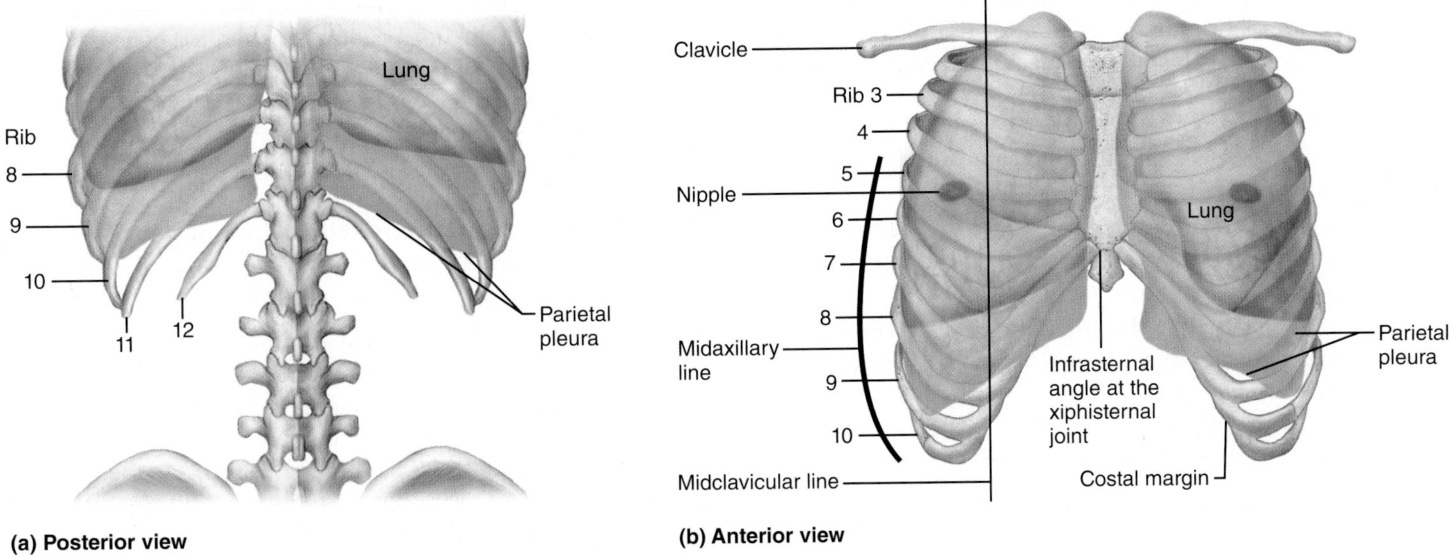

Figure 22.13 Position of the lungs and the pleural cavities in reference to the thoracic cage. The pleural cavity extends approximately two ribs below the inferior border of the lungs.

The space between the parietal and visceral pleurae is the **pleural cavity** (Figure 22.11d). The pleural cavity is filled with a thin film of *pleural fluid.* Produced by the pleurae, this lubricating fluid allows the lungs to glide without friction over the thoracic wall during breathing movements. The fluid also holds the parietal and visceral pleurae together, just as a film of oil or water would hold two glass plates together. The two pleurae can easily slide from side to side across each other, but their separation is strongly resisted. Consequently, the lungs cling tightly to the thoracic wall and are forced to expand and recoil as the volume of the thoracic cavity increases and decreases during breathing.

The pleurae also divide the thoracic cavity into three separate compartments—the central mediastinum and two lateral pleural compartments, each containing a lung (Figure 22.11d). This compartmentalization helps to prevent the moving lungs or heart from interfering with one another. Compartmentalization also limits the spread of local infections and the extent of traumatic injury.

Around all surfaces of the lung, with the exception of the inferior surface, the pleural cavity is a slitlike potential space. Inferiorly, the pleural cavity extends below the inferior border of the lungs (Figure 22.13). Posteriorly, the inferior margin of the parietal pleura lies adjacent to vertebra T_{12} near the posterior midline (Figure 22.13a) and runs horizontally across the back to reach rib 10 at the midaxillary line. Anteriorly, the parietal pleura ascends to rib 8 in the midclavicular line (Figure 22.13b) and to the level of the xiphisternal joint near the anterior midline. The inferior borders of the lungs are located two ribs superior to the pleural margin. Anteriorly, the lungs meet the pleural

CLINICAL APPLICATION

Pleurisy and Pleural Effusion Lung infections such as pneumonia produce inflammation of the pleura, called **pleurisy (pleuritis).** The rubbing together (friction) of the two inflamed pleural membranes produces a stabbing chest pain with each breath. Because the visceral pleura is relatively insensitive (see the discussion of visceral pain on p. 481), the pain of pleurisy actually originates from the parietal pleura only. If pleurisy persists, the inflamed pleurae may secrete excess pleural fluid, which then overfills the pleural cavity and exerts pressure on the lungs, thus hindering breathing movements.

Pleural effusion is a general term for the accumulation of one of various kinds of fluid in the pleural cavity. Although it characterizes many cases of pleurisy, it can also result from (1) hemorrhage of a damaged lung or lung vessel or (2) leakage of a serous fluid from the lung capillaries when either right ventricular failure or heart failure causes blood to back up in the pulmonary circuit.

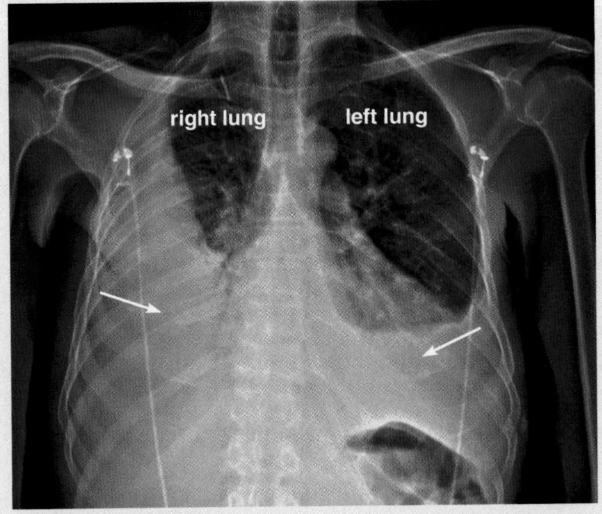

White arrows indicate areas of fluid accumulation below both lungs; right lung is collapsed.

margin near the xiphisternal joint. Surgeons and other clinicians must be aware of the inferior margin of the pleural cavities because cutting into a pleural cavity causes the lung to collapse.

check your understanding

☐ 9. Name the fissure that separates the superior and middle lobes of the right lung.

☐ 10. In what order, from superior to inferior, do the pulmonary vessels and main bronchus enter the left lung at the hilum?

☐ 11. Would a stab wound in the midclavicular line just above rib 7 puncture a lung? Would it puncture the pleural cavity?

(For answers, see Appendix B.)

VENTILATION

learning outcomes

- Explain the relative roles of the respiratory muscles, pleural cavity, and lung elasticity in the act of ventilation.
- Define surfactant, and explain its function in ventilation.
- Explain how the brain and peripheral chemoreceptors control the ventilation rate.

The Mechanism of Ventilation

Breathing, or **pulmonary ventilation,** consists of two phases: *inspiration* (inhalation), the period when air flows into the lungs, and *expiration* (exhalation), the period when gases exit the lungs. This section discusses the mechanical factors that promote this movement of air. (For detailed information on the *muscles* of ventilation, review Table 11.7 on pp. 291–292.)

Inspiration

The process of **inspiration** is easy to understand if you think of the thoracic cavity as an expandable container with a single entrance at the top: the tubelike trachea. Enlarging all dimensions of this container increases its volume and thus decreases the pressure within it (because pressure and volume in a closed container are inversely related). This decrease in internal gas pressure causes air to enter the container from the atmosphere, because gases always flow from areas of high pressure to areas of low pressure. During normal quiet inspiration, the inspiratory muscles—the diaphragm and intercostal muscles—function to increase the volume of the thorax. (Figure 22.14).

- **Action of the diaphragm.** When the dome-shaped diaphragm contracts, it moves inferiorly and flattens. As a result, the superior-inferior dimension of the thoracic cavity increases. Contraction of the diaphragm is stimulated by the *phrenic nerve.*
- **Action of the intercostal muscles.** The external intercostal muscles contract to raise the ribs. Lifting the ribs enlarges both the lateral dimensions of the thoracic cavity and the anterior-posterior dimensions. The intercostal muscles are innervated by the *intercostal nerves.*

The external and internal intercostal muscles also function together during quiet inspiration to stiffen the thoracic wall. Without this stiffening, the contraction of the diaphragm would result in a change of shape of the thorax but not a change in volume. Think about what would happen if the expandable container described above was made with rubber sides. As the container is enlarged by movement of its inferior surface, the flexible sides would collapse toward the center. A change of shape would result, but not a change of volume.

The lungs are surrounded by the pleural cavity. The air in the lungs is at atmospheric pressure and the pressure in the pleural cavity is less than atmospheric pressure, causing the lungs to expand and compress the pleural cavity against the thoracic wall (see Figure 22.13b). As the thoracic cavity enlarges by the action of the inspiratory muscles, the pleural cavity also enlarges, as does the volume of the lungs (Figure 22.14, inspiration). The increased volume results in a decrease in pressure in the pleural cavity and in the lungs. Because the lungs are open to the atmosphere via the bronchial tubes and trachea, when pressure in the lungs decreases, air from the atmosphere flows in.

For ventilation to occur, the pleural cavity must be intact. If the integrity of the pleural cavity is disrupted as a result of trauma or a disease process and the pressure differential between the lung and the pleural cavity is lost, the lung will collapse, and normal ventilation cannot occur.

CLINICAL APPLICATION

Collapsed Lung A lung will collapse if air enters the pleural cavity, a condition called **pneumothorax** (nu″mo-tho′raks; "air thorax"). The air breaks the seal of pleural fluid that holds the lung to the thoracic wall, allowing the elastic lung to collapse like a deflating balloon. Pneumothorax usually results from chest trauma or overexertion that raises intrathoracic pressure such that a lung pops like one pops a blown-up paper bag; it can also result from a wound that penetrates the thoracic wall or from a disease process that erodes a hole through the external surface of the lung. Pneumothorax is reversed surgically by closing the "hole" through which air enters the pleural cavity and then gradually withdrawing the air from this cavity using chest tubes. This treatment allows the lung to reinflate and resume its normal function.

Obstruction of a bronchus by a plug of mucus, an inhaled object, a tumor, or enlarged lymph nodes may also cause a lung to collapse; the collapse occurs as the air beyond the point of blockage is gradually reabsorbed into the pulmonary capillaries. Finally, pleural effusion (see p. 663) can cause collapse as the accumulating fluid compresses the lung.

Because the lungs are in completely separate pleural cavities, one lung can collapse without affecting the function of the other.

During deep or forced inspiration, additional muscles contract and further increase thoracic volume. The rib cage is elevated by the *sternocleidomastoid muscle* of the neck and by

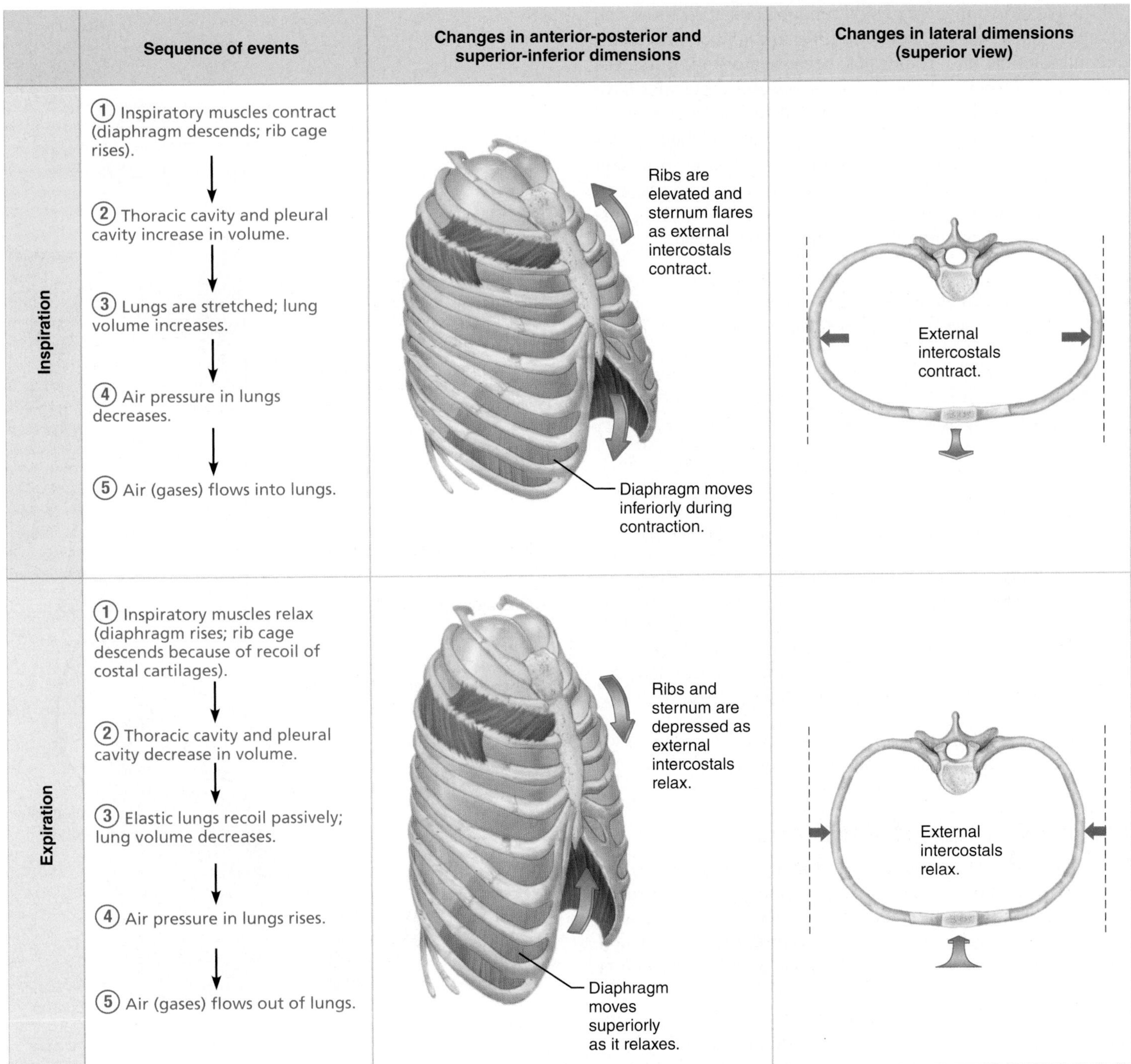

Figure 22.14 **Changes in thoracic volume and sequence of events during inspiration and expiration.**

the *scalenes* and *pectoralis minor* of the chest. Additionally, the *quadratus lumborum* fixes the 12th rib, resulting in a more powerful downward pull of the diaphragm, and the back extends as the thoracic curvature is straightened by the erector spinae muscles.

Expiration

Quiet **expiration** in healthy people is chiefly a passive process. As the inspiratory muscles relax, the rib cage drops under the force of gravity, and the relaxing diaphragm moves superiorly (Figure 22.14). At the same time, the many elastic fibers within the lungs recoil. The result is that the volumes of the thorax and lungs decrease simultaneously, increasing the pressure within the lungs and pushing the air out.

By contrast, *forced* expiration is an active process produced by contraction of muscles in the abdominal wall, primarily the *internal* and *external obliques* and the *transversus abdominis*. These contractions (1) increase the intra-abdominal pressure, forcing the diaphragm superiorly, and (2) sharply depress the rib cage, decreasing thoracic volume. The *internal intercostal muscles* and the *latissimus dorsi* also help to depress the rib cage.

In a healthy lung, the alveoli remain open at all times and do not collapse during exhalation. At first glance this seems to contradict the laws of physics, because a watery film coats the internal surfaces of the alveoli, and water molecules have a high attraction for one another (called *surface tension*) that should collapse the alveoli after each breath. Collapse does not occur because the alveolar film also contains surfactant secreted by the type II alveolar cells, which interferes with the attraction between water molecules, thereby reducing surface tension and enabling the alveoli to remain open.

CLINICAL APPLICATION

Respiratory Distress Syndrome Because pulmonary surfactant is not produced until the end of fetal life, its absence can have dire consequences for infants born prematurely. In such infants, the alveoli collapse during exhalation and must be completely reinflated during each inspiration, an effort that requires a tremendous expenditure of energy that can lead to exhaustion and respiratory failure. This condition, called **respiratory distress syndrome (RDS),** is responsible for one-third of all infant deaths. It is treated by using positive-pressure respirators to force air into the alveoli and keep them inflated between breaths and by administering natural or artificial surfactants. Even though such *surfactant therapy* has saved many lives since its introduction in 1990, many survivors still suffer from *chronic lung disease (bronchopulmonary dysplasia)* throughout childhood and beyond. This condition results from inflammatory injury to the respiratory zone, possibly caused by the high pressures the respirator exerts on the delicate lungs to distribute surfactant evenly.

Neural Control of Ventilation

The brain's most important respiratory center is in the reticular formation of the medulla oblongata. This center, called the ventral respiratory group (VRG) is a pacemaker whose neurons generate the basic ventilatory rhythm and rate with input from centers in the pons and dorsal medulla (Figure 22.15). Neurons from the medullary respiratory center stimulate the somatic motor neurons to the respiratory muscles: the phrenic nerve to the diaphragm and the intercostal nerves to the intercostal muscles. This basic ventilatory pattern can be modified by higher centers of the brain, such as the limbic system and hypothalamus (by which emotions influence breathing—as when people gasp), and the cerebral cortex (through which people have conscious control over the rate and depth of breathing).

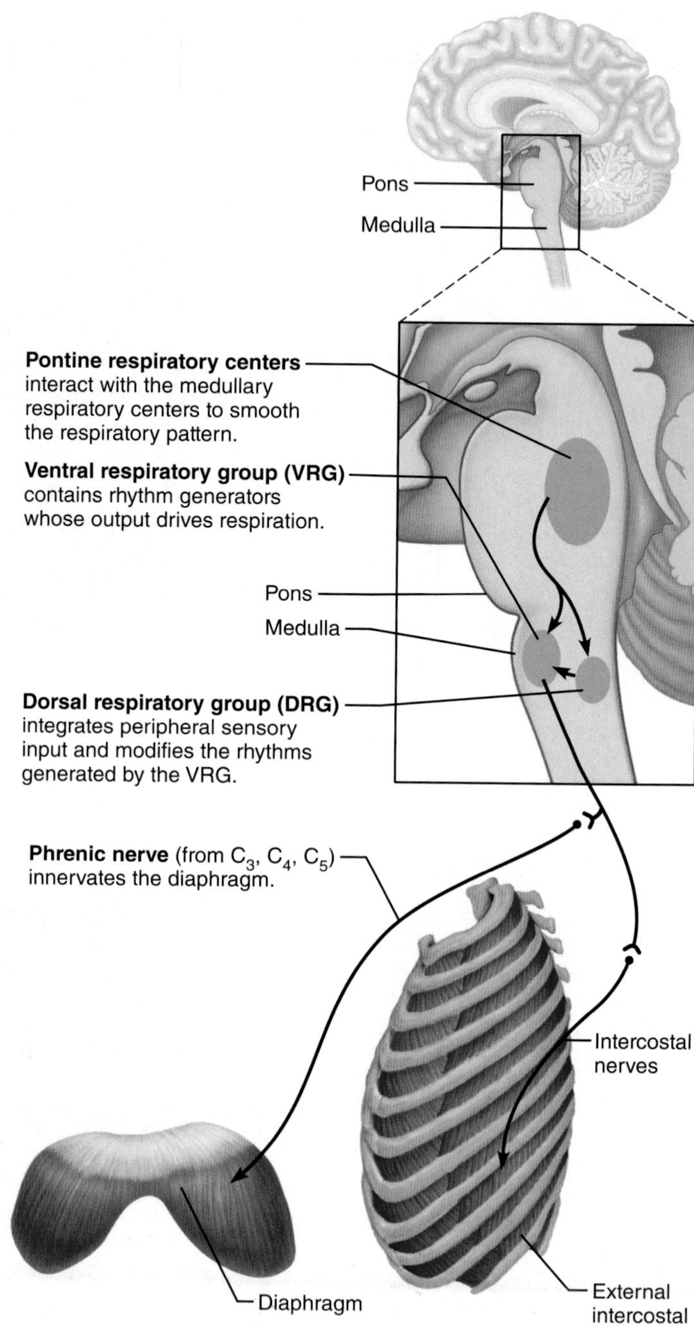

Figure 22.15 Respiratory centers in the brain stem. The synapses in the spinal cord with somatic motor neurons to the inspiratory muscles are shown schematically.

Although the medulla's respiratory center sets a baseline ventilatory rate, this rate is also modified by input from receptors that sense the chemistry of blood. These chemoreceptors respond to falling concentrations of oxygen, rising levels of carbon dioxide, or increased acidity of the blood by signaling the respiratory center to increase the rate and depth of breathing, which returns the blood gases to their normal concentrations. The chemoreceptors are of two types: *central chemoreceptors,* located mainly in the medulla, and *peripheral chemoreceptors,* the **aortic bodies** located on the aortic arch and the **carotid bodies** found in the fork of the common carotid artery (Figure 22.16). The aortic bodies send their sensory information to the medulla through the vagus nerves, whereas the carotid bodies send theirs through the glossopharyngeal (and perhaps the vagus) nerves (see Table 14.2 on pp. 441–442). In humans, it appears that the carotid bodies are the more important chemoreceptor for regulating respiration.

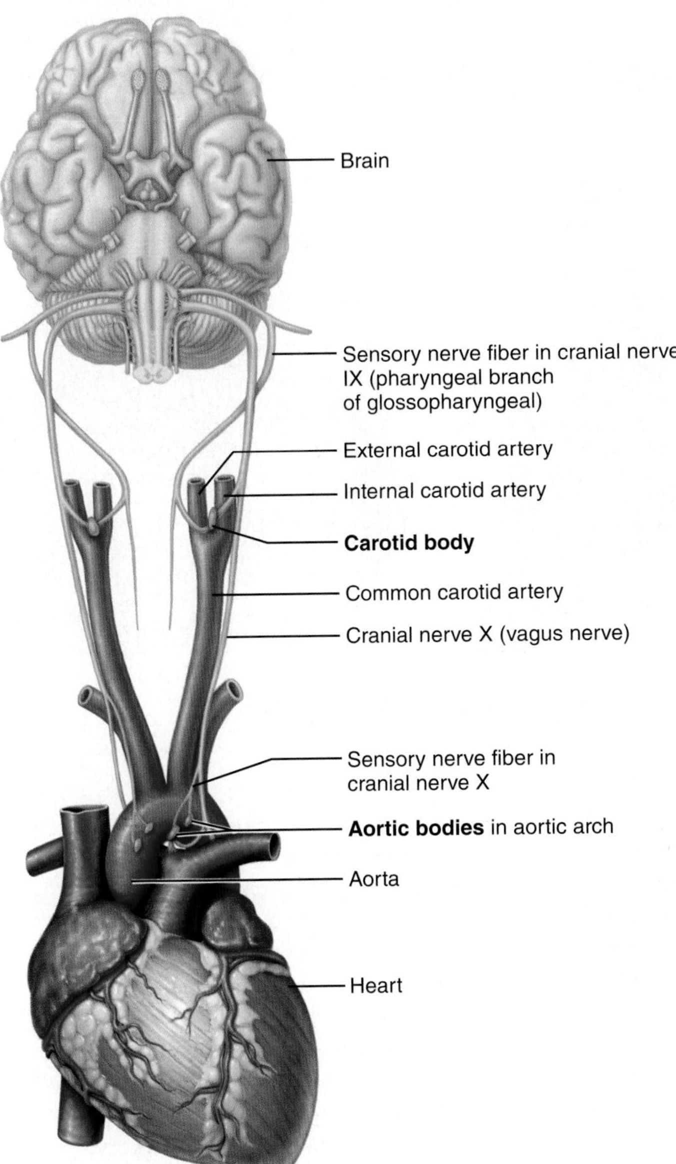

Figure 22.16 **Location and innervation of the peripheral chemoreceptors in the carotid and aortic bodies.** Also shown are the sensory pathways from these receptors through cranial nerves IX and X to the respiratory center in the medulla.

check your understanding

- ☐ 12. How does contraction of the diaphragm affect the volume of the thoracic cavity? How does this change of volume affect the pressure in the pleural cavity?
- ☐ 13. What will happen to the lung in the stab wound injury described in question 11 (p. 664)? Why is this type of injury called a "sucking chest wound"?
- ☐ 14. What accessory muscles may be used for inspiration in a person experiencing respiratory distress? What muscles are active during forced expiration?
- ☐ 15. Although the respiratory muscles are innervated by somatic motor neurons, and thus are under voluntary control, we do not have to think about breathing. Why is this so?

(For answers, see Appendix B.)

DISORDERS OF THE RESPIRATORY SYSTEM

learning outcome

▶ Consider the causes and consequences for respiration of asthma, cystic fibrosis, chronic bronchitis, emphysema, and lung cancer.

Because the respiratory system is open to the outside world and exposed to large volumes of air, it is highly susceptible to airborne microorganisms, pollutants, and irritants, leading to a remarkable number of common respiratory disorders. One of the most lethal respiratory diseases, lung cancer, is discussed in A CLOSER LOOK (p. 668).

Bronchial Asthma

Bronchial asthma affects about 7% of adults and 10% of children and is increasing in frequency. Asthma is a type of allergic inflammatory response that occurs in people who are hypersensitive to irritants in the air or to stress. Attacks may be triggered by inhaling substances to which the sufferer is allergic (dust mites, pollen, molds, bits of cockroach), by inhaling dust or smoke, by respiratory infections, by emotional upset, or by the mild shock of breathing cold air. Symptoms of asthma attacks include coughing, wheezing, and shortness of breath.

An asthma attack starts with an early phase, in which an allergen initiates the release of inflammation-mediating chemicals, such as histamine, from mast cells. These chemicals stimulate both contraction of the bronchial smooth musculature (bronchoconstriction) and secretion of mucus in these airways. Within several hours, a late phase develops as eosinophils, neutrophils, a certain type of helper T lymphocytes, and basophils accumulate in the bronchi and bronchioles. These leukocytes secrete additional inflammatory chemicals that damage the bronchial mucosa and further increase bronchoconstriction and secretion of mucus.

Until about 1990, bronchoconstriction was considered the primary symptom of asthma, and quick relief is still provided by drugs that either inhibit smooth-muscle contraction (bronchodilators) or inhibit the parasympathetic stimulation of such contraction (anticholinesterases). The realization that asthma is primarily an inflammatory disease has led to treatments using anti-inflammatory drugs (glucocorticoids and nonsteroidal anti-inflammatory agents). These drugs allow much better long-term management, leading to fewer asthma attacks and less damage to the airways.

Cystic Fibrosis

Cystic fibrosis (CF) is an inherited disease in which the functions of exocrine glands are disrupted throughout the body. Occurring mainly in people of European descent, it kills 1 of every 2500 Americans. CF affects the respiratory system by causing an accumulation of a viscous mucus in the respiratory passageway. This mucus clogs the respiratory tubes and acts as a feeding ground for bacteria, predisposing the child to early death due to respiratory infections.

Lung Cancer: The Facts Behind the Smoke Screen

Lung cancer accounts for fully one-third of all cancer deaths in the United States, and its incidence is increasing daily. The most common malignancy in both sexes, it is notoriously deadly. Most types of lung cancer are tremendously aggressive and metastasize rapidly and widely, and most cases cannot be diagnosed until they are far advanced. Whole-lung CT scans to detect early-stage tumors may improve treatment outcomes.

For years, Americans were remarkably unaware of the link between lung cancer and cigarette smoking, despite the fact that over 90% of lung cancer patients are smokers. Smoking even a single cigarette increases the heart rate, constricts peripheral blood vessels throughout the body, disrupts the flow of air in the lungs, and affects brain and mood. In adolescents, that first cigarette can trigger addiction. In a study of sixth graders in Massachusetts, of those who developed physical symptoms of dependence, 10% developed symptoms after their first cigarette and 25% within 2 weeks of intermittent cigarette use!

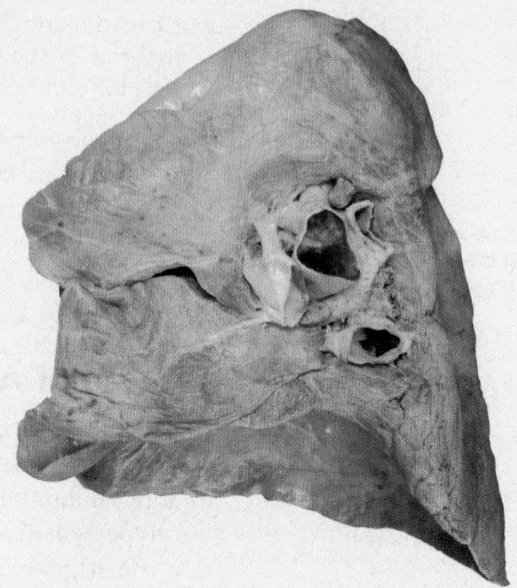

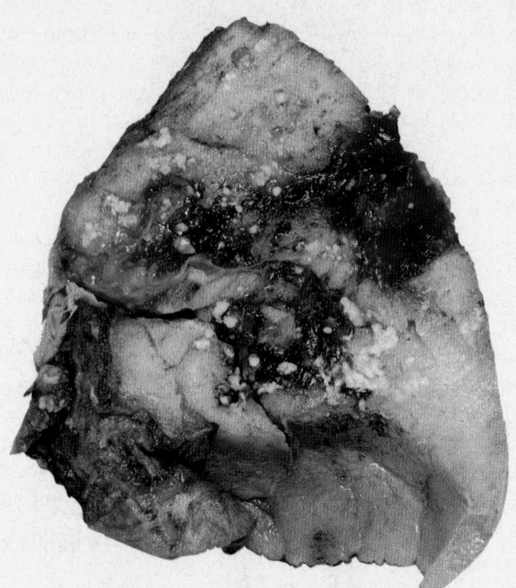

Medial view of right lung: healthy lung on left, cancerous lung on right.

Long-term smoking contributes to atherosclerosis and heart disease, strokes, cataracts, and osteoporosis. In addition to causing lung cancer, smoking raises the risk of cancers of the mouth, throat, esophagus, stomach, liver, pancreas, cervix, kidney, and bladder. Smoking contributes to one-fifth of all deaths in the United States, making it the single most preventable cause of death.

Ordinarily, sticky mucus and the action of cilia do a fine job of protecting the lungs from chemical and biological irritants, but smoking overwhelms these cleansing devices until they eventually become nonfunctional. Even though continuous irritation prompts the production of more mucus, smoking slows the movements of cilia that clear this mucus and depresses the activity of lung macrophages. Mucus pools in the lower respiratory tree and pulmonary illnesses, including pneumonia and COPD, become more frequent. However, it is the 15 or so carcinogens in tobacco smoke, including the highly addictive nicotine, that ultimately lead to lung cancer. These chemicals, plus the tars in tobacco, eventually cause the epithelial cells lining the bronchial tree to proliferate wildly and lose their characteristic structure. Ironically, the nicotine in tobacco smoke may be carcinogenic *because* it promotes cell survival. Normally, damaged cells either repair the damage or die, but nicotine appears to activate enzymes that disrupt the sequence of apoptosis.

The three most common types of lung cancer are (1) **squamous cell carcinoma** (25% to 30% of cases), which arises in the epithelium of the larger bronchi and tends to form masses that cavitate and bleed; (2) **adenocarcinoma** (40%), which originates in the peripheral areas of the lung as solitary nodules that develop from bronchial mucous glands and alveolar epithelial cells; and (3) **small cell carcinoma** (15% of cases, but rapidly increasing), which contains lymphocyte-like epithelial cells that originate in the main bronchi and grow aggressively in cords or small grapelike clusters within the mediastinum, a site from which metastasis is especially rapid.

The most effective treatment for lung cancer is complete removal of the diseased lung. However, removal is an option open to very few patients because metastasis has usually occurred by the time of diagnosis, and most patients' chances of survival are too poor to justify the surgery. In most cases, radiation therapy and chemotherapy are the only options, but most lung cancers are resistant to these treatments. Only small cell carcinoma responds to chemotherapy, but frequently it returns quickly and spawns brain tumors.

The good news is that quitting helps. While the incidence of lung cancer is 20 times greater in smokers than in nonsmokers, this ratio drops to 2:1 for former smokers who have not smoked in 15 years. In a British study, people who stopped smoking before age 35 lived full life spans, and quitting at any age prolonged survival. Overall, 48.8% of U.S. adults who ever smoked cigarettes have quit. Many succeeded by quitting "cold turkey," and others benefited from aids such as nicotine nasal sprays, patches, and inhalers.

In the 1980s, researchers isolated the gene responsible for cystic fibrosis (the *cftr* gene) and showed that a defect in this gene prevents the formation of a membrane channel protein that carries chloride ions out of epithelial cells. Attempts to cure CF by delivering nondefective copies of the *cftr* gene to the affected respiratory epithelial cells have not succeeded. Drug treatments that target the malformed membrane protein have led to progress. One drug approved by the U.S. Food and Drug Administration (FDA) in January 2012 (ivacaftor; trade name Kalydeco, and known during its development as VX-770) acts to prop open the blocked chloride channels, enabling chloride and water to exit the cell, thereby thinning the mucus in airways and improving lung function. This drug has shown significant success for a small percentage of CF patients. Modern antibiotics allow the average CF patient to survive until age 30, an increase in life span of 16 years since 1969. (For information on the nonrespiratory effects of CF, see p. 713.)

Chronic Obstructive Pulmonary Disease

Chronic obstructive pulmonary disease (COPD) is a category of disorders in which the flow of air into and out of the lungs is difficult or obstructed. It mostly refers to *chronic bronchitis* or *emphysema* (or both of these occurring together), and it is a major cause of death and disability in the United States. These two diseases share certain features: The patients almost always have a history of smoking; they experience difficult or labored breathing called **dyspnea** (disp-ne′ah; "bad breathing"); coughing and pulmonary infections occur frequently; and most COPD victims ultimately develop respiratory failure (Figure 22.17).

Chronic Bronchitis

In **chronic bronchitis,** inhaled irritants lead to a prolonged secretion of excess mucus by the mucosa of the lower respiratory passages and to inflammation and fibrosis (formation of scar tissue) of this mucosa. These responses obstruct the airways, severely impairing ventilation and gas exchange. Infections frequently develop because bacteria and viruses thrive in the stagnant pools of mucus. Coughing is especially persistent and productive. Patients with chronic bronchitis are sometimes called "blue bloaters" because lowered blood oxygenation often results in cyanosis and other signs of right heart failure, including edema. However, the degree of dyspnea is usually moderate compared to that of emphysema sufferers. So significant is cigarette smoking as a causative factor of chronic bronchitis that this disease would be an insignificant health problem if cigarettes were unavailable. Air pollution is another, although minor, causative factor.

Emphysema

Emphysema (em″fĭ-se′mah; "to inflate") is characterized by a permanent enlargement of the alveoli caused by a deterioration of the alveolar walls (Figure 22.18). The disease is most often associated with a smoking-related chronic inflammation of the lungs and increased activity of lung macrophages, whose lysosomal enzymes destroy the alveolar walls and down elastin. Chronic inflammation also leads to fibrosis, and the lungs become progressively less elastic, making expiration difficult and exhausting. For complex reasons, the bronchioles open during inspiration but collapse during expiration, trapping huge volumes of air in the alveoli. This enlarges the lung, leads to the development of an expanded "barrel chest," and flattens the diaphragm, which decreases ventilatory efficiency. Damage to the lung capillaries increases the resistance in the pulmonary vascular circuit, forcing the heart's right ventricle to enlarge through overwork.

COPD is routinely treated with bronchodilators and anti-inflammatory drugs (including glucocorticoids) in aerosol

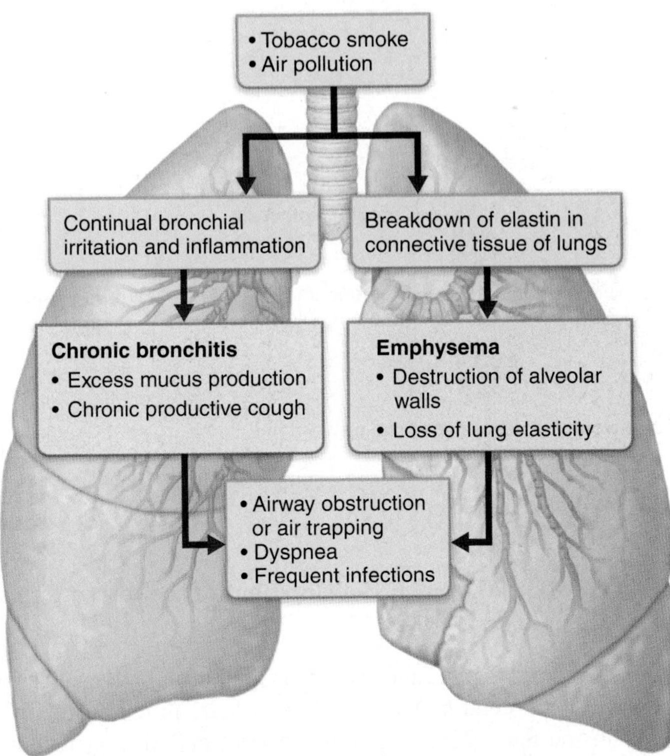

Figure 22.17 **The pathogenesis of chronic obstructive pulmonary disease (COPD).**

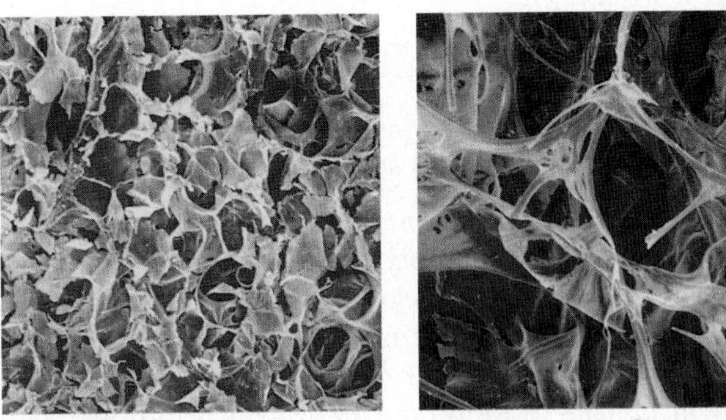

(a) SEM of alveoli from a normal lung (25×)

(b) SEM of alveoli from the lung of a patient with emphysema (25×)

Figure 22.18 **Alveolar changes in emphysema.** Scanning electron micrographs (SEMs) of alveoli at some magnification.

Frontonasal elevation
Olfactory placode
Stomodeum
(future mouth)

(a) 4 weeks: anterior superficial view of the embryo's head

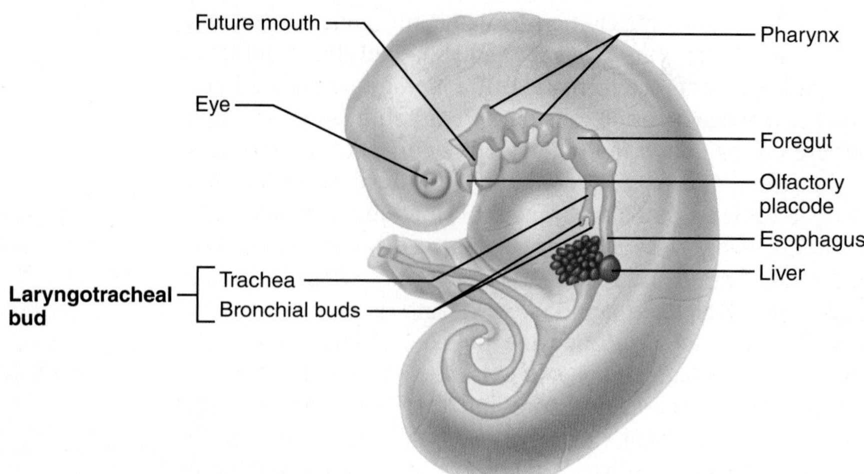

(b) 5 weeks: left lateral view of the developing lower respiratory passageway mucosae

Figure 22.19 **Embryonic development of the respiratory system.**

form. Severe dyspnea and lowered uptake of oxygen nearly mandates supplemental oxygen use. A treatment used initially in the late 1950s and reintroduced, called *lung volume reduction surgery,* has been performed on some emphysema patients. In this surgery, part of the grossly enlarged lung is removed to give the remaining lung tissue room to expand. This surgery has improved the breathing capacity and the quality of life in moderately ill patients, but it is less beneficial in patients with less severe emphysema.

☑ check your understanding

☐ **16.** Which of the described disorders results in damage to the respiratory zone?

☐ **17.** What causes "smoker's cough," the persistent cough that commonly afflicts long-term or heavy smokers?

(For answers, see Appendix B.)

THE RESPIRATORY SYSTEM THROUGHOUT LIFE

learning outcomes

- Trace the development of the respiratory system in the embryo and fetus.
- Describe the normal changes that occur in the respiratory system from infancy to old age.

Because embryos develop in a craniocaudal (head-to-tail) direction, the upper respiratory structures appear before the lower ones. By week 4, a thickened plate of ectoderm called the **olfactory placode** (plak′ōd; "plate") has appeared on each side of the future face (Figure 22.19a). These placodes quickly invaginate to form *olfactory pits* that form the nasal cavity, including the olfactory epithelium in its roof. The nasal cavity then connects with the future pharynx of the developing foregut, which forms at the same time.

The lower respiratory organs develop from a tubular outpocketing off the pharyngeal foregut called the **laryngotracheal bud** (Figure 22.19b). The proximal part of this bud forms the trachea, and its distal part branches repeatedly to form the bronchi and their subdivisions, including (eventually) the lung alveoli. The respiratory tubes, like the gut tube from which they arise, are lined by endoderm and covered by splanchnic mesoderm. The endoderm becomes the lining epithelium (and glands) of the trachea, bronchial tree, and alveoli. The splanchnic mesoderm gives rise to all other layers of the tracheal and bronchial walls (cartilage, smooth muscle, and all connective tissues) and to the stroma of the lungs.

The respiratory system reaches functional maturity relatively late in development. No alveoli appear until month 6 (24 weeks), and the type I alveolar cells do not attain their extreme thinness until the time of birth. Type II alveolar cells begin to produce surfactant by week 26. It is not until 26–30 weeks that a prematurely born baby can survive and breathe on its own. Infants born before this time are most severely threatened by respiratory distress syndrome resulting from inadequate production of surfactant.

During fetal life, the lungs are filled with fluid, and all respiratory exchange occurs across the placenta. At birth, the first breaths bring air into the lungs, and the alveoli inflate and begin to function in gas exchange. During the nearly 2 weeks it takes for the lungs to become fully inflated, the fluid that originally filled the lungs is absorbed into the alveolar capillaries.

At birth, only one-sixth of the final number of lung alveoli are present. The lungs continue to mature throughout childhood, and more alveoli are formed until young adulthood. Research has revealed that in individuals who begin smoking in their early teens, the lungs never completely mature, and those additional alveoli never form.

Under normal conditions, the respiratory system works so efficiently that people are not even aware of it. Most problems that occur are the result of external factors, such as viral or bacterial infections, irritants that trigger asthma in susceptible individuals, or obstruction of the trachea by a piece of food. For many years, tuberculosis and bacterial pneumonia (discussed in the "Related Clinical Terms" section at the end of this chapter) were the worst killers in the United States and Europe. Though antibiotics have decreased the mortality associated with these infections, they are still dangerous diseases. Influenza, a respiratory flu virus, killed at least 20 million people worldwide in 1918, and each year government health agencies keep a watchful eye on outbreaks caused by new viral strains. Despite the attention given to such dramatic illnesses, by far the most troublesome respiratory disorders are smoking-related COPD and lung cancer.

As humans age, the number of glands in the nasal mucosa declines, as does the flow of blood in this mucosa. Thus, the nose dries and crusts and produces a thickened mucus that makes one want to "clear the throat." In the elderly, the thoracic wall grows more rigid, and the lungs gradually lose their elasticity, resulting in a slowly declining ability to ventilate the lungs. The levels of oxygen in the blood may fall slightly while levels of carbon dioxide rise. Additionally, just as the overall efficiency of the immune system declines with age, many of the respiratory system's protective mechanisms become less effective—the activity of the cilia in its epithelial lining decreases, and the macrophages in the lungs become sluggish. The net result is that the elderly are at greater risk for respiratory infections, particularly pneumonia and influenza.

☑ check your understanding

☐ 18. Why do babies born earlier than 26 weeks experience severe respiratory distress?

(For the answer, see Appendix B.)

RELATED CLINICAL TERMS

Atelectasis (at″ĕ-lek′tah-sis) Collapse of the lung, either from airway obstruction or from the compression of pleural effusion. Can also refer to a failure of the lungs to inflate, as in premature infants.

Bronchoscopy (*scopy* = viewing) Use of a viewing tube to examine the internal surface of the main bronchi in the lung. The tube is inserted through the nose or mouth and guided inferiorly through the larynx and trachea. Forceps may be attached to the tip of the tube to remove trapped objects, take biopsy samples, or retrieve samples of mucus for examination.

Croup Disease in children in which viral-induced inflammation causes the air passageways to narrow; is characterized by coughing that sounds like the bark of a dog, hoarseness, and wheezing or grunting sounds during inspiration. Most cases resolve after a few days, but severe cases may require anti-inflammatory aerosols or a tracheotomy (see p. 655) to bypass the obstructed upper respiratory tubes.

Pneumonia An infectious inflammation of the lungs in which fluid accumulates in the alveoli. Of the over 50 known varieties, most are caused by viruses or bacteria (however, the type of pneumonia associated with AIDS, *Pneumocystis jirovecii* pneumonia, is caused by a fungus). Extremely common, pneumonia is the sixth most frequent cause of death in the United States because almost any severely ill person can develop it.

Sudden infant seath syndrome (SIDS) Unexpected death of an apparently healthy infant during sleep. Commonly called *crib death,* SIDS is the most frequent cause of death in infants under 1 year old. The cause is unknown, but it may reflect immaturity of the brain's respiratory control centers. Since 1992, a campaign to have babies sleep on their backs instead of on their bellies has led to a decline of 40% or more in the incidence of SIDS in the United States.

Tuberculosis (TB) A lung disease caused by the bacterium *Mycobacterium tuberculosis,* which is spread by coughing and enters the body in inhaled air. TB typically affects the lungs but can spread through lymphatic vessels to other organs. A massive inflammatory and immune response contains the primary infection within fibrous or calcified nodules in the lungs (tubercles), but the bacteria often survive, break out, and cause repeated infections. Symptoms of TB are coughing, weight loss, mild fever, and chest pain. Some strains of TB are now resistant to antibiotics, and TB cases are increasing in the United States. Current vaccines protect more than half of the children who receive them, but are not effective in adults.

CHAPTER SUMMARY

Take full advantage of the invaluable online practice and study tools for this chapter by going to the Study Area of MasteringA&P®. There you will find:

- Art Labeling Activities, Quizzes, Practice Tests, Memory Games
- Animal Organ Dissection, Cat Dissection and Bone Videos
- PAL Cadaver
- PAL Histology
- A&P Flix Animations
- Dynamic Study Modules

1. Respiration involves the delivery of O_2 to the body cells for use in energy production and the removal of the waste product CO_2. The respiratory processes are as follows: pulmonary ventilation, external respiration, cardiovascular transport of respiratory gases, internal respiration, and cellular respiration. Both the respiratory system and the cardiovascular system participate in respiration.

FUNCTIONAL ANATOMY OF THE RESPIRATORY SYSTEM (pp. 646–664)

2. The organs of the respiratory system include a conducting zone (nose to terminal bronchioles), which warms, moistens, and filters the inhaled air, and a respiratory zone (respiratory bronchioles to alveoli), where gas exchange occurs.

The Nose and the Paranasal Sinuses (pp. 647–651)

3. The nose and nasal cavity provide an airway for respiration and house the olfactory receptors.

4. The external nose is shaped by bone and by cartilage plates. The nasal cavity begins at the external nares and ends posteriorly at the posterior nasal apertures (choanae). Air-swirling conchae occupy its lateral walls. The paranasal sinuses drain into the nasal cavity.

5. The respiratory mucosa lines the nasal cavity and paranasal sinuses. Its epithelium, a pseudostratified ciliated columnar epithelium with goblet cells, is covered with a sheet of mucus that filters dust particles and moistens inhaled air. Its lamina propria contains glands that contribute to the mucus sheet and blood vessels that warm the air.

The Pharynx (p. 651)

6. The pharynx has three regions: The nasopharynx (behind the nasal cavity) is an air passageway; the oropharynx (behind the mouth) and the laryngopharynx (behind the larynx) are passageways for both food and air.

7. The soft palate moves superiorly to seal off the nasopharynx during swallowing. The tubal and pharyngeal tonsils occupy the mucosa that lines the nasopharynx. The oropharynx contains the lingual and palatine tonsils. The laryngopharynx opens into the larynx anteriorly and into the esophagus inferiorly.

The Larynx (pp. 651–654)

8. The larynx, or voice box, is the entryway to the trachea and lower respiratory tubes. The larynx has a skeleton of nine cartilages: the thyroid; the cricoid; the paired arytenoid, corniculate, and cuneiform cartilages; and the epiglottis.

9. The leaf-shaped epiglottis acts as a lid that prevents food or liquids from entering the lower respiratory channels. The larynx moves superiorly under this protective flap during swallowing.

10. The larynx contains the vocal folds (vocal cords). Exhaled air causes these folds to vibrate, producing the sounds of speech. The vocal folds extend anteriorly from the arytenoids to the thyroid cartilage, and muscles that move the arytenoids change the tension on the folds to vary the pitch of the voice. The pharynx, nasal cavity, lips, and tongue also aid in vocal articulation.

11. The main sensory and motor nerves of the larynx (and of speech) are the left and right recurrent laryngeal nerves, which are branches of the vagus nerves.

The Trachea (pp. 654–656)

12. The trachea, which extends from the larynx in the neck to the main bronchi in the thorax, is a tube reinforced by C-shaped cartilage rings, which keep it open. The trachealis muscle narrows the trachea, increasing the speed of airflow during coughing and sneezing.

13. The wall of the trachea contains several layers: a mucous membrane, a submucosa, a fibromusculocartilaginous layer, and an outer adventitia. The mucous membrane consists of an air-filtering pseudostratified ciliated columnar epithelium and a lamina propria rich in elastic fibers.

The Bronchial Tree (pp. 656–659)

14. The right and left main (primary) bronchi supply the lungs. Within the lungs, they subdivide into lobar (secondary) bronchi to the lobes, segmental (tertiary) bronchi to the bronchopulmonary segments, and finally to bronchioles and terminal bronchioles.

15. As the respiratory tubes become smaller, the cartilage in their walls is reduced and finally lost (in bronchioles); the epithelium thins and loses its air-filtering function (in the terminal bronchioles); smooth muscle becomes increasingly important; and elastic fibers continue to surround all of the tubes.

16. The terminal bronchioles lead into the respiratory zone, which consists of respiratory bronchioles, alveolar ducts, and alveolar sacs, all of which contain tiny chambers called alveoli. Gas exchange occurs in the alveoli, across the thin respiratory membrane. The respiratory membrane consists of type I alveolar cells, fused basement membranes, and capillary endothelial cells.

17. Alveoli also contain type II alveolar cells, which secrete surfactant, and freely wandering alveolar macrophages that remove dust particles that reach the alveoli.

The Lungs and Pleurae (pp. 659–664)

18. The lungs flank the mediastinum in the thoracic cavity. Each lung is suspended in a pleural cavity by its root (the vessels supplying it) and has a base, an apex, and medial and costal surfaces. The root structures enter the lung hilum.

19. The right lung has three lobes (superior, middle, and inferior lobes, as defined by the oblique and horizontal fissures); the left lung has two lobes (superior and inferior lobes, as defined by the oblique fissure). The lobes are divided into bronchopulmonary segments, which are supplied by segmental bronchi.

20. The lungs consist primarily of air tubes and alveoli, but they also contain a stroma of elastic connective tissue.

21. The pulmonary arteries carry deoxygenated blood and generally lie posterior to their corresponding bronchi within the lungs. The pulmonary veins, which carry oxygenated blood, tend to lie anterior to their corresponding bronchi. The lungs are innervated by sympathetic and parasympathetic fibers through the pulmonary plexus.

22. Pleurae are serous membranes. The parietal pleura lines the thoracic wall, diaphragm, and mediastinum; the visceral pleura covers the external surfaces of the lungs. The slitlike pleural cavity between the parietal and visceral pleura is filled with a serous fluid that holds the lungs to the thorax wall and reduces friction during breathing movements. Inferiorly, the pleural cavity extends below the inferior border of the lungs.

PAL Human Cadaver/Respiratory System

PAL Histology/Respiratory System

VENTILATION (pp. 664–667)

The Mechanism of Ventilation (pp. 664–666)

23. Ventilation consists of inspiration and expiration. Inspiration occurs when the diaphragm and intercostal muscles contract, increasing the volume of the thorax. As the intrathoracic pressure drops, pressure in the lungs also drops and air rushes into the lungs.

24. Expiration is largely passive, occurring as the inspiratory muscles relax. The volume of the thorax decreases and the lungs recoil elastically, raising pressure and expelling air from the lungs.

25. Surface tension of the alveolar fluid threatens to collapse the alveoli after each breath. This tendency is resisted by surfactant secreted by type II alveolar cells.

Neural Control of Ventilation (pp. 666–667)

26. The main respiratory control center is in the reticular formation of the medulla oblongata, although this center is influenced by input from the pons, limbic system, hypothalamus, and cerebral cortex.

27. Chemoreceptors monitor concentrations of respiratory gases and acid in the blood. The central chemoreceptors are neurons in the medulla, and the peripheral chemoreceptors are the carotid and aortic bodies.

DISORDERS OF THE RESPIRATORY SYSTEM (pp. 667–670)

28. Bronchial asthma is a chronic, allergic inflammatory condition.

29. Cystic fibrosis is an inherited disease in which the functions of the body's exocrine glands are disrupted. A viscous mucus accumulates in the bronchi, facilitating bacterial growth and leading to early death due to respiratory infection.

30. Chronic bronchitis, a COPD, is characterized by an excessive production of mucus in the lower respiratory passages, which impairs ventilation and leads to infection. Patients may become cyanotic as a result of low oxygen levels in the blood.

31. Emphysema, also a COPD, is characterized by permanent enlargement and destruction of alveoli. The lungs lose elasticity, and expiration becomes an active, exhaustive process.

32. Most lung cancers are caused by smoking. They are extremely aggressive, metastasize rapidly, and are most often fatal.

THE RESPIRATORY SYSTEM THROUGHOUT LIFE (pp. 670–671)

33. The nasal cavity develops from the olfactory placodes; the pharynx forms as part of the foregut; and the lower respiratory tubes grow from an outpocketing of the embryonic pharynx (laryngotracheal bud). The epithelium lining the pharynx and the lower respiratory tubes derives from endoderm. Mesoderm forms all other parts of these tubes and the lung stroma as well.

34. The respiratory system completes its development very late in the prenatal period: No alveoli or surfactant appear until month 6, and only one-sixth of the final number of alveoli are present at birth. Respiratory immaturity, especially a lack of surfactant, is a serious complication of premature birth.

35. With age, the nose dries, the thorax becomes more rigid, the lungs become less elastic, and ventilation capacity declines. Respiratory infections become more common in old age.

REVIEW QUESTIONS

Multiple Choice/Matching Questions

(For answers, see Appendix B.)

1. When the inspiratory muscles contract, (a) only the lateral dimension of the thoracic cavity increases, (b) only the anteroposterior dimension of the thoracic cavity increases, (c) the volume of the thoracic cavity decreases, (d) both the lateral and the anteroposterior dimensions of the thoracic cavity increase, (e) the diaphragm bulges superiorly.

2. The part of the respiratory mucosa that warms the inhaled air is the (a) pseudostratified epithelium, (b) vessels in the lamina propria, (c) type I alveolar cells, (d) cartilage and bone.

3. Which of the following statements about the vocal cords is *false*? (a) They are the same as the vocal folds. (b) They attach to the arytenoid cartilages. (c) Exhaled air flowing through the glottis vibrates them to produce sound. (d) They are also called the vestibular folds.

4. In both lungs, the surface that is the largest is the (a) costal, (b) mediastinal, (c) inferior (base), (d) superior (apex).

5. Match the proper type of lining epithelium from the key with each of the following respiratory structures.

Key: (a) stratified squamous (c) simple squamous
(b) pseudostratified columnar (d) stratified columnar

____ **(1)** nasal cavity

____ **(2)** nasopharynx

____ **(3)** laryngopharynx

____ **(4)** trachea and bronchi

____ **(5)** alveoli (type I cells)

6. Match the air tube in column B with the lung region in column A supplied by that air tube.

Column A	Column B
____ **(1)** bronchopulmonary segment	**(a)** main bronchus
____ **(2)** lobule	**(b)** lobar bronchus
____ **(3)** alveolar ducts and sacs	**(c)** segmental bronchus
____ **(4)** whole lung	**(d)** large bronchiole
____ **(5)** lobe	**(e)** respiratory bronchiole

7. The respiratory membrane consists of (a) type I alveolar cell, basement membranes, endothelial cell; (b) air, connective tissue, lung; (c) type II alveolar cell, macrophage, type I alveolar cell; (d) pseudostratified epithelium, lamina propria, capillaries.

8. The trachealis and the smooth muscle around the bronchi develop from which of the following embryonic layers? (a) ectoderm, (b) endoderm, (c) mesoderm, (d) neural crest.

9. A serous cell of a gland secretes (a) the slippery serous fluid in the body cavities, (b) mucus, (c) a watery lubricating fluid, (d) tissue fluid.

10. The function of type I alveolar cells is to (a) produce surfactant, (b) propel sheets of mucus, (c) remove dust particles through phagocytosis, (d) allow rapid diffusion of respiratory gases.

Short Answer Essay Questions

11. Trace the route of exhaled air from an alveolus to the external nares, naming all the structures through which the air passes. Then indicate which of these structures are in the respiratory zone and which are in the conducting zone.

12. The cilia lining the upper respiratory passages (superior to the larynx) beat inferiorly, whereas the cilia lining the lower respiratory passages (larynx and inferior) beat superiorly. What is the functional reason for this difference?

13. What is the function of the abundant elastin fibers that occur in the stroma of the lung and around all respiratory tubes from the trachea through the respiratory tree?

14. (a) Which tonsils lie in the fauces of the oropharynx? (b) Does the lingual tonsil lie near the superior part of the epiglottis? Explain.

15. (a) It is easy to confuse the hilum of the lung with the root of the lung. Define both structures, and contrast hilum and root. (b) Define the parietal pleura and the visceral pleura.

16. The three terms choanae, conchae, and carina are easily confused. Define each of them, clarifying the differences.
17. Briefly explain the anatomical reason why most men have deeper voices than boys or women.
18. Sketch a picture of the right and left lungs in anterior view, showing all the fissures and lung lobes, as well as the cardiac notch.
19. Explain the functions of the pseudostratified epithelium of the respiratory mucosa.
20. What is the function of the alveolar macrophages in the lung alveoli?

Critical Reasoning & Clinical Application Questions

1. (a) Two girls in a high school cafeteria were giggling over lunch, and milk accidentally sprayed out of the nostrils of one of them. Explain in anatomical terms why swallowed fluids can sometimes come out the nose. (b) A boy in the same cafeteria then stood on his head to show he could drink milk upside down without any of it entering his nasal cavity or nose. What prevented the milk from flowing downward into his nose?
2. A surgeon had to remove three adjacent bronchopulmonary segments from the left lung of a patient with tuberculosis. Even though almost half of the lung was removed, there was no severe bleeding, and relatively few blood vessels had to be cauterized (closed off). Why was this surgery so easy to perform?
3. While getting his 1-year-old daughter (who puts almost everything in her mouth) ready for bed, Mr. Gregoire could not find one of the small barrettes he had just removed from her hair. Two days later, she developed a cough and became feverish. What probably had happened to the barrette, and where (anatomically) would you expect to find it?
4. A taxi driver was carried into an emergency room after being knifed once in the left side of the thorax. The diagnosis was pneumothorax and a collapsed lung. (a) Explain exactly why the lung collapsed. (b) Explain why only one lung (not both) collapsed.
5. A man was driving to work when his car was hit broadside by a car that ran a red light. When freed from the wreckage, the man was deeply cyanotic, and his breathing had stopped. His heart was still beating, but his pulse was fast and thready. His head was cocked at a peculiar angle, and the emergency personnel said he seemed to have a fracture at the level of vertebra C_2. (a) How might these findings account for the cessation of breathing? (b) Define cyanosis (see Chapter 5), and indicate why the man was cyanotic.
6. After a 4-year-old child had surgery that successfully closed off a patent ductus arteriosus (see p. 622), the child's voice was hoarse, and an examination revealed that many muscles in the left half of the larynx were paralyzed. What had gone wrong during surgery to cause the paralysis?

The Digestive System 23

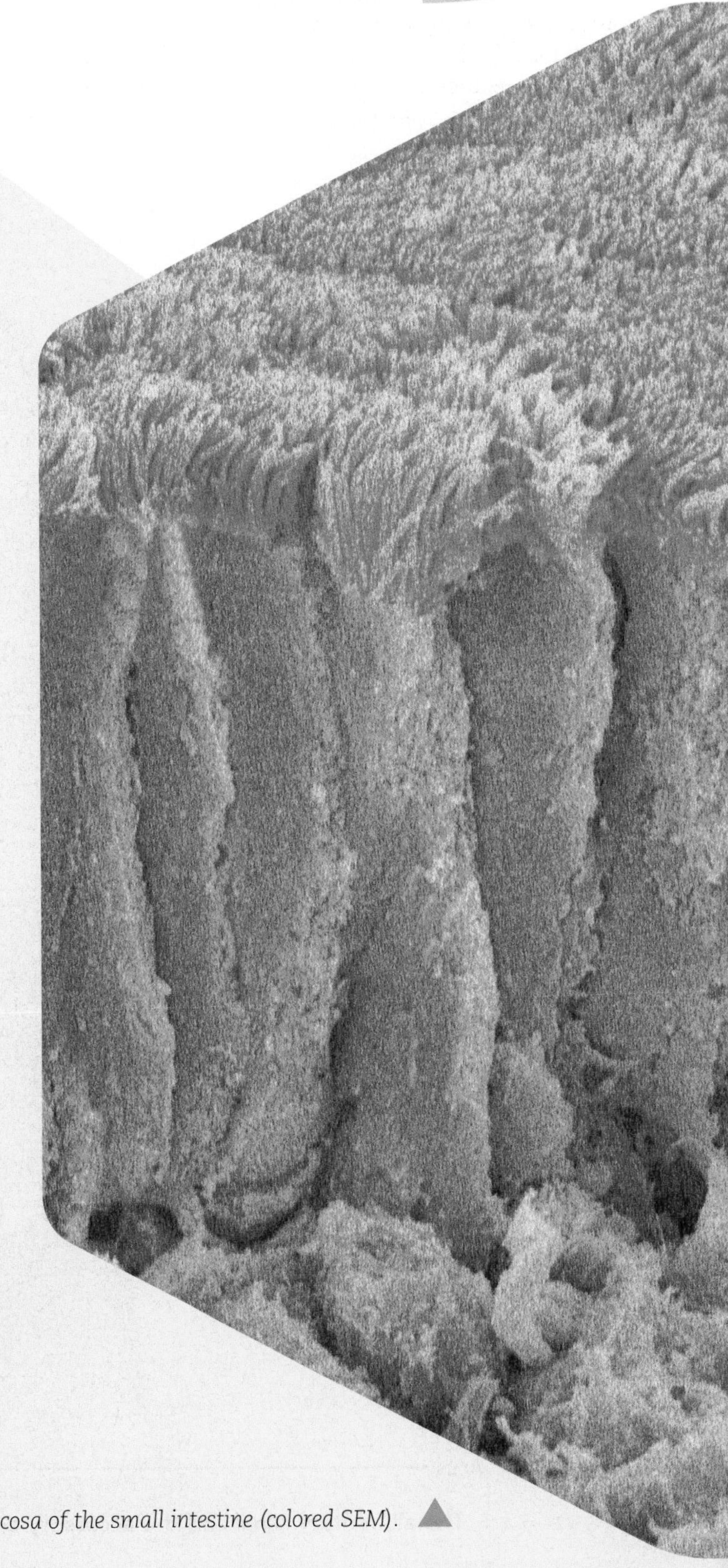

Microvilli (red) on villi (orange) from the mucosa of the small intestine (colored SEM). ▲

"You are what you eat": simplistic, but true! The nutrients the body uses to build its structures and to fuel its functions come from the food we ingest. Nutrition influences health status, mental and emotional state, and our overall sense of well-being. Proteins, complex carbohydrates, unsaturated fats, vitamins, minerals, and water are essential components of a healthy diet. The body system that converts the food we eat into units our body can absorb and use is the digestive system. This chapter explores structures of the body that take in food, break it into nutrient molecules, absorb these molecules into the circulatory system, and then eliminate the indigestible wastes.

OVERVIEW

learning outcomes

- Describe the overall function of the digestive system, and differentiate the alimentary canal from the accessory digestive organs.
- Draw the major subdivisions of the anterior abdominal wall, and identify the organs located in each region.
- Explain the location and function of the peritoneum and peritoneal cavity. Define mesentery.
- Differentiate between intraperitoneal and secondarily retroperitoneal digestive organs. Name the mesenteries associated with the intraperitoneal digestive organs.
- List the major processes that occur during digestion.

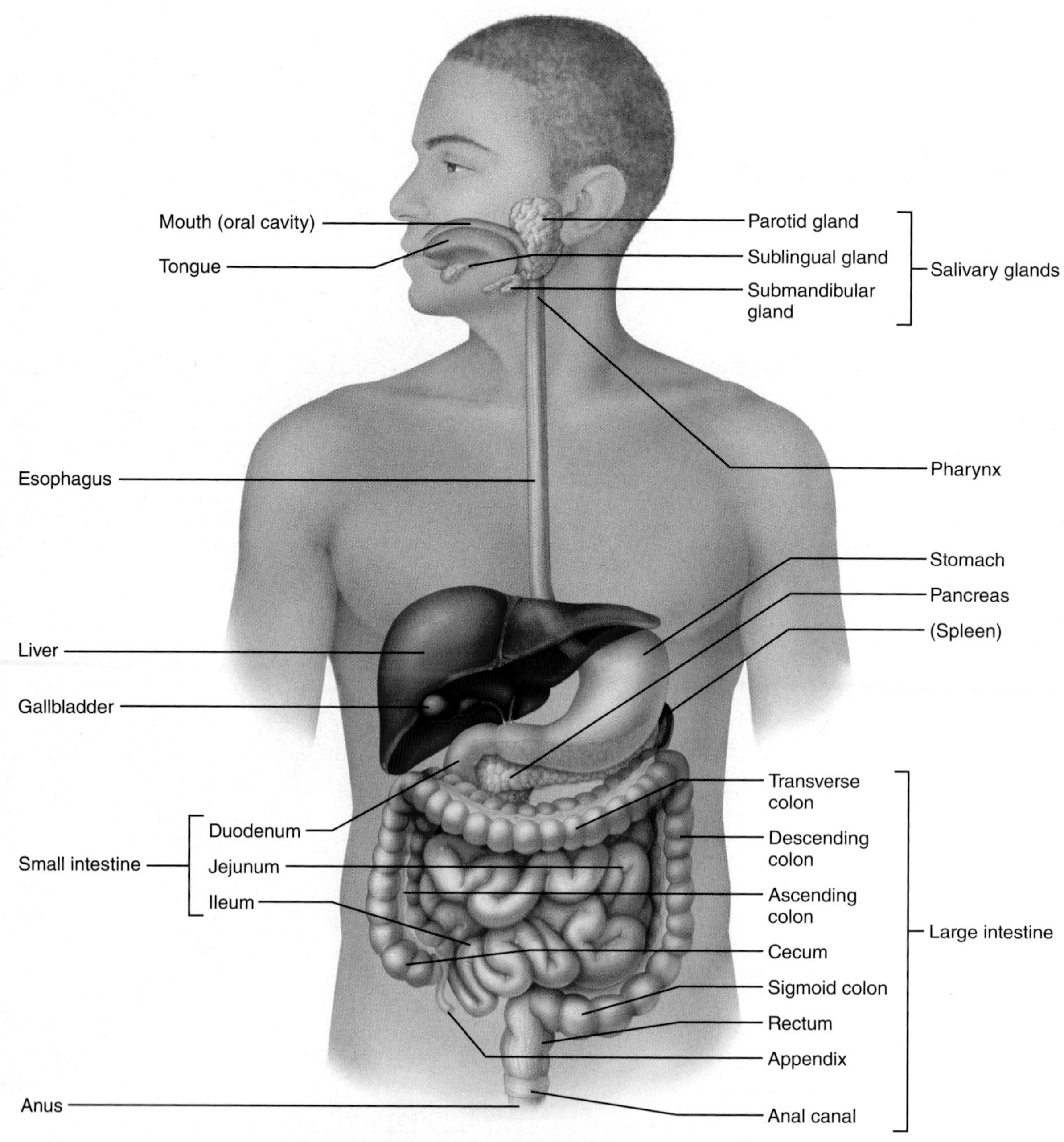

Figure 23.1 The alimentary canal and the accessory digestive organs.

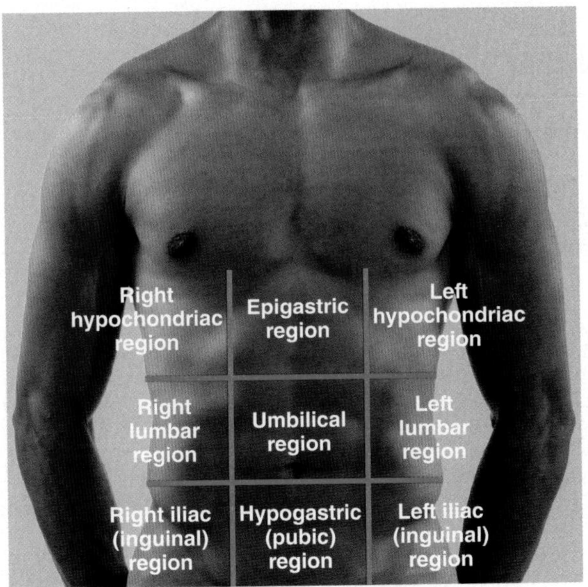

(a) Nine regions delineated by four planes

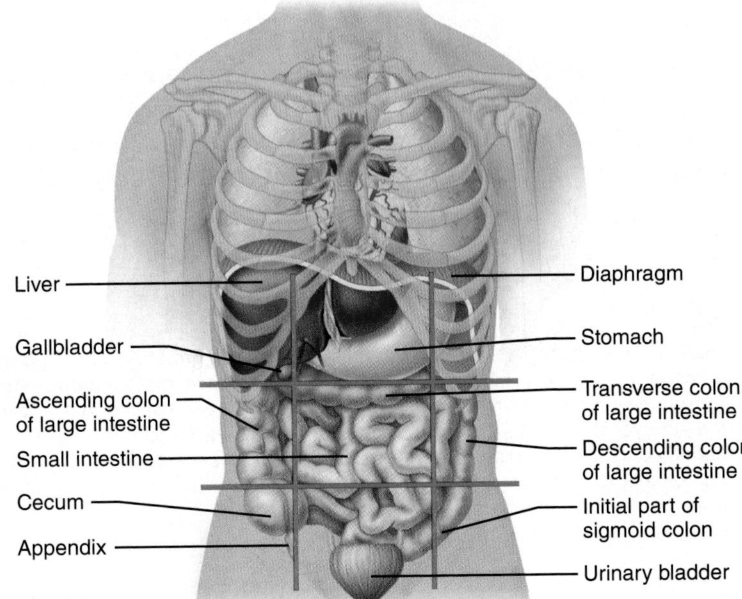

(b) Anterior view of the nine regions showing the superficial organs

Figure 23.2 **Divisions of the anterior abdominal wall.** The parasagittal planes are the midclavicular lines; the transverse planes are the subcostal plane superiorly and the transtubercle plane inferiorly.

The various organs of the digestive system (Figure 23.1) can be divided into two main groups: the *alimentary canal* (*aliment* = nourishment) and the *accessory digestive organs.*

The **alimentary canal,** also called the *gastrointestinal (GI) tract,* is the muscular digestive tube that winds through the body, extending from the mouth to the anus. The organs of the alimentary canal are the *mouth, pharynx, esophagus, stomach, small intestine* (small bowel), and *large intestine* (large bowel), the last of which leads to the terminal opening, or *anus.* In a cadaver, the alimentary canal is about 9 m (30 feet) long, but in a living person it is considerably shorter because of its muscle tone. Food material in the alimentary canal is technically considered to be *outside* the body because the canal is open to the external environment at both ends.

The **accessory digestive organs** are the *teeth* and *tongue,* plus the *gallbladder* and large digestive glands—the *salivary glands, liver,* and *pancreas*—that lie external to and are connected to the alimentary canal by ducts. The accessory digestive glands secrete saliva, bile, and digestive enzymes, all of which contribute to the breakdown of foodstuffs.

Abdominal Regions

Most digestive organs are contained in the abdominopelvic cavity, the largest division of the ventral body cavity. To help locate the positions of these abdominopelvic organs, clinicians typically divide the anterior abdominal wall into a number of regions.

In one scheme, four planes forming a pattern similar to a tic-tac-toe grid divide the abdominal wall into nine regions (Figure 23.2). The two parasagittal planes are the *midclavicular lines,* which extend inferiorly from the midpoint of each clavicle. The superior transverse plane is in the *subcostal* ("below the ribs") *plane* and connects the inferior points of the costal margins, whereas the inferior transverse plane is in the *transtubercular plane* and connects the tubercles (widest points) of the iliac crests. All of these bony landmarks can be felt on the body's surface.

- The superior three regions are the **right** and **left hypochondriac regions** (hī″po-kon′dre-ak; "deep to the cartilage") and the central **epigastric region** (ep′ĭ-gas′-trik; "superior to the belly").
- The middle three regions are the **right** and **left lumbar regions** (or *lateral regions*) and the central **umbilical region.**
- The inferior three regions are the **right** and **left iliac regions,** or **inguinal regions,** and the central **hypogastric region** ("inferior to the belly") (also called the **pubic region**).

Because many abdominal organs move, their positions within the abdominal grid are only approximate.

In a simpler scheme, a vertical and a horizontal line intersecting at the navel define four regions: the **right** and **left upper** and **lower quadrants** (see Figure 1.8, p. 13).

CLINICAL APPLICATION

Bowel Sounds To acquire the valuable information conveyed by **bowel sounds,** clinicians place a stethoscope in each of the four quadrants of the anterior abdominal wall. Normal bowel sounds, which result from the movement of gas and intestinal contents by peristalsis, are high-pitched gurgles that occur every 5–15 seconds. Less frequent bowel sounds can indicate a halt in intestinal activity, whereas loud sounds may indicate increased activity associated with inflammation, diarrhea, or other bowel disorders.

FOCUS Peritoneum and the Digestive System Organs

Figure 23.3

When fully developed, the digestive system organs are either intraperitoneal or secondarily retroperitoneal. These terms refer to organ location relative to the peritoneal cavity. Mesenteries connect the intraperitoneal organs to the body wall.

The Peritoneum and Peritoneal Cavity

Embryonic development

Folding of the embryo during week 4 of development forms the primitive gut, the inner tube. Surrounding the primitive gut is the embryonic coelom. In the abdomen the coelom forms the peritoneal cavity, a serous cavity lined by a serous membrane called the peritoneum.

Digestive organs develop from the primitive gut. They are covered externally by **visceral peritoneum** and surrounded by the **peritoneal cavity.** The outer lining of the peritoneal cavity is the **parietal peritoneum.**

In this early embryo, all digestive organs are **intraperitoneal,** surrounded by the peritoneal cavity. These organs are anchored to the dorsal and ventral body wall by two layers of peritoneum fused to form a **mesentery.**

Mesenteries provide a passageway for the blood vessels, lymphatic vessels, and nerves supplying the digestive organs. Mesenteries also anchor the digestive organs within the peritoneal cavity, store fat, and house lymphoid tissues that respond to ingested pathogens.

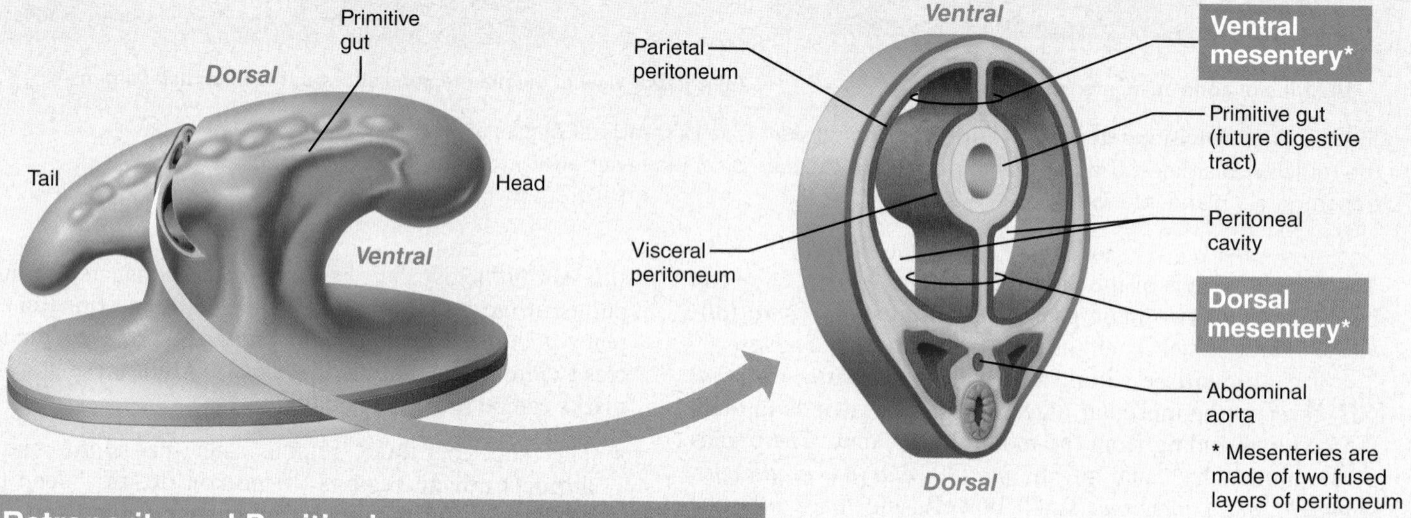

Retroperitoneal Positioning

Transition from fetal to adult structural arrangement

As the digestive tract elongates and rotates, some organs get pushed against the dorsal body wall. The **visceral peritoneum** of these organs fuses with the **parietal peritoneum** along the dorsal body wall. These organs lose their mesenteries and are located *behind* the peritoneal cavity, in a **secondarily retroperitoneal** location. Other abdominal organs, such as kidneys, ureters, and adrenal glands, develop outside the peritoneal cavity and so are in a retroperitoneal position from their beginnings.

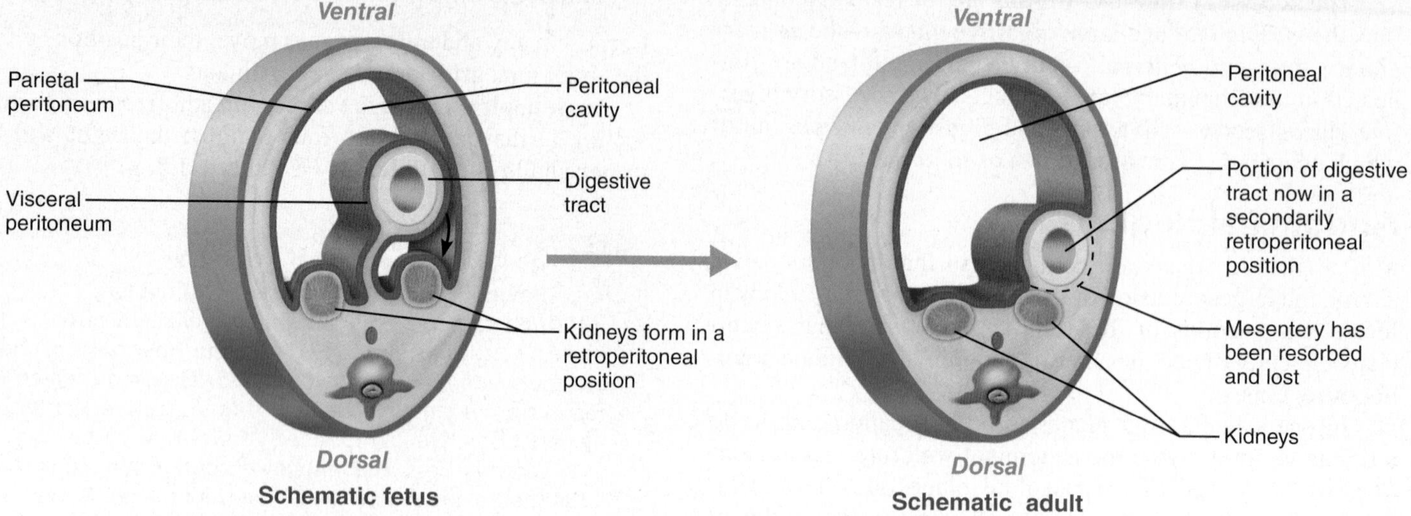

Mesenteries and Organ Position

Adult structures (sectional illustrations)

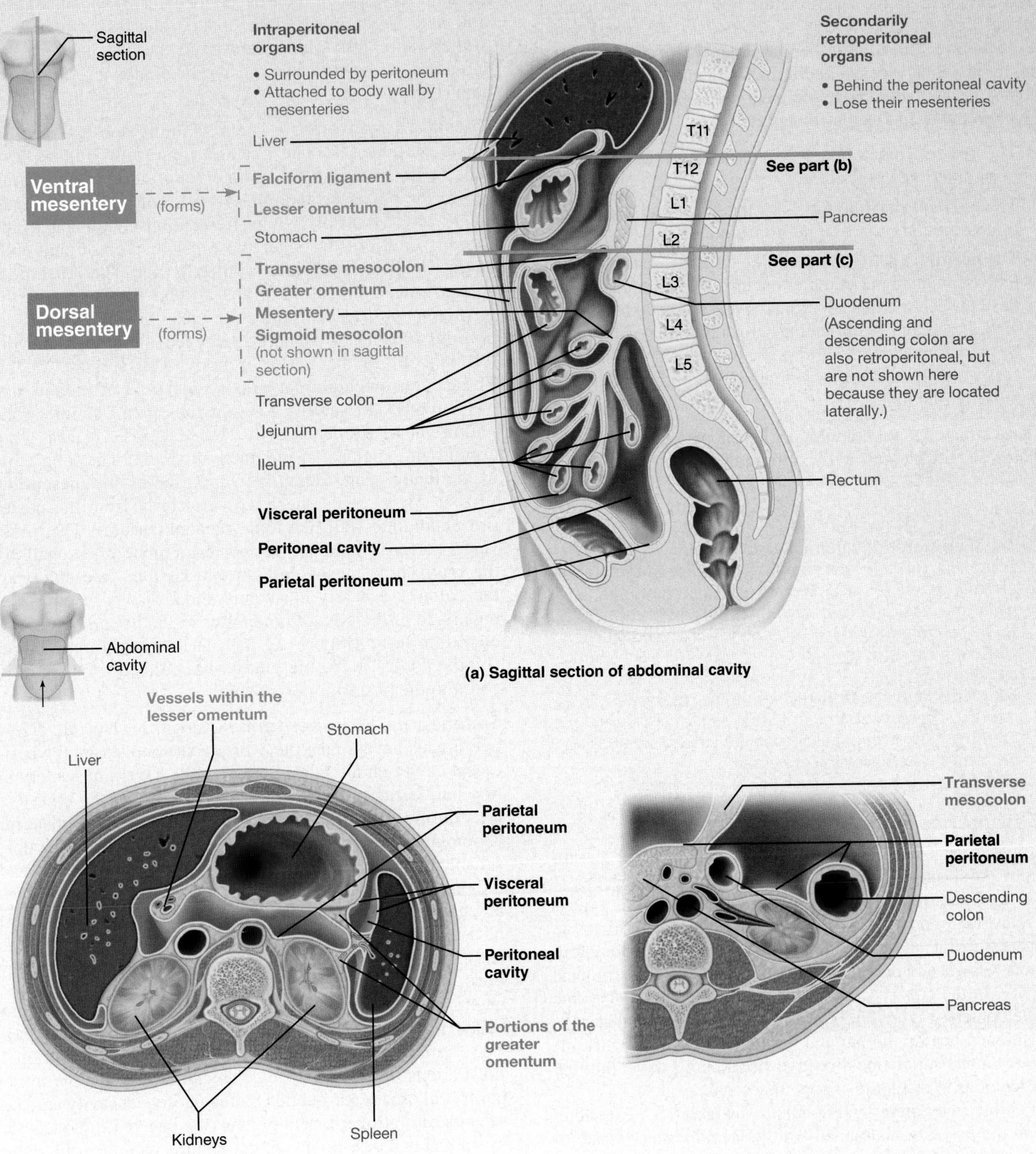

(a) Sagittal section of abdominal cavity

(b) Transverse section through abdomen at vertebra T_{12}

(c) Transverse section through abdomen at vertebra L_2 illustrating retroperitoneal position

The Peritoneal Cavity and Peritoneum

All divisions of the ventral body cavity contain slippery *serous membranes* (Chapter 1, p. 12), the most extensive of which is the **peritoneum** (per″i-to-ne′um) located in the abdominopelvic cavity. *Focus on the Peritoneum* (Figure 23.3) illustrates the development of the digestive organs and their asociated peritoneal structures. The digestive organs in the abdominopelvic cavity all develop surrounded by *peritoneum* and the *peritoneal cavity.*

- The **visceral peritoneum** covers the external surfaces of most digestive organs.
- The **parietal peritoneum** lines the body wall and is continuous with the visceral peritoneum.
- The **peritoneal cavity** is a slitlike potential space between the visceral and parietal peritoneum. The peritoneal cavity lies between the digestive organs and the abdominal body wall. It contains a lubricating serous fluid that is produced by the peritoneum and allows the digestive organs to glide easily along one another and along the body wall as they move during digestion.

These structures are homologous to the parietal and visceral pleura and the pleural cavity surrounding the lungs.

> **CLINICAL APPLICATION**
>
> **Peritonitis** Inflammation and infection of the peritoneum is called **peritonitis.** It can arise from a piercing wound to the abdomen, from a perforating ulcer that leaks stomach juices into the peritoneal cavity, or from poor sterile technique during abdominal surgery. Most commonly, however, it results from a burst appendix that leaks feces into the peritoneal cavity. Peritonitis that is widespread within the peritoneal cavity is a dangerous condition that can lead to death. Treatment involves intravenous antibiotics. Surgery may be needed to remove infected tissue.

Mesenteries

A **mesentery** (mes′en-ter″e) is a double layer of peritoneum—a sheet of two serous membranes fused together—that extends from the body wall to the digestive organs (Figure 23.3 and Figure 23.4). Mesenteries hold organs in place, store fat, and, most important, provide a route for circulatory vessels and nerves to reach the organs in the peritoneal cavity. In the cross-sectional illustration showing embryonic development (Figure 23.3), notice that the abdominal aorta is positioned behind the parietal peritoneum. The branches off the aorta that supply the digestive viscera must travel through mesenteries to reach the organs they serve.

Most mesenteries are *dorsal mesenteries,* extending from the posterior abdominal wall to the alimentary canal. In the superior abdomen, a *ventral mesentery* extends from the stomach and liver to the anterior abdominal wall. Note, some mesenteries are named "ligaments," although these peritoneal sheets have no resemblance and no relationship to the fibrous ligaments that interconnect bones.

Ventral Mesenteries The two ventral mesenteries are the falciform ligament and the lesser omentum. The **falciform ligament** (fal′sĭ-form; "sickle-shaped") binds the anterior aspect of the liver to the anterior abdominal wall and diaphragm (Figure 23.4a). The **lesser omentum** (o-men′tum; "fatty skin") runs from the liver to the lesser curvature of the stomach and the beginning of the duodenum (Figure 23.4b).

Dorsal Mesenteries The remaining mesenteries are dorsal mesenteries. The **greater omentum** (Figure 23.3 and 23.4a) connects the greater curvature of the stomach to the posterior abdominal wall, but in a very roundabout way: Anteriorly, it is tremendously elongated and extends inferiorly to cover the transverse colon and coils of the small intestine like a butterfly net. The left border of the greater omentum wraps around the spleen as the *gastrosplenic ligament* and continues dorsally as the *splenorenal ligament* (extending between the spleen and the left kidney) to the posterior body wall (Figure 23.3b). The greater omentum contains a great deal of fat and also has a remarkable ability to limit the spread of infections within the peritoneal cavity; for example, it can wrap around and enclose an inflamed appendix. The long coils of the jejunum and ileum are supported by the **mesentery** (Figure 23.4c). This sheet fans inferiorly from the posterior abdominal wall like long, pleated curtains. The transverse colon is held to the posterior abdominal wall by the **transverse mesocolon** (mez″o-ko′lon; "mesentery of the colon"), a nearly horizontal sheet that is fused to the underside of the greater omentum, so that it can be viewed only inferiorly (Figure 23.4c). The **sigmoid mesocolon** (Figure 23.4c) is the mesentery that connects the sigmoid colon to the posterior pelvic wall.

Secondarily Retroperitoneal Organs Not all digestive organs have a mesentery or are surrounded by the peritoneal cavity on all sides. Some organs have a mesentery at first but, because of the complex rotations of the digestive tract during development, they end up against the posterior abdominal wall (Figure 23.3). These organs fuse to the dorsal abdominal wall, in the process losing their mesentery and lodging behind the peritoneum (Figure 23.4d). Such organs are called **secondarily retroperitoneal** (*retro* = behind) because they are initially formed within the peritoneum but are located behind the peritoneum once they are fully developed. By contrast, the digestive organs that keep their mesentery and remain surrounded by the peritoneal cavity are called **intraperitoneal** or **peritoneal** organs (Table 23.1). The stomach is an example of such an organ. In sectional illustrations (Figure 23.3a–c), you can clearly see the intraperitoneal organs surrounded by the peritoneal cavity and the secondarily retroperitoneal organs (the pancreas, duodenum, ascending and descending colon, and the rectum) behind the parietal peritoneum along the posterior abdominal wall.

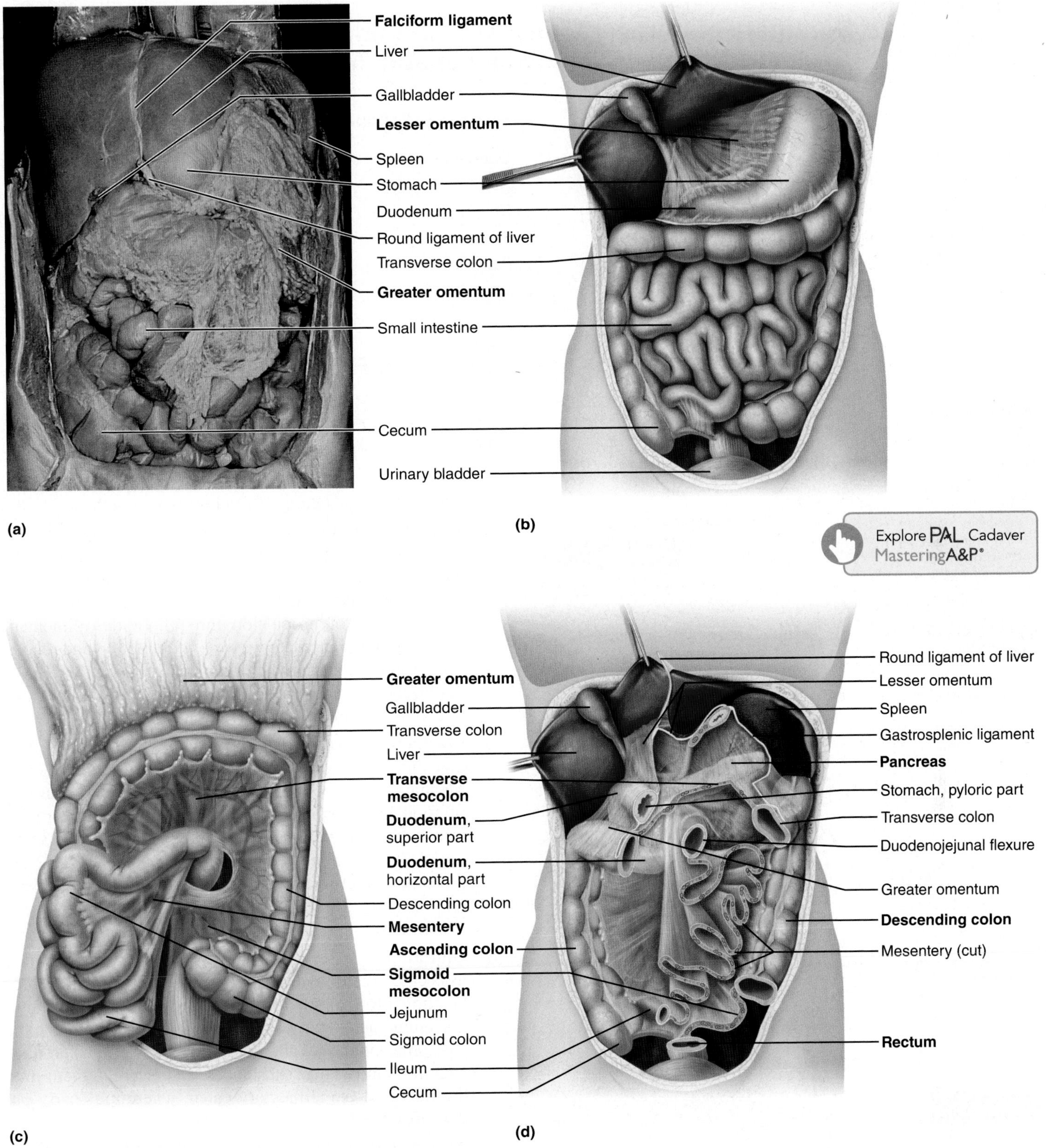

Falciform ligament
Liver
Gallbladder
Lesser omentum
Spleen
Stomach
Duodenum
Round ligament of liver
Transverse colon
Greater omentum
Small intestine
Cecum
Urinary bladder

(a)

(b)

Explore PAL Cadaver
MasteringA&P®

Greater omentum
Gallbladder
Transverse colon
Liver
Transverse mesocolon
Duodenum, superior part
Duodenum, horizontal part
Descending colon
Mesentery
Ascending colon
Sigmoid mesocolon
Jejunum
Sigmoid colon
Ileum
Cecum

Round ligament of liver
Lesser omentum
Spleen
Gastrosplenic ligament
Pancreas
Stomach, pyloric part
Transverse colon
Duodenojejunal flexure
Greater omentum
Descending colon
Mesentery (cut)
Rectum

(c)

(d)

Figure 23.4 **The mesenteries.** (**a**) The greater omentum, a dorsal mesentery, is in its normal position covering the abdominal viscera. (**b**) The lesser omentum attaches the lesser curvature of the stomach to the liver (the liver is lifted out of the way). The greater omentum has been removed, exposing the small and large intestines. (**c**) The greater omentum and transverse colon have been reflected superiorly to reveal the mesentery to the small intestine, the transverse mesocolon, and the sigmoid mesocolon. (**d**) Deep view of the peritoneal cavity illustrating the retroperitoneal organs. Stomach, jejunum, ileum, and transverse and sigmoid colon have been removed.

Table 23.1 Summary of Intraperitoneal and Secondarily Retroperitoneal Digestive Organs in the Abdomen and Pelvis

Intraperitoneal Organs (and their Mesenteries)	Secondarily Retroperitoneal Organs (Lack Mesenteries)
Liver (falciform ligament and lesser omentum)	Duodenum (almost all of it)
Stomach (greater and lesser omentum)	Ascending colon
Ileum and jejunum (mesentery proper)	Descending colon
Transverse colon (transverse mesocolon)	Rectum
Sigmoid colon (sigmoid mesocolon)	Pancreas

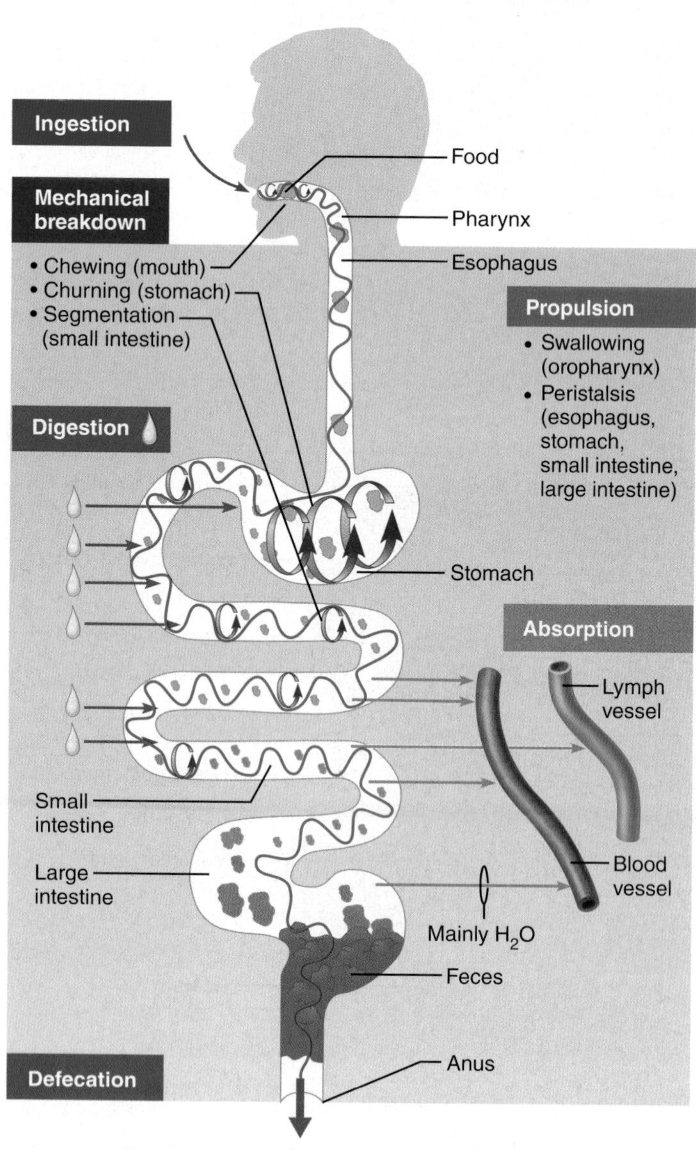

Figure 23.5 **Activities of the gastrointestinal tract.**

Digestive Processes

The organs of the digestive system perform the following six essential food-processing activities: ingestion, propulsion, mechanical breakdown, digestion, absorption, and defecation (Figure 23.5).

1. **Ingestion** is the taking of food into the mouth.
2. **Propulsion** is the movement of food through the alimentary canal. It includes swallowing, which is initiated voluntarily, and peristalsis, an involuntary process. **Peristalsis** (per″ĭ-stal′sis; "around contraction"), the major means of propulsion throughout the alimentary canal, involves alternate waves of contraction and relaxation of musculature in the organ walls (Figure 23.6a). Its net effect is to squeeze food from one organ to the next, but some mixing occurs as well.
3. **Mechanical breakdown** physically prepares food for digestion by enzymes by breaking it into smaller pieces. Mechanical processes include chewing, the churning of food in the stomach, and **segmentation,** the rhythmic local constrictions of the intestine (Figure 23.6b). Segmentation mixes food with digestive juices and increases the efficiency of nutrient absorption by repeatedly moving different parts of the food mass over the intestinal wall.
4. **Digestion** is a series of steps in which complex food molecules (carbohydrates, proteins, and lipids) are broken down to their chemical building blocks (simple sugars, amino acids, fatty acids, and glycerol). Glands in the gastrointestinal tract and in the accessory organs produce enzymes and other substances and secrete them into the lumen of the alimentary canal, where they carry out digestion.
5. **Absorption** is the transport of digested end products from the lumen of the alimentary canal into the blood and lymphatic capillaries located in the wall of the canal.
6. **Defecation** is the elimination of indigestible substances from the body as feces.

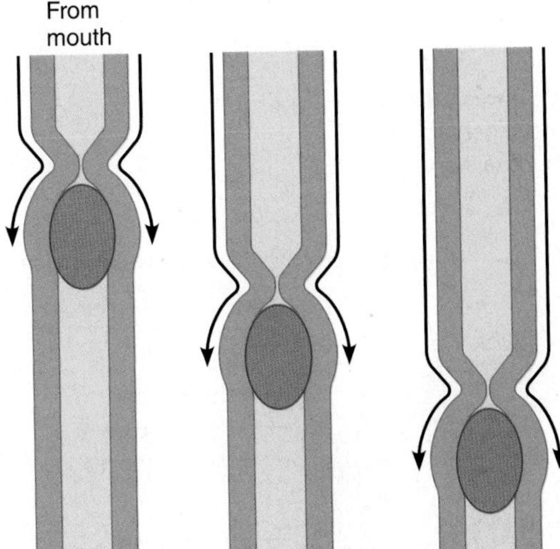

(a) Peristalsis: Adjacent segments of alimentary tract organs alternately contract and relax, moving food along the tract distally.

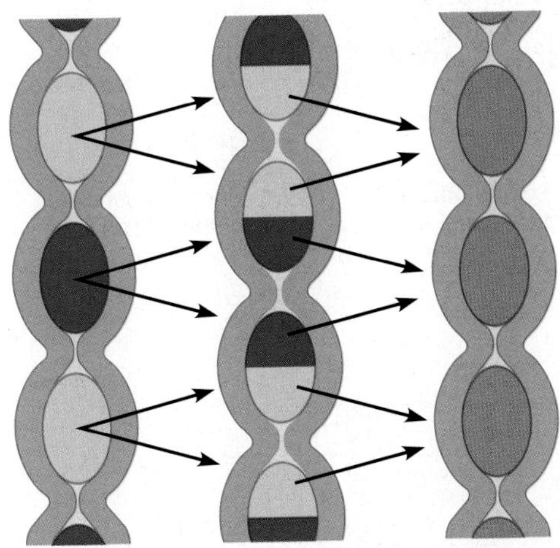

(b) Segmentation: Nonadjacent segments of alimentary tract organs alternately contract and relax, moving the food forward then backward. Food is mixed and slowly propelled.

Figure 23.6 **Peristalsis and segmentation.**

☑ check your understanding

- ☐ 1. Where in the alimentary canal does propulsion occur?
- ☐ 2. Differentiate the abdominal cavity from the peritoneal cavity. Which digestive system organs are located in the abdominal cavity but are not intraperitoneal?
- ☐ 3. Identify all the mesenteries that connect to each organ listed: (a) liver, (b) stomach, (c) sigmoid colon. For each, state whether it is a dorsal mesentery or a ventral mesentery.
- ☐ 4. Injury to the spleen or liver can cause extensive internal bleeding. Where would blood collect from such an injury?

(For answers, see Appendix B.)

ANATOMY OF THE ALIMENTARY CANAL

Histology

learning outcomes

- ▶ Describe the four layers of the wall of the alimentary canal.
- ▶ Describe the structure and appearance of smooth muscle fibers, and describe how smooth muscle cells are joined together to form a sheetlike tissue.
- ▶ Discuss the different stimuli that can initiate the contraction of smooth muscle tissue.
- ▶ Describe the innervation of the muscle and glands in the alimentary canal.

The walls of the alimentary canal, from the esophagus to the anal canal, have the same four tissue layers (Figure 23.7). In fact, most such layers occur in the hollow organs of the respiratory, urinary, and reproductive systems as well. From the lumen outward, these layers are the *mucosa, submucosa, muscularis externa,* and *serosa.*

The Mucosa

The innermost layer is the **mucosa,** or *mucous membrane.* More complex than other mucous membranes in the body, the typical digestive mucosa contains three sublayers: (1) a lining epithelium, (2) a lamina propria, and (3) a muscularis mucosae.

The lining **epithelium** abuts the lumen of the alimentary canal and performs many functions related to digestion, such as absorbing nutrients and secreting mucus. This epithelium is continuous with the ducts and secretory cells of the various digestive glands, most of which lie fully within the wall and are called *intrinsic glands.*

The **lamina propria** is a loose areolar or reticular connective tissue whose capillaries nourish the lining epithelium and absorb digested nutrients. The lamina propria contains most of the mucosa associated lymphoid tissue (MALT), which defends against invasion by bacteria and other microorganisms in the alimentary canal.

External to the lamina propria is the **muscularis mucosae,** a thin layer of smooth muscle that produces local movements of the mucosa. For example, the twitching of this muscle layer dislodges sharp food particles that become embedded in the mucosa.

The Submucosa

Just external to the mucosa is the **submucosa,** a layer of connective tissue containing major blood and lymphatic vessels and nerve fibers. Its rich vascular network sends branches to all other layers of the wall. Its connective tissue is a type intermediate between loose areolar and dense irregular—a "moderately dense" connective tissue. The many elastic fibers in the submucosa enable the alimentary canal to return to its shape after food material passes through it.

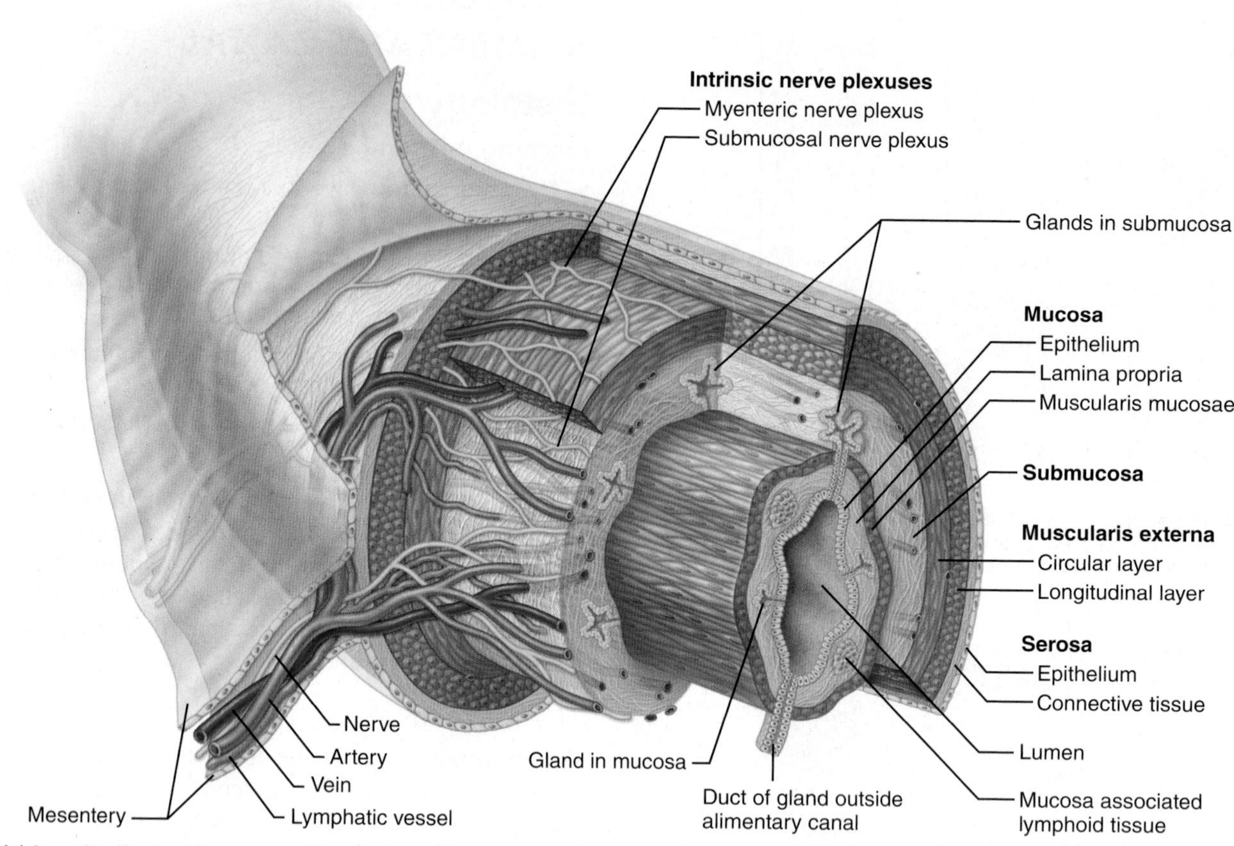

(a) Longitudinal and cross-sectional views through the small intestine

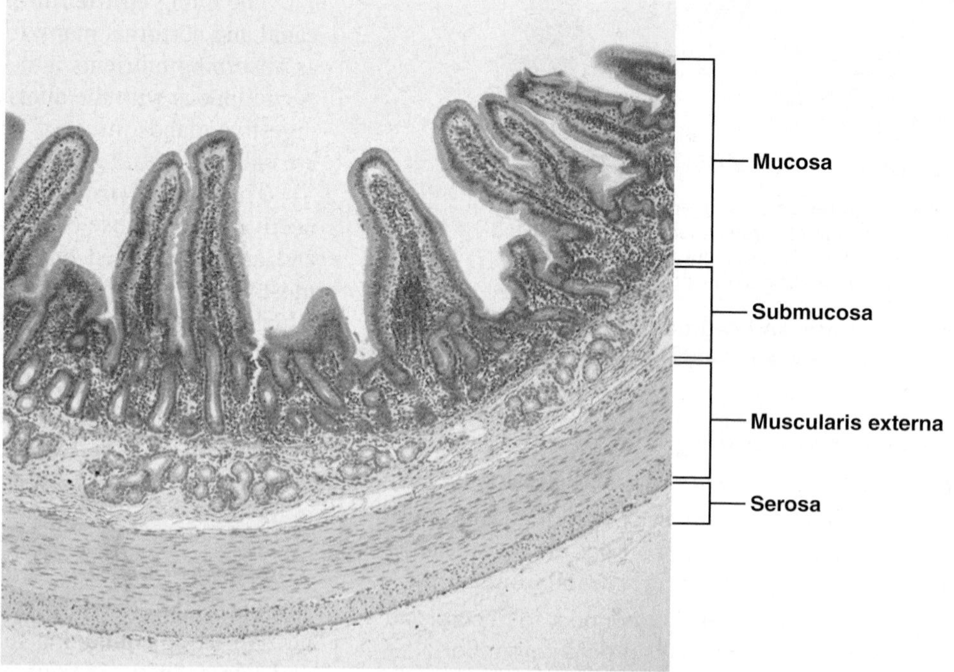

(b) Light micrograph cross section through the small intestine (85×)

Figure 23.7 **Histological layers of the alimentary canal.** The four basic layers in the wall are the mucosa, submucosa, muscularis externa, and serosa.

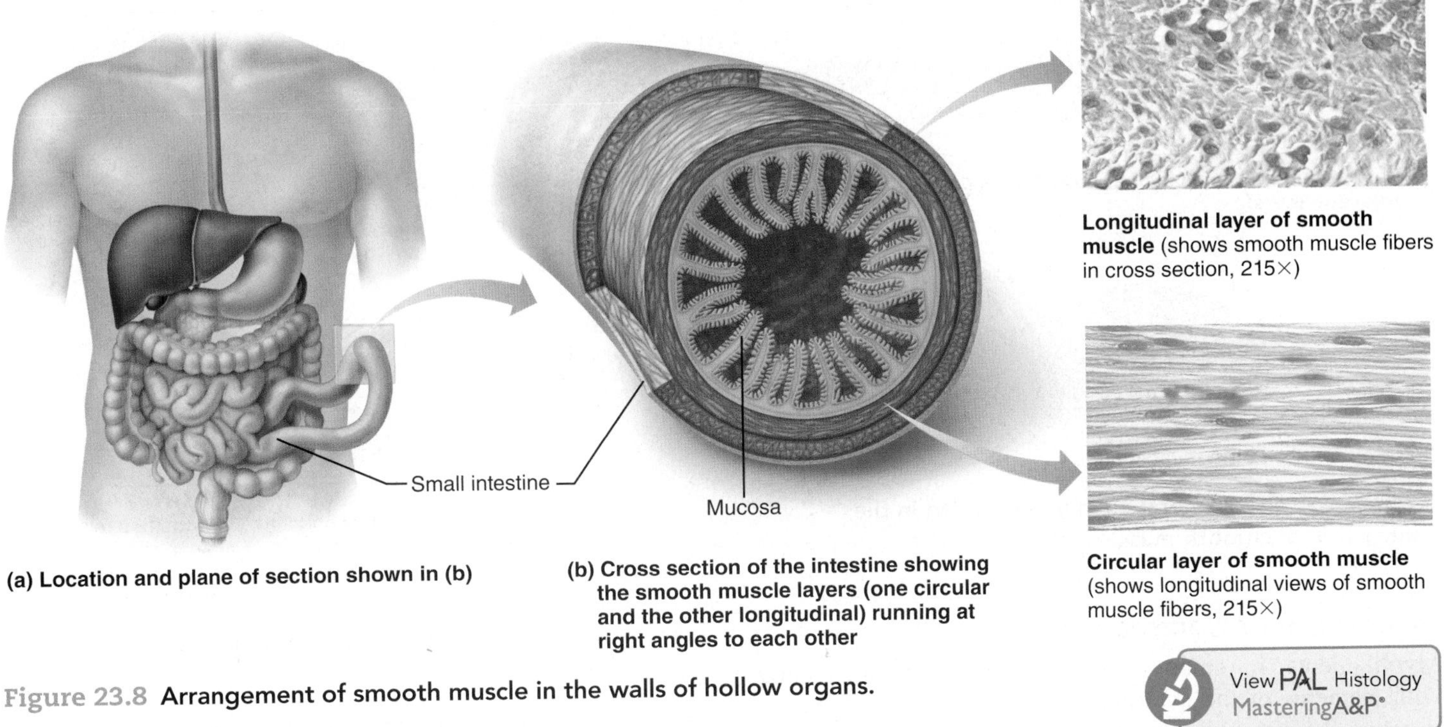

Figure 23.8 Arrangement of smooth muscle in the walls of hollow organs.

The Muscularis Externa

External to the submucosa is the **muscularis externa,** also simply called the *muscular layer*. Throughout most of the alimentary canal, this tunic consists of two layers of smooth muscle, an inner *circular layer* whose fibers orient around the circumference of the canal, and an outer *longitudinal layer* whose fibers orient along the length of the canal. Functionally, the circular layer squeezes the gut tube, and the longitudinal layer shortens it. Together, these layers are responsible for peristalsis and segmentation. In some places, the circular layer thickens to form sphincters that act as valves to prevent the backflow of food from one organ to the next.

The histological structure of smooth muscle, the mechanism of its contraction, and its innervation will be discussed in detail shortly.

The Serosa

The **serosa,** which is the visceral peritoneum, is the outermost layer of the intraperitoneal organs of the alimentary canal. Like all serous membranes (see p. 89), it is formed of a simple squamous epithelium (mesothelium) underlain by a thin layer of areolar connective tissue.

Parts of the alimentary canal that are not associated with the peritoneal cavity lack a serosa and have an *adventitia,* an ordinary fibrous connective tissue, as their outer layer. For example, the esophagus in the thorax has an adventitia that binds it to surrounding structures. Secondarily retroperitoneal organs (discussed on p. 680) have both a serosa and an adventitia—a serosa on the anterior side facing the peritoneal cavity and an adventitia on the posterior side embedded in the posterior abdominal wall.

Smooth Muscle

Most smooth muscle of the body is found in the walls of visceral organs such as the urinary bladder, uterus, and intestines (Figure 23.8). More specifically, smooth muscle has six major locations: the iris of the eye, and in the walls of the circulatory vessels, respiratory tubes, digestive tubes, urinary organs, and reproductive organs.

Structure of Smooth Muscle

Smooth muscle cells are called fibers because of their elongated shape. Each fiber tapers at its ends and has one centrally located nucleus (Figure 23.8b). Typically, these fibers have a diameter of 3–8 μm and a length ranging from 15 to 200 μm. They are separated from one another by a delicate connective tissue, the endomysium. In the walls of hollow viscera, the fibers are grouped into *sheets* of smooth muscle tissue. Most often two sheets are present, with their fibers oriented at right angles to each other. In the more externally located **longitudinal layer,** the muscle fibers run parallel to the long axis of the organ; in the deeper **circular layer,** the fibers run around the circumference of the organ. The circular layer constricts the hollow organ, and the longitudinal layer shortens the organ's length and enlarges its lumen. These muscle layers generate the alternate waves of contraction and relaxation that propel substances through the organ by peristalsis.

When viewed by light microscopy, smooth muscle fibers have no striations. Electron microscopy confirms that there are no sarcomeres. Interdigitating thick and thin filaments, however, are present. These contractile myofilaments lie nearly parallel to the long axis of the fiber, but at a slightly oblique angle, and fill much of the cell volume.

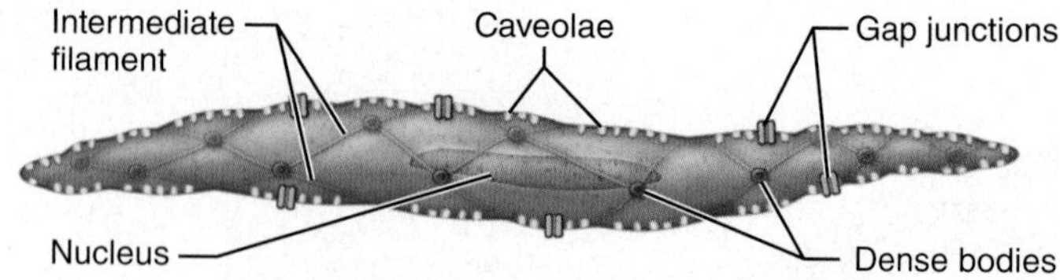

(a) Relaxed smooth muscle fiber (note that gap junctions connect adjacent fibers)

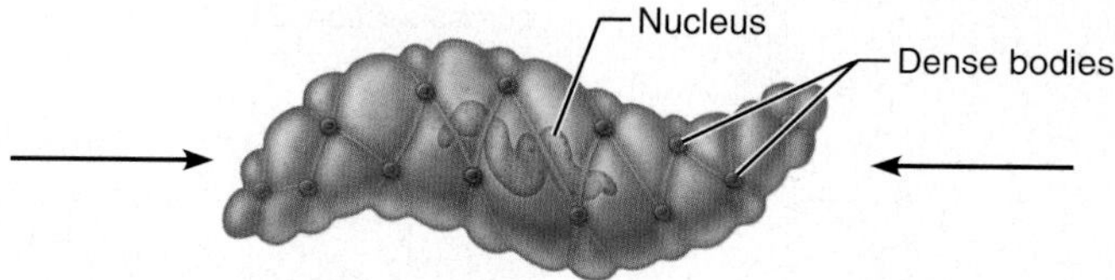

(b) Contracted smooth muscle fiber

Figure 23.9 **Cytoskeletal elements involved in the contraction of smooth muscle.**

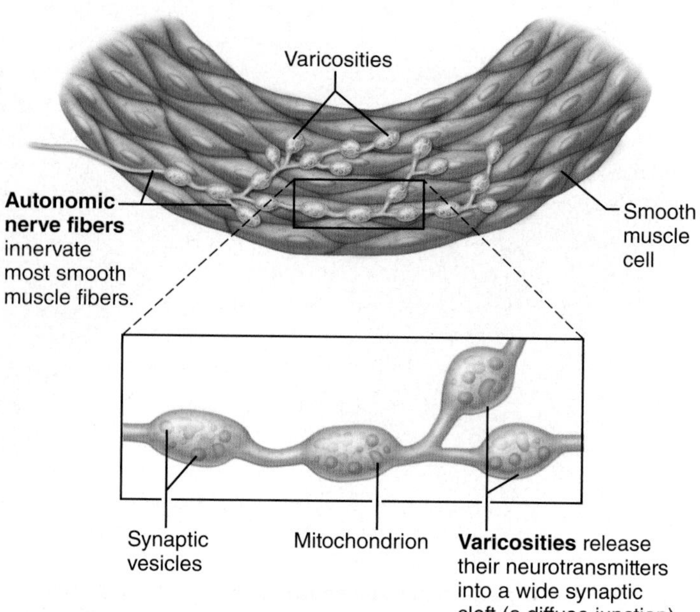

Figure 23.10 **Innervation of smooth muscle.**

Mechanism of Contraction

Myofilaments in smooth muscle operate by interacting with elements of the cytoskeleton (Figure 23.9a). Tension-resisting *intermediate filaments* extend through the cell in a lattice-like arrangement. Along these intermediate filaments lie **dense bodies** at regular intervals that anchor the thin filaments. These dense bodies correspond to the Z discs of skeletal muscle. Through this anchoring attachment, the sliding myofilaments shorten the muscle cell by pulling on the cytoskeleton (Figure 23.9b). The dense bodies attached to the sarcolemma also bind the muscle cell to the endomysium outside the cell and to adjacent cells, transmitting the contractile force to the surrounding connective tissue. This contributes to the synchronous contraction of most smooth muscle.

As in skeletal and cardiac muscle, the entry of Ca^{2+} into the sarcoplasm stimulates the smooth muscle fiber to contract. Some of these calcium ions enter from the extracellular fluid through tiny spherical infoldings of the sarcolemma, called *caveolae* (Figure 23.9a). These invaginations enclose bits of extracellular fluid, allowing high concentrations of Ca^{2+} to be sequestered close to the surface membrane. Ca^{2+} ions are also stored and released by an intracellular sarcoplasmic reticulum (not illustrated).

The contraction of smooth muscle is slow, sustained, and resistant to fatigue. Whereas smooth muscle takes 30 times longer to contract and relax than does skeletal muscle, it can maintain its contractile force for a long time without tiring. This is a valuable feature because the smooth muscle in the walls of the small arteries and visceral organs must sustain a moderate degree of contraction day in and day out without fatiguing. Because smooth muscle's energy requirements are low, mitochondria are not abundant in its fibers.

Innervation of Smooth Muscle and Glands

We do not have direct voluntary control over the contraction of our smooth muscle, because it is innervated by the autonomic nervous system. In general, *only a few smooth muscle fibers in each muscle sheet are innervated, and the impulse spreads through gap junctions between adjacent fibers.* Through such intercellular communication, the whole sheet contracts as a single unit. This arrangement is called *single-unit innervation.* There are exceptions: In the iris of the eye and the arrector pili muscles of the skin, every smooth muscle fiber is innervated individually, an arrangement called *multiunit innervation.* In addition, contraction of smooth muscle does not always require a nervous signal; contraction can be stimulated by stretching the muscle fibers or by hormones. The contraction of the muscular wall of the uterus during labor and delivery is an excellent example of both stretching and hormonal stimuli causing the contraction of smooth muscle.

The contacts between visceral motor neurons and the visceral effectors (smooth muscle, cardiac muscle, and glands) are much simpler than the elaborate neuromuscular junctions present in skeletal muscle (refer to Table 10.2, pp 254–255, for comparison). Near the smooth muscle or gland cells it innervates, a visceral motor axon swells into a row of knobs (varicosities) resembling beads on a necklace (Figure 23.10). These varicosities are the presynaptic terminals, which contain synaptic vesicles filled with neurotransmitter. Some of these axon terminals form shallow indentations on the membrane of the effector cell, but many remain a considerable distance from any cell. Because it takes time for neurotransmitters to diffuse across these wide synaptic clefts, visceral motor responses tend to be slower than somatic motor reflexes.

Enteric Nervous System The smooth muscle and glands of the alimentary canal are controlled largely by the independent **enteric nervous system** (*enteric* = gut). This "brain in the gut" is composed of over 100 million neurons—as many as in the entire spinal cord—and is located entirely within the wall of the alimentary canal. Within the wall, the *enteric neurons* form reflex arcs of sensory neurons,

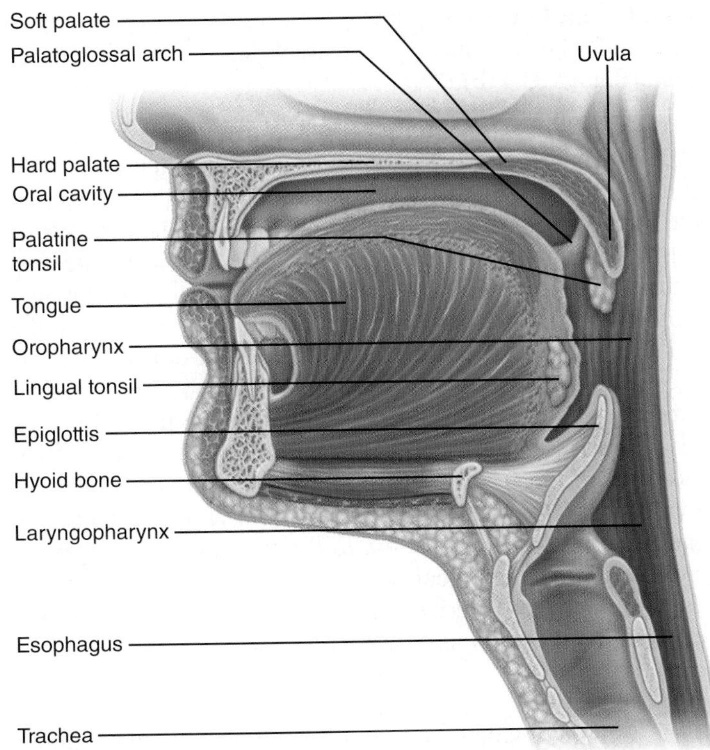

(a) Sagittal section of the oral cavity and pharynx

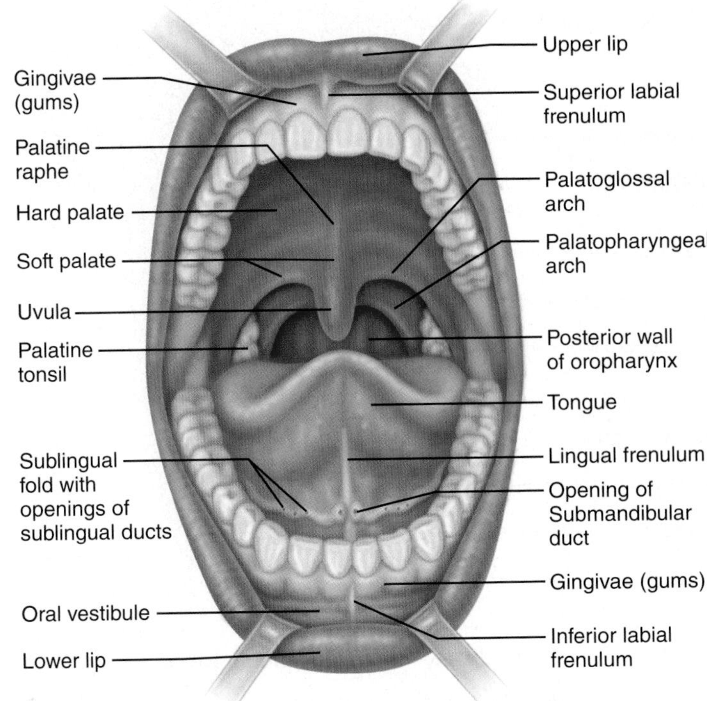

(b) Anterior view

Figure 23.11 Anatomy of the mouth.

interneurons, and motor neurons that control the muscular and secretory activity of the digestive organs in response to the contents of the lumen. The enteric neurons are located in two visceral nerve plexuses within the wall of the alimentary canal (Figure 23.7a).

- The **myenteric nerve plexus** (mi-en-ter′ik; "intestinal muscle") is in the muscularis externa between the circular and longitudinal layers. It innervates the muscularis externa to control peristalsis and segmentaion.
- The **submucosal nerve plexus** lies within the submucosa. Nerve fibers extend inward and signal the glands in the mucosa to secrete and the muscularis mucosae to contract.

Although the enteric nervous system can function independently, it is linked to and influenced by the central nervous system:

- Visceral sensory fibers carried in the vagus and splanchnic nerves transmit sensory stimuli from alimentary canal to the CNS.
- Visceral motor fibers from the classic autonomic nervous system (sympathetic and parasympathetic, Chapter 15) influence the activity of the enteric neurons. Postganglionic sympathetic fibers, preganglionic parasympathetic fibers, and postganglionic parasympathetic neurons synapse on the enteric neurons. Parasympathetic input stimulates digestive functions, increasing the activity of the smooth muscle and glands of the alimentary canal; sympathetic stimulation inhibitis digestive function.

check your understanding

☐ 5. Name the three sublayers of the mucosa. Which sublayer forms the instrinsic glands that produce digestive secretions?

☐ 6. Name the tissue layer of the alimentary canal that is responsible for peristalsis and segmentation.

☐ 7. Contrast smooth muscle to skeletal muscle, noting differences in cell shape, number and location of nuclei in each cell, the presence or absence of striations, innervation, the stimuli for contraction, and fatigue resistance.

(For answers, see Appendix B.)

The Mouth and Associated Organs

learning outcome

▶ Describe the gross and microscopic anatomy and the basic functions of the mouth, teeth, salivary glands, and pharynx.

The Mouth

Food enters the alimentary canal through the mouth, where it is chewed, manipulated by the tongue, and moistened with saliva. The **mouth,** or **oral cavity** (Figure 23.11), is a mucosa-lined cavity whose boundaries are the lips anteriorly, the cheeks laterally, the palate superiorly, and the tongue inferiorly. Its anterior opening is the **oral orifice.** Posteriorly, the mouth borders the fauces of the oropharynx (see p. 651).

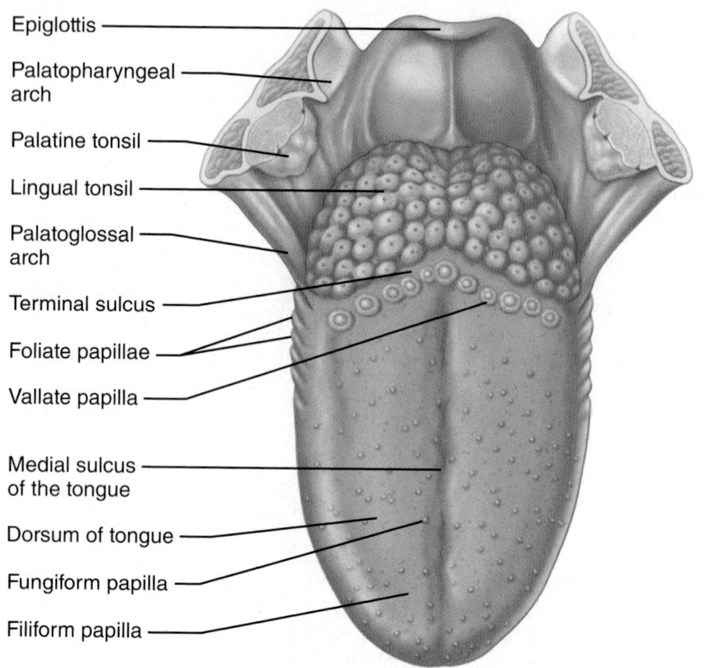

Figure 23.12 **The dorsal surface of the tongue.**

The mouth is divided into the oral vestibule and the oral cavity proper. The **oral vestibule** (ves′tĭ-būl; "porch") is the slit between the teeth and the cheeks (or lips). When you brush the outer surface of your teeth, your toothbrush is in the oral vestibule. The **oral cavity proper** is the region of the mouth that lies internal to the teeth.

Histology of the Mouth The walls of the oral cavity consist of just a few layers of tissue: an internal mucosa made of an epithelium and lamina propria only, a thin submucosa in some areas, and an external layer of muscle or bone. The lining of the mouth, a thick stratified squamous epithelium, protects it from abrasion by sharp pieces of food during chewing. On the tongue, palate, lips, and gums this epithelium may show slight keratinization, which provides extra protection against abrasion.

The Lips and Cheeks The **lips** (or **labia**) and the **cheeks,** which help keep food inside the mouth during chewing, are composed of a core of skeletal muscle covered by skin. Whereas the cheeks are formed largely by the buccinator muscles, the orbicularis oris muscle (p. 280) forms the bulk of the lips.

The lips are thick flaps extending from the inferior boundary of the nose to the superior boundary of the chin. The region of the lip where one applies lipstick or lands a kiss is called the *transition part,* a zone where the highly keratinized skin meets the oral mucosa. This region is poorly keratinized and translucent, so it derives its reddish color from blood in the underlying capillaries. Because the transition part lacks sweat and sebaceous glands, it must be moistened with saliva periodically to prevent drying and cracking. The **labial frenulum** (fren′u-lum; "little bridle of the lip") is a median fold that connects the internal aspect of each lip to the gum (Figure 23.11b).

The Palate The **palate,** which forms the roof of the mouth, has two distinct parts: the *hard palate* anteriorly and the *soft palate* posteriorly (Figure 23.11). The bony hard palate forms a rigid surface against which the tongue forces food during chewing. The muscular soft palate is a mobile flap that rises to close off the nasopharynx during swallowing. To demonstrate this action, try to breathe and swallow at the same time. The separation of the nasal and oral cavities by the palate is characteristic of all mammals and is essential for producing the suction necessary for suckling in infants (see discussion of cleft palate in Chapter 7, p. 180). Dipping inferiorly from the free edge of the soft palate is the fingerlike uvula. Laterally, the soft palate is anchored to the tongue by the **palatoglossal arches** and to the wall of the oropharynx by the **palatopharyngeal arches** (Figure 23.11b). These two folds form the boundaries of the fauces, the arched area of the oropharynx that contains the palatine tonsils.

The Tongue

The **tongue,** which occupies the floor of the mouth (Figure 23.11), is predominantly a muscle constructed of interlacing fascicles of skeletal muscle fibers. During chewing, the tongue grips food and constantly repositions it between the teeth. Tongue movements also mix the food with saliva and form it into a compact mass called a *bolus* (bo′lus; "lump"); then, during swallowing, the tongue moves posteriorly to push the bolus into the pharynx. In speech, the tongue helps to form some consonants (*k, d, t,* and *l,* for example). Finally, it houses most of the taste buds.

The tongue has both intrinsic and extrinsic muscle fibers. The *intrinsic muscles,* which are confined within the tongue and are not attached to bone, have fibers that run in several different planes. These intrinsic muscles change the shape of the tongue, for example rolling the tongue, but do not change its position. The *extrinsic muscles* extend to the tongue from bones of the skull and the hyoid bone (see pp. 282–283). These extrinsic muscles alter the position of the tongue: They protrude it, retract it, and move it laterally. The tongue is divided by a median septum of connective tissue, and both halves contain identical groups of muscles.

A fold of mucosa on the undersurface of the tongue, the **lingual frenulum** (Figure 23.11b), secures the tongue to the floor of the mouth and limits its posterior movements. Individuals in which the lingual frenulum is abnormally short or extends exceptionally far anteriorly are said to be "tongue-tied," and their speech is distorted because movement of the tongue is restricted. This congenital condition, called **ankyloglossia** (ang″kĭ-lo-glos′e-ah; "fused tongue"), is corrected surgically by snipping the frenulum.

The dorsal surface of the tongue (Figure 23.12) is covered with three major types of peglike projections of the mucosa: the filiform, fungiform, and vallate *papillae.* The terms *papillae* and *taste buds* are not synonymous; the fungiform and vallate papillae *contain* the taste buds.

The conical, pointed, and keratinized **filiform papillae** (fil′ĭ-form; "thread-shaped") roughen the tongue, enabling it to grasp and manipulate food during chewing. These smallest and most numerous papillae line up in parallel rows. They give the tongue's surface its whitish appearance.

The **fungiform papillae** (fun′jĭ-form; "mushroom-shaped"), which resemble tiny mushrooms, have a vascular core that gives them a red color. Although less abundant than filiform papillae, they are scattered widely over the tongue surface. Taste buds occur in the epithelium on the *tops* of these papillae.

Ten to twelve large **vallate papillae** (val′āt; "wall") line up in a V-shaped row bordering the posterior third of the tongue and directly anterior to a groove called the **terminal sulcus,** which marks the border between the mouth and pharynx. Each vallate papilla is surrounded by a circular ridge, from which it is separated by a deep furrow. Taste buds occupy the epithelium on the *sides* of these papillae (see Figure 16.1, p. 489).

The posterior third of the tongue, which lies in the oropharynx, not in the mouth, is covered not with papillae but with the bumpy *lingual tonsil.*

The Teeth

The **teeth** lie in sockets (alveoli) in the gum-covered margins of the mandible and maxilla. We masticate, or chew, by raising and lowering the mandible and by moving it from side to side while using the tongue to position food between the teeth. In the process, the teeth tear and grind the food, breaking it into smaller fragments.

Dentition and the Dental Formula

During their lifetime, humans have two sets of teeth, or *dentitions.* By age 21, the primary dentition, called the **deciduous teeth** (de-sid′-ū-us; "falling off"), has been replaced by the permanent dentition (Figure 23.13).

At about 6 months after birth, the lower central incisors become the first of the deciduous teeth to appear. Additional pairs of teeth erupt at varying intervals until all 20 deciduous teeth have emerged, by about 2 years of age. As the deep-lying **permanent teeth** enlarge and develop, the roots of the deciduous teeth are resorbed until these teeth loosen and fall out, typically between the ages of 6 and 12 years (Figure 23.13). Generally, by the end of adolescence, all permanent teeth have erupted except for the third molars (also called *wisdom teeth*), which emerge between the ages of 17 and 25 years. There are 32 permanent teeth in a full set, but in some people the wisdom teeth are either completely absent or fail to erupt.

> **CLINICAL APPLICATION**
>
> **Impacted Tooth** Instead of emerging normally, a tooth may remain embedded deep in the jawbone and push on the roots of the other teeth. The **impacted tooth** causes pressure and pain and must be removed by a dentist or oral surgeon. Wisdom teeth are the most commonly impacted teeth.

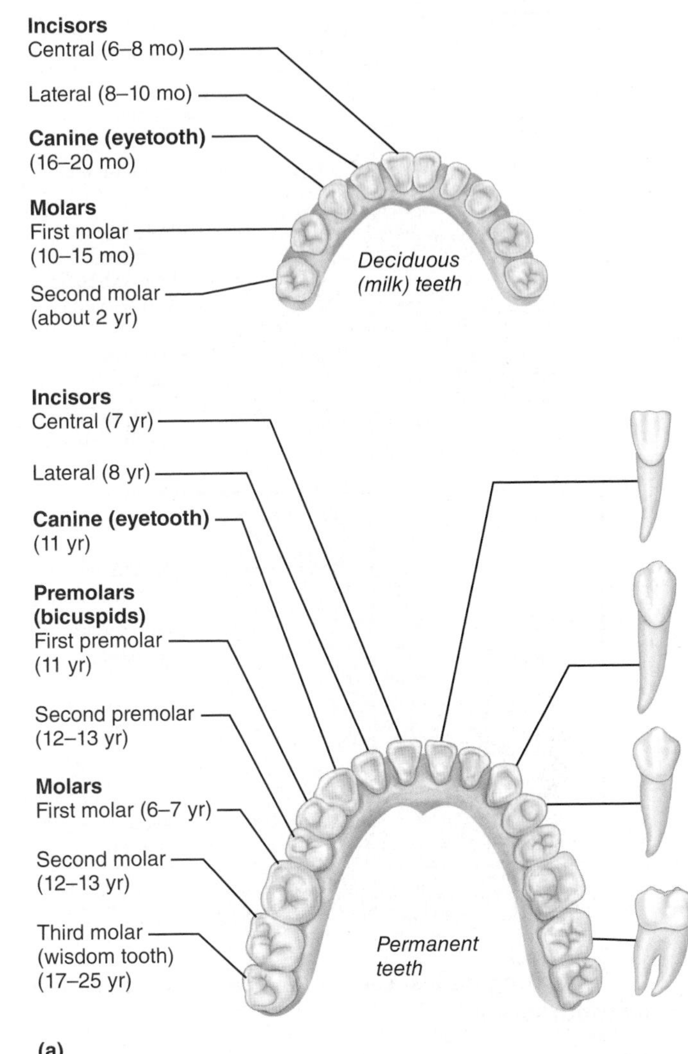

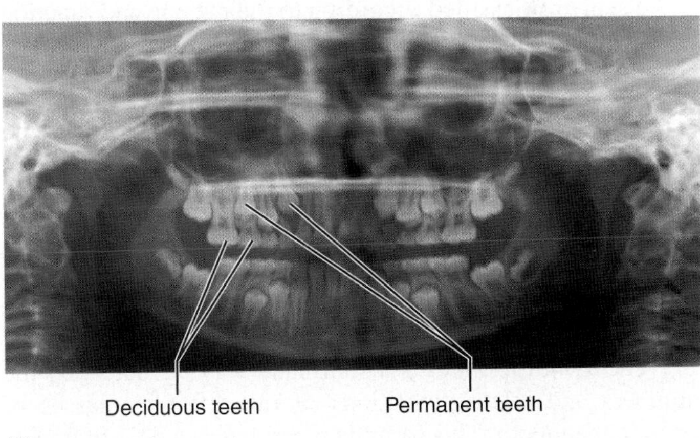

Figure 23.13 Human dentition. **(a)** Deciduous and permanent teeth of the lower jaw. Approximate ages at which the teeth erupt are shown in parentheses. Shapes of the individual teeth are shown at right. **(b)** X-ray image of the mouth of a 7-year-old child.

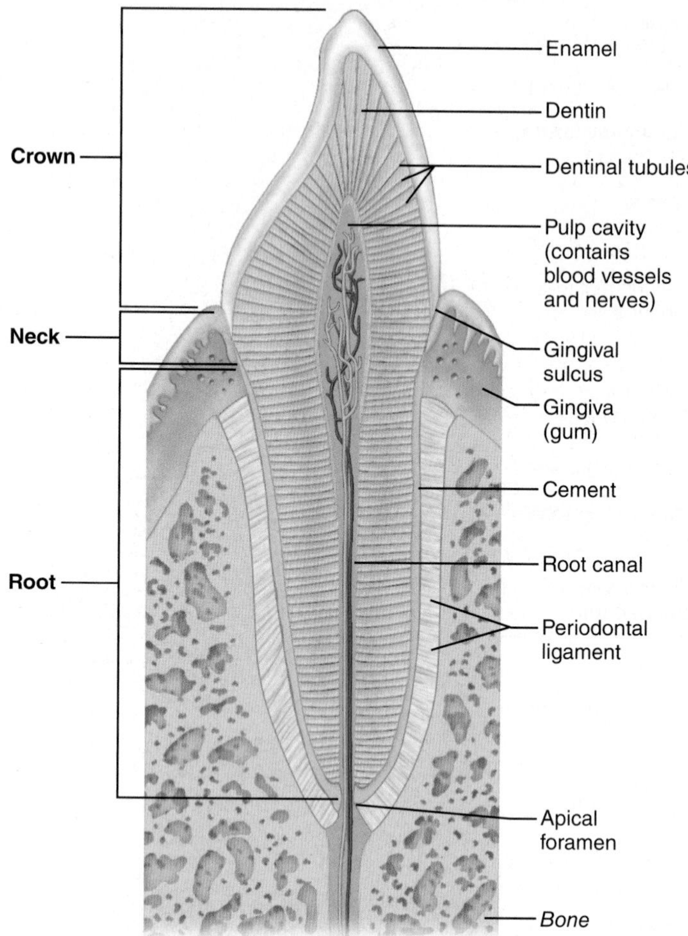

Figure 23.14 Longitudinal section of a canine tooth within its bony tooth socket.

Teeth are classified according to their shape and function as incisors, canines, premolars, or molars (Figure 23.13). The chisel-shaped **incisors** are adapted for nipping off pieces of food, and the cone-shaped **canines** (cuspids, eyeteeth) tear and pierce. The **premolars** (bicuspids) and **molars** have broad crowns with rounded *cusps* (surface bumps) for grinding food. The molars (literally, "millstones") have four or five cusps and are the best grinders. During chewing, the upper and lower molars lock repeatedly together, the cusps of the uppers fitting into the valleys in the lowers, and vice versa. This action generates tremendous crushing forces.

The *dental formula* is a shorthand way of indicating the numbers and relative positions of the different classes of teeth in the mouth. This formula is written as a ratio of uppers over lowers, for just half of the mouth (because right and left halves are the same). The total number of teeth is calculated by multiplying the dental formula by 2. The formula for the permanent dentition (two incisors, one canine, two premolars, and three molars) is written as follows:

$$\frac{2I, 1C, 2P, 3M}{2I, 1C, 2P, 3M} \times 2 \text{ (equals 32 teeth)}$$

The dental formula for the deciduous teeth is as follows:

$$\frac{2I, 1C, 2M}{2I, 1C, 2M} \times 2 \text{ (equals 20 teeth)}$$

Vessels and Nerves The upper teeth are innervated by the superior alveolar nerves, branches of the maxillary division of the trigeminal nerve (cranial nerve V). The lower teeth are innervated by the inferior alveolar nerves, branches of the mandibular division of the trigeminal nerve. The arterial supply for the teeth comes via the superior and inferior alveolar arteries, which are branches of the maxillary artery from the external carotid.

Tooth Structure Each tooth has two main regions, the exposed **crown** and the **root(s)** in the socket. These regions meet at the **neck** near the gum line (Figure 23.14). The surface of the crown, which bears the forces of chewing, is covered by a layer of **enamel,** the hardest substance in the body, that is 0.96–1.6 mm thick. Enamel lacks cells and vessels, and 99% of its mass consists of densely packed hydroxyapatite crystals (the same calcium salts found in bone) arranged in force-resisting rods or prisms oriented perpendicular to the tooth's surface.

Dentin, or **dentine,** underlies the enamel cap and forms the bulk of the tooth. This is a bonelike tissue with mineral and collagen components, but it is harder than bone and lacks internal blood vessels. Dentin contains unique radial striations called **dentinal tubules.**

The **pulp cavity,** in the center of the tooth, is filled with dental **pulp,** a loose connective tissue containing the tooth's vessels and nerves. Pulp supplies nutrients for the tooth's hard tissues and provides for tooth sensation. The part of the pulp cavity in the root is the **root canal.** The opening into the root canal at the tip of each root is the **apical foramen.** When a tooth is damaged by a blow or by a deep cavity, the pulp may die and become infected. In such cases, **root canal therapy** must be performed. In this procedure, all of the pulp is drilled out, and the pulp cavity is sterilized and filled with an artificial, inert material before the tooth is capped.

The external surface of the tooth root is covered by a calcified connective tissue called **cement.** Essentially a bone layer, cement attaches the tooth to the **periodontal ligament** (per″e-o-don′tal; "around the tooth"), or **periodontium.** This ligament anchors the tooth in the bony socket of the jaw. The periodontal ligament is continuous with the gum, or **gingiva** (jin-jĭ′vah), at the neck of the tooth.

Dental **cavities,** or **caries** (kar′ēz; "rottenness"), result from a gradual demineralization of the enamel and dentin by bacterial action. The decay process begins with the accumulation of **dental plaque,** a film of sugar, bacteria, and other debris that adheres to the teeth. Metabolism of the trapped sugars by the bacteria produces acids, which dissolve the calcium salts from the teeth. Once the salts have leached out, the remaining organic matrix is broken down by protein-digesting bacterial enzymes. Frequent brushing helps prevent tooth decay by removing plaque.

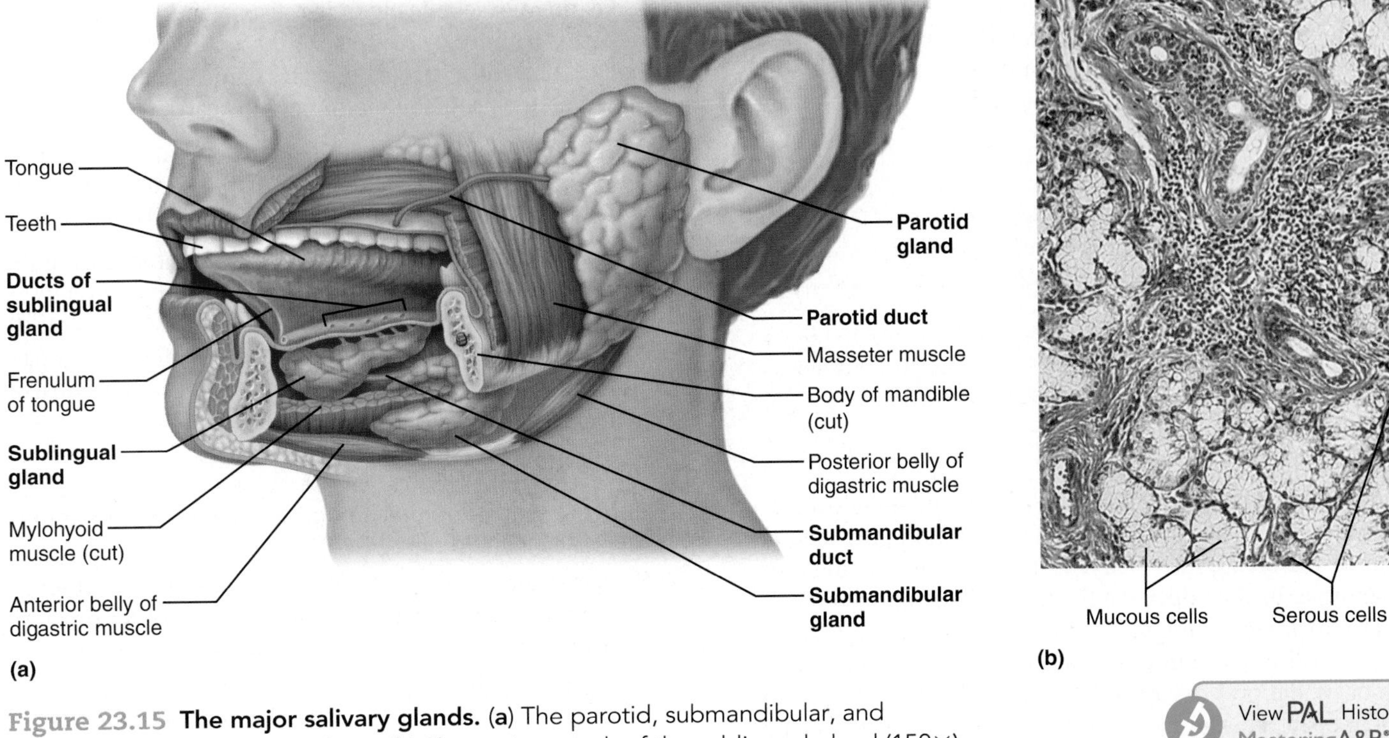

Figure 23.15 The major salivary glands. (**a**) The parotid, submandibular, and sublingual glands and their ducts. (**b**) Photomicrograph of the sublingual gland (150×), a mixed gland containing mostly mucous cells (white) with a few serous cells (purple).

View PAL Histology MasteringA&P®

Watch Cat Dissection MasteringA&P®

CLINICAL APPLICATION

Gum Disease Even more serious than tooth decay is the effect of plaque on the *gums*. As plaque accumulates around the necks of the teeth, the contained bacteria release toxins that irritate the gums and cause them to pull away from the teeth. The plaque calcifies into a layer called **calculus** (kal′ku-lus; "stone"), on which more plaque accumulates, further inflaming the gums. This condition, **gingivitis** (jin″jĭ-vi′tis), can be reversed if the calculus is removed, but if it is neglected the bacteria invade the periodontal tissues, forming pockets of infection that destroy the periodontal ligament and dissolve away the bone around the tooth. Called **periodontitis,** this condition begins around age 35 and eventually affects 75% of people. (Periodontitis often results in the loss of teeth and is the reason many people wear dentures.) Still, tooth loss is not inevitable. Even advanced cases of periodontitis can be treated by cleaning the infected pockets around the roots, then cutting and stitching the gums to shrink the pockets. Newly available gels, which are injected into the pockets, contain either antibiotics or substances that stimulate regeneration of the gum, cement, and periodontal ligament.

Dentists recommend that people floss between their teeth every day, beginning at a young age, because flossing removes plaque, thereby minimizing one's chances of developing periodontitis and losing one's teeth later in life.

The Salivary Glands

The **salivary glands** produce *saliva,* a complex mixture of water, ions, mucus, and enzymes that performs many functions: It moistens the mouth, dissolves food chemicals so that they can be tasted, wets food, and binds the food together into a bolus. Its enzymes, salivary amylase and lipase, begin the digestion of carbohydrates and fats. Saliva contains a bicarbonate buffer that neutralizes the acids that are produced by oral bacteria and that initiate tooth decay. Additionally, it contains bactericidal enzymes, antiviral substances, antibodies, and a cyanide compound, all of which kill harmful oral microorganisms. Saliva also contains proteins that stimulate the growth of beneficial bacteria to outcompete harmful bacteria in the mouth.

Minor (intrinsic) salivary glands are scattered within the mucosa of the tongue, palate, lips, and cheeks. Saliva from these glands keeps the mouth moist at all times. By contrast, *major (extrinsic) salivary glands,* which lie external to the mouth but connect to it through their ducts (Figure 23.15), secrete saliva only during eating or anticipation of a meal, causing the mouth to water. These paired major salivary glands are the *parotid, submandibular,* and *sublingual glands.*

The largest major gland is the **parotid** (pah-rot′id) **gland,** a compound acinar gland. True to its name (*par* = near; *otid* = the ear), it lies anterior to the ear, between the masseter muscle and the skin (Figure 23.15a). Its **parotid duct** runs parallel to the zygomatic arch, penetrates the muscle of the cheek, and opens into the mouth lateral to the second upper molar. Secretion of the parotid duct is stimulated by the glossopharyngeal nerve (cranial nerve IX). Because the branches of the facial nerve run through the parotid gland on their way to the muscles of facial expression, surgery on this gland can lead to facial paralysis.

CLINICAL APPLICATION

Mumps A virus that spreads from one person to another in the saliva causes **mumps**. Its dominant symptom is inflammation and swelling of the parotid gland. When a person with mumps opens the mouth or chews, movements of the mandible pull on the irritated parotid glands and the bulging masseter muscles compress these glands, causing pain.

The **submandibular gland,** a compound tubuloacinar gland, is about the size of a walnut. It lies along the medial surface of the mandibular body, just anterior to the angle of the mandible. Its duct opens in the floor of the mouth directly lateral to the tongue's lingual frenulum (see Figure 23.11b).

The **sublingual gland,** also a tubuloacinar gland, lies in the floor of the oral cavity, inferior to the tongue. Its 10 to 12 ducts open into the mouth, directly superior to the gland (Figure 23.15a). Both the sublingual and submandibular glands are innervated by the facial nerve (cranial nerve VII).

Microscopic Anatomy The secretory cells of the salivary glands are *serous cells,* which produce a watery secretion containing the enzymes and ions of saliva, and *mucous cells,* which produce mucus (Figure 23.15b). (Note, however, that because the serous cells in human salivary glands are now known to secrete a small amount of mucus as well, some scientists refer to these cells as *seromucous cells.*) The parotid glands contain only serous cells; the submandibular and minor salivary glands are mixed glands containing serous and mucous cells; and the sublingual glands, while also are mixed glands, contain mostly mucous cells. The smallest ducts of all salivary glands are formed by a simple cuboidal epithelium.

The Pharynx

From the mouth, swallowed food passes posteriorly into the **oropharynx** and then the **laryngopharynx** (Figure 23.11a), both of which are passageways for food, fluids, and inhaled air. The muscles of the neck and pharynx (Table 11.5, p. 284) contract in sequence to complete the swallowing process:

1. The *suprahyoid muscles* lift the larynx superiorly and anteriorly to position it beneath the protective flap of the epiglottis, thus closing the airway so food is not inhaled into the lungs.
2. The three *pharyngeal constrictor muscles*—superior, middle, and inferior (shown in Figure 11.11c, p. 286)—encircle the pharynx and partially overlap one another. Like three stacked, clutching fists, they contract from superior to inferior to squeeze the bolus into the esophagus. The pharyngeal muscles are skeletal muscles innervated by somatic motor neurons carried in the vagus nerve (cranial nerve X).
3. The *infrahyoid muscles* pull the hyoid bone and larynx inferiorly returning them to their original position.

The histology of the pharyngeal wall resembles that of the mouth: The mucosa of the oropharynx and laryngopharynx is lined by a stratified squamous epithelium, which protects against abrasion. (Swallowed food often contains rough particles after it is chewed.) The external muscular layer consists of the pharyngeal constrictors.

check your understanding

- ☐ 8. What type of epithelium forms the mucosa lining the oral cavity and pharynx?
- ☐ 9. Name the three major salivary glands. What nutrient macromolecules do the enzymes in saliva act on?
- ☐ 10. Which cranial nerve innervates the teeth? Which division of this nerve innervates the teeth in the upper jaw? Which innervates the teeth in the lower jaw?

(For answers, see Appendix B.)

The Esophagus

learning outcome

▶ Describe the gross and microscopic anatomy of the esophagus.

Gross Anatomy

The **esophagus** is a muscular tube that propels swallowed food to the stomach. Its lumen is collapsed when it is empty. The esophagus begins as a continuation of the pharynx in the midneck, descends through the thorax on the anterior surface of the vertebral column (see Figure 23.1), and passes through the *esophageal hiatus* (hi-a′tus; opening) in the diaphragm to enter the abdomen (see Figure 11.13b, p. 292). Its abdominal part, which is only about 2 cm long, joins the stomach at the **cardial orifice,** where a **cardiac sphincter** acts to close off the lumen and prevent regurgitation of acidic stomach juices into the esophagus. The only anatomical evidence of this sphincter is a minimal thickening of the smooth muscle in the wall. The edges of the esophageal hiatus in the diaphragm also help prevent regurgitation.

Microscopic Anatomy

Unlike the mouth and pharynx, the esophagus wall (Figure 23.16a) contains all four layers of the alimentary canal: *mucosa, submucosa, muscularis externa,* and *adventitia.* The following histological features are of interest:

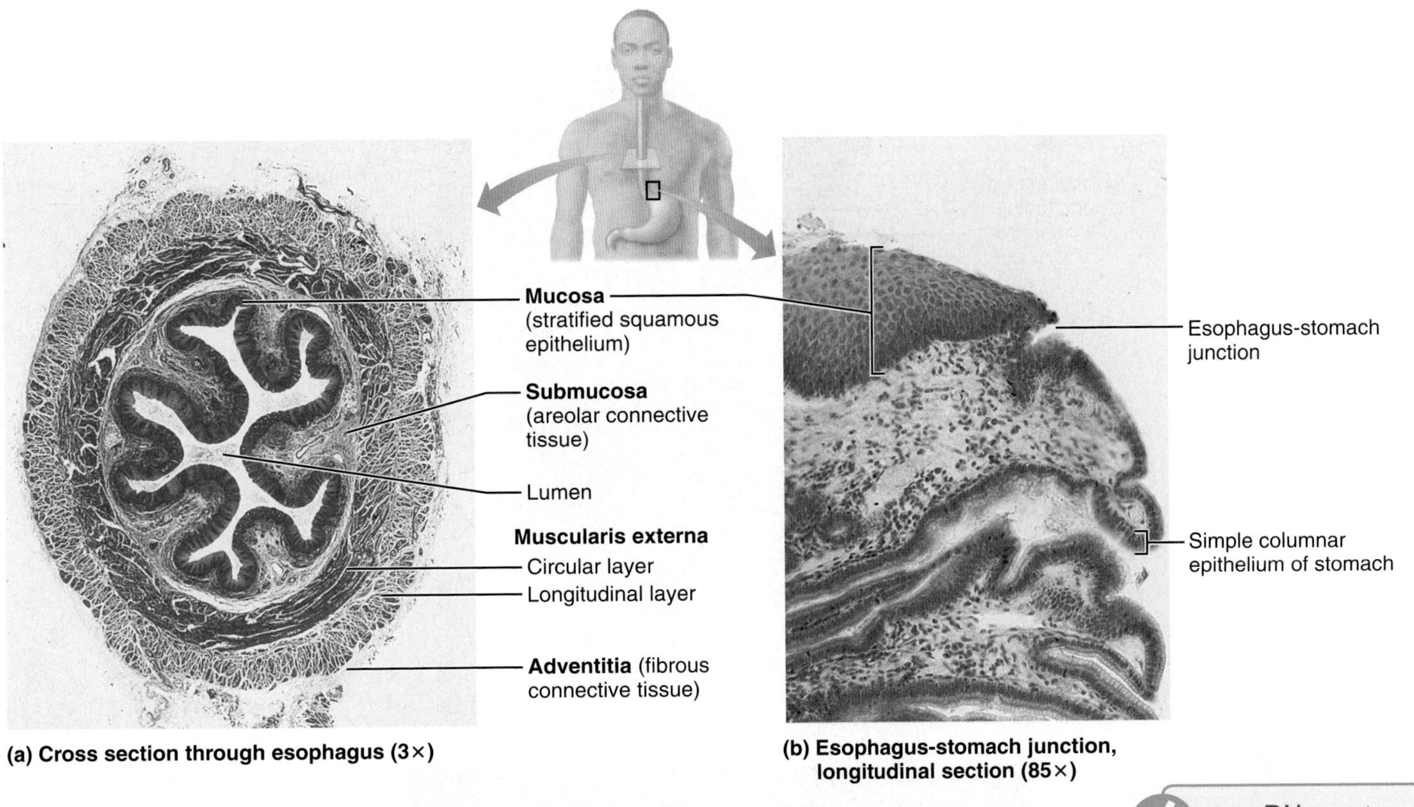

(a) Cross section through esophagus (3×)

(b) Esophagus-stomach junction, longitudinal section (85×)

Figure 23.16 **Microscopic structure of the esophagus.** In (b), notice the abrupt change between the stratified squamous epithelium lining the esophagus and the simple columnar epithelium lining the stomach.

View PAL Histology MasteringA&P®

- The mucosal epithelium is a nonkeratinized stratified squamous epithelium. At the junction of the esophagus and stomach, this thick, abrasion-resistant layer changes abruptly to the thin simple columnar epithelium of the stomach, which is specialized for secretion (Figure 23.16b).
- When the esophagus is empty, its mucosa and submucosa are thrown into longitudinal folds, but during passage of a bolus, these folds flatten out.
- The submucosa of the wall of the esophagus contains mucous glands, primarily compound tubuloalveolar glands, that extend to the lumen. As a bolus passes, it compresses these glands, causing them to secrete a lubricating mucus. This mucus helps the bolus pass through the esophagus.
- The muscularis externa consists of skeletal muscle in the superior third of the esophagus, a mixture of skeletal and smooth muscle in the middle third, and smooth muscle in the inferior third. This arrangement is easy to remember if the esophagus is viewed as the zone where the skeletal muscle of the mouth and pharynx gives way to the smooth muscle of the stomach and intestines.
- The most external esophageal layer is an adventitia, not a serosa, because the thoracic segment of the esophagus is not suspended in the peritoneal cavity.

CLINICAL APPLICATION

Hiatal Hernia and Gastroesophageal Reflux Disease In a **hiatal hernia,** the superior part of the stomach pushes through an enlarged esophageal hiatus into the thorax following a weakening of the diaphragmatic muscle fibers around the hiatus. Because the diaphragm no longer reinforces the action of the cardiac sphincter, the acidic stomach juices are persistently regurgitated, eroding the wall of the esophagus and causing a burning pain.

The regurgitation associated with hiatal hernia is just one form of **gastroesophageal reflux disease (GERD),** a condition that affects at least 4% of Americans. Most cases of GERD are due to abnormal relaxation or weakness of the cardiac sphincter (and probably of the sphincter mechanism of the esophageal hiatus as well). Symptoms include heartburn behind the sternum, regurgitation of stomach contents, and belching. The patient may aspirate the acids that are burped up, leading to hoarseness, coughing, and bronchial asthma. After persistent exposure to the acidic stomach contents, the lower esophagus develops ulcers, and the epithelium there becomes abnormal and precancerous, a condition called **Barrett's esophagus.** Treatment, which is usually successful, involves administering antacids and drugs that decrease the secretion of stomach acids. In severe cases, surgery is used to reinforce the cardiac sphincter in the lower esophagus.

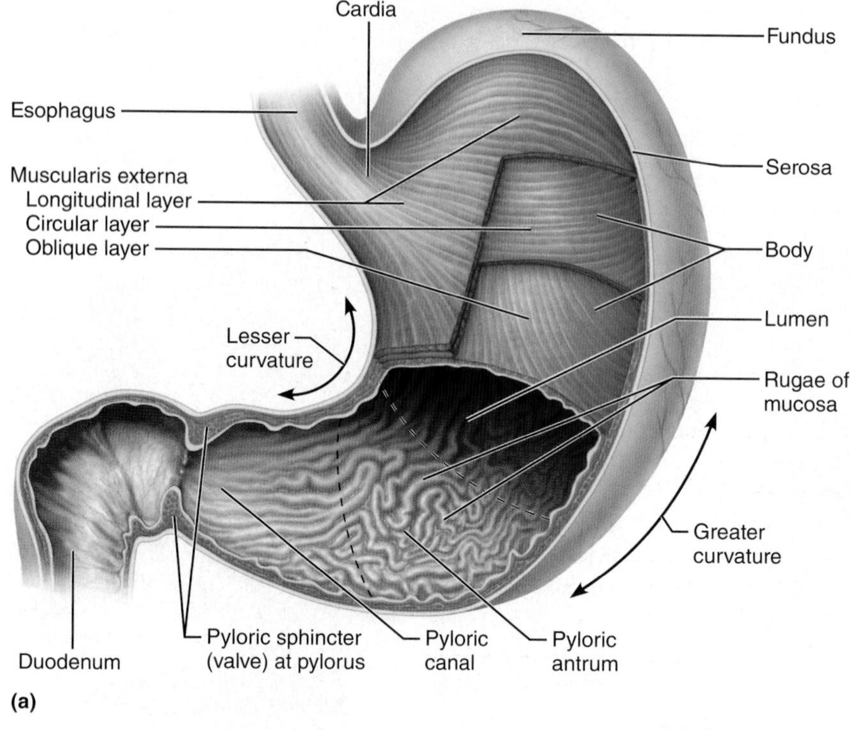

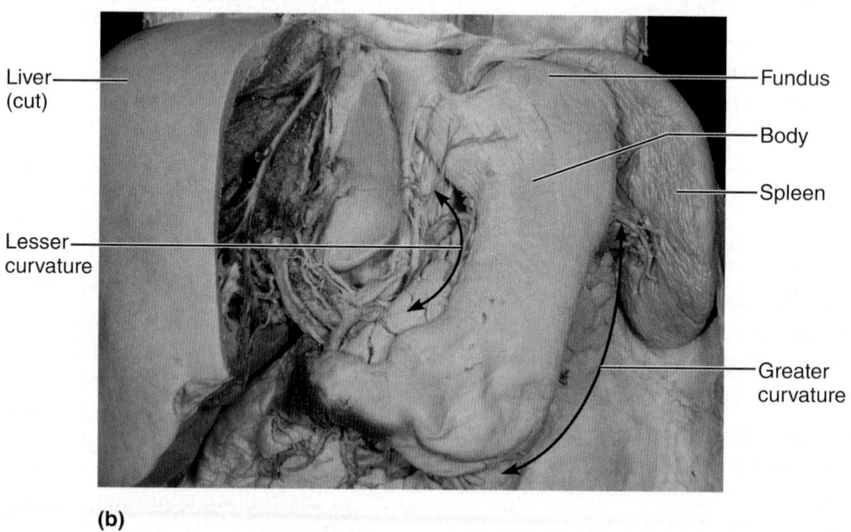

Figure 23.17 Gross anatomy of the stomach. (**a**) Basic regions of the stomach and gross internal anatomy. (**b**) Photograph of external aspect of stomach in situ.

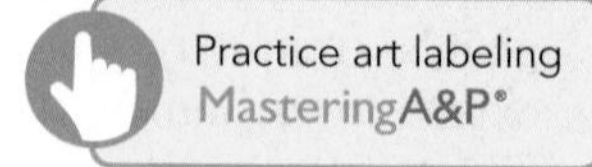

The Stomach

learning outcomes

- Describe the gross and microscopic anatomy of the stomach.
- Relate the histological modifications in the wall of the stomach to the digestive activities that occur there.

The J-shaped **stomach** (Figure 23.17), the widest part of the alimentary canal, is a temporary storage tank in which food is churned and turned into a paste called **chyme** (kīm; "juice"). The stomach also starts the breakdown of food proteins by secreting **pepsin,** a protein-digesting enzyme that can function only under acidic conditions, and hydrochloric acid, a strong acid that destroys many harmful bacteria in the food. Although most nutrients are absorbed in the small intestine, some substances are absorbed through the stomach, including water, electrolytes, and some drugs (aspirin and alcohol). Food remains in the stomach for roughly 4 hours.

Gross Anatomy

The stomach extends from the esophagus to the small intestine. The stomach lies in the superior left part of the peritoneal cavity, in the left hypochondriac, epigastric, and umbilical regions of the abdomen (see Figure 23.2). It is directly inferior to the diaphragm and anterior to the spleen and pancreas. Its upper part is hidden behind the left side of the liver. Although the stomach is anchored at both ends by esophageal and intestinal attachments, it is quite mobile in between. It tends to lie high and run horizontally in short, stout people (steerhorn stomach) and is elongated vertically in many tall, thin people (J-shaped stomach). When full, a J-shaped stomach may extend low enough to reach the pelvis!

The main regions of the stomach are described next (refer to Figure 23.17a). The **cardia** ("near the heart") is a ring-shaped zone encircling the cardial orifice at the junction with the esophagus. The **fundus,** the stomach's dome, is tucked under the diaphragm. The large midportion of the stomach, the **body,** ends at the funnel-shaped **pyloric part,** composed of the wider **pyloric antrum** ("cave") and the narrower **pyloric canal.** The pyloric part ends at the terminus of the stomach, the **pylorus** ("gatekeeper") containing the **pyloric sphincter,** which controls the entry of chyme into the intestine. The convex left surface of the stomach is its **greater curvature,** and the concave right margin is the **lesser curvature.** The greater and lesser omenta, mesenteries that connect to the stomach (p. 680), are named for their attachment to these curvatures.

The stomach's structure accounts for its great distensibility—it easily holds 1.5 liters of food and has a maximum capacity of about 4 liters (1 gallon). The internal surface of the empty stomach contains numerous longitudinal folds of mucosa called **rugae** (roo′ge; "wrinkles") (Figure 23.17), which flatten as the stomach fills; the resulting expansion in volume accommodates the increasing quantity of food within the stomach.

Vessels and Nerves The stomach is innervated by sympathetic fibers that derive from the thoracic splanchnic nerves by way of the celiac plexus, and by parasympathetic fibers that derive from the vagus. These autonomic fibers influence the activity of the enteric neurons in the stomach. The arteries to the stomach arise from the celiac trunk and include the right and left gastric, short gastric, and right and left gastroepiploic arteries; the corresponding veins drain into the portal, splenic, and superior mesenteric veins.

Microscopic Anatomy

The wall of the stomach has the typical layers of the alimentary canal (Figure 23.18). The lining epithelium of the *mucosa* ("surface epithelium" in the figure) is simple columnar epithelium and consists entirely of cells that secrete a coat of bicarbonate-buffered mucus. This mucus protects the stomach wall from the destructive effects of acid and pepsin in the lumen. This mucous secretion is necessary because the stomach mucosa is exposed to some of the harshest conditions in the entire alimentary canal. The oversecretion of stomach acid can result in peptic ulcers (see "Disorders of the Digestive System" on p. 712).

The surface of the stomach mucosa is dotted with millions of cup-shaped **gastric pits,** which open into tubular **gastric glands** (Figure 23.18b). Surface mucous cells invariably line the gastric pits, but the cells lining the gastric glands vary among the different regions of the stomach. In the pyloric and cardiac parts (not illustrated), the cells of the glands are primarily mucous cells. In the fundus and body, by contrast, the gastric glands contain three types of secretory cells: mucous neck cells, parietal (oxyntic) cells, and chief (zymogenic) cells (see Figure 23.18b–d and Figure 23.24, p. 705).

- **Mucous neck cells,** which occur in the upper ends, or necks, of the gastric glands, secrete a different type of mucus from that secreted by the surface cells. The specific function of this mucus is not known.
- **Parietal (oxyntic) cells,** which occur mainly in the middle regions of the glands, produce the stomach's hydrochloric acid (HCl) by pumping hydrogen and chloride ions into the lumen of the gland. Although parietal cells appear spherical when viewed by light microscopy, they actually have three thick prongs like those of a pitchfork. Many long microvilli cover each prong, providing a large surface area to enable rapid movement of H^+ and Cl^- out of the cells. The cytoplasm contains many mitochondria that supply the large amount of energy expended in pumping these ions. Parietal cells also secrete **gastric intrinsic factor,** a protein necessary for the absorption of vitamin B_{12} by the small intestine. The body uses this vitamin in the manufacture of red blood cells.
- **Chief (zymogenic) cells** occur mainly in the basal parts of the glands. Chief cells make and secrete the enzymatic protein **pepsinogen** (pep-sin′o-jen), which is activated to pepsin when it encounters acid in the apical region of the gland (Figure 23.18c). Chief cells have features typical of protein-secreting cells: a well-developed rough endoplasmic reticulum (rough ER) and Golgi apparatus, plus secretory granules in the apical cytoplasm. Chief cells also secrete gastric lipase, which functions in fat digestion.

At least two other epithelial cell types occur in the gastric glands but also extend beyond these glands:

- **Enteroendocrine cells** (en″ter-o-en′do-krin; "gut endocrine") are hormone-secreting cells scattered throughout the lining epithelium and glands of the alimentary canal. These cells (Figure 23.18) release their hormones into the capillaries of the underlying lamina propria. One of these hormones, *gastrin,* signals parietal cells to secrete HCl when food enters the stomach. Most enteroendocrine cells that produce gastrin are in the stomach's pyloric region.
- **Undifferentiated stem cells** (not illustrated) are located throughout the stomach, at the junction of the gastric glands and gastric pits. These cells divide continuously, replacing the entire lining epithelium of mucus-secreting cells every 3–7 days. Such rapid replacement is vital because these cells can survive for only a few days in the harsh environment of the stomach.

The *muscularis externa* of the stomach contains an additional layer of smooth muscle. Deep to the circular and longitudinal layers lies an innermost layer of smooth muscle fibers that run *obliquely* (Figure 23.17a). The circular and longitudinal layers churn and pummel food into smaller fragments. The oblique fibers jackknife the stomach into a V shape to move the chyme into the small intestine. The pyloric sphincter is a thickening of the circular layer.

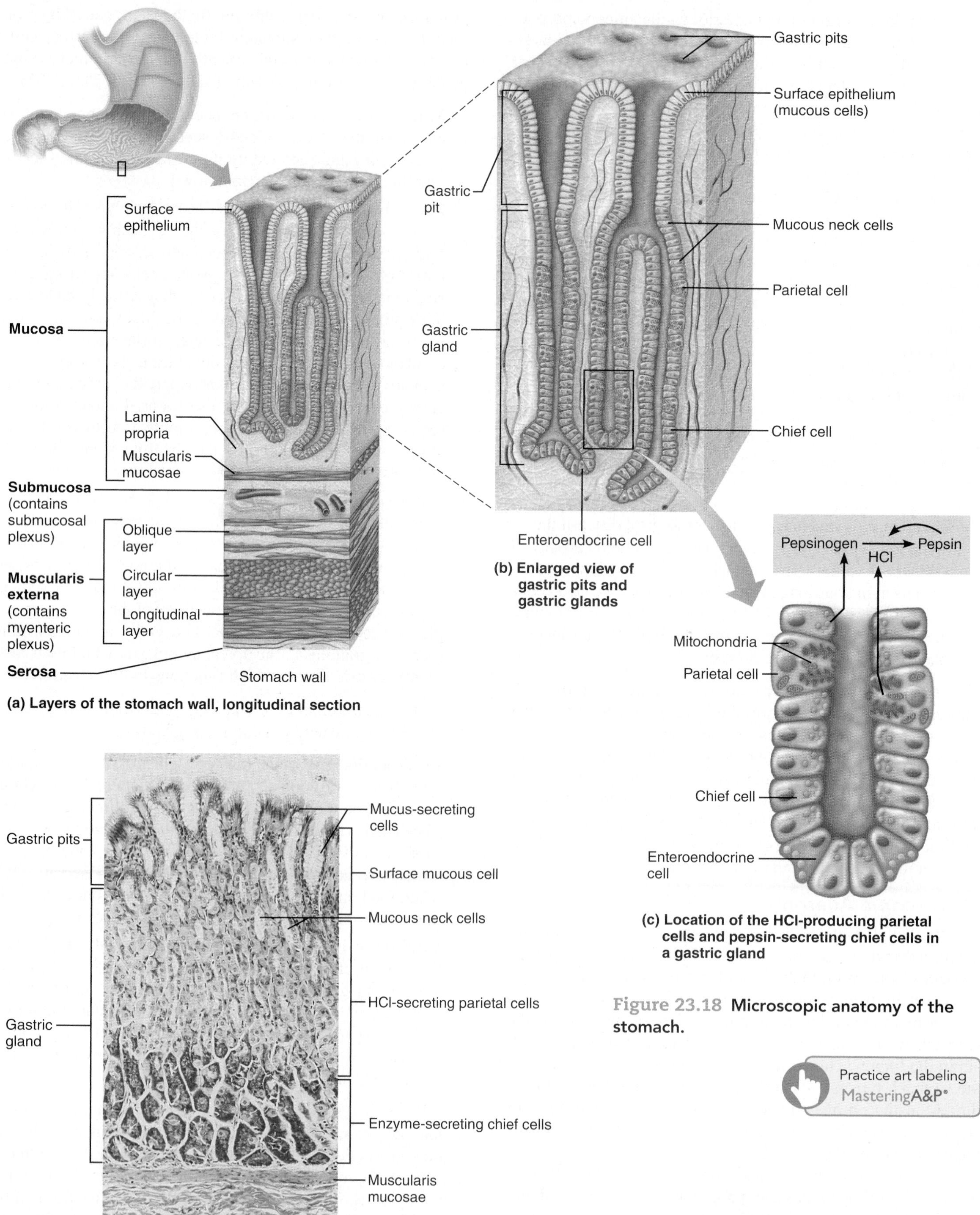

(a) Layers of the stomach wall, longitudinal section

(b) Enlarged view of gastric pits and gastric glands

(c) Location of the HCl-producing parietal cells and pepsin-secreting chief cells in a gastric gland

(d) Micrograph of the stomach mucosa, view similar to part (b) (110×)

Figure 23.18 Microscopic anatomy of the stomach.

✓ check your understanding

☐ **11.** How does the epithelial lining change from the esophagus to the stomach? What is unique about the muscularis externa in the stomach?

☐ **12.** Describe the location of the stomach in reference to the nine abdominal regions (Figure 23.2).

☐ **13.** What do chief cells produce? What do parietal cells produce? What do the surface cells that line the stomach produce?

(For answers, see Appendix B.)

The Small Intestine

learning outcomes

- ▶ Describe the gross and microscopic anatomy of the small and large intestine.
- ▶ Relate the histological modifications in the wall of the intestines to the digestive activities of these regions.

The **small intestine** is the longest part of the alimentary canal (see Figure 23.1) and the site of most enzymatic digestion and virtually all absorption of nutrients. Most digestive enzymes that operate within the small intestine are secreted not by the intestine, but by the pancreas. During digestion, the small intestine undergoes active segmentation movements, shuffling the chyme back and forth and thereby maximizing its contact with the nutrient-absorbing mucosa. Peristalsis propels chyme through the small intestine in about 3–6 hours.

Gross Anatomy

The small intestine is a convoluted tube that runs from the pyloric sphincter, in the epigastric region of the abdomen, to the first part of the large intestine, in the lower right quadrant. It is shorter in living people (2.7–5 meters) than in preserved cadavers (6–7 meters), where loss of muscle tone and the effects of preservatives have caused it to lengthen.

The small intestine has three subdivisions (see Figure 23.1): the **duodenum** (du″o-de′num; "twelve finger-widths long"), the **jejunum** (je-joo′num; "empty"), and the **ileum** (il′e-um; "twisted intestine"), which contribute 5%, almost 40%, and almost 60% of the length of the small intestine, respectively. Whereas most of the C-shaped duodenum lies secondarily retroperitoneal, the jejunum and ileum form sausagelike coils that hang from the posterior abdomen by the mesentery and are framed by the large intestine. The jejunum makes up the superior left part of this coiled intestinal mass, whereas the ileum makes up the inferior right part.

Even though the *duodenum* is the shortest subdivision of the small intestine, it has the most features of interest (Figure 23.19). It receives digestive enzymes from the pancreas via the *main pancreatic duct* and bile from the liver and gallbladder via the *bile duct*. These ducts enter the wall of the duodenum where they form a bulb called the **hepatopancreatic ampulla** (hep″ah-to-pan″kre-ah′tik am-pul′ah; "flask from the liver and pancreas"). This ampulla opens into the duodenum via a mound called the **major duodenal papilla**. Entry of bile and pancreatic juice into the duodenum is controlled by sphincters of smooth muscle that surround the hepatopancreatic ampulla and the ends of the pancreatic and bile ducts.

Vessels and Nerves The small intestine is innervated by enteric neurons. Parasympathetic fibers from the vagus and sympathetic fibers from the thoracic splanchnic nerves, both relayed through the superior mesenteric (and celiac) plexus, influence their activity. Arterial supply to the small intestine comes primarily via the superior mesenteric artery. The veins run parallel to the arteries and typically drain into the superior mesenteric vein; from there, the nutrient-rich venous blood drains into the hepatic portal vein, which carries it to the liver.

Microscopic Anatomy

As previously noted, nearly all nutrient absorption occurs in the small intestine, which is highly adapted for this function. In addition to the huge surface area for absorption provided by the small intestine's great length, several other structural features provide even more absorptive surface area, and serve other functions related to digestion as well.

Modifications for Absorption The wall of the small intestine has three structural modifications that amplify its absorptive surface enormously: *circular folds, villi,* and extensive *microvilli* (Figure 23.20). Because most absorption occurs in the proximal region of the small intestine, these specializations decrease in number toward the distal end.

The **circular folds,** or *plicae circulares* (pli′ke), are permanent, transverse ridges of the mucosa and submucosa (Figures 23.20a). They are nearly 1 cm tall. Besides increasing the absorptive surface area, these folds force the chyme to spiral through the intestinal lumen, slowing its movement and allowing time for complete absorption of nutrients.

Villi are fingerlike projections of the mucosa that give it a velvety texture, much like the soft nap of a towel. Over 1 mm high, and thus large enough to be seen with the unaided eye, villi are covered by a simple columnar epithelium made up primarily of **absorptive cells,** called **enterocytes,** specialized for absorbing digested nutrients (Figures 23.20b and c). Within the core of lamina propria in each villus is a network of blood capillaries and a wide lymphatic capillary called a **lacteal** (see p. 629). The end products of the digestion of carbohydrates and proteins enter the blood capillaries; absorbed fats enter the lacteals. The implications of this distinction are significant. The blood vessels that drain the small intestine carry absorbed nutrients to the liver via the hepatic portal system (Chapter 20, p. 613). Absorbed fats, however, do not go directly to the liver, but rather travel through the lymphatic vessels and empty into the venous system near the brachiocephalic vein (see Chapter 21, pp. 630–632). Thus, ingested and absorbed fat-soluble toxins, such as pesticides or herbicides, can circulate throughout the body before reaching the liver for detoxification.

Within the core of each villus is a slip of smooth muscle from the muscularis mucosae that allows the villus to move during digestion. These movements enhance absorption efficiency by increasing the amount of contact between the

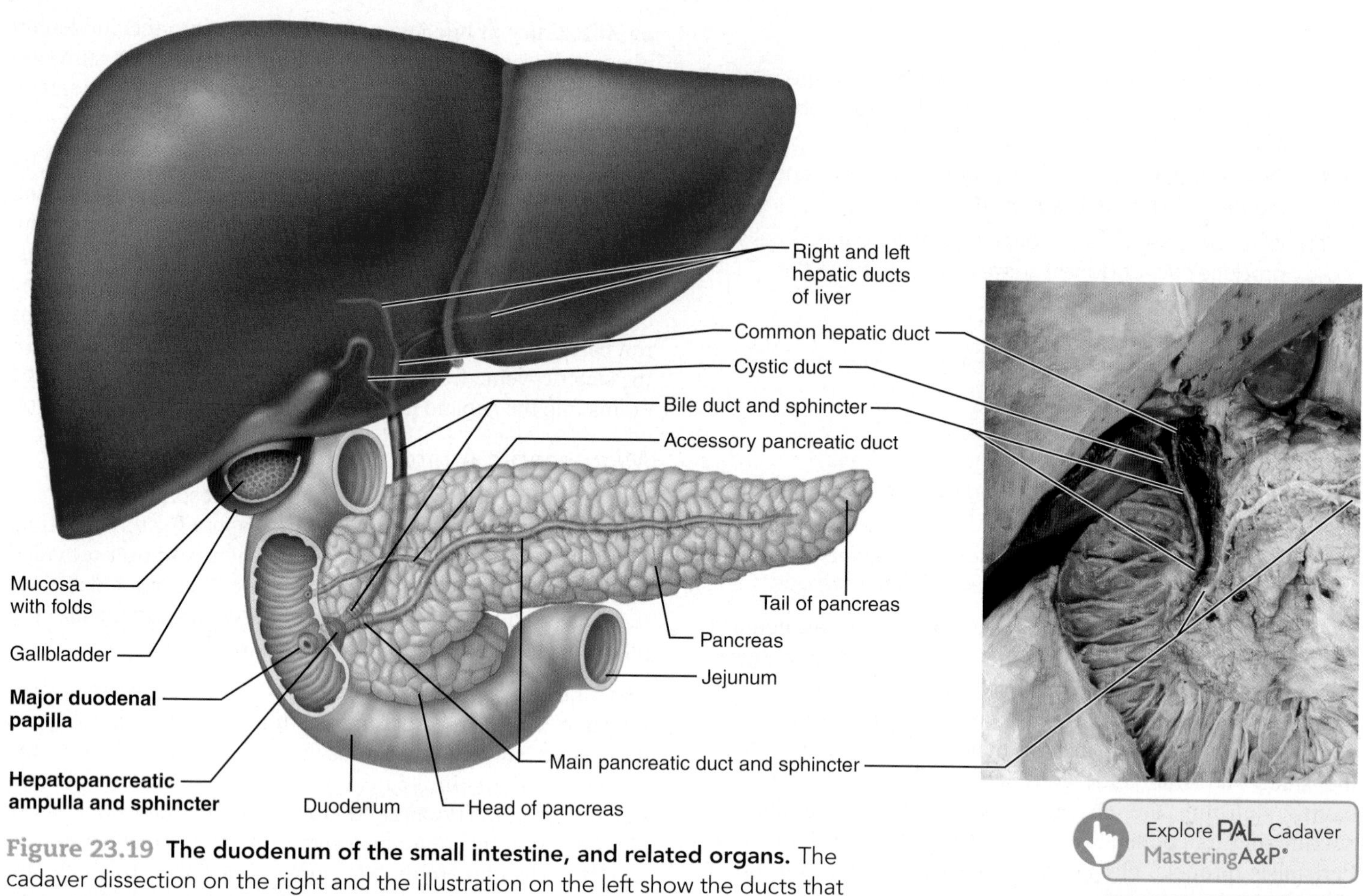

Figure 23.19 The duodenum of the small intestine, and related organs. The cadaver dissection on the right and the illustration on the left show the ducts that open into the duodenum from the pancreas, gallbladder, and liver.

villi and the nutrients in the intestinal lumen, and they also squeeze lymph through the lacteals.

The apical surfaces of the absorptive enterocytes have many **microvilli** (Figure 23.20b). Whereas such projections occur on most epithelial surfaces in the body, those in the small intestine are exceptionally long and densely packed. Besides amplifying the absorptive surface, the plasma membrane of these microvilli contains enzymes that complete the final stages of the breakdown of nutrient molecules.

The amount of absorptive surface in the small intestine is remarkable. Together, the circular folds, villi, and microvilli increase the intestinal surface area to about 200 square meters, equivalent to the floor area of an average two-story house!

Histology of the Wall All typical layers of the alimentary canal occur in the small intestine. The lining epithelium of the *mucosa,* which occurs not only on the villi but also on the intestinal surface between villi, contains the previously mentioned absorptive enterocytes plus some scattered goblet cells and enteroendocrine cells (see also Figure 23.24, p. 705):

- **Absorptive enterocytes** (Figure 23.20b and c). These cells contain many mitochondria because the uptake of digested nutrients is an energy-demanding process. They also contain an abundant endoplasmic reticulum, which assembles the newly absorbed lipid molecules into lipid-protein complexes called *chylomicrons* (ki″lo-mi′kronz). Once made, the chylomicrons enter the lacteal capillaries, so it is in this form that absorbed fat enters the circulation.
- **Goblet cells** (Figure 23.20b and c). These cells secrete onto the internal surface of the intestine a coat of mucus that lubricates the chyme and forms a protective barrier that prevents enzymatic digestion of the intestinal wall.
- **Enteroendocrine cells** (Figure 23.20b). The enteroendocrine cells of the duodenum secrete several hormones, which signal the gallbladder to release stored bile and the pancreas to secrete digestive enzymes and a bicarbonate-rich juice to neutralize the acidic chyme entering the duodenum.

Between the villi, the mucosa contains invaginations called **intestinal crypts** (Figure 23.20b). The epithelial cells that line these crypts secrete *intestinal juice,* a watery liquid that mixes with chyme in the intestinal lumen. Two other cell types are found within the intestinal crypts.

- **Undifferentiated epithelial cells** line the intestinal crypts and renew the mucosal epithelium by dividing rapidly and moving continuously onto the villi. These are among the most quickly dividing cells of the body, completely renewing the inner epithelium of the small intestine every 3–6 days. Such rapid replacement is necessary because individual

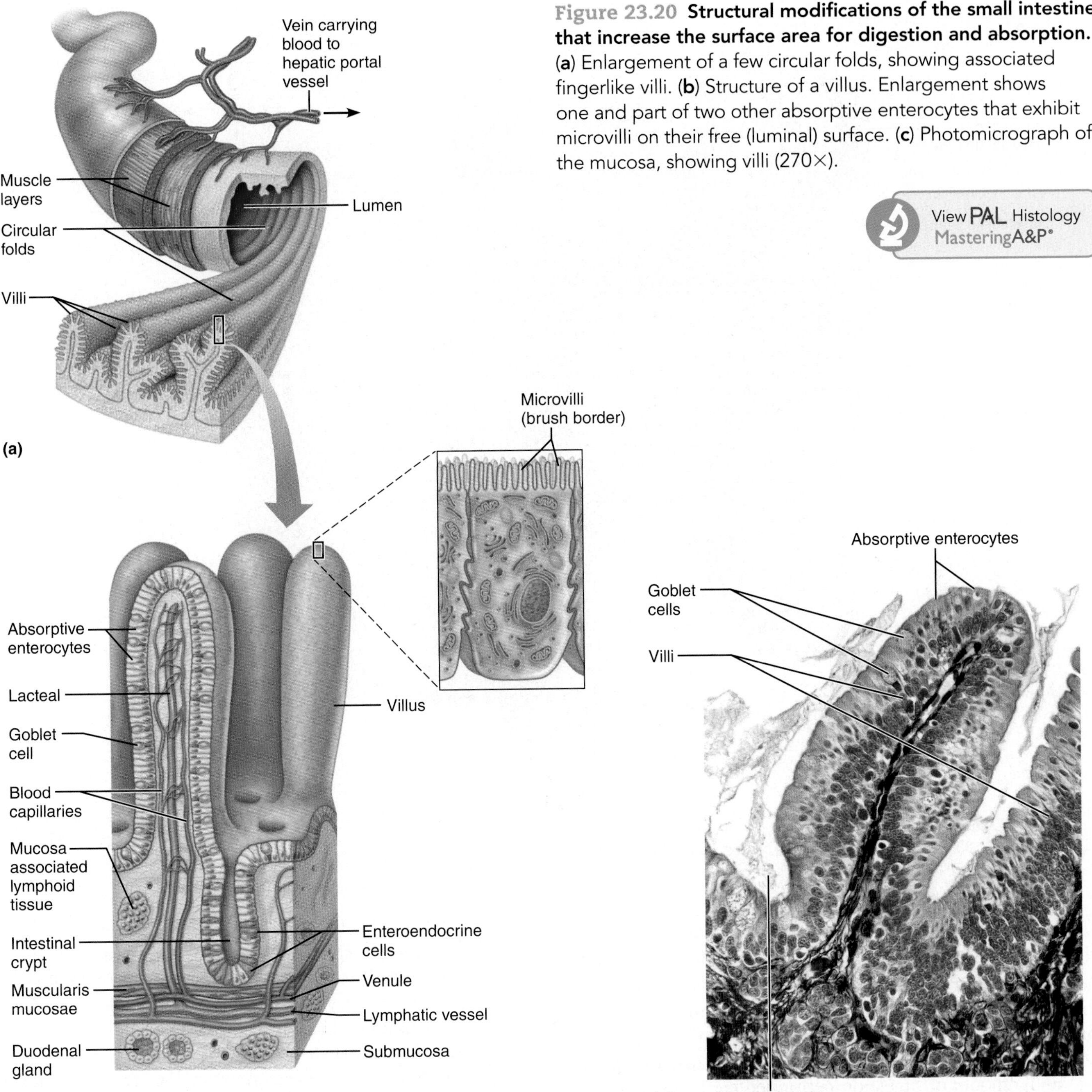

Figure 23.20 Structural modifications of the small intestine that increase the surface area for digestion and absorption. (**a**) Enlargement of a few circular folds, showing associated fingerlike villi. (**b**) Structure of a villus. Enlargement shows one and part of two other absorptive enterocytes that exhibit microvilli on their free (luminal) surface. (**c**) Photomicrograph of the mucosa, showing villi (270×).

View PAL Histology MasteringA&P®

epithelial cells cannot withstand the destructive effects of the digestive enzymes in the intestinal lumen for long.

- Mature **Paneth cells** (see Figure 23.24, p. 705) are found at the base of the crypt. These epithelial cells secrete enzymes that destroy certain bacteria and may help determine which kinds of bacteria live in the intestinal lumen. The permanent bacterial residents of the intestinal lumen, called the *intestinal flora,* manufacture some essential vitamins, which the intestines absorb. Vitamin K is one substance produced by the intestinal bacteria.

The small intestine contains many areas of lymphoid tissue. Mucosa associated lymphoid tissue (MALT) is found in the mucosal layer throughout the intestine, and *aggregated lymphoid nodules* (*Peyer's patches*) are located in the submucosa of the ileum (see p. 640).

The *submucosa* of the small intestine is a typical connective tissue. In the duodenum only, it contains a set of compound tubular **duodenal glands,** whose ducts open into the intestinal crypts (see Figure 23.20b). These glands secrete an alkaline, bicarbonate-rich mucus that helps neutralize the acidity of the chyme from the stomach and contributes to the protective layer of mucus on the inner surface of the small intestine.

The outer layers of the small intestine (*muscularis externa* and *serosa*) have no unusual features.

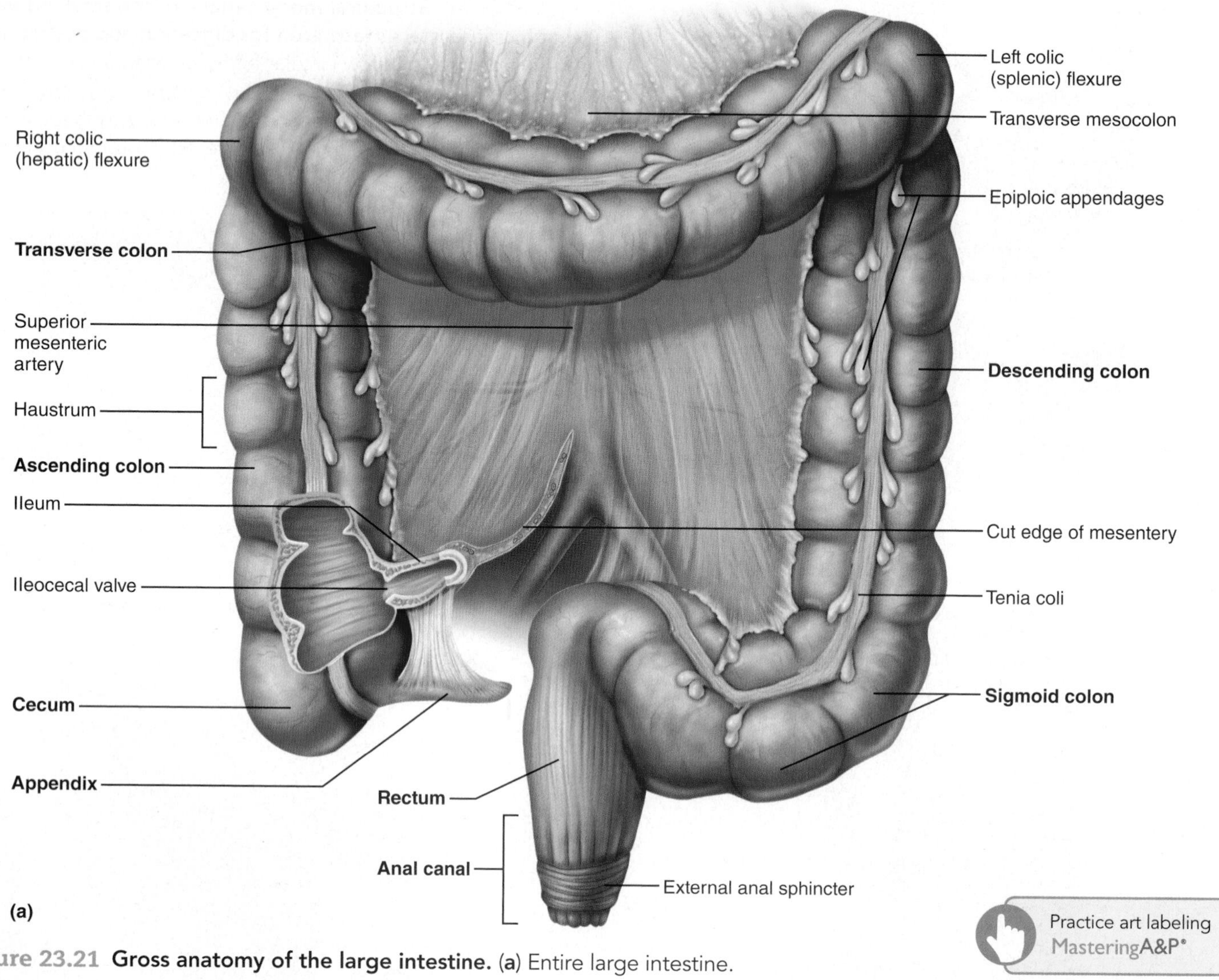

Figure 23.21 Gross anatomy of the large intestine. (a) Entire large intestine.

The Large Intestine

The **large intestine** is the last major organ in the alimentary canal (Figure 23.21). The material that reaches it is a largely digested residue that contains few nutrients. During the 12–24 hours that this residue remains in the large intestine, little additional breakdown of food occurs, except for the small amount of digestion performed by the many bacteria living there. Even though the large intestine absorbs these few remaining nutrients, its main function is to absorb water and electrolytes from the digested mass, resulting in semisolid feces. Propulsion through the large intestine is sluggish and weak, except for **mass peristaltic movements,** which pass over the colon a few times a day to force the feces powerfully toward the rectum.

Gross Anatomy

The large intestine frames the small intestine on 3½ sides, forming an open rectangle (see Figure 23.1). This organ, which is wider than the small intestine but less than half as long (1.5 meters), has the following subdivisions: *cecum, appendix, colon, rectum,* and *anal canal* (Figure 23.21a).

Over most of its length, the large intestine exhibits three special features: teniae coli, haustra, and epiploic appendages. **Teniae (taeniae) coli** (te′ne-e ko′li; "ribbons of the colon") are three longitudinal strips, spaced at equal intervals around the circumference of the cecum and colon. They are thickenings of the longitudinal layer of the muscularis externa, which is thin except at these sites. Because the teniae maintain muscle tone, they cause the large intestine to pucker into sacs, or **haustra** (haw′strah; "to draw up"). **Epiploic appendages** (ep″ĭ-plo′ik; "membrane-covered"), also called *omental appendices,* are fat-filled pouches of visceral peritoneum that hang from the intestine. Their significance is unknown.

The Cecum and Appendix The large intestine begins with the saclike **cecum** (se′kum; "blind pouch") in the right iliac fossa. The opening of the ileum of the small intestine into the cecum's medial wall is surrounded internally by the **ileocecal valve** (Figure 23.21a), which is formed by two raised edges of the mucosa. A sphincter in the distal ileum keeps the valve closed until there is food in the stomach, at which time the sphincter reflexively relaxes, opening the valve. As the cecum fills, its walls stretch, pulling the edges of the ileocecal valve together and closing the opening. This action prevents reflux of feces from the cecum back into the ileum.

The **appendix** is a blind tube that opens into the posteromedial wall of the cecum. Although almost always illustrated as hanging inferiorly, it more often lies "tucked up" posterior to the cecum in the right iliac fossa. The appendix has large masses of lymphoid tissue in its wall. Commonly considered a vestigial organ, current research proposes that the appendix functions as a safe haven for the beneficial bacteria that inhabit the large intestine. According to this theory, beneficial bacteria from the appendix can repopulate the gut following an infectious disease that causes diarrhea and flushes out the intestinal flora.

CLINICAL APPLICATION

Appendicitis Acute inflammation of the appendix, called **appendicitis,** results from a blockage that traps infectious bacteria within its lumen. The blockage often is caused by a lump of feces or by a virus-induced swelling of the lymphoid tissue of the appendix wall. Unable to empty its contents, the blocked appendix swells with the mucus it secretes, squeezing off its venous drainage and leading to ischemic necrosis and infection. If the appendix ruptures, bacteria and feces are released into the peritoneum, causing peritonitis.

Because the symptoms of appendicitis vary greatly, this condition is notoriously difficult to diagnose. Often, however, the first symptom is pain in the umbilical region, followed by loss of appetite, fever, nausea, vomiting, and relocalization of pain to the lower right quadrant of the abdominal surface. Palpation of this region that causes strong pain after the pressure is removed (so-called rebound tenderness) can indicate appendicitis.

Immediate surgical removal of the appendix, called **appendectomy** (ap″en-dek′to-me), is the usual treatment.

The Colon The **colon** (ko′lon) has several distinct segments (Figure 23.21a). From the cecum, the **ascending colon** ascends along the right side of the posterior abdominal wall in a secondarily retroperitoneal position and reaches the level of the right kidney, where it makes a right-angle turn, the **right colic flexure** (also called the **hepatic flexure** because the liver lies directly superior to it). From this flexure, the **transverse colon** extends intraperitoneal to the left across the peritoneal cavity. Directly anterior to the spleen, it bends acutely downward at the **left colic (splenic) flexure** and descends along the left side of the posterior abdominal wall again in a secondarily retroperitoneal position as the **descending colon.** Inferiorly, the colon becomes intraperitoneal and enters the true pelvis as the S-shaped **sigmoid colon** (*sigma* = the Greek letter corresponding to the letter *s*).

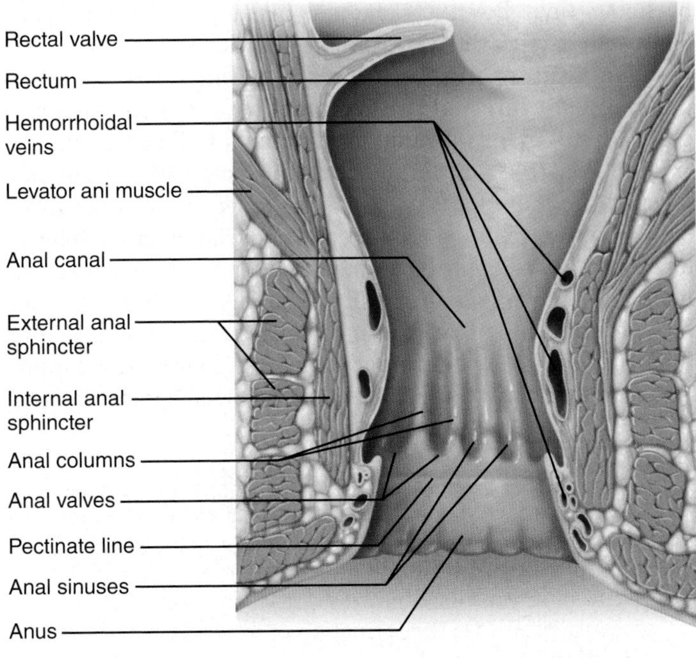

Figure 23.21 Gross anatomy of the large intestine, *continued.* **(b)** The inferior rectum and anal canal in frontal section.

The Rectum In the pelvis, the sigmoid colon joins the **rectum** (Figure 23.21a), which descends along the inferior half of the sacrum in a secondarily retroperitoneal position. The rectum has no teniae coli; its longitudinal muscle layer is complete and well developed, so that it can generate strong contractions for defecation. Even though the word *rectum* means "straight," the rectum actually has several tight bends. Internally, these bends are represented as three *transverse folds of the rectum,* or **rectal valves** (Figure 23.21b), which prevent feces from being passed along with flatus (gas).

The Anal Canal The last subdivision of the large intestine is the **anal canal** (see Figure 23.21b). About 3 cm long, it begins where the rectum passes through the levator ani, the muscle that forms the pelvic floor. A portion of the levator ani is responsible for maintaining the anorectal angle, an acute angle between the anus and the rectum that contributes to fecal continence. The anal canal lies entirely external to the abdominopelvic cavity in the perineum.

Internally, the superior half of the anal canal contains longitudinal folds of mucosa, the **anal columns.** These columns contain the terminal portions of the superior rectal artery and vein (the hemorrhoidal vessels). Neighboring anal columns join each other inferiorly at crescent-shaped transverse folds called **anal valves.** The pockets just superior to these valves are **anal sinuses,** which release mucus when they are compressed by feces, providing lubrication that eases fecal passage during defecation.

The horizontal line along which the anal valves lie is called the *pectinate* ("comb-shaped") *line.* Because the mucosa superior to this line is innervated by visceral sensory fibers, it is relatively insensitive to pain. Inferior to the pectinate line, however, the mucosa is sensitive to pain because it is innervated by somatic nerves.

The wall of the anal canal contains two sphincter muscles: an **internal anal sphincter** of smooth muscle and an **external anal sphincter** of skeletal muscle (Figure 23.21b). The former is a thickening of the circular layer of the muscularis, whereas the latter is a distinct muscle (shown in Figure 11.15 on p. 297). The external sphincter contracts voluntarily to inhibit defecation, whereas the internal sphincter contracts involuntarily, both to prevent feces from leaking from the anus between defecations and to inhibit defecation during emotional stress. During toilet training, children learn to control the external anal sphincter.

CLINICAL APPLICATION

Diverticulosis and Diverticulitis When the diet lacks fiber, the contents of the colon are reduced in volume, and the contractions of the circular muscle in the colon exert greater pressures on its wall. This pressure promotes the formation of multiple sacs called **diverticula** (dī″ver-tik′u-lah), which are small outward herniations of the mucosa through the colon wall. The resulting condition is termed **diverticulosis.** This condition arises most frequently in the sigmoid colon. Occurring in 30% to 40% of all Americans over age 50, and in half of those over 70, diverticulosis generally leads to nothing more than dull pain, although it may rupture an artery in the colon and produce bleeding from the anus. Increasing the amount of fiber in the diet generally relieves the symptoms.

In about 20% of diverticulosis cases, however, patients develop the more serious condition called **diverticulitis,** in which the inflamed diverticula become infected and may perforate, leaking feces into the peritoneal cavity. In serious cases, the affected region of the colon is removed by surgery and antibiotics are given to fight the peritonitis.

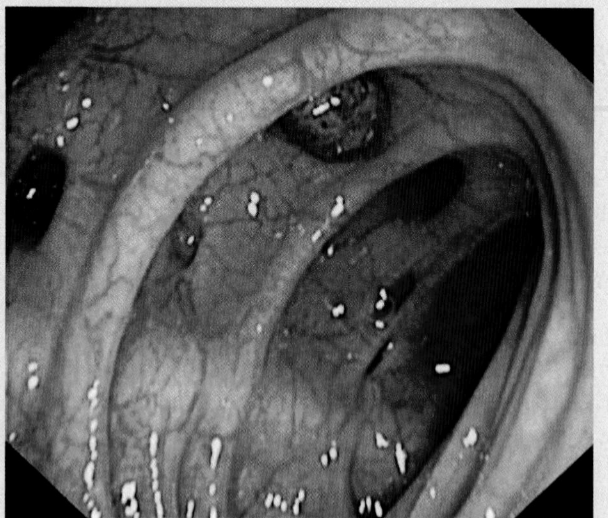

Diverticula in the wall of the sigmoid colon

Vessels and Nerves The first half of the large intestine—to a point two-thirds of the way along the transverse colon—is supplied by the superior mesenteric vessels. Its sympathetic innervation is from the superior mesenteric and celiac ganglia and plexuses, and its parasympathetic innervation is from the vagus nerve.

The distal half of the large intestine, up to the proximal portion of the rectum, is supplied by the inferior mesenteric vessels. The lower rectum and the anal canal are served by rectal branches of the internal iliac vessels. The sympathetic innervation of the distal half of the large intestine is via the inferior mesenteric and hypogastric plexuses, and the parasympathetic innervation is from the pelvic splanchnic nerves. As in other areas of the alimentary canal, autonomic innervation influences the intrinsic enteric neurons. The final part of the anal canal below the pectinate line is innervated by somatic nerves, such as the pudendal nerve.

CLINICAL APPLICATION

Hemorrhoids Varicose veins of the hemorrhoidal veins in the anal canal (see Figure 23.21b) are called **hemorrhoids;** they often result from straining to deliver a baby or to defecate. Because they are stretched and inflamed, the swollen veins throb and bulge into the lumen of the anal canal. Internal hemorrhoids occur above the pectinate line, and external hemorrhoids occur below the pectinate line. External hemorrhoids are itchier and more painful, but only internal hemorrhoids tend to bleed. About 75% of Americans develop hemorrhoids at some time in their lives.

Severe hemorrhoids are often treated by tying them at their base with small rubber bands, whereupon they wither and fall away. They can also be injected with a hardening agent or exposed to electricity or strong infrared light to coagulate the blood within them. These unpleasant-sounding treatments are simple and effective and have largely eliminated the difficult surgical removals that were formerly performed.

Defecation The rectum is usually empty and the anal sphincters contracted. When feces are squeezed into the rectum by mass peristaltic movements, the stretching of the rectal wall initiates the defecation reflex (**Figure 23.22**, ①). Mediated by the sacral spinal cord, this parasympathetic reflex signals the walls of the sigmoid colon and rectum to contract and the internal anal sphincter to relax (Figure 23.22, ②). If one decides to delay defecation, the reflexive contractions end, and the rectum relaxes. Another mass movement occurs a few minutes later, initiating the defecation reflex again—and so on, until one chooses to defecate (Figure 23.22, ③) or the urge to defecate becomes unavoidable.

During defecation, the musculature of the rectum contracts to expel the feces. This process is supplemented by the voluntary contraction of the diaphragm and the abdominal wall muscles, which increases intra-abdominal pressure, and of the levator ani muscle (diagrammed on p. 297), which lifts the anal canal superiorly, leaving the feces inferior to the anus and thus outside the body.

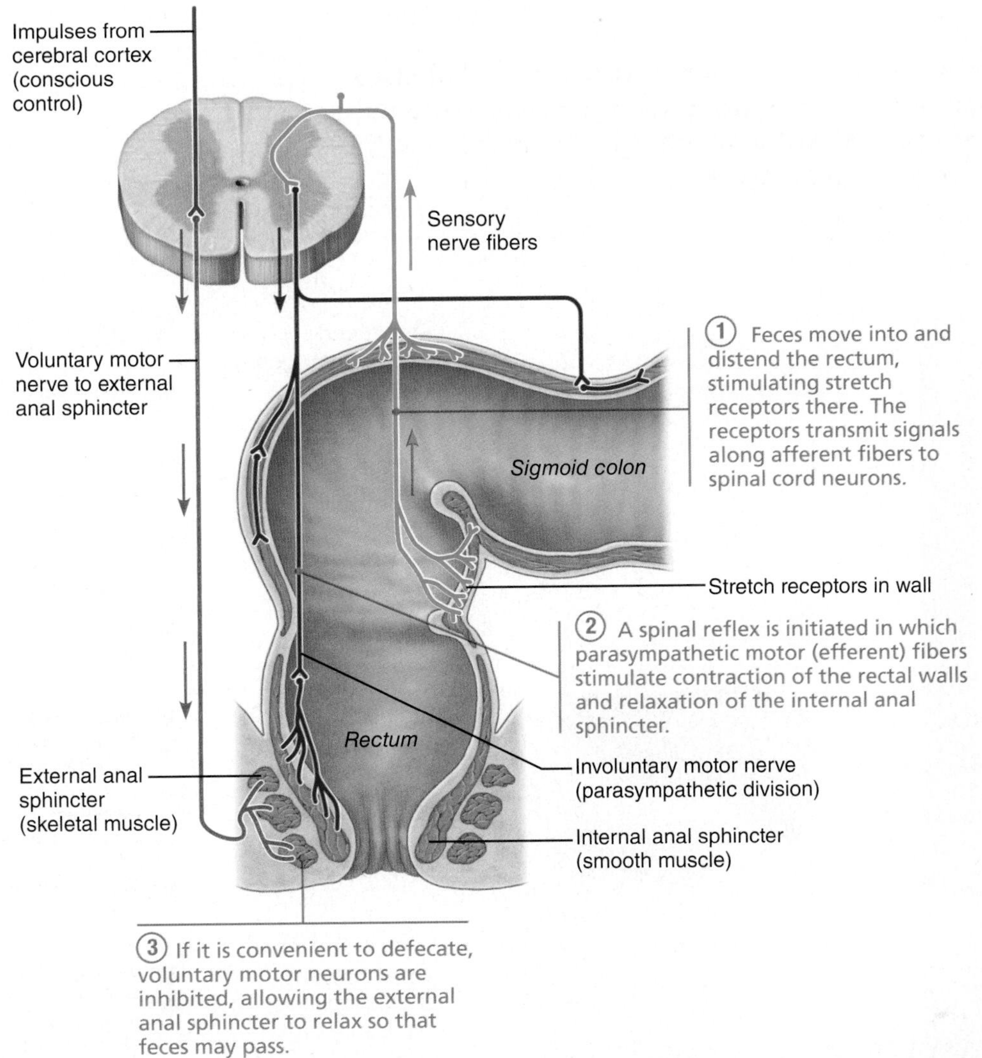

Figure 23.22 **Defecation reflex.**

Microscopic Anatomy

The wall of the large intestine (Figure 23.23) resembles that of the small intestine in some ways and differs from it in others.

- The *mucosal* epithelium of the colon is a simple columnar epithelium containing the same cell types as in the small intestine. Goblet cells are more abundant in the large intestine, for they secrete large amounts of lubricating mucus that eases the passage of feces toward the end of the alimentary canal. The absorptive cells, called **colonocytes**, take in water and electrolytes (see Figure 23.24, p. 705).
- Villi are absent, which reflects the fact that fewer nutrients are absorbed in the large intestine.
- Intestinal crypts are present as simple tubular glands containing many goblet cells. Undifferentiated stem cells occur at the bases of the intestinal crypts, and epithelial cells are fully replaced every week or so.

The other layers of the wall are rather typical. The lamina propria and *submucosa* contain more lymphoid tissue than occurs elsewhere in the alimentary canal, but this is not surprising, considering the extensive bacterial flora of the

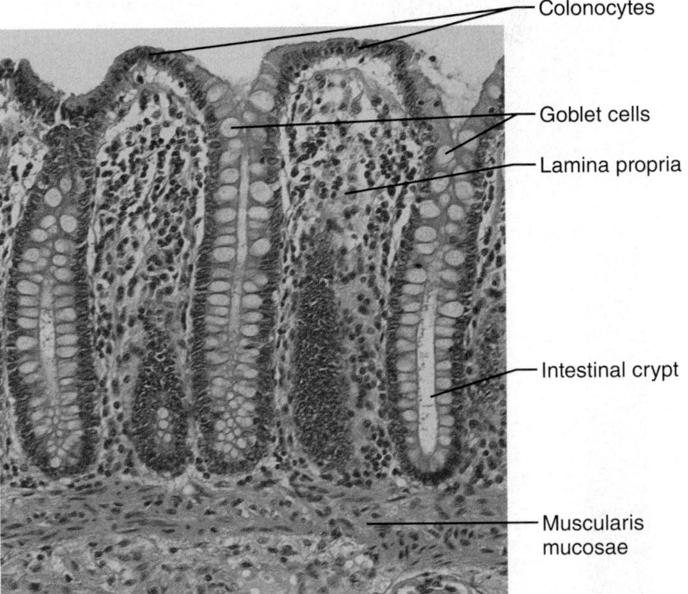

Figure 23.23 **The mucosa of the large intestine.** Photomicrograph. Note the abundance of goblet cells (130×).

Figure 23.24

As a burger travels through the alimentary canal, digestive secretions break down its components into absorbable units. The structure of the wall of the alimentary canal reflects the activities that occur at each region.

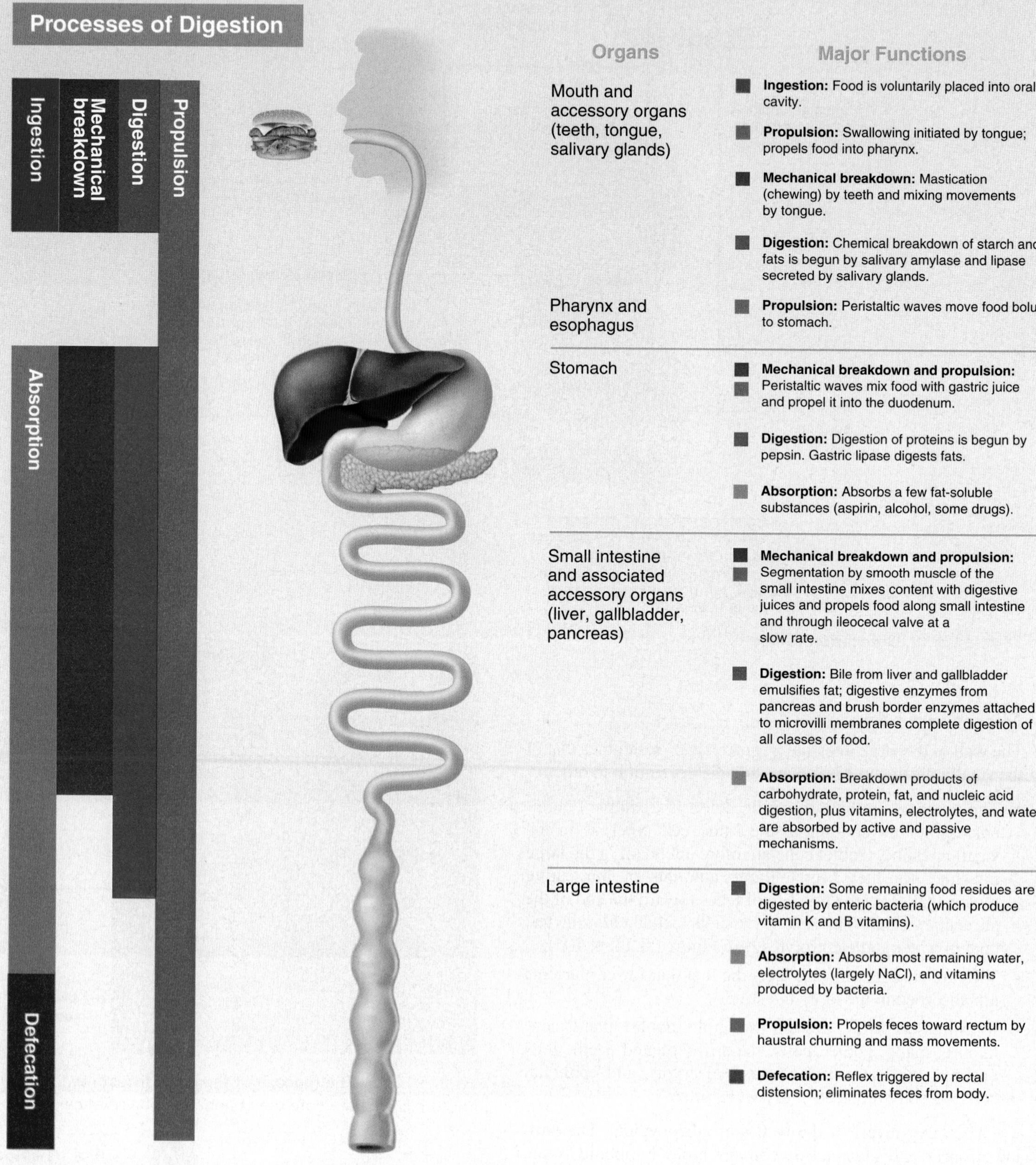

Histology of the Alimentary Canal

Cells in Mucosa

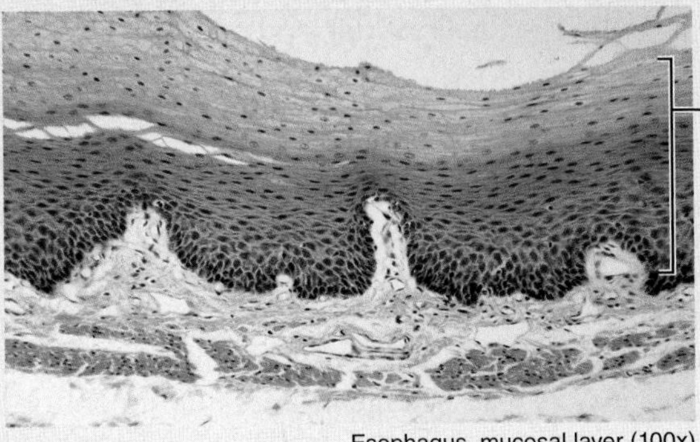

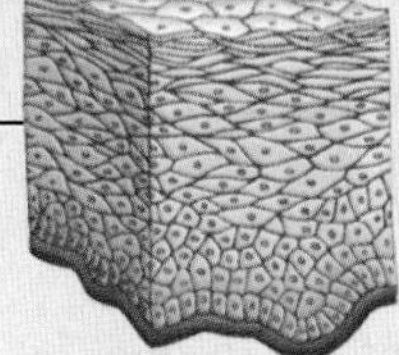

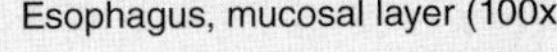

Esophagus, mucosal layer (100x)

Nonkeratinized stratified squamous epithelium
Protects underlying tissues

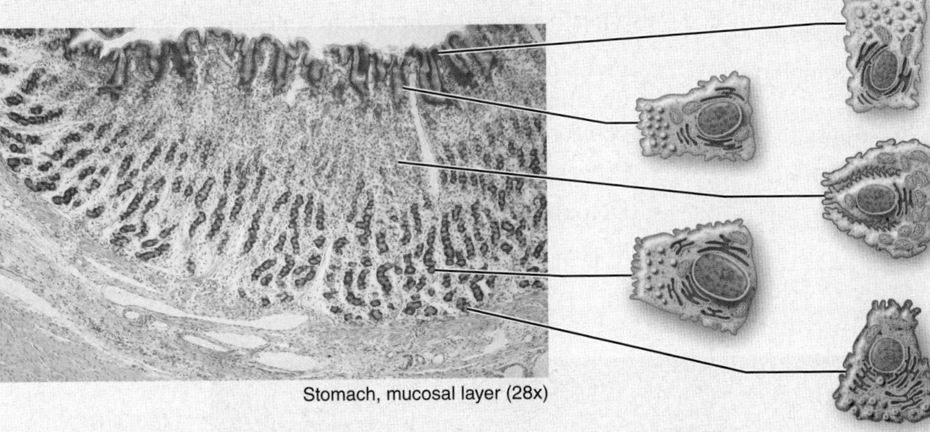

Stomach, mucosal layer (28x)

Surface mucous cell
Secretes mucus

Mucous neck cell
Secretes mucus

Parietal cell
Secretes HCl and gastric intrinsic factor

Chief cell
Secretes pepsinogen; begins protein digestion

Enteroendocrine cell
Secretes gastrin, which stimulates secretion by parietal cells

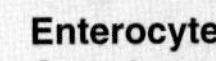

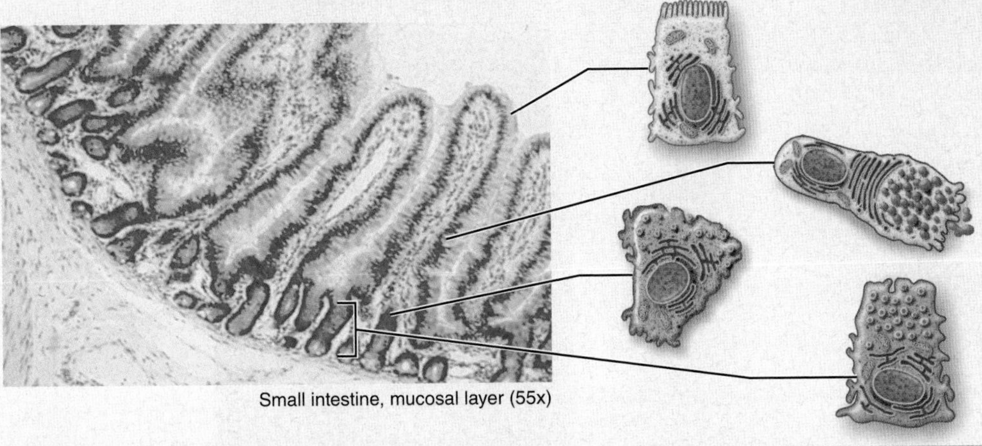

Small intestine, mucosal layer (55x)

Enterocyte
Completes digestion and absorbs nutrients across microvilli

Goblet cell
Secretes mucus

Enteroendocrine cell
Secretes secretin or cholecystokinin (CCK), which stimulates release of bile and pancreatic juice and inhibits stomach secretions

Paneth cell
Secretes substances that destroy bacteria

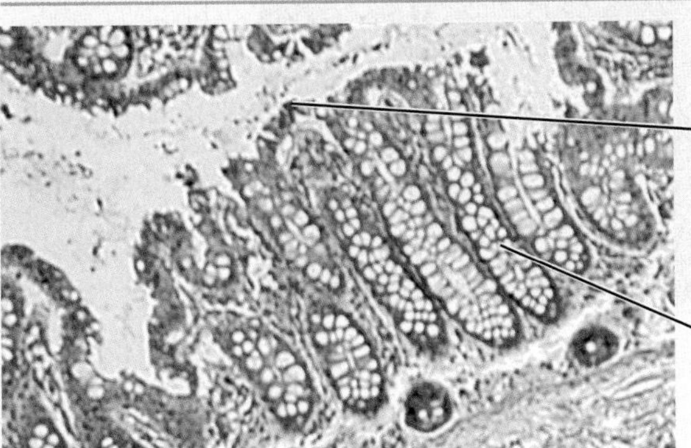

Large intestine, mucosal layer (115x)

Colonocyte
Absorbs water, electrolytes, and vitamins

Goblet cell
Secretes mucus

large intestine. The specializations of the *muscularis externa* and *serosa,* namely the teniae coli and epiploic appendages, were discussed in the section "Gross Anatomy" (p. 700).

The *anal canal* is a zone of epithelial transition in which the simple columnar epithelium of the intestine abruptly changes to stratified squamous epithelium near the level of the pectinate line. At the extreme inferior end of the anal canal, the mucosa merges with the true skin that surrounds the anus.

Focus on Digestive Processes (Figure 23.24) summarizes the digestive activities in each portion of the alimentary canal. The histology of the mucosa in each region is also illustrated, allowing a visual comparison of the distinctive features of this layer throughout the alimentary canal. Modifications of the mucosal layer most obviously differentiate the regions of the alimentary canal.

☑ check your understanding

- ☐ **14.** What is the typical life span for an intestinal epithelial cell? How are the cells of the epithelium replaced?
- ☐ **15.** Name all the parts of the large intestine, beginning with its junction with the ileum.
- ☐ **16.** Name the structures within the villus that receive absorbed nutrients. Which types of nutrients are absorbed into each structure?

(For answers, see Appendix B.)

ANATOMY OF THE ACCESSORY ORGANS

learning outcome

▶ Describe the gross and microscopic anatomy and functions of the liver, gallbladder, and pancreas.

The Liver

The **liver** is the largest gland in the body, weighing about 1.4 kg (3 pounds) in an average adult. Amazingly versatile, it performs over 500 functions. Its digestive function is to produce **bile,** a green alkaline liquid that is stored in the gallbladder and secreted into the duodenum. Bile salts emulsify fats in the small intestine; that is, they break up fatty nutrients into tiny particles, just as dish detergent breaks up a pool of fat drippings in a roasting pan. These smaller particles are more accessible to digestive enzymes from the pancreas. The liver also performs many metabolic functions. The liver:

- Picks up glucose from nutrient-rich blood returning from the alimentary canal and stores this carbohydrate as glycogen for subsequent use by the body.
- Processes fats and amino acids and stores certain vitamins.
- Detoxifies many poisons and drugs in the blood.
- Makes the blood proteins.

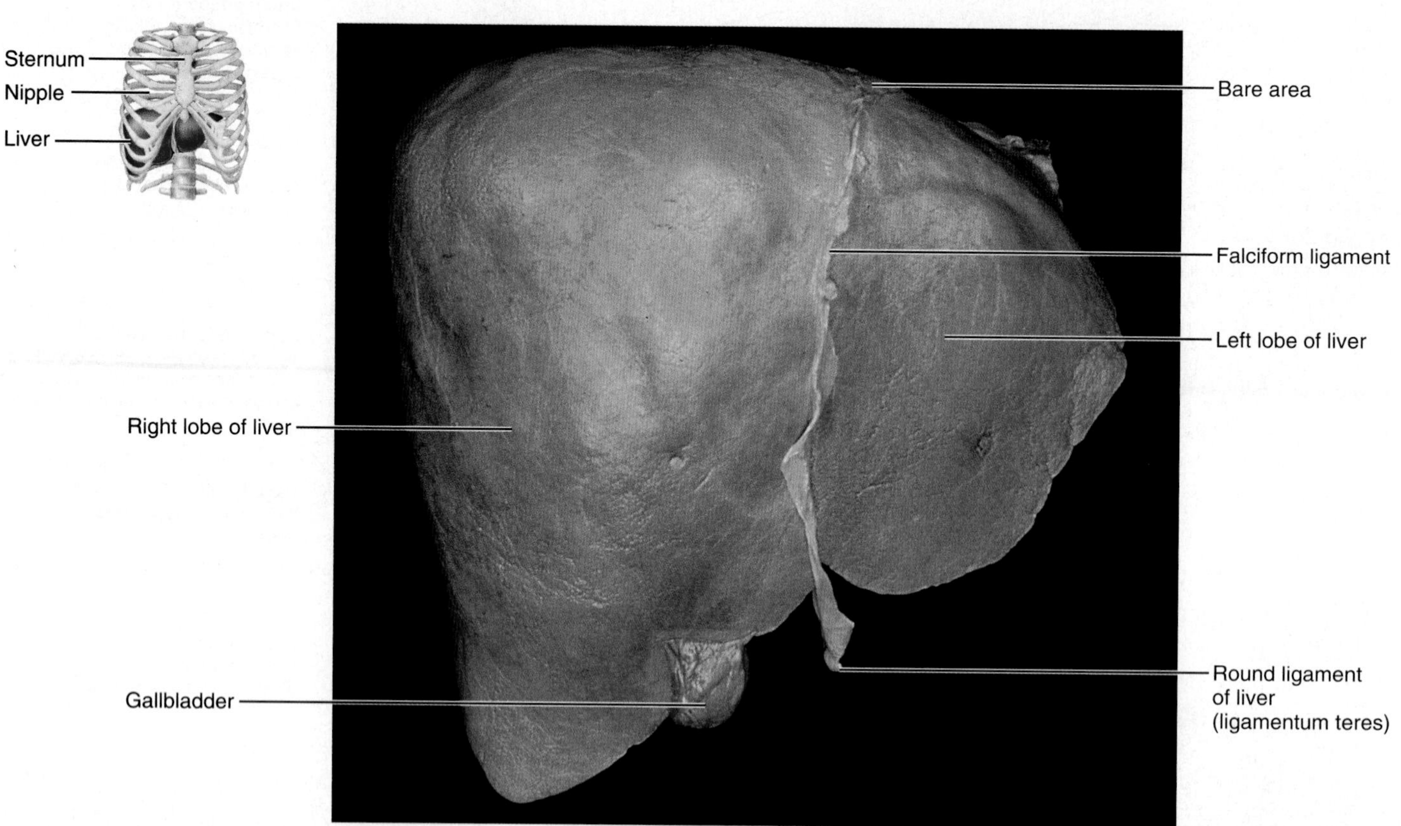

Figure 23.25 Anterior view of the liver.

Explore PAL Cadaver
MasteringA&P®

Almost all of these functions are carried out by a type of cell called a **hepatocyte** (hep′ah-to-sīt″), or simply a *liver cell.*

Gross Anatomy

The liver lies inferior to the diaphragm in the right superior part of the abdominal cavity (Figure 23.4), filling much of the right hypochondriac and epigastric regions and extending into the left hypochondriac region. It lies almost entirely within the rib cage, which protects this highly vascular organ from blows that could rupture it. The liver is shaped like a wedge, the wide base of which faces right and the narrow apex of which lies just inferior to the level of the left nipple.

The liver has two surfaces: the *diaphragmatic* and *visceral* surfaces (Figure 23.25 and Figure 23.26). The **diaphragmatic surface** faces anteriorly and superiorly, whereas the **visceral surface** faces posteroinferiorly. Even though most of the liver is covered with a layer of visceral peritoneum, the superior part, called the **bare area,** is fused to the diaphragm and is therefore devoid of peritoneum.

The liver has a **right lobe** and a **left lobe,** which traditionally were considered to be divided by the **falciform ligament** on the anterior part of the diaphragmatic surface (Figure 23.25) and the **fissure** on the visceral surface (Figure 23.26). The falciform ligament is a vertical mesentery that binds the liver to the anterior abdominal wall, and the fissure is a deep groove in the same sagittal plane as the falciform ligament. Two other lobes, the **quadrate lobe** and the **caudate lobe,** are visible on the visceral surface just to the right of the fissure. Long considered part of the right lobe, these lobes are now considered part of the left lobe, with which they share nerves and vessels.

Nerves and Vessels An important area near the center of the visceral surface is the **porta hepatis** (por′tah hep-ah′-tis; "gateway to the liver"), where most of the major vessels and nerves enter and leave the liver (Figure 23.26). The right and left branches of the *hepatic portal vein,* which carry nutrient-rich blood from the stomach and intestines, enter the porta hepatis, as do the right and left branches of the *hepatic artery proper* carrying oxygen-rich blood to the liver. The **right** and **left hepatic ducts,** which carry bile from the respective liver lobes, exit from the porta hepatis and fuse to form the **common hepatic duct,** which extends inferiorly toward the duodenum (shown in Figure 23.19). Autonomic nerves reach the liver from the celiac plexus and consist of both sympathetic and parasympathetic (vagal) fibers. Other important structures on the liver's visceral surface are the *gallbladder* and the *inferior vena cava,* which lie to the right of the quadrate and caudate lobes, respectively. The inferior vena cava receives the *hepatic veins* carrying blood out of the liver.

Several structures pass through the liver's fissure. Lying in the fissure's inferior half is the **round ligament of liver,** or **ligamentum teres** (*teres* = round). This cordlike ligament, the remnant of the umbilical vein in the fetus, ascends to the liver from the navel, within the inferior margin of the falciform ligament. Additionally, the superior half of the liver's fissure contains the **ligamentum venosum** (Figure 23.26), a cordlike remnant of the ductus venosus of the fetus (see pp. 620–623 for a review of fetal circulation).

Microscopic Anatomy

The liver contains over a million classical **liver lobules** (Figure 23.27a), each about the size of a sesame seed. Each lobule is shaped like a hexagonal (six-sided) solid and consists of plates of liver cells, or **hepatocytes,** radiating out from a **central vein** (Figure 23.27b and c). If you were to look at the top of a thick paperback book opened so wide that its two covers touched each other, you would have a rough model of the liver lobule, with the spreading pages representing the plates of hepatocytes, and the hollow cylinder formed by the rolled spine representing the central vein. The hepatocytes in each plate are organized like bricks in a wall.

At almost every corner of the lobule is a **portal triad** (tri′ad; "three"; Figure 23.27c). The portal triad contains three main vessels: a portal arteriole that is a branch of the hepatic artery, a portal venule that is a branch of the hepatic portal vein, and a **bile duct** (which carries bile away from the liver lobules). Note that the blood vessels bring both arterial and venous blood to the lobule. The arterial blood supplies the hepatocytes with oxygen, and blood from the portal vein delivers substances from the intestines for processing by the hepatocytes.

Between the plates of hepatocytes are large capillaries, the **liver sinusoids.** Near the portal triads, these sinusoids receive blood from both the portal arteriole and venule and carry this blood inward to reach the central vein (Figure 23.27c). From there, the central veins form tributaries (interlobular veins) that ultimately lead to the hepatic veins and then to the inferior vena cava outside the liver.

In the walls of the sinusoids are **stellate macrophages** (or *hepatic macrophages*), which destroy bacteria and other foreign particles in the blood flowing past them. Thus, even though microorganisms in the intestine may enter the intestinal capillaries, few of them make it past the liver. Besides cleansing the blood of microorganisms, the stellate macrophages also destroy worn out blood cells—as do macrophages of the spleen and bone marrow.

The liver sinusoids are lined by an exceptionally leaky, fenestrated endothelium (Figure 23.27c). Vast quantities of blood plasma pour out of the sinusoids, bathing the hepatocytes with fluid. Hepatocytes require proximity to such a large blood supply because so many of their functions depend on interactions with the fluid portion of blood.

Hepatocytes possess a large number of many different organelles that enable them to carry out their many functions:

- The abundant rough ER manufactures the blood proteins.
- The well-developed smooth ER helps produce bile salts and detoxifies bloodborne poisons.
- Abundant peroxisomes detoxify other poisons (including alcohol; the mechanism for this process is discussed on p. 32).
- The large Golgi apparatus packages the abundant secretory products from the ER.

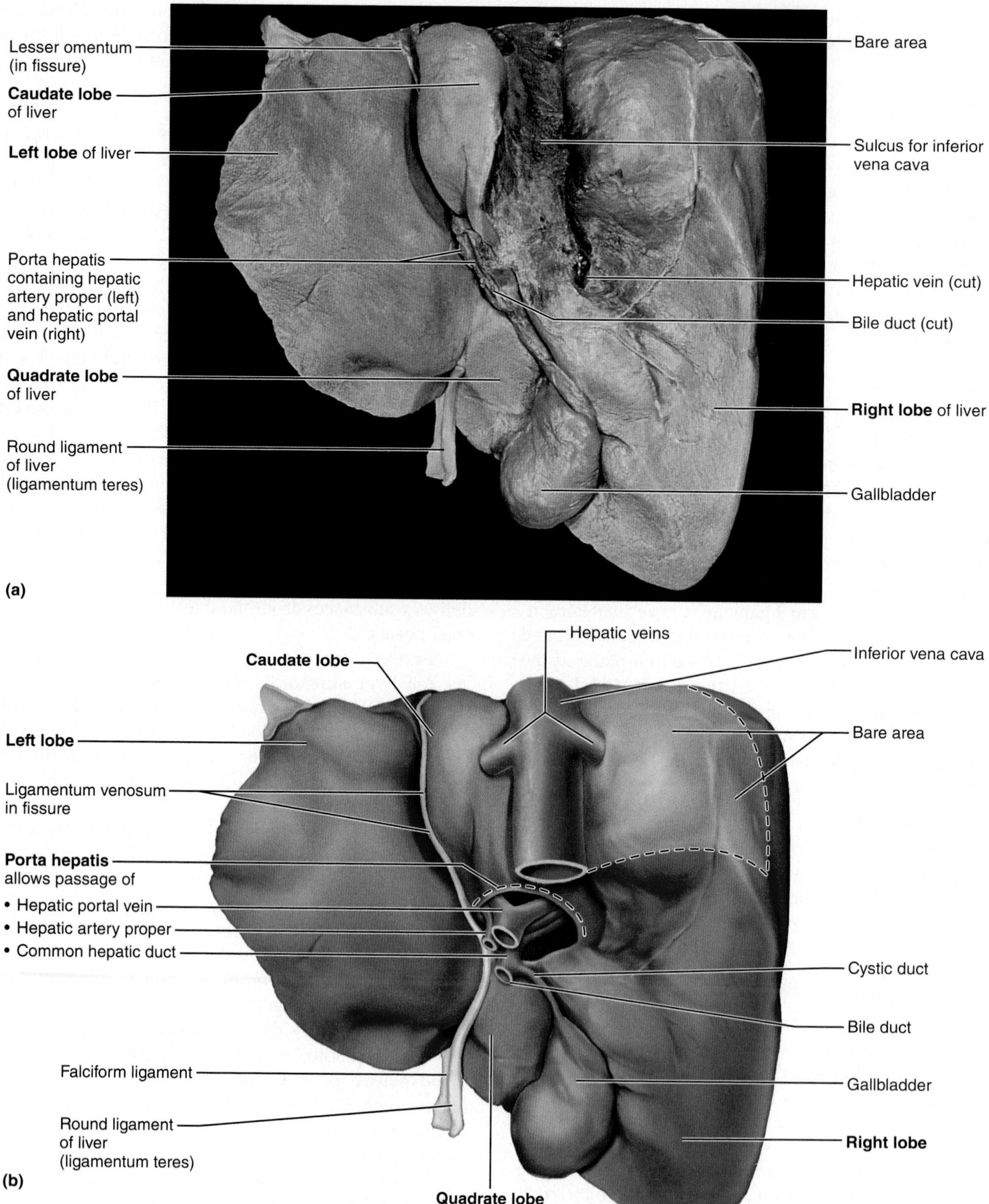

Lesser omentum (in fissure)
Caudate lobe of liver
Left lobe of liver
Porta hepatis containing hepatic artery proper (left) and hepatic portal vein (right)
Quadrate lobe of liver
Round ligament of liver (ligamentum teres)
Bare area
Sulcus for inferior vena cava
Hepatic vein (cut)
Bile duct (cut)
Right lobe of liver
Gallbladder
(a)

Hepatic veins
Caudate lobe
Inferior vena cava
Left lobe
Bare area
Ligamentum venosum in fissure
Porta hepatis allows passage of
• Hepatic portal vein
• Hepatic artery proper
• Common hepatic duct
Cystic duct
Bile duct
Falciform ligament
Gallbladder
Round ligament of liver (ligamentum teres)
Right lobe
Quadrate lobe
(b)

Figure 23.26 **Visceral surface of the liver (posteroinferior view).** (**a**) Photograph. (**b**) Illustration. Note the vessels and ducts that enter and leave the liver.

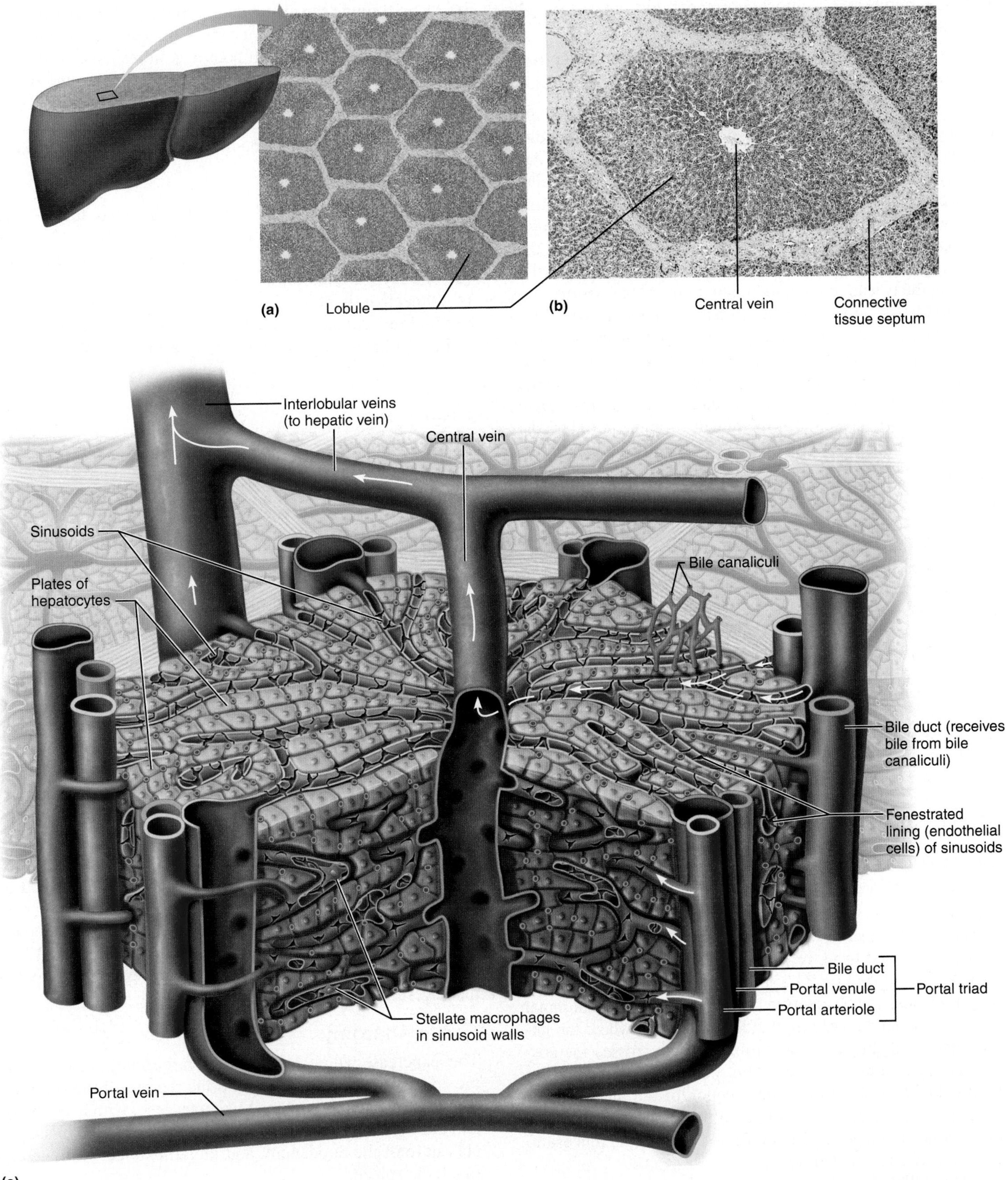

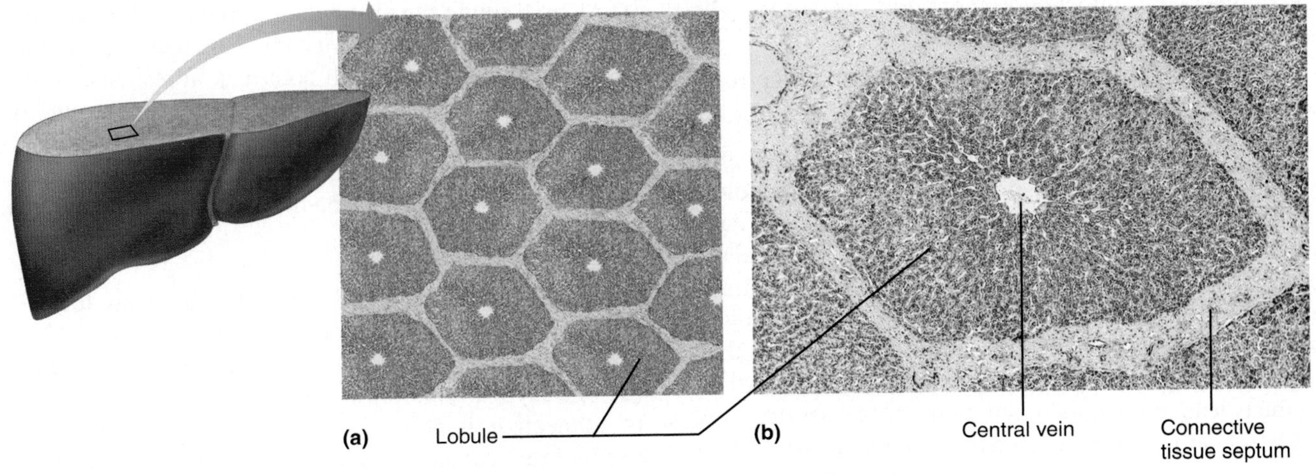

Figure 23.27 **Microscopic anatomy of the liver.** (**a**) Normal lobular pattern of the liver. (**b**) Enlarged view of one liver lobule. (**c**) Three-dimensional representation of a small portion of one liver lobule, showing the structure of sinusoids. Arrows indicate the direction of blood flow.

- Large numbers of mitochondria provide energy for all these processes.
- The numerous glycosomes store sugar, reflecting the role of hepatocytes in blood sugar regulation.

Collectively, hepatocytes produce about 500–1000 ml of bile each day. The secreted bile enters tiny intercellular spaces or channels, called **bile canaliculi** ("little canals"), that lie between adjacent hepatocytes (Figure 23.27c). These canaliculi carry bile outward through each lobule, emptying into the bile ducts in the portal triads. From there, the bile flows into progressively larger ducts, exiting the liver through the hepatic ducts at the porta hepatis. Beyond this, additional bile-carrying ducts lead to the duodenum, as described shortly.

Finally, hepatocytes have a great capacity for cell division and regeneration: Judging from experiments on laboratory animals, if half of a person's liver were removed, it would regenerate in a few weeks! Cellular replacement occurs through the division of mature hepatocytes and of *liver stem cells,* which are located near the bile ducts at the portal triads.

CLINICAL APPLICATION

Cirrhosis A progressive inflammation of the liver is called **cirrhosis** (sĭ-ro′sis; "orange-colored"); it usually results from chronic alcoholism. Even though the alcohol poisoned hepatocytes are continuously replaced, the liver's connective tissue regenerates faster, so the liver becomes fibrous and fatty, and its function declines. The scar tissue impedes the flow of blood through the liver, causing portal hypertension, elevation of blood pressure in the hepatic portal vessels. The patient may grow confused or comatose as toxins accumulate in the blood and depress brain functions. Besides alcoholism, other causes of cirrhosis include hepatitis (see "Viral Hepatitis" on p. 712) and autoimmune attack on the bile ducts.

The Gallbladder

Gross Anatomy

The **gallbladder** is a muscular sac, resting in a shallow depression on the visceral surface of the right lobe of the liver (Figure 23.26). It stores and concentrates bile produced by the liver. Its rounded head, or *fundus,* protrudes from the liver's inferior margin (Figure 23.25). The position of the fundus of the gallbladder can be identified superficially on the abdominal wall. It is located on the right side just deep to where the lateral margin of the rectus abdominis muscle crosses the costal margin of the rib cage. (This point is diagrammed on Figure 11.31, p. 336, at the intersection of the linea semilunaris and the costal margin.) Internally, a honeycomb pattern of mucosal foldings (see Figure 23.19) enables the mucosa to expand as the gallbladder fills.

The gallbladder's duct, the **cystic duct** (*cyst* = bladder; Figure 23.19), joins the common hepatic duct from the liver to form the **bile duct,** which empties into the duodenum. The liver secretes bile continuously, but sphincters at the end of the bile duct and at the hepatopancreatic ampulla are closed when bile is not needed for digestion. At these times, bile backs up through the cystic duct into the gallbladder for storage. When fatty chyme from a meal enters the duodenum, the gallbladder's muscular wall contracts in response to the hormone *cholecystokinin,* which is released from the enteroendocrine cells of the duodenum. The sphincters at the end of the duct system relax, and bile is expelled from the gallbladder through the cystic duct to the bile duct into the duodenum.

Microscopic Anatomy

Histologically, the wall of the gallbladder has fewer layers than the wall of the alimentary canal: (1) a mucosa consisting of a simple columnar epithelium and a lamina propria, (2) one layer of smooth muscle, and (3) a thick outer layer of connective tissue that is covered by a serosa wherever it is not in direct contact with the liver. The columnar cells of the lining epithelium concentrate the bile by absorbing some of its water and ions.

CLINICAL APPLICATION

Gallstones Bile is the normal vehicle in which cholesterol is excreted from the body, and bile salts keep the cholesterol dissolved within bile. Either too much cholesterol or too few bile salts can lead to the crystallization of cholesterol in the gallbladder, producing **gallstones** that can plug the cystic duct and cause agonizing pain when the gallbladder or its duct contracts. Gallstones are easy to diagnose because they show up well with ultrasound imaging. Treatments include administering drugs that might dissolve the stones, and *laparoscopic cholecystectomy* (ko″le-sis-tek′to-me), a minimally invasive surgical technique for removing the gallbladder. In this procedure, a viewing scope and surgical tools threaded through small holes in the anterior abdominal wall at the location of the fundus are used to excise the gallbladder. Any gallstones remaining in the common bile duct are vaporized using a laser.

The Pancreas

The **pancreas** ("all meat") is both an exocrine gland and an endocrine gland. In its endocrine function, the pancreas secretes two major hormones, insulin and glucagon, which lower and raise blood sugar levels, respectively (see Chapter 17 for details). Its exocrine function is to produce most of the enzymes that digest foodstuffs in the small intestine.

Gross Anatomy

The pancreas, which is secondarily retroperitoneal, lies in the epigastric and left hypochondriac regions of the abdomen (see Figure 23.2). It is shaped like a tadpole, with head, body, and tail regions (Figure 23.28a), its head lies in the C-shaped curvature of the duodenum, and its tail extends to the left to touch the spleen.

The **main pancreatic duct** extends through the length of the pancreas (Figure 23.19). This duct joins the bile duct to form the hepatopancreatic ampulla and empties into the duodenum at the major duodenal papilla. An **accessory pancreatic duct** lies in the head of the pancreas and either drains into the main duct or drains directly into the duodenum.

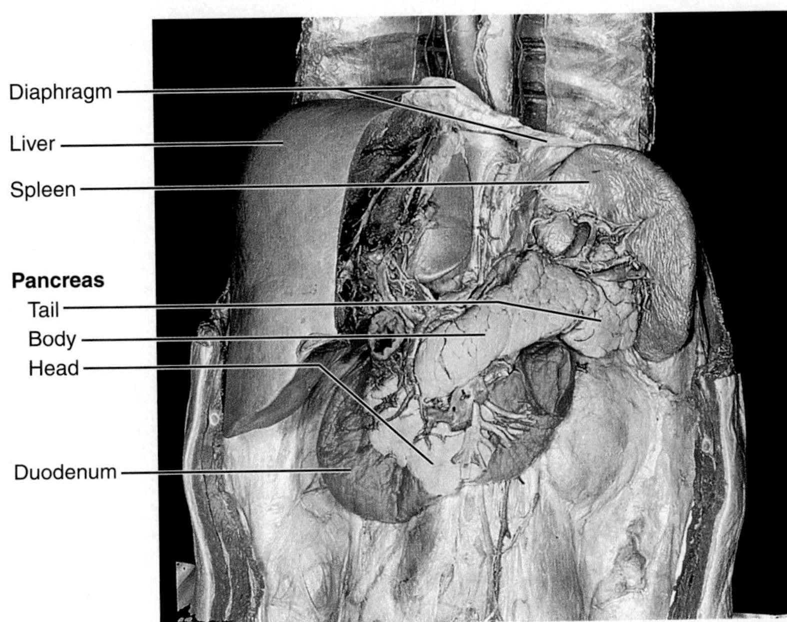

(a) Dissection illustrating the pancreas and its relationship to surrounding organs in the superior abdomen

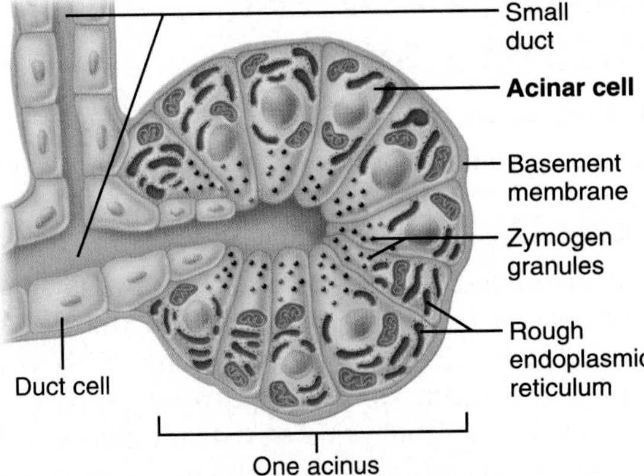

(b) Illustration of the pancreatic acinar cells

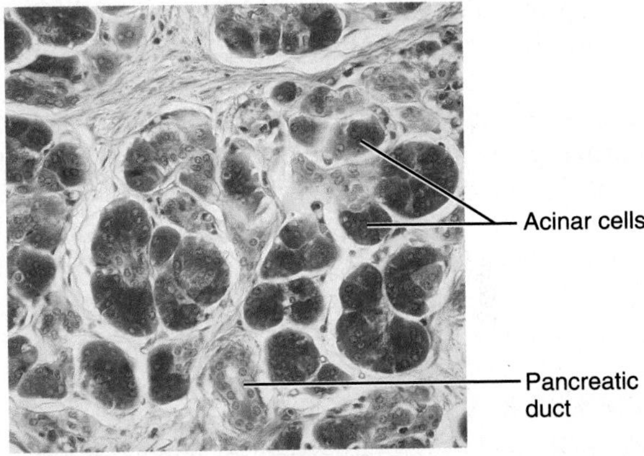

(c) Photomicrograph of the exocrine acinar cells of the pancreas (130×)

Figure 23.28 The gross and microscopic anatomy of the pancreas. In **(a)**, the stomach and much of the liver have been removed.

> **CLINICAL APPLICATION**
>
> **Pancreatitis** Inflammation of the pancreas, or **pancreatitis** (pan″kre-ah-tī′tis), usually accompanied by necrosis of pancreatic tissue, is a painful condition that is usually caused by the blockage of the pancreatic duct, either by gallstones or by an alcoholism-induced precipitation of protein. The blockage results in activation of the pancreatic enzymes in the pancreas instead of the small intestine. Pancreatitis can lead to nutritional deficiencies, diabetes, pancreatic infections, and, finally, to death through circulatory shock as the mediators of inflammation pour into the bloodstream from the damaged pancreas.

Nerves and Vessels The pancreas receives blood through branches of the hepatic, splenic, and superior mesenteric vessels. Its autonomic nerves are from the celiac plexus. Sympathetic input derives from the thoracic splanchnic nerves, whereas parasympathetic input is from the vagus nerve.

Microscopic Anatomy

Microscopically, the pancreas consists of many exocrine glands intermixed with fewer clusters of endocrine cells. These exocrine glands—compound acinar glands that open into the two large ducts like clusters of grapes attached to two main vines—will be considered first. The acini of these glands consist of serous **acinar cells** (Figure 23.28b and c), which make, store, and secrete at least 22 kinds of pancreatic enzymes capable of digesting the various categories of foodstuffs. The enzymes are stored in inactive form in intracellular secretory granules called **zymogen granules** (zī′mo-jen; "fermenting"). The acinar cells also contain an elaborate rough ER and Golgi apparatus, typical features of cells that secrete proteins. From each acinus, the secreted product travels through the pancreatic duct system to the duodenum, where the enzymes are activated. Furthermore, the epithelial cells that line the smallest pancreatic ducts secrete a bicarbonate-rich fluid that helps neutralize acidic chyme in the duodenum. Also among the duct cells are stem cells that form new acini and endocrine cells.

✓ check your understanding

☐ 17. Name the vessels and ducts that pass through the porta hepatis. Indicate what is found within each structure and whether its content goes into the liver or away from the liver.

☐ 18. Trace the path of blood drained from the digestive tract through a liver lobule.

☐ 19. Which cells in the pancreas produce and secrete digestive enzymes? Where do these secretions empty into the alimentary canal?

(For answers, see Appendix B.)

DISORDERS OF THE DIGESTIVE SYSTEM

learning outcome

▶ Describe some disorders of the digestive organs.

Peptic Ulcers

Ulcers are craterlike erosions of the mucosa in any region of the alimentary canal that is exposed to stomach secretions (Figure 23.29a). The majority of peptic ulcers occur in the pyloric region of the stomach (**gastric ulcers**) or in the duodenum of the small intestine (**duodenal ulcers**). The typical symptom of a peptic ulcer is a chronic burning sensation in the epigastric region usually 1–3 hours after a meal. Ulcers, traditionally thought to be caused by stress, are now known to be an infectious disease caused by a spiral-shaped, acid-resistant bacterium, *Helicobacter pylori* (Figure 23.29b). *H. pylori* binds to the gastric epithelium and induces over-secretion of acid and inflammation, which lead to ulcers. Approximately 20% of people in the United States under the age of 40 and 50% over age 60 harbor *H. pylori,* although most are either resistant or have a harmless strain. The bacterium is spread via fecal contamination of food or water. A simple 2-week regimen of antibiotics permanently cures peptic ulcers in most patients.

Intestinal Obstruction

Any hindrance to the movement of chyme or feces through the intestine is called **intestinal obstruction.** Most obstructions are *mechanical*—due to hernias of the bowel or twists that pinch the bowel shut, intestinal tumors or adhesions, or foreign objects lodged in the bowel. *Nonmechanical* obstruction, by contrast, is due to a halt in peristalsis. This can occur when movement of the intestine is inhibited by trauma or when the intestine is touched during surgery. About 85% of all obstructions occur in the small intestine; the remainder affect the large intestine. Common symptoms include cramps, vomiting, nausea, and failure to pass gas and feces.

Inflammatory Bowel Disease

Up to 2 of every 1000 people are affected by **inflammatory bowel disease,** a noncontagious, periodic inflammation of the intestinal wall characterized by chronic leukocyte infiltration of this wall. Symptoms include cramping, diarrhea, weight loss, and intestinal bleeding. Of the two subtypes of this condition, the form called *Crohn's disease* is the more serious, with deep ulcers and fissures developing along the entire intestine, but primarily in the terminal ileum. The other form, *ulcerative colitis,* is characterized by a shallow inflammation of the mucosa of the large intestine, mainly in the rectum. Although formerly thought to be a nervous condition, inflammatory bowel disease is now understood to be an abnormal immune and inflammatory response to bacterial antigens that normally occur in the intestine. Treatment involves adopting a special diet that is low in fiber and in dairy products, reducing stress, taking antibiotics, and, most effectively, administering anti-inflammatory and immunosuppressant drugs.

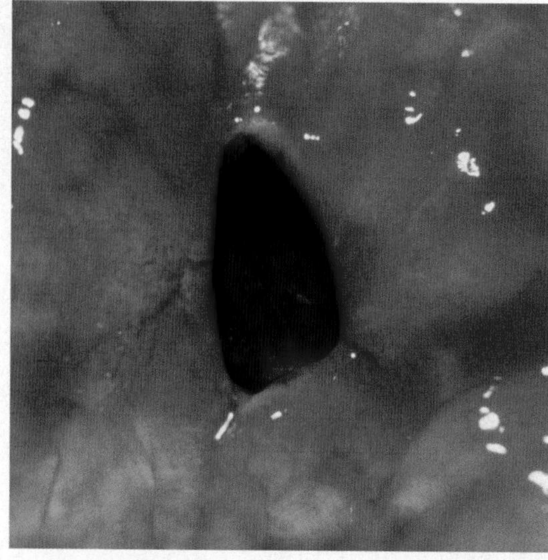

(a) A gastric ulcer lesion

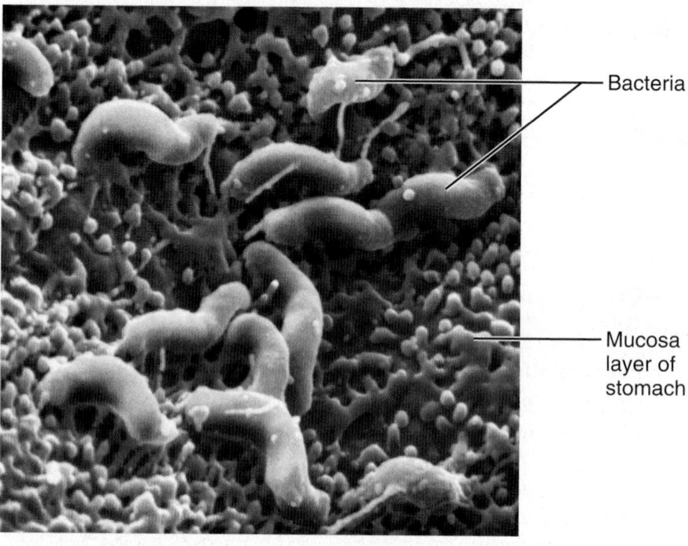

(b) *H. pylori* bacteria

Figure 23.29 **Peptic ulcers.**

Viral Hepatitis

Hepatitis, the general term for any inflammation of the liver, is largely of viral origin. Upon infection, most types of **viral hepatitis** lead to flulike symptoms and jaundice (yellow skin and mucous membranes, an indication that the liver is not removing bile pigments from the blood to make bile). The major types of hepatitis are A, B, C, and G.

Hepatitis A, spread by the fecal-oral route, often in contaminated food or water, is characterized by an acute infection without long-term damage followed by recovery and lifelong immunity. Treatments include administration of antibodies, and effective preventive vaccines are available.

Hepatitis B is transmitted via infected blood or body fluids, or from mothers to newborns at birth. Most infected individuals recover and gain immunity, but some develop chronic liver disease and eventually cirrhosis, with an increased likelihood of developing liver cancer. Many hepatitis B patients

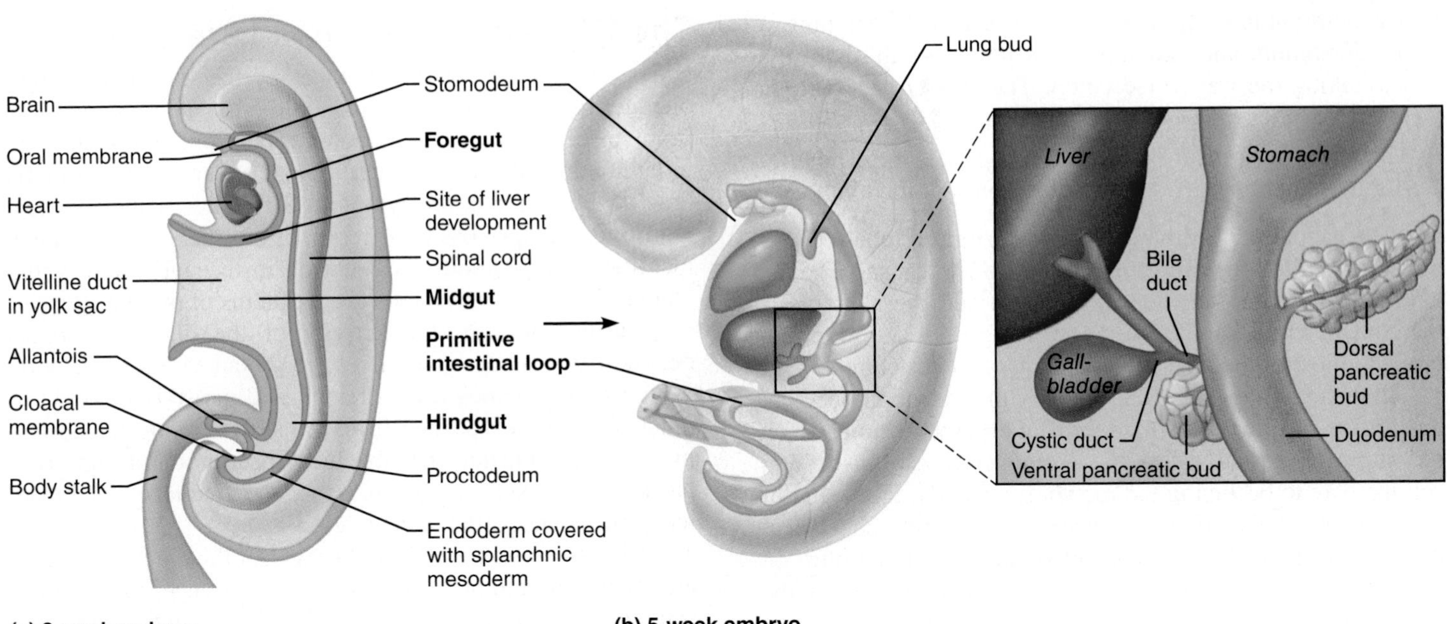

Figure 23.30 Development of the digestive system. (**a**) The primitive gut (composed of foregut, midgut, and hindgut) has formed. The midgut is still open and continuous with the yolk sac. The oral and cloacal membranes will be reabsorbed in a few weeks to form the oral and anal openings. (**b**) The liver and pancreas are budding off of the distal foregut. The pancreas forms from two pancreatic buds (ventral and dorsal) that later join.

can be helped with interferon (a substance that enhances the immune response against viruses) plus a combination of drugs that stops viral replication; an effective vaccine is also available.

Hepatitis C, like hepatitis B, is transmitted via body fluids and can lead to cirrhosis and liver cancer, but it has raised greater concern because it usually produces no short-term symptoms. As a result, it is difficult to diagnose, and many individuals do not know they are infected until they have spread the virus or develop serious symptoms, sometimes 20 years after infection. Some 4 million Americans have hepatitis C, and its spread is a serious health concern. No vaccine yet exists, but interferon and a drug that inhibits viral replication can help many patients.

Hepatitis G is as widespread as type C but seems to cause little liver damage.

Cystic Fibrosis and the Pancreas

Cystic fibrosis (described in more detail on p. 667) disrupts secretions in the respiratory system, but most of the body's other secretory epithelia are affected as well. The pancreas and intestinal glands, submandibular glands, and bile ducts in the liver all become blocked with thick secretions. The most serious is the effect on the pancreas, in which the clogged ducts prevent the pancreatic juices from reaching the small intestine. As a result, fats and other nutrients are not digested or absorbed, and the feces are bulky and fat laden. This problem can be treated by administering pancreatic enzymes with meals. Eventually, the pancreas may become a mass of cystically dilated ducts surrounded by dense fibrous tissue and no acini.

THE DIGESTIVE SYSTEM THROUGHOUT LIFE

learning outcome

▶ Explain how the digestive organs develop in the embryo, and define the foregut, midgut, and hindgut.

Embryonic Development

The alimentary canal originates when the flat embryo folds into the shape of a cylinder, enclosing a tubular part of the yolk sac within its body (Chapter 3, p. 54). This folding produces the *primitive gut,* a tube of endoderm that is covered by splanchnic mesoderm. The endoderm gives rise to the lining epithelium of the alimentary canal, the epithelial lining of the gut-derived organs (liver, pancreas, and gallbladder), and all the secretory cells of the digestive glands; the splanchnic mesoderm gives rise to all other layers in the wall of the alimentary canal and the gut-derived organs.

Initially, the middle region of the primitive gut is open to the yolk sac through the **vitelline duct** (vi-tel′in; "yolk") (Figure 23.30a). The vitelline duct is a key landmark that divides the embryonic gut into three basic regions: foregut, superior to the vitelline duct; midgut, open to the vitelline duct; and hindgut, inferior to the vitelline duct.

- The embryonic **foregut** develops into the first segment of the digestive system, from the pharynx to the point in the duodenum where the bile duct enters. The celiac trunk supplies blood to the abdominal forgut and its derivatives: the liver, gallbladder, and pancreas.

- The embryonic **midgut** becomes the segment beginning at the duodenum and extending to a point two-thirds of the way along the transverse colon. The superior mesenteric artery supplies the derivatives of the midgut.
- The **hindgut** forms the rest of the large intestine. The inferior mesenteric artery supplies the organs of the hindgut.

The caudal part of the early hindgut joins a tubelike outpocketing called the **allantois** (ah-lan′to-is; "sausage"). The expanded junction between the hindgut and the allantois is the **cloaca** (klo-a′kah; "sewer"), which gives rise to the rectum and most of the anal canal, among other structures.

In the mouth region of the embryo, the endoderm-lined gut touches the surface ectoderm to form an **oral membrane,** which lies in a depression called the **stomodeum** (sto″mo-de′um; "on the way to becoming the mouth"). Similarly, at the end of the hindgut, endoderm meets ectoderm to form the **cloacal membrane** in a pit called the **proctodeum** (prok″to-de′um; "on the way to becoming the anus"). The oral membrane lies at the future mouth-pharynx boundary (the fauces), and the cloacal membrane lies in the future anal canal, roughly where the pectinate line will occur. The oral and cloacal membranes are reabsorbed during month 2, thereby opening the alimentary canal to the outside.

During weeks 4 and 5, the embryonic gut starts to elongate, bend, and form outpocketings (Figure 23.30b). Salivary glands arise as outpocketings from the mouth; the pharynx develops four or five pairs of lateral *pharyngeal pouches* (see Figure 1.5b, p. 10); and the future lungs and trachea bud off the distal pharynx (see p. 670). A spindle-shaped enlargement of the abdominal foregut is the first sign of the stomach; the liver and pancreas arise as buds from the last part of the foregut; and the midgut elongates into the **primitive intestinal loop.** In months 2 and 3, this loop rotates and elongates to bring the intestines into their final positions.

CLINICAL APPLICATION

Developmental Abnormalities Failure of the vitelline duct to close completely can result in an outpocketing of the ileum, called **Meckel's diverticulum.** This is the most common developmental abnormality of the digestive system, occuring in 2% of the population, and is often asymptomatic. If the tissue within the diverticulum is of pancreatic or gastric origin, ulceration or bleeding may occur.

Abnormal rotation during development can cause the intestine to twist around itself, a condition called **volvulus,** which can disrupt the blood supply to the intestine, lead to death of the tissue in the affected portion, and cause intestinal blockage. Volvulus is treated by surgical removal of the affected portion of the gut.

The Digestive System in Later Life

Unless abnormal interferences occur, the digestive system operates through childhood and adulthood with relatively few problems. However, contaminated foods sometimes cause an inflammation of the alimentary canal called **gastroenteritis** (gas″tro-en″ĕ-ri′tis), the symptoms of which include nausea, vomiting, cramps, loss of appetite, or diarrhea. As previously mentioned, appendicitis is common in teenagers and young adults. Gallstones and ulcers are problems of middle age.

During old age, the activity of the digestive organs declines: Fewer digestive juices are produced; the absorption of nutrients becomes less efficient; and peristalsis slows. So much water is reabsorbed from the slow-moving fecal mass in the large intestine that the feces become hard and compacted. The result is a decrease in the frequency of bowel movements and, often, constipation.

Diverticulosis (p. 702) and cancer of the digestive organs are other common problems of the aged. Cancer of the stomach, colon, liver, or pancreas rarely exhibits early signs, and the cancer has often metastasized before the person seeks medical attention. These forms of cancer are deadly—colon cancer is the second leading cause of cancer deaths in the United States. However, if detected early, cancers of the digestive viscera are sometimes curable, colon and liver cancer more so than pancreatic or gallbladder cancer. The best advice is to have regular medical checkups. Half of all rectal cancers can be felt digitally during rectal exams, and nearly 80% of colon cancers can be seen during a colonoscopy (see pp. 99, 715). Evidence suggests that diets high in plant fiber decrease the incidence of colon cancer.

check your understanding

☐ 20. Why is it so important to wash your hands after using the restroom?

☐ 21. Which embryonic germ layer forms the epithelium of the mucosa? Which germ layer forms the submucosa and muscularis externa?

☐ 22. What blood vessel supplies the structures derived from the embryonic midgut?

(For answers, see Appendix B.)

RELATED CLINICAL TERMS

Anal fissure A longitudinal tear in the mucosa of the anal canal, often caused by the passage of hard, dry feces. Usually heals naturally, but most fissures that do not heal are in the posterior midline, which is poorly vascularized. Symptoms include pain and bleeding during defecation. Treatment includes using laxatives to soften the feces, glycerin suppositories, or in persistent cases, surgery.

Ascites (ah-si′tēz; *asci* = bag, bladder) Abnormal accumulation of serous fluid that has leaked out of peritoneal capillaries into the peritoneal cavity; may be caused by portal hypertension following liver cirrhosis or by heart or kidney disease. Excessive ascites causes visible bloating of the abdomen.

Endoscopy (en-dos′ko-pe; *endo* = inside; *scopy* = viewing) The viewing of the lining of a ventral body cavity or tubular organ with a flexible, tubelike device called an endoscope, which contains a lens and a light radiating from its tip. Endoscopes are used to view the internal surfaces of various parts of the alimentary canal, including the stomach (gastroscopy), the colon (colonoscopy), and the sigmoid colon (sigmoidoscopy). **Laparoscopy** (lap″ah-ros′ko-pe; "flank viewing") is the use of an endoscope inserted into the peritoneal cavity through the anterior abdominal wall, typically to assess the condition of the digestive organs and the pelvic reproductive organs in women.

Enteritis (*enteron* = intestine) Inflammation of the intestine, especially the small intestine.

Fecal microbial therapy This therapy, also called *fecal transplant*, is an experimental treatment for chronic intestinal inflammation caused by the bacterium *Clostridium difficile* ("*C. diff*"). This organism is resistant to most antibiotics and, as its name implies, is extremely diffecult to treat. Fecal microbial therapy transplants the gut bacteria from a healthy donor to the infected individual by delivering a sample of diluted feces from the donor into the recipient's colon. The transplanted feces restablishes the "normal" gut microbes in the recipient's colon. This treatment, while both simple and rather unusual, has been very successful in combating this bacterium.

Liver biopsy (bi′op-se; *bio* = living; *opsis* = vision, viewing) Removal from the liver of a small piece of living tissue, which is then examined for signs of disease. The puncturing needle is inserted through the seventh, eighth, or ninth intercostal space, in the right midaxillary line (straight inferiorly from the axilla) after the patient has exhaled as much air as possible. (Exhalation minimizes the chances that the needle will pierce the lung.)

Pyloric stenosis (*stenosis* = narrowing, constriction) Congenital condition in about 1 in 400 newborns, in which the pyloric sphincter of the stomach is abnormally constricted; the condition's characteristic sign, projectile vomiting, usually does not appear until the baby begins to eat solid food. This condition can usually be repaired surgically. Can also occur in adults through scarring caused by an ulcer or by a tumor that blocks the pyloric opening.

Rectocele (rek′to-sēl; *recto* = rectum; *cele* = sac) Condition in women in which the rectum pushes on the vagina and bulges into the posterior vaginal wall. Usually results from tearing of the supportive muscles of the pelvic floor during childbirth, which then allows the unsupported pelvic viscera to sink inferiorly. May also be associated with *rectal prolapse*, in which the rectal mucosa protrudes from the anus.

CHAPTER SUMMARY

Take full advantage of the invaluable online practice and study tools for this chapter by going to the Study Area of MasteringA&P®. There you will find:

- Art Labeling Activities, Quizzes, Practice Tests, Memory Games
- Animal Organ Dissection, Cat Dissection and Bone Videos
- PAL Cadaver
- PAL Histology
- A&P Flix Animations
- Dynamic Study Modules

OVERVIEW (pp. 676–683)

1. The digestive system includes the alimentary canal (mouth, pharynx, esophagus, stomach, and small and large intestines) and accessory digestive organs (teeth, tongue, salivary glands, liver, gallbladder, and pancreas).

Abdominal Regions (p. 677)

2. The nine regions of the anterior abdominal wall are defined by two horizontal planes (subcostal and transtubercular planes) and by the vertical midclavicular lines. There are one epigastric and two hypochondriac regions superiorly, one umbilical and two lumbar regions in the middle, and one hypogastric (pelvic) and two iliac (inguinal) regions inferiorly. The anterior abdominal wall can also be divided into four quadrants.

The Peritoneal Cavity and Peritoneum (pp. 678–682)

3. The serous membrane in the abdominopelvic cavity is the peritoneum, which has a parietal layer (on the internal surface of the body wall) and a visceral layer (on the viscera). The slit between the visceral and parietal peritoneum, the peritoneal cavity, contains slippery serous fluid, which decreases friction as the organs move.

4. A mesentery is a double layer of peritoneum that tethers the movable digestive organs to the body wall. Mesenteries also store fat and carry blood vessels and nerves.

5. The mesenteries associated with the intraperitoneal abdominal organs are (1) the ventral mesenteries, the falciform ligament and lesser omentum; and (2) the dorsal mesenteries, the greater omentum, mesentery proper, transverse mesocolon, and sigmoid mesocolon. Digestive organs that lack a mesentery and are fused to the posterior body wall are called secondarily retroperitoneal organs.

Digestive Processes (pp. 682–683)

6. The digestive system carries out the processes of ingestion, propulsion, mechanical breakdown, digestion, absorption, and defecation.

ANATOMY OF THE ALIMENTARY CANAL (pp. 683–706)

Histology (pp. 683–685)

7. The esophagus, stomach, and intestine share the same tissue layers: an inner mucosa (lining epithelium, lamina propria, and muscularis mucosae), a fibrous submucosa, a muscularis externa (circular and longitudinal layers), and an outer serosa (visceral peritoneum) or adventitia.

Smooth Muscle (pp. 685–687)

8. Smooth muscle fibers are elongated cells with tapering ends and one central nucleus. They have no striations or sarcomeres, but are filled with myofilaments that contract by the sliding filament mechanism. Intermediate filaments of the cytoskeleton run in lattice-like arrangement through the cell. Actin myofilaments attach to these filaments at dense bodies.

9. Smooth muscle fibers are most often arranged in circular and longitudinal sheets.

10. Smooth muscle contracts for extended periods at low energy cost and without fatigue.

11. Smooth muscle is innervated by involuntary nerve fibers, which usually contact only a few muscle fibers per sheet. The impulse that signals contraction usually spreads from fiber to fiber through gap junctions. This is called single-unit innervation.

12. Visceral motor neurons do not form elaborate neuromuscular junctions with the visceral muscle and glands they innervate. Their axon terminals (varicosities) may even end some distance from the effector cells.

13. The enteric nervous system is an intrinsic neural network within the wall of the alimentary canal. Enteric neurons located in the visceral nerve plexuses (myenteric and submucosal) control the muscular and secretory activities of the digestive organs. Visceral sensory, parasympathetic, and sympathetic fibers link the enteric nervous system with the CNS. Autonomic innervation influences the activity of the enteric neurons.

The Mouth and Associated Organs (pp. 687–692)

14. Food enters the alimentary canal through the mouth (oral cavity), which consists of an external oral vestibule and an internal oral cavity proper.

15. The mouth is lined by a stratified squamous epithelium, which resists abrasion by food fragments.

16. The lips and cheeks keep food inside the mouth during chewing. The red margin is the red part of the lips that borders the oral orifice.

17. The tongue is predominantly a mucosa-covered skeletal muscle. Its intrinsic muscles change its shape, and its extrinsic muscles change its position. Three classes of papillae occur on the tongue's superior surface: filiform papillae, which grip food during chewing, and fungiform and circumvallate papillae, which contain taste buds.

18. Saliva is produced by minor salivary glands in the oral mucosa and by three pairs of major salivary glands—the parotid, submandibular, and sublingual glands. The salivary glands are compound acinar or tubuloacinar glands containing varying amounts of serous and mucous cells.

19. Teeth tear and grind food in the chewing process (mastication). The 20 deciduous teeth begin to fall out at age 6 and are gradually replaced during childhood and youth by the 32 permanent teeth.

20. Teeth are classified as incisors, canines, premolars, and molars. Each tooth has an enamel-covered crown and a cement-covered root. The bulk of the tooth is dentin (dentine), which surrounds the central pulp cavity. A periodontal ligament (periodontium) secures the tooth to the bony tooth socket.

The Pharynx (p. 692)

21. During swallowing, food passes from the mouth into the oropharynx and laryngopharynx, which are lined by stratified squamous epithelium. The suprahyoid muscles elevate the larynx to protect the airway. The pharyngeal constrictor muscles squeeze food into the esophagus, and then the infrahyoid muscles depress the larynx returning it to its original position.

The Esophagus (pp. 692–693)

22. The esophagus descends from the pharynx through the posterior mediastinum and into the abdomen. There it joins the stomach at the cardial orifice, where a sphincter prevents superior regurgitation of stomach contents.

23. The esophageal mucosa contains a stratified squamous epithelium. The muscularis consists of skeletal muscle superiorly and smooth muscle inferiorly. The external layer of the esophagus is an adventitia.

The Stomach (pp. 694–697)

24. The J-shaped stomach churns food into chyme and secretes HCl and pepsin, which begins the breakdown of food proteins. Lying in the superior left part of the abdomen, its major regions are the cardia, fundus, body, pyloric part, and pylorus (containing the pyloric sphincter). Its right and left borders are the lesser and greater curvatures. When the stomach is empty, its internal surface exhibits rugae.

25. The internal surface of the stomach is lined by simple columnar epithelial cells that secrete mucus and dotted with gastric pits that lead into tubular gastric glands. Secretory cells in the gastric glands include pepsinogen-producing chief cells, parietal cells that secrete HCl, mucous neck cells, and enteroendocrine cells that secrete hormones (including gastrin).

26. The stomach wall is protected against self-digestion and acid by an internal coat of mucus and by the rapid regeneration of its lining epithelium.

The Small Intestine (pp. 697–699)

27. The small intestine is the main site of digestion and nutrient absorption. Its segments are the duodenum, jejunum, and ileum.

28. The duodenum lies secondarily retroperitoneally in the superior right quadrant of the abdomen. The bile duct and main pancreatic duct join to form the hepatopancreatic ampulla and empty into the duodenum through the major duodenal papilla.

29. The small intestine has four features that increase its surface area and allow rapid absorption of nutrients: its great length, circular folds of the mucosa, villi, and abundant microvilli on the absorptive cells. Absorbed nutrients enter capillaries in the core of the villi.

30. The epithelium lining the internal intestinal surface contains absorptive enterocytes, goblet cells, and enteroendocrine cells. Between the villi are the intestinal crypts, which secrete intestinal juice, continuously renew the lining epithelium, and contain Paneth cells.

31. Other special features of the small intestine are (1) aggregated lymphoid nodules in the ileum and (2) the duodenal mucous glands.

The Large Intestine (pp. 700–706)

32. The large intestine, which concentrates feces by absorbing water, forms an open rectangle around the small intestine. Its subdivisions are the cecum and appendix, colon (ascending, transverse, descending, and sigmoid), rectum, and anal canal. Special features on the external surface of the colon and cecum are teniae coli, haustra, and epiploic appendages.

33. The cecum lies in the right iliac fossa and contains the ileocecal valve. Attached to the cecum is the appendix, which contains abundant lymphoid tissue.

34. Near the end of the large intestine, the sigmoid colon enters the pelvis and joins the rectum. The rectum pierces the pelvic floor and joins the anal canal, which ends at the anus. Internally, the rectum contains the transverse rectal folds, and the anal canal contains anal columns, anal valves, the pectinate line, and anal sinuses.

35. The defecation reflex produces a contraction of the rectal walls when feces enter the rectum. During defecation, contraction of the diaphragm and abdominal muscles increases the intra-abdominal pressure and lifts the anal canal. Defecation can be inhibited by the external anal sphincters.

36. Like the small intestine, the large intestine is lined by a simple columnar epithelium with absorptive colonocytes and goblet cells and contains intestinal glands. It lacks villi, secretes more mucus than the small intestine, and contains more lymphoid tissue than the small intestine.

ANATOMY OF THE ACCESSORY ORGANS (pp. 706–711)

The Liver (pp. 706–710)

37. The liver performs many functions, including processing of absorbed nutrients, secreting fat-emulsifying bile, and removing toxins from the blood.

38. The liver lies in the superior right region of the abdomen. It has diaphragmatic and visceral surfaces and four lobes (right, left, quadrate, and caudate). The quadrate and caudate lobes are parts of the left lobe. The bare area lies against the diaphragm.

39. Most vessels enter or leave the liver through the porta hepatis. Other structures on the visceral surface are the inferior vena cava, hepatic veins, gallbladder, fissure, round ligament of liver (ligamentum teres), and ligamentum venosum.

40. Classical liver lobules are hexagonal structures consisting of plates of hepatocytes that are separated by blood sinusoids and converge on a central vein. Portal triads (venule branch of portal vein, arteriole branch of hepatic artery, bile duct) occur at most corners of the lobule.

41. Blood flows through the liver in the following sequence: through the branches of the hepatic artery and portal vein, through the sinusoids to supply the hepatocytes, and then to the central veins, hepatic veins, inferior vena cava, and heart. In the sinusoids, stellate cells (macrophages) remove bacteria and other foreign material from the blood.

42. Hepatocytes perform almost all liver functions: They manufacture blood proteins; detoxify poisons; metabolize glucose, fats, and amino acids; and produce bile. These liver cells secrete bile into the bile canaliculi. The bile proceeds to the bile ducts in the portal triads, to the hepatic ducts, and through the common hepatic duct to the gallbladder and duodenum.

The Gallbladder (p. 710)

43. The gallbladder, a green muscular sac, stores and concentrates bile. Its duct is the cystic duct. When fatty chyme enters the small intestine, the gallbladder squeezes bile into the bile duct and duodenum.

The Pancreas (pp. 710–711)

44. The tadpole-shaped pancreas is secondarily retroperitoneal. It runs horizontally across the posterior abdominal wall, between the duodenum and the spleen.

45. The pancreas is an exocrine gland containing many compound acinar glands that empty into the main and accessory pancreatic ducts. The serous acinar cells secrete digestive enzymes, and the smallest duct cells secrete an alkaline fluid. Both secretions are emptied into the duodenum.

46. The pancreas is also an endocrine gland, with hormone-secreting cells that regulate blood sugar.

PAL **Human Cadaver/Digestive System**

DISORDERS OF THE DIGESTIVE SYSTEM (pp. 712–713)

47. Disorders considered in this chapter include hiatal hernia of the esophagus, gastroesophageal reflux disease (p. 693), peptic ulcers, intestinal obstruction, inflammatory bowel disease, viral hepatitis, and the obstructive effect of cystic fibrosis on the pancreatic ducts.

THE DIGESTIVE SYSTEM THROUGHOUT LIFE (pp. 713–714)

Embryonic Development (pp. 713–714)

48. As the 3-week-old embryo assumes its cylindrical body shape, it encloses the primitive gut, a tube of endoderm covered by splanchnic mesoderm. The endoderm becomes the lining epithelium (and gland cells), and the splanchnic mesoderm gives rise to all other layers of the wall of the alimentary canal and the gut-derived organs.

49. The embryonic gut is divided into a foregut, midgut, and hindgut, which form distinct regions of the digestive system.

50. The primitive embryonic gut tube, straight at first, soon grows outpocketings (liver, pancreas, pharyngeal pouches), shows swellings (stomach, cloaca), and lengthens into a primitive intestinal loop. This loop rotates and elongates to bring the intestines into their final positions.

The Digestive System in Later Life (p. 714)

51. Various diseases may plague the digestive organs throughout life. Appendicitis is common in young adults; gastroenteritis and food poisoning can occur at any time; and ulcers and gallbladder problems increase in middle age.

52. The efficiency of digestive processes declines in the elderly, and constipation becomes common. Diverticulosis and cancers (such as colon and stomach cancer) appear with increasing frequency among older individuals.

REVIEW QUESTIONS

Multiple Choice/Matching Questions

(For answers, see Appendix B.)

1. Which of the following organs is secondarily retroperitoneal? (a) pharynx, (b) stomach, (c) ascending colon, (d) ileum.

2. The submucosal nerve plexus of the intestine (a) innervates the mucosa layer, (b) lies in the mucosa layer, (c) controls peristalsis, (d) contains only motor neurons.

3. For each organ in the left-hand column, select the type of epithelium that lines its lumen from the list in the right-hand column.

Organ	Epithelium types
____ **(1)** oral cavity	**(a)** simple squamous
____ **(2)** oropharynx	**(b)** simple cuboidal
____ **(3)** esophagus	**(c)** simple columnar
____ **(4)** stomach	**(d)** stratified squamous
____ **(5)** small intestine	**(e)** stratified cuboidal
____ **(6)** colon	

4. Match each of the mesenteries in the key with the appropriate description.

Key: (a) greater omentum (d) the mesentery
(b) lesser omentum (e) transverse mesocolon
(c) falciform ligament (f) sigmoid mesocolon

____ **(1)** connects the ileum and jejunum to the posterior abdominal wall

____ **(2)** connects anterior surface of the liver to the anterior abdominal wall

____ **(3)** connects the large intestine to the pelvic wall

____ **(4)** attaches to the greater curvature of the stomach; has the most fat

____ **(5)** runs from the stomach's lesser curvature to the fissure of the liver

____ **(6)** a mesentery of the large intestine that is fused to the underside of the greater omentum

5. The pointed type of tongue papilla that is *not* involved with taste reception is (a) filiform, (b) fungiform, (c) circumvallate, (d) dermal.

6. Which of the following statements about the gallbladder is *false*? (a) It makes bile. (b) Its duct is called the cystic duct. (c) It has a fundus lying inferior to the liver. (d) It has mucosal folds similar to the stomach rugae.

7. Which of the following correctly describe the flow of blood through the classical liver lobule and beyond? (More than one choice is correct.) (a) portal venule to sinusoids to central vein to hepatic vein to inferior vena cava, (b) porta hepatis to hepatic vein to portal venule, (c) portal venule to central vein to hepatic vein to sinusoids, (d) portal arteriole to sinusoids to central vein to hepatic vein.

8. The exocrine glands in the pancreas are (a) simple tubular, (b) simple acinar, (c) compound tubular, (d) compound acinar, (e) compound tubuloalveolar.

9. Which cell type occurs in the stomach mucosa, has three prongs, contains many mitochondria and many microvilli, and pumps hydrogen ions? (a) absorptive enteroyte, (b) parietal cell, (c) mucus-secreting cell, (d) chief cell, (e) mucous neck cell.

10. Which one of the following features is shared by both the small and large intestines? (a) intestinal crypts, (b) aggregated lymphoid nodules (Peyer's patches), (c) teniae coli, (d) haustra, (e) circular folds, (f) intestinal villi.

11. A digestive organ that has a head, body, and tail is the (a) pancreas, (b) gallbladder, (c) greater omentum, (d) stomach.

12. From the list of abdominal regions listed in column B, indicate the predominant region where each organ listed in column A is found.

Column A	Column B
____ **(1)** small intestine	**(a)** right hypochondriac
____ **(2)** liver	**(b)** left hypochondriac
____ **(3)** stomach	**(c)** right iliac
____ **(4)** ascending colon	**(d)** right lumbar
____ **(5)** cecum	**(e)** umbilical

13. Protein digestion begins (a) in the mouth by saliva, (b) in the stomach by pepsin, (c) in the duodenum by bile, (d) in the small intestine by intestinal secretions.

14. The calcified connective tissue that attaches the tooth to the periodontal ligament is the (a) pulp, (b) enamel, (c) dentine, (d) cement, (e) peridontium.

15. Which of the following statements about smooth muscle is *false*? (a) Smooth muscle cells are called fibers. (b) Most smooth muscle cells, like cardiac muscle cells, are joined by gap junctions. (c) Smooth muscle cells have a single, centrally located nucleus. (d) Contraction of smooth muscle is stimulated exclusively by involuntary nerves. (e) Smooth muscle tissue found in the wall of hollow organs is arranged in sheets.

16. Match the digestive organ listed in column B with the function listed in column A.

Column A	Column B
____ **(1)** produces bile	**(a)** salivary glands
____ **(2)** absorbs water	**(b)** esophagus
____ **(3)** churning occurs here	**(c)** stomach
____ **(4)** muscular tube connecting the laryngopharynx with the stomach	**(d)** small intestine
	(e) liver
____ **(5)** produces both endocrine and exocrine secretions	**(f)** gallbladder
____ **(6)** secretes a substance that initiates carbohydrate digestion	**(g)** pancreas
	(h) large intestine
____ **(7)** stores bile	
____ **(8)** segmentation occurs here	

17. The salivary gland that contains only serous cells is the (a) parotid gland, (b) submandibular gland, (c) sublingual gland, (d) minor salivary glands.

18. Use the key below to indicate the blood vessel that supplies arterial blood to each of the digestive organs listed.

Key: (a) celiac trunk
(b) superior mesenteric artery
(c) inferior mesenteric artery

____ **(1)** stomach	____ **(5)** ascending colon
____ **(2)** ileum	____ **(6)** rectum
____ **(3)** descending colon	____ **(7)** cecum
____ **(4)** liver	

Short Answer Essay Questions

19. Make a simple drawing of the organs of the alimentary canal, and label each organ. Also label the gross subparts of the stomach and intestines. Then add three more labels to your drawing—*salivary glands, liver,* and *pancreas*—and use arrows to show where each of these three glandular organs empties its secretion into the alimentary canal.
20. Name the layers of the wall of the alimentary canal. Describe the tissue composition and the major function of each layer. Then, identify the embryonic source of each layer.
21. (a) Write the dental formulas for both the deciduous and the permanent teeth. (b) Define *dental pulp,* and describe its location.
22. Bianca went on a trip to the Bahamas during spring vacation and did not study for her anatomy test scheduled early the following week. On the test, she mixed up the following pairs of structures: (a) the rectal valves and the anal valves, (b) pyloric part and pylorus of the stomach, (c) anal canal and anus, (d) villi and microvilli (in the small intestine), (e) hepatic vein and hepatic portal vein, (f) gastric pits and gastric glands. Help her by defining and differentiating between these sound-alike structures.
23. (a) Make a rough sketch of the visceral surface of the liver, and label the following five structures: fissure, porta hepatis, inferior vena cava, gallbladder, caudate lobe. (b) To follow up on part (a), it is easy to learn the structure of the visceral surface of the liver if you can find and mark the big H there. To mark the left vertical limb of the H, draw a line through the fissure. To mark the crossbar of the H, draw a horizontal line through the porta hepatis. Then, to mark the right limb of the H, draw a continuous vertical line through the inferior vena cava and gallbladder. Which two lobes are inside the H? Which limb of the H separates the liver's right and left lobes?
24. Name four structural features that increase the absorptive surface area of the small intestine.
25. (a) List the three major vessels of the portal triad, and describe the location of this triad with respect to the liver lobule. (b) Name three organelles that are abundant within hepatocytes, and explain how each of these organelles contributes to liver functions.
26. Of the three major salivary glands, which is the largest? Which one underlies the tongue? Which has only serous glands?
27. Trace the entire duct systems of the liver and pancreas. (Study these ducts in Figure 23.19, and then try to draw them from memory.)
28. On a diagram of the adult digestive tract (such as Figure 23.1), mark the boundaries between the derivatives of the embryonic foregut, midgut, and hindgut.
29. Where in the alimentary canal does chemical digestion begin for each of the three types of food: carbohydrates, proteins, and fats?
30. List all of the sphincters in the gastrointestinal tract and indicate the location of each.
31. Name the five processes of digestion and indicate all the locations in the gastrointestinal tract where each process occurs.

Critical Reasoning & Clinical Application Questions

1. This chapter describes schemes that divided the anterior abdominal wall into nine regions or four quadrants. List the regions that lie either entirely or partially within (a) the upper right quadrant and (b) the lower left quadrant.
2. A 21-year-old man with severe appendicitis did not seek treatment in time and died a week after his abdominal pain and fever began. Explain why appendicitis can quickly lead to death.
3. Duncan, an inquisitive 8-year-old, saw his grandfather's dentures soaking overnight in a glass of water. He asked his grandfather how his real teeth had fallen out. Assuming the grandfather remembered the events correctly but was not a medical expert and used layperson's terms, reconstruct the kind of story the man is likely to have told.
4. Eva, a middle-aged attorney, complains of a burning pain in the "pit of her stomach," usually beginning about 2 hours after eating and lessening after she drinks a glass of milk. When asked to indicate the site of pain, she points to her epigastric region. When her GI tract is examined by endoscopy, a gastric ulcer is visualized. What are the possible consequences if the ulcer goes untreated?
5. A doctor used an endoscope and located some polyps (precancerous tumors) in the wall of the large intestine of an elderly man. What is an endoscope?
6. The janitor who cleaned the anatomy lab had a protruding abdomen that looked like the biggest "beer-belly" the students had ever seen, even though the rest of his body did not look fat at all. Another janitor on the floor told some students that the man was a recovering alcoholic and that he had "over a hundred pounds of fluid in his belly." What was the man's probable condition? (See this chapter's Related Clinical Terms.)
7. Explain why cancer chemotherapies that stop the replication of cellular DNA throughout the body cause nausea, diarrhea, and vomiting.
8. From what embryonic layer (ectoderm, mesoderm, or endoderm) are the hepatocytes of the liver and the pancreatic acinar cells and islets derived?
9. Why is dehydration a concern for an individual with persistent diarrhea?
10. Rebound tenderness in the lower right quadrant of the abdomen is an indication of what disorder?

24 The Urinary System

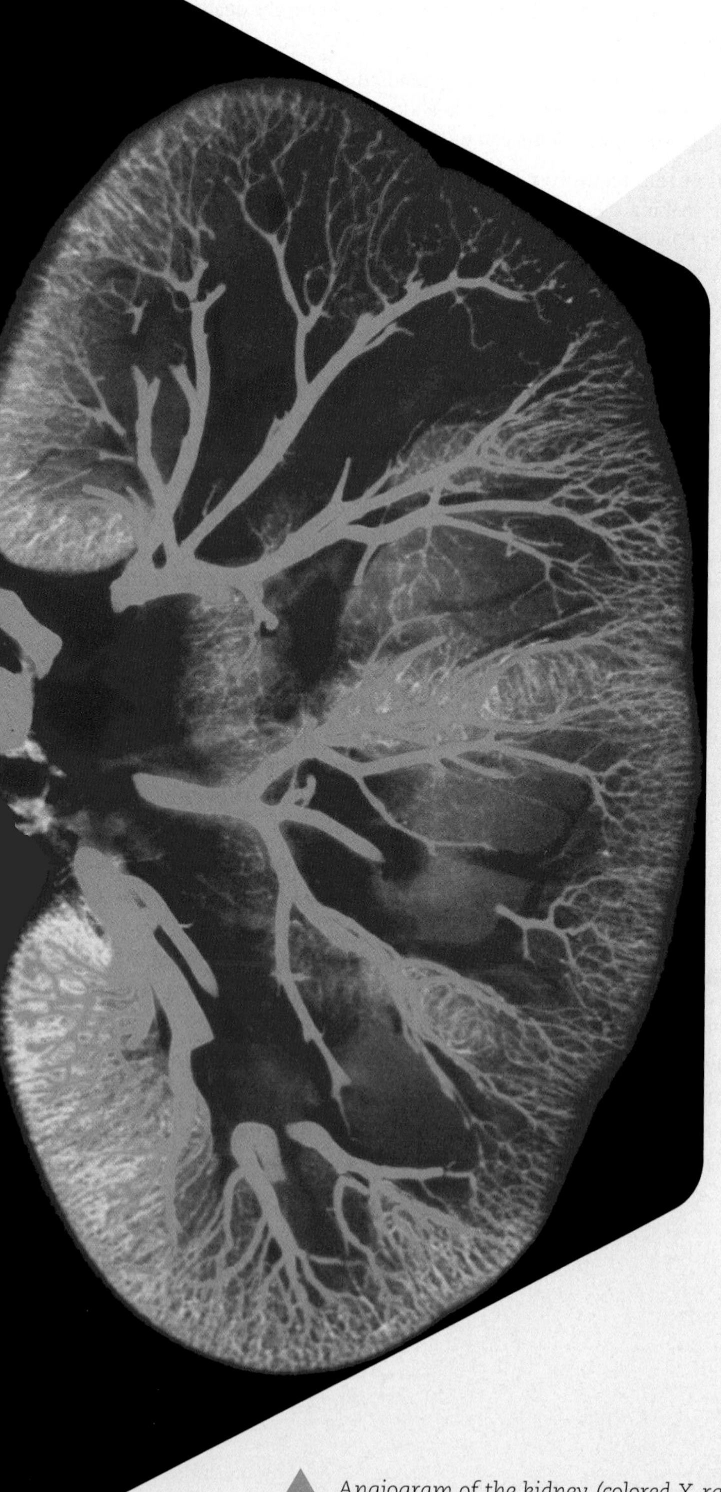

▲ *Angiogram of the kidney (colored X-ray image).*

The kidneys maintain the purity and chemical constancy of the blood and the other extracellular body fluids. Much like a water purification plant that keeps a city's water drinkable and disposes of its wastes, the kidneys are usually unappreciated until they malfunction and body fluids become contaminated. Every day, the kidneys filter many liters of fluid from the blood, sending toxins, metabolic wastes, excess water, and excess ions out of the body in urine while returning needed substances from the filtrate to the blood. The main waste products excreted in urine are three nitrogenous compounds: (1) *urea,* derived from the breakdown of amino acids during normal recycling of the body's proteins; (2) *uric acid,* which results from the turnover of nucleic acids; and (3) *creatinine* (kre-at′ĭ-nin), formed by the breakdown of creatine phosphate,

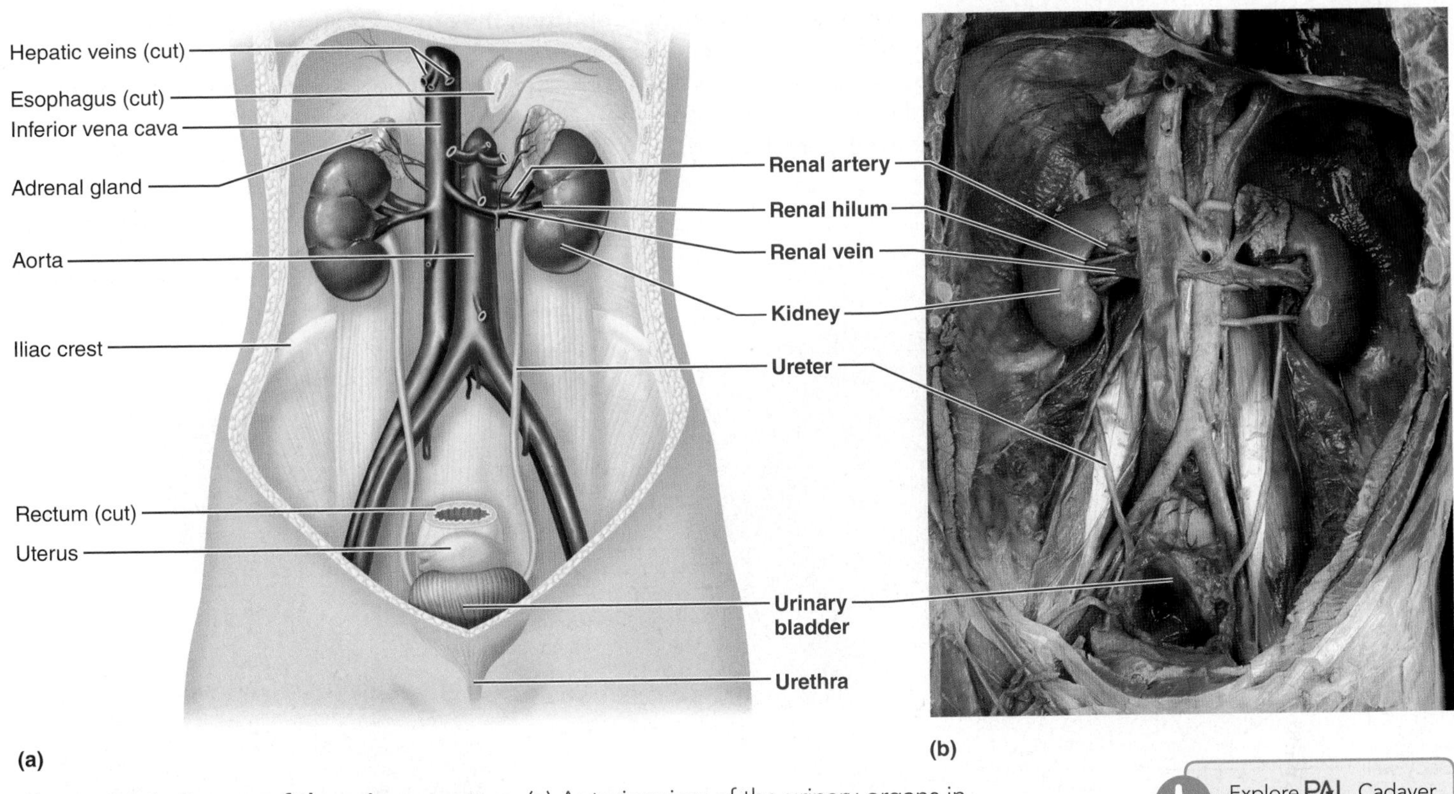

Figure 24.1 Organs of the urinary system. (**a**) Anterior view of the urinary organs in a female. (**b**) Dissection of a view similar to (**a**) in a male.

a molecule in muscle that stores energy for the manufacture of adenosine triphosphate (ATP). Although the lungs, liver, and skin also participate in excretion, the kidneys are the major excretory organs.

Disposing of wastes and excess ions is only one aspect of kidney function; the kidneys also regulate the volume and chemical makeup of the blood by maintaining the proper balance of water and salts and of acids and bases. The kidneys perform work that would be tricky for a chemical engineer, and they perform it efficiently most of the time.

Besides the urine-forming kidneys, the other organs of the urinary system (Figure 24.1) are the paired *ureters* (u-re′terz; "pertaining to urine"), the *urinary bladder,* and the *urethra* (u-re′thrah). The ureters are tubes that carry urine from the kidney to the bladder, a temporary storage sac for urine. The urethra is a tube that carries urine from the bladder to the body exterior.

KIDNEYS

Gross Anatomy of the Kidneys

learning outcomes

- ▶ Describe the location, coverings, and external gross anatomy of the kidney.
- ▶ Describe the internal gross anatomy and the macroscopic blood vessels of the kidney.

Location and External Anatomy

The red-brown, bean-shaped **kidneys** lie retroperitoneal (behind the parietal peritoneum) in the superior lumbar region of the posterior abdominal wall (Figure 24.2). They extend from the level of the 11th or 12th thoracic vertebra superiorly to the 3rd lumbar vertebra inferiorly, and thus they receive some protection from the lower two ribs (Figure 24.2b). The right kidney is crowded by the liver and lies slightly inferior to the left kidney. An average adult kidney is about 12 cm high, 6 cm wide, and 3 cm thick—the size of a large bar of soap. The lateral surface of each kidney is convex; the medial surface is concave and has a vertical cleft called the renal **hilum,** where vessels, ureters, and nerves enter and leave the kidney (see Figure 24.1). On the superior part of each kidney lies an adrenal (suprarenal) gland, an endocrine gland that is functionally unrelated to the kidney.

Several layers of supportive tissue surround each kidney (Figure 24.2a). A thin, tough layer of dense connective tissue called the **fibrous capsule** adheres directly to the kidney's surface, maintaining its shape and forming a barrier that can inhibit the spread of infection from the surrounding regions. Just external to the renal capsule is the **perirenal fat capsule** (per′e-re″nal; "around the kidney"), and external to that is an envelope of **renal fascia.** The renal fascia contains an

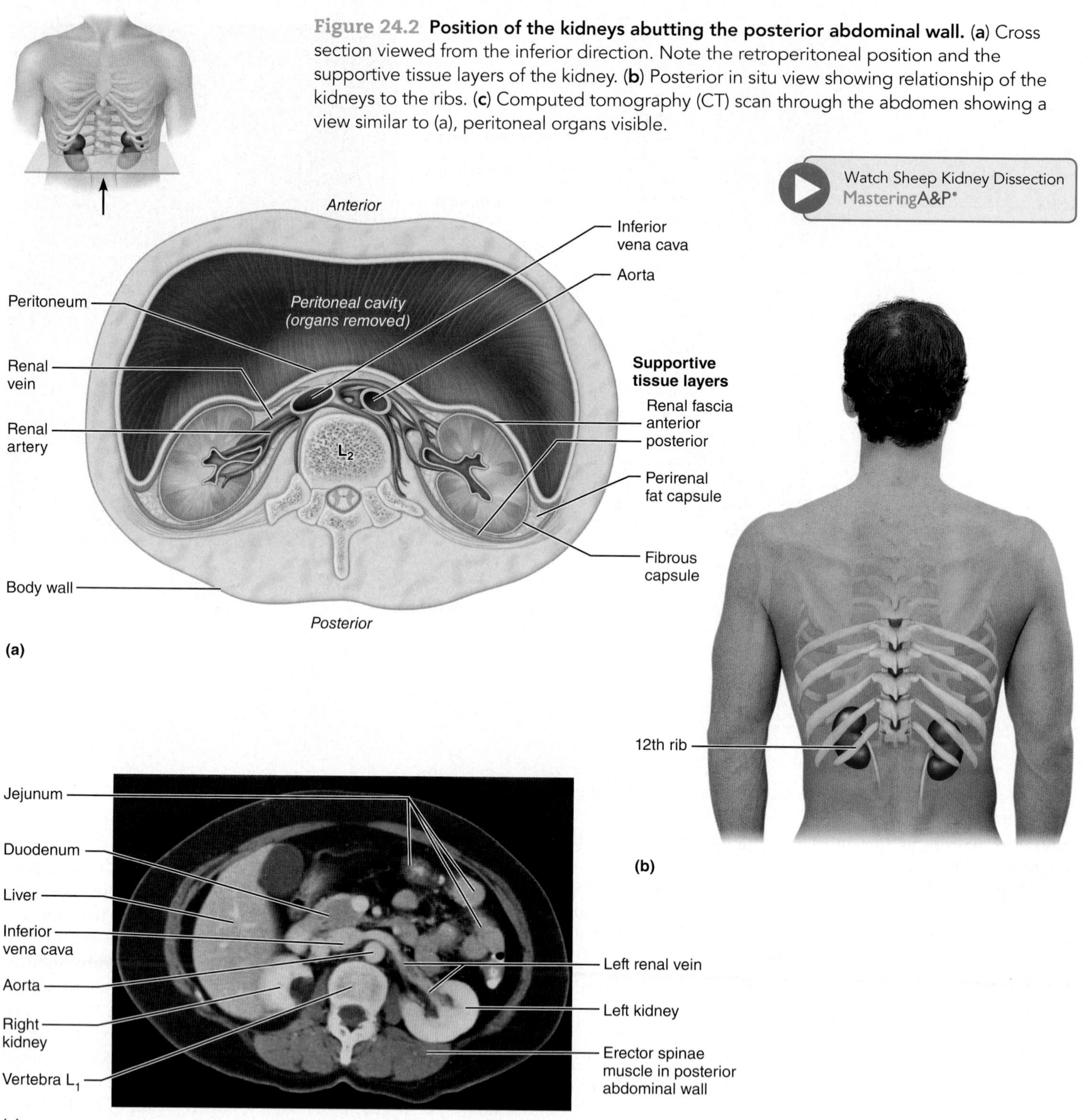

Figure 24.2 Position of the kidneys abutting the posterior abdominal wall. (**a**) Cross section viewed from the inferior direction. Note the retroperitoneal position and the supportive tissue layers of the kidney. (**b**) Posterior in situ view showing relationship of the kidneys to the ribs. (**c**) Computed tomography (CT) scan through the abdomen showing a view similar to (a), peritoneal organs visible.

external layer of fat, the *pararenal fat* (par′ah-re″nal; "near the kidney"). The perirenal and pararenal fat layers cushion the kidney against blows and help hold the kidneys in place.

The surgical approach to the kidney is usually through the posterolateral abdominal wall, where the kidney lies closest to the body surface (see Figure 24.2c). With this approach, very few muscles, vessels, or nerves need to be cut. However, the incision must be made inferior to the level of T_{12} to avoid puncturing the pleural cavity, which lies posterior to the superior third of each kidney. Puncturing the pleural cavity leads to pneumothorax and collapse of the lung (p. 664).

Internal Gross Anatomy

A frontal section through a kidney (Figure 24.3) reveals two distinct regions of kidney tissue: cortex and medulla. The more superficial region, the **renal cortex,** is light in color and has a granular appearance. Deep to the cortex is the darker **renal medulla,** which consists of cone-shaped masses called

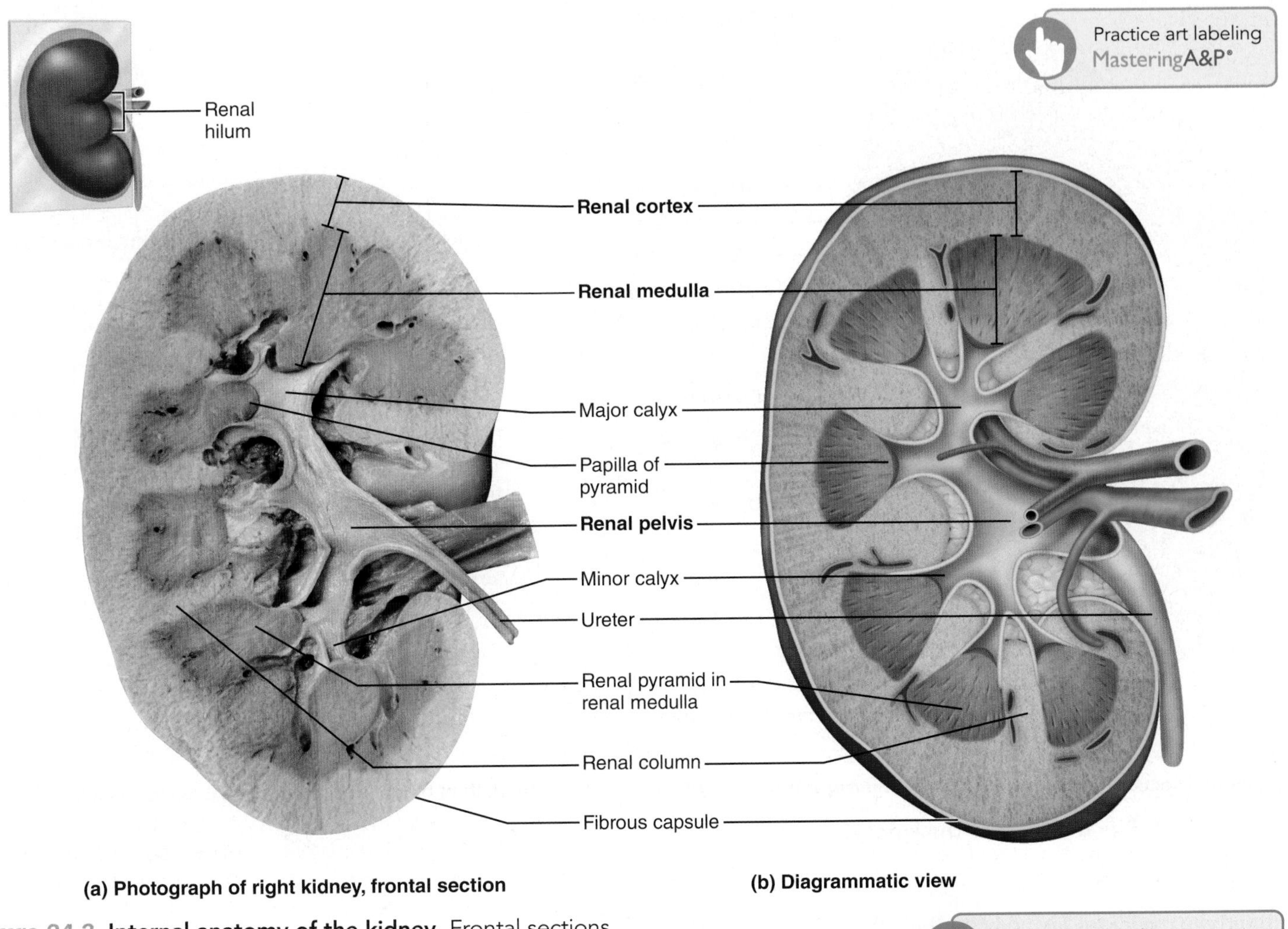

Figure 24.3 Internal anatomy of the kidney. Frontal sections.

renal pyramids. The broad base of each pyramid abuts the cortex, whereas the pyramid's apex, or papilla, points internally. The renal pyramids exhibit striations because they contain roughly parallel bundles of tiny urine-collecting tubules. The **renal columns,** inward extensions of the renal cortex, separate adjacent pyramids.

The human kidney is said to have *lobes,* each of which is a single renal pyramid plus the cortical tissue that surrounds that pyramid. There are 5 to 11 lobes and pyramids in each kidney.

The *renal sinus* is a large space within the medial part of the kidney opening to the exterior through the renal hilum. This sinus is actually a "filled space" in that it contains the renal vessels and nerves, some fat, and the urine-carrying tubes called the *renal pelvis* and *calices.* The **renal pelvis** (*pelvis* = basin), a flat, funnel-shaped tube, is simply the expanded superior part of the ureter. Branching extensions of the renal pelvis form two or three **major calices** (ca′lih-sēs; singular, **calyx**) each of which divides to form several **minor calices,** cup-shaped tubes that enclose the papillae of the pyramids (*calyx* = cup). The calices collect urine draining from the papillae and empty it into the renal pelvis; the urine then flows through the renal pelvis and into the ureter, which transports it to the bladder for storage.

CLINICAL APPLICATION

Pyelonephritis When an infection of the renal pelvis and calices, called *pyelitis* (pi″ĕ-li′tis; "inflammation of the renal pelvis"), spreads to involve the rest of the kidney as well, the result is **pyelonephritis** (pi″ĕ-lo-nĕ-fri′tis; *nephros* = kidney). Although it usually results from the spread of the fecal bacterium *Escherichia coli* from the anal region superiorly through the urinary tract, pyelonephritis also occurs when bloodborne bacteria lodge in the kidneys and proliferate there. In severe cases, the kidney swells and scars, abscesses form, and the renal pelvis fills with pus. Left untreated, the infected kidneys may be severely damaged, but timely administration of antibiotics usually achieves a total cure.

Gross Vasculature and Nerve Supply

Given that the kidneys continuously cleanse the blood, it is not surprising that they have a rich blood supply. Under normal resting conditions, about one-fourth of the heart's systemic output reaches the kidneys via the large **renal arteries.** These arteries branch at right angles from the abdominal aorta, between the first and second lumbar vertebrae (Figure 24.1 and Figure 20.12, p. 604). Because the aorta lies

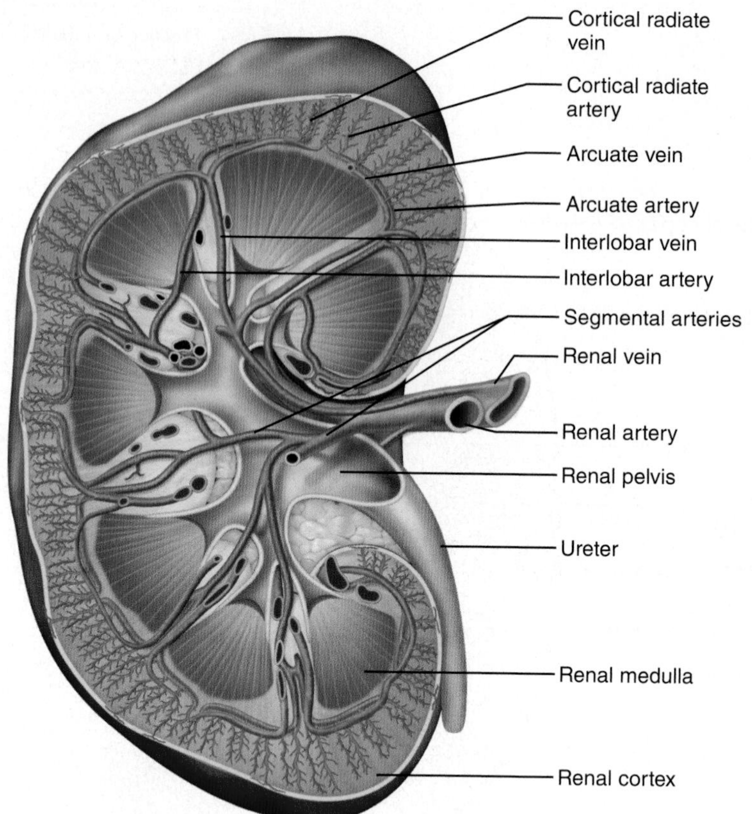

(a) Frontal section, posterior view, illustrating major blood vessels

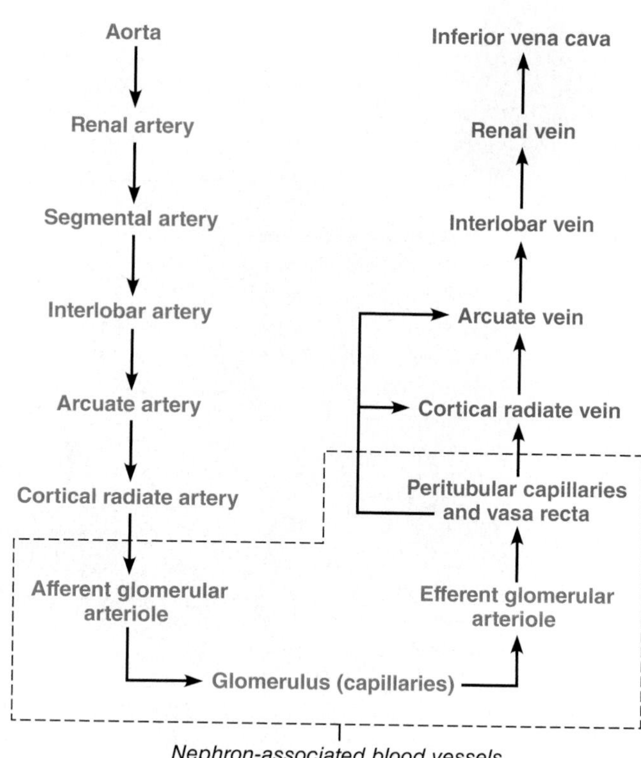

(b) Path of blood flow through renal blood vessels

Figure 24.4 Blood vessels of the kidney.

slightly to the left of the body midline, the right renal artery is longer than the left. As each renal artery approaches a kidney, it divides into five **segmental arteries** that enter the hilum (Figure 24.4). Within the renal sinus, each segmental artery divides into **interlobar arteries,** which lie in the renal columns between the renal pyramids.

At the medulla-cortex junction, the interlobar arteries branch into **arcuate arteries** (ar′ku-āt; "shaped like a bow"), which arch over the bases of the renal pyramids. Radiating outward from the arcuate arteries and supplying the cortical tissue are small **cortical radiate arteries.** More than 90% of the blood entering the kidney perfuses the cortex. These arteries give rise to the **glomerular arterioles,** which feed into the **peritubular capillaries** surrounding the tubules in the kidney (see p. 730 for further discussion of these microscopic vessels).

The *veins* of the kidney essentially trace the pathway of the arteries in reverse: Blood leaving the renal cortex drains sequentially into the **cortical radiate, arcuate, interlobar,** and **renal veins** (there are no segmental veins). The renal vein exits from the kidney at the hilum and empties into the inferior vena cava (Figure 24.1). Because the inferior vena cava lies on the right side of the vertebral column, the left renal vein is about twice as long as the right. Each renal vein lies anterior to the corresponding renal artery, and both blood vessels lie anterior to the renal pelvis at the hilum of the kidney.

The nerve supply of the kidney is provided by the *renal plexus,* a network of autonomic fibers and autonomic ganglia on the renal arteries. This plexus is an offshoot of the celiac plexus (see Figure 15.5, p. 474). The renal plexus is supplied by sympathetic fibers from the most inferior thoracic splanchnic nerve, the first lumbar splanchnic nerve, and other sources. These sympathetic fibers both control the diameters of the renal arteries and influence the urine-forming functions of the uriniferous tubules.

☑ check your understanding

☐ 1. The kidneys are located in the abdomen, but they are not within the peritoneum. Explain this distinction.

☐ 2. Describe the location of the kidneys in reference to the vertebrae. Why is the right kidney located more inferiorly than the left kidney?

☐ 3. Name the blood vessels that pass through the renal column.

(For answers, see Appendix B.)

Microscopic Anatomy of the Kidneys

learning outcomes

- ▶ Identify the portions of the nephron, and explain the role of each in the formation of urine.
- ▶ Describe the location and functions of the arterioles and capillaries surrounding the nephron.

The main structural and functional unit of the kidney is the **nephron** ("kidney"). More than a million of these tubules are crowded together within each kidney.

Before examining the detailed structure of the nephron, you will learn the mechanisms by which it produces urine.

Mechanisms of Urine Production

The nephron produces urine through three interacting mechanisms: filtration, resorption, and secretion (Figure 24.5). In **filtration,** a filtrate of the blood leaves the kidney capillaries and enters the renal tubule (arrow 1 in Figure 24.5). This filtrate resembles tissue fluid in that it contains all the small molecules of blood plasma. As it proceeds through the renal tubule, the filtrate is processed into urine by the mechanisms of resorption and secretion. During **resorption,** most of the nutrients, water, and essential ions are recovered from the filtrate and returned to the blood of capillaries in the surrounding connective tissue (arrow 2 in Figure 24.5). In fact, 99% of the volume of the renal filtrate is resorbed in this manner. As the essential molecules are reclaimed from the filtrate, the remaining wastes and unneeded substances contribute to the urine that ultimately leaves the body. Supplementing this passive method of waste disposal is the active process of **secretion,** which moves additional undesirable molecules into the tubule from the blood of surrounding capillaries (arrow 3 in Figure 24.5). Next you will explore the basic portions of the nephron and see how each contributes to the processes of filtration, resorption, and secretion.

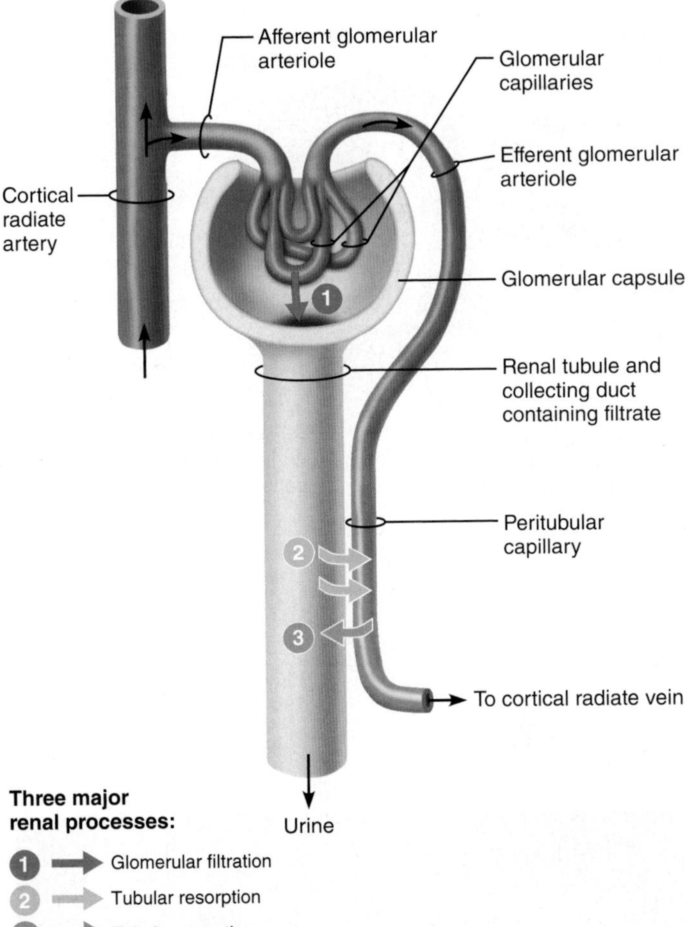

Figure 24.5 **Basic mechanisms of urine formation.** A single nephron is drawn schematically to illustrate the three processes involved in urine formation. A kidney contains more than a million nephrons acting in parallel.

Nephron Structure

Each **nephron** is composed of the *renal corpuscle* and a *renal tubule*. The renal tubule is divided into portions: the proximal tubule, the nephron loop (loop of Henle), the distal tubule, and the collecting duct. Throughout its length, the nephron is lined by a simple epithelium that is adapted for various aspects of the production of urine.

Renal Corpuscle The first part of the nephron, where filtration occurs, is the spherical **renal corpuscle** (Figure 24.6a). Renal corpuscles occur strictly in the cortex. They consist of a tuft of capillaries called a **glomerulus** (glo-mer′u-lus; "ball of yarn") surrounded by a cup-shaped, hollow **glomerular capsule** *(Bowman's capsule).* The glomerulus lies in its glomerular capsule like a fist pushed deeply into an underinflated balloon. This tuft of capillaries is supplied by an afferent arteriole and drained by an efferent arteriole. The endothelium of the glomerulus is fenestrated (has pores), and thus these capillaries are highly porous (Figure 24.6b), allowing large quantities of fluid and small molecules to pass from the capillary blood into the hollow interior of the glomerular capsule, the **capsular space.** This fluid is the filtrate that is ultimately processed into urine. Only about 20% of the fluid leaves the glomerulus and enters the capsular space; 80% remains in the blood within this capillary.

The external **parietal layer** of the glomerular capsule is a simple squamous epithelium (Figure 24.6a). It contributes to the structure of the capsule but plays no part in the formation of filtrate. By contrast, the capsule's **visceral layer** clings to the glomerulus and consists of unusual, branching epithelial cells called **podocytes** (pod′o-sītz; "foot cells"). The branches of the octopus-like podocytes (Figure 24.6b and c) end in **foot processes,** or *pedicels* ("little feet"), which interdigitate with one another as they surround the glomerular capillaries. The filtrate passes into the capsular space through thin clefts between the foot processes, called **filtration slits.**

The **filtration membrane,** the actual filter that lies between the blood in the glomerulus and the capsular space, consists of three layers (Figure 24.6d):

1. The fenestrated endothelium of the capillary. The capillary pores (fenestrations) restrict the passage of the largest elements such as blood cells.
2. The filtration slits between the foot processes of podocytes, each of which is covered by a thin **slit diaphragm.**
3. An intervening basement membrane consisting of the fused basal laminae of the endothelium and the podocyte epithelium. The basement membrane and slit diaphragm hold back all but the smallest proteins while letting through small molecules such as water, ions, glucose, amino acids, and urea.

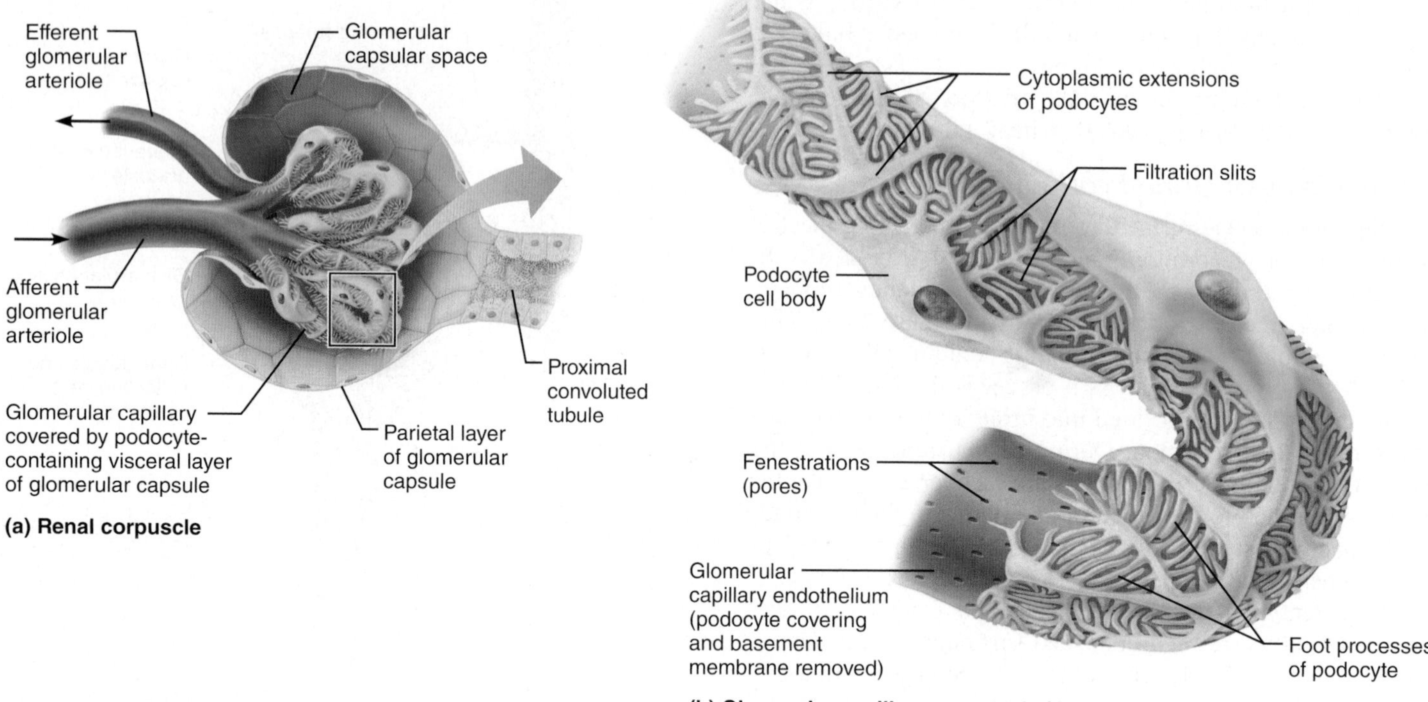

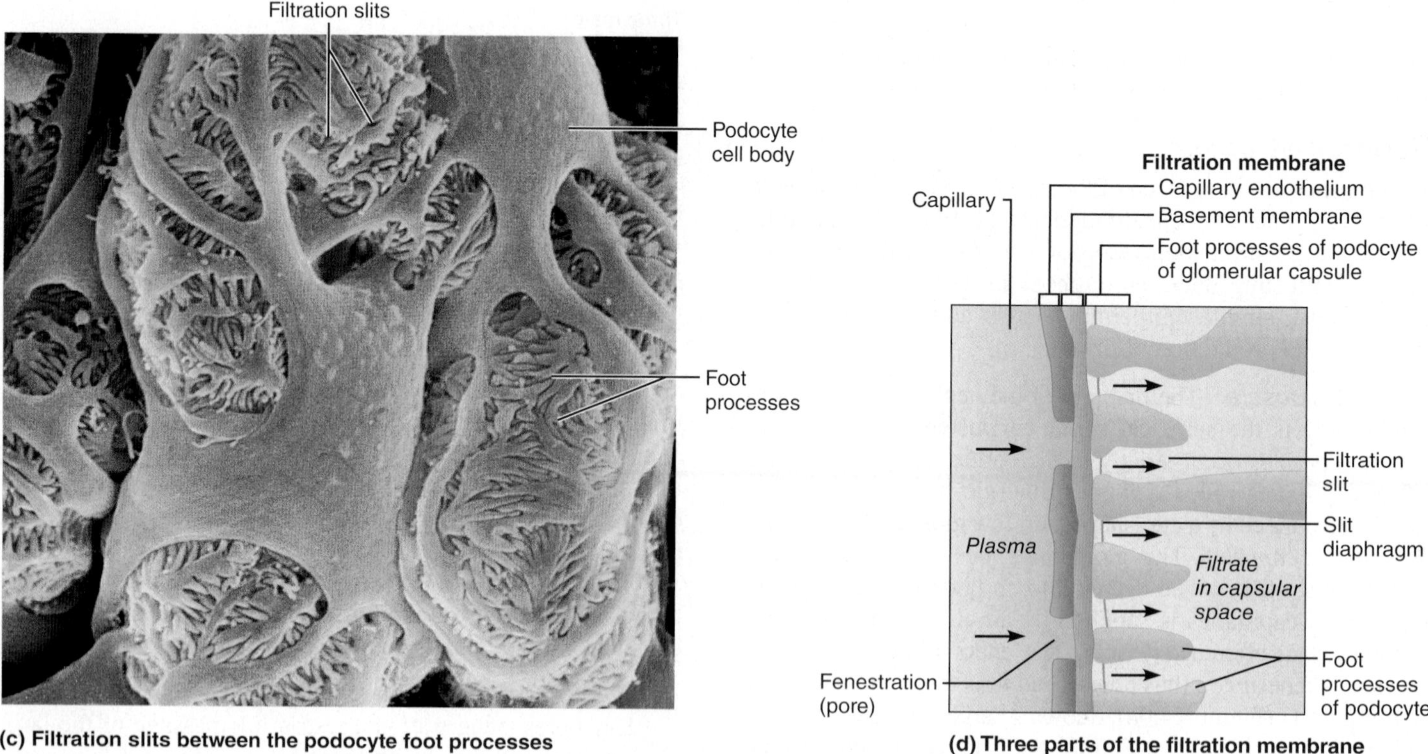

Figure 24.6 Renal corpuscle and the filtration membrane. (**a**) The renal corpuscle consists of a glomerulus surrounded by a glomerular capsule. (**b**) Enlargement of a glomerular capillary covered by the visceral (inner) layer of the glomerular capsule, composed of podocytes. Some podocytes and the basement membrane have been removed to show the fenestrations (pores) in the underlying capillary wall. (**c**) Scanning electron micrograph of the visceral layer clinging to the glomerular capillaries (artificially colored; 6500×). (**d**) Diagram of a section through the filtration membrane, showing its three layers.

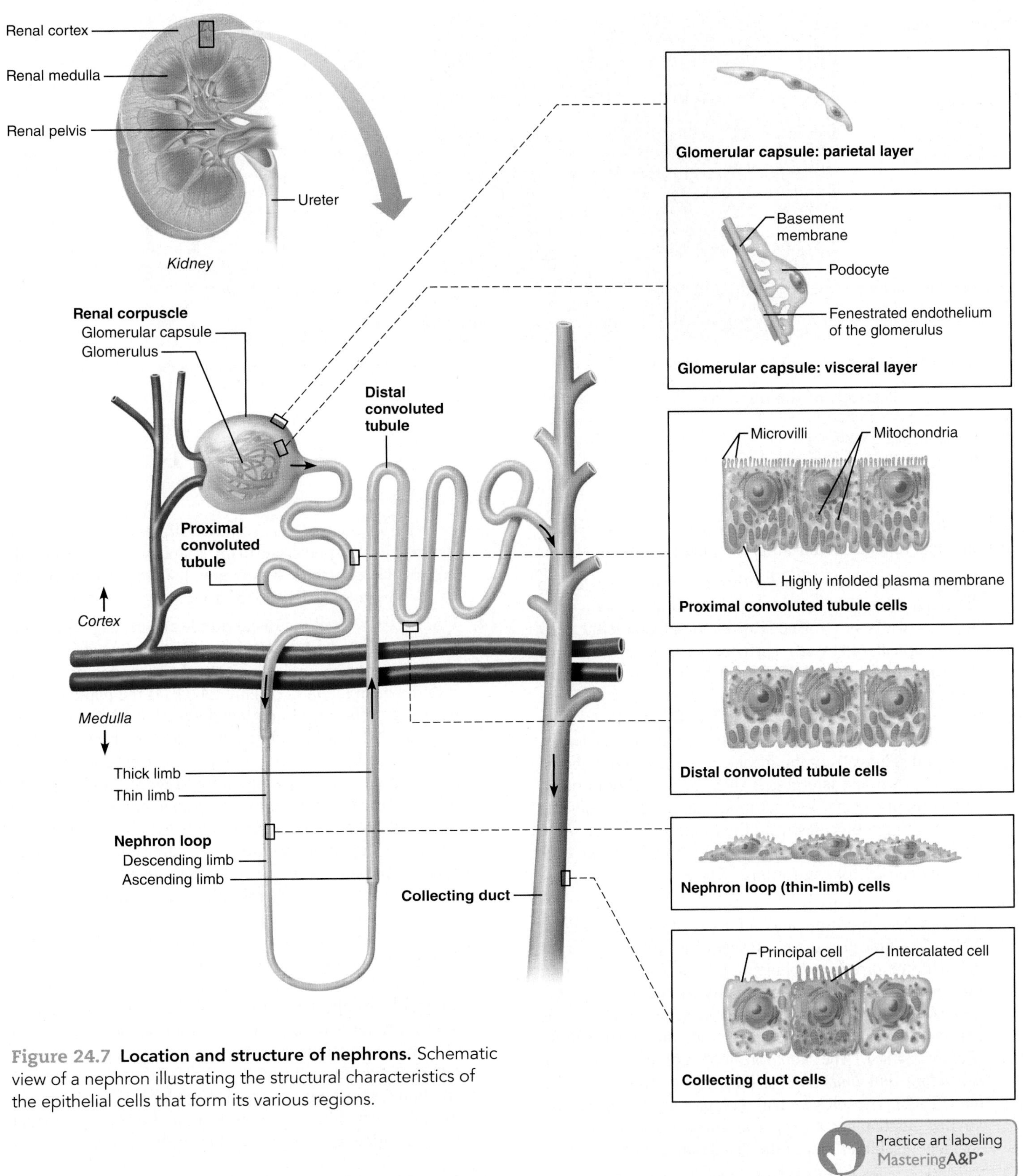

Figure 24.7 **Location and structure of nephrons.** Schematic view of a nephron illustrating the structural characteristics of the epithelial cells that form its various regions.

Renal Tubule After forming in the renal corpuscle, the filtrate proceeds into the long tubular section of the nephron (Figure 24.7), which begins as the elaborately coiled *proximal convoluted tubule,* makes a hairpin loop called the *nephron loop (loop of Henle),* winds and twists again as the *distal convoluted tubule,* and ends by joining a *collecting duct*. This meandering nature of the nephron increases its length and enhances its capabilities in processing the filtrate that flows through it. Each

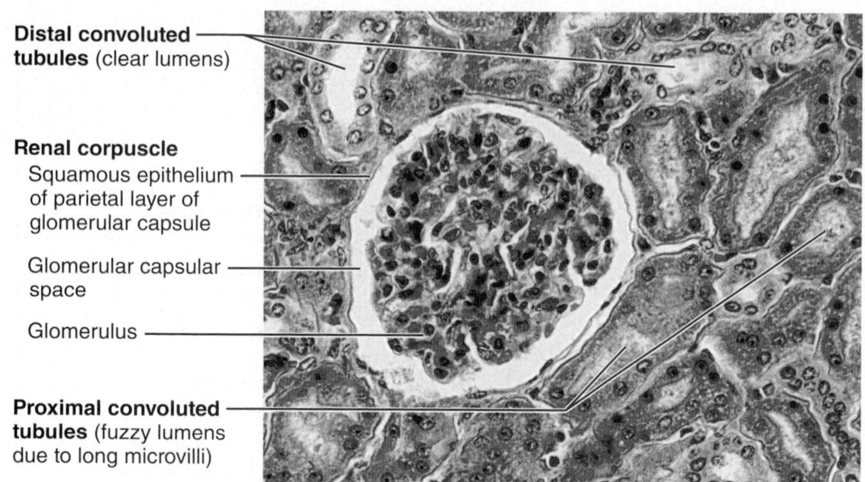

(a) Photomicrograph of renal cortical tissue (230×)

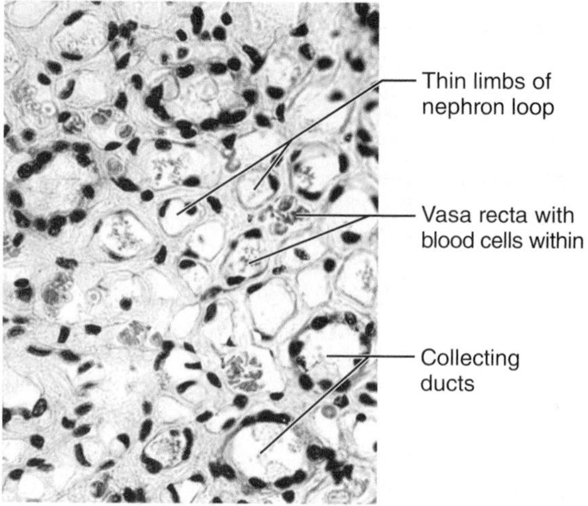

(b) Photomicrograph through the renal medulla (230×)

Figure 24.8 Histology of the nephron.

part of the nephron's tubular section has a unique cellular anatomy that reflects its filtrate-processing function.

1. The **proximal convoluted tubule,** confined entirely to the renal cortex, is most active in resorption and secretion. Its walls are formed by cuboidal epithelial cells whose apical (exposed) surfaces have long microvilli that seem to fill the tubule lumen with a "fuzz" in photo micrographs (Figure 24.7 and Figure 24.8a). These microvilli increase the surface area of these cells tremendously, maximizing their capacity for resorbing water, ions, and solutes from the filtrate. The plasma membrane on their basal and lateral cell surfaces is highly infolded and contains many ion-pumping enzymes responsible for resorbing molecules from the filtrate. The cells of the proximal tubule contain many mitochondria, which provide the energy for resorption.
2. The U-shaped **nephron loop** (loop of Henle), consists of a descending limb and an ascending limb (Figure 24.7). The first part of the **descending limb** is continuous with the proximal tubule and has a similar structure; the rest of the descending limb is the *descending thin limb* (*DTL*), the narrowest part of the nephron, with walls consisting of a permeable simple squamous epithelium (Figure 24.8b). The nephron loop continues into the *ascending thin limb* (ATL), joining the *thick ascending limb* (TAL). The thick ascending limb joins the distal convoluted tubule in the cortex. The cell structure of this thick segment resembles that of the distal convoluted tubule.
3. The **distal convoluted tubule,** like the proximal convoluted tubule, is confined to the renal cortex. It has walls of simple cuboidal epithelium and is specialized for the selective secretion and resorption of ions (Figure 24.7 and Figure 24.8a). The distal convoluted tubule is less active in resorption than the proximal tubule, and its cells do not have an abundance of absorptive microvilli. The cells of the distal tubule do have many mitochondria and infoldings of the basolateral membrane, features that are typical of all ion-pumping cells in the body.
4. Urine passes from the distal tubules of the nephrons into the **collecting ducts,** each of which receives urine from several nephrons and runs straight through the cortex into the deep medulla (Figure 24.7). At the papilla of the pyramid, adjacent collecting ducts join to form larger **papillary ducts,** which empty into the minor calices. Histologically, the walls of the collecting ducts consist of a simple cuboidal epithelium (Figure 24.8b), which thickens to become simple columnar epithelium in the papillary ducts. Most of these cells have few organelles, but a few, called *intercalated cells,* are rich in mitochondria and participate in resorption and secretion of ions.

The most important role of the collecting ducts is to conserve body fluids, a function they share with the distal tubules. When the body must conserve water, the posterior part of the pituitary gland secretes antidiuretic hormone (ADH, see Chapter 17, p. 528), which increases the permeability of the collecting ducts and distal tubules to water. As a result, water is resorbed from the filtrate in these tubules into the surrounding blood vessels, decreasing the total volume of urine produced. Alcohol inhibits the release of antidiuretic hormone, resulting in reduced water resorption from the renal tubules, the production of a copious amount of dilute urine, and the potential for dehydration.

Classes of Nephrons Although all nephrons have the structures just described, they are divided into two categories according to location (Figure 24.9). **Cortical nephrons,** which represent 85% of all nephrons, are located almost entirely within the cortex, with their nephron loops dipping only a short distance into the medulla. The remaining 15% are called

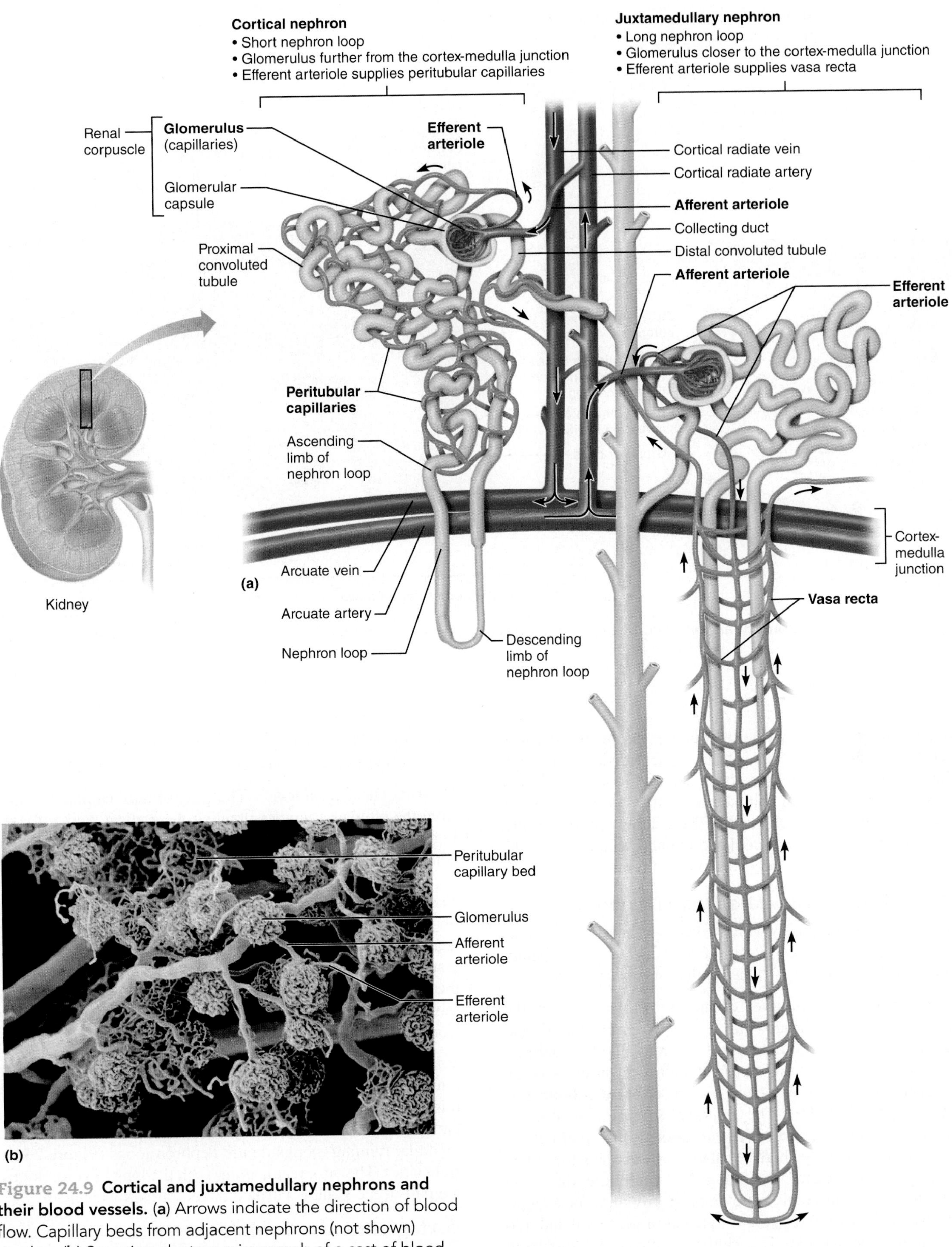

Figure 24.9 Cortical and juxtamedullary nephrons and their blood vessels. (**a**) Arrows indicate the direction of blood flow. Capillary beds from adjacent nephrons (not shown) overlap. (**b**) Scanning electron micrograph of a cast of blood vessels from the renal cortex (60×).

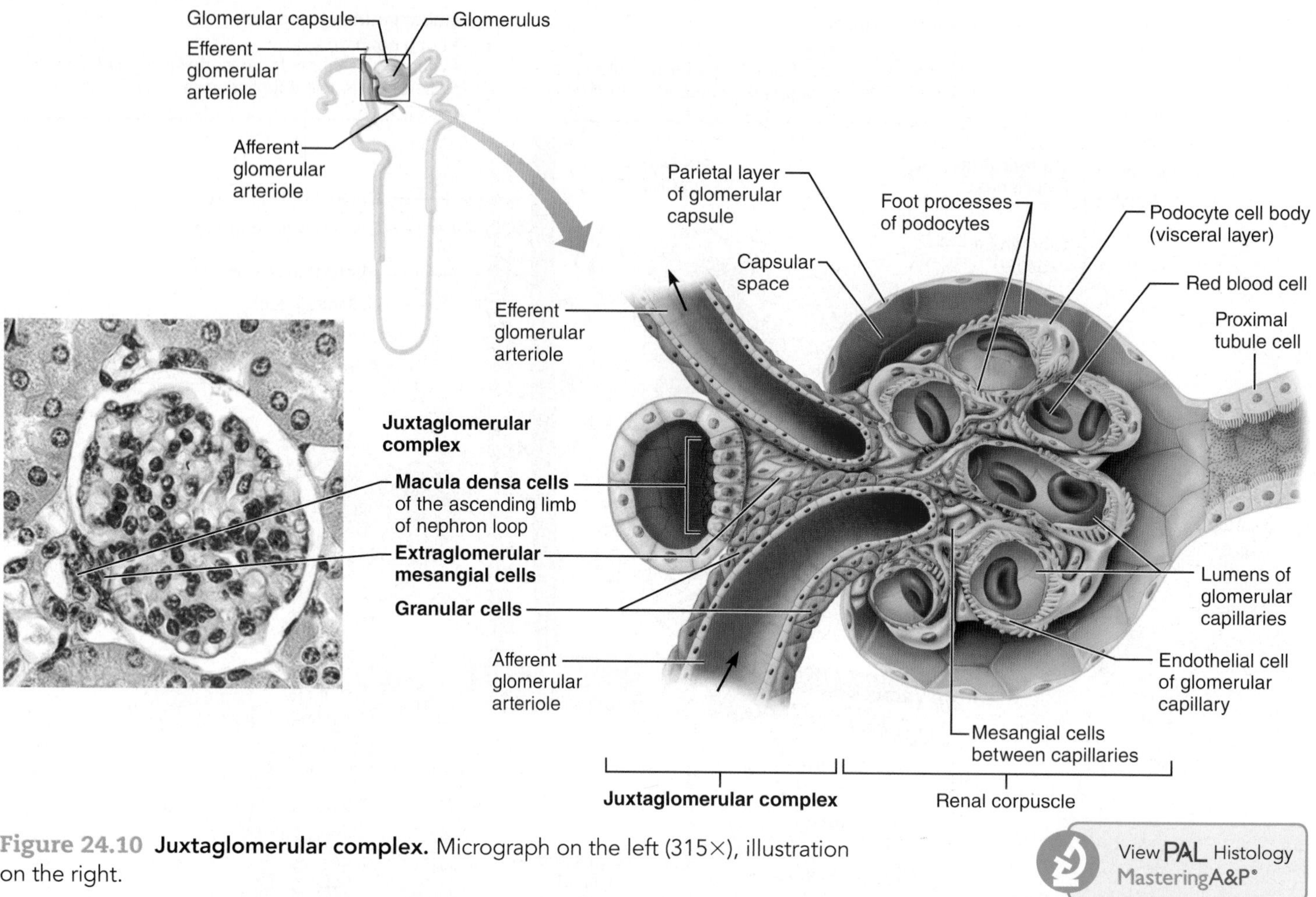

Figure 24.10 Juxtaglomerular complex. Micrograph on the left (315×), illustration on the right.

juxtamedullary ("near the medulla") **nephrons** because their renal corpuscles lie near the cortex-medulla junction. The juxtamedullary nephrons have nephron loops that deeply invade the medulla and thin segments that are much longer than those of cortical nephrons. These long nephron loops, in conjunction with nearby collecting ducts, contribute to the kidney's ability to produce a concentrated urine.

Blood Vessels Associated with Nephrons

Nephrons associate closely with two capillary beds, the *glomerulus* and the *peritubular capillaries* (Figure 24.9). Juxtamedullary nephrons also associate with the capillary-like *vasa recta.*

Glomeruli The capillaries of the glomerulus produce the filtrate that moves through the rest of the renal tubule and becomes urine. The glomerulus differs from all other capillary beds in the body: It is both fed and drained by arterioles—an **afferent glomeruler arteriole** and an **efferent glomeruler arteriole,** respectively. The *afferent arterioles* arise from the cortical radiate arteries that run through the renal cortex. Because arterioles are high-resistance vessels and the efferent arteriole is narrower than the afferent arteriole, the blood pressure in the glomerulus is extraordinarily high for a capillary bed and easily forces the filtrate out of the blood and into the glomerular capsule. The kidneys generate 1 liter (about 1 quart) of this filtrate every 8 minutes, but only 1% ends up as urine; the other 99% is resorbed by the uriniferous tubule and returned to the blood in the peritubular capillary beds.

Peritubular Capillaries The **peritubular capillaries** arise from the efferent arterioles draining the cortical glomeruli (see Figure 24.9a, left side). These capillaries lie in the interstitial connective tissue of the renal cortex, a loose areolar connective tissue that surrounds the renal tubules. The capillaries cling closely to the convoluted tubules and empty into nearby venules of the renal venous system. The peritubular capillaries are adapted for absorption: They are low-pressure, porous capillaries that readily absorb solutes and water from the tubule cells after these substances are resorbed from the filtrate. Furthermore, all molecules that are *secreted* by the nephrons into the urine are from the blood of nearby peritubular capillaries.

Vasa Recta In the deepest part of the renal cortex, efferent arterioles from the juxtamedullary glomeruli continue into thin-walled looping vessels called **vasa recta** (va′sah rek′tah; "straight vessels"). These hairpin loops descend into the medulla, running alongside the nephron loops (Figure 24.9a, right side). The vasa recta, along with the long nephron loops, are part of the kidney's urine-concentrating mechanism.

Juxtaglomerular Complex

The **juxtaglomerular complex** (juks″tah-glo-mer′u-lar; "near the glomerulus"), a structure that functions in the regulation of blood pressure, is an area of specialized contact between the terminal end of the thick ascending limb of the nephron loop and the afferent arteriole (Figure 24.10). Within the complex, the structures of both the tubule and the arteriole are modified.

The walls of the afferent and efferent glomerular arterioles contain **granular cells** *(juxtaglomerular cells),* modified smooth muscle cells with secretory granules containing a hormone called **renin** (re′nin; "kidney hormone"). The granular cells seem to be mechanoreceptors that secrete renin in response to falling blood pressure in the afferent arteriole.

The **macula densa** (mak′u-lah den′sah; "dense spot"), which is the terminal portion of the nephron loop adjacent to the granular cells, consists of tall, closely packed epithelial cells that act as chemoreceptors for monitoring solute concentrations in the filtrate. When solute concentrations in the filtrate fall below a certain level, the cells of the macula densa signal the granular cells to secrete renin. Renin initiates a sequence of chemical reactions in the blood (referred to as the *renin-angiotensin mechanism*) that eventually results in the secretion of the hormone aldosterone from the adrenal cortex. Aldosterone increases sodium (Na^+) resorption from the distal convoluted tubules, which increases blood-solute concentration. When sodium is resorbed, water follows along the osmotic gradient, causing blood volume, and most importantly, blood pressure, to rise. Caffeine and certain medications prescribed for hypertension act as diuretics, substances that increase the amount of urine excreted, by blocking the resorption of sodium from the distal convoluted tubules.

The mesangial cells (Figure 24.10) are irregularly shaped cells located around the base of the glomerular tuft. These cells show contractile properties that regulate blood flow within the glomerulus. The **extraglomerular mesangial cells** interact with cells of the macula densa and granular cells to regulate blood pressure. The details of this interaction are an area of ongoing research.

check your understanding

- ☐ 4. Which parts of the nephron are located in the renal cortex? Which parts are found in the medulla?
- ☐ 5. Which mechanism in the formation of urine involves the glomerular capillaries? Which mechanisms involve the peritubular capillaries?
- ☐ 6. What structure in the kidney functions in regulating blood pressure? How is the renal tubule modified in this region?

(For answers, see Appendix B.)

URETERS

learning outcome

- ▶ Describe the location, histology, and function of the ureters.

Gross Anatomy

Urine, collected from the renal calices and emptied into the renal pelvis, leaves the kidneys via the ureters. The **ureters** are slender tubes, about 25 cm (10 inches) long, that carry urine from the kidneys to the bladder (see Figure 24.1). Each ureter begins superiorly, at the level of L_2, as a continuation of the renal pelvis. From there, it descends retroperitoneal through the abdomen, enters the true pelvis by crossing the pelvic brim at the sacroiliac joint, enters the posterolateral corner of the bladder, and then runs medially within the posterior bladder wall before opening into the bladder's interior. This oblique entry into the bladder prevents backflow of urine from the bladder into the ureters, because any increase of pressure within the bladder compresses the bladder wall, thereby closing the distal ends of the ureters.

CLINICAL APPLICATION

Pyelography The radiographic procedure for examining the ureters and renal calices is called pyelography (pi″ĕ-log′rah-fe; "recording the renal pelvis"); the resulting image is called a **pyelogram** (see image). Radiologists introduce the X-ray contrast medium via one of two routes: either by injecting it into the ureters through a catheter in the bladder *(retrograde pyelography)* or by injecting it through a vein so that it reaches the ureters when excreted by the kidneys *(intravenous pyelography)*. Pyelography enables the clinician to examine the ureters along their course from the kidneys to the bladder as these tubes extend along an imaginary line defined by the transverse processes of the five lumbar vertebrae.

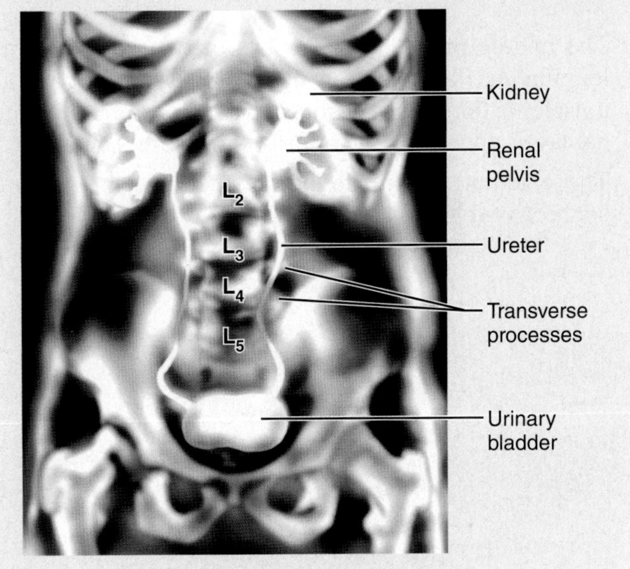

Microscopic Anatomy

The histological structure of the tubular ureters is the same as that of the renal calices and renal pelvis; the walls have three basic layers: a mucosa, a muscularis, and an adventitia (Figure 24.11).

- The lining **mucosa** is composed of a *transitional epithelium* that stretches when the ureters fill with urine (see p. 72 and Figure 24.13b) and a lamina propria composed of a stretchy, fibroelastic connective tissue containing rare patches of lymphoid tissue.

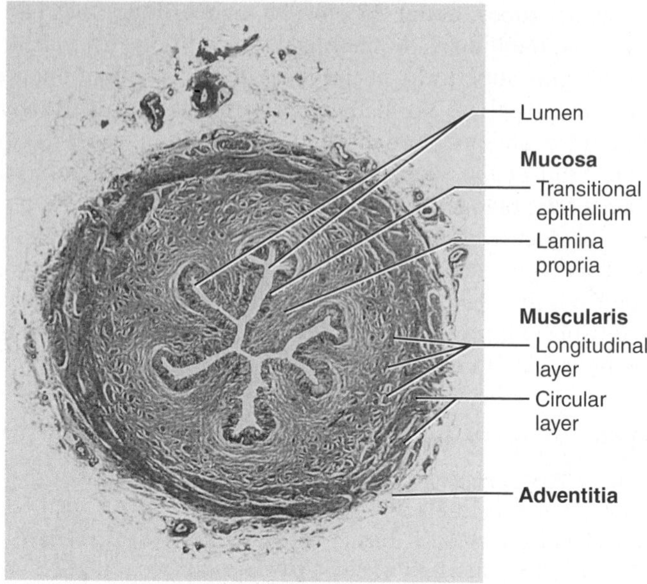

Figure 24.11 **Microscopic structure of the ureter, cross section (12×).** The mucosal folds, shown here in an empty ureter, stretch and flatten to accommodate large pulses of urine.

- The middle **muscularis** consists of two layers: an inner longitudinal layer and an outer circular layer of smooth muscle. A third layer of muscularis, an external longitudinal layer, appears in the inferior third of the ureter.
- The external **adventitia** of the ureter wall is a typical connective tissue.

The ureters play an active role in transporting urine. Distension of the ureter by entering urine stimulates its muscularis to contract, setting up peristaltic waves that propel urine to the bladder. This means that urine does *not* reach the bladder by gravity alone. Although the ureters are innervated by both sympathetic and parasympathetic nerve fibers, neural control of their peristalsis appears to be insignificant compared to the local stretch response of ureteric smooth muscle.

URINARY BLADDER

learning outcome

▶ Describe the shape, location, histology, and function of the bladder.

Gross Anatomy

The **urinary bladder,** a collapsible, muscular sac that stores and expels urine, lies inferior to the peritoneal cavity on the pelvic floor, just posterior to the pubic symphysis (Figure 24.12). In males, the bladder lies anterior to the rectum; in females, the bladder lies just anterior to the vagina and uterus.

A full bladder is roughly spherical and expands superiorly into the abdominal cavity; an empty bladder, by contrast, lies entirely within the pelvis. When full, the bladder becomes firm and can be palpated through the anterior abdominal wall just superior to the pubic symphysis. A bladder that can be palpated more than a few centimeters above this symphysis is dangerously full of urine and requires drainage by catheterization.

The empty bladder has the shape of an upside-down pyramid with four triangular surfaces and four corners, or angles (Figure 24.12a). The two *posterolateral angles* receive the ureters. At the bladder's *anterior angle* (or apex) is a fibrous band called the **urachus** (u′rah-kus; "urinary canal of the

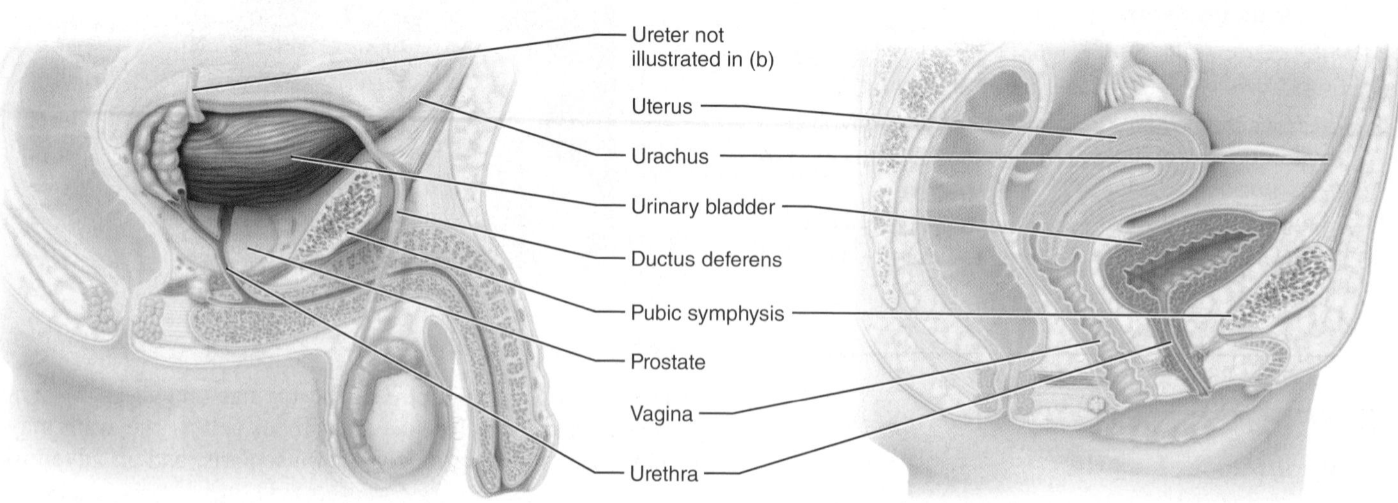

(a) Sagittal section through male pelvis, urinary bladder shown in lateral view

(b) Sagittal section through female pelvis

Figure 24.12 **Position of the urinary bladder in reference to the pelvic organs.**

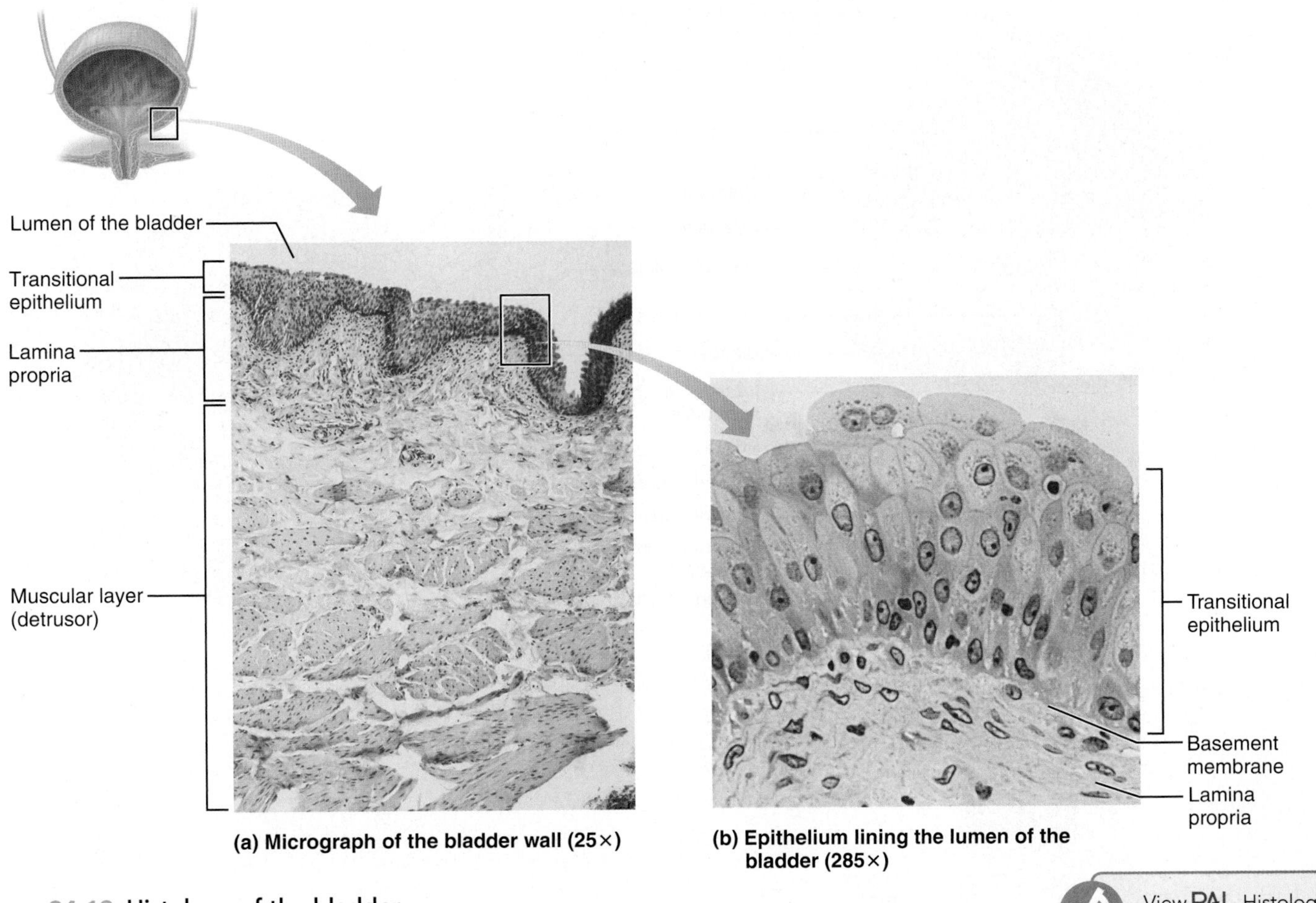

(a) Micrograph of the bladder wall (25×)

(b) Epithelium lining the lumen of the bladder (285×)

Figure 24.13 **Histology of the bladder.**

fetus"), the closed remnant of an embryonic tube called the allantois. The *inferior angle (neck)* drains into the urethra. In males, the prostate, a reproductive gland, lies directly inferior to the bladder, where it surrounds the urethra.

In the interior of the bladder, openings for both ureters and the urethra define a triangular region on the posterior wall called the **trigone** (tri′gon; "triangle") (see Figure 24.14). The trigone is of special clinical importance because infections tend to persist in this region.

Vessels and Nerves The arteries supplying the bladder are branches of the internal iliac arteries, primarily the *superior* and *inferior vesical arteries* (*vesical* = sac = the bladder) (see Figure 20.14 p. 606). Veins draining the bladder form a plexus on the bladder's inferior and posterior surfaces that empties into the internal iliac veins. Nerves extending to the bladder from the hypogastric plexus consist of parasympathetic fibers (ultimately from the pelvic splanchnic nerves), a few sympathetic fibers (ultimately from the lower thoracic and upper lumbar splanchnic nerves), and visceral sensory fibers.

Microscopic Anatomy

The wall of the bladder has three layers (Figure 24.13).

- A *mucosa* with a distensible transitional epithelium and a lamina propria forms the inner lining of the bladder. The mucosal lining contains folds, or **rugae,** that flatten as the bladder fills.
- A thick muscular layer called the **detrusor** (de-tru′sor; "to thrust out") forms the middle layer. This layer consists of highly intermingled smooth muscle fibers arranged in inner and outer longitudinal layers and a middle circular layer. Contraction of this muscle squeezes urine from the bladder during urination.
- On the lateral and inferior surfaces, the outermost layer is the *adventitia.* The superior surface of the bladder is covered by parietal peritoneum.

The bladder's great distensibility makes it uniquely suited for its function of storing urine. When there is little urine in it, the bladder collapses into its basic pyramidal shape, its walls thick and its mucosa thrown into rugae (Figure 24.14). But as urine accumulates, the rugae flatten, and the wall of the bladder thins as it stretches, allowing the bladder to store larger amounts of urine without a significant rise in internal pressure (at least until 300 ml has accumulated). A full adult bladder holds about 500 ml (or 1 pint) of urine, 15 times its empty volume.

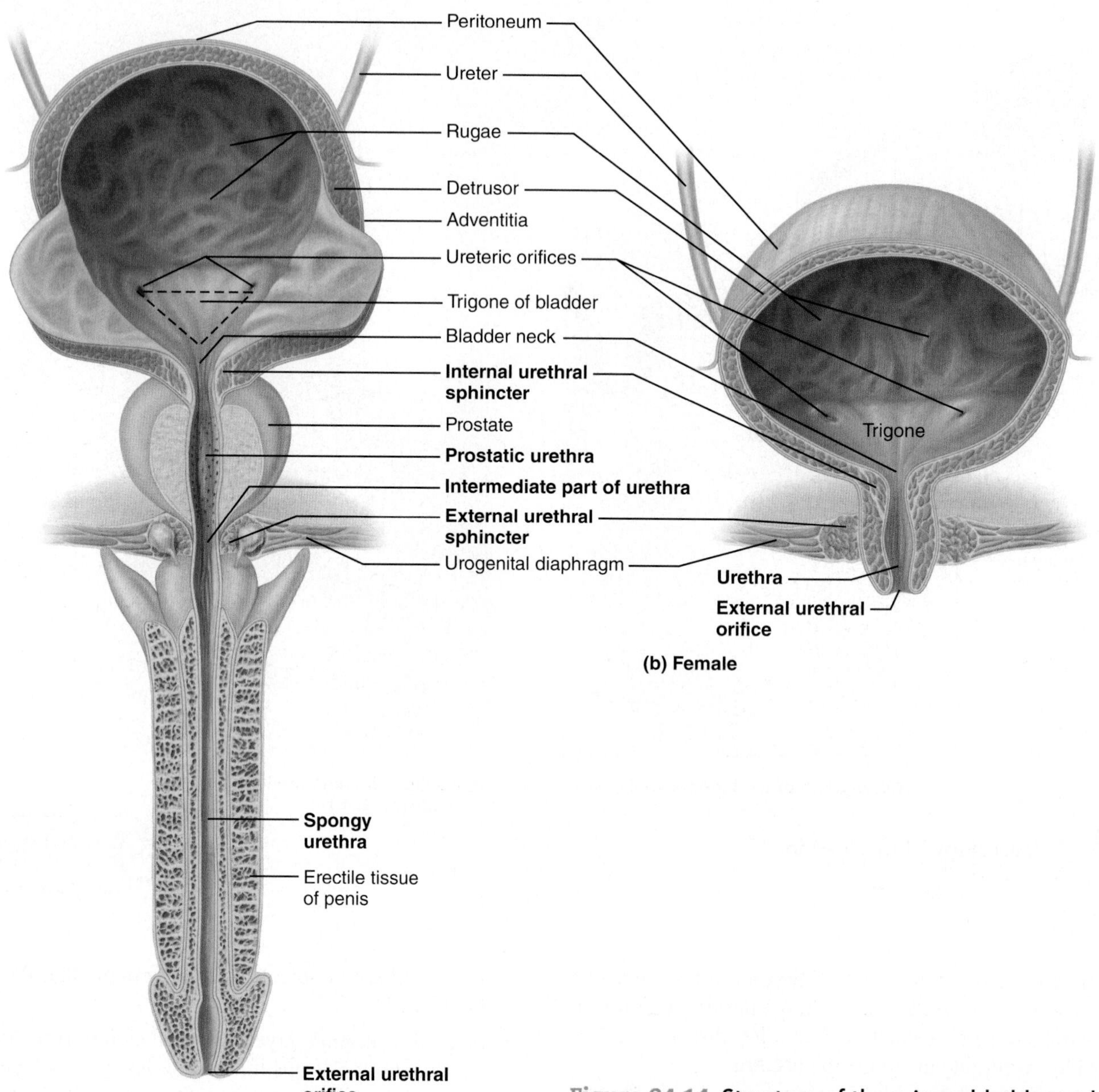

(a) Male. The long male urethra has three regions: prostatic, intermediate part, and spongy.

Figure 24.14 Structure of the urinary bladder and urethra. The anterior wall of the bladder has been cut away to reveal the position of the trigone.

URETHRA

learning outcomes

- Describe the structure and function of the urethra in both sexes.
- Define micturition, and explain its neural control.

The **urethra** is a thin-walled tube that drains urine from the bladder and conveys it out of the body (Figure 24.14). This tube consists of smooth muscle and an inner mucosa. In males, the muscle layer becomes very thin toward the distal end of the urethra. The lining epithelium changes from a transitional epithelium near the bladder to a stratified and pseudostratified columnar epithelium in midurethra (sparse in females), and then to a stratified squamous epithelium near the end of the urethra.

At the bladder-urethra junction, a thickening of the detrusor forms the **internal urethral sphincter.** This is an involuntary sphincter of smooth muscle that keeps the urethra closed when urine is not being passed and prevents dribbling of urine between voidings. A second sphincter, the **external urethral sphincter,** surrounds the urethra within the sheet of muscle called the *urogenital diaphragm.* This external sphincter is a skeletal muscle used to inhibit urination voluntarily until the proper time. The levator ani muscle of the pelvic floor also serves as a voluntary constrictor of the urethra. (For more information on the external urethral sphincter and levator ani muscles, see Table 11.9 on pp. 296–297).

The length and functions of the urethra differ in the two sexes. In females, the urethra is just 3–4 cm (1.5 inches) long and is bound to the anterior wall of the vagina by connective tissue (see Figure 24.12b). It opens to the outside at the

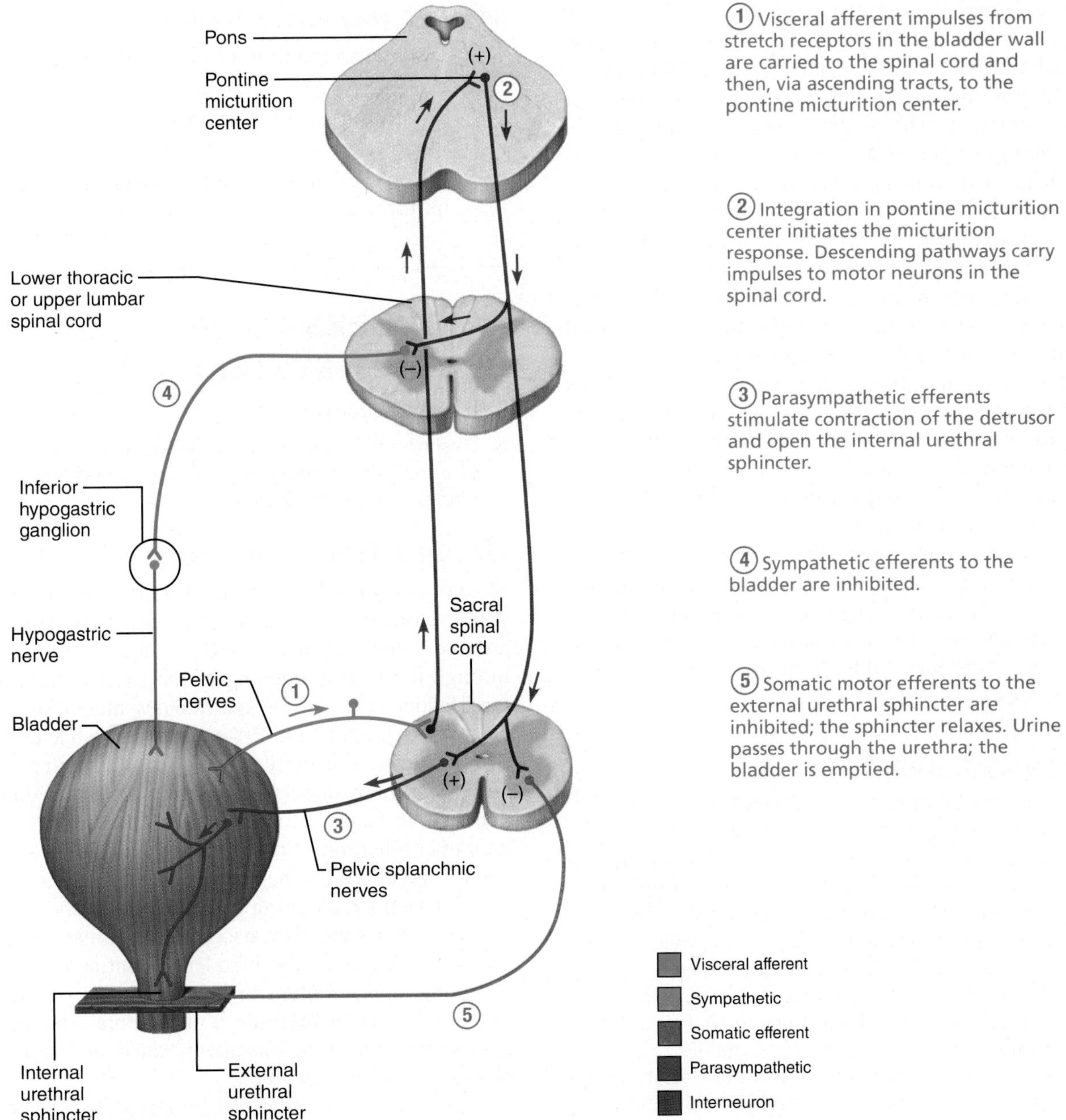

Figure 24.15 **Micturition.**

external urethral orifice, a small, often difficult-to-locate opening that lies anterior to the vaginal opening and posterior to the clitoris (shown in Figure 25.20, p. 766).

In males, the urethra is about 20 cm long (8 inches) and has three named regions: the **prostatic urethra,** which is about 2.5 cm long and runs in the prostate; the **intermediate part of the urethra,** or *membranous urethra,* which runs for about 2.5 cm through the membranelike urogenital diaphragm; and the **spongy urethra,** which is about 15 cm long, passes through the entire penis, and opens at the tip of the penis via the **external urethral orifice.** The male urethra carries ejaculating semen as well as urine (although not simultaneously) from the body (see Chapter 25, p. 750, for the male urethra's reproductive function).

MICTURITION

Micturition (mik″tu-rish′un), also called *voiding* or *urination,* is the act of emptying the bladder. It is brought about by the contraction of the bladder's detrusor muscle, assisted by the muscles of the abdominal wall, which contract to raise intra-abdominal pressure. Micturition is controlled by the brain and involves both autonomic and somatic pathways (Figure 24.15). As urine accumulates in the bladder, distension of the bladder wall activates stretch receptors, which send sensory impulses through visceral sensory neurons (Figure 24.15, ①) to the sacral region of the spinal cord, and then up to a micturition center in the dorsal part of the pons. Acting as an on/off switch, the neurons in the micturition center in the lower pons signal the parasympathetic neurons (Figure 24.15, ②) that stimulate contraction of the detrusor muscle, thereby emptying the bladder (Figure 24.15, ③). At the same time, the sympathetic pathways to the bladder (Figure 24.15, ④)—which would prevent micturition by relaxing the detrusor muscle—are inhibited. The somatic motor neurons to the external urethral sphincter are also inhibited, thus relaxing this voluntary muscle and allowing urine to pass through the urethra (Figure 24.15, ⑤).

The micturition center in the pons is heavily influenced by rostral brain regions, such as the inferior frontal region of the cerebral cortex (which enables a conscious decision that it is safe to micturate) and the anterior cingulate gyrus (involved in emotional evaluation of the urge to micturate).

The pons, cerebral cortex, and other parts of the CNS also *inhibit* micturition. They do so by (1) stimulating the somatic motor neurons to the external urethral sphincter, causing contraction of this muscle, and (2) activating sympathetic pathways that relax the detrusor and stimulate the internal urethral sphincter to contract.

The inability to control micturition is normal in babies who have not learned voluntarily to close their external urethral sphincter. Reflexive voiding occurs each time a baby's bladder fills enough to activate its stretch receptors, but the internal sphincter prevents dribbling of urine between voidings just as it does in adults.

CLINICAL APPLICATION

Micturition Dysfunctions **Urinary incontinence** is very common in the elderly, affecting half of those who are in nursing homes or are homebound. Its many causes can be divided into three main types: (1) **urge incontinence,** in which the detrusor has uncontrolled contractions; (2) **stress incontinence,** in which the urethral sphincter mechanisms malfunction, such that coughing and sneezing can force urine through these sphincters; and (3) **overflow incontinence,** in which the bladder overfills and urine dribbles from the urethra. Incontinence should not be viewed as an inevitable result of aging; up to 80% of the elderly with this condition can be helped by behavioral training, medications, or surgery.

Urinary retention, in which the bladder is unable to expel its contained urine, is common in patients recovering from surgeries involving general anesthesia, after which the detrusor apparently needs time to recover before resuming activity. It is also common in older men, because enlargement of the prostate squeezes the urethra. If urinary retention is prolonged, a slender rubber drainage tube called a **catheter** (kath′ĕ-ter; "thrust in") must be inserted through the urethra and into the bladder to drain the urine and prevent overfilling of the bladder. The straight, short urethra of females is much easier to catheterize than the long, curved urethra of males.

☑ check your understanding

- ☐ 7. Distinguish the ureter from the urethra.
- ☐ 8. Which type of epithelium lines the lumen of the ureter, the bladder, and the proximal portions of the urethra?
- ☐ 9. Which urethral sphincter is innervated by somatic motor neurons?

(For answers, see Appendix B.)

DISORDERS OF THE URINARY SYSTEM

learning outcome

▶ Describe the basic causes, symptoms, and treatment of urinary tract infections, kidney stones (calculi), and bladder and kidney cancers.

Urinary Tract Infections

Most **urinary tract infections (UTIs)** occur in sexually active young women. Intercourse drives bacteria from the vagina and the external genital region (and from the anus as well) through the nearby opening of the short urethra and toward the bladder. The use of spermicides magnifies the problem because they tend to kill normal resident bacteria and allow pathogenic fecal bacteria to colonize the vagina. *Escherichia coli,* a normal resident bacterium of the lower digestive tract that seldom causes problems there, produces 80% of all urinary tract infections. Sexually transmitted diseases, which are chiefly infections of the reproductive tract, can also inflame the urinary tract, clogging some urinary ducts.

The elderly are also susceptible to urinary tract infections due to weakness of the bladder, incontinence, poor bladder emptying, and retention of urine. The most common symptoms of a UTI in the elderly are mental changes and confusion, which can be mistakenly attributed to more serious disorders, such as dementia.

Overall, urinary tract infections occur in about 40% of women. Infection of the bladder, called **cystitis,** can spread superiorly to infect the ureters and kidneys, causing **pyelitis** and **pyelonephritis** (discussed on p. 723). Symptoms of urinary tract infections include a burning sensation during micturition, increased urgency and frequency of micturition, fever, and sometimes cloudy or blood-tinged urine. When the kidneys are involved, back pain and severe headache often occur as well. Most urinary tract infections are easily cured by antibiotics.

In men, urinary tract infections usually result from long-term catheterization of the penile urethra because it is difficult to keep indwelling catheters sterile.

Renal Calculi

In 12% of men and 5% of women in North America, calcium, magnesium, or uric acid salts in the urine crystallize and precipitate in the calices or renal pelvis, forming kidney stones, or **renal calculi** (kal′ku-li; "little stones"). Most calculi are smaller than 5 mm in diameter and thus can pass through the urinary tract without causing serious problems. The calculi, however, cause pain when they obstruct a ureter, thereby blocking the drainage of urine and increasing intrarenal pressure. The most severe pain results when the contracting walls of the ureter contact the sharp calculi during their periodic peristaltic contractions. Pain from kidney stones radiates from the lateral abdominal region to the anterior abdominal wall, and then perhaps to the groin. Calculi tend to lodge in three especially narrow regions of the ureters (see pyelogram on p. 731): (1) at the level of L_2, where the renal pelvis first narrows into the ureter; (2) at the sacroiliac joint, where the ureter enters the true pelvis; and (3) where the ureters enter the bladder. The clinician should be aware of these three points when searching for kidney stones on X-ray, CT, and ultrasound images.

Predisposing factors for calculi are (1) oversaturation of the renal filtrate with calcium ions, uric acid, or a substance called oxalate, which leads to the precipitation of crystals from the urine; (2) abnormal acidity or alkalinity of the urine; (3) dehydration; (4) blockage of urine flow in the urinary tract; and (5) bacterial infection, because at least some calculi precipitate around bacteria. Individuals with a history of kidney stones are encouraged to drink large volumes of water because this leads to a urine so dilute that salts will not precipitate and form calculi.

Surgical removal of calculi has been the traditional treatment, but now, surgery has been largely replaced by *extracorporeal shock wave therapy,* in which ultrasonic shock waves delivered from outside the patient's body break up the calculi.

Cancer of Urinary Organs

Bladder cancer, which accounts for about 3% of cancer deaths and is five times more common in men than in women, typically involves neoplasms of the bladder's lining epithelium. Blood in the urine is a common warning sign. This form of cancer may be induced by organic carcinogens that are deposited in the urine after being absorbed from the environment. Substances that have been linked to bladder cancer include tars from tobacco smoke, certain industrial chemicals, and some artificial sweeteners. Bladder cancer is usually lethal if it metastasizes. Surgical removal of the bladder and chemotherapy increase survival time significantly, to an average of 5 years after detection.

Kidney cancer—a cancer arising from the epithelial cells of either the renal tubules or the renal pelvis and calices—accounts for 2% of cancer deaths in the United States, but its incidence is rising. Risk factors for this type of cancer, which is twice as common in men as in women, include obesity, high blood pressure, and, perhaps, a high-protein diet. Most tumors are over 3 cm in diameter and have metastasized when detected (typically during a kidney examination using ultrasound imaging, CT, or MRI). In such cases the prognosis is poor, and the average survival time is only 12–18 months. Because renal cancers are resistant to radiation and chemotherapy, standard treatment is the surgical removal of the entire kidney and the associated adrenal gland and lymph nodes.

CLINICAL APPLICATION

Kidney Transplant A kidney transplant is the transfer of a functioning kidney from a donor to a recipient whose kidneys are failing. About 30% of transplanted kidneys come from living donors (usually a relative or spouse), and 70% come from cadavers (whose kidneys can be maintained for about 36 hours after death). Kidney transplants are more common than transplants of any other major organ. This procedure has a high rate of success (80% to 90% survival after 3 years if from a living donor, 70% if from a cadaver) and is comparatively easy to perform (only a few vessels need to be cut and rejoined). A single kidney is transplanted, because one is sufficient to carry out excretory functions. The failing kidney is usually left in place, and the new kidney is transplanted into the right iliac fossa of the pelvis, which has more room than the crowded lumbar region. As with other organ transplants, the recipient must take immunosuppressive drugs. Current efforts are directed toward improving long-term survival rates, which are diminished by chronic rejection and other factors.

THE URINARY SYSTEM THROUGHOUT LIFE

learning outcomes

- Trace the embryonic development of the urinary organs.
- List several effects of aging on the structure and function of the urinary system.

The embryo develops not just one pair of kidneys but three pairs, one after another, starting cranially and proceeding caudally. The three pairs are the pronephros, mesonephros, and metanephros. These different kidneys develop from the *urogenital ridges,* paired elevations of intermediate mesoderm on the dorsal abdominal wall (Figure 24.16). Of the three pairs of kidneys that form, only the last pair persists to become the adult kidneys. Initially, during week 4, the first kidney, or **pronephros** (pro-nef′ros), forms as a set of nephrons and then quickly degenerates. Although the pronephros is never functional and is gone by week 6, it sends a **pronephric duct** *(primary excretory duct)* to the cloaca, and this duct is used by the kidneys that develop later. As the second nephron system, called the **mesonephros** ("middle kidney"), claims the pronephric duct, this duct becomes the **mesonephric duct.** The nephrons of the mesonephros, in turn, degenerate after the third kidney, the **metanephros** (met″ah-nef′ros), becomes functional.

The metanephros is the definitive kidney (*metanephros* means "ultimate kidney"). It develops in the pelvic region. Starting in week 5, a hollow **ureteric bud** grows from the mesonephric duct into the urogenital ridge, inducing the mesoderm there to form the nephrons (Figure 24.16a). The ureteric bud, in turn, develops into the renal pelvis, calices, and collecting ducts; its unexpanded proximal part

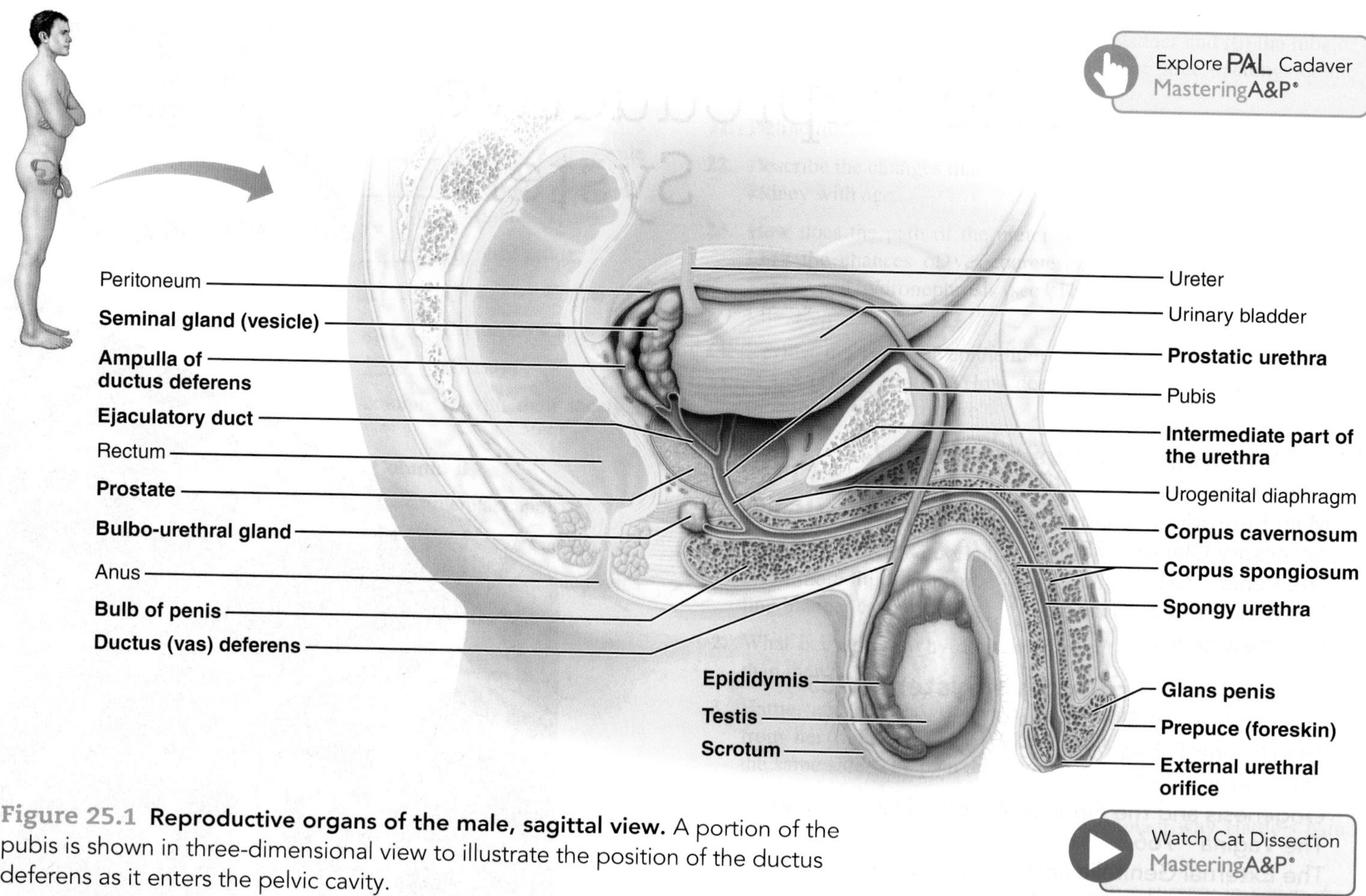

Figure 25.1 Reproductive organs of the male, sagittal view. A portion of the pubis is shown in three-dimensional view to illustrate the position of the ductus deferens as it enters the pelvic cavity.

Most organ systems of the body function almost continuously to maintain the well-being of the individual. The reproductive system, however, appears to "slumber" until puberty, after which it plays an important role in our adult lives. Although male and female reproductive organs are quite different, they share a common purpose—to produce offspring.

In both females and males, the sex organs, or **genitalia** (jen″ĭ-ta′le-ah), are divided into *primary* and *accessory* organs. The **primary sex organs,** or **gonads** (go′nadz; "seeds"), are the testes in males and the ovaries in females. Gonads produce the sex cells or **gametes** (gam′ēts)—the sperm in males and the ovum (egg) in females—that fuse to form a fertilized egg. All other genitalia in both sexes are **accessory sex organs,** including the internal glands and ducts that nourish the gametes and transport them toward the outside of the body, and the external genitalia.

Besides producing sex cells, the gonads also secrete *sex hormones* and therefore function as endocrine glands. Hormones are messenger molecules that travel through the bloodstream to signal various physiological responses in specific target organs. Sex hormones play vital roles in the development, maintenance, and function of all sex organs. Other hormones, secreted by the pituitary gland at the base of the brain, also influence reproductive functions.

The male reproductive system is considered first, followed by examination of the female reproductive system.

THE MALE REPRODUCTIVE SYSTEM

In the male reproductive system (Figure 25.1), the gonads are the sperm-producing *testes* (singular, *testis*), which lie in the scrotum. From the testes, sperm travel to the outside of the body through a system of ducts in the following order: the duct of the *epididymis,* the *ductus deferens,* the *ejaculatory duct,* and finally the *urethra,* which opens at the tip of the *penis.* The accessory sex glands, which empty their secretions into the sex ducts during ejaculation, are the *seminal glands, prostate,* and *bulbo-urethral glands.*

The Testes

learning outcome

▶ Describe the location, gross and microscopic structure, and function of the testes.

Location

The paired, oval **testes** (tes′tez; "witnesses"), or *testicles,* are located in the **scrotum** (skro′tum; "pouch"), a sac of skin and superficial fascia that hangs inferiorly external to the abdominopelvic cavity at the root of the penis (Figure 25.2). The scrotum is covered with sparse hairs. A septum in the midline divides the scrotum into right and left halves, providing one compartment for each testis.

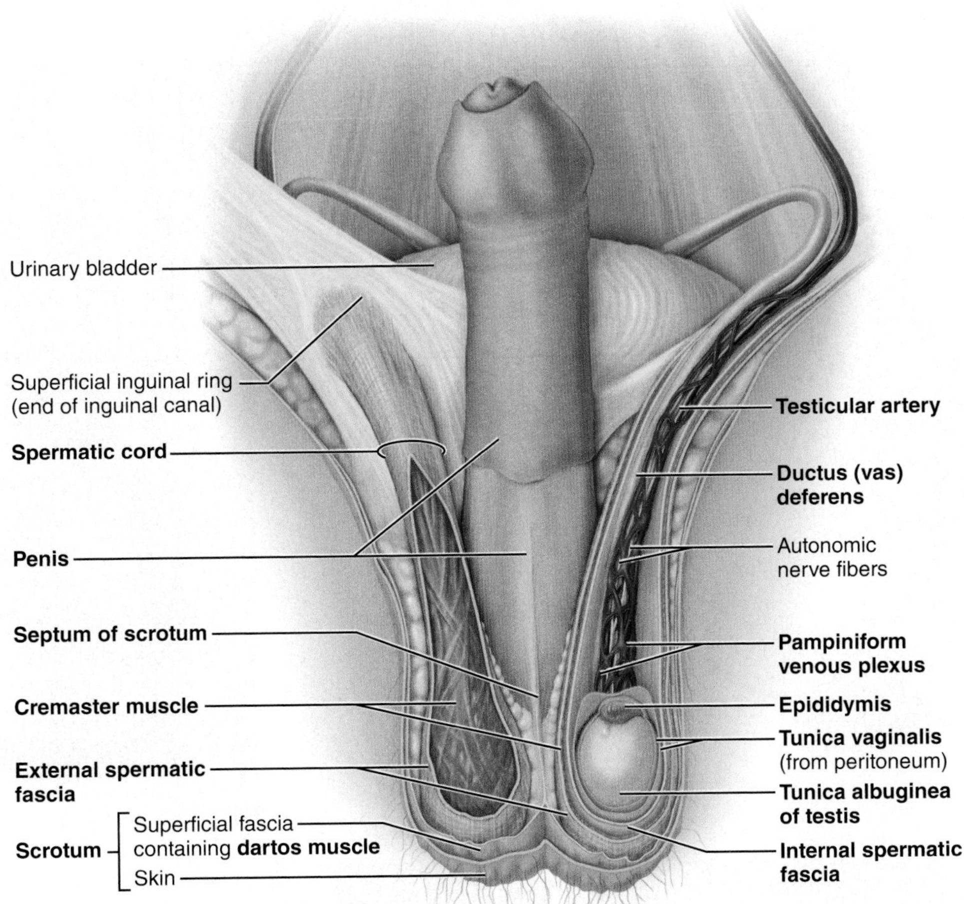

Figure 25.2 Relationships of the testis to the scrotum and spermatic cord (anterior view). The scrotum has been opened and its anterior wall removed.

As you will see (p. 777), the testes first develop deep in the posterior abdominal wall of the embryo and then migrate down into the scrotum, which is external to the body wall. Such a superficial location would seem to place a male's entire genetic heritage in a vulnerable position. However, because viable sperm cannot be produced at the core body temperature of 37°C, the scrotum's superficial position provides an environment that is about 3°C cooler, an essential adaptation.

Furthermore, the scrotum responds to changes in external temperature. Under cold conditions, the testes are pulled up toward the warm body wall, and the scrotal skin wrinkles to increase its thickness and reduce heat loss. These actions are performed by two muscles in the scrotum:

- The **dartos muscle** (dar′tos; "skinned"), a layer of smooth muscle in the superficial fascia, is responsible for wrinkling the scrotal skin.
- The **cremaster muscles** (kre-mas′ter; "a suspender"), bands of skeletal muscle that extend inferiorly from the internal oblique muscles of the trunk, are responsible for elevating the testes. Under hot conditions, these muscles relax, so the scrotal skin is flaccid and loose, and the testes hang low to increase the skin surface available for cooling (sweating); this also moves the testis farther away from the warm trunk.

Gross Anatomy

Each testis (Figure 25.3) averages about 2.5 cm (1 inch) in width and 4 cm in height. Within the scrotum, each testis is posterior to, and partially enclosed by, a serous sac called the **tunica vaginalis** (vaj-in-al′is; "ensheathing coat") (Figure 25.3a). This sac develops as an outpocketing of the abdominal peritoneal cavity that precedes the descending testes into the scrotum (see p. 779). The tunica vaginalis consists of a superficial parietal layer, an intermediate cavity containing serous fluid (a remnant of the peritoneal cavity), and a deeper visceral layer that hugs the surface of the testis. Thus, although the testes have descended into the scrotum, they are still retroperitoneal. Just deep to the visceral layer of the tunica vaginalis lies the **tunica albuginea** (al″bu-jin′e-ah; "white coat"), the fibrous capsule of the testis. Septal extensions of the tunica albuginea project inward to divide the testis into 250–300 wedge-shaped compartments called **lobules,** each containing one to four coiled **seminiferous tubules**

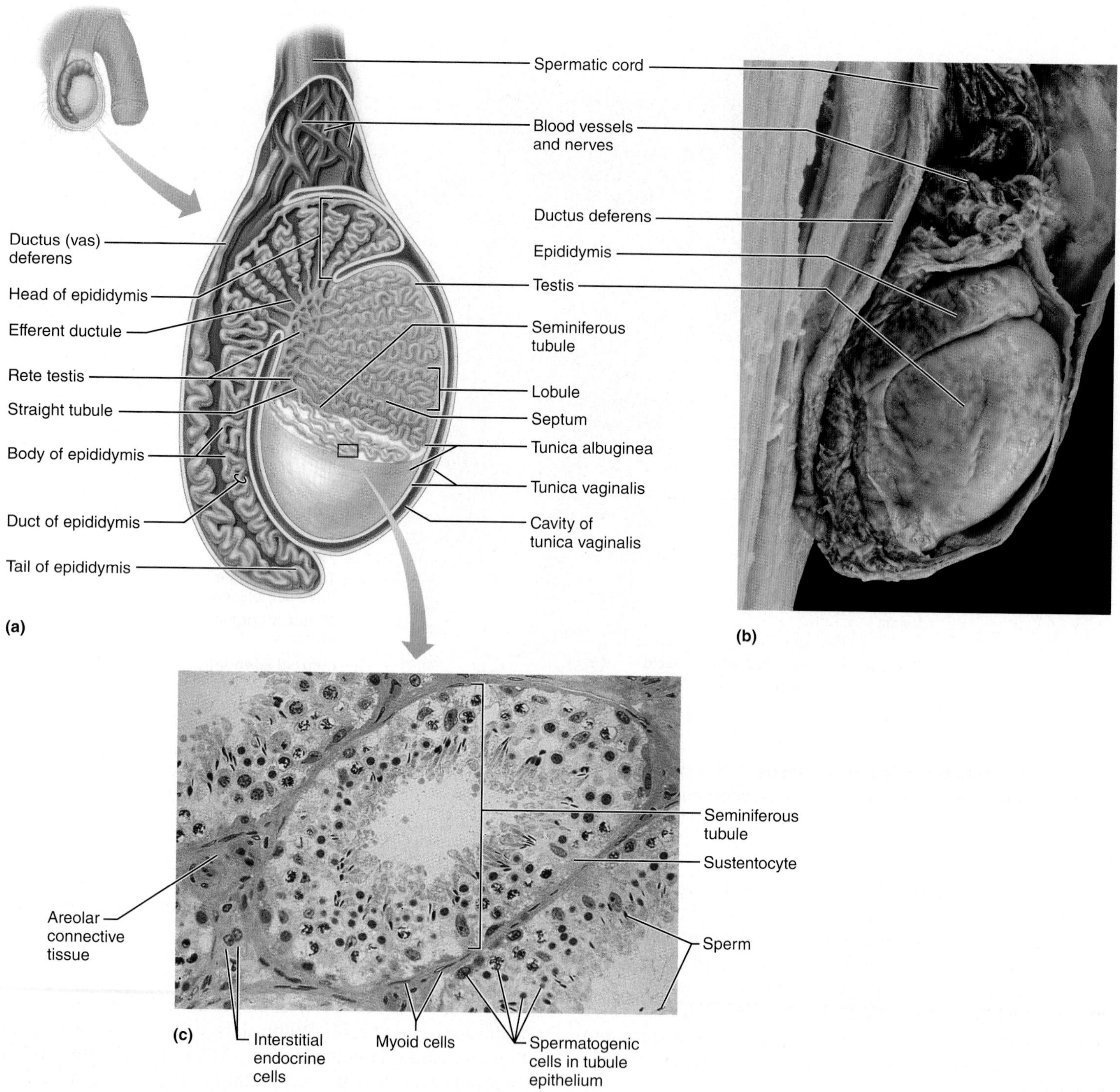

Figure 25.3 Structure of the testis. (**a**) Partial sagittal section through the testis and the epididymis. The anterior aspect is to the right. (**b**) External view of a testis from a cadaver; same orientation as in part (a). (**c**) Seminiferous tubule in cross section (250×). Note the spermatogenic (sperm-forming) cells in the tubule epithelium and the interstitial cells in the connective tissue between the tubules.

Explore PAL Cadaver MasteringA&P®

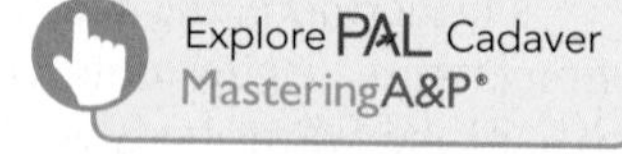

(sem″ĭ-nif′er-us; "sperm-carrying"), the actual "sperm factories." Most of the convoluted seminiferous tubules are looped like hairpins.

Posteriorly, the seminiferous tubules of each lobule converge to form a **straight tubule** that conveys sperm into the **rete testis** (re′te; "network of the testis"), a complex network of tiny branching tubes. The rete testis lies in the *mediastinum testis,* a region of dense connective tissue in the posterior part of the testis. From the rete testis, sperm leave the testis through about a dozen *efferent ductules* that enter the *epididymis,* a comma-shaped structure that hugs the posterior outer surface of the testis (Figures 25.1 and 25.3b).

Nerves and Vessels The testes receive their arterial blood from the long testicular arteries, which branch from the aorta in the superior abdomen (shown in Figure 20.12, p. 604). The testicular veins, which roughly parallel the testicular arteries in the posterior abdominal wall, arise from a venous network in the scrotum called the **pampiniform plexus** (pam-pin′ĭ-form; "tendril-shaped") (Figure 25.2). The veins of this plexus, which surround the testicular arteries like climbing vines, absorb heat from the arterial blood, cooling it before it enters the testes and thereby keeping the testes cool. This passive mechanism of heat transfer is referred to as *countercurrent heat exchange.* The testes are innervated by both divisions of the autonomic nervous system. The abundant visceral sensory nerves transmit impulses that result in agonizing pain and nausea when the testes are hit forcefully.

CLINICAL APPLICATION

Varicocele A **varicocele** is a varicose vein in the pampiniform plexus that disrupts venous return from the testicles. Varicoceles, found in 15% of men in the general population, are found in 40% of men being evaluated for fertility treatments and may cause sperm abnormalities.

It is worth emphasizing that the testicular vessels and nerves extend to the testis from approximately the level of L_2 on the posterior abdominal wall. In the embryo, the testes form in the superior lumbar region and thus receive their vessels from the abdominal aorta and inferior vena cava before descending into the scrotum.

Microscopic Anatomy

A histological section through a lobule of the mature testis (Figure 25.3c) reveals numerous *seminiferous tubules* separated from each other by an areolar connective tissue. The sperm-forming tubules consist of a thick stratified epithelium surrounding a hollow central lumen. The epithelium consists of spherical *spermatogenic* ("sperm-forming") *cells* embedded in columnar sustentocytes (supporting cells) (Figure 25.3c).

Spermatogenic Cells **Spermatogenic cells** are in the process of forming sperm, or *spermatogenesis* (sper″mah-to-jen′ĕ-sis), which begins at puberty. An adult male forms about 400 million sperm per day. The stem cells that form sperm are *spermatogonia* (sper″mah-to-go′ne-ah; "sperm seed"). Spermatogonia lie peripherally on the epithelial basal lamina. As these cells move inward toward the lumen, they differentiate sequentially into *primary* and *secondary spermatocytes, spermatids,* and finally *sperm.* This process takes approximately 75 days. (Spermatogenesis is described further on p. 752.)

Sustentocytes The spermatogenic cells are surrounded by **sustentocytes** (*Sertoli cells*) which extend from the basal lamina to the lumen of the seminiferous tubule (Figure 25.3c). Sustentocytes assist sperm production in many ways: They convey nutrients to the spermatogenic cells, they actively move these cells toward the tubule lumen, and they phagocytize the cytoplasm that is shed during sperm formation. Sustentocytes also secrete testicular fluid into the tubule lumen, which helps to push sperm through the tubule and out of the testes.

Myoid Cells Human seminiferous tubules are surrounded by several layers of smooth-muscle-like **myoid cells** (Figure 25.3c). Contracting rhythmically, these may help to squeeze sperm and testicular fluid through the tubules and out of the testis.

Interstitial Cells The loose connective tissue between the seminiferous tubules contains clusters of **interstitial cells,** or *Leydig* (li′dig) *cells* (Figure 25.3c), spherical or polygon-shaped cells that make and secrete the male sex hormones, or **androgens.** The main type of androgen secreted is testosterone. After it is secreted into the nearby blood and lymphatic capillaries, testosterone circulates throughout the body and maintains all male sex characteristics and sex organs. In fact, all male genitalia atrophy if the testes (and testosterone) are removed. Secretion of testosterone by the interstitial cells is controlled by **luteinizing** (loo′te-in-īz″ing) **hormone (LH),** a hormone from the anterior part of the pituitary. By stimulating testosterone secretion, LH controls testosterone's effects on the entire male reproductive system.

✓ check your understanding

☐ 1. Where are interstitial cells located? What is their function?

☐ 2. Distinguish the tunica albuginea from the tunica vaginalis.

(For answers, see Appendix B.)

Male Reproductive Ducts

learning outcome

▶ Describe the location, structure, and functions of the ducts and accessory glands of the male reproductive system.

As previously noted, sperm leaving the *seminiferous tubules* travel through the *straight tubules* and *rete testis* (Figure 25.3a), which are lined by a simple cuboidal epithelium. Sperm then leave the testis through the *efferent ductules,* which are lined by a simple columnar epithelium. Both ciliated epithelial cells and smooth musculature in the wall of the efferent ductules help move the sperm along while nonciliated cells absorb most of the testicular fluid. From the efferent ductules the sperm enter the duct of the epididymis.

The Epididymis

Sperm are not yet fully functional as they leave the testes. The **epididymis** (ep″ĭ-did′ĭ-mis; "beside the testis") is where sperm mature. It is a comma-shaped organ that arches over the posterior and lateral side of the testis (Figure 25.3a and b). The *head* of the epididymis contains the efferent ductules, which empty into the **duct of the epididymis,** a highly coiled duct that completes the head and forms all of the *body* and *tail* of this organ. With an uncoiled length of over 6 m (20 feet), the duct of the epididymis is longer than the entire intestine!

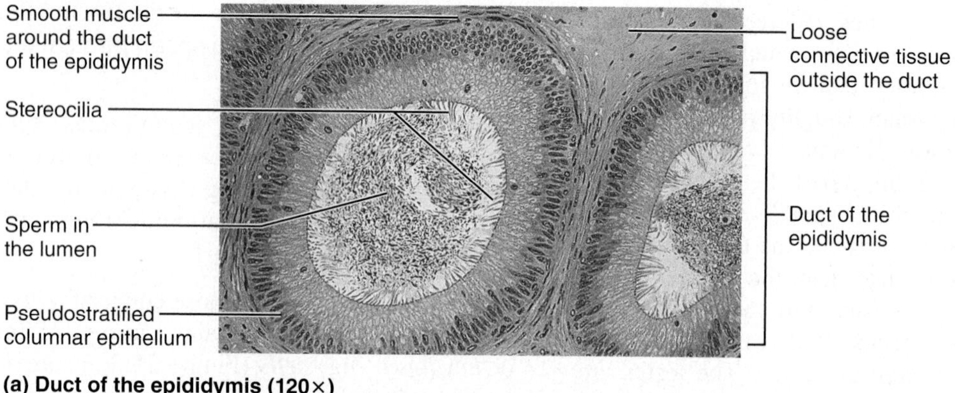

(a) Duct of the epididymis (120×)

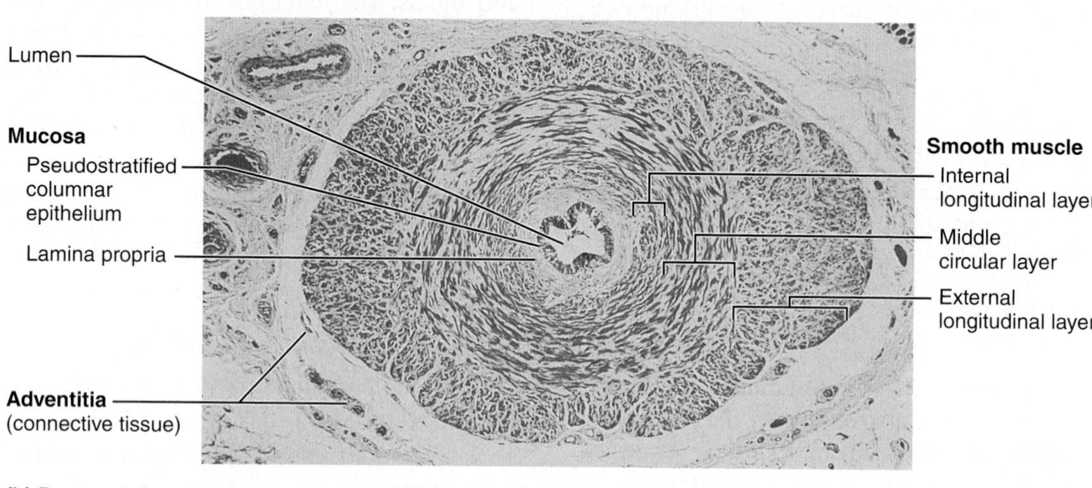

(b) Ductus deferens (45×)

Figure 25.4 Histology of the epididymis and ductus deferens. (**a**) Cross section through the coiled duct of the epididymis. (**b**) Cross section through the ductus deferens. Note the thick layer of smooth muscle.

Histologically, the duct of the epididymis is dominated by a tall, pseudostratified columnar epithelium (Figure 25.4a). The luminal surface of this epithelium bears tufts of long microvilli called **stereocilia** (ster″e-o-sil′e-ah), which are not cilia and do not move. Instead, they provide the tall epithelial cells with a vast surface area for reabsorbing testicular fluid and for transferring nutrients and secretions to the many sperm that are stored in the lumen of the epididymis. External to the epithelium lies a layer of smooth muscle.

The immature, nearly immotile sperm that leave the testis are moved slowly through the duct of the epididymis. During this journey, which takes about 20 days, the sperm gain the ability to swim and also the ability to fertilize the egg through the acrosome reaction (the events of fertilization occur in the female's uterine tube and are described on p. 768). These maturation processes are stimulated by proteins secreted by the epididymis epithelium. Sperm are ejaculated from the epididymis, *not* directly from the testes. At the beginning of ejaculation, the smooth muscle in the walls of the epididymis contracts, expelling sperm from the tail of the epididymis into the next segment of the duct system, the ductus deferens.

Sperm can be stored in the epididymis for several months, after which time they are phagocytized by the epithelial cells of the duct of the epididymis.

The Ductus Deferens

The **ductus deferens** (def′er-ens; "carrying away"), or *vas deferens,* stores and transports sperm during ejaculation. It is about 45 cm (18 inches) long. From the tail of the epididymis, the ductus deferens runs superiorly within the spermatic cord (discussed below), goes through the inguinal canal, pierces the anterior abdominal wall, and enters the pelvic cavity (Figure 25.1). From there the ductus deferens runs posteriorly along the lateral wall of the true pelvis, arches medially over the ureter, and descends along the posterior wall of the bladder. Its distal end expands as the **ampulla** ("flask") **of the ductus deferens** and then joins with the duct of the seminal gland to form the short **ejaculatory duct.** Each ejaculatory duct runs within the prostate, where it empties into the prostatic urethra.

The wall of the ductus deferens (Figure 25.4b) consists of:

- An inner mucosa with the same pseudostratified epithelium as that of the epididymis, plus a lamina propria.
- An extremely thick muscularis. During ejaculation, the smooth muscle in the muscularis creates strong peristaltic waves that rapidly propel sperm through the ductus deferens to the urethra.
- An outer adventitia of connective tissue.

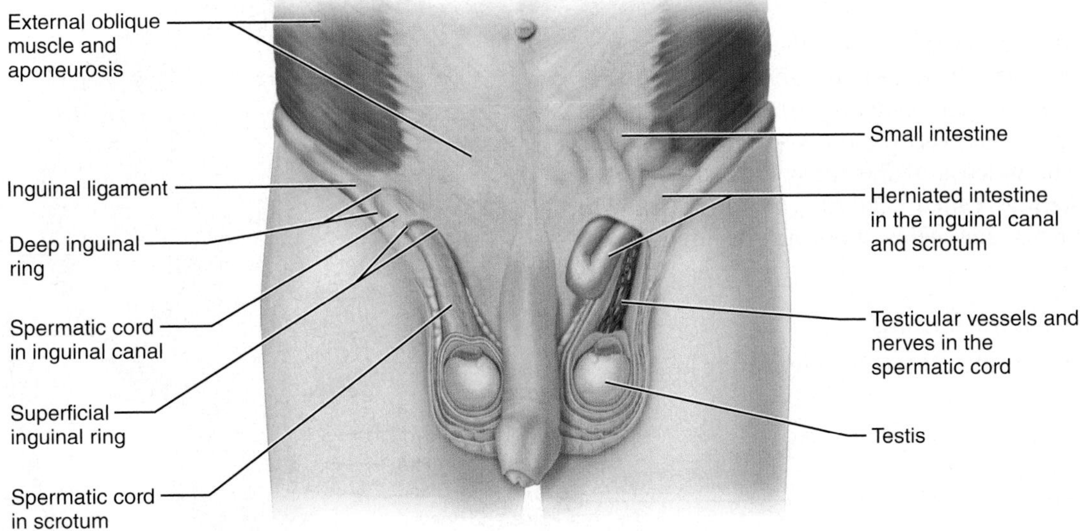

Figure 25.5 **Spermatic cord in the inguinal canal (at left) and an acquired inguinal hernia (at right).** The inguinal canal lies just deep to the aponeurosis of the external oblique muscle. An inguinal hernia forms a bulge below the skin in the groin. Here, a portion of the intestine is shown passing through the inguinal canal toward the scrotum.

CLINICAL APPLICATION

Vasectomy Some men take responsibility for birth control by having a **vasectomy** (vah-sek′to-me; "cutting the vas"). In this minor surgery, the physician makes a small incision into both sides of the scrotum, transects both ductus deferens, and then closes the cut ends, either by tying them off or by fusing them shut by cauterization. Although sperm continue to be produced, they can no longer exit the body and are phagocytized in the epididymis.

Some men who have had vasectomies wish to reverse them. New microsurgical techniques are making it easier to reopen and reattach the closed ends of the vas, and the rate of successful reversal (as judged by subsequent impregnation) is now over 50%. However, if sperm has leaked out of the closed end of the vas, the immune system responds with autoimmune reaction that destroys the sperm at the leakage site and in the testes and permanent sterility results.

Physicians performing a vasectomy are able to distinguish the ductus deferens from the testicular vessels that surround it because its muscular layer is so thick that it feels like a hard wire when squeezed between the fingertips.

The Spermatic Cord

The ductus deferens is the largest component of the **spermatic cord,** a tube of fascia that also contains the testicular vessels and nerves (Figure 25.2). The inferior part of the spermatic cord lies in the scrotum, and its superior part runs through the **inguinal canal,** an obliquely oriented trough in the anterior abdominal wall (Figure 25.5). The inguinal canal is partially formed by the inguinal ligament, which is the free inferior margin of the aponeurosis of the external oblique muscle (p. 294). The medial opening of the inguinal canal, the **superficial inguinal ring,** is a V-shaped opening in this aponeurosis. The canal runs laterally to the **deep inguinal ring,** an opening in the fascia deep to the abdominal muscle, the transversus abdominis, where the ductus deferens and testicular vessels enter the pelvic cavity.

CLINICAL APPLICATION

Inguinal Hernia The deep inguinal ring and the area of fascia just medial to it (the inguinal triangle) are weak areas in the abdominal wall; coils of intestine or the greater omentum can be forced anteriorly through these areas, into the inguinal canal, sometimes pushing all the way into the scrotum. The resulting condition, called an **inguinal hernia** (Figure 25.5), is often precipitated by lifting or straining, which raises intraabdominal pressure and forces the herniated elements through the body wall. Although most herniated elements pop back into the abdominal cavity whenever the pressure decreases, a few remain squeezed in the narrow inguinal canal, which can strangle their blood supply and lead to potentially fatal gangrene.

At least 2% of adult males develop inguinal hernias, which are surgically repaired by closing off the enlarged opening with surgical thread or a strong prosthetic patch. Increasingly popular is a minimally invasive procedure that uses a laparoscope to guide placement of a patch external to the parietal peritoneum on the inner side of the anterior abdominal wall.

Although composed of the same elements, the inguinal canals of *females* are considerably narrower than in males because they contain only a thin band called the round ligament of the uterus (described on pp. 758–759). As a result, inguinal hernias are far less common in females than in men.

The Urethra

The urethra in males carries sperm from the ejaculatory ducts to the outside of the body. Its three parts are the *prostatic urethra* in the prostate, the *intermediate part of urethra* in the urogenital diaphragm, and the *spongy urethra* in the penis (Figure 25.6a). The mucosa of the spongy urethra contains scattered outpocketings called *urethral glands* (not shown) that secrete a mucus that helps lubricate the urethra just before ejaculation.

check your understanding

- ☐ 3. What is meant by the phrase "sperm mature in the epididymis"?
- ☐ 4. Where is the ejaculatory duct located?

(For answers, see Appendix B.)

Accessory Glands

The accessory glands in males include the paired *seminal glands,* the single *prostate,* and paired *bulbo-urethral glands.* These glands produce the bulk of the **semen,** which is defined as sperm plus the secretions of the accessory glands and accessory ducts.

The Seminal Glands

The **seminal glands** or (**seminal vesicles**) lie on the posterior surface of the bladder (see Figures 25.1 and 25.6a). These hollow glands are about the shape and length of a finger (5 to 7 cm); however, because a seminal gland is pouched, coiled, and folded back on itself, its true (uncoiled) length is about 15 cm. Internally, the mucosa is folded into a honeycomb pattern of crypts and blind chambers and is lined by a secretory pseudostratified columnar epithelium. A thick layer of smooth muscle surrounds the mucosa. This muscle contracts during ejaculation to empty the gland. The external wall is composed of a fibrous capsule of dense connective tissue.

The secretion of the seminal glands, which constitutes about 60% of the volume of semen, is a viscous fluid that contains:

- A sugar called fructose and other nutrients that nourish the sperm on their journey
- Prostaglandins (pros″tah-glan′dinz), which stimulate contraction of the uterus to help move sperm through the female reproductive tract
- Substances that suppress the immune response against semen in females
- Substances that enhance sperm motility
- Enzymes that clot the ejaculated semen in the vagina and then liquefy it so that the sperm can swim out.

Seminal fluid also contains a yellow pigment that fluoresces under ultraviolet light. This feature enables the recognition of semen residues in criminal investigations of sexual assault.

As previously noted, the duct of each seminal gland joins the ductus deferens on the same side of the body to form an ejaculatory duct within the prostate. Sperm and seminal fluid mix in the ejaculatory duct and enter the prostatic urethra together during ejaculation.

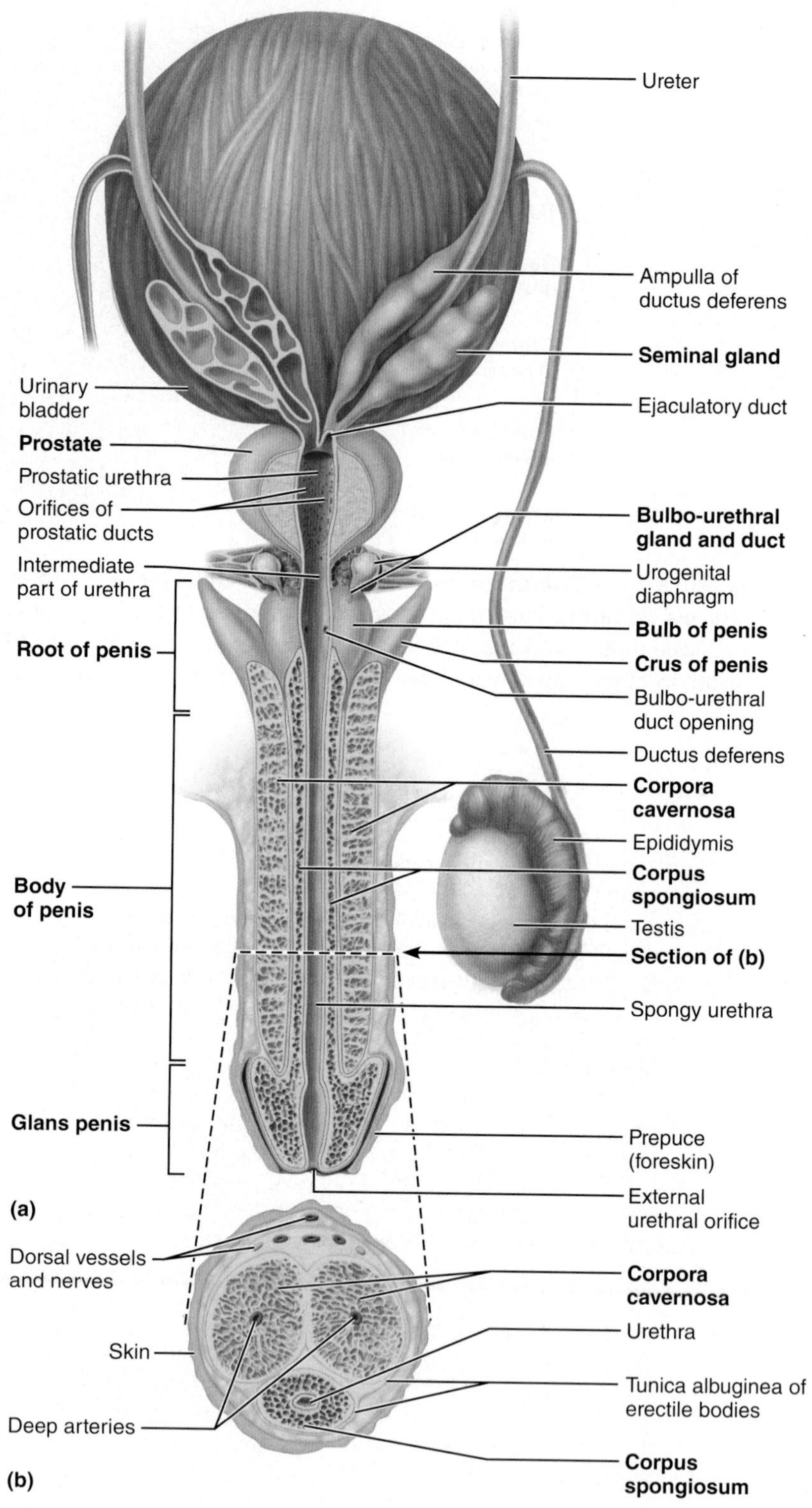

Figure 25.6 Male reproductive structures. (**a**) Posterior view showing a longitudinal (frontal) section through the penis. (**b**) Transverse section through the penis, dorsal aspect at top.

The Prostate

The **prostate** (pros′tāt), which is the size and shape of a chestnut, encircles the first part of the urethra just inferior to the bladder (Figure 25.6a). The prostate consists of 20–30 compound tubuloalveolar glands of three classes—*main, submucosal,* and *mucosal glands.* The glands are embedded in a mass of dense connective tissue and smooth muscle called the **fibromuscular stroma** and surrounded by a connective tissue capsule (Figure 25.7). The muscle of the stroma contracts during ejaculation to squeeze the prostatic secretion into the urethra.

The prostatic secretion constitutes about one-third of the volume of semen. It is a milky fluid that, like the secretion of the seminal glands, contains various substances that enhance sperm motility and enzymes that clot and liquefy ejaculated semen. One of the enzymes that liquefy semen is *prostate-specific antigen (PSA)*; measuring the levels of PSA in a man's blood is the most important method of screening for prostate cancer (discussed on p. 774).

In addition to its susceptibility to tumors, the prostate is also subject to infection in sexually transmitted diseases (STDs). **Prostatitis** (pros″tah-ti′tis), inflammation of the prostate, is the single most common reason that men consult a urologist.

CLINICAL APPLICATION

Benign Prostatic Hyperplasia A common, noncancerous tumor called **benign prostatic hyperplasia (BPH)** is characterized by uncontrolled growth in the deep, mucosal glands in the prostate and by proliferation of some of the nearby stroma cells. About 50% of men at age 50 and 80% of men at age 70 develop BPH. Enlargement of the prostate and contraction of its smooth musculature constrict the prostatic urethra, making micturition difficult. BPH can lead to urinary retention, continual dribbling of urine, urinary tract infections, and the formation of kidney stones.

Diagnosis of BPH begins with inquiries about the patient's urinary symptoms—whether he is hesitant to urinate, has a weak stream, must strain to urinate, feels that his bladder does not empty fully, has increased urinary frequency and urgency, and must urinate often at night. Next, to determine whether the prostate is enlarged, the physician performs a **digital rectal exam;** a finger inserted into the anal canal can feel the prostate because it lies just anterior to the rectum (as shown in Figure 25.1). The physician may also order a blood test to check the levels of PSA (which are only mildly elevated in BPH, generally less so than in prostate cancer). Treatment for BPH begins with drugs that inhibit the production or action of testosterone, which the tumor cells depend on, or drugs that relax the prostate's smooth muscle. If such treatment does not increase ease of urination, the prostate is either removed in a surgical procedure called *transurethral prostatectomy* or destroyed by heat from a microwave device or an electrode inserted into the urethra.

The Bulbo-urethral Glands

The **bulbo-urethral** (bul″bo-u-re′thral) **glands** are pea-sized glands situated inferior to the prostate, within the urogenital diaphragm (Figures 25.1 and 25.6a). These compound tubuloalveolar glands produce a mucus, some of which enters the spongy urethra when a man becomes sexually excited prior to ejaculation. This mucus neutralizes traces of acidic urine in the urethra and lubricates the urethra to smooth the passage of semen during ejaculation.

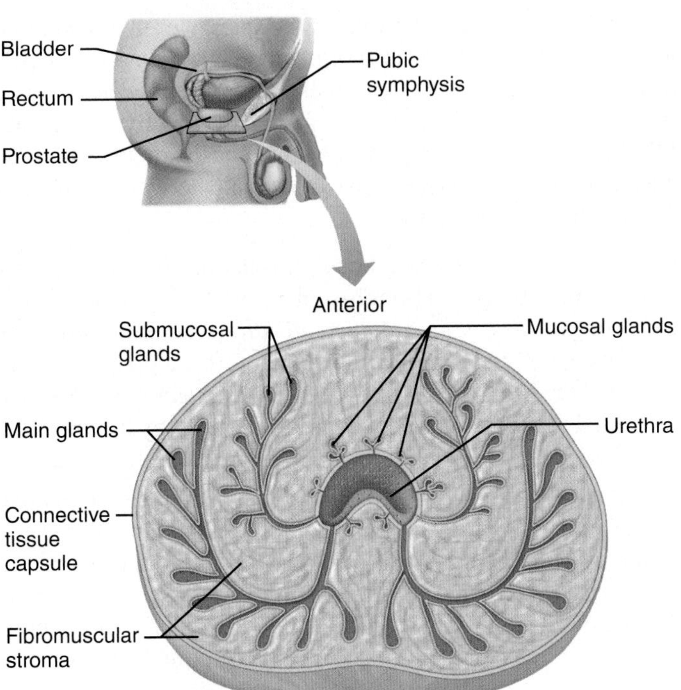

Figure 25.7 The prostate. Diagram of a transverse section through the prostate.

The Penis

learning outcome

- Describe the structure of the penis, and explain the mechanism of erection.

The **penis** ("tail"), the male organ of sexual intercourse, delivers sperm into the female reproductive tract (Figure 25.6). Together, it and the scrotum make up the external reproductive structures, or external genitalia, of the male. As we describe the structures within the penis, keep in mind that the dorsal and ventral surfaces of the penis are named in reference to the erect penis.

The penis consists of an attached **root** and a free **body** that ends in an enlarged tip called the **glans penis.** The skin covering the penis is loose and extends distally around the glans to form a cuff, called the **prepuce** (pre′pūs), or **foreskin.**

CLINICAL APPLICATION

Circumcision Over 60% of male babies in the United States undergo **circumcision** (ser″kum-sizh′un; "cutting around"), the surgical removal of the foreskin that is typically performed shortly after birth. Circumcision is a controversial procedure. Opponents charge that it is a medically unnecessary surgery that causes too much pain. Proponents contend that it reduces risk of penile cancer and infections of the glans, urethra, and urinary tract. Circumcision also reduces the risk for HIV infection.

Internally, the penis contains the spongy urethra and three long cylindrical bodies (corpora; singular, corpus) of erectile tissue (Figure 25.6). Each of these three **erectile bodies** is a thick tube covered by a sheath of dense connective tissue and filled with a network of partitions that consist of smooth muscle and connective tissue. This spongelike network, in turn, is filled with vascular spaces.

- The midventral erectile body surrounding the spongy urethra, the **corpus spongiosum** (kor′pus spun″je-o′sum; "spongy body"), is enlarged *distally* where it forms the glans penis, and *proximally* where it forms a part of the root called the **bulb of the penis.** This bulb is secured to the urogenital diaphragm and is covered externally by the sheetlike bulbospongiosus muscle (shown in Figure 11.15c, p. 297).
- The paired, dorsal erectile bodies, the **corpora cavernosa** (kav″er-no′sah; "cavernous bodies"), make up most of the mass of the penis. Their proximal ends in the root are the **crura** ("legs") **of the penis** (singular, **crus**); each crus is anchored to the pubic arch of the bony pelvis and is covered by an ischiocavernosus muscle (Figure 11.15c).

Vessels and Nerves

Most of the main vessels and nerves of the penis lie near the dorsal midline (Figure 25.6b). The sensory *dorsal nerves* are branches of the pudendal nerve from the sacral plexus, and the *dorsal arteries* are branches of the internal pudendal arteries from the internal iliac arteries. Two *dorsal veins* (superficial and deep) lie precisely in the dorsal midline and drain all blood from the penis. Finally, a *deep artery* runs within each corpus cavernosum. The autonomic nerves to the penis, which follow the arteries and supply the erectile bodies, arise from the inferior hypogastric plexus in the pelvis.

Male Sexual Response

The chief phases of the male sexual response are (1) erection of the penis, which allows it to penetrate the vagina, and (2) ejaculation, which expels semen into the vagina. *Erection* results from engorgement of the erectile bodies with blood. During sexual stimulation, parasympathetic innervation dilates the arteries supplying the erectile bodies, increasing the flow of blood to the vascular spaces within. The smooth muscle in the partitions in these bodies relaxes, allowing the bodies to expand as the blood enters them. As the erectile bodies begin to swell, they press on the small veins that normally drain them, slowing venous drainage and maintaining engorgement. The arrangement of the collagen fibers in the dense connective tissue outside the erectile bodies strengthens the penis during erection. Longitudinal fibers that run the length of the penis lie at right angles to circular fibers that form rings around the penile shaft. This is an optimal design for resisting bending forces, so the erect penis does not buckle or kink sharply during intercourse.

Whereas erection is largely under parasympathetic control, *ejaculation* is under sympathetic control. Ejaculation begins with a strong, sympathetically induced contraction of the smooth musculature throughout the reproductive ducts and glands, which squeezes the semen toward and into the urethra. Simultaneously, somatic contraction of the bulbospongiosus muscle of the penis rapidly squeezes the semen onward through the spongy urethra and out of the body.

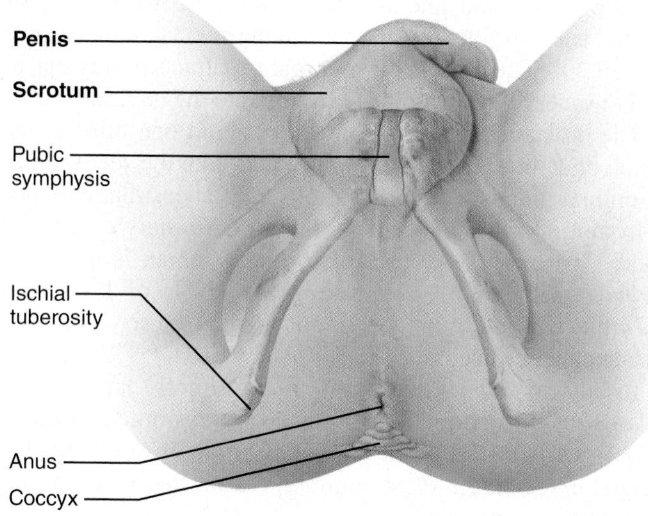

Figure 25.8 The male perineum.

The Male Perineum

The male **perineum** (per″ĭ-ne′um; "around the anus") contains the scrotum, the root of the penis, and the anus (Figure 25.8). More specifically, it is defined as the diamond-shaped area between the pubic symphysis anteriorly, the coccyx posteriorly, and the ischial tuberosities laterally. The floor of the perineum is formed by the muscles of the urogenital diaphragm and the superficial perineal space (Chapter 11, p. 296).

check your understanding

- ☐ 5. Which pair of accessory glands produces the majority of semen?
- ☐ 6. Which erectile tissue does the urethra pass through? What is this portion of the urethra called?

(For answers, see Appendix B.)

Spermatogenesis

learning outcome

▶ Outline the events of spermatogenesis.

Spermatogenesis, the formation of sperm, occurs within the seminiferous tubules of the testes throughout a man's life, from puberty until death (although the rate of sperm formation declines with advancing age). The process of spermatogenesis may be divided into three successive stages as spermatocytes move from the peripheral region of the seminiferous tubules to the lumen (Figure 25.9).

Stage 1: Formation of spermatocytes (top of Figure 25.9b and c). **Spermatogonia,** sperm stem cells, are located on the outer region of the seminiferous tubule on the epithelial basal lamina. These cells divide vigorously and continuously by mitosis. Each division forms two distinctive

Figure 25.9 Spermatogenesis (sperm formation).

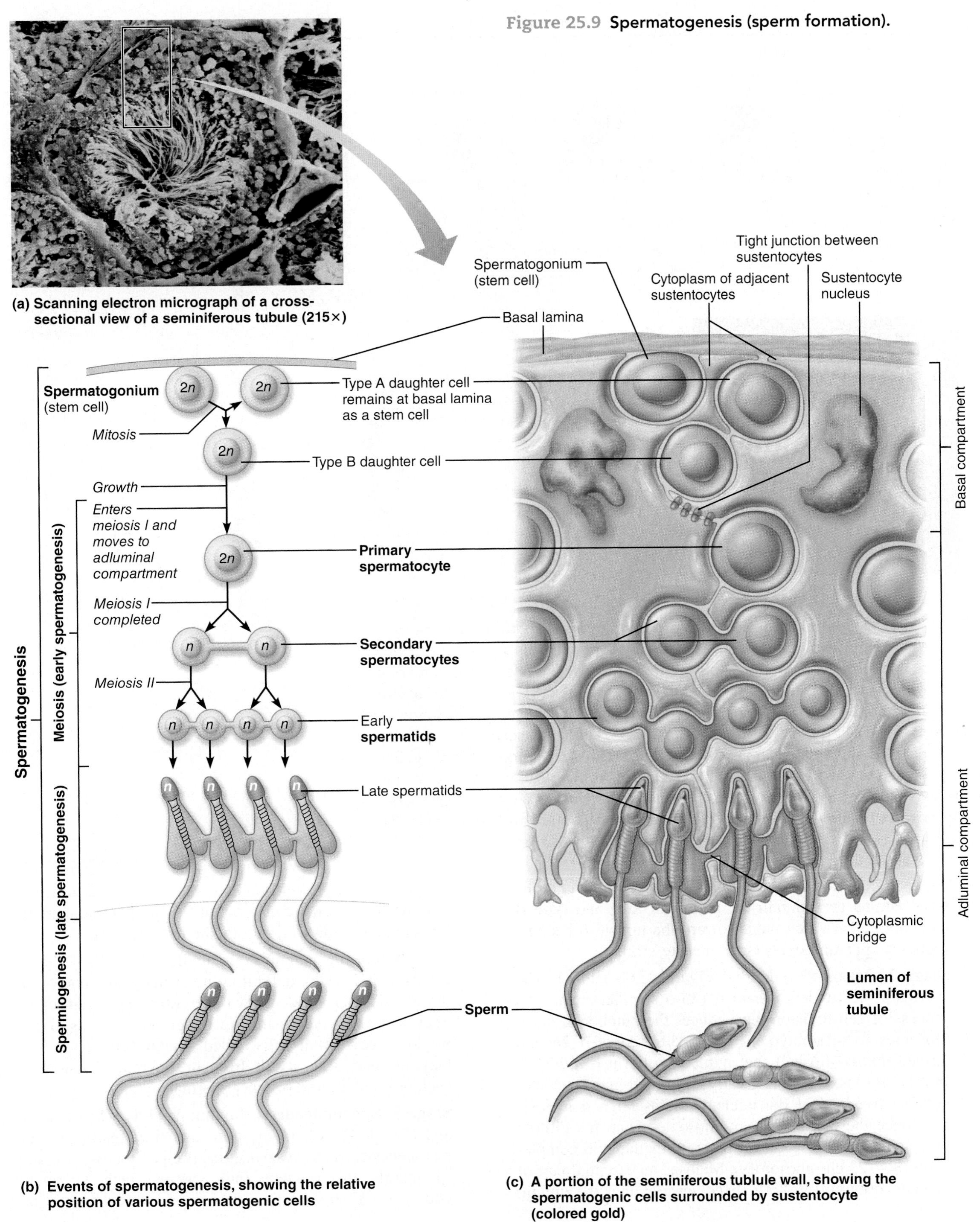

(a) Scanning electron micrograph of a cross-sectional view of a seminiferous tubule (215×)

(b) Events of spermatogenesis, showing the relative position of various spermatogenic cells

(c) A portion of the seminiferous tublule wall, showing the spermatogenic cells surrounded by sustentocyte (colored gold)

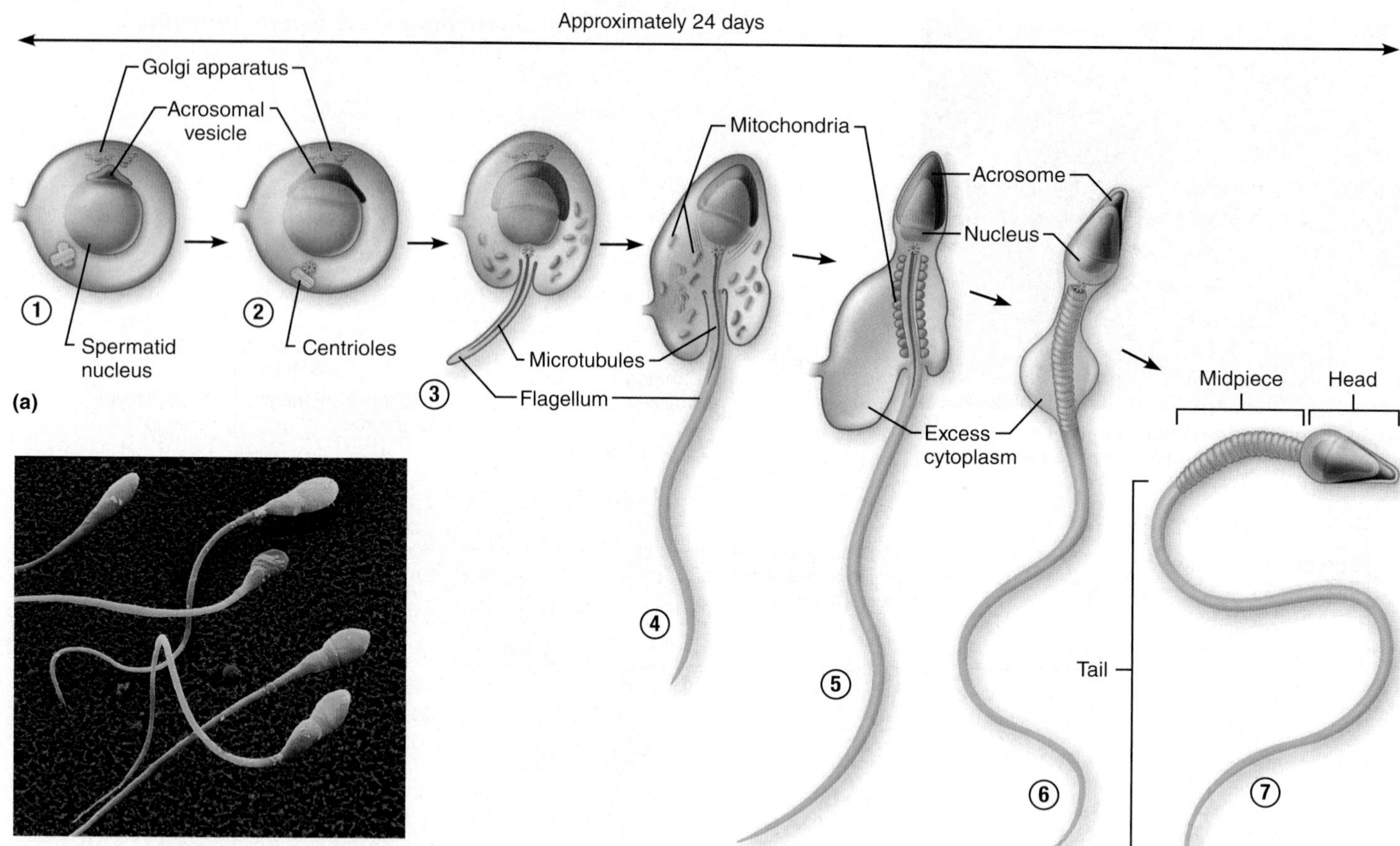

Figure 25.10 Spermiogenesis: Transformation of a spermatid into a sperm. (**a**) The steps in spermiogenesis are as follows: ① The Golgi apparatus produces vesicles that form the acrosome. ② The acrosome positions itself at the anterior end of the nucleus, and the centrioles move to the opposite end. ③ Microtubules assemble from a centriole and grow to form the flagellum that is the sperm tail. ④ Mitochondria multiply in the cytoplasm. ⑤ The mitochondria position themselves around the proximal core of the flagellum, and excess cytoplasm is shed from the cell. ⑥ Structure of an immature sperm that has just been released from a sustentocyte into the lumen of the seminiferous tubule. ⑦ A structurally mature sperm with a streamlined shape that allows active swimming. (**b**) Scanning electron micrograph of mature sperm (1,850×).

daughter cells: **type A** daughter cells, which remain at the basal lamina to maintain the germ cell line; and **type B** daughter cells, which move toward the lumen to become **primary spermatocytes** (sper″mah′to-sītz).

Stage 2: Meiosis (middle of Figure 25.9b and c). The spermatocytes undergo *meiosis* (mi-o′sis; "lessening"), a process of cell division that reduces the number of chromosomes found in typical body cells (denoted $2n$ and termed *diploid*) to half that number (n, termed *haploid*). In this process, which is greatly simplified here, two successive divisions of the nucleus, meiosis I and meiosis II, reduce the number of chromosomes from the diploid complement ($2n = 46 = 23$ pairs) to the haploid complement ($n = 23$ chromosomes). Meiosis, an essential part of gamete formation in both sexes, ensures that the diploid complement of chromosomes is reestablished at fertilization, when the genetic material of the two haploid gametes joins to make a diploid zygote, the fertilized egg.

Within the seminiferous tubules, the cells undergoing meiosis I are by definition the primary spermatocytes; these cells each produce two **secondary spermatocytes.** Each secondary spermatocyte undergoes meiosis II and produces two small cells called **spermatids.** Thus, four haploid spermatids result from the meiotic divisions of each original diploid primary spermatocyte.

Stage 3: Spermiogenesis (bottom of Figure 25.9b and c, and Figure 25.10). In the third and final phase of spermatogenesis, called *spermiogenesis* (sper″me-o-jen′ĕ-sis), spermatids differentiate into sperm. Each spermatid undergoes a streamlining process as it fashions a tail and sheds superfluous cytoplasm. The resulting **sperm cell** has a *head,* a *midpiece,* and a *tail.* The **head** of the sperm contains the nucleus with highly condensed chromatin surrounded by a helmetlike **acrosome** (ak′ro-sōm; "tip piece"), a vesicle containing enzymes that enable the sperm to penetrate and enter an egg. The **midpiece** contains mitochondria spiraled tightly around the core of the tail. The long **tail** is an elaborate flagellum. The mitochondria provide the energy needed for the whiplike movements of the tail that propel the sperm through the female reproductive tract. The newly formed sperm detach from the epithelium of the seminiferous tubule and enter the lumen of the seminiferous tubule (Figure 25.9c).

The *sustentocytes* of the tubule surround the spermatogenic cells (Figure 25.9c). These cells are bound to each other by tight junctions on their lateral surfaces and divide the seminiferous tubules into two compartments:

- The **basal compartment** extends from the basal lamina to the tight junctions and contains the spermatogonia and earliest primary spermatocytes.
- The **adluminal compartment** ("near the lumen") lies internal to the tight junctions and includes the more advanced spermatocytes and the lumen of the tubule.

The tight junctions between the sustentocytes constitute the blood testis barrier. This barrier protects the developing sperm from attack by the immune system. Sperm and late-stage spermatocytes produce unique membrane proteins, which the immune system identifies as foreign. If the immune system were exposed to the developing sperm, the resulting autoimmune response would destroy the gametes and result in sterility. Because the developing spermatogenic cells must cross the blood testis barrier on their way to the lumen, the tight junctions between sustentocytes allow these cells to pass into the adluminal compartment, much like canal locks that open to permit ships to pass.

Spermatogenesis is controlled by the stimulating action of two hormones: **follicle-stimulating hormone (FSH)** from the anterior pituitary gland and **testosterone,** the primary male sex hormone produced by the interstitial cells of the testes. In addition, secretions from the sustentocytes influence spermatogenesis: *Androgen-binding protein* concentrates testosterone near the spermatogenic cells, thus stimulating spermatogenesis, and *inhibin* inhibits FSH and slows the rate of sperm production.

✓ check your understanding

☐ 7. How many spermatids are formed from each primary spermatocyte?

☐ 8. What roles do sustentocytes play in sperm formation?

☐ 9. What are the components of the head of the sperm? Of the midpiece? Of the tail?

(For answers, see Appendix B.)

THE FEMALE REPRODUCTIVE SYSTEM

learning outcomes

- ▶ Describe the location, structure, and function of the ovaries.
- ▶ Describe the anatomy of the uterine tubes in terms of their function.
- ▶ Explain the location, regions, supportive structures, and layers of the uterus.

The female reproductive organs differ from those of males in several important ways. First, in addition to producing gametes (ova, or "eggs"), the female reproductive organs also prepare to support a developing embryo during pregnancy. Second, the female organs undergo changes according to a reproductive cycle, the **menstrual cycle,** which averages 28 days in length. By definition, the menstrual cycle is the female's monthly cycle as it affects *all* her reproductive organs. The primary female reproductive organs are the ovum-producing *ovaries* (Figure 25.11). The accessory ducts include the *uterine tubes,* where fertilization typically occurs; the *uterus,* where the embryo develops; and the *vagina,* which acts as a birth canal and receives the penis during sexual intercourse. The female external genitalia are referred to as the *vulva.* The milk-producing *mammary glands,* which are actually part of the integumentary system, are also considered here because of their reproductive function of nourishing the infant.

The Ovaries

Gross Anatomy

The paired, almond-shaped **ovaries,** located on the lateral sides of the uterus, measure about 3 cm by 1.5 cm by 1 cm. Each ovary lies against the bony lateral wall of the true pelvis, in the fork of the iliac vessels. The surfaces of the ovaries are smooth in young girls but after puberty become scarred and pitted from the monthly release of ova.

The ovaries, although they are retroperitoneal, are surrounded by the peritoneal cavity and held in place by mesenteries and ligaments derived from peritoneum. The mesentery of the ovary, the horizontal **mesovarium** (mez″o-va′re-um), is part of the **broad ligament,** a large fold of peritoneum that hangs from the uterus and the uterine tubes like a tent (Figure 25.12). The **suspensory ligament of the ovary,** a lateral continuation of the broad ligament, attaches the ovary to the lateral pelvic wall. Finally, the ovary is anchored to the uterus medially by the **ovarian ligament,** a distinct fibrous band enclosed within the broad ligament.

Vessels and Nerves The ovaries are served by the *ovarian arteries,* which are branches from the abdominal aorta, and by the *ovarian branches of the uterine arteries,* which arise from the internal iliac artery (Figure 25.12; see also Figures 20.12 and 20.14, pp. 604 and 606). The ovarian artery, veins, and nerves reach the ovary by traveling within the suspensory ligament and then through the mesovarium. The ovaries are innervated by both divisions of the autonomic nervous system.

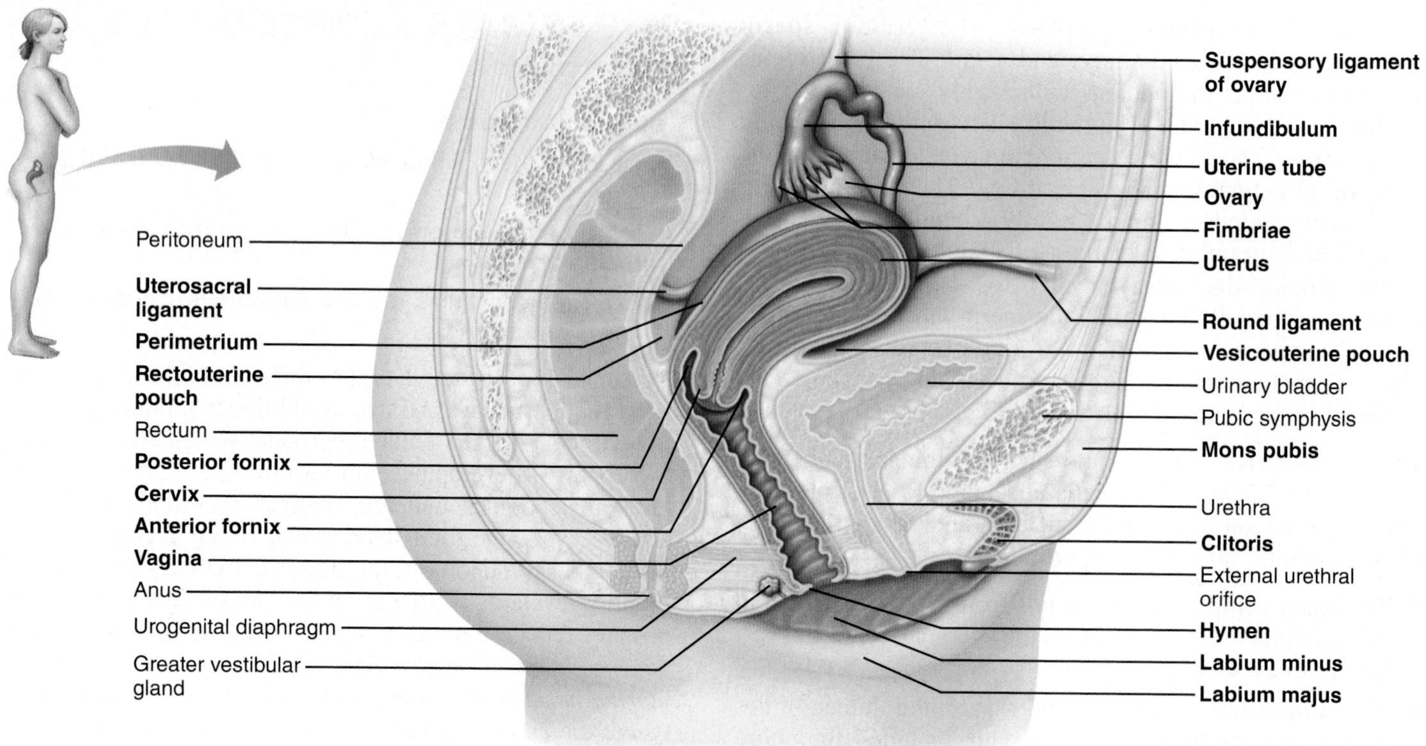

Figure 25.11 Internal organs of the female reproductive system, midsagittal section.

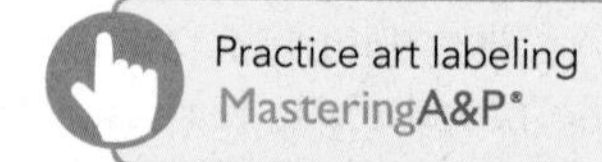

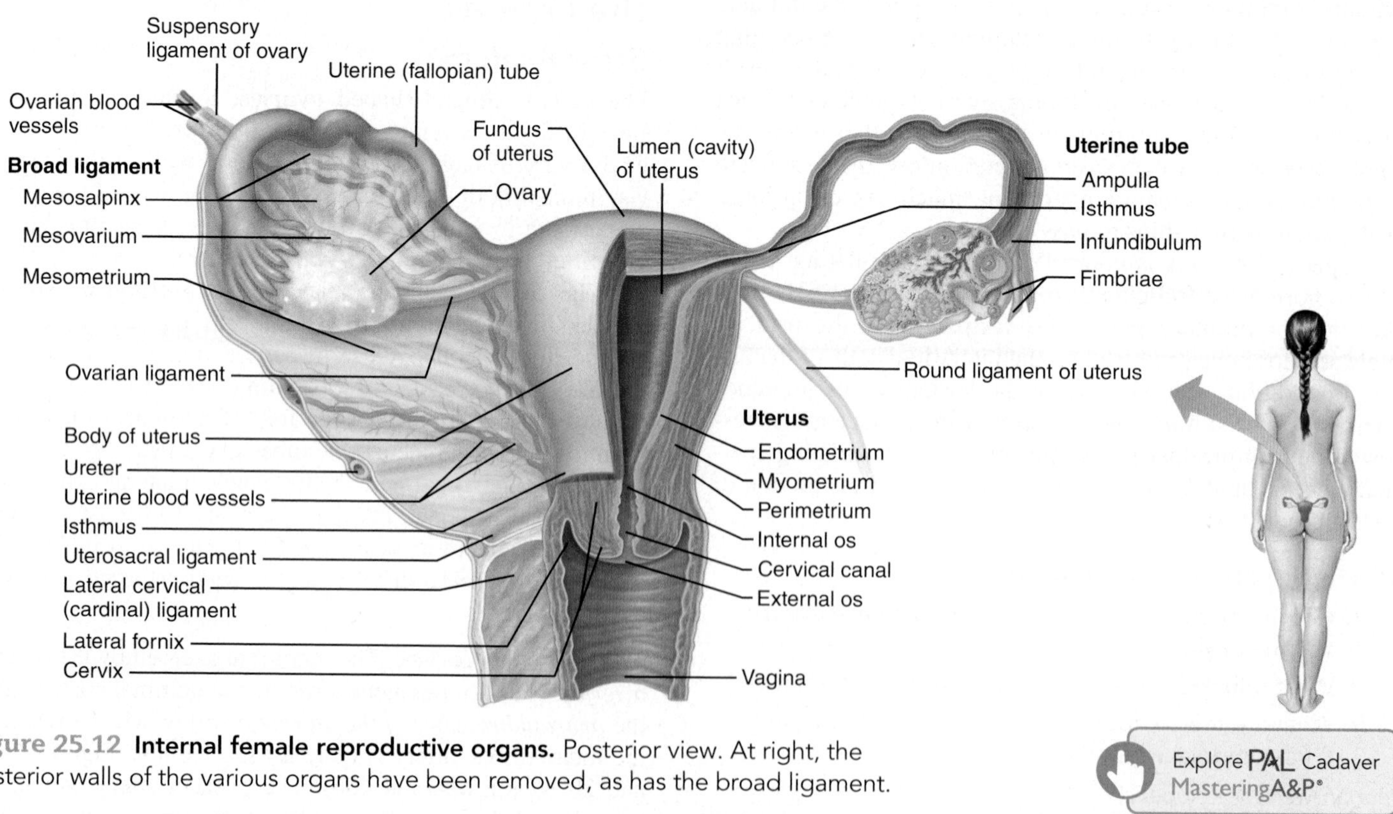

Figure 25.12 Internal female reproductive organs. Posterior view. At right, the posterior walls of the various organs have been removed, as has the broad ligament.

Explore PAL Cadaver MasteringA&P®

Watch Cat Dissection MasteringA&P®

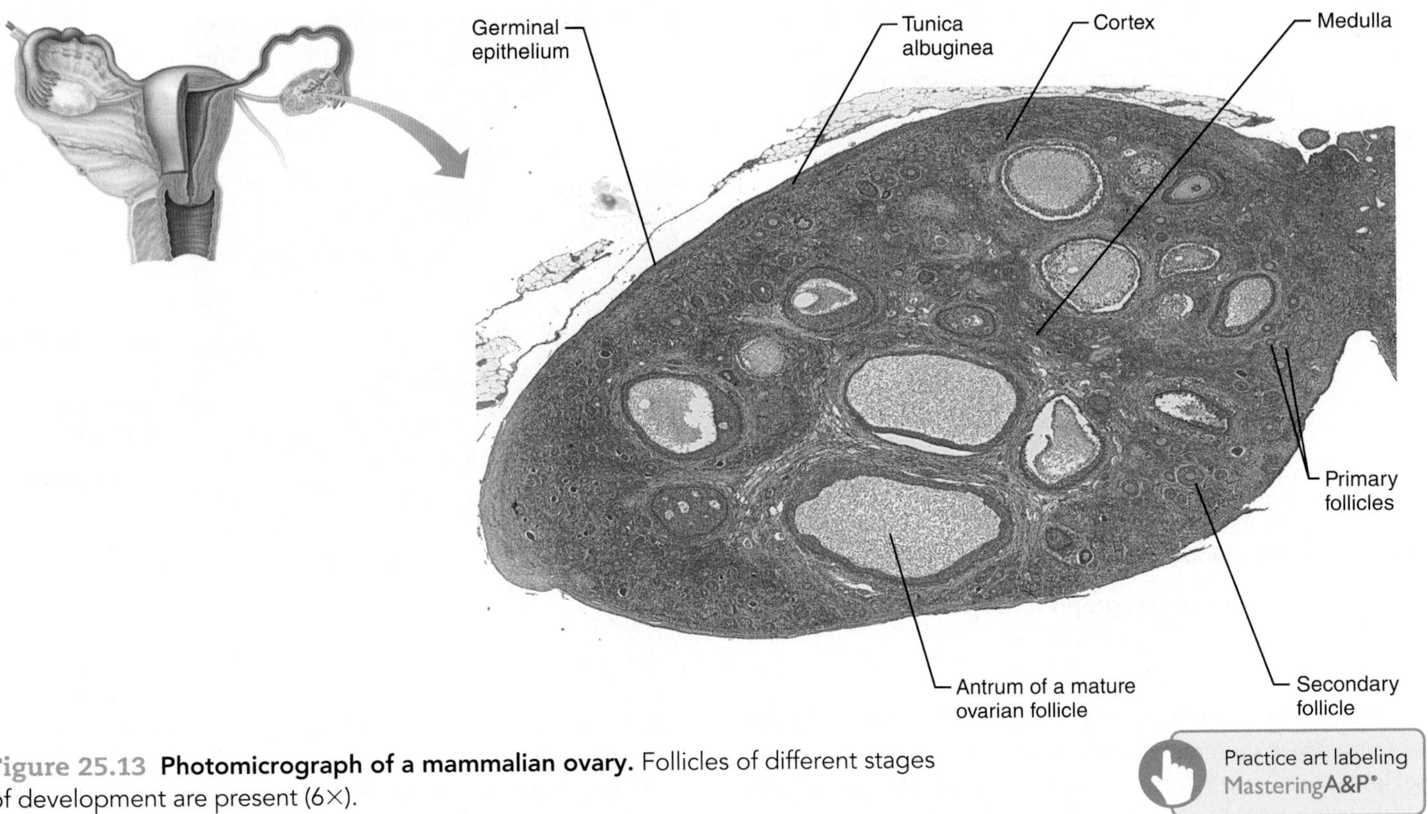

Figure 25.13 Photomicrograph of a mammalian ovary. Follicles of different stages of development are present (6×).

Practice art labeling MasteringA&P®

Microscopic Anatomy

The ovary is surrounded by a fibrous capsule called the **tunica albuginea** (Figure 25.13), which is much thinner than the tunica albuginea of the testis. The tunica albuginea is covered by a simple cuboidal epithelium called the **germinal epithelium.** Despite its name, this mesothelium does *not* germinate the ova. The main substance of the ovary is divided into an outer *cortex* and an inner *medulla.* The **ovarian cortex** houses the developing gametes, which are called **oocytes** (o′o-sīts; "egg cells") while in the ovary. All oocytes occur within saclike multicellular structures called **follicles** ("little bags"), which enlarge substantially as they mature. In human females, generally one, or occasionally two, follicles mature each month, thus only one (or two) ooctyes develop each month. The part of the cortex between the follicles is a cell-rich connective tissue. The deep **ovarian medulla** is a loose connective tissue containing the largest blood vessels, nerves, and lymphatic vessels of the ovary; these vessels enter the ovary through the *hilum,* a horizontal slit in the anterior ovarian surface where the mesovarium attaches.

The Uterine Tubes

The **uterine** (u′ter-in) **tubes,** also called **oviducts** or, more commonly, **fallopian tubes** (Figure 25.12), receive the ovulated oocyte and are the site for fertilization. Each uterine tube begins laterally near an ovary and ends medially, where it empties into the superior part of the uterus. The lateral region of the uterine tube, an open funnel called the **infundibulum** (in″fun-dib′u-lum; "funnel"), opens into the peritoneal cavity. The margin of the infundibulum is surrounded by ciliated, fingerlike projections called **fimbriae** (fim′bre-e; "fibers, a fringe") that drape over the ovary. Medial to the infundibulum is the expanded **ampulla** ("flask"), which forms half the length of the uterine tube and is the site where fertilization usually occurs. Finally, the medial third of the uterine tube is the **isthmus** (is′mus; "a narrow passage").

Unlike the male reproductive ducts, which are directly continuous with the tubules of the testes, the uterine tubes have little or no direct contact with the ovaries. Instead, each ovulated oocyte is released into the peritoneal cavity of the pelvis (where many oocytes are lost), and the uterine tube performs a complex series of movements to capture the oocyte: The infundibulum bends to cover the ovary while the fimbriae stiffen and sweep the ovarian surface; then the beating cilia on the fimbriae generate currents in the peritoneal fluid that carry the oocyte into the uterine tube, where it begins its journey toward the uterus.

The oocyte's journey is aided by peristaltic waves generated by sheets of smooth muscle in the wall of the uterine tube and by the beating cilia of the uterine tube's ciliated simple columnar epithelium (Figure 25.14). Alternating with the ciliated cells in this lining epithelium are tall, nonciliated cells that secrete substances that nourish the oocyte (or embryo) and facilitate fertilization.

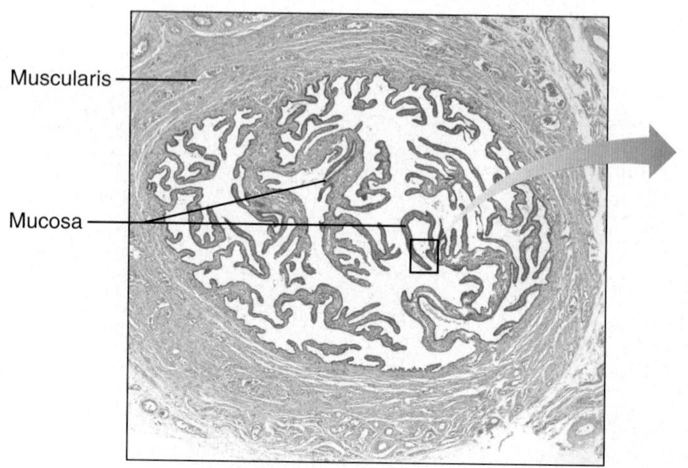

(a) Cross section through the ampulla (8×)

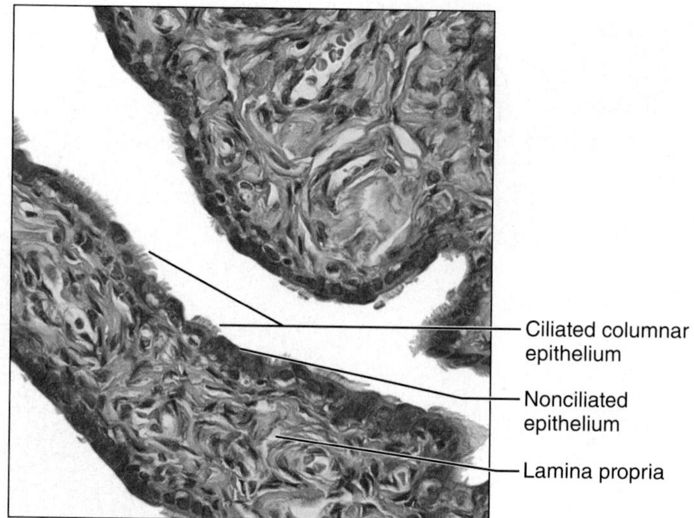

(b) Enlargement of the mucosa (170×)

Figure 25.14 Photomicrograph through the uterine tube.

Externally, the uterine tube is covered by peritoneum and supported by a short mesentery called the **mesosalpinx** (mez″o-sal′pinks), a part of the broad ligament (Figure 25.12). (*Mesosalpinx* means "mesentery of the trumpet," a reference to the uterine tube's trumpet shape; the root word *salpinx* is the basis for the terms **salpingectomy,** the surgical removal of a uterine tube, and **salpingitis,** inflammation of a uterine tube.)

CLINICAL APPLICATION

Pelvic Inflammatory Disease Widespread infection that originates in the vagina and uterus and spreads to the uterine tubes, ovaries, and ultimately the pelvic peritoneum is called **pelvic inflammatory disease (PID).** This condition, which occurs in about 10% of women in the United States, is usually caused by chlamydial or gonorrheal infection, but without exception other bacteria that infect the vagina are involved as well. Signs and symptoms include tenderness of the lower abdomen, fever, and a vaginal discharge. Even a single episode of PID causes infertility (due to scarring that blocks the uterine tubes) in 10% to 25% of cases unless halted within a few days. Therefore, patients are immediately given broad-spectrum antibiotics whenever PID is suspected.

The uterine tube is the site of 90% of all *ectopic pregnancies.* Ectopic pregnancy is the implantation of an embryo outside the uterus (see Clinical Application p. 771). Implantations in the uterine tube often result in rupture and hemorrhaging that threaten a woman's life.

The Uterus

The **uterus** (womb) lies in the pelvic cavity, anterior to the rectum and posterosuperior to the bladder (Figure 25.11). It is a hollow, thick-walled organ whose functions are to receive, retain, and nourish a fertilized egg throughout pregnancy. In a woman who has never been pregnant, the uterus is about the size and shape of a small, inverted pear, but it is somewhat larger in women who have had children. Normally, the uterus is tilted anteriorly, or **anteverted,** at the superior part of the vagina. However, in older women it is often inclined posteriorly, or **retroverted.** The major portion of the uterus is called the **body** (Figure 25.12). The rounded region superior to the entrance of the uterine tubes is the **fundus,** and the slightly narrowed region inferior to the body is the **isthmus.** Below this, the narrow neck of the uterus is the **cervix,** the inferior tip of which projects into the vagina. Containing much collagen, the cervix forms a tough, fibrous ring that keeps the uterus closed and the fetus within it during pregnancy.

The central lumen of the uterus is quite small (except in pregnancy). It is divided into the *cavity of the body* and the **cervical canal,** which communicates with the vagina inferiorly via the **external os** (*os* = mouth) and with the cavity of the body superiorly via the **internal os.** The mucosal lining of the cervical canal contains *cervical glands* that secrete a mucus that fills the cervical canal and covers the external os, presumably to block the spread of bacteria from the vagina into the uterus. Cervical mucus also blocks the entry of sperm—except at midcycle, when it becomes less viscous and allows the sperm to pass through.

Supports of the Uterus

Several ligaments and mesenteries help hold the uterus in place (Figure 25.15). The uterus is anchored to the lateral pelvic walls by the **mesometrium** (mez″o-me′tre-um; "mesentery of the uterus"), which is the largest division of the broad ligament. Inferiorly, the **lateral cervical (cardinal) ligaments** run horizontally from the uterine cervix and superior vagina to the lateral pelvic walls. These "ligaments" are thickenings of the fascia of the pelvis. The uterus is bound to

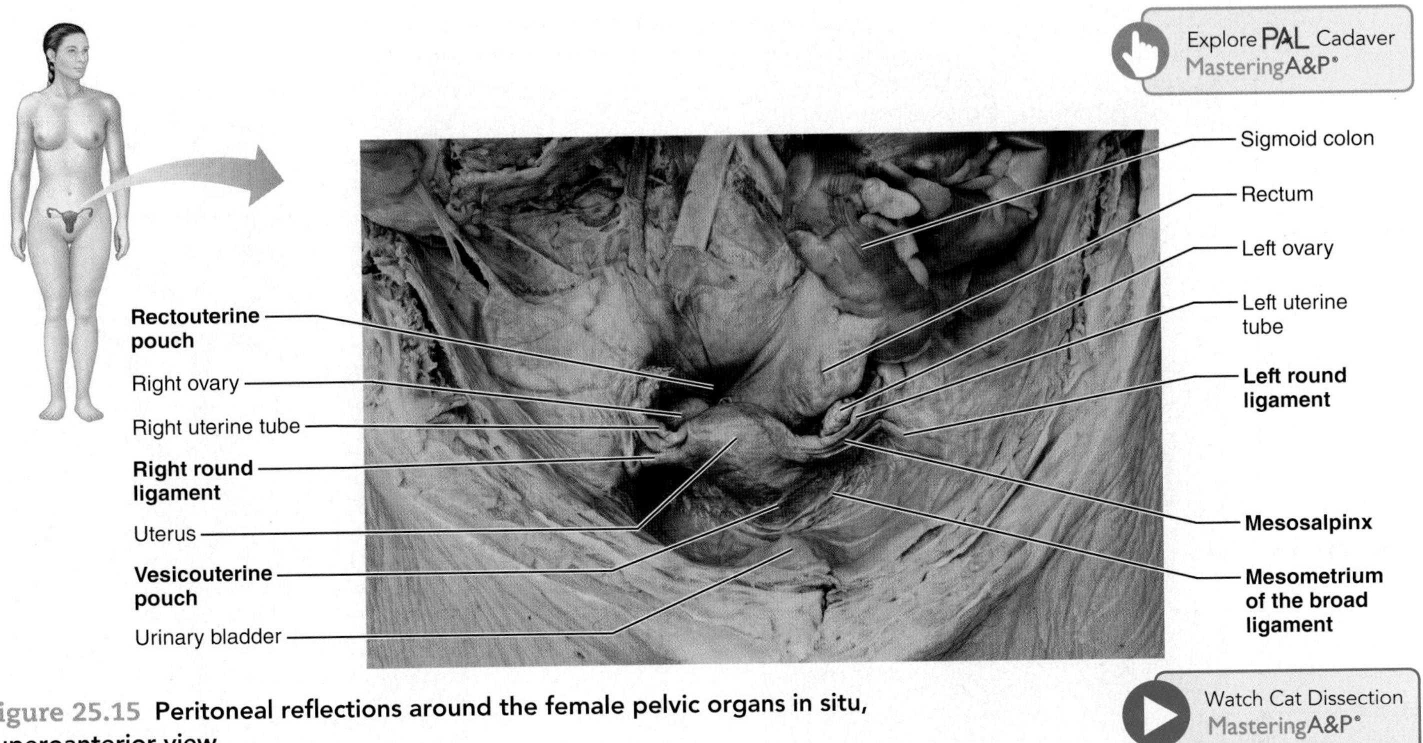

Figure 25.15 **Peritoneal reflections around the female pelvic organs in situ, superoanterior view.**

the anterior body wall by the paired **round ligaments of the uterus,** each of which starts as a continuation of the ovarian ligament on the superolateral aspect of the uterus, descends through the mesometrium and inguinal canal, and anchors in one of the labia majora of the external genitalia.

Despite the presence of these ligaments, the uterus is supported chiefly by the muscles of the pelvic floor—namely, the muscles of the urogenital and pelvic diaphragms (see pp. 296–297). If these muscles are torn during childbirth, the unsupported uterus may sink inferiorly, until the tip of the cervix protrudes through the external vaginal opening—a condition called **prolapse of the uterus.**

The undulating course of the peritoneum around and over the various pelvic organs produces several cul-de-sacs, or blind-ended peritoneal pouches (Figures 25.11 and 25.15). The *vesicouterine pouch* (ves″ĭ-ko-u′ter-in; "bladder to uterus") lies between the uterus and bladder, and the *rectouterine pouch* lies between the uterus and rectum. Because the rectouterine pouch forms the most inferior part of the peritoneal cavity, pus from intraperitoneal infections and blood from internal wounds can accumulate there.

The Uterine Wall

The wall of the uterus is composed of three basic layers (see Figure 25.12).

- The **perimetrium** (per″ĭ-me′tre-um; "around the uterus"), the outer serous membrane, is the peritoneum.
- The **myometrium** (mi″o-me′tre-um; "muscle of the uterus"), the bulky middle layer, consists of interlacing bundles of smooth muscle that contract during childbirth to expel the baby from the mother's body.
- The **endometrium** ("within the uterus") is the mucosal lining of the uterine cavity. It consists of a simple columnar epithelium containing secretory and ciliated cells underlain by a lamina propria connective tissue. If fertilization occurs, the embryo burrows into the endometrium and resides there for the rest of its development.

The endometrium has two chief layers: the *functional layer* and *basal layer* (Figure 25.16). The thick, inner **functional layer** undergoes cyclic changes in response to varying levels of ovarian hormones in the blood and is shed during menstruation (about every 28 days). The thin **basal layer** is not shed and is responsible for forming a new functional layer after menstruation ends. The endometrium contains straight tubular **uterine glands** that change in length as the endometrium thickens and thins.

The **uterine arteries,** which arise from the internal iliac arteries in the pelvis, ascend along the sides of the uterine body, send branches into the uterine wall (see Figure 25.12), and divide into **arcuate arteries** that course through the myometrium (Figure 25.16b). **Radial arteries** then reach the endometrium, where they give off **straight arteries** *(basal arteries)* to the basal layer and **spiral arteries** to the functional layer. The spiral arteries undergo degeneration and regeneration during each successive menstrual cycle, and they undergo spasms that cause the functional layer to shed during menstruation. Veins in the endometrium are thin-walled and form an extensive network, with occasional sinusoidal enlargements.

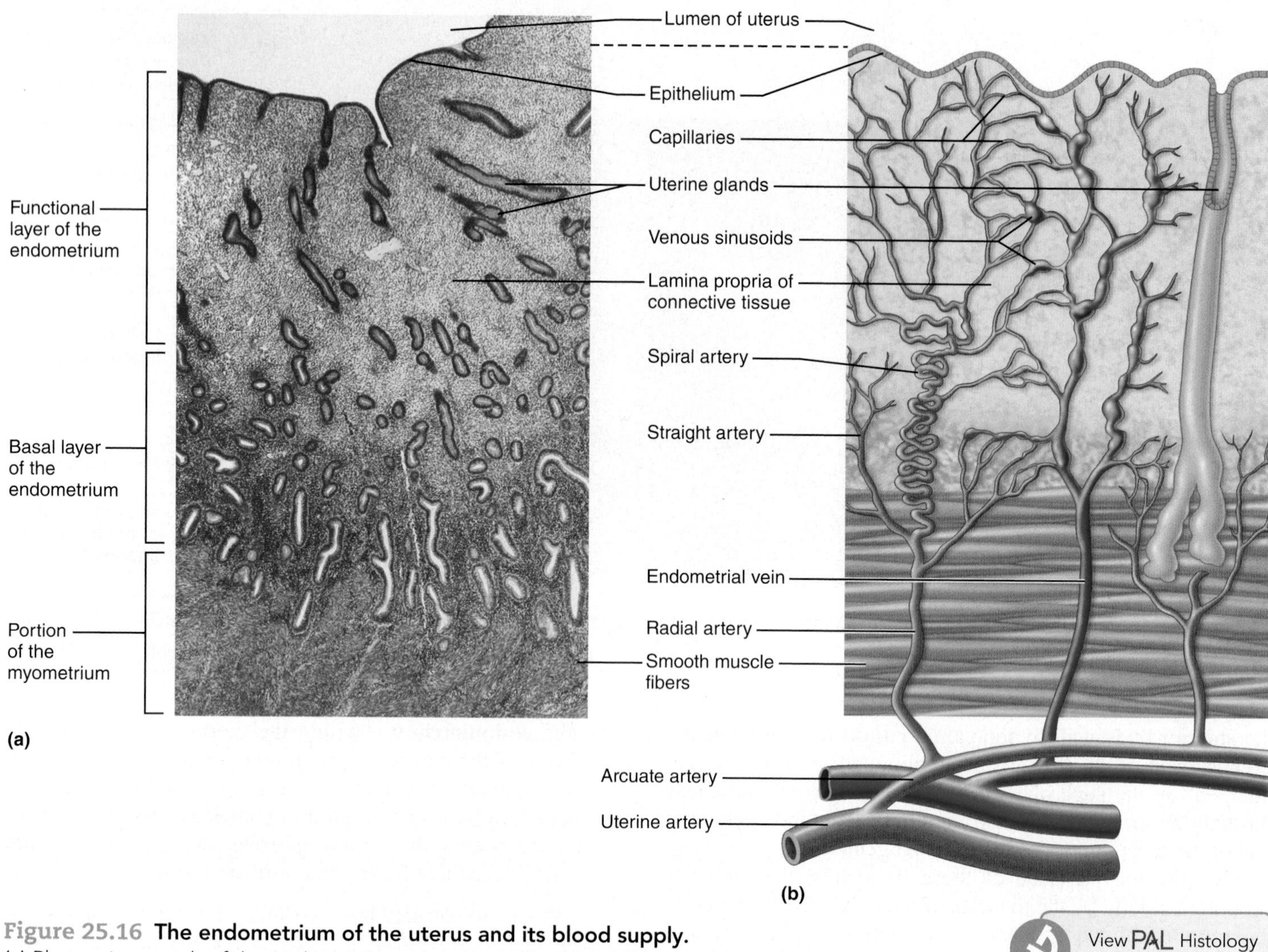

Figure 25.16 The endometrium of the uterus and its blood supply. (**a**) Photomicrograph of the endometrium, showing its basal and functional layers (30×). (**b**) Diagrammatic view of the endometrium, showing the straight arteries to the basal layer and the spiral arteries that serve the functional layer. Veins and venous sinuses are also shown.

View PAL Histology MasteringA&P®

check your understanding

☐ 10. What type of tissue lines the uterine tubes? How does this aid in "capturing" the ovulated oocyte?

☐ 11. Name the peritoneal supports of the female reproductive structures. To which organ does each attach?

☐ 12. Describe the location of each of the following structures: (a) fundus of the uterus, (b) infundibulum of the uterine tube, (c) endometrium.

(For answers, see Appendix B.)

Oogenesis and the Female Menstrual Cycle

learning outcomes

- ▶ Explain the phases of the ovarian cycle and the stages of oogenesis.
- ▶ Outline the three phases of the uterine cycle.

Oogenesis

Ova are produced in a process called **oogenesis** (o″o-jen′ĕ-sis; "egg generation") (Figure 25.17). Oogenesis, like spermatogenesis, includes the chromosome-reduction divisions of meiosis ($2n$ to n, as described on p. 754); however, there are distinct differences in the timing and outcome of these two processes.

- **Oogenesis takes many years to complete.** In the fetal period, stem cells called **oogonia** give rise to the female's lifelong supply of oocytes, which are arrested in an early stage of meiosis I around the time of birth. These **primary oocytes** remain "stalled" in meiosis I for decades, until they are ovulated from their follicle. Only in response to the influence of the LH surge that signals ovulation does a primary oocyte finish meiosis I and enter meiosis II as a **secondary oocyte**—but then it arrests again and does not finish meiosis II until a sperm penetrates its plasma membrane. Only after the completion of meiosis II is the egg technically called an **ovum.**
- **Oogenesis produces a single ovum.** Oogenesis typically produces four daughter cells: the large ovum and three

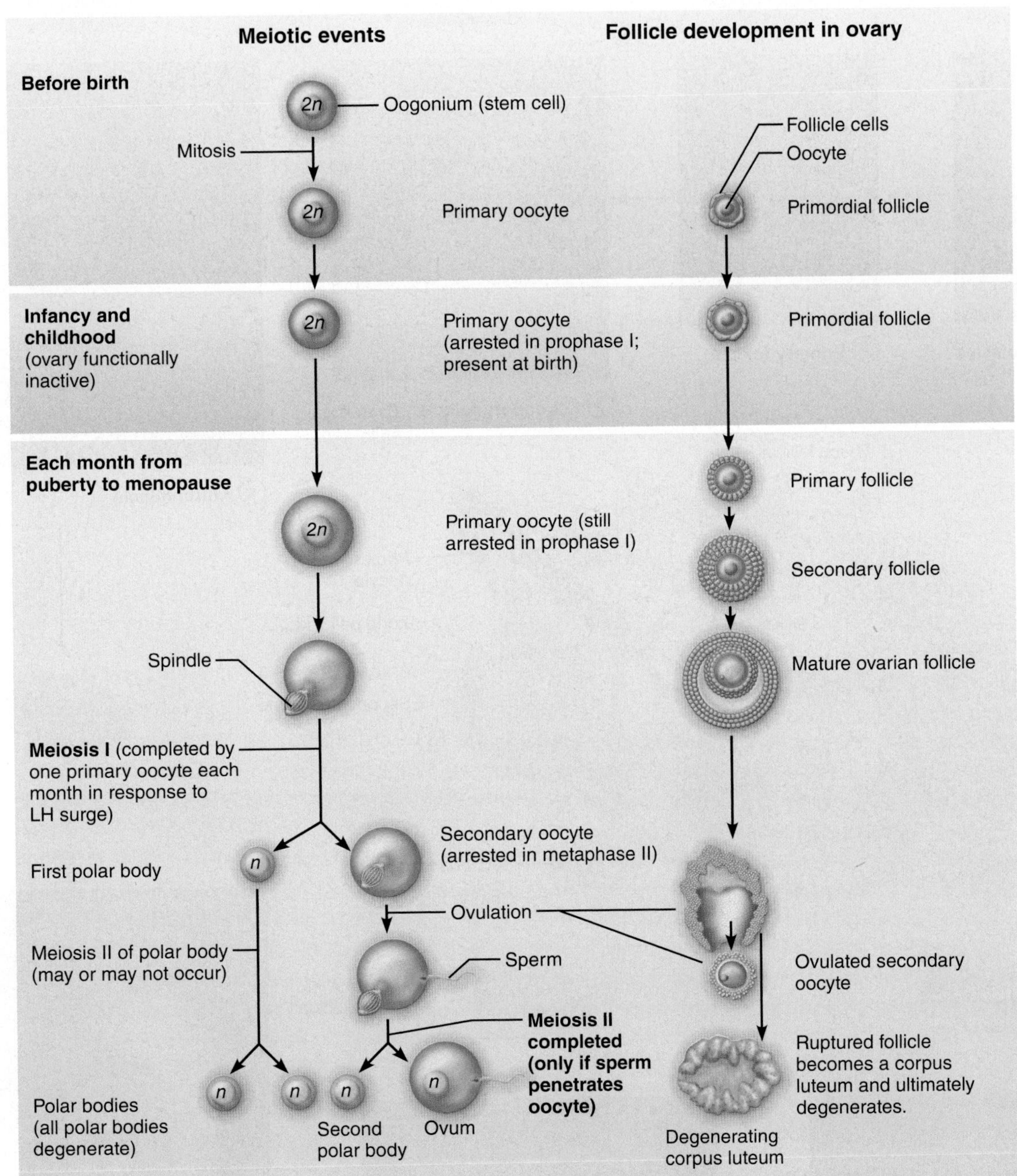

Figure 25.17 Oogenesis. Left, flowchart of the events of meiosis. Right, correlation with follicle development and ovulation in the ovary.

smaller cells called **polar bodies.** This occurs because during cell division most of the cytoplasm remains with a single daughter cell, the eventual ovum. The other three cells, the polar bodies, contain primarily DNA. The polar bodies degenerate quickly, without being fertilized or contributing to the developing embryo.

The monthly menstrual cycle of human females is due to hormonally induced cycling in both the ovary and the uterus. The *ovarian cycle* stimulates the development of ovarian follicles and the production of oocytes. The *uterine cycle* prepares the uterine wall for the implantation and nourishing of a fertilized ovum.

The Ovarian Cycle

The monthly **ovarian cycle** is the menstrual cycle as it relates to the ovary. It has three successive phases: the follicular phase, ovulation, and the luteal phase (Figures 25.18 and 25.19a and b).

Follicular Phase (First Half of the Ovarian Cycle)

From before birth to the end of a female's reproductive years, the ovarian cortex contains many thousands of follicles, most of which are **primordial follicles** (Figure 25.18 ①). Each of these primordial follicles consists of an oocyte surrounded by a single layer of flat supportive cells called **follicular cells.** At the start of each ovarian cycle, 6–12 primordial follicles start to grow, initiating the **follicular phase,** which lasts 2 weeks.

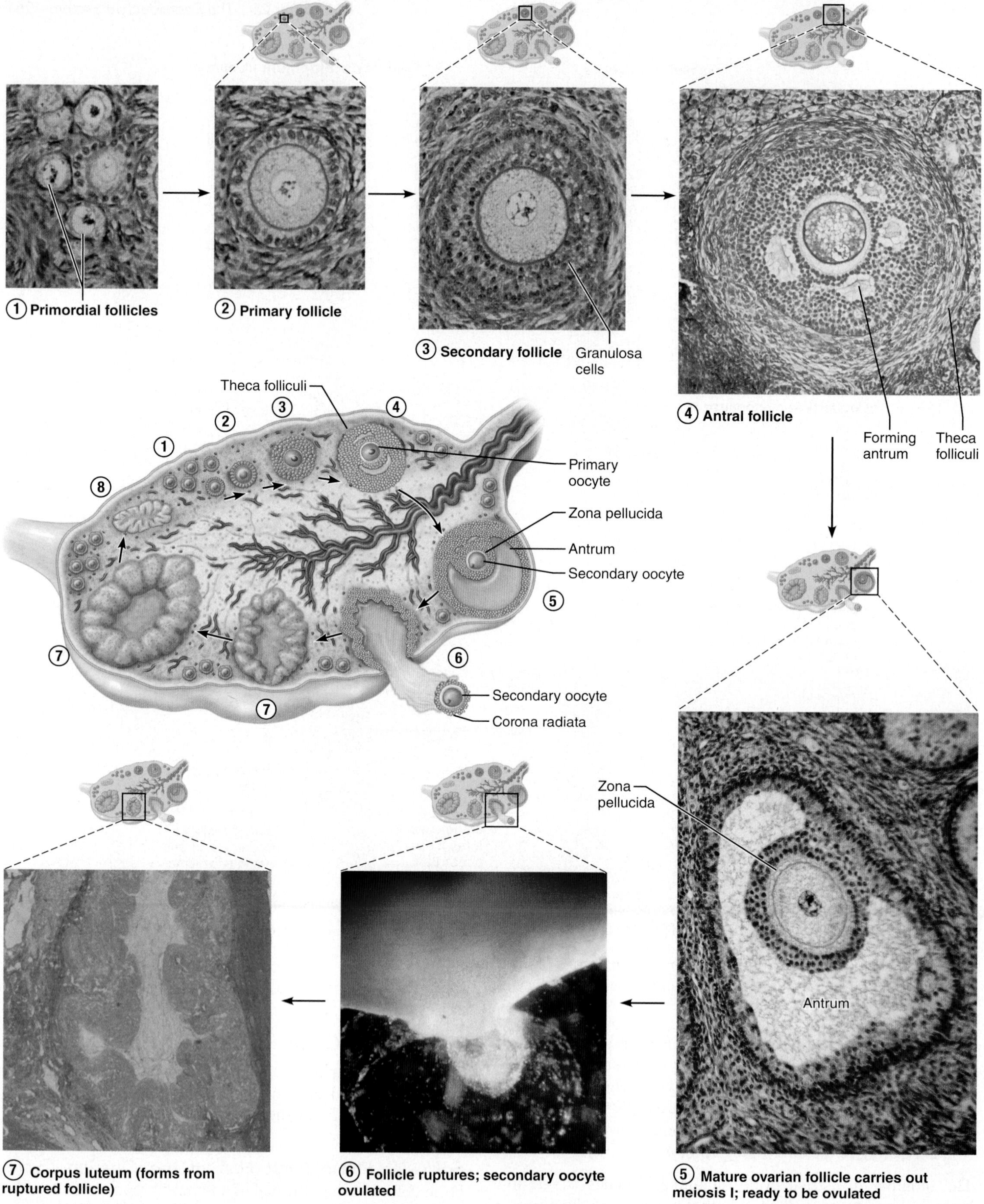

1 Primordial follicles
2 Primary follicle
3 Secondary follicle
Granulosa cells
4 Antral follicle
Forming antrum
Theca folliculi
Theca folliculi
1
2
3
4
8
Primary oocyte
Zona pellucida
Antrum
Secondary oocyte
5
6
7
7
Secondary oocyte
Corona radiata
Zona pellucida
Antrum
7 Corpus luteum (forms from ruptured follicle)
6 Follicle ruptures; secondary oocyte ovulated
5 Mature ovarian follicle carries out meiosis I; ready to be ovulated

Figure 25.18 The ovarian cycle. Schematic and microscopic views illustrate the development and fate of ovarian follicles. The numbers on the diagram of the ovary indicate the sequence of stages in follicle development, not the movement of a developing follicle within the ovary.

View PAL Histology MasteringA&P®

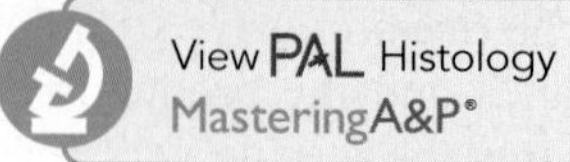

Beyond the smallest stages, growth is stimulated by follicle-stimulating hormone (FSH) secreted from the anterior part of the pituitary gland (Figure 25.19a).

When a primordial follicle starts to grow, its flat follicular cells become cuboidal, and the oocyte grows larger. Now the follicle is called a **primary follicle** (Figure 25.18, ②). Next, the follicular cells multiply to form a stratified epithelium around the oocyte (follicle ③ in the figure); from this point on, the follicular cells are called **granulosa cells.**

The oocyte develops a glycoprotein coat called the **zona pellucida** (zo′nah pĕ-lu′sid-ah; "transparent belt"), a protective shell that a sperm must ultimately penetrate to fertilize the oocyte. As the granulosa cells continue to divide, a layer of connective tissue called the **theca folliculi** (the′kah fo-lik′u-li; "box around the follicle") condenses around the exterior of the follicle (see follicle ④). The external cells of the theca are spindle-shaped and resemble smooth muscle cells; the internal cells of the theca are teardrop-shaped and secrete hormones. These theca cells are stimulated by luteinizing hormone (from the anterior pituitary) to secrete androgens (male sex hormones). The nearby granulosa cells, under the influence of follicle-stimulating hormone (also from the anterior pituitary), convert the androgens into the female sex hormones, **estrogens** (es′tro-jens). Once secreted into the bloodstream, the estrogens stimulate the growth and activity of all female sex organs. These estrogens also signal the uterine mucosa to repair itself after each menstrual period.

In the next stage of follicle development (follicle ④), a clear liquid gathers between the granulosa cells and coalesces to form a fluid-filled cavity called the **antrum** ("cave"); the follicle is now an **antral follicle.** The antrum expands with fluid until it isolates the oocyte, along with a surrounding coat of granulosa cells called a **corona radiata** (ko-ro′nah ra-de-ah′tah; "radiating crown"), on a stalk at the periphery of the follicle. Finally, one follicle attains full size (follicle ⑤ in Figure 25.18a), reaching a diameter of 2 cm (almost 1 inch). This **mature ovarian follicle** is ready to be ovulated.

Of the many follicles that grow in the follicular phase, most die and degenerate along the way, so that only one follicle each month completes the maturation process and expels its oocyte from an ovary for potential fertilization. Such expulsion, called *ovulation,* occurs just after the follicular phase ends, on about day 14 of the ovarian cycle.

Ovulation (Midpoint of the Ovarian Cycle) **Ovulation** occurs about halfway through each ovarian cycle (Figure 25.18, ⑥). In this process, one oocyte exits from one of the woman's two ovaries into the peritoneal cavity, and is swept into a uterine tube. The signal for ovulation is the sudden release of a large quantity of luteinizing hormone (LH) from the pituitary gland, just before day 14 (Figure 25.19a and b). In the process of ovulation, the ovarian wall over the follicle bulges, thins, and oozes fluid; this wall then ruptures and the oocyte exits, surrounded by its corona radiata. The forces responsible for this process are not fully understood, but they probably involve an enzymatic breakdown of the follicle wall, followed by a contraction of the muscle-like cells of the external layer of the theca that retracts the follicle and leaves the oocyte outside the ovary.

Luteal Phase (Second Half of the Ovarian Cycle) After ovulation, the part of the follicle that stays in the ovary collapses, and its wall is thrown into wavy folds (Figure 25.18, ⑦ and ⑧). This structure, now called the **corpus luteum** ("yellow body"), consists of the remaining granulosa and theca layers.

The corpus luteum is not a degenerative structure but an endocrine gland that persists through the second half, or **luteal phase,** of each ovarian cycle. It secretes estrogens and another hormone called **progesterone** (pro-jes′tĕ-rōn), which acts on the mucosa of the uterus, signaling it to prepare for implantation of an embryo (Figure 25.19b, c, and d). However, if there is no implantation, the corpus luteum dies after 2 weeks and becomes a scar called a **corpus albicans** (al′bĭ-kans; "white body"). The corpus albicans stays in the ovary for several months, shrinking until it is finally phagocytized by macrophages.

The Uterine Cycle

The **uterine cycle** is the menstrual cycle as it involves the endometrium (Figure 25.19d). Specifically, it is a series of cyclic phases that the endometrium undergoes month after month as it responds to changing levels of ovarian hormones in the blood (Figure 25.19c). These endometrial phases are closely coordinated with the phases of the ovarian cycle, which in turn are dictated by the pituitary hormones FSH and LH (Figure 25.19a). The uterine cycle has three phases:

1. The **menstrual phase** (days 1–5), in which the functional layer is shed
2. The **proliferative phase** (days 6–14), in which the functional layer rebuilds
3. The **secretory phase** (days 15–28), in which the endometrium prepares for implantation of an embryo

Many contraceptive methods function by altering the natural hormonal levels that regulate the ovarian and uterine cycles (see **A CLOSER LOOK** on p. 765).

CLINICAL APPLICATION

Endometriosis In the condition called **endometriosis** (en″do-me″tre-o′sis), fragments of endometrial tissue are present in the uterine tubes, on the ovary, and in the peritoneum of the pelvic cavity. The most widely favored explanation for this out-of-place endometrium is that a reflux of menstrual fluid spreads endometrial cells from the uterus through the uterine tubes. Because the uterine tubes open into the peritoneum, endometrial cells may also spread to the peritoneum. In endometriosis, the pain associated with menstruation is extreme because the endometrial fragments respond to the circulating ovarian hormones by building up endometrial tissue and bleeding. Blood accumulates in the pelvic cavity, forms cysts, and exerts pressure. Present in up to 10% of women of reproductive age, endometriosis causes one-third of all cases of infertility in women, when ectopic endometrial tissue covers the ovaries or blocks the uterine tubes. Treatment involves drugs that halt the secretion of estrogens and suppress menstruation, use of lasers to vaporize patches of ectopic endometrium, and hysterectomy.

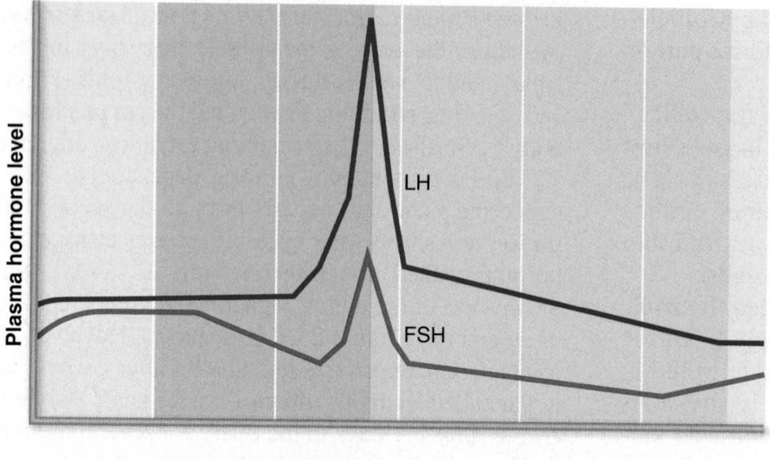

(a) Fluctuation of gonadotropin levels: Fluctuating levels of pituitary gonadotropins (follicle-stimulating hormone and luteinizing hormone) in the blood regulate the events of the ovarian cycle.

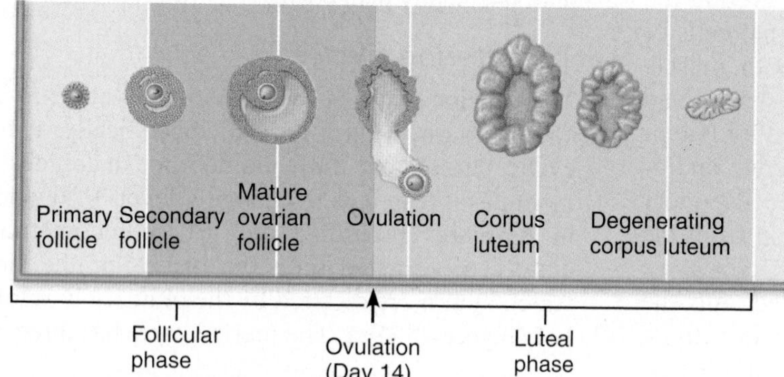

(b) Ovarian cycle: Structural changes in the ovarian follicles during the ovarian cycle are correlated with (d) changes in the endometrium of the uterus during the uterine cycle.

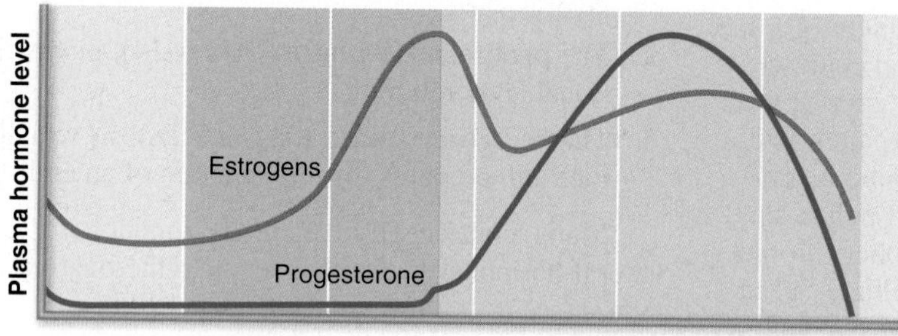

(c) Fluctuation of ovarian hormone levels: Fluctuating levels of ovarian hormones (estrogens and progesterone) cause the endometrial changes of the uterine cycle. The high estrogen levels are also responsible for the LH/FSH surge in (a).

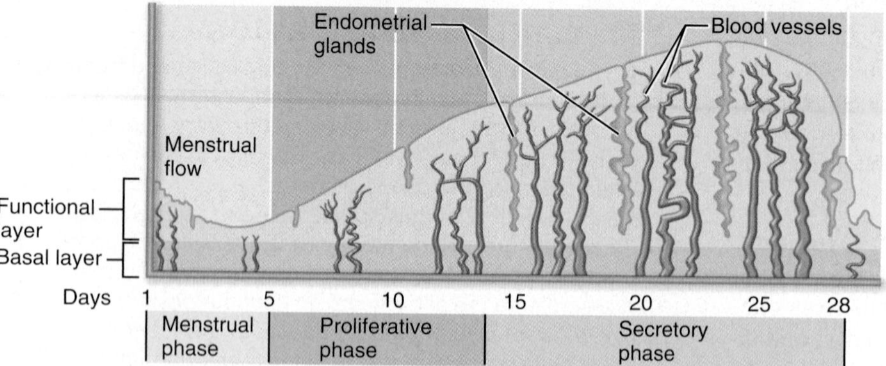

(d) The three phases of the uterine cycle

Menstrual: Shedding of the functional layer of the endometrium.

Proliferative: Rebuilding of the functional layer of the endometrium.

Secretory: Begins immediately after ovulation. Enrichment of the blood supply and glandular secretion of nutrients prepare the endometrium to receive an embryo.

Both the menstrual and proliferative phases occur before ovulation, and together they correspond to the follicular phase of the ovarian cycle. The secretory phase corresponds in time to the luteal phase of the ovarian cycle.

Figure 25.19 **The menstrual cycle: structural and hormonal changes.** The time bar near the bottom of the figure, reading Days 1 to 28, applies to all four parts of this figure.

A CLOSER LOOK

Contraception: To Be or Not To Be

The key to birth control is dependability. As the red arrows in the accompanying flowchart show, the birth control techniques currently available have many sites of action for blocking reproduction. Birth control methods can be categorized into three groups based on how each prevents pregnancy. **Behavioral methods** (coitus interruptus, rhythm method, abstinence) consist of altering behavior to prevent pregnancy. **Barrier methods** (condom, diaphragm or cervical cap, spermicides) prevent sperm and egg from meeting. **Hormonal methods** (pill, patch, intrauterine device or IUD, vaginal ring, implanted or injected hormonal agents, emergency contraceptive pill) alter the hormonal environment of the ovary and the uterus to prevent ovulation and/or implantation. Failure rates for behavioral methods, other than complete abstinence, are high, and these methods are not reliable. Barrier methods are quite effective, especially when used in combination, but they must be used consistently, during each sexual encounter. Hormonal methods are also highly effective, with a failure rate of less than 1%, and have a low risk of adverse effects in most women.

How do these hormonal interventions work? Many (the pill, the patch, the vaginal ring) contain a combined dose of estrogens and progestins (progesterone-like hormones). These hormones trick the hypothalamus-pituitary control of ovulation by keeping blood levels of these ovarian hormones relatively constant, as occurs during pregnancy. Ovarian follicles do not develop, ovulation ceases, and menstrual flow is much reduced.

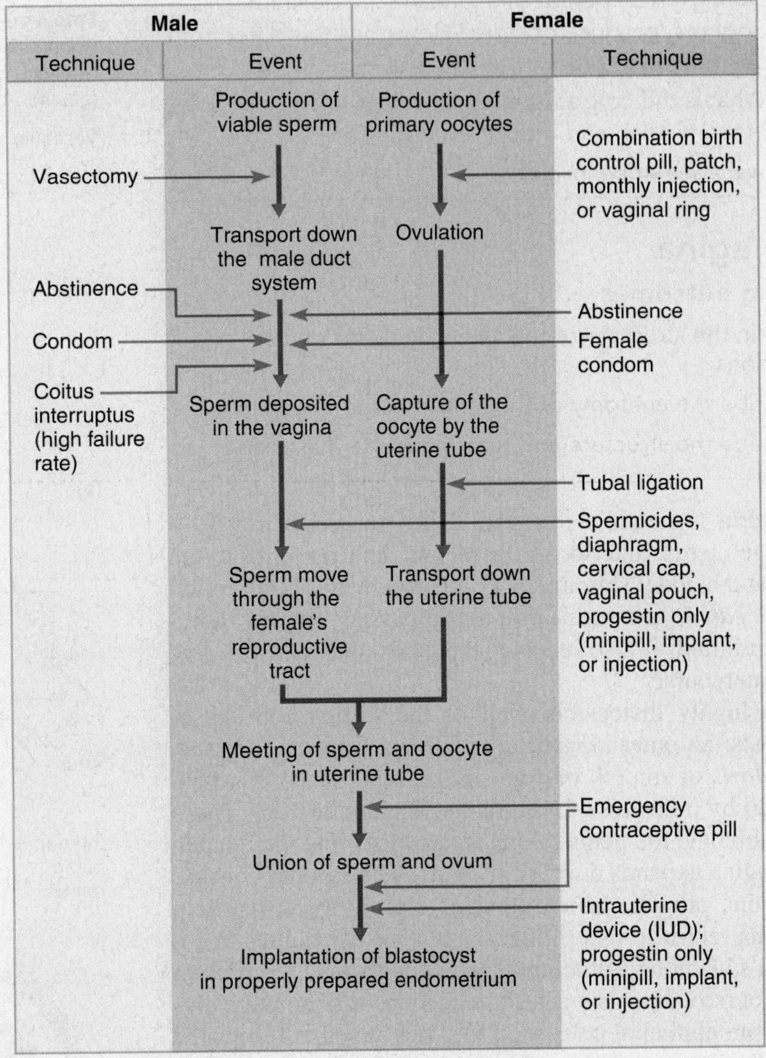

Mechanisms of contraception. Techniques or products that interfere with events from production of gametes to implantation are indicated by red arrows at the site of interference and act to prevent the next step from occurring.

Other hormonal methods use low doses of progestins only. These methods (intrauterine device, minipill, implanted or injected applications) cause the cervical mucus to thicken and thus block sperm entry into the uterus. They also disrupt the development of the uterine lining and prevent implantation. Higher doses of hormones found in emergency contraceptive pills or post-coital pills also disrupt the normal hormonal signals enough to prevent fertilization or implantation when taken within 3 days of unprotected intercourse.

A woman should choose a method of birth control within the context of her health and her commitment to use. No birth control method is effective if it is not used for each and every sexual encounter. Birth control also does not prevent sexually transmitted infection (STI). The only method that is 100% effective in birth control and in preventing STIs is total abstinence.

check your understanding

- 13. Which hormone stimulates follicle development in the ovary?
- 14. During which period of a female's life do stem cells, oogonia, produce oocytes? How does this differ from spermatocyte production in males?
- 15. What is the corpus luteum, and what role does it play in the ovarian and uterine cycles?

(For answers, see Appendix B.)

The Vagina

learning outcomes

- Explain the structure of the vagina in terms of its functions.
- Describe the anatomy of the female external genitalia.
- Describe the structure and location of the mammary glands.

The **vagina** ("sheath") is a thin-walled tube that lies inferior to the uterus, anterior to the rectum, and posterior to the urethra and bladder (Figure 25.11). The vagina is often called the *birth canal,* because it provides a passageway for delivery of an infant. It also receives the penis and semen during sexual intercourse.

The highly distensible wall of the vagina consists of three coats: an outer *adventitia* of fibrous connective tissue, a *muscularis* of smooth muscle, and an inner *mucosa,* which is marked by transverse folds (rugae) (Figure 25.12a). These ridges stimulate the penis during intercourse, and they flatten as the vagina expands during childbirth. The mucosa consists of a lamina propria, which contains elastic fibers that help the vagina return to its original shape after expanding, and a stratified squamous epithelium that can withstand the friction of intercourse and resist bacterial infection. Additionally, glycogen from epithelial cells shed into the lumen is fermented by beneficial resident bacteria into lactic acid; the resulting acidity discourages the growth of harmful bacteria, but it is also hostile to sperm.

Near the vagina's external opening, the **vaginal orifice,** the mucosa elaborates to form an incomplete diaphragm called the **hymen** (hi′men; "membrane") (Figure 25.20a). The hymen is vascular and tends to bleed when ruptured during the first sexual intercourse. However, its durability varies: In some females it is delicate and can rupture during a sports activity, the insertion of a tampon, or a pelvic examination; occasionally the hymen is so tough that it must be breached surgically if intercourse is to occur.

The recess formed where the widened superior part of the vagina encircles the tip of the cervix is called the **fornix** (Figure 25.12a). The posterior part of this recess, the posterior fornix, is much deeper than either the lateral or the anterior fornices (Figure 25.11).

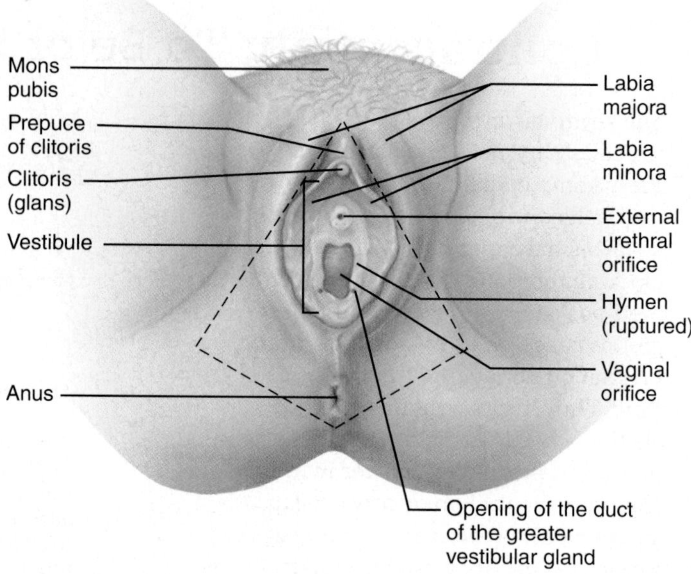

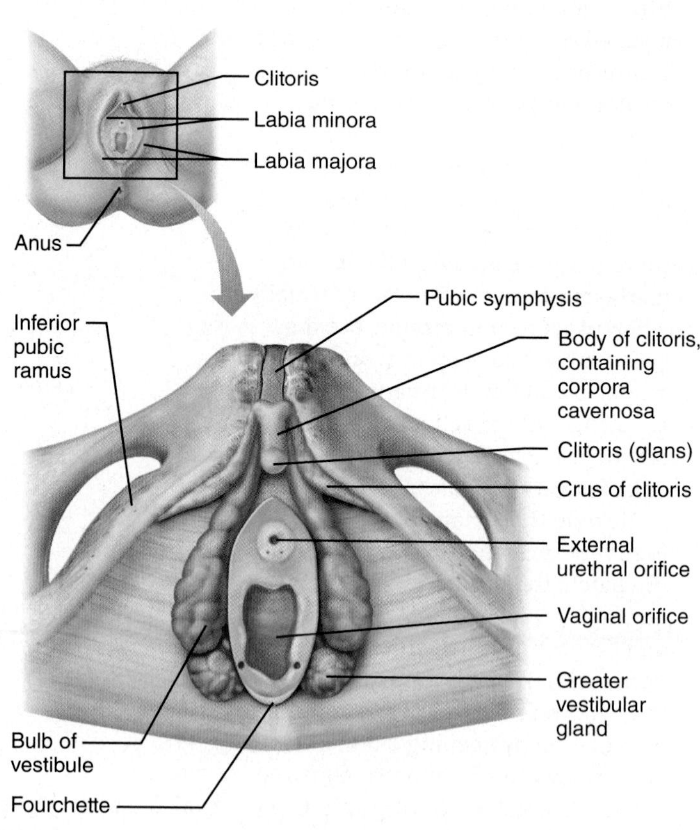

Figure 25.20 **The external genitalia (vulva) of the female. (a)** Superficial structures. The region enclosed by the dashed lines is the perineum. **(b)** Deep structures. The labia majora and associated skin have been removed to show the underlying erectile bodies. The associated superficial muscles are not illustrated (see Figure 11.15 on p. 297).

The External Genitalia and Female Perineum

The female reproductive structures that lie external to the vagina are the **external genitalia,** also called the **vulva** ("covering") or **pudendum** (pyoo-den′dum) (Figure 25.20a). These structures include the mons pubis, the labia, the clitoris, and structures associated with the vestibule.

The **mons pubis** (mons pu′bis; "mountain on the pubis") is a fatty, rounded pad overlying the pubic symphysis. After puberty, pubic hair covers the skin of this area. Extending posteriorly from the mons are two long, hair-covered, fatty skin folds, the **labia majora** (la′be-ah mah-jor′ah; "larger lips"), which are the female counterpart, or *homologue,* of the scrotum; that is, they derive from the same embryonic structure. The labia majora enclose two thin, hairless folds of skin called the **labia minora** (mi-nor′ah; "smaller lips"), which enclose a recess called the **vestibule** ("entrance hall") housing the external openings of the urethra and vagina. In the vestibule, the vaginal orifice lies posterior to the urethral orifice. Lateral to the vaginal orifice are the paired, pea-sized *greater vestibular glands* (Figure 25.20b), which lie deep to the posterior part of the labia on each side and secrete lubricating mucus into the vaginal orifice during sexual arousal, facilitating entry of the penis. At the extreme posterior point of the vestibule, the right and left labia minora come together to form a ridge called the *frenulum of the labia* ("little bridle of the lips") or the **fourchette** (foor-shet′).

Just anterior to the vestibule is the **clitoris** (klit′o-ris; "hill"), a protruding structure composed largely of erectile tissue that is sensitive to touch and swells with blood during sexual stimulation. The clitoris is homologous to the penis, having both a glans and a body (although there is no urethra within it). It is hooded by a fold of skin, the **prepuce of the clitoris,** formed by the anterior junction of the two labia minora (Figure 25.20a). As in the penis, the body of the clitoris contains paired **corpora cavernosa,** which continue posteriorly into **crura** that extend along the bony pubic arch (Figure 25.20b). The clitoris contains no corpus spongiosum. The homologous **bulbs of the vestibule** lie along each side of the vaginal orifice and directly deep to the bulbospongiosus muscle. During sexual stimulation, these erectile tissues engorge with blood and form the basis of the female sexual response.

The female **perineum** is a diamond-shaped region between the pubic arch anteriorly, the coccyx posteriorly, and the ischial tuberosities laterally (Figure 25.20a). In the exact center of the perineum, just posterior to the fourchette, lies the *central tendon* or *perineal body* (Figure 25.20b). This knob is the insertion tendon of most muscles that support the pelvic floor.

CLINICAL APPLICATION

Episiotomy Because of its proximity to the vagina, the central tendon is sometimes torn by an infant's head during childbirth. The resulting jagged tear may heal poorly, weakening the pelvic floor muscles and allowing the pelvic organs to sink inferiorly and the uterus to prolapse. To avoid these problems, an **episiotomy** (e-piz″e-ot′o-me; "cutting of the vulva") is performed in 50% to 80% of deliveries in the United States. In this procedure, the vaginal orifice is widened by a posterior cut through the fourchette at the time the baby's head appears at the vestibule. The posterior incision is made either straight through the central tendon or just lateral to it. After the birth, this clean incision is stitched together and allowed to heal. Episiotomies have generated much controversy. Opponents say that they are too often performed unnecessarily, that they are painful while healing, and that rigorous studies are needed to determine whether they are as effective as claimed.

The Mammary Glands

Mammary glands (breasts) are modified sweat glands that are present in both sexes but function only in lactating females (Figure 25.21) when they produce milk to nourish an infant in response to hormonal stimulation.

Embryonically, the mammary glands form as part of the skin, arising along bilateral lines that run between the axilla and groin on the lateral trunk of the embryo. These embryonic *milk lines* persist only in the midthorax, where the definitive breasts form. About 1 person in 500 develops extra (accessory) nipples or breasts, which can occur anywhere along the milk line.

The base of the cone-shaped breast in females extends from the second rib superiorly to the sixth rib inferiorly. Its medial border is the sternum, and its lateral boundary is the midaxillary line, a line extended straight inferiorly from the middle of the axilla. The muscles deep to the breast are the pectoralis major and minor and parts of the serratus anterior and external oblique. The main arteries to the breast are the lateral thoracic artery and cutaneous branches of the internal thoracic and posterior intercostals (see Figure 20.11 on p. 602). The lymphatic vessels draining the breast drain into *parasternal lymph nodes* (nodes along the internal thoracic arteries) and into the *axillary lymph nodes* in the armpit.

The **nipple,** the central protruding area from which an infant sucks milk, is surrounded by a ring of pigmented skin, the **areola** (ah-re′o-lah; "a small open area"). During nursing, large sebaceous glands in the areola produce an oily sebum that minimizes chapping and cracking of the skin of the nipple.

Internally, the mammary gland consists of 15 to 25 **lobes** (Figure 25.21), each of which is a distinct compound alveolar gland opening at the nipple. The lobes are

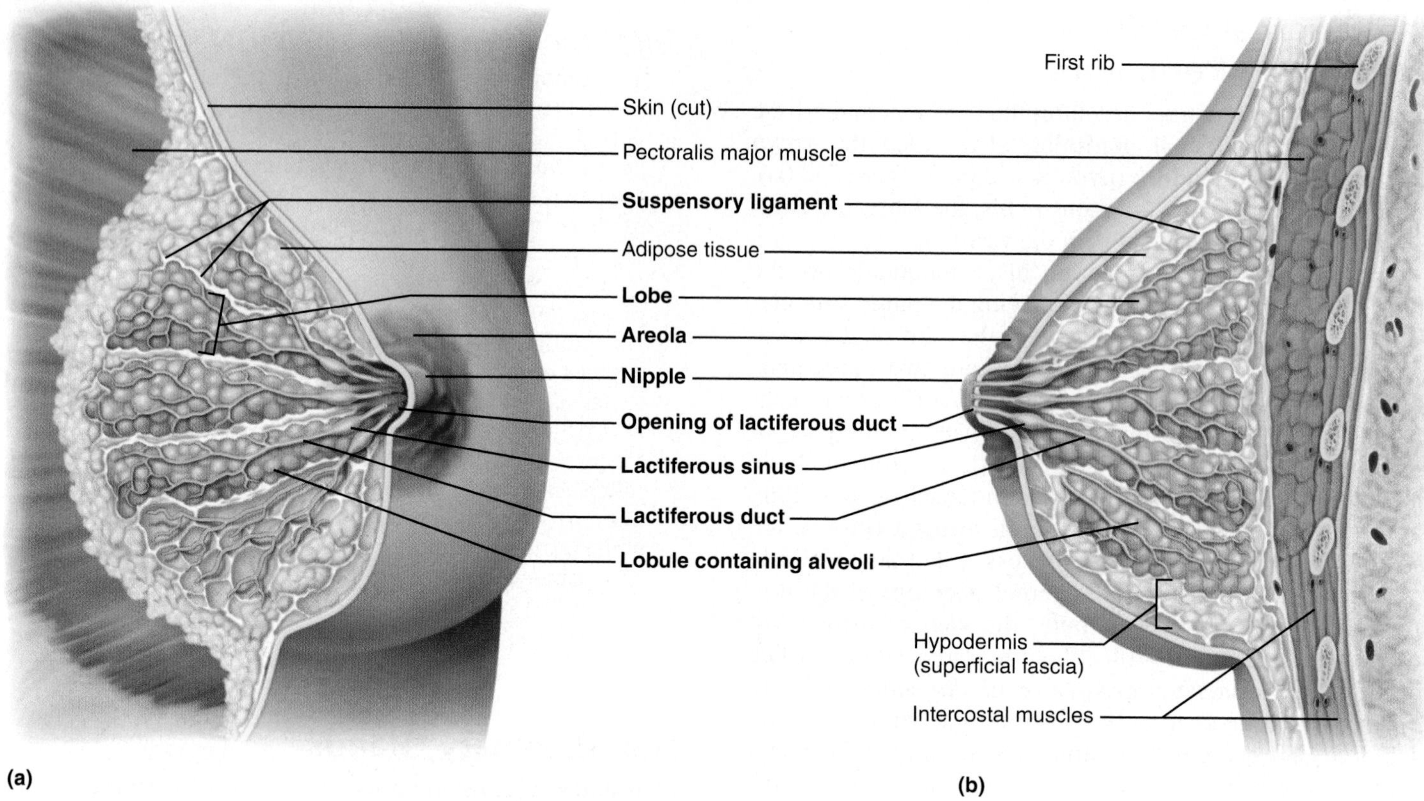

Figure 25.21 Structure of a lactating (milk-secreting) mammary gland. (**a**) Anterior view of a partly dissected breast. (**b**) Sagittal section of a breast.

separated from one another by a large amount of adipose tissue and strips of interlobar connective tissue that form **suspensory ligaments of the breasts.** These ligaments run from the underlying skeletal muscles to the overlying dermis and provide, as their name implies, support for the breasts. The lobes of the breast consist of smaller units called **lobules,** which are composed of tiny alveoli clustered like a bunch of grapes. The walls of the alveoli consist of a simple cuboidal epithelium of milk-secreting cells. From the alveoli, the milk passes through progressively larger ducts until it reaches the largest ducts, called **lactiferous ducts** (*lactiferous* = milk-carrying), which lie within and deep to the nipple. Just deep to the areola, each lactiferous duct has a dilated region called a **lactiferous sinus,** where milk accumulates during nursing.

The glandular structure of the breast is undeveloped in nonpregnant women. During childhood, a girl's breasts consist only of rudimentary ducts, as do the breasts of males throughout life. When a girl reaches puberty, these ducts grow and branch, but lobules and alveoli are still absent. (Breast enlargement at puberty is due to fat deposition.) The glandular alveoli finally form—from the smallest ducts—about halfway through pregnancy, and milk production starts a few days after childbirth.

Breast cancer, which usually arises from the duct system, is a leading cause of cancer deaths in women (see p. 775).

☑ check your understanding

- ☐ **16.** What are the male homologues of the following female genital structures: (a) labia majora, (b) clitoris, (c) bulb of the vestibule?
- ☐ **17.** When does the glandular portion of the breast develop in females?

(For answers, see Appendix B.)

PREGNANCY AND CHILDBIRTH

learning outcomes

- ▶ Describe the processes of implantation and placenta formation.
- ▶ Define a placental (chorionic) villus, and explain its functions.
- ▶ List the three phases of labor.

Pregnancy

Events Leading to Fertilization

Before fertilization can occur, sperm deposited in the vagina must swim through the uterus to reach the ovulated oocyte in the uterine tube. Sperm are known to have proteins called *olfactory receptors* that respond to chemical stimuli. It is postulated that the oocyte, its surrounding cells, or cells in

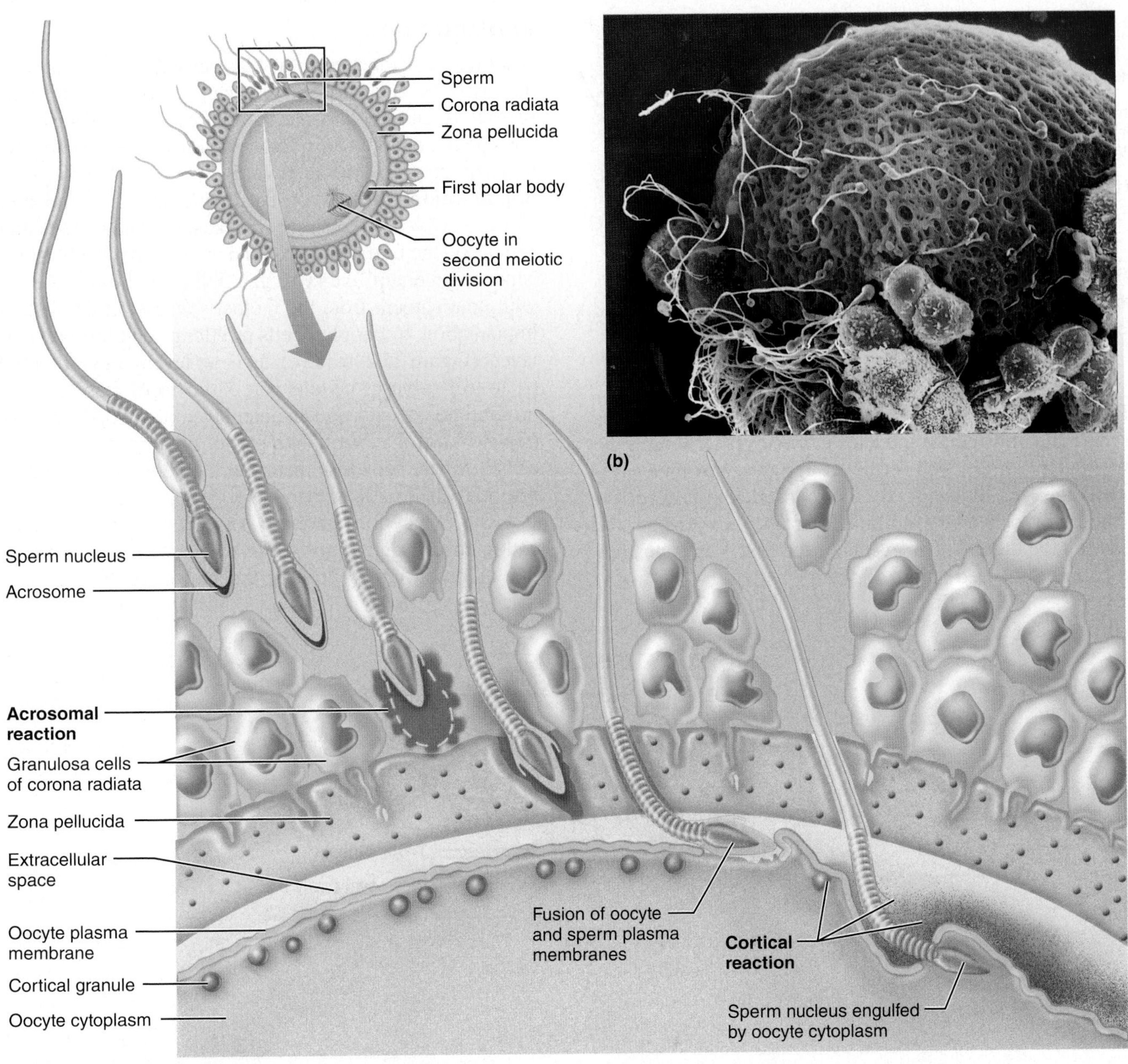

Figure 25.22 Events leading to fertilization: the acrosomal and cortical reactions. (**a**) Sperm penetration of an oocyte is depicted from left to right. A sperm passes between the cells of the corona radiata, binds to the zona pellucida, releases its acrosomal enzymes, and passes through the zona. Fusion of the plasma membrane of a single sperm with the oocyte membrane induces the cortical reaction. In this reaction, cortical granules in the oocyte release into the extracellular space enzymes that destroy the sperm receptors in the zona pellucida, thereby preventing entry of more than one sperm. (**b**) Scanning electron micrograph of human sperm surrounding an oocyte (artificially colored; 500×). The small, spherical cells are granulosa cells of the corona radiata.

the female reproductive tract release chemical signaling molecules that direct sperm.

Once sperm reach the oocyte, they bind to receptors on the zona pellucida around the oocyte and undergo the **acrosomal reaction,** the release of digestive enzymes from their acrosomes (Figure 25.22). These enzymes digest a slit in the zona, through which the sperm wriggles to reach the oocyte. The plasma membrane of a single sperm fuses with the plasma membrane of the oocyte, and the sperm nucleus is engulfed by the oocyte's cytoplasm. This fusion induces the **cortical reaction:** Granules in the oocyte secrete enzymes into the extracellular space beneath the zona pellucida. These enzymes alter the zona pellucida and destroy the sperm receptors, preventing any other sperm from binding to and entering the egg.

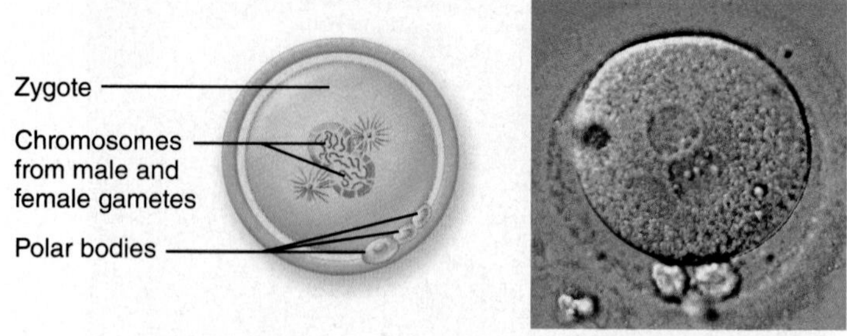

Figure 25.23 **Fertilization.** Illustration and micrograph of the merging of male and female chromosomes in an oocyte. Notice the polar bodies at the bottom of the oocyte.

Fertilization occurs at the moment the chromosomes from the male and female gametes come together within the ovum (Figure 25.23). The fertilized egg (zygote) then initiates cleavage (cell division). By the fourth day after fertilization, the embryo is in the multicellular *blastocyst* stage and has entered the uterus. (The events of the first 4 days postfertilization are discussed on pp. 48–49.)

Implantation

After reaching the uterus, the blastocyst floats freely in the uterine cavity for about 2 days, receiving nourishment from uterine secretions. During this time, it "hatches" by digesting part of the zona pellucida to create a large opening and squeezing through it. Then, about 6 days after fertilization, the blastocyst begins **implantation,** the act of burrowing into the endometrium (Figure 25.24a). At this time, the blastocyst consists of an *inner cell mass* (the future embryo) and an outer **trophoblast** ("nourishment generator"), a layer that will soon provide the embryo with nourishment from the mother's uterus. In the first step of implantation, trophoblast cells proliferate and form two distinct layers (Figure 25.24b and c): an inner layer, or **cytotrophoblast** (si″to-trof′o-blast; "cellular part of the trophoblast"), where cell proliferation occurs; and an outer layer, or **syncytiotrophoblast** (sin-sit″e-o-trof′o-blast; "the part of the trophoblast with fused cells"), where cells lose their plasma membranes and fuse into a multinuclear mass of cytoplasm.

The syncytiotrophoblast projects invasively into the endometrium and digests the uterine cells it contacts. As the endometrium is eroded, the blastocyst becomes deeply

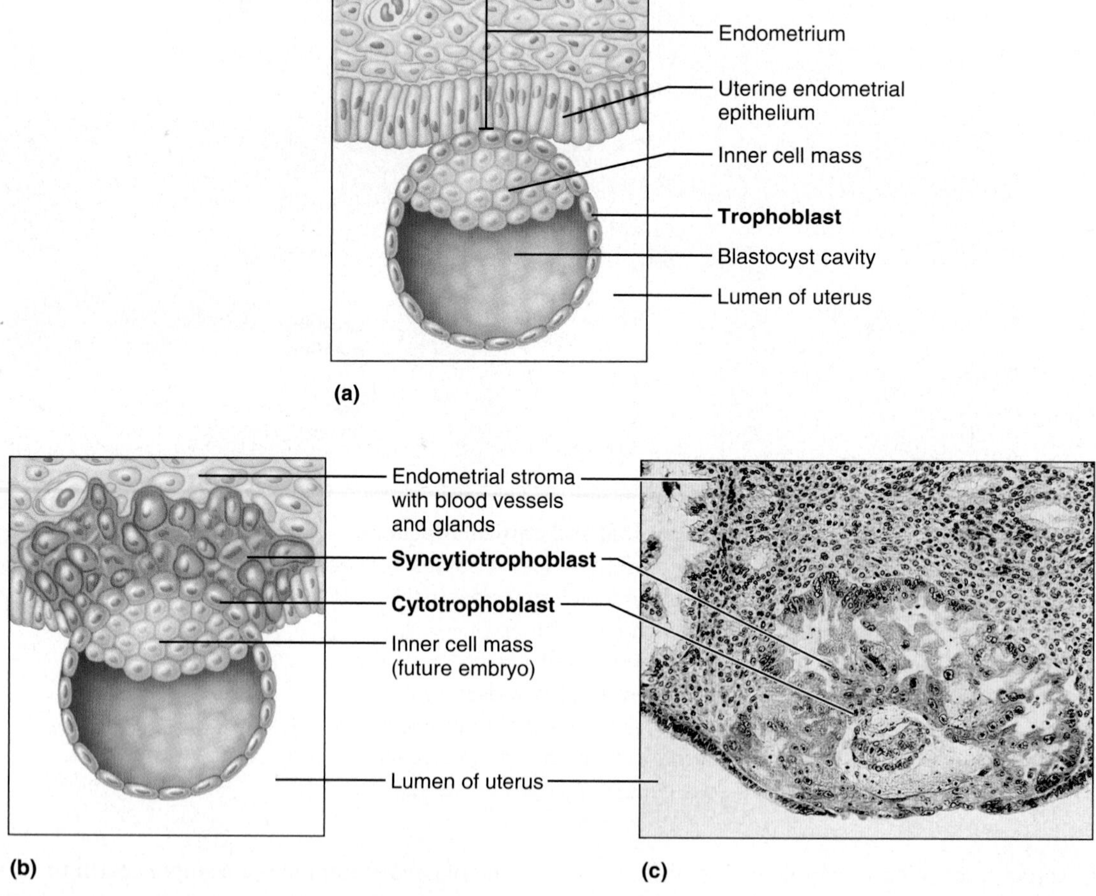

Figure 25.24 **Implantation of the blastocyst.** (**a**) Diagram of a blastocyst that has just adhered to the uterine endometrium (day 6). (**b**) Implanting embryo at a slightly later stage (day 7), showing the cytotrophoblast and syncytiotrophoblast. (**c**) Light micrograph of an implanting embryo (approximately day 9, 42×).

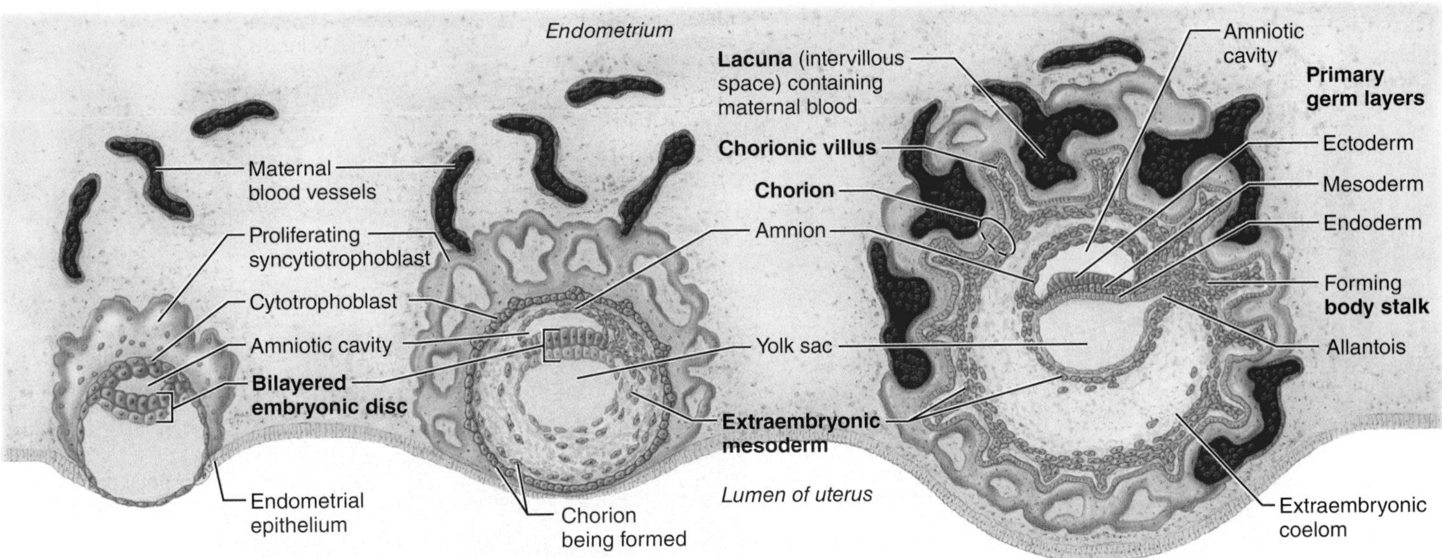

(a) Implanting 7½-day blastocyst. The syncytiotrophoblast is eroding the endometrium.

(b) 12-day blastocyst. Implantation is complete. Extraembryonic mesoderm is forming a discrete layer beneath the cytotrophoblast.

(c) 16-day embryo. Cytotrophoblast and associated mesoderm have become the chorion, and chorionic villi are elaborating. The body stalk forms the basis of the umbilical cord.

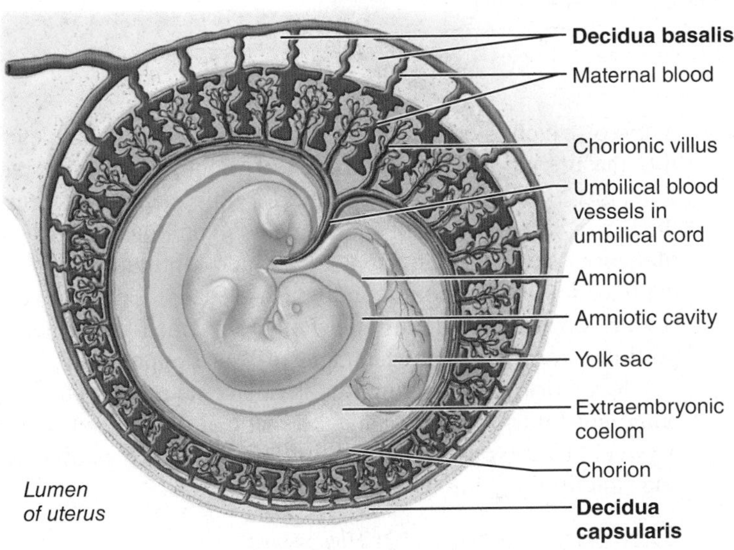

(d) 4½-week embryo. The decidua capsularis and decidua basalis are well formed. The chorionic villi lie in blood-filled intervillous spaces within the endometrium. The embryo is now receiving its nutrition via the umbilical vessels that connect it (through the umbilical cord) to the placenta.

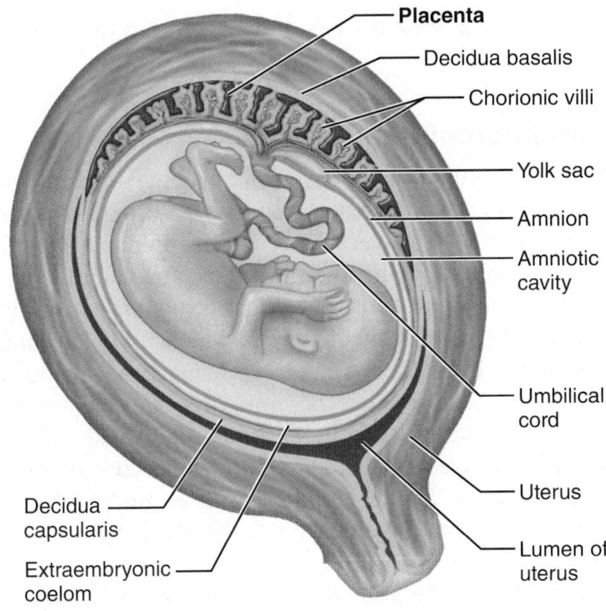

(e) 13-week fetus

Figure 25.25 **Placenta formation.**

embedded within this thick uterine lining (Figure 25.25a). Cleftlike spaces called **lacunae** then open within the syncytiotrophoblast and quickly fill with maternal blood that leaks from degraded endometrial blood vessels (Figure 25.25c). Now, 10–12 days after fertilization, embryo-derived tissues have made their first contact with maternal blood, their ultimate source of nourishment. Events now proceed toward the formation of the placenta, the organ that nourishes the developing fetus.

CLINICAL APPLICATION

Ectopic Pregnancy Implantation of an embryo in any other site than the uterus is called an **ectopic pregnancy** (*ecto* = "outside"). An ectopic pregnancy can occur on an intraperitoneal organ, such as the liver or spleen, or on the outside of the uterus, but it usually occurs in a uterine tube, a condition called a **tubal pregnancy.** Because the uterine tube is unable to establish a placenta or accommodate growth, it ruptures unless the condition is diagnosed early. If rupture occurs, internal bleeding may threaten the woman's life. For obvious reasons, ectopic pregnancies almost never result in live births.

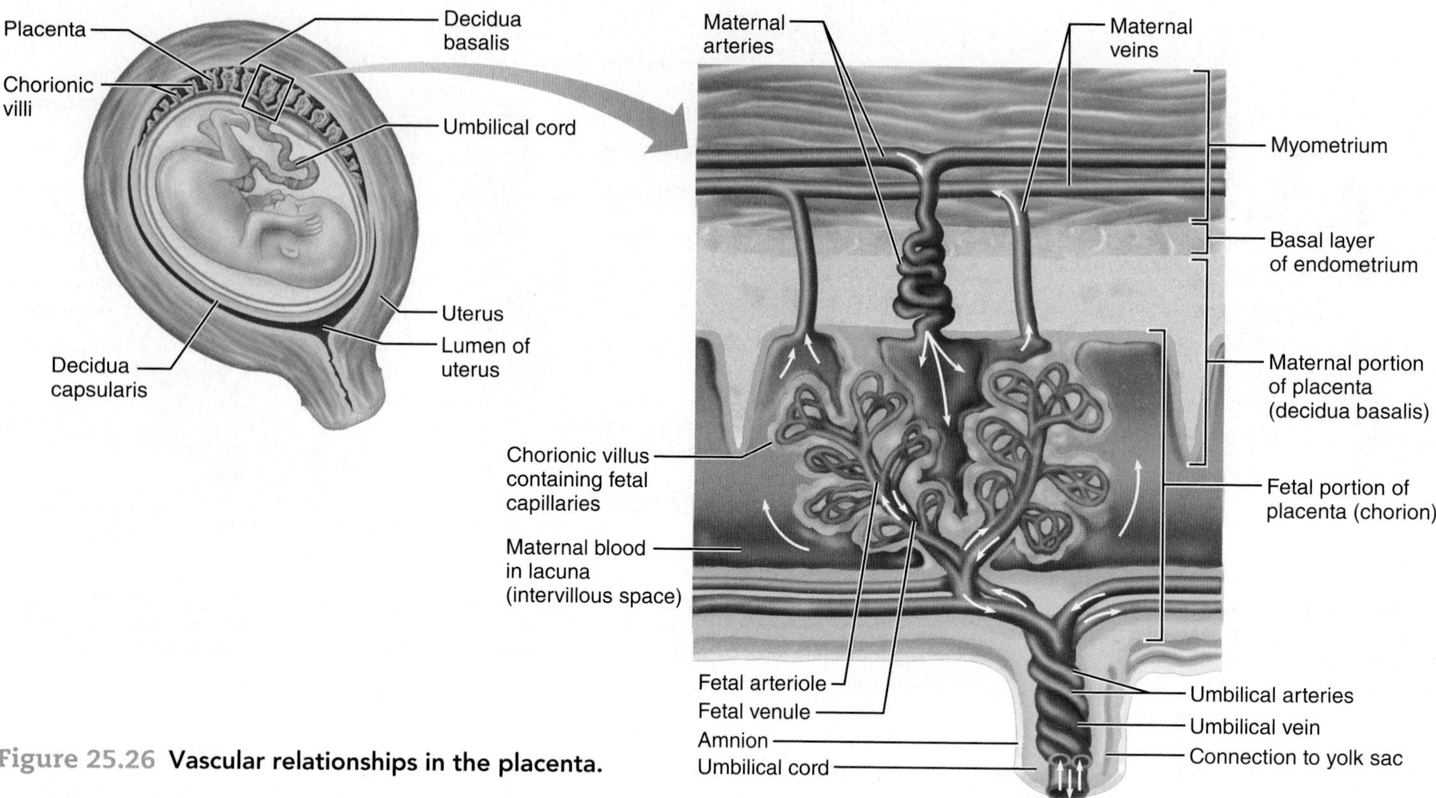

Figure 25.26 Vascular relationships in the placenta.

Formation of the Placenta

Both embryonic (trophoblastic) and maternal (endometrial) tissues contribute to the placenta. First, the proliferating trophoblast gives rise to a layer of **extraembryonic mesoderm** on its internal surface (Figure 25.25b and c). Together, the extraembryonic mesoderm and trophoblast layers are now called the **chorion** (ko′re-on; "membrane"), which folds into fingerlike **chorionic villi,** which contact the lacunae containing maternal blood. At this stage, the embryo's body connects to the chorion outside of it through a **body stalk** made of extraembryonic mesoderm. This stalk forms the core of the future umbilical cord, as the umbilical arteries and umbilical vein grow into it from the embryo's body. Over the next 2 weeks, the mesodermal cores of the chorionic villi develop capillaries that connect to the umbilical blood vessels growing from the embryo (Figure 25.25d). With the establishment of this vascular connection, the embryo's own blood now courses within the chorionic villi, very close to (but not mixing with) the maternal blood just external to these villi. Nutrients, wastes, and other substances begin diffusing between the two separate bloodstreams, across the villi. This exchange is occurring by the end of the first month of pregnancy.

Formation of the placenta is completed during months 2 and 3. The chorionic villi that lie nearest the umbilical cord grow in complexity, whereas those around the rest of the embryo regress and ultimately disappear. The part of the mother's endometrium adjacent to the complex chorionic villi and umbilical cord is called the **decidua basalis** (de-sid′u-ah), whereas the endometrium opposite this, on the uterine-luminal side of the implanted embryo, is the **decidua capsularis** (Figure 25.25d). The decidua capsularis expands to accommodate the growing embryo, which is called a fetus after month 2, and completely fills the uterine lumen by the start of month 4. At this latter time, the decidua basalis and chorionic villi together make up a thick, pancake-shaped disc at the end of the umbilical cord—the **placenta** ("cake") (Figure 25.25e)—which continues to nourish the fetus for 6 more months, until birth. Because it detaches and sloughs off from the uterus after birth, the name of its maternal portion—*decidua* ("that which falls off")—is appropriate.

It is worth emphasizing that the placenta is not given this name until it achieves its pancake shape at the start of month 4 (week 13), even though its chorionic villi are established and functioning long before this time.

Exchange Across the Placenta

The exchanges that occur across the chorionic villi between maternal and fetal blood are absolutely essential to prenatal life. These exchanges provide the fetus with nutrients and oxygen, dispose of its wastes, and allow the fetus to send hormonal signals to the mother. To travel between maternal and fetal blood, substances must pass through the *placental barrier,* which consists of all three layers of each chorionic villus—syncytiotrophoblast, cytotrophoblast, extraembryonic mesoderm (Figure 25.25c)—plus the endothelium of the fetal capillaries in the villi (Figure 25.26). The cytotrophoblast and mesoderm layers are soon reduced to mere traces, thereby thinning the barrier and allowing a very efficient exchange of molecules.

Many kinds of molecules pass across the chorionic villi of the placenta, by several mechanisms. Sugars, fats, and oxygen diffuse from mother to fetus, whereas urea and carbon dioxide diffuse from fetus to mother. Maternal antibodies are actively transported across the placenta and give the fetus some resistance to disease. The placenta is effective at blocking passage of most bacteria from mother to fetus, but many viruses (German measles virus, chicken pox virus, mononucleosis virus, and sometimes HIV) pass through, as do many drugs and toxins (alcohol, heroin, mercury). Because the placenta is not an effective barrier to drugs, the mother must minimize her intake of harmful substances during pregnancy to minimize the chances of birth defects (see Chapter 3).

The syncytiotrophoblast secretes many substances that reach the mother's blood and regulate the events of pregnancy. One of these substances is an enzyme that "turns off" the mother's T lymphocytes, which prevents her immune system from rejecting the fetus. Most of these substances, however, are hormones. Some of these hormones (progesterone and human chorionic gonadotropin, or HCG) maintain the uterus in its pregnant state, and others, secreted in the later stages of pregnancy (estrogens and corticotropin-releasing hormone), promote labor. The production of HCG begins shortly after the syncytiotrophoblast forms, a mere 1 week after fertilization; its presence in the urine of a pregnant woman is the basis of most home pregnancy tests.

> **CLINICAL APPLICATION**
>
> **Displaced Placenta Placenta previa** (pre′ve-ah; "appearing in front of") is a condition in about 1 in 200 pregnancies, in which the embryo implants in the inferior part—rather than in the usual superior part—of the uterine wall. The placenta may cover the internal os of the cervix. Placenta previa is usually associated with bleeding during the last 3 months of pregnancy as some of the placenta pulls away from the uterus. To limit the bleeding, bed rest is prescribed. Placenta previa differs from **placental abruption,** in which the placenta is in a normal position but becomes partly separated from the uterine wall before birth. This disorder, which is up to eight times more common than placenta previa, also produces vaginal bleeding during pregnancy. Placental abruption can interfere with fetal development by reducing the delivery of nutrients and oxygen to the fetus.

Childbirth

Parturition (par″tu-rish′un; "bring forth young"), the act of giving birth, occurs an average of 266 days after fertilization and 280 days after the last menstrual period. The events that expel the infant from the uterus are referred to as **labor,** the initiation of which is complex.

Labor involves three successive stages: the dilation, expulsion, and placental stages (Figure 25.27). The **dilation (first) stage** (Figure 25.27(1a) and (1b)) begins with the first regular contractions of the uterus and ends when the cervix is fully dilated (about 10 cm in diameter) by the baby's head. The dilation stage is the longest part of labor, lasting 6–12 hours or more.

The **expulsion (second) stage** (Figure 25.27(2)) lasts from full dilation to delivery, or actual childbirth. The uterine

(1a) **Early dilation.** Baby's head enters the pelvis; widest dimension is along left-right axis.

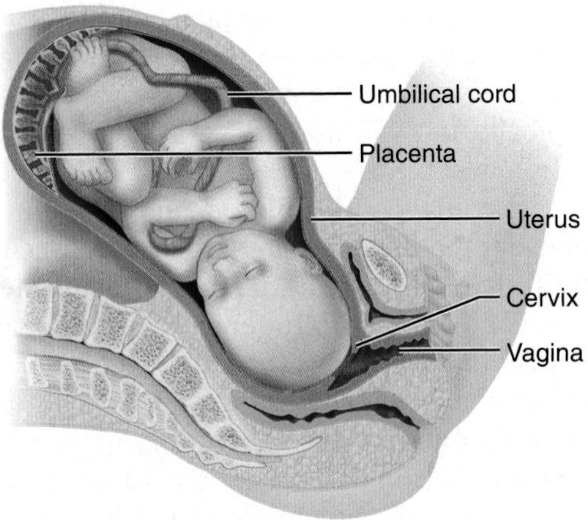

(1b) **Late dilation.** Baby's head rotates so widest dimension is in anteroposterior axis (of pelvic outlet). Dilation nearly complete.

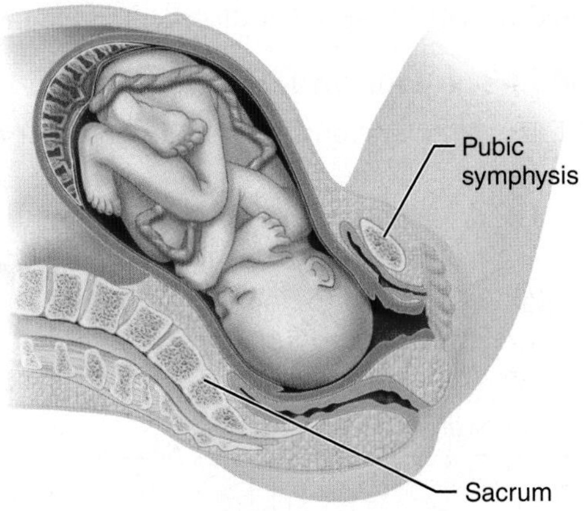

(2) **Expulsion.** Baby's head extends as it is delivered.

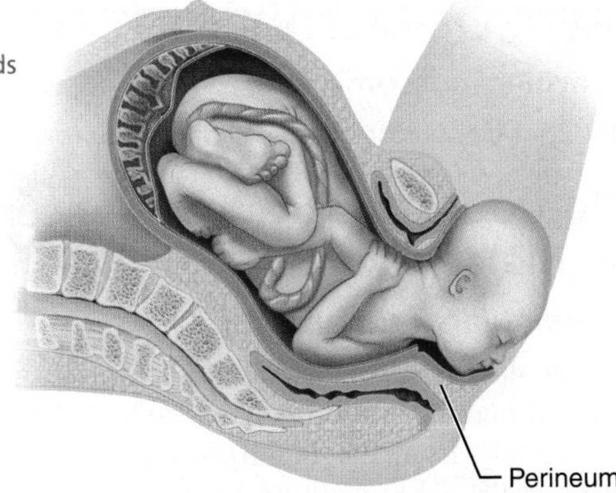

(3) **Placental stage.** After baby is delivered, the placenta detaches and is removed.

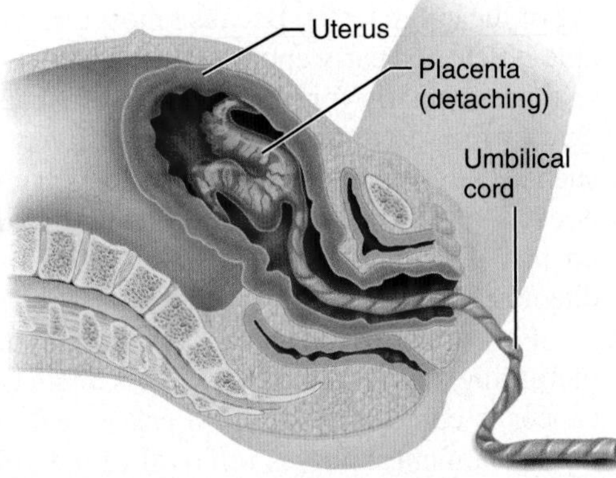

Figure 25.27 **Stages of labor.**

contractions become stronger during this stage, and the mother has an increasing urge to bear down with her abdominal muscles. Although this phase can take up to 2 hours, it typically lasts 50 minutes in a first birth and 20 minutes in subsequent births.

The **placental (third) stage** (Figure 25.27③), or the delivery of the placenta, is accomplished within 15 minutes after birth of the infant. Forceful uterine contractions that continue after birth compress the uterine blood vessels, which limits bleeding and causes the placenta to detach from the uterine wall. The placenta (now referred to as *afterbirth*) is easily removed by a slight tug on the umbilical cord. The blood vessels in the umbilical cord are then counted to verify the presence of two umbilical arteries and one umbilical vein because the absence of one umbilical artery is often associated with cardiovascular disorders in the infant.

✓ check your understanding

- ☐ 18. What prevents more than one sperm from fertilizing an ovum?
- ☐ 19. What part of the placenta is formed from maternal tissue? What part is formed from fetal tissue?
- ☐ 20. In addition to nourishing the developing fetus, the placenta also functions as an endocrine organ. What hormones does the placenta produce?

(For answers, see Appendix B.)

DISORDERS OF THE REPRODUCTIVE SYSTEM

learning outcome

▶ Consider various cancers of the reproductive organs.

Reproductive System Cancers in Males

Testicular Cancer

Testicular cancers, which affect approximately 1 of every 50,000 males, most often between ages 15 and 35, arise most commonly from the rapidly dividing, early-stage spermatogenic cells. The resulting tumors are firm, usually painless lumps, although diffuse pain may be associated with a swelling of the entire testis. Because most testicular cancers are curable if detected early enough, men are advised to examine their testicles regularly for lumps, which are readily felt through the skin of the scrotum. Although most lumps discovered during such self-examinations are not cancers but relatively harmless pockets of fluid called varicoceles and hydroceles (discussed on p. 747 and in "Related Clinical Terms," p. 780), proper diagnosis includes ruling out testicular cancers.

For unknown reasons, testicular cancer is becoming increasingly common. Despite its increased incidence, this cancer is cured in 95% of all cases, and death rates are falling. Treatment involves removal of the affected testis, radiotherapy, and chemotherapy.

Prostate Cancer

Prostate cancer is a slow-growing cancer that arises from the main glands in that organ. Although usually symptomless in the early stages, eventually the cancer can grow sufficiently to impinge on the urethra, and the flow of urine may be blocked. After reaching a certain size, prostate cancer metastasizes to the bony pelvis, pelvic lymph nodes, and axial skeleton, producing pain in the pelvis, the lower back, and the bones. Metastasis is most rapid in the relatively few men who develop prostate cancer before age 60.

Prostate cancer is the second most common cause of cancer death in men (after lung cancer), killing 3% of all men in the United States. Risk factors include a fatty diet and genetic predisposition; indeed, prostate cancer has a more prominent hereditary component than most other cancers.

Digital rectal examination and ultrasound imaging with a device inserted into the rectum can detect prostate cancer. A blood test measuring levels of prostate-specific antigen (PSA) can detect prostate cancer in the early stages. Broad use of the PSA test, however, is controversial. False positives occur in two-thirds of those tested (elevated PSA, but no cancer), and the PSA can miss cases of prostate cancer. When PSA levels are elevated, cancer is diagnosed by biopsy of the prostate. PSA screening decisions should be made jointly by the patient and provider based on the patient's history and an understanding of uncertainties and benefits associated with this screening test.

If the cancer is detected before metastasis, radical prostatectomy (complete removal of the prostate) or radiation therapy (including inserting rice-sized grains of radioactive material directly into the gland) is highly effective—85% of patients are cancer-free after 10 years. If metastasis has occurred, however, no effective control measures are available. Prostate cancer, however, usually grows so slowly that many patients die of other diseases or conditions related to old age. Overall, prostate cancer kills about 10% of the men who develop it.

Reproductive System Cancers in Females

Ovarian Cancer

Ovarian cancer typically arises from cells in the germinal epithelium covering the ovary. It affects 1.4% of women and is the fifth most common cause of cancer death in women. This cancer produces few symptoms until it enlarges sufficiently to produce either a feeling of pressure in the pelvis or changes in bowel or bladder habits. Metastasis occurs either via lymphatic and blood vessels or by direct spread to nearby structures.

Women who have the most ovulations over their lifetime (that is, women who have neither had children nor used oral contraceptives) have the highest risk for ovarian cancer, perhaps because the extensive cell division involved in repairing the site of a ruptured (ovulated) follicle each month predisposes the germinal epithelium cells to cancer.

Diagnosis may involve feeling a mass during a pelvic exam, visualizing it by using an ultrasound probe placed

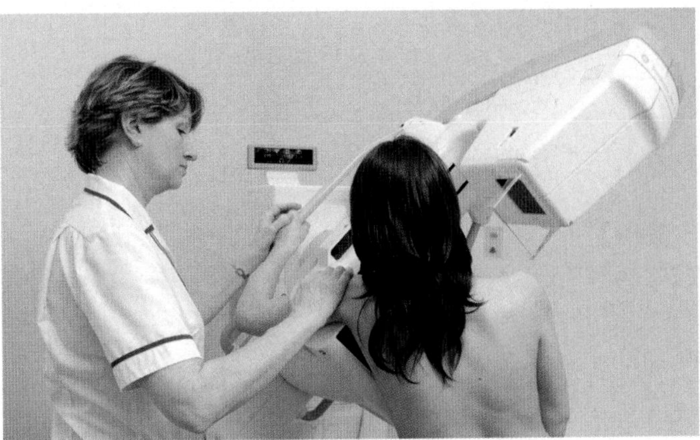

(a) Mammogram procedure

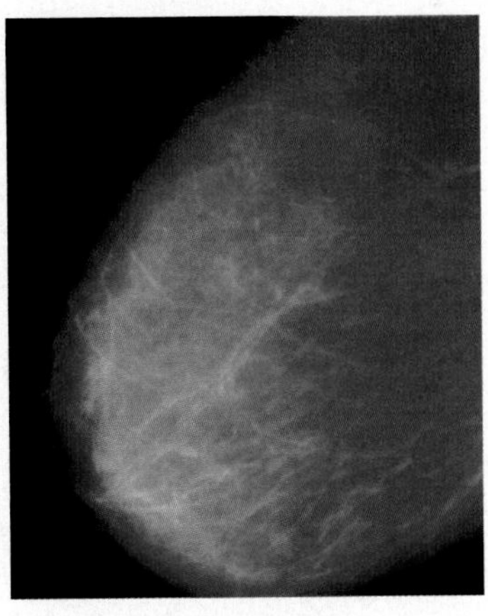

(b) Film of normal breast

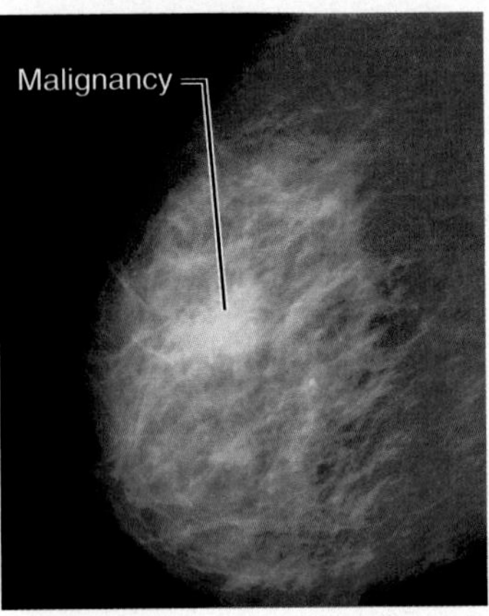

(c) Film of breast with tumor

Figure 25.28 **Mammograms.**

in the vagina, or using a blood test for a protein associated with ovarian cancer. Treatment is by surgical removal of the ovary, uterine tubes, and uterus, followed by radiotherapy and chemotherapy. Because early detection is difficult and the prognosis is poor if metastasis has occurred, the overall 5-year survival rate following ovarian cancer is under 40%.

Endometrial Cancer

Endometrial cancer, which arises from the endometrium of the uterus (usually from the uterine glands), is the fourth most common cancer of women (after lung, breast, and colorectal cancer); about 2% of all women develop it. The most important sign of endometrial cancer, bleeding from the vagina, allows early detection. Risk factors include obesity and postmenopausal estrogen-replacement therapy (because estrogens stimulate the growth and division of endometrial cells). Diagnosis typically involves the use of an ultrasound device inserted into the vagina to detect endometrial thickening, followed by an endometrial biopsy. Treatment involves removal of the uterus, commonly followed by pelvic irradiation. Because it is often detected early, endometrial cancer has a relatively high cure rate of 40% to 95%.

Cervical Cancer

Cervical cancer, which usually appears between the ages of 30 and 50 and occurs in about 1% of U.S. women, is a slow-growing cancer that arises from the epithelium covering the tip of the cervix. Ninety percent of cases are caused by sexually transmitted human papillomavirus (HPV). A **Papanicolaou (Pap) smear,** or *cervical smear test,* is the most effective way to detect this cancer in its earliest (precancerous) stage. In this screening test, some cervical epithelial cells are scraped off and examined microscopically for abnormalities. If cancerous or precancerous cells are confined to the cervix, they are excised or removed by freezing or a laser; in such cases survival rates are very high (over 95%). When the cancer has spread throughout the pelvic organs, radiation plus chemotherapy appears to produce 5-year survival rates of over 70%. A vaccine, marketed under the name Gardasil®, is now available to prevent infection by many types of HPV. Vaccination is encouraged for adolescent girls and young women to prevent this sexually transmitted infection and protect them from developing cervical cancer. The vaccine does not protect against all types of HPV that can cause cervical cancer, and thus regular Pap screening is still necessary. Because of the importance of early detection, women over the age of 18 (younger, if sexually active) are advised to have a Pap smear performed every 1–3 years.

Breast Cancer

Breast cancer, the second most common cause of cancer deaths in women and the killer of 3% of U.S. women, typically arises from the smallest ducts in the lobules of the breast. One of every eight women will develop breast cancer in her lifetime. Women are advised to perform monthly breast self-examinations to detect lumps in the breast, a change in breast shape, scaling on the nipple or areola, discharge from the nipple, or dimpling or an orange-peel texture on the breast.

Risk factors for breast cancer include a family history of the disease, late menopause, early onset of puberty, first live birth after age 30, and postmenopausal estrogen-replacement therapy. With the possible exception of family history, all these risk factors reflect an increased lifelong exposure to estrogens, which stimulate division of the duct cells and promote cancer.

Breast cancer is a rapidly spreading cancer that metastasizes from the breast through lymphatic vessels to axillary and parasternal lymph nodes, and then to the chest wall, lungs, liver, brain, and bones. If metastases reach the lungs or liver, average survival time is less than 6 months.

Screening techniques to enable early detection effectively reduce mortality from breast cancer. Women over age 40 are advised to obtain a **mammogram** (X-ray image of the breast) annually (Figure 25.28). Suspicious masses are biopsied and microscopically examined to determine whether they are cancerous. If one or more are, the axillary lymph nodes are removed and examined for cancer cells; the percentage of these axillary nodes that are affected is the best indicator of metastasis and the most reliable predictor of survival. **Radical mastectomy**—the removal of the entire affected

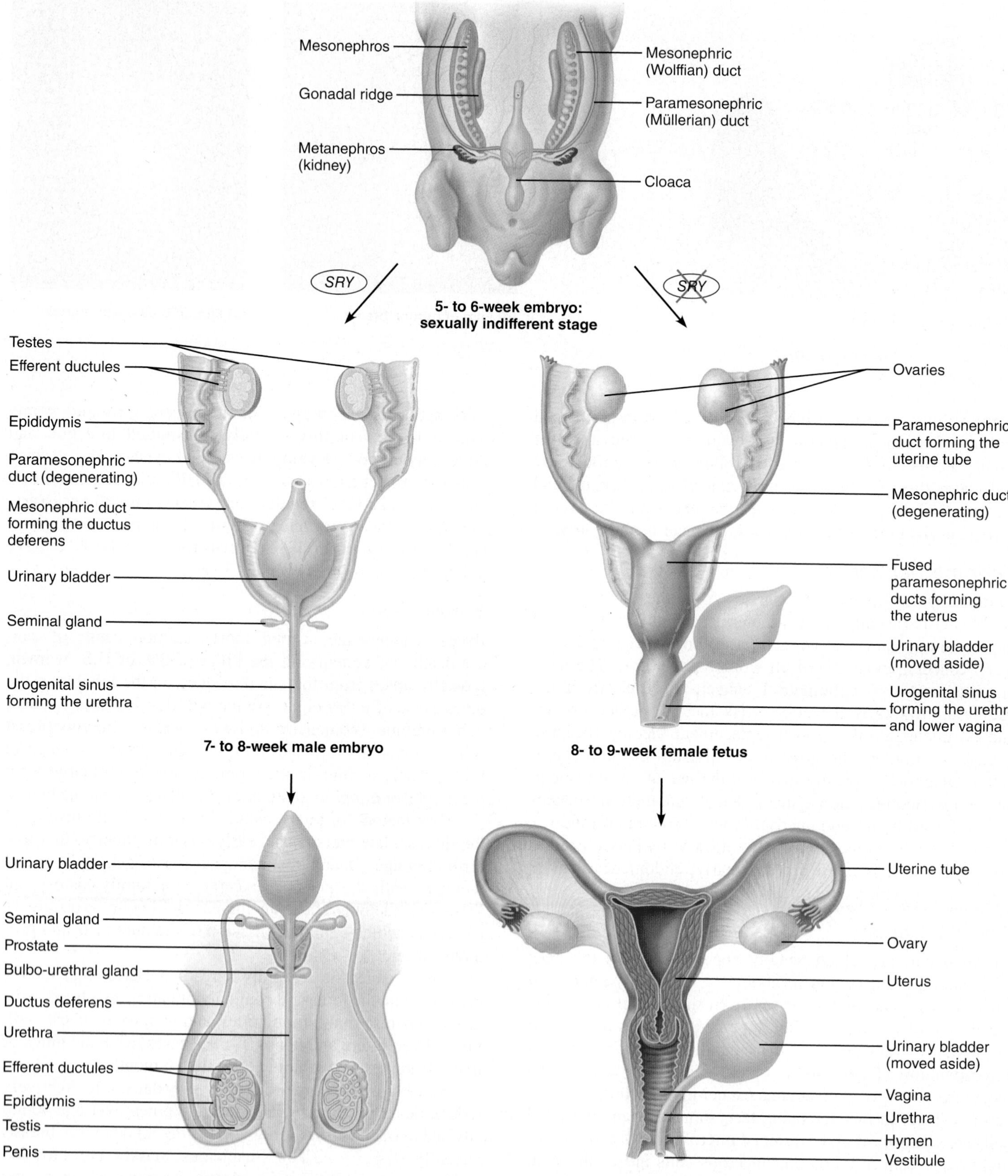

Figure 25.29 **Development of the internal reproductive organs in both sexes.**

breast plus all underlying muscles, fascia, and associated lymph nodes—is performed only in the most advanced cases. Current standard treatment consists of a **lumpectomy**—the removal of the cancerous mass plus a small rim of surrounding tissue—followed by radiation therapy and then administration of hormones (so-called selective estrogen replacement modulators, such as tamoxifen) and chemotherapy.

The importance of early detection through regular mammograms is apparent in the survival statistics. Among patients in which breast cancer has metastasized to almost all the axillary lymph nodes, the 5-year survival rate is about 30%, but when the axillary nodes are not involved, the 5-year survival rate is 78%.

THE REPRODUCTIVE SYSTEM THROUGHOUT LIFE

learning outcomes

- Compare and contrast the prenatal development of the male and female sex organs.
- Note the anatomical changes that occur during puberty and menopause.

Embryonic Development of the Sex Organs

The gonads of both males and females begin to develop during week 5 as masses of intermediate mesoderm called the **gonadal ridges** (Figure 25.29), which form bulges on the dorsal abdominal wall in the lumbar region, just medial to the mesonephros (an embryonic kidney; see p. 737). The **mesonephric (Wolffian) ducts,** the future male ducts, develop medial to the **paramesonephric (Müllerian) ducts,** the future female ducts. Both sets of ducts empty into the cloaca (future bladder and urethra). At this time, the embryo is said to be in the **sexually indifferent stage** because the gonadal ridge and ducts are structurally identical in both sexes.

In the week after the gonadal ridges appear, *primordial germ cells* migrate to them from the yolk sac and seed the developing gonads with germ cells destined to become male spermatogonia or female oogonia. After these germ cells are in place, the gonadal ridges differentiate into testes or ovaries, based on the presence or absence of a male-inducing signal, called SRY protein, that comes from the epithelium covering the ridge.

In *male* embryos, this differentiation begins in week 7 (Figure 25.29 left side). Testes cords (the future seminiferous tubules) grow from the gonadal surface into the inner part of the gonad, where they connect to the mesonephric duct via nephrons that become the efferent ductules. With further development, the mesonephric duct becomes the major male sex ducts: epididymis, ductus deferens, and ejaculatory duct. The paramesonephric ducts play no part in male development and soon degenerate.

In female embryos (Figure 25.29, right side), the process begins slightly later, in about week 8. The outer, or cortical, part of the immature ovaries forms the ovarian follicles. Shortly thereafter, the paramesonephric ducts differentiate into most structures of the female duct system—the uterine tubes, the uterus, and the superior part of the vagina—and the mesonephric ducts degenerate.

Like the gonads, the *external* genitalia develop from identical structures in both sexes (Figure 25.30). During the sexually indifferent stage, both male and female embryos have a small projection called the **genital tubercle** on their external perineal surface. The urogenital sinus of the cloaca (future urethra and bladder) lies deep to the tubercle, and the **urethral groove,** which serves as the external opening of the urogenital sinus, runs between the genital tubercle and the anus. The urethral groove is flanked laterally by the **urethral folds** *(genital folds)* and the **labioscrotal swellings.**

During week 8, the external genitalia begin to develop rapidly. In males (Figure 25.30b), the genital tubercle enlarges, forming most of the penis (the glans and the erectile bodies), and the urethral folds fuse in the midline to form the spongy urethra. The two labioscrotal swellings also join in the midline to become the scrotum. In females (Figure 25.30c), the genital tubercle becomes the clitoris, the unfused urethral folds become the labia minora, and the unfused labioscrotal swellings become the labia majora. The urethral groove persists in females as the vestibule.

CLINICAL APPLICATION

Hypospadias A condition called **hypospadias** (hi″po-spa′de-as) is the most common congenital abnormality of the male urethra. Occuring in nearly 1% of male births, it results from a failure of the two urethral folds to fuse completely, producing urethral openings on the undersurface of the penis. As a result, more urine exits from the underside of the penis than from its tip. This condition, which is becoming more common, is typically corrected surgically within the first 18 months of age.

Descent of the Gonads

In both sexes, the gonads originate in the dorsal body wall of the lumbar region and then descend caudally during the fetal period. In the male fetus, the testes descend toward the scrotum and are followed by their blood vessels and nerves (Figure 25.31a). The testes reach the pelvis near the inguinal region in month 3, when a fingerlike outpocketing of the peritoneal cavity, the **vaginal process,** pushes through the muscles of the anterior abdominal wall to form the inguinal canal. The testes descend no farther until month 7, when suddenly they pass quickly through the inguinal canal and enter the scrotum (Figure 25.31b). After the testes reach the scrotum, the proximal part of each vaginal process closes off, whereas its caudal part becomes the saclike tunica vaginalis (Figure 25.31c). In some boys, however, the vaginal process does not close at all, and this open path to the scrotum constitutes the route for an inguinal hernia into the vaginal process. Indeed, such **congenital inguinal hernias** are the most common hernias.

The causative mechanism of the descent of the testis is unknown, but most theories focus on a fibrous cord called

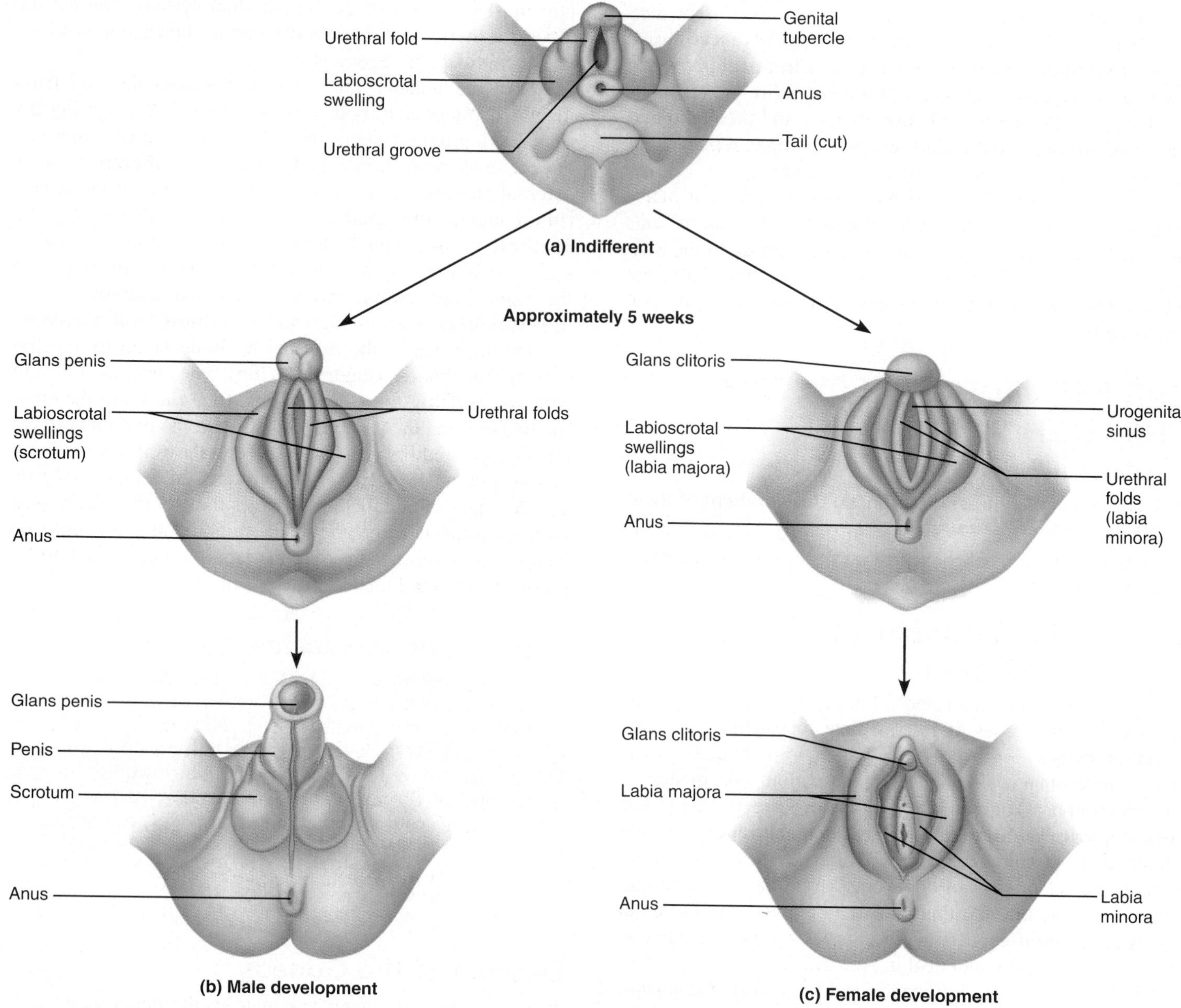

Figure 25.30 Development of homologous structures of the external genitalia in both sexes. The two pictures at the bottom of the figure show the fully developed perineal region.

the **gubernaculum** (gu″ber-nak′u-lum; "governor"), which extends caudally from the testis to the floor of the scrotal sac and is known to shorten (Figure 25.31). The final descent through the inguinal canal may be aided by an increase in intra-abdominal pressure that pushes the testis into the scrotum. In any case, this final descent through the inguinal canal is stimulated by testosterone.

Like the testes, the ovaries descend during fetal development, but only into the pelvis, where the tentlike broad ligament blocks any further descent. Each ovary is guided in its descent by a gubernaculum that is anchored in the labium majus and is homologous to the gubernaculum in the male (see Figure 25.31a). In the female, the inferior part of the gubernaculum becomes the round ligament of the uterus, and its superior part becomes the ovarian ligament (see Figures 25.12 and 25.15).

CLINICAL APPLICATION

Cryptorchidism A congenital condition called **cryptorchidism** (krip-tor′kĭ-dizm; *crypt* = hidden; *orcho* = testis) occurs in about 5% of newborn males in which one or both testes fail to descend into the scrotum. An undescended testis, which may be located in either the inguinal canal (typically) or in the pelvic cavity, is sterile because viable sperm cannot be produced at the higher temperature there. The interstitial cells of these testes, however, do secrete testosterone. Surgical correction within 6–18 months after birth is recommended to avoid damage to sperm-forming cells. Left uncorrected, undescended testes have an increased likelihood of developing testicular cancer.

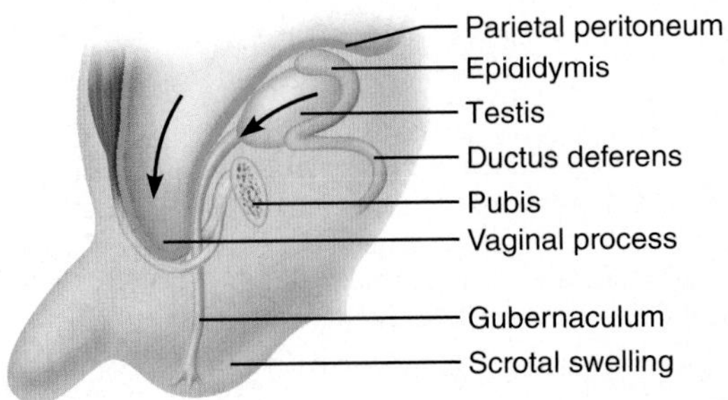

(a) 3-month fetus

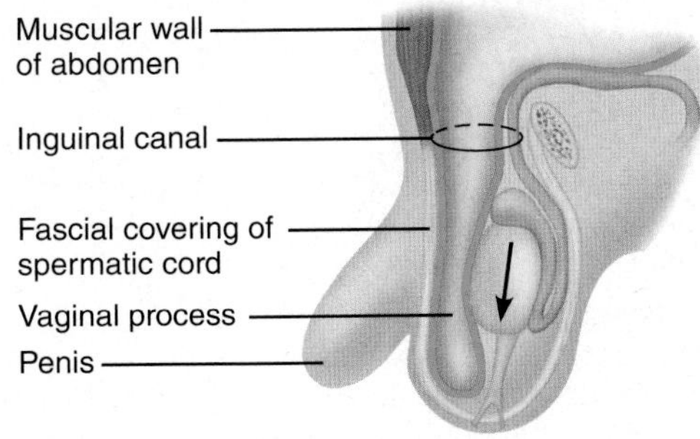

(b) 7-month fetus

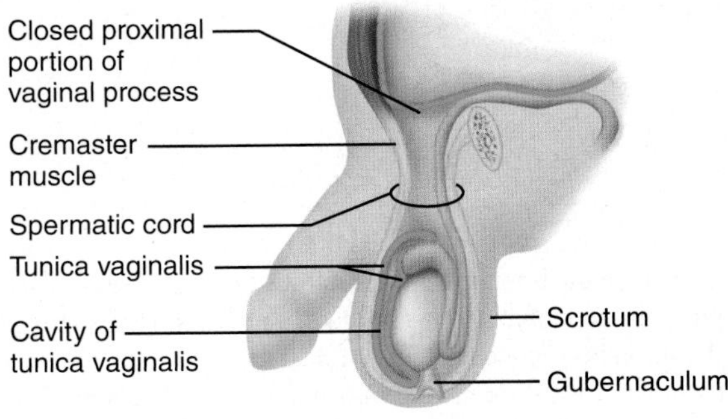

(c) At birth

Figure 25.31 **Descent of the testes.** Sagittal sections through the pelvis of the male fetus.

Puberty

Puberty is the period of life, generally between ages 10 and 15, when the reproductive organs grow to their adult size and reproduction becomes possible. These changes occur under the influence of rising levels of gonadal hormones, testosterone in males and estrogens in females. The events of puberty occur in the same sequence in all individuals, but the age when they occur varies widely.

In males, the first change is an enlargement of the testes and scrotum around age 13. This is followed by the appearance of pubic, axillary, and facial hair, which are **secondary sex characteristics**—features induced in the *nonreproductive* organs by sex hormones. Other secondary sex characteristics include the enlargement of the larynx that deepens the male voice and an increased oiliness of the skin that can lead to acne. The musculoskeletal system also increases in mass at puberty in a growth spurt that lasts up to 6 years. Sexual maturation is evidenced by the presence of mature sperm in the semen.

In females, the first sign of puberty is budding breasts, often apparent by age 11. *Menarche* (me-nar′ke; "month beginning"), the first menstruation, usually occurs 1–2 years later. Dependable ovulation and fertility await the maturation of pituitary hormonal controls, which takes nearly 2 more years. In the meantime, the estrogen-induced secondary sexual characteristics appear at around age 13, including an increase in subcutaneous fat, especially in the hips and breasts, and widening and lightening of the bones of the pelvic girdle, which are adaptations that facilitate childbirth. Furthermore, the ovaries secrete some androgens that signal the development of pubic and axillary hair, apocrine sweat glands, and oiliness of the skin. The estrogen-induced female growth spurt lasts from about age 12 until age 15 to 17.

Menopause

Most women, physiologically, reach their reproductive peak in their late 20s. Around age 35, the rate at which follicles degenerate in the ovaries accelerates. Later, when few follicles remain in the ovaries, ovulation and menstruation cease. This event, called **menopause,** normally occurs between ages 46 and 54. The cause of menopause is unclear: It may be induced by the exhaustion of follicles from the ovaries, or alternatively, it may be signaled by the brain.

After menopause, the ovaries stop secreting estrogens and testosterone. Deprived of estrogens, the reproductive organs and breasts start to atrophy. As the vagina becomes dry, intercourse may become painful, and vaginal infections become more common. The skin thins gradually, and bone mass declines, sometimes resulting in osteoporosis. Slowly rising blood cholesterol levels place postmenopausal women at risk for cardiovascular disorders.

There is no equivalent of menopause in men. Instead, starting at age 40, testosterone production begins a slow decline of about 1% per year. As a result, bone and muscle mass decrease gradually, as does sperm formation—although not enough to prevent some elderly men from fathering children. Erection becomes less efficient as the amount of connective tissue in the penile erectile bodies increases with age.

☑ check your understanding

- ☐ 21. What type of screening is most common for each cancer listed here: (a) prostate cancer, (b) cervical cancer, (c) breast cancer, (d) testicular cancer?
- ☐ 22. What do the embryonic mesonephric ducts form in a male embryo? What do the paramesonephric ducts form in a male embryo?
- ☐ 23. What are secondary sex characteristics in males and females, and what stimulates their development?

(For answers, see Appendix B).

RELATED CLINICAL TERMS

Fibroids Slow-growing, benign tumors in the wall of the uterus made of smooth muscle cells and fibrous connective tissue. Fibroids occur in about 20% of all women over 30. Their precise cause is unknown but may relate to an atypical response to estrogens. Small fibroids produce no symptoms, but large ones can cause pressure and pain in the pelvis, heavy menstrual periods, and infertility. A conservative treatment is myomectomy, removal of the fibroid from its capsule while preserving the rest of the uterus. However, if the fibroids are abundant, the whole uterus may require removal.

Hydrocele (hi′dro-sēl) (*hydro* = water; *cele* = sac) A swelling in the scrotum caused by an excessive accumulation of fluid in the cavity of the tunica vaginalis; may result from an infection or injury to the testis that causes the layers of the tunica vaginalis to secrete excess serous fluid. Most hydroceles are small and do not require treatment. Large hydroceles can be treated by first aspirating the fluid with a needle inserted into the cavity of the tunica vaginalis and then removing the serous lining of the tunica vaginalis.

Hysterectomy (his″tē-rek′to-me) (*hyster* = uterus; *ectomy* = cut out) Surgical removal of the uterus; a very common operation, typically performed to remove fibroids (25% to 30%), to stop abnormal uterine bleeding (20%), as a treatment for endometriosis (under 20%), and to relieve uterine prolapse (4% to 17%) and chronic pelvic pain (12% to 18%). The uterus can be removed either through the anterior abdominal wall or the vagina. In response to evidence that hysterectomies are performed too often, standardized criteria for deciding when they are appropriate have been developed by the American College of Obstetricians and Gynecologists.

Orchitis (or-ki′tis) (*orcho* = testis) Inflammation of the testes, sometimes caused by the mumps virus. If testicular infection spreads to the epididymis, it is called epididymo-orchitis.

Testicular torsion Twisting of the spermatic cord and testis within the scrotum, which can starve the testis of its blood supply and produce sudden urgent pain, swelling, and reddening of the scrotum. It can result from trauma or a spasm of the cremaster muscle and is most common in those males in whom the testis, instead of attaching to the posterior scrotal wall, hangs freely. The twist can be diagnosed by various imaging techniques, but surgery to undo it is often performed without waiting for imaging.

Tubal ligation Having one's uterine tubes "tied" surgically as a form of birth control. Usually involves threading a cutting instrument and a viewing tube (laparoscope) through a small incision in the anterior abdominal wall and then tying and/or cutting the uterine tubes. Tubal ligation usually is irreversible, although in some cases delicate microsurgery can successfully reopen or reattach the uterine tubes.

CHAPTER SUMMARY

Take full advantage of the invaluable online practice and study tools for this chapter by going to the Study Area of MasteringA&P®. There you will find:

- Art Labeling Activities, Quizzes, Practice Tests, Memory Games
- Animal Organ Dissection, Cat Dissection and Bone Videos
- PAL Cadaver
- PAL Histology
- A&P Flix Animations
- Dynamic Study Modules

1. The function of the reproductive system is to produce offspring. The gonads produce gametes (sperm or ova) and sex hormones. All other reproductive organs are accessory organs.

THE MALE REPRODUCTIVE SYSTEM (pp. 744–755)

The Testes (pp. 744–747)

2. The scrotum is the sac of skin and subcutaneous tissue that contains the testes. Housing the testes outside of the body cavity keeps the testes at a slightly lower temperature than the body, as required for viable sperm production. The dartos muscle, cremaster muscle, and pampiniform plexus help regulate the temperature of the testes in the scrotum.

3. Each testis, which is partly surrounded by a serous sac called the tunica vaginalis (derived from the peritoneum), is divided into many lobules that contain sperm-producing seminiferous tubules separated by areolar connective tissue.

4. The thick epithelium of the seminiferous tubule consists of spermatogenic cells (spermatogonia, primary and secondary spermatocytes, and spermatids), embedded in columnar sustentocytes. Spermatogenic cells move toward the tubule lumen as they differentiate into sperm by a process called spermatogenesis.

5. Sustentocytes (Sertoli cells) form the blood-testis barrier, nourish the spermatogenic cells, and move them toward the lumen. They also secrete testicular fluid (for sperm transport), androgen-binding protein, and the hormone inhibin.

6. Myoid cells are smooth-muscle–like cells that surround the seminiferous tubules and help to squeeze sperm through the tubules.

7. Interstitial (Leydig) cells, oval cells in the connective tissue between seminiferous tubules, secrete androgens (mostly testosterone) under the influence of luteinizing hormone from the pituitary gland.

Male Reproductive Ducts (pp. 747–750)

8. From the seminiferous tubules, sperm travel through the straight tubules and rete testes, then out of the testes through the efferent ductules into the epididymis.

9. The comma-shaped epididymis hugs the posterolateral surface of the testis. The duct of the epididymis, lined with a pseudostratified columnar epithelium with stereocilia, is where sperm gain the ability to swim and fertilize. Ejaculation begins with contraction of smooth muscle in the duct of the epididymis.

10. The ductus (vas) deferens extends from the epididymis in the scrotum to the ejaculatory duct in the pelvic cavity. During ejaculation, the thick layers of smooth muscle in its wall propel sperm into the urethra by peristalsis.

11. The fascia-covered spermatic cord contains the ductus deferens and the testicular vessels. It runs through the anterior abdominal wall from the scrotum to the superficial inguinal ring and then continues through the inguinal canal to the deep inguinal ring.

12. The urethra, which runs from the bladder to the tip of the penis, conducts semen and urine to the body exterior.

Accessory Glands (pp. 750–751)

13. The seminal glands (vesicles)—long, pouched glands posterior to the bladder—secrete a sugar-rich fluid that constitutes 60% of the ejaculate.

14. The prostate, surrounding the prostatic urethra, is a group of compound glands embedded in a fibromuscular stroma. Its secretion helps to clot, then to liquefy, the semen. Benign prostatic hyperplasia, which affects many men, arises from the deep glands in the prostate.

15. The bulbo-urethral glands in the urogenital diaphragm secrete mucus into the urethra before ejaculation to lubricate and neutralize this tube for the passage of semen.

The Penis (pp. 751–752)

16. The penis has a root, a body, and a glans covered by foreskin. Its main nerves and vessels lie dorsally. Most of its mass consists of vascular erectile bodies—namely, a corpus spongiosum and two corpora cavernosa. Engorgement of these bodies with blood causes erection.

The Male Perineum (p. 752)

17. In the male, the diamond-shaped perineum contains the scrotum, the root of the penis, and the anus.

Spermatogenesis (pp. 752–755)

18. The three stages of spermatogenesis are (1) formation of spermatocytes, (2) meiosis, and (3) spermiogenesis.

19. In spermiogenesis, round spermatids become tadpole-shaped sperm cells as an acrosome forms, a flagellum grows, the nucleus flattens as its chromatin condenses, and the superfluous cytoplasm is shed.

THE FEMALE REPRODUCTIVE SYSTEM (pp. 755–768)

20. The female reproductive system produces ova and sex hormones and houses, nourishes, protects, and delivers an infant. It undergoes a menstrual cycle of about 28 days.

The Ovaries (pp. 755–757)

21. The almond-shaped ovaries lie on the lateral walls of the pelvic cavity and are suspended by various mesenteries and ligaments.

22. The mesovarium (mesentery of the ovary) is part of a fold of peritoneum called the broad ligament. Other parts of the broad ligament are the mesosalpinx and mesometrium. Contained within the broad ligament are the suspensory ligament of the ovary and its continuation, the round ligament of the uterus.

23. Each ovary is divided into an outer cortex (follicles containing oocytes separated by connective tissue) and an inner medulla (connective tissue containing the main ovarian vessels and nerves). One follicle matures each month, releasing one oocyte from puberty to menopause.

The Uterine Tubes (pp. 757–758)

24. Each uterine tube extends from an ovary to the uterus. Its regions, from lateral to medial, are the infundibulum, ampulla, and isthmus. Fimbriae extending from the infundibulum create currents that help draw an ovulated oocyte into the uterine tube.

25. The wall of the uterine tube contains a muscularis layer and a folded inner mucosa with a simple columnar epithelium. Both the smooth muscle and ciliated epithelial cells help propel the oocyte toward the uterus.

The Uterus (pp. 758–760)

26. The hollow uterus, shaped like an inverted pear, has a fundus, a body, an isthmus, and a cervix. Cervical glands fill the cervical canal with a bacteria-blocking mucus.

27. The uterus is supported by the broad, lateral cervical, and round ligaments. Most support, however, comes from the muscles of the pelvic floor.

28. The uterine wall consists of the outer perimetrium, the muscular myometrium, and a thick mucosa called the endometrium. The myometrium functions to squeeze the baby out during childbirth. The endometrium consists of a functional layer, which sloughs off each month (except in pregnancy), and an underlying basal layer, which replenishes the functional layer.

Oogenesis and the Female Menstrual Cycle (pp. 760–766)

29. Oogenesis, the production of female gametes, starts before birth and takes decades to complete. The stem cells (oogonia) appear in the ovarian follicles of the fetus. Primary oocytes stay in meiosis I until ovulation occurs years later, and each secondary oocyte then stays in meiosis II until it is penetrated by a sperm.

30. The ovarian cycle is responsible for follicle maturation and ovulation. The uterine cycle prepares the uterus for implantation of a fertilized egg.

31. The ovarian cycle has three phases: the follicular phase, ovulation, and the luteal phase. The ovaries of a newborn female contain thousands of primordial follicles, each of which consists of an oocyte surrounded by a layer of flat follicular cells.

32. From puberty to menopause during the follicular phase of each monthly ovarian cycle (days 1–14), 6–12 follicles start maturing. The follicles progress from primordial follicles to primary follicles, to antral follicles, to mature ovarian follicles that are ready for ovulation. Beyond the smallest stages, this progression is stimulated by follicle-stimulating hormone (FSH) from the anterior part of the pituitary gland. Generally, only one follicle per month completes the maturation process.

33. The theca folliculi around a growing follicle cooperates with the follicle itself to produce estrogens from androgens under the stimulatory influence of luteinizing hormone (LH) and FSH. Estrogens stimulate the growth and activity of all female sex organs and signal the proliferative stage of the uterine cycle.

34. Upon stimulation by LH, ovulation occurs at midcycle (day 14), releasing the oocyte from the ovary into the peritoneal cavity.

35. In the luteal phase of the ovarian cycle (days 15–28), the ruptured follicle remaining in the ovary becomes a wavy corpus luteum that secretes progesterone and estrogens. Progesterone functions to maintain the secretory phase of the uterine cycle. If fertilization does not occur, the corpus luteum degenerates in about 2 weeks into a corpus albicans (scar).

36. The uterine cycle has three phases: the menstrual, proliferative, and secretory phases. The first two phases are a shedding and then a rebuilding of endometrium in the 2 weeks before ovulation. The third phase prepares the endometrium to receive an embryo in the 2 weeks after ovulation.

The Vagina (p. 766)

37. The vagina, a highly distensible tube that runs from the uterus to the body's exterior, receives the penis and semen during intercourse and acts as the birth canal.

38. The layers of the vaginal wall are an outer adventitia, a middle muscularis, and an inner mucosa consisting of an elastic lamina propria and a stratified squamous epithelium. The lumen is acidic.

39. The vaginal fornix is a ringlike recess around the tip of the cervix in the superior vagina.

The External Genitalia and Female Perineum (p. 767)

40. The female external genitalia (vulva) include the mons pubis, the labia majora and minora, the clitoris with its erectile bodies, and the vestibule containing the vaginal and urethral orifices. The mucus-secreting greater vestibular glands and the bulbs of the vestibule (erectile bodies) lie just deep to the labia.

41. The central tendon of the perineum lies just posterior to the vaginal orifice and the fourchette.

The Mammary Glands (pp. 767–768)

42. The mammary glands develop from the skin of the embryonic milk lines. Internally, each breast consists of 15–25 lobes (compound alveolar glands) that secrete milk. The lobes, which are subdivided into lobules and alveoli, are separated by adipose tissue and by supportive suspensory ligaments. The full glandular structure of the breast does not develop until the second half of pregnancy. Breast cancers usually arise from the duct system.

PAL **Human Cadaver/Reproductive System**

PREGNANCY AND CHILDBIRTH (pp. 768–774)

Pregnancy (pp. 768–773)

43. Fertilization usually occurs in the ampulla of a uterine tube. Many sperm must release enzymes from their acrosomes to penetrate the oocyte's zona pellucida. Entry of one sperm into the oocyte prevents the entry of others.

44. Once fertilized, the zygote divides to form a multicellular embryo, which travels to the uterus and implants in the endometrium. As it implants, its trophoblast divides into a cytotrophoblast and an invasive syncytiotrophoblast. Lacunae open in the syncytiotrophoblast and fill with blood from endometrial vessels, forming the first contact between maternal blood and embryonic tissue.

45. The placenta acts as the respiratory, nutritive, and excretory organ of the fetus and produces the hormones of pregnancy. It is formed both from embryonic tissue (chorionic villi) and from maternal tissue (endometrial decidua).

46. The placental exchange barrier consists of the layers of the chorionic villi plus the endothelium of the fetal capillaries inside those villi. Of these layers, the syncytiotrophoblast secretes the placental hormones. No direct mixing of maternal and fetal blood occurs across a healthy placenta.

Childbirth (pp. 773–774)

47. The three stages of labor are the dilation, expulsion, and placental.

DISORDERS OF THE REPRODUCTIVE SYSTEM (pp. 774–775, 777)

48. The most serious cancers of the reproductive organs are prostate cancer and breast cancer.

THE REPRODUCTIVE SYSTEM THROUGHOUT LIFE (pp. 776–779)

Embryonic Development of the Sex Organs (p. 776, 777)

49. The gonads of both sexes develop in the dorsal lumbar region from the gonadal ridges of intermediate mesoderm. The primordial germ cells migrate to the gonads from the endoderm of the yolk sac. The mesonephric ducts form most of the ducts and glands in males, whereas the paramesonephric ducts form the uterine tubes and uterus of the female.

50. The external genitalia arise from the genital tubercle (penis and clitoris), urethral folds (spongy urethra and labia minora), and labioscrotal swellings (scrotum and labia majora).

Descent of the Gonads (pp. 777–779)

51. The testes form in the dorsal abdomen and descend into the scrotum. The ovaries also descend, but only into the pelvis.

Puberty (p. 779)

52. Puberty is the interval when the reproductive organs mature and become functional. It starts in males with penile and scrotal growth and in females with breast development.

Menopause (pp. 779)

53. During menopause, ovarian function declines, and ovulation and menstruation cease.

REVIEW QUESTIONS

Multiple Choice/Matching Questions

(For answers, see Appendix B.)

1. Which of the following statements about the female vestibule is correct? (a) The vaginal orifice is the most anterior of the two orifices in the vestibule. (b) The urethra is between the vaginal orifice and the anus. (c) The anus is between the vaginal orifice and the urethra. (d) The urethral orifice is anterior to the vaginal orifice.

2. For each of the following hollow organs, choose the correct lining epithelium from the key. (Some linings may be used more than once.)

	Key:
____ (1) duct of the epididymis	(a) simple squamous
____ (2) uterine tube	(b) simple cuboidal
____ (3) ductus deferens	(c) simple columnar
____ (4) uterus	(d) pseudostratified columnar
____ (5) vagina	(e) stratified squamous

3. Which of the following structures produces the male sex hormones? (a) seminal glands, (b) seminiferous tubules, (c) developing follicles of the testes, (d) interstitial cells.

4. The uterine cycle can be divided into three continuous phases. Starting from the first day of the cycle, their order is (a) menstrual, proliferative, secretory; (b) menstrual, secretory, proliferative; (c) secretory, menstrual, proliferative; (d) proliferative, menstrual, secretory; (e) proliferative, secretory, menstrual.

5. Which of the following statements about luteinizing hormone and follicle-stimulating hormone is *false*? They are (a) secreted by the pituitary gland, (b) abbreviated as LH and FSH, (c) hormones with important functions in both males and females, (d) the sex hormones secreted by the gonads.

6. The hormone-secreting layer of the placental barrier is (a) the extraembryonic mesoderm, (b) the lacunae, (c) endothelium, (d) cytotrophoblast, (e) syncytiotrophoblast.

7. The function of the stereocilia in the duct of the epididymis and in the ductus deferens is (a) absorption and secretion, (b) beating forward to move sperm through, (c) initiating ejaculation, (d) beating backward to keep the sperm in these ducts for 20 days.

8. The broad ligament of the uterus in the female pelvis (a) is really a mesentery, (b) consists of the mesovarium and mesosalpinx but not the mesometrium, (c) is the same as the round ligament, (d) is the same as the lateral cervical ligament, (e) b and c.

9. The embryonic paramesonephric duct gives rise to the (a) uterus and uterine tubes, (b) ovary, (c) epididymis and vas deferens, (d) urethra.

10. Which of the following statements about the uterine cervix is *false*? (a) It dilates at the end of the first stage of labor. (b) The internal os opens into the vagina. (c) The cervical glands secrete mucus. (d) It contains the cervical canal.

11. If the uterine tube is a trumpet ("salpinx"), what part of it represents the wide, open end of the trumpet? (a) isthmus, (b) ampulla, (c) infundibulum, (d) flagellum.

12. The myometrium is the muscular layer of the uterus, and the endometrium is the (a) serosa, (b) adventitia, (c) submucosa, (d) mucosa.

13. What occurs during the proliferative phase of the uterine cycle? (a) The basal layer of the endometrium thickens. (b) The primary follicles undergo mitosis. (c) The myometrium thickens. (d) The functional layer of the endometrium thickens. (e) The primary spermatocytes form.

14. Which cells in the ovaries secrete hormones? (a) oocytes, (b) zona pellucida, (c) theca and granulosa cells, (d) corpus albicans.

15. Number the ducts through which sperm passes in order beginning with production in the seminiferous tubules (1) and ending with ejaculation out the external urethral orifice (10).

____ **(a)** seminiferous tubules

____ **(b)** epididymis

____ **(c)** straight tubule

____ **(d)** ejaculatory duct

____ **(e)** prostatic urethra

____ **(f)** rete testis

____ **(g)** spongy urethra

____ **(h)** ductus deferens

____ **(i)** intermediate part of urethra

____ **(j)** external urethral orifice

16. The hormones that stimulate the ovarian cycle are secreted by the ____________; the hormones that stimulate the uterine cycle are secreted by the ____________ (a) ovary / uterus, (b) pituitary gland / ovary, (c) follicle cells / functional layer, (d) interstitial cells / placenta, (e) ovary / pituitary gland.

17. Sustentocytes (a) nourish spermatocytes and assist in sperm production, (b) secrete testosterone, (c) secrete fluid that contributes to semen, (d) produce follicle-stimulating hormone.

18. Match the embryonic structures listed in column B with their adult derivatives listed in column A. Some answers may be used more than once.

Column A	Column B
____ **(1)** glans penis	**(a)** genital tubercle
____ **(2)** labia minora	**(b)** mesonephric duct
____ **(3)** epididymis and ductus deferens	**(c)** urethral folds
	(d) paramesonephric duct
____ **(4)** uterine tubes and uterus	**(e)** labioscrotal swellings
____ **(5)** spongy urethra	
____ **(6)** labia majora	

Short Answer Essay Questions

19. Describe the major structural and functional regions of a sperm cell.

20. Ryan was asked to trace the path of the ductus deferens. This response—that it runs from the scrotum directly up to the spongy urethra—was a common error, because the path of this duct is longer and more complex than he thought. (a) Trace the entire path of the ductus deferens. (b) To expand on part (a), trace the pathway of an ejaculated sperm from the epididymis all the way into the uterine tube in the female.

21. In menstruation, the functional layer is shed from the endometrium. Explain the hormonal and physical factors responsible for this shedding. (See Figure 25.19.)

22. (a) Outline the structural changes that a maturing follicle undergoes in the follicular stage of the ovarian cycle. (b) At what time in the ovarian cycle does ovulation occur?

23. Some anatomy students were saying that a man's bulbo-urethral glands (and the urethral glands) act like city workers who come around before a parade and clear the street of parked cars. What did they mean by this analogy?

24. A man swam in a cold lake for an hour and then noticed that his scrotum was shrunken and wrinkled. His first thought was that he had lost his testicles. What had really happened?

25. Indicate whether the following statement is true or false, and explain your answer: In the placenta, maternal and fetal blood come in direct contact with each other and intermix.

26. What is the blood testis barrier, and what is its function?

27. Tell the difference (if any) between each of the following pairs of sound-alike terms: (a) sperm and spermatocyte, (b) epididymis and duct of epididymis, (c) ovarian ligament and suspensory ligament of the ovary, (d) vulva and vestibule, (e) spermatogenesis and spermiogenesis.

28. What are the effects of follicle-stimulating hormone (FSH) and luteinizing hormone (LH) on both male and female reproductive structures?

29. Compare oogenesis with spermatogenesis. How do these processes differ (1) in the number of gametes produced from each primary oocyte or primary spermatocyte and (2) in the period during which mitosis of stem cells occurs?

Critical Reasoning & Clinical Application Questions

1. Why would a radical mastectomy affect the ability to adduct one's arms (as in doing push-ups)? Would a lumpectomy cause a similar deficit? Why or why not? (Try to remember the main muscle used in doing push-ups.)

2. Based on what you know about the sources of semen, explain the effect of a vasectomy on the volume of the ejaculate.

3. A 68-year-old man who has trouble urinating is given a rectal exam. What condition does he probably have, and what is the purpose of the rectal exam?

4. Gina gave birth to six children. During an examination, the doctor sees her cervix protruding from the vaginal orifice. What is Gina's condition, and what has caused it?

5. Why do doctors often recommend that men not wear tight underwear or tight jeans if they want to father children?

6. Explain how a woman with an ovary removed from one side and a uterine tube removed from the other side can get pregnant (exclude the possibility of in vitro fertilization).

7. In severe cases of endometriosis, endometrial tissue is found within the peritoneum. How can endometrial tissue from the uterus end up in the peritoneum?

8. Name three sets of arteries that must be cauterized during a simple mastectomy (removal of only breast tissue).

APPENDIX A

The Metric System

Measurement	Unit and Abbreviation	Metric Equivalent	Metric to English Conversion Factor	English to Metric Conversion Factor
Length	1 kilometer (km)	= 1000 (10^3) meters	1 km = 0.62 mile	1 mile = 1.61 km
	1 meter (m)	= 100 (10^2) centimeters	1 m = 1.09 yards	1 yard = 0.914 m
		= 1000 millimeters	1 m = 3.28 feet 1 m = 39.37 inches	1 foot = 0.305 m
	1 centimeter (cm)	= 0.01 (10^{-2}) meter	1 cm = 0.394 inch	1 foot = 30.5 cm 1 inch = 2.54 cm
	1 millimeter (mm)	= 0.001 (10^{-3}) meter	1 mm = 0.039 inch	
	1 micrometer (μm) [formerly micron (μ)]	= 0.000001 (10^{-6}) meter		
	1 nanometer (nm) [formerly millimicron (mμ)]	= 0.000000001 (10^{-9})meter		
	1 angstrom (Å)	= 0.0000000001 (10^{-10}) meter		
Area	1 square meter (m^2)	= 10,000 square centimeters	1 m^2 = 1.1960 square yards	1 square yard = 0.8361 m^2
			1 m^2 = 10.764 square feet	1 square foot = 0.0929 m^2
	1 square centimeter (cm^2)	= 100 square millimeters	1 cm^2 = 0.155 square inch	1 square inch = 6.4516 cm^2
Mass	1 metric ton (t)	= 1000 kilograms	1 t = 1.103 ton	1 ton = 0.907 t
	1 kilogram (kg)	= 1000 grams	1 kg = 2.205 pounds	1 pound = 0.4536 kg
	1 gram (g)	= 1000 milligrams	1 g = 0.0353 ounce 1 g = 15.432 grains	1 ounce = 28.35 g
	1 milligram (mg)	= 0.001 gram	1 mg = approx. 0.015 grain	
	1 microgram (μg)	= 0.000001 gram		
Volume (solids)	1 cubic meter (m^3)	= 1,000,000 cubic centimeters	1 m^3 = 1.3080 cubic yards	1 cubic yard = 0.7646 m^3
			1 m^3 = 35.315 cubic feet	1 cubic foot = 0.0283 m^3
	1 cubic centimeter (cm^3 or cc)	= 0.000001 cubic meter = 1 millimeter	1 cm^3 = 0.0610 cubic inch	1 cubic inch = 16.387 cm^3
	1 cubic millimeter (mm^3)	= 0.000000001 cubic meter		
Volume (liquids and gases)	1 kiloliter (kl or kL)	= 1000 liters	1 kl = 264.17 gallons	1 gallon = 3.785 L
	1 liter (l or L)	= 1000 milliliters	1 L = 0.264 gallon 1 L = 1.057 quarts	1 quart = 0.946 L
	1 milliliter (ml or mL)	= 0.001 liter	1 ml = 0.034 fluid ounce	1 quart = 946 ml
		= 1 cubic centimeter	1 ml = approx. $\frac{1}{4}$ teaspoon	1 pint = 473 ml
			1 ml = approx. 15–16 drops (gtt.)	1 fluid ounce = 29.57 ml 1 teaspoon = approx. 5 ml
	1 microliter (μl or μL)	= 0.000001 liter		
Time	1 second (s)	= $\frac{1}{60}$ minute		
	1 millisecond (ms)	= 0.001 second		
Temperature	Degrees Celsius (°C)		°F = $\frac{9}{5}$°C + 32	°C = $\frac{5}{9}$°F − 32

APPENDIX B

Answers to Check Your Understanding, Multiple Choice, and Matching Questions

Chapter 1

Check Your Understanding 1. Histology is the microscopic study of tissue structure; radiography is the study of internal body structures from X-ray and other imaging techniques. **2.** pathology: study of disease; hepatitis: inflammation of the liver; brachial: arm; leukocyte: white cell; pneumonia: condition of air. **3.** A tissue is a group of cells and associated extracellular material of similar structure with a common function. The four types of tissue are epithelial tissue, a covering or lining tissue; connective tissue, which functions in support and protection; muscle tissue, which produces movement; and nervous tissue, which allows for internal communication by electrical impulses. **4.** (a) The urinary system eliminates wastes and regulates water and ion balance. (b) The nervous system is a fast-acting control system. (c) The respiratory system supplies blood with oxygen and removes carbon dioxide. **5.** The liver lies inferior to the heart and extends laterally in the upper right quadrant of the abdomen. **6.** The outer tube shows evidence of segmentation. **7.** The parietal pleura is the outer layer of serous membrane that lines the pleural cavity. **8.** In a tissue stained with H&E, the cell nuclei are colored dark blue to purple. **9.** Scanning electron microscopy (SEM) produces three-dimensional images of surface features. **10.** (a) Ultrasound is best for examining the gallbladder in the ER. (b) Broken bones are best visualized using conventional X-ray imaging. (c) MRI is best for visualizing joint structure. (d) A CT scan is best for imaging internal organs in a trauma situation.

Review Questions 1. c **2.** (a) superior; (b) deep; (c) proximal; (d) lateral; (e) medial; (f) superior; (g) anterior **3.** (1) c; (2) e; (3) b; (4) a; (5) d; (6) f **4.** b **5.** (1) I; (2) I; (3) O; (4) O; (5) I; (6) O **6.** (1) D; (2) V; (3) D; (4) V; (5) V **7.** (a) 2; (b) 3; (c) 1; (d) 4 **8.** d **9.** (1) c; (2) a; (3) b; (4) d **10.** c **11.** e

Chapter 2

Check Your Understanding 1. The three regions of the cell are the plasma membrane, the cytoplasm, and the nucleus. **2.** The plasma membrane is composed of lipids (phospholipids), sugars (glycolipids and glycoproteins), and proteins. **3.** Water enters and leaves the cell via osmosis. **4.** Exocytosis carries large macromolecules out of the cell. **5.** Ribosomes and rough ER produce proteins; the Golgi apparatus packages proteins. **6.** The mitochondrion produces energy for cellular activity. **7.** A cell that specialized in phagocytosis would contain numerous lysosomes. **8.** The intermediate filament resists tension. **9.** The nucleolus produces ribosomal RNA and assembles the ribosomal subunits. **10.** Rough endoplasmic reticulum is continuous with the nuclear envelope. **11.** Extended chromatin is thin strings of chromatin that appear as beads (histones) on a string (DNA) in TEM. Extended chromatin is found in active regions of DNA where transcription is occurring. Condensed chromatin is thick, tightly coiled chromatin that is found in inactive regions of DNA. **12.** The cell spends most of its life in interphase. **13.** *Ana* = "apart"; the chromosomes move toward opposite poles of the cell during anaphase. *Meta* = "between, transition"; during metaphase, the chromosomes align in the midline of the cell. *Telo* = "the end"; in telophase, the nuclei of daughter cells reassemble, cytokinesis continues, cell division ends. **14.** Muscle cells have abundant actin and myosin microfilaments. **15.** Cells that produce and secrete hormones contain abundant rough ER, mitochondria, and secretory granules. **16.** According to the mitochondrial theory of aging, reducing caloric intake can increase life span.

Review Questions 1. a **2.** a **3.** b **4.** a **5.** (a) Golgi apparatus; (b) rough ER; (c) condensed; (d) smooth; (e) nucleosome; (f) microtubules; (g) mitochondrion; (h) mitochondrion; (i) ribosome or rough ER **6.** d **7.** b **8.** a **9.** (a) metaphase; (b) prophase; (c) telophase; (d) prophase; (e) anaphase; **10.** (a) microtubules; (b) intermediate filaments; (c) microtubules; (d) actin microfilaments; (e) intermediate filaments; (f) microtubules; (g) actin microfilaments **11.** (1) b; (2) g; (3) c or b; (4) d; (5) e; (6) f; (7) a **12.** b

Chapter 3

Check Your Understanding 1. The basic vertebrate body plan is established during the embryonic period. **2.** The digestive tract (stomach, intestines) forms from the inner tube, as do the liver, pancreas, and gallbladder. **3.** The kidneys are located posterior (or dorsal) to the peritoneal cavity. **4.** During gastrulation, cells from the epiblast invaginate (move inward) and form three separate germ layers. This process occurs during the third week of development. **5.** The notochord induces the formation of the neural tube. **6.** Somites, from the paraxial mesoderm, and intermediate mesoderm cluster into segments along the body axis. **7.** Ectoderm forms the outer covering of the embryo. **8.** Endoderm froms the inner lining of the inner tube. **9.** The sclerotome forms the vertebrae and ribs. **10.** Splanchnic lateral plate mesoderm forms the connective tissues, muscle tissues, visceral serosa of the respiratory and digestive tubes, and the heart and blood vessels. **11.** Most body organs, including the heart, are formed during the second month of embryonic development, weeks 5–8 of the embryonic period. Disruption of normal development during this period results in birth defects. **12.** Lungs are not fully developed until the end of the seventh month in utero. Babies born prior to this time commonly experience respiratory distress.

Review Questions 1. (1) a; (2) b; (3) b; (4) a **2.** b **3.** b **4.** c **5.** (1) c; (2) f; (3) i; (4) h; (5) a; (6) d; (7) b; (8) g; (9) e **6.** (1) c; (2) g; (3) d; (4) e; (5) f; (6) a; (7) b **7.** c **8.** d **9.** a, d **10.** e (Endoderm forms only the inner lining of the gut tube.) **11.** a **12.** a **13.** d **14.** a **15.** c **16.** d

Chapter 4

Check Your Understanding 1. The apical region is the free, unattached surface of the epithelium. It is easily identifiable in a micrograph as the region that abuts an open space, usually apparent as a white space. **2.** Larger-sized cells allow for more room for cellular machinery necessary for producing cellular secretions (ER, ribosomes, Golgi apparatus, mitochondria). Cuboidal epithelium is found in kidney tubules, glands, and the ovary surface. **3.** Diffusion is a distant-dependent process; columnar epithelium is too thick for efficient diffusion. Simple squamous epithelium is found in these locations. **4.** Goblet cells are unicellular exocrine glands that secrete mucin (mucus). **5.** endocrine **6.** Simple exocrine glands have unbranched ducts; the ducts of compound glands are branched. **7.** Microvilli are short extensions of the plasma membrane with a core of actin microfilaments that extend into the cytoskeleton. They increase the surface area of the plasma membrane. A cilium is a long extension of the plasma membrane composed of a core of microtubules arranged in pairs linked by motor proteins and anchored to a centriole at its base. The bending of a cilium produces movement of the fluid at the epithelial surface. **8.** Intermediate filaments extend across the cell and anchor at desmosomes on the opposite side. Linker proteins from the desmosomes link to adjacent cells. **9.** Other tissues that contain gap junctions are embryonic tissues, connective tissues, cardiac muscle, and smooth muscle. **10.** Epithelial tissues contain lots of cells of similar type closely packed together into sheets with little extracellular matrix. Connective tissues contain fewer cells of various types separated from each other by extensive extracellular material. **11.** The matrix of a connective tissue is composed of both ground substance and fibers. The ground substance is commonly a gel-like material that functions to hold fluid. **12.** Collagen fibers in connective tissues resist tension; the ground substance resists compression; and the elastic fibers allow for recoil. The blast cells (fibroblasts, chondroblasts, osteoblasts) initially produce the tissue matrix. Once the tissue is formed, the fibrocytes, chondrocytes, and osteocytes can secrete matrix components to keep the matrix intact and healthy. **13.** The fibers of loose connective tissues are separated from each other by ground substance. In dense connective tissue, the fibers are closely packed together. **14.** Ligaments and tendons are formed from dense regular connective tissue. The hypodermis is composed of adipose and loose areolar connective tissue. The connective tissue that underlies epithelia is loose areolar connective tissue. Reticular connective tissues contribute to the structure of lymph nodes. **15.** Collagen fibers are found in loose areolar, adipose, dense regular, and irregular connective tissues; hyaline cartilage, elastic cartilage, and fibrocartilage; and bone. **16.** Cells are located in lacunae in both cartilage and bone tissues. **17.** In cartilage, compressive forces are resisted by the gel-like ground substance. **18.** Most obvious way to distinguish smooth muscle tissue from stratified squamous epithelium is that smooth muscle tissue does not have an apical and basal region. **19.** Neurons function to transmit electrical impulses. **20.** *Striated* means "striped" and refers specifically to muscle tissue. Cardiac muscle and skeletal muscle are striated muscle tissue. *Stratified* means "layered" and is used in reference to epithelial tissues: stratified squamous, stratified cuboidal, and stratified columnar epithelia. **21.** Epithelia and some connective tissues (bone, areolar connective tissue, dense irregular connective tissue, blood) regenerate easily. Cardiac muscle and nervous tissue in brain and spinal cord do not regenerate. **22.** Scar tissue is located in the underlying connective tissue. **23.** Increased blood flow to an infected tissue and increased fluid leakage from capillaries serving the infected area cause the heat and swelling associated with infection. **24.** Epithelia form from all three embryonic germ layers: ectoderm, mesoderm, and endoderm. **25.** Cell division in cancerous cells is uncontrolled, and cells expand beyond their normal tissue boundaries.

Review Questions 1. c and e **2.** b **3.** (1) b; (2) f; (3) a; (4) d; (5) g **4.** a **5.** (a) cilia; (b) microvilli; (c) desmosome; (d) tight junction; (e) basement membrane **6.** (1) e; (2) a and c; (3) b; (4) c; (5) d **7.** (1) a; (2) a,b,c; (3) a; (4) c; (5) a **8.** d **9.** c **10.** b **11.** b **12.** (a) mesoderm; (b) mesoderm; (c) ectoderm

Chapter 5

Check Your Understanding 1. The word root *epi* means "over" or "above"; *hypo* means "below"; and *derm* means "skin." **2.** The five main functions of skin are protection, body temperature regulation, excretion, production of vitamin D, and sensory reception. **3.** The stratum basale, stratum spinosum, and stratum granulosum are the epidermal layers containing living cells. **4.** Cells in the epidermis function in protection (from UV radiation, water loss, chemicals, invading microorganisms, and scrapes and abrasions), production of vitamin D, and sensory reception. **5.** Thick skin contains an additional layer, the stratum lucidum, and has a much thicker stratum corneum. **6.** Dermal blood vessels dilate to release heat when the body is warm, and they constrict to prevent the loss of heat when the body is cold. **7.** (a) The papillary layer of the dermis is composed of loose areolar connective tissue. (b) The reticular layer is dense irregular connective tissue. (c) The hypodermis contains both loose areolar and adipose connective tissue. **8.** Fibroblasts in the dermis produce the collagen and elastic fibers of the dermis. These fibers give the skin strength, resilience, and recoil properties; the macrophages, mast cells, and white blood cells function to fight infections; adipocytes store energy. **9.** The large variation in skin color in humans is due to variation in the type and amount of melanin in the epidermal layer. **10.** A dark-skinned individual in northern latitudes would have concerns about vitamin D production. Greater amounts of melanin in darker skin reduces UV penetration. Because there is less sunlight in the winter months, dark skin produces less vitamin D than light skin. **11.** The keratin in hair and nails is a hard keratin: it is tougher and more durable, and its cells do not flake off. **12.** Hair is derived from the epidermal layer; hair follicles have both a dermal component, the peripheral connective tissue sheath, and an epidermal component, the epithelial root sheath. **13.** The cells of the medulla are the deepest layer of hair cells, the next layer of cells is the cortex, and the outermost layer is the cuticle. **14.** Chemotherapy stops the cell division of the cells in the hair matrix, causing hair to fall out, but it does not destroy the hair follicles. A severe burn, however, destroys hair follicles; thus, hair cannot regrow. **15.** Apocrine sweat glands do not cause acne. Acne is caused by sebaceous glands, which also become active at adolescence. **16.** Eccrine sweat glands, like most glands, are formed from cuboidal epithelium. In a micrograph, this epithelium may appear stratified because of the section. **17.** Eccrine glands function in temperature regulation (sweat causes evaporative cooling) and excretion (sweat contains metabolic wastes, salts, and water). **18.** Second-degree burns cause blisters. **19.** The life-threatening concerns associated with third-degree burns are fluid loss and the resulting circulatory shock, followed by risk of infection. Fluid loss is treated with intravenous fluids to maintain blood volume. To prevent infection, the damaged area is covered with either skin grafts or artificial skin, and intravenous antibiotics are given. **20.** Basal cell carcinoma is the most common type of skin cancer. Melanomas are the skin cancers that can be fatal. **21.** Appendages of the skin develop from ectoderm. The hypodermis develops from mesoderm. **22.** UV radiation degrades the collagen fibers and other components of the dermis.

Review Questions 1. c **2.** (1) e; (2) b or d; (3) a; (4) c; (5) d **3.** d **4.** d **5.** b **6.** a **7.** c **8.** b **9.** d **10.** d **11.** (1) b; (2) a; (3) a; (4) b; (5) d **12.** a **13.** (1) e; (2) d; (3) f; (4) b,g; (5) a; (6) g; (7) c **14.** c **15.** (1) d; (2) a; (3) e and c; (4) c; (5) d; (6) d and a; (7) d

Chapter 6

Check Your Understanding 1. The matrix of the three types of cartilage differs in the amount and type of fibers found in each. Hyaline cartilage has small collagen fibrils; elastic cartilage has both collagen and elastic fibers; and fibrocartilage contains thick bundles of collagen. **2.** Hyaline cartilage is most abundant. It is located in articular cartilages, respiratory passageways, costal cartilages of ribs, the embryonic skeleton, and the cartilages of the nose. **3.** Chondroblasts in the perichondrium produce new cartilage by appositional growth. **4.** Collagen fibers contribute to the strength and flexibility of bone, and the mineralized ground substance makes bone hard. **5.** Calcium and phosphate are stored in bone. These minerals are released by the action of osteoclasts. **6.** Osteoblasts secrete new bone matrix. Osteocytes are surrounded by bone matrix and function to repair the matrix and keep it healthy. **7.** Periosteum covers the external surfaces of bone (except for the articular areas); endosteum covers the internal surfaces of bone (including the central and perforating canals). Both membranes contain osteoblasts and osteoclasts. **8.** In a flat bone, compact bone is located on the external surfaces, and spongy bone is located internally. **9.** A condyle is a joint surface. A tubercle is an attachment site for muscle or ligament. A foramen is a hole in the bone that acts as a passageway for vessels or nerves. **10.** The central canal is in the middle of an osteon; the perforating canals connect the central canals of adjacent osteons. Both of these structures contain the vessels and nerves that supply the surrounding osteocytes. Canaliculi are small connections between lacunae that link adjacent osteocytes. **11.** Oxygen and nutrients diffuse through the spaces in the matrix, the canaliculi and lacunae of the inner lamellae, to reach the cells in the outermost lamella. **12.** A trabecula is a "beam" of bone composed of several lamellae and associated osteocytes. The bone tissue in a trabecula is not aligned in concentric rings around a central canal, as it is in an osteon. **13.** Most bones of the skull (except for parts of the base of the skull) and the clavicle are membranous bones. **14.** In a 6-month-old fetus, only the diaphysis of the long bone is ossified. **15.** The thickness of the epiphyseal plate does not change during childhood. As the long bone grows, bone is added to the diaphyseal end of the epiphyseal plate. **16.** Exercise stimulates bone deposition. Mechanical stress on the bone stimulates osteoblasts to produce additional bone matrix. **17.** Remodeling of the bony callus following a fracture makes the repaired area resemble the original unbroken bone. **18.** Comminuted and compression fractures are common in older individuals. **19.** Osteomalacia and rickets result from inadequate mineralization of bone. **20.** Activity of osteoclasts should be targeted to slow bone loss. **21.** Development of osteoporosis is best prevented during the bone-building years of adolescence and early adulthood. **22.** Bones begin to ossify by week 8 of embryonic development. Bone mass starts to decline around age 40. **23.** Estrogens aid in maintaining bone density. The precipitous estrogen decline in postmenopausal women increases their risk of bone loss.

Review Questions 1. e **2.** a **3.** e **4.** (1) a; (2) b; (3) a; (4) c; (5) a **5.** (1) T; (2) F; (3) T; (4) F; (5) F; (6) T; (7) T **6.** b **7.** c **8.** (1) a; (2) b; (3) c; (4) a; (5) b; (6) a **9.** a **10.** c **11.** (a) 3; (b) 2; (c) 4; (d) 1; (e) 5; (f) 6 **12.** b **13.** c **14.** b **15.** b **16.** a **17.** a **18.** (1) b,d; (2) a,d; (3) b; (4) c; (5) a,c

Chapter 7

Check Your Understanding 1. The frontal, ethmoid, and sphenoid bones form the anterior cranial fossa. **2.** The crista galli is on the ethmoid bone; the mastoid process is on the temporal bone; the nuchal line is on the occipital bone; the sella turcica is on the sphenoid bone; the supraorbital foramen is on the frontal bone; the zygomatic process is on the temporal bone. **3.** The four bones that articulate with the left parietal bone are the right parietal bone via the sagittal suture; the frontal bone via the coronal suture; the left temporal bone via the squamous suture; and the occipital bone via the lambdoid suture. **4.** The bones that articulate with the maxilla are the zygomatic, vomer, palatine, nasal, frontal, lacrimal, ethmoid, and inferior nasal concha. **5.** The hard palate is formed from the horizontal plates of the palatine bones and palatine processes of the maxillae. **6.** The alveolar processes are the superior margin of the mandible and the inferior margin of the maxillae that contain the teeth in the alveoli (sockets). **7.** Nasal conchae are formed from the ethmoid bone (superior and middle) and the inferior nasal conchae. These conchae create turbulence in the incoming air and function to warm, moisten, and filter the air. **8.** Cranial bones contributing to the orbit are the frontal, sphenoid, and ethmoid bones; the facial bones that contribute to the orbit are the lacrimal, maxillary, zygomatic, and palatine bones. **9.** Ethmoidal air cells and maxillary sinuses are located in the lateral walls of the nasal cavity. All sinuses are lined with a mucous membrane. **10.** The nucleus pulposus, the inner core of the intervertebral disc, expands under compression. The outer rings of fibrocartilage, the anulus fibrosus, function to resist twisting forces. **11.** The cervical curvature develops at 3 months, as the infant begins to hold up his head; the lumbar curvature develops around 1 year, when the baby begins to walk. **12.** The anulus fibrosus is thinnest posteriorly, but the posterior longitudinal ligaments prevent direct posterior herniation; thus, most herniations occur in the posterolateral direction. **13.** The superior articular process of a vertebra articulates with the inferior articular process of the vertebra just superior. **14.** All cervical vertebrae have transverse foramina. This is the most distinctive feature. Other possible answers are listed in Table 7.2. **15.** Thoracic vertebra T_{12} has costal facets on its vertebral body; lumbar vertebra L_1 does not. **16.** The fused spinous processes of the sacral vertebrae form the median sacral crest. **17.** The sternal angle is the junction between the manubrium and the body of the sternum. Rib 2 articulates here. **18.** The head of the rib articulates with the superior and inferior costal facets of the thoracic vertebrae. The tubercle of the rib articulates with the transverse costal facet. **19.** The frontal bone and mandible form as two individual bones that fuse during childhood. **20.** Age-related degenerative bone loss can cause kyphosis.

Review Questions 1. (1) b,g; (2) h; (3) d; (4) d,f; (5) e; (6) c; (7) a,b,d,h; (8) i **2.** c **3.** c **4.** a **5.** a **6.** (1) a; (2) c; (3) f; (4) b; (5) d; (6) e; (7) c, except C1 and C7; (8) d; (9) f; (10) e **7.** (1) e; (2) d; (3) b; (4) a; (5) c

Chapter 8

Check Your Understanding 1. The acromion of the scapula articulates with the clavicle. **2.** The only bony attachment between the pectoral girdle and the axial skeleton is the clavicle joining with the sternum and rib 1. **3.** The three fossae of the scapula are the supraspinous fossa, "above" the scapular spine; the infraspinous fossa, "below" the scapular spine; and the subscapular fossa, "beneath" the scapula on its anterior surface. **4.** The radius is on the lateral side of the forearm. **5.** The capitulum is on the humerus, and it articulates with the head of the radius. The trochlear notch is on the ulna, and it articulates with the trochlea of the humerus. The head of the ulna is on the ulna, and it articulates with the ulna notch of the radius. The radial notch is on the ulna, and it articulates with the head of the radius. **6.** The anatomical neck of the humerus is just inferior to the head; it is the area where the diaphysis joins with the proximal epiphysis. The surgical neck, the most frequently fractured region of the humerus, is inferior to the greater and lesser tubercles. **7.** The metacarpals are located in the palm of the hand. **8.** The pubis forms the anterior portion of the pelvic bone. **9.** When you sit, the ischial tuberosities carry your weight. **10.** When your hands are on your hips, your hands are resting on the iliac crests. **11.** The pubic arch is broader in females than in males; the greater sciatic notch is wider and more shallow in females than in males; and the sacrum is wider, shorter, and more curved in females. **12.** The acetabulum of the hip bone articulates with the head of the femur. **13.** The bony bumps on either side of your ankle are formed by the medial malleolus of the tibia and the lateral malleolus of the fibula. **14.** The lateral malleolus is on the fibula; the linea aspera is on the femur; the lesser trochanter is on the femur; the fibular notch is on the tibia; the talar shelf is on the calcaneus; and the tibial tuberosity is on the tibia. **15.** The anterior portion of the talus is the keystone of the medial longitudinal arch. The arches distribute weight anteriorly to the heads of the metatarsals (the "ball of the foot") and posteriorly to the calcaneus (heel). **16.** The skeleton in males becomes more robust at puberty. In females, the pelvis broadens. **17.** The three bones of the hip fuse together during childhood at the acetabulum. Splinting holds the femoral head in place so the acetabulum and surrounding ligaments can form normally.

Review Questions 1. e **2.** b **3.** d **4.** a **5.** c **6.** (1) g; (2) f; (3) b; (4) a; (5) b; (6) c; (7) d; (8) e **7.** (1) b; (2) c; (3) e; (4) a; (5) h; (6) e; (7) f **8.** c,d,f (The fibula has a styloid process, but it is not described in this text.) **9.** b **10.** c

Chapter 9

Check Your Understanding 1. A synarthrosis is an immovable joint. This is a functional joint classification. A syndesmosis is a fibrous joint where the adjoining bones are connected exclusively by ligaments. A synchondrosis is a cartilaginous joint where hyaline cartilage unites adjoining bones. Both syndesmosis and synchrondrosis are considered structural classifications. **2.** A symphysis joint contains both hyaline cartilage, which covers the articular joint surfaces, and fibrocartilage, which unites adjoining bones. Symphysis joints are found between the vertebral bodies (intervertebral discs) and joining the pubic bones (pubic symphysis). **3.** The six features of synovial joints are (1) articular cartilage, (2) a joint cavity, (3) an articular capsule, (4) synovial fluid, (5) reinforcing ligaments, and (6) nerves and vessels. **4.** An articular disc is composed of fibrocartilage, and it is located within the joint cavity. Articular cartilage is composed of hyaline cartilage and covers the adjoining surfaces of all synovial joints. **5.** Synovial fluid nourishes the cells of the articular cartilage and lubricates the joint surfaces. **6.** (a) The movements that occur at the elbow are flexion and extension. (b) The hip allows flexion/extension, abduction/adduction, medial and lateral rotation, and circumduction. (c) Dorsiflexion and plantar flexion occur at the ankle. (d) Rotation occurs at the atlantoaxial joint, as in turning your head to say "no". (e) The metacarpophalangeal joint can produce flexion/extension, abduction/adduction, and circumduction. **7.** (a) The elbow is a uniaxial hinge joint. (b) The hip is a multiaxial ball-and-socket joint. (c) The ankle is a uniaxial hinge joint. (d) The atlantoaxial joint is a uniaxial pivot joint. (e) The metacarpophalangeal joint is a biaxial condylar joint. **8.** Pronation is medial rotation of the forearm such that the palm faces posteriorly; supination is lateral rotation of the forearm such that the palm faces anteriorly. This movement occurs at the proximal and distal radioulnar joints. **9.** The articular disc in the sternoclavicular and the temporomandibular joints divide each joint cavity and enable multiple complex movements of each joint. **10.** The knee also contains an articular disc. **11.** The elbow is the most stable, and the shoulder is the least stable. **12.** The muscle tendons that cross the shoulder contribute most to its stability. **13.** The ulna forms part of the elbow. The radius forms part of the wrist joint. **14.** The intracapsular ligament of the hip is the ligament of the head of the femur. The intracapsular ligaments of the knee are the anterior and posterior cruciate ligaments. **15.** Forceful inversion of the ankle can sprain the lateral ligament. **16.** The medial and lateral menisci of the knee aid in joint stability, as do the extensive capsular, extracapsular, and intracapsular ligaments. Muscle tone in the quadriceps and hamstring muscles also contribute to stability of the knee. **17.** (a) Gouty arthritis is caused by crystallization of uric acid in the synovial membranes. (b) Osteoarthritis is the erosion of the articular cartilage. (c) Rheumatoid arthritis is an autoimmune response causing inflammation of the synovial membrane. **18.** A sprain or subluxation stretches the joint capsule and surrounding ligaments and makes the joint more susceptible to repeat injury. **19.** Synovial joints form by week 8 of fetal development.

Review Questions 1. (1) a,b; (2) a; (3) a; (4) a; (5) b; (6) c; (7) c; (8) a,b; (9) c **2.** e **3.** c **4.** e **5.** a **6.** (1) a,c; (2) b,d; (3) c; (4) b,d **7.** (1) g; (2) e; (3) a; (4) a,b,g; (5) d; (6) g; (7) h; (8) e,c (properly called *ulnar collateral ligament of wrist*); (9) g **8.** (1) c; (2) e; (3) b; (4) d; (5) d; (6) d; (7) c; (8) a; (9) a; (10) a

Chapter 10

Check Your Understanding 1. All muscle tissue contains the myofilaments actin and myosin, which generate contractile forces; the plasma membrane is called a sarcolemma; and the cytoplasm is called sarcoplasm. The functional properties of muscle tissue are (1) contractility, (2) excitability, (3) extensibility, and (4) elasticity. **2.** Cardiac muscle and skeletal muscle are striated. Cardiac muscle and smooth muscle are called visceral muscle, because they occur in the visceral organs and are not voluntarily controlled. Skeletal muscle and smooth muscle cells are called muscle fibers. **3.** Perimysium is the fibrous connective tissue that surrounds a fascicle. **4.** The origin of a muscle is the attachment site of the muscle that does not move when the muscle contracts. The insertion is the attachment site on the bone that does move when the muscle contracts. In the limbs, conventionally, the origin is the more proximal attachment, and the insertion is located distally. **5.** Smallest to largest: myofilament, sarcomere, myofibril, muscle fiber. Myofilaments are composed of the contractile proteins actin and myosin. In striated muscle, the myofilaments are arranged into sarcomeres, the smallest contractile unit in striated muscle. A myofibril is an organelle within a muscle cell made up of repeating sarcomeres. A muscle fiber is a muscle cell. **6.** The thick (myosin) myofilaments are limited to the A band. **7.** The terminal cistern stores Ca^{2+}. The T tubules carry the stimulus to contract from the surface sarcolemma to the deeper regions of the muscle fiber, initiating the release of Ca^{2+} from the terminal cistern. **8.** The myosin heads must attach to the overlapping thin filament to generate a contractile force. If there is no overlap, the myosin heads have nothing to attach to. **9.** The length of the I band shortens during muscle contraction. **10.** The neuromuscular junction is the connection between a single terminal bouton and a muscle fiber. A motor unit is a single motor neuron and *all* of the muscle fibers it innervates. **11.** Usain Bolt has a predominance of fast glycolytic muscle fibers in his lower limbs. **12.** Muscle tissue from an individual with Duchenne muscular dystrophy has muscle fibers of irregular size, and in many regions the muscle fibers are replaced with fat and connective tissue. **13.** Skeletal muscle fibers form from the fusion of multiple embryonic cells; thus, they are multinucleated. **14.** Weight-training exercise increases the size of muscle fibers and thus helps maintain muscle strength in older individuals.

Review Questions 1. c **2.** b **3.** (1) thin; (2) thick; (3) thick; (4) thin; (5) thick; (6) thin **4.** (1) yes; (2) no; (3) yes; (4) no **5.** (1) d; (2) a; (3) b; (4) e; (5) c **6.** c **7.** c **8.** a **9.** (1) b; (2) a; (3) c; (4) c; (5) b; (6) a; (7) a,c; (8) c; (9) c; (10) a,b,c

Chapter 11

Check Your Understanding 1. A muscle with fibers aligned along its long axis has a parallel arrangement of its fascicles. **2.** The power of a muscle depends on the total number of muscle fibers. Pennate muscles pack more muscle fibers into a unit area than parallel muscles; thus, they are able to generate greater force. **3.** Second class levers always operate at a mechanical advantage. First class levers operate at a mechanical advantage when the length of the effort arm is longer than the length of the load arm. **4.** The fulcrum of a skeletal/muscular lever system is the joint. **5.** Levers that operate at a mechanical disadvantage allow for greater speed and larger distance of movement for a given muscular effort. In addition, in these arrangements, muscles insert close to the joint, providing better stability to the joint. **6.** (a) The muscles that move the thigh are limb muscles. (b) The abdominal muscles are axial muscles. (c) The chewing muscles are pharyngeal arch or branchiomeric muscles. **7.** Muscles that develop on the dorsal side of the limb are extensor muscles. **8.** The biceps brachii and brachialis function as synergists. **9.** A thigh abductor crosses the hip on the lateral side of the joint. The antagonist muscles produce adduction. **10.** Muscles that cross on the posterior side of a joint generally produce extension. **11.** (a) *Latissumus dorsi* means "wide back," an excellent description of this superficial back muscle. (b) *Sternocleidomastoid* indicates that the origin of this muscle is at the sternum and clavicle and that its insertion is on the mastoid process of the temporal bone. (c) *Serratus anterior* refers to the sawlike appearance of this muscle, much like a serrated knife, and its location on the anterior/lateral thoracic wall. (d) The adductor magnus is the large adductor muscle of the thigh. **12.** The biceps brachii and brachialis are located in the anterior compartment of the arm. Their antagonist muscle group, the triceps brachii, is located in the posterior compartment of the arm. **13.** Muscles in the posterior compartment of the thigh extend the thigh and flex the leg. The quadriceps femoris group functions as an antagonist to these muscles. **14.** The gastrocnemius and soleus act in synergy as prime movers for plantar flexion. **15.** The diaphragm is the most important muscle for inspiration. The intercostal muscles, pectoralis minor, the scalenes, and sternocleidomastoid aid in forceful inspiration. The abdominal muscles aid in forceful expiration. **16.** The six movements of the shoulder joint are flexion, extension, abduction, adduction, medial rotation, and lateral rotation. The muscles involved in these movements are described in Table 11.11, **17.** The prime mover of dorsiflexion is the tibialis anterior. Dorsiflexion of the foot keeps the toes from dragging during the recovery phase of walking. **18.** The lesser gluteals shift the trunk from side to side during walking to keep the body upright and to keep the center of gravity over the limb that is on the ground. Without their action, the trunk would fall over toward the unsupported side when the limb is in the swing phase of gait. **19.** The linea alba runs down the anterior abdominal wall from the xiphoid process to the symphysis pubis. The external and internal oblique and the transversus abdominis muscles insert into the linea alba. **20.** The triangle of auscultation is bounded by the trapezius medially, the latissimus dorsi inferiorly, and the medial border of the scapula laterally. The area is used to listen for lung sounds. **21.** Intramuscular injections can be given in the deltoid, about 5 cm inferior to the greater tubercle of the humerus which form the rounded superior part of the shoulder; the gluteus medius, 5 cm superior from the midpoint between the posterior superior iliac spine and the greater trochanter; or midway down the vastus lateralis on the lateral thigh. **22.** (a) The supracristal line is in the lumbar region. (b) The linea semilunaris is in the abdomen. (c) The cubital fossa is on the anterior side of the elbow. (d) The median bicipital groove is on the medial side of the arm. **23.** The boundaries of the posterior triangle of the neck are the sternocleidomastoid muscle anteriorly, the trapezius posteriorly, and the clavicle inferiorly. The carotid pulse is located in the anterior triangle just anterior to the sternocleidomastoid and superior to the level of the larynx. **24.** In most individuals, the tendons of palmaris longus and flexor carpi radialis are apparent in the distal forearm when the person makes a fist.

Review Questions **1.** d **2.** c **3.** (1) c; (2) b; (3) e; (4) a; (5) d **4.** e **5.** c **6.** d **7.** c **8.** b **9.** b **10.** a **11.** d **12.** b **13.** b

Chapter 12

Check Your Understanding **1.** Afferent signals carry sensory information toward the CNS. **2.** The visceral motor (autonomic) division of the peripheral nervous system regulates contraction of cardiac muscle. **3.** (a) Pain from a pulled muscle is general somatic sensory. (b) Nausea is general visceral sensory. (c) Taste is a special visceral sensory sensation. **4.** The neuron cell processes that receive stimuli are the dendrites. **5.** When an electrical impulse in the presynaptic neuron reaches the terminal bouton, it stimulates the release of chemical neurotransmitters from this neuron. These neurotransmitters diffuse across the synaptic cleft, bind to receptor sites on the postsynaptic neuron (usually on the dendrite or cell body), and stimulate an electrical response in the postsynaptic neuron. In this way, the electrical signal is passed, via a chemical messenger, from one neuron to another. **6.** Most sensory neurons are unipolar (pseudounipolar) neurons. **7.** Multipolar neurons are the most abundant type of neuron. Motor neurons of the PNS are multipolar neurons. **8.** Oligodendrocytes produce myelin in the CNS, and Schwann cells produce myelin in the PNS. **9.** Astrocytes are common in regions where synapse occurs because they regulate ionic and neurotransmitter levels in neural tissue. **10.** Schwann cells do cover nonmyelinated axons in the PNS; however, these axons are not wrapped by concentric layers of the Schwann cell plasma membrane. Numerous nonmyelinated axons are embedded in a single Schwann cell (see Figure 12.7). **11.** The perineurium wraps a bundle of axons (nerve fibers) into a fascicle. **12.** Synapses in the CNS always occur in the gray matter as this is where neuron cell bodies are located in the CNS. **13.** The white matter of the CNS appears white because the majority of the nerve fibers in these regions are myelinated. **14.** The integration center occurs in the gray matter of the CNS at the synapse between the sensory neuron and motor neuron. **15.** The sensory impulse must be carried to the brain for processing the conscious sensation of pain. This processing takes more time than is required for processing the simple withdrawal reflex. **16.** A converging circuit contains multiple presynaptic neurons synapsing on a single postsynaptic neuron. **17.** After a peripheral injury, axons in the PNS can regrow within regeneration tubes formed by surviving Schwann cells. In the CNS, neuroglia do not guide regrowing axons and, in fact, secrete inhibitory chemicals that block neural regrowth. **18.** Myelin insulates nerve signals and increases the speed of impulse conduction. Loss of myelin in the CNS, as occurs with MS, interferes with the conduction of nerve signals. All of the symptoms described for MS indicate disruption of neural processing (balance disorders, muscle weakness, vision disturbance). **19.** The neuroblasts of the basal plate form motor neurons. **20.** Sensory neurons develop from neural crest tissue, which is located outside the developing neural tube.

Review Questions **1.** b **2.** (1) d; (2) b; (3) f; (4) c; (5) a; (6) a **3.** b **4.** (1) SS; (2) VS; (3) VM; (4) SM; (5) VS; (6) SS; (7) SM **5.** b **6.** b **7.** a **8.** c **9.** d **10.** (1) c; (2) b; (3) a **11.** b **12.** c **13.** d **14.** 2, 1, 3

Chapter 13

Check Your Understanding **1.** The telencephalon forms the cerebrum, and the mesencephalon forms the midbrain. **2.** Brain nuclei are clusters of neuronal cell bodies that form portions of the gray matter of the CNS. **3.** The third ventricle is connected to the fourth ventricle through the cerebral aqueduct, located in the midbrain. **4.** (a) The pyramids of the medulla are white fiber tracts connecting the cerebrum with the spinal cord. (b) The vestibular and cochlear nuclei are cranial nerve nuclei involved with innervation of the head. They receive sensory impulses from the cranial nerve that innervates the inner ear. (c) The cerebral peduncles are white fiber tracts in the midbrain connecting the cerebrum with the brain stem. (d) The periaqueductal gray matter is involved with the autonomic behaviors associated with the fight-or-flight response and mediates the response to visceral pain. Two cranial nerve nuclei are also located here. (e) The trigeminal nuclei are cranial nerve nuclei involved with sensory innervation of the face and motor innervation of the chewing muscles. **5.** The pyramids (a) and the vestibular and cochlear nuclei (b) are located in the medulla oblongata. The cerebral peduncles (c) and the periaqueductal gray matter (d) are in the midbrain. The trigeminal nuclei (e) are in the pons. **6.** The corpora quadrigemina are four brain nuclei located on the dorsal surface of the midbrain involved with auditory (inferior colliculi) and visual (superior colliculi) reflexes. **7.** The vermis connects the two cerebellar hemispheres. **8.** The flocculonodular lobe receives sensory information from the vestibular nuclei about equilibrium and head position. **9.** The superior cerebellar peduncles connect the midbrain to the cerebellum; the middle cerebellar peduncles connect to the pons; and the inferior cerebellar peduncles extend from the medulla to the cerebellum. **10.** Axonal fibers from the thalamic neurons extend to the cerebral cortex. **11.** The hypothalamus is the main visceral control center. Visceral control refers to regulation of the activities of the visceral organs, including autonomic motor activity and endocrine functions. **12.** The primary sensory cortex enables conscious awareness of a sensation. The sensory association area places the sensory stimulus into a context that gives meaning to the stimulus. **13.** The premotor cortex plans complex movements. The primary motor cortex signals the execution of these movements. **14.** Contralateral ("opposite side") projection refers to the crossing over of nerve fibers to the opposite side of the body, such that the right side of the body is controlled or perceived by the left side of the brain. **15.** The visual cortex and association area are located in the occipital lobe. Visual deficits would result from injury to this region. **16.** Projection fibers connect the cerebral cortex to more caudal regions of the CNS. **17.** The caudate nucleus is located just lateral to the lateral ventricles. **18.** The reticular nuclei receive input from all major ascending sensory tracts. These nuclei project to the cerebrum (via relays through the thalamus), thus maintaining alertness. **19.** The amygdala is the brain nucleus for processing and responding to fear. **20.** The falx cerebri lies in the longitudinal fissure. **21.** Cerebrospinal fluid (CSF) is produced by the choroid plexuses in the roof of the brain ventricles. CSF returns to the blood vascular system through the arachnoid granulations, which empty into the dural sinuses. **22.** Lumbar spinal nerves L_3–L_5 and sacral spinal nerves S_1–S_5 pass through the vertebral foramen of lumbar vertebra L_3. **23.** White matter provides the two-way conduction pathway between the brain and spinal cord. **24.** Somatic motor neurons to the muscles of the upper limb are located in the ventral horn of the lower cervical segments. Paralysis of these muscles would result from damage to this region of the spinal cord. **25.** Cerebrospinal fluid is located in the subarachnoid space, which is located deep to the arachnoid mater and superficial to the pia mater. **26.** Discriminative touch sensations are carried on the dorsal column pathways (fasciculus cuneatus and fasciculus gracilis). **27.** The dorsal column and spinothalamic pathways pass through the thalamus. The spinocerebellar pathway does not. **28.** The pyramidal pathways (lateral and ventral corticospinal) originate from the primary motor cortex. **29.** The ascending spinocerebellar tracts do not cross over. The descending vestibulospinal tract and some fibers of the reticulospinal tract do not cross over. These two descending pathways are not illustrated, but are discussed in Table 13.4. **30.** Most of the ascending and descending pathways project contralaterally. **31.** The thalamus and the somatosensory cortex are active in a newborn. The frontal lobe (the location of rational thought) is not fully developed until early adulthood.

Review Questions 1. (1) c; (2) f; (3) e; (4) g; (5) b; (6) e; (7) i; (8) a; (9) g; (10) h; (11) d **2.** b **3.** d **4.** (1) G; (2) W; (3) W; (4) G; (5) W; (6) G; (7) W **5.** a

Chapter 14

Check Your Understanding 1. Touch, pain, pressure, vibration, temperature, and proprioception from the somatic organs (skin, body wall, limbs) are general somatic sensations. Taste and smell are special visceral sensations. **2.** All spinal nerves carry both sensory and motor fibers, but some cranial nerves are exclusively sensory or primarily motor in function. **3.** Nociceptors respond to painful stimuli. These sensory receptors are free nerve endings. **4.** Proprioceptors are part of the somatic sensory system (outer tube). **5.** Encapsulated receptors are mechanoreceptors (respond to pressure, discriminative touch, vibration, and stretch). **6.** The glossopharyngeal nerve (CN IX), the vagus (CN X), and the accessory nerve (CN XI) pass through the jugular foramen. **7.** The optic nerve (CN II) carries the special somatic sensation (SSS) for vision, and the vertibulocochlear nerve (CN VIII) carries SSS for hearing and equilibrium. **8.** The trigeminal nerve (CN V) is the "great sensory nerve of the face." The facial nerve (CN VII) carries motor innervation to the muscles of the face. **9.** A lesion to the dorsal root of spinal nerve T_4 would result in sensory deficits from the muscles and skin of the anterior and posterior trunk in the region of thoracic vertebra T_4 and rib 4. There would be no motor deficits. A lesion to the dorsal ramus of T_4 would result in sensory and motor deficits in the skin and intrinsic back muscles in the vicinity of thoracic vertebra T_4. **10.** (a) Cervical nerve C_5 exits below cervical vertebra C_4. (b) Lumbar nerve L_3 exits below lumbar vertebra L_3. **11.** The diaphragm is innervated by spinal segments C_3, C_4, and C_5. Innervation from segments C_3 and C_4 would not be affected by an injury at C_5; therefore, independent breathing is possible with an injury in this location. **12.** The radial nerve innervates all of the extensor muscles of the arm and forearm. **13.** Sensory innervation from the medial portion of the hand would be lost, as would the ability to abduct and adduct the hand and extend the fingers (loss of motor innervation to the interossei and medial lumbricals). **14.** The muscles of the upper limb are innervated by nerves off the brachial plexus (C_5–T_1). Injury at C_5 would result in loss of function below the area of injury; thus, the muscles of the upper limb would be paralyzed. **15.** The femoral nerve innervates the muscles of the anterior compartment of the thigh; the obturator nerve innervates the muscles of the medial compartment of the thigh; and the sciatic nerve innervates the muscles of the posterior compartment of the thigh (tibial branch primarily). **16.** The nerves innervating the elbow are the musculocutaneous, radial, median, and ulnar nerves. **17.** Reviewing the dermatome map (Figure 14.15) shows that the skin on the anterolateral surface of the leg is innervated by spinal nerve L_5. Herniation at this level can result in loss of sensation at this dermatome. **18.** Shingles blisters at the level of the navel indicate viral infection in spinal nerve T_9 or T_{10}. **19.** (a) neuralgia = pain in a nerve; (b) paresthesia = beyond sensation, abnormal sensation; (c) neuropathy = disease of the nerves. **20.** Postpolio syndrome is not due to reactivation of the polio virus. It is thought that this syndrome results from the normal degeneration of motor neurons associated with age. Because the motor units of polio survivors are enlarged, loss of motor neurons results in noticeable loss of muscle function.

Review Questions 1. c **2.** b **3.** (1) d; (2) c; (3) f; (4) b; (5) e; (6) a **4.** (1) f; (2) i; (3) b; (4) g,h,j; (5) e; (6) i; (7) i; (8) k; (9) j; (10) c,d,f,k; (11) c **5.** (1) b,6; (2) d,8; (3) c,2; (4) c,5; (5) a,4; (6) a, 3 and 9; (7) a,7; (8) a,7; (9) d,1 **6.** c **7.** c **8.** c **9.** b **10.** a and e

Chapter 15

Check Your Understanding 1. Postganglionic axons of both the sympathetic and parasympathetic divisions are nonmyelinated. **2.** The sympathetic division of the ANS has been activated in this near-miss scenario. **3.** Sympathetic ganglia are located near the spinal cord and vertebral column (in the sympathetic trunk or collateral ganglia). Most parasympathetic ganglia are located in or near the target organ. **4.** Traditionally the ANS is considered a motor division; therefore, it does not include the visceral sensory fibers. **5.** Sacral spinal nerves (S_2–S_4) carry preganglionic parasympathetic fibers. **6.** (a) Vagal stimulation of the heart decreases heart rate. (b) Vagal stimulation of the small intestine increases motility and secretion. (c) Vagal stimulation of the salivary glands stimulates secretion. **7.** The preganglionic fibers synapse in ganglia located along the pathway of the trigeminal nerve (CN V). Postganglionic fibers then travel along CN V to reach their target organs. **8.** The white rami communicantes carry preganglionic outflow from the thoracic and lumbar regions of the cord into the sympathetic trunk. The gray rami communicantes are postganglionic fibers that exit off the sympathetic trunk onto a spinal nerve at each level of the sympathetic trunk. These postganglionic fibers are carried on spinal nerves to all regions of the body to innervate blood vessels, sweat glands, and arrector pili **9.** Sympathetic innervation to the abdominal organs decreases activity of the glands and muscles of the digestive system. **10.** The thoracic plexuses contain preganglionic parasympathetic fibers and postganglionic sympathetic fibers. The abdominal plexuses also contain preganglionic parasympathetic fibers and postganglionic sympathetic fibers. In addition, preganglionic sympathetic fibers travel through the abdominal plexuses (via the splanchnic nerves) to reach the collateral ganglia. **11.** Most visceral pain fibers follow the sympathetic pathway back to the CNS. **12.** A peripheral reflex does not involve the CNS at all. Some peripheral reflexes synapse in the sympathetic ganglia, and others occur entirely within the body wall of the target organ (enteric nervous system). **13.** The hypothalamus is the main control center of the ANS. **14.** Postganglionic neurons are formed from neural crest tissue. **15.** The parasympathetic division is deficient in achalasia of the cardia. Raynaud's disease is due to an exaggerated sympathetic vasoconstriction in response to cold.

Review Questions 1. d **2.** (1) S; (2) P; (3) P; (4) S; (5) S; (6) P; (7) S; (8) S; (9) P; (10) S; (11) P; (12) S; (13) S; (14) S; (15) S **3.** c **4.** d **5.** b **6.** d **7.** c **8.** a

Chapter 16

Check Your Understanding 1. The receptor cells for taste are epithelial cells. **2.** Olfactory receptors are bipolar neurons. These cells are replaced when damaged, making them among the few neurons that undergo replacement throughout adult life. **3.** The conjunctiva is a transparent mucous membrane that covers the inner surfaces of the eyelids and the anterior surface of the eye. **4.** The lateral rectus is not functioning if the eye is turned in medially. This muscle is innervated by the abducens nerve (CN VI). **5.** The cornea is part of the external fibrous layer of the eye, continuous with the sclera. It is a transparent layer covering the anterior one-sixth of the eye. The cornea is composed of dense connective tissue sandwiched between two layers of epithelium. It forms part of the light-bending apparatus of the eye. The choroid is part of the middle vascular layer of the eye. It is a highly vascular, pigmented membrane covering the posterior five-sixths of the eye. Blood vessels of the choroid nourish the eye, and its pigments absorb light. **6.** Macular degeneration diminishes the clarity of the visual field straight ahead. **7.** The aqueous humor is a clear watery fluid that fills the anterior segment. It supplies nutrients and oxygen to the avascular lens and cornea. It is constantly produced and drained, a process that maintains constant intraocular pressure. **8.** The lens becomes rounder when close objects are viewed. **9.** The resting eye is set for distance vision. Accommodation for viewing close objects requires elasticity of the lens. As we age, the lens becomes less elastic and cannot accommodate for near vision well. **10.** Axons from ganglion cells from the medial half of the retina extend through the right optic nerve and pass through the optic chiasma to the left optic tract. They synapse in the left lateral geniculate nucleus of the thalamus with neurons that extend via the optic radiation to the left primary visual cortex. **11.** The retina forms from the outpocketing of the neural tube. **12.** The tympanic membrane (eardrum) separates the external acoustic meatus from the middle ear cavity. **13.** The four openings into the middle ear cavity are the pharyngotympanic tube, the round window, the oval window, and the mastoid antrum. **14.** The malleus is the auditory ossicle that abuts the tympanic membrane. **15.** The membranous labyrinth is a continuous series of membranous sacs and ducts filled with endolymph that contain the sensory receptors for hearing and equilibrium. The bony labyrinth is the cavity within the petrous portion of the temporal bone that surrounds the membranous labyrinth. **16.** Perilymph fills the cavities of the bony labyrinth of the cochlea, the scala vestibuli, and the scala tympani. **17.** The vestibular membrane forms the roof of the cochlear duct; the basilar membrane forms the floor; and the tectorial membrane sits atop the sensory hair cells. Vibrations in the tectorial membrane stimulate the hair cells of the spiral organ. **18.** The maculae of the utricle and the saccule monitor stationary head position and linear acceleration. **19.** Sensorineural deafness results from damage to the cells of the spiral ganglion. **20.** The membranous labyrinth is formed from ectoderm. **21.** The vestibular nuclei of the medulla and the cerebellum are the major processing centers for input from the vestibular nerves.

Review Questions 1. a **2.** d **3.** c **4.** b **5.** b **6.** b **7.** e **8.** b **9.** c **10.** b **11.** d **12.** b

Chapter 17

Check Your Understanding 1. Hormones circulate throughout the body in the bloodstream. They leave the bloodstream at the capillaries and enter the tissue fluid. Although hormones encounter all types of cells, they influence the activity only of their target cells. **2.** Endocrine secretions are regulated via humoral stimuli (concentrations of specific substances in the blood); nervous stimuli (direct nerve innervation); or hormonal stimuli (hormones from one endocrine tissue influence the secretion of hormones from another endocrine tissue). **3.** Tropic hormones secreted by the pituitary regulate the secretion of hormones by other endocrine glands. The nontropic pituitary hormones act directly on nonendocrine target tissues. **4.** The hormones secreted by the posterior pituitary are produced in the neuron cell bodies located in the hypothalamus. The anterior pituitary hormones are produced by the cells of this gland. **5.** (a) The target organs for oxytocin are the uterus and the breast. (b) ACTH acts on the adrenal cortex. (c) FSH acts on the ovaries in females and the testes in males. (d) The target organ for ADH is the kidneys. **6.** A patient with hypothyroidism would not have concerns about blood calcium regulation because this disorder refers to low levels of thyroid hormone, not calcitonin levels. **7.** Aldosterone is produced in the outer layer, the zona glomerulosa, of the adrenal cortex. **8.** The adrenal medulla secretes the hormones norepinephrine (a sympathetic neurotransmitter) and epinephrine when stimulated by preganglionic sympathetic neurons. The cells of the adrenal medulla have no axonal processes, but rather release their secretions into the blood of nearby capillaries. Lastly, the cells of the adrenal medulla are derived from neural crest tissue, as are the postganglionic autonomic neurons. **9.** Glucocorticoid secretion is stimulated by ACTH, which is produced by the anterior pituitary. **10.** The endocrine cells of the pancreas are distributed throughout the pancreas in clusters of cells called the pancreatic islets. **11.** In a developing fetus, estrogens and progesterone are produced by the placenta. **12.** Addison's disease results from hyposecretion of the adrenal cortex. Typical symptoms are low blood pressure, dehydration, fatigue, and loss of appetite. Graves' disease is caused by hypersecretion of the thyroid. Symptoms are elevated metabolic rate, rapid heart rate, sweating, and weight loss. Cushing's disease results from hypersecretion of glucocorticoid hormones from the adrenal cortex. High blood glucose levels, lethargy, swollen face, and development of a "buffalo hump" are symptoms.

Review Questions **1.** a **2.** (1) c; (2) b; (3) g; (4) h; (5) d; (6) e; (7) f; (8) i; (9) a; (10) j; (11) b,h; (12) i **3.** a **4.** (1) a; (2) b; (3) a; (4) b; (5) b; (6) a **5.** (1) e; (2) a; (3) a and c; (4) d; (5) b **6.** d **7.** a **8.** d **9.** c **10.** (1) b,j; (2) a,g; (3) c,f; (4) d,h; (5) e,i **11.** a,c,d,e,g,h,j

Chapter 18

Check Your Understanding **1.** The percentage of blood volume composed of erythrocytes is called the hematocrit. Its normal value is around 45%. **2.** The three main plasma proteins are albumin (prevents water from leaving the bloodstream), globulins (antibodies and proteins that transport lipids, iron, and copper) and fibrinogen (involved with blood clotting). **3.** Erythrocytes and platelets do not contain nuclei. **4.** *Leukocyte* refers to all types of white blood cells; lymphoctyes are specific white blood cells that are part of the immune system. **5.** Wright's stain is composed of the acidic stain eosin, which binds to basic cellular elements and stains them pink, and the basic stain methylene blue, which binds to acidic cellular elements, staining them blue to purple in color. The cytoplasm of neutrophils takes up both stains, thus their name, "neutral." Eosinophils are "eosin loving," and the cytoplasmic granules stain pink. The granules in basophils ("basic loving") take up the basic stain methylene blue, and the cytoplasm of these cells is blue to purple in color. **6.** In adults, blood stem cells are located in the spongy bone of the axial skeleton and limb girdles and in the proximal epiphysis of the humerus and femur. **7.** The fibrous network within red bone marrow is formed from reticular connective tissue. **8.** Lymphocytes form from lymphoid stem cells. All other blood cells form from myeloid stem cells. **9.** Symptoms of anemia are fatigue, shortness of breath, and low body temperature. Low levels of erythrocytes or hemoglobin concentration result in poor oxygenation of body tissues. **10.** The spleen and the liver are the major fetal hematopoietic organs before month 7. **11.** The T cells in cord blood are immature and therefore less likely to cause tissue rejection.

Review Questions **1.** d **2.** a **3.** c **4.** a **5.** (a) 2; (b) 5; (c) 1; (d) 4; (e) 3 **6.** (1) c; (2) b; (3) a; (4) c; (5) a; (6) b; (7) f; (8) b; (9) a; (10) a; (11) a; (12) e **7.** d

Chapter 19

Check Your Understanding **1.** The right side of the heart receives and pumps deoxygenated blood. **2.** The apex of the heart is located on the left midclavicular line at intercostal space 5 (ICS 5). **3.** The epicardium is the visceral layer of the serous pericardium. **4.** (a) Pectinate muscles are located in both atria. (b) Papillary muscles are found in both ventricles. (c) The fossa ovalis is in the right atrium. (d) The trabeculae carneae are muscular ridges located in both ventricles. **5.** The superior vena cava, inferior vena cava, and coronary sinus are the three major veins that empty into the right atrium. (The anterior cardiac veins, small veins that drain the right ventricle, also empty into the right atrium.) **6.** From the aortic semilunar valve, blood enters into the aorta, the great artery that carries oxygenated blood to the systemic circuit. **7.** During ventricular systole, the AV valves are closed, and the semilunar valves are open. **8.** A stenotic valve has a narrowed opening due to fusion or stiffening of the cusps and thus doesn't open properly. An incompetent valve is a leaky valve that doesn't close completely. Both valve disorders reduce the efficiency of the heart. **9.** Gap junctions allow the stimulus to contract to be passed from one muscle cell to an adjacent cell. This linkage between muscle cells enables all the muscle cells in a heart chamber to contract at the same time. **10.** The "pacemaker" of the heart is the sinoatrial (SA) node. It is located in the right atrium and is composed of specialized cardiac muscle cells that initiate the electrical signal that sets heart rate. All other conducting system structures are composed of specialized cardiac muscle cells. **11.** Sympathetic innervation to the heart increases heart rate and increases strength of contraction. Parasympathetic innervation only influences heart rate; it decreases heart rate. **12.** The left coronary artery supplies blood to the left ventricle via two branches: The anterior wall receives blood from the left interventricular artery, or left anterior descending (LAD), and the posterior wall is supplied by the circumflex artery. **13.** Risk factors for coronary artery disease include high blood pressure, smoking, high cholesterol levels, diabetes, inactivity, and family history of heart disease. For lucky individuals, the first symptom is angina; for unlucky individuals, a fatal heart attack is often the first symptom. **14.** An opening in the interventricular septum would allow oxygenated and deoxygenated blood to mix. Because the left ventricle is stronger, more blood is shunted from the left ventricle to the right ventricle. **15.** The embryonic ventricle forms the left ventricle. The right ventricle forms from the bulbus cordis. **16.** Regular and vigorous exercise is the most important factor for maintaining a healthy heart throughout life.

Review Questions **1.** b **2.** d **3.** b **4.** c **5.** e **6.** a **7.** d **8.** d **9.** c **10.** b **11.** d

Chapter 20

Check Your Understanding **1.** Capillary walls are composed of a single layer of endothelial cells through which diffusion can occur readily. Capillary diameter is similar to that of a red blood cell (RBC), which maximizes the surface contact between the RBC and the capillary walls and thus maximizes gas diffusion. In addition, endothelial clefts and fenestrations contribute to capillary permeability, allowing fluid and dissolved molecules to move readily from the blood plasma to the tissue fluid. **2.** The thick smooth muscle of the tunica media and the internal and external elastic membranes that surround this layer function to keep blood pulsing through arteries. Venous blood flow is maintained by valves that prevent backflow, by normal body movement, and by the skeletal muscle pump. **3.** (a) Vasa vasorum are small vessels in the tunica externa of larger arteries that supply the outer layers of these large vessels. (b) Arterial anastomoses are junctions between two different arteries that supply a common organ. (c) Varicose veins form when the valves of a vein fail and blood pools in the vessel. (d) An artery is a vessel that carries blood away from the heart. **4.** The pulmonary artery is illustrated in blue because it is carrying poorly oxygenated blood (toward the lung to be oxygenated). This vessel is carrying blood away from the heart; thus, it is an artery. **5.** The brachiocephalic trunk supplies the head, neck, thorax, and upper limb on the right side. The left common carotid artery supplies the left portions of the head and neck, and the left subclavian artery supplies the thorax and upper limb on the left side. **6.** The four major vessels that supply blood to the brain are the right and left internal carotid arteries and the right and left vertebral arteries. The internal carotids branch off the common carotid arteries; the vertebral arteries branch off the subclavian arteries. **7.** (a) The femoral artery in the thigh; (b) the brachial artery in the arm; (c) the radial artery at the wrist; (d) the dorsalis pedis artery in the foot; (e) the common carotid artery in the neck. **8.** The celiac trunk supplies the stomach, liver, gallbladder, pancreas, spleen, and part of the small intestine. The superior mesenteric artery supplies the small intestine, the ascending colon, and part of the transverse colon. The inferior mesenteric artery supplies the distal portion of the transverse colon, the descending colon, and rectum. **9.** The hepatic portal vein drains the organs of the digestive tract, enters the liver, and empties into the liver sinusoids. The hepatic veins drain blood from the liver and empty into the inferior vena cava. **10.** (a) The cephalic vein is a superficial vein of the upper limb. (b) The popliteal vein is located on the posterior knee in the popliteal fossa. (c) The transverse sinus is in the head. It is one of the dural sinuses that drains the brain. (d) The saphenous vein is a superficial vein of the lower limb. (e) The azygos vein runs along the right posterior thoracic wall. **11.** The median cubital vein is a common site for withdrawing blood. The great saphenous vein is commonly used in coronary artery bypass surgery. **12.** Factors that contribute to atherosclerosis are smoking, a high-fat diet, and a sedentary lifestyle. **13.** The basement membrane of capillaries thickens in diabetes mellitus. This reduces the rate of the diffusion of gases and nutrients to the body tissues. **14.** The umbilical vein carries the most highly oxygenated blood in the fetus. **15.** The ductus arteriosus extends between the pulmonary trunk and the aorta. Blood in the pulmonary trunk is shunted to the aorta, and thus the pulmonary circuit is bypassed.

Review Questions **1.** d **2.** b **3.** b **4.** (1) intima; (2) externa; (3) externa; (4) externa; (5) media; (6) intima **5.** a **6.** b **7.** c **8.** b **9.** c **10.** d **11.** a **12.** d **13.** a **14.** b **15.** c **16.** a

Chapter 21

Check Your Understanding **1.** The endothelial cells of lymphatic capillaries have few intercellular junctions, and the edges of adjacent cells overlap, forming easily opened minivalves. Blood capillaries have tight junctions between endothelial cells. Intercellular clefts between adjacent cells and, in some cases, pores cause these vessels to be permeable. **2.** The sentinel node in a throat cancer patient would be a cervical node; in a breast cancer patient, it would be an axillary node; in a cervical cancer patient, it would be an iliac node. **3.** Intestinal trunk; thoracic duct; left internal jugular and left subclavian veins. **4.** Cytotoxic T lymphocytes, as well as all T lymphocytes, gain immunocompetence in the thymus. **5.** Infection caused by a bacterial agent will activate B lymphocytes. **6.** B lymphocytes are in the lymphoid follicles. Each lymphoid follicle is derived from the activation of a single B cell. Cell division occurs at the germinal center, producing new B cells that then migrate from the follicle to become plasma cells. **7.** In the red pulp of the spleen defective blood cells are phagocytized by macrophages. **8.** In the lymph nodes, T lymphocytes are located in the deeper portions of the cortex and in the medullary cords of the medulla. **9.** Aggregated lymphoid nodules are located in the ileum of the small intestine. **10.** The red lines indicate lymphangitis, inflammation of a lymphatic vessel due to infection. **11.** The thymus is derived from endoderm. The tonsils, spleen, and lymph nodes are all from mesoderm.

Review Questions 1. c **2.** c **3.** a **4.** b **5.** e **6.** b **7.** a,c **8.** c **9.** a **10.** d **11.** d

Chapter 22

Check Your Understanding 1. The nasal cavity, the nasal conchae, paranasal sinuses, and the nasopharynx, are covered with respiratory mucosa. This mucous membrane functions to warm, filter, and moisten inhaled air. **2.** External respiration is gas exchange between the blood and the air in the air sacs of the lungs. Internal respiration is gas exchange between the blood and the body tissues. Cellular respiration is the use of oxygen by cells to convert glucose into cellular energy (ATP). **3.** The pharynx, commonly called the throat, is located posterior to the nasal and oral cavities and the larynx. It extends inferiorly to the opening of the esophagus. The larynx, or voice box, is at the top of the trachea. It houses the vocal cords and functions as a gateway to prevent food from entering the airway. **4.** The inferior boundary of the nasopharynx is the uvula, that of the oropharynx is the epiglottis, and that of the laryngopharynx is the inferior edge of the cricoid cartilage. **5.** The vocal folds attach to the arytenoid cartilages posteriorly and extend anteriorly to the thyroid cartilage. The length and tension of the vocal cords influence pitch: short, tense vocal cords produce a higher pitch; longer vocal cords, as occur in males, produce a lower pitch. The force of the air passing through the vocal cords varies loudness of sound: greater force produces louder sound. Branches from the vagus nerve (CN X), the recurrent laryngeal nerves, innervate the muscles of the larynx that vary the tension in the vocal cords. **6.** The mucous membrane lining the trachea and large bronchi secretes mucus, which traps dust and foreign particles. The ciliated epithelium sweeps this dust-laden mucus up toward the pharynx, where it is swallowed. In the alveoli, alveolar macrophages remove foreign particles by phagocytosis. **7.** At the level of the bronchioles (tubules with a diameter of 1 mm or less) cartilage plates are no longer present in the tube wall. Elastin is found in the tubule walls throughout the bronchial tree, all the way down to and surrounding the alveoli. **8.** The respiratory membrane is composed of the endothelial cells of the pulmonary capillaries, the simple squamous epithelium (type I alveolar cells) of the alveolar wall, and the fused basal lamina of these tissues. **9.** The horizontal fissure separates the superior and middle lobes of the right lung. **10.** From superior to inferior, the pulmonary artery, the main bronchus, and the pulmonary vein enter the left lung at the hilum. **11.** A stab wound at rib 7 in the midclavicular line would not puncture a lung, but it would puncture the pleural cavity (see Figure 22.13). **12.** Contraction of the diaphragm increases the volume of the thoracic cavity. This volume change decreases the pressure in the pleural cavity, causing air to enter the lungs. **13.** If the pleural cavity is punctured in a stab wound, atmospheric air enters the pleural cavity—thus the "sucking" sound produced by this injury. The loss of the pressure differential between the lung and the pleural cavity results in collapse of the lung. **14.** The accessory inspiratory muscles include the scalenes, sternocleidomastoid, and pectoralis minor, which all aid in elevating the ribs. Primarily it is the abdominal muscles (the obliques and transversus abdominis) that are active during forced expiration. **15.** The respiratory centers of the brain stem (ventral respiratory group or VRG, with input from neurons in the pons and dorsal respiratory group) set the basic respiratory rhythm. The VRG neurons stimulate the somatic motor neurons that innervate the respiratory muscles. **16.** Emphysema destroys the alveolar walls in the respiratory zone. **17.** "Smoker's cough" is a symptom of chronic bronchitis, a COPD disease. Excess secretion of mucus, bronchial inflammation, and fibrosis in response to chronic exposure to smoke cause persistent and productive coughing. **18.** The lungs are not fully formed until late in fetal development. Specifically, the alveoli form at 24 weeks, and type II alveolar cells begin to produce surfactant at 26 weeks.

Review Questions 1. d **2.** b **3.** d **4.** a **5.** (1) b; (2) b; (3) a; (4) b; (5) c **6.** (1) c; (2) d; (3) e; (4) a; (5) b **7.** a **8.** c **9.** c **10.** d

Chapter 23

Check Your Understanding 1. Propulsion occurs throughout the alimentary canal, from the oral cavity to the anus. **2.** The abdominal cavity is the portion of the ventral body cavity extending from the diaphragm to the pelvis. The peritoneal cavity is the space between the parietal and visceral serous membranes: the parietal peritoneum and visceral peritoneum. The pancreas, duodenum, ascending colon, descending colon, and rectum are secondarily retroperitoneal. **3.** (a) The mesenteries to the liver are the falciform ligament and the lesser omentum, both ventral mesenteries. (b) The mesenteries attached to the stomach are the greater omentum, a dorsal mesentery, and the lesser omentum. (c) The mesentery to the sigmoid colon is the sigmoid mesocolon, a dorsal mesentery. **4.** Blood from a hemorrhaging spleen or liver would collect in the peritoneal cavity. **5.** From the lumen outward, the sublayers of the mucosa are the epithelium, lamina propria and muscularis mucosae. The intrinsic glands are formed from the epithelial layer. **6.** Peristalsis and segmentation result from contractions of the smooth muscle of the muscularis externa. **7.** The cells of smooth muscle are elongated with tapering ends, contain a single centrally located nucleus, and are nonstriated. Smooth muscle is innervated by the ANS and contracts in response to hormones, stretching, and nerve stimulation. It is extremely fatigue resistant. Skeletal muscle is composed of multinucleated, striated, cylindrically shaped cells. Skeletal muscle is innervated by somatic motor neurons and is less fatigue resistant than smooth muscle. **8.** Stratified squamous epithelium composes the mucosa of the oral cavity, oropharynx, and laryngopharynx. **9.** The three major salivary glands are the parotid, the submandibular, and the sublingual. Their secretions initiate the digestion of carbohydrates (salivary amylase) and fats (salivary lipase). **10.** The trigeminal nerve (CN V) innervates the teeth: the maxillary branch (CN V_2) to the teeth in the upper jaw; the mandibular branch (CN V_3) to the teeth in the lower jaw. **11.** The mucosal epithelium changes from stratified squamous in the esophagus to simple columnar in the stomach. The muscularis externa in the stomach has an additional deep layer, the oblique layer. **12.** The stomach is located in the left hypochondriac, epigastric, and umbilical regions of the abdomen. **13.** Chief cells produce pepsinogen and gastric lipase; parietal cells produce HCl; the surface epithelia secrete bicarbonate-buffered mucus. **14.** Intestinal epithelial cells live for 3–6 days. These cells are replaced by cells from the rapidly dividing undifferentiated epithelial cells that line the intestinal glands. **15.** The portions of the large intestine are the cecum, ascending colon, transverse colon, descending colon, sigmoid colon, rectum, and anal canal. **16.** The end products of carbohydrate and protein digestion are absorbed into the capillary network within the villus; digested fats are absorbed into the lacteal. **17.** The right and left hepatic arteries carry oxygen-rich blood into the liver; the right and left branches of the hepatic portal vein carry nutrient-rich blood into the liver; and the right and left hepatic ducts carry bile out of the liver. **18.** The path of blood from the digestive tract through the liver lobule is as follows: from the hepatic portal vein, to the portal venules in the portal triad at the periphery of a liver lobule, through the liver sinusoids, into the central vein of the lobule. **19.** The acinar cells of the pancreas produce and secrete digestive enzymes. These secretions empty into the duodenum. **20.** Many diseases are passed through fecal contamination of food and water. Washing your hands after using the restroom is an effective means to prevent the spread of infection. **21.** The mucosal epithelium is derived from endoderm. The submucosa and muscularis externa are derived from splanchnic lateral plate mesoderm. **22.** The superior mesenteric artery supplies the structures that develop from the embryonic midgut.

Review Questions 1. c **2.** a **3.** (1) d; (2) d; (3) d; (4) c; (5) c; (6) c **4.** (1) d; (2) c; (3) f; (4) a; (5) b; (6) e **5.** a **6.** a **7.** a,d **8.** d **9.** b **10.** a **11.** a **12.** (1) e; (2) a; (3) b; (4) d; (5) c **13.** b **14.** d **15.** d **16.** (1) e; (2) h; (3) c; (4) b; (5) g; (6) a; (7) f; (8) d **17.** a **18.** (1) a; (2) b; (3) c; (4) a; (5) b; (6) c; (7) b

Chapter 24

Check Your Understanding 1. The kidneys lie behind the parietal peritoneum, so they are not within the peritoneal cavity; they are retroperitoneal. They are, however, in the abdominal cavity. **2.** The kidneys extend from the eleventh or twelfth thoracic vertebra superiorly to the third lumbar vertebra inferiorly. The right kidney is slightly inferior to the left because the liver takes up much of the upper right abdominal quadrant. **3.** The interlobar arteries and veins pass through the renal columns. **4.** The renal corpuscle (glomerulus and glomerular capsule), and the proximal and distal tubules are located exclusively in the renal cortex. The nephron loop and the collecting duct extend into the medulla. **5.** Filtration occurs at the glomerular capillaries. Secretion and resorption occur across the peritubular capillaries. **6.** The juxtaglomerular complex functions in blood pressure regulation. The cells at the terminal portion of the nephron loop, the macula densa, are elongated where they abut the afferent arteriole. These cells monitor filtrate solute concentration and initiate a response when this concentration falls below a certain level. **7.** The ureter carries urine from the kidney to the urinary bladder. The urethra carries urine from the urinary bladder to the outside of the body. **8.** Transitional epithelium lines the ureter, bladder, and proximal urethra. **9.** The external urethral sphincter is innervated by somatic (voluntary) motor neurons. **10.** The urethra in females is much shorter than in males and is located close to the vaginal and anal openings. Bacteria from these regions can readily enter the urethra, particularly during sexual activity. **11.** All three pairs of kidneys—pronephros, mesonephros, and metanephros—form from intermediate mesoderm. **12.** A cloaca is the common opening of the digestive tract and the urinary and reproductive ducts.

Review Questions 1. d **2.** b **3.** c **4.** c **5.** d **6.** b **7.** e **8.** e **9.** a **10.** c **11.** b **12.** b **13.** (a) 1; (b) 6; (c) 12; (d) 3; (e) 7; (f) 10; (g) 2; (h) 5; (i) 9; (j) 11; (k) 4; (l) 8 **14.** (1) c; (2) d; (3) e; (4) b; (5) a

Chapter 25

Check Your Understanding 1. Interstitial cells are clustered in the loose connective tissue between the seminiferous tubules (thus their name). These cells secrete testosterone. **2.** The tunica albuginea is the fibrous capsule covering the testis. The tunica vaginalis is superficial to the tunica albuginea. It is a two-layered serous sac (a remnant of the peritoneum) that partially surrounds the testis. **3.** In the epididymis, sperm gain the ability to swim and the ability to fertilize the egg. **4.** The ejaculatory duct is located at the junction of the ampulla of the ductus deferens and the duct of the seminal gland. This duct passes through the prostate and empties into the prostatic urethra. **5.** Approximately 60% of the volume of semen is produced by the seminal glands. **6.** The urethra passes through the corpus spongiosum and thus this portion of the urethra is called the spongy urethra. **7.** Each primary spermatocyte produces four spermatids. **8.** Sustentocytes help form sperm by nourishing spermatogenic cells and moving these cells toward the tubule lumen. Secretions from sustentocytes influence testosterone and FSH levels and function either to stimulate or to inhibit sperm formation. Tight junctions between adjacent sustentocytes form the blood testis barrier, which protects sperm from attack by the immune system. **9.** The head of the sperm contains the nucleus and the acrosome; the midpiece contains mitochondria; and the tail is a flagellum. **10.** The uterine tubes are lined with ciliated and unciliated columnar epithelium. The beating of the cilia produces movement of the peritoneal fluid to draw the oocyte into the uterine tube. **11.** Peritoneal folds that support the female reproductive structures are the broad ligament composed of the mesometrium to the uterus, the mesosalpinx to the uterine tubes, and the mesovarium and suspensory ligament to the ovary. **12.** (a) The fundus of the uterus is the region superior to the entrance of the uterine tubes. (b) The infundibulum is the lateral funnel-shaped opening of the uterine tube. (c) The endometrium is the innermost layer of the uterine wall. **13.** Follicle-stimulating hormone (from the anterior pituitary) stimulates follicle development in the ovary. **14.** Oogonia differentiate into ooctyes during the fetal period in females. In males, spermatogonia differentiate into spermatoctyes throughout a male's adult life. **15.** The corpus luteum is the remnant of the ovarian follicle after ovulation. The corpus luteum secretes progesterone, which stimulates the secretory phase of the uterine cycle. **16.** (a) The female labia majora are homologous to the male scrotum. (b) The clitoris is homologous to the corpora cavernosa of the penis. (c) The bulb of the vestibule is homologous to the corpus spongiosum. **17.** The glandular portion of the breast develops midway through pregnancy. **18.** Once a sperm fuses to the plasma membrane of the ooctye, the cortical reaction prevents additional sperm from binding to and entering the egg. **19.** The decidua basalis is from the maternal endometrium; the chorionic villi are from the embryonic trophoblast. **20.** The placenta produces progesterone and human chorionic gonadotropin (HCG) throughout the pregnancy and produces estrogens and corticotropin-releasing hormone toward the end of gestation. **21.** (a) PSA tests are used to screen for prostate cancer. (b) Pap smears screen for cervical cancer. (c) Mammography screens for breast cancer. (d) Self-examination of the testicles is advised as screening for testicular cancer. **22.** In a male embryo, the mesonephric duct forms the efferent ductules, the epididymis, the ductus deferens, and the ejaculatory duct. The paramesonephric ducts degenerate in a male embryo. **23.** The testosterone-induced secondary sex characteristics in males are pubic, facial, and axillary hair; enlargement of the larynx; increased secretion of sebaceous glands; and increased skeletal and muscular mass. In females, estrogen-induced secondary sex characteristic include an increase in subcutaneous fat at the hips and breast and widening of the pelvic girdle.

Review Questions 1. d **2.** (1) d; (2) c; (3) d; (4) c; (5) e **3.** d **4.** a **5.** d **6.** e **7.** a **8.** a **9.** a **10.** b **11.** c **12.** d **13.** d **14.** c **15.** (a) 1; (b) 4; (c) 2; (d) 6; (e) 7; (f) 3; (g) 9; (h) 5; (i) 8; (j) 10 **16.** b **17.** a **18.** (1) a; (2) c; (3) b; (4) d; (5) c; (6) e

GLOSSARY

Pronunciation Key

The pronunciation key that is used throughout this book takes advantage of the fact that in scientific terms, long vowels usually occur at the end of syllables, whereas short vowels usually occur at the beginning or in the middle of syllables. Therefore:

1. When vowels are *unmarked,* they have a long sound when they occur at the end of a syllable, and a short sound when they are at the beginning or in the middle of a syllable. For example, in the word *kidney* (kid′ne), you automatically know that the middle 'i' is short and the terminal 'e' sound is long. In the word *renal* (re′nal), the 'e' is long, and the 'a' is short.
2. When a vowel sound violates the above rule, it is marked with a short or long symbol. Only those short vowels that come at the end of a syllable are marked with the short symbol, called a breve (˘); and only those long vowels that occur at the start or middle of a syllable are marked with a long symbol (‒). For example, in the word *pelvirectal* (pel″vĭ-rek′tal), all four vowels are short, but only the 'i' is marked as short, because it is the only vowel that falls at the end of a syllable. In *methane* (meth′ān), the long 'a' is marked as such because it does not fall at the end of a syllable.
3. Short 'a' is *never* marked with a breve. Instead, short 'a' at the end of a syllable is indicated by 'ah.' For example, in the word *papilla* (pah-pil′ah), each 'a' is short and pronounced as 'ah.'
4. In words that have more than one syllable, the syllable with the strongest accent is followed by a prime (′) mark, and syllables with the second strongest accent, where present, are indicated with a double prime (″). The unaccented syllables are followed by dashes. An example of these principles is the word *anesthetic* (an″es-thet′ik), in which *thet′* is emphasized most strongly, *an″* has a secondary emphasis, and the other two syllables are not spoken with any emphasis.

Abdomen (ab-do′men) Region of the body between the diaphragm and the pelvis.

Abduct (ab-dukt′) To move away from the midline of the body.

Absorption Process by which the products of digestion pass through the lining of the alimentary canal into the blood or lymph.

Accessory digestive organs Organs that contribute to the digestive process but are not part of the alimentary canal, including the tongue, teeth, salivary glands, pancreas, liver, and gallbladder.

Acetabulum (as″Ð-tab′u-lum) Cuplike cavity on lateral surface of hip bone that receives the femur.

Actin (ak′tin) A contractile protein in cells, especially abundant in muscle cells.

Action potential *See* Nerve impulse.

Acute Producing severe symptoms in the short term; rapidly developing.

Adduct (ah-dukt′) To move toward the midline of the body.

Adenohypophysis (ad″ĕ-no-hi-pof′ĭ-sis) The glandular part of the pituitary; also called the anterior lobe of the pituitary.

Adenoids (ad′ĕ-noids) The pharyngeal tonsil on the roof of the pharynx.

Adenosine triphosphate (ATP) (ah-den′o-sēn) Molecule in cells that stores and releases chemical energy for use in body cells.

Adipose (ad′ĭ-pōs) A loose connective tissue; fat.

Adrenal gland (ah-dre′nal) Hormone-secreting gland located superior to the kidney; consists of medulla and cortex areas; also called *suprarenal gland.*

Adrenaline (ah-dren′ah-lin) *See* Epinephrine.

Adrenocorticotropic hormone (ACTH) (ah-dre″no-kor″tĭ-ko-trop′ik) Hormone from the anterior pituitary that influences the activity of the adrenal cortex.

Adventitia (ad″ven-tish′e-ah) Outermost layer or covering of an organ; consists of connective tissue.

Aerobic (a′er-ōb-ik) Requiring oxygen.

Afferent (af′er-ent) Carrying to or toward; especially a nerve fiber that carries impulses toward the central nervous system; afferent neurons are sensory neurons.

Agonist (ag′o-nist) Muscle that bears primary responsibility for causing a certain movement; also called a *prime mover.*

AIDS (acquired immune deficiency syndrome) Disease caused by the human immunodeficiency virus (HIV): symptoms include severe weight loss, swollen lymph nodes, and many opportunistic infections that lead to death.

Aldosterone (al-dos′ter-ōn) Hormone secreted by the adrenal cortex that stimulates resorption of sodium ions and water from the kidney.

Alimentary canal (al″ĭ-men′tar-e) The digestive tube, extending from the mouth to the anus; its basic regions are the oral cavity, pharynx, esophagus, stomach, and small and large intestines.

Allantois (ah-lan′to-is) A tubular extension of the embryonic hindgut and cloaca; becomes the urachus, a fibrous cord attached to the adult bladder.

Alveolus (al-ve′o-lus) (1) One of the microscopic air sacs of the lungs; (2) a spherical sac formed by the secretory cells in a gland; also called an *acinus;* (3) the socket of a tooth.

Amino acid (ah-me′no) Organic compound containing nitrogen, carbon, hydrogen, and oxygen; building block of proteins.

Amnion (am′ne-on) Membrane that forms a fluid-filled sac around the embryo and fetus.

Amphiarthrosis (am″fe-ar-thro′sis) A slightly movable joint.

Anaerobic (an″a-er-o′bik) Not requiring oxygen.

Anastomosis (ah-nas″to-mo′sis) A union or joining of blood vessels or other tubular structures.

Androgen (an′dro-jen) A male sex hormone, the main example of which is testosterone.

Anemia (ah-ne′me-ah) Reduced oxygen-carrying capacity of the blood; results from too few erythrocytes or from abnormal hemoglobin.

Aneurysm (an′u-rizm) Blood-filled dilation of a blood vessel, caused by a weakening of the vessel wall.

Angina pectoris (an-ji′nah) Severe, suffocating chest pain caused by temporary lack of oxygen supply to heart muscle.

Antagonist (an-tag′o-nist) Muscle that reverses, or opposes, the action of another muscle.

Anterior The front of an organism, organ, or body part; the ventral direction.

Anterior lobe of pituitary *See* Adenohypophysis.

Antibody A protein molecule that is secreted by a plasma cell (a cell derived from an activated B lymphocyte) and that binds to an antigen in immune responses.

Antidiuretic hormone (ADH) (an″tĭ-di″u-ret′ik) Hormone produced by the hypothalamus and released by the posterior part of the pituitary gland (pars nervosa); stimulates the kidney to resorb more water.

Antigen (an′tĭ-jen) A molecule that is recognized as foreign by the immune system, activates the immune system, and reacts with immune cells or antibodies.

Anus (a′nus) The opening at the distal end of the alimentary canal.

Aorta (a-or′tah) Major systemic artery; arises from the left ventricle of the heart.

Apocrine gland (ap′o-krin) A type of sweat gland in the armpit and anal-genital regions; produces a secretion containing water, salts, proteins, and lipids. Becomes active at puberty.

Aponeurosis (ap″o-nu-ro′sis) Fibrous sheet connecting a muscle to the body part it moves.

Appendicular skeleton (ap″en-dik′u-lar) Bones of the limbs and limb girdles that are attached to the axial skeleton.

Apoptosis (ap″o-to′sis) Programmed cell death. Controlled cellular suicide that eliminates cells that are stressed, unneeded, excessive, or aged.

Aqueous humor (a′kwe-us) Watery fluid in the anterior segment of the eye.

Arachnoid mater (ah-rak′noid ma′ter) The weblike middle layer of the three meninges.

Areola (ah-re′o-lah) Circular, pigmented area of skin surrounding the nipple.

Arrector pili (ah-rek′tor) Tiny band of smooth muscle attached to each hair follicle; its contraction causes the hair to stand upright.

Arteriole (ar-te′re-ōl) A minute artery.

Arteriosclerosis (ar-te″re-o-sklĕ-ro′sis) Hardening of the arteries: any of a number of degenerative changes in the walls of arteries leading to a decrease in their elasticity. (*See* Atherosclerosis.)

Artery Vessel that carries blood away from the heart.

Arthritis (ar-thri′tis) Inflammation of joints.

Articular capsule (ar-tik′u-lar) The capsule of a synovial joint; consists of an outer layer of fibrous connective tissue and an inner synovial membrane.

Articulation Joint; point where two elements of the skeleton meet.

Atherosclerosis (ath″er-o″skle-ro′sis) Changes in the walls of large arteries involving the deposit of lipid plaques; the most common variety of arteriosclerosis.

Atlas First cervical vertebra.

Atria (a′tre-ah) Paired, superiorly located heart chambers that receive blood returning to the heart (singular: atrium).

Atrioventricular bundle (AV bundle) (a″tre-o-ven-trik′u-lar) Bundle of cardiac muscle cells that conducts impulses from the AV node to the walls of the right and left ventricles; located in the septum (wall) between the two ventricles of the heart; also called *bundle of His.*

Atrioventricular node (AV node) Specialized mass of conducting cells located in the interatrial septum of the heart.

Atrophy (at′ro-fe) Reduction in size or wasting away of an organ or tissue.

Auditory ossicles (os′ĭ-k′lz) The three tiny bones in the middle ear: malleus (hammer), incus (anvil), and stapes (stirrup).

Autonomic nervous system (ANS) (aw″to-nom′ik) General visceral motor division of the peripheral nervous system; innervates smooth and cardiac muscle, and glands.

Avascular (a-vas′ku-lar) Having no blood supply; containing no blood vessels.

Axial skeleton Portion of the skeleton that forms the central (longitudinal) axis of the body; includes the bones of the skull, the vertebral column, and the bony thorax.

Axilla (ak-sil′ah) Armpit.

Axis (1) Second cervical vertebra; (2) imaginary line about which a joint or structure rotates.

Axon Neuron process that carries impulses away from the cell body.

B cells Lymphocytes that oversee humoral immunity; they divide to generate plasma cells, which secrete antibodies; also called *B lymphocytes.*

Basal lamina A thin sheet of protein that underlies an epithelium.

Basal nuclei (ganglia) Areas of gray matter located deep within the white matter of the cerebral hemispheres; regulate certain aspects of movement.

Basement membrane A layer between an epithelium and the under lying connective tissue; consists of both a basal lamina and a network of reticular fibers.

Basophil (ba′so-fil) (1) Type of white blood cell whose cytoplasmic granules stain purple with the basic dyes in blood stains; mediates late stages of inflammation. (2) A gland cell in the anterior pituitary containing cytoplasmic granules that stain with basic dyes.

Benign (be-nīn) Not malignant; not life-threatening.

Biceps (bi′seps) Two-headed, especially applied to certain muscles.

Bile Greenish fluid secreted by the liver, stored in the gallbladder, and released into the small intestine; helps emulsify fats.

Biopsy (bi′op-se) Removing a piece of living tissue to examine it under a microscope. Usually done to diagnose a suspected disease condition.

Bipolar neuron Neuron with just two processes, which extend from opposite sides of the cell body.

Blastocyst (blas′to-sist) Stage of early embryonic development; a hollow ball of cells; the product of cleavage.

Blood brain barrier The feature that inhibits passage of harmful materials from the blood into brain tissues; reflects relative impermeability of brain capillaries.

Blood pressure Force exerted by blood against a unit area of the blood vessel walls; differences in blood pressure between different areas of the circulation provide the driving force for blood circulation.

Blood stem cell The cell type, present throughout life, from which all blood cells (erythrocytes, leukocytes, platelets) arise. Present in the bone marrow. It gives rise not only to blood cells, but also to mast cells, osteoclasts, and dendritic cells of the immune system. Also called *pluripotential hematopoietic stem cell.*

Bolus (bo′lus) A rounded mass of food prepared by the mouth for swallowing.

Bone remodeling Process involving bone formation and bone destruction in response to mechanical and hormonal factors.

Brain stem Collectively, the midbrain, pons, and medulla of the brain.

Bronchus Any of the air tubes of the respiratory tree between the trachea and bronchioles; bronchi enter and branch within the lungs.

Bursa A fibrous sac lined with synovial membrane and containing synovial fluid; occurs between bones and tendons (or other structures), where it acts to decrease friction during movement.

Calcaneal (Achilles) tendon Tendon that attaches the calf muscles to the heel bone.

Calcitonin (kal″sĭ-to′nin) Hormone released by the thyroid gland that promotes a decrease in calcium levels in the blood.

Calyx (ka′liks) A cuplike tributary of the pelvis of the kidney.

Cancer A malignant, invasive cellular tumor that has the capability of spreading throughout the body or body parts.

Capillary The smallest of blood vessels and the site of exchanges of molecules between the blood and the tissue fluid.

Carcinogen (kar-sin′o-jen) Cancer-causing agent.

Cardiac muscle Muscle tissue of the heart wall.

Cardiac sphincter The circular layer of smooth muscle at the junction of the esophagus and stomach; contracts to prevent reflux of stomach contents into the esophagus.

Cartilage (kar′tĭ-lij) White, semiopaque, resilient connective tissue. Three types, distinguished by fibers in extracellular matrix.

Caudal (kaw′dal) Literally, toward the tail; in humans, toward the inferior portion of the trunk.

Cecum (se′kum) The blind-ended pouch at the beginning of the large intestine.

Cell membrane *See* Plasma membrane.

Central nervous system (CNS) Brain and spinal cord.

Centriole (sen′tre-ōl) Barrel-shaped organelle formed of microtubules and located near the nucleus of the cell; active in cell division.

Cerebellum (ser″ĕ-bel′um) Brain region that is attached to the pons and smooths and coordinates body movements.

Cerebral aqueduct (sĕ-re′bral ak′wĕ-dukt″) The narrow cavity of the midbrain that connects the third and fourth ventricles.

Cerebral arterial circle A union of arteries at the base of the brain; also called *circle of Willis*.

Cerebral cortex The external, gray matter region of the cerebral hemispheres.

Cerebrospinal fluid (CSF) (ser″ĕ-bro-spi′nal) Clear fluid that fills the cavities of the central nervous system and surrounds the CNS externally; it floats, cushions and nourishes the brain and spinal cord.

Cerebrum (ser′ĕ-brum) The cerebral hemispheres (some authorities also include the diencephalon).

Cervix (ser′viks) The inferior, necklike part of the uterus (*cervix* = neck).

Chemoreceptor (ke″mo-re-sep′tor) Receptor sensitive to chemicals in solution.

Cholesterol (ko-les′ter-ol) A steroid lipid found in animal fats as well as in the plasma membranes of cells.

Chondroblast (kon′dro-blast) Actively mitotic form of a cartilage cell.

Chondrocyte (kon′dro-sīt) Mature form of a cartilage cell.

Chorion (ko′re-on) The outermost fetal membrane; helps form the placenta; technically, it consists of the trophoblast and the extraembryonic mesoderm.

Choroid plexus (ko′roid) A capillary-rich membrane on the roof of the brain that forms the cerebrospinal fluid; technically, it consists of pia mater and ependymal cells.

Chromatin (kro′mah-tin) Strands in the cell nucleus that consist of deoxyribonucleic acid (DNA) and histone proteins.

Chromosome (kro′mo-sōm) Barlike body of tightly coiled chromatin, visible during cell division; typical human cells have 46 chromosomes.

Chronic (kron′ik) Long-term; prolonged; not acute.

Chyme (kīm) Semifluid, creamy mass consisting of partially digested food and stomach juices.

Cilium (sil′e-um) Motile, hairlike projection from the apical surface of certain epithelial cells.

Circumduction (ser″kum-duk′shun) Movement of a body part so that it outlines a cone in space.

Cirrhosis (si-ro′sis) A chronic disease, particularly of the liver, characterized by an overgrowth of connective tissue, or fibrosis.

Cleavage An early embryonic stage consisting of rapid cell divisions without intervening growth periods; begins with a fertilized ovum and produces a blastocyst.

Cochlea (kok′le-ah) Snail-shaped chamber of the bony labyrinth in the internal ear; houses the receptor for hearing (spiral organ).

Cognition (kog-nish′un) All aspects of thinking, perceiving, and intentionally remembering and recalling information; the mental processes involved in obtaining knowledge.

Collagen fibers The strongest and most abundant fibrous component of connective tissues; composed of fibrils arranged in bundles. Has high tensile strength.

Collateral ganglia Ganglia of the sympathetic nervous system that are not part of the sympathetic trunk. Located anterior to the vertebral column (thus also called *prevertebral ganglia*) in the abdomen and pelvis; includes the celiac, superior mesenteric, inferior mesenteric, and inferior hypogastric ganglia. Contain postganglionic neurons that innervate the abdominopelvic organs.

Commissure A bundle of axons that crosses from one side of the central nervous system to the other.

Condyle (kon′dīl) A rounded projection at the end of a bone that articulates with another bone.

Cone cell One of the two types of photoreceptor cells in the retina of the eye; provides for color vision and sharp vision.

Congenital (kon-jen′ĭ-tal) Existing at birth.

Conjunctiva (kon″junk-ti′vah) Thin, protective mucous membrane that covers the white of the eye and the internal surface of eyelids.

Connective tissue A primary tissue; form and function vary widely, but all connective tissues contain a large amount of extracellular matrix; functions include support, holding tissue fluid, and protection from disease.

Constriction Narrowing of a blood or lymphatic vessel, or of an opening like the pupil. Often caused by the squeezing action of circular musculature. *See* Dilation.

Contraction The generation of a pulling force while shortening; this ability is highly developed in muscle cells.

Contralateral (kon″trah-lat′er-al) Concerning the opposite half of the body; when nerve fibers project contralaterally, they cross over to the opposite side of the body (right to left, or vice versa).

Cornea (kor′ne-ah) Transparent anterior portion of the eyeball.

Corona radiata (kŏ-ro′nah ra-de-ah′tah) (1) Crownlike arrangement of granulosa cells around an oocyte in an ovarian follicle after the appearance of an antrum; (2) crownlike arrangement of nerve fibers in the white matter of the cerebrum, radiating to and from every part of the cerebral cortex.

Coronal plane (kŏ-ro′nal) *See* Frontal plane.

Cortex (kor′teks) Outer region of an organ.

Corticosteroids (kor″tĭ-ko-ste′roids) Steroid hormones secreted by the adrenal cortex. Examples are cortisol, aldosterone, and some sex hormones.

Cortisol (kor′tĭ-sol) A glucocorticoid hormone produced by the adrenal cortex.

Cranial nerves The 12 pairs of nerves that attach to the brain.

Cutaneous (ku-ta′ne-us) Pertaining to the skin.

Cytokinesis (si″to-ki-ne′sis) Division of the cytoplasm that occurs after the cell nucleus has divided.

Cytoplasm (si′to-plazm) The part of a cell between the plasma membrane and the nucleus; contains many organelles.

Cytotoxic ($CD8^+$) T lymphocyte T lymphocyte that directly kills eukaryotic foreign cells, cancer cells, or virus-infected body cells; also called *killer T cell*.

Decussation A crossing of structures in the form of an X. Often applied to axons that cross the body midline from the left to the right side (or vice versa) of the central nervous system.

Deep Toward the inside; inner; internal.

Defecation (def′ĕ-ka-shun) Elimination of the contents (feces) of the bowel.

Dendrite (den′drīt) Neuron process that transmits signals toward the cell body and serves as receptive region of the neuron; most dendrites branch extensively.

Deoxyribonucleic acid (DNA) (de-ok″sĭ-ri″bo-nu-kle′ik) A nucleic acid found in all living cells; carries the organism's hereditary information.

Dermis The leathery layer of skin, deep to the epidermis; composed largely of dense irregular connective tissue.

Desmosome (des′mo-sōm) A cell junction composed of two disc-shaped plaques connected across the intercellular space; the most important junction for holding epithelial cells together.

Diabetes insipidus (di″ah-be′tēz) Disease characterized by passage of a large quantity of dilute urine plus intense thirst caused by inadequate release of, or kidney response to, antidiuretic hormone.

Diabetes mellitus Disease caused by deficient release or use of insulin; characterized by an inability of body cells to use glucose at a normal rate and by high glucose levels in blood and urine.

Diapedesis (di″ah-pĕ-de″sis) Active movement of white blood cells through the walls of capillaries and venules into the surrounding tissue.

Diaphragm (di′ah-fram) (1) Any partition or wall separating one area from another; (2) the muscular sheet that separates the thoracic cavity from the abdominopelvic cavity.

Diaphysis (di-af′ĭ-sis) Elongated shaft of a long bone.
Diarthrosis (di″ar-thro′sis) Freely movable joint; all synovial joints are diarthroses.
Diastole (di-as′to-le) Period during which the ventricles or atria of the heart relax.
Diencephalon (di″en-sef′ah-lon) The part of the forebrain between the cerebral hemispheres and the midbrain; includes the thalamus, hypothalamus, and third ventricle.
Diffusion (dĭ-fu′zhun) The spreading of particles in a gas or solution from regions of high particle concentration to regions of low concentration, with movement toward a uniform distribution of the particles.
Digestion Chemical and mechanical process of breaking down foodstuffs into molecules that can be absorbed.
Dilation Expansion or widening of a vessel, organ, or opening. (*See* Constriction.)
Distal (dis′tal) Away from the attached end of a structure, especially a limb.
Diverticulum (di″ver-tik′u-lum) A pouch or a sac in the walls of a hollow organ or structure.
Dorsal (dor′sal) Pertaining to the back; posterior.
Duct Canal or passageway, usually tubular.
Ductus (vas) deferens (def′er-ens) Extends from the epididymis to the urethra; propels sperm into the urethra via peristalsis during ejaculation.
Duodenum (du″o-de′num) First part of the small intestine.
Dura mater (du′rah ma′ter) Most external and toughest of the three membranes (meninges) covering the brain and spinal cord.
Ectoderm (ek′to-derm) Embryonic germ layer that forms both the outer layer of the skin (epidermis) and nervous tissues.
Edema (ĕ-de′mah) Abnormal accumulation of tissue fluid in the loose connective tissue; causes the affected body region to swell.
Effector (ef-fek′tor) Muscle or gland capable of being activated by motor nerve endings.
Efferent (ef′er-ent) Carrying away or away from, especially a nerve fiber that carries impulses away from the central nervous system; efferent neurons are motor neurons.
Elastin (e-las′tin) Main protein in elastic fibers of connective tissues; stretchable and resilient.
Embolus (embolism) (em′bo-lus) Any abnormal mass carried freely in the bloodstream; maybe a blood clot, bubbles of air, mass of fat, or clumps of cells.
Embryo (em′bre-o) The developing human from week 2 through week 8 after fertilization.
Endocarditis (en″do-kar-di′tis) Inflammation of the inner lining of the heart.
Endocardium (en″do-kar′de-um) The layer that lines the inner surface of the heart wall; consists of endothelium and areolar connective tissue.
Endocrine glands (en′do-krin) Ductless glands that secrete hormones into the blood.
Endocrine system Body system consisting of glands that secrete hormones.
Endocytosis (en″do-si-to′sis) Processes by which large molecules and particles enter cells; types are phagocytosis, pinocytosis, and receptor-mediated endocytosis.
Endoderm (en′do-derm) An embryonic germ layer that forms the lining and glands of the digestive and respiratory tubes.
Endometrium (en″do-me′tre-um) Mucous membrane lining the uterus.
Endomysium (en″do-mis′e-um) Thin connective tissue surrounding each muscle cell.
Endoplasmic reticulum (ER) (en″do-plaz′mik rĕ-tik′u-lum) A system of membranous envelopes and tubes in the cytoplasm of a cell; there are smooth and rough varieties.
Endosteum (en-dos′te-um) Layer of cells lining the internal surfaces of bone, specifically, lining the central canals of osteons and the medullary cavity, and covering the trabeculae of spongy bone.
Endothelium (en″do-the′le-um) The simple squamous epithelium that lines the walls of the heart, blood vessels, and lymphatic vessels.
Enzyme (en′zīm) A protein that acts as a biological catalyst to speed up chemical reactions.
Eosinophil (e″o-sin′o-fil) Granular white blood cell whose granules readily take up a pink dye called eosin; helps end allergy reaction and fights parasites.
Epidermis (ep″ĭ-der′mis) Superficial layer of the skin, composed of a keratinized stratified squamous epithelium.
Epididymis (ep″ĭ-did′ĭ-mis) Comma-shaped structure in the scrotum adjacent to the testis; contains a duct in which the sperm mature.
Epiglottis (ep″ĭ-glot′is) A leaf-shaped piece of elastic cartilage that extends from the posterior surface of the tongue to the larynx; covers the opening of the larynx during swallowing.
Epimysium (ep″ĭ-mis′e-um) Sheath of fibrous connective tissue surrounding a muscle.
Epinephrine (ep″ĭ-nef′rin) Chief hormone produced by the adrenal medulla; also called *adrenaline.*
Epiphyseal plate (ep″ĭ-fiz′e-al) Plate of hyaline cartilage at the junction of the diaphysis (shaft) and epiphysis (end) of most bones in the growing skeleton; provides growth in the length of the bone.
Epiphysis (e-pif′ĭ-sis) The end of a long bone, attached to the shaft.
Epithelium (ep″ĭ-the′le-um) A primary tissue that covers body surfaces and lines body cavities; its cells are arranged in sheets; also forms glands.
Equilibrium The sense of balance; measures both the position and the movement of the head.
Erythrocyte (ĕ-rith′ro-sīt) Red blood cell; when mature, an erythrocyte is literally a sac of hemoglobin (oxygen-carrying protein) covered by a plasma membrane.
Estrogens (es′tro-jens) Female sex hormones.
Exocrine glands (ek′so-krin) Glands that secrete onto body surfaces or into body cavities; except for the one-celled goblet cells, all exocrine glands have ducts.
Exocytosis (ek″so-si-to′sis) Mechanism by which substances are moved from the cell interior to the extracellular space; the main mechanism of secretion.
Expiration Act of expelling air from the lungs; exhalation.
Exteroceptor (ek″ster-o-sep′tor) Sensory end organ that responds to stimuli from the external world.
Extracellular Outside a cell.
Extracellular matrix (ma′triks) The material that lies between the cells in connective tissues; consists of fibers, ground substance, and tissue fluid.
Extrinsic (ek-strin′sik) Originating outside an organ or part.
Facet (fas′et) A smooth, nearly flat surface on a bone for articulation.
Fallopian tube (fah-lo′pe-an) *See* Uterine tube.
Fascia (fash′e-ah) Layers of fibrous connective tissue that cover and separate muscles and other structures. Superficial fascia is the fatty hypodermis of the skin.
Fascicle (fas′ĭ-k′l) Bundle of nerve or muscle fibers bound together by connective tissue.
Fenestrated (fen′es-tra-ted) Pierced with one or more small openings or pores.
Fertilization Fusion of the sperm and egg nuclei.
Fetus (fe′tus) Developmental stage lasting from week 9 of development to birth.
Fibroblast (fi′bro-blast) Young, actively mitotic cell that secretes the fibers and ground substance of connective tissue proper.
Fibrocyte (fi′bro-sīt) Mature fibroblast; maintains the matrix of connective tissue proper.
Filtration Passage of a solution or suspension through a membrane or filter, with the purpose of holding back the larger particles.

Fissure (fish′er) (1) A groove or cleft; (2) the deepest depressions or inward folds on the brain.

Fixator (fik-sa′tor) Muscle that immobilizes one or more bones, allowing other muscles to act from a stable base.

Flagellum (flah-jel′um) Long, whiplike extension of the plasma membrane of some bacteria and sperm cells; propels the cell.

Follicle (fol′ĭ-k′l) (1) Spherical structure in the ovary consisting of a developing egg cell surrounded by one or more layers of follicle cells; (2) colloid-containing structure in the thyroid gland.

Follicle-stimulating hormone (FSH) Hormone secreted by the anterior pituitary that stimulates the maturation of ovarian follicles in females and the production of sperm in males.

Foramen (fo-ra′men) Hole or opening in a bone or between body cavities.

Forebrain Rostral portion of the brain, consisting of the telencephalon and diencephalon.

Formed elements Blood cells (red and white cells and platelets).

Fossa (fos′ah) A depression, often a joint surface.

Fovea (fo′ve-ah) A pit.

Frontal (coronal) plane Vertical plane that divides the body into anterior and posterior parts.

Fundus (fun′dus) The base of an organ; that part farthest from the opening of an organ.

Funiculus (fu-nik′u-lus) (1) A cordlike structure; (2) a division of the white matter in the spinal cord.

Gallbladder Sac inferior to the right part of the liver; it stores and concentrates bile.

Gamete (gam′ēt) Sex cell; sperm or oocyte.

Gametogenesis (gam″ĕ-to-jen′ĕ-sis) Formation of gametes.

Ganglion (gang′gle-on; plural, *ganglia*) Collection of neuron cell bodies outside the central nervous system.

Gap junction A passageway between two adjacent cells; formed by transmembrane proteins called connexons.

Gene One of the biological units of heredity located in chromatin; transmits hereditary information; roughly speaking, one gene codes for the manufacture of one protein.

General Pertaining to sensory inputs or motor outputs that are *widely distributed* through the body rather than localized; opposite of *special.*

Germ layers Three cellular layers (ectoderm, mesoderm, and endoderm) that represent the early specialization of cells in the embryonic body and from which all body tissues arise.

Gestation (jes-ta′shun) The period of pregnancy; averages 280 days in humans.

Gland A structure whose cells are specialized for secretion.

Glial cells (gle′al) *See* Neuroglia.

Glomerular capsule (glo-mer′u-lar) Double-walled cup forming the initial portion of the nephron in the kidney; also called *Bowman's capsule.*

Glomerulus (glo-mer′u-lus) (1) A ball of capillaries forming part of the nephron in the kidney; forms a filtrate that will be modified into urine. (2) A cluster of complex synapses within the olfactory bulb between the axonal branches of the olfactory nerve and a dendritic process of a mitral cell.

Glottis (glot′is) The opening between the two vocal cords in the larynx.

Glucagon (gloo′kah-gon) Hormone secreted by alpha cells of the pancreatic islets; raises the glucose level of blood.

Glucocorticoids (gloo″ko-kor′tĭ-koids) Hormones secreted by the cortex of the adrenal gland; they increase the concentration of glucose in the blood and aid the body in resisting long-term stress.

Glucose (gloo′kōs) The principal blood sugar; the main sugar used by cells for energy.

Glycogen (gli′ko-jen) A long chain of glucose molecules; the main form in which sugar is stored in animal cells; glycogen takes the form of dense granules in the cytoplasm.

Goblet cells Individual mucus-secreting cells of the respiratory and digestive tracts.

Gonad (go′nad) Primary reproductive organ: the testis of the male or the ovary of the female.

Gonadotropins (go-nad″o-trōp′ins) Gonad-stimulating hormones secreted by the anterior pituitary: follicle-stimulating hormone and luteinizing hormone.

Gray matter Gray area of the central nervous system; contains neuron cell bodies and unmyelinated processes of neurons.

Ground substance The nonfiber part of the extracellular matrix of connective tissue. In many connective tissues it is a gel-like substance with large molecules that attract water and hold tissue fluid; in bone it is hard.

Growth hormone Hormone that stimulates growth of the body; secreted by the anterior pituitary; also called somatotropin and somatotropic hormone (SH).

Gustation (gus-ta′shun) Taste.

Gyrus (ji-rus) A ridge on the surface of the cerebral cortex.

Hair follicle Tubelike invagination of the epidermis of the skin from which a hair grows.

Haustra (hos′trah) Pouches (sacculations) of the colon.

Heart block Impaired transmission of impulses from atria to ventricles.

Heart murmur Abnormal heart sound (usually resulting from valve problems).

Helper ($CD4^+$) T cell A type of T lymphocyte that participates in the activation of other lymphocytes by secreting chemicals that stimulate newly activated lymphocytes to multiply.

Hematocrit (he-mat′o-krit) The percentage of total blood volume occupied by erythrocytes.

Hematoma (he″mah-to′mah) A mass of blood that has bled from blood vessels into the tissues.

Hematopoiesis (hem″ah-to-poi-e′sis) Blood cell formation; hemopoiesis.

Hemoglobin (he′mo-glo″bin) Oxygen-transporting protein in erythrocytes.

Hemopoiesis *See* Hematopoiesis.

Hemorrhage (hem′ŏ-rij) Bleeding.

Hepatic portal system (hĕ-pat′ik) The part of the circulation in which veins receive nutrients from capillaries in the stomach and intestines and carry these nutrients to capillaries in the liver; the liver cells then process the nutrients.

Hepatitis (hep″ah-ti′tis) Inflammation of the liver.

Hernia (her′ne-ah) Abnormal protrusion of an organ or body part through the containing wall of its cavity.

Hilum (hi′lum) A slit on the surface of an organ through which the vessels and nerves enter and leave; the spleen, lungs, kidneys, lymph nodes, and ovaries have prominent hilums.

Histamine (his′tah-mēn) Chemical substance that increases vascular permeability in the initial stages of inflammation.

Histology (his-tol′o-je) Branch of anatomy dealing with the microscopic structure of tissues, cells, and organs.

Holocrine gland (hol′o-krin) A gland in which entire cells break up to form the secretion product; sebaceous (oil) glands of the skin are the only example.

Hormones Messenger molecules that are released by endocrine glands and travel in the blood to regulate specific body functions.

Hypertension (hi″per-ten′shun) High blood pressure.

Hypertrophy (hi-per′tro-fe) Enlargement of an organ or tissue due to an increase in the size of its cells.

Hypodermis (hi″po-der′mis) The fatty layer deep to the skin; consists of adipose and areolar connective tissue; also called the *subcutaneous layer* and *superficial fascia.*

Hypophysis (hi-pof′ĭ-sis) The pituitary gland.

Hypothalamus (hi″po-thal′ah-mus) Inferior region of the diencephalon; visceral control center of the brain.

Ileum (il′e-um) Coiled terminal part of the small intestine, located between the jejunum and the cecum of the large intestine.

Immune system Organ system consisting of lymphocytes (T and B cells), lymphoid tissues, and lymphoid organs (lymph nodes, spleen, thymus, tonsils, and aggregated lymphoid nodules in the small intestine and appendix); responsible for antigen-specific defenses mounted by activated lymphocytes.

Immunity (ĭ-mu′nĭ-te) Ability of the body to develop resistance to specific foreign agents (both living and nonliving) that can cause disease.

Immunocompetence Ability of the body's immune system to recognize specific antigens.

Induction (ĭn-dŭk′shun) The influence exerted by a group of cells on the differentiation of adjacent cells or on the development of an embryonic structure.

Inferior (caudal) Below; toward the feet.

Inflammation (in″flah-ma′shun) A physiological response of the body to tissue injury; includes dilation of blood vessels and an increase in capillary permeability; indicated by redness, heat, swelling, and pain in the affected area.

Inguinal (ing′gwĭ-nal) Pertaining to the groin region.

Inner cell mass Accumulation of cells in the blastocyst from which the body of the embryo derives.

Innervation (in″er-va′shun) Supply of nerves to a body part.

Insertion Movable part or attachment of a muscle, as opposed to the muscle's origin.

Inspiration Drawing of air into the lungs; inhalation.

Insulin (in′su-lin) Hormone secreted by beta cells in pancreatic islets; it decreases blood glucose levels.

Integumentary system (in-teg″u-men′tar-e) The skin and its appendages (hairs, nails, and skin glands).

Intercalated discs (in-ter′kah-la″ted) Complex junctions that interconnect cardiac muscle cells in the wall of the heart.

Intercellular Between body cells.

Internal Deep to.

Internal capsule Band of white matter in the brain, between the basal ganglia and the thalamus.

Internal respiration Exchange of gases between blood and tissue fluid and between tissue fluid and cells.

Interneuron (in″ter-nu′ron) (1) Nerve cell that lies between a sensory neuron and a motor neuron in a reflex arc; (2) any nerve cell that is confined entirely within the central nervous system.

Interoceptor (in″ter-o-sep′tor) Nerve ending situated in a visceral organ; responds to changes and stimuli within the body's internal environment; also called *visceroceptor.*

Interstitial fluid (in″ter-stish′al) *See* Tissue fluid.

Intervertebral discs (in″ter-ver′tĕ-bral) The discs between the vertebrae of the spinal column; each consists of fibrous rings surrounding a springy core.

Intervertebral foramina (fo-ra′min-ah) Openings between the dorsal projections of adjacent vertebrae through which the spinal nerves pass.

Intracellular Within a cell.

Intraperitoneal Within the peritoneal cavity.

Invaginate To form an inpocketing or to grow inward. For example, an epithelium pushes inward, pocketlike, into the underlying connective tissue to form a gland during development.

Ion (i′on) Atom or molecule with a positive or negative electrical charge.

Ipsilateral (ip″sĭ-lat′er-al) Situated on the same side of the body; opposite of *contralateral.*

Ischemia (is-ke′me-ah) Local decrease in blood supply.

Jejunum (jĕ-joo′num) The coiled part of the small intestine that is located between the duodenum and ileum.

Joint Junction of two or more elements of the skeleton; an articulation.

Keratin (ker′ah-tin) Tension-resisting protein found in the epidermis, hair, and nails; keratin makes these structures tough and able to resist friction.

Labium (la′be-um) Lip.

Labyrinth (lab′ĭ-rinth) Bony cavities and membranes of the internal ear.

Lacrimal (lak′rĭ-mal) Pertaining to tears.

Lactation (lak-ta′shun) Production and secretion of milk.

Lacteal (lak′te-al) Lymphatic capillaries in the small intestine that take up lipids.

Lacuna (lah-ku′nah) Little depression or cavity; in bone and cartilage, each lacuna is occupied by a cell.

Lamina (lam′ĭ-nah) (1) A thin layer or flat plate; (2) the portion of a vertebra between the transverse process and the spinous process.

Laparoscopy (lap″ah-ros′ko-pe; "observing the flank") Examination of the peritoneal cavity and the associated organs with a laparoscope, a viewing device (endoscope) at the end of a thin tube that is inserted through the abdominal wall.

Larynx (lar′ingks) Cartilaginous organ located between the pharynx and trachea; contains the vocal cords; the voice box.

Lateral Away from the body midline.

Leukocyte (loo′ko-sīt) White blood cell; the five types of leukocytes are all involved in the defense against disease.

Ligament (lig′ah-ment) Band of dense regular connective tissue that connects bones.

Limbic system (lim′bik) A functional brain system involved in emotional and visceral responses; structurally, it includes medial portions of the cerebral cortex (septal nuclei, cingulate gyrus, hippocampal formation, and amygdala), the fornix, and parts of the diencephalon (hypothalamus and anterior thalamic nuclei).

Lumbar (lum′bar) Region of the back between the thorax and the pelvis.

Lumen (lu′men) The cavity inside a tube, blood vessel, or hollow organ.

Luteinizing hormone (LH) (loo′te-in-īz″ing) A hormone secreted by the anterior pituitary; in females, it aids maturation of follicles in the ovary and triggers ovulation; in males, it signals the interstitial cells of the testis to secrete testosterone.

Lymph (limf) The clear fluid transported by the lymphatic vessels.

Lymph node Bean-shaped lymphoid organ that filters and cleanses the lymph.

Lymphatic system (lim-fat′ik) Organ system consisting of lymphatic vessels, lymph nodes, and the lymphoid organs and tissues; drains excess tissue fluid and fights disease.

Lymphatics General term used to designate lymphatic vessels.

Lymphocyte (lim′fo-sīt) Agranular white blood cell that arises from bone marrow and becomes functionally activated in the lymphoid organs of the body; the main cell type of the immune system; each lymphocyte recognizes a specific antigen.

Lymphoid organs The organs of the lymphatic system that house lymphocytes and function in immunity; spleen, lymph nodes, tonsils, and thymus are the main examples.

Lymphoid tissue The main tissue of the immune system; a reticular connective tissue that houses and activates many lymphocytes.

Lyse (līz) To break up or disintegrate.

Lysosome (li′so-sōm) A membrane-bound, saclike cytoplasmic organelle that contains a wide variety of digestive enzymes.

Macrophages (mak′ro-fāj-es) The general phagocytic cells of the body, capable of engulfing and digesting a wide variety of foreign cells, particles, and molecules; present throughout the connective tissues of the body and especially abundant in lymphoid tissues of the immune system.

Malignant (mah-lig′nant) Life-threatening; pertains to neoplasms such as cancer that spread and lead to death.

Mammary glands (mam′ar-e) The breasts.

Mast cell A connective tissue cell containing secretory granules that initiate and mediate local inflammatory responses.

Mastication (mas″tĭ-ka′shun) Chewing.

Meatus (me-a′tus) A canal or opening.

Mechanoreceptor (mek″ah-no-re-sep′tor) Receptor sensitive to mechanical forces, such as touch, stretch, pressure, or vibration.

Medial Toward the midline of the body.
Median (me′de-an) In the midline of the body; midsagittal.
Mediastinum (me″de-ah-sti′num) Region of the thoracic cavity between the lungs; contains the heart, thoracic aorta, esophagus, and other structures.
Medulla (me-dul′ah) Middle or internal region of certain organs.
Medulla oblongata (ob″long-gah′tah) Inferior part of the brain stem.
Meiosis (mi-o′sis) A process of nuclear division that occurs during the production of the sex cells and reduces the chromosome number by half; results in the formation of haploid (n) cells.
Melanin (mel′ah-nin) Dark pigment formed by cells called melanocytes; imparts color to the skin and hair.
Memory lymphocytes T and B lymphocytes that provide for immunological memory (acquired, long-term immunity from diseases).
Meninges (mĕ-nin′jēz) Protective coverings around the brain and spinal cord; from external to internal, they are the dura mater, arachnoid mater, and pia mater.
Meningitis (men″in-ji′tis) Inflammation of the meninges.
Menstrual cycle (men′stroo-al) The changes in the female reproductive organs that occur every month (28 days, on the average).
Menstruation (men″stroo-a′shun) Menstrual phase of the uterine cycle; the periodic, cyclic discharge of blood and tissue from the lining of the female uterus in the absence of pregnancy.
Mesencephalon (mes″en-sef′ah-lon) Midbrain.
Mesenchyme (mes′eng-kīm) The type of embryonic tissue from which connective tissues and muscle tissues arise.
Mesenteries (mes′en-ter″ēz) Double-layered sheets of peritoneum that support most organs in the abdominopelvic cavity.
Mesoderm (mez′o-derm) The embryonic germ layer that gives rise to most structures in the body, including the skeleton, muscles, dermis, connective tissues, kidneys, and gonads.
Metabolic rate (mĕ-tah-bol′ik) Energy expended by the body per unit time.
Metabolism (mĕ-tab′o-lizm) Sum total of all the chemical reactions occurring in the cells of the body.
Metastasis (mĕ-tas′tah-sis) The spread of cancer from one body part or organ to another not directly connected to it.
Microvilli (mi″kro-vil′i) Immotile, cellular projections on the free surface of most epithelia; microvilli anchor sheets of mucus or increase surface area for absorption.
Micturition (mik″tu-rish′un) Urination, or voiding; emptying the bladder.
Midbrain Region of the brain stem that lies between the diencephalon and the pons.
Mineralocorticoids (min″er-al-o-kor′tĭ-koids) Steroid hormones secreted by the adrenal cortex that increase the resorption of sodium and water by the kidneys; the main mineralocorticoid is aldosterone.
Mitochondrion (mi″to-kon′dre-on) Cytoplasmic organelle that generates most adenosine triphosphate (ATP) for cellular activities; mitochondria are the cell's "power plants."
Mitosis (mi-to′sis) Division of the nucleus during the typical process of cell division, during which the chromosomes are distributed to the two daughter nuclei.
Mitral valve The valve on the left side of the heart between the left atrium and left ventricle. Also called the *bicuspid valve.*
Mixed nerve A nerve containing fibers of both sensory and motor neurons; all spinal nerves are mixed nerves.
Monocyte (mon′o-sīt) An agranular white blood cell, with a large nucleus that is often bent into the shape of a C; the largest of all blood cells; develops into a macrophage.
Motor neuron Nerve cell that signals muscle cells to contract or gland cells to secrete; also called *efferent neuron.*
Motor unit A motor neuron and all of the muscle cells it stimulates.
Mucosa (mu-ko′sah) *See* Mucous membranes.
Mucous membranes Moist membranes that line all tubular organs and body cavities that open to the exterior (digestive, respiratory, urinary, and reproductive tracts).
Mucus (mu′kus) A sticky, viscous fluid that covers many internal surfaces in the body; it consists of the protein mucin and a large amount of water.
Multipolar neuron (mul″tĭ-po′lar) A nerve cell that has more than two processes; most neurons are multipolar, having several dendrites and an axon.
Muscle fiber Either a smooth or skeletal muscle cell.
Muscle spindle Complex, spindle-shaped receptor in skeletal muscles that senses muscle stretch.
Muscle tone Continuous, low levels of contractile force produced by muscles that are not actively shortening.
Myelencephalon (mi″el-en-sef′ah-lon) Caudal part of the hindbrain; the medulla oblongata.
Myelin sheath (mi′ĕ-lin) Fatty insulating sheath that surrounds all but the thinnest nerve fibers; formed of the plasma membrane of neuroglia wrapped in concentric layers around the nerve fiber.
Myocardial infarction (mi″o-kar′de-al) Condition characterized by dead tissue areas in the myocardium of the heart; caused by interruption of blood supply to the area; also called *heart attack.*
Myocardium (mi″o-kar′de-um) Layer of the heart wall composed of cardiac muscle.
Myofibril (mi″o-fi′bril) Rodlike bundle of contractile myofilaments in the cytoplasm of a skeletal muscle cell; made of repeating segments called sarcomeres.
Myofilament (mi″o-fil′ah-ment) The contractile filaments in muscle cells; the two varieties are thick (myosin) filaments and thin (actin) filaments.
Myometrium (mi″o-me′tre-um) The thick layer of smooth muscle in the wall of the uterus.
Myosin (mi′o-sin) A contractile protein in cells, especially abundant in muscle cells.
Nares (na′rēz) The nostrils.
Necrosis (nĕ-kro′sis) Death of a cell or tissue caused by disease or injury.
Neoplasm (ne′o-plazm) *See* Tumor.
Nephron (nef′ron) Smallest functional unit in the kidney. Produces urine.
Nerve A collection of nerve fibers (long axons) in the peripheral nervous system.
Nerve fiber Any long axon of a neuron.
Nerve impulse A large, transient depolarization event, including polarity reversal, that is conducted along the plasma membrane of a nerve axon or muscle cell without diminishing in intensity. Also called an *action potential*.
Neural crest Embryonic tissue derived from ectoderm that migrates widely within the embryo and gives rise to sensory neurons, all nerve ganglia, melanocytes, and other structures.
Neuroglia (nu-rog′le-ah) Nonexcitable cells of neural tissue that support, protect, and insulate neurons; glial cells.
Neurohypophysis (nu″ro-hi-pof′ĭ-sis) The part of the pituitary gland that derives from the brain; contains the stalklike infundibulum and the posterior lobe of the pituitary.
Neurolemma (nu″ro-lem′ah) Cell membrane of a neuron.
Neuromuscular junction Location where the terminal bouton of a motor neuron meets a skeletal muscle fiber; this junction transmits the stimulus to contract from the neuron to the muscle fiber. Also called *motor end plate.*
Neuron (nu′ron) Cell of the nervous system specialized to generate and transmit electrical signals; a nerve cell.
Neurotransmitter (nu″ro-trans′mit-er) Chemical released by neurons that may, upon binding to receptors on neurons or effector cells, stimulate or inhibit them.
Neutrophil (nu′tro-fil) Most abundant type of white blood cell; a granulocyte specialized for destroying bacteria.
Nucleic acid (nu-kle′ik) The class of organic molecules that includes deoxyribonucleic acid (DNA) and ribonucleic acid (RNA).
Nucleolus (nu-kle′o-lus) A small, dark-staining body in the cell nucleus; represents parts of several chromosomes and manufactures the basic subunits of ribosomes.

Nucleus (nu′kle-us) (1) Control center of a cell; contains genetic material; (2) a cluster of neuron cell bodies in the brain.
Occipital (ok-sip′ĭ-tal) Pertaining to the area at the back of the head.
Occlusion (ŏ-kloo′zhun) Closure or obstruction.
Olfaction (ol-fak′shun) Smell.
Olfactory epithelium (ol-fak′to-re) A sensory receptor region in the superior lining of the nasal cavity; this epithelium contains olfactory neurons that respond to odors in the air.
Oocyte (o′o-sīt) Immature egg undergoing the process of meiosis.
Oogenesis (o″o-jen′ĕ-sis) Process of ovum (female gamete) formation.
Ophthalmic (of-thal′mik) Pertaining to the eye.
Optic (op′tik) Pertaining to the eye.
Optic chiasma (ki-az′mah) A cross-shaped structure anterior to the diencephalon of the brain, representing the point of crossover of half the axons of the optic nerve.
Organ A part of the body formed of two or more tissues and adapted to carry out a specific function; the stomach and biceps brachii muscle are examples of large organs, but many organs are smaller and simpler (sweat gland, hair follicle, muscle spindle).
Organ system A group of organs that work together to perform a vital body function; e.g., the nervous system.
Organelles (or″gah-nelz′) Small structures in the cytoplasm (ribosomes, mitochondria, and others) that perform specific functions for the cell. Nucleus is also an organelle.
Origin Attachment of a muscle that remains relatively fixed during muscular contraction.
Osmosis (oz-mo′sis) Diffusion of a solvent (water molecules) through a membrane from a dilute solution into a more concentrated one.
Ossification (os″ĭ-fĭ-ka′shun) Bone formation.
Osteoblast (os′te-o-blast″) A bone-forming cell.
Osteoclast (os′te-o-klast″) Large cell that resorbs or breaks down the bone matrix.
Osteocyte (os′te-o-sīt″) Mature bone cell, shaped like a spider with a body and long processes, that occupies a lacuna in the bone matrix.
Osteogenesis (os″te-o-jen′ĕ-sis) Process of bone formation.
Osteon (os′te-on) Tube-shaped unit in mature, compact bone; consists of concentric layers of bone lamellae surrounding a central canal; also called *Haversian system.*
Osteoporosis (os″te-o-po-ro′sis) Age-related condition (affects many elderly women) in which bones weaken as bone resorption outpaces bone deposition; the weakened bones break easily.
Ovarian cycle (o-va′re-an) Monthly cycle of follicle development in the ovaries, ovulation, and formation of the corpus luteum; the menstrual cycle as it involves the ovaries.
Ovary (o′var-e) Female sex organ in which ova (eggs) are produced; the female gonad in the pelvis.
Ovulation (o″vu-la′shun) Ejection of an egg (oocyte) from the ovary.
Ovum (o′vum) (1) General meaning: the female germ cell, or egg; (2) specific meaning: female germ cell after a sperm has entered it but before the sperm nucleus and the egg nucleus have fused.
Oxytocin (ok″sĭ-to′sin) Hormone produced by the hypothalamus and released by the posterior pituitary; it stimulates contraction of the uterus during childbirth and the ejection of milk during nursing.
Palate (pal′at) Roof of the mouth.
Palpation (pal-pa′shun) Using one's fingers to feel deep organs through the skin of the body surface.
Pancreas (pan′kre-as) Tadpole-shaped gland posterior to the stomach; produces both exocrine and endocrine secretions.
Pancreatic juice Bicarbonate-rich secretion of the pancreas containing enzymes for digestion of all food categories.
Parasympathetic division (par″ah-sim″pah-thet′ik) The division of the autonomic nervous system that oversees digestion, elimination, and glandular function; the resting and digesting division.
Parathyroid glands (par″ah-thi′roid) Small endocrine glands located on the posterior aspect of the thyroid gland.
Parathyroid hormone (PTH) Hormone secreted by the parathyroid glands; it increases the concentration of calcium ions in the blood.
Paresthesia An abnormal sensation of numbness, burning, or tingling.
Parietal (pah-ri′ĕ-tal) Pertaining to the walls of a cavity.
Pars distalis (parz dis-tal′is) The main division of the adenohypophysis of the pituitary gland; the anterior lobe of the pituitary.
Pars nervosa (ner-vo′sah) The region of the neurohypophysis of the pituitary gland from which hormones are secreted; the posterior lobe of the pituitary.
Pectoral (pek′tor-al) Pertaining to the chest.
Pectoral girdle Bones that attach an upper limb to the axial skeleton: clavicle and scapula.
Pelvic girdle The paired hip bones and sacrum, which attach the lower limbs to the axial skeleton.
Pelvis (pel′vis) Inferior region of the body trunk; contains the basin-shaped, bony structure called the bony pelvis.
Pepsin (pep′sin) Protein-digesting enzyme secreted by the stomach lining.
Pericardial cavity (per″ĭ-kar′de-al) Space between the parietal and visceral layers of pericardium. Contains a small amount of serous fluid.
Pericardium (per″ĭ-kar′de-um) Double-layered sac that encloses the heart and forms its superficial layer.
Perichondrium (per″ĭ-kon′dre-um) Membrane of fibrous connective tissue that covers the external surface of cartilages.
Perimysium (per″ĭ-mis′e-um) Connective tissue that surrounds and separates fascicles (bundles) of muscle fibers within a skeletal muscle.
Perineum (per″ĭ-ne′um) Region of the trunk superficial to the pelvic diaphragm and bounded by the pubic symphysis anteriorly, coccyx posteriorly, and ischial tuberosities laterally. Contains the anus, vulva (females), and scrotum (males).
Periosteum (per″e-os′te-um) Membrane of fibrous connective tissue that covers the external surface of bones of the skeleton.
Peripheral nervous system (PNS) Portion of the nervous system consisting of nerves and ganglia that lie outside the brain and spinal cord.
Peristalsis (per″ĭ-stal′sis) Progressive, wavelike contractions of smooth muscle that squeeze foodstuffs through the alimentary canal (or that move other substances through other body organs).
Peritoneal cavity (per″ĭ-to-ne′al) Space between the parietal and visceral layers of peritoneum. Contains a small amount of serous fluid.
Peritoneum (per″ĭ-to-ne′um) Serous membrane that lines the interior of the abdominopelvic cavity and covers the surfaces of the organs in this cavity.
Peritonitis (per″ĭ-to-ni′tis) Infection and inflammation of the peritoneum.
Peritubular capillaries Capillaries in the kidney that surround the proximal and distal convoluted tubules; active in resorption.
Phagocytosis (fag″o-si-to′sis) The process by which a cell forms cytoplasmic extensions to engulf foreign particles, cells, or macromolecules and then uses lysosomes to digest these substances.
Pharynx (far′ingks) Muscular tube extending from the region posterior to the nasal cavity to the esophagus; the "throat" part of the digestive tube.
Photoreceptors (fo″to-re-sep′tors) Specialized receptor cells that respond to light energy: rod cells and cone cells.
Pia mater (pi′ah ma′ter) Most internal and most delicate of the three membranes (meninges) covering the brain and spinal cord.
Pinocytosis (pin″o-si-to′sis) The process by which cells engulf extracellular fluids. Also called fluid-phase endocytosis.
Pituitary gland (pĭ-tu′ĭ-tār″e) A hormone-secreting, golf club–shaped structure that hangs inferiorly from the brain and performs a variety of endocrine functions, such as regulating the gonads, thyroid gland, adrenal cortex, lactation, and water balance. Also called the *hypophysis.*

Placenta (plah-sen′tah) Temporary organ formed from both fetal and maternal tissues that provides nutrients and oxygen to the developing fetus, carries away fetal waste molecules, and secretes the hormones of pregnancy; shed as the afterbirth when labor is over.
Plasma (plaz′mah) The nonliving, fluid component of blood, within which the blood cells are suspended.
Plasma cell Cell formed from the division of an activated B lymphocyte; secretes antibodies.
Plasma membrane Membrane that encloses cell contents; the external, limiting membrane of the cell.
Plasmalemma (plaz″mah-lem′ah) *See* Plasma membrane.
Platelet Cell fragment found in blood; plugs small tears in blood vessels and helps initiate clotting.
Pleura (ploo′rah) Serous membrane that lines the pleural cavity in the thorax and covers the external surface of the lung.
Pleural cavity (ploo′ral) Space between the parietal and visceral layers of pleura. Contains a small amount of serous fluid.
Plexus (plek′sus) A network of converging and diverging nerves or veins.
Plica (pli′kah) A fold.
Podocytes (pod′o-sīts) Octopus-shaped epithelial cells that surround the glomerular capillaries; they help produce and maintain the basement membrane (a filtration membrane in the kidney).
Pons The part of the brain stem between the midbrain and the medulla oblongata.
Portal system A system of vessels in which two capillary beds, rather than one, lie between the incoming artery and the outgoing vein of a body region; a portal vein lies between the two capillary beds. Examples are the hepatic portal system and the hypophyseal portal system.
Posterior Toward the back; dorsal.
Posterior lobe of pituitary *See* neurohypophysis.
Postganglionic neuron (post gang-gle-on′ik) Autonomic motor neuron that has its cell body in a peripheral autonomic ganglion and projects its axon to an effector.
Preganglionic neuron (pre″gang-gle-on′ik) Autonomic motor neuron that has its cell body in the central nervous system and projects its axon to a peripheral autonomic ganglion.
Prime mover Muscle that bears the major responsibility for a particular movement; agonist.
Process (1) Prominence or projection; (2) series of actions for a specific purpose; (3) projection from the cell body of a neuron.
Progesterone (pro-jes′tĕ-rōn) Hormone that prepares the uterus to receive the implanting embryo.
Pronation (pro-na′shun) Medial rotation of the forearm that causes the palm to face posteriorly.
Prone Refers to a body lying horizontally with the face downward.
Proprioceptor (pro″pre-o-sep′tor) Receptor that senses movement in the musculoskeletal system; more specifically, proprioceptors sense stretch in muscles, tendons, and joint capsules.
Protein (pro′tēn) A long chain of amino acids or several linked chains of amino acids; the amino acid chains have bent and folded (and often coiled) to give each protein a distinct shape.
Proximal (prok′sĭ-mal) Toward the attached end of a limb, or near the origin of a structure.
Pseudostratified (soo″do-strat′ĭ-fīd) Pertaining to an epithelium that appears to be stratified (consisting of more than one layer of cells) but is not; the cells vary in height, but all touch the base of the epithelium.
Puberty (pu′ber-te) Period of life when reproductive maturity is reached.
Pulmonary (pul′mo-ner″e) Pertaining to the lungs.
Pulmonary circuit System of blood vessels that serve gas exchange in the lungs: the pulmonary arteries, capillaries, and veins.
Pulse Rhythmic expansion and recoil of arteries resulting from the contraction of the heart; can be felt from outside the body.
Pupil Opening in the center of the iris through which light enters the eye.
Purkinje fibers (pur-kin′je) Long rows of modified cardiac muscle cells of the conduction system of the heart; also called *subendocardial branches.*
Pyloric region (pi-lor′ik) A funnel-shaped region of the stomach, just proximal to the pylorus.
Pylorus The distal, ring-shaped portion of the stomach that joins the small intestine and contains the pyloric sphincter muscle.
Ramus (ra′mus) Branch of a nerve, artery, or bone.
Raphe (ra′fe) A seam in the midline.
Receptor (1) Peripheral nerve ending, or complete cell, that responds to particular types of stimulus; (2) a cell component, usually a membrane protein, that binds to specific molecules to signal a certain response.
Reduction Restoring broken bone ends or dislocated bones to their original positions.
Reflex Automatic response to a stimulus.
Relay nucleus Any nucleus in the brain whose neurons receive signals from one region of the central nervous system and send this information to another region; within the relay nucleus, the information is organized and edited.
Renal (re′nal) Pertaining to the kidney.
Renin (re′nin) Hormone released by the kidneys that is involved with raising blood pressure, blood volume, and the sodium concentration in blood.
Respiratory system Organ system that carries out gas exchange; includes the nose, pharynx, larynx, trachea, bronchi, and lungs.
Rete (re′te) A network, often composed of nerve fibers or blood vessels.
Reticular cell (rĕ-tik′u-lar) A fibroblast in reticular connective tissue (in bone marrow, spleen, lymph nodes, and so on).
Reticular formation Functional system that runs through the core of the brain stem; involved in alertness, arousal, and sleep; also contains visceral centers that control heart rate, breathing rate, and vomiting; controls some body movements as well.
Reticulocyte (rĕ-tik′u-lo-sīt) Immature or young erythrocyte.
Retina (ret′ĭ-nah) Contains neural layer of the eyeball, including the photoreceptor cells for vision, and a supportive pigmented layer.
Retroperitoneal External or posterior to the peritoneum.
Rhinencephalon (ri″nen-sef′ah-lon) The part of the cerebrum that receives and integrates olfactory (smell) impulses.
Ribonucleic acid (RNA) (ri″bo-nu-kle′ik) Nucleic acid that contains the sugar ribose; acts in protein synthesis.
Ribosome (ri′bo-sōm) Cytoplasmic organelle on which proteins are synthesized.
RICE An acronym for rest, ice, compression, and elevation; standard treatment for injuries to muscles, tendons, and ligaments.
Rod cell One of the two types of photoreceptor cells in the retina of the eye.
Rostral Toward the nasal region or higher brain centers.
Rugae (roo′ge) Elevations or ridges, as in the mucosa of the stomach.
Sacral (sa′kral) Pertaining to the sacrum; the region in the midline of the buttocks.
Sagittal plane (saj′ĭ-tal) A vertical plane that divides the body or a body part into right and left portions.
Sarcolemma (sar″ko-lem′ah) The plasma membrane of a muscle cell.
Sarcomere (sar′ko-mēr) The smallest contractile unit of skeletal and cardiac muscle; the part of a myofibril between two Z discs; contains myofilaments composed mainly of contractile proteins (actin, myosin).
Sarcoplasm (sar′ko-plazm) The cytoplasm of a muscle cell.
Sarcoplasmic reticulum (SR) (sar′ko-plaz-mik rĕ-tik′u-lum) Specialized smooth endoplasmic reticulum of muscle cells; stores calcium ions.
Sclera (skle′rah) Outer fibrous layer of the eyeball.

Scrotum (skro′tum) The external sac that encloses the testes.
Sebaceous gland (se-ba′shus) Gland in the skin that produces an oily secretion called sebum.
Sebum (se′bum) The oily secretion of sebaceous glands.
Secretion (se-kre′shun) (1) The passage of material formed by a cell to its exterior; (2) cell product that is transported to the exterior of a cell.
Section A cut through the body (or an organ) along a particular plane; a thin slice of tissue prepared for microscopic study.
Semen (se′men) Fluid mixture containing sperm and secretions of the male accessory reproductive glands.
Semilunar valves (sem″ĭ-lu′nar) Valves at the base of the aorta and the pulmonary trunk that prevent blood from returning to the heart ventricles after ventricular contraction.
Seminiferous tubules (sem″ĭ-nif′er-us) Highly convoluted tubules within the testes that form sperm.
Sensory neuron Nerve cell that carries information received from sensory receptors; also called *afferent neuron.*
Serosa (se-ro′sah) *See* Serous membrane.
Serous cell Exocrine gland cell that secretes a watery product containing digestive enzymes.
Serous fluid (se′rus) A clear, watery lubricant produced by cells of a serous membrane.
Serous membrane Moist, slippery membrane that lines internal body cavities (pleural, pericardial, and peritoneal cavities) and covers visceral organs within these cavities; also called *serosa.*
Serum (se′rum) Amber-colored fluid that exudes from clotted blood plasma as the clot shrinks and then no longer contains clotting factors.
Sinoatrial (SA) node (si″no-a′tre-al) A collection of specialized cardiac muscle cells in the superior wall of the right atrium; pacemaker of the heart.
Sinus (si′nus) (1) Mucous-membrane-lined, air-filled cavity in certain bones of the face; (2) dilated channel for the passage of blood or lymph.
Sinusoid (si′nu-soid) An exceptionally wide, twisted, leaky capillary; large protein molecules or whole blood cells pass easily through the walls of sinusoids.
Skeletal muscle A striated muscle tissue composed of long, cylindrically shaped, multinucleated cells. Located in the muscles that attach to and move the skeleton.
Smooth muscle Musculature consisting of spindle-shaped, nonstriated muscle cells; present in the walls of most visceral organs.
Somatic (so-mat′ik) Pertaining to the region of the body that lies external to the ventral body cavity, including the skin, skeletal muscles, and the skeleton; opposite of *visceral.*
Somite (so′mīt) A mesodermal segment of the body of the embryo.
Special Pertaining to sensory inputs or motor outputs that are *localized* rather than being widespread through the body; opposite of *general.*
Special senses The senses whose receptors are confined to a small region rather than distributed widely through the body: taste, smell, vision, hearing, and equilibrium.
Spermatogenesis (sper″mah-to-jen′e-sis) The process by which sperm (male gametes) form in the testes; involves meiosis.
Sphincter (sfingk′ter) A muscle surrounding an opening; acts as a valve to close and open the orifice.
Spinal nerves The 31 pairs of nerves that attach to the spinal cord.
Spinal reflex A reflex mediated through the spinal cord.
Squamous (skwa′mus) Flat, platelike; pertaining to flat epithelial cells that are wider than they are tall.
Stenosis (stě-no′sis) Constriction or narrowing.
Steroids (ste′roids) Group of lipid molecules containing cholesterol and some hormones.
Stimulus (stim′u-lus) An excitant or irritant; a change in the environment that evokes a response.
Striated muscle (stri′āt-ed) Muscle consisting of cross-striated (striped) muscle fibers: skeletal and cardiac muscle.
Stroke Condition in which brain tissue is deprived of its blood supply, as in blockage of a cerebral blood vessel; also called *cerebrovascular accident.*
Stroma (stro′mah) The connective tissue framework of an organ.
Subcutaneous (sub″ku-ta′ne-us) Deep to the skin.
Sulcus (sul′kus) A groove.
Superficial Located close to or on the body surface; outer; external; opposite of *deep.*
Superficial fascia The hypodermis; the fatty layer just below the skin.
Superior Closer to the head; above.
Supination (soo″pĭ-na′shun) Lateral rotation of the forearm that causes the palm to face anteriorly.
Supine (soo′pīn) Refers to a body lying horizontally with the face upward.
Suprarenal gland (soo″prah-re′nal) *See* Adrenal gland.
Surfactant (ser-fak′tant) Detergent-like fluid secreted by type II alveolar cells lining the respiratory alveoli in the lungs; reduces the surface tension of water molecules, thus preventing collapse of the alveoli after each breath.
Suture (soo′cher) An immovable, fibrous joint; except at the jaw joint, all bones of the skull are united by sutures.
Sweat gland Tubular gland in the skin that secretes sweat (which cools the body).
Sympathetic division (sim″pah-thet′ik) Division of the autonomic nervous system that prepares the body to cope with danger or excitement; the fight-or-flight division.
Symphysis (sim′fĭ-sis) A joint in which the bones are connected by fibrocartilage.
Synapse (sin′aps) Specialized cell junction between two neurons, at which the neurons communicate.
Synaptic cleft Fluid-filled space at a synapse between neurons; also called *synaptic gap.*
Synarthrosis (sin″ar-thro′sis) Any immovable joint.
Synchondrosis (sin″kon-dro′sis) A joint in which bones are united by hyaline cartilage.
Syndesmosis (sin″des-mo′sis) A joint in which bones are united only by a ligament.
Synergist (sin′er-jist) Muscle that aids the action of a prime mover by contributing to the same movement or by stabilizing joints to prevent undesirable movements.
Synovial fluid (sĭ-no′ve-al) Fluid secreted by the synovial membranes of the freely movable joints of the body; lubricates the joint surfaces and nourishes the articular cartilages.
Synovial joint Freely movable joint with a cavity and a capsule. *See* Diarthrosis.
Systemic (sis-tem′ik) Pertaining to the whole body.
Systemic circuit System of blood vessels that carries oxygenated blood to the tissues throughout the body.
Systole (sis′to-le) Period during which the ventricles or the atria of the heart contract.
T cells Lymphocytes that mediate cellular immunity; include cytotoxic and helper T cells; also called *T lymphocytes.*
T tubule Extension of the muscle cell plasmalemma (sarcolemma) that protrudes deeply into the muscle cell.
Target cell A cell that is capable of responding to a hormone because it bears receptors to which the hormone can bind.
Taste buds Bulb-shaped sensory organs on and around the tongue that house the receptor cells for taste.
Telencephalon (tel″en-sef′ah-lon) Rostral division of the embryonic forebrain; develops into the cerebrum.
Tendon (ten′don) Cord of dense regular connective tissue that attaches muscle to bone.
Testis (tes′tis) Male primary sex organ that produces sperm; male gonad in the scrotum.

Testosterone (tes-tos′tĕ-rōn) *See* Androgen.
Thalamus (thal′ah-mus) An egg-shaped mass of gray matter in the diencephalon of the brain; consists of nuclei through which information is relayed to the cerebral cortex.
Thermoreceptor (ther″mo-re-sep′tor) Receptor sensitive to temperature changes.
Thoracic cage Skeletal structures that form the framework of the thorax; includes sternum, ribs, thoracic vertebrae, and costal cartilages.
Thoracic duct Large lymphatic duct that ascends anterior to the vertebral column; drains the lymph from up to three-fourths of the body (all except the body's superior right quarter).
Thorax (tho′raks) That portion of the body superior to the diaphragm and inferior to the neck.
Thymus (thi′mus) Organ of the immune system that is essential for the production of T cells (T lymphocytes); located in the anterior thorax.
Thyroid gland (thi′roid) Butterfly-shaped endocrine gland in the anterior neck; its main hormone (thyroid hormone) increases metabolic rate.
Tight junction A type of cell junction that closes off the intercellular space; also called a *zonula occludens.*
Tissue A group of similar cells (and extracellular material) that perform similar functions; primary tissues of the body are epithelial, connective, muscle, and nervous tissue.
Tissue fluid Watery fluid that, along with the molecules of ground substance, occupies the extracellular matrix of connective tissue; it is a filtrate of the blood containing all the small molecules of blood plasma; also called *interstitial fluid.*
Trabecula (trah-bek′u-lah) (1) Any one of the fibrous bands extending from the capsule to the interior of an organ; (2) a piece of the bony network in spongy bone.
Trachea (tra′ke-ah) The cartilage-reinforced air tube that extends from the larynx to the bronchi; windpipe.
Tract A collection of nerve fibers in the central nervous system having the same origin, destination, and function.
Transverse process One of a pair of projections that extend laterally from the neural arch of a vertebra.
Trauma (traw′mah) A wound, injury, or shock, usually caused by external forces.
Trochanter (tro-kan′ter) A large, somewhat blunt process on a bone. Located on the femur.
Trophoblast (trof′o-blast) External layer of cells in the blastocyst (early embryo); forms the embryo's contribution to the placenta.
Tropic hormone (tro′pik) A hormone that regulates the function of another endocrine organ; tropic hormones signal endocrine glands to secrete their own hormones.
Tropomyosin (trō-pō-mī′ō-sin) One of the regulator proteins of muscle contraction located on the thin myofilament. Forms a thin strand that spirals around the actin molecule, covering the binding sites for the thick myofilament, myosin.
Troponin (trō′pō-nin) One of the regulator proteins of muscle contraction located on the thin myofilament. Controls the interaction of actin with the thick myofilament, myosin. Has binding sites for calcium and tropomyosin.
Tubercle (too′ber-k′l) A nodule or small rounded process on a bone.
Tuberosity (too″bĕ-ros′ĭ-te) A broad process on a bone, larger than a tubercle.
Tumor An abnormal growth of cells; a swelling; a neoplasm; can be cancerous.
Tunica (too′nĭ-kah) A covering or coat; a layer or membrane of tissue.
Tympanic membrane (tim-pan′ik) The eardrum located between the outer and middle ear.
Ulcer (ul′ser) Erosion of the surface of an organ or tissue, such as a peptic ulcer in the wall of the stomach or small intestine.
Umbilical cord (um-bil′ĭ-kal) A cord that attaches to the navel before birth and connects the fetus to the placenta; contains the umbilical arteries and vein.
Umbilicus (um-bil′ĭ-kus) Navel; belly button.
Unipolar neuron (u″nĭ-po′lar) A sensory neuron in which a single short process projects from the cell body but divides like a T into two long processes (central process and peripheral process).
Urea (u-re′ah) The main nitrogen-containing waste excreted in urine.
Ureter (u-re′ter) Tube that carries urine from kidney to bladder.
Urethra (u-re′thrah) Tube that carries urine from the bladder to the exterior.
Uterine tube (u′ter-in) Tube through which the ovum travels to the uterus; also called *fallopian tube* and *oviduct.*
Uterus (u′ter-us) Hollow, thick-walled pelvic organ that receives the developing embryo; site where embryo/fetus develops; the womb.
Varicose vein A dilated vein filled with pooled blood resulting from the failure of the venous valve to prevent backflow in the vessel.
Vasa recta (va′sah rek′tah) Capillary-like blood vessels that supply the loops of Henle and collecting ducts in the medulla of the kidney.
Vasa vasorum Small blood vessels located in the tunica externa of large arteries that supply oxygen and nutrients to the outer layers of the vessel wall.
Vascularized (vas′ku-lar-īzd″) Having a blood supply; containing blood vessels.
Vasoconstriction (vas″o-kon-strik′shun) Narrowing of blood vessels, normally through the contraction of smooth muscle cells in the vessel walls.
Vasodilation (vas″o-di-la′shun) Relaxation of smooth muscle cells in the walls of blood vessels, causing the vessels to dilate (widen).
Vasomotor fibers (vas″o-mo′tor) Sympathetic nerve fibers that regulate the contraction of smooth muscle in the walls of blood vessels, thereby regulating the diameter of the vessels.
Vein Vessel that carries blood toward the heart.
Ventilation (ven″tĭ-la′shun) Breathing; consists of inspiration and expiration.
Ventral (ven′tral) Toward the front of the body; anterior.
Ventricles (1) Paired, inferiorly located heart chambers that function as the major blood pumps; (2) fluid-filled cavities of the brain.
Venule (ven′ūl) A small vein.
Vertebral column (ver′tē-bral) The spine or spinal column; formed of a number of bones called vertebrae, the discs between these vertebrae, and two composite bones (sacrum and coccyx).
Vesicle (ves′ĭ-k′l) A small, liquid-filled sac; also refers to the urinary bladder.
Vesicular follicle (vĕ-sik′u-lar fol′ĭ-k′l) Mature ovarian follicle; formerly called *Graafian follicle.*
Villus (vil′us) One of many fingerlike projections of the internal surface of the small intestine that together increase the surface area for nutrient absorption.
Viscera (visceral organs) (vis′er-ah) The organs within the ventral body cavity, including the stomach, bladder, heart, lungs, spleen.
Visceral Pertaining to the organs and structures within the ventral body cavity and to all smooth muscle and glands throughout the body; opposite of *somatic.*
Visceral muscle Smooth muscle and cardiac muscle.
Vitamins Organic compounds required by the body in minute amounts that generally must be obtained from the diet.
Vulva (vul′vah) The external genitalia of the female.
White matter White substance of the central nervous system; contains tracts of myelinated nerve fibers.
Yolk sac Embryonic sac that stores a tiny quantity of yolk and gives rise to the lining of the digestive tube; also gives rise to the primordial germ cells and the blood cells.
Zygote (zi′gōt) Fertilized egg.

Photo Credits

Chapter 1 Chapter Opener: Simon Fraser/Science Source. 1.2a–l: Pearson Education, Inc. 1.4.1, 1.8a: John Wilson White/Pearson Education, Inc. 1.4a: CNRI/Science Source. 1.4b: Scott Camazine/Science Source. 1.4c: James Cavallini/Science Source. 1.9a: Steve Downing/Pearson Education, Inc. 1.9b: Biophoto Associates/Science Source. 1.9c: CNRI/Science Source. 1.10a: Vanessa Vick/Science Source. 1.10b: MedImage/Science Source. 1.11.1: Blend Images/Alamy. 1.11.2: SGO/BSIP/Science Source. 1.12: Zephyr/Science Source. 1.13: Einat Even Sapir/Pearson Education, Inc. 1.14: Mangelli/Pearson Education, Inc. 1.15a: LADA/Science Source. 1.15b: National Library of Medicine. Table 1.1.1: Gustoimages/Science Source. Table 1.1.2: Scott Camazine/Science Source.

Chapter 2 Chapter Opener: MedImage/Science Source. 2.5b: Birgit Satir. 2.6b: CNRI/Science Source. 2.7b: Professors Pietro M. Motta/Tomonori Naguro/Science Source. 2.9: Dr. Gopal Murti/Science Source. 2.10c: Professors Pietro M. Motta/Tomonori Naguro/Science Source. 2.11a: Frank Solomon/J. Dinsmore/MIT.edu. 2.11b: Mary Osborn/Max Planck Institute of Neurobiology. 2.11c: Mark S. Ladinsky/Richard McIntosh. 2.12b, 2.13b.1, 2: Don W. Fawcett/Science Source. 2.13b.3: Aebi, Ueli et al. "The nuclear lamina is a meshwork of intermediate-type filaments." *Nature* 323 (9 October 1986): 560–564. doi:10.138/323560a0. 2.15a: Victoria E. Foe. 2.15b: GF Bahr. 2.17.1–6: Photographs by Dr. Conly L. Rieder, Wadsworth Center, Albany, New York 12201-0509.

Chapter 3 Chapter Opener: Dr. Yorgos Nikas/Science Source. 3.8d: Science Source. 3.11b: Lennart Nilsson Photography/TT Nyhetsbyrån. Table 3.1.1: CNRI/Science Source. Table 3.1.2: Garry Watson/Science Source. Table 3.1.3: Tissuepix/Science Source. Table 3.1.4: Petit Format/Science Source. A Closer Look: A. P. Streissguth/University of Washington.

Chapter 4 Chapter Opener: Biophoto Associates/Science Source. 4.3a, g: William Karkow/Pearson Education, Inc. 4.3b, f, h, 4.10d, k: Allen Bell/Pearson Education, Inc. 4.3c, d: Steve Downing/Pearson Education, Inc. 4.3e: Nina Zanetti/Pearson Education, Inc. 4.4a: SPL/Science Source. 4.7: Don W. Fawcett/Science Source. 4.10a, j: William Karkow/Pearson Education, Inc. 4.10b, h: Steve Downing/Pearson Education, Inc. 4.10c, g, i: Nina Zanetti/Pearson Education, Inc. 4.10e, f, l: Lisa Lee/Pearson Education, Inc. 4.11.1, 2, 10, 15, 17, 21: William Karkow/Pearson Education, Inc. 4.11.3–7, 9, 11–14, 16, 18–20, 22: Pearson Education, Inc. 4.11.8: Steve Downing/Pearson Education, Inc. 4.13a: Eric Graves/Science Source. 4.13b: Steve Downing/Pearson Education, Inc. 4.13c, 4.14: Lisa Lee/Pearson Education, Inc. A Closer Look: David Musher/Science Source.

Chapter 5 Chapter Opener: Steve Gschmeissner/Science Source. 5.2.Morton Publishing. 5.3a: William Karkow/Pearson Education, Inc. 5.4: Centers for Disease Control and Prevention. 5.5a: Clouds Hill Imaging Ltd./Science Source. 5.5c: John Wilson White/Pearson Education, Inc. 5.7b: Manfred Kage/Science Source. 5.7d: Steve Downing/Pearson Education, Inc. 5.8: Lisa Lee/Pearson Education, Inc. 5.9a: Scott Camazine/Science Source. 5.9b: Dr. M. A. Ansary/Science Source. 5.10a: Biophoto Associates/Science Source. 5.10b: P. Marazzi/SPL/Science Source. 5.10c: James Stevenson/ Science Source. A Closer Look: Phanie/Science Source.

Chapter 6 Chapter Opener: Michael Abbey/Science Source. 6.2: Nina Zanetti/Pearson Education, Inc. 6.5.1: Donald Gregory Clever/Pearson Education, Inc. 6.5.2: Steve Gschmeissner/Science Source. 6.6b: Larry DeLay/Pearson Education, Inc. 6.7c.1: Andrew Syred/Science Source. 6.7c.2: William Karkow/Pearson Education, Inc. 6.9a: Steve Gschmeissner/Science Source. 6.9b: Gwen V. Childs. 6.12a.Living Art Enterprises/Science Source. 6.12b: Deborah Vaughan, PhD. 6.14: CNRI/Science Source. 6.16a: SPL/Science Source. 6.16b: Professor Pietro M. Motta/Science Source. 6.17: Biophoto Associates/Science Source. 6.18: Scott Camazine/Science Source. Table 6.2.1: Custom Medical Stock Photo/Newscom. Table 6.2.2: Center for Disease Control and Prevention (CDC). Table 6.2.3: Brian Tidey/Pearson Education, Inc. Table 6.2.4: Dr. Frank Gaillard/Pearson Education, Inc. Table 6.2.5: William T. C. Yuh. Table 6.2.6: Hani Alsalam/Pearson Education, Inc. A Closer Look: PCN Photography/Alamy.

Chapter 7 Chapter Opener: SPL/Science Source. 7.4b: Michael Wiley/Pearson Education, Inc. 7.6b: Ralph T. Hutchings. 7.7b, 7.8: Pearson Education, Inc. 7.9b: Ralph T. Hutchings. 7.10: Pearson Education, Inc. 7.12: Michael Wiley/Pearson Education, Inc. 7.13c: Pearson Education, Inc. 7.15a: Living Art Enterprises, LLC/Science Source. 7.15b: BSIP/Science Source. 7.16a: Pearson Education, Inc. 7.22d: Creative Digital Visions/Pearson Education, Inc. 7.24b: Karen Krabbenhoft/Pearson Education, Inc. 7.25a: Pearson Education, Inc. 7.26: Center For Craniolfacial Anomalies. 7.27a: Princess Margaret Rose Orthopaedic Hospital/Science Source. 7.27b: Nordic Photos/SuperStock. 7.27c: Steve Gorton/Dorling Kindersley, Ltd. 7.CA.1: Neil Borden/Science Source. Table 7.2.7, 9: Larry DeLay/Pearson Education, Inc. Table 7.2.8: Creative Digital Visions/Pearson Education, Inc.

Chapter 8 Chapter Opener: Scott Camazine/Sue Trainor/Science Source. 8.5: John Wilson White/Pearson Education, Inc. 8.10c: M. J. Wiley/Pearson Education, Inc. 8.10d: Michael Wiley/Pearson Education, Inc. 8.12b: Elaine N. Marieb. 8.13a: Katherine Nimkin. 8.13b: Mediscan/Alamy. 8.CA.1: CMSP Custom Medical Stock Photo/Newscom. 8.CA.2: William Herring. 8.CA.3: David Nelson. 8.CA.4: Scott Camazine/Science Source. Table 8.2.3, 4: Larry Delay/Pearson Education, Inc. Table 8.2.5, 6: David L. Bassett Collection.

Chapter 9 Chapter Opener: CNRI/Science Source. 9.3b: Mark Nielsen/Pearson Education, Inc. 9.4: Robert A. Chase. 9.6, 9.7: John Wilson White/Pearson Education, Inc. 9.11b: Mark Nielsen/Pearson Education, Inc. 9.11e, 9.12c: Karen Krabbenhoft/Pearson Education, Inc. 9.14b: David L. Bassett Collection. 9.15f: Pearson Education, Inc. 9.17a: Benjamin Cummings/Pearson Education, Inc. 9.18: Elaine N. Marieb. 9.19a: Zephyr/Science Source. 9.19b: Living Art Enterprises/Science Source. 9.19c: Elaine N. Marieb. 9.20: P. Marazzi/Science Source.

Chapter 10 Chapter Opener: Sercomi/BSIP/Science Source. 10.1b: William Karkow/Pearson Education, Inc. 10.2: Biophoto Associates/Science Source. 10.4: Marian Rice. 10.6b, 10.7.1: Don W. Fawcett/Science Source. 10.7.2: Biophoto Associates/Science Source. 10.9b: Eric V. Graves/Science Source. 10.10: National Institutes of Health. 10.11a: Lisa Lee/Pearson Education, Inc. 10.11b: William Karkow/Pearson Education, Inc. Table 10.2.4: Eric Graves/Science Source. Table 10.2.5: Steve Downing/Pearson Education, Inc. Table 10.2.6: SPL/Science Source. A Closer Look: Corbis/SuperStock.

Chapter 11 Chapter Opener: David Gregs/Alamy. 11.9a: John Wilson White/Pearson Education, Inc. 11.9c, 11.11b: Karen Krabbenhoft/Pearson Education, Inc. 11.13c: David L. Bassett Collection. 11.14c: Karen Krabbenhoft/Pearson Education, Inc. 11.16b: Olga Malakhova/Pearson Education, Inc. 11.16d: Karen Krabbenhoft/Pearson Education, Inc. 11.17d: Mark Nielsen. 11.18b: Karen Krabbenhoft/Pearson Education, Inc. 11.19b: Creative Digital Visions/Pearson Education, Inc. 11.20: National Institutes of Health. 11.21: John Wilson White/Pearson Education, Inc. 11.23d: Karen Krabbenhoft/Pearson Education, Inc. 11.24: National Institutes of Health. 11.25b, 11.26b: Winston Charles Poulton/Pearson Education, Inc. 11.27d: Karen Krabbenhoft/Pearson Education, Inc. 11.28: Mark Nielsen. 11.29–11.33, 11.35–11.39: John Wilson White/Pearson Education, Inc. 11.CA.1: P. Marazzi/Science Source.

Chapter 12 Chapter Opener: Nancy Kedersha/Science Source. 12.4b: David M. Phillips/Science Source. 12.5c: Thomas Deerinck/NCMIR/Science Source. 12.7a: Don W. Fawcett/Science Source. 12.7b: Anthony M. Heape. 12.9b: Karen Krabbenhoft/Pearson Education, Inc. 12.9c: Nina Zanetti/Pearson Education, Inc. 12.14: Living Art Enterprises/Science Source. 12.17: Harkany, Tibor. "Hardwiring the Brain: Endocannabinoids Shape Neuronal Connectivity." *Science* 316, no. 5828 (2007, The American Association for the Advancement of Science): 1212–1216. 12.CA.1: Mediscan/Alamy.

Chapter 13 Chapter Opener: Innerspace Imaging/Science Source. 13.5: Pearson Education, Inc. 13.6d, 13.8, 13.9b, 13.11a: Karen Krabbenhoft/Pearson Education, Inc. 13.11b: Robert A. Chase. 13.11d: Winston Charles Poulton/Pearson Education, Inc. 13.12: WDCN/Univ. College London/Science Source. 13.17a: Karen Krabbenhoft/Pearson Education, Inc. 13.17b: From "A Stereoscopic Atlas of Human Anatomy," by David L. Bassett, M.D. 13.18b: Karen Krabbenhoft/Pearson Education, Inc. 13.22b: Robert A. Chase. 13.22c, 13.24b–d: Karen Krabbenhoft/Pearson Education, Inc. 13.26b: Lisa Lee/Pearson Education, Inc. 13.31: Patricia E. Phelps. 13.32: Biophoto Associates/Science Source. 13.CA.1: Ansary/Custom Medical Stock Photo/Newscom. A Closer Look: Claudio Villa/Getty Images.

Chapter 14 Chapter Opener: Steve Gschmeissner/Science Source. 14.4c, 14.9d: Karen Krabbenhoft/Pearson Education, Inc. 14.12a: David L. Bassett Collection. 14.14a: Pearson Education, Inc. 14.CA.2: Biophoto Associates/Science Source. Table 14.2.7: William Thompson/Pearson Education, Inc. A Closer Look: Science Source.

Chapter 15 Chapter Opener: Innerspace Imaging/Science Source. 15.6b: Karen Krabbenhoft/Pearson Education, Inc. 15.14, 15.CA.1: P. Marazzi/Science Source.

Chapter 16 Chapter Opener: Meckes Ottawa/Eye of Science/Science Source.

16.1c: Pearson Education, Inc. 16.4a: Richard Tauber/Pearson Education, Inc. 16.7b: David L. Bassett Collection. 16.9c: Lisa Lee/Pearson Education, Inc. 16.11: Dr. Charles Klettke/Pearson Education, Inc. 16.14b: Stephen Spector/Pearson Education, Inc. 16.19d: P. M. Motta/Univ. La Sapienza, Rome/ Science Source. 16.22b: I. M. Hunter-Duvar. 16.CA.1: Tea Potocnik/Getty Images. 16.CA.2, 16.CA.4: SPL/Science Source.

Chapter 17 Chapter Opener: Biophoto Associates/Science Source. 17.3d: Lisa Lee/Pearson Education, Inc. 17.5a: Karen Krabbenhoft/Pearson Education, Inc. 17.5b: Steve Downing/Pearson Education, Inc. 17.6c: Pearson Education, Inc. 17.7b, 17.9: Lisa Lee/Pearson Education, Inc. 17.10: General Photographic Agency/Getty Images. 17.11a: Ralph C. Eagle/Science Source. 17.11b: Scott Camazine/Science Source. 17.12: Dr. Charles B. Wilson. 17.16a: Lisa Lee/Pearson Education, Inc.

Chapter 18 Chapter Opener: Sercomi/ Science Source. 18.2a: National Cancer Institute/Science Source. 18.2b: William Karkow/Pearson Education, Inc. 18.4a–e: Pearson Education, Inc. 18.6: Oliver Meckes/Nicole Ottawa/ Science Source. 18.7b, c: William Karkow/Pearson Education, Inc. 18.9a: Cheryl Power/Science Source. 18.9b: Omikron/Science Source. A Closer Look: Cord Blood Registry.

Chapter 19 Chapter Opener: Quest/ Science Source. 19.2d: Mark Nielsen/ Shawn Miller. 19.5a, c, f: Karen Krabbenhoft/Pearson Education, Inc. 19.6b: Pearson Education, Inc. 19.6c: Lennart Nilsson Photography/ TT Nyhetsbyrån. 19.6d: Karen Krabbenhoft/Pearson Education, Inc. 19.9: John Wilson White/Pearson Education, Inc. 19.12a: Steve Downing/ Pearson Education, Inc. 19.13: Brenda Russell. 19.CA.1: Zotin Igor/Itar-Tass Photos/Newscom. 19.CA.2: Mediscan/ Alamy.

Chapter 20 Chapter Opener: 3D4 Medical. 20.1a: Ed Reschke. 20.2: Roger C. Wagner. 20.3: Steve Downing/ Pearson Education, Inc. 20.10c, 20.14c, 20.17b, 20.20b: Karen Krabbenhoft/ Pearson Education, Inc. 20.21b: Mark Nielsen. 20.23c: Karen Krabbenhoft/ Pearson Education, Inc. 20.24: Living Art Enterprises/LLC/Science Source. 20.CA.1: P. Marazzi/Science Source. A Closer Look 1: Sheila Terry/Science Source; 2: GJLP/Science Source.

Chapter 21 Chapter Opener: Steve Gschmeissner/Science Source. 21.3a, b: Biophoto Associates/Science Source. 21.4b: Karen Krabbenhoft/Pearson Education, Inc. 21.7: Victor Eroschenko/ Pearson Education, Inc. 21.9a: Karen Krabbenhoft/Pearson Education, Inc. 21.9b: Lisa Lee/Pearson Education, Inc. 21.10c: Mark Neilsen/Pearson Education, Inc. 21.10d: Victor Eroschenko/ Pearson Education, Inc. 21.11, 21.12: Biophoto Associates/Science Source. A Closer Look: Montagnier/Institut Pasteur/Science Source.

Chapter 22 Chapter Opener: CNRI/ Science Source. 22.2a: Jenny Thomas/ Pearson Education, Inc. 22.3a: David L. Bassett Collection. 22.4: Karen Krabbenhoft/Pearson Education, Inc. 22.5a: John Wilson White/Pearson Education, Inc. 22.5c: David L. Bassett Collection. 22.6a, b, 22.7c: CNRI/ Science Source. 22.7b: Nina Zanetti/ Pearson Education, Inc. 22.8b: London Scientific Films/Getty Images. 22.9b: Lisa Lee/Pearson Education, Inc. 22.10b: Motta/Macchiarelli/Science Source. 22.11a: Richard Tauber/ Pearson Education, Inc. 22.11b: David L. Bassett Collection. 22.11c: Karen Krabbenhoft/Pearson Education, Inc. 22.11e: National Institutes of Health. 22.18a, b: Centers for Disease Control and Prevention. 22.CA.1: Zephyr/ Science Source. A Closer Look 1: Karen Krabbenhoft/Pearson Education, Inc; 2: Mediscan/Alamy.

Chapter 23 Chapter Opener: Susumu Nishinaga/Science Source. 23.2a: John Wilson White/Pearson Education, Inc. 23.4a: David L. Bassett Collection. 23.7b: Pearson Education, Inc. 23.8b: Steve Downing/Pearson Education, Inc. 23.13b: Elaine N. Marieb. 23.15b: Science Photo Library/ Science Source. 23.16a: Biophoto Associates/Science Source. 23.16b: Lisa Lee/Pearson Education, Inc. 23.17b: Robert A. Chase. 23.18d: Biophoto Associates/Science Source. 23.19: Mark Nielsen. 23.20c: Steve Gschmeissner/ Science Source. 23.23: William Karkow/Pearson Education, Inc. 23.24.1: Steve Downing/Pearson Education, Inc. 23.24.2: Lisa Lee/Pearson Education, Inc. 23.24.3, 4: David G. King. 23.25, 23.26a, 23.27b: David L. Bassett Collection. 23.28a: Robert A. Chase. 23.28c: Lisa Lee/Pearson Education, Inc. 23.29a: CNRI/Science Source. 23.29b: Eye of Science/Science Source. 23.CA.1: Gastrolab/Science Source.

Chapter 24 Chapter Opener: Biophoto Associates/Science Source. 24.1b: Mark Nielsen. 24.2c: Ruedi Thoeni. 24.3a: Karen Krabbenhoft/ Pearson Education, Inc. 24.6c: P. Motta/M. Castellucci/SPL/Science Source. 24.8a: Lisa Lee/Pearson Education, Inc. 24.8b: Morton Publishing. 24.9b: Steve Gschmeissner/Science Source. 24.10: Morton Publishing. 24.11: Biophoto Associates/Science Source. 24.13a: Lisa Lee/Pearson Education, Inc. 24.13b: Pearson Benjamin Cummings. 24.CA.1: National Institutes of Health.

Chapter 25 Chapter Opener: P. M. Motta/J. Van Blerkom/Science Source. 25.3b: Pearson Education, Inc. 25.3c: Ed Reschke. 25.4a: Victor Eroschenko/ Pearson Education, Inc. 25.4b: Biophoto Associates/Science Source. 25.9a: P. M. Motta/Univ. La Sapienza Rome/Science Source. 25.10b: Juergen Berger/Science Source. 25.13: Lisa Lee/Pearson Education, Inc. 25.14a, b: Morton Publishing. 25.15: Winston Charles Poulton/Pearson Education, Inc. 25.16a: Pearson Education, Inc. 25.18.1–3: Science Pictures Ltd./ Science Source. 25.18.4: Biophoto Associates/Science Source. 25.18.5: Steve Downing/Pearson Education, Inc. 25.18.6: C. Edelman/La Vilette/ Science Source. 25.18.7: Lester V. Bergman/Corbis. 25.22b: Lennart Nilsson Photography/TT Nyhetsbyrån. 25.23b: CC Studios/Science Photo Library/Science Source. 25.24c: Allen C. Enders/Carnegie Collection. 25.28a: Mark Thomas/Science Source. 25.28b: Southern Illinois University/Science Source. 25.28c: Kings College Hospital/Science Source.

Illustration Credits

All illustrations by Imagineering STA Media Services unless otherwise noted.

Chapter 1 1.2: Vincent Perez/Wendy Hiller Gee.

Chapter 3 3.3, 3.7: Electronic Publishing Services, Inc./Imagineering STA Media Services.

Chapter 5 5.1, 5.3, 5.6: Electronic Publishing Services, Inc./Imagineering STA Media Services.

Chapter 6 6.8: Imagineering STA Media Services/Precision Graphics

Chapter 7 7.4a, 7.7a, b, 7.9a, b: Nadine Sokol.

Chapter 10 10.8, 10.13: Electronic Publishing Services, Inc.

Chapter 12 12.3, 12.5, 12.8, 12.10, 12.11, 12.15, Table 12.1.1, 3–6: Electronic Publishing Services, Inc.

Chapter 13 13.1–13.4, 13.6–13.8, 13.11, 13.13–13.23, 13.27–13.30: Electronic Publishing Services, Inc. 13.25: Marieb, Elaine N., Mitchell, Susan J., *Human Anatomy and Psychology Lab Manual, Main Version,* 8th ed., © 2008. Reprinted and electronically reproduced by permission of Pearson Education, Inc., Upper Saddle River, New Jersey.

Chapter 14 14.4: Electronic Publishing Services, Inc.

Chapter 15 15.2–15.4, 15.8, 15.11, 15.12: Electronic Publishing Services, Inc.

Chapter 16 16.2, 16.4, 16.5, 16.7, 16.12, 16.16–16.20, 16.22, 16.23: Electronic Publishing Services, Inc. 16.6, 16.21: Electronic Publishing Services, Inc./Imagineering STA Media Services.

Chapter 17 17.1: Electronic Publishing Services, Inc.

Chapter 19 19.2–19.6, 19.12, 19.14: Electronic Publishing Services, Inc.

Chapter 20 20.1, 20.5, 20.10–20.13, 20.15–20.19, 20.21, 20.23: Electronic Publishing Services, Inc. 20.8a Electronic Publishing Services, Inc. 20.8b Imagineering STA Media Services.

Chapter 22 22.1, 22.3, 22.5, 22.8, 22.10, 22.11, 22.16: Electronic Publishing Services, Inc.

Chapter 23 23.1, 23.4, 23.7–23.10, 23.15, 23.17–23.22: Electronic Publishing Services, Inc.

Chapter 24 24.1, 24.6, 24.7, 24.10, 24.12, 24.14: Electronic Publishing Services, Inc.

Chapter 25 25.1–25.3, 25.6, 25.13: Electronic Publishing Services, Inc.

INDEX

NOTE: Page numbers in **boldface** type indicate a major discussion. A *t* following a page number indicates tabular material, an *f* following a page number indicates a figure, and a *b* following a page number indicates a boxed feature.

WORD ROOTS: PREFIXES, SUFFIXES, AND COMBINING FORMS

Prefixes and Combining Forms

a-, an- *not, without* acardia, lack of a heart; anaerobic, in the absence of oxygen
ab- *off, from, away* abnormal, departing from normal
acou- *hearing* acoustics, the science of sound
acr-, acro- *at the apex, extremity; peak* acrodermatitis, inflammation of the skin of the extremities
ad- *to, toward* adorbital, toward the orbit
aden-, adeno- *gland* adeniform, resembling a gland in shape
aero- *air* aerobic respiration, oxygen-requiring metabolism
af- *toward* afferent neurons, which carry impulses to the central nervous system
alb- *white* corpus albicans of the ovary, a white scar tissue
amphi- *on both sides; of both kinds* amphibian, an organism capable of living both in water and on land
ana- *apart, up, again* anaphase of mitosis, when the chromosomes separate
angi- *vessel* angiitis, inflammation of a lymphatic vessel or blood vessel
ant-, anti- *against, opposite* anticoagulant, a substance that prevents blood coagulation
ante- *preceding; before* antecubital, in front of the elbow
ap-, api- *tip, extremity* apex of the heart
append- *hang to* appendicular skeleton
aqua-, aque- *water* aqueous solutions
areola- *open space* areolar connective tissue, a loose connective tissue
arthr-, arthro- *joint* arthropathy, any joint disease
artic- *joint* articular surfaces of bones, the points of connection
atri- *vestibule* atria, upper chambers of the heart
aut-, auto- *self* autogenous, self-generated
ax-, axi-, axo- *axis, axle* axial skeleton, axis of vertebral column
azyg- *unpaired* azygous vein, an unpaired vessel
basal *base* basal lamina of epithelial basement membrane
bi- *two* bicuspid, having two cusps
bio- *life* biology, the study of life and living organisms
blast- *bud, germ* blastocyte, undifferentiated embryonic cell
brachi- *arm* brachial plexus of peripheral nervous system supplies the arm
brev- *short* peroneus brevis, a short leg muscle
broncho- *the windpipe* bronchospasm, spasmodic contraction of bronchial muscle
bucco- *cheek* buccolabial, pertaining to the cheek and lip
capit- *head* decapitate, remove the head
carcin- *cras, cancer* carcinogen, a cancer-causing agent
cardi-, cardio- *heart* cardiotoxic, harmful to the heart
carneo- *flesh* trabeculae carneae, ridges of muscle in the ventricles of the heart
caud- *tail* caudal (directional term)
celi- *abdominal* celiac artery
cephal- *head* cephalometer, an instrument for measuring the head
cerebr- *brain, especially the cerebrum* cerebrospinal, pertaining to the brain and spinal cord
cervic, cervix *neck* cervical vertebrae, vertebrae located in the neck
chondr- *cartilage* chondrogenic, giving rise to cartilage
cili- *small hair* ciliated epithelium
circum- *around* circumnuclear, surrounding the nucleus
clavic- *key* clavicle, a "skeleton key"
co-, con- *with, together* concentric, common center, together in the center
coel- *hollow* coelom, the ventral body cavity
commis- *united* gray commissure of the spinal cord connects the two columns of gray matter
contra- *against* contraceptive, agent preventing conception
corn-, cornu- *horn* stratum corneum, outer layer of the skin composed of horny cells
corona *crown* coronal suture of the skull
corp- *body* corpse; corpus luteum, hormone-secreting body in the ovary
cort- *bark* cortex, the outer layer of the brain, kidney, adrenal glands, and lymph nodes
cost- *rib* intercostal, between the ribs
crani- *skull* craniotomy, a skull operation
crypt- *hidden* cryptomenorrhea, a condition in which menstrual symptoms are experienced but no external loss of blood occurs
cubit *the elbow* cubital fossa, depression on the anterior side of the elbow
cusp- *pointed* bicuspid and tricuspid, valves of the heart
cyst- *sac, bladder* cystitis, inflammation of the urinary bladder
cyt- *cell* cytology, the study of cells
de- from, *down, out* deactivation, becoming inactive
decid- *falling off* deciduous (milk) teeth
delta *triangular* deltoid muscle, roughly triangular in shape
den-, dent- *tooth* dens process of axis, tooth-shaped process
dendr- *tree, branch* dendrites, branches of a neuron
derm- *skin* dermis, deep layer of the skin
di- *twice, double* dimorphism, having two forms
dia- *across, separate, through, between* diaphragm, the wall through or between two areas
dors- *the back* dorsal; dorsum; dorsiflexion
duc-, duct *lead, draw* ductus deferens, structure that carries sperm from the epididymis into the urethra during ejaculation
dura- *hard* dura mater, tough outer meninx
dys- *difficult, faulty, bad* dyspepsia, disturbed digestion
ec-, ex-, ecto- *out, outside, from* excrete, to remove materials from the body
ef- *away* efferent nerve fibers, which carry impulses away from the central nervous system
en-, em- *in, into* encysted, enclosed in a cyst or capsule
endo- *within, inner* endoderm, innermost of the three embryonic layers
entero- *intestine* enterologist, one who specializes in the study of intestinal disorders
epi- *over, above* epidermis, outer layer of skin
erythr- *red* erythrocyte, red blood cell
eu- *well* euesthesia, a normal state of the senses

exo- *outside, outer layer* exophthalmos, an abnormal protrusion of the eye from the orbit
extra- *outside, beyond* extracellular, outside the body cells of an organism
extrins- *from the outside* extrinsic muscles of the tongue
fasci-, fascia- *bundle, band* superficial and deep fascia
fenestra- *window* fenestrated capillaries
fontan- *fountain* fontanelles of the skull
foram- *opening* foramen magnum of the skull
foss- *ditch* mandibular fossa of the skull
gangli- *swelling or knot* dorsal root ganglia of the spinal nerves
gastr- *stomach* gastrin, a hormone that influences gastric acid secretion
germin- *grow* germinal epithelium of the gonads
glosso- *tongue* glossopathy, any disease of the tongue
glute- *buttock* gluteus maximus, largest muscle of the buttock
gon-, gono- *seed, offspring* gonads, the sex organs
hema-, hemato-, hemo- *blood* hematocyst, a cyst containing blood
hemi- *half* hemiglossal, pertaining to one half of the tongue
hepat- *liver* hepatitis, inflammation of the liver
hetero- *different, other* heterosexuality, sexual desire for a person of the opposite sex
hist- *tissue* histology, the study of tissues
holo- *whole* holocrine glands, whose secretions are whole cells
hom-, homo- *same* homeoplasia, formation of tissue similar to normal tissue; homocentric, having the same center
hyal- *clear* hyaline cartilage, which has no visible fibers
hydr-, hydro- *water* dehydration, loss of body water
hyper- *excessive* hypertension, excessive tension
hypno- *sleep* hypnosis, a sleeplike state
hypo- *under, below, deficient* hypodermic, beneath the skin; hypokalemia, deficiency of potassium
hyster-, hystero- *uterus, womb* hysterectomy, removal ofthe uterus; hysterodynia, pain in the womb
ile- *intestine* ileum, the last portion of the small intestine
im- *not* impermeable, not permitting passage, not permeable
infra- *below, beneath* infraspinatus, the muscle located below the scapular spine
inter- *between* intercellular, between the cells
intra- *within, inside* intracellular, inside the cell
ipsi- *same* ipsilateral, on the same side
iso- *equal, same* isothermal, equal, or same, temperature
jugul- *throat* jugular veins, prominent vessels in the neck
kera- *horn* keratin, the water-repellent protein of the skin
labi-, labri- *lip* labial frenulum, the membrane which joins the lip to the gum
lact- *milk* lactiferous glands, milk-producing glands of the breast
lacun- *space, cavity, lake* lacunae, the spaces occupied by cells of cartilage and bone tissue
lamell- *small plate* concentric lamellae, rings of bone matrix in compact bone
lamina *layer, sheet* basal lamina, part of the epithelial basement membrane
lat- *wide* latissimus dorsi, a broad muscle of the back
later- *side* lateral (directional term)
leuko- *white* leukocyte, white blood cell
lingua- *tongue* lingual tonsil, adjacent to the tongue
lip-, lipo- *fat, lipid* lipophage, a cell that has taken up fat in its cytoplasm
luci- *clear* stratum lucidum, clear layer of the epidermis
lumen *light* lumen, center of a hollow structure
lut- *yellow* corpus luteum, a yellow, hormone-secreting structure in the ovary
lymph *water* lymphatic circulation, return of clear fluid to the bloodstream
macro- *large* macromolecule, large molecule
magn- *large* foramen magnum, largest opening of the skull
mal- *bad, abnormal* malfunction, abnormal functioning of an organ
mamm- *breast* mammary gland, breast
mast- *breast* mastectomy, removal of a mammary gland
medi- *middle* medial (directional term)
medull- *marrow* medulla, the middle portion of the kidney, adrenal gland, and lymph node
mega- *large* megakaryocyte, large precursor cell of platelets
meningo- *membrane* meningitis, inflammation of the membranes of the brain
meso- *middle* mesoderm, middle germ layer
meta- *after, between, change* metatarsus, the part of the foot between the tarsus and the phalanges
metro- *uterus* metroscope, instrument for examining the uterus
micro- *small* microscope, an instrument used to make small objects appear larger
mito- *thread, filament* mitochondria, small, filament-like structures located in cells
mono- *single* monospasm, spasm of a single limb
morpho- *form* morphology, the study of form and structure of organisms
multi- *many* multinuclear, having several nuclei
myelo- *spinal cord, marrow* myeloblasts, cells of the bone marrow
myo- *muscle* myocardium, heart muscle
narco- *numbness* narcotic, a drug producing stupor or numbed sensations
necro- *death* necrosis, tissue death
neo- *new* neoplasm, an abnormal growth
nephro- *kidney* nephritis, inflammation of the kidney
neuro- *nerve* neurophysiology, the physiology of the nervous system
noto- *back* notochord, the embryonic structure that precedes the vertebral column
ob- *reversed, against* obstruction, impeding or blocking up
oculo- *eye* monocular, pertaining to one eye
odonto- *tooth* orthodontist, one who specializes in proper positioning of the teeth in relation to each other
olfact- *smell* olfactory nerves
oligo- *few* oligodendrocytes, neuroglial cells with few branches
oo- *egg* oocyte, precursor of female gamete
ophthalmo- *eye* ophthalmology, the study of the eyes and related disease

orb- *circular* orbicularis oculi, muscle that encircles the eye
ortho- *straight, direct* orthopedic, correction of deformities of the musculoskeletal system
osteo- *bone* osteodermia, bony formations in the skin
oto- *ear* otoscope, a device for examining the ear
ov-, ovi- *egg* ovum, oviduct
oxy- *oxygen* oxygenation, the saturation of a substance with oxygen
pan- *all, universal* panacea, a cure-all
papill- *nipple* papillary muscles, muscles extending from the wall of the ventricles that anchor the atrioventricular valves
para- *beside, near, beyond* paraphrenitis, inflammation of tissues adjacent to the diaphragm
pariet- *a wall* parietal pleura, the layer of pleura against the thoracic wall
pect-, pectus *breast* pectoralis major, a large chest muscle
pelv- *a basin* pelvic girdle, which cradles the pelvic organs
peni- *a tail* penis, male copulatory organ
pep-, peps-, pept- *digest* pepsin, a digestive enzyme of the stomach; peptic ulcer
peri- *around* perianal, situated around the anus
phago- *eat* phagocyte, a cell that engulfs and digests particles or cells
phleb- *vein* phlebitis, inflammation of the veins
pia *tender* pia mater, delicate inner membrane around the brain and spinal cord
pin-, pino- *drink* pinocytosis, the engulfing of small particles by a cell
pleur- *side, rib* pleural serosa, the membrane that lines the thoracic cavity and covers the lungs
plex-, plexus *net, network* brachial plexus, the network of nerves that supplies the arm
pneumo- *lungs, air, wind* pneumothorax, air in the pleural cavity
pod- *foot* podiatry, the treatment of foot disorders
poly- *multiple* polymorphism, multiple forms
post- *after, behind* posterior, places behind (a specific) part
pre-, pro- *before, ahead of* prenatal, before birth
procto- *rectum, anus* proctoscope, an instrument for examining the rectum
pseudo- *false* pseudotumor, a false tumor
psycho- *mind, psyche* psychogram, a chart of personality traits
pulmo- *lung* pulmonary artery, which brings blood to the lungs
pyo- *pus* pyocyst, a cyst that contains pus
quad-, quadr- *four* quadriceps, four muscles located in the anterior thigh
re- *back, again* reinfect
rect- *straight* rectum; rectus abdominis, longitudinal muscle of the anterior abdominal wall
ren- *kidney* renal artery, arterial vessel to the kidney
retro- *backward, behind* retrogression, to move backward in development
ruga- *fold, wrinkle* rugae, the folds in the stomach, gallbladder, and urinary bladder
sagitt- *arrow* sagittal (directional term)
sarco- *flesh* sarcomere, unit of contraction in skeletal muscle
sclero- *hard* sclerodermatitis, inflammatory thickening and hardening of the skin
semi- *half* semicircular, having the form of half a circle
septo *a fence* interventricular septum in the heart
serrat- *saw* serratus anterior, a muscle of the chest wall that has a jagged edge
sinu,- *a hollow* sinuses of the skull
soma- *body* somatic nervous system
splanchn- *viscera* splanchnic nerve, autonomic supply to abdominal viscera
squam- *scale, flat* squamous epithelium, squamous suture of the skull
steno- *narrow* stenocoriasis, narrowing of the pupil
stroma *spread out* stroma, the connective tissue framework of some organs
sub- *below, under* sublingual, below the tongue
super- *above, over* superior, quality or state of being above others or a part
supra- *above, over, beyond* supracondylar, above a condyle
sym-, syn- *together, with* synapse, the region of communication between two neurons
synerg- *work together* synergists, muscles that work together to produce a specific movement
tachy- *rapid* tachycardia, abnormally rapid heartbeat
telo- *the end, complete* telophase, the end of mitosis
templ-, tempo- *time* temporal region of the skull, one of the first locations for hair to become gray, thus showing the passage of time or age
teres *round, smooth* ligamentum teres, also called the round ligament
therm- *heat* thermometer, an instrument used to measure heat
tissue *woven* tissue
toxic- *poison* antitoxic, effective against poison
trab- *beam, timber* trabeculae, spicules of bone in spongy bone tissue
trans- *across, through* transpleural, through the pleura
tri- *three* trifurcation, division into three branches
troph- *nourish* trophoblast, from which develops the fetal portion of the placenta
tuber- *swelling* tuberosity, a bump on a bone
vas- *a vessel, duct* vascular system, system of blood vessels
venter, ventr- *belly* ventral (directional term)
villus *shaggy hair* microvilli, which have the appearance of hair in light microscopy
viscero- *organs of the body cavity,* visceroinhibitory, inhibiting the movements of the viscera